The Giant Vesicle Book

The Giant Vesicle Book

Edited by

Rumiana Dimova
Carlos M. Marques

CRC Press is an imprint of the
Taylor & Francis Group, an **informa** business

CRC Press
Taylor & Francis Group
6000 Broken Sound Parkway NW, Suite 300
Boca Raton, FL 33487-2742

First issued in paperback 2022

© 2020 by Taylor & Francis Group, LLC
CRC Press is an imprint of Taylor & Francis Group, an Informa business

No claim to original U.S. Government works

ISBN 13: 978-1-498-75217-6 (hbk)
ISBN 13: 978-1-03-233789-0 (pbk)
DOI: 10.1201/9781315152516

This book contains information obtained from authentic and highly regarded sources. Reasonable efforts have been made to publish reliable data and information, but the author and publisher cannot assume responsibility for the validity of all materials or the consequences of their use. The authors and publishers have attempted to trace the copyright holders of all material reproduced in this publication and apologize to copyright holders if permission to publish in this form has not been obtained. If any copyright material has not been acknowledged please write and let us know so we may rectify in any future reprint.

Except as permitted under U.S. Copyright Law, no part of this book may be reprinted, reproduced, transmitted, or utilized in any form by any electronic, mechanical, or other means, now known or hereafter invented, including photocopying, microfilming, and recording, or in any information storage or retrieval system, without written permission from the publishers.

For permission to photocopy or use material electronically from this work, please access www.copyright.com (http://www.copyright.com/) or contact the Copyright Clearance Center, Inc. (CCC), 222 Rosewood Drive, Danvers, MA 01923, 978-750-8400. CCC is a not-for-profit organization that provides licenses and registration for a variety of users. For organizations that have been granted a photocopy license by the CCC, a separate system of payment has been arranged.

Trademark Notice: Product or corporate names may be trademarks or registered trademarks, and are used only for identification and explanation without intent to infringe.

Publisher's Note

The publisher has gone to great lengths to ensure the quality of this reprint but points out that some imperfections in the original copies may be apparent.

Library of Congress Cataloging-in-Publication Data

Names: Dimova, Rumiana (Scientist), editor. | Marques, Carlos (Scientist), editor.
Title: The giant vesicle book / edited by Rumiana Dimova, Carlos Marques.
Description: Boca Raton, FL : CRC Press, Taylor & Francis Group, [2020]
Identifiers: LCCN 2019006748 | ISBN 9781498752176 (hardback ; alk. paper) | ISBN 1498752179 (hardback ; alk. paper) | ISBN 9781315152516 (e-Book) | ISBN 1315152517 (e-Book)
Subjects: LCSH: Liposomes. | Lipid membranes--Biotechnology.
Classification: LCC QH602 .G53 2019 | DDC 571.6/55--dc23
LC record available at https://lccn.loc.gov/2019006748

Visit the Taylor & Francis Web site at
http://www.taylorandfrancis.com

and the CRC Press Web site at
http://www.crcpress.com

Contents

Preface

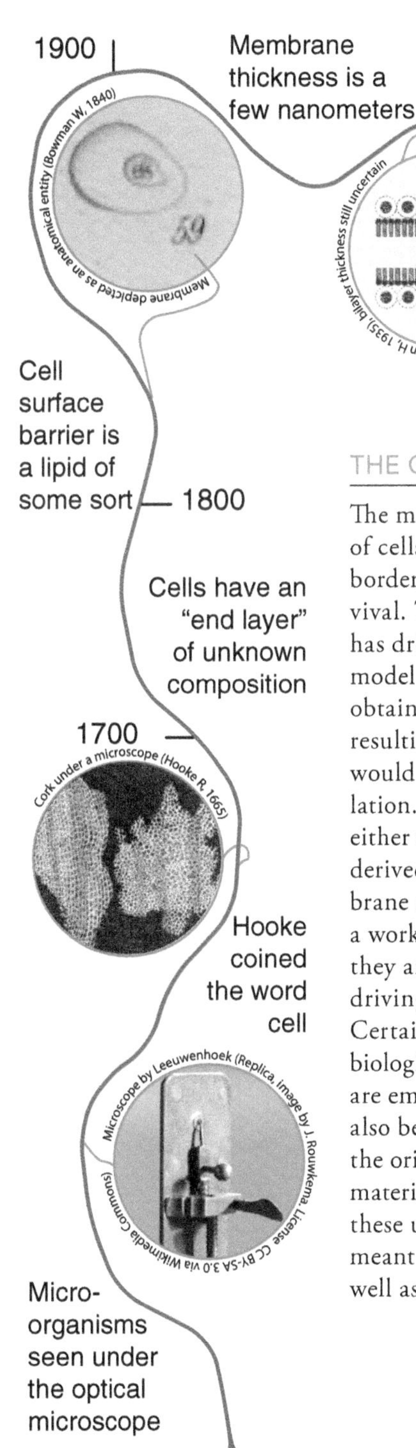

THE GIANT VESICLE BOOK: A PREFACE

The membrane is the first point of contact with the external environment of cells and membrane-bound organelles. It ensures their entity as an active border, being involved in (re)shaping, communication, protection and survival. The quest for unraveling the properties and interactions of membranes has driven efforts to develop model membrane systems. Arguably, the optimal model system to study the plasma membrane as a major organelle would be obtained by emptying the cell interior while modulating the composition of the resulting capsule of lipids, membrane proteins and other species; these capsules would be then directly visible under a microscope and amenable to manipulation. We cannot do that (yet), but we can use giant vesicles (10–100 μm), either synthetic, that is, prepared from a well-defined mixture of molecules, or derived from the plasma membrane of cells. Giant vesicles, the model membrane systems starring in this book, have been initially established and used as a workbench for studying basic properties of simple lipid bilayers. Nowadays, they are increasingly employed in biophysics to investigate the mechanisms driving biological processes at the level of the plasma and organelle membranes. Certainly, the widening acceptance of giant vesicles as a mimetic system among biologists is a good indicator of how successful this approach is. However, they are employed not only in the fields of biophysics and biology. Giant vesicles have also become important tools for scientists in physical chemistry, to understand the origin of life or in application-driven research such as in drug delivery or in material synthesis where the vesicles are used as micro-compartments. For all these undertakings, and for the many more to come, *The Giant Vesicle Book* is meant to be a road companion, a trusted guide for the first steps in this field, as well as a source for the information that experts require.

AIMS OF THE BOOK

This book provides

- A start-up package of theoretical and experimental information for newcomers in the field as well as specific protocols for establishing the required preparations and assays;
- Tips and instructions for carefully performing measurements with giant vesicles or for observing them, including pitfalls, and the corresponding theoretical background required to interpret the data;
- A description of approaches developed for investigating giant vesicles as well as brief overviews of previous studies implementing the described techniques;
- Ideas for further applications and discussions on the implications of the observed phenomena toward understanding membrane-related processes.

How to read the book

The book consists of 30 chapters grouped in five different parts dedicated to

- Part I: The making of (Chapters 1–4)
- Part II: Giant vesicles theoretically and *in silico* (Chapters 5–9)
- Part III: GUV-based techniques and what one can learn from them (Chapters 10–20)
- Part IV: GUVs as membrane interaction platforms (Chapters 21–25)
- Part V: GUVs as complex membrane containers (Chapters 26–30)

Every chapter has material highlighted within boxes providing a summary of specifically important content of the chapter or a description of protocols and assays. Every chapter also ends with a section discussing ideas for further applications or a vision for the future. At the end of the book, a set of four appendices provides additional information about the molecules employed or cited in the book: lipids, labeled lipids, detergents, and fluorescent probes.

The book is available both in print and as an eBook with links to chapters and external material including videos.

REFERENCES

Bowman W (1840) On the minute structure and movements voluntary muscle. *Philosophical Transactions of the Royal Society of London*, 130, 457–501.

Danielli JF, Davson, H (1935) A contribution to the theory of permeability of thin films. *Journal of Cellular and Comparative Physiology*, 5(4), 495–508.

Hooke R (1665) *Micrographia: Or Some Physiological Descriptions of Minute Bodies Made by Magnifying Glasses. With Observations and Inquiries Thereupon*. London, UK: J. Martyn and J. Allestry.

Sjöstrand FS, Andersson-Cedergren E, Dewey MM (1958) The ultrastructure of the intercalated discs of frog, mouse and guinea pig cardiac muscle. *Journal of Ultrastructure Research*, 1(3), 271–287.

Van Leeuwenhoek A (1677) Letter to Oldenburg dated 9 October 1676. *Philosophical Transactions*, 12, 831.

Acknowledgments

Peter Walde Pasquale Stano Ilya Levental Patricia Bassereau Petra Schwille
Reinhard Lipowsky Gerhard Gompper Petia M.Vlahovska J. Agudo-Canalejo Luis Bagatolli
David Needham Elisa Parra-Ortiz Andreas Janshoff Jochen Guck John H. Ipsen
Kheya Sengupta Sarah L. Keller Victor Steinberg Mireille Claessens Ana J. García-Sáez
Maurício S. Baptista Daniel A. Fletcher Karin A. Riske Brigitte Pépin-Donat Dennis E. Discher
J.-F. Le Meins Masayuki Imai Christine D. Keating Ilia Platzman Joachim P. Spatz

STORY OF THE BOOK

The story of this book started at the Biophysical Society meeting in San Francisco in 2014 when Lou Chosen had the idea of "a book that provides an overview on giant lipid vesicles, covering physical properties, formation, testing, and basic applications ... a new, up to date resource that would be important for the field and useful for scientists, practitioners, and students working at the intersection of life sciences, physics, chemistry, computing, and biomedical research." We thank Lou for pushing this through and for bearing with us all these years until the book was ready.

This was a time full of happy moments for us, drafting the idea and getting excited about it, approaching authors and getting them excited about it, seeing the first chapters trickle in: Thank you, Karin, Ilya, Sarah and Andreas, for being the first ones! This was certainly a time full of energizing and stirring moments—reading all these chapters and learning a lot—and at times getting frustrated with being late as well as breaking every possible deadline. This was a time of communication, with more than a thousand emails exchanged, of an absurd amount of appointments and hours spent teleconferencing. Particularly, this was also an intense time for our families, whether helping us roll the giant trampoline for the editors' picture or putting up with us reading, writing and correcting the chapters during family holidays. We cannot thank Yulia, Peter and the supporting tribes enough.

Thank you, Luis, Karin and Dennis, for being close to our targeted chapter length; thank you, Reinhard, for the book within the book. We did aim at having 30 chapters of 15 to 20 pages each, and we failed. We take our failure as a token of the extraordinary vitality of this field.

This work would not have come to light if more than 70 members of the biophysical community had not written, drawn, corrected and reviewed the 30 chapters of the book, and kindly accepted our infinite number of requests to bring the text and the figures to a common format. We do not apologize for being pushy and straightforward, but we do thank from our hearts all the leading authors and collaborators for their goodwill, enthusiasm, and readiness in their communication with us. It has been a pleasure working with you. It has also been a pleasure to be supported by our groups and close scientific environment, by all our collaborators that have test-read, reviewed and proofread the majority of the chapters. And finally, our special thanks to Vasil N. Georgiev, Albert Johner, Roland L. Knorr, Tom Robinson, Jan Steinkühler and Matthew Turner for kindly joining our reviewing odyssey.

COVER CONTENT

Images of giant vesicles under different experimental conditions, observed by optical microscopy techniques: epifluorescence, confocal, phase contrast, bright field, differential interference contrast, reflection-interference contrast microscopy and super-resolution (STED) microscopy. Courtesy of Pedro Aoki, Luis Bagatolli, Natalya Bezlyepkina, Tripta Bhatia, Rumiana Dimova, Susanne Fenz, Christopher Haluska, Roland L. Knorr, Yanhong Li, Rafael B. Lira, Yonggang Liu, Carlos M. Marques, Salome Pataraia, Karin A. Riske, André Schröder, Margarita Staykova, Jan Steinkühler, Vivien Walter, Georges Weber, Ziliang Zhao.

About the editors

Rumiana Dimova leads an experimental lab in biophysics at the Max Planck Institute of Colloids and Interfaces in Potsdam, Germany. She has been working with giant vesicles already during her MSc studies and was introduced to the magic of their preparation by M. Angelova, who established the well-known electroformation protocol for giant vesicle formation. Rumiana's PhD aimed at developing biophysical approaches to assess membrane rheological properties. Since then, she has been tackling a variety of open questions in cell membrane biophysiscs and synthetic biology while employing giant vesicles as a platform to develop new methods for the biophysical characterization of membranes and processes involving them. These studies have resulted in more than 100 peer-reviewed publications and book contributions. In 2017, she was nominated as the chair of the Membrane Structure and Assembly Subgroup of the Biophysical Society. Recently, she was also awarded the Emmy Noether Distinction for Women in Physics of the European Physical Society.

Carlos M. Marques, a CNRS senior scientist, founded the MCube group at the Charles Sadron Institute in Strasbourg, France, where he gears experimental and theoretical research toward the understanding of the physical properties of self-assembled lipid bilayers. Trained as a polymer theoretician, Carlos first got interested in membranes because they interact with polymers and published the first prediction for the membrane changes expected when polymers adsorb onto lipid bilayers. He then expanded the scope of his group to include experiments and numerical simulations and has now published over 130 scientific papers dealing with a variety of facets of soft condensed matter and, in particular, with lipid bilayers. He has published many papers based on research with giant vesicles, including the first study of lipid oxidation in GUVs, the discovery of the so-called PVA method for vesicle growth and the quantification of molecular adsorbed amounts from confocal microscopy images.

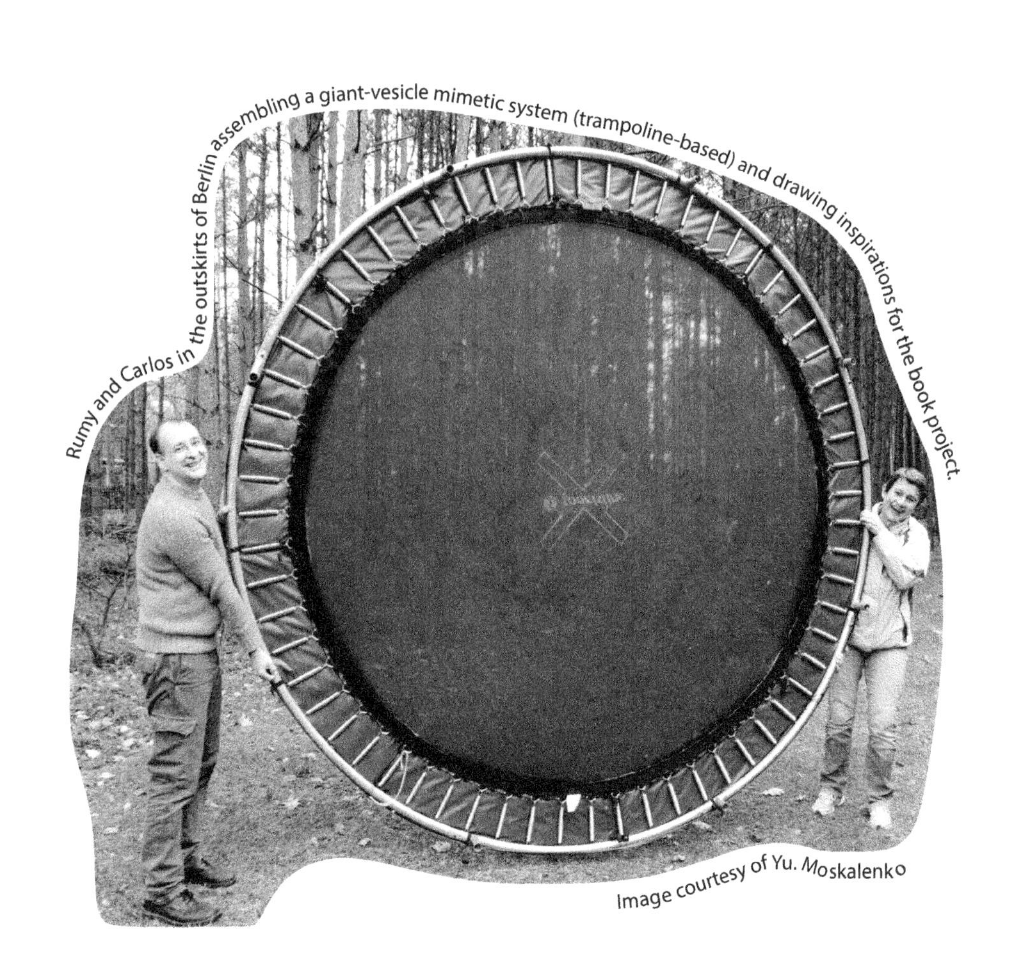

Rumy and Carlos in the outskirts of Berlin assembling a giant-vesicle mimetic system (trampoline-based) and drawing inspirations for the book project.

Image courtesy of Yu. Moskalenko

Contributors

Jaime Agudo-Canalejo
Theory & Bio-Systems Department
Max Planck Institute of Colloids and Interfaces
Potsdam, Germany

Thorsten Auth
Theoretical Soft Matter and Biophysics
Institute of Complex Systems and Institute for Advanced Simulation
Forschungszentrum Jülich
Jülich, Germany

Luis A. Bagatolli
Instituto de Investigación Médica Mercedes y Martín Ferreyra
INIMEC-CONICET-Universidad Nacional de Córdoba
Córdoba, Argentina

Mauricio S. Baptista
Institute of Chemistry, Department of Biochemistry
University of São Paulo
São Paulo, Brazil

Patricia Bassereau
Laboratoire Physico Chimie Curie, CNRS UMR168
Institut Curie
PSL Research University
and
Sorbonne Université
Paris, France

Lucia T. Benk
Department of Cellular Biophysics
Max Planck Institute for Medical Research
and
Department of Biophysical Chemistry
University of Heidelberg
Heidelberg, Germany

Tripta Bhatia
Department of Physics, Chemistry and Pharmacy
MEMPHYS–Center for Biomembrane Physics
University of Southern Denmark
Odense, Denmark
and
Max Planck Institute of Colloids and Interfaces
Potsdam, Germany

Matthew C. Blosser
Department of Chemical Engineering
University of Southern California
Los Angeles, California

Clément Campillo
LAMBE
Université Evry Val d'Essonne
Evry, France

David Christian
Biophysical Engineering Labs
University of Pennsylvania
Philadelphia, Pennsylvania

Mireille M. A. E. Claessens
Nanobiophysics Group
Faculty of Science and Technology
MESA+ Institute for Nanotechnology
University of Twente
and
MIRA Institute for Biomedical Technology and Technical Medicine
University of Twente
Enschede, the Netherlands

Gheorghe Cojoc
Biotechnology Center
Technische Universität Dresden
Dresden, Germany

Caitlin E. Cornell
Department of Chemistry
University of Washington
Seattle, Washington

Thi Phuong Tuyen Dao
Laboratoire de Chimie des Polymères Organiques UMR5 629 ENSCBP
University of Bordeaux, CNRS, Bordeaux INP
LCPO, UMR 5629
Pessac, France
and
Centro de Quimica-Fisica Molecular
Complexo Interdisciplinar
IST Universidade de Lisboa
Lisboa, Portugal

Kushal Kumar Das
Interfaculty Institute of Biochemistry
University of Tübingen
Tübingen, Germany

Ulysse Delabre
Laboratoire Ondes et Matière d'Aquitaine (LOMA)
University of Bordeaux, CNRS
LOMA, UMR 5798
Talence, France

Rumiana Dimova
Max Planck Institute of Colloids and Interfaces
Potsdam, Germany

Dennis E. Discher
Biophysical Engineering Labs
University of Pennsylvania
and
Pharmacological Sciences Graduate Group
University of Pennsylvania
Philadelphia, Pennsylvania

Dmitry A. Fedosov
Theoretical Soft Matter and Biophysics
Institute of Complex Systems and Institute for Advanced Simulation
Forschungszentrum Jülich
Jülich, Germany

Khalid Ferji
Laboratoire de Chimie des Polymères Organiques, UMR 5629
ENSCBP
University of Bordeaux, CNRS, Bordeaux INP
LCPO, UMR 5629
Pessac, France

Fabio Fernandes
Centro de Quimica-Fisica Molecular
Complexo Interdisciplinar
IST Universidade de Lisboa
Lisbon, Portugal

Daniel A. Fletcher
Department of Bioengineering & Biophysics
University of California
and
UC Berkeley/UC San Francisco Graduate Group in Bioengineering
and
Biological Systems and Engineering Division
Lawrence Berkeley National Laboratory
Berkeley, California

Johannes P. Frohnmayer
Department of Cellular Biophysics
Max Planck Institute for Medical Research
and
Department of Biophysical Chemistry
University of Heidelberg
Heidelberg, Germany

Ana J. García-Sáez
Interfaculty Institute of Biochemistry
University of Tübingen
Tübingen, Germany
and
Max Planck Institute for Intelligent Systems
Stuttgart, Germany

Matthias Garten
Laboratoire Physico Chimie Curie, CNRS UMR168
Institut Curie
PSL Research University
and
Sorbonne Université
Paris, France
and
Section on Integrative Biophysics
Division of Basic and Translational Biophysics
Eunice Kennedy Shriver National Institute of Child Health and Human Development
National Institutes of Health
Bethesda, Maryland

Antoine Girot
Laboratoire Ondes et Matière d'Aquitaine (LOMA)
University of Bordeaux, CNRS
LOMA, UMR 5798
Talence, France

Gerhard Gompper
Theoretical Soft Matter and Biophysics
Institute of Complex Systems and Institute for Advanced Simulation
Forschungszentrum Jülich
Jülich, Germany

Jochen Guck
Biotechnology Center
Technische Universität Dresden
Dresden, Germany

Barbara Haller
Department of Cellular Biophysics
Max Planck Institute for Medical Research
and
Department of Biophysical Chemistry
University of Heidelberg
Heidelberg, Germany

Allan Grønhøj Hansen
Department of Physics, Chemistry and Pharmacy
MEMPHYS–Center for Biomembrane Physics
University of Southern Denmark
Odense, Denmark

Tobias Härtel
Department of Cellular and Molecular Biophysics
Max Planck Institute of Biochemistry
Martinsried, Germany

Emmanuel Ibarboure
Laboratoire de Chimie des Polymères Organiques, UMR 5629
ENSCBP
University of Bordeaux, CNRS, Bordeaux INP
LCPO, UMR 5629
Pessac, France

Masayuki Imai
Department of Physics
Tohoku University
Sendai, Japan

John Hjort Ipsen
Department of Physics, Chemistry and Pharmacy
MEMPHYS–Center for Biomembrane Physics
University of Southern Denmark
Odense, Denmark

Rosângela Itri
Institute of Physics, Department of Applied Physics
University of São Paulo
São Paulo, Brazil

Jan-Willi Janiesch
Department of Cellular Biophysics
Max Planck Institute for Medical Research
and
Department of Biophysical Chemistry
University of Heidelberg
Heidelberg, Germany

Andreas Janshoff
Institute of Physical Chemistry
Georg-August-University of Goettingen
Goettingen, Germany

Christine D. Keating
Department of Chemistry
Pennsylvania State University
University Park, Pennsylvania

Sarah L. Keller
Department of Chemistry
University of Washington
Seattle, Washington

Sébastien Lecommandoux
Laboratoire de Chimie des Polymères Organiques, UMR 5629
ENSCBP
University of Bordeaux, CNRS, Bordeaux INP
LCPO, UMR 5629
Pessac, France

Jean-François Le Meins
Laboratoire de Chimie des Polymères Organiques, UMR 5629
ENSCBP
University of Bordeaux, CNRS, Bordeaux INP
LCPO, UMR 5629
Pessac, France

Michael Levant
Department of Physics of Complex Systems
Weizmann Institute of Science
Rehovot, Israel
and
Laboratoire de Physique
Ecole Normale Supérieure de Lyon
Lyon, France

Ilya Levental
Department of Integrative Biology and Pharmacology
University of Texas Health Science Center at Houston
Houston, Texas

Daniel Lévy
Laboratoire Physico Chimie Curie, CNRS UMR168
Institut Curie
PSL Research University
and
Sorbonne Université
Paris, France

Reinhard Lipowsky
Theory & Bio-Systems Department
Max Planck Institute of Colloids and Interfaces
Potsdam, Germany

Rafael B. Lira
Department of Theory and Bio-Systems
Max Planck Institute of Colloids and Interfaces
Potsdam, Germany

Joseph H. Lorent
Department of Integrative Biology and Pharmacology
University of Texas Health Science Center at Houston
Houston, Texas

Allyson M. Marianelli
Department of Chemistry
Pennsylvania State University
University Park, Pennsylvania

Carlos M. Marques
Institut Charles Sadron
Université de Strasbourg, CNRS
Strasbourg, France

Chaouqi Misbah
LIPHY
Université Grenoble Alpes
and
LIPHY
CNRS
Grenoble, France

Praful Nair
Biophysical Engineering Labs
University of Pennsylvania
Philadelphia, Pennsylvania

David Needham
Department of Mechanical Engineering and Material Science
School of Engineering
Duke University
Durham, North Carolina

Elisa Parra-Ortiz
Department of Physics, Chemistry, and Pharmacy
University of Southern Denmark
Odense, Denmark
and
Department of Pharmacy
University of Copenhagen
Copenhagen, Denmark

Brigitte Pépin-Donat
INAC-SPRAM CNRS
and
INAC-SPRAM
Université Grenoble Alpes
and
INAC-SPRAM
CEA
Grenoble, France

Ilia Platzman
Department of Cellular Biophysics
Max Planck Institute for Medical Research
and
Department of Biophysical Chemistry
University of Heidelberg
Heidelberg, Germany

Coline Prévost
Laboratoire Physico Chimie Curie, CNRS UMR168
Institut Curie
PSL Research University
and
Sorbonne Université
Paris, France

Manuel Prieto
Centro de Química-Física Molecular
Complexo Interdisciplinar
IST Universidade de Lisboa
Lisboa, Portugal

François Quemeneur
CYTODIAG
Biopôle Clermont-Limagne
Riom, France

Scott P. Rayermann
School of Interdisciplinary Arts and Sciences
Division of Sciences and Mathematics
University of Washington - Tacoma
Tacoma, Washington

Karin A. Riske
Departamento de Biofísica
Universidade Federal de São Paulo
São Paulo, Brazil

Olivier Sandre
Laboratoire de Chimie des Polymères Organiques, UMR5629
ENSCBP
University Bordeaux, CNRS, Bordeaux INP
LCPO, UMR 5629
Pessac, France

Eva M. Schmid
Department of Bioengineering & Biophysics
University of California
Berkeley, California

Petra Schwille
Department of Cellular and Molecular Biophysics
Max Planck Institute of Biochemistry
Martinsried, Germany

Kheya Sengupta
Aix-Marseille Université, CNRS, CINaM UMR 7325
13288 Marseille
Cedex 9, France

Mijo Simunovic
Laboratoire Physico Chimie Curie, CNRS UMR168
Institut Curie
PSL Research University
and
Sorbonne Université
Paris, France
and
The Rockefeller University
New York, New York

Ana-Sunčana Smith
PULS Group
Institut für Theoretische Physik and the Excellence Cluster:
Engineering of Advanced Materials
Universität Erlangen-Nürnberg
Erlangen, Germany
and
Group for Computational Life Sciences
Institute Rudjer Boskovic
Zagreb, Croatia

Joachim P. Spatz
Department of Cellular Biophysics
Max Planck Institute for Medical Research
and
Department of Biophysical Chemistry
University of Heidelberg
Heidelberg, Germany

Pasquale Stano
Department of Biological and Environmental Sciences and Technologies (DiSTeBA)
University of Salento
Lecce, Italy

Victor Steinberg
Department of Physics of Complex Systems
Weizmann Institute of Science
Rehovot, Israel

Fabrice Thalmann
Institut Charles Sadron
Université de Strasbourg, CNRS
Strasbourg, France

Begoña Ugarte-Uribe
Interfaculty Institute of Biochemistry
University of Tübingen
Tübingen, Germany
and
Department of Biochemistry and Molecular Biology
Biofisika Institute (UPV/EHU, CSIC)
University of the Basque Country (UPV/EHU)
Leioa, Spain

Petia M. Vlahovska
Engineering Sciences and Applied Math
Northwestern University
Evanston, Illinois

Peter Walde
Department of Materials
ETH Zürich
Zürich, Switzerland

Marian Weiss
Department of Cellular Biophysics
Max Planck Institute for Medical Research
and
Department of Biophysical Chemistry
University of Heidelberg
Heidelberg, Germany

List of boxes and protocols

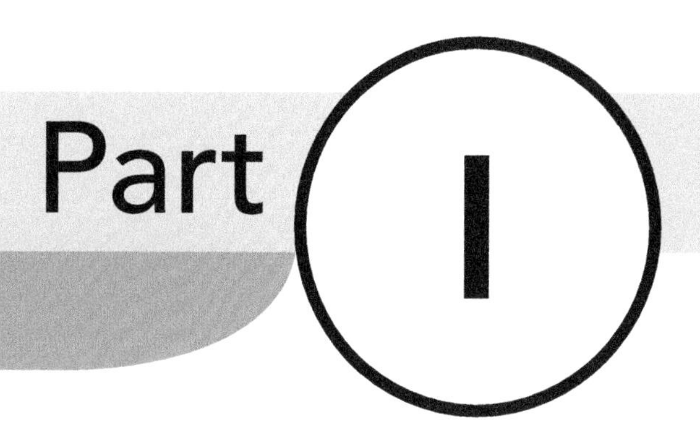

The making of

Preparation methods for giant unilamellar vesicles

Rumiana Dimova, Pasquale Stano, Carlos M. Marques, and Peter Walde

Lipids need your guidance for forming giant vesicles.

Contents

1.1 GIANT UNILAMELLAR VESICLES

1.1.1 THE BIRTH OF A NEW BIOPHYSICAL PLATFORM

"A drop of water or sucrose solution was then placed on the glass slide, and seemed to seep in between the phospholipid lamellae. The swollen phospholipid membranes formed long fingerlike projections attached at one end to the glass slide. Eventually these membranous structures became detached and rounded up to form vesicles" (Figure 1.1).

By reporting their first observations of the gentle swelling of the strata of bilayers of phosphatidylcholine (PC) phospholipids in an aqueous solution, Reeves and Dowben deliberately initiated a new approach to the science of lipid membranes (Reeves and Dowben, 1969). The community working with giant liposomes largely shares the motivations for such an approach because it allows studying of the behavior of *single* lipid bilayer assemblies of a cell size, under conditions where membrane composition and the influence of the environment are under strict control. Indeed, as the bimolecular leaflet proposition of Danielli and Davson (1935) became largely accepted, methods started to be developed to study the fatty bilayer. By the 1960s, thin films (also known as black films) formed by anchoring a mixture of organic solvent and lipids to a small plastic diaphragm, thus separating two aqueous regions, were already used to study bilayer permeability to ions (Mueller et al., 1963). Moreover, the limits of the approach singled out by Reeves and Dowben are still recognized today: The bilayer thus formed is not only in direct contact with the remaining organic solvent but also under considerable tension. A different route was taken by Alec Bangham, the "father of liposomes" and his collaborators and by others (Bangham and Horne, 1964; Bangham et al., 1965; Papahadjopoulos and Miller, 1967), who studied the structure and permeability of "multilamellar lipid spherules" then recognized as "fragments of tubular myelin figures which form spontaneously when dried lipids swell in an aqueous medium." The tubular structures were called "myelin figures" given that such types of tubular molecular assemblies were originally found to form from myelin (Neubauer, 1867), a phospholipid-rich heterogeneous waxy material that surrounds the axon of most of the nerve cells of vertebrates (Deber and Reynolds, 1991) (for example images, see Sakurai and Kawamura, 1984). The key contribution of Reeves and Dowben was then to control the spontaneous swelling conditions of the dry lipids such that one obtains "phospholipid vesicles of pre-determined composition several microns in diameter bounded by walls one or a few bilayers thick" (Reeves and Dowben, 1969).

Interest in these single-walled vesicles that could be visualized under an optical microscope grew rapidly over the next decade (Papahadjopoulos and Kimelberg, 1974), and prompted studies such as those of Helfrich, who investigated for the first time single bilayer melting at the optical length scales (Harbich et al., 1976).

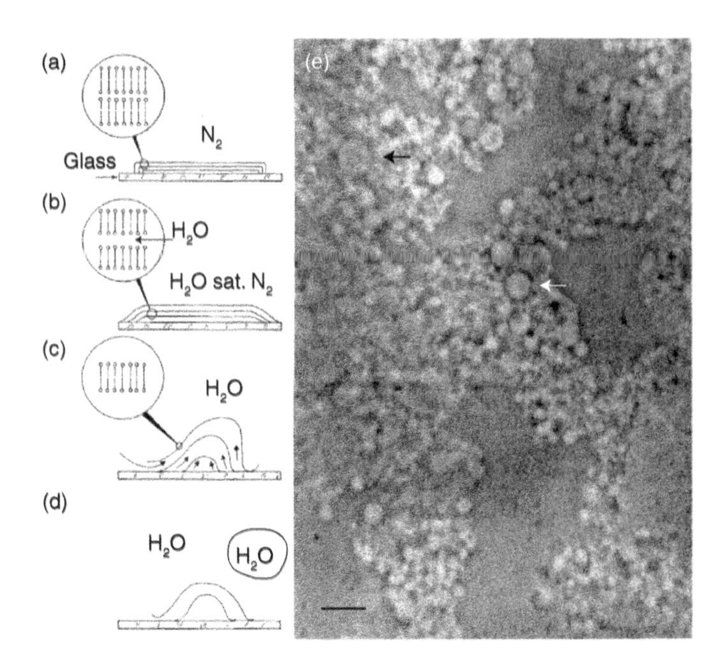

Figure 1.1 First light microscopy image showing giant vesicles—arrows in (e)—and the corresponding preparation method. (a) Lamellae of phospholipids swell and (b) become separated from one another in a moist atmosphere. When an aqueous solution is then added, (c) it runs in between the lamellae, which finally (d) become detached from the glass surface and round up into vesicles. (e) Photomicrograph of phosphatidylcholine vesicles prepared in 200 mM sucrose and suspended in 10% w/w gelatin to dampen Brownian movement. Scale bar: 10 µm. (From Reeves, J.P. and Dowben, R.M.: Formation and properties of thin-walled phospholipid vesicles. *Journal of Cellular Physiology*. 1969. 73. 49–60. Copyright Wiley-VCH Verlag GmbH & Co. KGaA. Reproduced with permission.)

However, only by the end of the 1970s was the term Giant Vesicles first used in a publication, an abstract of a Biophysical Society meeting (Murphy and Shamoo, 1978). However, the article at the root of citations carrying this designation (Harbich and Helfrich, 1979) did not cite Murphy and Shamoo. The name Giant Vesicles might thus have imposed itself as an evidence for several biophysical groups working with these fascinating objects (Luisi and Walde, 2010).

The science of single bilayer vesicles with a giant size that emerged at the end of the 1960s has since grown to a multidisciplinary community that *The Giant Vesicle Book* was written to serve. This first chapter of the book must obviously be a practical guide for the scientist wishing to prepare giant vesicles: It will start thus by introducing the founding spontaneous swelling strategies, but it will elaborate also on the large spectrum of the other methods that have since been developed to bring increased facility and improved control to the formation of giant unilamellar vesicles (GUVs).

1.1.2 GUVs, LIPIDS AND METHODS OF FORMATION

One can view phospholipid GUVs—also called giant unilamellar liposomes—as micrometer-sized, aqueous droplets that are separated from the bulk aqueous solution by a molecular bilayer. This bilayer is about 4–5 nm thick. As we will see below, depending on the method of preparation, the vesicle-confined volume can have the same or a different composition than the bulk aqueous solution. A single spherical 1,2-dioleoyl-*sn*-glycero-3-phosphocholine (DOPC) GUV with a diameter of 50 µm, for example, encloses a trapped volume of 65×10^{-15} m^3 (65 pL), see Appendix 1 of the book for structure and data on lipids. The boundary membrane measures 7.9×10^3 µm^2 assembled from 2.2×10^{10} DOPC molecules, as calculated on the basis of an average area per DOPC molecule of 0.72 nm^2.

Why and how do 22 billion molecules assemble into a structure of a well-defined shape, tens of thousand times larger than their size? The GUV bilayer self-assembles from amphiphiles having the proper balance of molecular dimensions of the hydrophobic and hydrophilic regions. The molecular structure of the amphiphile therefore determines the local structure of the assembly—whether it has a planar geometry, its thickness, the area per molecule—but the bilayer itself does not spontaneously form a GUV of a given size, as do for instance spherical or cylindrical micelles (Israelachvili, 2011). The methods developed for GUV formation are thus methods that allow for the spontaneous self-assembly of amphiphiles into a bilayer but that also constrain (we will often say guide) the bilayer such that it adopts the shape of a GUV.

It is worth stressing that the phase behavior of the bilayers self-assembled from amphiphiles is relevant for the formation of GUVs because most methods rely on self-assembled bilayers in their fluid state. The fluid phases occur for high enough temperatures, above the main-transition temperature (T_m), where the bilayer changes from a rigid low temperature gel phase with low molecular mobility to the high temperature fluid phase where membrane flexibility and high molecular mobility allow for shape control (Marsh, 2013). T_m values for all the lipids discussed in this book are given in Appendix 1.

Giant unilamellar vesicles are thus obtained by a guided assembly of the amphiphiles in their fluid state, that is, by using a certain method, by following a certain protocol. The details of this protocol—or method of preparation—depend on the type of amphiphiles used and on the experimental conditions such as the properties of the aqueous medium or the temperature. The giant vesicles are therefore only kinetically stable and not thermodynamically stable. If GUVs are fragmented into smaller vesicles by mechanical treatment, the GUVs do not spontaneously reform from the smaller vesicles. Nevertheless, the kinetic GUV stability may be so high that GUVs can exist over extended periods of time (at least several days).

GUVs represent a dispersed state of the lamellar phase of amphiphiles in an aqueous medium, independent of whether the amphiphiles are conventional biological phospholipids like 1-palmitoyl-2-oleoyl-*sn*-glycero-3-phosphocholine (POPC) or DOPC, mixtures of phospholipids and other biological molecules, or fully synthetic, non-natural amphiphiles of low or high molar mass. In the latter case, when GUVs are obtained from amphiphilic block copolymers, they are often called giant unilamellar "polymersomes," see Chapter 26 for details.

Conceptually, there are two main ways to obtain GUVs (Walde et al., 2010; Dimova, 2019). One starts with lamellae-forming amphiphiles in their dry—or nearly dry—state, often spread and deposited with the help of an organic solvent (e.g., chloroform) on a solid surface (e.g., silicate glass). The amphiphiles are then hydrated to form GUVs in a defined and controlled way by adding an aqueous solution. This is the founding "spontaneous swelling" method introduced above and extensively discussed in Section 1.2.1. The deposition of the amphiphiles and their hydration can be influenced in a positive way by an externally applied electric field if the amphiphiles are deposited on an electrically conductive solid surface (the so-called "electroformation" method, discussed in Section 1.2.3).

The second, main way for obtaining GUVs is based on the fact that GUVs are micrometer-sized aqueous droplets in a bulk aqueous solution, separated by a layer of amphiphiles surrounding the droplets. For obtaining such "water-in-water droplets" (i.e., GUVs), more easily producible, amphiphile-stabilized, micrometer-sized water-in-oil (w/o) droplets are first prepared, followed by the transformation of these droplets into GUVs, ideally with the retention of the droplet size and an elimination of all (or most) of the oil during the transformation process ("water droplet transfer" method, see Section 1.3.1).

Other ways for forming GUVs exist (see Section 1.3), but not every possibility is discussed here. In all cases, the methods are based on a controlled transformation of a certain aggregation state of lamellae-forming molecules into GUVs as a result of a particular procedure. The initial amphiphile system may also contain "helper molecules"—water-miscible solvents or other amphiphiles—that are then removed during the GUV formation process.

1.2 METHODS BASED ON VESICLE SWELLING ON SUBSTRATES

In these methods, one typically spreads a lipid solution in a volatile organic solvent as a film onto a substrate. Alternatively, when aiming at encapsulating water-soluble molecules within the GUVs, one can employ an approach used for small liposome preparation, namely the inverse-phase method. In this approach, the precursor film is spread from an inverted emulsion of water (with the target molecules) in an organic solvent containing the lipids (Mertins et al., 2009). The emulsion is typically sonicated to reduce the size of the water droplets and it acquires a bluish opalescent color. After spreading and gentle drying of the emulsion, the lipid film is then swollen with one of the methods described in Sections 1.2.1 through 1.2.3 in more detail. This method has been used for the encapsulation of various water-soluble proteins during GUV formation (Mertins et al., 2009; Weinberger et al., 2013).

1.2.1 THE SPONTANEOUS SWELLING (OR GENTLE HYDRATION) METHOD

We have seen that Reeves and Dowben (1969) were the first to demonstrate that instead of myelin figures, GUVs can be obtained in high yields from a dry film of bilayer-forming egg yolk PCs (or their mixtures with cholesterol) when deposited on a glass surface and then gently and slowly hydrated at room temperature in a controlled way with either distilled water or with an aqueous solution of non-electrolytes (see also Mueller et al., 1983). The authors realized that the formation of micrometer-sized phospholipid GUVs in high yields through this spontaneous swelling is only possible if the hydration is carried out in a specific and defined way. For some of the phospholipid samples investigated, swelling at an elevated temperature (35°C–40°C) was required. For successful GUV formation, the lipids must be in their fluid states, that is, the swelling has to be done at $T > T_m$ of the lipids. The driving factors for vesicle growth seems to be a combination of osmotic pressure, electrostatic interactions and the hydrophobic effect (Tsumoto et al., 2009).

Many different versions of the original method exist nowadays and we will summarize some of their features. A schematic illustration of one spontaneous swelling method, also called the gentle hydration method, is shown in Figure 1.2, where the main stages of vesicle growth and a version of the formation protocol are indicated. Details of the individual steps are given in Box 1.1.

There are various approaches and modifications of the spontaneous swelling method aimed at enhancing it. The next two sections will focus on using polymer-based substrates (Section 1.2.2) and electric fields (Section 1.2.3) to speed up and improve GUV growth. Here, we will summarize a

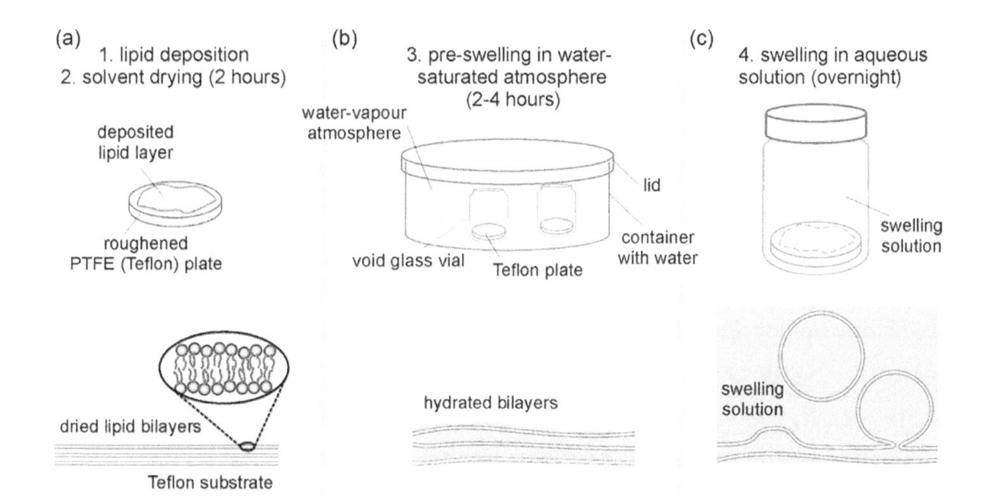

Figure 1.2 The sketch illustrates experimental steps during an exemplary spontaneous swelling protocol (top panel) with the corresponding stage of vesicle growth (lower panel). (a) Lipid film is spread onto a roughened PTFE (commonly referred to as Teflon) plate and dried from solvent. (b) The dried lipid film is then pre-swollen in a water-saturated atmosphere inside a closed container with water to facilitate the bilayer hydration and at a temperature above the main phase transition T_m of the lipids. (c) The pre-swollen lipid film finally becomes fully hydrated by the addition of the desired swelling solution onto the lipid-coated Teflon plate inside the glass vial. To avoid evaporation, the glass vial is securely sealed during overnight incubation. To minimize compositional variations within a batch, all steps need to be performed at a temperature where the lipid mixture is fully miscible. (Reproduced with permission from Kubsch, B et al., *J. Vis. Exp.*, 128, e56034, 2017.) An instructive video material of the preparation is available in the same reference.

Box 1.1 Protocol for GUV formation by the spontaneous swelling method

Example protocol for the formation of GUVs via the spontaneous swelling method as illustrated in Figure 1.2 and described by Kubsch et al. (2017). For the preparation, one needs a polytetrafluoroethylene (PTFE, commonly known as Teflon®) plate for lipid deposition of size ~1.5 × 1.5 cm and thickness of ~2 mm. The plate is roughened on one side with fine sandpaper. *TIP*: Teflon is lipophobic and will not be wetted by chloroform solutions if smooth. Both sides of the Teflon plate can be roughened to avoid confusion about the correct side for lipid deposition. Once roughened, the Teflon plate can be reused for new experiments after proper cleaning (Step 2).

Step 1. Prepare lipid stock at a total lipid concentration of 4 mM in chloroform or a chloroform:methanol mixture.
Step 2. Wash the Teflon plate and a glass vial (~15 mL in volume) with an aqueous solution of commercial dishwashing detergent, rinse thoroughly with water, ethanol, and chloroform, in that sequence, and then, finally, dry them.
Step 3. Using a glass syringe, deposit and evenly spread 10–15 μL of lipid stock onto the roughened side of the Teflon plate to create a uniform lipid film. Use the syringe needle to spread the solution if needed. *Note*: Plastic needles or pipette tips should not be used because polymeric material may be dissolved by the organic solvent and contaminate the sample.
Step 4. Place the Teflon plate with the deposited lipid film into the 15-mL glass vial and dry it in a desiccator for 2 h to remove the chloroform. This step, as well as the subsequent swelling, should be performed at a temperature at which the lipid is in the fluid state.
Step 5. After desiccation, take a sealable glass container (into which the 15-mL glass vial fits [Figure 1.2b]) and fill it with deionized water to a level that does not allow buoyancy to knock over the 15-mL glass vial (~1-cm water height suffices). Place the vial with the Teflon plate into the container. Cover the container so that a water-saturated atmosphere may emerge (next step) and place it in the oven at 60°C. *TIP*: Water condensation at the inner container walls is a good indication for successful water vapor saturation.
Step 6. Let the deposited lipid film pre-swell for 4 h. During this time, prepare the desired swelling solution. *TIP*: For vesicle preparations that can be conducted at a lower temperature (but still under conditions in which the lipids are in the fluid phase), this time can be extremely shortened—down to a few minutes if warm water-saturated nitrogen or argon is blown over the lipid film for the pre-swelling instead.
Step 7. Take the 15-mL glass vial out of the oven and, with a syringe connected to a 0.45-μm filter, add ~5 mL of the swelling solution into the vial to hydrate the lipid film. *TIP*: Pre-warming the syringe, filter, needle and solution is advisable to ensure fluidity and miscibility of the lipid film. Filtering the solution minimizes (in-)organic contamination with bacteria or salt precipitate, for example. Introduce the solution slowly, and afterward, avoid shaking the vial as much as possible.
Step 8. Seal the vial holding the hydrated lipid film on the Teflon plate to minimize evaporation and leave it undisturbed at 60°C overnight for final GUV swelling. The cluster of GUVs swollen and detached from the Teflon substrate results in a little (~1-mm) cloud-like clump visible to the naked eye (whitish when no fluorescence dye is used or colored otherwise). *TIP*: If proteins are present in the solution (i.e., for encapsulation in the GUVs), decrease the temperature for swelling.
Step 9. To harvest the vesicles, slowly cool the glass vial containing the GUV suspension to room temperature over a period of ~1 h (the cooling rate is particularly important if phase separation is expected because it alters the domain size [see also Knorr et al., 2018]). Cut ~1/10 of the pointy end of a piston pipette plastic tip for the cluster or aggregate of vesicles to fit through the orifice. Pipette up the cluster together with some volume of swelling solution and, if needed, resuspend in the desired isosmotic solution. The vesicle sample is then ready for observation.

few approaches that present slight modifications based on changes in the deposited lipid mixture, localized heating, sugars and some exotic inclusions:

1. *Charged lipids*: The spontaneous swelling method can be applied for the formation of GUVs in physiological solutions (Kubsch et al., 2017), but the yield is often low or the vesicles small (less than 10 μm in diameter). As demonstrated by Akashi et al. (1996), the growth under physiological conditions may be improved if the phospholipid mixtures contain 10–20 mol% charged lipids, such as phosphatidylserine or phosphatidylglycerol. The lipid charges provide electrostatic repulsion between bilayers and thus facilitate the GUV formation. However, higher fractions of charged lipids may lower the quality of the vesicles and result in defects (Rodriguez et al., 2005).
2. *Divalent ions*: The same group found that the growth of PC GUVs in the presence of divalent ions such as Ca^{2+} or Mg^{2+} in the concentration range of 1–30 mM also promotes vesicle swelling (Akashi et al., 1998).
3. *Sugars*: In the case of pure zwitterionic DOPC (no added charged lipids), the use of sugars, such as D-(–)-fructose, already present in the deposited lipid film, was shown to be beneficial for obtaining GUVs (with diameters of ≈10–40 μm) if prepared in the presence of high salinity buffers (Tsumoto et al., 2009). In this protocol, the lipid in chloroform solution is mixed with fructose dissolved in methanol, and the resulting solution is deposited and dried in a glass tube, followed by swelling in the physiological solution (Tsumoto et al., 2011). The presence of the sugars improves the swelling of the DOPC film, that is, the separation of the stacked DOPC bilayers, due to an increased interlamellar spacing in the dried films caused by an enhanced water movement toward the interlamellar space due to osmotic effects. Even though sugars are beneficial for GUV growth, one should be cautious about impurities, which have been shown to influence the phase state of the membrane and cause domains in quasi-single component vesicles (Knorr et al., 2018).

4. *Localized heating and pressure waves*: Spin-coated lipid films from charged and neutral lipids as well as from a complex lipid mixture have been shown to form good quality GUVs in both low and high ionic strength buffers when locally irradiated with an infrared laser beam guided through an optical fiber (Billerit et al., 2012). The local temperature was estimated to reach between 80°C and 90°C. Similar local heating of multilamellar vesicles was demonstrated by the same group to be able to produce single GUVs within a couple of minutes (Billerit et al., 2011); however, the produced vesicle remains attached to the multilamellar structure. Apart from heating, the use of pressure waves has also been shown to be useful for swelling vesicles in high-salinity buffers (Pavlič et al., 2011).
5. *Patterning*: The spontaneous swelling method (like all of the substrate-assisted methods) results in large variation in the vesicle size (typically in the range 1–150 μm). The use of surfaces with photolithographic templates as substrate has allowed some control over the vesicle size distribution (Howse et al., 2009), but this method has been applied only to polymer vesicles.
6. *Some exotic approaches*: The inclusion of photoactive compounds (with azobenzene groups) in the lipid deposited on the substrate has been shown to promote GUV formation within seconds after UV light irradiation (Shima et al., 2014). Here, as with localized heating and pressure waves, the GUVs remain attached to the starting cluster of lipids and photoactive compounds. To improve the efficiency of encapsulating macromolecules of interest in GUVs, polymeric crowding agents such as 10 wt% PEG2000 have been shown to increase the yield (Dominak and Keating, 2008). The swelling of GUVs on glass beads as a support has also been demonstrated to allow for encapsulating a minimal gene expression system (Nourian et al., 2012).

To conclude the section on spontaneous swelling of GUVs, we summarize some drawbacks of the method. First, it requires a *long time* for the preparation, on the order of a day or even longer. If the employed lipids are prone to oxidation, long times and elevated temperature may naturally results in lipid degradation (Morales-Penningston et al., 2010). Second, the *quality* of the vesicles (e.g., multilamellarity and the presence of defects such as tubes and smaller internal vesicles) is relatively poor and only a small fraction of the obtained vesicles are ideal (Rodriguez et al., 2005). Whether a vesicle is uni- or multilamellar can be roughly assessed with fluorescence microscopy (GUVs with two bilayers will be twice as bright as unilamellar ones) or by incorporating α-hemolysin (a pore-forming toxin) into the membrane and probing the exchange of a water-soluble dye with the vesicle exterior (e.g., Chiba et al., 2014) see Appendices 2 and 4 for information on fluorescent membrane markers and water-soluble dyes. Third, as aforementioned, a larger number of good quality GUVs (without defects) in physiological buffer can be achieved when the lipid mixture includes a small fraction of *charged lipids*. Fourth, the lipid composition of multicomponent vesicles may vary from vesicle to vesicle as suggested by Larsen et al. (2011), even though the later study involved the use of a relatively large fraction of fluorophores (a couple of mol%) and the preparation procedure included freeze–thawing cycles. It should be noted that the *variations in the membrane composition* across vesicles in the batch is probably pertinent to every method where lipids are spread on a solid substrate (because of the different affinities of the various lipids for the substrate [e.g., Steinkühler et al., 2018]) and may also result from uncontrolled budding off of a part of the GUV during solution handling.

1.2.2 GEL-ASSISTED SWELLING

This section will focus on approaches based on GUV growth similar to the spontaneous swelling method, except that the substrate is a polymer-based gel. The swelling in gel-assisted methods is enhanced by a buffer influx flow from below the bilayers deposited on a porous polymer layer. This trick leads to a significant speeding up of the swelling process, from days to practically minutes.

Gel-assisted GUV preparation was first achieved on agarose gels (Horger et al., 2009). The method relies on hydrating a film of lipids deposited on a preformed film of agarose and yields vesicles in a few minutes in a simple and straightforward way. It can be briefly summarized as follows: 1% (w/v) agarose solution is prepared in pure water above the polymer melting temperature. A small amount (e.g., ~200 μL) is spread on a clean cover glass placed on a heating plate at 80°C for ~15 min to ensure evaporation of excess water. Then, the lipid (e.g., 8–10 μL of a 3 mM lipid solution in chloroform) is spread over this agarose film and dried under vacuum. A chamber is assembled with a spacer around the deposit and filled with the swelling solution. Giant vesicles are observed to form within the first 20–30 min. Most of the obtained vesicles are unilamellar without any visible defects such as attached lipid debris or internal tubules.

This approach can be used with a wide range of lipid compositions and buffer conditions and has shown promising results in the efficient encapsulation of various biomolecules (Horger et al., 2009) as well as in producing GUVs with reconstituted proteins (Garten et al., 2015). It represents a very important step in the development of protocols for preparation of GUVs and has been the method of choice for producing GUVs in a number of studies (Tsai et al., 2011; Ikenouchi et al., 2012; Drucker et al., 2013; Hansen et al., 2013; Katayama et al., 2013; Lira et al., 2014; Saliba et al., 2014). However, agarose is left as a residual contamination in the formed vesicles, even though this has been reported to not change the molecular mobility of lipids in the bilayer (Horger et al., 2009; Lira et al., 2014). Agarose is also autofluorescent, which can bias fluorescence microscopy studies. Although GUVs could be obtained in high yields with this modified spontaneous swelling method, it was shown later on that the agarose molecules trapped inside during the vesicle formation lead to altered physicochemical properties of the GUVs (Lira et al., 2014). The same group also found that adding agarose to vesicles formed with other methods and subsequent jellifying can be effectively employed for GUV immobilization or encaging in the gel (Lira et al., 2016). Similarly, the immobilization can be achieved with agar gel (lysogeny broth agar) (Rampioni et al., 2018).

An improved version of the gel-assisted method replaced agarose with the chemically cross-linked polyvinyl alcohol (PVA) and circumvents polymer encapsulation because the lipids do not penetrate the PVA film, but, rather, they assemble into a bilayer stack on top of the matrix (Weinberger et al., 2013). Some traces of the polymer may still remain in the vesicles, resulting in changes in their membrane properties (Dao et al., 2017). The protocol is summarized in Box 1.2. GUV swelling has also been established on hydrogel forming polymer substrates made of chemically cross-linked polyacrylamide, but

Box 1.2 Protocol for GUV formation by PVA-assisted swelling
Example protocol for the formation of GUVs via the PVA-assisted swelling following (Weinberger et al., 2013; see also Stein et al., 2017).

Step 1. Prepare a 5% (w/w) PVA solution in water. The PVA should have a high molecular weight, typically above 150 kDa, such as the highest molecular weight of PVA available, for instance, from Sigma-Aldrich. Stir the PVA solution on a hot plate (or water bath) at around 90°C until the solution is clear. *TIP*: Supplementing the PVA solution with sugars (e.g., 50 mM sucrose) enhances the vesicle swelling but one has to consider that the sugar is also transferred to the swelling buffer (Step 5), thereby altering the final solution osmolarity.
Step 2. Clean (for better results, plasma clean) a 25 x 25 mm cover glass and deposit a layer of the PVA solution. The best results for easy harvesting of the GUVs after preparation are achieved by spin-coating the PVA solution, for example, spin-coat 150 μL of 5% PVA for 30 s at a speed of 1,200 rpm. To dry the PVA film, place the coated glass in an oven at 80°C for 30 min.
Step 3. Deposit 5–10 μL of 1 mg/mL lipid solution (in organic solvent) uniformly on the still-warm PVA film with a microsyringe. For well-separated GUVs and to facilitate the vesicle detachment at the end (Step 6), the lipid solution can be spin-coated and a small amount of lipids should be deposited on the PVA film (e.g., 5 μL for a 22-mm-diameter area).
Step 4. Dry the lipid-coated cover glass under vacuum for at least 2 h to evaporate the solvent from the dissolved lipid mixture.
Step 5. Form a chamber by, for example, gluing a ring onto the coated cover glass. Add the desired buffer solution (e.g., 1 mL). The vesicles form within seconds to several minutes. *TIP*: The swelling cannot be performed at elevated temperatures because PVA dissolves in the solution and becomes encapsulated in the vesicles. In practice, it works well for 1,2-dipalmitoyl-*sn*-glycero-3-phosphocholine (with 150 kDa PVA, one can work at up to 50°C without noticeable gel dissolution for up to half an hour), and it can also be used for 1,2-distearoyl-*sn*-glycero-3-phosphocholine if one makes the preparation in less than 15 min at 60°C (gel strong dissolution can usually be spotted visually, one should be aware of weak dissolution as temperature increases, contaminating the sample).
Step 6. To transfer the GUVs into an observation chamber, pipette out the vesicles with a cut tip or rinse the PVA substrate into the observation chamber. *TIP*: After swelling, the vesicles can be detached by gently pipetting up and down 200 μL of the buffer solution close above the glass surface. Gentle tipping against the bottom or the sidewalls of the chamber helps to detach the GUVs. Alternatively, fixing the swelling chamber to the bottom of a petri dish and letting it fall on the table upside-down from a height of 1 cm height helps as well.

this method was found to produce lower yields of GUV compared with swelling on agarose films (Horger et al., 2009). More recently, chemically cross-linked dextran–poly(ethylene glycol) was used as hydrogel substrate for GUV preparation under physiological ionic strength conditions (Mora et al., 2014). Dextran polymers (MW = 70 kDa) were cross-linked by poly(ethylene glycol) chains using Michael addition to simultaneously form the hydrogel and anchor it to a glass surface (Figure 1.3 provides more details on the protocol). Presumably, the anchored covalent hydrogel cannot be dissolved during the GUV formation, also at higher swelling temperatures, to potentially contaminate the lipid bilayer, as may be the concern with non-covalently crosslinked hydrogels. Modulating the degree of cross-linking within the hydrogel network results in tuning of the vesicle size distribution. It remains to be confirmed that the mechanical and thermodynamic properties of vesicles prepared following this protocol are not altered by the swelling procedure. Even though this method is presumably most efficient among the gel-assisted swelling methods for growing GUVs at higher temperatures and preventing the polymer from detaching from the gel, it is preparation-wise more demanding. And as this book goes to print, quite monodisperse membrane structures akin to giant vesicles were shown to form on micropatterned gel substrates, the pattern size of gels from cross-linked poly(*N*- isopropylacrylamide) controlling the size of the anchored giant vesicles (Schultze et al., 2019).

1.2.3 THE ELECTROFORMATION METHOD

Probably the most widely used approach for GUV preparation is the electroformation method (also called "electroswelling") developed by Angelova and Dimitrov (1986). It is very much related to the above swelling methods. In all cases, GUVs are obtained through a controlled hydration of bilayer-forming lipids that have been deposited as a thin film on a solid surface. In the case of the electroformation method, the hydration of the lipid film is influenced by an externally applied alternating current (AC) and, therefore, the lipids must be deposited on an electrically conductive surface. This conductive surface is usually either an indium tin oxide (ITO) coated glass that is separated from a second ITO-coated glass by a thin nonconductive spacer of about 0.3–1 mm thickness, or a 0.5- to 1-mm-thick platinum (Pt) wire that is separated from another Pt wire by a distance of about 3–5 mm (Angelova and Dimitrov, 1988; Angelova et al., 1992; Bucher et al., 1998; Angelova, 2000; Dimova et al., 2006; Méléard et al., 2009; Schmid et al., 2015). Titanium or stainless steel wires have also been used as electrodes (Ayuyan and Cohen, 2006; Pereno et al., 2017) as well as silicon substrates (Le Berre et al., 2008). One advantage of using ITO-coated glasses as electrodes is that the swelling area is larger and, thus, the yield is higher. For egg PC, electroformation was also shown to be feasible with lipid deposition on nonconductive substrates immersed between electrodes (Okumura et al., 2007). With time and use, ITO-coated plates may deteriorate, affecting the size of vesicles, but they can be also annealed to regain quality (for an annealing procedure, see Herold et al. 2012). Figure 1.4 shows images of homebuilt chambers based on ITO-coated glass electrodes and Pt electrodes. A simple chamber can be also improvised by tucking Pt wires down in a small petri dish or poking them through the lid of a microcentrifuge tube. Note, however, that long swelling times in plastic tubes should be avoided because they are known to release impurities into the samples (McDonald et al., 2008). A commercially available

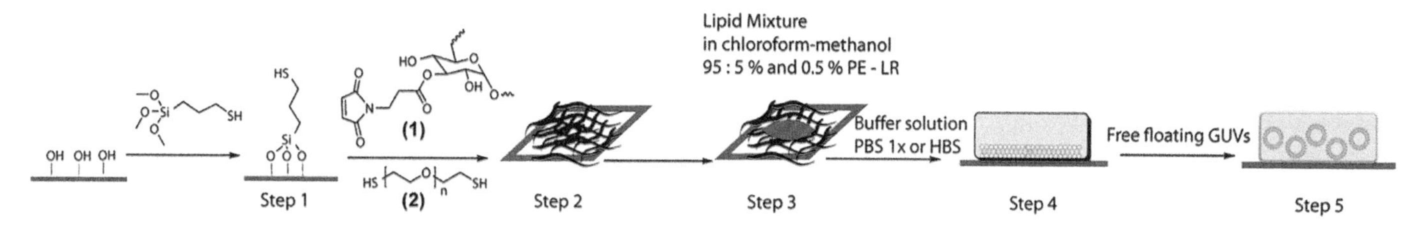

Figure 1.3 Formation of giant vesicles from dextran(ethylene glycol) hydrogels covalently cross-linked and anchored to the substrate. (Step 1) A reactive thiol moiety is introduced onto the glass slide surface with 3-mercaptopropyl trimethoxysilane. (Step 2) Thiol-coated microscope slides are cross-linked to the hydrogel by drop-casting 2 wt% dextran solution (1) and PEG dithiol (2) at various molar ratios at 40°C until a homogeneous hydrogel film is formed. (Step 3) The lipid mixture is deposited onto the hydrogel surface and the solvent evaporated in a vacuum oven for 30 min at 35°C or at room temperature under a gentle stream of nitrogen gas to prevent lipid oxidation. (Step 4) The hydrogel and lipid film on the glass slide are rehydrated in aqueous buffer solutions for 1–2 h to (Step 5) form free-floating vesicles. (Mora, N.L. et al., *Chem. Commun.*, 50, 1953–1955, 2014. Reproduced by permission of The Royal Society of Chemistry.)

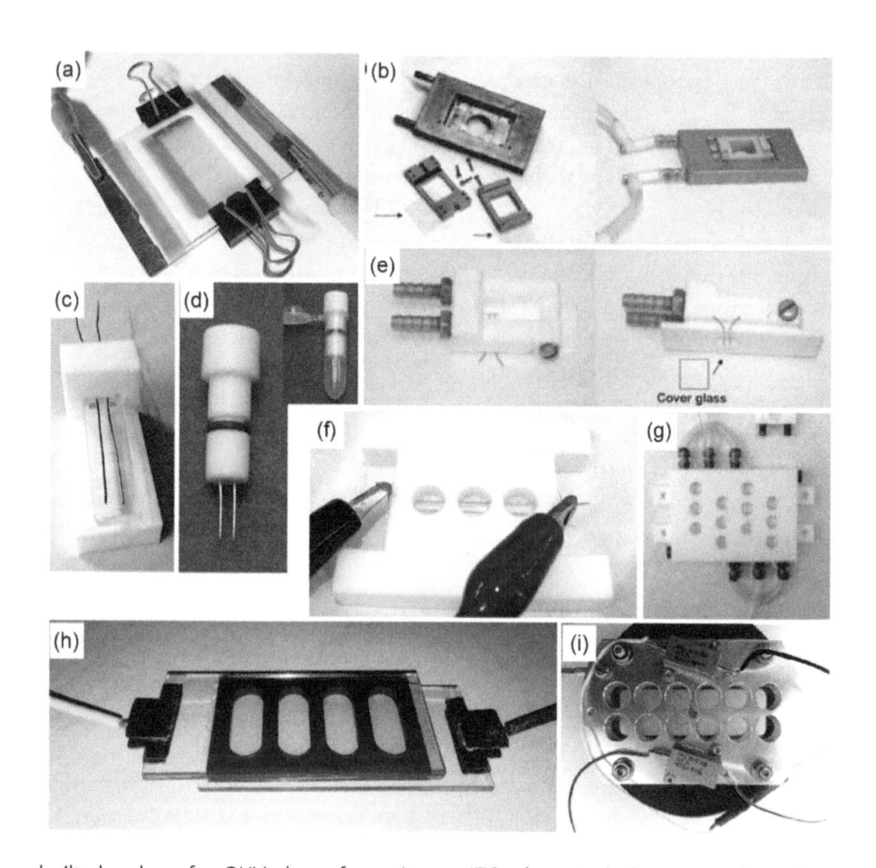

Figure 1.4 Examples of homebuilt chambers for GUV electroformation on ITO plates (a, b, h, i) or Pt electrodes (c–g). (a) The chamber is made out of two ITO-coated glasses and a Teflon frame (as a spacer) clamped together with office clamps (in this way, the use of silicone grease for sealing is avoided as well as the danger of introducing impurities). The ITO glasses are connected to an AC field generator using conductive copper tape and alligator clips. This setup was used, for example, in Georgiev et al. (2018). (Courtesy of V. Georgiev.) (b) A chamber to generate GUVs using ITO electrodes on plastic supports (indicated with arrows). The distance between electrode surfaces is 3 mm. The metal piece shown in the figure has a water circulation path to maintain the sample at the desired temperature. (c) A chamber formed by a cuvette with inserted Pt wires (diameter 1 mm, approximate spacing of 3–5 mm) and held by a Teflon frame; employed in Bouvrais et al. (2012). (Courtesy by T. Bhatia.) (d) Chamber with Pt wires where the cuvette is replaced by an Eppendorf Tube® (see inset); employed by Weinberger et al. (2017). (Courtesy of Vivien Walter and André Schroder.) (e) A chamber with Pt wires 3 mm apart. The chamber bottom is sealed by a standard microscope slide (glued, e.g., with epoxy J-B KWIK, J-B Weld, Sulphur Springs, TX). (f) A chamber milled out of a Teflon block with three wells (10 mm diameter, 5 mm depth) with Pt wires (0.5 mm in diameter, 3 mm apart) closed by bottom and top coverslips held in place with vacuum grease. (Reproduced with permission from Garten, M. et al., *J. Vis. Exp.*, 95, e52281, 2015). (g) An array of 12 individual chambers (each with volume of 600 mL) with circulating water for temperature control. The Pt wires diameter is 1 mm. (Images in (b, e, g) Reprinted from *Biophys J*, 90, Fidorra, M. et al., Absence of fluid-ordered/fluid-disordered phase coexistence in ceramide/POPC mixtures containing cholesterol, 4437–4451, Copyright 2006, with permission from Elsevier.) (h, i) Arrays of chambers produced by means of a spacer with holes (red in panel h) sandwiched between two ITO glasses. In (i), the chamber (top view) is attached to a heating device consisting of an aluminum plate with holes allowing for transmission microscopy observation and connected to resistors (yellow) and a temperature sensor; the setup was used by Sturzenegger et al. (2016). (Image in (i) courtesy of T. Robinson.)

electroformation setup is also available (and has already been used in several studies [e.g., O'Connor et al., 2018]) as well as a smartphone app for GUV production (Almendro Vedia et al., 2017). Before purchasing a commercial product for a particular project, it is worth considering the large possibility of preparing homebuilt setups (Figure 1.4) and the demand of increasing the membrane complexity as well as the variety of encapsulating buffers.

Under optimal conditions, homogeneous GUVs can be obtained with the electroformation method at room temperature—in pure water or in aqueous solutions of low ionic strength—from a number of different phospholipids with low T_m values or from many different lipid mixtures and native membranes (Méléard et al., 2009; Walde et al., 2010; Mikelj et al., 2013). The temperature at which the electroformation is performed must be above the T_m value of the lipids, that is, the lipids should be in their fluid state (Angelova, 2000; Shimanouchi et al., 2009; Baykal-Caglar et al., 2012). A typical protocol for obtaining GUVs with the electroformation method from DOPC in pure water is given in Box 1.3. Video material on the method and application of electroformation for the preparation of GUVs with reconstituted proteins is available (Garten et al., 2015; Schmid et al., 2015).

There are many variations of the procedure given in Box 1.3. They were elaborated, for example, for preparing GUVs in physiologically relevant aqueous solutions (10 mM Tris-HCl, pH 7.4, containing 250 mM NaCl) (see Pott et al., 2008; Méléard et al., 2009). Depending on the compositions of interest in terms of lipids and aqueous solution, optimal conditions must be found. One point concerns the thickness and the way the thin lipid film is initially formed on a conductive surface, that is, whether it is obtained by using a lipid solution made with an organic solvent (or a mixture of organic solvents) and by hand-spreading a drop of this solution on the surface (Box 1.3), by "spin-coating" of this solution (Estes and Mayer, 2005), or by using an aqueous suspension of large unilamellar vesicles (LUVs) of the lipids instead of lipids dissolved in an organic solvent (Pott et al., 2008), see also Chapter 3. In the case of spin-coating the lipid, a homogeneous coverage of the ITO-coated glass with a lipid film thickness of 25–50 nm turned out to give, in pure water, high yields of GUVs with diameters above 30 μm from egg yolk PC, POPC/1-palmitoyl-2-oleoyl-*sn*-glycero-3-phospho-(1′-*rac*-glycerol) sodium salt (POPG) (9:1, mol/mol), or pure POPG (Estes and Mayer, 2005), see Appendix 1 of the book for structure and data on these lipids. A variation of the AC electric field strength may have only a small (but significant) influence on the average size and size distribution of electroformed vesicles, as demonstrated in the case of GUVs formed from egg yolk PC in deionized water (Politano et al., 2010). Furthermore, the size of GUVs obtained by electroformation can be controlled to some extent by using a specifically modified or patterned conductive surface (Taylor et al., 2003a, 2003b; Le Berre et al., 2008).

One concern about using the electroformation method is that oxidized products may be formed in significant amounts during the formation of the GUVs if the lipids contain poly-unsaturated chains, for example, 1,2-dilinoleoyl-*sn*-glycero-3-phosphocholine (Breton et al., 2015). In this case, the physicochemical properties of the membranes of the GUVs obtained by electroformation are expected to be different from those of GUVs obtained from the same lipids with the spontaneous swelling method. For lipids with mono-unsaturated chains, like DOPC, the extent of oxidation is low if the applied voltage and the duration of

Box 1.3 Protocol for GUV formation by the electroformation method

Example protocol for the preparation of GUVs via the electroformation method by using either a chamber with two parallel Pt wires or ITO-coated glass plates and a spacer. (Example chambers are shown in Figure 1.4). Video material for similar protocols is available (Garten et al., 2015; Schmid et al., 2015).

Step 1. Clean the Pt electrodes and chamber or ITO plates and spacer by washing them with Millipore® water/ethanol/chloroform/ethanol/Millipore water. Polish electrodes with a clean tissue and wash them again with Millipore water. Blow-dry the spacer or chamber and electrodes, for example, with the help of high-pressure nitrogen gas.

Step 2. Prepare lipid stock at a total lipid concentration of 2 mM in chloroform or chloroform:methanol (e.g., 9:1, v/v) mixture. Using a glass syringe, deposit and evenly spread 2 μL of lipid stock onto the ITO plates or deposit small (<0.2 μL) droplets along the Pt wires (1 μL of solution is needed to form a series of drops along 1 cm of wire).

Step 3. Place the electrodes in a desiccator and dry for 2 h to remove the organic solvent. Afterward, assemble the chamber (example chambers are shown in Figure 1.4). One can use silicone (vacuum) grease and sealing paste. Alternatively, ITO-plate chambers can be assembled and held together by simple office clamps (Figure 1.4a), which avoids the use of grease (which may contaminate the sample).

Step 4. Slowly add the swelling solution until the chamber is filled. Avoid letting gas bubbles in. Avoid any rapid movement because this can strip the lipid film off the electrodes.

Step 5. Connect electrodes to an AC field generator. Set the frequency to 10 Hz/500 Hz sine wave for low-/high-salinity buffer, respectively, and use a multimeter to measure and adjust the voltage across the wires to 0.7 V/0.35 V root mean square (Vrms). For low-salinity buffers, approximately 2 h are sufficient; for high-salinity buffers, swelling takes overnight. Ramping voltage and frequency was reported to aid GUV formation for low-salinity buffers (Méléard et al., 2009). For example, one can start with 0.5 V, 10 Hz, increase the voltage each 30 min by 0.5 V until reaching 2.5 or 3 V, and leave the field on for another hour.

Step 6. The final step aimed at detaching the vesicles from the electrode often involves decreasing the voltage slowly back to 0.5 V and changing the frequency to 5 Hz for 10–20 min.

electric field application are kept at the typical values used for the formation of GUVs (Breton et al., 2015), see Box 1.3. For lipid mixtures containing charged lipids, the electroformation method was also shown to produce vesicles with an asymmetric distribution of the charged lipids across the bilayer unless the protocol was optimized (Steinkühler et al., 2018). Further tips are available in Section 1.5.

1.3 METHODS BASED ON ASSEMBLY FROM FLUID INTERFACES

Apart from preparing giant vesicles with the methods described in Section 1.2, there are a number of other procedures with which GUVs can be obtained (Walde et al., 2010; Matosevic, 2012; van Swaay and deMello, 2013; Patil and Jadhav, 2014; Stein et al., 2017). Some of them will be summarized in this section with particular emphasis on the key concept behind the method.

1.3.1 THE DROPLET-TRANSFER METHOD

The droplet-transfer method (some variations of which are also known as the method of the inverted emulsion, or the emulsion-transfer method) is based on the stepwise construction of vesicles by, first, preparing water droplets (suspended in a nonpolar solvent) surrounded by *one* (mono)layer of lipids, and, then, "enwrapping" the droplets by a *second* lipid layer to obtain the vesicle architecture (Figure 1.5). After the pioneering work in the 1970s (Träuble and Grell, 1971), commented on by Szoka and Papahadjopoulos (1980), the droplet-transfer method became popular with the reports by Weitz and collaborators in 2003 (Pautot et al., 2003b) and by Noireaux and Libchaber (2004). The procedure allows for control over the vesicle size because the size of water in oil (w/o) droplets is easily manipulated. The method is also rather fast; the preparation takes about 30 min.

In this section, we will refer to the aqueous phase that is used to prepare the droplet as the I-solution (inner solution), whereas the aqueous phase that will finally surround the GUVs will be referred to as the O-solution (outer solution) (Figure 1.5). The nonpolar organic solvent(s) used in the preparation will be simply referred to as "oil." The method uses lipids that can be solubilized in such oil. The lipid-containing oil is used in two instances: first, for preparing the w/o droplet or emulsion, where a certain volume of the I-solution is introduced in a certain volume of lipid-containing oil; second, for preparing the oil/water interface, here called the "intermediary phase," a certain volume of lipid-containing oil (equal to or different from the emulsion with respect to the lipid composition and oil nature) is overlaid onto a certain volume of the O-solution. As oil, one can use dodecane (Pautot et al., 2003a), mineral oil (Yamada et al., 2006, 2007; Hase et al., 2007), squalene (Pautot et al., 2003a), liquid paraffin (Fujii et al., 2014). Mixtures of solvents can be also used (Hu et al., 2011; Elani et al., 2013; Hamada et al., 2014). Because in many studies mixtures of hydrocarbons have been used (mineral oil or liquid paraffin), for comparative purposes it is important to verify the chemical composition of such oils, which are actually mixtures of linear, branched, and cyclic hydrocarbons. The droplet transfer can proceed spontaneously

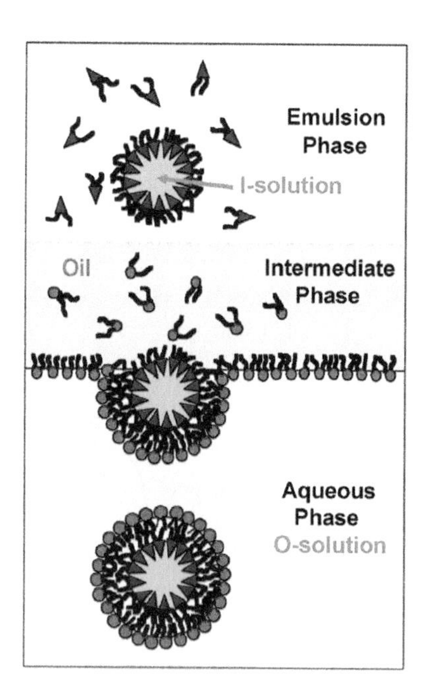

Figure 1.5 Schematic illustration of one variation of the droplet-transfer method as described by Pautot et al. (2003b); image not drawn to scale. An aqueous solution (I-solution, or inner solution) is emulsified in oil with lipid as the surfactant, forming a stable w/o emulsion. The aqueous solution that will receive the final vesicles (O-solution, or outer solution) is placed in a second vial, and lipid-saturated oil is poured on top of it. This second oil/water system is called here the "intermediate phase." A lipid monolayer forms at the oil/water interface. The water droplets are heavier than the oil that surrounds them and sediment into the second aqueous phase. It is common to include sucrose and glucose, respectively, in the inner and in the outer aqueous solutions so that the two solutions have different densities, yet are isotonic. As the w/o droplets pass through the interface where the second layer of lipid is sitting, the bilayer is completed and the final vesicles are formed. Note also that after passage, the oil/water interface is locally deprived of the lipid monolayer, which should re-form by lipid diffusion. (Reprinted with permission from Pautot, S. et al., 2003b. Engineering asymmetric vesicles. *Proc. Natl. Acad. Sci. USA*, 100, 10718. Copyright 2003b National Academy of Sciences, U.S.A.)

(under gravitational field) or be facilitated by applying a centrifugal force or hydrodynamic flow guided in a microfluidic device (Figure 1.6a–c).

Emulsion-based method

This method shares the starting point of using a w/o emulsion for forming LUVs with the reverse phase evaporation method (Szoka and Papahadjopoulos, 1978). The method consists of two major steps in preparing an emulsion and pouring it over the intermediate phase. Note that, depending on the stability of the emulsion, one might first need to prepare the vial containing the intermediate phase stratified above the O-solution (Figure 1.5). For lipid concentrations (in the intermediate phase) of ~200 μM, one typically waits 15–30 min for the lipid monolayer to form; during this time a preformed w/o emulsion might be destabilized. An example protocol is given in Box 1.4.

First step: Obtaining a good water-in-oil emulsion

A w/o emulsion is prepared by dispersion of a small amount of the aqueous I-solution in a lipid-containing oil (shaking, pipetting, vortexing, sonication, and so on can be used here).

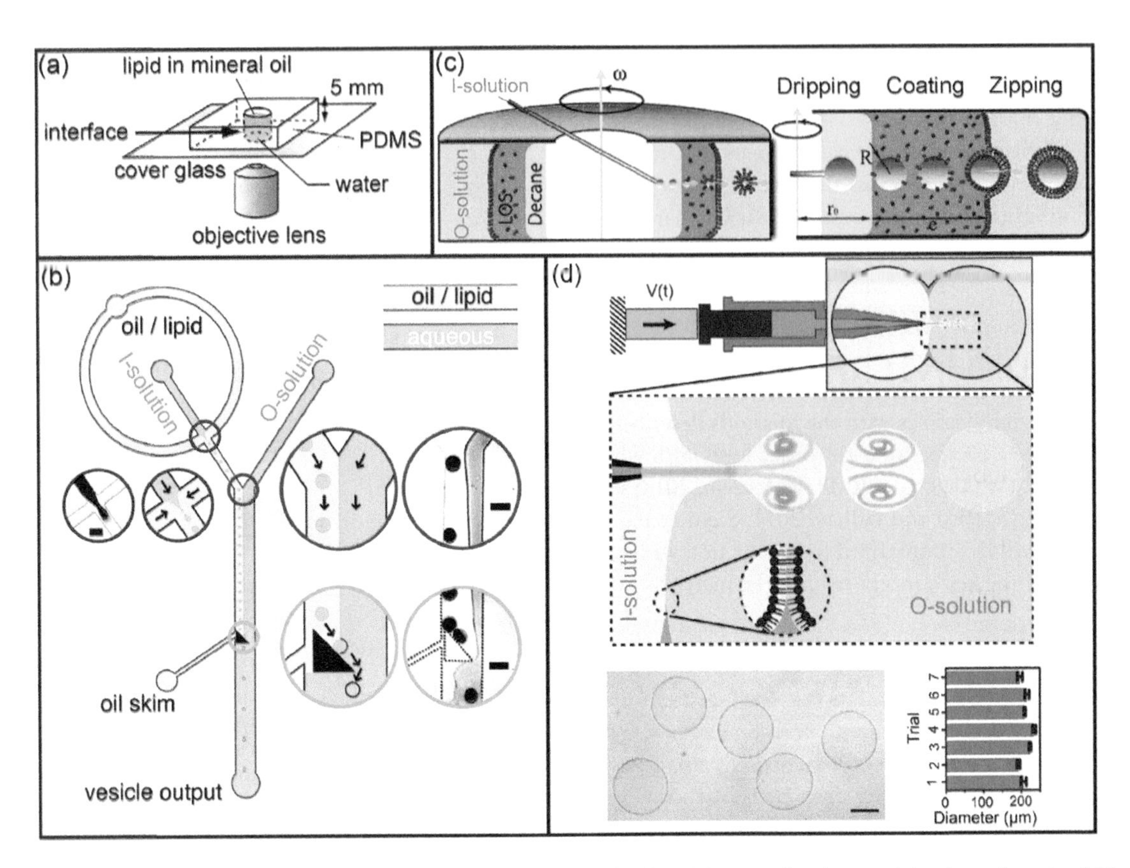

Figure 1.6 Schematic illustration of setups for different droplet-transfer methods. (a) A setup for the emulsion-based method. The observation chamber is made of poly(dimethylsiloxane) (PDMS) on a microscope coverslip. Water-in-oil microdroplets, initially in the oil phase, come through the oil/water interface to form GUVs in the lower aqueous phase and are observed from below. (Reprinted with permission from Yamada, A. et al., *Langmuir*, 22, 9824–9828, 2006. Copyright 2006 American Chemical Society.) (b) Circuit schematic and operation of a microfluidic-based GUV approach. The oil/lipid input is introduced at the top left, focusing the I-solution to generate uniform, lipid-stabilized droplets (purple insets). The droplet flow merges with an O-solution to form a lipid-stabilized oil/water interface adjacent to the droplet flow (red insets). Droplets impinge on a triangular post in the center of the channel, where the oil flow is skimmed while droplets are deflected along the hypotenuse of the post and traverse the interface, completing the lipid bilayer to form a unilamellar vesicle (green insets). Arrows indicate the flow field. (Micrograph scale bar: 100 μm). (Reproduced with permission from Matosevic, S. and Paegel, B.M., *J. Am. Chem. Soc.*, 133, 2798–2800, 2011.) (c) Schematic side view of the setup for the cDICE method. The capillary is fixed and the chamber is rotating at speed ω. The I-solution droplets drip off the capillary in a decane layer. The centrifugal force drives their flight through the lipid-in-oil solution (LOS) layer. The droplet surfaces become saturated with lipids and are then dispersed in the O-solution. (Abkarian, M. et al., *Soft Matter*, 7, 4610–4614, 2011. Reproduced by permission of The Royal Society of Chemistry.) (d) Formation of GUVs by microfluidic jetting. Schematic of the piezoelectric-driven microfluidic jetting device assembled with the planar lipid bilayer chamber (top). Close-up schematic of the vesicle-formation process, highlighting the interaction of a vortex ring structure with the planar lipid bilayer (middle). Separate, monodisperse, GUVs resulting from pulsed microfluidic jetting (scale bar: 100 μm); vesicle diameter over seven separate trials (bottom). (Reprinted with permission from Stachowiak, J.C. et al., 2008, Unilamellar vesicle formation and encapsulation by microfluidic jetting. *Proc. Natl. Acad. Sci. USA*, 105, 4697–4702. Copyright 2008 National Academy of Sciences, U.S.A.)

The I-solution can be a high-salinity buffer, but it is important that it be isotonic with the O-solution. To facilitate the detachment of the vesicles from the oil/water interface, Hamada et al. introduced the sugar gradient strategy (sucrose in the I-solution, glucose in the O-solution), which results in a density difference, allowing rapid sedimentation of the vesicles (Hamada et al., 2008). This is now considered a standard practice. In general, it is not necessary that the emulsion be stable for a long time because, once prepared, the emulsion is generally used in a short time (i.e., within several minutes). Note that lipids that do not allow emulsification or lead to unstable emulsions (unstable even in the first few minutes), will obstruct the procedure. Typically, PC lipids are used.

Second step: Transfer of w/o emulsion droplets to the external aqueous phase (O-solution)

The w/o emulsion is gently poured over the "intermediate phase"—another lipid-containing oil phase—stratified over an aqueous solution (the O-solution, isotonic to the I-solution) (Figures 1.5 and 1.6a). The lipids in this phase might be of the same type as the lipids used in the w/o emulsion, or different (in the latter case, asymmetric vesicles will be obtained). Also, the intermediate lipid oil solution can be the same (in most cases) or different (Pautot et al., 2003a, 2003b; Hu et al., 2011) from the oil used for preparing the w/o emulsion. The latter element is important for constructing asymmetric vesicles because it aids in preventing mixing of the lipids when the w/o emulsion is poured

Box 1.4 Protocol for GUV formation via the emulsion-transfer method

Example protocol for the formation of GUVs via the emulsion-transfer method following the procedure described by Fujii et al. (2014). A similar protocol is described by Rampioni et al. (2018), including scale-up parameters. Additional comments can be found in Stano (2019)". Useful video material is available (Natsume et al., 2017).

Step 1. Add 500 μL of liquid paraffin (e.g., Wako, Cat. No. 128-04375) to 100 μL of 25–100 mg/mL lipid solution in chloroform (this lipid will constitute the inner leaflet of the final GUVs) and vortex vigorously. Then, incubate the lipid–paraffin solution at 80°C for 30 min to evaporate the chloroform and let the solution cool to room temperature, see below panel (a) in the figure. Afterward, the lipid–paraffin solution can be stored at room temperature for about 1 week. *Note*: If the lipid–paraffin solution is white and turbid, water is present in the solution. Moisture should be eliminated from the lipid samples; alternatively, one could work in a nitrogen atmosphere.

Step 2. Transfer 250 μL of the lipid–paraffin solution to a single round-bottomed glass tube and add 25 μL of the I-solution (which includes 200 mM sucrose). Mix them by tapping the bottom of the glass tube. Vortex for 30 s to generate a w/o emulsion, see panel (b) *Note*: If the emulsion is not white and turbid, the I-solution has not been sufficiently emulsified. Tilt the glass tube horizontally and tap again. Extend the vortexing to 1 min. Finally, leave the emulsion for 10 min to equilibrate with the lipid monolayer assembly on the surface of a layer of droplets with lipids.

Step 3. Pour 150 μL of the O-solution (which includes 200 mM glucose) into a 1.5-mL test tube, and layer the emulsion mixture on top, see panel (c) *Note*: In this protocol, the intermediate phase is not used because the w/o emulsion is poured directly onto the O-solution. Alternatively, first overlay another portion of the lipid-containing oil solution (e.g., 150 μL) prepared in Step 1, on the O-solution, wait from 15 min to 3 h, then pour the w/o emusion over the lipid-containing oil solution stratified over the O-solution. For asymmetric vesicles, the two lipid-in-oil solutions, namely the one for preparing the intermediate phase and the one for preparing the w/o emulsion, must contain different lipids; details can be found in the report by Pautot et al. (2003b).

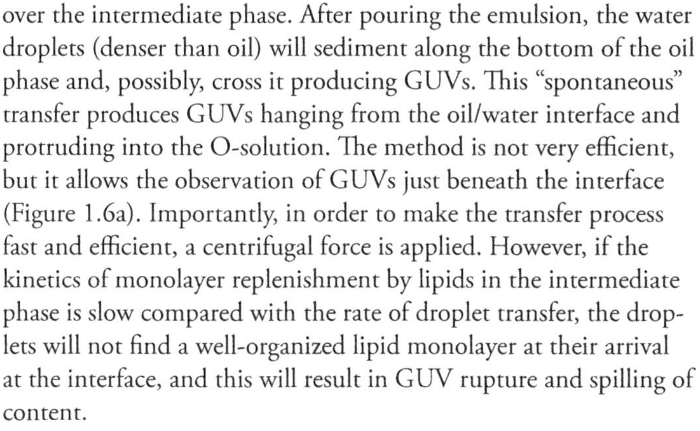

Step 4. Centrifuge the tube at 9,000g for 30 min, see panel (d) To collect the liposomes, first, open the test tube lid and carefully insert, from outside, a 21-gauge needle into the tube at the position where the liposome pellet is located, see panel (e) Next, remove the needle, close the lid, and collect the liposome suspension dripping from the hole into another test tube, see panel (f) *Note*: If the lid is pushed too tightly, the paraffin solution may flow out and contaminate the liposome suspension.

Step 5. Centrifuge the collected liposome suspension at 6,000g for 10 min. Discard the supernatant (which contains the emulsion droplets that failed to develop a lipid bilayer) and suspend the liposome pellet in 100 μL of the fresh O-solution. *Note*: If the liposome suspension is white and turbid, lipid particles contaminate the sample. Repeat this step and/or reduce the lipid concentration used in Step 1.

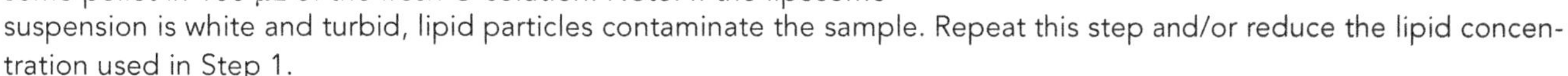

The figure is reprinted by permission from Springer Customer Service Centre GmbH: Springer Nature, Nature protocols (Fujii et al., 2014), Copyright 2014.

over the intermediate phase. After pouring the emulsion, the water droplets (denser than oil) will sediment along the bottom of the oil phase and, possibly, cross it producing GUVs. This "spontaneous" transfer produces GUVs hanging from the oil/water interface and protruding into the O-solution. The method is not very efficient, but it allows the observation of GUVs just beneath the interface (Figure 1.6a). Importantly, in order to make the transfer process fast and efficient, a centrifugal force is applied. However, if the kinetics of monolayer replenishment by lipids in the intermediate phase is slow compared with the rate of droplet transfer, the droplets will not find a well-organized lipid monolayer at their arrival at the interface, and this will result in GUV rupture and spilling of content.

An alternative approach to the emulsion-transfer method, in which an additional step in the preparation requires freezing of the aqueous phase (but not the oil phase), was reported and called the "lipid-coated ice-droplet hydration method" (Sugiura et al., 2008; Kuroiwa et al., 2012). This method requires first the preparation of well-defined, narrowly distributed, micrometer-sized, aqueous droplets stabilized by amphiphilic lipids and surfactant in hexane. This can be achieved by micro-channel emulsification. Subsequently, these w/o droplets are transformed into giant vesicles by first freezing the droplets at liquid nitrogen temperature (at which hexane remains liquid), followed by the removal of the hexane and coating of the frozen droplets with phospholipids (i.e., LUVs) to yield an aqueous suspension of giant vesicles. The vesicle sizes are, in general, smaller or around 10 μm in radius and the quality (e.g., unilamellarity) is not as good as that of vesicles produced with other methods, but the approach allows encapsulation of material. Similar to the water droplet-transfer method, the presence of remaining oil and, in this approach, also surfactant in the membrane of the giant vesicles obtained is an issue to consider.

Microfluidics-assisted approaches

To improve the monodispersity of the GUVs, various types of microfluidic devices utilizing the droplet-transfer method have been developed. In some examples, the w/o emulsions are generated in a microfluidic T-junction device, followed by transfer of the droplets through a second lipid monolayer via centrifugation in a tube (Hu et al., 2011). Matosevic and Paegel (2011) managed to integrate the full production process in one microfluidic device (Figure 1.6b). In this device, the emulsion droplets are formed using flow-focusing technology, and the phase transfer (from the oil phase to the aqueous phase) is stimulated using a triangular post in the microchannel. However, the amount of recovered vesicles is very low (5%) because the majority of the vesicles burst at the triangular post. A more practical and less complex version of this approach (avoiding the need of flow balancing in a complex chip as that shown in Figure 1.6b) relied on splitting the droplet production and vesicle generation steps into two microfluidic chips (Hu et al., 2011; see also Tan et al., 2006). A combination of a flow-focusing w/o droplet generator and a deeper channel to transfer the w/o droplets across the oil–water phase boundary allowed the formation of monodispersed GUVs with an asymmetric membrane (Karamdad et al., 2015). The rigidity of these bilayers was investigated, and the results suggested that the asymmetry affects the membrane mechanics (Karamdad et al., 2016). We will not discuss the microfluidics-assisted approaches further, but we will refer the reader to Chapter 30 of this book where one approach for preparing GUVs using a microfluidic system, as well as the preparation of microfluidic chips, are described in detail.

Capillary-generated droplets

This method, also called continuous droplet interface crossing encapsulation (cDICE) was developed by Abkarian et al. and involves the generation of droplets via a capillary and passing them through the oil–water interface by centrifugal forces of 4–100*g* (Abkarian et al., 2011) (Figure 1.6c). The method profits from a rotatory device that was designed in order to combine droplet formation and droplet transfer. Droplets of the I-solution drip continuously from a capillary into the lipid-containing oil phase (decane), radially overlaying the aqueous phase (O-solution). The droplets, whose frequency of appearance (typically 250 Hz) and size can be regulated by the rotational rate and by the capillary size, travel radially toward the oil/water interface and are eventually found in the O-solution. GUVs with narrow size distributions are obtained, with relatively high yield compared with microfluidics-based vesicle formation. Using this approach, 1-μm polystyrene beads, red blood cells, or actin filament bundles have been encapsulated into giant liposomes (Deek et al., 2018). The dynamics of lipid microdomains on GUVs formed with this method have also been studied (Blosser et al., 2016). However, the application of the method, as well as all other droplet-transfer methods, to multicomponent mixtures remains questionable because lipids might have different solubility in the organic phase as well as different partition coefficients between the interface and the organic phase (Blosser et al., 2016).

Examples of applications of the water-droplet transfer method

The water-droplet transfer method has been used successfully to prepare asymmetric vesicles (Xiao et al., 1998a, 1998b; Pautot et al., 2003a, 2003b; Hamada et al., 2008; Whittenton et al., 2008; Greco et al., 2012; Tahara et al., 2012; Elani et al., 2015) and for entrapping specific solutes with high efficiency. The latter is often used in research on the reconstitution of cellular functions—a field that is related to studies on the origins of life (in the construction of "protocell" models, [see Walde et al., 1994]), but that is currently converging with bottom-up synthetic biology (Stano et al., 2011, 2012).

The droplet-transfer method has been successfully used for the preparation of GUVs encapsulating enzymes (Carrara et al., 2012; Peters et al., 2015), cytoskeletal elements such as actin (Takiguchi et al., 2008, 2011; Pontani et al., 2009; Abkarian et al., 2011; Murrell et al., 2011; Campillo et al., 2013; Guevorkian et al., 2015), FtsZ/ZipA proteins (Cabré et al., 2013), dendronized polymer-enzyme conjugates (Grotzky et al., 2013), the potassium channel membrane protein (Yanagisawa et al., 2011) and the photosynthetic reaction center (Altamura et al., 2017), in correct orientation, paramagnetic nanoparticles (Toyota et al., 2012), microbeads (Abkarian et al., 2011; Natsume and Toyota, 2013; Saito et al., 2014) and whole cells (Abkarian et al., 2011). Additional studies have been reported on osmotic stress (Ohno et al., 2009; Fujiwara and Yanagisawa, 2014); on reconstituting (from outside) connexin 43 (Liu et al., 2013); on evaluating the Hofmeister effects (Hadorn et al., 2011); on creating hierarchical vesicles-inside-vesicles systems (Hadorn et al., 2013); and even on forming millimeter-sized vesicles (Kubatta and Rehage, 2009), but with limited stability.

1.3.2 GUV BLOWING

As we have seen in the first section, GUV formation started with spontaneous swelling. It is thus interesting that several decades later different versions of these films were used to prepare giant vesicles. Indeed, some groups (Funakoshi et al., 2007; Stachowiak et al., 2008) used thin films to prepare giant vesicles by blowing a focused solution jet on the planar bilayer, thus creating GUVs in a manner similar to the blowing of soap bubbles (Figure 1.6d). Although concerns about residual organic phase in the vesicle bilayers still exist (Kirchner et al., 2012), this family of techniques known as pulsed jetting has the great advantage of being very effective for encapsulation purposes. The original approach of Funakoshi et al. can be reproduced with modest laboratory equipment, but the positioning of the blowing nozzle is difficult and, because it is done by hand, not very reproducible. The more sophisticated and reliable method introduced by Stachowiak et al. requires the adaptation of an inkjet system and, thus, bears some intrinsic degree of technicity. A video tutorial for reproducing this method is available online (Coyne et al., 2014).

Although not a pulsed jetting technique, a method developed by Karlsson et al. (2002) to create networks of giant vesicles connected by membrane nanotubes, also relies on blowing (and puncturing) the lipid membrane by a thin (here a micropipette) tip. In this case, preformed giant vesicles with a lipid reservoir, attached to a substrate and free in solution, are micromanipulated and microelectronically fused, leading to a combination of lipid unilamellar microspheres and nanotubes containing a variety of solutes or macromolecules. These lipid unilamellar networks are the first step in the development of micro- or nanofluidic devices based on lipid bilayers.

Box 1.5 Useful tips to properly deal with the lipids for GUV formation

Tip 1: Chloroform is a common solvent for lipids, but it does not dissolve all types of lipids (the solution in the latter case would look milky). An alternative is to use chloroform/methanol 9:1 (v/v), diethyl ether/methanol 9:1 (v/v), or chloroform/diethyl ether/methanol 2:7:1 (v/v/v) or 4:5:1 (v/v/v). The last mixture of solvents is suitable for charged lipids. Phosphatidylglycerol is one example.

Tip 2: One should be aware that lipid solvents are volatile and that every time a vial of stock solution is open evaporation takes place, changing the lipid concentration. Thus, once a vial is open, one should work quickly and try to close it again as soon as possible.

Tip 3: Lipid stock solutions are to be stored in a freezer at −20°C. When they have to be used for GUV preparation, it is recommended to warm up the vial (e.g., with hands) before opening it because water may condense inside. Dried lipids may be stored in vials flushed over with nitrogen or argon to avoid oxidation.

Tip 4: When handling lipid solutions in organic solvents, always use glass (Hamilton) syringes rather than pipettes with plastic tips. Organic solvents such as chloroform dissolve plastic material and no matter how fast the transfer is, one has the danger of contaminating the lipid sample. Furthermore, because of the high vapor pressure of chloroform, measuring the exact volume of the lipid solution with the pipette will be jeopardized, which is crucial when aiming at preparing a mixed lipid membrane with a certain composition.

Tip 5: Lipids have a limited shelf life. A few approaches have been reported for assessing whether lipids have been strongly oxidized (Torchilin and Weissig, 2003), one of which is based on dissolving the lipid in ethanol and measuring the absorbance spectrum: Non-oxidized lipids do not absorb in the wavelength range 230–270 nm.

Tip 6: To avoid development of bacteria in GUV samples, one can employ a small amount (~1–3 mM) of sodium azide. One should be aware that this changes the solution conductivity.

Tip 7: Be aware that not using freshly prepared aqueous solutions for the vesicle preparation may be associated with changes in the solution conductivity and pH resulting from CO_2 absorption from the air.

1.4 DEALING WITH LIPIDS

Proper care needs to be exerted in dealing with GUV-forming lipids to ensure a successful preparation of the vesicles. Sometimes a little detail that one might not consider important can be the reason for a frustrating sequence of experiments leading to no vesicles or to vesicles with very poor properties. In Box 1.5, we summarize a few crucial tips to keep in mind when dealing with the lipids.

1.5 WHICH METHOD TO CHOOSE?

There are always at least three questions which should be addressed experimentally when reporting the preparation of GUVs using a particular (new) method: (i) Are the obtained vesicles (mainly) unilamellar? (ii) Are the vesicle membranes composed of the molecules one thinks that the membranes are composed of (e.g., is there presence of only negligible amounts of remaining oil; is the lipid composition of mixtures preserved in the obtained membrane)? (iii) Do the vesicles have the desired stability? Directly linked to the method of choice is the question whether there is an interest in preparing GUVs for experiments with—or making observations of—*individual GUVs* or a *population of GUVs*.

It is worth recalling here that GUVs represent a kinetically trapped state of lamellae-forming amphiphiles—or of mixtures of molecules, which are prone to form lamellar structures—in an aqueous solution. This is why, various methodologies (i.e., concepts) and protocols exist for obtaining GUVs through one of the different guided assembly processes. The method of choice depends very much on the type of amphiphiles used, on the composition of the aqueous solution, and on the application of the GUVs (Bagatolli et al., 2000; Morales-Penningston et al., 2010; Walde et al., 2010). For each method, there are advantages and disadvantages that become evident if one understands and critically questions the general principle behind the method and does not simply follow a specific recipe.

The electroformation method (Section 1.2.3) turned out to be very useful for gaining knowledge about the physics of phospholipid bilayers, for example their mechanical properties (Dimova et al., 2006; Dimova, 2019) (see also Chapters 11 through 17 of this book) and—in the case of lipid mixtures—the micrometer-sized domain formation if lipid-immiscibilities exist (Chapter 18). Depending on how the electroformed GUVs are obtained, one has to worry not only about vesicle-to-vesicle variations in terms of vesicle size and lamellarity (as in the case of all other methods as well) but also about differences in the lipid composition of the GUVs with respect to the lipid composition of the lipid mixtures used for the GUV preparation (Baykal-Caglar et al., 2012) and about the possible oxidation of polyunsaturated lipids that may be present. (See also Chapter 22 of this book.) For a multicomponent vesicle with charged lipids, one should be also aware of the possibility of leaflet asymmetry when using some protocols (Steinkühler et al., 2018).

If one is interested in using large populations of GUVs that are loaded with large amounts of water soluble molecules such as enzymes or nucleic acids, then a method based on the initial formation of micrometer-sized aqueous droplets in an organic solvent seems to be most appropriate (Section 1.3). In this case, the challenges are of, first, preparing a suitable and stable w/o-emulsion in which the aqueous droplets of appropriate buffer content host the desired molecules (e.g., enzymes) and then of transferring the emulsion droplets into a desired aqueous solution without extensive leakage of the trapped molecules from the droplets while retaining the enzyme activity in the case of trapped enzymes. Given that it is possible (i) that detectable amounts of organic solvent molecules (the "oil") are present within the membranes of the GUVs obtained by the water droplet-transfer method (Section 1.3.1) and (ii) that not all GUVs obtained with this method will be unilamellar, these two points must be considered if they are of importance for the type of study to be carried out with the vesicles. In any case, the water droplet-transfer method was found to be a convenient method for protein expression experiments inside giant vesicles (Noireaux and Libchaber, 2004; Fujii et al., 2014; Soga et al., 2014; Rampioni et al., 2018) whereby all the different required ingredients must be already present in the w/o emulsion from the very beginning.

To conclude, we should stress that as one starts to deal with giant vesicles and to learn how to observe them under a microscope, a "critical eye" easily develops for gauging the quality of vesicles—unilamellarity; absence of features such as bulges, tubes or inner vesicles; absence of defects that often appear as spots (dark in phase contrast and bright in fluorescence)—when, for instance, part of the lipids degrades during formation. Such a critical eye is of most help not only in evaluating the quality of published work but also as an incentive for all to produce GUVs of the utmost quality and to substantiate our claims with the beautiful images that quality GUVs generate.

LIST OF ABBREVIATIONS

AC	alternating current
cDICE	continuous droplet interface crossing encapsulation
DOPC	1,2-dioleoyl-*sn*-glycero-3-phosphocholine
DPH	1,6-dipenyl-1,3,5-hexatriene
DPPC	1,2-dipalmitoyl-*sn*-glycero-3-phosphocholine
DPPE	1 ,2-dipalmitoyl-*sn*-glycero-3-phosphoethanolamine
DSPC	1,2-distearoyl-*sn*-glycero-3-phosphocholine
ITO	indium tin oxide
LUV	large unilamellar vesicle
PC	phosphatidylcholine
PG	phosphatidylglycerol
POPC	1-palmitoyl-2-oleoyl-*sn*-glycero-3-phosphocholine
POPG	1-palmitoyl-2-oleoyl-*sn*-glycero-3-phospho-(1′-*rac*-glycerol) sodium salt
PS	phosphatidylserine
PTFE	polytetrafluoroethylene
PVA	polyvinyl alcohol
Tris	tris(hydroxymethyl)-aminomethane (= 2-amino-2-hydroxymethyl-propane-1,3-diol)
w/o	water-in-oil

REFERENCES

Abkarian M, Loiseau E, Massiera G (2011) Continuous droplet interface crossing encapsulation (cDICE) for high throughput monodisperse vesicle design. *Soft Matter* 7:4610–4614.

Akashi K, Miyata H, Itoh H, Kinosita K (1996) Preparation of giant liposomes in physiological conditions and their characterization under an optical microscope. *Biophys J* 71:3242–3250.

Akashi K, Miyata H, Itoh H, Kinosita K (1998) Formation of giant liposomes promoted by divalent cations: Critical role of electrostatic repulsion. *Biophys J* 74:2973–2982.

Almendro Vedia VG, Natale P, Chen S, Monroy F, Rosilio V, López-Montero I (2017) iGUVs: Preparing giant unilamellar vesicles with a smartphone and lipids easily extracted from chicken eggs. *J Chem Educ* 94:644–649.

Altamura E, Milano F, Tangorra RR, Trotta M, Omar OH, Stano P, Mavelli F (2017) Highly oriented photosynthetic reaction centers generate a proton gradient in synthetic protocells. *Proc Natl Acad Sci USA* 114:3837.

Angelova M, Dimitrov DS (1988) A mechanism of liposome electroformation. In: *Trends in Colloid and Interface Science II* (Degiorgio V, Ed.), pp. 59–67. Darmstadt, Germany: Steinkopff.

Angelova MI (2000) Liposome electroformation. In: *Giant Vesicles* (Luisi PL, Walde P, Eds.), pp. 27–36. Chichester, UK: John Wiley & Sons.

Angelova MI, Dimitrov DS (1986) Liposome electroformation. *Faraday Discuss* 81:303–311.

Angelova MI, Soléau S, Méléard P, Faucon F, Bothorel P (1992) Preparation of giant vesicles by external AC electric fields. Kinetics and applications. In: *Trends in Colloid and Interface Science VI* (Helm C, Lösche M, Möhwald H, Eds.), pp. 127–131. Darmstadt, Germany: Steinkopff.

Ayuyan AG, Cohen FS (2006) Lipid peroxides promote large rafts: Effects of excitation of probes in fluorescence microscopy and electrochemical reactions during vesicle formation. *Biophys J* 91:2172–2183.

Bagatolli LA, Parasassi T, Gratton E (2000) Giant phospholipid vesicles: Comparison among the whole lipid sample characteristics using different preparation methods—A two photon fluorescence microscopy study. *Chem Phys Lipids* 105:135–147.

Bangham AD, Horne RW (1964) Negative staining of phospholipids and their structural modification by surface-active agents as observed in the electron microscope. *J Mol Biol* 8:660–668.

Bangham AD, Standish MM, Watkins JC (1965) Diffusion of univalent ions across lamellae of swollen phospholipids. *J Mol Biol* 13:238–252.

Baykal-Caglar E, Hassan-Zadeh E, Saremi B, Huang J (2012) Preparation of giant unilamellar vesicles from damp lipid film for better lipid compositional uniformity. *Biochim Biophys Acta-Biomembr* 1818:2598–2604.

Billerit C, Jeffries GDM, Orwar O, Jesorka A (2012) Formation of giant unilamellar vesicles from spin-coated lipid films by localized IR heating. *Soft Matter* 8:10823–10826.

Billerit C, Wegrzyn I, Jeffries GDM, Dommersnes P, Orwar O, Jesorka A (2011) Heat-induced formation of single giant unilamellar vesicles. *Soft Matter* 7:9751–9757.

Blosser MC, Horst BG, Keller SL (2016) cDICE method produces giant lipid vesicles under physiological conditions of charged lipids and ionic solutions. *Soft Matter* 12:7364–7371.

Bouvrais H, Cornelius F, Ipsen JH, Mouritsen OG (2012) Intrinsic reaction-cycle time scale of Na^+,K^+-ATPase manifests itself in the lipid–protein interactions of nonequilibrium membranes. *Proc Natl Acad Sci* 109:18442–18446.

Breton M, Amirkavei M, Mir LM (2015) Optimization of the electroformation of giant unilamellar vesicles (GUVs) with unsaturated phospholipids. *J Membrane Biol* 248:827–835.

Bucher P, Fischer A, Luisi PL, Oberholzer T, Walde P (1998) Giant vesicles as biochemical compartments: The use of microinjection techniques. *Langmuir* 14:2712–2721.

Cabré EJ, Sánchez-Gorostiaga A, Carrara P, Ropero N, Casanova M, Palacios P, Stano P, Jiménez M, Rivas G, Vicente M (2013) Bacterial division proteins FtsZ and ZipA induce vesicle shrinkage and cell membrane invagination. *J Bio Chem* 288:26625–26634.

Campillo C, Sens P, Köster D, Pontani LL, Lévy D, Bassereau P, Nassoy P, Sykes C (2013) Unexpected membrane dynamics unveiled by membrane nanotube extrusion. *Biophys J* 104(6):1248–1256.

Carrara P, Stano P, Luisi PL (2012). Giant vesicles "colonies": A model for primitive cell communities. *ChemBioChem* 13(10):1497–1502.

Chiba M, Miyazaki M, Ishiwata S (2014) Quantitative analysis of the lamellarity of giant liposomes prepared by the inverted emulsion method. *Biophys J* 107:346–354.

Coyne CW, Patel K, Heureaux J, Stachowiak J, Fletcher DA, Liu AP (2014) Lipid bilayer vesicle generation using microfluidic jetting. *JoVE*:e51510.

Dao TPT, Fauquignon M, Fernandes F, Ibarboure E, Vax A, Prieto M, Le Meins JF (2017) Membrane properties of giant polymer and lipid vesicles obtained by electroformation and PVA gel-assisted hydration methods. *Colloids Surf Physicochem Eng Aspects* 533:347–353.

Deber CM, Reynolds SJ (1991) Central nervous system myelin: Structure, function, and pathology. *Clin Biochem* 24:113–134.

Deek J, Maan R, Loiseau E, Bausch AR (2018) Reconstitution of composite actin and keratin networks in vesicles. *Soft Matter* 14:1897–1902.

Dimova R (2019) Giant vesicles and their use in assays for assessing membrane phase state, curvature, mechanics, and electrical properties. *Annu Rev Bioph Biom* 48:93–119.

Dimova R, Aranda S, Bezlyepkina N, Nikolov V, Riske KA, Lipowsky R (2006) A practical guide to giant vesicles. Probing the membrane nanoregime via optical microscopy. *J Phys* 18:S1151–S1176.

Dominak LM, Keating CD (2008) Macromolecular crowding improves polymer encapsulation within giant lipid vesicles. *Langmuir* 24:13565–13571.

Danielli JF, Davson H (1935) A contribution to the theory of permeability of thin films. *J Cell Comp Physiol* 5:495–508.

Drucker P, Pejic M, Galla HJ, Gerke V (2013) Lipid segregation and membrane budding induced by the peripheral membrane binding protein annexin A2. *J Biol Chem* 288:24764–24776.

Elani Y, Gee A, Law RV, Ces O (2013) Engineering multi-compartment vesicle networks. *Chem Sci* 4:3332–3338.

Elani Y, Purushothaman S, Booth PJ, Seddon JM, Brooks NJ, Lawa RV, Ces O (2015). Measurements of the effect of membrane asymmetry on the mechanical properties of lipid bilayers. *Chem Commun* 51:6976–6979.

Estes DJ, Mayer M (2005) Electroformation of giant liposomes from spin-coated films of lipids. *Colloid Surf B* 42:115–123.

Fidorra M, Duelund L, Leidy C, Simonsen AC, Bagatolli LA (2006) Absence of fluid-ordered/fluid-disordered phase coexistence in ceramide/POPC mixtures containing cholesterol. *Biophys J* 90:4437–4451.

Fujii S, Matsuura T, Sunami T, Nishikawa T, Kazuta Y, Yomo T (2014) Liposome display for in vitro selection and evolution of membrane proteins. *Nat Protoc* 9:1578.

Fujiwara K, Yanagisawa M (2014) Generation of giant unilamellar liposomes containing biomacromolecules at physiological intracellular concentrations using hypertonic conditions. *ACS Synth Biol* 3(12):870–874.

Funakoshi K, Suzuki H, Takeuchi S (2007). Formation of giant lipid vesiclelike compartments from a planar lipid membrane by a pulsed jet flow. *J Am Chem Soc* 129(42):12608–12609.

Garten M, Aimon S, Bassereau P, Toombes GES (2015) Reconstitution of a transmembrane protein, the voltage-gated ion channel, KvAP, into giant unilamellar vesicles for microscopy and patch clamp studies. *J Vis Exp* 95:e52281.

Georgiev VN, Grafmüller A, Bléger D, Hecht S, Kunstmann S, Barbirz S, Lipowsky R, Dimova R (2018). Area increase and budding in giant vesicles triggered by light: behind the scene. *Adv Sci* 5:1800432.

Greco E, Quintiliani G, Santucci MB, Serafino A, Ciccaglione AR, Marcantonio C, Papi M et al. (2012) Janus-faced liposomes enhance antimicrobial innate immune response in Mycobacterium tuberculosis infection. *Proc Natl Acad Sci* 109:1360–1368.

Grotzky A, Altamura E, Adamcik J, Carrara P, Stano P, Mavelli F, Nauser T, Mezzenga R, Schluter AD, Walde P (2013) Structure and enzymatic properties of molecular dendronized polymer–enzyme conjugates and their entrapment inside giant vesicles. *Langmuir* 29(34):10831–10840.

Guevorkian K, Manzi J, Pontani LL, Brochard-Wyart F, Sykes C (2015) Mechanics of biomimetic liposomes encapsulating an actin shell. *Biophysi J* 109(12):2471–2479.

Hadorn M, Boenzli E, Hotz, PE (2011). A quantitative analytical method to test for salt effects on giant unilamellar vesicles. *Sci Rep* 1:168.

Hadorn M, Boenzli E, Sørensen, KT, De Lucrezia D, Hanczyc, MM, Yomo T (2013) Defined DNA-mediated assemblies of gene-expressing giant unilamellar vesicles. *Langmuir* 29(49):15309–15319.

Hamada S, Tabuchi M, Toyota T, Sakurai T, Hosoi T, Nomoto T, Nakatani K, Fujinami M, Kanzaki R (2014) Giant vesicles functionally expressing membrane receptors for an insect pheromone. *Chem Commun* 50:2958–2961.

Hamada T, Miura Y, Komatsu Y, Kishimoto Y, Vestergaard Md, Takagi M (2008) Construction of asymmetric cell-sized lipid vesicles from lipid-coated water-in-oil microdroplets. *J Phys Chem B* 112:14678–14681.

Hansen JS, Thompson JR, Helix-Nielsen C, Malmstadt N (2013) Lipid directed intrinsic membrane protein segregation. *J Am Chem Soc* 135:17294–17297.

Harbich W, Helfrich W (1979). Alignment and opening of giant lecithin vesicles by electric fields. *Zeitschrift für Naturforschung A* 34(9):1063–1065.

Harbich, W, Servuss, RM, Helfrich, W (1976). Optical studies of lecithin-membrane melting. *Phys Lett A* 57(3):294–296.

Hase M, Yamada A, Hamada T, Baigl D, Yoshikawa K (2007) Manipulation of cell-sized phospholipid-coated microdroplets and their use as biochemical microreactors. *Langmuir* 23:348–352.

Herold C, Chwastek G, Schwille P, Petrov EP (2012) Efficient electroformation of supergiant unilamellar vesicles containing cationic lipids on ITO-coated electrodes. *Langmuir* 28:5518–5521.

Horger KS, Estes DJ, Capone R, Mayer M (2009) Films of agarose enable rapid formation of giant liposomes in solutions of physiologic ionic strength. *J Am Chem Soc* 131:1810–1819.

Howse JR, Jones RAL, Battaglia G, Ducker RE, Leggett GJ, Ryan AJ (2009) Templated formation of giant polymer vesicles with controlled size distributions. *Nat Mater* 8:507–511.

Hu PC, Li S, Malmstadt N (2011) Microfluidic fabrication of asymmetric giant lipid vesicles. *ACS Appl Mater Interfaces* 3:1434–1440.

Ikenouchi J, Suzuki M, Umeda K, Ikeda K, Taguchi R, Kobayashi T, Sato SB, Kobayashi T, Stolz DB, Umeda M (2012) Lipid polarity is maintained in absence of tight junctions. *J Biol Chem* 287:9525–9533.

Israelachvili, J N (2011). *Intermolecular and Surface Forces*. Burlington, MA: Academic Press.

Karamdad K, Law RV, Seddon JM, Brooks NJ, Ces O (2015) Preparation and mechanical characterisation of giant unilamellar vesicles by a microfluidic method. *Lab Chip* 15:557–562.

Karamdad K, Law RV, Seddon JM, Brooks NJ, Ces O (2016) Studying the effects of asymmetry on the bending rigidity of lipid membranes formed by microfluidics. *Chem Commun* 52:5277–5280.

Karlsson M, Sott K, Davidson M, Cans AS, Linderholm P, Chiu D, Orwar O (2002). Formation of geometrically complex lipid nanotube-vesicle networks of higher-order topologies. *Proc Natl Acad Sci USA* 99(18):11573–11578.

Katayama S, Nakase I, Yano Y, Murayama T, Nakata Y, Matsuzaki K, Futaki S (2013) Effects of pyrenebutyrate on the translocation of arginine-rich cell-penetrating peptides through artificial membranes: Recruiting peptides to the membranes, dissipating liquid-ordered phases, and inducing curvature. *Biochim Biophys Acta-Biomembr* 1828:2134–2142.

Kirchner SR, Ohlinger A, Pfeiffer T, Urban AS, Stefani FD, Deak A, Lutich AA, Feldmann J (2012). Membrane composition of jetted lipid vesicles: A Raman spectroscopy study. *J Biophotonics* 5(1):40–46.

Knorr RL, Steinkühler J, Dimova R (2018) Micron-sized domains in quasi single-component giant vesicles. *Biochim Biophys Acta Biomembr* 1860:1957–1964.

Kubatta EA, Rehage H (2009) Characterization of giant vesicles formed by phase transfer processes. *Colloid Polym Sci* 287(9):1117–1122.

Kubsch B, Robinson T, Steinkühler J, Dimova R (2017) Phase behavior of charged vesicles under symmetric and asymmetric solution conditions monitored with fluorescence microscopy. *J Vis Exp* 128:e56034.

Kuroiwa T, Fujita R, Kobayashi I, Uemura K, Nakajima M, Sato S, Walde P, Ichikawa S (2012) Efficient preparation of giant vesicles as biomimetic compartment systems with high entrapment yields for biomacromolecules. *Chem Biodivers* 9:2453–2472.

Larsen J, Hatzakis NS, Stamou D (2011) Observation of inhomogeneity in the lipid composition of individual nanoscale liposomes. *J Am Chem Soc* 133:10685–10687.

Le Berre M, Yamada A, Reck L, Chen Y, Baigl D (2008) Electroformation of giant phospholipid vesicles on a silicon substrate: Advantages of controllable surface properties. *Langmuir* 24:2643–2649.

Lira RB, Dimova R, Riske Karin A (2014) Giant unilamellar vesicles formed by hybrid films of agarose and lipids display altered mechanical properties. *Biophys J* 107:1609–1619.

Lira RB, Steinkühler J, Knorr RL, Dimova R, Riske KA (2016) Posing for a picture: Vesicle immobilization in agarose gel. *Sci Rep* 6:25254.

Liu YJ, Hansen, GP, Venancio-Marques A, Baigl, D (2013) Cell-free preparation of functional and triggerable giant proteoliposomes. *ChemBioChem* 14(17):2243–2247.

Luisi PL, Walde P (Eds.) (2000) Giant vesicles, Perspectives in Supramolecular Chemistry. John Wiley & Sons Ltd., Chichester.

Marsh, D (2013). *Handbook of Lipid Bilayers*. Boca Raton, FL: CRC Press.

Matosevic S (2012) Synthesizing artificial cells from giant unilamellar vesicles: State-of-the art in the development of microfluidic technology. *Bioessays* 34:992–1001.

Matosevic S, Paegel BM (2011) Stepwise synthesis of giant unilamellar vesicles on a microfluidic assembly line. *J Am Chem Soc* 133:2798–2800.

McDonald GR, Hudson AL, Dunn SMJ, You HT, Baker GB, Whittal RM, Martin JW, Jha A, Edmondson DE, Holt A (2008) Bioactive contaminants leach from disposable laboratory plasticware. *Science* 322:917–917.

Méléard P, Bagatolli LA, Pott T (2009) Giant unilamellar vesicle electroformation: From lipid mixtures to native membranes under physiological conditions. In: *Methods in Enzymology*, Düzgüneş N (Ed). Vol 465 pp. 161–176. New York: Academic Press.

Mertins O, da Silveira NP, Pohlmann AR, Schroder AP, Marques CM (2009) Electroformation of giant vesicles from an inverse phase precursor. *Biophys J* 96:2719–2726.

Mikelj M, Praper T, Demič R, Hodnik V, Turk T, Anderluh G (2013) Electroformation of giant unilamellar vesicles from erythrocyte membranes under low-salt conditions. *Anal Biochem* 435:174–180.

Mora NL, Hansen JS, Gao Y, Ronald AA, Kieltyka R, Malmstadt N, Kros A (2014) Preparation of size tunable giant vesicles from cross-linked dextran(ethylene glycol) hydrogels. *Chem Commun* 50:1953–1955.

Morales-Penningston NF, Wu J, Farkas ER, Goh SL, Konyakhina TM, Zheng JY, Webb WW, Feigenson GW (2010) GUV preparation and imaging: Minimizing artifacts. *Biochim Biophys Acta* 1798:1324–1332.

Mueller P, Chien TF, Rudy B (1983) Formation and properties of cell-size lipid bilayer vesicles. *Biophys J* 44:375–381.

Mueller P, Rudin DO, Tien HT, Wescott WC (1963) Methods for the formation of single bimolecular lipid membranes in aqueous solution. *J Phys Chem* 67(2):534–535.

Murphy TJ, Shamoo AE (1978) Reconstitution of Ca^{2+}-Mg^{2+}-ATPase in giant vesicles. *Biophys J* 21:A27–A27.

Murrell M, Pontani LL, Guevorkian K, Cuvelier D, Nassoy P, Sykes C (2011). Spreading dynamics of biomimetic actin cortices. *Biophys J* 100(6):1400–1409.

Natsume Y, Toyota T (2013) Giant vesicles containing microspheres with high volume fraction prepared by water-in-oil emulsion centrifugation. *Chem Lett* 42(3):295–297.

Natsume Y, Wen H-i, Zhu T, Itoh K, Sheng L, Kurihara K (2017) Preparation of giant vesicles encapsulating microspheres by centrifugation of a water-in-oil emulsion. *J Vis Exp* 119:e55282.

Neubauer C (1867) Ueber das Myelin. *Zeitschrift für analytische Chemie* 6:189–195.

Noireaux V, Libchaber A (2004) A vesicle bioreactor as a step toward an artificial cell assembly. *Proc Natl Acad Sci USA* 101:17669–17674.

Nourian Z, Roelofsen W, Danelon C (2012) Triggered gene expression in fed-vesicle microreactors with a multifunctional membrane. *Angew Chem Int Ed* 51:3114–3118.

O'Connor D, Byrne A, Dolan C, Keyes TE (2018) Phase partitioning, solvent-switchable BODIPY probes for high contrast cellular imaging and FCS. *New J Chem* 42:3671–3682.

Ohno M, Hamada T, Takiguchi K, Homma M (2009) Dynamic behavior of giant liposomes at desired osmotic pressures. *Langmuir* 25(19):11680–11685.

Okumura Y, Zhang H, Sugiyama T, Iwata Y (2007) Electroformation of giant vesicles on a non-electroconductive substrate. *J Am Chem Soc* 129:1490–1491.

Papahadjopoulos D, Kimelberg, HK (1974) Phospholipid vesicles (liposomes) as models for biological membranes: Their properties and interactions with cholesterol and proteins. *Prog Surf Sci* 4:141–232.

Papahadjopoulos D, Miller N (1967). Phospholipid model membranes. I. Structural characteristics of hydrated liquid crystals. *Biochim Biophys Acta Biomembr* 135(4):624–638.

Patil YP, Jadhav S (2014) Novel methods for liposome preparation. *Chem Phys Lipids* 177:8–18.

Pautot S, Frisken BJ, Weitz DA (2003a) Production of unilamellar vesicles using an inverted emulsion. *Langmuir* 19:2870–2879.

Pautot S, Frisken BJ, Weitz DA (2003b) Engineering asymmetric vesicles. *Proc Natl Acad Sci USA* 100:10718.

Pavlič JI, Genova J, Popkirov G, Kralj-Iglič V, Iglič A, Mitov MD (2011) Mechanoformation of neutral giant phospholipid vesicles in high ionic strength solution. *Chem Phys Lipids* 164:727–731.

Pereno V, Carugo D, Bau L, Sezgin E, Bernardino de la Serna J, Eggeling C, Stride E (2017) Electroformation of giant unilamellar vesicles on stainless steel electrodes. *ACS Omega* 2:994–1002.

Peters RJ, Nijemeisland M, van Hest JC (2015) Reversibly triggered protein–ligand assemblies in giant vesicles. *Angewandte Chem* 127(33):9750–9753.

Politano TJ, Froude VE, Jing BX, Zhu YX (2010) AC-electric field dependent electroformation of giant lipid vesicles. *Colloid Surf B* 79:75–82.

Pontani LL, Van der Gucht J, Salbreux G, Heuvingh J, Joanny JF, Sykes C (2009). Reconstitution of an actin cortex inside a liposome. Biophys J 96(1):192–198.

Pott T, Bouvrais H, Meleard P (2008) Giant unilamellar vesicle formation under physiologically relevant conditions. *Chem Phys Lipids* 154:115–119.

Rampioni G, D'Angelo F, Messina M, Zennaro A, Kuruma Y, Tofani D, Leoni L, Stano P (2018) Synthetic cells produce a quorum sensing chemical signal perceived by Pseudomonas aeruginosa. *Chem Commun* 54:2090–2093.

Reeves JP, Dowben RM (1969) Formation and properties of thin-walled phospholipid vesicles. *J Cell Physiol* 73:49–60.

Rodriguez N, Pincet F, Cribier S (2005) Giant vesicles formed by gentle hydration and electroformation: A comparison by fluorescence microscopy. *Colloid Surf B* 42:125–130.

Saito AC, Ogura T, Fujiwara K, Murata S, Shin-ichiro MN (2014) Introducing micrometer-sized artificial objects into live cells: A method for cell–giant unilamellar vesicle electrofusion. *PloS One* 9(9):e106853.

Sakurai I, Kawamura Y (1984) Growth mechanism of myelin figures of phosphatidylcholine. *Biochim Biophys Acta-Biomembr* 777:347–351.

Saliba AE, Vonkova I, Ceschia S, Findlay GM, Maeda K, Tischer C, Deghou S et al. (2014) A quantitative liposome microarray to systematically characterize protein-lipid interactions. *Nat Methods* 11:47–50.

Schultze J, Vagias A, Ye L, Prantl E, Breising V, Best A, Koynov K, Marques CM and Butt HJ (2019) Preparation of Monodisperse Giant Unilamellar Anchored Vesicles Using Micropatterned Hydrogel Substrates. *ACS Omega* 4:9393–9399.

Schmid EM, Richmond DL, Fletcher DA (2015) Reconstitution of proteins on electroformed giant unilamellar vesicles. In: *Methods in Cell Biology* (Ross J, Marshall WF, Eds.), Vol. 128 pp. 319–338. Academic Press.

Shima T, Muraoka T, Hamada T, Morita M, Takagi M, Fukuoka H, Inoue Y et al. (2014) Micrometer-size vesicle formation triggered by UV light. *Langmuir* 30:7289–7295.

Shimanouchi T, Umakoshi H, Kuboi R (2009) Kinetic study on giant vesicle formation with electroformation method. *Langmuir* 25:4835–4840.

Soga H, Fujii S, Yomo T, Kato Y, Watanabe H, Matsuura T (2014) In vitro membrane protein synthesis inside cell-sized vesicles reveals the dependence of membrane protein integration on vesicle volume. *ACS Synth Biol* 3:372–379.

Stachowiak JC, Richmond DL, Li TH, Liu AP, Parekh SH, Fletcher DA (2008) Unilamellar vesicle formation and encapsulation by microfluidic jetting. *Proc Natl Acad Sci USA* 105:4697–4702.

Stano P (2019) Gene expression inside liposomes: from early studies to current protocols. *Chem Eur J* 25:7798 –7814.

Stano P, Carrara P, Kuruma Y, de Souza TP, Luisi PL (2011). Compartmentalized reactions as a case of soft-matter biotechnology: Synthesis of proteins and nucleic acids inside lipid vesicles. *J Mater Chem* 21(47):18887–18902.

Stano P, Rampioni G, Carrara P, Damiano L, Leoni L, Luisi PL (2012). Semi-synthetic minimal cells as a tool for biochemical ICT. *Biosystems* 109(1):24–34.

Stein H, Spindler S, Bonakdar N, Wang C, Sandoghdar V (2017) Production of isolated giant unilamellar vesicles under high salt concentrations. *Front Physiol* 8:63.

Steinkühler J, De Tillieux P, Knorr RL, Lipowsky R, Dimova R (2018) Charged giant unilamellar vesicles prepared by electroformation exhibit nanotubes and transbilayer lipid asymmetry. *Sci Rep* 8:11838.

Sturzenegger F, Robinson T, Hess D, Dittrich PS (2016) Membranes under shear stress: Visualization of non-equilibrium domain patterns and domain fusion in a microfluidic device. *Soft Matter* 12:5072–5076.

Sugiura S, Kuroiwa T, Kagota T, Nakajima M, Sato S, Mukataka S, Walde P, Ichikawa S (2008) Novel method for obtaining homogeneous giant vesicles from a monodisperse water-in-oil emulsion prepared with a microfluidic device. *Langmuir* 24:4581–4588.

Szoka F, Papahadjopoulos D (1978) Procedure for preparation of liposomes with large internal aqueous space and high capture by reverse-phase evaporation. *Proc Natl Acad Sci USA* 75:4194–4198.

Szoka F, Papahadjopoulos D (1980) Comparative properties and methods of preparation of lipid vesicles (Liposomes). *Annu Rev Biophys Bioeng* 9:467–508.

Tahara K, Tadokoro S, Kawashima Y, Hirashima N (2012) Endocytosis-like uptake of surface-modified drug nanocarriers into giant unilamellar vesicles. *Langmuir* 28:7114–7118.

Takiguchi K, Negishi M, Tanaka-Takiguchi Y, Homma M, Yoshikawa, K (2011) Transformation of actoHMM assembly confined in cell-sized liposome. *Langmuir* 27(18):11528–11535.

Takiguchi K, Yamada A, Negishi M, Tanaka-Takiguchi Y, Yoshikawa, K (2008). Entrapping desired amounts of actin filaments and molecular motor proteins in giant liposomes. *Langmuir* 24(20):11323–11326.

Tan Y-C, Hettiarachchi K, Siu M, Pan Y-R, Lee AP (2006) Controlled microfluidic encapsulation of cells, proteins, and microbeads in lipid vesicles. *J Am Chem Soc* 128:5656–5658.

Taylor P, Xu C, Fletcher PDI, Paunov VN (2003a) A novel technique for preparation of monodisperse giant liposomes. *Chem Commun* 14:1732–1733.

Taylor P, Xu C, Fletcher PDI, Paunov VN (2003b) Fabrication of 2D arrays of giant liposomes on solid substrates by microcontact printing. *Phys Chem Chem Phys* 5:4918–4922.

Torchilin V, Weissig V (2003) *Liposomes: A Practical Approach*. Oxford, UK: Oxford University Press.

Toyota T, Ohguri N, Maruyama K, Fujinami M, Saga T, Aoki I (2012). Giant vesicles containing superparamagnetic iron oxide as biodegradable cell-tracking MRI probes. *Anal Chem* 84(9):3952–3957.

Träuble H, Grell E (1971) Carriers and specificity in membranes. IV. Model vesicles and membranes. The formation of asymmetrical spherical lecithin vesicles. *Neurosci Res Program Bull* 9:373–380.

Tsai FC, Stuhrmann B, Koenderink GH (2011) Encapsulation of active cytoskeletal protein networks in cell-sized liposomes. *Langmuir* 27:10061–10071.

Tsumoto K, Oohashi M, Tomita M (2011) Monitoring of membrane collapse and enzymatic reaction with single giant liposomes embedded in agarose gel. *Colloid Polym Sci* 289:1337–1346.

Tsumoto K, Matsuo H, Tomita M, Yoshimura T (2009) Efficient formation of giant liposomes through the gentle hydration of phosphatidylcholine films doped with sugar. *Colloids Surf B* 68:98–105.

van Swaay D, deMello A (2013) Microfluidic methods for forming liposomes. *Lab Chip* 13:752–767.

Walde P, Cosentino K, Engel H, Stano P (2010) Giant vesicles: Preparations and applications. *ChemBioChem* 11:848–865.

Walde P, Goto A, Monnard PA, Wessicken M, Luisi, PL (1994) Oparin's reactions revisited: Enzymic synthesis of poly (adenylic acid) in micelles and self-reproducing vesicles. *J Am Chem Soc* 116(17):7541–7547.

Weinberger A, Tsai FC, Koenderink GH, Schmidt TF, Itri R, Meier W, Schmatko T, Schroder A, Marques C (2013) Gel-assisted formation of giant unilamellar vesicles. *Biophys J* 105:154–164.

Weinberger A, Walter V, MacEwan SR, Schmatko T, Muller P, Schroder AP, Chilkoti A, Marques CM (2017) Cargo self-assembly rescues affinity of cell-penetrating peptides to lipid membranes. *Sci Rep* 7:43963.

Whittenton J, Harendra S, Pitchumani R, Mohanty K, Vipulanandan C, Thevananther S (2008). Evaluation of asymmetric liposomal nanoparticles for encapsulation of polynucleotides. *Langmuir* 24(16):8533–8540.

Xiao Z, Huang N, Xu M, Lu Z, Wei Y (1998b). Novel preparation of asymmetric liposomes with inner and outer layer of different materials. *Chem Lett* 27(3):225–226.
Xiao Z, Xu M, Li M, Lu Z, Wei Y (1998a). Preparation of asymmetric bilayer-vesicles with inner and outer monolayers composed of different amphiphilic molecules. *Supramol Sci* 5(5–6):619–622.
Yanagisawa M, Iwamoto M, Kato A, Yoshikawa K, Oiki S (2011). Oriented reconstitution of a membrane protein in a giant unilamellar vesicle: experimental verification with the potassium channel KcsA. *J Am Chem Soc* 133:11774–11779.
Yamada A, Le Berre M, Yoshikawa K, Baigl D (2007) Spontaneous generation of giant liposomes from an oil/water interface. *ChemBioChem* 8:2215–2218.
Yamada A, Yamanaka T, Hamada T, Hase M, Yoshikawa K, Baigl D (2006) Spontaneous transfer of phospholipid-coated oil-in-oil and water-in-oil micro-droplets through an oil/water interface. *Langmuir* 22:9824–9828.

Preparation and properties of giant plasma membrane vesicles and giant unilamellar vesicles from natural membranes

Joseph H. Lorent and Ilya Levental

Seek simplicity and distrust it.

Alfred North Whitehead,

The Concept of Nature: The Tarner Lectures Delivered in Trinity College, Nov. 1919

Contents

2.1 INTRODUCTORY WORDS

Synthetic membrane model systems—including Langmuir monolayers, giant unilamellar vesicles (GUVs), and supported bilayers—have yielded wide-ranging insights into the phenomenology of biomembranes. One striking example is the detailed characterization of the liquid ordered phase (Ipsen et al., 1987) and liquid–liquid phase coexistence in sterol-containing membranes (Dietrich et al., 2001; Veatch and Keller, 2003), which converged with a plethora of biochemical and cellular observations (Brown and London, 1998; Simons and Vaz, 2004) to provide the physicochemical framework for the sweeping and controversial lipid raft hypothesis (Lingwood and Simons, 2010); Levental and Veatch, 2016. However, although synthetic membranes recapitulate many of the core features of biological membranes, several aspects are not fully represented. Most significantly, neither the lipid complexity of natural membranes nor their protein content and diversity are feasible to reproduce in synthetic models. In recent years, this niche has been filled by taking advantage of inherent cellular processes that produce large, intact plasma membrane vesicles that can serve as intermediate model systems, bridging the experimental flexibility and tractability of GUVs with the complexity of cellular membranes. This chapter will describe the most widely used such system—giant plasma membrane vesicles (GPMVs)—focusing on their isolation, properties, uses, and comparisons to alternative synthetic and natural models. Finally, we will describe methods to isolate lipids from natural sources, including GPMVs, for reconstitution of complex lipid mixtures into GUVs or other synthetic membrane systems.

2.2 COMPLEXITY OF BIOLOGICAL MEMBRANES

Although the genetic and protein complexity of living cells is widely appreciated, the detailed lipid repertoires of biological membranes have only recently become accessible through methodological advances in mass spectrometry (Shevchenko and Simons, 2010; Wenk, 2010). These observations have revealed that cellular lipidomes are much more complex and diverse than was previously assumed (Figure 2.1), opening a potentially large field of inquiry into why such complexity is required and how it is generated and regulated.

2.2.1 LIPID DIVERSITY AND PROTEIN CONTENT OF BIOLOGICAL MEMBRANES

The surprising complexity of mammalian lipidomes arises from combinatorial lipid synthesis, wherein independent arrangements of several possible hydrophilic headgroups, backbones, and hydrophobic chains can yield tens of thousands of distinct molecular species. In practice, most cells produce several hundred of these species at detectable abundances, with less than 100 species typically comprising >90% of the total lipidome (Gerl et al., 2012; Sampaio et al., 2011). Some of these species, such as the phosphoinositides (Janmey and Lindberg, 2004), have clear and unique functional roles in binding and regulating protein effectors. For most however, no species-specific functional role has been identified, leaving open the question of the purpose of such diversity.

In addition to lipids, biological membranes are composed of a plethora of proteins. Bioinformatic analyses estimate that ~30% of the mammalian proteome (~7,000 proteins) are integrated into membranes by hydrophobic transmembrane domains (Wallin and von Heijne, 1998). Together with the large number of proteins interacting with the membrane via posttranslational lipid anchors and direct lipid binding, proteins comprise approximately 50%–75% w/w of a typical membrane preparation (Mitra et al., 2004). Perhaps most relevant for membrane biophysics is that polypeptides are estimated to occupy ~20% of the area at the center of a membrane bilayer (Dupuy and Engelman, 2008), suggesting that these components are central determinants of physical phenotypes such as fluidity, rigidity, and phase separation.

2.2.2 DIVERSITY OF MEMBRANE COMPOSITIONS AND PROPERTIES

The diversity of membrane compositions in a single mammalian cell is well illustrated by considering the variety in cholesterol content across various intracellular organelles (Figure 2.2). The plasma membrane (PM) is typically the most cholesterol-rich membrane, containing up to 45 mol% (Gerl et al., 2012). In contrast, the endoplasmic reticulum (ER) is cholesterol poor, with ~5% at a healthy steady-state (Mesmin and Maxfield, 2009). It is believed that there is a gradual increase in cholesterol concentration through the biosynthetic pathway (i.e., ER < Golgi < PM), which is mirrored in a gradual decrease through the endocytic pathway (PM > early

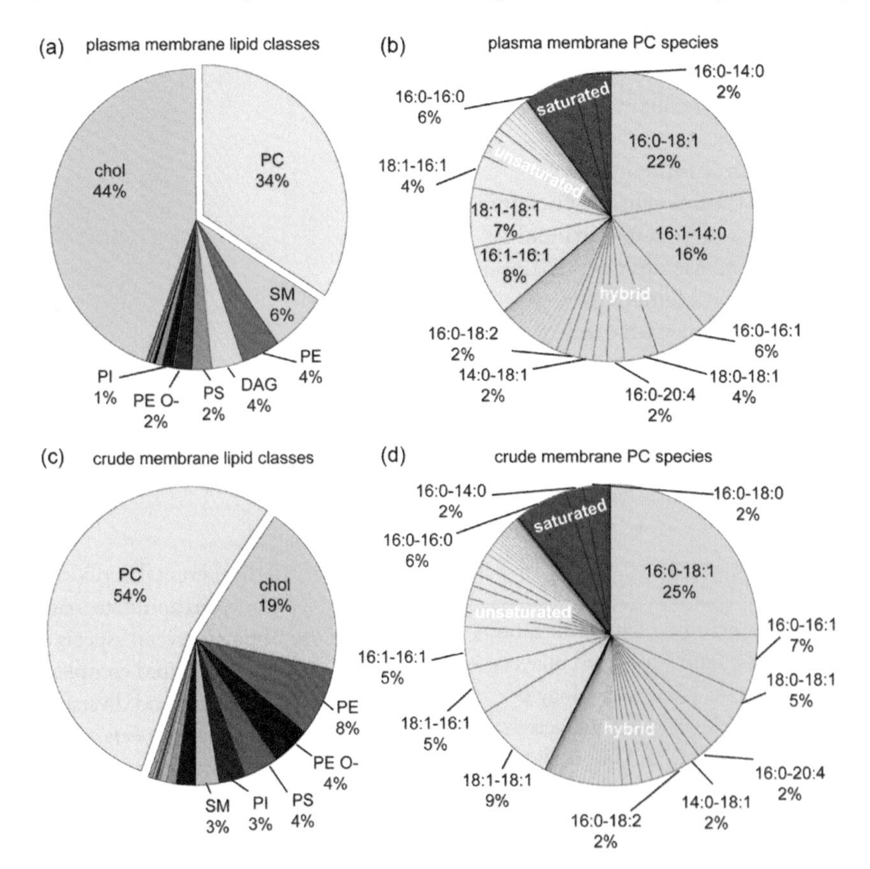

Figure 2.1 (a, b) Lipid diversity in plasma membrane versus (c, d) whole cell extracts from rat basophilic leukemia cells. Panels (a) and (c) display the percentage of main lipid species, and panels (b) and (d) the acyl chains of phosphatidylcholines.

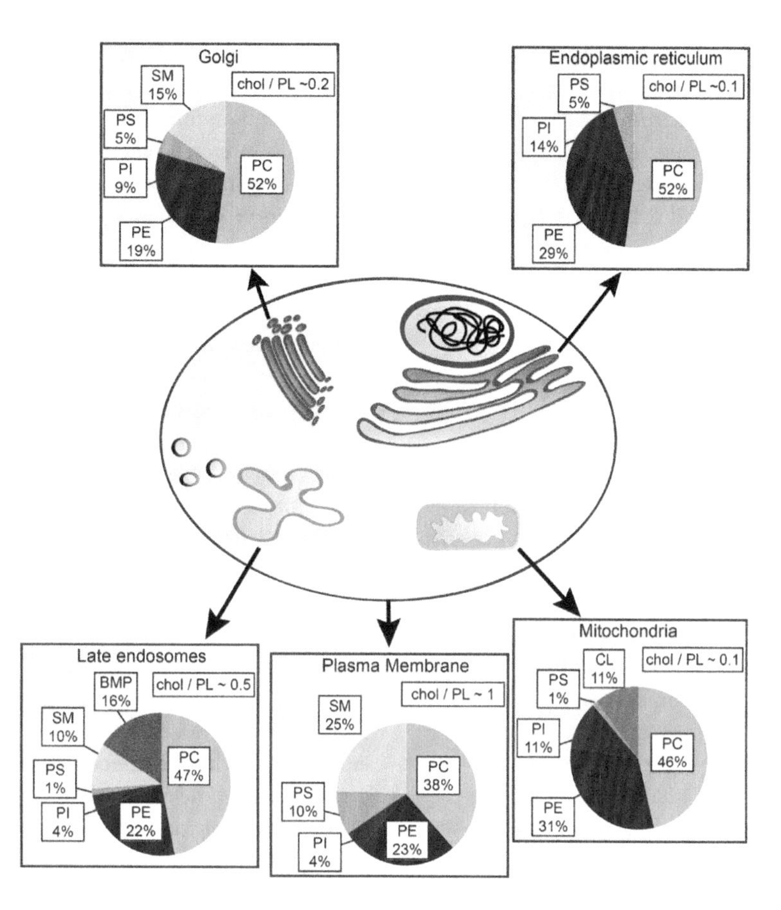

Figure 2.2 Major lipid compositions and cholesterol/phospholipid ratio of different cellular organelles. (Adapted from van Meer, G. et al., *Nat. Rev. Mol. Cell Biol.*, 9, 112–124, 2008.)

and late endosomes > lysosomes) (van Meer et al., 2008). Because cholesterol content is perhaps the major determinant of membrane order (Róg et al., 2009), fluidity, rigidity, and phase separation, it is likely that these properties also vary dramatically between intracellular membranes. These effects have not been widely explored in biological membranes, although an example is provided by the X-ray scattering measurements of membrane thickness showing clear differences between isolated cellular organelles (Mitra et al., 2004). Of course cholesterol is not the only lipid to vary between subcellular compartments; for example, glycosphingolipids are largely produced and modified in the Golgi and are, therefore, completely absent from the ER. Similarly, cardiolipin is believed to be exclusive to the inner mitochondrial membrane. These dramatic differences in membrane characteristics within a single cell likely extend to differences between cell/tissue types and interspecies differences, although these have only been anecdotally explored. Ultimately, these differences emphasize the need for careful consideration of the distinct compositional and biophysical phenotypes of specific membranes and the need for more extensive analysis of these phenotypes in various contexts (i.e., sources and manipulations).

2.3 ISOLATION, PROPERTIES, AND USES OF GIANT PLASMA MEMBRANE VESICLES

The capacity of cultured cells to shed large vesicles from their plasma membranes upon chemical treatment had been noted since the early 1960s (Belkin and Hardy, 1961). A number of chemicals can induce this behavior, most of them being cysteine-alkylating agents (Scott, 1976; Scott et al., 1979). Chemically induced blebbing produces highly PM-enriched preparations, and has been used to study the physical properties of the isolated PM (Dai and Sheetz, 1999; Tank et al., 1982). However, it was the discovery of liquid–liquid phase separation that prompted a renaissance in interest in cell-derived PM vesicles.

2.3.1 PHASE SEPARATION AND THE RAFT HYPOTHESIS

The lipid raft hypothesis has been one of the most influential and controversial in membrane biology since its introduction by Simons and Ikonen (1997). Essentially, the hypothesis posits lateral domains of distinct compositions and physical properties in cellular membranes, with these domains driven by weak, collective interactions between certain lipids and proteins in the membrane plane. The coexistence of liquid-ordered (Lo) and liquid-disordered (Ld) phases in synthetic biomimetic membranes provided the key physical underpinning for this hypothesis (Levental and Veatch, 2016). However, the major differences in lipid compositions and protein content between cellular membranes and synthetic models prevented definitive conclusions about the relevance of model membranes to cells. The question remained: Could liquid–liquid phase separation occur in a complex, protein-rich milieu, such as the plasma membrane? That question was answered definitively by seminal observations that showed liquid–liquid phase separation in PM vesicles derived from rat mast cells, termed

GPMVs (Baumgart et al., 2007b). Importantly, these phases sorted the native components of the PM, with some predicted lipid raft components being excluded from the unsaturated lipid-rich, presumably disordered phase. Later measurements confirmed the expected differences in membrane order (Kaiser et al., 2009; Levental et al., 2011) and diffusivity (Levental et al., 2009) between the two coexisting phases.

2.3.2 GPMV APPLICATIONS

These observations of Lo/Ld phase separation in GPMVs were a breakthrough for both conceptual and practical reasons. As noted above, direct observation of immiscible liquid phases in a complex, cell-derived membrane supports the central principle of the lipid raft hypothesis (though it should be noted that several caveats, discussed below, prevent the direct translation of this result to live cell PMs). As importantly, facile microscopic detection of coexisting phases allows investigation of a variety of phenomena related to the raft hypothesis, as well as general membrane features. The details for these analyses are described in Section 2.4 below.

Perhaps the most useful application of GPMVs is evaluation of partitioning of PM proteins between coexisting liquid phases. In GUVs, such experiments are complicated by the need for isolation or synthesis of the (often hydrophobic) polypeptide of interest, followed by reconstitution of the protein into a prefabricated membrane (protocols for this procedure are described in Chapter 3). GPMVs facilitate such measurements because the effort of producing and membrane-inserting the protein is undertaken by cellular machinery, with the isolated vesicles constitutively containing the protein of interest (provided it is trafficked to the PM). This method has been used extensively to explore the basis of ordered (i.e., raft) phase partitioning (Johnson et al., 2010; Sengupta et al., 2008), revealing a number of features including posttranslational lipidation (Levental et al., 2010), transmembrane domain length (Diaz-Rohrer et al., 2014) and surface area (Lorent et al., 2018) as key determinants. These three parameters comprise a predictive model for raft partitioning of single-pass transmembrane proteins (Lorent et al., 2018). Despite being highly predictive, other parameters like oligomerization or glycosylation likely also affect raft affinity (Lorent and Levental, 2015). Similar methods have been used to characterize the partitioning of lipid analogs and compare GUVs with GPMVs. Those comparisons revealed that fluorescent lipid analogs generally do not partition as predicted (e.g., sphingolipids partitioning to the more ordered raft phase) in GUVs, whereas the expected partitioning behavior is more accurately recapitulated in GPMVs (Sezgin et al., 2012).

Another common application of GPMVs is to investigate the properties of coexisting membrane domains in a biologically derived system. For example, the order of the coexisting phases has been compared between natural membranes and synthetic ones, revealing that GPMVs and other cell-derived systems have much smaller order differences between fluid domains than simple, synthetic GUVs (Kaiser et al., 2009). The differences between coexisting phases may underlie the temperature-dependent phase separation behavior observed in both synthetic and biological membrane models (Levental et al., 2011). Such phase separation temperature is important because it has been suggested to be related to the size and lifetime of nanodomains present at physiological temperatures (Veatch et al., 2008), thus there has been significant interest in agents that influence this temperature, and by supposition, the properties of domains *in vivo*. These factors include liquid anesthetics (Gray et al., 2013), bile acids (Zhou et al., 2013) and synthetic detergents (Levental et al., 2016), protein-lipid interactions (Johnson et al., 2010), lipid perturbations (Levental et al., 2009), cell cycle progression (Gray et al., 2015), dietary lipid feeding (Levental et al., 2016), and cellular differentiation (Levental et al., 2017).

2.3.3 COMPARISONS TO SYNTHETIC GUVs

GPMVs are somewhat smaller than GUVs (<10 μm), but otherwise appear microscopically similar using conventional imaging techniques (see Chapter 10). As in GUVs, liquid–liquid phase separation manifests as circular domains that exclude most fluorescent dyes not specifically designed for ordered phase partitioning (Sezgin et al., 2012). However, a number of important differences between the systems have been described; see Box 2.1. Most notable among these is the relatively small disparity between the coexisting ordered and disordered phases in GPMVs compared with GUVs (Kaiser et al., 2009). This difference is likely due to the specific lipid composition of the biological membranes, most notably the relative paucity of doubly unsaturated phospholipids like dioleoyl phosphatidylcholine (DOPC) (Gerl et al., 2012), which interact poorly with cholesterol and saturated lipids (Mannock et al., 2010; Róg et al., 2009). The majority of phospholipids in living membranes are so-called "hybrid lipids" with one saturated and one unsaturated acyl chain (e.g., 1-palmitoyl-2-oleoyl phosphatidylcholine (POPC); see Appendix 1 of the book for structure and data on this and other lipids) (Figure 2.1) and these are likely responsible for the relatively high order of the Ld phase in GPMVs. The consequence of the greater interdomain order discrepancy in GUVs is a packed Lo phase that excludes most lipid and protein components, even those with predicted raft affinity (Kahya et al., 2005; Sezgin et al., 2015; Shogomori et al., 2005).

An important consequence of the relatively small difference between phases in GPMVs is a relatively low miscibility transition temperature (T_{mix}; the temperature at which microscopic domains can be observed), which tends to be below room temperature, and can be as low as 5°C, depending on preparation conditions (Levental et al., 2011). Moreover, phase separation in GPMVs is often characterized by compositional fluctuations, indicative of a system near a critical point (Veatch et al., 2008). Similar behavior can be observed in GUVs with carefully chosen compositions (Veatch et al., 2007); however, it remains remarkable and unexplained how and why cells seemingly tune their highly complex PM compositions to be at or very near a compositional critical point, especially considering significant cell-to-cell heterogeneity (Levental and Veatch, 2016).

2.3.4 COMPARISONS TO LIVE CELL MEMBRANES

Experimentally, GPMVs offer a tractable experimental system to explore the physical phenotypes of biological membranes. Such phenotypes can include order, phase separation, and

Box 2.1 Important differences between GUVs, GPMVs and live cell membranes

	GUVs	GPMVs	LIVE CELL MEMBRANES
Lipids	Synthetic or extracted. Compositionally defined	Most plasma membrane lipids	See Figure 2.1
Proteins	Can be included at relatively low fractions. Contain usually only a small number of proteins	Contain most plasma membrane proteins. Exogenous constructs can be easily introduced (Box 2.4)	Organelle specific proteins
Asymmetry	Usually symmetric but protocols for the construction of asymmetric GUVs exist	Asymmetry is at least partially lost for lipids. Protein asymmetry has not been widely studied, but is likely preserved	The plasma membrane has asymmetric lipid and protein distribution
Phase separation	Nanoscopic/microscopic	Microscopic liquid–liquid at low temperature. Indirect evidence for nanoscopic domains at physiological temperature	No macroscopic domains. Nanoscopic domains are supported by extensive indirect data
	Large order difference between liquid phases	Small order difference between liquid phases	Not determined
	Most proteins are exclusively in Ld phase	Proteins partition between Lo/Ld phases	Not determined
Size	10–100 μm	Up to 10 μm	Organelle dependent
Cytoskeleton	No cytoskeleton	No cytoskeleton	Interacts with most/all organellar membranes

diffusivity, among a broad spectrum of other properties that have not yet been as widely characterized. GPMVs maintain the lipid complexity and protein content of native membranes, but are released from cytoskeletal and extracellular matrix components. Their most notable feature—microscopic liquid–liquid phase separation—is also the most striking difference between GPMVs and the PMs of live cells, which do not show microscopically discernable liquid domains. It remains unknown which of the many possible cellular factors is most responsible for this discrepancy, but physical predictions suggest that pinning a small number of membrane components can prevent large-scale phase separation (Machta et al., 2011; Yethiraj and Weisshaar, 2007). Thus, the presence of a relatively immobile membrane-associated scaffold, such as the cortical cytoskeleton or the extracellular matrix, could act as a mechanism to limit the size of domains in living cells (Arumugam et al., 2015; Honigmann et al., 2014).

Despite their utility, it is necessary to point out that while GPMVs are derived from live cells, they are not identical to the live cell PM, see Box 2.1. A number of important features of native membranes affected by GPMV isolation have been investigated, and these are unlikely to be the only caveats that must be taken into account when extending GPMV observations to living systems. For a recent extensive review of GPMV properties and caveats, see (Levental and Levental, 2015). The most important are (1) the loss of strict leaflet asymmetry, specifically of the anionic lipid phosphatidylserine (PS) (Baumgart et al., 2007b); (2) enzymatic modification/degradation of certain lipids (Keller et al., 2009); (3) loss of most assembled sub-membranous cytoskeleton; and (4) thermodynamic equilibrium (Gowrishankar et al., 2012).

2.4 PREPARATION OF GPMVs

2.4.1 ISOLATION OF GPMVs FROM CULTURED CELLS

Box 2.2 introduces a detailed protocol for inducing and isolating GPMVs from cultured cells. A representative image of GPMVs induced on a cell is given in Figure 2.3.

2.4.2 IMAGING GPMVs AND MEASUREMENT OF MISCIBILITY TRANSITION

Microscopic imaging of GPMVs is both simple and useful. These vesicles are comparable in size to GUVs and can be imaged with an inverted light/epifluorescence microscope (see Chapter 10). To observe membrane solubilization or changes in the shape of

Box 2.2 Isolation of GPMVs from mammalian cultured cells

Materials

- Mammalian cultured cells
- GPMV buffer: 10 mM HEPES, 100–150 mM NaCl, 2 mM $CaCl_2$, pH 7.4
- *N*-ethylmaleimide (NEM) stock solution: 1 M in ethanol (the solution is stable for 1 week at −20°C)
- Dithiothreitol (DTT) stock solution: 1 M in H_2O (the solution is stable for 1 week at −20°C)
- Paraformaldehyde (PFA) stock solution: 4% (wt/vol) in H_2O (stable for several months at −20°C)
- FAST DiO stock solution: 1,1′-dilinoleyl-3,3,3′,3′-tetramethylindocarbocyanine, 4-chlorobenzenesulfonate 0.5 mg/mL in dimethylsulfoxide (DMSO)

Notes:

- We have mainly isolated GPMVs from rat basophilic leukemia (RBL) cells and NIH 3T3 fibroblasts, both of which give robust and reproducible vesicle preparations. However, in our experience, GPMVs can be prepared from any mammalian cell line. Clearly, separation of GPMVs is much easier from cells that grow adherent to a culture surface rather than in suspension. Although suspension cells also produce large GPMV-like blebs, these often remain attached to the cells and require extra centrifugation steps to separate the vesicles from the cells.
- Extracellular calcium is required to induce GPMV formation, and this is likely related to the mechanism of GPMV formation (Keller et al., 2009).
- The osmolarity of the GPMV buffer can be reduced by decreasing NaCl concentration to 100 mM or lower. Lower extracellular osmolarity leads to the formation of larger GPMVs, suggesting the mechanism of formation is related to osmotic water influx into cells.
- A variety of chemicals can induce GPMV formation (Scott et al., 1979). Each likely contains a unique set of artifacts. The PFA/DTT combination is most commonly used because it is the most robust in terms of GPMV abundance and lack of cell detachment from the plate. However, there are several important caveats to this combination. First and most important, PFA is a chemical cross-linker that non-specifically reacts with and crosslinks proteins and lipids (Levental et al., 2011). Moreover, the DTT in this preparation likely affects the palmitoylation state of some proteins (Levental et al., 2010). Other chemicals (e.g., NEM) used for isolation obviate these issues, but phase separation is more difficult to induce and there are more detached cells.
- A variety of fluorescent membrane stains can be used to visualize GPMVs, ranging from highly non-raft preferring (e.g., FAST DiO) to raft-enriched (naphthopyrene) stains (Baumgart et al., 2007a). Importantly, such probes poorly recapitulate the partitioning of their native analogs (i.e., fluorescent sphingomyelin does not partition as native sphingomyelin) and probes do not necessarily partition identically in GPMVs and other model membranes (Sezgin et al., 2012)

Procedure

1. Grow adherent mammalian cells to 70% confluence in a 35-mm dish or 6-well plate (~1–1.5 × 10^6 cells for RBLs).
2. Wash cells two times with GPMV buffer.
3. Label cell membranes with fluorescent dye by adding 1 mL GPMV buffer containing FAST-DiO at 5 µg/mL (10 µL stock solution per mL of GPMV buffer) for 10 min at 4°C.
4. Wash cells four times with GPMV buffer.
5. Incubate cells with GPMV buffer supplemented with vesiculation chemicals for 30–60 min at 37°C.
   a. For NEM: Add 2 µL NEM stock solution to 1 mL GPMV buffer (final concentration: 2 mM).
   b. For PFA/DTT: Add 18 µL PFA stock solution and 2 µL DTT stock solution to 1 mL GPMV buffer (final concentrations: 25 mM PFA and 2 mM DTT).
6. Following incubation, verify the presence of GPMVs on a phase-contrast microscope. These should be present as both free-floating and cell-attached spheres near the bottom of the plate (Figure 2.3).
7. Decant or pipet the supernatant containing GPMVs and discard the cells on the plate.
8. If using suspension cells, it may be necessary to separate GPMVs from floating cells. For this, the preparation can be centrifuged at 100*g* for 3 min, which should pellet most cells while leaving some GPMVs in the supernatant. However, GPMV yield will be significantly reduced by this centrifugation because many GPMVs will also pellet.
9. To visualize GPMVs on a microscope, transfer the GPMV solution to a 1.5-mL conical microcentrifuge tube and incubate at 4°C for 15–30 min, then proceed to the protocol in Box 2.3.
10. To use GPMV for lipid analysis, proceed to the protocol in Box 2.5.

(Continued)

Box 2.2 (Continued) Isolation of GPMVs from mammalian cultured cells

Notes:

- The above conditions (incubation time, chemical concentrations) are reasonable starting points for a GPMV preparation; however, it may be necessary to modify them to optimize the preparation for any given cell type.
- An important consideration is the concentration of fluorescent dye, which should be kept at a minimum to avoid perturbation of membrane properties.
- Although detached cells in the GPMV preparation can be a nuisance, it is not necessary to remove them for microscopic experiments. However, for biochemical experiments (e.g., Western blotting or lipidomics) that require high PM enrichment, it is necessary to separate cells from GPMVs by centrifugation, as described in Box 2.3.
- Usually, GPMVs are visually stable after storage at 4°C for at least 1–2 days. However, for lipidomic analysis, the samples should be frozen or lyophilized immediately after preparation because lipid degradation/oxidation will increase over time.
- In a buffer, GPMVs quickly sink to the bottom of their container. This is helpful for concentrating the vesicles with a short incubation in a conical microcentrifuge tube. A concentrated vesicle suspension can then be pipetted directly from the bottom of the tube. Similarly, GPMVs will sink to the bottom of a visualization chamber/slide for observation on an inverted microscope.

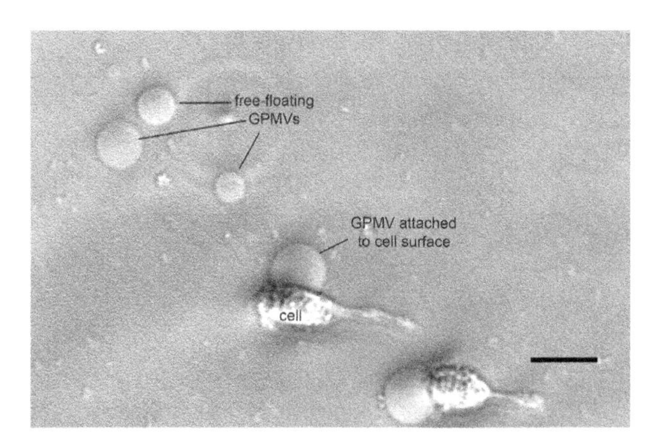

Figure 2.3 Free-floating and cell-attached GPMVs. Scale bar: 20 µm.

GPMVs, no fluorescent dye is required. The addition of fluorescent membrane dyes to GPMVs facilitates visualization of ordered/disordered phase separation. The best dyes are those with differential partitioning between phases which lead to a large contrast between phases (e.g., FAST-DiO partitions almost exclusively into the disordered phase in GPMVs). Phase separation in GPMVs is highly temperature dependent, with the T_{mix} denoting the temperature above which phase separation is no longer detectable by light microscopy (Chapter 18). This T_{mix} reflects the likelihood of the PM to phase separate and has been proposed to be related to the properties of raft domain *in vivo* (Veatch et al., 2008). Box 2.3 describes a protocol for visualization of phase separation in GPMVs on an inverted fluorescent microscope and presents an approach to determine the miscibility temperature as illustrated in Figure 2.4.

2.4.3 QUANTIFICATION OF LIPID AND PROTEIN PARTITIONING

Because they inherently include the proteins and lipids of the native cell membrane, using GPMVs to determine the partitioning of various membrane components between coexisting liquid phases is straightforward. In general, the component of interest is tagged with a fluorophore (for proteins, usually a genetically encoded tag such as green fluorescent protein; for lipids, typically analogs containing an organic fluorophore, but also specific lipid-binding protein domains), and the relative fluorescence intensity of that fluorophore between coexisting phases reflects the relative concentration of the component of interest. Raft and non-raft phases are distinguished as above, by a fluorescent marker with known partitioning characteristics (e.g., FAST-DiO always marks the disordered, non-raft phase) (Box 2.4).

2.4.4 GPMVs AS ISOLATED PLASMA MEMBRANE PREPARATION

Most methods used to isolate PMs from cultured cells are based on differential centrifugation (Maeda et al., 1983) or some specific biochemical property of the PM, such as enrichment of glycosylated material (Persson and Jergil, 1992). Such methods require high amounts of starting material to achieve suitable PM enrichments, and even then are often contaminated with intracellular organelle membranes. In contrast, GPMVs bleb specifically from the PM and appear free of internal membranes (Scott et al., 1979). Moreover, GPMVs can be easily and effectively separated from cells, yielding highly enriched PM preparations that are suitable for a variety of biochemical analysis (though we reiterate that GPMVs are not native PMs, with a variety of caveats associated with their isolation and preparation noted above). We describe here a method to prepare GPMVs for mass spectrometric analysis of their lipid composition (Box 2.5).

Box 2.3 Visualization of phase separation in GPMVs and determination of their miscibility transition temperature

Materials

- Bovine serum albumin (BSA) solution at 1 mg/mL in H_2O
- Small coverslips (22 × 22 mm)
- Large coverslips (22 × 44 mm)
- Vaseline or other paraffin wax
- Inverted fluorescence or confocal microscope
- Temperature controlled microscopy stage (e.g., Warner Instruments)

Notes:

- BSA coating of coverslips is used to prevent bursting of the vesicles on the coverslips in the microscopy chamber. Other hydrophilic coating agents such as gelatin and poly-L-lysine can also be used.

Procedure

1. Incubate small coverslips for 1 h at 4°C in BSA solution on a shaker.
2. Wash coverslips (22 × 22 mm) in water.
3. "Draw" a 1 × 1 cm square of paraffin wax in the center of the large (22 × 44 mm) coverslip using a wax-filled syringe and with a 20–200 µL pipet tip) (see figure below).

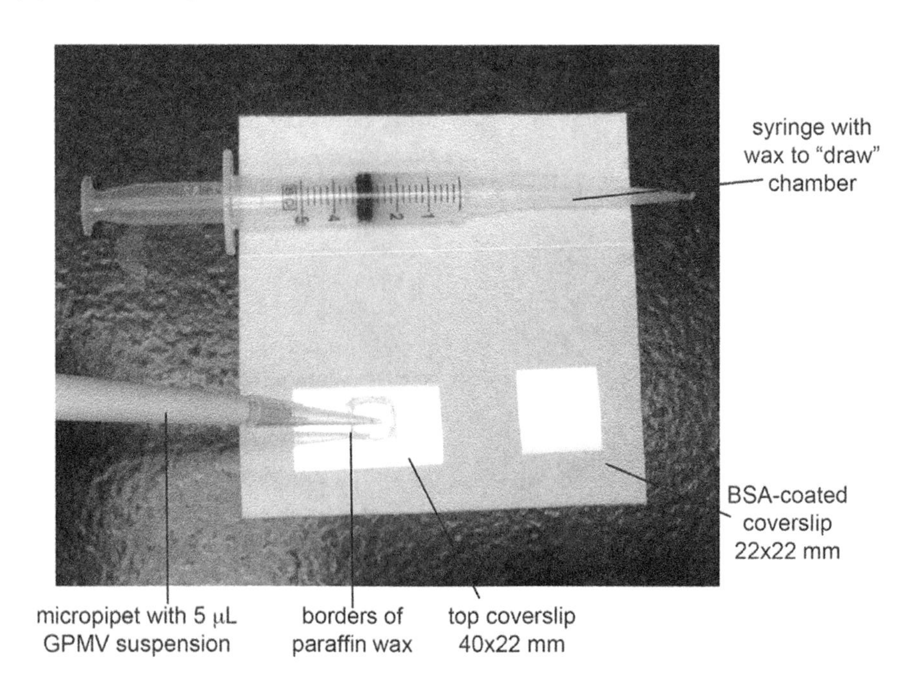

4. Add 5 µL of GPMV solution into square.
5. Put small BSA-coated coverslip on droplet of GPMV solution inside grease square to form a small wax-sealed imaging chamber.
6. Apply a small dot of wax to the other side of the large coverslip. (This will be used to affix the chamber to the microscope stage.)
7. Attach the chamber to the bottom of a temperature-controlled microscope stage, using the small dot of wax on the large coverslip. The small coverslip should be between the large coverslip and the imaging objective. In this configuration, the GPMVs will settle onto the BSA-coated small coverslip while the intimate contact between the large coverslip and the stage will facilitate efficient heat-transfer for temperature control of the chamber (see sketch on next page).

(Continued)

Box 2.3 (Continued) Visualization of phase separation in GPMVs and determination of their miscibility transition temperature

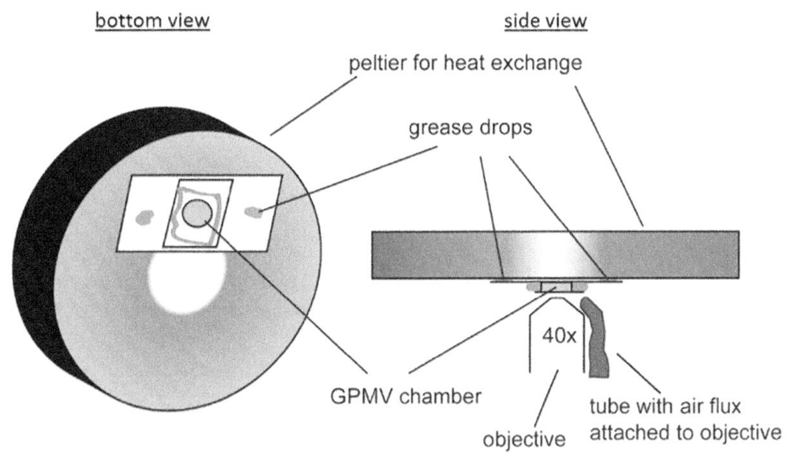

8. Set temperature to 4°C and observe GPMVs on an inverted microscope. Phase separation should be clearly observable as large circular domains on the surface of the vesicles (Figure 2.4).
9. Count ~50–100 GPMVs and determine the percentage of phase-separated vesicles.
10. Increase temperature by 3–5°C and repeat the counting procedure.
11. Repeat Step 10 until a large majority of vesicles (>90%) are no longer phase separated.
12. Plot the percentage of phase-separated vesicles as a function of temperature and fit to a sigmoid function. The temperature at which this fit gives 50% phase separation is defined as the T_{mix} (Figure 2.4).

Notes:

- The setup above is optimized for the determination of the miscibility temperature where accurate control of temperature is required. However, it is also useful for any experiments requiring observation of phase separation (e.g., measurement of the partitioning coefficient of lipids/proteins, as in Section 2.4.3) because the miscibility temperature of GPMVs is usually below room temperature.
- For accurate temperature control, we recommend using an air, rather than immersion, objective to prevent heat flux from the stage to the objective. High numerical aperture (up to 0.95) 40x air objectives provide sufficiently high magnification and resolution for most experiments.
- A significant problem for imaging cooled samples is condensation of ambient water vapor onto the sample. A simple solution is to gently stream dry air onto the sample, using the objective itself as a support for the tubing (as in the figure above).

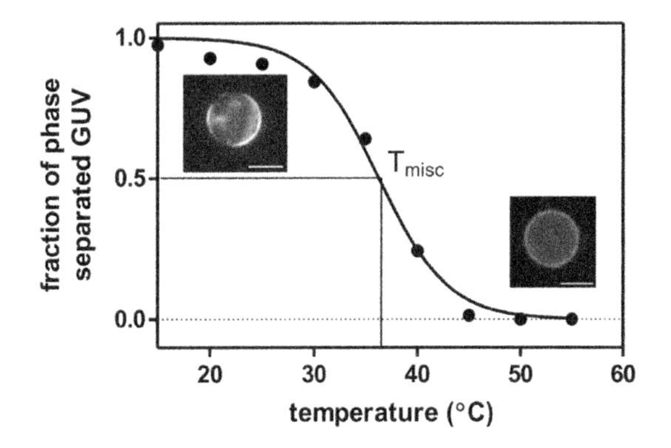

Figure 2.4 Determination of miscibility transition temperature (T_{mix}). The T_{mix} for this particular sample was ~37°C. However, most samples that we have encountered have significantly lower T_{mix}. Scale bar: 5 µm.

Box 2.4 Protocol for measuring lipid and protein phase partitioning

Materials

- Same as in Box 2.3, plus DNA construct of membrane protein (usually coupled to red fluorescent protein (RFP) when using FAST-DiO) or fluorescent lipid analog (e.g., NBD-sphingomyelin, see Appendix 2 of the book for structures and data on fluorescent lipid analogs) of component of interest

Notes:

- Any fluorescent moiety can be used to tag the component of interest, but care should be taken to avoid spectral overlap between it and the phase-specific dye.
- Although any transfection protocol that introduces a plasmid into cultured cells will work for the below procedure, it is important to be aware that common techniques (e.g., lipofection) are based on lipids and may interfere in unknown ways with the composition and/or properties of the GPMVs. For this reason, we prefer Nucleofection™, which uses electrical pulses to transiently perforate cells and introduce plasmids.
- As with all quantitative imaging experiments, it is important to consider artifacts associated with photo-oxidation, self-quenching, and non-radiative energy transfer (e.g., Förster resonance energy transfer (FRET)).
- If difficulties are encountered in imaging phase separation, for example, if a cooling device is unavailable, it is possible to raise the T_{mix} in GPMVs to above room temperature. We have observed that the bile acid deoxycholate at sub-lytic concentrations (~100–400 mM) can significantly increase phase separation temperature (Zhou et al., 2013), as can the synthetic detergent Triton™ X-100 at 0.001% (Levental et al., 2016), see Appendix 3 of the book for structure and data on this detergent. However, one must be cautious in interpreting the results obtained using such perturbations because their effects have not been clearly characterized.

Procedure

1. Transfect cells with plasmid or introduce fluorescent lipid analog (as in Step 3 in the procedure in Box 2.2).
2. Prepare GPMVs (see materials and procedure in Box 2.2).
3. Prepare GPMV slides/chambers (see materials and procedure in Box 2.2).
4. Observe GPMVs at a temperature below the miscibility transition, where most vesicles show phase separation (10°C typically works well for PFA/DTT GPMVs).
5. Find a single GPMV containing fluorescence from your component of interest, then image both the fluorescent channel containing the component of interest and the phase reference marker (i.e., FAST-DiO).
6. For each image, calculate the raft partition coefficient ($K_{p,raft}$) by measuring the background-subtracted fluorescence intensity of your component of interest in the raft/*Lo* (FAST-DiO–poor) and non-raft/*Ld* (FAST-DiO–rich) phase. $K_{p,raft} = I_{Lo}/I_{Ld}$ where I_{Lo} and I_{Ld} are the background-subtracted fluorescence intensities in the *Lo* and *Ld* phase, respectively (Figure 2.5).

Notes:

- We determine relative fluorescence intensity using the plot–profile (Analyze → Plot Profile) tool from the freely available software ImageJ, by profiling a line scan that passes through the two phases (Figure 2.5).
- Due to the inherent variability between individual cells, there can be significant variation in $K_{p,raft}$ measurements between individual GPMVs. We typically make measurements on at least 10 vesicles/experiment. Variability can also be reduced by careful maintenance of the cells, with attention to cell cycle, cell health, and construct expression.
- If exciting the fluorescent probe with linearly polarized light (e.g., laser excitation), it is important to consider the photoselection effect of certain membrane probes (Bagatolli, 2006), which will lead to higher excitation in the polarization plane. To avoid this artifact, a quarter-wave plate can be inserted into the light path to transform the linear polarized light into circular polarization.

The making of

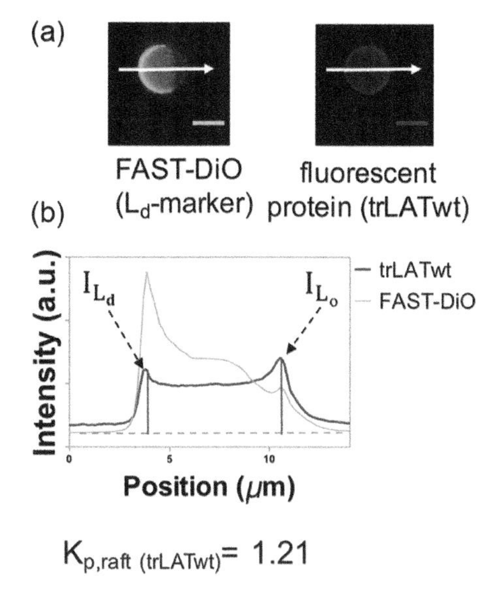

Figure 2.5 Determination of raft-partitioning coefficient ($K_{p,raft}$). (a) Epifluorescence microscopy images of a GPMV isolated from cells labeled with FAST-DiO (Ld marker) and expressing an RFP-tagged protein of interest. The arrow corresponds to the intensity profile seen in (b). Scale bar: 5 μm. (b) Intensity profile of fluorescent protein in GPMVs. The maxima correspond to the intensity in Ld and Lo phases, respectively.

Box 2.5 GPMV isolation for lipid and protein analysis

Materials

- The same materials as in Box 2.2
- Centrifuge (speed at least 20,000*g* for microcentrifuge tubes)
- NH_4HCO_3 (150 mM)

Notes:

- NH_4HCO_3 might be replaced by H_2O or other MS-suitable buffers.
- Microcentrifuge tubes coated to prevent leakage of phthalates that may interfere with subsequent mass spectrometric analysis are recommended.

Procedure

1. Isolate GPMVs using the procedure from Box 2.2, but starting from a 10-cm dish of densely cultured cells (~1–5 × 10^6 cells).
   a. Do not use PFA/DTT for biochemical preparations because these chemicals react with and crosslink proteins and lipids (especially amine-containing phospholipids). NEM is suitable for both mass spectrometry and/or Western blotting.
2. Centrifuge sample at 100*g* for 4 min to pellet cellular debris.
3. Transfer GPMV-rich supernatant to new microcentrifuge tube and centrifuge at 20,000*g* for 1 h at 4°C.
4. Discard supernatant.
5. Carefully wash the pellet without resuspending it with 2 mL of NH_4HCO_3 (150 mM), then discard supernatant.
6. Resuspend GPMVs in NH_4HCO_3. This preparation typically yields 50–100 μg of highly enriched PM, as determined by both mass spectrometric lipid analysis and Western blotting (Levental et al., 2016).

Notes:

- The efficiency and yield of the above preparation depends highly on cell type. Incubation times, vesicular concentration/type, and centrifugation speeds, for example, may need to be adjusted to optimize yield while minimizing contamination from intracellular membranes.

2.5 PREPARATION OF GUVs FROM NATURAL MEMBRANES

GPMVs maintain the complexity of native cellular PMs and allow quantification of membrane properties and protein partitioning in a near-native milieu; however, they are poorly controlled and subject to the variability and complexity inherent to biological systems. GUVs synthesized from pure lipid components obviate some of these drawbacks but fail to recapitulate some important aspects of native membranes. A potentially useful intermediate is the production of GUVs out of lipids extracted from natural membranes, as described for pulmonary surfactant lipids (Bernardino de la Serna et al., 2004), red blood cell (RBC) ghosts (Montes et al., 2007), neural synapses (Tulodziecka et al., 2016), yeast cells (Klose et al., 2010), and skin stratum corneum (Plasencia et al., 2007). Such GUVs are complex and biomimetic but have well-defined compositions (by analysis of lipid compositions) and are amenable to the suite of biophysical tools developed for model membrane studies. Notably, liquid–liquid phase separation has been directly observed in several such "natural" GUVs (see above references), with microsopic fluid domains persisting to physiological temperatures.

The most common extraction method for membrane lipids is described by Folch et al. (1957). Although the method is useful for most cell membranes, extraction efficiency is dependent on sample type (e.g., tissue or isolated cells). Further, although most lipid types are extracted by the Folch method, it has lower efficiency for highly hydrophilic lipids like glycolipids and phosphorylated inositols. Here we describe a slight modification on the Folch extraction, followed by GUV preparation from a crude cellular membrane extract, but the same general protocol applies for extracting/reconstituting GPMVs and other organelles (Box 2.6).

2.5.1 PREPARING GUVs FROM COMPLEX MIXTURES

Preparing GUVs from complex extracted lipid mixtures is similar to preparing GUVs from simple lipid mixtures (Chapter 1), although a few important points should be kept in mind. Lipid extracts usually contain a significant proportion of charged lipids, which means that electroformation protocols with higher frequencies (~500–1,000 Hz) should be used (Herold et al., 2012; Montes et al., 2007; Pott et al., 2008). Moreover, oxidation of polyunsaturated lipids could be a concern, so titanium wires rather than ITO slides are recommended (Ayuyan and Cohen, 2006). If practicable, the gentle hydration method may be preferred because it is amenable to charged lipids and avoids undue oxidation.

Notes:

- Even small quantities of oxidized lipids can significantly affect phase behavior in GUVs. Thus, for electroformation, Pt or Ti electrodes rather than ITO slides are recommended, as is the use of an antioxidant like *n*-propyl gallate (NPG) or butylated hydroxytoluene (BHT) (Ayuyan and Cohen, 2006).
- Lipid oxidation tests should be used to determine the amount of oxidation. A simple test is the ferrous oxidation–xylenol orange (FOX) assay to quantify peroxide levels (Ayuyan and Cohen, 2006).
- Buffered salt solutions may be preferable to sucrose because of the presence of charged lipids in extracts, however these require distinct electroformation conditions (Mikelj et al., 2013; Montes et al., 2007).

2.6 RED BLOOD CELL GHOSTS

An alternative isolated biological membrane model containing the full complement of biological lipids can be prepared directly from RBCs. RBC ghosts are prepared by osmotic or chemical lysis of

Box 2.6 Protocol for producing GUVs from lipid extracts

Materials

- Chloroform/methanol solution (2:1 v/v)
- $CaCl_2$ solution (4.5 mM)
- Glass syringe with needle (500 μL)
- Phosphate buffered saline (PBS pH = 7.4)
- Centrifuge
- Sonicator
- Ultrasonic bath
- Cells
- Chloroform resistant tubes (glass tubes)
- Methanol solution
- Analytical balance
- Gaseous nitrogen flow

Procedure

1. Start with ~10^6 cells for extraction.
2. Wash cells 3 times with PBS buffer.
3. Add 1 mL of PBS and scrape cells off plate into PBS suspension.
4. Weigh empty glass tube.

(Continued)

Box 2.6 (Continued) Protocol for producing GUVs from lipid extracts

5. Transfer cell suspension into glass tube and sonicate cell suspension for 5 s (with probe sonicator) to disrupt and homogenize cells.
6. Centrifuge cell suspension for 100,000*g* for 1 h to pellet crude cell membranes.
7. Discard supernatant and reweigh glass tube.
8. Calculate amount of material. The directions below are calculated for ~1 mg of membrane, a typical yield from 10^6 (70% confluent 10-cm dish of RBLs). The volumes below may need to be scaled up or down depending on isolated membrane amounts; however, the efficiency of Folch extraction is likely linear over a large range.
9. Add 1 volume (250 μL) of $CHCl_3$/MeOH solution.
10. Cover tube and put solution into ultrasonic bath for at least 10 min.
11. Add 0.2 volumes of $CaCl_2$ (4.5 mM), which will induce phase separation between organic and aqueous phases and present as either a milky solution when mixed or two clear, colorless liquids with a white filmy interface (largely proteins) when separated by centrifugation.
12. Vigorously vortex/agitate mixture for 10–20 min at room temperature.
13. Centrifuge mixture at 420*g* for 20 min to separate phases.
14. Remove upper (aqueous) phase and protein-rich interface using a glass syringe.
15. Transfer lower (organic) phase into a new glass vial using glass syringe.
16. If desired, this extract can be "washed" again by adding a methanol/$CaCl_2$ (1:1 v/v) solution and repeating Steps 12–15.
17. Evaporate organic solution under N_2 flow or speed vac.
18. Reconstitute lipids in organic solvent to desired concentration for GUVs production (typically 1–10 mg/mL).

Notes:

- A very significant concern for experiments involving extraction and reconstitution of cellular lipids is oxidation of the many unsaturated lipids present in such samples (Figure 2.1). Steps can be taken to reduce such oxidation, including using nitrogen (or another chemically inert gas) to dry the samples, keeping the samples at 4°C during extraction, degassing all buffers, and adding antioxidants; however, it is unlikely that these steps can fully prevent lipid oxidation, and even small amounts of such oxidized species can potentially affect experimental results (Ayuyan and Cohen, 2006).
- Efficiency of extraction may vary depending on the type of sample and the lipids of interest. Internal standards are necessary for quantitative reconstitution.
- After the low-speed centrifugation to separate the organic and aqueous phase, the intermediate white film contains most of the proteins from the preparation and can be used to quantify/normalize the protein content of the sample.
- The final concentration of the lipid solution can be determined by the inorganic phosphate assay (Bartlett, 1959). A simple way to semi-quantitatively determine the different lipid groups in a sample is by thin layer chromatography (Itoh et al., 1986; Leray et al., 1987). Unfortunately, these methods are not very sensitive and typically require much more material than is necessary to form GUVs. A simple alternative is the Amplex Red assay to quantify cholesterol concentration, from which the total lipid concentration can be estimated from measured cholesterol/phospholipid ratios for various membranes.
- Lipid extracts should be kept at −80°C in glass vials topped by Teflon® caps, which are resistant to chloroform. Prior to storage, lipid solutions should also be purged with N_2 or argon to eliminate oxygen and minimize oxidation.

erythrocytes, followed by resealing of the PM. Thus, these are essentially swelled erythrocyte PMs that are devoid of all non-membrane-associated intracellular material. There exist several established protocols for RBC ghost preparation (Schwoch and Passow, 1973), which we will not discuss in detail but point out important structural and compositional differences between ghost preparations and GUVs/GPMV.

Like other mammalian cells, native RBC possess a PM composed of hundreds of different lipid species with a typically high cholesterol/phospholipid ratio of almost 1 (compare to Figure 2.1) (Leidl et al., 2008; Rothman and Lenard, 1977). The fact that mature RBCs possess neither a nucleus, nor other organelles makes them excellent candidates for PM preparations. RBC ghosts, which lose most of their intracellular content during preparation, are even more specific. Also like other cells, the PM of erythrocytes has an asymmetric transbilayer lipid composition, with sphingomyelin and glycolipids enriched on the exoplasmic leaflet and negatively charged PS molecules on the cytoplasmic side (Harris et al., 2001; Rothman and Lenard, 1977). This asymmetry is lost, at least in part, during the preparation of ghosts (Harris et al., 2001; Steck et al., 1970). In this sense, ghosts are similar to GPMV, where PS appears also on the exoplasmic leaflet during preparation (Baumgart et al., 2007b), and GUVs, which usually do not possess an asymmetric membrane (though recent advances suggest asymmetric GUVs can be prepared [Cheng and London, 2009]). The abundance and nature of a membrane associated cytoskeleton in ghosts and GPMVs remains unclear. Neither preparation appears to have actin- or tubulin-based assemblies; however, spectrin is certainly present in RBC ghosts, where it may play an important biophysical and mechanical role. To our knowledge, the presence of spectrin in GPMVs has not been evaluated. Finally, RBC ghosts and GPMVs differ in their capacity for liquid–liquid phase separation, which is easily observed in GPMVs but not in RBC ghosts. The reason for this difference is currently unknown, but it may be related to the assembled cortical spectrin network in ghosts.

2.7 RAFTING INTO THE FUTURE

Although GPMVs have proven a useful model of an isolated mammalian PM, there remain a number of important caveats and limitations that prevent their unequivocal extrapolation to live cells. Perhaps the most significant are (1) PM lipid asymmetry; (2) lack of cytoskeleton/extracellular matrix; (3) undefined lipid compositions. Lipid asymmetry between opposing leaflets in the bilayer is a fundamental property of nearly all PMs, with mammalian cells expending significant energetic resources to prevent negatively charged lipids from flipping to the extracellular leaflet. GPMVs almost certainly retain some bilayer asymmetry since lipid-anchored proteins and glycosylated lipids are unlikely to flip across the bilayer, and transmembrane proteins probably retain their native orientation. However, strict asymmetry defined by exposure of PS is lost in GPMVs (Baumgart et al., 2007b), likely due to calcium-mediated activation of PM-resident scramblases. It is hoped that future technologies for isolation of GPMVs that retain the native lipid asymmetry of the PM will allow investigations of the biophysical consequences of this important property.

The interactions between membranes and cytoskeletal elements have become increasingly appreciated as key determinants of membrane function (Gowrishankar et al., 2012; Honigmann et al., 2014; Machta et al., 2011). Such interactions are almost certainly abrogated in GPMVs due to the isolation conditions; however, the components for cytoskeletal assembly are present in the lumen of GPMVs (which consists of cytoplasm, possibly somewhat diluted during blebbing), suggesting that it may be possible to re-activate their assembly.

The major advantage of GPMVs is that they allow access to microscopic liquid–liquid phase separation in a complex, biological milieu. However, up to now, this capacity has mostly been exploited in measurements of partitioning or phase separation that do not probe protein activity. An exciting direction would be the observation of some protein property, such as protein–protein binding affinity, enzyme activity, or channel physiology that is dependent on the lipid phase in which the protein is dissolved. Such experiments could be relatively straightforward, for example imaging the binding between a fluorescent receptor and a fluorescent ligand; if the two do not show identical partition coefficients, it would imply that the receptor preferentially binds ligand in one of the phases (as in Sezgin et al. (2015)).

Finally, although there exist some convincing hypotheses relating microscopic phase separation in GPMVs to ordered raft domains at physiological conditions (Veatch et al., 2008), the properties—and indeed the existence—of ordered domains in live cells remain controversial (Levental and Veatch, 2016). An important step toward resolving the controversy would be to investigate the putative domains remaining in GPMVs after phase separation is no longer microscopically observable. Super-resolution microscopy and spectroscopy suggests that this goal lies in the foreseeable future (Eggeling et al., 2009; Honigmann et al., 2014; Stone and Veatch, 2015).

LIST OF ABBREVIATIONS

BHT	butylated hydroxytoluene
BMP	bis(monoacylglycero)phosphate
BSA	bovine serum albumin
Chol	cholesterol
DAG	diacylglycerol
DMSO	dimethyl sulfoxide
DOPC	dioleoyl phosphatidylcholine
DTT	dithiothreitol
ER	endoplasmic reticulum
GFP	green fluorescent protein
GPMV	giant plasma membrane vesicle
HEPES	4-(2-hydroxyethyl)-1-piperazineethanesulfonic acid
NEM	N-ethylmaleimide
NPG	*n*-propyl gallate
PFA	paraformaldehyde
PC	phosphatidylcholine
PE	phosphatidylethanolamine
PI	phosphatidylinositol
PL	phospholipid
PM	plasma membrane
POPC	1-palmitoyl-2-oleoyl phosphatidylcholine
PS	phosphatidylserine
RBC	red blood cell
RBL	rat basophilic leukemia
RFP	red fluorescent protein
SM	sphingomyelin

GLOSSARY OF SYMBOLS

Lo	liquid ordered phase
I_{Lo}	fluorescence intensity in liquid ordered phase
Ld	liquid disordered phase
I_{Ld}	fluorescence intensity in liquid disordered phase
$K_{p,raft}$	raft partition coefficient
T_{mix}	miscibility transition temperature

REFERENCES

Arumugam S, Petrov EP, Schwille P (2015) Cytoskeletal pinning controls phase separation in multicomponent lipid membranes. *Biophys J* 108:1104–1113.

Ayuyan AG, Cohen FS (2006) Lipid peroxides promote large rafts: Effects of excitation of probes in fluorescence microscopy and electrochemical reactions during vesicle formation. *Biophys J* 91:2172–2183.

Bagatolli LA (2006) To see or not to see: Lateral organization of biological membranes and fluorescence microscopy. *Biochim Biophys Acta* 1758:1541–1556.

Bartlett GR (1959) Phosphorus assay in column chromatography. *J Biol Chem* 234:466–468.

Baumgart T, Hammond AT, Sengupta P, Hess ST, Holowka DA, Baird BA, Webb WW (2007b) Large-scale fluid/fluid phase separation of proteins and lipids in giant plasma membrane vesicles. *Proc Natl Acad Sci USA* 104:3165–3170.

Baumgart T, Hunt G, Farkas ER, Webb WW, Feigenson GW (2007a) Fluorescence probe partitioning between Lo/Ld phases in lipid membranes. *BBA—Biomembr* 1768:2182.

Belkin M, Hardy WG (1961) Relation between water permeability and integrity of sulfhydryl groups in malignant and normal cells. *J Biophys Biochem Cytol* 9:733–745.

Bernardino de la Serna J, Perez-Gil J, Simonsen AC, Bagatolli LA (2004) Cholesterol rules: Direct observation of the coexistence of two fluid phases in native pulmonary surfactant membranes at physiological temperatures. *J Biol Chem* 279:40715–40722.

Brown DA, London E (1998) Functions of lipid rafts in biological membranes. *Annu Rev Cell Dev Biol* 14:111–136.

Cheng HT, London E (2009) Preparation and properties of asymmetric vesicles that mimic cell membranes: Effect upon lipid raft formation and transmembrane helix orientation. *J Biol Chem* 284:6079–6092.

Dai J, Sheetz MP (1999) Membrane tether formation from blebbing cells. *Biophys J* 77:3363–3370.

Diaz-Rohrer BB, Levental KR, Simons K, Levental I (2014) Membrane raft association is a determinant of plasma membrane localization. *Proc Natl Acad Sci USA* 111:8500–8505.

Dietrich C, Bagatolli LA, Volovyk ZN, Thompson NL, Levi M, Jacobson K, Gratton E (2001) Lipid rafts reconstituted in model membranes. *Biophys J* 80:1417–1428.

Dupuy AD, Engelman DM (2008) Protein area occupancy at the center of the red blood cell membrane. *Proc Natl Acad Sci USA* 105:2848–2852.

Eggeling C, Ringemann C, Medda R, Schwarzmann G, Sandhoff K, Polyakova S, Belov VN et al. (2009) Direct observation of the nanoscale dynamics of membrane lipids in a living cell. *Nature* 457:1159–U1121.

Folch J, Lees M, Sloane Stanley GH (1957) A simple method for the isolation and purification of total lipides from animal tissues. *J Biol Chem* 226:497–509.

Gerl MJ, Sampaio JL, Urban S, Kalvodova L, Verbavatz JM, Binnington B, Lindemann D et al. (2012) Quantitative analysis of the lipidomes of the influenza virus envelope and MDCK cell apical membrane. *J Cell Biol* 196:213–221.

Gowrishankar K, Ghosh S, Saha S, Rumamol C, Mayor S, Rao M (2012) Active remodeling of cortical actin regulates spatiotemporal organization of cell surface molecules. *Cell* 149:1353–1367.

Gray EM, Diaz-Vazquez G, Veatch SL (2015) Growth conditions and cell cycle phase modulate phase transition temperatures in RBL-2H3 derived plasma membrane vesicles. *PLoS One* 10:e0137741.

Gray E, Karslake J, Machta BB, Veatch SL (2013) Liquid general anesthetics lower critical temperatures in plasma membrane vesicles. *Biophys J* 105:2751–2759.

Harris FM, Smith SK, Bell JD (2001) Physical properties of erythrocyte ghosts that determine susceptibility to secretory phospholipase A2. *J Biol Chem* 276:22722–22731.

Herold C, Chwastek G, Schwille P, Petrov EP (2012) Efficient electroformation of supergiant unilamellar vesicles containing cationic lipids on ITO-coated electrodes. *Langmuir* 28:5518–5521.

Honigmann A, Sadeghi S, Keller J, Hell SW, Eggeling C, Vink R (2014) A lipid bound actin meshwork organizes liquid phase separation in model membranes. *Elife* 3:e01671.

Ipsen JH, Karlstrom G, Mouritsen OG, Wennerstrom H, Zuckermann MJ (1987) Phase equilibria in the phosphatidylcholine-cholesterol system. *Biochim Biophys Acta* 905:162–172.

Itoh YH, Itoh T, Kaneko H (1986) Modified Bartlett assay for microscale lipid phosphorus analysis. *Anal Biochem* 154:200–204.

Janmey PA, Lindberg U (2004) Cytoskeletal regulation: Rich in lipids. *Nat Rev Mol Cell Biol* 5:658–666.

Johnson SA, Stinson BM, Go MS, Carmona LM, Reminick JI, Fang X, Baumgart T (2010) Temperature-dependent phase behavior and protein partitioning in giant plasma membrane vesicles. *Biochim Biophys Acta* 1798:1427–1435.

Kahya N, Brown DA, Schwille P (2005) Raft partitioning and dynamic behavior of human placental alkaline phosphatase in giant unilamellar vesicles. *Biochemistry* 44:7479–7489.

Kaiser HJ, Lingwood D, Levental I, Sampaio JL, Kalvodova L, Rajendran L, Simons K (2009) Order of lipid phases in model and plasma membranes. *Proc Natl Acad Sci USA* 106:16645–16650.

Keller H, Lorizate M, Schwille P (2009) PI(4,5)P2 degradation promotes the formation of cytoskeleton-free model membrane systems. *Chemphyschem* 10:2805–2812.

Klose C, Ejsing CS, Garcia-Saez AJ, Kaiser HJ, Sampaio JL, Surma MA, Shevchenko A, Schwille P, Simons K (2010) Yeast lipids can phase-separate into micrometer-scale membrane domains. *J Biol Chem* 285:30224–30232.

Leidl K, Liebisch G, Richter D, Schmitz G (2008) Mass spectrometric analysis of lipid species of human circulating blood cells. *Biochim Biophys Acta* 1781:655–664.

Leray C, Pelletier X, Hemmendinger S, Cazenave JP (1987) Thin-layer chromatography of human platelet phospholipids with fatty acid analysis. *J Chromatogr* 420:411–416.

Levental I, Grzybek M, Simons K (2011) Raft domains of variable properties and compositions in plasma membrane vesicles. *Proc Natl Acad Sci USA* 108:11411–11416.

Levental I, Lingwood D, Grzybek M, Coskun U, Simons K (2010) Palmitoylation regulates raft affinity for the majority of integral raft proteins. *Proc Natl Acad Sci USA* 107:22050–22054.

Levental I, Byfield FJ, Chowdhury P, Gai F, Baumgart T, Janmey PA (2009) Cholesterol-dependent phase separation in cell-derived giant plasma-membrane vesicles. *Biochem J* 424:163–167.

Levental KR, Levental I (2015) Giant plasma membrane vesicles: Models for understanding membrane organization. *Curr Top Membr* 75:25–57.

Levental KR, Lorent JH, Lin X, Skinkle AD, Surma MA, Stockenbojer EA, Gorfe AA, Levental I (2016) Polyunsaturated lipids regulate membrane domain stability by tuning membrane order. *Biophys J* 110(8):1800–1810.

Levental KR, Surma MA, Skinkle AD, Lorent JH, Zhou Y, Klose C, Chang JT, Hancock JF, Levental I (2017) ω-3 polyunsaturated fatty acids direct differentiation of the membrane phenotype in mesenchymal stem cells to potentiate osteogenesis. Sci Adv 3:eaao1193.

Lingwood D, Simons K (2010) Lipid rafts as a membrane-organizing principle. *Science* 327:46–50.

Lorent JH, Levental I (2015) Structural determinants of protein partitioning into ordered membrane domains and lipid rafts. *Chem Phys Lipids* 192:23–32.

Levental I, Veatch S. (2016) The continuing mystery of lipid rafts. J Mol Biol 428:4749–4764.

Lorent JH, Diaz-Rohrer B, Lin X, Spring K, Gorfe AA, Levental KR, Levental I (2017) Structural determinants and functional consequences of protein affinity for membrane rafts. Nat Commun 8:1219.

Machta BB, Papanikolaou S, Sethna JP, Veatch SL (2011) Minimal model of plasma membrane heterogeneity requires coupling cortical actin to criticality. *Biophys J* 100:1668–1677.

Maeda T, Balakrishnan K, Mehdi SQ (1983) A simple and rapid method for the preparation of plasma membranes. *Biochim Biophys Acta* 731:115–120.

Mannock DA, Lewis RN, McMullen TP, McElhaney RN (2010) The effect of variations in phospholipid and sterol structure on the nature of lipid-sterol interactions in lipid bilayer model membranes. *Chem Phys Lipids* 163:403–448.

Mesmin B, Maxfield FR (2009) Intracellular sterol dynamics. *Biochim Biophys Acta* 1791:636–645.
Mikelj M, Praper T, Demic R, Hodnik V, Turk T, Anderluh G (2013) Electroformation of giant unilamellar vesicles from erythrocyte membranes under low-salt conditions. *Anal Biochem* 435:174–180.
Mitra K, Ubarretxena-Belandia I, Taguchi T, Warren G, Engelman DM (2004) Modulation of the bilayer thickness of exocytic pathway membranes by membrane proteins rather than cholesterol. *Proc Natl Acad Sci USA* 101:4083–4088.
Montes LR, Alonso A, Goni FM, Bagatolli LA (2007) Giant unilamellar vesicles electroformed from native membranes and organic lipid mixtures under physiological conditions. *Biophys J* 93:3548–3554.
Persson A, Jergil B (1992) Purification of plasma membranes by aqueous two-phase affinity partitioning. *Anal Biochem* 204:131–136.
Plasencia I, Norlen L, Bagatolli LA (2007) Direct visualization of lipid domains in human skin stratum corneum's lipid membranes: Effect of pH and temperature. *Biophys J* 93:3142–3155.
Pott T, Bouvrais H, Meleard P (2008) Giant unilamellar vesicle formation under physiologically relevant conditions. *Chem Phys Lipids* 154:115–119.
Róg T, Pasenkiewicz-Gierula M, Vattulainen I, Karttunen M (2009) Ordering effects of cholesterol and its analogues. *Biochim Biophys Acta (BBA)—Biomembranes* 1788:97–121.
Rothman JE, Lenard J (1977) Membrane asymmetry. *Science* 195:743–753.
Sampaio JL, Gerl MJ, Klose C, Ejsing CS, Beug H, Simons K, Shevchenko A (2011) Membrane lipidome of an epithelial cell line. *Proc Natl Acad Sci USA* 108:1903–1907.
Schwoch G, Passow H (1973) Preparation and properties of human erythrocyte ghosts. *Mol Cell Biochem* 2:197–218.
Scott RE (1976) Plasma membrane vesiculation: A new technique for isolation of plasma membranes. *Science* 194:743–745.
Scott RE, Perkins RG, Zschunke MA, Hoerl BJ, Maercklein PB (1979) Plasma membrane vesiculation in 3T3 and SV3T3 cells. I. Morphological and biochemical characterization. *J Cell Sci* 35:229–243.
Sengupta P, Hammond A, Holowka D, Baird B (2008) Structural determinants for partitioning of lipids and proteins between coexisting fluid phases in giant plasma membrane vesicles. *Biochim Biophys Acta* 1778:20–32.
Sezgin E, Grzybek M, Buhl T, Dirkx R, Gutmann T, Coskun U, Solimena M, Simons K, Levental I, Schwille P (2015) Adaptive lipid packing and bioactivity in membrane domains. *PLoS One* 10(4):e0123930.
Sezgin E, Levental I, Grzybek M, Schwarzmann G, Mueller V, Honigmann A, Belov VN et al. (2012) Partitioning, diffusion, and ligand binding of raft lipid analogs in model and cellular plasma membranes. *Biochim Biophys Acta* 1818:1777–1784.
Shevchenko A, Simons K (2010) Lipidomics: Coming to grips with lipid diversity. *Nat Rev Mol Cell Biol* 11:593–598.
Shogomori H, Hammond AT, Ostermeyer-Fay AG, Barr DJ, Feigenson GW, London E, Brown DA (2005) Palmitoylation and intracellular domain interactions both contribute to raft targeting of linker for activation of T cells. *J Biol Chem* 280:18931–18942.
Simons K, Ikonen E (1997) Functional rafts in cell membranes. *Nature* 387:569–572.
Simons K, Vaz WL (2004) Model systems, lipid rafts, and cell membranes. *Annu Rev Biophys Biomol Struct* 33:269–295.
Steck TL, Weinstein RS, Straus JH, Wallach DF (1970) Inside-out red cell membrane vesicles: Preparation and purification. *Science* 168:255–257.
Stone MB, Veatch SL (2015) Steady-state cross-correlations for live two-colour super-resolution localization data sets. *Nat Commun* 6:7347.
Tank SW, Wu ES, Webb WW (1982) Enhanced molecular diffusibility in muscle membrane blebs; release of lateral constraints. J Cell Biol 92:207–212.
Tulodziecka K, Diaz-Rohrer BB, Farley MM, Chan RB, Di Paolo G, Levental KR, Waxham MN, Levental I (2016) Remodeling of the postsynaptic plasma membrane during neural development. *Mol Biol Cell* 27:3480–3489.
van Meer G, Voelker DR, Feigenson GW (2008) Membrane lipids: Where they are and how they behave. *Nat Rev Mol Cell Biol* 9:112–124.
Veatch SL, Cicuta P, Sengupta P, Honerkamp-Smith A, Holowka D, Baird B (2008) Critical fluctuations in plasma membrane vesicles. *ACS Chem Biol* 3:287–293.
Veatch SL, Keller SL (2003) Separation of liquid phases in giant vesicles of ternary mixtures of phospholipids and cholesterol. *Biophys J* 85:3074–3083.
Veatch SL, Soubias O, Keller SL, Gawrisch K (2007) Critical fluctuations in domain-forming lipid mixtures. *PNAS* 104:17650–17655.
Wallin E, von Heijne G (1998) Genome-wide analysis of integral membrane proteins from eubacterial, archaean, and eukaryotic organisms. Protein science: A publication of the *Protein Soc* 7:1029–1038.
Wenk MR (2010) Lipidomics: New tools and applications. *Cell* 143:888–895.
Yethiraj A, Weisshaar JC (2007) Why are lipid rafts not observed in vivo? *Biophys J* 93:3113–3119.
Zhou Y, Maxwell KN, Sezgin E, Lu M, Liang H, Hancock JF, Dial EJ, Lichtenberger LM, Levental I (2013) Bile acids modulate signaling by functional perturbation of plasma membrane domains. *J Biol Chem* 288:35660–35670.

Protein reconstitution in giant vesicles

Matthias Garten, Daniel Lévy, and Patricia Bassereau

Lipids and Proteins: the nutritious mix for your experiment.

Contents

3.1 INTRODUCTION

Membrane proteins are involved in all major cellular processes such as cell homeostasis, bioenergetics, cell division and communication. Nearly 25% of human genes encode for membrane proteins and about 60% of all drugs in use today target a membrane protein (Yıldırım et al., 2007). The protein class comprises peripheral proteins that transiently bind to the lipid membrane and integral (or transmembrane) proteins that contain single or several membrane spanning motives and that remain in the membrane.

When trying to understand their role in cellular functions or when designing, drugs it is often beneficial to isolate these proteins and study them in a well-controlled biomimetic membrane environment, such as in cell-sized giant unilamellar vesicles (GUVs). The process of transferring a membrane protein from its native membrane to a biomimetic environment is termed reconstitution and, in most cases, first results in the formation of small unilamellar vesicles with reconstituted proteins (so-called proteo-SUVs). For the sake of simplicity, we do not distinguish here SUVs (with a diameter smaller than 100 nm) from large unilamellar vesicles (LUVs) with a diameter larger than 100 nm up to 1 µm and we refer to both as SUVs. Giant unilamellar vesicles with reconstituted proteins (proteo-GUVs) correspond to proteo-giant unilamellar vesicles with a diameter larger than a few micrometers (Box 3.1). Proteo-SUVs are often essential to form protein-containing biomimetic membrane systems such as proteo-GUVs, as we will see in this chapter, but they are also used to deliver membrane proteins to supported lipid bilayers or black-lipid films.

Transmembrane proteins are generally obtained by overexpression in homologous or heterologous host cells, then extracted from the membrane and solubilized in detergent micelles that provide an amphiphilic environment similar to the lipid bilayer. Next, solubilized proteins are purified using affinity and size exclusion chromatography. Strategies of purification and use of detergents are not detailed in this chapter but readers can refer to reviews (Junge et al., 2008; Andréll and Tate, 2013; Zorman et al., 2015). Alternatively, native membrane preparations can be used as a source material, with all proteins naturally present in them (Keller et al., 1988; Montes et al., 2007; Dezi et al., 2013). The first reconstitutions of membrane proteins into proteo-SUV were performed in the late 1970s (Darszon et al., 1980), and the method is now well established in membrane biochemistry.

Box 3.1 General methods for membrane protein reconstitution in liposomes

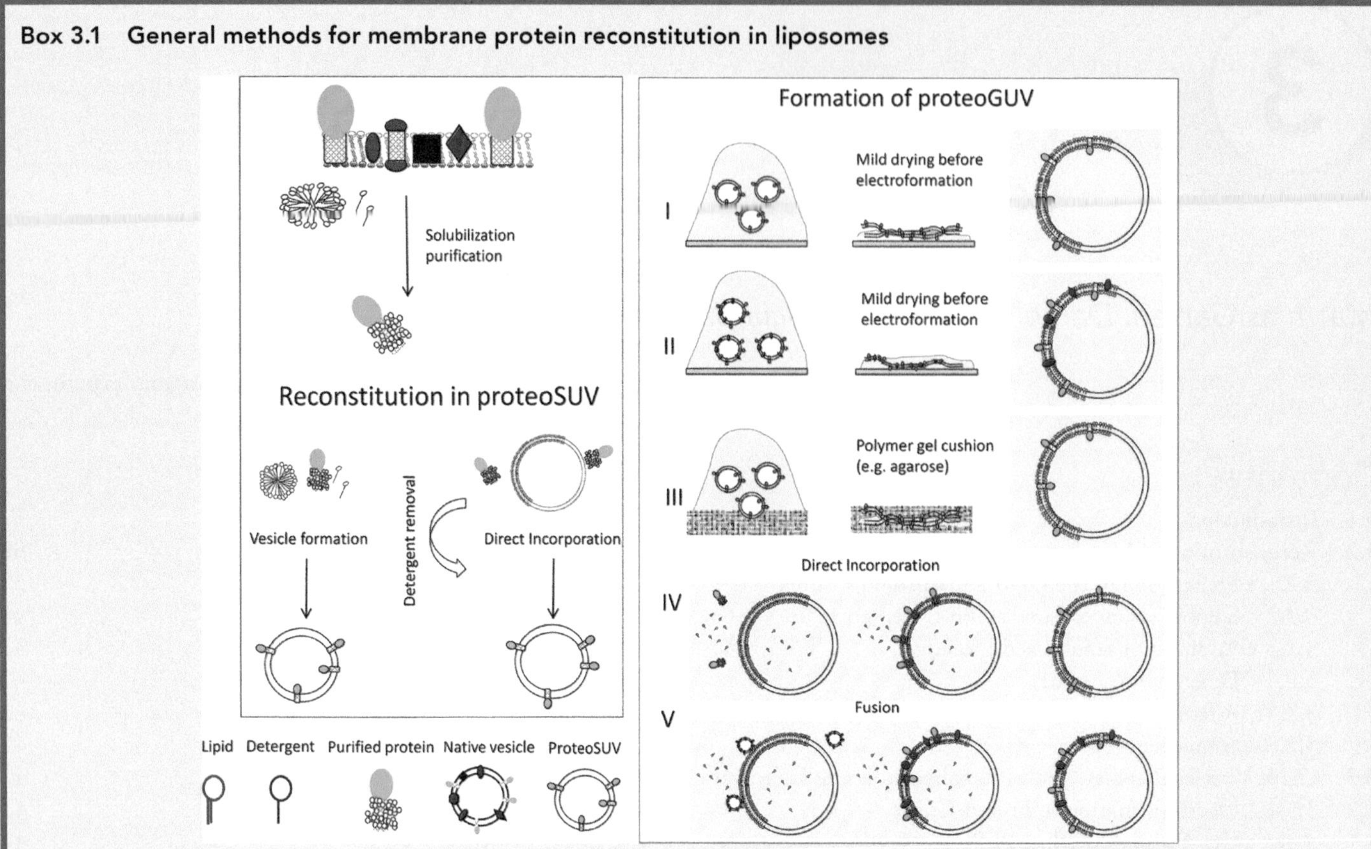

Methods for protein reconstitution are summarized in this box. The reconstitution of proteins into proteo-SUVs is shown in the panel outlined in red (left). The panel outlined in blue (right) summarizes methods for reconstitution of membrane proteins into proteo-GUVs. Electroformation from: (I) dried film of proteo-SUVs or (II) native vesicles (e.g., plasma membrane fragments obtained with a French Press). (III) Spontaneous swelling from a dried film of proteo-SUV on a polymer gel. (IV) Direct incorporation of solubilized membrane proteins in lipidic GUVs destabilized by a sub-solubilizing concentration of detergent followed by detergent removal. (V) Detergent-mediated fusion of native vesicles with lipidic GUVs, followed by detergent removal.

Hence, successful reconstitutions have been reported with prokaryotic and eukaryotic membrane proteins, constituted of β-sheets like porins or highly hydrophobic α-helices, with single and multiple helical spans, with large extramembranous domains, or with several subunits. Membrane proteins that are functionally coupled can be co-reconstituted to decipher their interactions. Reconstitution into SUVs is an obligatory step to assign the putative function of a newly purified membrane protein; it is also convenient for checking protein functionality before preparing GUVs. Extensive descriptions of the mechanism of reconstitution in SUVs and examples can be found in reviews (Rigaud et al., 1995; Rigaud and Lévy, 2003; Geertsma et al., 2008).

Since the late 1990s, successful strategies for the reconstitution of membrane proteins into GUVs have been developed (see Table 3.1 for an overview). In this chapter, after presenting protocols for membrane protein reconstitution into lipid bilayers to form proteo-SUVs, we will cover the two most commonly reported classes of proteo-GUV formation methods:

- Dehydration-rehydration
- Direct incorporation or fusion

Dehydration-rehydration techniques exploit the fact that dried lipid stacks tend to form separate bilayers upon rehydration. Correct control of the rehydration conditions allows GUV formation (Box 3.1 I-III, right panel). Direct incorporation or fusion methods utilize preformed lipidic GUVs. Membrane proteins are introduced *a posteriori* either by detergent-mediated direct incorporation or by fusion of vesicles, mediated by fusogenic peptides or by detergents (Box 3.1 IV-V, right panel).

3.2 RECONSTITUTION OF MEMBRANE PROTEINS IN SUVs (PROTEOLIPOSOMES)

In a large number of contemporary examples of proteo-GUV formation, proteo-SUVs serve as base material. In dehydration-rehydration methods, they are used to form the lipid film from which GUVs are swollen, and direct reconstitution methods can use SUVs to fuse them to GUVs. Details for the different steps of the preparation of proteo-SUVs are provided below (see also Box 3.1, left panel).

3.2.1 PREPARATION OF THE LIPID SUSPENSION IN BUFFER

In principle, any lipid or mixture of lipids can be used if reconstitution is performed at a temperature above the gel-to-fluid phase transition. Lipids solubilized in organic solvents are dried into a film from which a suspension of multilayered liposomes is formed after addition of a buffer. Fluorescent lipids can also be added for subsequent observation

Table 3.1 **Selection of membrane proteins reconstituted in GUVs and corresponding techniques (chronological order)**

RECONSTITUTION METHOD IN GUVs	PROTEIN STATE	PROTEIN	ASSAY FOR PROTEIN ACTIVITY	USE OF DETERGENT FOR GUV RECONSTITUTION	RECONSTITUTION BUFFER	REFERENCES
Spontaneous swelling from native membranes in organic solvent (I, no electric field)	Native membrane preparation	Bovine and squid rhodopsin, reaction centers from *R. sphaeroides*, beef heart cytochrome c oxidase, acetylcholine receptors from *T. californica*	Absorption of light with activity, pH changes, electrical recordings	No, organic solvents	Medium (25 mM)-to-low ionic strength	Darszon et al. 1980
Spontaneous swelling (I, no electric field)	Native membrane preparation	Diverse: rabbit skeletal muscle, rat brain synaptosomal membrane preparation, chloroplast envelopes	Electrical recordings	No	Physiological ionic strength	Keller et al. 1988
From a dried film of proteoliposomes (mix of II and IV)	Solubilized in organic solvent	Bacteriorhodopsin (BR)	Membrane fluctuations	No	Low ionic strength	Manneville et al. 1999
Direct by small fusogenic peptide (V)	Proteo-SUV	BR	H- pumping	No, fusogenic peptide	Physiological ionic strength	Kahya et al. 2001
From a dried film of proteo-liposomes (I)	Proteo-SUV	BR Ca-ATPase	H pumping ATP dependent Ca^{2+}-translocation	No	Low ionic strength	Girard et al. 2004
Dehydration-rehydration, electroformation (I)	Proteo-SUVs	SNARE complex	Test of binding specificity	No	Low ionic strength	Bacia et al. 2004
From a film of proteo-SUVs dried in the presence of sugars (protectants) (I)	Proteo-SUV	MscL LacS OppA	Tested in LUV not in GUVs	No	Low ionic strength	Doeven et al. 2005; Geertsma et al. 2008
From a dried film of native membranes (II)	Native membranes	Several: human erythrocyte ghosts	Not tested	No detergent	Physiological ionic strength	Montes et al. 2007
Direct mixing of purified proteins with GUVs (IV)	Solubilized in detergent	OmpF	Conductance in GUVs	OctylPoE below CMC	Low ionic strength	Kreir et al. 2008
Direct mixing of purified proteins with GUVs (IV)	Solubilized in detergent	MscL, MscS	Conductance in GUVs	DDM below CMC	Low ionic strength	Battle et al. 2009
Dehydration-rehydration, electroformation (I)	Proteo-SUV	Integrin $\alpha_{IIb}\beta_3$	Binding to fibrinogen	No	Low ionic strength	Streicher et al. 2009
Direct insertion into inverse emulsion GUVs (IV)	Solubilized in detergent	KcsA	pH sensitive K conductance	DDM at CMC	Physiological ionic strength	Yanagisawa et al. 2011

(Continued)

Table 3.1 (*Continued*) **Selection of membrane proteins reconstituted in GUVs and corresponding techniques (chronological order)**

RECONSTITUTION METHOD IN GUVs	PROTEIN STATE	PROTEIN	ASSAY FOR PROTEIN ACTIVITY	USE OF DETERGENT FOR GUV RECONSTITUTION	RECONSTITUTION BUFFER	REFERENCES
From a dried film of proteo-SUVs, electroformation and agarose gel (I and III)	Proteo-SUV	KvAP	Conductance in GUVs	No detergent	Physiological ionic strength	Aimon et al. 2011; Garten et al. 2015
Fusion of proteoliposomes to preformed monolayer (variation of V)	Proteo-SUV	SNARE Single span membrane proteins	SNARE- mediated Fusion of small vesicles to GUVs	No detergent required	Physiological ionic strength	Richmond et al. 2011
From a dried film of proteo-SUVs (I)	Proteo-SUV	VDAC	Electrical recording of GUV membrane transferred to a planar membrane	No detergent	Physiological ionic strength	Betaneli et al. 2012
Dehydration-rehydration, electroformation (I)	Proteo-SUV	Na^+,K^+–ATPase	ATPase activity	No detergent	Medium (30 mM) ionic strength	Bouvrais et al. 2012
Direct incorporation in GUVs (IV) Fusion of proteo-SUVs or native membranes to GUVs (V)	Solubilized in detergent proteo-SUV native membrane	BR FhuA BmrC/BmrD	H- pumping DNA transfer ATP-dependent Drug translocation	Detergent above CMC	Physiological and any ionic strengths	Dezi et al. 2013
Dehydration-rehydration, spontaneous swelling (I, no electric field)	Proteo-SUV	TRPV1 (capsaicin receptor)	Electrical recordings	No	Physiological ionic strength	Cao et al. 2013
Dehydration-rehydration, electroformation (I)	Proteo-SUV	Serotonin receptor	Single-channel activity	No	Physiological ionic strength	Hassaine et al. 2014
Direct mixing of purified proteins with GUVs (IV)	Lyophilized	Phospholamban	Electrical recordings	No	Physiological ionic strength	Smeazzetto et al. 2016
Dehydration-rehydration, electroformation (II)	Proteo-SUVs from native membranes	Aquaporin 0	Water permeability measurement	No	Low ionic strength	Berthaud et al. 2016
Fusion of proteo-SUVs to GUVs (V)	Proteo-SUV	Cytochrome bo_3-oxidase, F_1F_0 ATPase, cytochrome c oxidase, Na^+/H^+ antiporter NapA, proteorhodopsin	ATP-production in the co-reconstituted system	No	Physiological ionic strength	Biner et al. 2016

Note: Roman numerals in the first column refer to the techniques outlined in Box 3.1.

with fluorescent microscopy of proteo-GUVs. The fraction of fluorescent lipids must be kept below a maximum of 0.8%–1.0% to limit the effect of lipid photooxidation (Morales-Penningston et al., 2010) as well as the saturation of the fluorescence signal.

3.2.2 SOLUBILIZATION OF LIPIDS WITH DETERGENTS

During the next step, lipids are solubilized in detergent micelles. To that end, liposomes prepared as in Section 3.2.1 are mixed with a stock detergent solution. The concentration of detergent for the complete solubilization depends on the type of detergent, on its critical micelle concentration (CMC), and on the detergent/lipid ratio; for instance, 1 mL solution of egg phosphatidylcholine/egg phosphatidic acid (Egg PC/Egg PA, see Appendix 1 of the book for structure and data on these lipids) (9:1) at 4 mg/mL (~5 mM) is solubilized either with 10 mg/mL of Triton™ X-100 (see Appendix 3 for structure and data on this detergent), 12.5 mg/mL *n*-dodecyl β-D-maltoside (DDM) or 7.3 mg/mL (25 mM) octylglucoside (OG). More details on the detergent/lipid ratio for solubilization can be found in reports by Rigaud and Lévy (2003) and Lichtenberg et al. (2013). Solubilization takes only minutes with Triton X-100 and OG, but requires 2 h with DDM and other bulky-head detergents. It is worth noting that lipids can be solubilized with detergents different from the one used for the purification of the protein.

3.2.3 ADDITION OF SOLUBILIZED PROTEINS

The type of membrane protein usually does not interfere with the reconstitution process when the lipid-to-protein ratio (LPR) is high, for example, above LPR 20 w:w for a 100-kDa protein (~2,000 lipids/protein mol:mol). Proteins solubilized in detergent are added to the solubilized mixture of lipid/detergent and the mixture is equilibrated a few minutes before detergent removal. Reconstitution can be performed at 20°C or 4°C, depending on the stability of the proteins.

3.2.4 DETERGENT REMOVAL

An efficient way to remove detergents is by hydrophobic adsorption onto polystyrene beads, Bio-Beads™ (see reviews by Rigaud et al. 1995; Rigaud and Lévy 2003; Geertsma et al., 2008). Given that detergents are amphiphilic, all detergents can be removed by Bio-Beads. We and others have successfully used Bio-Beads to completely remove more than 30 high-CMC and low-CMC detergents. The resulting proteo-SUVs are unilamellar and their average sizes generally range between 50 and 200 nm. Detergents can be removed at any temperature. No protein adsorption and negligible lipid adsorption have been reported. The resulting proteo-SUVs have the lowest reported ionic permeability and sustained the formation of a pH gradient stable up to 2 pH units. Alternatively, detergents can be removed by dialysis using a dialysis bag with a cutoff of 14 kDa. The rate of detergent removal depends on its CMC: high-CMC (5–20 mM) and low-CMC detergents (0.1–1 mM) are removed within 24 h and over several weeks, respectively.

Bio-Beads are added to the lipid/detergent/protein mixture at a Bio-Beads/detergent ratio of 10 w:w. The solution is gently stirred. After 2 h, the same amount of Bio-Beads is added to the already present beads for 1 h, and eventually a third addition is made for another 1 h. After stirring is stopped, the Bio-Beads sediment and the SUV suspension are pipetted off. The formation of proteo-SUVs from a micellar solution can be followed using a UV spectrometer by recording the increase of the turbidity at 400 nm. Proteo-SUVs are unilamellar with a diameter of 100 nm. Proteoliposomes can be held at 4°C, and activity measurements are usually performed within 48 h, depending on the stability of proteins. (Box 3.2)

Box 3.2 Protocol for reconstitution of proteoliposomes
Example of BmrCD, a bacterial heterodimeric ABC transporter.

1. *Preparation of dried lipid films*
   First, 4 mg of lipids or lipid mixtures, for example, Egg PC/Egg PA (9/1) in an organic solvent ($CHCl_3$ or $CHCl_3$/MeOH 9/1), is dried in a 10-mL evaporation balloon with a rotavapor or under a N_2 flux, followed by 2 h drying with a vacuum pump. The dried film is resuspended by vortexing for 5 min in 1 mL HEPES 50 mM (pH 7.2), NaCl 150 mM, or any other buffer that stabilizes the purified proteins of interest.
2. *Solubilization of lipid suspensions*
   Next, 1 mL solution of Egg PC/Egg PA (9/1) is solubilized at 4 mg/mL with 12.5 mg/mL of Triton X-100 under gentle stirring at room temperature. The previously turbid suspension becomes transparent.
3. *Addition of BmrCD*
   BmrCD is a bacterial ABC transporter, 65 kDa, and overexpressed in *E. coli* (Galian et al., 2011). Proteins are purified and solubilized at 1 mg/mL in DDM 0.05%. An aliquot of 100 µL of protein solution is added to 1 mL of solubilized mixture of Egg PC/Egg PA at a lipid/protein ratio equal to 40 w:w and equilibrated under stirring for 15 min at room temperature.
4. *Bio-Bead preparation*
   Bio-Beads (Biorad, 50 mesh) are prepared ("washed") by adding 1–2 g beads for 30 min to 50 mL MeOH, followed by extensive rinsing with 50 mL distilled water to eliminate the MeOH. The washed beads are held at 4°C and used within the next 2 weeks. Bio-Beads are weighted as "wet beads," that is, a 1-mL aliquot of bead solution is pipetted from the stock solution and the water is removed with a pipette and finally with an absorbing Kimwipes® (lint-free) tissue.
5. *Detergent removal*
   Bio-Beads are weighed in a 500-µL microcentrifuge tube (Eppendorf®) and added directly to the reconstitution solution. For 1 mL of 4 mg/mL lipids solubilized with 10 mg/mL Triton X-100, add 100 mg Bio-Beads, which corresponds to a Bio-Bead/detergent ratio of 10 w:w. The solution is gently stirred. After 2 h, 100 mg Bio-Beads are added for 1 h, then eventually a third addition is made for 1 h. It is worth noting that volumes can be adapted, keeping the ratio of Bio-Beads to detergent constant, for example, for a 100-µL reconstitution volume, Bio-Beads are sequentially added: 10 mg (2 h)/10 mg (1 h)/10 mg (1 h). If reconstitution is performed at 4°C, you can adapt the amount of Bio-Beads and time of detergent removal according to Levy et al. (1990).

3.2.5 ALTERNATIVE METHOD: DIRECT INCORPORATION OF MEMBRANE PROTEINS IN PREFORMED SUVs

In the "direct incorporation" approach, solubilized proteins are added directly to preformed liposomes destabilized with sub-solubilizing detergent concentrations. This favors unidirectional insertion of membrane proteins into membranes. Direct insertion is mediated by sugar-based detergents, for example, OG or DDM, which creates specific defaults in the lipid bilayer. Proteins are inserted in a unique orientation into the membrane through their most hydrophobic domain, leaving their hydrophilic domain pointing outward (Rigaud et al., 1995; Geertsma et al., 2008).

Liposomes are first prepared using 1 mL lipid suspension at 5 mg/mL, further extruded though a 0.2-µm filter to decrease the amount of large multilamellar vesicles. Next, add DDM at 3.3 mg/mL, corresponding to a detergent/lipid ratio of 0.66 w:w to saturate the vesicles with DDM without solubilizing them. The mixture is equilibrated for 1 h. To reconstitute the protein, add the solubilized protein at a lipid/protein ratio of 40 w:w to the preformed liposomes. Then, incubate for 15 min to allow for protein insertion. Finally, remove the detergent by the addition of 60 mg Bio-Beads for 2 h.

3.2.6 SOME REMARKS

To verify that the protein is well preserved at this step of the reconstitution, it is recommended to perform functional assays on proteo-SUVs before forming proteo-GUVs. For instance, in the case of ion channels and transporters, this is classically done using electrophysiology on black lipid membranes fused with proteo-SUVs (Finol-Urdaneta et al., 2010) or with flow cytometry (Rusinova et al., 2014).

It is important to note, that for the dehydration-rehydration methods, as described in Section 3.3, in which the SUVs will be dehydrated, salt and other osmolite concentrations in the SUV buffer must be kept low (<5 mM). Buffers with higher osmolite concentrations prevent GUV formation, presumably due to the very high final osmolality in the dehydrated lipid film and its interaction with the swelling and/or formation of crystals that interfere with the lipid film formation.

3.3 GIANT VESICLES PREPARED FROM PROTEOLIPOSOMES BY DEHYDRATION-REHYDRATION TECHNIQUES

A classical technique to form GUVs is rehydration of a previously dehydrated lipid film (see Chapter 1). The first successful attempts to form giant proteo-liposomes were reported by dehydrating purified membranes from various sources and rehydrating them, resulting mostly in multi-lamellar membrane structures (Darszon et al., 1980; Criado and Keller, 1987). In contrast, giant liposome preparation by electroformation leads to a high yield of unilamellar vesicles (Angelova and Dimitrov, 1986; Méléard et al., 2009). This advantage is preserved for proteo-GUV formation when using electroformation protocols, at low salt (>5 mM total salt concentration) on indium tin oxide (ITO) electrodes (Girard et al., 2004) and at physiological salt concentrations (100–200 mM) on platinum (Pt) electrodes (Aimon et al., 2011) (Box 3.1, blue-outlined panel, I and II). More recently, an agarose-assisted swelling method has been adapted that allows a relative high yield of GUVs with a very homogeneous protein distribution while preserving the advantage of using physiological buffers (Garten et al., 2015; Horger et al., 2015) (Box 3.1, blue-outlined panel, III). In all cases, it is recommended to fluorescently label membrane proteins to eventually check their incorporation into the GUV membranes.

To illustrate reconstitution by electroformation on Pt-wires and gel-assisted swelling, we present the case of the archeal voltage-gated potassium channel from *Aeropyrum pernix* (KvAP) into GUVs (Garten et al., 2015).

3.3.1 ELECTROFORMATION OF PROTEO-GUVs

Swelling of GUVs in the presence of an electric field has the advantage of an increased yield of unilamellar vesicles. The electric field is applied via electrodes onto which the proteo-SUVs are dehydrated to form a lipid/protein film (Figure 3.1a). Just as for the protein-free GUVs in Chapter 1, two types of electrodes can be used, depending on the salinity of the medium: ITO-coated glass slides, for low salt (~5 mM) buffers, or Pt wires, which allow usage of physiological salt concentrations (~100 mM) (Méléard et al., 2009) and which may be important to preserve protein activity. After the dehydration step, the

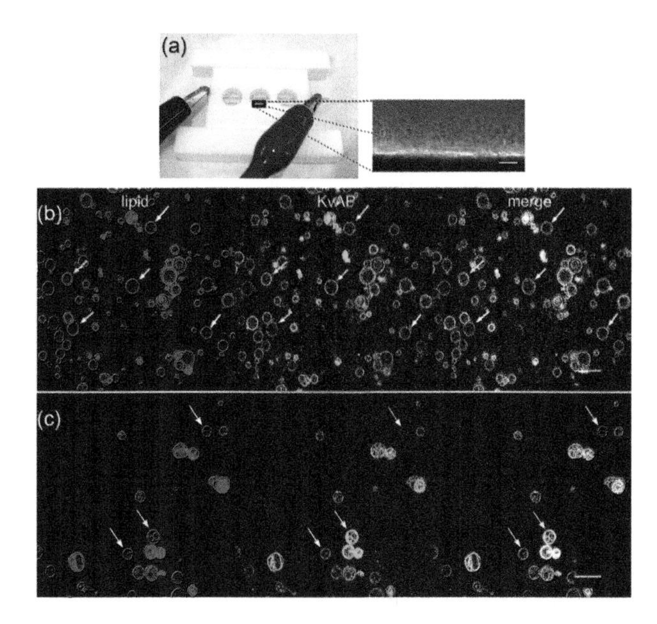

Figure 3.1 Proteo-GUV formation by electroformation. (a) Chamber used to grow the GUVs on platinum wires. Inset: GUVs on the wire. (b, c) Representative confocal images of proteo-GUVs (white arrows: KvAP, Egg PC:Egg PA 9:1) obtained from electroformation on platinum wires in (b) low (5 mM KCl)- and (c) high (100 mM KCl)-salt buffers. Yellow arrows point to bi- or multilamellar vesicles. Left (magenta): fluorescent lipid Texas Red® 1,2-dihexadecanoyl-*sn*-glycero-3-phosphoethanolamine (Texas Red DHPE), middle (green): Alexa488-labeled KvAP, right: overlay. Scale bars: 20 µm. ([a] Modified from Aimon, S., *PLoS One*, 6, e25529, 2011; a inset, b and c modified from Garten, M. et al., *J. Vis. Exp.*, 95, 52281, 2015.) (a, inset) Phase-contrast microscopy; (b, c) confocal microscopy (see Chapter 10).

GUVs are formed in the growth buffer with an applied electric field. Depending on the buffer salinity, protocols vary: For low-salt buffers, low-frequency fields are used: 1–3 h at 10 Hz, $V_{rms} = 0.7$ V (the root mean square $V_{rms} = V_{peak}/\sqrt{2}$); whereas high-salt buffers demand longer swelling times with higher frequencies: 12 h to overnight at 500 Hz, $V_{rms} = 0.35$ V, see Box 3.3. Ramping voltage and frequency has been reported to aid GUV formation (Méléard et al., 2009) but is not necessary in all cases (Garten et al., 2015). One may consider keeping the total output voltage below $V_{peak} = 1.1$ V to minimize the risk of sample oxidation. Note that the voltages given in the protocols correspond to voltages at the electrodes. We recommend measuring the voltage effectively applied to the electrodes to ensure good electrical connection and to compensate for the loss of electrical potential due to the impedance of the electroformation chamber.

The chamber used for KvAP-GUV formation is the same as that presented in Chapter 1. Note that a simple chamber can be improvised by poking the Pt wires through the lid of a glass vial or by tucking them down in a small petri dish.

Figures 3.1b and c show examples of resulting GUVs. Low-salt protocols (Figure 3.1b) give a higher yield for KvAP and a more homogeneous protein distribution among vesicles. High-salt protocols (Figure 3.1c) have the advantage of using physiological salt concentrations while still yielding enough GUVs for most experiments.

Box 3.3 Protocol for proteo-GUV electroformation on Pt wires

Example of the KvAP channel (Aimon et al., 2011; Garten et al., 2015). General advice on the choice of lipids and buffers can be found in Box 3.8.

- *Cleaning of the wires (~30 min)*
  A clean electroformation chamber is important for a successful GUV formation.
  1. Depending on your chamber design, remove glass windows and wipe off all sealant and grease.
  2. If possible, remove the wires. Rinse and scrub the chamber and the wires with ethanol and water, finish with acetone.
  3. Sonicate the chamber and wires in acetone, ethanol and deionized water for 5 min, scrubbing the wires between the steps. If your chamber wires are not held in place by sealant, insert them back in the chamber and scrub one last time with ethanol before the water sonication step.
  4. Dry the chamber under a stream of air or nitrogen. It is now ready for the lipid deposits.
- *Lipid deposit and dehydration (~35 min + SUV preparation time)*
  5. Prepare the KvAP-SUVs at 3 mg/mL following the recommendations in Section 3.2 of this chapter (see also Aimon et al., 2011). KvAP-GUVs can be prepared with protein concentrations as high as 1:10 (w:w) protein/lipid. The protein/lipid ratio in the SUVs eventually translates into the final concentration in the proteo-GUVs.
  6. Use a small pipette (<5 µL) to deposit many sub-microliter droplets onto the wire. Avoid contact between droplets, which would lead to coalescence and the formation of a single thick lipid film which is detrimental to GUV formation. About 1 µL SUV suspension can be deposited per 1-cm wire.
  7. Let the drops dry. Typically 30 min at room temperature work well for KvAP. Cover the chamber to avoid bleaching of fluorescent markers.
- *Swelling (1–3 h for low-salt, 12–16 h for high-salt buffers)*
  8. Assemble the chamber if necessary. The chamber shown in Figure 3.1a has a coverslip attached to the bottom with high-vacuum grease (Dow Corning) and the wire holes closed by sealant paste (Vitrex medical A/S, Denmark).
  9. Add growth buffer following the general considerations in this chapter. Avoid rapid movements of the buffer, which may detach the lipid film from the wire.
  10. Close the chamber. Tiny openings on the top are tolerable as long as the chamber is not put on absorbant tissue, which can soak liquid out of it.
  11. Apply voltage with a function generator. The detailed protocol depends on the salinity of the buffer. With low-salt buffers (<5 mM), proteo-GUVs generally tend to grow well at 10 Hz and 0.7 V_{rms} in 1–3 h. High-salt buffers (100–200 mM) need 500 Hz and 0.35 V_{rms} for 12–16 h. Check the voltage on the wires with a multimeter.
- *Harvest*
  If the chamber has a glass window, growth can be followed under a microscope. GUVs typically grow like a "bunch of grapes" (Figure 3.1a inset). GUVs on the wire are visible with phase-contrast or differential interference contrast microscopy. If no GUVs can be found, they may be on the other side of the wire. In any case, it is worth transferring the GUV preparation to an observation chamber to assess the yield.
  12. To pipette GUVs from the growth chamber, open it gently, keeping the wires covered with the growth buffer solution, that is, not exposing them to air.
  13. GUVs tend to remain attached to the wire and can be conveniently pipetted off. GUVs can be fragile and should be pipetted slowly with a wide pipette tip (~1-mm opening).
  14. GUVs can be transferred to a passivated glass vial tube and stored for up to 1 week at 4°C.
  15. To work with the GUVs under the microscope, they should be transferred to a chamber containing the observation buffer and a passivated coverslip at the bottom. The slide can be passivated by applying a β-casein solution (5 mg/mL in 20-mM HEPES [pH 7.4]; Sigma, ≥98% pure) for 5 min and rinsing thoroughly with deionized water.

3.3.2 GEL-ASSISTED SWELLING OF PROTEO-GUVs

Electroformation of GUVs in physiological salt buffers tends to have a low yield, and growth protocols take a long time to complete. The electric field may also cause lipid oxidation (Breton et al., 2015); thus, it may be desirable to avoid it for very fragile proteins. Gel-assisted swelling protocols have the best of both worlds: as a modification of spontaneous swelling methods, they do not rely on an electric field, and they yield a large number of GUVs under physiological salt conditions (Horger et al., 2009; Weinberger et al., 2013), faster (typically 30–60 min) and with a higher fraction of unilamellar vesicles than with conventional spontaneous swelling protocols. As for the electroformation methods, a lipid film is dehydrated. However, instead of applying an electric field to aid the swelling, the gel promotes the formation of GUVs (Figure 3.2a).

Agarose-assisted swelling has been successfully adapted to form proteo-GUVs (Garten et al., 2015; Horger et al., 2015) yielding GUVs with a very homogeneous protein distribution (Figure 3.2b and Box 3.4) A caveat lies in the gel itself, which can residually stay attached to the vesicle and change its properties (Lira et al., 2014).

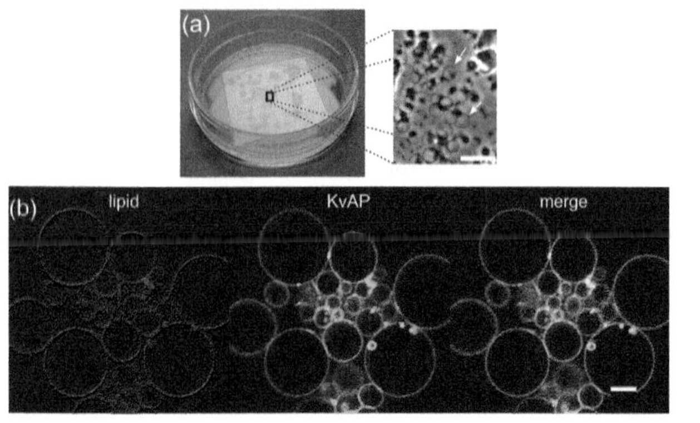

Figure 3.2 Agarose-assisted swelling of proteo-GUVs. (a) Petri dish holding an agarose-coated coverslip. Inset: GUV swelling on the agarose gel. Scale bar: 20 µm. (b) Example of proteo-GUVs (1,2-diphytanoyl-*sn*-glycero-3-phosphocholine [DPhPC see Appendix 1 of the book for structure and data on this lipid], KvAP) formed on agarose. Left (magenta): fluorescent lipid BODIPY® TR Ceramide, middle (green): Alexa488 labeled KvAP, right: overlay. Scale bar: 5 µm. ([a inset and b] Modified with permission from Garten, M. et al., *J. Vis. Exp.*, 95, 52281, 2015.) (a inset) Phase-contrast microscopy; (b) confocal microscopy (Chapter 10).

3.4 POST-INCORPORATION OF TRANSMEMBRANE PROTEINS INTO GIANT VESICLES

The two next methods differ from the previous ones in such that membrane proteins are directly inserted or fused to preformed lipidic GUVs in physiological buffers. Thus, steps that can be harmful for fragile membrane proteins like drying steps and incubation with potentially denaturing low-salt buffers are avoided (Dezi et al., 2013).

The strategy consists of first forming GUVs in the presence of mild detergents at sub-solubilizing concentrations (Box 3.1, blue-outlined panel, IV and V). For example, stable GUVs can be prepared in the presence of 100 µM *n*-dodecyl-β-D-thiomaltoside (DOTM) (with a CMC of 50 µM), whereas they solubilize at

Box 3.4 Protocol for agarose-aided swelling of proteo-GUVs

Example of the KvAP channel. General advice on the choice of lipids and buffers can be found in Box 3.8.

- *Prepare the agarose gel-coated coverslip (~30 min)*
  1. Prepare a 1% w:vol agarose solution by boiling it in a microwave twice and stirring (be careful, boiling retardation may lead to spraying of the hot gel upon touch). The solution can be used immediately or it can be stored for at least 1 month at 4°C and be reheated for later use.
  2. Plasma clean (high power, adjust vacuum so that the plasma is brightest; Harrick PDC-32G-2) for 45 s a small coverslip (e.g., 22 × 22 mm; Figure 3.2a) that will later fit into a small petri dish. Plasma aids spreading of the agarose solution on the glass. If no plasma cleaner is available, a piranha treatment (careful: highly toxic) may be considered to make the glass hydrophilic.
  3. Within 15 min after cleaning, apply ~200 µL warm agarose solution onto the slide. Remove excess solution by tilting the slide and touching a Kimwipes® tissue to it to draw off the solution.
  4. Let the slide dry for 30 min at 60°C.
- *Lipid deposit (~30 min)*
  5. Use a small pipette (<5 µL) to deposit sub-microliter droplets of KvAP-SUVs (Aimon et al., 2011) onto the agarose-coated slide. Avoid coalescence of the droplets to keep the lipid film thin. Small, non-contacting droplets work best. A small slide takes about 10–20 µL SUV solution.
  6. Let the drops dry in air until the water is evaporated. The evaporation can be expedited by placing the slide under a steam of air or nitrogen.
- *Swelling (30–60 min)*
  7. Place the slide in a small dish and carefully add growth buffer from the side so that the slide is covered. Avoid any movement of the dish that could detach the vesicles prematurely. The growth can be followed under a microscope (Figure 3.2a inset).
- *Harvest*
  8. To transfer the GUVs to the observation chamber, tap the dish a few times and drop it on the table from a height of ~1 cm to detach the vesicles. Aspirate a few microliters of the GUV suspension with a pipette with a large tip diameter (>1 mm).
  9. Add the suspension onto a passivated coverslip (see the electroformation protocol) in the observation buffer. Example GUVs are shown in Figure 3.2b.

200 μM DOTM. Electroformation of the GUVs in a sucrose solution is used to ensure formation of larger amounts of GUVs compared with growth in salt buffers. Moreover, GUVs can be electroformed with several lipids, including biologically important lipids such as phosphatidylinositol or cardiolipin. Then, either solubilized proteins are added to detergent-containing GUVs for direct incorporation (see Section 3.4.2) or purified native vesicles or proteo-SUVs are incorporated by detergent-mediated fusion with detergent-GUVs (see Section 3.4.3). Finally, detergent molecules are removed by the addition of Bio-Beads. Importantly, the presence of detergent during the process increases the ionic permeability of the GUV membrane and allows for the exchange of internal and external contents after mixing with any buffer. Thus, GUVs can be prepared with specific lipid compositions that, after detergent removal and buffer exchange, can have different internal and external ionic contents. Both protocols are detailed in the rest of this section.

3.4.1 FORMATION OF GUVs IN THE PRESENCE OF DETERGENT

The first step of these methods is to grow GUVs in the presence of detergent, for example, DOTM, from a dried mixed film of lipid/detergent. Given that detergent will equilibrate between the lipidic and aqueous phases after rehydration, the amount of detergent in the film should be calculated by also considering the volume of the growth chamber. In our experiments (Box 3.5), we chose a final DOTM concentration of 75 μM, which is above the CMC (50 μM) and below the solubilization of GUVs (200 μM).

3.4.2 DIRECT INCORPORATION OF TRANSMEMBRANE PROTEINS INTO GUVs

For direct incorporation of solubilized proteins in preformed GUVs, we recommend DOTM, a mild sugar-based detergent in which several membrane proteins are stable. Detergent concentration should remain higher than the CMC in order to keep the proteins solubilized during the incorporation step. The principle of direct incorporation into GUVs is to prepare mixtures of DOTM-GUVs at sub-solubilizing detergent concentrations but above the CMC, then to add solubilized proteins and remove the detergent after protein insertion. We applied this approach with bacteriorhodopsin (BR) a light-induced protein pump (Figure 3.3). When illuminated, the BR-GUVs could build up a large stable pH gradient of 0.8 pH units (acidic inside). The high pH gradient confirmed that BR proteins are collectively pumping protons in the same direction in an inside-out orientation, as also demonstrated after direct incorporation of BR in proteo-SUVs (Rigaud et al., 1988). This predominantly single orientation in GUVs represents an advantage of this method compared with dehydration-hydration methods, which produce GUVs with no favored directionality for the insertion. Moreover, the large amplitude of the proton gradient results also from the low passive proton permeability of BR-containing GUVs, confirming the extremely low residual detergent concentration at the end of the reconstitution process; this was further confirmed in an experiment in which GUVs subjected to a 1.5-pH unit jump returned to equilibrium only after 40 min (Dezi et al., 2013).

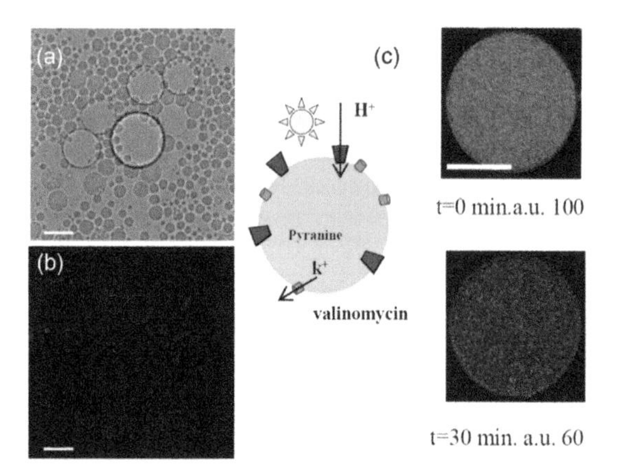

Figure 3.3 Direct incorporation of solubilized protein in GUVs. (a) Lipidic GUVs grown in the presence of 100 μM DOTM. (b) GUVs, labeled with Texas-Red® DHPE (red), after exchange to physiological buffer. (c) Incorporation of red-labeled BR in GUVs with encapsulated pyranine, a green pH probe. Pyranine fluorescence intensity in arbitrary units (a.u.). The central scheme represents the principle of the assay: Upon illumination at a wavelength higher than 500 nm, protons are pumped in by BR (red proteins), leading to an acidification of the internal volume and changing the fluorescence of pyranine. Valinomycin (blue) is used to release the membrane potential through K^+ counter-transport. Scale bars in (a, b): 10 μm; in (c): 5 μm. (a) Phase-contrast microscopy, (b, c) confocal microscopy (see Chapter 10).

Box 3.5 Protocol for growing GUV in the presence of detergent and in physiological buffers

1. *Prepare the lipid–detergent dried film (~60 min)*
   A mixture of 30 μg of lipid and 23 μg of DOTM is dried in a 600-μL chamber made of two facing ITO slides and a spacer (~3-mm thickness). Considering that 15 μL is spread onto each slide, prepare a lipid/DOTM solution with 1 mg/mL lipids (it may contain up to 0.5% w/w fluorescent marker) and 1.5 mM DOTM in $CHCl_3$/MeOH (9/1) (DOTM is highly soluble in organic solvents).
   Take 15 μL of lipid–detergent solution with a clean glass syringe and, keeping the needle attached and parallel to the glass surface, slowly spread it on each glass. Protect the solution from light with aluminum foil, and evaporate the solvent under vacuum for 60 min.
2. *Growth of GUVs from lipid–detergent dried film (~210 min)*
   Rehydration solution (600 μL) is usually a sucrose buffer at 400 mOsm with additional molecules if they need to be encapsulated into the GUVs such as soluble fluorescent probes, DNA, and substrates. Electroformation is performed over 210 min at 1.1 V_{peak}, 10 Hz on ITO slides.
3. *Exchange with physiological buffers (~ 10 min)*
   GUVs are equilibrated with the physiological buffer of the proteins by mixing one volume of GUVs with one volume of 2x concentrated buffer: a drop of 20 μL GUVs prepared in 75 μM DOTM and 400 mM sucrose is mixed in an Eppendorf tube with 20 μL of buffer (50-mM HEPES [pH 7], 500 mM KCl and 75 μM DOTM). This can be repeated with various buffers in parallel. These GUVs are used for the protocols in Box 3.6 and 3.7 and are stable when kept at room temperature or 4°C for 1 day.

Box 3.6 Protocol for direct incorporation of solubilized proteins in GUV
Example of BR (Dezi et al., 2013)

1. *Preparation of GUVs destabilized with sub-solubilizing concentration of DOTM*
DOTM-GUVs (e.g., Egg PC/Egg PA 9/1 mol/mol) are prepared as described in Box 3.5 in a sucrose buffer at 75 μM DOTM and equilibrated with 200-mM K_2SO_4, 20-mM PIPES, pH 7.5 supplemented with 150-μM pyranine, a pH sensitive probe. One volume (20–50 μL) of DOTM-GUVs is mixed with the same volume of 0–200 μM DOTM final concentration and gently mixed by hand over 30 min. Be careful to consider detergent added with the solubilized proteins. Avoid chloride-containing buffers because the passive permeability of the membrane to proton-associated chloride (such as HCl molecule) is high and would decrease the pH gradient generated by BR.
2. *Addition of solubilized BR*
BR prepared from the purple membrane of *H. salinarium* is solubilized in Triton X-100, then labeled with Alexa 546 and Triton X-100 is exchanged by chromatography to DOTM (Dezi et al., 2013). A 5-μL aliquot of BR solution is added to GUVs at 0.5–1 μM final concentration and incubated for 2–6 h.
3. *Detergent removal*
Detergent is removed by adding 5 mg Bio-Beads for 2 h. The amount of residual detergent can be estimated to be well below the nanomolar range (Levy et al., 1990).
4. *Pumping assay*
Five microliters of the sample are diluted 10–20 times in a solution of 200-mM glucose, 100-mM K_2SO_4, 10-mM PIPES, pH 7.8, and 0.01-μM valinomycin. Valinomycin is needed to perform a K^+ counter-transport required to release the electrical potential (positive inside) generated by the proton influx. Osmolarity is adjusted to the same value as the sample buffer. Before starting the observation, vesicles are sedimented for 10 min. Light-induced proton transport by BR is triggered by illuminating the sample using a 12-V, 100-W halogen lamp with a low wavelength cutoff at 500 nm and a heat filter.

Box 3.7 Protocol for fusion of native membranes to GUVs
Example of *E. coli* IMVs (Dezi et al., 2013)

1. *Preparation of fluorescent purified native membranes*
IMVs from *E. coli* expressing BmrC/D are purified by two passages through a French press (18,000 psi) according to (Galian et al., 2011). The vesicles are mostly unilamellar, with diameters ranging between 50 and 100 nm, with some small, open membrane fragments as seen by cryo-electron microscopy. For protein labeling, IMVs are diluted to a final total protein concentration of 3 mg/mL in 30 mM phosphate buffer (pH 8) and 150 mM KCl and then incubated with 1 mM Alexa 488-succinimidyl ester for 1 h at room temperature. After nonspecific labeling of the proteins present in the membrane, vesicles are passed through a Sephadex® column (PD10; Millipore), washed with 30 mM phosphate buffer (pH 8).
2. *Fusion of fluorescent IMVs with GUVs*
A 20 μL drop of GUVs prepared in 75 μM DOTM (Box 3.5) is mixed with 20 μL buffer (50-mM HEPES [pH 7], 500 mM KCl, 6 mM $MgCl_2$, and 100–300 μM Triton X-100). Next, a 2 μL drop of IMV suspension is added at a 1-mg/mL total protein final concentration. Note that the amount of detergent present during the fusion process is far below the solubilizing concentration for the native vesicles. GUVs are also not solubilized by the addition of Triton X-100, because detergent molecules partition between GUVs and native membranes. After a 2–16 h incubation at room temperature, without stirring, detergent is removed with 5 mg of Bio-Beads.
GUVs are initially prepared without fluorescent lipids but become fluorescent (green) due to the labeling of the fused IMVs (Figure 3.4).

3.4.3 FUSION OF NATIVE VESICLES OR PROTEOLIPOSOMES TO GUVs

This method is based on the property of some detergents, DDM, Triton X-100, and DOTM to induce the fusion of lipid SUVs or native membranes when added at sub-solubilizing concentrations (Kragh-Hansen et al., 1998; Urbaneja et al., 1988). It was adapted to the incorporation of membrane proteins in GUVs (Dezi et al., 2013). Practically, native membrane vesicles or proteo-SUVs are incubated with GUVs in the presence of detergent. Both membranes, the GUV and proteo-liposomes, are then destabilized and can fuse. Then, the detergent removal step is similar to the direct incorporation method (see Section 3.4.2) (Figure 3.4). One of the advantages when working with native vesicles is that proteins are incorporated in GUVs together with other components, lipids or protein partners that might be important for the function. Moreover, it does not require protein purification. This protocol has been successful with purified inverted inner membrane vesicles (IMVs, diameter 50–100 nm) of *E. coli* that contained the multidrug ABC transporters BmrC/D, or with chromatophores from photosynthetic bacteria; it has also worked well with proteo-SUVs, showing that fusion was not related to specific fusogenic components present in the membranes but, rather, to the detergents (Dezi et al., 2013).

Box 3.8 Hints for choosing buffers, solutions and lipids for proteo-liposomes

- *Dehydration* of a protein can result in its partial or total denaturation. To avoid complete dehydration, drying can be performed in a humid atmosphere (Girard et al., 2004). In addition, 1–2 mM sugars such as trehalose or sucrose, commonly used as cryo-protectants (Crowe et al., 1996) can be added to the SUV buffer to alleviate dehydration stress (Aimon et al., 2011). Keeping the time a protein is dehydrated minimal is critical for its stability. Note, however, that if the drying time is too short, that is, if rehydration is initiated while water is still visibly present (typically 10–15 min with the smallest attainable drops), the film does not form well and the lipids may simply be washed off the substrate, preventing GUV formation.
- *pH control:* Solutions must contain a pH buffer because protein function is impacted by pH. Buffers like HEPES, MOPS, or TRIS are common choices for maintaining a physiological pH.
- *Divalent ions*, such as Ca^{2+} and Mg^{2+}, are often avoided when working with GUVs because they tend to promote vesicle–vesicle adhesion. In addition, ~1 mM EDTA can be added to the solution to chelate divalent ions and prevent sticking.
- *Salts*: The salt concentration in the observation (external) buffer can be freely selected. Nevertheless, salt concentration in the growth buffer determines the technique to be used. Indium ITO slides and low AC frequency (~10 Hz) are used for low salt concentrations (>5 mM), whereas for physiological salt concentrations (~100–200 mM), either spontaneous swelling or high frequency (~500 Hz) on Pt wire are adapted.
- *Sedimentation of the GUVs:* It is classically facilitated by using external and internal buffers with different densities. For example, the growth buffer can contain sucrose, while the observation buffer may contain glucose, commonly in the range of 200–400 mM.
- *Matching osmolarities*: The growth and observation buffers must have matching osmolarities because GUVs are prone to lysis or to instabilities due to relatively small osmotic pressure differences (typically, a maximum of 10% difference is practically possible for a hyperosmotic external buffer, but only 1% for a hypoosmotic one). This can be achieved by measuring osmolarities with an osmometer and adding sucrose or glucose to the buffers until they match within about 1%. Note that for sucrose and glucose, 1 mM equals about 1 mOsm.
- *Lipids*: Many lipid types can be used to prepare proteo-GUVs, including neutral lipids such as phosphatidylcholine (PC), possibly complemented with phosphatidylethanolamine (PE) cholesterol, and/or charged lipids such as phosphatidylglycerol (PG), phosphatidic acid (PA) or phosphatidylserine (PS). To adjust the final lipid composition of the GUVs or dilute the proteins, protein-free SUVs can be mixed with proteo-SUVs. When forming the membrane stacks during dehydration, the two populations will fuse, yielding a mixed GUV population.

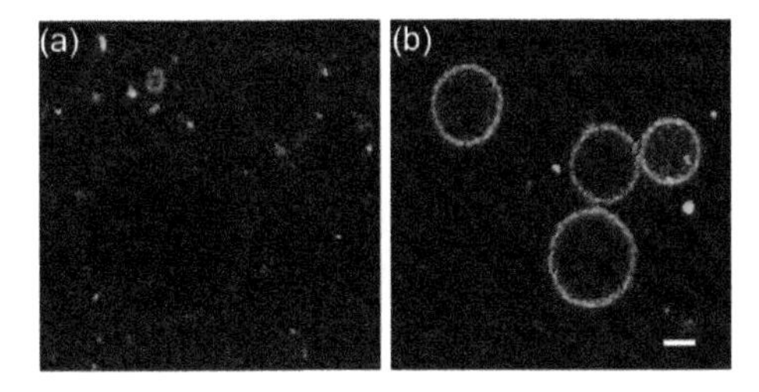

Figure 3.4 Incorporation of transmembrane proteins by fusion of native membranes with GUVs. Fusion of Alexa488-labeled inverted inner membrane vesicles of *E. coli* containing overexpressed BmrC/D, a multidrug resistance transporter, to non-labeled Egg PC–Egg PA GUVs (a) in the absence of Triton X-100, GUVs present but nearly unlabeled, and (b) in the presence of 300 μM Triton X-100. Scale bar: 5 μm. (a, b) Confocal microscopy (see Chapter 10).

3.5 SOME APPLICATIONS OF THE PROTEO-GUVs: PAST AND FUTURE POTENTIAL

In this chapter, we have listed different methods that can be used for the reconstitution of transmembrane proteins into GUVs. Gel-assisted swelling is probably the easiest and fastest method for proteo-GUV preparation, but it nevertheless produces GUVs with undifferentiated protein orientation and a higher fraction of multilamellar vesicles as compared with the direct/fusion method; the presence of gel remaining on the membrane is an additional drawback (Lira et al., 2014). Electroformation is an alternative to consider when gel-assisted swelling is unsuccessful and when protein orientation is not an issue (for instance, for ATPases, the presence of ATP in only the external medium guaranties that proteins activated have one orientation). Direct reconstitution or fusion allow unidirectional reconstitution and are expected to be milder methods for protein preservation. However, they require more preparation steps and care with detergent concentrations. Nevertheless, these post-reconstitution methods are very promising for more complex biomimetic experiments that require encapsulation of proteins inside the GUV lumen together with integral proteins in the GUV membrane (e.g., Lagny and Bassereau, 2015) although detergents might possibly induce leakage during the reconstitution process of some of the smallest components primarily trapped in the GUV and be required for the rest of the experiments.

Different types of transmembrane proteins have been reconstituted in GUVs, using, for a large part, the methods described in this chapter (a list is provided in Table 3.1). So far, they belong to pumps, channels (voltage-gated or mechanosensitive), transporters, adhesion, fusion or receptor protein classes (see Table 3.1 for the references). We believe that it is an obligatory step to check the functionality of proteins after reconstitution, as was done for many of the referenced proteins. We can reasonably expect that many more proteins will be reconstituted in the future, especially with the continuous development of biomimetic systems in biophysics and the blooming of synthetic biology.

Potential applications for proteo-GUVs are of various natures:

- *In biology:*

Transmembrane proteins often need both a well-defined membrane environment and specific "internal" and "external" bulk solutions to fully express their function. Some membrane proteins need external activation or a signal (e.g., ATP hydrolysis, photon absorption, voltage gating, mechanical triggering) to perform their function, whereas other membrane proteins do not (e.g., porins, adhesion, fusion proteins, some receptors). All these requirements are accessible when proteins are reconstituted in GUVs and their function can be studied in detail by various fluorescence microscopy techniques.

For example, we showed in this chapter how BR functionality was checked after reconstitution by demonstrating that BR-containing GUVs produce a large light-switchable transmembrane pH gradient (0.8-pH unit) across the membrane that remains stable over hours. Transfer of molecules can be followed using fluorescence-based assays (for examples, see Dezi et al., 2013), or water transport by monitoring aquaporin-GUV spreading (Berthaud et al., 2016). Ion channel functionality in GUVs is generally checked by electrophysiology and patch-clamping a part of the GUV with a glass micropipette (e.g., Battle et al., 2009; Aimon et al., 2011; Cao et al., 2013).

The role of specific lipids on the function of a protein can be also studied in detail: By reconstituting the capsaicin receptor, TRPV1, into giant vesicles, it was found that phosphoinositides, with the exception of PI(3,4,5)P3, inhibit channel opening (Cao et al., 2013). The aggregation of voltage-dependent anion channels (VDACs) of the outer mitochondrial membrane was shown to be disrupted by cardiolipin, a mitochondrial lipid (Betaneli et al., 2012). In patch-clamp experiments because membrane tension in the patch ranges from ~1 mN/m up to the lysis tension (~5–30 mN/m), mechanosensitive channels can be studied. It was thus shown that the activation of MscL, the mechanosensitive channel of large conductance (*E. coli*), is shifted to higher membrane tensions when phosphatidylethanolamine is a major membrane constituent (Moe and Blount, 2005). The study of larger membrane patches by patch clamp has been realized (Garten et al., 2017) and will allow measuring currents from an ensemble of proteins such as transporters or detecting gating charge movement. In addition, it will make membrane tension accessible in the full range from 0 to lysis tension while controlling the membrane potential.

Interestingly, because GUVs can sustain only very moderate stretching before lysis, reconstitution of mechanosensitive channels, or/and aquaporins into GUVs that mediate water fluxes through the membrane, would allow access to a rich membrane dynamics when GUVs are submitted to osmotic differences (Oglęcka et al., 2014) or to shear.

In contrast to giant plasma membrane vesicles (Chapter 2) that contain a plethora of proteins, only one type of membrane protein at a time was reconstituted in GUVs in this chapter. Many membrane complexes involved in energy production (e.g., ATP-synthase), photo-conversion (e.g., *light-harvesting complex*), or signaling (e.g., GPCRs) involve several transmembrane and/or membrane-associated proteins. Co-reconstitution of transmembrane proteins that are functionally coupled was only recently reported (Biner et al., 2016). A future challenge will be to study these membrane machineries.

Many organelles of cells including the nucleus, mitochondria and plastids have double membranes. To study the function of proteins embedded in these membranes it will be necessary to reconstitute double-membrane systems carrying multiple proteins. To that end, a combination of two different proteo-GUVs, proteo-GUVs and black lipid membranes or proteo-GUVs and supported lipid membranes can be useful.

The micrometer size of GUVs allows using fluorescence microscopy to tackle specific biological questions at the level of a single vesicle with unique advantages. The GUV's quasi-flat membrane can induce a different protein structure compared with the highly curved membrane of SUVs, ultimately modulating protein function (Tonnesen et al., 2014). Moreover, GUV size enables the observation of free diffusion over long distances. For example, measurements could determine whether protein–protein interactions in functionally coupled complexes result from short-distance specific forces or from long-distance physical forces and whether coupling is necessary to induce a protein response.

Another challenge will be to study the different oligomeric states of membrane proteins (such as the ABC transporter or GPCRs) and their variation during the function of proteins. Single-molecule FRET (fluorescence resonance energy transfer) approaches should provide a reliable measure of oligomeric state in a dilute, controlled system thanks to the GUV size.

- *In physics:*

Proteo-GUVs are very convenient systems for testing physical models on membranes containing inclusions and for better modeling of biomembranes. Using fluorescently labeled membrane proteins, confocal methods have been developed for GUVs to measure their respective protein surface fraction, allowing for quantitative physical experiments (Aimon et al., 2011).

In inhomogeneous membranes containing lipid domains, protein segregation into one phase type can be detected (Bacia et al., 2004), demonstrating the importance of the membrane organization for protein localization. Long-distance membrane-mediated interactions between inclusions also have been predicted. (For a recent review, see Yolcu et al., 2014). Protein clustering can be deduced from mobility measurements; for instance, for BR, it was shown that activity leads to protein aggregation at high protein surface fraction (Kahya et al., 2002) although the protein follows a standard 2D Brownian motion otherwise (Kahya et al., 2001). The Saffman–Delbrück model for protein diffusion in membranes has been extensively tested, also with GUVs (e.g., Ramadurai et al., 2009). *In vitro* experiments with proteo-GUVs have contributed to demonstrating the limits of this canonical model, in particular for proteins such as KvAP that are able to deform membranes locally (Quemeneur et al., 2014). The effects of membrane geometry and curvature can also be investigated with membrane nanotubes pulled from proteo-GUVs (Chapter 16), for example, on protein diffusion (Domanov et al., 2011) and on protein redistribution between flat and curved membranes (Aimon et al., 2014).

The reconstitution of a voltage-sensitive channel into GUVs connected to a lipid nanotube resembles the geometry of a neuron with its axon. Such a system inspires the possibility of studying action potential propagation *in vitro*. It could help to elucidate the role of ion channel cooperativity for action potential initiation (Naundorf et al., 2006) by establishing which geometry and channel density allows action potential generation. Eventually, this approach could be expanded to a network of GUVs linked by nanotubes (Davidson et al., 2003) to study a very basic neuronal network.

Reconstitution of active pumps or channels in GUVs has opened the way to study the physics of nonequilibrium membranes, also named "active membranes," first introduced in the pioneering work of Prost and Bruinsma (1996). Fluctuations of active membranes have also been studied (e.g., Bouvrais et al., 2012) for a recent review, see Lacoste and Bassereau (2014), which showed that protein activity has long-distance consequences and changes membrane mechanical properties such as bending rigidity and tension.

Eventually, primitive biomimetic cells can be built up containing adhesion molecules, allowing for testing models for cell spreading with biologically relevant adhesion molecules such as integrins (Streicher et al., 2009). The next challenge will be to reconstitute an actin cortex on the internal side of the GUV together with accessory proteins connecting integrins and actin to develop a more satisfactory model cell.

- *In biotechnology* (synthetic biology and bioinspired systems):

GUVs are promising systems for synthetic biology, where one of the ultimate goals is to develop a minimal functional cell integrating biological components in the lumen and in the membrane (Schwille, 2011; Lagny and Bassereau, 2015) (see Chapters 4 and 28–30). This "engineering" type of approach can not only help in understanding how cells work but can also, in the long term, have potential technical applications. The reconstitution of membrane proteins is an important part toward building up functional cell units. A very exciting perspective are the cell-free systems (such as the PURE [protein synthesis using recombinant elements] system of Shimizu et al., 2001) that are currently in a development phase: They could allow the production of membrane proteins in the lumen of the GUV and the direct incorporation of them into the membrane (Fenz et al., 2014; Sachse et al., 2014). Many other challenges remain, such as preparing membrane-limited compartments that can communicate or transfer components through transporters in their membrane (Elani et al., 2014), representing another step toward a self-sustained artificial cell.

As more and more reconstitution techniques are developed and transmembrane proteins are successfully reconstituted into GUVs, it may even become possible to build an engineered, fully functional model cell. Yet a dream project, it might one day allow unprecedented insight into understanding the complexity of life.

ACKNOWLEDGMENTS

This work was supported by the Intramural Program of the *Eunice Kennedy Shriver* National Institute of Child Health and Human Development (NICHD) of the National Institutes of Health (NIH). The P.B. group belongs to the CNRS consortium CellTiss; the D.L. and P.B. groups belong to Labex CelTisPhyBio (ANR-11-LABX0038) and to Paris Sciences et Lettres (ANR-10-IDEX-0001-02). The D.L. group is a member of the French National Research Infrastructure France-BioImaging (ANR10-INBS-04).

LIST OF ABBREVIATIONS

Bodipy-TRCer	BODIPY® TR Ceramide
BR	bacteriorhodopsin
$CHCl_3$	chloroform
CMC	critical micelle concentration
DDM	*n*-dodecyl β-D-maltoside
DOTM	*n*-dodecyl-β-D-thiomaltoside
DPhPC	1,2-diphytanoyl-*sn*-glycero-3-phosphocholine
EDTA	ethylenediaminetetraacetic acid
Egg PA	egg phosphatidic acid
Egg PC	egg phosphatidylcholine
IMV	inverted inner membrane vesicle
LUV	large unilamellar vesicle
MeOH	methanol
mol:mol	molar ratio
OG	octylglucoside
Proteo-GUV	giant unilamellar vesicle with reconstituted proteins
Proteo-SUV	small unilamellar vesicle with reconstituted proteins
SNARE	SNAP (Soluble NSF Attachment Protein) receptor
SUV	small unilamellar vesicle
Texas Red DHPE	Texas Red® 1,2-dihexadecanoyl-*sn*-glycero-3-phosphoethanolamine
w:vol	weight per volume ratio
w:w	weight to weight ratio

REFERENCES

Aimon S, Callan-Jones A, Berthaud A, Pinot M, Toombes GES, Bassereau P (2014) Membrane shape modulates transmembrane protein distribution. *Dev Cell* 28:212–218.

Aimon S, Manzi J, Schmidt D, Poveda Larrosa JA, Bassereau P, Toombes GES (2011) Functional reconstitution of a voltage-gated potassium channel in giant unilamellar vesicles. *PLoS One* 6:e25529.

Andréll J, Tate CG (2013) Overexpression of membrane proteins in mammalian cells for structural studies. *Mol Membr Biol* 30:52–63.

Angelova MI, Dimitrov DS (1986) Liposome electroformation. *Faraday Discuss Chem Soc* 81:303–311.

Bacia K, Schuette CG, Kahya N, Jahn R, Schwille P (2004) SNAREs prefer liquid-disordered over "Raft" (Liquid-ordered) domains when reconstituted into giant unilamellar vesicles. *J Biol Chem* 279:37951–37955.

Battle AR, Petrov E, Pal P, Martinac B (2009) Rapid and improved reconstitution of bacterial mechanosensitive ion channel proteins MscS and MscL into liposomes using a modified sucrose method. *FEBS Lett* 583:407–412.

Berthaud A, Quemeneur F, Deforet M, Bassereau P, Brochard-Wyart F, Mangenot S (2016) Spreading of porous vesicles subjected to osmotic shocks: The role of aquaporins. *Soft Matter* 12:1601–1609.

Betaneli V, Petrov EP, Schwille P (2012) The role of lipids in VDAC oligomerization. *Biophys J* 102:523–531.

Biner O, Schick T, Müller Y, von Ballmoos C (2016) Delivery of membrane proteins into small and giant unilamellar vesicles by charge-mediated fusion. *FEBS Lett* 590:2051

Bouvrais H, Cornelius F, Ipsen JH, Mouritsen OG (2012) Intrinsic reaction-cycle time scale of Na+,K+-ATPase manifests itself in the lipid–protein interactions of nonequilibrium membranes. *Proc Natl Acad Sci USA* 109:18442–18446.

Breton M, Amirkavei M, Mir LM (2015) Optimization of the electroformation of giant unilamellar vesicles (GUVs) with unsaturated phospholipids. *J Membr Biol* 248:827–835.

Cao E, Cordero-Morales JF, Liu B, Qin F, Julius D (2013) TRPV1 channels are intrinsically heat sensitive and negatively regulated by phosphoinositide lipids. *Neuron* 77:667–679.

Criado M, Keller BU (1987) A membrane fusion strategy for single-channel recordings of membranes usually non-accessible to patch-clamp pipette electrodes. *FEBS Lett* 224:172–176.

Crowe LM, Reid DS, Crowe JH (1996) Is trehalose special for preserving dry biomaterials? *Biophys J* 71:2087–2093.

Darszon A, Vandenberg CA, Schönfeld M, Ellisman MH, Spitzer NC, Montal M (1980) Reassembly of protein-lipid complexes into large bilayer vesicles: Perspectives for membrane reconstitution. *Proc Natl Acad Sci USA* 77:239–243.

Davidson M, Karlsson M, Sinclair J, Sott K, Orwar O (2003) Nanotube-vesicle networks with functionalized membranes and interiors. *J Am Chem Soc* 125:374–378.

Dezi M, Di Cicco A, Bassereau P, Levy D (2013) Detergent-mediated incorporation of transmembrane proteins in giant unilamellar vesicles with controlled physiological contents. *Proc Natl Acad Sci USA* 110:7276–7281.

Doeven MK, Folgering JHA, Krasnikov V, Geertsma ER, Van Den Bogaart G, Poolman B (2005) Distribution, lateral mobility and function of membrane proteins incorporated into giant unilamellar vesicles. *Biophys J* 88:1134–1142.

Domanov YA, Aimon S, Toombes GES, Renner M, Quemeneur F, Triller A, Turner MS, Bassereau P (2011) Mobility in geometrically confined membranes. *Proc Natl Acad Sci USA* 108:12605.

Elani Y, Law RV, Ces O (2014) Vesicle-based artificial cells as chemical microreactors with spatially segregated reaction pathways. *Nat Commun* 5:5305.

Fenz SF, Sachse R, Schmidt T, Kubick S (2014) Cell-free synthesis of membrane proteins: Tailored cell models out of microsomes. *Biochim Biophys Acta* 1838:1382–1388.

Finol-Urdaneta RK, McArthur JR, Juranka PF, French RJ, Morris CE (2010) Modulation of KvAP unitary conductance and gating by 1-alkanols and other surface active agents. *Biophys J* 98:762–772.

Galian C, Manon F, Dezi M, Torres C, Ebel C, Levy D, Jault JM (2011) Optimized purification of a heterodimeric ABC transporter in a highly stable form amenable to 2-D crystallization. *PLoS One* 6:e19677.

Garten M, Aimon S, Bassereau P, Toombes GES (2015) Reconstitution of a transmembrane protein, the voltage-gated ion channel, KvAP, into giant unilamellar vesicles for microscopy and patch clamp studies. *J Vis Exp* 95:52281.

Garten M, Mosgaard LD, Bornschlögl T, Dieudonné S, Bassereau P, Toombes GES (2017) Whole-GUV patch clamping *Proc. Natl. Acad. Sci. USA* 114: 328–333.

Geertsma ER, Nik Mahmood N a. B, Schuurman-Wolters GK, Poolman B (2008) Membrane reconstitution of ABC transporters and assays of translocator function. *Nat Protoc* 3:256–266.

Girard P, Pécréaux J, Lenoir G, Falson P, Rigaud J-L, Bassereau P (2004) A new method for the reconstitution of membrane proteins into giant unilamellar vesicles. *Biophys J* 87:419–429.

Horger KS, Estes DJ, Capone R, Mayer M (2009) Films of agarose enable rapid formation of giant liposomes in solutions of physiologic ionic strength. *J Am Chem Soc* 131:1810–1819.

Horger KS, Liu H, Rao DK, Shukla S, Sept D, Ambudkar SV, Mayer M (2015) Hydrogel-assisted functional reconstitution of human P-glycoprotein (ABCB1) in giant liposomes. *Biochim Biophys Acta* 1848:643–653.

Junge F, Schneider B, Reckel S, Schwarz D, Dötsch V, Bernhard F (2008) Large-scale production of functional membrane proteins. *Cell Mol Life Sci CMLS* 65:1729–1755.

Kahya N, Pecheur EI, de Boeij WP, Wiersma DA, Hoekstra D (2001) Reconstitution of membrane proteins into giant unilamellar vesicles via peptide-induced fusion. *Biophys J* 81:1464–1474.

Kahya N, Wiersma DA, Poolman B, Hoekstra D (2002) Spatial organization of bacteriorhodopsin in model membranes light-induced mobility changes. *J Biol Chem* 277:39304–39311.

Keller BU, Hedrich R, Vaz WLC, Criado M (1988) Single channel recordings of reconstituted ion channel proteins: An improved technique. *Pflüg Arch* 411:94–100.

Kragh-Hansen U, le Maire M, Moller JV (1998) The mechanism of detergent solubilization of liposomes and protein-containing membranes. *Biophys J* 75:2932–2946.

Kreir M, Farre C, Beckler M, George M, Fertig N (2008) Rapid screening of membrane protein activity: Electrophysiological analysis of OmpF reconstituted in proteoliposomes. *Lab Chip* 8:587.

Lacoste D, Bassereau P (2014) An update on active membranes. In: *Liposomes, Lipid Bilayers and Model Membranes: From Basic Research to Application*, p. 271. Boca Raton, FL: CRC Press, Taylor & Francis Group.

Lagny TJ, Bassereau P (2015) Bioinspired membrane-based systems for a physical approach of cell organization and dynamics: Usefulness and limitations. *Interface Focus* 5:20150038.

Levy D, Bluzat A, Seigneuret M, Rigaud JL (1990) A systematic study of liposome and proteoliposome reconstitution involving Bio-Bead-mediated Triton X-100 removal. *Biochim Biophys Acta* 1025:179–190.

Lichtenberg D, Ahyayauch H, Goni FM (2013) The mechanism of detergent solubilization of lipid bilayers. *Biophys J* 105:289–299.

Lira RB, Dimova R, Riske KA (2014) Giant unilamellar vesicles formed by hybrid films of agarose and lipids display altered mechanical properties. *Biophys J* 107:1609–1619.

Manneville J-B, Bassereau P, Lévy D, Prost J (1999) Activity of transmembrane proteins induces magnification of shape fluctuations of lipid membranes. *Phys Rev Lett* 82:4356–4359.

Méléard P, Bagatolli LA, Pott T (2009) Giant unilamellar vesicle electroformation from lipid mixtures to native membranes under physiological conditions. *Methods Enzymol* 465:161–176.

Moe P, Blount P (2005) Assessment of potential stimuli for mechano-dependent gating of MscL: Effects of pressure, tension, and lipid headgroups. *Biochemistry* 44:12239–12244.

Montes L-R, Alonso A, Goñi FM, Bagatolli LA (2007) Giant unilamellar vesicles electroformed from native membranes and organic lipid mixtures under physiological conditions. *Biophys J* 93:3548–3554.

Morales-Penningston NF, Wu J, Farkas ER, Goh SL, Konyakhina TM, Zheng JY, Webb WW, Feigenson GW (2010) GUV preparation and imaging: Minimizing artifacts. *Biochim Biophys Acta Biomembr* 1798:1324–1332.

Naundorf B, Wolf F, Volgushev M (2006) Unique features of action potential initiation in cortical neurons. *Nature* 440:1060–1063.

Oglęcka K, Rangamani P, Liedberg B, Kraut RS, Parikh AN (2014) Oscillatory phase separation in giant lipid vesicles induced by transmembrane osmotic differentials. *Elife* 3:e03695.

The making of

Prost J, Bruinsma R (1996) Shape fluctuations of active membranes. *EPL Europhys Lett* 33:321.

Quemeneur F, Sigurdsson JK, Renner M, Atzberger PJ, Bassereau P, Lacoste D (2014) Shape matters in protein mobility within membranes. *Proc Natl Acad Sci USA* 111:5083.

Ramadurai S, Holt A, Krasnikov V, van den Bogaart G, Killian JA, Poolman B (2009) Lateral diffusion of membrane proteins. *J Am Chem Soc* 131:12650–12656.

Richmond DL, Schmid EM, Martens S, Stachowiak JC, Liska N, Fletcher DA (2011) Forming giant vesicles with controlled membrane composition, asymmetry, and contents. *Proc Natl Acad Sci USA* 108:9431–9436.

Rigaud J-L, Lévy D (2003) Reconstitution of membrane proteins into liposomes. *Methods Enzymol* 372:65–86.

Rigaud JL, Paternostre MT, Bluzat A (1988) Mechanisms of membrane protein insertion into liposomes during reconstitution procedures involving the use of detergents. 2. Incorporation of the light-driven proton pump bacteriorhodopsin. *Biochemistry* 27:2677–2688.

Rigaud J-L, Pitard B, Levy D (1995) Reconstitution of membrane proteins into liposomes: Application to energy-transducing membrane proteins. *Biochim Biophys Acta Bioenerg* 1231:223–246.

Rusinova R, Kim DM, Nimigean CM, Andersen OS (2014) Regulation of ion channel function by the host lipid bilayer examined by a stopped-flow spectrofluorometric assay. *Biophys J* 106:1070–1078.

Sachse R, Dondapati SK, Fenz SF, Schmidt T, Kubick S (2014) Membrane protein synthesis in cell-free systems: From bio-mimetic systems to bio-membranes. *FEBS Lett* 588:2774–2781.

Schwille P (2011) Bottom-up synthetic biology: Engineering in a Tinkerer's world. *Science* 333:1252–1254.

Shimizu Y, Inoue A, Tomari Y, Suzuki T, Yokogawa T, Nishikawa K, Ueda T (2001) Cell-free translation reconstituted with purified components. *Nat Biotechnol* 19:751–755.

Smeazzetto S, Tadini-Buoninsegni F, Thiel G, Berti D, Montis C (2016) Phospholamban spontaneously reconstitutes into giant unilamellar vesicles where it generates a cation selective channel. *Phys Chem Chem Phys* 18:1629.

Streicher P, Nassoy P, Bärmann M, Dif A, Marchi-Artzner V, Brochard-Wyart F, Spatz J, Bassereau P (2009) Integrin reconstituted in GUVs: A biomimetic system to study initial steps of cell spreading. *Biochim Biophys Acta Biomembr* 1788:2291–2300.

Tonnesen A, Christensen SM, Tkach V, Stamou D (2014) Geometrical membrane curvature as an allosteric regulator of membrane protein structure and function. *Biophys J* 106:201–209.

Urbaneja MA, Goni FM, Alonso A (1988) Structural changes induced by Triton X-100 on sonicated phosphatidylcholine liposomes. *Eur J Biochem* 173:585–588.

Weinberger A, Tsai F-C, Koenderink GH, Schmidt TF, Itri R, Meier W, Schmatko T, Schröder A, Marques C (2013) Gel-assisted formation of giant unilamellar vesicles. *Biophys J* 105:154–164.

Yanagisawa M, Iwamoto M, Kato A, Yoshikawa K, Oiki S (2011) Oriented reconstitution of a membrane protein in a giant unilamellar vesicle: Experimental verification with the potassium channel KcsA. *J Am Chem Soc* 133:11774–11779.

Yıldırım MA, Goh K-I, Cusick ME, Barabási A-L, Vidal M (2007) Drug—target network. *Nat Biotechnol* 25:1119–1126.

Yolcu C, Haussman RC, Deserno M (2014) The effective field theory approach towards membrane-mediated interactions between particles. *Adv Colloid Interface Sci* 208:89–109.

Zorman S, Botte M, Jiang Q, Collinson I, Schaffitzel C (2015) Advances and challenges of membrane–protein complex production. *Curr Opin Struct Biol* 32:123–130.

Giant unilamellar vesicles with cytoskeleton

Tobias Härtel and Petra Schwille

Imagination is more important than knowledge.
Knowledge is limited; Imagination encircles the world

Albert Einstein

Contents

4.1 NEXT LEVEL OF COMPLEXITY IN A BOTTOM-UP APPROACH OF BUILDING CELL-LIKE COMPARTMENTS

Synthetic biology is the most recent variation of research and development in the life sciences. It is, on the one hand, the consequent application of cutting-edge molecular biology techniques to a new generation of biotechnology, culminating in the precise manipulations of single genes and large gene networks in cells and organisms. On the other hand, synthetic biology is a new angle under which biological systems are considered: modular architectures, whose parts can be dissected, categorized, and recombined, in a fully new fashion. This particular modular approach also spurs the idea of trying to construct biological functionality from the bottom-up, making the construction of a minimal cell with the smallest possible number of functional modules an attractive final goal (Schwille, 2015). However, even if the perspective of constructing a full cell, that is, a living entity, from scratch may be too distant, such bottom-up constructed systems are supposed to be better defined and easier to handle and characterize than the complex biological ones with their multitudes of lateral interactions and regulatory systems.

What the elementary set of functional modules is, for a system to start showing features of life, remains still a matter of debate. A presumable synthetic cell has been proposed to harvest, maintain a compositional homeostasis, as well as store information with the ability to mutate and evolve (Deamer, 2005). Obviously, a cell also needs a selective boundary to separate the inside (and biological identity) from the outside (source of nutrients and potentially hostile environment). It is thus no wonder that many recent strategies of creating protocell-like entities revolve around model membrane compartments. Giant unilamellar vesicles (GUVs), the heroes of this book, are obviously fantastic compartment model systems to start with, owing to their comfortable size of many tens to hundreds of micrometers, ideally suited to study the processes on the membranes and within with light microscopy and related techniques.

Many chapters in this book are consequently devoted to the possibility of reconstituting important biological machineries and processes in and on GUV membranes (see Chapters 2, 3, 16, 28 and 30). Of particular interest are obviously processes that lead to a deformation and restructuring of these vesicles, as in processes of endo- and exocytosis, as well as cell and organelle division. However, although a considerable number of exciting studies on vesicle membrane deformations has been published (see Chapters 16 and 23), it quite quickly becomes obvious how and why real cells have armored their delicate membranes with sugar or protein walls, coats and cytoskeletal structures: First, a membrane can faithfully contain and preserve a volume, but it cannot confer shape to a compartment. Also, the membrane offers flexibility and transformability, but hardly any mechanical stability, which is, however, ultimately required for most organisms to withstand a mostly hostile environment. Interactions between cytoskeletal filaments and the plasma membrane were researched using a spectrum of techniques focusing on the *in vitro* self-assembling filament network of GUVs (i.e., the actin-coated membrane), like optical tweezers and single-particle tracking experiments (Helfer et al., 2001a, 2001b). Thus, in devising a suitable compartment that could ideally be the basis for a minimal cell, a cytoskeleton- or cortex-like structure will likely have to be taken up to the canon of ultimately necessary modules for bottom-up biology. In addition to adding stability, the cytoskeleton is the key platform for an enormous variety of essential cellular functions and regulatory systems, interacting with and responding to environmental and internal triggers (Helfer et al., 2000).

In this chapter, we will give an overview of the many functions of the cytoskeleton, its composition and structure in various organisms. Afterward we will describe methods to anchor cytoskeletal filaments to membranes of supported lipid bilayers (SLBs) and GUVs by several strategies based on charge, amphipathic helices, streptavidin-biotin binding or by integral membrane proteins. This methodological part of the chapter will be followed by recent bottom-up applications referring to all kingdoms of live, for example, mimicries of actin cortices of eukaryotic cells, filamenting temperature-sensitive mutant A and Z (FtsZ-FtsA) filaments of *Escherichia coli*, and cell division ABC (CdvABC) membrane deformation as in the archaea *Sulfolobus acidocaldarius*. We will conclude the chapter with a summary and an outlook to the future of this interesting research direction.

4.2 FUNCTIONS OF THE CORTEX/CYTOSKELETON

We will start with an overview of composition and functionality of the cytoskeleton. The cytoskeleton can be considered as a complex network of interacting and interlinking protein filaments and tubules in the cytoplasm of cells in all three domains of life (Hardin et al., 2015). It is responsible for cell shape, motility of complete cells or the organelles within, and their mechanical interactions with the environment (Alberts et al., 2008). Furthermore, cells have to be able to reorganize their organelles and their cell shape under changing conditions like cell growth or division. (Alberts et al., 2008). The variability of the cytoskeleton depends on the specific features of three underlying protein families, which self-assemble to form three main types of filaments: microfilaments, microtubules, and intermediate filaments. These filaments have specific structural features, allowing them to assemble into an extremely dynamic filamentous network, able to expand or contract very rapidly. The single protein units of each filament vary between monomeric and polymeric states in dependence of a so-called critical concentration. This property allows the cell to rapidly control cytoskeletal structures and reinforcement of the membrane, deformation of membranes and changes of cell shape.

4.2.1 REINFORCEMENT OF MEMBRANE

The most obvious role of the cytoskeleton is that it confers reinforcement and stability to the cell membrane and allows cells to be correctly shaped, physically robust, but also flexible enough to support deformations by polymeric interactions (Geli and Riezman, 1998; Fletcher and Mullins, 2010). A dynamic interplay between actin, microtubules, and intermediate filaments also constitutes the cytoskeletal cortex underneath the membrane, where some crosslinking proteins exchange rapidly and the polymers themselves turn over on time scales of seconds to minutes (Pollard and Cooper, 2009). By these properties of the cortex and the fluidity of the membrane, the cell is able to deform fast or slowly such that the cell surface can be stiff or elastic. Even the cells of plants and fungi, despite being encased in a cell wall, use cytoskeletal polymers to adapt the shape of their compartments (Hussey et al., 2006). Furthermore, the cytoskeleton is part of a system that senses external forces applied to the cell as well as the mechanical properties of the cell's environment. This polymeric network can thus influence diverse aspects of cell function, including gene expression and differentiation (Discher et al., 2009).

4.2.2 CELL SHAPE INDUCTION/MAINTENANCE

The variety of eukaryotic and prokaryotic cell shapes reflects in a certain way the functions of the cell, confirming the structure–function relationship. Each cell type has evolved a shape that is best adjusted to its function and has also evolved regulation and maintenance of its shape. As already mentioned, polymers of actin filaments, microtubules, and intermediate filaments in various amounts and geometries are responsible for the mechanical properties and cell shapes, which are often critical to their functions (Heuser and Kirschner, 1980; Pollard and Cooper, 2009). Actin filaments are essential for mechanical structure and motility, microtubules for separating chromosomes and transporting larger cargo within the cell, and intermediate filaments for resisting mechanical forces (Pollard and Cooper, 2009).

4.2.3 DEFORMATION OF CELLS/MEMBRANES

Living cells are not just stiff "gift boxes of life," but dynamic and fluidic containers, most of them undergoing permanent shape deformations. These transformations are strictly regulated and catalyzed by certain proteins interacting with the local lipid environment (McMahon and Gallop, 2005; Schwille and Diez, 2009). In addition, force-inducing motor proteins are often involved in large-scale membrane transformations (Schwille and Diez, 2009). Strikingly, most of these "coating and/or membrane sculpting" peripheral proteins share structural motifs in terms of membrane-penetrating amphipathic helices or peptides (Farsad and De Camilli, 2003). *In vitro* reconstitution of cytoplasmic coats onto SLBs or GUVs has been recognized as a valuable tool for investigating specific sites and cues of coat binding in dependence of curvature, lipid asymmetry or fluidity in a minimal system (Schwille and Diez, 2009). In recent approaches, coat–membrane interactions were shown for, for example, the ADP ribosylation factor 1 (Arf1)-dependent assembly of coat protein complex I on GUVs (Manneville et al., 2008), and globotriaosylceramide (Gb3) (glycolipid)-binding B-subunit of bacterial Shiga toxin (Romer et al., 2007). It could also be shown by the use of different Bin/amphiphysin/Rvs (BAR) domain proteins that membranes can deform into protrusions or invaginations depending on the cooperativity of assembly or aggregation (Saarikangas et al., 2009). Further accomplishments, such as the *in vitro* reconstitution of cytoskeletal-like filaments of endosomal sorting complexes required for transport (ESCRT)-III in GUVs, showed membrane deformations leading to a budding event from the mother vesicle. This transformation can be induced bidirectionally depending on the localization of the protein assembly: either an out-budding event by intraluminal protein assembly, or budding into the vesicle if protein binding occurred from the outside (Barelli and Antonny, 2009; Wollert et al., 2009).

4.3 COMPOSITION AND STRUCTURE OF CORTEX/CYTOSKELETON

4.3.1 EUKARYOTES: THREE KINDS OF FILAMENTS

As mentioned in Section 4.2.2, the cytoskeleton confers shape and a manifold of structural functionalities to the cell, and it also increases the level of molecular crowding in the cytosol (Minton, 1992). Being essential for many processes at the cell periphery, cytoskeletal components interact extensively and intimately with cellular membranes (Doherty and McMahon, 2008). Most eukaryotic cells have three types of cytoskeletal filaments responsible for the cell's spatial organization and mechanical properties: actin filaments, microtubules, and intermediate filaments. Each of these cytoskeletal filament types exhibits its own geometry and intracellular distribution and is composed of different subunits (Alberts et al., 2008).

Actin filaments (also known as microfilaments) are two-stranded helical polymeric filaments formed by many units of the roughly 42-kDa heavy globular multifunctional protein actin. Actin itself is a monomeric subunit that may self-assemble into two different types of filaments: microfilaments (part of cytoskeleton) and thin filaments (part of muscle cells). They appear flexible but strong with an average diameter of 5–9 nm (Alberts et al., 2008). Microfilaments also serve as "rails" for attached myosin motor molecules. Myosin movement along F-actin filaments, so called actomyosin fibers, generates contractile forces in both muscle and non-muscle types (Gunning et al., 2015). The actin related protein 2/3 (Arp2/3) complex is a seven-subunit protein complex that simultaneously controls nucleation of polymerization and branching of filaments by creating a new nucleation core. This nucleation core activity of Arp2/3 is activated by members of the Wiskott-Aldrich syndrome family protein (WASP) family. Two models of actin filament branching are known so far: the "Side Branching Model" and "the Barbed End Branching model," of which the former model seems to be favored (Dayel and Mullins, 2004).

Microtubules are long, hollow cylinders made of a dimer of two globular proteins, α- and β-tubulin (about 50 kDa) (Weisenberg, 1972). These highly dynamic polymers can reach a length of 50 μm and an outer diameter of 24 nm, while the inner diameter is about 12 nm (Alberts et al., 2008). Microtubules are usually nucleated and organized by organelles called microtubule-organizing centers, which contain a third type of globular protein, γ-tubulin (Desai and Mitchison, 1997). The process of adding or removing tubulin monomers depends on the concentration of $\alpha\beta$-tubulin dimers in solution in relation to a so-called critical concentration. In the case of higher dimer concentration than the steady state level, the microtubule will polymerize and *vice versa,* the microtubule will shrink (Alberts et al., 2008). Microtubules largely make up the internal structures of cilia and flagella. Furthermore, they are essential for intracellular transport, movement of secretory vesicles, organelles, and intracellular macromolecules (Vale, 2003). During mitosis and meiosis, microtubules are involved in chromosome separation as main elements of the spindle apparatus (Walczak and Heald, 2008).

Intermediate filaments are string-like fibers with a diameter of about 10 nm and composed of a heterogeneous family of related proteins (intermediate filament proteins) sharing similar structural and sequence features encoded by about 70 genes (Herrmann et al., 2007; Alberts et al. 2008). Most types of intermediate filaments are located in the cytoplasm, but there is also a nuclear one, lamin. In contrast to microtubules and actin filaments, which contain a plus end and a minus end, intermediate filaments lack any polarity because of the anti-parallel structure of their tetramers, and are thus unable to support motility or directed intracellular transport. Cytoplasmic intermediate filaments do not show any treadmilling, but they are very dynamic (Helfand et al., 2004).

In several bottom-up reconstitution studies, the interaction between cytoskeletal filaments and motor proteins attached to membranes was used to mechanically investigate membrane transformations with the ultimate goal of obtaining a full division of membrane vesicles. The reconstitution of an interface mimicry of cortex-like structures attached to a membrane is required to investigate the activity of cytoskeletal motors on model membranes. The important transition from supported membrane assays to flexible membrane vesicles was made quite early, for example, by realizing the polymerization of an actin network within GUVs, but a defined attachment between the membrane and the filaments was still missing (Limozin and Sackmann, 2002). Several actin superstructures interacting with the plasma membrane exist at different cellular locations, for example: a highly branched, polymerizing actin network at the leading edge of migratory cells; stress fibers, which are attached to sites of adhesion; long actin filaments in filopodia; and actin-rich structures that are found at invaginations in endocytic and phagocytic structures. The reconstitution of actin-based motility was first attempted in phosphatidylinositol-4,5-bisphosphate (PIP_2)-containing GUVs with purified attachment components (Delatour et al., 2008). In a recent approach, neural–WASP (N-WASP) bound to PIP_2 and Arp2/3 was used to anchor an actin network to phase-separated GUVs (Liu and Fletcher, 2006) (see also Chapter 23). This will be extensively discussed in Section 4.4.

4.3.2 PROKARYOTIC FILAMENTS

The prokaryotic cytoskeleton is, like its eukaryotic counterpart, built up of different structural filaments (Bi and Lutkenhaus, 1991). Almost all cytoskeletal proteins of eukaryotes have analogs in the prokaryotic domain, and even proteins without any occurrence in eukaryotes could be discovered (Popp et al., 2010; Wickstead and Gull, 2011; Popp et al., 2012; Gunning et al., 2015). As in eukaryotic cells, cytoskeletal filaments play essential roles in cellular processes like cell division, motility, cell membrane and wall protection, shape determination, and polarity determination in various prokaryotes (Michie and Lowe, 2006; Shih and Rothfield, 2006). Because motor–filament systems that transport cargo over larger distances are usually not required due to the much smaller sizes of prokaryotic cells, motility and force are conferred by different mechanisms of a special class of cytoskeletal proteins, the so-called cytomotive filaments (Lowe and Amos, 2009) using nucleotide turnover to carry out directional polymerization dynamics. The most important ones will be briefly introduced.

Filamenting **t**emperature-**s**ensitive mutant **Z** (FtsZ) was the first discovered prokaryotic cytoskeletal protein. FtsZ polymerizes at mid-cell during cell division into a filamentous ring-like structure (Z ring), which constricts at the end of cytokinesis similar to the actin–myosin contractile ring in eukaryotes (Bi and Lutkenhaus, 1991). The Z ring is a highly dynamic structure made of several protofilaments extending and shrinking, acting as a cell division regulator. FtsZ is the first protein to assemble the septum during prokaryotic cytokinesis.

The **MinCDE** system is not strictly a cytoskeletal element, but a positioning device that places the divisome complex at mid-cell in several prokaryotes by inhibiting polymerization of FtsZ at the cell poles (Teather et al., 1974; de Boer et al., 1992; Kretschmer and Schwille, 2014). All three Min proteins (MinC, MinD, and MinE) oscillate together in a spatiotemporal manner to guarantee a time-averaged minimum of the inhibitor MinC in the mid-cell region to admit Z ring polymerization (Hu and Lutkenhaus, 1999; Raskin and de Boer, 1999).

MreB has a three-dimensional (3D) structure and polymerization activity highly related to eukaryotic actin and, thus, seems to be its bacterial analog. MreB polymerizes into a filamentous helical network at the membrane all along the cell periphery and seems to determine cell shape in mainly nonspherical bacteria (Kurner et al., 2004). Strikingly, a *mreB* deletion-mutant of *E. coli* that creates defective MreB proteins will be spherical instead of rod-like (Wachi and Matsuhashi, 1989). Here, locations and activity of peptidoglycan synthesizing enzymes are mediated by the stiff membrane touching MreB filaments, sculpting and strengthening the cell shape (Gitai, 2005).

Crescentin is a further analog to a eukaryotic cytoskeleton element. It has a rather high primary homology to intermediate filaments and an average diameter of about 10 nm, and thus lies within the diameter range of eukaryotic intermediate filaments (Ausmees et al., 2003). Similar to MreB, crescentin polymerizes into filaments from cell pole to pole along the inner membrane.

ParM is a cytoskeleton element of prokaryotes with a related structure to actin and the functionality of tubulin. Like eukaryotic tubulin, ParM polymerizes on both ends and features dynamic instability (Garner et al., 2004; Popp et al., 2012). In combination with ParR and the gene *parC*, ParM forms a plasmid separation system, whereat ParR is a DNA-binding protein and *parC* a repetitive binding site on the R1 plasmid. This binding happens on both sides of the ParM filament, leading to an extension of it and a simultaneous separation of the plasmids (Moller-Jensen et al., 2002).

Among the prokaryotic cytoskeletal filaments, it seems very promising that the Min system in combination with FtsZ could be the most reliable reconstituted system in GUVs for synthetic biology.

4.3.3 ARCHAEAL FILAMENTS

The development of a cytoskeleton in eukaryotic cells has been a milestone in evolution and is involved in uncountable numbers of intracellular processes like shape determination, phagocytosis or cell division. However, an actin-based cytoskeleton has been recently described in Archaea. The results of these investigations implicate further hints for an evolutionary relation between archaea and eukaryotes, focusing the attention on an ancestral archaeal actin gene for both domains (Bernander et al., 2011). Archaeal actin or any functional predecessor could thus have played an essential role in cellular processes at the origin of life and evolution of the eukaryotic lineage.

Crenactin is one archaeal actin ortholog that belongs to the conserved Arcade operon (the actin-related cytoskeleton in Archaea involved in shape determination). This operon has been found to be unique in the archaeal kingdom *Crenarchaeota*, especially in the orders *Thermoproteales* and *Candidatus Korarchaeum* (Ettema et al., 2011) and is also encoding additional cytoskeleton-associated proteins. Crenactin has the highest sequence similarity to all known eukaryotic actin homologs in comparison to bacterial cytoskeletal proteins (Yutin et al., 2009). All crenactin-expressing species are rod-shaped and compose membrane-spanning helical structures for shape determination like MreB in other prokaryotes (Ettema et al., 2011).

The **CdvABC** system is the cell division machinery in Crenarchaeota. CdvA assembles in a double helical structure and seems to be a functional analog to actin (Härtel and Schwille, 2014). Furthermore, CdvA recruits CdvB (a crenarchaeal ESCRT-III like protein), which polymerizes at the mid-cell and fissions into two separate daughter cells (Wollert et al., 2009; Wollert and Hurley, 2010). *Sulfolobus acidocaldarius* produces three additional CdvB paralogs to form 40-nm-wide membrane tubules, analogously to the eukaryotic system, after recruitment of CdvB to the cell division site by CdvA (Samson and Bell, 2009; Samson et al., 2011). Afterward, the CdvB division ring is disassembled by CdvC (the vacuolar protein sorting-associated protein 4 [Vps4]-like protein; AAA-type ATPase) in ATP-dependent oligomerization (Babst et al., 1998).

The CdvABC system seems to be a functional model system for membrane constriction as described above.

4.4 EXPERIMENTAL PROTOCOLS: HOW TO ANCHOR FILAMENTS

In the following sections, we will discuss possibilities to anchor membrane-transforming proteins to supported and freestanding lipid bilayers of defined composition, ideally being accessible from both sides and which can be visualized by optical microscopy (Rivas et al., 2014). The most frequently employed model system for such investigations is that of the GUVs, whose size (10–100 μm) is most convenient to investigate the dynamics and localization of membrane–protein interactions by fluorescence imaging and microspectroscopy (Garcia-Saez et al., 2010) (see also Chapter 21).

As mentioned in Section 4.2, cytoskeletal filaments have a severe influence on the stability of the cell membrane, on the determination of cell shape, and are heavily involved in membrane transformations. To investigate single cytoskeleton filaments or whole cortex networks on membranes, the unspecific or specific contact or interaction sites of these filaments with membrane lipids need to be established. In the following subsection, we will describe several protocols of various complexity to couple native or artificial filaments and cytoskeletal proteins to membranes: (1) simply by charge, (2) by streptavidin–biotin binding, (3) by amphipathic helices, (4) by physiological lipid-protein complexes, or (5) by integral membrane proteins. To illustrate the different anchoring strategies, different research applications, including eukaryotic, prokaryotic, and archaeal filaments, but also DNA origami will be presented as non–protein-based biological structure and function elements. The practical examples will be introduced as text boxes with detailed protocols that may be reproduced. Previous publications and results suggest that it is experimentally easier to polymerize scaffolds around a giant vesicle rather than inside (avoiding the need of introducing complex solutions in the GUV interior) as in artificial cytoskeleton. However, more recent advances have reported the formation of "active vesicles" where actin meshwork was reconstituted inside GUVs (Tsai and Koenderink, 2015) and further activity has been attributed by including molecular motors inside the vesicles (Schuppler et al., 2016).

4.4.1 EXAMPLE 1: CHARGE-INDUCED ATTACHMENT

In this example, membrane-scaffolding "bricks" were engineered from amphipathic DNA origami structures to mimic the function of coat-forming proteins like the inverse/Fes/CIP4 homology BAR family (I-/F BARF) (Czogalla et al., 2013, 2015). These structures have a flat membrane-binding interface supplied with cholesterol-derived anchors (Figure 4.1a). Lateral accessories of sticky oligonucleotides promote the interaction between DNA origami monomers leading to the formation of ordered arrays and stacks on the membrane. This tight, controlled binding of DNA origami arrays to the membrane is able to deform freestanding lipid membrane and membrane vesicles (Figure 4.1b). This application nicely shows how the anchoring of membrane coats can be accomplished simply by introducing complementary charges between lipids and polymers. Box 4.1 introduces the production of DNA origami scaffolds, preparation of GUVs (with 1,2-dioleoyl-*sn*-glycero-3-phosphocholine [DOPC see Appendix 1 of the book for structure and data on this lipid]) for scaffold anchoring and deformation experiments, fluorescence imaging, and fluorescence correlation and cross-correlation spectroscopy by laser scanning microscopy (LSM).

4.4.2 EXAMPLE 2: ATTACHMENT VIA STREPTAVIDIN-BIOTINYLATED LIPIDS

Cell shape transformations, such as during cytokinesis, require forces acting on membranes that are generated by the interactions between membrane attachment sites, actin filaments, and myosin motors. To investigate these interactions in a controlled setting,

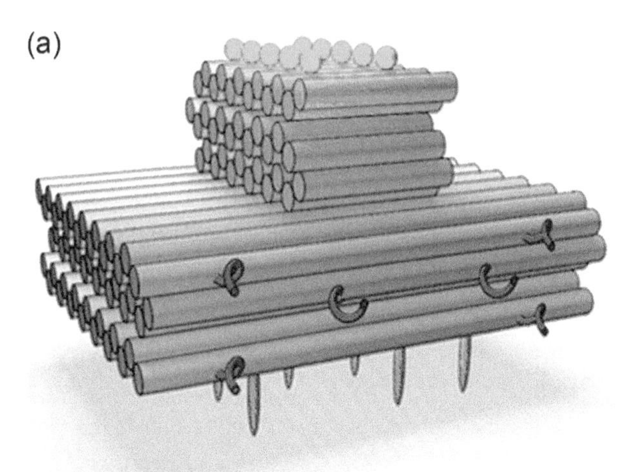

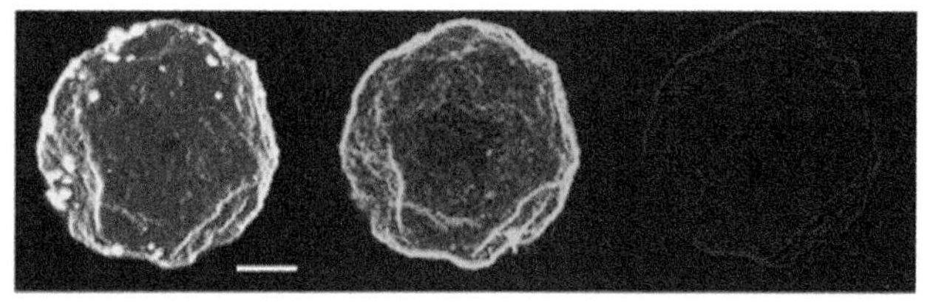

Figure 4.1 Amphipathic DNA origami nanoparticles can scaffold and deform lipid membrane vesicles. (a) Schematics of origami structure A: The helix bundles are shown as gray cylinders with "sticky" oligonucleotide overhangs shown in orange; the green circles represent bound fluorescent molecules (AlexaFluor 488). (b) A 3D-morphology projection in the equatorial plane of a GUV made of DOPC scaffolded with origami structures at a surface density of more than 10 particles/μm². The yellow channel (left) represents membrane labeled with fluorescent lipid analog DiI see Appendix 2 of the book for structure and data on this and other lipid dyes, green channel (middle) shows bound origami structure A, and the red channel shows bound origami structure B (AlexaFluor 647). (Adapted from Czogalla, A. et al.: Amphipathic DNA origami nanoparticles to scaffold and deform lipid membrane vesicles. *Angew. Chem.* 2015. 54. 6501–6505. Copyright Wiley-VCH Verlag GmbH & Co. KGaA. Reproduced with permission.)

Box 4.1 Protocol for production of DNA origami scaffolds and preparation of GUVs for scaffold anchoring and membrane vesicle deformation (Czogalla et al., 2015)

Materials and equipment

- Reverse-phase cartridge-purified oligonucleotides (Eurofins MWG Operon)
- Triethylene glycol (3′-cholesteryl-TEG) CPG (Glen Research)
- High-performance liquid chromatography (HPLC; Waters e2695 system)
- Electrospray ionization mass spectrometry (ESI-MS; Waters Acuity system)
- Terminal desoxyribonucleotidyl transferase (TdT) (Roche, Unterhaching, Germany)
- ChromaTide™ AlexaFluor® 488-5-dUTP (Thermo Fisher, Waltham, MA)
- ChromaTide AlexaFluor 647-5-dUTP (Thermo Fisher, Waltham, MA)
- Freeze 'N Squeeze™ DNA Gel Extraction Spin Columns (Bio-Rad, Munich, Germany)
- Transmission electron microscope FEI Morgagni 268D (FEI, Hillsboro, Oregon)
- MatriCal 384 MicroWell Plate (MatriCal, Spokane, USA)
- GUV buffer 10 mM HEPES pH7.5, 150 mM NaCl, 10 mM $MgCl_2$
- Lipid: DOPC (Avanti Polar Lipids, Alabaster, USA)
- Lipid dye: 1,1′-dioctadecyl-3,3,3′,3′-tetramethylindocarbocyanine perchlorate (DiI) fluorescent lipid analog (Invitrogen) added at 0.1 mol% of total lipid
- MatriCal 384 MicroWell Plate (MatriCal, Spokane, USA)
- LSM 780/CC3 or LSM 510/CC2 (Carl Zeiss, Germany)

DNA origami design, assembly and analysis

- Folded DNA origami structures can be designed via the *in silico* tool CaDNAno (Ke et al., 2009).
- The folding of single-stranded scaffold DNA can be performed as described by Douglas et al. (2009).
- The assembly reaction solution contains 10 nM of scaffold DNA, 75 nM of each oligonucleotide, 5 mM TrisHCl pH8.0, 1 mM EDTA, and 16 mM $MgCl_2$.
- This solution is heated to 80°C for 5 min and subsequently cooled slowly over 15 h by using an appropriate temperature ramp (Czogalla et al., 2015).
- Oligonucleotides are HPLC purified and their molecular weights are checked by ESI-MS.
- Fluorescent labels are introduced by modifying selected oligonucleotides enzymatically (i.e., by TdT) and ChromaTide AlexaFluor 488-5-dUTP or ChromaTide AlexaFluor 647-5-dUTP, as previously described (Kauert et al., 2011).
- 5 µM of each oligo is incubated together with 400 units of TdT in a 20-µL reaction according to the company's manual.
- For the preparation of staple oligonucleotides, a PolyGen 12-column synthesizer is used to synthesize at the 3′-end cholesteryl-modified oligonucleotides. The modification is moderated by 3′-cholesteryl-TEG CPG.
- The assembled origami structures are investigated by gel electrophoresis at 3.5 V/cm on a 1% agarose gel containing 50 mM Tris borate (pH8.3), 1 mM EDTA, and 14 mM $MgCl_2$.
- For subsequent measurements, origami objects are sliced out from the gel and purified using a commercial kit (e.g., Freeze'n Squeeze Kit).

Note: Avoid exposure to ethidium bromide and UV light during purification steps to reduce damage to assembled structures.

Note: A part fraction of the sample should run in an extra lane on the gel next to the main fraction. This extra lane will also be cut from the gel, stained with ethidium bromide and marked with a scalpel during subsequent UV illumination. These marks are then used to locate and excise the corresponding origami band of the main fraction.

Vesicle preparation and fluorescence imaging

- Production of GUVs composed of DOPC, labeled with 0.1 mol% DiI fluorescent lipid analog by electroformation in a polytetrafluoroethylene (PTFE) chamber with platinum (Pt) electrodes, as described in Chapter 1.
- Chamber should be filled with isosmotic sucrose solution during GUV formation.
- GUVs are pipetted into bovine serum albumin (BSA)–coated 384-microwell plates filled with iso-osmolar buffer.
- To maximize the base pairing between all complementary single stranded extensions on both constructs A (labeled with AlexaFluor 488) and B (labeled with AlexaFluor 647), an approximate 1:1 molar ratio between both constructs should be kept (10–100 pM in aqueous solution).
- Incubation of the mixture overnight at 4°C in a sealed 384-microwell plate.
- Confocal imaging is performed with an LSM equipped with C-Apochromat, 40 × 1.2W objective. An example of a vesicle deformed by the scaffolding is given in Figure 4.1b.

minimal *in vitro* systems like minimal actin cortexes (MACs) have been developed that allow a more quantitative understanding of cell cortex mechanics (Figure 4.2). In Box 4.2, the design of several bottom-up systems, with actin filaments attached to artificial membranes is presented (Heinemann et al., 2013; Vogel et al., 2013). Insights gained from these *in vitro* systems may help to uncover fundamental principles of how exactly actin-myosin-membrane interactions govern actin cortex remodeling for cell shape changes, and how this depends on specific membrane properties.

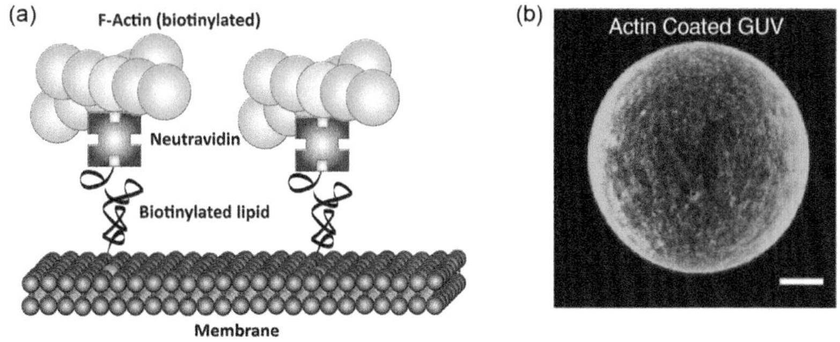

Figure 4.2 Bottom-up system based on actin immobilization on model membranes for quantitative understanding of the cell cortex mechanics. (a) A sketch for the anchoring of the actin filaments. (b) An actin-coated GUV. Scale bar: 10 μm. (Reprinted from *Curr. Opin. Chem. Biol.*, 22, Rivas, G. et al., Reconstitution of cytoskeletal protein assemblies for large-scale membrane transformation, 18–26, Copyright 2014, with permission from Elsevier.)

Box 4.2 Reconstitution of actin-based assemblies on supported lipid membranes and on giant vesicles (Vogel et al., 2013)

Materials and Equipment

- Rabbit skeletal muscle actin monomers (Molecular Probes)
- Biotinylated rabbit actin monomers (tebu-bio [Cytoskeleton; Offenbach, Germany])
- Neutravidin (Molecular Probes)
- Alexa-Fluor 488 Phalloidin (Molecular Probes)
- F-Buffer (50 mM KCl, 2 mM $MgCl_2$, 1 mM dithiothreitol (DTT), 1 mM ATP, 10 mM Tris-HCl pH7.5)
- Reaction buffer (50 mM KCl, 2 mM $MgCl_2$, 1 mM DTT, 10 mM Tris-HCl pH7.5)
- Lipids, for example, EggPC, DOPC, 1,2-dioleoyl-*sn*-glycero-3-phosphoglycerol (DOPG), *N*-palmitoyl-D-erythro-sphingosylphosphorylcholine (C16 SM), 1,2-distearoyl-*sn*-glycero-3-phosphoethanolamine-*N*-[biotinyl(polyethylene glycol)-2000] [DSPE-PEG(2000)-Biotin] from Avanti (Alabaster, AL) see Appendix 1 of the book for structure and data on the lipids; the fraction of the biotinylated lipid is typically in the range 0.01–1 mol%
- Lipid dyes, for example, Atto647N-1,2-dioleoyl-*sn*-glycero-3-phosphoethanolamine (Atto647N-DOPE see Appendix 2 of the book for structure and data on this lipid dye) from Atto-Tec (Siegen, Germany) added at 0.1 mol% of total lipid
- Perfusion chamber (tebu-bio [Cytoskeleton; Offenbach, Germany])
- Round glass slide (d = 24 mm, #1.5, Menzel Gläser; Thermo Fisher Scientific, Braunschweig, Germany)
- Coverslip holder (CoverSlipHolder, JPK; Berlin, Germany)
- Glass microscope slide (#1.5, Menzel Gläser; Thermo Fisher, Braunschweig, Germany)
- UV-curable glue 63 (Norland Products; Cranbury, NJ)

Actin preparation and labeling

- Mixture of actin monomers and biotinylated actin monomers in a ratio of 4:1.
- Polymerization of the mixture (39.6 μM) can be introduced in F-Buffer.
- Biotinylated actin filaments are stabilized by addition of Alexa-Fluor 488 Phalloidin. A final concentration of about 2 μM Alexa 488 phalloidin-labeled biotinylated actin filaments should be obtained.

Myosin preparation

- Myosin is purified from rabbit skeletal muscle tissue as described by Smith et al. (2007).
- Activity of myosin can be tested by a classical motility assay.
- Assembly of myosin filaments is obtained in myosin reaction buffer.

Minimal actin cortex preparation

- For neutravidin bridge between DSPE-PEG(2000)-Biotin and biotin actin filaments (Figure 4.2a), GUVs are incubated with 200 μL (2.5 ng/μL) neutravidin for 5 min.
- 20 μL of the GUV suspension should be pipetted into a cut 2.5-mL reaction tube (Eppendorf Tubes®) UV-glued onto a microscope slide filled with 2 mL of reaction buffer and Alexa 488 phalloidin-labeled actin filaments. Due to the higher density of the sucrose within the GUVs, they will sediment down and actin filaments will anchor to the neutravidin to form an actin filament capsule around the GUVs (Figure 4.2b).

4.4.3 EXAMPLE 3: ATTACHMENT VIA AMPHIPATHIC HELICES

During bacterial cell division, the cytoskeletal tubulin homolog protein FtsZ assembles into dynamic curved filaments to constitute the so-called Z ring. This Z ring is part of the complex divisome and appears to constrict during cytokinesis. For investigation of the key protein in bacterial cell division in a minimal setting, a membrane-targeted derivative of FtsZ (FtsZ-mts) was constructed by splicing an amphipathic helix to its C-terminal end (Osawa et al., 2009). After adding it to lipid vesicles (large multilamellar vesicles), FtsZ-mts was incorporated into the bilayer, where it formed multiple Z protorings, diffusing laterally bidirectionally along the liposomes and merging into brighter rings. These rings constricted the liposomes without further divisome proteins. To further investigate the assembly of the Z ring and its force generation, FtsZ-mts was attached to supported membranes as well as GUVs (Figure 4.3) and found to show a rich phenomenology of attachment and multimerization dynamics (Arumugam et al., 2014, 2015). Box 4.3 summarizes the essential assays.

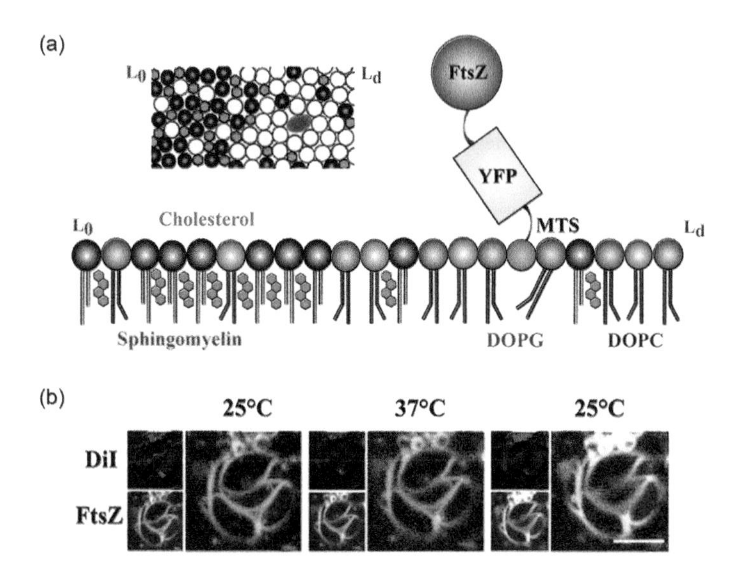

Figure 4.3 Cytoskeletal pinning controls phase separation in multicomponent lipid membranes. (a) A sketch of the external leaflet of a vesicle membrane composed of sphingomyelin (black), cholesterol (blue), DOPG (pink head groups), DOPC (purple head groups) and the orange circle represents the FtsZ-YFP-MTS construct, which is binding to the L_d phase. The insert in the top left corner shows a top view of the leaflet illustrating the lipid phase separation in liquid ordered (Lo) and liquid disordered (Ld) phase. (b) The presence of FtsZ-YFP-MTS (green channel) bound to the L_d phase of the SLB (purple channel; DiDC18 labeling) has no effect on phase transition temperature. (Reprinted from *Biophys. J.*, 108, Arumugam, S. et al., Cytoskeletal pinning controls phase separation in multicomponent lipid membranes, 1104–1113, Copyright 2015, with permission from Elsevier.)

Box 4.3 Protocol for attaching proteins via amphipathic helices as anchors (Arumugam et al., 2015)

Materials and Equipment

- TRIS buffer (50 mM Tris HCl, 1 mM EDTA, 50 mM KCl, 10% glycerol)
- Resource Q anion exchange column (Amersham Biosciences, Piscataway, NJ)
- HMKKG buffer (HEPES 50 mM, magnesium acetate 5 mM, potassium acetate 300 mM, potassium chloride 50 mM, and 10% glycerol, pH 7.8)
- *E. coli* lipid extract from Avanti Polar Lipids (Alabaster, AL)
- DiIC18(5) (Life technologies, Darmstadt, Germany)
- NanoWizard I (JPK Instruments, Berlin, Germany)
- Non-coated silicon cantilevers (DP15/HiRes-W/AIBS, MikroMasch, Spain)
- Silicon nitride cantilevers (MLCT/cantilever A, Veeco, CA)

FtsZ purification and assembly on supported lipid bilayers

- FtsZ-YFP-MTS is composed of a truncated version of *E. coli* FtsZ (amino acids 1-366), without the FtsA-binding site, combined with yellow fluorescent protein (YFP) (Venus; [Rekas et al., 2002]) and an amphipathic helix from *E. coli* MinD, a membrane-targeting sequence (MTS) (Osawa et al., 2009).
- FtsZ-YFP-MTS can be overexpressed in the vector pET11b in *E. coli* BL21 cells overnight at 16°C with isopropyl β-D-1-thiogalactopyranoside (IPTG) induction.

(Continued)

Box 4.3 (Continued) Protocol for attaching proteins via amphipathic helices as anchors (Arumugam et al., 2015)

- *E. coli* cells should be lysed by sonication in TRIS buffer, and FtSZ-YFP-MTS can be further enriched from the supernatant by 40% ammonium sulfate precipitation.
- FtsZ-YFP-MTS is resuspended and dialyzed against TRIS buffer, purified by a Resource Q anion exchange column, desalted and aliquoted in TRIS buffer.
- FtsZ polymers assembly is induced in HMKKG buffer.
- SLBs should be treated with the same buffer but with additional GTP for investigation of FtsZ interaction. FtsZ-YFP-MTS can be polymerized at a final concentration of 0.5 µM and 500 µM GTP.
- SLBs can be prepared by small unilamellar vesicles (SUVs) out of 4 mg/mL *E. coli* lipid extract with 0.1 mol% 1,1′-dioctadecyl-3,3,3′,3′-tetramethylindodicarbocyanine, 4-chlorobenzenesulfonate (DiDC18) sonicated in FtsZ assembly buffer at room temperature.
- SUV suspension should be diluted to a concentration of 0.5 mg/mL, added to the substrate or glass rods and incubated at 37°C.
- Bilayer formation on mica of 2.5-mM fused SUVs is triggered by adding $CaCl_2$.
- A washing step with FtsZ assembly buffer cleans supported lipid bilayer from unfused SUVs. All SLB should be prepared on cleaned glass substrates or glass rods from Whatman® GF/B filters.

Preparation of GUVs decorated with a FtsZ network

- GUVs composed of DOPC/DOPG/eSM/Chol (2.5:2.5:3:2 ratio at 1 mg/mL in chloroform), labeled with 0.1 mol% Fast DiI fluorescent lipid analog are prepared by electroformation in PTFE chamber with Pt electrodes as described in Chapter 1.
- GUVs are transferred to a BSA-coated chamber and sink to the bottom as a result of density differences.
- Half of the reaction solution is removed and replaced by polymerization buffer (50 mM 2-(*N*-morpholino)ethanesulfonic acid [MES], 50 mM KCL, 15 mM $MgCl_2$, pH 6.5).
- FtsZ-YFP-MTS is polymerized on GUVs at final concentration of 1 µM FtsZ-YFP-MTS and 0.5 mM guanosine-5′-[(α,β)-methyleno]triphosphate (GMPCPP).

4.4.4 EXAMPLE 4: ATTACHMENT VIA PERIPHERAL MEMBRANE PROTEINS

The actin cytoskeleton is involved in the organization of lipids and proteins in the plasma membrane. A minimal model for an actin-based cortex (i.e., a MAC) was described in Section 4.4.2. In a more physiological setting, the branched actin network is attached to lipid membranes by the lipid second messenger PIP_2 and the protein adaptor N-WASP. Phase-separated GUVs were used to investigate the polymerization of the actin network and its influence on the lipid phase separation. The PIP_2–N-WASP bridge between the lipid bilayer and the actin network behaves like a switch, and the presence of a preexisting actin network spatially influences the location of phase separation (Liu and Fletcher, 2006) (Figure 4.4). It can be shown that the dynamic, membrane-bound actin network can control location and dynamics of membrane domain formation and may actively contribute to membrane organization during cell processes like cell signaling (Section 4.2).

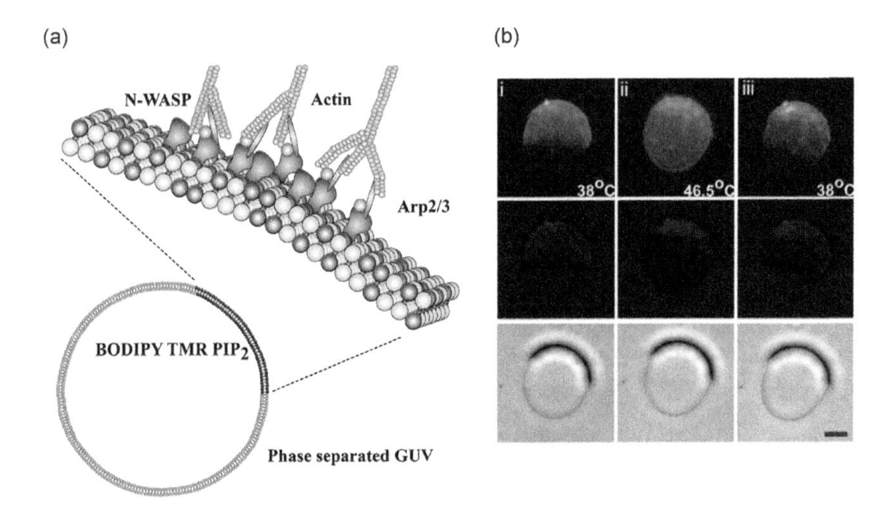

Figure 4.4 Assembly of actin networks on phase-separated GUVs. (a) A schematic presentation of the actin anchoring to the membrane. (b) Epifluorescence (top: fluo-DOPE middle: TMR-PIP_2) and phase-contrast images of a GUV (DPPC/DOPC/Chol/BODIPY TMR PIP_2) subjected to a temperature cycle. The vesicle is phase separated at low temperature and homogeneous at high temperature, but the actin network remains localized. Scale bar: 10 µM. (Reprinted from *Biophys. J.*, 91, Liu, A.P. and Fletcher, D.A., Actin polymerization serves as a membrane domain switch in model lipid bilayers, 4064–4070, Copyright 2006, with permission from Elsevier.)

Box 4.4 Protocol for protein attachment via peripheral protein bridges to giant vesicles (Liu and Fletcher, 2006)

Materials and equipment

- Lipids, such as dipalmitoylphosphatidylcholine (DPPC), DOPC, 1,2-dioleoyl-*sn*-glycero-3-phospho-L-serine (DOPS), Cholesterol (Alabaster, AL)
- 4,4-Difluoro-4-bora-3a,4a-diaza-*s*-indacene (BODIPY®) tetramethyrhodamine (TMR, see Appendix 4 of the book for structure and data on this dye) PIP_2 (Echelon Biosciences, Salt Lake City, UT)
- lis-1,2-Bis(diphenylphosphino)ethane (lis-DPPE), fluo-DOPE Avanti Polar Lipids (Alabaster, AL)
- Alexa Fluor 546 phalloidin (Invitrogen; Carlsbad, CA)
- LabTek II #1.5 8-well chamber (Thermo Fisher Scientific; Waltham, MA)
- Reaction buffer (20 mM HEPES pH7.5, 100 mM NaCl)
- Polymerization buffer (50 mM KCl, 2 mM $MgCl_2$, 5 mM Tris HCl pH7.4, 1 mM ATP, and 1 mM DTT).

Protein preparation

- His-tagged Rat ΔEVH1 N-WASP can be overexpressed and purified as described by Papayannopoulos et al. (2005).
- N-WASP activity is confirmed by bead motility assay using N-WASP-coated microspheres (Wiesner et al., 2003).
- Purification of actin from rabbit acetone dried powder is shown by Spudich and Watt (1971).
- Actin is labeled with rhodamine see Appendix 4 of the book for structure and data on this dye (Amann and Pollard, 2001).
- Arp2/3 complex can be purified from bovine brains (Egile et al., 1999).
- Arp2/3 complex activity is checked by pyrene actin polymerization assay (Welch et al., 1998).
- Protein purities are controlled by SDS-PAGE and protein yield is measured by protein absorbance at 280 nm.

Coating coverslips for vesicle observation

- Prepare a 1–2 mg/mL solution of BSA.
- Add an appropriate volume of BSA solution to your chamber.

Note: 200 µL per well in 8-well Nunc™ Lab-Tek™ II chamber is practicable.

- Incubate at room temperature for at least 30 min.
- Wash at least twice.

Note: BSA can bind to GUVs and promote their aggregation. It binds to GUVs containing phosphatidic acid (PA) at quite low concentrations. After coating the glass with 10 mg/mL BSA and just one rinsing step, it even clusters phosphatidylcholine/phosphatidylserine GUVs.

Actin-associated vesicles

- The GUVs composed of DPPC/DOPC/Chol/BODIPY TMR PIP_2 (2:1:30:0.6%), labeled with 0.5–1 mol% lis-DPPE or fluo-DOPE are prepared by electroformation on Pt electrodes at 60°C.
- PIP_2-containing vesicles should be incubated with N-WASP reaction buffer for 15 min.
- Add vesicles to a solution of actin and Arp2/3 complex.

Note: Final concentrations of 6 µM actin, 150 nM Arp2/3, 390 nM N-WASP.

- The sample are loaded into a flow cell with a glass slide and a cover glass with a 6-mm rim.
- Membrane-associated actin patches can be removed by adding 0.75 mg/mL proteinase K and incubating for 15 min at 37°C, whereas entangled actin filaments are found to be stable up to 50°C (Figure 4.4).

4.4.5 EXAMPLE 5: ATTACHMENT OF CYTOSKELETAL FILAMENTS VIA INTEGRAL MEMBRANE PROTEINS

Reconstitution of long actin filaments firmly attached to vesicle membranes from inside through trans-membrane proteins was first demonstrated by Merkle et al. (2008) (Figure 4.5). This system mimics the attachment of the cytoskeleton in erythrocytes. Two protocols were used to anchor actin filaments within GUVs. On the one hand, GUVs were made of isolated membrane fractions from porcine brain, a complex lipid composition with essential integral membrane proteins for anchoring. On the other hand, an enriched spectrin/ankyrin network was used to attach pre-polymerized actin filaments to the inner membrane of GUVs. This approach, as detailed in Box 4.5, validates GUVs as an important compartment system, within which multiprotein networks can be reconstituted and investigated in the presence of complex lipid mixtures, also by anchoring membrane-attached proteins to integral proteins.

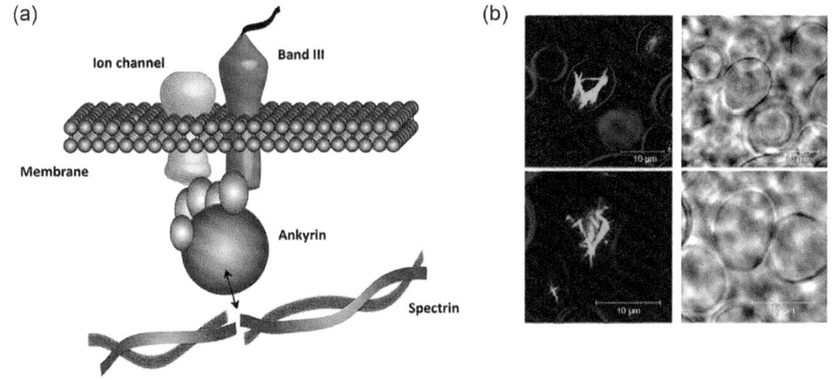

(a) (b)

Figure 4.5 Reconstitution and anchoring of cytoskeleton inside GUVs (Merkle et al. 2008). (a) A schematic presentation of protein components (ion channels, band III, Ankyrin, spectrin) required to anchor actin filaments to the membrane. (b) Confocal cross sections (left) and phase-contrast images (right) of GUVs prepared from porcine brain membrane extracts and labeled with DiDC18 (red) encapsulating actin filaments labeled with phaloidin-Alexa 488 (green) with the addition of a highly enriched fraction of spectrin/ankyrin proteins. (Adapted from Merkle, D. et al., *ChemBioChem*, 9, 2673–2681, 2008. With permission from John Wiley & Sons.)

Box 4.5 Protocol for attachment of cytoskeletal filaments via integral membrane proteins in vesicles (Merkle et al., 2008)

Material

- Reagents, salts protease inhibitors and buffers from Sigma Aldrich
- Antibodies from Abcam (Cambridge, UK)
- Secondary goat anti-mouse Cy5 conjugates from Jackson ImmunoResearch Laboratories (Suffolk, UK).
- DiD-C18, purified actin from rabbit skeletal muscle, calcium green and phalloidin conjugated to Alexafluor 488 from Molecular Probes (Invitrogen).
- Total porcine brain lipid extract (TBLE) from Avanti Polar Lipids (Alabaster, USA).

Isolation of porcine membrane fractions

- About 100 g of fresh porcine brain is prepared as described by Davis and Bennett (1986).
- Washing in PBS and homogenization in homogenization buffer (sodium phosphate, sodium azide and EDTA).
- Centrifugation 15 min at 900*g* and 45 min at 20,000*g*.
- Resuspending in homogenization buffer with sucrose and centrifugation 45 min at 30,000*g*.
- Pellet resuspending in NaOH (0.1 N) and extraction 30 min at 4°C.
- Centrifugation for 45 min at 30,000*g* through wash buffer (sodium phosphate (10 mM), sodium azide (5 mM), EGTA (0.2 mM), DTT (0.5 mM), 0.05% (v/v) Tween® 20 and 10% (w/v) sucrose.

Note: All buffers include phenylmethylsulfonyl fluoride (200 nM), pepstatin A, benzamidine (2 mM), leupeptin (100 μM), aprotonin (800 nM), chymostatin (100 μM), and antipain (100 μM).

Isolation of highly enriched spectrin/ankyrin pool

- Starting membrane pellet is extracted with wash buffer containing KCl (500 mM) for 30 min.
- Centrifugation 45 min at 30,000*g* and pellet extraction with 800-mM KCl wash buffer for 60 min.
- Spectrin and ankyrin remain in supernatant. Ammonium sulfate is added to the isolated supernatant at 4°C to 60% saturation, and the proteins are precipitated at −20°C.
- Centrifugation 15 min at 1,000*g*. Pellet resuspend and dialyze against 500 mM KCl buffer.
- Centrifugation 60 min at 100,000*g* and concentrate through a 300-kDa membrane.
- Purification by sepharose column.

GUV preparation from porcine brain fractions

- Production of GUVs by electroformation on Pt electrodes was performed as described in Chapter 1.
- Aliquots of porcine membrane fractions can be spotted evenly over indium tin oxide (ITO) coverslips.
- The GUVs preparation by using electroswelling for 2 h at 1.2 V
- DiDC18 (0.2 μL, 2 μM in chloroform) is added to the mixture in order to stain the lipid bilayer.
- Samples containing actin can be also treated with 1 unit of phalloidin–Alexa 488 conjugate.

4.5 APPLICATIONS OF EUKARYOTIC MEMBRANE–CYTOSKELETON SYSTEMS

4.5.1 CHANGE IN MEMBRANE DIFFUSION

In the biophysical study by Heinemann et al. (2013), the well-characterized MAC system (Figure 4.2 and Box 4.2) attached to a freestanding membrane was used to investigate the effect of the actin network on lateral diffusion of lipids and membrane proteins. Lateral diffusion of lipids and proteins within the cell membrane is essential for a large number of membrane-dependent processes. Employing fluorescence correlation spectroscopy (FCS) (see also Chapter 21), a marked relationship between the actin network density and the reduction in lipid and protein mobility was observed. It could also be determined that high actin densities influenced much more the mobility of bulky proteins than did lipid diffusion. This implies that membrane-bound cortical actin has a complex influence on the lateral diffusion, depending on the type and, accordingly, on the size of the diffusing objects. Moreover, the addition of myosin filaments allows to tune the diffusion rate in this minimal system, which is a potential mechanism to control lateral diffusion (Heinemann et al., 2013). Freestanding membranes (Heinemann and Schwille, 2011) were used to investigate the correlation of mobility reduction and actin density by point-FCS as mentioned by Heinemann and Schwille (2011). Complementing coarse-grain Monte Carlo simulations supported the analysis of the experimental FCS data.

4.5.2 CHANGE IN PHASE TRANSITION

In a recent series of theory and experiments, the response of membrane phase transition in glass- or mica-supported membranes on a membrane bound actin network was investigated, either by standard or by stimulated emission depletion super-resolution microscopy and FCS (Ehrig et al., 2011; Honigmann et al., 2014). The interplay with a cytoskeletal model system showed a significant effect on the lateral distribution and dynamics of lipids using a ternary lipid composition according to the so-called "canonical raft mixture" (Veatch and Keller, 2003a, 2003b), see Chapter 18 for membrane phase behavior. Experiments showed that in the presence of actin, a macroscopic effect on phase separation was missing (Honigmann et al., 2014). Furthermore, the alignment of liquid disordered (Ld) domains with actin fibers, leading to a channel-like domain structure, was confirmed by experiments as well as simulations (Ehrig et al., 2011; Fischer and Vink, 2011; Machta et al., 2011). Interesting findings from *in vivo* measurements could be revisited by this model system. Recent experiments showed that the type of lipid domain that will be stabilized by the artificial cytoskeleton depends on the phase-segregation properties of the pinning molecules. Cells could spatially sort their membrane components depending on the pinning species (Honigmann et al., 2014). This sorting mechanism also exists at physiological temperatures, which are above the phase-separation temperature. However, the pinning may also reduce or prevent phase separation within plasma membrane at lower temperatures. These effects are enhanced in the presence of membrane curvature. However, because the energy cost of lipid extraction by far exceeds that of membrane detachment from the support (Helm et al., 1991), it could be shown that in freestanding membranes, or in cell membranes, the curvature-coupling is likely stronger.

4.5.3 CONTRACTION OF ASTERS, SHAPE MODULATION

Contractile behavior of actomyosin network was investigated recently on phase-separated membranes by fluorescence recovery after photobleaching (FRAP) experiments with a total internal reflection fluorescence microscopy (TIRFM) setup (Greiss et al. in preparation), for more information on these techniques consult Chapters 18 through 20. Interference of aster formation with membrane domains can be explained by restricted diffusion of actin filaments at the phase boundary. Freely diffusing F-actin is pulled into bundles by ATP-hydrolyzing myofilaments, leading to a contractile network with local concentration hubs, the so-called asters. If two F-actin asters are separated by phase boundaries and pulled together, this leads to domain deformation events. The attachment to the phase boundaries tremendously influences the local distribution of the F-actin. Important parameters to pay attention to during the contraction of asters are the actin filament length and the average size of interspace regions between two domains. The motion of asters toward convex boundaries leads to minimized energy, whereas concave shapes behave *vice versa*, leading to an accumulation of actomyosin at the domain edges. It could also be concluded that the occurrence of asymmetric deformation is dependent on the geometry of the domains. Membrane deformation introduces a nonequilibrium state in the membrane, which changes the affinity for further lipids balancing the deformation. The rearrangement of the actin cortex by motor proteins is a basis for fully new approaches in order to study dynamic membrane organization under more physiological conditions. Raft clustering has been proposed to be induced by either hydrogen bonding via sphingomyelin or by protein crosslinking (Kusumi et al., 2004), but also by direct displacement of raft proteins attached to dynamic actin polymers. Reorganization of membrane proteins leads to a constant spatiotemporal synchronization of lipid/protein rafts and actin cortex remodeling. The friction coupling of lipids, as in supported bilayers, must be avoided to adapt this minimal system closer to the biological equivalent (Przybylo et al., 2006). Future experiments will include freestanding membranes, for example, in GUVs, for the assembly of a minimal actin cortex on phase-separated membranes. The boundary of lipid phases in freestanding membranes might promote the displacement of entire domains leading to the accumulation of phases at the aster formation site, unlike the focused deformation of phase boundaries in supported phase-separated membranes.

4.6 APPLICATIONS OF PROKARYOTIC MINIMAL CORTICES

4.6.1 DEFORMATION OF VESICLES

In the work by Cabre et al. (2013) permeable GUVs were used to investigate the influence of FtsZ and ZipA, two key constituents of the *E. coli* divisome, on membrane vesicles. In this study, vesicles were made permeable by adding α-hemolysin. (For more background and information, see Chapter 20 on membrane

The making of

permeability and Chapters 11, 14 and 15, which contain data on the mechanical properties of membranes.) Furthermore, it is known that FtsZ is an initial constriction force generator for the division step, and ZipA is an FtsZ anchor to the cytoplasmic membrane. This native ZipA anchoring was utilized to investigate polymerization effects of FtsZ on artificial vesicle membrane. ZipA associates with the inner membrane leaflet without GTP, whereas FtsZ remains soluble in the vesicle lumen. After the addition of GTP, the FtsZ polymerization starts, resulting in a shrinkage of the GUV (Cabre et al., 2013). Shorter and more flexible FtsZ polymers, attached to ZipA, appear to induce pulling forces, as, for instance, during membrane constriction. When added from outside of the vesicles, actin polymerization produces inward protrusions (Liu et al., 2008), whereas FtsZ polymerization leads to outward-oriented spikes (Osawa et al., 2009; Martos et al., 2012). A reduced invagination could be observed in dividing cells, in contrast to the strong shrinkage via FtsZ polymerization in ZipA containing vesicles. Structural support *in vivo* is guaranteed by peptidoglycan and the outer membrane, not being affected by the increase in ZipA. In the mentioned publication, it could be shown that the amount of ZipA when overproduced in *E. coli*, is similar to the limit for vesicle shrinkage. Recent studies (Cabre et al., 2013; Lopez-Montero et al., 2013) suggest that the force to constrict the cell envelope may modify the architecture of the cytoplasmic membrane. This plasticity of the membrane, supporting dilation of *E. coli* lipid vesicles in which ZipA is incorporated at both sides of the membrane (Lopez-Montero et al., 2013), may on the other hand be responsible for the fragility of the envelope and the breakdown of the permeability barrier. Membrane constriction is an essential stage in the cell division of modern cells and also an ancestral mechanism in the evolution into a peptidoglycan world. This *in vitro* system with GUVs revealed that a minimal divisome consisting of ZipA anchoring and FtsZ polymerization, reproduced inside of permeable vesicles may provide a faithful mimicry for bacterial cell division (Cabre et al., 2013).

4.6.2 CHANGES IN DIFFUSION AND PHASE SEPARATION

Comparable to what has been reported above on actin cortex mimicries, recent work by Arumugam et al. (2015) also revealed an influence of membrane-attached FtsZ networks on phase separation in GUV lipid membranes. In this approach a minimalistic system was designed with only one filamentous protein species (i.e., FtsZ) and four lipid species for creating a phase-separating membrane. The experiments showed that the membrane-attached cytoskeleton prevents large-scale domain formation in membranes and that the domain size is influenced by the density of the FtsZ network. This cytoskeleton mimicry thus suppresses large-scale phase separation and supports membrane heterogeneity to be functional even below phase transition temperatures. They also revealed that the interaction with the cytoskeleton maintains the phase separation in the membrane also above the transition temperature, as would be required for proper function of the cell membranes under hyperthermic conditions. The recombinant FtsZ with an amphiphilic membrane-targeting sequence self-assembles into a network attached to the membrane, which is connected to only one leaflet of the bilayer (Szeto et al., 2002, 2003). However, it could be shown that the interleaflet coupling may be strong enough for a synchronized phase separation in both leaflets of the bilayer. The presence of membrane-interacting components, like transmembrane proteins, membrane binding proteins, and cytoskeletal structures, thus preserves heterogeneity and furthermore stabilizes fluctuations over a broader range of compositions and temperature (Arumugam et al., 2015). Filaments may directly interact with the membrane or interact with transmembrane and/or cytoplasmic membrane-binding proteins. Such a minimal system thus nicely elucidates the specific effect of cytoskeletal pinning, through amphipathic helix insertion, on phase separation. Although in live cells, there are many more factors that may prevent lipid unmixing in the cell membrane, few of them may be so systematically tested and quantified for their specific relevance in a minimal model system.

4.7 APPLICATIONS WITH ARCHAEAL PROTEINS: MEMBRANE DEFORMATIONS BY CdvA-CdvB FILAMENTS

In spite of their obvious functional relationships to both prokaryotic but also eukaryotic cytoskeleton systems, the reconstitution of archaeal proteins into minimal model systems has only recently been attempted (Hartel and Schwille, 2014) (Figure 4.6). For investigation of potential membrane deformations by the CdvABC system discussed above, we set out to study the interaction of the three archaeal cell division proteins CdvA, CdvB (ESCRT-III like protein), and CdvC (Vps4-like protein) in a cell-free bottom-up approach following the rationale pioneered

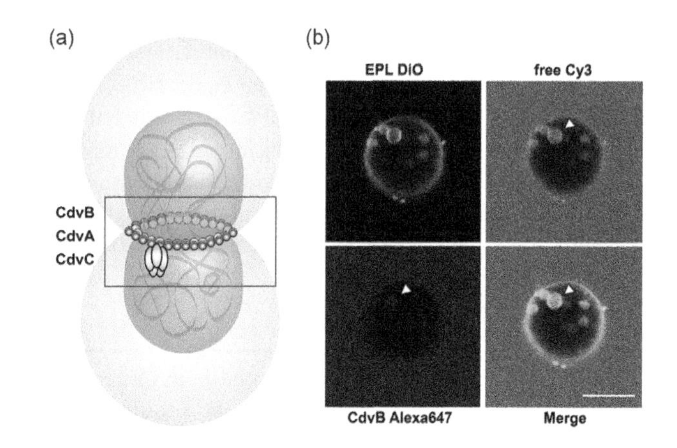

Figure 4.6 ESCRT-III–mediated cell division in *Sulfolobus acidocaldarius*—a reconstitution perspective (Hartel and Schwille, 2014). (a) CdvA (yellow) and DNA (green) build up double-helical structures horizontally to the cytokinesis region in *Sulfolobus acidocaldarius* (gray). CdvB (red) and paralogs interact with CdvA via its C-terminal winged helix-like domain and also bind to microtubule interaction and transport domain of the CdvC (cyan). The AAA-type ATPase CdvC regenerates the CdvB complex after cytokinesis. (b) ILV formation into EPL GUVs in presence of CdvB and CdvC. The GUV membrane is stained green with DiO. Red-labeled CdvB proteins (Alexa647), unstained CdvC proteins, and free Cy3 were added to the extraluminal buffer of the EPL GUVs. After ILV formation of the GUV membrane CdvB as well as the soluble marker Cy3 could be visualized in the vesicles (white arrowheads). Scale bar: 10 µm. (Adapted from Hartel, T. and Schwille, P., *Front. Microbiol.*, 5, 257, 2014.)

by Wollert et al. on the eukaryotic system (Wollert et al., 2009). Indeed, confocal microscopy revealed extraluminal membrane binding and inward directed intraluminal vesicles (ILV) formation into *Escherichia coli* polar lipids (EPL) GUVs in the presence of CdvB and CdvC (Hartel and Schwille, 2014) (Figure 4.6a).

Experiments with CdvABC proteins intraluminal in lipid vesicles were performed employing the droplet method for GUV formation (Pontani et al., 2009; van Swaay and deMello, 2013) (see also Chapter 1). In this approach, to study the function and interplay of Cdv-proteins, they were used either purified or by intraluminal *de novo* protein synthesis. Fluorescent recombinant proteins (C-terminally or N-terminally labeled with mCherry, Yellow1, or AmCyan1) could be visualized by confocal imaging. Strikingly, specific deformations of membrane vesicles after co-reconstitution of all three CdvABC proteins could be observed. Furthermore, the influence of electric charge, polarity, and lipid composition of the GUVs on curvature and abundance of budding and deformations events could be identified. Ongoing work is targeted toward further characterizing the function and interaction of the *crenarchaeal* ESCRT-III-like paralogs in membrane vesicles, to answer the question whether these paralogs are indeed essential for cell division in *Sulfolobus acidocaldarius* or whether they are only for fine-tuning. CdvABC-mediated proteo-cell division in such a minimal setting could help us explore the fundamental mechanics of membrane transformation independent of the very organism and nature of the membrane lipids (which are dramatically different in these archaea), and shed light on the phylogenetic evolution of the *crenarchaeal* division machinery. This may also answer the question why CdvABC-mediated cell division is mainly found in thermophilic and acidothermophilic *Crenarchaeota* and *Thaumarchaeota*. The linkage between archaeal cell division and ESCRT-III dependent biogenesis of multivesicular bodies in eukaryotes is another fascinating evolutionary aspect (Poole, 2006; Hanson et al., 2008). Future studies will investigate the variability of CdvB-mediated constriction, and facilitate CdvB based division of larger membrane compartments, to reveal its potential suitability as a truly "archetypical" cell division machinery to be assembled from the bottom-up (Hartel and Schwille, 2014).

4.8 SUMMARY AND OUTLOOK

Cytoskeletal and cortical structures in membrane vesicles may be relatively simple and well characterized bottom-up systems to investigate essential cellular processes like cell division, endocytosis, cell shape induction, and motility. In recent years, the reconstitution of eukaryotic, bacterial and archaeal cytoskeletal systems outside, but also inside, of GUVs have been mainly used to study biophysical properties, membrane–protein interactions, and to generally characterize protein activity on membrane transformation (as in Chapters 16 through 23). Thus, artificial cytoskeletons and cortices have meanwhile become the state-of-the-art in bottom-up reconstitution systems attempting to elucidate fundamental cellular phenomena beyond the simple membrane–protein system. As a striking example of how functionally significant such a minimal system can become, the cell division machinery of *E. coli* was reproduced *in vitro*, leading to self-organized, oscillating Min protein concentrations, spatially positioning the contractile Z ring to the middle of a cell-shaped compartment (Loose et al., 2008; Zieske and Schwille, 2013, 2014). Besides the actual septum positioning of *E. coli* by the MinCDE system (Teather et al., 1974; de Boer et al., 1989) the assembly and disassembly dynamics of FtsZ into filaments and proto-rings with curvature sensing and curvature inducing propensity could be quantitatively characterized (Arumugam et al., 2012, 2014). Several research groups worldwide are currently attempting the reconstitution and assembly of bacterial divisome, based on FtsZ, ZipA, FtsA, and the Min system in vesicles (Osawa et al., 2008; Jimenez et al., 2011; Cabre et al., 2013; Osawa and Erickson, 2013), with the final goal of accomplishing a self-organized controlled division of a membranous compartment with a minimal set of elements.

The insights gained from bottom-up *in vitro* experiments may also be used to better understand redundant properties in cellular systems when comparing them to reduced scenarios in minimal controllable systems (Vogel et al., 2013). As described in earlier subsections, the most popular model system so far is the reconstituted actin cortex on supported and /or freestanding membranes. The actin density in these systems can be varied by changing the concentration of the anchoring sites such as biotinylated lipids, the actin filaments, and the linking species neutravidin. Reconstitution on flat membranes, allows for observation by a large variety of modern biophysical techniques, such as TIRFM, FCS and atomic force microscopy (AFM). In the future, this minimalistic system could be gradually improved step by step, by adding ever more components to investigate specific functions of proteins or their ability to fine-tune the system. MACs could also be investigated in the presence of cytoplasmic cell extracts to closer mimic the *in vivo* system. Recent work has already accomplished to build MACs on the inside of liposomes (Pontani et al., 2009; Vogel and Schwille, 2012). Here, global parameters like liposome size can be varied by microfluidic techniques (Stachowiak et al., 2008), but also the shape of the liposomes will be of great relevance to be modified from spherical to elongated (Matosevic and Paegel, 2011).

Obviously, compared with the physiological situation, an artificial cell cortex or cytoskeleton is only one more layer of complexity added to the very minimal system consisting of membrane and (potentially) proteins inserted or attached to it. However, because many biochemical reaction systems are greatly dependent on the geometry of the respective cell volume, shape-inducing and stability-conferring elements are of crucial importance already for very simple cell mimicries. Thus, it is fair to say that membrane–cytoskeleton systems are presently the state-of-the-art behind which functional biophysical studies on reconstituted cellular systems should not fall behind. Moreover, it already becomes apparent that most essential factors that are involved in membrane and cell shape transformations actually do not only require the presence of an internal stabilizing element, but are also tightly coupled to (mostly) sugar-based cell walls and extracellular matrixes at the outside of the membrane. Future developments of faithful model systems with minimal complexity but maximal physiological relevance will thus have to encompass intelligent strategies of how to best mimic those external stabilizing elements.

LIST OF ABBREVIATIONS

ACM	actin-coated membranes
Arf1	ADP ribosylation factor 1
Arp2/3	actin related protein 2/3
ATP	adenosine triphosphate
BAR	bin amphiphysin rvs
BODIPY	4,4-difluoro-4-bora-3a,4a-diaza-*s*-indacene
BSA	bovine serum albumin
C16 SM	*N*-palmitoyl-D-erythro-sphingosylphosphorylcholine
Cdv	cell division protein
COPI	coat protein complex I
Cy	cyanine
DiDC18	1,1′-dioctadecyl-3,3,3′,3′-tetramethylindodicarbocyanine, 4-chlorobenzenesulfonate
Dil	1,1′-dioctadecyl-3,3,3′,3′-tetramethylindocarbocyanine perchlorate
DNA	deoxyribonucleic acid
DOPC	1,2-dioleoyl-*sn*-glycero-3-phosphocholine
DOPE	1,2-dioleoyl-*sn*-glycero-3-phosphoethanolamine
DOPG	1,2-dioleoyl-*sn*-glycero-3-phosphoglycerol
DOPS	1,2-dioleoyl-*sn*-glycero-3-phospho-L-serine
DPPC	dipalmitoylphosphatidylcholine
DPPE	1,2-bis(diphenylphosphino)ethane
DSPE-PEG (2000) Biotin	1,2-distearoyl-*sn*-glycero-3-phosphoethanolamine-N-[biotinyl(polyethylene glycol)-2000]
DTT	dithiothreitol
dUTP	2′-deoxyuridine, 5′-triphosphate
EDTA	ethylenediaminetetraacetic acid
Egg PC	Egg L-α-phosphatidylcholine
EGTA	ethylene glycol-bis(β-aminoethyl ether)-*N*, *N*,*N*′,*N*′-tetraacetic acid
EPL	*Escherichia coli* polar lipids
ESCRT	endosomal sorting complexes required for transport
ESI-MS	electrospray ionization mass spectrometry
F-BAR	Fes/CIP4 homology BAR
FCCS	fluorescence cross-correlation spectroscopy
FCS	fluorescence correlation spectroscopy
FRAP	fluorescence recovery after photobleaching
FtsA	filamenting temperature-sensitive mutant A
FtsZ	filamenting temperature-sensitive mutant Z
Gb3	globotriaosylceramide
GMPCPP	guanosine-5′-[(α,β)-methyleno]triphosphate
GTP	guanosine triphosphate
GUV	giant unilamellar vesicle
HEPES	4-(2-hydroxyethyl)-1-piperazineethanesulfonic acid
HPLC	high-performance liquid chromatography
I-BAR	inverse-BAR
IF	intermediate filament
ILV	intraluminal vesicles
IPTG	isopropyl β-D-1-thiogalactopyranoside
ITO	indium tin oxide
Ld	liquid disordered
Lo	liquid ordered
LSM	laser scanning microscopy
MAC	minimal actin cortex
MES	2-(*N*-morpholino)ethanesulfonic acid
MinCDE	minicell
MOTC	microtubule-organizing center
mts	membrane targeting sequence
MVB	multivesicular bodies
N-WASP	neural Wiskott-Aldrich syndrome protein
O/N	overnight
PIP_2	phosphatidylinositol-4,5-bisphosphate
Pt	platinum
PTFE	polytetrafluoroethylene
SDS-PAGE	sodium dodecyl sulfate polyacrylamide gel electrophoresis
SLB	supported lipid bilayer
STED	stimulated emission depletion
SUV	small unilamellar vesicles
TBLE	total porcine brain lipid extract
TdT	terminal desoxyribonucleotidyl transferase
TEG	triethylene glycol
TEM	transmission electron microscope
TIRFM	total internal reflection fluorescence microscopy
TMR	tetramethyrhodamine
Vps4	vacuolar protein sorting-associated protein 4
YFP	yellow fluorescent protein

REFERENCES

Alberts B, Johnson A, Lewis J, Raff M, Roberts K, Walter P (2008) *Molecular Biology of the Cell*–5th ed. Garland Science, New York.

Amann KJ, Pollard TD (2001) The Arp2/3 complex nucleates actin filament branches from the sides of pre-existing filaments. *Nature Cell Biol* 3:306–310.

Arumugam S, Chwastek G, Fischer-Friedrich E, Ehrig C, Monch I, Schwille P (2012) Surface topology engineering of membranes for the mechanical investigation of the tubulin homologue FtsZ. *Angewandte Chemie* 51:11858–11862.

Arumugam S, Petrasek Z, Schwille P (2014) MinCDE exploits the dynamic nature of FtsZ filaments for its spatial regulation. *Proceedings of the National Academy of Sciences of the United States of America* 111:E1192–1200.

Arumugam S, Petrov EP, Schwille P (2015) Cytoskeletal pinning controls phase separation in multicomponent lipid membranes. *Biophysical Journal* 108:1104–1113.

Ausmees N, Kuhn JR, Jacobs-Wagner C (2003) The bacterial cytoskeleton: An intermediate filament-like function in cell shape. *Cell* 115:705–713.

Babst M, Wendland B, Estepa EJ, Emr SD (1998) The Vps4p AAA ATPase regulates membrane association of a Vps protein complex required for normal endosome function. *EMBO J* 17:2982–2993.

Barelli H, Antonny B (2009) Cell biology: Detached membrane bending. *Nature* 458:159–160.

Bernander R, Lind AE, Ettema TJ (2011) An archaeal origin for the actin cytoskeleton: Implications for eukaryogenesis. *Communicative & Integrative Biology* 4:664–667.

Bi EF, Lutkenhaus J (1991) FtsZ ring structure associated with division in *Escherichia coli*. *Nature* 354:161–164.

Cabre EJ, Sanchez-Gorostiaga A, Carrara P, Ropero N, Casanova M, Palacios P, Stano P, Jimenez M, Rivas G, Vicente M (2013) Bacterial division proteins FtsZ and ZipA induce vesicle shrinkage and cell membrane invagination. *The Journal of Biological Chemistry* 288:26625–26634.

Czogalla A, Kauert DJ, Franquelim HG, Uzunova V, Zhang Y, Seidel R, Schwille P (2015) Amphipathic DNA origami nanoparticles to scaffold and deform lipid membrane vesicles. *Angewandte Chemie* 54:6501–6505.

Czogalla A, Petrov EP, Kauert DJ, Uzunova V, Zhang Y, Seidel R, Schwille P (2013) Switchable domain partitioning and diffusion of DNA origami rods on membranes. *Faraday Discussions* 161:31–43; discussion 113–150.

Davis JQ, Bennett V (1986) Association of brain ankyrin with brain membranes and isolation of active proteolytic fragments of membrane-associated ankyrin-binding protein(s). *The Journal of Biological Chemistry* 261:16198–16206.

Dayel MJ, Mullins RD (2004) Activation of Arp2/3 complex: Addition of the first subunit of the new filament by a WASP protein triggers rapid ATP hydrolysis on Arp2. *PLoS Biology* 2:E91.

de Boer P, Crossley R, Rothfield L (1992) The essential bacterial cell-division protein FtsZ is a GTPase. *Nature* 359:254–256.

de Boer PA, Crossley RE, Rothfield LI (1989) A division inhibitor and a topological specificity factor coded for by the minicell locus determine proper placement of the division septum in *E. coli*. *Cell* 56:641–649.

Deamer D (2005) A giant step towards artificial life? *Trends in Biotechnology* 23:336–338.

Delatour V, Helfer E, Didry D, Le KH, Gaucher JF, Carlier MF, Romet-Lemonne G (2008) Arp2/3 controls the motile behavior of N-WASP-functionalized GUVs and modulates N-WASP surface distribution by mediating transient links with actin filaments. *Biophysical Journal* 94:4890–4905.

Douglas SM, Marblestone AH, Teerapittayanon S, Vazquez A, Church GM, Shih WM (2009) Rapid prototyping of 3D DNA-origami shapes with caDNAno. *Nucleic Acids Res* 37:5001–5006.

Desai A, Mitchison TJ (1997) Microtubule polymerization dynamics. *Annual Review of Cell and Developmental Biology* 13:83–117.

Discher DE, Mooney DJ, Zandstra PW (2009) Growth factors, matrices, and forces combine and control stem cells. *Science* 324:1673–1677.

Doherty GJ, McMahon HT (2008) Mediation, modulation, and consequences of membrane-cytoskeleton interactions. *Annual Review of Biophysics* 37:65–95.

Egile C, Loisel TP, Laurent V, Li R, Pantaloni D, Sansonetti PJ, Carlier MF (1999) Activation of the CDC42 effector N-WASP by the Shigella flexneri IcsA protein promotes actin nucleation by Arp2/3 complex and bacterial actin-based motility. The Journal of cell biology 146:1319–1332.

Ehrig J, Petrov EP, Schwille P (2011) Near-critical fluctuations and cytoskeleton-assisted phase separation lead to subdiffusion in cell membranes. *Biophysical Journal* 100:80–89.

Ettema TJ, Lindas AC, Bernander R (2011) An actin-based cytoskeleton in archaea. *Molecular Microbiology* 80:1052–1061.

Farsad K, De Camilli P (2003) Mechanisms of membrane deformation. *Current Opinion in Cell Biology* 15:372–381.

Fischer T, Vink RL (2011) Domain formation in membranes with quenched protein obstacles: lateral heterogeneity and the connection to universality classes. *The Journal of Chemical Physics* 134:055106.

Fletcher DA, Mullins RD (2010) Cell mechanics and the cytoskeleton. *Nature* 463:485–492.

Garcia-Saez AJ, Carrer DC, Schwille P (2010) Fluorescence correlation spectroscopy for the study of membrane dynamics and organization in giant unilamellar vesicles. *Methods in Molecular Biology* 606:493–508.

Garner EC, Campbell CS, Mullins RD (2004) Dynamic instability in a DNA-segregating prokaryotic actin homolog. *Science* 306:1021–1025.

Geli MI, Riezman H (1998) Endocytic internalization in yeast and animal cells: Similar and different. *Journal of Cell Science* 111 (Pt 8):1031–1037.

Gitai Z (2005) The new bacterial cell biology: moving parts and subcellular architecture. *Cell* 120:577–586.

Gunning PW, Ghoshdastider U, Whitaker S, Popp D, Robinson RC (2015) The evolution of compositionally and functionally distinct actin filaments. *Journal of Cell Science* 128:2009–2019.

Hanson PI, Roth R, Lin Y, Heuser JE (2008) Plasma membrane deformation by circular arrays of ESCRT-III protein filaments. *The Journal of Cell Biology* 180:389–402.

Hardin J, Bertoni G, Kleinsmith LJ (2015). Becker's World of the Cell (8th ed.). New York: Pearson. pp. 422–446.

Hartel T, Schwille P (2014) ESCRT-III mediated cell division in Sulfolobus acidocaldarius—A reconstitution perspective. *Frontiers in Microbiology* 5:257.

Heinemann F, Schwille P (2011) Preparation of micrometer-sized free-standing membranes. *Chemphyschem: A European Journal of Chemical Physics and Physical Chemistry* 12:2568–2571.

Heinemann F, Vogel SK, Schwille P (2013) Lateral membrane diffusion modulated by a minimal actin cortex. *Biophysical Journal* 104:1465–1475.

Helfand BT, Chang L, Goldman RD (2004) Intermediate filaments are dynamic and motile elements of cellular architecture. *Journal of Cell Science* 117:133–141.

Helfer E, Harlepp S, Bourdieu L, Robert J, MacKintosh FC, Chatenay D (2000) Microrheology of biopolymer-membrane complexes. *Physical Review Letters* 85:457–460.

Helfer E, Harlepp S, Bourdieu L, Robert J, MacKintosh FC, Chatenay D (2001a) Viscoelastic properties of actin-coated membranes. *Physical Review E, Statistical, Nonlinear, and Soft Matter Physics* 63:021904.

Helfer E, Harlepp S, Bourdieu L, Robert J, MacKintosh FC, Chatenay D (2001b) Buckling of actin-coated membranes under application of a local force. *Physical Review Letters* 87:088103.

Helm CA, Knoll W, Israelachvili JN (1991) Measurement of ligand-receptor interactions. *Proceedings of the National Academy of Sciences of the United States of America* 88:8169–8173.

Herrmann H, Bar H, Kreplak L, Strelkov SV, Aebi U (2007) Intermediate filaments: from cell architecture to nanomechanics. *Nature Reviews Molecular Cell Biology* 8:562–573.

Heuser JE, Kirschner MW (1980) Filament organization revealed in platinum replicas of freeze-dried cytoskeletons. *The Journal of Cell Biology* 86:212–234.

Honigmann A, Sadeghi S, Keller J, Hell SW, Eggeling C, Vink R (2014) A lipid bound actin meshwork organizes liquid phase separation in model membranes. *ELife* 3:e01671.

Hu Z, Lutkenhaus J (1999) Topological regulation of cell division in *Escherichia coli* involves rapid pole to pole oscillation of the division inhibitor MinC under the control of MinD and MinE. *Molecular Microbiology* 34:82–90.

Hussey PJ, Ketelaar T, Deeks MJ (2006) Control of the actin cytoskeleton in plant cell growth. *Annual Review of Plant Biology* 57:109–125.

Jimenez M, Martos A, Vicente M, Rivas G (2011) Reconstitution and organization of *Escherichia coli* proto-ring elements (FtsZ and FtsA) inside giant unilamellar vesicles obtained from bacterial inner membranes. *The Journal of Biological Chemistry* 286:11236–11241.

Kauert DJ, Kurth T, Liedl T, Seidel R (2011) Direct mechanical measurements reveal the material properties of three-dimensional DNA origami. Nano Lett 11:5558–5563.

Ke Y, Douglas SM, Liu M, Sharma J, Cheng A, Leung A, Liu Y, Shih WM, Yan H (2009) Multilayer DNA origami packed on a square lattice. *J Am Chem* Soc 131:15903–15908.

Kretschmer S, Schwille P (2014) Toward spatially regulated division of protocells: Insights into the *E. coli* min system from in vitro studies. *Life* 4:915–928.

Kurner J, Medalia O, Linaroudis AA, Baumeister W (2004) New insights into the structural organization of eukaryotic and prokaryotic cytoskeletons using cryo-electron tomography. *Experimental Cell Research* 301:38–42.

Kusumi A, Koyama-Honda I, Suzuki K (2004) Molecular dynamics and interactions for creation of stimulation-induced stabilized rafts from small unstable steady-state rafts. *Traffic* 5:213–230.

Limozin L, Sackmann E (2002) Polymorphism of cross-linked actin networks in giant vesicles. *Physical Review Letters* 89:168103.

Liu AP, Fletcher DA (2006) Actin polymerization serves as a membrane domain switch in model lipid bilayers. *Biophysical Journal* 91:4064–4070.

Liu AP, Richmond DL, Maibaum L, Pronk S, Geissler PL, Fletcher DA (2008) Membrane-induced bundling of actin filaments. *Nature Physics* 4:789–793.

Loose M, Fischer-Friedrich E, Ries J, Kruse K, Schwille P (2008) Spatial regulators for bacterial cell division self-organize into surface waves *In Vitro*. *Science* 320:789–792.

Lopez-Montero I, Lopez-Navajas P, Mingorance J, Velez M, Vicente M, Monroy F (2013) Membrane reconstitution of FtsZ-ZipA complex inside giant spherical vesicles made of *E. coli* lipids: large membrane dilation and analysis of membrane plasticity. *Biochimica et Biophysica Acta* 1828:687–698.

Lowe J, Amos LA (2009) Evolution of cytomotive filaments: The cytoskeleton from prokaryotes to eukaryotes. *The International Journal of Biochemistry & Cell Biology* 41:323–329.

Machta BB, Papanikolaou S, Sethna JP, Veatch SL (2011) Minimal model of plasma membrane heterogeneity requires coupling cortical actin to criticality. *Biophysical Journal* 100:1668–1677.

Manneville JB, Casella JF, Ambroggio E, Gounon P, Bertherat J, Bassereau P, Cartaud J, Antonny B, Goud B (2008) COPI coat assembly occurs on liquid-disordered domains and the associated membrane deformations are limited by membrane tension. *Proceedings of the National Academy of Sciences of the United States of America* 105:16946–16951.

Martos A, Jimenez M, Rivas G, Schwille P (2012) Towards a bottom-up reconstitution of bacterial cell division. *Trends in Cell Biology* 22:634–643.

Matosevic S, Paegel BM (2011) Stepwise synthesis of giant unilamellar vesicles on a microfluidic assembly line. *Journal of the American Chemical Society* 133:2798–2800.

McMahon HT, Gallop JL (2005) Membrane curvature and mechanisms of dynamic cell membrane remodelling. *Nature* 438:590–596.

Merkle D, Kahya N, Schwille P (2008) Reconstitution and anchoring of cytoskeleton inside giant unilamellar vesicles. *ChemBioChem* 9:2673–2681.

Michie KA, Lowe J (2006) Dynamic filaments of the bacterial cytoskeleton. *Annual Review of Biochemistry* 75:467–492.

Minton AP (1992) Confinement as a determinant of macromolecular structure and reactivity. *Biophysical Journal* 63:1090–1100.

Moller-Jensen J, Jensen RB, Lowe J, Gerdes K (2002) Prokaryotic DNA segregation by an actin-like filament. *The EMBO Journal* 21:3119–3127.

Osawa M, Erickson HP (2013) Liposome division by a simple bacterial division machinery. *Proceedings of the National Academy of Sciences of the United States of America* 110:11000–11004.

Osawa M, Anderson DE, Erickson HP (2008) Reconstitution of contractile FtsZ rings in liposomes. *Science* 320:792–794.

Osawa M, Anderson DE, Erickson HP (2009) Curved FtsZ protofilaments generate bending forces on liposome membranes. *The EMBO Journal* 28:3476–3484.

Papayannopoulos V, Co C, Prehoda KE, Snapper S, Taunton J, Lim WA (2005) A polybasic motif allows N-WASP to act as a sensor of PIP(2) density. Mol Cell 17:181–191.

Pollard TD, Cooper JA (2009) Actin, a central player in cell shape and movement. *Science* 326:1208–1212.

Pontani LL, van der Gucht J, Salbreux G, Heuvingh J, Joanny JF, Sykes C (2009) Reconstitution of an actin cortex inside a liposome. *Biophysical Journal* 96:192–198.

Poole AM (2006) Did group II intron proliferation in an endosymbiont-bearing archaeon create eukaryotes? *Biology Direct* 1:36.

Popp D, Narita A, Lee LJ, Ghoshdastider U, Xue B, Srinivasan R, Balasubramanian MK, Tanaka T, Robinson RC (2012) Novel actin-like filament structure from clostridium tetani. *The Journal of Biological Chemistry* 287:21121–21129.

Popp D, Narita A, Ghoshdastider U, Maeda K, Maeda Y, Oda T, Fujisawa T, Onishi H, Ito K, Robinson RC (2010) Polymeric structures and dynamic properties of the bacterial actin AlfA. *Journal of Molecular Biology* 397:1031–1041.

Przybylo M, Sykora J, Humpolickova J, Benda A, Zan A, Hof M (2006) Lipid diffusion in giant unilamellar vesicles is more than 2 times faster than in supported phospholipid bilayers under identical conditions. *Langmuir* 22:9096–9099.

Raskin DM, de Boer PA (1999) MinDE-dependent pole-to-pole oscillation of division inhibitor MinC in *Escherichia coli*. *Journal of Bacteriology* 181:6419–6424.

Rekas A, Alattia JR, Nagai T, Miyawaki A, Ikura M (2002) Crystal structure of venus, a yellow fluorescent protein with improved maturation and reduced environmental sensitivity. *J Biol Chem* 277:50573–50578.

Rivas G, Vogel SK, Schwille P (2014) Reconstitution of cytoskeletal protein assemblies for large-scale membrane transformation. *Current Opinion in Chemical Biology* 22:18–26.

Romer W, Berland L, Chambon V, Gaus K, Windschiegl B, Tenza D, Aly MR et al. (2007) Shiga toxin induces tubular membrane invaginations for its uptake into cells. *Nature* 450:670–675.

Saarikangas J, Zhao H, Pykalainen A, Laurinmaki P, Mattila PK, Kinnunen PK, Butcher SJ, Lappalainen P (2009) Molecular mechanisms of membrane deformation by I-BAR domain proteins. *Current Biology: CB* 19:95–107.

Samson RY, Bell SD (2009) Ancient ESCRTs and the evolution of binary fission. *Trends Microbiol* 17:507–513.

Samson RY, Obita T, Hodgson B, Shaw MK, Chong PL, Williams RL, Bell SD (2011) Molecular and structural basis of ESCRT-III recruitment to membranes during archaeal cell division. *Mol Cell* 41:186–196.

Schuppler M, Keber FC, Kroger M, Bausch AR (2016) Boundaries steer the contraction of active gels. *Nature Communications* 7:13120.

Schwille P (2015) Jump-starting life? Fundamental aspects of synthetic biology. *The Journal of Cell Biology* 210:687–690.

Schwille P, Diez S (2009) Synthetic biology of minimal systems. *Critical Reviews in Biochemistry and Molecular Biology* 44:223–242.

Shih YL, Rothfield L (2006) The bacterial cytoskeleton. *Microbiology and Molecular Biology Reviews* 70:729–754.

Smith PG, Dreshaj A, Chaudhuri S, Onder BM, Mhanna MJ, Martin RJ (2007) Hyperoxic conditions inhibit airway smooth muscle myosin phosphatase in rat pups. Am J Physiol-Lung Cell Mol Physiol 292:L68–L73.

Spudich JA, Watt S (1971) The regulation of rabbit skeletal muscle contraction. I. Biochemical studies of the interaction of the tropomyosin-troponin complex with actin and the proteolytic fragments of myosin. The Journal of biological chemistry 246:4866–4871.

Stachowiak JC, Richmond DL, Li TH, Liu AP, Parekh SH, Fletcher DA (2008) Unilamellar vesicle formation and encapsulation by microfluidic jetting. *Proceedings of the National Academy of Sciences of the United States of America* 105:4697–4702.

Szeto TH, Rowland SL, Rothfield LI, King GF (2002) Membrane localization of MinD is mediated by a C-terminal motif that is conserved across eubacteria, archaea, and chloroplasts. *Proceedings of the National Academy of Sciences of the United States of America* 99:15693–15698.

Szeto TH, Rowland SL, Habrukowich CL, King GF (2003) The MinD membrane targeting sequence is a transplantable lipid-binding helix. *The Journal of Biological Chemistry* 278:40050–40056.

Teather RM, Collins JF, Donachie WD (1974) Quantal behavior of a diffusible factor which initiates septum formation at potential division sites in *Escherichia coli*. *Journal of Bacteriology* 118:407–413.

Tsai FC, Koenderink GH (2015) Shape control of lipid bilayer membranes by confined actin bundles. *Soft Matter* 11:8834–8847.

Vale RD (2003) The molecular motor toolbox for intracellular transport. *Cell* 112:467–480.

van Swaay D, deMello A (2013) Microfluidic methods for forming liposomes. *Lab on a Chip* 13:752–767.

Veatch SL, Keller SL (2003a) Separation of liquid phases in giant vesicles of ternary mixtures of phospholipids and cholesterol. *Biophysical Journal* 85:3074–3083.

Veatch SL, Keller SL (2003b) A closer look at the canonical "Raft Mixture" in model membrane studies. *Biophysical Journal* 84:725–726.

Vogel SK, Schwille P (2012) Minimal systems to study membrane-cytoskeleton interactions. *Current Opinion in Biotechnology* 23:758–765.

Vogel SK, Heinemann F, Chwastek G, Schwille P (2013) The design of MACs (minimal actin cortices). *Cytoskeleton* 70:706–717.

Wachi M, Matsuhashi M (1989) Negative control of cell division by mreB, a gene that functions in determining the rod shape of *Escherichia coli* cells. *Journal of Bacteriology* 171:3123–3127.

Walczak CE, Heald R (2008) Mechanisms of mitotic spindle assembly and function. *International Review of Cytology* 265:111–158.

Weisenberg RC (1972) Microtubule formation in vitro in solutions containing low calcium concentrations. *Science* 177:1104–1105.

Welch MD, Rosenblatt J, Skoble J, Portnoy DA, Mitchison TJ (1998) Interaction of human Arp2/3 complex and the Listeria monocytogenes ActA protein in actin filament nucleation. *Science* 281:105–108.

Wickstead B, Gull K (2011) The evolution of the cytoskeleton. *The Journal of Cell Biology* 194:513–525.

Wiesner S, Helfer E, Didry D, Ducouret G, Lafuma F, Carlier MF, Pantaloni D (2003) A biomimetic motility assay provides insight into the mechanism of actin-based motility. *J Cell Biol* 160:387–398.

Wollert T, Hurley JH (2010) Molecular mechanism of multivesicular body biogenesis by ESCRT complexes. *Nature* 464:864–869.

Wollert T, Yang D, Ren X, Lee HH, Im YJ, Hurley JH (2009) The ESCRT machinery at a glance. *Journal of Cell Science* 122:2163–2166.

Yutin N, Wolf MY, Wolf YI, Koonin EV (2009) The origins of phagocytosis and eukaryogenesis. *Biology Direct* 4:9.

Zieske K, Schwille P (2013) Reconstitution of pole-to-pole oscillations of min proteins in microengineered polydimethylsiloxane compartments. *Angewandte Chemie* 52:459–462.

Zieske K, Schwille P (2014) Reconstitution of self-organizing protein gradients as spatial cues in cell-free systems. *ELife* 3:e03949.

Giant vesicles theoretically and *in silico*

Understanding giant vesicles: A theoretical perspective

Reinhard Lipowsky

As simple as possible but not simpler.

Albert Einstein

Contents

5.1 INTRODUCTION AND OVERVIEW

The architecture of biological membranes is characterized by a wide range of length scales. On the μm scale, these membranes exhibit a unique combination of properties: (i) They form closed surfaces without edges; (ii) They are highly flexible and, thus, can easily adapt their shape to external perturbations; (iii) In spite of this flexibility, they provide robust and stable barriers between the different aqueous compartments; and (iv) In the cell, these compartments are continuously remodeled via membrane fusion and fission (or scission).

These properties arise from the specific molecular structure of these membranes. When viewed on the nm scale, each biomembrane consists of a specific mixture of many different lipids and membrane proteins which reflects the biological functions of this membrane. However, in spite of this chemical complexity, all biomembranes are organized according to the same universal principle: their basic building block is provided by a *bilayer of lipid molecules*. The latter molecules are essentially *insoluble* in the aqueous solution which ensures the stability of the membrane. In addition, these lipid bilayers are maintained in a *fluid* state which enables the membranes to adapt to external perturbations by remodeling of membrane composition, shape, and topology.

Many of the fascinating remodeling processes that have been found for biological membranes can also be observed for giant unilamellar vesicles (GUVs) that are formed by membranes with a relatively small number of molecular components. The theory described here will typically be compared to experimental observations on lipid vesicles but the same theory applies to vesicle membranes that are composed of lipids and membrane proteins.

One intriguing example for the remodeling of membrane shape is provided by the formation of membrane necks via budding, a crucial step of all endo- and exocytotic processes. Another example is provided by the formation of membrane nanotubes, highly curved membrane structures that protrude from weakly curved membrane segments. As far as the remodeling of composition is concerned, we now have a variety of lipid mixtures that can phase separate into two fluid phases, a liquid-ordered and a liquid-disordered phase. When we study this membrane phase separation in giant vesicles, we often observe large intramembrane domains that partition the vesicle membrane into a few membrane compartments with a lateral extension in the micrometer range. In addition, multicomponent membranes exposed to a heterogeneous environment form ambience-induced segments that can also differ in their molecular composition. One example for this type of segmentation is provided by vesicle membranes exposed to aqueous two-phase systems or water-in-water emulsions which exhibit several wetting morphologies. The interplay between ambience-induced segmentation and membrane phase separation leads to the confinement of phase separation to single membrane segments which represents a generic mechanism to suppress the formation of large intramembrane domains or rafts in cellular membranes.

The present chapter is organized as follows. The next two Sections 5.2 and 5.3 are introductory in nature: they describe basic aspects of biomembranes and provide an elementary view of membrane curvature. The relation between local curvature generation and spontaneous curvature is explained in Section 5.3.5. Different molecular mechanisms for local curvature generation are described in Box 5.1. Section 5.4 describes the theory of curvature elasticity for uniform membranes.[1] This theory is based on the *local* curvature-elastic properties of the membranes, but also takes into account that the ultralow lipid solubility and the osmotic conditions lead to *global* constraints on the membrane area and the vesicle volume. In fact, what makes this theory both appealing and challenging is this interplay between local and global membrane properties.

We will focus on the spontaneous curvature model but also discuss the modifications arising from area-difference-elasticity. On the one hand, the spontaneous curvature model is particularly attractive from a theoretical point of view because it depends only on a small number of curvature-elastic parameters. In fact, for membranes with a laterally uniform composition, the spontaneous curvature model involves only two such parameters, (i) the bending rigidity κ which describes the resistance of the membrane against bending deformations and (ii) the spontaneous curvature which provides a quantitative measure for the bilayer asymmetry of the membranes. On the other hand, the spontaneous curvature model is also sufficient to obtain a quantitative description for the behavior of many membranes of interest. Indeed, this model applies to all membranes with (at least) one molecular component such as cholesterol that undergoes frequent flip-flops between the two leaflets of the bilayer. Area difference elasticity is only relevant in the absence of flip-flops, i.e., when the number of molecules is separately conserved in each leaflet.

One striking consequence of curvature elasticity is the formation of closed membrane necks that represent narrow funnel-like membrane structures between two larger membrane segments. The stability of these necks depends on the relative magnitude of the neck curvature and the spontaneous curvature, which may contain a nonlocal contribution from area-difference-elasticity. These stability conditions for closed membrane necks can be reinterpreted as effective constriction forces generated by spontaneous curvature. Simple estimates show that sufficiently large spontaneous curvatures lead to the cleavage of the membrane necks and thus to complete membrane fission. The different aspects of membrane necks are summarized in Box 5.2.

Sections 5.5 and 5.6 are devoted to two striking morphologies formed by uniform membranes: (i) multi-sphere shapes that involve small spherical buds and (ii) membrane nanotubes that can be necklace-like or cylindrical. Section 5.7 describes the behavior of vesicles that interact with an adhesive and rigid surface. For simplicity, the latter section will focus on vesicle membranes with a laterally uniform composition but will also discuss adhesion of vesicles as an example for ambience-induced segmentation of membranes. A closely related subject, the behavior of adhesive nanoparticles in contact with membranes and vesicles, will be addressed in Chapter 8 of this book. The shapes and shape transformations of vesicles that contain two or multiple intramembrane domains are discussed in Section 5.8, and the wetting of membranes in contact with aqueous two-phase systems or water-in-water emulsions in Section 5.9. For partial wetting, the water-water interfaces exert capillary forces onto the membranes which then respond with strong shape deformations. On the nanometer scale, the membrane segments close to the three-phase contact line should be curved in a smooth manner and the capillary forces then lead to a complex force balance along this contact line which involves an intrinsic contact angle. On the micrometer scale, the membrane shapes exhibit kinks which define an apparent contact line and apparent contact angles. Experimental aspects of aqueous two-phase systems will be addressed in Chapter 29 of this book. Both membrane phase separation and membrane wetting leads to vesicle membranes that have a laterally nonuniform composition. At the end, we will briefly look at the consequences of curvature elasticity for membrane fusion and fission (or scission) of membranes, the two most important topological transformations of membranes.

Each of the different membrane systems discussed in Sections 5.7 through 5.9 involves one additional parameter: the adhesive strength W of substrate surfaces, the line tension λ of domain boundaries, and the interfacial tension $\Sigma_{\alpha\beta}$ between two different aqueous phases. Because all of these parameters can be measured or deduced from experimental observations, the theory leads to quantitative predictions. In fact, the theory described here leads to a large number of simple relationships between material parameters and geometric quantities which provide important checkpoints for the comparison between theory and experiment.

5.2 BIOMEMBRANES AND GIANT VESICLES: BASIC ASPECTS

Here and below, the term "biomembranes" will be used as an abbreviation for "biological and biomimetic membranes." These two types of membranes differ primarily in their chemical complexity. Biological or cellular membranes usually contain hundreds or even thousands of different lipid species and a large number of different membrane proteins. Biomimetic membranes as considered here have a much simpler composition with only a few molecular components but share one crucial physical property with biological membranes, namely their fluidity, which enables both types of membranes to undergo analogous remodeling processes. The simplest biomimetic membranes are provided by one-component lipid bilayers which have a molecular structure as in Figure 5.1.

[1] Here and below, a 'uniform membrane' is 'laterally uniform' and a 'uniform aqueous phase' is 'spatially uniform'.

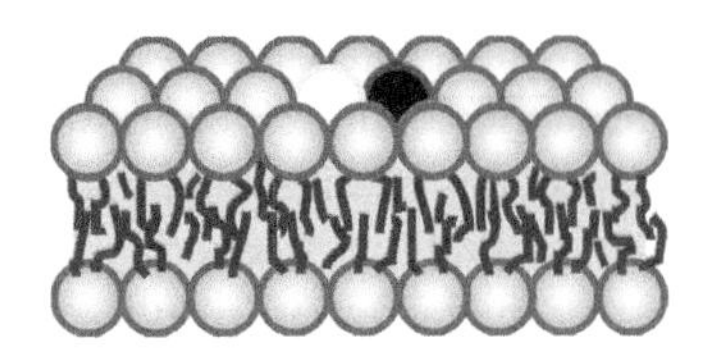

Figure 5.1 Lipid bilayer as the basic building block of all biomembranes. The lipid molecules are arranged into two monolayers or leaflets, with the lipid headgroups forming the two interfaces between the bilayer and the aqueous solutions. The thickness of the bilayer is 4 to 5 nm. For a *fluid* bilayer, each lipid molecule undergoes rapid lateral diffusion within the membrane. This diffusive process is based on the pairwise exchange of neighboring lipids (black and white) on the time scale of nanoseconds.

5.2.1 BIOMEMBRANES BASED ON LIPID BILAYERS

Essentially all biological membranes contain a single lipid bilayer as their basic building block. The importance of lipids was already realized by Langmuir and others at the beginning of the 20th century. This insight came from spreading experiments: the membranes were dissolved in a volatile organic solvent, the solution was spread on a water surface, and the solvent was evaporated. In this way, one obtains a lipid monolayer at the air–water interface. Such a technique was also used by Gorter and Grendel who extracted lipids from red blood cells (Gorter and Grendel, 1925; Robertson, 1960). They found that the area of the monolayer was approximately twice the area of the cell and proposed that the cell should be covered by a lipid bilayer. This proposal was confirmed, in the 1950s and 1960s, by imaging cross-sections of biomembranes via electron microscopy. Such electron microscopy images gave direct evidence that cell membranes are based upon a single bilayer and showed that these bilayers have a thickness of 4–5 nm (Robertson, 1959).

Electron microscopy studies also demonstrated that bilayers are already formed by a single species of phospholipid molecules (Bangham and Horne, 1964). Therefore, bilayers consisting of one or a few lipid components have become important model systems for biological membranes. Different bilayer systems have been developed and intensely studied, including multilamellar liposomes, black lipid membranes, solid-supported bilayers, and unilamellar vesicles. Giant unilamellar vesicles as considered here typically have a linear size of tens of micrometers and can be directly imaged in their fluid state using optical microscopy.

5.2.2 SEMI-PERMEABILITY AND OSMOTIC CONDITIONS

One basic function of biological membranes is that they partition space into separate aqueous compartments and represent effective barriers for the diffusion of ions and solute molecules from one compartment to another. These functions are also provided by lipid bilayers. When these bilayers form vesicles, they create an interior aqueous compartment that is well separated from the exterior solution. Indeed, the bilayers are permeable to small uncharged molecules such as H_2O, O_2, and CO_2 as well as H_3O^+ and OH^- ions, but do not allow the permeation of other ions or larger water-soluble molecules such as glucose and other monosaccharides. As a consequence, these solutes represent osmotically active "particles" and exert osmotic pressures onto the vesicle membranes. The experimental methods to measure the permeability of membranes are reviewed in Chapter 20 of this book.

The osmotic pressures depend on the solute concentrations in the interior and exterior solutions. If a vesicle membrane is exposed to different interior and exterior concentrations, the resulting osmotic pressure difference causes water to move through the membrane into the compartment with the higher solute concentration. First, consider a higher solute concentration in the exterior solution which leads to osmotic deflation of the vesicle. In this case, the water outflux reduces the vesicle volume until the interior particle concentration matches the exterior one and the osmotic pressure difference is close to zero. On the other hand, if we start with a higher solute concentration in the interior compartment, the volume of the vesicle is increased by osmotic inflation. However, this volume increase is truncated by the limited ability of the vesicle membrane to increase its area by mechanical stretching. Indeed, when a lipid bilayer is mechanically stretched, its area can only be increased by a few percent before it ruptures. Therefore, once the inflated vesicle has attained a spherical shape, further influx of water increases the membrane tension up to a limiting value at which the membrane ruptures and forms pores. These pores then provide an alternative pathway for the reduction of the osmotic pressure difference.

5.2.3 FLUIDITY OF BIOMEMBRANES

Another universal aspect of biological membranes is that they are maintained in a fluid state which is characterized by fast lateral diffusion of the molecules along the membrane. This membrane fluidity became generally accepted at the beginning of the 1970s as a result of three parallel developments. First, the lateral diffusion was probed by spin-labeled lipids (Kornberg and McConnell, 1971; Devaux and McConnell, 1972) and steroids (Sackmann and Träuble, 1972; Träuble and Sackmann, 1972) which led to lateral diffusion constants of the order of 1 μm^2 per second. Nowadays, the lateral diffusion of membrane molecules can be observed directly by fluorescence recovery after photobleaching (FRAP) (Almeida and Vaz, 1995) and by single particle tracking (Sako and Kusumi, 1994; Saxton and Jacobson, 1997; Fujiwara et al., 2002; Kusumi et al., 2005), two methods that have been applied to a large variety of biomimetic and biological membranes. These studies confirmed that the lateral diffusion constants of membrane molecules are indeed of the order of 1 μm^2 per second. A detailed discussion of both FRAP and single particle tracking as well as tables with diffusion constants for a variety of lipids can be found in Chapter 21 of this book.

Second, it has been realized that the observed shape transformations of red blood cells (Canham, 1970; Evans, 1974) and lipid vesicles (Helfrich, 1973; Deuling and Helfrich, 1976) are only possible if the membranes represent two-dimensional liquids. Indeed, these shape transformations change the curvature of the membranes in a smooth and continuous manner and would be impossible for solid-like or polymerized membranes. Particularly interesting shape changes are provided by budding processes in

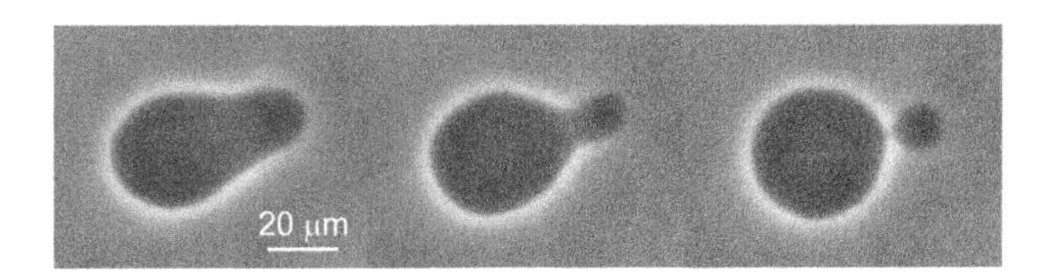

Figure 5.2 Formation of a spherical out-bud from a giant unilamellar vesicle (GUV) as observed by phase contrast microscopy. This budding process, which took about 5 s, proceeds in a smooth and continuous manner and provides direct evidence on the micrometer scale that the lipid membrane is in a fluid state on the molecular scale. (Reproduced with permission from Dimova, R. et al., A practical guide to giant vesicles: Probing the membrane nanoregime via optical microscopy, *J. Phys. Cond. Mat.*, 18, S1151–S1176, 2006, Institute of Physics)

which small spherical out- or in-buds are formed from larger mother vesicles. Out-buds point towards the exterior aqueous solution, in-buds towards the interior solution. One example for the formation of an out-bud is shown in Figure 5.2. Such a budding process provides direct evidence that the membrane is in a fluid state. The associated curvature elasticity of biomembranes has now been developed into a quantitative theory (Berndl et al., 1990; Seifert and Lipowsky, 1990; Seifert et al., 1991; Miao et al., 1991; Lipowsky, 1991; Miao et al., 1994; Döbereiner et al., 1997; Lipowsky, 2013; Liu et al., 2016; Lipowsky, 2018a) which will be described in this chapter.

Third, in 1972, a large body of observations on cellular membranes was integrated into the fluid mosaic model in which the membrane proteins are dispersed in a fluid bilayer of lipids (Singer and Nicolson, 1972). Whether the fluid mosaic model actually describes the supramolecular structure of cell membranes has been a matter of some debate. On the one hand, the endocytosis and exocytosis of cell membranes involves the formation of fluid domains that are enriched in membrane-anchored receptors and coat proteins and can be understood in terms of domain-induced budding (Lipowsky, 1992, 1993; Agudo-Canalejo and Lipowsky, 2015a).

On the other hand, it has also been proposed that cell membranes contain intramembrane domains, so-called rafts, that are enriched in certain lipids such as sphingomyelin and cholesterol (Simons and Ikonen, 1997). In spite of a large number of experimental studies, including superresolution microscopy methods such as stimulated emission depletion (STED) microscopy, it has not possible to obtain direct evidence for such rafts in cellular membranes. If these lipid rafts exist in mammalian cells, their diameter does not exceed 20 nm (Eggeling et al., 2009). The different experimental techniques used to search for such rafts have been critically reviewed by (Klotzsch and Schütz, 2013). One generic mechanism that explains the difficulty to observe membrane phase separation in cellular membranes is ambience-induced segmentation by the heterogeneous environment to which these membranes are exposed (Lipowsky, 2014b) as discussed in Section 5.8.5 below.

5.2.4 REMODELING OF COMPOSITION AND SHAPE

In general, the fluidity of biomembranes implies that these membranes can easily adapt to changes in their environment by remodeling their composition, shape, and topology. This multi-responsive behavior includes shape transformations of GUVs, membrane segmentation by laterally nonuniform environments such as adhesive surfaces, membrane phase separation, and the responses of GUVs to capillary forces arising from water-in-water droplets.

The remodeling of membrane composition in ternary lipid mixtures leads to the nucleation and growth of intramembrane domains that can be directly observed in the optical microscope, see Figure 5.3. Such domains, which demonstrate the coexistence of two (or more) lipid phases, have now been observed for a variety of membrane systems including giant vesicles (Dietrich et al., 2001; Veatch and Keller, 2003; Baumgart et al., 2003; Bacia et al., 2005; Riske et al., 2006; Dimova et al., 2007; Semrau et al., 2008), solid-supported membranes (Jensen et al., 2007; Garg et al., 2007; Kiessling et al., 2009), hole-spanning (or black lipid) membranes (Collins and Keller, 2008), as well as pore-spanning membranes (Orth et al., 2012). The phase diagrams of such three-component membranes have been determined using spectroscopic methods (David et al., 2009) as well as fluorescence microscopy of giant vesicles and X-ray diffraction of membrane stacks (Veatch et al., 2006; Vequi-Suplicy et al., 2010; Uppamoochikkal et al., 2010; Pataraia et al., 2014). The experimental aspects of lipid phase separation and domain formation are reviewed in more detail in Chapter 18 of this book.

Another particularly striking example for the remodeling of membrane shape that does not require membrane phase separation is provided by the spontaneous tubulation of GUVs (Li et al., 2011; Lipowsky, 2013; Liu et al., 2016). Two examples for the resulting pattern of nanotubes are displayed in Figure 5.4. In these examples, the vesicles respond to osmotic deflation by the formation of many nanotubes that emanate from the giant mother vesicle and protrude into the vesicle interior. As a result, highly curved membrane segments coexist with weakly curved segments even though the membrane has a laterally uniform composition. The nanotubes shown in Figure 5.4 were formed spontaneously, i.e., in the absence of external pulling forces. Another quite different mechanism for the formation of membrane

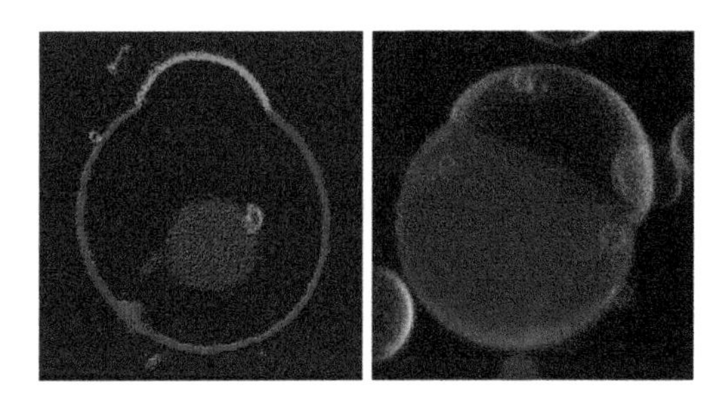

Figure 5.3 Remodeling of membrane composition can lead to domain-induced budding of vesicles as theoretically predicted in (Lipowsky, 1992, 1993; Jülicher and Lipowsky, 1993) and observed by fluorescence microscopy in (Baumgart et al., 2003; Riske et al., 2006): (left) Cross section through a vesicle that formed two domains after a decrease in temperature (Baumgart et al., 2003); and (right) Three-dimensional confocal scan of a two-domain vesicle that was formed by electrofusion. In both cases, the vesicle membrane is composed of dioleoyl phosphadityl choline (DOPC), sphingomyelin, and cholesterol (see Appendix 1 of the book for structure and data on these lipids) together with small concentrations of two fluorescent probes. (Reproduced with permission from Riske, K.A. et al., *Biophys. Rev. Lett.*, 1, 387–400, 2006. Copyright (c) 2006 World Scientific Publishing.)

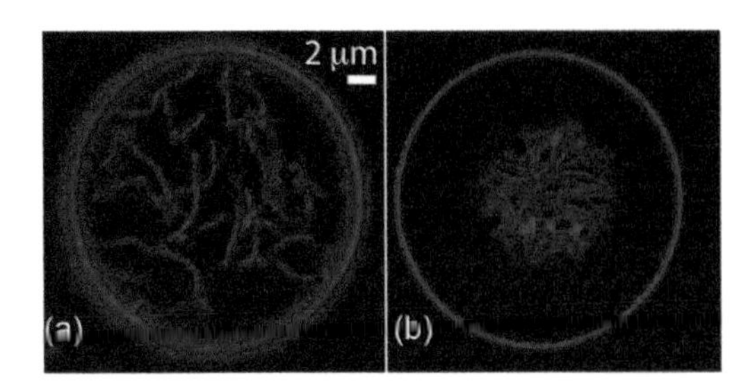

Figure 5.4 Remodeling of membrane shape can lead to complex patterns of flexible nanotubes. The nanotubes were formed by liquid-disordered membranes after the interior aqueous compartment separated into a PEG-rich and dextran-rich phase: (a) Disordered pattern corresponding to a vesicle membrane that is completely wetted by the PEG-rich phase; and (b) Layer of densely packed tubes corresponding to a membrane that is partially wetted by both aqueous phases. All tubes are connected to the outer vesicle membranes (red circles). In both images, the diameter of the tubes is below the diffraction limit of the confocal microscope but the tubes are theoretically predicted to be necklace-like and cylindrical in (a) and (b), respectively (Liu et al., 2016). (Reproduced with permission from Liu, Y. et al., *ACS Nano*, 10, 463–474, 2016. Copyright American Chemical Society.)

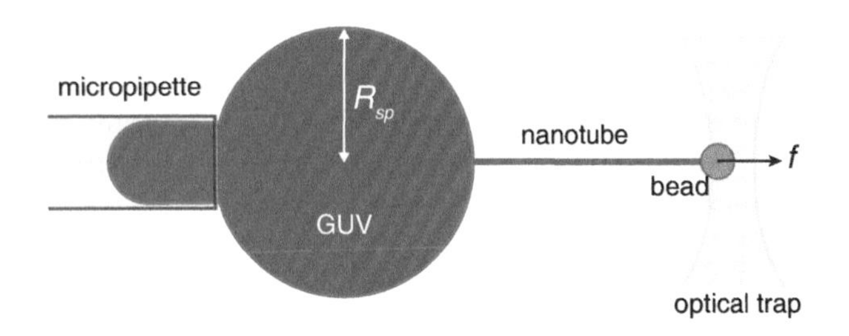

Figure 5.5 Pulling a membrane nanotube attached to a bead from a giant unilamellar vesicle (GUV) by an optical trap: The weakly curved GUV is aspirated by the micropipette, the right end of the strongly curved nanotube experiences the pulling force f arising from the optical trap. The latter force is typically of the order of 10 pN and can then generate tubes with a radius of 10–20 nm.

nanotubes is provided by external pulling forces that are locally applied to the membranes. A particularly instructive setup for the latter tubulation process is obtained if one aspirates a giant unilamellar vesicle in a micropipette and simultaneously applies a pulling force to a membrane-bound nanobead via magnetic tweezers (Heinrich and Waugh, 1996) or optical traps (Sorre et al., 2012), as schematically depicted in Figure 5.5.

The experimental methods that have been developed for GUVs composed of a few lipid components can also be applied to giant plasma membrane vesicles (GPMVs) or "blebs," which contain a wide assortment of different lipids and proteins, all oriented in the same way as in the original cell membrane. In spite of their chemical complexity, the membranes of GPMVs were found to phase separate into coexisting lipid phases (Baumgart et al., 2007; Veatch et al., 2008), in close analogy to ternary lipid mixtures. One cellular process that has been elucidated using GPMVs is the molecular recognition of "self" during phagocytosis by macrophages. This recognition process involves the binding of the immunoglobulin CD47, a ubiquitous "marker of self" protein, to the macrophage receptor SIRPα (Sosale et al., 2015). The adhesion of GPMVs with CD47 to SIRPα immobilized on a substrate surface revealed that the two proteins bind in a cooperative manner (Steinkühler et al., 2019), confirming previous theoretical studies (Weikl et al., 2009, 2016; Hu et al., 2013). Furthermore, it has also been observed that GPMVs form many nanotubes under deflation and that these tubulated vesicles exhibit rather unusual elastic properties (Steinkühler et al., 2018b).

5.2.5 STABILITY OF BILAYER MEMBRANES

In spite of their high flexibility, lipid membranes have a robust molecular architecture and maintain this architecture even under strong local deformations. One example is provided by force-induced tubulation as shown in Figure 5.5. Using this method, one can produce nanotubes or "tethers" with a radius of only 10 nm, which should be compared to the bilayer thickness of 4–5 nm (Sorre et al., 2012). Tubes of a similar width have also been generated by a slightly different setup in which the laser trap is replaced by another micropipette that grabs the nanobead (Hochmuth et al., 1982; Tian et al., 2009). However, in spite of the large curvature of these nanotubes, the tube membranes maintain their structural integrity and provide an efficient separation of the interior and exterior aqueous compartments. Detailed information about the experimental method to pull nanotubes from GUVs can be found in Chapter 16 of this book.

The stability of the bilayer structure reflects the ultralow solubility of phospholipids in water. One measure for this solubility is provided by the critical micelle concentration which represents both the concentration at which the lipids start to self-assemble into bilayers (instead of micelles) and the concentration of individual lipid molecules in the presence of bilayers. The critical micelle concentration of phospholipids decreases exponentially with their chain length, i.e., with the number of hydrocarbon groups per chain (Cevc and Marsh, 1987). The phospholipid dimyristoyl phosphatidyl choline (DMPC, see Appendix 1 of the book for structure and data on this and other lipids), for example, has the relatively short chain length of 14 hydrocarbon groups, but its critical micelle concentration is only $10^{-10.5}$ in mole fraction units or about 0.95 DMPC molecules per μm^3. When this lipid forms a giant unilamellar vesicle with a radius of 10 μm, the vesicle membrane consists of about 4×10^9 lipid molecules whereas the interior aqueous compartment of the vesicle contains only about 4×10^3 such molecules. Most biologically relevant phospholipids have a chain length that exceeds 14 hydrocarbon groups which implies an even lower critical micelle concentration. As a consequence, one can usually ignore any exchange of phospholipids between the bilayer membrane and the aqueous solutions and assume that the membrane contains a fixed number of such lipids.

5.2.6 POLYMORPHISM OF VESICLES

Because biomembranes are fluid, one might expect that their shape can be understood by analogy with liquid droplets. However, in the absence of external forces or constraints, a liquid droplet of a given volume always attains a spherical shape in order to minimize its interfacial area and, thus, its interfacial free energy. In contrast to liquid droplets, lipid vesicles can attain a large variety of different shapes such as discocytes, stomatocytes, and dumbbells. Furthermore, the vesicle may undergo shape transformations as one changes the osmotic conditions or the temperature. Because the lipid molecules are practically insoluble in water, the total number of lipid molecules within

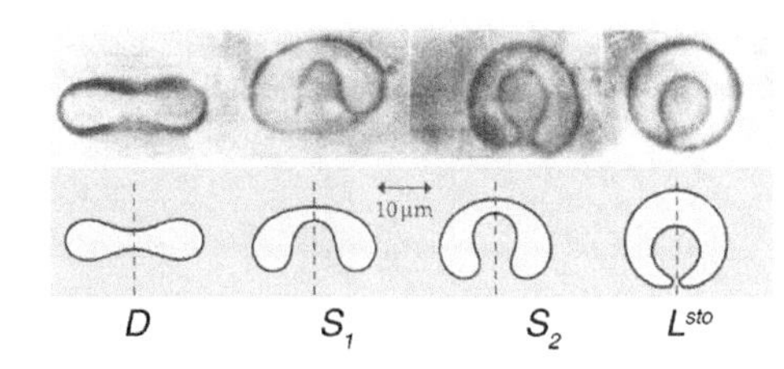

Figure 5.6 Temperature-induced shape transformation of a single vesicle: In this example, the vesicle starts from the initial shape of a discocyte (*D*) which is transformed, via the intermediate stomatocytes S_1 and S_2, into the limit shape L^{sto} consisting of two spheres. The small sphere of L^{sto} forms an in-bud that is connected to the large sphere via a closed membrane neck. The generation of a smooth spherical bud without any membrane folds again demonstrates the fluidity of the membrane. The top row displays images of phase contrast microscopy, the bottom row theoretical shapes with minimal curvature energy. (From Berndl, K. et al., *Europhys. Lett.*, 13, 659–664, 1990.)

the membrane is conserved during such shape transformations. In addition, at any given temperature, each lipid molecule tries to occupy a certain optimal area within the membrane. Furthermore, when exposed to external forces or constraints, lipid bilayers hardly change their area before they rupture. Therefore, the area of the vesicle membrane is conserved, to a very good approximation, during isothermal shape transformations arising, e.g., from osmotic deflation and inflation. The latter processes change the vesicle volume for fixed membrane area. In general, the volume of a vesicle can become arbitrarily small but cannot exceed the volume of a sphere.

Shape transformations can also be induced by temperature changes reflecting the different thermal expansivities of the lipid bilayer and the aqueous solution. When we increase the temperature by ΔT, the initial membrane area A_0 increases by $\Delta A = \alpha_A \Delta T A_0$ with $\alpha_A \simeq 2 \times 10^{-3}$/K for lipid bilayers. At the same time, the initial water volume V_0 increases by $\Delta V = \alpha_V \Delta T V_0$ with $\alpha_V \simeq 2 \times 10^{-4}$/K. When we apply these relations to a GUV, we find that an increase in temperature generates excess area of the membrane and reduces the volume-to-area ratio of the vesicle. One example for temperature-induced shape transformations is displayed in Figure 5.6.

The multi-responsive behavior of GUVs as illustrated by Figures 5.2 through 5.6 can be understood, in a quantitative manner, by the unusual curvature-elastic properties of the vesicle membranes. In the next two sections, we will first discuss the general concept of membrane curvature and then introduce the spontaneous curvature model for the description of curvature elasticity.

5.3 CURVATURE OF MEMBRANES

This section provides an elementary introduction into different aspects of curvature. It first emphasizes that membrane curvature emerges on nanoscopic scales and then describes basic concepts from differential geometry which include the two principal curvatures, the mean curvature, and the Gaussian curvature. Furthermore, one simple but important issue that is discussed in some detail is our convention for the sign of the principal curvatures, which can be positive or negative. At the end of this section, several molecular mechanisms for local curvature generation are briefly discussed and summarized in Box 5.1. Local curvature generation is intimately related to the preferred or spontaneous curvature of a membrane. The latter curvature can again be positive or negative. The present section is supplemented by Appendix 5.A on differential geometry.

5.3.1 EMERGENCE OF CURVATURE ON NANOSCOPIC SCALES

As shown in Figure 5.3 and Figure 5.6, vesicle shapes appear to be rather smooth when viewed under the optical microscope. Therefore, on the micrometer scale, membranes can be described as smoothly curved surfaces and then characterized by their curvature. However, this smoothness does not persist to molecular scales, i.e., when we resolve the molecular structure of a bilayer membrane as in Figure 5.7.

Because membranes are immersed in liquid water, each lipid and protein molecule undergoes thermal motion with displacements both parallel and perpendicular to the membrane. The perpendicular displacements represent molecular protrusions that roughen the two interfaces bounding the membrane. Therefore, in order to characterize a lipid/protein bilayer by its curvature, one has to consider small membrane patches and average over the molecular conformations within these patches. The minimal lateral size of these patches can be determined from the analysis of the bilayer's shape fluctuations and was found, from molecular dynamics simulations of a one-component lipid bilayer, to be about 1.5 times the membrane thickness, see Figure 5.7 (Goetz et al., 1999). For a membrane with a thickness of 4 nm, this minimal size is about 6 nm. Because such a membrane patch contains 80–100 lipid molecules, membrane curvature should be regarded as an emergent property arising from the collective behavior of a large number of lipid molecules.

The curvature just discussed applies to the midsurface of the bilayer membrane, i.e., to the surface between the two leaflets of the bilayer. Furthermore, for a membrane segment with midsurface area A and bending rigidity κ, curved conformations as in Figure 5.7 are only possible if the membrane is "tensionless" in the sense that the mechanical membrane tension is small compared to κ/A (Goetz and Lipowsky, 1998). For the example displayed in Figure 5.7, the latter tension scale is found to be $\kappa/A = 0.08$ mN/m.

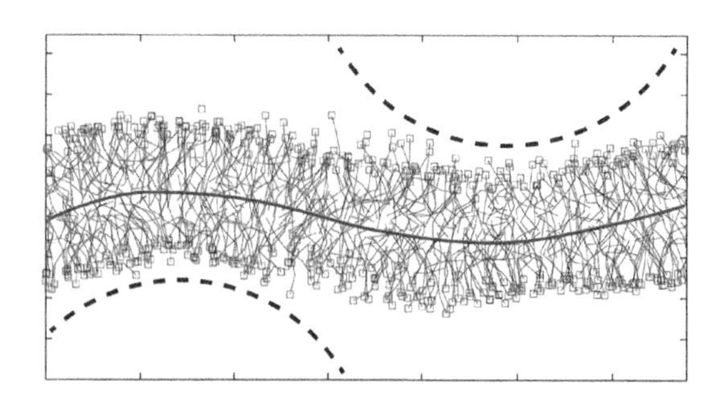

Figure 5.7 Emergence of membrane curvature on nanoscopic scales as observed in molecular dynamics simulations. The bilayer has a thickness of about 4 nm, the smallest curvature radius of its midsurface (red curve) is about 6 nm. For comparison, two circles (broken lines) with a radius of 6 nm are also displayed. (Reproduced from Goetz, R. et al., *Phys. Rev. Lett.*, 82, 221–224, 1999.)

5.3.2 MEAN AND GAUSSIAN CURVATURE

For each point on a smooth surface, we can construct a unit normal vector perpendicular to the membrane surface. Now, any plane that contains both the chosen point and this normal vector, defines a so-called normal section of the membrane surface, see Figure 5.8. The intersection between the surface and such a normal section defines a cross-sectional curve through the chosen point with a certain curvature C at this point. I will take this curvature to be positive if the cross-sectional curve bulges in the direction of the chosen normal vector as in Figure 5.8. This sign convention ensures that the cross-sectional curves on a sphere have positive curvature. Now, let us rotate the normal section around the normal vector. As a result of this rotation, the cross-sectional curve through the chosen point changes and so does the curvature C. As we change the rotation angle from 0 to 360 degrees, the latter curvature varies over a certain range as given by $C_{min} \leq C \leq C_{max}$. The two extremal values C_{min} and C_{max} define the principal curvatures, C_1 and C_2, at the chosen point. These principal curvatures correspond to the eigenvalues of the negative curvature tensor, see Appendix 5.A. Furthermore, for $C_1 \neq C_2$, the normal sections that contain the cross-sectional curves with $C = C_1$ and $C = C_2$ are always orthogonal to each other.

For fluid membranes as considered here, the molecules diffuse laterally along the membrane, which implies that the membrane surface should be described in terms of geometric quantities that do not depend on the choice of the surface coordinates, i.e., that are invariant under a reparametrization of the surface. Such quantities are provided, apart from a possible change of sign, by the principal curvatures C_1 and C_2 or equivalently by the mean curvature

$$M \equiv \frac{1}{2}(C_1 + C_2) \tag{5.1}$$

and the Gaussian curvature

$$G \equiv C_1 C_2. \tag{5.2}$$

The mean curvature is proportional to the trace of the curvature tensor whereas the Gaussian curvature is equal to its determinant (Appendix 5.A). Note that $C_1 = M - \sqrt{M^2 - G}$ and $C_2 = M + \sqrt{M^2 - G}$. Both expressions are always real-valued because $M^2 \geq G$.[2] Indeed, the latter inequality is equivalent to $(C_1 - C_2)^2 \geq 0$ and, thus, holds for any shape of the membrane segment. The equality $M^2 = G$ applies to spherical segments with $C_1 = C_2$.

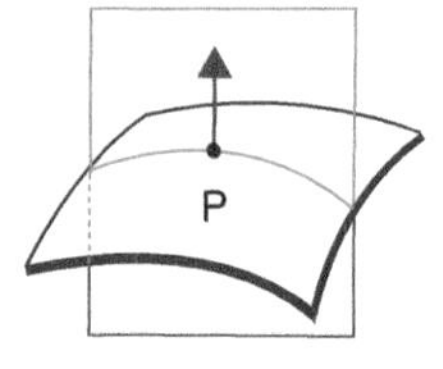

Figure 5.8 Normal section through membrane surface: Consider a point P of the membrane surface and the normal vector (arrow) at point P. A normal section is provided by any plane that contains both the point P and its normal vector. The intersection between the chosen normal section and the membrane surface defines a cross-sectional curve through point P. This curve has a certain curvature at point P. The latter curvature changes in a smooth manner as we rotate the normal section around the normal vector.

5.3.3 SIGN OF MEMBRANE CURVATURE

The mean curvature M is invariant under all orientation-preserving transformations of the surface coordinates, i.e., under all transformations that have a positive Jacobi determinant. The latter transformations do not affect the normal vectors of the membrane. However, we may also consider improper transformations of the surface coordinates which reverse the orientation of the normal vectors. A simple example of such an improper transformation A^1 is provided by a transposition of the two surface coordinates, i.e., by the transformation from (s^1, s^2) to $(\bar{s}^1 \equiv s^2, \bar{s}^2 \equiv s^1)$. The reversal of the normal vector implies that the principal curvatures change their sign and so does the mean curvature.

On the one hand, the reversal of the normal vectors provides a useful operation from a theoretical point of view because many physical properties of the membrane should not depend on our choice for the orientation of the normal vectors and must therefore be invariant under the reversal of these vectors. On the other hand, in order to avoid any ambiguity, we need a convention that always assigns a definite orientation to the normal vectors. For vesicle membranes as considered here, we can always distinguish between an interior and an exterior compartment and, thus, can always take the normal vectors to point towards the outer leaflet which is in contact with the exterior aqueous compartment, see Figure 5.9.

The sign of the mean curvature M depends on the sign of the principal curvatures C_1 and C_2. As explained before, each principal curvature is obtained from a certain normal section and taken to be positive if the corresponding cross-sectional curve bulges in the direction of the normal vector. If all cross-sectional curves of the membrane bulge into the direction of the normal vector as in Figures 5.8 and 5.9a, both C_1 and C_2 are positive which implies that the mean curvature M is positive as well.[3] Likewise, the mean

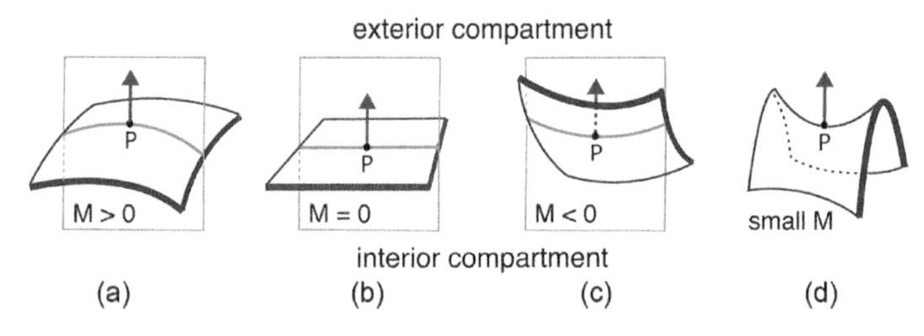

Figure 5.9 Sign convention for mean curvature M: (a) The mean curvature is *positive* if the membrane curves or bulges locally towards its outer leaflet in contact with the exterior compartment; (b) The mean curvature vanishes for a planar membrane; (c) The mean curvature is *negative* if the membrane curves or bulges locally towards its inner leaflet in contact with the interior compartment; and (d) If P is a saddle point, the two principal curvatures C_1 and C_2 have opposite sign and the mean curvature $M = \frac{1}{2}(C_1 + C_2)$ is small or even zero.

[2] The expressions for C_1 and C_2 imply that $C_1 = \dot{\psi}$ and $C_2 = \sin\psi / r$ for axisymmetric shapes parametrized by the tilt angle ψ and the radial coordinate r of the shape contour (Seifert et al., 1991).

[3] Choose local Cartesian coordinates (x,y,z) with the origin given by point $P = (0,0,0)$, normal vector $\hat{n} = (0,0,1)$, and the x-coordinate parallel to the normal section that contains the cross-sectional curve with the principal curvature $C_1 = C_{min}$. The cross-sectional curves within the normal sections with $y = 0$ and $x = 0$ are then described by $z \approx -C_1 x^2$ and $z \approx -C_2 y^2$ for small values of x and y.

curvature M is *negative* if all cross-sectional curves of the membrane bulge into the direction of the negative normal vector, see Figure 5.9c. At a saddle point of the membrane surface, the two principal curvatures have opposite signs and the mean curvature M can be positive or negative or even vanish, depending on the relative magnitude of the two principal curvatures, see Figure 5.9d.

5.3.4 CONSTANT-MEAN-CURVATURE SHAPES

In general, the principal curvatures and the mean curvature M are local quantities that vary along the membrane surface. Some particularly simple shapes are, however, characterized by constant mean curvature, i.e., all points on the surface have the same mean curvature, see Figure 5.10. Thus, a planar membrane has vanishing mean curvature, $M = 0$, whereas a sphere with radius R_{sp} has mean curvature $M = 1/R_{sp}$ and $M = -1/R_{sp}$ when its inner leaflet is in contact with the interior and the exterior solution, respectively. Likewise, a cylinder with radius R_{cy} has mean curvature $M = 1/(2R_{cy})$ when the enclosed volume of water belongs to the interior compartment and $M = -1/(2R_{cy})$ when this volume is connected to the exterior compartment. Another simple shape is a catenoid for which each point represents a saddle point with vanishing mean curvature $M = 0$ as depicted in Figure 5.10c.

Cylinders represent possible shapes for membrane nanotubes. Another tube morphology that has been observed are necklace-like tubes as shown in Figure 5.11a. The latter tubes consist of identical spheres connected by closed membrane necks. For spheres with radius R_{sp}, the necklace-like tube has mean curvature $M = 1/R_{sp}$ and $M = -1/R_{sp}$ when the enclosed volume of the tube is connected to the interior and exterior solution, respectively. A necklace-like tube consisting of spheres with radius R_{sp} can be continuously transformed into a cylindrical tube with radius $R_{cy} = \frac{1}{2} R_{sp}$, thereby preserving the value of the mean curvature. This transformation proceeds via a family of intermediate unduloids, all of which have the same mean curvature as the necklace-like and the cylindrical tube. The unduloids consist of lemon-like bulges connected by open necks, see the example in Figure 5.11b. Thus, during the constant-mean-curvature transformation, the closed necks of the necklace-like tube open up and the bulges of the necklace retract until the necks and the bellies have the same radius and form a cylindrical tube.

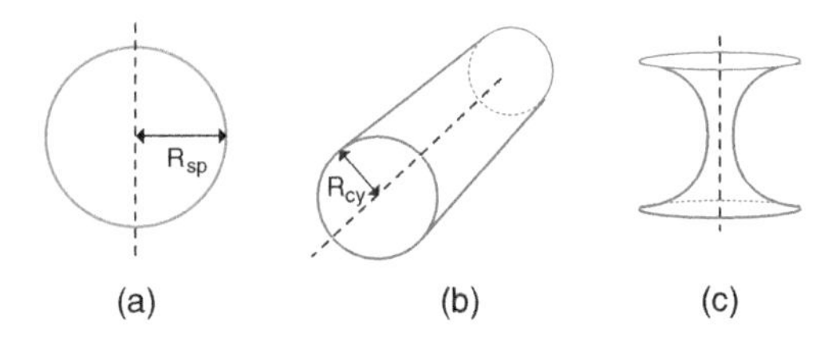

Figure 5.10 Simple membrane shapes with constant mean curvature M: (a) Sphere with radius R_{sp} and mean curvature $M = \pm 1/R_{sp}$; (b) Cylinder with radius R_{cy} and mean curvature $M = \pm 1/(2R_{cy})$; and (c) Catenoid with mean curvature $M = 0$. For spheres and cylinders, the sign of the mean curvature depends on whether the inner leaflet is in contact with the interior or exterior aqueous solution.

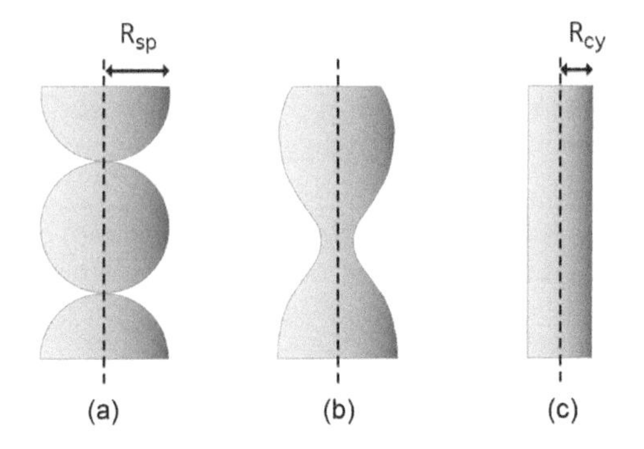

Figure 5.11 Three membrane tubes with different morphologies but the same constant mean curvature M: (a) Necklace-like tube consisting of identical spheres with radius $R_{sp} = 1/|M|$. The spheres are connected by closed membrane necks; (b) Unduloid with lemon-like bulges connected by open necks. The neck radius R_{ne} and the bulge radius R_{bu} are related to $|M|$ via $|M| = 1/(R_{ne} + R_{bu})$; and (c) Cylindrical tube with radius $R_{cy} = 1/(2|M|)$. (Reproduced from Lipowsky, R. *Biol. Chem.* 395, 253–274, 2014b. With permission of Walter de Gruyter GmbH & CO.KG.)

5.3.5 LOCAL CURVATURE GENERATION AND SPONTANEOUS CURVATURE

The simulation snapshot in Figure 5.7 displays a symmetric bilayer consisting of two leaflets that have the same molecular composition and are exposed to the same aqueous environment. Likewise, the cartoons in Figure 5.9 did not indicate any asymmetry between the two leaflets. In real systems, such symmetric bilayers are somewhat exceptional, but they provide a useful reference system because their elastic properties are governed by a single elastic parameter, the bending rigidity κ that provides the basic energy scale of membranes. For phospholipid bilayers, the latter scale is of the order of 10^{-19} J, which is about $20 k_B T$ at room temperature. For different lipid bilayers, the measured values of the bending rigidity vary by about an order of magnitude, see the corresponding tables in Chapters 11, 14, and 15 of this book.

Real bilayer membranes are typically asymmetric. This asymmetry can arise from a different lipid composition of the two leaflets as found in all biological membranes (van Meer et al., 2008; Fadeel and Xue, 2009). One prominent example is provided by the ganglioside GM1, a glycolipid that is abundant in all mammalian neurons (Aureli et al., 2016) and plays an important role in many neuronal processes and diseases (Schengrund, 2015). Furthermore, GM1 acts as a membrane anchor for various toxins, bacteria, and viruses such as the simian virus 40 (Ewers et al., 2010). The curvature generated by different leaflet concentrations of GM1 has been recently studied, both experimentally for giant vesicles (Bhatia et al., 2018; Dasgupta et al., 2018) and by simulations of molecular bilayers (Dasgupta et al., 2018; Sreekumari and Lipowsky, 2018; Miettinen and Lipowsky, 2019). Likewise, membrane proteins in biological membranes have a preferred orientation, which also contributes to their asymmetry. In addition, membranes can acquire such an asymmetry from their environment as provided by the exterior and interior aqueous compartments. Indeed, the membranes become asymmetric when these two compartments contain different concentrations of ions, small solutes such as sugar molecules, and/or proteins that form adsorption or depletion layers on the two leaflets of the bilayer membranes (Lipowsky and Döbereiner, 1998; Lipowsky, 2013; Rozycki and Lipowsky, 2015, 2016; Liu et al., 2016; Karimi et al., 2018; Ghosh et al., in preparation). Examples for mechanisms of local generation of membrane curvature are given in Box 5.1. Local curvature generation by proteins is reviewed in Chapter 23 of this book.

Box 5.1 Local generation of membrane curvature

Bilayer asymmetry and spontaneous curvature can be generated by a variety of molecular mechanisms as illustrated in this Box.

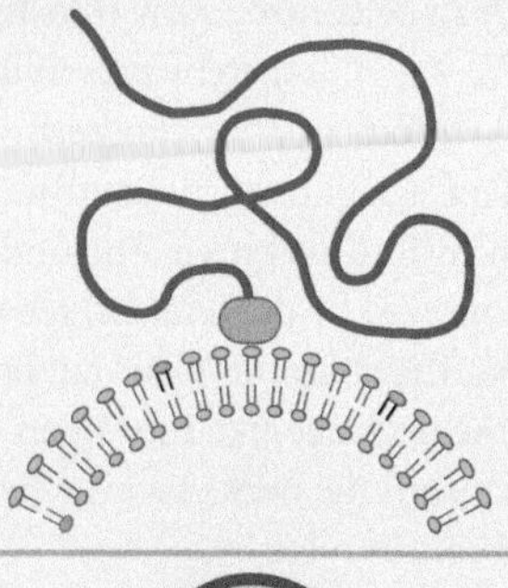

- A simple example is provided by a flexible polymer that is anchored with one of its ends to the membrane (Lipowsky, 1995; Nikolov et al., 2007).
- Such an anchored polymer generates curvature in order to increase its configurational entropy.

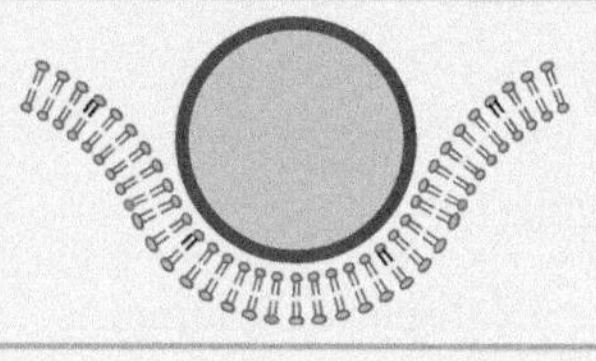

- Adhesive nanoparticles that are partially engulfed by the membrane act as scaffolds and impose their curvature onto this membrane, (Lipowsky and Döbereiner, 1998; Deserno, 2004; Agudo-Canalejo and Lipowsky, 2015a) see Chapter 8 of this book.

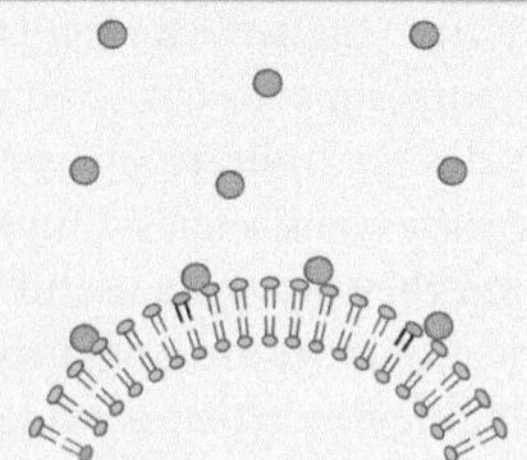

- Small adhesive solutes generate a substantial spontaneous curvature m as predicted theoretically (Lipowsky and Döbereiner, 1998; Lipowsky, 2013) and observed in molecular simulations (Rozycki and Lipowsky, 2015). For particles with a diameter of 1 nm and a concentration difference of 100 mM, adsorption leads to $m = \frac{1}{77nm}$.

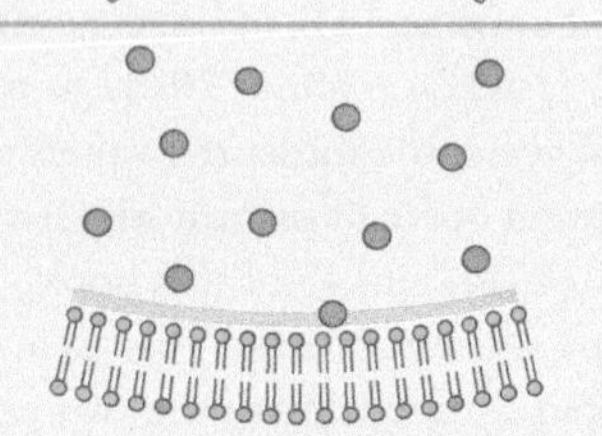

- Depletion layers of solutes induce a spontaneous curvature m of the opposite sign (Lipowsky and Döbereiner, 1998). This prediction has also been confirmed by recent molecular simulations (Różycki and Lipowsky, 2016). For particles with a diameter of 1 nm and a concentration difference of 100 mM, depletion leads to $m = -\frac{1}{270nm}$.

The case of divalent ions is controversial because two recent experimental studies on Ca^{2+} ions (Simunovic et al., 2015; Baumgart et al., 2017) led to different conclusions about the sign of the ion-induced spontaneous curvature.

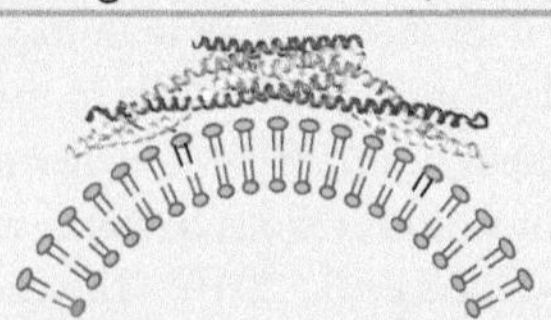

- N-BAR proteins such as amphiphysin (Takei et al., 1999; Peter et al., 2004) and endophilin (Farsad et al., 2001), F-BAR proteins such as pacsin/syndapin (Wang et al., 2009), and other proteins involved in endocytosis such as epsin (Ford et al., 2002) can bind to membranes and impose their curvature onto these membranes.

Membrane-binding proteins that act as scaffolds for the membrane shape are usually quite rigid. They can be regarded as adhesive nanoparticles with two characteristic properties: (i) their shape is typically nonspherical and often banana-like or convex-concave; and (ii) their surface contains a more or less complex pattern of adhesive and nonadhesive surface domains. Thus membrane-binding proteins that impose their shape onto the membrane can be regarded as nonspherical Janus-like nanoparticles.

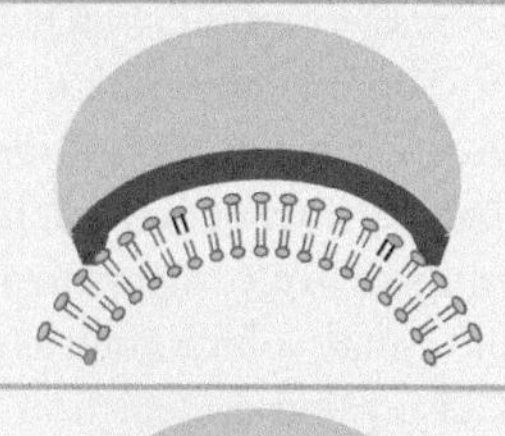

- If the *planar* membrane can bind to some of the adhesive surface domains (red) of the particle, the particle generates membrane curvature via an induced-fit mechanism.

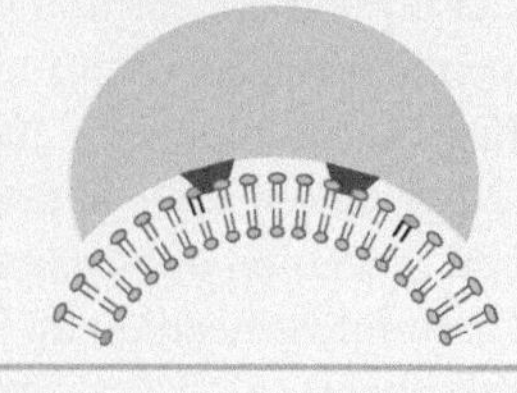

- If the adhesive surface domains (red) can only be reached by an appropriately *curved* membrane, the particle generates membrane curvature via conformational selection (Lipowsky, 2014b).

On length scales that exceed about twice the membrane thickness, the bilayer asymmetry can be described in terms of another curvature-elastic parameter, the spontaneous curvature m. In order to define the sign of m in an unambiguous manner, we use the same sign convention as for the mean curvature M, see Figure 5.9. Thus, we distinguish an interior from an exterior aqueous compartment and take the spontaneous curvature to be positive and negative if the membrane prefers to bulge towards the exterior and interior compartment, respectively. Note that, under the reversal of the normal vectors, the spontaneous curvature transforms in the same way as the mean curvature and, thus, changes sign.

If the membrane is decorated by many bound "particles," it will acquire a certain spontaneous curvature that depends both on the local particle-induced curvature and on the particle coverages for the two leaflets of the bilayer membrane (Breidenich et al., 2000; Lipowsky, 2002). Thus, if a single particle that is bound to the outer leaflet of an asymptotically flat bilayer generates the local, position-dependent mean curvature $M_{si}(s^1, s^2)$, the spontaneous curvature m is given by

$$m = I_{M,\mathrm{si}}(\Gamma_{\mathrm{ex}} - \Gamma_{\mathrm{in}}) \tag{5.3}$$

with the integrated mean curvature

$$I_{M,\mathrm{si}} \equiv \int \mathrm{d}A M_{\mathrm{si}}(s^1, s^2) \tag{5.4}$$

and the coverages Γ_{ex} and Γ_{in} which are equal to the numbers of particles bound to the outer and inner leaflets per unit area (Breidenich et al., 2000). In contrast to other elastic membrane parameters such as the bending rigidity or the area compressibility modulus, the spontaneous curvature can vary over more than three orders of magnitude, from the inverse size of giant vesicles, which is of the order of 1/(50 μm), to half the inverse membrane thickness, which is of the order of 1/(10 nm).

Inspection of the relationship Eq. 5.3 shows that the sign of the spontaneous curvature m is determined (i) by the sign of the integrated mean curvature $I_{M,si}$ induced by a single particle bound to the outer leaflet of the bilayer and (ii) by the sign of the difference $\Gamma_{ex} - \Gamma_{in}$ between the coverages of the outer and inner leaflets. Depending on the molar particle concentrations in the exterior and interior aqueous compartments, the sign of $\Gamma_{ex} - \Gamma_{in}$ can be positive or negative. Likewise, the sign of the integrated curvature $I_{M,si}$ can be positive or negative as well, reflecting different molecular interactions between the bound particle and the membrane. An anchored polymer, for example, generates a positive value of $I_{M,si}$ but this value becomes negative when all monomers of the polymer are strongly adsorbed onto the membrane (Breidenich et al., 2001, 2005). A negative sign of $I_{M,si}$ also applies if the particle is large and partially engulfed by the membrane.

As explained previously, we use two related conventions in order to define the sign of the local mean curvature of the membrane in an unambiguous manner. The first convention is that the normal vector of the membrane is taken to point towards the exterior compartment. The second convention is that we take the local mean curvature of the membrane to be positive if the membrane bulges in the direction of the normal vector. Therefore, the spontaneous curvature is taken to be positive as well if the membrane prefers to bulge towards the exterior solution, i.e., in the direction of the normal vector.

The intuitive notion that asymmetric membranes have a preferred curvature was originally discussed by Bancroft for surfactant monolayers in water-oil emulsions (Bancroft, 1913; Bancroft and Tucker, 1927) and was included by Frank as the so-called "splay term" in the curvature elasticity of liquid crystals (Frank, 1958). In the context of lipid bilayers, spontaneous curvature was first considered by Helfrich (1973), who introduced it in analogy to the splay term for liquid crystals. The corresponding curvature energy of the membrane is now known as the spontaneous curvature model (Seifert et al., 1991) which will be presented in the next section.

5.4 CURVATURE ELASTICITY OF UNIFORM MEMBRANES

This chapter describes the theoretical framework that has been crucial in order to understand the morphology of giant vesicles. This framework is based on membrane curvature and the associated elastic energy contributions. The theory also takes into account that the low lipid solubility and the osmotic conditions lead to important constraints on the membrane area and the vesicle volume. In fact, what makes this theory both appealing and challenging is the interplay between local and global membrane properties.

On the one hand, the shape of a membrane can be described locally by its mean and Gaussian curvatures. On the other hand, in the absence of topological transformations such as membrane fusion and fission, both the membrane area and the vesicle volume are essentially fixed which has a direct and strong influence on the local membrane behavior. The connection between local and global properties is provided by two quantities, the mechanical tension Σ within the membrane and the pressure difference ΔP across this membrane. For free vesicles, these two quantities cannot be measured experimentally. However, the theory described in this chapter provides explicit relations between Σ and ΔP and those quantities that are directly accessible to experimental observations.

Another intriguing aspect of the morphology of giant vesicles is the frequent observation of membrane necks that connect two larger membrane segments. One example is provided by the neck that connects the spherical bud to the mother vesicle in Figure 5.2, another example is provided by the shape L^{sto} in Figure 5.6. Theoretically, these necks were first discovered by numerical energy minimization (Seifert et al., 1991; Miao et al., 1991; Berndl et al., 1990) of vesicles with uniform membranes as considered in this section. The necks are interesting from a conceptual point of view because they lead to *local* relations between (i) geometric quantities that can be directly observed in the optical microscope and (ii) curvature-elastic parameters such as the spontaneous curvature.

This section focuses on the spontaneous curvature model which is theoretically appealing because it depends on a relatively small number of parameters. Indeed, uniform vesicle membranes involve two geometric quantities, the vesicle volume

V and the membrane area A, as well as two material parameters, the bending rigidity κ and the spontaneous curvature m introduced in Section 5.3.5. In fact, as shown below, the vesicle shapes depend only on two dimensionless parameters, the volume-to-area ratio proportional to $V/A^{3/2}$, also known as the reduced volume, and the dimensionless spontaneous curvature proportional to $mA^{1/2}$.

The spontaneous curvature model is based on an expansion in powers of the principal curvatures and should be reliable as long as these curvatures are small compared to the inverse membrane thickness. In addition, the spontaneous curvature model implicitly assumes that the area difference between the two leaflets can change via flip-flops of lipid molecules. While a phospholipid molecule may stay in the same leaflet for hours, a cholesterol molecule will, on average, flip-flop from one leaflet to the other within one second. Therefore, the spontaneous curvature model should provide a reliable description for bilayer membranes that contain cholesterol or another sterol. The latter membranes are of particular interest because they undergo phase separation into liquid-disordered and liquid-ordered phases, see Section 5.8 below and Chapter 18 of this book.

If all membrane components undergo relatively slow flip-flops, one should extend the spontaneous curvature model by adding a nonlocal term that depends on the quenched area difference between the two leaflets. This extension leads to the area-difference-elasticity model and to an effective spontaneous curvature as described at the end of this section.

The present section is supplemented by three appendices: Appendix 5.B on different topologies of vesicles; Appendix 5.D which explains the identity of the mechanical tension with the Lagrange multiplier for membrane area; and Appendix 5.E which describes the different variants of curvature models.

5.4.1 SPONTANEOUS CURVATURE MODEL

Curvature expansion of local curvature energy

Within the spontaneous curvature model, the curvature energy functional $\mathcal{E}_{\rm cu}\{S\}$ of a certain membrane shape S is provided by the area integral[4]

$$\mathcal{E}_{\rm cu}\{S\} = \int \mathrm{d}A\, \varepsilon_{\rm cu}(\underline{s}) \tag{5.5}$$

where $\varepsilon_{\rm cu}(\underline{s})$ represents a local energy density that varies smoothly with the two-dimensional surface coordinates $\underline{s} \equiv (s^1, s^2)$ used to parametrize the membrane surface via the three-dimensional vector $\vec{X}(\underline{s})$. When expressed in terms of these coordinates, the area element dA depends on the metric tensor g_{ij}, see Appendix 5.A , and has the form

$$\mathrm{d}A = \mathrm{d}s^1 \mathrm{d}s^2 \sqrt{g} \text{ with } g \equiv \det(g_{ij}) = g_{11} g_{22} - g_{12} g_{21}. \tag{5.6}$$

The local density $\varepsilon_{\rm cu}$ of the curvature energy should only depend on the principal curvatures C_1 and C_2. In addition, at any given point P of the membrane surface, this energy density must remain unchanged when we rotate the surface coordinates by $\pi/2$ which implies $\varepsilon_{\rm cu}(C_2, C_1) = \varepsilon_{\rm cu}(C_1, C_2)$. An expansion of $\varepsilon_{\rm cu}$ up to second order in the principal curvatures then leads to[5]

$$\varepsilon_{\rm cu}(C_1, C_2) \approx a_0 + a_1(C_1 + C_2) + a_2(C_1^2 + C_2^2) + a_3 C_1 C_2. \tag{5.7}$$

When this relation is expressed in terms of the mean curvature M and the Gaussian curvature G, we obtain

$$\varepsilon_{\rm cu} \approx 2\kappa(M - m)^2 + \kappa_G G \tag{5.8}$$

with the bending rigidity κ, the spontaneous curvature m, and the Gaussian curvature modulus κ_G.[6] As a result, the curvature energy functional has the form (Helfrich, 1973; Seifert et al., 1991)

$$\mathcal{E}_{\rm cu}\{S\} = \int \mathrm{d}A[2\kappa(M(\underline{s}) - m)^2 + \kappa_G G(\underline{s})] \tag{5.9}$$

which defines the spontaneous curvature model.

Vesicles without bilayer edges or pores

For a closed vesicle without bilayers edges or pores, the Gauss-Bonnet theorem of differential geometry implies

$$\int \mathrm{d}A G = 2\pi\chi = 2\pi(2 - 2\mathfrak{g}) \tag{5.10}$$

with the Euler characteristic χ and the topological genus $\mathfrak{g}$, which counts the number of handles, see Appendix 5.B. Thus, for a closed vesicle shape S and a uniform vesicle membrane, the spontaneous curvature model is defined by the curvature energy functional

$$\mathcal{E}_{\rm cu}\{S\} = \mathcal{E}_{\rm be}\{S\} + 2\pi\chi\kappa_G \tag{5.11}$$

with the bending energy functional

$$\mathcal{E}_{\rm be}\{S\} = 2\kappa \int \mathrm{d}A\,(M - m)^2. \tag{5.12}$$

When we evaluate the functionals $\mathcal{E}_{\rm cu}$ and $\mathcal{E}_{\rm be}$ for a certain shape S_o, we obtain the corresponding curvature and bending energies $E_{\rm cu} = \mathcal{E}_{\rm cu}\{S_o\}$ and $E_{\rm be} = \mathcal{E}_{\rm be}\{S_o\}$ for which we use normal capital letters E.

It is instructive to consider the behavior of the bending energy functional Eq. 5.12 under the reversal of the normal vectors. Thus, consider a certain shape S_o and map it onto another shape S_o' by reversing all normal vectors of its membrane surface. The mean curvature M of shape S_o is then transformed into the mean curvature $M'(\underline{s}) = -M(\underline{s})$ of shape S_o' which implies

[4] Here and below, large calligraphic letters such as $\mathcal{E}$ and $\mathcal{F}$ are used for functionals that map shapes into real numbers.

[5] Here and below, the symbol $\approx$ stands for 'asympotically equal' in a certain limit

[6] The constant term $a_0 - a_1^2/(4a_2)$ has been omitted.

Giant vesicles theoretically and in silico

$$\mathcal{E}_{be}(\{S'_o\}, m') = \mathcal{E}_{be}(\{S_o\}, m) \quad \text{for} \quad m' = -m, \tag{5.13}$$

i.e., the bending energy functional is invariant under a reversal of the normal vectors provided we reverse the spontaneous curvature m as well.

The bending energy functional $\mathcal{E}_{be}\{S\} \sim \int dA M^2$ of symmetric membranes with $m = 0$ has a long history in the calculus of variations. The quadratic expression in the mean curvature was first studied at the beginning of the 19th century by the French mathematician Germain in her theory of vibrating plates (Dalmédico, 1991). About a hundred years later, this expression played a prominent role in the work of the German mathematician Blaschke and his students, who were particularly interested in its invariance properties under conformal transformations. In the 1960s, the subject was studied in a systematic manner by the British mathematician Willmore, and the shapes that minimize $\int dA M^2$ are often referred to as Willmore surfaces (Willmore, 1982).

Separation of length scales

As described above, the spontaneous curvature model is based on the expansion of the curvature energy density in powers of the principal curvatures and includes all terms up to second order in these curvatures. This truncation of the curvature expansion at second order is clearly appropriate as long as the principal curvatures are much smaller than the inverse membrane thickness $1/\ell_{me} \simeq 1/(4 \text{ nm})$ as follows from the discussion in Section 5.3.1. Thus, the spontaneous curvature model should provide a reliable description for the shapes of giant vesicles as observed in the (conventional) optical microscope, which resolves membrane curvatures below 1/(300 nm). In fact, as explained in Appendix 5.C.1, the spontaneous curvature model is expected to be quite reliable up to principal curvatures of about 1/(80 nm). For more strongly curved membrane segments, third-order curvature terms may become important which involve two additional curvature-elastic parameters, see Appendix 5.C.1.

5.4.2 SPONTANEOUS TENSION

The bending energy functional as given by Eq. 5.12 attains its minimal value, $\mathcal{E}_{be} = 0$, when we consider shapes for which the mean curvature M is equal to the spontaneous curvature m. The expression Eq. 5.12 also implies that the bending rigidity κ represents a "spring constant" for deviations of the actual mean curvature M from the spontaneous curvature m of the membrane.

Real membranes experience a variety of constraints that necessarily lead to such deviations of M from m. One important constraint is provided by the size of the membrane. If the membrane area A is large compared to $4\pi/m^2$, which is the surface area of a sphere with radius $1/|m|$, the membrane cannot adapt its curvature to the spontaneous curvature by forming a single sphere but can do so, to a large extent, by forming a long cylinder with radius $R_{cy} = 1/(2m)$. Another important constraint arises from the osmotic conditions that determine the vesicle volume and, thus, the volume-to-area ratio, also known as the reduced volume. If the vesicle volume is increased by osmotic inflation, it will eventually attain a spherical shape with mean curvature $M = 1/R_{sp}$ that usually differs from the spontaneous curvature m of the vesicle membrane. In fact, for a giant spherical vesicle, the mean curvature $M = 1/R_{sp}$ can be very small compared to the absolute value $|m|$ of the spontaneous curvature. Likewise, supported lipid bilayers with $M = 0$ can have a large spontaneous curvature with magnitude $|m| \gg 0$. Whenever a large membrane segment of area A is forced to attain a mean curvature that is much smaller than the spontaneous curvature, the contribution of this segment to the bending energy obtained from Eq. 5.12 has the form $E_{be} \approx A\sigma$ with the spontaneous tension (Lipowsky, 2013)

$$\sigma \equiv 2\kappa m^2. \tag{5.14}$$

This tension represents the only tension scale that can be defined, apart from a dimensionless multiplicative factor, by the two parameters κ and m. Therefore, the spontaneous tension σ may be viewed as the intrinsic tension of curvature elasticity. If the membrane has a bending rigidity of about 10^{-19} J, a spontaneous curvature of 1/(20 μm) leads to a spontaneous tension of about 10^{-6} mN/m while a spontaneous curvature of 1/(20 nm) leads to a spontaneous tension of about 1 mN/m. Thus, in real membrane systems, the spontaneous tension can vary over six orders of magnitude, see the examples in Table 5.1.

5.4.3 GLOBAL AND LOCAL PARAMETERS

Volume and area as global control parameters

As explained in Section 5.2.2, lipid bilayers are permeable to water and small gas molecules but essentially impermeable to ions and solute molecules, see also Chapter 20 of this book. As a consequence, the vesicle volume is primarily determined

Table 5.1 **Spontaneous (or preferred) curvature m in units of 1/μm and associated spontaneous tension $\sigma = 2\kappa m^2$ in units of 2 mN/m for four different membrane systems where the bending rigidity was taken to have the typical value $\kappa \simeq 10^{-19}$ J.**

	SUGAR SOLUTIONS[a]	DNA STRANDS[b]	PEG/DEXTRAN SOLUTIONS[c]	BAR-DOMAIN PROTEINS[d]
m [1/μm]	0.01–0.1	0.1–1	3–10	10–50
σ [2 mN/m]	10^{-8}–10^{-6}	10^{-6}–10^{-4}	10^{-3}–10^{-2}	10^{-2}–0.5

[a] Döbereiner, H.G. et al., *Eur. Biophys. J.*, 28, 174–178, 1999.
[b] Nikolov, V. et al., *Biophys. J.*, 92, 4356–4368, 2007.
[c] Li, Y. et al., *Proc. Nat. Acad, Sci. USA*, 108, 4731–4736, 2011; Liu, Y. et al., *ACS Nano*, 10, 463–474, 2016.
[d] Peter, B.J. et al., *Science*, 303, 495–499, 2004; McMahon, H.T. and Gallop, J.L. *Nature*, 438, 590–596, 2005.

by the osmotic conditions and the temperature. Therefore, one convenient procedure to change the vesicle volume at constant temperature is via osmotic inflation and deflation. Osmotic deflation is limited by the attractive intermolecular forces that start to become important when different membrane segments come into close proximity. Thus, at very small volumes, different segments of the vesicle membrane may start to fold back onto themselves or to form local membrane stacks. On the other hand, osmotic inflation is limited by the available membrane area. Indeed, for a given membrane area A and the corresponding vesicle size

$$R_{ve} = \sqrt{A/(4\pi)}, \tag{5.15}$$

the vesicle volume V attains its maximal value when the vesicle has a spherical shape. Therefore, the vesicle volume satisfies the inequality

$$V \leq \frac{4\pi}{3} R_{ve}^3 = \frac{4\pi}{3}\left(\frac{A}{4\pi}\right)^{3/2}. \tag{5.16}$$

For constant temperature and lipid composition, the area A of the vesicle membrane is primarily determined by the number of lipid molecules within the membrane. Indeed, in the absence of external forces or constraints, the lipids attain a certain molecular area corresponding to their optimal packing density. In principle, the membrane area can be changed by a mechanical tension that acts to stretch the membrane. In practice, such a tension can increase the membrane area only by a few percent because the membrane starts to rupture for larger extensions of its area. Therefore, as long as the membrane does not rupture, the membrane area A should attain a constant value to a very good approximation.

For giant unilamellar vesicles, one can directly measure the vesicle volume V and the membrane area A. It is therefore rather natural from an experimental point of view to regard V and A as basic geometric parameters that determine the vesicle shape.

Dimensionless parameters of spontaneous curvature model

For closed vesicles, the Gaussian curvature modulus contributes a constant term to the curvature energy functional $\mathcal{E}_{cu}$ which is independent of the vesicle shape. We are then left with the bending energy functional $\mathcal{E}_{be}$ that depends on four (dimensionful) parameters: two material parameters, namely bending rigidity κ and spontaneous curvature m, as well as two geometric parameters, vesicle volume V and membrane area A. Furthermore, we can choose a basic energy and length scale. One convenient choice for these two scales is provided by the bending energy κ and the vesicle size R_{ve} as defined by Eq. 5.15.

For the latter choice, the dimensionless bending energy E_{be}/κ depends only on two dimensionless parameters: (i) the volume-to-area ratio or reduced volume of the vesicle

$$v \equiv \frac{V}{\frac{4\pi}{3} R_{ve}^3} = 6\sqrt{\pi}V/A^{3/2} \tag{5.17}$$

and (ii) the rescaled and dimensionless spontaneous curvature

$$\bar{m} \equiv mR_{ve} = m\sqrt{A/(4\pi)}. \tag{5.18}$$

In the following, we will often discuss the behavior of vesicles with a certain, fixed membrane area and, thus, with a fixed length scale R_{ve}. Deflation and inflation processes are then described by changes in the volume v for a certain value of the spontaneous curvature $\bar{m}$. Likewise, adsorption and desorption processes which affect the bilayer asymmetry are described by changes of the spontaneous curvature $\bar{m}$ for a fixed value of the volume v.

Scale transformations of vesicle shapes

The conclusions of the previous subsection can be understood from a somewhat different perspective if we study the behavior of the energy functional in Eq. 5.12 under scale transformations. As mentioned, the vesicle shape S can be described by a vector-valued function $\vec{X}(\underline{s})$ that depends on the two-dimensional surface coordinate $\underline{s}$. A scale transformation from the shape S to the new shape S' is then described by

$$\vec{X}(\underline{s}) \rightarrow \vec{X}'(\underline{s}) \equiv \zeta\,\vec{X}(\underline{s}) \quad \text{with a scale factor } \zeta > 0 \tag{5.19}$$

which implies the scale transformations

$$V \rightarrow V' = \zeta^3 V \quad \text{and} \quad A \rightarrow A' = \zeta^2 A \tag{5.20}$$

of vesicle volume and membrane area.

The bending energy functional $\mathcal{E}_{be}$ in Eq. 5.12 remains invariant under the scale transformation Eq. 5.19, i.e., $\mathcal{E}_{be}\{S'\} = \mathcal{E}_{be}\{S\}$ if we combine this transformation with the rescaling

$$m \rightarrow m' \equiv m/\zeta \tag{5.21}$$

of the spontaneous curvature.

Now, assume that we have minimized the energy functional and found the shape S_0 of minimal bending energy for a certain set of the (dimensionful) parameters V, A, κ, and m. Any slightly deformed shape, say S_1, will have a larger bending energy, i.e., $\mathcal{E}_{be}\{S_1\} > \mathcal{E}_{be}\{S_0\}$. This property remains valid if we compare the bending energies of the shapes S_0' and S_1' as obtained by rescaling both S_0 and S_1 with the same scale factor ζ, i.e., $\mathcal{E}_{be}\{S_1'\} > \mathcal{E}_{be}\{S_0'\}$ for any small deformation of S_0', provided we also rescale the spontaneous curvature according to Eq. 5.21. Therefore, the rescaled shape S_0' represents the shape of minimal bending energy for the parameters $\zeta^3 V$, $\zeta^2 A$, κ, and m/ζ.

The same conclusion can be drawn from the dimensionless parameters introduced in the previous subsection. Indeed, the dimensionless bending energy E_{be}/κ depends only (i) on the volume-to-area ratio $v \propto V/A^{3/2}$ and (ii) on the spontaneous curvature $\bar{m} = mR_{ve}$, both of which remain invariant under the combined scale transformation Eqs 5.20 and 5.21.

It is often instructive to consider the special case of a symmetric membrane with vanishing spontaneous curvature, $m = 0$. In this case, the energy functional Eq. 5.12 is invariant under the scale transformation of the vesicle geometry as described by Eq 5.20 and does not involve the rescaling of any material parameter. Thus, for $m = 0$, large and small vesicles have the same bending energy if they have the same shape.

5.4.4 LOCAL SHAPE EQUATION AND ENERGY BRANCHES

Constrained energy minimization

If we take the vesicle volume and the membrane area as control parameters, we are thus faced with the problem of minimizing the curvature energy functional as given by Eq. 5.11 for a given vesicle volume V and membrane area A. In principle, there are a variety of ways to tackle this minimization problem numerically.

Numerical minimization typically involves a discretization of the vesicle shape into a triangular mesh of membrane patches. Furthermore, in order to model the fluidity of the membrane, one has to choose a dynamic triangulation. The advantage of numerical minimization is that we do not have to make any simplifying assumptions about the vesicle shape. The disadvantage of such a numerical procedure is that we can only explore a limited region of the parameter space. Furthermore, numerical minimization methods becomes difficult whenever the vesicle shape involves narrow membrane necks or long tubes. As we will see further below, such somewhat exotic shapes are quite common for vesicle membranes.

In order to apply analytical approaches to the constrained minimization, we will now incorporate the area and volume constraints via Lagrange multipliers Σ and ΔP and consider the shape functional

$$\mathcal{F}\{S\} = -\Delta P \mathcal{V}\{S\} + \Sigma \mathcal{A}\{S\} + \mathcal{E}_{\mathrm{be}}\{S\} \quad (5.22)$$

where we have omitted the shape-independent term arising from the integrated Gaussian curvature. The two Lagrange multipliers have to be chosen in such a way that the volume functional $\mathcal{V}$ and the area functional $\mathcal{A}$ attain the values $\mathcal{V}\{S\} = V$ and $\mathcal{A}\{S\} = A$. Note that we again denote the functionals $\mathcal{F}$, $\mathcal{V}$, and $\mathcal{A}$ by large calligraphic letters and their numerical values for a certain shape by normal capital letters F, V, and A.

As shown in Appendix 5.D, the Lagrange multiplier Σ can be identified with the mechanical tension experienced by the uniform membrane. The latter identity can be derived by defining the overall elastic energy of the membrane to be the sum of its bending and stretching energy and by minimizing this overall elastic energy (Lipowsky, 2014a).

Euler-Lagrange or local shape equation

The first variation of the shape functional $\mathcal{F}\{S\}$ leads to the Euler-Lagrange equation

$$\Delta P = 2\Sigma M - 2\kappa \nabla_{\mathrm{LB}}^2 M - 4\kappa [M - m][M(M + m) - G] \quad (5.23)$$

with the Laplace-Beltrami operator ∇_{LB}^2 and the (local) Gaussian curvature G. When expressed in terms of the surface coordinates $\underline{s}$, the action of this operator onto a scalar function $f(\underline{s})$ has the explicit form

$$\nabla_{\mathrm{LB}}^2 f = \frac{1}{\sqrt{g}} \frac{\partial}{\partial s^k} \left(\sqrt{g}\, g^{kj} \frac{\partial}{\partial s^j} f \right) \quad (5.24)$$

with the inverse metric tensor $(g^{ij}) \equiv (g_{ij})^{-1}$ and an implicit summation over repeated indices (do Carmo, 1976). Note that the Euler-Lagrange Eq. 5.23 provides an explicit relation between the Lagrange multipliers ΔP and Σ with the mean and Gaussian curvatures, M and G, which describe the membrane shape locally. Therefore, the Euler-Lagrange equation represents a *local* shape equation.

The Euler-Lagrange Eq. 5.23 is equivalent to

$$\Delta P = 2\hat{\Sigma} M - 2\kappa \nabla_{\mathrm{LB}}^2 M - 4\kappa m M^2 - 4\kappa [M - m][M^2 - G] \quad (5.25)$$

with the total membrane tension

$$\hat{\Sigma} \equiv \Sigma + 2\kappa m^2 = \Sigma + \sigma \quad (5.26)$$

which represents the sum of the mechanical tension Σ and the spontaneous tension σ, where we identified the Lagrange multiplier Σ with the mechanical tension, see Appendix 5.D. Therefore, the only tension that enters the solution of the Euler-Lagrange equation is the total tension $\hat{\Sigma}$ that contains the spontaneous tension σ defined in Eq. 5.14.

For spontaneous curvature $m = 0$, the Euler-Lagrange Eq 5.23 assumes the simplified form

$$\Delta P = 2\Sigma M - 2\kappa \nabla_{\mathrm{LB}}^2 M - 4\kappa M[M^2 - G] \quad (m = 0) \quad (5.27)$$

which was derived by several mathematicians as reviewed in the monograph of Willmore (Willmore, 1982). It seems that the variation of the more general case with $m \neq 0$ was first considered by (Jenkins, 1977) who included both normal and tangential displacements of the membrane surface.[7] However, in order to derive the Euler-Lagrange Eq. 5.23, it is sufficient to include only normal displacements as shown by (Ou-Yang and Helfrich, 1989).

Energy branches of stationary shapes

The solutions of the Euler-Lagrange Eq. 5.23 represent the stationary shapes corresponding to local minima, saddle points, or local maxima of the bending energy. The physically relevant shapes are the local minima, which represent (meta)stable states, and the saddle points which provide the activation barriers between different (meta)stable states.

In practice, the combination of the Laplace-Beltrami operator and the nonlinearities in the principal curvatures C_1 and C_2, arising from the second and third power of the mean curvature $M = \frac{1}{2}(C_1 + C_2)$ and from the Gaussian curvature $G = C_1 C_2$, make the Euler-Lagrange Eq. 5.23 rather difficult to solve. As explained further below, much insight can be obtained for special shapes such as spheres, cylinders, and combinations

[7] The final result of the variational calculation by (Jenkins, 1977) contains one term that is cancelled by another, missing term.

thereof. For axisymmetric shapes, the partial differential Eq 5.23 is equivalent to a set of ordinary differential equations that can be solved numerically, e.g., by shooting methods. In this way, the regime of relatively small spontaneous curvatures m with $|\bar{m}|=|m|R_{ve}\lesssim 2$ has been studied in a systematic manner (Seifert et al., 1991).

These numerical solutions have shown that the stationary shapes form, in general, several branches for the same set of parameters as illustrated in Figure 5.12.[8] The latter figure displays the branches for vanishing spontaneous curvature $m = 0$. The different branches will now be labeled by the index j and the corresponding stationary shapes by S^j. Along branch j, the bending energy function

$$E_{be}(V,A;\kappa,m;j)=\mathcal{E}_{be}\{S^j\} \tag{5.28}$$

varies in a continuous manner as one changes one of the control parameters. When expressed in terms of the dimensionless parameters v and $\bar{m}=mR_{ve}$ as defined in Eqs 5.17 and 5.18, one obtains

$$E_{be}(V,A;\kappa,m;j)=8\pi\kappa\,\bar{E}(v,\bar{m};j), \tag{5.29}$$

see Figure 5.12. The corresponding shapes of minimal energy are displayed in Figure 5.13.

Pressure difference and membrane tension

In order to get further insight into the two Lagrange multipliers ΔP and Σ, it is useful to consider the shape energy

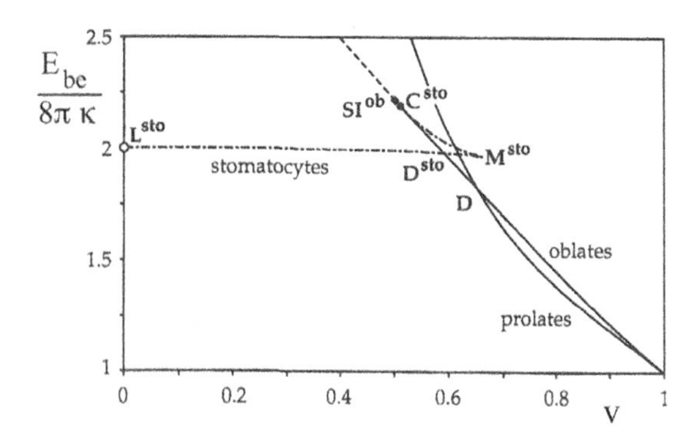

Figure 5.12 Dimensionless bending energy $\bar{E}_{be} = E_{be}/(8\pi\kappa)$ as a function of volume-to-area ratio v for spontaneous curvature $m = 0$: The sphere corresponds to the largest possible volume-to-area ratio $v = 1$. In the limit of small v, we obtain the limit shape L^{sto} of a stomatocyte consisting of two concentric spheres of (almost) equal size connected by a closed membrane neck. The two full lines emanating from the sphere correspond to (meta)stable prolates and oblates. The dashed-dotted line connecting the limit shape L^{sto} with the transition point D^{sto} corresponds to stable stomatocytes, the one between D^{sto} and M^{sto} to metastable stomatocytes, and the dashed-dotted line between M^{sto} and C^{sto} to the activation barriers between the oblates and the stomatocytes. (Reproduced from Seifert, U. et al., *Phys. Rev. A*, 44, 1182–1202, 1991.)

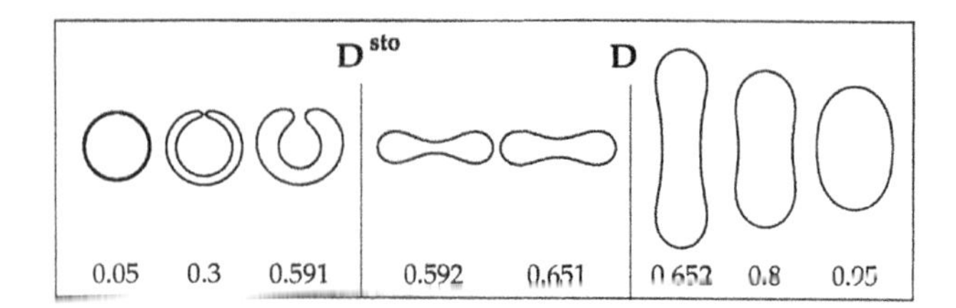

Figure 5.13 Axisymmetric shapes of a vesicle with constant area A and variable volume V as expressed in terms of the dimensionless volume v (bottom row) for spontaneous curvature $\bar{m} = 0$. (Reproduced from Seifert, U. et al., *Phys. Rev. A*, 44, 1182–1202, 1991.)

$$F(\Delta P,\Sigma;\kappa,m;j)\equiv-\Delta PV+\Sigma A+E_{be}(V,A;\kappa,m;j) \tag{5.30}$$

along a certain branch j of stationary shapes and to interpret this expression as the Legendre-transformed energy from the extensive variables V and A to the intensive variables ΔP and Σ. The formal structure of such a Legendre transformation, which plays an important role in thermodynamics, implies (Svetina and Zeks, 1989; Seifert et al., 1991; Miao et al., 1991; Seifert, 1997)

$$\Delta P=\left(\frac{dE_{be}(V,A;\kappa,m;j)}{dV}\right)_A \tag{5.31}$$

and

$$\Sigma=-\left(\frac{dE_{be}(V,A;\kappa,m;j)}{dA}\right)_V. \tag{5.32}$$

When we have several branches of stationary shapes for the same values of V and A, the derivatives on the right hand side of these relations will depend on the branch index j and so will the values of ΔP and Σ, compare Figure 5.12.

The relation Eq. 5.31 implies that the Lagrange multiplier ΔP is the pressure conjugate to the vesicle volume V and can, thus, be identified with the difference

$$\Delta P=P_{in}-P_{ex} \tag{5.33}$$

between the pressures P_{in} and P_{ex} within the interior and exterior compartments. In practise, these pressures are usually osmotic pressures but may also include hydrostatic pressures as imposed by a micropipette. The pressure difference ΔP is usually orders of magnitude smaller than the individual osmotic pressures P_{in} and P_{ex}. The relation Eq. 5.32 implies that the Lagrange multiplier Σ is the tension conjugate to the membrane area A. In fact, as previously mentioned, this tension can be identified with the mechanical tension experienced by the uniform membrane as shown in Appendix 5.D (Lipowsky, 2014a).

When expressed in terms of the dimensionless bending energy $\bar{E}_{be}=E_{be}/(8\pi\kappa)$, the general relations Eqs 5.31 and 5.32 for the pressure difference and the membrane tension can be rewritten in the form

$$\frac{\Delta P}{8\pi\kappa}=\left(\frac{dv}{dV}\right)_A\frac{\partial\bar{E}_{be}}{\partial v}=6\sqrt{\pi}\,\frac{1}{A^{3/2}}\frac{\partial\bar{E}_{be}}{\partial v} \tag{5.34}$$

[8] The 'branches' are really two-dimensional sheets over the $(v,\bar{m})$-plane.

or

$$\frac{\Delta P R_{ve}^3}{\kappa} = 6\frac{\partial \bar{E}_{be}}{\partial v} \quad (5.35)$$

and

$$\frac{\Sigma}{8\pi\kappa} = 9\sqrt{\pi}\frac{V}{A^{5/2}}\frac{\partial \bar{E}_{be}}{\partial v} - \frac{1}{4\sqrt{\pi}}\frac{m}{A^{1/2}}\frac{\partial \bar{E}_{be}}{\partial \bar{m}} \quad (5.36)$$

or

$$\frac{\Sigma R_{ve}^2}{\kappa} = 3v\frac{\partial \bar{E}_{be}}{\partial v} - \bar{m}\frac{\partial \bar{E}_{be}}{\partial \bar{m}}. \quad (5.37)$$

For vanishing spontaneous curvature, $\bar{m} = m = 0$, the second term in Eqs 5.37 and 5.36 vanishes which implies that both ΔP and Σ become proportional to the partial derivative $\partial \bar{E}_{be}/\partial v$. Inspection of Figure 5.12 shows that this derivative is negative along the prolate and oblate branch but close to zero along the stomatocyte branch. Thus, as we reduce the volume of a spherical vesicle with $m = 0$, the pressure difference ΔP and the membrane tension Σ are both negative along the prolate and oblate branches. A negative pressure difference $\Delta P = P_{in} - P_{ex}$ implies that the exterior osmotic pressure exceeds the interior one and that the pressure difference acts to compress the vesicle volume. A negative tension Σ implies that the membrane is slightly compressed compared to its optimal packing density. Along the stomatocyte branch, on the other hand, both the pressure difference and the membrane tension are close to zero.

A combination of the two relations Eqs 5.34 and 5.36 leads to

$$3\Delta PV - 2\Sigma A = 4\sqrt{\pi}\frac{\kappa m}{A^{1/2}}\frac{\partial \bar{E}_{be}}{\partial \bar{m}}, \quad (5.38)$$

independent of the derivative $\partial \bar{E}_{be}/\partial v$ which cancels out from this special combination of ΔP and Σ. In the absence of a spontaneous curvature, we then obtain the simple relation

$$3\Delta PV = 2\Sigma A \quad (m = 0). \quad (5.39)$$

We will see in the next subsection that the same relation also follows from special deformations (or variations) of the stationary shapes as provided by infinitesimal scale transformations.

5.4.5 GLOBAL SHAPE EQUATION

Now, consider a certain stationary shape S^j of the shape functional $\mathcal{F}$ as given by Eq. 5.22. The pressure difference ΔP and the tension Σ then have specific values as obtained from the partial derivatives in Eqs 5.31 and 5.32 along the corresponding branch that includes the chosen shape S^j. Small deformations of this shape can be described by membrane displacements $\vec{u}(\underline{s})$ which define the deformed shape S' via

$$\vec{X}(\underline{s}) \to \vec{X}'(\underline{s}) = \vec{X}(\underline{s}) + \varepsilon\vec{u}(\underline{s}) \quad \text{with} \quad |\varepsilon| \ll 1. \quad (5.40)$$

Because the shape S^j represents a local minimum or saddle point of the shape functional F, we know that

$$\mathcal{F}\{S'\} - \mathcal{F}\{S^j\} = O(\varepsilon^2) \quad \text{or} \quad \frac{d\mathcal{F}\{S'\}}{d\varepsilon}|_{\varepsilon=0} = 0. \quad (5.41)$$

A particular shape deformation is provided by the choice $\vec{u}(\underline{s}) = \vec{X}(\underline{s})$ which leads to the infinitesimal scale transformation

$$\vec{X}(\underline{s}) \to \vec{X}'(\underline{s}) = (1+\varepsilon)\vec{X}(\underline{s}). \quad (5.42)$$

This scale transformation implies that the area A and the volume V are transformed according to $A \to A' = (1 + \varepsilon)^2 A$ and $V \to V' = (1 + \varepsilon)^3 V$. Likewise the integrated mean curvature

$$I_M = \mathcal{I}_M\{S\} \equiv \int dA M \quad (5.43)$$

transforms according to

$$I_M = \mathcal{I}_M\{S\} \to I_M' = \mathcal{I}_M\{S'\} = (1+\varepsilon)I_M \quad (5.44)$$

while the integral $\int dA M^2$ remains unchanged. When applied to the explicit form of the shape functional $\mathcal{F}$, the condition Eq 5.41 leads to

$$-3\Delta PV + 2\hat{\Sigma}A - 4\kappa m I_M = 0 \quad (5.45)$$

with the total membrane tension $\hat{\Sigma} = \Sigma + 2\kappa m^2$ as in Eq. 5.26. For any stationary shape S^j, this equation provides an explicit connection between ΔP, $\hat{\Sigma}$ and the global geometric quantities V, A, and I_M. Therefore, Eq. 5.45 represents a *global* shape equation.

For $m = 0$, the global shape equation reduces to the relation Eq. 5.39. Furthermore, a combination of Eq. 5.45 with Eq. 5.38 leads to the expression

$$\frac{\partial \bar{E}_{be}}{\partial \bar{m}} = 2\bar{m} - \frac{I_M}{\sqrt{\pi A}} \quad (5.46)$$

for the partial derivative of the dimensionless bending energy $\bar{E}_{be}(v, \bar{m})$ with respect to the spontaneous curvature $\bar{m} = mR_{ve}$. Note that the integrated mean curvature I_M depends on the stationary shape S^j and, thus, on the spontaneous curvature $\bar{m}$.

5.4.6 VESICLE SHAPES WITH MEMBRANE NECKS

The numerical solutions of the Euler-Lagrange equations for axisymmetric shapes revealed that these shapes develop narrow membrane necks in certain regions of the parameter space and that these shapes approach limit shapes with closed necks. These necks provide information about the spontaneous curvature m as will be explained in the following subsections, see also Box 5.2 for a summary of necks for vesicle membranes with laterally uniform composition.

Neck closure condition

Let us consider a branch of stationary shapes S^{st} that represent local minima of the bending energy and, thus, solutions of the Euler-Lagrange Eq. 5.23. These shapes are smooth in the sense that the shape variable $\vec{X}(\underline{s})$ is twice differentiable with respect to the surface coordinates and that the mean curvature varies continuously along an arbitrary path on the membrane surface. For any point P on this surface and for any path through this point, we can thus define two mean curvature values, M_{P+} and M_{P-}, which represent the limiting values of the mean curvature as we approach the point P from the "left" and from the "right" along the chosen path. The continuous variation of M then implies that

$$M_P = \frac{1}{2}(M_{P+} + M_{P-}). \quad (5.47)$$

For a smooth surface, we could also use the more general expression $M_P = \zeta M_{P+} + (1-\zeta) M_{P-}$ with $0 \le \zeta \le 1$ corresponding to different weights for the left-sided and the right-sided limit. However, because the assignment of "left" and "right" is completely arbitrary, we want the expression to remain unchanged when we interchange "left" and "right," which implies $\zeta = 1/2$. We now interpret the expression Eq. 5.47 as an interpolation formula and extend it to closed necks, i.e., to points on the membrane surface at which the mean curvature develops a discontinuity. Thus, if the two membrane segments, 1 and 2, adjacent to the closed neck have the mean curvatures M_1 and M_2, we define the effective curvature of the closed neck by

$$M_{ne} \equiv \frac{1}{2}(M_1 + M_2). \quad (5.48)$$

This definition is analogous to the value $H(0) = \frac{1}{2}$ of the Heaviside step function $H(x)$ as obtained from smooth approximations for $H(x)$.

The numerical studies of membrane necks also showed that the neck closure makes no contributions to the bending energy. Because the energy density at the neck is given by

$$\mathcal{E}_{be}(M_{ne}) \equiv 2\kappa[M_{ne} - m]^2, \quad (5.49)$$

we conclude that the neck closes in such a way that

$$M_{ne} = \frac{1}{2}(M_1 + M_2) = m \quad \text{(neck closure)}. \quad (5.50)$$

It follows from this condition that the two membrane segments 1 and 2 have the same bending energy density, i.e., that

$$\mathcal{E}_{be}(M_1) = \mathcal{E}_{be}(M_2). \quad (5.51)$$

In fact, we could also start from the requirement that the bending energy density is continuous across the closed neck which leads to $M_1 - m = \pm(M_2 - m)$. For the root with the plus sign, we obtain the relation $M_1 = M_2$, i.e., a continuous variation of M and, thus, no neck but, for the root with the minus sign, we recover the neck closure condition Eq. 5.50.

The neck closure condition Eq. 5.50 has been confirmed for a large number of axisymmetric shapes as obtained by minimizing the bending energy numerically (Seifert et al., 1991). So far, necks between non-axisymmetric membrane segments have not been studied in a systematic manner but the continuity arguments given above also apply to such non-axisymmetric situations and then lead to the same closure condition.

Neck closure of membrane buds

It is instructive to apply the condition Eq. 5.50 to the neck closure of membrane buds as frequently observed in experiments. Two cases can be distinguished corresponding to in- and out-buds that point towards the interior and exterior compartment, respectively, see Figure 5.14.

First, consider spherical out-buds as shown in Figure 5.14a–c. For such a bud with radius R_2, the bud membrane adjacent to the neck has positive mean curvature $M_2 = 1/R_2$. The 1-segment on the other side of the neck must satisfy $M_1 \ge -M_2$ because the two membrane segments cannot intersect each other. Combining this geometric constraint with the neck closure condition Eq 5.50, we obtain the inequality

$$m = \frac{1}{2}(M_1 + M_2) \ge 0 \quad \text{(neck closure of out-bud)} \quad (5.52)$$

for the spontaneous curvature m. Thus, whenever we observe the neck closure of an out-bud, we can conclude that the spontaneous curvature must be positive or zero. Furthermore, for $m = 0$, neck closure of an out-bud implies $M_1 = -M_2$, i.e., the 1-segment partially engulfs the bud membrane in the vicinity of the neck. Therefore, for a 1-segment with mean curvature $M_1 > -M_2 = -1/R_2$, neck closure of an *out-bud* implies a *positive* spontaneous curvature.

Next, consider spherical in-buds as shown in Figure 5.14d–f. For a spherical in-bud with radius R_2, the bud membrane adjacent to the neck has negative mean curvature $M_2 = -1/R_2$. The 1-segment on the other side of the neck must satisfy $M_1 \le -M_2 = |M_2|$ because the two membrane segments should not intersect each other.

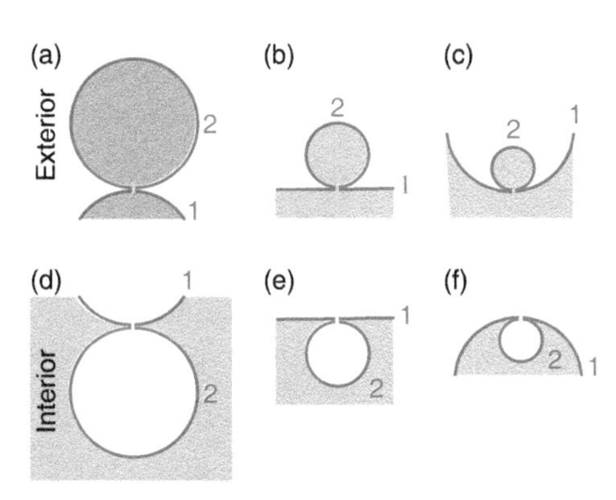

Figure 5.14 (a–c) Out-buds with closed necks, formed as limit shapes by a membrane with *positive* spontaneous curvature: The out-buds are filled with interior medium (gray) and point towards the exterior medium (white). The three membranes (blue) in (a–c) have the same spontaneous curvature $m > 0$ but differ in the mean curvatures of the 1- and 2-segments; (d–f) In-buds with closed necks, formed as limit shapes by a membrane with *negative* spontaneous curvature: The in-buds are filled with exterior medium (white) and point towards the interior medium (gray). The three membranes (blue) in (d–f) have the same spontaneous curvature $m < 0$ but differ in the mean curvatures of the two membrane segments. The two segments have mean curvature $M_1 = m$ and $M_2 = m$ in (a) and (d), $M_1 = 0$ and $M_2 = 2m$ in (b) and (e), and $M_1 = -m$ and $M_2 = 3m$ in (c) and (f).

A combination of the latter inequality with the neck closure condition Eq. 5.50 now leads to the condition

$$m = \frac{1}{2}(M_1 + M_2) \leq 0 \quad \text{(neck closure of in-bud)} \tag{5.53}$$

for the spontaneous curvature m. Thus, whenever we observe the neck closure of an in-bud, we can conclude that the spontaneous curvature must be negative or zero. For $m = 0$, neck closure of an in-bud now implies $M_1 = -M_2 = |M_2|$ as for the limit shape L^{sto} in Figure 5.12. Therefore, for a 1-segment with mean curvature $M_1 < |M_2| = 1/R_2$, neck closure of an *in-bud* implies a *negative* spontaneous curvature.

Stability of closed necks

The neck closure condition Eq. 5.50 applies to limit shapes as obtained from smooth solutions of the local shape Eq. 5.23 or the corresponding set of ordinary differential equations for axisymmetric shapes. One may also consider a closed neck and ask under what conditions this neck is locally stable. This problem has been addressed for axisymmetric vesicles consisting of two almost spherical vesicles that are connected by a narrow neck with radius R_{ne}. More precisely, these vesicle shapes consist of two spherical caps which are connected by two unduloid segments which form a membrane neck of radius R_{ne}. The shapes are parametrized in such a way that one can study the closure of the neck keeping the total membrane area constant. For vanishing neck radius R_{ne}, the shapes approach the two-sphere shapes Θ^{out} and Θ^{in}.[9] The two-sphere shape Θ^{out} consists of a sphere with radius R_1 and mean curvature $M_1 = 1/R_1$ connected, via a closed neck, to a spherical out-bud with radius $R_2 \leq R_1$ and mean curvature $M_2 = 1/R_2$ as in Figure 5.14a. The two-sphere shape Θ^{in} again consists of a sphere of radius R_1 and mean curvature $M_1 = 1/R_1$ but now connected, via a closed membrane neck, to a spherical in-bud with radius $R_2 \leq R_1$ and mean curvature $M_2 = -1/R_2$ as in Figure 5.14f. For small but nonzero R_{ne}, the bending energy of these vesicle shapes can then be expanded in powers of the neck radius R_{ne}.

If the two membrane segments 1 and 2 adjacent to the neck have positive mean curvatures as in Figure 5.14a, the bending energy is found to behave as (Fourcade et al., 1994)

$$E_{be}(R_{ne}) \approx E_{be}(0) - 4\pi\kappa(M_1 - m + M_2 - m)R_{ne} \quad \text{for small } R_{ne}. \tag{5.54}$$

On the other hand, if the 1-segment has positive mean curvature whereas the 2-segment has negative mean curvature as in Figure 5.14f, the bending energy has the asymptotic behavior (Lipowsky, 2014a)

$$E_{be}(R_{ne}) \approx E_{be}(0) + 4\pi\kappa(M_1 - m + M_2 - m)R_{ne} \quad \text{for small } R_{ne} \tag{5.55}$$

with a plus instead of a minus sign in front of the linear term. In both cases, the bending energy $E_{be}(0)$ of the two-sphere shapes Θ^{out} and Θ^{in}, which are characterized by vanishing neck radius $R_{ne} = 0$, does not involve any contribution from the neck itself.

The asymptotic behavior as given by Eq. 5.54 implies that the closed neck in Figure 5.14a, corresponding to an out-bud, is stable provided the average neck curvature M_{ne} satisfies

$$M_{ne} = \frac{1}{2}(M_1 + M_2) \leq m \quad \text{with } M_1 > 0 \text{ and } M_2 > 0 \tag{5.56}$$

but opens up if $M_{ne} > m$. The marginal case with $M_{ne} = m$ corresponds to the neck closure condition Eq. 5.52 with *positive* spontaneous curvature. Therefore, when a membrane with $m > 0$ forms a closed neck with $M_1 > 0$ and $M_2 > 0$ as in Figure 5.14a, this neck remains closed if the effective neck curvature M_{ne} decreases below the spontaneous curvature m.

On the other hand, the small R_{ne}-behavior in Eq. 5.55 implies that the closed neck of the in-bud in Figure 5.14f is stable provided

$$M_{ne} = \frac{1}{2}(M_1 + M_2) \geq m \quad \text{with } M_1 > 0 \text{ and } M_2 < 0 \tag{5.57}$$

but opens up if $M_{ne} < m$. Now, the marginal case with $M_{ne} = m$ corresponds to the neck closure condition Eq. 5.53 with *negative* spontaneous curvature. Therefore, when a membrane with $m < 0$ forms a closed neck with $M_1 > 0$ and $M_2 < 0$ as in Figure 5.14f, this neck remains stable if the effective neck curvature M_{ne} increases above the spontaneous curvature m, i.e., if the absolute value $|M_{ne}|$ of the effective neck curvature decreases below the absolute value $|m|$ of the spontaneous curvature.

The stability of a closed neck must not depend on our choice for the direction of the normal vectors. When we reverse the normal vectors, we change both the sign of the mean curvatures and the sign of the spontaneous curvature. Let us first apply this transformation to the neck configuration in Figure 5.14a which leads to the neck configuration in Figure 5.14d. The corresponding stability relation now becomes

$$M_{ne} = \frac{1}{2}(M_1 + M_2) \geq m \quad \text{with } M_1 < 0, M_2 < 0, \text{ and } m < 0. \tag{5.58}$$

Furthermore, if we reverse the normal vectors of the neck configuration in Figure 5.14f, we obtain the neck configuration in Figure 5.14c and the associated stability relation

$$M_{ne} = \frac{1}{2}(M_1 + M_2) \leq m \quad \text{with } M_1 < 0, M_2 > 0, \text{ and } m > 0. \tag{5.59}$$

In summary, we obtain essentially two different stability relations for the closed necks depicted in Figure 5.14. Closed necks with *non-negative* neck curvature M_{ne} can only exist for non-negative spontaneous curvature $m \geq 0$ and the neck curvature can then attain a value within the interval

$$0 \leq M_{ne} \leq m \quad \text{(out-bud, spontaneous curvature } m \geq 0) \tag{5.60}$$

[9] In the next Section 5.5 we will study such two-sphere vesicles in a systematic manner and distinguish limit shapes from persistent shapes. The two-sphere shapes Θ^{out} then correspond to the limit shapes L^{pea} and $L^{out}_{=}$ as well as to the persistent shapes Φ^{pea}. Likewise, the two-sphere shapes Θ^{in} represent both the limit shapes L^{sto} and $L^{in}_{=}$ as well as the persistent shapes Φ^{sto}.

which includes the neck configurations in Figure 5.14(a–c). The limiting case $M_{ne} = 0$ applies to an out-bud that is partially enclosed by the adjacent 1-segment of the mother vesicle whereas the equality $M_{ne} = m$ corresponds to the neck closure condition of the limit shape. An example for $M_{ne} = 0$ is provided by a discocyte with a membrane neck that connects the discocyte's north pole with mean curvature $M_1 < 0$ to a spherical out bud with mean curvature $M_2 = -M_1 > 0$.

Closed necks with *non-positive* neck curvature M_{ne}, on the other hand, can only exist for non-positive spontaneous curvature $m \leq 0$ and the neck curvature can then have a value within the interval

$$0 \geq M_{ne} \geq m \quad \text{(in-bud, spontaneous curvature } m \leq 0) \tag{5.61}$$

which includes the neck configurations in Figure 5.14(d–f). Now, the neck closure condition $M_{ne} = m$ and the enclosed bud condition $M_{ne} = 0$ provide lower and upper bounds for the range of possible M_{ne}-values.

Mismatch between neck curvature and spontaneous curvature

For a stably closed neck that satisfies the inequalities $M_{ne} < m$ and $m < M_{ne}$ in Eqs 5.60 and 5.61, the bending energy density $\varepsilon_{be} = 2\kappa[M_{ne} - m]^2$ as given by Eq. 5.49 does not vanish. The closed neck may just be considered as a curvature "defect" as discussed in Appendix 5.C.2. In the continuum description used here, this defect is point-like and has vanishing area which implies that its bending energy vanishes as well. The latter property is explicitly borne out in the derivation of the relations Eqs 5.54 and 5.55 because the energies $E_{be}(0)$ obtained for vanishing neck radius $R_{ne} = 0$ do not contain any contribution from the neck.

However, a large mismatch between the neck curvature and the spontaneous curvature as obtained for stable necks with $0 < M_{ne} = m$ and $m = M_{ne} < 0$ does have an important consequence for the morphology of the vesicle. Indeed, a sufficiently large mismatch leads to an effective, curvature-induced constriction force that cleaves the membrane neck and thus leads to membrane fission, see Section 5.5.4 below.

5.4.7 AREA DIFFERENCE ELASTICITY

As mentioned at the beginning of this section, the spontaneous curvature model provides a quantitative description for the morphology of vesicles as long as the membrane curvatures are large compared to the inverse membrane thickness. Thus, highly curved membrane structures such as nanobuds or nanotubes may involve higher order curvature terms as discussed in Appendix 5.E. In addition, the spontaneous curvature model implicitly assumes that the area difference between the two bilayer leaflets can change via fast flip-flops of at least one molecular membrane component. If flip-flops can be ignored on the experimentally relevant time scales, the spontaneous curvature model should be supplemented by an additional energy term as described in this subsection.

Nonlocal energy term for preferred area difference

The bending energy functional Eq. 5.12 represents the area integral over a *local* energy density. In general, the bending of a bilayer membrane consisting of two leaflets may be constrained in a *nonlocal* or *global* manner. Indeed, if the membrane molecules cannot undergo flip-flops between the two leaflets, the number of molecules are fixed within each leaflet and the quenched difference between these two numbers leads to a preferred area difference between these leaflets. This constraint was originally considered by Evans (1974), incorporated into the bilayer-coupling model by (Svetina and Zeks, 1989; Seifert et al., 1991), and generalized in terms of the area-difference-elasticity model (Miao et al., 1994; Döbereiner et al., 1997; Seifert, 1997).

The area difference ΔA between the area of the outer leaflet and the area of the inner leaflet is given by

$$\Delta A = 2 d_{mo} I_M \tag{5.62}$$

with the molecular length scale d_{mo}, which corresponds to the distance between the neutral surfaces of the two monolayers or leaflets, and the integrated mean curvature $I_M = \int dA M$ as in Eq 5.43. The area-difference-elasticity model is defined by the energy functional

$$\mathcal{E}_{ADE}\{S\} = \mathcal{E}_{be}\{S\} + \mathcal{D}_{ADE}\{S\} \tag{5.63}$$

with the local energy functional $\mathcal{E}_{be}\{S\}$ as defined by Eq. 5.12 corresponding to the spontaneous curvature model and the nonlocal area-difference-elasticity term (Miao et al., 1994; Döbereiner et al., 1997)

$$\mathcal{D}_{ADE}\{S\} = \frac{\pi\kappa_\Delta}{2 A d_{mo}^2}(\Delta\mathcal{A}\{S\} - \Delta A_0)^2$$
$$= \frac{2\pi\kappa_\Delta}{A}(\mathcal{I}_M\{S\} - I_{M,0})^2 \tag{5.64}$$

where $\Delta\mathcal{A}\{S\}$ represents the area difference of the vesicle shape S and $\mathcal{I}_M\{S\}$ the integrated mean curvature of this shape. The additional energy term $\mathcal{D}_{ADE}$ introduces two new parameters, the second bending rigidity κ_Δ and the integrated mean curvature $I_{M,0} = \Delta A_0 / 2d_{mo}$, corresponding to optimal molecular areas in both leaflets (Seifert, 1997). These molecular areas are, however, not accessible to current experimental methods and depend on the mechanical membrane tension. If the leaflets of a large spherical vesicle with radius R_{ve} had optimal molecular areas, we would obtain

$$I_{M,0} = \int dA \frac{1}{R_{ve}} = 4\pi R_{ve}. \tag{5.65}$$

Local and nonlocal spontaneous curvature

The stationary shapes with fixed membrane area A and fixed vesicle volume V are now more difficult to calculate because of the nonlocal character of the area-elasticity-difference but can be obtained using a two-step variational procedure, see Appendix 5.E. This procedure shows that all stationary shapes of the area-difference-elasticity model are also stationary shapes of the spontaneous curvature model with the shape functional $\mathcal{F}\{S\}$ as given by Eq. 5.22 and the effective spontaneous curvature (Döbereiner et al., 1997)

$$m_{eff} \equiv m + m_{nlo} \tag{5.66}$$

with the spontaneous curvature m, which is determined locally by the molecular interactions as considered in the previous subsections, and the nonlocal spontaneous curvature

$$m_{\text{nlo}} \equiv \pi \frac{\kappa_\Delta}{\kappa} \frac{I_{M,0} - \mathcal{I}_M\{S^j\}}{A} \tag{5.67}$$

which depends on the stationary shape S^j via the integrated mean curvature $\mathcal{I}_M\{S^j\}$.

As mentioned before, area-difference-elasticity is only relevant if the membrane contains no molecular components that undergo flip-flops on the experimentally relevant time scales. Therefore, as far as the effective spontaneous curvature m_{eff} is concerned, we need to distinguish two cases: (i) For relatively fast flip-flops of some membrane components such as cholesterol, we can ignore the nonlocal spontaneous curvature m_{nlo} which implies that the effective spontaneous curvature m_{eff} becomes equal to the spontaneous curvature m, i.e., the area-difference-elasticity model reduces to the spontaneous curvature model; and (ii) For relatively slow flip-flops of all molecular membrane components, we will, in general, have a nonlocal spontaneous curvature m_{nlo} contributing to the effective spontaneous curvature $m_{\text{eff}} = m + m_{\text{nlo}}$. In order to examine whether this nonlocal spontaneous curvature m_{nlo} is relevant for a given vesicle shape, we need to determine its magnitude and to compare it with the local spontaneous curvature m.

Generalized stability relations for membrane necks

The latter approach can be applied, in particular, to two-sphere shapes with closed membrane necks. The stability of these necks can also be examined for the area-difference-elasticity model using the shape parametrization described in Section 5.4.6. Thus, we again consider axisymmetric shapes with membrane necks, parametrized in such a way that they approach the two-sphere shapes Θ^{out} and Θ^{in} in the limit of small neck radii. As before, the two-sphere shape Θ^{out} consist of a sphere with a spherical out-bud and the two-sphere shape Θ^{in} of a sphere with a spherical in-bud. We now use the energy functional Eq. 5.63 of the area-difference-elasticity model to calculate the elastic energy of the vesicle shapes up to first order in the neck radius R_{ne}. One then finds that closed necks with positive curvature M_{ne} are stable if

$$0 < M_{\text{ne}} \leq m_{\text{eff}} = m + \pi \frac{\kappa_\Delta}{\kappa} \frac{I_{M,0} - I_M\{\Theta^{\text{out}}\}}{A} \tag{5.68}$$

(stable Θ^{out} shapes)

and necks with negative curvature M_{ne} are stable if

$$0 \geq M_{\text{ne}} \geq m_{\text{eff}} = m + \pi \frac{\kappa_\Delta}{\kappa} \frac{I_{M,0} - \mathcal{I}_M\{\Theta^{\text{in}}\}}{A} \tag{5.69}$$

(stable Θ^{in} shapes).

These stability conditions involve three different types of quantities: (i) the neck curvature, a purely geometric quantity that can be directly deduced from the two-sphere shapes; (ii) the local spontaneous curvature m, a material parameter determined by the molecular interactions, and (iii) the non-local spontaneous curvature m_{nlo} that depends both on the geometry of the shape via the integrated mean curvature and on the bending rigidity ratio κ_Δ/κ. In subsection 5.5.3 further below, we will discuss the consequences of the stability conditions Eqs 5.68 and 5.69 for multi-sphere vesicles.

5.5 MULTI-SPHERE SHAPES OF UNIFORM MEMBRANES

In this section, we will consider a variety of multi-sphere shapes for vesicles with uniform membranes, i.e., membranes that have laterally uniform compositions and curvature-elastic properties. This section should be considered as a case study which nicely illustrates the polymorphism and multi-responsive behavior of giant vesicles.

We will focus on multi-component membranes that contain at least one membrane component such as cholesterol that undergoes relatively fast flip-flops. As mentioned, these membranes are appealing from a theoretical point of view because we can study their shapes within the spontaneous curvature model which depends only on two dimensionless parameters, the volume-to-area ratio (or reduced volume) v and the (local) spontaneous curvature $\bar{m}$. These two parameters can be controlled experimentally, e.g., by the osmotic conditions and by the adsorption of small solutes. In addition, three-component membranes with cholesterol have been of particular interest recently because they can form liquid-ordered and liquid-disordered phases. For both types of intramembrane phases, multi-sphere shapes have indeed been observed experimentally (Liu et al., 2016).

We will start with the Euler-Lagrange equations for spherical shapes which reveal the coexistence of two different sphere radii. When combined with the stability relations for the individual spheres and for the closed necks, we obtain multi-sphere vesicles that consist of several spheres with two different radii. We first consider two-sphere shapes and show that these shapes can be found in extended regions of the $(v, \bar{m})$-plane and that these regions are bounded by two types of limit shapes. We also examine the changes of the morphology diagram when area difference elasticity is taken into account. We conclude that these changes are negligible both for large spontaneous curvatures and for small bud sizes.

Multi-sphere shapes consisting of more than two spheres will also be discussed. One interesting example is provided by one sphere with radius R_1 and N spherical buds with radius R_2, all connected by closed necks that have the same neck curvature. For $N > 1$, the morphology diagram exhibits a more complex bifurcation structure with two bifurcation points and three types of limit shapes. The multi-sphere shapes with $N > 1$ buds described in this section are intimately related to the necklace-like tubes with $N > 1$ spherules as considered in the next Section 5.6.

5.5.1 SPHERICAL VESICLES AND SPHERICAL SEGMENTS

We now specify the local shape Eq. 5.23, which represents the Euler-Lagrange equation of the bending energy functional, and the global shape Eq. 5.45, which follows from the invariance of the bending energy under infinitesimal scale transformations, to a spherical membrane segment with constant mean curvature $M = M_{sp}$. It turns out that *both shape equations lead to the same quadratic equation* for M_{sp} as given by

$$\Delta P = P_{in} - P_{ex} = 2\hat{\Sigma} M_{sp} - 4\kappa m M_{sp}^2 \quad (5.70)$$

with the total membrane tension $\hat{\Sigma} = \Sigma + \sigma$. For a symmetric bilayer membrane with $m = 0$, the relation Eq. 5.70 further simplifies and becomes

$$\Delta P = 2\Sigma M_{sp} \ (m = 0) \quad (5.71)$$

which has the same form as the Laplace equation for liquid droplets. The Euler-Lagrange Eq. 5.70 can be derived in a more intuitive manner if one parametrizes the spherical shape by its radius R_{sp} and minimizes the shape energy with respect to R_{sp} (Lipowsky, 2013).

It follows from Eqs 5.70 and 5.71 that each value of $M_{sp} = \pm 1/R_{sp}$ defines a straight M_{sp}-line in the $(\Sigma, \Delta P)$-plane. For $m = 0$, these M_{sp}-lines cover the whole $(\Sigma, \Delta P)$-plane. For $m \neq 0$, on the other hand, the straight M-lines do not cover the whole $(\hat{\Sigma}, \Delta P)$-plane as follows from the solution of the quadratic Eq. 5.70 which has the form

$$M_{1/2} = \frac{\hat{\Sigma}}{4\kappa m} \pm \left[\left(\frac{\hat{\Sigma}}{4\kappa m} \right)^2 - \frac{\Delta P}{4\kappa m} \right]^{1/2}. \quad (5.72)$$

Because the mean curvature must be real-valued, spherical segments are not possible for those values of $\hat{\Sigma}$ and ΔP for which the expression under the square root (or discriminant) becomes negative. Therefore, a certain choice of $\hat{\Sigma}$ and ΔP leads to spherical segments if

$$\Delta P \geq -\frac{\hat{\Sigma}^2}{4\kappa |m|} \quad \text{for} \quad m < 0 \quad (5.73)$$

and if

$$\Delta P \leq \frac{\hat{\Sigma}^2}{4\kappa m} \quad \text{for} \quad m > 0. \quad (5.74)$$

Along the parabolic boundaries $\Delta P = \hat{\Sigma}^2 / (4\kappa m)$ of these regions, we have only one solution as given by

$$M_1 = M_2 = \frac{\hat{\Sigma}}{4\kappa m} = \frac{\Sigma + 2\kappa m^2}{4\kappa m}. \quad (5.75)$$

For all other possible values of $\hat{\Sigma}$ and ΔP, we have two different solutions as in Eq. 5.72 with $M_1 \neq M_2$, corresponding to two different spherical segments. In general, the mean curvatures M_1 and M_2 may be positive or negative depending on the signs of the pressure difference ΔP, the membrane tension Σ, and the spontaneous curvature m.

Coexistence of two spherical segments

The two solutions M_1 and M_2 are characterized by the same values of the pressure difference ΔP and the mechanical tension Σ. Therefore, the two membrane segments can coexist for these values of ΔP and Σ. Vice versa, when we observe the coexistence of two spherical membrane segments with mean curvatures M_1 and M_2, we can use the two Euler-Lagrange equations to conclude that the membrane tension is given by

$$\Sigma = 2\kappa m (M_1 + M_2) - 2\kappa m^2 \quad (5.76)$$

and the pressure difference by

$$\Delta P = 4\kappa m M_1 M_2. \quad (5.77)$$

The coexistence of two spherical shapes is indeed observed when out- and in-buds are formed from larger mother vesicles as shown in Figure 5.2 through Figure 5.6 and discussed in more detail in the next subsection.

On the other hand, the coexistence of more than two spherical segments with pair-wise different mean curvatures M_i and M_j is not possible for a uniform membrane. Indeed, if $N \geq 3$ different types of spherical segments coexisted on the same vesicle, we would have N Euler-Lagrange equations of the form Eq. 5.70. When we now choose a pair of spherical segments with mean curvatures M_i and M_j, we obtain the relations Eqs 5.76 and 5.77 with M_1 and M_2 replaced by M_i and M_j. For fixed i, we can choose $N - 1$ different values for j and obtain $N - 1$ different relations of the form Eqs 5.76 and 5.77. These relations immediately imply that all mean curvatures M_j must be identical. Because we can repeat this procedure for each value of i, we conclude that the shape equations for spherical segments allow only two different values of the mean curvature to coexist for uniform membranes.

Multi-component membranes can lead to the coexistence of several lipid phases and several types of intramembrane domains that differ in their composition, see Section 5.8 below. For two types of domains, the membrane can form coexisting spherical segments with four different mean curvatures. In general, a membrane with K types of domains can form coexisting spherical segments with $2K$ different mean curvatures as follows from the Euler-Lagrange equations for the different membrane domains. This morphological complexity remains to be explored.

Stability of individual spheres

Now, consider a single sphere which experiences the pressure difference $P_{sp} = P_{sp,in} - P_{sp,ex}$ where $P_{sp,in}$ is the osmotic pressure acting within the volume enclosed by the sphere. The second variation of the shape functional shows that a sphere with radius R_{sp} and mean curvature $M = 1/R_{sp}$ is (locally) stable provided this pressure difference P_{sp} satisfies (Ou-Yang and Helfrich, 1989; Seifert et al., 1991; Miao et al., 1991)

$$P_{sp} > P_{sp}^{*+} \equiv \frac{4\kappa}{R_{sp}^3}(m R_{sp} - 3) \quad (M_{sp} = 1/R_{sp}) \quad (5.78)$$

When we reverse the normal vector of the sphere, we change the signs of both the mean curvature M and the spontaneous curvature m. For such an inverted sphere, we obtain the stability condition

$$P_{sp} > P_{sp}^{*-} \equiv \frac{4\kappa}{R_{sp}^3}(-m R_{sp} - 3) \quad (M_{sp} = -1/R_{sp}). \quad (5.79)$$

One example for an inverted sphere in real systems is provided by an in-bud protruding into a giant vesicle which is a possible shape for negative spontaneous curvature $m < 0$. The in-bud with radius $R_{sp} = R_2$ and mean curvature $M_2 = -1/R_2$ is attached to a spherical mother vesicle with radius $R_{sp} = R_1 \geq R_2$ and mean curvature $M_1 = 1/R_1$. In this case, the volume enclosed by the in-bud is a subvolume of the exterior solution. Therefore, the membrane of the in-bud experiences the pressure difference $P_{sp} = -\Delta P$ whereas the membrane of the mother vesicle is exposed to $P_{sp} = \Delta P$.

Because the mother vesicle and the in-bud experience two different pressure differences, the two spherical membrane segments are then governed by two different stability conditions. Indeed, using the stability relations Eqs 5.78 and 5.79 as well as the general expression Eq. 5.77 for ΔP, the spherical shape of the mother vesicle is found to be stable if

$$\Delta P = -\frac{4\kappa m}{R_1 R_2} > \frac{4\kappa}{R_1^3}(mR_1 - 3) \quad \text{or} \quad \frac{m}{R_2} < \frac{3 - mR_1}{R_1^2} \tag{5.80}$$

whereas the stability condition for the spherical in-bud has the form

$$-\Delta P = \frac{4\kappa m}{R_1 R_2} > \frac{4\kappa}{R_2^3}(-mR_2 - 3) \text{ or } \frac{m}{R_1} > \frac{-mR_2 - 3}{R_2^2} \tag{5.81}$$

At the critical pressures $P_{sp} = P_{sp}^{*\pm}$, the spherical shape undergoes a bifurcation which generates the branches of prolate and oblate shapes. For conventional spheres with $M_{sp} > 0$, the prolate shape has the lowest bending energy for small $|m|/M_{sp}$-values whereas the oblate shape represents the lower energy shape for sufficiently large negative values of m/M_{sp}, see the morphology diagram in Figure 5.16 (Seifert et al., 1991).

5.5.2 TWO-SPHERE VESICLES

Giant vesicles frequently form shapes that consist of two spheres connected by a narrow membrane neck. Within the spontaneous curvature model, such shapes arise quite naturally and can be reached by deflation of smoothly curved shapes. Two such limit shapes have been obtained from a systematic numerical study of axisymmetric shapes (Berndl, 1990; Seifert et al., 1991): the limit shapes L^{pea} with a spherical out-bud and the limit shapes L^{sto} with a spherical in-bud. These limit shapes represent two-sphere shapes and have the geometries displayed in Figure 5.15. The limit shapes L^{pea} are reached, for positive spontaneous curvature, by the deflation of pear-like vesicles, the limit shapes L^{sto} for negative spontaneous curvature by the deflation of stomatocytes, see the morphology diagram in Figure 5.16. Inspection of this diagram shows that these limit shapes are found along two lines within the $(v, \bar{m})$-plane.

Closer inspection of this morphology diagram also reveals that the deflation of a spherical vesicle with $v = 1$ and $\bar{m} > 0$ leads to a prolate-pear bifurcation before the limit shape L^{pea} is reached. Because the latter bifurcation is discontinuous and exhibits hysteresis, the experimental observation of the true limit shape will be facilitated if one studies both the deflation and the subsequent inflation of the GUV. Likewise, the deflation of a spherical vesicle with $v = 1$ and $\bar{m} < 0$ leads to an oblate-stomatocyte bifurcation before the limit shape L^{sto} is reached. The latter bifurcation is again discontinuous (Seifert et al., 1991).

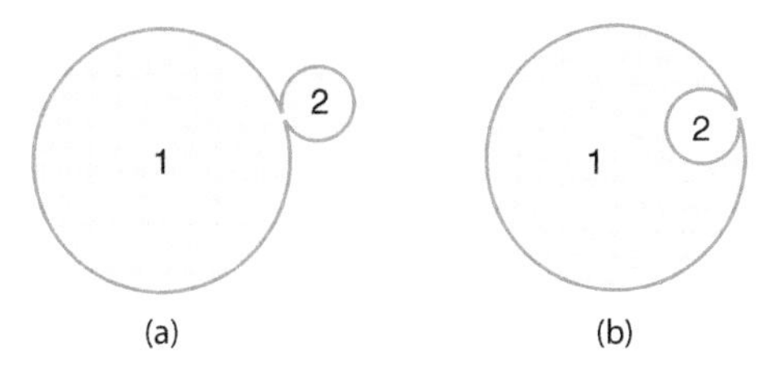

Figure 5.15 Geometry of shapes consisting of two spheres with radii $r_1 = R_1/R_{ve}$ and $r_2 = R_2/R_{ve} \leq r_1$ connected by a closed neck: (a) Two-sphere shape Θ^{out} with an out-bud and *positive* neck curvature $\bar{M}_{ne} = \frac{1}{2}(\frac{1}{r_1} + \frac{1}{r_2}) > 0$ which can only form for positive spontaneous curvature $\bar{m} \geq \sqrt{2}$ and (b) Two-sphere shape Θ^{in} with an in-bud and *non-positive* neck curvature $\bar{M}_{ne} = \frac{1}{2}(\frac{1}{r_1} - \frac{1}{r_2}) \leq 0$ which can only form for non-positive spontaneous curvature $\bar{m} \leq 0$. The stability of the membrane neck in (a) and (b) is governed by Eqs 5.60 and 5.61, respectively. $\bar{M}_{ne} = \bar{m}$, the shape Θ^{out} in (a) represents a limit shape L^{pea} as obtained by neck closure from a stationary pear-like shape of the Euler-Lagrange equation while it represents a persistent shape Φ^{pea} with a stably closed neck for $\bar{M}_{ne} < \bar{m}$. Likewise, the shape Θ^{in} may represent a limit shape L^{sto} as obtained by neck closure from a stationary stomatocyte or a persistent shape Φ^{sto}. The limit shapes are found along certain lines within the $(v, \bar{m})$-plane whereas the persistent shapes are stable within two-dimensional regions of this plane, see the morphology diagrams in Figures 5.16 and 5.17.

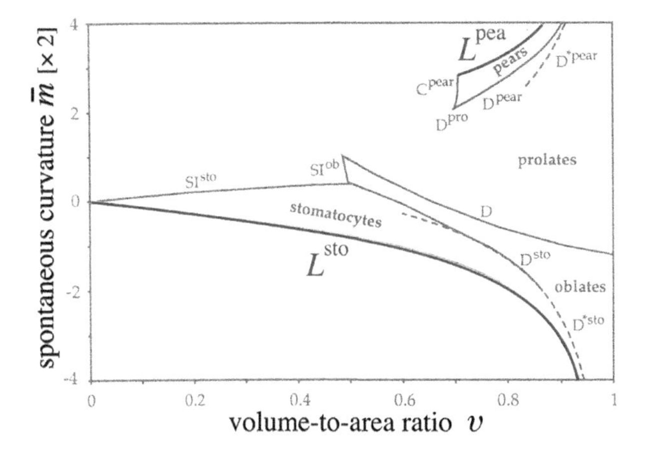

Figure 5.16 Morphology diagram as a function of volume-to-area ratio v and spontaneous curvature $c_0 \equiv 2\bar{m}$ which exhibits two lines of limit shapes. The limit shapes L^{pea} with an out-bud as in Figure 5.15a are found for $\bar{m} \geq \sqrt{2}s$ along the upper line which is truncated at the end point $(v_*^+, \bar{m}_*^+) = (1/\sqrt{2}, \sqrt{2})$ corresponding to two equal spheres. As we move along the L^{pea}-line by increasing the spontaneous curvature $\bar{m}$ and the volume-to-area-ratio v, the out-bud becomes smaller and smaller until the whole membrane area is taken up by the larger sphere. The limit shapes L^{sto} with an out-bud, see Figure 5.15b, are found for $\bar{m} \leq 0$ along the lower line which is truncated at the end point $(v_*^-, \bar{m}_*^-) = (0,0)$ corresponding to two nested spheres of equal size. As we move along the L^{sto}-line by decreasing $\bar{m} < 0$ and increasing v, the in-bud becomes smaller and smaller until the vesicle forms a single sphere with $v = 1$ (Berndl, 1990). (Reproduced from Seifert, U. et al., *Phys. Rev. A*, 44, 1182–1202, 1991; Berndl, K. *Formen Von Vesikeln Diplomarbeit*, Ludwig–Maximilians–Universität München, 1990.)

The following analysis of two-sphere vesicles involves several steps (Lipowsky, 2018b). First, the geometric properties of the two-sphere shapes lead to other types of limit shapes, $L_{=}^{out}$ and

$L^{in}_{=}$, consisting of two identical spheres. Second, the neck closure condition determines the limit shapes L^{pea} and L^{sto}. Finally, we must examine the stability of the two individual spheres in order to find instability lines at which the two-sphere vesicles transform into other types of shapes. We will also emphasize two-sphere vesicles with buds that have zero bending energy and consider the two sphere limit shapes obtained in the presence of area difference elasticity.

Geometric properties

The geometry of any two-sphere vesicle is determined by the radii R_1 and R_2 of the two spheres. In the following, we will consider vesicles with fixed area A and vesicle size $R_{ve} = \sqrt{A/(4\pi)}$ but variable volume V as controlled by the osmotic conditions. We then measure the radii of the two spheres in units of R_{ve} and define the dimensionless radii

$$r_1 \equiv R_1 / R_{ve} \quad \text{and} \quad r_2 \equiv R_2 / R_{ve}. \tag{5.82}$$

These two radii satisfy the implicit equations

$$r_1^2 + r_2^2 = \frac{A}{4\pi R_{ve}^2} = 1 \tag{5.83}$$

and

$$r_1^3 \pm r_2^3 = \frac{V}{\frac{4\pi}{3} R_{ve}^3} = v \tag{5.84}$$

where the plus and minus sign in Eq. 5.84 correspond to two-sphere shapes with an out- and in-bud, respectively. Therefore, the geometry of any two-sphere vesicle is determined by its area A and its volume V and depends only on the volume-to-area ratio v. As in Figure 5.15, we use the notation Θ^{out} and Θ^{in} for two-sphere shapes for which we have not examined the stability of their necks.

For a two-sphere vesicle with an in-bud, the radius r_2 of this bud must satisfy $r_2 \leq r_1$ because the membrane segments of the two spheres should not intersect. For a two-sphere vesicle with an out-bud, the shapes for $r_1 < r_2$ are identical with the shapes for $r_1 > r_2$. In order to avoid this degeneracy, we will impose the restriction $r_2 \leq r_1$ for out-buds as well. Because $r_1^2 + r_2^2 = 1$ as in Eq 5.83, the inequality $r_1 \geq r_2 = \sqrt{1 - r_1^2}$ implies

$$r_1 \geq \frac{1}{\sqrt{2}} \quad \text{and} \quad r_2 \leq \frac{1}{\sqrt{2}}. \tag{5.85}$$

The limiting cases with $r_2 = r_1 = 1/\sqrt{2}$ corresponds to two spheres with the same size and defines two other types of limit shapes, denoted by $L^{out}_{=}$ and $L^{in}_{=}$. The limit shape $L^{out}_{=}$ consists of two equal spheres with positive mean curvature whereas the limit shape $L^{in}_{=}$ consists of two nested spheres which have the same size but opposite mean curvatures. In addition, these limit shapes have the smallest possible volume of two-sphere vesicles as given by

$$\min(v) = v^{out}_{=} \equiv 1/\sqrt{2} \quad \text{for } L^{out}_{=} \tag{5.86}$$

and

$$\min(v) = v^{in}_{=} \equiv 0 \quad \text{for } L^{in}_{=}. \tag{5.87}$$

A related property of these limit shapes is that their neck curvatures have the smallest absolute values. When expressed in terms of the dimensionless neck curvature

$$\bar{M}_{ne} \equiv M_{ne} R_{ve} = \frac{1}{2}\left(\frac{1}{r_1} \pm \frac{1}{r_2}\right), \tag{5.88}$$

these minimal neck curvatures have the values

$$\min(\bar{M}_{ne}) = \sqrt{2} \quad \text{for } L^{out}_{=} \tag{5.89}$$

and

$$\min(|\bar{M}_{ne}|) = \max(\bar{M}_{ne}) = 0 \quad \text{for } L^{in}_{=}. \tag{5.90}$$

Neck closure and neck stability

A necessary prerequisite for a stable two-sphere vesicle is the stability of the closed neck connecting the two spheres. The stability of closed necks was already studied in subsection 5.4.6 where we distinguished neck closure from closed neck conditions. The closure condition has the dimensionless form

$$\bar{M}_{ne} = \frac{1}{2}\left(\frac{1}{r_1} \pm \frac{1}{r_2}\right) = \bar{m} \quad \text{(neck closure)} \tag{5.91}$$

where the plus and minus sign again corresponds to two-sphere shapes with out- and in-buds, respectively. In addition, the closed neck condition is given by

$$0 < \bar{M}_{ne} = \frac{1}{2}\left(\frac{1}{r_1} + \frac{1}{r_2}\right) < \bar{m} \quad \text{for out-buds} \tag{5.92}$$

and by

$$0 > \bar{M}_{ne} = \frac{1}{2}\left(\frac{1}{r_1} - \frac{1}{r_2}\right) > \bar{m} \quad \text{for in-buds.} \tag{5.93}$$

Limit shapes related to neck closure

The combination of the geometric relations Eqs 5.83 and 5.84 with the neck closure condition Eq. 5.91 determines the limit shapes L^{pea} and L^{sto}. When we eliminate the two radii from these three equations, we obtain the functional relationships

$$v = v^{pea}(\bar{m}) \quad \text{for the line of } L^{pea} \text{ shapes} \tag{5.94}$$

and

$$v = v^{sto}(\bar{m}) \quad \text{for the line of } L^{sto} \text{ shapes.} \tag{5.95}$$

The function $v^{pea}(\bar{m})$ has the explicit form (Seifert et al., 1991)

$$v = v^{\text{pea}}(\bar{m}) \equiv -\frac{1}{4\bar{m}^3} + \left(1 - \frac{1}{2\bar{m}^2}\right)\sqrt{1 + \frac{1}{4\bar{m}^2}} \quad \text{for} \quad \bar{m} \geq \sqrt{2}, \tag{5.96}$$

which behaves as

$$v^{\text{pea}}(\bar{m}) \approx 1 - \frac{3}{8\bar{m}^2} \quad \text{for large } \bar{m}. \tag{5.97}$$

The function $v^{\text{sto}}(\bar{m})$ has the same $\bar{m}$-dependence as $v^{\text{pea}}(\bar{m})$ but applies to $\bar{m} \leq 0$.

Both lines of limit shapes L^{pea} and L^{sto} extend down to the smallest possible volumes which they reach when both spheres have the same size. The corresponding values of the spontaneous curvature $\bar{m}$ are given by

$$\min(\bar{m}) = \bar{m}_{=}^{\text{out}} \equiv \sqrt{2} \quad \text{for } L^{\text{pea}} \tag{5.98}$$

and by

$$\max(\bar{m}) = \bar{m}_{=}^{\text{in}} \equiv 0 \quad \text{for } L^{\text{sto}}, \tag{5.99}$$

see the L^{pea} and L^{sto} lines in Figure 5.16. In the following subsection, the region of the morphology diagram with $\bar{m} > 0$, which contains the limit shapes L^{pea} and $L_{=}^{\text{out}}$, will be discussed in more detail.

Buds with zero bending energy

It is useful to distinguish another special case of budded vesicle shapes, denoted by Z^{out} and Z^{in}. The spherical buds of these shapes have radius $r_2 = 1/|\bar{m}|$ and thus zero bending energy. Because of the inequality $r_2 \leq 1/\sqrt{2}$ as in Eq. 5.85, we then have

$$r_2 = \frac{1}{|\bar{m}|} \leq \frac{1}{\sqrt{2}} \quad \text{and} \quad r_1 = \sqrt{1 - \frac{1}{\bar{m}^2}} \geq \frac{1}{\sqrt{2}}. \tag{5.100}$$

Both relations lead to the same inequality $|\bar{m}| \geq \sqrt{2}$ which implies

$$\bar{m} \geq \sqrt{2} \quad \text{for out-buds} \tag{5.101}$$

and

$$\bar{m} \leq -\sqrt{2} \quad \text{for in-buds.} \tag{5.102}$$

For these vesicles, the buds have vanishing bending energy and the whole bending energy is provided by the bending energy of the mother vesicle with radius r_1. The neck mean curvature is then given by

$$\bar{M}_{\text{ne}} = \bar{M}_{\text{ne}}^0(\bar{m}) \equiv \frac{1}{2}\left(\frac{1}{\sqrt{1 - 1/\bar{m}^2}} + \bar{m}\right) \tag{5.103}$$

which satisfies

$$\bar{M}_{\text{ne}}^0(\bar{m}) \leq \bar{m} \quad \text{for out-buds with } \bar{m} \geq \sqrt{2} \tag{5.104}$$

and

$$\bar{M}_{\text{ne}}^0(\bar{m}) \geq \bar{m} \quad \text{for in-buds with } \bar{m} \leq -\sqrt{2}. \tag{5.105}$$

Comparison with the stability relations as given by Eqs 5.92 and 5.93 then shows that the closed necks between the zero-energy buds and the mother vesicles are stable for both positive and negative spontaneous curvatures. For out-buds, the equality $\bar{M}_{\text{ne}} = \bar{M}_{\text{ne}}^0(\bar{m}) = \bar{m}$ describes the neck closure condition and applies to $r_1 = r_2 = 1/\bar{m}$, i.e., to the case of two identical spheres. This special morphology represents a limit shape for which the whole bending energy vanishes.

The volume of the two-sphere vesicles Z^{out} and Z^{in} with zero-energy buds is given by

$$v = v^{\text{zeb}} \equiv \left(1 - \frac{1}{\bar{m}^2}\right)^{3/2} \pm \frac{1}{\bar{m}^3} \tag{5.106}$$

where the plus and minus sign applies to out- and in-buds, respectively. This volume behaves as

$$v^{\text{zeb}} \approx 1 - \frac{3}{2\bar{m}^2} \quad \text{for large } |\bar{m} \tag{5.107}$$

which applies to both out-buds with $\bar{m} > 0$ and in-buds with $\bar{m} < 0$.

Morphology diagram for positive spontaneous curvature

As displayed in Figure 5.17, the morphology diagram for positive spontaneous curvature contains two lines of limit shapes, L^{pea} and $L_{=}^{\text{out}}$, that have a common end point at

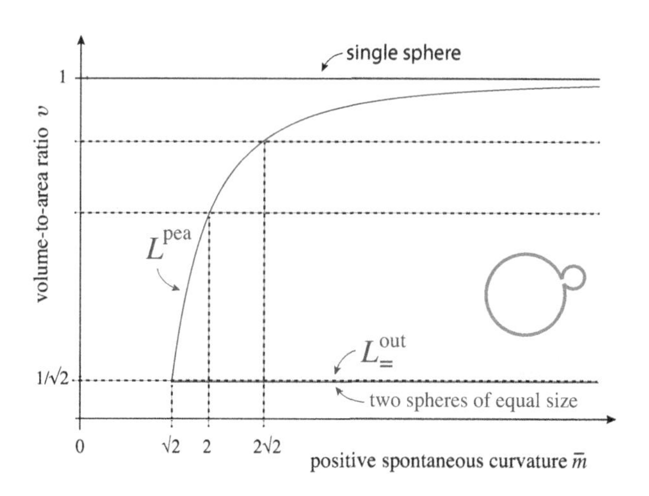

Figure 5.17 Morphology diagram for two-sphere vesicles with an out-bud (inset) and positive spontaneous curvature: Such vesicles have *positive* neck curvature and can be formed for spontaneous curvature $\bar{m} \geq \sqrt{2}$ as well as reduced volume v in the interval $1/\sqrt{2} \leq v \leq v^{\text{pea}}(\bar{m})$ corresponding to the shaded (yellow) region. The lower boundary of this region (horizontal line) is provided by the limit shapes $L_{=}^{\text{out}}$ that consist of two identical spheres and have the volume $v_{=}^{\text{out}} = 1/\sqrt{2}$, the upper boundary (curved line) by the limit shapes L^{pea} as described by $v = v^{\text{pea}}(\bar{m})$ in Eq. 5.96. The two boundary lines have a common end point at $(\bar{m}_{=}^{\text{out}}, v_{=}^{\text{out}}) = (\sqrt{2}, 1/\sqrt{2})$. When a limit shape L^{pea} is deflated for constant spontaneous curvature $\bar{m} > \sqrt{2}$, the larger sphere shrinks whereas the smaller sphere (or out-bud) grows transforming the limit shape L^{pea} into a persistent shape Φ^{pea} with neck curvature $\bar{M}_{\text{ne}} < \bar{m}$. The closed neck persists during further deflation until the lower limit shape $L_{=}^{\text{out}}$ with two identical spheres is reached. All two-sphere vesicles with the same volume v (broken horizontal lines) have the same neck curvature and the same shape but differ in their bending energy, see text. The upper broken line corresponds to the two-sphere geometry with $v = 0.941$ and $\bar{M}_{\text{ne}} = 2\sqrt{2}$, the intermediate broken line to $v = 0.871$ and $\bar{M}_{\text{ne}} = 2$.

$(\bar{m},v) = (\bar{m}_{=}^{\text{out}}, v_{=}^{\text{out}}) = (\sqrt{2}, 1/\sqrt{2})$. Thus, the limit shapes L^{pea} are located at

$$v = v^{\text{pea}}(\bar{m}) \quad \text{and} \quad \bar{m} \geq \bar{m}_{=}^{\text{out}} = \sqrt{2} \tag{5.108}$$

while the $L_{=}^{\text{out}}$ shapes are located at

$$v = v_{=}^{\text{out}} = 1/\sqrt{2} \quad \text{and} \quad \bar{m} \geq \bar{m}_{=}^{\text{out}} = \sqrt{2}. \tag{5.109}$$

Note that all $L_{=}^{\text{out}}$ shapes have the same geometry but differ in their bending energy which has the $\bar{m}$-dependent form

$$\bar{E}_{\text{be}}\{L_{=}^{\text{out}}\} = \frac{E_{\text{be}}\{L_{=}^{\text{out}}\}}{8\pi\kappa} = 2(1 - r_1\bar{m})^2 \quad \text{with } r_1 = 1/\sqrt{2}. \tag{5.110}$$

which vanishes for $\bar{m} = \bar{m}_{=}^{\text{out}} = \sqrt{2}$ and increases as $\sim \bar{m}^2$ for large $\bar{m}$.

Inspection of Figure 5.17 shows that the two lines of limit shapes enclose an extended region of two-sphere shapes, Φ^{pea}, with stably closed necks. This region can be entered by deflation of the L^{pea} shapes, by inflation of the $L_{=}^{\text{out}}$ shapes, and by increasing the spontaneous curvature of the L^{pea} shapes. All Φ^{pea} shapes that are produced by one of these processes are persistent in the sense that their necks remain stably closed during both deflation and inflation as well as under small changes of the spontaneous curvature.

Stability of individual spheres

A second requirement for the stability of two-sphere vesicles is the shape stability of both spheres. Thus, in order to examine the stability of the individual spheres, we now use the stability criterion Eq. 5.78 together with the pressure difference $P_{\text{sp}} = \Delta P$ and ΔP as given by Eq. 5.77. We then conclude that the spherical mother vesicle with radius R_1 is stable if

$$\Delta P = \frac{4\kappa m}{R_1 R_2} > \frac{4\kappa}{R_1^3}(mR_1 - 3) \quad \text{or} \quad \frac{\bar{m}}{r_2} > \frac{\bar{m}r_1 - 3}{r_1^2} \tag{5.111}$$

whereas the out-bud with radius R_2 is stable if

$$\Delta P = \frac{4\kappa m}{R_1 R_2} > \frac{4\kappa}{R_2^3}(mR_2 - 3) \quad \text{or} \quad \frac{\bar{m}}{r_1} > \frac{\bar{m}r_2 - 3}{r_2^2}. \tag{5.112}$$

Because the two radii $r_1 = R_1/R_{\text{ve}}$ and $r_2 = R_2/R_{\text{ve}}$ satisfy the geometric relation $r_1^2 + r_2^2 = 1$, we can express both stability relations in terms of a single radius, say r_2. One then finds that both individual spheres are stable for all limit shapes L^{pea} and $L_{=}^{\text{out}}$ as well as for the shapes Z^{out} with zero-energy buds. Furthermore, the larger sphere of the intermediate persistent shapes Φ^{pea} is always stable whereas the spherical out-bud may become unstable for sufficiently large values of the spontaneous curvature and a certain range of v-values. More precisely, the spherical out-bud with radius r_2 is stable if

$$r_2 - \frac{r_2^2}{\sqrt{1 - r_2^2}} < \frac{3}{\bar{m}} \tag{5.113}$$

and unstable if

$$r_2 - \frac{r_2^2}{\sqrt{1 - r_2^2}} > \frac{3}{\bar{m}}. \tag{5.114}$$

Therefore, the instability line between the stable and unstable out-buds follows from the solutions of the equation

$$r_2 - \frac{r_2^2}{\sqrt{1 - r_2^2}} = \frac{3}{\bar{m}}. \tag{5.115}$$

This equation has no solution for $\bar{m} < \bar{m}_{\text{ss}} = 13.29$, one solution for $\bar{m} = \bar{m}_{\text{ss}}$ and two solutions for $\bar{m} > \bar{m}_{\text{ss}}$.

Therefore, the out-buds of the persistent shapes Φ^{pea} are stable for $\bar{m} < \bar{m}_{\text{ss}}$ but become unstable for $\bar{m} \geq \bar{m}_{\text{ss}}$ and a certain $\bar{m}$-dependent range of v-values. At $\bar{m} = \bar{m}_{\text{ss}}$, the instability consists of the single point $(\bar{m}_{\text{ss}}, v_{\text{ss}}) = (13.29, 0.8259)$ which opens up into a parabola-like curve for $\bar{m} > \bar{m}_{\text{ss}}$. For large $\bar{m}$, the upper and lower branches of the parabola-like curve approach the Z^{out} line and the $L_{=}^{\text{out}}$ line, respectively. Because $\bar{m}_{\text{ss}} = 13.29$, this bifurcation structure is located outside of the $(\bar{m},v)$-region displayed in Figure 5.17.

Thus, we conclude that two-sphere vesicles with out-buds can be found in a large region of the morphology diagram for $\bar{m} > 0$. In particular, when we deflate a limit shape L^{pea} for $\sqrt{2} < \bar{m} < \bar{m}_{\text{ss}} \simeq 13.29$, we obtain a family of stable persistent shapes Φ^{pea} with decreasing neck curvatures $\bar{M}_{\text{ne}}$ until we reach the limit shape $L_{=}^{\text{out}}$ with the smallest possible neck curvature $\bar{M}_{\text{ne}} = \bar{m}_*^+ = \sqrt{2}$. Further deflation of the limit shape $L_{=}^{\text{out}}$ leads back to a dumbbell-like shape with an open neck.

5.5.3 MODIFICATIONS BY AREA DIFFERENCE ELASTICITY

So far, two-sphere vesicles have been discussed in the context of the spontaneous curvature model which depends on the locally generated spontaneous curvature m and assumes that one molecular component of the bilayer membrane can undergo frequent flip-flops between the two bilayer leaflets. It is instructive to see how the morphology diagram is changed when we consider bilayer membranes with slow flip-flops between the leaflets. In the latter situation, the area difference ΔA between the two leaflets is constrained as described by the nonlocal energy term in the area-difference-elasticity model, see the nonlocal expression in Eq. 5.64 that contributes to the energy functional Eq. 5.63 of this model.

As explained in Section 4.7.1, the shapes that minimize this energy functional also minimize the energy functional of the spontaneous curvature model as in Eq. 5.12, provided we use the effective spontaneous curvature $m_{\text{eff}} \equiv m + m_{\text{nlo}}$ as given by Eq. 5.66 which represents the sum of the local spontaneous curvature m and the nonlocal spontaneous curvature

$$m_{\text{nlo}} \equiv \pi \frac{\kappa_\Delta}{\kappa} \frac{I_{M,0} - \mathcal{I}_M\{S\}}{A}$$

as in Eq. 5.67. If the leaflets of a sphere with radius R_{ve} have optimal molecular areas, one has $I_{M,0} = 4\pi R_{ve}$ and the geometric factor of the nonlocal spontaneous curvature becomes

$$\frac{I_{M,0} - \mathcal{I}_M\{S\}}{A} = \frac{4\pi R_{ve} - \mathcal{I}_M\{S\}}{4\pi R_{ve}^2}. \quad (5.116)$$

Now, consider again the two-sphere vesicles Θ^{out} and Θ^{in} with radii R_1 and R_2 connected by a closed membrane neck as shown in Figure 5.15. The integrated mean curvature $\mathcal{I}_M$ of these shapes is given by

$$\mathcal{I}_M\{\Theta^{out}\} = 4\pi(R_1 + R_2) \quad \text{and} \quad \mathcal{I}_M\{\Theta^{in}\} = 4\pi(R_1 - R_2) \quad (5.117)$$

which leads to the geometric factors

$$\frac{4\pi R_{ve} - \mathcal{I}_M\{S\}}{4\pi R_{ve}^2} = \frac{1}{R_{ve}}(1 - r_1 \mp r_2) \quad (5.118)$$

and to the nonlocal spontaneous curvatures

$$\bar{m}_{nlo} = m_{nlo} R_{ve} = \pi \frac{\kappa_\Delta}{\kappa}(1 - r_1 \mp r_2) \quad (5.119)$$

where the minus and plus sign applies to out- and in-buds, respectively. The nonlocal spontaneous curvature involves the geometric factor

$$1 - r_1 \mp r_2 = 1 - r_1 \mp \sqrt{1 - r_1^2} \quad (5.120)$$

where we used the area relation $r_1^2 + r_2^2 = 1$. For the shape Θ^{out} with an out-bud, this expression is negative and bounded by

$$1 - \sqrt{2} \le 1 - r_1 - \sqrt{1 - r_1^2} \le 0 \quad \text{for } 0 \le r_1 \le 1 \text{ (out-bud)}. \quad (5.121)$$

For the shape Θ^{in} with an in-bud, on the other hand, the corresponding expression is positive and satisfies the bounds

$$0 \le 1 - r_1 + \sqrt{1 - r_1^2} \le \sqrt{2} + 1 \quad \text{for } 0 \le r_1 \le 1 \text{ (in-bud)}. \quad (5.122)$$

Therefore, the absolute value of the nonlocal spontaneous curvature satisfies the bounds

$$|\bar{m}_{nlo}| \le \pi(\sqrt{2} - 1)\frac{\kappa_\Delta}{\kappa} \quad \text{for } \Theta^{out} \quad (5.123)$$

and

$$|\bar{m}_{nlo}| \le \pi(\sqrt{2} + 1)\frac{\kappa_\Delta}{\kappa} \quad \text{for } \Theta^{in}. \quad (5.124)$$

These bounds can be used to estimate the relative magnitude of the nonlocal and local contributions to the spontaneous curvature, see further below.

When we include area-difference-elasticity, the stability conditions for the closed neck are given by Eqs 5.68 and 5.69 which imply the neck closure condition

$$\bar{M}_{ne} = \frac{1}{2}\left(\frac{1}{r_1} \pm \frac{1}{r_2}\right) = \bar{m} + \bar{m}_{nlo} = \bar{m} + \pi\frac{\kappa_\Delta}{\kappa}(1 - r_1 \mp r_2) \quad (5.125)$$

where the last equality follows from Eq. 5.119. In order to determine the location of the limit shapes L^{pea} and L^{sto} in the $(v, \bar{m})$-plane, we must now combine the neck closure relation Eq 5.125 with the geometric relations $r_1^2 + r_2^2 = 1$ and $r_1^3 \pm r_2^3 = v$. In general, the κ_Δ-term will shift the L^{pea}- and L^{sto}-lines in the $(v, \bar{m})$-plane, a shift that can be easily calculated for any value of κ_Δ/κ. For positive spontaneous curvature, for example, one then finds that the lines of limit shapes L^{pea} are shifted towards higher $\bar{m}$-values as we increase the rigidity ratio κ_Δ/κ. Furthermore, when we describe the shifted L^{pea} lines by $\bar{m}^{pea} = f(v)$. the function $f(v)$ develops a minimum for $\kappa_\Delta/\kappa > 1$.

In addition, we can draw some general conclusions about the morphology diagram when we include the area-difference-elasticity term proportional to κ_Δ. First, the limit shapes $L_=^{out}$ and $L_=^{in}$, consisting of two spheres with the same radius, are again located at $v = v_=^{out} = 1/\sqrt{2}$ for $\bar{m} > 0$ and at $v = v_=^{in} = 0$ for $\bar{m} < 0$ as follows from the two geometric relations alone. Therefore, the morphology diagram in the $(v, \bar{m})$-plane will always contain extended regions with (meta)stable two-sphere shapes as in Figure 5.17, irrespective of the value of κ_Δ/κ.

Second, we can conclude from the neck closure condition in Eq. 5.125 and from the bounds provided by Eqs 5.123 and 5.124 that the nonlocal contributions $\bar{m}_{nlo}$ arising from area difference elasticity can be neglected for sufficiently large local contributions $\bar{m}$. More precisely, we obtain from Eqs 5.125 and 5.123 that the nonlocal spontaneous curvature can be ignored for the shape Θ^{out} if the local spontaneous curvature is sufficiently large and positive with

$$\bar{m} \gg \pi(\sqrt{2} - 1)\frac{\kappa_\Delta}{\kappa} \quad \text{(out-bud)}. \quad (5.126)$$

Likewise, combining Eq. 5.125 with Eq. 5.124, we conclude that the nonlocal contribution can be ignored for the shape Θ^{in} if the local spontaneous curvature is large and negative with

$$\bar{m} \ll -\pi(\sqrt{2} + 1)\frac{\kappa_\Delta}{\kappa} \quad \text{(in-bud)}. \quad (5.127)$$

The ratio κ_Δ/κ of the bending rigidities is expected to be of the order of one (Döbereiner et al., 1997). Therefore, both for out- and for in-buds, the nonlocal contribution can be ignored compared to the local one if $|\bar{m}| \gg 1$ or $|m| \gg 1/R_{ve}$.

Finally, assume that we were able to measure the radii r_1 and r_2 of a vesicle during neck closure. We can then use the neck closure condition in Eq. 5.125 to estimate the local spontaneous curvature $\bar{m}$ via

$$\bar{m} = \frac{1}{2}\left(\frac{1}{r_1} \pm \frac{1}{r_2}\right) + \pi\frac{\kappa_\Delta}{\kappa}(r_1 \pm r_2 - 1) \quad (5.128)$$

where the plus and minus sign applies to an out- and in-bud, respectively. For small bud radius r_2, the radius $r_1 = \sqrt{1-r_2^2} \approx 1-r_2^2$. When we use this asymptotic equality in Eq. 5.128, we obtain the local spontaneous curvature which implies

$$\bar{m} \approx \frac{1}{2}\left(\frac{1}{r_1} \pm \frac{1}{r_2}\right) \pm \pi \frac{\kappa_\Delta}{\kappa} r_2 \quad \text{for small buds with } r_2 \ll 1. \quad (5.129)$$

The asymptotic behavior as given by Eq. 5.129 implies that the κ_Δ-term can also be ignored for sufficiently small buds. This behavior for small buds is consistent with the behavior for large spontaneous curvatures $\bar{m}$ because large $\bar{m}$ implies limit shapes with small buds.

The influence of area difference elasticity on two-sphere vesicles has been recently studied for giant vesicles that contained lipids with photoresponsive F-Azo groups and underwent light-induced budding (Georgiev et al., 2018). A theoretical analysis of the experimental data based on Eq. 5.128 showed that the spontaneous curvature can indeed be decomposed into a local and a nonlocal contribution, that all vesicles were governed by the same rigidity ratio κ_Δ/κ, and that the local spontaneous curvature $m = \bar{m}R_{ve}$ was about 1/(2.5 μm).

5.5.4 EFFECTIVE CONSTRICTION FORCES AND CLEAVAGE OF MEMBRANE NECKS

As explained in the previous subsections, the persistent shapes Φ^{pea} have the same geometry, for a given volume v, as the limit shapes L^{pea} but an increased spontaneous curvature $\bar{m}$ compared to the spontaneous curvature of L^{pea}. When expressed in terms of dimensionful variables, the spontaneous curvature m then satisfies the stability condition $m > M_{ne} = \frac{1}{2}(M_1 + M_2)$ for the closed necks of out-buds as in Eqs 5.60 and 5.92. Now, consider an explicit constriction force f that acts on the neck radius R_{ne}, which we take into account by adding the term fR_{ne} to the bending energy in Eq. 5.54.[10] We then obtain the generalized condition

$$f - 4\pi\kappa(M_1 + M_2 - 2m) > 0 \quad (5.130)$$

for a closed membrane neck which may be rewritten in the form

$$f + f_{eff}^{out} > 0 \quad (5.131)$$

with the effective constriction force

$$f_{eff}^{out} \equiv 4\pi\kappa(2m - M_1 - M_2) \geq 0 \quad \text{(out-buds with } m > 0\text{)}. \quad (5.132)$$

This constriction force vanishes when the neck satisfies the neck closure condition $M_1 + M_2 = 2m$.

Now, let us consider a persistent shape Φ^{pea} close to the line of limit shapes $L_=^{out}$ which consist of two identical spheres. These persistent shapes have a volume $v \gtrsim 1/\sqrt{2}$ and are characterized by two spheres with small mean curvatures M_1 and M_2, both of which are of the order of $\sqrt{2}/R_{ve}$. Furthermore, the individual spheres of these persistent shapes are stable up to fairly high m-values because the individual spheres of the limit shapes $L_=^{out}$ are stable for all values of m. If the spontaneous curvature m is large compared to both M_1 and M_2, the expression for the curvature-induced constriction force as given by Eq. 5.132 simplifies and becomes asymptotically equal to

$$f_{eff}^{out} \approx f_m^{out} \quad \text{with} \quad f_m^{out} \equiv 8\pi\kappa m \quad \text{for } 2m \gg M_1 + M_2, \quad (5.133)$$

where f_m^{out} represents the curvature-induced constriction force. Thus, for the bending rigidities $\kappa = 10^{-19}$ J and $\kappa = 4 \times 10^{-19}$ J, the spontaneous curvature $m = 1/(100\text{nm})$ generates the constriction forces $f_m^{out} \simeq 25$ pN and $f_m^{out} \simeq 100$ pN, respectively.

In the absence of flip-flops between the bilayer leaflets, we should include the effects of area-difference-elasticity as discussed in the previous subsection. In this case, the effective constriction force has the form

$$f_{eff}^{out} \equiv 4\pi\kappa(2m + 2m_{nlo}^{out} - M_1 - M_2) \geq 0 \quad (5.134)$$

with the nonlocal spontaneous curvature

$$m_{nlo}^{out} = \pi \frac{\kappa_\Delta}{\kappa} \frac{I_{M,0} - \mathcal{I}_M\{\Theta^{out}\}}{A} = \pi \frac{\kappa_\Delta}{\kappa} \frac{1 - r_1 - r_2}{R_{ve}} \quad (5.135)$$

as in Eq. 5.119. This term is negative, see Eq. 5.121, which implies that area-difference-elasticity acts to weaken the curvature-induced constriction forces for out-buds.

In-buds with closed necks are formed for negative spontaneous curvatures. In the latter case, we obtain the effective constriction force

$$f_{eff}^{in} \equiv 4\pi\kappa(M_1 + M_2 - 2m) > 0 \quad \text{(in-buds with } m < 0\text{)} \quad (5.136)$$

which behaves as $f_{eff}^{in} \approx f_m^{in}$ with the curvature-induced constriction force

$$f_m^{in} \equiv -8\pi\kappa m \quad \text{for } 2m \ll M_1 + M_2 < 0. \quad (5.137)$$

In the absence of molecular flip-flops between the bilayer leaflets, the effective constriction force is

$$f_{eff}^{in} \equiv 4\pi\kappa(M_1 + M_2 - 2m - 2m_{nlo}^{in}) \quad (5.138)$$

with the nonlocal spontaneous curvature

$$m_{nlo}^{in} = \pi \frac{\kappa_\Delta}{\kappa} \frac{I_{M,0} - \mathcal{I}_M\{\Theta^{in}\}}{A} = \pi \frac{\kappa_\Delta}{\kappa} \frac{1 - r_1 + r_2}{R_{ve}} \quad (5.139)$$

as in Eq. 5.119. This term is positive, see Eq. 5.122, which implies that area-difference-elasticity also acts to weaken the effective constriction forces for in-buds.

In the curvature models, a closed membrane neck is described by a point-like discontinuity of the membrane curvature. Because of the finite membrane thickness ℓ_{me}, the radius R_{ne} of the membrane neck is necessarily restricted to $R_{ne} \gtrsim \ell_{me}$. Therefore, strictly speaking, the above derivation of the effective constriction forces f_{eff}^{out} and f_{eff}^{in} implicitly assumed that $R_{ne} \gtrsim \ell_{me}$. However, we will now argue that these constriction forces may also be used to obtain a simple criterion for the cleavage of the membrane neck.

[10] The same approach has been used for the endocytosis and exocytosis of nanoparticles in (Agudo-Canalejo and Lipowsky, 2016).

Neck cleavage represents a topological transformation from a budded vesicle that has the same topology as a single sphere to a cleaved state with the topology of two spheres. The free energy difference between the budded and the cleaved state involves a contribution from the Gaussian curvature modulus κ_G, see Section 5.10 at the end of this chapter. Furthermore, this free energy difference depends strongly on the magnitude of the spontaneous curvature. For large values of $|m|$, the fission process is exergonic and reduces the free energy of the vesicle as explained in Section 5.10.3. Therefore, in the presence of a large spontaneous curvature, thermodynamics allows fission to occur spontaneously, i.e., without any free energy input from a chemical reaction such as ATP hydrolysis. How fast this exergonic process occurs depends, however, on the free energy barrier between the budded and the cleaved state of the vesicle membrane.

In order to cleave the membrane neck, we have to create two bilayer edges. For a neck with radius R_{ne}, these two bilayer edges have the combined length $4\pi R_{ne}$. The associated edge energy E_{ed} depends on the edge tension λ_{ed} and has the form

$$E_{ed} = 4\pi R_{ne}\lambda_{ed} \quad \text{with } R_{ne} \gtrsim \ell_{me} \tag{5.140}$$

where the latter inequality reminds us that the neck radius should exceed the membrane thickness ℓ_{me}. The edge energy provides a simple estimate for the free energy barrier between the budded and the cleaved state of the vesicle membrane. This barrier has to be overcome by the mechanical work $f_m R_{ne}$ expended by the curvature-induced constriction force $f_m = f_m^{out}$ or f_m^{in} from $R_{ne} = \ell_{me}$ to $R_{ne} = 0$. Therefore, we obtain the cleavage criterion $f_m R_{ne} \gg E_{ed}$ which is equivalent to

$$|m| \gg |m^{cl}| \equiv \frac{\lambda_{ed}}{2\kappa} \quad \text{for large } |m|. \tag{5.141}$$

This criterion predicts that the membrane neck is cleaved and undergoes fission if the absolute value $|m|$ of the spontaneous curvature is sufficiently large and exceeds the threshold value $|m^{cl}| = \lambda_{ed}/(2\kappa)$.

The main contribution to the edge tension λ_{ed} comes from the interface between the hydrophobic core of the bilayer and the aqueous solution. The corresponding interfacial tension Σ_{hc} may be reduced by a rearrangement of the head groups along the bilayer edge or by the adsorption of edge-active molecules. For an interfacial tension $\Sigma_{hc} \gtrsim 1$ mN/m and a thickness $\ell_{hc} \simeq 2$ nm of the hydrophobic core, we obtain the estimate $\lambda_{ed} = \Sigma_{hc}\ell_{hc} \gtrsim 2$ pN. Using the typical bending rigidity $\kappa = 10^{-19}$ J, neck cleavage requires the spontaneous curvature m to exceed the threshold value $|m^{cl}| \gtrsim 1/(100 \text{ nm})$. As we will see in Section 7.5 below, neck cleavage is further facilitated by the adhesion of membranes to solid substrates and nanoparticles.

Curvature-induced budding and fission has been recently observed in molecular dynamics simulations of nanovesicles (Ghosh et al., in preparation). In this case, the spontaneous curvature was generated by the adsorption of small solute particles. Combined budding and fission has also been observed experimentally for giant vesicles exposed to polyhistidine-tagged GFP proteins that were bound to certain lipid components within the vesicle membranes (Steinkühler et al., in preparation).

5.5.5 VESICLE SHAPES WITH SEVERAL BUDS

Let us now consider multi-sphere vesicles that consist of more than two spheres connected by more than one closed neck, see also Box 5.2. The Euler-Lagrange Eq. 5.70, which applies to all membrane segments of such a multi-sphere vesicle apart from the closed necks, implies that at most two different types of spheres with two distinct radii, $r_1 = R_1/R_{ve}$ and $r_2 = R_2/R_{ve}$, can coexist on the same vesicle.

These two radii are determined by the membrane area $A = 4\pi R_{ve}^2$, by the vesicle volume $V = v(4\pi/3)R_{ve}^3$, and by the numbers N_1 and N_2 of the two types of spheres. If both types of spheres have a positive mean curvature, the two radii r_1 and r_2 satisfy the geometric relations

$$N_1 r_1^2 + N_2 r_2^2 = 1 \tag{5.142}$$

and

$$N_1 r_1^3 + N_2 r_2^3 = v \quad (\bar{M}_1 > 0 \text{ and } \bar{M}_2 > 0). \tag{5.143}$$

If we define the volumes v_1 and v_2 of the individual spheres via

$$\frac{4\pi}{3}R_1^3 = v_1\frac{4\pi}{3}R_{ve}^3 \quad \text{and} \quad \frac{4\pi}{3}R_2^3 = v_2\frac{4\pi}{3}R_{ve}^3, \tag{5.144}$$

the relation Eq. 5.143 can be rewritten in the form

$$N_1 v_1 + N_2 v_2 = v. \tag{5.145}$$

Simple examples for such multi-sphere shapes with $N_1 = 1$ are shown in Figure 5.18a–c. If the r_1- and r_2-spheres have positive and negative mean curvature, respectively, multi-sphere shapes with $N_1 > 1$ are impossible because they would require different types of necks with positive and negative neck curvature. Therefore, we are left with $N_1 = 1$, i.e., one large sphere with N_2 in-buds as illustrated in Figure 5.18d. In the latter case, the second geometric relation Eq. 5.143 is replaced by

$$r_1^3 - N_2 r_2^3 = v_1 - N_2 v_2 = v \quad (\bar{M}_1 > 0 \text{ and } \bar{M}_2 < 0). \tag{5.146}$$

In contrast to these geometric relations, the stability relations for the membrane necks are local and do not depend on the sphere

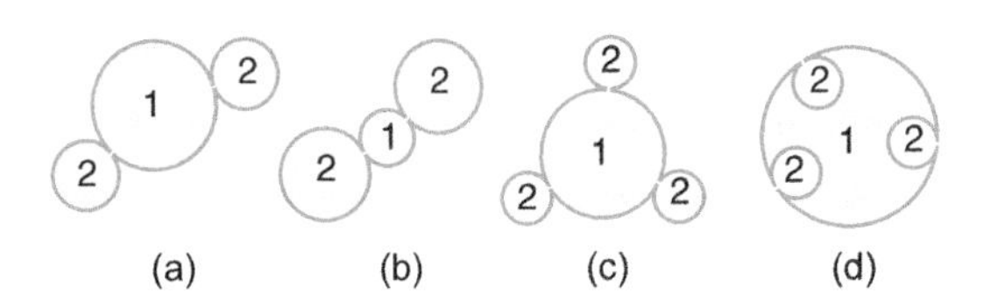

Figure 5.18 (a–c) Examples for vesicles consisting of 1 + N spheres with positive neck curvature: (a) Large r_1-sphere with two smaller r_2-spheres; (b) Small r_1-sphere with two larger r_2-spheres as observed in (Lipowsky and Dimova, 2003); (c) Large r_1-sphere with three smaller r_2-spheres; and (d) Example for a vesicle consisting of 1 + 3 spheres with negative neck curvature. For simplicity, all membrane necks have been placed in the plane of the figure. The positions of these necks are, however, arbitrary and can be shifted along the surface of the large sphere as long as the buds do not intersect each other.

numbers N_1 and N_2. Therefore, both the neck closure condition Eq. 5.91 as well as the closed neck conditions Eqs 5.92 and 5.93 are valid for arbitrary numbers N_1 and N_2 of r_1- and r_2-spheres, where we implicitly assume that these spheres do not intersect each other.

However, a multi-sphere vesicles built up from several r_1-spheres and r_2-spheres may exhibit different types of closed necks. Indeed, we can distinguish necks between two r_1-spheres from necks between two r_2-spheres and from necks between an r_1- and an r_2-sphere. These three types of necks have different neck curvatures $\bar{M}_{ne}$ as long as $r_1 \neq r_2$. In this section, we will focus on the simplest case in which all necks have the same curvature as in Figure 5.18 and again focus on the case with out-buds. Multi-sphere shapes with two types of necks will be discussed in the next section in the context of necklace-like tubes.

Multi-sphere vesicles with N out-buds

The simplest multi-sphere shapes with more than two spheres consist of one r_1-sphere and N r_2-spheres which are connected by N closed necks with the same neck curvature $\bar{M}_{ve}$. All examples in Figure 5.18 belong to this category. If the neck curvature is positive, the r_2-spheres form N out-buds of the r_1-sphere as in Figure 5.18a–c. The latter shapes lead to a morphology diagram with two bifurcation points B_*^+ and $B_\diamond^+$ as displayed in Figure 5.19.

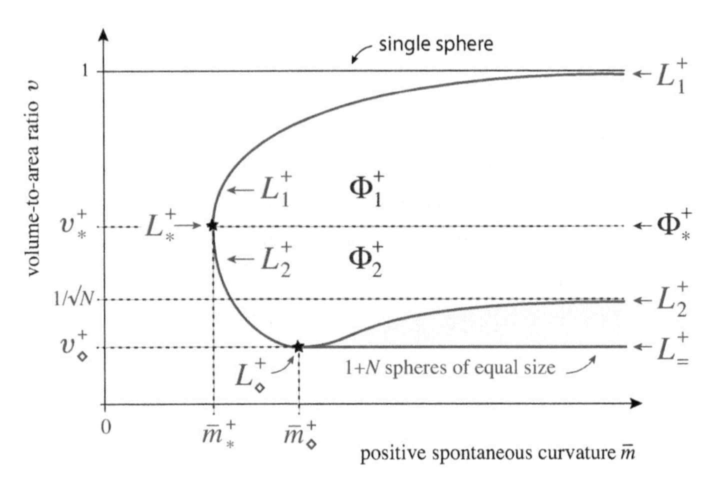

Figure 5.19 Morphology diagram for (1 + N)-sphere vesicles with positive spontaneous curvature: The vesicles consist of one r_1-sphere and N r_2-spheres as illustrated in Figure 5.18a–c. These vesicles are (meta)stable within the yellow (dark and light) parameter region bounded by three types of limit shapes (red lines), denoted by L_1^+, L_2^+, and $L_=^+$. The limit shapes L_1^+ and L_2^+ have variable neck curvature $\bar{M}_{ne} = \bar{m}$ whereas the limit shapes $L_=^+$ have the constant neck curvature $\bar{M}_{ne} = \sqrt{1+N}$. The limit shapes involve two types of bifurcation points (black stars). At the bifurcation point B_*^+ with coordinates $(\bar{m}, v) = (\bar{m}_*^+, v_*^+)$ as given by Eqs 5.147 and 5.148, the limit shape L_*^+ bifurcates into the shapes L_1^+ and L_2^+. The limit shape L_*^+ has a balanced geometry in the sense that the volume of the r_1-sphere is equal to the combined volume of all r_2-spheres. The same geometry applies to the persistent shapes Φ_*^+ along the horizontal broken line (blue) that emanates from the bifurcation point B_*^+. At the bifurcation point $B_\diamond^+$ with $(\bar{m}, v) = (\bar{m}_\diamond^+, v_\diamond^+) = (\sqrt{1+N}, 1/\sqrt{1+N})$ corresponding to the limit shape $L_\diamond^+$, the limit shapes $L_=^+$ bifurcate off from the line of L_2^+ shapes. The limit shapes $L_\diamond^+$ and $L_=^+$ consist of 1 + N spheres with the same size. The region between the $L_=^+$-line and the L_2^+-line with $\bar{m} > \bar{m}_\diamond^+$ (dark yellow) is special because two different Φ_2^+ shapes can be formed at each point within this region: one of these shapes is characterized by $r_1 > r_2$, the other by $r_1 < r_2$. Both shapes can be reached by inflation of the limit shape $L_=^+$ as illustrated in Figure 5.20 for N = 3.

Bifurcation of L_1^+ and L_2^+ shapes

Inspection of Figure 5.19 reveals that membranes with sufficiently small spontaneous curvatures do not form stable multi-sphere shapes. As we increase the spontaneous curvature, we encounter the bifurcation points B_*^+ at which a single multi-sphere shape, L_*^+, appears with spontaneous curvature

$$\bar{m} = \bar{m}_*^+(N) \equiv \frac{1}{2}(1+N^{1/3})^{3/2} \tag{5.147}$$

and volume

$$v = v_*^+(N) \equiv \frac{2}{(1+N^{1/3})^{3/2}}. \tag{5.148}$$

It is interesting to note that $\bar{m}_*^+(N) v_*^+(N) = 1$ for all values of N. The limit shape L_*^+ has a balanced volume in the sense that

$$v_1 = r_1^3 = N r_2^3 = N v_2 \quad \text{(balanced volume)}. \tag{5.149}$$

i.e., the volume of the r_1-sphere is equal to the combined volume of all r_2-spheres.

For $\bar{m} > \bar{m}_*^+(N)$, the limit shape L_*^+ bifurcates into two different branches of limit shapes, L_1^+ and L_2^+, as shown in Figure 5.19. For the upper branch with the limit shapes L_1^+, the volume v_1 of the r_1-sphere exceeds the combined volume $N v_2$ of the r_2-spheres. For the lower branch with the limit shapes L_2^+, on the other hand, the r_2-spheres dominate in the sense that $N v_2 > v_1$. Thus, the volume ratio

$$\rho_1 \equiv \frac{v_1}{N v_2} > 1 \quad \text{for the } L_1^+ \text{ shapes} \tag{5.150}$$

but

$$\rho_1 < 1 \quad \text{for the } L_2^+ \text{ shapes}. \tag{5.151}$$

As we move along the line of L_1^+-shapes by increasing the spontaneous curvature $\bar{m}$, both the total volume v and the volume ratio ρ_1 increase monotonically until the r_1-sphere has taken up the whole volume in the limit of large $\bar{m}$. More precisely, the volume $\mathcal{V}\{L_1^+\}$ of the L_1^+-shapes increases monotonically with increasing spontaneous curvature $\bar{m}$ and behaves as

$$\mathcal{V}\{L_1^+\} \approx \frac{4\pi}{3} R_{ve}^3 \left[1 - \frac{3N}{8\bar{m}^2} - \frac{N}{4\bar{m}^3}\right] \quad \text{for large } \bar{m}. \tag{5.152}$$

On the other hand, we can also move along the lower branch of the L_2^+-shapes by increasing L_2^+ which leads to a monotonic decay of the volume ratio ρ_1 until the N r_2-spheres have taken up the whole volume and $v \approx N r_2^3 \approx 1/\sqrt{N}$, see Figure 5.19.

As a consequence, the two limit shapes L_1^+ and L_2^+ look rather different for large $\bar{m}$. In this limit, the L_1^+-shapes consist of a large r_1-sphere and N small r_2-spheres with radii

$$r_1 \approx 1 \quad \text{and} \quad r_2 \approx \frac{1}{2\bar{m}} \quad (L_1^+, \text{large } \bar{m}). \tag{5.153}$$

In contrast, the L_2^+-shapes consist of a small r_1-sphere and N large r_2-spheres with radii

$$r_1 \approx \frac{1}{2\bar{m}} \quad \text{and} \quad r_2 \approx \frac{1}{\sqrt{N}} \quad (L_2^+, \text{large } \bar{m}). \tag{5.154}$$

For $N = 2$, these two limit shapes are illustrated in Figure 5.18a,b.

Bifurcation of $L_=^\pm$ from L_2^+ shapes

When we inspect the morphology diagram in Figure 5.19 more closely, we discover an additional complication related to the L_2^+-branch. In contrast to the volume ratio ρ_1 that decreases monotonically along this branch, the total volume $\mathcal{V}\{L_2^+\}$ of the L_2^+-shapes exhibits a minimum as a function of $\bar{m}$. At this minimum, the L_2^+-shape consists of 1 + N spheres of equal size with $r_1 = r_2 = 1/\sqrt{1+N}$, and provides the end point for the line of limit shapes $L_=^\pm$, see Figure 5.19. Therefore, the limit shape $L_\Diamond^+$ with

$$\bar{m} = \bar{m}_\Diamond^+(N) \equiv \sqrt{1+N} \quad \text{and} \quad v = v_\Diamond^+(N) \equiv \frac{1}{\sqrt{1+N}} \tag{5.155}$$

represents a second bifurcation point, $B_\Diamond^+$, at which the $L_=^\pm$ shapes split off from the L_2^+ shapes. Note that the limit shape $L_\Diamond^+$ is built up from 1 + N spheres of equal size with radius $r_1 = r_2 = 1/\sqrt{1+N} = 1/\bar{m}_\Diamond^+$. As a consequence, the bending energy vanishes for each of these spheres and, thus, for the whole limit shape $L_\Diamond^+$.

For $\bar{m} > \bar{m}_\Diamond^+(N)$, the volume $\mathcal{V}\{L_2^+\}$ increases again and behaves as

$$\mathcal{V}\{L_2^+\} \approx \frac{4\pi}{3} R_{ve}^3 \left[\frac{1}{\sqrt{N}} - \frac{3}{8\sqrt{N}\bar{m}^2} + \frac{1}{8\bar{m}^3} \right] \quad \text{for large } \bar{m}. \tag{5.156}$$

In contrast, all $L_=^\pm$ shapes have the same geometry and, thus, the same volume $\mathcal{V}\{L_=^\pm\} = (4\pi/3) R_{ve}^3 / \sqrt{1+N}$. The latter shapes are distinguished by their bending energies which depend on $\bar{m}$. It is again interesting to note that $\bar{m}_\Diamond^+(N) v_\Diamond^+(N) = 1$ for all values of N.

As shown in Figure 5.19, the lines of limit shapes L_1^+, L_2^+, and $L_=^\pm$ enclose an extended region of two-sphere shapes Φ^+ with stably closed necks. This region can be entered by deflation of L_1^+ or L_2^+ shapes, by inflation of L_2^+ or $L_=^\pm$ shapes, and by increasing the spontaneous curvature of L_1^+ or L_2^+ shapes. All Φ^+ shapes that are produced by one of these processes are persistent in the sense that their neck remains stably closed during both deflation and inflation as well as under small changes of the spontaneous curvature.

It is interesting to note that all bifurcation points B_*^+ and $B_\Diamond^+$ are located on the line $v = 1/\bar{m}$ within the $(\bar{m}, v)$-plane. Indeed, it follows from Eqs 5.147 and 5.148 that $v_*^+(N) = 1/\bar{m}_*^+(N)$ and from Eq. 5.155 that $v_\Diamond^+(N) = 1/\bar{m}_\Diamond^+(N)$ for all values of N. Furthermore, for large N, the $\bar{m}$-coordinates behave as $\bar{m}_*^+(N) \approx \frac{1}{2} N^{1/2}$ for the bifurcation points B_*^+ and as $\bar{m}_\Diamond^+(N) \approx N^{1/2}$ for the bifurcation points $B_\Diamond^+$ which implies that the points $B_\Diamond^+$ are more widely spaced compared to the points B_*^+.

Stability of individual spheres

When we apply the stability criterion Eq. 5.78 to examine the stability of the individual spheres, we find that both spheres are stable for all limit shapes L_1^+, L_2^+, and $L_=^\pm$. Furthermore, the larger sphere of the intermediate persistent shapes Φ^+ is always stable whereas the smaller sphere becomes unstable for sufficiently large values of the spontaneous curvature. The corresponding instability lines now follow from the solutions of the equation

$$r_2 - \frac{r_2^2}{\sqrt{1 - N r_2^2}} = \frac{3}{\bar{m}} \tag{5.157}$$

This equation has no solution for $\bar{m} < \bar{m}_{ss}(N)$, one solution for $\bar{m} = \bar{m}_{ss}(N)$ and two solutions for $\bar{m} > \bar{m}_{ss}(N)$. The critical value $\bar{m}_{ss}(N)$ for the instability of the small spheres is found to be $\bar{m}_{ss} = 14.3, 15.2$, and 19.6 for $N = 2, 3, 10$, respectively.

For large $\bar{m}$, the right hand side of Eq. 5.157 becomes small which implies the two asymptotic solutions

$$r_2 \approx \frac{3}{\bar{m}} \quad \text{and} \quad r_2 \approx \frac{1}{\sqrt{1+N}} \quad (\text{large } \bar{m}) \tag{5.158}$$

for the bud radius r_2. In the same limit, the reduced volume

$$v = r_1^3 + N r_2^3 = \left(1 - N r_2^2\right)^{3/2} + N r_2^3 \tag{5.159}$$

behaves as

$$v \approx 1 - \frac{27}{2} \frac{N}{\bar{m}^2} + 27 \frac{N}{\bar{m}^3} \quad \text{for } r_2 \approx 3/\bar{m} \tag{5.160}$$

and as

$$v \approx \frac{1}{\sqrt{1+N}} = v_\Diamond^+ \quad \text{for } r_2 \approx 1/\sqrt{1+N}. \tag{5.161}$$

Therefore, the two branches of the instability line approach the straight lines $v = 1$ and $v = v_\Diamond^+$ corresponding to a single sphere and to a multi-sphere consisting of (1 + N) spheres of equal size, respectively, compare Figure 5.19.

Along the instability line that approaches $v = v_\Diamond^+ = 1/\sqrt{1+N}$ for large $\bar{m}$, the N buds are smaller than or equal to the central sphere, i.e., $r_2 \leq r_1$, as illustrated in Figure 5.18a and Figure 5.20a. In contrast, the shapes along the L_2^+ line with $\bar{m} > \bar{m}_\Diamond^+$ are characterized by N buds that are larger than the central sphere, i.e., $r_2 > r_1$, as illustrated in Figure 5.18b and Figure 5.20c. As a consequence, the instability line that approaches $v = v_\Diamond^+ = 1/\sqrt{1+N}$ for large $\bar{m}$ does not cross the L_2^+ line obtained for $\bar{m} > \bar{m}_\Diamond^+$. Indeed, for the dark yellow region in Figure 5.19, we obtain a stack of two different sheets of (1 + N)-spheres, the two sheets being connected via the $L_=^\pm$ line. This bifurcation structure will be discussed in more detail in the next paragraphs.

Persistent shapes and deflation behavior

As for two-sphere vesicles, the geometry of the persistent shapes Φ^+ is fully determined by the volume v. Thus, if we consider any point $(\bar{m}_o, v_o)$ within the region bounded by the limit shapes, see the yellow region in Figure 5.19, the persistent shape at this point has the same geometry as the limit shape with the same volume $v = v_o$, i.e., as the limit shape obtained by projecting the point $(\bar{m}_o, v_o)$ parallel to the $\bar{m}$-axis onto the line of limit shapes. Using this constant-volume projection, we can then distinguish persistent shapes Φ_1^+, Φ_*^+, and Φ_2^+ which have the same geometry as the limit shapes L_1^+, L_*^+, and L_2^+, respectively.

Now, consider a point $(\bar{m}_o, v_o)$ within the region between the $L_=^+$-line and the L_2^+-line with $\bar{m} > \bar{m}_\Diamond^+$, corresponding to the dark yellow region in Figure 5.19. The constant-volume projection of this point onto the lines of limit shapes leads to two such shapes. One of these L_2^+ shapes is located at $\bar{m} < \bar{m}_\Diamond^+$ and characterized by $r_1 > r_2$ whereas the other L_2^+ shape is located at $\bar{m} > \bar{m}_\Diamond^+$ which implies $r_1 < r_2$. As a consequence, for each point $(\bar{m}_o, v_o)$ within the dark yellow region in Figure 5.19, we obtain two different persistent shapes Φ_2^+ with $r_2 < r_1$ and $r_2 > r_1$, respectively. Therefore, the dark yellow region in Figure 5.19 is characterized by a stack of two different sheets of shapes, two sheets that merge along the line of limit shapes $L_=^+$.

This two-sheet structure of the morphology diagram has interesting consequences for the deflation and inflation behavior of the multi-sphere vesicles considered here. Starting from a "balanced" persistent shape Φ_*^+, deflation eventually leads to a limit shape $L_=^+$, consisting of $1 + N$ spheres of equal size. Further deflation of the latter shape will open up the necks of the $L_=^+$-shapes. However, inflation of the $L_=^+$-shape will not necessarily lead back to the Φ_2^+-shapes that were obtained by deflation of the balanced Φ_*^+-shapes. Indeed, the whole $L_=^+$-line should be regarded as another bifurcation line from which two sheets of Φ_2^+-shapes emanate, both of which are accessible via inflation of the $L_=^+$-shapes. Inflation along one of these two sheets leads back to the balanced Φ_*^+-shapes, inflation along the other sheet leads to the limit shapes L_2^+ with $\bar{m} > \bar{m}_\Diamond^+$. This behavior is illustrated in Figure 5.20 for $N = 3$.

Out-buds with zero bending energy

The persistent shapes Φ^+ include the special shapes Z_N^{out} with N out-buds that have radius $r_2 = 1/\bar{m}$ and, thus, vanishing bending energy. The reduced volume of these latter shapes is given by

$$v = \left(1 - \frac{N}{\bar{m}^2}\right)^{3/2} + \frac{N}{\bar{m}^3} \approx 1 - \frac{3}{2}\frac{N}{\bar{m}^2} + \frac{N}{\bar{m}^2} \quad \text{for large } \bar{m}. \tag{5.162}$$

Therefore, the line of special shapes Z_N^{out} with zero-energy buds also approaches the straight line $v = 1$ for a single sphere. Comparison with Eqs 5.152 and 5.160 shows that the line of Z_N^{out} shapes is located between the line of limit shapes L_1^+ as described by Eq. 5.152 and the upper branch of the instability line for individual spheres as given by Eq. 5.160. As a consequence, the special shapes Z_N^{out} are stable for large $\bar{m}$. Furthermore, the line of Z_N^{out} shapes with zero-energy buds includes the limit shape $L_\Diamond^+$, see Figure 5.19, because

$$v = \left(1 - \frac{N}{\bar{m}^2}\right)^{3/2} + \frac{N}{\bar{m}^3} = \frac{1}{\sqrt{1+N}} = v_\Diamond^+ \quad \text{for } \bar{m} = \bar{m}_\Diamond^+ = \sqrt{1+N}. \tag{5.163}$$

In the latter case, the N out-buds have the same size as the mother vesicle which implies that the whole limit shape $L_\Diamond^+$ has vanishing bending energy as mentioned previously. Therefore, the Z_N^{out} shapes with zero-energy buds have stably closed necks connecting stable individual spheres, and the corresponding Z_N^{out} line in the morphology diagram emanates from the limit shape $L_\Diamond^+$ with $\bar{m} = \bar{m}_\Diamond^+ = \sqrt{1+N}$ and approaches the straight line $v = 1$, corresponding to a single sphere, for large $\bar{m}$.

Corrections arising from area-difference-elasticity

When we include area-difference-elasticity, the shapes with a large mother vesicle of radius r_1 and N spherical out-buds of radius r_2 generate the nonlocal spontaneous curvature

$$\bar{m}_{\text{nlo}} = \pi \frac{\kappa_\Delta}{\kappa}(1 - r_1 - N r_2) = \pi \frac{\kappa_\Delta}{\kappa}\left(1 - r_1 - \sqrt{N\left(1 - r_1^2\right)}\right) \tag{5.164}$$

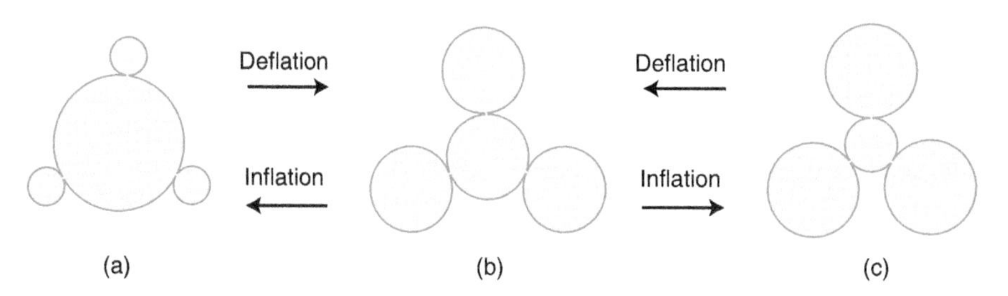

Figure 5.20 Three multi-sphere vesicles that can be transformed into each other by deflation or inflation. All three vesicles consist of a single r_1-sphere and three r_2-spheres: (a) Persistent shape Φ_1^+ with one large r_1-sphere with volume v_1 and $N = 3$ smaller r_2-spheres with combined volume $3v_2 < v_1$; (b) Limit shape $L_=^+$ for which the r_1-sphere and the three r_2-spheres have the same size; and (c) Persistent shape Φ_2^+ with a small r_1-sphere and three larger r_2-spheres. Deflation of Φ_1^+ in (a) leads, via an intermediate shape Φ_2^+, to $L_=^+$ as in (b) but inflation of $L_=^+$ can lead either back to (a) or to (c). Such deflation and inflation processes are possible for $\bar{m} > \bar{m}_\Diamond^+(3) = 2$, see Figure 5.19.

which generalizes Eq. 5.119 from $N = 1$ to $N \geq 1$. The last equality in Eq. 5.164 follows from the area relation $r_1^2 + N r_2^2 = 1$. The geometric factor in Eq. 5.164 is negative and bounded by

$$1-\sqrt{1+N} \leq 1-r_1-\sqrt{N\left(1-r_1^2\right)} \leq 0 \tag{5.165}$$

which implies that the absolute value of the nonlocal spontaneous curvature satisfies

$$|\bar{m}_{\rm nlo}| \leq \pi\left(\sqrt{1+N}-1\right)\frac{\kappa_\Delta}{\kappa}. \tag{5.166}$$

The neck closure condition is now given by

$$\bar{M}_{\rm ne} = \frac{1}{2}\left(\frac{1}{r_1}+\frac{1}{r_2}\right) = \bar{m}+\bar{m}_{\rm nlo} = \bar{m}+\pi\frac{\kappa_\Delta}{\kappa}(1-r_1-Nr_2) \tag{5.167}$$

Using the inequality in Eq. 5.166, we can ignore the nonlocal contribution $\bar{m}_{\rm nlo}$ in the neck closure condition as given by Eq 5.167 for

$$\bar{m} \gg \pi(\sqrt{1+N}-1)\frac{\kappa_\Delta}{\kappa} \tag{5.168}$$

which generalizes Eq. 5.126 for two-sphere shapes with $N = 1$ to arbitrary values of N.

Alternatively, we can consider the limit of small out-buds and, thus, small bud radii r_2. In this limit, the radius r_1 of the mother vesicle behaves as $r_1 \approx 1-\frac{1}{2}Nr_2^2$ for small Nr_2^2 as follows from the area relation $r_1^2 + Nr_2^2 = 1$. As a consequence, the neck closure condition in Eq. 5.167 leads to the local spontaneous curvature

$$\bar{m} \approx \frac{1}{2}\left(\frac{1}{r_1}+\frac{1}{r_2}\right)+\pi\frac{\kappa_\Delta}{\kappa}Nr_2 \quad \text{for small buds with } r_2 \ll 1/\sqrt{N} \tag{5.169}$$

which shows that we can ignore the κ_Δ term arising from area-difference-elasticity for small $r_2 \ll 1/\sqrt{N}$.

Mutual exclusion of out-buds

Because all out-buds or r_2-spheres are attached to the r_1-sphere, they may become closely packed when they reach a certain size. For $N = 2$ as shown in Figures 5.18a, b, the two r_2-spheres can become arbitrarily large without getting into contact. Therefore, mutual exclusion of the two r_2-spheres does not affect the morphology diagram in Figure 5.19. For $N = 3$, mutual exclusion of the three r_2-spheres starts to play a role when the radius of the r_2-spheres becomes sufficiently large compared to the radius of the r_1-sphere, compare Figure 5.20c. Indeed, the three r_2-spheres come into contact when $r_2 = \frac{\sqrt{3}}{2-\sqrt{3}}r_1 = 6.46r_1$ corresponding to the contact volume $v_{\rm co}^+ = 0.5712$ for $N = 3$. As a consequence, (1 + 3)-sphere shapes with $r_2 > r_1$ can no longer be formed for the volume range $0.5712 < v < 1/\sqrt{3} = 0.5774$.

In general, the mutual exclusion of the r_2-spheres acts to reduce the parameter region in which (1 + N)-sphere shapes can be formed for all $N \geq 3$. The corresponding contact volume $v_{\rm co}^+(N)$ decreases with increasing N. For $N = 12$, the r_2-spheres come into contact along the $L_=^{\pm}$-line where the r_2-spheres have the same size as the r_1-sphere. The corresponding contact volume $v_{\rm co}^+(12) = v_\lozenge^+(12) = 1/\sqrt{13} = 0.2774$. As a consequence, (1 + 12)-sphere shapes, for which each r_2-sphere is larger than the r_1-sphere can no longer be formed when we take mutual exclusion of the r_2-spheres into account. On the other hand, we can also conclude that the morphology diagram exhibits both bifurcation points B_*^+ and $B_\lozenge^+$ as well as the limit shapes L_1^+, L_2^+ with $r_1 > r_2$, and $L_=^{\pm}$ up to bud number $N = 11$. Thus, for $3 \leq N \leq 11$, the mutual exclusion of the out-buds will only affect the (1 + N)-spheres for which the bud radius r_2 exceeds the radius r_1 of the central sphere, as illustrated in Figure 5.20c for $N = 3$.

5.5.6 N-DEPENDENT ENERGY LANDSCAPE

Optimal bud number

In the previous subsections, we focused on the stability of different multi-sphere shapes and found certain stability regions within the $(\bar{m}, v)$-plane for each of these shapes. When we vary the spontaneous curvature $\bar{m}$ and the volume v within such a stability region, the bending energy of the corresponding multi-sphere shape changes smoothly and defines an energy surface over this region. Because the different stability regions overlap with each other in the $(\bar{m}, v)$-plane, we often find many energy surfaces stacked above one another, when we consider the vicinity of a certain point in the $(\bar{m}, v)$-plane. These energy surfaces of the multi-sphere shapes should be regarded as partial branches that supplement the branches of stationary solutions obtained from the Euler-Lagrange equations. Therefore, the overall energy landscape of the vesicle shapes is rather complex.

In order to determine the shape of lowest bending energy for given values of $\bar{m}$ and v, we need to compare the different branches of shapes. As an example, let us again consider multi-sphere shapes with N out-buds which have the dimensionless bending energy

$$\bar{E}_{\rm be} = (1-mR_1)^2 + N(1-mR_2)^2 = 1+N+\bar{m}^2-2\bar{m}(r_1+Nr_2) \tag{5.170}$$

where the radii r_1 and r_2 satisfy the geometric relations in Eqs 5.142 and 5.143 with $N_1 = 1$ and $N_2 = N$. When we minimize this bending energy with respect to N, we find the optimal bud number

$$N_{\rm opt} \approx \frac{2(1-v)}{3}\bar{m}^2 \quad \text{for large } \bar{m}. \tag{5.171}$$

For $N = N_{\rm opt}$, the radius of the out-buds has the value $r_2 \approx 1/\bar{m}$ which implies that shapes with an optimal bud number are identical with the shapes $Z_N^{\rm out}$ possessing N zero-energy out-buds. The asymptotic equality as given by Eq. 5.171 implies that the optimal number $N = N_{\rm opt}$ of out-buds increases with the spontaneous curvature $\bar{m}$ when we consider a fixed volume $v < 1$ as obtained by the osmotic deflation of a single sphere. The actual shape transition from a shape with N out-buds to a shape with $N + 1$ out-buds necessarily involves smooth vesicle shapes with open necks. For small values of N, the corresponding bifurcations have been calculated by numerical energy minimization in (Seifert et al., 1991; Liu et al., 2016). For large values of N, we need to consider sufficiently large GUVs with radius $r_1 \gg r_2 = 1/m$ so that we can ignore the mutual exclusion of the out-buds.

Box 5.2 Membrane necks of vesicles with laterally uniform composition

Membrane necks are funnel-like membrane structures that connect two different membrane compartments. The mean curvatures, M_1 and M_2, of the two membrane segments adjacent to the neck define the neck curvature $M_{ne} = \frac{1}{2}(M_1 + M_2)$ as introduced in Eq. 5.48.

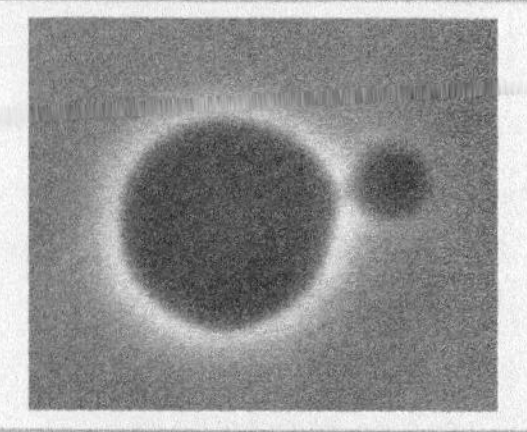

- GUV and out-bud connected by a narrow membrane neck (from Figure 5.2). The mother vesicle has the radius R_1 = 19 μm, the out-bud has the radius R_2 = 7.2 μm. The neck curvature M_{ne} then has the *positive* value $M_{ne} = \frac{1}{2}(\frac{1}{R_1} + \frac{1}{R_2}) = \frac{1}{10.4\,\mu m}$.

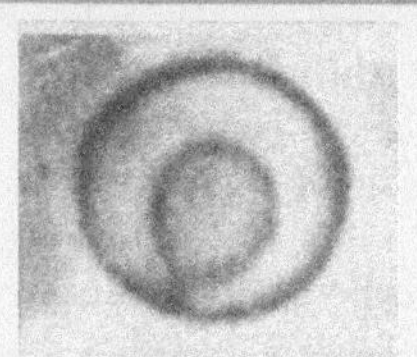

- GUV and in-bud connected by a narrow membrane neck (from Figure 5.6). The mother vesicle has the radius R_1 = 10.4 μm, the in-bud has the radius R_2 = 4.7 μm, which leads to the *negative* neck curvature $M_{ne} = \frac{1}{2}(\frac{1}{R_1} - \frac{1}{R_2}) = -\frac{1}{8.6\,\mu m}$.

When we observe the closure of a neck, the neck curvature M_{ne} is equal to the spontaneous curvature m_{eff}, which may include a non-local contribution from area-difference-elasticity as in Eq. 5.68. Thus, the observation of neck closure leads to an estimate for m_{eff}. Furthermore, sufficiently large values of m_{eff} lead to the cleavage of the membrane neck and thus to complete membrane fission, see Section 5.5.4.

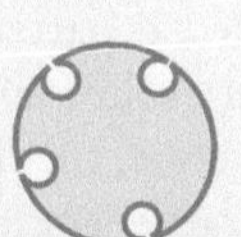

- A spherical vesicle may form several spherical buds with closed membrane necks. In equilibrium, all buds must have the same mean curvature as follows from the Euler-Lagrange Eq. 5.70 for spherical membrane segments. Therefore, the necks of all buds must have the same neck curvature M_{ne}.
- (Top) A vesicle with four out-buds and positive neck curvature.
- (Bottom) A vesicle with four in-buds and negative neck curvature.

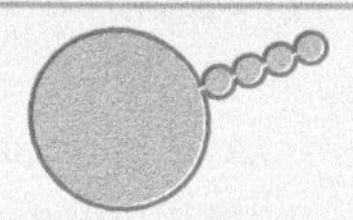

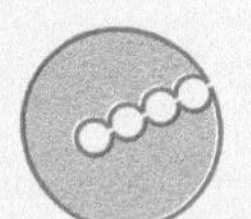

- (Top) A vesicle membrane with positive spontaneous curvature forming a necklace-like out-tube consisting of four out-beads with the same positive mean curvature.
- (Bottom) A vesicle membrane with negative spontaneous curvature forming a necklace-like in-tube consisting of four in-beads with the same negative mean curvature.
- In both cases, the neck curvature M_{ne} attains two different values (i) for the necks connecting the necklace-like tube with the mother vesicle and (ii) for the necks between two neighboring beads within the tube.

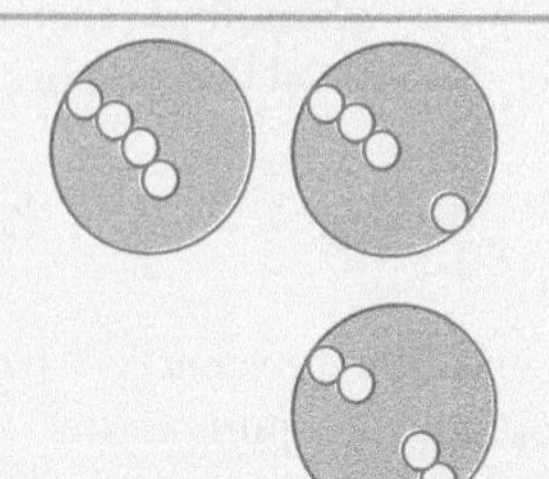

- Five different morphologies of a vesicle with four in-beads or in-buds of equal size. All five morphologies have the same membrane area, the same vesicle volume, the same integrated mean curvature, and the same bending energy. This degeneracy illustrates the morphological complexity of membranes, see Section 6.4 further below.
- Apart from the 4-bud morphology, all morphologies involve two types of necks that differ in their neck curvature.

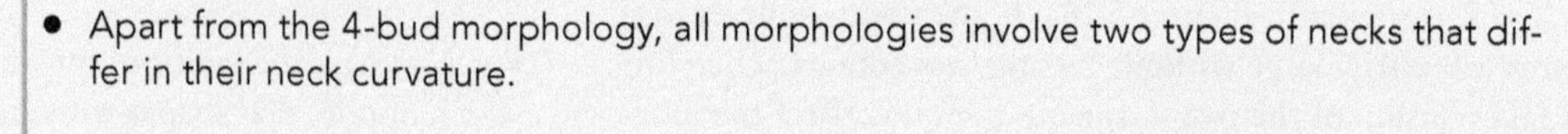

GUVs with buds and necklace-like tubes stabilized by membrane necks have some interesting properties. On the one hand, they provide aqueous subcompartments that could be used for the confinement of nanoparticles or microspheres. The closed necks represent diffusion barriers that can, however, be removed relatively easily, e.g., by osmotic inflation which leads to neck opening for all morphologies displayed in this box. On the other hand, the formation of many buds and necklace-like tubes provides an area reservoir to the mother vesicle which increases the vesicle's robustness against mechanical perturbation as shown by micropipette aspiration (Bhatia et al., 2018). The stability of membrane necks can be further enhanced by adhesion and constriction forces (Agudo-Canalejo and Lipowsky, 2016), see Chapter 8 of this book.

5.6 NANOTUBES OF UNIFORM MEMBRANES

Giant vesicles can spontaneously form long nanotubes that emanate from the vesicle membrane. Such a tubulation process provides direct evidence that the vesicle membrane has a relatively large spontaneous curvature m. In-tubes pointing towards the interior of the vesicle are formed for large negative m-values, see Figure 5.21, out-tubes pointing towards the exterior solution for large positive values of m. Therefore, a uniform membrane with constant spontaneous curvature will form either in-tubes or out-tubes but not both types of tubes simultaneously.

In general, in- and out-tubes differ in several important aspects. First, the in- and out-tubes are connected to different volume reservoirs: the in-tubes exchange volume with the exterior aqueous compartment, which represents an effectively unlimited volume reservoir, whereas the out-tubes exchange aqueous solution with the interior vesicle compartment. Second, the membranes of out- and in-tubes experience different osmotic pressure differences: the membrane of an out-tube is subject to the same pressure difference ΔP as the membrane of the large spherical segment whereas an in-tube feels the opposite pressure difference $-\Delta P$. Third, the membrane segments that form in- and out-tubes differ in the sign of their mean curvature which is negative for in-tubes and positive for out-tubes.

As shown in Figure 5.21, membrane nanotubes can have two different morphologies: necklace-like tubes consisting of small quasi-spherical beads connected by closed membrane necks as well as cylindrical tubes. From a theoretical point of view, necklace-like tubes represent multi-sphere vesicles with two types of necks whereas cylinders are governed by different shape equations. For cylindrical tubes, we include a pulling force that is applied locally to the tip of the tubes. For both tube morphologies, the mechanical tension is relatively small, reflecting the large area reservoir provided by the tubes, and the total membrane tension is dominated by the spontaneous tension, $\sigma = 2\kappa m^2$ (Lipowsky, 2013). At the end, we briefly discuss the transformation of necklace-like tubes into cylindrical ones, a transformation that occurs when the tube length has reached a certain critical value.

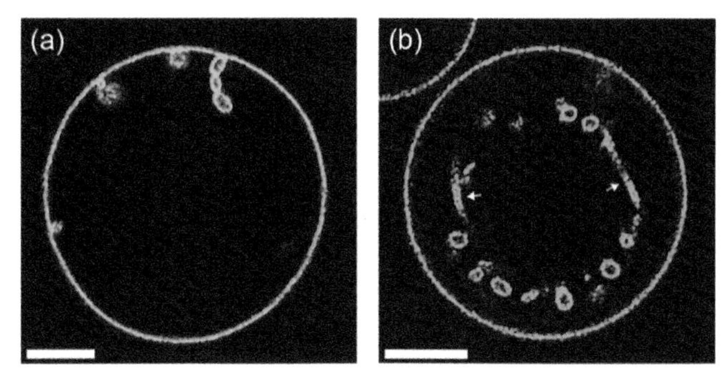

Figure 5.21 Giant vesicles with in-tubes, i.e., with membrane nanotubes that point towards the vesicle interior: (a) one necklace-like tube and several buds and (b) several necklace-like tubes and two cylindrical tube segments (white arrows). (Reproduced with permission from Liu, Y. et al., *ACS Nano*, 10, 463–474, 2016.)

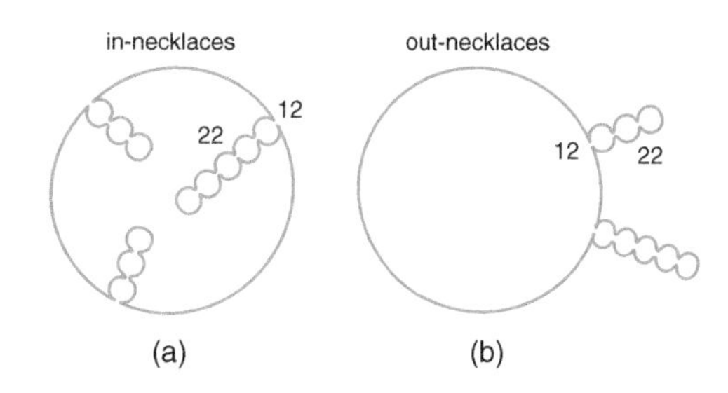

Figure 5.22 Necklace-like nanotubes consisting of spherules with radius r_2 emanating from a giant spherical vesicle with radius r_1. These shapes involve two different types of closed necks, 12-necks and 22-necks, that differ in their neck curvatures: (a) Necklace-like in-tubes with negative neck curvatures M_{12} and $M_{22} < M_{12}$ are formed for negative spontaneous curvature; and (b) Necklace-like out-tubes with positive neck curvatures M_{12} and $M_{22} > M_{12}$ require a membrane with positive spontaneous curvature.

5.6.1 NECKLACE-LIKE NANOTUBES

Necklace-like nanotubes as observed experimentally consist of identical quasi-spherical beads that are connected by closed membrane necks. One such necklace consisting of three beads is visible in Figure 5.21a. If one ignores thermally excited fluctuations, such a necklace can be described, in the context of curvature models, by a multi-sphere vesicle with two different types of closed necks as shown in Figure 5.22. Indeed, we now have to distinguish the necks between the large sphere and a necklace from the necks between two small spheres within the same necklace. In the following, we will use three terms for the sake of clarity. First, we will distinguish "buds" that are directly connected to the mother vesicle from "beads" that are connected to buds or other beads. Second, both buds and beads will be collectively called "spherules."

Necklace geometry and neck stability

Geometry of vesicle with necklace-like nanotubes. Consider a GUV consisting of a large spherical mother vesicle and one or several necklace-like nanotubes as displayed in Figure 5.22. The tubes contain a total number N of spherules. For a vesicle membrane with area A, we will again use the vesicle size $R_{ve} = \sqrt{A/(4\pi)}$ as the basic length scale and use the rescaled radii $r_1 = R_1/R_{ve}$ and $r_2 = R_2/R_{ve} < r_1$ of the mother vesicle and the spherules. These radii satisfy the relations

$$r_1^2 + N r_2^2 = 1 \tag{5.172}$$

corresponding to the total membrane area A and

$$r_1^3 \pm N r_2^3 = v \tag{5.173}$$

corresponding to the vesicle volume V where the plus and minus sign applies to out- and in-necklaces. Note that the same geometric relations apply to a GUV with N out- or in-buds as described in Section 5.5.

Stability of membrane necks. As mentioned, each necklace connected to a giant vesicle is characterized by two types of necks, 12- and 22-necks, see Figure 5.22. These two necks have two different neck curvatures as given by

$$M_{12} = \frac{1}{2}(M_1 + M_2) \tag{5.174}$$

and

$$M_{22} = \frac{1}{2}(M_2 + M_2) = M_2. \quad (5.175)$$

We will now examine the stability of these two types of necks. Out necklaces require positive spontaneous curvature $m > 0$ and are characterized by positive mean curvature $M_2 = 1/R_2$ of the spherules. For such a necklace, the 22-necks connecting two neighboring spherules, see Figure 5.22, are stable if the neck curvature M_{22}

$$0 < M_{22} = M_2 \le m \quad \text{(stable 22-neck of out-necklace).} \quad (5.176)$$

Furthermore, the stability condition for the 12-necks, connecting the mother vesicle with the out-necklace, has the form

$$M_1 + M_2 \le 2m \quad (5.177)$$

which follows from the stability condition Eq. 5.176 for the 22-necks because

$$M_1 + M_2 < 2M_2 \le 2m. \quad (5.178)$$

In-necklaces, on the other hand, can form for negative spontaneous curvature $m < 0$ and are characterized by negative mean curvature $M_2 = -1/R_2$ of the spherules. The stability condition for the 22-necks is now given by

$$m \le M_{22} = M_2 < 0 \quad \text{(stable 22-neck of in-necklace).} \quad (5.179)$$

Furthermore, the stability condition for the 12-necks, connecting the mother vesicle with the out-necklace, has the form

$$M_1 + M_2 \ge 2m \quad (5.180)$$

which follows from the stability condition Eq. 5.179 for the 22-necks because

$$M_1 + M_2 > 2M_2 \ge 2m. \quad (5.181)$$

Necklaces of zero-energy spherules

We now consider necklaces that consist of zero-energy spherules with radius $R_2 = 1/|m| \ll R_1$ and denote the shapes with N zero-energy spherules by L_N^{in} and L_N^{out}.[11] In contrast to the persistent shapes Z_N^{in} and Z_N^{out} with N zero-energy in- and out-buds as discussed in Section Multi-sphere vesicles with N out-buds, the shapes L_N^{in} and L_N^{out} are limit shapes because the closed 22-necks between neighboring spherules fulfill the neck closure condition $M_{22} = m$, compare Eqs 5.176 and 5.179.

For the limit shapes L_N^{in} and L_N^{out} with spherules of radius $r_2 = 1/|\bar{m}| \ll 1$, the mother vesicle has the radius

$$r_1 = r_1(N) = \sqrt{1 - \frac{N}{\bar{m}^2}} \quad (5.182)$$

and the volume is given by

$$v = v(N) = \left[1 - \frac{N}{\bar{m}^2}\right]^{3/2} \mp \frac{N}{|\bar{m}|^3}, \quad (5.183)$$

where the minus and plus sign applies to in- and out-necklaces, respectively.

Because the spherules have the radius $r_2 = 1/|\bar{m}|$, the in- and out-necklaces do not contribute to the bending energies of the L_N^{in} and L_N^{out} shapes. The latter energies are then equal to the bending energies of the mother vesicle with radius r_1 and mean curvature $\bar{M}_1 = 1/r_1$. These bending energies have the form

$$E_{\text{be}}(r_1) = 8\pi\kappa(1 - \bar{m}r_1)^2 \quad \text{for both in- and out-necklaces} \quad (5.184)$$

corresponding to $\bar{m} < 0$ and $\bar{m} > 0$, respectively. Using Eq. 5.182, the latter bending energy can be rewritten as

$$\bar{E}_{\text{be}}(r_1) = \frac{E_{\text{be}}(r_1)}{8\pi\kappa} = 1 + \bar{m}^2 - 2\bar{m}\sqrt{1 - \frac{N}{\bar{m}^2}} - N \quad (5.185)$$

which behaves as

$$\bar{E}_{\text{be}}(r_1) \approx (1 - \bar{m})^2 - N\left(1 - \frac{1}{\bar{m}}\right) \quad \text{for large } |\bar{m}|. \quad (5.186)$$

The first term of this expression represents the bending energy of a single sphere with spontaneous curvature $\bar{m}$. The second term proportional to N is negative for $\bar{m} < 0$ or $\bar{m} > 1$. Thus, for large negative or positive values of $\bar{m}$, the bending energies of the two limit shapes L_N^{in} and L_N^{out} decrease with increasing N. Therefore, these limit shapes provide possible low-energy pathways for the osmotic deflation of giant vesicles with large negative and large positive spontaneous curvatures, respectively.

The low-energy pathway provided by the sequence of L_N^{in} shapes has been studied in detail by numerical minimization of the shape functional $\mathcal{F}\{S\}$ in Eq. 5.22 (Liu et al., 2016). As a result, it was found that each limit shape L_N^{in} belongs to a different branch of (meta)stable shapes. When we start from such a limit shape with a certain value of N, an increase in vesicle volume via osmotic inflation leads to an opening of the necks and the necklaces then resemble unduloids as shown in Figure 5.23b, compare also Figure 5.28 further below. On the other hand, decreasing the vesicle volume by osmotic deflation does not open the closed necks connecting neighboring spherules but increases the radius of the spherules to $r_2 > 1/|\bar{m}|$, see Figure 5.23b. The corresponding metastable branch extends up to $r_2 = 3/|\bar{m}|$ at which point the spherules become unstable and undergo a sphere-prolate bifurcation.

5.6.2 DOMINANCE OF SPONTANEOUS TENSION

The mechanical equilibrium between the spherical mother vesicle and the spherules implies the two shape equations

$$\Delta P = 2\hat{\Sigma}M_{\text{sp}} - 4\kappa m M_{\text{sp}}^2 \quad \text{with } M_{\text{sp}} = M_1 = \frac{1}{R_1} \text{ or } M_2 = \pm\frac{1}{R_2}$$

[11] In (Liu et al., 2016), the shapes L_N^{in} have been denoted by $L^{[N]}$.

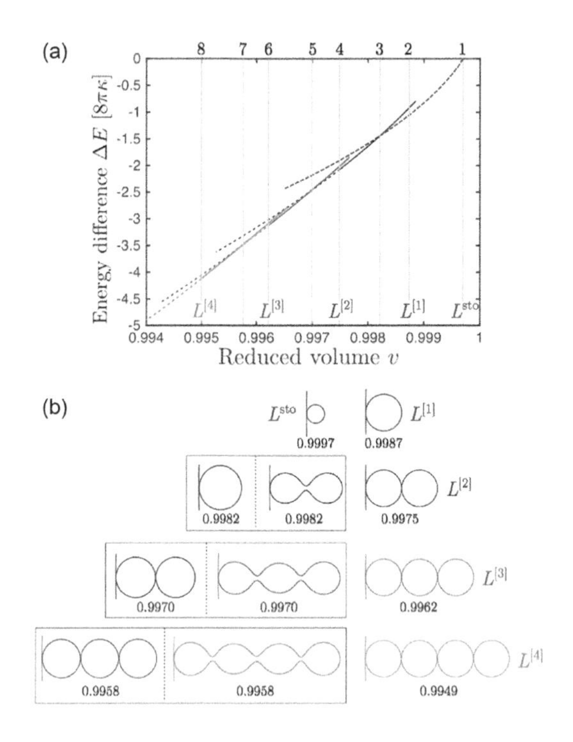

Figure 5.23 Osmotic deflation and inflation of a lipid vesicle with size $R_{ve} = 20.7$ μm and spontaneous curvature $m = -1/(599\text{nm})$: (a) Energy landscape of the vesicle as a function of the reduced volume v with the limit shapes $L^{[N]} \equiv L_N^{in}$. The energy difference ΔE describes the deflation-induced reduction in bending energy compared to the initial spherical vesicle, in units of $8\pi\kappa$. The eight vertical lines labeled from 1 to 8 (top) correspond to eight v-values obtained via eight discrete deflation steps; and (b) Tube shapes for the global energy minima at these eight v-values. The short vertical line on the left end of the tubes represents a short segment of the mother vesicle which is connected to each tube by a closed membrane neck. As we deflate the initial vesicle with $v = 1$, we move along the 1-necklace branch (red) that begins at the limit shape L^{sto} with bud radius $R_2 \approx 1/(2|m|)$ and $v = 0.9997$. After passing the shape $L^{[1]} = L_1^{in}$ with $R_2 = 1/|m|$ and $v = 0.9987$, we reach the reduced volume $v = 0.9982$ at which the 1-necklace branch crosses the 2-necklace branch (blue). For the latter v-value, a bud with radius $R_2 > 1/|m|$ coexists with a 2-necklace that has an open 22-neck. Further deflation leads to the 2-necklace $L^{[2]} = L_2^{in}$ with a closed neck at $v = 0.9975$ and, subsequently, to the 3-necklace branch (orange) and the 4-necklace branch (green). The dashed and solid segments of the free energy landscape in panel (a) correspond to tubes with closed and open necks, respectively. (Reproduced with permission from Liu, Y. et al., *ACS Nano*, 10, 463–474, 2016.)

as in Eq. 5.70 with the total membrane tension $\hat{\Sigma} = \Sigma + 2\kappa m^2$. Combining these two equations to eliminate the pressure difference ΔP, we obtain the mechanical tension

$$\Sigma = 2\kappa m(M_1 + M_2) - 2\kappa m^2 = 4\kappa m M_{12} - 2\kappa m^2 \quad (5.187)$$

where the first equality is equal to Eq. 5.76 and the second equality follows from the mean curvature M_{12} of the 12-neck as given by Eq. 5.174. Therefore, the mechanical tension Σ depends on the neck curvature M_{12} whereas the stability of the multi-sphere shape is determined by the neck curvature M_{22} of the 22-necks.

For the limit shapes L_N^{in} and L_N^{out}, the spherules have the mean curvature $M_2 = m$ and the mean curvature of the 12-necks is given by $M_{12} = \frac{1}{2}(M_1 + m)$. As a consequence, the mechanical tension in Eq. 5.187 becomes

$$\Sigma = 2\kappa m M_1 = \frac{\sigma}{m R_1} = \mp \frac{R_2}{R_1}\sigma \quad (5.188)$$

where the minus and plus sign applies to the limit shapes L_N^{in} and L_N^{out}, respectively. Because the radius R_1 of the mother vesicle is much larger than the radius $R_2 = 1/|m|$ of the spherules, the absolute value $|\Sigma|$ of the mechanical tension in Eq. 5.188 is much smaller than the spontaneous tension $\sigma = 2\kappa m^2$.

The limit shapes L_N^{in} and L_N^{out} represent the equilibrium shapes of the tubulated vesicle for certain vesicle volumes or, equivalently, for certain values of the membrane area

$$A_{nt} = A_{nt,N} \equiv N 4\pi/m^2 \quad \text{(limit shapes } L_N^{in} \text{ and } L_N^{out}) \quad (5.189)$$

stored in the tubes (Liu et al., 2016; Bhatia et al., 2018). Each of these limit shapes belongs to a whole branch of shapes, as illustrated for in-necklaces by the energy branches in Figures 5.23 and 5.24. The latter figure displays the bending energy landscape E_{nt} for the necklace-like tubes that grow as we reduce the volume of the GUV. The deflation process decreases the membrane area A_1 of the mother vesicle and increases the area A_{nt} stored in the tubes, for fixed total area $A = A_1 + A_{nt}$. The bending energy of the tubulated GUV is equal to $E_1 + E_{nt}$ where the bending energy E_1 of the mother vesicle is a monotonically decreasing function

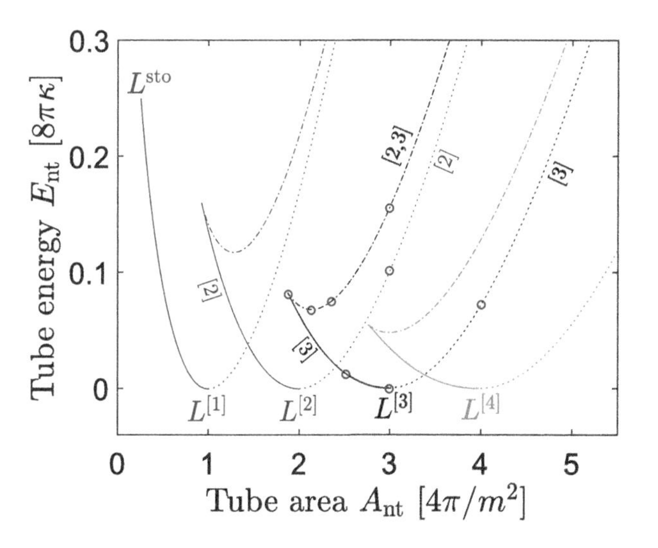

Figure 5.24 Energy landscape E_{nt} of a necklace-like nanotube protruding into a GUV as a function of membrane area A_{nt} stored in the tube. The size of the GUV is much larger than the width of the nanotube. The energy landscape is built up from a discrete set of [N]-branches with $N \geq 1$. The different branches are distinguished by different colors. Each [N]-branch attains its energy minimum for the limit shape $L^{[N]} = L_N^{in}$ which consists of N spherules with radius $R_2 = 1/|m|$ and area $4\pi/m^2$. When we deflate the limit shape $L^{[N]}$, i.e., when we reduce the vesicle volume for fixed membrane area, we move towards larger values of the tube area A_{nt} along the dotted lines which represent necklace-like tubes with N small spheres of radius $R_2 > 1/|m|$ and $N-1$ closed necks. When R_2 reaches the limiting value $R_2 = 3/|m|$, the spherules undergo a sphere-prolate bifurcation (outside of the figure). When we inflate the limit shape $L^{[N]} = L_N^{in}$, we move towards smaller values of A_{nt} along the full lines that represent necklace-like tubes with N bellies and $N-1$ open necks. The dash-dotted lines represent unstable necklace-like tubes corresponding to transition states [N, N + 1] between the (meta)stable [N] and [N + 1] states. The red circles mark the nanotube morphologies displayed in Figure 5.25. (From Bhatia, T. et al., *ACS Nano*, 12, 4478–4485, 2018.)

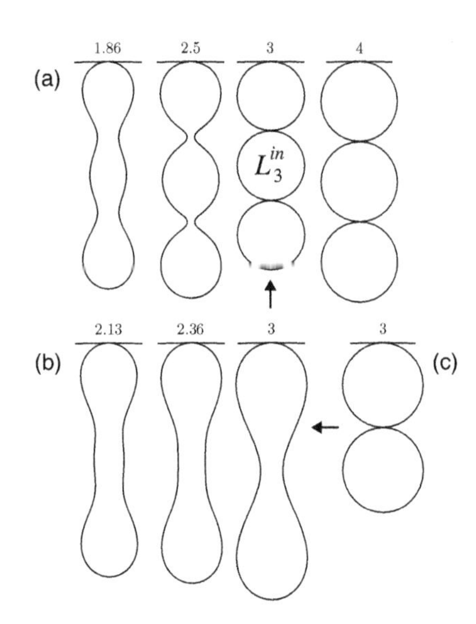

Figure 5.25 Morphologies of necklace-like nanotubes corresponding to the red circles in Figure 5.24. The number at the top of each tube represents the tube area A_{nt} in units of $4\pi/m^2$: (a) Four shapes along the (meta)stable [3]-branch. The shape with $A_{nt} = 1.86$ represents the bifurcation point between the [3]-branch and the unstable [2, 3]-branch of transition states. The shape with $A_{nt} = 3$ is the limit shape L_3^{in}; (b) Three shapes along the unstable [2, 3]-branch of transition states. The shape with $A_{nt} = 2.13$ is located at the energy minimum of the [2, 3]-branch, see Figure 5.24, the shape with $A_{nt} = 2.36$ separates transition states with three from those with two bellies; and (c) Metastable shape of the [2]-branch that decays into the limit shape $L^{[3]} = L_3^{in}$ via the rightmost transition state in (b) with $A_{nt} = 3$ (two arrows). (From Bhatia, T. et al., *ACS Nano*, 12, 4478–4485, 2018.)

of A_{nt}. Examples for the morphologies of the necklace-like tubes along several branches of the energy landscape are displayed in Figure 5.25.

Inspection of the energy landscape in Figure 5.24 reveals that the equilibrium shapes with the lowest bending energy E_{nt} are provided by short segments of the [N]-branches as obtained by slight deflation and slight inflation of the limit shapes L_N^{in}. Slight deflation of L_N^{in} reduces the vesicle volume and increases the area A_{nt} of the necklace-like tubes until we reach the intersection point of the [N]-branch with the [$N + 1$]-branch at tube area $A_{nt} = (N + \epsilon_N)4\pi/m^2$ with a dimensionless coefficient ϵ_N that satisfies $0 < \epsilon_N < 1$. We now consider the increase in tube area as given by

$$A_{nt} - A_{nt,N} \equiv \delta_N \frac{4\pi}{m^2} \quad \text{with } 0 \le \delta_N \le \epsilon_N \tag{5.190}$$

which leads to the mean curvature

$$M_2 = \frac{m}{\sqrt{1+\delta_N / N}} \approx m\left(1 - \frac{\delta_N}{2N}\right) \quad \text{for large } N. \tag{5.191}$$

The number N of spherules is directly related to the length L_{nt} of the necklace-like nanotubes via $L_{nt} = 2NR_2 = 2N/|M_2|$ which implies

$$M_2 \approx m + \frac{\delta_N}{L_{nt}} \quad \text{for large } L_{nt} \gg R_2, \tag{5.192}$$

i.e., for a tube length L_{nt} that is large compared to the radius R_2 of the spherules, with $0 < \delta_N < 1$. Using again the general expression for the mechanical tension Σ of necklace-like tubes as given by Eq. 5.187, we obtain

$$\Sigma \approx 2\kappa m\left(M_1 + \frac{\delta_N}{L_{nt}}\right) = \left(\frac{1}{mR_1} + \frac{\delta_N}{mL_{nt}}\right)\sigma \tag{5.193}$$

Therefore, the absolute value $|\Sigma|$ of a necklace-like tube is much smaller than the spontaneous tension σ if both the mother vesicle radius R_1 and the tube length L_{nt} are much larger than the small sphere radius $R_2 \approx 1/|m|$. In such a situation, the total membrane tension $\hat{\Sigma} = \Sigma + \sigma$ of a GUV with necklace-like nanotubes becomes

$$\hat{\Sigma} = \Sigma + \sigma \approx \left(\frac{1}{mR_1} + \frac{\delta_N}{mL_{nt}}\right)\sigma + \sigma \approx \sigma \quad (\text{large } |m|) \tag{5.194}$$

and is, thus dominated by the spontaneous tension σ. The small mechanical tension reflects the large area reservoirs as provided by the nanotubes. Indeed, when the tubulated vesicle is exposed to external forces or constraints, it can adapt to these perturbations, for fixed vesicle volume and membrane area, by simply shortening the nanotubes. This increased robustness of tubulated vesicles has been recently demonstrated by micropipette aspiration of tubulated GUVs (Bhatia et al., 2018).

5.6.3 MORPHOLOGICAL COMPLEXITY AND RUGGED ENERGY LANDSCAPE

As previously mentioned, the limit shapes L_N^{in} displayed in Figure 5.23 provide a low-energy pathway for the growth of a single necklace-like tube. The elongation of this tube from $L^{[N]}$ to $L^{[N+1]}$ proceeds via a sphere-prolate bifurcation. Inspection of the microscopy images displayed in Figure 5.21 and Figure 5.4 reveals however that giant vesicles can form much more complex shapes consisting of many buds and tubes. This morphological complexity emerges from the presence of a second low-energy pathway provided by the nucleation of another bud via an oblate-stomatocyte bifurcation (Liu et al., 2016). The competition of these two pathways—elongation of an existing bud or necklace and nucleation of another bud—can lead to many different morphologies (Lipowsky, 2018b).

In order to illustrate the morphological complexity, let us consider a monodisperse batch of vesicles with a certain spontaneous curvature m. These vesicles are now osmotically deflated by the same deflation steps as in Figures 5.23–5.25. As a result, we obtain the same sequence of vesicle volumes V_N that lead to the limit shapes L_N^{in}, but let us now include the possibilities (i) that the vesicle membrane can also form, at each step, a new bud and (ii) that the same deflation step can elongate any of the existing buds and necklaces. As a result, we obtain a complex sequence of morphologies as shown in Figure 5.26.

In Figure 5.26, all morphologies with the same number N of spherules have the same bending energy (Lipowsky, 2014a; Liu et al., 2016) and represent, in fact, the states of lowest bending energy for given area A and volume V_N. The N-bead morphologies

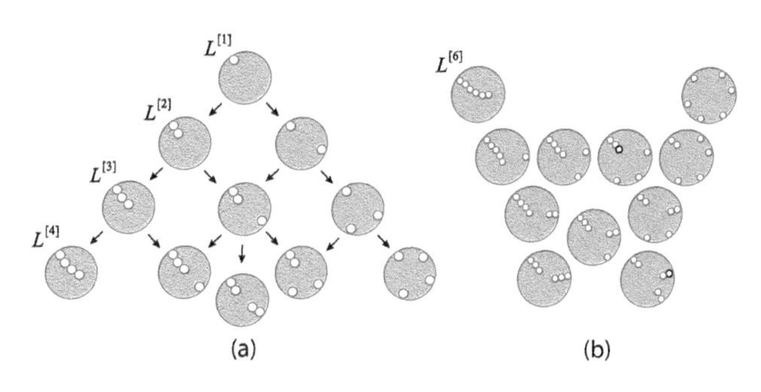

Figure 5.26 (a) The deflation of the limit shape $L^{[1]} = L_1^{in}$ in with a single in-bud (top) can lead to the shape $L^{[2]} = L_2^{in}$ with a necklace consisting of two spherules or to another shape with two in-buds. Further deflation steps (arrows) lead to an increasing number $|\Omega|$ of distinct N-bead morphologies which all have the same area, volume, and bending energy and represent, in fact, the states of lowest bending energy. Note that we have $|\Omega| = 5$ distinct morphologies with $N = 4$ spherules; and (b) For $N = 6$, the vesicle can attain $|\Omega| = 11$ different morphologies, all having the same volume, area, and bending energy as $L^{[6]} = L_6^{in}$. Neighboring morphologies differ in the location of only one bead and can be obtained by a "cut and paste" operation (Lipowsky, 2014a). In both (a) and (b), all contact zones between two spherical membrane segments contain a closed membrane neck which implies that all beads are filled with exterior solution (white).

differ, however, in the detailed arrangement of the spherules and belong to different energy branches that cross each other at volume $V = V_N$. Note also that all spherules connected to the same mother vesicle must have the same size. The latter feature follows directly from the Euler-Lagrange equation for uniform membranes because this equation allows only spherical segments with two different radii to coexist on the same vesicle.

What happens when we continue to deflate the vesicles displayed at the bottom of Figure 5.26a? It turns out that the number $|\Omega|$ of distinct N-spherule morphologies grows quite rapidly for $N > 4$. This is illustrated in Figure 5.26b by the $|\Omega| = 11$ distinct states of lowest bending energy for $N = 6$. Each of these 11 states has again the same area, volume, and bending energy. Therefore, we have 11 different branches of shapes that cross each other at volume $V = V_6$. For even larger values of N, the number $|\Omega|$ of distinct N-spherule morphologies grows exponentially with $\sqrt{N}$ as follows from known results about partitions in the sense of mathematical number theory. Furthermore, when we reach a certain volume V_N after the Nth deflation step, many n-spherule morphologies with $n < N$ can still exist as metastable states with larger spherule sizes. As a consequence, the energy landscape becomes more and more rugged as the volume decreases and the largest possible bead number N increases.

The morphological complexity described above has been recently studied experimentally by optical microscopy of giant vesicles (Bhatia et al., in preparation). These vesicles were exposed to aqueous solutions of two monosaccharides, sucrose and glucose. Varying the two sugar concentrations, one can independently change the volume-to-area ratio v and the spontaneous curvature m. As a result, a large variety of different morphologies has been observed, in agreement with the theoretical predictions.

5.6.4 CYLINDRICAL NANOTUBES

As shown in Figure 5.21b, the spontaneous tubulation of giant vesicles can also lead to cylindrical nanotubes. Cylindrical shapes are described by two shape equations, both of which differ from the shape equation for spherical shapes. In the next subsection, we will first derive the shape equations for cylinders. In the subsequent subsection, we will then combine the shape equations for cylinders and spheres in order to describe giant vesicles with cylindrical nanotubes.

5.6.5 SHAPE EQUATIONS FOR CYLINDRICAL TUBES

A cylindrical membrane segment is characterized by constant mean curvature $M = M_{cy}$ and vanishing Gaussian curvature $G = 0$. It then follows from the Euler-Lagrange Eq. 5.23 that the mean curvature M_{cy} satisfies the cubic equation

$$\Delta P = 2\Sigma M_{cy} - 4\kappa M_{cy}(M_{cy}^2 - m^2) = 2\hat{\Sigma} M_{cy} - 4\kappa M_{cy}^3 \qquad (5.195)$$

with the total membrane tension $\hat{\Sigma} = \Sigma + 2\kappa m^2$ as before. In contrast to spherical shapes, an infinitesimal scale transformation of cylindrical shapes leads to a global shape Eq. 5.45 that differs from the Euler-Lagrange Eq. 5.195. Indeed, the global shape equation has the form

$$3\Delta P = 8\hat{\Sigma} M_{cy} - 16\kappa m M_{cy}^2 \qquad (5.196)$$

for both in- and out-tubes. The Euler-Lagrange Eq. 5.195 and the global shape Eq. 5.196 can be derived in a more intuitive manner if one parametrizes the cylindrical shape by its radius R_{cy} and its length L_{cy} and minimizes the corresponding shape energy both with respect to R_{cy} and with respect to L_{cy} (Lipowsky, 2013).

We can now eliminate the term proportional to $\hat{\Sigma}$ by a combination of Eqs 5.195 and 5.196 which leads to the pressure difference

$$\Delta P = 16\kappa M_{cy}^2 (m - M_{cy}). \qquad (5.197)$$

When we insert the latter equation into Eq. 5.195, we obtain the total tension

$$\hat{\Sigma} = 8\kappa m M_{cy} - 6\kappa M_{cy}^2 \qquad (5.198)$$

and the mechanical tension

$$\Sigma = \hat{\Sigma} - 2\kappa m^2 = -6\kappa (M_{cy} - m)(M_{cy} - \frac{1}{3}m) \qquad (5.199)$$

as a function of mean curvature M_{cy}.

The two relations in Eqs 5.197 and 5.199 have two immediate consequences: (i) For fixed curvature-elastic parameters κ and $\overline{m}$, each possible value of M_{cy} leads to unique values of ΔP and Σ. Thus, as we vary the value of M_{cy}, we move along a certain line in the (Σ, ΔP)-plane; and (ii) Vice versa, for each point in the (Σ, ΔP)-plane, we find only a single solution for M_{cy}. Taken separately, both the cubic relationship Eq. 5.197 between the pressure difference ΔP and the mean curvature M_{cy} as well as the quadratic relationship Eq. 5.199 between the mechanical tension Σ and M_{cy} can lead to several solutions for M_{cy}. However, one cannot find two different values for M_{cy} that satisfy both relationships simultaneously. Therefore, these equations do not allow the coexistence of two cylinders with different radii.

5.6.6 SPONTANEOUS AND FORCE-INDUCED TUBULATION

To proceed, let us now consider a vesicle as shown in Figures 5.5 and 5.27 that has the shape of a large sphere with radius R_{sp} and a cylindrical tube with radius R_{cy} and length L_{cy}. As in the case of necklace-like tubes, we must distinguish cylindrical in-tubes as in Figure 5.27a from cylindrical out-tubes as in Figure 5.27b. To study the interplay of spontaneous and force-induced tubulation, a locally applied external force will be included that acts at the tip of the cylinder as shown in Figure 5.27. The force f is taken to be positive and negative if it points towards the exterior and interior aqueous solution, respectively, see Figure 5.27 (this convention is different from the one used in (Lipowsky, 2013), where f described the absolute value of the pulling force for both pulling directions). As shown in (Lipowsky, 2013), minimization with respect to R_{cy} and L_{cy} then leads to two equations that have the same form as Eqs 5.197 and 5.199 but with the spontaneous curvature m replaced by the composite curvature

$$m_{\rm com} \equiv m + \frac{f}{4\pi\kappa} \quad (5.200)$$

which represents the superposition of the spontaneous curvature m and the rescaled pulling force $f/(4\pi\kappa)$.

Next, we take into account that the cylindrical tubes emanate from a giant spherical vesicle as in Figure 5.27. The different membrane segments that form the tubes and the giant vesicle experience the same pressure difference ΔP and the same membrane tension $\hat{\Sigma}$. These two quantities are related to the mean curvature of the giant vesicle via the Euler-Lagrange equation

$$\Delta P = 2\hat{\Sigma} M_{\rm sp} - 4\kappa m M_{\rm sp}^2$$

as given by Eq. 5.70 with $M_{sp} = 1/R_{sp}$.

If we insert the expressions Eqs 5.197 and 5.199 for the cylinder, with m replaced by m_{com}, into the Euler-Lagrange Eq. 5.70 for the sphere, we obtain a cubic equation for the mean curvature M_{cy} which has the form (Lipowsky, 2013)

$$g(M_{\rm cy}) = 0 \quad (5.201)$$

with the polynomial

$$g(x) \equiv 4x^3 - \left(4m_{\rm com} + 3M_{\rm sp}\right)x^2 + 4m_{\rm com} M_{\rm sp} x - m M_{\rm sp}^2. \quad (5.202)$$

A cylindrical nanotube that emanates from a large mother vesicle must have a radius R_{cy} that is much smaller than the radius R_{sp} of the large mother vesicle. This separation of length scales is corroborated by the experimental observations, compare Figure 5.21b, and implies that the curvature $|M_{cy}| = 1/(2R_{cy})$ of the cylindrical tube is much larger than the curvature $M_{sp} = 1/R_{sp}$ of the giant vesicle. In this limit, the cubic equation Eq. 5.201 has the solution

$$M_{\rm cy} \approx m_{\rm com} - \frac{1}{4R_{\rm sp}} = m + \frac{f}{4\pi\kappa} - \frac{1}{4R_{\rm sp}} \quad \text{for } R_{\rm sp} \gg R_{\rm cy}. \quad (5.203)$$

Therefore, to leading order, the mean curvature of the cylindrical nanotube is equal to the composite curvature $m_{com} = m + f/(4\pi\kappa)$. For spontaneous tubulation with $f = 0$, the relation Eq. 5.203 also implies that the limit of large R_1/R_{cy} is equivalent to the limit of large $|m|R_1$ which is of the same order of magnitude as $|\bar{m}| = |m| R_{\rm ve}$.

Composite curvature and total membrane tension

Alternatively, we may also combine the Euler-Lagrange Eq 5.70 for the large sphere with the Euler-Lagrange Eq. 5.195 for the cylindrical nanotube to eliminate only the pressure difference. In the limit of giant vesicles, we then obtain the asymptotic equality

$$M_{\rm cy} \approx \pm\sqrt{\hat{\Sigma}/(2\kappa)} - \frac{1}{2R_{\rm sp}} \quad \text{for } |m| \gg 1/R_{\rm sp} \quad (5.204)$$

with the total membrane tension $\hat{\Sigma} = \Sigma + \sigma = \Sigma + 2\kappa m^2$ as in Eq. 5.26, where the plus and minus sign in Eq. 5.204 applies to out- and in-tubes, respectively. Note that the latter relation does not depend explicitly on the locally applied force f. A combination of the two asymptotic equalities Eqs 5.203 and 5.204 then leads to the relation

$$m_{\rm com} = m + \frac{f}{4\pi\kappa} \approx \pm\sqrt{\hat{\Sigma}/(2\kappa)} - \frac{1}{4R_{\rm sp}} \quad \text{for } R_{\rm sp} \gg R_{\rm cy} \quad (5.205)$$

between the spontaneous curvature m, the locally applied force f, and the total membrane tension $\hat{\Sigma} = \Sigma + \sigma$ which includes the spontaneous tension $\sigma = 2\kappa m^2$ and, thus, depends on the spontaneous curvature m as well.

It is also possible to pull both out- and in-tubes via an optical trap from the same aspirated GUV (Dasgupta and Dimova, 2014; Dasgupta et al., 2018). One can then measure the two forces f_{ex} and f_{in} that generate out- and in-tubes for the same aspiration

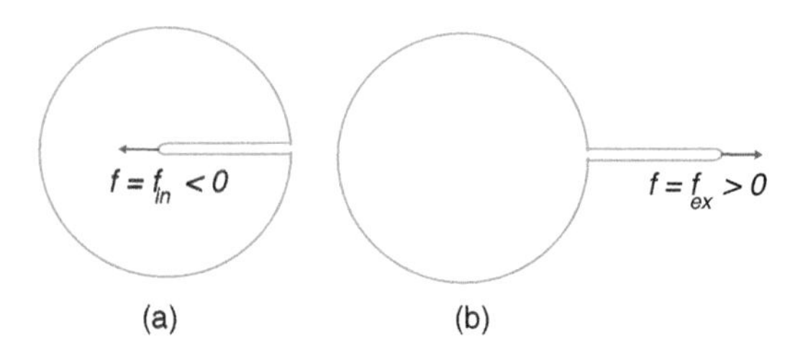

Figure 5.27 Giant vesicles with cylindrical nanotubes formed by spontaneous or force-induced tubulation: (a) Cylindrical in-tube in the presence of a pulling force $f = f_{in}$ that points towards the interior solution; and (b) Cylindrical out-tube in the presence of a pulling force $f = f_{ex}$ that points towards the exterior solution. The pulling forces f_{in} and f_{ex} are taken to be negative and positive, respectively.

pressure and, thus, for the same mechanical membrane tension Σ. Both cases are described by Eq. 5.205 with f replaced by f_{ex} for the plus sign and by f_{in} for the minus sign. The sum of these two relations leads to the simple expression

$$m \approx -\frac{f_{ex}+f_{in}}{8\pi\kappa} - \frac{1}{4R_{sp}} \quad (R_{sp} \gg R_{cy}) \tag{5.206}$$

for the spontaneous curvature m. The term $-1/(4R_{sp})$ represents again a small correction term because $|m| >> 1/R_{sp}$ as in Eq. 5.204. Therefore, one can determine the spontaneous curvature m by measuring the forces f_{ex} and f_{in}, irrespective of the membrane tension. For symmetric bilayers as studied in (Dasgupta and Dimova, 2014), the spontaneous curvature vanishes and the relation Eq. 5.206 implies that $f_{in} = -f_{ex}$. For GUVs containing a binary mixture of POPC and GM1, on the other hand, the out- and in-pulling forces, f_{ex} and f_{in}, were observed to have different magnitudes, i.e., $f_{in} \neq -f_{ex}$ which implies a nonzero spontaneous curvature (Dasgupta et al., 2018).

Total membrane tension and aspiration tension

The relationship between the composite curvature and the total membrane tension as given by Eq. 5.205 depends on the total membrane tension L_1^+. In some experimental studies of force-induced out-tubes, (Sorre et al., 2012; Simunovic et al., 2015) the relation in Eq. 5.205 was used with the total membrane tension $\hat{\Sigma}$ replaced by the aspiration tension Σ_{asp} as obtained from the spherical end cap of the membrane tongue within the micropipette. Thus, consider the membrane tongue of a GUV that is aspirated by a cylindrical micropipette with radius R_{pip}. The spherical end cap of this tongue has the mean curvature $M_{to} \leq 1/R_{pip}$ which increases initially from the value $M_{to} = 1/R_{ve}$, i.e., the mean curvature of the initial mother vesicle, up to $M_{to} = 1/R_{pip}$ and then remains constant during further aspiration. Thus, it is useful to distinguish *initial* aspiration with $1/R_{ve} < M_{to} < 1/R_{pip}$ from *prolonged* aspiration with $M_{to} = 1/R_{pip}$.

If the pressures within the interior vesicle compartment and within the pipette are denoted by P_{in} and P_{pip}, the spherical end cap of the tongue is then described by the shape equation

$$P_{in} - P_{pip} = 2\hat{\Sigma}M_{to} - 4\kappa m M_{to}^2 \tag{5.207}$$

as follows from Eq. 5.70 for spherical segments with M_{sp} replaced by M_{to}. In addition, the spherical mother vesicle with curvature radius R_{sp} and mean curvature $M_{sp} = 1/R_{sp}$ leads to the second shape equation

$$P_{in} - P_{ex} = 2\hat{\Sigma}M_{sp} - 4\kappa m M_{sp}^2$$

as in Eq. 5.70. Subtracting the latter equation from Eq. 5.207, we obtain the suction pressure

$$P_{ex} - P_{pip} = 2[M_{to} - M_{sp}]\left[\hat{\Sigma} - 2\kappa m(M_{to} + M_{sp})\right]. \tag{5.208}$$

Note that the suction pressure $P_{ex} - P_{pip}$ vanishes for $M_{to} = M_{sp}$ which corresponds to the initial contact between GUV and pipette.

Solving Eq. 5.208 for the total membrane tension $\hat{\Sigma}$, we obtain

$$\hat{\Sigma} = \Sigma_{asp} + \Delta\hat{\Sigma} \tag{5.209}$$

with the aspiration tension

$$\Sigma_{asp} \equiv \frac{P_{ex} - P_{pip}}{2(M_{to} - M_{sp})} \quad \text{for } M_{to} > M_{sp} \tag{5.210}$$

and the additional tension term

$$\Delta\hat{\Sigma} \equiv 2\kappa m(M_{sp} + M_{to}). \tag{5.211}$$

When the mean curvature M_{to} of the tongue's end cap has reached its maximal value $1/R_{pip}$, the aspiration tension and the additional tension term become

$$\Sigma_{asp} = \frac{(P_{ex} - P_{pip})R_{pip}}{2(1 - R_{pip}/R_{sp})} \tag{5.212}$$

and

$$\Delta\hat{\Sigma} = 2\kappa m(M_{sp} + 1/R_{pip}). \tag{5.213}$$

The expression in Eq. 5.212 has been widely used to obtain the aspiration tension from micropipette experiments by controlling the suction pressure $P_{ex} - P_{pip}$ and by measuring the pipette radius R_{pip} as well as the radius R_{sp} of the mother vesicle by optical microscopy. The approximation used in (Sorre et al., 2012; Simunovic et al., 2015; Dasgupta et al., 2018) was to ignore the additional tension term $\Delta\hat{\Sigma}$ and to replace the total tension $\hat{\Sigma}$ in Eq 5.205 by the aspiration tension Σ_{asp} as given by Eq. 5.212.

The accuracy of this approximation depends on the magnitude of the suction pressure and of the spontaneous curvature. As an example, let us consider a GUV membrane with bending rigidity $\kappa = 10^{-19}$ J and spontaneous curvature $m = \tilde{m}/\mu$m and let us assume that the GUV is aspirated by a micropipette of radius $R_{pip} = 3$ µm and then forms a larger spherical segment of radius $R_{sp} = 6$ µm. The additional tension term $\Delta\hat{\Sigma}$ then has the magnitude $2\kappa m(M_{sp} + 1/R_{pip}) = 0.1\tilde{m}$ µN/m which is equal to 1 µN/m for $\tilde{m} = 10$ or $m = 1/(100$ nm). This inaccuracy should be compared to the smallest values of the aspiration tension which are also of the order of 1 µN/m for the considered geometry, corresponding to the smallest accessible suction pressures of about 1 Pa. Therefore, we conclude that the additional tension term $\Delta\hat{\Sigma}$ should *not* be neglected if the spontaneous curvature is large and/or if the suction pressure is small.

Dominance of spontaneous tension

In the absence of locally applied pulling forces, the total tension $\hat{\Sigma} = \Sigma + \sigma$ of a cylindrical nanotube is given by the relation Eq 5.198, which depends on the bending rigidity κ, the spontaneous curvature m, and the tube's mean curvature M_{cy}. Inserting the asymptotic equality Eq. 5.203 for M_{cy} with $f = 0$ into Eq 5.198, the total tension becomes

$$\hat{\Sigma} \approx 2\kappa m^2 + \frac{\kappa m}{R_{sp}} = \sigma\left(1 + \frac{1}{mR_{sp}}\right) \quad \text{for large} \mid m \mid R_{sp} \qquad (5.214)$$

with the spontaneous tension $\sigma = 2\kappa m^2$. It then follows that the total membrane tension is again dominated by the spontaneous tension and that the mechanical tension $\Sigma = \hat{\Sigma} - \sigma$ behaves as (Lipowsky, 2013)

$$\Sigma \approx \frac{\kappa m}{R_{sp}} = \frac{1}{2mR_{sp}}\sigma = \pm\frac{R_{cy}}{R_{sp}}\sigma \quad \text{for large} \mid m \mid R_{sp} \qquad (5.215)$$

where the plus and minus sign applies to out- and in-tubes, respectively. Thus, in the limit of large R_{sp}/R_{cy} or large $R_{sp} \mid m \mid \simeq \mid \bar{m} \mid$ corresponding to large spherical segments or narrow tubes, the total tension $\hat{\Sigma}$ approaches the spontaneous tension σ whereas the mechanical tension goes to zero as $\Sigma \approx \kappa m/R_{sp}$.

It is interesting to note that the relation Eq. 5.215 is equivalent to $1/R_{sp} \approx \Sigma/(\kappa m)$. A combination of this latter relation with Eq 5.203 leads to

$$M_{cy} \approx m - \frac{\Sigma}{4\kappa m} \quad \text{for} \quad R_{sp} \gg R_{cy} \text{ and } f = 0. \qquad (5.216)$$

Thus, for fixed values of the curvature-elastic parameters κ and m, an increase in the mechanical membrane tension Σ leads to a reduction of $|M_{cy}|$ and, thus, to an increase in the tube radius R_{cy}. This conclusion, which applies to both $m > 0$ and $m < 0$, is somewhat counterintuitive but also follows from the quadratic expression Eq. 5.199 for the mechanical tension Σ as a function of M_{cy}. A closer look at this latter expression reveals that cylindrical tubes do not exist for

$$\Sigma > \Sigma_{max} \equiv \frac{2}{3}\kappa m^2 = \frac{\sigma}{3} \quad \text{(no cylindrical tubes).} \qquad (5.217)$$

Furthermore, starting from a cylinder with $M_{cy} = m$, corresponding to $\Sigma = 0$ and zero bending energy, an increase in the mechanical tension Σ decreases the mean curvature $|M_{cy}|$ and increases the cylinder radius $R_{cy} = 1/(2|M_{cy}|)$ until we reach $\Sigma = \Sigma_{max} = \sigma/3$ corresponding to a cylindrical tube with mean curvature $M_{cy} = 2|m|/3$ and radius $R_{cy} = 3/(4|m|)$.

5.6.7 NECKLACE-TO-CYLINDER TRANSFORMATIONS

As shown in Figure 5.21b, necklace-like and cylindrical nanotubes have been observed to coexist on the same vesicle. These observations can be understood from the competition of different energy contributions which favor necklace-like tubes below a certain critical tube length but cylindrical tubes above this length (Lipowsky, 2013; Liu et al., 2016). At the critical tube length, the necklace-like tube transforms into a cylindrical one. Such a transformation can proceed in a continuous manner *via* intermediate unduloids as shown in Figure 5.28.

The existence of a critical tube length can be understood intuitively from the following simple argument (Lipowsky, 2013). If the membrane has spontaneous curvature m, a necklace-like

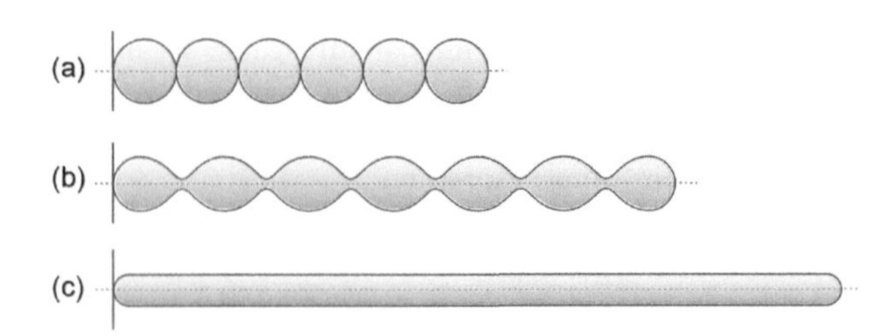

Figure 5.28 Low energy transformation of a necklace-like tube into a capped cylinder: All three tubes have the same surface area and, apart from the end caps, the same mean curvature M which is equal to the spontaneous curvature m. (a) Necklace-like tube L_6^{in} with vanishing bending energy consisting of six spherules connected by closed membrane necks. The spherules have the radius $R_2 = 1/|m|$ and mean curvature $M = -1/R_2 = m$; (b) Capped unduloid with neck radius R_{ne}, bulge radius R_{bu}, and mean curvature $M = -1/(R_{ne} + R_{bu}) = m$; and (c) Capped cylinder with radius $R_{cy} = 1/(2|m|)$ and mean curvature $M = -1/(2R_{cy}) = m$. The transformation of the sphere-necklace into the cylinder proceeds *via* a continuous family of intermediate unduloids. During this transformation, the tube volume is reduced by a factor 3/4. If we ignore the end caps of the unduloids in (b) and the cylinder in (c), both types of tubes have zero bending energy as does the necklace-like tube in (a). (Reproduced with permission from Liu, Y. et al., *ACS Nano*, 10, 463–474, 2016.)

tube consisting of spherules with radius $R_2 = 1/|m|$ connected by closed membrane necks has vanishing bending energy. For a cylindrical tube with radius $R_{cy} = 1/(2|m|)$, the main body of the cylinder also has vanishing bending energy but such a tube must be closed by two end caps which have the finite bending energy $2\pi\kappa$. Therefore, the bending energy of the end caps disfavors the cylindrical tube. On the other hand, the necklace-like tube has a larger volume compared to the cylindrical one and the osmotic pressure difference across the membranes acts to compress the tubes when they protrude into the interior solution within the vesicles. Therefore, such a tube can lower its free energy by reducing its volume which favors the cylindrical tube. The volume work is proportional to the tube length whereas the bending energy of the end caps is independent of this length. The competition between these two energies then implies that short tubes are necklace-like whereas long tubes are cylindrical.

The same conclusion is obtained by minimizing the bending energy of the whole vesicle membrane (Liu et al., 2016). One then finds that, for fixed vesicle volume and membrane area, the mother vesicle has a smaller bending energy when it forms a cylindrical tube and that this energy decrease of the mother vesicle overcompensates the bending energy increase from the end caps of the cylinder when the tube is sufficiently long. The critical tube length at which the necklace-like tube transforms into a cylindrical one is about three times the vesicle radius.

5.7 ADHESION OF VESICLES

When a vesicle is in contact with an adhesive substrate surface as in Figure 5.29, it can gain adhesion energy by spreading onto this surface but must then increase its bending energy to adapt its shape to the adhesive surface. For large vesicles, the adhesion energy must dominate because it is proportional to the contact area of the vesicle and thus grows quadratically with the size of the vesicle whereas the increased bending energy is concentrated

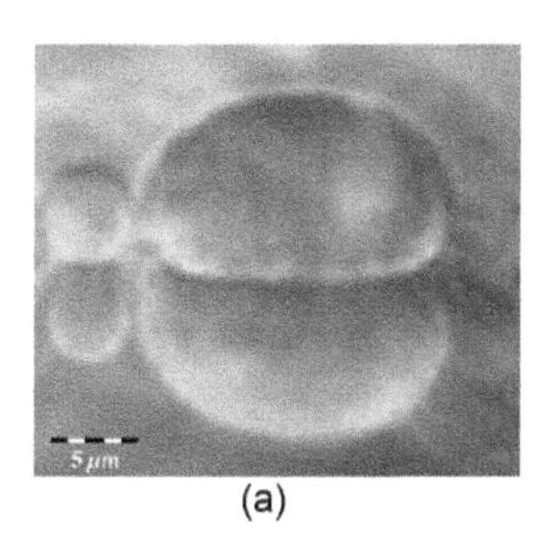

(a)

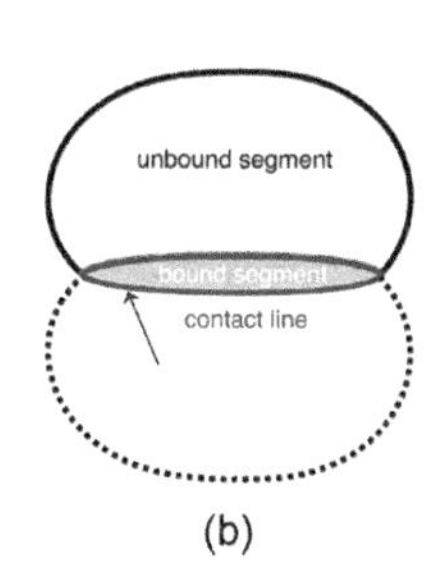

(b)

Figure 5.29 (a) Optical micrograph of two vesicles adhering to a pure glass surface that reflects the light and creates two mirror images; and (b) Shape of the larger vesicle consisting of a bound (gray region) and an unbound (white region) membrane segment. The two segments join along the contact line (red) which represents the boundary of the bound membrane segment. (Reproduced with permission from Gruhn, T. et al., *Langmuir*, 23, 5423–5429, 2007. Copyright 2007 American Chemical Society.)

along the contact line of the vesicle with the surface and thus grows only linearly with the size of the vesicle.

Within the contact area, the membrane experiences a variety of molecular forces. In order to study the overall shape of the adhering vesicle, one may ignore the molecular details and focus on the adhesive strength $|W|$ of the membrane-surface interactions which corresponds to the adhesion (free) energy per area (Seifert and Lipowsky, 1990). This coarse-grained description of the membrane-surface interactions in terms of the single parameter $|W|$ is consistent with the separation of length scales that has been used to construct the different curvature models.

Because the bound and the unbound membrane segments are exposed to different environments and, thus, to different molecular interactions, they can differ in their molecular composition and, thus, in their curvature-elastic properties (Rouhiparkouhi et al., 2013; Lipowsky et al., 2013; Lipowsky, 2014b). In order to reduce the number of parameters, we will first assume that this ambience-induced segmentation of the vesicle membranes can be ignored and that the bound and unbound membrane segments have the same curvature-elastic properties. Adhesion-induced segmentation of multi-component membranes will be discussed at the end of this section and at the end of Section 5.8.

Furthermore, we will again focus on the spontaneous curvature model which depends on only two dimensionless parameters, the volume v and the spontaneous curvature $\bar{m}$. When we parametrize the adhesion energy in terms of the dimensionless adhesive strength $|w|$ proportional to $|W|/\kappa$, vesicles adhering to planar surfaces are described by only three parameters. On the one hand, this parametrization is convenient from a theoretical point of view because it allows us to explore large regions of the parameter space. On the other hand, the additional parameter $|W|$ can be directly deduced from experimental observations of adhering vesicles. At the end of this section, more complex adhesion geometries will be briefly discussed corresponding to curved and/or chemically patterned substrate surfaces. The extension of the theory described here to the interactions of membranes with adhesive nanoparticles is described in Chapter 8 of this book. The experimental methods used to study the adhesion of GUVs are reviewed in Chapter 17.

5.7.1 INTERPLAY OF ADHESION AND BENDING

First, let us consider a planar substrate surface and focus on the competition between bending rigidity κ and adhesive strength $|W|$ for the simple case of a vesicle that is free to adapt its volume, corresponding to the osmotic pressure difference $\Delta P = 0$, and is bounded by a symmetric membrane with vanishing spontaneous curvature, $m = 0$. We are then left with only three dimensionful parameters, the membrane area A, the bending rigidity κ, and the adhesive strength $|W|$.

The non-adhering or free vesicle forms a spherical shape S_{fr} with bending energy $\mathcal{E}_{be}\{S_{fr}\} = 8\pi\kappa$. When the vesicle membrane spreads onto an adhesive surface, the vesicle attains the shape S_{ad} with contact area A_{bo} of the bound membrane segment and gains the adhesion energy

$$E_{ad} \equiv -|W|A_{bo}. \tag{5.218}$$

For a planar surface, this adhesion energy is the only energy contribution from the bound membrane segment. The unbound membrane segment, on the other hand, has to adapt its shape to the presence of the substrate surface which leads to the bending energy increase

$$\Delta E_{be} = \mathcal{E}_{be}\{S_{ad}\} - \mathcal{E}_{be}\{S_{fr}\} = 8\pi\kappa\Delta\bar{E}_{be}. \tag{5.219}$$

Adhesion is favored if

$$E_{ad} + \Delta E_{be} < 0 \quad \text{or} \quad 8\pi\kappa\Delta\bar{E}_{be} < |W|A_{bo}. \tag{5.220}$$

Because $\Delta\bar{E}_{be}$ is a dimensionless number, we can immediately conclude from this relation that the vesicle adheres to the surface if the adhesive strength $|W|$ is sufficiently large or if the bending rigidity κ is sufficiently small.

In general, the adhesion of vesicles involves three additional parameters: the osmotic conditions that determine the volume-to-area ratio, the spontaneous curvature m of asymmetric bilayers, and the mean curvature M_{bo} of the bound membrane segment arising from a curved adhesive surface. In order to take these additional parameters into account, we need a systematic theory based on an appropriate energy functional.

5.7.2 THEORY OF VESICLE ADHESION

The shape S of a vesicle that adheres to a rigid substrate surface can be decomposed into two membrane segments, a bound segment with shape S_{bo} in contact with the surface and an unbound segment with shape S_{un} not in contact with this surface. The total membrane area A can then be decomposed according to

$$A = A_{bo} + A_{un} = \mathcal{A}\{S_{bo}\} + \mathcal{A}\{S_{un}\} \tag{5.221}$$

where $A_{bo} = \mathcal{A}\{S_{bo}\}$ and $A_{un} = \mathcal{A}\{S_{un}\}$ are the partial areas of the bound and unbound membrane segments S_{bo} and S_{un}, respectively. In general, the two partial areas also depend on the shape of the adhesive surface. The combined bending and adhesion energy of the vesicle leads to the energy functional (Seifert and Lipowsky, 1990)

$$\mathcal{E}_{AV}\{S\} = 2\kappa \int dA\,(M-m)^2 + \mathcal{E}_{ad}\{S\} \qquad (5.222)$$

with the adhesion (free) energy functional

$$\mathcal{E}_{ad}\{S\} = -|W|\mathcal{A}\{S_{bo}\} \qquad (5.223)$$

where the subscript "AV" stands for "adhering vesicle." The first term on the right hand side of Eq. 5.222, which represents the bending energy functional of the spontaneous curvature model, can be decomposed into the bending energies of the unbound and the bound membrane segments according to

$$\mathcal{E}_{be}\{S\} = 2\kappa \int dA_{un}\,(M-m)^2 + 2\kappa \int dA_{bo}\,(M_{bo}-m)^2 \qquad (5.224)$$

where the mean curvature M_{bo} of the bound segment is imposed onto the latter segment by the shape of the rigid substrate.

The stationary states of the adhering vesicle are then obtained by minimizing the shape functional

$$\mathcal{F}_{AV}\{S\} = -\Delta P \mathcal{V}\{S\} + \Sigma \mathcal{A}\{S\} + \mathcal{E}_{AV}\{S\} \qquad (5.225)$$

with the constraints that $\mathcal{V}\{S\} = V$ and $\mathcal{A}\{S\} = A$ where V and A are the prescribed vesicle volume and membrane area as before. It is important to note that the value of the contact area A_{bo} of the bound membrane segment is not prescribed here which implies that the contact line is not pinned but free to find its optimal position.

Additional parameters related to adhesion

As before, it is again convenient to choose the vesicle size $R_{ve} = \sqrt{A/(4\pi)}$ as the basic length scale and the bending rigidity κ as the basic energy scale. The shape of the adhering vesicle then depends on the dimensionless volume $v = 6\sqrt{\pi}V/A^{3/2}$ and on the dimensionless spontaneous curvature $\bar{m} = mR_{ve}$, both of which also determine the shape of free vesicles. In addition, the adhering shape also depends on the dimensionless adhesion strength

$$|w| \equiv |W|R_{ve}^2/\kappa \qquad (5.226)$$

and on the dimensionless curvatures $\bar{M}_{bo} = M_{bo}R_{ve}$ that the substrate surface imposes on the bound membrane segment.

The simplest substrate geometry is provided by a planar surface with $M_{bo} = 0$ which reduces the parameter space to the three dimensionless parameters v, $\bar{m}$, and $|w|$. The next-to-simplest substrate geometry is obtained for constant-mean-curvature surfaces such as spherical surfaces or cavities. In the latter case, the mean curvature $\bar{M}_{bo}$ of the bound membrane segment is constant and the parameter space becomes four-dimensional. In the following subsections, we will first discuss the planar case and subsequently summarize the modifications arising from spherical surfaces and cavities.

5.7.3 VESICLES ADHERING TO PLANAR SURFACES

Contact curvature and contact mean curvature

For a planar substrate surface as in Figure 5.29, the bound membrane segment of the adhering vesicle is planar as well. We require the bound and the unbound membrane segments to join along the contact line in a smooth manner, i.e., that the two membrane segments have a common tangent plane or, equivalently, that the normal vector of the unbound membrane segment is also normal to the planar substrate along the contact line. In other words, the membrane shape should not exhibit any kink along the contact line. This geometric requirement is equivalent to the condition that the membrane has a finite bending energy (Seifert and Lipowsky, 1990).

Because the normal vector is required to vary continuously across the contact line, the principal curvature $C_{\parallel co}$ tangential to the contact line vanishes. In addition, the principal curvature $C_{\perp co}$ of the unbound membrane segment perpendicular to the contact line is given by

$$C_{\perp co} = \sqrt{2|W|/\kappa} \qquad (5.227)$$

as follows from the first variation of the shape functional Eq 5.225, both for axisymmetric (Seifert and Lipowsky, 1990) and for non-axisymmetric (Deserno et al., 2007) shapes. Therefore, the contact mean curvature becomes

$$M_{co} = \frac{1}{2}(C_{\parallel co} + C_{\perp co}) = \frac{1}{2}C_{\perp co} = \sqrt{|W|/(2\kappa)} \quad \text{(planar substrate).} \qquad (5.228)$$

Because the mean curvature of the bound segment vanishes, the mean curvature of the membrane jumps from $M = M_{co}$ to $M = 0$ when we cross the planar contact line.

It is interesting to note that the contact mean curvature M_{co} does not depend on the spontaneous curvature m, which is somewhat counterintuitive. This m-independence also applies when the vesicle adheres to a curved surface, see further below. However, the shape and the contact area of an adhering vesicle do depend quite significantly on the spontaneous curvature (Agudo-Canalejo and Lipowsky, in preparation).

One should also note that the principal curvature $C_{\perp co}$ jumps along the contact line from $C_{\perp co} = 0$ within the bound membrane segment to $C_{\perp co} = \sqrt{2|W|/\kappa}$ within the unbound segment. Likewise, as mentioned, the mean curvature jumps from $M = 0$ within the bound membrane segment to $M = M_{co}$ within the unbound segment. In the following sections, we will see that analogous curvature discontinuities are also present along domain boundaries separating two intramembrane domains and along three phase contact lines arising from membrane wetting.

Adhesion length

The contact mean curvature $M_{co} = \sqrt{|W|/(2\kappa)}$ as given by Eq 5.228 is a material parameter that directly encodes the competition between membrane bending as governed by the bending rigidity κ and membrane-surface adhesion as described by the adhesive strength $|W|$. For planar substrate surfaces as considered here, the inverse of the contact mean curvature is equal to the adhesion length

$$R_W \equiv \sqrt{2\kappa/|W|} = \sqrt{2/|w|}\,R_{ve}. \qquad (5.229)$$

Table 5.2 **Five combinations of lipid bilayers and adhesive materials, with estimates of the bending rigidity κ, the adhesive strength $|W|$, and the adhesion length R_W; see Appendix 1 of the book for structure and data on the lipids**

ADHESION REGIME	LIPID BILAYER	ADHESIVE MATERIAL	κ [10^{-19} J]	$\|W\|$ [mJ/m^2]	R_W [nm]
Strong	DMPC	Silica	0.8[a]	0.5–1[b]	13–18
Strong	EggPC	Glass	$\simeq 1$	0.15[c]	26
Intermediate	DMPC	Receptor-ligand	0.8[a]	0.03[d]	73
Weak	DOPC/DOPG	Coated glass	0.4[e]	3×10^{-4} [e]	510
Ultraweak	DOPC/DOPG	Glass	0.4[e]	10^{-5} [e]	2800

[a] Brüning, B.A. et al., *Biochim. Biophys. Acta*, 1838, 2412–2419, 2014.
[b] Anderson, T.H. et al., *Langmuir*, 25, 6997–7005, 2009.
[c] Schönherr, H. et al., *Langmuir*, 20, 11600–11606, 2004.
[d] Moy, V.T. et al., *Biophys. J.*, 76, 1632–1638, 1999.
[e] Gruhn, T. et al., *Langmuir*, 23, 5423–5429, 2007.

Depending on the lipid composition of the bilayer membrane and on the adhesive material, the adhesion length R_W can vary between about 10 nanometers for strong adhesion and a few micrometers for ultraweak adhesion as illustrated by the examples in Table 5.2. For the adhering vesicle displayed in Figure 5.29, the adhesion length was estimated to be 2.8 μm corresponding to the ultra-weak adhesion regime, see bottom row of Table 5.2. In this case, the contact curvature radius

$$R_{\perp co} \equiv 1/C_{\perp co} = \frac{1}{2} R_W = \sqrt{\kappa/(2|W|)} \qquad (5.230)$$

can be directly read off from the optical image displayed in Figure 5.29a.

When the adhesion length becomes of the order of 10 nanometer as in the first two rows of Table 5.2, we start to "see" the molecular structure of the lipid bilayers. As a consequence, higher-order curvature terms as discussed in Section C.1 may start to play a role. On the other hand, the estimates in the latter section also imply that we can certainly ignore such terms for $R_W \gtrsim 80$ nm.

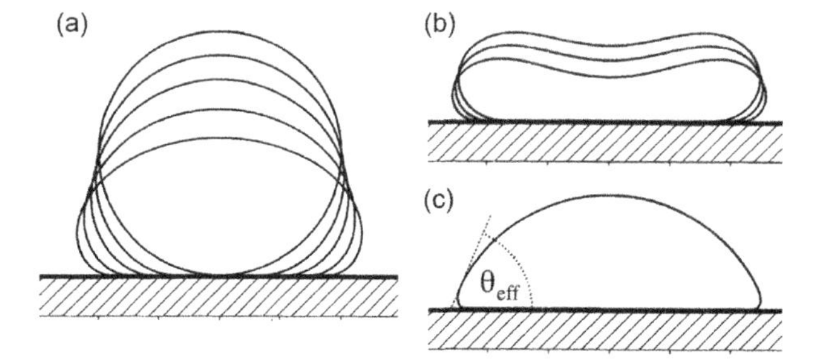

Figure 5.30 Vesicles with identical membrane area and vanishing spontaneous curvature adhering to substrate surfaces (shaded) with variable adhesive strength: (a) Vesicle shapes and five different values of the adhesive strength, $|w| = 2$, 2.9, 4.1, 6.4, and 10.2, in the absence of a volume constraint, corresponding to pressure difference $\Delta P = 0$. As $|w|$ decreases, so does the contact area of the bound membrane segment. The spherical shape with vanishing contact area is obtained for the finite value $|w| = 2$; (b) Adhering discocyte vesicles for different values of the adhesive strength $|w|$ and the pressure difference $\Delta P < 0$; and (c) In the strong adhesion regime with $|w| \gg 2$, the vesicle shape approaches a spherical cap, characterized by the effective (or apparent) contact angle θ_{eff}. (Reproduced from Seifert, U. and Lipowsky, R., *Phys. Rev.*, A 42, 4768–4771, 1990.)

Shapes of adhering vesicles

The shape of the unbound membrane segment of the adhering vesicle is obtained by solving the Euler-Lagrange Eq. 5.23 with the boundary condition as given by Eq. 5.228. If the shape is axisymmetric with respect to the normal vector of the planar surface, the Euler-Lagrange equation leads to a set of ordinary differential equations that can be solved numerically, see the examples in Figure 5.30 (Seifert and Lipowsky, 1990). In all panels of this figure, the membrane has the same area and the same bending rigidity as well as vanishing spontaneous curvature. In Figure 5.30a, we see the shapes of five vesicles that can freely adapt their volume corresponding to $\Delta P = 0$. The five vesicle shapes are obtained for five different values of the adhesive strength $|w|$.

Inspection of Figure 5.30a shows that the contact area of the bound membrane segment increases with increasing $|w|$ as one would expect intuitively. However, as we decrease the adhesive strength $|w|$, the contact area vanishes already at the threshold value

$$|w| = |w_{ad}| = 2 \quad (\Delta P = 0), \qquad (5.231)$$

corresponding to the spherical shape in Figure 5.30a. Thus, the vesicle starts to spread over the substrate surface provided (Seifert and Lipowsky, 1990)

$$|w| = |W| R_{ve}^2/\kappa > |w_{ad}| = 2 \quad \text{for } \Delta P = 0. \qquad (5.232)$$

The relation $|w| > 2$ is equivalent to the intuitive relations

$$R_{ve} > R_W \quad \text{or} \quad A|W| > 8\pi\kappa \quad (\Delta P = 0), \qquad (5.233)$$

i.e., the membrane starts to spread over the substrate surface when the vesicle size R_{ve} exceeds the adhesion length R_W. The latter criterion directly reflects the competition between the adhesive strength $|W|$ and the bending rigidity κ which favors and disfavors the onset of spreading, respectively.

The shapes in Figure 5.30a have been obtained for spontaneous curvature $m = 0$ but the threshold value $|w_{ad}| = 2$

or $|W_{ad}| = 2\kappa / R_{ve}^2$ should apply as long as the spherical shape of the free vesicle remains stable. Because a sphere with $\Delta P = 0$ is stable for $m < 3/R_{ve}$, the threshold value $|w_{ad}| = 2$ for the onset of spreading is expected to apply for this range of m-values as well. This expectation is confirmed by numerical energy minimization for axisymmetric shapes (Agudo-Canalejo and Lipowsky, in preparation). The latter calculations also show that the contact area increases with increasing spontaneous curvature $m > 0$ even though the contact mean curvature M_{co} does not depend on m.

If the vesicle volume is constrained by the osmotic conditions, the spreading of the vesicle membrane onto the adhesive surface sets in for (Lipowsky and Seifert, 1991)

$$|W| > |w_{ad}(v)| \kappa / R_{ve}^2 \tag{5.234}$$

where the dimensionless parameter $|w_{ad}|$ depends on the dimensionless volume v, approaches the value $|w_{ad}(v)| \approx 2$ for small $1 - v$, and stays of order one for arbitrary values of v. For an ensemble of vesicles with different sizes, the relation Eq. 5.234 implies that large vesicles with

$$R_{ve} > \sqrt{|w_{ad}(v)| \kappa / |W|} \quad \text{(bound vesicle)} \tag{5.235}$$

adhere to the adhesive surface whereas small vesicles do not. This difference in the size distribution of bound and free vesicles should be accessible to experiment.

General criterion for the onset of adhesion

The contact mean curvature M_{co} characterizes the membrane shape along the contact line between the bound and unbound membrane segment as described by Eq. 5.228. It turns out that this curvature also provides a general stability criterion for the onset of adhesion, i.e., for the initial spreading of the membrane onto the adhesive surface. This criterion is based on the comparison between the contact mean curvature M_{co} and the mean curvature M_{ms} of the membrane segment that comes initially into contact with the adhesive surface. Indeed, the membrane segment starts to spread onto the adhesive surface if (Agudo-Canalejo and Lipowsky, 2015a,b)

$$M_{ms} < M_{co} \quad \text{(onset of adhesion)}, \tag{5.236}$$

i.e., if the mean curvature M_{ms} of the adjacent membrane segment is smaller than the contact mean curvature M_{co}.

For a spherical vesicle with radius R_{ve}, all membrane segments have the same mean curvature, $M_{ms} = 1/R_{ve}$. Furthermore, for a planar surface as considered here, the contact mean curvature is given by $M_{co} = \sqrt{|W| / (2\kappa)}$ as in Eq 5.228. The general criterion Eq. 5.236 then assumes the form $|W| > 2\kappa / R_{ve}^2$ or $|w| > 2$ in agreement with the inequality Eq. 5.232. The general criterion for the onset of adhesion as given by Eq. 5.236 will be discussed further below for the adhesion of vesicles to spherical beads and cavities, and plays a prominent role for the engulfment of nanoparticles, see Chapter 8 of this book.

Strong adhesion regime and effective contact angle

The strong adhesion regime corresponds to the situation in which the adhesion length R_W is much smaller than the vesicle size, i.e.,

$$R_{ve} \gg R_W \quad \text{or} \quad |W| A \gg 8\pi\kappa \quad \text{or} \quad |w| \gg 2. \tag{5.237}$$

For a given value of the adhesion strength $|W|$, the strong adhesion regime corresponds to the limit of small bending rigidity κ. Thus, the limiting case $R_W/R_{ve} = 0$ can be obtained for a hypothetical membrane with vanishing bending rigidity $\kappa = 0$. In this limit, the shape functional Eq. 5.225 for the adhering vesicle reduces to

$$\mathcal{F}_{AV}\{S\} = -\Delta P \mathcal{V}\{S\} + \Sigma \mathcal{A}\{S\} - |W| \mathcal{A}\{S_{bo}\} \tag{5.238}$$

with the bound membrane segment S_{bo}. The shape functional in (5.238) is identical with the shape functional of a liquid droplet in contact with a planar surface (Lipowsky et al., 2005). This shape functional for $\kappa = 0$ is minimized by vesicle shapes which correspond to spherical caps in complete analogy to liquid droplets.

For $\kappa = 0$, the contact curvature radius R_{co} vanishes, and the vesicle forms a sharp "microscopic" contact angle with the surface along the contact line. For $\kappa > 0$ but small R_{co}/R_{ve}, the shape of the vesicle consists of a spherical cap, a strongly curved membrane segment along the contact line, and a bound membrane segment with area $A_{bo} < \frac{1}{2}A$. The strongly curved membrane segment has a mean curvature of the order of $M_{co} = (|W|/2\kappa)^{1/2}$ and provides the connection between the unbound spherical cap and the bound membrane segment. On length scales which are large compared to $1/M_{co}$, the adhering vesicle can be characterized by an effective (or apparent) contact angle θ_{eff} as in Figure 5.30c (Seifert and Lipowsky, 1990). The effective contact angle does not represent a material parameter but is determined by the spherical cap geometry and the volume-to-area ratio v via the geometric relation

$$v = 2\frac{[1-\cos(\theta_{eff})]^{1/2}[2+\cos(\theta_{eff})]}{[3+\cos(\theta_{eff})]^{3/2}}. \tag{5.239}$$

Furthermore, in the strong adhesion regime corresponding to the limit of large $|w|$, the combined bending and adhesion energy $\bar{E}_{AV} \equiv E_{AV}/(8\pi\kappa)$ of the vesicle can be expanded in powers of the dimensionless adhesive strength $|w|$ (Lipowsky and Seifert, 1991; Tordeux et al., 2002; Steinkühler et al., 2016). One then finds

$$\bar{E}_{AV} \approx -\frac{1+\cos\theta_{eff}}{2(3+\cos\theta_{eff})}|w| + 2\frac{1-\sin(\theta_{eff}/2)}{\sqrt{3+\cos\theta_{eff}}}\sqrt{|w|} \quad \text{for large } |w|. \tag{5.240}$$

When we rewrite this expression in terms of dimensionful parameters, we obtain

$$E_{AV} \approx -A_{bo}|W| + 8\sqrt{\pi}\frac{1-\sin(\theta_{eff}/2)}{\sqrt{3+\cos\theta_{eff}}}\sqrt{\kappa |W| A}. \tag{5.241}$$

The first-order term represents the adhesion energy of the bound membrane segment with area

$$A_{bo} = \frac{1+\cos\theta_{eff}}{3+\cos\theta_{eff}} A. \tag{5.242}$$

The second-order term in Eq. 5.241 is proportional to

$$\sqrt{\kappa |W| A} \sim R_{\perp co}\sqrt{A_{bo}}\,\kappa M_{co}^2 \tag{5.243}$$

where the right hand side represents an estimate for the bending energy of the strongly curved membrane segment close to the contact line because this segment has an area of the order of $R_{\perp co}\sqrt{A_{bo}}$ and the mean curvature M_{co}. Therefore, the second-order term can be regarded as a line energy term that depends, however, on the effective contact angle θ_{eff} and, thus, on the volume-to-area ratio v via the relation Eq. 5.239. In the absence of a volume constraint, i.e., for pressure difference $\Delta P = 0$, the strong adhesion regime leads to a pancake-like shape with $\theta_{eff} = 0$ and $A_{bo} = \frac{1}{2}A$. In this case, the expression Eq. 5.241 for the combined bending and adhesion energy simplifies and becomes

$$E_{AV} \approx -\frac{1}{2}A|W| + 4\sqrt{\pi}\sqrt{\kappa |W| A} \qquad \text{for large } |w| = |W|\, R_{ve}^2/\kappa. \tag{5.244}$$

5.7.4 MORE COMPLEX ADHESION GEOMETRIES

In the present subsection, we will discuss the contact mean curvature M_{co} for more complex adhesion systems as provided by curved surfaces and chemically patterned substrates.

Adhesion of vesicle to large spherical particle

When the vesicle adheres to a large spherical particle with radius R_{pa}, the bound membrane segment has the mean curvature $M_{bo} = -1/R_{pa}$ which implies the membrane curvature $C_{\parallel co} = -1/R_{pa}$ parallel to the contact line. Within the unbound membrane segment, the second principal curvature $C_{\perp co}$ perpendicular to the contact line is given by

$$C_{\perp co} = \sqrt{2|W|/\kappa} - 1/R_{pa} \tag{5.245}$$

as obtained by minimization of the bending energy (Seifert and Lipowsky, 1990). As a consequence, the contact mean curvature has the form

$$M_{co} = \frac{1}{2}(C_{\parallel co} + C_{\perp co}) = \left(\frac{|W|}{2\kappa}\right)^{1/2} - \frac{1}{R_{pa}} \tag{5.246}$$

or

$$M_{co} = \frac{1}{R_W} - \frac{1}{R_{pa}} \quad \text{(spherical particle of radius } R_{pa}) \tag{5.247}$$

where we used the definition of the adhesion length R_W as given by Eq. 5.229. The general criterion Eq. 5.236 for the onset of membrane adhesion now assumes the form (Agudo-Canalejo and Lipowsky, 2015a)

$$M_{ms} < M_{co} = \frac{1}{R_W} - \frac{1}{R_{pa}} \quad \text{(adhesion to spherical particle)} \tag{5.248}$$

where M_{ms} is the mean curvature of the membrane segment that comes initially in contact with the particle. The contact mean curvature is positive for large particles with $R_{pa} > R_W$ and negative for small particles with $R_{pa} < R_W$.[12]

Note that the principal curvature $C_{\perp co}$ and the mean curvature M are again discontinuous along the contact line. The principal curvature $C_{\perp co}$ jumps from the value $C_{\perp co} = -1/R_{pa}$ within the bound membrane segment to the value $C_{\perp co} = \sqrt{2|W|/\kappa} - 1/R_{pa}$ within the unbound membrane segment. In fact, the curvature discontinuity as given by $\sqrt{2|W|/\kappa}$ is independent of the particle size and thus applies also to the limit of a large R_{pa} corresponding to a planar surface. Likewise, as we move across the contact line, the mean curvature jumps from $M = -1/R_{pa}$ within the bound membrane segment to $M = M_{co} = 1/R_W - 1/R_{pa}$. Therefore, the discontinuity of the mean curvature is always equal to the inverse adhesion length, irrespective of the particle size R_{pa}.

Adhesion of vesicle to large spherical cavity

When the vesicle adheres to a large spherical cavity with radius R_{cav}, the bound membrane segment has the mean curvature $M_{bo} = 1/R_{cav}$ which also applies to the membrane curvature $C_{\parallel co}$ parallel to the contact line. The membrane curvature $C_{\perp co}$ perpendicular to the contact line is given by

$$C_{\perp co} = \sqrt{2|W|/\kappa} + 1/R_{cav} \tag{5.249}$$

as obtained by minimization of the energy functional. As a consequence, the contact mean curvature now has the form

$$M_{co} = \frac{1}{2}(C_{\parallel co} + C_{\perp co}) = \left(\frac{|W|}{2\kappa}\right)^{1/2} + \frac{1}{R_{cav}} \quad \text{(spherical cavity of radius } R_{cav}). \tag{5.250}$$

It now follows from the general adhesion criterion Eq. 5.236 that a membrane segment with mean curvature M_{ms} starts to adhere to the cavity wall if

$$M_{ms} < M_{co} = \frac{1}{R_W} + \frac{1}{R_{cav}} \quad \text{(adhesion to a spherical cavity)} \tag{5.251}$$

with the adhesion length R_W as defined by Eq. 5.229. Therefore, as we move across the contact line, the mean curvature now jumps from $M = 1/R_{cav}$ within the bound membrane segment to $M = M_{co} = 1/R_W + 1/R_{cav}$ within the unbound membrane segment, with the curvature discontinuity being again equal to $1/R_W$.

[12] The limiting case with $M_{ms} = M_{co}$ can be further elucidated for nanoparticles with $R_{pa} \ll R_{ve}$, see Eq. 5.257 below.

Adhesion of vesicle to chemically patterned surface

Finally, let us consider the adhesion of vesicles to a planar but chemically structured surface which contains two types of surface domains, D_1 and D_2. These two types of domains are characterized by two different adhesive strengths, W_1 and W_2, with $|W_2| < |W_1|$, i.e., the D_2 domain is less adhesive than the D_1 domain.

If the contact line of an adhering vesicle is located *within* the D_1 domain, the contact mean curvature is given by

$$M_{co}^{[1]} = \left(\frac{|W_1|}{2\kappa}\right)^{1/2}. \quad (5.252)$$

Likewise, for a contact line within the D_2 domain, the contact mean curvature is

$$M_{co}^{[2]} = \left(\frac{|W_2|}{2\kappa}\right)^{1/2} < M_{co,1}. \quad (5.253)$$

On the other hand, if a contact line segment (CLS) of the vesicle is pinned to the boundary between the two surface domains, the contact curvature radius $M_{co} = M_{co}^{pin}$ is not fixed but can vary within the range (Lipowsky et al., 2005)

$$M_{co}^{[2]} \le M_{co}^{pin} \le M_{co}^{[1]} \quad \text{(pinned CLS)}. \quad (5.254)$$

This freedom of the contact mean curvature M_{co}^{pin} along the boundaries of surface domains leads to transitions between different shapes of adhering vesicles (Lipowsky et al., 2005). One example is provided by a vesicle on a striped surface domain that is strongly adhesive and surrounded by another surface domain that is non-adhesive or only weakly adhesive. When the volume-to-area ratio v is close to a sphere, the adhering vesicle has a fairly compact shape and a relatively small contact area. During deflation, the vesicle then undergoes a morphological transition from this compact shape to a thin tube-like state with a large contact area.

5.7.5 ENDOCYTOSIS OF NANOPARTICLES

The adhesion of nanoparticles to cell membranes represents the first step for the process of endocytosis which is essential for the cellular uptake of such particles, see Chapter 8 of this book. In general, the endocytosis of a nanoparticle that comes into contact with the outer leaflet of the membrane consists of three steps: Onset of particle adhesion, spreading of the membrane over the particle surface until the particle is completely engulfed by the membrane, and cleavage (or scission) of the membrane neck connecting the completely engulfed particle with the mother membrane.

Completely engulfed particle

When a particle in contact with the outer leaflet becomes completely engulfed, the membrane forms a limit shape with a closed membrane neck. For this limit shape, the mean curvature M'_{ms} of the unbound membrane segment adjacent to the membrane neck satisfies the neck closure condition (Agudo-Canalejo and Lipowsky, 2015a)

$$M'_{ms} + M_{co} = M'_{ms} + \frac{1}{R_W} - \frac{1}{R_{pa}} = 2m \quad (5.255)$$

with the contact mean curvature M_{co} as given by Eq. 5.247. Comparison with the neck closure condition for spherical in- and out-buds as described by Eq. 5.50 and Figure 5.14 shows that the mean curvature of the bud is now replaced by the contact mean curvature M_{co} of the adhesive nanoparticle. Furthermore, the closed neck is stable provided

$$M'_{ms} + M_{co} - 2m \ge 0 \quad \text{(stable neck, endocytosis)}. \quad (5.256)$$

in close analogy to the case of an in-bud with a stably closed neck as described by Eq. 5.61.

The presumably simplest way to derive the neck closure condition in Eq. 5.255 is to require that the bending energy density of the membrane as given by $2\kappa(M - m)^2$, see Eq. 5.12, is continuous across the neck. The latter requirement implies $(M_{co} - m)^2 = (M'_{ms} - m)^2$ or $M_{co} - m = \pm(M'_{ms} - m)$. The root with the plus sign leads to $M_{co} = M'_{ms}$ and thus to a continuous variation of the mean curvature. The root with the minus sign, on the other hand, is equivalent to the neck closure condition in Eq. 5.255. In (Agudo-Canalejo and Lipowsky, 2016), the two relations in Eqs 5.255 and 5.256 have been derived in a systematic manner by calculating the free energy of certain membrane shapes with small neck radii R_{ne} and taking the limit of zero R_{ne}.

Energy landscape for small particles

In the limit of small particles with $R_{pa} \ll R_{ve}$, one can identify the mean curvature M'_{ms} of the unbound membrane segment adjacent to the closed neck for the completely engulfed particle with the mean curvature M_{ms} of the membrane segment that comes initially into contact with the particle, see Eq. 5.248 (Agudo-Canalejo and Lipowsky, 2015b). One can then explicitly calculate the local (free) energy landscape E as a function of the area fraction q of the particle surface that is covered by the vesicle membrane. The physically meaningful range of q-values corresponds to $0 \le q \le 1$. For small particles, the energy landscape is then found to have the simple quadratic form (Agudo-Canalejo and Lipowsky, 2017)

$$E(q) = E(0) + 16\pi\kappa R_{pa}[(M - M_{co})q + (m - M)q^2] \quad (5.257)$$

which depends on three parameters: the local mean curvature $M = M_{ms} = M'_{ms}$, the contact mean curvature M_{co}, and the spontaneous curvature m.

Local conditions for adhesion plus engulfment

Complete engulfment with a stable membrane neck corresponds to an energy landscape $E(q)$ that has a boundary minimum at $q = 1$. The latter criterion is equivalent to the stability condition in Eq. 5.256. Furthermore, the completely engulfed particle state represents the *global* minimum of this energy landscape when the three curvatures satisfy the inequalities

$$M_{co} \ge M \ge 2m - M_{co}. \quad (5.258)$$

The first inequality corresponds to the local criterion for the onset of adhesion, the second inequality to a completely engulfed particle with a stable membrane neck. Therefore, the inequalities in Eq. 5.258 imply both adhesion and complete engulfment of the nanoparticle.

Effective constriction forces

The stability relation as given by Eq. 5.256, which applies to a stably closed neck for the complete engulfment of a nanoparticle, can be generalized by including an external force $f > 0$ that acts to constrict the membrane neck. Such a force contributes the term fR_{ne} to the energy of the vesicle-particle system which is proportional to the neck radius R_{ne} (Agudo-Canalejo and Lipowsky, 2016). One then finds the stability relation

$$\frac{f}{4\pi\kappa} + M'_{ms} + M_{co} - 2m \geq 0 \tag{5.259}$$

which defines the effective constriction force

$$f_{eff}^{in} \equiv 4\pi\kappa(M'_{ms} + M_{co} - 2m) \quad \text{for endocytosis.} \tag{5.260}$$

For small M'_{ms}, i.e., for a weakly curved membrane of the mother vesicle, the effective constriction force behaves as

$$f_{eff}^{in} \approx f_W^{in} + f_m^{in} \tag{5.261}$$

with the adhesion-induced constriction force

$$f_W^{in} \equiv 4\pi\kappa\left(\frac{1}{R_W} - \frac{1}{R_{pa}}\right) \tag{5.262}$$

and the curvature-induced constriction force

$$f_m^{in} \equiv -8\pi\kappa m, \tag{5.263}$$

where f_m^{in} has the same form as in Eq. 5.137.

The final step of endocytosis corresponds to the cleavage (or scission) of the membrane neck. As explained in Section 5.4, the cleavage of a neck with radius R_{ne} leads to two bilayer edges and to a free energy barrier of the order of $4\pi R_{ne}\lambda_{ed}$ which depends on the edge tension λ_{ed}. To overcome this barrier, the effective constriction force must be sufficiently large and satisfy

$$f_{eff}^{in} \approx f_W^{in} + f_m^{in} \gg 4\pi\lambda_{ed}. \tag{5.264}$$

Inspection of Eq. 5.262 for the adhesion-induced constriction force f_W^{in} shows that this force facilitates neck cleavage for strong adhesion with $1/R_W \gg 1/R_{pa}$. Thus, even for a symmetric membrane with $m = 0$ and $f_m^{in} = 0$, strong adhesion with

$$f_W^{in} \gg 4\pi\lambda_{ed} \quad \text{or} \quad R_W \ll \frac{\kappa}{\lambda_{ed}} \tag{5.265}$$

leads to neck cleavage and, thus, to the release of the membrane-enclosed nanoparticle from the mother membrane. Using the typical value $\kappa = 10^{-19}$ J for the bending rigidity and the estimate $\lambda_{ed} \gtrsim 1$ pN for the edge tension, the inequality in Eq. 5.265 predicts neck cleavage for an adhesion length R_W that is small compared to 100 nm.

5.7.6 AMBIENCE-INDUCED SEGMENTATION

The membranes considered in the previous sections were taken to have a laterally uniform composition which implies laterally uniform curvature-elastic properties even if they contained several molecular components. However, when a multi-component membrane is in contact with an adhesive surface, different membrane components will typically experience different molecular interactions with this surface, which implies that the membrane-surface interactions can lead to an enrichment or depletion of the different components within the bound segment of the vesicle membrane. As a consequence, the bound membrane segment will, in general, differ in its composition from the unbound segment of the membrane which provides an example for ambience-induced segmentation of membranes as displayed in Figure 5.31a (Rouhiparkouhi et al., 2013; Lipowsky et al., 2013; Lipowsky, 2014b). For two-component membranes, this kind of segmentation has been theoretically studied in some detail, see Appendix 5.G.

The adhesion geometry in Figure 5.31a corresponds to a chemically uniform substrate surface which leads to only two membrane segments, one bound and one unbound segment. If the substrate surface is chemically patterned as in Figure 5.31b and consists of two chemically distinct surface domains, both of which are adhesive but differ in their adhesive strengths, the vesicle membrane is partitioned into three different segments, corresponding to two different bound segments and one unbound segment. An even more complex geometry is depicted in Figure 5.31c: three vesicle membranes that differ in their overall

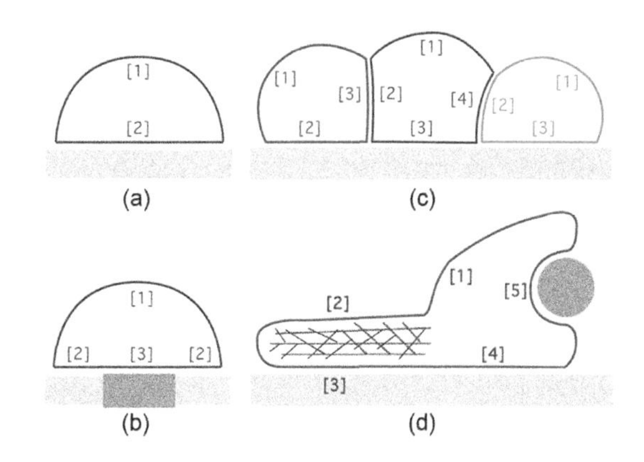

Figure 5.31 Ambience-induced segmentation of membranes that are exposed to different local environments: (a) Vesicle adhering to a planar, chemically uniform substrate surface; (b) Vesicle adhering to a planar and chemically patterned surface; (c) Cluster of three vesicles adhering to a planar, chemically uniform surface and to each other; and (d) Cartoon of a macrophage that moves along a solid surface and engulfs a small particle. The colors of the membranes represent their overall compositions. For each membrane, the numbers [k] = [1], [2], etc indicate the different ambience-induced membrane segments. Because of the different molecular interactions between the membrane components and the different environments, each membrane segment will, in general, have a molecular composition that differs from the overall composition. (From Lipowsky, R. *Biol. Chem.*, 395, 253–274, 2014.)

compositions and interact both with the solid support and with other membranes. In addition, Figure 5.31d displays, in a rather schematic manner, the outer cell membrane of a macrophage that moves along a solid surface, contains some cytoskeletal filaments, and engulfs a microparticle.

In all examples displayed in Figure 5.31, the different membrane segments, labeled by $[k] = 1, 2, ..., K$, can differ in their molecular composition which implies that they can also differ in their curvature-elastic properties. We are then led to consider membrane segments with different bending rigidities $\kappa^{[k]}$ and different spontaneous curvatures $m^{[k]}$. This approach has been recently applied to clathrin-dependent endocytosis which involves two membrane segments, corresponding to the presence and absence of the clathrin-containing protein coat (Agudo-Canalejo and Lipowsky, 2015a). The latter process is discussed in more detail in Chapter 8 of this book.

Ambience-induced segmentation of vesicle membranes has been recently observed for giant vesicles that adhere to planar electrodes (Steinkühler et al., 2016). The vesicles contained anionic lipids and adhered to the positively charged electrode at the bottom of the chamber. Using fluorescence quenching assays, the bound membrane segment was observed to have a different composition than the unbound segment, but, in contrast to naive expectations, only the outer leaflet of the bilayer membrane was affected and the bound segment of this latter segment was depleted of anionic lipids.

Ambience-induced segmentation will play an important role in the next two sections on membrane phase separation (Section 5.8) and membrane wetting (Section 5.9). Indeed, the interplay of ambience-induced segmentation and membrane phase separation (Section 5.8.5) confines the phase transition, for a given composition, to one of the membrane segments and each of these phase transitions occurs for a reduced range of compositions. In the case of wetting, the membranes are exposed to different aqueous phases that provide different local environments for these membranes, in close analogy to the adhesive substrate surfaces that have been discussed in the present section.

5.8 MEMBRANE PHASE SEPARATION AND MULTI-DOMAIN VESICLES

Biological and biomimetic membranes are fluid, contain several molecular components, and represent two-dimensional systems. As a consequence, the membranes should be able to undergo phase separation into two different liquid phases, in close analogy to phase separation of liquid mixtures in three dimensions. Membrane phase separation proceeds via the formation of intramembrane domains that differ in their molecular composition from the surrounding membrane matrix. The presence of domains implies the appearance of a new parameter, the line tension, which acts to shorten the domain boundaries (Lipowsky, 1992).

In the context of liquid droplets, the tension of the three-phase contact line, which was already considered by Gibbs, represents a relatively small correction term to the interfacial free energies that can be completely ignored on the micrometer scale. In contrast, the line tension associated with intramembrane domains has a rather strong effect on the

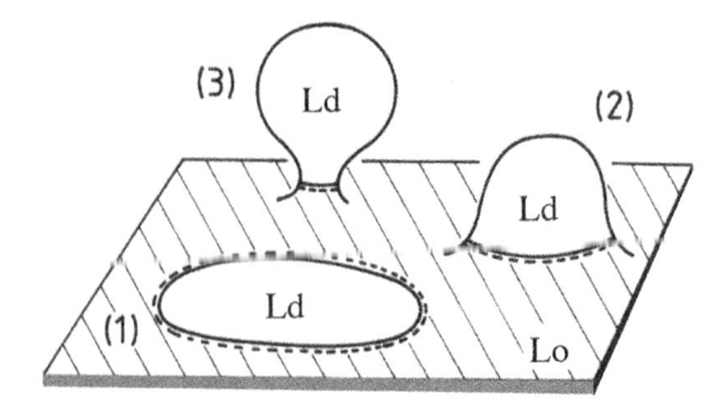

Figure 5.32 Domain-induced budding of a growing liquid-disordered (Ld) domain within an liquid-ordered (Lo) matrix: (1) Essentially flat Ld domain; (2) Partial Ld bud; and (3) Complete Ld bud. During the time evolution from (1) to (3) the domain boundary between the Ld domain and the Lo matrix shortens and the line energy of this boundary decreases continuously. In the following, the letters *a* and *b* will be used to indicate two coexisting fluid phases within the membranes. These membrane phases can be pure lipid phases or involve membrane proteins as well. (With kind permission Springer Science + Business Media: *J. Phys. II France*, Budding of membranes induced by intramembrane domains, 2, 1992, 1825–1840. Lipowsky, R.)

shape of membranes and vesicles. Indeed, the line tension of the domain boundaries can induce new types of shape transformations such as domain-induced budding, displayed in Figures 5.32 and 5.3. The latter process was first predicted theoretically (Lipowsky, 1992, 1993; Jülicher and Lipowsky, 1993) and then confirmed experimentally by optical microscopy of giant vesicles (Baumgart et al., 2003, 2005; Bacia et al., 2005; Dimova et al., 2007; Semrau et al., 2008).

At the beginnings of the 1990s, it was rather difficult to find experimental evidence for the coexistence of two fluid phases in membranes. This situation has now changed completely because many ternary lipid mixtures have been identified which exhibit two coexisting fluid phases, a liquid-ordered (Lo) and a liquid-disordered (Ld) phase. These lipid mixtures, which consist of a saturated lipid such as sphingomyelin, an unsaturated phospholipid, and cholesterol, form vesicles with several intramembrane domains. The intense experimental study of these mixtures was triggered by the proposal (Simons and Ikonen, 1997) that biological membranes contain intramembrane domains or rafts that are rich in sphingomyelin and cholesterol. In order to directly visualize the different domains formed in lipid vesicles, it was also crucial to find appropriate fluorescent probes that have a preference for one of the two fluid phases (Korlach et al., 1999; Dietrich et al., 2001; Veatch and Keller, 2003; Baumgart et al., 2003).

In this section, we will review the morphologies of multi-domain membranes and vesicles. We will consider multi-component membranes that consist of lipids and proteins and form two coexisting membrane phases, both of which are in a fluid state. Thus, the intramembrane domains could be pure lipid domains but they could also contain membrane proteins that participate in the phase separation. In the next subsection, the process of domain-induced budding as depicted in Figure 5.32 will be discussed. Second, the shape functional for two-domain vesicles will be described in some detail. The morphologies of these vesicles involve again closed membrane necks which are now governed by the interplay between the spontaneous curvatures of the two types of domains and the line tension of the domain boundary. In addition, the Gaussian curvature moduli of the two membrane domains also affect the vesicle shape and determine the relative position of domain boundary and membrane neck. If the two domains differ in their bending rigidities, this rigidity difference can stabilize multi-domain vesicles with more than two domains and thus truncate the phase separation process. Such multi-domain vesicles undergo morphological transitions which involve changes of both the vesicle shape and the domain pattern (Gutlederer et al., 2009; Hu et al., 2011). Finally, in Section 7.6, we will address the interplay between membrane phase separation and ambience-induced segmentation which acts to confine the phase separation to single membrane segments. The experimental methods to identify two coexisting fluid phases within the membranes of GUVs are reviewed in Chapter 18 of this book.

This section is supplemented by two appendices: Appendix 5.F on the matching conditions and curvature discontinuities along domain boundaries; and Appendix 5.G which discusses the interplay of segmentation and phase separation for two-component membranes.

5.8.1 BUDDING OF INTRAMEMBRANE DOMAINS

To be specific, let us consider a single Ld domain embedded in a larger Lo matrix as shown in Figure 5.32. Because the two phases differ in their molecular composition, they will also differ in their curvature-elastic parameters. First, the Ld phase is more flexible than the Lo phase. Second, the two phases will, in general, have different spontaneous curvatures. One mechanism that generates such a difference in preferred curvature is provided by adsorbate molecules with different affinities to the two phases. In addition, the domain boundary contributes a line (free) energy that is proportional to its length; the corresponding free energy per length defines the line tension λ (Lipowsky, 1992, 1993).

To simplify the notation, the Lo and Ld phases will now be denoted by the letters *a* and *b*. The Lo- or *a*-phase has the bending rigidity κ_a and the spontaneous curvature m_a. Likewise, the Ld- or *b*-domain has the bending rigidity κ_b and the spontaneous curvature m_b. We will first ignore possible contributions from the Gaussian curvature moduli which will be discussed further below.

In order to focus on the *b*-domain, let us further assume that the *a*-matrix is weakly curved and that its spontaneous curvature m_a can be ignored. After nucleation, the *b*-domain is weakly curved as well, see state (1) in Figure 5.32. The domain area A_b then grows by diffusion-limited aggregation. For a circular domain, the domain has the radius $L_b = \sqrt{A_b/\pi}$ which implies the domain boundary length $2\pi L_b$. The domain energy is then given by

$$E_{(1)} = 2\pi L_b \lambda + 2A_b \kappa_b m_b^2 \tag{5.266}$$

where the first term represents the line energy of the domain boundary and the second term the bending energy of the flat *b*-domain with spontaneous curvature m_b. If we transform the flat domain into a spherical bud connected to the *a*-matrix by a narrow membrane neck, see state (3) in Figure 5.32, we get essentially rid of the line energy. We now assume that the budding process is sufficiently fast and that we may ignore changes in the domain area A_b during this process. The bud then has the radius $R_b = \frac{1}{2}L_b$ and the energy

$$E_{(3)} = 8\pi\kappa_b(1 - R_b\,|\,m_b\,|)^2. \tag{5.267}$$

Budding is energetically favored for $E_{(3)} - E_{(1)} < 0$ or (Lipowsky, 1992)

$$L_b = 2R_b > \frac{4\xi_b}{1 + 4\xi_b\,|\,m_b\,|} \equiv L_{b,1} \quad \text{(bud energetically favored)} \tag{5.268}$$

with the invagination length

$$\xi_b \equiv \kappa_b / \lambda \tag{5.269}$$

This simple argument shows that the competition between bending and line tension leads to two regimes for the bud size, depending on the relative size of the invagination length ξ_b and the spontaneous curvature m_b. If the spontaneous curvature m_b is small compared to the inverse invagination length $1/\xi_b = \lambda/\kappa_b$, the budding process is dominated by the line tension, and the bud radius $R_b \approx 4\xi_b$. On the other hand, if the spontaneous curvature m_b is large, the budding process is dominated by this curvature and $R_b \sim 1/\,|\,m_b\,|$.

The argument just described ignores the stability of the closed neck between the *b*-bud and the weakly curved *a*-matrix. As discussed further below, such a neck is stable if

$$L_b = 2R_b > \frac{4\xi_b}{1 + 2\xi_b\,|\,m_b\,|} \equiv L_{b,2} \quad \text{(stability of closed neck).} \tag{5.270}$$

Comparison of the two criteria Eqs 5.268 and 5.270 indicates that the budding transition at $L_b = L_{b,1}$ occurs before the closed neck of the bud becomes stable at $L_b = L_{b,2} > L_{b,1}$. This conclusion is corroborated by systematic energy minimization calculations (Jülicher and Lipowsky, 1993, 1996) as described next.

5.8.2 THEORY OF TWO-DOMAIN VESICLES

When a vesicle membrane undergoes phase separation into two coexisting phases *a* and *b*, it will initially form many small *a*- and/or small *b*-domains which will then coarsen into larger domains.[13] In this subsection, we will consider the simplest situation in which the completion of this coarsening process leads to one large *a*-domain coexisting with one large *b*-domain. Further below, we will also discuss the possibility that the coarsening process is truncated and leads to an equilibrium state of a multi-domain vesicle with more than two domains.

Geometry and energetics of two-domain vesicles

Now, consider a vesicle of volume V that is bounded by a membrane with one *a* domain and one *b* domain. We can then decompose the vesicle shape S into three components: the shapes S_a and S_b of the two domains as well as the shape S_{ab} of the *ab* domain

[13] We focus here on the nucleation regime close to the binodal line of the membrane phase diagram. Further away from this line, the multi-component membrane phase separates via spinodal decomposition for which the description in terms of sharp domain boundaries does not apply.

boundary. The a and b domains have the surface areas A_a and A_b, respectively. The total area of the vesicle membrane is then given by

$$A = A_a + A_b = \mathcal{A}\{S_a\} + \mathcal{A}\{S_b\} \quad (5.271)$$

where $\mathcal{A}\{.\}$ denotes the area functional as before. The ab domain boundary with shape S_{ab} has a certain length, $\mathcal{L}\{S_{ab}\} = L_{ab}$ where $\mathcal{L}\{.\}$ denotes the length functional.

The energy of a two-domain vesicle can be decomposed into several contributions: the curvature energy of the a domain, the curvature energy of the b domain, and the line energy of the ab domain boundary. As for a GUV with a uniform or single-domain membrane, the curvature energies can be further decomposed into bending and Gaussian curvature contributions. The energy functional of the two-domain vesicle then has the form

$$\mathcal{E}_{2\text{Do}}\{S\} = \mathcal{E}_{\text{be}}\{S_a\} + \mathcal{E}_{\text{be}}\{S_b\} + \mathcal{E}_G\{S_a, S_b\} + \lambda \mathcal{L}\{S_{ab}\}. \quad (5.272)$$

The last term on the right hand side of this equation represents the contribution of the domain boundary which is proportional to the line tension λ (Lipowsky, 1992). Any stable domain pattern implies that the line tension λ has to be positive as will be assumed in the following. The energy functional

$$\mathcal{E}_G\{S_a, S_b\} \equiv \kappa_{Ga} \int \mathrm{d}A_a G + \kappa_{Gb} \int \mathrm{d}A_b G \quad (5.273)$$

represents the combined Gaussian curvature terms of both domains and depends on the Gaussian curvature moduli κ_{Ga} and κ_{Gb} of the a- and b-domains. Finally, the bending energy functionals $\mathcal{E}_{\text{be}}\{S_a\}$ and $\mathcal{E}_{\text{be}}\{S_b\}$ have the form

$$\mathcal{E}_{\text{be}}\{S_a\} = 2\kappa_a \int \mathrm{d}A_a (M - m_a)^2 \quad \text{and} \quad \mathcal{E}_{\text{be}}\{S_b\} = 2\kappa_b \int \mathrm{d}A_b (M - m_b)^2 \quad (5.274)$$

which generalizes the spontaneous curvature model for a uniform membrane to the case of two different domains. These energy functionals depend on the bending rigidities κ_a and κ_b as well as on the spontaneous curvatures m_a and m_b.

Shape functional for two-domain vesicles

The equilibrium shapes of a two-domain vesicle are obtained by minimizing the energy functional Eq. 5.272 for a certain volume $V = \mathcal{V}\{S\}$ and for certain areas A_a and A_b of the a- and b-domains. These three constraints can be taken into account by three Lagrange multipliers ΔP, Σ_a, and Σ_b. As a consequence, the shape functional of the two-domain vesicle has the form

$$\mathcal{F}_{2\text{Do}}\{S\} = -\Delta P \mathcal{V}\{S\} + \Sigma_a \mathcal{A}\{S_a\} + \Sigma_b \mathcal{A}\{S_b\} + \mathcal{E}_{2\text{Do}}\{S\}. \quad (5.275)$$

So far, a systematic minimization of this functional has been performed for axisymmetric vesicles using the shooting method (Jülicher and Lipowsky, 1993, 1996) and, to some extent, by numerical minimization of discretized membranes (Gutlederer et al., 2009; Hu et al., 2011). In these numerical studies, the spontaneous curvatures were taken to be relatively small. The same energy functional has also be used to calculate doubly-periodic bicontinuous shapes corresponding to "lattices of passages" (Góz'dz' and Gompper 1998).

Gaussian curvature energies

The energy functional of a two-domain vesicle contains the Gaussian curvature term $\mathcal{E}_G\{S_a, S_b\}$ as given by Eq. 5.273. If the two Gaussian curvature moduli κ_{Ga} and κ_{Gb} are equal, this term does not depend on the shape but only on the topology of the vesicle and is then given by

$$\mathcal{E}_G\{S_a, S_b\} = 2\pi\chi\kappa_G \quad \text{for } \kappa_{Ga} = \kappa_{Gb} = \kappa_G \quad (5.276)$$

where χ denotes the Euler characteristic of the whole vesicle, see Appendix 5.B. In the following, we will consider two-domain vesicles that have a spherical topology characterized by $\chi = 2$.

If the Gaussian curvature moduli of the a- and b-phases are different, however, the Gaussian curvature terms also make a shape-dependent contribution. Indeed, the Gaussian curvature term in Eq. 5.273 then becomes (Jülicher and Lipowsky, 1993, 1996)

$$\mathcal{E}_G\{S_a, S_b\} = -\Delta\kappa_G \oint \mathrm{d}l\, C_g + 2\pi(\kappa_{Ga} + \kappa_{Gb}). \quad (5.277)$$

with the difference

$$\Delta\kappa_G \equiv \kappa_{Ga} - \kappa_{Gb} \quad (5.278)$$

of the Gaussian curvature moduli. The first term on the right hand side of Eq. 5.277 is proportional to this difference $\Delta\kappa_G$ and to the line integral of the geodesic curvature C_g along the domain boundary. To obtain the correct sign of this term, the orientation of the line element $\mathrm{d}l$ has to be chosen in such a way that the line integral moves around the b-domain in a clockwise manner when one looks down onto this domain from the exterior solution. The line integral along the domain boundary implies that the first term on the right hand side of Eq. 5.277 depends on the shape S_{ab} of the domain boundary. In contrast, the second term on the right hand side of Eq 5.277 does not depend on the morphology of the vesicle but reflects its spherical topology. For $\kappa_{Ga} = \kappa_{Gb} = \kappa_G$, the first term vanishes and the second term reduces to $4\pi\kappa_G$ as in Eq. 5.276 with $\chi = 2$.

Euler-Lagrange or local shape equations

The first variation of the shape functional $\mathcal{F}_{2\text{Do}}\{S\}$ as given by Eq 5.275 leads to two Euler-Lagrange equations for the (local) mean curvature M and the (local) Gaussian curvature G within the membrane domains with shapes S_a and S_b. These equations have the form

$$\Delta P = 2\hat{\Sigma}_i M - 2\kappa_i \nabla^2_{\text{LB}} M - 4\kappa_i m_i M^2 - 4\kappa_i [M - m_i][M^2 - G] \quad (5.279)$$

with $i = a, b$, the total membrane tensions

$$\hat{\Sigma}_i \equiv \Sigma_i + 2\kappa_i m_i^2, \quad (5.280)$$

and the Laplace-Beltrami operator ∇^2_{LB}, generalizing the Euler-Lagrange Eq. 5.25 for a uniform membrane. When the two types of domains form spherical segments, the terms proportional to $M^2 - G$ vanish and we obtain two quadratic equations for the corresponding constant mean curvatures $M = M_a$ and $M = M_b$. Each of these quadratic equations can have up to two solutions which implies that the two-domain vesicles can form coexisting spherical segments with up to four different mean curvatures. One example is a two-domain vesicle with three closed membrane necks: one neck connects two membrane segment of *a* phase, one neck two membrane segments of *b* phase, and the third neck connects the *a* domain with the *b* domain. The latter neck is governed by a neck condition that includes the line tension of the domain boundary, see further below.

Matching conditions along the domain boundary

In addition to the two Euler-Lagrange Eqs 5.279, we need to impose appropriate matching conditions along the boundary between the two membrane domains. In the theoretical description considered here, we ignore the width of the *ab* domain boundary.[14] This simplification is justified when the linear size of the *a* and *b* domain is large compared to the boundary width, a condition that is usually fulfilled for the optically resolvable membrane domains of giant vesicles. Because we ignore the width of the domain boundary, the bending rigidity and the spontaneous curvature change abruptly as we cross this boundary. Nevertheless, we can still impose the physical requirement that the shapes of the two membrane domains meet "smoothly" along the domain boundary, i.e., that these shapes have a common tangent along this boundary, as explicitly shown for axisymmetric vesicle shapes (Jülicher and Lipowsky, 1996).

Even for axisymmetric vesicle shapes with smooth contours, the matching conditions turn out to be somewhat complex. Indeed, these matching conditions can lead to discontinuities along the domain boundary, both for the curvature and for the mechanical tension. For an axisymmetric vesicle, one of the principal curvatures, say C_1, is provided by the contour curvature. As described in Appendix 5.F, the contour curvature C_1 attains, in general, two different values C_{1b} and C_{1a} when we approach the domain boundary from the *b* and *a* domain, respectively. Defining the mean curvatures $M_a(s_1)$ and $M_b(s_1)$ at the *a*- and *b*-sides of the domain boundary, see Appendix 5.F, the curvature discontinuity can be written in the concise form

$$\kappa_a[M_a(s_1) - m_a] - \kappa_b[M_b(s_1) - m_b] = \frac{1}{2}(\kappa_{Gb} - \kappa_{Ga})C_2(s_1) \tag{5.281}$$

where $C_2(s_1)$ is the second principal curvature which is continuous across the domain boundary.

The curvature discontinuity also affects the difference $\Sigma_a - \Sigma_b$ of the mechanical tensions within the two membrane domains. In order to describe this tension difference, we use the parametrization of axisymmetric shapes as shown in Figure 5.33. Because of axisymmetry, the shape is determined by a one-dimensional

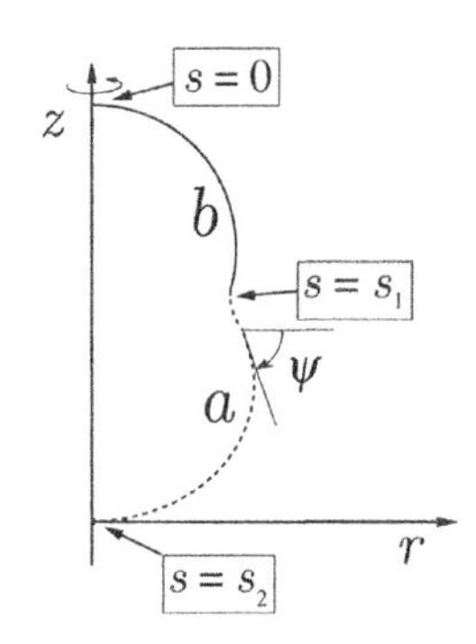

Figure 5.33 Contour of an axisymmetric vesicle with two domains, *a* (broken line) and *b* (full line). The contour is parametrized by the arc length *s*, the interval $0 \leq s < s_1$ corresponds to the *b*-domain and the interval $s_1 < s \leq s_2$ to the *a*-domain. The circular domain boundary is located at $s = s_1$. The shape of the contour is described by the radial coordinate $r = r(s)$ and the tilt angle $\psi = \psi(s)$ which varies from $\psi(s = 0) = 0$ at the north pole to $\psi(s = s_2) = \pi$ at the south pole.

contour which can be parametrized by the radial coordinate $r = r(s)$ and the tilt angle $\psi = \psi(s)$, both of which depend on the arc length *s* of the contour, see Appendix 5.F. The domain boundary is located at $s = s_1$ and the tension difference $\Sigma_a - \Sigma_b$ depends on the radius $r_1 \equiv r(s_1)$ of the circular domain boundary and the tilt angle $\psi_1 \equiv \psi(s_1)$ at this boundary. The tension difference then has the form

$$\Sigma_a - \Sigma_b = \lambda \frac{\cos\psi_1}{r_1} + \Delta_\Sigma \tag{5.282}$$

with Δ_Σ as given by the expression Eq. 5.17. The latter expression involves several terms and depends on the contour curvatures $C_{1a}(s_1)$ and $C_{1b}(s_1)$ and on the second principal curvature $C_2(s_1) = \sin\psi_1 / r_1$ at the domain boundary. If both membrane domains have identical curvature-elastic properties, the additional term Δ_Σ vanishes and we are left with the balance between the line tension λ and the mechanical tensions Σ_a and Σ_b within the two membrane domains. Finally, if the line tension λ vanishes as well, the mechanical tension within the *a*-domain is equal to the mechanical tension within the *b*-domain. The equality $\Sigma_a = \Sigma_b$ also holds for two domains with identical curvature-elastic properties if the radius $r_1 = r(s_1)$ of the domain boundary is a local minimum of $r(s)$ as in Figure 5.33, corresponding to the tilt angle $\psi_1 = \psi(s_1) = \pi/2$ and $\cos(\psi_1) = 0$. The latter situation applies to two membrane domains that have the same Gaussian curvature modulus, $\kappa_{Gb} = \kappa_{Ga}$, but is, in general, not valid for $\kappa_{Gb} \neq \kappa_{Ga}$, see last subsection of Section 5.8.3.

Parameters of two-domain vesicles

The morphology of two-domain vesicles depends on three geometric parameters, the vesicle volume *V* as well as on the partial areas A_a and A_b. Using again the vesicle size $R_{ve} = \sqrt{A/4\pi}$ as the basic length scale, we are left with two dimensionless parameters, the reduced volume $v \sim V/A^{3/2}$ with $A = A_a + A_b$ and $0 \leq v \leq 1$ as well as the area fraction

$$x_b \equiv \frac{A_b}{A_a + A_b} = \frac{A_b}{A} \tag{5.283}$$

[14] The width of the domain boundary is set by the correlation length for the compositional fluctuations. Far away from a critical demixing (or consolute) point, this correlation length will be comparable to the size of the lipid head groups while it becomes large compared to molecular length scales close to a critical point.

of the b-domain with $0 \leq x_b \leq 1$. The area fraction x_a of the a-domain in then given by $x_a = 1 - x_b$.

In addition, the morphology of two-domain vesicles depends on six curvature-elastic parameters: the spontaneous curvatures m_a and m_b, the bending rigidities κ_a and κ_b, the difference $\kappa_{Ga} - \kappa_{Gb}$ of the Gaussian curvature moduli, and the line tension λ. Using the bending rigidity κ_b as the basic energy scale, we obtain five dimensionless parameters: the dimensionless curvatures

$$\bar{m}_a \equiv m_a R_{\text{ve}} \quad \text{and} \quad \bar{m}_b \equiv m_b R_{\text{ve}}, \tag{5.284}$$

the rigidity ratios

$$\rho_\kappa \equiv \frac{\kappa_a}{\kappa_b} \quad \text{and} \quad \rho_G \equiv \frac{\Delta\kappa_G}{\kappa_b} = \frac{\kappa_{Ga} - \kappa_{Gb}}{\kappa_b}, \tag{5.285}$$

as well as the dimensionless line tension

$$\bar{\lambda} \equiv \frac{\lambda R_{\text{ve}}}{\kappa_b}. \tag{5.286}$$

The bending rigidity ratio ρ_κ is expected to be of order one. If we again identify the b- and the a-domains with the Ld and Lo phases of three-component lipid bilayers, the value $\rho_\kappa \simeq 4.5$ has been measured for a certain tie line within the two-phase coexistence region (Heinrich et al., 2010). The rigidity ratio ρ_G is also expected to be of order one. Two groups (Baumgart et al., 2005; Semrau et al., 2008) have compared the experimentally observed shapes of two-domain vesicles with those calculated from the theory reviewed here and developed in (Jülicher and Lipowsky, 1993, 1996). As a result, these groups obtained the estimates $\rho_G \simeq 3.9$ (Baumgart et al., 2005) and $1.1 \lesssim \rho_G \lesssim 2.5$ (Semrau et al., 2008).

An order of magnitude estimate of the line tension leads to the value $\lambda \simeq 10^{-11}$ N or 10 pN (Lipowsky, 1992). For the ternary lipid mixtures studied in (Baumgart et al., 2003, 2005; Semrau et al., 2008), the line tensions deduced from the experiments varied between 10^{-12} and 10^{-14} N, reflecting the vicinity of critical demixing points in these mixtures. For giant vesicles with a size R_{ve} between 10 and 50 μm, the dimensionless line tension $\bar{\lambda}$ then varies within the range $1 \lesssim \bar{\lambda} \lesssim 500$.

5.8.3 DOMAIN-INDUCED BUDDING OF VESICLES

The shape functional $\mathcal{F}_{\text{2Do}}\{S\}$ in Eq. 5.275 has been minimized in order to determine the equilibrium morphologies within the subspace of axisymmetric shapes (Jülicher and Lipowsky, 1993, 1996). As discussed in the previous subsection, these shapes depend on seven dimensionless parameters, two geometric and five material parameters. In order to illustrate the equilibrium morphologies of two-domain shapes, the next subsection describes the dependence of domain-induced budding on the volume-to-area volume v and on the line tension $\bar{\lambda}$, keeping all other parameters fixed. We will see that closed membrane necks play again a prominent role. The closure and the stability of these necks is governed by generalized neck conditions that depend on the line tension.

Budding controlled by osmotic conditions

We now consider a two-domain vesicle with area fraction $x_b = 0.1$, corresponding to a relatively small b-domain, and study the shape of this vesicle as a function of volume-to-area ratio v and line tension $\bar{\lambda}$. In order to reduce the dimension of the parameter space, the a- and b-domain are taken to have the same bending rigidity, $\kappa_a = \kappa_b$, and zero spontaneous curvatures, $m_a = m_b = 0$. Furthermore, we will also assume that the difference $\Delta\kappa_G$ between the Gaussian curvature moduli is small and can be ignored. We are then left with a 2-dimensional $(v, \bar{\lambda})$-section across the 7-dimensional parameter space. The corresponding morphology diagram is shown in Figure 5.34a.

This diagram contains two lines of limit shapes, L_{ss} and L_{ps}. The limit shapes L_{ss} have volume-to-area ratio $v = v_* = 0.885$ and line tension $\bar{\lambda} > \bar{\lambda}_* = 8.43$. These shapes consist of two spheres, a smaller b-sphere and a larger a-sphere that are connected by a closed neck. The domain boundary is located within this neck and has, thus, zero length. The a-sphere has radius $R_a = \sqrt{A_a / 4\pi}$ and mean curvature $M_a = 1/R_a$ while the b-sphere has radius $R_b = \sqrt{A_b / 4\pi}$ and mean curvature $M_b = 1/R_b$. Therefore, the geometry of the limit shapes L_{ss} is completely determined by the partial areas A_a and A_b. When we inflate one of the limit shapes L_{ss}, thereby increasing the

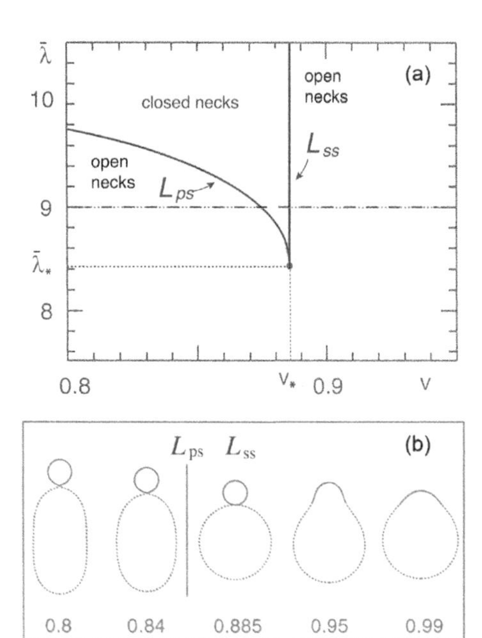

Figure 5.34 (a) Morphology diagram for two-domain vesicles as a function of reduced volume v and line tension $\bar{\lambda}$ and (b) Shapes of two-domain vesicles for $\bar{\lambda} = 9$ and variable v (bottom row), corresponding to the horizontal dashed line in (a). The b-domain covers the area fraction $x_b = 0.1$; both domains are taken to have the same bending rigidity and the same Gaussian curvature modulus as well as zero spontaneous curvatures. The limit shapes L_{ss} consist of two spheres, a larger sphere formed by the a-domain and a smaller sphere formed by the b-domain. The limit shapes L_{ps} consist of an a-prolate and a b-sphere. In (a), the two lines of limit shapes meet at the point $(v, \bar{\lambda}) = (v_*, \bar{\lambda}_*) = (8.43, 0.885)$. As we deflate a vesicle for $\bar{\lambda} > \bar{\lambda}_*$, we first reach the limit shape L_{ss}, at which the open neck closes, move across the shaded region (yellow) of persistent shapes with closed necks, and eventually reach the limit shape L_{ps}, at which the neck starts to open again. (Reproduced from Jülicher, F. and Lipowsky, R. *Phys. Rev. E*, 53, 2670–2683, 1996.)

volume-to-area ratio to $v > v_*$, the neck opens up and the domain boundary acquires a nonzero length.

The limit shapes L_{ps} are located at $v_{ps}(\bar{\lambda}) < v_*$ and again restricted to $\bar{\lambda} > \bar{\lambda}_*$, see Figure 5.34a. The latter shapes consist of an *a*-prolate and a *b*-sphere. The *b*-sphere of the limit shape L_{ps} is identical with the *b*-sphere of the limit shape L_{ss} and, thus, has the same radius $R_b = \sqrt{A_b / 4\pi}$. This *b*-sphere is connected to the pole of the *a*-prolate via a closed neck, and the domain boundary is again located within this neck. At its pole, the *a*-prolate has the mean curvature

$$M_a = \frac{\lambda}{2\kappa_b} - M_b = \frac{\lambda}{2\kappa_b} - \sqrt{4\pi / A_b}. \tag{5.287}$$

The latter relation represents an example for the neck closure condition of domain-induced budding, see further below. When we deflate one of the limit shapes L_{ps}, thereby decreasing the volume-to-area ratio to $v < v_{ps}$, the neck opens up and the domain boundary acquires a nonzero length.

Inspection of Figure 5.34a shows that the two lines of limit shapes, L_{ps} and L_{ss}, enclose an intermediate parameter regime in which all two-domain shapes have a closed neck. Now, assume that we move across this regime by inflation, thereby increasing the parameter v for fixed value of the line tension $\bar{\lambda} > \bar{\lambda}_*$. We start with a shape that has a volume-to-area ratio $v < v_{ps}(\bar{\lambda})$ and a slightly open neck, see Figure 5.34b. As we reach the limit shape L_{ps} by inflation, the neck closes and the two mean curvatures M_a and M_b adjacent to this neck fulfill the neck closure condition in Eq. 5.287. Further inflation does not affect the *b*-sphere but increases the volume of the *a*-prolate, thereby producing different persistent shapes Φ_{ps} with a closed neck. The volume of the *a*-prolate increases until it is transformed into an *a*-sphere. During this transformation, the mean curvature M_a at the pole of the *a*-prolate decreases continuously until it reaches the limiting value $M_a = \sqrt{4\pi / A_a}$ of the *a*-sphere. After this transformation, the two-domain vesicle forms the limit shape L_{ss}. Because the line tension forces the domain boundary to be located within the neck, a further increase in the vesicle volume necessarily leads to an open neck.

Neck closure and closed neck conditions

The $(v, \bar{\lambda})$-diagram discussed in the previous subsection, see Figure 5.34a, contains a large parameter region for which the shape of the two-domain vesicle involves a closed membrane neck. This abundance of necks is also obtained for other choices of the area fraction x_b, different values of the bending rigidities κ_a and κ_b, and nonzero values of the spontaneous curvatures m_a and m_b. In all of these cases, the domain boundary is again located within the neck provided the difference $\Delta\kappa_G$ of the Gaussian curvature moduli is small and can be neglected. Such *ab* necks that completely eliminate the domain boundary will now be considered in more detail.

Out-buds

If the *b*-domain forms an out-bud as in Figure 5.34b, the closed *ab*-neck is stable if the mean curvatures M_a and M_b of the *a*- and *b*-segments adjacent to the neck satisfy the relation (Jülicher and Lipowsky, 1993, 1996)

$$\kappa_a(M_a - m_a) + \kappa_b(M_b - m_b) \leq \frac{1}{2}\lambda \quad \text{for } \kappa_{Ga} = \kappa_{Gb}. \tag{5.288}$$

The equality sign of this relation provides the neck closure condition for the limit shapes, the inequality sign the closed neck condition. The relation in Eq. 5.288 for a domain-induced out-bud has been confirmed by numerical energy minimization for a large number of different parameter values. This relation can also be derived by parametrizing the shape of the two-domain vesicle in terms of membrane segments with constant mean curvature, compare Section Stability of closed necks. Recently, the neck closure condition corresponding to the equality sign in Eq. 5.288 has been shown to apply to non-axisymmetric shapes as well (Yang et al., 2017).

One should note that the matching condition along the domain boundary no longer applies when we reach a limit shape with a closed neck for which the domain boundary has zero length. Indeed, consider the simplest case of two membrane domains that have the same curvature-elastic parameters. In the latter case, the matching condition in Eq. 5.281 has the simple form $M_a = M_b$, corresponding to a continuous variation of the mean curvature across the domain boundary. In contrast, the limit shape is characterized by the neck closure condition in Eq. 5.288 which reduces to $M_a = \frac{\lambda}{2\kappa_b} - M_b$ when the two domains have the same curvature-elastic parameters. If we combined the latter relation with $M_a = M_b$, we would conclude that $M_a = M_b = \frac{\lambda}{4\kappa_b}$ which is, however, inconsistent with $M_b = \sqrt{4\pi / A_b}$ as in Eq. 5.287. The same conclusion follows also by inspection of the limit shape L_{ps} in Figure 5.34 which clearly shows that $M_a \neq M_b$.

In-buds

If the *b*-domain forms an in-bud with a closed *ab*-neck, this neck is stable if (Lipowsky, 2014b)

$$\kappa_a(M_a - m_a) + \kappa_b(M_b - m_b) \geq -\frac{1}{2}\lambda \quad \textit{for } \kappa_{Ga} = \kappa_G \tag{5.289}$$

This relation can again be derived by an appropriate hemisphere-unduloid parametrization of the vesicle shape or, alternatively, by changing the sign of all curvatures that appear in Eq. 5.288. Because the line tension of the domain boundary is necessarily positive, the right hand side of the inequality in Eq. 5.289 is always negative.

Special parameter values

It is instructive to consider some special cases of the neck closure condition corresponding to the equality in Eqs 5.288 and 5.289. If the *a*- and *b*-domains have the same lipid composition and, thus, the same curvature-elastic parameters, the line tension λ vanishes and the neck closure condition becomes $M_a + M_b = 2m$, corresponding to the neck closure relations Eqs 5.52 and 5.53 for a uniform membrane. For a weakly curved *a*-segment, a spherical *b*-bud then has the radius

$$R_b = \frac{1}{|M_b|} \approx \frac{1}{2|m|} \quad \text{(uniform membrane, weakly curved } a\text{-segment).} \tag{5.290}$$

Another simple case is provided by a weakly curved *a*-membrane characterized by a small spontaneous curvature $|m_a| \ll |m_b|$. In this case, the *b*-domain forms a spherical bud with radius

$$R_b = \frac{1}{|M_b|} \approx \frac{1}{|m_b + \lambda/(2\kappa_b)|} \quad \text{(weakly curved } a\text{-membrane, small } |m_a|). \tag{5.291}$$

Thus, depending on the relative size of the spontaneous curvature $|m_b|$ and the reduced line tension $\lambda/(2\kappa_b)$, the bud size may be dominated by spontaneous curvature or by line tension. For some ternary lipid mixtures, the measured line tension was found to be of the order of 10^{-12} N (Baumgart et al., 2005; Semrau et al., 2008). The bending rigidity κ_b has a typical value of the order of 10^{-19} J. Thus, in these systems, the inverse length scale $\lambda/(2\kappa_b) \simeq 1/(200\text{nm})$ which implies that the bud size is dominated by line tension with $R_b \approx 2\kappa_b/\lambda$ for $|m_b| \ll 1/(200 \text{ nm})$ and governed by spontaneous curvature with $R_b \approx 1/|m_b|$ for $|m_b| \gg 1/(200 \text{ nm})$.

Effect of Gaussian curvature moduli

In the previous subsection, it was tacitly assumed that the difference $\Delta\kappa_G = \kappa_{Ga} - \kappa_{Gb}$ between the Gaussian curvature moduli of the *a* and *b* domain can be ignored. This simplification will be valid as long as $\Delta\kappa_G$ is small compared to the bending rigidities κ_a and κ_b. For larger values of $\Delta\kappa_G$, this difference has a significant effect on the location of the domain boundary, see Figure 5.35.

For an axisymmetric shape as shown in the top figure of Figure 5.35, the shape contour can be parametrized by the arc length *s*, the radial coordinate *r*, and the angle ψ between the normal vector and the symmetry axis, see Figure 5.33. The Gaussian curvature contribution in Eq. 5.277 can then be expressed in terms of the tilt angle $\psi_1 = \psi(s_1)$ at the domain boundary and becomes (Jülicher and Lipowsky, 1993, 1996)

$$\mathcal{E}_G\{S_a, S_b\} = 2\pi(\kappa_{Ga} - \kappa_{Gb})\cos(\psi_1) \equiv E_G(\psi_1). \tag{5.292}$$

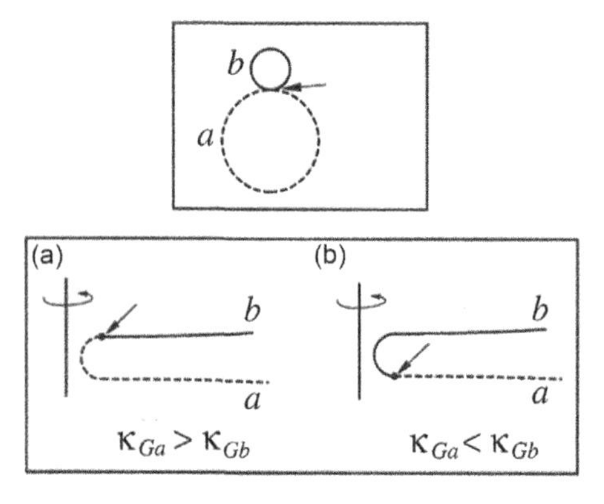

Figure 5.35 (Top) Side view of a vesicle that consists of a large *a* domain and a small *b* bud. The two domains are connected by a membrane neck which contains the *ab* domain boundary (arrow); (Bottom) More detailed view of the neck region which shows that the domain boundary position (arrows) depends on the relative size of the Gaussian curvature moduli κ_{Ga} and κ_{Gb} of the *a* and *b* domains. For $\kappa_{Ga} > \kappa_{Gb}$, the domain boundary is shifted towards the *b* bud. For $\kappa_{Ga} < \kappa_{Gb}$, this boundary is displaced towards the *a* domain. In both cases (a) and (b), the domain boundary is shifted out of the neck towards the domain with the smaller κ_G-value, and the neck is then formed by the domain with the larger Gaussian curvature modulus (Jülicher and Lipowsky, 1993, 1996). Such shifts of the domain boundaries have been experimentally observed by (Baumgart et al., 2005; Semrau et al., 2008). (Reproduced from Jülicher, F. and Lipowsky, R., *Phys. Rev. E*, 53, 2670–2683, 1996.)

If the domain boundary is located in the neck, i.e., at the closest point of the shape contour to the symmetry axis, the angle $\psi_1 = \pi/2$ and the energy term $E_G(\psi_1) = 0$.

Depending on the sign of $\kappa_{Ga} - \kappa_{Gb}$, the energy term E_G becomes negative as the domain boundary moves out of the neck towards the *b* or towards the *a* domain. If $\kappa_{Ga} > \kappa_{Gb}$, this term becomes negative for $\psi_1 > \pi/2$ which implies that the domain boundary prefers to move up towards the *b* domain as in Figure 5.35a. On the other hand, if $\kappa_{Ga} < \kappa_{Gb}$, E_G becomes negative for $\psi_1 < \pi/2$ which implies that the domain boundary prefers to move down towards the *a* domain, see Figure 5.35b. In both cases, the neck is then formed by the domain with the larger Gaussian curvature modulus.

The actual displacement of the domain boundary is limited by the line tension. Indeed, as the domain boundary moves out of the neck, the energy gain $|E_G(\psi_1)|$ arising from the Gaussian curvature terms is bounded by

$$|E_G(\psi_1)| \lesssim 2\pi|\kappa_{Ga} - \kappa_{Gb}| \quad \text{for any value of } \psi_1 \tag{5.293}$$

whereas the line energy of the domain boundary increases monotonically with the length of this boundary.

Such displacements of the domain boundaries away from the neck have indeed been observed experimentally for two-domain vesicles formed by ternary lipid mixtures (Baumgart et al., 2005; Semrau et al., 2008). Based on the observed location of the domain boundaries, the difference $\Delta\kappa_G = \kappa_{Ga} - \kappa_{Gb}$ in the Gaussian curvature moduli has been estimated to be $\Delta\kappa_G \simeq 3.9 \times 10^{-19}$ J in (Baumgart et al., 2005) and 3×10^{-19} J in (Semrau et al., 2008). So far, these values which are of the same order of magnitude as the bending rigidities represent the only experimentally deduced information about the Gaussian curvature moduli of lipid bilayers.

5.8.4 STABLE MULTI-DOMAIN PATTERNS

When we quench a vesicle membrane from the one-phase into the two-phase region, the phase separation process within the membrane starts with the formation of many small domains which then grow and merge into larger domains. Domain growth by coalescence, which is driven by the reduction in the line energy of the domain boundaries, has been observed both in computer simulations (Kumar et al., 2001) and in giant vesicle experiments (Veatch and Keller, 2003). If the line tension is sufficiently large, the coarsening process will often lead to complete phase separation and to two large membrane domains as studied in the previous subsections. However, if the two lipid phases differ in their bending rigidity, a multi-domain pattern with more than two domains can be energetically more favorable (Gutlederer et al., 2009; Hu et al., 2011). Some examples with 1 + 3 and 1 + 4 domains are displayed in Figure 5.36. Inspection of these figures shows that the more rigid *a*-domains are only weakly curved whereas the more flexible *b*-domains form the more strongly curved membrane segments. A reduction in the number of *b*-domains would reduce the line energy of these domains but, at the same time, increase the bending energy of the vesicle, and the bending energy increase outweighs the line energy reduction.

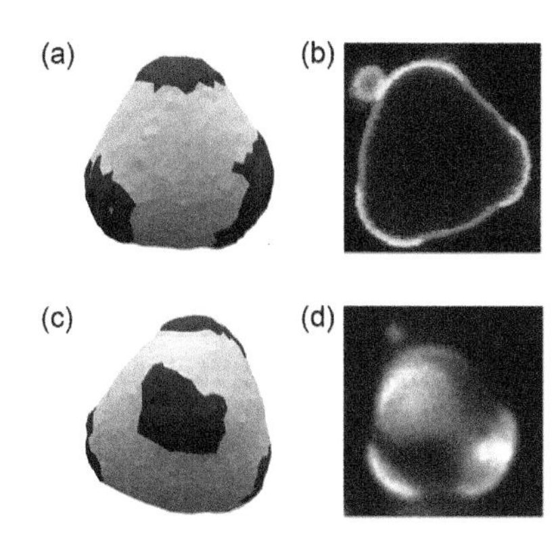

Figure 5.36 Multi-domain vesicles with two membrane domains that differ in their bending rigidities: (a, c) Snapshots from Monte Carlo simulations with (a) three and (c) four domains of the *b* phase (red) within a single domain of *a* phase (white); and (b, d) Corresponding images obtained by optical microscopy (Veatch and Keller, 2003; Gudheti et al., 2007). The *a* phase corresponds to the more rigid liquid-ordered phase, which forms a single, multiply-connected and weakly curved domain, whereas the *b* phase represents the more flexible liquid-disordered phase which forms three or four disconnected and more strongly curved domains. (Hu, J. et al., *Soft Matter*, 7, 6092–6102, 2011. Reproduced by permission of The Royal Society of Chemistry.)

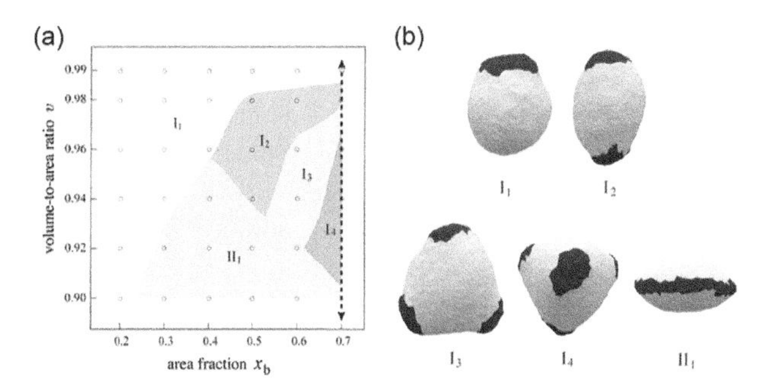

Figure 5.37 Morphological transitions of multi-domain vesicles that simultaneously change the vesicle shape and the domain pattern. (a) Morphology diagram as a function of area fraction x_b of the *b*-domains and volume-to-area ratio or reduced volume *v*. The diagram exhibits five different morphologies, labeled by I_1, I_2, I_3, I_4, and II_1 and depicted in (b). The dashed vertical line at $x_b = 0.7$ indicates a possible deflation/inflation trajectory; and (b) Sequence of vesicle morphologies and morphological transitions that the vesicle explores as we move along the dashed vertical line in (a). For each morphology, the white domain corresponds to the more rigid *a* or Lo phase, the red domains to the more flexible *b* or Ld phase. The multi-domain vesicle follows the sequence $I_1 \to I_2 \to I_3 \to I_4 \to II_1$ during deflation and the reverse sequence during inflation. All transitions $I_1 \leftrightarrow I_2 \leftrightarrow I_3 \leftrightarrow I_4 \leftrightarrow II_1$ break a spatial symmetry. Therefore, all of these transitions are discontinuous and exhibit hysteresis. The transitions from $I_{N+1} \to I_N$, as induced by inflation, are facilitated by the line tension and should thus be easier to observe experimentally. (Hu, J. et al., *Soft Matter* 7, 6092–6102, 2011. Reproduced by permission of The Royal Society of Chemistry.)

The shape energy of multi-domain vesicles with N_a and N_b domains is obtained by summing up the bending and Gaussian curvature energies over all $N_a + N_b$ domains and the line energies over all domain boundaries. The minimization of this shape energy has been performed both by solving the corresponding shape equations assuming certain symmetries of the domain patterns (Gutlederer et al., 2009) and by Monte Carlo simulations (Hu et al., 2011). As a result, the multi-domain vesicles are found to undergo new types of morphological transformations at which both the vesicle shape and the domain pattern are changed in a discontinuous manner. Presumably the simplest way to explore these morphological transitions is by changing the vesicle volume via osmotic deflation or inflation as illustrated in Figure 5.37.

Each vesicle morphology shown in Figure 5.37 is characterized by a different spatial symmetry: both with respect to the vesicle shape and with respect to the domain pattern. Therefore, all transitions that can be observed between these different morphologies are discontinuous and exhibit hysteresis. As we deflate the vesicle for fixed area fraction $x_b = A_b/(A_a + A_b)$, we can encounter the sequence of vesicle morphologies I_1, I_2, I_3, I_4, and II_1 displayed in Figure 5.37b. The corresponding transitions $I_N \to I_{N+1}$ involve the fission of N into $(N + 1)$ *b*-domains. Such a fission process has to overcome an energy barrier that involves longer domain boundaries and, thus, an increased line energy. In contrast, during inflation, the reverse transitions $I_{N+1} \to I_N$ lead to a reduction in the number of *b*-domains and are thus facilitated by the line tension. Therefore, it should be easier to experimentally observe these morphological transitions during inflation processes.

5.8.5 MEMBRANE PHASE SEPARATION AND AMBIENCE-INDUCED SEGMENTATION

As explained in Section 7.6 and illustrated in Figure 5.31, membranes are often exposed to different local environments which act to enrich or deplete certain molecular components of the membranes. As a result, the membranes are partitioned into several segments that can differ in their molecular composition. The interplay between this ambience-induced segmentation and membrane phase separation has some interesting consequences as shown theoretically for membranes consisting of two molecular components, see Appendix 5.G (Rouhiparkouhi et al., 2013; Lipowsky et al., 2013). First, the phase separation within the multi-component membrane is always spatially confined to a single segment as illustrated in Figure 5.38. Second, when the membrane is partitioned into K different membrane segments, we encounter K separate coexistence regions as we vary the membrane composition and/or the temperature. Third, the size of the coexistence regions, i.e., the range of compositions that exhibits two-phase coexistence, shrinks with increasing K. These generic properties have direct consequences for cell membranes.

The environment of a cell membrane is rather heterogeneous and the molecular interactions experienced by the different

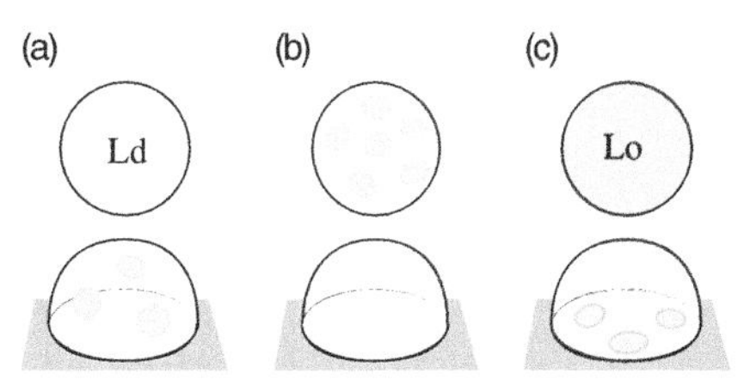

Figure 5.38 Multi-component vesicles with three different compositions. The top row displays the non-adhering vesicles with compositions that belong to (a) the liquid-disordered phase Ld (white), (b) the two-phase coexistence region, and (c) the liquid-ordered phase Lo (blue). The bottom row displays the same vesicles now adhering to a rigid surface or solid support. In the adhering state, membrane phase separation and domain formation can occur either in the bound or in the unbound segment but not in both segments simultaneously.

molecular components of the membrane change on nanoscopic scales. When we focus on the interactions with the cytoskeleton, we can distinguish at least two types of membrane segments, contact segments that interact with the cytoskeletal proteins and noncontact segments that do not experience such interactions (Sako and Kusumi, 1994; Saxton and Jacobson, 1997; Fujiwara et al., 2002; Kusumi et al., 2005). In addition, different contact segments are, in general, exposed to cytoskeletal structures that differ in their molecular composition of actin-binding proteins (Skau and Kovar, 2010; Michelot and Drubin, 2011) and non-contact segments involve additional supramolecular structures such as the protein scaffolds formed during clathrin-dependent endocytosis that have a lifetime in the range between 20 and 80s (Loerke et al., 2009; Cureton et al., 2012).

Thus, cell membranes are expected to be partitioned into many distinct membrane segments that are exposed to different local environments. If lipid phase domains form in such a cell membrane, this domain formation is necessarily restricted to one of the membrane segments and, thus, hard to detect (Lipowsky, 2014b). In the limiting case in which the environmental heterogeneities act as long-lived random fields on the cellular membranes, these heterogeneities would completely destroy the two-phase coexistence region, in analogy to the Ising model with random fields (Binder, 1983; Aizenman and Wehr, 1989; Fischer and Vink, 2011). This view is in agreement with experimental observation on membrane phase separation in giant plasma membrane vesicles (Baumgart et al., 2007; Veatch et al., 2008) because the latter vesicles have no cytoskeleton.

In contrast to lipid phase domains, the formation of intramembrane domains via the clustering of membrane proteins is frequently observed *in vivo*. One example is provided by clathrin-dependent endocytosis which can be understood as a domain-induced budding process that is governed by the membrane's spontaneous curvature. When the endocytic vesicles contain nanoparticles or other types of cargo, the uptake of this cargo becomes maximal at a certain, optimal cargo size (Agudo-Canalejo and Lipowsky, 2015a) as experimentally observed for the uptake of gold nanoparticles by HeLa cells (Chithrani et al., 2006; Chithrani and Chan, 2007) and discussed in more detail in Chapter 8 of this book. In general, protein-rich membrane domains or membrane domains induced by an extended protein coat should always undergo domain-induced budding as long as the lipid-protein domains remain in a fluid state. Recent examples are domain-induced budding processes arising from the clustering of Shiga toxin (Pezeshkian et al., 2016) and from the sequential adsorption of two types of ESCRT proteins (Avalos-Padilla et al., 2018).

5.9 WETTING OF MEMBRANES BY AQUEOUS DROPLETS

Aqueous two-phase systems, also called aqueous biphasic systems, have been used for a long time in biochemical analysis and biotechnology and are intimately related to water-in-water emulsions (Albertsson, 1986; Helfrich et al., 2002; Esquena, 2016). One prominent example are PEG-dextran solutions that undergo aqueous phase separation when the weight fractions of the polymers exceed a few percent. The corresponding interfacial tensions are ultralow, of the order of 10^{-6}–10^{-4} N/m, reflecting the vicinity of a critical demixing point in the phase diagram (Scholten et al., 2002; Liu et al., 2012; Atefi et al., 2014; de Freitas et al., 2016). The corresponding phase diagram is displayed in Figure 5.39 based on the experimental data in (Liu et al., 2012). As explained in the following section, aqueous two-phase systems and water-in-water emulsions also provide insight into the wetting behavior of membranes and vesicles. The experimental procedures used to encapsulate aqueous two-phase systems by GUVs are reviewed in Chapter 29 of this book.

In the experimental studies of phase separation of PEG-dextran solutions within GUVs, (Li et al., 2011; Liu et al., 2016) the GUV membranes were observed to form many nanotubes. More precisely, such tubes were formed by the membrane segments in contact with the PEG-rich aqueous phase. Thus, deflation of the PEG-dextran solutions led simultaneously to

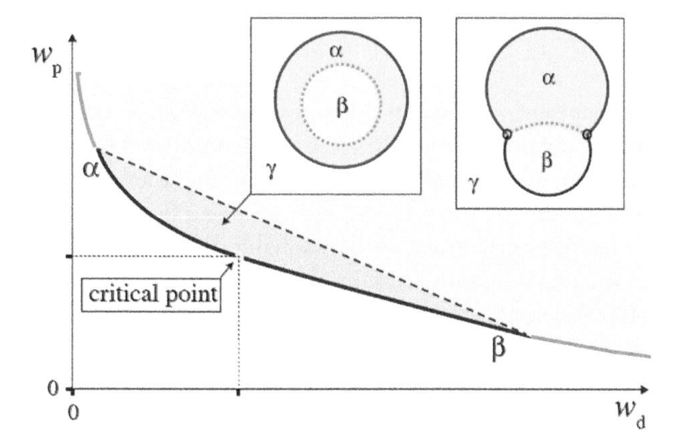

Figure 5.39 Phase diagram and membrane wetting behavior of aqueous PEG-dextran solutions as a function of the weight fractions w_p and w_d for PEG and dextran as determined experimentally in (Liu et al., 2012). For low weight fractions, the polymer mixture forms a spatially uniform aqueous phase corresponding to the one-phase region (white) in the phase diagram. The coexistence region of the PEG-rich phase α and the dextran-rich phase β contains two subregions, a complete wetting region (pink) close to the critical point and a partial wetting region (turquoise) further away from it. In the pink subregion, the membrane is completely wetted by the PEG-rich phase α which encloses the dextran-rich phase β. The corresponding wetting morphology is depicted in the left inset: the outer leaflet of the uniform vesicle membrane (red) is in contact with the exterior phase γ, the inner leaflet with the interior phase α but not with the interior phase β (gravitational effects arising from the different mass densities of the two phases have been ignored). In the turquoise subregion, the membrane is partially wetted by both phases as shown in the right inset: both interior phases α and β are now in contact with the vesicle membrane and induce two distinct membrane segments (red and purple). Within the phase diagram, the boundary between the complete and partial wetting subregions is provided by a certain tie line (red dashed line), the precise location of which depends on the lipid composition of the membrane. Along this tie line, the system undergoes a complete-to-partial wetting transition. The dashed tie-line partitions the binodal line into two line segments (red and blue). If one approaches the red segment of the binodal line from the one-phase region, a wetting layer of the α phase starts to form at the membrane and becomes mesoscopically thick as one reaches this line segment. No such layer is formed along the blue segment of the binodal line.

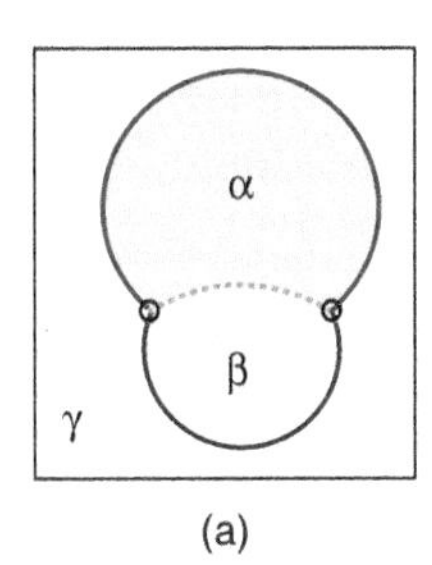

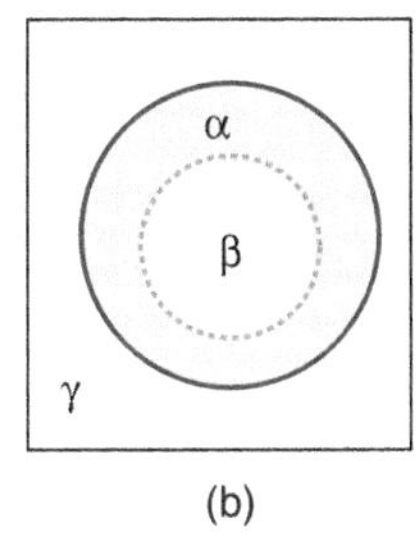

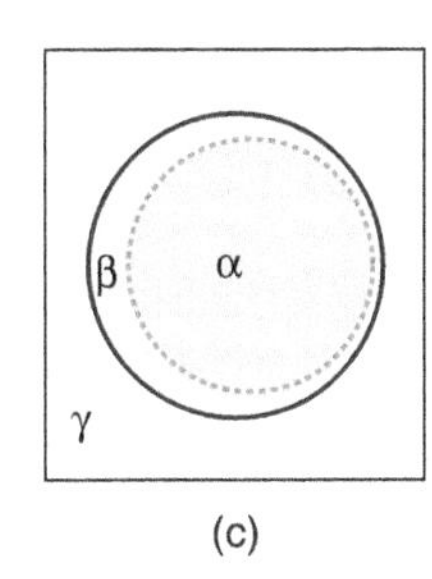

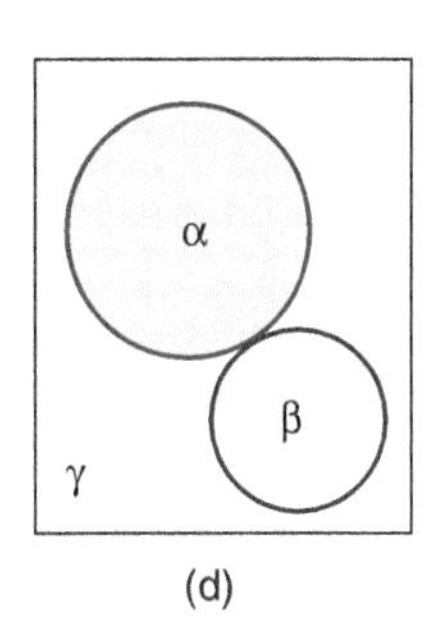

Figure 5.40 In-wetting morphologies arising from phase separation into two aqueous phases, α (yellow) and β (blue), *within* a giant vesicle. The vesicle is surrounded by the bulk liquid γ (white) which plays the role of an inert spectator phase. Red and purple segments of the vesicle membrane are in contact with the α and β droplets, respectively. The $\alpha\beta$ interfaces are depicted as dashed orange lines: (a) Partial wetting of the vesicle membrane by both the α and the β phase. This morphology involves a three-phase contact line (black circles). On the micrometer scale, the vesicle shape exhibits a kink along this contact line which directly reveals the capillary forces acting onto the vesicle membrane; (b) Complete wetting of the membrane by the α phase; (c) Complete wetting by the β phase; and (d) Special morphology for which the α and the β droplet are separated by a closed membrane neck. The latter morphology, which resembles complete wetting by the γ phase, is only possible if the membrane has a certain minimal area $A = A_{2sp}$ to enclose both spherical droplets completely, see Eq. 5.294.

both wetting and tubulation of the GUVs. However, wetting and tubulation should be regarded as two distinct and independent processes. First, nanotubes can be formed in the absence of aqueous phase separation as predicted theoretically for uniform membranes, see Section 5.6, and observed experimentally (Liu et al., 2016) for GUVs exposed to asymmetric PEG solutions without dextran. Second, membrane wetting is expected to always generate some spontaneous curvature but tubulation can only occur if the spontaneous curvature is sufficiently large compared to the inverse vesicle size as explained in Section 5.6. In the following subsections, we will first focus on wetting and ignore the possibility of tube formation. The additional aspects related to spontaneous tubulation will be addressed in a later subsection.

This section is supplemented by two Appendices: Appendix 5.H on wetting of two membraneless droplets and Appendix 5.I on out-wetting of membranes and vesicles by droplets that originate from the exterior solution.

5.9.1 DISTINCT IN-WETTING MORPHOLOGIES

Wetting phenomena arise in aqueous systems with three separate aqueous phases that will be denoted by α, β, and γ. In the presence of a GUV membrane, which separates the aqueous solution into an interior and exterior compartment, only two of these phases, say α and β, will be in chemical equilibrium and able to form two coexisting phases. We can then distinguish two different cases, out-wetting and in-wetting, depending on whether these coexisting phases are formed within the exterior or interior compartment. For out-wetting, the exterior solution undergoes aqueous phase separation into α and β droplets while the interior solution forms a spatially uniform γ phase. The γ phase does not participate in the wetting process and, thus, represents an inert spectator phase. For in-wetting, the interior solution separates into α and β droplets while the exterior solution forms a spatially uniform γ phase which again plays the role of an inert spectator phase.

In order to simplify the following discussion, I will focus in this section on the case of in-wetting. The case of out-wetting is considered in Appendix 5.I. In-wetting has been studied experimentally for PEG-dextran solutions, using two different methods to induce the phase separation within the GUVs: temperature changes (Helfrich et al., 2002; Long et al., 2008) and osmotic deflation (Li et al., 2008, 2011; Kusumaatmaja et al., 2009; Liu et al., 2016; Dimova and Lipowsky, 2016). After the phase separation has been completed, the vesicle contains two aqueous droplets consisting of the PEG-rich phase α and the dextran-rich phase β, which are both separated from the exterior phase γ by the GUV membrane.

In general, an aqueous solution with three distinct aqueous phases α, β and γ can form three different liquid-liquid interfaces, an $\alpha\beta$, an $\alpha\gamma$, and a $\beta\gamma$ interface. When the interior aqueous solution within the GUV undergoes aqueous phase separation as considered here, the membrane is partitioned into an $\alpha\gamma$ and a $\beta\gamma$ membrane segment. In principle, one can then distinguish four wetting morphologies: a partial wetting morphology which is characterized by a three-phase contact line and three distinct morphologies of complete wetting as depicted in Figure 5.40.

For the PEG-dextran solutions, complete wetting of the membrane by the β phase as in Figure 5.40c has not been observed. Complete wetting of the membrane by the PEG-rich phase α as in Figure 5.40b was observed close to the critical point of the PEG-dextran mixture, see pink region in Figure 5.39. Partial wetting as in Figure 5.40a was found further away from the critical point, see turquoise region in Figure 5.39. Deflation of the partial wetting morphologies should eventually lead to complete wetting of the $\alpha\beta$ interface by the γ phase, see Figure 5.40d. In the latter case, the GUV membrane consists of three segments: an $\alpha\gamma$ segment around the α droplet, a $\beta\gamma$ segment around the β droplet, and a membrane neck (or nanotube) connecting the $\alpha\gamma$ with the $\beta\gamma$ segment. The latter morphology is not possible if the volume-to-area ratio of the GUV is too large. Indeed, if the α and β droplets have the volumes V_α and V_β, they cannot be completely enclosed by the vesicle membrane if the membrane area A is too small and satisfies the inequality[15]

$$A < A_{2sp} \equiv (4\pi / 9)^{1/3}\left(V_\alpha^{2/3} + V_\beta^{2/3}\right). \tag{5.294}$$

[15] If the vesicle membrane forms nanotubes, the area A corresponds to the apparent area of the mother vesicle.

In the following, I will first focus on systems that fulfill the geometric constraint Eq. 5.294 and thus cannot attain the limit shape with $A = A_{2sp}$ in Figure 5.40d. The latter morphology will be discussed in Section 5.9.8 further below.

In the phase diagram of Figure 5.39, the complete and partial wetting subregions are separated by a certain tie line, at which the system undergoes a complete-to-partial wetting transition. The precise location of this tie line depends on the lipid composition of the membranes. So far, three compositions have been studied: binary lipid mixtures consisting of DOPC and GM1 (Li et al., 2008, 2011) as well as ternary mixtures containing DOPC, dipalmytoyl phosphatidyl choline (DPPC) and cholesterol (Liu et al., 2016). In general, the wetting transition along this tie line can be continuous or discontinous depending on the manner in which the contact angle vanishes as we approach the transition from the partial wetting regime. So far, the experimental data do not allow us to draw firm conclusions about the continuous or discontinuous nature of the transition.

A particularly interesting class of water-in-water droplets is provided by membraneless organelles and biomolecular condensates that have been discovered *in vivo* and are enriched in intrinsically disordered proteins such as FUS (Brangwynne et al., 2009). It has been recently shown that a FUS-rich droplet in contact with a lipid vesicle can attain three different wetting morphologies depending on the salt concentration in the exterior solution (Knorr et al., under review). First, the droplet may form a thin wetting layer that spreads over the whole vesicle membrane, corresponding to complete wetting by the FUS-rich phase. Second, the droplet may have a limited contact area with the vesicle membrane and can then be characterized by apparent contact angles. Third, the droplet may also avoid the contact with the membrane corresponding to dewetting of the FUS-rich phase.

5.9.2 FLUID-ELASTIC MOLDING OF MEMBRANES

The distinction between dewetting, partial wetting, and complete wetting as described in the previous subsection emphasizes the different morphologies of an aqueous droplet in contact with a vesicle membrane. Alternatively, we may also focus on the response of the membrane to such a droplet. This response reflects the fluid-elastic molding mechanisms by which the droplet shapes the membranes. These mechanisms involve the adhesion of the droplets to the membranes, the capillary forces that the $\alpha\beta$ interface exerts onto the membrane, as well as the bilayer asymmetry and curvature generation arising from the different aqueous phases in contact with the two membrane leaflets. For a large bilayer asymmetry and low membrane tension, the membrane forms nanobuds and nanotubes as observed for vesicle membranes in contact with PEG-dextran solutions (Li et al., 2011; Lipowsky, 2013; Liu et al., 2016).

The different molding mechanisms are governed by different fluid-elastic parameters. First of all, the contact areas between the different aqueous phases and the GUV membrane can be characterized by different adhesive strengths, $W_{\alpha\gamma}$ and $W_{\beta\gamma}$, which represent the adhesion free energies of the $\alpha\gamma$ and $\beta\gamma$ segments per unit area. If the α droplets are attracted towards the membrane, in a background of β phase, the corresponding affinity contrast $W_{\alpha\gamma} - W_{\beta\gamma}$ is negative. In such a situation, the α droplet tries to increase its contact area $A_{\alpha\gamma}$ with the membrane. However, an increase of the contact area $A_{\alpha\gamma}$ for fixed volume V_α usually implies an increase in the area $A_{\alpha\beta}$ of the $\alpha\beta$ interface and, thus, of the interfacial free energy $A_{\alpha\beta}\Sigma_{\alpha\beta}$ which is proportional to the interfacial tension $\Sigma_{\alpha\beta}$.

On the other hand, the α droplet can simultaneously increase the contact area $A_{\alpha\gamma}$ with the membrane and decrease the area $A_{\alpha\beta}$ of the $\alpha\beta$ interface when it is partially or completely engulfed by the membrane. Complete engulfment of the α droplet as depicted in Figure 5.40d is only possible if the membrane area A is sufficiently large and satisfies $A \geq A_{2sp}$ with the area threshold A_{2sp} as in Eq. 5.294. In general, complete engulfment of a liquid droplet by a vesicle membrane requires some area reservoir or, equivalently, a sufficiently small lateral stress Σ acting within the membrane. Vice versa, a large lateral stress as generated, e.g., by osmotic inflation reduces the contact area for partial wetting and suppresses engulfment.

The interfacial tension $\Sigma_{\alpha\beta}$ of an aqueous two-phase system or water-in-water emulsion can be very small and only of the order of 10^{-6}–10^{-5} N/m. In spite of these ultra-low tension values, the resulting capillary forces generate strong shape deformations of the vesicle membrane along the three-phase contact line. Indeed, when viewed with conventional optical resolution, the membrane shape exhibits an apparent kink along this contact line as schematically depicted in Figures 5.40a and 5.49a for partial in- and out-wetting, respectively.

Finally, both for in- and for out-wetting, the two leaflets of the different membrane segments are exposed to different aqueous solutions which implies that the membrane segments acquire a certain spontaneous curvature. For a sufficiently large spontaneous curvature, the membrane segment forms nanobuds and nanotubes as observed for giant vesicles in contact with phase-separated PEG-dextran solutions (Li et al., 2011; Lipowsky, 2013; Liu et al., 2016). In the latter case, the spontaneous curvature was generated by PEG adsorption which implies that the nanobuds and nanotubes were formed by the membrane segments $\alpha\gamma$ in contact with the PEG-rich phase, reflecting the more negative adhesive strength $W_{\alpha\gamma}$ of these segments.

5.9.3 THEORY OF VESICLE-DROPLET SYSTEMS

Basic assumptions about the composition of the vesicle membrane

As previously mentioned, multicomponent membranes exposed to two different aqueous solutions are partitioned into two segments that will, in general, differ in their molecular compositions. These different compositions reflect the different molecular interactions between the membrane molecules and the two aqueous phases. Membrane segmentation can also arise via two alternative mechanisms, (i) phase separation within the membrane as discussed in the previous Section 5.8 and (ii) curvature sorting, i.e., the preference of some membrane molecules for highly curved membrane segments.

In the present section, we consider membrane compositions that belong to the one-phase region when the vesicle membrane is exposed to a uniform aqueous environment provided by any of the three liquid phases α, β, and γ. Furthermore, to simplify the following discussion, we will assume that curvature sorting is negligible and can be ignored.[16] In such a situation, the different molecular compositions of the $\alpha\gamma$ and $\beta\gamma$ membrane segments are determined by the different molecular interactions of the membrane molecules with the two distinct aqueous phases, molecular interactions that will be described by the corresponding adhesion free energies.

Geometry of in-wetting morphologies

The in-wetting morphologies in Figure 5.40 involve one α and one β droplet enclosed by the vesicle membrane. It will be useful to decompose the corresponding shape S into several components. First, we define the shapes S_α and S_β of the two droplets with volumes

$$V_\alpha = \mathcal{V}\{S_\alpha\} \quad \text{and} \quad V_\beta = \mathcal{V}\{S_\beta\}. \tag{5.295}$$

The total volume of the vesicle is then given by

$$V = V_\alpha + V_\beta. \tag{5.296}$$

These volumes can be considered to be constant at constant temperature and fixed osmotic conditions. The two droplets are bounded by three surface segments: the $\alpha\beta$ interface between the α and the β droplet as well as two membrane segments, the $\alpha\gamma$ segment in contact with the α droplet and the $\beta\gamma$ segment exposed to the β droplet. The shapes of these three surfaces will be denoted by $S_{\alpha\beta}$, $S_{\alpha\gamma}$, and $S_{\beta\gamma}$, respectively. Their surface areas are then given by

$$A_{\alpha\beta} = \mathcal{A}\{S_{\alpha\beta}\}, \quad A_{\alpha\gamma} = \mathcal{A}\{S_{\alpha\gamma}\}, \quad \text{and} \quad A_{\beta\gamma} = \mathcal{A}\{S_{\beta\gamma}\}. \tag{5.297}$$

All three surface segments meet along the three-phase contact line which has the shape $S_{\alpha\beta\gamma}$ and the length

$$L_{\alpha\beta\gamma} = \mathcal{L}\{S_{\alpha\beta\gamma}\} \tag{5.298}$$

where $L\{.\}$ is the length functional as before.

The $\alpha\beta$ interface can adapt its area $A_{\alpha\beta}$ to changes in the droplet and membrane morphologies. As before, the total membrane area A will be taken to be constant at constant temperature. The vesicle-droplet system is then characterized by three geometric constraints as provided by the volumes V_α and V_β of the two droplets as well as the total membrane area A. In order to determine the morphology of the vesicle-droplet system, we will minimize the (free) energy of the system, taking these three constraints into account.

Different energetic contributions

The three surface segments and the contact line make different contributions to the total (free) energy of the vesicle-droplet system. One contribution arises from the interfacial tension $\Sigma_{\alpha\beta}$ of the interface between the two liquid phases α and β. The latter contribution is proportional to the interfacial area $A_{\alpha\beta}$ and given by

$$\Sigma_{\alpha\beta} A_{\alpha\beta} = \Sigma_{\alpha\beta}\mathcal{A}\{S_{\alpha\beta}\}. \tag{5.299}$$

The curvature elasticity of each membrane segment $j\gamma$ with $j = \alpha$ or β makes two contributions, a bending energy that depends on the bending rigidity $\kappa_{j\gamma}$ and the spontaneous curvature $m_{j\gamma}$ as well as a contribution from the Gaussian curvature modulus $\kappa_{G,j\gamma}$. In close analogy to the bending energy of a two-domain vesicle, see Eq. 5.274, the bending energy functional of a partially wetted membrane has the form

$$\mathcal{E}_{\rm be}^{\rm in}\{S_{\alpha\gamma}, S_{\beta\gamma}\} = \mathcal{E}_{\alpha\gamma}^{\rm in}\{S_{\alpha\gamma}\} + \mathcal{E}_{\beta\gamma}^{\rm in}\{S_{\beta\gamma}\} \tag{5.300}$$

with

$$\mathcal{E}_{j\gamma}^{\rm in}\{S_{j\gamma}\} = 2\kappa_{j\gamma}\int dA_{j\gamma}\left(M - m_{j\gamma}\right)^2 \quad \text{for } j = \alpha \text{ or } \beta \tag{5.301}$$

which depends on the (local) mean curvature M of the membrane. In addition, the Gaussian curvature energy functional is given by

$$\mathcal{E}_G\{S_{\alpha\beta\gamma}\} = (\kappa_{G,\alpha\gamma} - \kappa_{G,\beta\gamma})\oint dl\, C_g + 2\pi(\kappa_{G,\alpha\gamma} + \kappa_{G,\beta\gamma}), \tag{5.302}$$

where the first term involves the line integral over the geodesic curvature C_g along the three-phase contact line as follows from the Gauss-Bonnet theorem, see the analogous expression for two-domain vesicles in Eq. 5.277. To obtain the correct sign of this term, the orientation of the line element dl has to be chosen in such a way that the $\alpha\gamma$ segment is surrounded in a clockwise manner when one looks down onto this segment from the exterior phase γ. We will again focus on membrane compositions with (at least) one molecular species, such as cholesterol, that undergoes frequent flip-flops between the two leaflets. We can then ignore additional bending energy terms arising from area-difference elasticity as described by Eqs 5.63 and 5.64. Furthermore, as emphasized at the beginning of the present section, we will also assume that this multi-component membrane has no tendency to phase separate and has a laterally uniform composition when exposed to spatially uniform aqueous environments.

In addition, the molecular interactions between the aqueous droplets and the membrane lead to two additional contributions, the adhesion free energies of the droplets and the free energy of the three-phase contact line. The latter contribution is proportional to the length $L_{\alpha\beta\gamma}$ of the contact line and given by

$$\lambda_{\rm co} L_{\alpha\beta\gamma} = \lambda_{\rm co}\mathcal{L}\{S_{\alpha\beta\gamma}\} \tag{5.303}$$

with the contact line tension $\lambda_{\rm co}$. The latter line tension can be positive or negative in contrast to the line tension λ of a domain boundary, which must be positive to ensure the stability of the intramembrane domains. Finally, the adhesion free energies will now be discussed in some detail.

[16] In general, curvature sorting should be limited to highly curved membrane segments. For in-wetting morphologies as considered here, high curvatures can be present along the three-phase contact line. In addition, one type of membrane segment may form nanotubes (Li et al., 2011; Liu et al., 2016) which represent highly curved membrane segments as well.

Adhesion free energies of droplets

In order to determine the adhesion free energies of the droplets in contact with the vesicle membrane, we denote the outer and inner leaflet of the bilayer membrane by the subscript "ol" and "il," respectively, and view the leaflet-water interfaces as "walls" with different interfacial tensions, depending on whether they are exposed to the α or to the β phase.

To each shape S of the wetting morphology depicted in Figure 5.40a, we can define a reference system with the same shape but with both the α and β droplet replaced by γ phase. The intermolecular interactions between the leaflets and the adjacent γ phases then lead to the interfacial tensions $\Sigma_{\text{ol},\gamma}$ and $\Sigma_{\text{il},\gamma}$ of the corresponding leaflet-water interfaces and the combined interfacial free energy functional of both leaflet-water interfaces has the form

$$\mathcal{T}_{\gamma\gamma}\{S_{\alpha\gamma},S_{\beta\gamma}\} = (\Sigma_{\text{ol},\gamma} + \Sigma_{\text{il},\gamma})\big(\mathcal{A}\{S_{\alpha\gamma}\} + \mathcal{A}\{S_{\beta\gamma}\}\big). \quad (5.304)$$

On the length scale of several nanometers, we should be able to ignore the dependence of the interfacial tensions on the interfacial curvatures which implies that both leaflet-water interfaces are governed by the same interfacial tension

$$\Sigma_{\text{l}\gamma} \equiv \Sigma_{\text{ol},\gamma} = \Sigma_{\text{il},\gamma} \quad (5.305)$$

corresponding to the leaflet-water interfaces of a planar bilayer membrane.

If we now go back to the wetting morphology in Figure 5.40a, the interfacial free energy of the leaflet-water interfaces becomes

$$\mathcal{T}_{\alpha\beta}\{S_{\alpha\gamma},S_{\beta\gamma}\} = (\Sigma_{\text{l}\alpha} + \Sigma_{\text{l}\gamma})\mathcal{A}\{S_{\alpha\gamma}\} + (\Sigma_{\text{l}\beta} + \Sigma_{\text{l}\gamma})\mathcal{A}\{S_{\beta\gamma}\}. \quad (5.306)$$

The adhesion free energy functional $\mathcal{E}_{\text{ad}}$ of the α and the β droplet in contact with one of the bilayer leaflets is then defined by

$$\mathcal{E}_{\text{ad}}\{S_{\alpha\gamma},S_{\beta\gamma}\} \equiv \mathcal{T}_{\alpha\beta} - \mathcal{T}_{\gamma\gamma} = W_{\alpha\gamma}\,\mathcal{A}\{S_{\alpha\gamma}\} + W_{\beta\gamma}\,\mathcal{A}\{S_{\beta\gamma}\} \quad (5.307)$$

with the adhesion free energies per unit area, $W_{\alpha\gamma}$ and $W_{\beta\gamma}$, given by (Lipowsky, 2018a)

$$W_{\alpha\gamma} \equiv \Sigma_{\text{l}\alpha} - \Sigma_{\text{l}\gamma} \quad \text{and} \quad W_{\beta\gamma} \equiv \Sigma_{\text{l}\beta} - \Sigma_{\text{l}\gamma} \quad (5.308)$$

for the α and β droplet in contact with the inner bilayer leaflet. Thus, the system can be characterized by two adhesive strengths, $W_{\alpha\gamma}$ and $W_{\beta\gamma}$, in close analogy to (i) the adhesive strength W between a membrane and a substrate surface as discussed in Section 5.7 and to (ii) the adhesion of nanoparticles as described in Chapter 8 of this book. When the leaflet prefers the α phase over the γ phase, the adhesive strength $W_{\alpha\gamma} < 0$. Likewise, when the leaflet prefers the β phase over the γ phase, $W_{\beta\gamma} < 0$. The adhesive strength $W_{j\gamma}$ also represents the reversible work that has to be expended per unit area to replace the γ phase by the phase j with $j = \alpha,\beta$. In addition, we can also compare the adhesion of the α and β droplets to one of the leaflets without any reference to the γ phase. Thus, the reversible work per unit area to replace a droplet of β phase in contact with a leaflet by α phase is given by

$$W_{\alpha\beta} \equiv \Sigma_{\text{l}\alpha} - \Sigma_{\text{l}\beta} = W_{\alpha\gamma} - W_{\beta\gamma} \quad (5.309)$$

which is negative if the leaflet prefers the α phase over the β phase.

Energy functional for in-wetting

Now, let us collect the different terms described previously. As a result, we obtain the energy functional

$$\begin{aligned}\mathcal{E}^{\text{in}}_{\text{2Dr}}\{S\} \equiv{}& \Sigma_{\alpha\beta}\mathcal{A}\{S_{\alpha\beta}\} + \mathcal{E}^{\text{in}}_{\text{be}}\{S_{\alpha\gamma},S_{\beta\gamma}\} + \mathcal{E}_{\text{ad}}\{S_{\alpha\gamma},S_{\beta\gamma}\}\\ &+ \mathcal{E}_{\alpha\beta\gamma}\{S_{\alpha\beta\gamma}\}\end{aligned} \quad (5.310)$$

with the contact line contribution

$$\mathcal{E}_{\alpha\beta\gamma}\{S_{\alpha\beta\gamma}\} = \mathcal{E}_{G}\{S_{\alpha\beta\gamma}\} + \lambda_{\text{co}}\,\mathcal{L}\{S_{\alpha\beta\gamma}\}. \quad (5.311)$$

The subscript 2Dr stands for "two droplets" and the superscript "in" indicates that the energy functional $\mathcal{E}^{\text{in}}$ corresponds to in-wetting and should be distinguished from out-wetting. In fact, the only energy contribution that is different for in- and out-wetting is the one that arises from the bending energy $\mathcal{E}^{\text{in}}_{\text{be}}\{S\}$ of the two membrane segments, as described by Eq. 5.300, because the spontaneous curvatures change sign when we swap the α and β phases with the γ phase.

Shape functional for in-wetting

In addition to the different energetic contributions of the vesicle-droplet system, we have to take the constraints on the membrane area A and the droplet volumes V_α and V_β into account. The constraint on the membrane area A is implemented by the Lagrange multiplier Σ which can be identified with the lateral stress that acts to stretch (or compress) the membrane as explicitly shown for uniform membranes in Appendix 5.D. In addition, we have to enforce certain values for the volumes V_α and V_β of the α and β droplets. These volumes are determined by the pressures P_α, P_β, and P_γ within the three liquid phases α, β, and γ or, more precisely, by the pressure differences $P_\alpha - P_\gamma$ and $P_\beta - P_\gamma$. We are then led to study the stationary shapes (minima, maxima, and saddle points) of the shape functional

$$\mathcal{F}^{\text{in}}_{\text{2Dr}}\{S\} = (P_\gamma - P_\alpha)\mathcal{V}\{S_\alpha\} + (P_\gamma - P_\beta)\mathcal{V}\{S_\beta\} + \Sigma\mathcal{A}\{S\} + \mathcal{E}^{\text{in}}_{\text{2Dr}}\{S\} \quad (5.312)$$

where the last term $\mathcal{E}^{\text{in}}_{\text{2Dr}}\{S\}$ represents the energy functional for in-wetting as given by Eq. 5.310. Both the pressure differences $P_\gamma - P_\alpha$ and $P_\gamma - P_\beta$ as well as the lateral stress Σ will be used as Lagrange multipliers to fulfill the geometric constraints that the droplet volumes V_α and V_β as well as the total membrane area A have certain prescribed values.

Terms proportional to individual segment areas

The shape functional as given by Eq. 5.312 contains the term $\Sigma\mathcal{A}\{S\}$ which depends on the lateral membrane stress Σ and the adhesion term $\mathcal{E}_{\text{ad}}\{S\}$ as given by Eq. 5.307 which depends on the adhesive strengths of the two aqueous phases. When we combine these two terms, we obtain

$$\Sigma\mathcal{A}\{S\} + \mathcal{E}_{\text{ad}}\{S\} = \Sigma_{\alpha\gamma}\,\mathcal{A}\{S_{\alpha\gamma}\} + \Sigma_{\beta\gamma}\,\mathcal{A}\{S_{\beta\gamma}\} \quad (5.313)$$

with the mechanical segment tensions (Lipowsky, 2018a)

$$\Sigma_{\alpha\gamma} \equiv \Sigma + W_{\alpha\gamma} \text{ and } \Sigma_{\beta\gamma} \equiv \Sigma + W_{\beta\gamma}. \quad (5.314)$$

Thus, each segment tension $\Sigma_{j\gamma}$ depends both on the lateral membrane stress Σ and on the adhesive strength $W_{j\gamma}$. Individual vesicles from a given vesicle preparation are usually characterized by different Σ-values corresponding to different membrane areas and vesicle shapes. In contrast, the adhesive strength $W_{j\gamma}$ is determined by the molecular interactions across the leaflet-water interfaces and should have the same value for all GUVs from the same batch, assuming that their membranes have the same lipid-protein composition and are exposed to aqueous solutions with the same solute composition. As a consequence, the difference

$$\Sigma_{\alpha\gamma} - \Sigma_{\beta\gamma} = W_{\alpha\gamma} - W_{\beta\gamma} = W_{\alpha\beta} \quad (5.315)$$

of the two segment tensions is only determined by the adhesive strengths and should also have the same value for all GUVs from the same batch.

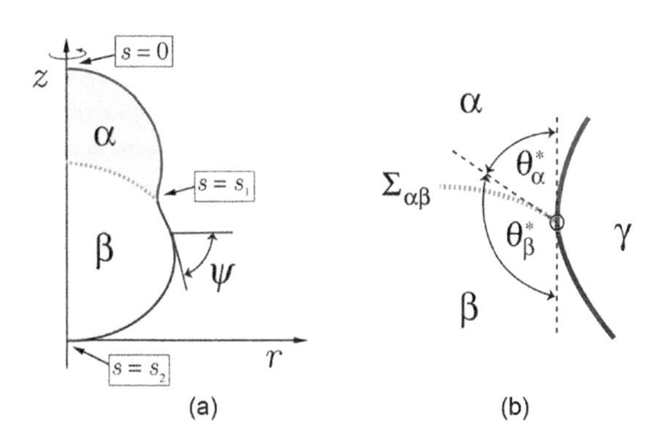

Figure 5.41 (a) Axisymmetric shape corresponding to partial in-wetting: As in Figure 5.33, the 2-dimensional shape of the membrane is uniquely determined by the 1-dimensional shape contour (red-purple) in the (r, z)-plane defined by the coordinate z along the symmetry axis and the radial coordinate r. The shape contour is parametrized by its arc length s, with the north and south pole of the vesicle being located at $s = 0$ and $s = s_2$, respectively, and the contact line at $s = s_1$. The angle ψ describes the tilt of the tangent vector at the shape contour from the horizontal r-direction; and (b) The $\alpha\gamma$ segment (red) and the $\beta\gamma$ segment (purple) meet at the contact line with a common tangent. The angles between this tangent and the tangent to the $\alpha\beta$ interface (dashed orange) represent the intrinsic contact angles θ^*_α and θ^*_β with $\theta^*_\alpha + \theta^*_\beta = \pi$.

5.9.4 SHAPE EQUATIONS AND MATCHING CONDITIONS

Shape equations for membrane segments

The first variation of the shape functional in Eq. 5.312 leads to two Euler-Lagrange or shape equations for the two membrane segments $\alpha\gamma$ and $\beta\gamma$, in close analogy to the shape Eqs 5.279 for two-domain vesicles. Indeed, the shape equations for the two membrane segments have the form

$$\begin{aligned} P_j - P_\gamma &= 2\hat{\Sigma}_{j\gamma} M - 2\kappa_{j\gamma} \nabla^2_{\mathrm{LB}} M - 4\kappa_{j\gamma} m_{j\gamma} M^2 \\ &\quad - 4\kappa_{j\gamma}[M - m_{j\gamma}][M^2 - G] \end{aligned} \quad (5.316)$$

with $j = \alpha, \beta$ and the total segment tensions

$$\hat{\Sigma}_{j\gamma} \equiv \Sigma_{j\gamma} + \sigma_{j\gamma} = \Sigma + W_{j\gamma} + \sigma_{j\gamma} \quad (5.317)$$

which include the spontaneous segment tensions

$$\sigma_{j\gamma} \equiv 2\kappa_{j\gamma} m^2_{j\gamma}. \quad (5.318)$$

As before, the ∇^2_{LB} symbol represents the Laplace-Beltrami operator, see Eq. 5.24, and G is the (local) Gaussian curvature. For the partial in-wetting morphologies depicted in Figure 5.40a, the pressure differences $P_\alpha - P_\gamma$ and $P_\beta - P_\gamma$ are positive.

Boundary or matching conditions for axisymmetric shapes

In addition to the shape equations for the two membrane segments, the first variation of the shape functional also leads to certain boundary or matching conditions for the two segments along the contact line. For axisymmetric vesicles as depicted in Figure 5.41, these matching conditions can be obtained by generalizing the corresponding conditions for two-domain vesicles as discussed in Section 5.8 and Appendix 5.F. Indeed, the axisymmetric shape shown in Figure 5.41 is quite similar to the one in Figure 5.33, the only difference is the presence of the two droplets α and β as well as the $\alpha\beta$ interface between these droplets. In Figure 5.41, the symmetry axis is again chosen to be the z-axis and the shape contour is again parametrized in terms of the arc length s, the radial coordinate $r = r(s)$, and the tilt angle $\psi = \psi(s)$. We can now directly use the matching conditions described in Appendix 5.F.1 if we substitute the domain indices b and a with the segment indices $\alpha\gamma$ and $\beta\gamma$, respectively.

The first variation of the shape functional with respect to the variable $\psi(s_1)$ is obtained by using the substitution $a \to \beta\gamma$ and $b \to \alpha\gamma$ in Eqs 5.2 and 5.3 which leads to the curvature discontinuity

$$\kappa_{\beta\gamma} C_1(s_1 + \varepsilon) - \kappa_{\alpha\gamma} C_1(s_1 - \varepsilon) = \delta\kappa C_2(s_1) + 2\kappa_{\beta\gamma} m_{\beta\gamma} - 2\kappa_{\alpha\gamma} m_{\alpha\gamma} \quad (5.319)$$

of the contour curvature C_1 along the three-phase contact line with the parameter

$$\delta\kappa \equiv \kappa_{\alpha\gamma} - \kappa_{\beta\gamma} + \kappa_{G,\alpha\gamma} - \kappa_{G,\beta\gamma}. \quad (5.320)$$

Note that the individual contour curvatures $C_1(s_1 + \varepsilon)$ and $C_1(s_1 - \varepsilon)$ are usually quite large compared to the orthogonal curvature $C_2(s_1)$ that satisfies $C_2(s_1) = \sin\psi(s_1) / r(s_1) \leq 1 / r(s_1)$. The discontinuity $C_1(s_1 + \varepsilon) - C_1(s_1 - \varepsilon)$ of the contour curvature vanishes if the two membrane segments have the same curvature-elastic properties, i.e., the same spontaneous curvature, bending rigidity, and Gaussian curvature modulus. The latter situation has been studied in (Kusumaatmaja et al., 2009) with the additional simplification that both membrane segments have zero spontaneous curvatures, i.e., $m_{\alpha\gamma} = m_{\beta\gamma} = 0$.

Balance between interfacial and segment tensions

A second boundary or matching condition is obtained from the first variation of the shape functional with respect to the variable $r_1 \equiv r(s_1)$ which represents the radius of the contact line. The resulting condition can be obtained from Eq. 5.F16, supplemented by one additional term arising from the interfacial tension $\Sigma_{\alpha\beta}$. We then obtain the balance condition (Lipowsky, 2018a)

$$\Sigma_{\beta\gamma} - \Sigma_{\alpha\gamma} = \Sigma_{\alpha\beta} \cos\theta_\alpha^* + \lambda_{co} \frac{\cos\psi_1}{r_1} + \Delta_{\Sigma,co} \quad (5.321)$$

with the intrinsic contact angle θ_α^* and the tilt angle $\psi_1 \equiv \psi(s_1)$, see Figure 5.41. The last term in Eq. 5.321 has the explicit form

$$\Delta_{\Sigma,co} = \frac{1}{2}\kappa_{\beta\gamma} \mathcal{Q}_{\beta\gamma}(s_1+\varepsilon) - \frac{1}{2}\kappa_{\alpha\gamma} \mathcal{Q}_{\alpha\gamma}(s_1-\varepsilon) \quad (5.322)$$

with the curvature-dependent terms

$$\mathcal{Q}_{j\gamma}(s) \equiv C_1^2(s) - [C_2(s) - 2m_{j\gamma}]^2 \quad \text{for } j = \alpha, \beta. \quad (5.323)$$

These relations describe the balance between the capillary forces arising from the interfacial tension $\Sigma_{\alpha\beta}$, the tensions $\Sigma_{\beta\gamma}$ and $\Sigma_{\alpha\gamma}$ of the two membrane segments, and the line tension λ_{co}. The additional term $\Delta_{\Sigma,co}$ in Eq. 5.321 arises from the different curvature-elastic properties of the two membrane segments. Indeed, the term $\Delta_{\Sigma,co}$ vanishes if the two membrane segments have the same curvature-elastic properties. In the latter case, the force balance condition Eqs 5.321 simplifies and becomes

$$\Sigma_{\beta\gamma} - \Sigma_{\alpha\gamma} = W_{\beta\gamma} - W_{\alpha\gamma} = \Sigma_{\alpha\beta} \cos\theta_\alpha^* + \lambda_{co} \frac{\cos\psi_1}{r_1} \quad (5.324)$$

which depends on the difference of the two adhesive strengths $W_{\beta\gamma}$ and $W_{\alpha\gamma}$, the interfacial tension $\Sigma_{\alpha\beta}$, and the contact line tension λ_{co}. Thus, if the vesicle membrane continued to have laterally uniform curvature-elastic properties even when it is partially wetted by the two aqueous droplets, the force balance along the contact line as described by Eq. 5.324 would involve neither the bending rigidity nor the spontaneous curvature of the membrane. For GUVs, the radius r_1 of the contact line is typically of the order of many micrometers. In such a situation, the term proportional to the line tension λ_{co} in (5.324) can be neglected which implies that the intrinsic contact angle θ_α^* depends only on two material parameters, the difference $W_{\beta\gamma} - W_{\alpha\gamma}$ of the two adhesive strengths and the interfacial tension $\Sigma_{\alpha\beta}$ of the water-water interface.

If the two membrane segments have different spontaneous curvatures but the same bending rigidities κ and the same Gaussian curvature moduli, the additional term $\Delta_{\Sigma,co}$ becomes

$$\Delta_{\Sigma,co} = 4\kappa[m_{\beta\gamma} - m_{\alpha\gamma}][M(s_1-\varepsilon) - m_{\alpha\gamma}] \quad (5.325)$$

with the mean curvature $M = \frac{1}{2}(C_1 + C_2)$ which satisfies, for $\kappa_{\alpha\gamma} = \kappa_{\beta\gamma} = \kappa$ and $\kappa_{G,\alpha\gamma} = \kappa_{G,\beta\gamma}$, the matching condition

$$M(s_1+\varepsilon) - m_{\beta\gamma} = M(s_1-\varepsilon) - m_{\alpha\gamma} \quad (5.326)$$

along the contact line as follows from Eq. 5.319. Thus, the discontinuity in the mean curvature, $M(s_1 + \varepsilon) - M(s_1 - \varepsilon)$, is now equal to the difference in the spontaneous curvatures, $m_{\beta\gamma} - m_{\alpha\gamma}$, and the additional term $\Delta_{\Sigma,co}$ is proportional to this discontinuity.

At present, both the curvature discontinuities and the additional term $\Delta_{\Sigma,co}$ that enters the force balance relation (5.321) cannot be used to analyze the shapes of GUVs because the local membrane curvatures along the contact line have not been resolved by optical microscopy. Therefore, these matching conditions will not be further pursued in the following. On the other hand, the experimental observations revealed one universal feature of the partial wetting morphologies for GUVs, namely that the shapes of the two membrane segments are very well described by spherical caps which is a direct consequence of the capillary forces exerted by the $\alpha\beta$ interface onto the vesicle membrane. Because the $\alpha\beta$ interface necessarily forms a spherical cap as follows from the classical Laplace equation, the partial wetting morphologies consist of three surface segments that form three spherical caps and meet along the three-phase contact line, as displayed in Figure 5.42.

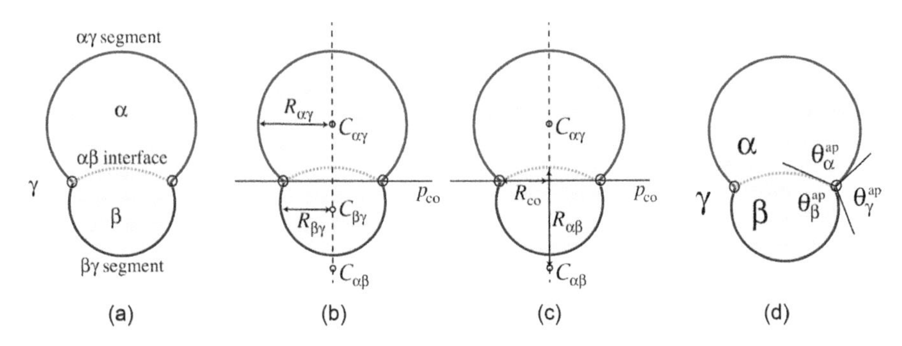

Figure 5.42 Cross-section of partial in-wetting morphology as observed experimentally: (a) Three spherical surface segments corresponding to the $\alpha\beta$ interface (dotted orange line) and to the two membrane segments $\alpha\gamma$ (red) and $\beta\gamma$ (purple). These three spherical caps meet along an apparent contact line (black circles); (b, c) The three-spherical-cap shape is determined by the curvature radii $R_{\alpha\gamma}$, $R_{\beta\gamma}$, and $R_{\alpha\beta}$ of the three spherical caps as well as by the contact line radius R_{co}. The three centers $C_{\alpha\gamma}$, $C_{\beta\gamma}$, and $C_{\alpha\beta}$ of the three spherical caps are located on the rotational symmetry axis (vertical dashed line). In order to obtain a unique shape, we also need to specify the locations of these cap centers relative to the contact line plane p_{co} (full horizontal line), see main text; and (d) At the contact line, the tangent planes to the three spherical surface segments define the three apparent contact angles θ_α^{ap}, θ_β^{ap}, and θ_γ^{ap} with $\theta_\alpha^{ap} + \theta_\beta^{ap} + \theta_\gamma^{ap} = 2\pi$. (From Lipowsky, R. *J. Phys. Chem. B*, 122, 3572–3586, 2018a.)

5.9.5 THREE-SPHERICAL-CAP SHAPES

Geometric relations for three spherical caps

From the optical microscopy images, we can directly deduce the curvature radii of the three spherical caps which will be denoted by $R_{\alpha\beta}$, $R_{\alpha\gamma}$, and $R_{\beta\gamma}$, respectively, see Figure 5.42b, c, and the centers of the spherical caps by $C_{\alpha\gamma}$, $C_{\beta\gamma}$, and $C_{\alpha\beta}$. We will again use the sign convention that all radii are always taken to be positive. Because the three spherical caps meet along the apparent contact line, the three cap centers $C_{\alpha\gamma}$, $C_{\beta\gamma}$, and $C_{\alpha\beta}$ are necessarily colinear. The straight line through these centers represents the axis of rotational symmetry for the three-spherical-cap shape corresponding to the vertical dashed line in Figure 5.42b,c. To obtain a certain three-spherical-cap shape, we also need to specify the radius R_{co} of the apparent contact line in addition to the curvature radii, see Figure 5.42c. In fact, the four length scales $R_{\alpha\beta}$, $R_{\alpha\gamma}$, $R_{\beta\gamma}$, and R_{co} are not quite sufficient to uniquely define the three-spherical-cap shape because we still need to specify (i) whether the two cap centers $C_{\alpha\gamma}$ and $C_{\beta\gamma}$ of the two membrane segments are located above or below the apparent contact line plane p_{co} as depicted by the horizontal full line in Figure 5.42b,c; and (ii) whether the cap center $C_{\alpha\beta}$ of the liquid-liquid interface is above or below this contact line plane corresponding to an $\alpha\beta$ interface that bulges towards the β or towards the α droplet.

For the example shown in Figure 5.42, the cap centers $C_{\alpha\gamma}$ and $C_{\beta\gamma}$ are located above and below the apparent contact line plane p_{co}, respectively. This location of the two cap centers implies that both membrane segments form spherical caps with an equator (or "belly"). In addition, the center $C_{\alpha\beta}$ of the $\alpha\beta$ interface is located below the plane p_{co} which implies that the $\alpha\beta$ interface bulges towards the α droplet corresponding to a pressure P_β in the β droplet that exceeds the pressure P_α in the α droplet. Keeping the four length scales fixed as well as the locations of the two cap centers $C_{\alpha\gamma}$ and $C_{\beta\gamma}$, we may also place the center $C_{\alpha\beta}$ above the contact line plane p_{co} which then leads to an $\alpha\beta$ interface that bulges towards the β droplet corresponding to $P_\alpha > P_\beta$.

We now introduce the sign convention that the mean curvature $M_{\alpha\beta}$ of the $\alpha\beta$ interface is *positive*, i.e.,

$$M_{\alpha\beta} = \frac{1}{R_{\alpha\beta}} > 0 \quad \text{for} \quad P_\alpha > P_\beta \tag{5.327}$$

and *negative* with

$$M_{\alpha\beta} = -\frac{1}{R_{\alpha\beta}} < 0 \quad \text{for} \quad P_\beta > P_\alpha. \tag{5.328}$$

With this sign convention, the classical Laplace equation for the $\alpha\beta$ interface assumes the form

$$P_\alpha - P_\beta = 2\Sigma_{\alpha\beta} M_{\alpha\beta} = \pm\frac{2\Sigma_{\alpha\beta}}{R_{\alpha\beta}} \tag{5.329}$$

where the plus and minus sign applies to $P_\alpha > P_\beta$ and $P_\beta > P_\alpha$, respectively.

Family of three-spherical-cap shapes with geometric constraints

As previously mentioned, the vesicle-droplet systems are characterized by three geometric constraints as provided by the droplet volumes V_α and V_β as well as by the total membrane area A. These three quantities can be expressed in terms of the four radii $R_{\alpha\beta}$, $R_{\alpha\gamma}$, $R_{\beta\gamma}$, and R_{co} which leads to three equations between the four radii. The solution of these three equations may be parametrized in terms of V_α, V_β, A, and a suitable reaction coordinate such as the apparent contact line radius R_{co}. As a result of this reparametrization, we obtain a one-parameter family of three-spherical-cap shapes that fulfill all three geometric constraints.

Apparent contact angles

Another set of geometric quantities that can be directly deduced from the optical microscopy images are the apparent contact angles θ_α^{ap}, θ_β^{ap}, and θ_γ^{ap}, with $\theta_\alpha^{ap} + \theta_\beta^{ap} + \theta_\gamma^{ap} = 2\pi$ introduced in Figure 5.42d. The sines of these angles can be expressed in terms of the three curvature radii and the apparent contact line radius R_{co}. In general, one has to distinguish several cases depending on the relative locations of the cap centers $C_{\alpha\gamma}$, $C_{\beta\gamma}$, and $C_{\alpha\beta}$ with respect to the contact line plane p_{co}. When these cap centers have the relative locations as in Figure 5.42b,c, corresponding to $P_\beta > P_\alpha$, we obtain the explicit relationships (Lipowsky, 2018a)

$$\sin\theta_\alpha^{ap} = \frac{R_{co}}{R_{\alpha\beta}R_{\alpha\gamma}}\left(\sqrt{R_{\alpha\beta}^2 - R_{co}^2} + \sqrt{R_{\alpha\gamma}^2 - R_{co}^2}\right), \tag{5.330}$$

$$\sin\theta_\beta^{ap} = \frac{R_{co}}{R_{\alpha\beta}R_{\beta\gamma}}\left(\sqrt{R_{\alpha\beta}^2 - R_{co}^2} - \sqrt{R_{\beta\gamma}^2 - R_{co}^2}\right), \tag{5.331}$$

with $R_{\alpha\beta} \geq R_{\beta\gamma}$ and

$$\sin\theta_\gamma^{ap} = \frac{R_{co}}{R_{\alpha\gamma}R_{\beta\gamma}}\left(\sqrt{R_{\alpha\gamma}^2 - R_{co}^2} + \sqrt{R_{\beta\gamma}^2 - R_{co}^2}\right). \tag{5.332}$$

If the two cap centers $C_{\alpha\gamma}$ and $C_{\beta\gamma}$ have the same locations as in Figure 5.42 but the cap center $C_{\alpha\beta}$ is moved to a location above the contact line plane p_{co}, corresponding to $P_\alpha > P_\beta$, these relations assume the slightly modified form

$$\sin\theta_\alpha^{ap} = \frac{R_{co}}{R_{\alpha\beta}R_{\alpha\gamma}}\left(\sqrt{R_{\alpha\beta}^2 - R_{co}^2} - \sqrt{R_{\alpha\gamma}^2 - R_{co}^2}\right) \tag{5.333}$$

with $R_{\alpha\beta} \geq R_{\alpha\gamma}$,

$$\sin\theta_\beta^{ap} = \frac{R_{co}}{R_{\alpha\beta}R_{\beta\gamma}}\left(\sqrt{R_{\alpha\beta}^2 - R_{co}^2} + \sqrt{R_{\beta\gamma}^2 - R_{co}^2}\right), \tag{5.334}$$

and

$$\sin\theta_\gamma^{ap} = \frac{R_{co}}{R_{\alpha\gamma}R_{\beta\gamma}}\left(\sqrt{R_{\alpha\gamma}^2 - R_{co}^2} + \sqrt{R_{\beta\gamma}^2 - R_{co}^2}\right). \tag{5.335}$$

The latter expression is identical with Eq. 5.332 but the first two expressions differ from Eqs 5.330 and 5.331 in the signs before the second square root.

These explicit relations between the sines of the apparent contact angles and the four radii directly demonstrate that the apparent contact angles are determined by the size and shape of the GUVs. In particular, all three angles change when we vary the apparent contact line radius R_{co}.

Angle-curvature relationship

Finally, using some trigonometric relations, it is not difficult to show that the curvature radii and the apparent contact angles satisfy the relation

$$\mp\frac{\sin\theta_\gamma^{ap}}{R_{\alpha\beta}} = \frac{\sin\theta_\alpha^{ap}}{R_{\beta\gamma}} - \frac{\sin\theta_\beta^{ap}}{R_{\alpha\gamma}} \tag{5.336}$$

where the minus and plus sign applies to an $\alpha\beta$ interface that bulges towards the α and the β droplet, respectively. The equalities in Eq. 5.336, which do not depend on the apparent contact line radius R_{co}, may be used to estimate the accuracy of the measured values for the curvature radii and apparent contact angles. When expressed in terms of the mean curvatures, the purely geometric relation (5.336) becomes

$$M_{\alpha\beta}\sin\theta_\gamma^{ap} = M_{\alpha\gamma}\sin\theta_\beta^{ap} - M_{\beta\gamma}\sin\theta_\alpha^{ap}. \tag{5.337}$$

Shape equations for spherical caps

When the membrane segments $\alpha\gamma$ and $\beta\gamma$ assume spherical cap shapes, the shape Eqs 5.316 assume the simplified form

$$P_j - P_\gamma = 2\hat{\Sigma}_{j\gamma} M_{j\gamma} - 4\kappa_{j\gamma} m_{j\gamma} M_{j\gamma}^2 \quad \text{with } j = \alpha, \beta \tag{5.338}$$

with the total segment tensions $\hat{\Sigma}_{j\gamma} \equiv \Sigma_{j\gamma} + \sigma_{j\gamma} = \Sigma + W_{j\gamma} + \sigma_{j\gamma}$ as in Eq. 5.317 and the spontaneous segment tensions $\sigma_{j\gamma} = 2\kappa_{j\gamma} m_{j\gamma}^2$ as in Eq. 5.318. The shape Eqs 5.338 can be rewritten in the more compact form

$$P_j - P_\gamma = 2\Sigma_{j\gamma}^{eff} M_{i\gamma} \quad \text{with} \quad j = \alpha, \beta \tag{5.339}$$

with the effective tensions

$$\Sigma_{j\gamma}^{eff} \equiv \hat{\Sigma}_{j\gamma} - 2\kappa_{j\gamma} m_{j\gamma} M_{j\gamma} = \Sigma_{j\gamma} + \sigma_{j\gamma} - 2\kappa_{j\gamma} m_{j\gamma} M_{j\gamma} \tag{5.340}$$

which depend on the mean curvatures $M_{j\gamma}$. Note that these shape equations now determine the *constant* mean curvatures $M_{\alpha\gamma}$ and $M_{\beta\gamma}$ of the two spherical membrane segments. Because both mean curvatures are necessarily positive, a positive value of $P_j - P_\gamma$ implies a positive value of the effective tension $\Sigma_{j\gamma}^{eff}$.

A linear combination of the Laplace Eq. 5.329 for the $\alpha\beta$ interface and the shape Eqs 5.339 for the two membrane segments can be used to eliminate the three pressure differences. As a result, we obtain the relation

$$\Sigma_{\alpha\beta} M_{\alpha\beta} = \Sigma_{\alpha\gamma}^{eff} M_{\alpha\gamma} - \Sigma_{\beta\gamma}^{eff} M_{\beta\gamma} \tag{5.341}$$

between the interfacial tension $\Sigma_{\alpha\beta}$ and the effective tensions $\Sigma_{\alpha\gamma}^{eff}$ and $\Sigma_{\beta\gamma}^{eff}$ experienced by the two membrane segments.

Relationship between tensions and angles

Using a combination of the geometric relation, Eq. 5.337, and the curvature-tension relation, Eq. 5.341, we can now eliminate the mean curvature $M_{\alpha\beta}$ of the $\alpha\beta$ interface which leads to the relationship (Lipowsky, 2018a)

$$M_{\alpha\gamma}\left(\frac{\Sigma_{\alpha\gamma}^{eff}}{\Sigma_{\alpha\beta}} - \frac{\sin\theta_\beta^{ap}}{\sin\theta_\gamma^{ap}}\right) = M_{\beta\gamma}\left(\frac{\Sigma_{\beta\gamma}^{eff}}{\Sigma_{\alpha\beta}} - \frac{\sin\theta_\alpha^{ap}}{\sin\theta_\gamma^{ap}}\right) \tag{5.342}$$

between the effective tensions, the apparent contact angles, and the mean curvatures of the $\alpha\gamma$ and $\beta\gamma$ membrane segments. It is important to note that the derivation of Eq. 5.342 was based (i) on the purely geometric relation, Eq. 5.337, which applies to three spherical caps that intersect along the apparent contact line and (ii) on the shape equations for the spherical membrane segments and the $\alpha\beta$ interface. In particular, this derivation did not make any assumptions about the mechanical balance of the interfacial and membrane tensions along the apparent contact line.

The relationship in Eq. 5.342 is reminiscent of the relation as given by Eq. 5.H7 in Appendix 5.H which applies to two membraneless droplets adhering to each other within a bulk liquid without a vesicle. The latter relation depends only on the contact angles and on the interfacial tensions, both of which represent material parameters. In contrast, the relationship in Eq. 5.342 for partial in-wetting of GUVs depends on several geometry-dependent parameters: (i) Explicitly on the mean curvatures $M_{\alpha\gamma} = 1/R_{\alpha\gamma}$ and $M_{\beta\gamma} = 1/R_{\beta\gamma}$ of the two membrane segments; (ii) Implicitly on these two curvatures via the effective tensions $\Sigma_{\alpha\gamma}^{eff}$ and $\Sigma_{\beta\gamma}^{eff}$; and (iii) On the apparent contact angles which are determined by the three-spherical-cap geometry as described in Eqs 5.330 to 5.332 for $P_\beta > P_\alpha$ and in Eqs 5.333 to 5.335 for $P_\alpha > P_\beta$.[17]

Parameter dependencies

On the other hand, many of the parameters that enter Eq. 5.342 can be determined experimentally. The interfacial tension $\Sigma_{\alpha\beta}$ represents a material parameter that can be obtained via experimental studies of macroscopic $\alpha\beta$ interfaces as demonstrated for PEG-dextran solutions in (Liu et al., 2012). In addition, the apparent contact angles and the mean curvatures can be obtained, for each vesicle-droplet couple, from optical microscopy experiments. It is less obvious how to determine the parameter combinations that enter the effective membrane tensions $\Sigma_{j\gamma}^{eff}$ as given by Eq. 5.340. These parameter combinations are the total segment tensions $\hat{\Sigma}_{j\gamma} = \Sigma + W_{j\gamma} + 2\kappa_{j\gamma} m_{j\gamma}^2$ as defined by Eq 5.317 and the combinations $\kappa_{j\gamma} m_{j\gamma}$ with $j = \alpha$ or β. Without prior knowledge about the bending rigidities and the spontaneous curvatures, these four parameter combinations should be regarded as unknowns that enter the relationship in Eq. 5.342 in a linear fashion. In order to determine four unknowns, we need four linearly independent equations.

[17] In both cases, the cap centers $C_{\alpha\gamma}$ and $C_{\beta\gamma}$ are located on different sides of the contact line plane p_{co}. Slightly different relations apply if these two cap centers are located on the same side of p_{co} which implies that one of the membrane segments attains a spherical cap without an equator.

To obtain such a set of equations, we might want to apply the relationship in Eq. 5.342 to four different vesicle-droplet couples as obtained from the same vesicle batch or the same preparation protocol. The four couples should then have the same composition of the vesicle membrane and the same composition of the different aqueous phases. As a consequence, all four vesicle-droplet couples should be characterized by the same interfacial tension $\Sigma_{\alpha\beta}$, the same adhesive strengths $W_{j\gamma}$, the same spontaneous curvatures $m_{j\gamma}$, and the same bending rigidities $\kappa_{j\gamma}$ because all of these quantities represent material parameters. However, the total segment tensions also include the overall lateral stress Σ that does *not* represent a material parameter but depends on the vesicle geometry and, thus, will vary from vesicle to vesicle even within the same batch. Therefore, if we applied the relationship in Eq. 5.342 to four different vesicle-droplet couples, the corresponding total segment tensions would involve four different stresses. As a consequence, each additional vesicle-droplet system would introduce one additional unknown as provided by the lateral stress experienced by the corresponding vesicle membrane.

To address this difficulty, two strategies can be pursued. First, we could consider GUVs with low lateral stresses Σ that fulfill the condition

$$|\Sigma| \ll W_{j\gamma} + 2\kappa_{j\gamma} m_{j\gamma}^2 \quad \text{for } j = \alpha \text{ or } \beta. \tag{5.343}$$

We could then ignore these stresses and estimate the total segment tensions by their asymptotic behavior

$$\hat{\Sigma}_{j\gamma} \approx W_{j\gamma} + 2\kappa_{j\gamma} m_{j\gamma}^2. \tag{5.344}$$

In the latter case, the total segment tensions would have the same values for all vesicle-droplet couples from the same batch. On the one hand, one would expect intuitively that the lateral stresses can be strongly reduced by osmotic deflation of the GUVs. On the other hand, the inequality in Eq. 5.343 involves two terms that may have different signs: the spontaneous tension $2\kappa_{j\gamma} m_{j\gamma}^2$ is always positive but the affinity strength $W_{j\gamma}$ will be negative when the membranes prefers the j phase over the γ phase. These two terms could cancel each other to a large extent, implying that the lateral stress must become ultralow in order to fulfill the inequality in Eq. 5.343.

A second strategy that does not involve any assumption about the magnitude of the lateral stress Σ is to consider several droplets adhering to the same vesicle. This strategy is described in the next paragraph.

Several droplets adhering to the same GUV

Thus, consider a situation in which several α droplets adhere to the interior leaflet of the same GUV membrane. These droplets coexist with one large β droplet inside the GUV. The different α droplets are labeled by $n = 1, 2, \ldots, N$. The vesicle membrane is then partitioned into $N + 1$ segments labeled by $n\gamma$ and $\beta\gamma$. The different $n\gamma$ segments experience the effective membrane tensions

$$\Sigma_{\alpha\gamma}^{(n)} = \Sigma + W_{\alpha\gamma} + \sigma_{\alpha\gamma} - 2\kappa_{\alpha\gamma} m_{\alpha\gamma} M_{\alpha\gamma}^{(n)} \tag{5.345}$$

where all parameters on the right hand side are independent of n apart from the mean curvatures $M_{\alpha\gamma}^{(n)}$ of the $n\gamma$ segments. For such a geometry, we obtain N relationships of the form

$$M_{\alpha\gamma}^{(n)} \left(\frac{\Sigma_{\alpha\gamma}^{(n)}}{\Sigma_{\alpha\beta}} - \frac{\sin\theta_{\beta}^{(n)}}{\sin\theta_{\gamma}^{(n)}} \right) = M_{\beta\gamma} \left(\frac{\Sigma_{\beta\gamma}^{\text{eff}}}{\Sigma_{\alpha\beta}} - \frac{\sin\theta_{\alpha}^{(n)}}{\sin\theta_{\gamma}^{(n)}} \right) \tag{5.346}$$

with $n = 1, 2, \ldots, N$. This set of equations can be rewritten in the form

$$\Upsilon_{\alpha\gamma}^{(n)} \equiv M_{\alpha\gamma}^{(n)} \left(\frac{\Sigma_{\alpha\gamma}^{(n)}}{\Sigma_{\alpha\beta}} - \frac{\sin\theta_{\beta}^{(n)}}{\sin\theta_{\gamma}^{(n)}} \right) + M_{\beta\gamma} \frac{\sin\theta_{\alpha}^{(n)}}{\sin\theta_{\gamma}^{(n)}} = M_{\beta\gamma} \frac{\Sigma_{\beta\gamma}^{\text{eff}}}{\Sigma_{\alpha\beta}} \tag{5.347}$$

where the last term is independent of the droplet label n. We then conclude that (Lipowsky, 2018a)

$$\Upsilon_{\alpha\gamma}^{(1)} = \Upsilon_{\alpha\gamma}^{(2)} = \ldots = \Upsilon_{\alpha\gamma}^{(N)}. \tag{5.348}$$

Therefore, from three different $n\gamma$ segments with three distinct mean curvatures $M_{\alpha\gamma}^{(n)}$ and, thus, three distinct expressions $\Upsilon_{\alpha\gamma}^{(n)}$, we obtain two linearly independent equations from which can deduce the two parameter combinations $(\Sigma + W_{\alpha\gamma} + \sigma_{\alpha\gamma})/\Sigma_{\alpha\beta}$ and $\kappa_{\alpha\gamma} m_{\alpha\gamma}/\Sigma_{\alpha\beta}$ for any value of Σ.

5.9.6 SHAPE FUNCTIONAL FOR THREE SPHERICAL CAPS

So far, we did not consider the force balance along the apparent contact line of the three spherical cap segments. We now address this force balance using a somewhat different approach. We start from the energy functional $\mathcal{E}_{2\text{Dr}}^{\text{in}}\{S\}$ and the shape functional $\mathcal{F}_{2\text{Dr}}^{\text{in}}\{S\}$ as given by Eqs 5.310 and 5.312 and apply these functionals to the three-spherical-cap shapes $S = S^{\text{sc}}$ which include the spherical cap shapes $S_{\alpha\gamma}^{\text{sc}}$ and $S_{\beta\gamma}^{\text{sc}}$ of the two membrane segments. The energy functional $\mathcal{E}_{2\text{Dr}}^{\text{in}}\{S\}$ then assumes the form

$$\mathcal{E}_{2\text{Dr}}^{\text{in}}\{S^{\text{sc}}\} = E^{\text{in}}(R_{\alpha\beta}, R_{\alpha\gamma}, R_{\beta\gamma}, R_{\text{co}}) \tag{5.349}$$

where the energy E^{in} represents an explicit function of the four variables $R_{\alpha\beta}$, $R_{\alpha\gamma}$, $R_{\beta\gamma}$, and R_{co}. The contributions from the Gaussian curvature energies and from the line tension are confined to the true contact line which is embedded in a highly curved membrane segment. These latter segment is lost when we use the three-spherical-cap approximation and replace the true by the apparent contact line. Therefore, we will now ignore these two energetic contributions. The energy function E^{in} then has the form

$$E^{\text{in}} = \sum_{j=\alpha,\beta} E_{j\gamma}^{\text{in}} \quad \text{with } E_{j\gamma}^{\text{in}} = W_{j\gamma} A_{j\gamma} + E_{j\gamma,\text{be}}^{\text{in}} \tag{5.350}$$

which consists of the adhesion free energies $W_{j\gamma}A_{j\gamma}$ and the bending energy contributions

$$E_{j\gamma,\text{be}}^{\text{in}} \equiv 2\kappa_{j\gamma} A_{j\gamma} \left(M_{j\gamma} - m_{j\gamma} \right)^2 \tag{5.351}$$

with $j = \alpha$ or β and constant mean curvatures $M_{j\gamma} = 1/R_{j\gamma}$.

Likewise, when we apply the shape functional in Eq. 5.312 to the three-spherical-cap shape S^{sc}, the resulting expression

$$\mathcal{F}^{\text{in}}\{S^{\text{sc}}\} = F^{\text{in}}(R_{\alpha\beta}, R_{\alpha\gamma}, R_{\beta\gamma}, R_{\text{co}}) \quad (5.352)$$

also becomes an explicit function F^{in} of the four radii. This function has the form

$$F^{\text{in}} = (P_\gamma - P_\alpha)V_\alpha + (P_\gamma - P_\beta)V_\beta + \Sigma(A_{\alpha\gamma} + A_{\beta\gamma}) + E^{\text{in}} \quad (5.353)$$

with the energy function E^{in} as given by Eq. 5.350. It will be convenient to rewrite this shape function according to

$$F^{\text{in}} = (P_\gamma - P_\alpha)V_\alpha + (P_\gamma - P_\beta)V_\beta + \Delta F^{\text{in}} \quad (5.354)$$

with

$$\Delta F^{\text{in}} \equiv \Sigma_{\alpha\beta} A_{\alpha\beta} + \Sigma_{\alpha\gamma} A_{\alpha\gamma} + \Sigma_{\beta\gamma} A_{\beta\gamma} + \sum_{j=\alpha,\beta} E^{\text{in}}_{j\gamma,\text{be}} \quad (5.355)$$

and the mechanical segment tensions $\Sigma_{j\gamma} = \Sigma + W_{j\gamma}$ as defined in Eq. 5.314.

In order to obtain a self-consistent description, we will now consider two limiting cases corresponding to small spontaneous curvatures and small bending energies as well as large spontaneous curvatures and large spontaneous tensions.

Small spontaneous curvatures and small bending energies

For membrane segment $j\gamma$, the regime of small spontaneous curvature will be defined by

$$(1 - |\eta|)M_{j\gamma} \le m_{j\gamma} \le (1 + |\eta|)M_{j\gamma} \quad (\text{small curvature } m_{j\gamma}) \quad (5.356)$$

with a dimensionless coefficient $|\eta| > 0$ of order one. For these $m_{j\gamma}$-values, the segment's bending energy $E^{\text{in}}_{j\gamma,\text{be}}$ satisfies the inequality

$$E^{\text{in}}_{j\gamma,\text{be}} = 2\kappa_{j\gamma}(M_{j\gamma} - m_{j\gamma})^2 A_{j\gamma} \le 2\kappa_{j\gamma} |\eta|^2 M^2_{j\gamma} A_{j\gamma}. \quad (5.357)$$

In terms of the curvature radius $R_{j\gamma}$ of membrane segment $j\gamma$, we obtain the squared mean curvature $M^2_{j\gamma} = R^{-2}_{j\gamma}$ and the segment area

$$A_{j\gamma} = 4\pi\zeta_{j\gamma} R^2_{j\gamma} \quad \text{with} \quad \zeta_{j\gamma} \equiv \frac{A_{j\gamma}}{4\pi R^2_{j\gamma}} \quad (5.358)$$

which implies the inequality

$$E^{\text{in}}_{j\gamma,\text{be}} \le 8\pi\kappa_{j\gamma} |\eta|^2 \zeta_{j\gamma} \quad \text{with} \quad 0 < \zeta_{j\gamma} < 1 \quad (5.359)$$

for the bending energy of the $j\gamma$ segment.

The small bending energy regime for the $j\gamma$ segment will now be defined by the condition that this energy is small compared to the interfacial free energy $\Sigma_{\alpha\beta} A_{\alpha\beta}$, i.e., by the condition

$$E^{\text{in}}_{j\gamma,\text{be}} \ll \Sigma_{\alpha\beta} A_{\alpha\beta} \quad (\text{small bending energy } E^{\text{in}}_{j\gamma,\text{be}}). \quad (5.360)$$

Using the two inequalities in Eq. 5.359, the condition in Eq. 5.360 can be fulfilled by

$$E^{\text{in}}_{j\gamma,\text{be}} \le 8\pi\kappa_{j\gamma} |\eta|^2 \ll \Sigma_{\alpha\beta} A_{\alpha\beta} \quad (5.361)$$

or

$$A_{\alpha\beta} \gg 8\pi |\eta|^2 \frac{\kappa_{j\gamma}}{\Sigma_{\alpha\beta}} \quad (5.362)$$

with the dimensionless coefficient $|\eta|$ of order one, see Eq 5.356.[18]

Thus, if the spontaneous curvature $m_{j\gamma}$ is small and satisfies the inequalities in Eq. 5.356 and if the interfacial area $A_{j\gamma}$ is large and satisfies the inequality in Eq. 5.362, we can ignore the bending energy $E^{\text{in}}_{j\gamma,\text{be}}$ of the membrane segment $j\gamma$ compared to the interfacial free energy $\Sigma_{\alpha\beta}A_{\alpha\beta}$. The energy contribution from this segment, see Eq. 5.350, then has the simple form

$$E^{\text{in}}_{j\gamma} \approx W_{j\gamma} A_{j\gamma}, \quad (5.363)$$

i.e., this contribution is dominated by the adhesion free energy between the membrane and the α or β droplet.

Large spontaneous curvatures and spontaneous tensions

For segment $j\gamma$, the regime of large spontaneous curvatures is defined by

$$|m_{j\gamma}| \gg |M_{j\gamma}| \quad (\text{regime of large } m_{j\gamma}). \quad (5.364)$$

In the latter regime, the bending energy $E^{\text{in}}_{j\gamma,\text{be}}$ of the $j\gamma$ segment becomes

$$E^{\text{in}}_{j\gamma,\text{be}} \approx 2\kappa_{j\gamma} m^2_{j\gamma} A_{j\gamma} = \sigma_{j\gamma} A_{j\gamma} \quad (5.365)$$

with the spontaneous tension $\sigma_{j\gamma}$ which implies the contribution

$$E^{\text{in}}_{j\gamma} \approx (W_{j\gamma} + \sigma_{j\gamma}) A_{j\gamma} \quad (5.366)$$

of the $j\gamma$ segment to the energy function E^{in} in Eq. 5.350.

Shape functions for special parameter regimes

If both segments belong to the small spontaneous curvature and small bending energy regime, the shape function ΔF^{in} as given by Eq. 5.355 simplifies and becomes $\Delta F^{\text{in}} = \Delta F^{\text{in}}_{s+s}$ with the area-dependent shape function

$$\Delta F^{\text{in}}_{s+s} \equiv \Sigma_{\alpha\beta} A_{\alpha\beta} + \Sigma_{\alpha\gamma} A_{\alpha\gamma} + \Sigma_{\beta\gamma} A_{\beta\gamma} \quad (\text{small} + \text{small regime}) \quad (5.367)$$

which depends on the mechanical segment tensions $\Sigma_{\alpha\gamma}$ and $\Sigma_{\beta\gamma}$. On the other hand, if both membrane segments belong to the

[18] The numerical value of $|\eta|$ was taken to be $|\eta| = 3/2$ in (Lipowsky, 2018a).

large spontaneous curvature regime, we obtain the shape function $\Delta F^{\text{in}} = \Delta F^{\text{in}}_{l+l}$ with

$$\Delta F^{\text{in}}_{l+l} \equiv \Sigma_{\alpha\beta} A_{\alpha\beta} + \hat{\Sigma}_{\alpha\gamma} A_{\alpha\gamma} + \hat{\Sigma}_{\beta\gamma} A_{\beta\gamma} \quad \text{(large + large regime)}. \tag{5.368}$$

with the total segment tensions $\hat{\Sigma}_{j\gamma} = \Sigma_{j\gamma} + \sigma_{j\gamma} = \Sigma + W_{j\gamma} + 2\kappa_{j\gamma} m^2_{j\gamma}$ as in Eq. 5.317.

Finally, if one membrane segment, say $\alpha\gamma$, has a large spontaneous curvature whereas the other membrane segment, $\beta\gamma$, has a small spontaneous curvature, the shape function becomes $\Delta F^{\text{in}} = \Delta F^{\text{in}}_{l+s}$ with

$$\Delta F^{\text{in}}_{l+s} \equiv \Sigma_{\alpha\beta} A_{\alpha\beta} + \hat{\Sigma}_{\alpha\gamma} A_{\alpha\gamma} + \Sigma_{\beta\gamma} A_{\beta\gamma} \quad \text{(large + small regime)}. \tag{5.369}$$

Note that we can obtain the shape function for the small-small regime from the shape function for the large-large regime by putting the spontaneous curvatures $m_{j\gamma}$ and, thus, the spontaneous segment tensions $\sigma_{j\gamma}$ equal to zero for both segments which implies that the total segment tensions $\hat{\Sigma}_{j\gamma}$ reduce to the mechanical segment tensions $\Sigma_{j\gamma}$. Likewise, we obtain the shape function for the large-small regime from the shape function of the large-large regime by putting the spontaneous tension $\sigma_{\beta\gamma}$ of the $\beta\gamma$ membrane segment equal to zero which leads to $\hat{\Sigma}_{\beta\gamma} = \Sigma_{\beta\gamma}$.

5.9.7 FORCE BALANCE ALONG APPARENT CONTACT LINE

Constrained energy minimization within the subspace of three-spherical-cap shapes then implies the four stationarity conditions (Lipowsky, 2018a)

$$\frac{\partial F^{\text{in}}}{\partial R_{\alpha\beta}} = 0, \quad \frac{\partial F^{\text{in}}}{\partial R_{\alpha\gamma}} = 0, \quad \frac{\partial F^{\text{in}}}{\partial R_{\beta\gamma}} = 0, \quad \text{and} \quad \frac{\partial F^{\text{in}}}{\partial R_{\text{co}}} = 0. \tag{5.370}$$

It is not difficult to show that the first condition $\partial F^{\text{in}}/\partial R_{\alpha\beta} = 0$ is equivalent to the classical Laplace Eq. 5.329 for the curvature radius $R_{\alpha\beta}$ of the $\alpha\beta$ interface. We should also require that the two stationarity relations $\partial F^{\text{in}}/\partial R_{\alpha\gamma} = 0$ and $\partial F^{\text{in}}/\partial R_{\beta\gamma} = 0$ lead back to the shape Eqs 5.339 for the curvature radii $R_{\alpha\gamma}$ and $R_{\beta\gamma}$ of the two membrane segments. The latter requirement is, however, not fulfilled in general but only for certain regions of the parameter space.

These special parameter regions include the small-small, large-large, and large-small regimes described in the previous subsection and defined by the shape functions ΔF^{in} in Eqs 5.367 to 5.369. All of these shape functions have the same form as the shape function ΔF_o for two membraneless droplets as given by Eq 5.H12 in Appendix 5.H when we substitute the interfacial tensions $\Sigma_{\alpha\gamma}$ and $\Sigma_{\beta\gamma}$ of the membraneless droplets by the mechanical or total tensions of the membrane segments. Using the same substitution in the force balance Eq. 5.H9 for membraneless droplets, we obtain the corresponding force balance conditions for the membrane-enclosed droplets.

As explained above, we can recover the small-small regime from the large-large regime by putting the spontaneous tensions of the two membrane segments equal to zero. Likewise, we can recover the large-small regime from the large-large regime by putting the spontaneous tension $\sigma_{\beta\gamma}$ equal to zero. Therefore, it is sufficient to consider the substitution in the force balance Eq. 5.H9 for the large-large regime. In the latter case, the interfacial tensions $\Sigma_{\alpha\gamma}$ and $\Sigma_{\beta\gamma}$ in Eq. 5.H9 for membraneless droplets have to be substituted by the total segment tensions $\hat{\Sigma}_{\alpha\gamma}$ and $\hat{\Sigma}_{\beta\gamma}$, respectively. As a result, we obtain the force balance conditions

$$\frac{\Sigma_{\alpha\beta}}{\sin\theta^{\text{ap}}_{\gamma}} = \frac{\hat{\Sigma}_{\alpha\gamma}}{\sin\theta^{\text{ap}}_{\beta}} = \frac{\hat{\Sigma}_{\beta\gamma}}{\sin\theta^{\text{ap}}_{\alpha}} \quad \text{(large-large regime)} \tag{5.371}$$

between the $\alpha\beta$ interface and the two membrane segments along the apparent contact line. These conditions are equivalent to the two linearly independent relationships

$$\frac{\hat{\Sigma}_{\alpha\gamma}}{\Sigma_{\alpha\beta}} = \frac{\sin\theta^{\text{ap}}_{\beta}}{\sin\theta^{\text{ap}}_{\gamma}} \quad \text{and} \quad \frac{\hat{\Sigma}_{\beta\gamma}}{\Sigma_{\alpha\beta}} = \frac{\sin\theta^{\text{ap}}_{\alpha}}{\sin\theta^{\text{ap}}_{\gamma}} \quad \text{(large-large regime)} \tag{5.372}$$

between the tensions and the contact angles (Lipowsky, 2013, 2014b). The force balance as given by Eq. 5.371 represents the law of sines for a triangle with the three sides $\Sigma_{\alpha\beta}$, $\hat{\Sigma}_{\alpha\gamma}$, and $\hat{\Sigma}_{\beta\gamma}$ as displayed in Figure 5.43b. For membraneless droplets, the corresponding triangle is displayed in Figure 5.48.

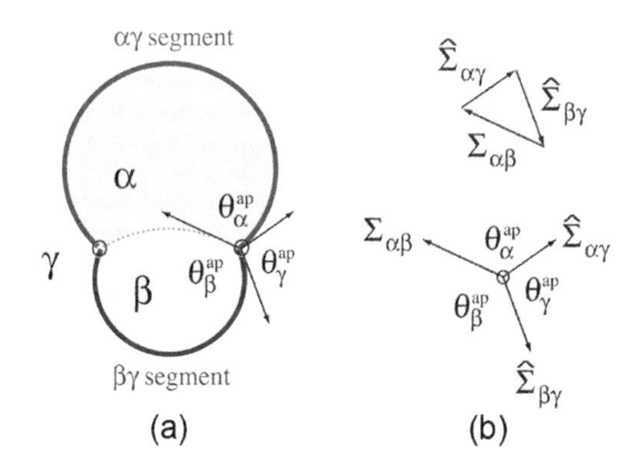

Figure 5.43 Force balance along the apparent contact line for small and large spontaneous curvatures: (a) Partial in-wetting morphology of vesicle (red, purple) enclosing two aqueous droplets of α (yellow) and β (blue) phase immersed in the exterior liquid γ (white). As in Figure 5.42, the membrane segments $\alpha\gamma$ (red) and $\beta\gamma$ (purple) form spherical caps that meet the $\alpha\beta$ interface (broken orange) along the apparent contact line (small black circles) where the three surface segments form the apparent contact angles $\theta^{\text{ap}}_{\alpha}$, $\theta^{\text{ap}}_{\beta}$, and $\theta^{\text{ap}}_{\gamma}$; and (b) Force balance between the interfacial tension $\Sigma_{\alpha\beta}$ as well as the total tensions $\hat{\Sigma}_{\alpha\gamma}$ and $\hat{\Sigma}_{\beta\gamma}$ of the two membrane segments as defined by Eq. 5.317. The three tensions form a triangle which implies the relations in Eqs 5.371 and 5.372 (Lipowsky, 2013, 2018a). The latter relations can be explicitly derived for three parameter regimes: (i) if both spontaneous curvatures are large as defined by Eq. 5.364, (ii) if both spontaneous curvatures are small and the interfacial area $A_{\alpha\beta}$ is sufficiently large as in Eqs 5.356 and 5.362, which implies $\hat{\Sigma}_{j\gamma} \approx \Sigma_{j\gamma}$; and (iii) for one small and one large spontaneous curvature.

Essentially the same force balance conditions apply to the small-small and large-small regimes. If the $\beta\gamma$ membrane segment belongs to the regime of small spontaneous curvature and small bending energy, the corresponding force balance conditions are obtained from those in Eqs 5.371 and 5.372 by replacing the total segment tension $\hat{\Sigma}_{\beta\gamma} = \Sigma_{\beta\gamma} + \sigma_{\beta\gamma}$ by the mechanical segment tension $\Sigma_{\beta\gamma}$. Likewise, the conditions for the small-small regime are obtained by replacing the total segment tensions $\hat{\Sigma}_{j\gamma}$ of both segments by the mechanical segment tensions $\Sigma_{j\gamma}$.

Difference of total segment tensions

Subtracting the two force balance relations in Eq. 5.372 from each other, we obtain the difference

$$\frac{\hat{\Sigma}_{\beta\gamma} - \hat{\Sigma}_{\alpha\gamma}}{\Sigma_{\alpha\beta}} = \frac{W_{\beta\gamma} - W_{\alpha\gamma} + \sigma_{\beta\gamma} - \sigma_{\alpha\gamma}}{\Sigma_{\alpha\beta}} = \Xi(\theta_\alpha^{\text{ap}}, \theta_\beta^{\text{ap}}, \theta_\gamma^{\text{ap}}) \quad (5.373)$$

with the function

$$\Xi(x, y, z) = \frac{\sin x - \sin y}{\sin z}, \quad (5.374)$$

as in Eq. 5.H19. Note that the overall lateral stress Σ, which depends on the vesicle geometry, drops out from the difference $\hat{\Sigma}_{\beta\gamma} - \hat{\Sigma}_{\alpha\gamma}$. As a consequence, Eq. 5.373 provides a relation between the apparent contact angles $\theta_\alpha^{\text{ap}}$, θ_β^{ap}, and $\theta_\gamma^{\text{ap}}$, the adhesive strengths $W_{\alpha\gamma}$ and $W_{\beta\gamma}$, and the spontaneous tensions $\sigma_{\alpha\gamma}$ and $\sigma_{\beta\gamma}$, i.e., between the apparent contact angles and material parameters. As shown in Appendix 5.H.4, the function $\Xi(\theta_\alpha^{\text{ap}}, \theta_\beta^{\text{ap}}, \theta_\gamma^{\text{ap}})$ satisfies the inequalities

$$-1 \le \Xi(\theta_\alpha^{\text{ap}}, \theta_\beta^{\text{ap}}, \theta_\gamma^{\text{ap}}) \le +1 \quad (5.375)$$

as follows from the triangle inequalities for the triangle in Figure 5.43b. The upper bound $\Xi = +1$ is obtained for the apparent contact angles $\theta_\alpha^{\text{ap}} = 0$ and $\theta_\beta^{\text{ap}} = \theta_\gamma^{\text{ap}} = \pi$, corresponding to complete wetting of the membrane by the α phase as in Figure 5.40b. The lower bound $\Xi = -1$ is obtained for the angles $\theta_\beta^{\text{ap}} = 0$ and $\theta_\alpha^{\text{ap}} = \theta_\gamma^{\text{ap}} = \pi$, corresponding to complete wetting of the membrane by the β phase as shown in Figure 5.40c.

Relation between apparent and intrinsic contact angles

For some special parameter regions, we can also obtain a simple relation between the apparent contact angles in Figure 5.42 and the intrinsic contact angle in Figure 5.41. We now consider two membrane segments that have essentially the same curvature-elastic properties which implies the simplified force balance

$$\Sigma_{\beta\gamma} - \Sigma_{\alpha\gamma} = \Sigma_{\alpha\beta} \cos\theta_\alpha^* \quad (5.376)$$

along the true contact line as described by Eq. 5.324 where we assumed a large contact line radius and ignored the term proportional to the line tension λ_{co}. Two membrane segments with the same curvature-elastic properties have the same spontaneous tensions. Therefore, the difference $\Sigma_{\beta\gamma} - \Sigma_{\alpha\gamma}$ between the mechanical tensions of the two segments is equal to the difference $\hat{\Sigma}_{\beta\gamma} - \hat{\Sigma}_{\alpha\gamma}$ between the total segment tensions. For small or large spontaneous curvatures, we then obtain

$$\Sigma_{\beta\gamma} - \Sigma_{\alpha\gamma} = \hat{\Sigma}_{\beta\gamma} - \hat{\Sigma}_{\alpha\gamma} = \Sigma_{\alpha\beta} \frac{\sin\theta_\alpha^{\text{ap}}}{\sin\theta_\gamma^{\text{ap}}} - \Sigma_{\alpha\beta} \frac{\sin\theta_\beta^{\text{ap}}}{\sin\theta_\gamma^{\text{ap}}} \quad (5.377)$$

where the second equality follows from Eq. 5.372. A combination of Eq. 5.377 with Eq. 5.376 then leads to the relation

$$\cos\theta_\alpha^* = \frac{\sin\theta_\alpha^{\text{ap}} - \sin\theta_\beta^{\text{ap}}}{\sin\theta_\gamma^{\text{ap}}} \quad (5.378)$$

between the intrinsic contact angle θ_α^* that is not accessible to conventional optical microscopy and the apparent contact angles that can be obtained from the microscopy images.

In (Kusumaatmaja et al., 2009), the relation in Eq. 5.378 was originally derived for the special case of vanishing spontaneous curvatures for both membrane segments, i.e., $m_{\alpha\gamma} = m_{\beta\gamma} = 0$, and was then used to analyze the shapes of vesicles that enclosed one PEG-rich and one dextran-rich droplet. Even though the apparent contact angles of these vesicles were quite different, the relation in Eq. 5.378 led to a fairly constant value for the intrinsic contact angle θ_α^*. Later experiments revealed, however, that the spontaneous curvatures $m_{\alpha\gamma}$ must be quite large because the $\alpha\gamma$ membrane segments in contact with the PEG-rich phase formed nanotubes, see Figures 5.4 and 5.21 corresponding to a spontaneous curvature of about 1/(125 nm) for the Ld phase and 1/(600 nm) for the Lo phase. (Li et al., 2011; Liu et al., 2016) Furthermore, the experimental data as well as molecular dynamics simulations provided strong evidence that this large spontaneous curvature was generated by asymmetric adsorption of PEG molecules. Therefore, it is tempting to assume that the spontaneous curvature $m_{\beta\gamma}$ of the $\beta\gamma$ membrane segments in contact with the dextran-rich phase was comparatively small. A small value of $m_{\beta\gamma}$ and a large value of $m_{\alpha\gamma}$ would justify the use of Eq 5.372 to describe the force balance along the apparent contact line but it would not justify the use of Eq. 5.376 to describe the force balance along the true contact line because the latter equation is based on the assumption that both membrane segments have essentially the same spontaneous curvature. On the other hand, if we assumed that the spontaneous curvature $m_{\beta\gamma}$ is large as well and comparable to $m_{\alpha\gamma}$, we could justify the use of both Eqs 5.372 and 5.376. Therefore, it would be rather valuable to determine the spontaneous curvature $m_{\beta\gamma}$ in an independent manner, e.g., by studying GUVs that are completely filled with the dextran-rich phase, corresponding to a point in the aqueous phase diagram of Figure 5.39 that is located on the binodal line between the partial wetting regime of the two-phase coexistence region and the uniform phase at high dextran concentrations, see lower blue segment of the binodal in Figure 5.39. Deflation of such a GUV will lead to budding for small spontaneous curvatures as in Section 5.5 or to tubulation for large spontaneous curvatures as in Section 5.6.

Membrane nanotubes for partial and complete wetting

As shown in Figure 5.39, the phase diagram of aqueous PEG-dextran solutions exhibits a two-phase coexistence region with both a complete wetting regime close to the critical point and a partial wetting regime further away from this point. The two wetting regimes are separated by a certain tie line corresponding to the dashed straight line in Figure 5.39. For complete wetting, the whole GUV membrane is exposed to the PEG-rich phase whereas, for partial wetting, only the $\alpha\gamma$ membrane is in contact with this aqueous phase. Therefore, in the complete and partial wetting regime, nanotubes were formed by the whole GUV membrane and the $\alpha\gamma$ membrane segment, respectively. Furthermore, for complete wetting, the tubes stayed away from the $\alpha\beta$ interface whereas they accumulated on this interface for partial wetting. In the latter case, the adhesion of the tubes to the $\alpha\beta$ interface lowers the (free) energy of the vesicle-droplet system as shown in (Liu et al., 2016). Each tube that adheres to the $\alpha\beta$ interface is in contact with both the α and the β phase and, thus, forms both an $\alpha\gamma$ and a $\beta\gamma$ membrane segment separated by a contact line parallel to the long tube axis. Along these microscopic contact lines, the angle between the $\alpha\beta$ interface and the $\alpha\gamma$ tube segments is again given by the intrinsic contact angle θ_α^* with the same local geometry as depicted in Fig. 5.41b, because the γ phase within the tubes is identical with the exterior aqueous phase.

If the $\alpha\gamma$ membrane segment forms nanotubes, the segment tension $\Sigma_{\alpha\gamma} = \Sigma + W_{\alpha\gamma}$ is small compared to the spontaneous tension $\sigma_{\alpha\gamma}$ of this segment (Lipowsky, 2013) as follows from the mechanical equilibrium between the highly curved tubes and the weakly curved spherical $\alpha\gamma$ segments, see the detailed discussion of this aspect in Section 5.6. The corresponding tension-angle relationship in Eq. 5.372 then assumes the simplified form

$$\frac{\sigma_{\alpha\gamma}}{\Sigma_{\alpha\beta}} = \frac{2\kappa_{\alpha\gamma} m_{\alpha\gamma}^2}{\Sigma_{\alpha\beta}} \approx \frac{\sin\theta_\beta^{\rm ap}}{\sin\theta_\gamma^{\rm ap}} \quad \text{(tubulated } \alpha\gamma \text{ segments)} \qquad (5.379)$$

which can be used to estimate the spontaneous curvature $m_{\alpha\gamma}$ from the apparent contact angles (Liu et al., 2016).

5.9.8 TWO-DROPLET VESICLES WITH CLOSED NECKS

For partial in-wetting, the vesicle membrane is in contact with two enclosed droplets, as displayed in Figure 5.43a. When we deflate such a two-droplet vesicle, it can decrease its interfacial energy by reducing the area $A_{\alpha\beta}$ of the $\alpha\beta$ interface. The corresponding energy gain is governed by $\Delta A_{\alpha\beta}\Sigma_{\alpha\beta}$ where $\Delta A_{\alpha\beta}$ is the change in interfacial area. Such a morphological change is, in fact, rather likely unless one of the membrane segments has a sufficiently large spontaneous curvature to form nanobuds and nanotubes. If the $\alpha\gamma$ segment forms nanotubes, for example, the energy gain is $\Delta A_{\alpha\gamma}\sigma_{\alpha\gamma}$ with the area $\Delta A_{\alpha\gamma}$ stored in the nanotubes and the spontaneous tension $\sigma_{\alpha\gamma} = 2\kappa_{\alpha\gamma} m_{\alpha\gamma}^2$. So, we expect that osmotic deflation of a partially wetted vesicle leads to a reduction of the interfacial area whenever $\sigma_{\alpha\gamma} \ll \Sigma_{\alpha\beta}$. This competition between different morphological pathways is more systematically described in Appendix 5.J for the special case of two-droplet vesicles with up-down symmetry.

Thus, in the absence of bud and tube formation, the area of the $\alpha\beta$ interface will eventually shrink to zero and the vesicle membrane will then form a closed membrane neck around this point-like interface as in Figure 5.40d. For such a morphology, which looks like the limit shape $L^{\rm pea}$ in Figure 5.15a but involves two different interior solutions α and β, the vesicle membrane has the area $A = A_{\rm 2sp} \propto V_\alpha^{2/3} + V_\beta^{2/3}$, which is determined by the volumes V_α and V_β of the two spherical droplets as in Eq. 5.294.

As described in Section 5.8.3 on domain-induced budding, spherical buds with closed necks are also formed by two-domain vesicles arising from lipid phase separation within multi-component membranes. Compared to such two-domain vesicles, the closed neck of a two-droplet vesicle is further stabilized by the formation of the $\alpha\beta$ interface during neck opening. If we assume an axisymmetric neck and ignore a possible difference of the Gaussian curvature moduli $\kappa_{G,\alpha\gamma}$ and $\kappa_{G,\beta\gamma}$, the contact line is located within the membrane neck and the contact line radius r_1 is equal to the neck radius $R_{\rm ne}$. Furthermore, because of the assumed axisymmetry, the neck-spanning $\alpha\beta$ interface has the shape of a spherical cap that meets the membrane along the circular contact line with the intrinsic contact angle θ_α^* of the α droplet, see Figure 5.41b. The free energy of the membrane neck then includes the interfacial free energy

$$\Sigma_{\alpha\beta} A_{\alpha\beta} = \frac{2\pi}{1+\sin\theta_\alpha^*} \Sigma_{\alpha\beta} R_{\rm ne}^2 \qquad (5.380)$$

which grows quadratically with increasing neck radius $R_{\rm ne}$. The bending energy of the vesicle membrane that consists of two membrane segments and forms an open neck of radius $R_{\rm ne}$ can be obtained from the corresponding expression for two-domain vesicles as derived in (Jülicher and Lipowsky, 1996). Adding the free energy of the contact line, we then obtain

$$E_{\rm be}(R_{\rm ne}) + 2\pi R_{\rm ne}\lambda_{\rm co} \approx E_{\rm be}(R_{\rm ne}=0) - 4\pi E_1 R_{\rm ne} \quad \text{for small } R_{\rm ne} \qquad (5.381)$$

with

$$E_1 \equiv \kappa_{\alpha\gamma}(M_{\alpha\gamma} - m_{\alpha\gamma}) + \kappa_{\beta\gamma}(M_{\beta\gamma} - m_{\beta\gamma}) - \frac{1}{2}\lambda_{\rm co}. \qquad (5.382)$$

The closure of the neck and the stability of the closed neck are governed by the behavior of the combined free energy $\Sigma_{\alpha\beta}A_{\alpha\beta} + E_{\rm be}(R_{\rm ne}) + 2\pi R_{\rm ne}\lambda_{\rm co}$ for small $R_{\rm ne}$. In the latter limit, the leading term is provided by the E_1-term in (5.381) because the interfacial free energy $\Sigma_{\alpha\beta}A_{\alpha\beta} \sim R_{\rm ne}^2$. Therefore, we obtain the stability criterion $E_1 \leq 0$ which is equivalent to

$$\kappa_{\beta\gamma}(M_{\beta\gamma} - m_{\beta\gamma}) + \kappa_{\alpha\gamma}(M_{\alpha\gamma} - m_{\alpha\gamma}) \leq \frac{1}{2}\lambda_{\rm co} \quad (\kappa_{G,\beta\gamma} \simeq \kappa_{G,\alpha\gamma}). \qquad (5.383)$$

The equality in Eq (5.383) describes the neck closure condition for limit shapes obtained from vesicle shapes with open necks whereas the inequality describes the stability of closed necks. Because the additional term arising from the $\alpha\beta$ interface is irrelevant in the limit of small neck radius $R_{\rm ne}$, the stability criterion in Eq. 5.383 has the same form as the corresponding criterion for two-domain vesicles as given by Eq. 5.288 with the line tension λ of the domain boundary replaced by the line tension λ_{co} of the three-phase contact line. It is important to note, however, that the stability condition in Eq. 5.383 has been obtained under the implicit assumption that the membrane neck is axisymmetric. The latter assumption is justified for a positive value of the contact line tension λ_{co} but may not apply to a negative value of λ_{co}. Indeed, recent molecular simulations have shown that a negative contact line tension can lead to a spontaneous symmetry breaking of the rotational symmetry and to a tight-lipped contact line (Satarifard et al., 2018).

5.9.9 NUCLEATION OF NANODROPLETS AT MEMBRANES

In general, phase separation in liquid mixtures may proceed via nucleation and growth of small droplets or via spinodal decomposition. In the nucleation regime, the droplets are formed by the minority phase and have to overcome a certain free energy barrier in order to grow. This barrier is reduced if a droplet is nucleated at an adhesive surface. For a rigid surface as provided by a tense membrane, the barrier reduction depends primarily on the contact angle of the droplet. For a flexible and deformable membrane, as considered here, the barrier may be further reduced by the elastic response of the membrane which can adapt its shape and composition to the molecular interactions with the droplet.

As in the previous subsections, we focus on phase separation of the interior aqueous solution into two coexisting liquid phases, α and β. For complete wetting of the vesicle membrane by the α phase, the intrinsic contact angle θ_α^* vanishes which implies that the phase separation starts via the formation of a thin α layer at the inner leaflet of the vesicle membrane, see pink subregion in Figure 5.39. For partial wetting, on the other hand, the intrinsic contact angle θ_α^* is finite, and the phase separation within the nucleation regime starts with nanodroplets of α phase that are formed at the inner membrane leaflet as shown in Figure 5.44a.

For such a small droplet, the intrinsic contact angle will be affected by the tension λ_{co} of the contact line, see Eqs 5.321 and 5.324. This contact line tension can be positive or negative, in contrast to the line tension of domain boundaries which is always positive. In fact, recent molecular simulation indicate that the contact line tension λ_{co} can be negative (Satarifard et al., 2018) which implies that it acts to decrease the contact angle θ_α^* of small droplets compared to larger ones.

After an α droplet as in Figure 5.44a has been formed, the $\alpha\gamma$ segment of the membrane in contact with this droplet is exposed to an asymmetric environment and can acquire an appreciable spontaneous curvature $m_{\alpha\gamma}$. In order to simplify the following discussion, let us assume that the spontaneous curvature $m_{\alpha\gamma}$ is large compared to the spontaneous curvature $m_{\beta\gamma}$ of the $\beta\gamma$ segment and that the latter curvature is small and can be ignored.

If the spontaneous curvature $m_{\alpha\gamma}$ is *negative* as in the case of PEG-dextran solutions that undergo phase separation within the vesicle interior, the membrane prefers to curve towards the inner leaflet and to form a spherical in-bud of radius R_γ that is filled with the exterior γ phase as in Figure 5.44b. As shown in this figure, all membrane segments adjacent to the closed neck are formed by the $\alpha\gamma$ membrane with spontaneous curvature $m_{\alpha\gamma}$. The membrane neck is then characterized by the condition $0 > M_{\text{ne}} = \frac{1}{2}(M_1 + M_2) \geq m_{\alpha\gamma}$ where M_1 and $M_2 = -1/R_\gamma$ are the mean curvatures of the two membrane segments 1 and 2 on the two sides of the neck. Because these two membrane segments have the same curvature-elastic properties, this stability condition is identical with Eq. 5.57 for uniform membranes, see also Figure 5.14(d–f) in Section 4.6. Inspection of Figure 5.44b reveals that the in-bud displaces some volume of α phase and *increases* the area of the $\alpha\beta$ interface which implies that the α droplet has to reach a sufficiently large volume before the in-bud becomes energetically favorable. After such an in-bud has been formed, the bud radius increases until the spherical shape becomes unstable and transforms into a short necklace-like tube as displayed in Figure 5.23.

On the other hand, if the droplet-induced curvature $m_{\alpha\gamma}$ is *positive*, the $\alpha\gamma$ membrane segment prefers to curve towards the outer leaflet of the vesicle membrane and to form a spherical out-bud of radius R_α that is filled with α phase as in Figure 5.44c. As shown in the latter panel, the two membrane segments adjacent to the neck of the out-bud are now provided by the $\alpha\gamma$ and the $\beta\gamma$ segments which have, in general, different spontaneous curvatures $m_{\alpha\gamma}$ and $m_{\beta\gamma}$. The formation of the out-bud reduces the free energy of the membrane-droplet system by (i) adapting the mean curvature of the $\alpha\gamma$ segment to its spontaneous curvature $m_{\alpha\gamma}$ and (ii) replacing the $\alpha\beta$ interface by a closed membrane neck which implies a strong reduction of the interfacial free energy. The corresponding neck condition is given by Eq. 5.383 if both membrane segments have essentially the same Gaussian curvature modulus, $\kappa_{G,\beta\gamma} \simeq \kappa_{G,\alpha\gamma}$. If the Gaussian curvature moduli are different, the vesicles may still form closed membrane necks but the domain boundaries are then shifted away from these necks and, thus, have a finite length, compare Figure 5.35.

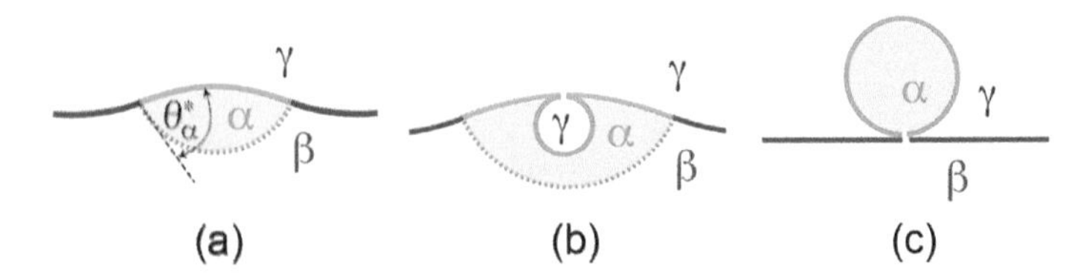

Figure 5.44 In-wetting: Nucleation and growth of an aqueous nanodroplet (yellow) consisting of α phase in contact with an aqueous β phase and the *inner* leaflet of a weakly curved vesicle membrane (blue/red) that separates the α and β phases from the exterior aqueous phase γ. The contact line with the $\alpha\beta$ interface (broken orange) divides the membrane into two segments, an $\alpha\gamma$ segment (blue) and a $\beta\gamma$ segment (red). Both segments are exposed to asymmetric aqueous environments which act to induce spontaneous curvatures $m_{\alpha\gamma}$ and $m_{\beta\gamma}$. Here, we focus on the case $m_{\beta\gamma} \simeq 0$ and $|m_{\alpha\gamma}| \gg m_{\beta\gamma}$: (a) Initially, the $\alpha\beta$ interface has the shape of a spherical cap and forms the intrinsic contact angle θ_α^* with the adjacent $\alpha\gamma$ segment (blue) of the membrane; (b) For *negative* values of $m_{\alpha\gamma}$, the $\alpha\gamma$ membrane segment prefers to form a spherical in-bud that is filled with exterior γ phase. The closure and stability of the in-bud's neck depends only on $m_{\alpha\gamma}$; and (c) For *positive* values of $m_{\alpha\gamma}$, the $\alpha\gamma$ membrane segment prefers to engulf the α droplet, in particular if the volume of the droplet matches the preferred bud size. Complete engulfment leads to a closed membrane neck that replaces the $\alpha\beta$ interface, thereby eliminating the contribution of this interface to the system's free energy.

5.10 TOPOLOGICAL CHANGES OF MEMBRANES

In the previous sections, we focused on processes that do not change the topology of the membranes. Now, let us briefly consider two important topology-transforming processes, membrane fusion and membrane fission (or scission). During membrane fusion, two separate membranes are combined into a single one; during fission, a single membrane is divided up into two separate ones. These processes are ubiquitous in eukaryotic cells: Both the outer cell membrane and the inner membranes of organelles act (i) as donor membranes that continuously produce vesicles via budding and fission and (ii) as acceptor membranes that integrate such vesicles via adhesion and fusion. One example for fission is provided by the closure of autophagosomes which are double-membrane organelles (Knorr et al., 2012, 2015).

5.10.1 FREE ENERGY LANDSCAPES

It is instructive to consider the free energy landscapes for fusion and fission as schematically depicted in Figure 5.45. Fusion is exergonic, if the free energy G_2 of the 2-vesicle state exceeds the free energy G_1 of the 1-vesicle state. In the opposite case with $G_1 > G_2$, fission is exergonic. Exergonic fusion or fission processes occur spontaneously but the kinetics of these processes is governed by the free energy barriers Δ between the 1-vesicle and the 2-vesicle state, see Figure 5.45. Because these barriers are typically large compared to k_BT, even exergonic fusion and fission processes will be rather slow unless coupled to other molecular processes that act to reduce these barriers. Indeed, in the living cell, the fusion and fission of biomembranes is controlled by membrane-bound proteins such as SNAREs and dynamin as will be discussed in later chapters of this book. It should also be emphasized that the free energy landscape may involve several barriers as has been observed in molecular dynamics simulations of tension-induced fusion (Grafmüller et al., 2007, 2009).

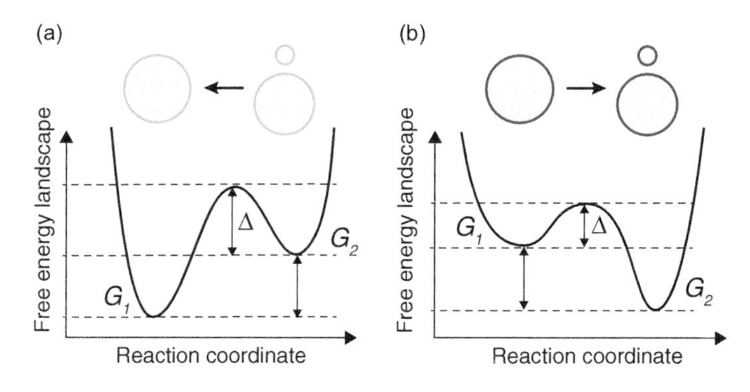

Figure 5.45 Free energy landscapes for membrane fusion and fission (or scission): (a) Schematic landscape for an exergonic fusion process. In this case, the free energy G_2 of the 2-vesicle state exceeds the free energy G_1 of the 1-vesicle state; and (b) Schematic landscape for an exergonic fission process. In the latter case the free energy G_1 of the 1-vesicle state is larger than the free energy G_2 of the 2-vesicle state. The cartoons (top row) show a 1-vesicle state on the left and a 2-vesicle state on the right; both states have the same membrane area. The small vesicle of the 2-vesicle state has the radius R_{ss} which is much smaller than the radius of the large vesicle. The blue membranes in (a) have a spontaneous curvature with magnitude $|m| \ll 1/R_{ss}$ whereas the red membranes in (b) have a large spontaneous curvature with $m \simeq 1/(2R_{ss})$. In both (a) and (b), the free energy difference $G_2 - G_1$ determines the direction in which the processes can proceed spontaneously (black arrows) while the kinetics of these processes is governed by the free energy barriers Δ.

Dependence on spontaneous curvature

The free energy difference $G_2 - G_1$ between the 2-vesicle and the 1-vesicle state can be estimated if one ignores energetic contributions arising from changes in volume and focuses on changes in curvature energy (Lipowsky, 2013). Because of the topological changes, we need to take the Gaussian curvature and the associated Gaussian curvature modulus κ_G into account. (Helfrich, 1973) Stability arguments indicate that $-2 < \kappa_G/\kappa < 0$ (Helfrich and Harbich, 1987). For the following considerations, it will be sufficient to use the rough estimate $\kappa_G \simeq -\kappa$ which is consistent with both experimental (Derzhanski et al., 1978; Lorenzen et al., 1986) and simulation (Hu et al., 2012) studies. A small spherical vesicle that is cleaved off from a donor membrane then changes the total curvature energy by a certain amount that can be used to estimate the free energy difference $G_2 - G_1$. It is important to note, however, that this change in curvature energy depends strongly on the magnitude of the spontaneous curvature.

5.10.2 EXERGONIC FUSION FOR SMALL *m*

Let us consider a 1-vesicle state corresponding to a spherical GUV that acts as the donor membrane and a 2-vesicle state obtained from this GUV by cleaving off a much smaller spherical vesicle, see top row of Figure 5.45. Both states have the same membrane area. The small vesicle of the 2-vesicle state has the radius R_{ss} which is taken to be much smaller than the radius of the GUV. We may then ignore any constraints on the vesicle volumes and assume that the large vesicle of the 2-vesicle state has a spherical shape as well. If the GUV membrane is uniform and the magnitude $|m|$ of its spontaneous curvature is much smaller than the inverse size, $1/R_{ss}$, of the small vesicle, the free energy difference between the 2-vesicle and 1-vesicle state is *positive* and given by

$$G_2 - G_1 = 8\pi\kappa + 4\pi\kappa_G \simeq +4\pi\kappa \quad \text{for } |m| \ll 1/R_{ss} \tag{5.384}$$

where the estimate $\kappa_G \simeq -\kappa$ has been used. In this case, the fission process is endergonic whereas the fusion process is exergonic, see the corresponding free energy landscape in Figure 5.45a. For the typical rigidity value $\kappa \simeq 20k_BT$, the relation Eq. 5.384 leads to the fairly large free energy difference $G_2 - G_1 \simeq +250k_BT$!

5.10.3 EXERGONIC FISSION FOR LARGE *m*

On the other hand, if the magnitude $|m|$ of the spontaneous curvature is large, the GUV can form a small spherical bud with radius $R_{ss} \simeq 1/(2|m|)$ as in Figure 5.45b as follows from the closed neck condition for the corresponding limit shapes L^{pea} and L^{sto} as discussed in Section 5.5.2. If this bud is cleaved off, the free energy difference between the resulting 2-vesicle state and the initial 1-vesicle state is now *negative* and given by

$$\begin{aligned} G_2 - G_1 &= 8\pi\kappa(1 - 2R_{ss}|m|) \\ &\quad + 4\pi\kappa_G \simeq 4\pi\kappa_G \simeq -4\pi\kappa \\ &\quad \text{for } R_{ss} \simeq 1/(2|m|). \end{aligned} \tag{5.385}$$

In the latter case, the fission process is exergonic and the fusion process is endergonic, corresponding to a free energy landscape as in Figure 5.45b. Now, the free energy difference $G_2 - G_1 \simeq -250k_BT$ for a typical value $\kappa \simeq 20k_BT$ of the bending rigidity.

Biological membranes often form intramembrane domains with an appreciable spontaneous curvature m_{do}. One example for this latter case is provided by clathrin dependent endocytosis which leads to membrane domains with a spontaneous curvature $m_{do} \simeq -1/(40\text{nm})$ (Agudo-Canalejo and Lipowsky, 2015a). Now, consider a GUV with a small membrane domain that has an appreciable spontaneous curvature m_{do} whereas the spontaneous curvature of the remaining GUV membrane is again negligible. The membrane domain can then form a small spherical bud of size $R_{ss} = 1/|m_{do}|$ as follows from the closed neck condition for domain-induced budding, see Eq. 5.291. If the latter bud is cleaved off, the free energy difference between the resulting 2-vesicle state and the initial 1-vesicle state is again negative and has the form

$$G_2 - G_1 = 8\pi\kappa(1 - 2R_{ss}\,|\,m_{do}\,|) + 4\pi\kappa_G - 4\pi\frac{\lambda}{|\,m_{do}\,|} \simeq -12\pi\kappa - 4\pi\frac{\lambda}{|\,m_{do}\,|} \quad (5.386)$$

where λ denotes the line tension of the domain boundary. Because this line tension has to be positive, the fission of a domain-induced bud is an exergonic process that leads to an even larger free energy gain $|G_2 - G_1| > 12\pi\kappa \gtrsim 750k_BT$ for bending rigidity $\kappa \simeq 20k_BT$.

5.11 SUMMARY AND OUTLOOK

This chapter addressed the multi-responsive behavior of giant vesicles from a theoretical point of view. Because the vesicle membranes are fluid, they can respond to external perturbation by remodeling both their shape and their local membrane composition. Two curvature-elastic parameters that play a prominent role in the whole chapter are the spontaneous curvature m, which provides a quantitative measure for bilayer asymmetry (Section 3.5), and the spontaneous tension $\sigma = 2\kappa m^2$, which provides the intrinsic tension scale of curvature elasticity (Section 5.4.2). If molecular flip-flops between the two leaflets of the bilayer membrane can be ignored, the spontaneous curvature becomes an effective spontaneous curvature m_{eff} that contains both a local and a non-local contribution, the latter arising from area-difference-elasticity, see Eqs 5.66 and 5.67.

All biomembranes are asymmetric in the sense that the two leaflets have different lipid compositions (Fadeel and Xue, 2009) and that the membrane proteins have a preferred orientation related to their biological function. It is important to realize that both lipids and membrane proteins as well as adsorbed solutes and anchored macromolecules can contribute to the spontaneous curvature as illustrated by the examples in Box 5.1. In fact, the framework of curvature elasticity as reviewed here applies to giant vesicles irrespective of the chemical nature of the molecular membrane components as long as the vesicle membranes are in a fluid state. Thus, these vesicles may be built up from different lipid components, membrane proteins, or other amphiphilic molecules such as diblock copolymers.

The shapes and shape transformations of membranes with laterally uniform curvature-elastic properties are governed by two dimensionless parameters, the volume-to-area ratio (or reduced volume) v and the spontaneous curvature $\bar{m} = R_{ve}m$. These two parameters can be controlled by changes in the osmotic conditions and by one of the curvature-generating mechanisms in Box 5.1. The resulting shape transformations often lead to budding and tubulation processes, which create nanobuds and nanotubes as described in Sections 5.5 and 5.6. The buds and tubes represent additional membrane compartments that are still connected to the mother vesicle via closed or narrow membrane necks. These necks are a direct consequence of curvature elasticity (Section 5.4.6, Figure 5.14) and can be used to deduce the spontaneous curvature from the GUV morphology as described in Box 5.2. The latter deduction is based on the *local* stability conditions for closed necks as given by Eqs 5.60 and 5.61 which relate the neck curvature to the spontaneous curvature. In the absence of flip-flops, one obtains the generalized stability conditions in Eqs 5.68 and 5.69. Sufficiently large values of m_{eff} lead to the cleavage of the membrane neck and thus to complete membrane fission, see Section 5.5.4.

In cell biology, the closure and cleavage of such membrane necks represents an essential step for many processes such as endo- and exocytosis, the secretion of giant plasma membrane vesicles (or "blebs") (Scott, 1976; Baumgart et al., 2007; Veatch et al., 2008; Keller et al., 2009) and outer membrane vesicles (Kulp and Kuehn, 2010; Schertzer and Whiteley, 2012) from eukaryotic and prokaryotic cells, as well as cytokinesis during cell division.

When a GUV undergoes spontaneous tubulation, the total membrane tension is dominated by the spontaneous tension as described by Eqs 5.193 and 5.215 for necklace-like and cylindrical nanotubes, respectively. Because the spontaneous tension is a material parameter, tubulated vesicles behave, to a large extent, like liquid droplets with a variable surface area and with an effective interfacial tension that is provided by the spontaneous tension σ. This droplet-like behavior, which reflects the area reservoir that the nanotubes provide for the mother vesicle, leads to an increased robustness against mechanical perturbations as has been recently demonstrated by micropipette aspiration and cycles of osmotic deflation and inflation (Bhatia et al., 2018).

Membrane nanotubes are also formed within eukaryotic cells and provide ubiquitous structural elements of many membrane-bound organelles such as the endoplasmic reticulum, the Golgi, the endosomal network, and mitochondria (Marchi et al., 2014; van Weering and Cullen, 2014; Westrate et al., 2015). These intracellular nanotubes are used for molecular sorting, signaling, and transport. Intercellular (or "tunneling") nanotubes formed by the plasma membranes of two or more cells provide long-distance connections for cell-cell communication, intercellular transport, and virus infections (Wang and Gerdes, 2015; He et al., 2010; Sowinski et al., 2008). It seems rather plausible to assume that these tubes are also generated by spontaneous curvature and/or locally applied forces but the relative importance of these two tubulation mechanisms remains to be elucidated.

Additional shape transformations of membranes and vesicles can be induced by adhesive surfaces as described in Section 5.7. The onset of adhesion is governed by the simple stability relation in Eq. 5.236 which depends on the adhesion length $R_W = \sqrt{2\kappa/|W|}$. This length can vary over several orders of magnitude as illustrated by the membrane-particle couples in Table 5.2. Analogous stability relations play an important role for the engulfment of nanoparticles by membranes as described in Chapter 8 of this book.

The adhesion of a vesicle to a rigid substrate or solid support leads to the segmentation of the vesicle membrane into a bound and unbound membrane segment. For multi-component vesicle membranes, these two segments can differ in their molecular composition and thus in their curvature-elastic properties when the vesicle membrane contains several molecular components, as explained in Section 7.6. Therefore, the adhesion of multi-component membranes provides a relatively simple example for ambience-induced segmentation. This kind of segmentation plays an important role for the adhesion of nanoparticles by membrane-anchored receptors (Agudo-Canalejo and Lipowsky, 2015a), see the more detailed discussion in Chapter 8 of this book.

Multi-component membranes can undergo phase separation into two fluid phases, a process that is now firmly established for a variety of three-component membranes as discussed in Section 2.4 and at the beginning of Section 5.8. Membrane phase separation leads to multi-domain vesicles, the shape of which is governed by the interplay between the curvature-elastic properties of the intramembrane domains and the line tension of the domain boundaries. One prominent example for this interplay is domain-induced budding, see Figure 5.32 and Section 5.8.3. Another example is provided by transformations between different patterns of intramembrane domains, which are coupled to drastic shape changes of the vesicles as illustrated in Figure 5.37.

Membrane phase separation of multi-component vesicles is strongly affected by ambience-induced segmentation of the vesicle membranes as explained in Section 5.8.5. Indeed, if the membrane is partitioned into several segments that differ in their molecular composition, membrane phase separation is only possible in one of the segments but not in several segments simultaneously. Because cellular membranes are exposed to rather heterogeneous environments, the associated segmentation acts to suppress the formation of intramembrane domains within such membranes. The latter mechanism explains the difficulty to detect lipid phase separation *in vivo*, in contrast to the large intramembrane domains frequently observed in multi-component lipid membranes.

Another interesting example for ambience-induced segmentation is provided by membranes and vesicles exposed to aqueous two-phase systems or water-in-water emulsions as described in Section 5.9. To simplify the discussion, Section 5.9 focused on aqueous phase separation within the GUVs which leads to the in-wetting morphologies displayed in Figure 5.40. Out-wetting morphologies arising from phase separation of the exterior aqueous solution are addressed in Appendix 5.I. For partial in-wetting as shown in Figure 5.40a, the interface between the two aqueous phases α and β exerts capillary forces onto the GUV membrane along the three-phase contact line. On the micrometer scale, these forces lead to apparent kinks of the membrane shapes. This response of the membranes to the capillary forces is quite remarkable because the interfacial tension of the $\alpha\beta$ interface is ultralow, of the order of 10^{-6}–10^{-4} N/m, reflecting the vicinity of a critical demixing point in the aqueous phase diagram.

However, the apparent kink of the membrane shape should not persist to the nanoscale because such a kink would imply a very large bending energy of the GUV membrane. Therefore, when viewed on the nanometer scale, the membrane should be smoothly curved, which implies the existence of an intrinsic contact angle as depicted in Figure 5.41. This angle is related to the difference of the segment tensions as given by the force balance Eq. 5.321. The latter equation also depends on the local curvatures of the two membrane segments at the contact line. At present, these curvatures cannot be determined experimentally which implies that the force balance Eq. 5.321 cannot be scrutinized by experiment.

On the other hand, the optical micrographs of the GUV shape showed that the two membrane segments in contact with the α and β droplets form spherical caps to a very good approximation. The extrapolation of these spherical cap shapes defines an apparent contact line and apparent contact angles as shown in Figures 5.42 and 5.43. The spherical cap geometry leads to the simplified shape Eqs 5.338 which imply the general relationship in Eq. 5.342. The latter relationship depends on the effective tensions and curvature radii of the two membrane segments as well as on the interfacial tension and the apparent contact angles. This relationship can be used to obtain the curvature-elastic parameters of the membrane segments from the observed wetting morphology.

For certain regions of the parameter space corresponding to small and large spontaneous curvatures, a simplified set of tension-angle relationships can be derived for the force balance along the apparent contact lines. For small spontaneous curvatures as defined by Eq. 5.356, the bending energies can be neglected compared to the interfacial free energy of the $\alpha\beta$ interface if the interfacial area $A_{\alpha\beta}$ is sufficiently large and satisfies the inequality in Eq. 5.362. In this parameter regime, we obtain the relationships in Eqs 5.371 and 5.372 which relate the total membrane tensions and the interfacial tension to the apparent contact angles, corresponding to the force triangle in Figure 5.43b. The same relationships apply to large spontaneous curvatures for which the bending energy is dominated by the spontaneous tension and behaves as in Eq. 5.365. If one of the membrane segments forms membrane nanotubes, one can ignore the mechanical tension within this segment compared to its spontaneous tension and use the simpler relationship in Eq. 5.379 to estimate the spontaneous curvature of the tubulated segment.

In the context of synthetic biology, GUVs are very attractive as possible microcompartments for the bottom-up assembly of artificial protocells (Walde et al., 2010; Fenz and Sengupta, 2012; Schwille, 2015; Weiss et al., 2018). One practical problem that has impeded research in this direction is the limited robustness of GUVs against mechanical perturbations. Very recently, this limitation has been overcome by two different strategies. One strategy is based on the formation of GUVs within emulsion droplets that support and stabilize the GUVs (Weiss et al., 2018), see also

Chapter 30 of this book. The other strategy uses the special properties of tubulated GUVs as discussed in Section 5.6. The nanotubes increase the robustness of the giant vesicles by providing a membrane reservoir for the mother vesicles which can then adapt their surface area to avoid membrane rupture (Bhatia et al., 2018). In the latter study, the increased robustness has already been demonstrated by micropipette experiments and by repeated cycles of osmotic deflation and inflation. Giant vesicles with membrane nanotubes will also tolerate other mechanical perturbations, arising, e.g., from the adhesion and engulfment of microparticles, in close analogy to cellular uptake via phagocytosis and pinocytosis, or in response to constriction forces that can lead to membrane fission and the formation of smaller membrane compartments. The latter process of artificial cytokinesis is an important objective for the bottom-up assembly of artificial protocells. Thus, both droplet-stabilized and tubulated GUVs provide new and promising modules for the bottom-up assembly of such artificial protocells.

ACKNOWLEDGMENTS

I thank all my collaborators for enjoyable and fruitful interactions and Jaime Agudo-Canalejo and Rumiana Dimova for detailed comments on earlier versions of this manuscript.

APPENDICES

5.A BRIEF EXCURSION INTO DIFFERENTIAL GEOMETRY

Any membrane shape S can be described in terms of two surface coordinates $\underline{s} \equiv (s^1, s^2)$ and a vector-valued function $\vec{X} = \vec{X}(\underline{s})$ that maps the surface coordinates into three-dimensional space (see, e.g., do Carmo, 1976). At any point P of the membrane surface, the tangent vectors $\vec{X}_i$ with $i = 1, 2$ and the normal vector $\hat{n}$ are then given by

$$\vec{X}_i = \frac{\partial \vec{X}}{\partial s^i} \quad \text{and} \quad \hat{n} = \frac{\vec{X}_1 \times \vec{X}_2}{|\vec{X}_1 \times \vec{X}_2|} \tag{5.A1}$$

where the symbol $\times$ denotes the vector product in three-dimensional space. The three vectors $\vec{X}_1$, $\vec{X}_2$, and $\hat{n}$ represent a right-handed trihedron at any point P of the membrane surface. Note that the normal vector $\hat{n}$ is a unit vector which is orthogonal to the plane spanned by the two tangent vectors. In general, the tangent vectors $\vec{X}_1$ and $\vec{X}_2$ are neither unit vectors nor orthogonal to each other. These tangent vectors define the metric tensor

$$g_{ij} = \vec{X}_i \cdot \vec{X}_j, \tag{5.A2}$$

where the symbol $\cdot$ denotes the scalar product. As we move along the membrane surface, the normal vector $\hat{n}$ is tilted and this tilt can be expressed in terms of the tangent vectors because the normal vector is a unit vector with $\hat{n} \cdot \hat{n} = 1$ and $\frac{\partial \hat{n}}{\partial s^i} \cdot \hat{n} = 0$. The tilt of the normal vector then defines the curvature tensor $h_i^{\,j}$ via

$$\frac{\partial \hat{n}}{\partial s^i} = -h_i^{\,j} \vec{X}_j \equiv -h_i^{\,1} \vec{X}_1 - h_i^{\,2} \vec{X}_2 \tag{5.A3}$$

where the second equation explains the summation over the repeated index j. The principal curvatures C_1 and C_2 discussed in Section 3.2 are the eigenvalues of the curvature tensor $-h_i^{\,j}$. This definition of the principal curvatures implies that a sphere is characterized by the principal curvatures $C_1 = C_2 > 0$.[19] Using the definition of the normal vector in Eq. 5.A1, we can express the first derivatives $\frac{\partial \hat{n}}{\partial s^i}$ of the normal vector via the second derivatives $\frac{\partial^2 \vec{X}}{\partial s^i \partial s^j}$ of the vector-valued function $\vec{X}(\underline{s})$. Therefore, in order to define the principal curvatures at a certain point on the membrane surface, the components of the vector $\vec{X}(\underline{s})$ that describes the membrane shape in the vicinity of this point must be sufficiently smooth and twice differentiable with respect to the surface coordinates s^i.

5.B TOPOLOGY OF VESICLES

Giant vesicles that do not experience external forces or constraints form closed membrane surfaces without pores or edges. In general, the topology of such a surface can be characterized by two related integers: (i) the number of handles, also known as the genus $\mathfrak{g}$ of the surface, and (ii) the Euler characteristic $\chi = 2 - 2\mathfrak{g}$. For any segmentation or partitioning of the membrane surface in terms of (curved) polygons, the Euler characteristic χ is equal to the number of polygons minus the number of edges plus the number of corners.

Three surfaces with genus $\mathfrak{g} = 0, 1$, and 2 are displayed in Figure 5.46: A surface with $\mathfrak{g} = 0$ and $\chi = 2$ is topologically equivalent to a sphere, a doughnut or torus is characterized by $\mathfrak{g} = 1$ and $\chi = 0$, and the Lawson surface with two handles has genus $\mathfrak{g} = 2$ and Euler characteristic $\chi = -2$. Furthermore, a set of several such surfaces has an Euler characteristic that is equal to the sum of the individual Euler characteristics.

Thus, a set of n spheres has the Euler characteristics $\chi = 2n$.

For a closed membrane surface without bilayer edges, the Gauss-Bonnet theorem implies that the integrated Gaussian curvature is given by $\int dAG = 2\pi\chi = 2\pi(2 - 2\mathfrak{g})$ as in Eq. 5.10. On the other hand, if the membrane surface has pores (or holes) that are bounded by bilayer edges, each edge makes a contribution to the integrated Gaussian curvature as given by

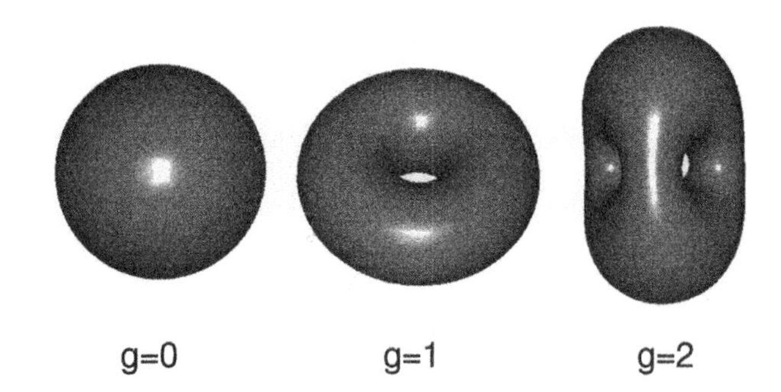

Figure 5.46 A sphere has no handle and genus $\mathfrak{g} = 0$; a torus has one handle and genus $\mathfrak{g} = 1$; the Lawson surface on the right has two handles and genus $\mathfrak{g} = 2$. The genus is a topological invariant and does not change for arbitrary shape deformations as long as we do not rupture or porate the surface.

[19] Most text books on differential geometry take the principal curvatures to be the eigenvalues of $h_i^{\,j}$ instead of $-h_i^{\,j}$. This conventional choice leads to $C_1 = C_2 < 0$ for a sphere.

$$\int dAG = -\oint dl C_g + 2\pi \quad (5.B1)$$

which depends on the line integral of the geodesic curvature C_g along the bilayer edge. In general, each bilayer edge will also contribute an edge energy which is proportional to the length of the edge.

Vesicles of genus $\mathfrak{g} = 2$ as illustrated by the rightmost shape in Figure 5.46 can undergo thermally excited shape transformations that correspond to conformal transformations of the vesicle shape, for which the vesicle volume, the membrane area, and the integrated mean curvature of the vesicle remain constant. This conformal diffusion in shape space was first predicted theoretically (Jülicher et al., 1993) and subsequently confirmed experimentally (Michalet and Bensimon, 1995).

5.C HIGHLY CURVED MEMBRANE SEGMENTS

5.C.1 HIGHER ORDER CURVATURE TERMS

As explained in Section 5.4.1, the spontaneous curvature model as defined by the curvature energy functional in Eq. 5.11 is obtained from a small curvature expansion up to second order in the principal curvatures C_1 and C_2 and ignores terms of higher order in these curvatures. These higher-order terms have the general form

$$\mathcal{E}_{\rm cu}^{p,q} \equiv \int dA \kappa_{p,q}[C_1^p C_2^q + C_2^p C_1^q] \quad \text{with} \quad p+q \geq 3. \quad (5.C1)$$

A somewhat different classification using symmetry arguments has been given by (Mitov, 1978).

A rough estimate for the magnitude of these terms can be obtained by dimensional analysis. The elastic parameter $\kappa_{p,q}$ has the dimension of energy multiplied by length to the power $p+q-2$. If we take the bending rigidity κ as the basic energy scale and the membrane thickness $\ell_{\rm me}$ as the basic molecular length, we obtain $\kappa_{p,q} \sim \kappa \ell_{\rm me}^{p+q-2}$. On the other hand, a vesicle with membrane area A has the overall size $R_{\rm ve} = \sqrt{A/(4\pi)}$. Therefore, dimensional analysis implies that the higher-order terms behave as

$$\mathcal{E}_{\rm cu}^{p,q} \sim \kappa (R_{\rm ve}/\ell_{\rm me})^{2-(p+q)} \quad (5.C2)$$

and decay to zero, in the limit of large $R_{\rm ve}/\ell_{\rm me}$, provided $p+q \geq 3$.

The estimate in Eq. 5.C2 indicates that, for $R_{\rm ve} \gtrsim 5\ell_{\rm me} \simeq 20$ nm, all higher-order terms with $p+q \geq 4$ should be negligible compared to the second-order terms of the spontaneous curvature model as given by Eq. 5.11. On the other hand, third-order terms with $p+q=3$ could make a significant contribution for $5\ell_{\rm me} \lesssim R_{\rm ve} \lesssim 20\ell_{\rm me} \simeq 80$ nm. The latter terms have the general form $C_1^3 + C_2^3$ and $C_1^2 C_2 + C_2^2 C_1$ and involve the additional elastic parameters $\kappa_{3,0}$ and $\kappa_{2,1}$.

The same conclusion applies to small spherical buds with radius $R_{\rm bud}$ and narrow cylindrical tubes with radius $R_{\rm tu}$. Thus, all higher-order terms should be negligible for $R_{\rm bud} \gtrsim 80$ nm and $R_{\rm tu} \gtrsim 80$ nm but third-order terms could make a significant contributions for smaller values of $R_{\rm bud}$ or $R_{\rm tu}$. In order to study the latter contributions in a systematic manner, molecular simulations should be rather useful.

5.C.2 MEMBRANE NECKS AS CURVATURE DEFECTS

As described in Section 5.4.6, closed membrane necks arise as limit shapes from the smooth solutions of the Euler-Lagrange or local shape equation. As the neck closes and the neck radius $R_{\rm ne}$ goes to zero, the adjacent membrane segment becomes highly curved because the curvature $1/R_{\rm ne}$ diverges. This divergence is truncated because the membrane curvature cannot exceed the inverse membrane thickness $1/\ell_{\rm me}$.

Taking the molecular structure of the bilayer membrane into account, this structure should be strongly perturbed in the vicinity of a closed neck and this perturbed molecular structure might lead to a finite "defect energy" $\delta E_{\rm ne}$ of the neck. A simple estimate of this latter energy can be obtained as follows. As explained in Section 5.3.1, curvature as a continuum concept emerges for membrane patches with a lateral dimension, say ℓ, that is about twice the membrane thickness. If we assume that the neck strongly perturbs the bilayer structure of a membrane patch of area ℓ^2, we obtain the estimate

$$\delta \bar{E}_{\rm ne} \equiv \frac{\delta E_{\rm ne}}{8\pi\kappa} = \frac{\ell^2}{4\pi}[M_{\rm ne} - m]^2 \quad (5.C3)$$

which behaves as

$$\delta \bar{E}_{\rm ne} \approx \frac{m^2 \ell^2}{4\pi} \quad \text{for large} \, |m| \gg |M_{\rm ne}|. \quad (5.C4)$$

This neck energy should be compared with the bending energy $\bar{E}_{\rm be}(R_2) = (1 - mR_2)^2$ of a spherical bud with radius R_2. Bud and neck then have the combined energy

$$\bar{E}_{\rm be} + \delta \bar{E}_{\rm ne} \approx m^2 R_2^2 \left(1 + \frac{\ell^2}{4\pi R_2^2}\right) \quad \text{for large} \, |m| \gg |M_{\rm ne}|. \quad (5.C5)$$

Thus, if we take $\ell \simeq 8$ nm, the correction term arising from the putative defect energy of the neck can be safely ignored for buds with radius $R_2 \gtrsim 40$ nm. In order to obtain a reliable estimate for smaller buds, molecular simulations should again be quite useful.

5.D MECHANICAL TENSION OF MEMBRANES

In this appendix, we consider vesicles with compressible membranes and determine their equilibrium shapes by minimizing the combined bending and stretching energy with respect to membrane area for fixed vesicle volume.

5.D.1 MECHANICAL TENSION AND STRETCHING ENERGY

In the absence of external forces or constraints, a bilayer membrane attains a certain optimal area $A_{\rm opt}$, which corresponds to the optimal packing of its molecules. The membrane experiences a tension, $\Sigma_{\rm st}$, when its area A is stretched and deviates from the optimal value $A_{\rm opt}$. This stretch tension can be expressed as

$$\Sigma_{\rm st}(A) = K_A \frac{A - A_{\rm opt}}{A_{\rm opt}} \quad (5.D1)$$

up to first order in $A - A_{opt}$, which defines the area compressibility modulus K_A. The stretch tension Σ_{st} must be smaller than the tension of rupture, Σ_{rup}. For lipid bilayers, the rupture tension Σ_{rup} is about two orders of magnitude smaller than the area compressibility modulus K_A and of the order of a few mN/m.

The work of stretching or compression, starting from the initial area $A = A_{opt}$, defines the stretching energy

$$E_{st}(A) = \int_{A_{opt}}^{A} dx \Sigma_{st}(x) = \frac{1}{2} K_A \frac{(A - A_{opt})^2}{A_{opt}}. \quad (5.D2)$$

For an arbitrary vesicle shape S, we define the stretching energy functional via

$$\mathcal{E}_{st}\{S\} \equiv E_{st}(\mathcal{A}\{S\}). \quad (5.D3)$$

5.D.2 COMBINED BENDING AND STRETCHING ENERGY

The total elastic energy of a compressible membrane, which consists of its combined bending and stretching energy, is now equal to

$$\mathcal{E}_{el}\{S\} \equiv \mathcal{E}_{be}\{S\} + \mathcal{E}_{st}\{S\}. \quad (5.D4)$$

The corresponding shape functional has the form

$$\mathcal{F}_{el}\{S\} = -\Delta P \mathcal{V}\{S\} + \mathcal{E}_{be}\{S\} + \mathcal{E}_{st}\{S\} \quad (5.D5)$$

where the pressure difference ΔP is used, as before, as a Lagrange multiplier to ensure that $\mathcal{V}\{S\} = V$.

5.D.3 TWO-STEP MINIMIZATION PROCEDURE

The minimization of the shape functional Eq. 5.D5 can be performed in two steps:

(i) First, we minimize the shape functional Eq. 5.22 for the spontaneous curvature model using the Lagrange multiplier tension Σ to enforce the membrane area $\mathcal{A}\{S\} = A$. As a result, we obtain the bending energy function

$$E_{be}(V, A; \kappa, m; j) = \mathcal{E}_{be}\{S^j\} \quad (5.D6)$$

as in Eq. 5.28, which represents the membrane's bending energy as a function of volume V and membrane area A along a branch of (meta)stable equilibrium shapes S^j. In general, we expect to find several branches of such shapes as illustrated in Figure 5.12 for vanishing spontaneous curvature, $m = 0$.

(ii) Second, we minimize the combined elastic energy functional $\mathcal{E}_{el} = \mathcal{E}_{be} + \mathcal{E}_{st}$ with respect to membrane area A for fixed volume V. Because the stretching energy is an explicit function of the membrane area, we can replace the minimization of the elastic energy functional $\mathcal{E}_{el}$ by the minimization of the elastic energy function

$$E_{el}(V, A) \equiv E_{be}(V, A) + E_{st}(A)$$
$$= E_{be}(V, A) + \frac{1}{2} K_A \frac{(A - A_{opt})^2}{A_{opt}}. \quad (5.D7)$$

The relation $(dE_{el}(V, A)/dA)_V = 0$ then determines the equilibrium value $A = A^{eq}$ of the membrane area via

$$K_A \frac{A^{eq} - A_{opt}}{A_{opt}} = -\left(\frac{dE_{be}(V, A^{eq})}{dA^{eq}}\right)_V. \quad (5.D8)$$

In this way, the minimization of the combination of bending and stretching energy has been reduced to the minimization of the bending energy functional alone, which determines the bending energy E_{be} as a function of V and A.

5.D.4 MECHANICAL TENSION

The relation as given by Eq. 5.D8 has a very simple physical interpretation. By definition, the left hand side of Eq. 5.D8 is equal to the stretch tension Σ_{st}, see Eq. 5.D1, whereas the right hand side of this equation corresponds to the relationship Eq. 5.32 which expresses the Lagrange multiplier tension Σ as the derivative of the bending energy with respect to membrane area A. Therefore, the relation Eq. 5.D8 is equivalent to

$$\Sigma_{st} = K_A \frac{A^{eq} - A_{opt}}{A_{opt}} = -\left(\frac{dE_{be}(V, A^{eq})}{dA^{eq}}\right)_V = \Sigma \quad (5.D9)$$

which reveals that the Lagrange multiplier tension Σ is, in fact, identical with the stretch tension Σ_{st}. The identity Eq. 5.D9 is not restricted to a specific form of the bending energy but holds for any such energy, when minimized for fixed vesicle volume and fixed membrane area. An analogous equation also holds for the bilayer coupling model (Svetina and Zeks, 1989), in which the bending energy function E_{be} depends on the volume V, membrane area A, as well as total mean curvature $I_M = \int dA M$, and the partial derivative on the right hand side of Eq. 5.D9 has to be taken at constant volume V and constant total mean curvature I_M.

5.E DIFFERENT VARIANTS OF CURVATURE MODELS

In this appendix, we will consider three variants of the curvature model: the spontaneous curvature (SC) model as studied in the main text, the bilayer coupling model, and the area-difference-elasticity model. Two general results will be shown explicitly: (i) all three models lead to the same stationary shapes of vesicles; and (ii) all stationary shapes of the area-difference-elasticity model are also stationary shapes of the spontaneous curvature model with an effective spontaneous curvature m_{eff}.

As in the main text, all functionals will be denoted by calligraphic letters. Thus, we again consider the geometric functionals $\mathcal{V}\{S\}, \mathcal{A}\{S\}, \Delta\mathcal{A}\{S\}$, and $\mathcal{I}_M\{S\}$ and denote their values for the stationary shapes $S = S^{st}$ by V, A, ΔA, and I_M.

5.E.1 BILAYER COUPLING (BC) MODEL

For the sake of clarity, it is convenient to start with the bilayer coupling (BC) model which is defined by the bending energy functional

$$\mathcal{E}_{BC}\{S\} \equiv 2\kappa \int dA\, M^2. \quad (5.E1)$$

In this model, one considers vesicle shapes with fixed volume V, fixed area A, and integrated mean curvature I_M where the latter quantity is proportional to area difference ΔA between the two leaflets of the bilayer membrane. The stationary shapes S_{BC}^{st} of this model follow from the first variation of the shape functional

$$\mathcal{F}_{BC}\{S\} = -P_{BC}\mathcal{V}\{S\} + \Sigma_{BC}\mathcal{A}\{S\} + Q_{BC}\mathcal{I}_M\{S\} + \mathcal{E}_{BC}\{S\}. \quad (5.E2)$$

The stationary shapes S_{BC}^{st} again form several branches labeled by j. The stationary shapes on branch j will be denoted by S_{BC}^{j}. The energies of these stationary shapes defines the energy functions as given by

$$E_{BC}(V, A, I_M; j) = \mathcal{E}_{BC}\{S_{BC}^{j}\} \quad (5.E3)$$

along the branch j. Interpreting the relation between the energy functional $\mathcal{E}_{BC}$ and the shape functional $\mathcal{F}_{BC}$ as a Legendre transformation, we obtain the relations

$$P_{BC} = \left(\frac{dE_{BC}}{dV}\right)_{A,I_M}, \quad \Sigma_{BC} = -\left(\frac{dE_{BC}}{dA}\right)_{V,I_M}, \quad (5.E4)$$

and

$$Q_{BC} = -\left(\frac{dE_{BC}}{dI_M}\right)_{V,A} \quad (5.E5)$$

for the three Lagrange multipliers P_{BC}, Σ_{BC}, and Q_{BC}.

The spontaneous curvature (SC) model studied in the main text is defined by the energy functional

$$\mathcal{E}_{SC}\{S\} \equiv \mathcal{E}_{be}\{S\} = \mathcal{E}_{BC}\{S\} - 4\kappa m\mathcal{I}_M\{S\} + 2\kappa m^2\mathcal{A}\{S\}. \quad (5.E6)$$

In this model, one considers vesicle shapes with fixed volume V and fixed area A. The stationary shapes S_{SC}^{st} of this model follow from the first variation of the shape functional

$$\mathcal{F}_{SC}\{S\} = -P_{SC}\mathcal{V}\{S\} + \Sigma_{SC}\mathcal{A}\{S\} + \mathcal{E}_{SC}\{S\} \quad (5.E7)$$

with $P_{SC} \equiv \Delta P$.

5.E.1.1 Identical stationary shapes in BC and SC models

A direct comparison of the two shape functionals $\mathcal{F}_{BC}$ and $\mathcal{F}_{SC}$ in Eqs 5.E2 and 5.E7 shows that these shape functionals are identical provided one chooses

$$P_{BC} = P_{SC}, \quad \Sigma_{BC} = \Sigma_{SC} + 2\kappa m^2, \quad \text{and} \quad Q_{BC} = -4\kappa m. \quad (5.E8)$$

or

$$P_{SC} = P_{BC}, \quad \Sigma_{SC} = \Sigma_{BC} - \frac{Q_{BC}^2}{8\kappa} \quad \text{and} \quad m = -\frac{Q_{BC}}{4\kappa}. \quad (5.E9)$$

As a consequence, the Euler-Lagrange equations of the two models are also identical. The Euler-Lagrange equation of the SC model has the form

$$P_{SC} = 2\Sigma_{SC}M - 2\kappa\nabla_{LB}^2 M - 4\kappa(M - m)[M(M + m) - G] \quad (5.E10)$$

as given by Eq. 5.23 with $P_{SC} \equiv \Delta P$ and $\Sigma_{SC} \equiv \Sigma$. The Euler-Lagrange equation of the BC model is obtained from Eq. 5.E10 by the parameter mapping Eq. 5.E9. Therefore, the stationary shapes of the SC model are also stationary shapes of the BC model and *vice versa* when we map the parameters of the two models according to Eq. 5.E8 or Eq. 5.E9, and we can identify the stationary shapes for each branch j, i.e., $S_{SC}^{j} = S_{BC}^{j}$, as well as the associated limit shapes.

5.E.2 AREA-DIFFERENCE-ELASTICITY MODEL

The bending energy functional of the area-difference-elasticity model as given by Eq. 5.63 can be rewritten in the form

$$\mathcal{E}_{ADE}\{S\} = \mathcal{E}_{BC}\{S\} - 4\kappa m\mathcal{I}\{S\} + 2\kappa m^2\mathcal{A}\{S\} + \mathcal{D}_{ADE}\{S\} \quad (5.E11)$$

with the nonlocal bending energy term $\mathcal{D}_{ADE}\{S\}$ as in Eq. 5.64.

In the area-difference-elasticity model, one again considers vesicle shapes with fixed volume V and fixed area A. In order to deal with the nonlocal character of $\mathcal{D}_{ADE}\{S\}$, it is useful to use a two-step variational procedure (Miao et al., 1994). In the first step, we determine the stationary shapes of Eq. 5.E11 for fixed volume V, fixed area A, *and fixed integrated mean curvature* I_M. These shapes are obtained from the first variation of the shape functional

$$\tilde{\mathcal{F}}_{ADE}\{S\} = -P\mathcal{V}\{S\} + \tilde{\Sigma}_{ADE}\mathcal{A}\{S\} + \tilde{Q}_{ADE}\mathcal{I}_M\{S\} + \mathcal{E}_{ADE}\{S\}. \quad (5.E12)$$

For fixed area A and fixed integrated mean curvature I_M, the energy functional in Eq. 5.E11 reduces to $\mathcal{E}_{ADE}\{S\} = \mathcal{E}_{BC}\{S\} + \text{const}$. Therefore, the stationary shapes $\tilde{S}_{ADE}^{j}$ of the ADE model for the given values of V, A, and I_M are identical with the stationary shapes S_{BC}^{j} of the BC model for the same values of V, A, and I_M and, thus, fulfill the same Euler-Lagrange equation as given by Eq. 5.E10 with the parameter mapping as in Eq. 5.E9. The energy function

$$\tilde{E}_{ADE}(V, A, I_M; j) = \mathcal{E}_{ADE}\{\tilde{S}_{ADE}^{j}\} \quad (5.E13)$$

is then equal to

$$\tilde{E}_{ADE} = E_{BC}(V, A, I_M; j) - 4\kappa m I_M + 2\kappa m^2 A + 2\pi\kappa_\Delta \frac{(I_M - I_{M,0})^2}{A}. \quad (5.E14)$$

Furthermore, the Lagrange multiplier $\tilde{Q}_{ADE}$ in Eq. 5.E12 fulfills the relation

$$\tilde{Q}_{ADE} = -\left(\frac{d\tilde{E}_{ADE}}{dI_M}\right)_{V,A} = Q_{BC} + 4\kappa m - 4\pi\kappa_\Delta \frac{I_M - I_{M,0}}{A} \quad (5.E15)$$

with Q_{BC} as in Eq. 5.E5).

In the second step of the variational procedure, we determine the values of the integrated mean curvature I_M that lead to extrema of the energy function $\tilde{E}_{ADE}$ for fixed volume V and fixed area A. These extrema follow from the condition

$$\left(\frac{d\tilde{E}_{\rm ADE}}{dI_M}\right)_{V,A} = 0. \qquad (5.E16)$$

Inserting this condition into Eq. 5.E15, we obtain the identity

$$Q_{\rm BC} = -4\kappa m + 4\pi\kappa_\Delta \frac{I_M - I_{M,0}}{A}. \qquad (5.E17)$$

Finally, we use the relation between $Q_{\rm BC}$ and the spontaneous curvature as given by Eq. 5.E8 with $m \equiv m_{\rm eff}$. As a result, we obtain the expression Eq. 5.66 for the effective spontaneous curvature $m_{\rm eff}$ of the equivalent spontaneous curvature model.

5.F DISCONTINUITIES ALONG DOMAIN BOUNDARIES

This appendix, which supplements Section 5.8 on multi-domain membranes and vesicles, describes the matching conditions for the domain shapes along a domain boundary in some detail. Even for axisymmetric vesicle shapes with smooth contours, these matching conditions turn out to be somewhat complex. Indeed, these matching conditions imply discontinuities along the domain boundary, both for the curvature and for the mechanical tension. In order to describe these discontinuities, we parametrize the contour of the axisymmetric shape by its arc length s starting from the north pole of the shape. We could then use cylindrical coordinates, r and z, to describe the vesicle shape but it is more convenient to use the coordinate r and the tilt angle ψ,[20] see Figure 5.33.

For an axisymmetric shape as in Figure 5.33, the two principal curvatures are given by

$$C_1 = \frac{d\psi}{ds} \equiv \dot{\psi} \quad \text{and} \quad C_2 = \frac{\sin\psi}{r} \qquad (5.F1)$$

where C_1 represents the curvature of the shape contour. The second principal curvature C_2 is continuous at the domain boundary with $s = s_1$ because both the tilt angle $\psi(s)$ and the coordinate $r(s)$ are continuous at this s-value. In contrast, the contour curvature C_1 can change discontinuously at the domain boundary.

5.F.1 CURVATURE DISCONTINUITIES

This discontinuity follows from the matching condition (Jülicher and Lipowsky, 1996)

$$\kappa_a\dot{\psi}(s_1+\varepsilon) - \kappa_b\dot{\psi}(s_1-\varepsilon) = \delta\kappa C_2(s_1) + 2\kappa_a m_a - 2\kappa_b m_b \qquad (5.F2)$$

with

$$\delta\kappa \equiv \kappa_b - \kappa_a + \kappa_{Gb} - \kappa_{Ga}. \qquad (5.F3)$$

We now introduce the notation

$$C_{1a}(s_1) \equiv \dot{\psi}(s_1+\varepsilon) \quad \text{and} \quad C_{1b}(s_1) \equiv \dot{\psi}(s_1-\varepsilon) \qquad (5.F4)$$

[20] The two variables ψ and r satisfy the relation $\frac{dr}{ds} = \cos\psi$, a condition that is incorporated into the variational calculation by a Lagrange parameter function (Seifert et al., 1991; Jülicher and Lipowsky, 1996).

for the contour curvatures and

$$M_a(s_1) \equiv \frac{1}{2}[C_{1a}(s_1) + C_2(s_1)] \quad \text{and}$$
$$M_b(s_1) \equiv \frac{1}{2}[C_{1b}(s_1) + C_2(s_1)] \qquad (5.F5)$$

for the mean curvatures at the two sides of the domain boundary. Using this notation, the matching condition Eq. 5.F2 can be rewritten as

$$\kappa_a[M_a(s_1) - m_a] - \kappa_b[M_b(s_1) - m_b] = \frac{1}{2}(\kappa_{Gb} - \kappa_{Ga})C_2(s_1). \qquad (5.F6)$$

The above matching conditions imply the discontinuity

$$C_{1a}(s_1) - C_{1b}(s_1) = \dot{\psi}(s_1+\varepsilon) - \dot{\psi}(s_1-\varepsilon) = \Delta_1 \qquad (5.F7)$$

of the contour curvature C_1 with

$$\Delta_1 \equiv \frac{\kappa_b - \kappa_a}{\kappa_a}C_{1b}(s_1) + \frac{\delta\kappa}{\kappa_a}C_2(s_1) + 2\frac{\kappa_a m_a - \kappa_b m_b}{\kappa_a} \qquad (5.F8)$$

as follows from Eq. 5.F2. Note that the discontinuity Δ_1 depends (i) on the contour curvature $C_{1b}(s_1)$ along the b-side of the domain boundary and (ii) on the second principal curvature $C_2(s_1)$ at this boundary. Rearranging the terms in Eq. 5.F7, we obtain the discontinuity

$$M_a(s_1) - M_b(s_1) = \frac{1}{2}\Delta_1 \qquad (5.F9)$$

of the mean curvature M. Note also that the curvature discontinuities as described by Eqs 5.F7 and 5.F9 depend only on *local* properties of the vesicle shape close to the domain boundary.

The matching conditions for the curvatures simplify when we consider two membrane domains for which some of the curvature-elastic parameters are identical. If both membrane domains have the same Gaussian curvature moduli, the expression Eq. 5.F8 becomes

$$\Delta_1 = 2\frac{\kappa_b - \kappa_a}{\kappa_a}M_{1b} + 2\frac{\kappa_a m_a - \kappa_b m_b}{\kappa_a} \quad (\kappa_{Gb} = \kappa_{Ga}) \qquad (5.F10)$$

and the matching condition Eq. 5.F6 attains the simple and concise form

$$\kappa_a[M_a(s_1) - m_a] = \kappa_b[M_b(s_1) - m_b] \quad (\kappa_{Gb} = \kappa_{Ga}). \qquad (5.F11)$$

If both domains have the same Gaussian curvature moduli and the same bending rigidity, the discontinuity Δ_1 becomes

$$\Delta_1 = 2(m_a - m_b) \quad (\kappa_{Gb} = \kappa_{Ga} \text{ and } \kappa_b = \kappa_a). \qquad (5.F12)$$

In this case, the curvature discontinuity is independent of the principal curvatures at the domain boundary and proportional to the difference $m_b - m_a$ of the spontaneous curvatures. Using the matching condition in the form Eq. 5.F11, we also obtain

$$M_b(s_1)-m_b = M_a(s_1)-m_a \quad \text{for } \kappa_{Gb}=\kappa_{Ga} \text{ and } \kappa_b=\kappa_a. \quad (5.F13)$$

Therefore, the deviation of the mean curvature from the spontaneous curvature is continuous across the domain boundary if the two membrane domains have the same Gaussian curvature modulus and the same bending rigidity. Likewise, the discontinuity simplifies to

$$\Delta_1 = 2\frac{\kappa_b-\kappa_a}{\kappa_a}\left[M_b(s_1)-m\right] \quad \text{for } m_b=m_a=m \text{ and } \kappa_{Gb}=\kappa_{Ga}. \quad (5.F14)$$

In the latter case, the curvature discontinuity is proportional to the difference $\kappa_b - \kappa_a$ of the bending rigidities and to the deviation $M_b(s_1) - m$ of the mean curvature $M_b(s_1)$ along the b-side of the domain boundary from the spontaneous curvature m. Finally, the curvature discontinuity Δ_1 vanishes if both membrane domains have the same curvature-elastic properties, i.e.,

$$\Delta_1 = 0 \quad \text{for } m_b=m_a, \kappa_b=\kappa_a, \text{ and } \kappa_{Gb}=\kappa_{Ga}. \quad (5.F15)$$

5.F.2 DIFFERENCE BETWEEN MECHANICAL TENSIONS

The discontinuity Δ_1 of the contour curvature C_1 at the domain boundary also affects the difference $\Sigma_b - \Sigma_a$ of the mechanical tensions within the two membrane domains. Using the results of (Jülicher and Lipowsky, 1996), one finds the tension difference

$$\Sigma_a - \Sigma_b = \lambda\frac{\cos\psi(s_1)}{r(s_1)} + \Delta_\Sigma \quad (5.F16)$$

with

$$\Delta_\Sigma \equiv \frac{1}{2}\kappa_a Q_a(s_1) - \frac{1}{2}\kappa_b Q_b(s_1) \quad (5.F17)$$

and

$$Q_j(s_1) \equiv C_{1j}^2(s_1) - [C_2(s_1)-2m_j]^2 \quad \text{for } j=a,b. \quad (5.F18)$$

It follows from the relations in Eqs 5.F2, 5.F3, and 5.F6 that the curvature discontinuities along the domain boundary depend on the difference $\kappa_{Gb} - \kappa_{Ga}$ of the Gaussian curvature moduli. Therefore, the expression Eq. 5.F17 for Δ_Σ implicitly depends on $\kappa_{Gb} - \kappa_{Ga}$ as well.

Inspection of the expression Eq. 5.F17 shows that Δ_Σ contains only two shape-independent terms as given by the spontaneous tensions $\sigma_j = 2\kappa_j m_j^2$ with $j = a, b$. Thus, we can decompose the expression Eq. 5.F17 according to

$$\Delta_\Sigma = -2\kappa_a m_a^2 + 2\kappa_b m_b^2 + \Delta_S = -\sigma_a + \sigma_b + \Delta_S \quad (5.F19)$$

with

$$\Delta_S \equiv \frac{1}{2}\kappa_a \tilde{Q}_a(s_1) - \frac{1}{2}\kappa_b \tilde{Q}_b(s_1) \quad (5.F20)$$

and

$$\tilde{Q}_j(s_1) = C_{1j}^2(s_1) - C_2^2(s_1) + 4C_2(s_1)m_j. \quad (5.F21)$$

The tension difference in Eq. 5.F16 can then be rewritten as

$$\hat{\Sigma}_a - \hat{\Sigma}_b = \lambda\frac{\cos\psi(s_1)}{r(s_1)} + \Delta_S. \quad (5.F22)$$

If both membrane domains have the same bending rigidity κ and the same Gaussian curvature modulus, the quantities Δ_Σ and Δ_S become

$$\Delta_\Sigma = 2\kappa(m_a-m_b)[M_a(s_1)-m_a+M_b(s_1)-m_b] \quad (\kappa_b=\kappa_a, \text{ and } \kappa_{Gb}=\kappa_{Ga}) \quad (5.F23)$$

and

$$\Delta_S = 2\kappa(m_a-m_b)[M_a(s_1)+M_b(s_1)] \quad (\kappa_b=\kappa_a, \text{ and } \kappa_{Gb}=\kappa_{Ga}). \quad (5.F24)$$

Note that $M_b(s_1) - m_b = M_a(s_1) - m_a$ according to Eq. 5.F13 for two domains with the same bending rigidity and the same Gaussian curvature modulus. Finally, if all curvature-elastic parameters of the two membrane domains are identical, the contour curvature is continuous across the domain boundary, see Eq. 5.F15, which implies $C_{1b}(s_1) = C_{1a}(s_1)$, $Q_b(s_1) = Q_a(s_1)$, and

$$\Delta_\Sigma = \Delta_S = 0 \quad \text{for } m_b=m_a, \kappa_b=\kappa_a, \text{ and } \kappa_{Gb}=\kappa_{Ga}. \quad (5.F25)$$

Therefore, in this case, the balance between the mechanical membrane tensions Σ_a and Σ_b within the two domains and the line tension γ of the domain boundary is described by

$$\Sigma_a - \Sigma_b = \lambda\frac{\cos\psi(s_1)}{r(s_1)} \quad \text{for } m_b=m_a, \kappa_b=\kappa_a, \text{ and } \kappa_{Gb}=\kappa_{Ga}. \quad (5.F26)$$

The minimization of the energy functional Eq. 5.272 also implies a third matching condition that describes a jump in $\ddot{\psi}$, i.e., in the first derivative of the contour curvature $C_1 = \dot{\psi}$ with respect to the arc length s.[21]

5.G SEGMENTATION AND PHASE SEPARATION OF TWO-COMPONENT MEMBRANES

The interplay of ambience-induced segmentation and phase separation of membranes has been theoretically studied in some detail for membranes with two lipid components, say l_a and l_b (Rouhiparkouhi et al., 2013; Lipowsky et al., 2013). If the membranes contains more than two components, we can single out one special component, denote this component by l_a, and combine all

[21] In order to derive this third matching condition, it is useful to start from the shape equation for $\ddot{\psi}$ within the two domains, see Eq. (A.13) in (Jülicher and Lipowsky, 1996), from which one can determine the quantity $\kappa_a\ddot{\psi}(s_1+\varepsilon) - \kappa_b\ddot{\psi}(s_1-\varepsilon)$. The latter quantity depends only on *local* properties of the vesicle shape close to the domain boundary.

other components into an effective second component l_b, thereby mapping a multi-component membrane onto a two-component one. To simplify the following discussion, we will ignore differences in the molecular areas of the two lipid components and take both molecular areas to be equal to A_l. If the membrane contains N_{la} lipids l_a and N_{lb} lipids l_b, the total membrane area A is then given by

$$A = \frac{1}{2}(N_{la} + N_{lb})A_l \tag{5.G1}$$

where the factor 1/2 takes into account that the bilayer membrane consists of two leaflets.

The membrane is exposed to K local environments that differ in their molecular compositions and thus partition the membrane into several segments distinguished by the superscript $[k]$ with $k = 1, 2, \ldots K$ as in Figure 5.31. The total membrane area A is then partitioned into the segmental areas $A^{[k]}$ with

$$A = A^{[1]} + A^{[2]} + \ldots + A^{[K]}. \tag{5.G2}$$

Furthermore, the total number of l_a and l_b molecules contained in segment $[k]$ is fixed and equal to $A^{[k]}/A_l$. Therefore, when one molecule diffuses from segment $[k]$ to a neighboring segment $[k']$, another molecule must diffuse from segment $[k']$ to segment $[k]$.

When a lipid molecule l_a or l_b is located in segment $[k]$, the molecular interactions with the adjacent environment $[k]$ lead to the interaction energies $U_{la}^{[k]}$ and $U_{lb}^{[k]}$, respectively, where effectively attractive interactions are described by negative values $U_{la}^{[k]} < 0$ and $U_{lb}^{[k]} < 0$. The enrichment or depletion of the two lipid species adjacent to environment $[k]$ is then determined by the relative affinity

$$\Delta U^{[k]} = U_{la}^{[k]} - U_{lb}^{[k]} \quad \text{within segment}[k], \tag{5.G3}$$

which is negative if environment $[k]$ prefers the l_a lipids and positive if this environment prefers the l_b lipids.

For a homogeneous environment with interaction energies $U_{la}^{[k]} = 0$ and $U_{lb}^{[k]} = 0$, the two lipid species have the chemical potentials μ_{la} and μ_{lb}. These chemical potentials are not independent because the lipid numbers N_{la} and N_{lb} are related via Eq. 5.G1. The membrane system is then described by the semi-grand canonical ensemble with the relative chemical potential (Lipowsky et al., 2013)

$$\Delta\mu \equiv \mu_{la} - \mu_{lb}. \tag{5.G4}$$

Within this statistical ensemble, the phase transition occurs along the line

$$\Delta\mu = \Delta\mu_*(T) \quad \text{for } T_t < T < T_c \tag{5.G5}$$

in the $(\Delta\mu, T)$ plane where T_t and T_c are the temperatures of the triple point and the critical demixing point, respectively. The function $\Delta\mu_*(T)$ is obtained from the free energy in the semigrand canonical ensemble and depends on all parameters that describe the interactions between the lipid components (Lipowsky et al., 2013).

When the membrane is now partitioned into several segments by the different local environments, the chemical potentials are shifted by the interaction energies $U_{la}^{[k]}$ and $U_{lb}^{[k]}$. Each segment $[k]$ is now characterized by the relative chemical potential

$$\Delta\mu^{[k]} \equiv \mu_{la} + U_{la}^{[k]} - (\mu_{lb} + U_{lb}^{[k]}) = \Delta\mu + \Delta U^{[k]}, \tag{5.G6}$$

which is equal to the relative chemical potential of the homogeneous system shifted by the relative affinity $\Delta U^{[k]}$. As a consequence, each segment $[k]$ undergoes a phase transition along the line

$$\Delta\mu^{[k]} = \Delta\mu + \Delta U^{[k]} = \Delta\mu_*(T) + \Delta U^{[k]} \quad \text{for } T_t < T < T_c, \tag{5.G7}$$

and the membrane consisting of K segments exhibits K phase transitions as shown in Figure 5.47. The transition lines for segment $[k + 1]$ and segment $[k]$ are separated by

$$\Delta\mu^{[k+1]} - \Delta\mu^{[k]} = U^{[k+1]} - \Delta U^{[k]} \equiv \Delta U_k \tag{5.G8}$$

with the affinity contrast ΔU_k between segment $[k + 1]$ and segment $[k]$.

In the canonical ensemble, the relative chemical potential $\Delta\mu$ is replaced by the mole fraction X_{la} of the l_a lipids with $0 \leq X_{la} \leq 1$. Each transition line within the $(\Delta\mu, T)$ phase diagram as displayed in Figure 5.47 is then mapped onto a coexistence region within the (X_{la}, T) phase diagram. Because the resulting K coexistence regions have to be accommodated, at each temperature T, within the interval $0 \leq X_{la} \leq 1$, the average width of a single coexistence region is necessarily smaller than $1/K$ and therefore decreases monotonically with increasing number K of distinct local environments.

5.H WETTING OF TWO MEMBRANELESS DROPLETS

Wetting of a vesicle membrane, arising from the aqueous phase separation within the vesicle, leads to two aqueous droplets enclosed by this membrane as depicted in the insets of

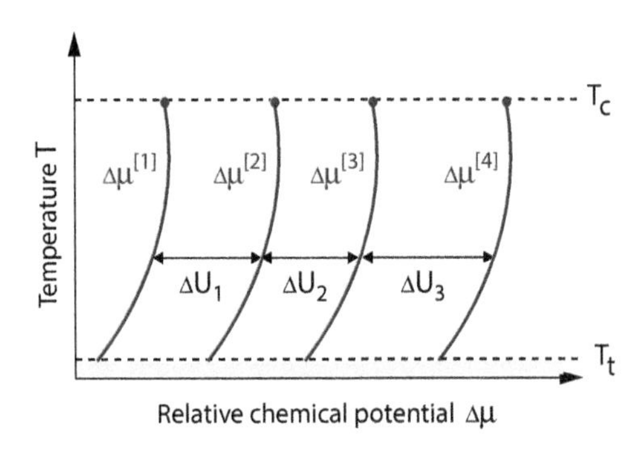

Figure 5.47 Phase diagram for a two-component membrane exposed to $K = 4$ different environments as a function of relative chemical potential $\Delta\mu$ and temperature T with $T_t < T \leq T_c$. Segment $[k]$ undergoes a phase transition along the demixing line $\Delta\mu = \Delta\mu^{[k]}$ as given by Eq. 5.G7. The demixing lines $\Delta\mu^{[k+1]}$ and $\Delta\mu^{[k]}$ are separated by the affinity contrast ΔU_k between segment $[k + 1]$ and segment $[k]$ as in Eq. 5.G8. Each demixing line has a critical point at $T = T_c$.

Figure 5.39. This appendix describes the analogous but somewhat simpler situation corresponding to the wetting of two droplets in the *absence* of the membrane. The two droplets consist of the two liquid phases α and β and are completely immersed into the bulk liquid phase γ, see Figure 5.48. The latter wetting system will now be discussed in some detail to reveal the similarities and differences compared to the wetting of membranes, see Section 9.3.

5.H.1 SPHERICAL GEOMETRY AND CONTACT ANGLES

The two droplets in Figure 5.48 consist of two aqueous phases, α and β, which are immersed into a third liquid phase γ. The geometry of such a droplet pair involves three interfaces: the $\alpha\gamma$ interface between the α droplet and the exterior γ phase; the $\beta\gamma$ interface between the β droplet and the γ phase; and the $\alpha\beta$ interface between the α and the β droplets. All three interfaces form spherical segments that meet at the three-phase contact line as shown in Figure 5.48a. The curvature radii of the three spherical radii are denoted by $R_{\alpha\gamma}$, $R_{\beta\gamma}$, and $R_{\alpha\beta}$ which are all taken to be positive.[22] Along the contact line, the tangent planes of the three interfaces form the three contact angles θ_α, θ_β, and θ_γ with $\theta_\alpha + \theta_\beta + \theta_\gamma = 2\pi$, see Figure 5.48b. It is not difficult to show that the shape consisting of three spherical segments implies the geometric relation

$$\pm\frac{\sin\theta_\gamma}{R_{\alpha\beta}} = \frac{\sin\theta_\alpha}{R_{\beta\gamma}} - \frac{\sin\theta_\beta}{R_{\alpha\gamma}} \tag{5.H1}$$

or

$$\pm\frac{1}{R_{\alpha\beta}} = \frac{\sin\theta_\alpha/\sin\theta_\gamma}{R_{\beta\gamma}} - \frac{\sin\theta_\beta/\sin\theta_\gamma}{R_{\alpha\gamma}} \tag{5.H2}$$

where the plus and minus sign corresponds to an $\alpha\beta$ interface that bulges towards the α and the β phase, respectively. Thus, the plus sign applies, in particular, to the geometry displayed in Figure 48a,b.

5.H.2 MECHANICAL EQUILIBRIUM BETWEEN INTERFACES

To proceed, let us consider the balance between the Laplace pressures and the interfacial tensions $\Sigma_{\alpha\gamma}$, $\Sigma_{\beta\gamma}$, and $\Sigma_{\alpha\beta}$ of the three interfaces. The mean curvatures $M_{\alpha\gamma} = 1/R_{\alpha\gamma}$ and $M_{\beta\gamma} = 1/R_{\beta\gamma}$ of the $\alpha\gamma$ and $\beta\gamma$ interfaces satisfy the two Laplace equations

$$\Delta P_{i\gamma} \equiv P_i - P_\gamma = 2\Sigma_{i\gamma}\, M_{i\gamma} = 2\Sigma_{i\gamma}/R_{i\gamma} > 0 \text{ for } i = \alpha, \beta. \tag{5.H3}$$

These equations are also valid when the two droplets are not in contact with each other and form two separate spheres immersed into the γ phase. For the partial wetting geometry, on the other hand, the mean curvature $M_{\alpha\beta} = \pm 1/R_{\alpha\beta}$ of the $\alpha\beta$ interface satisfies another Laplace equation as given by

$$P_\beta - P_\alpha = 2\Sigma_{\alpha\beta}\, M_{\alpha\beta} = \pm 2\Sigma_{\alpha\beta}/R_{\alpha\beta}. \tag{5.H4}$$

As before, the plus and minus sign corresponds to an $\alpha\beta$ interface that bulges towards the α and β phase, respectively. The pressure differences can be eliminated by a combination of all three Laplace equations which leads to the relationship

$$\pm\frac{\Sigma_{\alpha\beta}}{R_{\alpha\beta}} = \frac{\Sigma_{\beta\gamma}}{R_{\beta\gamma}} - \frac{\Sigma_{\alpha\gamma}}{R_{\alpha\gamma}} \tag{5.H5}$$

or

$$\pm\frac{1}{R_{\alpha\beta}} = \frac{\Sigma_{\beta\gamma}/\Sigma_{\alpha\beta}}{R_{\beta\gamma}} - \frac{\Sigma_{\alpha\gamma}/\Sigma_{\alpha\beta}}{R_{\alpha\gamma}}. \tag{5.H6}$$

between the three interfacial tensions.

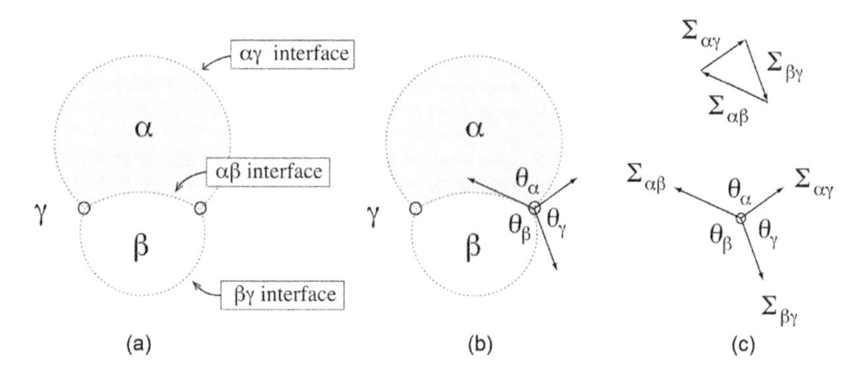

Figure 5.48 (a) Partial wetting of an α droplet (yellow) and a β droplet (blue) immersed in the liquid bulk phase γ (white). The two droplets are bounded by the $\alpha\gamma$, $\alpha\beta$, and $\beta\gamma$ interfaces. All three interfaces form spherical segments that meet at the three-phase contact line (small black circles); (b) Along the contact line, the tangent planes of the three interfaces form the three contact angles θ_α, θ_β, and θ_γ with $\theta_\alpha + \theta_\beta + \theta_\gamma = 2\pi$; and (c) The interfacial tensions $\Sigma_{\alpha\beta}$, $\Sigma_{\alpha\gamma}$, and $\Sigma_{\beta\gamma}$ pull at the contact line in the directions of the three tangent planes. In mechanical equilibrium, the three tensions must balance and add up to zero which implies that they form the sides of a triangle (upper panel) which is known as Neumann's triangle.

[22] Note that the $\alpha\beta$ interface may bulge towards the α phase as in Figure 5.48a or towards the β phase depending on the relative magnitude of the pressures within the α and β phases.

5.H.3 INTERFACIAL TENSIONS FROM CONTACT ANGLES

If we combine the relationship Eq. 5.H6 between the three tensions with the purely geometric relation Eq. 5.H2, we obtain

$$\frac{1}{R_{\alpha\gamma}}\left(\frac{\Sigma_{\alpha\gamma}}{\Sigma_{\alpha\beta}}-\frac{\sin\theta_\beta}{\sin\theta_\gamma}\right)=\frac{1}{R_{\beta\gamma}}\left(\frac{\Sigma_{\beta\gamma}}{\Sigma_{\alpha\beta}}-\frac{\sin\theta_\alpha}{\sin\theta_\gamma}\right). \quad (5.H7)$$

Both the interfacial tensions and the contact angles are material parameters that do not depend on the droplet geometry, provided the droplets are sufficiently large and we can ignore the contact line tension. As a consequence, the relation (5.7) can only hold for arbitrary values of the radii $R_{\alpha\gamma}$ and $R_{\beta\gamma}$, if the terms in the two parentheses vanish separately. Therefore, we conclude that

$$\frac{\Sigma_{\alpha\gamma}}{\Sigma_{\alpha\beta}}=\frac{\sin\theta_\beta}{\sin\theta_\gamma} \quad \text{`and} \quad \frac{\Sigma_{\beta\gamma}}{\Sigma_{\alpha\beta}}=\frac{\sin\theta_\alpha}{\sin\theta_\gamma} \quad (5.H8)$$

which relate the interfacial tensions to the contact angles. It is interesting to note that the derivation of Eq. 5.H8 was based (i) on the purely geometric relation Eq. 5.H2 for three spherical caps and (ii) on the Laplace Eqs 5.H3 and 5.H4 for the mechanical equilibrium of the spherical cap segments of the three interfaces away from the contact line. On the other hand, the relationships Eq. 5.H8 can also be derived from the force balance between the three interfacial tensions at the contact line. Indeed, in mechanical equilibrium, the three tensions must add up to zero which implies that these tensions form the sides of a triangle as shown in the upper panel of Figure 5.48c. In the literature on capillary forces, this triangle is known as Neumann's triangle (Rowlinson and Widom, 1989). The law of sines for triangles then leads to the equalities

$$\frac{\Sigma_{\alpha\beta}}{\sin\theta_\gamma}=\frac{\Sigma_{\alpha\gamma}}{\sin\theta_\beta}=\frac{\Sigma_{\beta\gamma}}{\sin\theta_\alpha} \quad (5.H9)$$

which are equivalent to the relations Eq. 5.H8.

It is instructive to rederive the force balance conditions as described by Eqs 5.8 or by the equivalent Eqs 5.9 using a variational approach. To do so, we start from the parametrization of the three-spherical-cap geometry in terms of the four radii $R_{\alpha\beta}$, $R_{\alpha\gamma}$, $R_{\beta\gamma}$, and R_{co} as described in Section 9.5.1 and consider the (free) energy of the three interfaces which has the form

$$E_o(R_{\alpha\beta},R_{\alpha\gamma},R_{\beta\gamma},R_{co})=\Sigma_{\alpha\beta}A_{\alpha\beta}+\Sigma_{\alpha\gamma}A_{\alpha\gamma}+\Sigma_{\beta\gamma}A_{\beta\gamma}. \quad (5.H10)$$

The three interfacial areas can be written as explicit functions of the four radii. To minimize this energy function for fixed droplet volumes V_α and V_β, we define the shape function

$$F_o(R_{\alpha\beta},R_{\alpha\gamma},R_{\beta\gamma},R_{co})\equiv(P_\gamma-P_\alpha)V_\alpha+(P_\gamma-P_\beta)V_\beta+\Delta F_o \quad (5.H11)$$

with

$$\Delta F_o\equiv E_o=\Sigma_{\alpha\beta}A_{\alpha\beta}+\Sigma_{\alpha\gamma}A_{\alpha\gamma}+\Sigma_{\beta\gamma}A_{\beta\gamma} \quad (5.H12)$$

where the two droplet volumes V_α and V_β are again explicit functions of the four radii. The stationary three-spherical-cap shapes are then obtained from

$$\frac{\partial F_o}{\partial R_{\alpha\beta}}=0,\quad \frac{\partial F_o}{\partial R_{\alpha\gamma}}=0,\quad \frac{\partial F_o}{\partial R_{\beta\gamma}}=0,\quad \text{and} \quad \frac{\partial F_o}{\partial R_{co}}=0. \quad (5.H13)$$

From these stationarity conditions, we recover the three Laplace Eqs 5.3 and 5.4 as well as the force balance Eqs 5.8 along the contact line.

5.H.4 TRIANGLE RzELATIONS FOR INTERFACIAL TENSIONS

Inspection of Figure 5.48c shows that the three contact angles are the exterior angles of the triangle formed by the three tensions. So far, it has been tacitly assumed that all three contact angles are neither zero nor equal to π. In fact, as one of the contact angle goes to zero, the two other angles must approach the limiting value π. In this limit, the interface of two phases is *completely* wet by the third phase. As an example, consider complete wetting of the $\beta\gamma$ interface by the α phase. In the latter case, the contact angle L_1^+ and the two other contact angles have the values $\theta_\beta=\theta_\gamma=\pi$. The α phase then forms a thin wetting layer between the β and the γ phases. In such a situation, one side of the tension triangle becomes equal to the sum of the two other sides and the triangle collapses.

For any triangle, the length of a given side must be smaller than or equal to the sum of the lengths of the two other sides. For the tension triangle in Figure 5.48c, the corresponding triangle relations are given by

$$\Sigma_{\alpha\gamma}\le\Sigma_{\beta\gamma}+\Sigma_{\alpha\beta},\quad \Sigma_{\beta\gamma}\le\Sigma_{\alpha\gamma}+\Sigma_{\alpha\beta} \quad \text{and} \quad \Sigma_{\alpha\beta}\le\Sigma_{\alpha\gamma}+\Sigma_{\beta\gamma}. \quad (5.H14)$$

It will be instructive to rewrite these relations in a somewhat redundant manner as given by

$$-\Sigma_{\alpha\beta}\le\Sigma_{\beta\gamma}-\Sigma_{\alpha\gamma}\le+\Sigma_{\alpha\beta}, \quad (5.H15)$$

$$-\Sigma_{\alpha\gamma}\le\Sigma_{\beta\gamma}-\Sigma_{\alpha\beta}\le+\Sigma_{\alpha\gamma} \quad (5.H16)$$

and

$$-\Sigma_{\beta\gamma}\le\Sigma_{\beta\gamma}-\Sigma_{\alpha\gamma}\le+\Sigma_{\beta\gamma} \quad (5.H17)$$

which provide lower and upper bounds for all tension differences. In fact, multiplying these inequalities by (−1), we obtain inequalities of the form $-\Sigma_{\alpha\beta}\le\Sigma_{\alpha\gamma}-\Sigma_{\beta\gamma}\le+\Sigma_{\alpha\beta}$ etc. Therefore, the difference between any two tensions is larger or equal to (−1) times the third tension and smaller or equal to (+1) times the third tension.

The inequalities in these triangle relations correspond to partial wetting while the equalities correspond to complete wetting.

As an example, consider the bounds for the tension difference $\Sigma_{\beta\gamma} - \Sigma_{\alpha\gamma}$ as given by Eq. 5.H15. Using the relations in Eq. 5.H8, we obtain the expression

$$\frac{\Sigma_{\beta\gamma} - \Sigma_{\alpha\gamma}}{\Sigma_{\alpha\gamma}} = \frac{\sin\theta_\alpha - \sin\theta_\beta}{\sin\theta_\gamma}. \quad (5.H18)$$

It will be convenient to define the function

$$\Xi(x, y, z) \equiv \frac{\sin x - \sin y}{\sin z} \quad (5.H19)$$

Combining the relations Eqs 5.H18 and 5.H15, we obtain the inequalities

$$-1 \leq \Xi(\theta_\alpha, \theta_\beta, \theta_\gamma) \leq +1 \quad (5.H20)$$

for the function Ξ that depends on all three contact angles. The lower bound

$$\Xi(\theta_\alpha = \pi, \theta_\beta = 0, \theta_\gamma = \pi) = -1 \quad (5.H21)$$

describes complete wetting of the $\alpha\gamma$ interface by the β phase whereas the upper bound

$$\Xi(\theta_\alpha = 0, \theta_\beta = \pi, \theta_\gamma = \pi) = +1 \quad (5.H22)$$

corresponds to complete wetting of the $\beta\gamma$ interface by the α phase.

5.I OUT-WETTING OF MEMBRANES AND VESICLES

In the main text, we focused on in-wetting morphologies of GUVs that arise from aqueous phase separation within the giant vesicles, see Figure 5.40. Wetting of vesicle membranes has also been observed when the vesicles were exposed to PEG-dextran solutions that underwent phase separation outside the GUVs (Li et al., 2012). The aqueous minority phase then forms droplets that can adhere to the vesicle membrane.

5.I.1 OUT-WETTING MORPHOLOGIES

The interaction of the membrane with one such droplet leads to several out-wetting morphologies as shown in Figure 5.49. The morphologies in Figure 5.49a and b have been observed for PEG-dextran solutions (Li et al., 2012). The morphology in Figure 5.49a corresponds to partial wetting of the vesicle membrane by the coexisting liquid phases α and β. This morphology is again characterized by a three-phase contact line that partitions the membrane into two segments. When viewed with optical resolution, the shape contour has an apparent kink at the contact line which should be replaced by a smoothly curved membrane segment when we look at this line with nanoscale resolution.

For partial out-wetting, the $\alpha\beta$ interface partitions the vesicle membrane into an $\gamma\alpha$ segment and a $\gamma\beta$ segment. At first sight, swapping the subscripts γ and α as well as γ and β for out-wetting compared to in-wetting morphologies might seem a bit pedantic but turns out to be important because of the spontaneous curvatures. These curvatures have a sign that is taken to be positive and negative if the membrane prefers to bulge towards the exterior and interior solution, respectively. Therefore, when we swap the interior and exterior solutions, the spontaneous curvature $m_{\gamma j}$ for out-wetting morphologies will differ from the spontaneous curvature $m_{j\gamma} = -m_{\gamma j}$ for in-wetting morphologies.

5.I.2 THEORY OF OUT-WETTING

Geometry of out-wetting morphologies

The out-wetting morphologies in Figure 5.49 involve the spectator phase γ inside the GUV as well as a single β droplet coexisting with the bulk phase α in the exterior solution. The shape S of the vesicle-droplet system can again be decomposed into several components. First, we define the shape S_γ of the interior β droplet, which is identical with the vesicle shape, and the shape S_β of the β droplet. The corresponding droplet volumes are denoted by

$$V_\gamma = \mathcal{V}\{S_\gamma\} \quad \text{and} \quad V_\beta = \mathcal{V}\{S_\beta\}. \quad (5.I1)$$

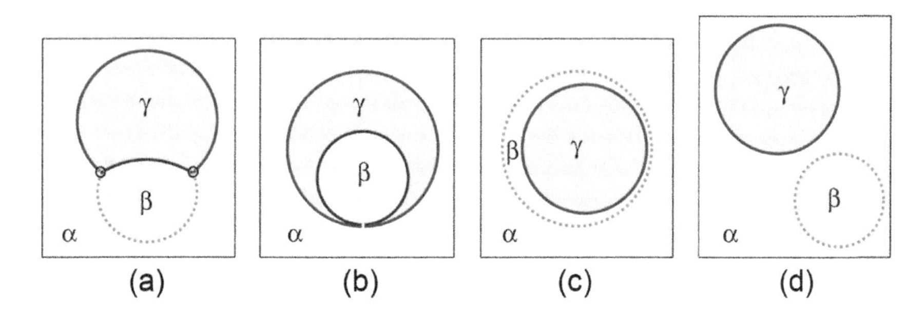

Figure 5.49 Out-wetting morphologies of giant vesicles arising from phase separation of the exterior solution into two aqueous phases, α (white) and β (blue). The vesicle is filled with the aqueous spectator phase γ. (yellow) The $\alpha\beta$ interfaces are depicted as dashed orange lines, the membrane segments in contact with the α and β droplets as red and purple lines, respectively: (a) Partial wetting of the vesicle membrane by α and β as observed on the micrometer scale. The apparent kink at the contact line (black circles) reveals the capillary forces that the $\alpha\beta$ interface exerts onto the vesicle membrane; (b) Special morphology for which the β droplet and the bulk phase α are separated by a closed membrane neck. This morphology resembles complete wetting by the γ phase and required a sufficiently small reduced volume v of the vesicle; (c) Complete wetting of the membrane by the β phase; and (d) Complete wetting by the α phase which leads to the release of the β droplet from the vesicle.

The vesicle volume is now identical with the volume of the γ droplet, i.e., $V = V_\gamma$. These volumes can be considered to be constant at constant temperature and fixed osmotic conditions. The two droplets are bounded by three surface segments: the $\alpha\beta$ interface between the β droplet and the aqueous bulk phase α as well as two membrane segments, the $\gamma\alpha$ segment in contact with the α phase and the $\beta\gamma$ segment exposed to the β droplet. The shapes of these three surfaces will be denoted by $S_{\alpha\beta}$, $S_{\gamma\alpha}$, and $S_{\gamma\beta}$, respectively, with surface areas

$$A_{\alpha\beta} = \mathcal{A}\{S_{\alpha\beta}\}, \quad A_{\gamma\alpha} = \mathcal{A}\{S_{\gamma\alpha}\}, \quad \text{and} \quad A_{\gamma\beta} = \mathcal{A}\{S_{\gamma\beta}\}. \quad (5.112)$$

The total surface area A of the vesicle membrane is then given by

$$A = A_{\gamma\alpha} + A_{\gamma\beta}. \quad (5.113)$$

All three surface segments meet along the three-phase contact line which has the shape $S_{\alpha\beta\gamma}$ and the length

$$L_{\alpha\beta\gamma} = \mathcal{L}\{S_{\alpha\beta\gamma}\}. \quad (5.114)$$

The $\alpha\beta$ interface can adapt its area $A_{\alpha\beta}$ to changes in the droplet and membrane morphologies. As before, the total membrane area A will be taken to be constant at constant temperature. The vesicle-droplet system is then characterized by three geometric constraints as provided by the volumes V_γ and V_β of the two droplets as well as the total membrane area A. In order to determine the morphology of the vesicle-droplet system, we will minimize the (free) energy of the system, taking these three constraints into account.

Adhesion free energies

The adhesion free energy per unit contact area between the outer leaflet of the vesicle membrane and the aqueous bulk phase α will be denoted by $W_{\gamma\alpha}$. Likewise, the adhesive strength $W_{\gamma\beta}$ describes the adhesion free energy per unit contact area between the outer leaflet of the vesicle membrane and the β droplet. The adhesion free energy of the vesicle-droplet system then has the form

$$E_{\rm ad} = W_{\gamma\alpha} A_{\gamma\alpha} + W_{\gamma\beta} A_{\gamma\beta} \quad (5.115)$$

corresponding to the adhesion free energy functional

$$\mathcal{E}_{\rm ad}\{S_{\gamma\alpha}, S_{\gamma\beta}\} = W_{\gamma\alpha} \mathcal{A}\{S_{\gamma\alpha}\} + W_{\gamma\beta} \mathcal{A}\{S_{\gamma\beta}\}. \quad (5.116)$$

We ignore any curvature-dependence of the adhesive strengths $W_{\gamma\alpha}$ and $W_{\gamma\beta}$ which leads to the identities

$$W_{\gamma\alpha} = W_{\alpha\gamma} \quad \text{and} \quad W_{\gamma\beta} = W_{\beta\gamma}, \quad (5.117)$$

i.e., the adhesive strengths $W_{\gamma\alpha}$ and $W_{\gamma\beta}$ for out-wetting are identical with the adhesive strengths $W_{\alpha\gamma}$ and $W_{\beta\gamma}$ for in-wetting as defined in Eq. 5.308. Therefore, the adhesion free energy functional for out-wetting has the same form as for in-wetting.

Mechanical, spontaneous, and total segment tensions

The adhesive strengths $W_{\gamma\alpha}$ and $W_{\gamma\beta}$ contribute to the mechanical tensions

$$\Sigma_{\gamma\alpha} = \Sigma + W_{\gamma\alpha} \quad \text{and} \quad \Sigma_{\gamma\beta} = \Sigma + W_{\gamma\beta} \quad (5.118)$$

of the two membrane segments where Σ is again the overall stress of the vesicle membrane arising from the constraint on the total membrane area. If the two segments have a spontaneous curvature, the weakly curved segments experience the spontaneous tension

$$\sigma_{\gamma\alpha} = 2\kappa_{\gamma\alpha} m_{\gamma\alpha}^2 \text{ and } \sigma_{\gamma\beta} = 2\kappa_{\gamma\beta} m_{\gamma\beta}^2. \quad (5.119)$$

The mechanical and the spontaneous segment tensions add up to the total segment tensions

$$\hat{\Sigma}_{\gamma\alpha} = \Sigma_{\gamma\alpha} + \sigma_{\gamma\alpha} \text{ and } \hat{\Sigma}_{\gamma\beta} = \Sigma_{\gamma\beta} + \sigma_{\gamma\beta} \quad (5.110)$$

which enter the shape equations for the two membrane segments $\gamma\alpha$ and $\gamma\beta$.

Shape functional for out-wetting

In close analogy to Eq. 5.312 for in-wetting, the shape functional for out-wetting has the form

$$\mathcal{F}_{\rm 2Dr}^{\rm out}\{S\} = (P_\alpha - P_\gamma)\mathcal{V}\{S_\gamma\} + (P_\alpha - P_\beta)\mathcal{V}\{S_\beta\} + \Sigma\mathcal{A}\{S\} + \mathcal{E}_{\rm 2Dr}^{\rm out}\{S\} \quad (5.111)$$

with the energy functional

$$\mathcal{E}_{\rm 2Dr}^{\rm out}\{S\} \equiv \Sigma_{\alpha\beta}\mathcal{A}\{S_{\alpha\beta}\} + \mathcal{E}_{\rm be}^{\rm out}\{S_{\gamma\alpha}, S_{\gamma\beta}\} + \mathcal{E}_{\rm ad}\{S_{\gamma\alpha}, S_{\gamma\beta}\} + \mathcal{E}_{\alpha\beta\gamma}\{S_{\alpha\beta\gamma}\}. \quad (5.112)$$

Compared to the energy functional for in-wetting, the energy functional for out-wetting differs only in the bending energy functional which has the form

$$\mathcal{E}_{\rm be}^{\rm out}\{S_{\gamma\alpha}, S_{\gamma\beta}\} = \sum_{j=\alpha,\beta} 2\kappa_{\gamma j} \int dA_{\gamma j} (M - m_{\gamma j})^2 \quad (5.113)$$

for out-wetting. As mentioned, the spontaneous curvatures $m_{\gamma j}$ for out-wetting and $m_{j\gamma}$ for in-wetting are different and related by

$$m_{\gamma j} = -m_{j\gamma}. \quad (5.114)$$

In contrast, the bending rigidities $\kappa_{\gamma j}$ for out-wetting are identical with the bending rigidities $\kappa_{\gamma j}$ for in-wetting.

5.1.3 THREE-SPHERICAL-CAP SHAPES

The out-wetting morphologies observed experimentally are well-described by three-spherical-cap shapes as depicted in Figure 5.49. The $\alpha\beta$ interface always forms a spherical cap with mean curvature $M_{\alpha\beta} = 1/R_{\alpha\beta} > 0$. Furthermore, when viewed on the micrometer scale as in Figure 5.49a, the two membrane segments $\gamma\alpha$ and $\gamma\beta$ also form two spherical caps with mean curvatures $M_{\gamma\alpha} = 1/R_{\gamma\alpha} > 0$ and $M_{\gamma\beta} = \pm 1/R_{\beta\gamma}$. These mean curvatures are governed by the shape equations

Giant vesicles theoretically and in silico

$$P_\gamma - P_j = 2\Sigma^{\rm eff}_{\gamma j} M_{\gamma j} \quad \text{with} \quad j = \alpha, \beta \tag{5.I15}$$

with the effective, curvature-dependent tensions

$$\Sigma^{\rm eff}_{\gamma j} \equiv \Sigma + W_{\gamma j} + \sigma_{\gamma j} - 2\kappa_{\gamma j} m_{\gamma j} M_{\gamma j} \tag{5.I16}$$

of the two membrane segments $\gamma\alpha$ and $\gamma\beta$. Because the mean curvature $M_{\gamma\alpha}$ of the membrane segment $\gamma\alpha$ in contact with the bulk phase α is necessarily positive, the effective tension $\Sigma^{\rm eff}_{\gamma\alpha}$ has the same sign as the pressure difference $P_\gamma - P_\alpha$. In contrast, the mean curvature $M_{\gamma\beta}$ of the membrane segment in contact with the β droplet may be positive or negative which implies that the effective tension $\Sigma^{\rm eff}_{\gamma\beta}$ need not have the same sign as the pressure difference $P_\gamma - P_\beta$.

In addition, we introduce three apparent contact angles $\theta^{\rm ap}_\gamma$, $\theta^{\rm ap}_\beta$, and $\theta^{\rm ap}_\alpha$ that open up towards the three liquid phases γ, β, and α. The tension-angle-curvature relationship for partial out-wetting is then given by

$$M_{\gamma\alpha}\left(\frac{\Sigma^{\rm eff}_{\gamma\alpha}}{\Sigma_{\alpha\beta}} - \frac{\sin\theta^{\rm ap}_\beta}{\sin\theta^{\rm ap}_\gamma}\right) = M_{\gamma\beta}\left(\frac{\Sigma^{\rm eff}_{\gamma\beta}}{\Sigma_{\alpha\beta}} - \frac{\sin\theta^{\rm ap}_\alpha}{\sin\theta^{\rm ap}_\gamma}\right). \tag{5.I17}$$

In contrast to in-wetting, the mean curvature $M_{\gamma\beta}$ of the $\gamma\beta$ membrane segment can now be negative corresponding to a $\gamma\beta$ segment that bulges towards the γ phase within the vesicle.

If several β droplets adhere to the exterior leaflet of a single GUV, we obtain several $\gamma\beta$ segments which we can distinguish by the label $n = 1, 2, \ldots, N$. These $\gamma\beta$ segments have the mean curvatures $M^{(n)}_{\gamma\beta}$ and experience the effective tensions $\Sigma^{(n)}_{\gamma\beta}$. We then obtain the relations $\Upsilon^{(1)}_{\gamma\beta} = \Upsilon^{(2)}_{\gamma\beta} = \ldots = \Upsilon^{(N)}_{\gamma\beta}$ with

$$\Upsilon^{(n)}_{\gamma\beta} \equiv M_{\gamma\alpha}\frac{\sin\theta^{(n)}_\beta}{\sin\theta^{(n)}_\gamma} + M^{(n)}_{\gamma\beta}\left(\frac{\Sigma^{(n)}_{\gamma\beta}}{\Sigma_{\alpha\beta}} - \frac{\sin\theta^{(n)}_\alpha}{\sin\theta^{(n)}_\gamma}\right), \tag{5.I18}$$

in close analogy to the relations as given by Eqs 5.347 and 5.348 for in-wetting. Thus, from three different $\gamma\beta$ segments with three distinct mean curvatures $M^{(n)}_{\gamma\beta}$, we can obtain the two parameter combinations $(\Sigma + W_{\gamma\beta} + \sigma_{\gamma\beta})/\Sigma_{\alpha\beta}$ and $\kappa_{\gamma\beta} m_{\gamma\beta}/\Sigma_{\alpha\beta}$ that determine the tension ratios $\Sigma^{(n)}_{\gamma\beta}/\Sigma_{\alpha\beta}$.

5.I.4 FORCE BALANCE ALONG APPARENT CONTACT LINE

In order to describe the force balance between the two membrane segments and the $\alpha\beta$ interface in a self-consistent manner, we consider again special parameter regimes in close analogy to the force balance for in-wetting morphologies. Thus, we can distinguish small-small, large-large, and large-small regimes for out-wetting as well.

Special parameter regimes

The relationships between the effective tensions and apparent contact angles as given by Eqs. 5.17 and 5.18 depend on the mean curvatures of the different membrane segments. We can again derive curvature-independent relationships if we consider membrane segments characterized by small spontaneous curvatures and small bending energies or large spontaneous curvatures and large spontaneous tensions. The corresponding shape function has the form

$$F^{\rm out} = (P_\alpha - P_\gamma)V_\gamma + (P_\alpha - P_\beta)V_\beta + \Delta F^{\rm out} \tag{5.I19}$$

where the area-dependent shape function $\Delta F^{\rm out}$ is somewhat different for the different regimes. If both membrane segments have large spontaneous curvatures, the area-dependent shape function $\Delta F^{\rm out}$ has the form

$$\Delta F^{\rm out}_{l+l} = \Sigma_{\alpha\beta} A_{\alpha\beta} + \hat{\Sigma}_{\gamma\alpha} A_{\gamma\alpha} + \hat{\Sigma}_{\gamma\beta} A_{\gamma\beta} \quad \text{for large-large regime.} \tag{5.I20}$$

Likewise, we obtain $\Delta F^{\rm out} = \Delta F^{\rm out}_{l+s}$ with

$$\Delta F^{\rm out}_{l+s} = \Sigma_{\alpha\beta} A_{\alpha\beta} + \Sigma_{\gamma\alpha} A_{\gamma\alpha} + \Sigma_{\gamma\beta} A_{\gamma\beta} \quad \text{for small-small regime.} \tag{5.I21}$$

and $\Delta F^{\rm out} = \Delta F^{\rm out}_{s+s}$ with

$$\Delta F^{\rm out}_{l+s} = \Sigma_{\alpha\beta} A_{\alpha\beta} + \Sigma_{\gamma\alpha} A_{\gamma\alpha} + \Sigma_{\gamma\beta} A_{\gamma\beta} \quad \text{for small-small regime.} \tag{5.I22}$$

Comparison with the area-dependent shape functions for in-wetting as given by Eqs 5.367–5.369 shows that, in all three regimes, the area-dependent shape function $\Delta F^{\rm out}$ for out-wetting is identical with the shape function $\Delta F^{\rm in}$ for in-wetting when we replace the segment labels $\gamma\alpha$ and $\gamma\beta$ by the segment labels $\alpha\gamma$ and $\beta\gamma$.

Force balance relations

Minimization of the shape function $\Delta F^{\rm out}$ with respect to the four curvature radii $R_{\gamma\alpha}$, $R_{\gamma\beta}$, and $R_{\alpha\beta}$ as well as with respect to the contact line radius $R_{\rm co}$ leads to the shape equations for the three spherical caps as well as to the force balance relations

$$\frac{\hat{\Sigma}_{\gamma\alpha}}{\Sigma_{\alpha\beta}} = \frac{\sin\theta^{\rm ap}_\beta}{\sin\theta^{\rm ap}_\gamma} \quad \text{and} \quad \frac{\hat{\Sigma}_{\gamma\beta}}{\Sigma_{\alpha\beta}} = \frac{\sin\theta^{\rm ap}_\alpha}{\sin\theta^{\rm ap}_\gamma}. \tag{5.I23}$$

More precisely, the latter relations describe the force balance along the apparent contact line if both membrane segments $\gamma\alpha$ and $\gamma\beta$ belong to the large spontaneous curvature regime. If the $\gamma\alpha$ segment belongs to the small spontaneous curvature and small bending energy regime, the total segment tension $\hat{\Sigma}_{\gamma\alpha}$ in Eq. 5.I23 is replaced by the mechanical segment tension $\Sigma_{\gamma\alpha}$. Likewise, if the $\gamma\beta$ segment belongs to the latter regime, the total segment tension $\hat{\Sigma}_{\gamma\beta}$ is replaced by the mechanical segment $\Sigma_{\gamma\beta}$.

5.J SYMMETRIC TWO-DROPLET VESICLES

In this appendix, we address the deflation of two-droplet vesicles that belong to the partial in-wetting regime as illustrated in Figure 5.40a. When such a vesicle is osmotically deflated, it may follow two distinct morphological pathways. The first pathway leads to the engulfment of both droplets by the membrane as in Figure 5.40d. The second pathway leads to the formation of many nanobuds and nanotubes as in Figure 5.21. In order to discuss the competition between these two morphological pathways, it is instructive to consider a simplified case corresponding to two-droplet vesicles with an up-down symmetry.

Up-down symmetric geometry and fluid-elastic parameters

Such a two-droplet vesicle contains one α and one β droplet, both of which have the same volume $V_\alpha = V_\beta \equiv V_1$. The two droplets are separated by a planar $\alpha\beta$ interface, corresponding to apparent contact angles $\theta_\alpha^{ap} = \theta_\beta^{ap}$. Likewise, the two membrane segments $\alpha\gamma$ and $\beta\gamma$ in contact with the α and β phase have the same areas $A_{\alpha\gamma} = A_{\beta\gamma} \equiv A_1$. Therefore, the vesicle volume V and the membrane area A are given by

$$V = V_\alpha + V_\beta = 2V_1 \text{ and } A = A_{\alpha\gamma} + A_{\beta\gamma} = 2A_1. \quad (5.J1)$$

Before deflation, the initial shape of the vesicle is taken to be a sphere with volume

$$V_{\text{ini}} = \frac{4\pi}{3} R_{\text{ve}}^3 \text{ with } R_{\text{ve}} = \sqrt{A/(4\pi)} \quad (5.J2)$$

which implies the reduced volume $v = v_{\text{ini}} = 1$.

In order to preserve the up-down symmetry during deflation, the two phases α and β and the two membrane segments $\alpha\gamma$ and $\beta\gamma$ are taken to have the same fluid-elastic parameters. Thus, both phases adhere to the membrane with the same adhesive strength $W_{\alpha\gamma} = W_{\beta\gamma} \equiv W$, and both membrane segments are characterized by the same bending rigidity $\kappa_{\alpha\gamma} = \kappa_{\beta\gamma} \equiv \kappa$ and the same spontaneous curvature $m_{\alpha\gamma} = m_{\beta\gamma} \equiv m$.

Regime of small spontaneous curvatures

First, consider the case of a small spontaneous curvature with $|m| \ll 1/R_{\text{ve}}$. The free energy E_{ini} of the initial spherical shape is then given by

$$E_{\text{ini}} \approx \Sigma_{\alpha\beta} A_{\alpha\beta} + 8\pi\kappa \text{ with } A_{\alpha\beta} = A/4 \text{ (small } |m|). \quad (5.J3)$$

Because both droplets have the same adhesive strength, the adhesion free energy WA does not depend on the shape of the vesicle and thus plays no role when we compare different vesicle morphologies.

For small $|m|$, the vesicle membrane cannot form any stable nanobuds or nanotubes and osmotic deflation from the initial volume V_{ini} to the volume

$$V_{\text{eng}} \equiv V_{\text{ini}}/\sqrt{2} \quad (5.J4)$$

leads to the engulfment of both droplets as in Figure 5.40d. Each droplet forms a sphere which is enclosed by the corresponding membrane segment. The two spherical segments are connected by a closed membrane neck which replaces the $\alpha\beta$ interface. Therefore, the free energy E_{eng} of this two-sphere shape does not involve any contribution from the interfacial tension $\Sigma_{\alpha\beta}$ and has the form

$$E_{\text{eng}} \approx 16\pi\kappa \quad \text{(small } |m|). \quad (5.J5)$$

arising from the bending energy of two spherical membrane segments.

The two-sphere shape without an $\alpha\beta$ interface has a lower free energy than the initial one-sphere shape if $E_{\text{eng}} - E_{\text{ini}} < 0$ which implies the inequalities

$$\Sigma_{\alpha\beta} > \frac{8\pi\kappa}{A} \quad \text{or} \quad A > \frac{8\pi\kappa}{\Sigma_{\alpha\beta}} \quad (5.J6)$$

for the interfacial tension and the membrane area. Therefore, for small $|m|$, deflation of the initial spherical vesicle leads to the two-sphere morphology without an $\alpha\beta$ interface for sufficiently large interfacial tension $\Sigma_{\alpha\beta}$ or sufficiently large membrane area A.

Regime of large spontaneous curvatures

For large spontaneous curvatures with $|m| \gg 1/R_{\text{ve}}$, the initial spherical vesicle with volume $V = V_{\text{ini}}$ has the free energy

$$E_{\text{ini}} \approx \left(\frac{1}{4}\Sigma_{\alpha\beta} + \sigma\right) A \quad \text{(large } |m|) \quad (5.J7)$$

which depends on the spontaneous tension $\sigma = 2\kappa m^2$. When we deflate this vesicle to obtain the smaller volume $V_{\text{eng}} = V_{\text{ini}}/\sqrt{2}$, the vesicle membrane may again engulf the two droplets completely, thereby replacing the $\alpha\beta$ interface by a closed membrane neck. The free energy E_{eng} of the latter shape is now given by

$$E_{\text{eng}} \approx \sigma A \quad \text{(large } |m|) \quad (5.J8)$$

which is smaller than E_{ini}. Therefore, the first morphological pathway which eliminates the $\alpha\beta$ interface always reduces the free energy of the vesicle-droplet system.

However, for large $|m|$, the deflated vesicle can also form nanobuds and nanotubes. To simplify the following discussion, the buds and tubes are built up from zero-energy spherules with radius $R_2 = 1/|m|$ as described in Sections 5.5 and 5.6. As a result of this second morphological pathway, the membrane forms a spherical mother vesicle with radius R_{mv} and N spherules of radius $1/|m|$ which are connected by closed membrane necks. The volume V_{tub} of this shape is given by

$$V_{\text{tub}} = \frac{4\pi}{3} R_{\text{mv}}^3 \pm \frac{4\pi}{3} \frac{N}{|m|^3} \quad (5.J9)$$

where the plus and minus sign corresponds to out- and in-spherules, respectively, and the conserved membrane area A can be decomposed according to

$$A = 4\pi R_{\text{mv}}^2 + 4\pi \frac{N}{m^2}. \quad (5.J10)$$

In order to compare the two morphological pathways of engulfment and tubulation, we now consider the same deflation depth in both cases corresponding to $V_{tub} = V_{eng} = V_{ini}/\sqrt{2}$ as in Eq 5.J4. The latter equality implies

$$\frac{4\pi}{3}R_{mv}^3 \pm \frac{4\pi}{3}\frac{N}{|m|^3} = \frac{4\pi}{3}\frac{R_{ve}^3}{\sqrt{2}}. \quad (5.J11)$$

In addition, the conservation of the membrane area leads to

$$A = 4\pi R_{ve}^2 = 4\pi R_{mv}^2 + \Delta A \quad (5.J12)$$

with the excess area

$$\Delta A = 4\pi \frac{N}{m^2} \quad (5.J13)$$

stored in the nanobuds and nanotubes. Thus, the area fraction φ stored in the N spherules is given by

$$\varphi \equiv \frac{\Delta A}{A} = \frac{N}{\bar{m}^2}. \quad (5.J14)$$

When expressed in terms of the dimensionless radius

$$r_{mv} \equiv R_{mv}/R_{ve} \quad (5.J15)$$

and the dimensionless spontaneous curvature $\bar{m} = mR_{ve}$, the two relationships in Eqs 5.11 and 5.12 attain the form

$$r_{mv}^3 \pm \frac{\varphi}{|\bar{m}|} = \frac{1}{\sqrt{2}} \quad \text{and} \quad r_{mv}^2 + \varphi = 1. \quad (5.J16)$$

These two equations determine the two unknown variables r_{mv} and φ in terms of $\bar{m}$. The solutions of these two equations have the asymptotic behavior

$$\varphi = \frac{N}{\bar{m}^2} \approx \frac{2}{3}\frac{\sqrt{2}-1}{\sqrt{2}} = 0.195 \quad (5.J17)$$

and

$$r_{mv} = \sqrt{1-\varphi} \approx 0.897 \quad \text{for large } |\bar{m}|. \quad (5.J18)$$

The asymptotic behavior for the area fraction $\varphi = N/\bar{m}^2$ also follows from Eq. 5.171 with $v = 1/\sqrt{2}$.

Because the spherules with radius $R_2 = 1/|m|$ do not contribute to the bending energy of the vesicle membrane, the tubulated vesicle has the free energy

$$E_{tub} \approx \Sigma_{\alpha\beta}\pi R_{mv}^2 + \sigma 4\pi R_{mv}^2 = (\frac{1}{4}\Sigma_{\alpha\beta} + \sigma)(A - \Delta A) \quad (5.J19)$$

which implies

$$E_{tub} - E_{ini} = -(\frac{1}{4}\Sigma_{\alpha\beta} + \sigma)\Delta A < 0 \quad (\text{large } |m|) \quad (5.J20)$$

as follows from the expression for E_{ini} in Eq. 5.J7. Therefore, the second morphological pathway induced by deflation also reduces the free energy of the vesicle-droplet system.

What remains to be done is to compare the free energies E_{tub} and E_{eng}, both of which are smaller than E_{ini}. Using Eqs 5.8 and 5.19, we obtain the free energy difference

$$E_{tub} - E_{eng} = \frac{1}{4}\Sigma_{\alpha\beta}A - (\frac{1}{4}\Sigma_{\alpha\beta} + \sigma)\Delta A. \quad (5.J21)$$

which is negative if

$$\varphi = \frac{\Delta A}{A} > \frac{\Sigma_{\alpha\beta}}{4\sigma + \Sigma_{\alpha\beta}} \quad (5.J22)$$

or

$$\sigma > \frac{1-\varphi}{4\varphi}\Sigma_{\alpha\beta}. \quad (5.J23)$$

Furthermore, the area fraction φ (as in Eq. 5.J23) attains the constant value $\frac{2}{3}\frac{\sqrt{2}-1}{\sqrt{2}} = 0.195$ for large $|m|$ as in Eq. 5.J17. Using this asymptotic behavior, we find that

$$E_{tub} < E_{eng} \quad \text{for} \quad \frac{\sigma}{\Sigma_{\alpha\beta}} > \frac{1+\sqrt{2}}{4(2-\sqrt{2})} = 1.030 \quad (\text{large } |m|). \quad (5.J24)$$

Therefore, the free energy E_{tub} of the tubulated vesicle is lower than the free energy E_{eng} of the vesicle with two completely engulfed droplets if the spontaneous tension σ is large compared to the interfacial tension $\Sigma_{\alpha\beta}$.

Spatial location of zero-energy spherules

In the previous discussion, we did not have to specify the spatial location of the spherules which may be attached to the two membrane segments or to other spherules within necklace-like tubes. Indeed, because the spherules have zero bending energy, the free energy E_{tub} depends only on the number N of the spherules but not on their spatial locations. In particular, for equal adhesive strengths $W_{\alpha\gamma} = W_{\beta\gamma}$ as considered above, an arbitrary number of N_α spherules can be in contact with the α phase which implies that $N_\beta = N - N_\alpha$ spherules are in contact with the β phase, extending the morphological complexity discussed in Section 5.6.4.

This degeneracy is, however, lifted if the adhesive strength $W_{\alpha\gamma}$ of the α droplet differs from the adhesive strength $W_{\beta\gamma}$ of the β droplet. If the α droplet is more adhesive than the β droplet, corresponding to $W_{\alpha\gamma} < W_{\beta\gamma}$, a spherule in contact with the α phase gains the adhesion free energy $(W_{\alpha\gamma} - W_{\beta\gamma})4\pi/|m|^2$ compared to a spherule in contact with the β phase. Therefore, if both membrane segments are still characterized by the same fluid-elastic parameters, the morphology with the lowest free energy is provided by N spherules that are all in contact with the α phase for $W_{\alpha\gamma} < W_{\beta\gamma}$.

GLOSSARY OF SYMBOLS

Symbols for membrane geometry and topology

A	membrane area
$\mathcal{A}\{S\}$	area functional of vesicle shape S
A_{bo}	membrane area bound to rigid substrate surface
ΔA	area difference between two leaflets of bilayer membrane
C_1, C_2	two principal curvatures of membrane surface
c	Euler characteristic, $\chi = 2 - 2\mathfrak{g}$
$\mathfrak{g}$	topological genus of vesicle, i.e., number of handles
G	determinant of metric tensor g_{ij}
g_{ij}	metric tensor, $g_{ij} = \vec{X}_i \cdot \vec{X}_j$
h_i^j	curvature tensor
G	Gaussian curvature of membrane surface, $G = C_1 C_2$
I_M	integrated mean curvature, $I_M = \int dAM$
$\mathcal{I}_M\{S\}$	integrated mean curvature functional of membrane shape S
ℓ_{me}	membrane thickness
M	mean curvature of membrane surface, $M = \frac{1}{2}(C_1 + C_2)$
M_1, M_2	mean curvature of two segments adjacent to a closed neck
M_{ne}	(effective) neck curvature, $M_{ne} = \frac{1}{2}(M_1 + M_2)$
$\bar{M}_{ne}$	dimensionless neck curvature, $\bar{M}_{ne} = M_{ne} R_{ve}$
$\hat{n}$	unit vector normal to membrane surface
ψ	tilt angle along the contour of an axisymmetric vesicle shape
r	radial coordinate for the contour of an axisymmetric vesicle shape
R_{cy}	radius of cylindrical membrane segment
R_{ne}	radius of membrane neck
R_{sp}	radius of spherical membrane segment
R_{ve}	vesicle size, $R_{ve} = \sqrt{A/(4\pi)}$, used as basic length scale
S	shape of vesicle
$\underline{s}$	two-dimensional surface coordinates, $\underline{s} \equiv (s^1, s^2)$, of membrane shape
$\vec{X}$	vector-valued function $\vec{X} = \vec{X}(\underline{s})$ in three dimensions
$\vec{X}_i$	two tangent vectors to membrane surface, $\vec{X}_i = \partial \vec{X} / \partial s^i$
V	volume of vesicle
$\mathcal{V}\{S\}$	volume functional of vesicle shape S
v	volume-to-area ratio or reduced volume, $v = 6\sqrt{\pi} V / A^{3/2}$

Symbols for curvature models of uniform membranes

A_{opt}	optimal membrane area corresponding to optimal molecular packing
E_{be}	bending energy
$\bar{E}_{be}$	dimensionless bending energy, $\bar{E}_{be} = E_{be} / (8\pi\kappa)$
$\mathcal{E}_{be}\{S\}$	bending energy functional of vesicle shape S
$\bar{\mathcal{E}}_{be}\{S\}$	dimensionless bending energy functional, $\bar{\mathcal{E}}_{be} = \mathcal{E}_{be} / (8\pi\kappa)$
$\mathcal{E}_{cu}\{S\}$	curvature energy functional of vesicle shape S
f	locally applied pulling force acting on a small membrane segment
f_{eff}^{in}	effective constriction force acting on the neck of an in-bud
f_{eff}^{out}	effective constriction force acting on the neck of an out-bud
$f_{ex} > 0$	pulling force pointing towards the exterior vesicle compartment
$f_{in} < 0$	pulling force pointing towards the interior vesicle compartment
f_m^{in}, f_m^{out}	constriction forces generated by spontaneous curvature
F	shape energy, $F = -\Delta PV + \Sigma A + E_{be}$
$\mathcal{F}\{S\}$	shape functional of vesicle shape S
$I_{M,0}$	integrated mean curvature of vesicle shape with an optimal area difference
κ	bending rigidity of membrane, used as basic energy scale
κ_Δ	second bending rigidity for area difference elasticity
κ_G	Gaussian curvature modulus
K_A	area compressibility modulus
λ_{ed}	line tension of bilayer edge
m	spontaneous curvature of bilayer membrane
$\bar{m}$	dimensionless spontaneous curvature, $\bar{m} = mR_{ve}$
m_{com}	composite curvature, $m_{com} = m + f / (4\pi\kappa)$
m_{eff}	effective spontaneous curvature, $m_{eff} = m + m_{nlo}$
m_{nlo}	nonlocal spontaneous curvature
P_{in}	osmotic pressure within interior compartment
P_{ex}	osmotic pressure within exterior compartment
ΔP	osmotic pressure difference across the membrane, $\Delta P = P_{in} - P_{ex}$
S^{st}	stationary shape, i.e., stationary solution of the Euler-Lagrange equation
σ	spontaneous tension, $\sigma = 2\kappa m^2$
Σ	mechanical membrane tension
$\hat{\Sigma}$	total membrane tension, $\hat{\Sigma} = \Sigma + \sigma$

Symbols for spheres and tubules (Sections 5.5 and 5.6)

$B_\star^+$,	bifurcation point for $(1+N)$-sphere vesicle at $(\bar{m}, v) = (\bar{m}_\star^+, v_\star^+)$
$B_\Diamond^+$	bifurcation point for $(1+N)$-sphere vesicle at $(\bar{m}, v) = (\bar{m}_\Diamond^+, v_\Diamond^+)$
L^{pea}	limit shape of two-sphere vesicle with $m > 0$
L^{sto}	limit shape of two-sphere vesicle with $m < 0$
$L_=^{out}$	limit shape of two-sphere vesicle consisting of two equal spheres
$L_=^{in}$	limit shape of two-sphere vesicle consisting of two nested spheres with equal radius
$L_\star^+$	limit shape of $(1+N)$-sphere vesicle at $B_\star^+$ with balanced volume, $v_1 = Nv_2$
L_1^+	limit shape of $(1+N)$-sphere vesicle dominated by r_1-sphere, $v_1 > Nv_2$
L_2^+	limit shape of $(1+N)$-sphere vesicle dominated by r_2-spheres, $v_1 < Nv_2$

$L_\Diamond^+$	limit shape of $(1+N)$-sphere vesicle at $B_\Diamond^+$ with $v_1 = v_2$
$L_=^+$	limit shape of $(1+N)$-sphere vesicle consisting of $(1+N)$ equal spheres
$L_{[N]}^{in}$	limit shape with in-necklaces containing N small spheres
$L_{[N]}^{out}$	limit shape with out-necklaces containing N small spheres
$\bar{m}_*^+$	$\bar{m}$-value of bifurcation point B_*^+, $\bar{m}_*^+ = \frac{1}{2}(1+N^{1/3})^{3/2}$
$\bar{m}_\Diamond^+$	$\bar{m}$-value of bifurcation point $B_\Diamond^+$, $\bar{m}_\Diamond^+ = \sqrt{1+N}$
$\bar{M}_{12}$	(effective) neck curvature of 12-neck of necklace-like tube
$\bar{M}_{22}$	(effective) neck curvature of 22-neck of necklace-like tube
N	number of r_2-spheres for $(1+N)$-sphere vesicles and necklace-like tubes
Φ^{pea}	persistent two-sphere vesicle with $m > 0$
Φ_*^+	persistent $(1+N)$-spheres, same geometry as L_*^+ shape but with $\bar{m} > \bar{m}_*^+$
Φ_1^+	persistent $(1+N)$-spheres, same geometries as L_1^+ shapes but with larger $\bar{m}$-values
Φ_2^+	persistent $(1+N)$-spheres, same geometries as L_2^+ shapes but with larger $\bar{m}$-values
r_1, r_2	dimensionless radii of two-sphere vesicle, $r_i = R_i/R_{ve}$
R_{cy}	radius of cylindrical membrane segment
R_{pip}	radius of cylindrical pipette
R_{sp}	radius of spherical membrane segment
R_1, R_2	two radii of two-sphere shape
ρ_1	volume fraction of large r_1-sphere, $\rho_1 = v_1/(Nv_2)$
ρ_2	volume fraction of N small r_2-spheres, $\rho_2 = Nv_2/v_1 = 1/\rho_1$
σ	spontaneous tension, $\sigma = 2\kappa m^2$
Σ	mechanical membrane tension
Σ_{asp}	aspiration tension as given by Eq. 5.210
$\hat{\Sigma}$	total membrane tension, $\hat{\Sigma} = \Sigma + \sigma$
Θ^{in}	two-sphere vesicle with an in-bud, unspecified neck condition
Θ^{out}	two-sphere vesicle with an out-bud, unspecified neck condition
v_1	dimensionless volume of single r_1-sphere, $v_1 = r_1^3$
v_2	dimensionless volume of single r_2-sphere, $v_2 = r_2^3$
v_*^+	v-value of bifurcation point B_*^+, $v_*^+ = 2/(1+N^{1/3})^{3/2}$
$v_\Diamond^+$	v-value of bifurcation point $B_\Diamond^+$, $v_\Diamond^+ = 1/\sqrt{1+N}$
v_{co}^+	smallest possible volume of $(1+N)$-sphere vesicle with mutual contacts of out-buds
v^{pea}	volume of limit shape L^{pea}
v^{sto}	volume of limit shape L^{sto}
$v_=^{out}$	volume of limit shape $L_=^{out}$ consisting of two equal spheres
$v_{[N]}^{in}$	volume of limit shape $L_{[N]}^{in}$ with in-necklaces of total length N
$v_{[N]}^{out}$	volume of limit shape $L_{[N]}^{out}$ with out-necklaces of total length N
Z^{in}	vesicle shape with one in-bud that has radius $r_2 = 1/\lvert\bar{m}\rvert$ and zero bending energy
Z_N^{in}	vesicle shape with N in-buds that have radius $r_2 = 1/\lvert\bar{m}\rvert$ and zero bending energy
Z^{out}	vesicle shape with one out-bud that has radius $r_2 = 1/\bar{m}$ and zero bending energy
Z_N^{out}	vesicle shape with N out-buds that have radius $r_2 = 1/\bar{m}$ and zero bending energy

Symbols for adhesion of vesicles (Section 5.7)

A_{bo}	area of bound membrane segment adhering to the substrate surface
A_{un}	area of unbound membrane segment not in contact with the surface
$C_{\parallel co}$	membrane curvature parallel to the contact line
$C_{\perp co}$	membrane curvature perpendicular to the contact line
E_{ad}	adhesion (free) energy
$\mathcal{E}_{ad}$	adhesion (free) energy functional
$\mathcal{E}_{AV}$	energy functional of adhering vesicle
f_W^{in}, f_W^{out}	effective constriction forces generated by adhesion
$\mathcal{F}_{AV}$	shape functional of adhering vesicle
l_1, l_2	two lipid species
M_{bo}	mean curvature of membrane segment bound to adhesive surface
M_{co}	contact mean curvature of unbound membrane segment
R_{be}	radius of spherical bead
$R_{\parallel co}$	membrane's curvature radius parallel to the contact line
$R_{\perp co}$	membrane's curvature radius perpendicular to the contact line
R_W	adhesion length, $R_W = \sqrt{2\kappa/\lvert W\rvert}$
S_{bo}	shape of bound membrane segment in contact with the adhesive surface
S_{un}	shape of unbound membrane segment not in contact with the adhesive surface
θ_{eff}	effective contact angle of adhering vesicle for strong adhesion
$\lvert W\rvert$	adhesion free energy density or adhesive strength
$\lvert w\rvert$	dimensionless adhesive strength, $\lvert w\rvert = \lvert W\rvert R_{ve}^2/\kappa$

Symbols for multi-domain vesicles (Section 5.8)

a, b	indices for different membrane phases
A_a, A_b	area of intramembrane domain formed by membrane phases a and b
$\Delta\kappa_G$	difference in Gaussian curvature moduli, $\Delta\kappa_G = \kappa_{Ga} - \kappa_{Gb}$
κ_a, κ_b	bending rigidities of a- and b-domains
κ_{Ga}, κ_{Gb}	Gaussian curvature moduli of a- and b-domains
$\mathcal{L}\{.\}$	length functional
L_{ab}	length of ab domain boundary, $L_{ab} = \mathcal{L}\{S_{ab}\}$
Ld,Lo	liquid-disordered and liquid-ordered phase of lipid mixtures
λ	line tension of domain boundary between intramembrane domains

$\bar{\lambda}$	dimensionless line tension, $\bar{\lambda} = \lambda R_{ve}/\kappa_b$
m_a, m_b	spontaneous curvatures of *a* and *b* domain
$\bar{m}_a, \bar{m}_b$	dimensionless spontaneous curvatures, $\bar{m}_a = m_a R_{ve}$
M_a	mean curvature of *a*-domain adjacent to closed neck
M_b	mean curvature of *b*-domain adjacent to closed neck
$M_a(s_1)$	mean curvature of *a*-domain along domain boundary of axisymmetric shape
$M_b(s_1)$	mean curvature of *b*-domain along domain boundary of axisymmetric shape
S_a, S_b	shapes of intramembrane domains consisting of membrane phases *a* and *b*
S_{ab}	shape of domain boundary between *a* and *b* domain
s_1	value of arc length *s* at the domain boundary of axisymmetric shape
Σ_a, Σ_b	mechanical membrane tensions in the *a* and *b* domains
x_a, x_b	area fractions of two-domain vesicles, $x_a = A_a/A$ and $x_b = A_b/A$

Symbols for wetting of membranes (Section 5.9)

α, β, γ	indices for different aqueous phases
$\alpha\beta$	index for interface between α and β phase
$\alpha\gamma$	index for membrane segment between α droplet and external phase γ
$\beta\gamma$	index for membrane segment between β droplet and external phase γ
$\mathcal{E}_{ad}$	adhesion free energy arising from the membrane-droplet interactions
$j\gamma$	index for $j\gamma$ membrane segment with $j = \alpha$ or β
$\kappa_{j\gamma}$	bending rigidity of $j\gamma$ membrane segment
λ_{co}	line tension of contact line
$m_{j\gamma}$	spontaneous curvature of $j\gamma$ membrane segment
$M_{j\gamma}$	mean curvature of $j\gamma$ membrane segment, three-spherical cap shape
r_1	radius of true contact line
$R_{\alpha\beta}$	curvature radius of the *ab* interface, three-spherical-cap shape
$R_{j\gamma}$	curvature radius of $j\gamma$ membrane segment, three-spherical-cap shape
R_{co}	radius of apparent contact line
$\sigma_{j\gamma}$	spontaneous tension of $j\gamma$ segment, $\sigma_{\alpha\gamma} = 2\kappa_{\alpha\gamma} m_{\alpha\gamma}^2$
Σ	overall lateral stress, Lagrange multiplier for the total membrane area
$\Sigma_{\alpha\beta}$	interfacial tension of $\alpha\beta$ interface
$\Sigma_{j\gamma}$	mechanical tension of $j\gamma$ segment, $\Sigma_{j\gamma} = \Sigma + W_{j\gamma}$
$\Sigma_{j\gamma}^{eff}$	effective membrane tension of $j\gamma$ segment, $\Sigma_{j\gamma}^{eff} = \hat{\Sigma}_{j\gamma} - 2\kappa_{j\gamma} m_{j\gamma} M_{j\gamma}$
$\hat{\Sigma}_{j\gamma}$	total membrane tension of $j\gamma$ segment, $\hat{\Sigma}_{j\gamma} = \Sigma_{j\gamma} + \sigma_{j\gamma}$
$\theta_\alpha^{ap}, \theta_\beta^{ap}, \theta_\gamma^{ap}$	apparent contact angles which depend on the vesicle geometry
θ_α^*	intrinsic contact angle between $\alpha\gamma$ membrane segment and $\alpha\beta$ interface
θ_β^*	intrinsic contact angle between $\beta\gamma$ membrane segment and $\alpha\beta$ interface
$\mathcal{T}_{\alpha\gamma}$	free energy of leaflet-water interfaces for partial in-wetting morphology
$\mathcal{T}_{\gamma\gamma}$	free energy of leaflet-water interfaces in contact with γ phase only
V_α, V_β	volumes of droplets formed by aqueous α and β phase
$W_{\alpha\gamma}$	adhesion (free) energy density of α phase replacing γ phase
$W_{\beta\gamma}$	adhesion (free) energy density of β phase replacing γ phase

REFERENCES

Agudo-Canalejo J, Lipowsky R (2015a) Critical particle sizes for the engulfment of nanoparticles by membranes and vesicles with bilayer asymmetry. *ACS Nano*, 9:3704–3720.

Agudo-Canalejo J, Lipowsky R (2015b) Adhesive nanoparticles as local probes of membrane curvature. *Nano Lett.*, 15:7168–7173.

Agudo-Canalejo J, Lipowsky R (2016) Stabilization of membrane necks by adhesive particles, substrate surfaces, and constriction forces. *Soft Matter*, 12:8155–8166.

Agudo-Canalejo J, Lipowsky R (2017) Uniform and janus-like nanoparticles in contact with vesicles: Energy landscapes and curvature-induced forces. *Soft Matter*, 13:2155–2173.

Agudo-Canalejo J, Lipowsky R (in preparation) Strong influence of spontaneous curvature on vesicle adhesion.

Aizenman M, Wehr J (1989) Rounding of first-order phase transitions in systems with quenched disorder. *Phys. Rev. Lett.*, 62:2503–2506.

Albertsson PA (1986) *Partition of Cell Particles and Macromolecules: Separation and Purification of Biomolecules, Cell Organelles Membranes, and Cells in Aqueous Polymer Two-Phase Systems and Their Use in Biochemical Analysis and Biotechnology.* 3 edition, New York: John Wiley & sons.

Almeida PFF, Vaz WLC (1995) Lateral diffusion in membranes. In Lipowsky R, Sackmann E (Eds.), *Structure and Dynamics of Membranes, Vol. 1A of Handbook of Biological Physics*. Amsterdam, the Netherlands: Elsevier.

Anderson TH, Min Y, Weirich KL, Zeng H, Fygenson D, Israelachvili JN (2009) Formation of supported bilayers on silica substrates. *Langmuir*, 25:6997–7005.

Atefi E, J. Adinin Mann J, Tavana H (2014) Ultralow interfacial tensions of aqueous two-phase systems measured using drop shape. *Langmuir*, 30:9691–9699.

Aureli M, Mauri L, Ciampa MG, Prinetti A, Toffano G, Secchieri C, Sonnino S (2016) GM1 Ganglioside: Past studies and future potential. *Mol. Neurobiol.*, 53:1824–1842.

Avalos-Padilla Y, Knorr RL, Javier-Reyna R, Garca-Rivera G, Lipowsky R, Dimova R, Orozco E (2018) The conserved ESCRT-III machinery participates in the phagocytosis of *Entamoeba histolytica*. *Front. Cell. Infect. Microbiol.*, 8:53. doi:10.3389/fcimb.2018.00053.

Bacia K, Schwille P, Kurzchalia T (2005) Sterol structure determines the separation of phases and the curvature of the liquid-ordered phase in model membranes. *PNAS*, 102:3272–3277.

Giant vesicles theoretically and *in silico*

Bancroft WD (1913) The theory of emulsification, V. *J. Phys. Chem.*, 17:501–519.
Bancroft W, Tucker C (1927) Gibbs on emulsification. *J. Phys. Chem.*, 31:1681–1692.
Bangham A, Horne R (1964) Negative staining of phospholipids and their structural modification by surface-active agents as observed in the electron microscope. *J. Mol. Biol.*, 8:660–668.
Baumgart T, Das S, Webb WW, Jenkins JT (2005) Membrane elasticity in giant vesicles with fluid phase coexistence. *Biophys. J.*, 89:1067–1080.
Baumgart T, Graber ZT, Shi Z (2017) Cations induce shape remodeling of negatively charged phospholipid membranes. *PCCP*, 19:15285–15295.
Baumgart T, Hammond AT, Sengupta P, Hess ST, Holowka DA, Baird BA, Webb WW (2007) Large-scale fluid/fluid phase separation of proteins and lipids in giant plasma membrane vesicles. *PNAS*, 104:3165–3170.
Baumgart T, Hess S, Webb W (2003) Imaging coexisting fluid domains in biomembrane models coupling curvature and line tension. *Nature*, 425:821–824.
Berndl K (1990) *Formen Von Vesikeln Diplomarbeit*, Ludwig–Maximilians–Universität München.
Berndl K, Käs J, Lipowsky R, Sackmann E, Seifert U (1990) Shape transformations of giant vesicles: Extreme sensitivity to bilayer asymmetry. *Europhys. Lett.*, 13:659–664.
Bhatia T, Agudo-Canalejo J, Dimova R, Lipowsky R (2018) Membrane nanotubes increase the robustness of giant vesicles. *ACS Nano*, 12:4478–4485.
Bhatia T, Dimova R, Lipowsky R in preparation Morphological complexity of giant vesicles exposed to asymmetric sugar solutions.
Binder K (1983) Random-field induced interface widths in ising systems. *Z. Phys. B*, 50:343–352.
Brangwynne CP, Eckmann CR, Courson DS, Rybarska A, Hoege C, Gharakhani J, Jülicher F, Hyman AA (2009) Germline P granules are liquid droplets that localize by controlled dissolution/condensation. *Science*, 324:1729–1732.
Breidenich M, Netz R, Lipowsky R (2000) The shape of polymer-decorated membranes. *Europhys. Lett.*, 49:431–437.
Breidenich M, Netz R, Lipowsky R (2001) Adsorption of polymers anchored to membranes. *Europ. Phys. J. E*, 5:403–414.
Breidenich M, Netz R, Lipowsky R (2005) The influence of non-anchored polymers on the curvature of vesicles. *Mol. Phys.*, 103:3160–3183.
Brüning BA, Prévost S, Stehle R, Steitz R, Falus P, Farago B, Hellweg T (2014) Bilayer undulation dynamics in unilamellar phospholipid vesicles: Effect of temperature, cholesterol and trehalose. *Biochim. Biophys. Acta*, 1838:2412–2419.
Canham P (1970) The minimum energy of bending as a possible explanation of the biconcave shape of the human red blood cell. *J. Theoret. Biol.*, 26:61–81.
Cevc G, Marsh D (1987) *Phospholipid Bilayers: Physical Principles and Models*, New York: John Wiley & Sons.
Chithrani BD, Chan WCW (2007) Elucidating the mechanism of cellular uptake and removal of protein-coated gold nanoparticles of different sizes and shapes. *Nano Lett.*, 7:1542–1550.
Chithrani BD, Ghazani AA, Chan WCW (2006) Determining the size and shape dependence of gold nanoparticle uptake into mammalian cells. *Nano Lett.*, 6:662–668.
Collins MD, Keller SL (2008) Tuning lipid mixtures to induce or suppress domain formation across leaflets of unsupported asymmetric bilayers. *PNAS*, 105:124–128.
Cureton DK, Harbison CE, Parrish CR, Kirchhausen T (2012) Limited transferrin receptor clustering allows rapid diffusion of canine parvovirus into clathrin endocytic structures. *J. Virol.*, 86:5330–5340.
Dalmédico A (1991) Sophie Germain. *Sci. Am.*, 265:117–122.
Dasgupta R, Dimova R (2014) Inward and outward membrane tubes pulled from giant vesicles. *J. Phys. D: Appl. Phys.*, 47:282001.
Dasgupta R, Miettinen M, Fricke N, Lipowsky R, Dimova R (2018) The glycolipid GM1 reshapes asymmetric biomembranes and giant vesicles by curvature generation. *Proc. Nat. Acad. Sci. USA*, 115:5756–5761.
David JH, Clair JJ, Juhasz J (2009) Phase equilibria in DOPC/DPPC-d62/Cholesterol mixtures. *Biophys. J.*, 96:521–539.
de Freitas RA, Nicolai T, Chassenieux C, Benyahia L (2016) Stabilization of water-in-water emulsions by polysaccharide-coated protein particles. *Langmuir*, 32:1227–1232.
Derzhanski A, Petrov AG, Mitov MD (1978) Molecular asymmetry and saddle-splay elasticity in lipid bilayers. *Ann. Phys.*, 3:297.
Deserno M (2004) Elastic deformation of a fluid membrane upon colloid binding. *Phys. Rev. E*, 69:031903.
Deserno M, Müller MM, Guven J (2007) Contact lines for fluid surface adhesion. *Phys. Rev. E*, 76:011605.
Deuling H, Helfrich W (1976) The curvature elasticity of fluid membranes: A catalogue of vesicle shapes. *J. Physique*, 37:1335–1345.
Devaux P, McConnell HM (1972) Lateral diffusion in spin-labeled phosphatidylcholine multilayers. *JACS*, 94:4475–4481.
Dietrich C, Bagatolli L, Volovyk Z, Thompson N, Levi M, Jacobson K, Gratton E (2001) Lipid rafts reconstituted in model membranes. *Biophys. J.*, 80:1417–1428.
Dimova R, Aranda S, Bezlyepkina N, Nikolov V, Riske K, Lipowsky R (2006) A practical guide to giant vesicles: Probing the membrane nanoregime via optical microscopy. *J. Phys. Cond. Mat.*, 18:S1151–S1176.
Dimova R, Lipowsky R (2016) Giant vesicles exposed to aqueous two-phase systems: Membrane wetting, budding processes, and spontaneous tubulation. *Adv. Mater. Interf.*, 4:1600451.
Dimova R, Riske KA, Aranda S, Bezlyepkina N, Knorr RL, Lipowsky R (2007) Giant vesicles in electric fields. *Soft Matter*, 3:817–827.
do Carmo M (1976) *Differential Geometry of Curves and Surfaces.* Englewood Cliffs: Upper Saddle River, NJ.
Döbereiner HG, Evans E, Kraus M, Seifert U, Wortis M (1997) Mapping vesicle shapes into the phase diagram: A comparison of experiment and theory. *Phys. Rev. E*, 55:4458–4474.
Döbereiner HG, Selchow O, Lipowsky R (1999) Spontaneous curvature of asymmetric bilayer membranes. *Eur. Biophys. J.*, 28:174–178.
Eggeling C, Ringemann C, Medda R, Schwarzmann G, Sandhoff K, Polyakova S, Belov VN, Hein B, von Middendorff C, Schönle A, Hell SW (2009) Direct observation of the nanoscale dynamics of membrane lipids in a living cell. *Nature*, 457:1159–1162.
Esquena J (2016) Water-in-water (W/W) emulsions. *Curr. Opin. Colloid Interface Sci.*, 25:109–119.
Evans E (1974) Bending resistance and chemically induced moments in membrane bilayers. *Biophys. J.*, 14:923–931.
Evans E, Needham D (1987) Physical properties of surfactant bilayer membranes: Thermal transitions, elasticity, rigidity, cohesion, and colloidal interactions. *J. Phys. Chem.*, 91:4219–4228.
Ewers H, Römer W, Smith AE, Bacia K, Dmitrieff S, Chai W, Mancini R et al., (2010) GM1 structure determines SV40-induced membrane invagination and infection. *Nature Cell Biol.*, 12:11–18.
Fadeel B, Xue D (2009) The ins and outs of phospholipid asymmetry in the plasma membrane: Roles in health and disease. *Crit. Rev. Biochem. Mol. Biol.*, 44:264–77.
Farsad K, Ringstad N, Takei K, Floyd SR, Rose K, Camilli PD (2001) Generation of high curvature membranes mediated by direct endophilin bilayer interactions. *J. Cell Biol.*, 155:193–200.
Fenz SF, Sengupta K (2012) Giant vesicles as cell models. *Integr. Biol.*, 4:982–995.

Fischer T, Vink RLC (2011) Domain formation in membranes with quenched protein obstacles: Lateral heterogeneity and the connection to universality classes. *J. Chem. Phys.*, 134:055106.

Ford MGJ, Mills IG, Peter BJ, Vallis Y, Praefcke GJK, Evans PR, McMahon HT (2002) Curvature of clathrin-coated pits driven by epsin. *Nature*, 419:361–366.

Fourcade B, Miao L, Rao M, Wortis M, Zia R (1994) Scaling analysis of narrow necks in curvature models of fluid lipid–bilayer vesicles. *Phys. Rev. E*, 49:5276–5286.

Frank FC (1958) I. Liquid crystals. on the theory of liquid crystals. *Discuss. Faraday Soc.*, 25:19–28.

Fujiwara T, Ritchie K, Murakoshi H, Jacobson K, Kusumi A (2002) Phospholipids undergo hop diffusion in compartmentalized cell membrane. *J. Cell Biol.*, 157:1071–1081.

Garg S, Rühe J, Lüdtke K, Jordan R, Naumann CA (2007) Domain registration in raft-mimicking lipid mixtures studied using polymer-tethered lipid bilayers. *Biophys. J.*, 92:1263–1270.

Georgiev VN, Grafmüller A, Bléger D, Hecht S, Kunstmann S, Barbirz S, Lipowsky R, Dimova R (2018) Area increase and budding in giant vesicles triggered by light: Behind the scene. *Adv. Sci.*, 5:1800432.

Ghosh R, Satarifard V, Grafmüller A, Lipowsky R in preparation Adsorption-Induced Budding and Fission of Nanovesicles.

Goetz R, Gompper G, Lipowsky R (1999) Mobilitiy and elasticity of self-assembled membranes. *Phys. Rev. Lett.*, 82:221–224.

Goetz R, Lipowsky R (1998) Computer simulations of bilayer membranes: Self-assembly and interfacial tension. *J. Chem. Phys.*, 108:7397–7409.

Gorter E, Grendel F (1925) On bimolecular layers of lipoids on the chromocytes of the blood. *J. Exp. Med.*, 41:439–443.

Góz'dz' W, Gompper G (1998) Composition-driven shape transformations of membranes of complex topology. *Phys. Rev. Lett.*, 80:4213–4216.

Grafmüller A, Shillcock J, Lipowsky R (2007) Pathway of membrane fusion with two tension-dependent energy barriers. *Phys. Rev. Lett.*, 98:218101.

Grafmüller A, Shillcock JC, Lipowsky R (2009) The fusion of membranes and vesicles: Pathway and energy barriers form dissipative particle dynamics. *Biophys. J.*, 96:2658–2675.

Gruhn T, Franke T, Dimova R, Lipowsky R (2007) Novel method for measuring the adhesion energy of vesicles. *Langmuir*, 23:5423–5429.

Gudheti MV, Mlodzianoski M, Hess ST (2007) Imaging and shape analysis of guvs as model plasma membranes: Effect of Trans dopc on membrane properties. *Biophys. J.*, 93:2011–2023.

Gutlederer E, Gruhn T, Lipowsky R (2009) Polymorpohism of vesicles with multi-domain patterns. *Soft Matter*, 5:3303–3311.

He K, Luo W, Zhang Y, Liu F, Liu D, Xu L, Qin L, Xiong C, Lu Z, Fang X, Zhang Y (2010) Intercellular transportation of quantum dots mediated by membrane nanotubes. *ACS Nano*, 6:3015–3022.

Heinrich M, Tian A, Esposito C, Baumgart T (2010) Dynamic sorting of lipids and proteins in membrane tubes with a moving phase boundary. *PNAS*, 107:7208–7213.

Heinrich V, Waugh RE (1996) A piconewton force transducer and its application to measurement of the bending stiffness of phospholipid membranes. *Ann. Biomed. Eng.*, 24:595–605.

Helfrich M, Mangeney-Slavin L, Long M, Djoko K, Keating C (2002) Aqueous phase separation in giant vesicles. *J. Am. Chem. Soc.*, *(JACS)* 124:13374–13375.

Helfrich W (1973) Elastic properties of lipid bilayers: Theory and possible experiments. *Z. Naturforsch.*, 28c:693–703.

Helfrich W, Harbich W (1987) Equilibrium configurations of fluid membranes. In Meunier J, Langevin D, Boccara N. (Ed.), *Physics of Amphiphilic Layers, Vol. 21 of Springer Proceedings in Physics*, 58–63. Berlin, Germany: Springer Berlin Heidelberg.

Hochmuth RM, Wiles HC, Evans EA, McCown JT (1982) Extensional flow of erythrocytemembrane: From cell body to elastic tether. *Biophys. J.*, 39:83–89.

Hu J, Weikl T, Lipowsky R (2011) Vesicles with multiple membrane domains. *Soft Matter*, 7:6092–6102.

Hu J, Lipowsky R, Weikl TR (2013) Binding constants of membrane-anchored receptors and ligands depend strongly on the nanoscale roughness of membranes. *Proc. Nat. Acad. Sci. USA*, 110:15283–15288.

Hu M, Briguglio JJ, Deserno M (2012) Determining the Gaussian curvature modulus of lipid membranes in simulations. *Biophys. J.*, 102:1403–1410.

Jenkins J (1977) Static equilibrium configurations of a model red blood cell. *J. Math. Biology*, 4:149–169.

Jensen MH, Morris EJ, Simonsen AC (2007) Domain shapes, coarsening, and random patterns in ternary membranes. *Langmuir*, 23:8135–8141.

Jülicher F, Lipowsky R (1993) Domain-induced budding of vesicles. *Phys. Rev. Lett.*, 70:2964–2967.

Jülicher F, Lipowsky R (1996) Shape transformations of inhomogeneous vesicles with intramembrane domains. *Phys. Rev. E*, 53:2670–2683.

Jülicher F, Seifert U, Lipowsky R (1993) Conformal degeneracy and conformal diffusion of vesicles. *Phys. Rev. Lett.*, 71:452–455.

Karimi M, Steinkühler S, Roy D, Dasgupta R, Lipowsky R, Dimova R (2018) Asymmetric ionic conditions generate large membrane curvatures. Nano Lett 18:7816–7821.

Keller H, Lorizate M, Schwille P (2009) PI(4,5)P2 Degradation promotes the formation of cytoskeleton-free model membrane systems. *ChemPhysChem*, 10:2805–2812.

Kiessling V, Wan C, Tamm LK (2009) Domain coupling in asymmetric lipid bilayers. *Biochim. Biophys. Acta*, 1788:64–71.

Klotzsch E, Schütz GJ (2013) A critical survey of methods to detect plasma membrane rafts. *Phil. Trans. R. Soc. B*, 368:20120033.

Knorr RL, Dimova R, Lipowsky R (2012) Curvature of double-membrane organelles generated by changes in membrane size and composition. *PLoS One*, 7:e32753.

Knorr RL, Franzmann T, Feeney M, Frigerio L, Hyman A, Dimova R, Lipowsky R (under review) Wetting and molding of membranes by biomolecular condensates.

Knorr RL, Lipowsky R, Dimova R (2015) Autophagosome closure requires membrane scission. *Autophagy*, 11:2134–2137.

Korlach J, Schwille P, Webb W, Feigenson G (1999) Characterization of lipid bilayer phases by confocal microscopy and fluorescence correlation spectroscopy. *Proc. Natl. Acad. Sci. USA*, 96:8461–8466.

Kornberg R, McConnell H (1971) Lateral diffusion of phospholipids in a vesicle membrane. *Proc. Nat. Acad. Sci. USA*, 68:2564–2568.

Kulp A, Kuehn MJ (2010) Biological functions and biogenesis of secreted bacterial outer membrane vesicles. *Annu. Rev. Microbiol.*, 64:163–184.

Kumar S, Gompper G, Lipowsky R (2001) Budding dynamics of multi-component membranes. *Phys. Rev. Lett.*, 86:3911–3914.

Kusumaatmaja H, Li Y, Dimova R, Lipowsky R (2009) Intrinsic contact angle of aqueous phases at membranes and vesicles. *Phys. Rev. Lett.*, 103:238103.

Kusumi A, Nakada C, Ritchie K, Murase K, Suzuki K, Murakoshi H, Kasai RS, Kondo J, Fujiwara T (2005) Paradigm shift of the plasma membrane concept from the two-dimensional continuum fluid to the partitioned fluid: High-speed single-molecule tracking of membrane molecules. *Annu. Rev. Biophys. Biomol. Struct.*, 34:351–378.

Li Y, Lipowsky R, Dimova. R (2008) Transition from complete to partial wetting within membrane compartments. *JACS*, 130:12252–12253.

Li Y, Lipowsky R, Dimova R (2011) Membrane nanotubes induced by aqueous phase separation and stabilized by spontaneous curvature. *Proc. Nat. Acad, Sci. USA*, 108:4731–4736.
Li Y, Kusumaatmaja H, Lipowsky R, Dimova R (2012) Wetting-induced budding of vesicles in contact with several aqueous phases. *J. Phys. Chem. B*, 116:1819–1823.
Lipowsky R (1991) The conformation of membranes. *Nature*, 349:475–481.
Lipowsky R (1992) Budding of membranes induced by intramembrane domains. *J. Phys. II France*, 2:1825–1840.
Lipowsky R (1993) Domain-induced budding of fluid membranes. *Biophys. J.*, 64:1133–1138.
Lipowsky R (1995) Bending of membranes by anchored polymers. *Europhys. Lett.*, 30:197–202.
Lipowsky R (2002) Domains and rafts in membranes: Hidden dimensions of selforganization. *J. Biological Phys.*, 28:195–210.
Lipowsky R (2013) Spontaneous tubulation of membranes and vesicles reveals membrane tension generated by spontaneous curvature. *Faraday Discuss.*, 161:305–331.
Lipowsky R (2014a) Coupling of bending and stretching deformations in vesicle membranes. *Adv. Colloid Interface Sci.*, 208:14–24.
Lipowsky R (2014b) Remodeling of membrane compartments: Some consequences of membrane fluidity. *Biol. Chem.*, 395:253–274.
Lipowsky R (2018a) The response of membranes and vesicles to capillary forces arising from aqueous two-phase systems and water-in-water emulsions. *J. Phys. Chem. B*, 122:3572–3586.
Lipowsky R (2018b) Curvature elasticity and multi-sphere morphologies. *J. Phys. D: Appl. Phys.*, 51:343001/22–24.
Lipowsky R, Brinkmann M, Dimova R, Franke T, Kierfeld J, Zhang X (2005) Droplets, bubbles, and vesicles at chemically structured surfaces. *J. Phys. Cond. Mat.*, 17:S537–S558.
Lipowsky R, Dimova R (2003) Domains in membranes and vesicles. *J. Phys. Cond. Mat.*, 15:S31–S45.
Lipowsky R, Döbereiner HG (1998) Vesicles in contact with nanoparticles and colloids. *Europhys. Lett.*, 43:219–225.
Lipowsky R, Rouhiparkouhi T, Discher DE, Weikl TR (2013) Domain formation in cholesterol/phospholipid membranes exposed to adhesive surfaces or environments. *Soft Matter*, 9:8438–3453.
Lipowsky R, Seifert U (1991) Adhesion of vesicles and membranes. *Mol. Cryst. Liq. Cryst.*, 202:17–25.
Liu Y, Agudo-Canalejo J, Grafmüller A, Dimova R, Lipowsky R (2016) Patterns of flexible nanotubes formed by liquid-ordered and liquid-disordered membranes. *ACS Nano*, 10:463–474.
Liu Y, Lipowsky R, Dimova R (2012) Concentration dependence of the interfacial tension for aqueous two-phase polymer solutions of dextran and polyethylene glycol. *Langmuir*, 28:3831–3839.
Loerke D, Mettlen M, Yarar D, Jaqaman K, Jaqaman H, Danuser G, Schmid SL (2009) Cargo and dynamin regulate clathrin-coated pit maturation. *PLoS Biol.*, 7:e1000057.
Long MS, Cans AS, Keating CD (2008) Budding and asymmetric protein microcompartmentation in giant vesicles containing two aqueous phases. *JACS*, 130:756–762.
Lorenzen S, Servuss RM, Helfrich W (1986) Elastic torques about membrane edges: A study of pierced egg lecithin vesicles. *Biophys. J.*, 50:565–572.
Marchi S, Patergnani S, Pinton *P* (2014) The endoplasmic reticulum-mitochondria connection: One touch, multiple functions. *Biochim. Biophys. Acta*, 1837:461–469.
McMahon HT, Gallop JL (2005) Membrane curvature and mechanisms of dynamic cell membrane remodelling. *Nature*, 438:590–596.
Miao L, Fourcade B, Rao M, Wortis M, Zia R (1991) Equilibrium budding and vesiculation in the curvature model of fluid lipid vesicles. *Phys. Rev. A*, 43:6843–6856.
Miao L, Seifert U, Wortis M, Döbereiner HG (1994) Budding transitions of fluid–bilayer vesicles: The effect of area–difference elasticity. *Phys. Rev. E*, 49:5389–5407.
Michalet X, Bensimon D (1995) Observation of stable shapes and conformal diffusion of genus 2 vesicles. *Science*, 269:666–668.
Michelot A, Drubin DG (2011) Building distinct actin filament networks in a common cytoplasm. *Curr. Biol.*, 21:R560–R569.
Miettinen M, Lipowsky, R (2019) Lipid bilayers with frequent flip-flops have tensionless leaflets. Nano Lett, in press.
Mitov M (1978) Third and fourth order curvature elasticity of lipid bilayers. *Compte Rendues de l'Academie bulgare des Sciences*, 31:513–515.
Moy VT, Jiao Y, Hillmann T, Lehmann H, Sano T (1999) Adhesion energy of receptor-mediated interaction measured by elastic deformation. *Biophys. J.*, 76:1632–1638.
Nikolov V, Lipowsky R, Dimova R (2007) Behavior of giant vesicles with anchored DNA molecules. *Biophys. J.*, 92:4356–4368.
Orth A, Johannes L, Römer W, Steinem C (2012) Creating and modulating microdomains in pore-spanning membranes. *Chem. Phys. Chem.*, 13:108–114.
Ou-Yang ZC, Helfrich W (1989) Bending energy of vesicle membranes: General expressions for the first, second and third variation of the shape energy and applications to spheres and cylinders. *Phys. Rev. A*, 39:5280–5288.
Pataraia S, Liu Y, Lipowsky R, Dimova R (2014) Effect of cytochrome c on the phase behavior of charged multicomponent lipid membranes. *Biochim. Biophys. Acta*, 1838:2036–2045.
Peter BJ, Kent HM, Mills IG, Vallis Y, Butler PJG, Evans PR, McMahon HT (2004) BAR domains as sensors of membrane curvature: The amphiphysin BAR structure. *Science*, 303:495–499.
Pezeshkian W, Gao H, Arumugam S, Becken U, Bassereau P, Florent JC, Ipsen JH, Johannes L, Shillcock JC (2016) Mechanism of shiga toxin clustering on membranes. *ACS Nano*, 11:314–324.
Riske KA, Bezlyepkina N, Lipowsky R, Dimova R (2006) Electrofusion of model lipid membranes viewed with high temporal resolution. *Biophys. Rev. Lett.*, 1:387–400.
Robertson JD (1959) The ultrastructure of cell membranes and their derivatives. *Biochem. Soc. Symp.*, 16:3–43.
Robertson JD (1960) The molecular structure and contact relationships of cell membranes. *Prog. Biophys. Biop. Ch.*, 10:343–418.
Rouhiparkouhi T, Weikl TR, Discher DE, Lipowsky R (2013) Adhesion-induced phase behavior of two-component membranes and vesicles. *Int. J. Mol. Sci.*, 14:2203–2229.
Rowlinson J, Widom B (1989) *Molecular Theory of Capillarity*. Oxford: Clarendon Press.
Rozycki B, Lipowsky R (2015) Spontaneous curvature of bilayer membranes from molecular simulations: Asymmetric lipid densities and asymmetric adsorption. *J. Chem. Phys.*, 142:054101.
Różycki B, Lipowsky R (2016) Membrane curvature generated by asymmetric depletion layers of ions, small molecules, and nanoparticles. *J. Chem. Phys.*, 145:074117.
Sackmann E, Träuble H (1972) Studies of the crystalline-liquid crystalline phase transition of lipid model membranes II: Analysis of electron spin resonance spectra of steroid labels incorporated into lipid membranes. *JACS*, 94:4492–4498.
Sako Y, Kusumi A (1994) Compartmentalized structure of the plasma membrane for receptor movements as revealed by a nanometer-level motion analysis. *J. Cell Biol.*, 125:1251–1264.
Satarifard V, Grafmüller A, Lipowsky R (2018) Nanodroplets at membranes create tight-lipped membrane necks via negative line tension. ACS Nano, 12:12424–12435.
Saxton MJ, Jacobson K (1997) Single-particle tracking: Applications to membrane dynamics. *Annu. Rev. Biophys. Biomol. Struct.*, 26:373–399.
Schengrund CL (2015) Gangliosides: Glycosphingolipids essential for normal neural development and function. *Trends Biochemical. Sci.*, 40:397–406.

Schertzer JW, Whiteley M (2012) A bilayer-couple model of bacterial outer membrane vesicle biogenesis. *MBIO*,3:e00297–e00311.

Scholten E, Tuinier R, Tromp RH, Lekkerkerker HNW (2002) Interfacial tension of a decomposed biopolymer mixture. *Langmuir*, 18:2234–2238.

Schönherr H, Johnson JM, Lenz P, Frank CW, Boxer SG (2004) Vesicle adsorption and lipid bilayer formation on glass studied by atomic force microscopy. *Langmuir*, 20:11600–11606.

Schwille P (2015) Jump-starting life? Fundamental aspects of synthetic biology. *J. Cell Biol.*, 210:687–690.

Scott RE (1976) Plasma membrane vesiculation: A new technique for isolation of plasma membranes. *Science*, 194:743–745.

Seifert U (1997) Configurations of membranes and vesicles. *Adv. Phys.*, 46:13–137.

Seifert U, Berndl K, Lipowsky R (1991) Shape transformations of vesicles: Phase diagram for spontaneous curvature and bilayer coupling model. *Phys. Rev. A*, 44:1182–1202.

Seifert U, Lipowsky R (1990) Adhesion of vesicles. *Phys. Rev. A*, 42:4768–4771.

Semrau S, Idema T, Holtzer L, Schmidt T, Storm C (2008) Accurate determination of elastic parameters for multi-component membranes. *Phys. Rev. Lett.*, 100:088101.

Simons K, Ikonen E (1997) Functional rafts in cell membranes. *Nature*, 387:569–572.

Simunovic M, Lee KYC, Bassereau P (2015) Celebrating soft matter's 10th anniversary: Screening of the calcium-induced spontaneous curvature of lipid membranes. *Soft Matter*, 11:5030–5036.

Singer S, Nicolson G (1972) The fluid mosaic model of the structure of cell membranes. *Science*, 175:720–731.

Skau CT, Kovar DR (2010) Fimbrin and tropomyosin competition regulates endocytosis and cytokinesis kinetics in fission yeast. *Curr. Biol.*, 20:1415–1422.

Sorre B, Callan-Jones A, Manzi J, Goud B, Prost J, Bassereau P, Roux A (2012) Nature of curvature coupling of amphiphysin with membranes depends on its bound density. *PNAS*, 109:173–178.

Sosale N, Rouhiparkouhi T, Bradshaw AM, Dimova R, Lipowsky R, Discher DE (2015) Cell rigidity and shape override CD47's "Self" signaling in phagocytosis by hyperactivating Myosin-II. *Blood*, 125:542–552.

Sowinski S, Jolly C, Berninghausen O, Purbhoo MA, Chauveau A, Köhler K, Oddos S, et al., (2008) Membrane nanotubes physically connect T cells over long distances presenting a novel route for HIV-1 transmission. *Nat. Cell Biol.*, 10:211–219.

Sreekumari A, Lipowsky R (2018) Lipids with bulky head groups generate large membrane curvatures by small compositional asymmetries. *J. Chem. Phys.*, 149: 084901.

Steinkühler J, Agudo-Canalejo J, Lipowsky R, Dimova R (2016) Modulating vesicle adhesion by electric fields. *Biophys. J.*, 111:1454–1464.

Steinkühler J, Różycki B, Alvey C, Lipowsky R, Weikl TR, Dimova R, Discher DE (2019) Membrane fluctuations and acidosis regulate cooperative binding of "marker of self" CD47 to macrophage receptor SIRP α. *J Cell Sci.*, 132:jcs216770.

Steinkühler J, Bhatia T, Lipowsky R, Dimova R (in preparation) Giant plasma membrane vesicles with nanotubes exhibit unusual elastic properties.

Steinkühler J, Knorr R, Zhao Z, Bhatia T, Bartelt S, Wegner S, Dimova R, Lipowsky R (in preparation) Controlled division of cell-sized vesicles by low densities of membrane-bound proteins.

Svetina S, Zeks B (1989) Membrane bending energy and shape determination of phospholipid vesicles and red blood cells. *Eur. Biophys. J.*, 17:101–111.

Takei K, Slepnev VI, Haucke V, Camilli PD (1999) Functional partnership between amphiphysin and dynamin in clathrin-mediated endocytosis. *Nat. Cell Biol.*, 1:33–39.

Tian A, Capraro BR, Esposito C, Baumgart T (2009) Bending stiffness depends on curvature of ternary lipid mixture tubular membranes. *Biophys. J.*, 97:1636–1646.

Tordeux C, Fournier JB, Galatola P (2002) Analytical characterization of adhering vesicles. *Phys. Rev. E*, 65:041912–1–041912–9.

Träuble H, Sackmann E (1972) Studies of the crystalline-liquid crystalline phase transition of lipid model membranes III: Structure of a steroid-lecithin system below and above the lipid-phase transition. *JACS*, 94:4499–4510.

Uppamoochikkal P, Tristram-Nagle S, Nagle JF (2010) Orientation of tie-lines in the phase diagram of DOPC/DPPC/Cholesterol model biomembranes. *Langmuir*, 26:17363–17368.

van Meer G, Voelker DR, Feigenson GW (2008) Membrane lipids: Where they are and how they behave. *Nature Rev.: Mol. Cell Biol.*, 9:112–124.

van Weering JRT, Cullen PJ (2014) Membrane-associated cargo recycling by tubule-based endosomal sorting. *Semin. Cell Dev. Biol.*, 31:40–47.

Veatch SL, Cicuta P, Sengupta P, Honerkamp-Smith A, Holowka D, Baird B (2008) Critical fluctuations in plasma membrane vesicles. *ACS Chem. Biol.*, 3:287–293.

Veatch SL, Gawrisch K, Keller SL (2006) Closed-loop miscibility gap and quantative tie-lines in ternary membranes containing diphytanoyl PC. *Biophys. J.*, 90:4428–4436.

Veatch S, Keller S (2003) Separation of liquid phases in giant vesicles of ternary mixtures of phospholipids and cholesterol. *Biophys. J.*, 85:3074–3083.

Vequi-Suplicy C, Riske K, Knorr R, Dimova R (2010) Vesicles with charged domains. *Biochim. Biophys. Acta*, 1798:1338–1347.

Walde P, Cosentino K, Engel H, Stano P (2010) Giant vesicles: Preparations and applications. *ChemBioChem*, 11:848–865.

Wang Q, Navarro VAS, Peng G, Molinelli E, Goh SL, Judson BL, Rajashankar KR, Sondermann H (2009) Molecular mechanism of membrane constriction and tubulation mediated by the F-BAR protein Pacsin/Syndapin. *PNAS*, 106:12700–12705.

Wang X, Gerdes HH (2015) Transfer of mitochondria via tunneling nanotubes rescues apoptotic PC12 cells. *Cell Death Differ.*, 22:1181–1191.

Weikl TR, Asfaw M, Krobath H, Różycki B, Lipowsky R (2009) Adhesion of membranes via receptor–ligand complexes: Domain formation, binding cooperativity, and active processes. *Soft Matter*, 5:3213–3224.

Weikl TR, Hu J, Xu GK, Lipowsky R (2016) Binding equilibrium and kinetics of membrane-anchored receptors and ligands in cell adhesion: Insights from computational model systems and theory. *Cell. Adh. Migr.*, 10:576–589.

Weiss M, Frohnmayer JP, Benk LT, Haller B, Janiesch JW, Heitkamp T, Börsch M, et al., (2018) Sequential bottom-up assembly of mechanically stabilized synthetic cells by microfluidics. *Nat. Mate.*, 17:89–95.

Westrate LM, Lee JE, Prinz WA, Voeltz GK (2015) Form follows function: The importance of endoplasmic reticulum shape. *Annu. Rev. Biochem.* 84:791–811.

Willmore T (1982) *Total Curvature in Riemannian Geometry.* Chichester, UK: Ellis Horwood.

Yang P, Du Q, Tu ZC (2017) General neck condition for the limit shape of budding vesicles. *Phys. Rev. E*, 95:042403.

Simulating membranes, vesicles, and cells

Thorsten Auth, Dmitry A. Fedosov, and Gerhard Gompper

The form is the outer expression of the inner content. [...]
Therefore one should not seek salvation in one form.

Wassily Kandinsky
The Blue Rider Almanac, 1912

Contents

6.1 INTRODUCTION

Amphiphilic molecules, in particular lipids, are the basic structural element of the membranes in biological cells. A biomembrane is typically composed of many different lipids, which gives a cell many opportunities to control membrane properties by adjusting the membrane composition. This can modify the spontaneous curvature and the bending rigidity, and even lead to phase separation and domain formation. In addition, a biological membrane contains a large number of trans-membrane proteins, which control the exchange of water, ions, and small molecules between the cell plasma and the extracellular space.

Vesicles are cells striped down to the minimum, a membrane enclosing a fluid volume. Vesicles are therefore ideal model systems to investigate the physical properties of many components of cells in isolation, without the full complexity of the cellular machinery. Because the systems are well defined, their properties can be analyzed and studied much more easily from a theoretical perspective. Although vesicles are relatively simple, they are still complex many-body systems, with lipids forming a bilayer due to the hydrophobic effect, a plethora of shapes and phase transformations, families of different genus, membranes with holes, and so on. Simulations therefore play a very important role in elucidating their equilibrium and dynamic properties. Here, simulation approaches range from the molecular scale—where the properties of lipids and membrane proteins are studied—over the supramolecular scale—where the self-assembly of lipids and their phase-behavior can be investigated—to the vesicle scale—where shapes and shape transitions, the effect of phase separation in the membrane and the internal fluid, and the deformations due to external forces and fluid flow are studied (see also Chapters 7, 15, and 19). Simulations are also important because the focus of the research is shifting from simple single-component to biologically more relevant multicomponent systems. An important example is red blood cells (RBCs), which have a cortical spectrin cytoskeleton attached to the lipid-bilayer membrane inside the cell. This gives the membrane a shear modulus. Moreover, RBCs do not fluctuate only due to thermal motion, but their fluctuations also have an active, metabolic component.

Because the physical effects in membranes and vesicles cover a large range of relevant length- and time-scales—from the quantum-mechanical behavior of single molecules and the hydrogen bonds between them (a few angstroms), the properties of single bilayer (a few nanometers), the behavior of small (100 nm) and giant (10 μm) vesicles, to the hydrodynamics of vesicles and cells under flow—that no single computer model can capture them all, compare Figure 6.1.

Therefore, several different models, which are suitable to study phenomena on a smaller range of length scales as illustrated in Box 6.1, have been developed over the last decades:

- **Atomistic Membrane Models**—On the microscopic scale, all-atom simulations are required, in which the positions of the atoms of all molecules as well as the interactions between them are taken into account explicitly. The interactions are sometimes treated quantum-mechanically, but are modeled in most cases using classical force fields. All-atom simulations are indispensable whenever the chemical structure of the participant molecules is relevant for the phenomena under investigation. For example, the functioning of membrane proteins that act as ion pumps can only be understood on the basis of such atomistic models. However, molecular–dynamics simulations of such models are restricted to a few thousand lipid molecules.
- **Coarse-Grained Membrane Models**—If the detailed chemical structure is not relevant but more generic properties of amphiphilic molecules are to be studied—such as the number of hydrocarbon tails per lipid, the chain length of the tails, or mixtures of two different amphiphiles—then a *coarse-grained description* can be used, in which several atoms are lumped into a single unit. These units are typically taken to be Lennard-Jones particles. In such a model, water becomes a Lennard-Jones fluid with attractive interactions, and amphiphilic molecules become short polymer chains with two kinds of monomers, with attractive or repulsive interactions with the solvent particles and the other monomers (den Otter and Briels, 2003; Goetz et al., 1999; Goetz and Lipowsky, 1998). The size of such a monomer is on the order of a few water molecules or CH_2 groups. Very similar models, with Lennard-Jones interactions replaced by linear "soft" potentials, have also been employed intensively in dissipative particle

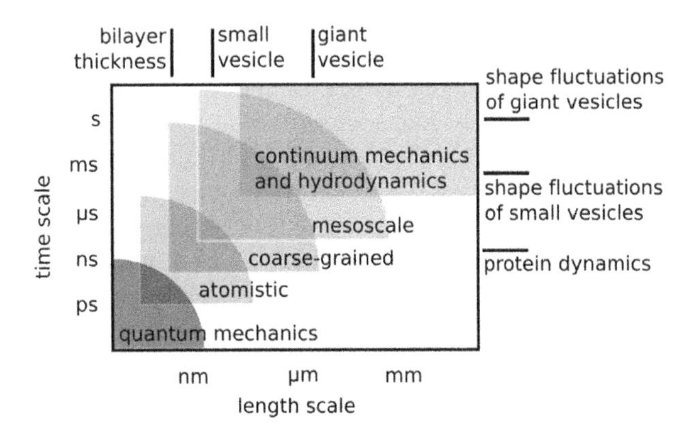

Figure 6.1 Characteristic time and length scales in amphiphile solutions. Physical phenomena occurring at the various scales are indicated. Different models and simulation techniques are required to capture the behavior at different scales. Their approximate ranges of validity are shown by the shaded regions.

Box 6.1 Membrane models on different length scales

MEMBRANE MODEL	CONSTITUENTS	MEMBRANE
Atomistic	Lipids, water molecules	
Coarse-grained	Coarse-grained lipids, water beads	
Solvent-free	Coarse-grained lipids	
Triangulated	Membrane vertices	
Meshless	Membrane beads	

Atomistic model (taken from avantilipids.com); simulation snapshot reprinted from Tieleman et al. (1997), with permission from Elsevier. Coarse-grained model; simulation snapshot reprinted from Boek et al. (2004), by permission of the Royal Society of Chemistry. Solvent-free bilayer model (Noguchi and Takasu, 2001a, 2001b); reprinted from Gompper and Noguchi (2006). Triangulated surface model (Gompper and Kroll, 1997, 2004); reprinted from Gompper and Noguchi (2006). Meshless membrane model; adapted with permission from Noguchi and Gompper (2006b). Copyrighted by the American Physical Society.

dynamics (DPD) simulations (Laradji and Sunil Kumar, 2004; Ortiz et al., 2005; Rekvig et al., 2004; Shillcock and Lipowsky, 2002; Venturoli et al., 2006).

- The coarse-grained modeling can be taken one step further by taking into account the different chemical nature (and electrical charge) of various head- and tail-groups (Marrink and Mark, 2003; Marrink et al., 2007). This allows progress from a more qualitative to a more quantitative description of membrane properties. Such models allow molecular dynamics simulations of few thousand lipids and make it possible to study the formation, structure, and dynamics of small phospholipid vesicles (Marrink et al., 2009; Marrink and Mark, 2003).
- **Solvent-Free Membrane Models**—The solvent in a coarse-grained model is required for two reasons. First, it is necessary to stabilize the bilayer structure due to the repulsion between the solvent and the amphiphile tails. Second, it mediates hydrodynamic interactions between different parts of the membrane. However, the simulation of the motion of solvent particles consumes a large fraction of the total simulation time. Therefore, *solvent-free membrane models* have been designed that work as well as the models with solvent when structural and thermodynamic properties are investigated. Additional interactions between amphiphiles have to be introduced in this case in order to mimic the hydrophobic interactions with the solvent (Brannigan and Brown, 2004; Cooke and Deserno, 2005; Cooke et al., 2005; Farago, 2003; Noguchi and Takasu, 2001a, 2001b). This approach is advantageous in the case of membranes in dilute solution because it reduces the number of molecules by orders of magnitude. However, the basic length scale is still on the order of magnitude of the size of the amphiphilic molecules.
- **Triangulated Surface Models**—The natural length scale of the previous two classes of membrane models is the size of the head group of a lipid molecule, that is, roughly 1 nm. This is far too small to describe phenomena on the scale of giant unilamellar vesicles (GUVs), capsules, and cells that have diameters on the order of 10 μm. In this case, a *continuum description* on the level of *elasticity theory* is required. In order to make such continuum models amenable to computer simulations, triangulated surfaces are often employed (Gompper and Kroll, 1997, 2004). The building blocks in such models correspond to membrane patches consisting of hundreds or thousands of lipid molecules. The main idea here is to connect membrane "nodes" (or "vertices") by a triangular network of bonds. The bond potentials are chosen so as to achieve a homogeneous distribution of vertices on the membrane. For fluid membranes, this requires a *dynamic triangulation* such that vertices can diffuse and flow within the membrane. For polymerized membranes, such as in capsules, a fixed connectivity represents the unbreakable bonding between neighboring molecules, and implies a shear elasticity of the membrane.
- **Meshless Membrane Models**—A different approach to discretize elasticity theory of a two-dimensional (2D) surface embedded in 3D space is to employ an ensemble of membrane nodes without connecting them to form a triangulated mesh. Meshless membrane models instead employ pairwise and multi-particle interactions to (i) achieve a roughly homogeneous density of nodes on the membrane and (ii) favor smoothly curved membrane conformations (Noguchi and Gompper, 2006b). The advantage of meshless membrane models is that open boundaries—which occur, for example, in membrane rupture—and topology changes—as in vesicle fusion—can be very easily simulated.

6.2 MEMBRANE MODELS AND SIMULATION TECHNIQUES

6.2.1 ATOMISTIC MEMBRANE MODELS

Simulations of molecular fluids based on classical force fields have made enormous progress in the last decades. The quality of the results depends crucially on the development of reliable force fields that allow a quantitative description of the collective behavior of many molecules. Force fields are often optimized for a special class of systems. For lipids, experimental reference data are employed that include specific structural properties of lipid bilayers, such as area per lipid, volume per lipid, bilayer thickness, order parameter for the lipid tail orientation, and headgroup hydration. Several such force fields have been developed and tested in recent years (Dickson et al., 2012, 2014; Lyubartsev and Rabinovich, 2011, 2016; Pastor and MacKerell, 2011; Skjevik et al., 2016). Although molecular dynamics (MD) simulations of such models are typically restricted to a few thousand lipid molecules, they have been successfully applied to study the spontaneous formation of a small dipalmitoylphosphatidylcholine (DPPC, see Appendix 1 of the book for structure and data on this lipid) vesicle in water (de Vries et al., 2004), and lipid vesicle fusion (Kasson et al., 2010; Knecht and Marrink, 2007).

Not only all-atomistic, but also some coarse-grained force fields retain chemical specificity. For lipid-bilayer membranes, a particularly popular coarse-grained force field is the MARTINI force field, which can be used to access larger system sizes and simulation times of several microseconds (Hsu et al., 2017; Wassenaar et al., 2015). Such a coarse-grained model can be obtained by transformation from an atomistic model or also directly, for example, using the CHARMM-GUI. After equilibration, the system can be transformed (back) to a relaxed atomistic model if required. In particular for complex multicomponent membranes, a chemically specific coarse-grained approach provides an efficient means for generating equilibrated atomistic models. However, the results on the coarse-grained level can also be interpreted directly without a subsequent atomistic simulation. For example, sorting of transmembrane helices into the liquid-disordered domains of phase-separated model membranes and their local lipidic environment has been studied using coarse-grained MARTINI simulations (Schäfer et al., 2011).

6.2.2 COARSE-GRAINED MEMBRANE MODELS

When the detailed chemical structure of the amphiphilic molecules is not important, a coarse-grained modeling is very useful, where groups of several atoms or molecules are described by only a single position vector. This is important, because

- It reduces the number of degrees of freedom and therefore allows the study of either a system over a longer time, or larger systems, or both.
- It emphasizes the universal aspects that are common to many different amphiphilic systems, independent of the detailed chemistry of a particular system.

In coarse-grained models, the solvent molecules are usually treated as spherical particles with attractive Lennard-Jones interactions,

$$U_{LJ}(r) = 4\varepsilon\left[\left(\frac{\sigma}{r}\right)^{12} - \left(\frac{\sigma}{r}\right)^{6}\right], \quad (6.1)$$

where σ is the (effective) hard-core radius. The amphiphilic molecules are modeled as short polymeric chains with head (H) and tail (T) particles, so that neighboring particles in the chain interact via the harmonic-spring potential

$$U_{chain}(r) = k_{chain}(r - \sigma)^2. \quad (6.2)$$

Different geometries of amphiphilic molecules are shown in Figure 6.2. Head particles mutually attract each other with the Lennard-Jones potential in Eq. (6.1), as well as head and solvent particles. Tail particles have a *repulsive* interaction with both the head and the solvent particles. This interaction can be conveniently described by a shifted and truncated Lennard-Jones potential

$$U_{LJ}(r) = \begin{cases} 4\in\left[\left(\frac{\sigma}{r}\right)^{12} - \left(\frac{\sigma}{r}\right)^{6}\right] + \in & \text{for } r < 2^{1/6}\sigma \\ 0 & \text{otherwise} \end{cases}$$

that has the advantage of being both continuous and differentiable at the cutoff $r = 2^{1/6}\sigma$. In this model, Newton's equation of motion for all particle positions can be solved by a MD simulation employing the velocity-Verlet algorithm (Allen and Tildesley, 1991).

An alternative approach to simulate coarse-grained membrane models is dissipative particle dynamics (DPD) (Laradji and Sunil Kumar, 2004; Shillcock and Lipowsky, 2002, 2005). An introduction to the DPD simulation technique can be found in Section 6.2.7. In this case, the Lennard-Jones interactions between two particle species i and j are replaced by the conservative forces

$$F_{ij}^{C} = a_{ij}\left(1 - r_{ij}/r_0\right)\hat{r}_{ij} \quad (6.3)$$

for interparticle distances, $r_{ij} < r_0$, and are zero otherwise. All conservative forces are taken to be *repulsive*. Water is slightly repelled from the amphiphile head and strongly repelled from the amphiphile tail, providing the hydrophobic interaction needed to form bilayers. The amphiphile head is hydrophilic and therefore strongly repelled from its tail. See Table 6.1 for an example of such interaction parameters (Table 6.1).

The coarse-grained membrane approach has been used to address a variety of questions recently, *inter alia* membrane self-assembly and structure (den Otter and Briels, 2003; Goetz et al., 1999; Goetz and Lipowsky, 1998; Illya et al., 2005; Ortiz et al., 2005; Shillcock and Lipowsky, 2002; Srinivas et al., 2004), the spectrum of thermal membrane fluctuations (den Otter and Briels, 2003; Goetz et al., 1999), phase diagrams of lipid bilayers (Kranenburg et al., 2003), pore formation in membranes (den Otter, 2005; Groot and Rabone, 2001; Tolpekina et al., 2004), domain-formation in multicomponent membranes (Laradji and Sunil Kumar, 2004), and membrane fusion (Li et al., 2005; Marrink and Mark, 2003; Müller et al., 2003; Shillcock and Lipowsky, 2005; Stevens et al., 2003).

6.2.3 SOLVENT-FREE MEMBRANE MODELS

Simulations of lipid membranes using MD require the calculation of the motion of a large number of water molecules in addition to the lipid molecules. To simulate a small patch of a membrane with an atomistic model, about 30 water molecules per lipid were found to be sufficient (Tieleman et al., 1997). However, much more water molecules are needed for simulations of vesicles because the formation of a vesicle (see Section 6.3.2) needs a large solvent volume to prevent membrane interactions through the periodic boundary conditions of the simulation box. Similarly, self-assembly of amphiphilic molecules in dilute solutions also requires a lot of water molecules.

In solvent-free models, the solvent is not taken into account explicitly. Instead, the hydrophobic effect is treated by an effective potential between amphiphilic molecules. This reduces the numerical cost of membrane simulations significantly. In particular, a solvent-free model is more efficient for simulations that require a large solvent space. A similar approach is also frequently used in simulations of protein folding (Feig and Brooks III, 2004).

The first solvent-free model was proposed by Drouffe et al. (1991). In this model, a lipid bilayer membrane consists of a single

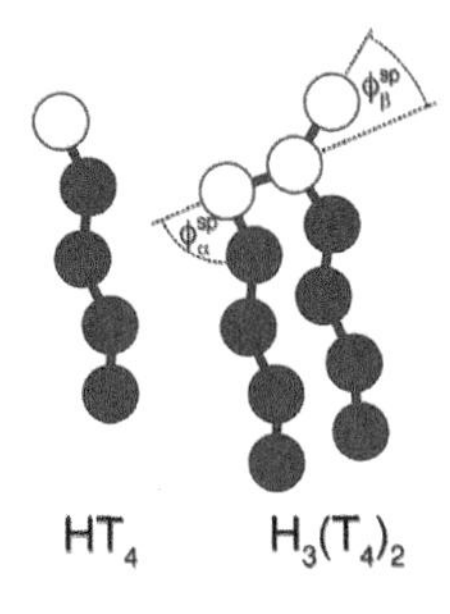

Figure 6.2 Typical amphiphilic molecules used in coarse-grained membrane models. The white particles (H) represent the head group, the blue particles (T) the tails. (Reprinted with permission from Goetz, R. and Lipowsky, R., *J. Chem. Phys.* 108: 7397–7409, 1998., Copyright 2009 by the American Physical Society.)

Table 6.1 **Two-particle conservative force parameters a_{ij} (in units of k_BT / r_0) and dissipative force parameters γ_{ij} (in units of $\sqrt{m_0 k_B T / r_0^2}$) for all particle pairs. k_BT is the thermal energy. Here, "H" denotes heads, "T" tails, and "W" solvent (water) particles**

INTERACTIONS	HH	TT	WW	HW	HT	TW
a_{ij}	25	25	25	35	50	75
γ_{ij}	4.5	4.5	4.5	4.5	9.0	20.0

Source: Shillcock, J.C. and Lipowsky, R. *J. Chem. Phys.*, 117, 5048, 2002.

layer of particles. The characteristic length scale is thus the same as for triangulated-membrane models and meshless membrane models discussed in Sections 6.2.4 and 6.2.5, respectively, both of which are indeed solvent-free models also. The particles of Drouffe et al. (1991) possess an orientational degree of freedom and interact with each other via three potentials: a soft-core repulsion, an anisotropic attraction, and a hydrophobic multibody interaction. The particles have been shown to self-assemble into membrane patches and vesicles. Recently, solvent-free models have also been developed to describe bilayer membranes, where the two monolayers are taken into account explicitly (Brannigan et al., 2006; Cooke and Deserno, 2005; Cooke et al., 2005; Farago, 2003; Noguchi and Takasu, 2001b). There are several variations of such bilayer models. An amphiphilic molecule is typically modeled as a rigid or flexible chain, which consists of one hydrophilic segment and two or three hydrophobic segments. The molecules interact with each other with pairwise (Brannigan et al., 2006; Cooke and Deserno, 2005; Cooke et al., 2005; Farago, 2003) or multibody (Drouffe et al., 1991; Noguchi and Takasu, 2001a, 2001b) potentials. One common feature is the requirement of an attractive potential between hydrophobic segments.

We introduce here one of the bilayer models (Noguchi and Takasu, 2001a, 2001b) in more detail. An amphiphilic molecule is modeled as one hydrophilic segment ($j = 1$) and two hydrophobic segments ($j = 2, 3$), which are separated by a fixed distance σ and are fixed on a line. Amphiphilic molecules ($i = 1, \ldots, N$) interact via a repulsive soft-core potential, U_{rep}, and an attractive "hydrophobic" potential, U_{hp}, so that the total interaction potential is given by

$$U_{am} = \sum_{i \neq i'} \sum_{j,j'} U_{rep}(|\mathbf{r}_{i,j} - \mathbf{r}_{i',j'}|) + \sum_{j=2,3} U_{hp}(\rho_{i,j}) \quad (6.4)$$

with

$$U_{rep}(r)/\varepsilon = \exp\{-20(r/\sigma - 1)\}. \quad (6.5)$$

Here, the prefactor 20 is chosen to obtain a very short interaction range of the repulsive interaction.

The multibody "hydrophobic" interaction is mimicked by a function of the local density of hydrophobic particles,

$$\rho_{i,j} = \sum_{i' \neq i} \sum_{j'=2,3} w_{\rho}(|\mathbf{r}_{i,j} - \mathbf{r}_{i',j'}|), \quad (6.6)$$

with the weight function $w_{\rho}(r) = 1/[\exp\{20(r/\sigma - 1.9)\} + 1]$. Thus, $\rho_{i,j}$ is the number of hydrophobic segments in a sphere with a radius of approximately 1.9σ. The multi-particle potential $U_{hp}(\rho)$ is then defined by

$$U_{hp}(\rho)/\varepsilon = \begin{cases} -0.5\rho & (\rho < \rho^* - 1) \\ 0.25(\rho - \rho^*)^2 - c & (\rho^* - 1 \leq \rho < \rho^*), \\ -c & (\rho^* \leq \rho) \end{cases} \quad (6.7)$$

where $c = 0.5\rho^* - 0.25$. The values $\rho^* = 10$ and 14 are used for $j = 2$ and 3 (the hydrophobic segments), respectively. At low density ($\rho < \rho^* - 1$), $U_{\mathrm{hp}}(\rho)$ acts as pairwise potential $-\varepsilon w_{\rho}(r)$. It is assumed that for $\rho > \rho^*$, the hydrophobic segments are shielded by hydrophilic segments from contact with solvent molecules and hydrophilic segments of other lipids. Thus, $U_{hp}(\rho)$ is constant at higher density ($\rho \geq \rho^*$). A similar "hydrophobic" potential is used in other solvent-free membrane and protein models. This multibody potential is employed in order to enhance the molecular diffusion in the membrane and to obtain a wide range of stability of a fluid phase.

This model is designed for simulations of fluid membranes, given that it has a wide temperature range where the fluid phase is stable, and a very low critical micelle concentration (CMC). The membrane properties can be varied easily by a modification of the model parameters and functional forms of the potentials. Other solvent-free models, with pair interactions only, have been used, for example, to study gel and crystalline phases (Brannigan et al., 2006; Cooke et al., 2005; Farago, 2003). However, the fluid phase can also be stabilized in models with pairwise interactions only by employing potentials with a broad attractive tail (Cooke and Deserno, 2005; Cooke et al., 2005). Thus, the solvent-free model can be adjusted depending on the type of physical problem under investigation. The use of density-dependent potentials seems to be advantageous in obtaining a wide parameter range where the membrane is fluid.

Solvent-free membrane models can be studied using Brownian dynamics and Monte Carlo simulations. In Brownian dynamics simulations, the motion of the jth segment of the ith molecule follows the underdamped Langevin equation,

$$m\frac{d^2\mathbf{r}_{i,j}}{dt^2} = -\zeta\frac{d\mathbf{r}_{i,j}}{dt} + g_{i,j}(t) - \frac{\partial U}{\partial \mathbf{r}_{i,j}}, \quad (6.8)$$

where m and ζ are the mass and the friction constant of the segments of molecules, respectively. $g_{i,j}(t)$ is a Gaussian white noise, which obeys the fluctuation–dissipation theorem

$$\langle g_{i,j}(t)\rangle = 0, \quad \langle g_{i,j}(t) g_{i',j'}(t')\rangle = 6k_B T\zeta\delta_{ii'}\delta_{jj'}\delta(t-t'). \quad (6.9)$$

Molecules in bilayers and vesicles diffuse laterally: the lateral diffusion constant is found to be $0.004\sigma^2/\tau_0$ at $k_B T/\varepsilon = 0.2$, where σ is the head-group diameter and $\tau_0 = \zeta\sigma^2/\varepsilon$ is a characteristic time scale (Noguchi and Takasu, 2001a, 2001b). The unit length σ corresponds to about 1 nm. The unit time, τ_0, can be estimated from the assumption that the lateral diffusion constant corresponds to that of phospholipids at 30°C, which is about 10^{-7} cm^2/s. This implies that τ_0 is about 1 ns.

Solvent-free bilayer models have been applied for a variety of studies, such as membrane fusion and fission (Noguchi and Takasu, 2001a, 2002), pore formation in membranes (Cooke and Deserno, 2005; Farago, 2003), the adhesion of a nanoparticle (Noguchi and Takasu, 2002), the fluid-gel phase transition (Brannigan et al., 2006; Cooke and Deserno, 2005), phase separation of lipids (Cooke et al., 2005), protein inclusions in membranes (Brannigan et al., 2006), and DNA-membrane complexes (Farago et al., 2006).

6.2.4 DYNAMICALLY TRIANGULATED SURFACES

The simulation of membranes and vesicles with characteristic sizes on the order of 100 nm–10 μm is impossible on the basis of a molecular model given that it would require an enormous number of lipid (and solvent) molecules. Therefore, on this level, a model is necessary in which the individual lipid molecules are no longer "visible". Instead, the membrane is described by a mathematical surface with an elastic energy that is most appropriate on these mesoscopic length scales (Gompper and Schick, 1994; Nelson et al., 2004; Safran, 1994). The shapes and fluctuations of the membrane are controlled by the curvature elastic energy (Canham, 1970; Helfrich, 1973) (see also Chapters 5 and 14)

$$H_{curv} = \int dS\left[\Sigma + 2\kappa(H - C_0)^2 + \bar{\kappa}K\right], \quad (6.10)$$

where the integral extends over the entire membrane surface. The shape of the membrane is expressed by

$$H = [c_1 + c_2]/2 \quad \text{and} \quad K = c_1c_2, \quad (6.11)$$

the mean and Gaussian curvature, respectively. Here, c_1 and c_2 are the two principal curvatures at each point of the membrane--the eigenvalues of the curvature tensor (Kreyszig, 1991). The parameters of the curvature energy are the membrane tension, Σ; the bending rigidity, κ; the saddle-splay modulus, $\bar{\kappa}$; and the spontaneous curvature C_0. These elastic constants of the membrane are the only place where the chemistry, the molecular architecture, and the interactions of the constituent lipid and protein molecules enter into this model.

In order to make this model suitable for simulations, the continuous surface has to be approximated by a network of vertices and bonds (Figure 6.3). A triangular network is usually used because it provides the most homogeneous and isotropic discretization of the surface (Gompper and Kroll, 1997). The simplest potential for the interaction of vertices that are connected by bonds is a *tethering potential*,

$$V(r) = \begin{cases} 0 & \text{if } r < l_0 \\ \infty & \text{otherwise} \end{cases} \quad (6.12)$$

which causes the particles to behave as tethered by a string. When hard spheres of diameter σ_0 are placed on the vertices and the bond lengths, ℓ_0, are restricted to be $\ell_0 \le \sqrt{3}\sigma_0$, the surface is *self-avoiding*, given that an arbitrary sphere does *not* fit through the holes of the network, so that no interpenetration of different parts of the network is possible.

The curvature energy can be discretized in different ways (Gompper and Kroll, 1996, 1997). The most commonly used form is (Kantor and Nelson, 1987a, 1987b)

$$E_b = \lambda_b \sum_{<ij>} (1 - \mathbf{n}_i \cdot \mathbf{n}_j) \quad (6.13)$$

where $\mathbf{n}_i$ and $\mathbf{n}_j$ are the normal vectors of neighboring triangles, and the sum runs over all pairs of neighboring triangles. The coupling constant, λ_b, in Eq. (6.13) is related to the bending

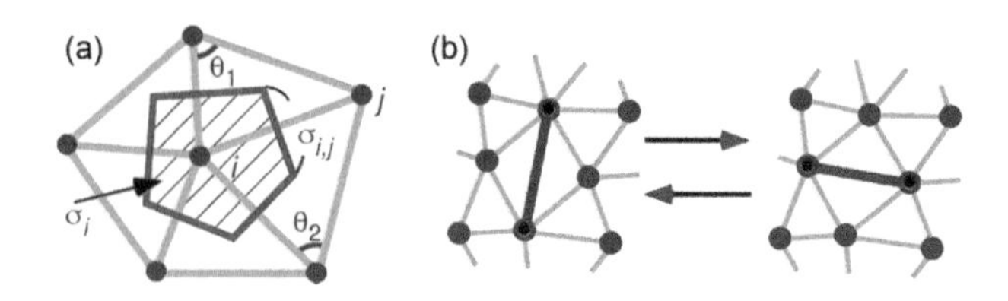

Figure 6.3 Geometry of triangulated surfaces. (a) Bond length and dual lattice, see text. (b) Bond flip in a dynamically triangulated membrane.

rigidity and saddle-splay modulus by $\kappa = \sqrt{3}\lambda_b/2$ and $\bar{\kappa} = -\kappa$, respectively (Gompper and Kroll, 1996; Seung and Nelson, 1988).

The discretization (6.13) of the curvature energy is not without problems, as discussed by Gompper and Kroll (1996). In particular, the discretization (Gompper and Kroll, 1996; Itzykson, 1986)

$$E_b = \frac{\kappa}{2}\sum_i \frac{1}{\sigma_i}\left\{\sum_{j(i)} \frac{\sigma_{i,j}\mathbf{r}_{i,j}}{r_{i,j}}\right\}^2 \quad (6.14)$$

has been found to give reliable results in comparison with the continuum expression (6.10). Here, the sum over $j(i)$ is over the neighbors of a vertex i that are connected by tethers. The bond vector between the vertices i and j is $\mathbf{r}_{i,j} = \mathbf{r}_i - \mathbf{r}_j$, and $r_{i,j} = |\mathbf{r}_{i,j}|$. The length of a bond in the dual lattice is $\sigma_{i,j} = r_{i,j}[\cot(\theta_1) + \cot(\theta_2)]/2$, where the angles θ_1 and θ_2 are opposite to bond ij in the two triangles sharing this bond, as illustrated in Figure 6.3. Finally, $\sigma_i = 0.25\sum_{j(i)}\sigma_{i,j}r_{i,j}$ is the area of the dual cell of vertex i.

Polymerized Membranes—Membranes in which neighboring particles are chemically linked together are called polymerized. Examples of such membranes are graphite monolayers, which are found in fullerenes, shells of artificial capsules obtained by cross-linking a polymer layer at the surface of a liquid droplet, and the polymer (protein) network attached at the inside of RBCs to the lipid membrane.

Triangulated surface models for polymerized membranes were first suggested and studied in the 1980s using Monte Carlo (Gompper and Kroll, 1991; Kantor and Nelson, 1987a, 1987b; Plischke and Boal, 1988; Vliegenthart and Gompper, 2006) and molecular dynamics (Abraham et al., 1989) simulations. Since then, the properties of triangulated surfaces of fixed triangulation have been investigated intensively (e.g., Gompper and Kroll 1997, 2004).

Fluid Membranes—For a study of fluid membranes, the connectivity of the membrane vertices cannot remain fixed during the simulation because otherwise a diffusion of vertices within the membrane is not possible. Therefore, dynamically triangulated surfaces (Boal and Rao, 1992; Ho and Baumgärtner, 1990; Kroll and Gompper, 1992) have to be used. The essential step of the dynamic triangulation procedure is shown in Figure 6.3b. Among the four vertices of two neighboring triangles, the "diagonal" bond is switched from one of the two possible positions to the other. This bond-switching is only allowed if the vertices remain connected to at least three neighbors after the switch. In addition, the distance between the newly connected vertices has to be smaller than the maximum bond length. This bond flip has the advantages (Gompper and Kroll, 1997, 2004):

- It is local, that is, only the vertices of two neighboring triangles are involved.
- It guarantees that the network retains its 2D connectivity during the whole simulation run.

Dynamically triangulated network models of fluid membranes have been applied in recent years to investigate a variety of systems and phenomena, such as phase separation and budding of two-component vesicles (Kohyama et al., 2003; Laradji and Sunil Kumar, 2004, 2005; Sunil Kumar et al., 2001), vesicles with membranes containing curvature-inducing nematogens and membrane tubulation (Ramakrishnan et al., 2013; Sreeja et al., 2015), defect scars on flexible vesicles with crystalline order (Kohyama and Gompper, 2007), the conformation of charged vesicles (Li et al., 2015), particle adhesion to vesicles (Bahrami et al., 2012; Šarić and Cacciuto, 2012a, 2012b, 2013), complex formation between a mixed fluid vesicle and a charged colloid (Fošnarič et al., 2009), vesicle adhesion to surfaces (Gruhn et al., 2007), sponge phases (Gompper and Kroll, 1998; Peltomäki et al., 2012), and vesicles in shear (Noguchi and Gompper, 2004, 2005a) and capillary flows (Noguchi and Gompper, 2005b; Noguchi et al., 2010).

Vesicles with the energy described in this subsection can be simulated using a Monte Carlo method, where one step consists of a random displacement of a randomly selected vertex. This step is accepted with the probability determined by the Boltzmann weight, determined by Eq. (6.13) or (6.14), In MD simulations, smooth bond potentials are usually employed, see Abraham et al. (1989).

6.2.5 MESHLESS MEMBRANE MODELS

The membrane conformation is described by the positions of N particles, which are the membrane "nodes". The particles either have no internal degrees of freedom (Noguchi and Gompper, 2006b) or can be characterized by an orientation vector (Drouffe et al., 1991). Models of membrane particles with orientation vector are very similar in spirit to the solvent-free models described in Section 6.2.3 above. Therefore, we focus here on meshless membrane models with particles without internal degrees of freedom, which can be understood as meshless discretization of the membrane shape and the curvature energy. These models are well suited to study, for example, vesicle dynamics accompanied by topological changes.

In the model of Noguchi and Gompper (2006b), the membrane particles interact with each other via potential

$$U = \varepsilon(U_{rep} + U_{att}) + k_\alpha U_\alpha, \tag{6.15}$$

which consists of a repulsive soft-core potential, U_{rep}, with a diameter, σ, an attractive potential, U_{att}, and a curvature potential, U_α. All three potentials only depend on the positions, r_i, of the particles. The curvature potential is based on the moving least-squares (MLS) method (Belytschko et al., 1996; Lancaster and Salkaskas, 1981). We briefly outline here the essential aspects of this simulation technique.

The MLS method is a least-squares fit of the membrane shape, weighted locally around each particle (Belytschko et al., 1996; Lancaster and Salkaskas, 1981; Noguchi and Gompper, 2006b). A Gaussian function is employed as a weight function (Noguchi and Gompper, 2006b)

$$w_{mls}(r_{i,j}) = \begin{cases} \exp\left(\dfrac{(r_{i,j}/\sigma)^2}{(r_{i,j}/r_{cc})^n - 1}\right) & (r_{i,j} < r_{cc}) \\ 0 & (r_{i,j} \geq r_{cc}) \end{cases} \tag{6.16}$$

where $r_{i,j}$ is the distance between particles i and j. This function is smoothly cut off at $r_{i,j} = r_{cc}$. Here, the parameters $n = 12$ and $r_{cc} = 3\sigma$ have been employed.

In the first-order MLS method, a plane is fitted locally to the particle positions by minimizing

$$\Lambda_1(r_i) = \frac{1}{w_0}\sum_j \{\mathbf{n}\cdot(\mathbf{r}_j - \mathbf{r}_0)^2\} w_{mls}(r_{i,j}), \tag{6.17}$$

where the sum is over all points (including i itself) and $w_0 \equiv \sum_j w_{mls}(r_{i,j})$ is a normalization factor. The normal vector $\mathbf{n}$ of the plane and the point $\mathbf{r}_0$ on the plane are fitting parameters. The minimum of Λ_1 is given by $\Lambda_1^{min} = \lambda_1$ when $\mathbf{r}_0$ is the weighted center of mass $\mathbf{r}_G = \sum_j \mathbf{r}_j\, w_{mls}(r_{i,j})/w_0$ and $\mathbf{n}$ is collinear with the eigenvector $\mathbf{u}_1$ of the lowest eigenvalue λ_1 of the weighted gyration tensor, $a_{\alpha\beta} = \sum_j (\alpha_j - \alpha_G)(\beta_j - \beta_G) w_{\text{mls}}(r_{i,j})$, where $\alpha, \beta = x, y, z$ and $\lambda_1 \leq \lambda_2 \leq \lambda_3$.

We now define the degree of deviation from a plane, the aplanarity, as

$$\alpha_{pl} = \frac{9D_w}{T_w M_w} = \frac{9\lambda_1\lambda_2\lambda_3}{(\lambda_1+\lambda_2+\lambda_3)(\lambda_1\lambda_2+\lambda_2\lambda_3+\lambda_3\lambda_1)}, \tag{6.18}$$

where D_w and T_w are determinant and trace of the weighted gyration tensor, respectively, and M_w is the sum of its three minors, $M_w = a_{xx}a_{yy} + a_{yy}a_{zz} + a_{zz}a_{xx} - a_{xy}^2 - a_{yz}^2 - a_{zx}^2$.

The aplanarity, α_{pl}, takes values in the interval $[0,1]$ and represents the degree of deviation from a plane. This quantity acts like λ_1 for $\lambda_1 \ll \lambda_2, \lambda_3$ because $\alpha_{pl} \simeq \lambda_1/(\lambda_2+\lambda_3)$ in this limit. Therefore, the curvature potential is defined as

$$U_\alpha = \sum_i \alpha_{pl}(r_i), \tag{6.19}$$

where $\alpha_{pl}(r_i) = 0$ when the ith particle has two or less particles within the cutoff distance $r_{i,j} < r_{cc}$. This potential increases with increasing deviation of the shape of the neighborhood of a particle from a plane, and favors the formation of quasi-2D membrane aggregates.

The particles interact with each other in the quasi-2D membrane surface via the potentials U_{rep} and U_{att}. These interaction potentials are necessary to obtain a homogeneous particle density in the membrane plane, and to avoid the membrane from rupturing and falling apart (Noguchi and Gompper, 2006b). The particles have an excluded-volume interaction via the repulsive potential

$$U_{rep} = \sum_{i<j} \exp\{-20(r_{i,j}/\sigma - 1) + B\} f_{cut}(r_{i,j}/\sigma), \tag{6.20}$$

with a cutoff function (Noguchi and Gompper, 2006b)

$$f_{cut}(s) = \begin{cases} \exp\left\{A\left(1+\dfrac{1}{(|s|/s_{cut})^n - 1}\right)\right\} & (s < s_{cut}) \\ 0 & (s \geq s_{cut}) \end{cases} \quad (6.21)$$

The factor A in Eq. (6.21) is determined such that $f_{cut}(s_{half}) = 0.5$. In Eq. (6.20), the parameters $n = 12$, $A = 1$, and $s_{cut} = 1.2$ were used by Noguchi and Gompper (2006b). The constant $B = 0.126$ has been introduced to satisfy $U_{rep}(\Sigma) = 1$.

The attractive interaction mimics the "hydrophobic" interaction. U_{att} is a potential of the local density of particles,

$$\rho_i = \sum_{j \neq i} f_{cut}(r_{i,j}/\sigma), \quad (6.22)$$

with the parameters $n = 12$ and $s_{cut} = s_{half} + 0.3$ in f_{cut} with s_{half} defined after Eq. (6.21). Here, ρ_i is the number of particles in a sphere whose radius is approximately $r_{att} = s_{half}\sigma$. The density-dependent attractive potential U_{att} is given by

$$U_{att} = \sum_i 0.25\ln\left[1+\exp\left\{-4(\rho_i - \rho^*)\right\}\right] - C \quad (6.23)$$

where $C = 0.25\ln\{1+\exp(4\rho^*)\}$. For $\rho_i < \rho^*$, the potential is approximately $U_{att} \simeq \rho_i s$ and therefore acts like a pair potential with $U_{att} \simeq \sum_{i<j} 2 f_{cut}(r_{i,j}/\sigma)$. For $\rho_i > \rho^*$, this function saturates to the constant $-C$. Thus, it is a pairwise potential with cutoff at densities higher than $\rho_i > \rho^*$. A convenient choice of parameters is $r_{att}/\sigma = 1.8$ and $\rho^* = 6$.

A detailed analysis of the fluctuation spectrum of a membrane (see Section 6.3.3 below) described by the meshless model reveals that the parameter k_α in Eq. (6.15) is proportional to the bending rigidity κ. Similarly, the analysis of pore opening in a membrane under tension shows that the line tension Γ of the membrane rim is proportional to the strength ε of the attraction potential in Eq. (6.15). Thus, two essential elastic moduli of the curvature Hamiltonian are well under control in the meshless membrane model (Noguchi and Gompper, 2006b).

A major advantage of meshless membrane models over dynamically triangulated surfaces is that topological changes are easily possible. However, because the models do not explicitly take into account solvent molecules, the volume of a vesicle cannot be kept constant. This is a disadvantage of this type of models. In addition, hydrodynamic interactions are not present. However, these interactions can be taken into account by combining a solvent-free model with a mesoscopic solvent technique, such as multi-particle collision dynamics (MPC) or DPD, as explained in Section 6.2.7.

6.2.6 CALCULATING SHAPES—FROM VESICLES TO CAPSULES AND RED BLOOD CELLS

Large vesicles can be simulated using a relatively small number of degrees of freedom employing meshless and triangulated membrane models. Although for dynamic simulations homogeneous discretizations are usually beneficial, very accurate results for energy minimization can be achieved by alternate refinement and minimization steps and by adapting the triangulation to the local membrane curvature (Berger and Colella, 1989). Therefore, membrane shapes with locally very different curvatures can be studied using energy minimization techniques that are challenging for dynamic simulation techniques, such as wrapping of ellipsoidal, cube-like, and rod-like nanoparticles (Dasgupta et al., 2013, 2014, 2017). Compared with other methods to calculate equilibrium shapes, such as numerical solution of shape equations (Kusumaatmaja et al., 2009; Seifert, 1997a; Smith et al., 2003) and phase-field models (Lowengrub et al., 2009), triangulation is a very versatile technique (Dasgupta et al., 2017; Gompper and Kroll, 2004). In particular for GUVs, the finite interface width that is used in phase-field models to describe the membrane requires large system sizes in order to obtain a realistic vesicle size compared with the membrane thickness. In experiments, a membrane thickness of 5 nm is negligible at the micrometer-length scale of the vesicle, which is well approximated by the mathematical surface used to model the membrane in triangulation techniques. Energy minimization using shape equations usually exploits the cylindrical symmetry of the system, which limits its range of applicability. Examples for systems that have been studied using these three techniques are shown in Figure 6.4.

Figure 6.5 illustrates the minimization process to obtain an oblate vesicle with reduced volume $v = 0.62$ modeled by a triangulated membrane using the freely available software package Surface Evolver (Brakke, 1992). The system is set up manually using only 8 vertices, 12 edges, and 6 faces in a cuboidal arrangement. An automatic initial refinement step adds additional vertices, edges, and facets (Figure 6.5a). A first energy minimization step using a steepest descent method deforms the shape of the initial structure (Figure 6.5b). In order to evolve the system toward the minimal energy state efficiently, it is important not to refine too much as long as the membrane shape is still far from the vesicle's equilibrium shape. After several refinement and minimization steps, energy and shape of the oblate vesicle are obtained (Figure 6.5c). Not only membrane mechanics, but also a volume and an area constraint for the vesicle can be taken into account. Here, the total membrane area is the sum of the areas of all triangles, whereas the volume can be calculated with the help of triple products of the position vectors of the membrane vertices and the two bond vectors that connect these vertices with neighboring membrane vertices. An example for a system that uses all these techniques is nanoparticle wrapping at vesicles (Yu et al., 2018).

Equilibrium shapes can also be calculated for more complex membranes than homogeneous lipid bilayers that are governed by bending rigidity only, such as for vesicles formed by a lipid bilayer membrane with a spontaneous curvature, e.g. induced by an area difference between the two monolayers that form the bilayer. Vesicle shapes have been shown to depend on both reduced volume of the cell and area difference (Seifert, 1997a). In the area–difference elasticity model, the preferred membrane curvature is taken into account by an additional energy contribution (Miao et al., 1994),

$$E_{ADE} = \frac{\pi k_{ADE}}{2Ah^2}(\Delta A - \Delta A_0)^2, \quad (6.24)$$

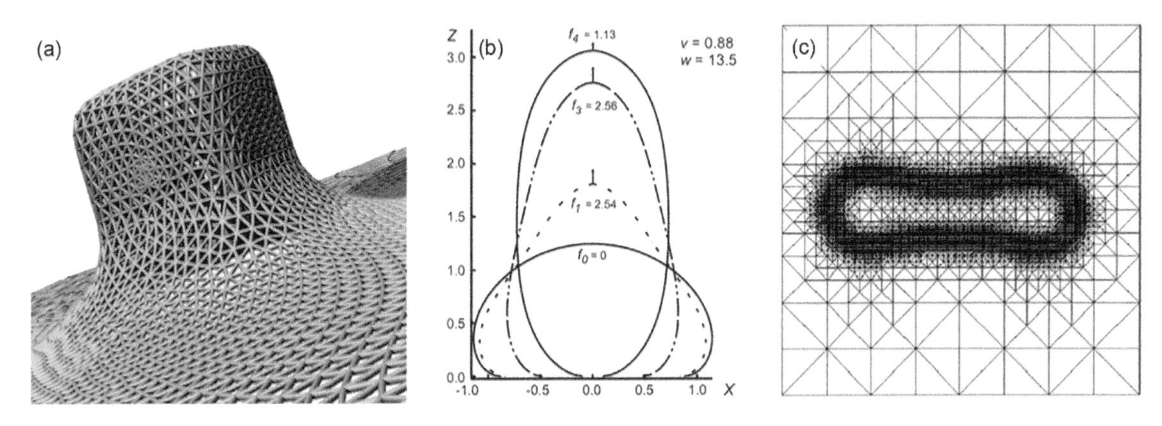

Figure 6.4 Equilibrium membrane shapes obtained using various energy minimization techniques. (a) A cube-like nanoparticle adhered with one face to a triangulated membrane. (Reprinted with permission from Dasgupta, S. et al., *Nano Lett.* 14, 687–693, 2014. Copyright 2014 American Chemical Society.) (b) Shapes of a vesicle with reduced volume v = 0.88 adhered to a planar substrate with adhesion strength w = 13.5 for various pulling forces, calculated using shape equations. (Reprinted with permission from Smith, A.-S. et al., *Europhys. Lett.*, 64, 281, 2003.) (c) Adaptively refined mesh for a vesicle shape calculation using a phase field model; the membrane shape can be extracted using isocontours. (Reprinted with permission from Lowengrub, J. S. et al., *Phys. Rev. E.*, 79, 031926, 2009. Copyright 2009 by the American Physical Society.)

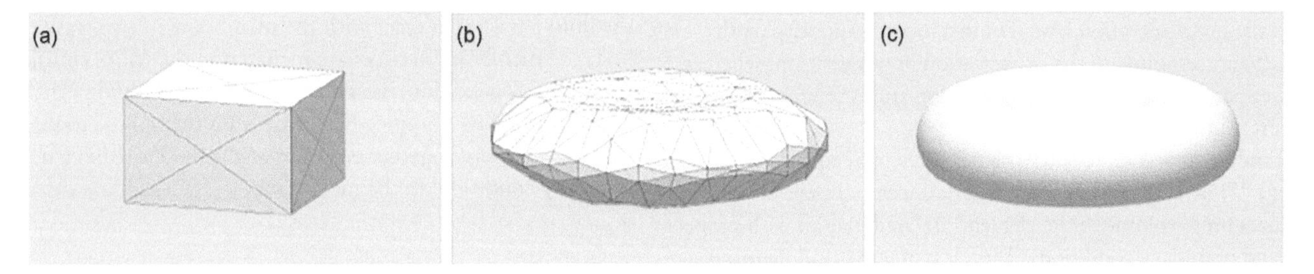

Figure 6.5 Vesicle shape determination via energy minimization using triangulated membranes. Starting from a cuboid (a), a first minimization step evolves the shape toward an oblate vesicle (b). The final shape of the oblate vesicle with reduced volume v = 0.62 is shown in (c). The calculations have been performed using Surface Evolver software.

where k_{ADE} is a constant on the order 1; A is the total membrane area of the vesicle; h is the thickness of the lipid bilayer; ΔA is the actual area difference between the two monolayers of the bilayer membrane; and ΔA_0 is the target area difference. Seifert (1997a) reviews calculations of vesicle shapes using shape equations in detail. Although the shapes of cylindrically symmetric vesicles, for which the membrane mechanics is governed by bending rigidity and spontaneous curvature only, are often calculated using shape equations, shapes of polymerized membranes and defect structures for crystalline membranes (Kohyama and Gompper, 2007; Seung and Nelson, 1988) can be obtained by using triangulated membranes with a shear modulus. The complex membrane of RBCs consists of a lipid bilayer supported by a network of entropic springs, the spectrin cytoskeleton (Auth et al., 2007a; Hansen et al., 1997; Ursitti and Wade, 1993; Zeman et al., 1990), see also subsections 6.3.7 and 6.3.8. It can be modeled using a fluid membrane with bending rigidity next to a polymerized membrane with shear modulus (Auth et al., 2007b). Using a membrane model that takes into account bending rigidity of the lipid bilayer membrane, shear elasticity of the cytoskeleton, and an area–difference elasticity, a large variety of experimentally observed shapes of RBCs can be reproduced (Lim et al., 2002).

In particular, spiculated echinocytic shapes of RBCs can only be obtained if the membrane is modeled with both bending and shear modulus (Figure 6.6a) (see Lim et al., 2002). For RBCs the ratio of shear to bending modulus is characterized by the Föppl-von Kármán number, $\Gamma = YD^2/\kappa = 2662$, with Young's modulus Y, bending rigidity κ, and diameter $D = \sqrt{A_{RBC}/\pi}$, where A_{RBC} is the surface area of the RBC (Fedosov et al., 2014b). Because of their thin spectrin cytoskeleton, for RBCs $Y \approx 4\mu$, where μ is the shear modulus of the membrane. An even stronger dominance of the shear elasticity is seen for capsules. Here, the shape is not only determined by its minimal energy, but also by the pathway and dynamics how the deformation has been achieved (Vliegenthart and Gompper, 2011). Figure 6.6b,c shows capsules that have

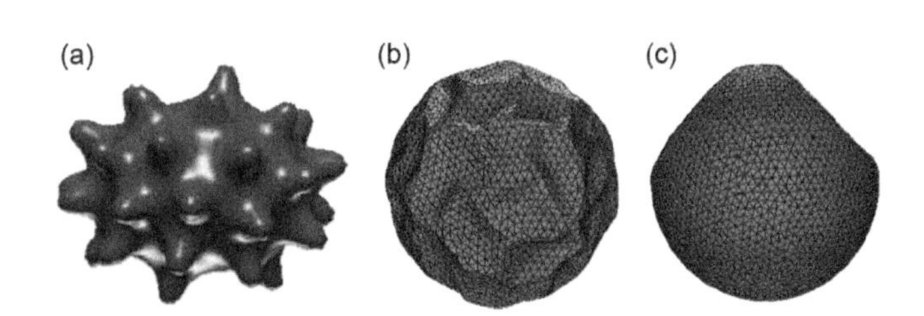

Figure 6.6 Shapes for membranes with shear modulus: (a) Echinocytic shape of an RBC obtained using energy minimization. (Reprinted with permission from Lim, G. et al., *Proc. Natl. Acad. Sci. USA*, 99, 16766–16769, 2002. Copyright 2002 National Academy of Sciences.) Shapes of an initially spherical shell that have been deflated by the same volume (b) quickly and (c) slowly obtained using Monte Carlo simulations. (Adapted from Vliegenthart, G.A. and Gompper, G. *New J. Phys.* 13, 045020, 2011. Copyright 2011, Institute of Physics.)

been compressed fast or slow using Monte Carlo simulations with the same volume change. Only the shape obtained with slow compression is close to a minimal-energy shape of the vesicle.

6.2.7 MODELING HYDRODYNAMICS

Vesicles, as well as cells, are typically studied in an aqueous environment. To describe their dynamics and the behavior under flow, hydrodynamics and hydrodynamic interactions have to be taken into account. Modeling fluid flow of a Newtonian solvent is often performed using the Navier-Stokes equation or its modifications (Wendt, 2009). For an incompressible fluid, the Navier-Stokes equation is given by

$$\begin{aligned} \frac{\partial \mathbf{u}}{\partial t} + (\mathbf{u}\cdot\nabla)\mathbf{u} &= -\frac{1}{\rho}\nabla p + \nu\nabla^2\mathbf{u}, \\ \nabla\cdot\mathbf{u} &= 0, \end{aligned} \tag{6.25}$$

where $\mathbf{u}$ is the local fluid velocity; ρ is the density; p is the pressure; and ν is the kinematic viscosity. These equations are derived using conservation laws. For instance, the upper part in Eq. (6.25) corresponds to the conservation of momentum, whereas the lower part represents mass conservation and is referred to as an incompressibility condition.

A standard approach to solve partial differential equations, such as Eq. (6.25) is to use various discretization techniques (e.g., finite difference, finite volume, finite element) in combination with proper initial and boundary conditions. This class of numerical methods is often referred to as computational fluid dynamics and represents well-established numerical techniques. However, in continuum approaches the inclusion of features present at the micro- and meso-scale (e.g., thermal fluctuations) is a nontrivial task.

Another class of efficient numerical approaches for modeling fluid dynamics includes particle-based Lagrangian methods, such as MD (Allen and Tildesley, 1991), DPD (Español and Warren, 1995; Hoogerbrugge and Koelman, 1992), MPC (Gompper et al., 2009; Malevanets and Kapral, 1999), and smoothed particle hydrodynamics (SPH) (Lucy, 1977; Monaghan, 2005). Microscopic modeling at the atomistic scale is often performed using MD, whereas the other aforementioned methods correspond to mesoscopic approaches. In these approaches, a fluid is represented by a number of particles that interact with each other through specified forces (in MD, DPD, and SPH) or collisions (in MPC). Through the conservation of local and global quantities, such as mass and momentum, all these methods provide proper hydrodynamic interactions at large enough length scales. Even though particle-based approaches are generally more expensive computationally than continuum techniques, they often allow a rather straightforward incorporation of desired micro- and mesoscopic features. This advantage often favors the use of particle-based methods in modeling complex fluids at the micro- and meso-scale over conventional computational fluid dynamics.

Due to the importance of particle-based approaches for simulations of the (hydro)dynamics of vesicles, we briefly describe the basic algorithms of two hydrodynamics techniques (i.e., DPD and MPC) in a little more detail. In DPD, the conservative forces between different particles i and j are assumed to be of the form

$$\mathbf{F}_{ij}^{C} = a_{ij}\left(1 - r_{ij}/r_0\right)\hat{\mathbf{r}}_{ij} \tag{6.26}$$

for distances $r_{ij} < r_0$ and zero otherwise. This guarantees that potentials are smooth, and relatively large time steps can be used in the integration of the equations of motion. Similarly, the dissipative friction forces are taken to be

$$\mathbf{F}_{ij}^{D} = \gamma_{ij}\left(1 - r_{ij}/r_0\right)^2\left(\mathbf{r}_{ij}\cdot\mathbf{v}_{ij}\right)\hat{\mathbf{r}}_{ij} \tag{6.27}$$

for distances $r_{ij} < r_0$ and zero otherwise, where $\mathbf{v}_{ij} = \mathbf{v}_i - \mathbf{v}_j$ is their relative velocity. Finally, there are thermal random forces that follow from the fluctuation–dissipation theorem (Español and Warren, 1995).

In MPC, the fluid consists of point particles that have *no* conservative interactions. The dynamics proceeds in two alternating steps. In the streaming step, particles move ballistically for a time interval δt. Then, all particles are sorted into the cells of a cubic lattice, which defines the collision environment. All fluid particles within one collision cell exchange momentum, for example, by a random exchange of momentum increments, such that the total momentum of each cell is conserved. The fluid particles interact with the membrane in two ways. First, the membrane vertices are included in the MPC collision procedure. Second, for triangulated surfaces, the fluid particles are scattered with a bounce-back rule from membrane triangles. These interactions together ensure that the fluid satisfies a no-slip boundary condition on the membrane.

6.3 APPLICATIONS

6.3.1 SELF-ASSEMBLY OF MICELLES AND BILAYERS

Amphiphilic molecules in aqueous solution self-assemble into a large variety of different structures (Gelbart et al., 1994; Gompper and Schick, 1994). The type of structures found depends very much on the amphiphile concentration, but also on the amphiphile architecture and environmental conditions, such as temperature and salt concentration.

At very small amphiphile concentrations, the amphiphiles are molecularly dispersed, because the translational entropy dominates over any interaction energy. Only when a minimal concentration—the *critical micelle concentration* (CMC) is exceeded, the amphiphiles aggregate into small droplets called *micelles*, in which the hydrocarbon tails are shielded from water contact by a layer of head groups. The typical size of a spherical micelle is, therefore, determined by the length of the amphiphilic molecules. In some systems, when the size of the head group is larger than the tail, micelles can grow into long cylindrical rods that are called cylindrical micelles. On the other hand, when the heads and tails of the amphiphiles have roughly the same size, micelles can grow into 2D bilayer patches. This can happen at still small amphiphile concentrations (above the CMC). In this case, the patch does not grow indefinitely in the lateral directions because the rim of the patch is energetically less favorable than the interior. This can be understood as a line tension of the rim. Because the rim energy grows linearly with the radius of the patch, at some point the flat bilayer becomes less favorable than a closed membrane shape or a vesicle, see Section 6.3.2 below. In contrast to micelles, vesicles can be much larger than the length of an amphiphile. At considerably higher amphiphile concentrations, micelles, bilayers, or vesicles can pack together to form 3D order phases, such as cubic micellar crystals, or lamellar phases in which bilayers form a stack in one direction.

Many aspects of this self-assembly process have been studied by simulations. For example, the formation of a bilayer from an initially random mixture of amphiphiles and water, as obtained from MD simulations of the coarse-grained Lennard-Jones model introduced in Section 6.2.2, is demonstrated in Figure 6.7. It shows the formation of a transient cylindrical micelle structure, which transforms after some time into a stable bilayer state. Note that due to the finite box size, the amphiphile concentration is rather large, so that this bilayer should be considered as a part of a lamellar phase.

6.3.2 VESICLE FORMATION

Amphiphilic molecules spontaneously form vesicles when the size of head and tail is approximately equal to favor bilayer formation, and the energy associated with the line tension of the membrane edge (where hydrophobic chains are exposed to water) exceeds the curvature energy of a spherical vesicle. This process can be studied with the solvent-free model described in Section 6.2.3. For an initially random spatial distribution of amphiphilic molecules, they first aggregate into small clusters, which have spherical or ellipsoidal shape, similarly as discussed for the coarse-grained membrane model in Section 6.3.1. These clusters then assemble into larger clusters and bilayer patches, which finally close into vesicles (Noguchi and Takasu, 2001b).

The transition from bilayer patches to vesicles is shown in Figure 6.8. The membrane first undulates due to thermal fluctuations, then deforms into a bowl-shaped conformation to reduce the length of the membrane edge, and finally closes into a spherical vesicle (Figure 6.8). The closed-bilayer vesicle is the equilibrium state under these conditions.

An interesting feature of solvent-free models is that they can easily be combined with a hydrodynamic simulation techniques for *structureless* solvents, that is, fluids that have no hydrophilic or hydrophobic interactions with the membrane particles, just a frictional interaction for their relative motion, as discussed in Sections 6.2.3 and 6.2.7. For example, a combination of the

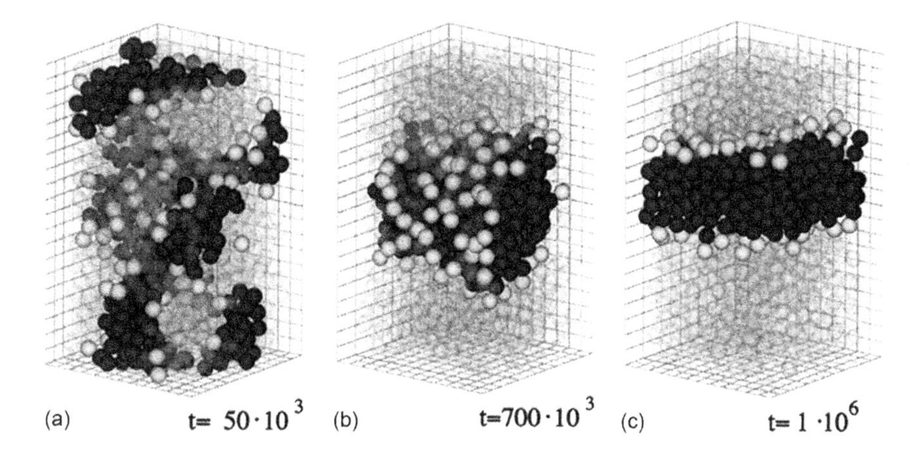

Figure 6.7 Self-assembly of a bilayer membrane in a mixture of HT_4 amphiphiles and solvent particles. The solvent particles are nearly transparent. The initial configuration, which is not shown, consists of a random mixture of 100 amphiphiles and 840 solvent particles. The configurations are snapshots that illustrate the time evolution of the structure starting with an early stage (a). After about 10^5 MD time steps, the amphiphiles form a cylindrical micelle (b), which spans the simulation box horizontally. This state is metastable for some time, before it transforms into a stable bilayer structure (c). (Reprinted with permission from Goetz, R. et al., *Phys. Rev. Lett.* 82, 221–224, 1999. Copyright 1999 by the American Physical Society.)

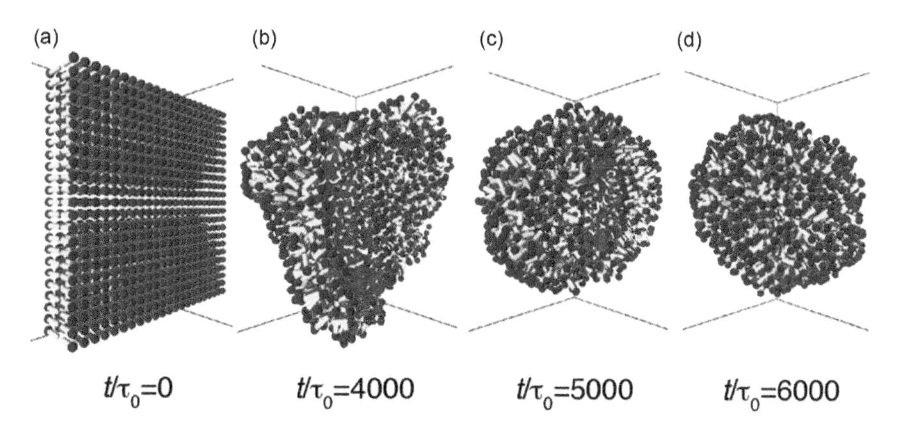

Figure 6.8 Snapshots of vesicle formation from (a) an initially planar membrane patch via (b, c) bowl-shaped structures, to (d) a complete vesicle, for $k_BT / \varepsilon = 0.5$ and $N = 1000$. Time is measured in units of $\tau_0 = \zeta\sigma^2 / \varepsilon$, where the mass $m = 1$ and the friction constant $\zeta = 1$ of segments are fixed, and ε is the strength of repulsive and hydrophobic interactions, see Eqs. 6.5 and 6.7. Red spheres and yellow cylinders represent the hydrophilic and hydrophobic segments of amphiphilic molecules, respectively. Vesicle closure is always observed for $k_BT / \varepsilon \leq 0.9$. (Reprinted from Gompper, G. and Noguchi, H. Coarse-grained and continuum models of membranes, *in* S. Blügel. et al. (Eds.), *Computational Condensed Matter Physics*, Vol. 32 of *Matter and Materials*, Forschungszentrum Jülich GmbH, Jülich, Germany, B.9, 2006.)

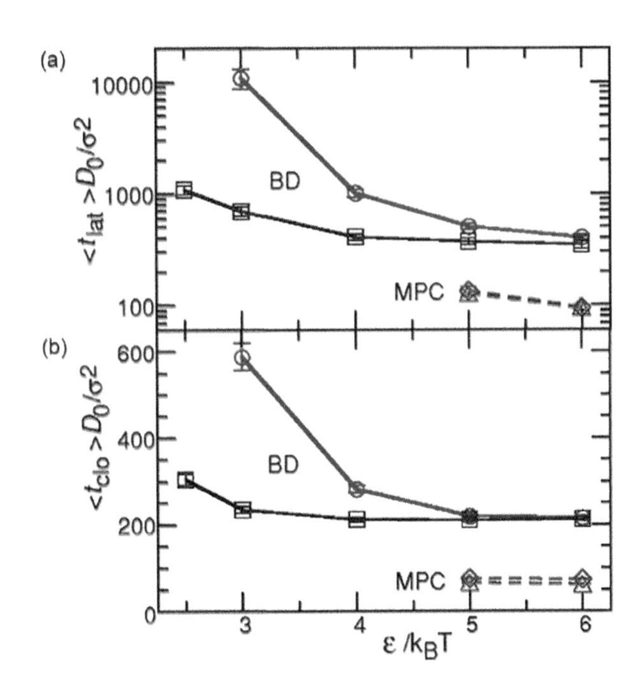

Figure 6.9 (a) Latency time, t_{lat}, and (b) closure time, t_{clo}, for a membrane of size N = 1000. The full lines with symbols represent data of BD simulations for $k_\alpha/k_BT = 5$ (squares) and $k_\alpha/k_BT = 10$ (circles). The dashed lines with symbols represent data of MPC simulations for $k_\alpha/k_BT = 10$ with higher (triangles) and lower (diamonds) solvent viscosity. (Reprinted from Noguchi, H. and Gompper, G. (2006a). Dynamics of vesicle self-assembly and dissolution, *J. Chem. Phys.*, 125, 164908, 2006a by the American Institute of Physics.)

meshless membrane model described in Section 6.2.5 and MPC fluid has been employed to study the (hydro)dynamics of vesicle formation (Noguchi and Gompper, 2006a).

Figure 6.9 shows both the latency time, t_{lat}, and the closure time, t_{clo}. Here, t_{lat} is defined as the last time before closure when the radius of gyration, R_g, crosses a threshold of about 90% of its value for the planar membrane disc; thus, the latency time is the time interval the membrane fluctuates thermally around the initial planar state before it starts its essentially deterministic path to a closed vesicle. The closure time, t_{clo}, is the subsequent time interval until closure is completed, that is, until R_g reaches its equilibrium value. For small line tension, $\Gamma \sim \varepsilon$ (compare Section 6.2.5), the latency time is much larger than the closure time, but the two times become comparable for large ε. Furthermore, both times level off at large ε. For the closure time, t_{clo}, the reason for this behavior is the simultaneous increase of line tension and membrane viscosity, which both depend approximately linearly on ε. Both t_{lat} and t_{clo} increase with increasing bending rigidity $\kappa \sim k_\alpha$ (compare Section 6.2.5) given that the resistance to curvature increases.

Figure 6.9 also compares Brownian dynamics and mesoscale hydrodynamics simulations. Hydrodynamics has two main effects. The first effect is that the initial stages of membrane closure are sped up because the characteristic time scale, $\tau \sim \eta R^3/\kappa$, of membrane fluctuations is shorter, and, as the membrane starts to bend into a bowl shape, the embedding fluid is set into motion, which is faster than the diffusive Brownian process. This results in shorter time scales t_{lat} and t_{clo}. The second effect is that as membrane closure is nearly complete, there is still some excess fluid volume inside the vesicle, which has to flow out through an increasingly narrow pore. This effect is difficult to see in the simulations of Noguchi and Gompper (2006a) because its observation depends on a very careful investigation of the final stage of vesicle closure. However, an effect of membrane viscosity is visible, which also causes a slowing down in the final stage of closure (Noguchi and Gompper, 2006a).

6.3.3 THERMAL MEMBRANE FLUCTUATIONS

Thermal fluctuations of the lipid molecules in a membrane lead to two types of thermal excitations (Figure 6.10b). On short-length scales, the lipid molecules are not perfectly aligned and do not have their heads all in the same plane but, rather, there are small vertical displacements between neighbors. These thermal motions are called *protrusion modes.* On length scales much larger than the bilayer thickness, there is a collective excitation where the whole membrane displays a wave-like deformation, which is called an *undulation mode.* The amplitudes are accessible experimentally, for example, by scattering techniques.

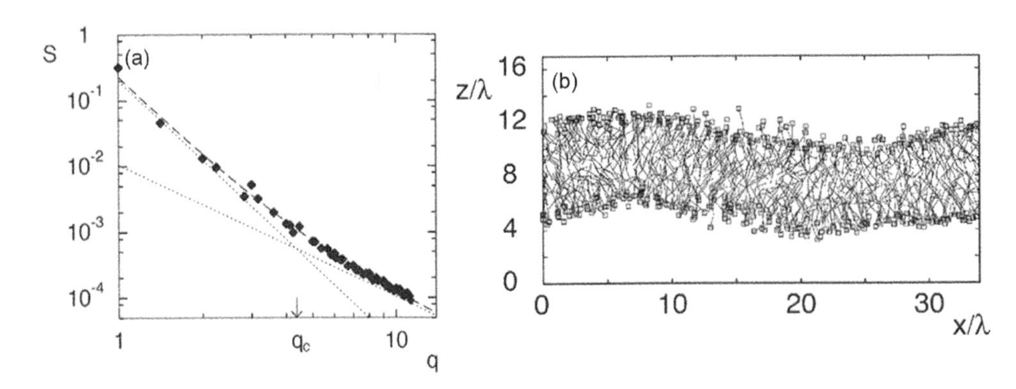

Figure 6.10 Fluctuation modes of thermally exited membrane deformations. (a) Fluctuation spectrum $S = \langle|h(\mathbf{q})|^2\rangle$ as function of the dimensionless wave number q. The largest wave number is determined by the box size and corresponds to $q = 1$. The dotted lines show the expected power-law behavior due to undulations (small q) and protrusions (large q), respectively. (b) Typical configuration of a bilayer membrane composed of 1,152 HT_4 amphiphiles. At small scales, individual molecules protrude from the bilayer. At large scales, the bilayer looks like an elastic, smoothly curved sheet. The basic length scale, λ, represents the range of the Lennard-Jones potential. (Reprinted with permission from Goetz, R. et al., *Phys. Rev. Lett.*, 82, 221–224, 1999. Copyright 1999 by the American Physical Society.)

In order to determine the spectrum of fluctuation modes of a membrane in simulations, a scalar height variable, $h(\mathbf{r})$, is introduced, which measures the deviation of the *local* position of the amphiphile head from a planar reference state (Monge parametrization). The theory of membrane fluctuations, for example, is well developed, as described and summarized in detail in Chapter 14. We mention here only those aspects of the theory that are relevant for the analysis of the simulation data.

In the Monge representation, the fluctuation spectrum is obtained from the correlation function

$$S(q) \equiv \langle | h(q) |^2 \rangle, \tag{6.28}$$

where

$$h(q) = \frac{1}{N} \sum_{i=1} h(r_i) \exp(iq \cdot r_i), \tag{6.29}$$

is the 2D Fourier-transform of the height-field, $h(r)$, with N being the number of amphiphiles.

Simulation results based on the coarse-grained Lennard-Jones model introduced in Section 6.2.2 are shown in Figure 6.10a. For small wave numbers q, the spectrum shows a q^{-4} decay, which is characteristic for surfaces that are governed by the *curvature elasticity*. The amplitude of this power law is the (inverse) bending rigidity, which can thereby be extracted from the simulations. This behavior should be compared with the spectrum of a interfaces governed by the surface tension (as the air–water interface), where the spectrum decays as q^{-2} for small wave numbers. The spectrum for large wave numbers, on the other hand, follows a q^{-2} power law. It is no coincidence that this is the same power law as for interfaces with interface tension given that the energy of the protrusion modes is proportional to the hydrophobic area exposed to the water when the amphiphiles "stick their head out" of the bilayer.

Taking measurements of the undulation spectrum of quasi-spherical vesicles is one of the standard experimental approaches to determine the bending rigidity of bilayers. In this case, the radial membrane displacements (from the center of the vesicle) are expanded in spherical harmonics, $Y_{lm}(\Omega)$, as

$$r(\Omega) = R_{A2}\left[1 + \sum_{l=0}^{l_M} \sum_{m=-l}^{l} u_{lm} Y_{lm}(\Omega)\right] \tag{6.30}$$

at the solid angle $\Omega = (\theta, \varphi)$. The spectrum of undulation modes is predicted to be

$$\langle | u_{lm} |^2 \rangle = \frac{k_B T}{\kappa} \frac{1}{(l+2)(l-1)[l(l+1)+Q]} \tag{6.31}$$

where $Q = 2(C_0 R_A)^2 - 4C_0 R_A + \Sigma R_A^2/\kappa$, with spontaneous curvature C_0. This result implies:

- The spectrum is governed by the bending rigidity for large l, and decays like $\kappa^{-1} l^{-4}$.
- The spectrum is governed by the "membrane tension" for small l, and decays like $\Sigma^{-1} l^{-2}$.
- The spontaneous curvature, C_0, cannot be measured in this approach because it only appears in combination with the membrane tension.

However, spontaneous curvature plays the key role in determining the morphology of biomembranes, lipid vesicles, and polymersomes; it is crucial for maintaining the spatial organization of, and traffic between, cellular organelles and the plasma membrane; and finally, it is believed that it controls the functional state of certain integral membrane proteins and membrane fusion competence. Therefore, it is very important to have a simple, straightforward procedure for the direct determination of the spontaneous curvature and bending modulus. Flicker spectroscopy of *nonspherical* vesicles avoids the shortcomings of analysis of undulations of quasi-spherical vesicles discussed above. By utilizing results of Monte Carlo simulations of dynamically triangulated vesicles, as introduced in Section 6.2.4, for a wide range of reduced volumes and spontaneous curvatures, the elastic parameters of the membrane can be extracted from experimental flicker spectroscopy data (Döbereiner et al., 2003).

In experiments, fluctuating prolate vesicles are stabilized by gravity—due to a small density difference of the solvent inside and outside the vesicle—at the bottom of a microchamber. The focal plane of a microscope is adjusted to include the long axis of the vesicle, and shape contours are recorded (Döbereiner et al., 1997) (Figure 6.11a). Choosing a coordinate system in which the x coordinate lies along the long axis of the vesicle, the 2D contours are then represented in polar coordinates as

$$r(\varphi) = r_0\left[1 + \sum_n (a_n \cos(n\varphi) + b_n \sin(n\varphi))\right]. \tag{6.32}$$

The mean values $\langle a_n \rangle$ describe the mean vesicle shape for oriented contours, $\langle b_n \rangle = 0$. The mean-square amplitudes $\langle (a_n - \langle a_n \rangle)^2 \rangle$ measure the thermal fluctuations of the vesicles about their mean shape.

The simulated vesicles are analyzed in the same way as the vesicles in experiments (Figure 6.11b). With increasing $\bar{c}_0 = C_0 R_A$, where $R_A = (A/4\pi)^{1/2}$ is the effective vesicle radius related to the membrane area, A, a transition from an oblate to a prolate shape is observed, which leads to a pronounced increase of $\langle a_2 \rangle$ and $\langle a_4 \rangle$. The oblate-to-prolate transition is reflected in a sharp peak of the fluctuations in a_2 (Figure 6.11c). By fitting the experimental data for the averages and variances of a_2, a_3, a_4, and a_5 to the simulation results, the parameters κ, $\bar{c}_0$, and v can be extracted simultaneously for a single vesicle (Döbereiner et al., 2003).

This method has been employed to measure electrostatically induced spontaneous curvature of 1-stearoyl-2-oleoyl-*sn*-glycero-3-phosphocholine (SOPC) vesicles, see Appendix 1 of the book for structure and data on this lipid. It has been suggested that a change in pH induces membrane curvature via the association of hydroxyl ions with the trimethyl-ammonium group of the phosphatidylcholine molecule (Lee et al., 1999). The results of the flicker analysis, as described above, are shown in Figure 6.11d.

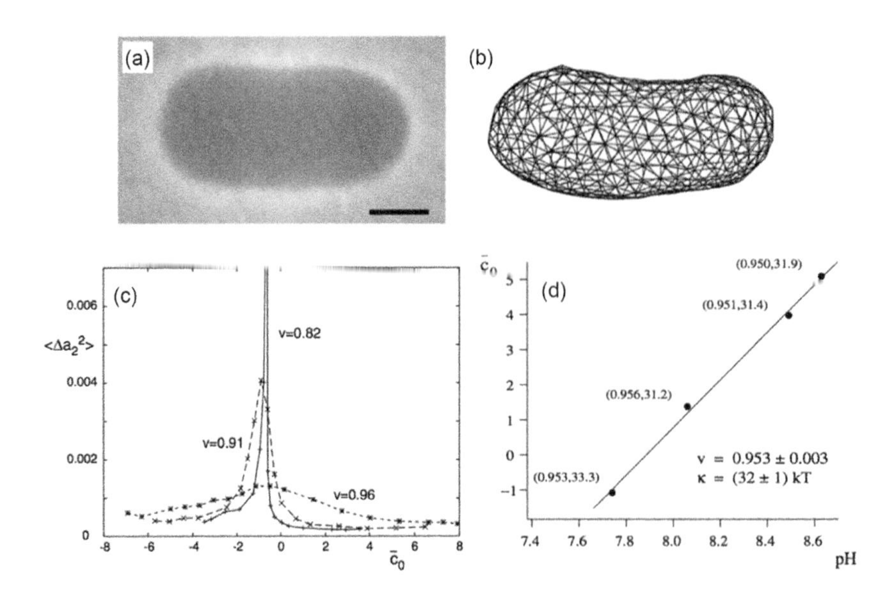

Figure 6.11 (a) Phase-contrast micrograph of a vesicle ($v = 0.828$) sedimented on a glass substrate. Scale bar: 5μm. (b) Simulation snapshot ($v = 0.825$, $\bar{c}_0 = -0.28$). (c) Simulated mean-square amplitude $\langle \Delta a_2^2 \rangle$ of shape fluctuations as a function of the effective spontaneous curvature $\bar{c}_0$. Note the peak at the prolate-to-oblate transition. Three different values of the reduced volume v are shown, as indicated. The other parameters in the simulations are $\kappa / k_BT = 25$, $k_{ADE} = 0.9$, and gravitational parameter $g = g_0 \Delta\rho R_S^4 / \kappa = 0.37$, where g_0 is the gravitational acceleration and $\Delta\rho$ is the excess mass density of the interior fluid. (d) Spontaneous curvature $\bar{c}_0 = C_0 R_A$ for $g = 0.8$. Note that the reduced volume and bending modulus, which are given in parentheses, (v, κ), remain constant. v, c_0, and κ are obtained simultaneously via comparison of the experimental data to the Monte Carlo simulations. (Reprinted with permission from Döbereiner, H.G. et al., *Phys. Rev. Lett.*, 91, 048301, 2003.Copyright 2003 by the American Physical Society.)

A strong change in spontaneous curvature at a constant bending modulus $\kappa = 32 k_B T$ is obtained. Indeed, the bending modulus of SOPC should not change considerably because electrostatic contributions to the elastic modulus are expected to be small. Note that the reduced volume of the vesicle is also found to be constant, as it should be. The large increase in the spontaneous curvature can be understood by considering the balance of the electrostatic free energy and the intrinsic bending energy of the membrane.

6.3.4 HIGH-GENUS VESICLES AND GAUSSIAN SADDLE-SPLAY MODULUS

Usually, vesicles with genus $g = 0$ that have spherical topology are investigated. However, also toroidal vesicles with genus $g = 1$ are experimentally observed, as well as vesicles with high genus and many handles. In addition, in biological cells organelles with high-genus membranes exist. For example, the nuclear membrane and the endoplasmic reticulum (ER) are multiply connected and together\ form complex shapes. The nucleus is wrapped by two lipid-bilayer membranes multiply connected by many lipid pores, and the ER can have a sponge-like structure. The genus of vesicles is controlled by the Gaussian saddle-splay modulus of the membrane, where large values for $\bar{\kappa}$ favor formation of high-genus vesicles. For a fixed genus, computer simulations of vesicles, based on triangulated surfaces as introduced in Section 6.2.4, do not have to take into account the value of $\bar{\kappa}$ because of the Gauss-Bonnet theorem: the integral over the Gaussian curvature of a closed vesicle depends only on its topology. Shape fluctuations are controlled by the bending rigidity κ and the area difference between the monolayers of the lipid bilayer. Figure 6.12 shows a collection of experimental and shapes figures for vesicles with genus $g \geq 1$ and finite values of the area–difference elasticity; all computer simulations have been performed using Monte Carlo simulations for triangulated membranes (Noguchi, 2015; Noguchi et al., 2015).

Because the Gaussian saddle splay-modulus of a membrane only affects the vesicle topology and not its shape fluctuations, its value is often not well known. Other systems besides vesicles with topology changes for which $\bar{\kappa}$ is important, are phase-separated multicomponent membranes (Allain et al., 2004; Das et al., 2009) and membranes with open edges (Hu et al., 2012, 2013). Such systems can therefore be used to determine the value of $\bar{\kappa}$ using experiments or computer simulations. Recently, MD simulations for partially bent circular membrane patches have been performed using coarse-grained lipid models (Hu et al., 2012, 2013). If membrane patch sizes are chosen such that an energy barrier between the initial patch and a closed vesicle exists, from multiple simulations the probability for closing can be measured and used to determine $\xi = \gamma R / (2\kappa + \bar{\kappa})$, where R is the initial radius and γ is the edge tension of the patch. For a particular system, κ can be extracted from analyzing the fluctuations as described in Section 6.3.3 or from simulations of membrane tethers (Harmandaris and Deserno, 2006; Shiba and Noguchi, 2011), the edge tension from pore formation (Farago, 2003; Marrink et al., 2004), and $\bar{\kappa}$ from the measurement of ξ. For DMPC membranes modeled using the MARTINI force field, $\gamma = 40.49 \pm 0.34\, pN$, $\kappa = (16.6 \pm 0.5) 10^{-20} J$, $\bar{\kappa} = (-17.3 \pm 1.0) 10^{-20} J$, and therefore $\bar{\kappa}/\kappa = -1/04 \pm 0.03$ have been extracted from simulation data (Hu et al., 2013).

6.3.5 COMPLEX MEMBRANES, INCLUSIONS, AND BUD FORMATION

Membranes of GUVs do not have to be homogeneous lipid-bilayer membranes, but can be multicomponent membranes that are more complex. For instance, curvature-inducing inclusions

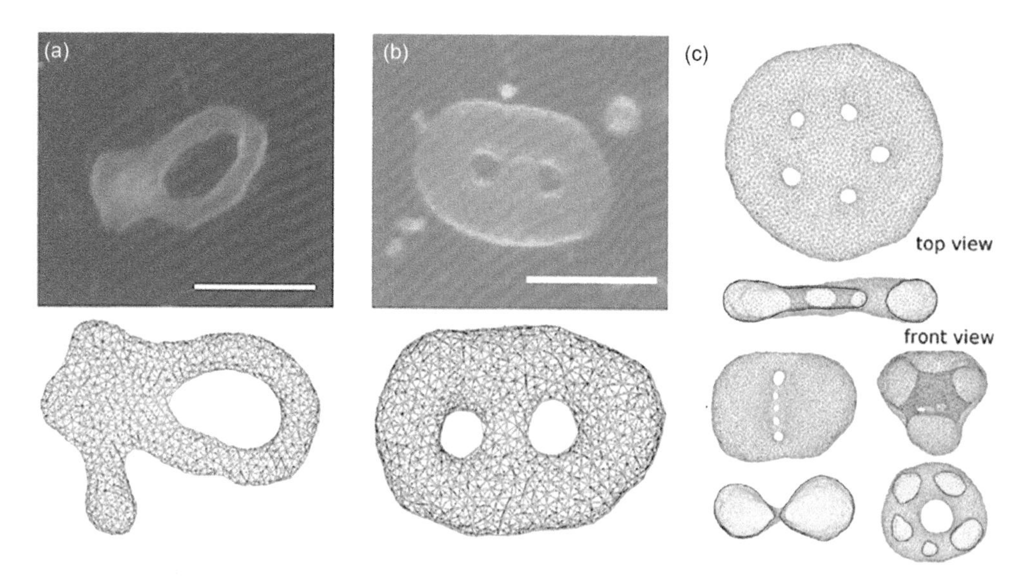

Figure 6.12 High-genus vesicles. Microscopy images and simulation snapshots for vesicles with (a) $g = 1$ and (b) $g = 2$. Scale bars: 10 μm. Adapted from Noguchi et al. (2015) with permission of the Royal Society of Chemistry. (c) Simulation snapshots for $g = 5$ vesicles. (Reprinted with permission from Noguchi, H. et al., *Soft Matter* 11, 193–20, 2015.)

such as proteins or spherical caps of viruses (Atilgan and Sun, 2007; Auth and Gompper, 2009; Reynwar et al., 2007) or polymers anchored only to one monolayer (Auth and Gompper, 2003, 2005; Bickel et al., 2000; Breidenich et al., 2000) induce an effective spontaneous curvature of the membrane. Although it is well known that curved inclusions on a planar membrane mutually repel each other (Goulian et al., 1993; Weikl, 2001), an effective attraction is found for curved inclusions on a vesicle. The optimal vesicle radius is (Auth and Gompper, 2009)

$$R \approx (\cos\alpha)/(\pi\sigma\sin^2\alpha)(1/r_i) \tag{6.33}$$

for a vesicle with

$$n \approx (4\cos^2\alpha)/(\pi\sigma\sin^4\alpha)(1/r_i^2) \tag{6.34}$$

inclusions. Here, r_i is the curvature radius of the inclusion, α is the opening angle of the spherical cap, and σ is the number density of inclusions on the total membrane area. This optimal radius corresponds to an effective spontaneous curvature for the inclusion-decorated membrane of $c_0 = 1/R$. The reasoning is that around each curved inclusion a catenoidal membrane patch with vanishing curvature energy forms. The energy of a vesicle covered with curved inclusions at low density is therefore $\mathcal{E} = 8\pi\kappa(1 - S_{cat}/S_{sph})$, where S_{cat}/S_{sph} is the area fraction of the vesicle that is covered with inclusions and catenoidal patches (Figure 6.13b). At optimal inclusion density, the vesicle is entirely covered by curved inclusions with their catenoidal patches (Figure 6.13c). Computer simulations of coarse-grained lipid bilayer membranes with spherical caps impressively show the dynamics of bud formation starting with planar membrane patches (Reynwar et al., 2007). Using Monte Carlo simulations with triangulated membranes, the deformation

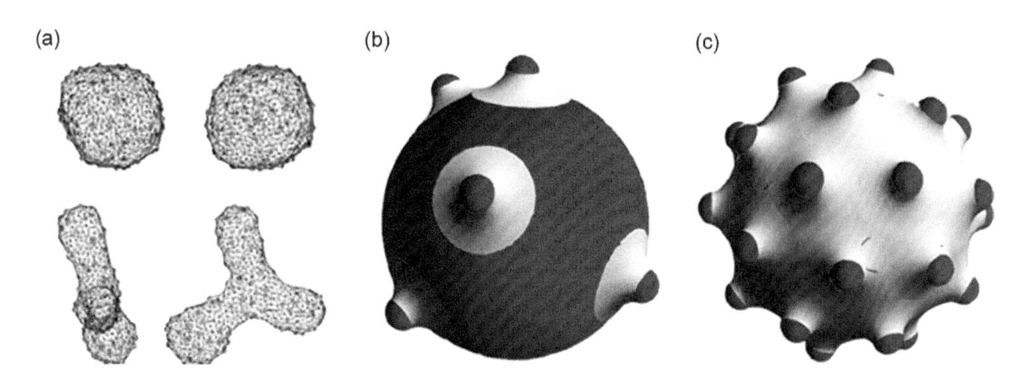

Figure 6.13 (a) A spherical vesicle with curved proteins modeled by a triangulated membrane and simulated using the Monte Carlo technique. The proteins are described by specific angle between the membrane normal and the adjacent edges of a triangle, the base length of these edges, and the angles between them. An initially spherical vesicle with 120 proteins (top) relaxes to a three-branched tubular structure. (Reprinted from Atilgan, E. and Sun, S.X., *J. Chem. Phys.*, 126, 095102, 2007. Copyright 2007 by the American Institute of Physics.) (b, c) Vesicle decorated with curved inclusions at (b) low and (c) optimal inclusion density. Around each inclusion, the membrane can be modeled by segments of the catenoid minimal surface (white). At optimal density, the bending energy of the vesicle vanishes. (Reprinted with permission from Auth, T. and Gompper, G. *Phys. Rev. E*, 80, 031901, 2009 Copyright 2009 y the American Physical Society.)

of an initially spherical vesicle into a three-armed star-shaped vesicle has been demonstrated, which is due to the effective spontaneous curvature induced by the inclusions (Figure 6.13a) and Atilgan and Sun (2007). Decoration of membranes by spherical caps that model partially-wrapped nanoparticles, together with shape predictions for the cells, have also been used to determine the effect of nanoparticle adhesion to red blood cells on the membrane properties from the GUVs decorated with curved proteins are a special case for GUVs with complex membranes. The large range of relevant length scales from several nanometers for the lipids and proteins to several micrometers for the GUVs makes such systems extremely challenging to be studied using a single simulation. Instead, a two-step approach can be applied where both length scales are decoupled. In a first step, effective curvature elastic properties of complex membrane are calculated using analytical calculations or computer simulations for small model systems, for example, membrane patches. In a second step, these effective curvature–elastic properties are used to simulate GUVs with the help of triangulated membranes that are well suited for the micrometer scale. Vice versa, the well-known dependence of RBC shapes on membrane spontaneous curvature has been used to extract effective spontaneous curvatures for RBC membranes with membrane-bound nanoparticles from the change of the RBC shape distribution upon exposure to particles (Barbul et al., 2018).

Common examples for complex membranes are biological membranes that usually contain a mixture of charged and uncharged lipids and proteins and that may be decorated by a glycocalyx. In particular, membranes decorated with polymers have been investigated in various studies (Auth and Gompper, 2003, 2005; Bickel et al., 2000; Bickel and Marques, 2002; Breidenich et al., 2000; Eisenriegler et al., 1996; Hanke et al., 1999; Hiergeist and Lipowsky, 1996; Lekkerkerker, 1990; Milner and Witten, 1988; Werner and Sommer, 2010; Yaman et al., 1997) (Figure 6.14a,b). Although the absolute value of the induced changes of the curvature–elastic constants due to certain components in complex membranes can often be tuned by adjusting their density, how they affect bending rigidity and Gaussian saddle-splay modulus is specific for each mechanism. Figure 6.14c shows the ratios $\Delta\kappa/\Delta\bar{\kappa}$ extracted for added charges (Lekkerkerker, 1990), polymers embedded into the membrane (Yaman et al., 1997), membrane-grafted ideal and self-avoiding linear polymer chains, polymer brushes, and star polymers (Auth and Gompper, 2003, 2005; Bickel et al., 2000; Bickel and Marques, 2002; Breidenich et al., 2000; Eisenriegler et al., 1996; Hanke et al., 1999; Hiergeist and Lipowsky, 1996; Milner and Witten, 1988; Werner and Sommer, 2010). In particular, star polymers allow for changing κ and $\bar{\kappa}$ independently by varying the functionality of the star.

6.3.6 VESICLES IN CAPILLARY FLOW

The deformability and dynamics of vesicles under flow through narrow channels and capillaries plays an important role in determining their behavior in microfluidic devices (Abreu et al., 2014; Barthés-Biesel, 2016). Vesicles can be simulated in capillary flow using a combination of the dynamically triangulated surface model (see Section 6.2.4) and mesoscale hydrodynamics simulation techniques (see Section 6.2.7). The triangulated-network model has to be slightly modified in order to combine it with one of the mesoscale hydrodynamics simulation techniques. Because the temporal evolution of the positions of the membrane vertices is determined by Newton's equation of motion, soft pairwise potentials have to be employed for the tether-bond and excluded volume. The volume V and surface area S of a vesicle are kept constant by constraint potentials. The membrane viscosity can be varied by changing the bond-flip rate, where the membrane viscosity increases with decreasing number of bond-flips per time step (Noguchi and Gompper, 2004, 2005a).

Simulation results for fluid and elastic (polymerized) vesicles in a cylindrical channel are displayed in Figure 6.15 for two different flow velocities (Noguchi and Gompper, 2005b). Simulations are performed for discocyte shapes (at rest) and the reduced volume $V^* = V/(4\pi R_A^3/3) = 0.59$, where $R_A = \sqrt{A/4\pi}$ is the effective vesicle radius. At this reduced volume, a biconcave discocyte is the

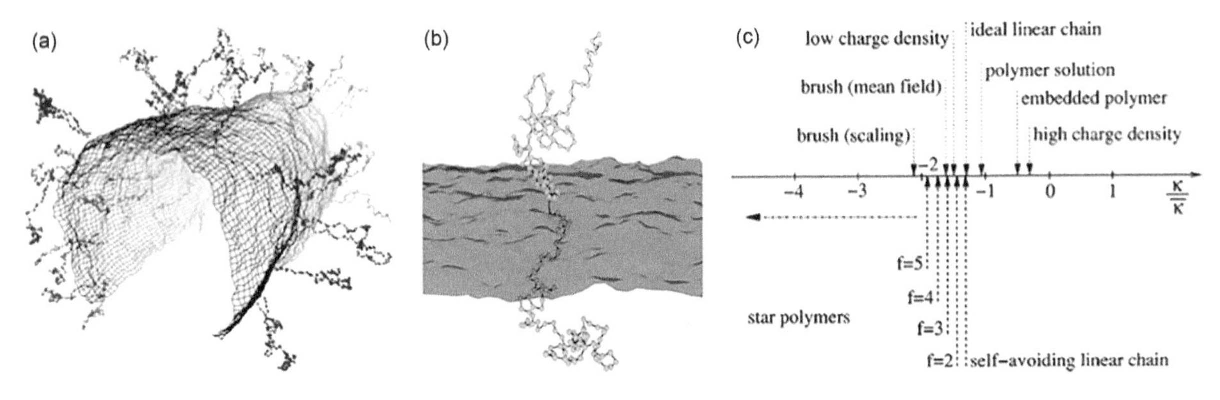

Figure 6.14 (a) Simulation snapshot of a tethered membrane with fixed connectivity decorated with end-grafted linear polymers with 64 monomers. (Reprinted with kind permission from Springer Sciences+Business Media: *Eur. Phys. J. E*, Polymer-decorated tethered membranes under good-and poor-solvent conditions, 31, 383–392, 2010, Werner, M. and Sommer, J.-U.) (b) Simulation snapshot of a symmetric diblock-copolymer attached to a fluid membrane. (Reprinted with permission from Auth, T. and Gompper, G. *Phys. Rev. E*, 72, 031904, 2005, Copyright 2005 by the American Physical Society.) (c) Various modifications change the elastic constants of a lipid bilayer membranes, κ and $\bar{\kappa}$. The ratio of the changes, $\Delta\kappa/\Delta\bar{\kappa}$, is specific for each mechanism (Auth and Gompper, 2003; Eisenriegler et al., 1996; Hanke et al., 1999; Hiergeist and Lipowsky, 1996; Lekkerkerker, 1990; Milner and Witten, 1988; Yaman et al., 1997). As indicated by the horizontal arrow, using star polymers a wide range of ratios $\kappa/\bar{\kappa}$ can be accessed. (Reprinted with permission from Auth, and Gompper, G. *Phys. Rev. E*, 68, 051801, 2003. Copyright 2003 by the American Physical Society.)

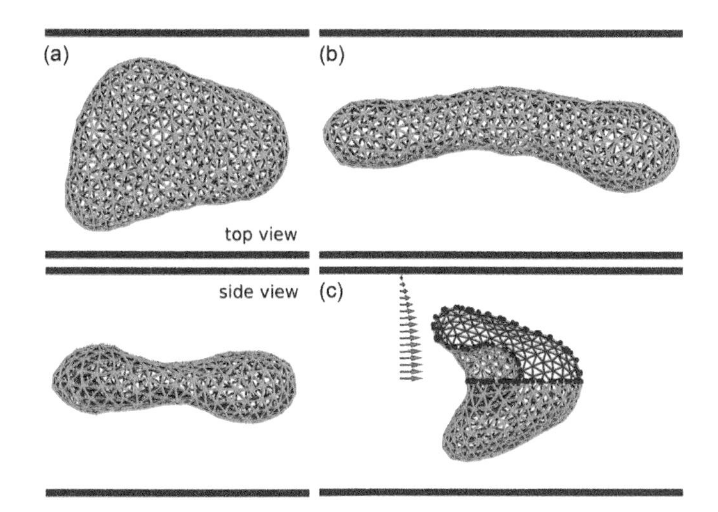

Figure 6.15 Snapshots of vesicles in capillary flow (from Noguchi, H. and Gompper, G. *Proc. Natl. Acad. Sci. USA* 102: 14159–14164, 2005b) for a capillary radius of $R_{cap} = 1.4R_0$ and membrane bending rigidity, $\kappa / k_BT = 20$. (a) Fluid vesicle with a discoidal shape at the mean fluid velocity, $v_m\tau / R_{cap} = 41$, both side and top views. The results are scaled with the intrinsic relaxation time, $\tau = \eta_0 R_{cap}^3 / k_BT$. (b) Fluid vesicle with a prolate shape at $v_m\tau / R_{cap} = 69$. (b) Elastic vesicle (RBC model) with a parachute shape at $v_m\tau / R_{cap} = 218$ (with shear modulus, $\mu R_0^2 / k_BT = 110$). The membrane consists of $N_{mb} = 500$ vertices. The blue arrows represent the velocity field of the solvent. The upper front quarter of the vesicle in (c) is removed to allow a look into the interior; the black circles indicate the lines where the membrane has been cut in this procedure. Thick black lines indicate the walls of the cylindrical capillary. (Reprinted from Noguchi, H. and Gompper, G., *J. Chem. Phys.*, 125, 164908 [1–13], 2006a; Noguchi, H. and Gompper, G. *Phys. Rev.*, *E* 73: 021903, 2006b.)

equilibrium vesicle shape, and a prolate ellipsoid and stomatocyte are metastable in the absence of flow (Seifert, 1997b); therefore, vesicle shapes should be very sensitive to flow at this particular reduced volume.

Under typical experimental conditions of capillary flows, the Reynolds number, $Re = \rho v_{ves} R_A / \eta_0$, is very small, typically $Re \simeq 10^{-2}$, where v_{ves} is the mean velocity of the vesicle. Therefore, parameters are chosen such that $Re < 1$ in all simulations. Both fluid and elastic vesicles retain their discoidal shapes in slow capillary flows (Figure 6.15a). The vesicles align their longest axis with the flow direction, even if their initial conformations are coaxial with the capillary. The discoidal shape is elongated in the flow direction and its front–rear symmetry is broken, but the biconcave dimples and the mirror symmetry with respect to the plane determined by the two eigenvectors of the gyration tensor with the largest eigenvalues are retained. For larger mean fluid velocity, a fluid vesicle transits into a prolate ellipsoidal shape (Figure 6.15b) given that this shape change reduces the flow resistance. An elastic vesicle transits into a parachute shape (Figure 6.15c) instead, because the shear elasticity of the membrane prevents the elongation of the vesicle into a prolate shape.

Even more interesting in comparison to homogeneous cylindrical (or rectangular) channels are structured channels, in which the channel cross section varies periodically along the channel. In such structured channels, the vesicle shape is no longer stationary, so the internal dynamics of a vesicle can be probed. This allows the exploration of the behavior of vesicles in more complex flow geometries, as they can be realized quite easily in modern microfluidic devices.

Snapshots of vesicles under flow through zig-zag channels are shown in Figure 6.16a. These snapshots already demonstrate that the vesicles deform periodically, as they move from wide to narrow regions of channel, and back (Noguchi et al., 2010). The reason is that the flow velocity is fast in the narrow parts, and slow in the wide parts of the channel. Therefore, the vesicle becomes elongated as it approaches the narrow parts, and shortened as it enters the wide parts. However, this behavior is only observed for flexible vesicles, with a small bending rigidity $\kappa^* \ll 1$, with $\kappa^* = \kappa L_y / (\eta R_V^3 v_m)$, where v_m is the mean flow velocity and R_V is the effective vesicle radius related to the vesicle volume. For stiff vesicles, with $\kappa^* \gg 1$, flow forces are too small to overcome the deformation energy costs. In this case, the vesicle adjusts to the compressional force when it enters the wide region by *tilting* its long axis away from the channel axis, as shown in Figure 6.16b. These results indicate that a lot of unexpected behaviors are to be discovered in complex flow geometries.

6.3.7 RED BLOOD CELLS—ACTIVE MEMBRANE FLUCTUATIONS

Red blood cells take the concept of vesicles as model systems one step further toward complex biological cells. These cells are still comparatively simple with a well-defined polymerized membrane attached to the lipid bilayer, the cortical spectrin cytoskeleton. Similarly to vesicles, RBCs are abundant and much easier to handle than cells with a 3D bulk cytoskeleton. A healthy human RBC has a biconcave shape with an average diameter of about 8 μm (Fung, 1993). Its membrane consists of a lipid bilayer with an attached cytoskeleton formed by a network of the protein spectrin linked by short filaments of actin. The membrane's shear elasticity supplied by the spectrin network constitutes the main difference between RBCs and GUVs. The presence of shear elasticity in an RBC membrane significantly affects its behavior and response to various external fields in comparison to vesicles.

One of the interesting measurements for RBCs or vesicles is membrane fluctuations, given that they should be directly associated with the membrane characteristics and properties of cytosol

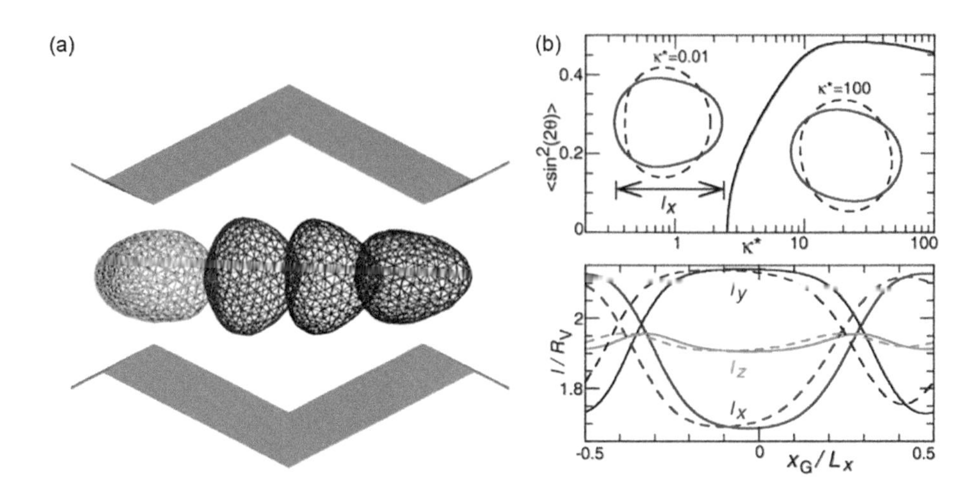

Figure 6.16 (a) Sequence of snapshots (at equal time intervals) of vesicles with reduced volume v = 0.96 and $R_A / L_y = 0.21$ (R_A is the effective vesicle radius related to the membrane area) moving through a structured microchannel, where L_y is the average channel width. (b) Dynamics of a quasi-spherical vesicle at reduced volume v = 0.988. Dependence of the tilt angle, θ, of a vesicle on κ^*, for $L_x / L_y = 4$, where L_x is the periodicity length along the channel and $a_y = 0.5$, which is the amplitude of the wall sawtooth, such that the maxima and minima are located at $y = (1 \pm a_y)L_y / 2$. Here, $\langle \sin^2(2\theta) \rangle$ describes the deviation from symmetric shape. The insets show sliced snapshots of vesicles in the x-y plane for $\kappa^* = 0.01$ and $\kappa^* = 50$. Solid and dashed lines indicate shapes of extremal elongation or tilt. In addition, maximum vesicle lengths in the x, y, and z directions as a function of the center-of-mass position x_G, for $\kappa^* = 0.01$ and $a_y = 0.5$. Solid and dashed lines represent results for $L_x / L_y = 4$ and $L_x / L_y = 16$, respectively. (Reprinted with permission from Noguchi, H. et al., *EPL*, 89, 28002 [1–6], 2010.)

and suspending media. RBC membrane fluctuations have been measured in a number of experiments including RBC edge flickering microscopy (Hale et al., 2009; Strey et al., 1995) and tracking of beads attached to the RBC (Amin et al., 2007; Betz et al., 2009). In contrast to fluctuation measurements on vesicles, see Section 6.3.3 the interpretation of these measurements for RBCs often leads to rather disparate outcomes. For instance, the results of measuring RBC edge fluctuations (Strey et al., 1995) have suggested a vanishing (or nearly negligible) effect of membrane shear elasticity, whereas experiments on RBC deformation with optical tweezers (Henon et al., 1999; Suresh et al., 2005) clearly identify a finite shear elasticity. Furthermore, the interpretation of fluctuation measurements by Betz et al. (2009) has resulted in very high (unrealistic) values for the effective viscosity of the fluid. These differences are likely to originate from the approximations used in analytical models that are mainly derived for planar lipid bilayer membranes such as the model in Eq. (6.31).

Recently, it has been recognized that cell activity (e.g., metabolic activity) through the consumption of ATP contributes to measured flickering for RBCs. The effect of ATP on membrane fluctuations has been investigated in a number of experiments (Betz et al., 2009; Boss et al., 2012; Evans et al., 2008; Park et al., 2010; Turlier et al., 2016; Tuvia et al., 1998) with contradicting outcomes. RBC fluctuations have been reported to depend on the viscosity of suspending media (Tuvia et al., 1998), which points toward out-of-equilibrium contributions. The studies with ATP depletion (Betz et al., 2009; Park et al., 2010) have shown that membrane fluctuations decrease; however, during the ATP depletion process there is no guarantee that RBCs are not subject to changes in membrane elasticity. In contrast, other investigations (Boss et al., 2012; Evans et al., 2008) have questioned the effect of ATP on measured flickering. Recent work (Turlier et al., 2016) has provided compelling evidence for cell activity by testing directly the fluctuation–dissipation relation, which is valid for any system in equilibrium. A violation of the fluctuation–dissipation relation has been shown using a setup illustrated in Figure 6.17a. In this setup, the three handle beads are held by a harmonic potential in simulations mimicking optical tweezers in experiments, whereas the probe bead is moved sinusoidally to determine the mechanical response function. The free fluctuations of the probe bead are measured separately in simulations and experiments. Simulations, based on the triangulated membrane model for RBCs (see Section 6.2.6), have closely mimicked experimental conditions (Turlier et al., 2016) and were used to quantitatively extract RBC membrane properties including shear elasticity, bending rigidity, and membrane viscosity. To simulate active processes, random active forces acting normally on membrane vertices were added. Figure 6.17b shows the power spectral density (PSD) of a passive system, where thermal fluctuations of the probe bead were monitored, and the corresponding system with the activity model. PSD is the Fourier transform of the trajectory of the probe bead. Clearly, certain frequencies are enhanced in case of an active process, which is in quantitative agreement with the experimental observations (Turlier et al., 2016). The combination of experiments, simulations, and theory has allowed the quantification of active fluctuations and the characterization of properties and kinetics of a cell's activity (Turlier et al., 2016).

6.3.8 RED BLOOD CELLS IN FLOW—CAPILLARY FLOW AND BLOOD RHEOLOGY

The behavior of RBCs in microcirculation plays an important role in blood flow resistance and in the cell partitioning within a microvascular network (Gompper and Fedosov, 2016). Therefore, a number of investigations have focused on the understanding of RBC deformation and dynamics in simple flows such as shear and Poiseuille flow (Fedosov et al., 2014b). Most of available experiments with single RBCs were performed in suspending fluids with a relatively high viscosity, often much larger than that of blood plasma. This allows the imposition of high enough fluid

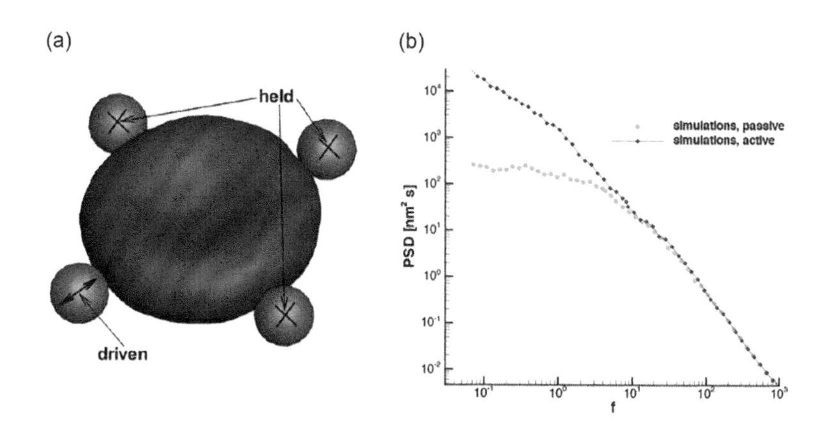

Figure 6.17 Simulations of RBC membrane fluctuations. (From Turlier, H. et al., *Nat. Phys.* 12, 513–519, 2016.) (a) Simulation setup mimicking the experimental conditions with an RBC and four beads attached. Three beads (marked by crosses) are used as handles via a harmonic potential, whereas the fourth probe bead (marked by a two-headed arrow) is either sinusoidally driven to measure the mechanical response or left free to monitor its fluctuations. (b) Comparison of the power spectral density (PSD) of a system with passive thermal fluctuations and the corresponding system with an activity generated by random kicks at the membrane.

stresses on RBCs, while keeping the corresponding shear rates under a certain limit that is often imposed by the experimental instruments used (e.g., rheometer). However, such conditions are very different from physiological conditions and may not properly reflect the behavior of RBCs in blood. We briefly review experimental and simulation results for the dynamics of RBCs in shear and Poiseuille flows, and emphasize the differences in RBC behavior in suspensions with different viscosities.

RBCs suspended in a relatively high-viscosity fluid (greater than about five times viscosity of water) show tumbling dynamics at low shear rates and tank-treading at high shear rates in Couette flow (Abkarian et al., 2007; Fischer, 2004; Tran-Son-Tay et al., 1984). The existence of the tumbling-to-tank-treading transition is attributed to an RBC minimum energy state such that a certain energy barrier has to be exceeded for an RBC to start the tank-treading motion. In the tank-treading state, an RBC also oscillates around the preferred inclination angle of tank-treading with a certain frequency and amplitude (Abkarian et al., 2007; Fedosov et al., 2010; Noguchi, 2009; Skotheim and Secomb, 2007). Recent experiments (Dupire et al., 2012) have identified another dynamics, RBC rolling, which occurs within the range of shear rates between RBC tumbling and tank-treading states.

Note that all these studies have been performed under the conditions where the viscosity of suspending media was larger than that of the RBC cytosol. However, under physiological conditions blood plasma has a viscosity about five times smaller than that of an RBC cytosol. Using a similar viscosity ratio, recent experiments (Vitkova et al., 2008) and simulations (Yazdani and Bagchi, 2011) have shown that a large enough viscosity contrast between inner and outer fluids suppresses the tank-treading motion of RBCs, leading to the preference for RBC tumbling. This behavior is qualitatively consistent with that for vesicles, where a transition from tank-treading to tumbling can be triggered by an increase in the viscosity contrast (Keller and Skalak, 1982). This indicates that membrane shear elasticity may play a secondary role for RBC dynamics at high enough viscosity contrast.

Similar to vesicles, RBCs in Poiseuille flow show a rich behavior, characterized by various shapes including parachutes and slippers

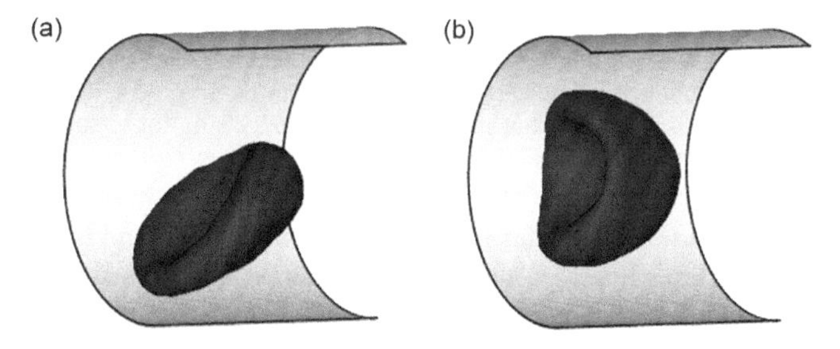

Figure 6.18 Simulation snapshots of an RBC in tube flow (from left to right): (a) an off-center slipper cell shape and (b) a parachute shape. (Fedosov, D.A. et al., *Soft Matter*, 10, 4258–4267, 2014. Reproduced by permission of the Royal Society of Chemistry.)

(Abkarian et al., 2008; Fedosov et al., 2014b; Kaoui et al., 2011; McWhirter et al., 2009; Tahiri et al., 2013; Tomaiuolo et al., 2009), as illustrated in Figure 6.18. Parachutes are characterized by a symmetric shape similar to a semi-spherical cap and they flow in the center of a tube without significant membrane motion. In contrast, slippers are nonsymmetric RBC shapes, where the membrane is subjected to a tank-treading motion. Thus, slippers are mainly differentiated from parachutes by an asymmetric shape and the membrane motion.

Recent simulations of vesicles and RBCs in two dimensions (2D) (where fluid and polymerized membranes cannot be distinguished) (Kaoui et al., 2011; Tahiri et al., 2013) have led to a phase diagram of various shapes including parachute, slipper, and a snaking dynamics, as function of RBC confinement and flow strength. The snaking dynamics is characterized by a wiggling motion of a discocyte shape near the tube center. Simulations of RBCs in three dimensions (3D) under flow in cylindrical channels (Fedosov et al., 2014b) (based on the models and techniques of Sections 6.2.4, 6.2.6, and 6.2.7) generate a similar diagram of RBC shapes in tube flow, which is qualitatively similar to the diagram in two dimensions. Figure 6.19 shows the RBC shape diagram in 3D for different flow rates and confinements. The flow rate is characterized by a nondimensional shear rate $\dot{\gamma}^*$, which is a product of the average shear rate (or pseudo-shear rate) and the characteristic relaxation time $\tau = \eta R^3/\kappa$ of an RBC (Fedosov et al., 2014b) The confinement χ is the ratio of an effective RBC

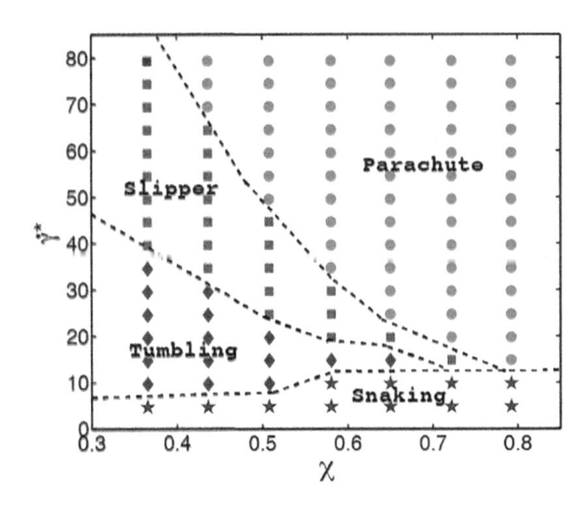

Figure 6.19 A phase diagram of RBC shapes in tube flow for the average membrane properties of a healthy RBC with a Föppl-von Kármán number $\Gamma = 2652$. Different dynamic states, depending on the flow strength characterized by a nondimensional flow rate $\dot{\gamma}^*$ and the confinement χ, are depicted by symbols: parachute (circles), slipper (squares), tumbling (diamonds) and snaking (stars) discocyte. The phase-boundary lines are drawn approximately to guide the eye. (Fedosov, D.A. et al., *Soft Matter*, 10, 4258–4267, 2014. Reproduced by permission of the Royal Society of Chemistry.)

diameter and the tube diameter. At strong confinements and high flow rates, parachutes are mainly found, whereas low confinements lead predominantly to off-center slippers. When the flow rate is small enough, off-center tumbling RBCs are found, which can be explained by the existence of the tumbling-to-tank-treading transition mentioned above for RBCs in shear flow. In contrast to the 3D model results, this region is absent in 2D simulations (Kaoui et al., 2011; Tahiri et al., 2013), because this transition cannot be captured by a 2D model. Another prominent difference between the phase diagrams in Figure 6.18 and in 2D simulations (Kaoui et al., 2011; Tahiri et al., 2013) is the existence of the "confined slipper" in 2D at high confinements, which is absent in 3D. Slippers at high confinements in 3D are hindered due to the cylindrical shape of a tube, which makes a confined slipper configuration energetically unfavorable because it would have to conform to the wall curvature. In microcapillary flow, changes in RBC membrane properties lead to a shift of boundaries between different RBC shapes and dynamics illustrated in Figure 6.18 (Fedosov et al., 2014b). Consequently, it should be possible to detect such changes based on the observation of RBCs in flow and simulations can provide the basis for a quantitative interpretation of these observations.

Cell deformation due to flow forces also has a strong effect on blood rheology. It is well known from various experiments (Chien, 1970; Chien et al., 1966; Popel and Johnson, 2005; Skalak et al., 1981) that whole blood in shear flow displays strong shear thinning. This effect is due to the attractive interaction between RBCs—probably due to attachment and bridging by fibrinogen (Baskurt and Meiselman, 2013; Neu and Meiselman, 2006; Rampling et al., 2004; Steffen et al., 2013)—which leads to the formation of large stacks of RBCs called rouleaux, and even networks of them, which imply a high shear viscosity. With increasing shear rate, these rouleaux break apart, and the viscosity decreases. Above a certain shear rate, approximately $1-10\,s^{-1}$, aggregation is no longer relevant, as can be seen from a comparison of "washed" blood, in which the blood plasma is replaced by another fluid of the same osmolarity, so that fibrinogen is absent, but the same viscosity is observed.

This behavior can be very well reproduced in simulations (Fedosov et al., 2011) using the triangulated-network model introduced in Sections 6.2.4 and 6.2.6. The attractive interaction is modeled by a short-range Morse potential. The only fitting parameter needed to reproduce the experimental results is the attraction strength, which is found by fitting (Fedosov et al., 2011). This attraction implies that tangential or sliding breakup requires a force in the range 3-5 pN. The magnitude of this force can be compared with the force needed for unfolding of proteins in single-molecule experiments, which is typically of order 100 pN. Therefore, the attraction between RBCs in whole blood is indeed very small. Further shear-thinning is found in the shear-rate range of $10-1000s^{-1}$, which is due to RBC deformation. The simulation results indicate that the cells at an intermediate shear rate of $50-100s^{-1}$ become more "spherical", in the sense that their size in three orthogonal directions becomes nearly equal, as in the case of a parachute-like or stomatocyte-like shape (Fedosov et al., 2011). For even higher shear rates around $1000s^{-1}$, they seem to elongate into an ellipsoid-like shape.

New experiments and simulations show an even more complex and interesting behavior (Lanotte et al., 2016). The stomatocyte-like shape at intermediate shear rates around $100–200s^{-1}$ is confirmed, but shown further to be a "rolling stomatocyte", which has a nearly time-independent shape with its symmetry axis pointing in the vorticity direction of the shear flow. Most spectacular, however, is a new dynamic shape occurring above a shear rate of about $500s^{-1}$, which has apparently been overlooked in previous studies. This is the "trilobe" shape displayed in Figure 6.20. Detailed studies of the shear viscosity reveal that both the shape and the dynamics of the trilobe contribute significantly to shear-thinning, with the shape yielding a factor 5, the dynamics another factor 2 in viscosity reduction.

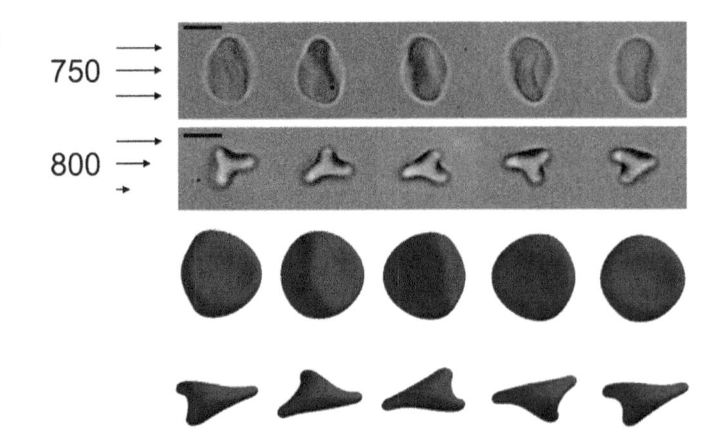

Figure 6.20 RBC dynamics in shear flow at high shear rates of about $750s^{-1}$, from both microfluidic experiments (top) and simulations (bottom). Snapshots parallel and perpendicular to the shear-gradient direction are shown, as indicated by the arrows on the left. (Adapted from Lanotte, L. et al., *Proc. Natl. Acad. Sci. USA*, 113, 13289–13294, 2016.)

It is important to emphasize that similar shapes have *not* been observed in simulations of vesicles in shear flow. Instead, vesicles in the regime of viscosity contrasts $\lambda \simeq 2$–10 display three types of shapes and dynamics (Lanotte et al., 2016): (i) squaring motion, where the angle between the longest axis and the flow direction undergoes discontinuous jumps over time, (ii) spontaneous parity breaking of the shape leading to cross-streamline migration, and (iii) tumbling of S-shaped vesicles. Thus, although vesicles are often taken as simple model systems of RBCs, extrapolations from the flow behavior of vesicles to RBCs have to be made with great care. The shear elasticity of the membrane can obviously make a significant difference.

6.4 CONCLUSIONS AND OUTLOOK

Models and simulation techniques for the entire range of length scales relevant for GUVs—from atomistic models for single lipids via coarse-grained molecular models for self-assembly and lipid organization in a membrane, to discretized continuum models for vesicle shapes—have been developed and applied in recent years. Developments of these techniques have extended the applicability domain for vesicle simulations toward more complex many-component systems that cannot be studied easily by analytical calculations. On the molecular scale, coarse-grained molecular models with chemical specificity have become available—independently from the increase in computational speed—for much larger systems. The combination of discretized continuum membrane models with mesoscopic hydrodynamic simulation techniques is nowadays successfully used to simulate vesicles and simple cells in flow, such as vesicle deformation in structured channels and blood flow in small capillaries. Finally, a combination of continuum membrane models with models for membrane proteins, the simulation of coupled fluid and polymerized membranes, and modeling of the interaction of membranes with cytoskeletal filaments allow the extension of bare vesicle simulations to biomimetic and biological systems. In combination with hydrodynamics, also the dynamics of such biomimetic systems can be accessed. This provides versatile tools and opens up exciting possibilities toward studying equilibrium and dynamic properties of multicomponent vesicular systems with passive and active components.

GLOSSARY OF SYMBOLS

A_v	area of unstressed vesicle
C_1	first principal curvature
C_2	second principal curvature
F_b	Free energy of bending
T	Temperature
k_BT	thermal energy
κ	bending modulus
$\bar{\kappa}$	saddle-splay modulus
P	pressure
t	time
t_{b0}	initial bilayer thickness uncompressed
V_v	volume of unstressed vesicle
Y	Young's modulus
Σ	membrane tension

ACRONYMS

ATP	adenosine triphosphate
DPD	dissipative particle dynamics
MD	molecular dynamics
MLS	moving least-squares method
MPC	multi-particle collision dynamics
RBC	red blood cell
SPH	smoothed particle hydrodynamics

REFERENCES

Abkarian, M., Faivre, M., Horton, R., Smistrup, K., Best-Popescu, C. A. and Stone, H. A. (2008). Cellular-scale hydrodynamics, *Biomed. Mater.* 3: 034011.

Abkarian, M., Faivre, M. and Viallat, A. (2007). Swinging of red blood cells under shear flow, *Phys. Rev. Lett.* 98: 188302.

Abraham, F. F., Rudge, W. E. and Plischke, M. (1989). Molecular dynamics of tethered membranes, *Phys. Rev. Lett.* 62: 1757–1759.

Abreu, D., Levant, M., Steinberg, V. and Seifert, U. (2014). Fluid vesicles in flow, *Adv. Colloid Interface Sci.* 208: 129–141.

Allain, J.-M., Storm, C., Roux, A., Amar, M. B. and Joanny, J.-F. (2004). Fission of a multiphase membrane tube, *Phys. Rev. Lett.* 93(15): 158104.

Allen, M. P. and Tildesley, D. J. (1991). *Computer Simulation of Liquids*, Clarendon Press, New York.

Amin, M. S., Park, Y.-K., Lue, N., Dasari, R. R., Badizadegan, K., Feld, M. S. and Popescu, G. (2007). Microrheology of red blood cell membranes using dynamic scattering microscopy, *Opt. Express* 15: 17001–17009.

Atilgan, E. and Sun, S. X. (2007). Shape transitions in lipid membranes and protein mediated vesicle fusion and fission, *J. Chem. Phys.* 126(9): 095102.

Auth, T. and Gompper, G. (2003). Self-avoiding linear and star polymers anchored to membranes, *Phys. Rev. E* 68(5): 051801.

Auth, T. and Gompper, G. (2005). Fluctuation spectrum of membranes with anchored linear and star polymers, *Phys. Rev. E* 72(3): 031904.

Auth, T. and Gompper, G. (2009). Budding and vesiculation induced by conical membrane inclusions, *Phys. Rev. E* 80(3): 031901.

Auth, T., Safran, S. and Gov, N. S. (2007a). Filament networks attached to membranes: Cytoskeletal pressure and local bilayer deformation, *New J. Phys.* 9(11): 430.

Auth, T., Safran, S. and Gov, N. S. (2007b). Fluctuations of coupled fluid and solid membranes with application to red blood cells, *Phys. Rev. E* 76(5): 051910.

Bahrami, A. H., Lipowsky, R. and Weikl, T. R. (2012). Tubulation and aggregation of spherical nanoparticles adsorbed on vesicles, *Phys. Rev. Lett.* 109(18): 188102.

Barbul, A., Singh, K., Horev- Azaria, L., Dasgupta, S., Auth, T., Korenstein, R., and Gompper, G. (2018). Nanoparticle-Decorated Erythrocytes Reveal That Particle Size Controls the Extent of Adsorption, Cell Shape, and Cell Deformability. *ACS Applied Nano Materials*, 1(8), 3785–3799.

Barthès-Biesel, D. (2016). Motion and deformation of elastic capsules and vesicles in flow, *Annu. Rev. Fluid Mech.* 48: 25–52.

Baskurt, O. K. and Meiselman, H. J. (2013). Erythrocyte aggregation: Basic aspects and clinical importance, *Clin. Hemorheol. Microcirc.* 53: 23–37.

Belytschko, T., Krongauz, Y., Organ, D., Fleming, M. and Krysl, P. (1996). Meshless methods: An overview and recent developments, *Comput. Methods Appl. Mech. Eng.* 139: 3.
Berger, M. J. and Colella, P. (1989). Local adaptive mesh refinement for shock hydrodynamics, *J. Comp. Phys.* 82(1): 64–84.
Betz, T., Lenz, M., Joanny, J.-F. and Sykes, C. (2009). ATP-dependent mechanics of red blood cells, *Proc. Natl. Acad. Sci. USA* 106: 15320–15325.
Bickel, T. and Marques, C. M. (2002). Scale-dependent rigidity of polymer-ornamented membranes, *Eur. Phys. J. E* 9(4): 349–352.
Bickel, T., Marques, C. and Jeppesen, C. (2000). Pressure patches for membranes: The induced pinch of a grafted polymer, *Phys. Rev. E* 62(1): 1124.
Boal, D. H. and Rao, M. (1992). Scaling behavior of fluid membranes in three dimensions, *Phys. Rev. A* 45: R6947–R6950.
Boek, E. S., den Otter, W. K., Briels, W. J. and Iakovlev, D. (2004). Molecular-dynamics simulation of amphiphilic bilayer membranes and wormlike micelles: A multi-scale modeling approach to the design of viscoelastic surfactant solutions, *Phil. Trans. R. Soc. Lond. A* 362: 1625–1638.
Boss, D., Hoffmann, A., Rappaz, B., Depeursinge, C., Magistretti, P. J., Van de Ville, D. and Marquet, P. (2012). Spatially-resolved eigenmode decomposition of red blood cells membrane fluctuations questions the role of ATP in flickering, *PLoS One* 7: e40667.
Brakke, K. A. (1992). The Surface Evolver, *Exp. Math.* 1(2): 141–165.
Brannigan, G. and Brown, F. L. H. (2004). Solvent-free simulations of fluid membrane bilayers, *J. Chem. Phys.* 120: 1059–1071.
Brannigan, G., Lin, L. C. L. and Brown, F. L. H. (2006). Implicit solvent simulation models for biomembranes, *Eur. Biophys. J.* 35: 104–124.
Breidenich, M., Netz, R. and Lipowsky, R. (2000). The shape of polymer-decorated membranes, *EPL (Europhys. Lett.)* 49(4): 431.
Canham, P. B. (1970). The minimum energy of bending as a possible explanation of the biconcave shape of the human red blood cell, *J. Theor. Biol.* 26: 61–81.
Chien, S. (1970). Shear dependence of effective cell volume as a determinant of blood viscosity, *Science* 168: 977–979.
Chien, S., Usami, S., Taylor, H. M., Lundberg, J. L. and Gregersen, M. I. (1966). Effects of hematocrit and plasma proteins on human blood rheology at low shear rates, *J. Appl. Physiol.* 21: 81–87.
Cooke, I. R. and Deserno, M. (2005). Solvent-free model for self-assembling fluid bilayer membranes: Stabilization of the fluid phase based on broad attractive tail potentials, *J. Chem. Phys.* 123: 224710.
Cooke, I. R., Kremer, K. and Deserno, M. (2005). Tunable generic model for fluid bilayer membranes, *Phys. Rev. E* 72: 011506.
Das, S., Jenkins, J. and Baumgart, T. (2009). Neck geometry and shape transitions in vesicles with co-existing fluid phases: Role of Gaussian curvature stiffness vs. spontaneous curvature, *EPL (Europhys. Lett.)* 86(4): 48003.
Dasgupta, S., Auth, T. and Gompper, G. (2013). Wrapping of ellipsoidal nano-particles by fluid membranes, *Soft Matter* 9(22): 5473–5482.
Dasgupta, S., Auth, T. and Gompper, G. (2014). Shape and orientation matter for the cellular uptake of nonspherical particles, *Nano Lett.* 14(2): 687–693.
Dasgupta, S., Auth, T. and Gompper, G. (2017). Nano- and microparticles at biological and fluid interfaces, *J. Phys. Condens. Matter* 29: 373003 [1–41].
de Vries, A. H., Mark, A. E. and Marrink, S. J. (2004). Molecular dynamics simulation of the spontaneous formation of a small DPPC vesicle in water in atomistic detail, *J. Am. Chem. Soc.* 126: 4488–4489.
den Otter, W. K. (2005). Area compressibility and buckling of amphiphilic bilayers in molecular dynamics simulations, *J. Chem. Phys.* 123: 214906.
den Otter, W. K. and Briels, W. J. (2003). The bending rigidity of an amphiphilic bilayer from equilibrium and nonequilibrium molecular dynamics, *J. Chem. Phys.* 118: 4712–4720.
Dickson, C. J., Madej, B. D., Skjevik, A. A., Betz, R. M., Teigen, K., Gould, I. R. and Walker, R. C. (2014). Lipid14: The Amber lipid force field, *J. Chem. Theory Comput.* 10: 865–879.
Dickson, C. J., Rosso, L., Betz, R. M., Walker, R. C. and Gould, I. R. (2012). GAFFlipid: A General Amber Force Field for the accurate molecular dynamics simulation of phospholipid, *Soft Matter* 8: 9617–9627.
Döbereiner, H.-G., Evans, E., Kraus, M., Seifert, U. and Wortis, M. (1997). Mapping vesicle shapes into the phase diagram: A comparison of experiment and theory, *Phys. Rev. E* 55: 4458–4474.
Döbereiner, H.-G., Gompper, G., Haluska, C., Kroll, D. M., Petrov, P. G. and Riske, K. A. (2003). Advanced flicker spectroscopy of fluid membranes, *Phys. Rev. Lett.* 91: 048301.
Drouffe, J. M., Maggs, A. C. and Leibler, S. (1991). Computer simulations of self-assembled membranes, *Science* 254: 1353–1356.
Dupire, J., Socol, M. and Viallat, A. (2012). Full dynamics of a red blood cell in shear flow, *Proc. Natl. Acad. Sci. USA* 109: 20808–20813.
Eisenriegler, E., Hanke, A. and Dietrich, S. (1996). Polymers interacting with spherical and rodlike particles, *Phys. Rev. E* 54(2): 1134.
Español, P. and Warren, P. (1995). Statistical mechanics of dissipative particle dynamics, *Europhys. Lett.* 30: 191–196.
Evans, J., Gratzer, W., Mohandas, N., Parker, K. and Sleep, J. (2008). Fluctuations of the red blood cell membrane: Relation to mechanical properties and lack of ATP dependence, *Biophys. J.* 94: 4134–4144.
Farago, O. (2003). "Water-free" computer model for fluid bilayer membranes, *J. Chem. Phys.* 119: 596–605.
Farago, O., Grønbech-Jensen, N. and Pincus, P. (2006). Mesoscale computer modeling of lipid-DNA complexes for gene therapy, *Phys. Rev. Lett.* 96: 018102.
Farutin, A. and Misbah, C. (2012). Squaring, parity breaking, and S tumbling of vesicles under shear flow, *Phys. Rev. Lett.* 109: 248106.
Fedosov, D. A., Caswell, B. and Karniadakis, G. E. (2010). A multiscale red blood cell model with accurate mechanics, rheology, and dynamics, *Biophys. J.* 98: 2215–2225.
Fedosov, D. A., Noguchi, H. and Gompper, G. (2014a). Multiscale modeling of blood flow: From single cells to blood rheology, *Biomech. Model. Mechanobiol.* 13: 239–258.
Fedosov, D. A., Pan, W., Caswell, B., Gompper, G. and Karniadakis, G. E. (2011). Predicting blood rheology in silico, *Proc. Natl. Acad. Sci. USA* 108: 11772–11777.
Fedosov, D. A., Peltomäki, M. and Gompper, G. (2014b). Deformation and dynamics of red blood cells in flow through cylindrical microchannels, *Soft Matter* 10: 4258–4267.
Feig, M. and Brooks III, C. L. (2004). Recent advances in the development and application of implicit solvent models in biomolecule simulations. *Current opinion in structural biology*, 14(2), 217–224.
Fischer, T. M. (2004). Shape memory of human red blood cells, *Biophys. J.* 86: 3304–3313.
Fošnarič, M., Iglič, A., Kroll, D. M. and May, S. (2009). Monte Carlo simulations of complex formation between a mixed fluid vesicle and a charged colloid, *J. Chem. Phys.* 131: 105103.

Fung, Y. C. (1993). *Biomechanics: Mechanical Properties of Living Tissues*, 2nd ed., Springer-Verlag, New York.

Gelbart, W. M., Ben-Shaul, A. and Roux, D. (Eds) (1994). *Micelles, Membranes, Microemulsions, and Monolayers*, Springer-Verlag, Berlin, Germany.

Goetz, R., Gompper, G. and Lipowsky, R. (1999). Mobility and elasticity of self-assembled membranes, *Phys. Rev. Lett.* 82: 221–224.

Goetz, R. and Lipowsky, R. (1998). Computer simulations of bilayer membranes: Self-assembly and interfacial tension, *J. Chem. Phys.* 108: 7397–7409.

Gompper, G. and Fedosov, D. A. (2016). Modeling microcirculatory blood flow: Current state and future perspectives, *WIREs Syst Biol Med.* 8: 157–168.

Gompper, G., Ihle, T., Kroll, D. M. and Winkler, R. G. (2009). Multi-Particle Collision Dynamics: A particle-based mesoscale simulation approach to the hydrodynamics of complex fluids, *Adv. Polym. Sci.* 221: 1–87.

Gompper, G. and Kroll, D. M. (1991). Fluctuations of a polymerized membrane between walls, *J. Phys. I France* 1: 1411–1432.

Gompper, G. and Kroll, D. M. (1996). Random surface discretizations and the renormalization of the bending rigidity, *J. Phys. I France* 6: 1305–1320.

Gompper, G. and Kroll, D. M. (1997). Network models of fluid, hexatic and polymerized membranes, *J. Phys.: Condens. Matter* 9: 8795–8834.

Gompper, G. and Kroll, D. M. (1998). Membranes with fluctuating topology: Monte Carlo simulations, *Phys. Rev. Lett.* 81: 2284–2287.

Gompper, G. and Kroll, D. M. (2004). Triangulated-surface models of fluctuating membranes, *in* D. R. Nelson, T. Piran and S. Weinberg (Eds.), *Statistical Mechanics of Membranes and Surfaces*, 2nd ed, World Scientific, Singapore, Chapter 12, pp. 359–426.

Gompper, G. and Noguchi, H. (2006). Coarse-grained and continuum models of membranes, *in* S. Blügel, G. Gompper, E. Koch, H. Müller-Krumbhaar, R. Spatschek and R. G. Winkler (Eds.), *Computational Condensed Matter Physics*, Vol. 32 of *Matter and Materials*, Forschungszentrum Jülich GmbH, Jülich, Germany, p. B.9.

Gompper, G. and Schick, M. (1994). Self-assembling amphiphilic systems, *in* C. Domb and J. Lebowitz (Eds.), *Phase Transitions and Critical Phenomena*, Vol. 16, Academic Press, London, UK, pp. 1–176.

Goulian, M., Bruinsma, R. and Pincus, P. (1993). Long-range forces in heterogeneous fluid membranes, *EPL (Europhys. Lett.)* 22(2): 145.

Groot, R. D. and Rabone, K. L. (2001). Mesoscopic simulation of cell membrane damage, morphology change and rupture by nonionic surfactants, *Biophys. J.* 81: 725–736.

Gruhn, T., Franke, T., Dimova, R. and Lipowsky, R. (2007). Novel method for measuring the adhesion energy of vesicles, *Langmuir* 23: 5423–5429.

Hale, J. P., Marcelli, G., Parker, K. H., Winlove, C. P. and Petrov, P. G. (2009). Red blood cell thermal fluctuations: Comparison between experiment and molecular dynamics simulations, *Soft Matter* 5: 3603–3606.

Hanke, A., Eisenriegler, E. and Dietrich, S. (1999). Polymer depletion effects near mesoscopic particles, *Phys. Rev. E* 59(6): 6853.

Hansen, J., Skalak, R., Chien, S. and Hoger, A. (1997). Influence of network topology on the elasticity of the red blood cell membrane skeleton., *Biophys. J.* 72(5): 2369.

Harmandaris, V. A. and Deserno, M. (2006). A novel method for measuring the bending rigidity of model lipid membranes by simulating tethers, *J. Chem. Phys.* 125(20): 204905.

Helfrich, W. (1973). Elastic properties of lipid bilayers: Theory and possible experiments, *Z. Naturforsch.* 28c: 693–703.

Henon, S., Lenormand, G., Richert, A. and Gallet, F. (1999). A new determination of the shear modulus of the human erythrocyte membrane using optical tweezers, *Biophys. J.* 76: 1145–1151.

Hiergeist, C. and Lipowsky, R. (1996). Elastic properties of polymer-decorated membranes, *J. Phys. (France) II* 6(10): 1465–1481.

Ho, J.-S. and Baumgärtner, A. (1990). Simulations of fluid flexible membranes, *Europhys. Lett.* 12: 295–300.

Hoogerbrugge, P. J. and Koelman, J. M. V. A. (1992). Simulating microscopic hydrodynamic phenomena with dissipative particle dynamics, *Europhys. Lett.* 19: 155–160.

Hsu, P.-C., Bruininks, B. M., Jefferies, D., Cesar Telles de Souza, P., Lee, J., Patel, D. S., Marrink, S. J., Qi, Y., Khalid, S. and Im, W. (2017). CHARMM-GUI Martini Maker for modeling and simulation of complex bacterial membranes with lipopolysaccharides, *J. Comput. Chem.* 38(27): 2354–2363.

Hu, M., Briguglio, J. J. and Deserno, M. (2012). Determining the Gaussian curvature modulus of lipid membranes in simulations, *Biophys. J.* 102(6): 1403–1410.

Hu, M., de Jong, D. H., Marrink, S. J. and Deserno, M. (2013). Gaussian curvature elasticity determined from global shape transformations and local stress distributions: A comparative study using the MARTINI model, *Faraday Discuss.* 161: 365–382.

Illya, G., Lipowsky, R. and Shillcock, J. C. (2005). Effect of chain length and asymmetry on material properties of bilayer membranes, *J. Chem. Phys.* 122: 244901.

Itzykson, C. (1986). Random geometry, lattices and fields, *in* J. Abad, M. Asorey and A. Cruz (Eds.), *Proceedings of the GIFT Seminar, Jaca 85*, World Scientific, Singapore, pp. 130–188.

Kantor, Y. and Nelson, D. R. (1987a). Crumpling transition in polymerized membranes, *Phys. Rev. Lett.* 58: 2774–2777.

Kantor, Y. and Nelson, D. R. (1987b). Phase transitions in flexible polymeric surfaces, *Phys. Rev. A* 36: 4020–4032.

Kaoui, B., Tahiri, N., Biben, T., Ez-Zahraouy, H., Benyoussef, A., Biros, G. and Misbah, C. (2011). Complexity of vesicle microcirculation, *Phys. Rev. E* 84: 041906.

Kasson, P. M., Lindahl, E. and Pande, V. S. (2010). Atomic-resolution simulations predict a transition state for vesicle fusion defined by contact of a few lipid tails, *PLoS Comput. Biol.* 6: e1000829.

Keller, S. R. and Skalak, R. (1982). Motion of a tank-treading ellipsoidal particle in a shear flow, *J. Fluid Mech.* 120: 27–47.

Knecht, V. and Marrink, S.-J. (2007). Molecular dynamics simulations of lipid vesicle fusion in atomic detail, *Biophys J.* 92: 4254–4261.

Kohyama, T. and Gompper, G. (2007). Defect scars on flexible surfaces with crystalline order, *Phys. Rev. Lett.* 98: 198101 [1–4].

Kohyama, T., Kroll, D. M. and Gompper, G. (2003). Budding of crystalline domains in fluid membranes, *Phys. Rev. E* 68: 061905 [1–15].

Kranenburg, M., Venturoli, M. and Smit, B. (2003). Molecular simulations of mesoscopic bilayer phases, *Phys. Rev. E* 67: 060901.

Kreyszig, E. (1991). *Differential Geometry*, Dover Publications, New York.

Kroll, D. M. and Gompper, G. (1992). The conformation of fluid membranes, *Science* 255: 968–971.

Kusumaatmaja, H., Li, Y., Dimova, R. and Lipowsky, R. (2009). Intrinsic contact angle of aqueous phases at membranes and vesicles, *Phys. Rev. Lett.* 103(23): 238103.

Lancaster, P. and Salkaskas, K. (1981). Surfaces generated by moving least square methods, *Math. Comput.* 37: 141–158.

Lanotte, L., Mauer, J., Mendez, S., Fedosov, D. A., Fromental, J.-M., Claveria, V., Nicoud, F., Gompper, G. and Abkarian, M. (2016). Red cells' dynamic morphologies govern blood shear thinning under microcirculatory flow conditions, *Proc. Natl. Acad. Sci. USA* 113: 13289–13294.

Laradji, M. and Sunil Kumar, P. B. (2004). Dynamics of domain growth in self-assembled fluid vesicles, *Phys. Rev. Lett.* 93: 198105.

Laradji, M. and Sunil Kumar, P. B. (2005). Domain growth, budding, and fission in phase-separating self-assembled fluid bilayers, *J. Chem. Phys.* 123: 224902.

Lee, J. B., Petrov, P. G. and Döbereiner, H.-G. (1999). Curvature of zwitterionic membranes in transverse pH gradients, *Langmuir* 15: 8543.

Lekkerkerker, H. (1990). The electric contribution to the curvature elastic moduli of charged fluid interfaces, *Physica A* 167(2): 384–394.

Li, J., Dao, M., Lim, C. T. and Suresh, S. (2005). Spectrin-level modeling of the cytoskeleton and optical tweezer stretching of the erythrocyte, *Biophys. J.* 88: 3707–3719.

Li, J., Zhang, H., Qiu, F., Yang, Y. and Chen, J. Z. Y. (2015). Conformation of a charged vesicle, *Soft Matter* 11: 1788–1793.

Lim, G., Wortis, M. and Mukhopadhyay, R. (2002). Stomatocyte-discocyte-echinocyte sequence of the human red blood cell: Evidence for the bilayer-couple hypothesis from membrane mechanics, *Proc. Natl. Acad. Sci. USA* 99: 16766–16769.

Lowengrub, J. S., Rätz, A. and Voigt, A. (2009). Phase-field modeling of the dynamics of multicomponent vesicles: Spinodal decomposition, coarsening, budding, and fission, *Phys. Rev. E* 79(3): 031926.

Lucy, L. B. (1977). A numerical approach to the testing the fission hypothesis, *Astronom. J.* 82: 1013–1024.

Lyubartsev, A. P. and Rabinovich, A. L. (2011). Recent development in computer simulations of lipid bilayers, *Soft Matter* 7: 25–39.

Lyubartsev, A. P. and Rabinovich, A. L. (2016). Force field development for lipid membrane simulations, *Biochimica et Biophysica Acta* 1858: 2483–2497.

Malevanets, A. and Kapral, R. (1999). Mesoscopic model for solvent dynamics, *J. Chem. Phys.* 110: 8605–8613.

Marrink, S. J., de Vries, A. H. and Mark, A. E. (2004). Coarse grained model for semiquantitative lipid simulations, *J. Phys. Chem. B* 108(2): 750–760.

Marrink, S. J., de Vries, A. H. and Tieleman, D. P. (2009). Lipids on the move: Simulations of membrane pores, domains, stalks and curves, *Biochim. Biophys. Acta* 1788: 149–168.

Marrink, S. J. and Mark, A. E. (2003). Molecular dynamics simulation of the formation, structure, and dynamics of small phospholipid vesicles, *J. Am. Chem. Soc.* 125: 15233–15242.

Marrink, S. J., Risselada, H. J., Yefimov, S., Tieleman, D. P. and de Vries, A. H. (2007). The MARTINI force field: Coarse grained model for biomolecular simulations, *J. Phys. Chem. B* 111: 7812–7824.

McWhirter, J. L., Noguchi, H. and Gompper, G. (2009). Flow-induced clustering and alignment of vesicles and red blood cells in microcapillaries, *Proc. Natl. Acad. Sci. USA* 106: 6039–6043.

Miao, L., Seifert, U., Wortis, M. and Döbereiner, H.-G. (1994). Budding transitions of fluid-bilayer vesicles: The effect of area-difference elasticity, *Phys. Rev. E* 49: 5389–5407.

Milner, S. and Witten, T. (1988). Bending moduli of polymeric surfactant interfaces, *J. Phys. (France)* 49(11): 1951–1962.

Monaghan, J. J. (2005). Smoothed particle hydrodynamics, *Rep. Prog. Phys.* 68: 1703–1759.

Müller, M., Katsov, K. and Schick, M. (2003). A new mechanism of model membrane fusion determined from Monte Carlo simulation, *Biophys. J.* 85: 1611–1623.

Nelson, D., Piran, T. and Weinberg, S. (Eds) (2004). *Statistical Mechanics of Membranes and Surfaces*, 2nd ed, World Scientific, Singapore.

Neu, B. and Meiselman, H. J. (2006). Depletion interactions in polymer solutions promote red blood cell adhesion to albumin-coated surfaces, *Biochimica et Biophysica Acta* 1760: 1772–1779.

Noguchi, H. (2009). Swinging and synchronized rotations of red blood cells in simple shear flow, *Phys. Rev. E* 80: 021902.

Noguchi, H. (2015). Shape transitions of high-genus fluid vesicles, *EPL (Europhys. Lett.)* 112(5): 58004.

Noguchi, H. and Gompper, G. (2004). Fluid vesicles with viscous membranes in shear flow, *Phys. Rev. Lett.* 93: 258102 [1–4].

Noguchi, H. and Gompper, G. (2005a). Dynamics of fluid vesicles in shear flow: Effect of membrane viscosity and thermal fluctuations, *Phys. Rev. E* 72: 011901 [1–14].

Noguchi, H. and Gompper, G. (2005b). Shape transitions of fluid vesicles and red blood cells in capillary flows, *Proc. Natl. Acad. Sci. USA* 102: 14159–14164.

Noguchi, H. and Gompper, G. (2006a). Dynamics of vesicle self-assembly and dissolution, *J. Chem. Phys.* 125: 164908 [1–13].

Noguchi, H. and Gompper, G. (2006b). Meshless membrane model based on the moving least-squares method, *Phys. Rev. E* 73: 021903.

Noguchi, H., Gompper, G., Schmid, L., Wixforth, A. and Franke, T. (2010). Dynamics of fluid vesicles in flow through structured microchannels, *EPL* 89: 28002 [1–6].

Noguchi, H. and Takasu, M. (2001a). Fusion pathways of vesicles: A Brownian dynamics simulation, *J. Chem. Phys.* 115: 9547.

Noguchi, H. and Takasu, M. (2001b). Self-assembly of amphiphiles into vesicles: A Brownian dynamics simulation, *Phys. Rev.* E 64: 041913.

Noguchi, H. and Takasu, M. (2002). Adhesion of nanoparticles to vesicles: A Brownian dynamics simulation, *Biophys. J.* 83: 299–308.

Noguchi, H., Sakashita, A. and Imai, M. (2015). Shape transformations of toroidal vesicles, *Soft Matter* 11(1): 193–201.

Ortiz, V., Nielsen, S. O., Discher, D. E., Klein, M. L., Lipowsky, R. and Shillcock, J. (2005). Dissipative particle dynamics simulations of polymersomes, *J. Phys. Chem.* B 109: 17708–17714.

Park, Y.-K., Best, C. A., Auth, T., Gov, N. S., Safran, S. A., Popescu, G., Suresh, S. and Feld, M. S. (2010). Metabolic remodeling of the human red blood cell membrane, *Proc. Natl. Acad. Sci. USA* 107: 1289–1294.

Pastor, R. W. and MacKerell, Jr., A. D. (2011). Development of the CHARMM force field for lipids, *J. Phys. Chem. Lett.* 2: 1526–1532.

Peltomäki, M., Gompper, G. and Kroll, D. M. (2012). Scattering of bicontinuous microemulsions and sponge phases, *J. Chem. Phys.* 136: 134708 [1–13].

Plischke, M. and Boal, D. (1988). Absence of a crumpling transition in strongly self-avoiding tethered membranes, *Phys. Rev. A* 38: 4943–4945.

Popel, A. S. and Johnson, P. C. (2005). Microcirculation and hemorheology, *Annu. Rev. Fluid Mech.* 37: 43–69.

Ramakrishnan, N., Sunil Kumar, P. B. and Ipsen, J. H. (2013). Membrane-mediated aggregation of curvature-inducing nematogens and membrane tubulation, *Biophys. J.* 104: 1018–1028.

Rampling, M. W., Meiselman, H. J., Neu, B. and Baskurt, O. K. (2004). Influence of cell-specific factors on red blood cell aggregation, *Biorheology* 41: 91–112.

Rekvig, L., Hafskjold, B. and Smit, B. (2004). Chain length dependencies of the bending modulus of surfactant monolayers, *Phys. Rev. Lett.* 92: 116101.

Reynwar, B. J., Illya, G., Harmandaris, V. A., Müller, M. M., Kremer, K. and Deserno, M. (2007). Aggregation and vesiculation of membrane proteins by curvature-mediated interactions, *Nature* 447(7143): 461–464.

Safran, S. A. (1994). *Statistical Thermodynamics of Surfaces, Interfaces, and Membranes*, Addison-Wesley, Reading, MA.

Šarić, A. and Cacciuto, A. (2012a). Fluid membranes can drive linear aggregation of adsorbed spherical nanoparticles, *Phys. Rev. Lett.* 108: 118101.

Šarić, A. and Cacciuto, A. (2012b). Mechanism of membrane tube formation induced by adhesive nanocomponents, *Phys. Rev. Lett.* 109: 188101.

Šarić, A. and Cacciuto, A. (2013). Self-assembly of nanoparticles adsorbed on fluid and elastic membranes, *Soft Matter* 9: 6677–6695.

Schäfer, L. V., de Jong, D. H., Holt, A., Rzepiela, A. J., de Vries, A. H., Poolman, B., Killian, J. A. and Marrink, S. J. (2011). Lipid packing drives the segregation of transmembrane helices into disordered lipid domains in model membranes, *Proc. Natl. Acad. Sci. U.S.A.* 108(4): 1343–1348.

Seifert, U. (1997a). Configurations of fluid membranes and vesicles, *Adv. Phys.* 46(1): 13–137.

Seifert, U. (1997b). Configurations of fluid membranes and vesicles, *Adv. Phys.* 46: 13–137.

Seung, H. S. and Nelson, D. R. (1988). Defects in flexible membranes with crystalline order, *Phys. Rev. A* 38: 1005–1018.

Shelley, J. C. and Shelley, M. Y. (2000). Computer simulations of surfactant solutions, *Curr. Opin. Colloid Interface Sci.* 5: 101–110.

Shiba, H. and Noguchi, H. (2011). Estimation of the bending rigidity and spontaneous curvature of fluid membranes in simulations, *Phys. Rev. E* 84(3): 031926.

Shillcock, J. C. and Lipowsky, R. (2002). Equilibrium structure and lateral stress distribution of amphiphilic bilayers from dissipative particle dynamics simulations, *J. Chem. Phys.* 117: 5048.

Shillcock, J. C. and Lipowsky, R. (2005). Tension-induced fusion of bilayer membranes and vesicles, *Nat. Mater.* 4: 225–228.

Skalak, R., Keller, S. R. and Secomb, T. W. (1981). Mechanics of blood flow, *J. Biomech. Eng.* 103: 102–115.

Skjevik, A. A., Madej, B. D., Dickson, C. J., Lin, C., Teigen, K., Walker, R. C. and Gould, I. R. (2016). Simulation of lipid bilayer self-assembly using all-atom lipid force fields, *Phys. Chem. Chem. Phys.* 18: 10573–10584.

Skotheim, J. M. and Secomb, T. W. (2007). Red blood cells and other nonspherical capsules in shear flow: Oscillatory dynamics and the tank-treading-to-tumbling transition, *Phys. Rev. Lett.* 98: 078301.

Smith, A.-S., Sackmann, E. and Seifert, U. (2003). Effects of a pulling force on the shape of a bound vesicle, *EPL (Europhys. Lett.)* 64(2): 281.

Sreeja, K. K., Ipsen, J. H. and Sunil Kumar, P. B. (2015). Monte Carlo simulations of fluid vesicles, *J. Phys. Condens. Matter* 27: 273104.

Srinivas, G., Disher, D. E. and Klein, M. L. (2004). Self-assembly and properties of diblock copolymers by coarse-grained molecular dynamics, *Nat. Mater.* 3: 638–644.

Steffen, P., Verdier, C. and Wagner, C. (2013). Quantification of depletion-induced adhesion of red blood cells, *Phys. Rev. Lett.* 110: 018102.

Stevens, M. J., Hoh, J. H. and Woolf, T. B. (2003). Insights into the molecular mechanism of membrane fusion from simulation: Evidence for the association of splayed tails, *Phys. Rev. Lett.* 91: 188102.

Strey, H., Peterson, M. and Sackmann, E. (1995). Measurement of erythrocyte membrane elasticity by flicker eigenmode decomposition, *Biophys. J.* 69: 478–488.

Sunil Kumar, P. B., Gompper, G. and Lipowsky, R. (2001). Budding dynamics of multicomponent membranes, *Phys. Rev. Lett.* 86: 3911–3914.

Suresh, S., Spatz, J., Mills, J. P., Micoulet, A., Dao, M., Lim, C. T., Beil, M. and Seufferlein, T. (2005). Connections between single-cell biomechanics and human disease states: Gastrointestinal cancer and malaria, *Acta Biomaterialia* 1: 15–30.

Tahiri, N., Biben, T., Ez-Zahraouy, H., Benyoussef, A. and Misbah, C. (2013). On the problem of slipper shapes of red blood cells in the microvasculature, *Microvasc. Res.* 85: 40–45.

Tieleman, D. P., Marrink, S. J. and Berendsen, H. J. C. (1997). A computer perspective of membranes: Molecular dynamics studies of lipid bilayer systems, *Biochim. Biophys. Acta* 1331: 235–270.

Tolpekina, T. V., den Otter, W. K. and Briels, W. J. (2004). Simulations of stable pores in membranes: System size dependence and line tension, *J. Chem. Phys.* 121: 8014–8020.

Tomaiuolo, G., Simeone, M., Martinelli, V., Rotoli, B. and Guido, S. (2009). Red blood cell deformation in microconfined flow, *Soft Matter* 5: 3736–3740.

Tran-Son-Tay, R., Sutera, S. P. and Rao, P. R. (1984). Determination of red blood cell membrane viscosity from rheoscopic observations of tank-treading motion, *Biophys. J.* 46: 65–72.

Turlier, H., Fedosov, D. A., Audoly, B., Auth, T., Gov, N., Sykes, C., Joanny, J.-F., Gompper, G. and Betz, T. (2016). Equilibrium physics breakdown reveals the active nature of red blood cell flickering, *Nat. Phys.* 12: 513–519.

Tuvia, S., Levin, S., Bitler, A. and Korenstein, R. (1998). Mechanical fluctuations of the membrane-skeleton are dependent on F-actin ATPase in human erythrocytes, *J. Cell Biol.* 141: 1551–1561.

Ursitti, J. A. and Wade, J. B. (1993). Ultrastructure and immunocytochemistry of the isolated human erythrocyte membrane skeleton, *Cell Motil. Cytoskeleton* 25(1): 30–42.

Venturoli, M., Sperotto, M. M., Kranenburg, M. and Smit, B. (2006). Mesoscopic models of biological membranes, *Phys. Rep.* 437: 1–54.

Vitkova, V., Mader, M.-A., Polack, B., Misbah, C. and Podgorski, T. (2008). Micro-macro link in rheology of erythrocyte and vesicle suspensions, *Biophys. J.* 95: L33–L35.

Vliegenthart, G. and Gompper, G. (2006). Forced crumpling of self-avoiding elastic sheets, *Nat. Mater.* 5: 216–221.

Vliegenthart, G. A. and Gompper, G. (2011). Compression, crumpling and collapse of spherical shells and capsules, *New J. Phys.* 13(4): 045020.

Wassenaar, T. A., Ingólfsson, H. I., Böckmann, R. A., Tieleman, D. P. and Marrink, S. J. (2015). Computational lipidomics with insane: A versatile tool for generating custom membranes for molecular simulations, *J. Chem. Theory Comput.* 11: 2144–2155.

Weikl, T. (2001). Fluctuation-induced aggregation of rigid membrane inclusions, *EPL (Europhys. Lett.)* 54(4): 547.

Wendt, J. F. (Ed.) (2009). *Computational Fluid Dynamics*, 3rd ed, Springer, Berlin, Germany.

Werner, M. and Sommer, J.-U. (2010). Polymer-decorated tethered membranes under good-and poor-solvent conditions, *Eur. Phys. J. E* 31(4): 383–392.

Yaman, K., Pincus, P., Solis, F. and Witten, T. (1997). Polymers in curved boxes, *Macromolecules* 30(4): 1173–1178.

Yazdani, A. Z. K. and Bagchi, P. (2011). Phase diagram and breathing dynamics of a single red blood cell and a biconcave capsule in dilute shear flow, *Phys. Rev. E* 84: 026314.

Yu, Q., Othman, S., Dasgupta, S., Auth, T., and Gompper, G. (2018). Nanoparticle wrapping at small non-spherical vesicles: curvatures at play. *Nanoscale*, 10(14), 6445–6458.

Zeman, K., Engelhard, H. and Sackmann, E. (1990). Bending undulations and elasticity of the erythrocyte membrane: Effects of cell shape and membrane organization, *Eur. Biophys. J.* 18(4): 203–219.

Theory of vesicle dynamics in flow and electric fields

Petia M. Vlahovska and Chaouqi Misbah

He who loves practice without theory is like the sailor who boards ship without a rudder and compass and never knows where he may cast.

Leonardo da Vinci

Contents

7.1 INTRODUCTION

The equilibrium and nonequilibrium behavior of giant unilamellar vesicles (GUVs) has been studied extensively in theory and simulations (Freund, 2014; Li et al., 2012; Seifert, 1997; Vlahovska et al., 2009c; Vlahovska, 2019). Vesicles display rich dynamics in external fields: multiple dynamical states in shear flow (tank-treading [TT], trembling [TR], tumbling [TB], swinging, squaring, parity-breaking) (Abkarian et al., 2007; Abreu et al., 2014; Deschamps et al., 2009; Farutin and Misbah, 2012b; Kantsler and Steinberg, 2005, 2006; Lebedev et al., 2008; Mader et al., 2006; Misbah, 2006; Vlahovska and Gracia, 2007; Vlahovska et al., 2011; Zabusky et al., 2011), asymmetric slipper-like shapes in Poiseuille flow (Abkarian et al., 2008; Coupier et al., 2012; Farutin and Misbah, 2011; Kaoui and Misbah, 2009), pearling and asymmetric dumbbell shapes in straining flows (Boedec et al., 2014; Kantsler et al., 2008; Narsimhan et al., 2014, 2015; Vlahovska, 2014; Zhao and Shaqfeh, 2013) or uniform electric fields (Sinha et al., 2013), drum-like "squared" shapes in direct current (DC) electric pulses (Dimova et al., 2009; Riske and Dimova, 2006; Salipante and Vlahovska, 2014; Vlahovska, 2015).

This chapter will provide a tutorial into the analytical modeling of the nonequilibrium dynamics of GUVs. Solutions for the deformation and motion of a nearly spherical vesicle are derived which illustrate the use of a formalism based on spherical harmonics. The results are applied to the analysis of vesicle dynamics in linear flows and vesicle response to electric pulses.

7.2 PROBLEM FORMULATION

The thickness of the lipid bilayer (ℓ_{me}, ~5 nm) is much smaller than the typical GUV size (radius, a, ~10 μm). Accordingly, the membrane can be treated as a two-dimensional (2D) surface embedded in a 3D space.

Let us consider a membrane with total area, A, enclosing fluid volume, V. The equilibrium shape can be nonspherical, characterized by a dimensionless excess area

$$\Delta = A / a^2 - 4\pi, \tag{7.1}$$

where the characteristic vesicle size, a, is defined by the radius of a sphere of the same volume, $a = (3V/4\pi)^{1/3}$. Another commonly used parameter to quantify the departure of the particle shape from a sphere is the reduced volume, $v = V / (\frac{4\pi}{3} R_0^3)$, where $R_0 = \sqrt{A/4\pi}$. For a sphere, $\Delta = 0$ and $v = 1$. The typical values of these parameters for a red blood cell (RBC), which has a biconcave disc shape, are $\Delta \sim 4$ and $v \sim 0.65$.

The encapsulated and suspending fluids are assumed incompressible and Newtonian with shear viscosities η^{in} and η^{ex}, conductivities σ^{in} and σ^{ex}, and permittivities ε^{in} and ε^{ex}, respectively. The mismatch in physical properties is characterized by the ratios

$$\chi = \frac{\eta^{in}}{\eta^{ex}}, \qquad \Lambda_\sigma = \frac{\sigma^{in}}{\sigma^{ex}}, \qquad \Lambda_\varepsilon = \frac{\varepsilon^{in}}{\varepsilon^{ex}}. \tag{7.2}$$

The membrane may also have different physical properties, viscosity (η^{me}), conductivity ($G_{me} = \sigma^{me} / \ell_{me}$), and capacitance ($C_{me} = \varepsilon^{me} / \ell_{me}$) than the embedding fluids. The corresponding dimensionless parameters are

$$\chi^{me} = \frac{\eta^{me} a}{\eta^{ex}}, \qquad g_m = \frac{\sigma^{me} a}{\sigma^{ex} \ell_{me}}, \qquad c_m = \frac{\varepsilon^{me} a}{\varepsilon^{ex} \ell_{me}}. \tag{7.3}$$

The vesicle is subject to interfacial forces generated by applied flow or electric field. In this chapter, we will develop a unified framework to treat the problem of vesicle response to an arbitrary distribution of interfacial forces. We will use a reference frame centered on and cotranslating with the vesicle (Figure 7.1 provides a definition of the configuration). We will consider linear flows with strain rate, $\dot{\varepsilon}$, and rotation, ω. Figure 7.2 illustrates the problem for the example of simple shear flow, $\omega = \dot{\varepsilon} = \dot{\gamma} / 2$,

$$\mathbf{v}^\infty(\mathbf{x}) = \dot{\varepsilon}\left(y\hat{\mathbf{x}} + x\hat{\mathbf{y}}\right) + \omega\left(y\hat{\mathbf{x}} - x\hat{\mathbf{y}}\right), \tag{7.4}$$

and quadratic flows with curvature, δ, such as the plane Poiseuille flow,

$$\mathbf{v}^\infty(\mathbf{x}) = \left(-\delta y^2 - 2\delta y_0 y\right)\hat{\mathbf{x}} - \mathbf{V}_p, \tag{7.5}$$

where y_0 is the vesicle distance to the centerline; and $\mathbf{V}_p$ is the vesicle velocity relative to the unperturbed background flow.

We will also analyze vesicle response to a uniform electric field $\mathbf{E} = E_0 \hat{\mathbf{y}}$.

7.2.1 FLUID MOTION AND FLUID–MEMBRANE COUPLING

For cell-sized vesicles, the Reynolds number is small and inertia effects are negligible. Accordingly, fluid velocity $\mathbf{v}^{(\alpha)}$ and pressure $p^{(\alpha)}$ of the interior ("(α) = in") and suspending ("(α) = ex") fluids obey the Stokes equations and the incompressibility condition

$$\nabla \cdot \mathbf{T}^{(\alpha)} = -\nabla p^{(\alpha)} + \eta^{(\alpha)} \nabla^2 \mathbf{v}^{(\alpha)} = \mathbf{0}, \quad \nabla \cdot \mathbf{v}^{(\alpha)} = 0, \tag{7.6}$$

where $\mathbf{T}$ is the bulk hydrodynamic stress tensor

$$\mathbf{T}^{(\alpha)} = -p^{(\alpha)}\mathbf{I} + \eta^{(\alpha)}\left[\nabla\mathbf{v}^{(\alpha)} + (\nabla\mathbf{v}^{(\alpha)})^\dagger\right]. \tag{7.7}$$

Here $\mathbf{I}$ denotes the unit tensor; and the superscript † denotes transpose. Note that the steady Stokes equations imply that the time scale over which the boundary configuration changes—for example, in oscillatory shear this would be the period of the oscillations—is much longer than the viscous time scale $\rho^{(\alpha)} a^2 / \eta^{(\alpha)}$, where $\rho^{(\alpha)}$ is the fluid density. Far from the vesicle, the flow field tends to the unperturbed external flow $\mathbf{v}^{ex} \to \mathbf{v}^\infty(\mathbf{x})$.

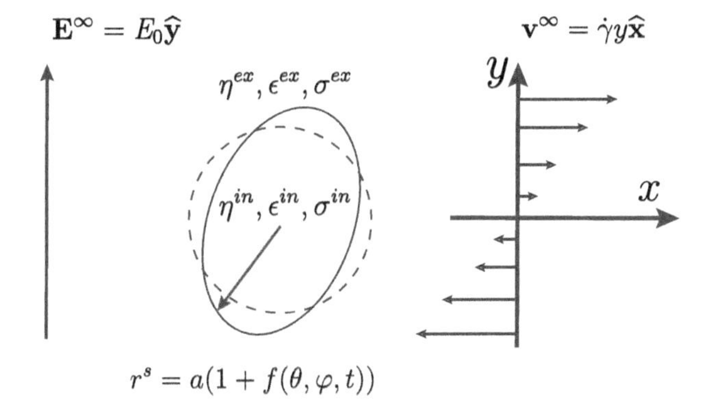

Figure 7.1 Sketch of the problem: a vesicle subjected to a shear flow and/or a uniform electric field. Vesicle shape is described by $r_s = a(1 + f(\theta, \varphi, t))$.

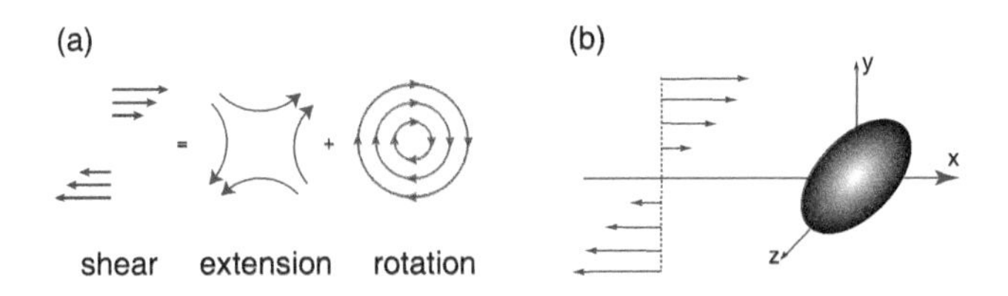

Figure 7.2 (a) Sketch of the straining and rotational components of the simple shear flow. (b) A soft particle deforms into an ellipsoid whose major axis does not align with the extensional axis of the flow. The angle between the ellipsoid major axis and the flow direction is ψ.

At the interface, the velocity is continuous, $\mathbf{v}^{\text{in}} = \mathbf{v}^{\text{ex}} = \mathbf{v}_{\text{me}}$, assuming a fluid-impermeable membrane and neglecting bilayer slip, that is, the possibility that the two leaflets of the lipid bilayer slide past each other (Schwalbe et al., 2010). The vesicle deformation is determined from the kinematic condition at the interface (Barthes-Biesel and Sgaier, 1985)

$$\frac{\partial r_s}{\partial t} = \mathbf{v}_{\text{me}} \cdot \mathbf{n}, \tag{7.8}$$

where $\mathbf{n}$ is the outward unit normal vector. Note that the vesicle can assume a steady TT shape (e.g., under shear) in which the deformation at a fixed Eulerian point is constant in time unlike that of a fixed material point because material elements rotate.

The stress balance at the interface is deduced from the interfacial transport of momentum (Edwards et al., 1991)

$$\mathbf{n} \cdot \left(\mathbf{T}^{\text{ex}} - \mathbf{T}^{\text{in}}\right) + \mathbf{f} = -\nabla_s \cdot \mathbf{P}_{\text{me}}, \tag{7.9}$$

where $\mathbf{f}$ is a surface excess force, for example, due to presence of surface charges or active processes, $\mathbf{P}_{\text{me}}$ is the 2D membrane stress tensor, $\nabla_s = \mathbf{I}_s \cdot \nabla$, and $\mathbf{I}_s = \mathbf{I} - \mathbf{nn}$ is the surface projection operator.

7.2.2 MEMBRANE MECHANICS

Self-assembled bilayers made of lipids or block copolymers are held together by non-covalent bonds, which allow for the molecules to rearrange freely. As a result, bilayer membranes behave as a viscous fluid.

Various constitutive laws are postulated to describe interfacial rheology (Edwards et al., 1991). For a Newtonian viscous interface, the dependence of the interfacial stress tensor on the rate of (inplane) interface deformation (i.e., shear or dilation/compression) is given by the Boussinesq-Scriven equation (Edwards et al., 1991)

$$\mathbf{P}_{\text{me}}^{BS} = \Sigma \mathbf{I}_s + \left(\eta_d^{\text{me}} - \eta^{\text{me}}\right)\left(\mathbf{I}_s : \mathbf{D}_s\right)\mathbf{I}_s + 2\eta^{\text{me}}\mathbf{D}_s, \tag{7.10}$$

where Σ is the surface tension; η^{me} and η_d^{me} are the shear and dilatational viscosities, respectively; and the rate-of-strain tensor is

$$\mathbf{D}_s = \frac{1}{2}\left[\nabla_s \mathbf{v}_{\text{me}} \cdot \mathbf{I}_s + \mathbf{I}_s \cdot \left(\nabla_s \mathbf{v}_{\text{me}}\right)^{\dagger}\right]. \tag{7.11}$$

Note that in general the velocity of the deforming membrane, $\mathbf{v}_{\text{me}}$, has normal and tangential components to the interface, $\mathbf{v}_{\text{me}} = v_n \mathbf{n} + \mathbf{v}_{||}$.

Resistance to out-of-plane deformation, that is, bending, is described by the Helfrich's model (Seifert, 1997), see also Chapter 5. It assumes that the energy cost for bending is quadratic in the mean curvature (the contribution due to the Gaussian curvature to the total energy is a topological invariant and therefore not considered in our analysis of vesicles that do not undergo break-up or poration). The corresponding stress tensor is (Capovilla and Guven, 2002; Maitre et al., 2012; Vlahovska et al., 2013)

$$\mathbf{P}_{\text{me}}^{H} = -2\kappa\left[-H^2\mathbf{I}_s - H\nabla_s\mathbf{n} + (\nabla_s H)\mathbf{n}\right], \tag{7.12}$$

where κ is the bending modulus; and the mean curvature is $H = -\frac{1}{2}\nabla_s \cdot \mathbf{n}$. For simplicity, we have assumed that the membrane has no spontaneous curvature, that is, there is no asymmetry in the lipid head and tail packing.

Taking the surface divergence of Eq. 7.12 yields the more familiar expression for the bending force density (Seifert, 1999) (see Appendix 7.A for the details)

$$\boldsymbol{\tau}^{B} = -\nabla_s \cdot \mathbf{P}_{\text{me}}^{H} = \kappa\left(4H^3 - 4KH + 2\nabla_s^2 H\right)\mathbf{n}, \tag{7.13}$$

where K is the Gaussian curvature. The equilibrium shapes of a vesicle (in absence of flow) satisfy the generalized Laplace's equation, which for this simplest membrane model is

$$p^{\text{in}} - p^{\text{ex}} = -2\Sigma_0 H + \kappa\left(4H^3 - 4HK + 2\nabla_s^2 H\right), \tag{7.14}$$

where Σ_0 is the membrane tension. Eq. 7.14 is a nonlinear equation for the surface profile and yields a plethora of solutions as a function the vesicle deflation (Seifert, 1997; Seifert et al., 1991).

The lipid bilayer is essentially shear-free (shear elastic modulus is zero), its area compressibility modulus is $K_A \sim 100$ mN/m; the resistance to stretching is much stronger than bending ($K_A a^2 / \kappa \gg 1$). Accordingly, the lipid bilayer is modeled as an area-incompressible fluid interface, $\nabla_s \cdot \mathbf{v}_{\text{me}} = 0$.

The viscous interfacial tractions derived from Eq. 7.10 (Edwards et al., 1991; Rahimi et al., 2013) for the 2D incompressible fluid are

$$\begin{aligned}\boldsymbol{\tau}^{BS} = \eta^{\text{me}}\Big[&2\mathbf{b} : \nabla\mathbf{v}_{\text{me}} - \nabla_s \times \nabla_s \times \mathbf{v}_{||} + K\mathbf{v}_{||} \\ &-2\left(\mathbf{b} - 2H\mathbf{I}_s\right)\cdot\nabla_s\left(\mathbf{v}_{\text{me}}\cdot\mathbf{n}\right) + 2H\left(\nabla_s\mathbf{v}_{\text{me}}\right)\cdot\mathbf{n}\Big],\end{aligned} \tag{7.15}$$

where $\mathbf{b} = -\nabla_s\mathbf{n}$.

7.2.3 NONDIMENSIONALIZATION AND CHOICE OF A SMALL PARAMETER FOR THE PERTURBATION ANALYSIS

Henceforth, all variables are nondimensionalized using the radius of a sphere with the same volume, a, the characteristic applied stresses, τ_c, and the properties of the suspending fluid (viscosity, η^{ex}, dielectric constant, ε^{ex}). In the case of shear flow, $\tau_c = \eta^{\text{ex}}\dot{\gamma}$

quadratic flows, $\tau_c = \eta^{ex}\delta a$, and in a uniform electric field with magnitude E_0, $\tau_c = \varepsilon^{ex} E_0^2$. Accordingly, the time scale is $t_c = \eta^{ex}/\tau_c$ and the velocity scale is $V_c = a\tau_c/\eta^{ex}$.

Assuming that the dominant dissipation occurs in the bulk fluids, forces exerted by the applied stresses distort the vesicle shape on a time scale

$$t_d = \frac{(1+\chi)\eta^{ex}}{\tau_c}. \tag{7.16}$$

Several relaxation mechanisms oppose the deformation. Bending stresses work to bring the shape back to its preferred curvature state; the corresponding time scale is

$$t_\kappa = \frac{(1+\chi)\eta^{ex} a^3}{\kappa}. \tag{7.17}$$

The factor $(1+\chi)$ in the above equations Eqs. 7.16 and 7.17 reflects the fact that the more viscous fluid controls the rate of deformation/relaxation. The strength of the relaxation mechanism that limits shape deformation imposed by the applied stresses is quantified by the corresponding dimensionless parameter: the bending number, similar to the capillary number commonly used for drops and capsules

$$B^{-1} = \frac{t_\kappa}{t_d} = \frac{\tau_c a^3}{\kappa}. \tag{7.18}$$

In flows with a rotational component, such as the shear flow (Figure 7.2), the rate of rigid-body rotation, t_r^{-1}, is proportional to the local shear rate,

$$t_r = \omega^{-1} \approx \dot{\gamma}^{-1}. \tag{7.19}$$

Vesicle rotation away from the extensional axis of the imposed flow effectively opposes particle stretching; the elongation produced by the straining flow is suppressed because within one period of rotation it gets convected toward the compressional axis. In linear flows, the rotation effect becomes important, that is, $t_r \le t_d$, for high-viscosity contrast particles

$$\chi = \frac{\eta^{in}}{\eta^{ex}} \approx \frac{t_d}{t_r}. \tag{7.20}$$

The interplay of these time scales leads to the complex dynamics of vesicles. Using the typical values for the material properties of a vesicle $\kappa \sim 10\, k_B T$, $a \sim 10\,\mu$m, we estimate that in a flow with shear rate $\dot{\gamma} = 1\,\text{s}^{-1}$, $B \sim 0.01$.

In the case of vesicles, in addition to the flow-related parameters, B and χ, there is a geometry-set one, the excess area, Δ (defined by Eq. 7.1). The smallest of these parameters (B, χ^{-1} or Δ) controls the magnitude of vesicle deformation, so in general there are several choices for a small parameter for a perturbation solution (see Rallison [1980] for an illuminating discussion about this issue in the "simpler" case of a drop.) In strong flows/electric fields, bending forces are negligible ($B \ll 1$) and deformation occurs via redistribution of excess area among various shape modes; accordingly, maximal deformation is set by the available excess area, which becomes the relevant small parameter (Misbah, 2006; Vlahovska and Gracia, 2007).

7.3 ASYMPTOTIC SOLUTION FOR SMALL DEFORMATIONS

The dynamics of a nearly spherical or a cylindrical vesicle are amenable to analytical solutions in the limit of small deviations from the initial configuration. In this section we consider the spherical geometry. The solutions for ellipsoidal and cylindrical vesicles (Boedec et al., 2014; Narsimhan et al., 2014; Rahimi et al., 2013) are conceptually similar.

We will address two problems: (1) How to determine the shape evolution of a vesicle subject to applied surface forces τ^s due to flow or electric field, and (2) what is the vesicle velocity relative to the background flow, that is, does a vesicle "go with the flow?"

7.3.1 PERTURBATION SCHEME

The instantaneous vesicle shape is parameterized relative to a reference unit sphere (i.e., sphere with a dimensionless radius 1) centered at $\mathbf{x}_p$, the particle center of mass (Figure 7.1)

$$\mathbf{x} = \mathbf{x}_p(t) + \big(1 + \varepsilon f(\theta,\varphi,t)\big)\hat{\mathbf{r}}, \tag{7.21}$$

where r, θ and φ are the radial distance, polar and azimuthal angles in a spherical coordinate system; $\hat{\mathbf{r}}$ is the unit radial vector. In general, the displacement of a material point, $\varepsilon(f(\theta,\varphi,t)\hat{\mathbf{r}} + \mathbf{u}_{||})$ has both radial and tangential (to a sphere) components; only the radial component, however, describes shape deformation. We will assume a nearly spherical vesicle such that $\varepsilon \ll 1$ is an appropriate small parameter. Then the flow about the vesicle can be found in terms of a perturbation expansion around a sphere

$$\mathbf{v} = \mathbf{v}^{(0)} + \varepsilon \mathbf{v}^{(1)} \ldots \tag{7.22}$$

The different terms in the expansion have physical meaning as follows. $\mathbf{v}^{(0)}$ corresponds to a flow around a sphere with no intrinsic interfacial stresses. $\mathbf{v}^{(1)}$ has two components linear in the shape deviation from sphere, f: (*i*) the flow driven by interfacial stresses trying to restore the unstressed membrane configuration in a quiescent fluid, and (*ii*) a disturbance in the applied flow produced by a viscous non-sphere described by deformed shape, $r_s = 1 + \varepsilon f$, with no intrinsic interfacial stresses.

Shape evolution

The shape evolution is determined from the kinematic condition Eq. 7.8

$$\mathbf{v}_{me} = \frac{d\mathbf{x}}{dt} = \mathbf{V}_p + \varepsilon\left(\frac{df}{dt}\hat{\mathbf{r}}\right), \tag{7.23}$$

where the material derivative is $d/dt = \partial/\partial t + \mathbf{v}_{me} \cdot \nabla$, and $\mathbf{V}_p = d\mathbf{x}_p/d$ is the particle translational velocity. The coordinate system and particle configuration is shown in Figure 7.1. Thus, we obtain for the radial displacement

$$\varepsilon \frac{\partial f}{\partial t} = \mathbf{v}_{me} \cdot \left(\hat{\mathbf{r}} - \varepsilon \nabla f\right) - \mathbf{V}_p \cdot \hat{\mathbf{r}} \quad \text{at} \quad r = 1. \tag{7.24}$$

Because translation does not lead to particle deformation (in the Stokes flow regime), the term $\mathbf{V}_p \cdot \hat{\mathbf{r}}$ is irrelevant in the case where we are interested in finding changes in shape. Inserting the flow decomposition Eq. 7.22 into the shape evolution equation Eq. 7.24 and expanding the velocity $\mathbf{v}$ in Taylor series about $r_s = 1 + \varepsilon f$ leads to

$$\varepsilon \frac{\partial f}{\partial t} = \mathbf{v}^{(0)}_{me} \cdot \hat{\mathbf{r}} + \varepsilon \left[\left(\mathbf{v}^{(1)}_{me} + f \frac{d\mathbf{v}^{(0)}}{dr} \right) \cdot \hat{\mathbf{r}} - \mathbf{v}^{(0)}_{me} \cdot \nabla f \right] + \ldots \quad \text{evaluated at} \quad r = 1. \tag{7.25}$$

Particle motion and cross-streamline migration

In Stokes flows, symmetry considerations dictate that a neutrally buoyant spherical particle will not migrate transversely to the local flow direction (although it can move with velocity different than the unperturbed flow velocity evaluated at the particle center). This result arises from the linearity of the Stokes equations and boundary conditions, and the symmetry of the problem under flow-reversal (Bretherton, 1962; Leal, 2007). However, such cross-stream drift may occur if the symmetry is lost, for example, by particle deformation in a shear gradient or in the presence of a wall (Leal, 1980).

Assuming that the particle velocity can be expanded in a similar manner as Eq. 7.22, and considering that translation does not give rise to deformation (i.e., $\partial f / \partial t = 0$), Eq. 7.24 leads to

$$0 = \left(-\mathbf{V}^{(0)}_p + \mathbf{v}^{(0)}_{me} \right) \cdot \hat{\mathbf{r}} + \varepsilon \left[-\mathbf{V}_p^{(1)} + \left(\mathbf{v}^{(1)}_{me} + f \frac{d\mathbf{v}^{(0)}}{dr} \right) \cdot \hat{\mathbf{r}} - \left(-\mathbf{V}^{(0)}_p + \mathbf{v}^{(0)}_{me} \right) \cdot \nabla f \right] \tag{7.26}$$

We will show in the next sections that at leading order, $O(\varepsilon)$ which corresponds to a sphere in flow, the particle can only move in the flow direction, although with a different velocity than the unperturbed flow. The velocity difference between the particle and the undisturbed flow velocity is $\mathbf{V}_p^{(0)}$. Only at next order, $O(\varepsilon^2)$, once the particle was deformed, the nonlinear term $(-\mathbf{V}_p^{(0)} + \mathbf{v}^{(0)}_{me}) \cdot \nabla f$ could lead to cross-streamline migration.

7.3.2 REPRESENTATION IN TERMS OF SPHERICAL HARMONICS

Particle geometry

Due to the spherical geometry of the problem, it is convenient to expand all variables in spherical harmonics (see Appendix 7.B for definitions). Let us rewrite the current position of a point on the interface, Eq. 7.21, in a coordinate system cotranslating with the particle

$$\mathbf{x} = \left(1 + \varepsilon f(\theta, \varphi, t)\right) \hat{\mathbf{r}} \tag{7.27}$$

$$f(\theta, \varphi, t) = \sum_{j=0, j \neq 1}^{\infty} \sum_{m=-j}^{j} f_{jm}(t) Y_{jm}(\theta, \varphi). \tag{7.28}$$

Here Y_{jm} are the scalar spherical harmonics, see Appendix 7.B for more details. The translational amplitudes f_{1m} are zero in the co-moving frame, and the particle velocity is determined from Eq. 7.37 below. The f_{00} amplitude is related to the other amplitudes because of conservation of particle volume and it can be shown (Seifert, 1999; Vlahovska et al., 2005) that

$$V = \frac{4\pi}{3} \left(1 + \varepsilon \frac{f_{00}}{\sqrt{4\pi}} \right)^3 + \varepsilon^2 \sum_{j \geq 2} \sum_{m=-j}^{j} f_{jm} f^*_{jm},$$

$$f_{00} = -\frac{\varepsilon}{\sqrt{4\pi}} \sum_{j \geq 2} \sum_{m=-j}^{j} f_{jm} f^*_{jm}, \tag{7.29}$$

where $f^*_{jm} = (-1)^m f_{j-m}$. Thus $V = 4\pi/3 + O(\varepsilon^2)$ and at linear perturbation order volume is conserved; one needs to worry about volume changes only when performing higher order perturbations (Barthès-Biesel and Acrivos, 1973; Danker et al., 2007; Vlahovska et al., 2005, 2009a).

A linear shape perturbation also preserves the total area, and the excess area, Δ, is

$$\Delta = A/a^2 - 4\pi = \int \frac{(1+\varepsilon f)^2}{\hat{r} \cdot \mathbf{n}} \sin\theta d\theta d\varphi - 4\pi$$

$$= \varepsilon^2 \sum_{jm} \frac{(j+2)(j-1)}{2} f_{jm} f^*_{jm} + O(\varepsilon^3), \tag{7.30}$$

where $\sum_{jm} \equiv \sum_{j \geq 2} \sum_{m=-j}^{j}$. The outward normal vector to a surface defined by a shape function $F = r - 1 - \varepsilon f(\theta, \varphi, t)$ is

$$\mathbf{n} = \frac{\nabla F}{|\nabla F|} = \hat{\mathbf{r}} - \varepsilon \sum_{jm} \sqrt{j(j+1)} f_{jm} \mathbf{y}_{jm0} + O(\varepsilon^2) \tag{7.31}$$

Accordingly, the mean curvature

$$H = -\frac{1}{2} \nabla \cdot \mathbf{n} = -\frac{1}{2} \left(\frac{2}{r} - \varepsilon \sum_{jm} f_{jm} \nabla_s^2 Y_{jm} \right)$$

$$= -1 - \frac{1}{2} \sum_{jm} \left(-2 + j(j+1) \right) f_{jm} Y_{jm} + O(\varepsilon^2) \tag{7.32}$$

where we used the fact that $\nabla \cdot \hat{\mathbf{r}} = 2/r = 2(1 - \varepsilon f) + O(\varepsilon^2)$ and $\nabla_s^2 Y_{jm} = -j(j+1) Y_{jm}$ on a unit sphere.

Flow

To solve for the flow, we adopt a basis of fundamental solutions of the Stokes equations in a spherical geometry, as listed in Appendix 7.D,

$$\mathbf{v}^{\text{ex}} = \mathbf{v}^{\infty} + \sum_{jmq} c^{-}_{jmq} \hat{\mathbf{v}}^{-}_{jmq}(r), \quad \mathbf{v}^{\text{in}} = \sum_{jmq} c^{+}_{jmq} \hat{\mathbf{v}}^{+}_{jmq}(r). \quad (7.33)$$

Summation over repeated indices is implied; q takes values 0, 1, and 2. The functions $\hat{\mathbf{v}}^{\pm}_{jmq}$ are vector solid spherical harmonics related to the harmonics in the Lamb solution. With respect to a sphere, $\hat{\mathbf{v}}^{\pm}_{jm2}$ is radial, whereas $\hat{\mathbf{v}}^{\pm}_{jm0}$ and $\hat{\mathbf{v}}^{\pm}_{jm1}$ are tangential; $\hat{\mathbf{v}}^{\pm}_{jm1}$ is surface-solenoidal ($\nabla_s \cdot \hat{\mathbf{v}}^{\pm}_{jm1} = 0$). The velocity coefficients $c^{\pm}_{jmq}$ are determined from the condition for velocity continuity and the stress balance. The far-field (imposed) flow is specified by $\mathbf{v}^{\infty} = c^{\infty}_{jmq} \hat{\mathbf{v}}^{+}_{jmq}$ and the coefficients for simple shear and plane Poiseuille flow are listed in Appendix 7.C.1.

Note that c^{-}_{1m1} is related to the torque exerted on the particle, c^{-}_{1m0} and c^{-}_{1m2}, to the force, and c^{-}_{2m0} and c^{-}_{2m2}, to the stresslet. For example, the $\mathbf{z}$ components of the torque, M, and the force, F, exerted on a particle are (Schmitz and Felderhof, 1982)

$$M_z = -\mathrm{i}\sqrt{24\pi}\, c^{-}_{101}, \quad F_z = -\sqrt{3\pi}\left(\sqrt{2}c^{-}_{100} + c^{-}_{102}\right). \quad (7.34)$$

Note that the torque and the force are nondimensionalized by $\tau_c a^3$ and $\tau_c a^2$, respectively.

Shape evolution

The interface deformation, Eq. 7.25, driven by a flow around a sphere is simply

$$\varepsilon \frac{\partial f_{jm}}{\partial t} = c^{+}_{jm2}. \quad (7.35)$$

The flow includes two components: flow about the sphere free of interfacial stresses, $\mathbf{v}^{(0)}$, and the flow driven by interfacial stresses on a deformed particle, trying to restore the unstressed particle shape, $\mathbf{v}^{(1)}$. The solution for the flow is straightforward at the leading order where the particle shape is a sphere: the velocity continuity reduces to $c^{+}_{jmq} = c^{-}_{jmq} + c^{\infty}_{jmq}$.

The shape evolution that is due to the perturbation in the applied flow by the deformed particle requires the evaluation of the nonlinear terms, $\mathbf{v}\cdot\nabla f$, and $f d\mathbf{v}/dr$. This requires evaluating of products of spherical harmonics, which is not trivial. In the case of a surface-incompressible membrane, the problem is a bit alleviated because $\nabla_s \cdot \mathbf{v} = 0$ and accordingly $\hat{\mathbf{r}} \cdot \mathrm{d}\mathbf{v}^{(0)}/\mathrm{d}r = 0$ on a sphere $r = 1$; the challenge of evaluating $\mathbf{v}^{(0)} \cdot \nabla f$, however, remains. In this chapter, we will focus on cases in which only the linear terms are relevant. We will consider only some very simple scenarios illustrating the effect of the nonlinear term. For example, if the particle rotates, the time derivative should be replaced by the Jaumann derivative. This situation arises when the particle is placed in shear flow or subjected to electric torque (e.g., the phenomenon of Quincke rotation observed with rigid particles and drops [Salipante and Vlahovska, 2013]). If the rotation rate is fast, $\omega \sim \varepsilon^{-1}$, the partial time derivative should be replaced by $D_t f_{jm} = \partial_t f_{jm} - \mathrm{i}m\omega f_{jm}$, in the case of clockwise rotation around the z axis.

Particle motion

In a fixed reference frame, the time-dependent vesicle position is specified by the particle center of mass

$$\mathbf{x}_p(t) = x_p(t)\hat{\mathbf{x}} + y_p(t)\hat{\mathbf{y}} + z_p(t)\hat{\mathbf{z}} = \sum_{m=-1}^{1} f_{1m}(t) Y_{1m} \hat{\mathbf{r}}, \quad (7.36)$$

where Y_{1m} are the scalar spherical harmonics of order one. The relation between the Cartesian and spherical harmonics representations of the particle position is $\mathbf{z}_p = f_{10}\sqrt{3/4\pi}$, $x_p = -(f_{11} - f_{1-1})\sqrt{3/8\pi}$, and $y_p = \mathrm{i}(f_{11} + f_{1-1})\sqrt{3/8\pi}$. Accordingly, the particle translational velocity, $\mathbf{V}_p = \mathrm{d}\mathbf{x}_p/\mathrm{d}t$, is

$$V_{p,z} = \sqrt{\frac{3}{4\pi}} \frac{\partial f_{10}}{\partial t}, \quad V_{p,x} = -\sqrt{\frac{3}{8\pi}} \frac{\partial}{\partial t}\left(f_{11} - f_{1-1}\right),$$
$$V_{p,y} = -\mathrm{i}\sqrt{\frac{3}{8\pi}} \frac{\partial}{\partial t}\left(f_{11} + f_{1-1}\right). \quad (7.37)$$

The equation for the particle velocity Eq. 7.26 shows that at leading order (a sphere)

$$\mathbf{V}^{(0)}_{\mathrm{p}} \cdot \hat{\mathbf{r}} = \mathbf{v}_{\mathrm{me}} \cdot \hat{\mathbf{r}} \Rightarrow \frac{\partial f_{1m}}{\partial t} = c^{+}_{1m2} \quad m = 0, \pm 1 \quad \text{at} \quad r = 1 \quad (7.38)$$

Combining Eqs. 7.37 and 7.38 yields

$$\mathbf{V}^{(0)}_{\mathrm{p}} = \sqrt{\frac{3}{8\pi}}\left[-(c^{+}_{112} - c^{+}_{1-12})\hat{\mathbf{x}} - \mathrm{i}(c^{+}_{112} + c^{+}_{1-12})\hat{\mathbf{y}} + \sqrt{2}c^{+}_{102}\hat{\mathbf{z}}\right]. \quad (7.39)$$

7.3.3 VELOCITY FIELD ABOUT A SPHERE

Here we show the velocity field corresponding to a stress distribution on the interface given by

$$\boldsymbol{\tau} = \sum_{jmq} \boldsymbol{\tau}^{s}_{jmq}. \quad (7.40)$$

The stresses can include "shape-distorting" tractions exerted by an applied flow, electric field, or active forces, τ^{a}, and "shape-restoring" viscoelastic membrane stresses, τ^{me},

$$\boldsymbol{\tau}^{s}_{jmq} = \boldsymbol{\tau}^{a}_{jmq} + \boldsymbol{\tau}^{\text{me}}_{jmq} \quad (7.41)$$

The expressions for the applied stresses are listed in Appendix 7.C and the membrane stresses are discussed in Section 7.3.4.

Applying the stress balance, $\tau^{\text{hd,ex}}_{jmq} - \chi\tau^{\text{hd,in}}_{jmq} = \tau^{s}_{jmq}$ (see Appendix 7.D for the definitions of the hydrodynamic tractions τ^{hd}), and the velocity continuity, $c^{+}_{jmq} = c^{-}_{jmq} + c^{\infty}_{jmq}$, conditions leads to

$$c^{+}_{jm2} = -\frac{j(1+j)\tau^{s}_{jm2} + 2\sqrt{j(1+j)}\,\tau^{s}_{jm0}}{2j^3 + 3j^2 + 4 + 4\left(j^2 + j - 2\right)\chi^{\text{me}} + \left(2j^3 + 3j^2 - 5\right)\chi}. \quad (7.42)$$

and

$$c^{+}_{jm1} = -\frac{1}{2+j+(j-1)\left(\chi+(j+2)\chi^{\text{me}}\right)}\tau^{s}_{jm1}. \quad (7.43)$$

The area-incompressibilty relates the amplitudes of the tangential and radial velocities, $2c^{+}_{jm2} = c^{+}_{jm0}\sqrt{j(1+j)}$.

7.3.4 INTERFACIAL STRESSES

Let us first start with the simple case of a drop. Its interface is governed by an isotropic surface tension γ. Deformation resulting in interfacial curvature H gives rise to capillary stress $-2\gamma H$. To linear order in the shape deviation from sphere, see Eq. 7.32, the restoring capillary stress (in dimensionless form) is

$$\tau^{\Sigma}_{jm0} = 0, \quad \tau^{\Sigma}_{jm1} = 0, \quad \tau^{\Sigma}_{jm2} = \quad Ca^{-1}\left(-2+j(j+1)\right)f_{jm}, \quad (7.44)$$

where $Ca = \tau_c a/\gamma$. The isotropic part in the capillary stress is omitted because it is irrelevant to particle dynamics and deformation.

In the case of a viscous area-incompressible interface, the stresses obtained from Eq. 7.15 are (Schwalbe et al., 2011)

$$\tau^{BS}_{jm0} = \chi^{\text{me}}(j-1)(j+2)c^{+}_{jm0},$$
$$\tau^{BS}_{jm1} = -\chi^{\text{me}}(j-1)(j+2)c^{+}_{jm1}, \quad \tau^{BS}_{jm2} = 0. \quad (7.45)$$

Note that the viscoelastic stresses due to solenoidal displacement and velocity fields (“$jm1$” type) are decoupled from the radial and tangential, “$jm2$” and “$jm0$” fields. Thus, to leading order “$jm1$” type forcing cannot excite “$jm2$” and $jm0$ flow and deformation, and vice versa.

To linear order, in the bending tractions Eq. 7.13 the terms H^3 and HK do not contribute, and only the term $2\nabla^2_s H$ is relevant. Taking the Laplacian of Eq. 7.32 (and using the fact that $\nabla^2_s Y_{jm} = -j(j+1)Y_{jm}$ results in

$$\tau^{B}_{jm0} = 0, \quad \tau^{B}_{jm1} = 0, \quad \tau^{B}_{jm2} = Bj(j+1)(j-1)(j+2)f_{jm}. \quad (7.46)$$

7.4 EXAMPLES: A QUASI-SPHERICAL VESICLE IN EXTERNAL FIELDS

A spherical vesicle with a surface-incompressible membrane cannot deform. Vesicles, however, although appearing spherical under the microscope, deform under stress. An example of such deformation in an electric field is shown in Figure 7.3.

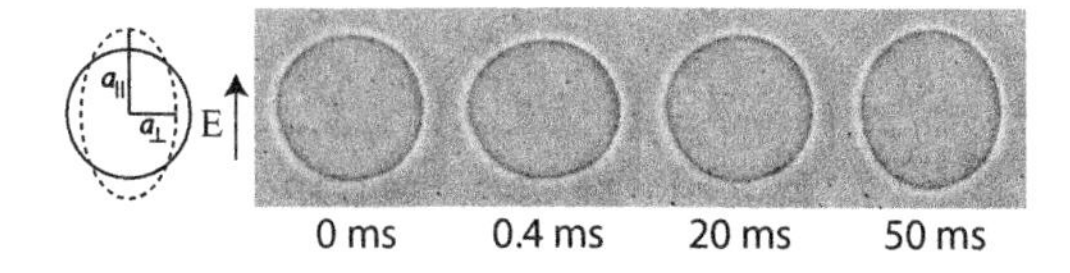

Figure 7.3 Deformation of a quasi-spherical vesicle upon application of a uniform DC electric field. (Salipante, P.F. and Vlahovska, P.M., *Soft Matter*, 10, 3386–3393, 2014. Reproduced by permission of The Royal Society of Chemistry.) Initial vesicle radius $a = 24.5$ µm.

The increase of apparent area comes from pulling area stored in suboptical membrane undulations. If the vesicle is nonfluctuating it has to be nonspherical at equilibrium in order to deform.

Because a drop is also spherical at equilibrium and its area increases under stress, it is tempting to model the vesicle as a drop, that is, a fluid particle with an extensible interface. However, the surface flows of a vesicle and a drop are fundamentally different: In the case of the lipid membrane, the flow is solenoidal, and the membrane tension depends on deformation. One consequence is that, in axisymmetric extensional flow or a uniform electric field, the fluid inside the vesicle is motionless, in contrast to the (surfactant-free, “clean”) drop. Note that even though the drop shape in a uniform electric field is stationary, the fluid inside flows in a toroidal flow pattern in the two drop hemispheres; the surface flow is either from the pole toward the equator or vice versa (depending on the fluid properties) (Taylor, 1966). In the case of vesicles, the tension gradients immobilize the interface and prevent surface flow. Next we review vesicle dynamics in linear flows. The analysis for the case of electric fields of a nonfluctuating, initially nonspherical (Schwalbe et al., 2011) or a fluctuating, quasi-spherical (Salipante and Vlahovska, 2014) vesicle follows the same steps.

7.4.1 DYNAMICS IN SIMPLE SHEAR FLOW: TANK TREADING AND TUMBLING

The shear flow excites the ellipsoidal shape modes $j = 2$. Inserting the expressions for the bending Eq. 7.46, tension Eq. 7.44 (with $Ca^{-1}_{\Sigma} = B\bar{\Sigma}_0$) and flow stresses Eq. 7.92 in Eqs. 7.42 and 7.35 we obtain

$$\frac{\partial f_{2m}}{\partial t} = -\mathrm{i}\frac{m}{2}\Lambda^{-1}\delta_{|m|2} + \mathrm{i}\frac{m}{2}f_{2m} - \varepsilon\Lambda^{-1}\sqrt{\frac{6}{5\pi}}\left[B(6+\bar{\Sigma}_0)f_{2m}\right]. \quad (7.47)$$

In Eq. 7.47 $\bar{\Sigma}_0$ is the isotropic part of the tension rescaled by the bending stress $\bar{\Sigma}_0 = \Sigma_0 a^2/\kappa$, $\delta_{|m|2}$ is the Kronecker delta function, and

$$\Lambda = \varepsilon\frac{(32+23\chi+16\chi^{\text{me}})}{4\sqrt{30\pi}} \quad (7.48)$$

The f_{20} and $f_{2\pm1}$ are slaved to the $f_{2\pm2}$ modes, which are excited by the imposed shear. Without loss of generality, we set $f_{2\pm1} = 0$; the $f_{2\pm1}$ modes describe deformations out of the shear plane, which are qualitatively captured by keeping only the f_{20} mode.

In Eq. 7.47, the tension, $\bar{\Sigma}_0$, needs to be determined self-consistently with deformation (Seifert, 1999). Under stress, a quasi-spherical vesicle deforms by pulling excess area stored in fluctuations. The area constraint requires that the total excess area (stored in fluctuations, Δ_f, and systematic deformation, $\bar{\Delta}$) is constant. Accordingly, $\Delta = \bar{\Delta} + \Delta_f$ is solved numerically at each time step to determine $\bar{\Sigma}_0$.

The area stored in fluctuations (Seifert, 1999) is

$$\Delta_f = \frac{k_B T}{2\kappa}\ln\left(\frac{j^2_{max} + j_{max} + \bar{\Sigma}_0}{6+\bar{\Sigma}_0}\right), \quad (7.49)$$

where $j_{max} = a / \ell_{me}$ is the ratio of the vesicle radius to the membrane thickness. The excess area corresponding to the systematic deformation is given by Eq. 7.30 as

$$\bar{\Delta} = \varepsilon^2 \sum_{jm} \frac{(j+2)(j-1)}{2} \bar{f}_{jm} \bar{f}^*_{jm} = 2\varepsilon^2 \left(2\bar{f}_{22}\bar{f}_{2-2} + \bar{f}^2_{20} \right), \quad (7.50)$$

where $\bar{f}_{jm}$ corresponds to a quasi-steady shape that satisfies Eq. 7.47 with $\partial f_{2m} / \partial t = 0$. If all excess area is to be engaged in the systematic deformation then $\varepsilon^2 \sim \Delta$, suggesting a choice of the small parameter, $\varepsilon = \Delta^{1/2} / 2$.

If the vesicle is not fluctuating, $\Delta_f = 0$, one can derive the explicit form for the tension using

$$\dot{\Delta} = 0 = \varepsilon^2 \sum_{jm} \frac{(j+2)(j-1)}{2} \left(\dot{f}_{jm} f^*_{jm} + f_{jm} \dot{f}^*_{jm} \right) \quad (7.51)$$

where the dot denotes time derivative. In shear flow, which involves only the $j = 2$ modes, this leads to

$$\dot{f}_{22} f_{2-2} + f_{22} \dot{f}_{2-2} + f_{20} \dot{f}_{20} = 0 \quad (7.52)$$

Inserting Eq. 7.47 in the above equation, using Eq. 7.74, and solving for the tension yields

$$\bar{\Sigma}_0 = -6 + \varepsilon \frac{iB^{-1}}{\Delta} \sqrt{\frac{10\pi}{3}} \left(f_{22} - f_{2-2} \right). \quad (7.53)$$

Note that with the choice of small parameter set by the excess area, that is, $\varepsilon = \Delta^{1/2} / 2$, $\bar{\Sigma}_0 \sim 1/\varepsilon$! Inserting Eq. 7.53 in Eq. 7.47 leads to a nonlinear evolution equation even at leading order, in contrast to droplets and initially spherical vesicles (Danker et al., 2007; Kaoui et al., 2009; Lebedev et al., 2008; Misbah, 2006; Schwalbe et al., 2010; Vlahovska and Gracia, 2007)

$$\dot{f}_{2m} = -i\frac{m}{2}\Lambda^{-1} + i\frac{m}{2} f_{2m} - \frac{1}{2}\Lambda^{-1}(f_{22} - f_{2-2}) f_{2m}, \quad m = \pm 2. \quad (7.54)$$

Because f_{20} is slaved to the $f_{2\pm2}$ modes, the evolution Eq. 7.54 can be used to determine $f_{2\pm2}$, whereas f_{20} is determined from the area constraint Eq. 7.74:

$$f_{20} = \left[2(1 - f_{22} f_{2-2}) \right]^{1/2} = \left[2(1 - R^2) \right]^{1/2}. \quad (7.55)$$

Instead of shape modes, the vesicle dynamics can be also described in terms of the orientation angle, ψ, and R, which measures the ellipticity of the vesicle contour in the x-y plane (Misbah, 2006), $f_{2\pm2} = R\exp(\mp 2i\psi)$. Eq. 7.54 takes the form (Box 7.1)

Analysis of the fixed points of the set of coupled nonlinear equations Eqs. 7.56 and 7.57 shows

- A stable fixed point ($R^* = 1, \psi^* = 1/2 \arccos \Lambda$). Physically, this corresponds to the TT state characterized by a steady inclination angle.

Box 7.1 Evolution equations for the orientation and shape of a nonfluctuating vesicle in a simple shear flow

- Inclination angle between the vesicle major axis and the flow direction

$$\dot{\psi} = -\frac{1}{2} + \frac{\Lambda^{-1}}{2R} \cos(2\psi). \quad (7.56)$$

- Asphericity of the vesicle contour in the shear plane

$$\dot{R} = \Lambda^{-1}(1 - R^2)\sin(2\psi). \quad (7.57)$$

- A closed orbit centered at ($\psi^* = 0, R^* = \Lambda^{-1}$). This corresponds to the TR state (Kantsler and Steinberg, 2006), also called vacillating-breathing (Misbah, 2006) and swinging (Noguchi and Gompper, 2007). In this motion, the mean inclination angle of the vesicle oscillates between $\pm\pi/4$ and the vesicle shape undergoes large deformations along the vorticity axis z.
- No equilibrium point. This corresponds to the TB state, where the mean inclination angle of the vesicle continuously increases, that is, the vesicle is flipping.

The TT fixed point loses stability at $\Lambda_c = 1$ and the vesicle starts tumbling. This corresponds to a critical viscosity ratio χ_c or membrane shear viscosity χ_c^{me}. These are obtained by setting Eq. 7.48 equal to 1 (and recalling that $\varepsilon = \sqrt{\Delta} / 2$)

$$\varepsilon \frac{(32 + 23\chi_c + 16\chi_c^{me})}{4\sqrt{30\pi}} = 1 \Rightarrow$$
$$\chi_c = \frac{8}{23}\left(\sqrt{\frac{30\pi}{\Delta}} - 2\left(2 + \chi^{me}\right) \right), \quad (7.58)$$
$$\chi_c^{me} = \sqrt{\frac{2\pi}{15\Delta}} - 2 - \frac{23}{16}\chi.$$

The TB and TR modes coexist within the leading-order theory (Kaoui et al., 2009; Lebedev et al., 2008), and the mode selection is determined by the initial conditions. If there is no deformation along the vorticity direction, that is, $f_{20} = 0$ at all times, Eq. 7.55 implies that R remains constant and equal to its maximum value 1. This situation resembles the classic model by Keller and Skalak (1982): the vesicle shape is a fixed ellipsoid and the vesicle dynamics is described only by the variations of the angle ψ (note, however, that unlike the Keller-Skalak solution, our velocity field is strictly area-incompressible). In this case the bifurcation is from TT to TB. If $f_{20} \neq 0$, the transition is from TT to TR. In the trembling mode, the vesicle undergoes periodic shape deformations along the vorticity direction. As a result, the vesicle appears to "breath" in the flow direction.

7.4.2 RHEOLOGY OF A VESICLE SUSPENSION

The effective stress, **T**, of a sheared dilute suspension, in which hydrodynamic interactions between the particles are negligible, is found from the individual particle stresslet, **S** (Kim and Karrila, 1991),

$$\mathbf{T}=\eta_{ex}\dot{\gamma}\left(2\Gamma^{s}+\phi\mathbf{S}\right), \quad (7.59)$$

where ϕ is the particle volume fraction, Γ^{s} is the symmetric part of the gradient velocity tensor describing the extensional component of the shear flow, $\Gamma^{s}\cdot\mathbf{x}=\frac{1}{2}(y,x,0)$. Rheological properties of interest are the vesicle contribution to the shear viscosity, S_{xy} and the normal stress differences, $N_1=T_{xx}-T_{yy}=\phi(S_{xx}-S_{yy})$ and $N_2=T_{yy}-T_{zz}=\phi(S_{yy}-S_{zz})$ (Danker and Misbah, 2007; Danker et al., 2008; Farutin and Misbah, 2012a; Vlahovska and Gracia, 2007). The suspension shear viscosity increases compared with the particle-free fluid viscosity as $\eta_{eff}=\eta_{ex}\left(1+\phi S_{xy}\right)$, where

$$S_{xy}=\frac{5}{2\left(23\hat{\chi}+32\right)}\left[23\hat{\chi}-16+48\left(R\sin(2\psi)\right)^{2}\right]. \quad (7.60)$$

$\hat{\chi}=\chi+16\chi^{me}/23$ is the effective viscosity ratio that accounts for both bulk and membrane viscosities, ψ and R are given by Eqs. 7.56 and 7.57, respectively. The normal stress is

$$N_1=-2N_2=\phi\frac{120}{(23\hat{\chi}+32)}R^{2}\sin\left(4\psi\right). \quad (7.61)$$

In the TT regime ($R=1$ and $\psi=1/2\arccos\Lambda$), the viscosity coefficient and normal stresses are

$$S_{xy}=\frac{5}{2}-\Delta\frac{(23\hat{\chi}+32)}{16\pi},\; N_1=-2N_2=\phi\frac{120}{(23\hat{\chi}+32)}\sin\left(2\Lambda\right) \quad (7.62)$$

The basic results for the rheology of dilute suspensions are summarized in Box 7.2. In the limit of a spherical vesicle, $\Delta=0$, and because a sphere with fixed area and volume in shear flow can only undergo rigid-body rotation, Einstein's result for a suspension of rigid spheres is recovered. The effective viscosity decreases with the increase of the excess area because the deformable vesicles elongate and thus offer less resistance to the flow. An increasing viscosity contrast also leads to a decrease in the effective viscosity because vesicles align better with the flow (the inclination angle ψ decreases with viscosity contrast). This expression for the effective viscosity of a suspension agrees surprisingly well with experimental data on rheology of dilute blood (Vitkova et al., 2008). In the TB regime, the suspension viscosity becomes time-dependent (Danker and Misbah, 2007; Danker et al., 2008; Ghigliotti et al., 2010), and its time-averaged value increases compared with the TT regime. Finally higher order evolution equations (Farutin et al., 2010) are necessary in order to have full agreement with the full numerical simulations (Biben et al., 2011).

Box 7.2 Rheology of a dilute suspension under steady shear flow

- **Solid spheres**

effective viscosity $\quad \eta_{eff}=\eta^{ex}\left(1+\frac{5}{2}\phi\right),$

normal stresses $\quad N_1=N_2=0 \quad (7.63)$

- **Spherical drops**

effective viscosity $\quad \eta_{eff}=\eta^{ex}\left(1+\frac{1+\frac{5}{2}\chi}{1+\chi}\phi\right),$

normal stresses $\quad N_1=N_2=0 \quad (7.64)$

- **Vesicles**

effective viscosity $\quad \eta_{eff}=\eta^{ex}\left(1+\left(\frac{5}{2}-\Delta\frac{(23\hat{\chi}+32)}{16\pi}\right)\phi\right),$

normal stresses $\quad N_1=-2N_2=\phi\sqrt{\frac{30\Delta}{\pi}} \quad (\Delta\ll1) \quad (7.65)$

7.4.3 WALL-INDUCED CROSS-STREAMLINE MIGRATION

A spherical particle in simple shear flow produces a symmetric disturbance velocity field and, therefore, in the low-Reynolds-number limit the particle does not drift relative to a bounding wall (Leal, 1980). Particle deformation breaks the symmetry and may lead to cross-streamline migration.

The leading order term in the far field of the disturbance velocity due to a force-free and torque-free particle is the stresslet, the symmetric and traceless force dipole. The boundary conditions at the wall can be satisfied by placing the stresslet hydrodynamic image on the opposite side (Kim and Karrila, 1991). Thus, a particle far from the wall moves with a velocity due to its corresponding image stresslet; in particular, the vesicle drift velocity, $V_{p,y}$, normal to a rigid wall is proportional to the stresslet component in the direction of the unit normal to the wall, that is S_{yy} (Smart and Leighton, 1990)

$$V_{p,y}=-\frac{3}{16d^{2}}S_{yy}=\frac{1}{d^{2}}\frac{3}{32}\frac{N_1}{\phi} \quad (7.66)$$

where d denotes the distance from the particle center to the wall.

For vesicles, the nonspherical rest shape provides a source of lift because the first normal stress difference is nonzero, see Eq. 7.61. In the TT regime, $\sin(\arccos\Lambda)\sim\varepsilon$ and hence $N_1\sim\varepsilon$ which shows that the more deflated the vesicle, that is the larger the excess area, the larger the migration velocity.

7.4.4 A VESICLE IN POISEUILLE FLOW

Here we illustrate the facility of the spherical harmonics approach by calculating vesicle slip in Poiseuille flow in just one line, in contrast to the tedious derivation with traditional approaches utilized for capsules (Helmy and Barthès-Biesel, 1982).

Analytical solutions are possible if we consider the limit of a quasi-steady (in addition to slightly perturbed) shape. Quasi-steadiness implies that the shape evolution occurs on a faster time scale than the vesicle migration. Eq. 7.39 leads to

$$\mathbf{V}_\mathrm{P} \cdot \hat{\mathbf{e}}_m = -\frac{2}{9}\left(\tau^s_{1m2} + \sqrt{2}\tau^s_{1m0}\right) \quad \text{only nonzero if } m = 0, \pm 1, \tag{7.67}$$

where $\hat{\mathbf{e}}_0 = \hat{\mathbf{z}}$ and $\hat{\mathbf{e}}_{\pm 1} = (\hat{\mathbf{x}} \pm i\hat{\mathbf{y}})/\sqrt{2}$. Poiseuille flow has a $j = 1$ component, see Eq. 7.90. The associated flow stresses cause the vesicle to lag the flow and its (dimensional) slip velocity is the same as for a rigid sphere $V_\mathrm{P} = -\delta/3$ in plane Poiseuille flow and $-2\delta/3$ in axisymmetric Poiseuille flow. The reason the vesicle behaves as a rigid sphere is that the $j = 1$ flow generates in-plane material point displacement and the associated tension gradients (which arise to preserve area-incompressibility) immobilize the vesicle surface.

Cross-streamline migration occurs at next order due to the interaction of the deformed shape and the perturbed flow. This problem has been solved analytically for vesicles (Danker et al., 2009), but the calculation is inconsistent because the velocity field in $\mathbf{v} \cdot \nabla f$ does not include the flow driven by membrane tension (which albeit linear in the shape deviation from a sphere is $O(1)$ because the membrane tension is $\sim 1/\varepsilon$ due to the total area constraint, see Eq. 7.53). For vesicles a consistent calculation has been presented recently (Farutin and Misbah, 2013). In general, this is a very complex analytical calculation, which is prone to errors, and surprisingly there is still no consensus even about the result for "simple" drops (Chan and Leal, 1979; Leal, 1980)!

7.4.5 OBLATE-PROLATE TRANSITION IN A UNIFORM DIRECT OR ALTERNATING CURRENT FIELD

The electrodeformation of GUVs made of artificial bilayer membranes provides fundamental insights into the electromechanics of biomembranes, namely the coupling of membrane shape and transmembrane potential (Dimova et al., 2009; Vlahovska, 2015). The steady shapes of vesicles in alternating current (AC) fields have been extensively studied experimentally (Aranda et al., 2008; Salipante et al., 2012) and theoretically (Nganguia et al., 2013; Vlahovska et al., 2009b; Yamamoto et al., 2010; Vlahovska, 2019). The behavior in DC pulses has been considered theoretically (Schwalbe et al., 2011; Zhang et al., 2013), but only limited experimental data is available due to problems with visualization and vesicle fragility (Riske and Dimova, 2005, 2006; Riske et al., 2009; Sadik et al., 2011; Salipante and Vlahovska, 2014). Only recently the theoretically predicted transition from an oblate to prolate ellipsoidal shape in the case of a quasi-spherical vesicle encapsulating solution less conducting than the suspending medium (Schwalbe et al., 2011) has been experimentally observed (Salipante and Vlahovska, 2014). The transition is detected by utilizing a two-step DC pulse in order to avoid electroporation and vesicle collapse.

Here, we show how to compute the steady and transient shapes of a quasi-spherical vesicle made of charge-free membrane in AC and DC uniform electric fields. The physical mechanisms of vesicle deformation in uniform electric field are summarized in Box 7.3. We will not go into details of how to solve the electrostatic problem (a sphere in a uniform electric field is a straightforward calculation [Schwalbe et al., 2011; Vlahovska, 2010; Vlahovska et al., 2009b]). We illustrate how to include the area increase due to ironing the suboptical fluctuations.

Box 7.3 Vesicle in a uniform electric field: Physical picture

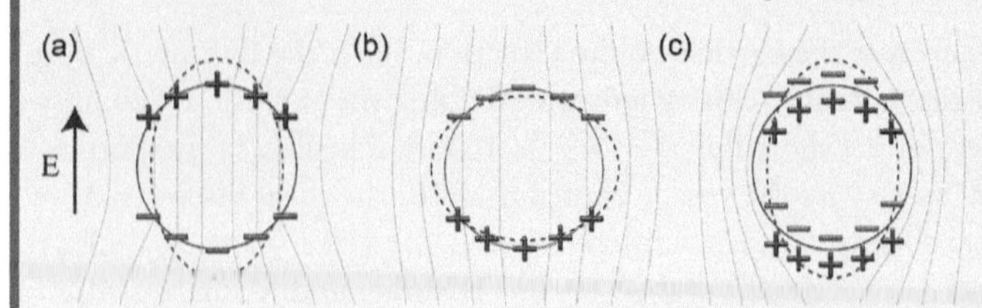

Application of an electric field leads to accumulation of ions at the membrane surfaces and the vesicle acts as a capacitor.

- At short times (intermediate frequencies), $t < t_\mathrm{me}$ ($\nu > \nu_c$ the membrane capacitor is short-circuited and there is charge imbalance between the inner and outer membrane surfaces. If the enclosed solution is more conducting than the suspending medium, $\Lambda_\sigma > 1$, the vesicle is pulled into an prolate ellipsoid. The polarization is reversed in the opposite case, $\Lambda_\sigma < 1$, and the vesicle deforms into an oblate ellipsoid.
- At long times (low frequencies) $t \gg t_\mathrm{me}$ ($\nu < \nu_c$) the membrane capacitor is fully charged and vesicle shape is an prolate ellipsoid at any Λ_σ. The dashed lines indicate the vesicle deformation.

The capacitor charging time (Grosse and Schwan, 1992; Kinosita et al., 1988; Schwan, 1989) and critical frequency (Yamamoto et al., 2010) are

$$t_\mathrm{me} = \frac{aC_\mathrm{me}}{\sigma^\mathrm{in}}\left(1 + \frac{\Lambda_\sigma}{2}\right),$$
$$\nu_c = \frac{\sigma^\mathrm{in}}{2\pi a C_\mathrm{me}}\left[(1-\Lambda_\sigma)(\Lambda_\sigma + 3)\right]^{-1/2}. \tag{7.68}$$

A vesicle with $\Lambda_\sigma < 1$ in a DC pulse may initially deform into an oblate spheroid but eventually adopts a prolate shape (Salipante and Vlahovska, 2014; Schwalbe et al., 2011). Similar transition in an AC field is readily observed by lowering the field (linear) frequency (Aranda et al., 2008; Vlahovska et al., 2009b) below the critical value ν_c.

We derive the evolution equation for vesicle deformation in response to an electric field in a similar manner to Section 7.4.1. A uniform electric field $\mathbf{E} = E_0\hat{\mathbf{z}}$ excites only $j = 2$, $m = 0$ (axisymmetric) mode at leading order. Hence, the vesicle shape is described by only one parameter, s, $r_s = 1 + sP_2(\cos\theta)$, where $P_2(\cos\theta)$ is the second-order Legendre polynomial. The aspect ratio (defined as the ratio of the axes parallel and perpendicular to the applied field) is then $D = (1 - s/2)/(1 + s)$.

$$\delta_m \frac{\partial s}{\partial t} = C(t) - \bar{\Sigma}_0 B \frac{24}{55 + 16\chi^\mathrm{me}} s(t), \tag{7.69}$$

where $\delta_m = t_\mathrm{me}/t_d$ is the ratio of the capacitor charging and electrohydrodynamic time scales, $t_d = \eta^\mathrm{ex}(1+\chi)/\varepsilon^\mathrm{ex}E_0^2$. For simplicity, we set $\chi = 1$ and $\Lambda_\varepsilon = 1$, because the inner and outer fluids are aqueous solutions with similar viscosity and permittivity. The forcing term due to the electric field is

$$C(t) = \frac{2\left[P_\mathrm{ex}(t) - 2\Theta(t - t_\mathrm{e})\right]^2 - 8P_\mathrm{in}(t)^2}{55 + 16\chi^\mathrm{me}}, \tag{7.70}$$

where $P_{ex}(t)$ and $P_{in}(t)$ are electric field coefficients defined as

$$P_{ex}(t) = \frac{(1-\Lambda_\sigma)\Theta(t-t_e)+\Lambda_\sigma \bar{V}_{me}(t)}{\Lambda_\sigma + 2},$$
$$P_{in}(t) = \frac{3\Theta(t-t_e)-2\bar{V}_{me}(t)}{\Lambda_\sigma + 2}. \tag{7.71}$$

$t_e = T / t_d$ is the dimensionless duration of the DC pulse. $\Theta(t-t_e) = 1$ if $t \le t_e$ and zero otherwise. The time dependence in the electric stress arises from the capacitor charging and discharging

$$V_{me}(t) = aE_0\bar{V}_{me}(t)\cos\theta, \tag{7.72}$$

$$\bar{V}_{me}(t) = \begin{cases} \frac{3}{2}\left(1-e^{-t/\delta_m}\right) & t \le t_e \\ \frac{3}{2}\left(1-e^{-t_e/\delta_m}\right)e^{-(t-t_e)/\delta_m} & t > t_e, \end{cases} \tag{7.73}$$

In Eq. 7.69, the tension Σ_0 needs to be determined self-consistently with deformation (Seifert, 1999) as explained in Section 7.4.1. Under stress, a quasi-spherical vesicle deforms by pulling excess area stored in fluctuations. The area constraint requires that the total excess area (stored in fluctuations, Δ_f, and systematic deformation, $\bar{\Delta}$) is constant. Accordingly, $\Delta = \bar{\Delta} + \Delta_f$ is solved numerically at each time step to determine Σ_0. The area stored in fluctuations is given by Eq. 7.49 and the excess area corresponding to the systematic deformation is

$$\bar{\Delta} = 8\pi\bar{s}^2 / 5 \tag{7.74}$$

where $\bar{s}$ corresponds to a quasi-steady shape that satisfies Eq. 7.69 with $\partial\bar{s}/\partial t = 0$.

7.4.6 DYNAMICS IN A COMBINED SHEAR FLOW AND UNIFORM ELECTRIC FIELD

Let us consider a vesicle subjected to a linear flow with strain-rate magnitude $\dot{\gamma}$ and a uniform DC electric field with magnitude E_0,

$$\mathbf{v}^\infty = \dot{\gamma}\, y\hat{\mathbf{x}}, \quad \mathbf{E}^\infty = E_0\,\hat{\mathbf{y}}. \tag{7.75}$$

Combining the analyses in Sections 7.4.1 and 7.4.5, we obtain for the evolution equations for the shape and orientation of a fluid membrane vesicle

$$\frac{\partial\psi}{\partial t} = -\frac{1}{2} - \frac{C''_{22}}{2R(t)}\cos(2\psi) - \frac{C'_{22}}{2R}\sin(2\psi]), \tag{7.76}$$

$$\frac{\partial R}{\partial t} = \left(1 - 4\frac{R^2}{\Delta}\right)\left\{C'_{22}\cos(2\psi) - C''_{22}\sin(2\psi)\right\}$$
$$-2C_{20}R\Delta^{-1}\left[\frac{\Delta}{2} - 2R^2\right]^{1/2}, \tag{7.77}$$

where $C_{22} = C'_{22} + iC''_{22}$ and

$$C_{2n} = C_{2n}^{shear} + Mn\,C_{2n}^{el}, \tag{7.78}$$

where the Mason number is the ratio of the electric and shear stresses

$$Mn = \frac{\varepsilon^{ex}E_0^2}{\eta^{ex}\dot{\gamma}}.$$

For the shear flow $C_{2n}^{shear} = -i\Lambda^{-1}$ Eq. 7.48. For an electric field in the $\mathbf{z}$ (vorticity) direction

$$C_{20}^{el} = \frac{4\sqrt{5\pi}}{5(32+23\hat{\chi})}\Xi(t), \quad C_{22} = 0, \tag{7.79}$$

and for an electric field in the y (velocity gradient) direction

$$C_{20}^{el} = \sqrt{\frac{2}{3}}C_{22}^{el} = -\frac{2\sqrt{5\pi}}{5(32+23\hat{\chi})}\Xi(t), \tag{7.80}$$

where

$$\Xi(t) = -4\Lambda_\varepsilon P_{in}^2 + (-2+P_{ex})^2$$
$$= \frac{1}{(\Lambda_\sigma+2)^2}\begin{bmatrix}\left(\Lambda_\sigma\bar{V}(t) - 3(\Lambda_\sigma+1)\right)^2 \\ -4\Lambda_\varepsilon\left(3-2\bar{V}(t)\right)^2\end{bmatrix}. \tag{7.81}$$

The amplitude of the transmembrane potential $\bar{V}(t)$ for a the general case of a conducting membrane is given by

$$\bar{V}_{me}(t) = \frac{3\Lambda_\sigma}{2\Lambda_\sigma + g_{me}(2+\Lambda_\sigma)}\left[1-\exp\left(-\frac{t}{\delta_m}\right)\right], \tag{7.82}$$

7.5 OUTLOOK

Vesicles are useful minimal models to study dynamics of cells—for example, RBCs and blood flow in the microcirculation (Li et al., 2012; Vlahovska et al., 2009c) or swimming micro-organisms and "ameboid" propulsion (exhibited by *Eutreptiella gymnastica*, which move via pronounced changes of the membrane shape) (Farutin et al., 2013; Wu et al., 2015). Analytical solutions are indispensable when it comes to understanding the basic physical mechanisms of vesicle dynamics and are useful to validate numerical simulations.

The theoretical analysis of vesicles is far from complete. Here we have considered a simple spherical geometry. Recently, progress has been made toward accounting for high-aspect ratio ellipsoidal shapes. A methodology similar to the one described here but utilizing ellipsoidal harmonics have been developed by Narsimhan et al. (2014) and applied to analyze vesicle stability in extensional flows. Very deflated vesicles do display behavior that cannot be captured by the nearly spherical formalism, for example, instability in extensional flow resulting in asymmetric dumbbell shapes (Narsimhan et al., 2014) and slipper shapes in unbounded quadratic flow (Kaoui and Misbah, 2009).

Other interesting problems that can be treated analytically include the far-field hydrodynamic interactions of vesicles and rheology of non-dilute suspensions; these questions have been studied only to a very limited extent for vesicles (Gires et al., 2012; Levant et al., 2012).

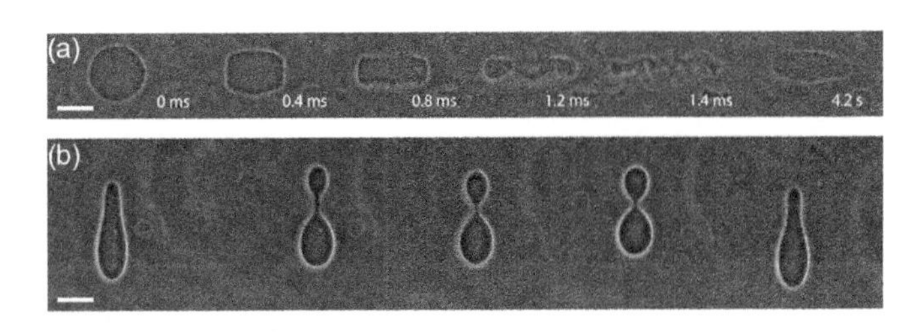

Figure 7.4 (a) Extreme deformation of a quasi-spherical vesicle upon application of a uniform DC electric field. (b) Time-dependent asymmetric dumbbell shaped vesicle in a uniform AC electric field. Scale bars: 10 μm. (Reproduced from Salipante, 2013. With permission.)

Dynamics of vesicles in electric fields is another widely open research area. Vesicles in electric field display many interesting behaviors: frequency- and time-dependent shapes, peculiar drum-like shapes with edges (Figure 7.4a), and poration to name a few (Aranda et al., 2008; Dimova et al., 2009; Riske and Dimova, 2005, 2006; Staykova et al., 2008; Vlahovska, 2010). Intriguingly, a deflated vesicle can also undergo an asymmetric dumbbell instability resembling the one in extensional flow (Figure 7.4b). The theoretical (Salipante and Vlahovska, 2014; Schwalbe et al., 2011; Seiwert et al., 2012) and numerical (Kolahdouz and Salac, 2015a, 2015b; McConnell et al., 2013, 2015a, 2015b; Veerapaneni, 2016) modeling of these systems is still at an early stage and presents exciting opportunities for research.

ACKNOWLEDGMENTS

P.M.V. acknowledges partial financial support by the National Science Foundation. C.M. acknowledges financial support from the Centre National d'Etudes Spatiales (CNES) and the European Space Agency (ESA).

APPENDICES

7.A SURFACE STRESS TENSOR

In this chapter, we follow the notation of Edwards et al. (1991) $\nabla_s = \mathbf{I}_s \cdot \nabla = e_\alpha \frac{\partial}{\partial x_\alpha}$, where e_α and x_α are the surface unit vectors and coordinates, respectively. This surface gradient derivative is not the covariant derivative on the surface as used in the works by Rahimi et al. (2013) and Guven and collaborators (Capovilla and Guven, 2002; Mueller, 2004).

Here we show how to obtain Eq. 7.13. First, we list some useful identities (Edwards and Wasan, 1988):

$$\nabla_s \cdot \mathbf{I}_s = 2H\mathbf{n}, \quad \nabla_s \cdot \nabla_s \mathbf{n} = -\left(4H^2 - 2K\right)\mathbf{n} \tag{7.83}$$

In taking the divergence of Eq. 7.12, the different terms transform as:

$$\begin{aligned}\nabla_s \cdot (-H\nabla_s \mathbf{n}) &= -\nabla_s H \cdot \nabla_s \mathbf{n} - H\nabla_s \cdot (\nabla_s \mathbf{n}) \\ &= -\nabla_s H \cdot \nabla_s \mathbf{n} + \left(4H^3 - 2HK\right)\mathbf{n}\end{aligned} \tag{7.84a}$$

$$\nabla_s \cdot ((\nabla_s H)\mathbf{n}) = \nabla_s^2 H\mathbf{n} + \nabla_s H \cdot (\nabla_s \mathbf{n}) \tag{7.84b}$$

$$\begin{aligned}-\nabla_s \cdot \left(H^2 \mathbf{I}_s\right) &= -2H\nabla_s H \cdot \mathbf{I}_s - H^2 \nabla_s \cdot \mathbf{I}_s \\ &= -2H\nabla_s H \cdot \mathbf{I}_s - 2H^3\mathbf{n}.\end{aligned} \tag{7.84c}$$

Adding the above equations leads to Eq. 7.13, with an extra term, $4H\nabla_s H \cdot \mathbf{I}_s$, which is, however, tangential to the surface and hence does not contribute to the normal bending forces.

7.B SPHERICAL HARMONICS

An extensive reference on spherical harmonics is a handbook by Varshalovich et al. (1988). Their properties in relation to the problems of particle microhydrodynamics are summarized in Bławzdziewicz et al. (2000).

The normalized spherical scalar harmonics are defined as

$$Y_{jm}(\theta,\varphi) = \left[\frac{2j+1}{4\pi}\frac{(j-m)!}{(j+m)!}\right]^{\frac{1}{2}} (-1)^m P_j^m(\cos\theta) e^{im\varphi}, \tag{7.85}$$

where $\hat{\mathbf{r}} = \mathbf{r}/r$, (r,θ,φ) are the spherical coordinates, and $P_j^m(\cos\theta)$ are the Legendre polynomials. The vector spherical harmonics are defined as

$$\mathbf{y}_{jm0} = \left[j(j+1)\right]^{-\frac{1}{2}} r\nabla_\Omega Y_{jm}, \mathbf{y}_{jm2} = \hat{\mathbf{r}} Y_{jm}, \mathbf{y}_{jm1} = -i\hat{\mathbf{r}} \times \mathbf{y}_{jm0} \tag{7.86}$$

where ∇_Ω denotes the angular part of the gradient operator. In spherical coordinates, the vector spherical harmonics that are tangential to a sphere are

$$\begin{aligned}\mathbf{y}_{jm0} &= \frac{1}{\sqrt{j(j+1)}}\frac{\partial Y_{jm}}{\partial\theta}\hat{\boldsymbol{\theta}} + \frac{im}{\sqrt{j(j+1)}}\frac{Y_{jm}}{\sin\theta}\hat{\boldsymbol{\varphi}} \\ \mathbf{y}_{jm1} &= -\frac{m}{\sqrt{j(j+1)}}\frac{Y_{jm}}{\sin\theta}\hat{\boldsymbol{\theta}} - \frac{i}{\sqrt{j(j+1)}}\frac{\partial Y_{jm}}{\partial\theta}\hat{\boldsymbol{\varphi}}\end{aligned} \tag{7.87}$$

For example

$$\begin{aligned}&\mathbf{y}_{200} = -\sqrt{\frac{15}{32\pi}}\sin(2\theta)\hat{\boldsymbol{\theta}}, \\ &\mathbf{y}_{101} = -i\sqrt{\frac{3}{8\pi}}\sin\theta\,\hat{\boldsymbol{\varphi}}, \quad \mathbf{y}_{202} = \frac{1}{8}\sqrt{\frac{5}{\pi}}[1+3\cos(2\theta)]\hat{\mathbf{r}}.\end{aligned} \tag{7.88}$$

7.C EXTERNAL STRESSES

7.C.1 FLOW

The linear flow defined by Eq. 7.4 is represented in the spherical harmonics basis as

$$c^{\infty}_{2\pm20} = \mp i2\dot{\varepsilon}\sqrt{\frac{\pi}{5}}, \quad c^{\infty}_{2\pm22} = \mp i2\dot{\varepsilon}\sqrt{\frac{2\pi}{15}}, \quad c^{\infty}_{101} = -i\omega\sqrt{\frac{8\pi}{3}}. \tag{7.89}$$

In the case of a simple shear flow $\dot{\varepsilon} = \omega = \dot{\gamma}/2$.

The unbounded quadratic shear flow written in a coordinate system centered at a distance, y_0, from the centerline, $\mathbf{v}^\infty = -\left[\dot{\gamma}y + \delta y^2\right]\hat{\mathbf{x}}$, is represented with coefficients

$$c^{\infty}_{3\pm30} = \mp\delta\sqrt{\frac{4\pi}{105}},\quad c^{\infty}_{3\pm32} = \mp\delta\sqrt{\frac{\pi}{35}},\quad c^{\infty}_{3\pm10} = \mp\delta\frac{2}{15}\sqrt{\frac{\pi}{7}},$$
$$c^{\infty}_{3\pm12} = \mp\delta\frac{1}{5}\sqrt{\frac{\pi}{21}},$$
$$c^{\infty}_{2\pm11} = \delta\frac{2}{3}\sqrt{\frac{\pi}{5}},\quad c^{\infty}_{1\pm10} = \pm\delta\frac{4}{5}\sqrt{\frac{\pi}{3}},\tag{7.90}$$
$$c^{\infty}_{1\pm12} = \pm\delta\frac{1}{5}\sqrt{\frac{2\pi}{3}}.$$

The shear part is the same as Eq. 7.4, with $\dot{\gamma} = 2\delta\, y_0$ being the local shear rate.

Axisymmetric Poiseuille flow $\mathbf{v}^{\infty} = -\left[\dot{\gamma}\, y + \delta(y^2 + \mathbf{z}^2)\right]\hat{\mathbf{x}}$ is given by

$$c^{\infty}_{300} = -\frac{8}{5}\sqrt{\frac{\pi}{21}},\quad c^{\infty}_{302} = -\frac{4}{5}\sqrt{\frac{\pi}{7}},\quad c^{\infty}_{100} = \frac{8}{5}\sqrt{\frac{2\pi}{3}},\tag{7.91}$$
$$c^{\infty}_{102} = \frac{4}{5}\sqrt{\frac{\pi}{3}}.$$

Similarly to the plane Poiseuille flow, the shear flow is as Eq. 7.4, with $\dot{\gamma} = 2\delta\, y_0$ being the local shear rate.

The forces due to flow are

$$\tau^{\infty}_{jm0} = (2j+1)[c^{\infty}_{jm0} - c^{\infty}_{jm2}\frac{3\sqrt{j(j+1)}}{j}\tag{7.92a}$$

$$\tau^{\infty}_{jm1} = c^{\infty}_{jm1}(j-1)\tag{7.92b}$$

$$\tau^{\infty}_{jm2} = \left[-c^{\infty}_{jm0}\frac{3\sqrt{j(j+1)}}{j} + c^{\infty}_{jm2}\frac{2j^2+j+3}{j}\right]\tag{7.92c}$$

7.C.2 ELECTRIC FIELD

A sphere in a uniform DC electric field $\mathbf{E} = E_0\hat{\mathbf{y}}$ experiences tangential tractions (Taylor, 1966; Vlahovska, 2011)

$$\tau^{E}_{220} = \tau^{E}_{2-20} = \sqrt{\frac{3}{2}}\tau^{E}_{200} = -\frac{9\sqrt{\frac{\pi}{5}}(\Lambda_{\sigma} - \Lambda_{\varepsilon})}{(\Lambda_{\sigma}+2)^2},\tag{7.93a}$$

and electric pressure

$$\tau^{E}_{222} = \tau^{E}_{2-22} = \sqrt{\frac{3}{2}}\tau^{E}_{202} = -\frac{3\sqrt{\frac{3\pi}{10}}\left(\Lambda_{\sigma}{}^2 - 2\Lambda_{\varepsilon} + 1\right)}{(\Lambda_{\sigma}+2)^2}.\tag{7.93b}$$

The isotropic component of the pressure, τ^{E}_{002}, is of no importance for shape dynamics because it is balanced by hydrostatic pressure. The above expressions are derived within the framework of the leaky dielectric model (Saville, 1997). The expressions for a uniform AC electric field can be found in (Torza et al., 1971; Vlahovska, 2011). Note that in electric fields, the characteristic stress $\tau_c = \varepsilon_{\text{ex}}E_0^2$, where E_0 is the magnitude of the applied electric field. For vesicles $\Lambda_{\varepsilon} = 1$ because both inner and outer fluids are typically aqueous solutions with similar permittivity.

7.D FUNDAMENTAL SET OF VELOCITY FIELDS, TRACTIONS, AND SOLUTION FOR THE FLOW $C^{\pm}_{jmq}$ AROUND A SPHERE

The velocity basis functions are

$$\hat{\mathbf{v}}^{-}_{jm0} = \frac{1}{2}r^{-j}\left(2 - j + jr^{-2}\right)\mathbf{y}_{jm0} + \frac{1}{2}r^{-j}\left[j(j+1)\right]^{\frac{1}{2}}\left(1 - r^{-2}\right)\mathbf{y}_{jm2},\tag{7.94a}$$

$$\hat{\mathbf{v}}^{-}_{jm1} = r^{(-j-1)}\mathbf{y}_{jm1},\tag{7.94b}$$

$$\hat{\mathbf{v}}^{-}_{jm2} = \frac{1}{2}r^{-j}(2-j)\left(\frac{j}{j+1}\right)^{\frac{1}{2}}\left(1 - r^{-2}\right)\mathbf{y}_{jm0} + \frac{1}{2}r^{-j}\left(j + (2-j)r^{-2}\right)\mathbf{y}_{jm2},\tag{7.94c}$$

$$\hat{\mathbf{v}}^{+}_{jm0} = \frac{1}{2}r^{j-1}\left(-(j+1) + (j+3)r^2\right)\mathbf{y}_{jm0} - \frac{1}{2}r^{j-1}\left[j(j+1)\right]^{\frac{1}{2}}\left(1 - r^2\right)\mathbf{y}_{jm2},\tag{7.95a}$$

$$\hat{\mathbf{v}}^{+}_{jm1} = r^{j}\, y_{jm1},\tag{7.95b}$$

$$\hat{\mathbf{v}}^{+}_{jm2} = \frac{1}{2}r^{j-1}(3+j)\left(\frac{j+1}{j}\right)^{\frac{1}{2}}\left(1 - r^2\right)\mathbf{y}_{jm0} + \frac{1}{2}r^{j-1}\left(j + 3 - (j+1)r^2\right)\mathbf{y}_{jm2}.\tag{7.95c}$$

On a sphere, $r = 1$, these velocity fields reduce to the vector spherical harmonics defined by Eq. 7.87

$$\hat{\mathbf{v}}^{\pm}_{jmq} = \mathbf{y}_{jmq}.\tag{7.96}$$

Hence the continuity of velocity becomes simply

$$c^{+}_{jmq} = c^{-}_{jmq} + c^{\infty}_{jmq}\tag{7.97}$$

The hydrodynamic tractions on a sphere due to the velocity fields Eqs. 7.94 and 7.95 are (Vlahovska and Gracia, 2007)

$$\tau^{\text{hd,in}}_{jm0} = (2j+1)c^{+}_{jm0} - 3\left(\frac{j+1}{j}\right)^{\frac{1}{2}}c^{+}_{jm2},\tag{7.98a}$$

$$\tau_{jm0}^{\mathrm{hd,ex}} = -(2j+1)c_{jm0}^{-} + 3\left(\frac{j}{j+1}\right)^{\frac{1}{2}} c_{jm2}^{-} \tag{7.98b}$$

$$\tau_{jm2}^{\mathrm{hd,ex}} = 3\left(\frac{j}{j+1}\right)^{\frac{1}{2}} c_{jm0}^{-} - \frac{4+3j+2j^2}{j+1} c_{jm2}^{-}, \tag{7.99a}$$

$$\tau_{jm2}^{\mathrm{hd,in}} = -3\left(\frac{j+1}{j}\right)^{\frac{1}{2}} c_{jm0}^{+} + \frac{3+j+2j^2}{j} c_{jm2}^{+} \tag{7.99b}$$

$$\tau_{jm1}^{\mathrm{hd,ex}} = -(j+2)c_{jm1}^{-}, \quad \tau_{jm1}^{\mathrm{hd,in}} = (j-1)c_{jm1}^{+}, \tag{7.100}$$

The stress balance on a sphere becomes

$$\tau_{jmq}^{\mathrm{hd,ex}} - \chi \tau_{jmq}^{\mathrm{hd,in}} = \tau_{jmq} \tag{7.101}$$

where $\tau = -\tau^{\infty} + \tau^{s}$ includes the applied (flow and/or electric) stresses, τ^{∞}, and interfacial stresses, τ^{s}, for example, bending, elastic, and surface-viscous stresses. Solving Eqs. 7.97 and 7.101 yields $c_{jmq}^{\pm}$.

GLOSSARY OF SYMBOLS

a	vesicle radius
C_{me}	membrane capacitance
Ca	capillary number
E	electric field
ℓ_{me}	membrane thickness
f_{jm}	shape deformation parameter
H	mean curvature
p	pressure
P_{me}	membrane stress tensor
t	time
T	bulk stress tensor
v	fluid velocity
V_{me}	transmembrane potential
Y_{jm}	spherical harmonic
η	dynamic viscosity
ρ	density
σ	conductivity
ε	dielectric permittivity
Λ_σ	conductivity ratio
Λ_ε	permittivity ratio
χ	viscosity ratio
Σ	membrane tension
κ	bending rigidity
Δ	excess area

REFERENCES

Abkarian M, Faivre M, Horton R, Smistrup K, Best-Popescu CA, Stone HA (2008) Cellular-scale hydrodynamics. *Biomed. Mater.* 3:034011.

Abkarian M, Faivre M, Viallat A (2007) Swinging of red blood cells under shear flow. *Phys. Rev. Lett.* 98:188302.

Abreu D, Levant M, Steinberg V, Seifert U (2014) Fluid vesicles in flow. *Adv. Colloid Interface Sci.* 208:129–141.

Aranda S, Riske KA, Lipowsky R, Dimova R (2008) Morphological transitions of vesicles induced by ac electric fields. *Biophys. J.* 95:L19–L21.

Barthès-Biesel D, Acrivos A (1973) Deformation and burst of a liquid droplet freely suspended in a linear shear field. *J. Fluid Mech.* 61:1–21.

Barthes-Biesel D, Sgaier H (1985) Role of membrane viscosity in the orientation and deformation of a spherical capsule suspended in shear flow. *J. Fluid. Mech.* 160:119–135.

Bławzdziewicz J, Vlahovska P, Loewenberg M (2000) Rheology of a dilute emulsion of surfactant-covered spherical drops. *Physica A* 276:50–80.

Biben T, Farutin A, Misbah C (2011) Three-dimensional vesicles under shear flow: Numerical study of dynamics and phase diagram. *Phys. Rev. E* 83:031921.

Boedec G, Jaeger M, Leonetti M (2014) Pearling instability of a cylindrical vesicle. *J. Fluid Mech.* 743:262–279.

Bretherton FP (1962) The motion of rigid particles in a shear flow at low Reynolds number. *J. Fluid Mech.* 14:284–304.

Capovilla R, Guven J (2002) Stresses in lipid membranes. *J. Phys. A* 35:6233–6247.

Chan PCH, Leal LG (1979) Motion of a deformable drop in a second-order fluid. *J. Fluid Mech.* 92:131–170.

Coupier G, Farutin A, Minetti C, Misbah C (2012) Shape diagram of vesicles in Poiseuille flow. *Phys. Rev. Lett.* 108:178106.

Danker G, Biben T, Podgorski T, Verdier C, Misbah C (2007) Dynamics and rheology of a dilute suspension of vesicles: Higher order theory. *Phys. Rev. E* 76:041905.

Danker G, Misbah C (2007) Rheology of a dilute suspension of vesicles. *Phys. Rev. Lett.* 98:088104.

Danker G, Verdier C, Misbah C (2008) Rheology and dynamics of vesicle suspension in comparison with droplet emulsion. *J. non-Newtonian Fluid. Mech.* 152:156–167.

Danker G, Vlahovska PM, Misbah C (2009) Vesicles in Poiseuille Flow. *Phys. Rev. Lett.* 102:148102.

Deschamps J, Kantsler V, Segre E, Steinberg V (2009) Dynamics of a vesicle in general flow. *Proc. Natl. Acad. Sci.* 106:11444–11447.

Dimova R, Bezlyepkina N, Jordo MD, Knorr RL, Riske KA, Staykova M, Vlahovska PM, Yamamoto T, Yang P, Lipowsky R (2009) Vesicles in electric fields: Some novel aspects of membrane behavior. *Soft Matter* 5:3201–3212.

Edwards DA, Brenner H, Wasan DT (1991) *Interfacial Transport Processes and Rheology* Butterworth-Heinemann, Boston, MA.

Edwards DA, Wasan DT (1988) Surface rheology II. The curved fluid surface. *J. Rheology* 32:447–472.

Farutin A, Biben T, Misbah C (2010) Analytical progress in the theory of vesicles under linear flow. *Phys. Rev. E* 81:061904.

Farutin A, Misbah C (2011) Symmetry breaking of vesicle shapes in poiseuille flow. *Phys. Rev. E* 84:011902.

Farutin A, Misbah C (2012a) Rheology of vesicle suspensions under combined steady and oscillating shear flows. *J. Fluid Mech.* 700:362–381.

Farutin A, Misbah C (2012b) Squaring, parity breaking, and S tumbling of vesicles under shear flow. *Phys. Rev. Lett.* 109:061922.

Farutin A, Misbah C (2013) Analytical and numerical study of three main migration laws for vesicles under flow. *Phys. Rev. Lett.* 110:108104.

Farutin A, Rafai S, Dysthe D, Duperray A, Peyla P, Misbah C (2013) Amoeboid swimming: A generic self-propulsion of cells in fluids by means of membrane deformations. *Phys. Rev. Lett.* 111:228102.

Freund JB (2014) Numerical simulation of flowing blood cells. *Annu. Rev. Fluid Mech.* 46:67–95.

Ghigliotti G, Biben T, Misbah C (2010) Rheology of a dilute two-dimensional suspension of vesicles. *J. Fluid Mech.* 653:489–518.

Gires PY, Danker G, Misbah C (2012) Hydrodynamic interaction between two vesicles in a linear shear flow: Asymptotic study. *Phys. Rev. E* 86:011408.

Grosse C, Schwan HP (1992) Cellular membrane potentials induced by alternating fields. *Biophys. J.* 63:1632–1642.

Helmy A, Barthès-Biesel D (1982) Migration of a spherical capsule freely suspended in an unbounded parabolic flow. *J. Mech. Theor. Appl.* 1:859–880.

Kantsler V, Segre E, Steinberg V (2008) Critical dynamics of vesicle stretching transition in elongational flow. *Phys. Rev. Lett.* 101:048101.

Kantsler V, Steinberg V (2005) Orientation and dynamics of a vesicle in tank-treading motion in shear flow. *Phys. Rev. Lett.* 95:258101.

Kantsler V, Steinberg V (2006) Transition to tumbling and two regimes of tumbling motion of a vesicle in shear flow. *Phys. Rev. Lett.* 96:036001.

Kaoui B, Farutin A, Misbah C (2009) Vesicles under simple shear flow: Elucidating the role of relevant control parameters. *Phys. Rev. E* 80:061905.

Kaoui B, Misbah C (2009) Why do red blood cells have asymmetric shapes even in a symmetric flow? *Phys. Rev. Lett.* 103:188101.

Keller SR, Skalak R (1982) Motion of a tank-treading ellipsoidal particle in shear flow. *J. Fluid Mech.* 120:27–47.

Kim S, Karrila SJ (1991) *Microhydrodynamics: Principles and Selected Applications* Butterworth-Heinemann, Boston, MA.

Kinosita Jr. K, Ashikawa I, Saita N, Yoshimura H, Itoh H, Nagayama K, Ikegami A (1988) Electroporation of cell membrane visualized under a pulsed laser fluorescence microscope. *Biophys. J.* 53:1015–1019.

Kolahdouz EM, Salac D (2015a) Dynamics of three-dimensional vesicles in dc electric fields. *Phys. Rev. E* 92:012302.

Kolahdouz EM, Salac D (2015b) Electrohydrodynamics of three-dimensional vesicles: A numerical approach. *SIAM J. Sci. Comput.* 37:B473–B494.

Leal LG (1980) Particle motions in a viscous fluid. *Annu. Rev. Fluid Mech.* 12:435–476.

Leal LG (2007) *Advanced Transport Phenomena.* Cambridge University Press, New York.

Lebedev VV, Turitsyn KS, Vergeles SS (2008) Nearly spherical vesicles in an external flow. *New J. Phys.* 10:043044.

Levant M, Deschamps J, Afik E, Steinberg V (2012) Characteristic spatial scale of vesicle pair interactions in a plane linear flow. *Phys. Rev. E.* 85:056306.

Li X, Vlahovska PM, Karniadakis GE (2012) Continuum- and particle-based modeling of shapes and dynamics of red blood cells in health and disease. *Soft Matter* 9:28–37.

Mader MA, Vitkova V, Abkarian M, Viallat A, Podgorski T (2006) Dynamics of viscous vesicles in shear flow. *Eur. Phys. J. E* 19:389–397.

Maitre E, Misbah C, Peyla P, Raoult A (2012) Comparison between advected-field and level-set methods in the study of vesicle dynamics. *Physica D* 241:1146–1157.

McConnell LC, Miksis MJ, Vlahovska PM (2013) Vesicle electrohydrodynamics in dc electric fields. *IMA J. Appl. Math.* 78:797–817.

McConnell LC, Miksis MJ, Vlahovska PM (2015a) Continuum modeling of the electric-field-induced tension in deforming lipid vesicles. *J. Chem. Phys.* 143:243132.

McConnell LC, Miksis MJ, Vlahovska PM (2015b) Vesicle dynamics in uniform electric fields: Squaring and breathing. *Soft Matter* 11:4840–4846.

Misbah C (2006) Vacillating breathing and tumbling of vesicles under shear flow. *Phys. Rev. Lett.* 96:028104.

Mueller M (2004) Theoretical examinations of interface mediated interactions between colloidal particles Ph.D. diss., MPI for Polymer Research.

Narsimhan V, Spann A, Shaqfeh ESG (2014) The mechanism of shape instability for a lipid vesicle in extensional flow. *J. Fluid Mech.* 750:144–190.

Narsimhan V, Spann AP, Shaqfeh ESG (2015) Pearling, wrinkling, and buckling of vesicles in elongational flows. *J. Fluid. Mech.* 777.

Nganguia H, Young YN, Vlahovska PM, Blawzdziewcz J, Zhang J, Lin H (2013) Equilibrium electro-deformation of a surfactant-laden viscous drop. *Phys. Fluids* 25:092106.

Noguchi H, Gompper G (2007) Swinging and tumbling of fluid vesicles in shear flow. *Phys. Rev. Lett.* 98:128103.

Rahimi M, DeSimone A, Arroyo M (2013) Curved fluid membranes behave laterally as effective viscoelastic media. *Soft Matter* 9:11033.

Rallison JM (1980) Note on the time-dependent deformation of a viscous drop which is almost spherical. *J. Fluid Mech.* 98:625–633.

Riske KA, Dimova R (2005) Electro-deformation and poration of giant vesicles viewed with high temporal resolution. *Biophys. J.* 88:1143–1155.

Riske KA, Dimova R (2006) Electric pulses induce cylindrical deformations on giant vesicles in salt solutions. *Biophys. J.* 91:1778–1786.

Riske KA, Knorr RL, Dimova R (2009) Bursting of charged multicomponent vesicles subjected to electric pulses. *Soft Matter* 5:1983–1986.

Sadik MM, Li JB, Shan JW, Shreiber DI, Lin H (2011) Vesicle deformation and poration under strong DC electric fields. *Phys. Rev. E.* 83:066316.

Salipante PF (2013) Electrohydrodynamics of simple and complex interfaces Ph.D. diss., Brown University.

Salipante PF, Knorr R, Dimova R, Vlahovska PM (2012) Electrodeformation method for measuring the capacitance of bilayer membranes. *Soft Matter* 8:3810–3816.

Salipante PF, Vlahovska PM (2013) Electrohydrodynamic rotations of a viscous droplet. *Phys. Rev. E* 88:043003.

Salipante PF, Vlahovska PM (2014) Vesicle deformation in dc electric pulses. *Soft Matter* 10:3386–3393.

Saville DA (1997) Electrohydrodynamics: The Taylor-Melcher leaky dielectric model. *Annu. Rev. Fluid Mech.* 29:27–64.

Schmitz R, Felderhof BU (1982) Creeping flow about a spherical particle. *Physica A* 113:90–102.

Schwalbe J, Vlahovska PM, Miksis MJ (2010) Monolayer slip effects on the dynamics of a lipid bilayer vesicle in a viscous flow. *J. Fluid Mech.* 647:403–419.

Schwalbe J, Vlahovska PM, Miksis MJ (2011) Vesicle electrohydrodynamics. *Phys. Rev E* 83:046309.

Schwalbe JT, Phelan FR, Vlahovska PM, Hudson SD (2011) Interfacial effects on droplet dynamics in poiseuille flow. *Soft Matter* 7:7797–7804.

Schwan HP (1989) Dielectrophoresis and rotation of cells In Neumann E, Sowers AE, Jordan CA, editors, *Electroporation and Electrofusion in Cell Biology*, pp. 3–21. Plenum Press, New York.

Seifert U (1997) Configurations of fluid membranes and vesicles. *Adv. Phys.* 46:13–137.

Seifert U (1999) Fluid membranes in hydrodynamic flow fields: Formalism and an application to fluctuating quasispherical vesicles. *Eur. Phys. J. B* 8:405–415.

Seifert U, Berndl K, Lipowsky R (1991) Shape transformations of vesicles: Phase diagram for spontaneous-curvature and bilayer-coupling models. *Phys. Rev. A* 44:1182–1202.

Seiwert J, Miksis MJ, Vlahovska PM (2012) Stability of biomimetic membranes in dc electric fields. *J. Fluid Mech.* 706:58–70.

Sinha KP, Gadkari S, Thaokar RM (2013) Electric field induced pearling instability in cylindrical vesicles. *Soft Matter* 9:7274–7293.
Smart JR, Leighton DT (1990) Measurement of the drift of a droplet due to the presence of a plane. *Phys. Fluids A* 3:21–28.
Staykova M, Lipowsky R, Dimova R (2008) Membrane flow patterns in multicomponent giant vesicles induced by alternating electric fields. *Soft Matter* 4:2168–2171.
Taylor GI (1966) Studies in electrohydrodynamics. I. Circulation produced in a drop by an electric field. *Proc. Royal Soc. A* 291:159–166.
Torza S, Cox R, Mason S (1971) Electrohydrodynamic deformation and burst of liquid drops. *Phil. Trans. Royal Soc. A* 269:295–319.
Varshalovich DA, Moskalev AN, Kheronskii VK (1988) *Quantum Theory of Angular Momentum*. World Scientific, Singapore.
Veerapaneni S (2016) Integral equation methods for vesicle electrohydrodynamics in three dimensions. *J. Comp. Phys.* 326:278–289.
Vitkova V, Mader M, Polack B, Misbah C, Podgorski T (2008) Micro-macro link in rheology of erythrocyte and vesicle suspensions. *Biophys. J.* 95:L33–L35.
Vlahovska P, Bławzdziewicz awzdziewicz J, Loewenberg M (2009a) Small-deformation theory for a surfactant-covered drop in linear flows. *J. Fluid Mech.* 624:293–337.
Vlahovska PM (2010) Non-equilibrium dynamics of lipid membranes: Deformation and stability in electric fields In Iglic A, editor, *Advances in Planar Lipid Bilayers and Liposomes, vol. 12*, pp. 103–146. Elsevier, Amsterdam, the Netherlands.
Vlahovska PM (2011) On the rheology of a dilute emulsion in a uniform electric field. *J. Fluid Mech.* 670:481–503.
Vlahovska PM (2014) Asymmetric shapes and pearling of a stretched vesicle. *J. Fluid Mech.* 754:1–4.
Vlahovska PM (2015) Voltage-morphology coupling in biomimetic membranes: Dynamics of giant vesicles in applied electric fields. *Soft Matter* 11:7232–7236.
Vlahovska PM (2019) Electrohydrodynamics of Drops and Vesicles. *Annual Review of Fluid Mechanics* 51:305–330.
Vlahovska PM, Barthes-Biesel D, Misbah C (2013) Flow dynamics of red blood cells and their biomimetic counterparts. *C. R. Physique* 14:451–458.
Vlahovska PM, Gracia R (2007) Dynamics of a viscous vesicle in linear flows. *Phys. Rev. E* 75:016313.
Vlahovska PM, Gracia RS, Aranda-Espinoza S, Dimova R (2009b) Electrohydrodynamic model of vesicle deformation in alternating electric fields. *Biophys. J.* 96:4789–4803.
Vlahovska PM, Podgorski T, Misbah C (2009c) Vesicles and red blood cells: From individual dynamics to rheology. *C. R. Physique* 10:775789.
Vlahovska PM, Young YN, Danker G, Misbah C (2011) Dynamics of a non-spherical microcapsule with incompressible interface in shear flow. *J. Fluid. Mech.* 678:221–247.
Wu H, Thiebaud M, Hu WF, Farutin A, Rafai S, Lai MC, Peyla P, Misbah C (2015) Amoeboid motion in confined geometry. *Phys. Rev. E* 92:050701.
Yamamoto T, Aranda-Espinoza S, Dimova R, Lipowsky R (2010) Stability of spherical vesicles in electric fields. *Langmuir* 26:12390–12407.
Zabusky NJ, Segre E, Deschamps J, Kantsler V, Steinberg V (2011) Dynamics of vesicles in shear and rotational flows: Modal dynamics and phase diagram. *Phys. Fluids* 23:041905.
Zhang J, Zahn JD, Tan W, Lin H (2013) A transient solution for vesicle electrodeformation and relaxation. *Phys. Fluids* 25:071903.
Zhao H, Shaqfeh ESG (2013) The shape instability of a lipid vesicle in a uniaxial extensional flow. *J. Fluid Mech.* 719:345–361.

Particle–membrane interactions

Jaime Agudo-Canalejo and Reinhard Lipowsky

There's lots of things that you can do alone, but it takes two to tango.

Louis Armstrong

Contents

8.1 INTRODUCTION AND OVERVIEW

Membranes are present in every living cell, and represent an essential component of life as we know it. They separate the inside from the outside of cells, and they form many smaller membrane-bounded compartments within the cell. Because of their fluidity, biological membranes are highly dynamic and can remodel their architecture in response to external and internal signals. In this chapter, we will be concerned with those remodeling processes that are involved in the entry of nanometer-sized particles into cells. Indeed, nanoparticles are increasingly used for targeted delivery of drugs to biological cells (Panyam and Labhasetwar, 2003; Singh and Lillard, 2009), reaching difficult targets such as tumors (Paciotti et al., 2004; Cho et al., 2008) or crossing the blood-brain barrier (Kreuter, 2001; Lockman et al., 2002). Novel magnetic nanoparticles are being developed as contrast agents in magnetic resonance imaging (Lee et al., 2007; Sun et al., 2008), and gold nanoparticles are being used in X-ray imaging and photothermal therapies (Hainfeld et al., 2006; Huang et al., 2006).

Nanoparticles are also more and more common in industrial processes, which has raised some concerns about their cytotoxicity. The world production of nanoparticles is projected to increase 25-fold in the period between 2008 and 2020

(Lewinski et al., 2008), with nanoparticles being extensively used in cosmetics, food, paints, powders, and surface treatments (Schmid and Riediker, 2008). Both *in vitro* and *in vivo* studies have found that nanoparticles can be cytotoxic when applied in high doses, with toxicity levels depending on nanoparticle size, shape, and surface chemistry (Fischer and Chan, 2007; Pan et al., 2007; Lewinski et al., 2008). A further particularly important class of nanoparticles is provided by viruses. Viral capsids typically range in size from 20 to 300 nm, and enter cells *via* the same pathways as artificial nanoparticles (Mercer et al., 2010). Insights into the nanoparticle pathways may thus help in creating new tools against viral infection.

In order to enter a cell, a particle must first cross the cellular membrane. This process is termed *endocytosis*, and starts with the adhesion and engulfment of the particle by the membrane, followed by membrane fission (Mukherjee et al., 1997; Sahay et al., 2010). The particle then ends up inside the cytoplasm, fully enclosed by the membrane. Several distinct endocytic pathways used by cells to take up nanoparticles have been described, depending on the molecular machinery involved. The most thoroughly studied pathway is clathrin-mediated endocytosis, which is assisted by the formation of a strongly curved protein coat on the inner side of the membrane (Mukherjee et al., 1997).

As demonstrated throughout this book, giant unilamellar vesicles (GUVs) are very important model systems that can provide useful insight into the structure and dynamics of cellular membranes. In this context, engulfment of rigid particles by GUVs (Dietrich et al., 1997; Koltover et al., 1999; Fery et al., 2003; Meinel et al., 2014; Strobl et al., 2014; van der Wel et al., 2016), smaller liposomes (Le Bihan et al., 2009; Michel et al., 2014), as well as polymer vesicles (Jaskiewicz et al., 2012) has been observed to occur spontaneously in experiments. In this way, spontaneous engulfment by GUVs may be seen as a membrane-driven mimetic mechanism of endocytosis in cells. Other experiments have observed adsorption of nanoparticles onto the GUV surface (Dinsmore et al., 1998; Natsume et al., 2010; Li et al., 2016; van der Wel et al., 2016), as well as particle-induced disruption of the membrane structure (Luccardini et al., 2006; Laurencin et al., 2010; Li and Malmstadt, 2013; Wei et al., 2015). Membrane disruption in GUVs illustrates one aspect of the cytotoxic capabilities of nanoparticles.

In this chapter, we consider the interaction between rigid particles and membranes, with particular emphasis on particle adhesion and engulfment (see also Chapters 9 and 25 for theoretical and experimental reviews of the interactions of membranes and polymers, which may be considered as soft particles). In Section 8.2, we will describe the basic particle properties that determine particle–membrane interactions and introduce the different processes arising from these interactions. Subsequent sections will be devoted to particle adhesion and engulfment. In Section 8.3, we introduce some basic aspects of particle engulfment, followed by a detailed theoretical description of the engulfment of rigid spherical particles in Section 8.4. In Section 8.5, we review some theoretical work on engulfment of nonspherical and deformable particles. Finally, we will review existing work on simultaneous engulfment of many particles, particle aggregation and membrane-mediated particle–particle interactions in Section 8.6, followed by some closing remarks and an outlook in Section 8.7.

8.2 DIFFERENT PROCESSES INDUCED BY PARTICLE–MEMBRANE INTERACTIONS

8.2.1 BASIC PARTICLE PROPERTIES: SIZE, SHAPE AND SURFACE CHEMISTRY

From a theoretical point of view, we would like to identify the basic principles underlying particle–membrane interactions. A first problem arises from the wide variety of particles that are used in applications. Indeed, the term "nanoparticle," as found in the scientific literature, encompasses an astonishing variety of particles, made from different materials and with widely differing characteristics, such as fullerenes (Kroto et al., 1985) and quantum dots (Alivisatos, 1996) as well as silver (Rai et al., 2009), gold (Daniel and Astruc, 2004) and silica (Slowing et al., 2008) nanoparticles.

As a first approximation, we may identify three particle properties of particular relevance for particle–membrane interactions (Albanese et al., 2012):

1. *Size*: Nanoparticles used in applications cover a wide range of sizes, from below 1 nm, in the case of spherical fullerenes (Kroto et al., 1985), up to a few hundreds of nanometers, in the case of large metallic (Daniel and Astruc, 2004; Dykman and Khlebtsov, 2014) or silica nanoparticles (Slowing et al., 2008). In experiments with model GUVs, latex particles as large as several micrometers have been used (Dietrich et al., 1997). Naturally, particles that are smaller than the membrane thickness, $\ell_{me} \approx 5$ nm, will interact with the membrane in very different ways than those that are of comparable size or much larger than the membrane thickness.
2. *Shape*: Commonly used particles are most often spherical but they can also have elongated quasi-one-dimensional shapes as in the case of carbon nanotubes (Odom et al., 1998) or gold nanorods (Jana et al., 2001) as well as flat disk-like shapes as in the case of graphene nanosheets. In all of these cases, it will be important whether one, two or all of the particle dimensions are larger or smaller than the membrane thickness, ℓ_{me}.
3. *Surface chemistry*: Hydrophilicity and hydrophobicity, as well as surface charge, strongly influence particle–membrane interactions. In addition, nanoparticles may be functionalized by coating them with lipids, polymers or specific protein ligands that can bind to receptors on the membrane surface (Verma and Stellacci, 2010; Monopoli et al., 2012; Mahmoudi et al., 2014). It is also possible to produce "patchy" particles with a nonhomogeneous surface, consisting of patches with different coatings or functionalization (Pawar and Kretzschmar, 2010; Poon et al., 2010). A particularly important kind of patchy particles are "Janus" particles, consisting of two distinct hemispherical patches or faces (Lattuada and Hatton, 2011; Li et al., 2016).

Depending on these properties, particles may (i) adsorb onto the membrane, (ii) incorporate into the hydrophobic core of the bilayer, (iii) translocate through the membrane, or (iv) be engulfed by the membrane.

8.2.2 PARTICLE ADSORPTION

Particles that are both hydrophilic—so that they do not incorporate into the bilayer membrane—and small—with a size comparable to the head group of the lipids—may adsorb onto the surface of the membrane (Lipowsky and Döbereiner, 1998; Lipowsky, 2013; Curtis et al., 2015; Różycki and Lipowsky, 2015). The latter process is illustrated by the molecular dynamics (MD) simulation snapshot in Figure 8.1a, which displays a hydrophilic particle of diameter 1 nm after adsorption onto a 1,2-dipalmitoyl-*sn*-glycero-3-phosphocholine membrane. Such "particles" may also be atomic ions or small molecules and solutes.

If a large number of such particles are present in the aqueous solution, they will adsorb onto the membrane surface with a certain coverage, Γ, proportional to the molar concentration, X, of particles in the solution. If the concentration of particles in the interior and exterior of the GUV are different, $X_{in} \neq X_{ex}$, the coverage of adsorbed particles will also differ between the interior and exterior surfaces of the membrane, $\Gamma_{in} \neq \Gamma_{ex}$. In this case, the asymmetric adsorption of particles onto the bilayer will generate a spontaneous curvature, m, in the membrane, with m as defined in Chapter 5, given by

$$m \approx \frac{k_B T}{4\kappa} \ell_{me}(\Gamma_{ex} - \Gamma_{in}) \quad (8.1)$$

This relation was first predicted theoretically by using a Langmuir adsorption model (Lipowsky and Döbereiner, 1998; Lipowsky, 2013) and later confirmed by MD simulations (Różycki and Lipowsky, 2015).

8.2.3 PARTICLE INCORPORATION

Hydrophobic particles such as spherical fullerenes, which have a size comparable to or smaller than the membrane thickness, ℓ_{me}, can be incorporated into the hydrophobic core of the bilayer membrane, as observed in many experiments (Hetzer et al., 1997; Gopalakrishnan et al., 2006; Bothun, 2008; Chen et al., 2010; Mai and Eisenberg, 2010; Rasch et al., 2010; Liu et al., 2013) and simulations (Ginzburg and Balijepalli, 2007; Wong-Ekkabut et al., 2008; D'Rozario et al., 2009; Curtis et al., 2015). A typical series of simulation snapshots is displayed in Figure 8.1b. Thus, very small hydrophobic nanoparticles behave just like hydrophobic molecules and can create "oily" pockets inside membranes, as observed in experiments (Hayward et al., 2006) and simulations (Greenall and Marques, 2012).

Some molecular simulations also found that hydrophilic particles that have a size comparable to the membrane thickness, $\ell_{me} \approx 5$ nm, and exhibit an attractive interaction with the lipid headgroups may be incorporated into the membrane (Figure 8.1c) (Noguchi and Takasu, 2002; Ginzburg and Balijepalli, 2007; Smith et al., 2007; Guo et al., 2013; Yue et al., 2014; Curtis et al., 2015). However, this type of particle incorporation has not been observed experimentally and may be suppressed in real systems. Indeed, in order to completely surround the particle and maximize the contact with the hydrophilic head groups, the membrane must change its topology and form an interior hydrophilic "pocket," with a line of T-junction defects around the particle (Figure 8.1c). This defect line involves a line energy proportional to the particle diameter. It is likely that the coarse-grained, "soft" potentials used in the simulations underestimate the energetic cost of such T-junction defects, which are expected to be rather large for real lipid bilayers.

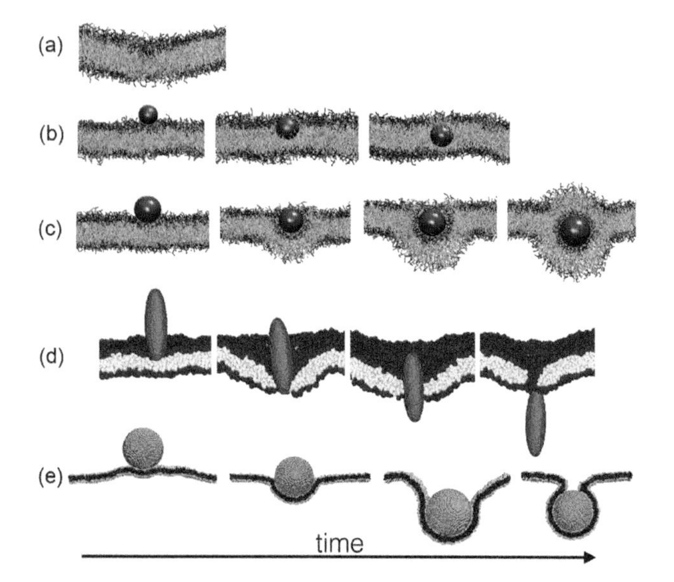

Figure 8.1 Different processes based on the interactions between particles and membranes. Time-series snapshots (cross sections) of molecular dynamics simulations of (a) adsorption, (b, c) incorporation, (d) translocation, and (e) engulfment of particles. The hydrophilic heads of the lipids are dark blue in (a–c), red in (d), and turquoise in (e); the hydrophobic tails of the lipids are turquoise in (a–c), yellow in (d), and black in (e). In (a), the particle is hydrophilic with radius $R_{pa} < \ell_{me}$ and adsorbs onto the membranes surface. In (b), the particle is hydrophobic with radius $R_{pa} < \ell_{me}$ and incorporates into the membrane core. In (c), the particle is hydrophilic with radius $R_{pa} \sim \ell_{me}$ and changes the topology of the bilayer. In (d), an elongated particle of a size comparable to ℓ_{me} and subject to an external force passes through the membrane. In (e), a hydrophilic particle of radius $R_{pa} \gg \ell_{me}$ is gradually engulfed but remains connected to the mother membrane by a small neck. ([a–c] Curtis, E.M. et al., *Nanoscale*, 7, 14505–14514, 2015. Reproduced by permission of The Royal Society of Chemistry; [d] Reprinted by permission from Macmillan Publishers Ltd. *Nat. Nanotechnol.*, Yang, K. and Ma, Y.-Q., 2010, copyright 2010; [e] Reprinted with permission from Smith, K.A. et al., *J. Chem. Phys.*, 127, 84703, 2007. Copyright 2007 by the American Institute of Physics.)

8.2.4 PARTICLE TRANSLOCATION

Lipid bilayers are permeable to small uncharged molecules such as H_2O, O_2, and CO_2 (Lodish et al., 2000) (see also Chapter 20), which can therefore pass through the membrane without disrupting the bilayer structure. One may then ask whether such a translocation process is also possible for larger, artificial nanoparticles.

In order to permeate the membrane, particles with a completely hydrophilic surface will have to cross a large energy barrier provided by the hydrophobic core of the bilayer. On the other hand, if the particle surface is completely hydrophobic, the particle will prefer to insert into the hydrophobic core as described in Section 8.2.3. It was first proposed, based on simulation studies, that nanoparticles with weakly bound amphiphilic ligands (Ding et al., 2012) or with mixed hydrophilic–hydrophobic surface domains (Li et al., 2012) may be able to pass through lipid bilayers more easily. As far as we know, there is no experimental evidence for such a translocation process facilitated by heterogeneous surfaces. However, recent experiments indicate that lipid-covered hydrophobic nanoparticles with a diameter of about 6 nm can translocate through membranes (Guo et al., 2016).

Nanoparticles can be forced to penetrate through membranes by external forces that push (or pull on) the particles. Thus, a carbon nanotube that is attached to an atomic force microscope can be pushed through cell membranes (Chen et al., 2007). In simulations, the minimal forces required to make particles of different size, shape and surface chemistry pass through a lipid bilayer have been measured (Yang and Ma, 2010; Li et al., 2012). An example is shown in Figure 8.1d, which displays simulation snapshots of an ellipsoidal particle that is moved through a bilayer by applying an external force to the particle's center of mass.

8.2.5 PARTICLE ENGULFMENT

In all the cases described so far, the size of the particles was comparable to or smaller than the membrane thickness, ℓ_{me}. As soon as the particles become larger (in all dimensions) than a few times the membrane thickness, incorporation or translocation become energetically unfavorable and the membrane will instead spread onto the particle and engulf it (Figure 8.1e) (Lipowsky and Döbereiner, 1998; Smith et al., 2007; Roiter et al., 2008) if the interaction between the hydrophilic surface of the membrane and the particle is attractive. The engulfment process ends with the particle fully covered by the membrane but still connected to the mother membrane by a small neck, as in the rightmost snapshot of Figure 8.1e. This neck might then break *via* membrane fission, which implies that the particle has been effectively transported from one side of the membrane to the other. In the case of a closed vesicle, we will distinguish between *endocytic* engulfment, for which the particle originates from the exterior compartment of the vesicle, and *exocytic* engulfment, for which the particle originates from the interior compartment.

In the rest of this chapter, we will focus on the process of particle engulfment for three main reasons. First, engulfment applies to the largest range of particle sizes: Whereas adsorption, incorporation or translocation require particles of a size comparable or smaller than the membrane thickness, engulfment can occur for particles ranging in size from a few nanometers to several micrometers. Second, the study of particle engulfment by model membranes provides fundamental insights into the process of endocytosis, the main pathway of particle entry into cells. Finally, from a theoretical perspective, particle engulfment can be understood in the context of the curvature elasticity theory of membranes as introduced in Chapter 5, in which the molecular details of the membrane are coarse-grained into two basic elastic parameters: the bending rigidity κ and the spontaneous curvature m. The processes of adsorption, incorporation or translocation, on the other hand, depend on the rearrangement of just a few lipids and are therefore dependent on the molecular details of the membrane. As a consequence, they are typically studied using simulations.

8.3 BASIC ASPECTS OF PARTICLE ENGULFMENT

8.3.1 CURVATURE-DOMINATED VERSUS TENSION-DOMINATED REGIMES

As explained in Chapter 5, the equilibrium shapes of free vesicles are governed by the curvature elasticity of membranes, as long as the vesicles are large compared with the membrane thickness, that is, with diameters larger than about 50 nm. In typical experimental conditions, a vesicle membrane has a fixed total number of lipid molecules, and therefore a fixed rest area, A. In addition, experiments are typically carried out in aqueous solutions that contain osmotically active particles, such as salts or sugars. In such conditions, the volume, V, enclosed by the vesicle is also fixed (Seifert et al., 1991; Seifert, 1997). The equilibrium shape of such a free vesicle then depends only on its volume-to-area ratio or reduced volume

$$v \equiv \frac{V}{\frac{4\pi}{3}\left(\frac{A}{4\pi}\right)^{3/2}} \tag{8.2}$$

and on its membrane spontaneous curvature, m. For $v = 1$, the vesicle has a spherical shape and is typically "tense," not displaying strong shape fluctuations. For volumes $0 < v < 1$, the vesicle will be "flaccid," with a nonspherical fluctuating shape. Note that we will often use the simple term "volume" as an abbreviation for reduced volume.

We can now distinguish two qualitatively different situations with respect to particle engulfment, depending on whether the vesicle has enough excess area to accommodate the particle or not. Suppose that we have a particle, not necessarily spherical, with volume V_{pa} and area A_{pa}. After completely engulfing the particle, the vesicle will enclose a volume $V \pm V_{pa}$, where the plus and minus signs apply to endocytic and exocytic engulfment respectively (Figure 8.2). The area of the remaining free membrane segment of the vesicle, not bound to the particle, will be $A - A_{pa}$. The new reduced volume of the vesicle after complete engulfment of the particle is then given by

$$v' \equiv \frac{V \pm V_{pa}}{\frac{4\pi}{3}\left(\frac{A - A_{pa}}{4\pi}\right)^{3/2}} \tag{8.3}$$

Because the new reduced volume v' always exceeds the original v, the vesicle becomes more spherical after a particle has

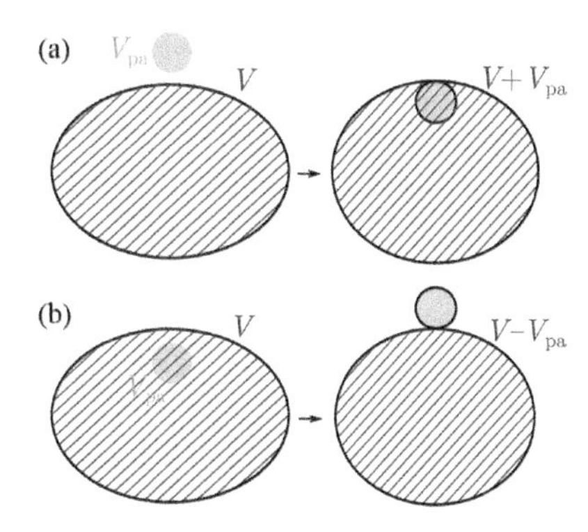

Figure 8.2 Endocytic (a) and exocytic (b) engulfment. In (a), the vesicle initially encloses a volume V (striped region) and engulfs a particle of volume V_{pa} originating from the exterior compartment. After engulfment, the effective volume enclosed by the mother vesicle includes the particle volume and is equal to $V+V_{pa}$ (striped region). In (b), the vesicle initially encloses a volume V (striped region), which includes the volume of the particle. After engulfment, the effective volume enclosed by the mother vesicle is $V-V_{pa}$ (striped region). In both (a) and (b), the area of the mother vesicle after engulfment is $A-A_{pa}$, where A is the area of the original vesicle and A_{pa} is the surface area of the particle.

been engulfed. We will now use the volume v' to distinguish a curvature-dominated from a tension-dominated regime. The curvature-dominated regime is defined by $v' \leq 1$, which implies that the vesicle has enough excess area to fully accommodate the particle. The curvature-dominated regime will be discussed in detail in Section 8.4 for the simplest case of a spherical particle. If, on the other hand, we find that $v' > 1$, the vesicle does not have enough excess area, and the vesicle membrane would need to stretch in order to accommodate the particle, leading to an increase in the mechanical tension of the membrane. This argument can naturally be extended to the engulfment of multiple particles: After complete engulfment of N particles, the new volume-to-area ratio v' of the vesicle is given by Eq. 8.3, with NV_{pa} and NA_{pa} instead of V_{pa} and A_{pa}.

As described in Chapter 5, the mechanical tension of the membrane can be related to the stretching of the membrane *via* $\Sigma = K_A \Delta A / A$, where K_A is the area compressibility modulus of the membrane and $\Delta A / A$ is the relative area stretching of the membrane with respect to the original (relaxed) vesicle membrane area. The energetics is then dominated by tension over bending if $\Sigma A_{pa} \gg \kappa$. In this tension-dominated regime, the equilibrium geometry of the system is dictated by a competition between particle–membrane adhesion and vesicle membrane tension, akin to the well-studied case of substrate wetting by droplets (Seifert and Lipowsky, 1990). Indeed, for vesicles that are initially close to spherical $v \simeq 1$, it was found by Dietrich et al. (1997) that the experimentally observed engulfment geometries could be explained in close analogy to the Young-Dupré equation for wetting. Furthermore, when the membrane was stretched beyond ~ 4% by the particle engulfment, that is, for $\Delta A / A \geq 0.04$, the membrane was observed to rupture. The intermediate case in which the bending and stretching contributions to the total energy are comparable was studied using an approximate numerical method by Deserno and Gelbart (2002).

8.3.2 GENERIC VERSUS SPECIFIC PARTICLE–MEMBRANE ADHESION

As we have seen, in order to engulf a particle, the membrane needs to bend and, in the tension-dominated regime, it also needs to stretch. Both of these processes are energetically unfavorable, so what constitutes, then, the driving force for engulfment? This driving force is provided by attractive interactions between the surface of the particle and the surface of the membrane, leading to favorable adhesion between the two. Generally, we can speak of two qualitatively different types of particle–membrane interactions: generic surface interactions; and specific interactions mediated by ligands on the particle that attach to corresponding receptors or "stickers" present on the membrane, thus forming receptor–ligand pairs; see Chapter 17 for an overview of adhesion of GUVs to substrates.

A thorough description of generic interactions in membranes can be found in the review by Lipowsky (1995). These interactions are always present as a combination of short-ranged repulsion arising from hydration forces and longer-ranged electrostatic and van der Waals forces that may be attractive or repulsive depending on the particle–membrane separation. These generic interactions are renormalized by membrane fluctuations in the proximity of the rigid particle (Lipowsky and Leibler, 1986). Due to their complexity, a detailed description of these forces lies outside the scope of this chapter. Suffice to say that, when adding all the contributions together, a typical attractive potential between the surface of the particle and the membrane will show (i) repulsion for short distances, (ii) a minimum at a certain distance and (iii) attraction at longer distances. Several such phenomenological effective potentials have been used in theoretical research on particle engulfment, such as a Morse potential (Raatz et al., 2014) or a Lennard-Jones type of potential (Šarić and Cacciuto, 2012).

In practice, the typical range of such generic attractive interactions is on the order of just a few nanometers, usually even smaller than the membrane thickness. In order to study the overall shape of the adhering membrane, one may ignore the molecular details and focus on the adhesive strength $|W_{gen}|$ of the membrane-surface interactions that corresponds to the adhesion (free) energy per area (Seifert and Lipowsky, 1990). This coarse-grained description of the membrane-surface interactions in terms of a single material parameter is in accordance with the separation of length scales that is used to describe the curvature elasticity of the membranes (see Chapter 5).

In addition to the generic interactions just discussed, biological membranes interact *via* membrane-anchored receptors and ligands (Lipowsky, 1996; Hu et al., 2013). Such specific interactions enable cellular membranes to selectively mediate binding of cells to other cells, an essential step in immune response or tissue development, as well as binding to small cargo that is to be internalized by the cell *via* endocytosis. In experiments, a commonly used pair of receptor–ligand molecules is provided by biotin and streptavidin, two molecules that can be easily used for the adhesion of particles and vesicles. The binding free energy, $|U|$, of such a receptor–ligand pair contains both enthalpic and entropic contributions. If the binding enthalpy of a single bond is $|H|$, its free energy can be estimated by $|U| = |H| - k_B T \ln(\rho_{lig}/\rho_0)$, where ρ_{lig} is the density of ligands on the particle surface and ρ_0 that of receptors on the membrane before contact with the particle (Gao et al., 2005). The second term represents the loss of translational entropy by the receptor–ligand bond. If this binding free energy is large compared with $k_B T$, all ligands will be bound to a receptor and the total energetic contribution of the specific interactions can be incorporated into a single adhesive energy per unit area of particle–membrane contact $|W_{spe}| \equiv |U| \rho_{lig}$.

In order to obtain a simple description of particle adhesion, we can combine both generic and specific attraction into a single parameter: the adhesive strength $|W| \equiv |W_{gen}| + |W_{spe}|$. In this way, the whole complexity of the adhesive interactions can be encoded into a single parameter, which can also be measured experimentally (see Table 5.2 in Section 5.8 of Chapter 5). The adhesion energy of a particle in contact with a membrane segment of area A_{bo} is then given simply by

$$\mathcal{E}_{ad} = -|W| A_{bo} \tag{8.4}$$

Such an approach was first used in the study of vesicle adhesion to planar substrates (Seifert and Lipowsky, 1990). Values for $|W|$ have been measured experimentally for many different systems, and can vary widely from 10^{-5} mJ/m² for 1,2-dioleoyl-*sn*-glycero-3-phosphocholine/1,2-dioleoyl-*sn*-glycero-3-phospho-rac-(1-glycerol) (DOPC/DOPG, see Appendix 1 of the book for structure and data on these lipids) membrane and glass surfaces (Gruhn et al., 2007), known as ultraweak adhesion, to strong adhesion with 0.5 mJ/m² for 1,2-dimyristoyl-*sn*-glycero-3-phosphocholine (DMPC, see Appendix 1 of the book for structure and data on this lipid) and silica surfaces (Anderson et al., 2009) (see Table 5.2 in Chapter 5).

8.4 ENGULFMENT OF RIGID SPHERICAL PARTICLES

8.4.1 ADHESION LENGTH

As mentioned in Section 8.3, in the curvature-dominated regime, spontaneous engulfment is the result of an interplay between membrane bending and particle adhesion: the gain in adhesion energy will be opposed by the energetic cost of bending the membrane around the particle. Such an energy balance was first considered by Lipowsky and Döbereiner (1998) for the case of a large vesicle (with zero spontaneous curvature) engulfing a small spherical particle. If the particle has radius R_{pa}, the completely engulfed particle will gain the adhesion energy $-|W| 4\pi R_{pa}^2$, which increases with particle size. On the other hand, it follows from the spontaneous curvature model (see Chapter 5) that the membrane segment bound to the particle has the bending energy $8\pi\kappa$ that is proportional to the bending rigidity κ and *independent* of the particle size. Ignoring the engulfment-induced changes in the bending energy of the unbound membrane segment, which should be negligible if the particle is much smaller than the vesicle, we find that complete engulfment is energetically favorable if $|W| 4\pi R_{pa}^2 > 8\pi\kappa$ or, equivalently,

$$R_{pa} > \sqrt{2\kappa / |W|} \equiv R_W. \tag{8.5}$$

Therefore, engulfment becomes energetically favorable for sufficiently large particles. The length scale R_W is an important material parameter of the system that we will call the *adhesion length*. Values of the adhesion length for different combinations of membrane composition and adhesive material are displayed in Table 5.2 of Chapter 5.

A limitation of the approach leading to Eq. 8.5 is that it simply compares the energy of the free and completely engulfed states but does not inform us about the stability of each state, that is, about the energy landscape of the engulfment process. Indeed, the transition from the free to the completely engulfed state may in principle occur (i) discontinuously, reflecting an energy barrier between the two states, or (ii) continuously, *via* partially engulfed states in which the membrane is bound to the particle but does not cover the particle completely. In order to distinguish these two cases, one has to (numerically) calculate the energy of the system along an appropriate reaction coordinate for the engulfment process, as will be done in Section 8.4.2.

8.4.2 ENERGY LANDSCAPES AND ENGULFMENT REGIMES

We will focus in this subsection on the engulfment of particles by vesicles with an axisymmetric geometry, as depicted in Figure 8.3. A possible reaction coordinate for this engulfment geometry is the *wrapping angle* ϕ, as defined in the figure. The wrapping angle can vary from $\phi = 0$, representing a free particle, to $\phi = \pi$, corresponding to a completely engulfed particle. For values $0 < \phi < \pi$ the particle is partially engulfed.

The restriction to axisymmetric geometries is due to the fact that, for each value of the wrapping angle ϕ, we need to find the shape of the unbound membrane segment that minimizes the bending energy of this segment while still satisfying the constraints on the total vesicle area A and enclosed volume V. A numerical solution of the corresponding Euler-Lagrange equation is only feasible for axisymmetric geometries, using a shooting method as described by Seifert et al. (1991) and Agudo-Canalejo and Lipowsky (2016). Numerical solutions therefore only allow us to describe engulfment *at the poles* of axisymmetric vesicles, although we will overcome this limitation using analytical considerations in Sections 8.4.3, 8.4.4 and 8.6.1. The numerical minimization of the membrane bending energy for non-axisymmetric geometries

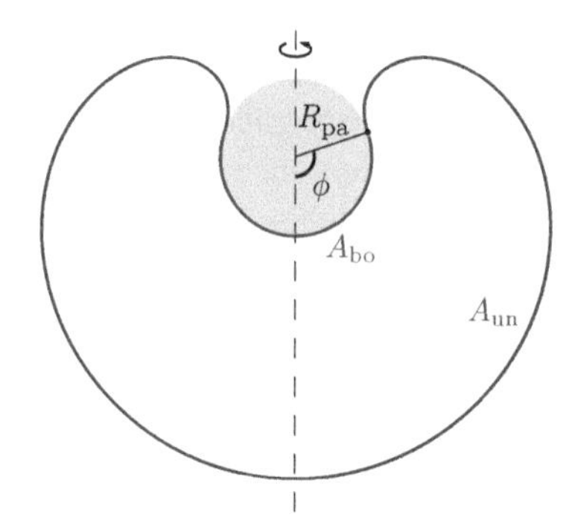

Figure 8.3 Vesicle membrane in contact with a spherical nanoparticle (gray) of radius R_{pa}. The vesicle shape is axially symmetric with respect to the vertical dashed line. The wrapping angle, ϕ, denotes the position of the contact line, which partitions the membrane into a bound (red) and an unbound (blue) segment. The wrapping angle varies from $\phi = 0$ for the onset of adhesion up to $\phi = \pi$ for the completely engulfed state. The bound and unbound membrane segment have the areas A_{bo} and $A_{un} = A - A_{bo}$, respectively.

requires the discretization or triangulation of these membranes, the shape of which can then be determined by energy gradient methods such as Monte Carlo simulations (see also Chapter 6).

Using the shooting method, we can obtain the energy of the unbound segment as a function of the wrapping angle, $\mathcal{E}_{\text{un}}(\phi)$, which comes exclusively from bending. The energy of the segment bound to the particle, on the other hand, can be calculated analytically and includes both bending and adhesion contributions that have the form

$$\mathcal{E}_{\text{bo}} = [-2\pi |W| R_{\text{pa}}^2 + 4\pi\kappa \left(1 \pm mR_{\text{pa}}\right)^2](1 - \cos\phi) \quad (8.6)$$

where the plus and minus signs apply to endocytic and exocytic engulfment, respectively. Notice that we have also included the possibility of a nonzero spontaneous curvature. For the case of zero spontaneous curvature $m = 0$, we recover the condition of energetically favorable engulfment as given by Eq. 8.5 by imposing that $\mathcal{E}_{\text{bo}} < 0$ in Eq. 8.6. The total energy, $\mathcal{E}(\phi)$, of the vesicle-particle system is then provided by the sum of the contributions from the bound and unbound segments, with $\mathcal{E}(\phi) = \mathcal{E}_{\text{un}}(\phi) + \mathcal{E}_{\text{bo}}(\phi)$.

We will now explore the energy landscapes for different sets of parameters. There are six parameters in the system: three material parameters, namely the bending rigidity of the membrane, κ; its spontaneous curvature, m; and the adhesive strength, $|W|$; and three geometric parameters, given by the membrane area, A; the enclosed volume, V; and the particle radius R_{pa}. By choosing the adhesive length, R_W, as a basic length scale and the bending rigidity as the basic energy scale, we are left with only four free parameters: the particle radius and the vesicle size, $R_{\text{ve}} \equiv \sqrt{A/4\pi}$, in units of R_W, the spontaneous curvature in units of R_W^{-1}, and the reduced volume of the vesicle, v, as defined in Eq. 8.2.

For particles that are several times smaller than the vesicle, that is, for $R_{\text{pa}} \lesssim 0.1 R_{\text{ve}}$, there are four qualitatively different types of energy landscapes or *engulfment regimes*, as shown in Figure 8.4: the free regime, $\mathcal{F}_{\text{st}}$, see (a); the partially engulfed regime, $\mathcal{P}_{\text{st}}$, see (b); the completely engulfed regime C_{st}, see (c); and the bistable regime $\mathcal{B}_{\text{st}}$, see (d–f). Which engulfment regime is present depends on the precise values of the four free parameters. In Figure 8.5, we show the engulfment regimes as a function of particle size and vesicle size, for an oblate vesicle with reduced volume $v = 0.98$ and three different values of the membrane spontaneous curvature, in the case of endocytic engulfment.

Particles that are much smaller than the adhesive length, R_W, as given by Eq. 8.5 are always found in the free regime, $\mathcal{F}_{\text{st}}$, in which the free state is stable and the completely engulfed state is unstable. Particles that are much larger than the adhesive length, on the other hand, are found in the completely engulfed regime, C_{st}, in which the free state is unstable but the completely engulfed state is stable. The most interesting behavior occurs for particles of intermediate size, on the order of the adhesion length: They may be either in a partially engulfed regime, $\mathcal{P}_{\text{st}}$—in which both the free and completely engulfed states are unstable and a partially engulfed state is stable—or in a bistable regime $\mathcal{B}_{\text{st}}$—in which both the free and completely engulfed states are stable and separated by an energy barrier. These four different engulfment regimes are separated from each other by the two lines L_{fr} and L_{ce} (Figure 8.5), which mark the stability limits of the free and completely engulfed state, respectively.

In the $\mathcal{P}_{\text{st}}$ regime, the contact area between the membrane and the particle changes in a continuous manner as we vary the particle radius, from a vanishingly small value at the instability line L_{fr} up to the total area of the particle at the instability line, L_{ce}. The system behavior is markedly different in the bistable regime, $\mathcal{B}_{\text{st}}$, in which the contact area changes abruptly or discontinuously as the particle size is varied. When, by increasing the particle size, the system crosses the instability line, L_{ce}, the free state remains stable and coexists with a metastable completely engulfed state until we cross the transition line, L_*, at which the free and completely engulfed states have the same free energy. The free state remains metastable until the system crosses the line

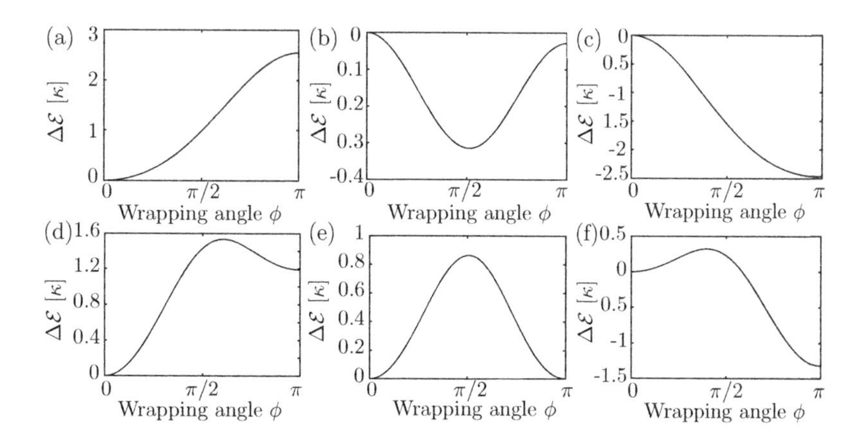

Figure 8.4 Free energy landscapes $\Delta\varepsilon(\phi) \equiv \varepsilon(\phi) - \varepsilon(0)$. (a) For the free regime, $\mathcal{F}_{\text{st}}$, the landscape is uphill, with a minimum at $\phi = 0$, which corresponds to the free state, and a maximum at $\phi = \pi$, which defines the completely engulfed state; (b) For the partial engulfment regime, $\mathcal{P}_{\text{st}}$, the landscape has maxima both at $\phi = 0$ and at $\phi = \pi$ and a minimum at an intermediate ϕ-value corresponding to a partially engulfed state; (c) For the complete engulfment regime, C_{st}, the landscape is downhill, with a minimum at $\phi = \pi$ and a maximum at $\phi = 0$; (d–f) Three landscapes within the bistable regime, $\mathcal{B}_{\text{st}}$, with two local minima at $\phi = 0$ and $\phi = \pi$ that are separated by a free energy barrier, implying that both the free and completely engulfed states are (meta)stable. In panels (d) and (f), the global minima (lowest energy states) are provided by the free and the completely engulfed state, respectively. Panel (e) corresponds to the transition line, L_* (see also Figure 8.5), at which both free and completely engulfed states have the same free energy.

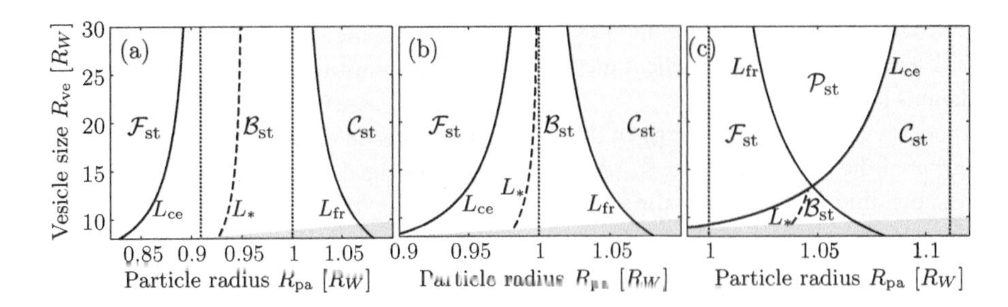

Figure 8.5 Different engulfment regimes for the endocytosis of a nanoparticle at the pole (see text) of an oblate vesicle with reduced volume $v = 0.98$. The three panels (a–c) correspond to the three values $m = -0.05 / R_W$, 0, $+0.05 / R_W$ of the spontaneous curvature. The different regimes are defined by the two instability lines, L_{fr} and L_{ce}, for the free and the completely engulfed state. The asymptotes of the two instability lines for large R_{ve} are indicated by vertical dotted lines. The bistable regimes, $\mathcal{B}_{st}$, contain the transition lines, L_* (dashed), at which the free and completely engulfed states coexist. For panel (c), the two instability lines L_{fr} and L_{ce} intersect. Close to the intersection point, the system is "multicritical" and reacts sensitively to small changes in both particle and vesicle size. The shaded area (gray) close to the x-axes indicates the size regime in which the vesicle does not have enough room to accommodate the completely engulfed particle, leading to a tension-dominated regime. (Reprinted from Agudo-Canalejo, J. and Lipowsky, R., *ACS Nano*, 9, 3704–3720, 2015a.)

L_{fr}, at which the free state becomes unstable and a particle in contact with the membrane would suddenly become completely engulfed. If we now reduce the particle size, the completely engulfed state would remain (meta)stable until we crossed the L_{ce}. The system therefore exhibits a marked hysteresis in this region.

As shown in Figure 8.5 for the case of endocytic engulfment, the intermediate size regime consists of a bistable regime for negative or zero values of the spontaneous curvature $m \leq 0$. For positive values of the spontaneous curvature (Figure 8.5c), we find again a bistable regime for small vesicles but a partially engulfed regime for large vesicles. In the latter case, the two instability lines cross each other. Around the intersection point, the system is "multicritical" in the sense that small changes in the particle and vesicle size will lead to large modifications of the system behavior.

As a particular example, let us consider GUVs made from DOPC/DOPG, and glass particles with a monolayer-coated surface (Gruhn et al., 2007). From Table 5.2 in Chapter 5, we see that the adhesion length corresponding to this system is $R_W \simeq 500$ nm. Particles with radii sufficiently smaller than R_W will remain free, whereas particles sufficiently larger than R_W should be completely engulfed. On the other hand, particles with radii in the vicinity of 500 nm are expected to explore the intermediate regimes of the engulfment diagrams in Figure 8.5. For GUVs with reduced volume $v = 0.98$, positive spontaneous curvature $m = +0.05 / R_W \approx 1/(10\,\mu\text{m})$, and a size around $R_{ve} = 30 R_W \approx 15\,\mu\text{m}$, we expect particles with radii between 510 and 540 nm to be partially engulfed (Figure 8.5c). Conversely, if we consider particles with a radius of about 510 nm, they should become partially engulfed by vesicles with sizes larger than $15\,\mu\text{m}$, but remain free in the presence of smaller vesicles. Furthermore, vesicles with sizes below $6\,\mu\text{m}$ will not lead to partial engulfment of intermediate-sized particles because they will display a bistable regime instead of the partially engulfed regime. We emphasize again, however, that due to the requirement of axisymmetry in numerical calculations, Figure 8.5 only describes the engulfment of a particle at one of the weakly curved poles of the oblate vesicle. This limitation is overcome below, where we provide analytical equations that can describe engulfment at any arbitrary point on the vesicle surface.

8.4.3 STABILITY RELATIONS

As mentioned above, the boundaries between the four different engulfment regimes are given by two instability lines, corresponding to the stability limits of the free and completely engulfed states. Interestingly, exact analytical expressions exist (Agudo-Canalejo and Lipowsky, 2015a) for the two instability lines, which are found to depend strongly on the local geometry of the membrane in contact with the particle.

The free, nonadhering state of the particle is unstable and the membrane starts to spread over the particle for particle radii

$$R_{pa} > \left[R_W^{-1} \mp M_{ms} \right]^{-1} \equiv R_{fr} \tag{8.7}$$

where M_{ms} is the mean curvature of the membrane before contact with the particle (Figure 8.6a), and the minus and plus signs apply to endocytic and exocytic engulfment respectively. The corresponding instability line L_{fr} is given by the equality $R_{pa} = R_{fr}$.

In the completely engulfed state, the particle is still connected to the original mother membrane by a closed neck. This state becomes unstable and the closed neck starts to open up if the particle is too small, with

$$R_{pa} < \left[R_W^{-1} \pm M'_{ms} \mp 2m \right]^{-1} \equiv R_{ce} \tag{8.8}$$

where now M'_{ms} is the mean curvature of the mother membrane at the position of the neck (Figure 8.6b), and the upper and lower signs apply again to endocytic and exocytic

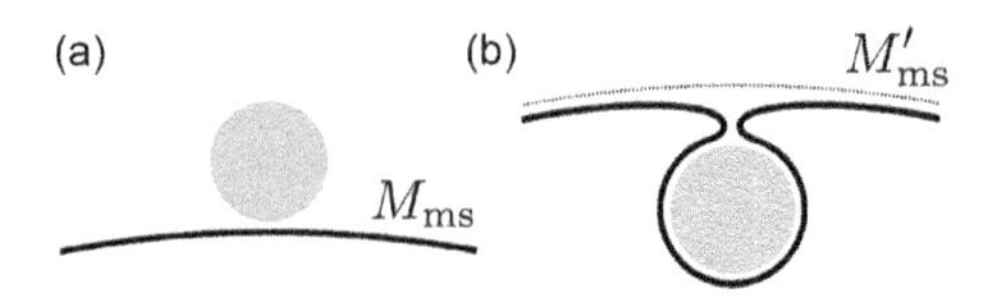

Figure 8.6 Definition of the mean curvatures of membrane segments relevant to the stability of free and completely engulfed particles. In (a), the mean curvature of the membrane before contact with the free particle is M_{ms}. In (b), the completely engulfed particle is still connected to the mother membrane by a narrow neck. The mean curvature of the mother membrane at the position of the neck is M'_{ms}.

engulfment respectively. The associated instability line L_{ce} is given by the equality $R_{pa} = R_{ce}$. Equation 8.8 can be viewed as a generalization of the closed neck condition for the budding of vesicles (see Chapter 5) to the case of "budding" around an adhesive particle (Agudo-Canalejo and Lipowsky, 2016), and has also been successfully applied to the interaction between ESCRT proteins and membranes (Agudo-Canalejo and Lipowsky, 2018).

Given that the two instability lines govern the engulfment behavior, the two relationships as provided by Eqs. 8.7 and 8.8 allow for a complete understanding of the engulfment process, without the need to perform numerical calculations. In particular, for the case of a planar membrane without spontaneous curvature, we have $M_{ms} = M'_{ms} = m = 0$, and we find that the free state becomes unstable and the completely engulfed state becomes stable when $R_{pa} > R_W$, therefore recovering the previously known result in Eq. 8.5. Moreover, by combining Eqs. 8.7 and 8.8, we find that they intersect whenever $m = (M_{ms} + M'_{ms})/2$. For convexly curved vesicles, such as the oblate vesicle with $v = 0.98$ in Figure 8.5, the mean curvatures M_{ms} and M'_{ms} are positive. This explains why the two instability lines do not intersect for negative or zero spontaneous curvature (Figure 8.5a and b) but do cross for positive spontaneous curvatures, as in Figure 8.5c. For the endocytic case, we find that $R_{fr} < R_{ce}$ if $m > (M_{ms} + M'_{ms})/2$, in which case particles of intermediate size are in the partially engulfed regime and the engulfment transition is continuous; whereas $R_{fr} > R_{ce}$ if $m < (M_{ms} + M'_{ms})/2$, in which case, particles of intermediate size are in the bistable regime and the engulfment transition is discontinuous. The opposite is true for exocytic engulfment, with the transition being discontinuous if $m > (M_{ms} + M'_{ms})/2$ and continuous otherwise.

In the particular case of a convex vesicle (so that M_{ms} and M'_{ms} are positive) with no spontaneous curvature, such a stability analysis predicts that endocytic engulfment is always discontinuous, whereas exocytic engulfment is always continuous. This was confirmed *via* numerical calculations in (Bahrami et al., 2016). Likewise, the predictions of the stability analysis in Eqs. 8.7 and 8.8 are also corroborated by the numerical results in (Góźdź, 2007; Cao et al., 2011). Furthermore, Eqs. 8.7 and 8.8 can be used to extract the material parameters of the system, such as the spontaneous curvature m and the adhesion length R_W, from experimental or computational studies of particle engulfment (see Box 8.1).

Box 8.1 Extracting material parameters from experimental or computational studies of particle engulfment

- As seen in Figure 8.4, the critical particle sizes R_{fr} and R_{ce} of the system are typically on the order of the adhesion length R_W. As shown in Table 5.2 of Chapter 5, values of the adhesion length can range from as little as 13 nm, for strong adhesion between DMPC and silica, to as much as 3 μm, for the ultraweak adhesion between DOPC/DOPG and glass. This implies that, for ultraweak adhesion, the different engulfment regimes should be accessible to optical microscopy experiments, whereas for strong adhesion the different engulfment states could be distinguished by cryoelectron or super-resolution microscopy. In the case of computer simulations, the strong adhesion regime should be accessible *via* coarse-grained MD simulations, whereas for ultraweak adhesion one could use Monte Carlo simulations.
- Suppose that, in experiments or simulations, the partially engulfed regime, $\mathcal{P}_{st}$, is observed for the endocytic case. This implies that, for this system, $m > (M_{ms} + M'_{ms})/2$, and engulfment proceeds continuously with increasing particle size in the endocytic case. We could then probe the system with particles of varying size, and record the two critical particle sizes, R_{fr} and R_{ce}, below and above which particles are free and completely engulfed, respectively. Simultaneously, one could easily measure from the experimental images or the simulation snapshots the corresponding values of M_{ms} and M'_{ms}. Using Eqs. 8.7 and 8.8, we can directly determine the values of the adhesion length R_W and the spontaneous curvature m as

$$R_W = R_{fr}\left(1 + R_{fr}M_{ms}\right)^{-1} \tag{8.9}$$

$$m = (R_{fr}^{-1} - R_{ce}^{-1} + M_{ms} + M'_{ms})/2 \tag{8.10}$$

- Suppose that, on the other hand, the partially engulfed regime, $\mathcal{P}_{st}$, is not observed for the endocytic case, implying that, for this system, $m < (M_{ms} + M'_{ms})/2$, and engulfment proceeds discontinuously with increasing particle size in the endocytic case. In this situation, only R_{fr} can be recorded, and therefore we would have access to the adhesion length R_W using Eq. 8.9 but not to the spontaneous curvature. However, we can now use the fact that, when $m < (M_{ms} + M'_{ms})/2$, engulfment proceeds continuously with increasing particle size in the exocytic case. Therefore, by repeating the same type of experiment or simulation, for the same particle–vesicle system, but this time for exocytic engulfment, we can now record the corresponding values of both R_{fr} and R_{ce}. The values of the adhesion length R_W and the spontaneous curvature m are then obtained as

$$R_W = R_{fr}\left(1 - R_{fr}M_{ms}\right)^{-1} \tag{8.11}$$

$$m = (R_{ce}^{-1} - R_{fr}^{-1} + M_{ms} + M'_{ms})/2 \tag{8.12}$$

8.4.4 ENERGY LANDSCAPES AND CURVATURE-INDUCED FORCES

The approach based on the stability of free and completely engulfed states presented in the previous section and Box 8.1 is very useful because it allows us to predict the fate of a particle coming into contact with a membrane, that is, whether the particle will remain free, be partially or completely engulfed or whether it will show bistability between the free and completely engulfed states. Unfortunately, the stability analysis does not provide us with any information about the height of the energy barriers for bistable regimes (Figure 8.4d–f) or about the binding energy and degree of engulfment for partially engulfed particles (Figure 8.4b).

Going beyond stability analysis, it is possible to develop an analytical approximation for the full energy landscapes experienced by particles coming into contact with vesicles, if we consider the limit of small particles with $R_{pa} \ll R_{ve}$. This analytical theory was developed in (Agudo-Canalejo and Lipowsky, 2017), through a systematic expansion of the free energy of the system to leading order in the particle-to-vesicle size ratio R_{pa}/R_{ve}. The free energy was found to behave as

$$\begin{aligned}\Delta\mathcal{E}(\phi) = 4\pi\kappa R_{pa}\,[2(R_{pa}^{-1} - R_{W}^{-1} \pm m)(1-\cos\phi) \\ \pm (M_{ms} - m)\sin^2\phi] \\ + O(\kappa R_{pa}^2 / R_{ve}^2)\end{aligned} \tag{8.13}$$

where the plus and minus signs correspond to endocytic and exocytic engulfment, respectively. We note that Eq. 8.13 depends on the local membrane curvature, M_{ms}, and therefore can be used to describe particle engulfment at any location on the surface of a vesicle, even non-axisymmetric locations.

Equation 8.13 reproduces the four kinds of energy landscapes displayed in Figure 8.4, that is, the four different engulfment regimes. Furthermore, it gives us access to the height of the energy barriers in bistable regimes and to the binding energy of partially engulfed particles. More precisely, according to Agudo-Canalejo and Lipowsky (2017), we find that the typical magnitude of barrier heights and binding energies is given by $4\pi\kappa R_{pa}\,|\,M_{ms} - m\,|$. Two important features should be noticed: First, the binding strength of partially engulfed particles as well as the size of the energy barriers in bistable regimes both increase with increasing particle size. Second, for a given particle size, the binding energy of a partially engulfed particle depends on the local curvature, M_{ms}.

More specifically, the binding of partially engulfed particles is more favorable for membrane segments with lower mean curvature in the case of endocytic engulfment, and for segments of higher mean curvature in the case of exocytic engulfment. This dependence of the binding energy of partially engulfed particles on the local mean curvature leads to curvature-induced forces that pull partially engulfed particles toward regions of lower or higher membrane curvature, in the case of endocytic or exocytic engulfment, respectively (Agudo-Canalejo and Lipowsky, 2017). Because particles with a chemically uniform surface as described here are only partially engulfed if their size and adhesiveness is in the right range (Figure 8.5), these curvature-induced forces can be most conveniently explored in experiments with partially adhesive Janus-like particles, which will be described in Section 8.6.3.

Finally, it is worth pointing out that the description of particle engulfment in the present section and Sections 8.4.2 and 8.4.3 is only strictly valid for particles that are many times smaller than the vesicle, with $R_{pa} \lesssim 0.1 R_{ve}$. On the other hand, if a particle is much larger than the vesicle, with $R_{pa} \gg R_{ve}$, we will recover the adhesion behavior of vesicles to planar substrates described in Chapter 5. Particles of a size comparable to the vesicle size thus represent an intermediate regime between these two limits. For particles that are smaller than the vesicle but still relatively large, with $R_{pa} \gtrsim 0.1 R_{ve}$, deviations from the ideal engulfment behavior described above lead to *satellite minima*, partially engulfed states with wrapping angles close to $\phi \approx 0$ or $\phi \approx \pi$ that are metastable and can coexist with free and completely engulfed states (Agudo-Canalejo and Lipowsky, 2015a, 2017).

8.4.5 ADHESION-INDUCED SEGREGATION OF MEMBRANE COMPONENTS

Biological membranes are always composed of a mixture of lipids and proteins (van Meer et al., 2008), and this compositional complexity can be mimicked in GUVs with multicomponent membranes. If a particle comes into contact with such a membrane, it is expected to preferentially attract and/or repel certain membrane components (Lipowsky et al., 2013) leading to adhesion-induced segregation of the membrane components. For example, if a charged particle comes into contact with a membrane that contains lipids of the opposite charge, these lipids are likely to be enriched in the membrane segment that is bound to the particle.

If the particle-bound segment has a different lipid composition, it will also have different elastic properties compared with the unbound segment. Thus, we will now consider the case in which the bound segment has bending rigidity, κ_{bo}, and spontaneous curvature, m_{bo}, whereas the unbound segment has a different bending rigidity, κ, and spontaneous curvature, m. The energy of such a segregated membrane is equal to the energy of a uniform membrane with bending rigidity, κ, and spontaneous curvature, m, provided we replace the adhesive strength, $|W|$, by the effective adhesive strength

$$W_{eff} = |W| + \frac{2\kappa}{R_{pa}^2}\left(1 \pm m R_{pa}\right)^2 - \frac{2\kappa_{bo}}{R_{pa}^2}\left(1 \pm m_{bo} R_{pa}\right)^2 \tag{8.14}$$

as follows from the difference between the elastic parameters of the unbound and bound segments. As before, the plus and minus signs apply to endocytic and exocytic engulfment, respectively.

It turns out that the stability relations for the new system are now again given by Eqs. 8.7 and 8.8 provided we replace the adhesive length, R_W, by $\sqrt{2\kappa / W_{eff}}$. It is important to note that this substitution introduces a nonlinear dependence on the particle size. For sufficiently negative values of the spontaneous curvature of the bound segment, one now finds regions of the parameter space in which only intermediate-sized particles can be completely engulfed (Agudo-Canalejo and Lipowsky, 2015a). The latter parameter dependence for complete engulfment explains the nonmonotonic size dependence of the uptake of gold nanoparticles by HeLa cells as observed experimentally (Chithrani and Chan, 2007).

We note that, in general, there will also be an energetic contribution proportional to the length of the contact line that separates the particle-bound and the unbound segments of the membrane. The associated line tension may be positive or negative. Positive line tensions will favor bistable regimes, because they provide an energy barrier for the engulfment process. Negative line tensions, on the other hand, will favor partially engulfed states, as well as completely engulfed states with non-axisymmetric 'tight-lipped' membrane necks, as found for the engulfment of nanodroplets (Satarifard et al., 2018).

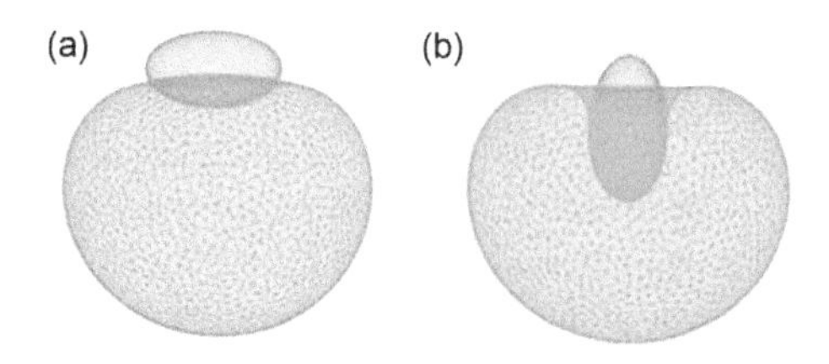

Figure 8.7 Engulfment of a prolate ellipsoidal particle, from energy minimization of a triangulated vesicle. In (a), the particle area fraction bound to the membrane is small with $A_{bo}/A_{pa} = 0.35$, and the particle preferentially adheres *via* its weakly curved surface segment. In (b), the particle area fraction bound to the membrane is large with $A_{bo}/A_{pa} = 0.67$ and the particle has reoriented with its strongly curved tip pointing toward the vesicle interior. (Bahrami, A.H., *Soft Matter*, 9, 8642, 2013. Reproduced by permission of The Royal Society of Chemistry.)

8.5 ENGULFMENT OF COMPLEX PARTICLES

8.5.1 NONSPHERICAL PARTICLES

Although the most commonly used nanoparticles in experiments and applications are spherical, it is also possible to produce nonspherical particles. Such particles may for example be ellipsoidal, cubic, or rod-like. In principle, the engulfment of nonspherical particles should not be conceptually different from that of spherical particles because it can still be understood as a competition between particle–membrane adhesion and membrane bending. In practice, however, the computations become more difficult. First of all, the interaction of nonspherical particles with membranes typically leads to non-axisymmetric configurations. Therefore, an exact solution to the problem of finding the minimum energy shapes of the vesicle as was done in Section 8.4.2 is not feasible, and approximate techniques such as energy minimization of discretized membranes (Bahrami, 2013; Dasgupta et al., 2013, 2014; Yi et al., 2014) or MD simulations (Vácha et al., 2011; Huang et al., 2013) have to be employed. These techniques are much more computationally expensive. Second, whereas the engulfment process of spherical particles could be described by a single reaction coordinate (Figures 8.3 and 8.4), the engulfment of nonspherical particles requires at least two such coordinates in order to describe the orientation of the particle with respect to the membrane.

It is precisely the rotation of elongated particles during engulfment that has been more thoroughly studied in theoretical and simulation work. These elongated particles tend to first attach to the membrane *via* their weakly curved surface segments but are then completely engulfed at their strongly curved tips. This engulfment-mediated rotation was first described by Bahrami (2013) and can be understood in terms of bending considerations. Initially, the membrane prefers to bind to the weakly curved side because it can gain the same amount of adhesion energy with a smaller cost in bending (Figure 8.7a). Close to complete engulfment, on the other hand, an elongated particle lying on its flat side would require that the membrane has already spread around both of its strongly curved tips, whereas if it stands against the membrane on its tip, the membrane can avoid wrapping around one of the strongly curved tips (Figure 8.7b). This conformational transition from "lying down to standing up" has also been observed in MD simulations (Huang et al., 2013).

8.5.2 DEFORMABLE PARTICLES AND FLUID DROPLETS

So far, we have considered particles that are rigid, that is, that do not change their shape as a result of their interaction with the membrane. A natural way to add complexity to the system would be to consider deformable particles. A simple example of a deformable particle that could be engulfed by a GUV would be a smaller lipid vesicle, which would have a bending rigidity comparable to that of the GUV, or a small polymer vesicle, whose bending rigidity would be on the order of 10 times that of the GUV (see Chapter 26). Another example of a deformable particle that may be engulfed by a GUV could be a fluid droplet, such as those arising in GUVs containing aqueous two-phase systems (see Chapter 29). Deformable particles may also be achieved by grafting a polymer brush onto rigid particles.

The particular case corresponding to the engulfment of a smaller fluid vesicle by a larger fluid membrane was considered in (Yi et al., 2011). In this case, the adhesion energy needed for complete engulfment increases as one decreases the bending rigidity of the smaller vesicle. This suppression of completely engulfed states is a consequence of a preference for partially engulfed states. Indeed, small vesicles that are less rigid than the engulfing membrane will tend to spread onto the larger membrane without deforming it.

A similar prominence of partially engulfed states is found for the engulfment of fluid droplets, as was studied in (Kusumaatmaja and Lipowsky, 2011; Satarifard et al., 2018). Small droplets again have a tendency to spread onto the membrane. In analogy to the engulfment of rigid particles, however, it is found that for large droplets, the partially engulfed state can become unstable and undergo a discontinuous transition to a completely engulfed state.

8.6 ENGULFMENT OF MULTIPLE PARTICLES

8.6.1 ENGULFMENT PATTERNS

We have until now described the engulfment of single particles by vesicles. It is, however, of practical relevance and theoretical interest to consider the case of simultaneous engulfment of many particles by a single vesicle. In Section 8.4, we described how the engulfment behavior of a single spherical particle is highly dependent on the local curvature of the engulfing vesicle, as described by the (in)stability conditions for the free and completely engulfed particle states in Eqs. 8.7 and 8.8. When we consider a vesicle with a complex, nonspherical shape, these instability conditions vary continuously as we move along the membrane surface. In particular, we can then take advantage of the local nature of these conditions to study the engulfment of many particles at non-axisymmetric locations of the vesicle shape.

Depending on the local curvature of the vesicle at the point of contact with the particle, this particle will either remain free, be partially or completely engulfed, or exhibit bistability between the free and the completely engulfed states. These four engulfment regimes can therefore coexist on the surface of a single vesicle, forming different engulfment patterns when a vesicle is exposed to many particles (Figure 8.8).

For a vesicle membrane with a laterally uniform composition, up to three different engulfment regimes can be simultaneously present on a single vesicle, if the particles are much smaller than the vesicle, which implies that the vesicle-particle system can form 10 possible engulfment patterns (Agudo-Canalejo and Lipowsky, 2015b). These patterns depend strongly on the vesicle shape, as defined by its reduced volume, on the spontaneous curvature, as well as on the particle size and adhesiveness. Therefore, small variations of any of these four parameters lead to morphological transitions between different engulfment patterns.

Completely engulfed particles in bistable, $\mathcal{B}_{st}$, or completely engulfed, C_{st}, segments are connected to the mother vesicle by a very narrow neck, and therefore create only a local deformation of the membrane that costs no energy. As a consequence, these particles cannot "feel" their surrounding membrane curvature gradient and will be subject to diffusive motion, unaware of each other. However, if these completely engulfed particles diffuse from the $\mathcal{B}_{st}$ or C_{st} segments into $\mathcal{F}_{st}$ or $\mathcal{P}_{st}$ segments, they will completely or partially detach from the vesicle membrane, respectively. Once in a $\mathcal{P}_{st}$ segment, partially engulfed particles can "feel" the surrounding curvature gradient and will be subject to curvature-mediated forces, see Section 8.4.4, as well as membrane-mediated particle–particle interactions that may lead to aggregation of particles. The latter processes will be discussed in the following two subsections.

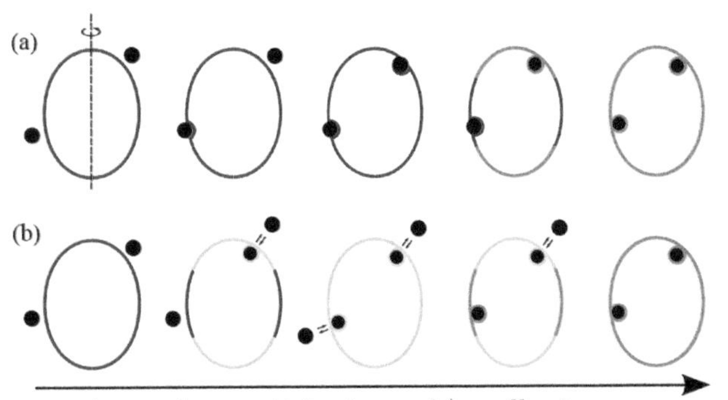

Figure 8.8 Different engulfment patterns of nanoparticles (black) on a prolate vesicle. The spontaneous curvature of the vesicle membrane is positive in (a) and negative in (b). The patterns involve coexistence of the four different engulfment regimes along separate segments of the same vesicle: Free segments $\mathcal{F}_{st}$ with no engulfment (red) and bistable segments $\mathcal{B}_{st}$ with activated engulfment and release (yellowish orange), as well as segments decorated by partially engulfed, $\mathcal{P}_{st}$ (blue), and completely engulfed, C_{st} (green), particles. A change in particle size or adhesiveness leads to continuous morphological transitions between these patterns. (Reprinted from Agudo-Canalejo, J. and Lipowsky, R., *Nano Lett.*, 15, 7168–7173, 2015b.)

8.6.2 PARTICLE AGGREGATION INTO MEMBRANE TUBES

Aggregation of partially engulfed particles into membrane tubes has been observed in two independent simulation studies using triangulated membranes and multiple spherical particles (Bahrami et al., 2012; Šarić and Cacciuto, 2012). As was clarified later (Bahrami et al., 2014; Raatz et al., 2014), this cooperative engulfment process turns out to be energetically favorable, compared with the individual engulfment of each particle, when the range of the interaction potential is sufficiently large. For potential ranges on the order of a few percent of the particle radius, these tubular structures are found to be very stable against thermal fluctuations. For a typical adhesion potential range of 1 nm, tubular aggregates should be favorable for particles with radii on the order of tens of nanometers.

8.6.3 JANUS PARTICLES AND MEMBRANE-MEDIATED INTERACTIONS

So far, we have considered particles that have a chemically uniform surface, that is, particles with a single value of the adhesive strength, $|W|$. As discussed in Sections 8.4 and 8.6.1, such particles are found in partially engulfed states only for a narrow region of the parameter space that includes the particle size and adhesive strength, as well as the vesicle shape and size, and the membrane spontaneous curvature and bending rigidity. Particles that are too small or too large will instead be found in free or completely engulfed states. As a consequence, it is a challenging task to explore the membrane-mediated interactions that arise between partially engulfed particles with a chemically uniform surface.

A better option to study such membrane-mediated interactions is provided by Janus or "patchy" particles, the surface of which contains both a strongly adhesive and a non-adhesive patch. If the adhesive strength, $|W|$, of the sticky patch is sufficiently large, with $|W|R_{pa}^2 \gg \kappa$, the membrane will always spread over this patch until it reaches the boundary between the adhesive and the nonadhesive patches, where it becomes "pinned." This pinning is caused by the abrupt change of the contact curvature as one crosses the boundary (Lipowsky et al., 2005). As a consequence, Janus or patchy particles will always be found in a partially engulfed state, independent of the size and shape of the vesicle or the membrane spontaneous curvature, and provide perfect candidates for the study of membrane-mediated particle–particle interactions.

Partially engulfed particles deform the surrounding membrane (Figure 8.3). If two identical Janus particles are sitting on the same vesicle, their deformation fields will overlap, and the shape as well as the bending energy of the membrane will depend on the distance between the two particles. This distance-dependence of the bending energy necessarily translates into a membrane-mediated force between the two particles, as they will tend to move in the direction that minimizes the bending energy of the membrane.

For two identical partially engulfed cylindrical particles that both lie on the same side of the membrane parallel to their symmetry axis, it has been known for a long time that the membrane-mediated interaction between the two particles is repulsive (Weikl, 2003; Müller et al., 2005). For spherical particles, the behavior is more subtle. Particles whose adhesive patch corresponds to less than half of the particle, that is, which are partially engulfed up to wrapping angles $\phi < \pi / 2$ will experience repulsive membrane-mediated interactions (Weikl et al., 1998; Reynwar and Deserno, 2011). Particles whose adhesive patch is larger than half the particle, corresponding to partial engulfment with wrapping angle $\phi > \pi / 2$ will experience long-range repulsion, mid-range attraction and short-range repulsion (Reynwar and Deserno, 2011).

8.7 SUMMARY AND OUTLOOK

In this chapter, we described some recent theoretical insights on particle–membrane interactions. For engulfment of rigid spherical particles, in Section 8.4, we showed how the engulfment process is governed by the stability of the free and completely engulfed states of the particle, as described by Eqs. 8.7 and 8.8. These two analytical conditions provide a concise but detailed description of the engulfment behavior, by defining the boundaries between the four relevant engulfment regimes (Figure 8.5). This type of stability analysis can then be used to describe complex engulfment patterns when a single vesicle is exposed to many small particles, as was shown in Section 8.6.1.

A limitation of the stability analysis is that we cannot access the height of the energy barriers present in a bistable, $\mathcal{B}_{st}$, regime nor the binding energy or equilibrium wrapping angle of partially engulfed states in a partially engulfed, $\mathcal{P}_{st}$, regime. Obtaining these latter quantities typically requires numerical computations such as those described in Section 8.4.2. However, in the limit in which the particles are much smaller than the vesicle, we can obtain these quantities analytically, see Section 8.4.4. Most importantly, we find that partially engulfed particles can "feel" the local membrane curvature and experience curvature-mediated forces toward regions of lower or higher membrane curvature, depending on whether the particles originate from the exterior (endocytic engulfment) or interior (exocytic engulfment) of the vesicle.

For more complex particles, such as nonspherical and deformable particles, a unifying theoretical framework such as the one we have for rigid spherical particles is still lacking. It is attractive to hypothesize that the stability relations for the free and completely engulfed states in Eqs. 8.7 and 8.8 will continue to hold for nonspherical particles, provided that we substitute the particle radius, R_{pa}, by the inverse local mean curvature of the particle at the point of contact with the membrane, $1 / M_{pa}$, which varies along the surface of the particle. This hypothesis is plausible because the two instability lines depend only on the local curvature of the vesicle membrane at the point of contact with the particle. The stability relation of the free particle state would then imply that the minimal adhesive strength required for binding to the membrane at a weakly curved region of the particle is lower than at a strongly curved region. The same would be true for the minimal adhesive strength required for a stable closed neck in a completely engulfed state, that is, closed necks are more stable for weakly curved segments of the particle surface. However, whereas for spherical particles the numerical results showed that these two instability lines determine the system behavior, with a partially engulfed state being stable only if the free and completely engulfed states are unstable (Figures 8.4 and 8.5), the same cannot be said for nonspherical particles. In the latter case, the energy landscapes are multidimensional and can display coexistence of several partially engulfed states, or of partially engulfed states with free or completely engulfed states.

All theoretical results described in this chapter are accessible to experiment. In particular, the theoretical predictions about engulfment patterns of vesicles with nonspherical shapes, as described in Section 8.6.1, could be studied by exposing lipid vesicles to fluorescently labeled particles. Vesicle membrane segments corresponding to the free, $\mathcal{F}_{st}$, or bistable, $\mathcal{B}_{st}$, regimes should appear dark, whereas completely engulfed, $\mathcal{C}_{st}$, segments should show a homogeneous fluorescence. For membrane segments that belong to a partially engulfed, $\mathcal{P}_{st}$, regime, curvature-mediated interactions will lead to more complex behavior. Such experimental studies of engulfment patterns would also provide a direct test of the stability relations in Eqs. 8.7 and 8.8. Furthermore, using partially adhesive Janus-like particles in contact with deflated vesicles of nonspherical shape, it should be possible to experimentally study the curvature-induced forces described in Section 8.4.4.

It will also be interesting to directly probe the bistable regime, $\mathcal{B}_{st}$, in which both the free and completely engulfed particle states are (meta)stable, using experiments with optical tweezers. Indeed, we expect that, in a bistable regime, a particle in contact with the membrane will not be spontaneously engulfed. However, if the particle is pushed against the membrane by the tweezers, it should become completely engulfed, and it should remain engulfed even if the force exerted by the tweezers is switched off. In order to be released, the particle will have to be pulled out of the vesicle using the tweezers once again. The minimal forces required to push and pull particles into and out of the engulfing membrane are governed by the height of the energy barrier for engulfment in the bistable regime and by the energy difference between the free and completely engulfed states (Figure 8.4d–f).

Other experimental challenges that will prove interesting, and which will require an expansion of the theoretical ideas described in this chapter, involve the interaction of particles with more complex vesicles, such as (i) vesicles with phase-separated membranes and (ii) vesicles with nanotubes, both described in Chapter 5. In the case of phase-separated vesicles, the particles should be preferentially engulfed by liquid-disordered domains rather than by liquid-ordered domains, because liquid-disordered membranes have a lower bending rigidity, unless this effect is overcompensated by a stronger adhesion of the particles to the liquid-ordered domains. Furthermore, the positive line tension of the domain boundaries should favor particle engulfment, because the engulfment process will tend to reduce the length of the contact line between domains. In the case of vesicles with nanotubes, the nanotubes should act as an area reservoir that can be used to engulf particles if these are sufficiently adhesive in the same way that neutrophils can extract area from membrane reservoirs during phagocytosis of many large particles (Herant et al., 2005). Whereas spherical vesicles without nanotubes cannot engulf particles without having their membrane stretched and ruptured, see Section 8.3.1, spherical vesicles with nanotubes should be able to engulf a large number of particles.

LIST OF ABBREVIATIONS

MD molecular dynamics
DOPC 1,2-dioleoyl-*sn*-glycero-3-phosphocholine
DOPG 1,2-dioleoyl-*sn*-glycero-3-phospho-rac-(1-glycerol)
DMPC 1,2-dimyristoyl-*sn*-glycero-3-phosphocholine

GLOSSARY OF SYMBOLS

A surface area of vesicle
A_{pa} surface area of particle
A_{bo} area of membrane segment bound to the particle
A_{un} area of the unbound vesicle membrane segment
ΔA increase in membrane area due to stretching
$\mathcal{B}_{st}$ bistable regime of engulfment
$\mathcal{C}_{st}$ completely engulfed regime
$\mathcal{E}_{ad}$ adhesion energy
$\mathcal{E}_{bo}$ energy of the particle-bound vesicle membrane segment
$\mathcal{E}_{un}$ energy of the unbound vesicle membrane segment
$\mathcal{F}_{st}$ free regime of engulfment
Γ coverage by adsorbate
H binding enthalpy of a receptor-ligand pair
κ membrane bending rigidity
κ_{bo} bending rigidity of the bound membrane segment
K_A area compressibility modulus
k_B Boltzmann's constant
ℓ_{me} membrane thickness
L_{fr} stability limit of the free particle state
L_{ce} stability limit of the completely engulfed particle state
L_* transition line in the bistable regime
m membrane spontaneous curvature
m_{bo} spontaneous curvature of the bound membrane segment
M_{ms} mean curvature of the vesicle membrane at the point of contact with the particle
M'_{ms} mean curvature of the vesicle membrane at the position of the narrow neck adjacent to a completely engulfed particle
M_{pa} local mean curvature of particle surface
$\mathcal{P}_{st}$ partially engulfed regime
ϕ wrapping angle
R_{pa} radius of spherical particle
R_{ve} effective radius of vesicle defined by $R_{ve} \equiv \sqrt{A/4\pi}$
R_W adhesion length defined by $R_W \equiv \sqrt{2\kappa/|W|}$
R_{fr} critical particle radius above which the free state is unstable
R_{ce} critical particle radius below which the completely engulfed state is unstable
ρ_{lig} density of ligands on the particle surface
ρ_0 density of receptors on the membrane
Σ mechanical tension of the membrane
T temperature
U binding free energy of a receptor-ligand pair
V volume of vesicle
V_{pa} volume of particle
v reduced volume (volume-to-area ratio) of the vesicle
v' effective reduced volume of vesicle after particle engulfment
W_{gen} adhesion (free) energy per area contributed by generic interactions
W_{spe} adhesion (free) energy per area contributed by specific interactions
W total adhesion (free) energy per area
W_{eff} effective adhesion energy per area, see Eq. 8.14.
X molar concentration of adsorbates in solution

REFERENCES

Agudo-Canalejo J, Lipowsky R (2015a) Critical particle sizes for the engulfment of nanoparticles by membranes and vesicles with bilayer asymmetry. *ACS Nano* 9:3704–3720.

Agudo-Canalejo J, Lipowsky R (2015b) Adhesive nanoparticles as local probes of membrane curvature. *Nano Lett* 15:7168–7173.

Agudo-Canalejo J, Lipowsky R (2016) Stabilization of membrane necks by adhesive particles, substrate surfaces, and constriction forces. *Soft Matter* 12:8155–8166.

Agudo-Canalejo J, Lipowsky R (2017) Uniform and Janus-like nanoparticles in contact with vesicles: Energy landscapes and curvature-induced forces. *Soft Matter* 13:2155–2173.

Agudo-Canalejo J, Lipowsky R (2018) Domes and cones: Adhesion-induced fission of membranes by ESCRT proteins. PLOS Comp Biol 14:e1006422.

Albanese A, Tang PS, Chan WCW (2012) The effect of nanoparticle size, shape, and surface chemistry on biological systems. *Annu Rev Biomed Eng* 14:1–16.

Alivisatos AP (1996) Semiconductor clusters, nanocrystals and quantum dots. *Science* 271:933–937.

Anderson TH, Min Y, Weirich KL, Zeng H, Fygenson D, Israelachvili JN (2009) Formation of supported bilayers on silica substrates. *Langmuir* 25:6997–7005.

Bahrami AH (2013) Orientational changes and impaired internalization of ellipsoidal nanoparticles by vesicle membranes. *Soft Matter* 9:8642.

Bahrami AH, Lipowsky R, Weikl TR (2012) Tubulation and aggregation of spherical nanoparticles adsorbed on vesicles. *Phys Rev Lett* 109:188102.

Bahrami AH, Lipowsky R, Weikl TR (2016) The role of membrane curvature for the wrapping of nanoparticles. *Soft Matter* 12:581–587.

Bahrami AH, Raatz M, Agudo-Canalejo J, Michel R, Curtis EM, Hall CK, Gradzielski M, Lipowsky R, Weikl TR (2014) Wrapping of nanoparticles by membranes. *Adv Colloid Interface Sci* 208:214–224.

Bothun GD (2008) Hydrophobic silver nanoparticles trapped in lipid bilayers: Size distribution, bilayer phase behavior, and optical properties. *J Nanobiotechnol* 6:1–10.

Cao S, Wei G, Chen JZY (2011) Transformation of an oblate-shaped vesicle induced by an adhering spherical particle. *Phys Rev E* 84:50901.

Chen X, Kis A, Zettl A, Bertozzi CR (2007) A cell nanoinjector based on carbon nanotubes. *Proc Natl Acad Sci USA* 104:8218–8222.

Chen Y, Bose A, Bothun GD (2010) Controlled release from bilayer-decorated magnetoliposomes via electromagnetic heating. *ACS Nano* 4:3215–3221.

Chithrani BD, Chan WCW (2007) Elucidating the mechanism of cellular uptake and removal of protein-coated gold nanoparticles of different sizes and shapes. *Nano Lett* 7:1542–1550.

Cho K, Wang X, Nie S, Chen ZG, Shin DM (2008) Therapeutic nanoparticles for drug delivery in cancer. *Clin Cancer Res* 14:1310–1316.

Curtis EM, Bahrami AH, Weikl TR, Hall CK (2015) Modeling nanoparticle wrapping or translocation in bilayer membranes. *Nanoscale* 7:14505–14514.

D'Rozario RSG, Wee CL, Wallace EJ, Sansom MSP (2009) The interaction of C60 and its derivatives with a lipid bilayer via molecular dynamics simulations. *Nanotechnology* 20:115102.

Daniel MC, Astruc D (2004) Gold nanoparticles: Assembly, supramolecular chemistry, quantum-size-related properties, and applications toward biology, catalysis, and nanotechnology. *Chem Rev* 104:293–346.

Dasgupta S, Auth T, Gompper G (2013) Wrapping of ellipsoidal nanoparticles by fluid membranes. *Soft Matter* 9:5473.

Dasgupta S, Auth T, Gompper G (2014) Shape and orientation matter for the cellular uptake of nonspherical particles. *Nano Lett* 14:687–693.

Deserno M, Gelbart WM (2002) Adhesion and wrapping in colloid-vesicle complexes. *J Phys Chem B* 106:5543–5552.

Dietrich C, Angelova M, Pouligny B (1997) Adhesion of latex spheres to giant phospholipid vesicles: Statics and dynamics. *J Phys II* 7:1651–1682.

Ding HM, Tian W De, Ma YQ (2012) Designing nanoparticle translocation through membranes by computer simulations. *ACS Nano* 6:1230–1238.

Dinsmore AD, Wong DT, Nelson P, Yodh AG (1998) Hard spheres in vesicles: Curvature-induced forces and particle-induced curvature. *Phys Rev Lett* 80:409–412.

Dykman LA, Khlebtsov NG (2014) Uptake of engineered gold nanoparticles into mammalian cells. *Chem Rev* 114:1258–1288.

Fery A, Moya S, Puech PH, Brochard-Wyart F, Mohwald H (2003) Interaction of polyelectrolyte coated beads with phospholipid vesicles. *C R Phys* 4:259–264.

Fischer HC, Chan WCW (2007) Nanotoxicity: The growing need for in vivo study. *Curr Opin Biotechnol* 18:565–571.

Gao H, Shi W, Freund LB (2005) Mechanics of receptor-mediated endocytosis. *Proc Natl Acad Sci USA* 102:9469–9474.

Ginzburg VV., Balijepalli S (2007) Modeling the thermodynamics of the interaction of nanoparticles with cell membranes. *Nano Lett* 7:3716–3722.

Gopalakrishnan G, Danelon C, Izewska P, Prummer M, Bolinger PY, Geissbühler I, Demurtas D, Dubochet J, Vogel H (2006) Multifunctional lipid/quantum dot hybrid nanocontainers for controlled targeting of live cells. *Angew Chemie Int Ed* 45:5478–5483.

Góźdź WT (2007) Deformations of lipid vesicles induced by attached spherical particles. *Langmuir* 23:5665–5669.

Greenall MJ, Marques CM (2012) Hydrophobic droplets in amphiphilic bilayers: A coarse-grained mean-field theory study. *Soft Matter* 8:3308.

Gruhn T, Franke T, Dimova R, Lipowsky R (2007) Novel method for measuring the adhesion energy of vesicles. *Langmuir* 23:5423–5429.

Guo R, Mao J, Yan L-T (2013) Unique dynamical approach of fully wrapping Dendrimer-like soft nanoparticles by lipid bilayer membrane. *ACS Nano* 7:10646–10653.

Guo Y, Terazzi E, Seemann R, Fleury JB, Baulin VA (2016) Direct proof of spontaneous translocation of lipid-covered hydrophobic nanoparticles through a phospholipid bilayer. *Sci Adv* 2:e1600261.

Hainfeld JF, Slatkin DN, Focella TM, Smilowitz HM (2006) Gold nanoparticles: A new X-ray contrast agent. *Br J Radiol* 79:248–253.

Hayward RC, Utada AS, Dan N, Weitz DA (2006) Dewetting instability during the formation of polymersomes from block-copolymer-stabilized double emulsions. *Langmuir* 22:4457–4461.

Herant M, Heinrich V, Dembo M (2005) Mechanics of neutrophil phagocytosis: Behavior of the cortical tension. *J Cell Sci* 118:1789–1797.

Hetzer M, Bayerl S, Camps X, Vostrowsky O, Hirsch A, Bayed TM (1997) Fullerenes in membranes: Structural and dynamic effects of lipophilic C60 derivatives in phospholipid bilayers. *Adv Mater* 9:913–917.

Hu J, Lipowsky R, Weikl TR (2013) Binding constants of membrane-anchored receptors and ligands depend strongly on the nanoscale roughness of membranes. *Proc Natl Acad Sci* 110:15283–15288.

Huang C, Zhang Y, Yuan H, Gao H, Zhang S (2013) Role of nanoparticle geometry in endocytosis: Laying down to stand up. *Nano Lett* 13:4546–4550.

Huang X, El-Sayed IH, Qian W, El-Sayed MA (2006) Cancer cell imaging and photothermal therapy in the near-infrared region by using gold nanorods. *J Am Chem Soc* 128:2115–2120.

Jana NR, Gearheart L, Murphy CJ (2001) Wet chemical synthesis of high aspect ratio cylindrical gold nanorods. *J Phys Chem B* 105:4065–4067.

Jaskiewicz K, Larsen A, Schaeffel D, Koynov K, Lieberwirth I, Fytas G, Landfester K, Kroeger A (2012) Incorporation of nanoparticles into polymersomes: Size and concentration effects. *ACS Nano* 6:7254–7262.

Koltover I, Rädler JO, Safinya C (1999) Membrane mediated attraction and ordered aggregation of colloidal particles bound to giant phospholipid vesicles. *Phys Rev Lett* 82:1991–1994.

Kreuter J (2001) Nanoparticulate systems for brain delivery of drugs. *Adv Drug Deliv Rev* 47:65–81.

Kroto HW, Heath JR, O'Brien SC, Curl RF, Smalley RE (1985) C60: Buckminsterfullerene. *Nature* 318:162–163.

Kusumaatmaja H, Lipowsky R (2011) Droplet-induced budding transitions of membranes. *Soft Matter* 7:6914–6919.

Lattuada M, Hatton TA (2011) Synthesis, properties and applications of Janus nanoparticles. *Nano Today* 6:286–308.

Laurencin M, Georgelin T, Malezieux B, Siaugue JM, Ménager C (2010) Interactions between giant unilamellar vesicles and charged core-shell magnetic nanoparticles. *Langmuir* 26:16025–16030.

Le Bihan O, Bonnafous P, Marak L, Bickel T, Trépout S, Mornet S, De Haas F, Talbot H, Taveau J-C, Lambert O (2009) Cryo-electron tomography of nanoparticle transmigration into liposome. *J Struct Biol* 168:419–425.

Lee J-H, Huh Y-M, Jun Y, Seo J, Jang J, Song H-T, Kim S et al. (2007) Artificially engineered magnetic nanoparticles for ultra-sensitive molecular imaging. *Nat Med* 13:95–99.

Lewinski N, Colvin V, Drezek R (2008) Cytotoxicity of nanoparticles. *Small* 4:26–49.

Li N, Sharifi-Mood N, Tu F, Lee D, Radhakrishnan R, Baumgart T, Stebe KJ (2016) Curvature-driven migration of colloids on tense lipid bilayers. *Langmuir* 33:600–610.

Li S, Malmstadt N (2013) Deformation and poration of lipid bilayer membranes by cationic nanoparticles. *Soft Matter* 9:4969.

Li Y, Li X, Li Z, Gao H (2012) Surface-structure-regulated penetration of nanoparticles across a cell membrane. *Nanoscale* 4:3768–3775.

Lipowsky R (1995) Generic interactions of flexible membranes. In: *Handbook of Biological Physics* (Lipowsky R, Sackmann E, Eds.), pp. 521–602. Amsterdam, the Netherlands: Elsevier Science BV.

Lipowsky R (1996) Adhesion of membranes via anchored stickers. *Phys Rev Lett* 77:1652–1655.

Lipowsky R (2013) Spontaneous tubulation of membranes and vesicles reveals membrane tension generated by spontaneous curvature. *Faraday Discuss* 161:305–331.

Lipowsky R, Brinkmann M, Dimova R, Franke T, Kierfeld J, Zhang X (2005) Droplets, bubbles, and vesicles at chemically structured surfaces. *J Phys Condens Matter* 17:S537–S558.

Lipowsky R, Döbereiner H-G (1998) Vesicles in contact with nanoparticles and colloids. *Europhys Lett* 43:219–225.

Lipowsky R, Leibler S (1986) Unbinding transition of interacting membranes. *Phys Rev Lett* 56:2541–2544.

Lipowsky R, Rouhiparkouhi T, Discher DE, Weikl TR (2013) Domain formation in cholesterol–phospholipid membranes exposed to adhesive surfaces or environments. *Soft Matter* 9:8438.

Liu J, Lu N, Li J, Weng Y, Yuan B, Yang K, Ma Y (2013) Influence of surface chemistry on particle internalization into giant unilamellar vesicles. *Langmuir* 29:8039–8045.

Lockman PR, Mumper RJ, Khan MA, Allen DD (2002) Nanoparticle technology for drug delivery across the blood-brain barrier. *Drug Dev Ind Pharm* 28:1–13.

Lodish H, Baltimore D, Berk A, Zipursky SL, Matsudaira P, Darnell J (2000) *Molecular Cell Biology* (Freeman WH, Ed.)., 4th ed. New York: Scientific American Books New York.

Luccardini C, Tribet C, Vial F, Marchi-Artzner V, Dahan M (2006) Size, charge, and interactions with giant lipid vesicles of quantum dots coated with an amphiphilic macromolecule. *Langmuir* 22:2304–2310.

Mahmoudi M, Meng J, Xue X, Liang XJ, Rahman M, Pfeiffer C, Hartmann R et al. (2014) Interaction of stable colloidal nanoparticles with cellular membranes. *Biotechnol Adv* 32:679–692.

Mai Y, Eisenberg A (2010) Controlled incorporation of particles into the central portion of vesicle walls. *J Am Chem Soc* 132:10078–10084.

Meinel A, Tränkle B, Römer W, Rohrbach A (2014) Induced phagocytic particle uptake into a giant unilamellar vesicle. *Soft Matter* 10:3667–3678.

Mercer J, Schelhaas M, Helenius A (2010) Virus entry by endocytosis. *Annu Rev Biochem* 79:803–833.

Michel R, Kesselman E, Plostica T, Danino D, Gradzielski M (2014) Internalization of silica nanoparticles into fluid liposomes: Formation of interesting hybrid colloids. *Angew Chemie Int Ed* 53:12441–12445.

Monopoli MP, Åberg C, Salvati A, Dawson KA (2012) Biomolecular coronas provide the biological identity of nanosized materials. *Nat Nanotechnol* 7:779–786.

Mukherjee S, Ghosh RN, Maxfield FR (1997) Endocytosis. *Physiol Rev* 77:759–803.

Müller MM, Deserno M, Guven J (2005) Interface-mediated interactions between particles: A geometrical approach. *Phys Rev E* 72:1–17.

Natsume Y, Pravaz O, Yoshida H, Imai M (2010) Shape deformation of giant vesicles encapsulating charged colloidal particles. *Soft Matter* 6:5359.

Noguchi H, Takasu M (2002) Adhesion of nanoparticles to vesicles: A Brownian dynamics simulation. *Biophys J* 83:299–308.

Odom TW, Huang J-L, Kim P, Lieber CM (1998) Atomic structure and electronic properties of single-walled carbon nanotubes. *Nature* 391:62–64.

Paciotti GF, Myer L, Weinreich D, Goia D, Pavel N, McLaughlin RE, Tamarkin L (2004) Colloidal gold: A novel nanoparticle vector for tumor directed drug delivery. *Drug Deliv* 11:169–183.

Pan Y, Neuss S, Leifert A, Fischler M, Wen F, Simon U, Schmid G, Brandau W, Jahnen-Dechent W (2007) Size-dependent cytotoxicity of gold nanoparticles. *Small* 3:1941–1949.

Panyam J, Labhasetwar V (2003) Biodegradable nanoparticles for drug and gene delivery to cells and tissue. *Adv Drug Deliv Rev* 55:329–347.

Pawar AB, Kretzschmar I (2010) Fabrication, assembly, and application of patchy particles. *Macromol Rapid Commun* 31:150–168.

Poon Z, Chen S, Engler AC, Lee H Il, Atas E, Von Maltzahn G, Bhatia SN, Hammond PT (2010) Ligand-clustered "patchy" nanoparticles for modulated cellular uptake and in vivo tumor targeting. *Angew Chemie Int Ed* 49:7266–7270.

Raatz M, Lipowsky R, Weikl TR (2014) Cooperative wrapping of nanoparticles by membrane tubes. *Soft Matter* 10:3570–3577.

Rai M, Yadav A, Gade A (2009) Silver nanoparticles as a new generation of antimicrobials. *Biotechnol Adv* 27:76–83.

Rasch MR, Rossinyol E, Hueso JL, Goodfellow BW, Arbiol J, Korgel BA (2010) Hydrophobic gold nanoparticle self-assembly with phosphatidylcholine lipid: Membrane-loaded and Janus vesicles. *Nano Lett* 10:3733–3739.

Reynwar BJ, Deserno M (2011) Membrane-mediated interactions between circular particles in the strongly curved regime. *Soft Matter* 7:8567.

Roiter Y, Ornatska M, Rammohan AR, Balakrishnan J, Heine DR, Minko S (2008) Interaction of nanoparticles with lipid membrane. *Nano Lett* 8:941–944.

Różycki B, Lipowsky R (2015) Spontaneous curvature of bilayer membranes from molecular simulations: Asymmetric lipid densities and asymmetric adsorption. *J Chem Phys* 142:54101.

Sahay G, Alakhova DY, Kabanov AV (2010) Endocytosis of nanomedicines. *J Control Release* 145:182–195.

Šarić A, Cacciuto A (2012) Mechanism of membrane tube formation induced by adhesive nanocomponents. *Phys Rev Lett* 109:188101.

Satarifard V, Grafmüller A, Lipowsky R (2018) Nanodroplets at membranes create tight-lipped membrane necks via negative line tension. ACS Nano 12:12424–12435.

Schmid K, Riediker M (2008) Use of nanoparticles in Swiss industry: A targeted survey. *Environ Sci Technol* 42:2253–2260.

Seifert U (1997) Configurations of fluid membranes and vesicles. *Adv Phys* 46:13–137.

Seifert U, Berndl K, Lipowsky R (1991) Shape transformations of vesicles: Phase diagram for spontaneous-curvature and bilayer-coupling models. *Phys Rev A* 44:1182–1202.

Seifert U, Lipowsky R (1990) Adhesion of vesicles. *Phys Rev A* 42:4768–4771.

Singh R, Lillard JW (2009) Nanoparticle-based targeted drug delivery. *Exp Mol Pathol* 86:215–223.

Slowing II, Vivero-Escoto JL, Wu C-W, Lin VS-Y (2008) Mesoporous silica nanoparticles as controlled release drug delivery and gene transfection carriers. *Adv Drug Deliv Rev* 60:1278–1288.

Smith KA, Jasnow D, Balazs AC (2007) Designing synthetic vesicles that engulf nanoscopic particles. *J Chem Phys* 127:84703.

Strobl FG, Seitz F, Westerhausen C, Reller A, Torrano AA, Bräuchle C, Wixforth A, Schneider MF (2014) Intake of silica nanoparticles by giant lipid vesicles: Influence of particle size and thermodynamic membrane state. *Beilstein J Nanotechnol* 5:2468–2478.

Sun C, Lee JSH, Zhang M (2008) Magnetic nanoparticles in MR imaging and drug delivery. *Adv Drug Deliv Rev* 60:1252–1265.

Vácha R, Martinez-Veracoechea FJ, Frenkel D (2011) Receptor-mediated endocytosis of nanoparticles of various shapes. *Nano Lett* 11:5391–5395.

van der Wel C, Vahid A, Šarić A, Idema T, Heinrich D, Kraft DJ (2016) Lipid membrane-mediated attraction between curvature inducing objects. *Sci Rep* 6:32825.

van Meer G, Voelker DR, Feigenson GW (2008) Membrane lipids: Where they are and how they behave. *Nat Rev Mol Cell Biol* 9:112–124.

Verma A, Stellacci F (2010) Effect of surface properties on nanoparticle-cell interactions. *Small* 6:12–21.

Wei X, Jiang W, Yu J, Ding L, Hu J, Jiang G (2015) Effects of SiO_2 nanoparticles on phospholipid membrane integrity and fluidity. *J Hazard Mater* 287:217–224.

Weikl TR (2003) Indirect interactions of membrane-adsorbed cylinders. *Eur Phys J E* 12:265–273.

Weikl TR, Kozlov MM, Helfrich W (1998) Interaction of conical membrane inclusions: Effect of lateral tension. *Phys Rev E* 57:10.

Wong-Ekkabut J, Baoukina S, Triampo W, Tang I-M, Tieleman DP, Monticelli L (2008) Computer simulation study of fullerene translocation through lipid membranes. *Nat Nanotechnol* 3:363–368.

Yang K, Ma Y-Q (2010) Computer simulation of the translocation of nanoparticles with different shapes across a lipid bilayer. *Nat Nanotechnol* 5:579–583.

Yi X, Shi X, Gao H (2011) Cellular uptake of elastic nanoparticles. *Phys Rev Lett* 107:98101.

Yi X, Shi X, Gao H (2014) A universal law for cell uptake of one-dimensional nanomaterials. *Nano Lett* 14:1049–1055.

Yue T, Zhang X, Huang F (2014) Membrane monolayer protrusion mediates a new nanoparticle wrapping pathway. *Soft Matter* 10:2024–2034.

Theory of polymer–membrane interactions

Fabrice Thalmann and Carlos M. Marques

And in this way, we will be as precise as required.

Pedro Nunes

in De Crepusculis, 1542 (Lisbon) describing the nonius, his invention

Contents

9.1 BILAYERS AND POLYMERS ARE INTIMATE OLD FRIENDS

Polymers are ubiquitous in natural and industrial systems of self-assembled bilayers. Also known as macromolecules, polymers are high molecular weight species made by the covalent binding of units called monomers. In a living cell, macromolecules are by weight the most abundant carbon-containing molecules: DNA and RNA, proteins and polysaccharides make up to 24% of a bacteria cell weight and account for a comparable fraction in the much larger animal cells (Lodish et al., 2000; Alberts et al., 2008). In cosmetics, pharmaceutics or detergency most formulations of membrane solutions have polymers added for performance, processing, conditioning or delivery (Lasic and Papahadjopoulos, 1998; Lasic and Barenholz, 1996; Vandepas et al., 1994). The presence of polymers in, on, or in the vicinity of a membrane changes not only the structure and viscosity of the liquid media where the membranes evolve: it modifies the properties of the membrane itself and its interactions with the environment. Giant unilamellar vesicles (GUVs) are systems of choice for studying such changes, and this book devotes several chapters to polymer-related matters in giant vesicles. Chapter 25 reviews the experimental techniques and conditions under which the consequences of the interactions between polymers and GUVs can be studied, whereas Chapters 26 and 27 review aspects of GUV formation with membranes made from or containing polymers. Chapter 5 also briefly discusses some consequences of polymer–membrane interactions for membrane curvature. In this chapter, we introduce basic theoretical concepts for understanding, quantifying and predicting membrane transformations induced by the presence of macromolecules. We first set the stage in the next section by classifying the different modes of interaction between macromolecules and self-assembled bilayers, and by discussing the most relevant control parameters. The following sections will follow this classification, presenting polymer concepts required to the theoretical treatment of the different situations. We will finish this chapter by discussing the key challenges ahead for the theoretical treatment of polymer–membrane interactions.

9.2 HOW DO POLYMERS INTERACT WITH MEMBRANES?

The theoretical understanding of the interactions between polymers and GUV membranes requires concepts from both one- and two-dimensional fluctuating objects or, otherwise stated, requires that the physical descriptions of both polymers and membranes are brought together. Experimentally there is however an important practical asymmetry: measurements made at the optical scales collect information on the membrane state but not on detailed polymer conformations that are, except for the rare cases involving large DNA molecules, inaccessible to the optical techniques. Thus, the most useful predictions for comparison with experiments on GUVs are those computing the polymer contributions to the effective membrane behavior.

From the point of view of the polymers, the presence of the bilayers introduces an interface, and thus an external field that will modify the state of the chains. From the point of view of the bilayer, the presence of the polymers will not only transform the membrane into a new effective membrane with different elastic properties or membrane permeability: the polymers might also induce strong shape transformations, change the structure of the membrane fluctuations and ultimately even compromise self-assembly viability. In this section, we first recall how polymers behave in the presence of interfaces and then discuss in general terms how effective elastic constants and spontaneous curvature arise for membranes in solutions containing other macromolecular species.

9.2.1 POLYMERS AND MEMBRANES: THE POLYMER VIEWPOINT

Polymers at interfaces have been studied for a long time, well before the interactions between polymers and bilayers started to be scrutinized. Boxes 9.1 and 9.2 summarize the main bulk and interfacial situations that a linear polymer chain might experience. As we will see in the next sections below, the theoretical framework developed to describe such situations is key to understand theoretically polymer–membrane interactions.

Box 9.1 Bulk states of polymers

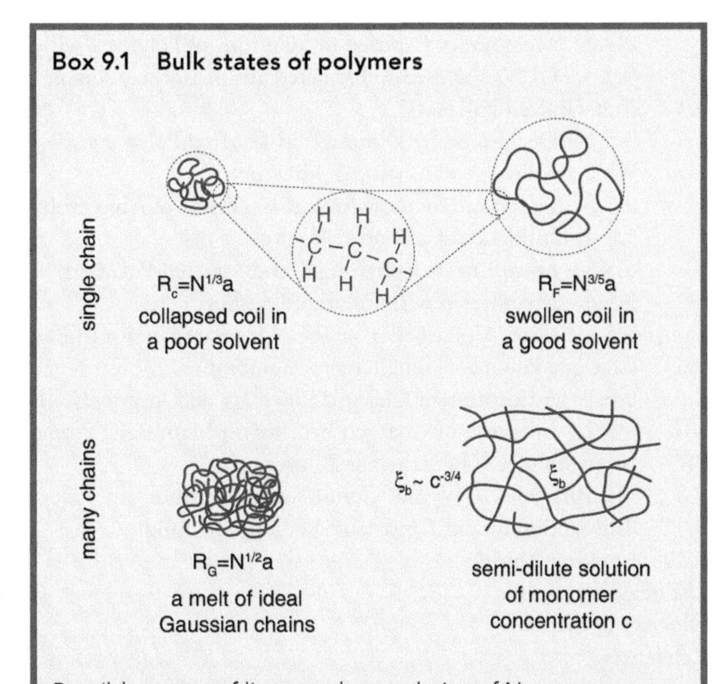

Possible states of linear polymer chains of N monomers in a bulk solution. At low enough concentrations chains do not interact, and their single chain behavior depends on the forces between monomers. Attraction leads to chain collapse, whereas repulsion leads to chain swelling. At higher concentrations chain–chain interactions lead to polymer screening: at small scales chains are swollen, but behave at large scales as ideal Gaussian polymers. Screening in polymer melts dominates the interactions and the chains are ideal at all scales.

Box 9.2 Polymers at interfaces: Penetrability, depletion and adsorption

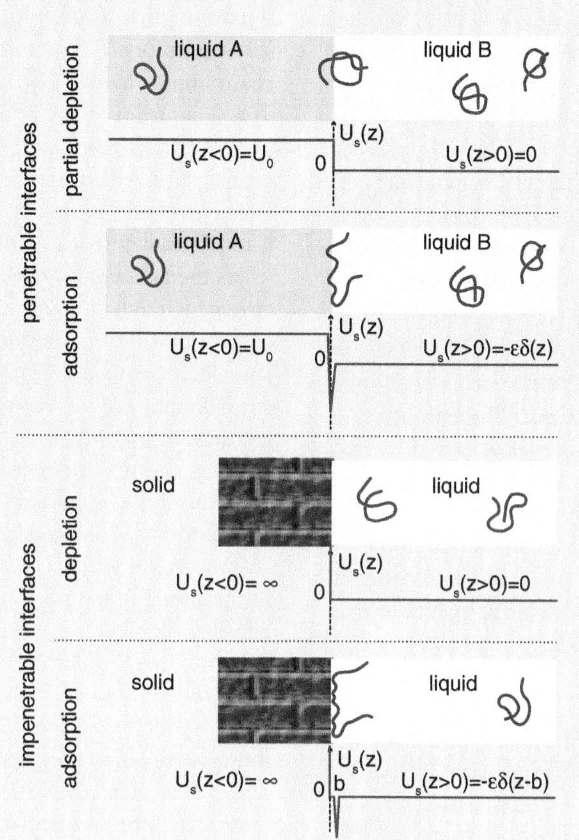

Different states of linear polymer chains at penetrable and impenetrable interfaces, with corresponding simple mathematical representations of the surface potentials $U_s(z)$.

Possible states of a chain in the bulk solution

For a chain in solution, two factors are important to know: (i) whether or not the chain is well solubilized and (ii) whether the polymer concentration c is larger or smaller than the so-called crossover concentration $c^\star$ above which different chains overlap.

The effective forces between the monomers determine the state of the polymer in solution. If they are repulsive, a single polymer in solution will be a swollen coil, with conformations well described by self-avoiding walks (SAW). SAW statistics predict that the average size of the polymer R_F scales with the number of monomers N as $R_F \simeq N^\nu a$ where $\nu \simeq 3/5$ and a is the Kuhn length, comparable to the monomer size when the chain is flexible. If the effective interactions between the monomers are attractive, the chain collapses into a dense globule containing little or no solvent. The dimensions of the polymer in this case are much smaller, they follow $R_c \simeq N^{1/3}a$. Transitions from a swollen to a collapsed state can be induced by temperature changes, often by increasing the temperature in aqueous solvents, by adding a cosolvent or by any other means that makes monomer–monomer interactions effectively attractive. For conditions (e.g., for temperatures) intermediate between the swollen and the collapsed states, attractions are sufficient to cancel the excluded volume between monomers but not enough to collapse the chain. At this particular point named the θ-point, the chain behaves like a random walk and it follows Gaussian statistics with $R_G = N^{1/2}a$.

The concentration of chains determines whether individual chains behave in an independent manner or if they interact significantly with other chains. For swollen chains, if the number of monomers per unit volume c is smaller than $c^\star = N / R_F^3$, the solution is dilute, and each chain with N monomers behaves independently from the others. Above $c^\star$, the chains interpenetrate and interact. The resulting solution is semi-dilute, it can be described by a crossover length, ξ_b, also called the blob size. For distances between monomers smaller than the blob size, the segments of the chains obey SAW statistics, whereas for distances larger than ξ_b the interactions between different chains screen excluded volume effects and the "chains of blobs" follow Gaussian statistics. In poor solvent conditions, similar considerations apply: A system of many chains makes a melt, where the excluded volume interactions are fully screened and where chains obey Gaussian statistics.

Possible states of a chain at the interface

From a fundamental point of view, the presence of a surface or an interface changes the forces felt by the monomers: The state of the chain is no longer only determined by the interactions between monomers but also by the surface potential.

Penetrable versus impenetrable interfaces. Although lipid bilayers are very efficient in preventing crossing by macromolecular species, they are after all liquid thin layers and they can be penetrated at least partially by the molecules to which they are exposed. It is thus conceptually important to distinguish penetrable from impenetrable interfaces.

Box 9.2 summarizes the different combinations of surface penetrability and surface–polymer interactions that are commonly found. The simplest case of an impenetrable interface is that of an infinitely repulsive flat surface such as a solid surface exposed to a solution of swollen homopolymers (homopolymers are polymers composed of a single type of monomers); here the net effect on the polymer is to forbid the presence of any monomer on say $z \leq 0$ if we assume that the surface is perpendicular to the z axis. The surface potential $U_s(r)$ can, in this case, be simply described as $U_s(z \leq 0) = \infty$; $U_s(z > 0) = 0$, it results in the complete depletion of the polymer from the interface. However, if one considers a liquid–liquid interface, like that, for instance, between water and an immiscible oil, the polymer can, in principle, be in both parts of the interface, albeit in different energetic states. Here a realistic potential could be described as $U_s(z \leq 0) = U_0$; $U_s(z > 0) = 0$, U_0 corresponding to the penalty of bringing a monomer from the phase above $z = 0$ to the phase below it. In this case, the chain has a finite probability of having simultaneously some of its monomers in the region $z > 0$ and others in $z < 0$.

Adsorption in good solvents. In many cases, the surface forces are not purely repulsive, but also exhibit an attractive component. For impenetrable surfaces, a simple theoretical approximation for such potential is given by $U_s(z \leq 0) = \infty$; $U_s(z > 0) = -\varepsilon\delta(z - b)$, where ε measures the strength of the attraction and b the distance from the wall where the attraction described by a delta function is felt. The result for the polymer is that above some attraction strength, ε_c, the polymer is in an adsorbed state, with some of its monomers attached to the surface, chain loops between the attracted monomers and chain tails at the two polymer extremities. A similar situation can occur for penetrable surfaces where the potential is written as $U_s(z \leq 0) = U_0$; $U_s(z > 0) = -\varepsilon\delta(z - b)$. An important difference with impenetrable surfaces is that adsorbed states exist for $\varepsilon > 0$ and, of course, the adsorbed state of the chain implies having loops and tails in both sides of the interface. The structure of the adsorbed polymer depends also on the bulk polymer concentration. For very dilute conditions, only a few chains cover the surface, without mutual interactions between different chains. In this case, the thickness of the adsorbed chain D is much smaller that its bulk radius, R_F, see Box 9.3.

As the bulk concentration increases, the surface becomes fully covered with chains, and different chains can compete for the same adsorption sites. As a result, the thickness of the adsorbed polymer layer increases until it reaches a polymer size, $D = R_F$. The internal structure of the adsorbed polymer layer is at a given height, z, identical to that of a semi-dilute solution, with a screening length $\xi(z) = z$ and thus a z-dependent concentration $c(z) \sim z^{-4/3}$; here the semi-dilute interfacial structure is exposed to a dilute polymer solution. If the bulk concentration increases such that $c > c^*$, the adsorbed layer keeps its self-similar structure but the thickness of the adsorbed layer is now given by $\xi(D) = \xi_b$. The difference between the adsorbed layer and the bulk becomes more tenuous because some of the chains have a fraction of their monomers in the adsorbed zone and others in the bulk semi-dilute solution. These aspects are pictorially summarized in Box 9.3, which also shows end-grafted polymer layers that we now introduce.

Grafted layers in good solvents. Many chains of practical interest are not homopolymers, but are composed of different

Box 9.3 Polymers at interfaces: From dilute to crowded

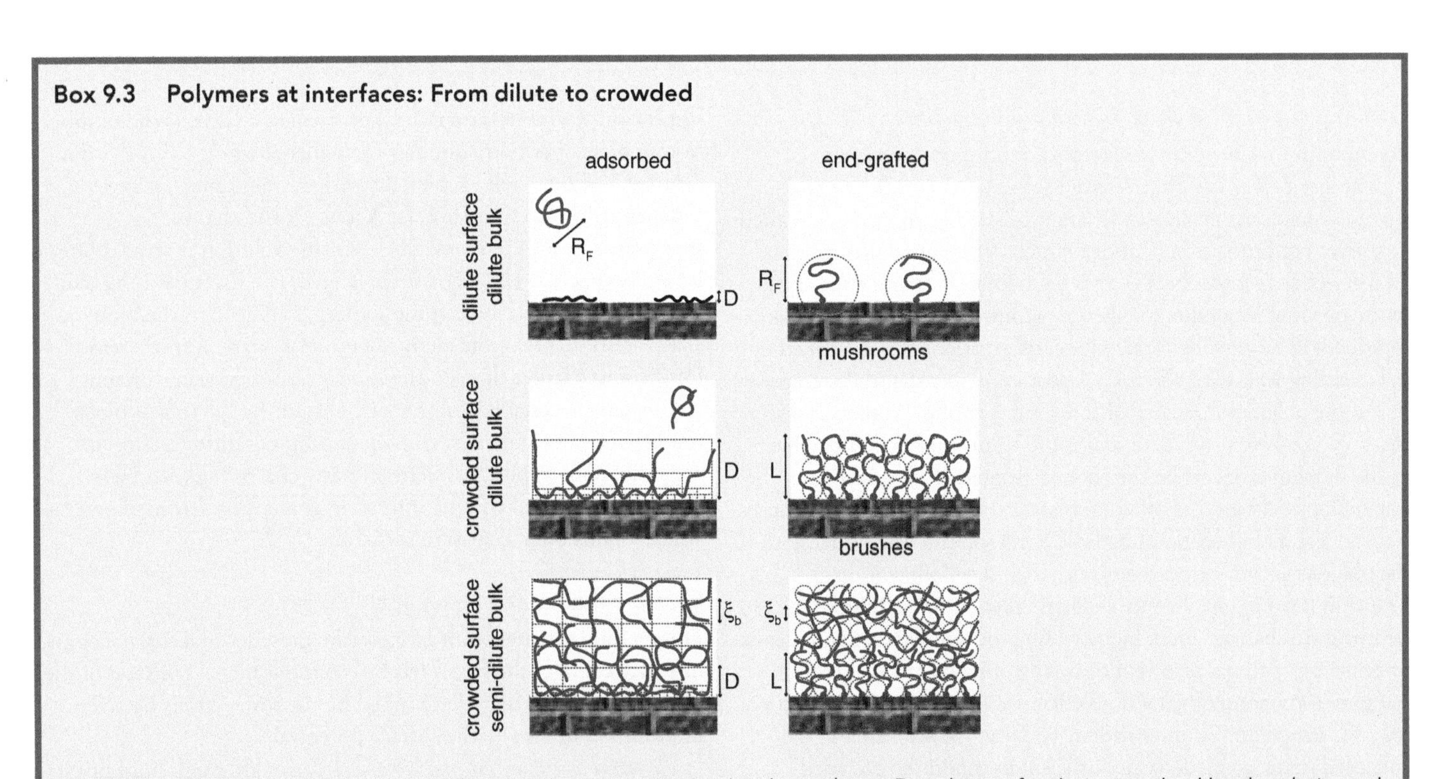

Attraction of all monomers to the interface results in an adsorbed polymer layer. Repulsion of a chain attached by the chain end leads to a end-grafted polymer layer. Very-dilute polymer solutions result in dilute, very thin adsorbed layers; however, even a dilute solution of polymer can lead to a fully covered surface. One refers to end-grafted layers of sparsely grafted chains as mushroom layers, whereas grafting densities high enough for the chains to interact build up the so-called polymer brushes. Semi-dilute solutions impose an outer osmotic pressure at the layer edges, influencing the structure of the polymer layer.

monomers, and many situations can arise that combine surface repulsion to some of the monomers and surface attraction for others. Arguably the most well-studied case is that of polymer end-grafting but adsorption of heteropolymers has also been the focus of intense work. For end-grafted polymers, a monomer or a group of monomers at one of the chain extremities has a different chemical nature and experiences a strong attraction to the surface, the attraction being considered as infinity when the extremity is chemically bound to the surface by, say, a covalent bond. The state of the end-grafted chains is mainly determined by the surface grafting density σ. When $\sigma \ll R_F^{-2}$ the end-grafted chains are far from each other, and only feel monomer–monomer and surface–monomer repulsive forces. One pictorially refers to this state as the mushroom-state, where the vertical extension of the end-grafted chain is of the same order of, but slightly larger than, its bulk coil size $D \simeq R_F$. In the opposite limit where $\sigma \gg R_F^{-2}$, one has the so-called brush layer where the different chains interact strongly, resulting in stretched chain conformations and a layer thickness $D \gg R_F$.

The particular case of θ solvents. Despite the scarcity of practical situations strictly corresponding to θ conditions where monomer attraction and repulsion exactly compensate, a considerable amount of work has been dedicated to the study of Gaussian chains interacting with interfaces. The most fundamental reason for this is that Gaussian chain models can be in many case exactly solved, providing an analytical benchmark for further developments accounting for chain excluded volume, charges and other factors. For θ solvents, the distinctions laid out in the previous paragraphs with respect to surface interactions also apply, but the compensation of monomer attraction and repulsion does not directly imply that the chains can be taken as Gaussian, and the problem requires a more subtle treatment of semi-dilute conditions such as those in dense adsorbed of grafted layers.

Adsorbed and grafted chains in poor solvents. All the above situations, where one or all monomers can interact attractively or repulsively with the interface can of course occur for all the possible qualities of the solvent. For very poor solvents where the solvent is completely excluded from the collapsed globules, attractive interactions with the interfaces lead to the formation of dense interfacial polymer layers, a situation comparable to that of the wetting of the surface by a second immiscible solvent. Wetting considerations play a lesser role for end-grafted dense layers where chain stretching dictates the interfacial dense layer properties.

9.2.2 MEMBRANES AND POLYMERS: THE MEMBRANE POINT OF VIEW

The wide variety of shapes and topologies observed in GUVs can be understood within the framework of membrane elasticity, as discussed in Chapter 5. In this section, we discuss how the parameters controlling membrane elasticity are *on average* modified by the presence in the surrounding solution of different species interacting with the membrane. If the interactions of the macromolecular species with the membrane are spatially localized, the effects on membrane stretching modulus, bending rigidity or spontaneous curvature cannot be described by average modifications only but assume also a local and spatially correlated character. This will be discussed in Section 9.5.2.

Average modifications of the membrane bending parameters

The membrane energy density associated with curvature deformation modes—see also Chapter 5—can be written as

$$h(C_1,C_2)=\frac{\kappa}{2}\left(C_1+C_2-2C_{\rm spo}\right)^2+\bar{\kappa}\,C_1C_2\,, \quad (9.1)$$

where C_1 and C_2 are the local principal curvatures that describe the shape of the membrane, and C_{spo} is the local spontaneous curvature, an intrinsic property of each membrane system, the spontaneous curvature, C_{spo}, vanishes for symmetric membranes. κ and $\bar{\kappa}$ are the membrane elastic constants that have dimensions of energy. The curvature rigidity, κ, controls, for instance, the amplitude of thermal fluctuations. Experimental κ values for fluid bilayers are found in the range of one to several tens of k_BT units. The Gaussian curvature rigidity, $\bar{\kappa}$, plays an important role in the determination of the membrane topology. Indeed, according to the Gauss-Bonnet theorem, the total energy contribution from the Gaussian curvature is a topological constant: $\int dS\left(C_1C_2\right)^{-1}=4\pi(1-g)$, where g is the surface topological genus ($g=0$ for a sphere, $g=1$ for a torus...). $\bar{\kappa}$ can thus be described as a chemical potential conjugated to surface topology.

For a bare membrane in a solution without other species, the energy density can be easily computed for simple geometries. The energy density of a sphere of radius $R=C_1^{-1}=C_2^{-1}$ is

$$\Delta h_s(R)=-\frac{4\kappa C_{spo}}{R}+\frac{2\kappa+\bar{\kappa}}{R^2}\,, \quad (9.2)$$

where $\Delta h_s(R)=h_s(R)-h_s(\infty)$. For a cylinder, $C_1=R^{-1}$ et $C_2=0$, one has

$$\Delta h_c(R)=-\frac{2\kappa C_{spo}}{R}+\frac{\kappa}{2R^2}\,. \quad (9.3)$$

It is worth noting that there is a factor two between the linear terms of the cylinder and sphere energy densities, a general geometric feature that holds for all energy densities with analytical expansions in R^{-1}.

In the vicinity of a membrane, the interactions of the suspended macromolecular species with the interface leads to a modification of the free energy of the system proportional to the area of the interface. When the surface is slightly curved, such excess surface energy can be expanded in powers of $R^{-1} \ll 1$ both for spheres and for cylinders. The renormalization of the elastic curvature constants can then be obtained by comparing the first two terms of the expansion of the excess surface energy density to the bending energy density. In the following, we consider only bare (not exposed to polymers) symmetric membranes for which the spontaneous curvature vanishes, C_{spo}; results for cases where $C_{spo} \neq 0$ can be obtained in a similar manner. Formally, the expansion of the excess energy density for a sphere ($i=s$) or a cylinder ($i=c$) is

$$\gamma_i(R)=\gamma_i^0\left(\frac{1}{R}=0\right)+\frac{\mu_i}{R}+\frac{\nu_i}{R^2}\,. \quad (9.4)$$

It is worth noting than one must have $\gamma_c^0 = \gamma_s^0 = \gamma^0$ because both cylinders and spheres reduce to flat planes in the limit of infinite curvature radius. This coefficient measures the excess surface free energy of a flat wall exposed to a macromolecular solution. The curvature-dependent terms of the excess energy density of a membrane in contact with the macromolecular solution is

$$\Delta h_s(R) + \Delta\gamma_s(R) = -\frac{4\kappa\Delta C_{spo}}{R} + \frac{2\kappa_{eff} + \bar{\kappa}_{eff}}{R^2}, \quad (9.5)$$

$$\Delta h_c(R) + \Delta\gamma_c(R) = -\frac{2\kappa\Delta C_{spo}}{R} + \frac{\kappa_{eff}}{2R^2}, \quad (9.6)$$

for the sphere and the cylinder, respectively, with $\Delta\gamma_i(R) = \gamma_i(R) - \gamma^0$. Exposure of one side only of the membrane to the particle surface interaction induces a spontaneous curvature that can be written as

$$\Delta C_{spo} = \frac{-\mu_s}{4\kappa} = \frac{-\mu_c}{2\kappa}. \quad (9.7)$$

A positive contribution indicates a tendency of the membrane to spontaneously bend toward the solution. When the membrane is embedded in the solution, with both sides exposed to the same polymer solution, the spontaneous curvature vanishes because the excess surface energy is the sum of the two contributions $\Delta\gamma_i = \Delta\gamma_i(R) + \Delta\gamma_i(-R)$. In this chapter, we will define the corrections to the elastic bending constants for this symmetric situation:

$$\Delta\kappa = \kappa_{eff} - \kappa = 4\nu_c, \quad (9.8)$$

$$\Delta\bar{\kappa} = \bar{\kappa}_{eff} - \bar{\kappa} = 2\nu_s - 8\nu_c. \quad (9.9)$$

Average modification of the membrane stretching modulus

The considerations described above for the curvature modes of a lipid bilayer assume implicitly that the membrane has a fixed area, a good approximation for situations where a GUV is only subjected to thermal random forces or to other external forces not exceeding a fraction of mN.m^{-1}. The lipid bilayer has however a finite stretching modulus K_A (also known as area compressibility modulus), that can be easily measured on GUVs, for instance, by micropipette techniques, see Chapter 11. The stretching modulus reflects the energy cost to change the area per lipid of the bilayer, an energetic cost that can be increased or decreased by interaction of the membrane with a macromolecular environment. Here we follow a simple description (Marsh et al., 2003) of bilayer stretching (or compression) elasticity to illustrate how one can theoretically compute polymer-induced corrections to K_A. The local structure of the bilayer is described by the area per lipid a_ℓ, and the energy per lipid F_ℓ is described as

$$F_\ell = \gamma a_\ell + \frac{c_0}{a_\ell}, \quad (9.10)$$

where the first term is the cohesive energy originating from the cost of the interfacial energy γ between the hydrophobic core of the bilayer and the solution ($\gamma \sim 40$ mN.m^{-1}), and the second term represents the lipid–lipid repulsion. Minimizing Eq. 9.10 with respect to a_ℓ provides the equilibrium lipid area a_ℓ^0 and the bare membrane stretching modulus K_A^0

$$a_\ell^0 = \sqrt{\frac{c_0}{\gamma}},$$

$$K_A^0 = 2a_\ell \frac{\partial^2 F_\ell}{\partial a_\ell^2}\Big|_{a_\ell = a_\ell^0} = 4\gamma, \quad (9.11)$$

where the factor 2 in the definition of K_A^0 accounts for the two leaflets of the bilayer. Two situations now need to be distinguished.

For chemical equilibrium conditions, where the macromolecules exchange freely between the surface and the surrounding solution, changes in the area of the membrane will not impact the excess surface energy γ^0, which is thus independent of a_ℓ. In this case, one has simply

$$F_\ell = \gamma a_\ell + \frac{c_0}{a_\ell} + \gamma^0 a_\ell,$$

$$a_\ell = \sqrt{\frac{c_0}{\gamma + \gamma^0}},$$

$$K_A = K_A^0 + 4\gamma^0. \quad (9.12)$$

Note that equilibrium cases such as those corresponding to depletion have positive values of γ^0 and are thus expected to decrease the area per lipid and to increase the area compressibility. Conversely, equilibrium adsorption that has negative γ^0 values is expected to increase the area per lipid and to decrease the stretching modulus.

For situations such as those with membranes decorated with end-grafted polymers, the excess surface energy γ^0 is a function of the area per lipid a_ℓ, and one needs to minimize

$$F_\ell = \gamma a_\ell + \frac{c_0}{a_\ell} + \gamma^0(a_\ell)a_\ell, \quad (9.13)$$

and recompute K_A. In general, for the situations of interest in this chapter such as those discussed in Section 9.5.3 below, $\gamma^0(a_\ell)a_\ell$ is a positive, decreasing function of a_ℓ, leading to equilibrium areas per lipid larger than those of the pure membrane. The sign of the correction to K_A^0 depends on the details of the function $\gamma^0(a_\ell)$ and needs to be computed for each particular case.

These approaches have been extensively used in the literature to determine the *average* modifications of the elastic parameters for many experimentally relevant situations involving polymers but also other molecules or suspended particles. The case of depleted colloidal suspensions of sphere and rods that we now describe is of particular interest because it illustrates in a simple manner the methods and the concepts of this approach while still leading to nontrivial results.

9.3 DEPLETION OF RIGID PARTICLES: RODS AND SPHERES

Depletion is arguably the simplest interaction between a macromolecule or a suspended particle and an interface see also Box 9.2. The potential of interaction vanishes away from the surface into the solution, and assumes a very high value otherwise. This results in the impenetrable character of the surface and on the depletion of the macromolecule or the particle away from the surface. Depletion phenomena was first studied by (Asakura and Oosawa, 1954) and it is of great importance for predicting the stability of colloidal suspensions because the attractive nature of the depletion forces between colloids can lead to flocculation. In this section, we consider the simple case of hard spheres and cylinders depleted by a membrane, and we will analyze polymer depletion in the next section. We show first using scaling arguments that spherical particles do not significantly change the membrane elastic constants but that rods do. Exact values of the effective elastic constants are then derived for the limit of dilute suspensions.

In a solution of hard spheres of radius r_0 and number density ρ_b, the typical scale for the energy density is $k_BT\rho_b$. This is also the value of the depletion pressure pushing two flat surfaces together. The order of magnitude of the excess surface energy due to depletion is thus $\Delta\gamma \simeq k_BT\rho_b r_0$. Typical values for the bare (without depletion) interface energies are of the order $\gamma \simeq k_BT/a^2$, where a is a microscopic size. For a characteristic value $a \sim 1$ nm, γ is on the order of a few tens of mN/m. The corrections due to the depletion of spherical particles are thus a factor $(a/r_0)^2$ lower than typical surface energies even at volume fractions $\phi = \rho_b(4\pi/3)r_0^3$ on the order of unity. Corrections to the curvature modulus are on the order of $\Delta\kappa \simeq k_BT\rho_b r_0^3$: Even for the largest possible volume fractions, this contribution is only on the order of k_BT, a value at the lower end of the range $[1-20]\ k_BT$ for most bare elastic constants. Contrary to the weak changes expected for elastic constants, changes in spontaneous curvature due to exposure to asymmetric depletion solutions can reach experimentally relevant values on the order of $\Delta C_{spo} \sim 1/r_0(k_BT/\kappa)\rho_b r_0^3$, as further discussed in Chapter 5.

In a rod solution with rod number density, ρ_b, the upper concentration limit of the isotropic solution is the Onsager concentration, $\rho_b^\star = 4.2L_r^{-2}\varnothing^{-1}$, where L_r is the rod length and $\varnothing$ is the rod diameter (Onsager, 1949) The contributions to the interfacial tension are now of the order $\Delta\gamma \simeq k_BT\rho_b L_r$, but even for rod diameters on the order of the microscopic length, a, the contribution to the interfacial tension of rod solutions at the Onsager concentration is still a factor (a/L_r) smaller than typical interfacial tension values. However, modifications of the elastic constants are here on the order of $\Delta\kappa \simeq k_BT\rho_b L_r^3$, a factor $(L_r/\varnothing)$ larger than k_BT. Therefore, even rather rigid phospholipid membranes with elastic constants as large as 20 k_BT may have their rigidities substantially modified at low rod concentrations where rod–rod interactions are still negligible.

Following the approach described by Yaman et al. (1997a, 1997b), we consider an ideal gas of hard spheres or of hard rods of length L_r in the presence of flat and curved surfaces that repel the rods. For rods, we parameterize the possible conformations by the center of mass coordinates $\vec{r}$, and by two angles specifying the rod direction, $\omega \equiv (\theta,\phi)$, whereas for spheres, the conformational space depends only on the position of the center of mass. The relevant potential describing the thermodynamics of the system is written as

$$F[\rho(r,\omega)] = \int d\mathbf{r}\, d\omega\, \rho(\mathbf{r},\omega)(k_BT\ln(v\rho/e) - (\mu_b - U_s(\mathbf{r},\omega))), \tag{9.14}$$

where ρ is the local concentration, v a normalization volume, μ_b is the solution chemical potential and $U_s(r,\omega)$ is the interaction potential between the rods and the surface. Functional minimization of F with respect to $\rho(r,\omega)$ gives the equilibrium density profile. After integration over the angular degrees of freedom, the excess surface energy is given by

$$\Delta\gamma = \frac{1}{S}\left(F[\rho(z)] - F[\rho(z\to\infty)]\right) = k_BT\int dz[\rho_b - \rho(z)]J(z,R), \tag{9.15}$$

where S is the surface area; z the perpendicular distance from the surface; and $J(z,R)$ is the appropriate Jacobian for the geometry. For flat surfaces $J(z,R)=1$, for the outside of a cylinder of radius R, $J(z,R)=1+z/R$, and for the outside of a sphere $J(z,R)=(1+z/R)^2$. Equation (9.15) holds for both spherical and rodlike particles provided that the corresponding concentration profile is known.

9.3.1 DEPLETION OF HARD SPHERES

In the case of hard sphere solutions—see also Figure 9.1a—where there is no coupling between angular conformation and curvature, one has ($U_s(z<r_0)=\infty$, $U_s(z<r_0)=0$) and ($\rho(z<r_0)=0$, $\rho(z>r_0)=\rho_b$). To second order in curvature, Eq. 9.15 leads to the following corrections to the interfacial energy

$$\Delta\gamma_s = k_BT\rho_b r_0\left(1+\frac{r_0}{R}+\frac{r_0^2}{3R^2}\right),$$

$$\Delta\gamma_c = k_BT\rho_b r_0\left(1+\frac{r_0}{2R}\right). \tag{9.16}$$

The excess free energy resulting from exposing a flat surface to a hard sphere suspension is thus exactly $\gamma^0 = k_BT\rho_b r_0$. From Eq. 9.12, this implies that the stretching modulus of the bilayer increases according to

$$K_A = K_A^0 + 4k_BT\rho_b r_0 \tag{9.17}$$

For a measurement where two flat surfaces are brought together to contact, thus excluding all colloids from the gap between the surfaces, the associated adhesion energy is simply twice the excess energy for one surface (Lekkerkerker and Tuinier, 2011)

$$W_a = 2k_BT\rho_b r_0 \tag{9.18}$$

This can be understood easily from a *gedankenexperiment* where two flat, infinitely thin surfaces are immersed in a colloidal suspension far from each other, thus creating four depletion layers. If the surfaces are now immersed together at contact, they create only two depletion layers. The adhesion energy, which is the difference between the energy cost for the two different immersions is thus twice the cost of the energy per unit surface required to creating a depletion layer.

The renormalized values of the elastic constants can be directly obtained from Eqs. (9.8) and (9.9). If the membrane is exposed on both sides to a bead solution there is no spontaneous curvature, the modifications to the elastic constants read:

$$\Delta\kappa = 0\,, \tag{9.19}$$

$$\Delta\bar{\kappa} = \frac{1}{2\pi}k_BT\rho_b\frac{4\pi r_0^3}{3}. \tag{9.20}$$

A different nanobead concentration on both sides of the membrane will lead to a spontaneous curvature, as further discussed in Chapter 5. The maximum effect is obtained when one side only of the membrane is exposed to a colloidal concentration ρ_b, leading to a predicted spontaneous curvature toward the colloidal suspension side

$$\Delta C_{\text{spo}} = -\frac{1}{r_0}\frac{3}{16\pi}\frac{k_BT}{\kappa}\rho_b\frac{4\pi r_0^3}{3}. \tag{9.21}$$

9.3.2 DEPLETION OF HARD RODS

The depletion interaction of infinitely thin rods and curved walls was first considered by (Auvray, 1981), then by (Yaman et al., 1997a). These last authors have in particular determined the contribution of the depletion interactions to the elastic constants of the membranes.

The geometrical constraints depend on the convexity of the surface (Figure 9.1b). Eq. 9.15 gives, in this case, the following values for the surface energy (Yaman et al., 1997b):

$$\Delta\gamma_{\text{out}} = k_BT\rho_b\frac{L_r}{4} \tag{9.22}$$

when the rods interact with a convex surface and

$$\Delta\gamma_{\text{in}} = k_BT\rho_b\frac{L_r}{4}\left(1-\alpha\frac{L_r^2}{R^2}\right) \tag{9.23}$$

when the rods interact with concave spherical ($\alpha = 1/12$) or cylindrical ($\alpha = 1/32$) surfaces. The asymmetry of the expressions for convex and concave surfaces shows that a membrane will spontaneously bend toward a rod solution even if there is no spontaneous curvature in the traditional sense.

The area compressibility of the bilayer is also here increased according to Eq. 9.12

$$K_A = K_A^0 + k_BT\rho_bL_r \tag{9.24}$$

When two membranes immersed in a solution and brought into a flat contact, the induced adhesion energy is

$$W_a = k_BT\rho_b\frac{L_r}{2} \tag{9.25}$$

which has a scaling form similar to Eq. 9.18 but with a different numerical factor (Lekkerkerker and Tuinier, 2011). For a membrane immersed in a rod solution one extracts the new elastic constant values

$$\Delta\kappa = \frac{-1}{64}k_BT\rho_bL_r^3 = -k_BT\frac{\rho_b}{\rho_b^\star}\frac{1}{15.2}\frac{L_r}{\varnothing}, \tag{9.26}$$

$$\Delta\bar{\kappa} = \frac{1}{96}k_BT\rho_bL_r^3 = k_BT\frac{\rho_b}{\rho_b^\star}\frac{1}{22.9}\frac{L_r}{\varnothing}. \tag{9.27}$$

κ decreases and $\bar{\kappa}$ increases, with an amplitude $L_r/\varnothing$ times larger that k_BT as anticipated: a rod solution might in principle destabilize a fluid membrane by lowering its curvature rigidity. A prediction for the changes in vesicle shape induced by a colloidal rod suspension was also made, in particular for the phase boundary between prolate and oblate GUV shapes (Groh, 1999). It was found that rods outside the vesicles favor oblate shapes, whereas prolate shapes are favored for rods inside. If rods are present on both sides the equilibrium shapes and the boundary of the prolate to oblate transition remain unchanged.

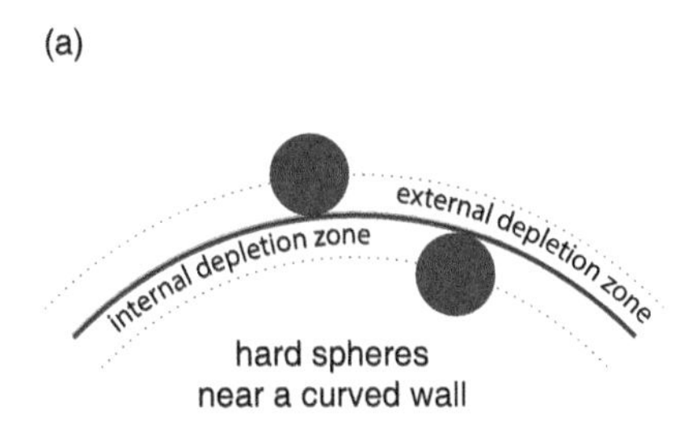

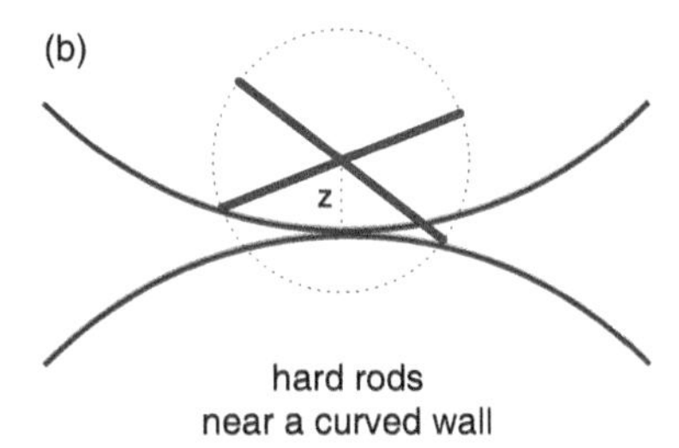

Figure 9.1 Hard spheres and hard rods close to curved surfaces. (a) For hard spheres the center of mass cannot access a zone close to the wall, named the depletion zone. The volume of the depletion zone is smaller in the concave part of the wall. (b) The center of mass of infinitely thin rods can get arbitrarily closer to the wall from the convex side, but not from the concave side creating an essential asymmetry in the response to curvature of membranes immersed in a rod suspension. For a given height, z, of the center of mass, rods on the concave side loose also more angular conformational entropy.

9.3.3 GIANT UNILAMELLAR VESICLES AND DEPLETED COLLOIDS

Several experiments involving depleted colloids and giant vesicles have been reported in the literature. Dinsmore et al. encapsulated small colloidal spheres of 42 nm in a giant vesicle of 1-stearoyl-2-oleoyl-*sn*-glycero-3-phosphocholine (SOPC, see Appendix 1 of the book for structure and data on this lipid) (Dinsmore et al., 1998) at a volume fraction $\phi \simeq 0.30$, along with a small number of larger spheres of 237 nm radius, see also Figure 9.2. The small spheres act in this case as a depletant for the large spheres inducing adhesion at the membrane interface, with a large adhesion being promoted in the regions of higher membrane curvature, as clearly seen in Figure 9.2c. By encapsulating increasing concentrations of polystyrene beads of 1-μm diameter in GUVs of 1,2-dioleoyl-*sn*-glycero-3-phosphocholine (DOPC, see Appendix 1 of the book for structure and data on this lipid), Natsume et al. observed diverse shape changes of the vesicle shape that could be rationalized in terms of the modifications of the vesicle free energy due to the formation of the colloid depletion layers (Figure 9.3) (Natsume et al., 2010). Also, protocols of GUV formation that allow for the encapsulation of large volume fractions of beads (Natsume and Toyota, 2013) have lead to studies on how deflated GUV membranes can interact with the dense encapsulated colloidal suspension and induce, for instance, ordered arrays of colloids and membrane facets (Natsume and Toyota, 2016). Although the theoretical predictions for the influence of colloidal depletion on the elastic parameters of membranes have not yet been directly tested, it is likely that the current development of more precise methods for the control of vesicle size and for the control of encapsulation will lead in a near future to more quantitative comparisons between theory and experiment.

9.4 MEMBRANES IN SOLUTIONS OF NONIONIC POLYMERS

By the end of the eighties, it had become clear that the addition of polymers to membrane systems lead to noticeable effects (Kekicheff et al., 1984; Seki and Tirrell, 1984; Ringsdorf et al., 1988; Rupert et al., 1988), seen also in pioneering experiments on GUVs (Evans and Needham, 1988; Decher et al., 1989; Cates, 1991). Following seminal work of Cantor that had computed diblock copolymer contributions to the bending rigidities of interfaces (Cantor, 1981), a number of theoretical contributions emerged about the influence of end-grafted (Milner and Witten, 1988) and adsorbed polymers (de Gennes, 1990; Brooks et al., 1991a, 1991b) on the bending and spontaneous curvature of lipid bilayers. Unusually, the effects on membranes of end-grafted and adsorbed/depleted polymers where discussed from the onset for chains with excluded volume, the less realistic but analytically tractable case of ideal (Gaussian) chains started to be discussed in the following years (Podgornik, 1993; Lipowsky, 1995; Eisenriegler et al., 1996). This section will be devoted to the discussion of the effects of adsorbed and depleted chains, the influence of end-grafted chains will be studied in Section 9.5.

9.4.1 MEMBRANES EXPOSED TO DEPLETED AND ADSORBED IDEAL POLYMERS

We now discuss effects from ideal Gaussian chains first because they provide exact results that can be taken as a benchmark for further, more realistic developments. An ideal Gaussian polymer is described by the positions, $\mathbf{R}_n$, of its $N+1$ monomers, $n = 0,1,2...N$ and

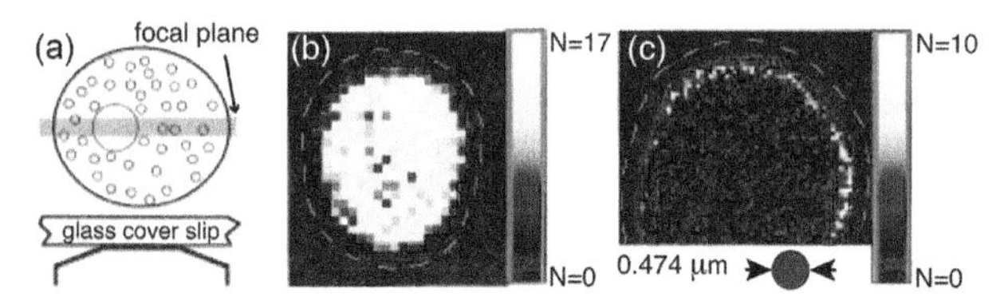

Figure 9.2 (a) Cartoon of the 600-nm thick slice through a SOPC giant vesicle imaged with an optical microscope. The in-plane positions of a large colloidal sphere encapsulated in the GUV was measured when it was in focus. (b) Probability distribution of a single 0.237-mm-radius polystyrene sphere inside a vesicle (without additional small spheres). The white dashed line is the edge of the vesicle, and the colored points indicate the number of times, N, the center of the sphere was observed in a bin located at a given point. There were 2,000 events and the bins were 130 × 130 nm. The sphere simply diffused freely throughout all of the available space. (c) Same as in (b), but with a vesicle that also contained small spheres (spheres volume fraction $\phi \simeq 0.30$, radius of small spheres $\simeq 0.042$ μm). There were 2,300 events and the bin size was 65 nm. The large sphere was clearly attracted to the vesicle wall, especially where the vesicle was most curved. (Reprinted with permission from Dinsmore, A. et al., *Phys. Rev. Lett.*, 80, 409, 1998. Copyright 1998 by the American Physical Society.)

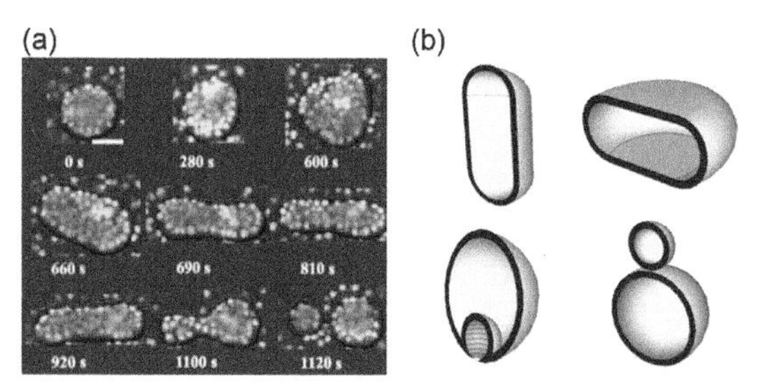

Figure 9.3 GUVs encapsulating depleted colloids. (a) Shape deformation pathway of a DOPC GUV encapsulating polystyrene beads with 1 μm diameter. The transformation is induced by applying a osmotic pressure difference between the inside and outside of the vesicle. Scale bar: 5 μm, it applies to all images. (b) Geometrical models (cross-section images) of different observed shapes. Dark gray region indicates internal depletion zone of the membrane surface with a thickness r_0. (From Natsume, Y. et al., *Soft Matter* 6, 5359–5366, 2010.)

by the potentials for the N links between neighboring monomers, $U_0 = k_BT/(2a^2)\Sigma_1^N(\mathbf{R}_n - \mathbf{R}_{n-1})^2$, with a the monomer size, k_B the Boltzmann constant, and T the absolute temperature. This chain model lacks excluded volume interactions between monomers, but it can accommodate additive external potentials of the form $U_{ext} = \Sigma_0^N u(\mathbf{R}_n)$ such as those resulting from the presence of an interface. The statistical properties of the polymer chain are often described by the restricted partition function also called Green function or chain propagator $G_N(\mathbf{R}_0,\mathbf{R}_n)$, which measures the conditional probability of finding one chain extremity at position $\mathbf{R}_n$, knowing that the other chain end is at $\mathbf{R}_0$. Under the potential U_{ext}, the propagator $G_N(\mathbf{R}_0,\mathbf{R}_N)$ obeys the Edwards equation

$$\left(\frac{\partial}{\partial N} - \frac{a^2}{6}\nabla^2 + \frac{U_{ext}(\mathbf{R}_N)}{k_BT}\right)G_N(\mathbf{R}_0,\mathbf{R}_N) = \delta(N)\delta(\mathbf{R}_0 - \mathbf{R}_N) \tag{9.28}$$

subjected also to the boundary conditions appropriate to the geometry where the chain is embedded. A free chain in an infinite space, $U_{ext} = 0$, is described by the Gaussian propagator

$$G_N^0(\mathbf{R}_0,\mathbf{R}_N) = \left(\frac{3}{2\pi a^2 N}\right)^{3/2} \exp\left\{-\frac{3(\mathbf{R}_N - \mathbf{R}_0)^2}{2a^2N}\right\} \tag{9.29}$$

and solutions of Eq. 9.28 for many common geometries such as spheres, cylinders or cones are available from textbooks (Carslaw and Jaeger, 1959) when U_{ext} vanishes everywhere except at the interface, in which case the potential can be translated into a boundary condition. Computing for a single chain the free-energy changes, $\Delta\mathcal{F}$, induced by the presence of an interface requires evaluating the chain partition function, Z, and is thus achieved by solving Eq. 9.28 and integrating out the remaining degrees of freedom. For chains without any fixed extremity, as it is the case for depleted or adsorbed polymers, one has

$$\Delta\mathcal{F} = -k_BT\ln\frac{\mathcal{Z}}{\mathcal{Z}_0} = -k_BT\ln\frac{\iint d\mathbf{R}_N\, d\mathbf{R}_0\, G_N(\mathbf{R}_0,\mathbf{R}_N)}{\iint d\mathbf{R}_N\, d\mathbf{R}_0 G_N^0(\mathbf{R}_0,\mathbf{R}_N)} \tag{9.30}$$

Effects of a solution of ideal depleted chains

For ideal chains depleted by impenetrable cylinders and spheres, Eq. 9.28 is to be solved in the corresponding geometry with $U_{ext} = 0$ in the bulk and the boundary conditions $G_N(\mathbf{R}_0 = \mathbf{r}_S,\mathbf{R}_N) = G_N(\mathbf{R}_0,\mathbf{R}_N = \mathbf{r}_S) = 0$ with $\mathbf{r}_S$ a vector describing the spherical or cylindrical surfaces (Podgornik, 1993; Eisenriegler et al., 1996; Hiergeist and Lipowsky, 1996). At a bulk chain concentration, ρ_b, the surface free energy obtained from Eq. 9.30 then reads

$$\Delta\gamma_c = k_BT\rho_b\frac{2}{\sqrt{\pi}}R_g\left(1 + \frac{\sqrt{\pi}}{4}\frac{R_g}{R} - \frac{1}{12}\frac{R_g^2}{R^2}\right),$$

$$\Delta\gamma_s = k_BT\rho_b\frac{2}{\sqrt{\pi}}R_g\left(1 + \frac{\sqrt{\pi}}{2}\frac{R_g}{R}\right), \tag{9.31}$$

with $R_g = Na^2/6$ the radius of gyration of the ideal polymer chain. Eq. 9.31 have a similar structure to Eq. 9.16, showing that depleted polymer solutions can qualitatively be considered as depleted colloidal particles with a radius comparable to the radius of gyration. However, as the comparison of the two set of relations also show, there is not a perfect mapping between the two systems given that it is not possible to rescale the colloidal radius in Eq. 9.16 (e.g., with $r_0 \to 2R_g/\sqrt{\pi}$) to reproduce Eq. 9.31. The adhesion energy between two flat membranes is in this case

$$W_a = k_BT\rho_b\frac{4}{\sqrt{\pi}}R_g \tag{9.32}$$

which has a scaling form similar to Eq. 9.18 but with a different numerical factor (Lekkerkerker and Tuinier, 2011).

The stretching modulus—see Eq. 9.12—increases by

$$K_A = K_A^0 + k_BT\rho_b\frac{8}{\sqrt{\pi}}R_g \tag{9.33}$$

whereas the effects on the membrane parameters follow

$$\Delta\kappa = -\frac{2}{3\sqrt{\pi}}k_BT\rho_bR_g^3,$$

$$\Delta\bar{\kappa} = \frac{4}{3\sqrt{\pi}}k_BT\rho_bR_g^3,$$

$$\Delta C_{spo} = -\frac{1}{R_g}\frac{1}{4}\frac{k_BT}{\kappa}\rho_bR_g^3. \tag{9.34}$$

Exposure of the membrane to a solution of ideal chains on one side induces a spontaneous curvature *toward* the polymer solution, a decrease in bending rigidity and an increase of the Gaussian elastic constant. For chain concentrations smaller or comparable to the crossover threshold $\rho_bR_g^3 \lesssim 1$ (note that $c = N\rho_b$ is at $c^\star$ when $\rho_bR_g^3 = 1$) the modifications of the elastic constants $\Delta\kappa$ and $\Delta\bar{\kappa}$ are negligible. However, contrary to colloidal suspensions, one can reach polymer concentration values much above the threshold, where strong modifications of the elastic constants are to be expected. This will be further discussed in Section 9.4.2 where predictions for chains with excluded volume will be given.

To conclude this paragraph it is worth stressing that the influence of the degree of impenetrability of the surface on the modifications of the membrane parameters has not yet been clarified. Given that for fully penetrable surfaces no effects are expected, a possible scenario is that the values of the membrane parameters monotonically decrease from those in Eq. 9.34 as a function of the degree of penetrability of the interface. However, as we will now see, finely tuning the interaction parameter between the ideal polymer and the surface can lead to a more complex behavior—and see also the comments at the end of this section on methods based on fluctuating interfaces.

Effects of a solution of ideal adsorbed chains

When the surface is not purely repulsive to the polymer, it is convenient to describe the effects of partial repulsion or attraction by an extrapolation length D defined by

$$\frac{1}{G_N(\mathbf{R}_0 = \mathbf{r}_S,\mathbf{R}_N)}\mathbf{n}.\nabla G_N(\mathbf{R}_0 = \mathbf{r}_S,\mathbf{R}_N) = -\frac{1}{2D} \tag{9.35}$$

where n is a vector normal to the surface pointing to the solution. This can be seen as the boundary condition corresponding to a short-range potential of strength $\varepsilon \sim a / D$. Note that as defined here one has surface attraction for $D > 0$ and surface repulsion for $D < 0$. Although such boundary condition does not reproduce all the features of microscopic potentials such as those represented in Box 9.2, it has not only the advantage of providing somewhat tractable results for ideal chains but it allows also tuning surface interactions from pure repulsion to strong attraction with a single parameter. This approach was first pursued in curved surfaces by Podgornik (1993), revisited later by several authors (Eisenriegler et al., 1996; Hiergeist et al., 1996) and corrected by Eisenriegler et al. (1996). Despite the potential of the method, corrections to the elastic parameters for arbitrary values of D were not yet obtained, likely due to the algebraic complexity of implementing the calculations up to second order in curvature. An explicit expansion of the free energy to first order in curvature allowed (Breidenich et al., 2005) to discuss spontaneous curvature effects, the central result can be written as

$$\Delta C_{\text{spo}} = -\frac{1}{R_g}\frac{1}{4}\frac{k_BT}{\kappa}\rho_b R_g^3\, g\left(\frac{R_g}{2D}\right)$$

$$g(x) = \left(1 + \frac{4}{\sqrt{\pi}}\frac{1}{x} - 2\frac{1}{x^2}\left(\text{erfc}(-x)\exp(x^2) - 1\right)\right), \quad (9.36)$$

where $erfc(x)$ is the complementary error function $erfc(x) = 1 - 2/\sqrt{\pi}\int_0^x du \exp(-u^2)$. In the limit of very strong repulsion ($1/D \rightarrow -\infty$) one has $g(x \rightarrow -\infty) \rightarrow 1$, and the result for the spontaneous curvature in Eqs. 9.34 is recovered. However, in this model, the absence of interactions ($1/D \rightarrow 0$) does not correspond to a vanishing spontaneous curvature($g(x \rightarrow 0) \rightarrow -1$). Moreover, this approach predicts that for finite adsorption strengths the membrane should bend away from the adsorbed chain, a result at odds with other approaches that we now describe.

9.4.2 MEMBRANES EXPOSED TO SOLUTIONS OF POLYMERS WITH EXCLUDED VOLUME

Predictions based on the ideal representations of polymers lead most often to analytical results but misrepresent the real nature of polymer chains where monomers experience excluded volume interactions. Accounting for excluded volume effects was a central driving force leading to the development of modern polymer physics (de Gennes, 1979; Des Cloizeaux and Jannink, 2010). In this section, we will present several methods required for tackling adsorption of polymers with excluded volume in curved geometries and summarize the main results obtained in this field.

The contribution of adsorbed polymers to the elastic constants of fluid bilayers was first qualitatively studied by de Gennes (1990), who provided the corrected scaling behavior as a function of adsorption strength and chain length but did not compute the amplitude nor the sign of the corrections. A systematic approach at the mean-field and scaling levels was initiated by Brooks et al. (1991a, 1991b). We will describe below the mean field-approach that allows for a closed treatment of excluded volume effects but does not account for the strong fluctuating nature of polymer systems, and then a scaling approach that correctly predicts the scaling forms of the corrections but without exact numerical pre-factors. For completeness, we will first report results obtained from group renormalization methods on the effects of excluded volume on the depletion of single polymer chains.

Effects from single depleted chains with excluded volume

Ideal chain models predict that the mean-square average of the end-to end distance of polymers obeys $R_e = N^{1/2}a$, a puzzle for polymer scientists in the middle of the 20th century who systematically observed exponents larger than 0.5, and closer to 0.6. The origin of this discrepancy was understood by Flory (1949), who estimated the swelling of the polymer chains due to monomer–monomer repulsions. It followed many decades of an intense theoretical activity where methods from statistical physics where applied the describe single- and many-chain polymer systems. In the 1960s, Edwards starts to apply to polymers self-consistent field methods inspired by condensed matter theory, laying the basis for modern polymer physics theory. A powerful insight by de Gennes (1972) pointing to the formal analogy between the statistical nature of polymer chains and certain classes of critical phenomena brought to the field of polymers many theoretical tools developed for describing phase transitions. Renormalization group methods proved particularly useful. In this class of approaches, one recognizes that long-distance correlations in polymer systems become irrelevant above four dimensions ($d = 4$) and therefore expand in $\epsilon = 4 - d$. For instance, for the Flory exponent ν describing chain dimensions $R_F = N^{\nu}a$, one obtains to second order in ϵ, $\nu = 1/2(1 + \epsilon/8 + 15\epsilon^2/256)$ predicting $\nu = 0.592$ for $\epsilon = 1$, close to the Flory estimation of $\nu = 0.6$ and only a few parts per thousand off the more precise admitted value of $\nu = 0.588$ obtained by re-summation of a perturbation series (Brézin et al., 1977).

Applying these methods to single chains with excluded volume depleted by curved surfaces, Hanke et al. (1999), obtained to first order in ε corrections to the membrane parameters given by

$$\Delta\kappa = -\frac{2}{3\sqrt{\pi}}k_BT\rho_b R_{gF}^3(1 - 0.0713\epsilon),$$

$$\Delta\bar{\kappa} = \frac{4}{3\sqrt{\pi}}k_BT\rho_b R_{gF}^3(1 - 0.177\epsilon),$$

$$\Delta C_{\text{spo}} = -\frac{1}{R_{gF}}\frac{1}{4}\frac{k_BT}{\kappa}\rho_b R_{gF}^3(1 - 0.131\epsilon). \quad (9.37)$$

with R_{gF} in these expressions being given by $R_{gF}^2 = N^{2\nu}a^2/6$, compare with Eq. 9.34. Excluded volume effects thus preserve the scaling forms and the signs of the corrections to the bare elastic constants given by ideal chains. The amplitude of the effects can also be larger because as N increases the reduction of the pre- factors is compensated by the increase of the radius of the chain with respect to its ideal value.

Mean-field theory for a polymer solution in contact with a membrane

Reversible adsorption of polymer chains can be described at the mean-field level by an energy density that is a functional of the local monomer volume fraction ϕ through the order parameter $\psi = \phi^{1/2}$ (de Gennes, 1981). Hereafter a is the size of a monomer, c the monomer number density and $\phi = ca^3$. At fixed chemical potential the Cahn-de Gennes free energy can be written as

$$F[\psi] = -k_B T \frac{\varepsilon}{a^2} \int dS \psi^2$$

$$+ \frac{k_B T}{a^3} \int dV \left(\frac{a^2}{6} (\nabla \psi)^2 + \frac{1}{2} \tilde{v} (\psi^2 - \psi_b^2)^2 \right). \quad (9.38)$$

The first integral describes the direct interaction with the surface, the dimensionless parameter ε represents the contact energy between the monomers and the wall. According to the sign of ε, this energy functional can describe adsorption or depletion. The gradient term is related to the chain connectivity and the last term in the functional describes volume excluded interactions. $\tilde{v} = v / a^3$ is the excluded volume parameter, related to the Flory interaction parameter, χ, through $v = (1 - 2\chi) a^3$. ψ_b is the bulk value of the order parameter.

Functional minimization of the energy density (9.38) provides an Euler-Lagrange differential equation for the profile

$$\frac{a^2}{6} \nabla^2 \psi - \tilde{v} \psi^3 + \tilde{v} \psi_b^2 \psi = 0 , \quad (9.39)$$

with boundary conditions

$$\frac{1}{\psi} \frac{\partial \psi}{\partial n} \Big|_{\text{surf}} = \frac{-1}{2D} , \quad (9.40)$$

where n is the normal to the surface. This condition defines the extrapolation length $D = a / (12\varepsilon)$ that characterizes the strength of adsorption ($\varepsilon > 0$) or depletion ($\varepsilon < 0$). The extrapolation length is in principle also a function of the surface curvature, but this dependence can be neglected in the limit of infinitely narrow wall potentials. A second characteristic length of the problem is the mean-field correlation length, $\xi_b = a / (3vc)^{1/2}$, introduced by Edwards (1965). For concentrated or semidilute solutions, ξ_b determines the longest decay of the profile that reaches the bulk value for $z \ll \xi_b$. As explained above, for distances larger than ξ_b the excluded volume interactions are screened and the polymer chain correlations exhibit a Gaussian behavior.

The solution of Eq. 9.39 is available for flat surfaces for all values of D, that is, for the whole range of adsorption and depletion strengths. Insertion of the solution in Eq. 9.38 and integrating gives the excess surface energy for flat surfaces

$$\gamma^0 = \frac{4}{3} \Pi_b \xi_b \left(1 - \frac{3}{4} \frac{c_s^{1/2}}{c_b^{1/2}} - \frac{1}{4} \frac{c_s^{3/2}}{c_b^{3/2}} \right) \quad (9.41)$$

where c_s is the value of monomer concentration at the surface determined by the surface potential strength ε or equivalently by the extrapolation length D

$$\left(\frac{c_s^{1/2}}{c_b^{1/2}} - \frac{c_b^{1/2}}{c_s^{1/2}} \right) = \frac{1}{2} \frac{\xi_b}{D}, \quad (9.42)$$

and $\Pi_b = k_B T v / 2 c_b^2$ is the osmotic pressure of the solution. For infinite repulsion ($D \to 0^-$ or $c_s = 0$) one gets thus the adhesion strength induced by the depletion of a polymer solution

$$W_a = \frac{8}{3} \Pi_b \xi_b \sim k_B T c_b^{3/2} \quad (9.43)$$

Note that γ^0 is positive for depletion and negative for adsorption leading thus, according to Eq. 9.12, to an increase of the area compressibility for depletion and a decrease for adsorption

$$K_A = K_A^0 + \frac{16}{3} \Pi_b \xi_b \qquad \text{for depletion and}$$

$$K_A = K_A^0 - \frac{k_B T}{a^2} \frac{16}{v} \varepsilon^3 \qquad \text{for adsorption.} \quad (9.44)$$

Changes induced by polymer depletion and adsorption are thus of the same order of magnitude as the increase or the reduction of interfacial tensions. Although in practice strong perturbations from depleted polymers are only expected above the crossover concentration c*, where the high viscosity of the polymer solutions make evaluations difficult, perturbations from strong adsorption can in principle easily reach finite fraction values of the bare stretching modulus.

Brooks et al. solved the Euler-Lagrange Eq. (9.39) in the limit of weak adsorption $\xi_b \ll D$ for cylindrical and spherical boundary conditions. The expansion of the excess surface energy leads to the following modifications of the membrane parameters

$$\Delta\kappa = -\frac{9}{8} \varepsilon^2 k_B T c \, \xi_b^3 ,$$

$$\Delta\bar{\kappa} = \frac{3}{4} \varepsilon^2 k_B T c \, \xi_b^3 ,$$

$$\Delta C_{spo} = \frac{-3}{8} \varepsilon^2 \frac{1}{\xi_b} \frac{k_B T}{\kappa} c \, \xi_b^3 , \quad (9.45)$$

showing that weak polymer adsorption or depletion softens the membranes and increases their tendency to form structures of high genus number. Similarly to the other equilibrium cases discussed before, here also the curvature energy is lowered if the surface bends *toward* the solution, both for the adsorption and the depletion cases. The scaling form of the corrections bares some similarities with the results on single chains in Eq. 9.34, with the role of the chain length being here played by the correlation length of the semi-dilute solution.

The authors in (Brooks et al., 1991a) solved also numerically the mean field, strong adsorption limit $\xi_b \gg D$, for which

Clement and Joanny (1997) and Skau and Blokhuis (2002, 2003) provided later analytical solutions. For mean-field strong adsorption these authors obtained the following expressions

$$\Delta\kappa = \frac{-8}{9}\varepsilon\, k_B T a c \xi_b^2$$

$$\Delta\bar{\kappa} = \frac{4}{3}\varepsilon\, k_B T a c \xi_b^2$$

$$\Delta C_{\text{spo}} = -\varepsilon^2 \frac{1}{\xi_b}\frac{k_B T}{\kappa} c\, \xi_b^3, \qquad (9.46)$$

For strong adsorption the corrections increase, becoming independent of the bulk polymer concentration, c (note that $c\xi_b^2$ is a constant), but preserve the sign of the effects both for the elastic constants and for the spontaneous curvature.

The sign of the modifications induced by the adsorbed polymers does not trivially follow intuition, and less systematic approaches have lead to different conclusions (Kim and Sung, 2001). The results in the papers by Brooks et al. (1991a, 1991b) have puzzled many authors because for covalently end-grafted polymers it is known, both theoretically (Milner and Witten, 1988) and experimentally (Evans and Rawicz, 1997), that the bending rigidity increases and the membrane bends *away* from the polymers (see also Section 9.5). One might thus speculate that if an adsorbed polymer is irreversibly attached to the membrane, the effects should be similar to those of end-grafted chains. This hypothesis was first tested at the mean-field level in (Clement and Joanny, 1997) and further discussed by (Skau and Blokhuis, 2002). These authors computed changes in membrane parameters induced by irreversible adsorbed polymers under different types of global constraints on the total adsorbed amount of polymer. Although the values of $\Delta\kappa$, $\Delta\bar{\kappa}$ and ΔC_{spo} are under these conditions marginally reduced, they keep the same sign. It is thus likely that a single global constraint on the adsorbed amount of the adsorbed layer still allows for enough relaxation degrees of freedom in the layer structure to remain close to equilibrium adsorption conditions. More drastic changes are to be expected as more strict constraints are imposed, by maintaining, for instance, the polymer loop distribution, effectively keeping irreversibly attached to the surface all adsorbed monomers. Such studies, of a more complex nature, were not yet undertaken.

Scaling results for a polymer solution in contact with a membrane

The mean-field approach presented in the previous paragraphs is known to correctly predict the main trends of solutions of polymers with excluded volume, but to overlook fluctuations effects, which for polymer solutions are crucial. For instance, the solution correlation length naturally arising in mean-field theories scales as $\xi_b \sim c^{-1}$, in contrast with appropriate scaling $\xi_b \sim c^{-3/4}$ (de Gennes, 1979), see also Box 9.1. Actually, a mean-field polymer theory describes mean-field effects in four dimensions, and its results need to be appropriately reproduced into three dimensions. For the adsorption and depletion of polymer solutions, de Gennes (1981) has proposed to replace the mean-field energy density in Eq. 9.38 by a corresponding scaling energy density which minimization produces the correct scaling behavior for the adsorbed and depleted polymer layers. The pre-factors of such energy functional are however not known, although they could in principle be computed from a comparison between the predictions from this energy density for solution properties such as the correlation length or the osmotic pressure and the corresponding predictions from renormalization theories. Interestingly, in the scaling approach, the adhesion strength between two surfaces follows the same power law with polymer concentration $W_a \simeq \Pi\xi_b \simeq k_B T c_b^{9/4} c_b^{-3/4} = k_B T c_b^{3/2}$. Brooks et al. extended this approach to curved surfaces and computed the corresponding modification of the membrane parameters, leading to results qualitatively similar to mean-field in the limit of weak adsorption, and to

$$\Delta\kappa \simeq -k_B T \log\left(\frac{\xi_b}{D}\right)$$

$$\Delta\bar{\kappa} \simeq k_B T \log\left(\frac{\xi_b}{D}\right)$$

$$\Delta C_{spo} \simeq -\frac{1}{D} k_B T \kappa, \qquad (9.47)$$

for strong adsorption ($D \ll \xi_b$). Eq 9.47 reproduce the scaling predictions in (de Gennes, 1990) for the elastic constants, but also provide the sign of the variations which is similar to the mean-field predictions. However, the logarithmic dependences of the corrections on the ratio ξ_b / D can be, at the crossover concentration $c^\star$ where $\xi_b = R_F$, much larger than the predictions from mean-field theory. One can thus anticipate, for the strong adsorption of large chains of polymerization degree N, a reduction of the membrane bending rigidity and an increase of the Gaussian elastic constant of the order $\log N$. The spontaneous curvature is in this limit very strong and depends only on the adsorption length D.

Methods based on fluctuating interfaces

Most of the studies on the influence of adsorbed and depleted polymer solutions on the membrane constitutive parameters are based on the approach explained and detailed above: polymer solution properties are computed close to a cylinder and a sphere and the excess surface energies of the two geometries are then expanded in powers of the curvatures up to second order. There is however an alternative pathway based on the calculation of the polymer configurations and free-energies close to an undulating surface. This allows to extract the modification of the surface energy in powers of the undulating mode, q, thus determining bending rigidity modifications, $\Delta\kappa$, from the fourth order term, q^4 (but not Gaussian rigidity modifications, $\Delta\bar{\kappa}$).

Polymer adsorption was first considered close to fluctuating interfaces by (Hone et al., 1987; Ji and Hone, 1988), without extracting, however, parameters relevant for membranes. This method was then used by several authors (Garel et al., 1995; Sung and Oh, 1996; Laradji, 1999; Sung and Lee, 2004) with conflicting results. Although Garel et al. (1995) and Sung and Oh (1996) found that the polymer adsorption does not change the bending constants of the membrane, Langevin simulations presented by Laradji (1999) on adsorbed penetrable surfaces and calculations on attractive impenetrable surfaces (Sung and Lee, 2004) found also a reduction in κ in qualitative agreement with the mean-field treatments described above in this section.

9.4.3 GUVs EXPOSED TO DEPLETED AND ADSORBED POLYMERS

GUVs have played a pioneering role in elucidating and quantifying the mechanisms of polymer–membrane interactions. Although the majority of work has focused, as we will see below in Sections 9.5 and 9.6, on membranes decorated by end-grafted polymers and on charged polymers, micromanipulation of GUVs has provided a quantitative test of the predictions for the adhesion strength induced by depleted polymers (Evans et al., 1996), see also Figure 9.4. GUVs have also been used as qualitative tests for the destabilizing power of hydrophilic polymers adsorbed on vesicles (Zhang et al., 2004) or, in a more quantitative manner, to explore the role of depleted polymer concentration and chain length on vesicle fusion and shape control (Terasawa et al., 2012; Okano et al., 2018), as shown in Figure 9.5.

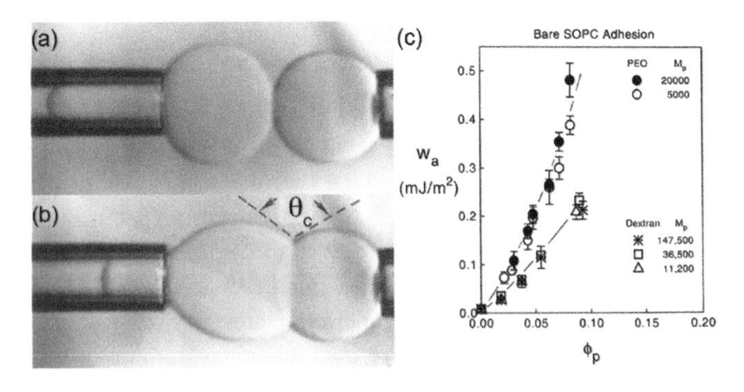

Figure 9.4 Micromanipulated GUVs for measuring adhesion of membranes in a depleted polymer solution. (a) Two GUVs are brought together under a given tension controlled by the suction pressure on the micropipettes. (b) Release of the tension on one of the micropipettes (here the left one) leads to an increase of the adhesive contact between the GUVs. Measure of the contact angle as a function of the applied tension gives the adhesion energy. (c) Adhesion energy for GUVs in solutions of polyethylene oxide and dextran. The results confirm the prediction $W_a \simeq \Pi\xi_b \simeq k_BTc_b^{3/2}$. (Reprinted with permission from Evans, E. et al., *Langmuir* 12, 3031–3037, 1996. Copyright 1996 American Chemical Society.)

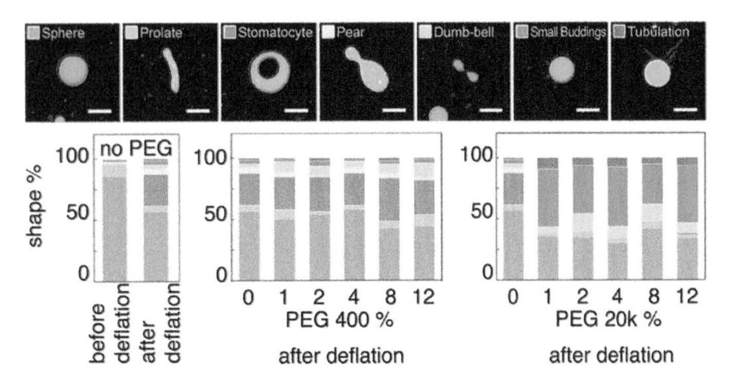

Figure 9.5 Representative images of seven shape categories of GUVs (membrane in red) containing a nonadsorbed fluorescently labeled PEG (green), after deflation, see Appendices 2 and 4 of the book for fluorescent lipids and dyes. Scale bar: 10 μm. Also shown the percentages of the shapes obtained after deflation, see Appendices 2 and 4 of the book for fluorescent lipids and dyes in different (w/w) concentrations of PEG (green) of two molecular weights. More than 100 vesicle images in each condition were used for analysis. (Adapted with permission from Okano, T. et al., *ACS Synth. Biol.*, 7, 739–747, 2018. Copyright 2018 American Chemical Society.)

9.5 END-GRAFTED POLYMERS

The discovery of Stealth® liposomes by Liposome Technology Inc. (Needham et al., 1992), the work at Unilever on the stabilization of multilamellar detergent dispersions (Van de Pas and Buytenhek, 1992) and the growing interest about biomimetic systems for surface recognition (Ringsdorf et al., 1988) triggered a series of fundamental studies on membrane systems with end-grafted polymers. In all these systems the membrane is modified by a coating of hydrophilic polymers anchored in the membrane. In stealth liposome technology, designed to increase liposome survival in the circulatory system by avoiding immune recognition, the coating is achieved by polyethylene glycol (PEG)-lipids, a family of phospholipids bearing a PEG chain covalently bound to the phospholipid head. In detergent formulations and in biomimetic systems it is common to use hydrophobically modified polymers, that is, hydrophilic polymers with terminal or side groups that can insert in the hydrophobic region of the bilayer. Typical hydrophobic groups include alkyl chains of different lengths (Decher et al., 1989).

In this section, we will discuss the considerable body of theory that was developed to study the effects of end-grafted polymers on lipid bilayers. Although we discuss its different facets, one should keep in mind that these concepts deal with the *polymer* influence on the bilayers. As we have seen, polymers are attached to the bilayers by *anchors*, and the anchors themselves can potentially change the bilayer properties to different degrees. Anchor changes are better described by the theories dealing with mixtures of lipids (Wolff et al., 2011) or with inclusions in lipid bilayers (Aranda-Espinoza et al., 1996; Netz, 1997). Comparison between theories presented here and experiments must therefore be exerted with care, balancing the relative importance of the *polymer* and of its *anchor(s)*.

Let us also stress, before discussing below more technical approaches, that the scaling forms of the corrections from end-grafted polymers to the membrane constitutive parameters can be easily computed. If γ^0 is the surface free energy of the end-grafted polymer layer for flat surfaces, then one expects to second order in curvature, curvature corrections of the form $\Delta\gamma = \gamma^0(1 + c_1R_l / R + c_2R_l^2 / R^2)$, where c_1 and c_2 are numerical constants and R_l the relevant length for the layer: $R_l = R_G = N^{1/2}a$ for Gaussian chains, $R_l = R_F = N^{3/5}a$ for chains with excluded volume and $R_l = L \simeq N\sigma^{1/3}$ for brushes, with L the brush thickness. For instance, at low grafting density, in the mushroom regime—see Box 9.3—before chain–chain interactions play a dominant role, the excess free energy per unit area is simply proportional to the density, $\gamma^0 \sim k_BT\sigma$. Sparse end-grafted polymers do not contribute to the stretching modulus K_A and the corrections to the elastic constants behave as $\Delta\kappa \sim \Delta\bar{\kappa} \sim k_BT\sigma R_l^2$, a scaling form independent of the solvent quality; the corrections to the spontaneous curvature like $\Delta C_{spo} \sim k_BT\sigma R_l / \kappa$. For brushes in a good solvent $\gamma_0 \simeq k_BTN\sigma^{11/6}$ and $\Delta\kappa \sim \Delta\bar{\kappa} \sim \Delta\gamma_0 L^2 \sim k_BTN^3\sigma^{5/2}$; $\Delta C_{spo} \sim \gamma^0 L / \kappa \sim k_BTN^2\sigma^{13/6} / \kappa$. However, details of the polymer layer such as chain architecture, solvent quality and others do matter, as they determine the value and the sign of the prefactors, as we will discuss below.

This section will first present global and local effects of a membrane decorated with a mushroom layer, that is, with a coverage of diluted end-grafted polymers, for which a range of exact and numerical results have been obtained for ideal and excluded volume chains. The section will then discuss the more dense cases of membrane coverage by brushes and presents some experimental results obtained on these systems by using GUVs.

9.5.1 DECORATION BY DILUTED END-GRAFTED POLYMERS: GLOBAL EFFECTS OF A MUSHROOM LAYER

A number of exact results can be obtained for ideal end-grafted chains. Because ideal chains do not interact with each other, the global effect of end-grafting σ chains per unit surface is given by $\Delta\gamma = \sigma\Delta\mathcal{F}$ with $\Delta\mathcal{F}$ the free-energy excess of end-grafting one chain to the surface, given by

$$\Delta\mathcal{F} = -k_BT\ln\frac{\mathcal{Z}}{\mathcal{Z}_0} = -k_BT\ln\frac{\int d\mathbf{R}_N\ G_N(\mathbf{R}_0 = r_{S'},\mathbf{R}_N)}{\int d\mathbf{R}_N\ G_N^0(\mathbf{R}_0 = r_{S'},\mathbf{R}_N)} \tag{9.48}$$

where $G_N(\mathbf{R}_0,\mathbf{R}_N)$ is solution of Eq. 9.28 in the corresponding flat, spherical or cylindrical geometry with $U_{ext} = 0$ in the bulk and the boundary conditions $G_N(\mathbf{R}_0 = \mathbf{r}_S,\mathbf{R}_N) = G_N(\mathbf{R}_0,\mathbf{R}_N = \mathbf{r}_S) = 0$, with $\mathbf{r}_S$ a vector describing the surfaces and $\mathbf{r}_{S'} = \mathbf{r}_S + a\mathbf{n}$ a vector describing the possible grafting positions, usually a monomer distance a away from the surface ($\mathbf{n}$ is here the normal to the surface). Note that compared with Eq. 9.30, there is one integration less in Eq. 9.64 in the partitions functions of the chains due to grafting. If grafting does not freeze the lateral degrees of freedom of the chain-end, one would need to supplement the surface excess energy $\Delta\gamma$ with the corresponding surface translational entropy. The green functions $G_N(\mathbf{R}_0,\mathbf{R}_N)$ are the same as those corresponding to the case of depleted Gaussian chain in Section 9.4.1. Technically, one expands Eq. 9.64 by keeping one end at one monomer distance from the spherical or cylindrical surface leading to

$$\Delta\gamma_c = k_BT\sigma\log\frac{\sqrt{\pi}R_g}{a} - k_BT\sigma\frac{\sqrt{\pi}}{2}\frac{R_g}{R} + k_BT\sigma\frac{2+\pi}{8}\left(\frac{R_g}{R}\right)^2,$$

$$\Delta\gamma_s = k_BT\sigma\log\frac{\sqrt{\pi}R_g}{a} - k_BT\sigma\sqrt{\pi}\frac{R_g}{R} + k_BT\sigma\frac{\pi}{2}\left(\frac{R_g}{R}\right)^2, \tag{9.49}$$

from where—see Section 9.2.2—one gets the corrections

$$\Delta\kappa = \frac{2+\pi}{2}k_BT\sigma R_g^2,$$

$$\Delta\bar{\kappa} = -2k_BT\sigma R_g^2,$$

$$\Delta C_{spo} = \frac{1}{R_g}\frac{\sqrt{\pi}}{4}\frac{k_BT}{\kappa}\sigma R_g^2. \tag{9.50}$$

as computed by (Marques and Fournier, 1996; Hiergeist and Lipowsky, 1996; Eisenriegler et al., 1996) and also found by numerical simulations that also provided finite N corrections (Auth and Gompper, 2003). Membranes are thus predicted to bend away from end grafted chains ($\Delta C_{spo} > 0$), to increase the bending modulus $\Delta\kappa > 0$ and to decrease the Gaussian rigidity $\Delta\bar{\kappa} < 0$. This is the reverse effect of adsorbed or depleted polymers and does correspond, at least for the spontaneous curvature and for the bending rigidity, to what simple intuition would predict. Indeed, by bending away from the polymer chain, the membrane increases the chain conformational space and thus its entropy and, by irreversibly grafting chains on both sides of the membrane, one effectively increases its thickness and thus its rigidity. Note however that the values of the corrections, even at the densest of the mushroom regime when $\sigma R_g^2 \sim 1$, are only on the order of k_BT, a minor contribution to the rigidity of most phospholipid bilayers, around $20k_BT$. As we will see below, corrections in the brush regime are much larger and can significantly contribute to the membrane rigidity.

9.5.2 INFLUENCE OF CHAIN ARCHITECTURE

We now consider four different architectures with the same total monomer number, schematically drawn in Figure 9.6. For simplicity we will designate "hair" a polymer chain grafted by a single extremity, "siamese" molecule a chain grafted by the median monomer, "loop" molecule when both extremities are anchored, and "gemini" a double chain connected to the same bola amphiphile. The same contributions to the elastic moduli are obtained for hairs, siamese and gemini molecules (Marques and Fournier, 1996; Hiergeist and Lipowsky, 1996; Bickel, 2001; Bickel and Marques, 2006) and are thus given by Eq. 9.50, whereas the spontaneous curvatures are architecture dependent. It vanishes obviously for gemini molecules, whereas for siamese molecules it is given by

$$\Delta C_{spo} = \frac{1}{R_g}\frac{\sqrt{2\pi}}{4}\frac{k_BT}{\kappa}\sigma R_g^2. \tag{9.51}$$

with the same sign but slightly larger than the corresponding value for hairs, see Eq. 9.50. Note that the siamese architecture is the first step toward a star polymer. The global effects of star polymers on membranes where also simulated and computed (Auth and Gompper, 2003), they are better described by techniques adapted to account for excluded volume, see Section 9.5.4 below.

The case of the loops is perhaps less obvious. First, corrections with the opposite sign are found for the elastic moduli

$$\Delta\kappa = -k_BT\sigma R_g^2, \tag{9.52}$$

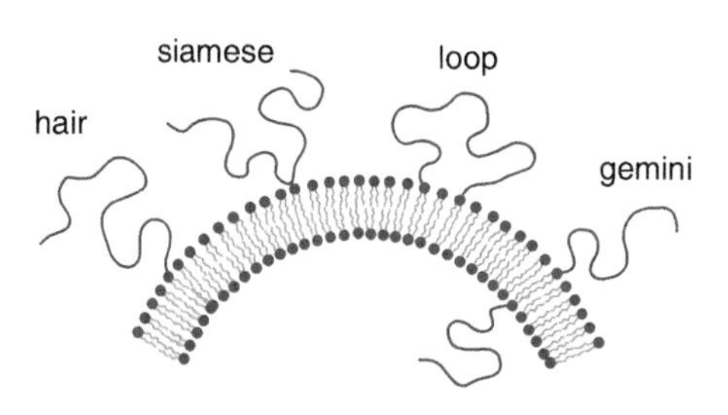

Figure 9.6 Different possible architectures of the end-grafted chains.

$$\Delta\bar{\kappa} = 2k_BT\sigma R_g^2 \,. \tag{9.53}$$

End-grafted loops behave thus as adsorbed polymers with respect to the corrections to the elastic constants, reducing the bending rigidity and favoring saddle-splay membrane conformations. However, contrary to adsorbed polymers, no spontaneous curvature is induced by loop polymers because for this case (Lipowsky et al., 1998)

$$\Delta C_{\text{spo}} = 0 \,. \tag{9.54}$$

When the anchors are close to each other, the polymer forms an anchored ring and the membrane bends away from the chain. On the other hand, if both ends are far apart, the polymer is in a stretched state and then pulls on the membrane as shown in Figure 9.7. These two competing effects happen to compensate exactly if the anchors are free to probe all the possible configurations.

To better understand this point, let us consider a loop attached on a sphere. The spontaneous curvature is in fact a function of the (horizontal) distance, ℓ, between anchoring points (Bickel, 2001)

$$\Delta C_{\text{spo}}(\ell) = \frac{1}{R_g}\frac{\sqrt{\pi}}{4}\left(1 - \frac{\ell^2}{2R_g^2}\right)\frac{k_BT}{\kappa}\sigma R_g^2 \,, \tag{9.55}$$

so that it vanishes at separation, $\ell = \sqrt{2}R_g$. Remarkably, it can be shown that the average of this quantity over all separations cancels out exactly

$$\langle \Delta C_{\text{spo}} \rangle = \int_0^\infty d\ell \mathcal{P}(\ell)\Delta C_{\text{spo}}(\ell) \propto$$

$$\int_0^\infty d\ell \left(1 - \frac{\ell^2}{2R_g^2}\right)\exp\left(-\frac{\ell^2}{4R_g^2}\right) = 0 \,. \tag{9.56}$$

9.5.3 LOOPS AND SIAMESE MOLECULES: EFFECT OF ANISOTROPY

As pointed out in the previous section, loop molecules present some intriguing features. The analysis can be refined and extended to the more general situation where the surface is locally described by its two principal radii of curvature R_1 and R_2. We focus here more closely on the coupling between particle anisotropy and local surface conformation. This question has

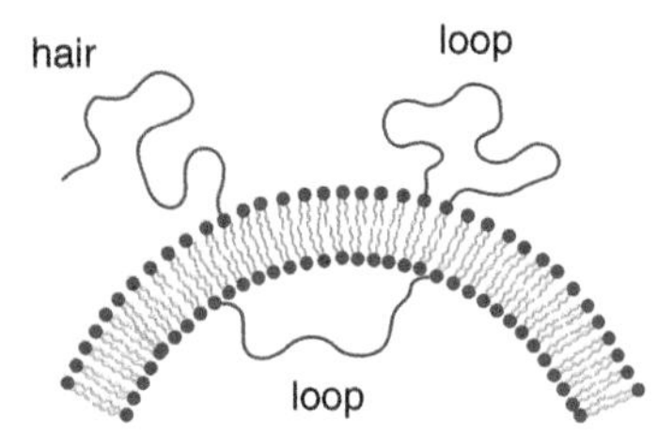

Figure 9.7 A chain grafted by its extremity has more conformational entropy when the grafting surface is bent away from the chain. The same applies for loops when the grafting points are close to each other. However, when the grafting points are far away, the chain entropy is increased if the surface bends toward the chain.

been studied quite generally in (Fournier, 1996), and the results have been applied to both siamese molecules (Marques and Fournier, 1996) and loop polymers (Bickel, 2001).

We first focus on loop molecules, and assume that the (projected) distance ℓ between anchors is fixed. Experimentally, this configuration could in principle be achieved by connecting the chain ends by a rigid spacer. The orientation of the molecule is then described by the vector $\ell = (x - x', y - y', 0)$, and we call θ the angle between and, for instance, the axis of principal curvature $1/R_1$. The discussion is restricted to the situation where both extremities are anchored symmetrically onto the surface, $x' = -x$, $y' = -y$, and $z = z' = -(R_1^{-2}\cos^2\theta + R_2^{-2}\sin^2\theta)\ell^2/4$. The partition function of the chain can be evaluated following a perturbative method developed in (Podgornik, 1993). Up to the first order, the partition function is (Bickel, 2001)

$$\mathcal{Z}(\ell) = \mathcal{Z}^{(0)}\exp\left(-\frac{\ell^2}{4R_g^2}\right)\times$$

$$\left\{1 + \frac{\sqrt{\pi}R_g}{2R_1}\left(1 - \frac{\ell^2\cos^2\theta}{R_g^2}\right) + \right.$$

$$\left.\frac{\sqrt{\pi}R_g}{2R_2}\left(1 - \frac{\ell^2\sin^2\theta}{R_g^2}\right)\right\}, \tag{9.57}$$

with $\mathcal{Z}^{(0)}$ the partition function of the chain with both extremities grafted at the same point on a flat surface. The polymer contribution $\Delta\mathcal{F} = -k_BT\ln\mathcal{Z}(\ell)$ to the free energy takes the following form

$$\Delta\mathcal{F} = \Delta\mathcal{F}_0 + \lambda\left(\frac{1}{R_1} + \frac{1}{R_2}\right) + \mu\left(\frac{1}{R_1} - \frac{1}{R_2}\right)\cos 2\theta \,, \tag{9.58}$$

where the coefficients λ and μ are given by

$$\lambda = -k_BT\frac{\sqrt{\pi}R_g}{2}\left(1 - \frac{\ell^2}{2R_g^2}\right) \quad \text{and} \quad \mu = k_BT\frac{\sqrt{\pi}\ell^2}{4R_g} \,. \tag{9.59}$$

The anisotropy of the spontaneous curvature μ/λ is vanishingly small for ℓ on the order of the size of a monomer. However, μ/λ is of order unity when ℓ become comparable to the polymer size. Note that this ratio eventually diverges for $\ell = \sqrt{2}R_g$.

Anisotropic polymer architectures thus promote different spontaneous curvatures in the direction parallel or perpendicular to their orientation in the plane of the membrane. It has been shown in general that, whereas isotropic inclusions yield no spontaneous mean curvature when symmetrically adsorbed in a bilayer, anisotropic inclusions yield a spontaneous deviatoric bending $R_1^{-1} - R_2^{-1} \neq 0$ by orienting at right angle across the bilayer (Fournier, 1996). The deviatoric contribution to the free energy (9.58) renormalizes the bending moduli according to (Fournier, 1996)

$$\Delta\kappa = -\frac{1}{2}\Delta\bar{\kappa} = -\frac{\sigma\mu^2}{k_BT} \,. \tag{9.60}$$

For siamese molecules, this contribution is (Fournier, 1996)

$$\Delta\kappa = -\frac{9\pi}{16} k_B T \sigma R_g^2 , \quad (9.61)$$

and it is found for the loops (Bickel, 2001)

$$\Delta\kappa = -\frac{\pi}{16} k_B T \sigma R_g^2 \frac{\ell^4}{R_g^4} , \quad (9.62)$$

The effect of anisotropic Gaussian inclusions is then to decrease the bending modulus κ and to increase the Gaussian modulus $\bar{\kappa}$. This can lead to interesting applications because the deviatoric contribution to the bending energy induces, above a concentration threshold, an "egg-carton" instability in flat membranes and a vesicle instability yielding long wormlike shapes (Fournier, 1996). Also interesting might be the possibility of changing the solvent quality given that excluded volume effects can change the sign of the siamese molecule contribution to the bending rigidity, as shown by Monte Carlo simulations (Auth and Gompper, 2005).

9.5.4 INFLUENCE OF EXCLUDED VOLUME

The effects of excluded volume on the corrections to the elastic parameters have been evaluated by Monte Carlo numerical simulations (Auth and Gompper, 2003). If we define R_{gF} as in Eq. 9.37 the corrections read

$$\Delta\kappa = 0.935 \times \frac{2+\pi}{2} k_B T \sigma R_{gF}^2 ,$$

$$\Delta\bar{\kappa} = -0.922 \times 2 k_B T \sigma R_{gF}^2 ,$$

$$\Delta C_{\text{spo}} = 0.928 \times \frac{1}{R_{gF}} \frac{\sqrt{\pi}}{4} \frac{k_B T}{\kappa} \sigma R_{gF}^2 . \quad (9.63)$$

and a direct comparison with the ideal chain results Eq. 9.34 can be made. Note that as for the depletion case Eq. 9.37, the amplitudes are reduced by less than 10%, but the overall effects with excluded volumed chains are larger for large enough chains, due to the swelling of the polymers $R_{gF}/R_g > 1$.

Excluded volume effects are also essential when describing the effect of dense architectures such as star polymers. A star polymer attached to a membrane consists of f chains (the f arms of the star) attached to the same point on the membrane. The mutual repulsion of the arms leads to corrections from this grafted architecture much stronger than those of single chains (Auth and Gompper, 2003). Numerical simulations and calculations point to an effect on the bending rigidity per arm that increases almost linearly with arm number. Interestingly, the correction to the Gaussian rigidity is much smaller, a vanishing correction is even predicted from simple models.

9.5.5 LOCAL EFFECTS FROM SINGLE END-GRAFTED POLYMERS

The interaction between one end-grafted polymer and a lipid bilayer has obviously a local character: the probability of interaction between the chain monomers and the membrane is much larger close to the grafting point than away from it. Because the bilayer is a flexible surface, this will translate into nontrivial membrane deformations that are the result of the force balance between the entropic forces exerted by the chain on the membrane and the elastic bending restoring forces. Intuition for the possible existence of nontrivial bent shapes of the membranes decorated with one end-grafted polymer was first acquired by (Lipowsky, 1995) who computed the free-energy of chains end-grafted to spheres, cones and catenoids. Amongst these three shapes, cones have the smallest energetic cost when the polymer free energy and the bilayer bending cost are combined. The full determination of the bilayer deformation profile became possible a few years later (Breidenich et al., 2000, 2001; Bickel et al., 2000b, 2001) when the pressure that end-grafted chains exert on the grafting surface was first computed.

The pressure can be extracted from the evaluation of the energy difference associated with a virtual small displacement $\zeta(x,y)$ of the grafting surface. For ideal chains, this is achieved by solving first Edwards Eq. 9.28 with $U(z>0)=0$ and the boundary condition $G_N^{\zeta}(R_0 \equiv (x,y,\zeta(x,y)), R_N) = 0$ and then integrating the remaining degrees of freedom

$$\Delta F = -k_B T \ln \frac{\int d\mathbf{R}_N \, G_N^{\zeta}(\mathbf{R}_0 \equiv (0,0,a), \mathbf{R}_N)}{\int d\mathbf{R}_N \, G_N^{\zeta=0}(\mathbf{R}_0 \equiv (0,0,a), \mathbf{R}_N)} = \int dx dy \; p(x,y)\zeta(x,y) , \quad (9.64)$$

thus showing that the work of displacement of $\zeta(x,y)$ is achieved by a pressure field $p(r=\sqrt{x^2+y^2})$ given by

$$p(r) = \frac{1}{2\pi} \frac{k_B T}{(r^2+a^2)^{3/2}} \left(1 + \frac{r^2+a^2}{2R_g^2}\right) \exp\left\{-\frac{r^2+a^2}{4R_g^2}\right\} , \quad (9.65)$$

Evaluation of the influence of excluded volume effects on the pressure profile is more challenging. A first step can be made by the ansatz that the pressure must be related to the chain monomer concentration $c(r)$ in the vicinity of the wall. For ideal chains this can be shown to be the case with

$$p(r) = k_B T \frac{a^2}{12} \frac{\partial^2 c}{\partial z^2}(r,0) \quad (9.66)$$

with $r=\sqrt{x^2+y^2}$. Qualitatively, the pressure can be associated with an ideal gas pressure caused by the concentration of monomers at a distance $z = a/\sqrt{6}$ from the wall

$$p(r) = k_B T c(r, z = \frac{a}{\sqrt{6}}) \quad (9.67)$$

The ansatz is comforted also by work on two-dimensional SAW polymers on a semi-infinite square lattice where it can be shown (Jensen et al., 2013) that

$$p(r) = -k_B T \ln(1 - c(r, z=1)) \quad (9.68)$$

with here $p(r)$ and $c(r)$ in dimensionless (lattice) units. For three dimensional cases the concentration of a SAW polymer can be computed by MC simulation, and one finds that its value close to the wall follows the same scaling law as the ideal chain (Bickel et al., 2001). It is also worth stressing that the pressure is also predicted to follow the same scaling form for end-grafted stars, but with an increased amplitude dependent on the arm number, $p(r) \simeq f^{3/2} k_B T r^{-3}$ (Breidenich et al., 2000; Bickel et al., 2001).

When the polymer is end-grafted to a lipid bilayer, a deformation, $h(r)$, of the membrane will follow, that obeys

$$\kappa \Delta_r \Delta_r h(r) + p(r) = 0 \tag{9.69}$$

with $\Delta_r = \frac{1}{r}\frac{d}{dr} r \frac{d}{dr}$ the radial Laplace operator. For the ideal pressure Eq. 9.65 the deformation can be exactly computed (Bickel et al., 2001), and it is displayed in Figure 9.8b. The deformation close to the grafting point has a conical shape

$$h(r) \underset{r \to 0}{\simeq} -\left(\frac{k_B T}{\kappa}\right) \frac{r}{2\pi} \tag{9.70}$$

with a slope inversely proportional to the membrane bare rigidity, thus confirming the original intuition in (Lipowsky, 1995). This is also confirmed by Monte Carlo simulations, shown in Figure 9.8c and 9.8d (Breidenich et al., 2000). The amplitude of the deformation is also predicted to be amplified by a factor $f^{3/2}$ when the pressure is exerted by polymer stars.

9.5.6 MEMBRANES WITH POLYMER BRUSHES

When the chain grafting density σ is larger than the crossover density $\sigma^\star \simeq R_F^{-2}$, the end grafted chains cannot be seen any longer as isolated mushrooms sparsely covering the surface, they rather form a continuous dense layer commonly known as a *brush*, where interactions between different neighboring chains are important, see also Box 9.3. The importance of brushes for sterical stabilization of colloids, for controlling lubrication and also for providing stealth properties to liposomes (Needham et al., 1992) triggered sustained efforts to describe theoretically and numerically these polymer layers. At the scaling level, the first descriptions of the structure of the polymer brushes where provided by Alexander (1977) and de Gennes (1980). The equilibrium brush layer thickness L can be obtained by balancing the stretching energy of the chains with the excluded volume repulsion between monomers. In a Flory-like calculation (de Gennes, 1979) the chain elasticity is taken as deriving from a Gaussian chain $F_{el} \simeq k_B T L^2 / Na^2$ and the excluded volume interactions are treated at the mean-field level, as proportional to the square average monomer density $F_{ex} \simeq v c^2 L / \sigma = v(N\sigma / L)^2 L / \sigma = v N^2 \sigma / L$. The layer thickness is derived by minimizing $F_{el} + F_{ex}$ with respect to L, leading to $L \sim Na(\sigma a^2)^{1/3}$, and to the corresponding mean-field free energy per unit area $\Delta\gamma \sim k_B T a^{-2} N(\sigma a^2)^{5/3}$ (Alexander, 1977).

Alexander approach assumes that all chain ends lie at the outer edge of the brush, which implies a step-like monomer density fraction profile $c(0 < z < L) = N\sigma / L \simeq \sigma^{2/3}$. The existence of a parabolic—thus non-steplike—profile and a continuous distribution of chain ends was shown by several authors (Semenov, 1985; Milner et al., 1988a, 1988b) within the so-called mean-field strong-stretching approximation, where the vertical component of the chain trajectory is computed without fluctuations and excluded volume interactions are accounted for at the mean-field level. From the point of view of the scaling forms of the chain height and of the free-energy, this is equivalent to the Alexander approach, but the constants associated with the scaling laws can now be exactly computed. Within this approach (Milner and Witten, 1988) and later others (Wang and Safran, 1991; Birshtein and Zhulina, 1997), have evaluated the surface energy for spherical and cylindrical geometries and thus provided corrections to the different membrane parameters

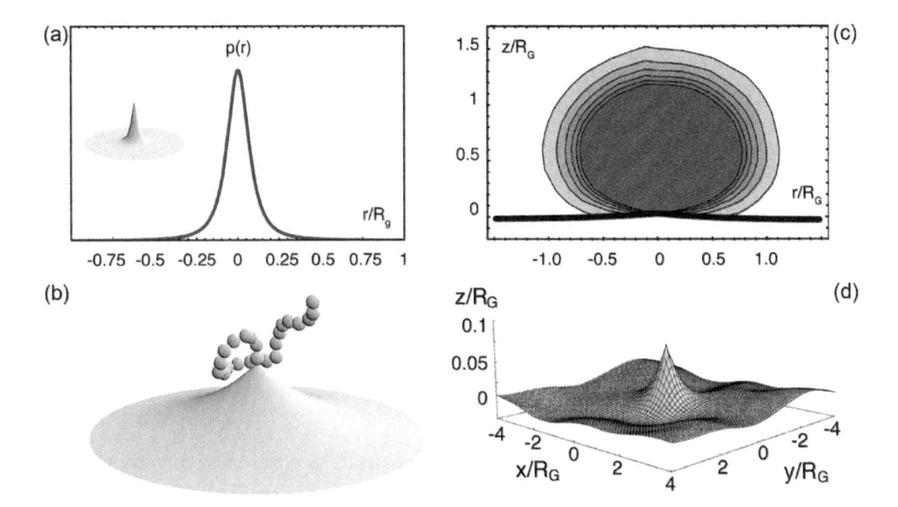

Figure 9.8 Pressure from an end-grafted polymer and corresponding membrane deformation. (a) Pressure in arbitrary units as a function of distance r from grafting point, in $R_g = N^{1/2}a / \sqrt{6}$ units. The inset highlights the radial symmetry of the function. (b) Induced deformation on a free bilayer, the shape close to the grafting point is well described by a cone. (Reprinted with permission from Bickel, T. et al., *Phys. Rev. E*, 62, 1124–1127, 2000b. Copyright 2000 American Physical Society). (c) Monomer density profile from a Monte Carlo simulation of 10^7 steps. Distances here are measured in end-to-end distance units $R_G = N^{1/2}a$. (d) Average membrane deformation extracted from the same simulations where the membrane bare rigidity is $\kappa = k_B T$. (Adapted with permission from Breidenich, M. et al., *EPL*, 49, 431, 2000. Copyright 2000 EDP Sciences.)

$$\Delta\kappa = \frac{9}{64}\left(\frac{12}{\pi^2}\right)^{1/3} k_BTN^3\tilde{v}^{4/3}(\sigma a^2)^{7/3} = 0.15k_BTN^3\tilde{v}^{4/3}(\sigma a^2)^{7/3},$$

$$\Delta\bar{\kappa} = -\frac{3}{35}\left(\frac{12}{\pi^2}\right)^{1/3} k_BTN^3\tilde{v}^{4/3}(\sigma a^2)^{7/3} = -0.09k_BTN^3\tilde{v}^{4/3}(\sigma a^2)^{7/3},$$

$$\Delta C_{\text{spo}} = \frac{3}{32}\frac{k_BT}{\kappa a}N^2\tilde{v}(\sigma a^2)^2 = 0.09\frac{k_BT}{\kappa a}N^2\tilde{v}(\sigma a^2)^2. \quad (9.71)$$

where $\tilde{v}$ is the dimensionless excluded volume, predicting that the membrane bends away from a brush grafted on one of its sides, and that symmetric membranes with one brush grafted on each side display zero spontaneous curvature and are significantly more rigid. As for mushrooms, brushes are also found to reduce the Gaussian elastic modulus and thus to promote surfaces with low genus numbers.

Several authors (Hristova and Needham, 1994; Bivas et al., 1998; Marsh, 2001; Marsh et al., 2003) have provided predictions for the variations of the stretching modulus of the membranes. Based on the simple bilayer model of Eq. 9.13 and on the brush mean-field result $\gamma^0(a_\ell) = (9/10)(\pi^2/12)^{1/3}N\tilde{v}^{2/3}x^{5/3}a_\ell^{-1}(a^2/a_\ell)^{2/3}$, with $x = \sigma a_\ell$ the molar fraction of the end-grafted polymers, one predicts an increase of area per lipid and a decrease of the stretching modulus. The control parameter of the variations is the ratio β of the brush energy density to the bare stretching modulus $\beta = \gamma^0(a_\ell^0)/K_A^0$. Given that typically the membrane does not stand an increase of area per lipid larger than 5%, one can find the equilibrium values of Eq. 9.13 for $\beta \ll 1$

$$a_\ell = a_\ell^0(1 + 4/3\beta)$$

$$K_A = K_A^0(1 - 4/9\beta). \quad (9.72)$$

An area increase of 5% can then be caused by a value $\beta = 0.04$ with a corresponding decrease of the stretching modulus of less than 2%.

It is known that if a mean-field estimation does predict accurately the scaling behavior of the brush thickness, it does not provide the correct scaling form for the free energy (de Gennes, 1979). Excluded volume correlations, similar to those properly describing semi-dilute polymer solutions—see Box 9.1—need to be considered for interfacial polymer layers, as recognized earlier (de Gennes, 1976). Extending (Alexander, 1977) step profile approach to properly account for excluded volume effects, (de Gennes, 1980) described the end-grafted chains as strings of blobs whose size ξ is set by the grafting density, $\xi \sim \sigma^{-1/2}$. Each blob containing $g_\xi \sim (\xi/a)^{5/3}$ monomers, the brush thickness is found to follow the same scaling law $L \sim (N/g_\xi)\xi \sim Na(\sigma a^2)^{1/3}$, but with a corresponding surface energy density $\gamma^0 \sim (N/g_\xi)k_BT\sigma \sim k_BTa^{-2}N(\sigma a^2)^{11/6}$. A scaling picture for curved geometries was first elaborated by (Daoud and Cotton, 1982) who computed the structure of star polymers, and later adapted to spherical and cylindrical surfaces by several authors (Hristova and Needham, 1994; Hiergeist and Lipowsky, 1996; Marsh, 2001; Marsh et al., 2003). Based on this picture, one finds also small modifications of the stretching modulus

$$a_\ell = a_\ell^0(1 + 5/3\beta)$$

$$K_A = K_A^0(1 - 4/18\beta). \quad (9.73)$$

with $\beta = \gamma^0(a_\ell^0)/K_A^0$ comparing the scaling brush surface density and the bare membrane stretching modulus, but also significant corrections to the bending parameters

$$\Delta\kappa = \frac{65}{72}k_BTN^3(\sigma a^2)^{5/2} = 0.9k_BTN^3(\sigma a^2)^{5/2},$$

$$\Delta\bar{\kappa} = -\frac{5}{18}k_BTN^3(\sigma a^2)^{5/2} = -0.3k_BTN^3(\sigma a^2)^{5/2},$$

$$\Delta C_{\text{spo}} = \frac{5}{24}\frac{k_BT}{\kappa a}N^2(\sigma a^2)^{13/6} = 0.2\frac{k_BT}{\kappa a}N^2(\sigma a^2)^{13/6}, \quad (9.74)$$

qualitatively similar to Eq. 9.71 but with slightly different exponents and numerical factors, we reproduce here numerical factors from (Hiergeist and Lipowsky, 1996). Although scaling approaches provide accurate scaling exponents, they do not provide reliable values for the prefactors. Numerical factors of Eq. 9.74 should thus be taken with care, and only regarded as a useful tool to compare the relative values of the corrections. It must also be pointed that these scaling approaches do not relax the fixed-end constraint, and thus overstate the value of the free-energy. Relaxation of this constraint within a scaling approach was computed (Milner et al., 1988a), but not extended to curved surfaces.

Numerical methods based on Monte-Carlo, molecular dynamics or dissipative particle dynamics simulations (Laradji, 2002; Thakkar and Ayappa, 2010; Wu et al., 2013) or on self-consistent field theory calculations (Szleifer and Carignano, 1996; Birshtein et al., 2008; Lei et al., 2015) validate the strong increase of bending corrections with chain length and surface coverage, but differ in details such as the numerical factors or even the exact value of the power-law dependences with these quantities. For instance, by coupling the effect of polymer coverage and lipid area expansion (Szleifer et al., 1998) conclude that the combination $\Delta\kappa + \Delta\bar{\kappa}/2$, which is relevant for determining liposome formation, actually decreases as the end-grafting density increases.

9.5.7 GIANT UNILAMELLAR VESICLES DECORATED BY END-GRAFTED POLYMERS

Micropipette aspiration experiments on GUVs provided the first striking results of bending rigidity increase by decoration of a lipid bilayer with end-grafted polymers (Evans and Rawicz, 1997). In experiments where hydrophobically modified polymers are added to liposomes or GUVs, the insertion of the hydrophobic anchor perturbs the symmetry of the bilayer and leads to strong perturbations that are not caused by the polymer alone (Frette et al., 1999). Even when the bilayers are formed with the anchored polymer, the nature of the anchor, usually a few carbons-long alkyl chain, does not insure a permanent grafting. In the experiments of Evans and Rawicz (1997) the polymers were covalently bonded to 1,2-stearoyl-*sn*-glycero-3-phosphocholine (DSPC, see Appendix 1 of the book for structure and data on this lipid) lipids that were mixed with digalactosyl diacylglycerol (DGDG, see Appendix 1 of the book for structure and data on this lipid), a lipid leading to a bilayer of bending rigidity $\kappa \simeq 11$ k_BT. Figure 9.9 displays the results of the measurements of the corrections to the bending constant. Importantly, they confirmed

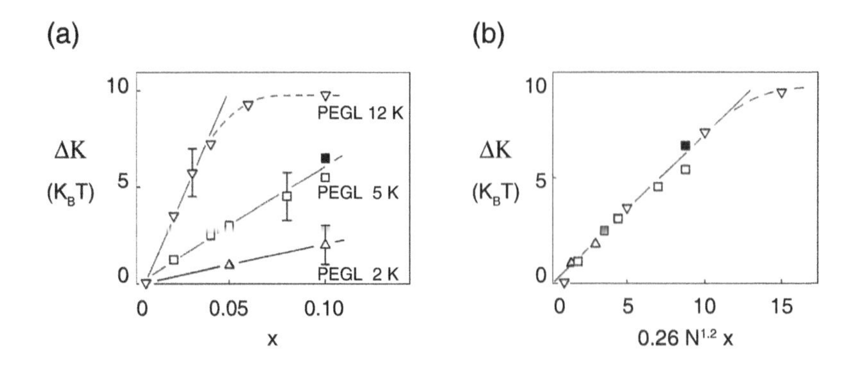

Figure 9.9 Bending rigidity increases ΔK relative to naked (DSPE+DSPG) GUV bilayers produced by the addition of grafted PEG polymer chains to the membrane, measured by micropipette aspiration. (a) Measurements for 10 vesicles at each composition are plotted as averages (in units of thermal energy) versus mole fraction x of PEG-lipid (PEGL) in DGDG bilayers. Typical standard deviations are shown by brackets. In (a) and (b) triangles represent DGDG + PEGL 2000, squares represent DGDG + PEGL 5000, and wedges (upside-down triangles) represent DGDG + PEGL 12000. DGDG + PEGL 12000 vesicles appear to have reached saturation above $x \sim 0.05$ (indicated by the dashed curve) with phase separation for higher concentrations of the PEG lipid. (b) Added stiffness ΔK plotted as a function of dimensionless surface density, $R_F^2\sigma \equiv 0.26\, N^{1.2}x$, for all lengths of grafted PEG polymer. (Adapted with permission from Evans, E. Rawicz, W. *Phys. Rev. Lett.*, 79, 2379. Copyright 1997 APS.)

that the end-grafted polymers bring a significant contribution to the bending elasticity, they also revealed that the polymers bring a strong perturbation to the membrane, to the point where the formation of GUVs could not be achieved with molar fractions above a few percent. The stability of bilayers with PEG-anchored polymers was further discussed by several authors, showing that for too large polymer content micellar mixed aggregates are the preferred self-assembly structures (Szleifer et al., 1998).

Although different techniques can be used to test the theoretical predictions for the corrections induced by end-grafted polymers (Gompper et al., 2001; Endo et al., 2000, 2001; Appell et al., 2004), work with GUVs and λ –phage DNA, a long polymer of 20 μm contour length allows such studies to be performed at the optical length scales. This was first shown by (Nikolov et al., 2007) who studied DOPC GUVs with end-anchored λ –phage DNA. Figure 9.10 shows the system and the main results of these

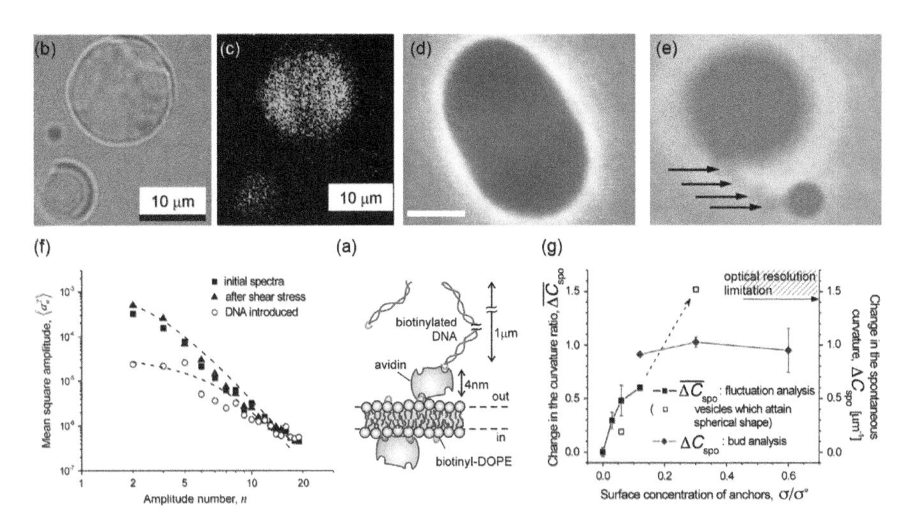

Figure 9.10 Corrections to the bending rigidity and spontaneous curvature of DOPC GUVs with end-anchored λ –phage DNA molecules. (a) Anchoring of a polymer (biotinylated λ –phage DNA) to a membrane already containing the anchoring sites such as avidin attached to a biotinylated membrane. (b) and (c) Vesicles immobilized on a glass surface in the presence of fluorescently labeled DNA. The anchor density is $0.6\,\sigma^*$, with $\sigma^* = 0.75\ \mu m^{-2}$. In panel (b) the vesicles are observed in transmitted light with a 100X oil immersion objective. The image in panel (c) shows the fluorescence signal from the same vesicles. (d) a vesicle with surface coverage $0.3\,\sigma^*$ and reduced volume $v = 0.975$. Scale bar: 5 μm. (e) Multiple budding observed after the DNA solution is introduced in the observation chamber. The bud formed first (diameter~4.1 μm) is connected to the mother vesicle (diameter~12.7 μm) with a necklace of vesicle pearls (diameter~1.4 μm). The necklace and the mother vesicle are out of focus but the pearl locations are shown with arrows. (f) Mean-square amplitudes $< a_n^2 >$ for three different 10-min fluctuation measurements as described in Chapter 14: Initial fluctuation spectra, shear stress test, and after adding DNA. The surface concentration of anchors is $0.03\,\sigma^*$. The first amplitude cannot be distinguished from zero and is not shown. The dashed lines are standard fits further described in Chapter 14. (g) Spontaneous curvature as a function of surface densities σ / σ^* in the mushroom regime as obtained from fluctuation spectroscopy (solid and open squares), left axis, and bud analyses (diamonds), right axis. The dimensionless spontaneous curvature ratio $\overline{\Delta C_{spo}} = < a_3^2 > / < a_2^2 >$ is proportional to the induced spontaneous curvature. The open-square symbols indicate two cases when the vesicle attained a spherical shape and no saturation in the curvature ratio could be observed. Error bars indicate scattering over several vesicles. The optical resolution limit for bud detecting is indicated by the arrow. (Reprinted from *Biophys. J.*, Nikolov, V. et al., Behavior of giant vesicles with anchored DNA molecules, 92, 4356–4368, 2007 with permission from Elsevier.)

studies. By evaluating the modifications induced by end-anchored DNA molecules on the fluctuation spectra of the vesicles, the authors have shown that the elastic bending constant is increased by a factor comparable to that predicted theoretically, and that a spontaneous curvature develops with the sign predicted by the theory. Complementary information confirming the development of a finite spontaneous curvature was also extracted by the onset of the formation of vesicle buds.

9.6 CHARGED POLYMERS AND CHARGED MEMBRANES

Charges are ubiquitous in aqueous solutions containing membranes and polymers. In the living, phospholipid head-groups such as phosphatidylserine or phosphatidylglycerol are negatively charged, synthetic cationic lipid-heads such as trimethylammonium propane and others have also been synthesized, and are presently used for transfection formulations. Charged polymers, commonly known as polyelectrolytes, can also carry positive or negative charged groups.

The presence of charges can potentially induce strong modifications in the behavior of membrane and polymers, due to the long range nature of the charge–charge mutual forces. In practice however the presence of a finite concentration of charged species in solution, due to the polymers or membranes themselves or to added salt, screens the forces between the charges. A finite interaction amplitude is only kept for distances smaller than the Debye length λ_D, at larger distances the forces vanish exponentially. For instance, for the physiologically relevant molar concentration of 100 mM of a monovalent salt in water the ionic strength is $I = 0.1$ M, and the Debye length of the solution at room temperature is $\lambda_D = 0.304 / \sqrt{I}$ nm $\simeq 1$ nm. The electrostatic interactions are under these conditions only relevant at a local scale; large polymers will behave under physiological conditions as nonionic polymers but with charge-renormalized excluded volume interactions.

For membranes, the presence of charges increases the bending rigidity and decreases the Gaussian modulus, a phenomenon that has been widely studied (Lekkerkerker, 1989; Harden et al., 1992). The electrostatic corrections to the membrane bare rigidity $\Delta\kappa_e, \Delta\bar{\kappa}_e$ depend on the surface number density of the charges σ_e and on the solution conditions through the Debye length λ_D. One has, for instance, for small surface charge density and high salt concentration (small λ_D) the moderate correction $\Delta\kappa_e = 3\pi k_B T \ell_B \lambda_D^3 \sigma_e^2 / 2$, whereas for high charge density and low salt (large λ_D) the rigidity can increase noticeably $\Delta\kappa_e = k_B T \lambda_D / (2\pi \ell_B)$, with ℓ_B the Bjerrum length, which is the distance at which the electrostatic energy of the interaction between two charges equals the thermal energy $k_B T$. For water $\ell_B = 0.7$ nm.

Exposing a charged membrane to a charged polymer can thus lead to a rich behavior, depending on the respective charges of the polymer and of the membranes, see also Chapter 25. In this section, we will consider the most relevant cases for both adsorbed and end-grafted polymers. For polymer adsorption, we will recall results for charged polymers in a salt solution interacting also by nonelectrostatic, short-range forces, with a surface carrying an opposite charge. For end-grafted polyelectrolytes, we will consider the effect of the charges from the polymer only, the grafting surface being neutral.

9.6.1 ADSORPTION OF CHARGED POLYMERS ON OPPOSITELY CHARGED MEMBRANES

The physics of the adsorption of hydrophilic polyelectrolytes on an oppositely charged surface is now well understood (Shafir and Andelman, 2007). The polymer backbone is taken as a water soluble polymer by itself, with a short-range interaction with the surface described as in the sections above by an adsorption length D. Recall that according to the convention of this chapter when D is small and positive the interaction is strongly attractive, and for D small and negative the short-range interaction is purely repulsive. The polymer is also decorated by charges, distributed randomly along its backbone; $1/N \leq f_e \leq 1$ measures the fraction of charged monomers.

For low salt conditions ($I = 0.1$ mM, $\lambda_D = 30$ nm) and a relatively strongly charged polymer ($f_e = 0.5$), Figure 9.11 shows the corrections to the bare membrane moduli. As the polymer strongly adsorbs on the surface it overcompensates surface charges, the net charge thus changing sign. In the limit of very strong adsorption one reaches the strongly charged limit of ionic membranes with $\Delta\kappa_e = k_B T \lambda_D / (2\pi \ell_B) \sim 7\ k_B T$ for the conditions of the figures.

Figure 9.12 shows the effect of surface charge on the corrections to the bare membrane moduli at low salt concentrations of strongly depleted polymers. As the figure shows, in this case, exposure of the membrane to the depleted polyelectrolyte solution strongly reduces the effects of the membrane charges on the elastic moduli.

The strongest effects of polyelectrolyte adsorption occur at low salt concentrations, where charge–charge interactions are not strongly screened. For higher salt concentrations, when $\lambda_D < D$, the salt screens both the surface and the polymer charges, and one crosses over to the behavior of nonionic polymers described in Section 9.4 above.

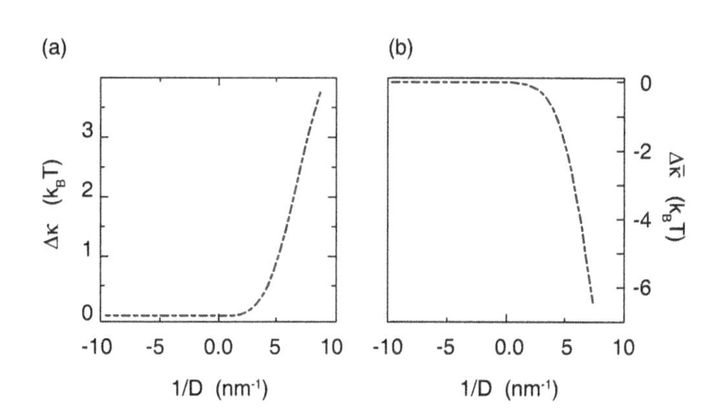

Figure 9.11 Corrections to the membrane bare (a) bending modulus and (b) Gaussian modulus, of a weakly charged ($\sigma_e = 0.1\ \text{nm}^{-2}$) membrane by a depleted ($D < 0$) or adsorbed ($D > 0$) aqueous polyelectrolyte solution—see also Section 9.4 for the meaning of D—at low salt concentrations ($I = 0.1$ mM, $\lambda_D = 30$ nm). Here half of the polymer monomers are charged, the monomer size is 0.5 nm, the polymer excluded volume is 0.05 nm^3 and the monomer concentration is 10^{-5} nm^{-3}. (From Shafir, A. and Andelman, D. *Soft Matter*, 3, 644–650, 2007. Reproduced by permission of the Royal Society of Chemistry.)

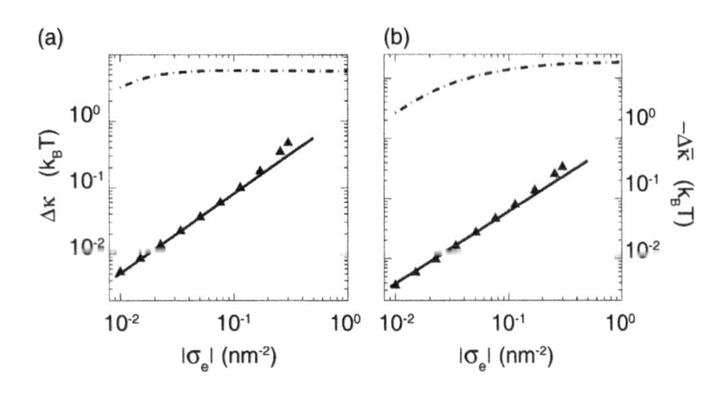

Figure 9.12 Corrections to the membrane bare (a) bending modulus and (b) Gaussian modulus of a strongly depleted ($D = -0.005$ nm) aqueous polyelectrolyte solution at low salt concentrations ($I = 0.1$ mM, $\lambda_D = 30$ nm). Here half of the polymer monomers are charged ($f_e = 0.5$), the monomer size is 1 nm, the polymer excluded volume is 0.05 nm^3 and the monomer concentration is 10^{-5} nm^{-3}. The triangular symbols are the numerically calculated corrections for these conditions, the solid lines scale as $|\sigma_e|^\beta$ with $\beta = 1.2$. The dashed-dotted line is the numerically calculated corrections for the case of an ionic solution with no polymers, with same ionic strength. The addition of polyelectrolytes can, in this case, reduce the curvature moduli significantly. (Adapted from Shafir, A. and Andelman, D. *Soft Matter*, 3, 644–650, 2007. Reproduced by permission of the Royal Society of Chemistry.)

9.6.2 INFLUENCE OF CHARGED BRUSHES ON MEMBRANE PROPERTIES

Several authors have studied end-grafted polyelectrolytes in curved geometries (Zhulina et al., 2006), and their influence on the membrane elastic parameters (Victorov, 2012; Lei et al., 2015). For a charged brush with monomer charge fraction f_e, in contact with a salt solution characterized by the Debye length λ_D, the strong stretching calculation (Milner and Witten, 1988) can be performed to give the corrections

$$\Delta\kappa = 0.15 k_B T N^3 (\sigma a^2)^{7/3} \times 4 \left(\frac{4\pi}{3} \frac{\ell_B f_e^2 \lambda_D^2}{a^3} \right)^{4/3},$$

$$\Delta\bar{\kappa} = -0.09 k_B T N^3 (\sigma a^2)^{7/3} \times 39 \left(\frac{4\pi}{3} \frac{\ell_B f_e^2 \lambda_D^2}{a^3} \right)^{4/3}, \quad (9.75)$$

which have been written here for easy comparison with Eq. 9.71. As expected, electrostatics increases the amplitude of the corrections, by an excluded volume-like term. When the quantity of salt is very high (very small λ_D) or the charge very low $f_e \ll 1$, the electrostatic excluded volume becomes smaller than the intrinsic volume of the monomers and Eq. 9.75 will reduce to the neutral brush result with excluded volume (Wang and Safran, 1991), the correction factor being replaced by $\tilde{v}^{4/3}$.

The strong-stretching treatment leading to Eq. 9.75 implies the linearization of the nonlinear (Poisson-Boltzman) electrostatic equations describing this surface and polymer geometry, which holds for small surface potentials, but is not valid otherwise. By applying a self-consistent field theory to this problem and thus fully accounting for the nonlinear nature of Poisson-Boltzmann equations (Lei et al., 2015), found results similar to those of Eq. 9.75 but with different power laws and numerical factors. For instance, for a solution ionic strength $I = 0.1$ M they find $\Delta\kappa = 0.74 f_e^2 \sigma^2 N^{2.8}, \Delta\bar{\kappa} = -0.35 f_e^2 \sigma^2 N^{2.8}$.

9.7 INSERTION OF A POLYMER IN THE BILAYER

A lipid bilayer can be seen as a thin hydrophobic slab (2–3 nm wide) surrounded by two interfacial regions (Nagle and Tristram-Nagle, 2000). Besides being hydrophobic, the inner bilayer region is anisotropic, with the lipid chains showing nematic ordering (Seelig and Seelig, 1974; Cevc and Marsh, 1987). In addition, many single component lipids or lipid mixtures display a well marked gel-fluid melting transition. In the low temperature gel phase, the close-packed aligned chains do not easily allow for the penetration of foreign material, except possibly in the interleaflet gap. It is only in the fluid, disordered state that the solubility of foreign molecules is expect to reach a significant value. Cholesterol can be considered as being exceptional in this respect, as it penetrates and transforms the gel phase into a new dense liquid-ordered (Lo) structure.

Disregarding anisotropy, small molecules can only be found in the bilayer core if they are sufficiently hydrophobic. This propensity of small solute to partition into the bilayer core is commonly characterized by the octanol-water partitioning coefficient P_{ow} defined as $\log P_{\text{ow}} = \log_{10}(c_{\text{oil}} / c_{\text{water}})$, obtained from the equilibrium concentrations of the solute dissolved in a diphasic octanol-water mixture. It is therefore expected that monomers and short oligomers could be found in the bilayer core provided they behave as hydrophobic molecules. The confinement of a large macromolecule within the inner bilayer comes at a significant entropic cost. The polymer center of mass sees its possible location strongly restricted, and the internal polymer conformations loose one spatial dimension. It is therefore expected on general grounds that the solubility of macromolecules decreases with increasing molecular weight.

If we now assume that a chain is hydrophobic enough to see the bilayer core as a favorable place, comes the problem of bringing the polymer to this region. With giant vesicles as a target, the most natural strategy would be to co-solubilize lipids and polymer in the appropriate volatile solvent, and then proceed with vesicle formation and growth. The polymer glass transition temperature is a serious issue as vesicle growth requires enough fluidity of its constitutive components. An alternative way consists in inserting the monomers into the vesicle bilayer and polymerize them *in-situ* (Jung et al., 2000; Krafft et al., 2001). In the paper by Krafft et al. (2001), for instance, it was reported the successful polymerization of poly(isodecyl acrylate) in egg extracted phosphatidyl choline (eggPC, see Appendix 1 of the book for structure and data on this lipid) liposomes, with an estimated molecular weight comprised between 10^6 and 10^7 g.mol^{-1}. These large and ramified macromolecules were found in a segregated globular state. If nothing opposes *in-situ* polymerization in giant vesicles, this has yet to be accomplished.

The insertion of molecules from the water region to the inner part of the bilayer and the translocation of these molecules across the bilayer are related problems, as insertion can be seen as the first step of a translocation pathway. Molecular dynamics simulations offer a useful framework, but are likely to fail to accurately describe the insertion mechanism for all but the smallest molecules, due to kinetic activation barriers. Therefore the customary approach for dealing with the insertion of large molecules relies on biasing, for example, umbrella sampling, weighted histograms and metadynamics. All these procedures rely on a potential of mean force (pmf) that reflects the thermodynamic cost of maintaining the molecule at a given distance from the bilayer center. They are all subject to equilibration issues and subjective choice of the collective reaction coordinates (Bochicchio et al., 2015; Shinoda, 2016). When successfully implemented, the numerical simulations provides a pmf of thermodynamical (water-bilayer partitioning) and dynamical (transition state model for the passive permeation rate) interest. As a rule, hydrophobic compounds coming from the solvent side encounter first a barrier in the bilayer interfacial region, followed by a deep minimum located in the inner bilayer.

Molecular dynamics simulations overcome the insertion problem by preparing initial configuration with polymers already inserted in the membrane, or close to it. In this case, one gains insights on the lipid organization around the polymer chains and the lateral partitioning in the diverse membrane microenvironments (e.g., fluid, gel, Lo domains). Several authors investigated the conformation of polystyrene (PS), polypropylene (PP) and polyethylene (PE) in the framework of Martini coarse-grained simulations (Barnoud et al., 2014; Rossi et al., 2014; Rossi and Monticelli, 2014; Bochicchio et al., 2017). PS and PP were found to disperse well in fluid bilayers, whereas PE remained as localized aggregates. The behavior of PP and PS is different as far as ternary "raft forming" model mixtures is concerned. PS was found to move preferentially into liquid disordered (Ld) regions, and stabilize the domains. PP was found to deplete cholesterol from the Lo domains, and destabilize the domains. A decrease in the elastic and bending curvature moduli was also predicted, moderate for PP and stronger for PS. Both predictions would be amenable to testing using GUVs, provided that insertion of the polymer was feasible.

Lattice Monte-Carlo simulations are not subjected to the same kinetic limitations as classical molecular dynamics. Werner et al. developed an original lattice bond-fluctuation model (BFM) for bilayer simulations(Sommer et al., 2012; Werner et al., 2012; Werner and Sommer, 2015). Despite suffering from a few restrictions, this BFM implements a realistic local dynamics comparable to coarse-grained molecular models. Benefiting from a massively parallel implementation on graphical processing units (GPUs), the BFM was shown to simulate the translocation of polymer chains of varying hydrophobicity across a bilayer without the need of biasing (Figure 9.13). For instance, polymer chains of about 5 kDa were simulated for about 10^7 Monte-Carlo steps. Werner et al. demonstrated the existence of an optimal hydrophobic/hydrophilic ratio that minimizes the pmf barriers. Under these conditions, the insertion and translocation kinetics of polymers is fastest. The same authors applied their model to block copolymers comprising sequences of hydrophilic and hydrophobic monomers, to further improve hydrophobic matching. The model makes predictions as far as designing efficient polymers for passive membrane insertion and translocation. For instance, weak polyelectrolytes whose hydrophilic character depends on external pH buffer conditions are good candidates for insertion or translocation across the bilayers.

Self-consistent mean field (SCMF) models usefully complement the aforementioned simulation approaches. These models are realistic numerical implementation of the Edwards self-consistent field approach for inhomogeneous polymer solutions. In the simplest case, lipids are treated as diblock oligomeric chains in a z-dependent self-consistent field, and polymer chains fit naturally into the theoretical picture. In addition, SCMF give directly access to thermodynamical

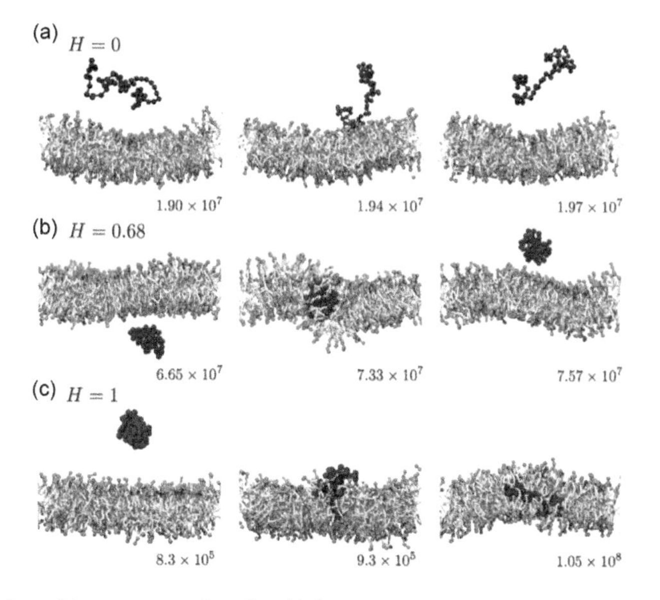

Figure 9.13 Three different scenarios for a 64-monomer–chain lipid bilayer encounter depending on the hydrophobicity parameter H of the monomers (estimated polymer chain mass of about 5000 Da). (a) An hydrophilic chain ($H = 0$) cannot penetrate into the bilayer. (b) For the optimal value $H = 0.68$ of the parameter, the chain penetrates into the bilayer but is eventually randomly released into one of the two surrounding water regions. (c) An hydrophobic chain ($H = 1$) penetrates into the bilayer and remains trapped into the interleaflet region. Predictions from the BFM model of Werner et al. (2012). (Werner, M. et al., *Soft Matter*, 8, 11714–11722, 2012. Reproduced by permission of The Royal Society of Chemistry.)

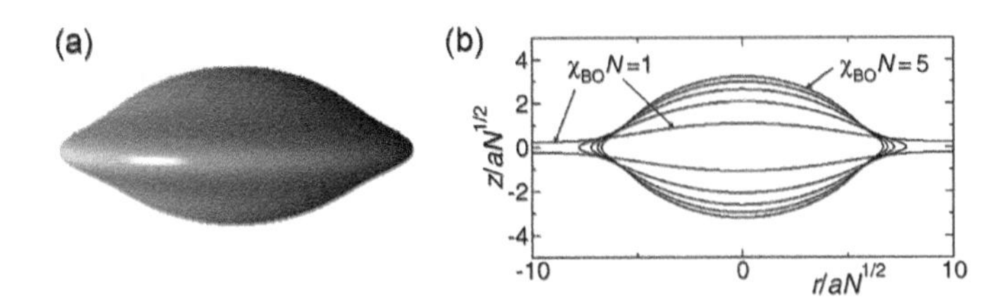

Figure 9.14 Embedment of polymers in a lipid bilayer. (a) Shape of a segregated oil droplet confined between the two leaflets of a phospholipid bilayer. (b) The bulge thickness increases as the molecular mass of the oil molecules increases. (Greenall, M.J. and Marques, C.M. *Soft Matter*, 8, 3308–3314, 2012. Reproduced by permission of The Royal Society of Chemistry.)

quantities, such as the Gibbs free-energy, and makes it possible to compare different structural organizations. Greenall and Marques investigated within this framework the distribution of oil molecules in a fluid bilayer (Greenall and Marques, 2012). Oil molecules are linear alkanes (paraffins) that inherit fluidity from the surrounding membrane state. Depending on the relative monomer affinity, the oil component can either distribute into the interstitial region, or form droplets of finite radius (Figure 9.14). Both limits can be understood in terms of wetting angle. This suggests that, putting aside issues of chain connectivity, the spreading of a polymeric material within the bilayer interior is controlled by the wetting angle of the monomer. The presence of insoluble triglycerides and sterol esters droplets confined between two phospholipid monolayers in some organelles has been experimentally reported and subject to theoretical modeling (Deslandes et al., 2017).

Altogether, experimental evidence regarding polymer chains inserted in bilayers is scarce, even more when considering the special case of GUVs. Theoretical investigations suggests a number of directions for designing polymers susceptible of interacting and inserting into the membrane. Molecular dynamics simulations have made predictions on the lateral partitioning into lipid domains, as well as regarding changes in the elastic properties. Giant vesicles experiments could possibly address both issues.

9.8 THE BILAYER AS A POLYMER CONFINEMENT MEDIUM

GUV can be used for investigating the adhesion of membranes onto solid substrates, with the possibility of functionalizing both the adhesive surface and the lipid bilayer. Adhesion can be monitored by means of reflection interference contrast methods (Radler and Sackmann, 1993), which measure the elevation of a membrane above a glass surface with subwavelength precision. Of particular interest is the adhesion of bilayers onto sparse grafted polymer brushes, as this reveals the repulsive potential of single confined chains.

Experiments performed on 20 μm long λ-phage double stranded DNAs demonstrated that semiflexible chains (persistence length $\ell_p = 50$ nm) can be efficiently pinned below a properly functionalized bilayer (Hisette et al., 2008; Nam et al., 2010).

Most flexible polymer chains cannot be fluorescently labeled as easily as DNA, and the repulsive potential of a dilute polymer brush is harder to detect. Predictions were made for the external pressure exerted by an ideal chain grafted on a surface acting on a confining bilayer (Thalmann et al., 2011). As anticipated, a flexible ideal chain exerts an osmotic repulsion on the opposing surface of range R_g, namely a position dependent pressure field:

$$p(r;d) = \frac{\pi k_B T}{2d_0^3}\Gamma(0, r^2/4R_g^2), \tag{9.76}$$

with d_0 the elevation of the confinement surface above the grafting point and r the lateral distance with respect to the chain grafting point and $\Gamma(0,x) = \int_x^\infty dy e^{-y}/y$ the incomplete Gamma function. The expression 9.76 is valid for κ, with a the monomer size.

Treating the membrane as an athermal elastic medium subject to nonspecific short-range surface adhesion, it is possible to determine quantitatively the shape of the bulge created by a polymer inserted at the origin, assuming a Monge representation $(r, d(r))$ of the elevation of the membrane, a low tilt limit $(d'(r) \ll 1)$ and a "Derjaguin approximation" of quasi-planar gap between the membrane and the surface. Under these assumptions, the mechanical balance between the grafted chain osmotic pressure and the curvature (κ) and tension (Σ) contributions reads:

$$\kappa \Delta\Delta d(r) - \Sigma \Delta d(r) = p(r; d_0), \tag{9.77}$$

with $d_0 = d(0)$ the central elevation of the membrane and $\Delta d(r)$ the Laplacian. One must specify in addition the boundary conditions obeyed by the elevation $d(r)$. For instance, in the case of a nonspecific, reversible adhesion on the surface, a boundary condition $\kappa \Delta d|_{d=0} = 2W_a$ involving the bending modulus and the adhesive energy per unit area W_a has to be fulfilled. The self-consistent shape of shallow membrane deformations can be computed numerically within this framework (Thalmann et al., 2011).

In the general case, membrane shapes can be classified according to the ratio d_0/L_b, with d_0 the bulge elevation above the surface and L_b the lateral extension (radius) of the bubble. Situations of strong adhesion leading to mushroom like conformations of the polymer chains $d_0/L_b \sim 1$ results from minimizing a scaling free-energy expression, adding curvature, adhesion, tension and confinement entropy (see state diagram in Figure 9.15):

$$F = \kappa\left(\frac{d_0}{L_b}\right)^2 + W_a L_b^2 + \Sigma d_0^2 + T\left(\frac{R_g}{d_0}\right)^2 \tag{9.78}$$

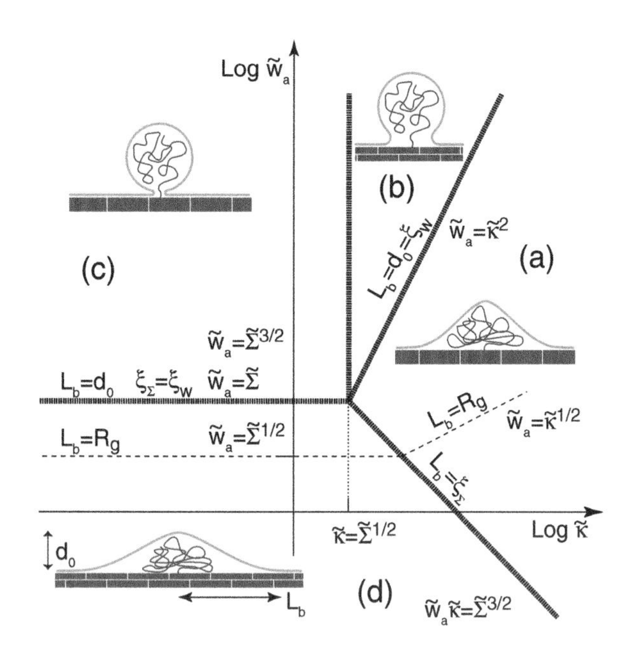

Figure 9.15 State diagram of a polymer-tensionless membrane system as a function of the curvature modulus κ (horizontal axis) and adhesion energy W_a (vertical axis), or more specifically of the dimensionless modulus $\tilde{\kappa} = \kappa / (k_B T)$ and adhesion energy $\tilde{W}_a = W_a R_g^2 / (k_B T)$ (Thalmann et al., 2011). Other parameters are the dimensionless membrane tension $\tilde{\Sigma} = \Sigma R_g^2 / (k_B T)$ and the typical lengths $\xi_\Sigma = \sqrt{\kappa / \Sigma}$ and $\xi_W = \sqrt{\kappa / W_a}$. (a) Shallow bubbles at large κ. (b) Balloon conformation for moderate κ and large adhesion W_a. (c) Mushroom polymer wrapped by the membrane at low κ, large W_a. (d) Shallow, large bulges at low adhesion and low curvature modulus. (Reprinted with permission from Thalmann, F. et al., *Phys. Rev. E*, 83, 061922, 2011, Copyright 2011 by the American Physical Society.)

These scaling results can be extended to the case of chains in good solvent. Different scaling laws for swollen chains, supported by Monte-Carlo simulations, can be found in (Su and Chen, 2013).

The case of a single grafted polymer, confined under a cylindrical piston-shaped surface constitutes an interesting problem in polymer physics, the escape transition (Subramanian et al., 1996; Dimitrov et al., 2009). It is known that for a certain range of gap size, piston radius and chain gyration radius, the polymer conformation can switch discontinuously between two competing conformational states. The first one is a homogeneously confined mushroom, whereas the second state is composed of a stretched stem escaping the piston, followed by a much less confined coil end (Figure 9.16).

A interesting question is whether GUV confinement could create the conditions of an escape transition. Nam et al. observed experimentally escaped DNA chains (Nam et al., 2010), but this was the outcome of a markedly out-of-equilibrium adhesion kinetics mechanisms, and is not conclusive as far as equilibrium escape transition is concerned. Williams and MacKintosh (1995) investigated the situation of a polymer chain confined under a smooth curved piston with prescribed analytical profile. Their scaling analysis, based on a flower blob picture predicts that there should not be any discontinuous transition for a parabolic convex shape. It would be interesting, however, to carry out a more systematic analysis of the escape transition in the presence of a spherically shaped piston, which starts with parabolic profile, and ends as a vertical object.

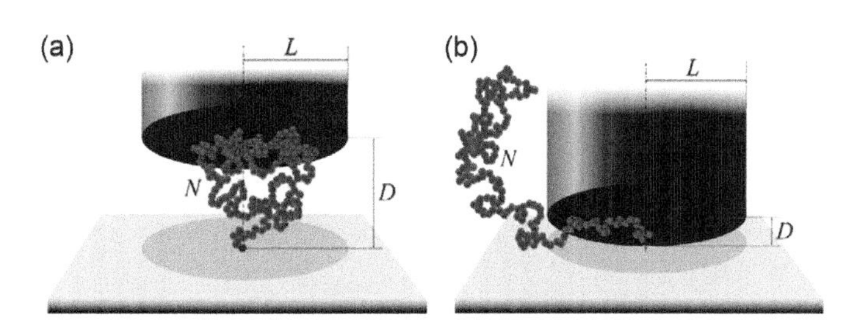

Figure 9.16 Escape transition, according to Dimitrov et al. (2009). (a) confined mushroom. (b) stem-flower conformation. (With kind permission from Springer Science+Business Media: Dimitrov, D.I. et al., Eur. *Phys. J. E*, The escape transition of a polymer: A unique case of non-equivalence between statistical ensembles. 29, 2009, 9–25)

9.9 THE SECTION FOR THE IMPATIENT: LOOK UP TECHNIQUE

Box 9.4 Macromolecular environments and their effects on membranes
$\Delta C_{spo} < 0$ the membrane bends *toward* the macromolecules
$\Delta C_{spo} > 0$ the membrane bends *away* from macromolecules
Weak: difficult to measure. Medium: can be measured. Strong: easily measured.

NONIONIC SYSTEMS				
ENVIRONMENT	PARAMETERS	STRENGTH	EQUATIONS	REFERENCES
Depletion of hard spheres	$\Delta\kappa < 0$ $\Delta\bar{\kappa} > 0$ $\Delta C_{spo} < 0$ $\Delta K_a > 0$	Weak Weak Medium Weak	9.19 9.19 9.19 9.17	Yaman et al. (1997b) Lipowsky et al. (1998)
Depletion of hard rods	$\Delta\kappa < 0$ $\Delta\bar{\kappa} > 0$ $\Delta C_{spo} < 0$ $\Delta K_a > 0$	Medium Medium Medium Medium	9.26 9.26 9.26 9.24	Yaman et al. (1997b) Yaman et al. (1997a)
Depletion of ideal chains	$\Delta\kappa < 0$ $\Delta\bar{\kappa} > 0$ $\Delta C_{spo} < 0$ $\Delta K_a > 0$	Weak Weak Medium Weak	9.34 9.34 9.34 9.33	Podgornik (1993) Eisenriegler et al. (1996) Hiergeist and Lipowsky (1996)
Adsorption of ideal chains	$\Delta\kappa$ $\Delta\bar{\kappa}$ $\Delta C_{spo} <> 0$ ΔK_a	Unknown Unknown Medium Unknown	 9.36 	Breidenich et al. (2005)
Depletion of diluted excluded volume chains	$\Delta\kappa < 0$ $\Delta\bar{\kappa} > 0$ $\Delta C_{spo} < 0$ $\Delta K_a > 0$	Weak Weak Medium Weak	9.37 9.37 9.37	Hanke et al. (1999)
Depletion of a polymer solution: mean-field	$\Delta\kappa < 0$ $\Delta\bar{\kappa} > 0$ $\Delta C_{spo} < 0$ $\Delta K_a > 0$	Medium Medium Medium Weak	9.45 9.45 9.45 9.44	Brooks et al., (1991a) Brooks et al. (1991b) Skau and Blokhuis (2002)
Adsorption of a polymer solution: mean-field	$\Delta\kappa < 0$ $\Delta\bar{\kappa} > 0$ $\Delta C_{spo} < 0$ $\Delta K_a < 0$	Medium Medium Medium Weak	9.46 9.46 9.46 9.44	Brooks et al. (1991a) Brooks et al. (1991b) Clement and Joanny (1997) Skau and Blokhuis (2002)
Adsorption of a polymer solution: scaling	$\Delta\kappa < 0$ $\Delta\bar{\kappa} > 0$ $\Delta C_{spo} < 0$ $\Delta K_a < 0$	Medium Medium Medium Weak	9.47 9.47 9.47	de Gennes (1990) Brooks et al. (1991a) Brooks et al. (1991b)

(Continued)

Box 9.4 (Continued) Macromolecular environments and their effects on membranes

$\Delta C_{spo} < 0$ the membrane bends *toward* the macromolecules

$\Delta C_{spo} > 0$ the membrane bends *away* from macromolecules

Weak: difficult to measure. Medium: can be measured. Strong: easily measured.

NONIONIC SYSTEMS				
ENVIRONMENT	PARAMETERS	STRENGTH	EQUATIONS	REFERENCES
End-grafted ideal chains	$\Delta\kappa > 0$ $\Delta\bar{\kappa} < 0$ $\Delta C_{spo} > 0$ $\Delta K_a = 0$	Weak Weak Medium no effect	9.50 9.50 9.50	Marques and Fournier (1996) Hiergeist and Lipowsky (1996) Eisenriegler et al. (1996)
End-grafted ideal loops	$\Delta\kappa < 0$ $\Delta\bar{\kappa} > 0$ $\Delta C_{spo} = 0$ $\Delta K_a = 0$	Weak Weak no effect no effect	9.52 9.52 9.52	Lipowsky et al. (1998) Bickel (2001)
Sparsely end-grafted excluded volume chains	$\Delta\kappa > 0$ $\Delta\bar{\kappa} < 0$ $\Delta C_{spo} > 0$ $\Delta K_a = 0$	Weak Weak Medium no effect	9.63 9.63 9.63	Auth and Gompper (2003)
Mean-field brushes	$\Delta\kappa > 0$ $\Delta\bar{\kappa} < 0$ $\Delta C_{spo} > 0$ $\Delta K_a < 0$	Strong Strong Strong Medium	9.71 9.71 9.71 9.72	Milner and Witten (1988) Wang and Safran (1991) Birshtein and Zhulina (1997) Marsh et al. (2003)
Scaling brushes	$\Delta\kappa > 0$ $\Delta\bar{\kappa} < 0$ $\Delta C_{spo} > 0$ $\Delta K_a < 0$	Strong Strong Strong Medium	9.74 9.74 9.74 9.73	Hristova and Needham (1994) Hiergeist and Lipowsky (1996) Marsh et al. (2003) Wu et al. (2013)
IONIC SYSTEMS				
Strong adsorption of charged polymers	$\Delta\kappa > 0$ $\Delta\bar{\kappa} < 0$ ΔC_{spo} ΔK_a	Strong Strong Unknown Unknown	Figure 9.11 Figure 9.11	Shafir and Andelman (2007)
Membrane repulsion of charged polymers	$\Delta\kappa < 0$ $\Delta\bar{\kappa} > 0$ ΔC_{spo} ΔK_a	Strong Strong	Figure 9.12 Figure 9.12	Shafir and Andelman (2007)
Charged brushes	$\Delta\kappa > 0$ $\Delta\bar{\kappa} < 0$ $\Delta C_{spo} > 0$ ΔK_a	Strong Strong Strong	9.75 9.75	Zhulina et al. (2006) Victorov (2012) Lei et al. (2015)

9.10 LOOKING AHEAD

Bilayers and polymers are intimate old friends, and their intimate friendship is here to stay. After all, both in natural systems and in man-made formulations, polymers and lipids, should we say polymers with lipids and lipids with polymers, seem to be the basic one and two-dimensional bricks making of the large majority of self-assembly structures.

So where does the theoretical understanding of the interactions between lipid bilayers and polymers stand, three decades after the first contributions started to be made? As in many other fields, the answer can be given according to two different gauges. The first is the internal consistency of the theoretical framework that has developed, pursuing both questions that have not been answered yet or questions that became reachable and answerable due to the development of new theoretical and numerical tools. The second is how effective and successful is the contribution of the theory to the explanation and quantification of the physical phenomena that is here at stake. We will discuss now the first question, and address the second later.

The simplest, less realistic, but fully tractable model for polymers is the ideal chain, and we have presented above many of the exact results that such approach allows. For the simplest surface fields with an attractive or repulsive component normal to the surfaces but otherwise homogeneous, most of the results for simple chain architectures are now known, a few gaps only remain to be filled. For instance, to our knowledge, the contribution to the membrane parameters from the conformation of a Gaussian chain adsorbed on a curved surface, in equilibrium with the bulk solution, has not been yet computed.

If on a large enough length scale a single component lipid bilayer appears as a homogeneous surface, it has been understood for a long time that surface heterogeneity can be important, at small enough length scales but also for many of the bilayers of practical interest that are composed of several types of lipids and other molecules. Surface chemical heterogeneity (Andelman and Joanny, 1991, 1993), and the coupling of polymer interactions with lipid composition and lipid phase behavior (Duan et al., 2013; Wu et al., 2013) has been investigated for flat membranes, both theoretically and by numerical simulations, but their consequences for the elastic parameters of the membrane still remain to be studied.

Arguably, the most counterintuitive result from this body of theory, is the *reduction* of bending rigidity following polymer adsorption and the tendency of the bilayer to bend *toward* the polymer in this case. As it was shown in Section 9.4, this does not change if a global constraint is imposed on the total adsorbed amount. Understanding how one crosses over from this "hard to understand" case to the more intuitive situation of end-grafted polymer layers, requires imposing a given structure for the loop distribution of the layer. As this chapter recalled, even ideal loops reveal a surprising behavior given that their effect depends on the distance between its two attachment points and on the associated probability distribution. Clarifying these questions for real chains would certainly be a significant contribution to the field.

An interesting output of the studies on the interactions between polymers and membranes, was the realization that the interactions have a local, inhomogeneous character and that a polymer acts as a nanoscopic device that applies a pressure field on the membrane. This is an interesting physical concept that might help in the future not only to rationalize results from experiments or numerical simulations, but perhaps to help also as a designing principle for new systems.

Although several polymer architectures have already been explored, the emergence of synthetic polymers with a protein-like controlled structure, with well-defined sequences of chemical composition and thus charge, solubility or rigidity, raises potential new challenges for their theoretical treatment. Maybe here again the ideal chain models will be a first step where chain conformation could be easily dealt with, whereas developing methods to account for the effect of monomer encoding on membrane behavior.

At this point the authors of this chapter must ask, before closing it, how much have these theoretical developments been important for GUV systems, and what questions have the GUV systems raised for polymers and membranes systems that are worth exploring. Although it would be pretentious pretending that we have any complete answer, we are tempted to provide a few lines of our candid opinion.

The first important success of GUVs as a physical system, was to allow measuring the physical parameters of the lipid bilayer. This has been pursued with a certain success in the presence of grafted polymers, much less experiments exist on GUVs that measure modifications induced by adsorbed or depleted polymers. Clearly, along this direction, experiments, and probably numerical simulations, will have the next say because the theoretical predictions do show that the effects are challenging to measure.

Also, one cannot but remark the sheer difference in emphasis between the experimental body of work where the interactions between charged polymers and charged membranes are the most studied, see, for instance, Chapter 25, and the theoretical developments where charged systems constitute only a fraction of the literature, see Box 9.4. There is certainly here room for theoretical development.

But stronger changes under way might concern how these questions are dealt with theoretically, both at the larger (optical) and at nanometric scales. For the large scales, where the basic membrane information is encoded by its shape and dynamics, the large majority of theoretical work focused on the changes to the constitutive parameters of the membranes, leaving a large gap between the predictions and what might actually happen to a GUV exposed to polymers. This has started to some extent to be tackled by computer simulations where the effects of a single flexible (Sun et al., 2006) or rigid (Guo et al., 2009; Li and Abel, 2018) end-grafted polymers on a vesicle shape have been predicted. Improved simulation methods and computer power now allows to consider finite amount of grafted chains (Werner and Sommer, 2010; Wu et al., 2013), but the computer-accessible vesicle dimensions are still far from the sizes of GUVs, a size window where methods such as those discussed in Chapter 5 of this book show their usefulness. However, GUVs also provide information about the smaller scales, through measurements of permeability, membrane polarity, area change or charge, to only quote a few membrane properties that polymers strongly modify. Here too, numerical methods became an important contributor (Werner et al., 2012; Troiano et al., 2017), but will not replace theoretical approaches. The road is thus opened for a deeper integration of polymer science into modeling efforts that were successfully used to describe phase separation, membrane poration and other phenomena.

LIST OF ABBREVIATIONS

BFM	bond-fluctuation model
DGDG	digalactosyl diacylglycerol
DNA	deoxyribonucleic acid
DOPC	1,2-dioleoyl-*sn*-glycero-3-phosphocholine
DSPC	1,2-stearoyl-*sn*-glycero-3-phosphocholine
Lo	liquid ordered
Ld	liquid disordered
MCS	Monte-Carlo steps
PE	polyethylene
PEG	polyethylene glycol
PP	polypropylene
PS	polystyrene
PEO	polyethylene oxide
RICM	reflection interference contrast microscopy
RNA	ribonucleic acid
SAW	self-avoiding random walk
SCMF	self-consistent mean-field
SCFT	self-consistent field theory
SOPC	1-stearoyl-2-oleoyl-*sn*-glycero-3-phosphocholine

GLOSSARY OF SYMBOLS

a	monomer size
a_ℓ	area per lipid
a_ℓ^0	area per lipid of the bare membrane
c	number concentration of monomers
c_s	number concentration of monomers at the surface
c_0	strength of interaction between lipids
c^*	crossover concentration between dilute and semi-dilute regimes
C_1, C_2	local principal curvatures
C_{spo}	spontaneous curvature
D	thickness of an interfacial polymer layer
d_0	elevation of a membrane above a substrate
f	number of arms of a polymer star
f_e	fraction of charged monomers in a polymer chain
g	surface topological genus
g_ξ	number of monomers in a blob of size ξ
G_N	Green function or propagator
Δh	membrane excess elastic energy
I	ionic strength of the solution
K_A	stretching modulus
K_A^0	stretching modulus of the bare membrane
k_B	Boltzmann constant
ℓ	distance between anchoring points of a loop
ℓ_B	Bjerrum length
ℓ_p	persistence length
L	thickness of a polymer brush
L_b	lateral extension of a membrane bulge
L_r	length of a rod
N	polymerization index
$p(r)$	pressure applied by the polymer on its grafting surface
r_0	radius of a hard sphere
R	radius of curvature
R_c	average radius of a collapsed polymer in poor solvent
R_F	average radius of a polymer in good solvent
R_G	average radius of a Gaussian chain
R_g	radius of gyration of an ideal chain
R_{gF}	radius of gyration of an excluded-volume chain
$\mathbf{R}_n$	position of the *n*th monomer of a chain
T	temperature
U_0	connectivity potential of a Gaussian chain
U_a	adsorption surface potential
U_s	impenetrable surface potential
U_l	penetrable surface potential
v	monomer excluded volume
$\tilde{v}$	dimensionless monomer excluded volume
W_a	nonspecific membrane surface adhesion energy
$\mathcal{Z}$	partition function
γ	cohesive energy of the hydrophobic part of the lipid layer
γ^0	zero order term of expansion of γ_c or γ_s in powers of $1/R$
γ_c	excess surface energy density for a cylinder
γ_s	excess surface energy density for a sphere
ΔF	energy difference for a polymer chain in the presence of a surface
$\Delta\gamma_c$	excess surface energy density for a cylinder w.r.t. a flat surface
$\Delta\gamma_s$	excess surface energy density for a sphere w.r.t. a flat surface
$\Delta\kappa$	correction to the bending rigidity
$\Delta\bar{\kappa}$	correction to the Gaussian rigidity
$\Delta\kappa_e$	electrostatic correction to the bending rigidity
$\Delta\bar{\kappa}_e$	electrostatic correction to the Gaussian rigidity
ε	strength of the surface potential
κ	bending rigidity
$\bar{\kappa}$	Gaussian rigidity
κ_{eff}	bending rigidity accounting for polymer influence
$\bar{\kappa}_{eff}$	effective Gaussian rigidity accounting for polymer influence
λ_D	Debye length
μ_c	first order coefficient of expansion of γ_c in powers of $1/R$
μ_s	first order coefficient of expansion of γ_s in powers of $1/R$
ν_c	second order coefficient of expansion of γ_c in powers of $1/R$
ν_s	second order coefficient of expansion of γ_s in powers of $1/R$
ξ_b	"blob"-size, screening length of a semi-dilute solution
Π_b	osmotic pressure of a polymer solution
ρ_b	bulk number density of colloids or polymer chains

ρ_b	Onsager concentration for isotropic-nematic transition
σ	surface number density of end-grafted chains
σ_e	surface number density of surface charges
Σ	membrane tension
ϕ	monomer volume fraction
χ	Flory interaction monomer
$\varnothing$	rod diameter

REFERENCES

Alberts B, Jonhson A, Lewis J, Raff M, Roberts K, Walter P (2008) *Molecular Biology of the Cell* Garland Science, Taylor and Francis Group, New York.

Alexander S (1977) Adsorption of chain molecules with a polar head: A scaling description. *Journal de Physique (Paris)* 38:983–987.

Andelman D, Joanny JF (1991) On the adsorption of polymer solutions on random surfaces: The annealed case. *Macromolecules* 24:6040–6042.

Andelman D, Joanny JF (1993) Polymer adsorption on surfactant monolayers and heterogeneous solid surfaces. *Journal de Physique II* 3:121–138.

Appell J, Ligoure C, Porte G (2004) Bending elasticity of a curved amphiphilic film decorated with anchored copolymers: A small angle neutron scattering study. *Journal of Statistical Mechanics: Theory and Experiment* 2004:P08002.

Aranda-Espinoza H, Berman A, Dan N, Pincus P, Safran S (1996) Interaction between inclusions embedded in membranes. *Biophysical Journal* 71:648–656.

Asakura S, Oosawa F (1954) On interaction between two bodies immersed in a solution of macromolecules. *The Journal of Chemical Physics* 22:1255–1256.

Auth T, Gompper G (2003) Self-avoiding linear and star polymers anchored to membranes. *Physical Review E* 68:051801.

Auth T, Gompper G (2005) Fluctuation spectrum of membranes with anchored linear and star polymers. *Physical Review E* 72:031904.

Auvray L (1981) Solutions de macromolécules rigides: Effets de paroi, de confinement et d'orientation par un écoulement. *Journal de Physique* 42:79–95.

Barnoud J, Rossi G, Marrink SJ, Monticelli L (2014) Hydrophobic compounds reshape membrane domains. *PLoS Computational Biology* 10:1–9.

Bickel T (2001) Interactions polymères-membranes: Une approche locale Ph.D. diss., Université Louis Pasteur-Strasbourg I.

Bickel T, Jeppesen C, Marques C (2001) Local entropic effects of polymers grafted to soft interfaces. *The European Physical Journal E* 4:33–43.

Bickel T, Marques C (2006) Entropic interactions in soft nanomaterials. *Journal of Nanosciences and Nanotechnology* 6:2386–2395.

Bickel T, Marques C, Jeppesen C (2000a) Grafted polymers are miniaturized pressure tools. *Comptes Rendus de l'Académie des Sciences-Series IV-Physics* 1:661–664.

Bickel T, Marques C, Jeppesen C (2000b) Pressure patches for membranes: The induced pinch of a grafted polymer. *Physical Review E* 62:1124–1127.

Birshtein T, Iakovlev P, Amoskov V, Leermakers F, Zhulina E, Borisov O (2008) On the curvature energy of a thin membrane decorated by polymer brushes. *Macromolecules* 41:478–488.

Birshtein TM, Zhulina EB (1997) The effect of tethered polymers on the conformation of a lipid membrane. *Macromolecular Theory and Simulations* 6:1169–1176.

Bivas I, Winterhalter M, Meleard P, Bothorel P (1998) Elasticity of bilayers containing PEG lipids. *EPL (Europhysics Letters)* 41:261.

Bochicchio D, Panizon E, Ferrando R, Monticelli L, Rossi G (2015) Calculating the free energy of transfer of small solutes into a model lipid membrane: Comparison between metadynamics and umbrella sampling. *The Journal of Chemical Physics* 143:144108.

Bochicchio D, Panizon E, Monticelli L, Rossi G (2017) Interaction of hydrophobic polymers with model lipid bilayers. *Scientific Reports* 7:6357.

Breidenich M, Netz R, Lipowsky R (2001) Adsorption of polymers anchored to membranes. *The European Physical Journal E* 5:403–414.

Breidenich M, Netz RR, Lipowsky R (2000) The shape of polymer-decorated membranes. *EPL (Europhysics Letters)* 49:431.

Breidenich M, Netz RR, Lipowsky R (2005) The influence of non-anchored polymers on the curvature of vesicles. *Molecular Physics* 103:3169–3183.

Brézin E, Le Guillou JC, Zinn-Justin J (1977) Perturbation theory at large order. ii. role of the vacuum instability. *Physical Review D* 15:1558.

Brooks J, Marques C, Cates M (1991a) The effect of adsorbed polymer on the elastic moduli of surfactant bilayers. *Journal de Physique II* 1:673–690.

Brooks J, Marques C, Cates M (1991b) Role of adsorbed polymer in bilayer elasticity. *EPL (Europhysics Letters)* 14:713.

Cantor R (1981) Nonionic diblock copolymers as surfactants between immiscible solvents. *Macromolecules* 14:1186–1193.

Carslaw H, Jaeger J (1959) *Conduction of Heat in Solids: Oxford Science Publications* Oxford, Clarendon, TX.

Cates M (1991) Model membranes: Playing a molecular accordion. *Nature* 351:102–102.

Cevc G, Marsh D (1987) *Phospholipid Bilayers. Physical Principles and Models* John Wiley & Sons, New York.

Clement F, Joanny JF (1997) Curvature elasticity of an adsorbed polymer layer. *Journal de Physique II* 7:973–980.

Daoud M, Cotton J (1982) Star shaped polymers: A model for the conformation and its concentration dependence. *Journal de Physique* 43:531–538.

de Gennes P (1972) Exponents for the excluded volume problem as derived by the Wilson method. *Physics Letters A* 38:339–340.

de Gennes P (1976) Scaling theory of polymer adsorption. *Journal de Physique* 37:1445–1452.

de Gennes P (1980) Conformations of polymers attached to an interface. *Macromolecules* 13:1069–1075.

de Gennes P (1990) Interactions between polymers and surfactants. *Journal of Physical Chemistry* 94:8407–8413.

de Gennes Pd (1981) Polymer solutions near an interface. Adsorption and depletion layers. *Macromolecules* 14:1637–1644.

de Gennes PG (1979) *Scaling Concepts in Polymer Physics* Cornell University Press, Ithaca, NY.

Decher G, Kuchinka E, Ringsdorf H, Venzmer J, Bitter-Suermann D, Weisgerber C (1989) Interaction of amphiphilic polymers with model membranes. *Die Angewandte Makromolekulare Chemie* 166:71–80.

Des Cloizeaux J, Jannink G (2010) *Polymers In Solution: Their Modelling and Structure* OUP, Oxford, UK.

Deslandes F, Thiam AR, Forêt L (2017) Lipid droplets can spontaneously bud off from a symmetric bilayer. *Biophysical Journal* 113:1–4.

Dimitrov, D. I., Klushin, L. I., Skvortsov, A., Milchev, A., Binder, K. (2009) The escape transition of a polymer: A unique case of non-equivalence between statistical ensembles. *The European Physical Journal E* 29:9–25.

Dinsmore A, Wong D, Nelson P, Yodh A (1998) Hard spheres in vesicles: Curvature-induced forces and particle-induced curvature. *Physical Review Letters* 80:409.

Duan X, Zhang R, Li Y, Shi T, An L, Huang Q (2013) Monte Carlo study of polyelectrolyte adsorption on mixed lipid membrane. *The Journal of Physical Chemistry B* 117:989–1002.

Edwards SF (1965) The statistical mechanics of polymers with excluded volume. *Proceedings of the Physical Society* 85:613.

Eisenriegler E, Hanke A, Dietrich S (1996) Polymers interacting with spherical and rodlike particles. *Physical Review E* 54:1134.

Endo H, Allgaier J, Gompper G, Jakobs B, Monkenbusch M, Richter D, Sottmann T, Strey R (2000) Membrane decoration by amphiphilic block copolymers in bicontinuous microemulsions. *Physical Review Letters* 85:102.

Endo H, Mihailescu M, Monkenbusch M, Allgaier J, Gompper G, Richter D, Jakobs B, Sottmann T, Strey R, Grillo I (2001) Effect of amphiphilic block copolymers on the structure and phase behavior of oil–water-surfactant mixtures. *The Journal of Chemical Physics* 115:580–600.

Evans E, Klingenberg D, Rawicz W, Szoka F (1996) Interactions between polymer-grafted membranes in concentrated solutions of free polymer. *Langmuir* 12:3031–3037.

Evans E, Needham D (1988) Attraction between lipid bilayer membranes in concentrated solutions of nonadsorbing polymers: Comparison of mean-field theory with measurements of adhesion energy. *Macromolecules* 21:1822–1831.

Evans E, Rawicz W (1997) Elasticity of "fuzzy" biomembranes. *Physical Review Letters* 79:2379.

Flory PJ (1949) The configuration of real polymer chains. *The Journal of Chemical Physics* 17:303–310.

Fournier JB (1996) Nontopological saddle-splay and curvature instabilities from anisotropic membrane inclusions. *Physical Review Letters* 76:4436.

Frette V, Tsafrir I, Guedeau-Boudeville M, Jullien L, Kandel D, Stavans J (1999) Coiling of cylindrical membrane stacks with anchored polymers. *Physical Review Letters* 83:2465–2468.

Garel T, Kardar M, Orland H (1995) Adsorption of polymers on a fluctuating surface. *EPL (Europhysics Letters)* 29:303.

Gompper G, Endo H, Mihailescu M, Allgaier J, Monkenbusch M, Richter D, Jakobs B, Sottmann T, Strey R (2001) Measuring bending rigidity and spatial renormalization in bicontinuous microemulsions. *EPL (Europhysics Letters)* 56:683.

Greenall MJ, Marques CM (2012) Hydrophobic droplets in amphiphilic bilayers: A coarse-grained mean-field theory study. *Soft Matter* 8:3308–3314.

Groh B (1999) Vesicles in solutions of hard rods. *Physical Review E* 59:5606–5612.

Guo K, Wang J, Qiu F, Zhang H, Yang Y (2009) Shapes of fluid vesicles anchored by polymer chains. *Soft Matter* 5:1646–1655.

Hanke A, Eisenriegler E, Dietrich S (1999) Polymer depletion effects near mesoscopic particles. *Physical Review E* 59:6853.

Harden J, Marques C, Joanny J, Andelman D (1992) Membrane curvature elasticity in weakly charged lamellar phases. *Langmuir* 8:1170–1175.

Hiergeist C, Lipowsky R (1996) Elastic properties of polymer-decorated membranes. *Journal de Physique II* 6:1465–1481.

Hiergeist C, Indrani V, Lipowsky R (1996) Membranes with anchored polymers at the adsorption transition. *EPL (Europhysics Letters)* 36:491.

Hisette ML, Haddad P, Gisler T, Marques CM, Schroder AP (2008) Spreading of bio-adhesive vesicles on DNA carpets. *Soft Matter* 4:828–832.

Hone D, Ji H, Pincus P (1987) Polymer adsorption on rough surfaces. 1. Ideal long chain. *Macromolecules* 20:2543–2549.

Hristova K, Needham D (1994) The influence of polymer-grafted lipids on the physical properties of lipid bilayers: A theoretical study. *Journal of Colloid and Interface Science* 168:302–314.

Jensen I, Dantas WG, Marques CM, Stilck JF (2013) Pressure exerted by a grafted polymer on the limiting line of a semi-infinite square lattice. *Journal of Physics A: Mathematical and Theoretical* 46:115004.

Ji H, Hone D (1988) Polymer adsorption on rough surfaces. 2. Good solvent conditions. *Macromolecules* 21:2600–2605.

Jung M, van Casteren I, Monteiro MJ, van Herk AM, German AL (2000) Pulsed-laser polymerization in compartmentalized liquids. 1. polymerization in vesicles. *Macromolecules* 33:3620–3629.

Kekicheff P, Cabane B, Rawiso M (1984) Macromolecules dissolved in a lamellar lyotropic mesophase. *Journal of Colloid and Interface Science* 102:51–70.

Kim YW, Sung W (2001) Membrane curvature induced by polymer adsorption. *Physical Review E* 63:041910.

Krafft MP, Schieldknecht L, Marie P, Giulieri F, Schmutz M, Poulain N, Nakache E (2001) Fluorinated vesicles allow intrabilayer polymerization of a hydrophobic monomer, yielding polymerized microcapsules. *Langmuir* 17:2872–2877.

Laradji M (1999) Polymer adsorption on fluctuating surfaces. *EPL (Europhysics Letters)* 47:694.

Laradji M (2002) Elasticity of polymer-anchored membranes. *EPL (Europhysics Letters)* 60:594.

Lasic DD, Barenholz Y (1996) *Handbook of Nonmedical Applications of Liposomes* CRC Press, Boca Raton, FL.

Lasic DD, Papahadjopoulos D (1998) *Medical Applications of Liposomes* Elsevier, Amsterdam, the Netherlands.

Lei Z, Yang S, Chen EQ (2015) Membrane rigidity induced by grafted polymer brush. *Soft matter* 11:1376–1385.

Lekkerkerker H (1989) Contribution of the electric double layer to the curvature elasticity of charged amphiphilic monolayers. *Physica A: Statistical Mechanics and its Applications* 159:319–328.

Lekkerkerker HN, Tuinier R (2011) *Colloids and The Depletion Interaction*, Vol. 833 Springer, Dordrecht, the Netherlands.

Li B, Abel SM (2018) Shaping membrane vesicles by adsorption of a semiflexible polymer. *Soft Matter* 14:185–193.

Lipowsky R (1995) Bending of membranes by anchored polymers. *EPL (Europhysics Letters)* 30:197.

Lipowsky R, Döbereiner HG, Hiergeist C, Indrani V (1998) Membrane curvature induced by polymers and colloids. *Physica A: Statistical Mechanics and its Applications* 249:536–543.

Lodish H, Berk A, Zipursky SL, Matsudaira P, Baltimore D, Darnell J (2000) Molecular Cell Biology, WH Freeman, New York, 4th ed.

Marques CM, Fournier JB (1996) Deviatoric spontaneous curvature of lipid membranes induced by siamese macromolecular cosurfactants. *EPL (Europhysics Letters)* 35:361.

Marsh D (2001) Elastic constants of polymer-grafted lipid membranes. *Biophysical Journal* 81:2154–2162.

Marsh D, Bartucci R, Sportelli L (2003) Lipid membranes with grafted polymers: Physicochemical aspects. *Biochimica et Biophysica Acta (BBA)-Biomembranes* 1615:33–59.

Milner S, Witten T (1988) Bending moduli of polymeric surfactant interfaces. *Journal de Physique* 49:1951–1962.

Milner S, Witten T, Cates M (1988b) A parabolic density profile for grafted polymers. *EPL (Europhysics Letters)* 5:413.

Milner ST, Witten TA, Cates ME (1988a) Theory of the grafted polymer brush. *Macromolecules* 21:2610–2619.
Nagle JF, Tristram-Nagle S (2000) Structure of lipid bilayers. *Biochimica and Biophysica Acta* 1469:159–195.
Nam G, Hisette M, Sun Y, Gisler T, Johner A, Thalmann F, Schröder A, Marques C, Lee N (2010) Scraping and stapling of end-grafted DNA chains by a bioadhesive spreading vesicle to reveal chain internal friction and topological complexity. *Physical Review Letters* 105:88101.
Natsume Y, Pravaz O, Yoshida H, Imai M (2010) Shape deformation of giant vesicles encapsulating charged colloidal particles. *Soft Matter* 6:5359–5366.
Natsume Y, Toyota T (2013) Giant vesicles containing microspheres with high volume fraction prepared by water-in-oil emulsion centrifugation. *Chemistry Letters* 42:295–297.
Natsume Y, Toyota T (2016) Asymmetrical polyhedral configuration of giant vesicles induced by orderly array of encapsulated colloidal particles. *PLoS One* 11:e0146683.
Needham D, McIntosh T, Lasic D (1992) Repulsive interactions and mechanical stability of polymer-grafted lipid membranes. *Biochimica et Biophysica Acta (BBA)-Biomembranes* 1108:40–48.
Netz RR (1997) Inclusions in fluctuating membranes: Exact results. *Journal de Physique I* 7:833–852.
Nikolov V, Lipowsky R, Dimova R (2007) Behavior of giant vesicles with anchored DNA molecules. *Biophysical Journal* 92:4356–4368.
Okano T, Inoue K, Koseki K, Suzuki H (2018) Deformation modes of giant unilamellar vesicles encapsulating biopolymers. *ACS Synthetic Biology* 7:739–747.
Onsager L (1949) The effects of shape on the interaction of colloidal particles. *Annals of the New York Academy of Sciences* 51:627–659.
Podgornik R (1993) Polymer-boundary surface interactions and bilayer curvature elasticity. *EPL (Europhysics Letters)* 21:245.
Radler J, Sackmann E (1993) Imaging optical thicknesses and separation distances of phospholipid vesicles at solid surfaces. *Journal de Physique II France* 3:727–748.
Ringsdorf H, Schlarb B, Venzmer J (1988) Molecular architecture and function of polymeric oriented systems: Models for the study of organization, surface recognition, and dynamics of biomembranes. *Angewandte Chemie International Edition in English* 27:113–158.
Rossi G, Barnoud J, Monticelli L (2014) Polystyrene nanoparticles perturb lipid membranes. *The Journal of Physical Chemistry Letters* 5:241–246 PMID: 26276207.
Rossi G, Monticelli L (2014) Modeling the effect of nano-sized polymer particles on the properties of lipid membranes. *Journal of Physics: Condensed Matter* 26:503101.
Rupert LA, Engberts JB, Hoekstra D (1988) Effect of poly(ethylene glycol) on the Ca^{2+}-induced fusion of didodecyl phosphate vesicles. *Biochemistry* 27:8232–8239.
Seelig A, Seelig J (1974) Dynamic structure of fatty acyl chains in a phospholipid bilayer measured by deuterium magnetic resonance. *Biochemistry* 13:4839–4845 PMID: 4371820.
Seki K, Tirrell DA (1984) pH-dependent complexation of poly (acrylic acid) derivatives with phospholipid vesicle membranes. *Macromolecules* 17:1692–1698.
Semenov A (1985) Contribution to the theory of microphase layering in block-copolymer melts. *Zh. Eksp. Teor. Fiz* 88:1242–1256.
Shafir A, Andelman D (2007) Bending moduli of charged membranes immersed in polyelectrolyte solutions. *Soft Matter* 3:644–650.
Shinoda W (2016) Permeability across lipid membranes. *Biochimica et Biophysica Acta (BBA): Biomembranes* 1858:2254–2265.
Skau K, Blokhuis E (2002) Mean-field theory for polymer adsorption on curved surfaces. *The European Physical Journal E* 7:13–22.
Skau K, Blokhuis E (2003) Polymer adsorption on curved surfaces: Finite chain length corrections. *Macromolecules* 36:4637–4645.
Sommer JU, Werner M, Baulin VA (2012) Critical adsorption controls translocation of polymer chains through lipid bilayers and permeation of solvent. *EPL (Europhysics Letters)* 98:18003.
Su YC, Chen JZY (2013) Budding transition of a self-avoiding polymer confined by a soft membrane adhering onto a flat wall. *Soft Matter* 9:570–576.
Subramanian G, Williams D, Pincus P (1996) Interaction between finite-sized particles and end grafted polymers. *Macromolecules* 29:4045–4050.
Sun M, Qiu F, Zhang H, Yang Y (2006) Shape of fluid vesicles anchored by rigid rod. *The Journal of Physical Chemistry B* 110:9698–9707.
Sung W, Lee S (2004) The soft-mode instability of a membrane induced by strong polymer adsorption. *EPL (Europhysics Letters)* 68:596.
Sung W, Oh E (1996) Membrane fluctuation and polymer adsorption. *Journal de Physique II* 6:1195–1206.
Szleifer I, Carignano M (1996) Tethered polymer layers. *Advances in Chemical Physics: Polymeric Systems* 94:165–260.
Szleifer I, Gerasimov OV, Thompson DH (1998) Spontaneous liposome formation induced by grafted poly (ethylene oxide) layers: Theoretical prediction and experimental verification. *Proceedings of the National Academy of Sciences* 95:1032–1037.
Terasawa H, Nishimura K, Suzuki H, Matsuura T, Yomo T (2012) Coupling of the fusion and budding of giant phospholipid vesicles containing macromolecules. *Proceedings of the National Academy of Sciences* 109:5942–5947.
Thakkar FM, Ayappa K (2010) Investigations on the melting and bending modulus of polymer grafted bilayers using dissipative particle dynamics. *Biomicrofluidics* 4:032203.
Thalmann F, Billot V, Marques CM (2011) Lipid bilayer adhesion on sparse DNA carpets: Theoretical analysis of membrane deformations induced by single-end-grafted polymers. *Physical Review E* 83:061922.
Troiano JM, McGeachy AC, Olenick LL, Fang D, Liang D, Hong J, Kuech TR, Caudill ER, Pedersen JA, Cui Q, Geiger FM (2017) Quantifying the electrostatics of polycation–lipid bilayer interactions. *Journal of the American Chemical Society* 139:5808–5816.
Van de Pas J, Buytenhek C (1992) The effects of free polymers on osmotic compression, depletion flocculation and fusion of lamellar liquid-crystalline droplets. *Colloids and Surfaces* 68:127–139.
Vandepas J, Olsthoorn T, Schepers F, Devries C, Buytenhek C (1994) Colloidal effects of anchored polymers in lamellar liquid-crystalline dispersions. *Colloids and Surfaces A-Physicochemical and Engineering Aspects* 85:221–236.
Victorov A (2012) Curvature elasticity of a weak polyelectrolyte brush and shape transitions in assemblies of amphiphilic diblock copolymers. *Soft Matter* 8:5513–5524.
Wang ZG, Safran S (1991) Curvature elasticity of diblock copolymer monolayers. *The Journal of Chemical Physics* 94:679–687.
Werner M, Sommer JU (2010) Polymer-decorated tethered membranes under good-and poor-solvent conditions. *The European Physical Journal E* 31:383–392.
Werner M, Sommer JU (2015) Translocation and induced permeability of random amphiphilic copolymers interacting with lipid bilayer membranes. *Biomacromolecules* 16:125–135 PMID: 25539014.

Werner M, Sommer JU, Baulin VA (2012) Homo-polymers with balanced hydrophobicity translocate through lipid bilayers and enhance local solvent permeability. *Soft Matter* 8:11714–11722.

Williams D. R. M, MacKintosh F. C. (1995) Polymer mushrooms compressed under curved surfaces. *Journal de Physique II France* 5:1407–1417.

Wolff J, Marques CM, Thalmann F (2011) Thermodynamic approach to phase coexistence in ternary phospholipid-cholesterol mixtures. *Physical Review Letters* 106:128104.

Wu H, Shiba H, Noguchi H (2013) Mechanical properties and microdomain separation of fluid membranes with anchored polymers. *Soft Matter* 9:9907–9917.

Yaman K, Jeng M, Pincus P, Jeppesen C, Marques C (1997a) Rods near curved surfaces and in curved boxes. *Physica A* 247:159–182.

Yaman K, Pincus P, Marques C (1997b) Membranes in rod solutions: A system with spontaneously broken symmetry. *Physical Review Letters* 78:4514.

Zhang L, Peng T, Cheng SX, Zhuo RX (2004) Destabilization of liposomes by uncharged hydrophilic and amphiphilic polymers. *The Journal of Physical Chemistry B* 108:7763–7770.

Zhulina E, Birshtein T, Borisov O (2006) Curved polymer and polyelectrolyte brushes beyond the Daoud-Cotton model. *The European Physical Journal E* 20:243–256.

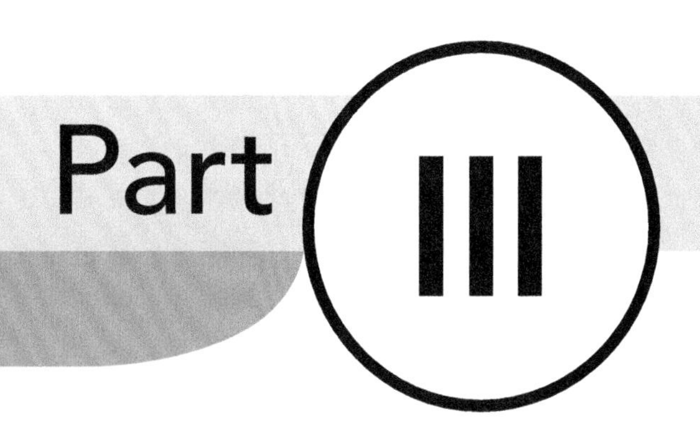

GUV-based techniques and what one can learn from them

Application of optical microscopy techniques on giant unilamellar vesicles

Luis A. Bagatolli

FAITH is a fine invention
For gentlemen who see;
But microscopes are prudent
In an emergency!

Emily Dickinson
Poems, Second Series; Eds. Higginson and Todd (1891), Roberts Brothers

Contents

10.1 INTRODUCTION

Since the end of the 1960s, major research efforts have been presented to study physical aspects of membranes in artificial model systems assembled in the laboratory. Among different options, the most popular membrane model systems are liposomes (Lasic and Papahadjopoulos, 1995; Bagatolli, 2009). Physical studies involving liposomes usually use an array of different experimental techniques—fluorescence or infrared spectroscopy, electron paramagnetic resonance, NMR, differential scanning calorimetry, X-ray diffraction and neutron scattering to mention a few—producing relevant mean physical parameters on the basis of data generally collected from *bulk dispersions* of sub-micrometer sized vesicles. These types of studies lack, however, spatially resolved information at the level of *single vesicles*, which is easily accessible from optical microscopy experiments on giant unilamellar vesicles (GUVs).

GUVs arose as versatile freestanding membrane models for performing microscopy experiments, particularly because their size (mean diameter between ~5 to 100 μm) is much bigger that the resolution limit of a regular optical microscope (~250 nm radial due to the diffraction limit of light). Moreover, GUVs of different compositions are relatively easy to produce and handle (see Part II of this book). Over the past 30 years, the study of GUVs using different optical microscopy techniques resulted in numerous relevant results on various membrane-related events at the level of single vesicles. Examples of these studies are membrane phase behavior (Veatch and Keller, 2005; Bagatolli, 2006; Heberle and Feigenson, 2011 [Chapter 18]) effects of membrane proteins (Fischer et al., 2000; Bacia et al., 2004; Kahya et al., 2005; Bouvrais et al., 2012 [Chapters 3, 16 and 24]), membrane mechanical properties (Bagatolli and Needham, 2014 [Chapters 11 through 16]) and enzymatic effect of lipases (Wick et al., 1996; Sanchez et al., 2002; Stock et al., 2012). GUVs are also objects of great interest in the field of the origin of life, where they can act as small bioreactors compartmentalizing distinct chemicals in order to perform in vitro studies of some relevant cellular processes (Luisi, 2000 [Chapter 28]).

This chapter is intended to provide some practical guidelines for performing experiments on GUVs using different optical microscopy techniques. In addition, it also discussed the pros and cons of distinct microscopy methods currently used to study membranes at the level of single vesicles.[1] Although many studies have been reported on GUVs using different microscopy-based strategies (several examples are discussed in Parts IV, V and VI of this book) only a few relevant cases will be presented in this chapter. Here, I thus focus more on the capabilities of the different experimental approaches and techniques than on the specific scientific aims of particular studies.

10.2 LABEL-FREE OPTICAL MICROSCOPY TECHNIQUES

10.2.1 BRIGHT-FIELD, PHASE-CONTRAST AND DIFFERENTIAL INTERFERENCE CONTRAST METHODS

There are several microscopy techniques suitable for performing experiments with GUVs. Some of these options require very specialized, expensive, and not easily accessible instrumentation. However, there are simple (and cheap) choices available to start the exploration of these model membrane systems. Examples include regular bright-field microscopes or microscopes equipped for either phase-contrast or differential interference contrast (DIC) methods, which are commonly used in laboratories focused on cell biology. Unlike fluorescence microscopy (discussed in the coming sections), these techniques do not require adding external labels in the membrane.

Although the observation of GUVs using regular bright-field microscopy is doable, acquisition of suitable images is technically challenging because the contrast obtained by the GUV membrane in aqueous media is too weak to be optimally detected. DIC or phase-contrast methods can easily circumvent this problem. These methods are complementary techniques capable of producing high-contrast images of transparent biological phases that do not ordinarily affect the amplitude of visible light waves passing though the specimen. A fundamental distinction between DIC and phase-contrast is the optical basis upon which images are formed by the complementary techniques (Murphy et al. 2009). The image intensity values observed in phase-contrast depend on the magnitude of the optical path length of the specimen, where very dense regions (with a high refractive index) appear darker than the background. Features of a specimen that have a refractive index less than the surrounding medium on the other hand are rendered much brighter. The situation is quite different for DIC, where optical path length gradients are primarily responsible for introducing contrast into the images. Steep gradients in path length produce very good contrast, and images display a pseudo three-dimensional (3D) relief shading that is a characteristic of this technique. On the contrary, areas having very narrow optical path slopes, such as those produced in very flat specimens, cause a marginal contrast, which is undistinguished from the background. Overall, DIC requires more technical skills to configure and operate compared with phase-contrast. Microscopes equipped with phase-contrast are far easier to align and operate and can be utilized quite effectively by relatively inexperienced microscopists (Murphy et al. 2009). For a comparison among GUV images obtained by these three methods (bright-field, phase-contrast and DIC, respectively), see Box 10.1.

Applications of the microscopy techniques described in this section are numerous, particularly using phase-contrast or Nomarsky optics (which is related to DIC). These techniques have been used to

Box 10.1 What do GUVs look like under a microscope?
The images in this Box illustrate the most common techniques for observing GUVs with diameters in the range 20–40 μm. Notice that most digital cameras only detect intensity, not color. Images: courtesy of A. Schröder and V. Walter.

	Bright-field imaging. Without experience, it might be difficult to notice the vesicle when looking directly through the eyepiece, but it is easy to detect with a camera.
	Phase-contrast conditions. Easy to detect also through the eyepiece, particularly if the GUV solution has been prepared with different optical densities inside and outside the vesicles, as, for instance, with osmotically matched sucrose and glucose solutions.
	Differential interference contrast. Slightly more difficult to detect through the eyepiece than under phase-contrast. The light distribution creates the impression of 3-D illumination, here from the bottom right.
	Epi-fluorescence. The membrane was labeled with a fluorescent dye. Notice the large spreading of the light. The light intensity away from the membrane is a function of the size of the vesicle, smaller vesicles appear "filled" with more light than the larger ones. In the eyepiece the GUV appears colored.
	Laser scanning confocal fluorescence cross section. It is a scanned image, i.e., built pixel by pixel, within a certain wavelength (color) range. The confocal images are generally reconstructed as colored images, but the intrinsic information of each scanned image is colorless.
	3D reconstruction by laser scanning confocal fluorescence. A simple 3D reconstruction of a GUV from 20 confocal slices.

[1] I believe that this information is relevant to best choose how to perform the experiments: Recall that the use of very fancy and sophisticated equipment does not necessarily guarantee better experimental results.

assist measurements of membrane mechanical properties using giant vesicles or to follow changes in the shape of GUVs under specific conditions (Wick et al., 1996; Döbereiner et al., 1997; Dimova et al., 2006; Haluska et al., 2012; Bagatolli and Needham, 2014). These particular topics are carefully addressed in other chapters of this book (see, e.g., Chapters 11 through 16, 22, 24). Among other examples, quantitative phase-contrast measurements have been used to determine membrane permeability using GUVs (Peterlin et al., 2009; Mertins et al., 2014 [Chapter 22]) or morphological aspects of GUVs containing archaebacterial lipids (Sustar et al., 2012). In the same line, quantitative DIC measurements have been recently reported to ascertain the lamellarity of giant vesicles (McPhee et al., 2013). These studies are interesting examples of quantitative label-free optical microscopy studies on GUVs.

10.2.2 PRACTICAL TIPS ABOUT OBSERVATION OF GUVs USING INVERTED MICROSCOPES

Concerning the way that the specimen is visualized in a microscope, there are two configurations that are commonly used, called upright and inverted. With inverted microscopes, the samples are observed from below—that is, the objective is placed under the sample—whereas with upright microscopes, samples are observed from above. Traditionally, inverted microscopes are used for life science research because gravity makes samples sink to the bottom of a holder in aqueous media, aiding better access to the specimen. For the very same reason the inverted configuration is preferred over the upright to observe GUVs. In addition, inverted microscopes also offer better possibilities to adapt special custom-built chambers on the microscope stage to visualize the vesicles (see Chapter 1).

Depending on the technical characteristics of the system employed to prepare GUVs, it is possible to directly follow the growth of these vesicles using the aforementioned microscopy techniques. For example, particular devices can be constructed for electroforming GUVs (Part II of this book [Fidorra et al., 2006]) that allow visualization of the surface of the electrode where GUVs are forming (platinum or glass indium tin oxide–coated electrodes) (Figure 10.1). Alternatively, the vesicles can be prepared in similar chambers that do not allow direct observation in the microscope. In this last case, GUVs are transferred after electroformation to a glass slide or special chamber for observation under the microscope (e.g., the 8-well plastic chamber from Lab-Tek® Brand Products, Naperville, IL, is an excellent option). When transfer is desired it is very convenient to generate a density difference between the interior of the vesicle and the surrounding medium, that is, by preparing GUVs in an aqueous solution containing sucrose (from 50 to 200 mOsM) and then transferred aliquots of the GUV suspensions into an iso-osmolar solution containing glucose (200 μL of glucose + 50 μL of the GUVs in sucrose, for example). The density difference between the interior and exterior of the GUVs caused by the presence of the sugars induces the vesicles to sink to the bottom of the chamber, and within a few minutes, the vesicles are ready to be observed using an inverted microscope. Notice that the difference in density between these two solutions also aids to obtain much better contrast for imaging using the microscopy methods described above. In addition, GUV immobilization can also be performed under the above-mentioned conditions to avoid the effect of the drifting of vesicles during imaging. This can be relevant for acquisition of image stacks using laser scanning systems (see Sections 3.3 and 3.4) or performing fluctuation correlation spectroscopy (FCS) experiments (Section 3.5; see also Chapter 21). Simple immobilization techniques use 0.5% w/v agarose gels (Lira et al., 2016) or millimolar concentrations of calcium or magnesium salts. A more involved strategy uses the avidin–biotin complex (the strongest known non-covalent interaction between a protein and ligand, $K_d = 10^{-15}$ M) as a linker to immobilize GUVs to the coverslip in the observation chamber. The bond formation between biotin and avidin is very rapid and, once formed, is unaffected by extremes of pH or temperature and organic solvents and other denaturing agents. This procedure is accomplished by labeling first the organic solvent lipid solution with a very small fraction of biotinylated lipids (0.1 mol % of the total lipids; e.g., 1,2-dioleoyl-*sn*-glycero-3-phosphoethanolamine-*N*-biotinyl from Avanti) prior to preparation of the GUVs. The vesicles are then added to an observation chamber where the coverslip surface has been coated with avidin (Stock et al., 2012). Chemically neutral immobilization techniques also exist, based on microfluidics devices where flow forces hold the vesicle still, albeit with a possible slight deformation (Nuss et al., 2012; Robinson et al., 2013).

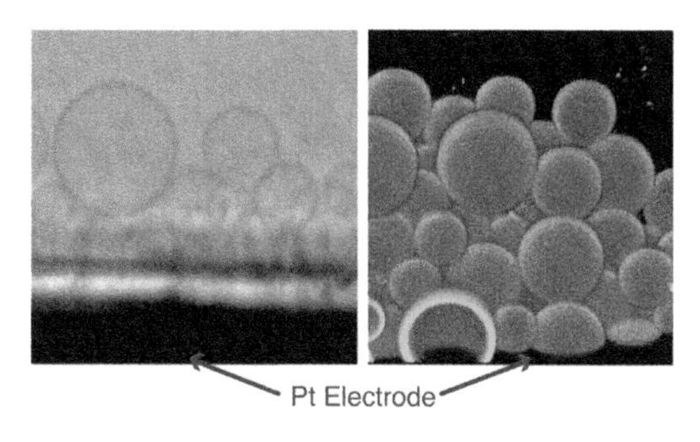

Figure 10.1 Bright-field (left) and fluorescence confocal (3D reconstruction; right) images of GUVs growing on platinum wires.

10.3 FLUORESCENCE MICROSCOPY TECHNIQUES

The main advantage in using fluorescence microscopy in membranous systems over more classical experimental approaches is clear: the sensitivity and flexibility of a microscope with the addition of fluorescence spectroscopy allows the collection of spatially resolved information. Ultimately, this latter information bridges membrane morphology with the dynamical and structural information obtained at the molecular level using fluorescence spectroscopy (e.g., lipid mobility, membrane hydration). There are several fluorescence microscopy methodologies available to study membranes (either for planar supported membranes or GUVs). Some of them will be discussed in the coming subsections, particularly regarding applications in GUVs. As a first step, I will start with a brief overview on fluorescent membrane probes, which are crucial actors for performing this type of experiment.

10.3.1 FLUORESCENCE PROBES

These molecules, called also "fluorophores" or "fluorescent probes," are required in fluorescence microscopy experiments to create the necessary contrast needed to produce an image. Although all fluorescent molecules can be used to create contrast, there are special probes whose fluorescence parameters display a remarkable sensitivity to the properties of the microenvironment where they are inserted (e.g., polarity, solvent relaxation). This feature has been largely exploited to obtain additional quantitative information on GUVs displaying membrane lateral heterogeneity (as is the case of the probe 6-dodecanoyl-2-dimethylaminonaphthalene (LAURDAN, see Appendix 2 of the book for structure and data on this probe) (Bagatolli, 2006 [see Section 10.3.4 below]). Because we are particularly interested in labeling membranes, an important requirement is to select an amphipathic (or hydrophobic) fluorescent molecule as a reporter, that is, fluorescent molecules with selective partition membranes.

There are a large number of fluorophores commercially available for tagging membranes. For practical reasons, I will arbitrarily divide them into two families. The first contains those probes where excitation is feasible by one photon excitation mode, particularly in the visible light range. Generally, these fluorescent molecules present a chemical structure similar to lipids, that is, a polar head plus two methylene chains of variable size and unsaturation. The fluorescent moiety can be attached either to the lipid chains or the polar head group. Examples of these probes are, amphiphilic derivatives of ATTO® dyes, rhodamine, fluorescein, dialkylcarbocyanine (DiI, DiO), dialkylaminostyryl (DiA), coumarins and BODIPY®, as well as perylene and naphtopyrene[2] (see Figure 10.2a for representative examples). These probes have been commonly used in many wide-field fluorescence and laser scanning confocal fluorescence microscopy (LSCFM) studies of GUVs (particularly those studies focused on the coexistence of equilibrium thermodynamic phases [see Korlach et al., 1999; Bagatolli and Gratton, 2000a; Kahya et al., 2003; Veatch and Keller, 2005; Bagatolli, 2006; Baumgart et al., 2007]). A general characteristic of these fluorophores is that they show *uneven partition* into coexisting membrane regions.

Occasionally, the partition properties of this family of fluorophores have been used as criteria to assign lipid phases in GUVs displaying lipid phase coexistence (Bagatolli, 2006). Even though an extensive characterization of the partition of several fluorescent dyes *on selected lipid mixtures* was reported (Baumgart et al., 2007; Sezgin et al., 2012), the aforementioned criterion to define the nature of a lipid phase is risky. Changes in the partition properties of the probes on different membrane regions are highly dependent *on the local chemical composition of the membrane domain and not on the phase state* (Bagatolli and Gratton, 2000a, 2000b; Bagatolli, 2006; Juhasz et al., 2010, 2012). If fluorescent images are used to assign lipid phases, a practical solution to strengthen the data interpretation is the use of FCS (Section 3.5, see also Chapter 21 [Korlach et al., 1999; Kahya et al., 2003]). With this approach, the diffusion of the probe can be measured and related to the nature of the existing lipid phases. Alternative measurements of fluorescence lifetime (de Almeida et al., 2007) or fluorescence polarization (anisotropy) (Ariola et al., 2009) under the microscope are also robust alternatives to explore membrane phase state because these parameters are responsive to the lateral packing of the membrane.

The second family of fluorescent probes mostly comprises UV-excited fluorescent molecules. In fact, the vast majority of fluorescence spectroscopy studies on membranes largely exploit fluorescent parameters obtained by these probes, such as fluorescence lifetime, fluorescence polarization and emission spectral information. Examples are 1,6-diphenyl-hexa-1,3,5-triene (DPH, see Appendix 2 of the book for structure and data on this probe), 1-(4-(trimethylamino)phenyl)-6-phenylhexa-1,3,5-triene (TMA-DPH, see Appendix 2 of the book for structure and data on this probe), (9*Z*,11*E*,13*E*,15*Z*)-octadeca-9,11,13,15-tetraenoic acid (parinaric acid, see Appendix 2 of the book for structure and data on this probe), LAURDAN (Figure 10.2b), 6-propionyl-2-dimethylaminonaphthalene (PRODAN, see Appendix 2 of the book for structure and data on this probe), and pyrene (Lentz et al., 1976b, 1976a; Jones and Lentz, 1986; Parasassi et al., 1990; Mateo et al., 1991; Garda et al., 1994; Parasassi et al., 1994; Velez et al., 1995; Krasnowska et al., 1998; Bagatolli, 2013). Notably, these probes have not been fully exploited in fluorescence microscopy experiments (a clear exception is LAURDAN, however, see Section 10.3.4). Two main reasons for the dearth of such studies are (i) the uneasy access for researchers to the advanced microscopy techniques required for performing the experiments, that is, fluorescence lifetime imaging microscopy (FLIM) or

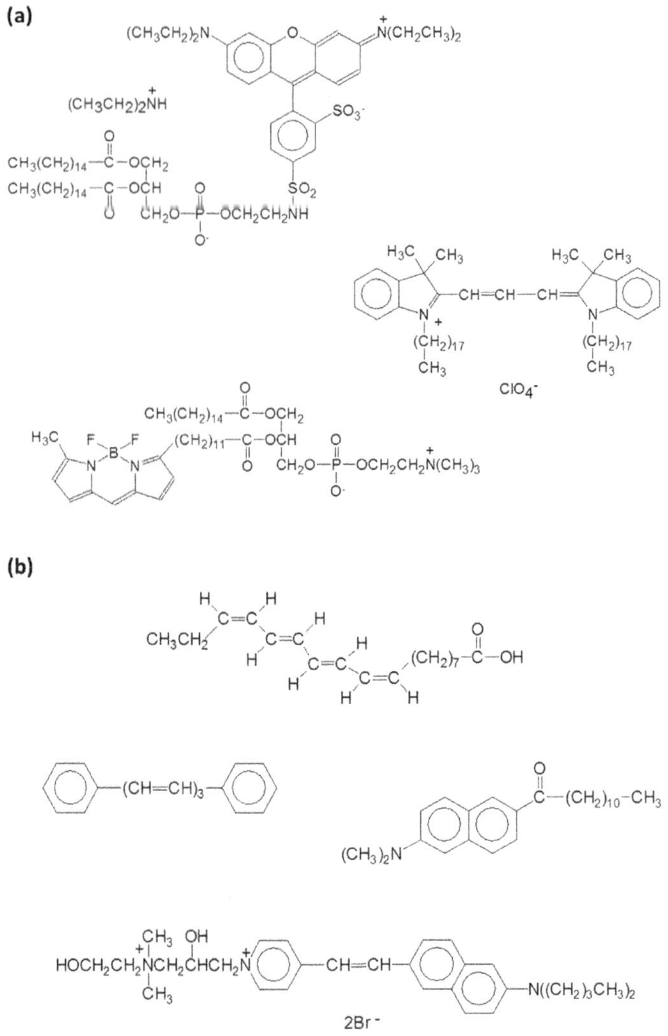

Figure 10.2 Chemical structure of representative fluorescent probes: (a) rhodamine-DPPE (top, see Appendix 2 of the book for structure and data on this probe), DiIC$_{18}$ (center), BODIPY-PC (bottom). (b) parinaric acid (top), DPH and LAURDAN (center panel, from left to right), and di-4-ANEPPDHQ (bottom).

[2] Notice that some of these hydrophobic fluorescent moieties can themselves be used to label membranes without further chemical modification.

polarization fluorescence microscopy (expensive and specialized equipment is required along with significant user expertise), and (ii) it is practically challenging for performing fluorescence microscopy experiments via one photon excitation (using wide-field or confocal fluorescence microscopy) with UV-fading fluorescent probes given that the extent of photobleaching is high and that it is often technically difficult to obtain reliable fluorescence images. An alternative solution for using these probes is to use multiphoton excitation fluorescence microscopy (see Section 10.3.4).

One of the most remarkable features of this family of probes (particularly for those that keep a fatty acid like structure, e.g., LAURDAN, parinaric acid) is the fact that in membranes displaying phase coexistence (i.e., liquid disordered/liquid ordered [l_d/l_o] or l_d/solid ordered [s_o]), the probes are present in the two phases. This allows simultaneous correlation of the probe's fluorescence parameters (changes in polarization or lifetimes, spectral shift) with the local physical properties of distinct membrane regions. Notice that this is different from what was discussed above in connection to the first family of fluorescent probes. Other polarity-sensitive membrane probes used in fluorescence microscopy experiments are 3-hydroxyflavone derivatives (M'Baye et al., 2008), 3-hydroxychromone (3HC) dyes (which are also sensitive to electrostatic effects [Demchenko et al., 2009]), 6-dodecanoyl-2-[*N*-methyl-*N*-(carboxymethyl) amino]naphthalene (C-LAURDAN) (Kim et al., 2007) and di-4-ANEPPDHQ (Jin et al., 2006 [chemical structure included in Figure 10.2b]). These probes show different emission spectra depending on the membrane's phase state, also showing a quasi-uniform distribution between coexisting phases.

Potential harmful effects of fluorescent reporters on the properties of membranes must be taken into account when GUV/fluorescence microscopy experiments are performed. For example, it was recently shown that although the mechanical properties of LAURDAN (or 1,1′-dioctadecyl-3,3,3′,3′-tetramethylindocarbocyanine perchlorate [$DiIC_{18}$]) -labeled 1-hexadecanoyl-2-(9Z-octadecenoyl)-*sn*-glycero-3-phosphocholine (POPC, see Appendix 1 of the book for structure and data on this lipid) membranes are not affected compared with membranes devoid of probes, these properties change when the probe 1,2-dihexadecanoyl-*sn*-glycero-3-phosphoethanolamine (rhodamine-DPPE) is used (Bouvrais et al., 2010). This phenomenon has been associated with the photo-oxidation of unsaturated lipids caused by the fluorescence probe, which can generate reactive oxygen species after prolonged exposure to the excitation source (see Chapter 22 of this book). In addition, it has been reported that specific fluorescent probes such as the case of 1,1′-didodecyl-3,3,3′,3′-tetramethylindocarbocyanine perchlorate ($DiIC_{12}$, see Appendix 2 of the book for structure and data on this probe) and 3,3′-dioctadecyloxacarbocyanine perchlorate ($DiOC_{18}$, see Appendix 2 of the book for structure and data on this probe) raise the miscibility transition temperature in dispersions of ternary mixtures containing cholesterol (Veatch et al., 2007). All these problems can be avoided if proper care is taken during the preparation procedures and acquisition of the images (Morales-Penningston et al., 2010). Control experiments designed to check for the presence of oxidative species are highly recommended if GUVs contain unsaturated lipid species, which are very susceptible to oxidation.

10.3.2 WIDE-FIELD FLUORESCENCE MICROSCOPY

This technique, which can be easily combined with phase-contrast or DIC, allows for easily achieving fluorescence images of GUVs labeled with different fluorophores. Wide-field fluorescence microscopy permits very rapid acquisition using cameras, allowing for live imaging of the specimen. This last feature, which is not available in scanning confocal fluorescence microscopes, warrants high temporal resolution to explore dynamical processes occurring in a GUV membrane (e.g., an enzymatic action).

An important difference with respect to more sophisticated fluorescence techniques—such as laser scanning confocal and multiphoton excitation fluorescence microscopy—is the inability of wide-field fluorescence microscopy to produce optical sectioning of the specimen (see Box 10.1 for a comparison). This limitation is reflected in a relatively important contribution of out-of-focus fluorescence in the images. This situation can be disadvantageous if many vesicles are in close contact or if one is interested in membrane budding or lipid tubes emerging from a GUV membrane. This problem could be somewhat solved by using deconvolution routines after image acquisition (e.g., deblurring [Sibarita, 2005]).

In connection to the last paragraph of the Section 10.3.1, it is important to recall that wide-field fluorescence microscopy uses relatively high-power excitation sources (generally mercury lamps), illuminating a relatively big field of view. Prolonged illumination of labeled GUVs containing unsaturated lipid species may result in undesired photo-oxidation effects mediated by fluorescent probes (particularly some of those probes belonging to the first family of probes; see Section 10.3.1). In this particular case, controlled time of exposure (or power) to the excitation lamp are highly recommended. In addition, although doable, the use of UV-excited fluorophores under wide-field fluorescence microscopy can be sometimes problematic. The high energy delivered by the UV excitation light (particularly from mercury lamps) can easily induce photobleaching of the fluorophores, making it difficult to perform time-lapse experiments or repeated images of the same GUV. The use of UV diodes with simultaneous control of the excitation light's exposure time on the sample can partially overcome this problem.

Wide-field fluorescence microscopy has been regularly applied to visualize single GUVs displaying membrane lateral heterogeneity (see Chapter 18). In fact, several studies exploring and characterizing domain coexistence in different lipid mixtures were reported using this technique (Veatch and Keller, 2002, 2003, 2005). Some of these studies explored phase separation in lipid mixtures generally composed of unsaturated phospholipids, single or natural mixtures of sphingomyelin (SM; *N*-hexadecanoyl-D-*erythro*-sphingosylphosphorylcholine, see Appendix 1 of the book for structure and data on this lipid) and cholesterol (although in some cases SM has been replaced by 1,2-dihexadecanoyl-*sn*-glycero-3-phosphocholine [DPPC, see Appendix 1 of the book for structure and data on this lipid]) (Veatch and Keller, 2002, 2003, 2005) or mixtures containing other sphingolipids (e.g., cerebrosides) (Lin et al., 2007). A representative example of a wide-field fluorescence image obtained in GUVs composed of an equimolar mixture of 1,2-dioctadecenoyl-*sn*-glycero-3-phosphocholine (DOPC, see Appendix 1 of the book for structure and data on this lipid), SM, and cholesterol is shown in Figure 10.3. Notice that this mixture displays liquid immiscibility (Dietrich et al., 2001) (see Chapter 18).

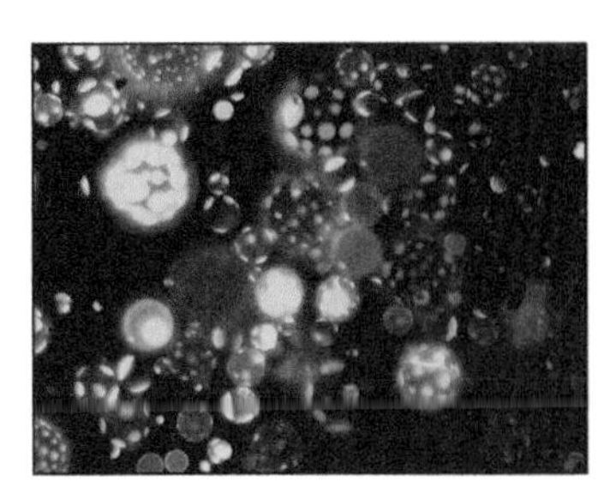

Figure 10.3 Wide-field fluorescence microscopy images of GUVs composed of DOPC/sphingomyelin/cholesterol (1:1:1 mol) showing coexistence of liquid ordered and liquid disordered phases. The fluorescent probe is rhodamine-DPPE (0.1% mol). The largest vesicles in the image have a diameter of about 50 μm.

To finalize this section, it is worth mentioning that this technique is very suitable for experiments in planar supported bilayers or Langmuir films (Brewer et al., 2010, 2017) because the specimen itself confines the fluorescence to a single plane of illumination.

10.3.3 LASER SCANNING CONFOCAL FLUORESCENCE MICROSCOPY

LSCFM is currently a valuable tool for a broad range of scientific research in the biological and medical sciences for imaging thin optical sections in living and fixed specimens ranging in thickness up to 100 micrometers (Claxton et al., 2006). The basic concept of confocal microscopy was originally developed by Marvin Minsky in the mid-1950s (Minsky, 1988). The basic key to the confocal approach is the use of spatial filtering techniques (pinholes) to eliminate out-of-focus fluorescence light in specimens whose thickness exceeds the immediate plane of focus. Thus, unlike wide-field fluorescence microscopy, this technique allows sectioning of the specimen of interest (Claxton et al., 2006), see Box 10.1 for a comparison. This capability allows for acquiring different sections of the specimen along the axial direction (a "z stack") allowing its 3D reconstruction using digital image analysis. An example illustrating this principle is shown in Figure 10.4 for a fluorescently labeled GUV.

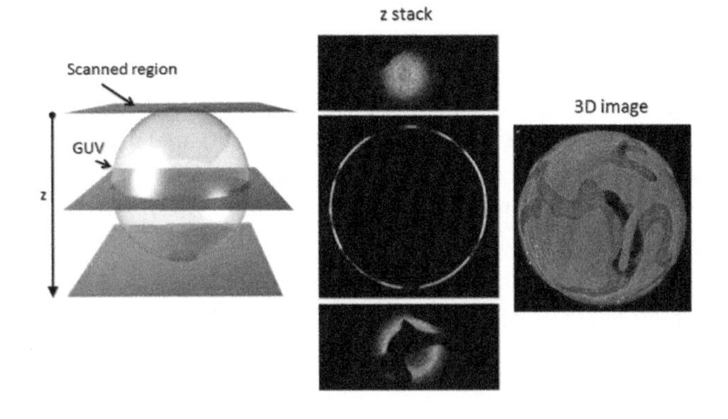

Figure 10.4 Laser scanning fluorescence confocal microscopy of a GUV (left), showing selected sections (fluorescent images) along the axial direction (middle). The stack of fluorescent image is used to reconstruct a 3D picture of the GUV (right). Notice the presence of two distinct regions (domains) in the lateral plane of the membrane. GUV composition is POPC/ceramide 7:3 mol, labeled with $DiIC_{18}$ (0.1 mol %); the vesicle diameter is 35 μm.

In LSCFM, the image of an extended specimen is generated by scanning the focused beam across a defined area in a raster pattern controlled by two high-speed oscillating mirrors driven with galvanometer motors (the scanner). Unlike wide-field fluorescence microscopy, the confocal image of a specimen is reconstructed, point by point, from emission photon signals by the photomultiplier and accompanying electronics, yet never exists as a real image that can be observed through the microscope eyepieces (Claxton et al., 2006). Temporal data can be also gathered using LSCFM either from time-lapse experiments conducted over extended periods (where it is recommended to use an appropriate instrument setup to avoid photobleaching of the probes) or through real-time image acquisition in smaller frames for short periods of time. However, it is important to note that the time required to obtain a frame using LSCFM is primarily limited by how fast the excitation source (i.e., a laser) is scanned on the region of interest. This situation generally limits the time of acquisition of an image—say the equatorial plane of a giant vesicle of ~30 μm diameter—from one to a few seconds depending on the fluorophore quantum yield and the detection efficiency of the set up used to measure (e.g., detectors, filters). Regarding experiments with GUVs, this limitation can in some cases cause difficulties in properly acquiring a z stack, particularly if there is drift of the vesicle during imaging. Systematic drifts can be corrected by available software, but a better manner of dealing with this problem is to immobilize the vesicles onto the surface of the coverslip using the avidin–biotin strategy discussed in Section 10.2.2. In addition, this problem can be worked out by using another related microscopy technique such as spinning disk confocal microscopy. This confocal methodology, which uses a Nipkow disk containing an array of pinholes and microlenses instead of point scanners, allows acquisition of several frames per second (for details, see Toomre et al., 2012) and constitutes a convenient choice when rapid acquisition of image stacks is desired. Alternatively, light sheet fluorescence microscopy (or selective plane illumination fluorescence microscopy [SPIM]) has recently emerged as an attractive option for dynamic experiments in GUVs (Loftus et al., 2013). In this last paper, the authors used this technique to perform accurate measurements of membrane bending moduli from giant vesicles. In light sheet fluorescence microscopy, a laser is formed into a thin sheet that excites fluorophores in one plane of the sample, the emission from which is imaged onto a camera using a perpendicular objective lens (i.e., the technique has sectioning effect capabilities [Keller et al., 2008]). This allows for fast 3D imaging, being also applicable to freely suspended GUVs (Loftus et al., 2013). SPIM was recently implemented in microscopes with conventional sample geometry (i.e., inverted epifluorescence microscopes), in a new method called sideSPIM and applied to fast track lipid domains on time series of 3D images of GUVs showing coexistence of l_d and l_o phases (Hedde et al., 2017) (Figure 10.5). The authors were able to image 3D stacks of 60 planes at 1.25 stacks/s. The exposure time for a single plane was 10 ms, resulting in 600 ms for all 60 planes plus a 200-ms long overhead for repositioning of the piezo stage at the starting position. The authors also showed that after fifty 3D stacks (acquired in ~15 s), the photobleaching was negligible (Hedde et al., 2017).

Another interesting feature of LSCFM is the capability to simultaneously acquire images using different fluorescent probes (practically, two to four colors), which require the combination of

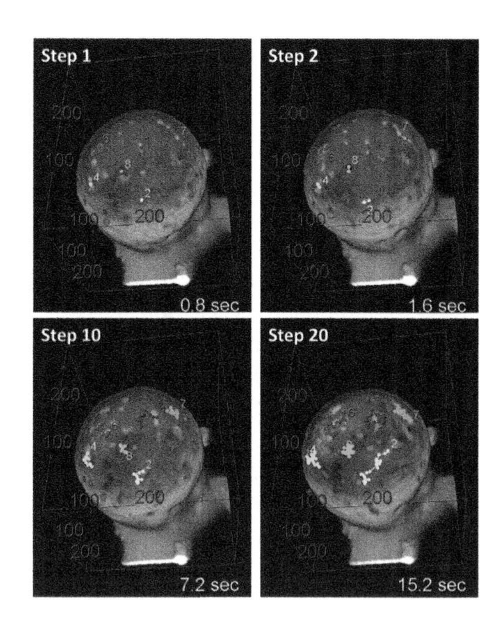

Figure 10.5 3D imaging of GUVs on top of a transparent support obtained with sideSPIM. The rendered fluorescence images are obtained approximately every 800 ms. The $DiIC_{18}$ labeled GUV are composed of a ternary mixture 1:1:1 mol (DOPC, DPPC, and cholesterol) displaying coexistence of liquid ordered/liquid disordered phases. For practical reasons, only four images of the whole time series are shown in the figure. The GUV images also include the trajectory of the free diffusing liquid ordered domains (which excluded $DiIC_{18}$) at different times. The GUV has been immobilized using the biotin-avidin strategy discussed in Section 10.2.2.

multiple excitation laser sources. In addition, commercially available LSCFM systems also permit spectral imaging, that is, acquisition of the emission spectra of fluorophores at the level of single pixels in the image (Claxton et al., 2006). Even though the resolution obtained in LSCFM is somewhat enhanced with respect to conventional wide-field fluorescence (Claxton et al., 2006), it is still in the range of few hundred nanometers (approximately 250 nm radial and 600 nm axial under optimal conditions, that is, using high numerical aperture objectives).

A fairly large number of studies have been reported using LSCFM on GUVs; some of them are addressed in other chapters of this book. For example, the sectioning effect provided by this technique was exploited to perform a time-lapse "leakage" experiment of fluorophores entrapped into GUVs, providing a way of disentangling membrane-destabilization mechanisms mediated by membrane-active peptides (Ambroggio et al., 2005; Tamba and Yamazaki, 2005; Hasper et al., 2006; Henriques et al., 2007 [see Chapter 24]). I will concentrate on one specific example regarding thermodynamic studies of membranes performed in my laboratory using GUVs (see next sub section below). For additional examples, the reader can refer to some of the chapters included in Parts II, IV through VI of this book.

Equilibrium thermodynamic studies on GUVs displaying phase coexistence

It was not until 1999–2001 that seminal papers appeared in the literature using LSCFM or multiphoton excitation fluorescence microscopy to reveal membrane lateral heterogeneity in GUVs composed of single phospholipids, phospholipid binary mixtures, and ternary lipid mixtures containing phospholipids, sphingolipids and cholesterol at different temperatures and composition (Bagatolli and Gratton, 1999; Korlach et al., 1999; Bagatolli and Gratton, 2000a, 2000b; Dietrich et al., 2001; Feigenson and Buboltz, 2001). These papers showed for the first time images of different micrometer-sized lipid domains in bilayers, including dynamic information from the coexisting membrane regions. Particular features of these domains were their shape (e.g., elongated, flower shape, snowflake-like, circular) (Figure 10.6a) and their size dependence on lipid composition and temperature (Bagatolli and Gratton, 1999; Korlach et al., 1999; Bagatolli and Gratton, 2000a, 2000b; Dietrich et al., 2001; Feigenson and Buboltz, 2001). Many subsequent studies exploited wide-field fluorescence microscopy or LSCFM approaches to study the lateral structure of several lipid mixtures. However, a major challenge has been and still is to take the study of GUVs beyond the stage of "pretty pictures." Exploiting advanced digital image analysis of GUVs can assist in achieving this goal.

In fact, novel image analyses have been recently introduced to examine data from GUVs obtained using LSCFM experiments.

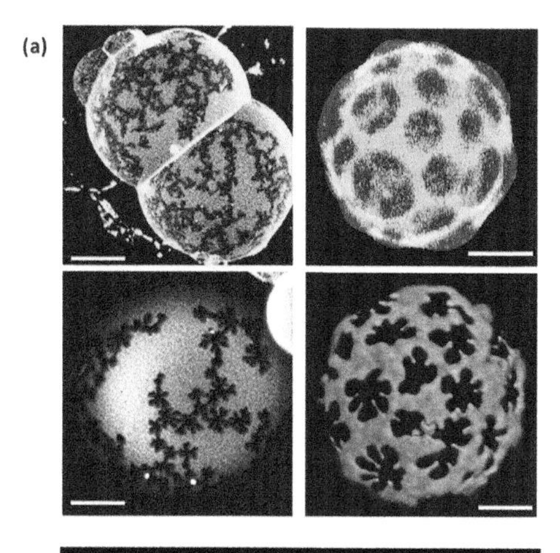

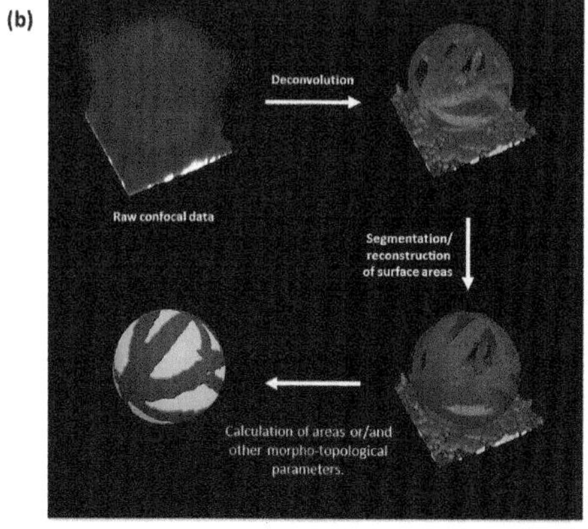

Figure 10.6 (a) Confocal 3D images of GUVs displaying membrane domains. Compositions are: human skin stratum corneum lipid extracts, probe: $DiIC_{18}$ (Plasencia et al., 2007), top left; native pig lung surfactant membranes, probes: $DiIC_{18}$ (red) and Bodipy-PC (green) (Bernardino de la Serna et al., 2004), top right; polar lipid fraction E (PLFE) from the thermoacidophilic archaebacteria Sulfolobus acidocaldarius, probe: Laurdan (Bagatolli et al., 2000c), bottom left; eggSM/egg ceramide (7:3 mol), probe: $DiIC_{18}$ (Sot et al., 2006), bottom right. Scale bars: 20 µm. (b) Sequential image analysis leads to the deconvolution and segmentation of the liquid disordered (green) and the solid ordered (red) phase areas in 2D-slices and to the reconstruction of surface areas in 3D-stacks. The GUV is composed of a DLPC/DPPC mixture (3:7 mol) (for details, see Fidorra et al., 2009). Vesicle diameter: 30 µm.

These procedures allow *quantitative information* on morpho-topological parameters from GUVs displaying domain coexistence. For example, Fidorra et al. (2009) reported a novel analytical procedure for measuring the surface areas of coexisting lipid domains in GUVs. The method is based on 3D image processing of confocal images of GUVs labeled with two fluorescent probes that differentially label the coexisting membrane areas. This procedure involves deconvolution and further segmentation of the images obtained (z stacks of GUVs), followed by 3D reconstruction of the surface of the GUVs, providing information on domain area and perimeter at the level of single vesicles (Figure 10.6b). Measurement of area fractions at different compositions allowed scrutiny of the thermodynamic lever rule from an already known phase diagram. In the work of Fidorra et al. (2009), the lever rule was validated for GUVs composed of DLPC/DPPC mixtures using domain-area information. Importantly, these experiments first confirmed a correspondence between the presence of membrane domains and s_o and l_d equilibrium thermodynamic phases in GUVs. In a comparable way, Juhasz et al. (2009) showed that the lever rule also applies to GUVs composed of DOPC/DPPC/cholesterol using fluorescence microscopy, showing a fairly good correlation with tie lines obtained by NMR in the l_d/ l_o phase coexistence region. Recently, Husen et al. (2012a) demonstrated that the data obtained from LSCFM can be independently analyzed (Husen et al., 2012b) to obtain information on the orientation and length of tie lines in DOPC/DPPC/cholesterol mixtures, without the need of recourse to tie-line information obtained by other methods. This idea has been also explored by Bezlyepkina et al. (2013) for mixtures of DOPC/eggSM/cholesterol. All these findings support the fact that GUVs experiments using LSCFM can be quantitatively exploited for equilibrium thermodynamic studies of freestanding membranes.

10.3.4 LASER SCANNING MULTIPHOTON EXCITATION FLUORESCENCE MICROSCOPY

The application of laser scanning multiphoton excitation fluorescence microscopy in biology was introduced by Denk et al. (1990). This technique still constitutes one of the most promising and fastest developing areas in biological and medical imaging at the optical resolution level (Diaspro, 2002). Multiphoton excitation is a nonlinear process in which a fluorophore absorbs two (or three) infrared photons simultaneously. In the case of two-photon excitation, each photon provides half the energy required for excitation via one photon (1/3 for the case of three-photon absorption). The high photon densities required for multiphoton absorption are achieved by focusing a high peak power laser light source on a diffraction-limited spot through a high numerical aperture objective (Denk et al., 1990; So et al., 1996; Master et al., 1999). Therefore, unlike LSCFM, excitation does not occur in the areas above and below the focal plane because of insufficient photon flux. This phenomenon allows for a sectioning effect without the aid of pinholes, that is, the excitation process provides inherent spatial resolution (Denk et al., 1990; So et al., 1996; Master et al., 1999).

The benefits of multiphoton excitation includes improved background discrimination, reduced photobleaching of fluorophores, minimal photodamage to living cell specimens, high penetration in thick specimens (up to 1 mm thick) and excitation of multiple fluorescent probes using a single excitation wavelength (Denk et al., 1990; So et al., 1995; So et al., 1996; Master et al., 1999; Diaspro, 2002). Practically speaking, the resolution of this technique is comparable to that obtained in LSCFM. In addition, multiphoton excitation also operates by focusing the excitation light across a defined area using scanners. Therefore, all considerations mentioned in the previous section regarding limited time of acquisitions apply as well for this technique. The most common excitation sources for multiphoton excitation are Ti-Saphire™ tunable pulsed lasers (which are infra-red high power lasers that operate in the range of 700 to 1,000 nm, with a pulse width of about 100 fs and a repetition rate of 80–100 MHz).

As mentioned in Section 10.3.2, the use of UV-excited probes using wide field fluorescence microscopy (also LSCFM) is not easy to accomplish. Instead, the characteristics of multiphoton excitation simply allow the use of these fluorescent probes in a microscope. For example UV-excited probes have been applied to explore lateral heterogeneity in model membranes (GUVs experiments [for a comprehensive review, see Bagatolli, 2006]), cellular membranes (Sanchez et al., 2012; Golfetto et al., 2015), enzymatic reactions by lipases (Sanchez et al., 2002; Stock et al., 2012) and the effect of the insertion of peptides in lipid membranes (Fahsel et al., 2002; Janosch et al., 2004). Most of these applications rely on the use of LAURDAN as a fluorescent probe. This fluorescent molecule shows an exquisite sensitivity to solvent dipolar relaxation, which in turn is sensitive to the packing of lipids in membranes (Bagatolli, 2006, 2013).

Brief overview of LAURDAN two-photon excitation imaging in GUVs

Although there are already comprehensive reviews about this topic (e.g., Bagatolli and Gratton, 2001; Bagatolli, 2006, 2013), this section aims to provide key technical details to perform fluorescence measurements using LAURDAN. The main advantages of LAURDAN over many of the fluorescence probes presented in Section 10.3.1 are (i) the negligible contribution of the probe from water (its partition to the membrane is highly favored), (ii) the even partition of LAURDAN in membranes displaying lateral heterogeneity and (iii) the sensitivity of this probe to lipid packing (50 nm emission shift in a s_o to l_d phase transition [Parasassi et al., 1990]). This sensitivity is due to dipolar relaxation processes, which occur during the lifetime of the probe (nanoseconds), caused by the presence of water molecules with restricted mobility in the region where LAURDAN is located in the membrane (Bagatolli, 2013). Most of the applications of LAURDAN in membranes studies rely on measurements of the generalized polarization (GP) function. The GP function was introduced by Parasassi et al. (1990) as an analytical method to quantitatively determine the relative amount of coexisting phases in a membrane and to study their temporal fluctuations. The GP function was originally defined as

$$GP = \frac{I_B - I_R}{I_B + I_R} \tag{10.1}$$

where I_B and I_R are the measured fluorescence intensities under conditions in which wavelengths (or a band of wavelengths) corresponding to the blue and red side of the probe's emission

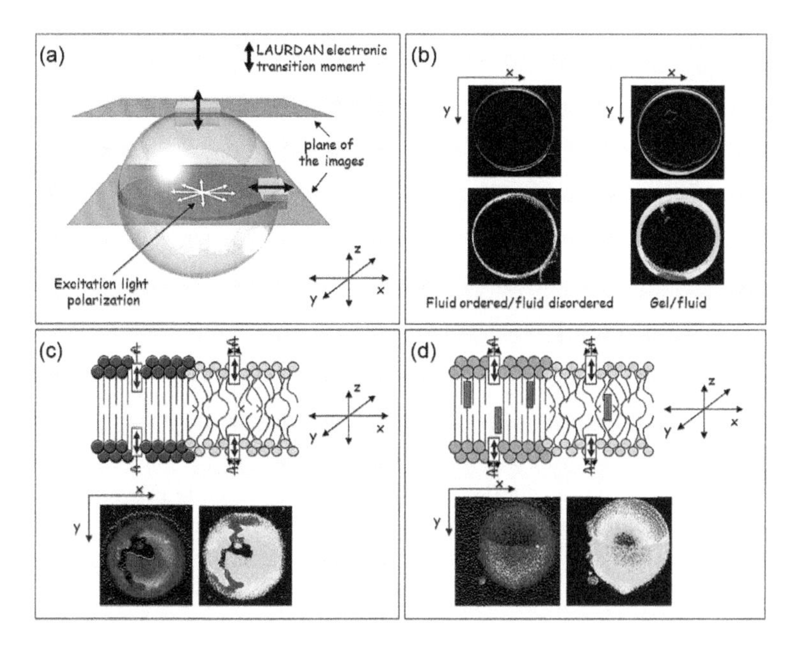

Figure 10.7 The photoselection effect is dictated by the relative orientation of the LAURDAN electronic transition moment with respect to the polarization plane of the excitation light. The last is also influenced by the lipid bilayer phase state (see text). (a) Sketch of the GUV and the position of the LAURDAN electronic transition moment; (b) the photoselection effect does not operate in the equatorial region of the GUV independent of the phase coexistence scenario; (c) gel/fluid phase coexistence observed at the polar region of the GUV composed of ceramide/POPC 1:5 mol (fluorescence intensity and GP), notice the strong photoselection effect observed in the gel phase area with respect to the fluid phase; (d) fluid ordered/fluid disordered phase coexistence observed at the polar region of the GUV (fluorescence intensity and GP) composed of DOPC/SM/cholesterol 1:1:1 mol. The photoselection effect in the fluid ordered phase is less pronounced compared with that observed in the gel phase (compare panels c and d). (Adapted from Bagatolli, L.A., *Biochim. Biophys. Acta*, 1758, 1541–1556, 2006.)

spectrum (B [440 nm] and R [490 nm], respectively) are both observed using a given excitation wavelength (360 nm and 780 nm for one- and two-photon excitation, respectively). This definition corresponds to the classical fluorescence polarization definition (Jameson et al., 2003) if B and R represent two different orientations of the observation polarizers. The advantage of the GP for the analysis of the spectral properties of LAURDAN (Bagatolli, 2013) is related to the well-known properties of the classical polarization function, which contains information on the interconversion between different "states." In the classical fluorescence polarization definition, the states correspond to different orientations of the emitting dipole with respect to the laboratory axis. In the case of the GP the states correspond with the extent of solvent relaxation nearby the probe.

The GP function shows very characteristic values depending on the membrane's phase state. For instance, in the s_o phase, the observed GP values are around 0.5–0.6 (the extent of solvent relaxation in the nanosecond regime is very low), whereas for the l_d phase, it is around 0.1 or below (the extent of solvent relaxation in the nanosecond regime is high). The l_o phase instead shows intermediate values depending on the cholesterol molar fraction in the membrane, which in turn is related to the hydration properties of the membrane (Dietrich et al., 2001).

To measure the GP in the microscope, it is necessary to split the LAURDAN fluorescence signal into two channels. Each channel is coupled with a band-pass filter that selects the fluorescence emission of the blue and red sides of the LAURDAN emission spectrum (Sanchez et al., 2007). Generally, a correction factor (called G-factor) is necessary to compensate potential differences between the two detection channels (Brewer et al., 2010). Another important point to consider when imaging GUVs using LAURDAN is the so-called photoselection effect. The photoselection effect arises from the fact that only those fluorophores that have electronic (absorption) transition moments aligned parallel or nearly so to the plane of polarization of the excitation light are excited, that is, the excitation efficiency is proportional to the cosine to the fourth power (because of the two photon excitation) of the angle between the transition moment of the probe and the polarization plane of the excitation light (Bagatolli, 2006, 2013). This effect depends on the position of the probe's transition moment relative to the plane of polarization of the excitation light (Figure 10.7a). At *the equatorial region* of the vesicle, circularly polarized excitation light[3] allows excitation of all LAURDAN molecules with the same efficiency, independent of the lipid phase scenario present in the vesicle (one single phase or phase coexistence), that is, the probe's electronic transition moment is always parallel to the polarization plane. This allows calculation of the GP without the influence of the photoselection effect (as seen in Figure 10.7b for coexistence of s_o/l_d and l_o/l_d phases). Instead, at *the polar region* of the vesicle, the dominant signal is that coming from the more fluid part of the bilayer (Figure 10.7c). This effect is dictated by the fact that a component of LAURDAN's electronic transition moment is always parallel to the excitation polarization plane (because of the relatively low lipid order). This last phenomenon

[3] Because the laser light is linearly polarized, a quarter wave plate is necessary to circularly polarize the excitation light. Linearly polarized light can also affect the fluorescence signal observed in the GUV equatorial region (see Bagatolli, 2006, for details).

does not take place in the s_o phase because the high lipid lateral order precludes the wobbling movement of LAURDAN, Figure 10.7c. In the case of l_o/l_d phase coexistence, a component of LAURDAN's transition moment (parallel to the excitation light polarization plane) will be present in both phases at the polar region of the GUV, allowing collection of the fluorescence signal from these two regions. In particular, the l_o phase shows a lower impact of the photoselection effect compared with s_o phase regions (compare Figure 10.7c and d). Interestingly, the photoselection effect per se allows qualitative information about lipid phases directly from the intensity images (Bagatolli, 2006) and the spatial orientation of fluorophores in the membrane (Bagatolli and Gratton, 2000b; Bernchou et al., 2009).

To finalize this Part I would like to briefly comment on a new way to analyze LAURDAN fluorescence based on the Fourier transformation of the probe's emission spectrum. This new method, called spectral phasor analysis, is emerging as a powerful tool to analyze data from cellular and model membranes (Golfetto et al., 2015; Malacrida et al., 2015). The Fourier transformation of LAURDAN spectra allows one to determine, with a simple visual inspection, the presence of complex interactions in the LAURDAN/membrane system without the assumption of a particular model as required by the classical GP function (Golfetto et al., 2015; Malacrida et al., 2015).

Presence of lipid domains in specialized biological membranes

An interesting example that uses multiphoton excitation fluorescence microscopy to evaluate the potential existence of coexisting thermodynamic phases in biological membranes pertains to skin stratum corneum lipid membranes. These membranes are composed of saturated ceramides with very long chains, cholesterol, and long-chain free fatty acids (lacking glycerophospholipids and SM). LAURDAN two-photon excitation fluorescence microscopy experiments using excised pig, mice, and human skin show that these membranes display a gel-like character, being—to the best of my knowledge—the only functional biological membrane exhibiting this type of stiff organization (Carrer et al., 2008; Bloksgaard et al., 2012a, 2012b; Iwai et al., 2012). Hydrated bilayers (free standing giant structures) composed of lipid mixtures extracted from human skin were directly visualized using confocal and multiphoton excitation fluorescence microscopy techniques (Plasencia et al., 2007). At skin physiological temperatures (28°C–32°C), the state of these hydrated bilayers corresponds microscopically (radial resolution limit 300 nm) to a single gel phase at pH 7. However, coexistence of two distinct micrometer-sized gel-like lipid domains is observed between pH 5 and 6 (Figure 10.6a, top left), and no fluid phase is observed at the pH range explored (5–8). This observation suggests that the proton activity gradient existing in the stratum corneum could distinctly influence the physical properties of the extracellular lipid matrix, impacting membrane lateral structure and stability (Plasencia et al., 2007). Local equilibrium conditions may be asserted in this system because a slow molecular turnover is expected after the lipids reach the stratum corneum (upon secretion from lamellar bodies contained in specialized cells). Similar studies using this methodology have been performed in GUVs composed of native pulmonary surfactant membranes where liquid immiscibility has been observed at physiological temperatures and related to the functional state of the material (Bernardino de la Serna et al., 2004, 2009, 2013).

10.3.5 FLUCTUATION CORRELATION SPECTROSCOPY-BASED TECHNIQUES

FCS based techniques have been extensively applied to study the lateral diffusion of fluorescent probes in models and biological membranes (Korlach et al., 1999; Ruan et al., 2004; Sanchez and Gratton, 2005; Chiantia et al., 2006; Kahya and Schwille, 2006; Machan and Hof, 2010). Given that this book contains a full chapter fully dedicated to the principles of FCS (providing also examples in GUVs and practical tips about the experiments; see Chapter 21), I will briefly state some generalities about some related methods and provide only a few examples of applications in GUVs. As a starting point, it is important to mention that fluctuation correlation spectroscopy can be performed both using LSCFM or multiphoton excitation fluorescence microscopy setups. Although FCS exploits fluorescence intensity temporal fluctuations acquired at a single point, there are other techniques that exploit fluorescence fluctuations obtained along trajectories or even correlating different pixels in images obtained by laser scanning fluorescence microscopy techniques. For example, it is possible to perform a very fast scan of the excitation beam along a trajectory (e.g., a circular shape) to obtain temporal fluctuations of the specimen of interest. This method, called scanning-FCS, was successfully applied in GUVs to study membrane–protein interactions (Ruan et al., 2004) and to explore the *temporal evolution* of domains in membranes (Celli et al., 2008). For example, this last study found that just below the l_d to l_d/s_o phase transition temperature in DPPC/DLPC mixtures, coupling between the two leaflets of the bilayer was observed to begin within the first 5 min after the onset of phase separation (Celli et al., 2008). Other study using scanning-FCS allowed detecting sub-resolution structural fluctuations in membranes composed of three different single lipid species displaying l_d phase. This was achieved by measuring fluctuation of the LAURDAN GP function on the scale of a few pixels (50-nm pixels) (Celli and Gratton, 2010). This result was interpreted as evidence of an underlying microscale structure of the membrane in which water is not uniformly distributed at the micrometer scale (Celli and Gratton, 2010) supporting the presence of dynamic heterogeneity in a single phase membrane near a phase transition.

Alternatively, the scanning process used to generate an image in a regular laser scanning fluorescence microscopy system can also be exploited to measure temporal fluctuations of fluorescence. This method, called raster imaging correlation spectroscopy (RICS) (Brown et al., 2008), has been successfully applied to study diffusion of membrane probes in models (i.e., GUVs) and natural membranes (Gielen et al., 2009). RICS offers a much wider dynamic range compared with FCS, fluorescence recovery after photobleaching and single-particle tracking separately (which are alternative ways of measuring diffusion of probes; see Chapter 21), allowing also for spatial mapping of dynamic properties.

Finally, studies in model (planar membranes) and cell membranes combining FCS with stimulated emission depletion (STED) microscopy were reported a few years ago (Eggeling et al., 2009; Honigmann et al., 2013; Mueller et al., 2013). STED microscopy is a super-resolution fluorescence microscopy technique that improves the resolution obtained in a regular fluorescence microscope (wide-field, LSCFM, multiphoton) to about tens of nanometers. STED operates using a scanning system that combined two laser sources; one to excite the fluorophore as in regular confocal microscopy and the other to deplete the probe fluorescent emission in specific regions of the sample while leaving a center small focal spot active to emit fluorescence (Hell, 2007). The combination of STED with FCS allowed monitoring diffusion in membranes with a better spatial resolution than the techniques described above. By performing STED-FCS experiments in cellular plasma membranes, Eggeling et al. reported that, unlike phosphoglycerolipids, sphingolipids and glycosylphosphatidylinositol-anchored proteins are transiently (approximately 10–20 ms) trapped in cholesterol-mediated molecular complexes dwelling within <20-nm diameter areas (Eggeling et al., 2009).

10.4 APPLICATIONS OF SECOND HARMONIC GENERATION AND COHERENT ANTI-STOKES RAMAN SCATTERING MICROSCOPY ON GUVs

As discussed at the beginning of Section 10.3.4, molecular excitation of fluorophores by the absorption of two or more photons can often be advantageous for imaging as in the case of multiphoton excitation fluorescence microscopy. Nonlinear microscopy can, however, be extended to the use of multi-harmonic light wherein the energy of incident photons instead of being absorbed by a molecule is scattered via a process of harmonic up-conversion (Moreaux et al., 2001). This last physical process is the ground operating in second harmonic generation (SHG) microscopy. Particular molecules displaying a non-centrosymmetrical symmetry are necessary to produce this phenomenon. Although there are naturally occurring molecules suitable for performing SHG microscopy experiments (collagen for example), in the particular case of membranes the use of probes is required. An example of an SHG probe is di-4-ANEPPDHQ (also used as a fluorescence probe) (Figure 10.2b) where SHG images where acquired in GUVs displaying l_o/l_d phase coexistence (Jin et al., 2005). Another example is the dye Di-6-ASPBS that has been shown to be useful in detecting close interaction between membranes, as well as flip-flop dynamics (Moreaux et al., 2001). A characteristic of SHG is the wavelength of the scattered signal, which is half of the wavelength used to excite the specimen. The microscope setup to perform second harmonic imaging is very similar to that for multiphoton fluorescence microscopy. For instance, Ti-Saphire high-power pulsed lasers are used as excitation sources and the use of a scanner is required to produce an image (Moreaux et al., 2001).

Another emerging nonlinear imaging technique useful to image lipid membranes is called coherent anti-Stokes Raman scattering (CARS) microscopy. CARS is a dye-free microscopy technique that images structures by displaying the characteristic intrinsic vibrational contrast of their molecules. In the case of lipids, the strong resonant Raman signal of the C–H stretching vibration is exploited. CARS is generated by two tightly co-focused beams (pump and Stokes) that are scanned over the sample so that the image is created point-by-point using filtered out CARS signal. Due to the nonlinear nature of CARS, an appreciable amount of signal is generated only in a small focal volume (typically less than 1 μm in diameter), which allows for high lateral and axial resolution (Toytman et al., 2009). CARS images have been performed in giant unilamellar vesicles (Potma and Xie, 2003). These experiments have also provided, for example, information on intermembrane distance beyond the diffraction-limited resolution of the microscope.

10.5 COMBINING FLUORESCENCE MICROSCOPY OF RUPTURED GUVs WITH ATOMIC FORCE MICROSCOPY

One of the limitations of optical imaging of membranes is the inability to easily access to the nanometer-length scale. This situation is due to the diffraction limit of light. Although several papers reported on high-resolution images obtained using super-resolution techniques (e.g., STED or other techniques such as photoactivated localization microscopy [PALM], or stochastic optical reconstruction microscopy [STORM]) of proteins in cells (e.g., actin cytoskeleton), to the best of my knowledge there are no images performed in GUVs using these methods (although some STED images were reported in planar supported membranes [see Honigmann et al., 2013]). In the particular case of STED, problems generated by the high power delivered by the lasers on individual GUVs may prevent proper image acquisition.

High-resolution images of membranes from GUVs where recently reported by combining fluorescence imaging with atomic force microscopy (AFM) (Bhatia et al., 2014). Specifically, this paper reports on the phenomenon of dynamic heterogeneity in membranes. These authors devised a methodology to fix and image (using both confocal fluorescence microscopy and AFM) dynamic fluid domain patterns of GUVs composed of DOPC/DPPC/cholesterol, which display l_o/l_d coexistence. Briefly, the individual GUVs were transferred to a solid support and rapidly ruptured using a pulse of Mg^{2+}. This procedure generates planar bilayer patches (Figure 10.8a), which are taken to represent a fixed state of the freestanding membrane, where lateral domain structures are kinetically trapped. High-resolution images of domain patterns (corresponding to the l_o and l_d co-existence region in the phase-diagram of ternary lipid mixtures) are revealed by AFM[4] scans of the membrane patches (Figure 10.8b). Using this strategy the authors demonstrated not only that macroscopic phase separation—as known from fluorescence images—superimposed fluctuations in the form of nanoscale domains of the l_o and l_d phases, but also that the size of the fluctuating domains increases as the composition approaches the critical point (Bhatia et al., 2014). Interestingly, this dynamic heterogeneity is also detected even deep in the l_o/l_d phase coexistence

[4] It is important to recall that AFM permits membrane images with much superior x–y and z (axial) resolution than optical microscopy, in the order of very few nanometers; see also Chapter 12 of this book for more details and applications of this technique.

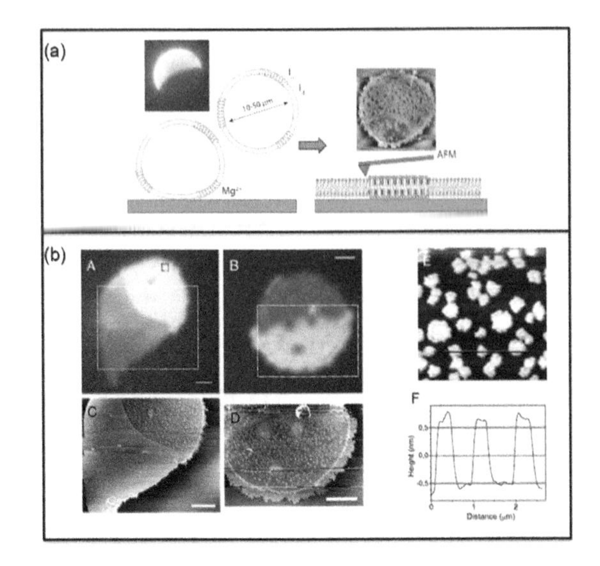

Figure 10.8 (a) Sketch indicating the deposition of fluorescently labeled giant vesicles onto a solid support. The membrane patches are subsequently imaged using fluorescence microscopy and AFM. (b) Small-scale domain patterns coexisting with macroscopic liquid disordered and liquid ordered phase separation in two patches with composition DOPC:DPPC:cholesterol (3:5:2 mol). wide-field fluorescence microscopy images (A, B) and matching AFM topography images (C, D, E). The regions of the AFM images are indicated with squares in panel labels A and B. The topography profile in (F) corresponds to the line (red) in (E). Scale bar: 10 µm. (Adapted from Bhatia, T. et al., *Biochim. Biophys. Acta*, 1838, 2503–2510, 2014.)

Box 10.2 Comprehensive summary of different microscopy techniques used to image GUVs

IMAGING METHOD	LABEL	IMAGE ACQUISITION (SPEED)	SECTIONING EFFECT	APPLICATIONS IN GUVs
Bright-field	No	Fast (ms)[a]	No	Observation of GUVs during/after preparation
Phase-contrast	No	Fast (ms)[a]	No	Observation of GUVs during/after preparation, permeability experiments, membrane deformation
Differential interference contrast	No	Fast (ms)[a]	No	Observation of GUVs during/after preparation, permeability experiments, membrane deformation, lamellarity
Wide-field fluorescence	Yes	Fast (ms)[a]	No	Observation of GUVs during/after preparation, permeability experiments, membrane domains, membrane deformation
Scanning confocal fluorescence	Yes	Slow (s)	Yes	Membrane domains, permeability, lipid–protein interactions, membrane deformation
Spinning disk fluorescence	Yes	Fast (ms)[a]	Yes	Lateral domains, permeability, lipid–protein interactions, membrane deformation
Light sheet illumination fluorescence	Yes	Fast (ms)[a]	Yes	Vesicle fluctuations analysis (so far)
Scanning multiphoton excitation fluorescence	Yes	Slow (s)	Yes	Membrane domains, permeability, lipid–protein interactions, membrane deformation
Second harmonic generation	Yes	Slow (s)	Yes	Membrane domains, flip-flop dynamics
CARS	No	Slow (s)	Yes	Intermembrane distance (so far)
Fluorescence based fluctuations techniques (FCS, RICS, scanning FCS)	Yes	N/A	Yes	Membrane dynamics (lateral diffusion of membrane domains, lipids and proteins). Broad temporal range (nanoseconds to minutes).

[a] Fast acquisition is possible using cameras. Resolution in all these techniques is limited by the diffraction limit of light. References for applications are provided in the chapter's main text.

region of the mixture. Importantly, this work demonstrated an excellent agreement between the area fraction of domains in intact GUVs and the membrane patches (i.e., using the LSCFM approach discussed in Section 3.3), supporting the assumption that the thermodynamic state of the membrane remains stable (Bhatia et al., 2014). This interesting approach is not limited to specific lipid compositions and could also potentially help to uncover lateral structure information in highly complex membranes.

10.6 CONCLUSIONS AND PERSPECTIVES

The different methods discussed in this chapter provide a broad range of possibilities to perform membrane related experiments using GUVs (see Box 10.2 for a comprehensive summary).

The spatially resolved information obtained with optical microscopy at the level of single vesicles is one of the most interesting advantages of these methods, allowing information that covers several length (restrained to hundreds of nanometers by the difraction limit of light) and time scales, that is, from morphological changes at the level of single vesicles to supramolecular or even molecular events in the lateral plane of the membrane. In particular, fluorescence methods offer a great variety of applications, not solely based in fluorescence intensity measurements but also when other fluorescence parameters are acquired. Examples are fluorescence lifetime or fluorescence polarization leading to other microscopy techniques not discussed here (e.g., FLIM [de Almeida et al., 2007], fluorescence polarization microscopy [Ariola et al., 2009]). Emerging techniques such as CARS (label free) and SHG open new alternatives to explore membrane related events using GUVs. Perhaps one of the interesting challenges of optical microscopy-based methodologies is to provide images of membranes below the diffraction limit of light (which limits the resolution of all the techniques discussed in this chapter to ~250 nm). Although several super-resolution techniques are available (e.g., STED, STORM, PALM) it is obvious that these methods still require further optimization for imaging freestanding membrane models such as GUVs.

ACKNOWLEDGMENTS

The author would like to thank Dr. Roberto Stock for the critical reading of the manuscript, Dr. Rosangela Itri and Gustavo Scanavachi for providing Figure 10.4, and Dr. Leonel Malacrida for providing Figure 10.5.

LIST OF ABBREVIATIONS

AFM	atomic force microscopy
C-LAURDAN	6-dodecanoyl-2-[N-methyl-N-(carboxymethyl)amino]naphthalene
CARS	coherent anti-Stokes Raman scattering
DIC	differential interference contrast microscopy
$DiIC_{12}$	1,1′-didodecyl-3,3,3′,3′-tetramethylindocarbocyanine perchlorate
$DiIC_{18}$	1,1′-dioctadecyl-3,3,3′,3′-tetramethylindocarbocyanine perchlorate
$DiOC_{18}$	3,3′-dioctadecyloxacarbocyanine perchlorate
DOPC	1,2-dioctadecenoyl-*sn*-glycero-3-phosphocholine
DPH	1,6-diphenyl-hexa-1,3,5-triene
DPPC	1,2-dihexadecanoyl-*sn*-glycero-3-phosphocholine
DSC	differential scanning calorimetry
EPR	electron paramagnetic resonance
FCS	fluorescence correlation spectroscopy
FLIM	fluorescence lifetime imaging
FRAP	fluorescence recovery after photobleaching
LAURDAN	6-dodecanoyl-2-dimethylaminonaphthalene
LSCFM	laser scanning confocal fluorescence microscopy
LUVs	large unilamellar vesicles
MLVs	multilamellar vesicles
NMR	nuclear magnetic resonance
PALM	photoactivated localization microscopy
Parinaric acid	(9*Z*,11*E*,13*E*,15*Z*)-octadeca-9,11,13,15-tetraenoic acid
POPC	1-hexadecanoyl-2-(9Z-octadecenoyl)-*sn*-glycero-3-phosphocholine
PRODAN	6-propionyl-2-dimethylamino-naphthalene
Rhodamine-DPPE	1,2-dihexadecanoyl-*sn*-glycero-3-phosphoethanolamine
RICS	raster image correlation spectroscopy
SHG	second harmonic generation
SM	*N*-hexadecanoyl-D-*erythro*-sphingosylphosphorylcholine
SPIM	selective plane illumination microscopy
SPT	single particle tracking
STED	stimulated emission depletion
STORM	stochastic optical reconstruction microscopy
SUVs	small unilamellar vesicles
TMA-DPH	1-(4-(trimethylamino)phenyl)-6-phenylhexa-1,3,5-triene
UV	ultraviolet

GLOSSARY OF SYMBOLS

GP	generalized polarization
I_B	fluorescence intensity at 440 nm
I_R	fluorescence intensity at 490 nm
l_o	liquid ordered phase
l_d	liquid disordered phase
s_o	solid ordered phase

REFERENCES

Ambroggio EE, Separovic F, Bowie JH, Fidelio GD, Bagatolli LA (2005) Direct visualization of membrane leakage induced by the antibiotic peptides: Maculatin, citropin, and aurein. *Biophys J* 89:1874–1881.

Ariola FS, Li Z, Cornejo C, Bittman R, Heikal AA (2009) Membrane fluidity and lipid order in ternary giant unilamellar vesicles using a new BODIPY-cholesterol derivative. *Biophys J* 96:2696–2708.

Bacia K, Schuette CG, Kahya N, Jahn R, Schwille P (2004) SNAREs prefer liquid-disordered over "raft" (liquid-ordered) domains when reconstituted into giant unilamellar vesicles. *J Biol Chem* 279:37951–37955.

Bagatolli LA (2006) To see or not to see: Lateral organization of biological membranes and fluorescence microscopy. *Biochim Biophys Acta* 1758:1541–1556.

Bagatolli LA (2009) Lipid membrane technology. In: *Encyclopedia of Applied Biophysics* (Bohr H, ed), pp. 711–740. Berlin, Germany: Wiley-VCH Verlag GmbH & Co.

Bagatolli LA (2013) LAURDAN fluorescence properties in membranes: A journey from the fluorometer to the microscope. In: *Fluorescent Methods to study Biological Membranes* (Mely Y, Duportail G, eds), pp. 3–36. Heiderlberg, New York, Dordrecht, London: Springer.

Bagatolli LA, Gratton E (1999) Two-photon fluorescence microscopy observation of shape changes at the phase transition in phospholipid giant unilamellar vesicles. *Biophys J* 77:2090–2101.

Bagatolli LA, Gratton E (2000a) A correlation between lipid domain shape and binary phospholipid mixture composition in free standing bilayers: A two-photon fluorescence microscopy study. *Biophys J* 79:434–447.

Bagatolli LA, Gratton E (2000b) Two photon fluorescence microscopy of coexisting lipid domains in giant unilamellar vesicles of binary phospholipid mixtures. *Biophys J* 78:290–305.

Bagatolli LA, Gratton E (2001) Direct observation of lipid domains in free standing bilayers using two-photon excitation fluorescence microscopy. *J Fluoresc* 11:141–160.

Bagatolli LA, Gratton E, Khan TK, Chong PL (2000c) Two-photon fluorescence microscopy studies of bipolar tetraether giant liposomes from thermoacidophilic archaebacteria Sulfolobus acidocaldarius. *Biophys J* 79:416–425.

Bagatolli LA, Needham D (2014) Quantitative optical microscopy and micromanipulation studies on the lipid bilayer membranes of giant unilamellar vesicles. *Chem Phys Lipids* 181:99–120.

Baumgart T, Hunt G, Farkas ER, Webb WW, Feigenson GW (2007) Fluorescence probe partitioning between Lo/Ld phases in lipid membranes. *Biochim Biophys Acta* 1768:2182–2194.

Bernchou U, Brewer J, Midtiby HS, Ipsen JH, Bagatolli LA, Simonsen AC (2009) Texture of lipid bilayer domains. *J Am Chem Soc* 131:14130–14131.

Bezlyepkina N, Gracia RS, Shchelokovskyy P, Lipowsky R, Dimova R (2013) Phase diagram and tie-line determination for the ternary mixture DOPC/eSM/cholesterol. *Biophys J* 104:1456–1464.

Bhatia T, Husen P, Ipsen JH, Bagatolli LA, Simonsen AC (2014) Fluid domain patterns in free-standing membranes captured on a solid support. *Biochim Biophys Acta* 1838:2503–2510.

Bloksgaard M, Bek S, Marcher AB, Neess D, Brewer J, Hannibal-Bach HK, Helledie T et al. (2012b) The acyl-CoA binding protein is required for normal epidermal barrier function in mice. *J Lipid Res* 53:2162–2174.

Bloksgaard M, Svane-Knudsen V, Sorensen JA, Bagatolli L, Brewer J (2012a) Structural characterization and lipid composition of acquired cholesteatoma: A comparative study with normal skin. *Otol Neurotol* 33:177–183.

Bouvrais H, Cornelius F, Ipsen JH, Mouritsen OG (2012) Intrinsic reaction-cycle time scale of Na^+,K^+-ATPase manifests itself in the lipid-protein interactions of nonequilibrium membranes. *Proc Natl Acad Sci USA* 109:18442–18446.

Bouvrais H, Pott T, Bagatolli LA, Ipsen JH, Meleard P (2010) Impact of membrane-anchored fluorescent probes on the mechanical properties of lipid bilayers. *Biochim Biophys Acta* 1798:1333–1337.

Brewer J, Bernardino de la Serna J, Wagner K, Bagatolli LA (2010) Multiphoton excitation fluorescence microscopy in planar membrane systems. *Biochim Biophys Acta* 1798:1301–1308.

Brewer J, Thoke HS, Stock RP, Bagatolli LA (2017) Enzymatic studies on planar supported membranes using a widefield fluorescence LAURDAN Generalized Polarization imaging approach. *Biochim Biophys Acta* 1859: 888–895

Brown CM, Dalal RB, Hebert B, Digman MA, Horwitz AR, Gratton E (2008) Raster image correlation spectroscopy (RICS) for measuring fast protein dynamics and concentrations with a commercial laser scanning confocal microscope. *J Microsc* 229:78–91.

Carrer DC, Vermehren C, Bagatolli LA (2008) Pig skin structure and transdermal delivery of liposomes: a two photon microscopy study. *J Control Release* 132:12–20.

Celli A, Beretta S, Gratton E (2008) Phase fluctuations on the micron-submicron scale in GUVs composed of a binary lipid mixture. *Biophys J* 94:104–116.

Celli A, Gratton E (2010) Dynamics of lipid domain formation: Fluctuation analysis. *Biochim Biophys Acta* 1798:1368–1376.

Chiantia S, Ries J, Kahya N, Schwille P (2006) Combined AFM and two-focus SFCS study of raft-exhibiting model membranes. *Chemphyschem* 7:2409–2418.

Claxton NS, Fellers TJ, Davidson MW (2006) Microscopy, confocal. In: *Encyclopedia of Medical Devices and Instrumentation*. John Wiley & Sons, Ltd.

de Almeida RF, Borst J, Fedorov A, Prieto M, Visser AJ (2007) Complexity of lipid domains and rafts in giant unilamellar vesicles revealed by combining imaging and microscopic and macroscopic time-resolved fluorescence. *Biophys J* 93:539–553.

de la Serna JB, Hansen S, Berzina Z, Simonsen AC, Hannibal-Bach HK, Knudsen J, Ejsing CS, Bagatolli LA (2013) Compositional and structural characterization of monolayers and bilayers composed of native pulmonary surfactant from wild type mice. *Biochim Biophys Acta* 1828:2450–2459.

de la Serna JB, Oradd G, Bagatolli LA, Simonsen AC, Marsh D, Lindblom G, Perez-Gil J (2009) Segregated phases in pulmonary surfactant membranes do not show coexistence of lipid populations with differentiated dynamic properties. *Biophys J* 97:1381–1389.

de la Serna JB, Perez-Gil J, Simonsen AC, Bagatolli LA (2004) Cholesterol rules: Direct observation of the coexistence of two fluid phases in native pulmonary surfactant membranes at physiological temperatures. *J Biol Chem* 279:40715–40722.

Demchenko AP, Mely Y, Duportail G, Klymchenko AS (2009) Monitoring biophysical properties of lipid membranes by environment-sensitive fluorescent probes. *Biophys J* 96:3461–3470.

Denk W, Strickler JH, Webb WW (1990) Two-photon laser scanning fluorescence microscopy. *Science* 248:73–76.

Diaspro A (2002) Confocal and two photon microscopy. In: *Foundations, Aplications and Advances*. New York: Wiley-Liss.

Dietrich C, Bagatolli LA, Volovyk ZN, Thompson NL, Levi M, Jacobson K, Gratton E (2001) Lipid rafts reconstituted in model membranes. *Biophys J* 80:1417–1428.

Dimova R, Aranda S, Bezlyepkina N, Nikolov V, Riske KA, Lipowsky R (2006) A practical guide to giant vesicles. Probing the membrane nanoregime via optical microscopy. *J Phys Condens Matter* 18:S1151–S1176.

Döbereiner HG, Evans E, Kraus M, Seifert U, Wortis M (1997) Mapping vesicle shapes into the phase diagram: A comparison of experiment and theory. *Phys Rev E* 55:4458–4474.

Eggeling C, Ringemann C, Medda R, Schwarzmann G, Sandhoff K, Polyakova S, Belov VN et al. (2009) Direct observation of the nanoscale dynamics of membrane lipids in a living cell. *Nature* 457:1159–1162.

Fahsel S, Pospiech EM, Zein M, Hazlet TL, Gratton E, Winter R (2002) Modulation of concentration fluctuations in phase-separated lipid membranes by polypeptide insertion. *Biophys J* 83:334–344.

Feigenson GW, Buboltz JT (2001) Ternary phase diagram of dipalmitoyl-PC/dilauroyl-PC/cholesterol: nanoscopic domain formation driven by cholesterol. *Biophys J* 80:2775–2788.

Fidorra M, Duelund L, Leidy C, Simonsen AC, Bagatolli LA (2006) Absence of fluid-ordered/fluid-disordered phase coexistence in ceramide/POPC mixtures containing cholesterol. *Biophys J* 90:4437–4451.

Fidorra M, Garcia A, Ipsen JH, Hartel S, Bagatolli LA (2009) Lipid domains in giant unilamellar vesicles and their correspondence with equilibrium thermodynamic phases: A quantitative fluorescence microscopy imaging approach. *Biochim Biophys Acta* 1788:2142–2149.

Fischer A, Oberholzer T, Luisi PL (2000) Giant vesicles as models to study the interactions between membranes and proteins. *Biochim Biophys Acta* 1467:177–188.

Garda HA, Bernasconi AM, Brenner RR (1994) Possible compensation of structural and viscotropic properties in hepatic microsomes and erythrocyte membranes of rats with essential fatty acid deficiency. *J Lipid Res* 35:1367–1377.

Gielen E, Smisdom N, vandeVen M, De Clercq B, Gratton E, Digman M, Rigo JM, Hofkens J, Engelborghs Y, Ameloot M (2009) Measuring diffusion of lipid-like probes in artificial and natural membranes by raster image correlation spectroscopy (RICS): Use of a commercial laser-scanning microscope with analog detection. *Langmuir* 25:5209–5218.

Golfetto O, Hinde E, Gratton E (2015) The Laurdan spectral phasor method to explore membrane micro-heterogeneity and lipid domains in live cells. *Methods Mol Biol* 1232:273–290.

Haluska CK, Baptista MS, Fernandes AU, Schroder AP, Marques CM, Itri R (2012) Photo-activated phase separation in giant vesicles made from different lipid mixtures. *Biochim Biophys Acta* 1818:666–672.

Hasper HE, Kramer NE, Smith JL, Hillman JD, Zachariah C, Kuipers OP, de Kruijff B, Breukink E (2006) An alternative bactericidal mechanism of action for lantibiotic peptides that target lipid II. *Science* 313:1636–1637.

Heberle FA, Feigenson GW (2011) Phase separation in lipid membranes. *Cold Spring Harb Perspect Biol* 3(4):a004630.

Hedde PN, Malacrida L, Arahr S, Siryapron A, Gratton E (2017) sideSPIM—selective plane illumination based on a conventional inverted microscope. *Biomed Opt Express* 8(8):3918–3937.

Hell SW (2007) Far-field optical nanoscopy. *Science* 316:1153–1158.

Henriques ST, Quintas A, Bagatolli LA, Homble F, Castanho MA (2007) Energy-independent translocation of cell-penetrating peptides occurs without formation of pores. A biophysical study with pep-1. *Mol Membr Biol* 24:282–293.

Honigmann A, Mueller V, Hell SW, Eggeling C (2013) STED microscopy detects and quantifies liquid phase separation in lipid membranes using a new far-red emitting fluorescent phosphoglycerolipid analogue. *Faraday Discuss* 161:77–89; discussion 113–150.

Husen P, Arriaga LR, Monroy F, Ipsen JH, Bagatolli LA (2012a) Morphometric image analysis of giant vesicles: A new tool for quantitative thermodynamics studies of phase separation in lipid membranes. *Biophys J* 103:2304–2310.

Husen P, Fidorra M, Hartel S, Bagatolli LA, Ipsen JH (2012b) A method for analysis of lipid vesicle domain structure from confocal image data. *Eur Biophys J* 41:161–175.

Iwai I, Han H, den Hollander L, Svensson S, Ofverstedt LG, Anwar J, Brewer J et al. (2012) The human skin barrier is organized as stacked bilayers of fully extended ceramides with cholesterol molecules associated with the ceramide sphingoid moiety. *J Invest Dermatol* 132:2215–2225.

Jameson DM, Croney JC, Moens PD (2003) Fluorescence: basic concepts, practical aspects, and some anecdotes. *Methods Enzymol* 360:1–43.

Janosch S, Nicolini C, Ludolph B, Peters C, Volkert M, Hazlet TL, Gratton E, Waldmann H, Winter R (2004) Partitioning of dual-lipidated peptides into membrane microdomains: Lipid sorting vs peptide aggregation. *J Am Chem Soc* 126:7496–7503.

Jin L, Millard AC, Wuskell JP, Clark HA, Loew LM (2005) Cholesterol-enriched lipid domains can be visualized by di-4-ANEPPDHQ with linear and nonlinear optics. *Biophys J* 89:L04–L06.

Jin L, Millard AC, Wuskell JP, Dong X, Wu D, Clark HA, Loew LM (2006) Characterization and application of a new optical probe for membrane lipid domains. *Biophys J* 90:2563–2575.

Jones ME, Lentz BR (1986) Phospholipid lateral organization in synthetic membranes as monitored by pyrene-labeled phospholipids: Effects of temperature and prothrombin fragment 1 binding. *Biochemistry* 25:567–574.

Juhasz J, Davis JH, Sharom FJ (2010) Fluorescent probe partitioning in giant unilamellar vesicles of "lipid raft" mixtures. *Biochem J* 430:415–423.

Juhasz J, Davis JH, Sharom FJ (2012) Fluorescent probe partitioning in GUVs of binary phospholipid mixtures: implications for interpreting phase behavior. *Biochim Biophys Acta* 1818:19–26.

Juhasz J, Sharom FJ, Davis JH (2009) Quantitative characterization of coexisting phases in DOPC/DPPC/cholesterol mixtures: Comparing confocal fluorescence microscopy and deuterium nuclear magnetic resonance. *Biochim Biophys Acta* 1788:2541–2552.

Kahya N, Brown DA, Schwille P (2005) Raft partitioning and dynamic behavior of human placental alkaline phosphatase in giant unilamellar vesicles. *Biochemistry* 44:7479–7489.

Kahya N, Schwille P (2006) Fluorescence correlation studies of lipid domains in model membranes. *Mol Membr Biol* 23:29–39.

Kahya N, Scherfeld D, Bacia K, Poolman B, Schwille P (2003) Probing lipid mobility of raft-exhibiting model membranes by fluorescence correlation spectroscopy. *J Biol Chem* 278:28109–28115.

Keller PJ, Schmidt AD, Wittbrodt J, Stelzer EH (2008) Reconstruction of zebrafish early embryonic development by scanned light sheet microscopy. *Science* 322:1065–1069.

Kim HM, Choo HJ, Jung SY, Ko YG, Park WH, Jeon SJ, Kim CH, Joo T, Cho BR (2007) A two-photon fluorescent probe for lipid raft imaging: C-laurdan. *Chembiochem* 8:553–559.

Korlach J, Schwille P, Webb WW, Feigenson GW (1999) Characterization of lipid bilayer phases by confocal microscopy and fluorescence correlation spectroscopy. *Proc Natl Acad Sci USA* 96:8461–8466.

Krasnowska EK, Gratton E, Parasassi T (1998) Prodan as a membrane surface fluorescence probe: Partitioning between water and phospholipid phases. *Biophys J* 74:1984–1993.

Lasic DD, Papahadjopoulos D (1995) Liposomes revisited. *Science* 267:1275–1276.

Lentz BR, Barenholz Y, Thompson TE (1976a) Fluorescence depolarization studies of phase transitions and fluidity in phospholipid bilayers. 1. Single component phosphatidylcholine liposomes. *Biochemistry* 15:4521–4528.
Lentz BR, Barenholz Y, Thompson TE (1976b) Fluorescence depolarization studies of phase transitions and fluidity in phospholipid bilayers. 2 two-component phosphatidylcholine liposomes. *Biochemistry* 15:4529–4537.
Lin WC, Blanchette CD, Longo ML (2007) Fluid-phase chain unsaturation controlling domain microstructure and phase in ternary lipid bilayers containing GalCer and cholesterol. *Biophys J* 92:2831–2841.
Loftus AF, Noreng S, Hsieh VL, Parthasarathy R (2013) Robust measurement of membrane bending moduli using light sheet fluorescence imaging of vesicle fluctuations. *Langmuir* 29:14588–14594.
Lira RB, Steinkühler J, Knorr RL, Dimova R, Riske, K (2016) Posing for a picture: Vesicle immobilization in agarose gel. *Sci Rep* 6:25254.
Luisi PL (2000) Why Giant Vesicles? In: *Giant Vesicles* (Luisi PL, Walde P, eds), pp. 3–10. Chichester, New York, Weinheim: John Wiley and Son.
M'Baye G, Mely Y, Duportail G, Klymchenko AS (2008) Liquid ordered and gel phases of lipid bilayers: Fluorescent probes reveal close fluidity but different hydration. *Biophys J* 95:1217–1225.
Machan R, Hof M (2010) Lipid diffusion in planar membranes investigated by fluorescence correlation spectroscopy. *Biochim Biophys Acta* 1798:1377–1391.
Malacrida L, Gratton E, Jameson DM (2015) Model-free methods to study membrane environmental probes: A comparison of the spectral phasor and generalized polarization approaches. *Methods Appl Fluoresc* 3:047001.
Master BR, So PTC, Gratton E (1999) Multiphoton excitation microscopy and spectroscopy of cells, tissues and human skin in vivo. Fluorescent and luminescent probes for biological activity. In: *Fluorescent and Luminescent Probes*, 2nd Edition, pp. 414–432. New York: Academic press.
Mateo CR, Lillo MP, Gonzalez-Rodriguez J, Acuna AU (1991) Lateral heterogeneity in human platelet plasma membrane and lipids from the time-resolved fluorescence of trans-parinaric acid. *Eur Biophys J* 20:53–59.
McPhee CI, Zoriniants G, Langbein W, Borri P (2013) Measuring the lamellarity of giant lipid vesicles with differential interference contrast microscopy. *Biophys J* 105:1414–1420.
Mertins O, Bacellar IO, Thalmann F, Marques CM, Baptista MS, Itri R (2014) Physical damage on giant vesicles membrane as a result of methylene blue photoirradiation. *Biophys J* 106:162–171.
Minsky M (1988) Memoir on Inventing the confocal scanning microscopy. *Scanning* 10:128–138.
Morales-Penningston NF, Wu J, Farkas ER, Goh SL, Konyakhina TM, Zheng JY, Webb WW, Feigenson GW (2010) GUV preparation and imaging: Minimizing artifacts. *Biochim Biophys Acta* 1798:1324–1332.
Moreaux L, Sandre O, Charpak S, Blanchard-Desce M, Mertz J (2001) Coherent scattering in multi-harmonic light microscopy. *Biophys J* 80:1568–1574.
Mueller V, Honigmann A, Ringemann C, Medda R, Schwarzmann G, Eggeling C (2013) FCS in STED microscopy: Studying the nanoscale of lipid membrane dynamics. *Methods Enzymol* 519:1–38.
Murphy DB, Spring KR, Davidson MW (2009) Comparison of phase contrast and DIC microscopy. In: *Microscopy Resource Center*, https://www.olympus-lifescience.com/en/microscope-resource/primer/techniques/dic/dicphasecomparison/.
Nuss H, Chevallard C, Guenoun P, Malloggi F (2012) Microfluidic trap-and-release system for lab-on-a-chip-based studies on giant vesicles. *Lab Chip* 12:5257–5261.
Parasassi T, De Stasio G, d'Ubaldo A, Gratton E (1990) Phase fluctuation in phospholipid membranes revealed by Laurdan fluorescence. *Biophys J* 57:1179–1186.
Parasassi T, Di Stefano M, Loiero M, Ravagnan G, Gratton E (1994) Influence of cholesterol on phospholipid bilayers phase domains as detected by Laurdan fluorescence. *Biophys J* 66:120–132.
Peterlin P, Jaklic G, Pisanski T (2009) Determining membrane permeability of giant phospholipid vesicles from a series of videomicroscopy images. *Meas Sci Technol* 20:055801.
Plasencia I, Norlen L, Bagatolli LA (2007) Direct visualization of lipid domains in human skin stratum corneum's lipid membranes: Effect of pH and temperature. *Biophys J* 93:3142–3155.
Potma EO, Xie XS (2003) Detection of single lipid bilayers with coherent anti-Stokes Raman scattering (CARS) microscopy. *J Raman Spectrosc* 34:642–650.
Robinson T, Kuhn P, Eyer K, Dittrich PS (2013) Microfluidic trapping of giant unilamellar vesicles to study transport through a membrane pore. *Biomicrofluidics* 7:044105.
Ruan Q, Cheng MA, Levi M, Gratton E, Mantulin WW (2004) Spatial-temporal studies of membrane dynamics: Scanning fluorescence correlation spectroscopy (SFCS). *Biophys J* 87:1260–1267.
Sanchez SA, Gratton E (2005) Lipid—Protein interactions revealed by two-photon microscopy and fluorescence correlation spectroscopy. *Acc Chem Res* 38:469–477.
Sanchez SA, Tricerri MA, Gratton E (2012) Laurdan generalized polarization fluctuations measures membrane packing micro-heterogeneity in vivo. *Proc Natl Acad Sci USA* 109:7314–7319.
Sanchez SA, Bagatolli LA, Gratton E, Hazlett TL (2002) A two-photon view of an enzyme at work: Crotalus atrox venom PLA2 interaction with single-lipid and mixed-lipid giant unilamellar vesicles. *Biophys J* 82:2232–2243.
Sanchez SA, Tricerri MA, Gunther G, Gratton E (2007) LAURDAN generalized polarization: From cuvette to microscope. In: *Modern Research and Educational Topics in Microscopy* (Méndez-Vilas A, Díaz J, eds), pp. 1007–1014. Badajoz, Spain: Formatex Research Center.
Sezgin E, Levental I, Grzybek M, Schwarzmann G, Mueller V, Honigmann A, Belov VN et al. (2012) Partitioning, diffusion, and ligand binding of raft lipid analogs in model and cellular plasma membranes. *Biochim Biophys Acta* 1818:1777–1784.
Sibarita JB (2005) Deconvolution microscopy. *Adv Biochem Eng Biotechnol* 95:201–243.
So PTC, French T, Yu WM, Berland KM, Dong CY, Gratton E (1995) Time resolved fluorescence microscopy using two-photon excitation. *Bioimaging* 3:49–63.
So PTC, French T, Yu WM, Berland KM, Dong CY, Gratton E (1996) Two-photon fluorescence microscopy: Time-resolved and intensity imaging. In: *Fluorescence Imaging Spectroscopy and Microscopy* (Wang XF, Herman B, eds), pp. 351–374. New York: John Wiley and Sons.
Sot J, Bagatolli LA, Goni FM, Alonso A (2006) Detergent-resistant, ceramide-enriched domains in sphingomyelin/ceramide bilayers. *Biophys J* 90:903–914.
Stock RP, Brewer J, Wagner K, Ramos-Cerrillo B, Duelund L, Jernshoj KD, Olsen LF, Bagatolli LA (2012) Sphingomyelinase D activity in model membranes: structural effects of in situ generation of ceramide-1-phosphate. *PLoS One* 7:e36003.

Sustar V, Zelko J, Lopalco P, Lobasso S, Ota A, Poklar Ulrih N, Corcelli A, Kralj-Iglic V (2012) Morphology, biophysical properties and protein-mediated fusion of archaeosomes. *PLoS One* 7:e39401.

Tamba Y, Yamazaki M (2005) Single giant unilamellar vesicle method reveals effect of antimicrobial peptide magainin 2 on membrane permeability. *Biochemistry* 44:15823–15833.

Toomre DK, Langhorst MF, Davidson MW (2012) Introduction to spinning disk confocal microscopy. In: *Education in Microscopy and Digital Imaging.* http://zeiss-campus.magnet.fsu.edu/articles/spinningdisk/introduction.html.

Toytman I, Simanovskii D, Palanker D (2009) On illumination schemes for wide-field CARS microscopy. *Opt Express* 17:7339–7347.

Veatch SL, Keller SL (2002) Organization in lipid membranes containing cholesterol. *Phys Rev Lett* 89:268101.

Veatch SL, Keller SL (2003) Separation of liquid phases in giant vesicles of ternary mixtures of phospholipids and cholesterol. *Biophys J* 85:3074–3083.

Veatch SL, Keller SL (2005) Seeing spots: Complex phase behavior in simple membranes. *Biochim Biophys Acta* 1746:172–185.

Veatch SL, Leung SS, Hancock RE, Thewalt JL (2007) Fluorescent probes alter miscibility phase boundaries in ternary vesicles. *J Physi Chem B* 111:502–504.

Velez M, Lillo MP, Acuna AU, Gonzalez-Rodriguez J (1995) Cholesterol effect on the physical state of lipid multibilayers from the platelet plasma membrane by time-resolved fluorescence. *Biochim Biophys Acta* 1235:343–350.

Wick R, Angelova MI, Walde P, Luisi PL (1996) Microinjection into giant vesicles and light microscopy investigation of enzyme-mediated vesicle transformations. *Chem Biol* 3:105–111.

Mechanic assays of synthetic lipid membranes based on micropipette aspiration

Elisa Parra-Ortiz and David Needham

Well known, entropic effects are endemic to biomembranes and their aqueous environment; e.g., reduction in entropy of the water that reluctantly forms an interface with lipid hydrocarbon chains, confinement of the chain configurations by tight packing in the interface and even suppression of collective shape fluctuations when mechanically or osmotically stressing a vesicle, all of which are principal factors in the elastic properties of membranes

Evan Evans

Contents

In this chapter, the micropipette manipulation technique is presented as a useful and wide-spread tool to assess the mechanical and thermomechanical properties of single or pairs of giant vesicles with different levels of sophistication. These micromechanical experiments are used to establish the nature of the material structure of lipid membranes: Based on the study of well-defined chemical compositions, this approach can inform about the chemical state and origin of the material properties observed for these membranes, as well as identifying the colloidal forces involved in the chemical affinity or interaction between membrane surfaces. Here, we will present the different micromechanical studies that can be performed on giant unilamellar vesicles (GUVs) by using the micropipette technique, first giving a detailed explanation on the basic experimental setup and its possible optimizations and automations. These mechanical experiments will be explained by introducing the fundamental and methodological aspects and will include measurements on bending modulus, area expansivity or compressibility, tensile strength (for fluid-phase lipid bilayers), yield shear and shear viscosity (in the case of solid-phase bilayers) and on the thermomechanical behavior of the different phases. In addition, examples of different non-purely mechanical micropipette experiments will be also presented and connected to the corresponding chapters in this book, such as molecular exchange between the solution and the membrane, adhesion between pairs of vesicles and including colloidal attraction and repulsion and ligand–receptor bonds, hemifusion and complete fusion. Finally, a short outlook on the explored combinations between the micropipette manipulation techniques and other experimental approaches will be addressed, and some ideas for future research will be given.

11.1 HISTORICAL OVERVIEW

Prior to micromechanical experiments on giant vesicles, micropipette manipulation techniques were first motivated by, and developed for, the study of biological cells. In this introductory section, we will give a historical perspective that briefly reviews the extensive work done in this area, including experimental developments, measurements, analyses and modeling of biological cells, in particular red and white blood cells. These studies then led to the pioneering micropipette studies developed and performed by Evans and coworkers on GUVs.

11.1.1 CELL STUDIES

Motivated by their new conception of the mechanism of cell division, "the expanding membrane theory," the first attempts to use micropipette manipulation to interrogate the properties of cells and their membranes was in a series of three papers by Mitchison and Swann (1954a, 1954b, 1955). The authors developed a method of measuring the properties of the cell membrane with an instrument they called the "cell elastimeter" (Mitchison and Swann, 1954a), consisting of a glass micropipette filled with water and connected by rubber tubing to a small movable reservoir of water. Using a microscope and a micromanipulator, the pipette was brought up to the cells (eggs of sea urchins) and the reservoir was then lowered slightly with a micrometer screw. This created a small suction onto the egg surface at the end of the pipette, where it formed a seal. By lowering the reservoir again, the cell surface was progressively aspirated by the pipette, the cell deformation was measured directly under the microscope, and the negative hydrostatic pressure was measured on the micrometer screw. The second paper (Mitchison and Swann, 1954b) described the application of this method to the unfertilized sea urchin egg and showed that the cell membrane behaved as a relatively rigid structure of appreciable thickness. In the third paper (Mitchison and Swann, 1955), they described measurements of the stiffness and, indirectly, the internal pressure of the sea urchin egg from fertilization to the second interphase.

Interested in the problem of the shape of the red cell, 10 years later, Rand and Burton modified the cell elastimeter technique for determining the resistance to deformation, or stiffness, of the red cell membrane and the pressure gradient across the cell wall, including its viscoelastic breakdown (Rand and Burton, 1964). Subsequently, several different techniques and experimental systems, like osmotic swelling (Fung and Tong, 1968) and stretching adherent red cells (Hochmuth and Mohandas, 1972), yielded a broad range of values for the material constants of elastic and viscous deformations, for example, from 10^3 to 10^7 mN/m^2 for the elasticity modulus. In attempts to rationalize these discrepancies, more sophisticated mechanical models of the red cell membrane were introduced in the early 1970s by both Skalak (1973) and Evans (1973a, 1973b). Evans unified a new material concept for the red cell membrane by analyzing two micromechanical experiments on red blood cells: fluid shear deformation of point-attached red cells, and micropipette aspiration of red cell discocytes. What followed was a period of intense activity perfecting the micropipette technique, including much of the micromechanical analyses necessary to interpret the application of strain deformations and the resulting elastic, bending and shear stresses that could be induced by a single micropipette (Evans and Hochmuth, 1976, 1978). These ground-breaking studies culminated in the seminal book by Evans and Skalak (1980) *Mechanics and Thermodynamics of Biomembranes.* Thus, for almost 50 years these micropipette techniques have provided a unique ability to apply well-defined stresses to study dilation, shear and bending modes of membrane and cellular deformation, including, for instance, the characterization of the material behavior of individual erythrocytes (Evans and Hochmuth, 1978), leukocytes (Meiselman et al., 1984) and cancer cells (Needham, 1991) in terms of elastic moduli and viscous coefficients, as well as intersurface interactions (Evans, 1980).

11.1.2 GIANT UNILAMELLAR VESICLE STUDIES

Starting in the 1980s, the technique was adapted and developed by Evans and Kwok (1981, 1982) and then by Evans and Needham (1987) to study individual GUVs of various lipid compositions. This experimental approach involved the use of Hoffman Modulation Contrast (HMC) optics in a regular bright-field microscope in order to give a more well-defined, optical-gradient image of the 4-nm-thick membrane. In this chapter, we will mostly focus on this particular imaging approach, whereas those readers interested in other imaging strategies of GUVs are referred to Chapter 10. These micromechanical techniques have largely characterized the composition–structure–property relationships for the lipid bilayer membranes, including the influence of hydrocarbon chain length, degree of unsaturation, cholesterol (Chol, see Appendix 1 of the book for structure and data on this lipid) and lipid phases on bending modulus, membrane elasticity, tensile strength or water permeability (Evans and Rawicz, 1990; Rawicz et al., 2000, 2008; Evans et al., 2003; Evans and Smith, 2011).

A particularly historical and newly insightful perspective on the mechanics and thermodynamics of lipid biomembranes were given recently by Evans et al. (2013), emphasizing "*the inherent softness of fluid-lipid biomembranes and the important entropic restrictions that play major roles in the elastic properties of vesicle bilayers.*"

11.2 EXPERIMENTAL SETUP FOR MICROPIPETTE MANIPULATION

In brief, the micropipette manipulation technique is centered around an inverted microscope where one or more glass micropipettes are mounted directly on the microscope stage plate via three-dimensional (3D) micromanipulators, used to fix and position them with submicrometer precision (Olbrich, 1997; Needham and Zhelev, 2000; Heinrich and Rawicz, 2005). Control over micropipette suction pressure is in the range of a microatmosphere (μatm) to tenths of an atmosphere (atm) (0.1 to 10^4 Pa) and is achieved by a water height difference between two tanks equipped with micrometer-driven displacement. Positive and negative pressures are recorded by in-line water filled pressure transducers. All experiments are monitored in real time, along with the temperature and pressure data, and recorded using digital cameras (Box 11.1).

Going into further detail and level of sophistication, the micropipette setup would consist of three major subsystems, shown in the figure of Box 11.1, that can: (I) apply and measure pressure in the micropipette; (II) visualize in real time the micromanipulation and membrane deformation; and (III) record, control and run the experiments with custom-written software. The different elements will be described in detail in Sections 11.2.1 through 11.2.4; the automation of the micropipette setup and its optimization to achieve fast and precise tension ramps and high-resolution video acquisition and analysis was comprehensively reported by Heinrich and Rawicz (2005).

11.2.1 MICROPIPETTE AND CHAMBER

The GUVs are pressurized using micrometric glass pipettes, typically fabricated from 0.75-mm outer diameter, 0.4-mm inner diameter borosilicate glass capillaries (A-M Systems). The capillaries are first pulled with a pipette puller (for instance, a vertical puller such as Model 720 from David Kopf Instruments, or a more advanced horizontal puller such as Model P-97 from Sutter Instruments, Inc.). This process gives two pulled glass needles. In order to create the all-important open tip of the micropipette, the use of the following two-step forging process developed by Evans and coworkers is strongly recommended; it ensures perfectly sharp and clean pipette tips of the desired diameter (between 1 and 50 μm). The process is illustrated in Figure 11.1. A commercial microforge (like MF-900, Narishige, shown in Figure 11.1a) is employed using a modification: a small amount of low-melting point glass (for instance, vitreous low-temperature powder glasses from Ferro Corp.) is placed on a platinum wire connected to the heater, which can be seen in Figure 11.1b and c, forming a glass bead that can be heated to temperatures above its melting point by applying an electric current through the wire. The pulled pipette tip is then inserted into the molten glass bead, and the molten glass flows over the outside of the capillary. Once the bead is resolidified with the capillary embedded, it provides a fracture point to cut the pipette tip. Then, in order to get sharp, flat tips, a second forging step is carried out, allowing the molten glass to flow inward inside the glass capillary through its now open tip. When the glass meniscus reaches the desired cutting point, the heater is turned off, once again solidifying the glass bead. By simply tapping the microforge or gently retracting the capillary, it results in a sharp cut of the pipette tip at the meniscus position, the

Box 11.1 Basic elements of the micropipette aspiration setup

- Borosilicate capillaries
- Pipette puller and microforge for micropipette fabrication
- Vibration isolation table
- Water-filled pressurization system (Group I, Panel a): (1) tube for external pressure application; (2) air-tight syringe; (3) syringe pump for advanced setups, which can be also computer-controlled; (4) three-way valve; (5) main water reservoir connected to the pipette; (6) reference water reservoir open to the atmosphere; (7) two-way stop valve; (8) water-filled differential pressure transducer
- Sample visualization and micromanipulation (Group II, Panel a): (9) experimental chamber, connected to bath circulator for temperature-controlled experiments; (10) inverted microscope with a proper optical system (e.g., HMC optics, differential interference contrast microscopy [DIC], phase-contrast); (11) standard B/W side-port camera; (Panel b) 3D micromanipulator with micrometers to fix and position the micropipette on the microscope stage
- Computers with interfaces to the camera and imaging software and to the pressure transducer and/or syringe pump for advanced setups (Group III, Panel a)

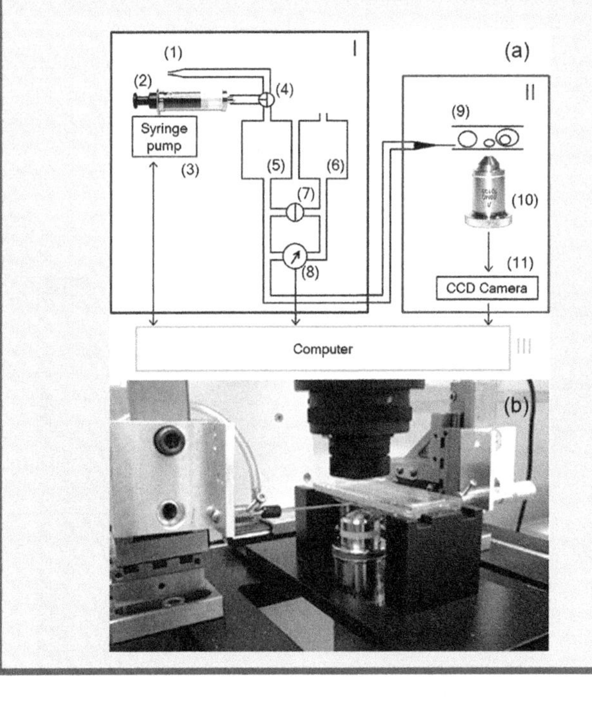

desired diameter. An actual image of the forged tip taken through the microforge eyepiece is shown in Figure 11.1c, where the final pipette diameter was approximately 10 μm. The choice of the right micropipette tip diameter for the experiments is important and depends on the target vesicle size and experiment to be carried out. Vesicles that are too large and with excess area would tend to neck in and pinch off at low pressures inside narrow micropipettes, whereas vesicles that are too small can slightly expand (depending on the their excess area) and be completely aspirated. Thus, using a pipette with an inner diameter of around one-third of the vesicle size would be the optimum choice for measuring, for example, the elastic or bending modulus.

The pipette is then backfilled with the same aqueous solution as the one in which the GUVs will be suspended and tested in the microchamber. This is done by using MicroFil syringe needles (World Precision Instruments, Inc.), carefully avoiding air bubbles inside the micropipette. The pipette is inserted into a holding chuck that will be then fixed to the micromanipulator on the microscope stage plate (see figure in Box 11.1, Panel b). In order to monitor it properly and get accurate measurements, the pipette must be oriented perpendicular to the optical axis of the microscope. In addition, a fine control of its position in the three dimensions is required. For these reasons, the pipette is mounted on a 3-axis stage (such as Model M-461-XYZ-M, Newport Corp.), rigidly attached to the microscope plate, and equipped with either manual micrometer heads (Newport Corp.) or motorized micrometers (Thermo Oriel at Spectra Physics) that control the pipette position with submicrometer precision. In the case of the motorized micrometers, their motion can be controlled with a pneumatic or 3D game joystick, or directly controlled by the computer in order to get a more automatized setup.

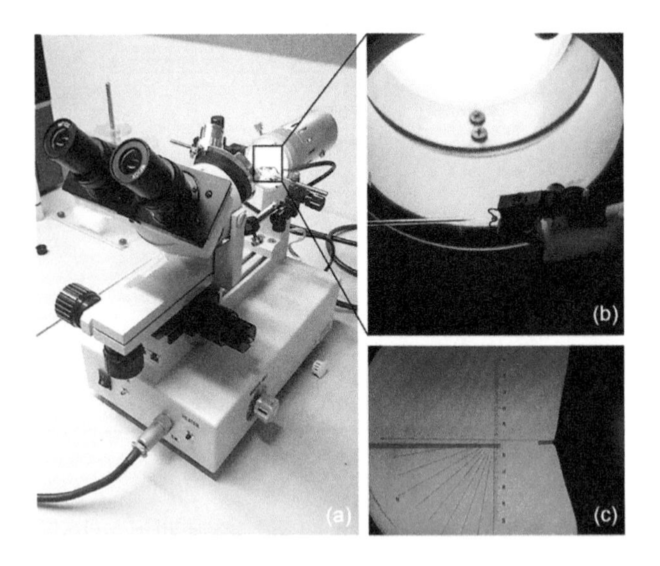

Figure 11.1 Forging process of the glass micropipettes. (a) A MF-900 microforge, consisting of a microscope and 10x eyepieces, horizontal 3D micromanipulator to hold the pipette, back lamp, electric heater and foot switch (not shown in the image). (b) Detail of the micropipette tip held on the micromanipulator, inserted into the low-melting point glass bead formed at the platinum wire. (c) View through the microforge eyepiece of an already forged micropipette tip (~10 μm diameter, left), where the tip leftover is trapped into the solidified bead (right). See main text for further details on the forging method.

For experiments carried out at room temperature, the basic experimental chamber can simply consist of two glass coverslips separated by a polytetrafluoroethylene (PTFE) or silicon spacer, open at one side to allow the insertion of the pipette. A two-sided open chamber is required in order to perform experiments with two or more micropipettes, built using two parallel spacers that would hold the sample in between. Another alternative is to use hollow tubing made of borosilicate glass (such as the miniature hollow rectangle capillaries, ID 20 × 4 mm, wall 0.4 mm, from VitroCom), cut to the desired length and mounted on a metal frame to be placed at the microscope stage. A temperature-controlled chamber, consisting of a metal frame that both holds the sample cell and a water circuit connected to a cooling/heating circulating bath, would allow carrying out experiments between 15°C and 70°C. When working at high temperatures, water-insoluble liquids, such as oils or long-chain alkanes, can be used to "close" the chamber openings and avoid water evaporation. To work at either low or high temperatures, an external device can be implemented to control the chamber humidity and hinder condensation or evaporation, respectively. In addition, a two-cell chamber system, holding two parallel cells opened at both sides, can be employed for experiments involving transfer of GUVs between two different media. This is the method used, for instance, in water permeability studies (Olbrich et al., 2000) and other experiments involving the effect of exposing the GUVs to different environments. Prior to GUV experiments, the sample cell and micropipette walls need to be pre-coated in order to prevent lipid adhesion to glass surfaces. This can be done by incubating the cell and pipette tip for 5–10 min in a bovine serum albumin (BSA) or casein 0.1% solution, and then rinsing it before placing the GUV suspension.

11.2.2 PRESSURE APPLICATION AND MEASUREMENT

The basic control over micropipette suction pressure is simply achieved by hydrostatics. The pressure is imposed by a water height difference between two water tanks that is adjusted by micrometers, and is continuously monitored by a liquid–liquid pressure transducer.

An optimum setup would involve a water-filled manometer based on two water reservoirs, equipped with a sensitive micrometer-driven displacement, a syringe pump (manual or automatic), and a coarser external pressure control, such as that shown in Figure 11.2. The micropipette is connected to the main water reservoir (M in Figure 11.2) of the pressure system, and pressure can be applied inside the pipette in two main ways: by lowering the position of the main water reservoir with respect to the vertical position of the reference reservoir, open to the atmosphere (R in Figure 11.2), or by evacuating the air volume above the water in the main reservoir with the syringe control. An extra tube can be attached to the air valve (AV in Figure 11.2) of the main reservoir in order to exert an extra positive or negative pressure with a second syringe or mouthpiece, very convenient in some cases, for instance to clean the pipette tip from lipid membranes blocking it.

The pressure difference between the two partly filled water reservoirs is measured by a pressure transducer (such as Model DP15, Validyne Engineering; T in Figure 11.2) assembled with a No. 32 diaphragm, optimal for the typical pressure range

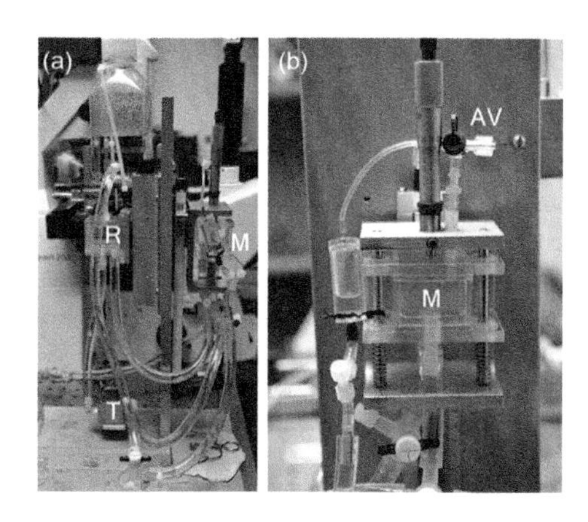

Figure 11.2 The water-filled pressurization system. (a) Side view of the system for micropipette pressurization, including the two water reservoirs, reference reservoir, *R*, and main reservoir, *M*, tubing connections and valves and the pressure transducer, *T*. (b) Detail of the main water reservoir, *M*, with the air valve, *AV*, open.

($0-10^4$ Pa). The transducer is connected to a demodulator (Model CD23, also from Valydine) that can be interfaced via a standard peripheral component interconnect (PCI) board with the data acquisition software in the computer in more advanced setups. The transducer and demodulator are routinely calibrated by adjusting the zero pressure and by applying a known pressure difference using an external manometer. In addition, the system should always be zeroed at the height at which the measurement will be performed. This is done by adjusting the vertical position of the two water-filled chambers until the pressure at the pipette tip is zero (for instance, by looking at the in- or outward flow of suspended particles through the pipette tip). The transducer and all the tubing of the pressure system are assembled and carefully filled with distilled and degassed water in order to avoid air bubbles in the system, which is crucial for a precise pressure control and measurement.

11.2.3 VIDEO MICROSCOPY AND DATA ACQUISITION

In order to visualize and control the GUV preparation and the micropipette, an inverted microscope is used (Axiovert 200, Carl Zeiss Microimaging, Inc.), ideally equipped with HMC optics (Modulation Optics, Inc.). This optical system produces a high-contrast and high-resolution image, converting optical gradients of the specimen, such as refraction indexes, into intensity variations due to three special components added to the microscope: the HMC objective, containing a three-region filter called modulator; the HMC condenser, consisting of a lens system and slits that correspond to the HMC objective, whose image must be aligned to the modulator; and the HMC polarizer, which is rotated to control the image contrast by varying the background intensity and partial coherence of the illumination (Hoffman and Gross, 1975). Thus, in order to maximize the image quality of a GUV sample, the vesicles are routinely prepared in sucrose solutions and dissolved in equiosmolar glucose solutions prior to observation (see Chapter 1 for an overview on the different methods). This provides a sharp gradient of the refraction index between the inside and the outside of the GUV, together with a density difference that settles the vesicles at the bottom of the chamber. These osmolarities are also used to trap water and oppose the suction pressure to prevent hydraulic transport of the volume because some experiments such as area dilation require constant volume. The usual osmolarities of the sucrose and glucose solutions are ~200 mOsm/kg; however, using lower values (~100 mOsm/kg) might be advisable for bending rigidity measurements due to the effect of sugars on this mechanical property (see Section 11.3.2). Other imaging systems, such as phase-contrast and DIC as well as epifluorescence and confocal microscopy can be employed. The reader is referred to Chapter 10 and to the recent review by Bagatolli and Needham (2014) for more details. A standard B/W analog camera is attached to a side port of the microscope in order to get a real-time recording of the microscope image, and even achieve an extra magnification of the image with a 12.5× or 16× eyepiece mounted on the camera tube (in case that the microscope objective is 20× this will be required to get the proper magnification). An actual image of a single GUV aspirated by a 10-μm micropipette and visualized with the HMC-equipped microscope is shown in Figure 11.3. The dimensions of the image are calibrated by taking an image of a standard stage micrometer (10-μm divisions) and using this in the subsequent digital analyses in order to get a precise measurement of the relevant geometric parameters of the system: the vesicle radius, R_{ve}; the projection length of the aspirated membrane portion of the vesicle inside the pipette, L_p; and the inner pipette radius, R_{pip}. The measurement of this last parameter is hampered by the diffraction patterns and slight curvature of the pipette inner walls, but it should be measured in a careful and systematic way because most measurements are very sensitive to it. One of the most reliable methods consists of using a second conical needle as a probe, the exact dimensions of which have been measured by scanning electron microscopy, or by simply relying on conventional optics as follows: the needle is inserted inside the micropipette, so its outer diameter is determined at the pipette entrance with a higher accuracy than the pipette inner diameter; it requires a second micromanipulator under the microscope to fix and align the probe (Heinrich and Rawicz, 2005).

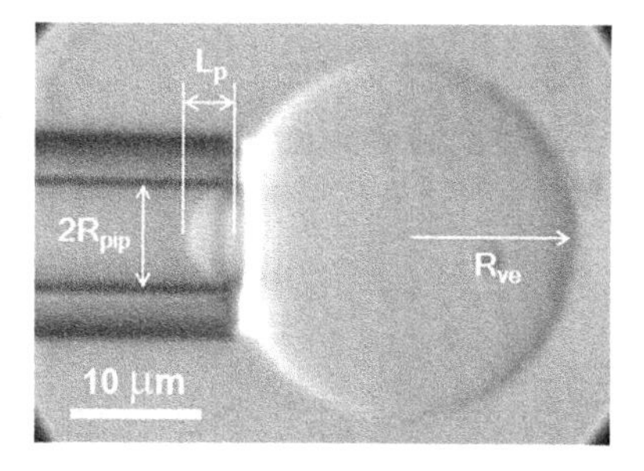

Figure 11.3 Image of a micropipette-aspirated GUV obtained with the Hoffman Modulation Contrast system. The vesicle has a sucrose solution inside and an equiosmolar glucose solution outside, creating a difference in refractive index for better visualization. The main geometrical parameters of the experiments are indicated: the micropipette inner diameter, $2R_{pip}$; the projection length of the vesicle aspirated by the pipette, L_p, and the vesicle radius, R_{ve}. (From Parra-Ortiz, E., Effects of pulmonary surfactant proteins SP-B and SP-C on the physical properties of biological membranes, in *Department Applied Physics III*, Madrid, Spain, Complutense University, 2013.)

11.2.4 FURTHER OPTIMIZATION AND AUTOMATION OF DATA COLLECTION

Different modifications of the basic setup can be performed in order to achieve a higher level of control and automation (Heinrich and Rawicz, 2005). As mentioned in Section 11.2.1, the pipette can be mounted on the 3-axis stage equipped with motorized micrometers (Thermo Oriel at Spectra Physics), providing a finer control of the pipette position by using a standard 3D game joystick. Regarding the pressure control, the creation of a negative pressure in the main reservoir with respect to the reference reservoir can be completely automated by using, for instance, a bidirectional and computer-controlled syringe pump (Model PHD 2000 Infuse/Withdraw High Force, Harvard Apparatus, Inc.). In the case of those experiments that require precise aspiration ramps, custom-written computer software can be then used to set the pump's desired suction pressure and pressurization speed. In addition, the experiments can be recorded by using high-speed cameras for a much higher time resolution than the standard 20–30 fps (for instance, SensiCam cooled digital 12-bit charge-coupled device (CCD) camera systems from the Cooke Corp.). They are equipped with an external cooling fan to minimize vibrations, and their fast frame rate mode allows real-time image acquisition up to 1,500 fps. However, this requires using a small region of interest. For example, the whole vesicle geometry can be determined just from the axial intensity profile of the pipette–vesicle system shown in Figure 11.3, that is, the imaginary horizontal line that crosses the image along its axis of symmetry. Thus, a region of the image of 32 full-length (640 pixels) video lines centered at the pipette middle axis can be recorded and subsequently binned into one single line, right before reading out from the camera in order to improve the signal-to-noise ratio and the transfer speed, resulting in a 1D axial intensity profile such as the ones shown in Figure 11.4. Here, the relevant diffraction patterns of the system can be easily identified, including the pipette entrance, the vesicle edge out of the pipette, and the inner vesicle projection aspirated by the pipette. Other imaging systems or modes of observation, such as phase-contrast, DIC or fluorescence image, will give different gray intensity profiles, depending on how the characteristic patterns of the image are visualized (see Chapter 10 for more details). Increasing aspiration pressures will be translated into longer projection lengths L_p of the GUV inside the pipette, and the precise moment of vesicle rupture can be identified as the disappearance of the projection edge and subsequent displacement and aspiration of the outer edge into the pipette. Because the field of view has to be reduced at such extent, a supplementary video system can be implemented for continuous visualization on a second monitor by using a standard B/W side-port camera. This intensity profiles can be then acquired at the highest sampling rate and recorded as a function of the aspiration pressure and can be further analyzed with custom-written software to calculate the geometrical changes of the vesicle in terms of area and volume as a function of pressure. The reader is referred to a more detailed description, published in 2005 by Heinrich and Rawicz (2005), for the possible ways to automate the setup to obtain advanced data analysis of the vesicle intensity profiles.

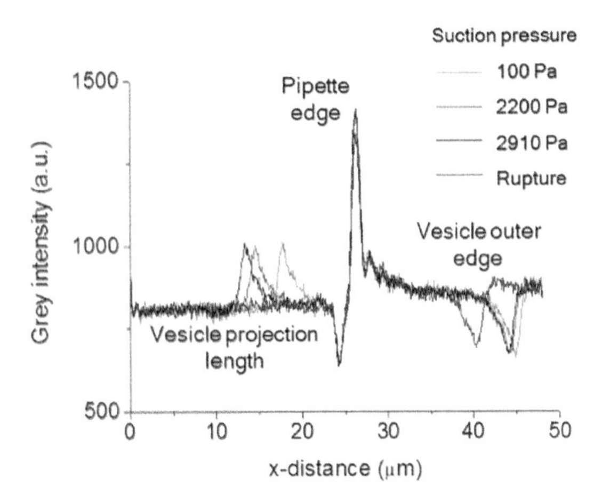

Figure 11.4 Intensity profiles corresponding to different times of a vesicle aspiration experiment. The characteristic signatures of the diffraction patterns of the vesicle (inner and outer edges) and pipette entrance are identified in the graph. The profiles correspond to the initial moment of low initial pressure (lightest gray line); to increasing suction pressures that cause the projection length to increase; and the eventual rupture (red line), in which the projection edge of the vesicle is broken and thus disappears. (Modified from Parra-Ortiz, E., Effects of pulmonary surfactant proteins SP-B and SP-C on the physical properties of biological membranes, in *Department Applied Physics III*, Madrid, Spain, Complutense University, 2013.)

11.3 MICROMECHANICS OF SINGLE GIANT UNILAMELLAR VESICLES

In this main section of the chapter, the analysis of GUV deformations resulting from mechanical stress applied by micropipette aspiration will be described. Much of this work has been previously reviewed (Evans and Needham, 1987; Needham and Zhelev, 1996; Needham and Zhelev, 2000). The section will focus on the very soft mode of membrane-bending deformation, where the thermal fluctuations of a single bilayer can be suppressed by very low pipette suction pressures; the stiffer resistance to area deformation associated with the area expansivity, revealed by the application of much larger suction pressures that gradually expand the bilayer until it fails in tension, giving a direct measure of the membrane tensile strength; the yield shear and shear viscosity for bilayers below their main transition temperature, when these solid membranes now resist in-plane shear; and the thermomechanical behavior of the bilayer and changes in molecular area due to phase transitions. Finally, other mechanical studies using micropipettes will be briefly presented, including measurements in conjunction with shear to develop a micro-rheometer of giant vesicles, measuring flow field effects on the membrane behavior; and studies on the line tensions that occur at membrane pore edges and between distinct lipid domains, on the spontaneous curvature of single giant vesicles, and on nanotube pulling from GUVs.

11.3.1 CONSTITUTIVE EQUATIONS

Analysis of micropipette suction pressure combined with vesicle and pipette geometry give membrane stresses and resulting shape changes as a series of constitutive equations. They provide direct measures of the coefficients related to membrane deformations. By collapsing all the forces in and on all the faces of a 3D cube (Evans and Skalak, 1980), the basic mechanics for the analyses of micropipette experimentation provides three independent shape changes for thin membrane deformations: *bending, area expansion or dilation*, and *in-plane shear*, explained in Box 11.2. They can be represented by simple shape changes of local surface elements: (i) bending or curvature change at constant rectangular shape, $\Delta C = \Delta(1/R)$; (ii) area change (dilation or condensation), $\alpha = \Delta A/A_0$; and (iii) in-plane extension or surface shear at constant surface density, $\lambda_e = L/L_0$, being the ratio between the length of a unit element of the stressed, L, and the non-stressed bilayer, L_0. These independent deformations are caused by the application of forces that exert specific actions on membrane elements as indicated in the figure in Box 11.2: bending moments, B, proportional to changes in membrane curvatures, ΔC; mean membrane tension, Σ, proportional to the fractional area dilation or condensation, α; and surface shear, τ_s, proportional to in-plane extension or shear strain, λ_e (Evans and Skalak, 1980; Evans and Needham, 1987). Together, these modes characterize the deformation and rate of deformation of the lipid bilayer. For a liquid lipid bilayer, the two relevant elastic relations are area dilation and bending. Membranes above their phase transition temperature by definition cannot support elastic shear but do have a low viscosity that is on the order of a two-poise oil. Below their transition, solid or gel phase lipid bilayers do support elastic shear stress and shear viscosity, displaying a yield shear and shear viscosity for applied shear stresses above that yield, characteristic of a Bingham

Box 11.2 Modes of deformation of an infinitely thin 2D membrane

- Based on the original analysis by (Evans and Skalak, 1980) and subsequently developed in (Evans and Needham, 1987), the mechanics of a cube can be collapsed into an infinitely thin 2D membrane. Normal and in-plane stresses lead to the following deformations: bending, a change in membrane curvature for a constant area, where B are the bending moments; area dilation, $\alpha = A/Ao$, an isothermal change in membrane area induced by isotropic membrane stress, Σ; and shear strain, an in-plane extension under constant membrane area caused by surface shear stress, t_s.

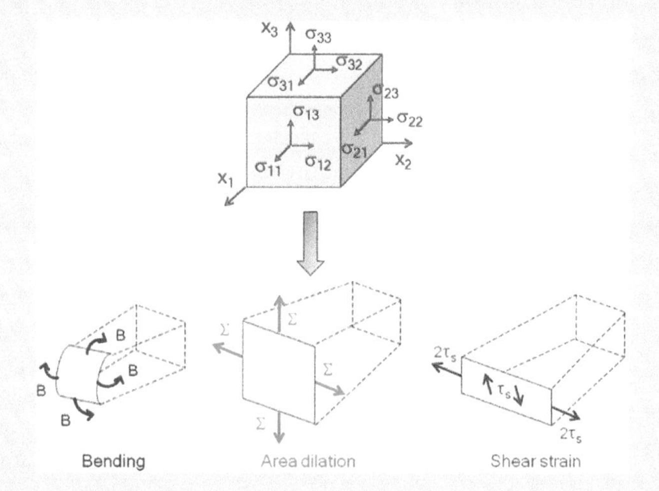

- Membrane bending is characterized by the *bending rigidity*, κ, which is the ratio of the change in membrane bending moment, ΔB, to changes in total membrane curvature, ΔC:

$$\Delta B = \kappa \Delta C \tag{11.1}$$

- At a given position in the membrane, curvature change is the change in the principal radii of curvature R_1 *and* R_2:

$$\Delta C = \Delta\left(1/R_1 + 1/R_2\right) \tag{11.2}$$

- The area dilation is characterized by the isothermal *area expansivity modulus*, K_A, given by the ratio between the fractional change in membrane area, α, and the isotropic membrane stress, $\Delta\Sigma$:

$$\Delta\Sigma = K_A \alpha \tag{11.3}$$

- Although liquid bilayers, by definition, do not support elastic shear, solid (or gel) phase lipid bilayers do support shear stress elastically, and so, for this solid bilayer state, shear deformation is characterized by the *surface shear rigidity*, μ, for a membrane element, which is the ratio between surface shear stress, τ_s, and shear deformation, e_s:

$$\tau_s = 2\mu e_s \tag{11.4}$$

- The shear deformation, e_s, is given by the in-plane extension, λ_e, as

$$e_s = \left(\lambda_e^2 - \lambda_e^{-2}\right)/2 \tag{11.5}$$

plastic (Evans and Needham, 1987), which is also true for lipid monolayers on gas microparticles (Kim et al., 2003).

Such independent deformations are produced by the application of external forces to the membrane elements, such as the hydrostatic pressure provided by the micropipette aspiration of a single lipid vesicle. For micropipette suction pressures, ΔP, in the range of a microatmosphere to tenths of an atmosphere (μatm to tenths of atm; i.e., 0.1 to 10^4 Pa), the force at the pipette tip, F_{pip}, can be estimated from the pipette cross-sectional area, A_{pip}, as follows:

$$F_{pip} = \Delta P . A_{pip} \tag{11.6}$$

Therefore, for a typical 8-μm diameter pipette, it is possible to deliver forces as small as 5 pN up to 750 nN. At the lowest end, this is equivalent to the force developed by molecular motors, like kinesin "walking" along a microtubule. It is also possible to forge a 2-μm diameter pipette, which delivers a force of only 0.3 pN, just over twice the force needed to stretch double-strand DNA to 50% relative extension. We will see that this range of forces is the same as that involved in membrane bending, showing just how soft the lipid bilayer actually is.

In order to provide the tension part of the analysis of a micropipette experiment of a single vesicle, the pipette suction pressure has to be related to the radii of curvature of the vesicle and assume that the tension, Σ, is constant around the whole geometry, that is, no work is put into the curvature around the pipette mouth. Applying the Young-Laplace equation to both the membrane cap aspirated by the pipette and the outer vesicle membrane, the pressure differences produce a homogeneous membrane tension, Σ, scaled by their respective radii of curvature, R_{pip} and R_{ve}, and the pressures inside the vesicle, P_{ve}, inside the pipette, P_{pip}, and in the microchamber, P_o, as follows:

$$P_{ve} - P_{pip} = \frac{2\Sigma}{R_{pip}} \; ; \; P_{ve} - P_o = \frac{2\Sigma}{R_{ve}}$$

The subtraction to eliminate P_{ve} gives the micropipette suction pressure $\Delta P = P_{pip} - P_o$, which simply yields to the following expression for the membrane tension, Σ:

$$\Sigma = \frac{\Delta P \cdot R_{pip}}{2\left(1 - R_{pip} / R_{ve}\right)} \tag{11.7}$$

Notice that this expression applies only for $L_p > R_{pip}$ because for smaller L_p the radius of the spherical cap inside the micropipette is not R_{pip}. Furthermore, the deformation exerted by this mechanical stress is detected as changes in vesicle area, ΔA, in which the pipette offers a unique sensitivity: The area change is linearized as changes in length of the membrane projection aspirated by the pipette, ΔL. From simple geometry, the change in area is given by

$$\Delta A = 2\pi R_{pip} \Delta L \left(1 - R_{pip} / R_{ve}\right) \tag{11.8}$$

Then, the fractional change in membrane area is obtained from the ratio with the area of the non-stressed vesicle, A_0: $\alpha = \Delta A / A_0$.

Therefore, proportionalities between intensive forces and static deformations describe the three modes of deformation—bending, area expansion and shear—and are used to characterize the membrane's material properties and give rise to first-order constitutive relations. Experimental details involved in the study of these three modes using the micropipette technique will now be explained in further detail in Sections 11.3.2 through 11.3.4.

11.3.2 BENDING

Under very low suction pressures created by the pipette, the thermal fluctuations of a single bilayer can be suppressed, and the bending rigidity of the GUV can be accurately determined (Evans and Needham, 1987). Figure 11.5 shows a simple micropipette experiment in which a pipette suction, at the limit of control and resolution (2 μatm, or 0.2 Pa), is applied to a single lipid vesicle. Using a very simple approach, the bending modulus, κ, can be calculated from the suction pressure, ΔP, required to bend the membrane into the micropipette entrance to just one pipette radius (R_{pip}), and it is approximately given by the following equation (Evans and Needham, 1987):

$$\kappa \sim \frac{\Delta P \cdot R_{pip}^3}{8} \tag{11.9}$$

Below this level, the thermal undulations of the single bilayer are dominant in the GUV shape and mechanics. Thus, with a suction pressure of only 0.2 Pa, a membrane tension on the order of 2.5×10^{-4} mN/m is being applied, and the value calculated for the bending rigidity for this lipid bilayer is 1.6×10^{-18} N.m. This is actually an energy unit, and because 1 N.m = 1 J, the energy required to bend the membrane is 1.6×10^{-18} J (1.6 attoJoules). The value

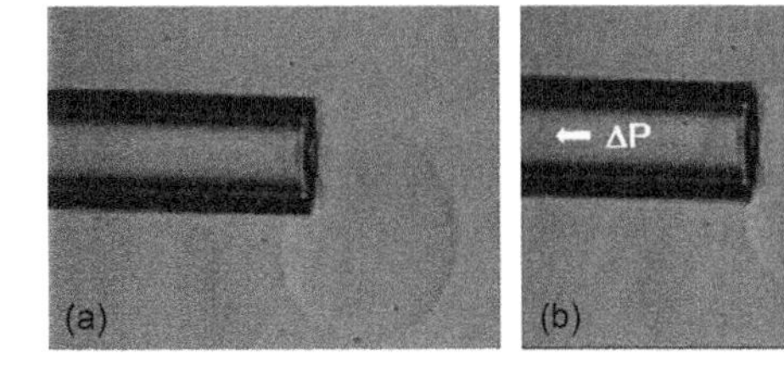

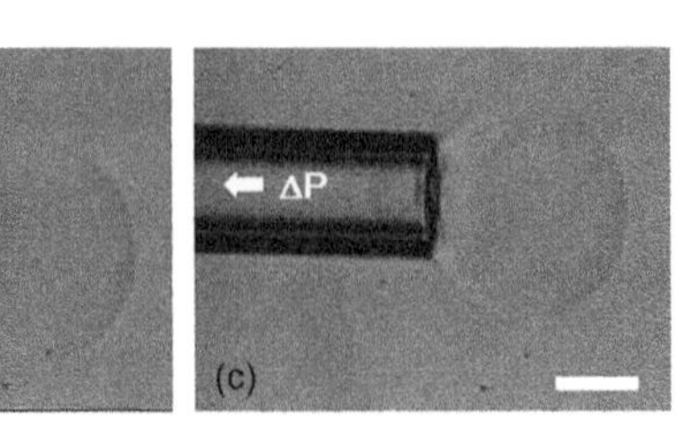

Figure 11.5 Experiment to determine membrane bending rigidity. (a) GUV at the pipette tip with zero applied suction pressure; (b) application of a low suction pressure starts aspirating the GUV; (c) application of a slightly higher $\Delta P \sim 0.2$ Pa bends the membrane into the micropipette and suppresses the thermal undulations. Scale bar: 10 μm. (Reprinted from *Chem. Phys. Lipids*, 181, Bagatolli, L.A. and Needham, D., Quantitative optical microscopy and micromanipulation studies on the lipid bilayer membranes of giant unilamellar vesicles, 99–120, Copyright 2014, with permission from Elsevier.)

measured by this technique was actually found to be an overestimate, when other more sensitive methods were developed and employed.

Noting that there was a small but discernable soft area expansion at the beginning of every area dilation measurement, Evans and Rawicz (1990) developed a new method to measure the bending modulus by gently pulling out the thermal undulations stored in the membrane. They showed that, when increasing the pressurization of a GUV from extremely low levels of tension, below a certain threshold, expansion is largely due to extending membrane undulations, and higher tensions would create actual lipid–lipid area change of the bilayer, shown in Figure 11.6a (Evans and Rawicz, 1990). This phenomenon in the low-tension regime is due to restrictions in the thermal fluctuations, an entropy-driven tension characteristic of the bending rigidity of the condensed-fluid membrane. These bending experiments need to be performed in a very careful way, applying small ΔP values (from 0.1 up to 2–3 × 10^2 Pa) and ΔP increments (30–50 Pa), in order to study the tension range between 0.001–0.5 mN/m. Each step should be held steady for several seconds to ensure a stationary state, and a stepwise course of pressure reduction should be used to verify reversibility (Rawicz et al., 2000). In addition, it is advisable to first prestress each vesicle under a tension of ~0.5 mN/m to ensure that small hidden surface defects are pulled into the membrane, and then release the pressure and drop it to ~0.1 Pa prior to each bending experiment (Rawicz et al., 2000; Vitkova et al., 2004). The new measurement gave a value for the bending modulus of 0.9 × 10^{-18} N·m (or J) for the common phospholipid 1-stearoyl-2-oeloylphosphatidylcholine (SOPC). This energy (5.6 eV) is of the same order than the bond-dissociation energy of a single C-H covalent bond (415.0 kJ/mol = 4.3 eV/bond), and is two orders of magnitude above the thermal energy at room temperature, $k_BT = 4{\cdot}10^{-21}$ J.

In these soft-expansion experiments, a more accurate way to obtain κ was therefore proposed from the study of the area expansion response against varying levels of tension, according to the following equations (Evans and Rawicz, 1990; Rawicz et al., 2000):

$$\alpha = \frac{k_BT}{8\pi\kappa}\, ln\left(1+\frac{c\Sigma A}{\kappa}\right)+\frac{\Sigma}{K_A} \qquad (11.10)$$

Low-tension regime:

$$ln(\Sigma/\Sigma_0) \approx (8\pi\kappa/k_BT)\cdot\alpha \qquad (11.11)$$

where c is a constant of ~0.1, which depends on the type of modes of surface undulations (spherical harmonics or plane waves); and $\alpha = \Delta A/A_0$ is the fractional change in membrane area obtained from the ratio with the area of the non-stressed vesicle, A_0. The tension, Σ, is exerted by hydrostatic pressure with the pipette and was given by Eq. 11.7, constituting the isotropic tensile stress that effectively expands the membrane. Therefore, κ will be obtained from the low-tension regime, given by the slope of the logarithm of membrane tension, Σ, versus area expansion, α, and, as will be explained in Section 11.3.3, the area expansivity modulus, K_A, can be readily obtained from the high-tension regime (Figure 11.6b), considering the impact of smoothing thermal undulations in the apparent area expansion.

Thus, using this approach, sensitive measurements (Evans and Rawicz, 1990; Waugh et al., 1992; Zhelev et al., 1994; Rawicz et al., 2000) gave κ values between (0.1 – 40) × 10^{-18} N·m (between ~10 and 10^4 k_BT) for different fluid bilayer compositions, a wide range of values that critically depend on chain length, degree of unsaturation, and especially cholesterol contents (Rawicz et al., 2000; Rawicz et al., 2008). In two tables shown at the end of Section 11.3.3, more specific κ values are listed for different bilayer compositions in two tables.

More recently, a careful work done by Henriksen and Ipsen aimed to conceal the observed differences between these results and the shape fluctuation method, where the bending rigidities where consistently higher than those obtained by micropipette aspiration experiments (Henriksen and Ipsen, 2004). They proposed that these discrepancies arise from the

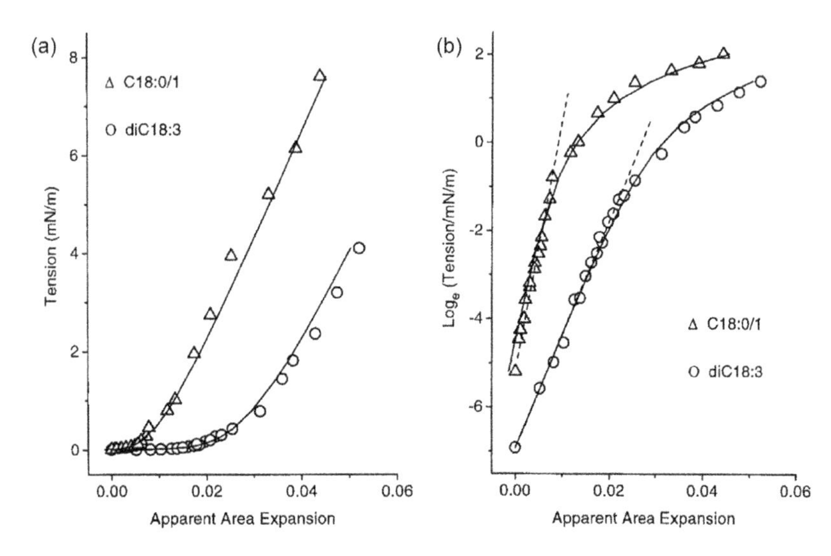

Figure 11.6 Experimental curves showing the full range of membrane tension versus apparent area change in both logarithmic and linear scales. (a) At low tensions, a fluctuation-dominated regime appears with a slope of $8\pi\kappa/k_BT$ in the logarithmic representation. (b) At high tensions, a crossover to a different regime occurs, with a slope of K_A in the linear plot. Vesicles of two different membrane compositions were tested, SOPC (C18:0/1) and dilinolenoylphosphatidylcholine (DLnPC; diC18:3). (Reprinted from *Biophys. J.*, 79, Rawicz, W. et al., Effect of chain length and unsaturation on elasticity of lipid bilayers, 328–339, Copyright 2000, with permission from Elsevier.)

effects of membrane elasticity in the low-tension regime, which are neglected in the conventional interpretation of micropipette data, and introduced a new approach that considers the full tension range in the analysis of both membrane bending and area expansivity (Henriksen and Ipsen, 2004). Further details about the shape fluctuation method can be found in Chapter 14.

Not surprisingly, because it is so soft, membrane bending rigidity can be largely affected by different additives and membrane inclusions, and some micropipette studies have been performed in that direction (Dimova, 2014). For instance, GUVs exposed to mono- and oligosaccharide solutions undergo a strong decrease in their bending rigidity, being around a 4-fold reduction for SOPC vesicles in the presence of 0.3-M sucrose compared with pure water (Vitkova et al., 2006). In addition to the measurements of Evans and Rawicz (Evans and Rawicz, 1990; Rawicz et al., 2008), where 1:1 SOPC:Chol membranes had an increased bending modulus of 2.46×10^{-18} N·m, more recent studies also show a cholesterol effect on membrane stiffness in a lipid-specific manner, suggesting that saturated acyl chains would have a greater interaction with cholesterol (as we will see later in the area dilation expansion section, Section 11.3.3), so their mixed bilayers show systematically higher rigidities than those with unsaturated chains (Pan et al., 2008, 2009; Gracià et al., 2010). Other interesting studies have also addressed the modulation of bending rigidity by certain peptides, determined by their impact on membrane thickness through the mismatch between lipid chains and the hydrophobic segments of membrane peptides (Dimova, 2014). For instance, a strong decrease in bending rigidity has been measured both by micropipette aspiration and shape fluctuation analysis in the presence of the membrane-thinning HIV fusion peptide FP23 (Shchelokovskyy et al., 2011; Dimova, 2014).

All these data collected on membrane bending underline the fact that, as thin deformable sheets that surround every cell on the planet, lipid bilayers are extremely flexible materials and can be easily deformable just through thermal fluctuations in the surrounding medium.

11.3.3 AREA EXPANSIVITY AND TENSILE STRENGTH

When much larger suction pressures are sequentially applied to a single GUV, $\Delta P \sim 1\text{–}2$ kPa, the bilayer is gradually expanded until it eventually fails, and the area expansion at failure is only ~3%; however, this can be over twice as large, at over twice the tensile strength, when stress is applied much more quickly (Evans et al., 2003). The ability of the membrane to accommodate to area changes under varying lateral mechanical stress can be analyzed from the relationship between mechanical tension, Σ, and the fractional change in membrane area, α, based on Eq. 11.3. The tension, Σ, that the pipette exerts on the membrane by applying a hydrostatic pressure, given by Eq. 11.7, constitutes the isotropic tensile stress that expands the membrane. Thus, as shown in Figure 11.7, analyses of the pipette (R_{pip}) and vesicle (R_{ve}) radii and the change in projection length, ΔL, provide an accurate measurement on the increase in membrane area by using Eq. 11.8, which then give relationships for these tensile stress and area-strain parameters in a first-order approximation.

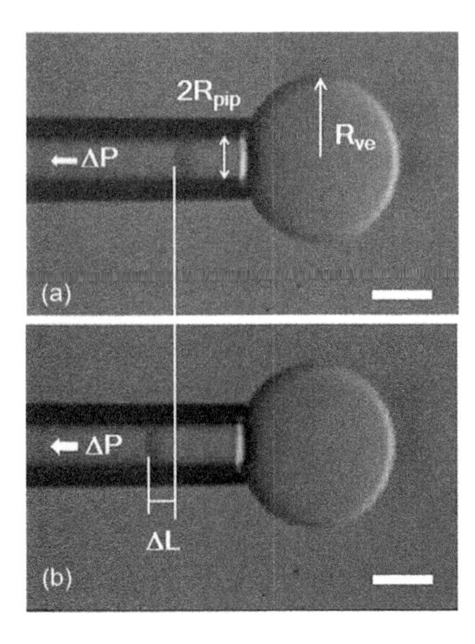

Figure 11.7 Experiment to determine membrane area expansivity. A SOPC GUV (25-μm diameter) is aspirated into the micropipette (8-μm inner diameter) and ready for the expansion experiment: (a) application of an intermediate suction pressure $\Delta P \sim 1$ *kPa* producing ~1.5% area change; and (b) further expansion just before tensile failure at $\Delta P \sim 2$ *kPa* that corresponds to an area change of ~3%. Scale bar: 10 μm. (Reprinted from *Chem. Phys. Lipids*, 181, Bagatolli, L.A. and Needham, D., Quantitative optical microscopy and micromanipulation studies on the lipid bilayer membranes of giant unilamellar vesicles, 99–120, Copyright 2014, with permission from Elsevier.)

The typical aspiration of a single 30-μm-diameter vesicle by an 8-μm-diameter pipette at suction pressures of around 0.1 kPa represents the application of forces ~10^2 nN. Using Eqs. 11.7 and 11.8, it can be calculated that a suction pressure of 1 kPa produces a tension of ~2.3 mN/m and a corresponding area change of ~1.5% for a membrane lipid composition of SOPC. When carried out in small area increments, the elastic area expansion modulus is obtained from the slope of this series of membrane tension, Σ, plotted against the corresponding area change, α. For this particular experiment done in SOPC GUV, under steady-state conditions (holding ΔP in each step for several seconds), K_A was measured to be ~190 mN/m.

Nevertheless, as it was shown in Section 11.3.2, a more sophisticated model predicts a significant impact of thermal undulations on area expansion under the full range of increasing tensions. Thus, the slope of Σ versus α only reflects an apparent expansivity modulus, K_{app}, rather than the actual K_A given that the applied tension is also employed to smooth the thermal bending fluctuations in the high-tension regime. An average bending modulus κ can be then employed to correct the apparent area changes under high tensions as it follows:

$$\alpha_{corr} \equiv \alpha - \frac{k_B T}{8\pi\kappa} ln\left(\Sigma / \Sigma_0\right) = \frac{\Sigma}{K_A} \tag{11.12}$$

Therefore, the applied membrane tension, Σ, can be plotted as a function of the measured values of area expansion, $\alpha = \Delta A / A_0$, obtained from Eq. 11.8 and corrected this way, α_{corr}, and a linear trend is predicted, whose slope will be directly the area expansivity, K_A. The K_A values calculated for fluid bilayers following this method are significantly higher than the apparent expansivity moduli, K_{app}, and show a little variation with chain length and unsaturation, although they are strongly affected by the cholesterol contents, especially in the case of phospholipids containing saturated acyl chains (Rawicz et al., 2000, 2008).

By using a more sophisticated micropipette setup, described in Section 11.2.4 and more thoroughly by Heinrich and Rawicz (2005), it is possible to obtain information about different kinetic regimes occurring upon membrane expansion and eventual rupture. Very precise linear tension ramps, ranging between 10^{-2} and 10^{2} mN/m/s, can be applied to GUVs and obtain accurate information on membrane area expansivity and rupture from nonequilibrium measurements. When performed at medium-high loading rates (above ~1 mN/m/s), the area expansion tests reveal a certain deviation from linearity, especially noticeable for the largest deformations (Rawicz et al., 2000, 2008). As presented in Figure 11.8, the experimental data of membrane tension as a function of area expansion (corrected from thermal undulations) deviates from linearity for area changes larger than 2%–3%. In order to explain this deviation, a new model was proposed by Evans and coworkers (Rawicz et al., 2000), the so-called *polymer brush model*, based on the classic theory of Flory for polymers (Flory and Volkenstein, 1969). In this model, each monolayer is considered as a collection of extended, structureless polymer chains, and the surface pressure in the fluid bilayer is dominated by confinement of chain entropy, neglecting van der Waals attraction between chains and specific headgroup interactions. It predicts an inverse cubic relationship between the membrane tension, Σ, and the area change, α, given by the following expression:

$$\Sigma = 2\Pi_0\left[1 - 1/(1+\alpha)^3\right] \tag{11.13}$$

where Π_0 is the monolayer surface pressure in the absence of mechanical tension. From thermodynamic minimization of

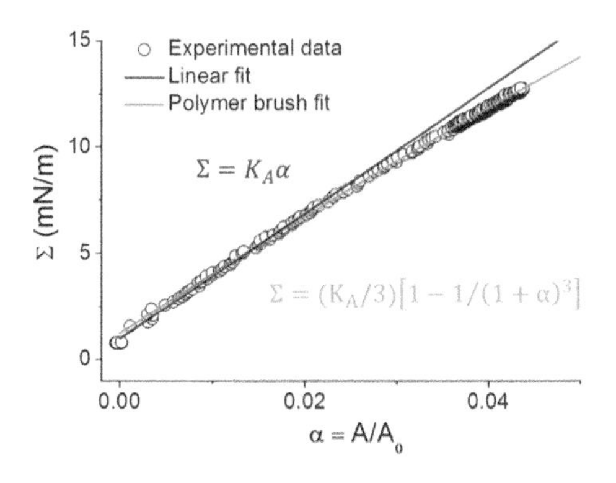

Figure 11.8 Example of an area expansion test done at a fast tension ramp. Collected data for one POPC GUV aspirated at increasing suction pressures (5 mN/m/s) is represented by open circles and fitted in two different ways: using a standard linear fit (Eq. 11.3, red line) and using the polymer brush model (Eq. 11.13, green line). Notice that at large tensions, bilayers undergo a certain degree of hyper-stretching before rupture. Area expansion moduli obtained from each fitting were the following: K_A(polymer brush) = 293.16 mN/m; K_A(linear) = 263.47 mN/m. (Adapted from Parra-Ortiz, E., Effects of pulmonary surfactant proteins SP-B and SP-C on the physical properties of biological membranes, in *Department of Applied Physics III*, Madrid, Spain, Complutense University, 2013.)

the free energy with respect to lipid surface density, Π_0 is equal to the interfacial energy density for exposing hydrocarbons to water, γ_{hcw}: $\Pi_0 = \gamma_{hcw}$ (Evans and Skalak, 1980). Following the assumption that surface pressure only depends on chain extension, the elastic modulus, K_A, is given by monolayer surface pressure as follows:

$$K_A = \left(\frac{\partial \Sigma}{\partial \alpha}\right) = 6\Pi_0 \tag{11.14}$$

Consequently, the model predicts that surface pressure and elastic modulus are governed by the interfacial energy density, which is affected by the chemical properties of the acyl chains, such as the unsaturation degree, discarding polar interactions at the headgroup level. As was mentioned before, the nonlinearity of the process of membrane expansion is especially noticeable in hyper-stretched vesicles, aspirated at fast tension ramps (above ~1 mN/m/s), being able to reach surface tensions higher than 10 mN/m before vesicle rupture. This can be achieved thanks to an optimized setup such as the one described in Section 11.2.4, using a computer-controlled syringe pump able to reach sufficiently fast tension ramps. By this way, it is possible to examine both lipid area elasticity and rupture process over a 2-fold greater tension range than in other studies, which reveals the anharmonic character of lipid chain interactions in GUVs (Rawicz et al., 2000, 2008).

From these results, it is clear that the lipid bilayer membrane is incredibly soft and expandable. However, it can only be expanded to a few percent before it fails in tension. As shown in Figure 11.7b, increasing the suction pressure to ~2 kPa brings the vesicle to the onset of tensile rupture at Σ_{rup} ~ 6 mN/m and only ~3% area dilation. The incorporation of cholesterol is a very effective way to increase this resistance to expansion. When taken to the limits of composition, that is, 50 mol% Chol in a bilayer composed of a long (diarachidonoylphosphatidylcholine [diC20]) saturated chain phospholipid, the elastic modulus become K_A ~ 10^3 mN/m and the tensile strength Σ_{rup} ~ 40 mN/m (Needham and Zhelev, 2000), equivalent to the compressibility and strength of bulk hydrocarbons and polyethylene. Polyunsaturated lipid bilayers show prominent changes in rupture strength depending on the number of double bonds in the acyl chain, finding that two or more unsaturations make the bilayer around one-third to one-half as strong as those with monounsaturated or saturated chains (Olbrich et al., 2000), which directly correlates with water permeability (explained in more detail in Chapter 20).

Also, in a kinetic sense, the strengths of molecular anchoring and material cohesion in fluid–lipid membranes increase with the rate of force and tension loading, found to be a function of the logarithm of the loading rate. Using a range of force rates, Evans and coworkers confirmed this by pulling single molecules from membranes with force rates from 1 to 10^4 pN/s (Evans and Ludwig, 2000) and rupturing giant membrane vesicles with tension rates from 10^{-2} to 10^2 mN/m/s (Evans et al., 2003). At a loading rate of ~10^{-2} mN/m/s, the tensile strength for an SOPC membrane was measured to be ~10 mN/m; but at 100 times greater loading rate (10 mN/m/s), the tensile strength had doubled to 20 mN/m. They concluded that, "under dynamic loading, rupture events below the peak in the tension distribution are likely to be governed by collective behavior, whereas beyond the peak events are likely to reflect microscopic peculiarities." The rupture kinetics of whole GUVs aspirated at a wide range of tension ramps were further analyzed in greater detail, from which quantitative images of the energy landscapes controlling the thermally activated process of membrane rupture were obtained by examining classical nucleation theory (Evans and Smith, 2011). From this sophisticated analysis, the authors were even able to estimate that the sizes of the holes, created by tension and responsible for bilayer breakdown, were on the order of nanometers, thus remaining in the molecular scale. Therefore, membrane rupture is a complex process in which many factors interplay: the hydrophobic and van der Waals interactions, balanced by acyl chain entropy confinement, forms a rather stable structure that requires a large energy to create a hole, which is a thermally activated process that also depends crucially on the applied mechanical tension.

To summarize Sections 11.3.2 and 11.3.3, some of the reported values of bending rigidity κ, apparent area expansivity, K_{app}, area expansivity corrected from thermal undulations, K_A, and rupture tensions, Σ_{rup}, obtained by micropipette techniques for selected lipid compositions are listed in Table 11.1, showing the effect of acyl chain lengths and unsaturation in these parameters. The effect of cholesterol contents and temperature in K_A and Σ_{rup} of GUVs prepared from binary and ternary mixtures is shown in Table 11.2.

11.3.4 YIELD SHEAR AND SHEAR VISCOSITY

Although liquid bilayers, by definition, do not support elastic shear, solid (or gel) phase lipid bilayers at temperatures below their main acyl melting transition do support shear stress elastically and have a measureable shear viscosity. When the acyl chains are crystallized, initial suction pressures for partially deflated GUVs (for instance, obtained by suspending in a hypertonic medium) are dominated by the surface shear rigidity or yield shear μ. For this experiment, we therefore need to work below the transition temperature of the corresponding lipid composition. The particular experiment shown in Figure 11.9 was performed on a single dimyristoylphosphatidylcholine (DMPC, see Appendix 1 of the book for structure and data on this lipid) GUV at 13°C, being that its T_m = 24°C (Evans and Needham, 1987). To start the experiment, the GUV is fully aspirated into the pipette and shaped into a solid, "nose-like" replica of the pipette-aspirated geometry, equivalent to that depicted in Figure 11.9d. This replica is carefully ejected, the GUV rotated 180° in suspension and re-aspirated under low suction pressure again, and the vesicle is now in the starting position for a yield shear and shear viscosity experiment. For these solid-phase bilayers, there is a pressure threshold, P_0, below which the solid membrane is elastically deformed by suction but only up to a projection into the pipette of one pipette radius (Figure 11.9a), beyond which it yields and flows as a plastic material (Figure 11.9b and c). This threshold can be then determined by obtaining the suction pressure necessary to achieve a projection length, L_p, equal to one R_{pip} inside the pipette. This threshold is proportional to the surface shear rigidity or yield shear, μ, following the expression (Evans and Needham, 1987):

$$P_0 \sim 8\mu \, ln\left(R_{ve} / R_{pip}\right) / R_{pip} \qquad (11.15)$$

Table 11.1 **Mechanical parameters for fluid PC bilayers of varying chain lengths and degrees of unsaturation measured with the micropipette technique, see Appendix 1 of the book for structure and data on these lipids**

GUV COMPOSITION	κ [10^{-19} J]	K_{app} [mN/m]	K_A [mN/m]	Σ_{rup} [mN/m]
DMPC (diC14:0)	0.56 ± 0.06 (29°C)[a]	150 ± 14 (29°C)[a]	234 ± 23 (29°C)[a] 290 ± 6 (15°C)[a]	~2.5 (Eq, ± 15°C)[b] 12 ± 3 (2 mN/m/s, 15°C)[c]
SOPC (C18:0/1)	0.90 ± 0.06 (18°C)[a]	208 ± 10 (18°C)[a]	235 ± 14 (18°C)[a] 290 ± 17 (32°–35°C)[c]	10 (2 mN/m/s, 18°C)[d] 9 ± 2 (0.1 mN/m/s, 21°C)[e]
DOPC (diC18:1_{c9})	0.85 ± 0.10 (18°C)[a]	237 ± 16 (18°C)[a]	310 ± 20 (15°C)[c] 265 ± 18 (18°C)[a]	10 ± 3 (2 mN/m/s, 15°C)[c] 9.9 ± 2.6 (0.1 mN/m/s, 21°C)[e]
SLoPC (C18:0/2)	0.46 ± 0.07 (18°C)[a]	193 ± 17 (18°C)[a]	241 ± 22 (18°C)[a]	4.9 ± 1.6 (0.1 mN/m/s, 21°C)[e]
DLoPC (diC18:2)	0.44 ± 0.07 (18°C)[a]	190 ± 18 (18°C)[a]	247 ± 21 (18°C)[a]	9.9 ± 2.6 (0.1 mN/m/s, 21°C)[e]
DLnPC (diC18:3)	0.38 ± 0.04 (18°C)[a]	159 ± 19 (18°C)[a]	244 ± 32 (18°C)[a]	3.1 ± 1.0 (0.1 mN/m/s, 21°C)[e]
DAPC (diC20:4)	0.44 ± 0.05 (18°C)[a]	183 ± 8 (18°C)[a]	250 ± 10 (18°C)[a]	2.3 ± 0.6 (Eq, 15°C)[f]
DEPC (diC22:1)	1.20 ± 0.15 (21°C)[a]	244 ± 8 (21°C)[a]	263 ± 10 (21°C)[a]	17 (2 mN/m/s, 18°C)[d]

Note: The values of bending modulus, κ, apparent area expansivity, K_{app}, area expansivity corrected from thermal undulations, K_A, and mechanical tensions upon rupture, Σ_{rup}, are listed; the corresponding experimental conditions such as temperature and tension loading rate are also shown. DAPC, diarachidonoylphosphatidylcholine; DEPC, dierucoylphosphatidylcholine; DLoPC, dilinoleoylphosphatidylcholine; DOPC, dioleylphosphatidylcholine; Eq, equilibrium conditions for Σ_{rup} determination (stepwise experiment); SLoPC, 1-stearoyl-2-linoleoylphosphatidylcholine.

[a] Rawicz et al. (2000).
[b] Needham et al. (1988).
[c] Rawicz et al. (2008).
[d] Evans et al. (2003).
[e] Olbrich et al. (2000).
[f] Needham and Nunn (1990).

Table 11.2 **Elasticity and strength of binary and ternary lipid mixtures measured with the micropipette technique**

	K_A [mN/m]		Σ_{rup} [mN/m]	
GUV COMPOSITION	15°C	32°C–35°C	15°C	32°C–35°C
Binary				
DOPC/Chol 1:1	890 ± 64	870 ± 141	19 ± 4	16 ± 2
SOPC/Chol 1:1	1,985 ± 330	1,130 ± 110	26 ± 3	21 ± 3
SM/Chol 1:1	3,327 ± 276	2,193 ± 209	33 ± 4	26 ± 5
Ternary				
DOPC/SM/Chol 1:1:1	655 ± 128	610 ± 61	15 ± 3	12 ± 3
SOPC/SM/Chol 1:1:1	1,725 ± 300	880 ± 130	21 ± 3	17 ± 3
SOPC/SM/Chol 1:1:2	2,188 ± 331	1,377 ± 172	26 ± 3	—

Source: Rawicz, W., et al., *Biophys. J.*, *94*, 4725–4736, 2008.
Note: The values of area expansivity corrected from thermal undulations, K_A, and mechanical tensions upon rupture, Σ_{rup}, are listed, measured at tension loading rates of 2 mN/m/s and two different temperatures.
Abbreviation: SM, sphingomyelin.

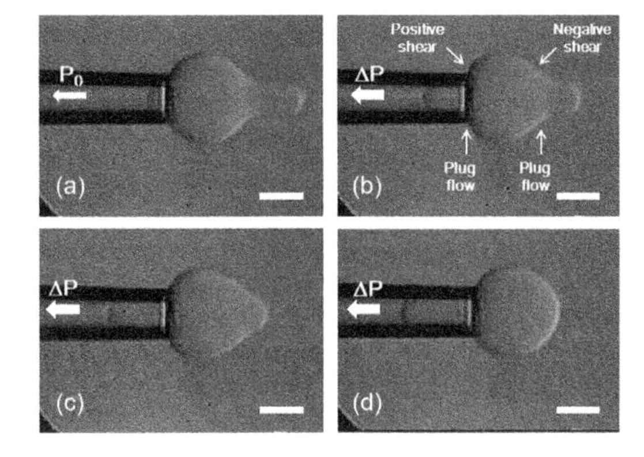

Figure 11.9 Example of a yield shear and shear viscosity experiment. It is done on a solid-phase GUV (here, DMPC at 13°C) and consists on several steps. First, the GUV is aspirated into the pipette and shaped into a "nose," a solid replica of the pipette-aspirated geometry. This replica is carefully ejected, the GUV is rotated 180° and re-aspirated under low suction pressure again; (a) the threshold pressure, P_0, is obtained as the ΔP that makes $L_p = 1 \cdot R_{pip}$, which is determined by the yield shear, μ; (b) a $\Delta P \sim 6$ times excess of yield, P_0, is applied, causing the membrane to flow, undergoing positive and negative shear; (c) a constant $\Delta P \sim 6 \cdot P_0$ creates plug flow of the vesicle projection into the pipette, whose rate of entry depends on the surface shear viscosity, η_s; and (d) the solid vesicle is completely re-aspirated into the micropipette, creating a new "nose." Scale bar: 10 μm. (Reprinted with permissions from Evans, E. and Needham, D., *J. Phys. Chem.*, 91, 4219–4228. Copyright 1987 American Chemical Society. and Reprinted from *Chem. Phys. Lipids*, 181, Bagatolli, L.A. and Needham, D., Quantitative optical microscopy and micromanipulation studies on the lipid bilayer membranes of giant unilamellar vesicles, 99–120, Copyright 2014, with permission from Elsevier.)

Furthermore, to measure and observe the shear viscous response of the solid bilayer, the suction pressure is increased to ~6 times beyond this P_0 yield point. The vesicle is then deformed continuously and flows into the micropipette at a steady rate, exhibiting positive and negative shear as depicted in Figure 11.9b. The "nose" of the vesicle and the projection inside the micropipette undergo simple plug flow as shown in Figure 11.9c; then, this excess pressure is maintained constant until the solid vesicle membrane is completely re-aspirated into the micropipette (Figure 11.9d). The rate of entry $\dot{L}$ of the solid-bilayer flowing into the pipette is determined by the surface shear viscosity, η_s:

$$\dot{L} \sim \Delta P \cdot R_{pip}^2 / \left[4\eta_s \, ln\left(R_{ve} / R_{pip} \right) \right] \quad (11.16)$$

Notice that here, ΔP denotes the difference between the suction pressure and the threshold, $\Delta P = P - P_0$. Therefore, the membrane viscosity can be obtained by representing $\dot{L}$ as a function of the logarithmic term.

The results of this particular experiment done on DMPC GUVs showed that solid-phase phospholipid membranes display Bingham plastic material behavior, that is, it behaves as a rigid body at low stresses but flows as a highly viscous fluid at high stress (Evans and Needham, 1987). As discussed there, "yield shear and shear viscosity reflect density and mobility of crystal defects (grain boundaries and/or intragrain dislocations) in the solid bilayer membranes." In its rippled solid phase $P_{\beta'}$,[1] yield shear values for DMPC go from 0.001 to 0.003 mN/m by reducing its temperature just below its T_m (from 20°C down to 13°C).

[1] The prime superscript identifies lipid phases where the acyl chains are tilted with respect to the normal to the layer surface.

This structure can be removed or prevented from formation by applying a moderate lateral stress (Needham and Evans, 1988). Upon crossing DMPC solid-solid phase transition from $P_{\beta'}$ to $L_{\beta'}$ there is a corresponding increase in the shear rigidity μ by a factor of 10, up to 0.036 mN/m. Similarly, the surface shear viscosity in flow η_s increases from close to zero in the liquid phase L_α (liquid bilayers have a viscosity similar to a 2 poise oil), covering 0.003–0.026 mN·s/m over the same temperature range. These surface shears represent equivalent bulk shear rigidities from 250 to 750 N/m² (0.25 to 0.75 MPa), reaching 9 MPa in the $L_{\beta'}$ phase; and the surface shear viscosity ranges from 750 to 6,500 N·s/m² (0.75 to 6 kPa·s). These values are comparable to yields for polypropylene of ~20–80 MPa, and plastic viscosities of 6 kPa·s at relatively low shear rates. Once again this hydrocarbon membrane, despite being only two molecules thick, behaves much like common hydrocarbon materials.

11.3.5 THERMAL TRANSITIONS AND THERMOMECHANICAL BEHAVIOR

The micropipette technique also allows studies to be made of thermal transitions in single GUVs. As is well described by Nagle and others, the partial specific volume for pure lipids has a direct temperature dependence (increases with increasing temperature), best represented as the coefficient of thermal expansion α_T given by (1/V)(dV/dT), which is around 10^{-3}°C^{-1} for DPPC in gel phase (Nagle and Wilkinson, 1982; Wiener et al., 1988). At the main acyl melting transition for DPPC centered at 41.6°C, there is an anomalous specific volume change of 0.0383 mL/g, which corresponds to an increase in bilayer volume of ~1170–1220 Å³/molecule, that is, 4% (Tristram-Nagle and Nagle, 2004), and a peak in the coefficient of expansion. This peak gradually disappears with increasing cholesterol (Melchior et al., 1980), consistent with the disappearance of the specific enthalpy of the transition measured by differential scanning calorimetry (Heimburg, 1998). Thus, the inclusion of cholesterol not only stiffens lipid bilayer membranes, it also abolishes all main acyl melting transitions. These effects have also been seen and quantified by micropipette experiments on GUVs.

The increase in specific volume upon melting is made up of a decrease in the average length of each lipid, detected as a thickness change of ~25% (from 54 Å to 40 Å), and an increase in area per molecule of ~60% (from 40 Å² to 65 Å²) (Tristram-Nagle and Nagle, 2004). If a free-standing GUV membrane goes through its melting transition, this large change in area means that the membrane will simply bud off and form attached daughter vesicles given the presence of thermal undulations. However, the micropipette can support the membrane with low suction pressures, and hold the single vesicle while it undergoes its main crystallization phase transition, T_m. Moreover, although not able to measure the bilayer thickness, these structural transition effects are manifest in micropipette experiments that measure the area of a GUV.

Micropipette experiments devised and conducted on DMPC by first Evans and Kwok (1982) and then Needham and Evans (1988), measured the area changes below, at, and above the phase transition, together with the structure and mechanical properties of the liquid crystalline and gel phases, including a mechanical model for the ripple phase, $P_{\beta'}$ (Evans and Kwok, 1982; Needham and Evans, 1988). When subjected to hyperosmotic conditions, the vesicle volume can be reduced and so produces excess area over a sphere of the same volume, allowing the membrane to contract in area through the transition (from above the transition) as it freezes and then back up through the transition (melting) with all membrane supported. For DMPC under low stress, the $P_{\beta'}$ rippled phase was allowed to form, and the area change of the GUV membrane as it was cooled through the transition region was measured to be 22%. By measuring the area of the vesicle versus temperature, this experiment also determined the thermal area expansivities of the membrane as: $4.2 \pm 0.20 \times 10^{-3}$°C^{-1} above the transition at 29°C; $5.8 \pm 0.42 \times 10^{-3}$°C^{-1} in the $P_{\beta'}$ phase at 16°C; and 3.0×10^{-3}°C^{-1} in the L_b phase below the pretransition at 8°C, matching the dilatometry measurements for membrane volume for DPPC (Melchior et al., 1980).

This technique has been also applied to phospholipid mixtures (Evans and Needham, 1986) and phospholipid:Chol mixtures (Evans and Needham, 1986; Needham et al., 1988). For lipid mixtures, in the example for SOPC:1-palmitoyl-2-oleoylphosphatidylethanolamine (POPE) (Evans and Needham, 1986) shown in Figure 11.10, by reducing the temperature of the bathing chamber from 18°C to 15°C for a GUV composed of the phospholipid mixture SOPC:POPE 40:60, the vesicle undergoes a phase transition from liquid L_α to solid L_β. The area change between both states can be readily measured using Eq. 11.8, and for this lipid mixture it was found to be around 26% of the L_α vesicle area.

Applied levels of membrane tension should be low or moderate (~0.1–2 mN/m) so as to be far below the rupture threshold. Usual cooling/heating rates are between 0.1°C and 10°C/min: Low rates are used for more accurate measurements. However, for heating and cooling rates of 0.5°C/min, some hysteresis in the area versus temperature curves was observed (Evans and Needham, 1986). For phospholipid:Chol mixtures, the micropipette experiments correlated exactly with the progressive reduction in excess enthalpy and dilatometry of the main acyl chain melting transitions (Evans and Needham, 1986; Needham and Evans, 1988), again measured

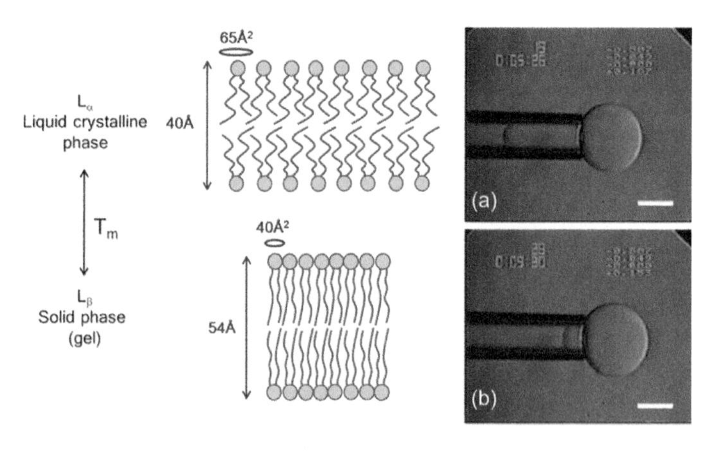

Figure 11.10 Example of a phase transition experiment through area changes. It was done applying a fixed ΔP on a SOPC:POPE 40:60 GUV: (a) vesicle in liquid crystalline phase L_α by setting the chamber temperature at 18.7°C (higher than its T_m, 17°C in this case); and (b) the same vesicle after solidification, at 15.3°C in its solid L_β phase, where the membrane thickness is larger and the area per molecule smaller so that the total membrane area is reduced. At the melting transition, a lipid bilayer typically goes from a thickness and area per molecule of 54 Å and 40 Å², to 40 Å and 65 Å², respectively. Scale bar: 10 μm.

for DPPC:Chol (Melchior et al., 1980). Membrane area changes were gradually reduced and eventually abolished as the cholesterol content was increased from 0 to 50 mol%.

Therefore, this approach can be potentially used to construct, inform or complement phase diagrams and tie lines for different lipid compositions in terms of areas per molecule or mechanical properties (see Chapter 18 for further information on these topics).

11.3.6 OTHER MECHANICAL STUDIES USING MICROPIPETTES

Micropipette-based approaches have been also used for different studies related to membrane mechanics, taking advantage of its ability to manipulate single GUVs and operate and adapt to different kinds of experiments. For instance, micropipettes have been used in conjunction with shear to develop a micro-rheometer of giant vesicles, measuring flow field effects on the membrane behavior (Fa et al., 2004). Oscillatory shear motions of different amplitudes and frequencies were applied by a moving plate, yielding to different shear rates that were found to reduce the amount of membrane area stored in fluctuations. These results would introduce a shear flow contribution to the relationships between area change, mechanical tension and bending modulus described in Section 11.3.2.

Line tension is another interesting mechanical parameter that has been studied by micropipette manipulation. This term is used for describing two intrinsically different processes: on the one hand, the formation and expansion of membrane pores by application of external stresses; and on the other hand, the lateral segregation between lipid domains that occurs in membranes of phase-separated mixtures. In both cases, the line tension corresponds to the energy per unit length of the bilayer contour at the pore/domain edge, and ultimately controls the kinetics of opening and closure of the new phase (the aqueous pore or phase-separated domain, respectively). As for the pore line tensions, electromechanical experiments were performed combining electroporation and mechanical control by micropipette aspiration to create hydrophilic pores under different voltages and membrane tensions. The critical voltage for electropermeabilization was found to decrease for increasing membrane tensions (Needham and Hochmuth, 1989). In later experiments, micrometer-sized pores were formed in single GUVs using electroporation and opposed by a far-field tension applied by the micropipette, which balanced the tendency for the pore to close, and allowed a direct determination of pore sizes and line tensions. There, the line tensions were determined for two lipid compositions, SOPC and SOPC with 50 mol% cholesterol, and the obtained values for single bilayers were 0.92×10^{-11} N and 3.05×10^{-11} N (Zhelev and Needham, 1993). In Chapter 15, the interested reader can find further details on the application of electric fields on GUVs. Regarding the line tension that arises between lipid domains in complex membrane compositions, direct measurements have been performed by the micropipette technique combined with fluorescence microscopy, obtaining values in the range between 0.5–3 pN for different GUV compositions containing saturated, unsaturated phospholipids, and cholesterol (Tian et al., 2007). Based on these approaches, a theoretical framework was also developed to understand vesicle shape transitions in ternary lipid mixtures modulated by bending stiffness and line tensions (Das et al., 2008). The reader is referred to Chapters 2 and 18 for further information about complex lipid compositions, lipid domains and phase diagrams on GUVs.

In addition, micropipette aspiration has been employed to assess spontaneous curvature in single giant vesicles, being able to connect curvature, excess area, phase separation, and membrane tension with nanotube formation and stability: Deflated vesicles would form stable nanotubes only in the presence of positive spontaneous curvature, whereas the tubes can be retracted back into the vesicle by increasing membrane tensions (Li et al., 2011). Further applications of micropipettes on nanotube pulling from GUVs can be found in Chapter 16; this is a method that has been proposed to characterize membrane mechanics and dynamics such as internal friction, and to study the effect of curvature on lipids and membrane proteins.

11.4 MOLECULAR EXCHANGE AND INTERACTIONS BETWEEN PAIRS OF VESICLES

In addition to the measurement of purely membrane mechanical properties, the versatility of the micropipette technique allows for the analysis of a range of other composition–structure–property relationships for single and pairs of GUVs. In Sections 11.4.1 through 11.4.3, some of these applications will be presented, including the following:

1. Molecular exchange from solution, where lipid bilayers are shown to exchange small amounts of other lipids and surfactants with their surrounding milieu
2. Accurate mechanical control and measurement of adhesion between pairs of vesicles
3. Positioning and mechanical control of the membrane tension to study fusion and hemifusion between vesicles.

The reader can find other applications of the micropipette manipulation methods in Chapters 15, 16, 20 and 24, where glass micropipettes are employed to micromanipulate single or pairs of vesicles to bring them together, to pull nanotubes from the vesicle surface under controlled membrane tensions, to measure volume changes in order to determine water permeability coefficients, and to measure area changes related to the incorporation of peptides and detergents, respectively.

11.4.1 MOLECULAR EXCHANGE FROM SOLUTION

Using a delivery pipette directed at a single vesicle, molecular exchange experiments with small water-soluble molecules like lysolipids (Needham and Zhelev, 1995; Needham et al., 1997) and bile acids (Evans et al., 1995) have shown the propensity of the membrane to take up these molecules from the surrounding solution, creating a corresponding membrane area change as the molecules become incorporated in the bilayer. The experiment that exposes a single GUV to a solution of lysolipid uses three micropipettes positioned around the microscope stage. As shown in Figures 11a–d, one pipette is used to hold the vesicle (right), and the other two to deliver test and bathing solutions (left). Using this method, it was shown that a water-soluble lysolipid (monoloeloylphosphatidylcholine [MOPC, see Appendix 1 of the book for structure and data on this lipid]) can partition into a liquid lipid bilayer membrane (SOPC) and actually expand its area

(Needham et al., 1997). This area expansion, measured at a low, constant suction pressure of ~0.2 mN/m to avoid GUV rupture, is readily measured from the vesicle geometry outside and inside the micropipette and can be used to quantify the lysolipid incorporation to the bilayer. For MOPC tested at its critical micelle concentration (CMC) of 1 μM, the initial maximum area change (Figure 11.11c in comparison with Figure 11.11a) was ~3%, which occurred in ~200 s of MOPC solution exposure. As shown in Figure 11.11d, when the lower flow pipette is repositioned to deliver the bathing solution, the lysolipid is readily washed out as the surrounding solution is exchanged for a MOPC-free media. This change in area, $\Delta A/A_o$ (calculated from Eq. 11.8), can be then converted to molar percentage of MOPC from the known areas per molecule of A_{MOPC} (35 Å^2) and A_{SOPC} (67 Å^2) by using the following equation:

$$\text{mol}\%\,MOPC_{bilayer} = \frac{\Delta A}{A_0}\frac{A_{SOPC}}{A_{MOPC}}\cdot 100 \qquad (11.17)$$

The uptake and wash-out kinetics are shown in Figure 11.11e, and this particular experiment showed that lysolipid saturates the outer monolayer at a concentration of ~6 mol% (Needham et al., 1997).

Although exposure to a surfactant at or below its CMC simply promotes adsorption into the bilayer, as shown in Figure 11.11f, when a GUV was exposed to a 100-μM MOPC solution (100 times in excess of its CMC), the vesicle area rapidly increased and the vesicle membrane fails within 20 s. MOPC was rapidly partitioned into the outer monolayer of the bilayer at a rate (0.2 s^{-1}), which was much faster than its ability to cross the bilayer by transmembrane flip-flop (0.0019 s^{-1}) (Needham and Zhelev, 1995). This resulted in rupture of the bilayer corresponding to an uptake of 16 mol% (Needham et al., 1997), consistent with previous studies in which MOPC was exchanged with egg phosphatidylcholine (PC) vesicles (Needham and Zhelev, 1995). The inference is that a rapid uptake into the outer monolayer induces a tensile failure before flip-flop can relax the tension. Interestingly, the presence of a bound polyethylene glycol (PEG) layer can inhibit micelle transfer into the bilayer and so prevent membrane failure in the presence of excess surfactants. As shown in Figure 11.11f, when PEG-lipid (DSPE-PEG750) was included in the bilayer, the presence of 20 mol% PEG-lipid completely prevented rupture even when the vesicle was exposed to 100 μM MOPC (Needham et al., 1997). With 20 mol% PEG-lipid incorporated in the bilayer, the level of uptake was found to be essentially identical to that for a vesicle exposed to the CMC of MOPC, 3 μM. This result implied that the presence of a saturating amount (20 mol%) of PEG-lipid in the lipid bilayer decreased the transport of micelles to the vesicle surface and therefore eliminated the partitioning of MOPC micelles into the bilayer at elevated bulk solution concentrations that would otherwise dissolve the vesicle. The initial inference is that

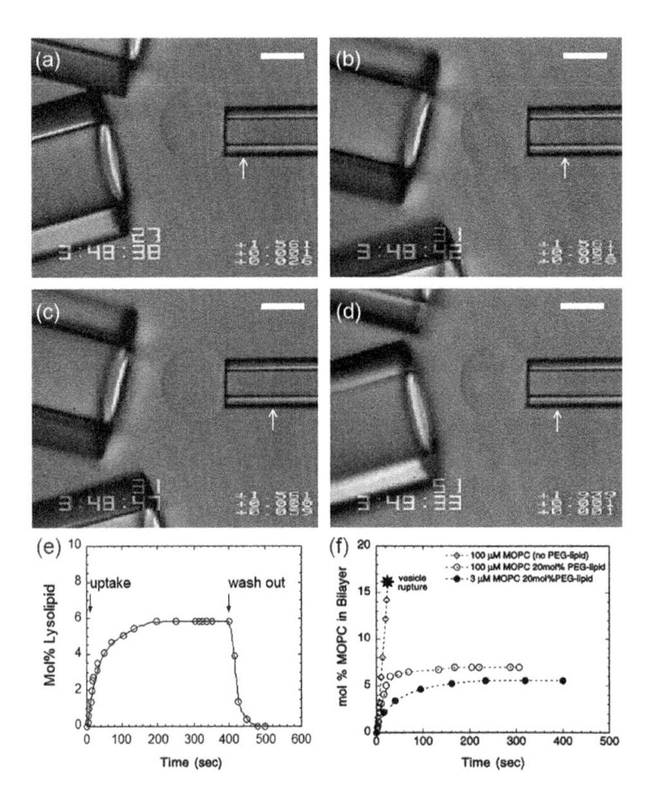

Figure 11.11 Example of a molecular exchange experiment. It was done for SOPC GUV exposed to an aqueous solution of the lysolipid MOPC. (a) GUV exposed to the bathing solution from the bottom flow pipette; (b) injection of a solution of MOPC from the top flow pipette, leading to lysolipid insertion that creates a change in total membrane area, measured from the increase in L_p; (c) maximum area change under lysolipid flow (1 μM); and (d) re-establish the bathing solution flow and wash out the lysolipid from the membrane, measured as a reduction in L_p. The arrows indicate the location of the edge of the vesicle projection inside the pipette. Scale bar: 10 μm. Images adapted from (Bagatolli and Needham, 2014), with permission from Elsevier. (e) Graph of uptake and desorption of MOPC by single SOPC vesicles after exposure to 1-μM MOPC solutions. (f) Exposure of single SOPC vesicles, with and without 20 mol% PEG-lipid, to a flow of MOPC solution (3 and 100 μM). Without PEG-lipid in the bilayer, exposure to 100 μM MOPC causes rapid expansion and rupture of the membrane. With 20 mol% PEG-lipid, uptake in 100 μM MOPC is essentially the same as that observed at the MOPC CMC (3 μM) and bilayers are stable. (Reprinted from *Biophys. J.*, 73, Needham, D. et al., Exchange of monooleoylphosphatidylcholine as monomer and micelle with membranes containing poly (ethylene glycol)-lipid, p. 2615, Copyright 1997, with permission from Elsevier.)

only monomeric species of MOPC can pass through the thin (25 Å) polymer layer (Kenworthy et al., 1995), even when there is a large excess concentration of micelles in the bathing medium.

In other micropipette solution-exchange experiments, instead of replacing the surrounding medium of the GUV using a flow pipette, the GUV was exposed to the adsorbing molecules by direct transfer to the final suspension, using dual-chamber assemblies on the microscope stage and a transfer capillary (or larger pipette) to protect the vesicle on its way through the air gap (Evans et al., 1995; Olbrich et al., 2000; Shi and Baumgart, 2015). For instance, the partitioning of the trihydroxy bile acid cholylglycine (CG) into a bilayer was studied with this transfer technique and a thermodynamic theory was developed, based on entropy of mixing and activity, both contributed by mechanical tension (Evans et al., 1995). With this more complex ionic surfactant, it was found that the energy barriers for uptake can arise from the electrical charges present on the acid group and restriction of bile acid movement across the bilayer. These measurements of area change also yielded important molecular scale properties for CG, such as the area per molecular complex in the bilayer (~60 A^2) and the ratio of aggregate size in the bilayer to that in aqueous solution (possibly monomers in the bilayer to dimers in solution at low ionic strength). They also showed a pronounced and reversible softening of the bilayers even at low CG concentrations, and the bilayer area elastic expansion modulus approached zero as the CG concentration in the bathing medium approached its CMC of 12 mM (Evans et al., 1995). More recent molecular exchange experiments using also more advanced setups have studied the interactions of micropipette-aspirated GUVs with amphiphilic proteins, namely endophilin N-terminal BAR (Bin/Amphiphysin/Rvs-homology) domains, by using a similar, well-controlled transfer method combined with confocal microscopy and optical tweezers: the coupling between membrane tension and density of curvature-inducing proteins was addressed, obtaining a membrane shape stability diagram consistent with a thermodynamic curvature instability model (Shi and Baumgart, 2015).

Therefore, these molecular-exchange experiments, made by virtue of being able to hold a single GUV under low and finite tension and to position additional micropipettes to deliver solutions of the exchangeable materials or to transfer the vesicle to different suspensions, allow an accurate quantification of the adsorption and desorption of membrane-active molecules, informing about the kinetics of the process and yielding even molecular scale properties. The reader is referred to Chapter 24 for further details on the interactions of GUVs with peptides or detergents and other experimental approaches to characterize them.

11.4.2 ADHESION BETWEEN PAIRS OF VESICLES

When manipulated in pairs and brought into tension-controlled contact, two GUVs are shown to be subject to the same range of attractive and repulsive colloidal interactions as many other colloidal particles, including: van der Waals attraction, limited by a very short range hydration repulsion; the variable-range power law of electrostatic repulsion (see Chapters 5 and 17 for detailed information on GUV adhesion measurements and the theory behind it); the presence of nonadsorbing polymers like PEG or dextrans that are then excluded from any adherent gap and exert an additional attractive stress in the contact by a depletion flocculation mechanism; and steric repulsive barriers due to the presence of lipid-anchored aqueous polymers like PEG (directly connected to Chapter 9, which describes the theory of polymer–membrane interactions). GUVs are also subjected to intimate mixing when manipulated into contacts that reduce the hydration barrier and allow membrane–membrane fusion or hemifusion under certain circumstances.

As first outlined and analyzed by Evans, an experimental procedure can be used to measure the interfacial free energy density for the adhesion of the membrane of large vesicles to other surfaces when the membrane force resultants are dominated by isotropic tension (Evans, 1980). A single GUV is aspirated by a micropipette with sufficient suction pressure to form a spherical segment outside the pipette. The vesicle is then brought into close proximity of the surface to be tested (e.g., another vesicle, but it could also be any microscopic surface), the suction pressure is reduced to permit adhesion, and the new equilibrium configuration is established. The mechanical analysis of the equilibrium shape provides the interfacial free energy density for the surface affinity.

Thus, by manipulating two GUVs into initial contact and then allowing one to spread on the other, the first measurement of vesicle–vesicle adhesion was for van der Waals interactions (Evans and Metcalfe, 1984). The adhesion of vesicles is promoted by the attractive potential between surfaces, and it is opposed by the mechanical rigidity of the membrane (Evans, 1980). In this experiment, vesicles are first slightly dehydrated so that long L_ps are produced inside the pipette. As shown for two SOPC vesicles in Figure 11.12, the right-hand vesicle is pressurized until the flaccid membrane becomes a rigid, spherical adhesion surface (~50–100 Pa). The adherent vesicle (left hand) is aspirated at low suction so it remains slightly flaccid (~20 Pa). Then, it is brought into initial contact with the adhesion surface (Figure 11.12a), and reductions in its membrane tension allows the vesicle to spread in discrete (equilibrium) steps (Figure 11.12b) until, as shown in Figure 11.12c, it is completely spread at an applied pressure value of 1–10 Pa, yet still held at one R_{pip} projection length in the left-hand pipette. Reversibility can be verified by observing the decrease in contact area as the suction is increased again. Therefore, the extent of adhesion is controlled *via* tension in the adherent vesicle membrane, where the membrane tension can be readily calculated with Eq. 11.7.

The mechanical equilibrium translates into the familiar Young-Dupré equation, which relates the free energy potential for membrane assembly of adhesion per unit area W (defined as minus the free energy potential at the minimum) to the membrane tension and to the contact angle, θ, between the vesicles at the contact zone (Evans, 1980; Evans and Needham, 1987), which in equilibrium is:

$$W = \Sigma_{eq}\left(1-\cos\theta\right) \tag{11.18}$$

Adhesion tests for bilayers composed of neutral PC (egg PC, DMPC, SOPC, and SOPC:POPE mixtures) in the L_α state yield comparable values of W between 0.01 and 0.015 mJ/m^2 (Evans and Needham, 1986, 1987), consistent with the van der Waals attractive energy potential limited by hydration repulsion at a gap of ~25 Å as measured by X-ray diffraction on multibilayer systems (Lis et al., 1982). Vesicles containing the less-hydrated phospholipid POPE exhibit much stronger adhesion, exhibiting

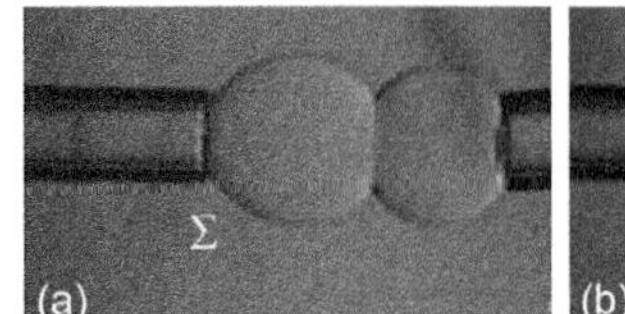

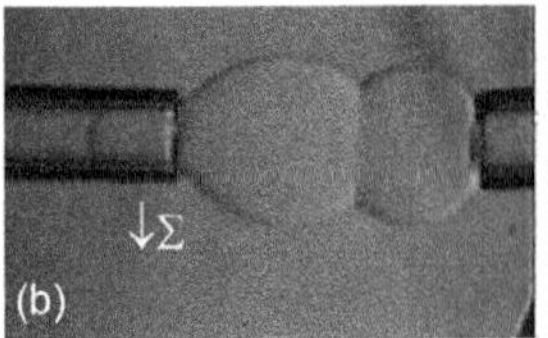

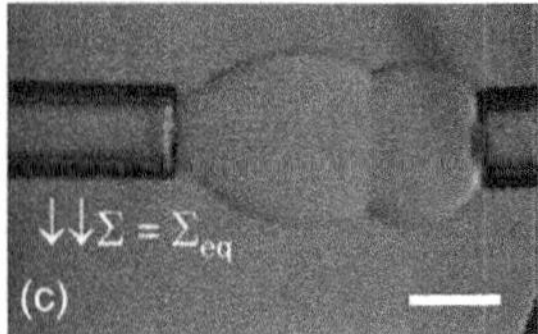

Figure 11.12 Example of a vesicle–vesicle adhesion experiment. (a) Starting position, where the two vesicles are brought into initial contact. (b) The left-hand vesicle is allowed to spread on the right-hand vesicle surface by stepwise reductions in membrane tension Σ. (c) The two vesicles are in equilibrium spread adherent contact when the left-hand vesicle is held at Σ_{eq}, for $L_p = 1 \cdot R_{pip}$. Scale bar: 10 μm. (Reprinted with permissions from Evans, E. and Needham, D., *J. Phys. Chem.*, 91, 4219–4228. Copyright 1987 American Chemical Society.)

contact energies an order of magnitude greater (0.12–0.15 mJ/m^2) consistent with the optimum fit of a theoretical prescription for the interlamellar stress to data again consistent with the smaller interbilayer gaps of ~11 Å (Evans and Needham, 1986). Even larger free energy potentials (0.22 mJ/m^2) are measured for adhesion between bilayers of the glycolipid digalactosyldiglyceride (DGDG , see Appendix 1 of the book for structure and data on this lipid) where sugar headgroups are now exerting additional attractive interactions (Evans and Needham, 1987).

Specific adhesion between GUVs can also be driven by ligand–receptor bonds, where membrane–membrane adhesion involves specific molecular binding and cross-bridging reactions. The mechanics of membrane–membrane adhesion and separation were developed by Evans for the case of discrete, kinetically trapped cross bridges (Evans, 1985). The results yielded specific values of the macroscopic tension applied to the membrane in the plane region away from the contact zone and the microscopic angle at the edge of the contact zone. The attraction between bilayers in concentrated solutions of nonadsorbing polymers has been also carefully analyzed, by exposing GUVs to concentrated solutions of dextrans, finding that it is due to the osmotic pressure reduction at the midpoint of the gap between surfaces arising from polymer depletion (Evans and Needham, 1988). Later, interbilayer adhesion has been studied with micropipette techniques for single GUVs containing a certain ligand manipulated and presented to their corresponding receptors. For instance, the biotin–avidin interaction at lipid vesicle surfaces was used to mimic ligand–receptor binding, receptor-mediated adhesion and macromolecule aggregation, including the influence of a surface grafted polymer, PEG750 (Noppl-Simson and Needham, 1996). Single vesicles were manipulated in solutions of fluorescently labeled avidin to measure the binding rate of avidin to a biotinylated vesicle as a function of biotin concentration at the surface. When incorporating a certain amount of PEG on the bilayer surface, the rate of avidin adsorption was found to be four times less with 2 mol% PEG750 than for the unmodified surface, and 10 mol% PEG completely inhibited binding of avidin to biotin for a 2-min incubation. Then, using two pipettes, a direct vesicle–vesicle adhesion test can be also performed, similar to that shown previously in Figure 11.12. By controlling the vesicle membrane tension, this adhesion test provided a direct measure of the spreading pressure of the biotin–avidin–biotin cross bridges in the contact zone. From mechanical equilibrium, an expression equivalent to Eq. 11.18 can be derived, where the work of adhesion will be equal to the excess spreading pressure of cross-bridged avidin (Noppl-Simson and Needham, 1996). Assuming ideality, this spreading pressure gives the concentration of avidin cross-bridges in the contact zone. Once adherent, the membranes failed in tension before they could be peeled apart. This vesicle–vesicle adhesion experiment, with a lower tension limit of 0.01 mN/m, provided a useful method to measure the spreading pressures and therefore colligative properties of a range of membrane-bound macromolecules.

From this fairly involved experimental procedure and analytical approach, GUVs and micropipettes have been proven as unique and versatile systems for the study and evaluation of a range of colloid and surface phenomena at and between surfaces (Evans, 1980; Evans and Metcalfe, 1984; Evans and Needham, 1986, 1987, 1988; Evans et al., 1996), including the following:

- van der Waals attraction limited by hydration (plus thermal) repulsion for neutral bilayers;
- The classic Derjaguin-Landau-Verwey-Overbeek (DLVO) interaction of van der Waals attraction limited by electrostatic repulsion for charged bilayers;
- Depletion flocculation caused by nonadsorbing polymers (like PEG and dextrans) in the bathing solution, limited again by hydration repulsion, that can overcome electrostatic repulsion;
- Steric repulsion offered by grafted PEG polymer at the vesicle surfaces that oppose the inherent weak van der Waals attraction;
- The delicate balance between depletion flocculation of PEG in solution and the steric repulsion of the same molecule grafted to the bilayer surfaces;
- And the breakdown of membranes in contact (van der Waals versus hydration) producing membrane fusion.

These last studies regarding membrane hemifusion and fusion using micropipettes will be described in the next section, and all of them are intimately related to Chapters 9, 17 and 25, to which the reader is referred for further information.

11.4.3 FUSION AND HEMIFUSION

When manipulated into close contacts and under certain conditions, the adhesion between pairs of GUVs can lead to intimate lipid mixing and eventually to membrane merging in a total or partial way (named fusion or hemifusion, respectively). For instance, the processes of bilayer adhesion and hemifusion mediated by factors such as low pH, osmotic depletion by polymers, and fusogenic peptides has been characterized by careful micromanipulation of two GUVs, based on the method described in Section 11.4.2 and fluorescence imaging (Sun et al., 2011). Micropipette manipulation has been also used to study fusion and electrofusion, inducing them either by a ligand–mediated mechanism or by applying electric fields, respectively (Haluska et al., 2006; Bezlyepkina et al., 2013). Therefore, this methodology has allowed detailed observations and good control on the different stages of the fusion process, including adhesion, hemifusion, pore and neck formation and total merging, proposing specific mechanisms for this process until certain biologically relevant conditions.

11.5 OUTLOOK: CONCLUSIONS AND COMBINATION BETWEEN THE MICROPIPETTE TECHNIQUE AND OTHER APPROACHES

To summarize, in this chapter we have described the development and use of several micropipette techniques and analyses for GUVs, from the classical, pioneering studies done by Evans and coworkers to some of the latest optimizations and developments, demonstrating its usefulness and versatility to study the two molecule-thick lipid membrane. Based on the models and data for GUVs accumulated over the past 40 years, these micropipette manipulation techniques show the lipid bilayer to be an elastic material that is very soft, with compressibilities between those of a bulk liquid and a gas; is stiffened considerably by the inclusion of cholesterol to levels equivalent to polyethylene; as a solid material, it shows the yield shear and shear viscosity of a Bingham plastic; displays a 25% change in area when taken through its main acyl chain freezing transition; is permeable to water in relation to its compliance; and can exchange small amounts of other lipids and surfactants with its surrounding milieu. When manipulated in pairs, GUVs have also been found to be subject to the same range of attractive and repulsive colloidal interactions as many other particles, including: van der Waals attraction, limited by a very short range hydration repulsion and the variable-range power law of electrostatic repulsion; steric repulsive barriers, due to the presence of bound aqueous polymers like PEG; depletion flocculation by the same molecules when free in surrounding solution; and intimate mixing when manipulated into contacts that reduce the hydration barrier and allow membrane–membrane fusion under certain conditions. However, many other studies and combinations with different experimental and theoretical approaches can be still explored, especially to understand the mechanical properties of increasingly complex, native or quasi-native membrane systems, including for instance protein complexes and other cellular machinery, such as the protocells described in Chapter 28 and other giant vesicles directly prepared from natural membranes or extracts (see Chapter 2).

Furthermore, the combination of the micropipette technique with high-resolution fluorescence microscopy may offer important information about the correlation between phase behavior, domain separation, and mechanical properties, as described and discussed in a recent review article (Bagatolli and Needham, 2014). Several studies have already merged fluorescence microscopy with micromanipulation of single or pairs of GUVs and obtained interesting results. For instance, the group of Patricia Bassereau generated tubes from GUVs using molecular motors, combining a micropipette used for GUV manipulation and to set the membrane tension and optical tweezers to extract tubes and measure forces (Roux et al., 2002). In addition, these authors used this technical combination to explore whether lipid sorting can be mediated by membrane curvature in cholesterol-containing GUVs showing phase coexistence (Roux et al., 2005). The reader is referred to Chapter 16 for further details on these experiments. The combination of the micropipette technique and fluorescence confocal imaging has been also done by the group of Tobias Baumgart, which proposed a strategy to measure line tension in phase separated vesicles of various compositions (Tian et al., 2007) and developed a theoretical framework to understand vesicle shape transitions in ternary lipid mixtures modulated by bending stiffness and line tensions (Das et al., 2008). Electroporation and micropipette techniques have been also combined, where the mechanical tension applied to the GUV could be adjusted to keep the pores formed by electric pulses open from tenths of a second to several seconds, providing also a platform to measure line tensions at the pore edge (Zhelev and Needham, 1993), as already mentioned in Section 11.3.6, a combination that will be further explained in Chapter 15.

The micropipette technique provides a unique ability to characterize the mechanics and deformation of single GUVs in terms of well-defined material constants, including the elastic area expansivity, bending modulus and elastic yield and viscosity in shear for solid phase membranes, allowing a high control on vesicle geometry, suction pressure, membrane tension, surrounding conditions such as temperature, pH and controlled exposure to different molecules in the bathing medium. However, further improvements on spatial and time resolution, the evolution of the technique to more optimized and automatized micropipette setups able to acquire more and more precise information, together with the potential combinations between these improved setups with other experimental and theoretical approaches, still remains to be fully exploited.

In this chapter, we have described in detail the basic instrument together with possible optimizations and automations, hopefully encouraging any interested student or researcher to implement these improvements and apply them to the understanding of new GUV systems, such as proteolipid vesicles containing membrane proteins or giant vesicles directly formed from natural membrane extracts, systems that are still largely unexplored with this technique and not totally understood.

LIST OF ABBREVIATIONS

Chol cholesterol
CMC critical micelle concentration
CG cholylglycine (bile acid)
DAPC diarachidonoylphosphatidylcholine
DEPC dierucoylphosphatidylcholine
DGDG digalactosyldiglyceride
DLnPC dilinolenoylphosphatidylcholine
DLoPC dilinoleoylphosphatidylcholine
DMPC dimyristoylphosphatidylcholine
DOPC dioleylphosphatidylcholine
eq equilibrium
HMC Hoffman Modulation Contrast optics
MOPC monoloeloylphosphatidylcholine
pip pipette
POPE 1-palmitoyl-2-oleoylphosphatidylethanolamine
PEG polyethylene glycol

PC	phosphatidylcholine
SLoPC	1-stearoyl-2-linoleoylphosphatidylcholine
SOPC	1-stearoyl-2-oeloylphosphatidylcholine
SM	sphingomyelin
ve	vesicle

GLOSSARY OF SYMBOLS

A_0	area of the non-stressed vesicle
A_{pip}	pipette cross-sectional area
α	fractional change in membrane area (dilation or expansion)
α_T	coefficient of thermal expansion
ΔA	vesicle area change
ΔL	change in length of the vesicle projection in the pipette
ΔP	micropipette suction pressure
B	bending moments
C	membrane curvature
C_1, C_2	principal curvatures of the membrane
e_s	surface shear deformation
η_s	surface shear viscosity
F_{pip}	force applied at the pipette tip
K_A	area expansivity or elastic modulus
K_{app}	apparent area expansivity modulus
k_B	boltzmann constant:
κ	bending rigidity
$\dot{L}$	rate of entry of the vesicle projection length inside the pipette
L_α	liquid crystalline phase
L_β	solid/gel phase
L_p	projection length of the vesicle inside the pipette
λ_e	in-plane extension or surface shear at constant surface density
μ	surface shear rigidity or membrane shear modulus
$P_{\beta'}$	ripple phase
P_0	monolayer surface pressure in the absence of mechanical tension
R_{pip}	pipette radius
R_{ve}	vesicle radius
Σ	membrane tension
Σ_{rup}	membrane tension at rupture
T_m	phase transition temperature
τs	surface shear stress
θ	contact angle
W	free energy potential of adhesion per unit area

REFERENCES

Bagatolli LA, Needham D (2014) Quantitative optical microscopy and micromanipulation studies on the lipid bilayer membranes of giant unilamellar vesicles. *Chemistry and Physics of Lipids* 181:99–120.

Bezlyepkina N, Gracià R, Shchelokovskyy P, Lipowsky R, Dimova R (2013) Phase diagram and tie-line determination for the ternary mixture DOPC/eSM/cholesterol. *Biophysical Journal* 104:1456–1464.

Das S, Tian A, Baumgart T (2008) Mechanical stability of micropipet-aspirated giant vesicles with fluid phase coexistence. *Journal of Physical Chemistry B* 112:11625–11630.

Dimova R (2014) Recent developments in the field of bending rigidity measurements on membranes. *Advances in Colloid and Interface Science* 208:225–234.

Evans E (1973a) A new material concept for the red cell membrane. *Biophysical Journal* 13:926.

Evans E (1973b) New membrane concept applied to the analysis of fluid shear-and micropipette-deformed red blood cells. *Biophysical Journal* 13:941.

Evans E (1980) Analysis of adhesion of large vesicles to surfaces. *Biophysical Journal* 31:425.

Evans E (1985) Detailed mechanics of membrane-membrane adhesion and separation. II. Discrete kinetically trapped molecular cross-bridges. *Biophysical Journal* 48:185–192.

Evans E, Heinrich V, Ludwig F, Rawicz W (2003) Dynamic tension spectroscopy and strength of biomembranes. *Biophysical Journal* 85:2342–2350.

Evans E, Hochmuth R (1976) A solid-liquid composite model of the red cell membrane. *Journal of Membrane Biology* 30:351–362.

Evans E, Hochmuth R (1978) Mechanochemical properties of membranes. *Current Topics in Membranes & Transport* 10:1.

Evans E, Klingenberg D, Rawicz W, Szoka F (1996) Interactions between polymer-grafted membranes in concentrated solutions of free polymer. *Langmuir* 12:3031–3037.

Evans E, Kwok R (1982) Mechanical calorimetry of large dimyristoylphosphatidylcholine vesicles in the phase transition region. *Biochemistry* 21:4874–4879.

Evans E, Ludwig F (2000) Dynamic strengths of molecular anchoring and material cohesion in fluid biomembranes. *Journal of Physics: Condensed Matter* 12:A315.

Evans E, Metcalfe M (1984) Free energy potential for aggregation of giant, neutral lipid bilayer vesicles by Van der Waals attraction. *Biophysical Journal* 46:423–426.

Evans E, Needham D (1986) Giant vesicle bilayers composed of mixtures of lipids, cholesterol and polypeptides. Thermomechanical and (mutual) adherence properties. *Faraday Discussions of the Chemical Society* 81:267–280.

Evans E, Needham D (1987) Physical properties of surfactant bilayer membranes: Thermal transitions, elasticity, rigidity, cohesion and colloidal interactions. *Journal of Physical Chemistry* 91:4219–4228.

Evans E, Needham D (1988) Attraction between lipid bilayer membranes in concentrated solutions of nonadsorbing polymers: Comparison of mean-field theory with measurements of adhesion energy. *Macromolecules* 21:1822–1831.

Evans E, Rawicz W (1990) Entropy-driven tension and bending elasticity in condensed-fluid membranes. *Physical Review Letters* 64:2094.

Evans E, Rawicz W, Hofmann A (1995) Lipid bilayer expansion and mechanical disruption in solutions of water-soluble bile acid. In: *Falk Symposium*, pp 59–59: Kluwer Academic Publication.

Evans E, Rawicz W, Smith B (2013) Concluding remarks back to the future: Mechanics and thermodynamics of lipid biomembranes. *Faraday Discussions* 161:591–611.

Evans E, Skalak R (1980) *Mechanics and Thermodynamics of Biomembranes*. Boca Raton, FL: CRC Press.

Evans E, Smith B (2011) Kinetics of hole nucleation in biomembrane rupture. *New Journal of Physics* 13:095010.

Fa N, Marques C, Mendes E, Schröder A (2004) Rheology of giant vesicles: A micropipette study. *Physical Review Letters* 92:108103.

Flory P, Volkenstein M (1969) Statistical mechanics of chain molecules. *Biopolymers* 8:699–700.

Fung Y, Tong P (1968) Theory of the sphering of red blood cells. *Biophysical Journal* 8:175.

Gracià R, Bezlyepkina N, Knorr R, Lipowsky R, Dimova R (2010) Effect of cholesterol on the rigidity of saturated and unsaturated membranes: Fluctuation and electrodeformation analysis of giant vesicles. *Soft Matter* 6:1472–1482.

Haluska C, Riske K, Marchi-Artzner V, Lehn J-M, Lipowsky R, Dimova R (2006) Time scales of membrane fusion revealed by direct imaging of vesicle fusion with high temporal resolution. *Proceedings of the National Academy of Sciences* 103:15841–15846.

Heimburg T (1998) Mechanical aspects of membrane thermodynamics. Estimation of the mechanical properties of lipid membranes close to the chain melting transition from calorimetry. *Biochimica et Biophysica Acta (BBA)—Biomembranes* 1415:147–162.

Heinrich V, Rawicz W (2005) Automated, high-resolution micropipet aspiration reveals new insight into the physical properties of fluid membranes. *Langmuir* 21:1962–1971.

Henriksen J, Ipsen J (2004) Measurement of membrane elasticity by micro-pipette aspiration. *The European Physical Journal E* 14:149–167.

Hochmuth R, Mohandas N (1972) Uniaxial loading of the red-cell membrane. *Journal of Biomechanics* 5:501–509.

Hoffman R, Gross L (1975) Modulation Contrast Microscope. *Applied Optics* 14:1169–1176.

Kenworthy A, Hristova K, Needham D, McIntosh T (1995) Range and magnitude of the steric pressure between bilayers containing phospholipids with covalently attached poly(ethylene glycol). *Biophysical Journal* 68:1921–1936.

Kim D, Costello M, Duncan P, Needham D (2003) Mechanical properties and microstructure of polycrystalline phospholipid monolayer shells—Novel solid nanoparticles. *Langmuir* 19:8455–8466.

Kwok R, Evans E (1981) Thermoelasticity of large lecithin bilayer vesicles. *Biophysical Journal* 35:637.

Li Y, Lipowsky R, Dimova R (2011) Membrane nanotubes induced by aqueous phase separation and stabilized by spontaneous curvature. *Proceedings of the National Academy of Sciences* 108:4731–4736.

Lis L, McAlister M, Fuller N, Rand R, Parsegian V (1982) Interactions between neutral phospholipid bilayer membranes. *Biophysical Journal* 37:657–665.

Meiselman H, Lichtman M, LaCelle P (1984) White cell mechanics: Basic science and clinical aspects. In: *Symposium Held at the Kroc Foundation*. Santa Barbara, CA: AR Liss.

Melchior D, Francis J, Scavitto J, Steim J (1980) Dilatometry of dipalmitoyllecithin-cholesterol bilayers. *Biochemistry* 19:4828–4834.

Mitchison J, Swann M (1954a) The mechanical properties of the cell surface I. The cell elastimeter. *Journal of Experimental Biology* 31:443–460.

Mitchison J, Swann M (1954b) The mechanical properties of the cell surface II. The unfertilized sea-urchin egg. *Journal of Experimental Biology* 31:461–472.

Mitchison J, Swann M (1955) The mechanical properties of the cell surface III. The sea-urchin egg from fertilization to cleavage. *Journal of Experimental Biology* 32:734–750.

Nagle J, Wilkinson D (1982) Dilatometric studies of the subtransition in dipalmitoylphosphatidylcholine. *Biochemistry* 21:3817–3821.

Needham D (1991) Possible role of cell cycle-dependent morphology, geometry, and mechanical properties in tumor cell metastasis. *Cell Biophysics* 18:99–121.

Needham D, Evans E (1988) Structure and mechanical properties of giant lipid (DMPC) vesicle bilayers from 20. degree. C below to 10. degree. C above the liquid crystal-crystalline phase transition at 24. degree. C. *Biochemistry* 27:8261–8269.

Needham D, Hochmuth R (1989) Electro-mechanical permeabilization of lipid vesicles. Role of membrane tension and compressibility. *Biophysical Journal* 55:1001.

Needham D, McIntosh T, Evans E (1988) Thermomechanical and transition properties of dimyristoylphosphatidylcholine/cholesterol bilayers. *Biochemistry* 27:4668–4673.

Needham D, Nunn R (1990) Elastic deformation and failure of lipid bilayer membranes containing cholesterol. *Biophysical Journal* 58:997.

Needham D, Stoicheva N, Zhelev D (1997) Exchange of monooleoylphosphatidylcholine as monomer and micelle with membranes containing poly (ethylene glycol)-lipid. *Biophysical Journal* 73:2615.

Needham D, Zhelev D (1995) Lysolipid exchange with lipid vesicle membranes. *Annals of Biomedical Engineering* 23:287–298.

Needham D, Zhelev D (1996) The mechanochemistry of lipid vesicles examined by micropipet manipulation techniques. *Surfactant Science Series* 62:373–444.

Needham D, Zhelev D (2000) Use of micropipet manipulation techniques to measure the properties of giant lipid vesicles. In: *Perspectives in Supramolecular Chemistry: Giant Vesicles* (Walde PLLaP, ed), pp. 102–147. Chichester, UK: John Wiley & Sons.

Noppl-Simson D, Needham D (1996) Avidin-biotin interactions at vesicle surfaces: Adsorption and binding, cross-bridge formation, and lateral interactions. *Biophysical Journal* 70:1391–1401.

Olbrich K (1997) Water permeability and mechanical properties of unsaturated lipid membranes and sarcolemmal vesicles. PhD Thesis, In: Department of Mechanical Engineering and Materials Science; Duke University, Durham, NC.

Olbrich K, Rawicz W, Needham D, Evans E (2000) Water permeability and mechanical strength of polyunsaturated lipid bilayers. *Biophysical Journal* 79:321–327.

Pan J, Tristram-Nagle S, Kučerka N, Nagle J (2008) Temperature dependence of structure, bending rigidity, and bilayer interactions of dioleoylphosphatidylcholine bilayers. *Biophysical Journal* 94:117–124.

Pan J, Tristram-Nagle S, Nagle J (2009) Effect of cholesterol on structural and mechanical properties of membranes depends on lipid chain saturation. *Physical Review E* 80:021931.

Parra-Ortiz E (2013) Effects of pulmonary surfactant proteins SP-B and SP-C on the physical properties of biological membranes. PhD Thesis, In: Department of Applied Physics III, Complutense University (Madrid, Spain); URL: https://eprints.ucm.es/23506/.

Rand R, Burton A (1964) Mechanical properties of the red cell membrane: I. Membrane stiffness and intracellular pressure. *Biophysical Journal* 4:115.

Rawicz W, Olbrich K, McIntosh T, Needham D, Evans E (2000) Effect of chain length and unsaturation on elasticity of lipid bilayers. *Biophysical Journal* 79:328–339.

Rawicz W, Smith B, McIntosh T, Simon S, Evans E (2008) Elasticity, strength, and water permeability of bilayers that contain raft microdomain-forming lipids. *Biophysical Journal* 94:4725–4736.

Roux A, Cappello G, Cartaud J, Prost J, Goud B, Bassereau P (2002) A minimal system allowing tubulation with molecular motors pulling on giant liposomes. *Proceedings of the National Academy of Sciences* 99:5394–5399.

Roux A, Cuvelier D, Nassoy P, Prost J, Bassereau P, Goud B (2005) Role of curvature and phase transition in lipid sorting and fission of membrane tubules. *The EMBO Journal* 24:1537–1545.

Shchelokovskyy P, Tristram-Nagle S, Dimova R (2011) Effect of the HIV-1 fusion peptide on the mechanical properties and leaflet coupling of lipid bilayers. *New Journal of Physics* 13:025004.

Shi Z, Baumgart T (2015) Membrane tension and peripheral protein density mediate membrane shape transitions. *Nature Communications* 6:5974.

Skalak R (1973) Modelling the mechanical behavior of red blood cells. *Biorheology* 10:229.

Sun Y, Lee C-C, Huang H (2011) Adhesion and merging of lipid bilayers: A method for measuring the free energy of adhesion and hemifusion. *Biophysical Journal* 100:987–995.
Tian A, Johnson C, Wang W, Baumgart T (2007) Line tension at fluid membrane domain boundaries measured by micropipette aspiration. *Physical Review Letters* 98:208102.
Tristram-Nagle S, Nagle J (2004) Lipid bilayers: Thermodynamics, structure, fluctuations, and interactions. *Chemistry and Physics of Lipids* 127:3–14.
Vitkova V, Genova J, Bivas I (2004) Permeability and the hidden area of lipid bilayers. *European Biophysics Journal* 33:706–714.
Vitkova V, Genova J, Mitov M, Bivas I (2006) Sugars in the aqueous phase change the mechanical properties of lipid mono-and bilayers. *Molecular Crystals and Liquid Crystals* 449:95–106.
Waugh R, Song J, Svetina S, Zeks B (1992) Local and nonlocal curvature elasticity in bilayer membranes by tether formation from lecithin vesicles. *Biophysical Journal* 61:974.
Wiener M, Tristram-Nagle S, Wilkinson DA, Campbell L, Nagle J (1988) Specific volumes of lipids in fully hydrated bilayer dispersions. *Biochimica et Biophysica Acta (BBA)-Biomembranes* 938:135–142.
Zhelev D, Needham D (1993) Tension-stabilized pores in giant vesicles: determination of pore size and pore line tension. *Biochimica et Biophysica Acta (BBA)-Biomembranes* 1147:89–104.
Zhelev D, Needham D, Hochmuth R (1994) A novel micropipet method for measuring the bending modulus of vesicle membranes. *Biophysical Journal* 67:720.

Atomic force microscopy of giant unilamellar vesicles

Andreas Janshoff

I suppose it is tempting, if the only tool you have is a hammer, to treat everything as if it were a nail

Abraham Maslow
Toward a Psychology of Being

Contents

12.1 INTRODUCTION

The mechanical properties of cells play a pivotal role in many biological processes comprising adhesion, migration, differentiation, cell division, embryogenesis and tumorigenesis (Fletcher et al., 2010). Typically, cellular elasticity is assessed using indentation experiments in which a probe of defined geometry locally deforms the cell. Response of cells to this site-specific deformation permits calculation of elastic moduli depending on the chosen viscoelastic model. It was found that the elastic properties of cells originate mainly from the plasma membrane firmly attached to a thin but contractile cortex composed of cross-linked actin filaments associated with myosin motors (Fletcher et al., 2010; Hoffman et al., 2009; Janmey et al., 2007; Pietuch et al., 2013a; Pollard et al., 2003, 2009; Stricker et al., 2010). The intricate nature of the cellular cortex and the inevitable presence of cytosolic components and organelles, in particular the nucleus, however, prevents a quantitative assessment of the elastic properties. Therefore, model membranes were frequently employed to substantially reduce complexity while still capturing the essential physical properties of the plasma membrane/F-actin cortex (Evans et al., 1997; Fenz et al., 2012). Giant unilamellar vesicles (GUVs) have proven to be among the most versatile model membranes to mimic the plasma membrane cortex and investigate the mechanical properties of cells (Pautot et al., 2003; Richmond et al., 2011) (see Chapter 4). Often, basic mechanical properties of lipid bilayers were inferred from micropipette suction experiments (Needham et al., 1988, 1990; Rawicz et al., 2000) (see Chapter 11), flicker spectroscopy (Esposito et al., 2007; Häckl et al., 1998; Pécréaux et al., 2004) (see Chapter 14), and also atomic force microscopy (Dieluweit et al., 2010).

Atomic force microscopy (AFM) not only permits obtaining high resolution topographic images of surfaces but also allows exerting and measuring forces ranging from few piconewtons to several micronewtons at defined spots. These experiments generate so-called force–indentation curves that can be interpreted in terms of stress–strain relationships considering the geometry of indenter and sample. These classes of experiments give access to various aspects of membrane mechanics such as bending moduli, area compressibility, prestress and lysis tension. It is safe to assume that vesicles can be described as fluid-filled capsules with a thin wall and low water permeability. Therefore, deformation of a spherical vesicle inevitably leads to bending and, more importantly, also to stretching of the bilayer that usually dominates at larger strains.

Notably, different ways exist to describe the indentation experiments performed on liposomes. Frequently, models based on Hertzian contact mechanics adapted to the indenter geometry are used (Brochu et al., 2008; Liang et al., 2004). These approaches, which are often the model of choice to describe cellular mechanics measured in AFM experiments, assume that the capsules behave like a solid, homogeneous continuum and therefore provide a single parameter to describe the mechanics of the material, the Young's modulus (Hertz, 1882; Sneddon, 1965) Although this is a convenient way to analyze the deformation at low strain, its underlying assumptions are clearly unfulfilled in the context of membranes, ignoring the shell-like structure of liposomes and cells. In particular, deeper indentation corresponding also to larger lateral strain cannot be captured by conventional contact mechanics models. Besides contact mechanics, more realistic models also exist that consider the two dimensional nature of vesicles or, more generally, liquid-filled capsules (Bando et al., 2013; Fery et al., 2007; Schaefer et al., 2013; Vella et al., 2012). The corresponding theoretical models employ shell mechanics, showing that bending governs the mechanical response at low strain smaller than the thickness of the shell, whereas at larger strain, nonlinear contributions from area dilatation of the shell rule, especially if the enclosed volume is conserved. If volume conservation does not hold, bending at larger strain adopts a square root dependence (Fery et al., 2007). The rigorous treatment of elastic shells is very involved. It requires computing the exact shape of the liposome during indentation, which can become difficult because the contact of the fluid membrane with the indenter depends on the depth of the penetration producing a moving boundary condition. However, limiting cases such as point-load forces or parallel plate compression have been considered in the past (Bando et al., 2013; Schaefer et al., 2013; Vella et al., 2012).

In AFM experiments, two main indenter geometries are usually employed, conical tips and spheres (colloidal probes). Conical tips are the most frequently used because this geometry is also suitable for imaging of the specimen by raster scanning of the surface. In this chapter, indentation and compression experiments on adherent giant liposomes are described and simple solutions to infer their elasticity are discussed.

12.2 ATOMIC FORCE MICROSCOPY

12.2.1 GENERAL SETUP

With the development of the atomic force microscope in the 1980s, an instrument became available with force resolution down to the piconewton regime (Binnig et al., 1986). Paired with its ability to permit lateral scanning beyond the optical diffraction limit and unprecedented vertical resolution, the method quickly became relevant in research fields requiring *in situ* imaging of surfaces with nanometer resolution. The versatility of this method is also based on the large number of contrast mechanisms comprising topography, elastic moduli, friction coefficients, molecular recognition units, surface charges and energy dissipation.

Since its invention by Binnig, Quate and Gerber, AFM has evolved from a predominantly imaging technique to a method allowing mechanical manipulation of soft matter with precise force feedback (Binnig et al., 1986; Butt et al., 2005). The cantilever of the microscope acts as a sensor for the local interaction between tip and sample (Figure 12.1).

12.2.2 FORCE CURVES

By approaching and withdrawing the tip of the force microscope from the surface a so-termed force–distance curve can be monitored by detecting the cantilever deflection providing the force depending on the piezo movement (Butt et al., 2005). The applied

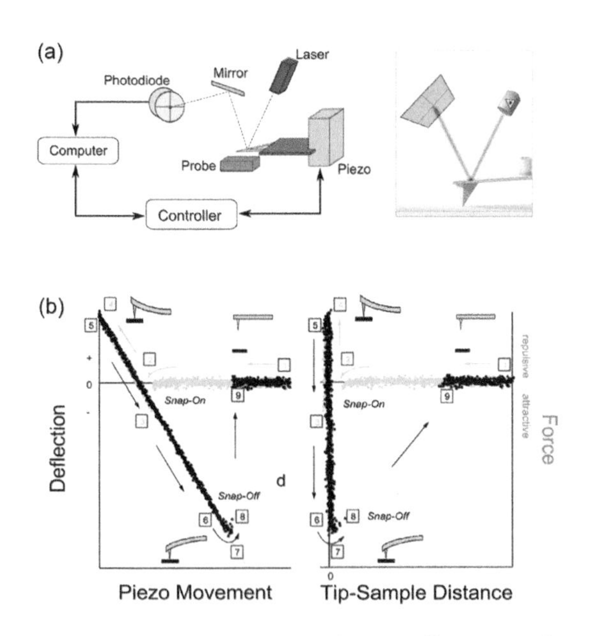

Figure 12.1 AFM measurements. (a) Schematic illustration of a typical AFM setup. The deflection of the cantilever is typically read out by reflection of a laser beam. (b) Transformation of experimentally recorded force curves into force–distance curves.

force, f, can be deduced from the cantilever deflection, z_c, multiplied by the spring constant, k_c, according to Hooke's law:

$$f = k_c z_c. \quad (12.1)$$

Usually one is, however, interested in the force as a function of tip-sample separation, z_{ts}, or *distance*, which is calculated by adding the deflection to the z-piezo position, z_p:

$$z_{ts} = z_p + z_c \quad (12.2)$$

Note that attractive forces produce a negative cantilever deflection. Eqs 12.1 and 12.2 allow transforming the raw data ($z_c(z_p)$) into meaningful force distance curves (Figure 12.1b). In most experiments, a force ramp is applied at constant velocity, v, so that $z_p = vt$ with time, t. In the contact part with a hard surface, $z_p = -z_c$ and $z_{ts} = 0$.

Commercial instruments usually measure the deflection of the cantilever with the optical lever technique in which a laser beam is reflected from the backside of the end of the cantilever and monitored by a position sensitive detector. If the tip experiences a force, the cantilever bends and the laser beam is reflected at an angle proportional to the change of the slope at the end of the cantilever. For rectangular beams the change of the slope is proportional to its deflection.

The cantilever's geometry and material is pivotal for the performance of the AFM. Generally, commercial cantilevers are made of silicon or silicon nitride in a rectangular or a triangular shape. Usually, triangular-shaped cantilevers are used if high lateral forces are expected and the torsion of the cantilever needs to be minimized, whereas rectangular cantilevers are more suitable for parallel plate compression of liposomes due to their larger area at the tip. Force resolution is ultimately limited by thermal noise of the cantilever, which depends on the spring constant depending on the material properties and dimensions. For a cantilever with rectangular cross section, the spring constant can be computed from

$$k_c = \frac{Ewt_c^3}{4L^3} \quad (12.3)$$

with w, the width; L, the length; and t_c, the thickness of the cantilever. E is the Young's modulus of the material. High force sensitivity is accomplished with low spring constants (long and thin). Calibration of the cantilever is most frequently achieved by computing the thermal noise using the equipartition theorem ($\frac{1}{2}k_B T = \frac{1}{2}\langle z_c^2 \rangle$) (Butt et al., 1995; Hutter et al., 1993). Note that $k_b T$ is 4.11 pN nm.

12.3 CONTACT MECHANICS

A useful starting point for the physical description of membranes is continuum elasticity theory, in which deformation is described in term of strain and stress. Strain, as a dimensionless quantity, comprises the linear spatial variation of the displacement vector, whereas the stress tensor as a force per unit area contains the internal forces that lead to the solid's equilibrium. Because all solid materials deform if not infinitely rigid and in contact with another solid, the contact part of the force curve is usually not linear, except for a cylindrical punch, but this must be appropriately described by a contact model. In the case of a spherical indenter and a planar elastic sample ignoring surface forces (adhesion), the force indentation curve is described by (Hertz, 1882)

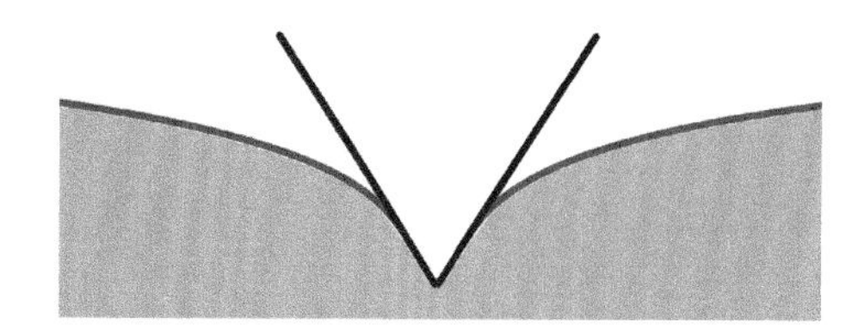

Figure 12.2 Deformation of a soft surface (elastic half space) with a conical indenter.

$$f = \frac{2E\sqrt{R}}{3(1-\nu^2)}\delta^{\frac{3}{2}}, \quad (12.4)$$

where δ is the indentation depth; E the Young modulus of the soft material neglecting a possible deformation of the tip; R the tip radius; and ν the Poisson ratio. The indentation problem of the linear elastic half-space deformed by indenters of arbitrary shape has been solved by Sneddon employing Hankel transformation that allow to readily obtain expressions for force indentation curves. In case of a conical indenter, Sneddon found (Figure 12.2) (Sneddon, 1965)

$$f = \frac{2E\tan\theta}{\pi(1-\nu^2)}\delta^2, \quad (12.5)$$

whereas for a flat-ended cylindrical punch, a linear force indentation curve is obtained:

$$f = 2Ea\delta, \quad (12.6)$$

with a the contact radius; and θ the half opening angle of the cone. Notably, approximate solutions exist that include adhesion such as the Johnson-Kendall-Roberts (JKR) model (Johnson et al., 1971). The JKR model describes indentation with a sphere (approximated by a paraboloid) in the presence of adhesion forces at the contact. The JKR model is a good approximation of the load–indentation relationship for soft samples, large tip radii and large energy of adhesion. The model explains persistent attractive forces upon retraction of the cantilever and thereby explains the formation of a "neck" while pulling away from the surface. Because contact models usually assume a continuous elastic half space, they fail to describe the deformation of thin shells that enclose a liquid. In the next section, membrane mechanics is briefly reviewed and a model proposed that is more suitable to describe force indentation curves obtained from probing giant liposomes.

12.4 MEMBRANE MECHANICS

The mechanical properties of membranes can be described by continuum elasticity theory to determine the free energy changes in response to arbitrary membrane deformations. Albeit treated in

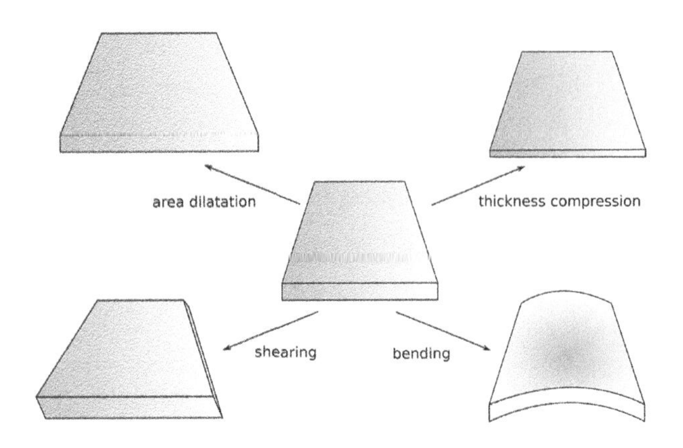

Figure 12.3 Schematic illustration of prominent membrane deformations.

great detail in Chapter 5, it is instructive to review the dominant types of membrane deformations that occur when GUVs are subject to various loading conditions. Figure 12.3 shows four basic classes of membrane deformations encompassing alteration of the membrane thickness, membrane area, bending, and shearing (Boal, 2012; Deserno, 2015; Seifert, 1997).

12.4.1 MEMBRANE AREA CHANGES—STRETCHING

Stretching along the sides of a membrane sheet generates a lateral tension, Σ (force per length, not to be confused with stress), through area dilatation. In equilibrium, the external tension will be balanced by the internal stress in the membrane. The free energy change associated with area dilatation F_a can be inferred from Hookes law in two dimensions:

$$F_a = \frac{K_A}{2}\frac{\Delta A^2}{A_0} \tag{12.7}$$

with ΔA, the change in membrane area from the equilibrium area A_0; and K_A the area compressibility modulus with units of energy per unit area. Typical values of K_A fall in the range of 0.1–0.5 N/m.

12.4.2 MEMBRANE BENDING

The by far most important membrane deformation—due to its extremely low modulus—is bending. Canham and Helfrich derived the elastic free energy of a mechanically deformed membrane based on curvature considerations (Deserno, 2015). Helfrich's expression for the total free bending energy, F_b, of a symmetric membrane (no spontaneous curvature) reads

$$F_b = \oint \left\{ \frac{1}{2}\kappa (C_1 + C_2)^2 + \bar{\kappa} C_1 C_2 \right\} dA. \tag{12.8}$$

The Hamiltonian has two moduli, κ and $\bar{\kappa}$, termed the bending modulus and the saddle splay modulus (Gauss rigidity), respectively. C_1 and C_2 are the two principle curvatures. Typical values for the bending rigidity, κ, are 10–50 k_BT for fluid membranes. The integral has to be performed over the whole bilayer surface. The resistance of a membrane or more general thin sheet of material to bending deformation heavily depends on the thickness, t_b, of the sheet. Classical plate theory predicts a cubic dependence of κ on plate thickness. The bilayer has only a thickness of about 5 nm, giving rise to extremely low values for κ. Because the bending modulus of a fluid bilayer is only a few k_BT, thermally excited undulations occur in unstressed membranes that generate dynamic repulsion between flickering vesicles and a solid wall, preventing adhesion. The nature of this force is entropic because more and more degrees of freedom are lost if long-wavelength modes are gradually frozen upon contact with the surface. These flickering modes are largely ironed out during adhesion and compression of liposomes.

12.4.3 MEMBRANE COMPRESSION

Changes in membrane thickness from t_{b0} to t_b are usually only relevant on small length scales because the modulus is typically rather large. Thickness changes sometimes also occur spontaneously in gel phase lipids due to insertion of small chain alcohols in a process referred to as interdigitation. The energy cost associated with compression of the lipid bilayer, F_c, is to first order

$$F_c = \frac{K_C}{2}\int \left(\frac{t_b - t_{b0}}{t_{b0}} \right)^2 dA \tag{12.9}$$

with K_C the thickness compressional modulus ($\approx 60 k_BT/\text{nm}^2$). Compression of a bilayer plays a role in force experiment where solid supported lipid bilayers are indented. The sharp tip of an AFM cantilever usually leads to a breakthrough—visible as a mechanical instability—if the applied forces exceed the so-called breakthrough limit. This limit depends on the radius of the indenter and the nature of the lipids. Unsaturated lipids are usually associated with lower breakthrough forces (Kuenneke et al., 2004).

12.4.4 MEMBRANE SHEARING

Shear deformations force the lipids to displace relative to each other, whereas the area per molecule is conserved. Therefore, a lipid bilayer can only resist this deformation if the relative positions of its constituent molecules are fixed. However, fluid membranes are liquid-like and are therefore unable to resist shear deformations—the shear modulus is essentially zero. Shear becomes, however, relevant in the context of coupling the membrane to the cytoskeleton because it occurs in a native plasma membrane/cortex shell.

12.5 MODELING GUV MECHANICS PROBED WITH AN AFM

12.5.1 GENERAL CONSIDERATIONS

In a typical AFM indentation experiment, the deformation of a GUV formed by a fluid lipid bilayer can be in-plane stretching and out-of-plane bending. As detailed above, membranes are characterized by a low resistance to bending and shearing so that area dilatation, which is very energy costly, is avoided and vesicles preferentially deform by pure bending. Generally, however, in spherical shells such as GUVs, stretching cannot be avoided upon deformation. Because GUVs are filled with liquid and display only a limited water permeability of the membrane (see Chapter 20), the lumen is largely maintained during deformation because the liquid inside is incompressible. As a consequence, volume conservation inevitably leads to in-plane stretching of the membrane upon deformation. General stretching of the shell is by far more energy costly than bending, albeit bending of shells involves stretching. Stretching and compression generated through bending is usually neglected for small deflections and only important if curvature is extremely large or thickness increased as in the case of a cortex attached to the lipid bilayer. Analytical solutions for the deformation of shells can only be obtained for simple indenter geometries. The normal displacement of the pole (Figure 12.4), under point loading with force, f, as a function of indentation depth, δ, is given by Reissner (Fery et al., 2007; Reissner, 1946a,b):

$$f = \frac{4Et^2}{R_v\sqrt{3(1-\nu^2)}}\delta \tag{12.10}$$

where R_v is the radius of the vesicle. This approximation is only valid for extremely small indentation depths ($\delta < t$). Bending is limited to a small dimple formed around the pole, where the force is concentrated. The size of the dimple, d, is obtained from minimizing the free energy composed of stretching and bending energies yielding for the dimple size (Fery et al., 2007):

$$d \propto \sqrt{t_b R_v}\,. \tag{12.11}$$

Hence, a reduction of the shell thickness, t_b, results in stronger confinement of the local deformation by bending.

If the enclosed volume is variable, it has been shown that "leaky" capsules indented by a point load force display a square root dependence on indentation depth under the assumption that the deformation energy is localized on the rim of the formed dimple ($f \propto \delta^{\frac{1}{2}}$), whereas stretching usually obeys a cubic dependency on indention ($f \propto \delta^3$), as found also in GUV experiments (Fery et al., 2007). In GUV and also cell experiments, prestress is usually stronger, obscuring the effect of bending at low strains.

In the following, we ignore the fact that at the tip of the conical indenter at $r = 0$ curvature becomes infinite because in reality the tip has a finite curvature around 20–60 nm at the end. Besides, the end of the tip is readily wrapped by membrane in indentation experiments so that bending is only an offset to the energy penalty (the cone widens with increased penetration depth). Apart from the tip region, the adhesion of the GUV to the substrate also produces a large curvature at the contact region. Although the contact region increases during indentation, the contribution to the energy functional does not explain a nonlinear increase of force with indentation depth. It is also instructive to compare energy associated with area dilatation (stretching) and bending for a limiting case. Stretching energy, F_{area}, relates to bending energy, F_{bend}, for point load forces roughly as $\frac{F_{area}}{F_{bend}} \propto \left(\frac{R_v}{t_b}\right)^2$ (Fery et al., 2007). Although the bilayer is extremely thin, $t_b \approx 5$nm, the radius of the liposome, R_v, is on the order of several micrometers. Hence, bending contributions are small compared with area dilatation and are therefore often ignored in indentation experiments. In the next section, we will exemplarily derive a tension-based model that permits computing of force indentation curves with a conical indenter. The indenter geometry will later be generalized. The model relies on the fact that volume conservation forces the GUV to dilate its shell laterally upon indentation and thereby produce a lateral tension. The deviation from a spherical shape requires a larger area

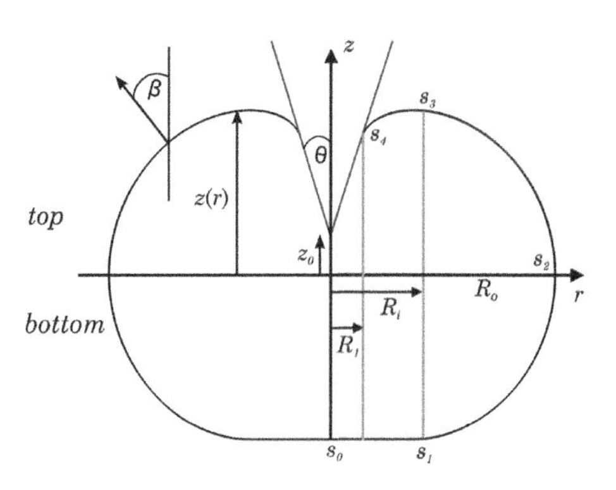

Figure 12.4 Schematic illustration and parametrization of a liposome subject to indentation with a conical indenter. (Schaefer, E. et al., *Soft Matter*, 11, 4487–4495, 2015.).

compared with the unstressed vesicle. In contrast to the aforementioned tension model, many studies on small nanometer-sized liposomes employ contact mechanics models such as the Hertz model to describe the mechanical properties of adhered vesicles (Liang et al., 2004). For egg-phosphatidylcholine (egg-PC) liposomes, the authors found Young's moduli on the order of MPa. This value is close to what is found for the compression modulus K_C of a fluid lipid bilayer (≈ 10MPa). In these experiments, it is highly unlikely that volume conservation still holds.

12.5.2 INDENTATION OF A GUV WITH A CONICAL INDENTER

The most frequently used indenter geometry is the conical (including pyramidal) shape because this type of geometry also allows imaging of the surface and performing site-specific indentation experiments. Figure 12.4 illustrates the envisioned geometry of a spherical liposome subject to indentation with a conical indenter. The shape of the deformed vesicle should be axisymmetric if the vesicle is poked along the axis of symmetry and the initial radius of the spherical vesicle prior to compression is R_v. The contact region with the flat substrate extends from $s_0 \rightarrow s_1$. The free contour ranges from $s_1 \rightarrow s_4$, with the largest radius, R_0, at s_2 and the greatest height, z, at s_3. The contour is parameterized by the angle β between the surface normal and the z-direction.

The following treatment is partly based on the work of Yoneda (1964), Evans and Skalak (1980), Bando et al. (2013), Bando and Oiso (2013) and Sen et al. (2005). The goal is to compute force indentation curves from the deformed shape. Central assumptions are negligible bending stiffness, uniform tension and constant volume (Schaefer et al., 2015).

Contour of the vesicle

The shape of the indented liposome can be computed under the assumption that the enclosed volume is fixed and pressure across the membrane conserved. Pressure P relates to tension Σ according to Young-Laplace's law:

$$\Delta P = \Sigma\left(C_1 + C_2\right). \tag{12.12}$$

Considering a small line element, ds, of the meridian at an arbitrary point on the contour, in which dr is the projection of ds on the r-axis ($dr = ds\cos\beta$), we find that $r = C_2^{-1}\sin\beta$ and $ds = C_1^{-1}d\beta$. Elimination of ds gives

$$C_1 = \frac{d\beta}{ds} = \frac{d\beta}{dr}\cos\beta = \frac{du}{dr}. \quad (12.13)$$

$$C_2 = \frac{1}{r}\sin\beta = \frac{u}{r}. \quad (12.14)$$

with $u = \sin\beta$. Small angles allow the approximation of $\sin d\beta \approx d\beta = C_1 ds$. As a consequence, Eq. 12.12 can be written as a differential equation:

$$\frac{\Delta P}{\Sigma} = \frac{du}{dr} + \frac{u}{r}. \quad (12.15)$$

Because $\frac{\Delta P}{\Sigma}$ is constant, we can integrate Eq. 12.15 to find

$$u_i(r) = A_i r + \frac{B_i}{r}. \quad (12.16)$$

with $i = 1,2,3$ referring to the corresponding regions of the free contour ($s_1 \to s_2 (i=1)$, $s_2 \to s_3 (i=2)$, $s_3 \to s_4 (i=3)$). For each of the regions, appropriate boundary conditions have to be fulfilled. A_1 and B_1 correspond to the unbound region $i = 1$ ranging from $s_1 \to s_2$. The following boundary conditions hold:

$$\beta = \frac{\pi}{2} \text{ at } r = R_o$$

$$\beta = 0 \text{ at } r = R_i. \quad (12.17)$$

where R_i is the contact radius of the GUV formed with the flat substrate at the bottom, and R_o the equatorial radius of the deformed liposome (Figure 12.4). From Eqs. 12.16 and 12.17 we obtain,

$$A_1 = \frac{R_o}{R_o^2 - R_i^2} \quad (12.18)$$

$$B_1 = \frac{-R_i^2 R_o}{R_o^2 - R_i^2} = -A_1 R_i^2. \quad (12.19)$$

In region $i = 2$ ($s_2 \to s_3$), the free contour obeys the boundary conditions (Pietuch et al., 2013b; Sen et al., 2005):

$$\beta = \frac{\pi}{2} \text{ at } r = R_o \quad (12.20)$$

$$\beta = 0 \text{ at } r(s_3). \quad (12.21)$$

Therefore,

$$A_2 = \frac{R_o}{R_o^2 - r(s_3)^2} \quad (12.22)$$

$$B_2 = \frac{-r(s_3)^2 R_o}{R_o^2 - r(s_3)^2} = -A_2 r(s_3)^2. \quad (12.23)$$

Because the contour is continuous at R_o, $r(s_3) = R_i$ holds, and therefore also $A_1 = A_2$ and $B_1 = B_2$, that is, the free contour from $s_1 \to s_2$ and $s_2 \to s_3$ are mirror-inverted. A_3 and B_3 for region $i = 3$ ($s_3 \to s_4$) that reaches up to the contact with the indenter at R_1 are obtained from the following boundary conditions:

$$\beta = 0 \text{ at } r = R_i \quad (12.24)$$

$$\beta = -\left(\frac{\pi}{2} - \theta\right) \text{ at } r = R_1, \quad (12.25)$$

leading to

$$A_3 = \frac{R_1 \sin\left(\frac{\pi}{2} - \theta\right)}{R_i^2 - R_1^2} \quad (12.26)$$

$$B_3 = -A_3 R_1^2 - R_1 \sin\left(\frac{\pi}{2} - \theta\right). \quad (12.27)$$

where θ is the half opening angle of the indenter. Once the radii R_o, R_i, and R_1 are found, the free contour corresponding to the regions ($s_1 \to s_2$ using $u_1(r)$, $s_2 \to s_3$ using $u_2(r) = u_1(r)$, and $s_3 \to s_4$ using $u_3(r)$) can be readily obtained from integrating

$$\frac{dz}{dr} = \tan\beta = \frac{u(r)}{\sqrt{1 - u(r)^2}}. \quad (12.28)$$

The remaining contour is defined by the boundaries, a flat substrate at the bottom and the conical indenter from the top.

The task is now to find expressions for R_o, R_1, and R_i depending on the distance between the tip of the indenter and the flat base plate at the bottom z_0. Three conditions apply to an indented GUV that permit computing force–indentation curves ($f(\delta)$). The next section describes how to find a set of parameters R_o, R_1, and R_i at a given force.

Volume constraint

First, volume changes during compression can be ignored because no hysteresis is found in compression experiments of GUVs. Permeability of water across the lipid bilayer is low compared with the time scale (~ 1s) of a single force curve (Boroske et al., 1981) (see also Chapter 20). The volume of the sphere prior to indentation is denoted as V_v and the volume of the indented liposome as V_{ind}. Therefore, the condition of volume conservation is

$$V_v = \frac{4}{3}\pi R_v^3 = V_{ind}. \quad (12.29)$$

The indented liposome is a solid of revolution, which facilitates the integration to obtain the volume V_{ind}

$$V_{ind} = \int_{R_i}^{R_o} \left(\frac{u_1(r)\pi r^2}{\sqrt{1 - u_1(r)^2}} + \frac{u_3(r)\pi r^2}{\sqrt{1 - u_3(r)^2}} \right) dr - \pi R_i^2 z(R_i) + \int_{R_1}^{R_i} \frac{u_3(r)\pi r^2}{\sqrt{1 - u_3(r)^2}} dr - \frac{\pi R_1^3}{3\tan\theta} \quad (12.30)$$

with $z(R_i) = \int_{R_i}^{R_o} \frac{u_1(r)}{\sqrt{1 - u_1(r)^2}} dr$.

Forces balance

The key assumption is that the only source of the restoring force to indentation is the in-plane tension, $\Sigma = \Sigma_0 + K_A \frac{\Delta A}{A_v}$, due to area dilatation. $\Delta A = A_{ind} - A_v$ denotes the difference between the actual area A_{ind} and the initial area prior to compression A_v. Membrane tension arises due to adhesion of the liposome, also referred to as prestress. The force balance of the top part of the liposome in the z-direction is

$$f = 2\pi(R_1 \sin(\pi/2-\theta) + R_1^2 A_3)\left(\Sigma_0 + K_A \frac{A_{ind} - A_v}{A_v}\right), \quad (12.31)$$

which is the second condition, whereas force equilibrium at the bottom part is the third condition (Schaefer et al., 2013)

$$f = \Delta P\, \pi R_i^2 = 2\pi R_i^2 A_1\left(\Sigma_0 + K_A \frac{A_{ind} - A_v}{A_v}\right). \quad (12.32)$$

Surface area A_{ind} of the vesicle

The area, A_v, prior to indentation is $4\pi R_v^2$, whereas the actual area, A_{ind}, can be divided into two surfaces of revolution, the top, A_{ind}^{top}, and bottom part, A_{ind}^{bottom}, of the liposome according to Figure 12.4:

$$A_{ind}^{bottom} = \pi R_i^2 + 2\pi \int_{R_i}^{R_o} \frac{r}{\sqrt{1-u_1(r)^2}} dr \quad (12.33)$$

$$A_{ind}^{top} = 2\pi \int_{R_i}^{R_o} \frac{r}{\sqrt{1-u_1(r)^2}} dr + 2\pi \int_{R_1}^{R_i} \frac{r}{\sqrt{1-u_3(r)^2}} dr + \frac{\pi R_1^2}{\sin(\theta)}. \quad (12.34)$$

Indentation depth

The indentation depth in the center at $r = 0$ is readily obtained from

$$\delta = 2R_v - \left(2\int_{R_i}^{R_o} \frac{u_1(r)}{\sqrt{1-u_1(r)^2}} dr + \int_{R_1}^{R_i} \frac{u_3(r)}{\sqrt{1-u_3(r)^2}} dr - \frac{R_1}{\tan\theta}\right). \quad (12.35)$$

The contour in region $s_1 \rightarrow s_3$ corresponds to the first integral, whereas the contour along the path $s_3 \rightarrow s_4$ is represented by the second integral.

Procedure to compute shape and force response

The procedure summarized in Box 12.1 permits computing the shape of the indented liposome and the corresponding force indentation curves (Bando and Oiso, 2013).

In a nutshell, the three parameters R_1, R_i, and R_o are obtained for a given force by solving the system of nonlinear equations comprising force balances (Eqs. 12.31 and 12.32) and volume constraint (Eq. 12.29). Once the three parameters are obtained, the corresponding indentation depth can be calculated.

Figure 12.5 shows a number of simulated force indentation curves and contour plots of a GUV subject to indention with a conical indenter. The impact of K_A and Σ_0 on the force response of a liposome is shown in Figure 12.5a,b. Although an increase in K_A results in a steeper slope at large strain, an increasing the prestress Σ_0 leads only to stiffening at low strain. Figure 12.5c shows how the contact radius with the flat substrate, R_i, and the contact radius with the indenter, R_1, increase with indention depth. R_1 follows a rather linear trend, as one would expect for wetting of a cone with an unstressed membrane, whereas R_i rapidly grows at low indention depth.

In Figure 12.5d,f, the influence of indenter geometry on the expected force indention curves is shown. A blunt indenter squeezes the liposome into a more pancake-like geometry, producing larger radii R_o, R_i and R_1, whereas sharper indenters reach deeper inside the vesicle. Figure 12.5e shows the shape of a GUV at different forces, illustrating that flattening of the shell and deeper penetration of the indenter take place so that the minimal surface area is reached. Notably, rupture of membranes consisting of two phospholipid leaflets occurs at an area dilatation ($\frac{\Delta A}{A_v}$) of merely 2%–5%, ultimately limiting the largest possible indentation depth.

Box 12.1 Basic procedure for computing shape and force response

1. A starting value for the externally applied force, f, is assigned.
2. The system of Eqs. 12.29, 12.31, and 12.32 is numerically solved for the radii R_1, R_i, and R_o to provide the contour of the indented vesicle by integrating Eq. 12.35.
3. The corresponding indentation depth, δ, is calculated from Eq. 12.35.
4. The initial force value is increased or decreased by a given increment and the previous set of radii (R_1, R_i, and R_o) used as new starting values.

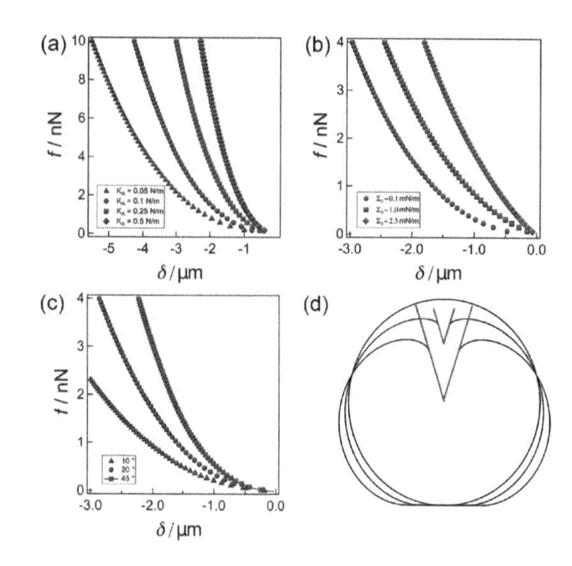

Figure 12.5 Simulated force indentation curves and GUV contour plots. (a) Calculated force indentation curves demonstrating the influence of the area compressibility modulus K_a on the force response of a GUV to indentation with a conical indenter. The following parameters were used: $\Sigma_0 = 0.1$mN/m; $R_v = 10\mu$m; $\theta = 18°$. (b) Impact of prestress Σ_0 on the force indentation curves using the identical set of parameters and $K_A = 0.1$N/m. (c) Force indentation curves as a function of half-opening angle of the conical indenter with the following parameters: $\Sigma_0 = 0.1$mN/m; $R_v = 10\,\mu$m; $K_A = 0.1$N/m used in (b). (d) Shape of the GUVs as a function of applied force (0 nN, 10 nN, and 100 nN); parameters as in (c).

12.5.3 PARALLEL PLATE COMPRESSION OF A GUV

Parallel plate compression is obtained by setting $\theta = 90^\circ$ corresponding to a horizontal plate acting as the indenter. This angle ensures a fully symmetric situation, in which $s_3 = s_4$. The restoring force maximizes compared with a sharp tip and leads to smallest bending contributions. With an AFM, this situation requires some adoption because an inherent tilt of $\sim 10^\circ$ is used for regular cantilevers with a sharp tip. Usually the base plate is skewed to compensate for this tilt (Schaefer et al., 2013).

12.5.4 INDENTATION OF A GUV WITH A SPHERICAL INDENTER

The only modification to the treatment above for any arbitrary indenter geometry is to adapt the boundary conditions for region 3 ($s_3 \rightarrow s_4$) and adjust the corresponding integrals for the volume and the surface area. Consequently, we obtain

$$\beta = 0 \quad \text{at} \quad r = R_i \tag{12.36}$$

$$\beta = -\arctan\left(\frac{R_1}{\sqrt{R_p^2 - R_1^2}}\right) \quad \text{at} \quad r = R_1, \tag{12.37}$$

as boundary conditions for spherical indenters with radius, R_p. Figure 12.6 shows the shapes of GUVs compressed with a force of 150 nN using three different indenter geometries. The shape does not depend on the elastic properties.

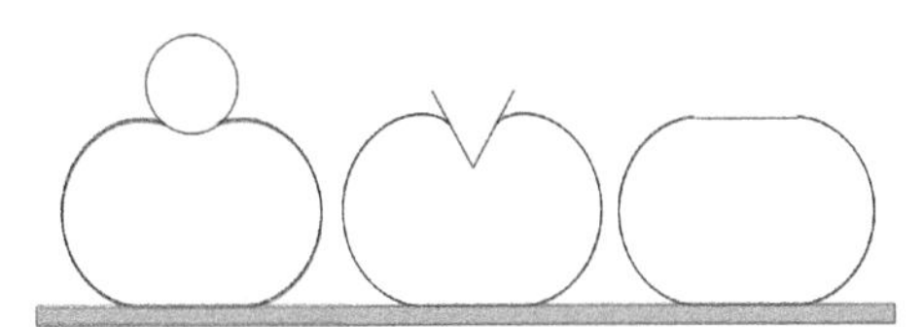

Figure 12.6 Shape of a GUV compressed with 150 nN using different indenter geometries.

12.6 INDENTATION OF SESSILE GUVs

In order to precisely measure the mechanical properties of GUVs, the liposomes need to adhere to the surface strongly enough that they will not displace upon contact with the probe but gently enough so that a continuous enlargement of the adhesion area does not occur. Moreover, the position of the probe needs to be carefully adjusted to ensure that the GUV is hit precisely in the center or compressed between two plates in the absence of tilt. If these conditions are not met, the vesicle might displace and escape from the central indentation, generating systematically lower apparent moduli.

12.6.1 PREPARATION OF GUVs FOR AFM MEASUREMENTS

The preparation of GUVs for AFM measurements usually follows the protocol of electroformation detailed in Chapter 1 (Dimitrov et al., 1987). A critical step in performing AFM experiments on GUVs is a proper surface functionalization that permits immobilization of the vesicle at the surface but at the same time prevents the vesicle from flattening or even spreading on the surface. Therefore, a specialized protocol is used to ensure proper surface functionalization to gently adhere vesicles. In brief, clean, activated glass slides were first incubated in an avidin solution followed by deposition of casein in order passivate the surface. After addition of the vesicle solution the Mg^{2+} ion concentration was increased to 2 mM to ensure sufficient adhesion of the vesicles on the surface (Schaefer et al., 2013, 2015). Note that Mg^{2+} ions might actually affect the membrane phase behavior (see Chapter 18).

12.6.2 AFM EXPERIMENTS

Conical and pyramidal indenters

Box 12.2 summarizes the experimental conditions for performing experiments on GUVs with an AFM.

Box 12.2 Basic steps for an AFM experiment on a GUV, see Appendix 1 of the book for structure and data on these lipids

A typical experimental setup used for indentation experiments is shown in panel A (Schaefer et al., 2015).

1. **Vesicle preparation:** GUVs can be prepared by electroformation as described in Chapter 1. An example protocol follows here: 8 µL of 1 mg/mL lipid (DOPC/1,2-dioleoyl-*sn*-glycero-3-phosphoethanolamine [DOPE]/A23187/DOPE-Bio [60:30:5:5]) dissolved in chloroform are deposited on indium tin oxide slides. The chamber is filled with 300 µl of buffer consisting of Tris-HCl (2 mM), $MgCl_2$ (0.5 mM), ATP (0.2 mM), DTT (0.25 mM), and sucrose (50 mM) (pH 7.5). For actin-containing vesicles, 5–7 µM actin monomers and 0.5–2 µM Alexa Fluor 488 ™actin are added. A waveform generator is used to apply a peak-to-peak voltage of $\sim$2.4 V at 70 Hz.
2. **Sample preparation and surface functionalization:** Glass slides are activated in $NH_4OH/H_2O_2/H_2O$ (1:1:5, *v/v*) solution heated to 75°C for 20 min, resulting in the formation of a thin hydrophilic SiO_2 layer. This substrate is first incubated in an avidin solution (1 µM) for 30 min, subsequently followed by deposition of casein (100 µM, wafer incubated for 30 min). Afterward, the sample is washed with buffer and 40 µL vesicle solution is then added. After 10 min, the Mg^{2+} ion concentration should increase to at least 2 mM. The buffer solution is supplemented with glucose solution to reach iso-osmolar conditions. Actin polymerization is achieved using the ionophore A23187 embedded in the membrane to enable Mg^{2+} influx into the vesicle, which initiates actin polymerization.
3. **Compression of GUV with an AFM:** Force compression curves are recorded using a conventional AFM such as the JPK NanoWizard (JPK Instruments, Berlin, Germany). Silicon nitride AFM probes with spring constants of approximately 0.03 N/m

(Continued)

Box 12.2 (*Continued*) Basic steps for an AFM experiment on a GUV, see Appendix 1 of the book for structure and data on these lipids

are suitable. For parallel plate compression, it is necessary to compensate for the device-specific 10-11° tilt of the cantilever with respect to the sample surface. The AFM can be placed on an inverse fluorescence microscope to monitor compression.

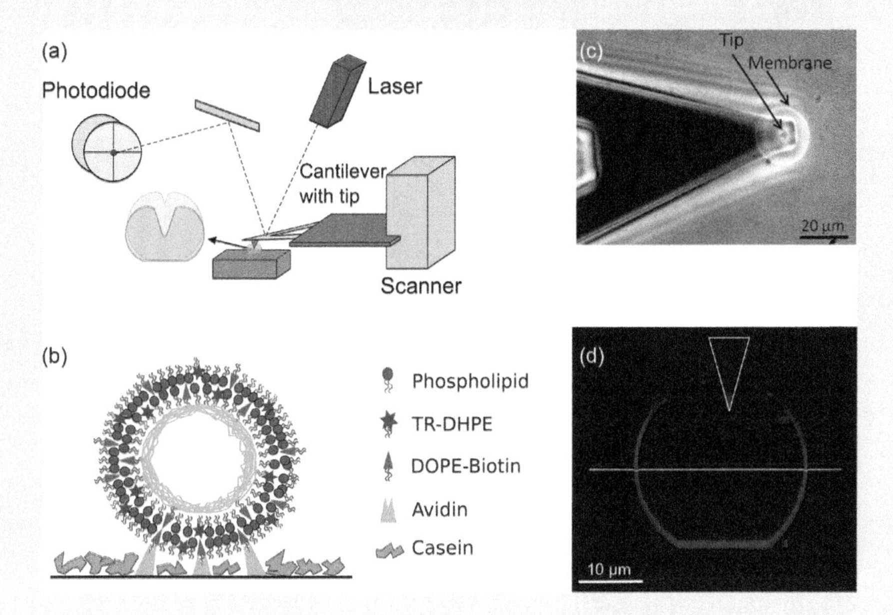

In panel (a), a conventional atomic force microscope with square-based pyramidal tips mounted on an inverted optical microscope is shown. The inverted microscope is used to position the AFM-cantilever on the liposome. If a confocal microscope with motorized z-stage is available z-stack can be recorded to visualize both indenter and GUV during deformation. The tip height is typically between 2.5–8 µm, tip radius in between 20 and 60 nm and the nominal spring constant of the cantilever lies in between 0.01 and 0.1 N/m. Panel (b) shows how immobilization of vesicles is achieved by adding small amounts of biotinylated phospholipids prior to electroformation to link the liposome gently to the surface functionalized with avidin and passivated with casein to prevent spreading of the GUVs. Vesicles are labeled with a red fluorophor (Texas Red) tagged to a phospholipid. Bright-field and confocal images of an adhered vesicles are shown in panel (c) demonstrating that only a small contact zone is formed with the glassy substrate. Also the cantilever is visible and the tip (arrow) is placed over the center of the liposome prior to indentation experiment. Panel (d) displays Z-stacks of the sessile liposome recorded with a confocal microscope (Olympus FluoView, FV1000) placed under the AFM prior to indentation at load force (overlay) $f = 2$nN. Indentation or compression experiments do not show a pronounced hysteresis, which confirms our most important assumption that the volume does not change during indentation. Moreover, viscous losses are too small to appear in force distance curves. Indeed giant liposomes can be continuously compressed without losing volume (Schaefer et al., 2013).

The area compressibility modulus, K_A, of membranes determines the amount of elastic energy required to laterally stretch or compress a lipid bilayer. It is an intrinsic property of the lipid bilayer and is related to the surface tension, γ, of the interface between the aqueous phase and the aliphatic chains of the phospholipids ($K_A \approx 4\gamma$). The bending modulus of the bilayer, κ, can also be inferred from the area compressibility modulus through ($K_A \approx \kappa t^{-2}$), with t the thickness of the bilayer. Albeit the bending modulus of the bilayer is extremely small, on the order of few $k_B T$, the associated area compressibility modulus suggests a laterally almost inextensible material. The prestress in the sessile liposome can be largely attributed to adhesion and the associated area dilatation (Murrell et al., 2011; Schaefer et al., 2013; Schwarz and Safran, 2013; Seifert, 1997). Because the liposomes change their shape from a sphere in solution to a truncated sphere upon adhesion, their surface area increases in order to keep the enclosed volume constant. This increase in surface area essentially generates a finite membrane tension ($\Sigma_0 = K_A \frac{A_{ad} - A_v}{A_v}$), the largest contribution to the prestress Σ_0. Prestress obscures effects from bending that should also occur at low strain.

Blunt indenters and parallel plate compression

Measuring the mechanical response of sessile liposomes to site-specific indentation with an atomic force microscope produces a number of challenges that may compromise an accurate assessment of elastic properties. One way to avoid some of the problems associated with sharp AFM tips is to use other types of indenters such as a sphere glued to cantilevers or parallel plate compression with tipless cantilevers (Figure 12.7). Because the elastic properties of membrane are nonlocal the size of the indenter does not matter in terms of accuracy. On the one hand, using sharp, point-like indenters like those frequently employed in AFM imaging with only a few nanometers of tip radius readily puncture the membrane by creating defects in the contact zone. On the other hand, larger indenters such as spherical probes with a size in the micrometer regime need

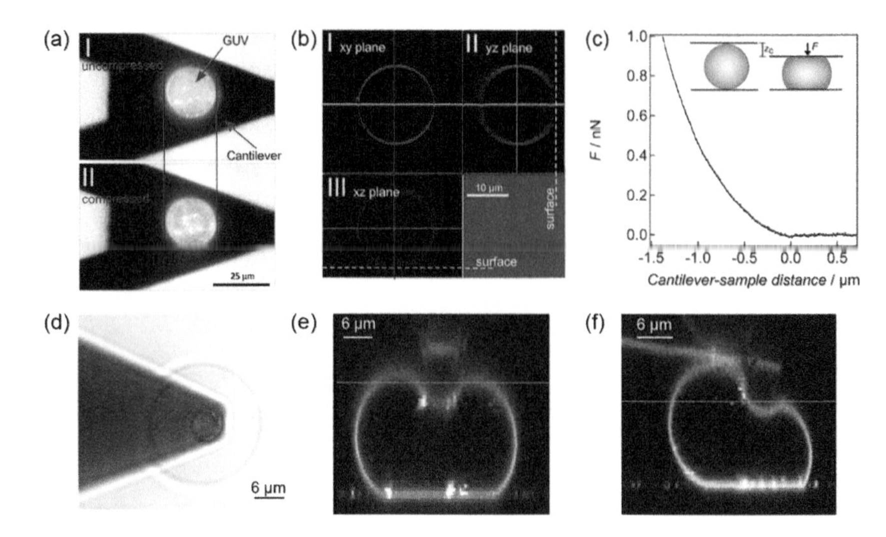

Figure 12.7 (a) Two optical micrographs taken from the bottom (fluorescence combined with bright-field) showing compression of two fluorescently labeled giant liposomes between a tipless cantilever and a glassy substrate. (b) Confocal laser scanning micrograph showing the contour of a sessile POPC vesicle. (c) Typical force compression curve. (d-f) Confocal images showing deformation of a sessile GUV with a spherical indenter.

to be placed even more precisely in the center of liposome to prevent lateral movement of the object due to emergence of lateral forces and violation of axisymmetry. Tipless cantilevers that are compensated for the inherent tilt angle of the cantilever by skewing the sample to almost the same degree essentially allow for parallel plate compression experiments of adhered vesicles (Figure 12.7a,b) (Schaefer et al., 2013). Parallel plate compression largely prevents the vesicle to displace under force. Moreover, the restoring force to compression are much larger than vesicles experience if subjected to a point-load producing. Another advantage is computation of the contour during compression and a negligible contribution from bending arises that would otherwise complicate the theoretical description considerably. As a consequence, the K_A values found in these experiments are larger than in the case of sharp indenters.

GUV adhered to a avidin-coated coverslip. (Figure 12.7c) Typical force compression curve obtained from squeezing a 1-palmitoyl-2-oleoyl-*sn*-glycero-3-phosphocholine (POPC) vesicle between cantilever and substrate ($R_v = 6$ µm). (Figure 12.7d) GUV probed with a spherical indenter (view from the bottom) (Figure 12.7e) Confocal laser scanning micrograph of the *x*-*z* plane during central indentation with 4 nN load force. (Figure 12.7f) Confocal image of a vesicle not hit in the center therefore dislocating from the normal axis (images courtesy of T. Kliesch).

Figure 12.7f illustrates what happens if a GUV is not centrally indented as shown in Figure 12.7d,e. Due to the appreciable adhesion the GUV dislocates and thereby compromises axisymmetry of the problem. Computation of elastic membrane properties becomes very complicated.

12.6.3 TETHER PULLING

Generally, approach and retraction curves performed on giant liposomes match each other because viscous losses are too small to be measured by AFM. However, membrane tethers are frequently pulled out of the GUV, which additionally allows assessing the bending modulus of the membrane from the plateau force, f_{pl}, with

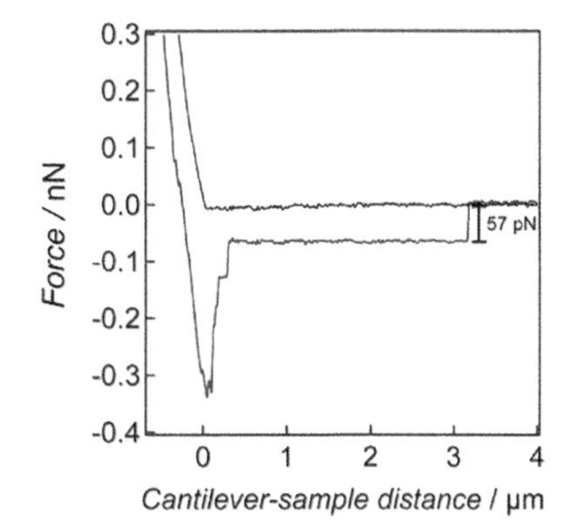

Figure 12.8 GUV compression with tipless cantilevers or blunt indented. Typical compression (top) and retraction curve (bottom) displaying the formation of a membrane tether. The force plateau with respect to the baseline at zero force allows the extraction of the membrane tension generated by adhesion to the surface.

respect to the baseline (Bo and Waugh, 1989; Cuvelier et al., 2005; Kocun et al., 2012; Powers et al., 2002) (see also Chapter 16):

$$f_{pl} = 2\pi\sqrt{2\Sigma_0\kappa} \tag{12.38}$$

with κ the bending modulus of the membrane. Alternatively, one can also compare tension values from indentation with those from tether pulling because the bending modulus of the membrane is known. This procedure allows one to validate the theoretical approach used to describe the indentation experiments. Figure 12.8 shows a typical force curve in which a tether is formed upon retraction of the cantilever. Both cantilevers equipped with a tip or tipless cantilever frequently generate tethers upon retraction.

12.6.4 GUVs WITH AN INTERNAL ACTIN CORTEX

There is a widespread interest in the use of giant vesicles as a platform for producing artificial cells to study adhesion, mechanics and dynamics as a function of composition and complexity (see Chapter 4). In particular, inclusion of cytoskeleton filaments is of great interest in order to realize active matter in confined geometry. The presence of myosin motors and actin filaments powered by ATP enables one to build a minimal model of a motile cell (Murrell et al., 2011; Tsai et al., 2011; Vogel et al., 2013a,b). The question therefore is, what happens mechanically if the liposome's shell is reinforced with an inner layer of actin—an artificial cortex? It is well known that the cortex plays a pivotal role for cellular mechanics and also cell shape during migration. Especially, the cortical tension is generated by the contractility of the cortex due to the presence of myosin II motors. Sackmann and coworkers were among the first to assemble thin actin shells in giant liposomes and elucidate their mechanical properties by recording thermal membrane undulations with optical microscopy (Häckl et al., 1998). Actin-filled liposomes have also been prepared by gentle hydration (Honda et al., 1999), electroformation, (Häckl et al., 1998) inkjet electroformation, (Stachowiak et al., 2008, 2009) and the inverted emulsion method (Pontani et al., 2009). The use of oil in the latter method has the disadvantage of interfering with mechanical measurements by residual oil partitioned in the membrane–phase. Koenderink and coworkers successfully generated GUVs filled with an actomyosin gel using "gentle hydration" of lipids on an agarose hydrogel (Carvalho et al., 2013; Tsai et al., 2011). Both Sackmann as well as Koenderink and coworkers report only a small contribution of the actin shell to the elastic properties of the membrane using membrane undulation monitoring (Häckl et al., 1998; Tsai et al., 2011). Sykes and coworkers investigated the spreading behavior of giant liposomes equipped with an actin shell (Murrell et al., 2011). They found that early spreading of actin-filled GUVs can be described by distinct power laws depending on the homogeneity of the actin cortex mirroring the spreading behavior of living eukaryotic cells. Recently, Guevorkian et al. used hydrodynamic tube extrusion to study the viscoelasticity of liposome with an internal actin cortex (Guevorkian et al., 2015). They found that the elastic as well as the viscous behavior of the membrane heavily depends on the presence of the actin cortex.

Previously, we investigated how the presence of an actin cortex changes the mechanical response of GUVs to compression between two parallel plates or indentation with a conical indenter (Schaefer et al., 2013, 2015). Figure 12.9 shows a representative set of force indentation experiments using a pyramidal indenter acting on two vesicles, one in the presence (squares) and one in the absence of actin (circles). The red and green lines are fits according to the aforementioned tension model. Although the area compressibility modulus increases here substantially, the effect of actin on GUV mechanics is less pronounced, on average, because many liposomes with an actin cortex do not show an altered elastic response compared with GUVs without actin. In some cases, however, membrane theory fails to describe the elastic behavior of the significantly stiffer cortex of liposomes with an actin shell. Force distance curves become more linear with a steeper slope. With increasing thickness of the actin cortex, bending becomes more relevant. It is conceivable that the additional actin shell forms a composite with the inner leaflet of membrane, which causes further stiffening of the structure. Depending on the thickness of the shell and coupling of the cortex to the bilayer, this effect might be detectable by force compression experiments. The area compressibility modulus of the composite shell itself might be strongly increased due to electrostatic interactions leading to cross-linking of phospholipids at the interface between filaments and inner leaflet. These cross-links would foster a larger apparent K_A and thereby explain the observed stiffening.

12.7 CONCLUSIONS AND OUTLOOK

GUVs are the ideal starting material to mimic cellular mechanics, ranging from adhering and subsequent spreading of cells on a surface to deformation with an atomic force microscope. Many studies exist in which an AFM is used to indent adherent cells in order to assess the elastic properties of living cells. In most cases, however, simple contact models based on Hertzian mechanics are used to describe the experimental force–distance curves. The physics of the intricate shell comprising an active contractile cortex attached to a fluid lipid bilayer is frequently ignored, and the cell is represented by an elastic continuum instead. GUVs allow the approach of this problem in a systematic manner by increasing complexity in a tailored bottom-up approach. This approach allows bringing together theory and experimentation in a more defined way while working with living cells. In particular, advancements in cortex mimics allow the study of the viscoelastic properties of active matter in a confined

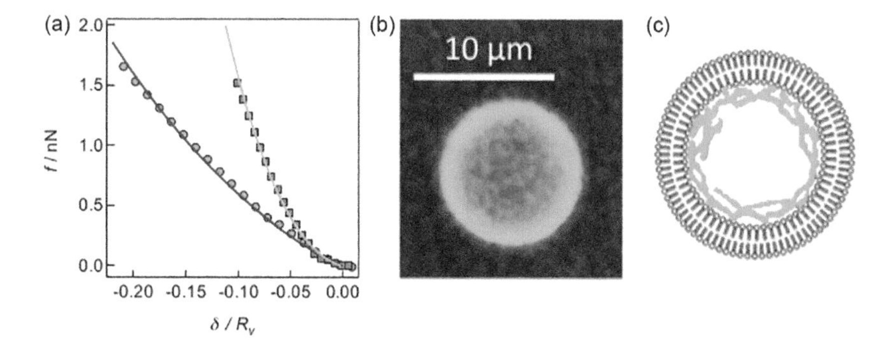

Figure 12.9 AFM on GUVs with internal actin shell. (a) Typical experimental force indentation curve of GUVs with an actin shell (filled squares) subject to fitting of the tension-based model (solid line) resulting in $\Sigma_0 = 0.53 \pm 0.02$ mN/m and $K_A = 0.434 \pm 0.008$ N/m. The plots show force f as a function of the dimensionless indentation δ / R_v to account for the two different radii of the two vesicles. Fixed parameters for modeling the actin-filled GUV: $R_v = 10.25\ \mu m$; $\theta = 18^\circ$. (b) Fluorescence image of a typical GUV with an actin shell (green dye: Alexa Fluor 488 actin). (c) Scheme illustrating the envisioned actin shell assembled inside the vesicle.

environment. The viscoelastic properties of active systems are of particular interest because living cells are paradigmatic out of equilibrium systems, driven by ATP consumption. Long-term memory fluctuation give rise to a new fluctuation–dissipation theorem that relates the response or creep function to the mean square displacement of the probe. Here, GUVs provide an excellent model system to study rheological properties of active matter as a function of cross-linker density, lipid composition, degree of actin polymerization and presence of myosin motors. The challenges in the future will be to reconstitute the necessary set of proteins or peptide with defined orientation and maintained functionality.

LIST OF ABBREVIATIONS

AFM	atomic force microscopy
DOPE	1,2-dioleoyl-*sn*-glycero-3-phosphoethanolamine
ITO	indium tin oxide
JKR	Johnson-Kendall-Roberts
POPC	1-palmitoyl-2-oleoyl-*sn*-glycero-3-phosphocholine

GLOSSARY OF SYMBOLS

A_0	initial membrane area in the unextended state
A_{ind}	surface area of the indented vesicle
A_v	surface area of the unstressed vesicle
C_1	first principle curvature
C_2	second principle curvature
E	Young's modulus
F_a	free energy associated with area dilatation
F_b	free energy of bending
F_c	free energy of compression
K_A	area compressibility modulus
K_C	bilayer thickness
L	length of the cantilever
P	pressure
R_1	contact radius of the vesicle with the AFM tip
R_i	contact radius of the vesicle with the substrate
R_o	radius of the vesicle at the equator
R_v	radius of the vesicle prior to indentation
R	radius of the AFM tip
V_{ind}	volume of the indented vesicle
V_v	volume of the unstressed vesicle
ΔA	change in membrane area
Σ_0	prestress of vesicle
Σ	lateral tension—force per unit length
$\bar{\kappa}$	saddle splay modulus
δ	indentation depth
κ	bending modulus
v	Poisson's ratio, the negative ratio of transverse to axial strain
θ	half opening angle of the cone representing the AFM tip
a	contact radius
d	size of the indentation dimple where bending is limited to
f	force
k_BT	thermal energy
k_c	spring constant of the cantilever
t_{b0}	initial bilayer thickness uncompressed
t_b	bilayer thickness
t_c	thickness of the cantilever
t	time
v	velocity of the z-piezo
w	width of the cantilever
z_p	z-piezo movement
z_{ts}	tip-sample separation

REFERENCES

Bando K, Ohba K, Oiso Y (2013) Deformation analysis of microcapsules compressed by two rigid parallel plates. *J. Biorheol.* 27:18–25.

Bando K, Oiso Y (2013) Indentation analysis of microcapsule with initial stretch. *J. Biomech. Sci. Eng.* 8:268–277.

Binnig G, Quate CF, Gerber C (1986) Atomic force microscope. *Phys. Rev. Lett.* 56:930–933.

Bo L, Waugh RE (1989) Determination of bilayer membrane bending stiffness by tether formation from giant, thin-walled vesicles. *Biophys. J.* 55:509–517.

Boal D (2012) *Mechanics of the Cell.* Cambridge, UK: Cambridge University Press.

Boroske E, Elwenspoek M, Helfrich W (1981) Osmotic shrinkage of giant egg-lecithin vesicles. *Biophys. J.* 34:95–109.

Brochu H, Vermette P (2008) Young's moduli of surface-bound liposomes by atomic force microscopy force measurements. *Langmuir* 24:2009–2014.

Butt HJ, Jaschke M (1995) Calculation of thermal noise in atomic force microscopy. *Nanotechnology* 6:1–7.

Butt HJ, Jaschke M (2005) Force measurements with the atomic force microscope: Technique, interpretation and applications. *Surf. Sci. Rep.* 59:1–152.

Carvalho K, Tsaid F-C, Lees E, Voituriez R, Koenderink GH, Sykes C (2013) Cell-sized liposomes reveal how actomyosin cortical tension drives shape change. *Proc. Natl. Acad. Sci. USA* 110:16456–16461.

Cuvelier D, Derényi I, Bassereau P, Nassoy P (2005) Coalescence of membrane tethers: Experiments, theory, and applications. *Biophys. J.* 88:2714–2726.

Deserno M (2015) Fluid lipid membranes: From differential geometry to curvature stresses. *Chem. Phys. Lipids.* 185:1–45.

Dieluweit S, Csiszár A, Rubner W, Fleischhauer J, Houben S, Merkel R (2010) Mechanical properties of bare and protein-coated giant unilamellar phospholipid vesicles. A comparative study of micropipet aspiration and atomic force microscopy. *Langmuir* 26:11041–11049.

Dimitrov DS, Angelova MI (1987) Lipid swelling and liposome formation on solid surfaces in external electric fields. *Prog. Colloid. Polym. Sci.* 73:48–56.

Esposito C, Tian A, Melamed S, Johnson C, Tee S-Y, Baumgart T (2007) Flicker spectroscopy of thermal lipid bilayer domain boundary fluctuations. *Biophys. J.* 93:3169–3181.

Evans E, Rawicz W (1997) Elasticity of fuzzy biomembranes. *Phys. Rev. Lett.* 79:2379–2382.

Evans E, Skalak R (1980) *Mechanics and Thermodynamics of Biomembranes.* Boca Raton, FL: CRC Press.

Fenz SF, Sengupta K (2012) Giant vesicles as cell models. *Integr. Biol.* 4:982–995.

Fery A, Weinkamer R (2007) Mechanical properties of micro- and nanocapsules: Single-capsule measurements. *Polymer* 48:7221–7235.

Fletcher DA., Mullins RD (2010) Cell mechanics and the cytoskeleton. *Nature* 463:485–492.

Guevorkian K, Manzi J, Pontani LL, Brochard-Wyart F, Sykes C (2015) Mechanics of biomimetic liposomes encapsulating an actin shell. *Biophys. J.* 109:2471–2479.

Häckl W, Bärmann M, Sackmann E (1998) Shape changes of self–assembled actin bilayer composite membranes. *Phys. Rev. Lett.* 80:1786–1789.

Hertz HR (1882) On contact between elastic bodies [Ueber die Beruehrung fester elastischer koerper]. *J. Reine. Angew. Math.* 94:156–171.

Hoffman BD, Crocker JC (2009) Cell mechanics: Dissecting the physical responses of cells to force. *Annu. Rev. Biomed. Eng.* 11:259–288.

Honda M, Takiguchi K, Ishikawa S, Hotani H (1999) Morphogenesis of liposomes encapsulating actin depends on the type of actin-crosslinking. *J. Mol. Biol.* 287:293–300.

Hutter JL, Bechhoefer J (1993) Calibration of atomic–force microscope tips. *Rev. Sci. Instrum.* 64:1868–1873.

Janmey PA, McCulloch CA (2007) Cell mechanics: integrating cell responses to mechanical stimuli. *Annu. Rev. Biomed. Eng.* 9:1–34.

Johnson KL, Kendall K, Roberts AD (1971) Surface energy and the contact of elastic solids. *Proc. Roy. Soc. A* 324:301–313.

Kocun M, Janshoff A (2012) Pulling tethers from pore spanning bilayer: towards simultaneous determination of local bending modulus and lateral tension of membranes. *Small* 8:847–851.

Kuenneke S, Krueger D, Janshoff A (2004) Scrutiny of the failure of lipid membranes as a function of headgroups, chain length and lamellarity measured by scanning force microscopy. *Biophys. J.* 86:1545–1553.

Liang X, Mao G, Ng KYS (2004) Mechanical properties and stability measurement of cholesterol-containing liposome on mica by atomic force microscopy. *J. Adv Coll Int Sci* 278:53–62.

Murrell M Pontani LL Guevorkian K, Cuvelier D, Nassoy P, Sykes C (2011) Spreading dynamics of biomimetic actin cortices. *Biophys. J.* 100:1400–1409.

Needham D, Evans E (1988) Structure and mechanical properties of giant lipid (DMPC) vesicle bilayers from 20°C below to 10°C above the liquid crystal-crystalline phase transition at 24°C. *Biochemistry* 27:8261–8269.

Needham D, Nunn RS (1990) Elastic deformation and failure of lipid bilayer membranes containing cholesterol. *Biophys. J.* 58:997–1009.

Pautot S, Frisken BJ, Weitz DA (2003) Engineering asymmetric vesicles. *Proc. Natl. Acad. Sci. USA* 100:10718–10721.

Pécréaux J, Doebereiner H-G, Prost J, Joanny J-F, Bassereau P (2004) Refined contour analysis of giant unilamellar vesicles. *Eur. Phys. J. E* 13:277–290.

Pietuch A, Brückner BR, Janshoff A (2013a) Membrane tension and homeostasis of epithelial cells through surface area regulation in response to osmotic stress. *BBA—Mol. Cell. Res.* 1833:712–722.

Pietuch A, Brueckner BR, Fine T, Mey I, Janshoff A (2013b) Elastic properties of cells in the context of confluent cell monolayers: impact of tension and surface area regulation. *Soft Matter* 9:11490–11502.

Pollard TD, Borisy GG (2003) Cellular motility driven by assembly and disassembly of actin filaments. *Cell* 112:453–465.

Pollard TD, Cooper JA (2009) Actin, a central player in cell shape and movement. *Science* 326:1208–1212.

Pontani LL, van der Gucht J, Salbreux G, Heuvingh J, Joanny JF, Sykes C (2009) Reconstitution of an actin cortex inside a liposome. *Biophys. J.* 96:192–198.

Powers TR, Huber G, Goldstein RE (2002) Fluid-membrane tethers: Minimal surfaces and elastic boundary layers. *Phys. Rev. E* 65:041901–041912.

Rawicz W, Olbrich KC, McIntosh T, Needham D, Evans E (2000) Effect of chain length and unsaturation on elasticity of lipid bilayers. *Biophys. J.* 79:328–339.

Reissner E (1946a) Stresses and small displacements of shallow spherical shells I. *J. Math. Phys.* 25:80–85.

Reissner E (1946b) Stresses and small displacements of shallow spherical shells II. *J. Math. Phys.* 25:279–300.

Richmond DL, Schmid EM, Martens S, Stachowiak JC, Liska N, Fletcher DA (2011) Forming giant vesicles with controlled membrane composition, asymmetry, and contents. *Proc. Natl. Acad. Sci. USA* 108:9431–9436.

Schaefer E, Vache M, KLiesch TT, Janshoff A (2015) Mechanical response of adherent giant liposomes to indentation with a conical AFM-tip. *Soft Matter* 11:4487–4495.

Schaefer E., Kliesch T-T, Janshoff A (2013) Mechanical properties of giant liposomes compressed between two parallel plates: Impact of artificial actin shells. *Langmuir* 29:10463–10474.

Schwarz US, Safran SA (2013) Physics of adherent cells. *Rev. Mod. Phys.* 85:1327–1381.

Seifert U (1997) Configurations of fluid membranes and vesicles. *Adv. Phys.* 46:13–137.

Sen, S.; Subramanian S, Discher DE (2005) Indentation and adhesive probing of a cell membrane with AFM: Theoretical model and experiments. *Biophys. J.* 89:3203–3213.

Sneddon IN (1965) The relation between load and penetration in the axisymmetric Boussinesq problem for a punch of arbitrary profile. *Int. J. Eng. Sci.* 3:47–57.

Stachowiak JC, Richmond DL, Li TH, Brochard-Wyart F, Fletcher DA (2009) Inkjet formation of unilamellar lipid vesicles for cell-like encapsulation. *Lab. Chip.* 9:2003–2009.

Stachowiak JC, Richmond DL, Li TH, Liu AP, Parekh SH, Fletcher DA (2008) Unilamellar vesicle formation and encapsulation by microfluidic jetting. *Proc. Natl. Acad. Sci. USA* 105:4697–4702.

Stricker J, Falzone T, Gardel ML (2010) Mechanics of the F-actin cytoskeleton. *J. Biomech.* 43:9–14.

Tsai FC, Stuhrmann B, Koenderink GH (2011) Encapsulation of active cytoskeletal protein networks in cell-sized liposomes. *Langmuir* 27:10061–10071.

Vella D, Ajdari A, Vaziri A, Boudaoud A (2012) The indentation of pressurized elastic shells: From polymeric capsules to yeast cells. *J. R. Soc. Interface* 68:448–455.

Vogel SK, Heinemann F, Chwastek G, Schwille P (2013a) The Design of MACs (Minimal Actin Cortices). *Cytoskeleton* 70:706–717.

Vogel SK, Petrasek Z, Heinemann F, Schwille P (2013b) Myosin motors fragment and compact membrane-bound actin filaments. *eLife* 2:e00116.

Yoneda M (1964) Tension at the surface of sea-urchin egg: a critical examination of Cole's experiment. *J. Exp. Biol.* 41:893–906.

Manipulation and biophysical characterization of giant unilamellar vesicles with an optical stretcher

Gheorghe Cojoc, Antoine Girot, Ulysse Delabre, and Jochen Guck

To stretch or to heat—that is the question.

Contents

13.1 INTRODUCTORY WORDS

This chapter describes how to manipulate giant unilamellar vesicles (GUVs) with optical traps, and especially with a dual-beam laser trap (DBLT)—called an optical stretcher, to extract their mechanical and thermodynamic properties. After a brief general introduction to optical traps, we will present the basic principles of the optical stretcher that enable the deformation of vesicles without any contact. A critical choice for the study of vesicles with optically induced stresses is the choice of the laser wavelength. If it is not well selected, the absorption of laser light can induce heating effects that can be detrimental. We will explain how to use the optical stretcher with an appropriate laser source, either to deform and extract mechanical properties of GUVs or to heat GUVs intentionally in order to perform some thermodynamic characterization. When several wavelengths are combined in one setup, it is also possible to manipulate vesicles optically and to control their temperature almost independently.

13.1.1 HISTORICAL INTRODUCTION TO OPTICAL TRAPS

In 1957, Hendrick C. Van de Hulst wrote, "Radiation pressure will never be relevant in the lab (only extraterrestrial affairs) because there is no light source available that could create the light intensities required" (van de Hulst, 1957). However, only a few years later, in 1960, Theodore H. Maiman was credited with the invention of the first working laser. This opened the door to utilizing radiation pressure in the lab for the optical manipulation of micrometer-sized objects in so-called optical traps. The early pioneer of optical manipulation was Arthur Ashkin at Bell Labs in the early 1970s. He was the first to demonstrate that microscopic objects can be manipulated with light. He went on to demonstrate the first levitation of objects with one vertical laser beam and to create a stable trap using two counter-propagating laser beams—the first DBLT (Ashkin, 1970). Ashkin, 16 years after the DBLT, went on to also invent optical tweezers by utilizing the forces produced by a single, focused laser beam (Ashkin et al., 1986). Optical tweezers have since been employed in many ways as a biophysical tool, whereas the DBLT has been scarcely used in membrane biophysics (Dietrich et al., 1997; Velikov et al., 1997, 1999; Dimova et al., 2000, 2002). Later, in 1993, based on Ashkin's early dual-beam work, Mara Prentiss and co-workers created the first dual-beam fiber trap using optical fibers to deliver the two beams (Constable et al., 1993). With this significantly simplified experimental setup, they were successfully able to trap polystyrene spheres as well as yeast cells. This renaissance of DBLTs set off their use also for a diverse set of biophysical and biological experiments (Guck et al., 2005; Franze et al., 2007; Lautenschläger et al., 2009; Kreysing et al., 2014; Ekpenyong et al., 2017)—including the manipulation and characterization of GUVs (Solmaz et al., 2012; Piñón et al., 2013; Delabre et al., 2015; Huff et al., 2015). Optical forces and their application in biology in general have been well described elsewhere (Svoboda and Block, 1994) so that we limit our discussion here to the aspects essential for GUV manipulation.

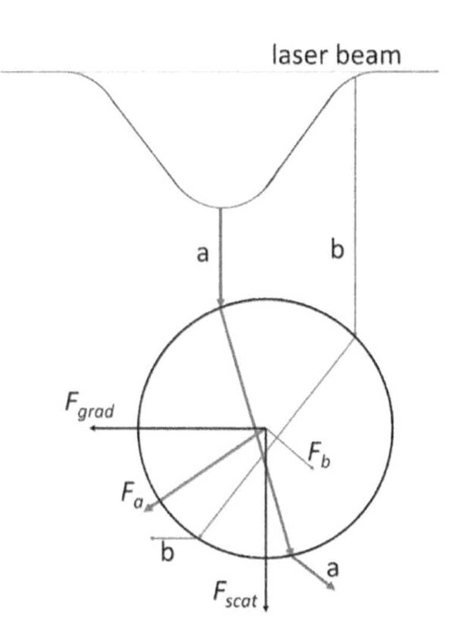

Figure 13.1 Geometry of two single rays that are incident at different angles with respect to the laser axis and giving rise to gradient and scattering forces F_{grad} and F_{scat}. Light intensity increases from b to a.

13.1.2 OPTICAL TWEEZERS

In single-beam gradient traps, known as optical tweezers, a tightly focused Gaussian laser beam generates strong electromagnetic fields that are capable of trapping an object against gravity and thermal motion (Ashkin, 1970; Ashkin et al., 1986). The electromagnetic net force is often decomposed into two components: scattering and gradient forces. The scattering force always pushes the particles away from the light source and can be explained by the scattering of the incident light. In contrast, the gradient force is directed along the field gradient, in most cases toward regions of highest field intensity. If the refractive index of the particle is lower than that of the surrounding medium, then the direction of the gradient force reverses and the particle is pushed away from highest field intensity. Both scattering and gradient forces stem from the interaction of light with the boundary of two mediums with different refractive indices. A sketch of the forces exerted on the spherical object is shown in Figure 13.1. The forces are due to the change in momentum of the focused ray as it passes through the sphere.

A variety of optical configurations and beam profiles, both in intensity and phase, have been used to manipulate a range of micrometer-sized particles (Molloy and Padgett, 2002; Franke-Arnold and Allen, 2008). In the last two decades optical tweezers have been employed to characterize GUVs: membrane viscoelasticity (Helfer et al., 2001; Mizuno et al., 2001), viscosity (Dimova et al., 1999, 2000), adhesion (Dietrich et al., 1997) or pearling instabilities (Bar-Ziv et al., 1998) and, in Chapter 16 of this book, there is an extensive description of how one can use optical tweezers to mechanically characterize GUVs by pulling a membrane tube out of a giant vesicle. Unfortunately, the small extent of the laser beams focus in optical tweezers poses limitations on the types of particles that can be trapped because the high field intensities near the focus may cause photodamage to sensitive samples (Kamm et al., 2010; Whyte et al., 2010), more about advantages and disadvantages of single beam traps can be found in Chapter 16.

13.1.3 DUAL-BEAM LASER TRAPS—OPTICAL STRETCHER

The DBLT is another kind of optical trap consisting of two non-focused, counter-propagating laser beams. This DBLT precedes optical tweezers by almost two decades. In 1993, the DBLT was used for the first time with optical fibers and successfully trapped polystyrene spheres as well as yeast cells (Constable et al., 1993). This led to the development of lensed fibers for trapping beads (Lyons and Sonek, 1995) and even a four-fiber divergent optical trap (Sidick et al., 1997). These techniques have shown the ability to trap and move objects using a variety of configurations and various wavelengths of light, but object deformation was generally viewed as a negative side effect. An idea that was ripe for exploration was the use of these forces to deform and measure soft materials in a controlled manner. The term optical stretcher was invented by Jochen Guck and Josef Käs at the University of Texas at Austin when they discovered that the DBLT could be used to stretch out red blood cells (Guck et al., 2000). Since that time, this deformation technique, used to measure the viscoelastic properties of a number of cell types and to utilize cell deformability as a cell marker (Guck et al., 2005; Lautenschläger et al., 2009), has evolved from a slow, open setup into a high-throughput microfluidic assembly (Lincoln et al., 2004, 2007a, 2007b; Bellini et al., 2012; Faigle et al., 2015; Nava et al., 2015). Recently, several groups have employed DBLTs to successfully stretch GUVs (Solmaz et al., 2012; Piñón et al., 2013; Delabre et al., 2015; Huff et al., 2015). The advantages of DBLTs include lower intensities on the trapped objects, a well-defined stress profile, no requirement of handles, decoupling from imaging optics, and the possibility to trap cells and cell clusters up to 100 μm in diameter.

13.2 OPTICAL STRETCHING BASICS

13.2.1 WORKING PRINCIPLE

Consisting of two counter-propagating, diverging Gaussian beams emerging from single-mode optical fibers, the optical stretcher is capable of trapping and deforming individual objects without establishing mechanical contact. It is noteworthy that the trapped particle is not compressed between the scattering forces from the two incident laser beams, as might be expected, but, rather counter-intuitively, elongated along the beam axis (Guck et al., 2000, 2001). An illustration of the trapping and deformation of a spherical particle in the optical stretcher is shown in Figure 13.2.

As the trapping optics are completely separated from the microscope optics, the optical stretcher becomes an extremely versatile tool for the micromanipulation of biological samples (Kemper et al., 2010)—and relatively high-throughput microfluidic particle delivery device (~150 particles/h) (Lincoln et al., 2007a, 2007b).

13.2.2 PHYSICS BACKGROUND

Now that the general idea of trapping and deforming cells with laser beams has been introduced, we can look at which forces are involved. Each beam can be well approximated by a Gaussian profile that begins to diverge once it exits the optical fiber. The factors needed to describe the beam width or beam waist (the distance at which the intensity falls to $1/e^2$), w, as a function of distance are the laser wavelength in vacuum, λ_0, the index of refraction of the medium, n_{med}, and the initial beam width where the beam exits the fiber, w_0. This is defined by

$$w(z) = w_0\sqrt{1+\left(\frac{z\lambda_0}{\pi n_{med} w_0^2}\right)^2} \tag{13.1}$$

where z is the distance from the fiber end. The typical z distance in such experiments is from 60 to 100 μm. For specific wavelengths (and using dedicated fibers), 808 and 1070 nm, the beam has an initial radius of $w_0 = 2.2$ μm and $w_0 = 3.1$ μm, respectively. With these values fixed, and considering n_{med} typically close to water ($n_{med} = 1.335$), the radius of the beam varies only as a function of the distance from the fiber (Figure 13.3). Thus, by changing the distance between the fiber ends, the trap width,

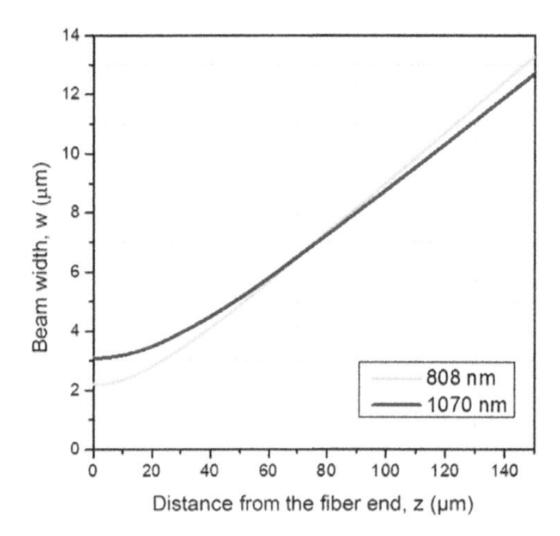

Figure 13.3 Laser beam width called "waist," as function of distance from fiber end and wavelength.

and with that the size of the particles that can be trapped and manipulated can be set to almost any value. Practical considerations limit the sizes of objects typically manipulated in optical stretchers to 1–50 μm diameter.

When this beam is incident on a dielectric object of differing index of refraction than the medium, there is a resulting net force on the object. This force can be calculated using ray optics by representing a Gaussian beam as individual rays incident on a spherical object, as shown in Figure 13.4a.

When a light ray passes through an interface where there is a change in the index of refraction, there will be a normal force exerted on the interface toward the region of lower refractive index (Ashkin and Dziedzic, 1973; Guck et al., 2000; Casner and Delville, 2001). This will act to balance the change in momentum of the light as it travels from one medium to the next. So, for objects with a higher index of refraction than the surrounding medium, whenever light passes through the surface there will be a normal outward force.

This results in a net force, which is generally described in terms of two perpendicular components, called the scattering force, F_{scat}, and the gradient force, F_{grad}, (Figure 13.4b). Although gradient and scattering forces do not have to be perpendicular to each other in general, in a paraxial, non-focused beam, the intensity gradient along the axis can be ignored. F_{scat} is due to the

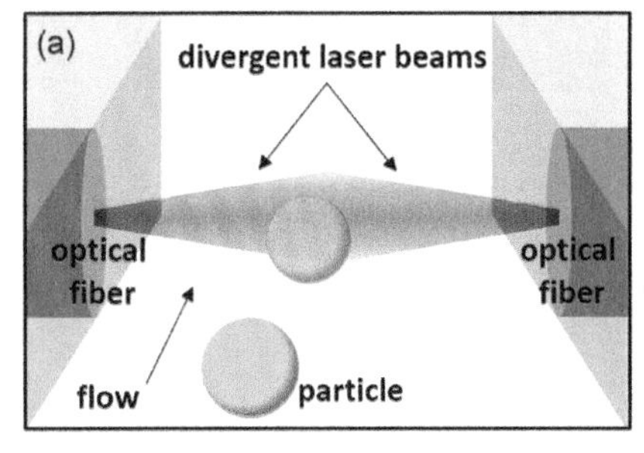

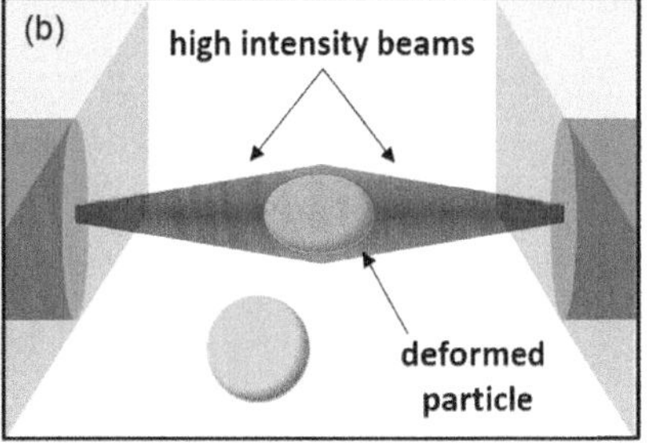

Figure 13.2 Schematic of the serial trapping and stretching of particles in the optical stretcher. (a) Particles yellow are delivered to the trapping region by means of a microfluidic channel. Low-intensity Gaussian laser beams (red) emanating from two opposing optical fibers initially capture the object by subjecting it to scattering and gradient forces directed toward the center of the optical trap. (b) Once the object is stably trapped, the laser intensity is increased so that the resulting optical surface forces are strong enough to deform the object. (Adapted from Guck, J. et al., *Biophys J*, 88, 3689–3698, 2005.)

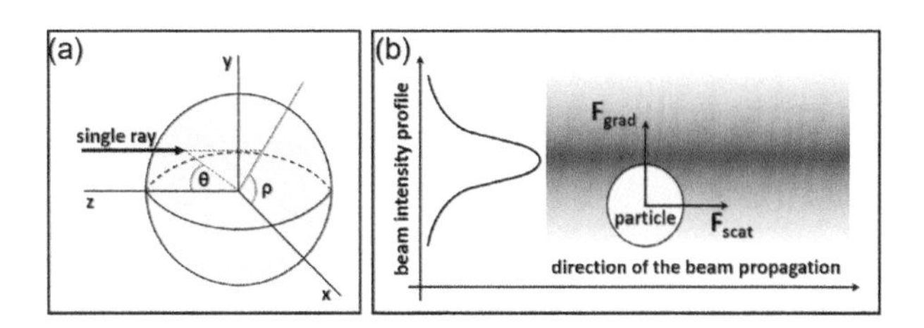

Figure 13.4 (a) Single ray incident on a sphere and (b) net forces on a spherical object with a higher index of refraction than the surrounding medium. (Adapted from Lincoln, B., The microfluidic optical stretcher, PhD thesis, University of Leipzig, 2006.)

forward momentum of the scattered light, which is removed from the laser beam and transferred to the scattering object, whereas F_{grad} is due to the symmetry breaking of the scattered momentum when the object is not centered on the laser axis. The scattering force points in the direction of the beam propagation and has the magnitude (Roosen and Imbert, 1976; Askhin, 1992)

$$F_{scat} = \frac{n_{med}}{c} \int_0^{\frac{\pi}{2}} d\theta \int_0^{2\pi} d\varphi I(\theta,\varphi,\rho_0)\rho^2 \sin(\theta)\cos(\theta)\times \left\{1+R_F\cos(2\theta)-\frac{T_F^2\left[\cos(2\theta-2r)+R_F\cos(2\theta)\right]}{1+R_F^2+2R_F\cos(2r)}\right\} \tag{13.2}$$

whereas the gradient force is directed toward (or away from, depending on the relative refractive index of the object and the surrounding medium) the beam axis and has the magnitude (Roosen and Imbert, 1976; Ashkin, 1992)

$$F_{grad} = \frac{n_{med}}{c} \int_0^{\frac{\pi}{2}} d\theta \int_0^{2\pi} d\varphi I(\theta,\varphi,\rho_0)\rho^2 \sin^2(\theta)\sin(\varphi)\times \left\{1+R_F\sin(2\theta)-\frac{T_F^2\left[\cos(2\theta-2r)+R_F\sin(2\theta)\right]}{1+R_F^2+2R_F\cos(2r)}\right\} \tag{13.3}$$

where

$$I(\theta,\varphi,\rho_0)=\frac{2P}{\omega(d)^2\pi}\exp\left[-\frac{2\rho^2\sin^2(\theta)+\rho_0^2-2\rho_0\rho\sin(\theta)\sin(\varphi)}{\omega(d)^2}\right] \tag{13.4}$$

Here, ρ is the sphere's radius; ρ_0 is the distance of the center of the sphere from the beam axis; P is the laser power; c is the speed of light in vacuum; and r is the refractive angle. R_F and T_F are the Fresnel intensity reflection and transmission coefficients, which are functions of the meridional and azimuthal angles, θ and φ, respectively as shown in Figure 13.4a. They are given by

$$R_F = \frac{R_{\parallel}+R_{\perp}}{2} \tag{13.5}$$

$$T_F = \frac{T_{\parallel}+T_{\perp}}{2} \tag{13.6}$$

where $R_{\parallel}$ and $T_{\parallel}$ represent the relative magnitudes of the fields before and after reflection and transmission for the transverse electric (TE) of electric field, and $R_{\perp}$ and $T_{\perp}$ represent the intensity of the reflection and transmission for the transverse magnetic (TM) of electric field (Roosen and Imbert, 1976). For nonpolarized beams, the average may be taken as shown in Eqs. 13.5 and 13.6 (R_F, T_F). (For explicit formulas, see Jackson, 1962).

When the object is centered on the beam axis, the net gradient force is zero due to symmetry, and only the scattering force remains. So, to create a stable trap, two identical beams can be aligned opposing each other so that the scattering forces from each beam cancel and the object is held stationary at the center point between them. Despite the fact that the object is held in place, there is still a steady flux of light passing through the object. Because the change in momentum (and thus the force) occurs at the interface, a stress is applied locally, where stress $= \frac{\text{force}}{\text{unit area}}$. The key difference using cells instead of rigid objects such as beads is the fact that these surface forces can additionally deform the object. Thus, although retaining the trapping characteristics of the body forces, a stress profile can be calculated over the surface of the object by considering all of the rays from an incident Gaussian beam (Guck et al., 2000) (Figure 13.5a).

The stress profile can be well approximated (Guck et al., 2001) by

$$\sigma(\theta)=\sigma_0 cos^2(\theta) \tag{13.7}$$

if one accounts only for the lowest-order dependence in θ. The peak stress, σ_0, is defined as the maximal stress along the beam axis ($\theta = 0$) from the combined effect of the two central rays passing through the object, and is given by

$$\sigma_0 \approx \frac{n_{med}I}{c}\left(2-R(\theta=0)+R(\theta=0)^2\right)\left(\frac{n_{cell}}{n_{med}}-1\right). \tag{13.8}$$

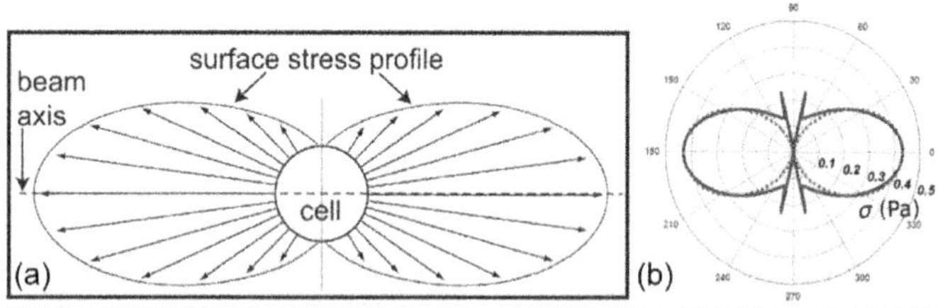

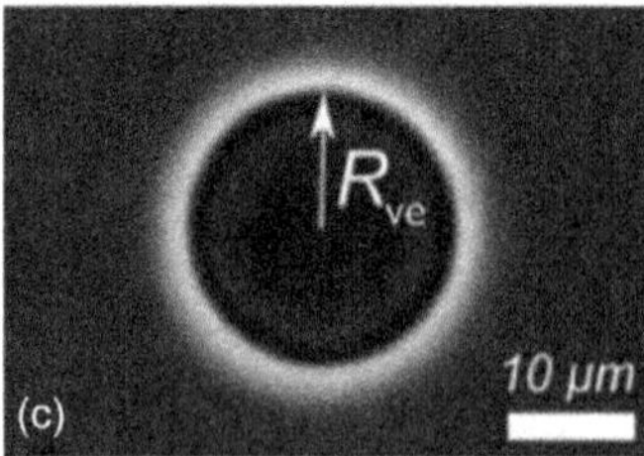

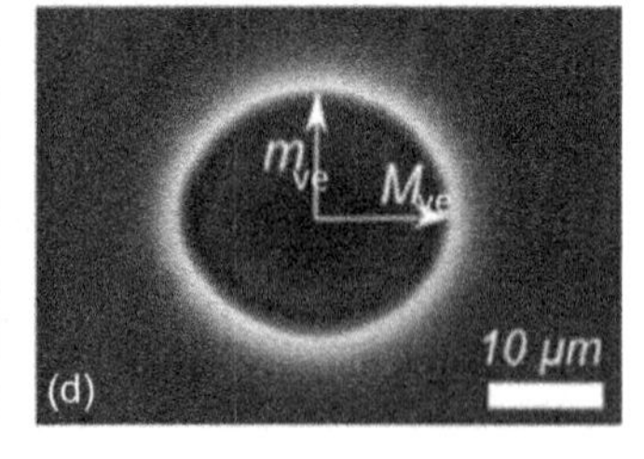

Figure 13.5 The optical stress on the surface deforms a compliant trapped object, such as a GUV. (a) Sketch of the surface stress profile. (b) Comparison between ray optics calculations of the optical stress (full line) and cos² approximation (dashed line) for a vesicle with radius R_{ve} = 10.6 μm and a beam radius w = 9.0 μm. (c) Vesicle trapped in an optical stretcher. (d) Vesicle deformed by an optical stretcher. (Delabre, U. et al., *Soft Matter*, 11, 6075–6088, 2015. Reproduced by permission of The Royal Society of Chemistry.)

This equation is an approximation because focusing of the beam by the presence of the object itself (which acts as a spherical lens) has not been accounted for. Figure 13.5 shows a comparison between the optical stress calculated based on ray optics and from the $\cos^2$ approximation. The resulting deformation (Figure 13.5c and d) can be tracked in the form of an axial elongation as a function of time. A fit to this shape can then be used to determine the mechanical properties of the trapped object, such as a vesicle, as will be discussed in the next section.

13.2.3 SPECIFICS FOR VESICLES: THE REFRACTIVE INDEX DIFFERENCE

In the case of vesicles, it is possible to control the inner and outer buffer solutions to modify the refractive index difference. The higher the refractive index difference is, the higher the optical stress is at a given power (see Eq. 13.8). When vesicles are prepared by electroformation, for example, the refractive index of the inner buffer solution can be adjusted with the sucrose concentration, whereas the refractive index of the outer buffer solution can be adjusted with the glucose solution. Here we exploit the fact that the refractive indices of the two sugar solutions increase differently with increasing concentration and osmolarity. The refractive indices of the buffers can be measured with an Abbe refractometer, for instance. For 1,000 mOsm of osmolarity of the inner sucrose solution and outer glucose solution, we measured $n_{\text{sucrose}} = 1.370$ and $n_{\text{glucose}} = 1.355$. Vesicles are then well adapted for optical characterization studies. However, high osmolarity solutions have proved to modify the mechanical properties of vesicles (Genova et al., 2006; Solmaz et al., 2013; Dimova, 2014), so there is a balance between maximizing stress for deformation and minimizing such adverse effects, which has to be optimized for the specific question at hand.

13.3 BUILDING AN OPTICAL STRETCHER

One of the main differences between various optical stretcher (OS) setups is the way in which particles, vesicles or cells are handled and delivered to the trap. In this section, we describe the two main OS configurations: open OS and closed OS (single-channel).

Open OS: Both optical fibers are aligned longitudinally against a straight edge, or backstop (Figure 13.6a), attached to a glass slide. A simple backstop can be obtained by gluing a thin, cylindrical object, such as another optical fiber or a glass capillary, onto the coverslip (see Lincoln, 2006). More advanced versions can be made by photolithography. The fiber positions can be adjusted by means of a three-axis translation stage onto which the fibers are mounted. Once the particle is placed into the gap between the fibers, the laser is supplied with power to initiate the trapping and deformation process.

Closed OS: A number of particles are serially delivered into the trapping region with the help of a glass microcapillary arranged perpendicular to the optical fibers. Laser light reflections by the capillary walls are minimized using index-matching gel. The physical separation of the optical and microfluidic elements has the advantage that the performance of the fibers is not affected by pollution from the particle suspension and that the capillary can be maintained and cleaned separately.

Box 13.1 Components required to build an optical stretcher setup

1. Single mode fiber laser/s.
2. Remove the polyacrylamide cover from the fibers using a fiber stripper. Clean them using paper tissues and ethanol. Use a fiber cleaver to perfectly cut the fibers.
3. Place the fibers on a coverslip having a polymer pattern. The polymer structures help with aligning the fibers.
4. Clamp the fibers on an aluminum stage to hold them in place.
5. To deliver the particle into the trap, build a gravity flow system using microfluidic tubing and a glass capillary. When working in open optical stretcher (OS) configuration the particle delivery can be achieved using three-dimensional (3D) manipulator, pressure pump and glass pipettes.
6. To avoid back reflections of the laser beam (when working in closed OS configuration), use a gel, with similar refractive index as the glass capillary and the fiber core, to fill the space between the fibers and the glass capillary.

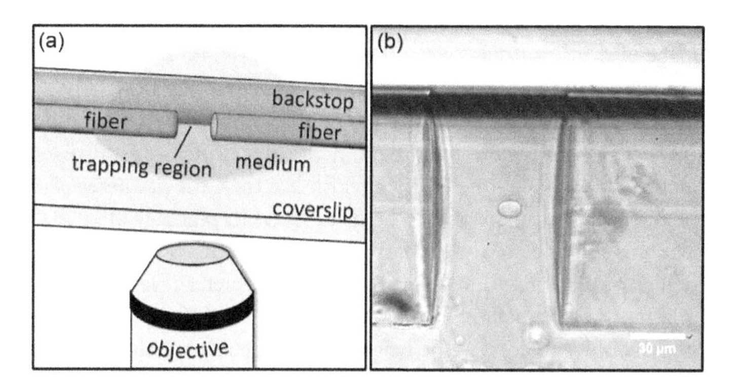

Figure 13.6 (a) The most basic optical stretcher setup. Two opposing fibers are aligned against a backstop fixed to a glass slide that is mounted on a microscope. A drop of medium envelopes the trap region. Particles are delivered by pipetting small amounts of additional particle suspension into the drop, after which individual particles can be trapped as they settle toward the bottom. (b) A phase-contrast image of an open sètup. This image also shows the distance between the optical fiber ends, which is required to calculate the size of the trap for each measurement.

Based on the same principle as the closed OS described here, another type of OS, multiple-channel OS, can be built, for the delivery of different drugs (Lincoln, 2006), for example, or the separation of cells after the measurement (Faigle, 2015). We are not describing this last OS setup in this chapter, but we discuss pitfalls and tips for building open and single channel OS setups.

13.3.1 OPEN SETUP

When deciding to perform experiments with an open configuration, the OS must be assembled and mounted onto an inverted microscope prior to each experiment. For clean and efficient working conditions, it is necessary that the optical fibers are freshly cleaned and cleaved, and a new, clean coverslip is used. Fibers are "peeled off" by the protection jacket, wiped with laboratory tissue and ethanol and cleaved. Fiber alignment is crucial for the counter-propagating beams to form a stable trap. It has been observed that a transverse offset of one of the fibers by a few micrometers can cause the cell to be pushed back and forth by the two beams. A stable trap, where the object is held stationary, can consistently be achieved by way of a backstop against which optical fibers (Figure 13.6a), held by three axis translation stages, are bent at a high enough angle that they become well aligned against it. Particles in suspension are transferred by a standard laboratory pipette to the region of the trap, which is located at the midpoint between the two fibers (Figure 13.6b). As the particles settle, each fiber can be translated along the backstop in an attempt to move the trap into their path. Pulling the fibers back from each other actually helps to create a little flow directed toward the trap to suck in a cell that is settling off to one side. The ability to modify the distance between the fiber ends also represents an advantage when the stress profile needs to be changed or when the laser power is a limitation. Typical distances between fibers ends used are from 100 to 250 μm. Once a cell is trapped and the fiber location is fixed, an image of the distance between the fibers ends should be acquired for later determination of the actual stress applied (Figure 13.6b).

An alternative simple method to get the optical fibers well and stably aligned is to constrain them in all three dimensions. Five micro-fabricated channels with a width less than the diameter of the fibers (125 μm) are able to hold the fibers in precisely aligned positions provided that the channel made from SU-8 is sufficiently high (25–40 μm) with respect to the fiber diameter (Figure 13.7a). Different width of channels, ranging from 90 to 110 μm, offer the possibility to control the height of the trap. Channels are created in SU-8 photoresist deposited on glass substrates with standard photolithography techniques (Lorenz et al., 1997). Glass substrates could be either thin coverslips or thicker microscopy slides. Coverslips are to be preferred if very good optical imaging quality of the trapped particle, fluorescence imaging or additional optical manipulation with optical tweezers is important. Microscopy slides provide easier handling and better mechanical stability during the experiment. For a fiber diameter of 125 μm, one can optimize the channel width and resist thickness to 100 and 25 μm, respectively. These parameters were determined taking into account ease of fiber placement, stable alignment, and volume of space underneath the trapping region. Care has to be taken that the bonding between SU-8 and the coverslip is tight. This can be made sure by hard post-baking. Otherwise an aqueous medium can creep underneath the photoresist and cause liftoff and, ultimately, fiber misalignment during the experiment (Figure 13.8).

Once the glass substrates with the SU-8 structures are prepared, the further setup can proceed under a stereoscope. Fibers are gently placed into the appropriate channels one at a time and temporarily clamped with screw nuts (Figure 13.7b). Because fiber alignment is critical, care must be taken to ensure that the fiber cores are coaxial. Once both fibers are in place and clamped, the fibers can be glued to the SU-8 coated coverslip with drops of nail polish deposited with a syringe needle. Fibers and the nail polish are usually left untouched for about 20 min, to allow the nail polish to dry. Fiber alignment can be double checked with a higher magnification objective, and reassembled if necessary.

To avoid coverslip bending during operation, which would misalign the optical fibers, the entire chip is placed on a custom microscope stage (technical drawings available upon request) and tightened down with an aluminum clamp (Figure 13.9a).

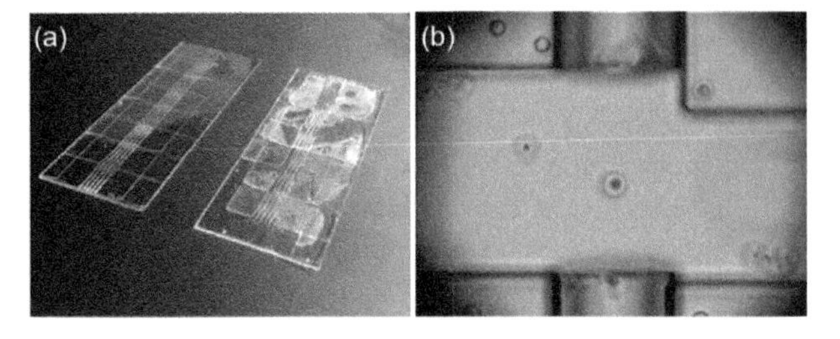

Figure 13.8 (a) Insufficient bonding between glass substrate and SU-8 can lead to liftoff. Left: Unused patterned chip. The different grooves shown have different widths in order to be able to find the optimal size. Right: Patterned chip after submersion in deionized water for 2 h. (b) Liftoff removes structures in the chip and leads to fiber misalignment as evidenced by the structure in the upper right corner of the image, which has moved downward. (Reproduced from Faigle, C., Optical stretcher: Towards a cell sorter based on high-content analysis, PhD thesis, Technical University of Dresden, 2015; Courtesy of Christoph Faigle.)

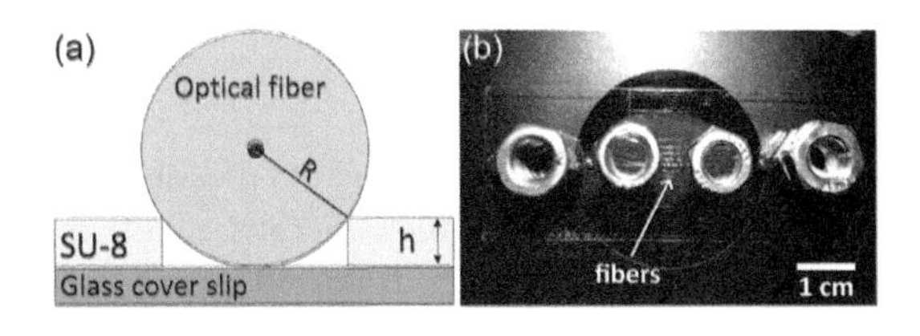

Figure 13.7 (a) Illustration of an alternative optical fiber alignment geometry. Two blocks of SU-8 photoresist create a channel in which the fiber can sit. The drawing shows the ideal case where the optical fiber rests on the glass and is simultaneously in contact with both blocks of SU-8. Because this can hardly ever be achieved perfectly, it is recommended to increase the height of, or reduce the width between, the SU-8 blocks so that the fiber only rests on the SU-8. (b) Image of the fibers aligned under a stereoscope, clamped with screw nuts, and glued with quick-drying nail polish (red dots).

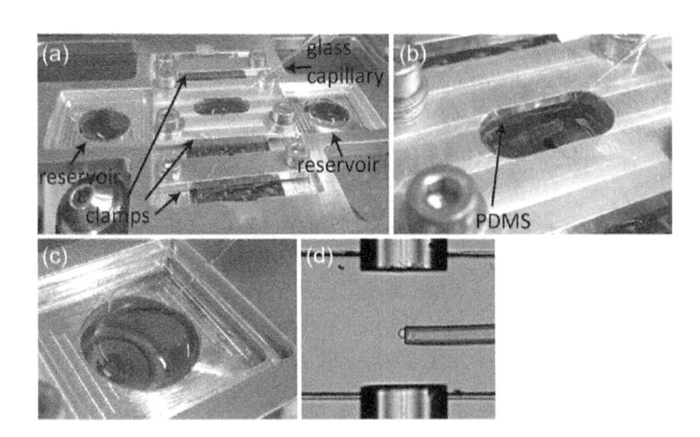

Figure 13.9 Open OS and pipette manipulation. (a) The open OS stage. (b) Zoom-in of the trapping region. (c) Zoom-in of one of the two side reservoirs. (d) Particles can be delivered into the trap by means of a glass capillary. (Reproduced from Faigle, C., Optical stretcher: Towards a cell sorter based on high-content analysis, Ph.D. thesis, Technical University of Dresden, 2015; Courtesy of Christoph Faigle.)

The aluminum clamp does not come in direct contact with the glass, but a polydimethylsiloxane (PDMS) spacer is placed in between (Figure 13.9b). Both the PDMS spacer (3–5 mm height) and aluminum have a hole to permit the access of a glass pipette to the trapping region (Figure 13.9a and b). When tightening the whole device, a leakage-free well is formed and one can perform experiments for about 1.5 h. The evaporation during this time is minimal. For longer experiments, evaporation could be avoided by adding a thin film of silicone oil to cover the droplet surface.

A more efficient delivery of particles into the trap than merely relying on sedimentation into the trap from a suspension placed over the general trapping region is the targeted delivery using a micropipette. For this, particles can be picked up from reservoir regions located on either side of the SU-9 patterned glass substrate (Figure 13.9c) with a micropipette (e.g., 20-μm inner diameter) connected to a pressure-driven air syringe. With the help of an electronic micropipette holder, particles can be transported by moving the pipette along the stage of the microscope and delivered directly into the laser beam of the optical trap (Figure 13.9d).

13.3.2 CLOSED SETUP

When a higher throughput of particles to be measured in a single experiment is desired, the closed OS is a better choice than the open setup. Its construction, first described by Lincoln et al. in 2007 (Lincoln et al., 2007a, 2007b) and since in use fairly unaltered, is slightly more involved than the open setup. The major difference is that it features a closed microfluidic system to deliver particles to the trap region. In its simplest form, a square glass microcapillary serves as the flow channel and the laser beams are sent through the capillary walls to trap and stretch the cells inside (Figure 13.10a and b). The fact that the optical fibers are separated from the particle suspension has the advantage that the fiber ends do not collect debris, which could affect beam properties. The glass capillary should be externally cleaned with ethanol prior to being placed perpendicular with the fibers. The perpendicular alignment of both fibers and the capillary is ensured by an SU-8 pattern similar to that of the open setup with an additional gap running perpendicular to the one holding the fibers. To reduce reflections of the laser light as it enters the capillary wall, an index-matching gel is used. When applying the gel, it is important to ensure that no air bubbles or debris are present between the fiber ends and the capillary. As for the open OS, a slab of PDMS, a coverslip and a clamp are used to tighten down the fibers and the microcapillary (Figure 13.10a and c). Once assembled, a closed setup can be reused for many experiments easily for up to a year. Rinsing with ethanol and deionized water after each experiment is usually sufficient to keep the microcapillary clean and free of clogging. The particle suspension is introduced into the square capillary via graphite ferrules, and its flow regulated by the height difference between two communicating reservoirs. It should be noted that the use of a closed setup requires the availability of a sufficiently large number of objects (10^5 is a useful order of magnitude) in a volume of about 1 mL.

13.3.3 CHOICE OF LASERS

Different kinds of laser sources are available for optical trapping in general. One important aspect is the power of the laser light because it determines the forces that can be applied. Although a few tens of milliwatts per optical fiber are sufficient to trap a particle in a DBLT, 0.1–1.0 W/fiber should be used for deforming particles ranging from GUVs to cells. The choice of trapping power results from a compromise between two effects: On the one hand, it should be high enough to trap and offset gravitational sedimentation, but on the other hand, it should be low enough to avoid deformation at trapping power. Otherwise, high-molecular weight polymers, such as polyethylene glycol or methylcellulose, can also be used to match density while keeping a high enough refractive index contrast. What is more important in the context of trapping and deforming GUVs is the choice of the right wavelength because absorption of the light by the medium in and around the GUV, or the GUV material itself, will cause varying amounts of heating. Table 13.1 gives the absorption coefficient of water and the resulting maximum temperature in the trap for various common wavelengths. Considering conveniently available laser sources, 808 nm is a good wavelength to deform objects while avoiding heating effects, whereas 1,480 nm is highly absorbed by water and can be used to heat the aqueous medium even without any optical stress. An intermediate choice is 1,064 nm, which is a classic wavelength in optical trapping. However, if 1,064-nm laser sources are standard products and can be found as fiber lasers, it is much more difficult to find 808 nm laser sources with high power. This is the main reason why an Ytterbium-doped fiber laser operating at a wavelength of 1,064 nm with a maximum power of 2–10 W has been the laser of choice for most optical stretching studies. The light is generated inside a single-mode fiber with sufficiently high power and the laser features turnkey operation and fast rise times

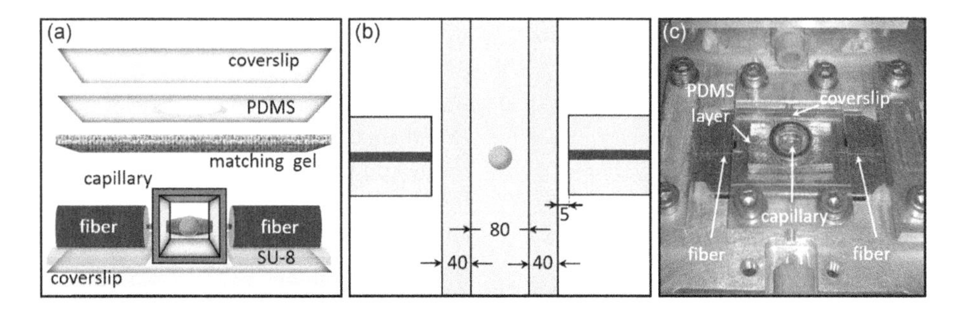

Figure 13.10 Assembly of a closed setup. (a) 3D rendering, (b) 2D schematic diagram of the optical trapping region including relevant dimensions in micrometers, and (c) picture of the closed optical stretcher.

Table 13.1 **Absorption coefficients and temperature increases in an optical stretcher setup for various laser wavelengths**

Wavelength (nm)	808	1,064	1,480
Absorption coefficient (cm^{-1})	0.019	0.1458	23.24
$\Delta T/P$ (°K/W)	1.7	13	2.072

Source: Ebert, S. et al., *Opt. Express*, 15, 15493–15499, 2007.
Note: It should be noted that the only value actually measured was for 1,064 nm in a closed, microfluidic setup. Whereas the absorption, and thus the heating, should scale linearly with the absorption coefficient, and the temperature increase per power reported here should be fairly accurate in a closed setup, the temperatures in an open setup could well be lower due to better heat dissipation.

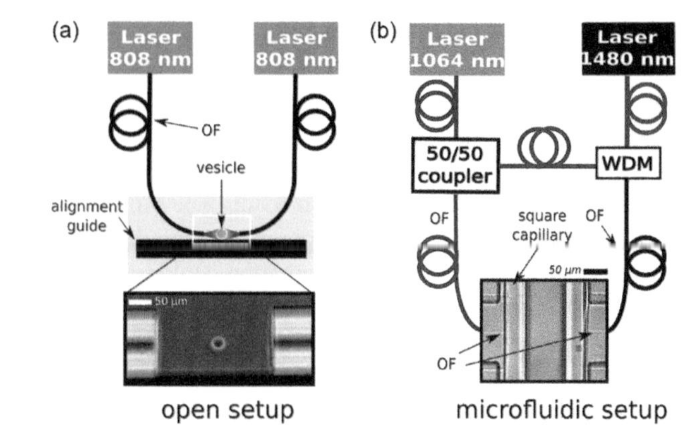

Figure 13.11 Optical stretcher setups used for characterization of vesicles. (a) Sketch of the open setup used for mechanical characterization of vesicles composed of two 808 nm laser sources. (b) Sketch of the closed, microfluidic setup used for thermodynamic investigation of vesicles. Two different wavelengths were used to almost independently deform and tune the temperature of the vesicle. The 1,064-nm laser fiber is split into two optical fibers with a 50/50 coupler. One 1,064-nm optical fiber is mixed with a 1,480-nm laser beam with a wavelength division multiplexer. Here the 1,480-nm laser beam has been coupled only into one 1,064-nm optical fiber, but coupling into both 1,064-nm optical fibers is also possible. The 1,064- and 1,064/1,480-nm optical fibers are aligned perpendicular to a square capillary of the closed microfluidic setup. (Delabre, U. et al., *Soft Matter*, 11, 6075–6088, 2015. Reproduced by permission of The Royal Society of Chemistry.)

(directly controlled by the current of the pump diode). Either two identical lasers are used for the two sides or a single laser with sufficient power, whose output fiber is spliced to a 50/50 fiber coupler in order to have two fibers with identical power for the two sides. Light powers approaching 1 W at 808 nm are possible with a tapered diode laser, whose astigmatic output beam has to be coupled into a single-mode fiber, which then serves as a spatial filter and ensures a Gaussian profile of the output beam for optical stretching. This coupling requires delicate free-beam to fiber coupling using a cylindrical lens, which causes the output power to be sensitive to mechanical and thermal drift, especially when switching between low and high power operation. A feedback-controlled power adjustment can be used to reduce output power drifts. As 1,480-nm sources, fiber-coupled laser diodes operating at 1,480 nm with a maximum power of 200 mW are commercially available. More flexibility in wavelength (650–1,100 nm) at sufficient output power can be obtained by using a tunable continuous wave titanium Ti-Sapphire laser, but these are more costly and their free-beam output still needs to be coupled into single-mode fibers, so this option does not provide the convenience of a turnkey operation. Very recently, also 780-nm fiber-coupled lasers with 1-W output power have become available, which could become a future standard in optical trapping and stretching of temperature sensitive objects.

Before inserting the fibers into the setups, the power at the end of each fiber has to be measured accurately and balanced if necessary with a fiber attenuator in order to ensure that each fiber carries 50% of the total power. It should be noted that powers quoted in this chapter refer to the output powers at the end of the fibers. This should be very close to the actual power also hitting the trapped object because the only interface is that from glass (fiber end or capillary wall) to water, which at normal incidence leads to losses of less than 0.2%.

13.3.4 SETUPS USED FOR VESICLE CHARACTERIZATION

For vesicle characterization, two different setups are presented below: an open setup and a closed microfluidic setup as illustrated in Figure 13.11. As described later, the first one (Figure 13.11a) will be used for the mechanical characterization of vesicles and is a standard open setup. The second setup makes use of a combination of two different wavelengths to deform and tune temperature almost independently and will be used for thermodynamic characterization of vesicles. In order to estimate the power inside the setups, shooting experiments with polystyrene beads can be performed as described by Guck et al. (2001) and Ferrara et al. (2011).

13.4 DEFORMATION OF VESICLES WITHOUT HEATING: MECHANICAL INVESTIGATION

In this section, we describe how to analyze the deformation of vesicles inside an optical stretcher, which generally follows the experiments reported in (Delabre et al., 2015). In order to only mechanically deform the vesicles and to minimize heating, the open 808-nm setup described in the preceding chapter has been used.

13.4.1 PREPARATION OF VESICLES

Vesicles are typically be prepared by electroformation as described in Chapter 1 of this book. A particular aspect of trapping and manipulating vesicles with an optical stretcher is that there has to be a difference in refractive index between the inside and the outside. To achieve this, vesicles can be prepared in a sucrose solution, which then constitutes the inner buffer, and then transferred by 20× dilution into an outer buffer solution of glucose. If glucose and sucrose concentrations are around 1,000 mOsm of osmolarity, then refractive indices, for example, measured with an Abbe refractometer, are $n_{sucrose} = 1.370$ and $n_{glucose} = 1.355$, respectively.

GUV-based techniques and what one can learn from them

13.4.2 EXPERIMENTS

Vesicles with radii from typically few micrometers (~4 μm) up to few tens of micrometers can be easily trapped with the optical stretcher. The only limitations depend on the fiber–fiber distance and the applied power. When a spherical and floppy vesicle is found, it can be moved inside the optical trap with a micropipette to be trapped in an 808-nm optical stretcher. When the vesicle is then deformed at high power, the vesicle appears to deform very quickly as shown in Figure 13.12a–c. The deformation is essentially into a prolate spheroid for small optical stress as illustrated in Figure 13.12a and b. The maximum deformation is constant for a given stretching power, without any obvious time dependence, and increases linearly with applied laser power (Figure 13.12c–d). In addition, the deformation is reversible—that is, no hysteresis is observed when the power is reduced. This reversibility is the signature of the absence of significant heating as discussed later.

13.4.3 THEORY

From a theoretical point of view, in order to extract the mechanical properties of vesicles (bending modulus, κ, and tension, Σ,), the shape of the vesicle is first decomposed in spherical harmonics (as described by Eqs. 14.4 and 14.5 in Chapter 14 of this book):

$$R(\theta,\phi) = R_{\text{ve}}\left(1+\sum_{l\geq 0}^{l_{\max}}\sum_{m=-l}^{l} u_{lm}Y_{lm}(\theta,\phi)\right) \tag{13.9}$$

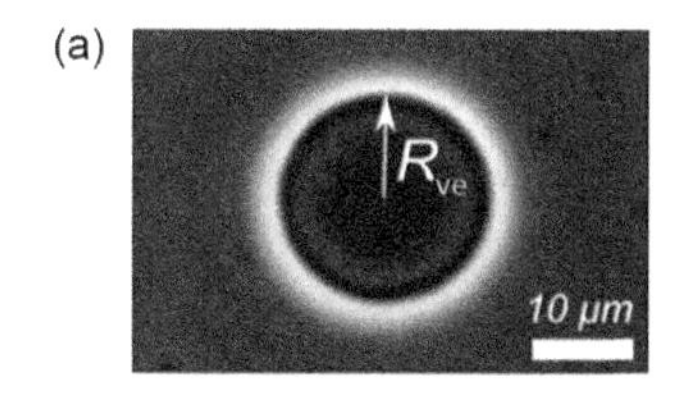

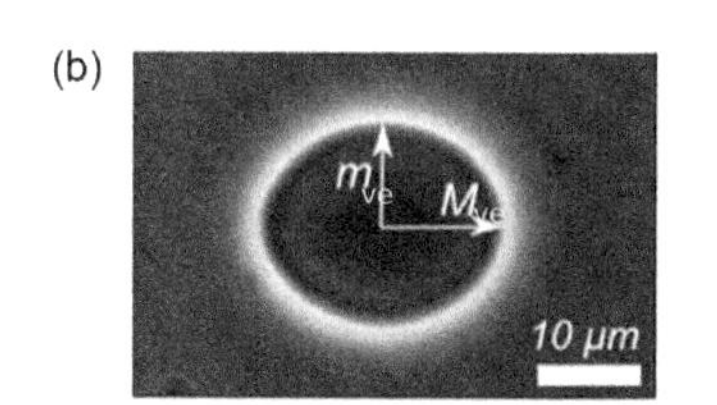

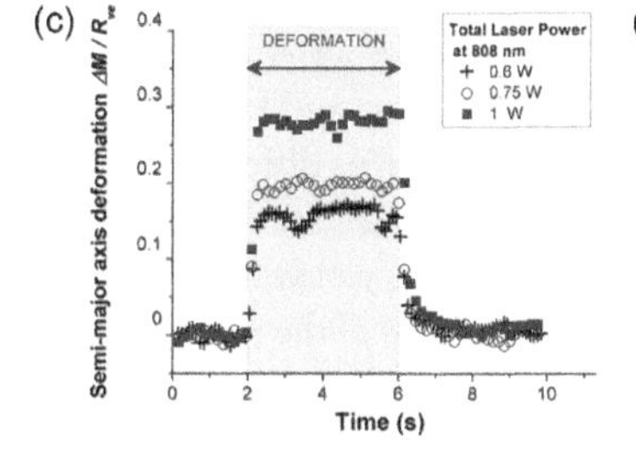

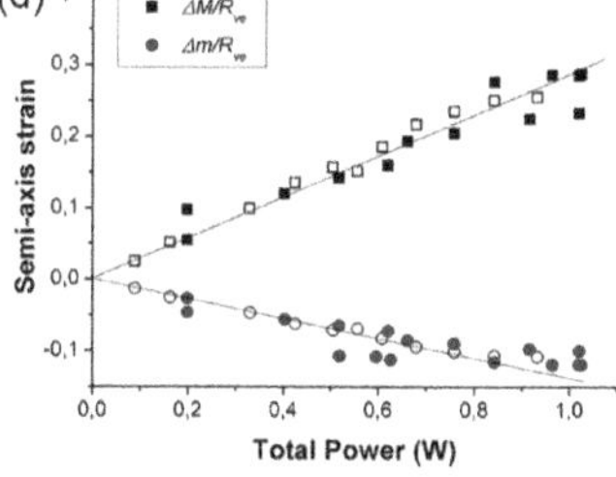

Figure 13.12 Vesicle deformation in an 808-nm optical stretcher without heating. (a) When the laser power is low ($P = 0.1$ W per fiber), the vesicle is trapped and still spherical. (b) At higher power, the vesicle is deformed into a prolate spheroidal shape with the major axis along the fiber axis. R_{ve} is the mean radius of the undeformed vesicle, m_{ve} and M_{ve} are the semi-minor axis and the semi-major axis of the stretched vesicle, respectively. Here, $w = 9$ μm and $R_{ve} = 10.6$ μm ($\Delta M / R_{ve} = 0.15$). (c) Major axis strain for various applied powers during a step-stress experiment. (d) Semi-major axis strain and semi-minor axis strain for two series (solid and open symbols) of experiments using the same vesicle. No hysteresis is observed and the volume is constant during deformation. Results shown are obtained with vesicles composed of a mixture used by Pontani et al. (2009): egg PC phospholipids, 1,2-dioleoyl-*sn*-glycero-3-[(N-(5-amino-1-carboxypentyl)iminodiacetic acid)succinyl] (nickel salt (DOGS-NTA-Ni, see Appendix 1 of the book for structure and data on these lipids) phospholipids and cholesterol in molar ratio (58:5:37). (Delabre, U. et al., *Soft Matter*, 11, 6075–6088, 2015. Reproduced by permission of The Royal Society of Chemistry.)

where Y_{lm} are the normalized spherical harmonic functions and their u_{lm} associated amplitudes. The radius of the vesicle, R_{ve}, is related to the constant volume of the vesicle as $V = \frac{4\pi R_{ve}^3}{3}$. The total area, A_{ve}, of the vesicle is assumed to be fixed and can be expressed by $A_{ve} = (4\pi + \Delta)R^2$ where Δ, the dimensionless excess area, is related to the difference between apparent area and true area. This area constraint, as discussed in more details elsewhere (Seifert, 1995, 1999), is taken into account in the free energy as a surface energy using the membrane tension, Σ, as a Lagrangian multiplier associated with the area of the vesicle.

Basically, the optical stretcher pulls out the microscopic membrane fluctuations to induce a global deformation of the vesicle. In this sense, for small laser power, the vesicle membrane is not stretched. The deformation in the optical stretcher can be deduced from the minimization of the free energy and from the area constraint. The free energy is composed of three terms: the bending energy, the surface energy and the deformation energy due to the optical stress. Using the $\cos^2$ approximation ($\sigma = \sigma_0 \cos^2\theta$) to characterize the optical stress as described in Section 13.2.2 of this chapter, the deformation energy can be decomposed in spherical harmonics as well, which enables to write the total free energy as (Delabre et al., 2015)

$$F = \frac{\kappa}{2}\sum_{l,m} E_l \left|u_{lm}\right|^2 - \kappa\left(\frac{2\sigma_0 R_{ve}^3}{3\kappa}\right)\sqrt{4\pi/5}\, u_{20} \tag{13.10}$$

The mean shape of a vesicle in the optical stretcher can now be deduced from the energy minimization and as there is only one mode with a nonzero mean amplitude (which is the prolate spheroidal one), the amplitude of this mode is

$$u_{20} = \frac{\sqrt{16\pi/5}\,\sigma_0 R_{ve}^3}{72\kappa + 12\Sigma R_{ve}^2 + 2\sigma_0 R_{ve}^3} \tag{13.11}$$

Note that the effective tension, Σ, depends implicitly on the optical stress due to the area constraint. For small laser power, this effective tension should be close to the initial tension, Σ_0, which can be estimated analytically depending on the considered regime (entropic regime, tense regime or prolate regime) (Seifert, 1995). This expression can be linearized for small optical stress ($\sigma_0 R_{ve} \ll 6\Sigma_0$):

$$u_{20} = \frac{\sqrt{16\pi/5}\,\sigma_0 R_{ve}^3}{72\kappa + 12\Sigma R_{ve}^2} \tag{13.12}$$

Experimentally, u_{20} is directly related to the major and minor axis strains, $\Delta M = M_{ve} - R_{ve}$ and $\Delta m = m_{ve} - R_{ve}$, by

$$u_{20} = \frac{2}{3}\sqrt{\frac{4\pi}{5}}\left(\frac{\Delta M}{R_{ve}} - \frac{\Delta m}{R_{ve}}\right). \tag{13.13}$$

In addition, in the low peak stress regime, it is possible to find an approximation for the area constraint as described by Delabre et al. (2015)

$$\frac{\Delta A}{A} = \frac{k_B T}{2\kappa}\ln\left(\frac{\Sigma}{\Sigma_0}\right), \tag{13.14}$$

which is a standard equation also found in the papers by Kummrow and Helfrich (1991) and De Haas et al. (1997) as also described in Chapter 11 of this book.

To determine the bending modulus, it seems possible to calculate the tension from the Law of Laplace using the geometrical shape of the vesicle as done in (Solmaz et al., 2012, 2013). However, the effective tension, Σ, cannot be rigorously determined by the Law of Laplace Because fluctuations and bending rigidity should be taken into account (Vitkova et al., 2004). The second possibility, described below, makes use of the definition of the vesicle tension as the Lagrangian multiplier of vesicle area. Using Eq. 13.13 and the fact that the excess area $\frac{\Delta A}{A} = 2u_{20}^2$ (Seifert, 1999; Delabre et al., 2015), it is possible to eliminate the effective tension in Eq. 13.14, to obtain

$$\sigma_0 = \sqrt{\frac{5}{4\pi}}u_{20}\left(\frac{36\kappa}{R_{ve}^3} + \frac{6\Sigma_0}{R_{ve}}\exp\left(\frac{4\kappa u_{20}^2}{k_B T}\right)\right) \quad (13.15)$$

where k_B is the Boltzmann constant and T is temperature. Eq. 13.15 is the global equation that can be tested experimentally to find the bending modulus and the initial tension where u_{20} is evaluated experimentally (with Eq. 13.13) and the peak stress σ_0 calculated numerically from experimental parameters (with Eq. 13.8) as described by Guck et al. (2000) and explained in Box 13.2. This formula is expected to work at low and intermediate power given that a saturation regime exists at high power (Delabre et al., 2015), and as long as the vesicle shape is well fitted by an ellipse.

Box 13.2 Measurement and analysis of vesicle deformation

1. Find a floppy vesicle that exhibits visual membrane fluctuations.
2. Aspirate the vesicle into the micropipette and move it to the trapping region.
3. Trap it in the optical stretcher and check that the vesicle is trapped at the center of the dual optical trap. A few tens of milliwatt laser power should be sufficient for trapping.
4. Measure the vesicle radius, R_{ve}, after at least 2 s to be sure that the vesicle is trapped correctly.
5. Increase the beam power to somewhere between 0.1 and 1.0 W (total power) to deform it for at least 2 s, and record the final deformation. The deformations should look similar to what is shown in Figure 13.12 and should reach steady-state.
6. Repeat Step 5 to record the deformation of the same vesicle for several applied powers.
7. For each final deformation, measure major axis strain, ΔM, and minor axis strain, Δm, (Figure 13.5), and then calculate u_{20} with Eq. 13.13.
8. Calculate σ_0 with Eq. 13.8 for each applied power.
9. Plot σ_0 versus u_{20} and fit the data with Eq. 13.15 to get the bending modulus and initial tension.
10. Repeat this procedure for several vesicles with various radii to get a mean value of the bending modulus.

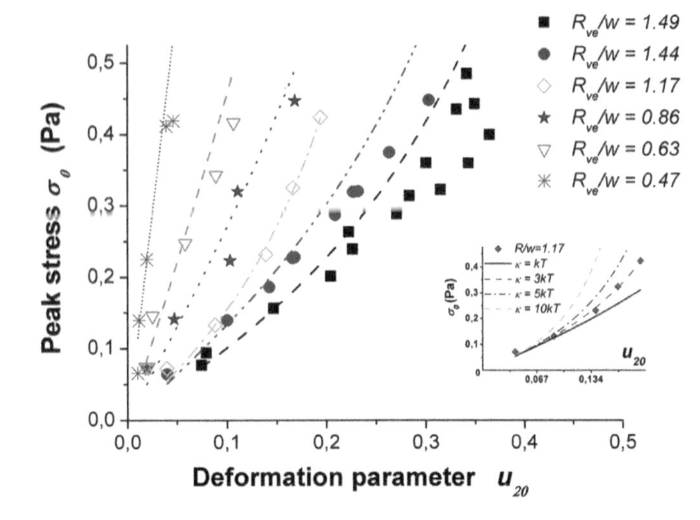

Figure 13.13 Comparison of experimental data with the theoretical model (Eq. 13.15) for a vesicle deformed in an 808-nm optical stretcher for different ratios R_{ve}/w with $w = 9$ µm for all the data sets. A fitting procedure (using Eq. 13.15) allows to extract the values of bending moduli, κ, and the initial effective tension Σ_0. The insert shows the effect of, κ, on the peak stress evolution for a given ratio R_{ve}/w following Eq. 13.15 with $\Sigma_0 = 3.810^{-6}$ N/m. Discrepancies between data and fit (Eq. 13.15) at high peak stress might result from shape deviations of the vesicle relative to an ellipse shape. (Delabre, U. et al., *Soft Matter*, 11, 6075–6088, 2015. Reproduced by permission of The Royal Society of Chemistry.)

13.4.4 COMPARISON BETWEEN EXPERIMENTS AND THEORY

The comparison of the experimental data for egg phosphatidylcholine (PC) vesicles with the deformation model is given in Figure 13.13.

The fitted values of bending moduli are between 1 and 7.5 k_BT, with a mean value of 2.5 k_BT, whereas the fitting values for Σ_0 are between 3 and 7.5 10^{-6} N/m, which is a good order of magnitude for flaccid vesicles. An example of the sensitivity of the model to fitting parameters is shown in the insert of Figure 13.13. The values obtained here are slightly lower than the values reported in literature, which are generally around 6–18 k_BT (Kummrow and Helfrich, 1991) and 10–11 k_BT (Vlahovska et al., 2009). This difference may indicate that the simple $\cos^2$ approximation to describe the optical stress is not appropriate to fully describe the shape of vesicles in the optical stretcher for any ratio between the radius of the vesicle and the waist of the laser, R_{ve}/w. Another possible explanation might be the high osmolarity of solutions used in the experiment to increase refractive index contrast. It has been reported that the bending modulus tends to decrease with osmolarity (Genova et al., 2006; Solmaz et al., 2013), which could explain the smaller values obtained here.

13.5 DEFORMATION OF VESICLES WITH HEATING: THERMODYNAMICAL INVESTIGATION

In this part, we show how it is possible to induce heating of the trapped vesicle with the optical stretcher thanks to an appropriate choice of laser wavelength. This enables some thermodynamic measurements or to combine deformation and thermodynamic studies on vesicles.

13.5.1 A DUAL WAVELENGTH OPTICAL STRETCHER SETUP

An ideal choice would be to combine an 808-nm laser with a 1,480-nm laser in the same optical fiber. This would enable one to independently tune the mechanical stress (with the 808-nm laser, which only causes minimal heating) and the temperature (with the 1,480-nm laser at very low power, which does not contribute any significant mechanical stress). However, a single-mode optical fiber at 1,480 nm is not single-mode for 808 nm light, which would excite higher order modes and the output beam would not be suitable for trapping and controlled deformation. To overcome this problem, we make use of the fact that the single-mode fiber for 1,064-nm light still permits sufficient power transmission for heating with 1,480-nm light. In addition, at a given power, the heating, directly related to the absorption coefficient, is much stronger at 1,480 nm than at 1,064 nm (Table 13.1). A combination of 1,064/1,480-nm sources could then be used to both deform (mainly with 1,064 nm) and heat vesicles (mainly with the 1,480 source). We used a combined 1,064/1,480-nm optical stretcher setup as described in Figure 13.11 to deform and heat pure lipid vesicles (dipalmitoyl-*sn*-glycero-3-phosphocholine [DPPC] and 1,2-distearoyl-*sn*-glycero-3-phosphocholine vesicles [DSPC], see Appendix 1 of the book for structure and data on these lipids) below and above their melting temperature.

13.5.2 EXPERIMENTS

The vesicles were prepared by electroformation as described in Section 13.4 of this chapter with similar sucrose and glucose solutions. Figure 13.14a shows a typical deformation of a DPPC vesicle in a combined 1,064/1,480-nm optical stretcher where the 1,480-nm laser power is kept constant and the 1,064-nm laser power is increased step by step. It is then possible to tune the temperature applied to the vesicle accurately while applying an optical stress. The measurement of the local temperature inside this double beam optical trap can be achieved using confocal microscopy combined with temperature sensitive fluorescent probes as described by Ebert et al. (2007). Here, at low 1,064-nm power, the temperature is below the melting temperature of DPPC (T_m = 41.6°C); therefore, the DPPC vesicle is in the gel phase and no deformation can be detected.

When the 1,064-nm power is at 240 mW, a small deformation appears along the fiber axis. At higher laser power (i.e., higher temperature) the deformation is extended to the whole vesicle and results in a "cigar" shape. The associated time dependent deformations are presented in Figure 13.14b and show complex behaviors. The most interesting feature is presented in Figure 13.14c where the final deformation strain for DPPC (T_m = 41.6°C) and DSPC vesicles (T_m = 54.4°C) is reported for various 1,064-nm laser powers, that is, various temperatures. It appears that the melting transition between the gel phase and the liquid phase can be quantitatively detected thanks to a huge increase in deformation for both types of vesicles. Actual temperatures in the trap can be measured directly by fluorescence ratio thermometry, as explained in (Ebert et al., 2007).

13.5.3 NONTRIVIAL DEFORMATION OF VESICLES WITH HEATING EFFECTS

In this part, we will show that laser-induced heating can have nontrivial effects on vesicles especially if vesicles are composed of a mixture of lipids. Nontrivial effects are observed, for example, when egg

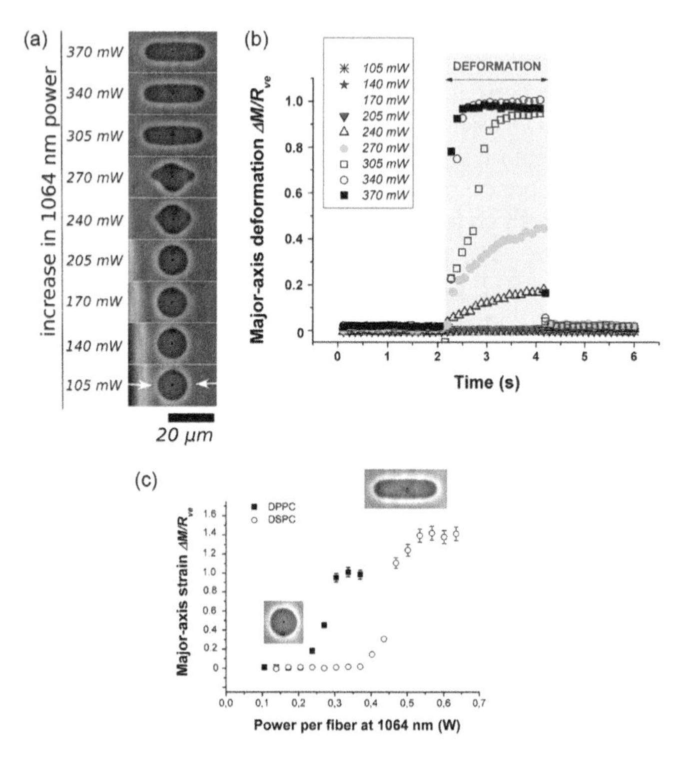

Figure 13.14 (a) Deformation of a DPPC vesicle in a 1,064/1,480-nm optical stretcher. Above 240 mW (1,064-nm power), a steady deformation starts to appear at the center of the vesicle, which is initially in the gel phase (images are acquired 2 s after the beginning of the deformation). (b) Major axis strain of a DPPC vesicle as a function of time for several 1,064-nm power values. (c) Major axis strain for DPPC and DSPC vesicles for 8 mW at 1,480 nm for DPPC and 11 mW at 1,480 nm for DSPC. (Delabre, U. et al., *Soft Matter*, 11, 6075–6088, 2015. Reproduced by permission of The Royal Society of Chemistry.)

PC vesicles similar to the ones used in Section 13.4 are deformed in a 1,064-nm optical stretcher. The absorption for 1,064-nm laser is around 13°C/W, which is much higher than for 808-nm (only 1.7°C/W). Contrary to the deformation of egg PC vesicles with an 808 nm optical stretcher (see Section 13.4), Figure 13.15 shows that the major axis strain changes drastically during the stretching. Evidently, such a decreasing strain under constant stress defies any mechanical characterization as a conventional material. Given that this effect was absent at 808 nm, heating due to the 1,064-nm laser seems to be the likely cause of this phenomenon. It is important to note that such phenomenon was not present for pure DPPC and DSPC phospholipid vesicles. There is also strong hysteresis in the deformation when a single vesicle is deformed several times. This hysteresis might result from the hidden area stored in membrane defects and Vitkova et al. (2004) have shown that prestressing of the vesicles is needed before such measurements to pull out the hidden area stored in defects so they do not contribute to the overall deformation. However, the microscopic origins of these phenomena related to the temperature increase remain to be understood, but a reorganization of the lipid mixture promoted by the increase in temperature might be at its core.

13.6 DISCUSSION

Laser-induced heating can be a problem for a rheological characterization of vesicle properties because it can lead to unwanted (and hard to understand) effects, as shown in Section 13.5.3. However, it can be turned into an advantage if the temperature can be controlled with a specific laser and specific points in the phase diagram can be assessed, as illustrated in Section 13.5.2. This chapter has presented a mechanical characterization of vesicles with an 808-nm optical stretcher, which enables deformation of vesicles without significantly heating effects even at high laser power. The theoretical description presented here allows determining the bending modulus and the initial tension of the vesicles from a fitting procedure. The use of an optical stretcher to trap and deform vesicles without any contact presents several advantages, especially compared with contact methods (e.g., micropipette aspiration) that could alter the physicochemical properties of the vesicle surface and obscure measurements due to adhesion. Further, optical stretching leads to axisymmetric, regular shapes during deformation, which are easily amenable to theoretical description. In conjunction with microfluidic delivery, optical stretching also enables a fast characterization of tens of vesicles in a short amount of time as shown in (Solmaz et al., 2013; Delabre et al., 2015). On the other hand, an open setup is more flexible and can be interesting for specific studies with fewer vesicles to characterize, for example. Importantly, the optical characterization of vesicles can only be performed for vesicles with a refractive index contrast between the inner medium and the outer medium. In our case, the contrast was adjusted with glucose and sucrose solutions. Even if the concentrations used here were quite high to induce a high optical stress, the refractive index contrast does not need to be high if enough laser power is used.

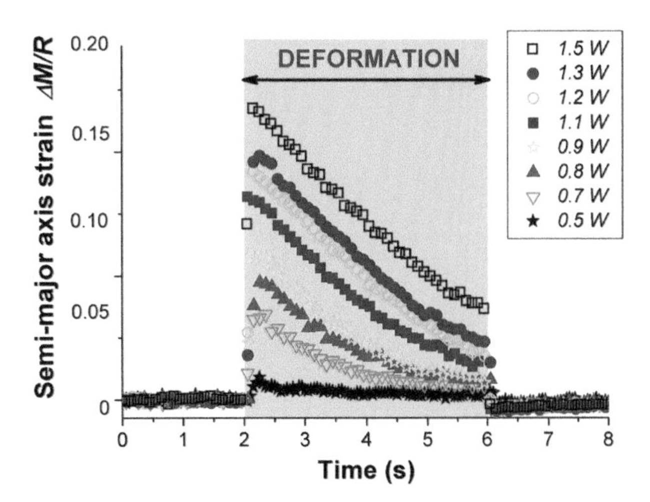

Figure 13.15 Major axis deformation of egg PC vesicles with a 1,064-nm optical stretcher. The deformation decreases during the stretching period at constant stress, possibly due to a possible microscopic reorganization triggered by the heating. (Delabre, U. et al., *Soft Matter*, 11, 6075–6088, 2015. Reproduced by permission of The Royal Society of Chemistry.)

13.7 OUTLOOK

Although most experiments with optical stretchers to date have tested biological cells, there is now an increasing number of studies also utilizing it for the mechanical and thermodynamic investigation of GUVs. Because DBLTs, such as the optical stretcher can be operated completely independent of the imaging modality, the combination of optical stretcher and confocal microscopy can, for instance, give access to the overall vesicle shape during deformation in three dimensions. Future studies might well include the use of fluorescence microscopy to monitor phase separation of lipid mixtures or protein clustering due to the induced variable membrane curvatures. In addition, optical tweezers could be employed simultaneously, for example, to control the tension of the vesicle by pulling out tethers (see Chapter 16) or inducing nonsymmetric curvatures. Using periodic optical stress signals, it can also be interesting to perform rheometric characterization of vesicles depending on the membrane composition but also on the inner composition of the vesicle (e.g., polymer solution, reconstituted actin cortex, bulk hydrogels). The use of optical stretchers for vesicle characterization, despite obvious advantages, is still in its infancy and many interesting and obvious experiments are well within reach.

LIST OF ABBREVIATIONS

CW	continuous wave
DBLT	dual-beam laser trap
DOGS-NTA-Ni	1,2-dioleoyl-*sn*-glycero-3-[(N-(5-amino-1-carboxypentyl)iminodiacetic acid)succinyl] (nickel salt)
DPPC	dipalmitoyl-*sn*-glycero-3-phosphocholine
DSPC	1,2-distearoyl-*sn*-glycero-3-phosphocholine
OS	optical stretcher
PC	phosphatidylcholine
PEG	polyethylene glycol
PDMS	polydimethylsiloxane
WDM	wavelength division multiplexer

GLOSSARY OF SYMBOLS

A	area of the vesicle
Δ	dimensionless excess area
Δm	minor axis strain
ΔM	major axis strain of the vesicle
F_{grad}	gradient force
F_{scat}	scattering force
κ	bending moduli
λ_0	laser wavelength in vacuum
m_{ve}	minor axis of the vesicle
M_{ve}	major axis of the vesicle
n_{med}	refractive index of the medium
R_{ve}	radius of the vesicle
σ	optical stress
σ_0	maximum of the optical stress
Σ	vesicle tension
Σ_0	initial vesicle tension
T_m	melting temperature of pure lipid vesicle
V	vesicle volume
ω	waist of the laser beam at vesicle position
ω_0	waist of the laser beam at the exit of the fiber

REFERENCES

Ashkin A (1970) Acceleration and trapping of particles by radiation pressure. *Phys Rev Lett* 24:156.

Ashkin A (1992) Forces of a single-beam gradient laser trap on a dielectric sphere in the ray optics regime. *Optics Lett* 61:569–582.

Ashkin A, Dziedzic JM (1973) Radiation pressure on a free liquid surface. *Phys Rev Lett* 30:139.

Ashkin A, Dziedzic JM, Bjorkholm JE, Chu S (1986) Observation of a single-beam gradient force optical trap for dielectric particles. *Optics Lett* 11:288–290.

Bar-Ziv R, Moses E, Nelson P (1998) Dynamic excitations in membranes induced by optical tweezers. *Biophys J* 75:294–320.

Bellini N, Bragheri F, Cristiani I, Guck J, Osellame R, Whyte G (2012) Validation and perspectives of a femtosecond laser fabricated monolithic optical stretcher. *Biomed Optics Express* 3:2658–2668.

Casner A, Delville JP (2001) Giant deformations of a liquid-liquid interface induced by the optical radiation pressure. *Phys Rev Lett* 87:054503.

Constable A, Kim J, Mervis J, Zarinetchi F (1993) Demonstration of a fiber-optical light-force trap. *Optics Lett* 18:1867–1869.

De Haas KH, Blom C, Van den Ende D, Duits M (1997) Deformation of giant lipid bilayer vesicles in shear flow. *Phys Rev E* 56:7132–7137.

Delabre U, Feld K, Crespo E, Whyte G, Sykes C, Seifert U, Guck J (2015) Deformation of phospholipid vesicles in an optical stretcher. *Soft Matter* 11:6075–6088.

Dietrich C, Angelova M, Pouligny B (1997) Adhesion of latex spheres to giant phospholipid vesicles: Statics and dynamics. *J Phys II* 7:1651–1682.

Dimova R (2014) Recent developments in the field of bending rigidity measurements on membranes. *Adv Coll Interf Sci* 208:225–234.

Dimova R, Dietrich C, Hadjiisky A, Danov K, Pouligny B (1999) Falling ball viscosimetry of giant vesicle membranes: Finite-size effects. *Eur Phys J B* 12:589–598.

Dimova R, Pouligny B, Dietrich C (2000) Pretransitional effects in dimyristoylphosphatidylcholine vesicle membranes: Optical dynamometry study. *Biophys J* 79:340–356.

Dimova R, Seifert U, Pouligny B, Förster S, Dobereiner H-G (2002) Hyperviscous diblock copolymer vesicles. *Eur Phys J E* 7:241–250.

Ebert S, Travis K, Lincoln B, Guck J (2007) Fluorescence ratio thermometry in a microfluidic dual-beam laser trap. *Opt Express* 15:15493–15499.

Ekpenyong AE, Töpfner N, Fiddler C, Herbig M, Li W, Summers C, Guck J, Chilvers ER (2017) Mechanical deformation induces depriming of neutrophils. *Sci Adv* 3:e1602536.

Faigle C (2015) Optical Stretcher: Towards a cell sorter based on high-content analysis. PhD thesis. Technical University of Dresden.

Faigle C, Lautenschlager F, Whyte G, Homewood P, Martin-Badosa E, Guck J (2015) A monolithic glass chip for active single-cell sorting based on mechanical phenotyping. *Lab Chip* 15:1267–1275.

Ferrara L, Baldini E, Minzioni P, Bragheri F, Ferrara L, Baldini E, Minzioni P et al. (2011) Experimental study of the optical forces exerted by a Gaussian beam within the Rayleigh range. *J Optics* 13:1–7.

Franke-Arnold S, Allen L (2008) Advances in optical angular momentum. *Laser Photon Rev* 2:299–313.

Franze K, Grosche J, Skatchkov SN, Schinkinger S, Foja C, Schild D, Uckermann O, Travis K, Reichenbach A, Guck J (2007) Müller cells are living optical fibers in the vertebrate retina. *Proc Natl Acad Sci USA* 104:8287–8292.

Genova J, Zheliaskova A, Mitov MD (2006) The influence of sucrose on the elasticity of SOPC lipid membrane studied by the analysis of thermally induced shape fluctuations. *Coll Surf A* 282:420–422.

Guck J, Ananthakrishnan R, Hamid M, Moon TJ, Cunningham CC, Käs JA (2001) The optical stretcher: A novel laser tool to micromanipulate cells. *Biophys J* 81:767–784.

Guck J, Ananthakrishnan R, Moon J, Cunningham CC, Käs JA (2000) Optical deformability of soft biological dielectrics. *Phys Rev Lett* 84:5451–5454.

Guck J, Schinkinger S, Lincoln B, Wottawah F, Ebert S, Romeyke M, Lenz D et al. (2005) Optical deformability as an inherent cell marker for testing malignant transformation and metastatic competence. *Biophys J* 88:3689–3698.

Helfer E, Harlepp S, Bourdieu L, Robert J (2001) Viscoelastic properties of actin-coated membranes. *Phys Rev E* 63:021904–021913.

Huff A, Melton CN, Hirst LS, Sharping JE (2015) Stability and instability for low refractive-index-contrast particle trapping in a dual-beam optical trap. *Biomed Optics Express* 6:3812–3819.

Jackson JD (1962) *Classical Electrodynamics*. New York: Wiley.

Kamm R, Lammerding J, Mofrad M (2010) Cellular nanomechanics. In: *Springer Handbook of Nanotechnology* (Bushan B, Ed.), pp. 1171–1200. Berlin, Germany: Springer.

Kemper B, Langehanenberg P, Höink A, Bally von G, Wottowah F, Schinkinger S, Guck J et al. (2010) Monitoring of laser micromanipulated optically trapped cells by digital holographic microscopy. *J Biophotonics* 3:425–431.

Kreysing M, Ott D, Schmidberger MJ, Otto O, Schürmann M, Martin-Badosa E, Whyte G, Guck J (2014) Dynamic operation of optical fibres beyond the single-mode regime facilitates the orientation of biological cells. *Nat Commun* 5:1–6.

Kummrow M, Helfrich W (1991) Deformation of giant lipid vesicles by electric fields. *Phys Rev A* 44:8356.

Lautenschläger F, Paschke S, Schinkinger S, Bruel A, Beil M, Guck J (2009) The regulatory role of cell mechanics for migration of differentiating myeloid cells. *Proc Natl Acad Sci USA* 106:15696–15701.

Lincoln B (2006) The microfluidic optical stretcher. PhD thesis. University of Leipzig.

Lincoln B, Erickson HM, Schinkinger S, Wottawah F, Mitchell D, Ulvick S, Bilby C, Guck J (2004) Deformability-based flow cytometry. *Cytometry A* 59A:203–209.

Lincoln B, Schinkinger S, Travis K, Wottawah F, Ebert S, Sauer F, Guck J (2007a) Reconfigurable microfluidic integration of a dual-beam laser trap with biomedical applications. *Biomed Microdevices* 9:703–710.

Lincoln B, Wottawah F, Schinkinger S, Ebert S, Guck J (2007b) High-throughput rheological measurements with an optical stretcher. In: *Methods in Cell Biology, Cell Mechanics* (Wang, YL, Discher, DE, Eds.), pp. 397–423. Amsterdam, the Netherlands: Elsevier Inc.

Lorenz H, Despont M, Fahrni N (1997) SU-8: A low-cost negative resist for MEMS. *J Micromech Microeng* 7:121–124.

Lyons ER, Sonek GJ (1995) Confinement and bistability in a tapered hemispherically lensed optical fiber trap. *Appl Phys Lett* 66:1584–1586.

Mizuno D, Kimura Y, Hayakawa R (2001) Electrophoretic microrheology in a dilute lamellar phase of a nonionic surfactant. *Phys Rev Lett* 87:088104.

Molloy JE, Padgett MJ (2002) Lights, action: Optical tweezers. *Contemp Phys* 43:241–258.

Nava G, Bragheri F, Yang T, Minzioni P, Osellame R, Cristiani I, Berg-Sørensen K (2015) All-silica microfluidic optical stretcher with acoustophoretic prefocusing. *Microfluid Nanofluidics* 19:837–844.

Piñón TM, Castelli AR, Hirst LS, Sharping JE (2013) Fiber-optic trap-on-a-chip platform for probing low refractive index contrast biomaterials. *Appl Optics* 52:2340–2345.

Pontani L-L, van der Gucht J, Salbreux G, Heuvingh J, Joanny J-F, Sykes C (2009) Reconstitution of an actin cortex inside a liposome. *Biophys J* 96:192–198.

Roosen G, Imbert C (1976) Optical levitation by means of two horizontal laser beams: A theoretical and experimental study. *Phys Lett A* 59:6–8.

Seifert U (1995) The concept of effective tension for fluctuating vesicles. *Z Phys B Condensed Matter* 97:299–309.

Seifert U (1999) Fluid membranes in hydrodynamic flow fields: Formalism and an application to fluctuating quasispherical vesicles in shear flow. *Eur Phys J B* 8:405–415.

Sidick E, Collins SD, Knoesen A (1997) Trapping forces in a multiple-beam fiber-optic trap. *Appl Optics* 36:6423–6433.

Solmaz ME, Biswas R, Sankhagowit S (2012) Optical stretching of giant unilamellar vesicles with an integrated dual-beam optical trap. *Biomed Optics Express* 3:2419–2427.

Solmaz ME, Sankhagowit S, Biswas R, Mejia CA, Povinelli ML, Malmstadt N (2013) Optical stretching as a tool to investigate the mechanical properties of lipid bilayers. *RSC Adv* 3:16632–16638.

Svoboda K, Block SM (1994) Biological applications of optical forces. *Annu Rev Biophys Biomol Struct* 23:247–285.

van de Hulst HC (1957) *Light Scattering by Small Particles*. New York: Dover Publications.

Velikov K, Danov K, Angelova M, Dietrich C, Pouligny B (1999) Motion of a massive particle attached to a spherical interface: Statistical properties of the particle path. *Coll Surf A* 149:245–251.

Velikov K, Dietrich C, Hadjiisky A, Danov K, Pouligny B (1997) Motion of a massive microsphere bound to a spherical vesicle. *Europhys Lett* 40:405–410.

Vitkova V, Genova J, Bivas I (2004) Permeability and the hidden area of lipid bilayers. *Eur Biophys J* 33:706–714.

Vlahovska PM, Gracia RS, Aranda-Espinoza S, Dimova R (2009) Electrohydrodynamic model of vesicle deformation in alternating electric fields. *Biophys J* 96:4789–4803.

Whyte G, Lautenschläger F, Kreysing MK, Boyde L, Ekpenyong AE, Delabre U, Chalut K, Franze K, Guck J (2010) Dual-beam laser traps in biology and medicine: When one beam is not Enough. In: *Proceedings of SPIE—The International Society for Optical Engineering* (Dholakia K, Spalding GC, Eds.), pp. 77620G–1–77620G–6. SPIE.

14 Vesicle fluctuation analysis

John Hjort Ipsen, Allan Grønhøj Hansen, and Tripta Bhatia

An experiment is a question which science poses to Nature,
and a measurement is the recording of Nature's answer.

Max Planck

Contents

14.1 INTRODUCTION

Vesicle fluctuation analysis (VFA) is a simple experimental technique for extracting information about membrane physical properties by analyzing the conformational fluctuations of giant vesicles visible by optical microscopy. Detailed analysis of membrane conformational fluctuations reveal information about material properties of the membrane, in particular the bending rigidity (κ), but it has a range of additional applications. VFA provides a unique possibility for analyzing the consequences of the cooperative phenomena occurring at mesocopic length scales that are contained within the membrane fluctuations at macroscopic length by testing theoretical predictions of phenomenological models and extracting estimates of model parameters. The background and principles for VFA of quasi-spherical giant unilamellar vesicles (GUVs) will be introduced. The minimal requirements for establishing an experimental facility will be discussed along with the key elements of the image analysis and software for data treatment. Finally, some applications and future possibilities of VFA are given.

14.1.1 A BRIEF HISTORY OF MEMBRANE FLICKERING

The flicker phenomena of the red blood cell (RBC) has probably been observed by many microscopists since the first description by Jan Swammerdam in 1658 (Swammerdam, 2016); however, it was first reported in 1890 by Browicz (1890), who described the irregular motion of the RBC shape. Although he thought of it as a pathological condition, the phenomenon itself does not originate from living processes given that it persists with temperature changes. Rather, flicker was attributed to the *molecular motion* and possibly to the artifacts of the observation conditions (Cabot, 1901). The improved understanding of the metabolic processes in cells from the 1940s made it possible to observe how the RBC's flicker is influenced by perturbation from the active processes. In 1949, Pulvertaft made systematic studies that excluded possible artifacts as the cause of the flickering (Pulvertaft, 1949). He showed that various poisons may impede the flicker, and although *Brownian motion* was mentioned, he concluded that metabolism is the probable origin of flicker. This view was confirmed by Blowers et al. (1951) in a very extensive study. In the 1960s, the bilayer nature of membranes with proteins embedded got a solid experimental foundation (Sjöstrand et al., 1958) as a fluid amphiphilic interface with vanishing surface tension (Schulman and Montagne, 1961). By the theoretical works of Canham and Helfrich, it became clear in the beginning of the 1970s that bilayer conformations in equilibrium are mainly controlled by curvature elastic energy and the mechanical constraints on the membrane (Canham, 1970; Helfrich, 1973). For a closed vesicle, Helfrich's interfacial free energy is given by

$$\mathcal{F}_{\mathrm{Hel}} = \Sigma A_{\mathrm{ve}} + \frac{\kappa}{2}\oint_{A_{\mathrm{ve}}} \mathrm{d}A(C_1 + C_2 - 2m)^2 \qquad (14.1)$$

where A_{ve} is the surface area of the vesicle; and Σ is the surface tension, which is vanishing for a freestanding membrane. C_1 and C_2 are the local principal curvatures, whereas the previously mentioned κ is the mean curvature elastic constant. If there is an asymmetry between the two monolayers of the membrane, the bilayer may have a spontaneous curvature (m). Many features of the RBC conformations can be described as elastic properties of a tension-free fluid interface.

The consequences of Helfrich's theory on the equilibrium conformational fluctuations were analyzed by Brochard et al. in 1975–1976 (Brochard and Lennon, 1975; Brochard et al., 1976), which paved the way for interpreting the flicker phenomena of the RBC as Brownian excitations of the membrane shape. The same phenomena was apparently also observed for synthetic vesicles. An important outcome of the modeling was the first ever estimate of κ for the RBC membranes using VFA. In the following decades, numerous methods for the measurement of κ based on the analysis of thermal membrane fluctuations were proposed and some of them have been established, for example, tubular VFA (Servus et al., 1976), reflection interference contrast microscopy (Zilker et al., 1987), NMR (Struppe et al., 1997), X-ray diffraction (Pan et al., 2008) and quasi-spherical VFA. The last two above-mentioned techniques have become the preferred methods of choice for estimating κ. Derived effects of the membrane fluctuations can also be used to estimate κ, for example, measurements of the vesicle entropic elasticity by micropipette aspiration techniques (Evans and Rawicz, 1990; Chapter 11). In this chapter, we focus on the quasi-spherical VFA, which is established in several laboratories (Center of Biomembrane Physics, Denmark; Bulgarian Academy of Sciences, Bulgaria; Max Planck Institute of Colloids and Interfaces, Potsdam; Institut Curie, Paris; Institut des Sciences Chimiques de Rennes, Rennes). For the analysis of quasi-spherical vesicle shape fluctuations, their dynamics is approximated by the Helfrich's theory for small configurational deviations around a spherical geometry (Schneider et al., 1984; Peterson, 1985; Milner and Safran, 1987). Experimentally, this technique was strongly advanced in the late 1980s with the works of Bivas et al. (1987) and evolved as a tool for the measurement of κ of membranes in a Bulgarian-French collaboration (Faucon et al., 1989; Méléard et al., 1998) in the 1990s. VFA techniques are still under continuous development (Döbereiner et al., 1999; Pécréaux et al., 2004; Drabik et al., 2016).

Although the membrane flicker phenomena of synthetic vesicles found a satisfactory explanation as thermal equilibrium fluctuations of a fluid, elastic interface, the fluctuations of RBC membranes, and other living membranes have become a subject for renewed ongoing debate. Detailed theoretical and experimental analysis of active membranes shows that many aspects of membrane fluctuations caused by enzymatic active processes have striking similarities with equilibrium fluctuations (Bouvrais, 2012). Therefore, the historical quest between the active and the non-active origin of membrane flickering of living membranes must currently be settled with a compromise where both types of processes must be considered as important (Yoon et al., 2009; Gov and Safran, 2005).

14.1.2 THE PROBLEMS WITH MEASURING BENDING RIGIDITY (κ)

As described above, the various attempts to model the membrane flickering phenomena led to methods for measuring κ from the analysis of equilibrium fluctuations of the membrane. It is of major interest to characterize this material parameter because the relevance of membrane shape and elasticity in a multitude of biophysical phenomena; however, in practice it is nontrivial to measure κ. Already a first glimpse of Eq. 14.1 reveals that for a freestanding, symmetric bilayer vesicle, $\Sigma = 0, m = 0$, the curvature elastic energy is scale invariant, independent of the size of the membrane. This makes the membrane shape highly flexible with thermal-excitations visible at macroscopic length scales, making VFA possible with optical microscopy. With typical values of $\kappa \sim 10 k_{\mathrm{B}}T$, membrane shape deformations are energetically inexpensive and thus easily influenced by external perturbations. Furthermore, if the membrane is subject to a confining potential, for example, in a multilamellar system (Struppe et al., 1997) or close to a solid support, it requires very accurate models of the potential to pull out a reliable estimate of κ from the fluctuations analysis. The thermodynamic, non-extensive character of the membrane bending elasticity problem posses a number of additional practical difficulties in analyzing membrane-shape deformations (Peliti, 1996). However, thermal-configurational fluctuations generate a small entropic tension, giving a thermodynamic, extensive contribution to the interfacial free energy that is measurable, for example, by micropipette aspiration techniques (Evans and Rawicz, 1990). This technique has been very popular in obtaining bending rigidities. However, as discussed by Henriksen and Ipsen (2004) the method requires extreme care in the measurements and data analysis to avoid artifacts from the measurement setup. VFA provides a method where an estimate of κ can be extracted from the observations of shape fluctuations of freestanding giant vesicles. Still, this technique requires great care in sample preparation (McDonald et al., 2008) and data analysis to achieve reliable estimates of κ given that even very small amounts of buffer components (Rowat et al., 2005) or probes (Bouvrais et al., 2010) can influence the results (see Section 14.4). Some of these problems may well be underlying the systematic differences found in estimates of κ obtained using different experimental techniques (Niggemann et al., 1995; Nagle, 2013). Therefore, in this chapter we primarily present VFA as a method to obtain spatial and temporal correlation functions for the shape fluctuations of GUVs. The estimation of κ and other material parameters are discussed in Section 14.4 on the applications of the technique. More comprehensive reviews on the measured bending rigidities are found in (Marsh, 2013; Dimova, 2014; Vitkova and Petrov, 2013).

14.2 THE MEASUREMENT

The measurements in VFA of quasi-spherical GUVs consists of two parts: (1) live recordings of the contour fluctuations of the membrane obtained from optical microscopy, and (2) construction of spatial and temporal correlation functions to characterize membrane's shape fluctuations from the recordings. The similar procedure can also be applied for fluctuation analysis of prolate-shaped vesicles (Döbereiner et al., 1999).

14.2.1 TIME AND LENGTH SCALES

Before establishing experimental equipment to capture live time series of fluctuating vesicles and software for their analysis, it is useful to make some elementary considerations about the length and time scales that are important in the problem. The typical size of GUVs adequate for recordings are in the range of radii R_{ve} $5-50$ μm. Let us consider a patch of lateral extension, L, of a freestanding vesicle. Applying the equipartition theorem on Eq. 14.1 the average deviation from the planar configuration in the patch is given by $h \sim \sqrt{k_B T / \kappa}\, L$ (Lipowsky, 1995). Thus for typical values of $\kappa \sim 10 k_B T$, we have $h \leq L$. The optical resolution is a limit on resolving any length scales below $L_{opt} \simeq 250$ nm in the membrane configurations. The equilibrium dynamics of the membrane patch with over-damped relaxation is characterized by a time scale, $\tau_L \sim \eta L^3 / \kappa$ (Brochard and Lennon, 1975), where η is the viscosity. Thus at L_{opt}, the membrane relaxes at a millisecond (ms) time scale, whereas it is in the range of seconds for $L \sim R_{ve}$. Therefore, it is important that the image acquisition is fast, with a camera integration time of milliseconds to prevent smearing of the configurational details due to membrane motion. On the other hand, nothing is gained with an even faster camera integration time. The length scales ranges from L_{opt} up to R_{ve} and the resolvable relaxation times span from 10 ms to 10 s. The upper limit is set mainly by the fact that if a vesicle is very large with long relaxation times in minutes, it will make the experimental data collection unreasonably long (i.e., hours) in order to gather sufficient data statistics. The measurement duration depends on camera integration time, and a shorter camera integration time results in faster measurement recordings (Gracià et al., 2010).

14.2.2 VESICLE PREPARATION

In principle, any technique for the preparation of GUVs can be applied for VFA (as described in Chapter 1), provided sufficient control of the composition of membrane and solvent can be maintained. Our preferred GUV preparation method is the electroformation (or electroswelling) technique (Angelova and Dimitrov, 1986), which gives a high yield of large unilamellar, quasi-spherical vesicles. In its simplest version, lipids are dissolved in chloroform at a concentration of around 0.5 mg/mL to prepare GUVs of radius 1–100 μm in pure water and in low-salt buffers (~ mM) using electroswelling. Alternatively, electroswelling can be performed on small unilamellar vesicles (SUVs) in high-salt or physiological buffers (Pott et al., 2008). Electroswelling technique has proven to be highly suited for VFA with reconstituted proteins (Bouvrais, 2012) or peptides (Bouvrais et al., 2008). Recently, we developed an electroswelling technique for the preparation of GUVs at physiological buffer and temperature conditions (Bhatia et al., 2015, 2016; Bhatia et al., 2016a,b) that can be used to reconstitute transmembrane proteins into GUVs of complex mixtures.

14.2.3 INSTRUMENTATION

Microscope

The microscope plays a central role in VFA given that the basis for the measurement is a long live recording of the configurational fluctuations of a single GUV. If the quality of the images are high and the recording is uninterrupted, the information content is maximal. The only relevant information from the images is the instantaneous membrane conformation in the equatorial plane of the vesicle, whereas all other features should be suppressed to avoid artifacts and disturbances in the subsequent image analysis. Various microscopy techniques (Chapter 10) tested for this purpose are bright-field, phase-contrast, Hoffman Modulation Contrast, wide-field fluorescence, and confocal fluorescence microscopy. We find that standard phase-contrast microscopy is by far the best technique for meeting the above criteria and, to some extent, confocal fluorescence microscopy. So, any standard inverted microscope with phase-contrast optics is applicable.

A typical phase-contrast image of a quasi-spherical GUV is shown in panel (a) (of Box 14.1), in which the vesicle's surface boundary (contour) appears dark due to the differences in the refractive index between the membrane and the solvent. Due to this high contrast, the position of the vesicle's membrane at the equatorial plane can be detected with a very high accuracy. Maximum contrast is achieved at the equator where the optical axis is tangential to the membrane. Any small vertical movements of the vesicle body can be seen as a blur in the vesicle's contour and the focal plane can be adjusted thereafter. The resolution of

Box 14.1 Series of steps for determining the contour of a GUV, see Appendix 1 of the book for structure and data on these lipids

Phase-contrast image of a 1,2-dioleoyl-*sn*-glycero-3-phosphocholine/cholesterol (DOPC/chol; 3:2) GUV containing the transmembrane ion-pump Na/K-ATPase is shown in panel (a). The vesicle's contour at the equatorial plane appears dark on a bright background (solvent). GUVs are prepared by electroswelling in physiological buffer composed of 200 mM sucrose, 30 mM NaCl, 2 mM $MgCl_2$, 30 mM histidine at pH 7.

Given a video sequence of *N* frames containing preferably only one vesicle in focus, as illustrated in panel (a), one can start by picking the center and an approximate radius of the vesicle by following the steps written in the block diagram shown in panel (b) as described below.

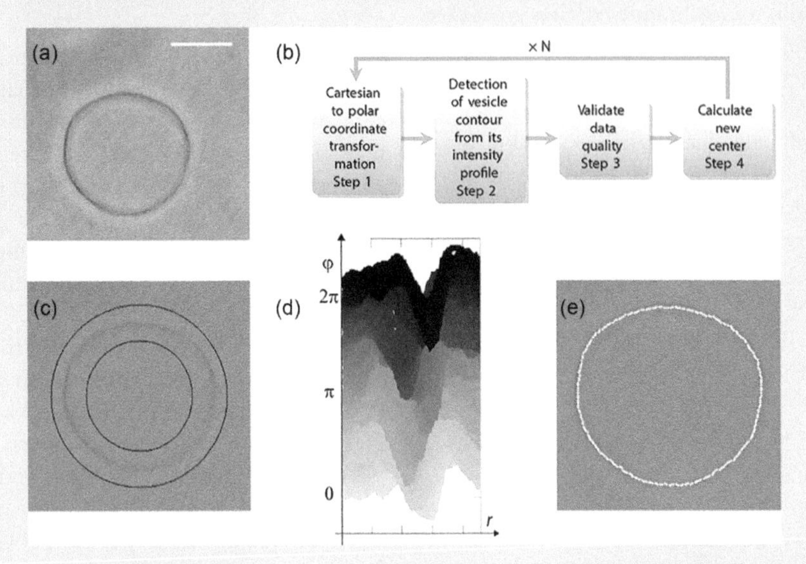

1. The coordinate transformation in step 1 in panel (b) is carried out with respect to the vesicle center and the effect is shown in panel (c) and panel (d). We draw the circles shown in panel (c) to indicate the circular region within which we anticipate to find the vesicle's contour. φ is the azimuthal angle defined at the equatorial plane of the vesicle. We used phase-contrast microscopy for VFA, and therefore the contour detection is done by simply identifying the minimum intensities for each φ as shown in panel (d). Polar-transformed intensity profiles are shown in panel (d), where *r* runs between the two circles.
2. The resulting contour is shown in panel (e). Instead of choosing the minimum intensity, a better estimate of the contour location is obtained by fitting to the relative intensity profile. This brings us to Step 3 of panel (b) where the obtained contour is examined based on certain criteria; for instance, pixel outliers originating from irregular images should be discarded.
3. The newly segmented contour serves as a base for the following consecutive frames. In general, the center of a selected GUV must be determined in each individual frame. However, given that the translational motion between frames is almost absent, the disadvantages with this method are minimal.
4. To assist the contour segmentation, a number of filters can be applied to the images. In panel (d) we have used a contrast limited adaptive histogram equalization (CLAHE) filter (Karel, 1994) in order to increase the image contrast, and a Gaussian filter thereafter for smoothening of the image.

a phase-contrast microscope is similar to that of a fluorescence microscope, and we cannot resolve the membrane's spatial undulations smaller than $L_{\text{opt}} \sim 0.61\lambda/\text{NA}$, where λ is the wavelength of light and NA is the numerical aperture of the objective. The detection of the position of the vesicle's contour at the equatorial plane is best achieved using a well-aligned phase-contrast microscope. In a fluorescence microscope, the contour resolution is reduced due to fluctuations of the fluorophore's absorption and emission transition dipole moments and fluorescence signal coming off the equatorial plane. For phase-contrast microscopy, we have mostly used a 40x objective, and a phase-plate corresponding to a 40x phase-ring. First, the Köhler illumination is achieved and then the objective phase plate is aligned with the condenser annulus to provide maximum contrast in the specimen. If solvent inside the GUV is different from the solvent outside then an additional contrast can be detected. Furthermore, GUVs have the advantage that they settle at either the top or the bottom of the observation chamber due to differences in the density of the solvent. However, it could hamper the analysis of the shape fluctuations because vesicles also deformed due to gravity (Henriksen and Ipsen, 2002).

Camera

To capture live recordings of membrane fluctuations, a camera must be mounted at the microscope. A standard charge-coupled device (CCD) video camera with 25 fps is sufficient for VFA. In the interlaced mode, the camera integration time is few milliseconds for each half-frame and the time between each half frame is 20 ms. Both of these times can be reduced substantially with a modern complementary metal-oxide-semiconductor (CMOS) video camera, which has much more flexibility. The stack of images are stored as 256 gray-level bitmaps (BMPs), TIFFs or in another raw image format for further analysis.

14.2.4 VESICLE CONTOUR DETECTION AT THE EQUATORIAL PLANE

Operative

Before starting the experiment, the microscope must be aligned as described above. A freely floating quasi-spherical GUV of radius 10–100 μm in the observation chamber is identified. In order to find the equatorial plane of the vesicle, the focal plane of the microscope is adjusted manually in small steps until the vesicle's contour appears sharp. Once a suitable focal plane is found, the recording of the images can start. The vertical movements of the vesicle in the observation chamber result in a blurred image of the vesicle's contour. Therefore, the vesicle must be observed while recording, so that small adjustments can be performed quickly to bring the vesicle's equatorial plane back into focus and to keep the recording uninterrupted. Moreover, motion of the GUV in the horizontal plane can lead to the vesicle going out of the field of view, which has to be adjusted accordingly. Sometimes, a flow in the observation chamber can lead to a lateral motion of the vesicles that can be avoided by keeping the chamber horizontally flat and by sealing the open ends of the observation chamber. The recording is closed after 10–60 min, depending on the size of the vesicle and problems to be investigated.

Software

The analysis of the flicker video sequence can be divided into four separate steps:

1. Segmentation of the vesicle contour
2. Construction of the angular correlation function from the segmented contour
3. Decomposition of the angular correlation function into Legendre modes
4. Comparison of the Legendre modes to a model of choice.

This section focuses mostly on the first step, whereas the next steps will become more apparent in the following section. Image segmentation is offered by various open source software distributions (a good and well-known example is FIJI, http://fiji.sc/Fiji), and can be quite helpful as a good starting point. However, for more specific needs and efficiency, one might need to do some coding on one's own. We will walk through the main ideas of a vesicle's contour segmentation in this section. This approach will lead to a natural way of decomposing a vesicle's contour in Legendre modes, which is essential for the next section. Before we start, we will take a brief moment to consider that a computer recognizes the intensities stored in an image. With a few simple mathematical operations, it is possible to use a computer to read gradients of intensities (edges) or even textures (shapes) present in an image. To begin with, MATLAB® has built-in functions that are easy to use, including an Image Processing Toolbox™. For advanced programming options, one can use CUDA™, using which 10–100 times faster runtime can be gained compared with MATLAB®. CUDA™ programming is applicable for massively parallelized problems on graphical processing units (GPUs or graphics cards), which are particularly beneficial for the image analysis. Furthermore, later versions of MATLAB® have some CUDA-based functionality that could be used for prototyping. Regardless of the programming language chosen, in Box 14.1 is shown a schematic walk-through for one of the many possible ways to segment a vesicle's contour.

14.3 ESTIMATION OF THE CORRELATION FUNCTION

The selection of a quasi-spherical GUV establishes an easily obtainable system with a well-defined overall geometry to study membrane configurational fluctuations. The spherical geometry makes it possible to calculate a correlation function for equatorial contour fluctuations. This correlation function can be related to configurational correlations of the whole vesicle and thus can be compared with model predictions. Figure 14.1 shows a fluctuating vesicle whose contour at the equatorial plane is highlighted. Each point on this contour can be labeled by polar coordinates (r,φ). $\overline{r}(t)$ is the instantaneous averaged contour radius. The average vesicle size, R_{ve}, can be chosen as $\overline{r}(t)$ or the time averaged measure of this quantity (see Section 14.3.2).

14.3.1 EQUATORIAL CONTOUR CORRELATION FUNCTION

The measured equatorial contours can now be analyzed and hopefully physical properties can be extracted. We expect that the observed contour fluctuations are the results of cooperative phenomena within the membrane and its environment. Correlation functions serve as the natural quantity for describing such behavior. The angular correlation function,

$$\Xi(\gamma,t)=\frac{1}{R_{ve}^2}\frac{1}{2\pi}\int_0^{2\pi}\mathrm{d}\varphi[r(\varphi+\gamma)r(\varphi)-\overline{r}^2(t)] \qquad (14.2)$$

is directly obtainable from the measured contours (Bivas et al., 1987) and depends on angle φ and time (t). $\Xi(\gamma,t)$ expresses how correlated the contour deviations are for contour positions separated by an angle γ. In Figure 14.2, the calculated $\Xi(\gamma,t)$ as obtained from a vesicle's contour, is shown. By definition, $\Xi(\gamma,t)$ (Eq. 14.2) is symmetric around $\gamma=\pi$. In the numerical calculation of $\Xi(\gamma,t)$, it is crucial to keep track of the numerical accuracy given that the finer details are easily washed away by uncontrolled numerical errors.

Although this instantaneous $\Xi(\gamma,t)$ is of interest for characterizing a single contour, its time average, $\langle\Xi(\gamma)\rangle$, is expected to contain general information about the processes governing membrane-shape fluctuations. For example, from the ergodic

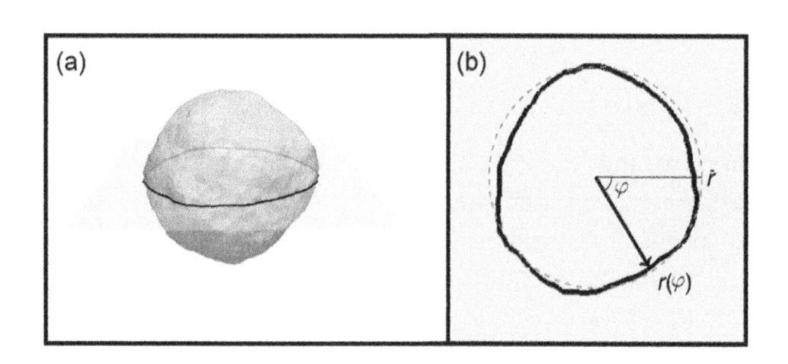

Figure 14.1 (a) Computer model of a quasi-spherical vesicle with highlighted equatorial plane. (b) The corresponding contour where the dashed line indicates mean shape of the contour. $\overline{r}$ is the corresponding mean radius.

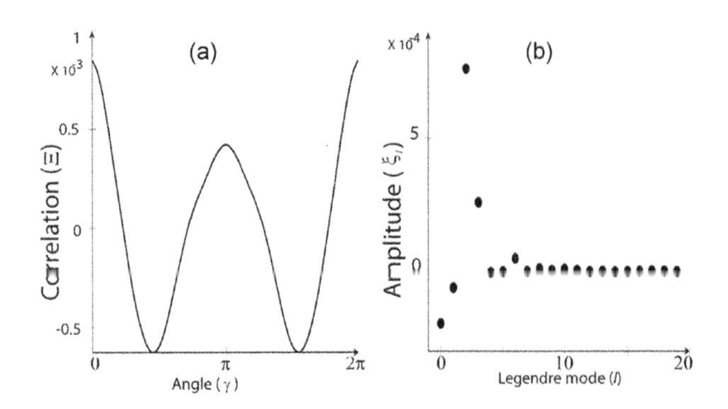

Figure 14.2 (a) Angular correlation function for the contour shown in panel (e) of Box 14.1. (b) The corresponding Legendre decomposition amplitudes.

theorem, we expect that for an equilibrium system the thermal average can be expressed as a time average over long time, t_1,

$$\langle \Xi(\gamma) \rangle_{\text{eq}} = \langle \Xi(\gamma) \rangle = \frac{1}{t_1} \int_0^{t_1} dt' \, \Xi(\gamma, t') \tag{14.3}$$

Similarly, for homogeneous nonequilibrium systems, the time-averaged correlation functions form the natural measures for physical characterization.

14.3.2 MODE ANALYSIS OF SPATIAL CORRELATIONS

As described in Section 14.2, the outcome of the measurements is a time series of equatorial contour positions for a quasi-spherical GUV that can be characterized in terms of the angular correlation functions. The next challenge is to understand how these contour fluctuations relate to the configurational fluctuations of the whole vesicle. The quasi-spherical geometry of a vesicle invites the formulation of its surface undulations in the spherical coordinate system as given below.

$$\vec{X}(\vartheta, \varphi, t) = R_{\text{ve}}[(1 + u(\vartheta, \varphi, t)] \cdot \vec{n}(\vartheta, \varphi) \tag{14.4}$$

where R_{ve} is the average vesicle radius; $\vec{n}(\vartheta, \varphi)$ is the positions on the unit sphere; and $u(\vartheta, \varphi, t)$ is the local deviation of the vesicle configuration relative to the spherical mean shape at time t. In Eq. 14.4 it is assumed that $u \ll 1$ and that no overhangs are present. Furthermore, the natural mode decomposition of u in the spherical background geometry is in terms of the spherical harmonics analogous with the Fourier decomposition in a flat space:

$$u(\vartheta, \varphi, t) = \sum_{l \geq 0} \sum_{n=-l}^{l} u_{ln}(t) Y_{ln}(\vartheta, \varphi) \tag{14.5}$$

The first two modes, $l = 0, 1$, adjust the radius and the center of mass respectively, and the modes for $l \geq 2$ capture the shape deformations of the vesicle. Therefore, the experimentally obtained contour configurations have a simple identification $r(\varphi, t) = R_{\text{ve}}\left(1 + u(\frac{\pi}{2}, \varphi, t)\right)$, and the angular-correlation function can be expressed in terms of u_{ln} and $Y_{ln}\left(\frac{\pi}{2}, \varphi\right)$:

$$\Xi(\gamma, t) = \sum_{l \geq 0} \sum_{l' \geq 0} \sum_{n \neq 0} u_{ln}(t) u^*_{l'n}(t) Y_{ln}(\frac{\pi}{2}, \gamma) Y^*_{l'n}(\frac{\pi}{2}, 0) \tag{14.6}$$

For most problems of interest $\langle u_{ln} u^*_{l'n} \rangle = \langle | u_{ln} |^2 \rangle \delta_{ll'} = \langle | u_{l0} |^2 \rangle \delta_{ll'}$ (see Section 14.4). Then mean value of $\Xi(\gamma, t)$ takes the following form:

$$\begin{aligned} \langle \Xi(\gamma) \rangle &= \sum_{l \geq 0} \langle | u_{l0} |^2 \rangle \sum_{n=-l}^{l} Y_{ln}(\frac{\pi}{2}, \gamma) Y^*_{ln}(\frac{\pi}{2}, 0) \\ &= \sum_{l \geq 0} \langle \xi_l \rangle P_l(\cos(\gamma)) \end{aligned} \tag{14.7}$$

where

$$\langle \xi_l \rangle = \frac{2l+1}{4\pi} \langle | u_{l0} |^2 \rangle \text{ for } l \geq 2 \tag{14.8}$$

and $P_l(x)$ are the Legendre Polynomials. To obtain Eq. 14.7, the addition theorem of spherical harmonics has been employed. So, a Legendre Polynomial decomposition of $\langle \Xi(\gamma) \rangle$ provides us with information about the spatial correlations in the vesicle's fluctuations expressed by $\langle | u_{l0} |^2 \rangle$. In the determination of $\langle \xi_l \rangle$ from experiments, $\langle \xi_0 \rangle$ and $\langle \xi_1 \rangle$ will be dominated by the fluctuations in the radius and center of mass, which are artifacts of the numerical contour detection and can be left out of consideration (Faucon et al., 1989). Because the averaging is performed over time, it is most practical to form the Legendre polynomial transform for each contour.

$$\xi_l(t) = \frac{2l+1}{2} \int_{-1}^{1} d\cos(\gamma) \Xi(\gamma, t) P_l(\cos(\gamma)). \tag{14.9}$$

$\xi_l(t)$ can be calculated numerically using the fast Legendre transform technique and the recursive relations connecting Legendre polynomials. Again, the control with the numerical error is essential for a good result. $\xi_l(t)$ is a stochastic quantity fluctuating around $\langle \xi_l \rangle$, as shown in Figure 14.3.

The estimation of $\xi_l(t)$ from microscopy images also makes it possible to check if the l modes are independent, as predicted in many theoretical model calculations, that is, $\langle \xi_l \xi_{l'} \rangle = \langle \xi_l \rangle \langle \xi_{l'} \rangle$ (Bouvrais, 2012).

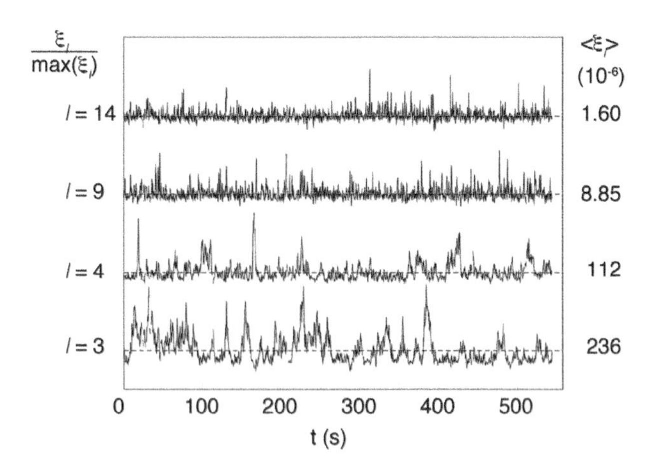

Figure 14.3 For the GUV shown in Box 14.1, time series of selected ξ_l are shown. ξ_l is scaled with respect to the maximum value of ξ_l for the respective mode. The dashed lines represent the average for corresponding ξ_l, and the values are displayed at the right vertical axis.

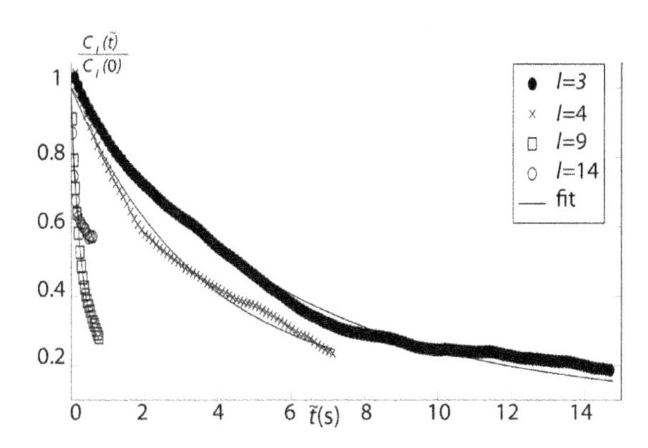

Figure 14.4 The autocorrelation functions $C_l(\tilde{t})$ are shown for the vesicle shown in Box 14.1 for a few modes. The full lines show single exponential fits to the data. The relaxation time (τ_l) is estimated from the fits and are given by $(\tau_3) = 11.4$ s, $(\tau_4) = 7.0$ s, $(\tau_9) = 0.53$ s, and $(\tau_{14}) = 0.22$ s.

14.3.3 TEMPORAL CORRELATIONS

The estimation of $\xi_l(t)$ also makes it possible to study temporal correlations in the vesicle shape fluctuations by the autocorrelation functions $C_l(\tilde{t}) = \langle \xi_l(t)\xi_l(t+\tilde{t})\rangle$. For systems with decoupled modes, $C_l(\tilde{t})$ is related to the correlations in $u_{lm}(t)$ as given below:

$$C_l(\tilde{t}) = \langle \xi_l(t)\xi_l(t+\tilde{t})\rangle$$

$$= \left(\frac{2l+1}{4\pi}\right)^2 \langle |u_{l0}(t)|^2 |u_{l0}(t+\tilde{t})|^2\rangle \text{ for } l \geq 2. \quad (14.10)$$

Numerically $C_l(\tilde{t})$ can be generated directly from the time series $\xi_l(t)$ by establishment of its Fourier transformation $\tilde{\xi}_l(\omega)$ and the reverse Fourier transform of the power spectrum $\tilde{\xi}_l(\omega)\tilde{\xi}_l(-\omega)$. Figure 14.4 shows single exponential decay of $C_l(\tilde{t})$ for few values of l.

14.4 APPLICATIONS

VFA is an experimental technique to produce spatiotemporal correlation functions for the conformational fluctuations at length scales larger than L_{opt}. These data first become truly useful if they can reveal information about physical mechanisms and give estimates of the model parameters. Some example of such applications will be given with emphasis on the estimation of κ.

14.4.1 PHENOMENOLOGICAL MODEL OF MEMBRANES

Because the observation length scales for VFA are much larger than the molecular length scales, the relevant theoretical membrane models are coarse-grained phenomenological models describing the cooperative long-distance behavior of the membrane. The Helfrich free energy (Eq. 14.1) is one such model, where the local principle curvatures, C_1 and C_2, are order parameters describing the local surface geometry. κ and m are material constants that can be influenced by many processes in the membrane and its environment. In some cases, such influences can be caught in a more extensive phenomenological model. A common class of modifications of Eq. 14.1 for interpreting VFA data takes the following form:

$$\mathcal{F} = \oint_{A_{\text{ve}}} dA\left(\frac{\kappa}{2}(C_1+C_2)^2 - w(\varphi_+ - \varphi_-)(C_1+C_2) + f(\varphi_+,\varphi_-)\right) \quad (14.11)$$

where φ_- and φ_+ are local scalar fields attributed to a density in the lower and upper monolayers in the bilayer. $f(\varphi_+,\varphi_-)$ is the free-energy lateral density for the φ_- and φ_+ fields (examples will be given below). The term $w(\varphi_+ - \varphi_-)$ represents the local preferred curvature due to asymmetry between the monolayers and induces a coupling between the densities and the curvature, which is also reflected in the fluctuations. w is the associated coupling constant. In 1986, Leibler (1986) suggested that such coupling may have a dramatic effect on the conformational fluctuations and leads to curvature instabilities of the membrane. Eq. 14.11 contains a number of model parameters that will appear in a model prediction for the correlation functions, for example, $\langle u_{lm}(t)u^*_{lm}(0)\rangle_{\text{eq}}$. Therefore, VFA provides a possibility to estimate or put some bounds on these model parameters. Another type of extension of Helfrich's model is to include frictional effects such as the viscous damping (Brochard and Lennon, 1975) or intermonolayer friction (Seifert and Langer, 1993; Evans and Yeung, 1994; Merkel et al., 1989; Shchelokovskyy et al., 2011), which is reflected in the relaxation times of the temporal angular correlation function (Miao et al., 2002). Finally, Helfrich's theory can be extended to describe nonequilibrium phenomena where enzymatic processes in the membrane are modeled as simple fluctuating forces on the membrane with parameters that will appear in the calculation of $\langle u_{lm}(t)u^*_{lm}(0)\rangle$ (Ramaswamy et al., 2000; Lomholt, 2006; Loubet et al., 2012). Thus, VFA has the capacity to verify a broad class of phenomenological models by analyzing the fluid membrane fluctuations and to estimate the model parameters. Some caution must be taken with such estimates: Although the natural lower length scales at which a phenomenological model like Eq. 14.11 (and its model parameters) are defined in nanometers, the observation length scale in VFA is above L_{opt}. In some cases, the model parameters may be highly dressed or renormalized at long length scales, for example, when the membrane becomes very soft (Hansen et al., 1998).

14.4.2 MEASURING BENDING RIGIDITY (κ)

Most applications of VFA involve the measurement of κ based on equilibrium fluctuations of the vesicle shape. The basis for this estimation is expansion of Eq. 14.1 for small deviations, u, from the spherical configuration. By use of Eqs. 14.4 and 14.5, $\mathcal{F}_{\text{Hel}}$ becomes (Faucon et al., 1989).

$$\mathcal{F}_{\text{Hel}} \simeq \Sigma 4\pi R^2 + \frac{1}{2}\sum_{l\geq 0}\sum_{m=-l}^{l} E_l(\kappa, m, \Sigma_0)\,|u_{lm}|^2 \quad (14.12)$$

where $E_l(\kappa, m, \Sigma) = \kappa(l-1)(l+2)\left(l(l+1)+\tilde{\Sigma}\right)$; and $\tilde{\Sigma} = \left(\frac{\Sigma R_{\text{ve}}^2}{\kappa} + 2mR_{\text{ve}} + \frac{m^2R_{\text{ve}}^2}{2}\right)$ is the reduced tension. Note that the spontaneous curvature (m) appears in the reduced tension (see Chapter 5). In this approximation, Eq. 14.1 appears as a sum of independent contributions from the amplitudes u_{ln}. The Boltzmann probability distribution, $\sim \exp(-\frac{1}{k_B T}\mathcal{F}_{\text{Hel}})$, is a simple product of Gaussian distributions in u_{ln} variables. The application of the equipartition theorem on Eq. 14.12 gives

$$\langle u_{ln} u^*_{l'n'} \rangle_{\text{eq}} = \frac{k_B T}{E_l(\kappa, m, \Sigma)} \delta_{ll'} \delta_{nn'} \quad (14.13)$$

κ and $\tilde{\Sigma}$ can be estimated from a vesicle recording by using Eq. 14.9 and fitting it to Eqs. 14.8 and 14.13. The fit to the model is found by minimizing the χ^2-function:

$$\chi(\kappa, \tilde{\Sigma}) = \sum_{l=2}^{l_{\max}} \left(\frac{\langle \xi_l \rangle - \langle \xi_l \rangle_{\text{eq}}(\kappa, \tilde{\Sigma})}{\text{var}(\xi_l)} \right)^2 \quad (14.14)$$

where $\langle \xi_l \rangle_{\text{eq}}(\kappa, \tilde{\Sigma})$ is the theoretical expression for $\langle \xi_l \rangle_{\text{eq}}$. It is important to scale with the variance in ξ_l (Figure 14.3) since there are orders of magnitudes difference between $\langle \xi_l \rangle$ values for different l values. The errors in the estimate are derived from the covariance matrix (the second derivative of the χ^2-function) between κ and Σ at the estimated values of κ and Σ. The maximum l-value ($l_{\max}$) in the fit must be chosen such that the associated length scale is larger than the optical resolution $l_{\max} < R_{\text{ve}} / L_{\text{opt}}$. Furthermore, fitting beyond the $l_{\max}$ values where the numerical error is significant in the calculation of $\langle \xi_l \rangle$ (in Section 14.3) must be avoided. Figure 14.5 shows an example of such a fit. The reduced tension $\tilde{\Sigma}$ varies strongly between the vesicles reflecting different areas and volumes of the GUVs typically in a measurable range (-6) up to 100. High $\tilde{\Sigma}$ values indicate that the vesicle is tension dominated for the most observable part of the membrane fluctuations, which leads to high error in the determination of κ. Therefore, it is desirable to analyze GUVs with low $\tilde{\Sigma}$ (having visual membrane undulations). It takes some practice to spot such vesicles in the experiments. The variation in the estimated κ is less among vesicles and a simple average of the estimated values weighted with the variance, provides a statistically reliable estimate of the κ in a measurement. Besides variation in ξ_l^2 due to physical fluctuations, systematic errors from instrumentation and analysis are unavoidable, for example, composition of membrane, camera, and contour detection.

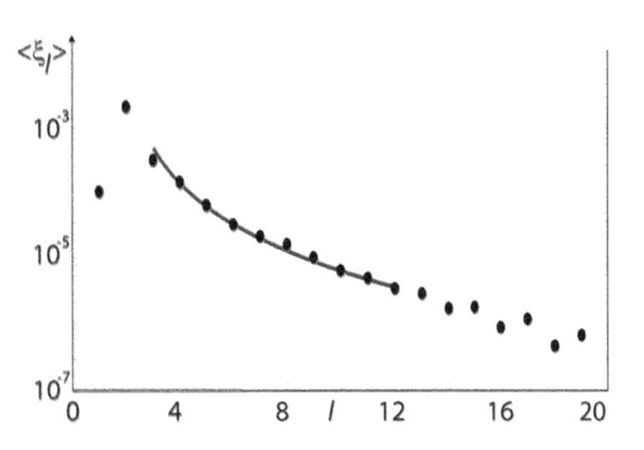

Figure 14.5 For the GUV shown in panel (a) of Box 14.1, we determine κ by fitting $\langle \xi_l \rangle$ to $\langle \xi_l \rangle_{\text{eq}}(\kappa, \tilde{\Sigma})$. The line shows the fit to the dotted experimental data. We get $\kappa = 25.76 \pm 0.79$ in the units of $k_B T$ and $\tilde{\Sigma} = -6.22 \pm 2.39$.

Recently, an alternative method for estimating κ from VFA data was introduced (Méléard et al., 2011) based on direct analysis of the equilibrium probability distributions of $\Xi(\gamma)$. Theoretical analysis of Gaussian fluctuations in u_{ln} from Eq. 14.11 shows that the probability distribution in the Fourier amplitudes of $\Xi(\gamma)$ is single exponential (easy to fit) dependent on κ and $\tilde{\Sigma}$. Besides, a reduced error in the κ determination, it also provides a simple test of the Gaussian character of the fluctuations in u_{ln}.

14.4.3 MEASURING RELAXATION TIMES

To model the temporal aspects of the thermal fluctuations the mechanical restoring forces for a quasi-spherical GUV, (as derived from Eq. 14.12) must be balanced with the frictional forces and the random forces originating from thermal excitations at a low Reynolds number. For the simplest case, with the membrane mechanics described by Eq. 14.12. It can be formulated as the Langevin equation (Miao et al., 2002) given below.

$$\frac{\eta}{\Gamma_l} \dot{u}_{ln}(t) = -\frac{\kappa}{R_{\text{ve}}^3} E_l u_{ln}(t) + \zeta_{ln}^{\text{th}}(t) \quad (14.15)$$

where $\Gamma_l = \frac{l(l+1)}{4l^3+6l^2-1}$; η is the viscosity of solvent causing the friction; and $\zeta^{\text{th}}(t)$ is the random, Gaussian thermal noise, fixed by the fluctuation–dissipation theorem.

$$\langle \zeta_{ln}^{\text{th}}(t) \rangle = 0, \langle \zeta_{ln}^{\text{th}}(t) \zeta_{l'n'}^{\text{th}}(t') \rangle = \frac{2\eta}{\Gamma_l} \frac{k_B T}{R^3} \delta_{ll'} \delta_{nn'} \delta(t - t') \quad (14.16)$$

The average is performed with respect to Gaussian distribution for $\zeta_{ln}^{\text{th}}(t)$. The solution of Eq. 14.15 becomes an extension of Eq. 14.13

$$\langle u_{ln}(0) u_{l'n'}(t) \rangle = \frac{k_B T}{E_l} \exp\left(-\frac{t}{\tau_l^{MS}} \right) \delta_{ll'} \delta_{nn'} \quad (14.17)$$

The decay time $\tau_l^{MS} = \frac{\eta R_{\text{ve}}^3}{E_l \Gamma_l}$ is the Milner-Safran relaxation time (Milner and Safran, 1987) with the characteristic l^{-3} behavior for flaccid vesicles. For Gaussian fluctuating variables, the Wick's theorem applies to calculate the temporal correlation in the variable $\xi_l(t)$ as given below,

$$\frac{\langle \xi_l(t) \xi_l(0) \rangle - \langle \xi_l \rangle^2}{\langle \xi_l \rangle^2} = \left(\frac{\langle u_{ln}(0) u_{ln}(t) \rangle}{\langle u_{ln}^2 \rangle} \right)^2 \quad (14.18)$$

14.4.4 LOCAL CURVATURE OF MEMBRANE INCLUSIONS

Many membrane inclusions give a local curvature imprint as described by Eq. 14.11. For a low concentration of inclusions in the membrane, the potential, $f(\varphi_+, \varphi_-)$, in Eq. 14.11 has the ideal gas form given by

$$f(\varphi_+, \varphi_-) = k_B T [\varphi_+ (\ln(a\varphi_+) - 1) + \varphi_- (\ln(a\varphi_-) - 1)] \quad (14.19)$$

where $\varphi_\pm$ are the lateral densities of the inclusions in the two monolayers. For a symmetric bilayer, the densities $\langle \varphi_+ \rangle_{\text{eq}} = \langle \varphi_- \rangle_{\text{eq}} = \varphi_0$. A stability analysis of Eq. 14.11 shows that the measured effective bending rigidity (κ_{eff}) is strongly modified:

$$\kappa_{\text{eff}} = \kappa - \frac{4w^2}{k_B T}\varphi_0. \tag{14.20}$$

Thus, κ_{eff} declines linearly with the increasing content of the inclusions with the slope as given by $(4w^2 / k_B T)$. This gives a possibility of estimating the magnitude of w by VFA. For lysolipids present in the 1-palmitoyl-2-oleoyl-*sn*-glycero-3-phosphocholine (POPC) bilayers (Henriksen et al., 2010), it was found that the short-chain lysolipids cause a curvature instability such that at some critical concentration φ_0^c, κ_{eff} vanishes and, for $\varphi_0 > \varphi_0^c$, GUVs do not form. In addition, for antimicrobial peptides, it has been possible to identify w. For Magainin II, a strong reduction in κ_{eff} is found for small φ_0, whereas it is stabilized at a low value of κ_{eff} at higher φ_0 (Bouvrais et al., 2008). This behavior can be understood from a simple extension of Eq. 14.19 that includes lateral and trans-monolayer coupling.

14.4.5 BENDING RIGIDITY AND THE MAIN TRANSITION

Most single lipid component membranes display several thermotropic phase transitions involving many degrees of freedom. The most pronounced phase transition is the *main* chain melting transition, which for many biologically interesting lipids take place in a temperature range that is accessible for VFA. The measurement of the κ above the main transition for saturated phosphatidylcholine (PC) lipids probably represents the first application of VFA for the analysis of complex properties of membranes (Fernandez-Puente et al., 1994). The measured κ shows a dramatic decrease close to the main transition temperature (T_m). A simple interpretation of the phenomenon can be written in terms of Eq. 14.11 such as

$$f(\varphi_+, \varphi_-) = \frac{K_A}{2}\left[(\varphi_+ - \varphi_0)^2 + (\varphi_- - \varphi_0)^2\right] \tag{14.21}$$

where $\varphi_\pm$ are the lateral molecular densities; and K_A is lateral compressibility modulus. The transition has pseudo-critical properties with a pronounced increase in the compressibility $(1/K_A)$ due to density fluctuations (Evans and Kwok, 1982; Ipsen et al., 1990). From the analysis of Eq. 14.11, an effective κ can be extracted (Hønger et al., 1994),

$$\kappa_{\text{eff}} = \kappa - \frac{2w^2}{K_A}. \tag{14.22}$$

Eq. 14.21 predicts a decrease in κ_{eff} close to T_m and gives a rough estimate of w (note that the physical origin of the w in Sections 4.4 and 4.5 are completely different). The effect of the bilayer softening is also seen significantly in the swelling experiments involving multilamellar membrane stacks around T_m (Hønger et al., 1994).

14.4.6 EFFECT OF MEMBRANE ADDITIVES

The effect of various membrane additives have been analyzed by VFA for many membrane systems. It is beyond the scope of this chapter to review these results, but a few lessons can be drawn. κ depends both on the lipid and the additive. For example, the polyprenyl lipid chains of prenylated peptides seems to reduce κ of the fluid 1,2-Dimyristoyl-sn-glycero-3-phosphorylcholine (DMPC) bilayers in water, whereas in the presence of charged peptides, κ is strongly enhanced (Rowat et al., 2004). Qualitatively, this can be understood by extending Eq. 14.1 to include membrane electrostatics. So, both the hydrophobic and the hydrophilic moieties of the peptides play a role for κ. VFA is particularly useful for the measurement of κ in membranes containing partitioning additives given that both the micropipette aspiration and the X-ray diffraction techniques fail. For hyper-swelling additives like triglycerides, VFA is possibly the only technique for the measurement of κ (Pakkanen et al., 2011). An important finding is that many commonly used fluorescent membrane probes affect κ (Bouvrais et al., 2010), so great care must be taken in their usage in fluorescence microscopy. In some cases, a clear trend on how the membrane additives influence κ have been identified. Sterols are known to have an ordering effect on membranes with saturated and mono-unsaturated lipids, whereas for double-saturated DOPC, hardly any change in κ is observed (Gracià et al., 2010). In a study of the mechanical and chain ordering properties of POPC bilayers with three different sterols, it was shown that both κ and K_A are completely determined by the chain ordering capacity of the sterols (Henriksen et al., 2006).

14.4.7 EFFECT OF SOLVENT ADDITIVES

VFA is also particularly well suited for the studies of the effects of solvent additives on the membrane mechanics given that issues about their partitioning into membranes (as in the micropipette aspiration technique) or into the interlamellar space (as in the X-ray diffraction measurements) can be excluded. Sugars are commonly used solvent components in the studies of vesicle properties. VFA studies indicate that sugars are not inert to the membrane properties and a reduction in κ is reported (Vitkova et al., 2006; Genova et al., 2006, 2007, 2010), whereas data from X-ray diffraction measurements disputes this finding (Nagle, 2013; Nagle et al., 2015, 2016). Similarly, standard buffers used in biochemical and cell biological studies are found to affect the measured κ for POPC membranes (Bouvrais et al., 2014) (see Table 14.1), indicating that the salt composition of the solvent is an important factor (Bouvrais et al., 2010). Furthermore, GUVs prepared from lipid extracts of whole smaller invertebrates have showed that the measured values of κ (Bouvrais et al., 2013) are affected by environmental pollutants.

Table 14.1 **Bending rigidities of POPC GUVs in the presence of 10 mM of various buffers (Bouvrais et al., 2014)**

BUFFER	$\kappa[k_B T]$
mM HEPES	39.09 ± 0.70
mM histidine	40.61 ± 0.44
mM MES	33.07 ± 0.43
mM MOPS	34.70 ± 0.74
mM PIPES	37.98 ± 1.00

Note: MES, 2-(*N*-morpholino)ethanesulfonic acid; MOPS, 3-(*N*-morpholino)propanesulfonic acid; PIPES, piperazine-*N*, *N*-bis(2-ethanesulfonic acid).

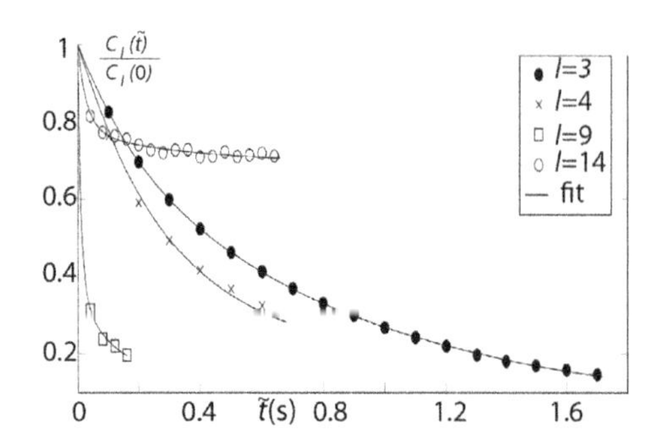

Figure 14.6 For a DOPC/chol (3:2) GUV with active Na/K-ATPase, $C_l(\tilde{t})$ is shown for a few modes. The full lines show a double-exponential fit to $C_l(\tilde{t})$ with two relaxation times, τ_{l1} and τ_{l2}, where $\tau_{l2} \sim 0.6$s. The estimated value of $\kappa = 18.57 +/-0.94$ in units of k_BT.

14.4.8 ACTIVE MEMBRANES

VFA was first developed as an outcome of the description of the RBC flicker phenomena and since then it has been a standard technique for the studies of RBC membrane conformations (Fricke et al., 1986). VFA has recently been applied to GUVs with reconstituted Na-K-ATPase (Bouvrais, 2012). For DOPC/chol (3:2) membranes, the measured κ is increased in the presence of non-active proteins. Interestingly, the estimated spectrum of configurational fluctuations of the membrane with the active ion pumps has the same form as found for equilibrium fluctuations, that is, ξ_l display mode decoupling and $\langle \xi_l \rangle$ can be well fitted to the form Eqs. 14.8 and 14.13, but the effective κ value is significantly lowered. However, the temporal correlations are strongly altered. $\langle \xi_l(t)\xi_l(0) \rangle$ shows a double-exponential relaxation time as shown in the Figure 14.6. The first relaxation time shows the characteristic Milner-Safran behavior and the second relaxation time (which is mode independent) shows a value close to the turn-over time of the active ion pump. So, apparently a molecular time scale propagates through the system up to the longest length scales in the system. These are exciting prospects for the applications of VFA to the active membrane systems.

14.5 DISCUSSION AND CONCLUSION

In this section, we summarize some of the possibilities and challenges with the VFA technique.

14.5.1 ADVANTAGES AND DISADVANTAGES WITH VFA

Both the advantages and problems with the use of VFA have been discussed in this chapter, but we will here summarize some of the main points. Besides an inverted light microscope with phase-contrast optics, a camera, a computer for data collection and their analysis and a minimal wet lab, the only requirement is a good understanding of physics and physical chemistry principles. VFA gives the possibility of analyzing the conformational fluctuations of individual free-standing and probe-free GUVs and of testing theoretical model predictions by estimating model parameters, in particular κ. This program works well for simple model membranes with a single fluid phase. It is particularly useful for the studies of membrane elastic properties and membrane stability in the presence of partitioning membrane inclusions or solutes where other methods fail. VFA as presented here also has a number of limitations. The method does not work well for phase-separating membranes or membranes with heterogeneities at optically resolvable length scales (Semrau et al., 2008). Similarly, the membrane fluctuations are difficult to resolve in molecularly crowded solvent conditions, for example, for biomembranes.

14.5.2 OUTLOOK

The setup for VFA of quasi-spherical vesicles described in this chapter has become a standard technique for the measurement of κ and other model parameters. It must be expected that it will find wider applications and importance as the interest in the interplay between membrane elasticity, membrane organization, and protein activity grows. However, there are important properties of the vesicles that cannot be caught from quasi-spherical vesicles, for example, the spontaneous curvature (m) in Eq. (14.1). Döbereiner et al. have introduced a technique to obtain m from the contour fluctuations of prolate vesicles stabilized by gravity using a sugar gradient, which furthermore induces a spontaneous curvature (Döbereiner et al., 1999). With a little further development, this technique has an obvious potential in the measurement of m for asymmetric membranes and in analyzing the effect of active intermonolayer transport proteins like ABC-transporters and flippases. Similarly, applications for tubular vesicles in VFA (Servus et al., 1976) have been limited. However, many membrane-coating proteins form naturally tubular membranes, and the analysis of their shape fluctuations can reveal information about the in-plane elastic constants induced by the protein coat (Ramakrishnan et al., 2013). It is also likely that the rapid developments in super-resolution optical microscopy techniques will make it possible to enhance the observation length scales below L_{opt}. Good luck with the VFA!

ACKNOWLEDGMENTS

We thank the Danish Council for Independent Research | Natural Sciences (FNU), grant 95-305-23443, for the financial support.

LIST OF ABBREVIATIONS

AC	alternating current
BAS	Bulgarian Academy of Sciences, Bulgaria
BMP	bitmap image file
CCD	charge-coupled device
chol	cholesterol
CI-Paris	Institut Curie, Paris
CLAHE	contrast limited adaptive histogram equalization
CMOS	complementary metal-oxide-semiconductor
DOPC	1,2-dioleoyl-*sn*-glycero-3-phosphocholine
HEPES	4-(2-hydroxyethyl)-1-piperazineethanesulfonic acid
ISCR	Institut des Sciences Chimiques de Rennes, Rennes
MEMPHYS	Center of Biomembrane Physics, Denmark
MES	2-(*N*-morpholino)ethanesulfonic acid
MOPS	3-(*N*-morpholino)propanesulfonic acid
MPIKG	Max Planck Institute of Colloids and Interfaces, Potsdam
NA	numerical aperture
PIPES	piperazine-*N*,*N*-bis(2-ethanesulfonic acid)

POPC	1-palmitoyl-2-oleoyl-*sn*-glycero-3-phosphocholine
RBC	red blood cell
SUV	small unilamellar vesicle
TIFF	Tagged Image File Format
VFA	vesicle fluctuation analysis

GLOSSARY OF SYMBOLS

A	area
A_{ve}	vesicle surface area
C_1, C_2	principal curvatures
C_l	temporal correlation
$\mathcal{F}_{Hel}$	Helfrich free energy
g	metric tensor
h	deviation from planar configuration
k_B	Boltzmann constant
κ	membrane bending rigidity
L	length
L_{opt}	optical resolution limit
l	spherical harmonic degree
m	membrane spontaneous curvature
n	spherical harmonic order
$\vec{n}$	surface normal vector
P_l	Legendre polynomial
$r(\varphi)$	contour radius as function of angle
$\bar{r}(t)$	averaged contour radius
Σ	membrane surface tension
T	temperature n
T_m	main chain melting transition temperature
τ_L	equilibrium time scale
η	solvent viscosity
φ, γ	Azimuthal angle
ϑ	polar angle
Ξ	angular correlation function
ξ_l	angular correlation amplitude
$t, \tau, \tilde{t}$	time
u	local vesicle deviation
Y_{ln}	spherical harmonic Laplace equation
V	volume
ω	angular frequency (Fourier)

REFERENCES

Angelova MI, Dimitrov DS (1986) Liposomes electroformation. *Faraday Discuss. Chem. Soc.* 81:303–311.

Bhatia T, Cornelius F, Brewer J, Bagatolli LA, Simonsen AC, Ipsen JH, Mouritsen OG (2016) Spatial distribution and activity of Na^+/K^+-ATPase in lipid bilayer membranes with phase boundaries. *Biochim. Biophys. Acta. Biomembr.* 1858:1390–1399.

Bhatia T, Cornelius F, Ipsen JH (2016a) Capturing sub-optical dynamic structures in the lipid bilayer of free-standing giant unilamellar vesicles. *Nature Protocols* 12:1563.

Bhatia T, Cornelius F, Ipsen JH (2016b) Reconstitution of transmembrane protein Na +/K +-ATPase in giant unilamellar vesicles of lipid mixtures involving PSM, DOPC, DPPC and cholesterol at physiological buffer and temperature conditions. *Nature Protocols/Protocol Exchange/protocols/4571 (Online).*

Bhatia T, Husen P, Brewer J, Bagatolli LA, Hansen PL, Ipsen JH, Mouritsen OG (2015) Preparing giant unilamellar vesicles of complex lipid mixtures on demand: mixing small unilamellar vesicles of compositionally heterogeneous mixtures. *Biochim. Biophys. Acta. Biomembr.* 1848:3175–3180.

Bivas I, Hanusse P, Bothorel P, Lalanne J, Aguerre-Chariol O (1987) An application of the optical microscopy to the determination of the curvature elastic modulus of biological and model membranes. *J. Phys. (France)* 48:855–867.

Blowers R, Clarkson EM, Maizels M (1951) Flicker phenomenon in human erythrocytes. *J. Physiol.* 113:228–239.

Bouvrais H (2012) Bending rigidities of lipid bilayers: Their determination and main inouts in biophysical studies. In *Advances in Planar Lipid Bilayers*, pp. 1–67. Elsevier, Amsterdam, the Netherlands.

Bouvrais H, Duelund L, Ipsen JH (2014) Buffers affect the bending rigidity of model lipid membranes. *Langmuir* 30:13–16.

Bouvrais H, Garvik OS, Pott T, Méléard P, Ipsen JH (2010) Mechanics of POPC bilayers in presence of alkali salts. *Biophys. J.* 98:272a.

Bouvrais H, Holmstrup M, Westh P, Ipsen JH (2013) Analysis of the shape fluctuations of reconstituted membranes using guvs made from lipid extracts of invertebrates. *Biol. open* 2:373–378.

Bouvrais H, Méléard P, Pott T, Jensen KJ, Brask J, Ipsen JH (2008) Softening of POPC membranes by magainin. *Biophys. Chem.* 137:7–12.

Bouvrais H, Pott T, Bagatolli LA, Ipsen JH, Méléard P (2010) Impact of membrane-anchored fluorescent probes on the mechanical properties of lipid bilayers. *Biochim. Biophys. Acta. Biomembr.* 1798:1333–1337.

Brochard F, Gennes PGD, Pfeuty P (1976) Surface tension and deformations of membrane structures: relation to two-dimensional phase transitions. *J. Phys. (France)* 37:1099–1104.

Brochard F, Lennon JF (1975) Frequency spectrum of the flicker phenomenon in erythrocytes. *J. Phys. (France)* 36:1035–1047.

Browicz T (1890) Further observation of motion phenomena on red blood cells in pathological states. *Zbl. med Wissen.* 28:625–627.

Cabot RC (1901) A guide to the clinical examination of the blood. Longmans, Green and Co., 4 edition.

Canham PB (1970) The minimum energy of bending as a possible explanation of the biconcave shape of the human red blood cell. *J. Theor. Biol.* 26:61–76.

Dimova R (2014) Recent developments in the field of bending rigidity measurements on membranes. *Adv. Colloid Interface Sci.* 208:225–234.

Döbereiner HG, Selchow O, Lipowsky R (1999) Spontaneous curvature of fluid vesicles induced by trans-bilayer sugar asymmetry. *Eur. Biophys. J.* 28:174–8.

Drabik D, Przybyło M, Chodaczek G, Iglič A, Langner M (2016) The modified fluorescence based vesicle fluctuation spectroscopy technique for determination of lipid bilayer bending properties. *Biochim. Biophys. Acta* 1858:244–252.

Evans E, Kwok R (1982) Mechanical calorimetry of large dimyristoylphosphatidylcholine vesicles in the phase transition region. *Biochemistry* 21:4874–4879.

Evans E, Rawicz W (1990) Entropy-driven tension and bending elasticity in condensed-fluid membranes. *Phys. Rev. Lett.* 64:2094–2097.

Evans E, Yeung A (1994) Hidden dynamics in rapid changes of bilayer shape. *Chem. Phys. Lipids* 73:39–56.

Faucon JF, Mitov MD, Méléard P, Bivas I, Bothorel P (1989) Bending elasticity and thermal fluctuations of lipid membranes. theoretical and experimental requirements. *J. Phys. (France)* 50:2389–2414.

Fernandez-Puente L, Bivas I, Mitov MD, Méléard P (1994) Temperature and chain length effects on bending elasticity of phosphatidylcholine bilayers. *Europhys. Lett.* 28:181–186.

Fricke K, Wirthensohn K, Laxhuber R, Sackmann E (1986) Flicker spectroscopy of erythrocytes. *Eur. Biophys. J.* 14:67–81.

Genova J, Zheliaskova A, Mitov MD (2006) The influence of sucrose on the elasticity of sopc lipid membrane studied by the analysis of thermally induced shape fluctuations. *Colloids Surf. A Physicochem. Eng. Asp.* 282:420–422.

Genova J, Zheliaskova A, Mitov MD (2007) Monosaccharides (fructose, glucose) and disaccharides (sucrose, trehalose) influence the elasticity of sopc membranes. *Optoelectron. Adv. Mat.* 9:427–430.

Genova J, Zheliaskova A, Mitov MD (2010) Does maltose influence on the elasticity of sopc membrane? *J. Phys. Conf. Ser.* 253:012063–1–012063–6.

Gov NS, Safran S (2005) Red blood cell membrane fluctuations and shape controlled by ATP-induced cytoskeletal defects. *Biophys. J.* 88:1859–1874.

Gracià RS, Bezlyepkina N, Knorr RL, Lipowsky R, Dimova R (2010) Effect of cholesterol on the rigidity of saturated and unsaturated membranes: fluctuation and electrodeformation analysis of giant vesicles. *Soft Matter* 6:1472–1482.

Hansen PL, Miao L, Ipsen JH (1998) Fluid lipid bilayers: intermonolayer coupling and its thermodynamic manifestations. *Phys. Rev. E* 58:2311–2324.

Helfrich W (1973) Elastic properties of lipid bilayers: Theory and possible experiments. *Z. Naturforsch. C* 28:693–703.

Henriksen JR, Andresen TL, Feldborg LN, Duelund L, Ipsen JH (2010) Understanding detergent effects on lipid membranes: A model study of lysolipids. *Biophys. J.* 98:2199–2205.

Henriksen JR, Ipsen JH (2002) Thermal undulations of quasi-spherical vesicles stabilized by gravity. *Eur. Phys. J. E.* 9:365–374.

Henriksen J, Ipsen J (2004) Measurement of membrane elasticity by micro-pipette aspiration. *Eur. Phys. J. E.* 14:149–167.

Henriksen J, Rowat AC, Brief E, Hsueh YW, Thewalt JL, Zuckermann MJ, Ipsen JH (2006) Universal behavior of membranes with sterols. *Biophys. J.* 90:1639–1649.

Hønger T, Mortensen K, Ipsen JH, Lemmich J, Bauer R, Mouritsen OG (1994) Anomalous swelling of multilamellar lipid bilayers in the transition region by renormalization of curvature elasticity. *Phys. Rev. Lett.* 72:3911–4.

Ipsen JH, Jørgensen K, Mouritsen OG (1990) Density fluctuations in saturated phospholipid bilayers increase as the acyl-chain length decreases. *Biophys. J.* 58:1099–1107.

Karel Z (1994) Contrast limited adaptive histogram equalization In Heckbert PS, editor, *Graphics Gems IV*, pp. 474–485. Academic Press Professional, Inc., San Diego, USA.

Leibler S (1986) Curvature instability in membranes. *J. de Physique* 47:507–516.

Lipowsky R (1995) Generic interactions of flexible membranes In *Handbook of Biological Physics*, pp. 521–602. Elsevier BV, Amsterdam, the Netherlands.

Lomholt MA (2006) Fluctuation spectrum of quasispherical membranes with force-dipole activity. *Phys. Rev. E* 73:061914–1–061914–9.

Loubet B, Seifert U, Lomholt MA (2012) Effective tension and fluctuations in active membranes. *Phys. Rev. E* 85:031913–1–031913–8.

Marsh D (2013) Handbook of lipid bilayers In *Handbook of Biological Physics*. CRC Press, Boca Raton, FL.

McDonald GR, Hudson AL, Dunn SMJ, You HT, Baker GB, et. al RMW (2008) Bioactive contaminants leach from disposable laboratory plasticware. *Science* 322:917.

Méléard P, Gerbeaud C, Bardusco P, Jeandaine N, Mitov MD, Fernandez-Puente L (1998) Mechanical properties of model membranes studied from shape transformations of giant vesicles. *Biochimie* 80:401–413.

Méléard P, Pott T, Bouvrais H, Ipsen JH (2011) Advantages of statistical analysis of giant vesicle flickering for bending elasticity measurements. *Eur. Phys. J. E.* 34:1–14.

Merkel R, Sackmann E, Evans E (1989) Molecular friction and epitactic coupling between monolayers in supported bilayers. *J. de Physique* 50:1535–1555.

Miao L, Lomholt MA, Kleis J (2002) Dynamics of shape fluctuations of quasi-spherical vesicles revisited. *Eur. Phys. J. E.* 9:143–160.

Milner ST, Safran SA (1987) Dynamical fluctuations of droplet microemulsions and vesicles. *Phys. Rev. A* 36:4371–4379.

Nagle JF (2013) Introductory lecture: Basic quantities in model biomembranes. *Faraday Discuss.* 161:11–29.

Nagle JF, Jablin MS, Tristram-Nagle S (2016) Sugar does not affect the bending and tilt moduli of simple lipid bilayers. *Chem. Phys. Lipids* 196:76–80.

Nagle JF, Jablin MS, Tristram-Nagle S, Akabori K (2015) What are the true values of the bending modulus of simple lipid bilayers? *Chem. Phys. Lipids* 185:3–10.

Niggemann G, Kummrow M, Helfrich W (1995) The bending rigidity of phosphatidylcholine bilayers—dependences on experimental-method, sample cell sealing and temperature. *J. Phys. II* 5:413–425.

Pakkanen KI, Duelund L, Qvortrup K, Pedersen JS, Ipsen JH (2011) Mechanics and dynamics of triglyceride-phospholipid model membranes: Implications for cellular properties and function. *Biochim. Biophys. Acta. Biomembr.* 1808:1947–1956.

Pan J, Tristram-Nagle S, Kučerka N, Nagle JF (2008) Temperature dependence of structure, bending rigidity, and bilayer interactions of dioleoylphosphatidylcholine bilayers. *Biophys. J.* 94:117–124.

Pécréaux J, Döbereiner HG, Prost J, Joanny JF, Bassereau P (2004) Refined contour analysis of giant unilamellar vesicles. *Eur. Phys. J. E.* 13:277–290.

Peliti L (1996) Fluctuating geometries in statistical mechanics and field theory, session LXII Vol. 62 of *Les Houches Summer School Proceedings*.

Peterson MA (1985) Shape fluctuations of red blood cells. *Mol. Cryst. Liq. Cryst.* 127:159–186.

Pott T, Bouvrais H, Méléard P (2008) Giant unilamellar vesicle formation under physiologically relevant conditions. *Chem. Phys. Lipids.* 154:115–9.

Pulvertaft RJV (1949) Vibratory movement in the cytoplasm of erythrocytes. *J. Clin. Path.* 2:281–283.

Ramakrishnan N, Kumar PBS, Ipsen JH (2013) Membrane-mediated aggregation of curvature-inducing nematogens and membrane tubulation. *Biophys. J.* 104:1018–1028.

Ramaswamy S, Toner J, Prost J (2000) Nonequilibrium fluctuations, traveling waves, and instabilities in active membranes. *Phys. Rev. Lett.* 84:3494–7.

Rowat AC, Hansen PL, Ipsen JH (2004) Experimental evidence of the electrostatic contribution to membrane bending rigidity. *Europhys. Lett.* 67:144–9.

Rowat AC, Keller D, Ipsen JH (2005) Effects of farnesol on the physical properties of DMPC membranes. *Biochim. Biophys. Acta.* 1713:29–39.

Schneider MB, Jenkins JT, Web WW (1984) Thermal fluctuations of large quasi spherical bimolecular phospholipid vesicles. *J. Phys.* 45:1457–1472.

Schulman JH, Montagne JB (1961) Formation of microemulsions by amino alkyl alcohols. *Ann. N.Y. Acad. Sci.* 92:366–371.

Seifert U, Langer SA (1993) Viscous modes of fluid bilayer membranes. *Europhys. Lett.* 23:71–76.

Semrau S, Idema T, Holtzer L, Schmidt T, Storm C (2008) Accurate determination of elastic parameters for multicomponent membranes. *Phys. Rev. Lett.* 100:088101–1–088101–4.

Servus RM, Harbich W, Helfrich W (1976) Measurement of the curvature-elastic modulus of egg lecithin bilayers. *Biochim. Biophys. Acta.* 436:900–903.

Shchelokovskyy P, Tristram-Nagle S, Dimova R (2011) Effect of the HIV-1 fusion peptide on the mechanical properties and leaflet coupling of lipid bilayers. *New J. Phys.* 13:025004–1–025004–16.

Sjöstrand FS, Andersson-Cedergren E, Dewey MM (1958) The ultrastructure of the intercalated discs of frog, mouse and guinea pig cardiac muscle. *J. Ultrastruct. Res.* 1:271–287.

Struppe J, Noack F, Klose G (1997) NMR study of collective motions and bending rigidity in multilamellar system of lipid and surfactant bilayers. *Z. Naturforsch. C* 52:681–694.

Swammerdam J (2016) In Porter R, editor, *The Hutchinson Dictionary of Scientific Biography (Helicon science).* Hodder Arnold H&S.

Vitkova V, Genova J, Mitov MD, Bivas I (2006) Sugars in the aqueous phase change the mechanical properties of lipid mono-and bilayers. *Mol. Cryst. Liq. Cryst.* 449:95–106.

Vitkova V, Petrov AG (2013) Lipid bilayers and membranes: material properties. *Adv. Planar Lipid Bilayers Liposomes* 17:89–138.

Yoon Y, Hong H, Brown A, Kim DC, Kang DJ, Lew VL, Cicuta P (2009) Flickering analysis of erythrocyte mechanical properties: Dependence on oxygenation level, cell shape, and hydration level. *Biophys. J.* 97:1606–1615.

Zilker A, Engelhardt H, Sackmann E (1987) Dynamic reflection interference contrast (RIC) microscopy: A new method to study surface excitations of cells and to measure membrane bending elastic moduli. *J. Phys. (France)* 48:2139–2151.

Using electric fields to assess membrane material properties in giant unilamellar vesicles

Rumiana Dimova and Karin A. Riske

Let's electrify the membrane.

Contents

15.1 INTRODUCTORY WORDS

"Cells can be funny. Try to grow them with a slightly wrong recipe, and they turn over and die. But hit them with an electric field strong enough to knock over a horse, and they do things to justify international meetings, fill a sizeable book, and lead one to speak of an entirely new technology for cell manipulation."[1] Indeed, for a long time, cells have been the subject of different manipulation protocols involving electric fields as exemplified by cell alignment, deformation, permeabilization and hybridization. Efforts to understand these effects have long been developed on model systems. In this chapter, we will see how a seminal understanding of the effects of electrical fields on lipid membranes was obtained by electrifying giant unilamellar vesicles (GUVs).

The phenomenon of electroporation or electropermeabilization is used for introducing into the cell various molecules to which the membrane is otherwise impermeable (Tekle et al., 1994; Teissie et al., 2005). Because of its efficiency, this method, in combination with chemotherapy (electrochemotherapy), has become an established approach for treatment of carcinoma, melanoma and connective tissue cancer (Heller et al., 1999; Gothelf et al., 2003; Nuccitelli et al., 2006; Calvet et al., 2014) and holds great promise for gene therapy (Golzio et al., 2002). Electrofusion is of particular interest because of its wide use in cell biology and biotechnology as a means for cell hybridization (Zimmermann, 1986). Modulation of the transmembrane potential is associated with membrane deformation (flexoelectricity) and is involved in a number of membrane processes (Petrov, 2002, 2006), as, for example, in electromotility of the outer hair cells (Raphael et al., 2000; Spector et al., 2006). Finally, electric fields are also extensively employed for microbial inactivation in food processing (Jeyamkondan et al., 1999;

[1] Adrian Parsegian, from the foreword to "Electroporation and electrofusion in cell biology," eds. E. Neumann, A. E. Sowers & C. A. Jordan, Plenum, New York, 1989.

Toepfl et al., 2006) and water cleaning (Vernhes et al., 2002). (For reviews on the effects of electric fields on cells, consult e.g., Robinson, 1985; Zimmermann, 1986; Zimmermann and Neil, 1996; Weaver, 2003; Teissie et al., 2005.)

Most of the above applications rely on the effects of electric fields on the lipid bilayer, which acts as an insulator and can withstand charge separation to a certain extent. The first studies using model bilayer systems were mainly performed on black lipid membranes, a system in which the applied voltage and the composition of the aqueous phase of both sides of the membrane can be easily controlled and the membrane conductance directly measured (e.g., Abidor et al., 1979). Studies on lipid vesicles followed and provided further knowledge on these electrical phenomena (Teissie and Tsong, 1981; Neumann et al., 1998; Kakorin et al., 2003). The clear advantages of using GUVs to study the effects of electric fields on lipid bilayers were quite soon acknowledged and extensively explored in the last decades to, among others, characterize membrane material properties (e.g., Needham and Hochmuth, 1989; Zhelev and Needham, 1993; Tekle et al., 2001; Riske and Dimova, 2005, 2006; Dimova et al., 2007; Dimova et al., 2009; Portet et al., 2009; Dimova, 2010; Dimova, 2011; Portet et al., 2012), investigate electroporation as means to transport (macro)molecules through the impermeable membrane (Golzio et al., 2010; Breton et al., 2012), and understand and exploit membrane electrofusion (Haluska et al., 2006; Riske et al., 2006; Yang et al., 2009; Shirakashi et al., 2012; Robinson et al., 2014; Saito et al., 2014).

Subjecting GUVs to electric fields offers a way to study the associated effects in a systematic and controlled way, and, as will be discussed in this chapter, to develop methods for assessing the membrane material properties and to manipulate GUVs without direct mechanical contact. In the last decade, a number of experimental tools were developed in our laboratory to pull, squeeze, and even tear apart vesicles for the sake of learning something about the membrane. Squeezing GUVs with alternating current (AC) fields allows assessing the bending rigidity (Section 15.3.1), the membrane capacitance (Section 15.3.2) and the vesicle excess area (Section 15.4.1). By poking and making holes in the membrane using direct current (DC) pulses, one can measure the pore edge tension (Section 15.3.3) and assess membrane rheological properties (Section 15.3.4). Occasionally, vesicles can be even "electrocuted" (vesicle bursting) to find out how stable their membrane is (Section 15.3.3). The combination of AC fields and DC pulses allow also investigating the fusion of two vesicles, which turns out to be a handy way of transforming GUVs into microreactors (Section 15.4.4) or creating multicomponent membranes with precisely known composition on which tie lines can be deduced (Section 15.4.3). All in all, electrifying the membrane is very useful and can be rewarding.

This chapter is organized as follows. We will first introduce some important equations describing the behavior of membranes and vesicles in electric fields and will then show their use to extract membrane mechanical, electrical and rheological properties. We will also demonstrate ways to manipulate GUVs via electric fields and discuss some particular applications. The purpose of the chapter is to share the methods that were developed, make available details on experimental protocols and offer tips for executing the proposed approaches successfully.

Because constructing chambers for electromanipulation (Section 15.5) is easy and the required equipment (AC field and/or DC pulse generators) is not demanding and already partially available in labs employing the electroformation method for GUV preparation, it is obvious that the application of the methods described in this chapter can be widely employed.

15.2 SOME EQUATIONS

The response of membranes to electric fields involves dynamic physical processes occurring on different time scales. The lipid bilayer is impermeable to ions and in the presence of an electric field free charges accumulate on both membrane surfaces. Hence, the vesicle membrane acts as a capacitor, which charges on a time scale (Schwan, 1985; Grosse and Schwan, 1992):

$$t_c = R_{ve} C_{me} \left(\frac{1}{\sigma_{in}} + \frac{1}{2\sigma_{ex}} \right) \tag{15.1}$$

where R_{ve} is the vesicle radius; C_{me} is the membrane capacitance; and σ_{in} and σ_{ex} are the conductivities of the solutions inside and outside the vesicle, respectively. We can estimate the capacitor charging time, $t_c \cong 10$ μs, for conditions corresponding to experiments on giant vesicles in 1 mM NaCl, namely, $\sigma_{in} \cong \sigma_{ex} \cong$ 10 mS/m, $C_{me} \cong 0.01$ F/m^2, and $R_{ve} \cong 10$ μm.

When exposed to relatively moderate AC fields, (deflated) quasi-spherical vesicles adopt elliptical shapes—prolate or oblate (with axis of symmetry along the field direction), depending on the field frequency and solution conductivities (see Box 15.1) (Dimova et al., 2007; Aranda et al., 2008); for a detailed morphological diagram of the vesicle shapes (see Dimova et al. 2009). Relatively high AC fields (>3 kV/m)[2] can lead to electroporation of giant vesicles (Harbich and Helfrich, 1979). In the regime of low field strengths, the tension is weak, resulting in pulling out the membrane undulations. The dependence of the tension on the field strength in this regime can be exploited to deduce the membrane bending rigidity by measuring the vesicle deformation (Kummrow and Helfrich, 1991; Niggemann et al., 1995; Gracià et al., 2010) as introduced in Section 15.3.1. The membrane tension of a vesicle exposed to an electric field can be obtained from a balance between the electric stresses and the Laplace pressure at the poles and the equator as proposed earlier (Kummrow and Helfrich, 1991; Niggemann et al., 1995) and elaborated in more detail by Vlahovska et al. (2009) and Yamamoto et al. (2010), see also Chapter 7. Roughly, the membrane tension, Σ, can be expressed as a function of the principal curvatures, C_1 and C_2, taken either at the equator (*equ*) or at the pole (*pol*) of the vesicles (Gracià et al., 2010):

$$\Sigma = \frac{g \varepsilon_w E^2}{\left(C_1 + C_2\right)_{equ} - \left(C_1 + C_2\right)_{pol}} \tag{15.2}$$

where ε_w is the dielectric constant of water; E is the field strength far from the vesicle; and g is a dimensionless parameter, which is

[2] In this chapter, the field strength is given either in kilovolts per meter or in volts per centimeter (1 kV/m = 10 V/cm) for the sake of easy comparison with the original publications.

GUV-based techniques and what one can learn from them

a function of the field frequency, the electrical properties of the membrane and those of the solutions inside and outside the vesicle (Gracià et al., 2010). The vesicle deformation can be employed to deduce the membrane tension for a range of field frequencies.

The charges accumulating on both sides of the bilayer in the presence of the electric field give rise to a transmembrane potential (Kinosita et al., 1988):

$$\Psi_{me}(t) = 1.5 R_{ve} E |\cos\theta| \left[1 - \exp(-t/t_c)\right] \tag{15.3}$$

where θ is the tilt angle between the electric field and the surface normal (see also Box 15.1); t is time; and t_c is the charging time as defined in Eq. 15.1. Above some electroporation threshold, the transmembrane potential, Ψ_{me}, cannot be further increased and the membrane becomes permeable to ions. The electroporation phenomenon can also be understood in terms of a stress in the bilayer. The transmembrane potential, Ψ_{me}, induces an effective electrical tension, Σ_{el}, as defined by the Maxwell stress tensor (Abidor et al., 1979; Needham and Hochmuth, 1989; Riske and Dimova, 2005):

$$\Sigma_{el} = \varepsilon_{me} \frac{l}{2 l_e^2} \Psi_{me}^2 \tag{15.4}$$

where l is the total bilayer thickness; $l \approx 4$ nm; l_e is the dielectric thickness; $l_e \approx 2.7$ nm for lecithin bilayers (Nagle and Tristram-Nagle, 2000); and ε_{me} is the membrane permittivity, $\varepsilon_{me} \approx 2\varepsilon_0$. For vesicles with some initial tension, Σ_0, the total tension reached during the pulse is a sum of Σ_0 and Σ_{el}. The vesicle ruptures if the total membrane tension exceeds the lysis tension, which is on the order of 5–10 mN/m (Olbrich et al., 2000), see also Table 15.1. This tension increase corresponds to building up a certain critical transmembrane potential, Ψ_c. The critical transmembrane potential for cell membranes and vesicles is $\Psi_c \approx 1$V (e.g., Needham and Hochmuth, 1989; Tsong, 1991; Weaver and Chizmadzhev, 1996) but decreases for taut (non-fluctuating) vesicles that exhibit an appreciable tension (Riske and Dimova, 2005; Portet et al., 2009). The critical transmembrane potential characterizes the membrane stability and can be directly deduced from electroporation of GUVs. Following the approach of Schwan (Neumann et al., 1989), the critical transmembrane potential can be assessed from the maximal pore size, $r_{por,max}$, in the membrane (Portet and Dimova, 2010):

$$\Psi_c = 1.5 E \sqrt{R_{ve}^2 - r_{por,max}^2}\,. \tag{15.5}$$

Fast digital microscopy of GUV electroporation can be used to measure the critical transmembrane potential and, thus, to characterize the membrane stability (Portet and Dimova, 2010).

The lifetime of pores depends on the edge tension, which represents the energy penalty (per unit length of pore circumference) for reorganizing the lipids in order to shield the exposed hydrophobic core of the membrane. The edge tension of lipid membranes is on the order of a few tens of picoNewtons (Harbich and Helfrich, 1979; Portet and Dimova, 2010) and can be measured on GUVs by following the time dependence of pore closure; see Section 15.3.3 for details. The pore dynamics typically consists of four stages: growing, stabilization at some maximal pore radius, slow decrease in pore size and fast closure (e.g., data in Figure 15.3). The third stage of slow pore closure is used to determine the membrane edge tension, applying the dependence derived by Brochard-Wyart et al. (2000):

$$R_{ve}^2 \ln\left(r_{por}\right) = -\frac{2\lambda_{por}}{3\pi\eta} t + const \tag{15.6}$$

where r_{por} is the pore radius (see also Box 15.1); λ_{por} denotes the edge tension; η is the viscosity of the aqueous medium; and the constant, *const*, depends on the maximal pore radius reached. For more detailed analysis of the pore dynamics taking into account both membrane and aqueous viscosities, see Ryham et al. (2011).

Table 15.1 **Typical values (order of magnitude) for the characteristic properties of lipid membranes in fluid and gel phase and of polymersome bilayers as deduced from measurements on GUVs**

MEMBRANE MATERIAL PROPERTY	FLUID-PHASE LIPID MEMBRANES	GEL-PHASE LIPID MEMBRANES	POLYMER MEMBRANES
Bending rigidity ($k_B T$)[a]	20	350	35–400[b]
Stretching elasticity (mN/m)[c]	200	850	120; 470
Shear surface viscosity (N.s/m)[d]	3–7 × 10^{-9}	Diverges	2 × 10^{-6}
Critical poration potential (V)[e]	1	10	4–9
Lysis tension (mN/m)[f]	5–10	>15	20–30
Pore edge tension (pN)[g]	10–50	—	10–50

[a] Data from Seifert and Lipowsky (1995); Discher et al. (1999); Dimova et al. (2000); Rawicz et al. (2000); Dimova et al. (2002); Mecke et al. (2003); Bermudez et al. (2004); Dimova (2014); the value for gel-phase lipid membranes corresponds to the bending rigidity of dimyristoylphosphatidylcholine (DMPC) measured at temperature about 5° below the main phase transition temperature of the lipid.
[b] The values reported for polymersomes depend on the diblock copolymer used (and the respective membrane thickness) and can vary strongly in different references.
[c] Data from Needham and Evans (1988); Discher et al. (1999); Rawicz et al. (2000); Dimova et al. (2002).
[d] Data from Dimova et al. (1999, 2000, 2002).
[e] Data from Aranda-Espinoza et al. (2001); Knorr et al. (2010).
[f] Data from Evans and Needham (1987); Olbrich et al. (2000); Aranda-Espinoza et al. (2001); Dimova et al. (2002).
[g] Data from Bermúdez et al. (2003); Portet and Dimova (2010).

Box 15.1 Examples for various effects of electric fields on GUVs

In quasi-spherical vesicles, AC fields can induce elliptical deformations—prolate or oblate, depending on the field frequency and solution conductivity—and can be used to align vesicles in pearl chains (Section 15.4.2) or lift them a few micrometers above the chamber bottom. Inhomogeneous AC fields can trigger flows in the membrane and around the vesicle (Section 15.4.5). DC pulses applied to GUVs in the absence of salt in their exterior also deform the vesicles into prolates; however, in the presence of salt, spherocylindrical shapes can be observed with an aspect ratio depending on the conductivity conditions; the snapshots of spherocylindrical vesicles in this Box are reproduced from Riske and Dimova (2006), with permission from Elsevier. Above the poration threshold, micrometer-sized pores (macropores) can be observed. When two neighboring vesicles are porated in their contact zone, fusion occurs.

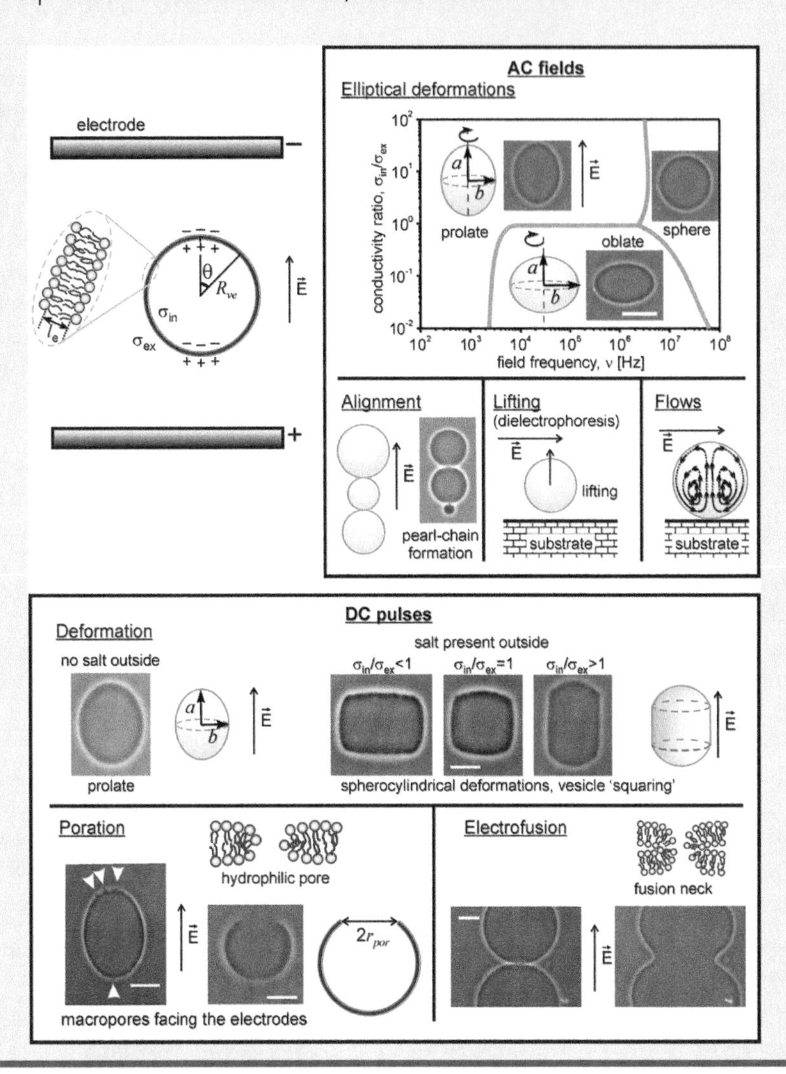

15.3 HOW TO MEASURE MEMBRANE PROPERTIES BY EXPOSING GUVs TO ELECTRIC FIELDS

This section will introduce approaches to deduce the bending rigidity, membrane capacitance and membrane stability from exposing vesicles to electric fields and observing their response and relaxation. The details about experimental chambers and tips for successful measurements are given in Section 15.5.

15.3.1 ASSESSING BENDING RIGIDITY FROM VESICLE ELECTRODEFORMATION

Vesicle electrodeformation as a means to deduce the bending rigidity of membranes has been introduced by Helfrich and coworkers (Kummrow and Helfrich, 1991; Niggemann et al., 1995) and later further developed by Gracià et al. (2010).

In this approach, a selected vesicle is subjected to an AC electric field of frequency in the range of 25–300 kHz and increasing strength; note that at relatively low field frequencies around

1–2 kHz and strong fields, one may observe the formation of large pores of several micrometers (Harbich and Helfrich, 1979), or protrusions of long membrane tubules (Antonova et al., 2016). The shape deformation of the vesicle is recorded. The induced membrane tension can be evaluated using Eq. 15.2. The full theoretical description of the force densities acting on the vesicle membrane has been derived by Yamamoto et al. (2010). The forces depend quadratically on the applied electric field strength, E. All other parameters influencing the force densities (such as permittivities, inner and outer vesicle radii and membrane thickness, conductivities and field frequency) are constant during the experiment. The membrane bending rigidity is deduced by applying very mild tensions to the membrane, for which the logarithmic dependence of the area change as a function of tension holds (see Chapters 5 and 11). Because of this logarithmic dependence, all system parameters contribute only as a constant term to the change in area ΔA_{ve} (Fricke and Dimova, 2016):

$$\alpha \equiv \frac{\Delta A_{ve}}{A_{ve}^{0}} = \frac{k_B T}{8\pi\kappa} \ln\left[\frac{E^2}{\left(C_1+C_2\right)_{equ} - \left(C_1+C_2\right)_{pol}}\right] + const \quad (15.7)$$

where A_{ve}^{0} is the initial surface area of the vesicle; k_B the Boltzmann constant; T the temperature; and the curvatures at the equator and pole are $C_{1,equ} = 1/b$, $C_{2,equ} = b/a^2$, $C_{1,pol} \equiv C_{2,pol} = a/b^2$, with a and b be being the short and long semi-axes of a prolate vesicle. Thus, simply plotting the logarithmic term as a function of the relative area change yields the bending rigidity from the slope of the data. In this way, the cost of extensive calculations is strongly simplified and only the applied electrical voltage and the two semi-axes of the deformed vesicles must be obtained experimentally (for the expressions for the area change of a prolate ellipsoidal vesicle see Eq. 15.10). Figure 15.1 shows an example for data collected on a vesicle made of palmitoyloleoylphosphatidylcholine (POPC), see Appendix 1 of the book for structure and data on this lipid. At weak electric fields, the vesicle deformation—and thus the area increase and membrane principle curvatures—is poorly resolved because of optical resolution, resulting in an apparent threshold electric field, which induces the deformation.

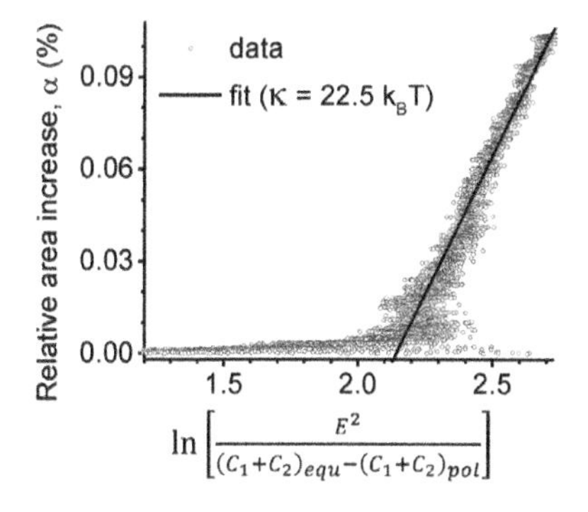

Figure 15.1 Measuring the bending rigidity from vesicle electrodeformation. Change in the area of a POPC vesicle at 40°C as a function of applied electric field strength (the field strength is measured in V/m units and the curvatures in 1/m). From the slope of the data (solid line), one obtains the bending rigidity following Eq. 15.7. (From Fricke, N., and Dimova, R. *Biophys. J.*, 111, 1935–1945, 2016.)

15.3.2 MEASURING MEMBRANE CAPACITANCE FROM ELECTRODEFORMATION

If the solution inside a vesicle is less conducting than the suspending medium, the vesicle changes its shape from a prolate to oblate ellipsoid with increasing frequency, that is, at low frequencies the vesicle elongates in the field direction, whereas at higher frequencies the vesicle long semi axis is perpendicular to the applied field (see Box 15.1). The transition frequency, ν_c, depends on the membrane charging time (Vlahovska et al., 2009; Peterlin, 2010; Yamamoto et al., 2010) and has the following form:

$$\nu_c = \frac{\sigma_{in}}{2\pi R_{ve} C_{me}}\left[\left(1-\frac{\sigma_{in}}{\sigma_{ex}}\right)\left(\frac{\sigma_{in}}{\sigma_{ex}}+3\right)\right]^{-1/2} \quad (15.8)$$

Thus, knowing the conductivities of the solutions inside and outside the vesicles and after measuring the vesicle size, one can infer the membrane capacitance from the dependence of the prolate–oblate transition frequency, ν_c, on the inverse vesicle diameter (Salipante et al., 2012). The measurement itself requires a frequency sweep, which can be quickly and easily done under phase-contrast observation with either home-made or commercially available chambers, and typical function generators for electroswelling of GUVs. The method is noninvasive because it does not require any direct contact with the membrane. It can be applied to lipid vesicles and polymersomes (see Chapter 26). Figure 15.2a shows exemplary data for such a frequency sweep on a polymersome used to measure the critical frequency, ν_c, and Figure 15.2b summarizes the dependence of the prolate–oblate transition frequencies on vesicle size: the slopes of the linear fits (with zero intercept) yield the membrane capacitance following Eq. 15.8.

15.3.3 PORE EDGE TENSION AND VESICLE STABILITY

Upon poration, the lipid molecules in the bilayer reorient so that their polar heads can line the pore walls and form a hydrophilic pore (Litster, 1975), see Box 15.1. The energy penalty per unit length for this reorganization is described by the edge tension, which emerges from the amphiphilic nature of the lipids. It also gives rise to a force driving the closure of transient pores. A few

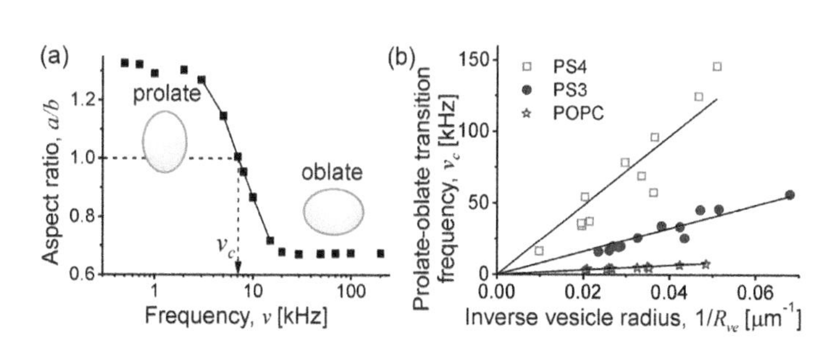

Figure 15.2 Capacitance measurements. (a) Frequency-dependent deformation of a polymersome used to determine the critical frequency, ν_c, of prolate-oblate transition. (b) Transition frequencies measured on vesicles with different sizes and compositions: Polymersomes made of two different diblock copolymers (PS4 and PS3) and GUVs made of POPC. (Adapted from Salipante, P.F. et al., *Soft Matter*, 8, 3810–3816, 2012. Reproduced by permission from the Royal Society of Chemistry.)

experimental methods have been developed to directly assess this physical quantity in cells (Chang and Reese, 1990; Chang, 1992) and in GUVs (Harbich and Helfrich, 1979; Zhelev and Needham, 1993; Karatekin et al., 2003; Puech et al., 2003). Here, we present an approach based on the electroporation of giant vesicles and on the observation of the pore closure with a fast digital imaging (Portet and Dimova, 2010). The analysis of the process of pore dynamics was further developed by Ryham et al. (2011), showing that the solution viscosity slows down the pore dynamics. The process can be followed under phase-contrast microscopy with a high-speed digital camera (the acquisition speed was typically above 1,000 fps). The electroporation can be induced by applying electric pulses of 5 ms duration and field strength in the range of 20–80 kV/m. In phosphatidylocholine (PC) vesicles, these field parameters induce micrometer-sized pores (macropores).

After recording the event, the time dependence of the pore radius is extracted from image analysis (either manual or software-automated). The dynamics of pore closure is then examined in view of Eq. 15.6. In practice, one only has to consider the linear part of $R_{ve}^2 \ln(r_{por})$ as a function of time in the period corresponding to the slow closure stage. Linear fit of this part is characterized by a slope, β, and the edge tension, λ_{por}, is estimated from the relation $\lambda_{por} = -(3/2)\pi\eta\beta$. Figure 15.3a illustrates the analysis performed on an L-α-phosphatidylcholine (egg PC) vesicle. Measurements on many vesicles yielded the average value of λ_{por} = 14.3 pN for the edge tension of such membranes. The example shown in Figure 15.3a is an ideal case. A significant amount of data may need to be discarded in the following situations: (i) During deformation and relaxation, deflated vesicles may exhibit strong out-of-sphere deformations. (ii) Pore closure might be physically obstructed, which is evidenced by additional steps in the pore-closing curve. (iii) Vesicle multilamellarity may also affect this dependence. (iv) Other vesicles located close to the examined one may affect the relaxation. (vi) Application of subsequent pulses on the same vesicle can drastically affect the relaxation because of defective membrane resealing (Portet et al., 2009; Mauroy et al., 2012). Additional tips are included in Section 15.5.

An alternative (and very rough) approach for deducing the edge tension in membranes was reported by Riske and Dimova (2005). Instead of following in detail the relaxation dynamics of the closing macropores, one can simply measure their lifetime, t_{por}, and maximum size, $r_{por,max}$. The pore edge tension can be then deduced from a linear fit with a slope proportional to η_s / λ_{por} (Sandre et al., 1999; Riske and Dimova, 2005), where η_s is the membrane shear surface viscosity. An example for data analyzed in this way is given in Figure 15.3b.

We emphasize that the dynamics discussed above applies only to fluid vesicles. A few features regarding the response of gel-phase membranes are discussed in Section 15.3.5 and an example of a porated gel-phase vesicle is shown in Figure 15.3c.

In fluid vesicles, occasionally DC pulses can induce macropores that remain stably open or even never reseal followed by vesicle destabilization and collapse. Stable and long-lived pores can be observed in the presence of cone-shaped molecules (such as Tween 20; see Appendix 3 of the book for structure and data on this molecule) (Karatekin et al., 2003; Rodriguez et al., 2006), or in the presence of hydrogel polymers (agarose) physically hindering pore closure (Lira et al., 2014). Similarly, when DC pulses are applied to GUVs containing a large fraction of negatively charged lipids, such as palmitoyl oleoyl phosphatidylglycerol (POPG) or palmitoyl oleoyl phosphatidylserine (POPS), vesicle burst/collapse is often observed (Riske et al., 2009a); see Appendix 1 of the book for structure and data on the lipids. After the end of the pulse, macropores can be detected, but instead of resealing as in the

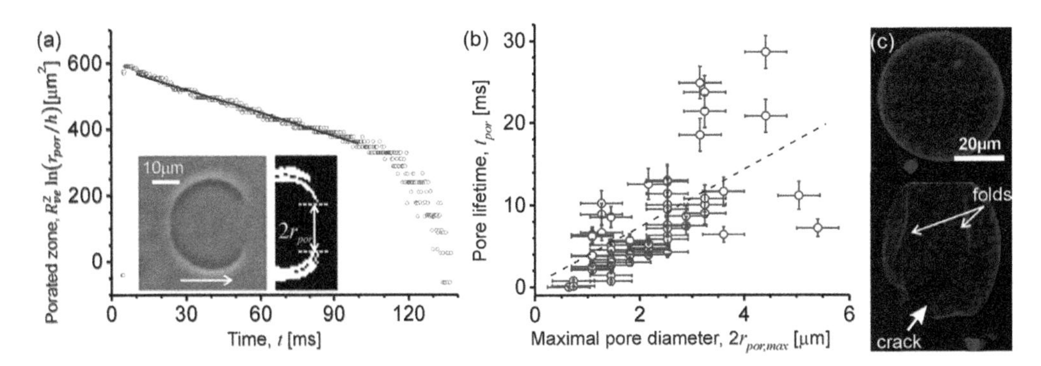

Figure 15.3 Pores in fluid and gel vesicles. (a) Evolution of the porated region in an egg PC vesicle as characterized by $R_{ve}^2 \ln(r_{por}/h)$ as a function of time, t; see Eq. 15.6 (note that to avoid plotting a dimensional value in the logarithmic term, we have introduced h = 1 μm). The open circles are experimental data and the solid line is a linear fit, whose slope yields the edge tension λ_{por}. The inset shows a raw image (left) of a porated vesicle 50 ms after being exposed to an electric pulse with duration of 5 ms and amplitude of 50 kV/m; see also Movie 15.1 (movies captions at the end of chapter). The field direction is indicated with an arrow. To the right is an enhanced and processed image of the vesicle half facing the cathode. The inner white contour corresponds to the location of the membrane. The pore diameter, $2r_{por}$, is schematically indicated. Adapted from Portet and Dimova (2010) with permission from Elsevier. (b) Lifetime of macropores, t_{por}, as a function of their maximal size, $2r_{por,max}$, measured on 15 different vesicles made of egg PC, data adapted from Riske and Dimova (2005). The dashed line has a slope η_s / λ_{por} (see text for details). (c) Fluorescently labeled dipalmitoylphosphatidylcholine (DPPC) vesicle at room temperature (in the gel phase) observed with confocal microscopy. Before the pulse (upper image), the vesicle has a spherical shape. After applying a pulse with field strength of 6 kV/cm and duration of 300 μs, the vesicle cracks open and the membrane folds as indicated by the arrows. (Adapted from Knorr, R.L. et al., *Soft Matter*, 6, 1990–1996, 2010. Reproduced by permission of the Royal Society of Chemistry.)

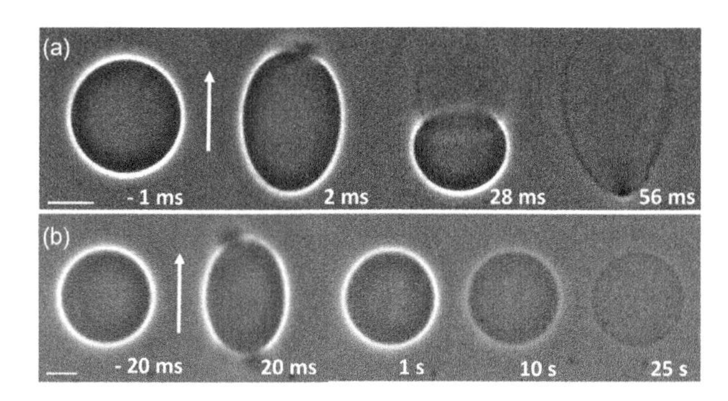

Figure 15.4 Destabilization of vesicles containing charged lipids. Image sequences of two GUVs composed of POPC:POPG 1:1 after a DC pulse (3 kV/cm, 150 µs; the field direction is indicated by the white arrows). The time stamps are relative to the beginning of the pulse. (a) Vesicle burst. The pore opens continuously and the vesicle restructures into a tubular network; see also Movie 15.2 (movies captions at the end of chapter). (b) Contrast loss after poration. The macropore reseals within tens of milliseconds, but long-lived submicroscopic pores persist and the vesicle loses the original sugar asymmetry seconds after the pulse. Scale bars: 20 µm.

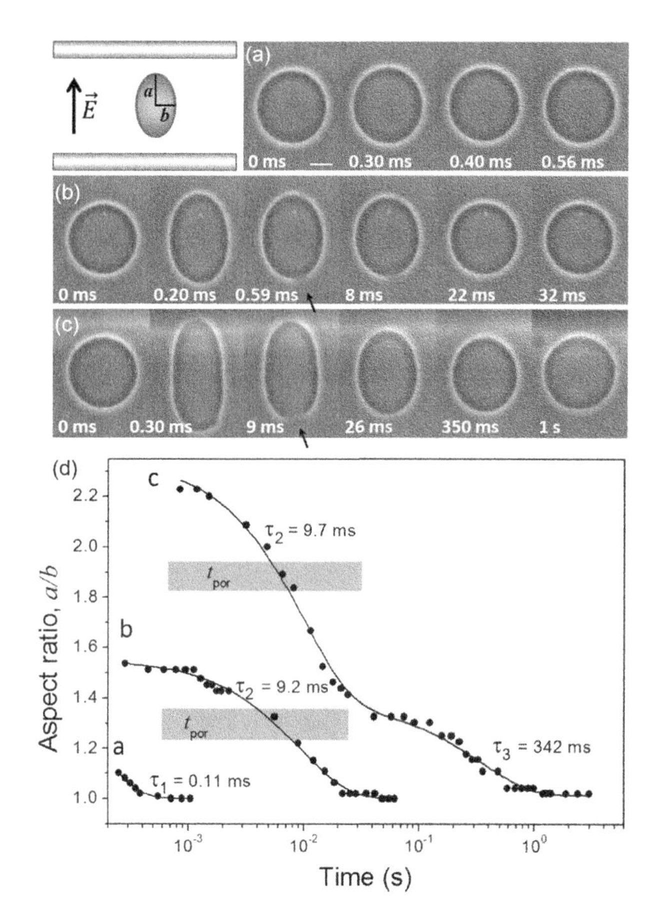

Figure 15.5 Vesicle relaxation dynamics after application of DC pulses. The cartoon illustrates the chamber with two electrodes, the direction of the electric field and a prolate GUV with its main axes *a* and *b*. The sequences show images of the same GUV experiencing three pulses applied with 2–5 min intervals between them: (a) 1 kV/cm, 250 µs; (b) 3 kV/cm, 100 µs; (c) 3 kV/cm, 200 µs. The time stamps show the time after the beginning of the pulse, the black arrows point to macropores. Scale bar: 10 µm. The graph in (d) shows the aspect ratio *a* / *b* of the vesicle after the maximum deformation is achieved. The points represent measurements and the curves are exponential fits with characteristic times indicated. The shaded areas in pulses (b) and (c) indicate the period within which macropores are optically detected. (Adapted and reprinted from *Biophys. J.*, Riske, K.A., and Dimova, R., Electro-deformation and poration of giant vesicles viewed with high temporal resolution, 1143–1155, copyright 2005, with permission from Elsevier.)

case of pure PC membranes (Figure 15.3a), the pores often continue to open and the whole vesicle collapses into a network of lipid tubes, as shown in Figure 15.4a; see also Movie 15.2 (movies captions at the end of chapter). In cases in which the macropores do reseal, many GUVs are found to lose their optical contrast (sugar asymmetry) within several seconds after the pulse, revealing that the membrane integrity is not fully restored and that submicrometer-sized pores remain open (Figure 15.4b). The frequency for the occurrence of vesicle burst and of long-lasting permeable states increases with the fraction of anionic lipids in the membrane. Therefore, the presence of negatively charged lipids has an impact on membrane stability, most probably by reducing the edge tension significantly (Lira et al., manuscript in preparation). In general cases, the occurrence of vesicle burst and imperfect pore resealing induced by DC pulses indicates a low stability of the membrane.

15.3.4 MEMBRANE RHEOLOGICAL AND MECHANICAL PROPERTIES DEDUCED FROM VESICLE RELAXATION DYNAMICS

As discussed in the previous sections and summarized in Box 15.1, electric DC pulses induce vesicle deformation into prolate or spherocylindrical shapes, depending on experimental conditions (Riske and Dimova, 2005, 2006). After the end of the pulse, the vesicle relaxes, usually into a spherical or quasi-spherical shape. The characteristic relaxation times depend on the bilayer properties and on the involved processes. Thus, the relaxation dynamics can be used to obtain information about membrane characteristics. Here, we will focus on conditions in which the DC pulse induces prolate deformation (the conductivity inside is higher than the one outside the vesicle and no salt is present in the external medium). The vesicle deformation is represented by the aspect ratio between the two semi-axes, a/b, of the prolate vesicle (see Box 15.1).

The relaxation dynamics of lipid GUVs after the application of DC pulses (1–3 kV/cm magnitude, 50–300 µs duration) can be studied using a fast camera acquisition (the system used by Riske and Dimova [2005] allowed achieving a time resolution of 30 µs). Three different relaxation times, τ_1, τ_2 and τ_3, can be detected in the vesicle relaxation depending on the specific conditions as follows (the examples given below refer to GUVs made of egg PC). For relatively weak DC pulses, which do not induce poration (see Eq. 15.5 for the poration threshold), only a mild deformation ($a/b < 1.1$) is induced, and the vesicle quickly relaxes back to the initial shape after the end of the pulse with a characteristic time, $\tau_1 \sim 0.1$ ms (Figure 15.5a). Such fast relaxation can only be resolved when using fast digital imaging. When DC pulses, strong enough to reach the poration threshold of the membrane, are applied to spherical vesicles, the deformation induced is stronger (a/b up to ~1.6) and macropores with diameters between 1 and 5 µm may open. The pores can be clearly visualized with phase-contrast microscopy when glucose/sucrose symmetry is

Box 15.2 Membrane relaxation times after GUV deformation induced by DC pulses
GUVs exposed to DC pulses relax back to their initial (quasi-)spherical shape with three distinct characteristic times. Using dimensional analysis, these relaxation times can be associated with different membrane properties and occurring processes:

Stretching relaxation, τ_1 ~ 0.1 ms. Subporation pulses are able to stretch the membrane due to the Maxwell stress imposed on the bilayer and the corresponding tension builds up (see Eq. 15.4) but the poration threshold is not reached. After the end of the pulse, the vesicle relaxes back by releasing the acquired membrane tension through shear in the membrane. Thus, τ_1 is related to (i) the membrane tension, Σ_{me}, acquired at the end of the pulse (typical values are $\Sigma_{me} \approx 5$ mN/m, right below the lysis tension; see Table 15.1); (ii) the relative area increase, α, of the stretched vesicle, which for the deformed vesicles is on the order of 1%; and (iii) the shear surface viscosity of the bilayer, which is on the order of 5×10^{-9} N.s/m (see Table 15.1 and Dimova et al. [2000]), such that

$$\tau_1 \sim \frac{\eta_s}{\alpha \Sigma_{me}} \approx 0.1 \text{ ms}$$

Pore-resealing relaxation, τ_2 ~10 ms. The time to reseal macropores, t_{por}, was measured for several egg PC GUVs (see shaded region in Figure 15.5) and was found to be $t_{por} \approx 4$–20 ms, depending on pore size. Thus, the second characteristic time, τ_2, is related to pore closing time, which is set by the pore edge tension, $\lambda_{por} \approx 10$ pN (see Section 15.3.3 and Table 15.1). Then, for typical pore radius, $r_{por} \approx 2$ µm:

$$\tau_2 \sim \frac{\eta_s r_{por}}{\lambda_{por}} \approx 1 \text{ ms}$$

Shape fluctuation relaxation, τ_3 ~ 1 s. This long relaxation time is observed only for vesicles with excess area, and increases from 0.1 s to 3 s with increasing excess area. This slow time reflects the dynamics of volume displacement around the deformed vesicle. It is associated with shape fluctuations and entails contributions from the medium viscosity η (of sucrose/glucose solutions), the membrane bending rigidity, $\kappa \approx 10^{-19}$ J (see Table 15.1), and the vesicle reduced volume ($v \approx 0.95$–0.99 for deflated quasi-spherical vesicles), such that

$$\tau_3 \sim \frac{4\pi\eta R_{ve}^3}{3\kappa}\frac{1}{1-v} \approx 1 \text{ s}$$

employed (see region with broken halo indicated with black arrows in Figure 15.5b). After the pulse, the vesicle deformation relaxation is about two orders of magnitude slower (τ_2 ~10 ms) than when no macropores are present and corresponds to typical times required for pore resealing (t_{por}). Note that submicroscopic pores may also be present but are not detected optically. When the pulses are applied to GUVs with some excess area (seen by their shape fluctuation) or if the vesicle lost significant volume after macroporation, a third and much longer relaxation time can be detected, τ_3 ~ 0.1–1 s, depending on the excess area available (Figure 15.5c). Box 15.2 describes the three characteristic relaxation times in terms of bilayer properties and processes involved.

Several theoretical studies were motivated by the experimental results on vesicle deformation induced by electric fields; some are discussed by Vlahovska (2015) (see also Chapter 7). To mention a few of them, the degree of vesicle deformation as a function of field strength and conductivity ratio was modeled by Sadik et al. (2011), where strong (even excessive) vesicle deformation with aspect ratios, a / b, up to 10 was observed. The vesicle relaxation dynamics was later considered using a droplet-based model (Zhang et al., 2013; Yu et al., 2015), which yielded a way of deducing the initial tension of the vesicle and the membrane bending rigidity (Yu et al., 2015). Theoretical modeling of the vesicle deformation and relaxation in DC pulses have demonstrated that the time dependence of the vesicle shape can be used to measure the membrane viscosity and capacitance (Salipante and Vlahovska, 2014). Theoretical description of the unusual spherocylindrical deformations observed in vesicles in the presence of salt (Riske and Dimova, 2006) was also recently developed, and the flat and curved regions in the vesicle interpreted as coexistence of porated and nonporated regions (Salipante et al., 2015).

As shown above, the study of vesicle relaxation dynamics after electrodeformation and electroporation allows for assessing the membrane material properties. From a practical viewpoint, the membrane viscosity and membrane edge tension influence the vesicle mechanical response after electroporation on a timescale that can be assessed with conventional fast cameras (time resolution of 10 ms is sufficient). Such an approach was used to investigate the effect of residual agarose entrapped in GUVs grown in hybrid films of lipids and agarose (see preparation protocols in Chapter 1) on the vesicle mechanical response (Lira et al., 2014). In that case, the mean value of the relaxation time, τ_2, doubled and many vesicles showed prolonged pore lifetimes (t_{por} ~ seconds) when agarose was present, suggesting that the polymer physically hinders pore resealing. This example demonstrates that analysis of vesicle relaxation dynamics after electroporation can be a powerful method for investigating membrane response to external forces under different circumstances (e.g., membrane composition, interaction with molecules, effect of additives in the medium).

15.3.5 A NOTE ABOUT GEL-PHASE VESICLES AND POLYMERSOMES

The mechanical, rheological and electrical properties of membranes in the gel phase and polymersomes differ significantly from those of fluid lipid membranes. Some of them are summarized in Table 15.1. These differences introduce new features in the response of gel-phase membranes to electric fields. Because of the high bending rigidity, gel-phase vesicles relax much faster after electrodeformation induced by electric pulses (Knorr et al., 2010). The critical transmembrane potential leading to poration of gel-phase vesicles and polymersomes is several times higher compared with that of fluid membranes (Aranda-Espinoza et al., 2001; Knorr et al., 2010) (one of the reasons being the larger membrane thickness), which implies that pulses of much higher field strength are needed to porate them. The resealing of macropores in polymersomes is slowed down because of the higher membrane viscosity (Dimova et al., 2002), and is arrested in gel-phase membranes (Knorr et al., 2010). Upon poration, gel-phase GUVs develop cracks that do not reseal within minutes (Figure 15.3c).

15.4 ASSESSING OVERALL VESICLE PROPERTIES AND MANIPULATION OF GUVs

In the previous sections, we illustrated several methods that can be applied to assess certain membrane material properties. Furthermore, electric fields can be easily employed to also manipulate giant vesicles either to assess hidden parameters such as their total area (sometimes stored in tubes or, in what one might call "defects," or in fluctuations) or to move two vesicles close together and fuse them. This section will present a few protocols as examples.

15.4.1 MEASURING VESICLE EXCESS AREA

GUVs in the presence of AC fields deform into prolate and oblate shapes with their principal axes oriented parallel to the electric field. The degree of deformation, measured from the aspect ratio, a/b, depends not only on the magnitude of the field applied (as shown in Section 15.3.1) but also on the vesicle excess area available. Electric fields of moderate strength (~0.2 kV/cm) are able to pull the vesicle excess area but are not strong enough to stretch the bilayer at a molecular level. Such fields can be used to estimate the vesicle surface area, which is usually difficult to measure from the projection of (deflated) vesicles in the focal plane. This approach offers an alternative to employing micropipette aspiration (see Chapter 11) to measure the actual vesicle area. Depending on the field frequency and conductivity ratio, the AC field can induce prolate and oblate deformation (see Box 15.1) (Aranda et al., 2008; Dimova et al., 2009). For the estimation of the area increase, the prolate shape is preferred because less deformation due to gravity and the proximity of the coverslip occurs. To induce prolate shapes, a small amount of salt (~0.2 mM NaCl) can be added to the sucrose solution in which the GUVs are grown. This ensures that the conductivity inside the vesicle is higher than the external isotonic glucose solution. From the aspect ratio, a/b, of the prolate, the vesicle surface area, A_{ve}, and volume, V_{ve}, can be easily obtained:

$$A_{ve} = 2\pi b\left(b + a\frac{\sin^{-1}\epsilon}{\epsilon}\right), V_{ve} = \frac{4\pi}{3}ab^2, \quad (15.9)$$

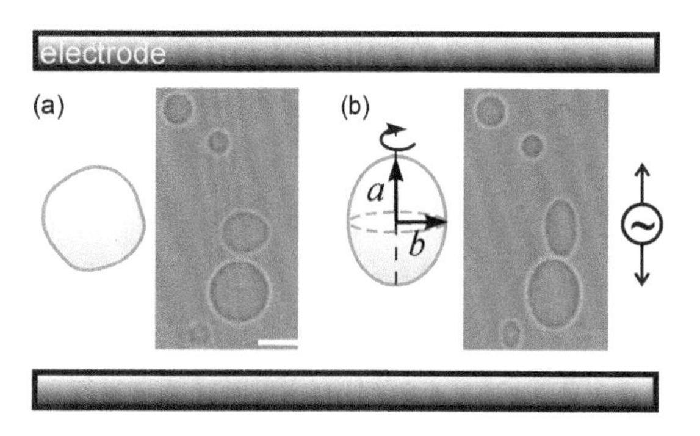

Figure 15.6 Employing AC fields to measure the vesicle surface area. (a) In the absence of electric field, vesicles with (not-so-large) excess area adopt quasi-spherical shapes. Scale bar: 15 µm. (b) In the presence of AC field of the right frequency and provided $\sigma_{in}/\sigma_{ex} > 1$, a prolate shape is induced, with the main axis parallel to the electric field. The degree of deformation is characterized by the aspect ratio a/b, which can be used to assess the vesicle excess area.

where ϵ, given by $\epsilon^2 = 1-(b/a)^2$, is the ellipticity. More importantly, the area increase of the prolate ellipsoid relative to a spherical vesicle with the same volume can be calculated from

$$\alpha = \frac{1}{2}\sqrt[3]{\frac{a}{b}}\left(\frac{b}{a} + \frac{\sin^{-1}\epsilon}{\epsilon}\right) \quad (15.10)$$

Figure 15.6 shows a cartoon and representative images of the prolate shape deformation induced by an AC field on deflated GUVs. The degree of deformation varies from vesicle to vesicle and depends on the excess area available as can be seen in the images in Figure 15.6b. Applications of this method to measure area increase can be found in papers by Riske et al. (2009b), Mattei et al. (2015), Georgiev et al. (2017), and Lira et al. (2019) and in Chapters 22 and 24.

Prerequisites and some insights for performing these measurements include the following: (i) The membrane should be impermeable to ions and in the fluid phase to allow vesicle deformation in response to AC fields. (ii) For strongly deflated GUVs (with a lot of excess area), the deformation may become too large and the vesicles may attain pointed or noneliptical shapes (the vesicles may even bud off at the poles facing the electrodes). In this case, Eqs. 15.9 and 15.10 cannot be applied to correctly assess the membrane area and will, at best, give only a rough estimate. (iii) The presence of charged lipids might influence the vesicle response to the field. (iv) When using this method to measure the area increase caused by external agents (e.g., molecules added as in Chapter 24, illumination as in Chapter 22), it is advisable to first measure the vesicle area in the absence of the perturbation to avoid contributions to the area increase resulting from unfolding of membrane stored as buds and invaginations (for example originating from asymmetry of buffers or leaflet composition [Karimi et al. 2018; Dasgupta et al. 2018]), sometimes termed as hidden area (Vitkova et al., 2004). (v) For long exposure to AC fields (minutes), GUVs tend to align and form pearl-chains (see the next section). More tips are given in Section 15.5.

15.4.2 BRINGING VESICLES TOGETHER BY AC FIELDS

GUVs are usually found as single objects separated in the observation chamber. Due to their large size and slow diffusion,

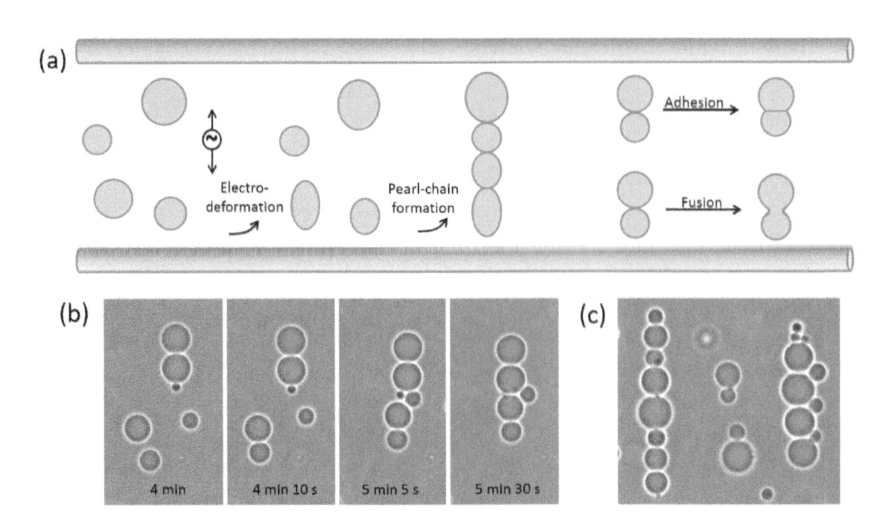

Figure 15.7 Vesicle alignment induced by AC field. (a) The cartoon shows a few AC field effects: First the GUVs with excess area are deformed and then, within seconds to minutes, the GUVs align as a pearl-chain. (b) Sequence of images showing GUVs after applying an AC field (0.2 kV/cm, 1 kHz) for the duration indicated on each snapshot. (c) Many GUVs aligned after several minutes-long exposure to the AC field.

the chances of GUV encounters are low. In some applications, vesicle–vesicle contact is required when, for instance, the aim is to study vesicle adhesion and/or fusion. Therefore, ways to manipulate GUVs to bring them together are sometimes useful. With micropipette aspiration (Chapter 11), one can easily bring selected GUVs into contact, but this requires the necessary setup. An alternative way is to apply an AC field (typically 0.2 kV/cm amplitude and 1 kHz–1 MHz frequency) to a population of GUVs. Electric-field alignment of particles (dielectrophoresis) has been known for decades (Zimmermann and Vienken, 1982; Takashima and Schwan, 1985). Similarly, the AC field induces alignment of GUVs along the field as in a pearl-chain (Figure 15.7), which results from the nonuniform field distribution is caused by neighboring vesicles. This procedure can be done in the presence of molecules that induce membrane adhesion, or a strong DC pulse can be applied afterward to induce vesicle fusion (see Sections 15.4.3 and 15.4.4).

Some requirements and tips for applying AC fields to bring two vesicles together include the following (see also Section 15.5): (i) The concentration of GUVs should be chosen such that it is not too high—to avoid yielding excessively long pearl-chains—or too low—otherwise the vesicles are too far apart to align. For typical GUV electroformation protocols on indium tin oxide plates (see Chapter 1), 10–20 × dilution is usually fine. (ii) The AC field can induce elliptical deformation, depending on the excess area of the GUV (see the preceding section). If GUVs with too much excess area are used, they will align as prolates and their poles might be deformed (squeezed) upon pearl-chain formation. (iii) If the salt concentration is too high (>10 mM), long exposure to electric fields might induce sample heating and electrochemical deposition on the electrodes.

15.4.3 USING ELECTROFUSION TO MAKE MULTICOMPONENT VESICLES WITH PRECISELY KNOWN COMPOSITION AND TO MEASURE TIE LINES

Electrofusion has been previously employed (along with ligand-mediated fusion) to investigate the timescales of opening of the fusion neck (Haluska et al., 2006). Employing fast digital imaging has lead to the conclusion that the opening of the fusion necks is very fast, with an average expansion velocity of centimeters per second. This velocity indicates that the initial formation of a single fusion neck can be completed in a few hundred nanoseconds, consistent with simulation data (Shillcock and Lipowsky, 2005). From a more practical viewpoint, electrofusion of GUVs can be used to initiate targeted mixing of either their encapsulated content (see next section) or their membranes. Of course, a necessary condition is that the membranes of both vesicles are fluid. In this Section, we will describe how electrofusion can be employed to create vesicles with precisely known composition and how these vesicles can be used to deduce tie lines in the phase diagram of ternary mixtures (tie lines show the composition of domains in phase separated vesicles, see Chapter 18).

The most obvious way to obtain multicomponent vesicles is to prepare them from lipid mixtures (Chapter 1). However, in this way, the composition of the different vesicles in a batch can vary drastically depending on the individual vesicle history. For example, before observation, a phase-separated vesicle may have budded in the region of one of the domains and the two daughter vesicles would then have attained compositions that are different from the composition of the mother vesicle. This is obvious from the observation that sometimes, in samples with compositions belonging to the region of coexistence of two fluid phases (see Chapter 18), one finds vesicles that are not phase separated; for more exact characterization based on measuring the area fractions of different domains within one batch, see Bezlyepkina et al. (2013). Particularly strong deviations in the vesicle composition are observed for multicomponent lipid mixtures that are not fully miscible at the temperature of observation (Tian et al., 2009; Vequi-Suplicy et al., 2010; Bezlyepkina et al., 2013). To overcome this problem, an alternative means of arriving at a specific vesicle composition can be used based on producing vesicles with domains via electrofusion of two vesicles made of two different fully miscible lipid mixtures (Riske et al., 2006; Bezlyepkina et al., 2013). This approach is illustrated in Figure 15.8. Two populations of vesicles made of compositions

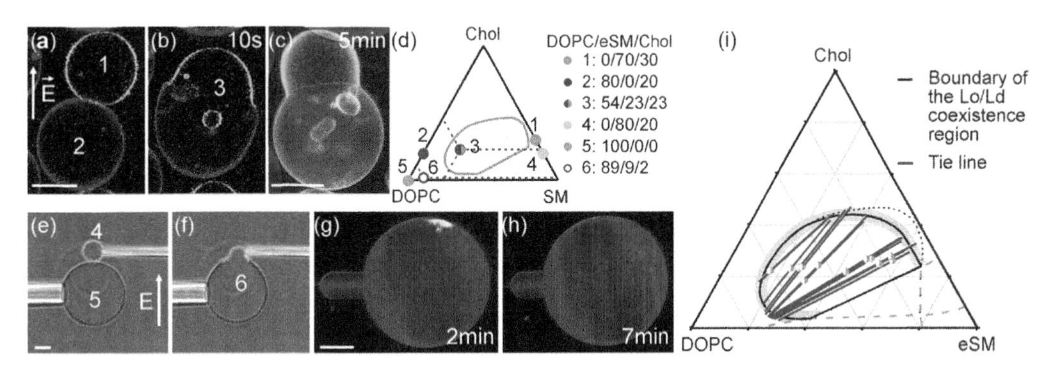

Figure 15.8 Electrofusion of single- or two-component vesicles as a way to create three-component vesicles with precisely controlled composition. (a–c) Fusion of two freely suspended vesicles observed with confocal microscopy: (a and b) cross sections and (c) a 3D projection. Vesicle 1 in panel (a) is composed of eSM/Chol (70:30). Vesicle 2 is made of DOPC/Chol (80:20). The two vesicles were subjected to an electric pulse (400 kV/m, 150 μs; the arrow indicates the field direction) and fused to form Vesicle 3 shown in (b) and (c). The time after applying the electric pulse is indicated in the upper-right corners. (d) Phase diagram and compositions of the vesicles in the images. The numbered compositions in the Gibbs triangle correspond to the numbered vesicles in (a), (b), (e) and (f). (e–f) Electrofusion of two vesicles brought into contact via micropipettes, as observed with an overlay of phase-contrast images and confocal cross sections (e and f) or 3D projections (g and h). The initial vesicles differ significantly in size. Vesicle 4 is made of eSM/Chol (80:20) and has a radius of 17 μm. Vesicle 5 is composed of DOPC and has a radius of 52 μm. After application of an electric pulse (250 kV/m, 100 μs) the vesicles fuse to form Vesicle 6, which is located in the single-phase region (d). The lipids mix quickly after the fusion, as shown in image (h), which was taken 7 min after the pulse. All scale bars: 20 μm. (i) Examples for tie lines deduced for the phase diagram of DOPC/eSM/Chol at 23°C. Half-solid circles in gray indicate the compositions of the fused vesicles whose images were used to locate tie lines (red) within the Lo/Ld coexistence region. The latter region is indicated by the solid black curve. (Reprinted from *Biophys. J.*, Bezlyepkina, N. et al., Phase diagram and tie-line determination for the ternary mixture DOPC/eSM/Cholesterol, 1456–1464, Copyright 2013, with permission from Elsevier.)

1 and 2 are labeled differently and are mixed (Figure 15.8d). Two vesicles of different compositions can be brought together either by applying an AC field (Figure 15.8a; see also Section 15.4.2) or by means of micropipette aspiration (Figure 15.8e–h). In the latter case, a custom-made chamber was made as illustrated in Figure 15.11e, allowing for pipette entry between the electrodes. Then, a strong pulse (above the poration thresholds of each membrane) is applied and a three-component vesicle is obtained. Measuring the areas of the fused domains right after fusion allows estimation of the exact composition of the fused vesicle, provided that data on the molecular area of the respective lipids is available. The areas of the two domains change over time after fusion as the lipids are allowed to mix and redistribute according to the tie line going through the composition of the fused vesicle. The mixing may proceed for several minutes before equilibrium is reached (on the order of 10–30 min, depending on the lipid diffusivity) (Bezlyepkina, 2012). After equilibration, one can measure the areas of the domains of such a vesicle and use these values to deduce the tie line passing through the compositional point of the vesicle in the phase diagram. Box 15.3 lists the major steps in this approach; we refer the reader to Bezlyepkina et al. (2013) for the details. A necessary condition is the availability of data about the phase boundaries of the lipid ordered/lipid disordered (Lo/Ld) coexistence region and about the approximate molecular areas of the membrane species in the two different phases. Example tie lines deduced for the dioleoylphosphatidylcholine/egg sphingomyelin/cholesterol (DOPC/eSM/Chol) system using this approach are given in Figure 15.8i.

15.4.4 FUSING VESICLES WITH DIFFERENT ENCAPSULATED REACTANTS: GUVs AS MICROREACTORS

Electrofusion of two GUVs in order to initiate content mixing reactions has been employed previously (Chiu et al., 1999). The example that we will consider here demonstrates how GUV electrofusion can be applied for the synthesis of quantum-dot-like CdS nanoparticles in closed compartments (Yang et al., 2009; Yang and Dimova, 2011). The approach consists of mixing two vesicle populations, one loaded with Na_2S and labeled with one fluorescent dye (red), the other loaded with $CdCl_2$ and labeled differently (green). The vesicle external media is almost free of Na_2S or $CdCl_2$ (in practice, the vesicles are strongly diluted). An AC field is applied to align the vesicles in the direction of the field due to dielectric screening (see Section 15.4.2). We then locate a red-and-green vesicle couple and apply a DC pulse strong and long enough to porate each of the vesicles (typically, pulses of 50–200 kV/m field strength and 150–300 μs duration are sufficient); see also Movie 15.3 (movies captions at the end of chapter). The steps of this protocol are schematically illustrated in Figure 15.9a. The product, in this case, quantum-dot-like CdS nanoparticles (with sizes between 4 and 8 nm, as determined from transmission electron microscopy [Yang et al., 2009]), is visualized under laser excitation as a fluorescent bright spot in the fusion zone (Figure 15.9b). Obviously, this protocol provides us with a visualizing analytical tool to follow the reaction kinetics with high temporal sensitivity. This method could be especially suitable for the online monitoring of ultrafast physicochemical processes such as photosynthesis, enzyme catalysis and photopolymerization, which usually require complex and abstracted spectroscopy techniques.

15.4.5 "STIRRING" THE LIPIDS IN A VESICLE USING AC FIELDS

Electric fields can be used to trigger flows in the vesicle (Staykova et al., 2008), and in this way mix the lipids (in a similar way, electric fields can drive flows in droplets). This can be achieved by exposing the vesicles to nonhomogeneous AC fields. In practice, the applied

Box 15.3 Flowchart of the approach of deducing tie lines in the Lo/Ld coexistence region following Bezlyepkina et al. (2013)

The approach is based on fusing two single- or two-component vesicles with precisely known composition and measuring areas of domains.

Step 1. Prepare two samples of vesicles belonging to the shoulders of the Gibbs triangle. The compositions could be either single- or two-component mixtures (like the two examples given in Figure 15.8a–b), and the lipids should be fully miscible at the temperature of observation. Use fluorescent labels, which partition differentially in the two phases (Lo and Ld); see also Chapters 10 and 18.
Step 2. Mix the two vesicle populations and apply an AC field to bring vesicles together; see Section 15.4.2. Select a couple of vesicles of different composition (different color) and apply a DC pulse strong enough to porate each of the membranes.
Step 3. Immediately after fusion, take a snapshot of the newly formed vesicle and measure the areas of the domains. From literature data on areas per molecule (in the respective environment), calculate the number of molecules in each domain and, in this way, determine the exact composition of the new electrofused vesicle.
Step 4. Allow 30 min for equilibration and measure domain areas again.
Step 5. Draw a hypothetical tie line through the composition point of the vesicle (see figure in box).
Step 6. From the intercepts with the binodal, obtain hypothetical domain compositions.
Step 7. From hypothetical domain compositions and literature molecular areas, calculate hypothetical domain areas.
Step 8. Compare hypothetical domain areas with domain areas measured in Step 3.
Step 9. If areas do not match, return to Step 5. If they do, then your hypothetical tie line is the real one.

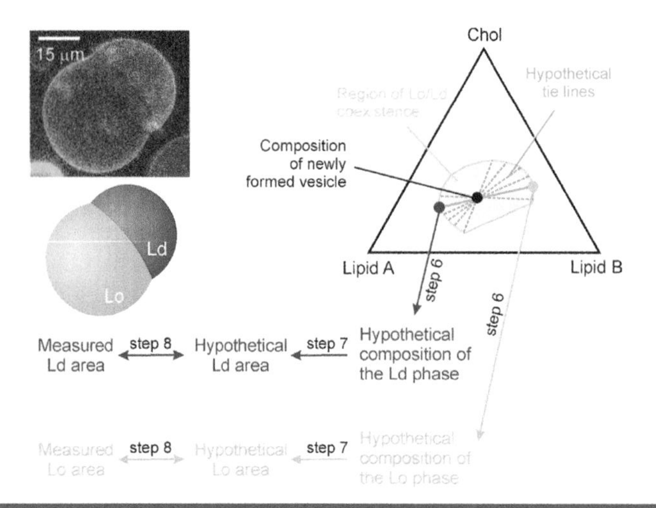

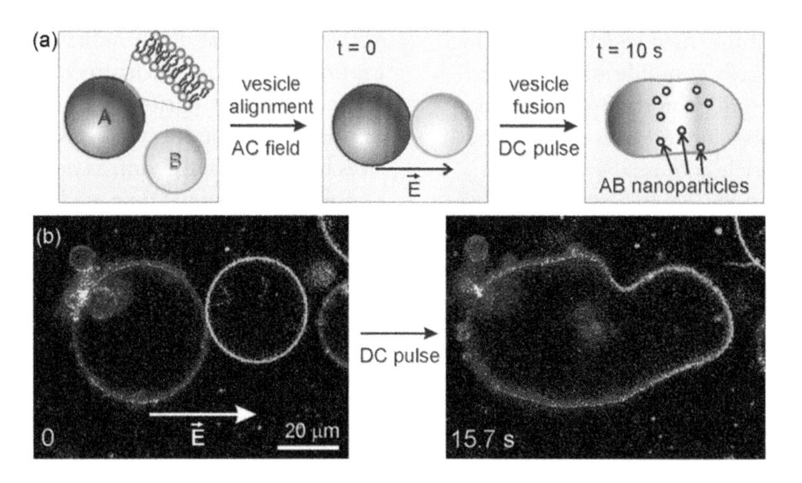

Figure 15.9 Electrofusion of giant vesicles as a method for nanoparticle synthesis. (a) Schematic illustration of the electrofusion protocol: Two populations of vesicles containing reactant A or B are mixed (in A- and B-free environment). The vesicles are subjected to an AC field to align them in the direction of the field and bring them close together. A DC pulse initiates the electrofusion of the two vesicles and the reaction between A and B proceeds to the formation of nanoparticles encapsulated in the fused vesicle. (b) Confocal cross sections of eggPC vesicles loaded with 0.3 mM Na_2S (red) and 0.3 mM $CdCl_2$ (green) undergoing electrofusion. The direction of the field is indicated in the first snapshot. After fusion (right), fluorescence from the product is detected in the interior of the fused vesicle; see Movie 15.3 (movies captions at the end of chapter). The time after applying the pulse is indicated on the micrographs. (From Yang, P. et al., Nanoparticle formation in giant vesicles: Synthesis in biomimetic compartments. *Small*, 2009. 5:2033–2037. Copyright Wiley-VCH Verlag Gmbh & Co. KGaA. Reproduced with permission.)

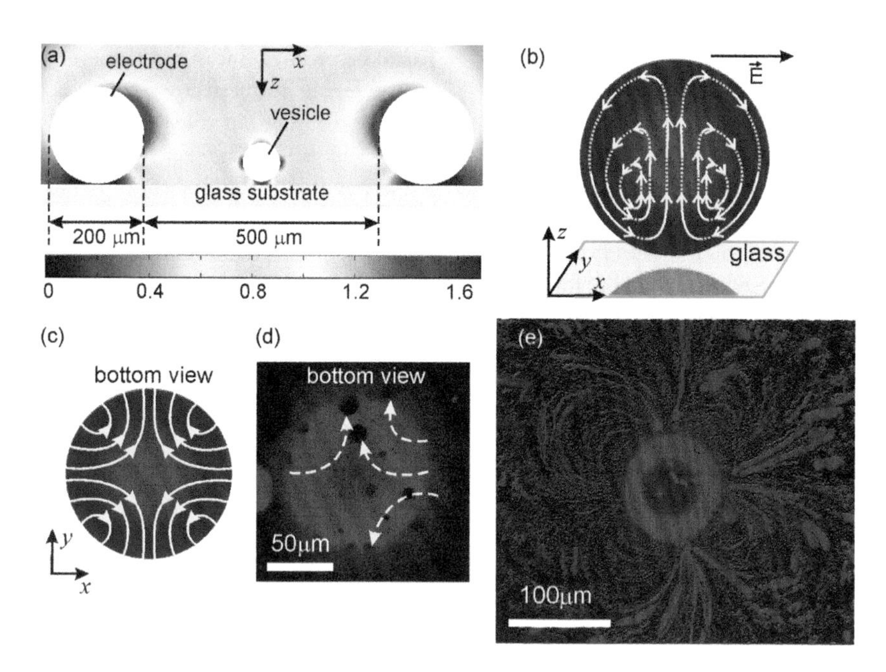

Figure 15.10 Generation of flows in GUVs by means of inhomogeneous electric fields. (a) Electric field distribution at 100 kHz in a cross section of the chamber, passing through the center of the vesicle. The vesicle (40 μm in radius) is located at 8 μm above the glass. The media conductivity is 300 μS/cm. The field inside the vesicle is not calculated. The data are rescaled with the strength of a field, which would be induced between two parallel planar electrodes at a distance of 500 μm. (b, c) Sketches of the side and bottom view of the vesicle with the flow lines. The length of the arrows in (b) roughly corresponds to the flow velocity. (d) Confocal micrograph (open pinhole) illustrating the membrane flow at the bottom part of a giant vesicle (~150 μm in diameter, DOPC/DPPC/Chol 48:32:20, labeled with 0.1 mol% dipalmitoylphosphatidylethanolamine-*N*-(lissamine rhodamine B) (Rh-DPPE; see Appendix 2 of the book for structure and data on this lipid dye), room temperature) induced by an AC field (360 V/cm, 80 KHz), at external conductivity of 250 μS/cm. The yellow dashed arrows indicate the trajectories of selected domains. Reproduced from Staykova et al. (2008), with permission from the Royal Society of Chemistry. See also Movie 15.4 (movies captions at the end of chapter). (e) Time-averaged image illustrating the movement of fluorescent microparticles in four quadrants around a GUV (central red circle) exposed to an asymmetric field. (Courtesy of Margarita Staykova.)

fields in experimental chambers are never perfectly homogeneous. The cause for this could be the presence of other vesicles in the vicinity of the target GUV under observation, or the proximity of a surface, for example, the chamber bottom (to which the vesicles sediment if they possess, for example, sucrose/glucose in/out asymmetry). In the latter case, the inhomogeneity in field distribution around the vesicle can drive flows in the membrane. Figure 15.10 shows an example of a numerically calculated field distribution around a vesicle located near the bottom of an Eppendorf™ electrofusion chamber (Figure 15.10a) and exposed to an AC field.

The field induces circular patterns of lipid transport in the membranes (Figure 15.10b–d and Movie 15.4; movies captions at the end of chapter). Such electroosmotic flows occur in the vicinity of polarizable objects and are driven by the displacement of the free induced surface charges by the lateral component of the electric field. Hydrodynamically, the membrane behaves as a two-dimensional (2D) incompressible fluid. To keep the surface area constant, the lipid bilayer develops tension under forcing. Thus, even weakly inhomogeneous AC fields may induce a pronounced membrane flow in giant lipid vesicles. The flow can be visualized by the displacement of fluorescently labeled lipid domains in phase-separated vesicles. The flow velocity reaches about 30 μm/s and depends on the field conditions and on the location within the vesicle surface. In all experiments, the lipid transport is faster at the bottom part of the vesicle. The lipid dynamics on the vesicle surface is organized in four symmetric quadrants, each extending from the lower to the top part of the vesicle, as illustrated in Figure 15.10c. The motion follows concentric closed trajectories with the highest velocity at the periphery of each quadrant and at the vesicle bottom.

From the field distribution, one can calculate the lateral electric stress (surface force density) on the membrane as the product of the field-induced free charge density at the vesicle surface and the tangential component of the electric field. The inhomogeneous electric field distribution for vesicles close to the glass breaks the symmetry in the electric stress distribution between the lower and the upper hemispheres of the vesicle. As a result, a nonuniform and nonsymmetric membrane tension builds up. It triggers lipid flow toward the regions of highest tension, analogously to Marangoni flows. The gradients in the tension are strongest in the regions facing the electrodes. There the membrane starts flowing downward. The flows are sufficient to allow fission of large domains in the vesicles (Staykova et al., 2008).

The flows on the vesicle membrane are coupled to flows in the vesicle exterior and interior. The movement around the vesicle can be visualized using fluorescent microparticles outside the vesicle producing beautiful images as seen in the time-averaged snapshots in Figure 15.10 (unpublished results). The flow inside the vesicle can be visualized in a similar manner or by encapsulating droplets of aqueous two-phase system (see Chapter 29), as demonstrated by Dimova et al. (2009).

15.5 EXPERIMENTAL CHAMBERS FOR STUDYING GUVs EXPOSED TO ELECTRIC FIELDS AND TIPS FOR SUCCESSFUL EXPERIMENTS

The experimental chamber needed for exploring the effects of electric fields on vesicles is simple. Indeed, any chamber for vesicle electroformation on platinum (Pt) wires (see Chapter 1) can be employed as long as the wires are located close to the chamber bottom so that heavy vesicles do not sediment away from the zone between the electrodes. Figure 15.11 shows images of chambers that are either purchased or home built. The commercially available Eppendorf™ electrofusion chamber, Figure 15.11a, consists of a polytetrafluoroethylene (Teflon™) frame confined from below by a glass plate through which observation is possible. A pair of parallel electrode wires (92 μm in radius, Pt) are fixed at the lower glass at a distance of 0.2 or 0.5 mm between the electrodes. After loading with the vesicle solution, the chamber is closed from above with another coverslip. A simple home-built chamber can be assembled from a glass slide and a coverslip, whereby two parallel adhesive copper strips can be used as electrodes (Figure 15.11d). The chamber can be glued with heated Parafilm®.

Electrodes made of Pt wires are preferred because they are more inert and present no danger of releasing oxidation species in the solution, which is even more important for long experiments. In any home-built chamber, the distance between the electrodes has to be precisely measured (for every measurement, if the electrodes are not perfectly parallel) because it affects the field strength measured in volts per meter.

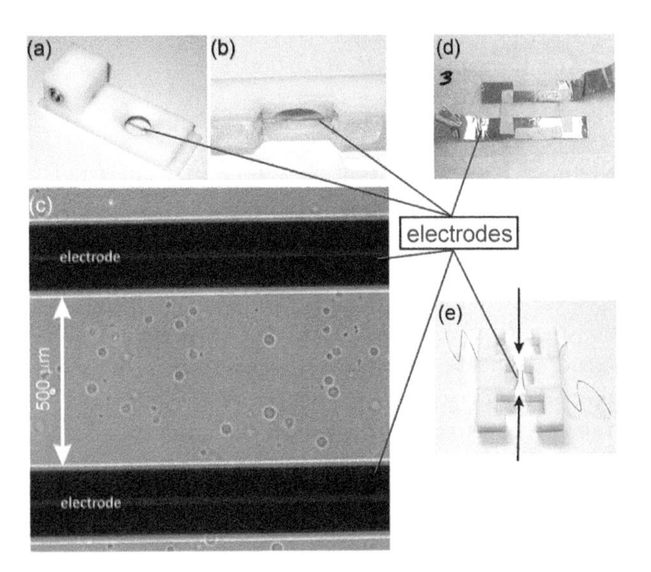

Figure 15.11 Chambers for experiments with GUVs exposed to electric fields. (a) Commercial electrofusion chamber of Eppendorf™ (top view) with 0.5-mm spacing between the electrodes (the inner diameter of the chamber is ~13 mm). (b) Eppendorf™ chamber modified for experiments at a confocal microscope (view from below): The lower part of the Teflon™ frame is cut out to allow replacing the original glass with slides of thickness 0.17 mm. (c) Microscopy image (phase-contrast) of a vesicle suspension between the electrodes of an Eppendorf™ chamber. (d) Home-built chamber made of glass coverslips, adhesive copper tapes (3M™, Cergy-Pontoise, France) as electrodes stuck 5 mm apart, and Parafilm® as a spacer. (e) Home-made chamber made of Teflon™ and Pt wires (without the covering glass slides on top and below) used for electrofusion with micropipette assistance as in Section 15.4.3. The compartments out of the inter-electrode regions are used to introduce the vesicles. The arrows point to the sides from where the two micropipettes are inserted.

As field generators for delivering strong DC pulses, one can use the Eppendorf™ Multiporator (50–300 μs duration of the pulse, and 50–300 V output voltage) and the pulse generator of β tech GHT_Bi500 (5 μs–50 ms duration of the pulse, and up to 500 V output voltage).

Below, we summarize some tips for measurements of membrane properties on vesicles in electric fields:

- One should select vesicles located in the middle between the two electrodes, otherwise the field inhomogeneity (Figure 15.10) might influence the vesicle response and movement of the whole vesicle might be observed.
- In capacitance measurements, frequency sweeps with increasing and decreasing frequency should not show hysteresis. If they do, this might be an indication that the frequency has been changed too fast.
- In bending rigidity measurements, voltage sweeps with increasing and decreasing field strength should not show hysteresis. If they do, this might be an indication that the voltage has been changed too fast or that the vesicle had some defects. One should also measure the real voltage in the solution, which might be different from the output of the instrument (particularly important for solutions with higher conductivity).
- If the vesicles are very tense, no deformation in weak AC fields will be detected. One could consider deflating the GUVs with hypertonic solutions, but, in this case, the change in the conductivities should be also accounted for.
- In measurements where fast temporal resolution is required (edge tension, relaxation dynamics), to ensure appropriate illumination, strong enough to be detected by the camera, one can employ either stroboscopic illumination or a Hg lamp. Because the whole process of pore formation and resealing takes less than a second, heating of the sample because of the Hg lamp is negligible. In general, the sample should not be continually illuminated for more than 30 s.
- To enhance the image quality and facilitate the detection of the vesicle and pores, one can employ sucrose/glucose (in/out) asymmetry in the vesicles. For measuring the bending rigidity, however, one should consider the effect of sugars see Dimova (2014) and also Chapter 14. For these measurements as well as for capacitance measurements, the vesicles should not be "too heavy," otherwise gravity might affect the deformation (Kraus et al., 1995). To ensure sufficient optical contrast, osmolarities on the order of 50 mOsm may be appropriate; see also Nikolov et al. (2007) for more precise estimates depending on the vesicle size. For measurements in pure water, if optically possible (e.g., when using fluorescence microscopy for observing the vesicles), one should have in mind that the vesicle volume is not conserved.
- When consecutive pulses are applied to the same vesicle, enough time for complete relaxation (30–60 seconds) should be given between the pulses. The membrane, in particular at the poles, might not have relaxed fully, leading to tubular protrusions caused by the subsequent pulses (Portet et al., 2009). One should be also aware that the internal conductivity in the vesicle might change with every following pulse when the membrane is transiently porated.

15.6 FINAL WORDS

Manipulation of GUVs via electric fields offers an easy way to assess a number of material properties of the membrane, elastic, electrical and rheological. We have demonstrated this on very simple and predominantly single-component membranes. Surely, as a next step, the effect of various lipid components can be explored. For example, by using vesicle electrodeformation we have investigated the effect of cholesterol on the membrane bending rigidity and found that, contrary to earlier understanding, increasing fractions of cholesterol do not lead to membrane stiffening of DOPC bilayers and that the effect is rather specific to the lipid geometry (Gracià et al., 2010). We have also explored the effect of composition on membrane stability by applying electric pulses to GUVs as discussed in Section 15.3.3, and found that vesicles composed of charged lipids rupture easily (Riske et al., 2009a). These are just two examples of how the approaches presented in this chapter can be employed to reveal the role of various membrane components. Application of weak DC fields holds promise for investigating vesicle adhesion to a substrate where the adhesion strength can be directly modulated by the field (Steinkühler et al., 2015, 2016).

The membrane compositions that we have explored so far are very simple (up to three components for the electrofusion experiments), but the behavior of more complex membranes appear to exhibit not very different behavior. For example, when giant vesicles made of lipid extract from red blood cells are exposed to electric fields, their response can be characterized like that of two-component (phosphatidylcholine–phosphatidylglycerol, PC–PG) mixtures (Riske et al., 2009a). Of course, this will probably not be the case of membranes doped with proteins. An interesting system to start with could be cell blebs or giant plasma membrane vesicles (see Chapter 2) or the simpler version of vesicles with one or a couple of reconstituted protein types (see Chapter 3). In any case, we want to convey the idea that setting up such experiment is easy and probably possible in most labs that already possess vesicle electroformation chambers with Pt wires as electrodes.

Exploring the effect of electric fields on model membranes is not only of academic interest. As discussed in the introduction, the phenomenon of electroporation or electropermeabilization is used for introducing various molecules in the cell and for cell hybridization, whereas flexoelectricity is involved in a number of membrane processes. However, understanding of membrane deformation and stability in electric fields remains elusive, and it is obvious that experiments on basic biomimetic systems such as giant vesicles will help clarify many aspects of membrane behavior.

MOVIES

The Movies can be found under http://www.crcpress.com/9781498752176.

Movie 15.1: Opening and closing of a macropore in a giant vesicle subjected to a DC pulse as shown in Figure 15.3a. Analysis of the pore dynamics is used to deduce the membrane edge tension (Portet and Dimova, 2010).

Movie 15.2: Bursting and collapse of a giant vesicle made of POPC:POPG 1:1 exposed to a DC pulse (150 V, 150 μs) (Riske et al., 2009a). The movie is courtesy of R.B. Lira.

Movie 15.3: Employing electrofusion as a means to establish giant vesicles as microreactors (Yang et al., 2009). The movie shows the fusion between NaCl-loaded (red) and $CdCl_2$-loaded (green) vesicles; snapshots of which are presented in Figure 15.9. The event is displayed about 10 times faster than in reality (the real time duration of the movie is about 90 s).

Movie 15.4: Inhomogeneous AC fields drive surface flows on the membrane of giant vesicles (Staykova et al., 2008). The movie shows the lower part of a vesicle prepared from 5.33:2.66:2 DOPC:DPPC:Chol. An AC field at a fixed strength of 260 V/cm and for three different frequencies (500 kHz, 1 MHz, and 3 MHz) was applied. The conductivity of the external medium is 360 μS/cm and the diameter of the vesicle is approximately 120 μm. For the low frequency (500 KHz), the direction of the membrane flow is as described in Figure 15.10. At 1 MHz, the flow almost ceases, and at 3 MHz, its direction reverses. Between the field frequency switches, the vesicle is not exposed to an AC field. The movie is sped up 8.4 times.

ACKNOWLEDGMENTS

We thank all our coworkers who helped us, enriching our understanding of vesicles in electric fields (in alphabetical order) and also for the fun time together:[3] S. Aranda, N. Bezlyepkina, D. Dudas, N. Fricke, R. S. Gracià, R. L. Knorr, H. Lin, R. Lipowsky, Y. Li, R. B. Lira, T. Portet, M. Staykova, P. Salipante, P. Shchelokovskyy, P. Vlahovska, T. Yamamoto, P. Yang. We thank O. Sandre for carefully reading the text and for the useful discussions. K.A.R. acknowledges the financial support of Fundação de Amparo à Pesquisa do Estado de São Paulo (FAPESP).

LIST OF ABBREVIATIONS

AC	alternating current
Chol	cholesterol
DC	direct current
DMPC	dimyristoylphosphatidylcholine
DOPC	dioleoylphosphatidylcholine
DPPC	dipalmitoylphosphatidylcholine
eggPC	L-α-phosphatidylcholine (Egg, Chicken)
eSM	egg sphingomyelin
PC	phosphatidylcholine
PG	phosphatidylglycerol
POPC	palmitoyloleoylphosphatidylcholine
POPG	palmitoyloleoylphosphatidylglycerol
POPS	palmitoyloleoylphosphatidylserine
Rh-DPPE	dipalmitoylphosphatidylethanolamine-N-(lissamine rhodamine B)

[3] One of the authors of this chapter was surprised one day to find a note on her office door typed in large letters in electric green: "electrofocused." The note still hangs there and the author remains "electrofocused."

GLOSSARY OF SYMBOLS

A_{ve}	vesicle surface area
A_{ve}^0	initial surface area of the vesicle
α	relative area increase of a vesicle
C_1 and C_2	principal curvatures
C_{me}	membrane capacitance
E	electric field strength
ε_0	vacuum permittivity
ε_{in}, ε_{ex}	dielectric constants of solutions inside and outside the vesicle
ε_{me}	membrane permittivity
ε_w	dielectric constant of water
k_B	the Boltzmann constant
κ	bending rigidity
l	total bilayer thickness
l_e	dielectric thickness
η	viscosity of the aqueous medium
η_s	shear surface viscosity of the membrane
θ	tilt angle between the electric field and the surface normal of the vesicle
λ_{por}	pore edge tension
ν	field frequency
ν_c	field frequency for prolate-oblate transition
r_{por}	pore radius
$r_{por,max}$	maximal pore size
R_{ve}	vesicle radius
σ_{in} and σ_{ex}	conductivities of the solutions inside and outside the vesicle
Σ	membrane tension
Σ_0	initial membrane tension
Σ_{el}	effective electrical tension
t	time
t_c	capacitor charging time
t_{por}	macropore resealing time
T	temperature
τ_1	stretching relaxation time
τ_2	pore-resealing relaxation time
τ_3	shape fluctuation relaxation time
Ψ_c	critical transmembrane potential
Ψ_{me}	transmembrane potential
v	vesicle reduced volume
V_{ve}	vesicle volume

REFERENCES

Abidor IG, Arakelyan VB, Chernomordik LV, Chizmadzhev YA, Pastushenko VF, Tarasevich MR (1979) Electrical breakdown of bilayer lipid-membranes. 1. Main experimental facts and their qualitative discussion. *Bioelectrochem Bioenerg* 6:37–52.

Antonova K, Vitkova V, Meyer C (2016) Membrane tubulation from giant lipid vesicles in alternating electric fields. *Phys Rev E* 93:012413.

Aranda S, Riske KA, Lipowsky R, Dimova R (2008) Morphological transitions of vesicles induced by alternating electric fields. *Biophys J* 95:L19–L21.

Aranda-Espinoza H, Bermudez H, Bates FS, Discher DE (2001) Electromechanical limits of polymersomes. *Phys Rev Lett* 8720:4.

Bermúdez H, Aranda-Espinoza H, Hammer DA, Discher DE (2003) Pore stability and dynamics in polymer membranes. *EPL* (Europhysics Letters) 64:550.

Bermudez H, Hammer DA, Discher DE (2004) Effect of bilayer thickness on membrane bending rigidity. *Langmuir* 20:540–543.

Bezlyepkina N (2012) Domain formation in model lipid membranes induced by electrofusion of giant vesicles. PhD thesis, Potsdam University.

Bezlyepkina N, Gracià RS, Shchelokovskyy P, Lipowsky R, Dimova R (2013) Phase diagram and tie-line determination for the ternary mixture DOPC/eSM/Cholesterol. *Biophys J* 104:1456–1464.

Breton M, Delemotte L, Silve A, Mir LM, Tarek M (2012) Transport of siRNA through lipid membranes driven by nanosecond electric pulses: An experimental and computational study. *J Am Chem Soc* 134:13938–13941.

Brochard-Wyart F, de Gennes PG, Sandre O (2000) Transient pores in stretched vesicles: Role of leak-out. *Physica A* 278:32–51.

Calvet CY, Famin D, André FM, Mir LM (2014) Electrochemotherapy with bleomycin induces hallmarks of immunogenic cell death in murine colon cancer cells. *Oncoimmunology* 3:e28131.

Chang DC (1992) Structure and dynamics of electric field-induced membrane pores as revealed by rapid-freezing electron microscopy. In: *Guide to Electroporation and Electrofusion* (Chang DC, Chassy BM, Saunders JA, Sowers AE, Eds), pp 9–27. San Diego, CA: Academic Press.

Chang DC, Reese TS (1990) Changes in membrane-structure induced by electroporation as revealed by rapid-freezing electron-microscopy. *Biophys J* 58:1–12.

Chiu DT, Wilson CF, Ryttsen F, Stromberg A, Farre C, Karlsson A, Nordholm S, Gaggar A, Modi BP, Moscho A, Garza-Lopez RA, Orwar O, Zare RN (1999) Chemical transformations in individual ultrasmall biomimetic containers. *Science* 283:1892–1895.

Dasgupta R, Miettinen MS, Fricke N, Lipowsky R, Dimova R (2018) The glycolipid GM1 reshapes asymmetric biomembranes and giant vesicles by curvature generation. *Proc Natl Acad Sci USA*, 115:5756–5761.

Dimova R (2010) Electrodeformation, electroporation, and electrofusion of cell-sized lipid vesicles. In: *Advanced Electroporation Techniques in Biology and Medicine* (Pakhomov AG, Miklavcic D, Markov M, Eds), pp 97–122. Boca Raton, FL: CRC Press.

Dimova R (2011) Membrane electroporation in high electric fields. In: *Bioelectrochemistry. Fundamentals, Applications and Recent Developments* (Alkire RC, Kolb DM, Lipkowski J, Eds), pp 335–367. Weinheim, Germany: Wiley-VCH Verlag GmbH & Co. KGaA.

Dimova R (2014) Recent developments in the field of bending rigidity measurements on membranes. *Adv Colloid Interface Sci* 208:225–234.

Dimova R, Bezlyepkina N, Jordo MD, Knorr RL, Riske KA, Staykova M, Vlahovska PM, Yamamoto T, Yang P, Lipowsky R (2009) Vesicles in electric fields: Some novel aspects of membrane behavior. *Soft Matter* 5:3201–3212.

Dimova R, Dietrich C, Hadjiisky A, Danov K, Pouligny B (1999) Falling ball viscosimetry of giant vesicle membranes: Finite-size effects. *Eur Phys J B* 12:589.

Dimova R, Pouligny B, Dietrich C (2000) Pretransitional effects in dimyristoylphosphatidylcholine vesicle membranes: Optical dynamometry study. *Biophys J* 79:340–356.

Dimova R, Riske KA, Aranda S, Bezlyepkina N, Knorr RL, Lipowsky R (2007) Giant vesicles in electric fields. *Soft Matter* 3:817–827.

Dimova R, Seifert U, Pouligny B, Forster S, Dobereiner HG (2002) Hyperviscous diblock copolymer vesicles. *Eur Phys J E* 7:241–250.

Discher BM, Won YY, Ege DS, Lee JCM, Bates FS, Discher DE, Hammer DA (1999) Polymersomes: Tough vesicles made from diblock copolymers. *Science* 284:1143–1146.

Evans E, Needham D (1987) Physical-properties of surfactant bilayer-membranes: Thermal transitions, elasticity, rigidity, cohesion, and colloidal interactions. *J Phys Chem* 91:4219–4228.

Fricke N, Dimova R (2016) GM1 softens POPC membranes and induces the formation of micron-sized domains. *Biophys J* 111:1935–1945.
Georgiev V, Grafmüller A, Bléger D, Hecht S, Kunstmann S, Barbirz S, Lipowsky R, Dimova R (2018) Area increase and budding in giant vesicles triggered by light: Behind the scene. *Adv Sci* 5:1800432.
Golzio M, Escoffre JM, Portet T, Mauroy C, Teissie J, Dean DS, Rols MP (2010) Observations of the mechanisms of electromediated DNA uptake: From vesicles to tissues. *Curr Gene Ther* 10:256–266.
Golzio M, Teissié J, Rols M-P (2002) Direct visualization at the single-cell level of electrically mediated gene delivery. *Proc Natl Acad Sci U S A* 99:1292–1297.
Gothelf A, Mir LM, Gehl J (2003) Electrochemotherapy: Results of cancer treatment using enhanced delivery of bleomycin by electroporation. *Cancer Treat Rev* 29:371–387.
Gracià RS, Bezlyepkina N, Knorr RL, Lipowsky R, Dimova R (2010) Effect of cholesterol on the rigidity of saturated and unsaturated membranes: Fluctuation and electrodeformation analysis of giant vesicles. *Soft Matter* 6:1472–1482.
Grosse C, Schwan HP (1992) Cellular membrane-potentials induced by alternating-fields. *Biophys J* 63:1632–1642.
Haluska CK, Riske KA, Marchi-Artzner V, Lehn JM, Lipowsky R, Dimova R (2006) Time scales of membrane fusion revealed by direct imaging of vesicle fusion with high temporal resolution. *Proc Natl Acad Sci U S A* 103:15841–15846.
Harbich W, Helfrich W (1979) Alignment and opening of giant lecithin vesicles by electric-fields. *Z Naturforsch, A: Phys Sci* 34:1063–1065.
Heller R, Gilbert R, Jaroszeski MJ (1999) Clinical applications of electrochemotherapy. *Adv Drug Del Rev* 35:119–129.
Jeyamkondan S, Jayas DS, Holley RA (1999) Pulsed electric field processing of foods: A Review. *J Food Prot* 62:1088–1096.
Kakorin S, Liese T, Neumann E (2003) Membrane curvature and high-field electroporation of lipid bilayer vesicles. *J Phys Chem B* 107:10243–10251.
Karatekin E, Sandre O, Guitouni H, Borghi N, Puech PH, Brochard-Wyart F (2003) Cascades of transient pores in giant vesicles: Line tension and transport. *Biophys J* 84:1734–1749.
Karimi M, Steinkühler J, Roy D, Dasgupta R, Lipowsky R, Dimova R (2018) Asymmetric ionic conditions generate large membrane curvatures. *Nano Lett* 18:7816–7821.
Kinosita K, Ashikawa I, Saita N, Yoshimura H, Itoh H, Nagayama K, Ikegami A (1988) Electroporation of cell-membrane visualized under a pulsed-laser fluorescence microscope. *Biophys J* 53:1015–1019.
Knorr RL, Staykova M, Gracia RS, Dimova R (2010) Wrinkling and electroporation of giant vesicles in the gel phase. *Soft Matter* 6:1990–1996.
Kraus M, Seifert U, Lipowsky R (1995) Gravity-induced shape transformations of vesicles. *EPL* (Europhysics Letters) 32:431.
Kummrow M, Helfrich W (1991) Deformation of giant lipid vesicles by electric-fields. *Phys Rev A* 44:8356–8360.
Lira RB, Dimova R, Riske KA (2014) Giant unilamellar vesicles formed by hybrid films of agarose and lipids display altered mechanical properties. *Biophys J* 107:1609–1619.
Lira RB, Robinson T, Dimova R, Riske KA (2019) Highly efficient protein-free membrane fusion: a giant vesicle study, *Biophys J* 116:79–91.
Litster JD (1975) Stability of lipid bilayers and red blood-cell membranes. *Phys Lett A* 53:193–194.
Mattei B, Franca ADC, Riske KA (2015) Solubilization of binary lipid mixtures by the detergent triton X-100: The role of cholesterol. *Langmuir* 31:378–386.
Mauroy C, Portet T, Winterhalder M, Bellard E, Blache M-C, Teissié J, Zumbusch A, Rols M-P (2012) Giant lipid vesicles under electric field pulses assessed by non invasive imaging. *Bioelectrochemistry* 87:253–259.
Mecke KR, Charitat T, Graner F (2003) Fluctuating lipid bilayer in an arbitrary potential: Theory and experimental determination of bending rigidity. *Langmuir* 19:2080–2087.
Nagle JF, Tristram-Nagle S (2000) Structure of lipid bilayers. *Biochim Biophys Acta* 1469:159–195.
Needham D, Evans E (1988) Structure and mechanical-properties of giant lipid (DMPC) vesicle bilayers from 20-Degrees-C Below to 10-Degrees-C above the Liquid-crystal crystalline phase-transition at 24-Degrees-C. *Biochemistry* 27:8261–8269.
Needham D, Hochmuth RM (1989) Electro-mechanical permeabilization of lipid vesicles: Role of membrane tension and compressibility. *Biophys J* 55:1001–1009.
Neumann E, Kakorin S, Toensing K (1998) Membrane electroporation and electromechanical deformation of vesicles and cells. *Faraday Discuss* 111:111–125.
Neumann E, Sowers AE, Jordan C (1989) *Electroporation and Electrofusion in Cell Biology*. New York: Plenum Press.
Niggemann G, Kummrow M, Helfrich W (1995) The bending rigidity of phosphatidylcholine bilayers: Dependences on experimental-method, sample cell sealing and temperature. *J Phys II* 5:413–425.
Nikolov V, Lipowsky R, Dimova R (2007) Behavior of giant vesicles with anchored DNA molecules. *Biophys J* 92:4356–4368.
Nuccitelli R, Pliquett U, Chen X, Ford W, Swanson RJ, Beebe SJ, Kolb JF, Schoenbach KH (2006) Nanosecond pulsed electric fields cause melanomas to self-destruct. *Biochem Biophys Res Commun* 343:351–360.
Olbrich K, Rawicz W, Needham D, Evans E (2000) Water permeability and mechanical strength of polyunsaturated lipid bilayers. *Biophys J* 79:321–327.
Peterlin P (2010) Frequency-dependent electrodeformation of giant phospholipid vesicles in AC electric field. *J Biol Phys* 36:339–354. doi:10.1007/s10867-010-9187-3.
Petrov AG (2002) Flexoelectricity of model and living membranes. *Biochim Biophys Acta* 1561:1–25.
Petrov AG (2006) Electricity and mechanics of biomembrane systems: Flexoelectricity in living membranes. *Anal Chim Acta* 568:70–83.
Portet T, Dimova R (2010) A new method for measuring edge tensions and stability of lipid bilayers: Effect of membrane composition. *Biophys J* 99:3264–3273.
Portet T, Febrer FCI, Escoffre JM, Favard C, Rols MP, Dean DS (2009) Visualization of membrane loss during the shrinkage of giant vesicles under electropulsation. *Biophys J* 96:4109–4121.
Portet T, Mauroy C, Demery V, Houles T, Escoffre JM, Dean DS, Rols MP (2012) Destabilizing giant vesicles with electric fields: An overview of current applications. *J Membr Biol* 245:555–564.
Puech PH, Borghi N, Karatekin E, Brochard-Wyart F (2003) Line thermodynamics: Adsorption at a membrane edge. *Phys Rev Lett* 90:128304.
Raphael RM, Popel AS, Brownell WE (2000) A membrane bending model of outer hair cell electromotility. *Biophys J* 78:2844–2862.
Rawicz W, Olbrich KC, McIntosh T, Needham D, Evans E (2000) Effect of chain length and unsaturation on elasticity of lipid bilayers. *Biophys J* 79:328–339.
Riske KA, Bezlyepkina N, Lipowsky R, Dimova R (2006) Electrofusion of model lipid membranes viewed with high temporal resolution. *Biophys Rev Lett* 1:387–400.
Riske KA, Dimova R (2005) Electro-deformation and poration of giant vesicles viewed with high temporal resolution. *Biophys J* 88:1143–1155.

Riske KA, Dimova R (2006) Electric pulses induce cylindrical deformations on giant vesicles in salt solutions. *Biophys J* 91:1778–1786.

Riske KA, Knorr RL, Dimova R (2009a) Bursting of charged multicomponent vesicles subjected to electric pulses. *Soft Matter* 5:1983–1986.

Riske KA, Sudbrack TP, Archilha NL, Uchoa AF, Schroder AP, Marques CM, Baptista MS, Itri R (2009b) Giant vesicles under oxidative stress induced by a membrane-anchored photosensitizer. *Biophys J* 97:1362–1370.

Robinson KR (1985) The responses of cells to electrical fields: A Review. *J Cell Biol* 101:2023–2027.

Robinson T, Verboket PE, Eyer K, Dittrich PS (2014) Controllable electrofusion of lipid vesicles: Initiation and analysis of reactions within biomimetic containers. *Lab Chip* 14:2852–2859.

Rodriguez N, Cribier S, Pincet F (2006) Transition from long- to short-lived transient pores in giant vesicles in an aqueous medium. *Phys Rev E* 74:061902.

Ryham R, Berezovik I, Cohen FS (2011) Aqueous viscosity is the primary source of friction in lipidic pore dynamics. *Biophys J* 101:2929–2938.

Sadik MM, Li JB, Shan JW, Shreiber DI, Lin H (2011) Vesicle deformation and poration under strong dc electric fields. *Phys Rev E* 83:066316.

Saito AC, Ogura T, Fujiwara K, Murata S, Nomura SM (2014) Introducing micrometer-sized artificial objects into live cells: A method for cell-giant unilamellar vesicle electrofusion. *PLoS One* 9:e106853.

Salipante PF, Knorr RL, Dimova R, Vlahovska PM (2012) Electrodeformation method for measuring the capacitance of bilayer membranes. *Soft Matter* 8:3810–3816.

Salipante PF, Shapiro ML, Vlahovska PM (2015) Electric field induced deformations of biomimetic fluid membranes. *Procedia IUTAM* 16:60–69.

Salipante PF, Vlahovska PM (2014) Vesicle deformation in DC electric pulses. *Soft Matter* 10:3386–3393.

Sandre O, Moreaux L, Brochard-Wyart F (1999) Dynamics of transient pores in stretched vesicles. *Proc Natl Acad Sci U S A* 96:10591–10596.

Schwan HP (1985) Dielectric properties of cells and tissues. In: *Interactions between electromagnetic fields and cells* (Chiabrera A, Nicolini C, Schwan HP, Eds), pp 75–97. New York: Plenum Press.

Seifert U, Lipowsky R (1995) Morphology of vesicles. In: *Structure and Dynamics of Membranes (Handbook of Biological Physics)* (Lipowsky R, Sackmann E, Eds), pp 403–463. Amsterdam, the Netherlands: Elsevier.

Shillcock JC, Lipowsky R (2005) Tension-induced fusion of bilayer membranes and vesicles. *Nat Mater* 4:225–228.

Shirakashi R, Sukhorukov VL, Reuss R, Schulz A, Zimmermann U (2012) Effects of a pulse electric field on electrofusion of giant unilamellar vesicle (GUV)-Jurkat Cell (Measurement of fusion ratio and electric field analysis of pulsed GUV-Jurkat Cell). *J Therm Sci Tech-Jpn* 7:589–602.

Spector AA, Deo N, Grosh K, Ratnanather JT, Raphael RM (2006) Electromechanical models of the outer hair cell composite membrane. *J Membrane Biol* 209:135–152.

Staykova M, Lipowsky R, Dimova R (2008) Membrane flow patterns in multicomponent giant vesicles induced by alternating electric fields. *Soft Matter* 4:2168–2171.

Steinkühler J, Agudo-Canalejo J, Lipowsky R, Dimova R (2015) Variable adhesion strength for giant unilamellar vesicles controlled by external electrostatic potentials. *Biophys J* 108:402a.

Steinkühler J, Agudo-Canalejo J, Lipowsky R, Dimova R (2016) Modulating vesicle adhesion by electric fields. *Biophys J* 111:1454–1464.

Takashima S, Schwan HP (1985) Alignment of microscopic particles in electric-fields and its biological implications. *Biophys J* 47:513–518.

Teissie J, Golzio M, Rols MP (2005) Mechanisms of cell membrane electropermeabilization: A minireview of our present (lack of ?) knowledge. *Biochim Biophys Acta* 1724:270–280.

Teissie J, Tsong TY (1981) Electric-Field induced transient pores in phospholipid-bilayer vesicles. *Biochemistry* 20:1548–1554.

Tekle E, Astumian RD, Chock PB (1994) Selective and asymmetric molecular-transport across electroporated cell-membranes. *Proc Natl Acad Sci U S A* 91:11512–11516.

Tekle E, Astumian RD, Friauf WA, Chock PB (2001) Asymmetric pore distribution and loss of membrane lipid in electroporated DOPC vesicles. *Biophys J* 81:960–968.

Tian AW, Capraro BR, Esposito C, Baumgart T (2009) Bending stiffness depends on curvature of ternary lipid mixture tubular membranes. *Biophys J* 97:1636–1646.

Toepfl S, Mathys A, Heinz V, Knorr D (2006) Review: Potential of high hydrostatic pressure and pulsed electric fields for energy efficient and environmentally friendly food processing. *Food Rev Int* 22:405–423.

Tsong TY (1991) Electroporation of cell-membranes. *Biophys J* 60:297–306.

Vequi-Suplicy CC, Riske KA, Knorr RL, Dimova R (2010) Vesicles with charged domains. *Biochimica et Biophysica Acta (BBA)—Biomembranes* 1798:1338–1347.

Vernhes MC, Benichou A, Pernin P, Cabanes PA, Teissié J (2002) Elimination of free-living amoebae in fresh water with pulsed electric fields. *Water Res* 36:3429–3438.

Vitkova V, Genova J, Bivas I (2004) Permeability and the hidden area of lipid bilayers. *Eur Biophys J Biophy Letters* 33:706–714.

Vlahovska PM (2015) Voltage-morphology coupling in biomimetic membranes: Dynamics of giant vesicles in applied electric fields. *Soft Matter* 11:7232–7236.

Vlahovska PM, Gracia RS, Aranda-Espinoza S, Dimova R (2009) Electrohydrodynamic model of vesicle deformation in alternating electric fields. *Biophys J* 96:4789–4803.

Weaver JC (2003) Electroporation of biological membranes from multicellular to nano scales. *IEEE T Dielect El In* 10:754–768.

Weaver JC, Chizmadzhev YA (1996) Theory of electroporation: A review. *Bioelectrochem Bioenerg* 41:135–160.

Yamamoto T, Aranda-Espinoza S, Dimova R, Lipowsky R (2010) Stability of spherical vesicles in electric fields. *Langmuir* 26:12390–12407.

Yang P, Dimova R (2011) Nanoparticle synthesis in vesicle microreactors. In: *Biomimetic Based Applications* (George A, ed), pp 523–552. Rijeka: InTech.

Yang P, Lipowsky R, Dimova R (2009) Nanoparticle formation in giant vesicles: Synthesis in biomimetic compartments. *Small* 5:2033–2037.

Yu M, Lira RB, Riske KA, Dimova R, Lin H (2015) Ellipsoidal relaxation of deformed vesicles. *Phys Rev Lett* 115: 128303.

Zhang J, Zahn JD, Tan WC, Lin H (2013) A transient solution for vesicle electrodeformation and relaxation. *Phys Fluids* 25: 071903.

Zhelev DV, Needham D (1993) Tension-stabilized pores in giant vesicles: Determination of pore-size and pore line tension. *Biochim Biophys Acta* 1147:89–104.

Zimmermann U (1986) Electrical breakdown, electropermeabilization and electrofusion. *Rev Physiol Biochem Pharmacol* 105:175–256.

Zimmermann U, Neil GA (1996) *Electromanipulation of Cells.* Boca Raton, FL: CRC Press.

Zimmermann U, Vienken J (1982) Electric field-induced cell-to-cell fusion. *J Membr Biol* 67:165–182.

16 Creating membrane nanotubes from giant unilamellar vesicles

Coline Prévost, Mijo Simunovic, and Patricia Bassereau

Pulling tubes from GUVs looks like pulling hair
but is a lot less painful… or maybe it isn't.

Contents

16.1 INTRODUCTION

The relatively large size of giant unilamellar vesicles (GUVs) makes them amenable for micromanipulation, which means that we can directly change the shape of the membrane while visualizing the effect with optical microscopy-based methods. The shape that mostly interests us here is a tube. How can one create a tube from a vesicle? If you press on a piece of paper, you deform it, leaving sharp corners. Pressing on lipid membranes does not leave corners because **membranes are fluid** and so they will minimize and smooth out their surface. They can also be strongly bent. So if you pinch a membrane and push it toward you or away from you, the membrane will form a tube between where it is pinched and the underlying membrane surface, to minimize its stretching. By contrast, if you did the same to a balloon—which does not have a fluid surface—it would pop! Practically, pulling tubes out of GUVs typically forms tiny tubes, with radii ranging between about 10 nm to a few hundred nanometers, and so we call them nanotubes.

The aim of this chapter is to describe methods of creating membrane nanotubes by pulling them from the surface of GUVs using external forces of different origins. Because the GUV diameter is several orders of magnitude larger than that of the pulled tube, it acts as a quasi-flat and quasi-infinite lipid reservoir for the nanotube. So far, this method has seen two main applications: (1) measuring the equilibrium and dynamical mechanical properties of membranes and, more recently, (2) characterizing cellular processes occurring at curved membrane interfaces.

The method was originally designed to measure the elasticity of the membrane of red blood cells. In the initial implementation, cells were adhered to a glass surface and tubules were pulled by using a fluid flow (Hochmuth et al., 1973). It was later modernized by introducing a micropipette to hold a vesicle (or a cell) and by applying a localized pulling force to extrude a nanotube (Hochmuth and Evans, 1982; Hochmuth et al., 1982). Most applications today are based on the latter approach, and they are the focus of this chapter.

First, we will summarize the theoretical basis that underlies membrane tube-pulling experiments. Then, we will describe the most widely used tube-pulling setup employing micropipette aspiration techniques, optical tweezers and confocal microscopy, highlighting experimental details and difficulties surrounding the experiment. Next, we will describe alternative methods of pulling membrane nanotubes and their applications. Finally, we will outline the past and future applications of the tube-pulling assay.

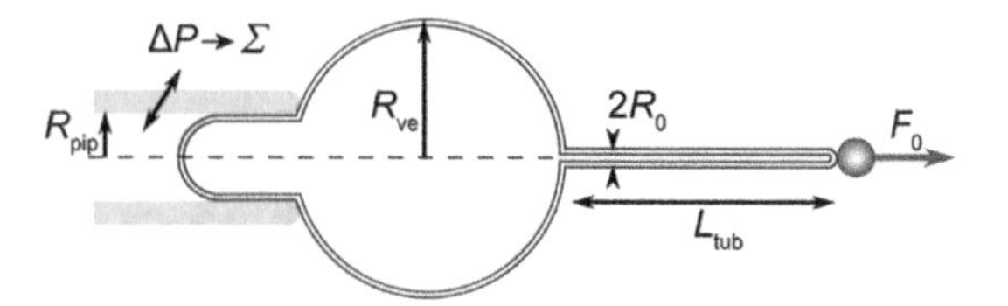

Figure 16.1 The tube-pulling assay. A GUV is aspirated in a micropipette (left) with an aspiration pressure (pressure difference between the pipette and the experimental chamber), ΔP. The aspiration pressure directly sets membrane tension (see Eq. 16.6). A membrane nanotube pulled from the body of the vesicle is held by an external force, F_0. This is often achieved by sticking a micrometer-sized bead opposite of the pipette and using it as a handle to pull on the membrane. Practically, the radii of the GUV and of the tube differ by two to three orders of magnitude, with R_{ve} ~ 10m and R_{tub} ~ 10–200 nm.

16.2 PHYSICS OF A MEMBRANE NANOTUBE CONNECTED TO A GUV

16.2.1 STATICS

The mechanics of tube formation has been extensively studied by different groups (Evans and Yeung, 1994; Heinrich and Waugh, 1996; Svetina et al., 1998; Derenyi et al., 2002; Powers et al., 2002). We give the main results below.

The total free energy of a membrane surface (neglecting Gaussian and spontaneous curvatures) is written (Canham, 1970; Helfrich, 1973)

$$\mathcal{F} = \int dA_{me}\left(\frac{\kappa}{2}(2M)^2 + \Sigma\right) \tag{16.1}$$

Here M is the local mean curvature; κ is the bending modulus (or bending rigidity); and Σ is the membrane tension. M is defined in each point of the surface by $M = \frac{1}{2}\left(\frac{1}{R_1} + \frac{1}{R_2}\right)$, where R_1 and R_2 are the two principal radii of curvature at this point. Refer to Chapter 5 for an extensive description of the mechanical properties of lipid membranes.

We consider the system where a nanotube is pulled from a GUV aspirated in a micropipette by application of a point force F (Figure 16.1). Including the work of the pressure forces and of the external pulling force F, the general expression of the free energy of the system is given by

$$\mathcal{F} = \int dA\frac{\kappa}{2}(2M)^2 + \Sigma A - \Delta PV - FL_{tub} \tag{16.2}$$

where A and V are the vesicle area and volume, respectively; ΔP is the pressure difference at the membrane interface; and L is the length of the tube. Under these experimental conditions, the shape transformations associated with pulling the tube occur at constant surface tension and pressure: The membrane projection inside the pipette—which we call the tongue—serves as a reservoir of lipids buffering the variations of area in the rest of the system, and the pressure is maintained via osmosis. Moreover, it can be shown that the pressure has a sub-leading contribution to the energy, and this term is usually disregarded (Evans and Yeung, 1994; Derenyi et al., 2002; Powers et al., 2002) although recent theories show that pressure may play an important part in determining the stability and fluctuations of tubes (Monnier et al., 2010).

The free energy in Eq. 16.2 also neglects the nonlocal curvature elasticity (Waugh et al., 1992; Evans and Yeung, 1994). This term accounts for the asymmetric stretching of both monolayers upon membrane bending (as the tube forms and extends, the outer monolayer is expanded while the inner monolayer is compressed). However, this contribution is only sizeable when the GUV and tube have comparable area (e.g., in the case of very long tubes). Therefore, it can be neglected in most typical experimental conditions.

The energy of the tube alone, $\mathcal{F}_{tub}$, can be written as

$$\mathcal{F}_{tub} = 2\pi R_{tub}L\left[\frac{\kappa}{2R_{tub}^2} + \Sigma\right] - FL \tag{16.3}$$

where R_{tub} is the tube radius. This expression tells us that the equilibrium radius results from the competition between bending rigidity and surface tension, with the former working to expand the tube and the latter working to constrict it.

The equilibrium radius R_0 and force F_0 are obtained through minimization of $\mathcal{F}_{tub}$ with respect to R_{tub} and L_{tub}, respectively (Waugh and Hochmuth, 1987; Evans and Yeung, 1994; Heinrich and Waugh, 1996; Derenyi et al., 2002)

$$R_0 = \sqrt{\frac{\kappa}{2\Sigma}} \tag{16.4}$$

$$F_0 = 2\pi\sqrt{2\kappa\Sigma} \tag{16.5}$$

The force does not depend on the length of the tube, which is again a consequence of the fact that the material needed to extend the tube is taken from the reservoir (the vesicle tongue)

(Hochmuth et al., 1982). Applying the Laplace law to the hemispherical cap of the tongue and to the spherical portion of the vesicle outside the pipette, one gets (Evans et al., 1976)

$$\Sigma = \frac{\Delta P R_{pip}}{2\left(1 - R_{pip}/R_{ve}\right)} \tag{16.6}$$

where R_{pip} and R_{ve} are the radii of the pipette and vesicle respectively, and ΔP is the hydrostatic pressure difference between the chamber pressure and the pipette pressure. Here we recommend Chapter 11 for a comprehensive review on the micropipette aspiration method.

Eq. 16.5 for the force can be used to measure the bending rigidity of membranes with specific lipid compositions in experiments where the tension is controlled by micropipette aspiration and the force is measured using optical or magnetic tweezers (Heinrich and Waugh, 1996; Cuvelier et al., 2005). In addition, combining Eqs. 16.4 and 16.5, one obtains

$$R_0 = \frac{F_0}{4\pi\Sigma} \tag{16.7}$$

which can be used to measure the radius of the tube. We will come back to these relations in Section 3.

A vesicle subjected to an axisymmetric extension initially assumes a catenoid shape in the vicinity of the point of application of the force. For larger deformations, a first-order shape transition occurs, resulting in the coexistence of a quasi-spherical vesicle and a tube. At the transition the pulling force overshoots before converging to F_0. The magnitude of the overshoot has been calculated to be ~ $0.13F_0$ for an applied point force. This reflects the mechanical instability of the linear deformation, which leads to the nucleation of the tube concomitant with the relaxation of the body of the vesicle to a more spherical shape (Figure 16.2) (Derenyi et al., 2002; Powers et al., 2002; Rossier et al., 2003; Koster et al., 2005; Ashok and Ananthakrishna, 2014). In reality, much larger overshoots are observed due to the finite contact area of the membrane on the bead that is used to pull on it, and the force overshoot scales linearly with the radius of the patch (Koster et al., 2005). This point has practical implications, as we will see later.

Although it is limited to simple systems, the theory outlined above is conceptually enlightening, and the basis for the theoretical description of more complex systems. A case of great biological relevance is the asymmetrical insertion of proteins or protein motifs in the membrane, or the binding to the membrane of proteins with a curvature preference. It is often modeled by a membrane spontaneous curvature (Helfrich, 1973; Markin, 1981; Leibler, 1986) such that the bending energy becomes

$$\varepsilon_{be} = \frac{\kappa}{2}\left(M - C_{pr}\left(\phi_{pr}\right)\right)^2 \tag{16.8}$$

where ϕ_{pr} is the fractional protein coverage and $C_{pr}\left(\phi_{pr}\right)$ is the spontaneous curvature of the membrane upon protein binding. This approach has been applied to the theoretical description of curvature–protein coupling in tube experiments, yielding modified equations for the radius and the force. Several studies have assumed a linear relationship between spontaneous curvature and fractional protein coverage (Marcerou et al., 1984; Leibler, 1986; Campelo et al., 2008; Sorre et al., 2012; Aimon et al., 2014): $C_{pr}(\phi) = \bar{C}_{pr}\phi_{pr}$, where $\bar{C}_{pr}$ is the intrinsic spontaneous curvature of the protein. There are variations to this model, including nonlinear curvature/composition coupling (Zhu et al., 2012) and composition-dependent bending rigidity (Aimon et al., 2014). Protein–protein interactions have been explicitly included in the paper by Zhu et al. (2012). More theoretical details about modeling membranes with spontaneous curvature can be found in Chapter 5.

16.2.2 DYNAMICS OF PULLING AND MEASURE OF FRICTION

We now briefly introduce the dynamical properties of membrane tubes. These properties are probed in rheology measurements where the length of the tube is varied at different rates while monitoring the force response. Such experiments reveal the friction forces in the system, in particular, in cells, the friction between the plasma membrane (PM) and the underlying cytoskeleton (Sheetz, 2001; Borghi et al., 2003; Brochard-Wyart et al., 2006).

However, dynamical force responses are observed in artificial systems as well, although mainly for large pulling speed and/or displacement. In a seminal theoretical study, Evans and Yeung (1994) uncovered the viscous force arising from the relative motion of the monolayers at the neck of the tube (interlayer drag). For very long tubes (greater than hundreds of micrometers), friction against the solvent overcomes the interlayer drag and causes a gradient of tension and a nonuniform radius along the length of the tube (Rossier et al., 2003). Nevertheless, these effects are small for short tubes extended at moderate speed (~10 μm, 1 μm/s). However, dynamical effects arise from even low levels of impurities in

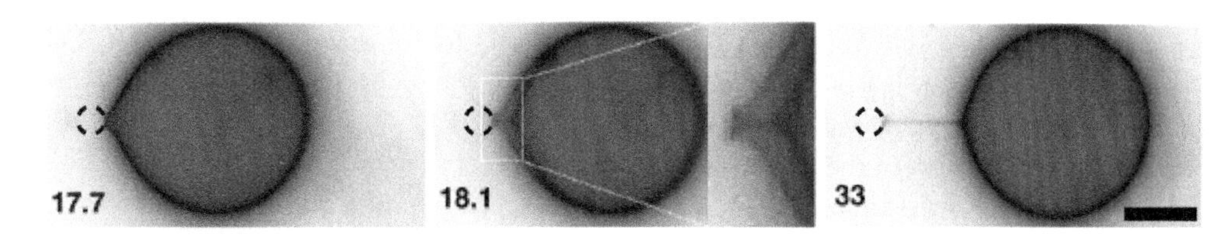

Figure 16.2 Inverted contrast fluorescence images showing the evolution of the shape of the vesicle upon pulling. In the middle image, the signal was integrated over 160 ms such that overlapping images of the vesicle immediately before and after the nucleation of the tube can be seen. The zoom in the inset highlights the relaxation of the body of the vesicle once the tube has nucleated. The dashed circle highlights the position of the bead. The adhesion patch can be seen at the junction between the bead and the membrane. The numbers in the bottom left corner indicate the time after the start of pulling in seconds. Scale bar: 10 μm. (Adapted from Koster, G. et al., *Phys. Rev. Lett.*, 94, 068101, 2005.)

the membrane, which increase the effective friction coefficient (Campillo et al., 2013). Similar effects are also present, when pulling tubes from cells (Datar et al., 2015).

16.3 PULLING NANOTUBES WITH OPTICAL TWEEZERS

Before going into the description of a sophisticated tube-pulling setup (combining micromanipulation, optical trapping and confocal imaging), we point out that, depending on the application, optical tweezers and a confocal microscope can be dispensable. There are several suitable techniques to pull a tube from a GUV (see Section 16.4), and fluorescence microscopy is not required to study mechanical or dynamical properties of membranes for instance (when fluorescence microscopy is required, confocal imaging would be the technique of choice because in epifluorescence the GUV appears much brighter than the tube, resulting in poor signal-to-noise ratio). However, in this section we describe an optimal system in the sense that we can extract the greatest amount of information from the measurements.

The micromanipulation of vesicles is achieved with a micropipette that not only controls the position of the vesicle, but also sets membrane tension through its aspiration pressure. Moreover, using a force transducer (i.e., magnetic or optical tweezers) offers the possibility of measuring the force exerted by the tubule (Heinrich and Waugh, 1996; Cuvelier et al., 2005). Last, by coupling these elements with a confocal microscope, it is possible to visualize the nanotube, deduce its radius, and evaluate, for example, the effect of curvature on the redistribution of lipids and proteins between GUV and nanotube (Callan-Jones and Bassereau, 2013).

16.3.1 MICROMANIPULATION

Micromanipulation of a vesicle can be done in a homemade experimental chamber. The chamber should be open at its sides, so that the micropipette can be inserted with its tip approximately parallel to the coverslip. Figure 16.3c shows a homemade experimental chamber positioned on the stage of an inverted microscope, with one micropipette inserted at each of its sides. A second micropipette can be used to microinject a protein solution near the vesicle (Morlot et al., 2012; Sorre et al., 2012), to hold a bead for tube pulling (Hochmuth et al., 1982; Zhu et al., 2012) if no optical tweezers are available, or in experiments where two tubes are pulled from the same vesicle to create a Y-junction (Cuvelier et al., 2005). The chamber is made of two parallel coverslips about 3 cm long, 1 cm wide, and positioned 1 mm apart.

In preparation for an experiment, the back of the micropipette is inserted into a metal holder connected via plastic tubing to a water tank with adjustable vertical position (Figure 16.3b), which allows for hydrostatic pressure control (recall that membrane tension is set by the pressure difference between the pipette and the chamber, Eq. 16.6). The flow can be switched on and off with a valve. Using a three-way valve and connecting one inlet to a syringe will facilitate purging the circuit when necessary.

After filling the chamber with solution, the micropipette is inserted until its tip is seen through the microscope. The displacement of the micropipette is typically achieved with a mechanical

(a)

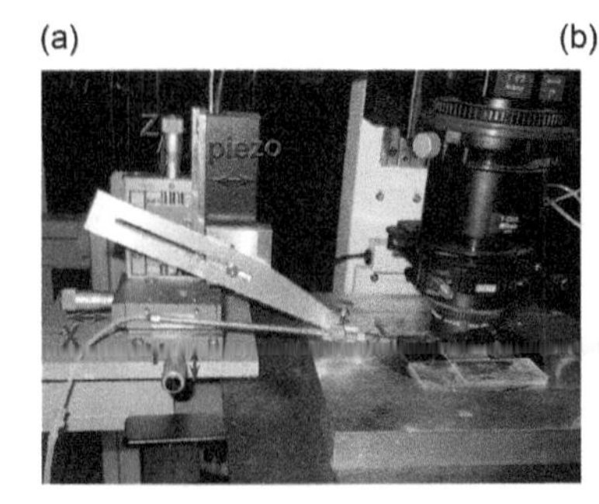

(b)

(c)

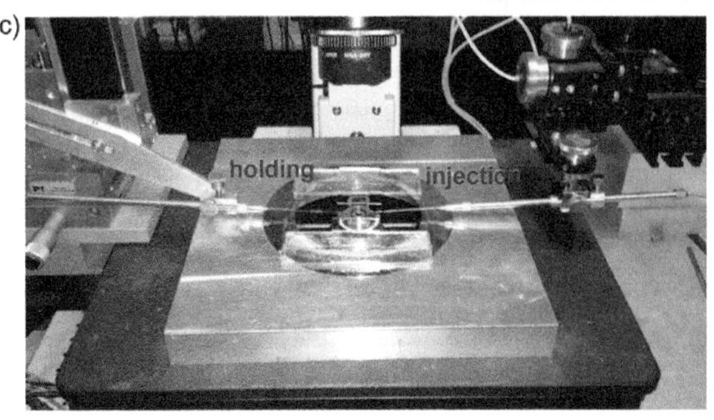

Figure 16.3 An example of an experimental setup. (a) The micromanipulation block. The block is attached to the side of the microscope stage. It is made up of three mechanical micromanipulators assembled to achieve three-dimensional displacements. In addition, a piezoelectric actuator is integrated to achieve more precise displacement along the X-axis (which is the direction of tube pulling). The micropipette holder is back-connected to the water circuit (which ends in the water tank), and clamped to the micromanipulation unit through a custom part. (b) The water tank. It is convenient to use two micromanipulators: One is used to set the zero position (#1) and the other to adjust membrane tension (#2). In this way the displacement relative to the zero-position can directly be read on the second manipulator. (c) The experimental chamber is made up of two coverslips separated by ~1-mm spacers. The pipettes are inserted in between the coverslips. One is used to manipulate and set the tension of the GUVs; the other is used here for protein microinjection. The displacement of the second pipette is controlled via a hydraulic micromanipulator.

or hydraulic micromanipulator. A more precise way of controlling the movement of the pipette is with a piezoelectric actuator, which is used for applications that require more control of the tube length or of the pulling speed, for example, for measuring membrane rheological properties (Bornschlogl et al., 2013; Campillo et al., 2013).

For GUV aspiration, micropipettes are generally made such that their opening is ~5 μm in diameter. We refer the reader to Chapter 11 for information on the specifics of the micropipette aspiration technique, including micropipette fabrication.

16.3.2 OPTICAL TWEEZERS

Laser-based trapping and application to manipulation of small objects was pioneered by Ashkin (Ashkin, 1970; Ashkin et al., 1986; Neuman and Nagy, 2008). It uses the force arising from the transfer of momentum from a light beam to an interacting particle to trap and manipulate micrometer-sized objects. In addition, an optical trap (or optical tweezers) can be used to apply calibrated forces in the pN range, and to measure such forces. For the purpose of tube pulling, it is used to trap small beads (~3 μm in size, optimum for these experiments in term of trapping and force achieved) that provide the handle to pull the tube.

Optical trapping is achieved by focusing a laser beam with an objective of high numerical aperture (NA > 1.2) within the experiment chamber. The high NA objective tightly focuses the incoming beam, resulting in a steep intensity gradient at the focus. In these conditions the gradient force, the force that pulls the object toward the center of the trap, exceeds the scattering force, which pushes the objects in the direction of the light, resulting in efficient trapping (Neuman and Block, 2004; Nieminen et al., 2007). Strong trapping also requires slightly overfilling the objective back aperture to ensure that the light converges as a tight, diffraction-limited spot (Neuman and Block, 2004). This is achieved by expanding the beam before coupling it into the microscope optical path. In the most typical configuration, the trapping laser beam is introduced in the optical path of the microscope before the objective, such that the objective is used for both imaging and focusing the trapping beam. Alternatively two separate objectives can be used (Heinrich et al., 2010a).

When a tube is pulled, it slightly displaces the bead from its equilibrium position. For a range of displacements, the force exerted by the tube is proportional to the displacement of the bead in the trap. The force is then given by

$$F = k_{OT}\Delta x \qquad (16.9)$$

where k_{OT} is the stiffness of the optical trap; and Δx is the position of the center of mass of the bead relative to its equilibrium position. A typical value of k_{OT} suitable for tube pulling is ~100 pN/μm. A simple way of measuring the position of the bead is through video tracking. The bright-field image is recorded on a video camera and the digitalized images are analyzed by a program that finds the center of mass of the bead on each frame.

Describing optical tweezers in further detail is beyond the scope of this book. For a step-by-step protocol for implementing a custom-built optical trap on a fluorescence microscope, refer to Lee et al. (2007); see Neuman and Block (2004) for a review of the various techniques of bead position detection and trap stiffness calibration. Readers interested in the subject can also refer to Chapter 13 of this book, which describes another kind of optical trap, the dual-beam laser trap, and its applications as an optical stretcher.

16.3.3 CONFOCAL MICROSCOPY

Depending on membrane tension, the tube diameter is in the range of 15 to a few hundred nanometers, which is below or very close to the diffraction limit. Therefore, it is hardly visible with differential interference contrast or phase-contrast microscopy. The use of confocal fluorescence microscopy has several advantages: (1) the tube can be visualized (Figure 16.4); (2) changes in shape and integrity can be followed (e.g., under the action of proteins the tube may become flexible, or split into coexisting segments of different radii, or it may fission, see Sections 5.1 and 5.2); (3) the radius can be measured after appropriate calibration (see Section 3.5); and (4) enrichment/depletion of proteins or lipids in the tube can be quantified. To visualize the membrane, the composition is typically doped with 0.1%–1% molar fraction of a fluorescent derivative of a lipid (often phosphatidylethanolamine). Proteins are labeled with a spectrally distinct fluorophore. Refer to Chapter 10 for details on the advantages and disadvantages of using fluorescence microscopy for the visualization of GUVs.

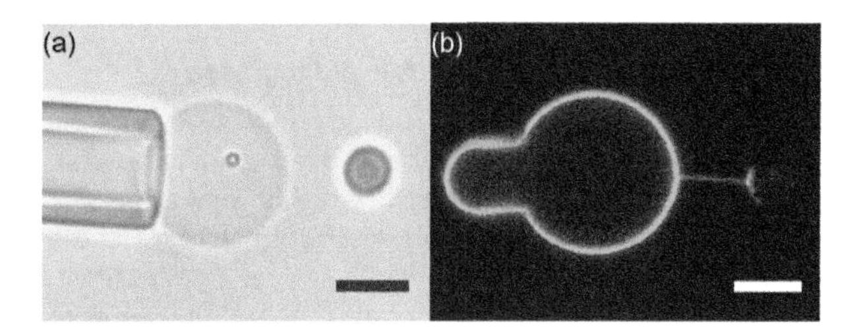

Figure 16.4 Microscopy images of the system. (a) Bright-field image showing the pipette, GUV and bead. The tube is below the resolution limit and is not visible on this image. Scale bar: 5 μm. (b) Confocal slice of the same GUV showing the tube connecting the GUV and the bead. The membrane contains 0.5% of the fluorescent tracer BODIPY® TR ceramide. Scale bar: 5 μm.

16.3.4 THE EXPERIMENTAL PROTOCOL

Tube pulling

Box 16.1 summarizes the practical details of tube pulling.

Microinjection

To study the binding of molecules (such as proteins) as a function of curvature, another micropipette can be introduced in the same way as the aspiration pipette. The injection pipette is filled with the molecule solubilized in an experimental buffer with the same osmolarity as the experimental solution. To be very precise, the injection buffer osmolarity should be slightly increased to compensate for the initial intentional evaporation of the experimental buffer. If dealing with precious samples, such as proteins, a few microliters are sucked into the tip of the pipette and the rest of the pipette then back-filled with oil to prevent mixing. The diameter of the injection pipette is not critical, but a larger opening will facilitate the filling step. An inner diameter of 5–10 μm is usually fine. When the pipette is immersed into the solution (typically just before sealing the chamber), it is important to set the pressure of the injection pipette close to zero (in the same way as for the aspiration pipette) to prevent the protein from leaking out into the chamber.

When the experimental system is ready, the injection pipette is brought close to the vesicle and the tube and the solution inside is injected at low pressure (low enough to prevent effects due to flow). Keep in mind that the effective concentration of the injected molecule in the vicinity of the vesicle is going to be lower than the concentration inside the pipette due to dilution (Sorre et al., 2012; Simunovic et al., 2015). Fluorescence calibration can be made to measure the dilution rate. In principle, a commercial microinjector could be used, but they are usually designed for cells and do not allow for a slow delivery, and the flow cannot be completely stopped.

Box 16.1 Flowchart of the tube-pulling assay

Preparation

- Grow GUVs containing a fraction of a biotinylated lipid (~0.05 molar %), and of a lipid conjugated to a fluorophore (~0.5 molar %). Remember to use a solution with approximately equal osmolarity to the experimental solution to prevent osmotic shock. The experimental buffer should contain at least a few millimoles of salt (e.g., NaCl) to ensure biotin–streptavidin interaction.
- Fill the experimental chamber and the holding pipette with β-casein (dissolved at ~5 mg/mL in buffer). **Critical step:** β-casein adheres to the hydrophilic glass surface and prevents nonspecific adhesion of the GUVs to the chamber surface and to the pipette.
- Insert the pipette in its holder with the valve open to avoid introducing air bubbles in the circuit. Position the pipette in the chamber, with its tip close to the bottom surface.
- Replace the casein solution in the chamber with the experimental buffer. Add a few microliters of the solutions of GUVs and of streptavidin-coated beads (the final concentration of beads in the chamber should be about 0.110^{-3}% w/v or less). **Critical step:** it is important that not too many of either is added to the chamber as they will be attracted to the center of the optical trap and disrupt the measurement.
- Let the beads and GUVs settle. Let the experimental buffer evaporate. **Critical step:** this raises the osmolarity of the external buffer by ~1 mOsm/min in the experimental chamber described above at 21°C, causing water to flow out of the GUVs to equilibrate the osmolarities. As a result, the GUVs deflate (become floppy), creating excess area that is optically visible as undulations of the membrane. This process can take anywhere from a few minutes to an hour (corresponding to a decrease in the volume of the solution of up to 25% approximately). It is important to work with floppy vesicles because (1) it is easier to pull a tube from a low-tension vesicle (Koster et al., 2003) and (2) the relation between tension and aspiration pressure (Eq. 16.6) holds only if the length of the aspired part of the vesicle is larger than pipette diameter.
- Once the GUVs are floppy, seal the edges of the chamber with mineral oil. **Critical step:** this prevents further evaporation, and therefore any pressure variations in the chamber.

Measurement

1. "Zero" the height of the water tank: Observe the movement of a bead close to the opening of the pipette while adjusting the height of the tank, until the motion of the bead reveals the absence of flow in the vicinity of the pipette. This procedure should be done at the height where the bead will be trapped during the experiment (typically ~15–20 μm above the surface so that the vesicle does not sit on the bottom of the chamber, but the trapping power—which decreases with height—remains high). **Critical step:** the zero-position can drift in time if the oil does not ideally seal the chamber, so it is advisable to readjust it before each measurement.
2. Find a floppy GUV and aspirate it into the pipette. The length of the GUV tongue must exceed the radius of the pipette. GUV membrane might be pre-stretched to eliminate possible preexisting membrane reservoirs, but this is not a requirement. To avoid losing the GUV while exploring the chamber, move the pipette away from the surface.
3. Look for a bead. Bring the pipette back in focus and trap the bead in the tweezers. Record a bright-field movie to get the equilibrium position of the bead in the trap.
4. Decrease the aspiration pressure as much as possible to minimize membrane tension while keeping the vesicle in the micropipette. Carefully bring the GUV in contact with the bead to establish biotin–streptavidin bonding, then gently pull it back. **Critical step:** precisely how long to keep the GUV and bead in contact depends on many factors, primarily on bead coating efficiency and on lipid membrane composition (it takes between a few milliseconds and a few seconds). If the contact lasts too long, the adhesion surface, and thus the force to initiate the tube, may overcome the strength of the optical trap (Koster et al., 2003).
5. Increase the aspiration pressure to recover the tongue. Slightly move the pipette such that the tube lies along the *X*-axis (pipette axis). Adjust the focus to the equator of the GUV. Start ramping up the tension.
6. At each step, record the tank height (for tension measurement, Eq. 16.6), a bright-field movie (for force measurement, Eq. 16.9), and take a fluorescent image of the vesicle.

16.3.5 MEASUREMENTS

Tension. The tension is obtained from Eq. 16.6, with $\Delta P = \rho g h$, ρ being the density of water, g the gravitational acceleration, and h the vertical position of the water tank relative to the zero-position. R_{pip} and R_{ve} are obtained from the confocal images of the aspirated vesicle.

Force. In the elastic range of the optical trap, the force exerted by the tube is given by Eq. 16.9. As mentioned, the equilibrium position is measured for each bead before pulling the tube. For pure lipid membranes, the membrane bending rigidity can readily be obtained by linear regression of the force curve (F^2 versus Σ) using Eq. 16.5 (see a typical force curve in Figure 16.5a).

Radius. For pure symmetric lipid membranes (same composition of the inner and the outer leaflets, same internal and external buffers), far from demixing transitions, Eq. 16.7 applies. In this case, knowing the force and tension, it is possible to measure the radius of the tube (see an example in Figure 16.5b). In addition, this expression can be used to calibrate the fluorescence of lipid tracers in the tube, which is proportional to the tube radius (this is not necessarily true close to demixing points where curvature-induced fluorophore sorting can occur [Sorre et al., 2009]). As long as the tube is thin enough to be contained within the focal volume (in the transverse direction), the fluorescence of each pixel along the tube will

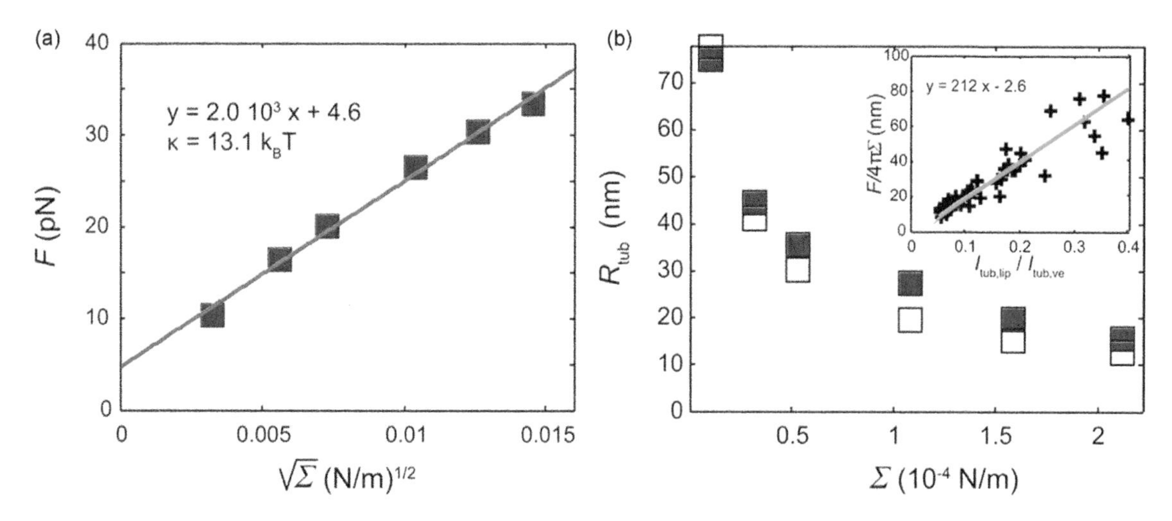

Figure 16.5 Assessing the mechanical properties of GUVs. (a) A typical force curve for a naked GUV with composition 57% egg phosphatidylcholine, 8% phosphatidylinositol 4,5-bisphosphate, 15% cholesterol, 10% dioleoylphosphatidylserine, 10% dioleoylphosphatidylethanolamine, complemented with 0.5% BODIPY® TR ceramide and 0.1% DSPE-PEG biotin, see Appendices 1 and 2 of the book for structure and data on these lipids and lipid dyes. In these conditions, there is a simple relationship between force, tension and bending rigidity (Eq. 16.5). The bending rigidity is obtained by fitting Eq. 16.5 to the tension–force data. Note the offset of the curve. See Box 16.2 for tentative explanations for this effect (b) Tube radius as a function of membrane tension for the same GUV. The radius was calculated from the force (see Eq. 16.7, open squares), or from fluorescence measurements (see Eq. 16.11, solid squares), using $R_{cal} = 212$nm. For these GUVs, both measurements yield similar values. In the presence of proteins, or in the proximity of a lipid–lipid demixing transition, only the radius inferred from fluorescence measurement is reliable, as Eq. 16.7 may not apply. Inset: calibration of the tube fluorescence. Pulling experiments were done on pure lipid membranes. The radius of the tube (as given by Eq. 16.7) is plotted against the ratio of fluorescence between the tube and vesicle. The calibration factor ($R_{cal} = 212$nm) is obtained by linear fit to the data (N = 11 GUVs).

be proportional to the number of fluorophores per unit length, hence to the area of the tube per unit length:

$$I_{\text{tub,lip}} \propto \frac{A_{\text{tub}}}{L_{\text{tub}}} \propto \frac{2\pi R_{\text{tub}} L_{\text{tub}}}{L_{\text{tub}}} \propto R_{\text{tub}} \qquad (16.10)$$

where $I_{\text{tub, lip}}$ is fluorescence intensity of lipids in the tubule; A_{tub} is the area of the tubule; and L_{tub} is its length. There are several suitable ways of quantifying fluorescence. The method chosen is not critical as long as it is used consistently. One possibility is to measure the intensity profile in the direction perpendicular to the tube and take $I_{\text{tub, lip}}$ as the peak intensity. The same method can be applied to quantify the fluorescence of the vesicle, working on a small portion of the image where the membrane is approximately straight.

The proportionality constant depends on the type of fluorophore and its concentration in the vesicle. However, by taking the ratio of the fluorescence of tube to the fluorescence of the vesicle, these dependences cancel out, giving

$$R_{\text{tub}} = R_{\text{cal}} \frac{I_{\text{tub,lip}}}{I_{\text{ve,lip}}} \qquad (16.11)$$

where R_{cal} is a calibration constant independent of the type of fluorophore and its concentration; and $I_{\text{ve, lip}}$ is the fluorescence intensity of lipids in the vesicle. To measure R_{cal}, it is advisable to use GUVs with simple lipid composition, for example, egg phosphatidylcholine with ~0.5% molar fraction of a fluorescent lipid, for which Eq. 16.7 applies. The quantity $\frac{F_0}{4\pi\Sigma}$ is plotted as a function of $\frac{I_{tub,lip}}{I_{ve,lip}}$ (each point corresponds to one tension step). This is repeated for several vesicles. Fitting a linear function to the data yields the proportionality constant R_{cal} (Figure 16.5b inset). Note that depending on the fluorophore used, a correction factor may need to be introduced to correct for polarization effects (Sorre et al., 2012; Aimon et al., 2014). Knowing R_{cal}, the radius of the tube can be calculated under all experimental conditions (see an example in Figure 16.5b).

Working with proteins

With the tube-pulling assay, it is possible to quantify the redistribution of proteins in the presence of the highly curved tube (curvature-induced sorting) as well as how protein insertion or binding affects the mechanics of the membrane.

- Sorting (S): It refers to the ratio between the densities of a component (e.g., protein) in the tube and in the GUV: $S = \frac{\phi_{\text{pr,tub}}}{\phi_{\text{pr,ve}}}$. Experimentally, it is given by the ratio of protein fluorescence in the tube and GUV, normalized by the ratio of lipid fluorescence to account for geometrical effects. Thus,

$$S = \frac{I_{\text{tub,pr}} \, / \, I_{\text{ve,pr}}}{I_{\text{tub,lip}} \, / \, I_{\text{ve,lip}}} \qquad (16.12)$$

where $I_{\text{tub,pr}}$ and $I_{\text{ve,pr}}$ are the fluorescence due to the labeled protein in the tube and in the vesicle, respectively; and $I_{\text{tub,lip}}$ and $I_{\text{ve,lip}}$ are the fluorescence due to the lipid tracer. S is a measure of the relative enrichment of protein in the highly curved region, with $S > 1$ when the protein concentration is higher in the tube than in the vesicle.

- In addition, this assay offers two independent ways for quantifying how protein insertion or binding affects the mechanics of the membrane:

(1) By measuring the radius of the tube, it is possible to detect a constriction or a dilation of the tube (Sorre et al., 2012; Prevost et al., 2015). (2) By measuring the tube force, it is possible to detect protein insertion or binding. Indeed, proteins that bend

Box 16.2 Tips to resolve known issues in tube-pulling experiments

1. **Nonspecific adhesion.** A glass surface is very hydrophilic, so the very strong adhesion of GUVs to glass would burst them and spread them on the glass surface. Adhesive interactions between a GUV and a micropipette will increase the tension in the membrane in a way that cannot be directly accounted for and may prevent the GUVs from becoming "floppy." A common and cheap glass passivating reagent is bovine serum albumin; however, it is a lipid transporter and thus should be avoided when working with GUVs. Purified β-casein from bovine milk (>98%) works very well. Another possibility is to graft all glass surfaces with polyethylene glycol (PEG) (Perret et al., 2002). The protocol is much more involved than coating with β-casein, but should be considered if the protein or the GUVs adhere too strongly to the glass.
2. **Offset in the force curve.** Often, a linear interpolation of the force versus the square root of membrane tension gives a slight positive offset, usually smaller than 10 pN in the absence of proteins (Prevost et al., 2015; Simunovic et al., 2015), indicating an effective negative spontaneous curvature. This observation likely results from a combination of (1) an imbalance in ionic strengths inside and outside the GUV (which would screen the lipid–lipid repulsions on one side more), (2) direct interaction of ions with the lipids (Levental et al., 2008; Wang et al., 2012; Simunovic et al., 2015), (3) some adsorption to the micropipette despite passivation (which increases the apparent tension).
3. **No bead-membrane attachment.** Various sugars, including glucose, inhibit biotin–streptavidin bonds (Houen and Hansen, 1997). We advise using <50 mM glucose (or any other sugar) in the external buffer.
4. **Choice of buffers.** When choosing experimental solutions, there are several key considerations. (1) It is recommended to buffer the solutions to prevent pH changes, especially when using proteins. There is a potential, but unconfirmed, danger that HEPES buffers degrade over time and affect lipids. We recommend using, for example, Tris buffer. (2) The buffers need to have a near-physiological ionic strength, especially the ones containing the proteins. Usually, NaCl or KCl is used in the range of 100–150 mM. Keep in mind that GUVs can be grown in solutions of high ionic strength if grown on a platinum-wire instead on indium-tin oxide plates (Meleard et al., 2009; Aimon et al., 2011; Prevost et al., 2015). (3) The buffers inside and outside the GUV need to have the same osmolarity, which is often achieved by adding sucrose or glucose respectively. (4) The density inside the GUV needs to be higher than the outside so that GUVs sediment to the bottom of the chamber. Moreover, the refractive index between the two solutions should be sufficiently different to observe GUVs in bright-field. Both are achieved by using sucrose inside the GUV and salt or glucose (or a combination of both) outside.
5. **Fluorescent probe.** When choosing the fluorescent marker for the protein, it is important to choose one that does not interact with the membrane. For example, Alexa fluorophores were shown to be inert, whereas several Atto dyes showed high interactivity with the membrane (Hughes et al., 2014).

membranes reduce the force required to hold a tube. This effect has been observed for several proteins so far (Roux et al., 2010; Sorre et al., 2012; Prevost et al., 2015; Renard et al., 2015). Note that similar experiments can, in principle, be done with lipids, provided that a curvature-insensitive fluorescent lipid is used as a reference, such as BODIPY® FLC5-GM1, Texas Red® DHPE, or BODIPY® TR ceramide (Sorre et al., 2009).

16.4 PULLING NANOTUBES WITHOUT OPTICAL TWEEZERS

16.4.1 SECOND MICROPIPETTE

Although optical tweezers provide a very sophisticated way of manipulating membrane geometry and measuring membrane mechanics, building and calibrating this setup may be too laborious. One way of replacing the optical trap is simply by holding the streptavidin-coated bead with another micropipette. The approach is identical to pulling a tubule described in the previous section, except that in this case the bead is held in place at the exit of the second pipette with aspiration pressure and not in the optical trap. Essentially, we are able to measure all quantities as in the setup using optical tweezers, except that we cannot infer the membrane force, considering that the force against the micropipette is significantly greater than the pN forces required to maintain a membrane tubule. The realization of the assay is still very powerful. If coupled with a confocal microscope, it allows us to study the coupling of membrane curvature with, say, membrane-bound proteins (Capraro et al., 2010; Heinrich et al., 2010a; Zhu et al., 2012), see also Chapter 23.

16.4.2 GRAVITATIONAL FORCE

Another alternative to the optical trap is using gravity to extrude a tubule from a GUV (Bo and Waugh, 1989). In the assay, a bead of known size and density is attached to an aspired vesicle and is allowed to sink in the chamber, pulling with it a membrane tether. The bead stops sinking once the buoyancy force cancels the membrane force. Therefore, we can control the equilibrium length of the tether by adjusting the aspiration pressure (implicitly, membrane tension). The buoyancy force, F_{by}, is given by the following expression:

$$F_{by} = \frac{4}{3}\pi R_{bd}{}^{3} g\left(\rho_{bd} - \rho_{fl}\right) \tag{16.13}$$

where R_{bd} is the radius of the bead; ρ_{bd} and ρ_{fl} are the densities of the bead and of the fluid, respectively; and g is the gravitational acceleration. In the original article, glass beads with diameters of 10–30 μm were used. The opposing membrane force, as previously, is the sum of contributions from bending and stretching (i.e., in-plane tension). We can calculate the tubule radius either from the aspiration pressure (Eq. 16.4) or using confocal microscopy. Therefore, this assay provides the same measurements as when using optical tweezers, albeit with much lower precision.

16.4.3 FLOW

In another method, developed by Rossier, Brochard-Wyart and coworkers, tubules are extruded by sticking vesicles to beads or nanopillars dispersed on the chamber surface then applying a flow. The flow can be created using, for example, a microfluidic system (Borghi et al., 2003; Rossier et al., 2003) (Figure 16.6a). An important benefit of the method is that it does not require micromanipulation of vesicles, allowing much faster and higher throughput measurements; however, determining the precise force is more challenging. Unlike in previous experiments, the tension in the vesicle is not externally controlled. As the tubule is initially pulled, both the tension and the tubule length increase until reaching equilibrium. This can be monitored with standard epifluorescence microscopy. The stationary length of the tubule, L_∞, was derived as (Borghi et al., 2003)

$$L_\infty = \tau_0 U \ln \frac{U_{\text{fl}}}{U_{\text{th}}} \quad (16.14)$$

where τ_0 is an extrusion characteristic time; U_{fl} is the flow velocity (calculated based on how the flow is generated in the system); and U_{th} is the threshold flow velocity required to pull a tubule. τ_0 can be calculated as

$$\tau_0 = \frac{3k_{\text{B}}TR_{\text{ve}}{}^3\eta_{\text{fl}}}{2\pi\kappa} \quad (16.15)$$

where k_B is the Botzmann constant, T is temperature and η_{fl} is the viscosity of the fluid (Figure 16.6a).

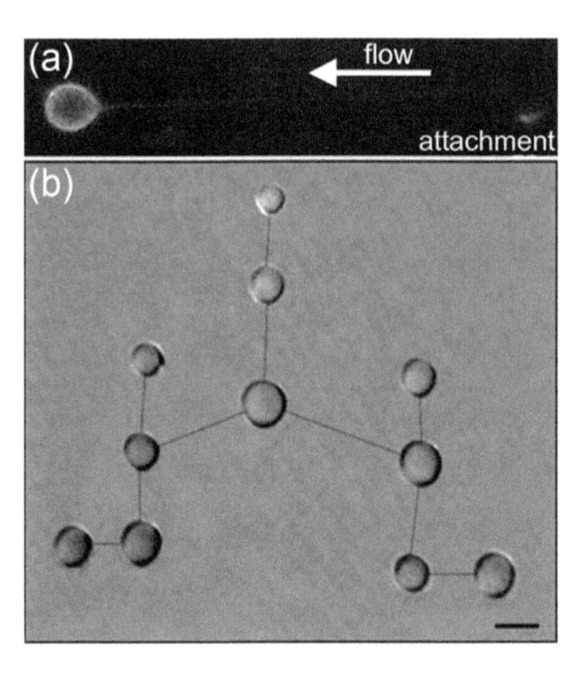

Figure 16.6 Alternative methods for pulling tubes. (a) Flow applied to a GUV, which is stuck to a surface-bound object, extracts a tubule. Scale bar: 20 µm. (Reprinted with permission from Rossier, O et al., *Langmuir*, 19, 575–584, 2003. Copyright 2003 by the American Chemical Society.) (b) Engineering GUV–tube networks by a series of micromanipulations and microinjections. Scale bar: 10 µm. (Reprinted by permission from Macmillan Publishers Ltd. *Nature*, Karlsson, A. et al., 2001a, copyright 2001.)

This versatile method can be applied for studying the mechanical properties of membranes and the dynamics of tube extrusion and, potentially, for exploring how proteins interact with curved membranes.

16.4.4 MICROMANIPULATION SCULPTING

Orwar and coworkers have developed a micromanipulation technique to create a network of tubules connecting multiple vesicles by membrane sculpting. In the assay, a vesicle is spontaneously formed from a lipid reservoir and remains attached to it. This lipid reservoir is crucial for providing the lipids for more vesicles while keeping membrane tension below lysis. Next, a micropipette is penetrated into this vesicle using an electric pulse, then rapidly pulled away, followed by an injection of the buffer (Karlsson et al., 2001b, 2002; Jesorka et al., 2011). This process initially creates a tubule between the pipette and the vesicle, while the injection inflates another vesicle at the pipette through a Marangoni effect (Dommersnes et al., 2005), thus forming two connected vesicles. If using multilamellar vesicles, the movements can be repeated several times, creating a network of vesicles interconnected with membrane tubules (Jesorka et al., 2011) (Figure 16.6b). The method is potentially a useful tool in engineering a nanoscopic membrane network, which can be used as a synthesis scaffold (Lizana et al., 2008; Jesorka et al., 2011) or to study more complex transport phenomena in a biological context. However, there are too many uncontrolled variables in the experiment—such as adhesion to the surface, in-plane tension—which directly control membrane geometry and mechanics, making it unsuitable for quantitative studies on membrane nanotubes.

16.4.5 MOLECULAR MOTORS

In cells, membrane tubules may be created by the action of molecular motors or actin polymerization. Mimicking this phenomenon, two groups have developed an assay in which a molecular motor kinesin attaches to a GUV (via biotin–streptavidin bonds) then, powered by ATP hydrolysis, it walks along the polymerized tubulin on the glass surface, pulling with it a tubule from the vesicle (Roux et al., 2002; Koster et al., 2003; Leduc et al., 2004; Shaklee et al., 2008; Leduc et al., 2010). Similarly to the previous assay, this approach cannot be used for quantitative measurements of membrane force or tension; however, it is very powerful in studying the action of molecular motors in an environment under much higher control than the cell.

16.4.6 TUBULATION BY OSMOTIC DEFLATION

The methods described so far employ an external force to deform the membrane. Tubules can also be produced by osmotically deflating GUVs. If exposing the GUVs to a solution of higher osmolarity than its interior, the water will escape via osmosis and the total volume will reduce. The excess area for the new smaller volume will be converted into tubules (Li et al., 2011; Liu et al., 2016) whereby the asymmetry in the compositions of solutions inside and outside of GUVs gives rise to spontaneous curvature stabilizing the tubes (Lipowsky, 2013).

This method is quite simple and general and it can be used to produce tubules from liquid-disordered but also liquid-ordered membranes (Liu et al., 2016). Potentially, it could be used to study some membrane shape transitions and to quickly determine the spontaneous curvature effect of adhering proteins or particles.

16.5 PAST AND POTENTIAL FUTURE APPLICATIONS OF NANOTUBES

In the past years, nanotube pulling, in combination with fluorescence confocal microscopy, has become a popular tool to investigate biological processes where membrane curvature is an important parameter as well as more basic questions related to membrane physics (Baumgart et al., 2011; Callan-Jones and Bassereau, 2013; Morlot and Roux, 2013; Bassereau et al., 2014; Dasgupta and Dimova, 2014). We list here some of the applications that have already been implemented and some others that could be considered.

16.5.1 CURVATURE/COMPOSITION COUPLING

- ***Lipids:*** It is well known that cellular membranes have different lipid compositions from organelle to organelle, but also between the two leaflets of the same membrane (van Meer et al., 2008). Notably, the concentration of sterols and sphingolipids increases along the secretory pathway. This suggests the existence of lipid sorting mechanisms able to prevent the mixing of the membranes lipid content during vesicular trafficking (Callan-Jones and Bassereau, 2013). Considering that transport intermediates are highly curved, it was hypothesized that lipids could be sorted in and out of these structures according to their structural properties: Lipid species that tend to form rigid membranes would be excluded, whereas those that tend to form flexible membranes would be enriched within transport intermediates. However, it was predicted and verified experimentally using the tube-pulling assay that such effect is undetectable in the case where the membrane composition is far from lipid demixing transitions (Sorre et al., 2009; Tian et al., 2009). In contrast, using fluorescent lipid derivatives with distinct partitioning behavior between lipid phases and tube force measurements, Sorre et al. (2009) demonstrated that lipids are effectively sorted when the membrane composition is close to a lipid demixing transition. In addition, using multiphase GUVs, Heinrich et al. (2010a) studied the dynamics of lipid sorting. In particular, they found that when a tube is pulled from the more rigid phase, a domain of the less rigid phase nucleates at the GUV–tube junction and progressively invades the tube.
- ***Proteins:*** A growing list of proteins are known to be sensitive to membrane curvature, and to deform membranes at high concentration. They include BAR (Bin/Amphiphysin/Rvs) domain proteins, proteins with amphipathic helices and dynamin. The GUV–tube geometry is well adapted to probe the influence of curvature on protein distribution, and several studies have used this approach (Ambroggio et al., 2010; Capraro et al., 2010; Heinrich et al., 2010b; Roux et al., 2010; Morlot et al., 2012; Sorre et al., 2012; Zhu et al., 2012; Ramesh et al., 2013; Aimon et al., 2014; Knorr et al., 2014; Wu and Baumgart, 2014; Prevost et al., 2015). As described in Section 2.5, several manifestations of curvature–protein coupling can be detected and quantified with the tube-pulling assay. Curvature-induced sorting is calculated from fluorescence measurements (Figure 16.7a). In addition, several proteins have been found to affect the mechanics of the tube in ways that can be quantified with the assay: (1) by modifying the radius with respect to the "bare membrane" radius given by Eq. 16.4, and (2) by decreasing the tube force compared with the force given by Eq. 16.5. The force can even vanish at relatively high protein concentrations (and low

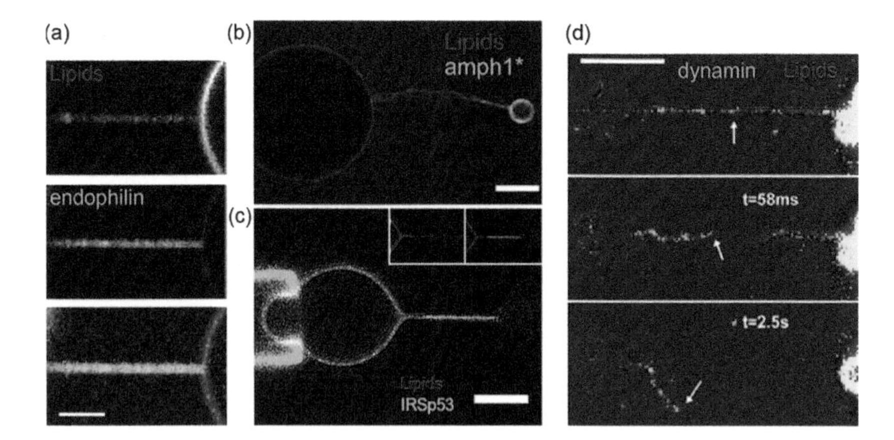

Figure 16.7 Several manifestations of composition/curvature coupling. (a–d) All proteins were labeled with Alexa 488 (green-emitting dye). The GUVs contained 0.3%–1% of a lipid labeled with a red-emitting dye (magenta in c). (a) Curvature-induced protein sorting. The N-BAR domain of endophilin A1 was added to the GUVs before pulling the tube. Lipid (top), protein (middle), and merged (bottom) channels. Scale bar: 3 µm. (Reprinted from *Biophys. J.*, 102, Zhu, C. et al., Nonlinear sorting, curvature generation, and crowding of endophilin N-BAR on tubular membranes, 1837–1845, Copyright 2012, with permission from Elsevier.) (b) Scaffolding. Full length amphiphysin 1 (an N-BAR domain protein) was microinjected near the GUV after pulling the tube. Merged green and red fluorescence channels. Scale bar: 5 µm. (Reprinted with permission from Sorre, B. et al., 2012. Nature of curvature coupling of amphiphysin with membranes depends on its bound density. *Proc. Nat. Acad. Sci. USA*, 109, 173–178. Copyright 2012 National Academy of Sciences, USA.) (c) Phase coexistence. The GUVs were grown in a solution containing the I-BAR domain of IRSp53. The experimental buffer contained a high concentration of salt to screen the attractive electrostatic interaction between the protein and the outer leaflet of the membrane. Merged green and red fluorescence channels. Inset: individual red and green channels. Scale bar: 5 µm. (Reprinted with permission from Prevost, C. et al., 2015. IRSp53 senses negative membrane curvature and phase separates along membrane tubules. *Nat. Commun.*, 6, 8529. Copyright 2015 National Academy of Sciences, USA.) (d) Scission. Dynamin 1 was injected together with GTP near the GUV after pulling the tube. Merged green and red fluorescence channels. Arrows indicate the location where scission occurs, Scale bar: 5 µm. (Reprinted from *Cell*, 151, Morlot, S. et al., Membrane shape at the edge of the dynamin helix sets location and duration of the fission reaction, 619–629, Copyright 2012, with permission from Elsevier.)

enough tensions), which is reflected by the fact that the tube becomes flexible (Figure 16.7b). Finally, a recent study has uncovered the existence of an original type of phase separation induced by curvature/concentration coupling in the tube (Prevost et al., 2015) (Figure 16.7c).

It is important to note that peripheral membrane proteins (see also Chapter 23), whether present in bulk or microinjected near the GUV, are exposed to positive curvature. The sign of the curvature is arbitrarily defined with respect to the membrane side that the protein solution faces: positive if it faces the convex side, and negative if it faces the concave side. As a matter of fact most proteins are sensitive to positive membrane curvature, but there are a few known exceptions, notably I-BAR (I stands for inverse) domain proteins. In this case, the protein solution has to be encapsulated inside the GUVs before performing the assay, which was achieved in a recent study by Prevost et al. (2015).

The combined experimental and theoretical efforts highlight that the membrane curvature itself (in the absence of explicit protein–protein attractions) can provide cues for membrane-remodeling events, such as in endocytosis. Future work will likely focus on exploring how the structure of the protein impacts its ability to sort, induce curvature, and impose a mechanical effect on the membrane. Further theoretical analysis will attempt to unveil the precise physical forces that underlie protein–protein interactions on the membrane (whether it is mostly driven by membrane fluctuations or curvature or by protein structure and direct interactions). The advancement in super-resolution microscopy also opens possibilities to explore the structure of protein assemblies on curved membrane at a very high level of detail.

16.5.2 SCISSION

The tube-pulling assay was used to study the mechanism of scission, a dynamic process in which the topology of the membrane is changed to form two separate membrane structures. Scission takes place in many cellular processes, such as in endocytosis where it permits the entry of the incoming cargo into the cell's lumen. By pulling tubes from a GUV with a composition that gives rise to phase separation on the tube, it has been shown that scission can be driven simply by line tension (Allain et al., 2004). The assay has also been used to study scission in the context of endocytosis by injecting scission proteins near the vesicle and investigating the minimum components required for breakage of tubules, namely dynamin (Morlot et al., 2012) and endophilin (Renard et al., 2015) (Figure 16.7d). The assay will likely prove valuable in addressing the problem of scission induced by the family of endosomal sorting complexes required for transport (ESCRT), which play an essential role in many membrane-remodeling phenomena, such as the budding of HIV (Hurley and Hanson, 2010; McCullough et al., 2013). This assay is ideal for unraveling the precise mechanism of ESCRT-induced scission, a burning question in the field.

16.5.3 TUBE COALESCENCE

When two tubes are pulled from the same GUV (the simplest way to do so is to hold one bead with the optical tweezers and the other with a second micropipette), and the GUV is moved away from the beads, the tubes coalesce when the angle between them reaches a certain value (Cuvelier et al., 2005). The resulting geometry is a Y junction, with a unique tube continued by two branches connected to the beads. By measuring the force and angle at coalescence, it is possible to determine the bending rigidity of the membrane in the absence of any control of the tension. In addition, the region connecting both tubes displays a negative Gaussian curvature (saddle shape). Although this type of curvature is presumably common within cellular membranes (e.g., neck of budding vesicles), its influence on the localization and diffusion of proteins has not been investigated.

16.5.4 INFLUENCE OF GEOMETRY ON PROTEIN DIFFUSION

In addition to being highly curved, lipid tubes are a naturally occurring instance of geometrically confined membranes. Saffman and Delbrück theoretically investigated the diffusion of particles embedded in a membrane patch of finite size, and predicted a slower diffusion for smaller patch sizes (Saffman and Delbruck, 1975). In the case of a cylindrical geometry, the diffusion of such particles was predicted to be slower for narrower tubes (Daniels and Turner, 2007). This prediction was tested for both lipids and a transmembrane protein with the tube-pulling assay (Domanov et al., 2011). The diffusion of each of these species was measured by single particle tracking of quantum dots attached to them, whereas the tube radius was controlled by adjusting the aspiration pressure. The results were consistent with the theoretical predictions. Tubes pulled from GUVs may thus represent a good system to study such processes as the diffusion of receptors in the tubular neck of neuronal spines and how it is influenced by their morphology.

16.5.5 PULLING TUBES FROM CELLS

As a last illustration of the multiple applications of membrane tube pulling, we switch from artificial systems to cells. In the case of tubes pulled from the PM of cells, it has essentially been used for years for probing the membrane tension of the PM rather than consequences of membrane curvature. Indeed, the tube force reflects the effective tension of the membrane, which is the sum of the PM tension and of the density of PM-cytoskeleton adhesion energy (Sheetz, 2001). Therefore, although it is in general difficult to discriminate between both contributions, tube-pulling experiments have been invaluable in probing the role of PM tension in processes such as exocytosis (Gauthier et al., 2011) or motility (Batchelder et al., 2011), the role of caveolae in buffering changes in PM tension (Sinha et al., 2011), as well as the coupling between the cytoskeleton and the PM (Dai and Sheetz, 1999; Bornschlogl et al., 2013).

16.5.6 OUTLOOK

The main strengths of the tube-pulling assay are (1) the simple, well-controlled geometry of the tube, combined with the sensitive microscopy-based detection methods allow to perform highly quantitative measurements; (2) the development of theoretical descriptions over the past decades provides a solid basis to interpret these measurements in terms of membrane mechanical properties and the interplay between external components (most notably proteins) and highly curved membranes.

Nevertheless, the assay suffers from technical pitfalls and low throughput. Some experience is required to attain the highest degree of control over the experimental parameters. The number of data points (number of tubes pulled) per experiment usually does not exceed 5–10. A significant improvement of the technique would be to develop a higher throughput configuration. On this perspective, the most promising approach may be microfluidics, where vesicles can be immobilized in parallelized arrays of traps. Microfluidics further offers the possibility of switching the external conditions (injecting proteins, rinsing). Relatively simple implementations of this idea have been achieved in the past (Waugh, 1982; Rossier et al., 2003). In these cases, tubes were pulled from GUVs using a liquid flow. These assays could be improved by combining them with confocal fluorescence microscopy. A more challenging approach would be to combine microfluidics with multiple-beam optical trapping to benefit from the precision of the force measurement. Implementing aspiration within microfluidics devices may also be technically very challenging.

Another technical challenge has been to investigate the coupling between proteins and negative membrane curvature. There are several methods to encapsulate protein solutions in GUVs, but most of them require using oil that eventually contaminates the lipid bilayer and affects its mechanical properties. In a recent study, proteins were encapsulating in GUVs by forming the GUVs in the presence of the protein solution, and detaching the proteins adsorbed on the outer layer with salt (Prevost et al., 2015). However, this method only works for proteins that bind to membranes solely through electrostatic interactions, and at relatively low protein concentrations. Therefore, new encapsulation methods allowing for a more systematic study of the interplay between proteins and negative curvature still need to be developed.

The tube-pulling assay is particularly well adapted to study the coupling between membrane curvature and protein concentration. In particular, several proteins with BAR domains, which are elongated in shape and promote the formation of tubules at high protein coverage on membranes, have been studied with this assay. Until now, most theoretical descriptions of the tubule–protein interplay have assumed that protein binding induces an isotropic spontaneous curvature. However, the spontaneous curvature generated by these protein domains is intrinsically anisotropic. Anisotropic spontaneous curvatures have been considered in recent theoretical studies (e.g., Iglic et al., 2007; Ayton et al., 2009; Ramakrishnan et al., 2013; Simunovic et al., 2013; Noguchi, 2014; Walani et al., 2014; Schweitzer and Kozlov, 2015), but such a description has not yet been applied directly to the analysis of the tube-pulling experiments. Taking this effect into account will strengthen the interpretation of the measurements obtained with the assay and generalize the analysis.

Finally, for complex membranes, in particular cell membranes, although pulling experiments do not present particular technical issues, their interpretation still requires further theoretical modeling to include various possible contributions to the tube force and a comprehensive description of cell membrane tension.

LIST OF ABBREVIATIONS

OT	optical tweezers
PE	phosphatidylethanolamine
PEG	polyethylene glycol
PM	plasma membrane

GLOSSARY OF SYMBOLS

A_{me}	area of membrane
C_{pr}	spontaneous curvature of the membrane upon protein binding
$\overline{C_{pr}}$	intrinsic spontaneous curvature of the protein
ΔP	pressure difference between the pipette and the experimental chamber
Δx	displacement of the bead in the trap with respect to its equilibrium position
ν_{be}	bending energy of the membrane
F	tube pulling force
F_0	equilibrium tube pulling force
$\mathcal{F}$	free energy of a membrane surface
$\mathcal{F}_{tub}$	free energy of the tube
ϕ_{pr}	fractional protein coverage of the membrane
h	height of water tank with respect to zero-position (zero difference of pressure)
k_{OT}	stiffness of the optical trap
κ	membrane bending rigidity
L_{tub}	length of the tube
M	local mean curvature
P	pressure difference across the GUV membrane
R_0	equilibrium radius of the tube
R_1, R_2	principal radii of curvature
R_{cal}	proportionality factor between tube radius and ratio of tube to vesicle fluorescence
R_{pip}	radius of the pipette
R_{tub}	radius of the tube
R_{ve}	radius of the GUV
S	sorting ratio
Σ	membrane tension
V	volume of the GUV

REFERENCES

Aimon S, Callan-Jones A, Berthaud A, Pinot M, Toombes GE, Bassereau P (2014) Membrane shape modulates transmembrane protein distribution. *Developmental Cell* 28:212–218.

Aimon S, Manzi J, Schmidt D, Poveda Larrosa JA, Bassereau P, Toombes GE (2011) Functional reconstitution of a voltage-gated potassium channel in giant unilamellar vesicles. *PLoS One* 6:e25529.

Allain JM, Storm C, Roux A, Ben Amar M, Joanny JF (2004) Fission of a multiphase membrane tube. *Physical Review Letters* 93:158104.

Ambroggio E, Sorre B, Bassereau P, Goud B, Manneville JB, Antonny B (2010) ArfGAP1 generates an Arf1 gradient on continuous lipid membranes displaying flat and curved regions. *The EMBO Journal* 29:292–303.

Ashkin A (1970) Acceleration and trapping of particles by radiation pressure. *Physical Review Letters* 24:156–159.

Ashkin A, Dziedzic JM, Bjorkholm JE, Chu S (1986) Observation of a single-beam gradient force optical trap for dielectric particles. *Optics Letters* 11:288.

Ashok B, Ananthakrishna G (2014) Dynamics of intermittent force fluctuations in vesicular nanotubulation. *The Journal of Chemical Physics* 141:174905.

Ayton GS, Lyman E, Krishna V, Swenson RD, Mim C, Unger VM, Voth GA (2009) New insights into BAR domain-induced membrane remodeling. *Biophysical Journal* 97:1616–1625.

Bassereau P, Sorre B, Levy A (2014) Bending lipid membranes: Experiments after W. Helfrich's model. *Advances in Colloid and Interface Science* 208:47–57.

Batchelder EL, Hollopeter G, Campillo C, Mezanges X, Jorgensen EM, Nassoy P, Sens P, Plastino J (2011) Membrane tension regulates motility by controlling lamellipodium organization. *Proceedings of the National Academy of Sciences of the United States of America* 108:11429–11434.

Baumgart T, Capraro BR, Zhu C, Das SL (2011) Thermodynamics and mechanics of membrane curvature generation and sensing by proteins and lipids. *Annual Review of Physical Chemistry* 62:483–506.

Bo L, Waugh RE (1989) Determination of bilayer membrane bending stiffness by tether formation from giant, thin-walled vesicles. *Biophysical Journal* 55:509–517.

Borghi N, Rossier O, Brochard-Wyart F (2003) Hydrodynamic extrusion of tubes from giant vesicles. *EPL (Europhysics Letters)* 64:837.

Bornschlogl T, Romero S, Vestergaard CL, Joanny JF, Van Nhieu GT, Bassereau P (2013) Filopodial retraction force is generated by cortical actin dynamics and controlled by reversible tethering at the tip. *Proceedings of the National Academy of Sciences of the United States of America* 110:18928–18933.

Brochard-Wyart F, Borghi N, Cuvelier D, Nassoy P (2006) Hydrodynamic narrowing of tubes extruded from cells. *Proceedings of the National Academy of Sciences of the United States of America* 103:7660–7663.

Callan-Jones A, Bassereau P (2013) Curvature-driven membrane lipid and protein distribution. *Current Opinion in Solid State & Materials Science* 17:143–150.

Campelo F, McMahon HT, Kozlov MM (2008) The hydrophobic insertion mechanism of membrane curvature generation by proteins. *Biophysical Journal* 95:2325–2339.

Campillo C, Sens P, Koster D, Pontani LL, Levy D, Bassereau P, Nassoy P, Sykes C (2013) Unexpected membrane dynamics unveiled by membrane nanotube extrusion. *Biophysical Journal* 104:1248–1256.

Canham PB (1970) The minimum energy of bending as a possible explanation of the biconcave shape of the human red blood cell. *Journal of Theoretical Biology* 26:61–81.

Capraro BR, Yoon Y, Cho W, Baumgart T (2010) Curvature sensing by the epsin N-terminal homology domain measured on cylindrical lipid membrane tethers. *Journal of the American Chemical Society* 132:1200–1201.

Cuvelier D, Derenyi I, Bassereau P, Nassoy P (2005) Coalescence of membrane tethers: Experiments, theory, and applications. *Biophysical Journal* 88:2714–2726.

Dai J, Sheetz MP (1999) Membrane tether formation from blebbing cells. *Biophysical Journal* 77:3363–3370.

Daniels DR, Turner MS (2007) Diffusion on membrane tubes: A highly discriminatory test of the Saffman-Delbruck theory. *Langmuir* 23:6667–6670.

Dasgupta R, Dimova R (2014) Inward and outward membrane tubes pulled from giant vesicles. *Journal of Physics D: Applied Physics* 47:282001.

Datar A, Bornschlogl T, Bassereau P, Prost J, Pullarkat PA (2015) Dynamics of membrane tethers reveal novel aspects of cytoskeleton-membrane interactions in axons. *Biophysical Journal* 108:489–497.

Derenyi I, Julicher F, Prost J (2002) Formation and interaction of membrane tubes. *Physical Review Letters* 88:238101.

Domanov YA, Aimon S, Toombes GE, Renner M, Quemeneur F, Triller A, Turner MS, Bassereau P (2011) Mobility in geometrically confined membranes. *Proceedings of the National Academy of Sciences of the United States of America* 108:12605–12610.

Dommersnes P, Orwar O, Brochard-Wyart F, Joanny J (2005) Marangoni transport in lipid nanotubes. *EPL (Europhysics Letters)* 70:271.

Evans E, Yeung A (1994) Hidden dynamics in rapid changes of bilayer shape. *Chemistry and Physics of Lipids* 73:39–56.

Evans EA, Waugh R, Melnik L (1976) Elastic area compressibility modulus of red cell membrane. *Biophysical Journal* 16:585–595.

Gauthier NC, Fardin MA, Roca-Cusachs P, Sheetz MP (2011) Temporary increase in plasma membrane tension coordinates the activation of exocytosis and contraction during cell spreading. *Proceedings of the National Academy of Sciences of the United States of America* 108:14467–14472.

Heinrich M, Tian A, Esposito C, Baumgart T (2010a) Dynamic sorting of lipids and proteins in membrane tubes with a moving phase boundary. *Proceedings of the National Academy of Sciences of the United States of America* 107:7208–7213.

Heinrich MC, Capraro BR, Tian A, Isas JM, Langen R, Baumgart T (2010b) Quantifying membrane curvature generation of amphiphysin N-BAR domains. *The Journal of Physical Chemistry Letters* 1:3401–3406.

Heinrich V, Waugh RE (1996) A piconewton force transducer and its application to measurement of the bending stiffness of phospholipid membranes. *Annals of Biomedical Engineering* 24:595–605.

Helfrich W (1973) Elastic properties of lipid bilayers: Theory and possible experiments. *Zeitschrift fur Naturforschung Teil C: Biochemie, Biophysik, Biologie, Virologie* 28:693–703.

Hochmuth RM, Evans EA (1982) Extensional flow of erythrocyte membrane from cell body to elastic tether. I. Analysis. *Biophysical Journal* 39:71–81.

Hochmuth RM, Mohandas N, Blackshear PL, Jr. (1973) Measurement of the elastic modulus for red cell membrane using a fluid mechanical technique. *Biophysical Journal* 13:747–762.

Hochmuth RM, Wiles HC, Evans EA, McCown JT (1982) Extensional flow of erythrocyte membrane from cell body to elastic tether. II. Experiment. *Biophysical Journal* 39:83–89.

Houen G, Hansen K (1997) Interference of sugars with the binding of biotin to streptavidin and avidin. *Journal Immunol Methods* 210:115–123.

Hughes LD, Rawle RJ, Boxer SG (2014) Choose your label wisely: Water-soluble fluorophores often interact with lipid bilayers. *PloS One* 9.

Hurley JH, Hanson PI (2010) Membrane budding and scission by the ESCRT machinery: It's all in the neck. *Nature Reviews Molecular Cell Biology* 11:556–566.

Iglic A, Slivnik T, Kralj-Iglic V (2007) Elastic properties of biological membranes influenced by attached proteins. *Journal of Biomechanics* 40:2492–2500.

Jesorka A, Stepanyants N, Zhang H, Ortmen B, Hakonen B, Orwar O (2011) Generation of phospholipid vesicle-nanotube networks and transport of molecules therein. *Nature Protocol* 6:791–805.

Karlsson A, Karlsson R, Karlsson M, Cans AS, Stromberg A, Ryttsen F, Orwar O (2001a) Networks of nanotubes and containers. *Nature* 409:150–152.

Karlsson M, Sott K, Cans A-S, Karlsson A, Karlsson R, Orwar O (2001b) Micropipet-assisted formation of microscopic networks of unilamellar lipid bilayer nanotubes and containers. *Langmuir* 17:6754–6758.

Karlsson M, Sott K, Davidson M, Cans AS, Linderholm P, Chiu D, Orwar O (2002) Formation of geometrically complex lipid nanotube-vesicle networks of higher-order topologies. *Proceedings of the National Academy of Sciences of the United States of America* 99:11573–11578.

Knorr RL, Nakatogawa H, Ohsumi Y, Lipowsky R, Baumgart T, Dimova R (2014) Membrane morphology is actively transformed by covalent binding of the protein Atg8 to PE-lipids. *PLoS One* 9:e115357.

Koster G, Cacciuto A, Derenyi I, Frenkel D, Dogterom M (2005) Force barriers for membrane tube formation. *Physical Review Letters* 94:068101.

Koster G, VanDuijn M, Hofs B, Dogterom M (2003) Membrane tube formation from giant vesicles by dynamic association of motor proteins. *Proceedings of the National Academy of Sciences of the United States of America* 100:15583–15588.

Leduc C, Campas O, Joanny JF, Prost J, Bassereau P (2010) Mechanism of membrane nanotube formation by molecular motors. *Biochimica et Biophysica Acta* 1798:1418–1426.

Leduc C, Campas O, Zeldovich KB, Roux A, Jolimaitre P, Bourel-Bonnet L, Goud B, Joanny JF, Bassereau P, Prost J (2004) Cooperative extraction of membrane nanotubes by molecular motors. *Proceedings of the National Academy of Sciences of the United States of America* 101:17096–17101.

Lee WM, Reece PJ, Marchington RF, Metzger NK, Dholakia K (2007) Construction and calibration of an optical trap on a fluorescence optical microscope. *Nature Protocols* 2:3226–3238.

Leibler S (1986) Curvature instability in membranes. *Journal de Physique* 47:507–516.

Levental I, Cebers A, Janmey PA (2008) Combined electrostatics and hydrogen bonding determine intermolecular interactions between polyphosphoinositides. *Journal of the American Chemical Society* 130:9025–9030.

Li Y, Lipowsky R, Dimova R (2011) Membrane nanotubes induced by aqueous phase separation and stabilized by spontaneous curvature. *Proceedings of the National Academy of Sciences of the United States of America* 108:4731–4736.

Lipowsky R (2013) Spontaneous tubulation of membranes and vesicles reveals membrane tension generated by spontaneous curvature. *Faraday Discuss* 161:305–331.

Liu Y, Agudo-Canalejo J, Grafmuller A, Dimova R, Lipowsky R (2016) Patterns of flexible nanotubes formed by liquid-ordered and liquid-disordered membranes. *ACS Nano* 10:463–474.

Lizana L, Bauer B, Orwar O (2008) Controlling the rates of biochemical reactions and signaling networks by shape and volume changes. *Proceedings of the National Academy of Sciences of the United States of America* 105:4099–4104.

Marcerou JP, Prost J, Gruler H (1984) Elastic model of protein-protein interaction. *Nuovo Cimento D* 3:204–210.

Markin VS (1981) Lateral organization of membranes and cell shapes. *Biophysical Journal* 36:1–19.

McCullough J, Colf LA, Sundquist WI (2013) Membrane fission reactions of the mammalian ESCRT pathway. *Annual Review of Biochemistry* 82:663–692.

Meleard P, Bagatolli LA, Pott T (2009) Giant unilamellar vesicle electroformation from lipid mixtures to native membranes under physiological conditions. *Methods in Enzymology* 465:161–176.

Monnier S, Rochal SB, Parmeggiani A, Lorman VL (2010) Long-range protein coupling mediated by critical low-energy modes of tubular lipid membranes. *Physical Review Letters* 105:028102.

Morlot S, Galli V, Klein M, Chiaruttini N, Manzi J, Humbert F, Dinis L, Lenz M, Cappello G, Roux A (2012) Membrane shape at the edge of the dynamin helix sets location and duration of the fission reaction. *Cell* 151:619–629.

Morlot S, Roux A (2013) Mechanics of dynamin-mediated membrane fission. *Annual Review of Biophysics* 42:629–649.

Neuman KC, Block SM (2004) Optical trapping. *The Review of Scientific Instruments* 75:2787–2809.

Neuman KC, Nagy A (2008) Single-molecule force spectroscopy: Optical tweezers, magnetic tweezers and atomic force microscopy. *Nature Methods* 5:491–505.

Nieminen TA, Knoner G, Heckenberg NR, Rubinsztein-Dunlop H (2007) Physics of optical tweezers. *Methods in Cell Biology* 82:207–236.

Noguchi H (2014) Two- or three-step assembly of banana-shaped proteins coupled with shape transformation of lipid membranes. *EPL-Europhys Letters* 108.

Perret E, Leung A, Morel A, Feracci H, Nassoy P (2002) Versatile decoration of glass surfaces to probe individual protein–protein interactions and cellular adhesion. *Langmuir* 18:846–854.

Powers TR, Huber G, Goldstein RE (2002) Fluid-membrane tethers. Minimal surfaces and elastic boundary layers. *Physical Review E, Statistical, Nonlinear, and Soft Matter Physics* 65:041901.

Prevost C, Zhao H, Manzi J, Lemichez E, Lappalainen P, Callan-Jones A, Bassereau P (2015) IRSp53 senses negative membrane curvature and phase separates along membrane tubules. *Nature Communications* 6:8529.

Ramakrishnan N, Sunil Kumar PB, Ipsen JH (2013) Membrane-mediated aggregation of curvature-inducing nematogens and membrane tubulation. *Biophysical Journal* 104:1018–1028.

Ramesh P, Baroji YF, Reihani SN, Stamou D, Oddershede LB, Bendix PM (2013) FBAR syndapin 1 recognizes and stabilizes highly curved tubular membranes in a concentration dependent manner. *Scientific Reports* 3:1565.

Renard HF, Simunovic M, Lemiere J, Boucrot E, Garcia-Castillo MD, Arumugam S, Chambon V et al. (2015) Endophilin-A2 functions in membrane scission in clathrin-independent endocytosis. *Nature* 517:493–496.

Rossier O, Cuvelier D, Borghi N, Puech PH, Derényi I, Buguin A, Nassoy P, Brochard-Wyart F (2003) Giant vesicles under flows: Extrusion and retraction of tubes. *Langmuir* 19:575–584.

Roux A, Cappello G, Cartaud J, Prost J, Goud B, Bassereau P (2002) A minimal system allowing tubulation with molecular motors pulling on giant liposomes. *Proceedings of the National Academy of Sciences of the United States of America* 99:5394–5399.

Roux A, Koster G, Lenz M, Sorre B, Manneville JB, Nassoy P, Bassereau P (2010) Membrane curvature controls dynamin polymerization. *Proceedings of the National Academy of Sciences of the United States of America* 107:4141–4146.

Saffman PG, Delbruck M (1975) Brownian motion in biological membranes. *Proceedings of the National Academy of Sciences of the United States of America* 72:3111–3113.

Schweitzer Y, Kozlov MM (2015) Membrane-mediated interaction between strongly anisotropic protein scaffolds. *PLoS Computational Biology* 11.

Shaklee PM, Idema T, Koster G, Storm C, Schmidt T, Dogterom M (2008) Bidirectional membrane tube dynamics driven by nonprocessive motors. *Proceedings of the National Academy of Sciences of the United States of America* 105:7993–7997.

Sheetz MP (2001) Cell control by membrane-cytoskeleton adhesion. *Nature Reviews Molecular Cell Biology* 2:392–396.

Simunovic M, Lee KY, Bassereau P (2015) Celebrating Soft Matter's 10th anniversary: Screening of the calcium-induced spontaneous curvature of lipid membranes. *Soft Matter* 11:5030–5036.

Simunovic M, Mim C, Marlovits TC, Resch G, Unger VM, Voth GA (2013) Protein-mediated transformation of lipid vesicles into tubular networks. *Biophysical Journal* 105:711–719.

Sinha B, Koster D, Ruez R, Gonnord P, Bastiani M, Abankwa D, Stan RV et al. (2011) Cells respond to mechanical stress by rapid disassembly of caveolae. *Cell* 144:402–413.

Sorre B, Callan-Jones A, Manneville JB, Nassoy P, Joanny JF, Prost J, Goud B, Bassereau P (2009) Curvature-driven lipid sorting needs proximity to a demixing point and is aided by proteins. *Proceedings of the National Academy of Sciences of the United States of America* 106:5622–5626.

Sorre B, Callan-Jones A, Manzi J, Goud B, Prost J, Bassereau P, Roux A (2012) Nature of curvature coupling of amphiphysin with membranes depends on its bound density. *Proceedings of the National Academy of Sciences of the United States of America* 109:173–178.

Svetina S, Zeks B, Waugh RE, Raphael RM (1998) Theoretical analysis of the effect of the transbilayer movement of phospholipid molecules on the dynamic behavior of a microtube pulled out of an aspirated vesicle. *European Biophysics Journal: EBJ* 27:197–209.

Tian A, Capraro BR, Esposito C, Baumgart T (2009) Bending stiffness depends on curvature of ternary lipid mixture tubular membranes. *Biophysical Journal* 97:1636–1646.

van Meer G, Voelker DR, Feigenson GW (2008) Membrane lipids: Where they are and how they behave. *Nature Reviews Molecular Cell Biology* 9:112–124.

Walani N, Torres J, Agrawal A (2014) Anisotropic spontaneous curvatures in lipid membranes. *Physical Review E, Statistical, Nonlinear, and Soft Matter Physics* 89:062715.

Wang YH, Collins A, Guo L, Smith-Dupont KB, Gai F, Svitkina T, Janmey PA (2012) Divalent cation-induced cluster formation by polyphosphoinositides in model membranes. *Journal of the American Chemical Society* 134:3387–3395.

Waugh RE (1982) Surface viscosity measurements from large bilayer vesicle tether formation. II. Experiments. *Biophysical Journal* 38:29–37.

Waugh RE, Hochmuth RM (1987) Mechanical equilibrium of thick, hollow, liquid membrane cylinders. *Biophysical Journal* 52:391–400.

Waugh RE, Song J, Svetina S, Zeks B (1992) Local and nonlocal curvature elasticity in bilayer membranes by tether formation from lecithin vesicles. *Biophysical Journal* 61:974–982.

Wu T, Baumgart T (2014) BIN1 membrane curvature sensing and generation show autoinhibition regulated by downstream ligands and PI(4,5)P2. *Biochemistry* 53:7297–7309.

Zhu C, Das SL, Baumgart T (2012) Nonlinear sorting, curvature generation, and crowding of endophilin N-BAR on tubular membranes. *Biophysical Journal* 102:1837–1845.

Measuring giant unilamellar vesicle adhesion

Kheya Sengupta and Ana-Sunčana Smith

Dates? Figs? Sticky things onna stick?

Terry Prachett
C.M.O.T. Dibbler in Small Gods

Contents

17.1 INTRODUCTORY WORDS

Stickiness is essential for all kinds of life. Indeed, the importance of adhesion in biology cannot be overstated. Many bacteria adhere, for example, to form biofilms as part of their life cycle. Multicellular organisms are made of adhesive cells in tissues. Viruses and other infectious organisms need to adhere to their target cells in order to infect them and in return, adhesion is a crucial component of the body's immune response. What makes cells stick is very different from, for example, how blobs of modeling clay stick together. The latter depends essentially on van der Waal's attraction to provide adhesion, whereas in cells, adhesion is mediated by specific proteins. These proteins reside on the membrane of the cell. Consequently, cell adhesion is intimately linked with the physics and physical chemistry of membranes (see reviews Fenz and Sengupta [2012]; Sackmann and Smith [2014]; Smith and Sackmann [2009] and the references therein).

In molecular and cell biology, the focus has often been on identification of the molecules involved in adhesion and on their possible mechanosensitivity at the molecular or cellular scale, often focusing on active processes. In contrast, soft matter physics provides model reference systems amenable to rigorous theoretical treatment (Schwarz and Safran, 2013). It is challenging but fruitful to combine insights from such simplified model systems with molecular specificity and activity of real cells. For example, just the fact that the molecules reside on flexible two-dimensional (2D) membranes has an enormous impact on the way they participate in adhesion. It is exactly this coupling that can be addressed with adhesion of giant unilamellar vesicles (GUVs), which can capture certain crucial aspects of cell adhesion (Fenz and Sengupta, 2012; Sackmann and Smith, 2014).

In this chapter, we review the traditional as well as new techniques used to observe, quantify and manipulate adhesion. We first discuss the basics of cell adhesion in nature, with the view of defining the limits of GUV technology *vis-à-vis* cell architecture. Next we briefly discuss the biophysics of membrane and GUV adhesion at different scales (see Chapter 5 for a detailed view). We then detail the various options available for choosing the ingredients and fabrication methods for adhesive GUVs. Finally, we describe measurement and analysis techniques at different length scales, integrating seminal results. We end with a discussion of technological applications and a concluding overview.

17.2 CONTEXT

17.2.1 CELL ADHESION AND GUV MIMICS

The cell membrane "interfaces" cell adhesion, which is effected by the formation of "specific" biochemical bonds between proteins embedded in the cell membrane and their counterparts, which may be either on the membrane of another cell or embedded on the extracellular matrix (see Lipowsky and Sackmann [1995] for an early review). The outer plasma membrane of the cell is composed essentially of different kinds of lipids and cholesterol (see Chapter 2), with hundreds of different proteins attached and/or embedded. Due to this composition, the cell membrane is essentially a 2D fluid, about 5-nm thick and several hundreds of micrometers in lateral extent. The cell membrane is heterogeneous, complex and is being constantly remodeled (Case and Waterman, 2015; Iskratsch et al., 2014; Saha et al., 2015; Saxton and Jacobson, 1997). GUV membranes mimic cell membranes in a highly simplified way and are usually made of two-tailed phospholipids arranged as a bilayer, with additional cholesterol, glycolipids, and/or embedded proteins to capture one or more of the essential properties of the cell membrane. In recent years, the creation of increasingly complex membranes has been attempted (see Chapters 2–4 and 30), including in the context of adhesion.

Cell adhesion molecules (CAMs) are proteins that mediate adhesion and link the cell membrane to another surface. They are called

Box 17.1 GUVs can be used to mimic certain aspects of cell adhesion

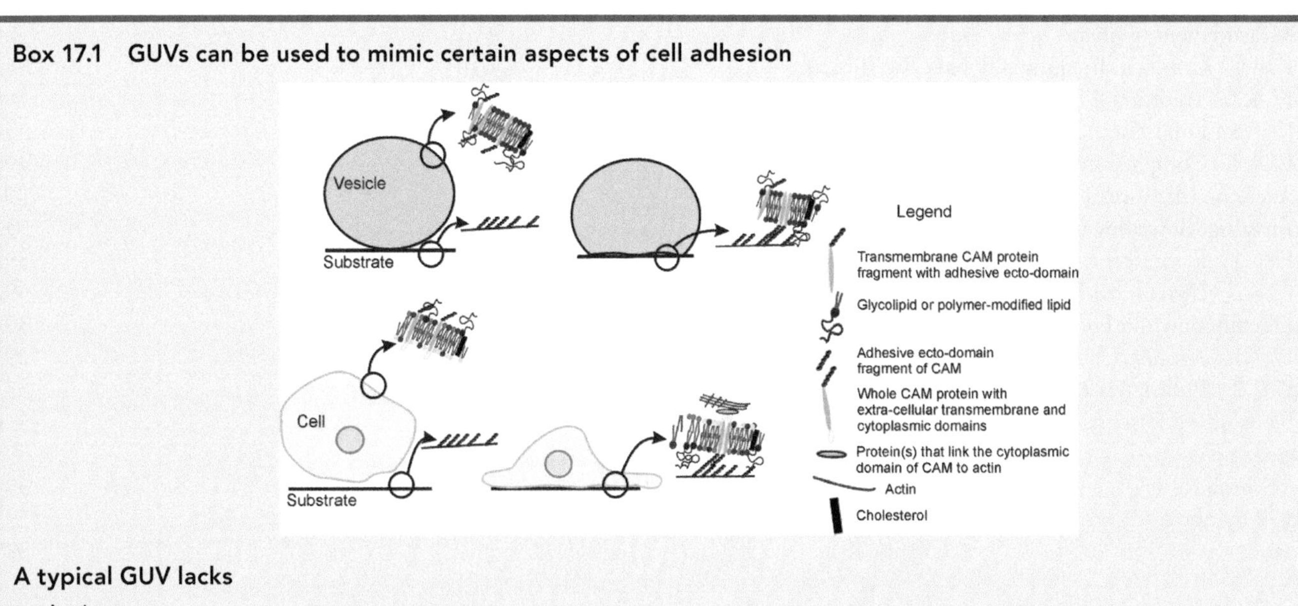

A typical GUV lacks

- Actin cortex
- Asymmetry between leaflets seen in cells
- Post-adhesive outside-in signaling that could initiate changes in molecular organization or even composition
- Possibility of coupling the adhesion molecules to actin

GUVs do have

- A membrane that is soft and flexible
- Molecular diffusion and clustering on the membrane
- Potential to carry appropriate cell adhesion molecules
- Polymers on the surface to mimic the glycocalyx
- Possibility of mimicking out-of-equilibrium membrane fluctuations

receptors and are often transmembrane proteins: selectins, integrins and cadherins being ubiquitous examples in the context of mammalian cells (Alberts et al., 2000). These molecules generally undergo diffusion and/or transport in the plane of the cell membrane (Singer and Nicolson, 1972), at least prior to adhesion. On ligation to their complementary ligand, receptors tend to form mobile and labile clusters, which may eventually form very well-defined micrometer-scale structures (Brasch et al., 2012; Geiger et al., 2001).

In general, the outer, extracellular part of a cell adhesion molecule is the domain responsible for forming the specific bond with its ligand, and the inner, cytoplasmic part connects to the cell interior and may participate in outside-in signaling as well as organization of actin. Adhesion molecules are force sensitive at the molecular level (Evans, 2001; Leckband and de Rooij, 2014; Merkel, 2001; Schiavo et al., 2012). The bond they form often becomes weaker under force (Merkel et al., 1999). Less often the bonds may respond in the opposite manner, becoming stronger under a pulling force (Marshall et al., 2003; Thomas et al., 2008).

In addition to interacting *via* the specific receptor-ligand links discussed above, generic or nonspecific forces of physical origin, like Coulomb forces or van der Waal's attraction, or entropic forces originating from the presence of cell surface polymers, are always present (Sackmann and Smith, 2014; Smith and Sackmann, 2009). Fluctuations of thermal or active origin are thought to play a role in adhesion, both for cells (Pierres et al., 2009; Zidovska and Sackmann, 2006) and in model membranes (Fenz et al., 2011; Smith et al., 2008). Another potentially important physical effect that is only beginning to be considered at the cell* or GUV (Fenz et al., 2011) level is crowding.

A potentially important player, often overlooked in cell adhesion studies is the glycocalyx (Paszek et al., 2014; Robert et al., 2006). It consists of long sugar molecules facing the extracellular space, and one of its functions is thought to be prevention of generic or unspecific interaction between cells. This function is fulfilled though steric hindrance† as well as through repulsive forces of entropic origin (De Gennes, 1979)‡. Hyaluronan is a ubiquitous and rather special cell surface sugar molecule that may be up to several micrometers long (see Chapter 25). In the context of cell adhesion, it clearly has a repulsive role but intriguingly may even participate as a ligand in cell adhesion (Cohen et al., 2006), extending toward another cell bearing its receptor long before traditional CAMs on the two surfaces are close enough to interact. Knowledge of the role of the glycocalyx in cell adhesion is still limited; GUV-based mimetic studies (Limozin and Sengupta, 2007; Sengupta and Limozin, 2010) may be important and relevant in this context.

The intracellular domain of an adhesion molecule usually gets connected to the actin cytoskeleton, which can then exert mechanical forces through the adhesion domains (see, for example, the reviews by Geiger et al. [2009] and Salbreux et al. [2012] or the book by Boal [2012]). Although building this full machinery in a GUV is still a distant dream, forces can still be applied in an artificial manner (Smith et al., 2008). In addition, the membrane itself is coupled to the actin cortex of the cell, a phenomenon that has been successfully mimicked in GUVs (Limozin et al., 2005; Limozin and Sackmann, 2002; Takiguchi et al., 2008) (see also Chapter 4). Clearly, in the context of cell adhesion, the mechanics of the entire membrane–cortex shell needs to be considered. Recent mechanical studies on cell adhesion hint that the cortex may be responsible, at least partly, for the contractile forces applied by cells (Oakes et al., 2014). Intriguingly, adherent GUVs have been shown to also be able to transmit such forces to their substrate (Caorsi et al., 2016; Murrell et al., 2014).

Biological systems in general and cell adhesion in particular is very complex, and therefore meaningful physical models have to be selective and focus on specific phenomena that can be first quantified in experiments and then be integrated into theoretical models that are necessarily partial. As outlined above, GUVs partly capture some of the essential characteristics of cell adhesion. Unlike liquid droplets, but like cells, GUVs are delimited by a membrane. Although the mechanics of GUVs is very different from mechanics of cells at the global scale, the GUV membrane locally provides the opportunity to couple protein reaction and diffusion to the membrane mechanics. GUVs can provide conditions to study adhesion-induced domain formation, influence of fluctuations on adhesion, the consequences of diffusion and a host of other membrane related phenomena (Fenz et al., 2017).

Clearly, GUVs cannot capture all aspects of cells that are pertinent for adhesion; see Box 17.1. The presence of specific conditions outside the cell can trigger a chain of biochemical reactions (called a signaling cascade) inside it that may lead to a particular biological activity of the cell. In cells, adhesion typically sets off such a signaling cascade that may lead, for example, to migration of the cell. A process like active migration cannot yet be realistically mimicked in GUVs.§

17.2.2 GUV ADHESION: TYPICAL EXPERIMENTS

From a purely practical point of view, use of GUVs rather than smaller vesicles for adhesion studies has the simple advantage that the GUVs can be followed in real time by optical microscopy. The early work of Evans (1980) set out the means to measure adhesion energy through imaging GUV deformation, either in GUV–GUV or in GUV–surface adhesion scenarios, giving rise to the two possible classes of experiments (Figure 17.1), which have since then often developed independently.

In a GUV–GUV scenario, the GUVs are either floating in solution and interact freely (Hadorn and Hotz, 2010) or are held by micropipettes (Kenworthy et al., 1995; Noppl-Simson and Needham, 1996; Prechtel et al., 2002) (see Bagatolli and Needham [2014] for a recent review) (see also Chapter 11). In the former case, often many GUVs adhere together forming bunches

* Crowding effects in cells are invoked, for example, to explain certain aspects of T-cell adhesion (Shaw and Dustin, 1997) or cadherin-mediated adhesion (Matre and Heisenberg, 2013), but direct experimental studies are rare (Biswas et al., 2015), if not absent

† Since long and bulky sugar molecules occupy space, nothing else can occupy the same space.

‡ The polymer-like sugars are caught between two surfaces and therefore the configurations accessible to them become restricted, giving rise to entropic repulsion.

§ GUV translation has indeed been reported (Solon et al., 2006), but the mechanism—driven by the transfer of charges—is very different from that of cell migration, which is based on active processes, including actin polymerization.

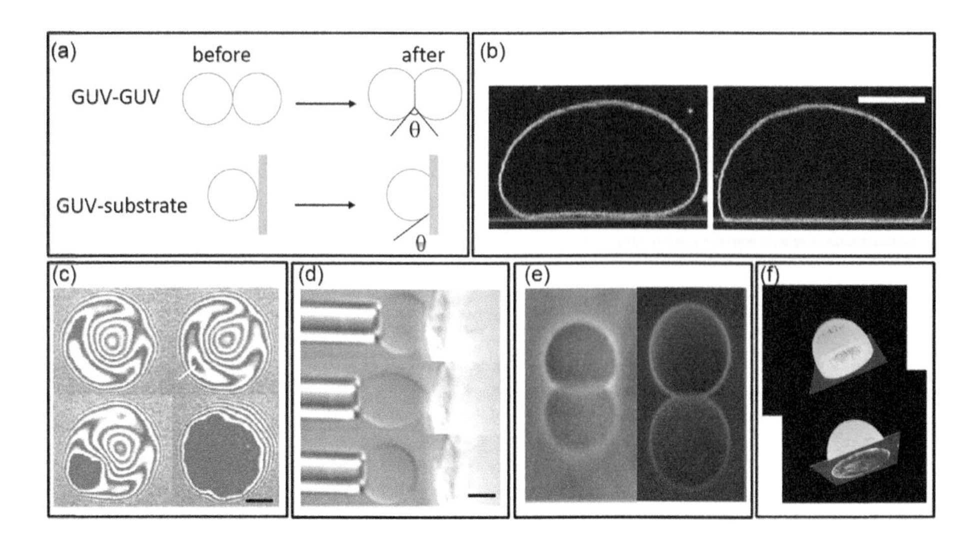

Figure 17.1 GUV–GUV versus GUV–substrate adhesion: In most studies of GUV adhesion, the vesicle adheres either to another GUV or to a well-defined surface. In both cases, the overall shape, the contact angle θ and the size of the contact zone are the important parameters to measure. (a) Schematic of GUV–GUV or GUV–substrate measurement. (b) Confocal microscopy images of a GUV before (left) and after (right) adhesion. Image courtesy of Rumiana Dimova, after (Steinkühler et al., 2016). (c) Interference contrast reflection microscopy pictures of a GUV undergoing adhesion. The dark patch in the contact zone is a result of specific binding, and grows with time. (Reproduced from Boulbitch et al., 2001, with permission from *Biophysical Journal*.) (d) Unbinding of a vesicle specifically adhered to a cell, manipulated by the micropipette technique and observed in phase-contrast. (Reproduced from Prechtel et al., 2002, with permission from *Physical Review Letters*.) (e) Example of a flaccid (left) and a tense (right) vesicle, resting on glass and imaged using an inclined microscope that images the GUV as well as its reflection on glass. (Courtesy of Annie Viallat; After Vézy, C. et al., *Soft Matter*, 3, 844–851, 2007.) (f) 3D rendering of a strongly adhered GUV (biotin–streptavidin mediated) seen at two different angles. (Courtesy of Carlos Marques and Pedro Aoki.)

or clumps, making fine observation difficult. In the latter case, the act of holding the GUV with micropipettes makes it impossible to study the case of very low tension.

Early work on GUV–substrate adhesion was inspired by the spreading of pure liquids and polymer solutions on solids (Feder et al., 1995). In a typical experiment, the GUVs either sediment down or, in rare cases, are buoyed up, making gravitational forces a relevant player. Typically, the substrate is a glass coverslip, treated in order to prevent strong unspecific interaction with the membrane which, if not controlled may dominate the adhesion process and may be strong enough to burst the vesicle. The slide can also be functionalized with specific proteins that may either be immobilized through chemical or physical means or linked to a supported lipid bilayer (SLB). In all cases, even after passivation, a residual interaction with the underlying glass may be present. However, importantly, surface-sensitive techniques like reflection interference contrast microscopy (RICM) and total internal reflection microscopy (TIRF-M) become possible. As will be discussed in Section 17.4, this has enabled probing of adhesion much beyond simple shape calculations. Note that the substrate does not always have to be some flat surface: Particle–GUV adhesion is an example.

The main advantage of using a GUV–GUV system is that the diffusion of lipids and proteins, especially transmembrane proteins is not hindered in any way, and it occurs on a finite surface that is of a similar size as the cell surface. In the GUV–substrate system, the diffusion of both lipids and proteins are severely retarded due to friction generated by the substrate (Merkel et al., 1989). Note, however, that diffusion is also hindered on a cell. In both systems, the nonspecific attraction between the adhesive surfaces may need to be screened for certain applications (see Section 17.2.3). In general, this is low and can be screened relatively easily (Kenworthy et al., 1995; Noppl-Simson and Needham, 1996) for GUV–GUV systems; a more elaborate procedure may be needed for GUV–substrate systems (Albersdörfer et al., 1997; Limozin and Sengupta, 2007; Lorz et al., 2007). Finally, the choice of a GUV–substrate system may allow more control over the organization of ligands on the second interacting surface (Monzel et al., 2012, 2016; Pi et al., 2015, 2013).

17.2.3 GUV ADHESION: BRIEF DESCRIPTION

As in the case of cells, in addition to the specific ligand/receptor bond formation, other forces of physicochemical origin are also important for model membrane adhesion. An important contribution comes from the repulsion due to the presence of the glycocalyx in cells, which in vesicles is mimicked by incorporation of lipids carrying a short passive polymer chain (typically polyethylene glycol). In addition, when the two surfaces approach, depending on the physical chemistry of the specific lipids and the medium used, other physical forces, such as van der Waals, Coulomb and hydration forces emerge. In addition, the GUV membranes are usually soft, with a bending modulus of several k_BT. Typically, GUVs are prepared under hypo-osmotic conditions leading to excess membrane area exhibiting thermal fluctuations (so-called Helfrich undulations; see also Chapter 14) that contribute to repulsive forces. The generic forces of physical origin, collectively called the "unspecific" forces, can be mathematically described in terms of an effective interaction potential. To fully describe the interaction of model membranes with a hard surface or another membrane, both specific and unspecific interactions need to be taken into account.

Modern understanding of membrane mechanics dates back to the work by Helfrich (1973), who used insights from the L_α liquid crystalline phases to understand membrane deformations.

The most relevant deformation mode for model membranes, including for GUVs, is bending. This results from the fact that lipids are virtually incompressible, shear-related deformations are not sustained because the membrane is fluid in 2D, and stretching costs are energetically prohibitive because stretching exposes the hydrophobic chains to water. Furthermore, membranes are fragile because the lateral cohesion between lipid molecules is weak.

In addition to bending deformations, a membrane is defined by its tension. The tension in a lipid membrane arises not from stretching of the material of the bilayer but from unfolding of the area "hidden" due to fluctuation modes (see Chapter 5 for a deeper discussion). Changing the tension also costs energy. Finally, especially in the context of adhesion, an interaction potential representing adhesive or repulsive interactions between the two participating surfaces contributes to the free energy.

Adhesion of vesicles was originally studied at the macroscopic level where the shape of the vesicle is determined through minimization of the Hamiltonian, which contains the bending, tension and adhesion terms. The optimal shape arises from the minimization of the Hamiltonian with the constraints on the total enclosed volume and area of the GUV. This strategy gives insight into the interplay between fully homogenized weak adhesion and elasticity. However, in the context of protein-mediated interactions, shape deformations occur at energy scales that are orders of magnitude smaller than the energy contributions due to the specific adhesion contacts (Smith and Seifert, 2007). This is also an important difference between GUVs and cells; in the latter, the membrane is a composite structure and shape deformations involve the much stiffer cytoskeleton.

In the context of specific adhesion, a very successful strategy in linking the effects of the membrane and the protein binding emerges from considering energy cost for bending, membrane tension (Σ) and the interaction potential (Servuss and Helfrich, 1989) in the Monge representation.* In this case, the Hamiltonian

$$\mathcal{H} = \int_A d\mathbf{x}\left[\frac{\kappa}{2}(\nabla^2 h(\mathbf{x}))^2 + \frac{\Sigma}{2}(\nabla h(\mathbf{x}))^2 + \frac{\gamma}{2}(h(\mathbf{x}) - \langle h(\mathbf{x})\rangle)^2\right], \quad (17.1)$$

allows for the investigation of the local coupling between the membrane and proteins. Here, κ is the bending modulus, $\mathbf{x} \equiv (x, y)$ is the lateral position of the membrane, $h(\mathbf{x})$ is the distance from the surface (or from the other membrane) at the position x and γ is related to the curvature of the interaction potential in a harmonic approximation. When specific adhesion is present, an additional term is suitably introduced to account for the energy of bond formation (Smith and Sackmann, 2009). In alternative approaches, the nonspecific adhesion term is omitted (Weikl et al., 2009).

In a typical experiment, a GUV is allowed to sediment onto a substrate. The bottom of the vesicle flattens and forms a contact zone where the membrane is close to the substrate and interacts with it. For high enthalpy and abundant linkers, numerous bonds quickly fill the adhesion zone, the vesicle is further deformed such that the contact zone, now converted to an adhesion zone, expands until the cost of elastic deformation and increasing tension balances the gain in adhesion energy (Figure 17.1). However, in most cases, the equilibrium is determined by a balance of entropy and enthalpy of the binders. The change in free energy ΔF arises from the enthalpy due to bond formation and the change in entropy of the binding pairs (Fenz et al., 2011; Smith et al., 2008; Smith and Seifert, 2005):

$$\Delta F = E_a N_b + k_B T \ln \Omega \quad (17.2)$$

where E_a is the enthalpy of bond formation of a single bond, N_b is the average number of bonds formed during adhesions and Ω represents the loss in the number of possible configurations of the system. Ω consists of several terms accounting for the contribution of permuting (i) free receptors, (ii) free ligands and (iii) bound receptor–ligand pairs, coupled to the appropriate statistical system. Hence, the equilibrium state in GUV–SLB adhesion experiments is significantly different for the cases where the ligands and receptors are (i) mobile or immobile (ii) binders diffuse in a vesicle of a finite size, or in the unstructured SLB, that for all practical purposes serves as a reservoir of binders at a constant chemical potential. Typically, the number of receptors and ligands are known because the GUVs are prepared with these system parameters and the free energy can be minimized with respect to N_b to determine the equilibrium bond configuration. In more sophisticated approaches, E_a is determined from the final adhesion state, because the fluctuations in the membrane change between the unbound state and the adhered state of the GUV (Fenz et al., 2017).

17.2.4 DETERMINANTS OF EQUILIBRIUM AND DYNAMICS

According to the above discussion, and depending on the characteristics of the GUV adhesion system, different scenarios are possible (De Gennes et al., 2003; Fenz and Sengupta, 2012; Sackmann and Smith, 2014) (see Box 17.2). Furthermore, at different length scales, different measurables may be relevant.

At the scale of the vesicle, the global shape is the measurable and, as seen in Section 17.2.4, is determined by minimization of Eq. 17.1 (Seifert and Langer, 1993). The adhesion can be induced by nonspecific forces or by specific linkers or, as is usually the case, by a combination of the two. For strong adhesion induced either by nonspecific attraction or by strong and abundant specific linkers, the shape is typically a truncated sphere (Evans, 1980; Seifert and Langer, 1993). This case of strong adhesion formed the basis of most early (Albersdörfer et al., 1997; Guttenberg et al., 2001; Nardi et al., 1998) and some recent experimental studies (Gruhn et al., 2007; Limozin and Sengupta, 2007; Nam and Santore, 2007a, 2007b; Shimobayashi et al., 2015; Steinkühler et al., 2016). (See also Table 5.2 in Chapter 5.)

For GUV–substrate adhesion, an important scenario is where gravity and nonspecific weak repulsive forces conspire to produce a relatively shallow minimum high above the substrate (Boulbitch et al., 2001; Marx et al., 2002; Sackmann and Bruinsma, 2002). Note that electrostatic interactions are strongly screened because the GUV is under water. The overall vesicle shape is determined as usual from Eq. 17.1: The spheroidal vesicle develops a flattened base—the contact zone (Limozin and Sengupta, 2007; Marx

* The Monge representation parameterizes vertical displacements as a function of the in-plane spatial coordinates and is most suitable for describing "small" deformations where overhangs are absent.

Box 17.2 Equilibrium and dynamics

Equilibrium: For nonspecific adhesion or at high linker concentration and for strong linkers, the equilibrium state is determined by balance of adhesion energy and vesicle deformation/tension. At low receptor concentration and/or affinity, the equilibrium state is determined by balance of entropy and enthalpy.

Dynamics: For nonspecific adhesion and for abundant and strong linkers with very fast reaction rate, the adhesion dynamics is dominated by hydrodynamic effects. However, for specific binding with abundant linkers, the reaction rate may be the determinant factor. For specific binding with sparse linkers, which is the most biologically relevant case, the kinetics is dominated by competition between diffusion and reaction rates.

et al., 2002; Sengupta and Limozin, 2010). The location and the width of the minimum of the potential determine at what height the flattened basal membrane is held and how much it can fluctuate. In typical experiments, this distance may be as high as several hundred nanometers, but the corresponding fluctuation amplitude may be relatively low at a few tens of nanometers (Monzel et al., 2012; Schmidt et al., 2014). In the contact zone, the GUV membrane is available for participating in adhesion without further macroscopic deformation of the GUV.

In the absence of other attractive forces, the membrane may live indefinitely in this state. However, if a deeper minimum is present closer to the substrate due to attractive specific or nonspecific forces, the membrane eventually shifts fully or partially into the deeper minimum (Bruinsma et al., 2000). Importantly, in most experimental scenarios, there is a barrier for this transition, which determines the nucleation time and size (Bihr et al., 2012; Limozin and Sengupta, 2009).

The parts of the membrane that transit to the deep adhesive minimum are together called the adhesion zone. Two cases can be identified: Case 1: the adhesion zone remains restricted within the initial contact zone, and Case 2: the adhesion zone goes beyond the initial contact zone. In the latter case, the vesicle is said to "spread" due to the formation of bonds. The size of the adhesion zone is determined by the extent to which the GUV can spread, given the constrains on GUV volume and membrane area. The energy balance is governed by balance of adhesion and deformation with tension and osmotic pressure being determined through constraints of fixed area and pressure (Lipowsky, 1991; Lipowsky and Sackmann, 1995).

In case of specific adhesion, the distance between the adherent surfaces is determined by the length of the linkers (Fenz et al., 2009). In case of nonspecific adhesion, or in the presence of glycocalyx mimetic polymers, this may be more complicated (Limozin and Sengupta, 2007). In case of weak adhesion or inhomogeneous substrates, the overall shape may be of a truncated spheroid (Seifert, 1997) with an irregularly shaped base (Limozin and Sengupta, 2007). The later is usually a sign that the GUV is still not dominated by tension, and the shape is a result of a projected membrane surface, rather than the total available unwrinkled membrane area.

At the meso-scale of the membrane and the linker molecules, the measurables are the molecular distribution and membrane topography. Membrane fluctuations and configuration of polymers may also play a role: The former can now be measured with great accuracy (Betz and Sykes, 2012; Monzel et al., 2015, 2012, 2016), but the latter is yet to be experimentally measured in the context of membrane adhesion.

The dynamics, for adhesion driven by nonspecific interactions or for strong bonds and abundant linker, is determined by hydrodynamic dissipation (Sengupta and Limozin, 2010). For the case of low abundance of linkers, it is determined by diffusion (Bihr et al., 2015; Boulbitch et al., 2001; Cuvelier and Nassoy, 2004; Lorz et al., 2007; Streicher et al., 2009). In case of low affinity linkers, the rate is expected to be reaction limited (Bihr et al., 2015; Boulbitch et al., 2001; Reister-Gottfried et al., 2008). Accessibility of the biding site of the ligands at the molecular level due to molecular conformation or due to limitations on diffusion induced by crowding (Fenz et al., 2011; Schmidt et al., 2015, 2014) may modify the picture.

17.3 INGREDIENTS

In general, a central challenge in GUV experiments is to identify and reproduce essential physical and biochemical constraints (Richmond et al., 2011)—this is particularly true for adhesion studies. Therefore, each of the essential ingredient of an adhesive GUV needs to be chosen carefully. The (de)merits of GUV–GUV and GUV–substrate adhesion were discussed in Section 17.2.2. Here we discuss the choice of molecular ingredients that are mostly common to both approaches.

Molecular specificity plays an important role in adhesion of cells. This aspect can be experimentally mimicked in GUV systems by incorporation of the protein of interest, the receptor, in the GUV, and its counterpart, the ligand, onto another surface. In case of GUV–GUV adhesion, the second surface is the other GUV. In case of GUV–substrate adhesion, the ligands can be either directly grafted on the surface or incorporated into a supported membrane. Such a configuration has been extensively used (see Fenz and Sengupta [2012] for a review), including by us (Fenz et al., 2011, 2009; Smith et al., 2008), and has yielded insight into the interplay of different actors in membrane adhesion: for example, reaction rates and diffusion times in determining GUV adhesion dynamics (Bihr et al., 2014).

17.3.1 MATRIX LIPIDS

The main component of the GUV membrane is usually phospholipids, see Appendix 1 of the book for structure and data on the lipids here below. The lipid of choice has often been 1,2-dioleoyl-*sn*-glycero-3-phosphocholine (DOPC) or 1-stearoyl-2-oleoyl-*sn*-glycero-3-phosphocholine (SOPC). The choice of unsaturated-chained lipids ensures that the GUV membrane is fluid even at relatively low ambient temperatures. A mixture of 1,2-dimyristoyl-*sn*-glycero-3-phosphocholine (DMPC) and cholesterol has also been extensively used. All these molecules are commercially available from multiple sources. A potential disadvantage of using DOPC is that a double bond is vulnerable to oxidation and having two double bonds may render the lipid particularly fragile and necessitate specific precautions. The choice of the matrix lipid of course determines the bending modulus of the GUV membrane, DOPC membranes being the softest among those discussed above. Diblock copolymer-based GUVs, called polymersomes, are less fragile and more rigid and have already been used in the context of adhesion (Nam and Santore, 2007a, 2007b) (see Chapter 26). The different methods to prepare GUVs are described in detail in Chapter 1.

17.3.2 LINKERS

In some of the earliest experimental systems, the adhesion was induced unspecifically, and charged systems were expected to provide simple means of inducing adhesion (Nardi et al., 1998).* To achieve this, charged surfactants or lipids are usually used; typically, a small percentage is included in the lipid matrix (Gruhn et al., 2007; Steinkühler et al., 2016).

A relatively simple means of inducing specific adhesion is by grafting the linker of interest on lipids by modification of the headgroup. A very popular choice is the biotin–avidin pair (Albersdörfer et al., 1997; Cuvelier and Nassoy, 2004; Feder et al., 1995; Fenz et al., 2011, 2009; Monzel et al., 2012). This link is purely artificial because *in vivo*, there is no known situation in which biotin comes in contact with avidin. With an enthalpy of 30 kT per bond, it is also a very strong link compared with other natural CAMs. In an avidin†–biotin system, one of the adhesive surfaces (GUV or substrate) carries the biotin ligands, and the other is functionalized with avidin, which itself may be bound using biotin. This system is easy to handle and the molecules are commercially available.

Another popular choice is the NTA-His-tag pair, NTA is grafted on a lipid by modification of its headgroup (Dorn et al., 1998; Schmitt et al., 1994) and the histidine tag (typically hexahistidin) is on the protein of interest. E-cadherin mediated adhesion has been studied using NTA-his6 linkers (Fenz et al., 2017, 2009; Puech et al., 2006). Other systems used in the past include Arg-Gly-Asp (RGD) (Guttenberg et al., 2000, 2001; Hu et al., 2000; Marchi-Artzner et al., 2003; Smith et al., 2008; Streicher et al., 2009) and SyalLewisX (Bihr et al., 2014; Lorz et al., 2007; Reister-Gottfried et al., 2008; Smith et al., 2006), binding to integrins and E-selectins, respectively, typically physisorbed on the surface to mimic cell–tissue adhesion. In both cases, the molecules were locally synthesized. Though RDG modified lipids are now commercially available, they have, to our knowledge, not yet been used in the context of GUV adhesion.

As discussed in Section 17.2, most CAMs are transmembrane proteins: An aspect that was mimicked in early studies using a bacterial protein contact site A, which resembles cadherins, was used for both GUV–substrate (Bruinsma et al., 2000; Kloboucek et al., 1999) and GUV–GUV (Maier et al., 1997) adhesion studies. The ubiquitous adhesion protein, integrin, was first incorporated into bilayers and used extensively in the context of GUV adhesion in the laboratory of Erich Sackmann (Goennenwein et al., 2003; Smith et al., 2008). Other groups have also successfully incorporated integrins into GUVs or bilayers in the context of GUV adhesion (Brüggemann et al., 2014; Streicher et al., 2009). These systems have the advantage that the protein remains fully functional and mobile in the membrane of interest, which reproduces more realistically the establishment of cell–cell contacts. Protein reconstitution in GUVs is the subject of Chapter 3.

An alternative to linking through proteins, is to use purely artificial links. DNA is becoming a popular choice. It is either used in a biotinylated form and is linked *via* avidin (Hadorn and Hotz, 2010) or comes with a sterol moiety that can insert into the membrane *via* hydrophobic interactions (Parolini et al., 2015; Shimobayashi et al., 2015). The advantage of using DNA is that it can be designed, in terms of length, affinity and flexibility in a way that proteins cannot. Recently, other chemical linkers have been used with the view of technological applications (Ushiyama et al., 2015).

17.3.3 MIMICKING OTHER MEMBRANE-ASSOCIATED CELLULAR FEATURES

An essential cell–surface feature commonly mimicked in GUV adhesion is the glycocalyx, which *in vivo* prevents undesirable cell–cell contacts. This is usually achieved by including lipids with headgroups bearing a polyethylene glycol (PEG) chain into the GUV and/or the substrate membrane (Albersdörfer et al., 1997; Evans and Rawicz, 1997; Feder et al., 1995; Kenworthy et al., 1995; Lorz et al., 2007; Needham et al., 1992) (Figure 17.2). More recently, DNA has been used as a glycocalyx mimic (Hisette et al., 2008; Moreira et al., 2003; Nam et al., 2010). Arguably, a more physiological choice is hyaluronan, which has been used as a repulsive polymer cushion in GUV–substrate adhesion studies (Limozin and Sengupta, 2007; Sengupta and Limozin, 2010). In all cases, the artificial gylcocalyx blocks some of the nonspecific interactions, especially if it is constructed from polymers longer than the binders themselves. However, any protein or large lipid headgroup that is not bound can act as a repeller (Lorz et al., 2007; Smith and Seifert, 2007).

In the last two decades, significant effort has gone into construction of artificial cytoskeletons mimicked in GUVs with actin networks either outside (Helfer et al., 2001; Liu et al., 2008) or inside (Cortese et al., 1989; Limozin et al., 2003, 2005; Limozin and Sackmann, 2002; Takiguchi et al., 2008). Despite this success, these vesicles have not yet been combined with adhesion proteins, mainly because of the technical

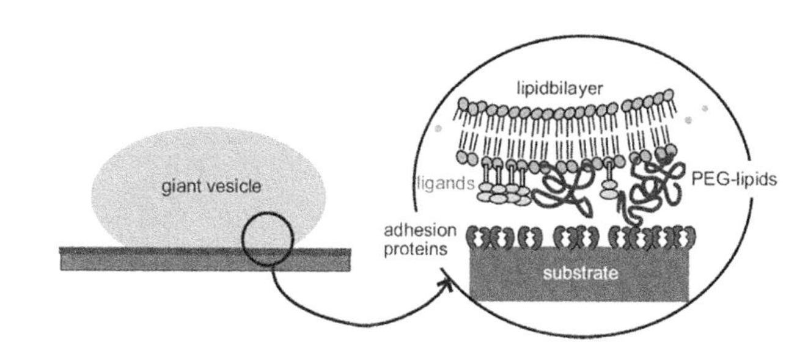

Figure 17.2 Typical composition of a basic adherent vesicle, with the lipid bilayer carrying ligands and repellers. (From Marx, 2003; Courtesy of Stephanie Goennenwein.)

* See also (Frostad et al., 2014) where multilamellar vesicles of size comparable to GUVs were used.

† The word "avidin," as used here, includes its derivatives—streptavidin and neutravidin. In fact, avidin itself, being charged at pH 7, is a poor choice.

complexity of the system. Likewise, vesicles filled with artificial polymer gels have been synthesized (Campillo et al., 2009; Viallat et al., 2004), but they also remain unexploited in adhesion studies.

17.3.4 CHOICE OF GUV PREPARATION TECHNIQUE AND OF THE COUNTER-SURFACE

As discussed in Section 17.2.2, the GUV under study may adhere to another GUV suitably functionalized or to a substrate. Electroswelling remains the most popular method for GUV fabrication (see Chapter 1). Although different types of fluidics-based techniques open up considerable flexibility and ease of preparation given that they invariably rely on the presence of an oil-like phase that ideally should not be incorporated into the GUV, there is always a risk that a few molecules of the oil phase does get incorporated into the hydrophobic region of the GUV membrane. This in turn may change the membrane mechanical properties and impact the interpretation of adhesion assays. New gel-assisted swelling protocols (Weinberger et al., 2013) hold out the hope for a technique that is more flexible and easier to set up than electroswelling, at the same time being as clean. These are, however, only at the beginning of being tested in the context of adhesion (Aoki et al., 2015).

GUV–substrate adhesion studies, the substrate plays an important role and must be prepared as carefully as the GUV. The different options are shown in Figure 17.3. Typically, the substrate is supported on a glass coverslip, to take advantage of advanced surface imaging techniques (see [Limozin and Sengupta, 2009] for a review). The linkers may be either physisorbed or chemically grafted to the surface. Typically, the former results in random orientation, which may compromise the access to the adhesion domain of the CAM and, therefore, wherever possible, the latter should be preferred. In either case, the ligands are necessarily immobilized. An alternative is to graft the CAM (or the suitable adhesion moiety of the molecule/ligand) on a supported bilayer. Recently, supported bilayers with immobilized or mobile ligands with variable mobility were created in the context of cell adhesion (Dillard et al., 2014; Hsu et al., 2012). Molecules are either attached to the headgroup of a lipid (see discussion under linkers) or they may be transmembrane proteins embedded in the bilayer. In the latter case, the bilayer may need to be supported on a polymer cushion to ensure mobility (Goennenwein et al., 2003; Smith et al., 2008).

17.4 MEASUREMENT OF SHAPE

17.4.1 CONFOCAL IMAGING: GLOBAL GUV SHAPE

GUVs have been extensively studied using confocal microscopy, for example, in the context of phase separation (Baumgart et al., 2007) (see Chapter 10). Adhesion is also studied with confocal or other fluorescence techniques (Beales et al., 2011; Sakuma et al., 2008; Sarmento et al., 2012; Shindell et al., 2015). This is particularly suitable for imaging domain formation and phase separation for both proteins and lipidic phases. Measurements of shape can be used to determine the effective adhesion strength (Smith et al., 2006). Recently, confocal imaging of the vesicle shape was performed in conjunction with imaging of the adhesion disc (Parolini et al., 2015).

17.4.2 TRADITIONAL RICM: ADHESION AREA AND MEMBRANE CURVATURE

RICM and the closely related interference reflection microscopy were originally introduced to observe adhesion of cells (Curtis, 1964; Gingell and Todd, 1979), and are now used extensively in studies of cell adhesion (Limozin and Sengupta, 2009). Cells, however, are optically complex and it was quickly realized that the precision of RICM in cells may be limited (Verschueren, 1985). Later, RICM was applied with great success to simpler objects like colloidal beads and GUVs (Rädler et al., 1995). It was developed and refined, in particular in the laboratory of Erich Sackmann, by accounting for the divergence of the incident light beam (Wiegand et al., 1998) and the nonlocal contributions to the image formation associated with the spatial and temporal variations of the curvature of nonplanar interfaces (Kühner and Sackmann, 1996). Early work on GUV using RICM (Albersdörfer et al., 1997; Feder et al., 1995; Goennenwein et al., 2003; Kloboucek et al., 1999; Marchi-Artzner et al., 2003) forms the basis of the majority of the GUV–substrate adhesion studies to date (Cuvelier and Nassoy, 2004; Fenz et al., 2009, 2011; Puech et al., 2006; Sengupta et al., 2006; Smith et al., 2008; Streicher et al., 2009). The basic principles of this technique are summarized in Box 17.3, for a practical guide, see Limozin and Sengupta (2009).

In its early application to GUV adhesion, RICM was used to measure the thickness of an adherent and burst GUV membrane (Feder et al., 1995). Soon afterward, it was used to not only measure the growth of the adhesion zone but also the shape of the vesicle rim as it curves away from the substrate (Albersdörfer et al., 1997). The reconstruction of this shape was used to determine the contact angle between the membrane and

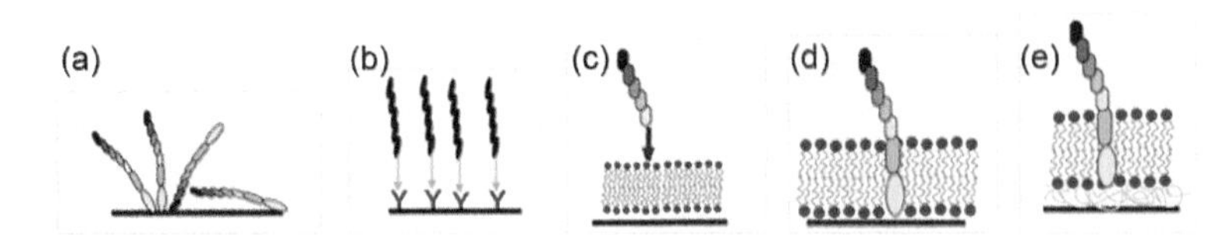

Figure 17.3 Example of different ways to functionalize the substrate. Increasingly sophisticated strategies are seen going from left to right: (a) unspecific physisorbtion of a CAM on glass, (b) well-oriented CAMs chemically grafted on glass, (c) CAMs linked to headgroups of lipids in an SLB, (d) transmembrane CAM incorporated into an SLB, (e) transmembrane CAM incorporated into an SLB and supported on a polymer cushion.

Box 17.3 Reflection interference contrast microscopy

RICM is arguably the most popular method for measuring GUV adhesion. Because "measuring" adhesion effectively implies measuring interaction between surfaces, the intersurface distance is a very important parameter for adhesion studies. Interference is the obvious technique of choice when small distances need to be measured. In the case of GUV adhesion, the challenge is to do so under water, with poor reflectivity and therefore often poor signal-to-noise ratio. In RICM of GUV–substrate adhesion, light reflected from the GUV–buffer interface interferes with light reflected from the substrate–buffer interface and is imaged with an objective. The intensity level is related to the distance between the GUV surface and the substrate through a sinusoidal relationship.

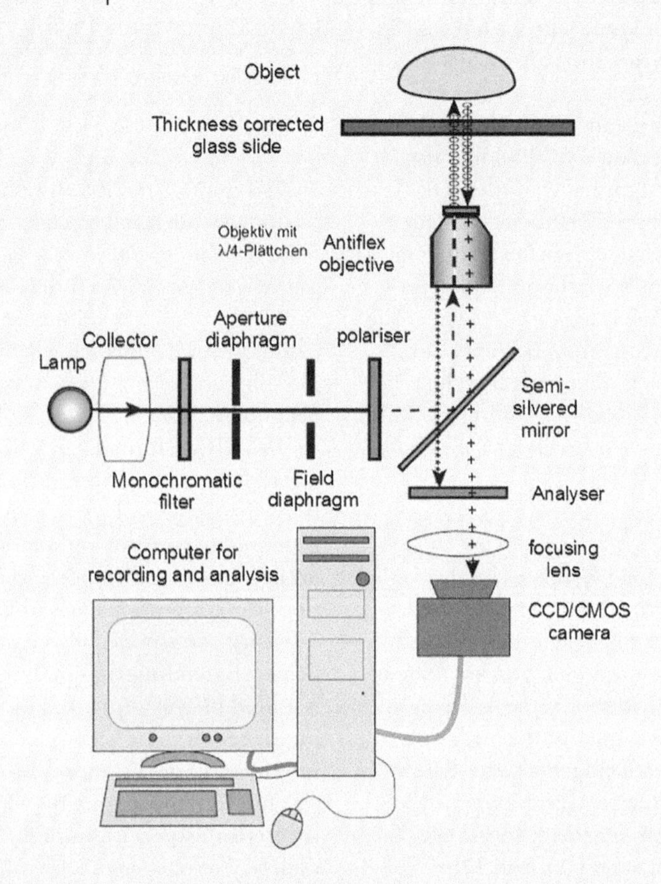

A basic RICM set up. White light from a semi-coherent source (usually a mercury lamp) is rendered monochromatic and polarized. It passes through an antiflex objective containing a quarter-wave plate, rendering the light circularly polarized. Light is reflected from the glass–buffer interface (not shown for clarity) and from the buffer–object interface (shown here). Note that in complex objects like cells, there may be more reflections to account for. The reflected light passes back through the quarter-wave plate and is turned back into plane-polarized light, but with the polarization reversed. The analyzer is set such that it cuts off all light except that which passed twice through the objective. This procedure, called the "antiflex" technique, enhances contrast and is essential for quantitative treatment of the data. (Image courtesy of Barbara Feneberg.)

the substrate. At equilibrium, the tension should be given by the Young-Duprè law:

$$W = \Sigma(1 - \cos\theta) \tag{17.3}$$

where W is the adhesion energy; Σ is the interface tension; and θ is the contact angle. However, because curving the membrane costs energy, unlike the case of liquid droplet, the GUV–substrate contact cannot be infinitely sharp. Bruinsma pointed out that the effect of the finite bending modulus results in a "rim" of width Λ around the adhesion disc, where the membrane configuration differs from that predicted for a liquid droplet (Bruinsma, 1995; Sackmann and Bruinsma, 2002). He proposed that this effect can be absorbed into an effective line tension term. The total energy gained (or lost) in the adhesion process is then a sum of a tension term ($\Sigma \cdot$ change in area of the membrane), an adhesion term ($W \cdot$ area of the adhesion disc), a bending energy term and the line tension term. In the Bruinsma formalism, it is assumed that the line tension term is decoupled from the rest, and can be minimized separately from the rest. Further assumptions are that the GUV radius R is much larger than the radius of the adhesion disc and that $\Sigma >> \frac{\kappa}{R^2}$. Under these assumptions, minimization of the free energy (first a minimization without the line tension term, and then a separate minimization of the line-tension) shows that $\Lambda = \sqrt{\frac{\kappa}{\Sigma}}$, yielding $W = \frac{\kappa}{\Lambda^2}(1-\cos(\theta))$.

This formulation proved to be very useful for analysis of RICM images because Λ can be easily estimated from standard RICM images, simply by detecting the location of the fringes and assigning the correct height to each fringe (Albersdörfer et al., 1997; Sengupta and Limozin, 2010). However, it must be emphasized that the Bruinsma formalism is valid only for strong adhesion, a circular adhesion disc, and moderately tense vesicles. Caution must be exercised to ensure that the experiments fulfill all the assumptions of the model.

In an alternative formulation, Seifert and Lipowsky worked with contact curvature (Seifert and Lipowsky, 1990) (see also Chapter 5), thus eliminating the need for many of the *ad hoc* assumptions in the Bruinsma formalism. The main drawback of the Seifert-Lipowsky approach is that a precise experimental measure of the contact curvature is difficult. In particular, for RICM analysis, because the membrane is highly curved near the substrate, the optical perturbation arising from this curvature needs to be accounted for. When this is done correctly, the two formalisms give the same results (see supporting information in Sengupta and Limozin, 2010).

17.4.3 DUAL-WAVE RICM AND RICM WITH MULTILAYERS

One major drawback of the conventional RICM technique is that the information about the phases of the beams reflected by the various interfaces of the GUV and/or substrate is not known. As a result, absolute distances cannot be measured, because there is an ambiguity of a phase factor $\lambda/2n$ (where λ is the wavelength of the light used and n is the refractive index of the medium). This drawback was overcome by introducing dual-wave RICM (DW-RICM), which compares the interferograms obtained simultaneously with two different wavelengths (Monzel et al., 2009; Schilling et al., 2004; Sengupta and Limozin, 2010). In other words, an additional periodicity and boundary condition is introduced by observing a second wavelength, as can be seen in Figure 17.4 This enables measurement of quasi-instantaneous absolute distance of a membrane above a planar surface, with nanometric precision in the vertical direction (Monzel et al., 2012).*

The DW-RICM technique was initially developed for colloidal beads and was then applied to vesicles (Schilling et al., 2004). In GUVs, an additional consideration arises from the fact that it is bound by a membrane of finite thickness. Therefore, light is reflected both from the outer and the inner surface of the membrane. Thus, three interfering beams must now be taken into account to describe the interference pattern. Introduction of multiple reflections simply introduces a phase factor in the cosine function connecting the intensity and height (Limozin and Sengupta, 2007). As a result, the traditional RICM analysis underestimates the membrane–substrate distance (h) by an amount that depends on the values of the inner and outer buffer refractive indices. When the refractive index of the inner buffer is sufficiently high, the minimum of the intensity corresponds to a h of almost zero. However, when it decreases, the minimum of the intensity occurs at a nonzero h (Fenz et al., 2009; Limozin and Sengupta, 2007).

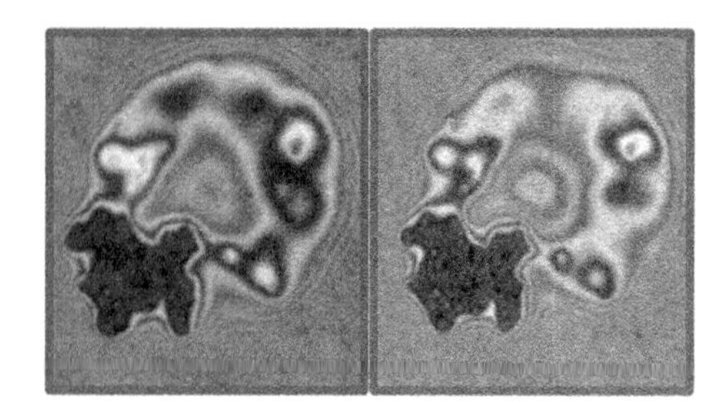

Figure 17.4 Example of a GUV undergoing adhesion, imaged in two colors: 561 nm (left) and 488 nm (right). Note the different fringe pattern arising from the difference in illumination wavelength.

The RICM formalism, which takes into account the refractive indexes of the various layers, has important implications for analysis of data from vesicles. Results show that it is definitely required for quantitative measurement of membrane–substrate distance. Even for a qualitative interpretation, the conventional assumption that the minimum of the intensity corresponds to the minimum of the height is insufficient. Extending this analysis to the case where the substrate is a lipid bilayer supported on glass, implies that an extra interface has to be accounted for. Using this technique, the intermembrane distance for biotin–avidin mediated binding was found to be 7 ± 1 nm, which compares very well with the theoretical expectations of 8 nm (Fenz et al., 2009). Thus, an accuracy of one nanometer can be achieved using this formalism for non-fluctuating membranes. With this insight, RICM was used to show that in E-cadherin mediated adhesion of artificial membranes, the cadherin is bound *via* the two outermost ectodomains (Fenz et al., 2009).

17.4.4 COMPARISON WITH TIRF-M

The main alternative to RICM for mapping the membrane topography has been TIRF-M (see also Chapter 21) and its variations. It has been used for colloidal beads, GUVs and cells (Brodovitch et al., 2013). For the bead case, TIRF is indeed as accurate as RICM, provided that the bead is close to the coverslip: far from the slide, DW-RICM is the method of choice. Of course, instead of a bead, a droplet many be followed in RICM. In case of GUVs and cells, the major advantage of RICM is that it is label free and can be easily implemented in a standard microscope with a florescence lamp. Furthermore, the recent developments sketched above (including fluctuation analysis), makes RICM the current method of choice for GUVs. Figure 17.5 shows an example where GUV spreading dynamics was analyzed not only in terms of spreading area but also the time evolution of the moving edge. TIRF-M is of course able to map the location of proteins, which RICM cannot do.

* For adherent vesicles, where the signal can be integrated over several time frames, the precision of DW-RICM is same as that of single-wave RICM, that is to say 1 nm (Fenz et al., 2009). For a fluctuating membrane, single-wavelength RICM can be applied only if the fluctuation amplitude is rather low (Smith et al., 2008, 2006). The accuracy of DW-RICM for a strongly fluctuating membrane is 5 nm (Monzel et al., 2012).

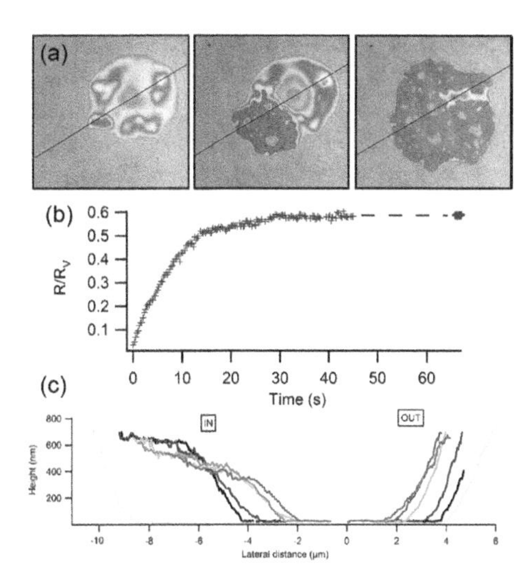

Figure 17.5 Spreading of a GUV on a substrate through nonspecific attraction. Presence of a polymer on the substrate slows down the process, so that imaging with good signal-to-noise ratio becomes possible, which in turn enables analysis not only of the area evolution (middle, R is the radius of the adhesion zone and R_V that of the GUV), but also reconstruction of the moving edge of the spreading vesicle (bottom, reconstruction along the red line indicated on the RICM snap-shots). (Adapted from Sengupta, K. and Limozin, L., *Phys. Rev. Lett.*, 104, 088101, 2010.)

17.5 ACCURATE MEASURE OF FLUCTUATIONS

For a soft membrane like that of lipidic GUVs, fluctuations are intimately linked to adhesion (Reister-Gottfried et al., 2008; Weikl et al., 2009). On the one hand, they give rise to a repulsive contribution to the nonspecific potential (see discussion in Chapter 5), and on the other hand, they control the binding and unbinding rates of CAMS, which could be determined theoretically (Bihr et al., 2012, 2015) and experimentally (Bihr et al., 2014). Consequently, the formation of adhesion domains is controlled by fluctuations, from the early, nucleation stages (Bihr et al., 2012; Fenz et al., 2011), to the macroscopic growth of the domains of ligand-receptor bonds (Bihr et al., 2015; Fenz et al., 2011; Schmidt et al., 2015). For example, it was shown that at the location of a new domain being constructed, the membrane fluctuations show an increase just before they are frozen (Fenz et al., 2011) while the separation between membrane is continuously decreasing. This kind of hot-spot could be explained in terms of a nontrivial dependence of the membrane roughness on the density of bonds (Figure 17.6c) (Fenz et al., 2011; Reister et al., 2011). It is now possible to model this entire process with great accuracy in Monte Carlo simulations that capture the essential physics, from the onset of adhesion to the relaxation into a steady state, and is able to handle the entire vesicle carrying about 10^6 binders with nanometric and microsecond resolution over several seconds (Bihr et al., 2015). Very importantly, direct mapping to analogous experiments can be performed (Schmidt et al., 2015).

Moreover, fluctuations are totally damped inside an adhesion patch full of densely packed bonds. This suppression of fluctuations has a consequence for the energetics of the system and also provides an alternative way to detect adhesion (Fenz et al., 2011;

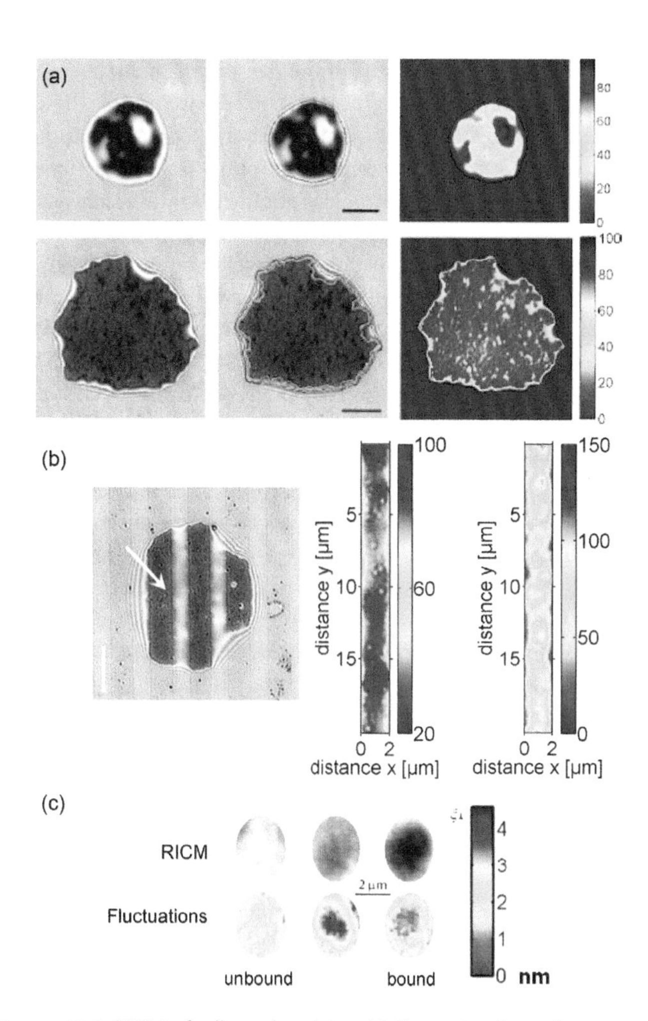

Figure 17.6 RICM of adhered vesicles. (a) Example of membrane topography reconstruction with RICM, considering reflection from both faces of the GUV membrane. Color bar indicates membrane to surface distance. Scale bar: 10 μm. The GUV is either fluctuating at a considerable distance above the substrate (top row) or adhered (bottom row). Note that darker flakes in the RICM image (bottom left) are in fact at a higher height (bottom right). (From Fenz, 2008. Image courtesy of Susanne Fenz.) (b) RICM image (scale bar: 5 μm) and reconstructed average height in one on one of the linear nonadherent regions, and corresponding fluctuation map (color scale for both in nm) for GUVs partially adhered to line patterns on a substrate. (From Monzel, 2012, courtesy of Cornelia Monzel.) (c) Zoom-in on a patch of membrane in the contact zone between a GUV carrying biotin ligands and interacting with an SLB carrying avidin receptors. Top: RICM snap-shots, Bottom: Fluctuation maps. In RICM, the patch progressively becomes darker with time (left to right) whereas the fluctuations are seen to increase (middle panel) before being frozen (left panel).

Smith et al., 2008). Experimentally, membrane fluctuations are not easy to measure, the main challenge being to develop techniques capable of measuring very small displacements at very high speed, and preferably over a large area and long time. Scattering techniques have given access to fluctuations in membrane stacks but are not relevant for GUV adhesion. Interestingly, fluctuations are thought to be relevant for cell adhesion (Biswas et al., 2015; Perez et al., 2008; Pierres et al., 2008; Zidovska and Sackmann, 2006).

DW-RICM as described above is a reliable way to quantify fluctuations (Monzel et al., 2009, 2012; Schmidt et al., 2014; Sengupta and Limozin, 2010). Figure 17.6b shows a membrane partly adhering specifically, *via* biotin–avidin bonds, to a pattern of lines. The nonadhered parts of the membrane fluctuate and their instantaneous configuration can be reconstructed. These can be used to create a pixel-by-pixel average height map and a fluctuation map, where the amplitude of fluctuation is defined as the standard deviation of the membrane height in each pixel. Fluctuations of a membrane close to a substrate were used to quantify the membrane–substrate interaction potential (γ in Eq. 17.1), which was found to be highly asymmetric (Schmidt et al., 2014) and to widen with decreasing membrane tension (Monzel et al., 2012). The accuracy in height is about 5 nm, in the lateral direction it is given by the Rayleigh limit, about 200 nm and time resolution is limited by the capacity of the camera to about 30 ms. Although DW-RICM is a very powerful technique for quick imaging of fluctuations, the relatively poor lateral space and time resolution means that some of the modes are not detected. These difficulties were partially circumvented with a novel technique called dynamic optical displacement spectroscopy (DODS) (Monzel et al., 2015), which can detect fluctuations with 20 nm lateral and 10 μs time resolution at a given position, anywhere on the GUV membrane, including within the adhesion zone.

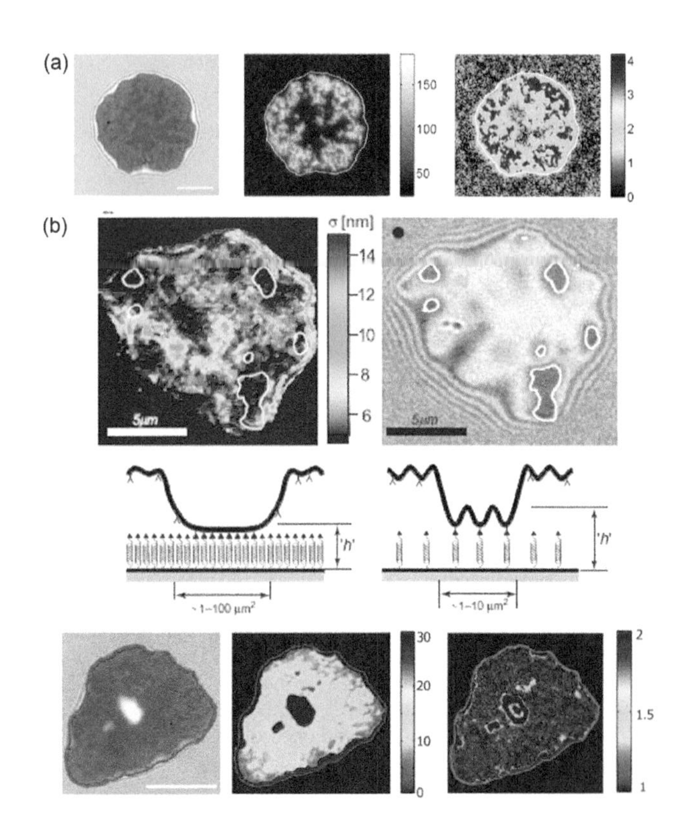

Figure 17.7 Advanced RICM approaches. (a) Dy-RICM (left), corresponding raw fluorescence image (middle) and color-coded fluorescence map, in arbitrary units (right), of an adherent GUV. The biotin-carrying GUV adheres to an SLB, carrying biotin which is further functionalized with fluorescent-labeled avidin receptors. The receptors diffuse into the adhesion zone and are immobilized, thus leading to accumulation under the vesicle. Crowding at the rim prevents further filling of the interior of the adhesion zone, resulting in an annular arrangement of receptors. Scale bar: 10 μm. (Courtesy of Susanne Fenz.) (b) Fluctuation maps as indicator of adhesion. Fluctuation maps (top left and bottom right) and RICM snapshots (top right and bottom left) for integrin/RGD system (top) and biotin–avidin system (bottom). Top row: In addition to the dark patches seen in RICM (picked-out in white), extra black patches show up in the fluctuation map (Smith et al., 2008). The patches that are dark in both RICM and fluctuation maps correspond to adhesion patches with dense bond configuration (case A, middle row). The patches that are dark in the fluctuation map, indicating frozen fluctuations, but are nevertheless gray or white in RICM images, correspond to dilute bonds (case B, middle row). Bottom row: height map (middle) constructed from RICM image (left) shows a heterogeneous height distribution. The membrane is nevertheless adhered everywhere in the contact zone except in the central bubble and at the edges, as evidenced by the fluctuation map. (From Fenz, 2008; Courtesy of Susanne Fenz.)

17.6 IMAGING MOLECULES AND BONDS

17.6.1 FLUORESCENCE TECHNIQUES

As already briefly discussed in Section 17.4 in the context of confocal microscopy, fluorescence labeling can be useful to detect the position of binders. Given that, unlike in cellular systems, typically, the molecules of interest are close to the surface (and do not usually leak into the GUV interior), epifluorescence microscopy is enough and TIRF-M is not necessary. Using labeled receptors in a biotin–avidin system, it was shown that the receptors diffuse into the adhesion zone and get immobilized, thus raising the local receptor concentration (Fenz et al., 2009), and, unexpectedly, at low receptors concentrations, incomplete filling and crowding results in annular structures (Fenz et al., 2011) (Figure 17.7a). One weakness of this approach is that free and bound receptors cannot be distinguished. Recently, bond density was directly quantified using Förster Resonance Energy Transfer (FRET) in a DNA-mediated system (Shimobayashi et al., 2015).

17.6.2 DYNAMICAL-RICM

As seen above, the accuracy of RICM makes it possible to record and analyze the temporal fluctuations of a membrane very precisely. This ability is exploited in dynamical-RICM (Dy-RICM) to detect complex and suboptical features on soft surfaces (Fenz et al., 2011, 2009; Smith et al., 2010, 2008). In Dy-RICM, the fluctuations in the interface–substrate distance, as determined from a time resolved sequence of RICM images, serve as the indicator of membrane binding. Thus, as discussed above, the key to Dy-RICM analysis is construction of a fluctuation map, which is simply a pixel-by-pixel image of the fluctuation amplitude of the soft interface. Given that, at a given location, the fluctuation amplitude is directly related to the curvature of the free energy potential in which the interface resides, such a map contains information about the confinement or binding at each pixel.

Figure 17.7b shows an example where, in addition to the conventional dark domains in RICM at a low height that correspond to micrometer size dense aggregates of bonds, Dy-RICM is able to identify dilute bond-clusters that would have escaped detection in conventional RICM.

Dy-RICM should ideally be used with two colors, but as long as the fluctuations are confined to one branch of the RICM intensity-height relation (see Limozin and Sengupta [2009] for a recapitulation), a single color analysis is acceptable. However, posttreatment verification of this assumption is imperative. Importantly, the most significant source of noise in a typical RICM experiment is the shot noise of the camera. The shot noise is proportional to the square root of the detected intensity; thus, a stationary membrane will seem to fluctuate more if its height coincides with the maximum of intensity compared with one where the intensity is minimum (Fenz, 2008). Therefore, to detect full suppression of fluctuations, it is important to normalize the fluctuations with respect to the shot noise. Using this kind of normalization, in addition to the dilute bonds already discussed above, very small bond clusters can be visualized using dy-RICM (Smith et al., 2010, 2008).

17.7 DE-ADHESION AND FORCE

Along with adhesion, de-adhesion is a process of extreme importance for normal cell functioning. However, de-adhesion of cells has not been studied as much, particularly from a biophysical point of view and hence is a less well-understood process. Nevertheless, several mechanisms for the control of de-adhesion have been previously identified. In particular, the control of the local adhesion strength and inducement of de-adhesion can be established through the manipulation of the extracellular matrix, for example, during tissue growth. Another possibility is by proteolytic decomposition of adhesion macromolecules. Like adhesion, de-adhesion is regulated through both biochemical and mechanical means. Hence, micropipette experiments in which lateral tension is increased are suitable to induce de-adhesion.

17.7.1 AGONIST/COMPETITION INDUCES DE-ADHESION

Competitive binders (agonists) provide a very useful tool to study the regulation of cell morphology or the polarity of protein distribution in the plasma membrane by cell adhesion. Antagonists, in the form of antibodies, were used to induce de-adhesion in a E-selectin/SiaL-X GUV system (Figure 17.8a). The ability of the antibodies to induce de-adhesion of GUVs pre-adhered by means of multiple ligand-receptor bonds was demonstrated (Smith et al., 2006). Intriguingly, such a de-adhesion, though to a lesser extent, is also seen in biotin–avidin mediated adhesion (Fenz et al., 2011) (Figure 17.8b) when pure biotin is added as competitor. This is not expected because the biotin–avidin bond is considered to be too strong to have a thermally driven on/off dynamics. In fact, at the limit of low density, the bonds are effectively weakened simply by the virtue of residing on a membrane (Fenz et al., 2011).

This behavior is well predicted by a theory that uses Langmuir absorption to model the effect of the antagonist. Accordingly, if the antagonist acts on a densely packed domain of bound CAM molecules, it produces a lateral pressure on to the domain. If the domain spans the entire vesicle, and the pressure is sufficiently high, the immediate response will be a decrease of the contact zone size. At some point this lateral pressure is balanced by the spreading pressure of the vesicle, at which point the antagonist penetrates the contact zone without changing its size.

Once in the contact zone, the antagonist blocks free receptors within domains, including receptors that would usually be only transiently free, due to the statistical turn-over of bonds. The efficiency of the antagonist is regulated by its binding affinity and concentration. More specifically, the affinity regulates the critical density that is necessary to induce unbinding of ligand-receptor bonds. Consequently, even antagonists that have a lower affinity for the receptor can induce at least partial unbinding if present in sufficient quantities. However, although a significant change in the state of adhesion can be obtained, due to the sigmoid dependence of the number of ligand-receptor bonds as a function of the antagonist concentration, very large concentrations of antagonist may be necessary to induce the full unbinding. This result is very important for understanding the competitive processes occurring on membranes.

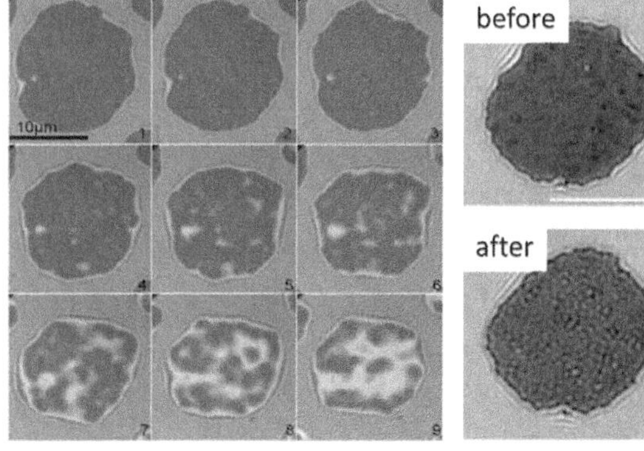

Figure 17.8 Competitor-induced de-adhesion in E-selectin/SyaL-X (a) or biotin–avidin (b) system. (a) E-selectin is on the surface and SyaL-X in the GUV, the bonds are weak (enthalpy in solution of about 10 k_BT) and the competitor agonist is an antibody that binds readily to the E-selectin. GUV-to-surface bonds are easily broken as evidenced by the decrease in black adhered zones seen in RICM. (Reproduced from PhD thesis of B. Feneberg, with permission of the author [Lorz, 2003].) (b) The avidin is on the surface and biotin on the GUV, the bonds are relatively strong (enthalpy in solution of 35 k_BT) and the competitor and the ligands is the same (biotin). In spite of this, the biotin in solution eventually displaces the biotin bound to the GUV as seen in the RICM images (appearance of white patches). Scale bar: 10 μm. (Reproduced with permission from PhD thesis of Susanne Fenz [Fenz, 2008].)

17.7.2 PULLING VIA MAGNETIC TWEEZERS

Vertical unbinding forces can be applied to adhered GUVs *via* a two-pole magnetic tweezers (Lorz, 2003; Marx, 2003; Smith and Sackmann, 2009). In this set up, paramagnetic or ferromagnetic beads are manipulated by subjecting them to an inhomogeneous magnetic field. A constant force (rather than a deformation) is applied. The heart of the tweezers consists of a solenoid formed by winding copper coils around two opposing arms of a rectangular soft-iron core, which has a small gap of few millimeters. Passing electric current through the coils gives rise to a magnetic field in the gap. The pole pieces (formed by the iron-core next to the gap) are shaped such that they provide a strong magnetic field gradient in the vertical direction at the position of the sample (Figure 17.9). The whole setup is fixed on an optical microscope. The observation chamber is mounted on an *x-y* stage and sits in between the poles. A paramagnetic bead, typically less than 2 μm diameter, and attached covalently to the vesicle, acts as the force transducer and transmits the upward lift force F_{pull} generated as a result of magnetic field gradient to the vesicle. For a given sample-to-pole distance, the pulling force is a function of the current through the solenoid. A protocol for calibrating the force on the vesicle as a function of the current was developed and the setup needs to be calibrated for each sample-to-pole distance that is used (Smith et al., 2008) (Figure 17.9).

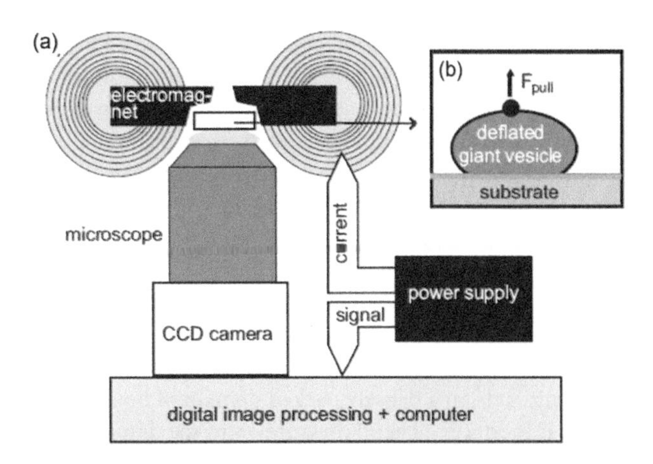

Figure 17.9 Schematic view of a vertical magnetic tweezers mounted on an inverted microscope. At the measuring chamber the field gradient is vertical. The current through the coils of the magnet is controlled by the power supply that produces a pulse-like signal. The digital unit controls the coil-current and simultaneously records images. (Adapted from PhD thesis of Stephanie Goennenwein [Marx, 2003], with permission of the author.)

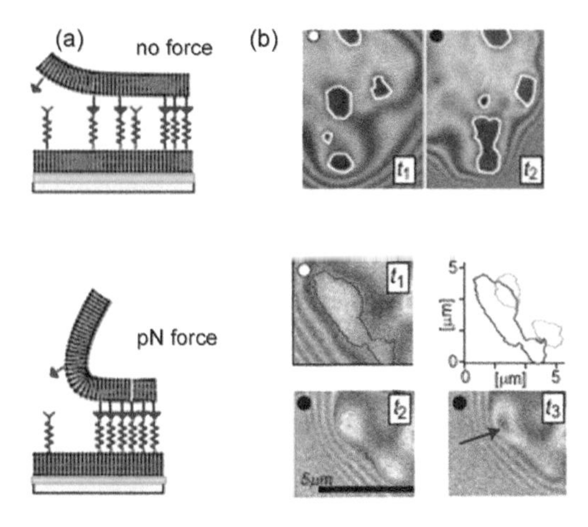

Figure 17.10 Evolution of bonds under force. (a) Schematic depiction of the transmission of force through the GUV membrane under pulling exerted using magnetic tweezers. Dilute bond domains are compacted under force and dragged toward the interior of the contact zone as seen in RICM snapshots in the bottom right panels. (b) Growth of existing adhesion patches that are dark in RICM. Bottom panel shows the patches with and without force. The domains are dragged inward and compacted under force. (After Smith, A.S. et al., *Proc. Natl. Acad. Sci. U.S.A.*, 105, 6906–6911.)

In the context of adhered vesicles, previous studies investigated the effect of the application of a picon-Newton external force at the apical unbound part of the vesicle with magnetic tweezers (Guttenberg et al., 2000). This force is transmitted to the adhered area through the membrane, and consequently, changes in the contact zone can be observed. This early effort was complemented with the modeling of the elastic response of vesicles through a shape equation (Smith et al., 2003, 2004) which was later coupled to the thermodynamic response of the bonds (Smith and Seifert, 2005) Consequently, the response of the shape was experimentally confirmed and the passive growth of adhesion under force was first demonstrated for E-selectin binding (Smith et al., 2004; Smith and Seifert, 2005). In the case of E-selectin, the bonds were immobile, and the growth of adhesion domains required unbinding and rebinding.

Somewhat later, a new mechanism for force resistance was suggested as a first step in the mechanosensing of cells, in a combined experimental and theoretical study of integrin mediated adhesion in vesicles (Smith et al., 2008). In this case, integrins maintained their mobility on the substrate, and the bonds could slide along the bilayers and were compacted into a dense array (Figure 17.10). The overall adhered area increased under repeated force application/release cycles. This could be understood in terms of a thermodynamic model that accounts for not only the bond energy but also the overall entropy of binders and bonds in the system. This behavior is reminiscent of mechanotransduction seen in living cells. In numerous studies in the last decade, living cells, when subjected to an external force have been shown to strengthen the adhesion. Whereas no specific force transduction protein has yet been identified, it is generally believed that such (a) protein(s) will be identified in the future. Furthermore, the presence of the cytoskeleton is thought to be indispensable for mechanotransduction. In the model system, however, adhesion strengthening under force was detected in a cytoskeleton-free system arising purely out of the physics of the two membranes. We believe that this result calls for a rethinking on the putative mechanism of the on-set of mechanotransduction.

17.8 CONCLUDING DISCUSSIONS

Throughout this chapter, we illustrated experiments where GUV adhesion was used first and foremost for gaining a deeper understanding of the physics of membrane and cell adhesion. Compared with cells, GUV systems offer more specific and robust control on physical parameters and hence are the preferred choice for testing concepts and theoretical predictions. Adhesive vesicles also serve as cell adhesion benchmarks for newly devised experimental techniques.

In the pharmaceutical context, understanding adhesion is a key step in designing novel drug carriers. In fact, today there are several FDA-approved drugs based on vesicles delivery devices. Because GUVs are rather fragile, smaller vesicles are typically used. Specific targeting of pathological areas for treatment of diseases is typically achieved by incorporating ligands onto the carrier-vesicle surface allowing for their surface-attachment to the relevant cells or to the cellular microenvironments. Optimization of this step can clearly benefit from the understanding of basic principles of adhesion that have been revealed with GUVs. Furthermore, with new supra-vesicular systems, GUVs may be directly used as carriers (Hadorn and Hotz, 2010).

Another innovative use of adherent GUVs is as a sensor: for example, to monitor photooxidation (Aoki et al., 2015) or the activity of aquaporins (Berthaud et al., 2016). In both cases, the change of the adhesive state is, with a help of a model, translated into activity, either of light or of a particular protein.

Adhesion is the first step in fusion and there is a host of past and on-going research that looks at transient adhesion leading to hemi or full fusion. Similarly, many early works on lipid transfer through contact between membranes touched on the adhesion problem indirectly. However, because a fine measurement of adhesion is not the goal of these experiments, we have not included them here in this chapter. Another class of problems that are indirectly related to adhesion is the question of adhesion-induced pore-opening, which we have discussed wherever the study contributed significantly to advancing our understanding of adhesion.

In terms of technical development, the most important challenge at the moment is the simultaneous imaging of bonds and the membrane. Although new super-resolution techniques can improve the resolution of imaging, that is far from being accurate enough for imaging individual bonds. The most appropriate technique to image the membrane remains RICM. Imaging of fluctuations can be complemented by other techniques, including the newly invented technique of DODS (Monzel et al., 2016). The most interesting future technical advancement to our mind will be coupling of RICM to powerful fluorescence based techniques, including FRET, FCS, DODS and super-resolution.

The study of GUV adhesion has often been justified by making connection with cell recognition processes. Although GUVs are indeed invaluable models for certain specific aspects of cell adhesion, they are also fascinating systems on their own right. By coupling a reaction-diffusion system to membrane mechanics, GUVs represent a complex soft matter system where statistical mechanics has to be combined with elasticity and hydrodynamics to gain full understanding. The challenge for the experimentalist is twofold: On the one hand, (s)he must identify the specific phenomena to study and choose the relevant ingredients, and on the other hand, (s)he must devise measurements and analysis that provide clear and well-quantified data. Whenever these challenges have been overcome, GUVs have provided valuable insight into membrane physics as well as biology.

ACKNOWLEDGMENTS

We thank our mentors Erich Sackmann, Udo Seifert and Rudi Merkel for their insights and on-going collaborations. We also thank our teams, colleagues, and collaborators who contributed to the work presented here, especially Barbara Lorz, Stefanie Goennenwein, Susanne Fenz, Cornelia Monzel, Daniel Smith, Timo Bihr and Laurent Limozin. K.S. thanks Arnaud Hemmerle for careful reading of the manuscript.

LIST OF ABBREVIATIONS

GUV	giant unilamellar vesicles
CAM	cell adhesion molecule
RICM	reflection interference contrast microscopy
DW-RICM	dual-wave reflection interference contrast microscopy
dy-RICM	dynamical reflection interference contrast microscopy
TIRF-M	total internal reflection fluorescence microscopy
FRET	Förster resonant energy transfer
DOPC	1,2-dioleoyl-*sn*-glycero-3-phosphocholine
SOPC	1-Stearoyl-2-oleoyl-*sn*-glycero-3-phosphocholine
DMPC	1,2-dimyristoyl-*sn*-glycero-3-phosphocholine
SLB	supported lipid bilayer

GLOSSARY OF SYMBOLS

E_a	bond energy
F	free energy
$h(x)$	membrane surface distance
T	temperature
W	adhesion energy density
x	spatial coordinate
k_B	boltzmann constant
γ	strength of membrane–surface interaction potential
κ	membrane bending modulus
λ	wavelength of light
Λ	adhesion rim width
Σ	membrane tension
θ	contact angle
n	refractive index

REFERENCES

Albersdörfer A, Feder T, Sackmann E (1997) Adhesion-induced domain formation by interplay of long-range repulsion and short-range attraction force: A model membrane study. *Biophysical Journal* 73:245–257.

Alberts B, Lewis J, Bray D (2000) *Molecular Biology of the Cell*, Garland Science, Taylor & Francis Group, UK.

Aoki P, Schroder A, Constantino C, Marques C (2015) Bioadhesive giant vesicles for monitoring hydroperoxidation in lipid membranes. *Soft Matter* 11:5995–5998.

Bagatolli LA, Needham D (2014) Quantitative optical microscopy and micromanipulation studies on the lipid bilayer membranes of giant unilamellar vesicles. *Chemistry and Physics of Lipids* 181:99–120.

Baumgart T, Hammond AT, Sengupta P, Hess ST, Holowka DA, Baird BA, Webb WW (2007) Large-scale fluid/fluid phase separation of proteins and lipids in giant plasma membrane vesicles. *Proceedings of the National Academy of Sciences* 104:3165–3170.

Beales PA, Nam J, Vanderlick TK (2011) Specific adhesion between DNA-functionalized Janus vesicles: Size-limited clusters. *Soft Matter* 7:1747–1755.

Berthaud A, Quemeneur F, Deforet M, Bassereau P, Brochard-Wyart F, Mangenot S (2016) Spreading of porous vesicles subjected to osmotic shocks: The role of aquaporins. *Soft Matter* 12:1601–1609.

Betz T, Sykes C (2012) Time resolved membrane fluctuation spectroscopy. *Soft Matter* 8:5317.

Bihr T, Fenz S, Sackmann E, Merkel R, Seifert U, Sengupta K, Smith AS (2014) Association rates of membrane-coupled cell adhesion molecules. *Biophysical Journal* 107:L33–L36.

Bihr T, Seifert U, Smith AS (2012) Nucleation of ligand-receptor domains in membrane adhesion. *Physical Review Letters* 109:258101.

Bihr T, Seifert U, Smith AS (2015) Multiscale approaches to protein-mediated interactions between membranes relating microscopic and macroscopic dynamics in radially growing adhesions. *New Journal of Physics* 17:083016.

Biswas KH, Hartman KL, Yu Ch, Harrison OJ, Song H, Smith AW, Huang WY, Lin WC, Guo Z, Padmanabhan A et al. (2015) E-cadherin junction formation involves an active kinetic nucleation process. *Proceedings of the National Academy of Sciences* 112:10932–10937.

Boal D, (2012) *Mechanics of the Cell*, Cambridge University Press, Cambridge, UK.

Boulbitch A, Guttenberg Z, Sackmann E (2001) Kinetics of membrane adhesion mediated by ligand-receptor interaction studied with a biomimetic system. *Biophysical Journal* 81:2743–2751.

Brasch J, Harrison OJ, Honig B, Shapiro L (2012) Thinking outside the cell: How cadherins drive adhesion. *Trends in Cell Biology* 22:299–310.

Brodovitch A, Bongrand P, Pierres A (2013) T lymphocytes sense antigens within seconds and make a decision within one minute. *The Journal of Immunology* 191:2064–2071.

Brüggemann D, Frohnmayer JP, Spatz JP (2014) Model systems for studying cell adhesion and biomimetic actin networks. *Beilstein Journal of Nanotechnology* 5:1193–1202.

Bruinsma R (1995) Adhesion and rolling of leukocytes: A physical model In *Proc. NATO Adv. Inst. Phys. Biomater. NATO ASI Ser*, Vol. 332, pp. 61–75.

Bruinsma R, Behrisch A, Sackmann E (2000) Adhesive switching of membranes: Experiment and theory. *Physical Review E* 61:4253–4267.

Campillo CC, Schroder AP, Marques CM, Pépin-Donat B (2009) Composite gel-filled giant vesicles: Membrane homogeneity and mechanical properties. *Materials Science and Engineering: C* 29:393–397.

Caorsi V, Lemiere J, Campillo C, Bussonnier M, Manzi J, Betz T, Plastino J, Carvalho K, Sykes C (2016) Cell-sized liposome doublets reveal active tension build-up driven by acto-myosin dynamics. *Soft Matter* 12:6223–6231.

Case LB, Waterman CM (2015) Integration of actin dynamics and cell adhesion by a three-dimensional, mechanosensitive molecular clutch. *Nature Cell Biology* 17:955.

Cohen M, Kam Z, Addadi L, Geiger B (2006) Dynamic study of the transition from hyaluronan-to integrin-mediated adhesion in chondrocytes. *The EMBO Journal* 25:302–311.

Cortese JD, Schwab B, Frieden C, Elson EL (1989) Actin polymerization induces a shape change in actin-containing vesicles. *Proceedings of the National Academy of Sciences* 86:5773–5777.

Curtis A (1964) The mechanism of adhesion of cells to glass a study by interference reflection microscopy. *The Journal of Cell Biology* 20:199–215.

Cuvelier D, Nassoy P (2004) Hidden dynamics of vesicle adhesion induced by specific stickers. *Physical Review Letters* 93:228101.

De Gennes PG, Puech PH, Brochard-Wyart F (2003) Adhesion induced by mobile stickers: A list of scenarios. *Langmuir* 19:7112–7119.

De Gennes PG (1979) *Scaling Concepts in Polymer Physics*, Cornell University Press, Ithaca, NY.

Dillard P, Varma R, Sengupta K, Limozin L (2014) Ligand-mediated friction determines morphodynamics of spreading T cells. *Biophysical Journal* 107:2629–2638.

Dorn IT, Neumaier KR, Tamp R (1998) Molecular recognition of histidine-tagged molecules by metal-chelating lipids monitored by fluorescence energy transfer and correlation spectroscopy. *Journal of the American Chemical Society* 120:2753–2763.

Evans E, Rawicz W (1997) Elasticity of "fuzzy" biomembranes. *Physical Review Letters* 79:2379.

Evans E (1980) Analysis of adhesion of large vesicles to surfaces. *Biophysical Journal* 31:425.

Evans E (2001) Probing the relation between force-lifetime-and chemistry in single molecular bonds. *Annual Review of Biophysics and Biomolecular Structure* 30:105–128.

Feder TJ, Weissmüller G, Žekš B, Sackmann E (1995) Spreading of giant vesicles on moderately adhesive substrates by fingering: A reflection interference contrast microscopy study. *Physical Review E* 51:3427.

Fenz SF, Bihr T, Merkel R, Seifert U, Sengupta K, Smith AS (2011) Switching from ultraweak to strong Adhesion. *Advanced Materials* 23:2622–2626.

Fenz S (2008) Cell-cell adhesion mediated by mobile receptor-ligand pairs: A biomimetic study. PhD thesis, Bonn University, Bonn, Germany.

Fenz SF, Bihr T, Schmidt D, Merkel R, Seifert U, Sengupta K, Smith AS (2017) Membrane fluctuations mediate lateral interaction between cadherin bonds. *Nature Physics* 13:906.

Fenz SF, Merkel R, Sengupta K (2009) Diffusion and intermembrane distance: Case study of avidin and e-cadherin mediated adhesion. *Langmuir* 25:1074–1085.

Fenz SF, Sengupta K (2012) Giant vesicles as cell models. *Integrative Biology: Quantitative Biosciences from Nano to Macro* 4:982–995.

Fenz SF, Smith AS, Merkel R, Sengupta K (2011) Inter-membrane adhesion mediated by mobile linkers: Effect of receptor shortage. *Soft Matter* 7:952.

Frostad JM, Seth M, Bernasek SM, Leal LG (2014) Direct measurement of interaction forces between charged multilamellar vesicles. *Soft Matter* 10:7769–7780.

Geiger B, Bershadsky A, Pankov R, Yamada KM (2001) Transmembrane crosstalk between the extracellular matrix and the cytoskeleton. *Nature Reviews Molecular Cell Biology* 2:793–805.

Geiger B, Spatz JP, Bershadsky AD (2009) Environmental sensing through focal adhesions. *Nature Reviews Molecular Cell Biology* 10:21–33.

Gingell D, Todd I (1979) Interference reflection microscopy. A quantitative theory for image interpretation and its application to cell-substratum separation measurement. *Biophysical Journal* 26:507.

Goennenwein S, Tanaka M, Hu B, Moroder L, Sackmann E (2003) Functional incorporation of intergins into solid supported membrane on ultrathin films of cellulose: Impact on adhesion. *Biophysical Journal* 85:646–655.

Gruhn T, Franke T, Dimova R, Lipowsky R (2007) Novel method for measuring the adhesion energy of vesicles. *Langmuir* 23:5423–5429.

Guttenberg Z, Bausch A, Hu B, Bruinsma R, Moroder L, Sackmann E (2000) Measuring ligand-receptor unbinding forces with magnetic beads: Molecular leverage. *Langmuir* 16:8984–8993.

Guttenberg Z, Lorz B, Sackmann E, Boulbitch A (2001) First-order transition between adhesion states in a system mimicking cell-tissue interaction. *EPL (Europhysics Letters)* 54:826.

Hadorn M, Hotz PE (2010) DNA-mediated self-assembly of artificial vesicles. *PLoS One* 5:e9886.

Helfer E, Harlepp S, Bourdieu L, Robert J, MacKintosh F, Chatenay D (2001) Viscoelastic properties of actin-coated membranes. *Physical Review E* 63:021904.

Helfrich W (1973) Elastic properties of lipid bilayers: Theory and possible experiments. *Zeitschrift für Naturforschung, Teil C* 28:693–703.

Hisette ML, Haddad P, Gisler T, Marques CM, Schröder AP (2008) Spreading of bio-adhesive vesicles on DNA carpets. *Soft Matter* 4:828–832.

Hsu CJ, Hsieh WT, Waldman A, Clarke F, Huseby ES, Burkhardt JK, Baumgart T (2012) Ligand mobility modulates immunological synapse formation and T cell activation. *PLoS One* 7:e32398.

Hu B, Finsinger D, Peter K, Guttenberg Z, Bärmann M, Kessler H, Escherich A, Moroder L, Böhm J, Baumeister W et al. (2000) Intervesicle cross-linking with integrin αIIbβ3 and cyclic-RGD-lipopeptide. A model of cell-adhesion processes. *Biochemistry* 39:12284–12294.

Iskratsch T, Wolfenson H, Sheetz MP (2014) Appreciating force and shape—The rise of mechanotransduction in cell biology. *Nature Reviews Molecular Cell Biology* 15:825.

Kenworthy AK, Hristova K, Needham D, McIntosh TJ (1995) Range and magnitude of the steric pressure between bilayers containing phospholipids with covalently attached poly (ethylene glycol). *Biophysical Journal* 68:1921.

Kloboucek A, Behrisch A, Faix J, Sackmann E (1999) Adhesion-induced receptor segregation and adhesion plaque formation: A model membrane study. *Biophysical Journal* 77:2311–2328.

Kühner M, Sackmann E (1996) Ultrathin hydrated dextran films grafted on glass: Preparation and characterization of structural, viscous, and elastic properties by quantitative microinterferometry. *Langmuir* 12:4866–4876.

Leckband D, de Rooij J (2014) Cadherin adhesion and mechanotransduction. *Annual Review of Cell and Developmental Biology* 30:291–315.

Limozin L, Bärmann M, Sackmann E (2003) On the organization of self-assembled actin networks in giant vesicles. *The European Physical Journal E* 10:319–330.

Limozin L, Sengupta K (2007) Modulation of vesicle adhesion and spreading kinetics by hyaluronan cushions. *Biophysical Journal* 93:3300–3313.

Limozin L, Sengupta K (2009) Quantitative reflection interference contrast microscopy (RICM) in soft matter and cell adhesion. *Chemphyschem* 10:2752–2768.

Limozin L, Roth A, Sackmann E (2005) Microviscoelastic moduli of biomimetic cell envelopes. *Physical Review Letters* 95:178101.

Limozin L, Sackmann E (2002) Polymorphism of cross-linked actin networks in giant vesicles. *Physical Review Letters* 89:168103.

Lipowsky R (1991) The conformation of membranes. *Nature* 349:475–481.

Lipowsky R, Sackmann E (1995) *Structure and Dynamics of Membranes: I. From Cells to Vesicles/II. Generic and Specific Interactions*, Elsevier, Amsterdam, the Netherlands.

Liu AP, Richmond DL, Maibaum L, Pronk S, Geissler PL, Fletcher DA (2008) Membrane-induced bundling of actin filaments. *Nature Physics* 4:789–793.

Lorz B (2003) Etablierung eines Modellsystems der Zelladhäsion über spezifische Bindungen geringer Affinität. PhD thesis, TU-Munich, München, Germany.

Lorz BG, Smith AS, Gege C, Sackmann E (2007) Adhesion of giant vesicles mediated by weak binding of sialyl-Lewisx to E-selectin in the presence of repelling poly (ethylene glycol) molecules. *Langmuir* 23:12293–12300.

Maier CW, Behrisch A, Kloboucek A, Merkel R (1997) Adhesion of lipid membranes mediated by electrostatic and specific interactions In *MRS Proceedings*, Vol. 489, p. 107. Cambridge University Press, Cambridge, UK.

Maitre JL, Heisenberg CP (2013) Three functions of cadherins in cell adhesion. *Current Biology* 23:R626–R633.

Marchi-Artzner V, Lorz B, Gosse C, Jullien L, Merkel R, Kessler H, Sackmann E (2003) Adhesion of Arg-Gly-Asp (RGD) peptide vesicles onto an integrin surface: Visualization of the segregation of rgd ligands into the adhesion plaques by fluorescence. *Langmuir* 19:835–841.

Marshall BT, Long M, Piper JW, Yago T, McEver RP, Zhu C (2003) Direct observation of catch bonds involving cell-adhesion molecules. *Nature* 423:190–193.

Marx S (2003) Generic and specific cell adhesion: Investigations of a model system by micro-interferometry. PhD thesis, TU-Munich, München, Germany.

Marx S, Schilling J, Sackmann E, Bruinsma R (2002) Helfrich repulsion and dynamical phase separation of multicomponent lipid bilayers. *Physical Review Letters* 88:138102.

Merkel R, Nassoy P, Leung A, Ritchie K, Evans E (1999) Energy landscapes of receptor–ligand bonds explored with dynamic force spectroscopy. *Nature* 397:50–53.

Merkel R, Sackmann E, Evans E (1989) Molecular friction and epitactic coupling between monolayers in supported bilayers. *Journal de Physique* 50:1535–1555.

Merkel R (2001) Force spectroscopy on single passive biomolecules and single biomolecular bonds. *Physics Reports* 346:343–385.

Monzel C, Fenz S, Merkel R, Sengupta K (2009) Probing biomembrane dynamics by dual-wavelength reflection interference contrast microscopy. *Chemphyschem* 10:2828–2838.

Monzel C, Schmidt D, Kleusch C, Kirchenbüchler D, Seifert U, Smith AS, Sengupta K, Merkel R (2015) Measuring fast stochastic displacements of bio-membranes with dynamic optical displacement spectroscopy. *Nature Communications* 6:8162.

Monzel C (2012) Analyses of adhesion topography and fluctuations in bio-membranes by advanced optical microscopy. PhD thesis, Bonn University, Bonn, Germany.

Monzel C, Fenz SF, Giesen M, Merkel R, Sengupta K (2012) Mapping fluctuations in biomembranes adhered to micropatterns. *Soft Matter* 8:6128.

Monzel C, Schmidt D, Seifert U, Smith AS, Merkel R, Sengupta K (2016) Nanometric thermal fluctuations of weakly confined biomembranes measured with microsecond time-resolution. *Soft Matter* 12:4755–4768.

Moreira A, Jeppesen C, Tanaka F, Marques C (2003) Irreversible vs. reversible bridging: When is kinetics relevant for adhesion? *EPL (Europhysics Letters)* 62:876.

Murrell MP, Voituriez R, Joanny JF, Nassoy P, Sykes C, Gardel ML (2014) Liposome adhesion generates traction stress. *Nature Physics* 10:163–169.

Nam G, Hisette ML, Sun YL, Gisler T, Johner A, Thalmann F, Schröder AP, Marques CM, Lee NK (2010) Scraping and stapling of end-grafted DNA chains by a bioadhesive spreading vesicle to reveal chain internal friction and topological complexity. *Physical Review Letters* 105:088101.

Nam J, Santore MM (2007a) The adhesion kinetics of sticky vesicles in tension: The distinction between spreading and receptor binding. *Langmuir* 23:10650–10660.

Nam J, Santore MM (2007b) Adhesion plaque formation dynamics between polymer vesicles in the limit of highly concentrated binding sites. *Langmuir* 23:7216–7224.

Nardi J, Bruinsma R, Sackmann E (1998) Adhesion-induced reorganization of charged fluid membranes. *Physical Review E* 58:6340.

Needham D, McIntosh T, Lasic D (1992) Repulsive interactions and mechanical stability of polymer-grafted lipid membranes. *Biochimica et Biophysica Acta (BBA)-Biomembranes* 1108:40–48.

Noppl-Simson DA, Needham D (1996) Avidin-biotin interactions at vesicle surfaces: Adsorption and binding, cross-bridge formation, and lateral interactions. *Biophysical Journal* 70:1391–1401.

Oakes PW, Banerjee S, Marchetti MC, Gardel ML (2014) Geometry regulates traction stresses in adherent cells. *Biophysical Journal* 107:825–833.

Parolini L, Mognetti BM, Kotar J, Eiser E, Cicuta P, Di Michele L (2015) Volume and porosity thermal regulation in lipid mesophases by coupling mobile ligands to soft membranes. *Nature Communications* 6:5948.

Paszek MJ, DuFort CC, Rossier O, Bainer R, Mouw JK, Godula K, Hudak JE, Lakins JN, Wijekoon AC, Cassereau L et al. (2014) The cancer glycocalyx mechanically primes integrin-mediated growth and survival. *Nature* 511:319–325.

Perez TD, Tamada M, Sheetz MP, Nelson WJ (2008) Immediate-early signaling induced by E-cadherin engagement and adhesion. *The Journal of Biological Chemistry* 283:5014–5022.

Pi F, Dillard P, Alammeddin R, Benard E, Ozerov I, Charrier A, Limozin L, Sengupta K (2015) Functional organized organic nano-dots: A versatile platform for manipulating and imaging whole cells on surfaces. *Nano Lett* 15:5178–5184.

Pi F, Dillard P, Limozin L, Charrier A, Sengupta K (2013) Nanometric protein-patch arrays on glass and polydimethylsiloxane for cell adhesion studies. *Nano Lett* 13:3372–3378.

Pierres A, Benoliel AM, Touchard D, Bongrand P (2008) How cells tiptoe on adhesive surfaces before sticking. *Biophysical Journal* 94:4114–4122.

Pierres A, Monnet-Corti V, Benoliel AM, Bongrand P (2009) Do membrane undulations help cells probe the world? *Trends in Cell Biology* 19:428–433.
Prechtel K, Bausch A, Marchi-Artzner V, Kantlehner M, Kessler H, Merkel R (2002) Dynamic force spectroscopy to probe adhesion strength of living cells. *Physical Review Letters* 89:028101.
Puech PH, Askovic V, De Gennes PG, Brochard-Wyart F (2006) Dynamics of vesicle adhesion: Spreading versus dewetting coupled to binder diffusion. *Biophysical Reviews and Letters* 01:85–95.
Rädler JO, Feder TJ, Strey HH, Sackmann E (1995) Fluctuation analysis of tension-controlled undulation forces between giant vesicles and solid substrates. *Physical Review E* 51:4526–4536.
Reister E, Bihr T, Seifert U, Smith AS (2011) Two intertwined facets of adherent membranes: Membrane roughness and correlations between ligand–receptors bonds. *New Journal of Physics* 13:025003.
Reister-Gottfried E, Sengupta K, Lorz B, Sackmann E, Seifert U, Smith AS (2008) Dynamics of specific vesicle–substrate adhesion: From local events to global dynamics. *Physical Review Letters* 101:208103.
Richmond DL, Schmid EM, Martens S, Stachowiak JC, Liska N, Fletcher DA (2011) Forming giant vesicles with controlled membrane composition, asymmetry, and contents. *Proceedings of the National Academy of Sciences* 108:9431–9436.
Robert P, Limozin L, Benoliel AM, Bongrand P (2006) Gycocalyx regulation of cell adhesion. *Principles of Cellular Engineering: Understanding the Biomolecular Interface* p. 143.
Sackmann E, Bruinsma RF (2002) Cell adhesion as wetting transition? *Chemphyschem: A European Journal of Chemical Physics and Physical Chemistry* 3:262–269.
Sackmann E, Smith AS (2014) Physics of cell adhesion: Some lessons from cell-mimetic systems. *Soft Matter* 10:1644–1659.
Saha S, Anilkumar AA, Mayor S (2015) Gpi-anchored protein organization and dynamics at the cell surface. *Journal of Lipid Research* jlr–R062885.
Sakuma Y, Imai M, Yanagisawa M, Komura S (2008) Adhesion of binary giant vesicles containing negative spontaneous curvature lipids induced by phase separation. *The European Physical Journal E* 25:403–413.
Salbreux G, Charras G, Paluch E (2012) Actin cortex mechanics and cellular morphogenesis. *Trends in Cell Biology* 22:536–545.
Sarmento M, Prieto M, Fernandes F (2012) Reorganization of lipid domain distribution in giant unilamellar vesicles upon immobilization with different membrane tethers. *Biochimica et Biophysica Acta (BBA)-Biomembranes* 1818:2605–2615.
Saxton MJ, Jacobson K (1997) Single-particle tracking: Applications to membrane dynamics. *Annual Review of Biophysics and Biomolecular Structure* 26:373–399.
Schiavo VL, Robert P, Limozin L, Bongrand P (2012) Quantitative modeling assesses the contribution of bond strengthening, rebinding and force sharing to the avidity of biomolecule interactions. *PLoS One* 7:e44070.
Schilling J, Sengupta K, Goennenwein S, Bausch AR, Sackmann E (2004) Absolute interfacial distance measurements by dual-wavelength reflection interference contrast microscopy. *Physical Review E* 69:021901.
Schmidt D, Bihr T, Fenz S, Merkel R, Seifert U, Sengupta K, Smith AS (2015) Crowding of receptors induces ring-like adhesions in model membranes. *Biochimica et Biophysica Acta (BBA)-Molecular Cell Research* 1853:2984–2991.
Schmidt D, Monzel C, Bihr T, Merkel R, Seifert U, Sengupta K, Smith AS (2014) Signature of a nonharmonic potential as revealed from a consistent shape and fluctuation analysis of an adherent membrane. *Physical Review X* 4:021023.
Schmitt L, Dietrich C, Tampe R (1994) Synthesis and characterization of chelator-lipids for reversible immobilization of engineered proteins at self-assembled lipid interfaces. *Journal of the American Chemical Society* 116:8485–8491.
Schwarz US, Safran SA (2013) Physics of adherent cells. *Reviews of Modern Physics* 85:1327–1381.
Seifert U (1997) Configurations of fluid membranes and vesicles. *Advances in Physics* 46:13–137.
Seifert U, Langer SA (1993) Viscous modes of fluid bilayer membranes. *Europhysics Letters* 23:71–76.
Seifert U, Lipowsky R (1990) Adhesion of vesicles. *Physical Review A* 42:4768–4771.
Sengupta K, Aranda-Espinoza H, Smith L, Janmey P, Hammer D (2006) Spreading of neutrophils: From activation to migration. *Biophysical Journal* 91:4638–4648.
Sengupta K, Limozin L (2010) Adhesion of soft membranes controlled by tension and interfacial polymers. *Physical Review Letters* 104:088101.
Servuss RM, Helfrich W (1989) Mutual adhesion of lecithin membranes at ultralow tensions. *Journal De Physique* 50:809–827.
Shaw AS, Dustin ML (1997) Making the t cell receptor go the distance: A topological view of t cell activation. *Immunity* 6:361–369.
Shimobayashi S, Mognetti BM, Parolini L, Orsi D, Cicuta P, Di Michele L (2015) Direct measurement of DNA-mediated adhesion between lipid bilayers. *Physical Chemistry Chemical Physics* 17:15615–15628.
Shindell O, Mica N, Ritzer M, Gordon V (2015) Specific adhesion of membranes simultaneously supports dual heterogeneities in lipids and proteins. *Physical Chemistry Chemical Physics* 17:15598–15607.
Singer S, Nicolson GL (1972) The fluid mosaic model of the structure of cell membranes. *Day and Good Membranes and Viruses in Immunopathology* pp. 7–47.
Smith AS, Fenz SF, Sengupta K (2010) Interferring spatial organization of bonds within adhesion clusters by exploiting fluctuations of soft interfaces. *Europhysics Letters* 89:28003.
Smith AS, Sackmann E, Seifert U (2003) Effects of a pulling force on the shape of a bound vesicle. *Europhysics Letters (EPL)* 64:281–287.
Smith AS, Sengupta K, Goennenwein S, Seifert U, Sackmann E (2008) Force-induced growth of adhesion domains is controlled by receptor mobility. *Proceedings of the National Academy of Sciences of the United States of America* 105:6906–6911.
Smith AS, Lorz BG, Goennenwein S, Sackmann E (2006) Force-controlled equilibria of specific vesicle-substrate adhesion. *Biophysical Journal* 90:L52–L54.
Smith AS, Sackmann E (2009) Progress in mimetic studies of cell adhesion and the mechanosensing. *ChemPhysChem* 10:66–78.
Smith AS, Sackmann E, Seifert U (2004) Pulling tethers from adhered vesicles. *Physical Review Letters* 92:208101.
Smith AS, Seifert U (2005) Effective adhesion strength of specifically bound vesicles. *Physical Review E* 71:061902.
Smith AS, Seifert U (2007) Vesicles as a model for controlled (de-)adhesion of cells: A thermodynamic approach. *Soft Matter* 3:275–289.
Solon J, Streicher P, Richter R, Brochard-Wyart F, Bassereau P (2006) Vesicles surfing on a lipid bilayer: Self-induced haptotactic motion. *Proceedings of the National Academy of Sciences* 103:12382–12387.
Steinkühler J, Agudo-Canalejo J, Lipowsky R, Dimova R (2016) Modulating vesicle adhesion by electric fields. *Biophysical Journal* 111:1454–1464.
Streicher P, Nassoy P, Bärmann M, Dif A, Marchi-Artzner V, Brochard-Wyart F, Spatz J, Bassereau P (2009) Integrin reconstituted in guvs: A biomimetic system to study initial steps of cell spreading. *Biochimica et Biophysica Acta (BBA)-Biomembranes* 1788:2291–2300.

Takiguchi K, Yamada A, Negishi M, Tanaka-Takiguchi Y, Yoshikawa K (2008) Entrapping desired amounts of actin filaments and molecular motor proteins in giant liposomes. *Langmuir* 24:11323–11326.
Thomas WE, Vogel V, Sokurenko E (2008) Biophysics of catch bonds. *Biophysics* 37.
Ushiyama A, Ono M, Kataoka-Hamai C, Taguchi T, Kaizuka Y (2015) Induction of intermembrane adhesion by incorporation of synthetic adhesive molecules into cell membranes. *Langmuir* 31:1988–1998.
Verschueren H (1985) Interference reflection microscopy in cell biology: Methodology and applications. *Journal of Cell Science* 75:279–301.
Vézy C, Massiera G, Viallat A (2007) Adhesion induced non-planar and asynchronous flow of a giant vesicle membrane in an external shear flow. *Soft Matter* 3:844–851.
Viallat A, Dalous J, Abkarian M (2004) Giant lipid vesicles filled with a gel: Shape instability induced by osmotic shrinkage. *Biophysical Journal* 86:2179–2187.
Weikl TR, Asfaw M, Krobath H, Rozycki B, Lipowsky R (2009) Adhesion of membranes via receptor-ligand complexes: Domain formation, binding cooperativity, and active processes. *Soft Matter* 5:13.
Weinberger A, Tsai FC, Koenderink GH, Schmidt TF, Itri R, Meier W, Schmatko T, Schröder A, Marques C (2013) Gel-assisted formation of giant unilamellar vesicles. *Biophysical Journal* 105:154–164.
Wiegand G, Neumaier KR, Sackmann E (1998) Microinterferometry: Three-dimensional reconstruction of surface microtopography for thin-film and wetting studies by reflection interference contrast microscopy (RICM). *Applied Optics* 37:6892–6905.
Zidovska A, Sackmann E (2006) Brownian motion of nucleated cell envelopes impedes adhesion. *Physical Review Letters* 96:048103.

Phase diagrams and tie lines in giant unilamellar vesicles

Matthew C. Blosser, Caitlin E. Cornell, Scott P. Rayermann, and Sarah L. Keller

Don't worry; it's just a phase.

Contents

18.1 INTRODUCTION

In many types of GUVs composed of mixtures of lipids and sterols, a dramatic change occurs when the temperature of the system decreases: The membrane spontaneously demixes into two or more coexisting phases. This phenomenon has attracted the attention of researchers from a wide range of fields. From a biological perspective, the ease with which model lipid membranes phase separate into two liquid phases lends credence to proposals and observations that lipids in cell membranes are poised to demix (Lingwood et al., 2008; Veatch et al., 2008; Rayermann et al., 2017). Demixing of cell membranes is particularly relevant with respect to its possible effects on protein behavior. In model systems, proteins partition differently into the two membrane phases, and protein activities are affected by the local lipid composition of the membrane (Keller et al., 1993; Cornea and Thomas, 1994; Cornelius, 2001; Kahya et al., 2005; Levental et al., 2010; Lin and London, 2013). From a physical perspective, coexisting phases in lipid membranes provide an ideal quasi two-dimensional (2D) system in which to test fundamental theories as in Chapter 5 of this book.

A phase diagram is a record of which phases are observed at each temperature and at each ratio of lipid species in the membrane. Because general concepts of phase separation are the subject of textbooks (e.g., Hillert, 2008), because other authors have previously compiled ternary phase diagrams of lipid membranes (Marsh, 2009), and because we have previously published a review on binary and ternary phase diagrams of lipid membranes (Veatch and Keller, 2005b), this chapter focuses narrowly on how to experimentally identify coexisting phases within GUVs, and on how to measure tie lines. Tie lines contain information about the composition of the different membrane phases. A tie line is a segment drawn within a phase diagram at constant temperature for a membrane that demixes into two phases. The two end points of the tie line fall at the two ratios of lipid species that comprise the two phases. Most of the illustrations we draw upon in this chapter are from work by our group and our collaborators; other examples are readily found throughout the literature.

Why are phase diagrams and tie lines of GUVs important? Phase diagrams are important because they empower a researcher to quantitatively measure how phase separation in membranes is affected by physical attributes relevant to cells.

Some examples include cross-linking of membrane components as discussed in Chapter 23 (Kahya et al., 2005; Liu and Fletcher, 2006; Hammond et al., 2007; Garbes Putzel and Schick, 2009; Safouane et al., 2010), membrane bending and curvature as discussed in Chapter 16 (Seifert, 1993; Harden et al., 2005; Roux et al., 2005; Parthasarathy et al., 2006; Das et al., 2008; Semrau and Schmidt, 2009; Tian et al., 2009; Ursell et al., 2009; Camley et al., 2010; Maleki and Fried, 2013), membrane shear (Garbes Putzel et al., 2011; Blosser et al., 2015), membrane tension (Portet et al., 2012; Uline et al., 2012), membrane charge (Mengistu et al., 2010; Vequi-Suplicy et al., 2010; Blosser et al., 2013), adhesion as discussed in Chapter 17 (Rozovsky et al., 2005; Gordon et al., 2008; Lipowsky et al., 2013; Zhao et al., 2013), photooxidation as discussed in Chapter 22 (Roux et al., 2005; Ayuyan and Cohen, 2006), fluorescent probes as discussed in Chapter 10 (Veatch et al., 2007a), and establishment of membrane asymmetry to construct membranes with different lipid compositions in their inner and outer faces (Collins and Keller, 2008; Garbes Putzel and Schick, 2008a; May, 2009; Williamson and Olmsted, 2015).

Tie lines are important because they allow researchers to specify the lipid composition of each phase at each temperature. To illustrate the importance of both temperature and composition, consider a single GUV membrane that phase separates into coexisting liquid-ordered (L_o) and liquid-disordered (L_d) phases. At a given temperature, the acyl chains of the lipids in the L_o phase are more ordered than the acyl chains of the lipids in the L_d phase. However, if we compare an L_o phase at a high temperature to an L_d phase at a low temperature, the L_o lipids may be *less* ordered than the L_d lipids. Similarly, if we compare the L_o phase of a GUV with a certain overall lipid composition (a term that encompasses both the types of lipids and the ratio of the lipids) to the L_d phase of a GUV with a different overall composition, lipids in an L_o phase may be *less* ordered than lipids in an L_d phase. Specifying the lipid ratio within each membrane phase enables researchers to compare results among equivalent samples.

18.2 IDENTIFYING COEXISTING PHASES IN GUVs

Microscopy of GUVs is one of several powerful tools for identifying lipid phase behavior. Inferring membrane phases from micrometer-scale observations requires familiarity with the characteristics of each phase.

18.2.1 AN INTUITIVE ANALOGY ABOUT DEMIXING OF LIQUID PHASES

To gain an intuition about the demixing of a membrane into two liquid phases and about tie lines, consider the following analogy. Imagine that you are suddenly transported into the middle of a crowded party in a large, open room where nobody previously knew each other. As people move, they jostle each other. As people laterally diffuse across the entire room, they talk with each other, which results in either favorable or unfavorable interactions. People involved in favorable interactions assemble into groups. Individuals periodically break off from their group and walk in a random path across the room, perhaps joining another group. If interactions between individuals in a group are very weak (on the order of k_BT), then there is only a tiny difference between the composition of the individuals inside the group and outside the group. When the interactions are sufficiently weak, all individuals in a group will disperse on a time scale much shorter than the duration of the party. Because this type of group exists only temporarily, it is termed a "fluctuation." On the other hand, if the interactions between individuals are strong, then groups persist for the duration of the party (although individuals continually join and leave their groups). In this case, each group is a "domain" and the composition of the people in a group is termed a "phase." The longer the tie line for this system, the larger the difference between the composition of individuals inside a group and outside that group. When two domains of the same phase (i.e., two groups containing the same types of people) are jostled into contact with each other, they smoothly merge into one large group.

18.2.2 FLUORESCENCE MICROSCOPY TO IDENTIFY LIQUID PHASES

Box 18.1 contains schematics of fluorescence micrographs of GUVs that correspond to some major points in the analogy. The vesicles contain different ratios of—at minimum—three lipids (namely, a phospholipid that melts at a high temperature, a phospholipid that melts at a low temperature, and cholesterol). Initially, the GUV is at an elevated temperature at which all lipids within the membrane are liquid and mix uniformly. At this high temperature, mixing entropy overwhelms interactions between different lipid types. As temperature decreases, lipid–lipid interactions gradually dominate over the mixing entropy. Small domains of one liquid phase nucleate within the initially uniform membrane.

In experiments using fluorescence microscopy (see Chapter 10), domains are visualized via the differential partitioning of a dye-labeled lipid, as in Figure 18.1. Labeled lipids or hydrophobic environmental probes (Bagatolli, 2006; Klymchenko and Kreder, 2014; Sezgin et al., 2014) are incorporated into GUVs at trace amounts, typically less than 1 mol%. The primary requirement for a fluorescently labeled lipid is that it preferentially partitions between phases, giving rise to a contrast. The majority of labeled lipids partition preferentially into the least ordered membrane phase, even when the fluorophore is attached to a lipid that would be expected to partition preferentially into a more ordered membrane phase (Silvius, 2003; Baumgart et al., 2007; Sezgin et al., 2012b; Klymchenko and Kreder, 2014). For example, fluorescently labeled cholesterol and saturated lipids typically partition to L_d phases rather than L_o phases, as do tail-labeled sphingolipids. Further illustrating the point that the designation of "ordered" and "disordered" is relative rather than absolute, the same fluorescent probe can partition to different phases under different conditions. For example, several probes partition to the ordered phase of giant plasma membrane vesicles (see Chapter 2) but partition to the disordered phase of model GUVs produced from ternary mixtures of dioleoyl-phosphatidylcholine (DOPC), brain sphingomyelin, and cholesterol (Sezgin et al., 2012b), see Appendix 1 of the book for structure and data on lipids. In a different membrane system, a single probe has been found to switch its phase preference after cross-linking (Kahya et al., 2005).

To characterize which phases are present in GUV membranes, it is useful to examine domain morphology and

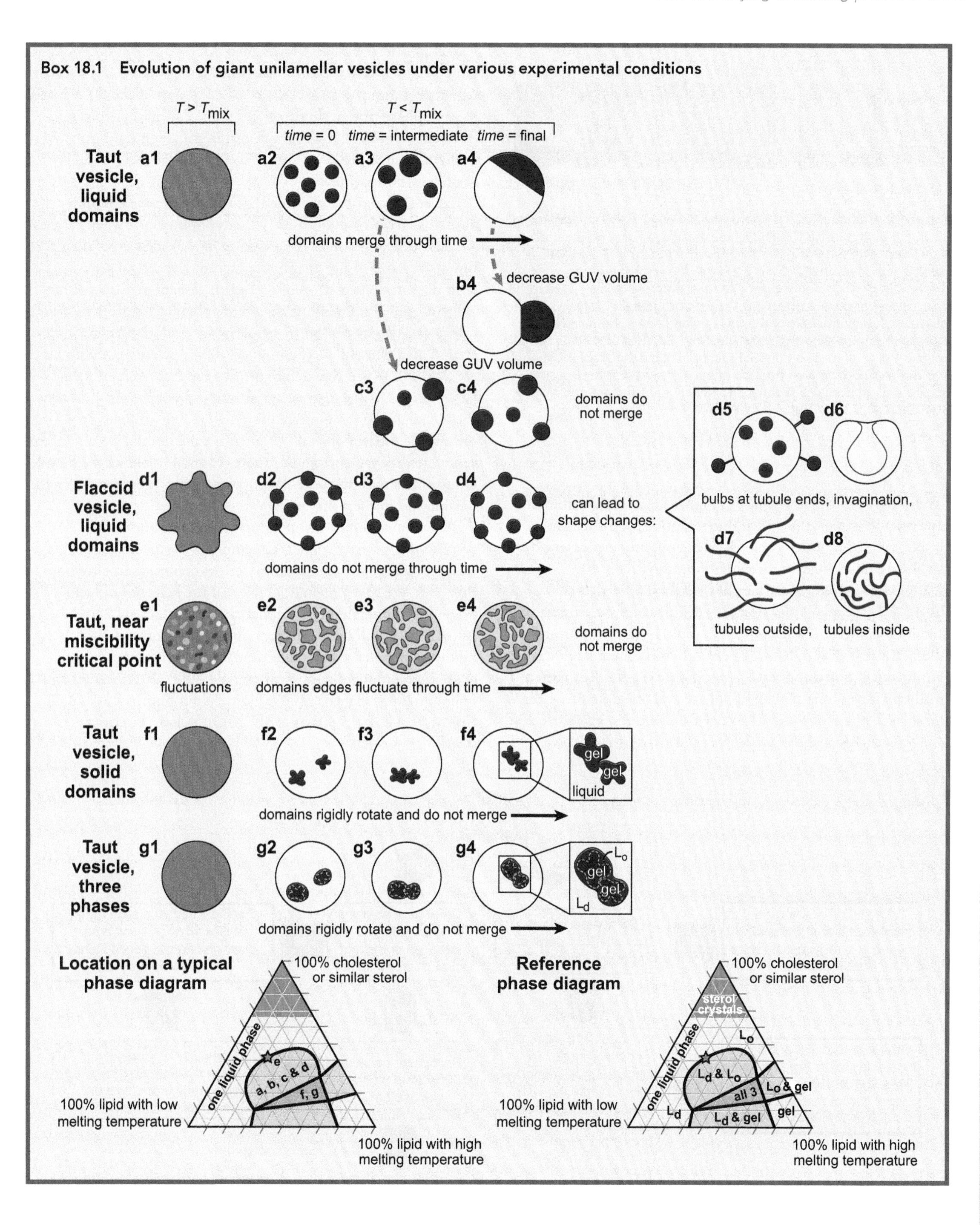
Box 18.1 Evolution of giant unilamellar vesicles under various experimental conditions
T > Tmix
T < Tmix
time = 0 time = intermediate time = final
Taut vesicle, liquid domains
a1 a2 a3 a4
domains merge through time
b4
decrease GUV volume
decrease GUV volume
c3 c4
domains do not merge
Flaccid vesicle, liquid domains
d1 d2 d3 d4
domains do not merge through time
can lead to shape changes:
d5 d6
bulbs at tubule ends, invagination,
d7 d8
tubules outside, tubules inside
e1 Taut, near miscibility critical point
e2 e3 e4
fluctuations
domains edges fluctuate through time
domains do not merge
Taut vesicle, solid domains
f1 f2 f3 f4
gel
gel
liquid
domains rigidly rotate and do not merge
Taut vesicle, three phases
g1 g2 g3 g4
Lo
gel
gel
Ld
domains rigidly rotate and do not merge
Location on a typical phase diagram
100% cholesterol or similar sterol
one liquid phase
e
a, b, c & d
f, g
100% lipid with low melting temperature
100% lipid with high melting temperature
Reference phase diagram
100% cholesterol or similar sterol
sterol crystals
Lo
one liquid phase
Ld & Lo
all 3
Lo & gel
Ld
Ld & gel
gel
100% lipid with low melting temperature
100% lipid with high melting temperature

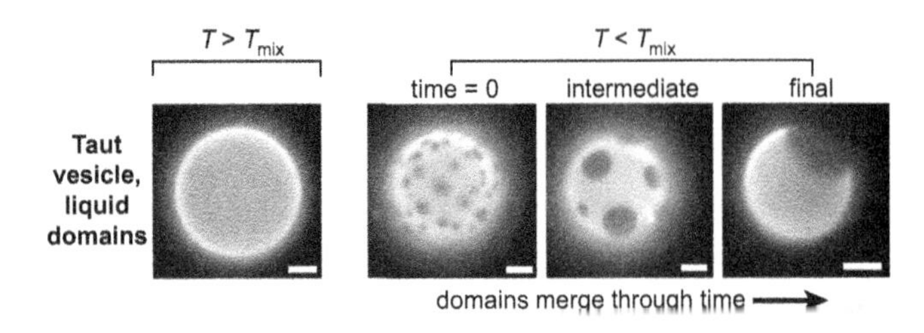

Figure 18.1 Representative micrographs of taut GUVs composed of 50/20/30 diphytanoylPC/DPPC/cholesterol, labeled with 0.8 mol% Texas Red® DPPE (see Appendix 2 for structure of this dye), which partitions preferentially to the L_d phase. At temperatures above T_{mix}, the vesicles are uniformly mixed. At temperatures below T_{mix}, two distinct liquid phases are observed. As time progresses, domains of each phase coalesce until only one L_d domain and one L_o domain remain, corresponding to schematic a2–a4 in Box 18.1. The time scale for full coarsening is typically on the time scale of minutes to tens of minutes, and depends on the area fraction of each phase and the diffusion coefficients of domains (Veatch and Keller, 2003; Juhasz et al., 2009; Stanich et al., 2013). Scale bars: 20 µm.

dynamics, as in the flow chart in Figure 18.2. Liquid domains are typically circular, for the same reason that 3D liquid droplets are spherical. When GUV membranes demix into coexisting liquid phases, Brownian motion drives diffusion of the domains across the surface of the GUV (Dietrich et al., 2001). When these liquid domains collide, they quickly and smoothly coalesce (Samsonov et al., 2001; Veatch and Keller, 2003; Stanich et al., 2013). If the GUV membrane is sufficiently taut, then all domains of each liquid phase eventually merge until only one domain of the L_d phase and one domain of the L_o phase persist, as shown in a2–a4 of Box 18.1 and in Figure 18.1.

Researchers often choose to publish micrographs of GUVs as in panels a2–a3 (rather than in the final state of panel a4) because readers can immediately estimate the fraction of the GUV's area in each phase from a single micrograph. Estimating the area fractions in a single micrograph can be challenging if only one domain of each phase persists and if the unlabeled phase has very low fluorescence.

Exceptions exist to the rule that micrometer-scale liquid domains merge through time. One example occurs when vesicles are flaccid. Flaccid vesicles are said to have "excess area" such that the surface area of the membrane exceeds the area of a sphere with the same enclosed volume. In this case, domains are kinetically trapped from merging due to membrane curvature as in c3–c4 and d2–d4 of Box 18.1 (Rozovsky et al., 2005; Yanagisawa et al., 2007; Semrau and Schmidt, 2009; Ursell et al., 2009; Hu et al., 2011). When the membrane has even more excess area, larger shape changes can occur, as in d5–d8 of Box 18.1. When these changes have a preferred orientation (e.g., tubules grow only toward the interior of a GUV), then the membrane has a nonzero spontaneous curvature and/or more material in one of the monolayer leaflets than the other. More commonly, tubules grow both outside and inside a single GUV.

Another example in which micrometer-scale domains do not merge occurs when the membrane composition and temperature are near a miscibility critical point, as in e1–e4 of Box 18.1. We have previously reviewed the behavior of membranes near critical points (Honerkamp-Smith et al., 2009). Above the critical temperature, noncircular composition fluctuations are observed through time in the membrane (Veatch et al., 2007b; Honerkamp-Smith et al., 2008, 2012; Inaura and Fujitani, 2008; Haataja, 2009; Honerkamp-Smith et al., 2009; Connell et al., 2013; Davis et al., 2013). Below the critical

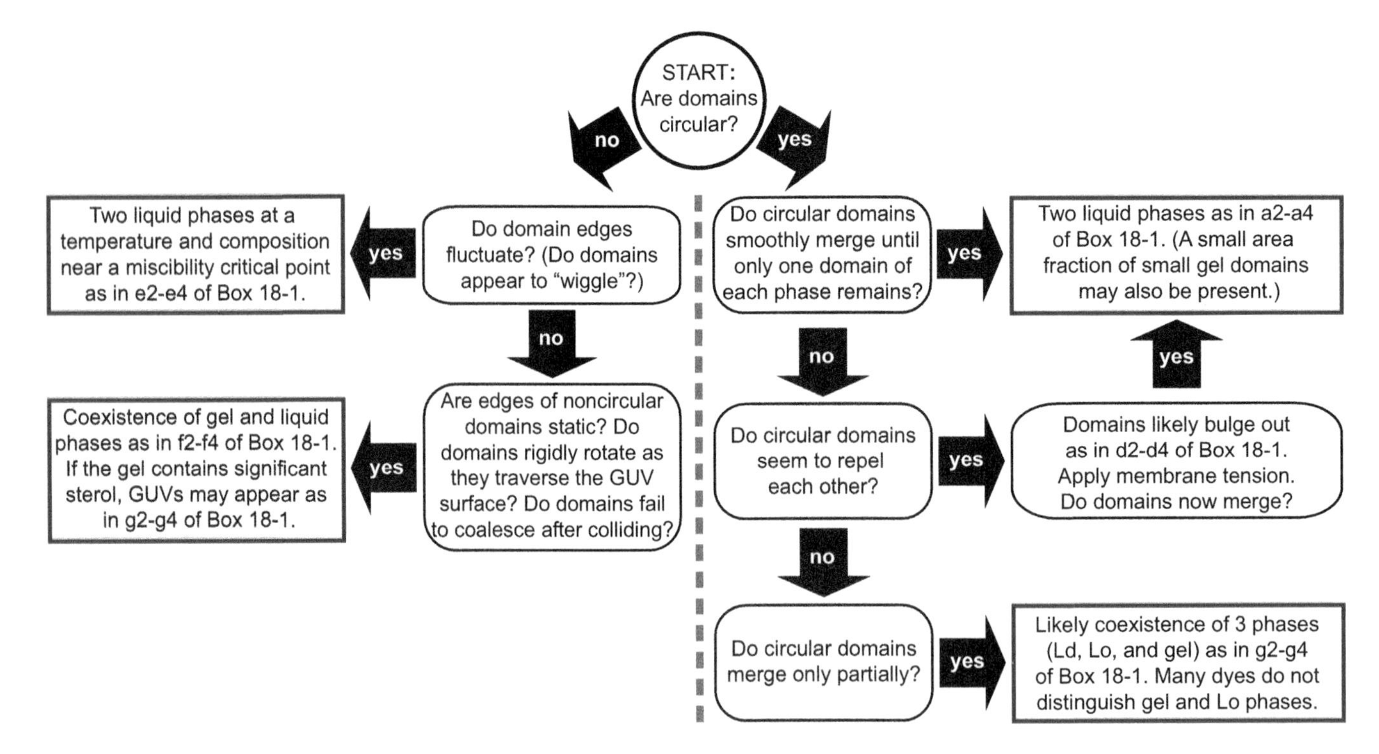

Figure 18.2 Identification of coexisting membrane phases in GUVs by fluorescence microscopy.

GUV-based techniques and what one can learn from them

temperature, domains boundaries fluctuate through time such that domains appear to wiggle (Esposito et al., 2007; Tian et al., 2007; Honerkamp-Smith et al., 2008). The characteristic length scale of these fluctuations varies with temperature. Near the critical point, the fractions of the GUV's surface area covered by L_o phase and L_d phase membranes are nearly equal. The same phenomena have been observed in giant plasma membranes (Veatch et al., 2008), introduced in Chapter 2.

18.2.3 ONLY TWO COEXISTING LIQUID PHASES HAVE BEEN OBSERVED IN GUVs

To date, a maximum of two coexisting *liquid* phases have been observed in a model lipid membrane at equilibrium, even when the membrane contains three or more lipid components. For example, only two fluorescence levels are observed in the GUVs in Figure 18.1. This observation also holds in complex systems with a very large number of lipid types, such as giant plasma membrane vesicles in Chapter 2 (Veatch et al. 2008; Levental et al., 2011; Sezgin et al., 2012a). The Gibbs Phase Rule states that more coexisting phases are possible in this system, but are not required.

18.2.4 FLUORESCENCE MICROSCOPY TO IDENTIFY COEXISTING GEL AND LIQUID PHASES

Gel phases are typically identified within GUV membranes as domains with noncircular shapes that do not change or relax over time. For many binary lipid systems, phase boundaries and tie lines derived from observation of GUVs can be compared with a decades-long history of data in which tie lines have been deduced from X-ray scattering of multilamellar vesicles (e.g., Furuya and Mitsui, 1979). Canonical examples of noncircular gel domains in GUV membranes appear widely in the literature (Korlach et al., 1999; Bagatolli and Gratton, 2000; Feigenson and Buboltz, 2001; Bagatolli, 2006; Zhao et al., 2007b). Whereas liquid domains quickly relax to circular shapes in order to minimize line tension, reorganization of gel domains is not observed on time scales accessible to most experiments due to slow lipid diffusion rates and intermolecular interactions that may result in lipid tilt with respect to the bilayer normal. Some gel phase domains have sharp, faceted edges, whereas others appear more curved and lobed like the two gel domains in panel f2 of Box 18.1, especially if the GUV membrane contains cholesterol.

Gel domains nucleate within a liquid GUV membrane during a temperature quench. Individual gel domains are larger, and therefore easier to identify by microscopy, when the rate of cooling is slower, such that fewer gel domains are nucleated. In contrast to gel domains, liquid domains coarsen predominantly through a process of collision and coalescence (as in a2–a4 of Box 18.1) to become several micrometers in diameter within seconds to minutes (Veatch and Keller, 2003; Stanich et al., 2013), whether the rate of cooling is fast or slow.

Because some gel domains assume a circular shape, particularly if they contain significant fractions of cholesterol, observation of domains over time is a useful technique for determining membrane phases by fluorescence microscopy. When two gel domains come into contact through diffusion across the surface of a GUV, they do not merge. They may adhere to each other as shown in panels f2–f4 in Box 18.1, but the aggregate as a whole does not reorganize into a new shape on the time scale of experiments.

GUV membranes that contain three phases (L_o, L_d, and gel) can be challenging to identify by fluorescence microscopy for two reasons. First, the partitioning of most (but not all) lipidic dyes between the L_o and gel phases is indistinguishable. Second, in all reported instances in which all three phases coexist within a membrane, gel domains are found within L_o domains (Fidorra et al., 2006), as depicted in g2 of Box 18.1. If a significant fraction of the membrane is in the L_o phase, then the difference between a domain that is entirely in an L_o phase and one that harbors a gel phase may not be obvious in a single micrograph.

To identify gel phases within L_o phases, one can evaluate how domains change shape (or not) after they collide. If two colliding L_o domains each contains a domain of gel phase, then after the collision, the two gel domains will not smoothly merge. Instead, the two gel domains will lie side-by-side, making an asymmetric structure. This structure will be coated by a single, merged L_o domain, as in g2–g4 of Box 18.1. If the total area of L_o phase is much larger than the total area of gel phase, then the merged domain in panel g3–g4 of Box 18.1 will appear circular, and it will be difficult to completely exclude the presence of gel phase. Complementary GUV imaging techniques, such as fluorescence lifetime imaging microscopy (FLIM) (Margineanu et al., 2007; Haluska et al., 2008), or the use of environmental probes such as Laurdan (Fidorra et al., 2006) can also be useful in identifying phases. Complementary techniques using geometries other than GUVs (e.g., nuclear magnetic resonance or electron paramagnetic resonance on multilamellar vesicles) produce excellent determinations of the lipid ratios and temperatures at which all three phases coexist (Veatch et al., 2007b; Davis et al., 2009; Ionova et al., 2012).

18.2.5 TREASURE MAPS: GIBBS PHASE DIAGRAMS

Phase diagrams provide a map of the lipid compositions and temperatures where particular phase behavior is found. A common procedure for mapping a phase diagram for a GUV system is to (1) produce GUVs of a known composition; (2) decrease the temperature; (3) identify the transition temperature at which a new phase appears, and identify the type of phase using the flowchart in Figure 18.2; and (4) repeat for many different lipid compositions.

Figure 18.3 shows how to read a ternary phase diagram of a GUV membrane. The area within the triangle represents all possible combinations of the three types of lipids in the membrane. For ease of visualization, imagine that the three lipids are called yellow, magenta, and cyan. Each vertex corresponds to a pure membrane; a membrane made of 100% yellow lipids appears at the top vertex. Points along each edge of the triangle correspond to binary mixtures of two lipid types. For example, the left edge of the triangles in Figure 18.3 corresponds to binary mixtures of yellow and cyan lipids (with no magenta). A point halfway along the left edge is green, an equimolar mixture of yellow and cyan. Any point in the interior of the triangle is a ternary mixture, where the mole fractions of each component sum

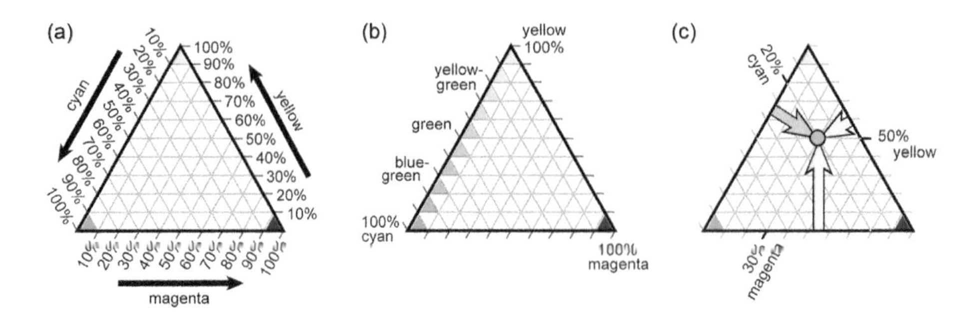

Figure 18.3 Mapping a ternary lipid composition within a phase diagram. (a) Vertices of a ternary (Gibbs) phase diagram denote that GUVs are made from only one lipid species. Here, the three lipid species are named yellow, cyan, and magenta. (b) Binary mixtures of two of the lipid species appear along the edges of the triangle. (c) Points in the interior of the triangle denote ternary mixtures of all three species. The tan point shown is composed of 50% yellow, 20% cyan, and 30% magenta.

to 100%. For example, the tan color in Figure 18.3 corresponds to a mixture of 50% yellow, 30% magenta, and 20% cyan. It is made by mixing the colors in the yellow, magenta, and cyan arrows that are perpendicular to the edges corresponding to 0% yellow, 0% magenta, and 0% cyan, respectively.

Figure 18.4 shows a schematic phase diagram determined by fluorescence microscopy for GUVs made from a ternary mixture of a lipid with a high melting temperature, a lipid with a low melting temperature, and cholesterol. The temperature is constant and is below the highest melting temperature of any single lipid in the system. Essential features of the diagram include a region of coexisting L_o and L_d phases in the interior of the triangle, gel–liquid coexistence along two of the binary axes, and a triangular region of coexisting gel, L_o and L_d phases. These general features are the same for a broad range of lipid systems; the exact shape of the coexistence regions are sensitive to molecular details of the lipids and surrounding buffer.

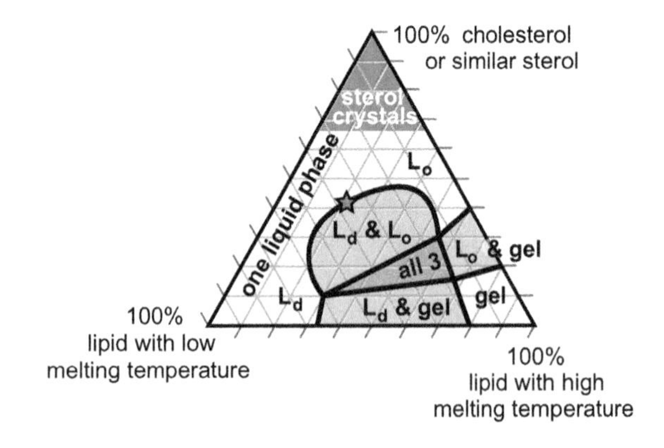

Figure 18.4 Schematic of a typical phase diagram found by fluorescence microscopy of GUVs and by nuclear magnetic resonance of multilamellar vesicles at a single temperature. The GUVs were made from mixtures of a lipid with a high melting temperature, a lipid with a low melting temperature, and cholesterol or a similar sterol. The gray region overlays mole fractions of cholesterol that exceed its solubility limit in membranes composed of phosphatidylcholine lipids (Huang et al., 1999). Solubility limits of other sterols can be much lower (Stevens et al., 2010). Within the "all 3" region, L_o, L_d, and gel phases coexist. The miscibility critical point is denoted by a star. Phase boundaries in the schematic are loosely modeled after three previously published phase diagrams (Veatch and Keller, 2005a, 2005b; Veatch et al., 2007b). In Box 18.1, domain morphologies are identified on this same phase diagram.

For lipid ratios that fall outside of the gel and coexistence regions shown in the phase diagram, the GUV membrane is in one liquid phase. Of course, physical properties of the membrane vary with lipid mole fraction within this region. For example, a bilayer with 10 mol% cholesterol will behave differently from one with 40 mol% cholesterol, even if no phase transition is encountered along a path from 10 mol% to 40 mol% cholesterol. Membranes with small fractions of cholesterol and small fractions of lipids with high melting temperatures are typically termed L_d phase. Those with high fractions of both are termed L_o phase. Within the one-phase liquid region of the diagram, the L_d and L_o phases cannot be distinguished, except by establishing an arbitrary definition. Illustrating that the designations of L_d and L_o phases are relative, there is no specific line or transition dividing the L_d and L_o phases within the "one liquid phase" region of Figure 18.4. The ability to traverse from one phase to another continuously by following a path around a critical point is a standard feature of miscibility phase diagrams. One reason that the determination of tie lines is important is that the region of the phase diagram with short tie lines is likely to yield GUVs exhibiting interesting critical phenomena (Veatch et al., 2007b; Honerkamp-Smith et al., 2008, 2009, 2012; Inaura and Fujitani, 2008; Haataja, 2009; Connell et al., 2013; Davis et al., 2013).

The phase boundaries change with temperature. Most notably, all of the coexistence regions shrink as temperature increases. A complete description of the membrane's phase behavior over a temperature range is straightforward to collect in the laboratory by producing a series of GUV samples, each made of a different ratio of lipids, and observing which phases are present in the membrane as temperature is varied. Measuring the temperatures at which membrane phases change (the transition temperatures) in this way determines the phase behavior over a range of temperatures. Measuring shifts in transition temperatures is a sensitive method of measuring perturbations in phase behavior. The opposite tactic, to hold temperature constant and vary the lipid ratio of GUVs *in situ* in a known, measureable way, is much more difficult experimentally.

18.3 IDENTIFYING TIE LINES IN GUVs

For a membrane that demixes into two phases, end points of tie lines identify the ratio of lipids in each of the phases, from which the partitioning of each lipid species into the phases can be calculated. For example, if the L_o tie line end point is at 60 mol% dipalmitoyl-phosphatidylcholine (DPPC), and the L_d end point is at 20 mol% DPPC, then the ordered phase is three times more enriched in DPPC than the disordered phase, and DPPC is said to partition preferentially to the ordered phase. Moving along a single tie line does not change these compositions within each phase. Instead, the amounts of the two phases change linearly with the position along the tie line, so that at one end point of the tie line all lipids are in the ordered phase, at the other end point all lipids are in the disordered phase, and at the midpoint half of the lipids are in each phase. This is known as the lever rule. An example of how one might deduce tie lines in a particular scenario appears in Chapter 15.

Knowledge of tie lines gives insight into what drives phase separation. For example, the initial observation that unsaturated lipids with disordered acyl chains partition strongly into the L_d phase and that saturated lipids with highly ordered chains partition strongly into the L_o phase (Veatch and Keller, 2002) inspired a theory that demixing of L_o and L_d phases is driven by a difference in lipid acyl chain order parameters (Garbes Putzel and Schick, 2008b). Because the bulk properties of each membrane phase (e.g., bending rigidity and thickness) do not change along a tie line, measurements of these properties can be performed at any point on the line, from 0% to 100% of the L_o (or L_d) phase. For instance, it is often convenient to perform measurements of bulk properties on single-phase vesicles whose lipid composition corresponds to an end point of the tie line (e.g., Heftberger et al., 2015). Similarly, by moving along a known tie line, the same phases can be interrogated at different area fractions of the membrane (e.g., Blosser et al., 2015; Bleecker et al., 2016b).

Inspection of the area fractions of the two membrane phases of a demixed GUV quickly yields qualitative tie lines by applying three rules: (1) Tie lines cannot cross at a single temperature; (2) they terminate in vesicles that are entirely one phase (i.e., at the boundary of the two-phase region); and (3) the mole fraction of each phase changes monotonically along the tie line. Approximate tie lines can be found by mapping a trail of lipid ratios on a Gibbs phase triangle for which each phase occupies half of the GUV membrane area, and then drawing perpendicular lines to this trail, terminating at the boundary of the two-phase region (Figure 18.5). This approach results in the conclusion that the primary difference between L_o and L_d phases in a GUVs of DOPC/DPPC/cholesterol is in the ratio of lipids with high melting temperature (e.g., saturated and long acyl chains) to lipids with low melting temperature (e.g., unsaturated, methylated, or short acyl chains) (Veatch and Keller, 2003). Additional constraints can be made based on other features of the phase diagram. For example, the lengths of tie lines approach zero as the tie lines approach a critical point. As another example, the edges of a triangular region in which three phases coexist (L_o, L_d, and gel) correspond to the first tie lines within the adjoining two-phase regions.

The process of extracting tie lines from GUV images can be made fully quantitative only by (1) imaging of the entire spherical surface of a free-floating GUVs and then (2) translating area fractions of the two phases into mole fractions. Depending on the type of microscopy used, different experimental details are relevant. When a confocal microscope or an inverted microscope is used, GUVs typically rest on other GUVs or on a surface. When they do so, interactions between the GUV and the neighboring surface can induce the formation of domains that would not otherwise have been present (Gordon et al., 2008; Zhao et al., 2013). Alternately, when an upright microscope is used, individual GUVs are typically free-floating; these GUVs can diffuse too quickly to be accurately imaged by confocal microscopy. In order to assess the area fraction of each phase for a single GUV, either 2D images of the top and bottom of the GUV must be correctly mapped onto 3D surfaces (Veatch et al., 2008; Stanich et al., 2013), simple geometric rules must be employed (Juhasz et al., 2009; Bleecker et al., 2016a), or 3D reconstructions of slices much be achieved (Fidorra et al., 2009).

Because the area per lipid varies with lipid species (and the surrounding environment), a vesicle with equal area fractions of L_o and L_d phases will generally have unequal mole fractions of the two phases. Translating area fractions of the two phases into mole

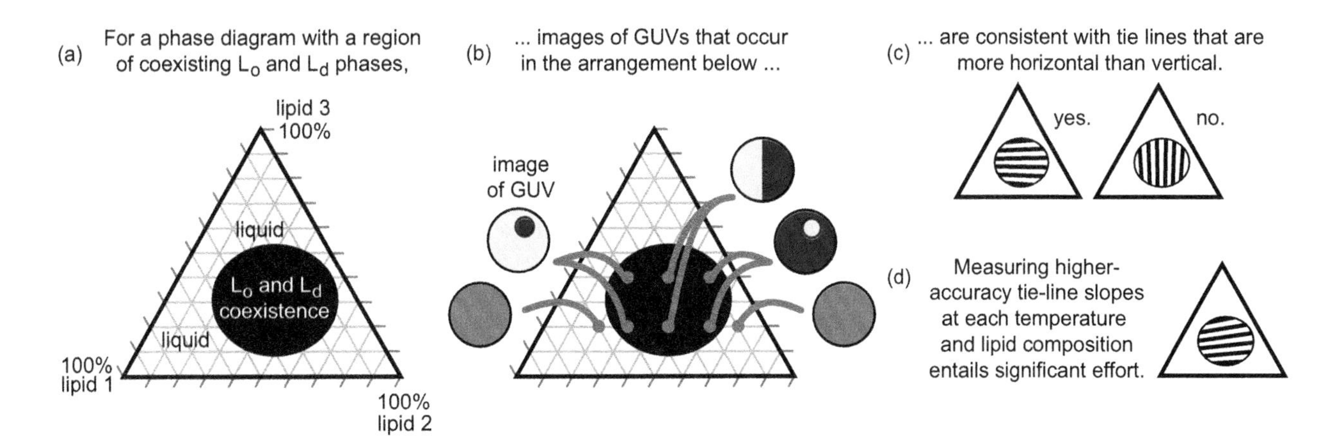

Figure 18.5 Inspection of fluorescence micrographs of phase-separated GUVs at different compositions give estimates of tie line directions. This approach it applies to any membrane composed of three lipid types with coexisting phases as in Panel *a*. The GUVs sketched in Panel *b* correspond to a case in which lipid 1 has a low melting temperature, lipid 2 has a high melting temperature, and lipid 3 is a sterol. The sketched GUVs give rise to approximated tie lines that are more horizontal than vertical (Panel c), but high resolution of the tie lines requires other methods (Panel d).

fractions is particularly challenging in ternary systems containing cholesterol because of the "area condensation effect" whereby cholesterol nonlinearly decreases the area per molecule of the other lipids in the bilayer membrane. Area condensation values are known for only a few binary mixtures of cholesterol and a phosphatidylcholine (PC) (Hung et al., 2007; Kučerka et al., 2007, 2009, 2011; Pan et al., 2009; Gallová et al., 2010; Heftberger et al., 2014; Litz et al., 2016). Phospholipid area condensations due to ergosterol are much smaller than those due to cholesterol (Hung et al., 2016).

In order to make the problem of using images of GUVs to estimate tie lines tractable, some researchers have approximated that the area fraction of each phase is equal to the mole fraction of each phase (Husen et al., 2012; Bezlyepkina et al., 2013; Khadka et al., 2015). This approximation is valid when the areas per lipid in the phases are sufficiently similar, differing by less than ~10%. In many ternary lipid systems, the average area per lipid in the L_o and L_d phases can differ by much more (e.g., Heftberger et al., 2015). Another approach, which has been applied to L_o and L_d phases, is to approximate the area per lipid in coexisting liquid phases as the weighted average of literature values for each species (Bezlyepkina et al., 2013). To account for area condensation, the areas are varied with cholesterol according to data from binary mixtures of phospholipid and cholesterol. This approach is valid when the areas per lipid of the coexisting phases are similar to the closest binary compositions. Wide use of this approach is limited by the fact that data for area per lipid have been published for so few binary lipid membranes to date. As noted by others (Fidorra et al., 2009), it is especially important to consider molecular areas when investigating GUV membranes containing coexisting liquid and gel phases, in which the areas per lipid differ significantly.

To summarize the discussion above, without previous knowledge of molecular areas in each phase, analysis of tie lines from images of GUVs currently presents several challenges. As a result, techniques of quantitatively determining tie lines from images of GUVs are not yet generalizable to previously unexplored lipid systems nor capable of high accuracy. Nevertheless, the use of images to approximate the location of tie lines within phase diagrams is extremely valuable because the resulting information can alert researchers about which lipid compositions and temperatures will be interesting to probe by more time-consuming and expensive methods, which often employ membranes in geometries other than GUVs.

For example, tie lines estimated by assessing area fractions of GUVs (Veatch and Keller, 2003; Fidorra et al., 2009; Husen et al., 2012; Bezlyepkina et al., 2013) can be compared with quantitative tie line results gleaned from techniques that use multilamellar vesicles, such as electron paramagnetic resonance (EPR) (Chiang et al., 2005; Ionova et al., 2012) and nuclear magnetic resonance (NMR) (Veatch et al., 2004, 2008; Veatch and Keller, 2005b; Davis et al., 2009; Yasuda et al., 2015). Figure 18.6 shows an example from Veatch et al. (2006) of a tie line determined by NMR of multilamellar vesicles overlain on a phase diagram determined by fluorescence microscopy of GUVs. The tie line was found by deconvolving distinct, superimposed NMR spectra for lipids in each phase. An outstanding attribute of NMR is that it distinguishes lipids in the L_o phase from lipids in a gel phase, resulting in an accurate determination of the three-phase region (Veatch et al., 2007b). X-ray scattering also uses multilamellar systems, either as vesicles (Heftberger et al., 2015) or as flat bilayer stacks at less than full hydration (Uppamoochikkal et al., 2010). X-ray scattering profiles can be used to find compositions of phases, and hence tie lines, with the caveat that the process generates a significant number of fitting parameters. Neutron scattering uses unilamellar vesicles, but they must be small (~60 nm) (Heberle et al., 2013). Similarly, fluorescence resonance energy transfer has been used to probe phase boundaries and partition coefficients in smaller vesicles (Feigenson and Buboltz, 2001; Heberle et al., 2010).

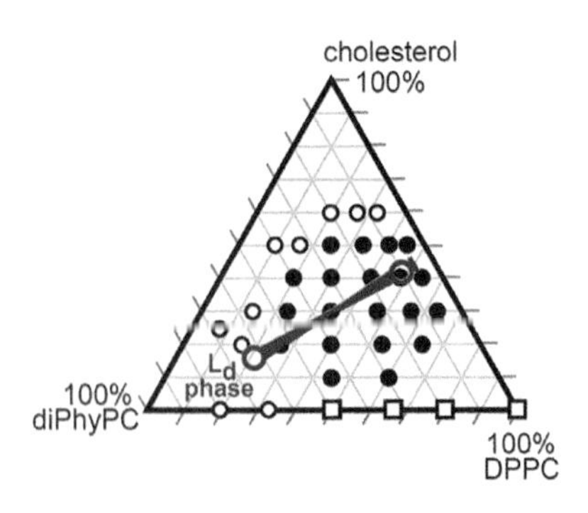

Figure 18.6 A ternary phase diagram for GUVs composed of diphytanoylPC (diPhyPC) lipids, DPPC lipids, and cholesterol. Data are extracted from Figure 7d of Veatch et al. (2006). Solid circles denote coexisting liquid phases. Open circles denote one liquid phase: L_o on the right and L_d on the left. Open squares denote coexistence of solid and liquid phases. The diagonal line overlain on the phase diagram is a tie line determined by nuclear magnetic resonance. Calculated experimental uncertainties in the tie line tilt are represented by the splay of the line. Uncertainties in the end point location on the left side of the diagram are on the order of the size of the large open circle and on the right are shown by an error bar.

In addition to using intact GUVs to assess area fractions, GUVs can be ruptured on a solid surface and interrogated by coherent anti-Stokes Raman scattering imaging (Li et al., 2005; Li and Cheng, 2008) or atomic force microscopy (Khadka et al., 2015). The first technique currently requires a stationary membrane so that images can be averaged over many frames; the second requires a flat membrane. To find tie lines or to comment on molar fractions, these works have approximated mole fractions from area fractions. As noted above, this assumption process relies on complementary data and/or introduces significant error in many systems, especially for coexisting gel and liquid phases. Imaging via stimulated Raman scattering holds promise as a more quantitative technique because of its automatic removal of nonresonant background signals. Geometries like total internal reflection have been used to enhance signals of Raman spectra (Lee and Bain, 2005).

Atomic force microscopy of ruptured GUVs has also been used to draw conclusions about phase diagrams and tie lines determined by other methods (Bleecker et al., 2016b). In general, spectrometric methods are promising for membrane geometries other than GUVs that yield high signals (Silvius, 2005), but can be difficult to interpret if the signal changes nonlinearly with lipid composition as in (Zhao et al., 2007a). Multiphoton microscopy has been performed for one tie line to date (Farkas and Webb, 2010). Tie lines can also be found by mass spectrometry of freeze-dried supported lipid bilayers or lipid monolayers, although the technique is not widely available (Kraft et al., 2006; Zheng et al., 2007; Lozano et al., 2013).

18.4 EXPERIMENTAL CAVEATS

The fundamental procedure for determining phase diagrams and tie lines for GUV membranes is conceptually straightforward. However, as in every field, there are many ways in which an experiment can yield poor results. This section presents general guidelines as well as specific advice about how useful data can be gathered and reported.

18.4.1 THREE BROAD THEMES

Researchers' opinions about which experimental factors are the most crucial to control in order to minimize error while measuring GUV transition temperatures for phase diagrams are well documented, and occasionally in disagreement (Veatch and Keller, 2005b; Goñi et al., 2008; Morales-Penningston et al., 2010). A theme that will recur throughout this section is that data are most meaningful when they are accompanied by estimates of experimental uncertainties and of known systematic errors, independent of which methods were used to generate those data. Just as very poor data can be collected under nominally ideal experimental conditions, very strong data can be collected under conditions that have not been optimized in every attribute.

A second theme is that raw data and open source code are more broadly useful (although less easy to immediately comprehend) than processed data. It is exciting that researchers are now producing phase diagrams of the same GUV systems using different experimental techniques (e.g., Ionova et al., 2012). In the last few years, reproducibility of data has become a subject of intense discussion among scientists, funding agencies, and policy makers. One good outcome of this situation is that many publishers now allow extensive compilations of raw data to be attached as supplementary information to publications. Comparison of that raw data can help researchers decide if discrepancies between phase diagrams lie in differences in sample treatment or in differences in data analysis. A variety of ways exist to publish and archive original data analysis methods and computer code, some of which are required by publishers (Ince et al., 2012). In our own laboratory, the single most challenging hurdle to overcome in disseminating computer code is the reluctance of the program's author to release code that is less than beautiful, for fear of being judged harshly by programmers with more experience. As a community, when we review manuscripts, we can request inclusion of open source code in supplementary info and in open software archives (e.g., the Open Bioinformatics Foundation at https://open-bio.org), and we can ask journals to adopt software sharing policies like, for example, those of PLOS (http://journals.plos.org/plosone/s/materials-and-software-sharing).

A third theme is that every technical hint in Section 18.4.3 is less important than a researcher's critical eye and willingness to discard a sample and start the experiment over. To date, the following vignette has repeated itself for nearly every researcher who has joined our laboratory. On the first day in the lab, the new researcher follows a well-established, written protocol for producing electroformed GUVs from ternary lipid mixtures. First-day researchers (and principal investigators who have fallen out of practice) require an entire day to prepare a sample. The resulting first-day GUV solution contains lipid aggregates and/or tubules in every field of view of the microscope. If imaging is postponed until the following day, the results are even worse. Overall, measured miscibility transition temperature of the first-day GUVs rarely matches published values, and the experimental uncertainty is large, independent of the type of lipids used to produce the GUVs. The new researcher repeats the same protocol daily for a period that can last a few weeks. Over this time, the abundance of aggregates and tubules in the resulting samples drops precipitously, the experimental uncertainty narrows dramatically, transition temperatures become reproducible, and sample preparation time drops by a factor of three to four. In an attempt to help other researchers master the technique faster, our lab has conducted tests to isolate which single step in the protocol contributes most to reproducibility, but found no clear conclusion (Joan Bleecker and Morgan McGuinness, unpublished results). Even experienced, well-practiced researchers occasionally produce poor samples that are immediately recognizable by their high incidence of non-GUV structures. Seasoned researchers are more productive and produce consistently reproducible results because they produce extra samples every time in case they discover that one of the samples is rife with lipid aggregates and tubules, so should be discarded. They also discard GUV solutions 1 h after they are made. For these experts, a new GUV solution is only 2 h away.

The details of this vignette will vary from laboratory to laboratory, but the message is universal: Reproducible results require both a proper protocol and an experienced and discerning researcher. Until all GUVs are made following some universal scientific protocol language (Sadowski et al., 2015) by standardized robots that achieve reproducible outcomes with small experimental uncertainties, clear presentation of those uncertainties and of raw data will help researchers assess their and each other's results.

18.4.2 INTRODUCTION TO SPECIFIC ADVICE AND PITFALLS

The following section compiles technical hints that are particularly pertinent for researchers who are compiling phase diagrams of GUVs. These hints were discovered largely by trial and error by the community. Discussion of GUV preparation methods complements Chapter 1, and discussion of imaging methods complements Chapter 10.

Experimental problems can be especially challenging to recognize when vesicles form and undergo phase separation at a well-defined temperature that is shifted from the true transition temperature. The list of topics below is not exhaustive, and some appear in and/or are complemented by existing resources in the literature (Veatch and Keller, 2005b; Goñi et al., 2008; Morales-Penningston et al., 2010). The list below highlights which protocols we would check first when confronted with results that do not make sense, as well as the list of first suggestions that we would offer to a colleague encountering problems in the lab.

18.4.3 GUV PREPARATION METHODS

Compositional Variation—All methods of producing GUVs result in some variation in lipid ratio from vesicle to vesicle. As a result, every sample containing more than one GUV will produce a distribution of measured physical quantities such as transition temperature (Veatch and Keller, 2005b) or area fractions (Bezlyepkina et al., 2013) centered at the "true" value, which is the transition temperature or area fraction of the average lipid composition of the population of GUVs. This variation is exacerbated when domains or tubes bud off from GUVs. The way that the community has reported experimental uncertainties in transition temperatures has evolved over the past decade. In the case of a GUV membrane demixing into coexisting L_o and L_d phases, the percentage of phase-separated vesicles varies sigmoidally with temperature, as in Figure 18.7, passing through 50% at the true transition temperature (Veatch et al., 2008; Levental et al., 2009, 2011; Gray et al., 2013). Current best practice is to fit the sigmoidal curve with 95% confidence bounds given by counting statistics and to quote the uncertainty as the width of these bounds when 50% of GUVs have phase separated (Gray et al., 2013). This uncertainty is smaller than the full width of the sigmoid, and shrinks as the number of vesicles in the sample set grows. The width of the sigmoid is affected both by the vesicle-to-vesicle variation in lipid ratio and by how sensitive the transition temperature is to changes in composition. That sensitivity varies across the phase diagram. Phase diagrams recorded over many temperatures resemble topographic maps. When a hiker walks across the top of a mountain with a broad, flat top, her elevation changes little even though her latitude and longitude change significantly. When that hiker encounters the steep side of the mountain, a small change in latitude or longitude results in a large change in elevation. Similarly, a small change in lipid ratio along a steep edge of the phase boundary results in a large experimental uncertainty. If a GUV sample produces a much larger range of transition temperatures than expected, then all subsequent data will be of low quality. The most expedient solution is to discard the sample and make new samples until the experimental uncertainty is consistently minimized.

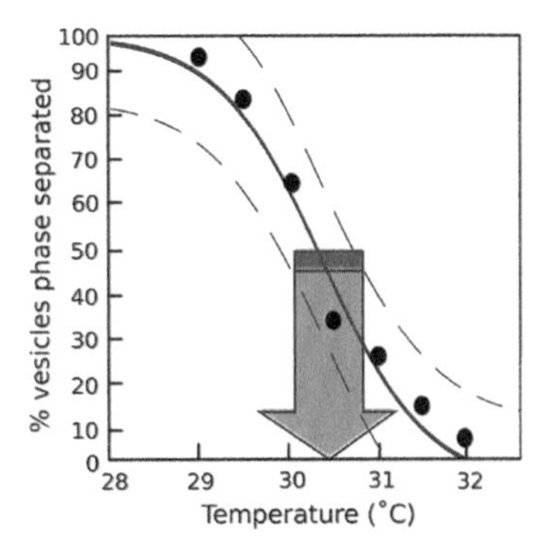

Figure 18.7 The black points, fit to a sigmoidal curve, represent the percentage of phase separated vesicles within a population of vesicles, versus temperature. All vesicles have the same nominal composition because they come from the same sample. The vesicles were formed from a mixture of 35/35/30 DOPC:DPPC:cholesterol and labeled with 0.8 mol% Texas Red DPPE. The distribution of transition temperatures (the width of the sigmoid) is due to the distribution of vesicle lipid compositions. Dashed lines represent 95% confidence intervals from counting statistics (Gray et al., 2013). The blue arrow points down to the miscibility transition temperature, where the width of the arrow is the experimental uncertainty.

Fabrication temperature and film thickness—When using electroformation or gentle hydration, lipid stock solutions should be mixed, spread onto substrates, and hydrated at a temperature significantly higher (~10°C) than the highest main chain transition temperature in the system (Veatch and Keller, 2005b). This ensures that all lipids are well mixed when vesicles are formed. Failure to do so can result in systematic errors in GUV composition. Similar systematic errors occur when lipid films used to produce GUVs by electroformation are too thin (Veatch and Keller, 2005b).

Probes and impurities—The addition of probes and impurities, especially hydrophobic and amphiphilic molecules, can shift transition temperatures in GUV membranes. To avoid the introduction of plasticizers to membranes, chloroform solutions should be stored in and handled with glass or polytetrafluoroethylene. Detergents should be thoroughly rinsed from all glassware. Glassware cleaned with acid should also be thoroughly rinsed. Some common solvents (e.g., ethanol) have been shown to alter membrane miscibility transition temperatures (Gray et al., 2013; Machta et al., 2016; Cornell et al., 2017). Fluorescent probes that strongly partition into one of two coexisting phases within a membrane also shift miscibility transition temperatures (Veatch et al., 2007a). The probe Laurdan is generally thought to partition relatively equally between L_o and L_d phases and so presents a good alternative to probes that partition strongly (Bagatolli, 2006; Kaiser et al., 2009). A caveat is that determining partitioning of any dye is nontrivial because it is difficult to disentangle aggregation (Kaiser et al., 2009), photoselection (Kim et al., 2007), and other photophysical effects of the membrane on the dye from the dye's concentration. There are almost certainly cases in which C-Laurdan, a version of Laurdan with a carboxylic acid moiety (Kim et al., 2007), does not partition equally between L_o and L_d phases. For example, when C-Laurdan is incorporated into giant plasma membrane vesicles or plasma membrane spheres, the difference in C-Laurdan intensities between the two phases is striking, whereas the difference in C-Laurdan signals between the L_o and L_d phases (the "difference in generalized polarization") is slight (Kaiser et al., 2009; Sezgin et al., 2014, 2015).

Membrane Tension and Osmotic Pressure—Electroformation typically produces taut GUVs, which become flaccid over time. An increase in surface tension produces small changes in GUV miscibility temperatures of only a few degrees at most (Portet et al., 2012; Uline et al., 2012). In contrast, a decrease in surface tension can give rise to modulated phases or microemulsions (Cornell et al., 2018) and can have an enormous effect on the rate of domain coarsening, as illustrated in the schematic of Box 18.1. Because evaporation can occasionally occur in electroformation or gentle hydration chambers, even when they appear sealed, confirmation of osmotic pressures via direct measurement is prudent.

Solutions Containing Sugars—GUVs filled with sucrose solutions are commonly diluted into glucose solutions in order to provide a density difference. This difference is used to sink vesicles to the bottom of an experimental chamber, to provide contrast for differential interference contrast microscopy, and to control osmotic pressure. The presence of these sugars usually does not significantly change miscibility transition temperatures of GUV membranes. An important known exception is in lipids with a net charge, for which the effect of sugars on membrane transition temperatures is large (Blosser et al., 2013).

Solutions Containing Divalent Cations—Divalent cations (e.g., Ca^{2+}) have a large effect on vesicle stability and phase behavior, especially in vesicles containing charged lipids (Shimokawa et al., 2010). Measureable effects persist down to micromolar concentrations. Buffers that nominally contain only monovalent salt can contain divalent impurities, and these impurities can have a significant effect that is often larger than the effect of the monovalent salt (Vequi-Suplicy et al., 2010; Blosser et al., 2013). To achieve ionic buffers without divalent cations, a chelator such as EDTA should be included in the solution.

18.4.4 IMAGING GUVs

Photooxidation

Exposure to light can induce chemical changes in lipid molecules. Photooxidation changes the composition of GUVs by degrading the original lipids and by creating products that act as additional components. This shift in membrane composition can result in a shift in membrane phase behavior. For GUVs far from a miscibility phase transition, well within a region of the phase diagram with coexisting L_o and L_d phases, a hallmark of photooxidation is the nucleation of new small domains within preexisting larger domains as in Figure 18.8. The effect is equivalent to a slow temperature quench (Stanich et al., 2013). The process of photooxidation in lipid membranes has elicited significant attention in the literature, with researchers focusing both on how to minimize photooxidation (Ayuyan and Cohen, 2006; Morales-Penningston et al., 2010) and on how to intentionally harness it to induce or probe new membrane phase behavior (Roux et al., 2005; Zhao et al., 2007b), as described more fully in Chapter 22.

Whether photooxidation is perceived as a blight or a boon, it is most helpful to the scientific community if researchers quantify effects of light exposure in their experiments. For example, for GUVs composed of ternary mixtures of cholesterol and entirely saturated phospholipids, it is trivial to reduce the "measured drift in transition temperature due to photooxidation to at most –0.09°C per exposure" (Honerkamp-Smith et al., 2008), which represents an inconsequential effect for most experiments. One way of quantitatively describing the effects of photooxidation is to quote the shift in GUV transition temperature from the beginning to the end of the experiment, for example, by repeating a measurement on the same sample. Other ways to quantitatively describe those effects are to quote the number of visible domains nucleated (as in Figure 18.8) per unit time at a constant temperature or to quote the change in area fraction over the course of an experiment at constant temperature. Given that domains that have

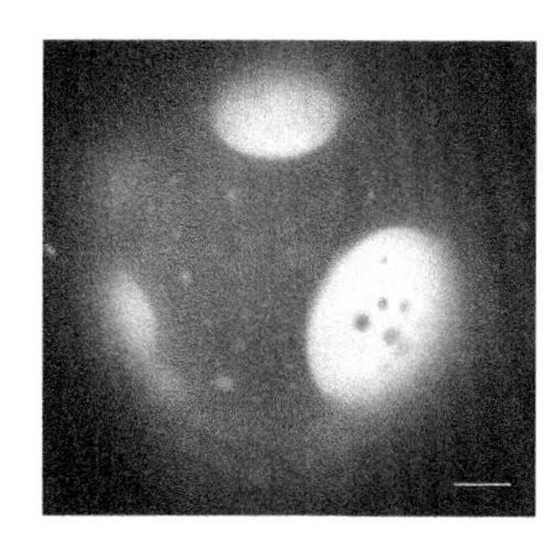

Figure 18.8 A representative micrograph of a photooxidized GUV composed of 35/35/30 DOPC:DPPC:cholesterol labeled with 0.8 mol% Texas Red DPPE. Small, bright L_d domains are visible within the darker L_o region, and small, dark L_o domains are visible within large L_d domains. The large domains formed in the period after the initial temperature quench, and the small domains nucleated after an increase in the membrane's miscibility transition temperature, caused by photooxidation. Scale bar: 20 μm.

recently nucleated are small, the former value is a much more sensitive measurement than the latter. For a population of GUVs in which the distribution of area fraction values is unusually narrow (~1%), quantifying the degree of photooxidation by measuring the change in area fraction over time produces a value that can readily be compared between laboratories.

For GUVs in which all phospholipid acyl chains are saturated, as in the example above, light exposure *decreases* miscibility transition temperatures in GUVs. For membranes comprising at least one unsaturated phospholipid, light *increases* the same transition temperature, and the magnitude of the shift is larger. Lipids with polyunsaturated acyl chains are exquisitely vulnerable to photooxidation. In addition, the degree of photooxidation depends on the concentration and species of fluorophore in the system as well as the concentration of oxygen dissolved in the buffer (Veatch et al., 2007a). Depending on the mechanism of photooxidation, a sample exposed to light will continue to photooxidize after the light is shut off, and products of the photooxidation reaction can diffuse to regions outside the original area of illumination.

Because the degree of photooxidation is influenced by so many factors (e.g., the type of lipid, the type of probe, the type of solvent or substrate, and the degree of oxygen exposure), there are no universally recommended values for the maximum concentration of probe that should be incorporated in a GUV or the maximum number of photons per voxel that should be employed to image a GUV. A more broadly useful approach is to determine those maximum values for each type of experiment in each laboratory, by assessing whether new domains nucleate or whether transition temperatures shift.

There are several strategies for mitigating the effects of photooxidation (Morales-Penningston et al., 2010). The simplest is to minimize the exposure to light. This includes (a) illuminating any individual sample for only a short time by frequently replacing the sample with new aliquots from the same vesicle preparation; (b) decreasing overall light levels, for example, through the use of neutral density filters in a standard fluorescence microscope or through the use of lower laser intensity in a confocal microscope (especially for final measurements of transition temperatures in a new sample after an estimated transition temperature has been found in a sacrificial sample using higher light levels); (c) illuminating samples only when imaging, for example, by using a shutter and short exposures; and (d) illuminating only the area of interest, for example, by restricting illumination with an aperture or by using a light sheet microscope in order to illuminate only the focal plane through a GUV (Loftus et al., 2013). Other strategies include employing complementary imaging techniques such as differential interference contrast microscopy that does not require fluorescent probes, degasing aqueous solutions, preparing and imaging samples in an oxygen-free glove bag, and adding antioxidants to GUV solutions (which can themselves cause a shift in transition temperatures [Zhao et al., 2007b]). The use of indium tin oxide–coated slides during electroformation of GUVs has been reported both to contribute to oxidation of lipids (Ayuyan and Cohen, 2006) and to not be the dominant cause of shifts in membrane transition temperatures (Zhao et al., 2007b).

Chemical assays can be used to verify the purity of lipid samples, but the common, semi-quantitative, method of thin layer chromatography typically registers the presence of lipid contaminants only when their concentrations rise to a few mole percent. This level of sensitivity is not high enough for researchers to confidently assert that photooxidation has not significantly shifted phase boundaries in lipid membranes. As with any assay of impurities, a prudent approach is to intentionally add known amounts of impurities to a sample in order to gauge the concentration of impurity at which the assay registers a shift in the signal. Standardized versions of oxidized lipids are currently available commercially.

18.4.5 BILAYERS IN NONSPHERICAL SHAPES

Budding and Tubulation—Phase separated vesicles are vulnerable to tubulation or budding over time. Daughter vesicles will have compositions roughly corresponding to tie line end points for the temperature at which the budding occurred, which can cause a population of GUV membranes to artifactually appear to be in a single phase or to yield an artifactual transition temperature. Minimizing the period of time over which vesicles are stored at a temperature below the transition temperature reduces problems associated with GUV budding. A GUV undergoing liquid–liquid phase separation, as in Figure 18.1, typically produces micrometer-scale, observable domains within tens of seconds of passing through the transition temperature. To verify gel–liquid coexistence, a gradual temperature quench may be required to observe large-scale gel domains (Jørgensen and Mouritsen, 1995; de Almeida et al., 2002). Repeatedly cycling through the transition temperature can exacerbate budding and tubulation.

Substrate interactions—Contact between GUVs and a substrate or other vesicles can shift transition temperatures and the spatial arrangement of domains. As usual, a best practice is to simply state the magnitude of the shift so that other researchers can assess if that offset will affect their own experiments. For example, miscibility transition temperatures of free floating GUVs and of supported bilayers produced by rupture of the same GUVs on a substrate differ by less than 5°C (Blosser et al., 2015) for membranes of diphytanoylPC (diPhyPC), DPPC, and cholesterol. Similar studies found a shift in gel–liquid (main phase) transition temperatures of ≤2°C (Yang and Appleyard, 2000; Yarrow et al., 2005). When GUVs are in close contact with a surface, L_o domains tend to gather in the membrane at the contact area, such that the area fraction of domains on the remainder of the membrane appear depleted of L_o domains (Gordon et al., 2008; Zhao et al., 2013).

18.5 PARTING THOUGHTS

Phase transitions in GUVs are a beautiful system to study, both in the scientific sense of beautiful phenomena and theory and in the aesthetic sense of beautiful images. Movies of miscibility phase transitions and of coalescence of domains in GUV membranes are captivating. Moreover, the phenomena that the movies illustrate are important in their biological applications.

Our hope is that as the community of researchers who study GUVs and phase separation grows, so too will the array of tools available to quantitatively assess the phase state and lipid composition of vesicles and of individual domains. Spectroscopic techniques in particular seem poised for a leap forward to enable high-accuracy measurements of GUV lipid compositions and tie lines. In principle, tie lines can be determined by any probe that is sensitive to local composition. Because compositions along a tie line are linear combinations of the end point compositions, with weights given by the lever rule, finding the correct, unique tie line orientation that produces a linear change in signal is theoretically straightforward. In practice, few experimental methods have proven sufficiently sensitive, non-perturbing, and feasible using current technologies. We look forward to watching the technology mature. Similarly, as sensitivities of NMR and EPR spectroscopies increase and as fluxes at X-ray scattering facilities increase, we can imagine a time when spectra and wide angle X-ray scattering profiles from GUVs might be evaluated to extract tie lines and to map phase diagrams. Improved tools will help us all provide rigorous estimates of experimental uncertainties and systematic errors so that others in the community can assess the validity and relevance of our results and can build theories that rest upon those results.

ACKNOWLEDGMENTS

During preparation of this chapter, the authors were supported in whole or in part by the National Institute of General Medical Sciences of the National Institutes of Health award 1F32GM115236 (to M.C.B.) and T32GM008268 (to C.C.), and by the National Science Foundation award MCB1402059 (to S.P.R. and S.L.K.). The authors thank Sarah Veatch for her comments on the manuscript, Ilya Levental for his C-Laurdan expertise and Dan Fu for his spectroscopy expertise.

LIST OF ABBREVIATIONS

DOPC	dioleoyl-phosphatidylcholine
DPPC	dipalmitoyl-phosphatidylcholine
DPPE	dipalmitoyl-phosphatidylethanolamine
EDTA	ethylenediaminetetraacetic acid
NMR	nuclear magnetic resonance
PC	phosphatidylcholine
PTFE	polytetrafluoroethylene
TLC	thin layer chromatography

GLOSSARY OF SYMBOLS

k_B Boltzmann's constant
L_d liquid disordered phase
L_o liquid ordered phase
t time
T temperature
T_{mix} miscibility transition temperature

REFERENCES

Ayuyan, A.G. & Cohen, F.S. (2006) Lipid peroxides promote large rafts: Effects of excitation of probes in fluorescence microscopy and electrochemical reactions during vesicle formation. *Biophys. J.*, **91**, 2172–2183.

Bagatolli, L.A. & Gratton, E. (2000) A correlation between lipid domain shape and binary phospholipid mixture composition in free standing bilayers: A two-photon fluorescence microscopy study. *Biophys. J.*, **79**, 434–447.

Bagatolli, L.A. (2006) To see or not to see: Lateral organization of biological membranes and fluorescence microscopy. *BBA-Biomembranes*, **1758**, 1541–1556.

Baumgart, T., Hunt, G., Farkas, E.R., Webb, W.W. & Feigenson, G.W. (2007) Fluorescence probe partitioning between Lo/Ld phases in lipid membranes. *Biochim. Biophys. Acta*, **1768**, 2182–2194.

Bezlyepkina, N., Gracia, R.S., Shchelokovskyy, P., Lipowsky, R. & Dimova, R. (2013) Phase diagram and tie-line determination for the ternary mixture DOPC/eSM/cholesterol. *Biophys. J.*, **104**, 1456–1464.

Bleecker, J.V., Cox, P.A. & Keller, S.L. (2016b) Thickness differences of bilayers not simply related to thickness differences between Lo and Ld phases. *Biophys. J.*, **110**, 2305–2308.

Bleecker, J.V., Cox, P.A., Foster, R.N., Litz, J.P., Blosser, M.C., Castner, D.G. & Keller, S.L. (2016a) Thickness mismatch of coexisting liquid phases in noncanonical lipid bilayers. *J. Phys. Chem. B*, **120**, 2761–2770.

Blosser, M.C., Honerkamp-Smith, A.R., Han, T., Haataja, M. & Keller, S.L. (2015) Transbilayer colocalization of liquid domains explained via measurement of strong coupling parameters. *Biophys. J.*, **109**, 2317–2327.

Blosser, M.C., Starr, J.B., Turtle, C.W., Ashcraft, J. & Keller, S.L. (2013) Minimal effect of lipid charge on membrane miscibility phase behavior in three ternary systems. *Biophys. J.*, **104**, 2629–2638.

Camley, B.A., Esposito, C., Baumgart, T. & Brown, F.L.H. (2010) Lipid bilayer domain fluctuations as a probe of membrane viscosity. *Biophys. J.*, **99**, L44–L46.

Chiang, Y.-W., Zhao, J., Wu, J., Shimoyama, Y., Freed, J.H. & Feigenson, G.W. (2005) New method for determining tie-lines in coexisting membrane phases using spin-label ESR. *Biochim. Biophys. Acta*, **1668**, 99–105.

Collins, M.D. & Keller, S.L. (2008) Tuning lipid mixtures to induce or suppress domain formation across leaflets of unsupported asymmetric bilayers. *Proc. Natl. Acad. Sci. USA*, **105**, 124–128.

Connell, S.D., Heath, G., Olmsted, P.D. & Kisil, A. (2013) Critical point fluctuations in supported lipid membranes. *Faraday Discuss.*, **161**, 91–111.

Cornea, R.L. & Thomas, D.D. (1994) Effects of membrane thickness on the molecular dynamics and enzymatic activity of reconstituted Ca-ATPase. *Biochemistry*, **33**, 2912–2920.

Cornelius, F. (2001) Modulation of Na, K-ATPase and Na-ATPase activity by phospholipids and cholesterol. I. Steady-state kinetics. *Biochemistry*, **40**, 8842–8851.

Cornell, C.E., McCarthy, N.L.C., Levental, K.R., Levental, I., Brooks, N.J. & Keller, S.L. (2007) n-Alcohol length governs shift in Lo-Ld mixing temperatures in synthetic and cell-derived membranes. *Biophys. J.*, **113**, 1200–1211.

Cornell, C.E., Skinkle, A.D., He, S., Levental, I. Levental, K.R. & Keller, S.L. (2018) Tuning length scales of small domains in cell-derived membranes and synthetic model membranes. *Biophys. J.*, **115**, 690–701.

Das, S., Tian, A. & Baumgart, T. (2008) Mechanical stability of micropipet-aspirated giant vesicles with fluid phase coexistence. *J. Phys. Chem. B*, **112**, 11625–11630.

Davis, J.H., Clair, J.J. & Juhasz, J. (2009) Phase equilibria in DOPC/DPPC-d62/cholesterol mixtures. *Biophys. J.*, **96**, 521–539.

Davis, J.H., Ziani, L. & Schmidt, M.L. (2013) Critical fluctuations in DOPC/DPPC-d62/cholesterol mixtures: 2H magnetic resonance and relaxation. *J. Chem. Phys.*, **139**, 045104.

de Almeida, R.F.M., Loura, L.M.S., Fedorov, A. & Prieto, M. (2002) Nonequilibrium phenomena in the phase separation of a two-component lipid bilayer. *Biophys. J.*, **82**, 823–834.

Dietrich, C., Bagatolli, L.A., Volovyk, Z.N., Thompson, N.L., Levi, M., Jacobson, K. & Gratton, E. (2001) Lipid rafts reconstituted in model membranes. *Biophys. J.*, **80**, 1417–1428.

Esposito, C., Tian, A., Melamed, S., Johnson, C., Tee, S.-Y. & Baumgart, T. (2007) Flicker spectroscopy of thermal lipid bilayer domain boundary fluctuations. *Biophys. J.*, **93**, 3169–3181.

Farkas, E.R. & Webb, W.W. (2010) Multiphoton polarization imaging of steady-state molecular order in ternary lipid vesicles for the purpose of lipid phase assignment. *J. Phys. Chem. B*, **114**, 15512–15522.

Feigenson, G.W. & Buboltz, J.T. (2001) Ternary phase diagram of dipalmitoyl-PC/Dilauroyl-PC/cholesterol: Nanoscopic domain formation driven by cholesterol. *Biophys. J.*, **80**, 2755–2788.

Fidorra, M., Duelund, L., Leidy, C., Simonsen, A.C. & Bagatolli, L.A. (2006) Absence of fluid-ordered/fluid-disordered phase coexistence in ceramide/POPC mixtures containing cholesterol. *Biophys. J.*, **90**, 4437–4451.

Fidorra, M., Garcia, A., Ipsen, J.H., Härtel, S. & Bagatolli, L.A. (2009) Lipid domains in giant unilamellar vesicles and their correspondence with equilibrium thermodynamic phases: A quantitative fluorescence microscopy imaging approach. *BBA-Biomembranes*, **1788**, 2142–2149.

Furuya, K. & Mitsui, T. (1979) Phase transitions in bilayer membranes of dioleoyl-phosphatidylcholine/dipalmitoyl-phosphatidylcholine. *J. Phys. Soc. Jpn.*, **46**, 611–616.

Gallová, J., Uhríková, D., Kučerka, N., Teixeira, J. & Balgavy, P. (2010) Partial area of cholesterol in monounsaturated diacylphosphatidylcholine bilayers. *Chem. Phys. Lipids*, **163**, 765–770.

Garbes Putzel, G. & Schick, M. (2008a) Phase behavior of a model bilayer membrane with coupled leaves. *Biophys. J.*, **94**, 869–877.

Garbes Putzel, G. & Schick, M. (2008b) Phenomenological model and phase behavior of saturated and unsaturated lipids and cholesterol. *Biophys. J.*, **95**, 4756–4762.

Garbes Putzel, G. & Schick, M. (2009) Theory of raft formation by the cross-linking of saturated or unsaturated lipids in model lipid bilayers. *Biophys. J.*, **96**, 4935–4940.

Garbes Putzel, G., Uline, M.J., Szleifer, I. & Schick, M. (2011) Interleaflet coupling and domain registry in phase-separated lipid bilayers. *Biophys. J.*, **100**, 996–1004.

Goñi, F.M., Alonso, A., Bagatolli, L.A., Brown, R.E., Marsh, D., Prieto, M. & Thewalt, J.L. (2008) Phase diagrams of lipid mixtures relevant to the study of membrane rafts. *Biochim. Biophys. Acta*, **1781**, 665–684.

Gordon, V.D., Deserno, M., Andrew, C.M.J., Egelhaaf, S.U. & Poon, W.C.K. (2008) Adhesion promotes phase separation in mixed-lipid membranes. *EuroPhys. Lett.*, **84**, 48003.

Gray, E., Karlslake, J., Machta, B.B. & Veatch, S.L. (2013) Liquid general anesthetics lower critical temperatures in plasma membrane vesicles. *Biophys. J.*, **105**, 2751–2759.

Haataja, M. (2009) Critical dynamics in multicomponent lipid membranes. *Phys. Rev. E*, **80**, 020902.

Haluska, C.K., Schröder, A.P., Didier, P., Heissler, D., Duportail, G., Mély, Y. & Marques, C.M. (2008) Combining fluorescence lifetime and polarization microscopy to discriminate phase separated domains in giant unilamellar vesicles. *Biophys. J.*, **95**, 5737–5747.

Hammond, A.T., Heberle, F.A., Baumgart, T., Holowka, D., Baird, B. & Feigenson, G.W. (2007) Crosslinking a lipid raft component triggers liquid ordered-liquid disordered phase separation in model plasma membranes. *Proc. Natl. Acad. Sci. USA*, **102**, 6320–6325.

Harden, J.L., MacKintosh, F.C. & Olmsted, P.D. (2005) Phase coexistence in mixed fluid membranes. *Phys. Rev. E*, **72**, 011903.

Heberle, F.A., Petruzielo, R.S., Pan, J., Drazba, P., Kučerka, N., Standaert, R.F., Feigenson, G.W. & Katsaras, J. (2013) Bilayer thickness mismatch controls domain size in model membranes. *J. Am. Chem. Soc.*, **135**, 6853–6859.

Heberle, F.A., Wu, J., Goh, S.L., Petruzielo, R.S. & Feigenson, G.W. (2010) Comparison of three ternary lipid bilayer mixtures: FRET and ESR reveal nanodomains. *Biophys. J.*, **99**, 3309–3318.

Heftberger, P., Kollmitzer, B., Heberle, F.A., Pan, J., Rappolt, M., Amenistch, H., Kučerka, N., Katsaras, J. & Pabst, G. (2014) Global small-angle X-ray scattering data analysis for multilamellar vesicles: The evolution of the scattering density profile model. *J. Appl. Crystallogr.*, **47**, 173–180.

Heftberger, P., Kollmitzer, B., Rieder, A.A., Amenitsch, H. & Pabst, G. (2015) In situ determination of structure and fluctuations of coexisting fluid membrane domains. *Biophys. J.*, **108**, 854–862.

Hillert, M. (2008) *Phase Equilibria, Phase Diagrams and Phase Transformations*. Cambridge University Press, Cambridge, UK.

Honerkamp-Smith, A.R., Machta, B.B. & Keller, S.L. (2012) Experimental observations of dynamic critical phenomena in a lipid membrane. *Phys. Rev. Lett.*, **108**, 1–5.

Honerkamp-Smith, A.R., Veatch, S.L. & Keller, S.L. (2008) Line tensions, correlation lengths, and critical exponents in lipid membranes near critical points. *Biophys. J.*, **95**, 236–246.

Honerkamp-Smith, A.R., Veatch, S.L. & Keller, S.L. (2009) An introduction to critical points for biophysicists: Observations of compositional heterogeneity in lipid membranes. *Biochim. Biophys. Acta*, **1788**, 53–63.

Hu, J., Weikl, T.R. & Lipowsky, R. (2011) Vesicles with multiple membrane domains. *Soft Matter*, **7**, 6092.

Huang, J., Buboltz, J.T. & Feigenson, G.W. (1999) Maximum solubility of cholesterol in phosphatidylcholine and phosphatidylethanolamine bilayers. *BBA-Biomembranes*, **1417**, 89–100.

Hung, W.-C., Lee, M.-T., Chen, F.-Y. & Huang, H.W. (2007) The condensing effect of cholesterol in lipid bilayers. *Biophys. J.*, **92**, 3960–3967.

Hung, W.-C., Lee, M.-T., Chung, H., Sun, Y.-T., Chen, H., Charron, N.E. & Huang, H.W. (2016) Comparative study of the condensing effects of ergosterol and cholesterol. *Biophys. J.*, **110**, 2026–2033.

Husen, P., Arriaga, L.R. & Bagatolli, L.A. (2012) Morphometric image analysis of giant vesicles: A new tool for quantitative thermodynamics studies of phase separation in lipid membranes. *Biophys. J.*, **103**, 2304–2310.

Inaura, K. & Fujitani, Y. (2008) Concentration fluctuation in a two-component fluid membrane surrounded with three-dimensional fluids. *J. Phys. Soc. Jpn.*, **77**, 114603.

Ince, D.C., Hatton, L. & Graham-Cumming, J. (2012) The case for open computer programs. *Nat. Protoc.*, **482**, 485–488.

Ionova, I.V., Livshits, V.A. & Marsh, D. (2012) Phase diagram of ternary cholesterol/palmitoylsphingomyelin/palmitoyloleoyl-phosphatidylcholine mixtures: Spin-label EPR study of lipid-raft formation. *Biophys. J.*, **102**, 1856–1865.

Jørgensen, K. & Mouritsen, O.G. (1995) Phase separation dynamics and lateral organization of two-component lipid membranes. *Biophys. J.*, **95**, 942–954.

Juhasz, J., Sharom, F.J. & Davis, J.H. (2009) Quantitative characterization of coexisting phases in DOPC/DPPC/cholesterol mixtures: Comparing confocal fluorescence microscopy and deuterium nuclear magnetic resonance. *BBA-Biomembranes*, **1788**, 2541–2552.

Kahya, N., Brown, D. & Schwille, P. (2005) Raft partitioning and dynamic behavior of human placental alkaline phosphatase in giant unilamellar vesicles. *Biochemistry*, **44**, 7479–7489.

Kaiser, H.-J., Lingwood, D., Levental, I., Sampaio, J.L., Kalvodova, L., Rajendran, L. & Simons, K. (2009) Order of lipid phases in model and plasma membranes. *Proc. Natl. Acad. Sci. USA*, **106**, 16645–16650.

Keller, S.L., Bezrukov, S.M., Gruner, S.M., Tate, M.W., Vodyanoy, I. & Parsegian, V.A. (1993) Probability of alamethicin conductance states varies with nonlamellar tendency of bilayer phospholipids. *Biophys. J.*, **65**, 23–27.

Khadka, N.K., Ho, C.S. & Pan, J. (2015) Macroscopic and nanoscopic heterogeneous structures in a three-component lipid bilayer mixtures determined by atomic force microscopy. *Langmuir*, **31**, 12417–12425.

Kim, H.M., Choo, H.-J., Jung, S.-Y., Ko, Y.-G., Park, W.-H., Jeon, S.-J., Kim, C.H., Joo, T. & Cho, B.R. (2007) A two-photon fluorescent probe for lipid raft imaging: C-Laurdan. *ChemBioChem*, **8**, 553–559.

Klymchenko, A.S. & Kreder, R. (2014) Fluorescent probes for lipid rafts: From model membranes to living cells. *Chem. Biol.*, **21**, 97–113.

Korlach, J., Schwille, P., Webb, W.W. & Feigenson, G.W. (1999) Characterization of lipid bilayer phases by confocal microscopy and fluorescence correlation spectroscopy. *Proc. Natl. Acad. Sci. USA*, **96**, 8461–8466.

Kraft, M.L., Weber, P.K., Longo, M.L., Hutcheon, I.D. & Boxer, S.G. (2006) Phase separation of lipid membranes analyzed with high-resolution secondary ion mass spectroscopy. *Science*, **313**, 1948–1951.

Kučerka, N., Gallová, J., Uhríková, D., Balgavy, P., Bulacu, M., Marrink, S.-J. & Katsaras, J. (2009) Areas of monounsaturated diacylphosphatidylcholines. *Biophys. J.*, **97**, 1926–1932.

Kučerka, N., Nieh, M.-P. & Katsaras, J. (2011) Fluid phase lipid areas and bilayer thicknesses of commonly used phosphatidylcholines as a function of temperature. *BBA-Biomembranes*, **1808**, 2761–2771.

Kučerka, N., Pencer, J., Nieh, M.-P. & Katsaras, J. (2007) Influence of cholesterol on the bilayer properties of monounsaturated phosphatidylcholine unilamellar vesicles. *Eur. Phys. J. E*, **23**, 247–254.

Lee, C. & Bain, C.D. (2005) Raman spectra of planar supported lipid bilayers. *BBA-Biomembranes*, **1711**, 59–71.

Levental, I., Byfield, F.J., Chowdhury, P., Gai, F., Baumgart, T. & Janmey, P.A. (2009) Cholesterol-dependent phase separation in cell-derived membrane vesicles. *Biochem. J.*, **424**, 163–167.

Levental, I., Grzybek, M. & Simons, K. (2010) Greasing their way: Lipid modifications determine protein association with lipid rafts. *Biochemistry*, **49**, 6305–6316.

Levental, I., Grzybek, M. & Simons, K. (2011) Raft domains of variable properties and compositions in plasma membrane vesicles. *Proc. Natl. Acad. Sci. USA*, **108**, 11411–11416.

Li, L. & Cheng, J.-X. (2008) Label-free coherent anti-Stokes Raman scattering imaging of coexisting lipid domains in single bilayers. *J. Phys. Chem. B*, **112**, 1576–1579.

Li, L., Wang, H. & Cheng, J.-X. (2005) Quantitative coherent anti-Stokes Raman scattering imaging of lipid distribution in coexisting domains. *Biophys. J.*, **89**, 3480–3490.

Lin, Q. & London, E. (2013) Altering hydrophobic sequence lengths shows that hydrophobic mismatch controls affinity for ordered lipid domains (rafts) in the multitransmembrane strand protein perfringolysin O. *J. Biol. Chem.*, **288**, 1340–1352.

Lingwood, D., Ries, J., Schwille, P. & Simons, K. (2008) Plasma membranes are poised for activation of raft phase coalescence at physiological temperature. *Proc. Natl. Acad. Sci. USA*, **105**, 10005–10010.

Lipowsky, R., Rouhiparkouhi, T., Discher, D. & Weikl, T.R. (2013) Domain formation in cholesterol-phospholipid membranes exposed to adhesive surfaces or environments. *Soft Matter*, **9**, 8438–8453.

Litz, J.P., Thakkar, N., Portet, T. & Keller, S.L. (2016) Depletion with cyclodextrin reveals two populations of cholesterol in model lipid membranes. *Biophys. J.*, **110**, 635–645.

Liu, A.P. & Fletcher, D.A. (2006) Actin polymerization serves as a membrane domain switch in model lipid bilayers. *Biophys. J.*, **91**, 4064–4070.

Loftus, A.F., Noreng, S., Hsieh, V.L. & Parthasarathy, R. (2013) Robust measurement of membrane bending moduli using light sheet fluorescence imaging of vesicle fluctuations. *Langmuir*, **29**, 14588–14594.

Lozano, M.M., Liu, Z., Sunnick, E., Janshoff, A., Kumar, K. & Boxer, S.G. (2013) Colocalization of the ganglioside GM1 and cholesterol detected by secondary ion mass spectrometry. *J. Am. Chem. Soc.*, **135**, 5620–5630.

Machta, B.B., Gray, E., Nouri, M., McCarthy, N.L.C., Gray, E.M., Miller, A.L., Brooks, N.J. & Veatch, S.L. (2016) Conditions that stabilize membrane domains also antagonize n-alcohol anesthesia. *Biophys. J.*, **111**, 537–545.

Maleki, M. & Fried, E. (2013) Multidomain and ground state configurations of two-phase vesicles. *J. R. Soc. Interface*, **10**, 20130112.

Margineanu, A., Hotta, J.-I., Van der Auweraer, M., Ameloot, M., Stefan, A., Beljonne, D., Engelborghs, Y., Herrmann, A., Müllen, K., De Schryver, F.C. & Hofkens, J. (2007) Visualization of membrane rafts using a perylene monoimide derivative and fluorescence lifetime imaging. *Biophys. J.*, **93**, 2877–2891.

Marsh, D. (2009) Cholesterol-induced fluid membrane domains: A compendium of lipid-raft ternary phase diagrams. *Biochim. Biophys. Acta*, **1788**, 2114–2123.

May, S. (2009) Trans-monolayer coupling of fluid domains in lipid bilayers. *Soft Matter*, **5**, 3148–3156.

Mengistu, D.H., Bohinc, K. & May, S. (2010) A model for the electrostatic contribution to the pH-dependent nonideal mixing of a binary charged-zwitterionic lipid bilayer. *Biophys. Chem.*, **150**, 112–118.

Morales-Penningston, N.F., Wu, J., Farkas, E.R., Goh, S.L., Konyakhina, T.M., Zheng, J.Y., Webb, W.W. & Feigenson, G.W. (2010) GUV preparation and imaging: Minimizing artifacts. *Biochim. Biophys. Acta*, **1798**, 1324–1332.

Pan, J., Tristram-Nagle, S. & Nagle, J.F. (2009) Effect of cholesterol on structural and mechanical properties of membranes depends on lipid chain saturation. *Phys. Rev. E*, **80**, 021931.

Parthasarathy, R., Yu, C.-H. & Groves, J.T. (2006) Curvature-modulated phase separation in lipid bilayer membranes. *Langmuir*, **22**, 5095–5099.

Portet, T., Gordon, S.E. & Keller, S.L. (2012) Increasing membrane tension decreases miscibility temperatures; an experimental demonstration via micropipette aspiration. *Biophys. J.*, **103**, L35–L37.

Rayermann, S.P., Rayermann, G.E., Cornell, C.E., Merz, A.J. & Keller, S.L. (2017) Hallmarks of reversible separation of living, unperturbed cell membranes into two liquid phases. *Biophys. J.*, **113**, 2425–2432.

Roux, A., Cuvelier, D., Nassoy, P., Prost, J., Bassereau, P. & Goud, B. (2005) Role of curvature and phase transition in lipid sorting and fission of membrane tubules. *EMBO J.*, **24**, 1537–1545.

Rozovsky, S., Kaizuka, Y. & Groves, J.T. (2005) Formation and spatiotemporal evolution of periodic structures in lipid bilayers. *J. Amer. Chem. Soc.*, **127**, 36–37.

Sadowski, M.I., Grant, C. & Fell, T.S. (2015) Harnessing QbD, programming languages, and automation for reproducible biology. *Trends Biotech.*, **34**, 214–227.

Safouane, M., Berland, L., Callan-Jones, A., Sorre, B., Römer, W., Johannes, L., Toombes, G.E.S. & Bassereau, P. (2010) Lipid cosorting mediated by shiga toxin induced tubulation. *Traffic*, **11**, 1519–1529.

Samsonov, A.V., Mihalyov, I. & Cohen, F.S. (2001) Characterization of cholesterol-sphingomyelin domains and their dynamics in bilayer membranes. *Biophys. J.*, **81**, 1486–1500.

Seifert, U. (1993) Curvature-induced lateral phase segregation in two-component vesicles. *Phys. Rev. Lett.*, **70**, 1335–1338.

Semrau, S. & Schmidt, T. (2009) Membrane heterogeneity—From lipid domains to curvature effects. *Soft Matter*, **5**, 3129–3364.

Sezgin, E., Kaiser, H.-J., Baumgart, T., Schwille, P., Simons, K. & Levental, I. (2012a) Elucidating membrane structure and protein behavior using giant plasma membrane vesicles. *Nat. Protoc.*, **7**, 1042–1051.

Sezgin, E., Levental, I., Grzybek, M., Schwarzmann, G., Mueller, V., Honigmann, A., Belov, V.N. et al. (2012b) Partitioning, diffusion, and ligand binding of raft lipid analogs in model and cellular plasma membranes. *Biochim. Biophys. Acta*, **1818**, 1777–1784.

Sezgin, E., Sadowski, T. & Simons, K. (2014) Measuring lipid packing of model and cellular membranes with environment sensitive probes. *Langmuir*, **30**, 8160–8166.

Sezgin, E., Waithe, D., Bernadino de la Serna, J. & Eggeling, C. (2015) Spectral imaging to measure heterogeneity in membrane lipid packing. *ChemPhysChem*, **16**, 1387–1394.

Shimokawa, N., Hishida, M., Seto, H. & Yoshikawa, K. (2010) Phase separation of a mixture of charged and neutral lipids on a giant vesicle induced by small cations. *Chem. Phys. Lett.*, **496**, 59–63.

Silvius, J.R. (2003) Fluorescence energy transfer reveals microdomain formation at physiological temperatures in lipid mixtures modeling the outer leaflet of the plasma membrane. *Biophys. J.*, **85**, 1034–1045.

Silvius, J.R. (2005) Lipid microdomains in model and biological membranes: How strong are the connections? *Quarterly Rev. Biophys.*, **38**, 373–383.

Stanich, C.A., Honerkamp-Smith, A.R., Garbes Putzel, G., Warth, C.S., Lamprecht, A.K., Mandal, P., Mann, E., Hua, T.-A.D. & Keller, S.L. (2013) Coarsening dynamics of domains in lipid membranes. *Biophys. J.*, **105**, 444–454.

Stevens, M.M., Honerkamp-Smith, A.R. & Keller, S.L. (2010) Solubility limits of cholesterol, lanosterol, ergosterol, stigmasterol, and β-sitosterol in electroformed lipid vesicles. *Soft Matter*, **6**, 5882–5890.

Tian, A., Capraro, B.R., Esposito, C. & Baumgart, T. (2009) Bending stiffness depends on curvature of ternary lipid mixture tubular membranes. *Biophys. J.*, **97**, 1636–1646.

Tian, A., Johnson, C., Wang, W. & Baumgart, T. (2007) Line tension at fluid membrane domain boundaries measured by micropipette aspiration. *Phys. Rev. Lett.*, **98**, 208102.

Uline, M.J., Schick, M. & Szleifer, I. (2012) Phase behavior of lipid bilayers under tension. *Biophys. J.*, **102**, 517–522.

Uppamoochikkal, P., Tristram-Nagle, S. & Nagle, J.F. (2010) Orientation of tie-lines in the phase diagram of DOPC/DPPC/cholesterol model biomembranes. *Langmuir*, **26**, 17363–17368.

Ursell, T.S., Klug, W.S. & Phillips, R. (2009) Morphology and interaction between lipid domains. *Proc. Natl. Acad. Sci. USA*, **106**, 13301–13306.

Veatch, S.L. & Keller, S.L. (2002) Organization in lipid membranes containing cholesterol. *Phys. Rev. Lett.*, **89**, 1–4.

Veatch, S.L. & Keller, S.L. (2003) Separation of liquid phases in giant vesicles of ternary mixtures of phospholipids and cholesterol. *Biophys. J.*, **85**, 3074–3083.

Veatch, S.L. & Keller, S.L. (2005a) Miscibility phase diagrams of giant vesicles containing sphingomyelin. *Phys. Rev. Lett.*, **94**, 148101.

Veatch, S.L. & Keller, S.L. (2005b) Seeing spots: Complex phase behavior in simple membranes. *Biochim. Biophys. Acta*, **1746**, 172–185.

Veatch, S.L., Cicuta, P., Sengupta, P., Honerkamp-Smith, A.R., Holowka, D. & Baird, B. (2008) Critical fluctuations in plasma membrane vesicles. *ACS Chem. Biol.*, **3**, 287–293.

Veatch, S.L., Gawrisch, K. & Keller, S.L. (2006) Closed-loop miscibility gap and quantitative tie-lines in ternary membranes containing diphytanoyl PC. *Biophys. J.*, **90**, 4428–4436.

Veatch, S.L., Leung, S.S.W., Hancock, R.E.W. & Thewalt, J.L. (2007a) Fluorescent probes alter miscibility phase boundaries in ternary vesicles. *J. Phys. Chem. B*, **111**, 502–504.

Veatch, S.L., Polozov, I.V., Gawrisch, K. & Keller, S.L. (2004) Liquid domains in vesicles investigated by NMR and fluorescence microscopy. *Biophys. J.*, **86**, 2910–2922.

Veatch, S.L., Soubias, O., Keller, S.L. & Gawrisch, K. (2007b) Critical fluctuations in domain-forming lipid mixtures. *Proc. Natl. Acad. Sci. USA*, **104**, 17650–17655.

Vequi-Suplicy, C.C., Riske, K.A., Knorr, R.L. & Dimova, R. (2010) Vesicles with charged domains. *Biochim. Biophys. Acta*, **1798**, 1338–1347.

Williamson, J.J. & Olmsted, P.D. (2015) Kinetics of symmetry and asymmetry in a phase-separating bilayer membrane. *Phys. Rev. E*, **92**, 052721.

Yanagisawa, M., Imai, M., Masui, T., Komura, S. & Ohta, T. (2007) Growth dynamics of domains in ternary fluid vesicles. *Biophys. J.*, **92**, 115–125.

Yang, J. & Appleyard, J. (2000) The main phase transition of mica-supported phosphatidylcholine membranes. *J. Phys. Chem. B*, **104**, 8097–8100.

Yarrow, F., Vlugt, T.J.H., van der Eerden, J.P.J.M. & Snel, M.M.E. (2005) Melting of a DPPC lipid bilayer observed with atomic force microscopy and computer simulation. *J. Cryst. Growth*, **275**, e1417–e1421.

Yasuda, T., Tsuchikawa, H., Murata, M. & Matsumori, N. (2015) Deuterium NMR of raft model membranes reveals domain-specific order profiles and compositional distribution. *Biophys. J.*, **108**, 2502–2506.

Zhao, J., Wu, J. & Veatch, S.L. (2013) Adhesion stabilizes robust heterogeneity in supercritical membranes at physiological temperature. *Biophys. J.*, **104**, 825–834.

Zhao, J., Wu, J., Heberle, F.A., Mills, T.T., Klawitter, P., Huang, G., Costanza, G. & Feigenson, G.W. (2007a) Phase studies of model biomembranes: Complex behavior of DSPC/DOPC/cholesterol. *Biochim. Biophys. Acta*, **1768**, 2764–2776.

Zhao, J., Wu, J., Shao, H., Kong, F., Jain, N., Hunt, G. & Feigenson, G.W. (2007b) Phase studies of model biomembranes: Macroscopic coexistence of Lα+Lβ, with light-induced coexistence of Lα+Lo phases. *Biochim. Biophys. Acta*, **1768**, 2777–2786.

Zheng, L., McQuaw, C.M., Ewing, A.G. & Winograd, N. (2007) Sphingomyelin/phosphatidylcholine and cholesterol interactions studied by imaging mass spectrometry. *J. Am. Chem. Soc.*, **129**, 15730–15731.

19 Vesicle dynamics in flow: An experimental approach

Victor Steinberg and Michael Levant

Simple object with complex dynamics

Contents

19.1 INTRODUCTION

Understanding the rheology of complex fluids, including biofluids such as blood, remains a great challenge, and the progress here requires detailed studies of the dynamics of a single micro-object, or cell in the case of biofluids. Each cell is bounded by a membrane, which consists of a phospholipid bilayer. The bilayer membrane is responsible for various physiological cell functions and contributes to the cell material properties. It displays bending elasticity and resistance to stretching and compression: the material properties that control the shape of the cellular bilayer under fluid motion. Because the red blood cell (RBC) has a complex structure consisting of a lipid bilayer with incorporated proteins coupled to the spectrin network, simple theoretical models are considered such as capsules, objects enclosed by a shell with stretching elasticity and packed with a fluid, liquid droplets with uniform surface tension, and incompressible fluid membranes with bending rigidity surrounding a fluid droplet. The latter, a microscopic deformable object of arbitrary form is called a vesicle, which is used as a physical model to study the dynamical behavior of more complex biological cells and, in particular, of RBCs.

A giant unilamellar vesicle (GUV) is a droplet of a viscous fluid encapsulated by a lipid bilayer membrane and suspended into a fluid of either the same or different viscosity as the inner one. Thermodynamically the lipid membrane is very stable. Under external stress in flow, lipid molecules (Luisi and Walde, 2000) can freely move in the membrane plane and diffuse and rearrange as a response to the planar stress. A necessary condition here is that the membrane is in the liquid phase, above the main transition temperature (see Chapter 18). It should be emphasized that hydrodynamics (or rheology) of a vesicle suspension is significantly influenced by vesicle dynamics, which in turn is drastically altered by a molecularly thick liquid membrane. It means that a molecular-scale object (membrane) strongly affects large-scale hydrodynamic scales. This is a rather unusual case for a hydrodynamic coarse-grained description at the scale of micrometers.

There are two main geometrical constraints of a vesicle membrane due to its physical properties: conservation of the vesicle volume and membrane surface area. The former means that the vesicle membrane is impermeable, at least on the timescale of the experiment, whereas the latter means that the membrane stretching can be neglected because it is a two-dimensional (2D) incompressible fluid. Indeed, the stretching elasticity is two to three orders of magnitude larger than bending rigidity (for a theoretical description of vesicles, see Chapter 5). However, it can be shown that despite the two constraints, a vesicle's shape is not uniquely defined; shape deformations are permitted and occur under external perturbations. Moreover, the constraints have far-reaching consequences on vesicle dynamics and membrane surface tension, which is determined by the external stresses exerted on a vesicle and can reach any number, positive as well as negative. The membrane surface tension plays the same role as the pressure in incompressible fluids: It is adjusted to external stresses to ensure the local membrane incompressibility.

The goal of this chapter is to present a review of the experimental and theoretical developments occurred during the last 15 years in vesicle dynamics in various hydrodynamic flows, together with detailed descriptions of new techniques, methods and approaches in experiment and data analysis.

19.2 DYNAMICS OF A VESICLE IN A LINEAR FLOW

19.2.1 SHORT REVIEW OF THEORETICAL AND EXPERIMENTAL STUDIES OF VESICLE DYNAMICS IN A CHANNEL FLOW

A first phenomenological theory of vesicle dynamics was initiated by experimental observations of two types of motion, namely tank-treading (TT) and tumbling (TU), of the red blood cell (RBC) in Couette and Poiseuille flows (Schmid-Schönbein and Wells, 1969; Goldsmith and Marlow, 1972). In the TT regime, a vesicle subjected to a shear flow acquires a stationary mean inclination angle θ with a membrane TT. In the TU regime, the main vesicle axis rotates continuously with a membrane TT. The theory was based on two basic assumptions that drastically simplified the problem: (i) fixed ellipsoidal shape, and (ii) a simplified *ad hoc* velocity field on the membrane surface, which violates the local incompressibility constraint and used energy considerations, leading to the following equation of motion:

$$d\theta/dt = \mathrm{a} + \mathrm{b}\cos 2\theta \tag{19.1}$$

Here $\mathrm{a} = -\dot{\gamma}/2$ is the mean rotation velocity due to the shear flow vorticity $\omega = \dot{\gamma}/2$, and $\mathrm{b}(\lambda, \Delta)$ is proportional to the shear rate $\dot{\gamma}$. The expression for $\mathrm{b}(\lambda, \Delta)$ depends on the details of a specific model (see Keller and Skalak, 1982), and is a function of the viscosity contrast, $\lambda = \eta_{in}/\eta_{out}$, and of the total excess area, $\Delta \equiv A/R^2 - 4\pi$, which is directly related to the ratio of the ellipsoidal axis. Here η_{in} and η_{out} are the dynamic viscosities of the inner and outer fluids, respectively, A is the vesicle surface area, and R is the effective vesicle radius related to its volume via $V = (4/3)\pi R^3$. In spite of the model simplicity an analysis of Eq. (1), even without knowledge of the specific expression for b, shows the presence of two types of motion: (i) at $|\mathrm{a/b}|<1$, a steady TT regime takes place with θ given by $\cos(2\theta) = -\mathrm{a/b}$; and (ii) at $|\mathrm{a/b}|>1$ a time-dependent TU regime is found, whereas a/b is independent of $\dot{\gamma}$. The latter was verified by measuring θ at various $\dot{\gamma}$, though large scatter hindered a quantitative test of the $\theta(\Delta)$ prediction (Abkarian and Viallat, 2005). Further on, an extended phenomenological model that takes into account vesicle shape deformations, r, obtained from heuristic arguments was suggested by Noguchi and Gompper (2004, 2005).

A new analytical approach to the vesicle dynamics in a shear flow at $\lambda = 1$ was introduced by Seifert (1999), where thermal fluctuations were also taken into account. Later on it was extended by Misbah (2006) and Vlahovska and Gracia (2007) by taking into account r, which leads to two coupled nonlinear equations at arbitrary λ for θ and r within the same approximations as in the paper by Seifert (1999). The latter used an approximation of the second-order spherical harmonics for a quasi-spherical vesicle ($\Delta << 1$), neglecting higher-order couplings between deformations and flow field and expanding the geometrical quantities as well as the Helfrich free energy (Helfrich,1973) around a sphere up to the second order. Then for TT motion at $\sqrt{\Delta}/2h << 1$, where $h = 60\sqrt{\frac{2\pi}{15}}/(32+23\lambda)$, one obtains an expression for θ independent of $\chi \equiv \dot{\gamma}\,\eta_{out}R^3/\kappa$, where χ is the capillary number and κ is the bending elasticity

$$\theta \cong \frac{1}{2}\arctan\sqrt{-1+\frac{4h^2}{\Delta}} \approx \frac{\pi}{4} - \frac{23\lambda+32}{240}\sqrt{\frac{15\Delta}{2\pi}} \tag{19.2}$$

which provides at $\theta = 0$, the expression for the transition line $\lambda_c(\Delta)$ from TT to TU. The TU regime, in a general case of the vesicle deformations, significantly differs from that in the Keller-Skalak (KS) theory due to a nonlinear coupling between θ and r, although under a shape-preserving assumption, the dynamic TU equation is naturally the same as in the paper by Keller and Skalak (1982) with a specific expression for the coefficient. Furthermore, a new type of an oscillatory motion, coined as a vacillating-breathing (VB) mode, is found and characterized by vesicle oscillations around the flow direction, whereas the short and long vesicle axes show a breathing motion (Misbah, 2006). The distinctive features of the VB mode are its independence of χ and coexistence with the TU mode, where each of them is chosen by initial conditions.

A quantitative experimental test of the theoretical and numerical predictions (Keller and Skalak, 1982; Seifert, 1999; Kraus et al., 1996) became possible due to technological achievements that we now describe. A significant improvement of the spatial resolution of a vesicle contour is achieved by adding a fluorescent lipid to enhance image contrast and by using fluorescence microscopy. The development of vesicle production technology with different λ, the use of microfluidic technology for channel fabrication, and flow control combined with a quantitative characterization of the velocity field allow to significantly simplify experiments and improve experimental control. Further, collecting large sample statistics in the laboratory frame and capturing images of a vesicle for sufficiently long time in a frame moving with a vesicle velocity drastically reduced error bars in the characterization of vesicle geometry and the inclination angle θ. Finally, the development of a software package under MATLAB® for image processing and control of various stepping motors lead to the automatization of the experiment and to fast image processing. As the result, the average $\langle\theta\rangle$ and the rms fluctuations of the inclination angle $\delta\theta$ in TT, the temporal variations of θ and shape deformations in TU, and a newly discovered intermediate time-dependent regime, coined trembling (TR), were characterized (Kantsler and Steinberg, 2005, 2006) (see Box 19.1). To summarize, the following conclusions are made: (i) the dependence $\langle\theta(\Delta,\lambda)\rangle$ in TT in a wide range of Δ and λ and scatter in θ due to thermal noise are described quantitatively well by theory (Seifert, 1999; Misbah, 2006; Vlahovska and Gracia, 2007), although at sufficiently small θ and large Δ and λ, long tails in the $\theta(\Delta)$ dependencies are observed (Kantsler and Steinberg, 2006) in a sharp contrast with the theory (Box 19.1a,b); (ii) the TT-TU transition curve is fitted surprisingly well by the solution of the KS theory in spite of the long tails in $\theta(\Delta)$; (iii) strong shape deformations are observed in TU in a contradiction to the KS theory assumptions; (iv) the newly observed TR state (Kantsler and Steinberg, 2006) differs qualitatively from the VB dynamics (Misbah, 2006; Vlahovska and Gracia, 2007).

19.2.2 EXPERIMENTAL TECHNIQUES TO STUDY VESICLE DYNAMICS IN A MICRO-CHANNEL FLOW

Experimental setup of a micro-channel flow

Measurements of vesicle dynamics were conducted in a shear flow of the near-wall region at a mid-height of a micro-channel via epi-fluorescent microscopy (Figure 19.1). Straight micro-channels of $380 \times 250\ \mu m^2$ and $250 \times 150\ \mu m^2$ rectangular cross sections with a Poiseuille's velocity profile were fabricated

Box 19.1 Vesicle dynamics in a channel flow (see also Movies 19.1, 19.2 and 19.3)

Vesicles subjected to a shear flow can exhibit several types of motion. In an earlier observed TT motion, the vesicle vertical section is almost elliptical with the ellipse main axis at a mean inclination angle, $\langle\theta\rangle$ (e.g., fig. 1b [Kantsler and Steinberg, 2005]). The mean inclination angle, $\langle\theta\rangle$, as a function of the excess area Δ with no viscosity contrast ($\lambda = 1$) but with low (upper plot) and high (lower plot) viscosities demonstrates a strong scatter reduction at higher $\eta_{in} = \eta_{out} = 10^{-2}$ Pa.s (see fig. 1b). An importance of the second control parameter λ, besides Δ, is demonstrated in fig. 1c, where λ varies from 1 up to 9.1 and strongly affects $\langle\theta\rangle$. In a TU motion, the vesicle main axis rotates in its vertical plane (see second raw of snapshots in fig. 1e) with the normalized angular velocity $\omega_n = \dot{\gamma}^{-1}d\theta/dt$, which shows strong deviations from a constant value (upper plot) accompanied by strong shape deformations (lower plot) (see fig. 1d). A first raw of snapshots in fig. 1e illustrates an intermediate motion (i.e., TR), and a lower plot shows that ω_n is periodic in θ for TU in contrast to TR, where the range of θ is limited. In Box 19.1 figures (a) and (b) reprinted from Kantsler and Steinberg (2005); (d) and (e) are reprinted from Kantsler and Steinberg (2006); and (c) is reprinted from Kantsler (2007) by a permission of Dr. V. Kantsler.

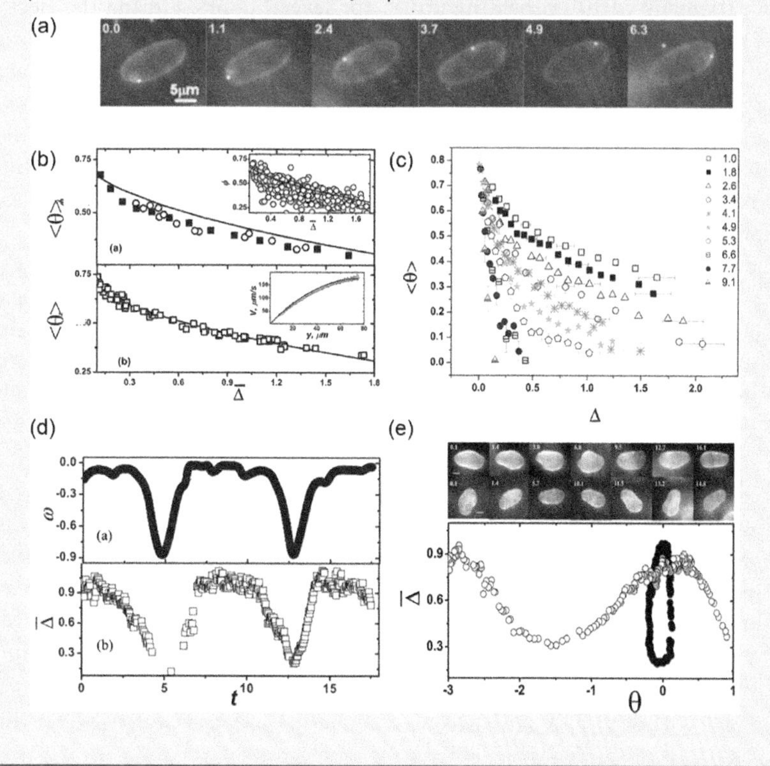

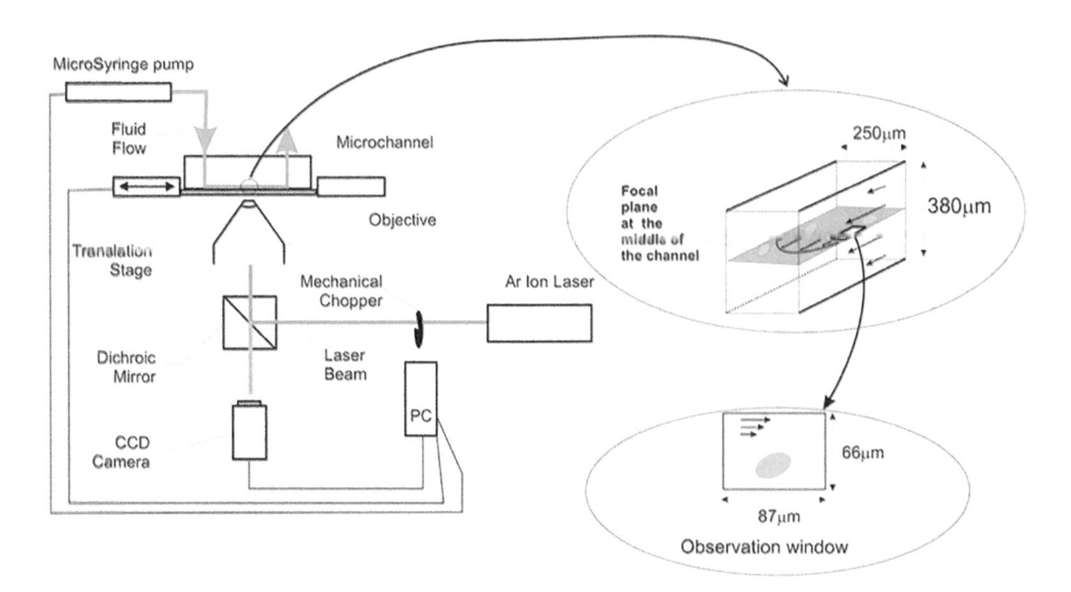

Figure 19.1 Experimental setup of a micro-channel flow. (Reprinted from Kantsler, V. et al., *Europhys. Lett.*, 82, 58005, 2008b.)

in an elastomer poly(dimethylsiloxane) (PDMS) by a soft lithography (Xia and Whitesides, 1998) and used for the experiments in a shear flow (Kantsler and Steinberg, 2005, 2006). An observation area of 87 × 66 μm² was captured by Mintron MTV-12V1® interlaced charge-coupled device (CCD) camera at a rate of 25 fps and digitized via Ellips Rio® frame grabber. A mechanical chopper was synchronized with the CCD camera and installed in the Ar-ion laser beam path to reduce exposure time down to ~1 ms, thus reducing photo-bleaching and allowing for the capture of a vesicle image with velocity up to hundreds of micrometers per second without noticeable image smearing. The micro-channel could be also moved in the flow direction by computer-controlled motorized translation stage to trace a vesicle in its reference frame up to 40 s providing up to 1,000 measurements with a single vesicle. A dilute suspension of deflated vesicles was driven through a micro-channel by a computer-controlled micro-syringe pump at a constant flow rate. The horizontal velocity profile in a mid-plane was measured by particle tracking velocimetry (PTV) technique with 1-μm fluorescent beads at a distance from the wall between 5 and 45 μm, where the deviation of the shear rate from the constant was less than ±5% on the vesicle size. The cross-stream vesicle drift resulting from the lift force in the near wall region was at least three orders of magnitude smaller than a horizontal velocity, so the overall drift in a vesicle position during the entire experiment was less than 3 μm.

Vesicle preparation

Vesicles of 10–100 μm diameter with either the same or different fluids outside and inside vesicles are prepared by electroformation (Angelova et al., 1992) (see also Chapter 1). Most of the experiments described in the present chapter used a solution consisting either of pure dioleoyl-phosphatidylcholine (DOPC, see Appendix 1 of the book for structure and data on this lipid; Sigma) dissolved in 9:1 v/v chloroform/methanol solvent (1.5 mg total lipids/mL solvent) or of 85% DOPC and 15% fluorescent phosphatidylcholine (NBD-PC®, Molecular Probes, see Appendix 2 of the book for structure and data on this probe) dissolved in the same solvent (1.8 mg/mL). The NBD-PC fluorescent lipid is added to enhance the image contrast and so to improve the spatial resolution of the vesicle contour. About 10 μL of a lipid solution is uniformly spread on an indium tin oxide (ITO)–coated glass electrode (Sigma-Aldrich; 75 × 25 × 1.1 mm, surface resistivity 70–100 Ω/sq). After solvent evaporation, the electrode is covered by a second ITO electrode separated by a 1-mm polytetrafluoroethylene (Teflon®) spacer, and the gap is filled with the aqueous solution. Sinusoidal AC voltage 3 $V_{p\text{-}p}$ peak-to-peak at 10Hz is applied to the electrodes, which are placed on an orbital shaker for several hours. During the electroformation, the lipids form bilayer sheets, which then detach from each other as a result of the shaking and swell under the action of an applied voltage. Finally, the swollen bilayers form vesicles containing the same fluid as the external solution (the viscosity contrast $\lambda = 1$). The yield of GUVs prepared by this procedure is 0.1%–1% v/v, with a wide size and shape distribution, but mostly they are larger than 5 μm.

For higher viscosity experiments sucrose-in-water solutions with the addition of high-molecular-weight dextran (molecular weight 5–20 × 10⁵) are used to reach high inner fluid viscosities up to about 20 mPa·s and more. Then viscosity measurements of sucrose–water and dextran solutions are conducted in a rheometer at stabilized temperature of T > 23°C to ensure fluidity of the membrane. The vesicle production decreases with increasing solvent viscosity, and various "exotic" structures such as tubes and vesicle-in-vesicle structures are more likely to appear in viscous solutions (Levant and Steinberg, 2014). To control the volume-to-surface ratio by inflation and deflation, the vesicles are introduced into a slightly hypertonic aqueous (glucose or sucrose) solution prior to an experiment. Vesicles in sucrose solution are perfectly density matched. The density difference between sucrose and glucose solutions at 200 mM is $\Delta\rho/\rho = 0.012$, which implies a small buoyancy force.

To achieve the viscosity contrasts, $\lambda > 1$, one uses a gentle centrifugation to wash out the outer fluid of higher viscosity and to replace it with a glucose–water solution (5.5% w/w) of the same osmolarity and slightly less density. Sedimentation is performed at an acceleration of 15–40*g* for 30–60 min, causing vesicles with $R \geq 5$ μm to form a sediment. Finally, an outer fluid consisting of 5.5% glucose–water solution with an appropriate amount of dextran is added to get the desired viscosity contrast and to compensate gravity. To adjust the excess area Δ to a desired value to produce a flaccid vesicle, one applies an osmotic deflation by using solution of higher sugar concentration.

Quantitative characterization of vesicle dynamics: Large statistics and long observation time

As one learned from a comparison of theoretical predictions on the vesicle TT and TU dynamical regimes (Keller and Skalak, 1982; Kraus et al., 1996; Seifert, 1999; Beaucourt et al., 2004; Misbah,2006; Vlahovska and Gracia, 2007) with experimental data collected from limited number of vesicles (Abkarian et al., 2002; Abkarian and Viallat, 2005; Mader et al., 2006), one needs to drastically improve the accuracy and spatial resolution of the detection of the vesicle contour and of its angle θ. The main progress here is achieved by collecting large sample statistics realized in two ways: by tracking an ensemble of more than 500 vesicles in the laboratory frame at $\lambda < \lambda_c$ and in a frame moving with a velocity of a single captured vesicle for sufficiently long time (up to 40 s) at the grabbing rate of 25 fps provided up to 1,000 measurements on a single vesicle in the TU motion at $\lambda > \lambda_c$. Such approach allows getting the average inclination angle $<\theta>$ and its dependence on Δ with high accuracy (Box 19.1). This was not achievable in previous experiments because they were carried out on only a few vesicles, and the resulting scatter in the data caused by thermal noise hinder a quantitative comparison with theory.

Image analysis of vesicle shape

Due to the relatively fast flow in a micro-channel, the reconstruction of an experimental 3D vesicle shape is extremely difficult, in particular if one wants to collect large statistics. On the contrary, a sample of the 2D cross section provides the principal axes of the approximated elliptical shape $2L$ and $2B$, and thus also, the Taylor deformation parameter $D = (L - B)/(L + B)$ as well as R and Δ (Seifert, 1999). Detection of a vesicle membrane with a high precision is based on its high fluorescent intensity, and a vesicle contour is determined by the maximum intensity of the corresponding grayscale profile along the radial direction. Thus, the maximum accuracy reached by this approach is less than 0.1 μm (Kantsler et al., 2007; Deschamps et al., 2009; Levant and Steinberg, 2012; Levant et al., 2014; Levant and Steinberg, 2014). The subsequent analysis included elliptical approximation of the vesicle contour and determination of $2L$ and $2B$, D, $\Delta = 32\pi D^2/15$, the ellipse center, and θ (see Pecreaux et al., 2004; Kantsler and Steinberg, 2005, 2006). Determination of 3D vesicle surface area and radius is done via the Knud-Thomsen approximate formula: $A \approx 4\pi\{[(L)^p(B)^p + Z^p(B)^p + (L)^pZ^p]/3\}^{1/p}$ and $R = (LBZ)^{1/3}$, where $Z = (L + B)/2$ and $p = 1.607$.

Phase diagram of vesicle dynamical regimes and the control parameters

After the discovery of the TR state, the central questions being addressed theoretically, numerically, and experimentally are the following. (i) Can the transition between different dynamical regimes of a vesicle be induced by varying the shear rate as pointed out in the experiment (Kantsler and Steinberg, 2006)? (ii) What is the structure of the phase diagram of the three vesicle dynamical regimes? (iii) What is the role of thermal noise?

Contrary to the KS phenomenological model (Keller and Skalak, 1982), where the TT-TU transition line is determined by only two controlled parameters λ and Δ, both the modified phenomenological (Noguchi and Gompper, 2007) and analytical (Danker and Misbah, 2007a; Vlavovska and Gracia, 2007; Kaoui et al., 2009; Vlahovska et al., 2009; Farutin et al., 2010; Biben et al., 2011; Farutin and Misbah, 2012) models use three control parameters: λ, Δ, and χ, thus also considering vesicle shape deformations. This implies that the corresponding phase diagram of the vesicle dynamical regimes in a shear flow should be presented in the 3D parameter space of λ, Δ, and χ. Thus, for each Δ three separate regions of the TT, TU and TR (VB) regimes exist and their locations depend on χ and λ values, being shifted relatively to each other for each Δ.

On the other hand, an analytical theory for the phase diagram of the vesicle dynamical regimes in a plane linear flow was developed by Lebedev et al. (2007, 2008). This is a natural extension of the theory made for quasi-spherical vesicles (Seifert, 1999; Olla, 2000; Misbah, 2006; Vlahovska and Gracia, 2007) that takes into account a third-order expansion term in the Helfrich free energy. The main prediction of the theory is the phase diagram in a space of only two control parameters, S and Λ, as the result of self-similar solution of two equations for θ and ψ, the ratio of the ellipsoid main axes of the vesicle shape (Lebedev et al., 2007, 2008). The expressions for the control parameters are the following:

$$S = \frac{14\pi}{3\sqrt{3}}\frac{s\eta r_0^3}{\kappa\Delta},\quad \Lambda = \frac{4(1+23\lambda/32)}{\sqrt{30\pi}}\frac{\omega}{s}\sqrt{\Delta},\qquad (19.3)$$

where the strain rate is $s = \sqrt{tr(s_{ik})}$, s_{ik} is the symmetric strain tensor, and ω_j is the vorticity vector defined via the velocity gradient as $\partial_i V_k = s_{ik} + \varepsilon_{ikj}\omega_j$, and ω is the vorticity. A shear flow is a special case of a general (or linear) flow with $\omega = s = \dot{\gamma}/2$.

The main goal of an experimental test is first of all to verify whether two control parameters indeed are sufficient to characterize the phase diagram of three vesicles dynamical states. As argued by the authors (Deschamps et al., 2009a, 2009b), getting reliable results for a quantitative comparison with theory requires new experimental techniques and the improvement of the existing ones. The previous approach obtaining values of λ, Δ, R and uniformity of the shear rate results in an error in the values of S and Λ up to 100% even with rather high spatial resolution in vesicle visualization. Such errors would hinder a quantitative test of theory. These technical improvements are described below in Subsection 19.2.3.

The first result of the experiment on the phase diagram of the vesicle dynamical states in the (S,Λ) coordinates in a shear flow is realized in a planar Couette flow geometry and presented in

Box 19.2 (Deschamps et al., 2009a) together with the theoretical results of (Lebedev et al., 2007, 2008). The main conclusion is that the experimental data show qualitatively the same topological structure of the phase diagram in the (S,Λ) plane, as predicted by Lebedev et al. (2007, 2008), with the data on vesicles of different Δ distributed rather randomly across the plane. It proves the independence of the location of the transition lines as well as the width of the TR region on Δ in a wide range of S, Λ and Δ. Undoubtedly it is a surprising result due to the fact that the theory is developed in the $\Delta << 1$ approximation, with vesicle shape perturbations limited by the second-order spherical harmonics, and by neglecting thermal noise, whereas in the experiment vesicles are found with $\Delta \in [0.2;2.2]$, with strong shape deformations that contained even and odd modes, and in the presence of the thermal fluctuations.

A new approach to study not only vesicle dynamics, but also a wide variety of micro-objects—including biologically related ones—in a linear flow is based on the continuous scanning the entire phase diagram for a single vesicle at the fixed value of λ by varying parameters ω/s and s (Deschamps et al., 2009b). The latter causes the changing S and Λ. To explore the whole (*S*, Λ) plane, vesicles with various values of *R* and Δ are loaded one by one, individually characterized, and observed in the TT, TR, and TU regimes, which distinct locations identify the phase diagram transition lines in the (*S*, Λ) plane for a wide range of Δ (see Box 19.2b,d). Thus the phase diagram determined in such a way is found in a qualitative

Box 19.2 Dynamical states of vesicles in plane Couette shear flow and linear flow in a microfluidic four-roll mill (see also Movie 19.4)

In shear as well as linear flows, vesicles display TT, TU, and TR motions. Examples are presented in fig. 2a as snapshots for a vesicle with radius $R = 6.19$ μm and excess area $\Delta = 0.82$ in a linear flow: upper, TT: $\omega/s = 1.7$, $s = 0.22\ s^{-1}$, $S = 6.8$, $\Lambda = 1.09$; middle, TU: $\omega/s = 5.3$, $s = 0.07\ s^{-1}$, $S = 2.17$, $\Lambda = 3.4$; lower, TR: $\omega/s = 3.1$, $s = 0.12\ s^{-1}$, $S = 3.7$, $\Lambda = 1.99$. Similar dynamics in a linear flow for a single vesicle with $R = 14.4$ μm and $\Delta = 0.64$ are shown as θ and D temporal dependencies in three dynamical regimes due to variations of ω/s and s (fig. 2b). Figure 2c presents phase diagrams of the vesicle dynamical states in a shear flow: symbols green, TU; red, TR; blue, TT. Filled squares, $\Delta \in [0\text{–}0.55]$; open squares, $\Delta \in [0.55\text{–}0.8]$; filled circles, $\Delta \in [0.8\text{–}1.05]$; open circles, $\Delta \in [1.05\text{–}1.25]$; filled triangles, $\Delta \in [1.25\text{–}2]$; and fig. 2d shows the phase diagram in a linear flow: symbols green, TU; red, TR; blue, TT. Filled squares, $\Delta \in [0\text{–}0.8]$; open squares, $\Delta \in [0.8\text{–}1.3]$; filled circles, $\Delta \in [1.3\text{–}1.8]$; open circles, $\Delta \in [1.8\text{–}2.6]$. Gray bands are guide for the eye. Dotted, dashed, and solid black lines are the theoretical boundaries between TT, TU, and TR, respectively (Lebedev et al., 2007, 2008). The black arrows indicate the experimental path. In Box 19.2 figures (a), (b), and (d) are reprinted from Deschamps J., Kantsler V., Segre E., and Steinberg V. (2009b); and figure (c) is reprinted from Deschamps J., Kantsler V., and Steinberg V. (2009a).

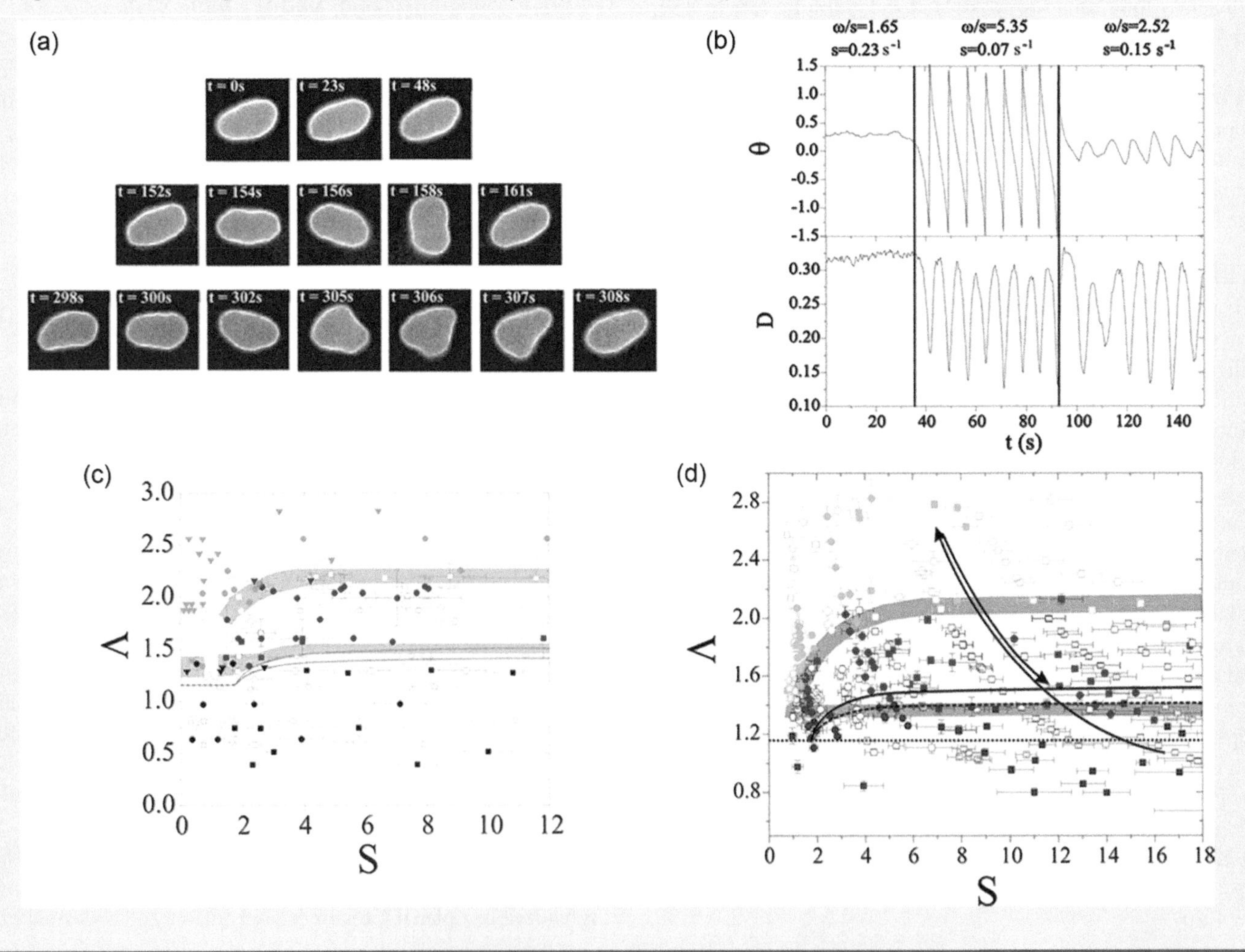

agreement with theory (Lebedev et al., 2007, 2008). However, the puzzle remains: what is the reason for the prevailing self-similar solution up to $\Delta \sim O(1)$ and in the presence of thermal noise?

19.2.3 EXPERIMENTAL TECHNIQUES AND DATA ANALYSIS TO STUDY A PHASE DIAGRAM OF VESICLE DYNAMICAL STATES IN PLANE COUETTE AND LINEAR FLOWS

Experimental setup of a plane Couette flow geometry

A plane Couette flow apparatus mounted on top of an inverted epi-fluorescent microscope includes two parallel microscope coverslips separated by a Teflon spacer to maintain a constant gap of d = 111 μm, and sliding on a coverslip (photo in Figure 19.2 and schematics in Figure 19.3a,b). One coverslip is driven by the computer-controlled close-loop high resolution linear DC-Mike® actuator (M-230.10, PI, Germany) coupled via a pulley to the second coverslip to bring it into a relative motion. PTV measurements show that the deviation of the measured $\dot{\gamma}$ from the expected $\dot{\gamma}_e = 2V/d$ is with an error below 4% (Figure 19.3d) that is

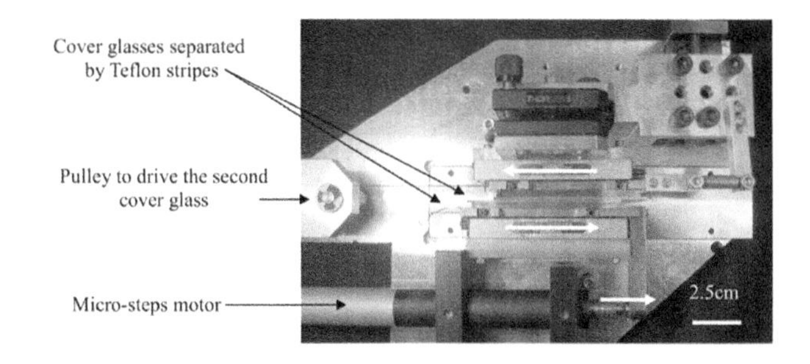

Figure 19.2 Photo of a plane Couette flow apparatus (view from above). (Reprinted from Kantsler, V., Hydrodynamics of Fluid Vesicles, PhD thesis, Weizmann Institute of Science, Rehovot, 2007b. With permission.)

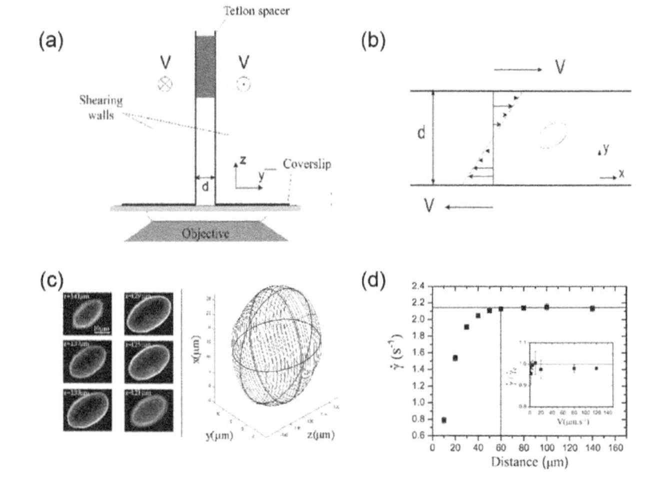

Figure 19.3 Schematics of the plane Couette flow device: (a) Side view. (b) View from below. (c) Reconstruction of 3D vesicle ellipsoidal shape: left two columns, slices captured at different z and each one is elliptically approximated; on the right, 3D ellipsoid approximation of a vesicle via stacking elliptical slices gives $R = 14.56 \pm 0.03$ μm, $\Delta = 0.49 \pm 0.04$. (d) Calibration plot of a plane Couette flow for $V = 120$ μm/s corresponding to $\dot{\gamma} = 2.138 \pm 0.01\,s^{-1}$ versus the expected value $\dot{\gamma}_e = 2.1521/s$. Inset: $\dot{\gamma}/\dot{\gamma}_e$ versus V. (Reprinted from Deschamps, J. et al., *Phys. Rev. Lett.*, 102, 118105, 2009a.)

significantly below the error in $\dot{\gamma}$ found in the near-wall shear flow. Observation of a vesicle is carried out above 60 μm, where $\dot{\gamma}$ is saturated (Figure 19.3d).

Improvements in measurements of experimental parameters

As pointed out above, the existing techniques to measure vesicle geometrical and physical parameters result in significant errors in S and Λ that would hinder a quantitative test of theory. To reduce an error in λ, the sedimentation velocity, V_{sed}, of each vesicle *in situ* in the Couette flow device is measured prior to experiments. It was related to the difference between the inner ρ_{in} and outer ρ_{out} densities of the fluids. By considering a vesicle as a hard ellipsoid of effective radius R, which is a function of three ellipsoid axes (Happel and Brenner, 1983) moving under gravity in a viscous flow one has $\rho_{in} - \rho_{out} = 9/2\,(\eta_{out} V_{sed}/gR^2)$. Then by using a relation $\lambda = f(\rho_{in} - \rho_{out})$, one obtains the expression for the viscosity contrast: $\lambda = F(V_{sed})$. Applying this expression for each vesicle, one finds the error below 8% for the expected λ in the range from 1 to 7.5.

Another new aspect of the improvement is 3D *in situ* reconstruction of a shape of each vesicle that provides precise values of its R and Δ. To conduct the measurement of a vesicle during its either TT or TU at low $\dot{\gamma}$ the Zeiss Acroplan® 63X and NA1.2 objective with submicrometer focal depth is scanned up or down by a step of 1 μm. The scan is performed sufficiently fast (<2 s) compared to vesicle dynamics to avoid any large fluctuations in a size and a position of the vesicle (Figure 19.3c). Ellipses arisen from a fit of each slice are stacked according to their heights with correction of the position of each slice on V_{sed} and the displacement in the (x, y) plane during the motion (Figure 19.3c). R and Δ were obtained from the fit of a vesicle by 3D ellipsoid. Thus, R varied from 4.5 up to 14 μm is measured with an error below 3.5%, whereas $\Delta \in [0.2, 2.2]$ with an error is less than 16% (Deschamps et al., 2009a).

Experimental setup of four-roll mill microfluidic device

In a linear flow, by changing the parameter ω/s from 0 to ∞, one can continuously pass from elongation to shear and to rotational flows. This scenario can be realized in a microfluidic analog of the four-roll mill (Hudson et al., 2004; Lee et al., 2007; Deschamps et al., 2009b) first suggested by Taylor (1934). The "heart" of this device is a dynamical trap—a cylindrical structure with ~400 μm in diameter and ~300 μm in height, where the object(s) of interest can be trapped far from boundaries for a long time observation compared to the characteristic vesicle timescales, where the flow is described by a planar linear flow velocity field. The flow inside the trap is driven by a hydrostatic pressure P_0, and ω is roughly set by the initial P_0 and the trap geometry, whereas s is varied continuously by a variation of the pressure drop across the channel ΔP in a typical range of $\Delta P/P_0 \sim 10\%$. The pressures P_0 and ΔP are not measured directly but via the fluid height h_0 and Δh (Figure 19.4a). The outlets of the four-roll mill are connected to a common pressure reservoir P_{out}. The device is fabricated in PDMS that is cast from a mold via soft lithography (Xia and Whitesides, 1998) (Figure 19.4b,c). The inlet resistance is an important parameter, which affects the sensitivity of the flow

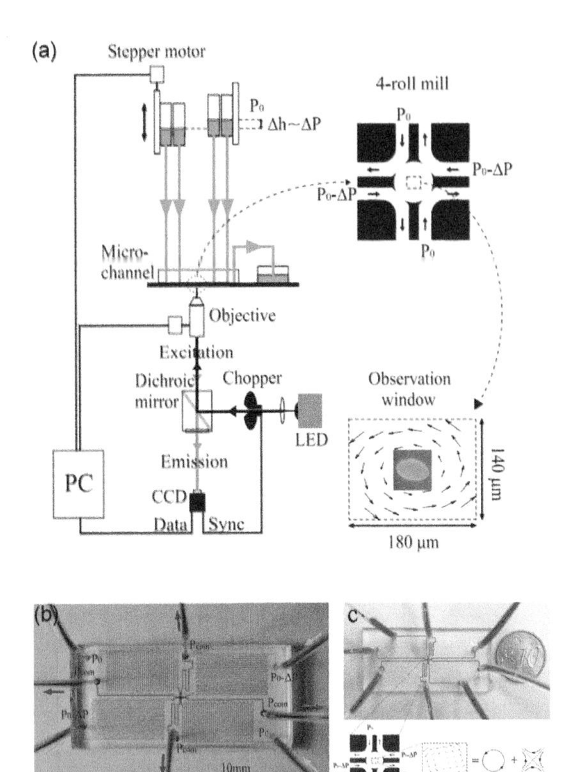

Figure 19.4 (a) Experimental setup with a four-roll mill microfluidic device. (Reprinted from Levant, M, and Steinberg, V., *Phys. Rev. Lett.* 109, 268103, 2012b.) (b) High-resistance four-roll mill micro-device for low viscosity fluids. (From Levant, M., Hydrodynamics of vesicles and red blood cells. PhD thesis, Weizmann Institute of Science, Rehovot, 2015.) (c) Low resistance four-roll mill micro-device for high viscosity fluids. (Reprinted from Levant, M and Steinberg, V., *Phys. Rev. Lett.* 112, 138106, 2014b.)

to small perturbations and so dynamical range of the flow rates in the trap. For example, the flow control in the trap by the Δh variation for a low viscosity fluid becomes unstable, if the flow resistance of the device is not adjusted properly according to fluid viscosity (Figure 19.4b and c).

The flow inside the four-roll mill is pre-calibrated to determine the dependence of ω and s on h_0 and Δh by using PTV as presented in Figure 19.5. The linearity of the velocity field and parabolic velocity profile in the vertical plane are also verified separately and the rms deviations from the simulations do not exceed 5% (see Levant, 2015). Thus, due to significant reduction of errors in the linearity of the velocity field and so in ω and s, the mean errors in S and Λ can be brought down to 9% and 3%, respectively.

Image processing

The main steps of image processing are presented in Figure 19.6. A set of morphological functions to fill an image and to clean it from noise and boundary effects are used in MATLAB® (**imfill, imareaopen, imborderclear** [MATLAB, 2003]). Then a Fourier analysis of a vesicle shape perturbations is made. The radial position of the membrane at each polar angle φ at a given time, $r(\varphi,t)$ is determined in the frame of reference of the vesicle via a maximum of intensity variations along r (Kantsler, 2007; Deschamps et al., 2009b; Levant and Steinberg, 2012). Another effective and more accurate approach of a ridge directed edge detection is suggested recently and used for colloid particles but is also applicable for vesicles (Afik, 2015). For each image up to 500 discrete positions are sampled along the contour. To quantitatively analyze vesicle dynamics and shape deformations, the dimensionless shape deformations $u(\varphi,t) = r(\varphi,t)/r_0(t)-1$ at $0 \leq \varphi \leq 2\pi$ and

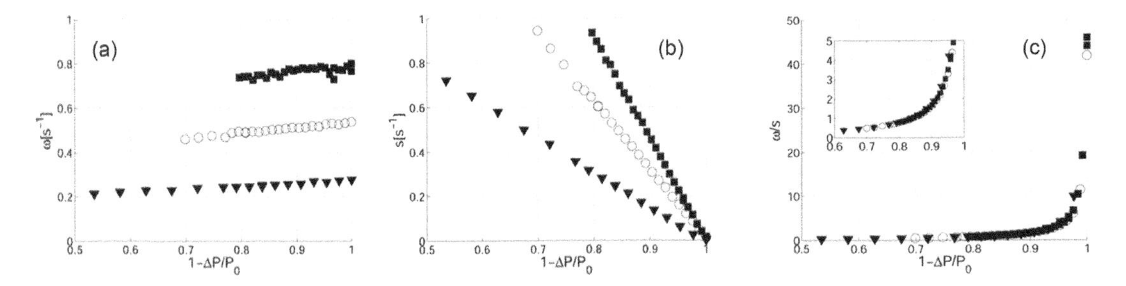

Figure 19.5 Flow calibrations for high-resistance channels with water $\eta = 1$ mPa.s. Typical dependence of (a) ω; (b) s; and (c) ω/s on 1-$\Delta P/P_0$ for three measured values P_0~h_0 (squares, circles, and triangles). The slopes of the linear dependence of s and ω versus 1-$\Delta P/P_0$ as a function of P_0 are not shown, whereas the ω/s function of 1-$\Delta P/P_0$ does not depend on P_0. (Reprinted from Levant, M., Hydrodynamics of vesicles and red blood cells. PhD thesis, Weizmann Institute of Science, Rehovot, 2015.)

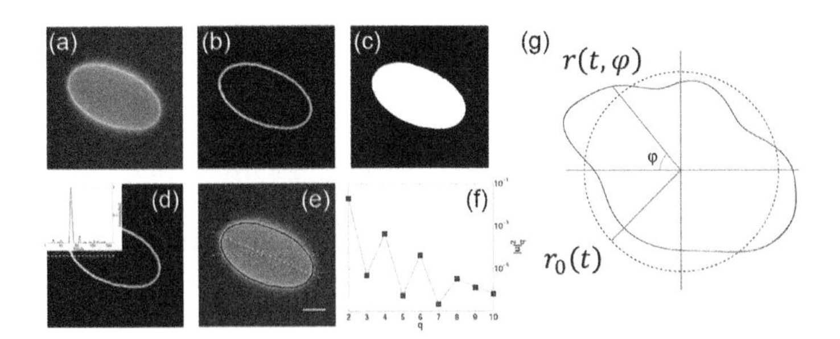

Figure 19.6 Main steps in image processing: (a) an original image; (b) the image after spectral filtering (band pass filter used is from the PTV package [From Crocker, J, C and Grier, D, G., *J. Colloid Interface Sci.* 179, 298, 1996]); (c) binary image; (d) detection of the image edge via the intensity radial profile maximum along line $\varphi = \pi$ as shown in inset (green dashed line); (e) the resulting contour overlapped on the original image; (f) Fourier decomposition of the contour at (e) of vesicle in TT with odd modes strongly suppressed. (Reprinted from Levant, M., Hydrodynamics of vesicles and red blood cells. PhD thesis, Weizmann Institute of Science, Rehovot, 2015.)

$r_0(t) = < r(\varphi,t)>_\varphi$ is Fourier decomposed via fast Fourier transform: $u(\varphi,t) = \sum^n_{q=1} u_q(t)e^{-iq\varphi}$. The inclination angle θ is calculated from the autocorrelation function of the time sequence of the angle of the second mode u_2.

19.3 VESICLE DYNAMICS IN AN ELONGATION FLOW

19.3.1 STATIONARY ELONGATION FLOW, STRETCHING TRANSITION AND PEARLING INSTABILITY

Biological membranes often develop tubular structures and form highly dynamic tubular networks, which play a key role in many biological processes (Davis and Sowinski, 2008; Sowinski et al., 2008). Membrane nanotubes are transient long-distance connections between cells that can facilitate intercellular communication but they can also contribute to pathologies by directing the spread of viruses (Davis and Sowinski, 2008; Sowinski et al., 2008).

An experiment on a floppy tubular shape vesicle with an initial length-to-diameter ratio $L_0/2R_0 > 4.2$ conducted in a hyperbolic flow carried out in a cross-slot microfluidic channel flow (see Section 19.3.3) exhibits a transition to a dumbbell shape at the critical extension rate $\dot{\varepsilon}_c$ and at $\dot{\varepsilon} \geq \dot{\varepsilon}_c$ a further transition into a transient pearling state is found (Kantsler et al., 2008a). The vesicle stretching for $L_0/2R_0 < 4.2$ is not observed due to a divergence of $\dot{\varepsilon}_c$ at small $L_0/2R_0$. Close to the transition at $\dot{\varepsilon} \leq \dot{\varepsilon}_c$ only a slight steady vesicle stretching is observed, whereas at $\dot{\varepsilon} \geq \dot{\varepsilon}_c$ a tether becomes unstable, and a sequence of conformation transitions to transient pearling configurations with different number of beads due to a different order of involved unstable modes is detected, as seen in Box 19.3.

The tubule-to-dumbbell transition is a continuous one accompanied by a critical slowing down of the dynamics and enhanced fluctuations close to $\dot{\varepsilon}_c$, similar to a coil–stretch transition of flexible polymers (Gerashchenko and Steinberg, 2008). Conformational fluctuations of a vesicle membrane define its configuration entropy and therewith an entropy-driven tension, analogously to a linear flexible polymer, whose elasticity is also entropy driven. Then a balance between the Stokes drag on a tubule $\sim\eta\dot{\epsilon}L_0^2/2/ln(L_0/2R_0)$ and the entropic force $\sim\kappa/2R_0$ provides the scaling relation between the critical extension rate and the tubule aspect ratio as $\dot{\epsilon}_c\tau \sim ln(L_0/2R_0)/(L_0/2R_0)$, where $\tau = R_0^2L_0\eta/\kappa$ agrees well with the tubule relaxation (retraction) time found experimentally (Kantsler et al., 2008a). Moreover, the data and the scaling relation also agree well with the numerical simulations (Narsimhan et al., 2014).

Another experiment on an intermediate-aspect-ratio vesicle shows qualitatively different dynamics in an elongation flow, when $\dot{\varepsilon}$ exceeds the critical value (see Spjut, 2010; Narsimhan et al., 2014). A breaking of mirror symmetry with a consequent transition into an unsteady, asymmetric dumbbell separated by a long tether is found (Box 19.3c). As illustrated in Box 19.3c, recent simulations get the same phenomenon and show that at $L_0/2R_0$ below the experimentally found value

Box 19.3 Vesicle dynamics in stationary elongation flow: Stretching, dumbbell and pearling instabilities (see also Movies 19.5 and 19.6)
Tubular vesicles exhibit a number of responses when exposed to elongation flow as shown in (a): stretching, dumbbell mode instability, second mode instability-pearling, multiple mode instability. Figure (a) is reprinted from Kantsler V., Segre E., Steinberg V. (2008a) by a permission of Cambridge University Press. Numerical simulations present the shapes of a tubular vesicle under pearling (b) in a rather close agreement with (a) (other shapes not shown). Figure (b) is reprinted from Narsimhan V, Spann A, Shaqfeh ESG. (2015) by a permission of Cambridge University Press. Prolate vesicles can exhibit stretching and budding in elongation flow during instability shown in upper row of (c) (Spjut, 2010). Numerical simulations correctly reproduce the shape. Figure (c) is reprinted from Zhao H. and Shaqfeh ESG. (2013) by a permission of Cambridge University Press.

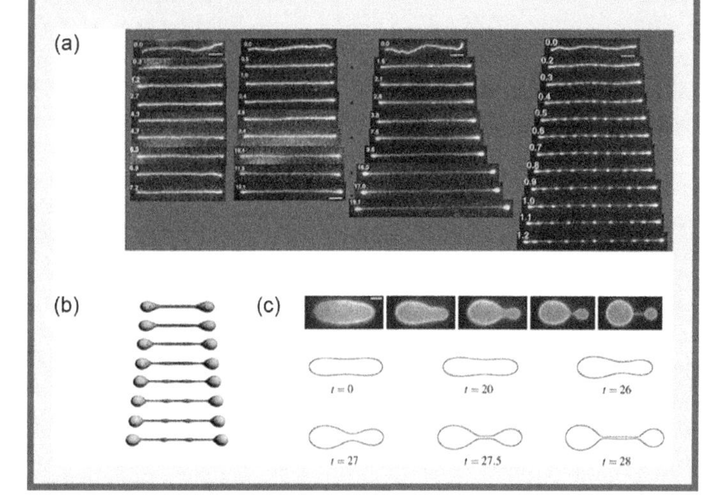

~4.2, the growth rate of the first odd (asymmetric) mode becomes positive, and the curvature energy is the mostly responsible for the destabilization, similar to the Rayleigh-Plateau phenomenon responsible for the capillary breakup of a viscous thread (Zhao and Shaqfeh, 2013a; Narsimhan et al., 2014, 2015). At $L_0/2R_0 > 4.2$, the numerical simulations show a symmetric tubule extension with a consequent transition to pearling dynamics in a good agreement with the experiment (Narsimhan et al., 2014, 2015).

19.3.2 TIME-DEPENDENT ELONGATION FLOW AND WRINKLING INSTABILITY

If the direction of the elongation flow is suddenly reversed, the new transient phenomenon of vesicle wrinkling characterized by the appearance of higher-order modes is observed (Kantsler et al., 2007). In a stationary flow the higher-order excitation modes are suppressed due to their high energetic contribution. The latter is seen from the second-order expansion of the Helfrich free energy functional for a membrane in a shape perturbation $u(x, y)$ presented in the Fourier space (Helfrich, 1973)

$$F^{(2)} = \frac{1}{2}\sum_{k}(\kappa k^4 + \sigma k^2)|u_k|^2 \qquad (19.4)$$

where σ is the Lagrange multiplier defined by the surface area conservation constraint in a contrast to the surface tension, which is the physical constant. So at $\sigma > 0$, higher k-modes are energetically less favorable and unrealizable, whereas at $\sigma < 0$, the modes with $k < \sqrt{|\sigma|/\kappa}$ become unstable and energetically favorable. A possible realization of $\sigma < 0$ occurs for a vesicle under compression due to the direction-reversed elongation flow (Box 19.4a). The experiment is conducted in the same cross-slot channel at $\dot{\varepsilon} = 0.05$–$10\ \mathrm{s}^{-1}$ that corresponds to the wider range of the dimensional strain rate $\chi = \dot{\varepsilon}\eta_{\mathrm{out}}R^3/\kappa$ between 2 and 340. Then the initially stretched vesicle is temporary compressed after a fast flow reversal on the timescale shorter than $\chi/\dot{\varepsilon}$ (Figure 19.7a). It is found that in strong flows at $\chi > \chi_c$ a vesicle developed high-order membrane deformation modes called wrinkles, as seen on the snapshots in Box 19.4b, where the data from Kantsler et al. (2007) are shown together with numerical simulations from Liu and Li (2014). The wrinkling instability is closely related to the Euler buckling instability due to compression (Landau and Lifschitz, 1987). Wrinkling of thin films is frequently observed in nature and in everyday life as a result of either external stretching of extensible materials or compression of inextensible films (Walter et al., 2001; Cerda et al., 2002; Cerda and Mahadevan, 2003; Cerda et al., 2004; Finken and Seifert, 2006; Huang et al., 2007).

Box 19.4 Vesicle wrinkling in elongation flow (see also Movie 19.7)

Flow reversal in an elongation flow leads to a vesicle compression that can cause vesicle wrinkling (a). (Reprinted from Levant M, Abreu D, Seifert U, Steinberg V. [2014].) The wrinkling instability exhibited by the experimental snapshots in a bottom row of (b) (Kantsler et al., 2007) is reproduced in 2D numerical simulations presented in a top row of (b). (Reprinted from Liu K. and Li S. (2014) by a permission of John Wiley & Sons.)

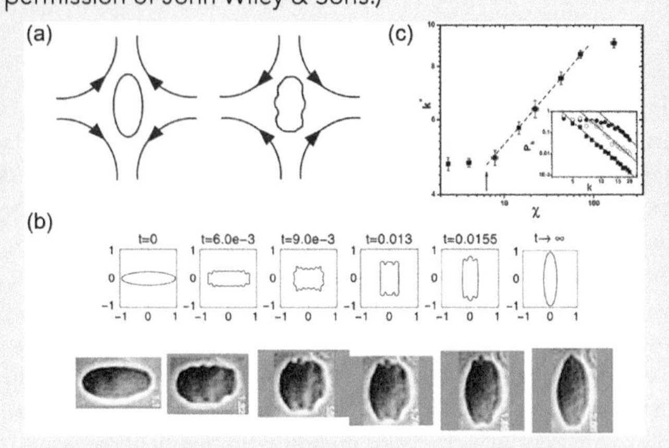

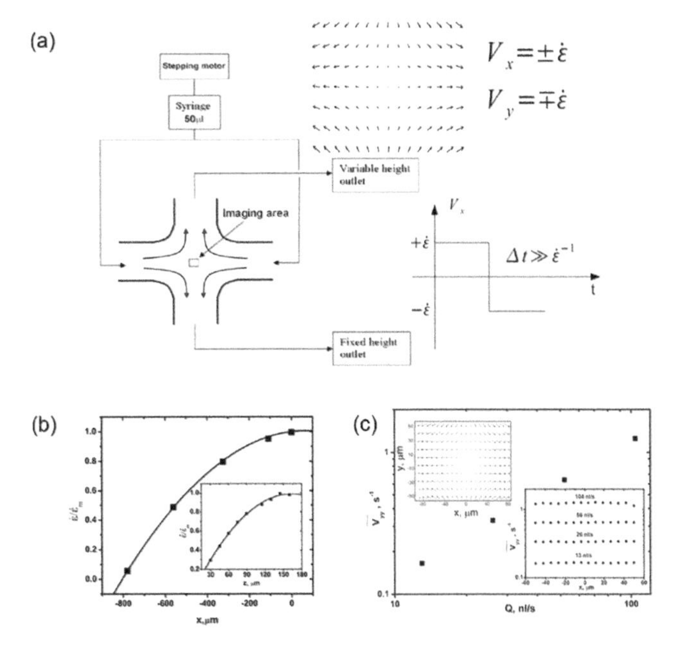

Figure 19.7 (a) Schematics of setup for stationary and transient hyperbolic flows. (b) Velocity gradient measurements as a function of distance from a stagnation point (x, y) = 0. Inset: Velocity gradient measurements at the stagnation point as a function of vertical distance z from the coverslip. Solid lines are the fits with polynomial functions of the second order. (c) Velocity gradient measurements in the field of view as a function of flow rate Q. Lower inset: Velocity gradient measurements in the field of view as a function of x-coordinate for different flow rates. Upper inset: Averaged velocity vector field. (Reprinted from Kantsler, V., Hydrodynamics of fluid vesicles, PhD thesis, Weizmann Institute of Science, Rehovot, 2007b. With permission.)

High-order modes due to the wrinkling instability are characterized via time-averaged power spectra of the vesicle shape perturbations, $|u_k|^2$ for different χ, as shown in the inset in Box 19.4c. At $\chi < \chi_c$, a power spectrum of thermal noise with $|u_k|^2 \sim k^{-4}$ is found (Faucon et al., 1989), whereas at $\chi > \chi_c$ a sharp transition from the flat at smaller k to the k^{-4} spectrum is observed. Thus, χ_c is defined as the instability onset, above which the modes with $k \geq 3$ are excited. More precise determination of $\chi_c = 6.5 \pm 0.8$ and scaling dependence $k^* \sim \chi^{1/4}$ at $\chi \geq \chi_c$ based on the dependence of spectrally averaged wave number $k^* = \sqrt{\sum_{k=3}^{19} k^2 |u_k|^2 / \sum_{k=3}^{19} |u_k|^2}$, where $k \leq 19$ limited by the spatial resolution of a vesicle contour, are found in a fair agreement with numerical simulations (Turitsyn and Vergeles, 2008).

19.3.3 EXPERIMENTAL TECHNIQUES

Technology of preparation of tubular vesicles

Nanotubes (tethers or tubular vesicles) first observed in biology as the result of self-assembly are produced by a variety of techniques such as various tether-pulling using optical tweezer (Dai and Sheetz, 1999), mechanical micropipette (Evans, 1994), and electrical microelectrode (Karlsson et al., 2001) manipulations, and by a very effective extrusion through a micro-sized aperture (Dittrich et al., 2006). However, here we review only a hydrodynamic mechanism of stretching and conformal transformations of free tubular vesicles. A tether stretching by a hydrodynamic (electro-osmosis induced) flow is reported by Rossier et al. (2003) and fabrication of tubules with a high aspect ratio by microfluidic tweezing is described by West et al. (2008). In Kantsler et al. (2008a), vesicles being prepared by a standard electroformation technique are subjected to a strong elongation flow. In this way, a sufficient amount of tubular shape vesicles in a wide range of large Δ are obtained.

A hyperbolic flow setup and velocity field

Measurements of the vesicle dynamics by (Kantsler et al. (2007, 2008a) are conducted in the vicinity of the stagnation point at the middle height of a cross-slot micro-channel of 500 μm wide and 320 μm high for the elongation rate $\dot{\varepsilon} = 0.01$–$3\ s^{-1}$ for stationary and $\dot{\varepsilon} = 0.05$–$10\ s^{-1}$ for transient hyperbolic flow (Figure 19.7a) via epi-fluorescent microscopy. In the case of a transient flow, a switching time from $V_x, V_y = (+\dot{\varepsilon},-\dot{\varepsilon})$ to $V_x, V_y = (-\dot{\varepsilon},+\dot{\varepsilon})$ is much shorter than $\chi/\dot{\varepsilon}$ s (Figure 19.7a) and as a result a vesicle undergoes a relaxation from one stretched state at $D = D_{sat}$ and $\varphi = 0$ to another at $D = D_{sat}$ and $\varphi = \pi/2$. PTV measurements of the velocity gradients reveal the deviations of $(\Delta\dot{\varepsilon})_{x,y,z}/\dot{\varepsilon}$ across the observation window and z-direction less than 5% (Figure 19.7a). The same errors are found for the reduced shear rate (Figure 19.7c) (Kantsler, 2007). To hold a vesicle at the stagnation point during the experiment, ΔP is controlled manually within [0.1–200] Pa on the outlets.

19.4 ROLE OF THERMAL FLUCTUATIONS IN VESICLE DYNAMICS

19.4.1 SHORT REVIEW OF THE EXPERIMENTAL AND THEORETICAL DEVELOPMENTS

The role of thermal noise in the vesicle dynamics in a flow depends on a dynamical regime. In equilibrium, thermal noise plays a minor role in particular at $\Delta < 1$ due to smallness of the ratio $k_BT/\kappa \approx 0.05$ (Faucon et al., 1989). However, even in the TT motion regime, thermal fluctuations cause considerable scattering in the θ measurements at $\chi \leq 1$ and low η, as discussed in Section 2.1 (Kantsler and Steinberg, 2005). The experimental data on the dependence of rms fluctuations of θ on the normalized shear rate (Kantsler and Steinberg, 2005) and the independence on λ are found in a good quantitative agreement with theory (Seifert, 1999).

In contrast, the role of thermal noise in the TR and TU dynamics, in spite of its smallness, is decisive. This paradox raises the following questions: Is it possible to clarify the role of thermal noise and to single out the characteristic features that lead to its enhancement in the TR dynamics? How one can analyze and quantify noise amplification in TR compared with the TT motion?

The first clear effect of thermal fluctuations is observed in numerical simulations of the KS model with thermal noise (Noguchi and Gompper, 2004, 2005a) in the dependence of θ on Δ and λ in the vicinity of the TT-to-TU transition. The smoothing of the TT-to-TR transition is further observed experimentally for intermediate values of λ and presented in Box 19.1c (Kantsler and Steinberg, 2006). However, the most striking manifestation of the role of thermal noise is observed in a significant contribution of higher-order odd modes into the TR regime (Deschamps et al., 2009a, 2009b; Zabusky et al., 2011; Levant et al., 2014) (see Box 19.5a) that makes it very different from numerical simulations without noise of the VB mode (Biben et al., 2011) (Box 19.5a) and similar to TR (Messlinger et al., 2009; Abreu and Seifert, 2013; Abreu et al., 2014a) with thermal noise (Box 19.5a).

Box 19.5 Role of thermal noise in vesicle dynamics

The most prominent manifestation of role of thermal noise in vesicle dynamics was observed in TR. Both theory and 3D numerical simulations without thermal noise shown in a middle row of (a) found the intermediate motion as pure oscillations of a vesicle of a symmetrical shape consisting only even modes (VB). (Reprinted from Biben T, Farutin A, Misbah C. [2011].) Adding thermal noise in 2D numerical simulations presented in a bottom row of (a) results in asymmetric vesicle shapes with a broken symmetry of even modes. (Reprinted from Messlinger S, Schmidt B, Noguchi H, and Gompper G. [2009].) The latter is much closer to vesicle snapshots obtained experimentally and shown in a top row of (a). (Reprinted from Levant M. and Steinberg V. [2012].) The strong vesicle shape deformation occurred in TR due to thermal noise amplification by a virtual wrinkling instability (see on the right of (a) a schematic reproduced from Levant M, Abreu D, Seifert U, Steinberg V. [2014]).

A stochastic nature of the vesicle shape perturbations and their strong amplifications is illustrated in shape spectra of two subsequent cycles in TR of a vesicle with $R = 15$ μm, $\Delta = 0.66$ at $S = 60.7$ and $\Lambda = 1.8$ shown on their left in (b), in probability distribution functions (PDFs) of fluctuations $\delta|u_2|$, $\delta\theta$, where $\delta|u_q| = |u_q| - \langle|u_q|\rangle$ and $\delta\theta = \theta - \langle\theta\rangle$, at two extreme values of (S, Λ) out of four presented in the insets: (70.3,0.84) and (60.7,1.8) in (c), and in trajectories in the (D,θ) plane (upper) and the autocorrelation functions of θ(t) (lower) in TT and TR at (S,Λ) = (51.2,1.18) and (53.6,1.56), respectively, in (d). (Figure (b), (c), (d) is reprinted from Levant M. and Steinberg V. [2012].)

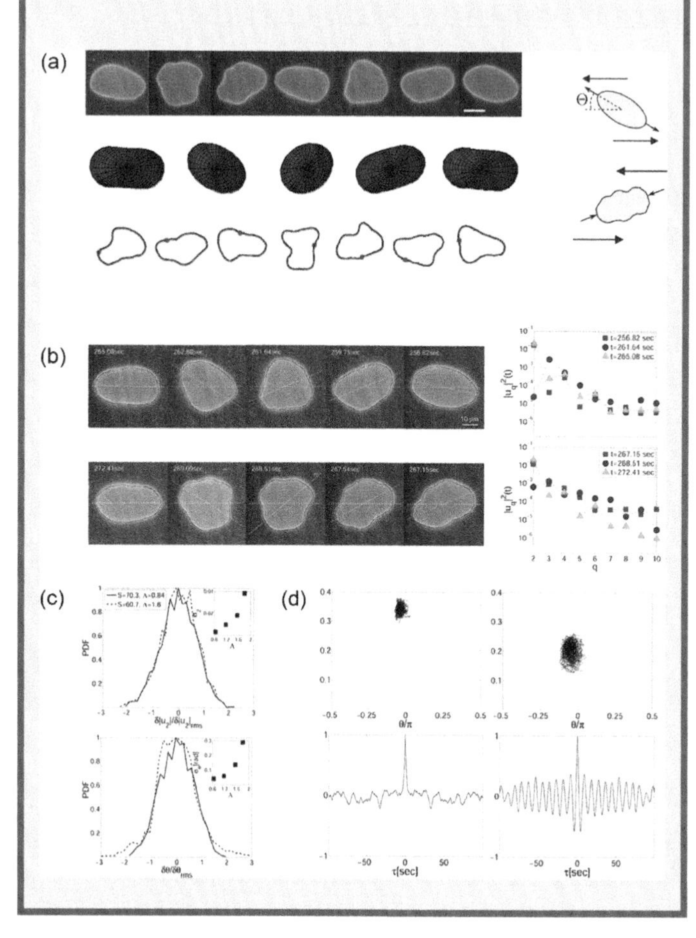

A novel noise amplification mechanism was identified (Levant and Steinberg, 2012) in experiments conducted in the dynamical trap with a high spatial resolution in vesicle imaging and with long time series compared to the characteristic period of vesicle dynamics, a condition necessary to collect sufficient statistics (up to 8,250 images for a run). In the compression part of the first cycle at $\theta < 0$ (Box 19.5a), the triangular vesicle shape becomes predominant, and the third-order mode exceeds by two orders of magnitude the second-order $|u_2|^2$ one, which varies up to three orders of magnitude during the cycle (Box 19.5b). In the subsequent cycle (Box 19.5b), the compression results in predominantly concave perturbations of short wavelengths that further leads to the amplification of high-order modes up to 10th order. This symmetry breaking of the even modes dictated by the $r \to -r$ flow symmetry reveals the most fundamental discrepancy with the theoretical models (Levant and Steinberg, 2012). A comparison of PDFs of $\delta|u_2|/\delta|u_2|_{rms}$ and $\delta\theta/\delta\theta_{rms}$ (Box 19.5c) for several (S, Λ) indicates a continuous increase of the PDF variances σ_2 and σ_θ with Λ up to 20 times in the TR region compared to thermal noise. The latter occurs in the vicinity of the wrinkling instability during the compression at $\theta < 0$ in each TR cycle, similar to the wrinkling instability in transient hyperbolic flow, though a crucial difference exists in an arbitrary vesicle shape on each cycle in TR. A presentation of the trajectories of the vesicle dynamical states is another way to quantify them (see Box 19.5d, top plots). However, in the case of the strong noise amplification, the identification of TT in the vicinity of TT-TR transition and TR above is hindered by the noise that smeared the fixed point in the former and the limit cycle in the latter. Then the corresponding correlation functions of $\theta(t)$ $C(\tau) = <\theta(t+\tau)\,\theta(t)>/\sigma^2_\theta$ demonstrate a clear difference between them (see Box 19.5d, bottom plots).

19.4.2 EXPERIMENTAL AND NUMERICAL STOCHASTIC APPROACHES

Further experiments and numerical simulations based on the stochastic model lead to the observation of a wrinkling instability that occurs at very high strain stresses in TR (Levant et al., 2014). The quantitative agreement between experiment and theory turns out to be quite remarkable (Figure 19.8). The key ingredient in the success of the experiment is a high spatial resolution in vesicle visualization and longtime observation of the TR vesicle dynamics provided by the dynamical trap. As illustrated in Figure 19.8a, the time evolution of the second and fourth modes together with even k^*_{even} and odd k^*_{odd} parts of the spectral averaged wave number $k^* = \sqrt{\sum_{k=3}^{16} k^2 |u_k|^2 / \sum_k |u_k|^2}$ of the vesicle shape deformations (similar to the analysis done in Section 19.3.2) is presented. One finds that k^*_{even} shows quasi-regular dynamics with clearly visible peaks anti-correlated with $|u_4|$ with a period of about half that of $|u_2|$, whereas the dynamics of k^*_{odd}s is rather erratic (Figure 19.8a). This observation is in a good accord with the simulations (Figure 19.8b) and is explained by the fact that the symmetric even modes are associated with the dynamics, whereas the odd modes are associated with the thermal noise amplification mechanism.

A quantitative characterization of the increasing contribution of the higher-order modes in the vesicle shape deformations for larger S follows from PDFs of k^* and time averaged power spectra of the vesicle shape at different S. Using PDFs of k^*, the normalized probability of k^* to be above a certain number of odd modes n as a function of S serves as a parameter to determine the wrinkling transition (Figure 5 in (Levant et al., 2014)). Thus, the novel stochastic approach in both modeling and analysis of the experimental data reveals the onset of the wrinkling instability and its characterization; however, a complete quantitative agreement with experiment requires full scale numerical simulations.

19.5 HYDRODYNAMIC INTERACTION OF VESICLES AND DYNAMICS OF VESICLES UNDER CONFINEMENT

19.5.1 FROM ONE TO MANY VESICLES: HYDRODYNAMIC INTERACTIONS IN VESICLE SUSPENSIONS

As the concentration of vesicles in a suspension increases, a hydrodynamic interaction between vesicles becomes significant. A direct evidence of the effect of a single vesicle

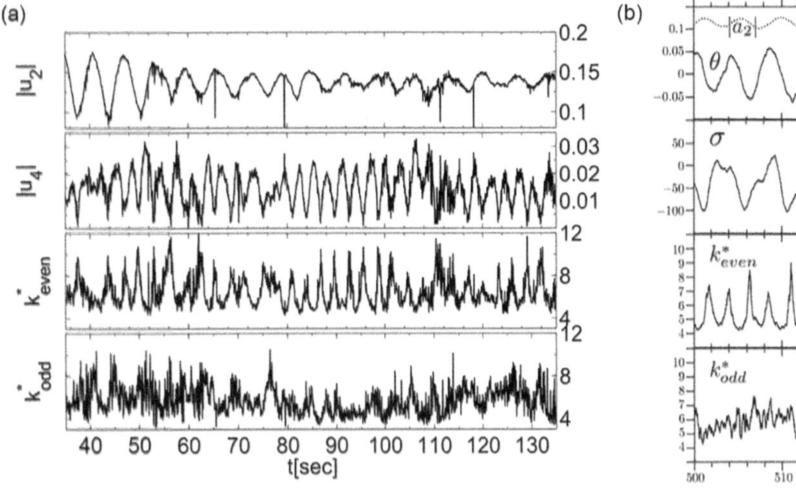

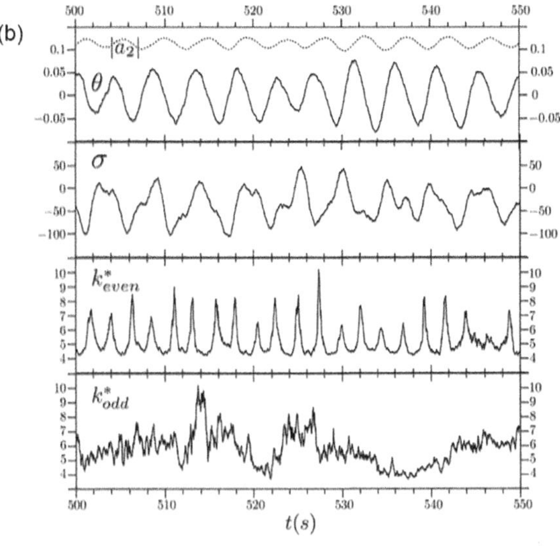

Figure 19.8 (a) Experimental data on TR dynamics at S = 1147, Λ = 1.78, Δ = 1.27. (b) Simulations of TR dynamics at the same S,Λ,Δ. (Reprinted from Levant, M. et al., *EPL*, 107, 28001, 2014a.)

undergoing TT on a shear flow via induced velocity disturbance field δV was studied in a microfluidic four-roll mill device (Levant et al., 2012; Afik et al., 2016). Time-lapsed velocity field images were obtained via 2D micro-particle image velocimetry (μPIV) filtered by a Laplace filter using Gaussian second derivatives (Jones et al., 2001) and processed using Gpiv (van der Graaf, 2008) at interrogation windows of 32 × 32 pxl (8.6 × 8.6 μm²) with 50% overlap. Then the δV amplitude was found to be significant up to four vesicle radii (Figure 19.9). An impressive agreement between experiments and numerical simulations is found (Afik et al., 2016). A direct collision between two vesicles and long-range hydrodynamic interaction mediated by the flow field show a significant impact on the vesicles' shape and orientation, which in turn contributes to rheology of a vesicle suspension. A typical experiment to visualize the vesicle interactions in a flow conducted in a controlled way in the four-roll mill is displayed in Figure 19.10 (see also Movie 19.8), where dynamical parameters such as θ and a distance between the vesicles in a pair can be measured in a controlled way and compared to theory and numerical simulations.

Then σ_θ, the rms of the θ deviation $\delta\theta = \theta - <\theta>$ from its mean value $<\theta>$, of a vesicle in TT as a function of d_{min}/R, where d_{min} is the minimal distance between spherical and deflated vesicles in a pair (Figure 19.10a), is used as the parameter evaluating the interaction strength (Figure 19.11b). Indeed, a correlation between changes of θ and vesicle shear stress variations is found in numerical calculations (Zhao and Shaqfeh, 2013b). The experiments reveal that σ_θ becomes significant at 0.02–0.03 rad corresponding to $d_{min}/R \approx 3.2$–3.7 (Figure 19.11b) obtained for the spherical and deflated vesicles in the pair

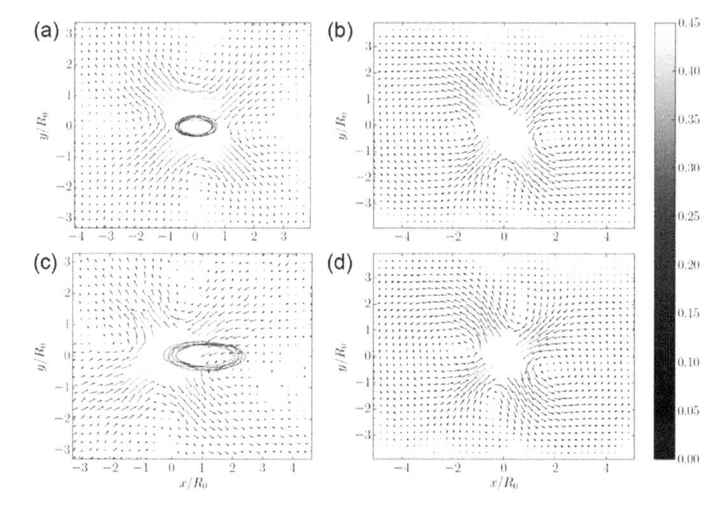

Figure 19.9 Time-averaged velocity disturbance fields δV: (a) a quasi-spherical vesicle with $R = 17$ μm, experiment; (b) the same at $\Delta = 0.05$, numerical simulations; (c) a vesicle with $R = 16.6$ μm, $\Delta = 0.39$, experiment; (d) a vesicle with $\Delta = 0.40$, numerical simulations. The black lines indicate the trajectory of the vesicle center of mass. (The plots (a) and (c) are adapted from Levant, M. et al., *Phys. Rev. E*, 85, 056306, 2012a. The figure is reprinted from Afik, E. et al., *EPL*, 113, 38003, 2016.)

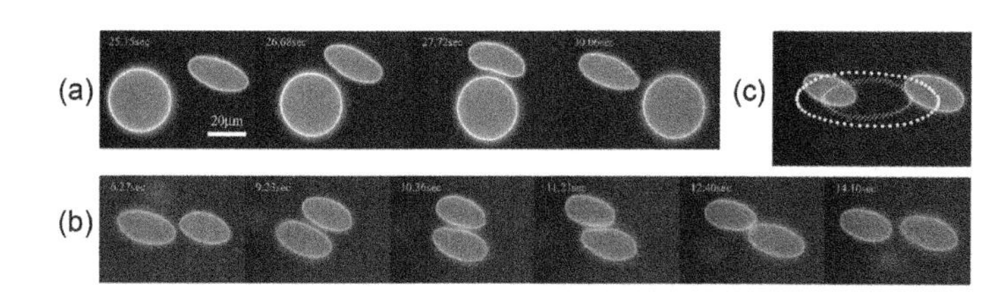

Figure 19.10 Snapshots of vesicle pair interactions: (a) $\Delta \approx 0$; 1.0, $R_1 = 13.2$, $R_2 = 9.2$ μm, $\omega/s = 1.33$; (b) $\Delta_1 \approx 0.88$, $\Delta_2 \approx 0.77$, $R_1 = 10.1$ μm, $R_2 = 11.9$ μm, $\omega/s = 1.52$; (c) trajectories of interacting at vesicles in a pair $\omega/s = 1.52$. (Reprinted from Levant, M. et al., *Phys. Rev.*, *E* 85, 056306, 2012a.) Characteristic spatial scale of vesicle pair interactions in a plane linear shear flow.

(Levant et al., 2012). In the case of two deflated vesicles in a pair (Figure 19.10b), the interaction strength is evaluated via the peak $C_{max}(\tau)$ of the normalized cross-correlation function $C(\tau) = <\delta\theta_1(t)\delta\theta_2(t+\tau)>_t/\sigma_{\theta 1}\sigma_{\theta 2}$ (Figure 19.11a,c) as a function of d_{min}/R providing a similar result as above (Figure 19.11d). These findings allow determining the volume fraction $\varphi \approx 0.1$, below which vesicle interactions are negligible and a vesicle suspension can be considered as dilute (Levant et al., 2012) that is consistent with numerical findings (Zhao and Shaqfeh, 2013b).

In a dilute vesicle suspension, the viscosity η_{eff} is slightly higher than the viscosity of the solvent η_{out}. The viscosity measurements in cone-plate geometry at a constant rotation are carried out with about 1% accuracy on a rheometer Rheolyst AR 1000-N and in a capillary of 1-mm diameter on a viscometer Villastic-3 at 22.8 ± 0.05°C and each measurement was averaged on several vesicle batches (Kantsler et al., 2008b). The contribution of the vesicles can be measured by the relative viscosity η_{eff}/η_{out}, which depends on the size and shape of the vesicles and the vesicle volume fraction. For a dilute suspension of quasi-spherical vesicles in TT in a shear flow one expects $\eta_{eff}/\eta_{out} = 1 + 5/2 - \Delta(23\lambda + 32)/16\pi$, where the last term is a correction to the Einstein formula for hard spheres due to vesicles (Danker and Misbah, 2007a; Danker et al., 2007b; Vlahovska and Gracia, 2007; Vergeles, 2008). Moreover, at the transition to TU/TR the η_{eff}/η_{out} dependence changes a trend from decrease to growth with λ. This nonmonotonic behavior of η_{eff}/η_{out} with the minimum at the TT-TU/TR transition is verified in several numerical simulations for dilute (Ghigliotti et al., 2010; Zhao and Shaqfeh, 2011; Zhao et al., 2011) and non-dilute (Thiebaud and Misbah, 2013; Zhao and Shaqfeh, 2013b) vesicle suspensions and in an experiment (Vitkova et al., 2008), though the latter is conducted in non-dilute suspensions and with a wide distribution of vesicles in R and Δ. Another experiment carried out also in a non-dilute regime contradicts the predictions at $\lambda < 1$, where η_{eff}/η_{out} is found to increase with λ at $\lambda < 1$ (Kantsler et al., 2008b). This dependence is also observed in numerical simulations (Lamura and Gompper, 2013; Kaoui et al., 2014) due to thermal noise and confinement. Thus, the contradiction in the dependence of η_{eff}/η_{out} on λ still remains unresolved experimentally due to lack of technology to produce uniform size and shape vesicles.

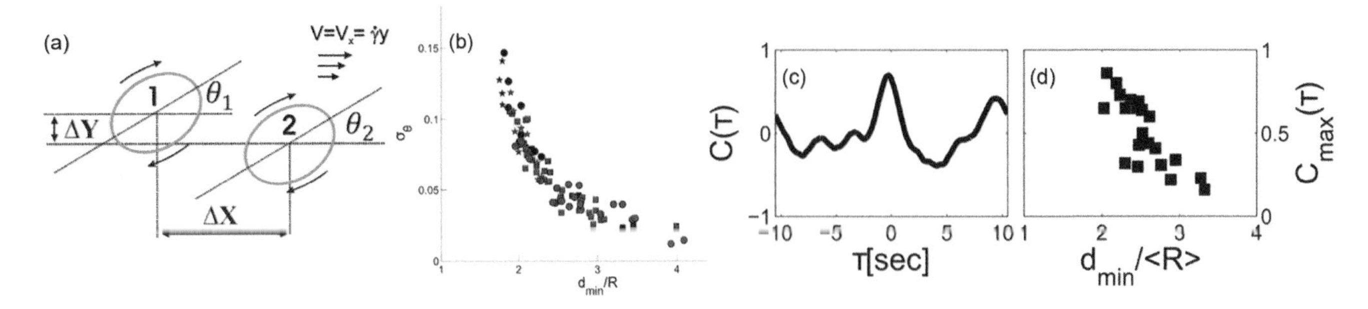

Figure 19.11 (a) Schematic of vesicle interaction in TT in a non-dilute suspension; (b) σ_θ versus d_{min}/R over many cycles for nine vesicle pairs; (c) $C(\tau)$ versus τ; (d) $C(\tau)_{max}$ versus d_{min}/R. (Reprinted from Levant, M. et al. *Phys. Rev. E*, 85, 056306, 2012a.)

19.5.2 ROLE OF CONFINEMENT AND CHANNEL GEOMETRY IN VESICLE DYNAMICS

The first experimental observation on the influence of a confinement on vesicle dynamics was observed at sufficiently large φ in a shear flow: the TT-TU/TR transition is shifted up significantly, rms velocity fluctuations growth with $\dot{\gamma}$, and rms fluctuations of θ exceeds the thermal values up to an order of magnitude (Kantsler et al., 2008b). Similar effects are found in numerical simulations for a single vesicle under confinement (Zhao and Shaqfeh, 2011; Zhao et al., 2011; Gires et al., 2012; Kaoui et al., 2012) and for a non-dilute vesicle suspension under shear (Zhao and Shaqfeh, 2013b; Kaoui et al., 2011; Lamura and Gompper, 2013).

In capillaries with a diameter comparable to a vesicle size, vesicles align with the flow axis and adopt either bullet-like or parachute shapes depending on Δ and a flow velocity, as reported by Vitkova et al. (2004) and shown in Figure 19.12 that qualitatively agree with the vesicle shapes obtained in simulations (Noguchi and Gompper, 2005b). The larger the confinement, the lower the ratio of the vesicle-to-flow velocity. Vesicles flowing at the center of the channel form clusters due to hydrodynamic interactions if they come close enough to each other (Ghigliotti et al., 2012), similar to RBC clusters observed experimentally (Tomaiuolo et al., 2012).

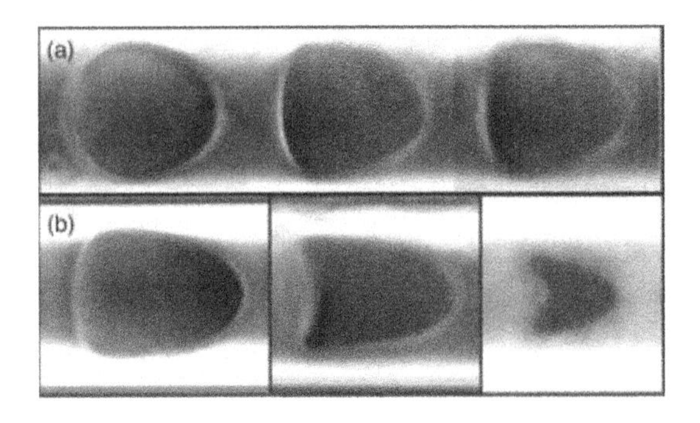

Figure 19.12 Snapshots of vesicles in a capillary with diameter close to $2R$ (flow from left to right. (a) Deformations of a vesicle with $R = 20.2$ µm, $\Delta = 0.14$, $\lambda = 0.71$ at $U = 36$, 292, and 491 µm/s. (b) Bullet and parachute shapes observed for vesicles with $R = 20.9$, 18.8, and 8.7 µm; $\Delta = 0.46$, 1.7, and 1.99; $\lambda = 0.8$, 0.67, and 0.4; $U = 219$, 541, and 505 µm/s. (Reprinted from Vitkova, V. et al., *Europhys. Lett.*, 68, 398–404, 2004.)

The dynamics of vesicles flowing in confined microchannels with oscillating width are studied both experimentally and theoretically (Noguchi, 2010; Noguchi et al., 2010; Braunmueller et al., 2011). For fast flows at $\chi >> 1$ and a wider channel ($L_y = 75$ µm), a vesicle shape oscillates between almost elliptical and bullet-like at $\Delta \leq 0.91$ (Figure 19.13, two left panels; see also Figure 6.16 and Chapter 6 for more details on the simulation methods), in contrast to an unbounded Poiseuille flow, where theory predicts a coexistence of bullet- and parachute-like (Danker et al., 2009). At $\Delta \geq 0.91$, the vesicles of asymmetric slipper-like shapes are observed in both experiment and simulations (Kaoui et al., 2009; Noguchi, 2010; Noguchi et al., 2010) and similar to that found in RBCs (Fung, 1993). For vesicles shifted off the central line in a wider channel at $\chi \geq 1$, oscillations in vesicle orientation and shape were observed (Figure 19.13, right panel).

19.5.3 DYNAMICS OF A COMPOUND VESICLE IN A LINEAR FLOW

A compound vesicle is a unilamellar vesicle encapsulating either a solid particle (Veerapaneni et al., 2011) or another GUV (Kaoui et al., 2013; Levant and Steinberg, 2014) suspended in a membrane-bounded fluid. It is useful to model the dynamics of cells having a complex internal structure such as muscle cells. The simulations considering a solid inner particle of either spherical or elliptical shape show that new dynamical features in a shear flow appear due to a hydrodynamic interaction between the inclusion and the membrane (Veerapaneni et al., 2011). For example, the transition from TT to TU can occur in the absence of any viscosity contrast but at some critical value of the filling factor. Moreover, a vesicle can swing if the enclosed particle is nonspherical. A compound vesicle including another vesicle instead of a solid particle leads to much richer vesicle dynamics observed for GUVs but experimentally less traceable due to larger number of the control parameters (Kaoui et al., 2013). Experimental observations of such a compound vesicle in a linear flow were realized in the four-roll mill device (Levant and Steinberg, 2014). It is found that although a compound vesicle can undergo the same TT, TR, and TU regimes as GUVs, a new swinging motion of the inner vesicle exists in the TR regime of the main GUV in accord with simulations (Kaoui et al., 2013). In addition, the inner and outer vesicles can be simultaneously

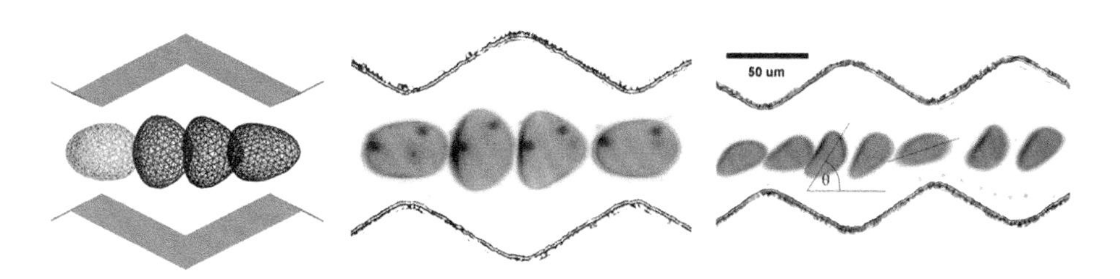

Figure 19.13 Left two panels: Snapshots of vesicle with $\Delta = 0.35$, $R/L_y = 0.207$, $L_y = 75$ µm, $\chi = 25$ (right, experiment; left, simulations for $\chi = 12.5$). Right panel: Snapshots of orientation oscillations of a vesicle in a wide channel with $L_y = 75$ µm, $R = 24$ $L_y = 75$ µm, $\chi = 75.8$. (Reprinted from Noguchi, H. et al., *EPL*, 89, 28002, 2010b.)

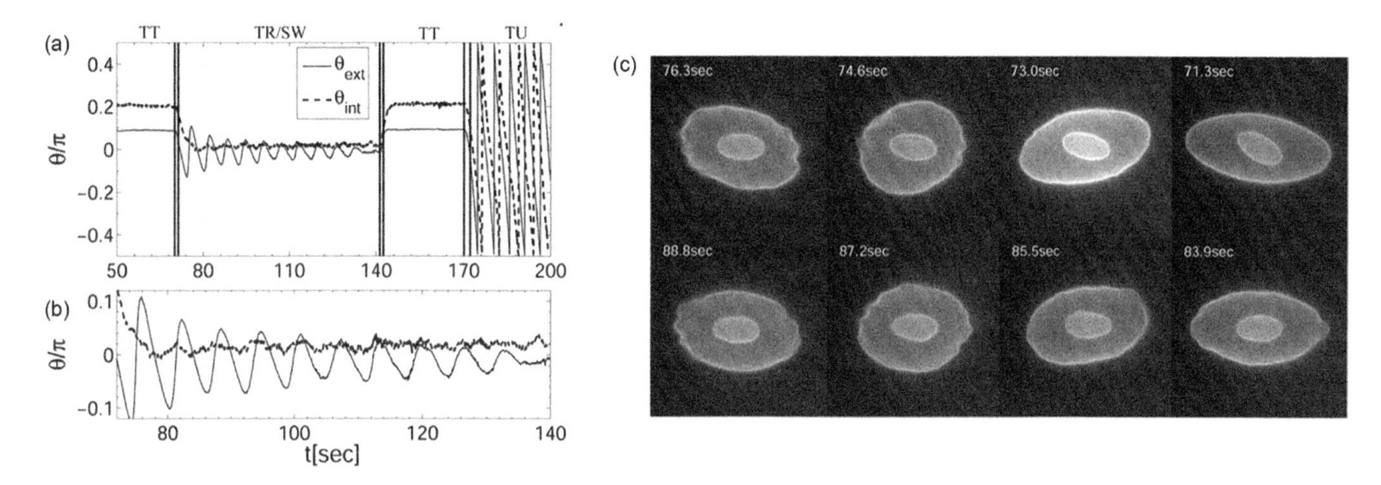

Figure 19.14 (a) Time evolution of θ_{ext} and θ_{int} of external and internal vesicles with the volume fraction 0.35, $\Delta_{ext} = 1.04$, and $\Delta_{int} = 0.73$ in consecutively TT, TR, TU regimes due to variation of the control parameters Λ and S: TT (1.03; 1270), TR(1.93; 675), TU (4.83; 270) regimes. (b) Zoom-in presentation of (a) showing the swinging regime. (c) Snapshots of two subsequent cycles in TR with clearly visible wrinkles. (Reprinted from Levant, M. and Steinberg, V., *Phys. Rev. Lett.*, 112, 138106, 2014b.)

found in different dynamical regimes and either synchronized or unsynchronized depending on the filling factor (Figure 19.14). Whether this model can mimic the dynamical behavior of muscle cells or whether one needs to take into account an additional non-Newtonian nature of a muscle cell and its nuclei (consisting of, e.g., fibers, organelles, cytoplasm) remains an open question.

19.5.4 RELAXATION OF A STRETCHED VESICLE AND AN OSCILLATORY FLOW

A vesicle deformed by an external flow will relax back to its equilibrium shape once the flow is turned off. For a vesicle relaxing from a tubular back to a rest shape in an extensional flow as described in Section 19.3.1, a relaxation time, τ, of several seconds is found (inset in Figure 5 of Kantsler et al. [2008a]) and close to one in a plane linear flow during changing the control parameters inside the same or between different vesicle dynamical regimes (Deschamps et al., 2009b). The same order of magnitude relaxation time has been found for a vesicle deformed by a strong uniform flow while trapped by optical tweezers (Foo et al., 2004) or directly deformed by optical tweezers (Zhou et al., 2011). This relaxation time also depends on the initial tension of the vesicle, as shown experimentally for a vesicle point-attached to a solid substrate or to a moving particle (Rossier et al., 2003). In all these experiments, the relaxation times are at least an order of magnitude smaller than estimated from a model for quasi-spherical vesicles (Lebedev et al., 2007, 2008) calling for further improvement of the model.

Effects of an external oscillatory flow on vesicles are studied in rather limited extent theoretically (Noguchi, 2010), numerically (Farutin and Misbah, 2012), and experimentally (Fa et al., 2004), in contrast to RBCs, which experience *in situ* oscillatory blood flow and where detailed numerical and experimental investigations are devoted to this subject. Regarding the vesicle dynamics, both theory and simulations show complex dynamical regimes (Noguchi, 2010) and resonances are expected (Farutin and Misbah, 2012), though no single relevant experiment has been conducted. The only experiment where the oscillatory flow effect on a membrane tension is tested using a micropipette technique (Evans and Needham, 1987). It was designed to measure the vesicle bending elasticity κ (Figure 19.15) (Fa et al., 2004). It is found that the oscillatory flow significantly suppresses membrane fluctuations due to increase of the membrane tension. The latter is revealed by the decrease of the effective bending elasticity, which dependence on $\dot{\gamma}$ is well described by the empirical expression $\kappa(\dot{\gamma}) = \kappa(0)/(1 + \tau_{0}\dot{\gamma})$, where the relaxation time $\tau_0 = 143$ *s* is obtained from the best fit (Fa et al., 2004). The result is strikingly different from the effect of a shear flow discussed above in details, where the excess area is stored in higher-order harmonics at $\chi \equiv \tau\dot{\gamma} >> 1$ in TT and the fluctuations are even strongly amplified in the vicinity of the TT-TR transition and in TR. The experimentally determined τ_0 turns out to be of the same order as the longest relaxation time for vesicle shape deformation, τ, discussed in the previous paragraph, which contradicts to an estimate made by Fa et al. (2004). A similar suppression of fluctuations due to a

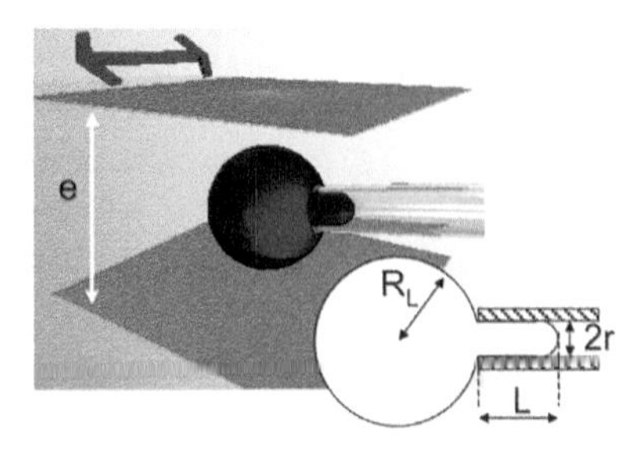

Figure 19.15 A setup of a microscopic rheometer. The membrane is sucked into the micropipette acquiring the geometry of a cylinder capped by a hemisphere. The top surface moves perpendicular to the micropipette axis shown by red arrows. (Reprinted from Fa, N., *Phys. Rev. Lett.*, 92, 108103, 2004.)

periodical flow has been reported in related membrane systems such as smectic lyotropics and sponge phases (Yamamoto and Tanaka, 1995). The empirical expression for $\kappa(\dot{\gamma})$ is strongly reminiscent of the expression for a physical parameter of a thermodynamic system subjected to a periodic adiabatic perturbation in a weak deviation from equilibrium in a slow relaxation process (Landau and Lifschitz, 1987). However, the phenomenological theory suggested about 80 years ago should be properly applied to the system presented.

19.6 OUTLOOK

A huge progress in the experimental, theoretical and numerical investigations of vesicle dynamics was made during the last decade. Thus the dynamics of a single vesicle in various flows are rather well understood now, in some cases even in quantitative details. Many experimental findings have been successfully explained by both analytical theory and simulations. Progress was achieved thanks to numerous novel techniques, technologies and approaches in experiments, data analysis, theoretical modeling and numerical simulations, as summarized in this review. However, fundamental problems still remain unresolved in particular in respect to the role of thermal noise. Moreover, more complex problems such as the dynamics and sedimentation of a single vesicle in a viscoelastic flow, which is directly relevant to biological applications, still remain untouched.

MOVIES

The Movies can be found under http://www.crcpress.com/9781498752176

Movie 19.1: Dynamics of a vesicle in the TT motion ($R = 14.4$ μm, $\Delta = 0.64$).

Movie 19.2: Dynamics of a vesicle in the TU motion ($R = 14.4$ μm, $\Delta = 0.64$).

Movie 19.3: Dynamics of a vesicle in the TR motion ($R = 14.4$ μm, $\Delta = 0.64$).

Movie 19.4: Movie presents three regimes of vesicle dynamics, when the control parameters are changed ($R = 14.4$ μm, $\Delta = 0.64$) (Deschamps et al., 2009b).

Movie 19.5: Dynamics of a tubular vesicle in a steady elongation flow via the tubule-to-dumbbell transition ($L_0/2R_0 = 8.8$).

Movie 19.6: Dynamics of a tubular vesicle in a steady elongation flow during the pearling transition ($L_0/2R_0 = 47$).

Movie 19.7: Dynamics of a vesicle in a transient elongation flow-wrinkling transition ($R = 14.4$ μm, $\Delta \approx 1$).

Movie 19.8: Dynamics of interaction of spherical and deflated vesicles in a pair, which snapshots are presented in Figure 19.10a. The plot in the upper left corner shows the time evolution of the inclination angle of the deflated vesicle ($R = 11$ μm, $\Delta \approx 1$) (Levant and Steinberg, 2012).

ACKNOWLEDGMENTS

We are grateful to our collaborators in experiments V. Kantsler, J. Deschamps, E. Afik, E. Segre, N. Zabusky and in theory V. Lebedev, K. Turitsyn, S. Vergeles, A. Lamura, D. Abreu, and U. Seifert in the endeavor to understand quantitatively vesicle dynamics in various flows. The authors acknowledge financial support from the Israel Scientific Foundation.

LIST OF ABBREVIATIONS

KS	Keller-Skalak
PDF	probability distribution function
RBC	red blood cell
SW	swinging
TT	tank-treading
TR	trembling
TU	tumbling
VB	vacillating-breathing

GLOSSARY OF SYMBOLS

A	area
D	Taylor deformation factor of vesicle- $D = (L-B)/(L+B) = \sqrt{15\Delta/2\pi}$
2B	vesicle short axis
2L	vesicle long axis
f	frequency
F	force
k	wave number
k*	spectrally averaged wave number
k_B	Boltzmann constant
P	pressure
ΔP	pressure difference
R	effective radius of vesicle defined as $R=(3V/4\pi)^{1/3}$
s	strain rate, $s=\sqrt{tr(s_{ij}^2)}$
s_{ij}	2D strain matrix
S	dimensionless parameter of theory related to χ
t	time
T	temperature
V	volume of vesicle
v	velocity (scalar)
$\dot{\gamma}$	shear rate
$\dot{\varepsilon}$	elongation rate
ρ	density
Δρ	density difference
Σ	shear rate normalized by k_BT: $\varsigma = \dot{\gamma}\eta_{out}R^3/k_BT$

η	dynamic viscosity
Δ	vesicle excess area; $\Delta = A/R^2 - 4\pi$
θ	vesicle inclination angle
$\delta\theta$	rms fluctuation amplitude of θ: $\delta\theta = <(\theta-\theta 0)^2>^{1/2}$
κ	bending rigidity of membrane
λ	η_{in}/η_{out} -vesicle viscosity contrast
Λ	dimensionless parameter of theory related to λ and ω/s
σ	Lagrangian multiplyer defined by surface area constraint
τ	characteristic timescale of vesicle relaxation
χ	normalized shear rate, analog of capillary number $\chi = \dot{\gamma}\eta_{out}R^3/\kappa$
ω	vorticity $\omega = \lvert\omega_k\rvert$
ω_k	angular velocity vector

REFERENCES

Abkarian M, Lartigue C, Viallat A. (2002) Tank-treading and unbinding of deformable vesicles in shear flow: Determination of the lift force. *Phys. Rev. Lett.* 88: 068103.

Abkarian M and Viallat A. (2005) Dynamics of vesicles in a wall-bounded shear flow. *Biophys. J.* 89: 1055–1066.

Abreu D. (2014b) Vesicles in flow: Role of thermal fluctuations. PhD thesis, University of Stuttgart, Germany.

Abreu D, Levant M, Steinberg V, Seifert U. (2014a) Fluid vesicles in flow. *Adv. Colloid Interface Sci.* 208: 129.

Abreu D and Seifert U. (2012) Effect of thermal noise on vesicles and capsules in shear flow. *Phys. Rev. E* 86: 010902.

Abreu D and Seifert U. (2013) Noisy nonlinear dynamics of vesicles in flow. *Phys. Rev. Lett.* 110: 238103.

Afik E. (2015) Robust and highly performant ring detection algorithm for 3D particle tracking using 2D microscope imaging. *Sci. Rep.* 5: 13584; ridge directed ring detector https://github.com/eldad-a/ridge-directed-ring-detector (2014).

Afik E, Lamura A, Steinberg V. (2016) Long-range hydrodynamic effect due to a single vesicle in linear flow. *EPL* 113: 38003.

Angelova, M. I, Soleau, S, Meleard, P, Faucon, J, Bothorel, P. (1992) Preparation of giant vesicles by external AC electric fields. *Prog. Colloid Polym. Sci.* 89: 127–133.

Beaucourt J, Rioual F, Seon T, Biben T, Misbah C. (2004) Steady to unsteady dynamics of a vesicle in a flow. *Phys. Rev. E* 69: 011906.

Biben T, Farutin A, Misbah C. (2011) Three-dimensional vesicles under shear flow: Numerical study of dynamics and phase diagram. *Phys Rev E* 83: 031921.

Braunmueller S, Scmid L, Franke T. (2011) Dynamics of red blood cells and vesicles in microchannels of oscillating width. *J. Phys. Condens. Matter* 23: 184116.

Cerda E and Mahadevan L. (2003) Geometry and physics of Wrinkling. *Phys. Rev. Lett.* 90: 074302.

Cerda E and Mahadevan L, Pasini JM. (2004) The elements of draping. *PNAS* 101: 1806–1810.

Cerda E, Ravi-Chandar K, Mahadevan L. (2002) Wrinkling of an elastic sheet under tension. *Nature* 419: 579.

Crocker J. C and Grier D. G. (1996) Methods of digital video-microscopy for colloidal studies. *J. Colloid Interface Sci.* 179: 298. And particle tracking routine: http://physics.nyu.edu/grierlab/software.html

Dai J and Sheetz M. P. (1999) Membrane tether formation from blebbing cells. *Biophys. J.* 77, 3363–3370.

Danker G, Biben, Podgorski TT, Verdier C, Misbah C. (2007b) Dynamics and rheology of a dilute suspension of vesicles: Higher order theory. *Phys. Rev. E* 76: 041905.

Danker G and Misbah C. (2007a) Rheology of a dilute suspension of vesicles. *Phys. Rev. Lett.* 98: 088104.

Danker G, Vlahovska PM, Misbah C. (2009) Vesicles in Poiseuille flow. *Phys. Rev. Lett.* 102: 148102.

Davis D and Sowinski S. (2008) Membrane nanotubes: Dynamic long-distance connections between animal cells, *Nat. Rev. Mol. Cell Bio* 9: 431.

Deschamps J, Kantsler V, Segre E, Steinberg V. (2009b) Dynamics of a vesicle in general flow. *Proc. Natl. Acad. Sci. U.S.A.* 106: 11444–11447.

Deschamps J, Kantsler V, Steinberg V. (2009a) Phase diagram of single vesicle dynamical states in shear flow. *Phys. Rev. Lett.* 102: 118105.

Dittrich P. S, Heule M, Renaud P, Manz A. (2006) On-chip extrusion of lipid vesicles and tubes through microsized apertures. *Lab Chip* 6, 488–493.

Evans E and Needham D. (1987) Physical properties of surfactant bilayer membranes: Thermal transitions, elasticity, rigidity, cohesion and colloidal interactions. *J. Phys. Chem.* 91: 4219.

Evans E and Yeung A. (1994) Hidden dynamics in rapid changes of bilayer shape. *Chem. Phys. Lipids* 73, 39–56.

Fa N, Marques CM, Mendes E, Schroder AP. (2004) Rheology of giant vesicles: A micropipette study. *Phys. Rev. Lett.* 92: 108103.

Farutin A, Biben T, Misbah C. (2010) Analytical progress in the theory of vesicles under linear flow. *Phys. Rev. E* 81: 061904.

Farutin A and Misbah C. (2012) Rheology of vesicle suspensions under combined steady and oscillating shear flows. *J. Fluid Mech.* 700: 362–381.

Faucon JF, Mitov MD, Meleard P, Bivas I, Bothorel P. (1989) Bending elasticity and thermal fluctuations of lipid membranes. Theoretical and experimental requirements. *J. Phys. France* 50: 2389–2414.

Finken R and Seifert U. (2006) Wrinkling of microcapsules in shear flow. *J. Phys. Condens. Matter* 18: L185.

Foo J-J, Chan V, Liu K-K. (2004) Shape recovery of an optically trapped vesicle: effect of flow velocity and temperature. *IEEE Trans Nanobioscience* 3: 96–100.

Fung Y-C. (1993) Biomechanics: Mechanical properties of living Tissues. Springer-Verlag, New York.

Gerashchenko S and Steinberg V. (2008) Critical slowdown in polymer dynamics near the coil-stretch transition in elongation flow. *Phys. Rev. E* 78: 040801 (R).

Ghigliotti G, Biben T, Misbah C. (2010) Rheology of a dilute two-dimensional suspension of vesicles. *J. Fluid Mech.* 653: 489–518.

Ghigliotti G, Selmi H, El Asmi L, Misbah C. (2012) Why and how does collective red blood cells motion occur in the blood microcirculation? *Phys Fluids* 24: 101901.

Gires PY, Danker G, Misbah C. (2012) Hydrodynamic interaction between two vesicles in a linear shear flow: Asymptotic study. *Phys. Rev. E* 86: 011408.

Goldsmith HL and Marlow J. (1972) Flow behavior of erythrocytes. I. Rotation and deformation in dilute suspensions. *Proc. Royal Soc. London B* 182: 351–384.

Happel J and Brenner H. (1983) Low Reynolds number hydrodynamics. Martinus Nijhoff Publishers.

Helfrich W. (1973) Elastic properties of lipid bilayers: Theory and possible experiments. *Z. Nat.* 28: C693–C703.

Huang J et al. (2007) Capillary wrinkling of floating thin polymer films. *Science* 317: 650.

Hudson SD, Phelan Jr FR, Handler MD. (2004) Microfluidic analog of the four-roll mill. *Appl. Phys. Lett.* 85: 335–337.

Jones E, Oliphant T, Peterson P. et al. (2001) *SciPy: Open source scientific tools for Python,* http://www.scipy.org/.

Kantsler V. (2007) Hydrodynamics of Fluid Vesicles. PhD thesis, Weizmann Institute of Science, Rehovot.

Kantsler V, Segre E, Steinberg V. (2007) Vesicle dynamics in time-dependent elongation flow: Wrinkling instability. *Phys. Rev. Lett.* 99: 178102.

Kantsler V, Segre E, Steinberg V. (2008a) Critical dynamics of vesicle stretching transition in elongation flow. *Phys. Rev. Lett.* 101: 048101.

Kantsler V, Segre E, Steinberg V. (2008b) Dynamics of interacting vesicles and rheology of vesicle suspension in shear flow. *Europhys. Lett.* 82: 58005.

Kantsler V and Steinberg V. (2005) Orientation and dynamics of a vesicle in tank-treading motion in shear flow. *Phys. Rev. Lett.* 95: 258101.

Kantsler V and Steinberg V. (2006) Transition to tumbling and two regimes of tumbling motion of a vesicle in shear flow. *Phys. Rev. Lett.* 96: 036001.

Kaoui B, Farutin A, Misbah C. (2009) Vesicles under simple shear flow: Elucidating the role of relevant control parameters. *Phys. Rev. E* 80: 061905.

Kaoui B, Harting J, Misbah C. (2011) Two-dimensional vesicle dynamics under shear flow: effect of confinement. *Phys. Rev. E* 83: 066319.

Kaoui B, Jonk R. J. W, Harting J. (2014) Interplay between microdynamics and macrorheology in vesicle suspensions. *Soft Matter* 10, 4735–4742.

Kaoui B, Krüger T, Harting J. (2012) How does confinement affect the dynamics of viscous vesicles and red blood cells? *Soft Matter* 8: 9246.

Kaoui B, Krüger T, Harting J. (2013) Complex dynamics of a bilamellar vesicle as a simple model for leukocytes. *Soft Matter* 9: 8057–8061.

Karlsson M, Sott K, Cans A.-S, Karlsson A, Karlsson R, Orwar O. (2001) Micropipet-assisted formation of microscopic networks of unilamellar lipid bilayer nanotubes and containers. *Langmuir* 17, 6754–6758.

Keller SR and Skalak R. (1982) Motion of a tank-treading ellipsoidal particle in a shear flow. *J. Fluid Mech.* 120: 27–47.

Kraus M, Wintz W, Seifert U, Lipowsky R. (1996) Fluid vesicle in shear flow. *Phys. Rev. Lett.* 77: 3685–3688.

Lamura A and Gompper G. (2013) Dynamics and rheology of vesicle suspensions in wallbounded shear flow. *EPL* 102: 28004.

Landau LD and Lifschitz EM. (1987) Fluid Mechanics. Elsevier, Oxford, UK.

Lebedev V, Turitsyn K, Vergeles S. (2007) Dynamics of nearly spherical vesicles in an external flow. *Phys. Rev. Lett.* 99: 218101.

Lebedev V, Turitsyn K, Vergeles S. (2008) Nearly spherical vesicles in an external flow. *New J. Phys.* 10: 043044.

Lee JS, Dylla-Spears R, Teclemariam NP, Muller SJ. (2007) Microfluidic four-roll mill for all flow types. *Appl. Phys. Lett.* 90: 074103.

Levant M. (2015) Hydrodynamics of vesicles and red blood cells. PhD thesis, Weizmann Institute of Science, Rehovot.

Levant M, Abreu D, Seifert U, Steinberg V. (2014) Wrinkling instability in vesicle dynamics in linear flow. *EPL* 107: 28001.

Levant M, Deschamps J, Afik E, Steinberg V. (2012) Characteristic spatial scale of vesicle pair interactions in a plane linear shear flow. *Phys. Rev. E* 85: 056306.

Levant M and Steinberg V. (2012) Amplification of thermal noise through vesicle dynamics. *Phys. Rev. Lett.* 109: 268103.

Levant M and Steinberg V. (2014) Complex dynamics of compound vesicles in linear flow. *Phys. Rev. Lett.* 112: 138106.

Liu K and Li S. (2014) Nonlinear simulations of vesicle wrinkling. *Math. Met. Appl. Sci.* 37: 1093.

Luisi PL and Walde P. (Eds). (2000) Giant Vesicles. John Wiley & Sons, New York

Mader MA, Vitkova V, Abkarian M, Viallat A, Podgorski T. (2006) Dynamics of viscous vesicles in shear flow. *Eur. Phys. J. E* 19: 389–397.

Messlinger S, Schmidt B, Noguchi H, and Gompper G. (2009) Dynamical regimes and hydrodynamic lift of viscous vesicles under shear. *Phys. Rev. E* 80: 021902.

Misbah C. (2006) Vacillating breathing and tumbling of vesicles under shear flow. *Phys. Rev. Lett.* 96: 028104.

Narsimhan V, Spann A, Shaqfeh ESG. (2014) The mechanism of shape instability for a vesicle in extensional flow. *J. Fluid Mech.* 750: 144–190

Narsimhan V, Spann A, Shaqfeh ESG. (2015) Pearling, wrinkling, and buckling of vesicles in elongational flows. *J. Fluid Mech.* 777: 1–26.

Noguchi H. (2010) Dynamics of fluid vesicles in oscillatory shear flow. *J Phys. Soc. Japan* 79: 024801.

Noguchi H. and Gompper G. (2004) Fluid vesicles with viscous membranes in shear flow. *Phys. Rev. Lett.* 93: 258102.

Noguchi H and Gompper G. (2005a) Dynamics of fluid vesicles in shear flow: Effect of membrane viscosity and thermal fluctuations. *Phys. Rev. E* 72: 011901.

Noguchi H and Gompper G. (2005b) Shape transitions of fluid vesicles and red blood cells in capillary flows. *Proc. Natl. Acad. Sci. U.S.A.* 102: 14159.

Noguchi H and Gompper G. (2007) Swinging and tumbling of fluid vesicles in shear flow. *Phys. Rev. Lett.* 98: 128103.

Noguchi H, Gompper G, Schmid L, Wixforth A, Franke T. (2010) Dynamics of fluid vesicles in flow through structured microchannels. *EPL* 89: 28002.

Olla P. (2000) The behavior of closed inextensible membranes in linear and quadratic shear flows. *Physica A* 278: 87–106.

Pecreaux J, Doebereiner H-G, Prost J, Joanny J-F, Bassereau P. (2004) Refined contour analysis of giant unilamellar vesicles. *Eur. Phys. J. E* 13: 277–290.

Rossier O, Cuvelier D, Borghi N, Puech PH, Derényi I, Buguin A, et al. (2003) Giant vesicles under flows: extrusion and retraction of tubes. *Langmuir* 19: 575–584.

Seifert U. (1999) Fluid membranes in hydrodynamic flow fields: Formalism and an application to fluctuating quasi-spherical vesicles in shear flow. *Eur. Phys. J. B* 8: 405–415.

Schmid-Schönbein H and Wells R. (1969) Fluid drop-like transition of erythrocytes under shear. *Science* 165: 288–291.

Sowinski S, Jolly C, Berninghausen O, Purbhoo MA, Chauveau A, Köhler K, et al. (2008) Membrane nanotubes physically connect T cells over long distances presenting a novel route for HIV-1 transmission. *Nat. Cell. Biol.* 10: 211.

Spjut JE. (2010) Trapping, deformation, and dynamics of phospholipid vesicles. MS thesis, University of California, Berkeley. Chap. 3; and Spjut JE, Muller S. unpublished.

Taylor G. I. (1934) The formation of emulsions in definable fields of flow. *Proc. R. Soc. London A* 146: 501.

Thiebaud M and Misbah C. (2013) Rheology of a vesicle suspension with finite concentration: A numerical study. *Phys. Rev. E* 88: 062707.

Tomaiuolo G, Lanotte L, Ghigliotti G, Misbah C, Guido S. (2012) Red blood cell clustering in Poiseuille microcapillary flow. *Phys Fluids* 24: 051903.

Turitsyn K and Vergeles S. (2008) Wrinkling of vesicles during transient dynamics in elongational flow. *Phys. Rev. Lett.* 100: 028103.

van der Graaf G. (2008) Gpiv, open source software for particle image velocimetry, http://gpiv.sourceforge.net/.

Veerapaneni SV, Young Y-N, Vlahovska PM, Bławzdziewicz J. (2011) Dynamics of a compound vesicle in shear flow. *Phys. Rev. Lett.*106:158103.

Vergeles SS. (2008) Rheological properties of a vesicle suspension. *JETP* 87: 511–515.

Vitkova V, Mader MA, Podgorski T. (2004) Deformation of vesicles flowing through capillaries. *Europhys. Lett.* 68: 398–404.

Vitkova V, Mader MA, Polack B, Misbah C, Podgorski T. (2008) Micro–macro link in rheology of erythrocyte and vesicle suspensions. *Biophys. J.* 95: L33–L35.

Vlahovska P and Gracia R. (2007) Dynamics of a viscous vesicle in linear flows. *Phys. Rev. E* 75: 016313.

Vlahovska PM, Podgorski T, Misbah C. (2009) Vesicles and red blood cells in flow: From individual dynamics to rheology. *C R Phys* 10: 775–789.

Walter A, Rehage H, Leonhard H. (2001) Shear induced deformation of microcapsules: Shape oscillations and membrane folding. *Colloid Surf.A* 183–185: 123–132.

West J, Manz A, Dittrich P. S. (2008) Lipid nanotubule fabrication by microfluidic tweezing. *Langmuir* 24, 6754–6758.

Xia YN and Whitesides GM. (1998) Soft lithography. *Annu. Rev. Mater. Sci.* 28: 153.

Yamamoto J and Tanaka H. (1995) Shear effects on layer undulation fluctuations of a hyperswollen lamellar phase. *Phys. Rev. Lett.* 74: 932

Zabusky NJ, Segre E, Deschamps J, Kantsler V, Steinberg V. (2011) Dynamics of vesicles in shear and rotational flows: Modal dynamics and phase diagram. *Phys. Fluids* 23: 041905.

Zhao H and Shaqfeh ESG. (2011) The dynamics of a vesicle in simple shear flow. *J. Fluid Mech.* 674: 578–604.

Zhao H and Shaqfeh ESG. (2013a) The shape stability of a lipid vesicle in a uniaxial extensional flow. *J. Fluid Mech.* 719: 345–361.

Zhao H and Shaqfeh ESG. (2013b) The dynamics of a non-dilute vesicle suspension in a simple shear flow. *J. Fluid Mech.* 725: 709–731.

Zhao H, Spann AP, Shaqfeh ESG. (2011) The dynamics of a vesicle in a wall-bound shear flow. *Phys Fluids* 23: 121901.

Zhou H, Gabilondo BB, Losert W, van de Water W. (2011) Stretching and relaxation of vesicles. *Phys. Rev. E* 83: 011905.

Membrane permeability measurements

Begoña Ugarte-Uribe, Ana J. García-Sáez, and Mireille M. A. E. Claessens

A brain like a sieve

Contents

20.1 MEMBRANE PERMEABILITY

Membranes separate cells and cell organelles from the outside world and allow for the creation and maintenance of chemically different environments. However, at the same time cells need to communicate with the outside world. For communication, exchange of materials is required. Although proteins control most of this exchange, some molecules are transported passively through the lipid membrane. In this chapter, several methods to explore and quantify the passive permeability of membranes using giant unilamellar vesicle (GUV) model systems will be described.

The intrinsic permeability, p_s, of a membrane to a solute is defined in the phenomenological expression for the flux J; the number of solute molecules N that cross a unit area A per unit time t:

$$J = p_s\left(c_{in} - c_{ex}\right) \tag{20.1}$$

By convention, fluxes are negative when the solute moves from the exterior to the interior of the GUVs, that is, $J < 0$ when $c_{in} < c_{ex}$. This permeability depends both on the properties of the membrane and on the solute molecules that are transported. In the simplest cases, the value of p_s reflects the diffusion coefficient of the solute molecules in the hydrophobic environment of the membrane and the thickness of the bilayer. In these cases, a solute molecule is thought to dissolve in the membrane from one side, diffuse over the thickness of the bilayer to the other side and leave the membrane. How many of the solute molecules are dissolved in the membrane is given by the partition coefficient B, a dimensionless quantity that characterizes the distribution of the solute between the hydrocarbon region of the lipid bilayer and the water phase. Taking into account this partition coefficient, the permeability is given by $p_s = BD/l_{me}$, where D is the diffusion coefficient of the solute molecules in the hydrocarbon phase, and l_{me} is the thickness of the lipid bilayer. The scaling relation between p_s and BD was verified in experiments using black lipid membranes made of phospholipids (Finkelstein, 1987). In these experiments the permeability to water was determined to be relatively high (10^{-4} m/s). Much lower values of p_s are found for charged and large hydrophilic molecules. Examples of membrane permeability values for different molecules and different membrane compositions are given in Tables 20.1 through 20.3.

Using black lipid membranes to study membrane permeability has the disadvantage that solvents and a solid support are required to prepare freestanding lipid bilayers. Traces of solvents in the membrane and the solid support itself can have a large effect on the permeability measurements. When using vesicles, one does not face these problems. However, in bulk experiments with submicrometer-sized vesicles, artifacts may arise from vesicle polydispersity and if the membrane permeability depends on curvature the results obtained with small unilamellar vesicles (SUVs) or large unilamellar vesicles (LUVs) cannot be applied to flat bilayers. Moreover, by looking at the bulk response, one loses the information on stochastic processes and on the possible existence of vesicle populations with different permeation properties. Therefore, video microscopy, in which the transport of molecules over the membrane of single GUVs is followed in time provides an attractive alternative to permeability experiments with black lipid membrane or LUV/SUV solutions.

Table 20.1 **Overview of the coefficient p_w for membrane permeability with respect to water obtained from measurements on GUVs, see Appendix 1 of the book for structure and data on these lipids**

LIPID(S)	P_W (10^{-6} M/S)			METHOD AND REFERENCES
Egg lecithin	41			Microscopy of GUVs in hypertonic solution (Boroske et al., 1981)
SOPC (C18:0/1)	28 ± 6			Micropipette aspiration (Olbrich et al., 2000)
OSPC (C18:1/0)	30 ± 2			
DOPC (diC18:1$_{c9}$)	42 ± 6			
DEIPC (diC18:1$_{t9}$)	30 ± 5			
DPSPC (diC18:1$_{c6}$)	35 ± 4			
SLPC (C18:0/2)	49 ± 6			
DLPC (diC18:2)	91 ± 24			
DLnPC (diC18:3)	146 ± 26			
EggPC/dicetyl-phosphate/Chol	10 ± 10			Microscopy of GUVs in hypertonic solution (Bernard et al., 2002)
DOPG/Chol (80/20)	15.3 ± 3.4			Microscopy of GUVs in hypertonic solution (Claessens et al., 2008)
DOPG/Chol (60/40)	6.6 ± 1.5			
	15°C	**30°C**	**35°C**	Micropipette aspiration (Rawicz et al., 2008)
DOPC	25.9 ± 3.1	56 ± 9	70 ± 6	
SOPC	19.7 ± 3.3	34 ± 7	44.6 ± 6	
DOPC/Chol	5.8 ± 0.6	—	—	
SOPC/Chol	1.5 ± 0.2	6.4 ± 1.3	12.3 ± 3	
SM/Chol	0.36 ± 0.1	3.9 ± 1	5.9 ± 2	
DOPC/SM/Chol	3.4 ± 0.6	9.3 ± 1.2	19.1 ± 5	
SOPC/SM/Chol	1.2 ± 0.1	4.3 ± 0.3	8.9 ± 1.4	
SOPC/SM/Chol	1.5 ± 0.2	—	4.8 ± 0.8	

Note: DEIPC, 1,2-elaidoyl-*sn*-glycero-3-phosphocholine; DLPC, 1,2-dilinoleoyl-*sn*-glycero-3-phosphocholine; DOPC, 1,2-dioleoyl-*sn*-glycero-3-phosphocholine; DPSPC, 1,2-dipetroselinoleoyl-*sn*-glycero-3-phosphocholine; OSPC, 1-oleoyl-2-stearoyl-*sn*-glycero-3-phosphocholine; SLPC, 1-stearoyl-2-linoleoyl-*sn*-glycero-3-phosphocholine; SOPC, 1-stearoyl-2-oleoyl-*sn*-glycero-3-phosphocholine.

Table 20.2 **Overview of the coefficient p_s for membrane permeability with respect to weak acids obtained from measurements on GUVs, see Appendix 1 of the book for structure and data on these lipids**

LIPIDS	ACIDS	P_S (10^{-6} M/S)	METHOD AND REFERENCES
DOPC/DPPC/Chol (1/1/1)	CH_2O_2 (Formic acid)	3.6 (3.3–4.0)	Spinning disc confocal microscopy and modeling (Li et al., 2011)
DOPC/DPPC/Chol (1/1/1)	$C_2H_4O_2$ (Acetic acid)	6 (5.7–6.3)	
DOPC/DPPC/Chol (1/1/1)	$C_3H_6O_2$ (Propionic acid)	19 (16–21)	
DOPC/DPPC/Chol (1/1/1)	$C_4H_8O_2$ (Butyric acid)	72 (69–75)	
DOPC/DPPC/Chol (1/1/1)	$C_5H_{10}O_2$ (Pentanoic acid)	800 (660–980)	
DOPC/DPPC/Chol (1/1/1)	$C_6H_{12}O_2$ (Hexanoic acid)	2300 (1770–3330)	

Note: DPPC, 1,2-dipalmitoyl-*sn*-glycero-3-phosphocholine.

Table 20.3 **Overview of the coefficient p_s for membrane permeability with respect to hydrophilic molecules obtained from measurements on GUVs, see Appendix 1 of the book for structure and data on these lipids**

LIPID(S)	MOLECULE STUDIED	P_S (10^{-6} M/S)	METHOD AND REFERENCES
POPC	TMR ligand	5×10^{-2}	FCM (Nishimura et al., 2014)
POPC	AF488 (740 Da)	1.7×10^{-2} (prepared by transfer) 1.4×10^{-2} (prepared by swelling)	
POPC	Glycerol	1.7×10^{-4}	Video-microscopy (Peterlin et al., 2009)
POPC	Urea	0.013 ± 0.001	Osmotic swelling (Peterlin et al., 2012)
POPC	0.1 M Glycerol 0.11 M Glycerol 0.2 M Glycerol	0.0053 ± 0.006 0.074 ± 0.006 0.019 ± 0.006	
POPC	1.1 M Ethylene glycol 0.2 M Ethylene glycol	0.046 ± 0.006 0.085 ± 0.01	
DPPC/DOPC/Chol (1/1/1)	PEG-4-NBD	$1.13 \pm 0.08 \times 10^{-1}$	Scanning disc confocal microscopy (Li et al., 2010)
DPPC/DOPC/Chol (1/1/1)	PEG-8-NBD	$2.04 \pm 0.17 \times 10^{-3}$	
DPPC/DOPC/Chol (1/1/1)	PEG-12-NBD	$2.27 \pm 0.21 \times 10^{-4}$	

20.2 PERMEABILITY TO WATER

The permeability of membranes to water is relatively high. Osmotic gradients across the membrane therefore result in transport of water. In hypotonic solutions, where the osmolarity of the exterior solution is lower than in the interior of the vesicle, the influx of water results in a volume increase. Such vesicle swelling experiments can, at known water permeability, be used to determine the elastic modulus of membranes (Rivers and Williams, 1990); membrane elasticity will oppose swelling. However, although the permeability for water is much higher than for other solutes even a small permeability for the solute will slow down the increase in vesicle volume. In membrane elasticity experiments using swelling, impermeability with respect to solutes is therefore a prerequisite (Rivers and Williams, 1990; Degier, 1993). Because GUV membranes cannot support large hypotonic gradients, swelling will eventually result in vesicle lysis. To induce water flux and quantify the water permeability of a membrane, GUVs are typically exposed to a hypertonic solution. Because vesicles behave as ideal osmometers (Degier, 1993), the loss of vesicle volume can be followed with video microscopy techniques and subsequently used to quantify the water permeability, p_w.

20.2.1 GUV SHAPE CHANGES

In some cases, the changes in volume can be visualized and quantified in a relatively easy manner. GUVs composed of lipid extracts from rye plasma membranes vesiculate upon osmotic contraction (Steponkus and Lynch, 1989). GUVs of egg phosphatidylcholine (eggPC) were observed to retain their spherical shape during osmotic shrinkage (Boroske et al., 1981) and develop spherical membrane invaginations (Bernard et al., 2002). The volume reduction of GUVs made of 1,2-di-oleoyl-*sn*-glycero-3-phospho-(1′-*rac*-glycerol) (DOPG) and cholesterol (Chol) was observed to be accompanied by the formation of daughter vesicles inside the shrinking giant (Claessens et al., 2008). When during osmotic shrinkage the shape of the GUVs remains approximately spherical and the size distribution of the daughter vesicles is narrow, the decrease in the outer vesicle radius can be used to determine the water permeability of the membrane (Figure 20.1a,c). To be able to obtain p_w, the decrease in outer radius of single vesicles has to be followed in time and the size of the internal daughter vesicles has to be determined.

To enable accurate tracking of the vesicle contour and measure the radius, the optical contrast has to be improved. In experiments using phase-contrast microscopy a refractive index asymmetry is created for this purpose. In the GUV preparation procedure, the dried lipid film is therefore often hydrated in a sucrose containing solution after which the GUVs are diluted in an equiosmolar glucose or electrolyte buffer (Chapter 1). Alternatively, fluorescence microscopy can be used to track the radius, in this case optical contrast is created by doping the membrane of interest with fluorescently labeled lipids (Chapter 10). Considering the sensitivity of p_w to the properties of the hydrophobic phase of the bilayer a small percentage of headgroup-labeled lipids may be preferable over tail labeled lipids. However, one should keep in mind that both the creation of a sucrose gradient and the use of fluorescently labeled lipids have their own disadvantages. The density difference provided by sucrose may result in shape changes and affect the fluctuation spectrum of the GUVs (Chapter 14), whereas fluorescence microscopy requires larger acquisition times (Dimova et al., 2006).

Besides improving the optical contrast, incorporating sucrose to create a density difference between the GUV interior and exterior solution has an additional advantage; it causes the GUVs to sediment, which makes them more easily accessible for long time imaging (as further discussed in Chapter 1). Although sedimentation facilitates long time imaging of vesicles, it is by itself not enough to stabilize the position of the vesicle in convective currents that may arise when a hypertonic solution is added to deflate the vesicles. It may therefore be necessary to anchor the vesicle to the imaged surface of the microscopy chamber. This can be achieved, for example, by doping the GUVs with biotinylated lipids and immobilizing

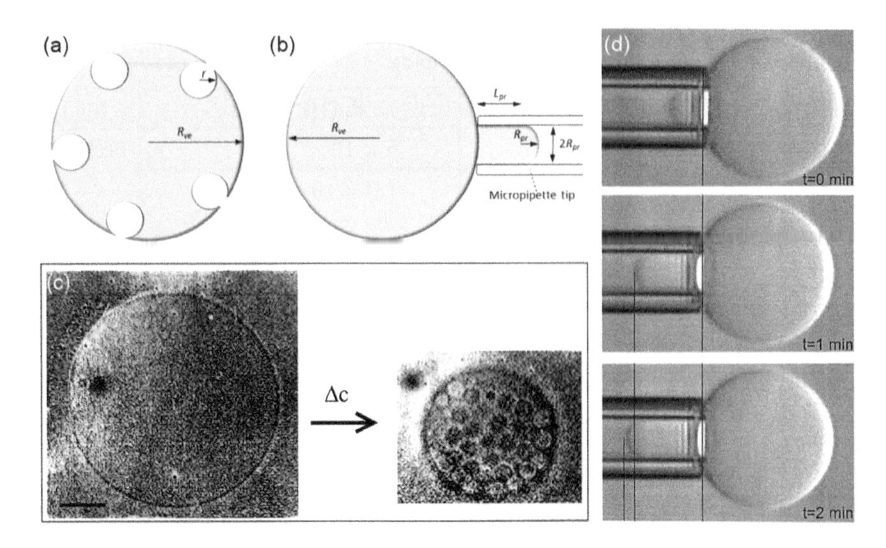

Figure 20.1 The transport of water through GUV membranes in hypertonic osmotic gradients results in loss of vesicle volume. The time course of these volume changes can be used to obtain the water permeability of the membrane, p_w. (a) Under certain conditions, volume loss leads to daughter vesicles with a narrow size distribution inside the vesicle. Then, p_w can be estimated from the change in vesicle radius. (b) Micropipette aspiration experiments on deflating vesicles provide a more general and elegant way to obtain p_w (see also Chapter 11). For this purpose, the geometry of the aspirated vesicle in conjunction with the established osmotic gradient at a fixed pipette pressure have to be taken into account. (c) Phase-contrast image of a GUV in isotonic solution (left), and the same vesicle (at the same magnification) 15 min after a controlled osmotic shock of glucose with $\Delta c = 250$ mM. The osmotic shock has resulted in the formation of daughter vesicles as sketched in (a). (Reprinted from *Biochim Biophys Acta Biomembr*, 1567, Bernard, A.L. et al., Raspberry vesicles, 1–5, Copyright 2002, with permission from Elsevier.) Scale bar: 10 μm. (d) Video micrographs of a single bilayer vesicle during the course of deflation in a water permeation experiment as sketched in (b). The increase in aspiration length inside the pipette is related to the reduction in vesicle volume. (Reprinted from *Biophys J*, 79, Olbrich, K. et al., Water permeability and mechanical strength of polyunsaturated lipid bilayers, 321–327, Copyright 2000, with permission from Elsevier.)

them on a streptavidin coated glass surface (Apellaniz et al., 2010; Su et al., 2016). In the simplest approach, streptavidin is immobilized on UV-O_3–cleaned glass coverslip by incubating a 1 μg/mL streptavidin in phosphate-buffered saline (PBS) overnight at +4°C in a humidified chamber. These coverslips are washed with PBS and incubated with GUVs containing up to 1% biotinylated lipids for at least half an hour. The unbound GUVs can subsequently be gently washed away. Because the streptavidin coating procedure is not very efficient the adhesion of the GUVs is not very strong. In addition, the biotinylated lipids have significant flexibility and therefore the tension imposed on the membrane by this immobilization method is minimal.

However, strong adhesion may cause the vesicles to become truncated spheres, see also Chapter 17. This shape change may significantly complicate the interpretation of experiments. Therefore, alternatively, special chambers can be constructed in which the convection from thermal and concentration gradients is broken. Bernard et al. used a microscopy chamber that was compartmentalized by an aluminum oxide membrane with a pore size of 0.2 μm (Bernard et al., 2002). In this configuration the vesicle position in the bottom part of the chamber remained stationary even after the solution of glucose was added to the top. Microfluidic systems in which GUVs can be trapped, studied in real time and released (Nuss et al., 2012) also provide an attractive alternative to chemical immobilization methods. Another alternative method for vesicle immobilization was recently reported, where the GUVs are caged in a meshwork of agarose (Lira et al., 2016).

When the equiosmolar solution outside the vesicle is replaced by a hypertonic solution, water withdraws from the vesicle. In bright-field microscopy images eggPC GUVs are observed to remain approximately spherical during shrinkage (Figure 20.2a). The radius of the GUV $R_{ve}(t)$ decreased linearly with time (Figure 20.2b). To obtain p_w from this decrease in $R_{ve}(t)$ we rewrite the expression for the flux in terms of the difference in molar concentration of the interior and exterior solution Δc:

$$J = \frac{1}{A}\frac{dV}{dt} = p_w v_w \Delta c \tag{20.2}$$

where v_w represents the molar volume of water; dV/dt the change in vesicle volume per unit time; and Δc the osmotic gradient over the membrane. We assume that the total vesicle membrane area A remains constant and equal to $A = 4\pi R_0^2$ where R_0 is the initial GUV radius.

In the simplest case, when the GUVs retain their spherical shape and the changes in osmotic gradient due to shrinkage can be neglected, Eq. 20.2 can be rewritten as (Boroske et al., 1981):

$$\frac{dR}{dt} = p_w v_w \Delta c \tag{20.3}$$

The observed linear relation between the change in R_{ve} and t thus gives direct access to p_w (Figure 20.2b). However, phase-contrast microscopy experiments indicate that the GUVs do not remain simple spheres. In eggPC/dicetyl-phosphate/Chol (Bernard et al., 2002) and DOPG vesicles containing 0%, 20% or 40% Chol (Claessens et al., 2008) the decrease in GUV radius coincides with the appearance of N smaller similarly sized vesicles in the interior face of the GUV membrane (Figure 20.1). The appearance of these small daughter vesicles has consequences for the determination of p_w. There are several factors that have to be taken into account (Bernard et al., 2002; Claessens et al., 2008). The first factor is the

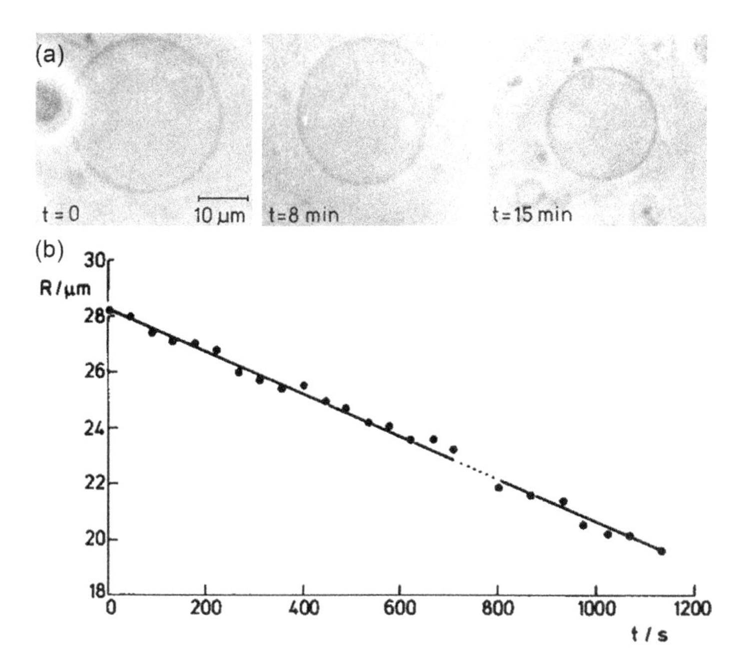

Figure 20.2 (a) Bright-field images of osmotic shrinkage of a GUV induced by a concentration difference of $\Delta c = 15$ mM. The shapes represent an initial spherical state ($t = 0$), an unstable state ($t = 8$ min), and another spherical state ($t = 15$ min). (b) Radius of an osmotically shrinking spherical vesicle with an initial radius $R_0 = 28.2$ µm as a function of time. The drawn line represents a linear fit, the dotted portion indicating a period of shape instability. The applied initial osmotic gradient was $\Delta c = 10$ mM. (Reprinted from *Biophys. J.*, 34, Boroske, E. et al., Osmotic shrinkage of giant egg-lecithin vesicles, 95–109, Copyright 1981, with permission from Elsevier.)

change in volume. Assuming that the membrane area is conserved, that is, $R_0^2 = R_{ve}(t)^2 + N(t)r^2$, $V(t)$ is a function of the radius of the mother and daughter vesicles, R_{ve} and r:

$$\frac{dR}{dt} = \frac{p_w v_w}{\left(R_{ve}(t)^2 + 2/3rR_{ve}(t)\right)} \left(\frac{c_0 R_0^3}{\left(R_{ve}(t)^3 + rR_{ve}(t)^2 - rR_0^2\right)} - c_b \right) \tag{20.4}$$
$$(1-\alpha)R_{ve}(t)^2 + \alpha R_0^2)$$

The value of r in this expression can be determined experimentally or calculated from the final R/R_0 ratio (Claessens et al., 2008). A second complicating factor is that osmotic shrinkage causes the gradient over the membrane Δc to decrease. This change in the osmotic gradient can be accounted for. Because $c(t) = c_0\left(V_0/V(t)\right)$, and $\Delta c = c(t) - c_0$, Δc can be calculated at any time using Eq. 20.4. The assumption that the membrane area over which permeation takes place is conserved may also not be correct. Experiments indicate that the osmotic deflation experiment is not completely reversible. This indicates that some daughter vesicles may detach from the mother GUV or are not readily available for inflation (Claessens et al., 2008). In that case the area that is accessible for water transport at any time is given by: $A(t) = 4\pi[R_{ve}(t)^2 - (\alpha R_0^2 + R_{ve}(t)^2)]$. In this expression α represents the area fraction of vesicles that remain continuous with the vesicle bilayer. Because the daughter vesicles entrap the hypertonic solution from the outside they are considered osmotically inactive once they are released.

Taking into account all these factors Eq. 20.2 can be rewritten as

$$\frac{dR}{dt} = \frac{p_w v_w}{\left(R_{ve}(t)^2 + 2/3rR_{ve}(t)\right)} \left(\frac{c_0 R_0^3}{\left(R_{ve}(t)^3 + rR_{ve}(t)^2 - rR_0^2\right)} - c_b \right) \tag{20.5}$$
$$(1-\alpha)R_{ve}(t)^2 + \alpha R_0^2)$$

in which c_b is the final concentration in- and outside of the vesicle. Equation 20.5 can be solved analytically but this results in a lengthy implicit relation. Fitting this equation to the derivative of the observed size decrease to obtain p_w and α is therefore more convenient. Whereas the initial slope of R versus time curves is sensitive to p_w, the value of α can be obtained from the $R(t)$ dependence at higher values of t. Using this method the permeability of DOPG vesicles was observed to decrease with increasing cholesterol content (Figure 20.3). Whereas for DOPG vesicles containing 20% Chol, one finds $p_w = 15.3 \pm 3.4$ µm/s, its value is decreased to $p_w = 6.6 \pm 0.5$ µm/s when the membrane contained 40% Chol (Claessens et al., 2008). Using the linear relation between the decrease in vesicle radius and time at short time scales, yields $p_w = 10 \pm 1$ µm/s for eggPC/dicetyl-phosphate/Chol GUVs (Bernard et al., 2002) (Table 20.1). The value obtained for α can be later confirmed in a control reswelling experiment.

20.2.2 MICROPIPETTE ASPIRATION EXPERIMENTS

Micropipette aspiration experiments (see also Chapter 11) provide a more elegant but technically more demanding method to determine the water permeability of membranes (Ramahaleo et al., 1999; Olbrich et al., 2000; Rawicz et al., 2008). When aspirated GUVs are placed into a hyperosmotic environment, the volume loss at constant membrane area can be directly followed using video microscopy. As a result of the efflux of water, the membrane projection length L_{pr} inside the micropipette will increase in time

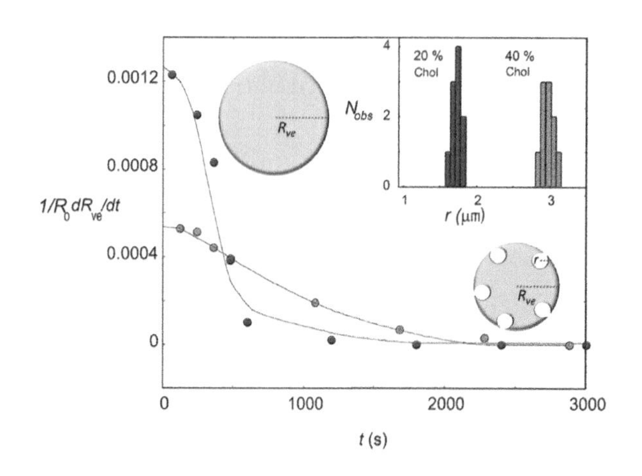

Figure 20.3 The measured $1/R_0\, dR_{ve}/dt$ as a function of time for GUVs composed of DOPG/20% Chol (●) and DOPG/40% Chol (O) at an initial osmotic gradient of 200 mM. The data are fitted with Eq. 20.5, which results in an average $p_w = 15.3 \pm 3.4$ µm/s and 6.6 ± 1.5 µm/s, respectively. The average value for α is 0.14 ± 0.4 and did not change with the membrane composition. The inset shows the histogram of daughter vesicle radii for DOPG GUVs with 20% or 40% Chol. (Reprinted from *Biochim Biophys Acta Biomembr*, 1778, Claessens, M.M.A.E. et al., Osmotic shrinkage and reswelling of giant vesicles composed of dioleoylphosphatidylglycerol and cholesterol, 890–895, Copyright 2008, with permission from Elsevier.)

(Figure 20.1b,d). Accurate video tracking of the projection edge, enabled by the creation of optical contrast between the GUV interior and exterior, makes a discrimination of <0.1% relative change in vesicle area or volume possible (Olbrich et al., 2000).

To determine p_w, individual GUVs are aspirated by a micropipette at a suction pressure that sets the membrane tension in the range of 0.5–1 mN/m (Olbrich et al., 2000; Rawicz et al., 2008). The tension should be chosen such that the thermal fluctuations (see Chapter 14) are ironed out but actual stretching of the membrane is prevented (see Chapter 11, Figure 11.8). At low membrane tension, area hidden in thermal fluctuations causes the projection length to increase faster than expected based on the water permeability of the membrane (Vitkova et al., 2004). High membrane tension may result in the formation of membrane pores (Vitkova et al., 2004) and collapse (lysis) of the GUV. For appropriate intermediate tensions the shape of the vesicle outside the pipette is perfectly spherical and the projection length of the vesicle inside the pipette is limited. Such an aspirated vesicle is subsequently transferred to a solution with ~10% higher solute concentration (Olbrich et al., 2000). Higher or lower osmolarities result in linearly proportional faster or slower permeation rates with the same value for p_w. To prevent contact with air during transfer between chambers with different osmolarity, the aspirated vesicle can be maneuvered into a larger micropipette (Olbrich et al., 2000; Rawicz et al., 2008). Once transferred, the higher solute concentration around the vesicle drives water transport over the membrane until a new osmotic equilibrium is reached. During the osmotic deflation experiment the vesicle is held fixed under a small suction pressure to control the bilayer tension (~1 mN/m). The changes in vesicle volume ΔV thus occur at a constant area, A, and can be calculated from the changes in the projection length, ΔL_{pr}, inside the pipette. The change in volume can be approximated by (Olbrich et al., 2000)

$$\Delta V \sim -\pi R_{pr}\left(R_{ve} - R_{pr}\right)\Delta L_{pr} \tag{20.6}$$

where R_{pr} is the inner radius of the micropipette; and R_{ve} is the radius of the spherical part of the GUV (Figure 20.1b). In this case, Eq. 20.2 can be rewritten to

$$\frac{dV^*}{dt} = -p_w c_\infty v_w \left(\frac{A}{V_\infty}\right)\left[\frac{\left(V^* - 1\right)}{V^*}\right] \tag{20.7}$$

With $V^* = V / V_\infty$ and c_∞ is the final osmolarity of the solution. The decrease in dimensionless vesicle volume relative to the initial dimensionless volume V_0^* at $t = 0$ thus follows the equation:

$$\left(V^* - 1\right)e^{V^*} = \left(V_0^* - 1\right)e^{-kt + V^*} \tag{20.8}$$

where $k = -p_w c_\infty v_w (\frac{A}{V_\infty})$. Nonlinear fitting of this equation to the experimentally observed decrease in the dimensionless volume with time gives access to the time constant, $1/k$. From the time constant associated with volume change to a new equilibrium, the apparent change in p_w can be calculated (Figure 20.4). Using this method, the membrane permeability to water, p_w, was also observed to decrease with increasing cholesterol content (Rawicz et al., 2008). Because the water permeability is mainly limited by the solubility of water in the hydrocarbon part of the bilayer, it is expected to be a function of phospholipid (poly)unsaturation. For mono- and di-mono-unsaturated phosphatidylcholine (PC) bilayers only a modest variation in p_w from ~30 to 40 µm/s was observed, with subtle variations between different positions of the double bonds and *cis* or *trans* configurations (Table 20.1).

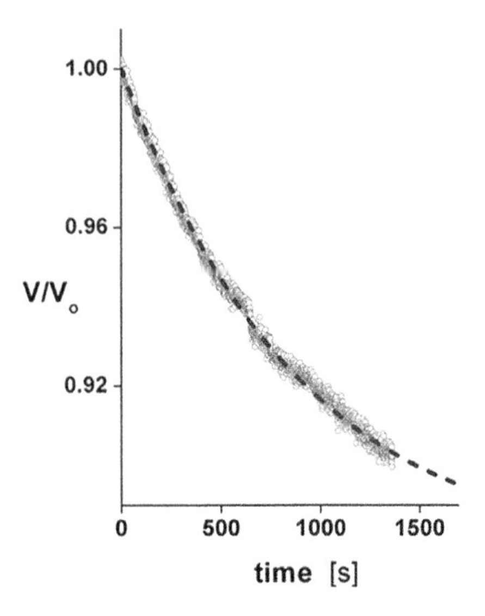

Figure 20.4 Example of an osmotic deflation experiment from a 1/1/1 1-stearoyl-2-oleoyl-*sn*-glycero-3-phosphocholine (SOPC)/SM/Chol GUV measured at 15°C after transfer from 205 mOsm solution to 236 mOsm. Superposed is the single parameter fit of Eq. 20.7 defined by the permeation rate, which yields the coefficient p_w for water permeability. (Reprinted from *Biophys J*, 94, Rawicz, W. et al., Elasticity, strength, and water permeability of bilayers that contain raft microdomain-forming lipids, 4725–4736, Copyright 2008, with permission from Elsevier.)

The introduction of two or more *cis*-double bonds in a chain does have a dramatic effect on the apparent water permeability. It was observed to rise from ~50 µm/s for c18:0/2 to 90 µm/s for diC18:2 and 150 µm/s for diC18:3 (Olbrich et al., 2000) (Table 20.1). The measurements of water permeability were found to scale exponentially with the reduced temperature $T_r = \left(T - T_m\right) / T_m$ where T is the temperature at which the experiment was performed and T_m is the gel-to-liquid crystalline phase transition temperature (Olbrich et al., 2000). This correlation supports the idea that the increase of free volume due to thermal expansion above the main gel–liquid crystalline transition of the bilayer is a major factor in water transport. This is consistent with the partition of solutes in the hydrocarbon region of the bilayer being strongly affected by chain ordering, which diminishes progressively with T_r in bilayers (Xiang and Anderson, 1997). Generally membrane water permeability seems to correlate with the rupture tension of the membranes (Rawicz et al., 2008). However, the apparent water permeability of ternary mixtures of PC/sphingomyelin (SM)/Chol containing microdomains deviates from this trend; it is low compared to membranes with similar rupture tension (Rawicz et al., 2008) (Table 20.1). This low rupture tension is not well understood but does not support the idea that there is considerable water leakage at micro domain boundaries.

Although technically more demanding, the micropipette aspiration is generally the preferred method to study membrane water permeability. This method does not suffer from additional shape changes that have to be accounted for, and more importantly this method is much more versatile. It additionally allows the determination of the elastic moduli and lysis tension of the membrane and thereby provides a more complete picture of the physical properties of the lipid bilayer.

20.3 PERMEABILITY TO OTHER MOLECULES

The partition coefficient B of molecules increases with their hydrophobicity and more lipophilic molecules are therefore thought to cross membranes more readily. The idea that the membrane permeability of a molecule increases with its hydrophobicity is known as Overton's rule. There has been quite some discussion in the literature on the validity of this rule (Grime et al., 2008a, 2008b; Missner et al., 2008; Li et al., 2010, 2011). To clarify how molecular structure relates to the membrane permeability of molecules, accurate membrane permeability measurements are required. Although the transport of water can be measured directly from the associated changes in volume, the quantification of the influx of other molecules is often more challenging. Osmotic swelling experiments have been reported to give sensitive measurements of the membrane permeability of solutes (Peterlin et al., 2012). The influx of fluorophores can be directly visualized in (fluorescence) microscopy or spectroscopy experiments, but to quantify the transport of other molecules over the GUV membrane reporter molecules are required (Figure 20.5).

20.3.1 pH-SENSITIVE DYES

To follow the transport of weak acids over the GUV membrane, pH sensitive dyes can be used as reporter molecules. For this

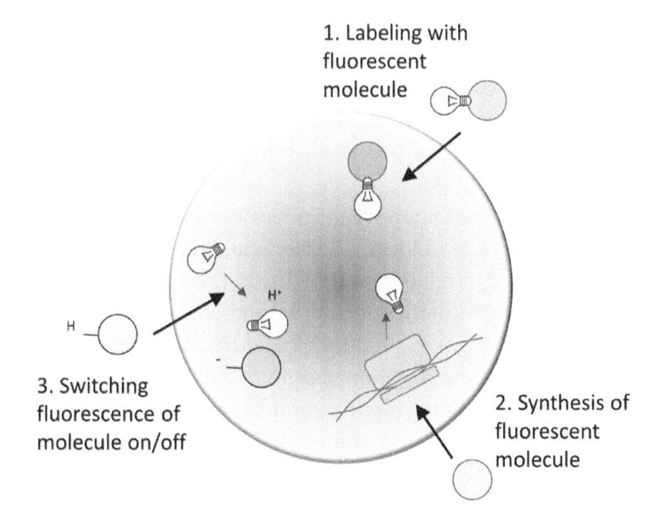

Figure 20.5 Fluorescent molecules are convenient reporters to visualize the transport of molecules into single GUVs: (1) The permeation of fluorescently labeled molecules, represented by yellow light bulbs coupled purple spheres, into GUVs can be directly visualized. (2) To follow the permeability of molecules that are required for translation of DNA into proteins (light blue sphere), a translation system can be reconstituted in GUVs. Synthesis of a fluorescent protein (light bulb) can thus be used as a readout for membrane permeation of the required molecule (light blue sphere). (3) By encapsulating pH-sensitive dyes in GUVs (light bulb), the transport of weak acids over the membrane can be followed in time. The on and off state of the fluorophore are depicted by a yellow and blue light bulb, respectively.

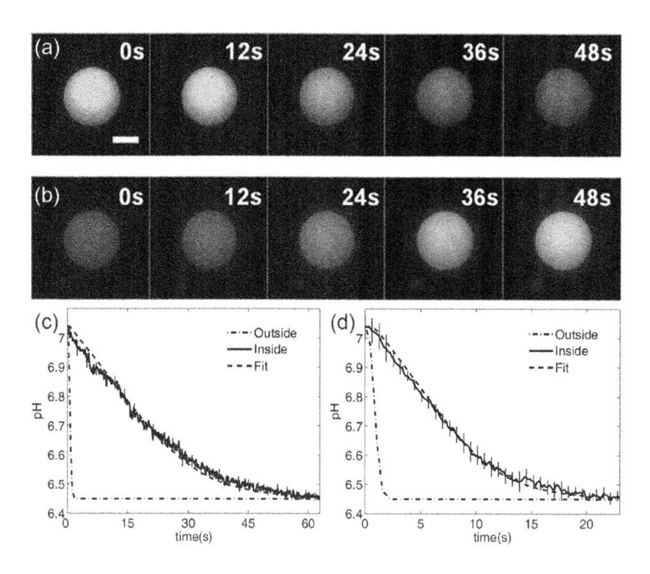

Figure 20.6 Time series of scanning disc confocal microscopy showing the fluorescence intensity change inside GUVs encapsulating the pH sensitive dye fluorescein coupled to a 40 kD dextran when acetic acid is transported (a) into and (b) out of a GUV. Scale bar: 20 μm. Ex 491 nm, Em 525 nm. (c, d) Solution pH change inside and outside the GUV with time for (c) acetic acid, (d) propionic acid. The solid line and the dot-dashed line represent the pH change inside and outside GUV. The dashed lines are finite difference modeling fitting results. (Reprinted from *Biophys J*, 101, Li, S. et al., Imaging molecular transport across lipid bilayers, 700–708, Copyright 2011, with permission from Elsevier.)

purpose, Li et al. encapsulated a 40 kD fluorescein-dextran in GUVs by rehydrating the lipid film used for GUV production in a 1 mg/mL 40 kDa fluorescein-dextran in 2 mM HEPES buffer containing 200 mM sucrose at pH 7.4 (Li et al., 2011). Using spinning-disc confocal microscopy, the speed with which the pH gradient changes upon addition of the acid was followed in real time and the time evolution of the concentration profile was determined (Figure 20.6). The transport of acids over the lipid bilayer was observed to be a fast phenomenon. Depending on the acid, it can take seconds to a minute to equilibrate the acid concentration (Li et al., 2011). Keep in mind that in fluorescence microscopy experiments it is important to take into account possible photobleaching of the fluorophores. As for other long term experiments, anchoring of the GUVs to the surface of the sample chamber may be required. Ideally the experiments are performed in microfluidic chambers. Such chambers allow changing the buffer and make it possible to create a flow enhancing the exchange. Moreover, with microfluidic systems influx and efflux of molecules can be followed in the same vesicle. By monitoring, in a microfluidic system, the pH changes caused by the flux of weak acids into GUVs that encapsulate fluorescein, it was shown that the change in pH occurs faster for the more lipophilic acids (Li et al., 2010, 2011).

A problem with permeability studies is that although you can have well-mixed bulk fluids on either side of the membrane, the region directly adjacent to the membrane is stagnant due to non-slip boundary conditions. In this so called unstirred boundary layer there is no mixing by convection and the diffusive transport dominates. Although the bulk concentrations on either side of the membrane can be easily measured, the concentration directly adjacent to the membrane determines the flux over the membrane. To circumvent the problems associated with this unknown concentration, Li et al. experimentally measured how the concentration profiles inside the GUVs changed with time. By fitting the sets of concentration profiles with a diffusion equation based finite difference model that was developed to simulate the experimental process p_s was obtained (Figure 20.6c,d) (Li et al., 2011) (Table 20.2).

20.3.2 FLUORESCENT REPORTER MOLECULES

To measure the permeability of GUV membranes to other molecules than weak acids, fluorescent reporter molecules have been used. Some of them are known to be fluorescent per se that is, fluorescent proteins such as green fluorescent protein (GFP) and allophycocyanin (APC); specific proteins and dextrans labeled with fluorescent dyes; free fluorescent dyes such as Alexa Fluor (AF)™ and ATTO™ dyes, whereas others become more fluorescent when they are bound to their substrate. For example, the fluorescence of the nucleic acid dye YO-PRO-1 increases significantly once it binds DNA. To study the membrane permeability of the YO-PRO-1 dye, DNA was injected into single GUVs. The controlled injection of soluble and insoluble substances into GUVs using borosilicate glass microneedles was first described by Wick et al. (1996) and Bucher et al. (1998). To inject 180 fL using a microinjection needle with an inner diameter of 100 nm and an outer diameter of 205 nm, a pressure of 2000 hPa had to be applied over 10 s for Ficoll 70 and 1 s for oleic anhydride (Wick et al., 1996). After the injection of DNA into the GUVs, 1 µM YO-PRO-1 was added to the outside solution and subsequently the increase in the fluorescence intensity of the DNA/YO-PRO-1 complex (Ex/Em = 491/507 nm) inside the GUV was followed in time. The permeability of a 1-palmitoyl-2-oleoyl-*sn*-glycero-3-phosphocholine (POPC) membrane to this dye could thus be established (Fischer et al., 2000), see Table 20.3. In addition, the permeability of POPC GUV membranes to fluorescein diphosphate was determined in a similar way but now using an enzymatic assay (Table 20.3). Exposure to the enzyme alkaline phosphatase causes the fluorescein diphosphate to dephosphorylate. The final hydroxylation product is highly fluorescent (Ex/Em = 485/528 nm). By exposing GUVs that were microinjected with enzyme to fluorescein diphosphate containing solutions, the increase in fluorescence intensity could be followed in time due to substrate transport into the vesicle interior (Fischer et al., 2000). However, transport of these charged molecules was not observed in experiments with LUVs, suggesting that the GUVs membranes might not be intact or be more fragile, and calling for careful control of membrane integrity. In their experiments Fischer et al. inspected the integrity of the bilayers by confirming that there was no leakage of DNA, RNA or fluorescent nucleotides from the GUVs.

Besides microscopy experiments, flow cytometry (FCM) has also been used to probe the permeation of molecules through POPC GUV membranes at the single vesicle level (Nishimura et al., 2014) (Table 20.3). FCM can be used to measure the properties of a large population of particles (i.e., GUVs, cells). Unilamellar vesicles can be discriminated from multilamellar ones by evaluating the scattering properties (i.e., the forward-scattered light and the side-scattered light are proportional to the size and the internal complexity of the particle, respectively). In this particular case, a solution containing the GUVs encapsulating a fluorescent reporter molecule was directed into a thin stream so that all the GUVs could pass in a single file and be detected individually (Nishimura et al., 2012). In general the permeability of single GUV membranes to polar and charged molecules was low compared to nonpolar molecules. However, for several charged molecules at least two distinct GUV populations could be discriminated. Although most vesicles were impermeable to small charged molecules, approximately 10% of the GUVs were permeable to nucleotides, lysine and arginine (Nishimura et al., 2014). Hence, this finding highlights the importance of single vesicle experiments on large vesicle populations.

To address the permeability of GUVs in FCM experiments, several approaches were used to report on the influx of molecules that are also suited for fluorescence microscopy. Nucleotides and (most) amino acids are not fluorescent, their diffusion into or out of a GUV can therefore not be directly observed. To circumvent this problem Nishimura and colleagues introduced the PURE system in their POPC GUVs and used it to produce GFP inside the GUVs. The PURE system contains all the components necessary for GFP synthesis in a cell-free manner. Even when encapsulated in GUVs, GFP production commenced in most vesicles. In agreement with this observation, with GUV volumes ranging from 1 to 100 fL, only 0.01% of the GUVs were calculated to lack one or more components. An additional membrane impermeable dye (i.e., R-phycoerythrin) was incorporated inside the vesicles to establish that the membrane of the non-GFP producing GUVs was intact (Nishimura et al., 2012). By depleting the PURE system in the GUVs of one of the molecules necessary for GFP production and adding it to the outside solution, the transport of this molecule over the membrane could be followed by measuring the GFP fluorescence intensity (Nishimura et al., 2014).

To determine the size limit of the permeable molecules, poly-dA oligonucleotides with different lengths were labeled with biotin and AF 350 at the 5′ and 3′ ends, respectively. These labeled oligonucleotides were subsequently added to streptavidin containing vesicles and the fraction of GUVs containing the constructs was measured as a function of the oligonucleotide length (Nishimura et al., 2014).

However, confocal fluorescence microscopy shows important advantages over FCM. During the last years, recent improvements in software development have led to the efficient analysis of permeabilization kinetics of a high number of GUVs at a single vesicle level (Hermann et al., 2014). In addition, microscopy provides additional information about possible morphological changes in vesicles during the permeabilization process that could be related to the formation of membrane domains and deformations (e.g., highly curved structures) (Lorent et al., 2014; Sankhagowit et al., 2014).

20.4 ALTERATION OF MEMBRANE PERMEABILITY: PORE FORMING PROTEINS AND PEPTIDES

The interaction of amphipathic molecules with membranes often interferes with the integrity and permeability of membranes. This membrane disruption can involve several mechanisms including pore formation and membrane thinning. It results in an all-or-none or graded loss of probes entrapped in the vesicles (see Chapter 24). In general, these membrane disruptive molecules switch between soluble and membrane-inserted conformations. The best studied examples include toxins and antimicrobial peptides, for example, Maganin 2, Equinatoxin II, peptides derived from the membrane-proximal external region of the HIV fusion glycoprotein gp41 subunit, and the bee venom mellitin (Tamba and Yamazaki, 2005; Schoen et al., 2008; Apellaniz et al., 2010; Kokot et al., 2012), proteins involved in mitochondrial permeabilization during apoptosis such as Bcl-2 proteins and derived peptides, for example, Bcl-xLΔCt, Baxα5 peptide, Bax and BakΔC21 (Garcia-Saez et al., 2009; Fuertes et al., 2010; Bleicken et al., 2013a, 2013b), oligomers of proteins involved in protein aggregation diseases such as α-synuclein or $A\beta$ (van Rooijen et al., 2010b), and unconventional secreted proteins directly passing membranes, for example, fibroblast growth factor 2 (Steringer et al., 2012). The mechanisms of membrane disruption and resulting content loss are not always well understood. Experiments on the permeabilization of single GUVs complement observations with SUVs, LUVs and supported lipid bilayers and can improve our understanding of the physical mechanisms responsible for membrane damage (Lee et al., 2008; Schoen et al., 2008; Apellaniz et al., 2010; Fuertes et al., 2010; van Rooijen et al., 2010b; Kokot et al., 2012). Stochasticity of release onset, heterogeneities in solute encapsulation and the existence of mixed mechanisms within the sample are a few of the parameters that can be obtained from studies on a population of single GUVs that are not easily accessible in bulk studies on SUVs or LUVs.

To assess if GUVs are permeabilized by specific molecules, the GUVs are produced, by electro-swelling or another method, in a buffer without fluorophores after which the GUV containing solution is diluted with an equiosmolar solution containing fluorophores such as the AF dyes (Schoen et al., 2008; Apellaniz et al., 2010; Fuertes et al., 2010). Alternatively the vesicles can be made in the presence of a fluorophore such as 8-hydroxypyrene-1,3,6-trisulfonic acid (HPTS), after which the outside fluorescence is quenched by a small molecule such as *p*-xylene-bis(*N*-pyridinium bromide) (DPX) (van Rooijen et al., 2010b). The influx of the fluorophore or quencher can subsequently be followed in time using confocal microscopy (Figure 20.7). The distribution of filling degrees for individual GUVs is associated with all-or-none, graded or mixed vesicle permeabilization mechanisms (Apellaniz et al., 2010). The distribution of lag-times associated with the refilling of individual vesicles provides information on the stochasticity of the permeabilization process (Schoen et al., 2008; Apellaniz et al., 2010; Fuertes et al., 2010). Although for many of these systems the influx of fluorophores or quenchers can be followed in time with high time resolution, it is often difficult to calculate p_s from this data. Multiple processes may happen at the same time and pore formation can be dynamic, resulting in ill-defined or time dependent value for p_s. As an alternative the membrane flux is therefore quantified. To allow for this quantification, it has been ensured that the fluorescence intensity scales with the concentration of fluorophore at the concentrations used. To determine J, the fluorescence intensity inside a GUV $I(t)$ at any time is normalized with respect to the fluorescence at time zero $I(0)$ and to the fluorescence outside the vesicle $I_{ex}(t)$. Because the volume of the solution outside the vesicle is large compared to the volume of the vesicles, $I_{ex}(t)$ is expected to be constant. Photobleaching of the fluorophores may however slightly affect $I_{ex}(t)$. Using the expression for the volume flux J (20.2) the concentration of fluorophore inside the vesicle with radius R is predicted to change as:

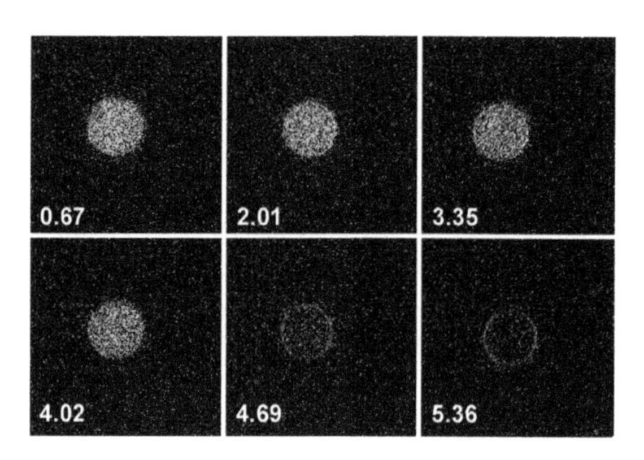

Figure 20.7 Confocal microscopy images of α-synuclein oligomer induced dye leakage from POPG GUVs. The GUVs are filled with the dye HPTS (green) and the quencher DPX was present on the outside of the vesicles. The GUV membrane was stained with 1,2-dioleoyl-*sn*-glycero-3-phosphoethanolamine (DOPE)-Rhodamine (red), see Appendix 2 of the book for structure and data on this probe. The consecutive images (time stamp in seconds) show the kinetics of dye efflux from a single GUV. (Figure reproduced from van Rooijen, B.D. et al., *Plos One*, 5, e14292, 2010b.)

$$c_{\text{in}}(t) = 1 - e^{-\frac{3J}{R}t}. \tag{20.9}$$

The slope of a plot of $-\frac{R}{3}\ln(1-c(t))$ directly gives the volume flux J (Schoen et al., 2008). Analysis of the flux induced by equine-toxin II and Baxα5 shows broad distributions in J (Schoen et al., 2008; Fuertes et al., 2010). These broad distributions may arise from differences in the size or number of pores or defects per vesicle. Differences in Laplace pressure could cause the same effect. To obtain better insights into physical mechanisms behind the experimental observation on pore formation, comparison with more complex modeling approaches that take into account changes in the distribution of pore sizes and numbers can be useful (Kokot et al., 2012).

Besides establishing the kinetics of the leakage process, GUV experiments can also address aspects concerning pore size and dynamics. For this purpose the entry of fluorophores with different sizes at different time points can be investigated (Apellaniz et al., 2010; Fuertes et al., 2010). Dyes can be coupled to specific proteins or high molecular weight molecules such as dextrans of different sizes (van Rooijen et al., 2010b; Bergstrom et al., 2013; Bleicken et al., 2013a, 2013b). The minimum pore size can be estimated by using solution Fluorescence Correlation Spectroscopy on these fluorescent reporters in order to quantify their diffusion coefficient and their hydrodynamic radius (see Chapter 21) (Bleicken et al., 2013a). Experiments with differently sized fluorophores have shown that Baxα5-induced pores in GUVs initially facilitate the influx of both AF 555 (M_w = 1,250 Da) and fluorescein labeled dextran (M_w = 10 kD). However, the entry rate of fluorescein labeled dextran decreased with time, resulting in incomplete filling of the GUVs. The smaller AF 555 dye was able to fully equilibrate over the membrane. This indicated that the pores shrunk over time, no longer allowing the entry of the larger labeled dextran (Fuertes et al., 2010). This was confirmed with photobleaching experiments; the entrapped labeled dextran could no longer exchange with the outside solution resulting in complete photobleaching of the vesicle interior (Fuertes et al., 2010). In contrast, both full Bax protein and BakΔC21 formed large, stable pores via all-or-none mechanisms and these pores were large enough to allow the passage of cytochrome c-AF488 (Cyt c) and APC into the lumen of GUVs (Bleicken et al., 2013a, 2013b). The size of these pores also evolved with time and surprisingly, it depended on protein concentration. Altogether, these results revealed that proapoptotic Bcl-2 proteins such as Bax and Bak tend to form pores tunable in size, in agreement with their proteolipid nature.

Moreover, a time delay between the addition of similar sized but spectrally different fluorophores allows distinguishing transient from permanent pores. After the equilibration of AF 555 over GUV membranes containing Baxα5 induced pores, AF 488 was added to the outside solution. The latter fluorophore equilibrated with the membrane interior with a delay. The membranes that were permeable to the first dye were also permeable to the second and even to a third dye. On the other hand, the vesicles that were initially nonpermeable remained sealed for the successive dyes. Interestingly, in the presence of Baxα5, the pores formed were not transient and the GUVs remained in the porated state (Fuertes et al., 2010). Similarly, AF 488 and AF 633 were also used as first and second dye to test the permeability effect of full Bcl-2 proteins. In case of full Bax, it showed a permeability behavior similar to Baxα5, whereas the antiapoptotic Bcl-2 protein Bcl-xL induced only transient permeability alterations (Bleicken et al., 2013a, 2016).

20.4.1 SINGLE VESICLE APPROACH: EXPERIMENTAL DESIGN AND DATA ANALYSIS

A set of methods have been recently implemented to study the membrane-permeabilizing activity of pore forming proteins and peptides (or other molecules of interest) at the single molecule level (Bleicken and Garcia-Saez, 2014). This strategy is based on the use of reduced systems composed of purified proteins/peptides and GUVs, together with fluorescence confocal microscopy. This allows for the study of membrane binding and permeabilization mechanisms of individual vesicles, providing an unprecedented degree of detail as well as a direct observation of the process under the microscope.

Here, we explain four different experimental approaches to derive detailed information about protein binding and membrane permeabilization (Figure 20.8). First, the membrane binding experiment is designed to analyze membrane association of the protein/peptide of interest (POI) into GUVs with different lipid composition (Box 20.1). Second, the mechanism of membrane permeabilization by molecules of interest (and size of individual pores) can be characterized in great detail by quantifying the degree of filling after a defined incubation time (i.e., "all-or-none" versus "graded" type) (Box 20.2), the stability of the permeabilized state (i.e., transient versus permanent membrane permeabilization) (Box 20.3), and the kinetics of filling for the individual vesicles in the GUV population (i.e., total permeabilized area and its progression over time) (Box 20.4).

In all cases, the observation chamber needs to be incubated with blocking solution around 30 min before adding the sample to it. The blocking solution can be prepared with casein or

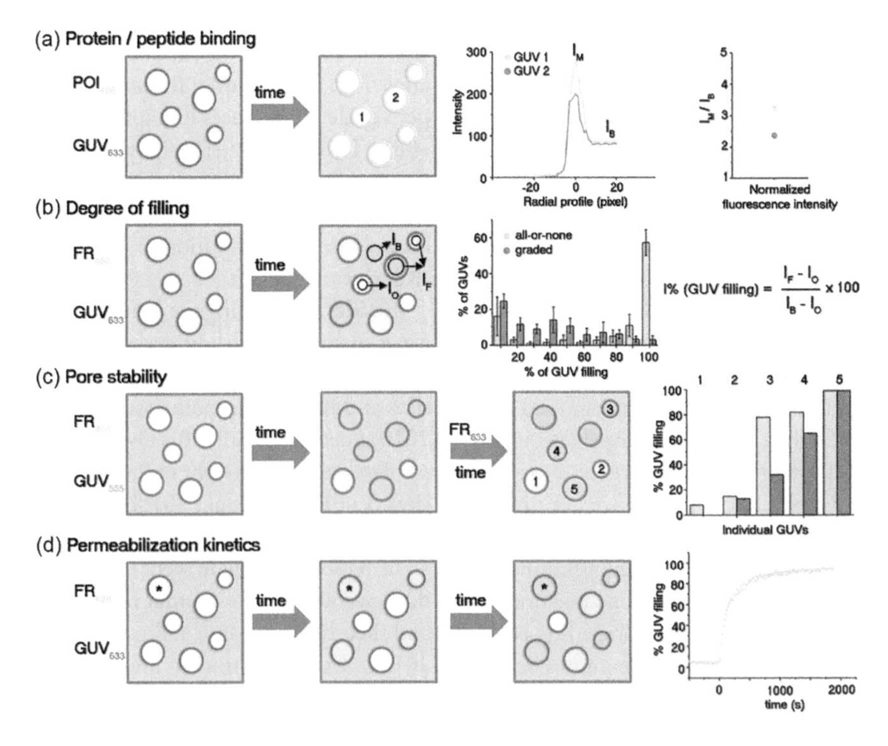

Figure 20.8 Schematic drawing and quantitative analysis examples of four different experimental setups. (a) Protein/peptide binding: radial profiles and intensity ratio after normalization of two representative GUVs. (b) Degree of filling: graph of analyzed data for "all-or-none" and graded examples. (c) Pore stability: filling of individual GUVs for both dyes. (d) Permeabilization kinetics: filling of one individual GUV (asterisk) over time. FR, fluorescent reporter; I, offset determination; I_B, background determination; I_F, GUV-filling; I_M, intensity at the membrane; POI, protein/peptide of interest. Color code: blue (488 channel); orange (633 channel); green (merge).

Box 20.1 Assessing protein/peptide binding to membranes

- For this experiment at least three test samples are prepared: (a) GUVs in buffer, (b) protein/peptide of choice in buffer, (c) all three components together. The controls are required in order to detect and remove potential channel cross talk during the setting up of the confocal microscope.
- Optimize the microscope setup so that the membrane channel shows no or minimal cross talk with the protein/peptide channel. This step is critical in the experiment as the cross talk can be detected as false protein/peptide binding. For example, GUV labeling with 1,1′-dioctadecyl-3,3,3′,3′-tetramethylindodicarbocyanine (DiD, see Appendix 2 of the book for structure and data on this probe) dye (<0.05%) is recommended when the POI is labeled with Alexa 488. Incubate at least 5–15 min to allow the GUVs to sediment at the bottom of the chamber.
- Image the samples, making sure that enough unilamellar vesicles are acquired to obtain statistically relevant results (optimally, in the order of hundreds). Due to possible protein/dye binding to the glass, images should not be taken very close to the glass in order to avoid high fluorescence signals near it. In addition, artificial vesicle structures due to glass binding could also occur in that area. Make sure that representative areas are shown. In this way, it is recommended to take images at different distances from the glass to decide which distance works best. Similar distances should also be used in all samples.
- Data analysis can be carried out manually with Image J or semi-automatically with the program "GUV detector" (Hermann et al., 2014). For manual analysis, the radial profile plug-in needs to be installed in Image J.

Box 20.2 Estimating the degree of GUVs filling

- In this experiment, the concentration of the external fluorescent reporter should be optimized (i.e., in general 50–500 nM of external fluorescent reporter is used). Here, the pore forming protein/peptide is unlabeled and its permeabilization activity is studied by analyzing the degree of filling of the GUVs (i.e., the entrance of the fluorescent reporter to the lumen of the GUVs is analyzed). For example, DiD-labeled GUVs are a good choice when Alexa 488-labeled fluorescent reporters are selected. A variation of the "degree of filling method" consists into working with big and small size markers in a three color experiment. Thus, additional information about the pore size can be achieved at the same time. To avoid cross talk problems, size markers with excitation maxima corresponding to the shortest (e.g., 488 nm) and the longest (e.g., 633 nm) wavelength laser lines are normally used, whereas the membrane dye corresponds to the laser line in between (e.g., DiI dye at 561 nm).
- A negative control is needed (i.e., the small and big size fluorescent reporters with the GUVs in the absence of the pore-forming protein/peptide) to analyze any basal permeabilization of the GUVs.
- Incubate the GUVs at least for 5–15 min before imaging to allow the vesicles to settle. Set the optimal incubation time for imaging. The timing is critical in this experiment because visualization should be performed in or close to equilibrium. It is also important to use always the same incubation time for all samples. Image enough unilamellar vesicles per reaction to have statistically relevant results. It is recommended to analyze 200–600 GUVs per reaction.
- Data analysis can be carried out semi-automatically with the program "GUV detector" (Hermann et al., 2014). Here, the data can be displayed as percentage of nonpermeabilized or permeabilized vesicles, or as percentage of GUV filling by each fluorescent reporter per individual vesicle.
- The example below shows the effect of Bcl-2 proteins: cBid-induced Bax permeabilization in GUVs composed of mitochondrial membrane lipids after 90 min of incubation with 10 nM cBid and 20 nM Bax. Images of each fluorescence channel (top) and the corresponding quantitative analysis (bottom) as the degree of filling of individual GUVs for Cyt c and APC. The hydrodynamic radius of each reporter molecule: Cyt c, 1.15 ± 0.18 nm and APC, 3.69 ± 0.4 nm (Bleicken et al., 2013a).

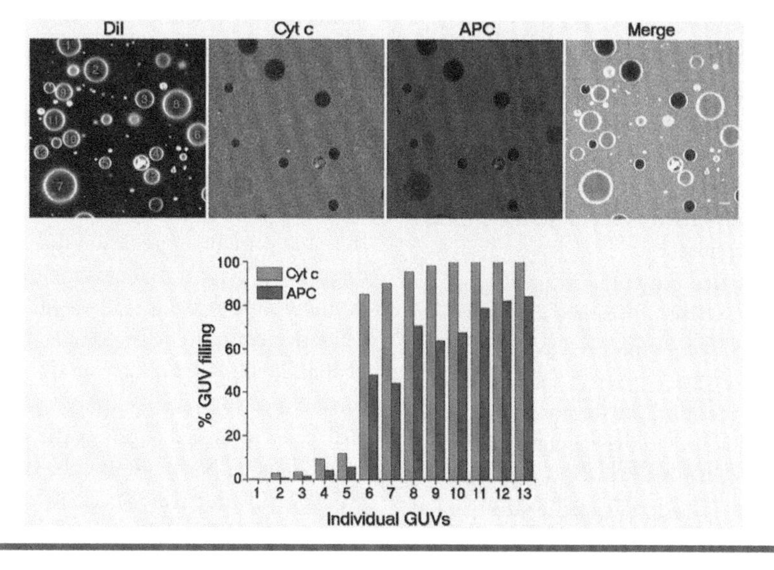

Box 20.3 Determining the pore stability

- In this experiment fluorescent reporters with excitation maxima corresponding to the shortest (e.g., 488 nm) and the longest (e.g., 633 nm) wavelength laser lines are normally used, whereas the membrane dye corresponds to the laser line in between (e.g., DiI dye at 561 nm).
- As in the protocol in Box 20.2, add one of the fluorescent reporters and the nonlabeled protein/peptide into the wells. Then, add DiI-labeled GUVs and incubate the GUVs at least for 5–15 min before imaging to allow the vesicles to settle. Here, it is critical to optimize the experimental settings so that the second fluorescent reporter is added under conditions close to equilibrium (i.e., the opening of new pores after adding the second fluorescent reporter would be problematic, making the data noisy). It has to be taken into account that the repeated opening and closure of pores in a vesicle over a long period of time cannot be distinguished from a pore that remains continuously open during the experiment. Both scenarios lead to the same situation.
- Add the second fluorescent reporter to the well and incubate for another 10–30 min to allow equilibration. It is very important to mix the sample carefully because even gentle mixing could lead to GUV destruction.
- Image 200–600 vesicles per reaction well in order to do statistics. Data analysis can be carried out semi-automatically with the program "GUV detector" (Hermann et al., 2014). For example, the data can be displayed as percentage of GUV-filling by each fluorescent reporter per individual vesicle (Figure B20.2.1).
- From this experiment, interesting information can be obtained about several processes: how many GUVs are permeabilized and the nature of the pores (i.e., transient or stable pores). Stable pores allow the passage of both fluorescent reporters, whereas transient pores close before the second fluorescent reporter is added to the well.

Box 20.4 Assessing the permeabilization kinetics

- This experiment is performed in a similar manner as the "degree of filling" and "pore stability" experiments. However, the sample is incubated with the GUVs in the well to let the vesicles settle down and avoid vesicle diffusion. The length of this incubation will be affected by how fast the molecule of interest permeabilized the GUV population. In addition, only one reaction well per experiment can be imaged because it is a time lapse experiment.
- Set up the microscope to perform a time lapse experiment in order to acquire one image every 10–30 s in a long period of time (e.g., 60–120 min). It is recommended to select a region in the sample with many GUVs in the field of view (e.g., around 20 or more vesicles).
- Mathematical fitting of the data can be done in order to estimate the initial and relaxed total permeabilized area of each vesicle, as well as the relaxation time (Fuertes et al., 2010; Hermann et al., 2014).

bovine serum albumin and avoids GUV spread onto the glass surface of the observation chamber due to unspecific adsorption. In addition, detailed information for GUV formation is given in Chapters 1 through 4. For these experiments, GUVs are generated in a sucrose solution in order to allow the vesicles to sediment in the observation chamber.

For data collection, the confocal microscope needs to be equipped with laser lines and detectors suitable for the fluorescent molecules and a water objective. The function "tile scan" to visualize a larger sample area is often helpful. For example, an LSM710 microscope with a C-Apochromat 40 × 1.2 water immersion objective (Zeiss), and Ar-ion (488 nm), HeNe (561 nm) and HeNe (633 nm) lasers is suitable for these experiments. A spectral beam guide is also used to separate the emitted photons from the different fluorophores.

Data analysis can be easily carried out with the program "GUV detector" (Hermann et al., 2014). This program has been developed for the automated detection of GUVs on digital microscopy images, providing information about the GUV size, the radial intensity at the membrane and the degree of permeabilization. The tool is suited to analyze all experiments presented here and it decreases the analysis time by 70%–90%. A free version of the program can be downloaded from the author's website.

20.5 SUMMARY AND OUTLOOK

Assessing permeability at the single GUV level has several advantages. Working with GUVs allows for discrimination of populations of vesicles with different permeation properties and it makes possible to study the stochasticity of permeation processes. These parameters are not easily accessible in conventional bulk measurements but are essential to develop a mechanistic understanding of membrane permeability. Studying membrane permeation of a population of single GUVs has greatly contributed to establishing the physical and molecular mechanisms responsible for the observed flux of molecules.

Understanding how membrane permeation of molecules with different properties works is of importance for several fields. In many protein aggregation diseases including Parkinson's disease and type II diabetes mellitus, early protein aggregates or membrane associated aggregation is thought to permeabilize cellular membranes (Engel et al., 2008; van Rooijen et al., 2009, 2010a, 2010b; Chaudhary et al., 2014). The mechanism(s) of this permeation process are not well understood, which makes it difficult to design drugs against these diseases. In liposome based drug delivery systems the conflicting need for limiting drug leakage and effective release at the target site requires control over membrane permeation in space and time. In this respect, release triggers that change membrane permeation properties are investigated and used. Current liposome based controlled release systems make use of the difference in the concentrations of protein or pH in the tissue of interest but remote triggers such as ultrasound, heat or light are also promising (Bibi et al., 2012). Although the liposome based drug delivery systems are much smaller than GUVs, GUV studies may contribute to unraveling the triggered release mechanisms. Ultimately this should result in a better control over the release. Control over membrane permeation has also been important in several vesicle based minimal cell models (Chapter 28). In these models, proteins or even the whole protein production machinery have been reconstituted in GUVs (Chapter 3). The function(ing) of these proteins may depend on the environmental conditions, for example, ionic strength, pH. Control over permeability of ions may thus allow the switching of properties of the encapsulated proteins. Actin polymerization and the interaction with of some actin binding proteins are very sensitive to the presence of specific ions such as Ca^{2+} and Mg^{2+}. The incorporation of ionophores in GUV membranes gave control over the influx of ions and thus over the polymerization and organization of GUV encapsulated actin in the absence and presence of actin binding proteins (Limozin and Sackmann, 2002; Limozin et al., 2003). Another example in which membrane permeation was controlled involves the reconstruction of bacteriorhodopsin and ATP-synthase in vesicles. By simultaneously reconstituting both proteins in a membrane, the symbiotic relationship between light-driven pumping of protons, the rotation of the ATP synthase and production of ATP could be demonstrated (Steinberg-Yfrach et al., 1998). To further our understanding of the physical mechanism of cellular organization and dynamics and to make minimal vesicle based cell models, control over in- and efflux of specific molecules will continue to be important. Challenges that involve (control over) membrane permeability in minimal cell models include: area to volume changes upon cell division, waste removal and the formation of specialized membrane sub-compartments. However, gaining control over membrane properties is not the only reason why membrane permeability studies will remain important in the coming decade. Considering that membranes are believed to have played an essential role in the transfer of substrates into primitive cells, understanding membrane permeability may help us to unravel the still enigmatic early steps in the evolution of primitive cells (Nishimura et al., 2014).

LIST OF ABBREVIATIONS

AF	Alexa Fluor dye
APC	allophycocyanin
ATP	adenosine triphosphate
Chol	cholesterol
Cyt c	cytochrome c
DiD	1,1′-dioctadecyl-3,3,3′,3′-tetramethylindodicarbocyanine (lipophilic dye)
DEIPC	1,2-elaidoyl-*sn*-glycero-3-phosphocholine
DLPC	1,2-dilinoleoyl-*sn*-glycero-3-phosphocholine
DOPC	1,2-dioleoyl-*sn*-glycero-3-phosphocholine
DOPE	1,2-dioleoyl-*sn*-glycero-3-phosphoethanolamine
DOPG	1,2-di-oleoyl-*sn*-glycero-3-phospho-(1′-rac-glycerol)
DPPC	1,2-dipalmitoyl-*sn*-glycero-3-phosphocholine
DPSPC	1,2-dipetroselinoleoyl-*sn*-glycero-3-phosphocholine
DPX	*p*-xylene-bis(*N*-pyridinium bromide)
EggPC	egg phosphatidylcholine
FCM	flow cytometry
FR	fluorescence reporter
GFP	green fluorescent protein
HPTS	8-hydroxypyrene-1,3,6-trisulfonic acid
LUV	large unilamellar vesicle
OSPC	1-oleoyl-2-stearoyl-*sn*-glycero-3-phospho-choline
PC	phosphatidylcholine
POI	protein/peptide of interest
POPC	1-palmitoyl-2-oleoyl-*sn*-glycero-3-phosphocholine
SM	sphingomyelin
SLB	supported lipid bilayer
SLPC	1-stearoyl-2-linoleoyl-*sn*-glycero-3-phosphocholine
SOPC	1-stearoyl-2-oleoyl-*sn*-glycero-3-phosphocholine
SUV	small unilamellar vesicle

GLOSSARY OF SYMBOLS

α	area fraction of vesicles that remain continuous with the vesicle bilayer
A	area
B	partition coefficient of molecule between water and the hydrocarbon region of the lipid bilayer
c	concentration
c_0	initial concentration
c_∞	final osmolarity
c_{in}	number concentration inside the vesicle
c_{ex}	number concentration in the exterior solution
D	diffusion coefficient
Ex	excitation wavelength
Em	emission wavelength
$I(t)$	fluorescence intensity at time t
I_B	background determination
I_F	fluorescence intensity (GUV-filling)
I_M	fluorescence intensity at the membrane
I_O	offset determination
I_{ex}	fluorescence intensity outside the vesicle
J	flux
l_{me}	membrane thickness
Mw	molecular weight
p_s	permeability to molecule s
p_w	permeability to water
r	radius of daughter vesicles
R_0	initial vesicle radius
R_{pr}	inner radius of the micropipette
R_{ve}	vesicle radius
t	time
v_w	molar volume of water
V	volume of vesicle
V_0	initial volume of the vesicle
V_∞	final equilibrium volume of the vesicle

REFERENCES

Apellaniz B, Nieva JL, Schwille P, Garcia-Saez AJ (2010) All-or-none versus graded single-vesicle analysis reveals lipid composition effects on membrane permeabilization. *Biophys J* 99:3619–3628.

Bergstrom CL, Beales PA, Lv Y, Vanderlick TK, Groves JT (2013) Cytochrome c causes pore formation in cardiolipin-containing membranes. *Proc Natl Acad Sci USA* 110:6269–6274.

Bernard AL, Guedeau-Boudeville MA, Jullien L, di Meglio JM (2002) Raspberry vesicles. *Biochim Biophys Acta Biomembr* 1567:1–5.

Bibi S, Lattmann E, Mohammed AR, Perrie Y (2012) Trigger release liposome systems: Local and remote controlled delivery? *J Microencapsul* 29:262–276.

Bleicken S, Garcia-Saez AJ (2014) New biophysical methods to study the membrane activity of Bcl-2 proteins. *Methods Mol Biol* 1176:191–207.

Bleicken S, Hofhaus G, Ugarte-Uribe B, Schroder R, Garcia-Saez AJ (2016) cBid, Bax and Bcl-xL exhibit opposite membrane remodeling activities. *Cell Death Dis* 7:e2121.

Bleicken S, Landeta O, Landajuela A, Basanez G, Garcia-Saez AJ (2013a) Proapoptotic Bax and Bak proteins form stable protein-permeable pores of tunable size. *J Biol Chem* 288:33241–33252.

Bleicken S, Wagner C, Garcia-Saez AJ (2013b) Mechanistic differences in the membrane activity of Bax and Bcl-xL correlate with their opposing roles in apoptosis. *Biophys J* 104:421–431.

Boroske E, Elwenspoek M, Helfrich W (1981) Osmotic shrinkage of giant egg-lecithin vesicles. *Biophys J* 34:95–109.

Bucher P, Fischer A, Luisi PL, Oberholzer T, Walde P (1998) Giant vesicles as biochemical compartments: The use of microinjection techniques. *Langmuir* 14:2712–2721.

Chaudhary H, Stefanovic AND, Subramaniam V, Claessens MMAE (2014) Membrane interactions and fibrillization of alpha-synuclein play an essential role in membrane disruption. *Febs Lett* 588:4457–4463.

Claessens MMAE, Leermakers FAM, Hoekstra FA, Stuart MAC (2008) Osmotic shrinkage and reswelling of giant vesicles composed of dioleoylphosphatidylglycerol and cholesterol. *Biochim Biophys Acta Biomembr* 1778:890–895.

Degier J (1993) Osmotic behavior and permeability of liposomes. *Chem Phys Lipids* 64:187–196.

Dimova R, Aranda S, Bezlyepkina N, Nikolov V, Riske KA, Lipowsky R (2006) A practical guide to giant vesicles. Probing the membrane nanoregime via optical microscopy. *J Phys-Condens Matter* 18:S1151–S1176.

Engel MFM, Khemtemourian L, Kleijer CC, Meeldijk HJD, Jacobs J, Verkleij AJ, de Kruijff B, Killian JA, Hoppener JWM (2008) Membrane damage by human islet amyloid polypeptide through fibril growth at the membrane. *Proc Natl Acad Sci USA* 105:6033–6038.

Finkelstein A (1987) *Water Movement through Lipid Bilayers, Pores and Plasma Membranes: Theory and Reality*. New York: John Wiley & Sons.

Fischer A, Oberholzer T, Luisi PL (2000) Giant vesicles as models to study the interactions between membranes and proteins. *Biochim Biophys Acta Biomembr* 1467:177–188.

Fuertes G, Garcia-Saez AJ, Esteban-Martin S, Gimenez D, Sanchez-Munoz OL, Schwille P, Salgado J (2010) Pores formed by Bax alpha 5 relax to a smaller size and keep at equilibrium. *Biophys J* 99:2917–2925.

Garcia-Saez AJ, Ries J, Orzaez M, Perez-Paya E, Schwille P (2009) Membrane promotes tBID interaction with BCL(XL). *Nat Struct Mol Biol* 16:1178–1185.

Grime JMA, Edwards MA, Rudd NC, Unwin PR (2008a) Quantitative visualization of passive transport across bilayer lipid membranes. *Proc Natl Acad Sci USA* 105:14277–14282.

Grime JMA, Edwards MA, Unwin PR (2008b) Reply to Missner et al.: Timescale for passive diffusion across bilayer lipid membranes. *Proc Natl Acad Sci USA* 105:E124–E124.

Hermann E, Bleicken S, Subburaj Y, Garcia-Saez AJ (2014) Automated analysis of giant unilamellar vesicles using circular Hough transformation. *Bioinformatics* 30:1747–1754.

Kokot G, Mally M, Svetina S (2012) The dynamics of melittin-induced membrane permeability. *Eur Biophys J* 41:461–474.

Lee MT, Hung WC, Chen FY, Huang HW (2008) Mechanism and kinetics of pore formation in membranes by water-soluble amphipathic peptides. *Proc Natl Acad Sci USA* 105:5087–5092.

Li S, Hu P, Malmstadt N (2010) Confocal imaging to quantify passive transport across biomimetic lipid membranes. *Anal Chem* 82:7766–7771.

Li S, Hu PC, Malmstadt N (2011) Imaging molecular transport across lipid bilayers. *Biophys J* 101:700–708.

Limozin L, Barmann M, Sackmann E (2003) On the organization of self-assembled actin networks in giant vesicles. *Eur Phys J E* 10:319–330.

Limozin L, Sackmann E (2002) Polymorphism of cross-linked actin networks in giant vesicles. *Phys Rev Lett* 89:168103.

Lira RB, Steinkuhler J, Knorr RL, Dimova R, Riske KA (2016) Posing for a picture: Vesicle immobilization in agarose gel. *Sci Rep* 6:25254.

Lorent J, Lins L, Domenech O, Quetin-Leclercq J, Brasseur R, Mingeot-Leclercq MP (2014) Domain formation and permeabilization induced by the saponin alpha-hederin and its aglycone hederagenin in a cholesterol-containing bilayer. *Langmuir* 30:4556–4569.

Missner A, Kuegler P, Antonenko YN, Pohl P (2008) Passive transport across bilayer lipid membranes: Overton continues to rule. *Proc Natl Acad Sci USA* 105:E123–E123.

Nishimura K, Matsuura T, Nishimura K, Sunami T, Suzuki H, Yomo T (2012) Cell-free protein synthesis inside giant unilamellar vesicles analyzed by flow cytometry. *Langmuir* 28:8426–8432.

Nishimura K, Matsuura T, Sunami T, Fujii S, Nishimura K, Suzuki H, Yomo T (2014) Identification of giant unilamellar vesicles with permeability to small charged molecules. *RSC Adv* 4:35224–35232.

Nuss H, Chevallard C, Guenoun P, Malloggi F (2012) Microfluidic trap-and-release system for lab-on-a-chip-based studies on giant vesicles. *Lab Chip* 12:5257–5261.

Olbrich K, Rawicz W, Needham D, Evans E (2000) Water permeability and mechanical strength of polyunsaturated lipid bilayers. *Biophys J* 79:321–327.

Peterlin P, Arrigler V, Haleva E, Diamant H (2012) Law of corresponding states for osmotic swelling of vesicles. *Soft Matter* 8:2185–2193.

Peterlin P, Jaklic G, Pisanski T (2009) Determining membrane permeability of giant phospholipid vesicles from a series of videomicroscopy images. *Meas Sci Technol* 20:055801.

Ramahaleo T, Morillon R, Alexandre J, Lassalles JP (1999) Osmotic water permeability of isolated protoplasts. Modifications during development. *Plant Physiol* 119:885–896.

Rawicz W, Smith BA, McIntosh TJ, Simon SA, Evans E (2008) Elasticity, strength, and water permeability of bilayers that contain raft microdomain-forming lipids. *Biophys J* 94:4725–4736.

Rivers RL, Williams JC (1990) Effect of solute permeability in determination of elastic-modulus using the vesicular swelling method. *Biophys J* 57:627–631.

Sankhagowit S, Wu SH, Biswas R, Riche CT, Povinelli ML, Malmstadt N (2014) The dynamics of giant unilamellar vesicle oxidation probed by morphological transitions. *Biochim Biophys Acta* 1838:2615–2624.

Schoen P, Garcia-Saez AJ, Malovrh P, Bacia K, Anderluh G, Schwille P (2008) Equinatoxin II permeabilizing activity depends on the presence of sphingomyelin and lipid phase coexistence. *Biophys J* 95:691–698.

Steinberg-Yfrach G, Rigaud JL, Durantini EN, Moore AL, Gust D, Moore TA (1998) Light-driven production of ATP catalysed by F0F1-ATP synthase in an artificial photosynthetic membrane. *Nature* 392:479–482.

Steponkus PL, Lynch DV (1989) The behavior of large unilamellar vesicles of rye plasma-membrane lipids during freeze thaw-induced osmotic excursions. *Cryo-Letters* 10:43–50.

Steringer JP, Bleicken S, Andreas H, Zacherl S, Laussmann M, Temmerman K, Contreras FX et al. (2012) Phosphatidylinositol 4,5-bisphosphate (PI(4,5)P2)-dependent oligomerization of fibroblast growth factor 2 (FGF2) triggers the formation of a lipidic membrane pore implicated in unconventional secretion. *J Biol Chem* 287:27659–27669.

Su QP, Du WQ, Ji QH, Xue BX, Jiang D, Zhu YY, Lou JZ, Yu L, Sun YJ (2016) Vesicle Size Regulates Nanotube Formation in the Cell. *Sci Rep* 6:24002.

Tamba Y, Yamazaki M (2005) Single giant unilamellar vesicle method reveals effect of antimicrobial peptide magainin 2 on membrane permeability. *Biochem* 44:15823–15833.

van Rooijen BD, Claessens M, Subramaniam V (2010a) Membrane interactions of oligomeric alpha-synuclein: Potential role in Parkinson's disease. *Curr Protein Pept Sci* 11:334–342.

van Rooijen BD, Claessens MM, Subramaniam V (2009) Lipid bilayer disruption by oligomeric alpha-synuclein depends on bilayer charge and accessibility of the hydrophobic core. *Biochim Biophys Acta* 1788:1271–1278.

van Rooijen BD, Claessens MMAE, Subramaniam V (2010b) Membrane permeabilization by oligomeric alpha-synuclein: In search of the mechanism. *PLoS One* 5:e14292.

Vitkova V, Genova J, Bivas I (2004) Permeability and the hidden area of lipid bilayers. *Eur Biophys J* 33:706–714.

Wick R, Angelova MI, Walde P, Luisi PL (1996) Microinjection into giant vesicles and light microscopy investigation of enzyme-mediated vesicle transformations. *Chem Biol* 3:105–111.

Xiang TX, Anderson BD (1997) Permeability of acetic acid across gel and liquid-crystalline lipid bilayers conforms to free-surface-area theory. *Biophys J* 72:223–237.

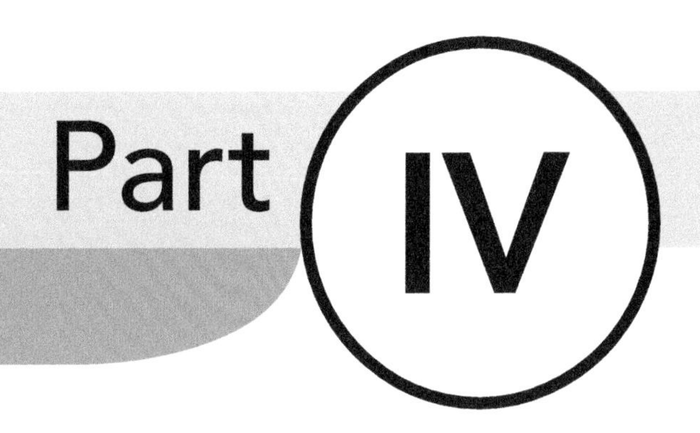

GUVs as membrane interaction platforms

Lipid and protein mobility in giant unilamellar vesicles

Begoña Ugarte-Uribe, Kushal Kumar Das, and Ana J. García-Sáez

Diffusing alone or interacting with others?

Contents

21.1 INTRODUCTION

Biological membranes play an essential role in cell shape, structure and function. They act not only as a barrier between the inner and outer aqueous environment of a cell and between cell organelles, but also as a suitable milieu for folding and activity of a number of proteins (Vereb et al., 2003; Chandler, 2005; Marguet et al., 2006). Biological membranes are dynamic in nature and formed by amphipathic lipid molecules, where membrane proteins can diffuse (Singer and Nicolson, 1972). The intraorganellar, transversal and lateral heterogeneity of membrane bilayers, together with their complex diffusion patterns, has been a subject of intense research over many decades. This had led to the proposal of several models of nonrandom molecular distribution, such as lipid rafts, micro- and nanodomains, and confinement zones, including the "pickets and fences" model (Kusumi et al., 2004).

One of the most relevant parameters related to the organization of biological membranes is the lateral diffusion coefficient of membrane lipids and proteins because it is directly linked to the membrane fluidity and structure. Hence, the characterization of the dynamic behavior of components within a membrane can provide useful information about the organization of that membrane.

Current fluorescence-based technological developments have enabled the study of membrane organization, dynamics and interactions. Due to the high complexity of cellular membrane systems, the use of artificial model systems such as giant unilamellar vesicles (GUVs) provides a biochemically well-defined strategy to gain a better knowledge about the presence and size of membrane domains, lipid–protein and protein–protein interactions within the membrane, and additional factors that can influence the lateral mobility of lipids (Sheetz et al., 1980; Kusumi et al., 1993, 2004; Dietrich et al., 2002; Bacia et al., 2004; Falck et al., 2004; Garcia-Saez et al., 2009). In this chapter, we focus on fluorescence-based approaches involved in broad understanding of the dynamic properties of biological membranes that can be combined with the use of GUVs as artificial model membranes. We describe the principles and applications of fluorescence correlation spectroscopy (FCS), fluorescence recovery after photobleaching (FRAP) and single-molecule imaging (SMI) (Subburaj et al., 2013; Unsay and Garcia-Saez, 2013; Cosentino et al., 2015; Das et al., 2015; Hermann et al., 2015). As an example, Table 21.1 shows an overview of the timescale of processes involved in biological membrane dynamics and the temporal resolution of a number of microscopy techniques (Alessandrini and Facci, 2014).

Table 21.1 **Examples of the timescale of (A) biological membrane dynamics and (B) the temporal resolution of microscopy techniques, where timescale represents the characteristic time for the specific biomolecular events to occur and time resolution represents the temporal sensitivity of these techniques**

(A) BIOLOGICAL MEMBRANE DYNAMICS	TIMESCALE
Protein lateral diffusion (per 0–1 μm)	10^{-6}–10^{-3} s
Lipid flip-flop	10^{0}–10^{3} s
Lipid lateral diffusion (per 0.01–1 μm)	10^{-9}–10^{-6} s
(B) MICROSCOPY TECHNIQUES	TIME RESOLUTION
FCS	10^{-6}–10^{-3} s
FRAP	10^{-3} ms–10^{0} s
SMI	10^{-9}–10^{-3} s

Source: Alessandrini, A. and Facci, P., *Soft Matter*, 10, 7145–7164, 2014.

21.2 SELECTION OF PROBES FOR STUDYING MEMBRANE DYNAMICS

Advanced microscopy techniques have been established as important physical methods to study the structure and dynamics of biological membranes. Most biological molecules and structures are not intrinsically fluorescent in spectral ranges that are useful for detection and need to be labeled with fluorescent dyes. The requirement for a high signal-to-noise ratio in single-molecule detection in membranes depends not only on the optical setup but also on the fluorophore that is linked to the molecule of interest in order to achieve a successful detection. Several factors have to be taken into account such as the nature of the fluorophore itself and its photophysical properties (i.e., the number and energy distribution of photons available for detection including excitation and emission spectra, molar absorptivity or extinction coefficient, fluorescence quantum yield, and photobleaching rate). The most common fluorescent probes are fluorescent proteins, quantum dots and organic dyes.

When fluorescent proteins are selected as fluorescent probes absolute specificity is assumed because the labeling occurs at the genetic level. They are biocompatible with many organisms and less phototoxic than other probes. Over the last years, protein biotechnology has enabled the generation of a number of fluorescent proteins covering a broad range of the visible spectrum (Shaner et al., 2007). The first member of this family is GFP, a green fluorescent protein composed of 238 amino acid residues (26.9 kDa) and isolated from the jellyfish *Aequorea victoria* (Tsien, 1998). Site-directed mutagenesis of this protein has led to the generation of cyan and yellow variants (e.g., CFP and YFP) with shifted excitation and emission spectra (Siegel et al., 2000). Spectral coverage has also been extended to the red range by the use of DsRed fluorescent protein isolated from *Discosoma* coral with further modifications to avoid undesirable characteristics related to slow maturation and oligomerization (Campbell et al., 2002). Although they have been applied successfully in many studies (Shaner et al., 2005), they are bulky and could affect the structure and the function of the molecule of interest. Moreover, they show worse photophysical properties in terms of brightness and photostability than quantum dots and organic fluorophores.

Quantum dots are crystals consisting of a spherical core of the semiconductor cadmium selenide surrounded by a zinc sulfide shell, which is in turn surrounded by a hydrophilic polymer surface coating (Michalet et al., 2005). The hydrophilic coating confers water solubility and functional groups can be incorporated by linkage to biomolecules, such as biotin, streptavidin and antibodies, which specifically bind to the molecule of interest (Wegner and Hildebrandt, 2015). These inorganic fluorescent probes are extremely bright and resistant to photobleaching. Thus, they provide high signal-to-noise ratio (up to 25) and can be detected individually, even by conventional wide-field illumination. In addition, they have a broad excitation spectra and a narrow emission wavelength range that depends on the core size. Therefore, the use of various quantum dots of variable size provides the whole visible spectrum coverage with high spectral resolution, allowing multicolor imaging. However, some drawbacks in their properties are related to their blinking properties and their large size, ranging between 10 and 20 nm, because they can be several times larger than the molecule to which they are attached and this could directly affect the functionality of the molecule of interest. In addition, they tend to aggregate and monovalent labeling is difficult to achieve, although some progress has been made toward reducing multivalent conjugation (Farlow et al., 2013).

Organic dyes are a good alternative to quantum dots and fluorescent proteins. They exhibit a conjugated π-electron system for absorption in the visible or near-visible spectrum. The most important advantage is their small size (<1 nm), allowing their linkage to a large variety of biomolecules, such as proteins, lipids, nucleic acids, and sugars (Goncalves, 2009; Wysocki and Lavis, 2011). They offer a wider spectral range, higher stability and quantum yield compared with fluorescent proteins (but less than quantum dots). Similarly to quantum dots, these dyes are linked through chemical conjugation and do not ensure absolute labeling efficiency. However, in contrast to quantum dots, they can offer monovalent ligation and they can be attached to the molecule of interest using well-established conjugation protocols. In fact, further development has been made with the introduction of different functional groups for labeling (i.e., *N*-hydroxysuccinimide [NHS] ester, tetrafluorophenyl [TFP] ester, maleimide, hydrazine, carbodiimide) in order to increase water solubility and photostability. These fluorophores can be classified into groups such as xanthene (e.g., fluorescein, rhodamine), cyanine (e.g., dialkylcarbocyanines, ATTO and Alexa series of dyes), naphthalene (e.g., dansyl, 6-dodecanoyl-2-dimethylaminonaphtalene [LAURDAN]), oxazine (e.g., Nile red), BODIPY® (i.e., a boron-containing class of fluorophores called 4,4-difluoro-4-bora-3a,4a-diaza-s-indacene-3-pentanoyl), and perylene dyes (see also Chapter 10), see Appendix 2 of this book for structures and data on fluorescent dyes.

21.3 FLUORESCENCE CORRELATION SPECTROSCOPY

FCS is a technique with high spatiotemporal resolution, which uses fluorescence fluctuations rather than absolute fluorescence intensity. It detects fluctuations arising from the diffusion of individual molecules passing in and out through a tiny, sub-femtoliter detection volume, which is usually the focal volume of a confocal microscope (Schwille et al., 1999). As the detection volume is tiny, FCS can be considered as a method with single-molecule sensitivity. Thus, this technique provides improved sensitivity when compared with FRAP, and allows working with nanomolar concentration of fluorophores. FCS has long been used to characterize lipid and cell membrane organization (Korlach et al., 1999; Bacia et al., 2004). In addition, the recent characterization of membrane nanodomains has been achieved thanks to the combination of this technique with super-resolution techniques, such as stimulated emission depletion (STED) microscopy (Eggeling et al., 2009). Due to its intensive applications to biological membranes, we discuss the basic principle and applications of FCS for studying membrane dynamics.

FCS was first established in the 1970s and technically improved in the following years (Magde et al., 1972, 1974; Rigler et al., 1993; Eigen and Rigler, 1994). Originally, FCS has been implemented in home-built microscopes but nowadays there are many commercial FCS setups available on the market. Some examples of the FCS setups currently available on the market are the TCP SP8 (Leica Microsystems Ltd.) and the laser scanning microscope (LSM) 710 with ConfoCor3 (or the LSM 780) (Carl Zeiss, Inc.). In the case of PicoQuant systems, FCS can in principle be coupled to any confocal microscope. The aforementioned setups combine a confocal microscopy system with FCS capabilities and have user-optimized platforms. They include several lasers for excitation and several detection channels with fiber-coupled avalanche photodiodes (APDs) or new generation imaging detectors with photon-quantification capabilities. In general, an FCS setup consists of an inverted confocal microscope equipped with a high-numerical aperture water objective like the 40× NA 1.2 UV-VIS-IR C-Apochromat water-immersion objective from Zeiss (Jena, Germany) to reduce aberrations. A dichroic mirror reflects the laser light that is then focused by the microscope objective to a spot of 0.3–0.5 μm diameter (of the measurement volume or focal volume). Then, the sample emits fluorescence light, which passes the dichroic mirror and the emission filter in order to remove residual Rayleigh and Raman scattered light. In addition, a pinhole in the image plane enhances the axial resolution (e.g., a pinhole of 1 Airy unit is usually used for better detection efficiency). Alternatively, an optical fiber of the same diameter can be used. For detection and fluorescence trace correlation, these setups are equipped with single-photon sensitivity detectors and a hardware or software correlator, respectively. It is essential to record photon arrival times with adequate time resolution. If the commercial setup selected does not allow it, an alternative is to use the photon mode of the hardware correlator Flex 02-01D (Bridgewater, NJ, USA, www.correlator.com). For example, this should be taken into account for an FCS variant called scanning FCS (SFCS) (see Section 21.3.1).

In general, FCS utilizes the fluorescence fluctuations generated by the diffusion of fluorescently tagged particles through a femtoliter sized measurement volume. The signal is recorded using highly sensitive detectors, as discussed above, as a stream of single photon arrival times, which corresponds to the raw data in FCS (Box 21.1). The signal is used to generate a function of fluorescence intensity versus time, which is then convoluted to generate the autocorrelation curve with the help of hardware or software correlator. The autocorrelation analysis on which FCS is based performs statistical analysis to the fluorescence fluctuations over time. In this way, it correlates a signal at time t with the same signal after a lag time $t+\tau$ and takes the temporal average. This temporal autocorrelation measures the self-similarity of the signal with itself over time. The following expression depicts the autocorrelation function:

$$G(\tau)=\frac{\langle \delta F(t)\cdot \delta F(t+\tau)\rangle}{\langle F(t)\rangle^2} \tag{21.1}$$

where $G(\tau)$ is the autocorrelation function; $F(t)$, is the overall fluorescence signal, also called fluorescence trace, from the confocal volume as a function of time; τ is the lag time or correlation time; and the angular brackets stand for averaging over time. $\delta F(t)$ is the fluorescence fluctuation at time t, defined as a deviation of the signal from its average over time $\delta F(t)=F(t)-\langle F(t)\rangle$ (Medina and Schwille, 2002; Garcia-Saez and Schwille, 2008; Sezgin and Schwille, 2011).

The autocorrelation analysis of the fluctuations depends on the concentration and diffusion coefficients. From these parameters, properties like association and dissociation constant and rate constant can be calculated.

The shape of the measurement volume or the focal volume has a large influence on the analysis of FCS data. This volume is described mathematically by the point spread function (PSF), which is often approached to an ellipsoid with few nanometers in diameter and micrometers along the optical axis (Figure 21.1) (Hess and Webb, 2002). Thus, it is important that the real shape of the PSF or measurement volume is verified and calibrated every time before starting the sample measurements as deviations from it would introduce artifacts in the FCS analysis. This is achieved by using known standard dyes or species with known diffusion coefficients under the same experimental conditions (Box 21.2). As an example, Table 21.2 gives an overview over some dyes that are commonly used as references in FCS measurements (see Appendices 1 and 2 of this book for structure and data on the lipids and fluorescent dyes).

In FCS the spatial intensity profile of a focused laser beam is referred to as the PSF of the instrument. The PSF is critical for FCS measurements as its size and shape characterizes the fluctuations of fluorescence molecules diffusing through the optical beam and thus the fluorescence intensity correlation function. The shape of the

Box 21.1 Principle of FCS

Panel a shows a schematic representation of a confocal FCS setup. Fluorescence fluctuations as shown in panel b are recorded by detectors with single-photon sensitivity (like APDs) in their respective channel (shown in green). The fluctuations are temporally autocorrelated (see panel c) to measure self-similarity of the signal over time and then fitted with a model function (here, a 3D diffusion fitting is used for FCS in solution). (Adapted from Das et al., 2015.)

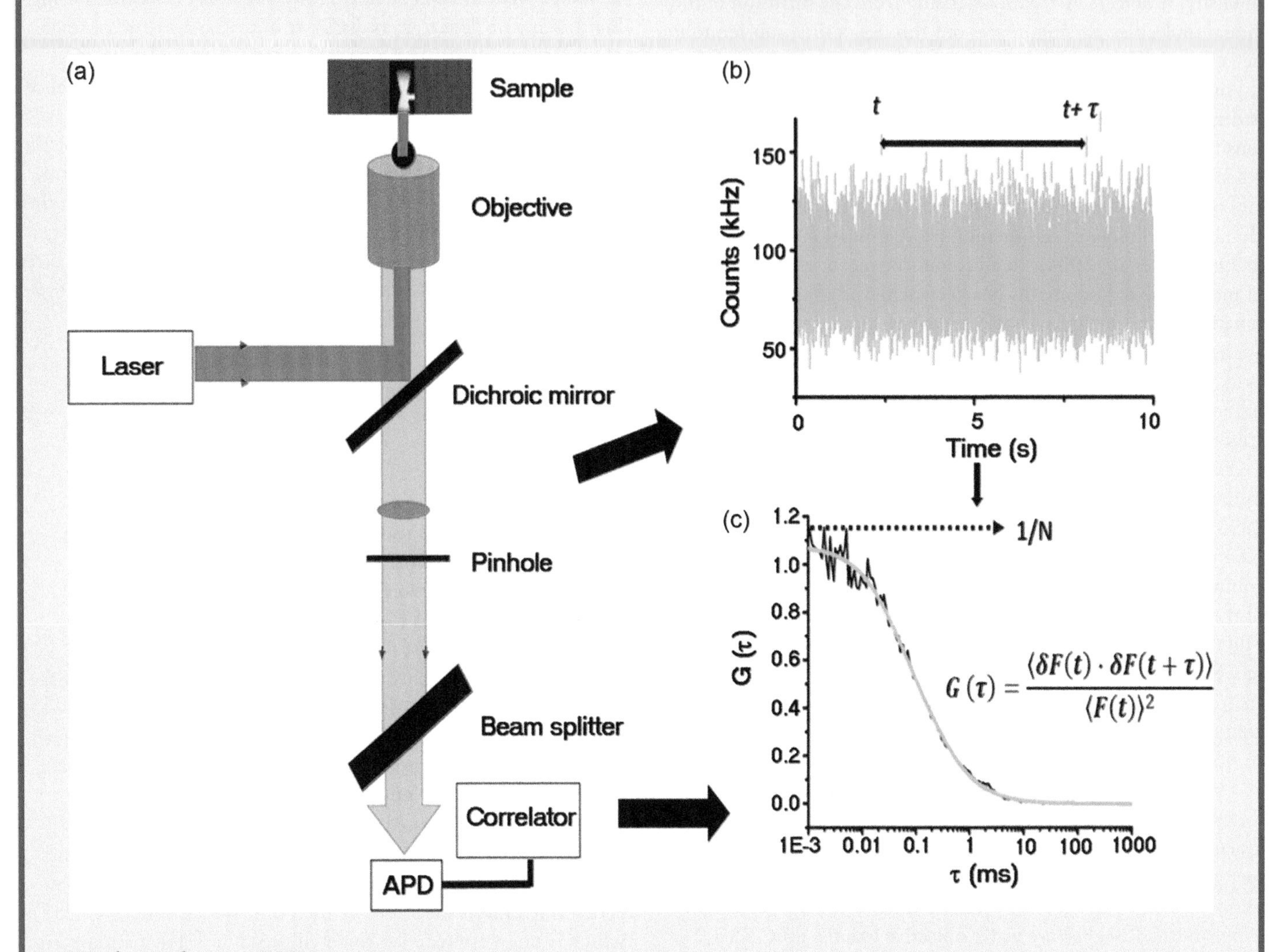

- Tips for performing FCS measurements:
- The measurement time should be 10,000 times higher than the residence time or diffusion time of the species in the detection or focal volume (τ_D). For example, in a sample containing green ($\tau_D = 15-30\mu s$) and red ($\tau_D = 40-60\mu s$) fluorophores, the slowest diffusion time in this case would be for the latter fluorophore (60 µs). Therefore, the measuring time during the experiment should be at least 0.6 s. According to this, we could keep the measuring time to 1 s. However, for better statistics we could keep the measuring time to 10 s.
- In membranes, the diffusion is slower than in solution. For example, if the diffusion time for one fluorophore is 1 ms, then the measuring time should be at least 10 s. For better statistics we could measure longer (e.g., 100 s).
- The above measuring times are hypothetical times related to free fluorophores. Nonetheless, the diffusion time can vary depending on the size of the molecule of interest (e.g., lipids, proteins) and the viscosity of the sample.

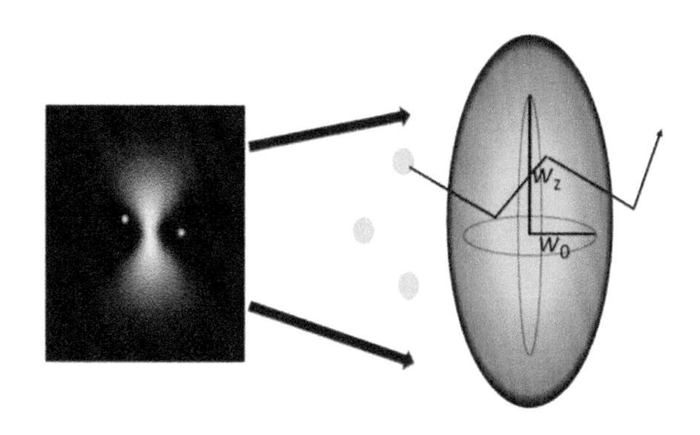

Figure 21.1 Schematic representation of the detection volume or focal volume, which is described mathematically by the PSF. In confocal microscopy with small pinholes (around one Airy unit), the PSF is approximated by a 3D Gaussian shape. The dimensions w_0 and w_z define the structural parameter or the geometry of the measurement volume.

Table 21.2 **Diffusion coefficients of (A) fluorophores, (B) fluorescent proteins, and (C) lipidic dyes. The free fluorophores are commonly used in FCS calibration measurements, see Appendices 1, 2, and 4 of the book for structure and data on lipids and probes.**

(A) FLUOROPHORES (IN WATER)	DIFFUSION COEFFICIENT IN SOLUTION (μm²/s)		REFERENCES
Rhodamine 6G	426 at 22.5°C/414 ± 1 at 25°C		Culbertson et al. (2002) and Petrasek and Schwille (2008)
Rhodamine B	427 ± 4 at 25°C		Culbertson et al. (2002)
Fluorescein	425 ± 1 at 25°C		Culbertson et al. (2002)
eGFP	95 at 22.5°C		Petrasek and Schwille (2008)
Alexa 488	435 at 22.5°C		Petrasek and Schwille (2008)
Alexa 546	341 at 22.5°C		Petrasek and Schwille (2008)
Atto 655	426 ± 8 at 25°C		Dertinger et al. (2007)
(B) PROTEINS (IN BUFFER)	DIFFUSION COEFFICIENT IN SOLUTION (μm²/s)	DIFFUSION COEFFICIENT IN GUVs (μm²/s)	REFERENCES
Bcl-xL ΔCt Alexa 488	78 ± 10 at 22.5°C	4.8 ± 0.7 at 22.5°C	Garcia-Saez et al. (2009)
tBID Alexa 647	143 ± 9 at 22.5°C	5 ± 0.3 at 22.5°C	Garcia-Saez et al. (2009)
cBid Alexa 647	96 ± 11 at 22.5°C	~5 at 22.5°C	Unpublished data
BclxL FL Alexa 488	40 at 22.5°C	~4.9 at 22.5°C	Unpublished data
Bax Atto 488	60 at 22.5°C	4.1 ± 0.1 at 22.5°C	Subburaj et al. (2015a), Unpublished data
(C) LIPIDIC DYES (IN Ld PHASE)	GUV: LIPID COMPOSITION	DIFFUSION COEFFICIENT (μm²/s)	REFERENCES
DiI-C20	DLPC	4.4 ± 0.9 at 23°C	Schwille et al. (1999)
DiI-C20	DLPC	3.0 ± 0.6 at 25°C	Korlach et al. (1999)
DiI-C18	DOPC/SM/Chol (2/2/1)	6.1 ± 0.5 at 20°C	Kahya and Schwille (2006)
DiI-C18	DOPC/SM/Chol (2/2/1)	2.5 ± 0.2 at 20°C	Carrer et al. (2008)
DiI-C18	DOPC/DSPC (1/1)	6.5 ± 0.4 at 20°C	Scherfeld et al. (2003)
DiI-C18	DOPC/SM/Chol (5/5/2)	5.1 ± 0.4 at 20°C	Scherfeld et al. (2003)
DiI-C18	DOPC/SM/Chol (1/1/1)	1.4 ± 0.1 at 20°C	Scherfeld et al. (2003)
DiI-C20	DLPC/DPPC (3/2)	5 ± 1 at 25°C	Scherfeld et al. (2003)
DiI-C18	POPC	7 ± 3 at 25°C	Gielen et al. (2009)

Note: Chol, cholesterol; DLPC, dilauroyl phosphatidylcholine; DOPC, dioleoyl phosphatidylcholine; DPPC, dipalmitoyl phosphatidylcholine; DSPC, distearoyl phosphatidylcholine; Ld, liquid disordered phase; POPC, palmitoyl oleoyl phosphatidylcholine; SM, sphingomyelin.

excitation PSF is dependent on optics and the overfill factor of the back aperture of the objective. The focused beam is approximated by a Gaussian-Lorentzian (GL) shape. Thus, the PSF for two photon excitation is given by a GL function (Dertinger et al., 2007):

$$PSF_{GL}(\rho,z) = \frac{w_0^4}{w^4(z)\exp\left\langle -4\frac{\rho^2}{w^2(z)}\right\rangle} \tag{21.2}$$

$$w^2(z) = w_o^2\left\langle 1+\frac{z^2}{z_r^2}\right\rangle \tag{21.3}$$

$$z_r = \frac{\pi w_o^2}{\lambda} \tag{21.4}$$

where w is the beam waist; and ρ is the photon detected position. PSF is expressed in cylindrical coordinates with a radial waist (w_0) and an axial waist (z_r) that depends on the excitation wavelength (λ). Alternative FCS schemes that use different detection volumes, such as two-photon FCS, should take into account the corresponding shape of the focal volume in the analysis of the data.

When the autocorrelation curves are fitted with adequate model equations, physical parameters can be obtained. These models take into account the excitation profile, size and shape of the confocal volume, the molecular brightness, and the fluorophore concentration as a function of position and time. In Table 21.3 commonly used examples of models are shown and detailed derivations are found in previous publications (Magde et al., 1972; Petrov and Schwille, 2008). These models also need to consider the photophysical processes, such as blinking and triplet state, which can affect the molecular brightness. In case of blinking, the fluorescent molecules have a dark and a bright state depending on their chemical environment. For example, this phenomenon can occur when the excitation of the fluorophore is affected by pH. In addition, excited fluorescent molecules can undergo a transition from singlet to metastable triplet state. In this case, molecules remain dark at the timescale of a few microseconds and then the triplet state relaxes and eventually emits a photon.

Diffusion in three dimensions (3D) is fitted with a mathematical model function to obtain parameters such as N and D, which stand for the average number of fluorescent particles in the detection area and their diffusion coefficient, respectively. The model function describing 3D Brownian diffusion is expressed as:

$$G_{3D}(\tau) = \frac{1}{N}\left(1+\frac{\tau}{\tau_D}\right)^{-1}\left(1+\frac{\tau}{S^2\tau_D}\right)^{-\frac{1}{2}} \tag{21.5}$$

where G is the autocorrelation function; N is the average number of fluorescent particles; τ_D is the residence time or diffusion time of the species in the detection or focal volume; and S is the structure parameter. S corresponds to the aspect ratio of the focal volume: $S = w_z/w_0$, where w_z and w_0 are the extensions of the focal volume in axial (z-axis) and radial directions, respectively.

Table 21.3 **FCS fitting models**

DIFFUSION TYPES	FITTING MODEL FUNCTION
3D diffusion	$G_{3D}(\tau) = \frac{1}{N}\left(1+\frac{\tau}{\tau_D}\right)^{-1}\left(1+\frac{\tau}{S^2\tau_D}\right)^{-\frac{1}{2}}$
3D diffusion for two components	$G_{3D+2C}(\tau) = \frac{1}{N_{total}}\frac{q_f^2Y_fG_{3Df}(\tau)+q_s^2Y_sG_{3Ds}(\tau)}{\left(q_fY_f+q_sY_s\right)^2}$
3D diffusion with triplet	$G_{3D+T}(\tau) = \left[1+\frac{T_R}{1-T_R}exp\left(-\frac{\tau}{\tau_{TR}}\right)\right]G_{3D}(\tau)$
2D diffusion[a]	$G_{2D}(\tau) = \frac{1}{N}\left(1+\frac{\tau}{\tau_D}\right)^{-1}$
2D with elliptical-Gaussian profile[b]	$G_{2DG}(\tau) = \frac{1}{N}\left(1+\frac{\tau}{\tau_D}\right)^{-\frac{1}{2}}\left(1+\frac{\tau}{S^2\tau_D}\right)^{-\frac{1}{2}}$

Note: The 3D diffusion model is used for measurement is solution with one population. The 3D diffusion model for two components is used for the measurement in solution with two different populations: for example, when there is free dye within the protein sample or when the protein shows two populations depending on their size (e.g., monomeric and oligomeric). The 3D diffusion model with triplet is used when the sample shows some contribution caused by transition into the triplet state
G, autocorrelation curve; N, average number of particles in the detection volume; q, molecular brightness of the f (fast) and s (slow) diffusing components and Y refers to their molar fraction; T_R, fraction of the fluorophores in the triplet state within the detection volume; $S = w_z/w_0$, structure parameter; τ, lag time; τ_D, diffusion time; τ_{TR}, triplet time. The terms introduced to correct for two components and triplet state are also valid for 2D diffusion (note that the diffusion in membranes is slower than the triplet relaxation time, and thus, this phenomenon may be negligible).

[a] This model is used when the membrane is oriented perpendicular to the optical axis (z-axis) as in supported lipid bilayers.

[b] This model is used when the membrane is oriented parallel to the optical axis as in GUVs.

The average number of particles, N, is calculated from the amplitude of the autocorrelation curve because:

$$G(0) = \frac{1}{N} \tag{21.6}$$

The fluorophore concentration can also be derived from N, by using the following formula: $N = CV_{eff}$. Where V_{eff} is the effective detection volume because in confocal FCS a collimated Gaussian laser beam is focused by an objective with high numerical aperture into the sample. Here, in 3D Brownian diffusion the correlation function is given by a 3D Gaussian detection volume and this effective volume is given as: $V_{eff} = \pi^{3/2}w_0^2w_z$.

The average amount of time that a fluorescent particle stays in the detection volume is represented as τ_D. This parameter is related to the diffusion coefficient (D) in the following expression:

$$\tau_D = \frac{w_0^2}{4D} \tag{21.7}$$

For spherical particles undergoing Brownian motion in a homogeneous viscous solution, the diffusion coefficient (D) is given by the Stokes-Einstein relation:

$$D = \frac{k_B T}{6\pi\eta R_h} \quad (21.8)$$

where k_B is the Boltzmann constant; T is the temperature; η is the viscosity of the solution; and R_h is the hydrodynamic radius of the particle. Thus, the diffusion coefficient is a parameter dependent on the temperature of the system, the viscosity of the medium, and the particle size (Magde et al., 1972; Haustein and Schwille, 2007; Petrov and Schwille, 2008). It should be taken into account that this applies only to bulk diffusion and not diffusion on the membrane.

In case of GUVs, FCS can be used by placing the detection volume at the membrane at the top or bottom poles of the GUVs. However, diffusion in membranes is slower than in solution, resulting in longer measurement times, which can lead to photobleaching, membrane fluctuations and instabilities (Enderlein et al., 2004, 2005; Tcherniak et al., 2009). To overcome these problems, the following section describes an alternative approach that scans the membrane perpendicular to the equatorial plane of the GUVs.

21.3.1 SCANNING FLUORESCENCE CORRELATION SPECTROSCOPY

Scanning FCS (SFCS) is a useful technique especially for free standing membranes such as GUVs, where the focal volume is scanned with a constant velocity along a defined path while recording the fluorescence intensity (Ries and Schwille, 2006; Ries et al., 2009). When applied to GUVs, the scan path is oriented perpendicular to the membrane plane at the vesicle equator (Figure 21.2). By selecting this orientation, the focal volume is repeatedly scanned along the linear path and crosses the membrane only at specific time points. Thus, the residence time of the fluorophores in the detection volume is reduced, minimizing the probability of photobleaching. Consequently, long measuring times become feasible. The latter issue is very important because long measurement times need to be performed in membranes by a high number of subsequent scans. In addition, SFCS also solves problems associated with membrane fluctuations, because small membrane movements during the acquisition can be corrected. To optimize data quality, larger GUVs (>20 μm diameter) should be selected and the scanning path should be placed at the equatorial plane, where the membrane is virtually flat. The advantage of choosing linear SFCS in membrane studies is that it uses the scanning unit of standard fluorescence microscopes and can be easily used on most commercial setups.

The arrival times of the emitted photons can be recorded and grouped according to the scan rate of the LSM by binning. As an example, if the measurement time is 300 s with a scan rate of 1.5 ms, grouping the data in 1.5 ms segments results in about 200,000 scans. The next step is to count the number of photons arriving at a particular position in the scan over time. In this way, a higher number of photons results in a higher intensity in a scan position. The fluorescence intensity is then represented in a pseudo-image, where the horizontal and the vertical axes stand for the scan position and the scan number, respectively (Figure 21.2). In the pseudo-image, the region where the detection volume passes through the membrane is clearly shown as a trace of high intensity values. Because of thermal membrane undulations and drifts, this trace may not be straight vertical and the membrane region may appear at a different position from scan to scan. However, this can be corrected by using aligning the scans with respect to the position of maximum of the membrane contributions, which

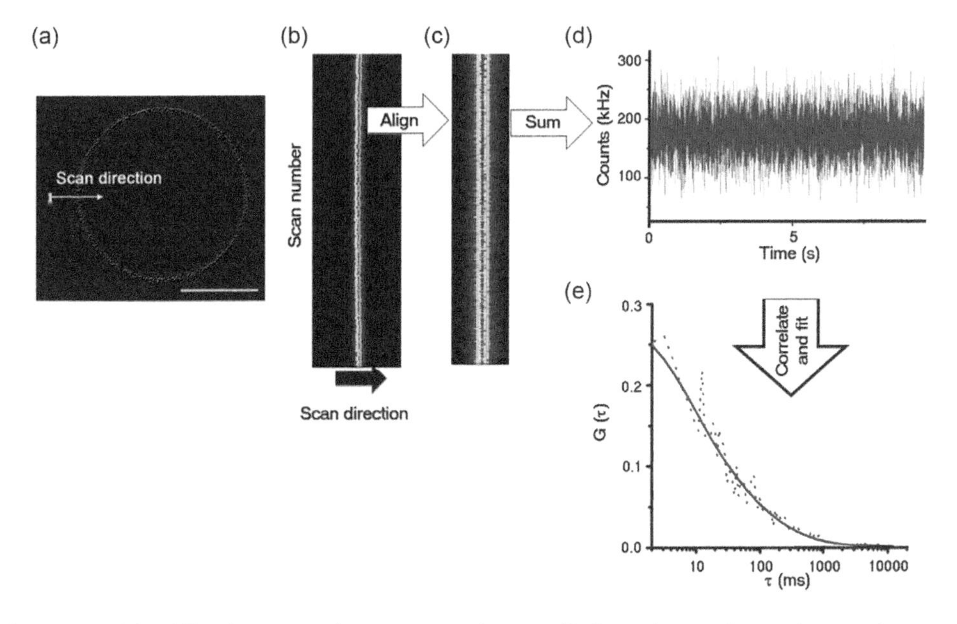

Figure 21.2 Principle of scanning FCS. (a) The detection volume is scanned perpendicular to the membrane plane, at the equatorial plane of the GUV. Scale bar: 10 μm. (b) Contribution of the membrane in fluorescence fluctuations can be observed with the line scans extending each other (scan width, 17.2 μm; number of scans: 158,730). (c) Membrane movements caused by instabilities are corrected by aligning the membrane for all scans considering a scan width that contains the fluorescence signal. (d) The photons detected in each line scan are summed up for the fluctuations in the fluorescence intensity. (e) Fluctuations are temporally autocorrelated and fitted with diffusion models (here, a 2D diffusion fitting is used for FCS in membranes). More information is detailed in Box 21.3. (Adapted from Unsay, J.D. and Garcia-Saez, A.J., *Methods Mol. Biol.*, 1033, 185–205, 2013.)

is usually with software (Figure 21.2). By summing up the intensity of the membrane contributions per each scan, one creates a discrete fluorescence intensity trace that can be autocorrelated as in point FCS and described using 2D diffusion of one component in a Gaussian elliptical detection volume (Ries and Schwille, 2006):

$$G(\tau)=\frac{1}{N}\left(1+\frac{\tau}{\tau_D}\right)^{-\frac{1}{2}}\left(1+\frac{\tau}{\tau_D S^2}\right)^{-\frac{1}{2}} \tag{21.9}$$

As discussed above, S is the structure parameter that stands for the aspect ratio of the focal volume: $S = w_z / w_0$. This can be obtained by performing a calibration experiment of a well-characterized fluorescent dye (see Table 21.2). Then, the autocorrelation curve of the calibration measurement is fitted with a simple 3D diffusion model to obtain N, τ_D and S. However, it has to be taken into account that this calibration is performed in solution and therefore the refraction index is different from that of membranes, thereby affecting the estimation of the detection volume.

Other advanced FCS techniques can be combined with SFCS to study interactions of two differently labeled molecules (two-color fluorescence cross correlation spectroscopy; two-color FCCS) or to provide calibration-free measurements (two-focus FCCS) (Ries and Schwille, 2006; Bacia and Schwille, 2007; Dertinger et al., 2008) (Figure 21.3). Two-color FCCS is an appropriate technique to investigate molecule binding by determining the concentration of complexes formed by two species (Garcia-Saez et al., 2009). In this regard, this technique provides information about the interaction of two labeled species. If the species interact they will diffuse together through the detection volume, generating simultaneous fluorescence fluctuations and positive cross correlation (CC) amplitude (Schwille et al., 1997). Thus, the higher the amplitude of cross correlation percentage is, the larger the extent of complex formation.

In this setup, the fluorophores are excited by two lasers in the same focal volume and the fluorescence traces of both channels are recorded. The excitation can be performed alternately to avoid cross talk (alternating excitation or pulse interleaved excitation), that is, only one laser excites the fluorophores at any time point and thus the signal from respective green or red channel is collected separately. On the other hand, in two-focus FCCS, the detection volume is scanned along two parallel paths. The possibility of obtaining the distance between the paths with high precision spares the determination of the shape of the focal volume. This can be done with one single laser if the focal volume alternately scans along two parallel paths. These methods can be combined for calibration-free binding and diffusion measurements (Ries et al., 2010). In both techniques, two different fluorescence traces are cross-correlated by the following expression (Ries and Schwille, 2008):

$$G_{gr}(\tau)=\frac{\langle \delta F_g(t)\cdot \delta F_r(t+\tau)\rangle}{\langle F_g(t)\rangle \cdot \langle F_r(t)\rangle} \tag{21.10}$$

Here, F_g and F_r are the fluorescence signal from the green and the red channels, respectively.

To obtain physical parameters, the theoretical model functions are fitted to the measured cross-correlation curves. The theoretical cross-correlation function for two-focus FCCS is given by:

$$G_{gr}(\tau)=G_g(\tau)\cdot G_r(\tau)\cdot \exp\left(-\frac{d^2}{4D\tau+w_0^2}\right) \tag{21.11}$$

In this equation, $G_g(\tau)$, $G_r(\tau)$ and d stand for the autocorrelation curves of the individual traces and the distance between the scan paths, respectively.

The resolution of FCS to distinguish distinct diffusing species is limited to a diffusion time difference of 1.6 fold, which corresponds to approximately six-fold change in mass between two molecules (Meseth et al., 1999; Sezgin and Schwille, 2011). The cross-correlation percentage provides information about the concentration of bound species gr (C_{gr}) (i.e., complex of green-and red-labeled molecules) and the unbound molecules in the green (C_g) and red (C_r) channels using the following equation (Ries and Schwille, 2008) (Box 21.3):

$$\%CC=\frac{G_{gr}(0)}{G_r(0)}=\frac{C_{gr}}{C_{gr}+C_g} \quad \text{or} \quad \%CC=\frac{G_{gr}(0)}{G_g(0)}=\frac{C_{gr}}{C_{gr}+C_r} \tag{21.12}$$

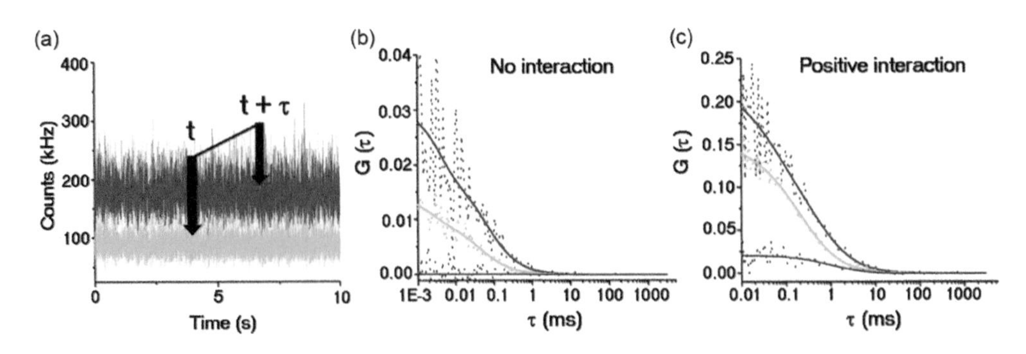

Figure 21.3 Two-color FCCS. (a) Fluorescence fluctuations in two channels is cross-correlated to find similarity between the two signals (red and green channels). (b) Particles do not interact when the cross correlation amplitude is near zero. (c) Particles showing positive cross correlation due to complex formation when the amplitude is above negative control levels. In (b–c), dots stand for auto (green/red) and cross (blue) correlation curves and lines for fitting curves. (Adapted from Das, K.K. et al., Microscopy of model membranes: Understanding how Bcl-2 proteins mediate apoptosis, in: *Advances in Planar Lipid Bilayers and Liposomes*, vol. 21 [Iglic, A. et al., eds], pp. 64–97, Academic Press, 2015.)

Box 21.2 Protocol for microscope alignment and calibration

The FCS setup needs to be properly aligned with a calibration measurement before use in order to optimize conditions during data acquisition. The calibration is also necessary to calculate the parameters related to the shape of the detection volume (w_0 and S). One should use a dye with similar spectral characteristics as the dye that will be measured in the membrane in your study. As an example, we will consider Alexa 488, which is a well-known dye in the green spectrum. Some other examples are listed in Table 21.2.

- Before the measurement, switch on the excitation lasers around 2 h to guarantee stabilization of the light source.
- Place 200 µL of 10 nM calibration dye solution in an observation chamber already preblocked with blocking solution (normally, a solution containing casein or bovine serum albumin). We usually use 8-well chambered coverglass system of 0.13–0.17 mm thickness.
- Place the observation chamber above the objective with the proper immersion medium.
- Set up the FSC configuration of the microscope that best fits the spectral properties under the study.
- Adjust the laser power that gives maximum counts per molecule without significant photobleaching during the acquisition time. Usually, this is around 0.3%–1%, depending on the molecule and sample as well as the sensitivity of the detectors for photon counts.
- Position the focal volume inside the solution at least 100 µm above the coverslip surface. FCS measurements are sensitive to distortions in the detection volume. These distortions may arise from improper axial positioning, differences in refractive index along the beam path, coverslip thickness and other effects (Enderlein et al., 2005; Ries and Schwille, 2008). Hence, it is very important to ensure that calibration measurements are performed away from the coverslip surface.
- Set the size of the pinhole to 40 µm or 1 Airy unit and align it with respect to the beam path. For this, select the x and y positions of the pinhole that give the highest fluorescence count rate or counts/molecule.
- Depending on the setup, it may be necessary to adjust the position of the collimator to achieve maximum count rate or counts/molecule.
- Adjust the correction collar of the objective to correct for the glass thickness of the observation chamber in order to maximize the count rate or counts/molecule.
- Perform the FCS measurement with your calibration dye solution. For instance, three acquisitions of 20 s each.
- Save the FCS data for further analysis. Depending on the software selected for data analysis, it may be necessary to export the data in ASCII format.
- Finally, fit an appropriate model function to the measured autocorrelation curve and estimate the diffusion time (τ_D) and the structure parameter (S). Generally, the autocorrelation curve of the calibration measurement is fitted with the 3D diffusion model with Gaussian detection to estimate these two parameters. Because the diffusion coefficient of the dye is known and can be obtained from the literature, the waist radius w_0 can be calculated using the equation (21.7). This should be assessed before performing measurements on samples containing particles with unknown characteristics. In addition, it has to be taken into account that different microscope setups will have different detection volume parameters (w_0 and S) because there is no exact true value for these parameters. In two-color experimental setups, the detection volume is wavelength dependent. This means that the linear dimensions are first-order proportional to the excitation wavelength, resulting in a focal volume, differing by a factor of 1.58 for the laser line 488 nm and 568 nm (Weidemann et al., 2002). For an unknown probe, it is always important to calibrate the FCS setup with the same or similar fluorophores that are tagged to the molecules of interest. It is recommended to keep a record of these calibration measurements to test for consistency of measurements over several months. In this way, eventual problems can be detected in the microscope setup.

Box 21.3 Protocol for data acquisition using SFCS on GUVs

First of all, you need to set up the light path, for example, to use APDs as the detector in your imaging mode of the microscope. If this is not possible, you need to perform first the GUV selection and positioning of the focal volume in the normal imaging mode, and then change to APD detection for the FCS measurement. This problem is solved when the same detector is used for imaging and FCS. Next, the following sequential steps need to be followed:

- Set the laser power to 0.3%–1%.
- Place the observation chamber containing the sample on the sample holder of the microscope. The observation chamber needs to be incubated with blocking solution around 30 min before adding the sample to it. The blocking solution can be prepared with 2 mg/mL casein or bovine serum albumin and avoids GUV spread onto the glass surface of the observation chamber. In addition, detailed information is given in Section II (Chapters 1 through 4) for GUV formation.
- Focus in the sample and select a uniform GUV. It is important to make sure that only truly unilamellar, tense (no visible fluctuations), homogeneous in intensity and immobile vesicles are selected. In our system we do not immobilize the GUVs (we wait until the GUVs filled with sucrose settle down in the chamber). However, one can immobilize the GUVs by different methods (e.g., by tethering to a surface, optical trapping, micropipette aspiration, microfluidics, agarose). Indeed, GUV immobilization with 0.25%–0.5% w/v agarose is shown to be an appropriate approach because the lateral

(Continued)

Box 21.3 (Continued) Protocol for data acquisition using SFCS on GUVs

displacement of the vesicles is completely suppressed without altering the lateral diffusion of the lipids (Lira et al., 2016). Depending on the lipid composition, GUVs with sizes ranging between 20 and 300 μm can be obtained with the electroformation method. Selecting larger vesicles is more suitable for FCS measurements because they show less curvature in the detection volume and cause less photobleaching problems. Unfortunately, it may be difficult to get large enough vesicles when using proteoliposomes.

- Once a suitable GUV is selected, place it to the center of your imaging field and zoom in to find the focal plane of the equator of the GUV (Figure 21.2). The usual scanning mode of a typical LSM scans in the horizontal ($x - y$) plane and SFCS measurements need to be performed perpendicular to the ideally flat membrane. Thus, the equator of the GUV is the best region for these measurements because it shows the focal plane where the GUV appears largest and exhibits zero curvature on the vertical plane. Otherwise, performing the scan elsewhere will result in a non-perpendicular scanning, and may distort the autocorrelation curves during analysis.
- Align the scanning path of the microscope, placing the GUV rim at the center and perpendicular to the scanned line. We generally use a zoom of 12 and acquire 32 pixel in x and 1 or 2 pixel in y (for one focus or two focus SFCS, respectively) in order to have a pixel size of 0.03 μm. In case of two focus SFCS, we use an angle of 45° for the measurement because in our microscope these conditions provide the closest to two parallel lines. This condition should be checked depending on the selected microscope.
- Select the scanning rate of your microscope at maximum speed (unidirectional, no averaging) and the analysis time. Accurate FCS data are achieved when appropriate time resolution for the fluorescent species diffusion is taken into account (Ries and Schwille, 2008; Tcherniak et al., 2009). Hence, it is advisable to measure a time period that is at least of the order of magnitude of 10,000 times longer than the predicted diffusion time of the fluorophore under study.
- Then one can start a time-lapse experiment of desired duration if the microscope software allows saving the photon arrival times. If this is not possible, one could use the photon mode of the hardware correlator. In case of using a hardware correlator, open the software for the photon mode and set up the parameters such as the analysis time and data storage (for one or two channels). Perform continuous scanning on the selected area by using the imaging mode of your microscope and start acquisition on the hardware correlator. Once the desired time of the measurement has elapsed, stop the experiment (i.e., stop acquisition and continuous scanning).

21.3.2 AUTOCORRELATION OF SFCS DATA

This section is focused on processing the raw data in the form of photon arrival times and on performing the autocorrelation. In our case, we use a home-written MATLAB® program (Figure 21.2). The process is briefly described in the following steps (Ries and Schwille, 2006):

- Using the data of the photon arrival times, bin the photon streams in bins of 100 ns to 5 μs depending on the scan rate. In theory, the bin size is determined by photon count rate, but in any case it should be much smaller compared with the scan rate.
- Arrange the photon beam according to the scan rate as a matrix/pseudo-image such that every row stands for one line scan.
- To smooth out small intensity irregularities along the fluorescence trace, every line scan can be convoluted with an averaging filter.
- Every line scan shows a maximum corresponding to the position of the membrane. To correct for membrane movement, shift all the scan lines to align this maximum value.
- Fit a Gaussian function to each line scan. For each line scan, sum up only the values between -2.5σ and 2.5σ to get one intensity value. When this is done for all line scans, a discrete fluorescence intensity trace over time is created that can be correlated using multi-tau algorithm (Magatti and Ferri, 2001).

21.3.3 DATA ANALYSIS AND FITTING A MODEL FUNCTION TO THE AUTOCORRELATION CURVE

Finally, a theoretical model has to be fitted to the experimental autocorrelation curve in order to obtain the physical parameters. This can be done by opening the autocorrelation curve with FCS-specific software (usually implemented in commercial systems) or with a program that has mathematical fitting option, such as MATLAB or Origin®. In our case, we use a home-written MATLAB program to fit model functions to FCS autocorrelation curves. Detailed information about the process is given in the following steps:

- After opening the autocorrelation curve with a suitable fitting program, plot the autocorrelation curves and discard those with distorted shapes (the latter ones usually result from major instabilities). Qualitatively, the shape of the curve is related to the process causing the fluorescence fluctuations, indicating the diffusion characteristics of the particles. Among others, it depends on the type of particle motion. Thus, the curve decay gives us information about the diffusion time of the molecules: a steep decay is typical for transport phenomena whereas slower decay stands for random Brownian or, alternatively, anomalous diffusion. In addition, the amplitude of the curve is a function of the particle area concentration (e.g., larger amplitude corresponds to lower concentration).
- For quantitative analysis, the autocorrelation curves have to be fitted with a proper model function using a nonlinear least-squares fitting algorithm (Table 21.3). A plot of the fitting residuals provides information about the fit quality. Successful fitting of the data will give us the value of the diffusion time (τ_D) and the number of particles (N) in the focal area. Other parameters such as the diffusion coefficient (D) can then be derived. In case of two-focus SFCCS, absolute measurements of fluorophore concentration and diffusion coefficient are obtained by fitting the data with the equation (21.11).

21.4 FLUORESCENT RECOVERY AFTER PHOTOBLEACHING

FRAP is a powerful biophysical technique that is commonly used to study the dynamic processes that govern the organization and structure of living cells. Among others, this methodology has been used to provide information about protein diffusion and interaction (Reits and Neefjes, 2001; Carrero et al., 2003; Kenworthy et al., 2004), lipid fluctuations (Goodwin et al., 2005; Goodwin and Kenworthy, 2005; Kenworthy, 2007; Wong and Wessel, 2008), vesicle transport (Smith et al., 2003; Tagawa et al., 2005), cell adhesion (Zhang et al., 2010), and organelle dynamics (Mitra and Lippincott-Schwartz, 2010). In addition, the constant progress in mathematical interpretation of data and the implementation of confocal laser-scanning microscopes allow a continuous development of this growing field (Braga et al., 2004; Kang et al., 2010).

The principle of FRAP is based on the irreversible photobleaching of a region of interest (ROI) bearing the fluorescent particle and the fluorescence recovery of the bleached region over time. This recovery takes place as a result of diffusion of fluorescent particles from the surrounding non-bleached areas, known as mobile fraction (M_f), whereas those that cannot interchange between the two regions are known as immobile fraction (I_f). This fluorescence recovery needs to be recorded with a low laser power in order to avoid additional photobleaching and results in an FRAP curve where the recovery speed and the particle fraction diffusion can be measured. Thus, FRAP analysis can provide many parameters including the size of diffusion particles, the internal organization of the cells and the binding degree of particles (Reits and Neefjes, 2001; Kang and Kenworthy, 2008; McNally, 2008). In case of GUVs, this technique can be used to characterize the diffusion of lipids, lipidic probes and proteins that bind to the membrane. In this regard, the diffusion of these molecules can also be analyzed depending on the lipid phase(s) present in the GUVs (i.e., liquid disordered phase, liquid ordered phase, gel phase).

A typical graphic representation of an FRAP measurement is shown in Figure 21.4. Here, the changes in fluorescence intensity of the bleached region are plotted over time.

First, there is a maximum or initial fluorescence intensity level (I_i), that corresponds to the pre-bleached step. This step is necessary for data normalization. After the photobleaching step, the signal decays to a minimum (I_0) and later it recovers exponentially to a constant value (I_∞). After the curve has been normalized and corrected for imaging-related bleaching in a control region, it is fitted to a nonlinear model to obtain parameters such as the half-time recovery ($\tau_{1/2}$) (i.e., the point when half of the fluorescence recovery has taken place) and the mobile (M_f) and immobile (I_f) fractions, see their definitions in the caption of Figure 21.4. The simplest one is an exponential equation:

$$\frac{I_t - I_0}{I_i - I_0} = M_f\left(1 - e^{-t/\tau}\right) \quad (21.13)$$

Then, the diffusion coefficient (D) can be calculated:

$$D = \frac{r_0^2 \gamma}{4\tau_{1/2}} \quad (21.14)$$

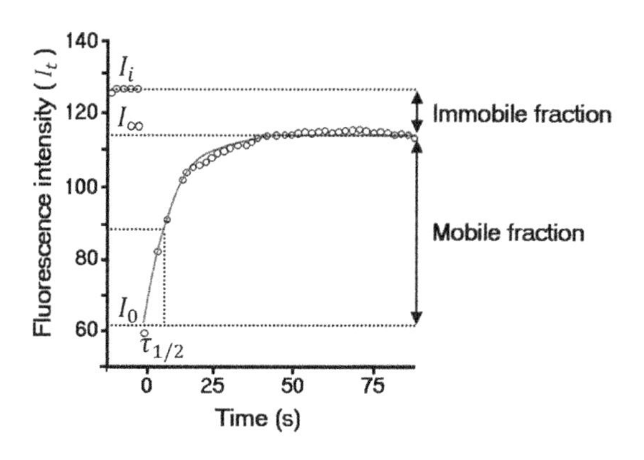

Figure 21.4 A typical FRAP curve. The initial intensity at the pre-bleached state (I_i) drops to a minimum (I_0) after fluorescence photobleaching and it recovers exponentially to a constant value (I_∞). $\tau_{1/2}$ stands for the half-time recovery, and the mobile and immobile fractions can be computed from $M_f = \frac{I_\infty - I_0}{I_i - I_0}$ and $I_f = 1 - M_f$, respectively. The solid curve corresponds to the data fitting. (Adapted from Subburaj, Y. et al., Membrane dynamics: Fluorescence correlation spectroscopy, in: *Encyclopedia of Analytical Chemistry: Applications, Theory and Instrumentation*, pp. 1–29, John Wiley & Sons, 2013.)

Here, r_0 is the radius of the excitation spot, $\tau_{1/2}$ is the half-time recovery, and γ is the bleaching parameter, resulting from the ratio between the nominal radius of the bleached area defined by the user (r_n) and the effective bleached radius from the post-bleached profile (r_e) (Sezgin and Schwille, 2011; Kang et al., 2012):

$$\gamma = \frac{r_n}{r_e} \quad (21.15)$$

This approach accounts for diffusion occurring already during photobleaching. In Table 21.4, some FRAP fitting models are listed.

21.4.1 APPLICATION OF FRAP ON GUVs

GUVs are a suitable model system for FRAP experiments because different regions within the GUV can be selected for this purpose. Useful GUV regions for FRAP experiments are the bottom and the top areas of the GUV (similar to supported lipid bilayers), the equatorial zone and regions from tubes generated by motor proteins or formed by tube pulling experiments (see Chapter 16). In fact, the latter regions have been successfully described in a number of reports in order to determine the role of membrane curvature in protein sorting and mobility (Ambroggio et al., 2010; Roux et al., 2010; Zhu et al., 2012; Aimon et al., 2014; Fossati et al., 2014; Lira et al., 2014). On the other hand, the equatorial zone of the GUV can also be selected to perform FRAP experiments in the GUV lumen in order to characterize encapsulated material (Lira et al., 2014). Figure 21.5 shows an overview of the GUV membrane regions that can be chosen for FRAP experiments. Due to polarization problems that are visible at the equatorial plane of the GUV (Figure 21.5), one should select similar regions along the angular position of the equator for appropriate comparison.

Next, we describe some consecutive steps and recommendations to perform FRAP experiments on GUVs based on general FRAP protocols:

- Use a confocal LSM with appropriate laser and filter sets for your study.
- For molecular mobility assays, set the scanning parameters including mode, zoom, pixel size, scan speed, scan number

Table 21.4 **FRAP fitting models**

TYPE OF FITTING MODEL	FUNCTION	REFERENCES
Diffusion	$I_t = \dfrac{\left[I_0 + I_\infty\left(\dfrac{t}{\tau_{1/2}}\right)\right]}{\left[1+\left(\dfrac{t}{\tau_{1/2}}\right)\right]}$	Feder et al. (1996) and Phair et al. (2004)
Diffusion (accounting for diffusion during photobleaching)	$I_t = I_i\left(1-\dfrac{K}{1+\gamma^2+2t/\tau_D}\right)M_f + \left(1-M_f\right)I_0$	Kang et al. (2012)
Chemical interaction dominant	$I_t = \gamma_0 + Ae^{-t/\tau_1}$	Phair et al. (2004)

Note: K, bleaching depth parameter; γ_0, plateau of the curve at bleaching time point; τ_D, characteristic diffusion time.

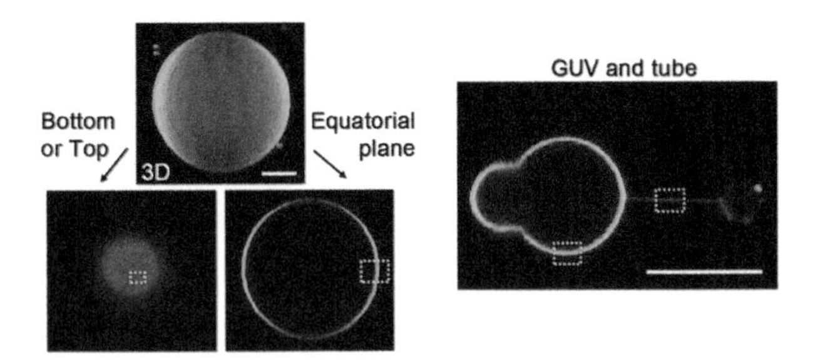

Figure 21.5 Application of FRAP on GUVs. GUV membrane areas where FRAP can be performed (shown as white dotted boxes): bottom or top part (similar to supported lipid bilayers), equatorial plane and tube generated by tube pulling experiments. Scale bars: 10 µm.

of averaging (1), pinhole (open), laser intensity (0.5%–10%) and detector gain. Verify the set of parameters by doing single scans each time after adjusting a parameter. Depending on the system, these settings may be altered to improve imaging.

- Define the ROI where bleaching will occur and recovery will be followed (Figure 21.5, white dotted boxes). To have an accurate quantification, it is better to select circular ROIs and if possible small enough to approximate the profile to a Gaussian function, but not too small to obstruct the analysis. In this way, the ROI should not be more than 30%–50% of the total signal area, so that the fluorescence recovery in a given time from the unbleached zone (50%–70%) can be easily detected. For molecular mobility assays, it is recommended to use an ROI just enough to cover the zone of interest. Additional control ROIs must be selected that will not be bleached. These ROIs are collected to correct the curves for background fluorescence, photofading, and loss of fluorescence as a result of the bleaching and imaging (Day et al., 2012). It is preferable to have the same size and shape for the photobleaching and control ROIs.
- Standardize bleaching conditions. Photobleaching is achieved using the same laser that is used for scanning, but in this case maximum power is required. For FRAP the ideal bleach is the one that reduces the ROI intensity by 60%–80% but does not cause photodamage. Then, it is recommended to bleach at the hard and fast. Thus, it is preferable to use high laser intensity with few iterations and short time (i.e., how long the bleaching laser is in the region of interest and how many times it bleaches). This is especially important for fluorescent molecules with a fast turnover rate (i.e., recovery kinetics in milliseconds) because the fluorescence recovery cannot occur while the region is being bleached. Moreover, speed is the key in FRAP experiments rather than a pretty image. Hence, the gain should be set high and the exposure time short (this parameter limits the speed of data collection). Alternatively, the laser power could be slightly increased.
- After the bleaching conditions are optimized, set the time interval and the total imaging time. In general, images should be acquired for 10–50 times longer than the half-life or until the signal in the bleached area reaches a plateau. The frame rate is also important and one needs to sample twice as fast as the measured parameter is changing.
- Acquire and save the time series to be analyzed. Briefly, for data processing a few recommendations have to be taken into account using control ROIs: background subtraction from the total image and photodamage correction. The curve is then normalized and fitted to a nonlinear model in order to obtain the half-time recovery ($\tau_{1/2}$), the mobile (M_f) and immobile (I_f) fractions, and the diffusion coefficient (D) by using Eqs. 21.13 and 21.14 (see also Figure 21.4 and Table 21.4).

21.5 SINGLE-MOLECULE IMAGING

Over the past few years, essential tools have been developed in the fields of biophysics and cell biology to study the dynamics of biological processes. Among them, single-molecule techniques provide crucial information that is averaged out in other traditional ensemble methods (Weiss, 1999). Whereas FCS and FRAP approaches can only provide information about the tendency of molecular behavior averaged over all molecules under study, single-molecule imaging (SMI) is capable of determining individual characteristics of a molecule that may be faded in the whole averaging inherent in bulk studies (Axelrod et al., 1976; Kusumi et al., 2004).

In general, a number of individual molecules within a cellular process come together and interact with each other in order to transmit information and respond to the environmental cues. Therefore, it is relevant to gain more insight into the motion of these molecules in order to achieve a better understanding of their function and regulation in the cells. Nevertheless, this issue becomes very complicated taking into account that the molecular behavior may be inhomogeneous, even finding molecules of a single species interacting stochastically with different molecules or cellular structures in a certain local milieu.

The method of single molecule detection provides combined spatial and temporal resolution information on single events (Jaiswal and Simon, 2007). Here, total internal reflection fluorescence (TIRF) microscopy is used for imaging because of its ability to limit the illumination of the sample up to a thickness of 200 nm (Schneckenburger, 2005). Briefly, the total internal reflection phenomenon occurs when the light beam passes from a medium with a high refractive index into a medium with a low refractive index. The light will bend and travel along the interface if the incident angle is smaller or equal to a critical angle. When the incident angle is higher than the critical one, the light will turn back into the high refractive medium and only a short-range electromagnetic disturbance, known as evanescent wave, will pass into the low refractive medium. Because the evanescent field intensity decreases exponentially with the distance from the interface, high-contrast images of the near-membrane area can be acquired with a maximum penetration depth of 100–200 nm (Figure 21.6). Hence, the signal-to-noise ratio is drastically improved, reducing out-of-focus fluorescence and photodamage. Moreover, the use of an objective with a high numerical aperture makes this approach extremely suitable for measurements with a low number of photons, as in the case of single molecule detection (Axelrod, 2001, 2008).

In general, an SMI experiment that analyzes the trajectories of the molecules (also known as single-particle tracking) is based on the following steps (Cosentino et al., 2015):

- First, a single fluorescent molecule needs to be detected. This requires working in a single-molecule regime, which means that only a low amount of fluorescent particles is needed. Its intensity is then fitted with a known distribution function (2D Gaussian fit), thereby providing precise information about its x and y position.
- The detection of molecules is repeated temporally in order to determine their trajectories.

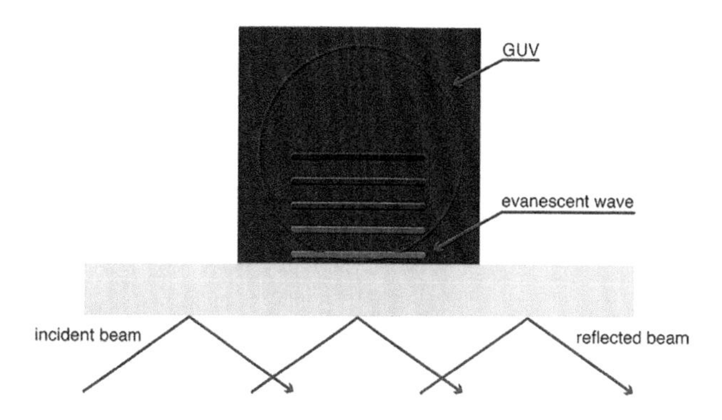

Figure 21.6 Principle of TIRF microscopy and its application on GUVs. TIRF occurs when the incident angle (defined respect to the normal to the interface), is higher than a critical angle. The excitation beam is reflected off the coverslip–sample interface and an evanescent field is generated in the sample that excites fluorescent molecules up to around 100–200 nm above the glass surface. In this scheme, the evanescence wave scope is shown from bright (best) to dark (none) waves. Thus, only the GUV part that is close to the interface is illuminated with TIRF microscopy.

- These trajectories are correlated over time (i.e., displacement of the molecule is calculated in consecutive frames), resulting in a plot of the mean square displacement (MSD) versus time lag. At this point the generated curve can be fitted with one of the theoretical diffusion models (e.g., Brownian, confined, anomalous or directed diffusion models). From the fitting one can get information about its mode of motion and quantitative parameters such as its diffusion coefficient (Figure 21.7).

21.5.1 APPLICATION OF SINGLE-MOLECULE IMAGING ON GUVs AND INCURRING CHALLENGES

This technique is mainly used to study mode of motion and interaction of molecules at the cell surface or in supported lipid bilayers (Rassam et al., 2015; Subburaj et al., 2015b). Unfortunately, GUVs are not suitable because only the GUV region that is at close contact with the coverslip is illuminated with this approach and the trajectory of molecules diffusing to upper planes cannot be detected. An example of this drawback is shown in Figure 21.8. Alternatively to TIRF microscopy, single-molecule experiments have been reported

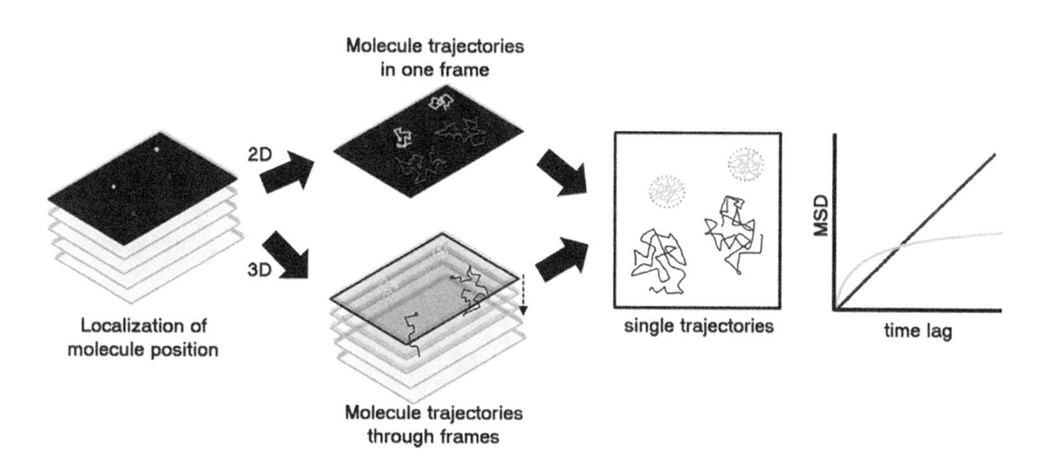

Figure 21.7 Scheme of single-molecule imaging (SMI) for one or multiple frames. The images are acquired with single molecule sensitivity in 2D (only one frame) or 3D (in different frames). After image acquisition, the position of the molecules can be calculated. The tracking of the molecules allows calculation of single trajectories and of the MSD with time. The curve shape is dependent on the motion type. Green curve, confined motion; magenta curve, Brownian motion.

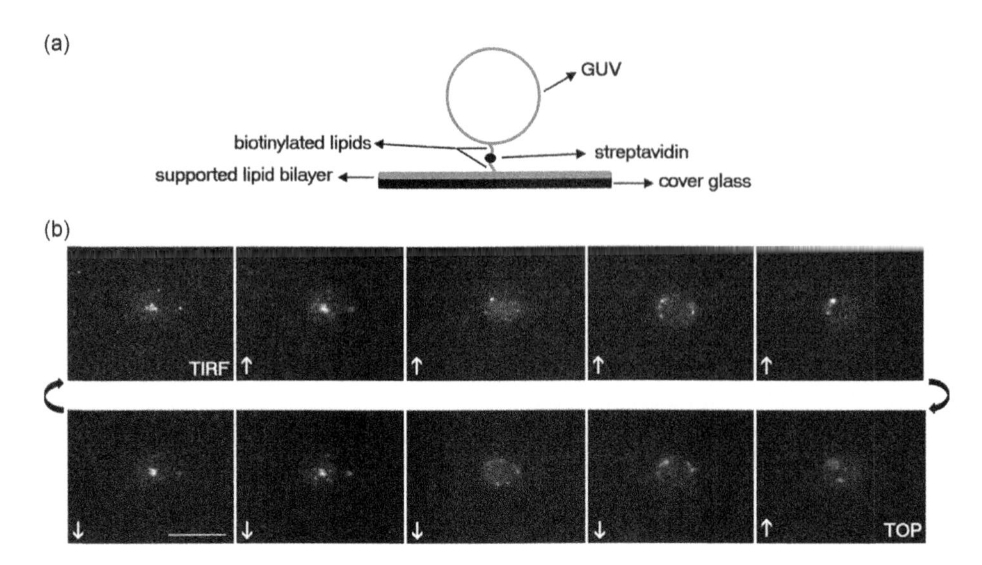

Figure 21.8 SMI on GUVs. (a) Schematic representation of the sample. The GUV is immobilized onto the preformed supported lipid bilayer with streptavidin through biotinylated lipids. (b) SMI on an individual GUV from the interface to the top of the vesicle surface. Only the top left image is taken in TIRF illumination range (the *z*-axis resolution is around 100–200 nm); the rest of the images are taken in epifluorescence (up to the top of the vesicle). High-contrast images of the near-membrane area can be acquired only with TIRF microscopy, where the signal-to-noise ratio is drastically improved, reducing out-of-focus fluorescence and photodamage. White arrows, plane movement in the *z*-axis; black arrows, the imaging direction. Here, the molecules under study tend to tether membranes and they appear as a concentrated spot at the contact surface between the GUV and the supported lipid bilayer. Scale bar: 10 μm.

in 80 μm diameter-GUVs using wide-field microscopes with an axial resolution of around 1 μm (Ciobanasu et al., 2009). In this case, the authors measured the radius of the GUVs at the equator plane by employing a membrane fluorescent dye and a grid with known dimensions within the eyepiece. Then, they shifted the focus into the sample by a distance corresponding to the GUV radius to reach the top pole of the spherical GUV. Single-molecule signals could be detected in this plane by fluorescence mode, where the region on the GUV top had a diameter of around 14 μm.

Although several attempts have been made in the last years to follow 2D and 3D single- or multiple-molecule trajectories in cells (Levi et al., 2003; Ragan et al., 2006) and on GUVs (Ciobanasu et al., 2009; Spille et al., 2015), its application is still limited to a certain region of the membrane. Therefore, new approaches should be developed in order to overcome this limitation.

21.6 FUTURE OUTLOOK

Using advanced microscopy techniques, we are only starting to unravel the complex relationships between molecule interactions and their functionality in cellular processes. The single-molecule approaches described in this chapter are powerful techniques that offer the possibility of studying membrane dynamics and gain information on the properties and motion of molecules. The easy accessibility and maturation of them during the last years will continue to make single-molecule analysis a major tool for dynamic processes that occur at the membrane level. However, not all the aforementioned methods are suitable when GUVs are selected as model systems. Thus, the choice of one technique over the other depends mainly on the specific biological question to be addressed.

LIST OF ABBREVIATIONS

APD	avalanche photodiode
BODIPY	4,4-difluoro-5,7dimethyl-4-bora-3a,4a-diaza-s-indacene-3-pentanoyl
CC	cross correlation
CFP	cyan fluorescent protein
Chol	cholesterol
DLPC	dilauroyl phosphatidylcholine
DOPC	dioleoyl phosphatidylcholine
DPPC	dipalmitoyl phosphatidylcholine
DSPC	distearoyl phosphatidylcholine
DsRed	*Dicosoma striata* red fluorescent protein
FCS	fluorescence correlation spectroscopy
FCCS	fluorescence cross correlation spectroscopy
FRAP	fluorescence recovery after photobleaching
GFP	green fluorescent protein
GL	Gaussian-Lorentzian
LAURDAN	6-dodecanoyl-2-dimethylaminonaphtalene
Ld	liquid disordered phase
LSM	laser scanning microscope
MSD	mean square displacement
NHS	*N*-hydroxysuccinimide
POPC	palmitoyl oleoyl phosphatidylcholine
PSF	point spread function
ROI	region of interest
SFCS	scanning FCS
SM	sphingomyelin
SMI	single molecule imaging
TFP	tetrafluorophenyl
TIRF	total internal reflection fluorescence
YFP	yellow fluorescent protein

GLOSSARY OF SYMBOLS

C	concentration
C_g	concentration of unbound molecules in the green channel
C_{gr}	concentration of bound molecules
C_r	concentration of unbound molecules in the red channel
D	diffusion coefficient
$\delta F(t)$	fluorescence fluctuation at time t
F_g	fluorescence signal from the green channel
F_r	fluorescence signal from the red channel
$F(t)$	overall fluorescence signal
$G(\tau)$	autocorrelation function
$G_{2D}(\tau)$	autocorrelation function for 2D diffusion
$G_{2DG}(\tau)$	autocorrelation function for 2D diffusion with elliptical-Gaussian profile
$G_{3D}(\tau)$	autocorrelation function for 3D diffusion
$G_{3D+2C}(\tau)$	autocorrelation function for 3D diffusion with two components
$G_{3D+T}(\tau)$	autocorrelation function for 3D diffusion with triplet
$G_g(\tau)$	autocorrelation function for the green channel
$G_{gr}(\tau)$	cross correlation function
$G_r(\tau)$	autocorrelation function for the red channel
γ	bleaching parameter
γ_0	plateau of the curve at bleaching time point
I_0	minimum intensity after photobleaching
I_∞	intensity recovery to a constant
I_f	immobile fraction
I_i	initial intensity at the pre-bleach state
K	bleaching depth parameter
k_B	Boltzmann constant
λ	excitation wavelength
$M_f x$	mobile fraction
N	number of particles
η	viscosity
ρ	photon detected position
q_f	molecular brightness of the fast diffusing component
q_s	molecular brightness of the slow diffusing component
r_0	radius of the excitation spot
r_e	effective bleached radius from the post-bleached profile
r_n	nominal radius of the bleached area defined by the user
R_h	hydrodynamic radius of the particle
S	structure parameter
T	temperature
T_R	fraction of the fluorophores in the triplet state
τ	lag or correlation time
$\tau_{1/2}$	half-time recovery
τ_D	diffusion time
τ_{TR}	triplet time
V_{eff}	effective volume
Y_f	molar fraction of the fast diffusing component
Y_s	molar fraction of the slow diffusing component
w	beam waist
w_0	extension of the focal volume in radial direction
w_z	extension of the focal volume in axial direction

REFERENCES

Aimon S, Callan-Jones A, Berthaud A, Pinot M, Toombes GE, Bassereau P (2014) Membrane shape modulates transmembrane protein distribution. *Developmental Cell* 28:212–218.

Alessandrini A, Facci P (2014) Phase transitions in supported lipid bilayers studied by AFM. *Soft Matter* 10:7145–7164.

Ambroggio E, Sorre B, Bassereau P, Goud B, Manneville JB, Antonny B (2010) ArfGAP1 generates an Arf1 gradient on continuous lipid membranes displaying flat and curved regions. *The EMBO Journal* 29:292–303.

Axelrod D (2001) Selective imaging of surface fluorescence with very high aperture microscope objectives. *Journal of Biomedical Optics* 6:6–13.

Axelrod D (2008) Chapter 7: Total internal reflection fluorescence microscopy. *Methods in Cell Biology* 89:169–221.

Axelrod D, Koppel DE, Schlessinger J, Elson E, Webb WW (1976) Mobility measurement by analysis of fluorescence photobleaching recovery kinetics. *Biophysical Journal* 16:1055–1069.

Bacia K, Scherfeld D, Kahya N, Schwille P (2004) Fluorescence correlation spectroscopy relates rafts in model and native membranes. *Biophysical Journal* 87:1034–1043.

Bacia K, Schwille P (2007) Practical guidelines for dual-color fluorescence cross-correlation spectroscopy. *Nature Protocols* 2:2842–2856.

Braga J, Desterro JM, Carmo-Fonseca M (2004) Intracellular macromolecular mobility measured by fluorescence recovery after photobleaching with confocal laser scanning microscopes. *Molecular Biology of the Cell* 15:4749–4760.

Campbell RE, Tour O, Palmer AE, Steinbach PA, Baird GS, Zacharias DA, Tsien RY (2002) A monomeric red fluorescent protein. *Proceedings of the National Academy of Sciences of the United States of America* 99:7877–7882.

Carrer DC, Schmidt AW, Knolker HJ, Schwille P (2008) Membrane domain-disrupting effects of 4-substitued cholesterol derivatives. *Langmuir: The ACS Journal of Surfaces and Colloids* 24:8807–8812.

Carrero G, McDonald D, Crawford E, de Vries G, Hendzel MJ (2003) Using FRAP and mathematical modeling to determine the in vivo kinetics of nuclear proteins. *Methods* 29:14–28.

Chandler D (2005) Interfaces and the driving force of hydrophobic assembly. *Nature* 437:640–647.

Ciobanasu C, Harms E, Tunnemann G, Cardoso MC, Kubitscheck U (2009) Cell-penetrating HIV1 TAT peptides float on model lipid bilayers. *Biochemistry* 48:4728–4737.

Cosentino K, Bleicken S, Garcia-Saez AJ (2015) Analysis of membrane-protein complexes by single-molecule methods. *Pumps, Channels and Transporters: Methods of Functional Analysis* 183:269–297.

Culbertson CT, Jacobson SC, Michael Ramsey J (2002) Diffusion coefficient measurements in microfluidic devices. *Talanta* 56:365–373.

Das KK, Unsay JD, Garcia-Saez AJ (2015) Microscopy of model membranes: Understanding how Bcl-2 proteins mediate apoptosis. In: *Advances in Planar Lipid Bilayers and Liposomes*, vol. 21 (Iglic, A. et al., eds), pp. 64–97. Academic Press.

Day CA, Kraft LJ, Kang M, Kenworthy AK (2012) Analysis of protein and lipid dynamics using confocal fluorescence recovery after photobleaching (FRAP). *Current Protocols in Cytometry* Chapter 2, 62:2–19.

Dertinger T, Loman A, Ewers B, Muller CB, Kramer B, Enderlein J (2008) The optics and performance of dual-focus fluorescence correlation spectroscopy. *Optics Express* 16:14353–14368.

Dertinger T, Pacheco V, von der Hocht I, Hartmann R, Gregor I, Enderlein J (2007) Two-focus fluorescence correlation spectroscopy: A new tool for accurate and absolute diffusion measurements. *Chemphyschem: A European Journal of Chemical Physics and Physical Chemistry* 8:433–443.

Dietrich C, Yang B, Fujiwara T, Kusumi A, Jacobson K (2002) Relationship of lipid rafts to transient confinement zones detected by single particle tracking. *Biophysical Journal* 82:274–284.

Eggeling C, Ringemann C, Medda R, Schwarzmann G, Sandhoff K, Polyakova S, Belov VN, Hein B, von Middendorff C, Schonle A, Hell SW (2009) Direct observation of the nanoscale dynamics of membrane lipids in a living cell. *Nature* 457:1159–1162.

Eigen M, Rigler R (1994) Sorting single molecules: Application to diagnostics and evolutionary biotechnology. *Proceedings of the National Academy of Sciences of the United States of America* 91:5740–5747.

Enderlein J, Gregor I, Patra D, Dertinger T, Kaupp UB (2005) Performance of fluorescence correlation spectroscopy for measuring diffusion and concentration. *Chemphyschem: A European Journal of Chemical Physics and Physical Chemistry* 6:2324–2336.

Enderlein J, Gregor I, Patra D, Fitter J (2004) Art and artefacts of fluorescence correlation spectroscopy. *Current Pharmaceutical Biotechnology* 5:155–161.

Falck E, Patra M, Karttunen M, Hyvonen MT, Vattulainen I (2004) Lessons of slicing membranes: Interplay of packing, free area, and lateral diffusion in phospholipid/cholesterol bilayers. *Biophysical Journal* 87:1076–1091.

Farlow J, Seo D, Broaders KE, Taylor MJ, Gartner ZJ, Jun YW (2013) Formation of targeted monovalent quantum dots by steric exclusion. *Nature Methods* 10:1203–1205.

Feder TJ, Brust-Mascher I, Slattery JP, Baird B, Webb WW (1996) Constrained diffusion or immobile fraction on cell surfaces: A new interpretation. *Biophysical Journal* 70:2767–2773.

Fossati M, Goud B, Borgese N, Manneville JB (2014) An investigation of the effect of membrane curvature on transmembrane-domain dependent protein sorting in lipid bilayers. *Cellular Logistics* 4:e29087.

Garcia-Saez AJ, Ries J, Orzaez M, Perez-Paya E, Schwille P (2009) Membrane promotes tBID interaction with BCL(XL). *Nature Structural & Molecular Biology* 16:1178–1185.

Garcia-Saez AJ, Schwille P (2008) Fluorescence correlation spectroscopy for the study of membrane dynamics and protein/lipid interactions. *Methods* 46:116–122.

Gielen E, Smisdom N, vandeVen M, De Clercq B, Gratton E, Digman M, Rigo JM, Hofkens J, Engelborghs Y, Ameloot M (2009) Measuring diffusion of lipid-like probes in artificial and natural membranes by raster image correlation spectroscopy (RICS): Use of a commercial laser-scanning microscope with analog detection. *Langmuir: The ACS Journal of Surfaces and Colloids* 25:5209–5218.

Goncalves MS (2009) Fluorescent labeling of biomolecules with organic probes. *Chemical Reviews* 109:190–212.

Goodwin JS, Drake KR, Rogers C, Wright L, Lippincott-Schwartz J, Philips MR, Kenworthy AK (2005) Depalmitoylated Ras traffics to and from the Golgi complex via a nonvesicular pathway. *The Journal of Cell Biology* 170:261–272.

Goodwin JS, Kenworthy AK (2005) Photobleaching approaches to investigate diffusional mobility and trafficking of Ras in living cells. *Methods* 37:154–164.

Haustein E, Schwille P (2007) Fluorescence correlation spectroscopy: Novel variations of an established technique. *Annual Review of Biophysics and Biomolecular Structure* 36:151–169.

Hermann E, Ries J, Garcia-Saez AJ (2015) Scanning fluorescence correlation spectroscopy on biomembranes. *Methods in Molecular Biology* 1232:181–197.

Hess ST, Webb WW (2002) Focal volume optics and experimental artifacts in confocal fluorescence correlation spectroscopy. *Biophysical Journal* 83:2300–2317.

Jaiswal JK, Simon SM (2007) Imaging single events at the cell membrane. *Nature Chemical Biology* 3:92–98.

Kahya N, Schwille P (2006) How phospholipid-cholesterol interactions modulate lipid lateral diffusion, as revealed by fluorescence correlation spectroscopy. *Journal of Fluorescence* 16:671–678.

Kang M, Day CA, DiBenedetto E, Kenworthy AK (2010) A quantitative approach to analyze binding diffusion kinetics by confocal FRAP. *Biophysical Journal* 99:2737–2747.

Kang M, Day CA, Kenworthy AK, DiBenedetto E (2012) Simplified equation to extract diffusion coefficients from confocal FRAP data. *Traffic* 13:1589–1600.

Kang M, Kenworthy AK (2008) A closed-form analytic expression for FRAP formula for the binding diffusion model. *Biophysical Journal* 95:L13–L15.

Kenworthy AK (2007) Fluorescence recovery after photobleaching studies of lipid rafts. *Methods in Molecular Biology* 398:179–192.

Kenworthy AK, Nichols BJ, Remmert CL, Hendrix GM, Kumar M, Zimmerberg J, Lippincott-Schwartz J (2004) Dynamics of putative raft-associated proteins at the cell surface. *The Journal of Cell Biology* 165:735–746.

Korlach J, Schwille P, Webb WW, Feigenson GW (1999) Characterization of lipid bilayer phases by confocal microscopy and fluorescence correlation spectroscopy. *Proceedings of the National Academy of Sciences of the United States of America* 96:8461–8466.

Kusumi A, Koyama-Honda I, Suzuki K (2004) Molecular dynamics and interactions for creation of stimulation-induced stabilized rafts from small unstable steady-state rafts. *Traffic* 5:213–230.

Kusumi A, Sako Y, Yamamoto M (1993) Confined lateral diffusion of membrane receptors as studied by single particle tracking (nanovid microscopy). Effects of calcium-induced differentiation in cultured epithelial cells. *Biophysical Journal* 65:2021–2040.

Levi V, Ruan Q, Kis-Petikova K, Gratton E (2003) Scanning FCS, a novel method for three-dimensional particle tracking. *Biochemical Society Transactions* 31:997–1000.

Lira RB, Dimova R, Riske KA (2014) Giant unilamellar vesicles formed by hybrid films of agarose and lipids display altered mechanical properties. *Biophysical Journal* 107:1609–1619.

Lira RB, Steinkuhler J, Knorr RL, Dimova R, Riske KA (2016) Posing for a picture: Vesicle immobilization in agarose gel. *Scientific Reports* 6:25254.

Magatti D, Ferri F (2001) Fast multi-tau real-time software correlator for dynamic light scattering. *Applied Optics* 40:4011–4021.

Magde D, Elson EL, Webb WW (1974) Fluorescence correlation spectroscopy. II. An experimental realization. *Biopolymers* 13:29–61.

Magde D, Webb WW, Elson E (1972) Thermodynamic fluctuations in a reacting system—Measurement by fluorescence correlation spectroscopy. *Physical Review Letters* 29:705.

Marguet D, Lenne PF, Rigneault H, He HT (2006) Dynamics in the plasma membrane: How to combine fluidity and order. *The EMBO Journal* 25:3446–3457.

McNally JG (2008) Quantitative FRAP in analysis of molecular binding dynamics in vivo. *Methods in Cell Biology* 85:329–351.

Medina MA, Schwille P (2002) Fluorescence correlation spectroscopy for the detection and study of single molecules in biology. *BioEssays: News and Reviews in Molecular, Cellular and Developmental Biology* 24:758–764.

Meseth U, Wohland T, Rigler R, Vogel H (1999) Resolution of fluorescence correlation measurements. *Biophysical Journal* 76:1619–1631.

Michalet X, Pinaud FF, Bentolila LA, Tsay JM, Doose S, Li JJ, Sundaresan G, Wu AM, Gambhir SS, Weiss S (2005) Quantum dots for live cells, in vivo imaging, and diagnostics. *Science* 307:538–544.

Mitra K, Lippincott-Schwartz J (2010) Analysis of mitochondrial dynamics and functions using imaging approaches. *Current Protocols in Cell Biology* Chapter 4, Unit 4, 25:21–21.

Petrasek Z, Schwille P (2008) Precise measurement of diffusion coefficients using scanning fluorescence correlation spectroscopy. *Biophysical Journal* 94:1437–1448.

Petrov EP, Schwille P (2008) State of the art and novel trends in fluorescence correlation spectroscopy. *Springer Ser Fluores* 6:145–197.

GUVs as membrane interaction platforms

Phair RD, Gorski SA, Misteli T (2004) Measurement of dynamic protein binding to chromatin in vivo, using photobleaching microscopy. *Methods in Enzymology* 375:393–414.
Ragan T, Huang H, So P, Gratton E (2006) 3D particle tracking on a two-photon microscope. *Journal of Fluorescence* 16:325–336.
Rassam P, Copeland NA, Birkholz O, Toth C, Chavent M, Duncan AL, Cross SJ et al. (2015) Supramolecular assemblies underpin turnover of outer membrane proteins in bacteria. *Nature* 523:333–336.
Reits EA, Neefjes JJ (2001) From fixed to FRAP: Measuring protein mobility and activity in living cells. *Nature Cell Biology* 3:E145–E147.
Ries J, Chiantia S, Schwille P (2009) Accurate determination of membrane dynamics with line-scan FCS. *Biophysical Journal* 96:1999–2008.
Ries J, Petrasek Z, Garcia-Saez AJ, Schwille P (2010) A comprehensive framework for fluorescence cross-correlation spectroscopy. *New Journal of Physics* 12:113009.
Ries J, Schwille P (2006) Studying slow membrane dynamics with continuous wave scanning fluorescence correlation spectroscopy. *Biophysical Journal* 91:1915–1924.
Ries J, Schwille P (2008) New concepts for fluorescence correlation spectroscopy on membranes. *Physical Chemistry Chemical Physics: PCCP* 10:3487–3497.
Rigler R, Mets U, Widengren J, Kask P (1993) Fluorescence correlation spectroscopy with high count rate and low-background—Analysis of translational diffusion. *European Biophysics Journal* 22:169–175.
Roux A, Koster G, Lenz M, Sorre B, Manneville JB, Nassoy P, Bassereau P (2010) Membrane curvature controls dynamin polymerization. *Proceedings of the National Academy of Sciences of the United States of America* 107:4141–4146.
Scherfeld D, Kahya N, Schwille P (2003) Lipid dynamics and domain formation in model membranes composed of ternary mixtures of unsaturated and saturated phosphatidylcholines and cholesterol. *Biophysical Journal* 85:3758–3768.
Schneckenburger H (2005) Total internal reflection fluorescence microscopy: Technical innovations and novel applications. *Current Opinion in Biotechnology* 16:13–18.
Schwille P, Korlach J, Webb WW (1999) Fluorescence correlation spectroscopy with single-molecule sensitivity on cell and model membranes. *Cytometry* 36:176–182.
Schwille P, Meyer-Almes FJ, Rigler R (1997) Dual-color fluorescence cross-correlation spectroscopy for multicomponent diffusional analysis in solution. *Biophysical Journal* 72:1878–1886.
Sezgin E, Schwille P (2011) Fluorescence techniques to study lipid dynamics. *Cold Spring Harbor Perspectives in Biology* 3:a009803.
Shaner NC, Patterson GH, Davidson MW (2007) Advances in fluorescent protein technology. *Journal of Cell Science* 120:4247–4260.
Shaner NC, Steinbach PA, Tsien RY (2005) A guide to choosing fluorescent proteins. *Nature Methods* 2:905–909.
Sheetz MP, Schindler M, Koppel DE (1980) Lateral mobility of integral membrane proteins is increased in spherocytic erythrocytes. *Nature* 285:510–511.
Siegel RM, Chan FK, Zacharias DA, Swofford R, Holmes KL, Tsien RY, Lenardo MJ (2000) Measurement of molecular interactions in living cells by fluorescence resonance energy transfer between variants of the green fluorescent protein. *Science's STKE: Signal Transduction Knowledge Environment* 2000:pl1.
Singer SJ, Nicolson GL (1972) The fluid mosaic model of the structure of cell membranes. *Science* 175:720–731.
Smith AJ, Pfeiffer JR, Zhang J, Martinez AM, Griffiths GM, Wilson BS (2003) Microtubule-dependent transport of secretory vesicles in RBL-2H3 cells. *Traffic* 4:302–312.
Spille JH, Kaminski TP, Scherer K, Rinne JS, Heckel A, Kubitscheck U (2015) Direct observation of mobility state transitions in RNA trajectories by sensitive single molecule feedback tracking. *Nucleic Acids Research* 43:e14.
Subburaj Y, Cosentino K, Axmann M, Pedrueza-Villalmanzo E, Hermann E, Bleicken S, Spatz J, Garcia-Saez AJ (2015a) Bax monomers form dimer units in the membrane that further self-assemble into multiple oligomeric species. *Nature Communication* 6:8042.
Subburaj Y, Ros U, Hermann E, Tong R, Garcia-Saez AJ (2015b) Toxicity of an alpha-pore-forming toxin depends on the assembly mechanism on the target membrane as revealed by single molecule imaging. *The Journal of Biological Chemistry* 290:4856–4865.
Subburaj Y, Salvador-Gallego R, García-Sáez AJ (2013) Membrane dynamics: Fluorescence correlation spectroscopy. In: *Encyclopedia of Analytical Chemistry: Applications, Theory and Instrumentation*, pp. 1–29: John Wiley & Sons.
Tagawa A, Mezzacasa A, Hayer A, Longatti A, Pelkmans L, Helenius A (2005) Assembly and trafficking of caveolar domains in the cell: caveolae as stable, cargo-triggered, vesicular transporters. *The Journal of Cell Biology* 170:769–779.
Tcherniak A, Reznik C, Link S, Landes CF (2009) Fluorescence correlation spectroscopy: Criteria for analysis in complex systems. *Analytical Chemistry* 81:746–754.
Tsien RY (1998) The green fluorescent protein. *Annual Review of Biochemistry* 67:509–544.
Unsay JD, Garcia-Saez AJ (2013) Scanning fluorescence correlation spectroscopy in model membrane systems. *Methods in Molecular Biology* 1033:185–205.
Vereb G, Szollosi J, Matko J, Nagy P, Farkas T, Vigh L, Matyus L, Waldmann TA, Damjanovich S (2003) Dynamic, yet structured: The cell membrane three decades after the Singer-Nicolson model. *Proceedings of the National Academy of Sciences of the United States of America* 100:8053–8058.
Wegner KD, Hildebrandt N (2015) Quantum dots: Bright and versatile in vitro and in vivo fluorescence imaging biosensors. *Chemical Society Reviews* 44:4792–4834.
Weidemann T, Wachsmuth M, Tewes M, Rippe K, Langowski J (2002) Analysis of ligand binding by two-colour fluorescence cross-correlation spectroscopy. *Single Molecules* 3:49–61.
Weiss S (1999) Fluorescence spectroscopy of single biomolecules. *Science* 283:1676–1683.
Wong JL, Wessel GM (2008) FRAP analysis of secretory granule lipids and proteins in the sea urchin egg. *Methods in Molecular Biology* 440:61–76.
Wysocki LM, Lavis LD (2011) Advances in the chemistry of small molecule fluorescent probes. *Current Opinion in Chemical Biology* 15:752–759.
Zhang X, Tee YH, Heng JK, Zhu Y, Hu X, Margadant F, Ballestrem C, Bershadsky A, Griffiths G, Yu H (2010) Kinectin-mediated endoplasmic reticulum dynamics supports focal adhesion growth in the cellular lamella. *Journal of Cell Science* 123:3901–3912.
Zhu C, Das SL, Baumgart T (2012) Nonlinear sorting, curvature generation, and crowding of endophilin N-BAR on tubular membranes. *Biophysical Journal* 102:1837–1845.

22 Shining light on membranes

Rosângela Itri, Carlos M. Marques, and Mauricio S. Baptista

There is a crack in everything, that's how the light gets in.
Leonard Cohen

Contents

22.1 MEMBRANES AND LIGHT

The optical absorption spectrum of lipids or of lipid membranes does not display any peculiar feature in the visible wavelength range, from 400 to 700 nm. Indeed, the notable peaks of the absorption spectrum of a typical lipid are either in the UV-region, where double bonds absorb in the UVC region[1] around 200 nm, or in the near infrared region where the motion of the C-H bonds gives rise to several strong absorption peaks above 900 nm (Kuksis, 2012). Lipid membrane systems in general and giant unilamellar vesicles (GUVs) in particular can, therefore, be easily studied by techniques employing visible light, such as light microscopy: A GUV can be observed under a light microscope for many hours without being destroyed, transformed or even without suffering any visible degradation. Given such premises, one might wonder why a chapter about the interactions between light and lipid membranes is at all needed in this book … that is, until actual experiments would be performed under a microscope with realistic GUV systems. Phospholipids, cholesterol, proteins and other membrane-forming molecules that are fluorescently labeled—see book Appendix 1—for visualization, solutions with fluorophores added for measuring permeability and other properties, biomimetic constructs, including light-sensitive proteins embedded or interacting with the bilayer … arguably most practical GUV systems strongly absorb light and bear thus the potential of inducing membrane transformations when exposed to visible light, by mechanisms that we discuss in this chapter. In the simplest cases, the membrane does not suffer any alterations from the illumination, yet it is still crucial as a support for the photoactivity of the hosted protein; we will see as an example in the next section how GUVs are essential platforms for understanding the photoreceptor activity of bacteriorhodopsin (BR). In most cases, however, the light activity of the hosted molecules induces a deep transformation of the membrane. We will refer to alterations where the lipid chemical structure in not modified, as physical transformations of the membrane. Section 22.3 describes two typical cases of such transformations where light-induced conformational changes of molecules embedded in the bilayer lead to pore formation or to complete destruction of the GUVs. The most dramatic perturbations induced by light correspond to chemical transformations of the lipids themselves, either (i) by a direct reaction with the photosensitizer (PhS)—a photosensitive molecule that becomes reactive by absorption of a photon—or (ii) indirectly by reactions with species activated by the PhS—for instance, with singlet oxygen ($^1O_2$). These aspects, discussed in Section 22.4, are important for many metabolic mechanisms. Indeed, lipid chemical transformations

[1] UVA: 315–400 nm; UVB: 280–315 nm; UVC: 100–280 nm.

induced by exposure to light have been shown to play an important role in diseases such as Parkinson's and Alzheimer's, atherosclerosis, diabetes, aging, and carcinogenesis and in physiological processes involved in the immunologic response and in the generation of signaling molecules and hormones (Ames et al., 1993). Photoinduced lipid transformations are also key to many practical applications. Section 22.5 will introduce applications in medicine for photodynamic therapy (PDT), in dermo-cosmetics for membrane protection and in pharmaceutics for liposome-based drug delivery.

22.2 GUVs AS A SUPPORT IN UNDERSTANDING PHOTORECEPTORS

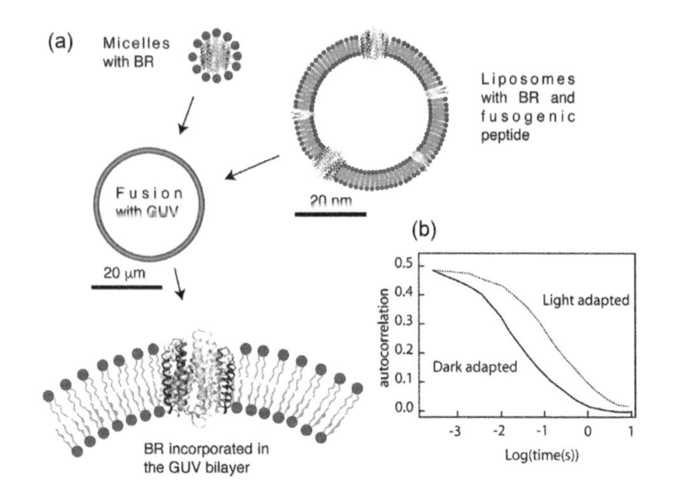

Figure 22.1 (a) Different strategies to incorporate membrane proteins in GUVs: fusion of liposomes containing BR and a fusogenic peptide, and direct transfer to the GUV from micelles solubilizing the protein. (b) FCS autocorrelation curves for Alexa Fluor 488™-labeled BR (protein-to-lipid ratio 1:40 w/w) diffusing in GUV in the dark-adapted (solid curve) and light-adapted states, after 4 min of illumination (dashed curve). (Adapted from Kahya, N. et al., *J. Biol. Chem.*, 277, 39304–39311, 2002.)

How do we humans see the world, and how do plants manage to convert solar into chemical energy? These fundamental processes of light are membrane-based photoinduced reactions that eventually cause charge separation, changes in protein conformation, proton gradients and several other dark reactions that follow. Membranes are key components in these processes because they are responsible for keeping concentration gradients generated by the photoinduced events. The elucidation of several of these mechanisms was made possible by experiments using vesicles and, more recently GUVs, as experimental platforms, as we will see in this section.

The first technical challenge to use GUVs as platforms for studying photoreceptors in membranes is the incorporation into the bilayer of large membrane proteins in their native conformation. Such a challenge has been successfully tackled and several strategies are now available (Fenz and Sengupta, 2012), as further discussed in Chapter 3 of this book. In this section, we discuss the specifics of membrane protein incorporation for light receptors, for which several techniques were developed and tested. Generally speaking, reconstitution of proteins in GUVs may be done (Figure 22.1a) either by solubilizing purified proteins in detergent micelles that eventually transfer the protein to the bilayer or by fusion of proteoliposomes or small liposomes containing the purified protein (Kahya, 2010).

BR is a microbial rhodopsin that shares some similarities with the vertebrate rhodopsins, the pigments responsible for vision. Both BR and rhodopsin are membrane proteins with a characteristic motif of seven transmembrane helices with a covalently bound retinal group, which undergoes isomerization reaction after light absorption. Rhodopsin is a G protein-coupled receptor and BR is a proton pump (van der Horst and Hellingwerf, 2004). BR monomers reconstituted in liposomes at low lipid-to-protein ratios retain the proton pumping activity (Dencher and Heyn, 1979). However, because in the bacterial membrane BR is organized in hexagonal lattices of trimers, one might wonder if the proton pumping activity of the protein depends on its spatial organization. As we will now see, GUVs provide an important platform to address this question.

A method that allows incorporation of BR in GUVs (Kahya et al., 2001) relies on using LUVs with small fusogenic peptides, decorated with BR by dialysis with detergent and fused with GUVs by the action of the peptide (Figure 22.1a). It can be shown by fluorescence correlation spectroscopy (FCS) that the incorporated BR presents unrestricted Brownian motion with a diffusion coefficient of 1.2 $\mu m^2 s^{-1}$ (Kahya et al., 2001). Such value is equivalent to the one measured in the 1970s in the membranes of isolated frog and mudpuppy rods (Poo and Cone, 1974). GUVs prepared with different concentrations of BR, in different optical states (photoactivated and dark-state), show also that the shift from monomeric/trimeric BR to larger oligomeric forms has no effect on the proton pumping activity of BR (Kahya et al., 2002). Interestingly, FCS performed on GUVs with incorporated BR reveals that the photoactivation of the photoreceptor increases its tendency to form higher-order aggregates, as shown from the large shift in the FCS autocorrelation curves shown in Figure 22.1b. BR oligomerization following photoactivation may thus be of importance for facilitating signal transduction and for regulating proton transport and other biological properties (Kahya et al., 2002).

BR-containing GUVs can also be prepared by using detergent solubilization, to display photoinduced generation of pH gradients and transmembrane potential that remain stable for several hours after light exposition (Dezi et al., 2013).

The sensory microbial rhodopsins, for example, sensory rhodopsin II (SRII) of *Natronomonas pharaonis* (NpSRII), have a receptor motif of the rhodopsins that work in close contact with a transducer that is not the G coupled protein but instead is a cognate transducer (HtrII of Natronomonas pharaonis [NpHtrII]). The type of complex formed between SRII and the transducer protein of SRII (HtrII) can be understood with the help of GUVs and time-resolved fluorescence techniques (Kriegsmann et al., 2009a, 2009b). The two-dimensional diffusion coefficient of separately diffusing proteins is twice as large for the transducer ($D = 4.1 \times 10^8$ $cm^2 s^{-1}$) as for the photoreceptor with

($D = 2.2 \times 10^8$ cm^{2}s^{-1}). In GUVs having both proteins incorporated, there is a significantly smaller diffusion coefficient, indicating larger diffusing units and, therefore, intermolecular protein binding (Kriegsmann et al., 2009a, 2009b). Intermolecular binding between the photoreceptor NpSRII and its NpHtrII transducer is extremely strong, with heterodimeric complexes being present in GUVs with concentrations as low as one thousand membrane proteins of each type.

22.3 PHOTOINDUCED PHYSICAL TRANSFORMATIONS

Many molecules can undergo photoinduced conformational changes. A well-known example is provided by the cationic surfactant azobenzene-modified trimethylammonium bromide surfactant (azoTAB) (Figure 22.2a). At room temperature in the dark, the molecule adopts a stable *trans* configuration where the surfactant tail is straight. When irradiated in the UVA region at 365 nm, *trans* azo-TAB photoisomerizes into the *cis* configuration, with a bent, more polar tail. The *cis* configuration is stable in the dark for many hours, and can be reverted to the *trans* configuration by irradiation in the blue region at 480 nm. Azo-TAB dissolved in the solution inserts into the lipid membrane (Diguet et al., 2012). GUVs are stable in 1 μM solutions of azo-TAB, a concentration small enough to avoid membrane perturbations due to area increase or lipid dissolution as described in Chapter 24. However, when *trans* to *cis* isomerization is promoted by UVA irradiation, the GUVs can burst, depending on the lipid composition. Fluid membranes resist well the configuration changes of the surfactant, but more rigid bilayers incorporating lipids in the gel phase burst under the irradiation (Figure 22.2b). Interestingly, the *cis* to *trans* isomerization of azo-TAB does not perturb the membrane. Milder effects of *cis* to *trans* isomerization can be obtained with other azo-containing molecules such the water-soluble *ortho*-tetrafluoroazobenzene that induces reversible area changes and modifies the membrane curvature (Georgiev et al., 2017).

Groups undergoing *cis-trans* photo-isomerization can also be directly included in the hydrophobic tails of the molecules composing the bilayers. In this case, the stability of the bilayer structure is directly perturbed by the configurational changes of the tails under irradiation. A striking example is provided by giant polymersomes with an asymmetric bilayer assembled from two different diblock copolymers (Mabrouk et al., 2009). The internal leaflet (Figures 22.3a,b) is composed of a PEG-*b*-PBD copolymer, which is a common polymersome-forming polymer (see also Chapter 26). The outer leaflet is assembled from a PEG-*b*-PMAazo444, a liquid-crystalline copolymer. Under illumination, the configurational changes of the PMAazo444 block increase the area per copolymer of the outer leaflet leading to burst with an interesting dynamics associated to the rolling neck associated with the bursting pore (Figure 22.3c).

Irradiation of GUVs incorporating photosensitive molecules can also be used to study transient pore formation, without bursting the vesicles. This was achieved by adding the fluorescent marker di$_6$ASP-BS to 1,2-dioleoyl-*sn*-glycero-3-phosphocholine (DOPC, see Appendix 1 of the book for structure and data on this and other lipids in this Chapter) GUVs (Figure 22.4a) (Sandre et al., 1999; Karatekin et al., 2003a, 2003b). As displayed in Figure 22.4b, under illumination by blue light, the vesicles become tense and open one pore.

By working with water–glycerol mixtures, which control the viscosity of the solution, it is possible to study the opening and closing dynamics of the pores, and to measure the pore line tension (see also Section 15.3.3 of Chapter 15). Although the driving force for pore opening seems to be the membrane tension, the mechanism of tension generation is not clear. A possible explanation is that DOPC lipids become oxidized by the probe and are extracted from the membrane, which would make this an example of mixed physical and chemical transformations. How chemical transformations are induced by light is comprehensively discussed in the next section.

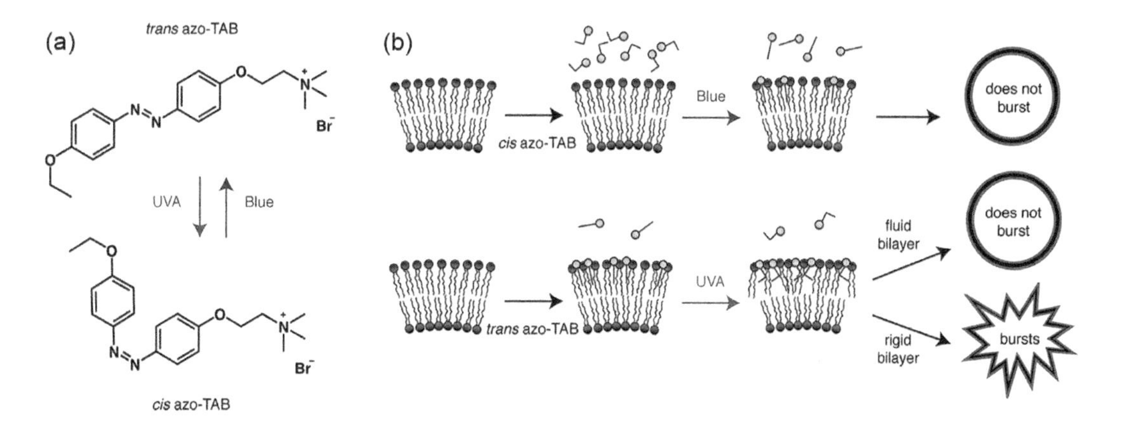

Figure 22.2 Effect of azo-TAB on giant vesicles. (a) The two isomers of azo-TAB and the typical wavelengths for photo-isomerization, courtesy of Dr. Y. Moskalenko. (b) *cis* azo-TAB does not insert in the membrane and *cis-trans* isomerization does not affect membrane stability. *Trans* azo-TAB inserts in the membrane: *trans-cis* isomerization leads to GUV bursting when the membrane is not fluid enough (presence of solid domains in the absence of cholesterol or high fraction of liquid-ordered (L_o) phase at high cholesterol concentration). (Reprinted with permission from Diguet, A. et al., *J. Am. Chem. Soc.*, 134, 4898–4904, 2012. Copyright 2012 American Chemical Society.)

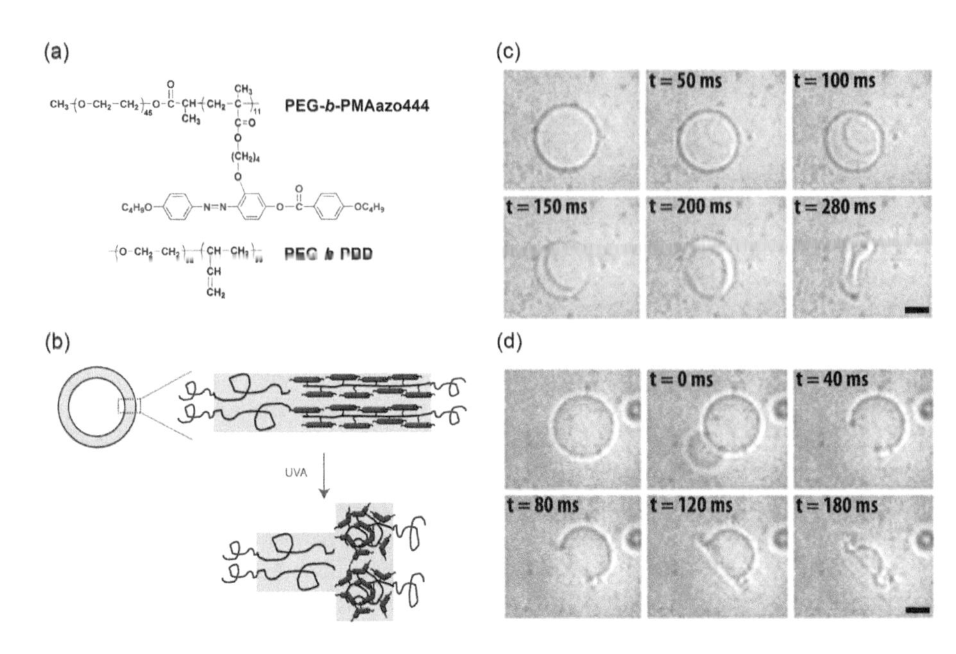

Figure 22.3 Bursting of polymersomes containing Azo groups. (a) Chemical structures of the two copolymers, PEG-*b*-PBD and PEG-*b*-PMAazo444. (b) Schematic representation of a bilayer made from the two copolymers. PEG-*b*-PBD copolymer is in a coil–coil state. PEG-*b*-PMAazo444 copolymer has a rod-like nematic conformation. Under UVA illumination, Azo isomerization induces a conformational change of the polymer into a disordered state, increasing the area per molecule of the PEG-*b*-PMAazo444. (c) Snapshots of a polymersome bursting under UVA illumination. The first image shows the vesicle before illumination. Time $t = 0$ corresponds to pore nucleation. The other images show the same vesicle as the pore grows. Scale bar: 5 µm. (d) Here the image at time $t = 0$ also corresponds to pore nucleation, and the expulsion of sucrose solution is clearly visible at the lower left of the vesicle. The other images display pore growth and clearly show outward curling. Scale bar: 5 µm. (Reprinted with permission from Mabrouk, E. et al., 2009. Bursting of sensitive polymersomes induced by curling. *PNAS*., 106, 7294–7298. Copyright 2009, National Academy of Sciences.)

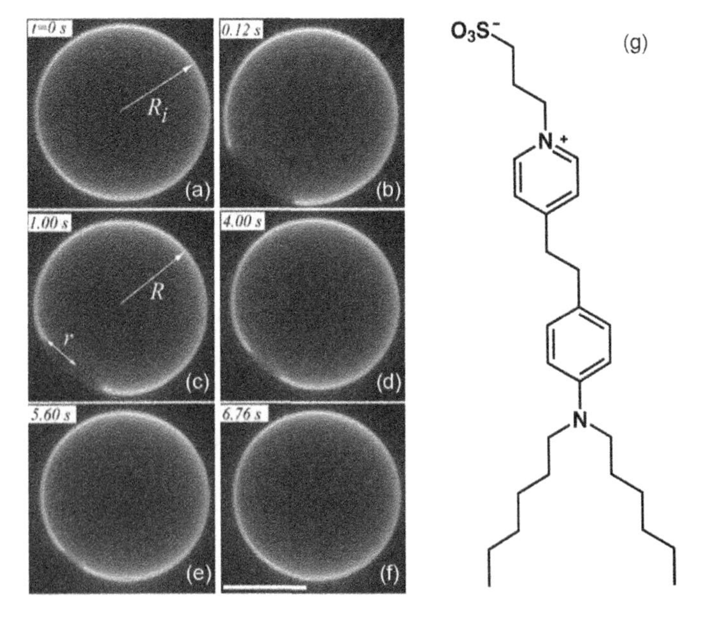

Figure 22.4 Pore opening in DOPC GUVs containing di$_6$ASP-BS, see Appendix 2 of the book for further data on this probe. When a tense vesicle with initial radius R_i (a) is illuminated further, tension builds up and the vesicle finally responds by the sudden opening of a pore, which reaches its maximum size very rapidly (b). The pore size r_{por} then decreases slowly until complete resealing (c–f). In this example, the vesicle radius, R, is approximately 10 µm and the bilayer contains 20 mol% cholesterol. (Reprinted from *Biophys. J.*, 84, Karatekin, E. et al., Cascades of transient pores in giant vesicles: Line tension and transport, 734–1749, Copyright 2003a, with permission from Elsevier). (g) Chemical structure of the dye di$_6$ASP-BS. (Courtesy of Y. Moskalenko.)

22.4 PHOTOINDUCED CHEMICAL TRANSFORMATIONS

Photoactive molecules, photoreceptors, chromophores … for all PhSs, the generation of excited states by light absorption always induces an increase in reactivity and possibly damages biological structures by photooxidation (Foote, 1968). A clear example of this is the damage caused by light in photosynthetic organisms, that is, those that depend on light to survive (Long et al., 1994). Photosynthetic reaction centers produce measurable amounts of $^1O_2$ (Uchoa et al., 2008).

$^1O_2$, produced by the reaction of the PhS triplet state with molecular oxygen, can react with biomolecules and induce damage by many mechanisms. This indirect mechanism of photosensitized oxidation has been called type II (Foote, 1968). The other main type of photosensitized oxidation is the direct reaction of the PhS triplet state with the biological target, a mechanism that has been generically called type I. As a matter of fact, $^1O_2$ and the triplet species of PhSs are the main agents damaging the skin under sun irradiation in addition to those involved in treating cancer by PDT (Chiarelli-Neto et al., 2014; Bacellar et al., 2015). The comprehensive understanding of oxidative damage in membranes is thus fundamental for proposing better methods for PDT as well as for defining better strategies to protect the skin from the sun.

In order to investigate in a controlled manner how lipid oxidation impacts the properties of biomimetic lipid membranes, GUVs can be photooxidized under a microscope. This can be achieved by decorating the GUV membrane with embedded or anchored PhSs or by exposing the GUVs to a solution of water-soluble PhS. These different methods result in different localizations of the PhS with respect to membrane leaflets (see also Box 22.1). For water-soluble PhS, GUVs can be brought into contact with the PhS after GUV formation or during GUV formation, resulting in PhS localized outside the vesicle or on both sides of the membrane, respectively. Many water-soluble PhSs also display some affinity for the membrane, depending on their partition coefficient (Bacellar et al., 2014). Hydrophobic enough PhS can generally be dissolved with the lipid itself before GUV formation and will stay embedded in the membrane. Irradiation can be performed with different types of light sources and photoinduced morphological transitions can be followed by video microscopy (Heuvingh and Bonneau, 2009; Riske et al., 2009; Kerdous et al., 2011; Mertins et al., 2014; Weber et al., 2014; Sankhagowit et al., 2014).

High levels of lipid oxidation usually lead to membrane destabilization (Caetano et al., 2007; Cwiklik and Jungwirth, 2010), but formation of the first oxidation product of $^1O_2$ reaction with a double bond, that is, a lipid hydroperoxide (see also Box 22.1) does not promote leakage of sugars (Riske et al., 2009; Weber et al., 2014). The oxidation level that a lipid bilayer can withstand before disruption depends on the membrane composition and on the lipid chemical transformations imposed by the oxidation.

Box 22.1 Photosensitization of GUV membranes

Photosensitizer localization

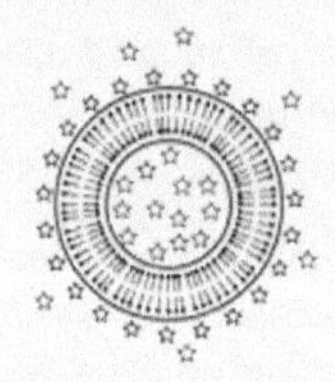

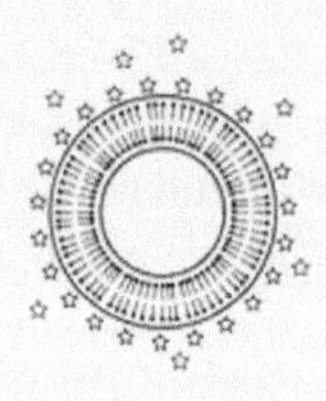

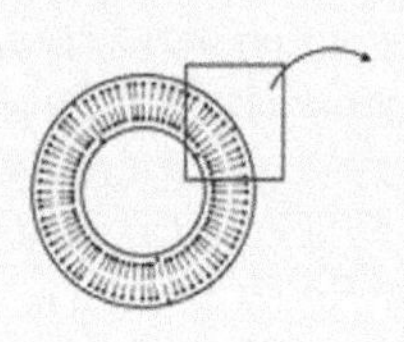

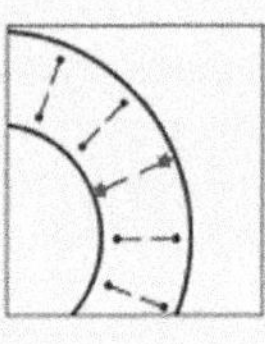

By using different PhSs—hydrophilic, hydrophobic, carrying a hydrophobic anchor or covalently bound to a lipid—it is possible to control the localization of the PhS (red stars) with respect to the bilayer: in the solution outside and inside the vesicle or adsorbed in the bilayer (left), in the solution outside (middle), permanently attached to the membrane (right).

Photosensitized oxidation of lipids

$^1O_2$ is first generated by a reaction between the photoactivated triplet state of the PhS and dissolved molecular oxygen O_2. $^1O_2$ then reacts with the lipid double bond, leading to an organic hydroperoxide group inserted at one of the carbons that hold the double bond (Stage I) and shifting the double bond by one carbon along the chain. The hydroperoxides accumulate in the membrane and can progress to form other radicals such as alkoxyls or peroxyls. These radicals in turn can feed peroxidation chain reactions in the presence of metals or by direct reactions with the PhS triplet state. Most dramatically, this gives rise to oxidized lipids with shortened alkyl chains (Stage II) that can strongly compromise the bilayer structure (Figure adapted from Sankhagowit et al., 2014). Details of this process were recently described by mass spectroscopic analysis (Bacellar et al., 2018).

Photo induced shape transformations

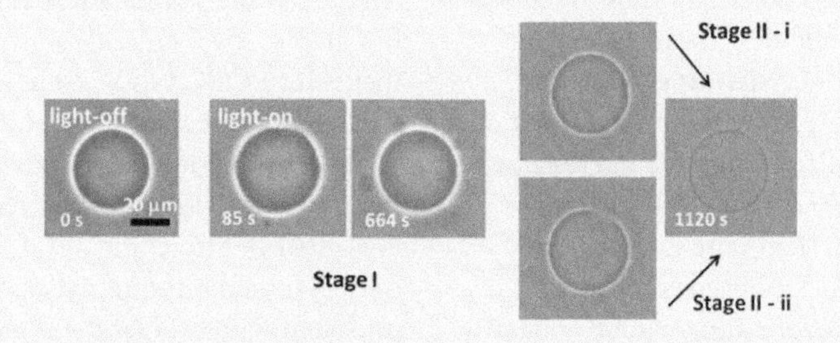

The formation of hydroperoxides increases the area per lipid and induces strong fluctuations, bud formation and other characteristic phenomena related to an increased area at constant volume (Stage I). The GUV then settles in a tenser shape and the membrane permeability increases (Stage II–i). Sometimes, micrometer-sized transient pores can also be observed (Stage II–ii).

Membrane composition can be easily tuned in GUVs, whereas the formed byproducts depend on the mechanisms of photosensitization, type I or type II PhS (Itri et al., 2014) associated with the biological environment (Bacellar et al., 2018).

Box 22.1 summarizes several facets of photosensitization in GUVs, here we provide a more detailed description of the different aspects of this phenomenon. A typical time evolution of photoinduced oxidation of 1-palmitoyl-2-oleoyl-*sn*-glycero-3-phosphocholine (POPC) GUVs dispersed in a glucose solution containing methylene blue as PhS is shown in Box 22.1 under continuous irradiation at wavelength of 665 nm (Mertins et al., 2014). Observation is made in the phase-contrast mode (see also Chapter 10), enhanced by the difference in optical index between the sucrose solution inside and the glucose solution outside the vesicle, obtained during GUV formation as explained in Chapter 1. The kinetics of the different alterations is obviously a function of the PhS concentration and localization and of the light intensity (Caetano et al., 2007; Mertins et al., 2014). These factors determine the density and distribution of $^1O_2$ and, therefore, the output of the oxidation reactions between $^1O_2$ and the unsaturated bonds. Under the microscope, one can easily observe how giant vesicles first respond to the light excitation with large fluctuations and formation of membrane buds, a result of membrane area increase caused by the addition of hydroperoxide groups at the position of the unsaturated double bond in the acyl chain (Wong-ekkabut et al., 2007; Riske et al., 2009; Itri et al., 2014, Sankhagowit et al., 2014). Interestingly, photosensitization of polymersomes mediated by $^1O_2$ also results in an area increase that can be ascribed to the hydroperoxidation of the double bounds of the polymer chain (Mabrouk et al., 2010).

It is worth stressing that the increase in the area per lipid should be accompanied by a decrease of both bilayer thickness and order parameter in the oxidized *sn*-2 tails. Although these changes have been predicted by molecular dynamic simulations (Guo et al., 2016; Siani et al., 2016), they cannot be accessed directly by GUV observation; they could instead be assessed on lamellar stacks by neutron or X-rays scattering techniques or by spectroscopic techniques such as NMR or infrared (IR). Other alterations such as lateral area increase as well as changes in the membrane bending and elastic moduli can be quantitatively determined by using GUVs as we describe below in Sections 22.3.1 and 22.3.2.

After exhibiting large fluctuations and bud formation, as the light irradiation continues, GUVs eventually display a tenser spherical shape, followed by an increase in membrane permeability as observed by loss of contrast (see also Box 22.1). Loss of contrast results from the formation of pores, allowing sugar exchange between the GUV interior and exterior (Mertins et al., 2014). Even though contrast loss can be achieved without the formation of any pore large enough to be visualized under the microscope, thus pointing to a majority of pores of suboptical dimensions, sometimes the opening of a large transient pore (see also Box 22.1) is also observed (Mertins et al., 2014; Sankhagowit et al., 2014).

Increased vesicle rigidity and pore formation can be understood from the tail scission of lipid molecules, resulting into shortened acyl chains capped with alcohol and aldehyde groups as probed by NMR (Sankhagowit et al., 2014). In particular, it has been shown by reverse-phase liquid chromatography and mass spectrometry that the main oxidized products from POPC obtained under oxidative stress conditions are hydroperoxides and aldehydes (Reis et al., 2005). Truncated lipids present different lipid packing with respect to their nonoxidized counterparts allowing pore stabilization in the membrane (Jurkiewicz et al., 2012). Interestingly, it has been shown by GUVs experiments that membrane damage is delayed in the presence of high levels of cholesterol (Kerdous et al., 2011). Further, the kinetics of pore formation in GUVs depends on the lipid species (Mertins et al., 2014), being faster for DOPC, which has two unsaturated chains than for POPC, which contains only one unsaturated chain. The oxidation kinetics can be quantitatively assessed by following the decrease of the vesicle contrast under light irradiation. A simple analysis of the GUVs morphological transitions under irradiation (Sankhagowit et al., 2014) also confirms that hydroperoxide formation is a faster process than the production of truncated lipids with associated loss of permeability. Indeed, the mechanism of membrane leakage has been recently shown to depend on the accumulation of trucanted lipid aldehydes formed in the initial steps of the photosensitized oxidation by contact-depend (Type I) reactions between the PhSs and both the lipid double bonds and the lipid hydroperoxides (Bacellar et al., 2018). The process of membrane permeation and pore size determination will be further discussed in Section 22.4.3.

Eventually, massive oxidation promotes severe damage of the membrane. Bursting, for instance, can be observed for DOPC GUVs dispersed in a methylene blue solution under irradiation, a likely consequence of the formation of a high number of truncated DOPC tails (Caetano et al., 2007). Indeed, lipids with one or two shorter lipid tails spontaneously assemble into spherical micelles instead of bilayers. A high number of these oxidized molecules eventually compromise the bilayer stability, leading to vesicle burst.

22.4.1 INCREASE IN AREA OF THE HYDROPEROXIDIZED BILAYER

One might be tempted to determine the maximum increase of surface area promoted by lipid photooxidation in GUVs by using vesicle electrodeformation as described in Chapter 15 (Riske et al., 2009; Mertins et al., 2014). Briefly, an alternating electrical field (alternating current [AC]) of 10 V amplitude at the frequency of 1 MHz is applied to vesicles formed in a sucrose solution containing a small amount of salt (0.5 mM NaCl) and diluted in a salt-free glucose solution to ensure higher conductivity of the internal GUV solution. This is known to induce a prolate shape deformation (Aranda et al., 2008) from which an excess area can be extracted (see Eqs. 15.9 and 15.10 in Chapter 15). Figure 22.5 displays an example of such experiments for POPC GUVs containing a porphyrin-based molecule anchored in the membrane. However, despite its apparent simplicity, this method fails to reveal the full area generated by the hydroperoxidation insertion reaction (Riske et al., 2009; Mertins et al., 2014). Stretching of GUVs by optical tweezers is another potential method to reveal area increase even though it leads to underestimated values (Sankhagowit et al., 2014).

One of the most efficient methods to reveal the new excess area resulting from hydroperoxidation is micropipette pulling (Weber et al., 2014). As further discussed in Chapter 11, a microcapillary glass is brought into contact with the GUV, and a small suction

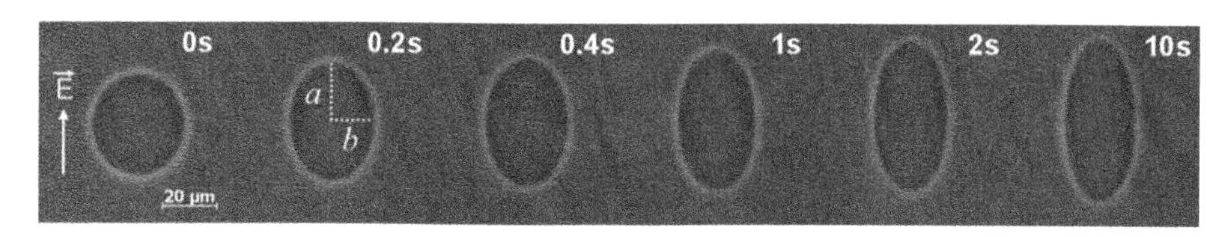

Figure 22.5 Effect of irradiation of POPC GUVs containing 3 mol% of porphyrin-based PhS observed under phase-contrast, displaying area increase (Riske et al., 2009). The prolate shape is a result of the continuous application of an AC field of 10 V amplitude and 1 MHz frequency. The irradiation time is shown on the top of each snapshot. Scale bar: 20 µm for all snapshots. Irradiation of the samples was performed using the HBO 103 W Hg lamp of an inverted microscope Axiovert 200™, Carl Zeiss™, with a 400-nm excitation filter. (Reprinted from *Biophys. J.*, 97, Riske, K.A. et al., Giant vesicles under oxidative stress induced by a membrane-anchored photosensitizer, 1362–1370, Copyright 2009, with permission from Elsevier.)

pressure is applied in order to partially bring a sphero-cylindrical membrane section of length L_p into the capillary. By keeping the pressure, and thus the membrane tension, at an adequate level (~0.5 mN/m) one can unfold the area fraction hidden in the fluctuation modes without elastically stretching the membrane. Under oxidation by $^1O_2$ the number of hydroperoxidized lipids increases, giving rise to an excess of membrane surface area and a consequent increase of L_p. From L_p and the vesicle and inner capillary diameters one can compute the excess surface area, as detailed in Chapter 11. For POPC and DOPC GUVs, this gives an excess area of 15%–19%, respectively. Similar values can also be obtained by a simple method developed recently (Aoki et al., 2015) that involves affixing the GUVs onto a strongly adhesive substrate. As the excess area is produced under illumination, the contact area of the GUV with the substrate increases. Here also, a simple measurement of the geometric parameters of the experiment allows computing the excess area (see also Box 22.2).

Interestingly, hydroperoxidation of a lipid with one unsaturation in each tail, such as DOPC, does not lead to doubling the excess area that is achieved with the hydroperoxidation of a lipid such as POPC, which has only one unsaturated tail. This reveals that the increase in area is a function of several factors, as it can be better understood from coarse-grained numerical simulations (see Chapter 6) of an hydroperoxidized bilayer (Guo et al., 2016; Siani et al., 2016). Indeed, a first contribution that amounts to 30%–50% of excess area can be attributed to the increase of volume of the lipid tail due to the insertion of two oxygen atoms. The second important contribution depends on the equilibrium distribution of the hydroperoxide group across the membrane thickness and thus on the average configuration of the lipid tails. Roughly speaking, the migration of the hydroperoxidized groups to the surface of the bilayer extends the space that the tail occupies laterally and by volume conservation reduces the bilayer thickness (De Rosa et al., 2018).

22.4.2 CHANGES IN MEMBRANE BENDING AND ELASTIC MODULI

Changes in the mechanical properties of the hydroperoxidized bilayer can in principle be evaluated by any of the available techniques, such as micropipette aspiration, fluctuation analysis and others, but the best technique might depend on the preparation conditions of the GUVs (Fournier et al. 2001). For instance, one can foresee using the fluctuation analysis method described in Chapter 14 for GUVs prepared with pre-hydroperoxidized lipids, but not by *in situ* oxidation because the emission of buds perturbs the measurement conditions. Figure 22.6 shows the typical curves of a micropipette suction experiment for a nonoxidized DOPC vesicle and one containing 33% of DOPC hydroperoxide (Weber, 2012) along with the measured bending and stretching moduli k and K_A (see also Chapter 11). For the nonoxidized vesicle, values of $k = 16 \pm 1$ K_BT and $K_A = 200 \pm 14$ mN m^{-1} were obtained, in good agreement with data reported by (Rawicz et al., 2000). For partially oxidized membranes generated by PhS photoactivation, $k = 10 \pm 1$ K_BT and $K_A = 100 \pm 6$ mN m^{-1} were determined (Weber, 2012). A fully hydroperoxidized bilayer of POPC has a stretching modulus of 50 mN m^{-1} (Weber et al., 2014), fourfold smaller than that evaluated for nonoxidized POPC GUVs. This can be rationalized as a decrease in hydrophobicity of the bilayer core due to the insertion of hydroperoxides, and it compares favorably with numerical simulation results (Guo et al., 2016) (Figure 22.7).

22.4.3 MEMBRANE PERMEABILIZATION

Membrane permeability of GUVs formed with mixtures of oxidized and nonoxidized lipids can be studied by the techniques described in Chapter 20. For instance, confocal imaging combined with a microfluidic setup has been applied to evaluate the permeability of GUVs made from a mixture of 1-palmitoyl-2-linoleoyl-*sn*-glycero-3-phosphocholine (PLinPC), 1,2-dimyristoyl-*sn*-glycero-3-phosphocholine (DMPC), cholesterol and an increasing amount of the oxidized form of PLinPC (POxnoPC), a commercially available oxidized lipid that has an aldehyde group in the end of its truncated *sn*-2 chain (Runas and Malmstadt, 2015). If one first exposes a GUV without POxnoPC to a glucose solution containing fluorescein-dextran of 40 kDa and 2,000 kDa molecular weights, one finds that the fluorescent probe transport across the membrane is low, with a measured permeability on the order of 1.5×10^{-8} m.s^{-1}. When 2.5%–10% of oxidized species are added, the permeability becomes one order of magnitude larger.

Changes in the lipid bilayer permeability can be easily noticed by visual inspection of GUVs under phase-contrast and epifluorescence and thus followed also under *in situ* oxidation. One can, for instance, analyze changes of contrast following oxidation due to the formation of oxidized lipid species (see Box 22.1). Figure 22.8 shows, as an example, the experimentally measured contrast decay as a function of irradiation time, t, for GUVs composed of POPC:POPG (8:2) dispersed in a glucose solution containing 10 µM of methylene blue. As one can observe, the contrast between the vesicle and the surrounding solution fades away so that differences between the inside and outside media become barely observable at the end of the process (Figure 22.8a). Using standard software such as ImageJ, one can quantify the

Box 22.2 Measuring excess area by GUV adhesion on a substrate

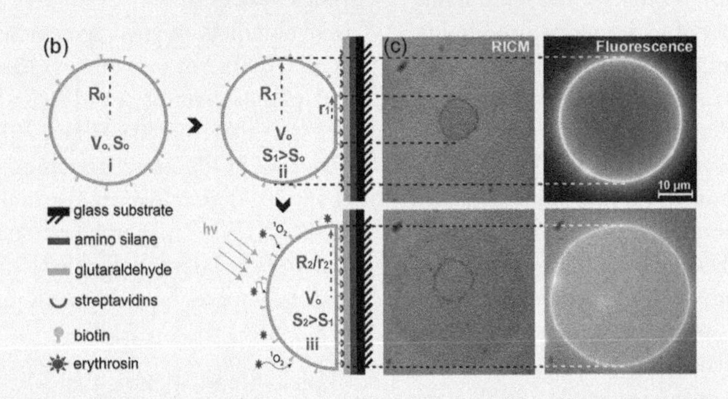

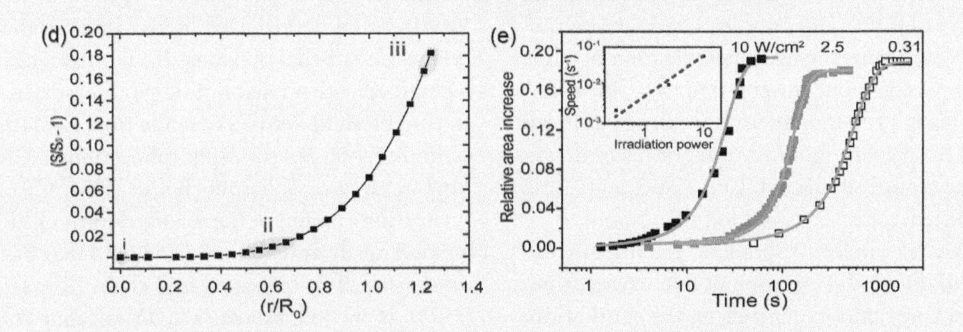

The principle of this method relies on first adhering GUVs on a substrate (a). Here this is achieved as shown in (b) by adding biotinylated lipids to the GUV and by functionalizing the substrate with streptavidin, with a method described by Aoki et al. (2015). See Chapters 10 and 17 for comparable procedures. Once the vesicle has adhered to the substrate in the presence of a PhS (in this case, erythrosine), the vesicle is irradiated and one observes an increase of the contact area between the vesicle and the substrate, up to a maximum value when changes stop. In this particular case, the radius r of the contact disk of the GUV with the substrate was measured by reflection interference contrast microscopy (RICM) (see Chapter 17 for a description of the technique) and the radius R of the equator circle by epifluorescence as shown in (c). From the recorded values of r and R, one can compute the volume, V, and the area, S, of the vesicle from $V = \frac{2}{3}\pi R^3\left(1+\frac{3}{2}\left(1-\frac{r^2}{R^2}\right)^{\frac{1}{2}} - \frac{1}{2}\left(1-\frac{r^2}{R^2}\right)^{\frac{3}{2}}\right)$ and $S = \pi r^2 + 2\pi R^2\left(\left(1-\frac{r^2}{R^2}\right)^{\frac{1}{2}}+1\right)$. Note that under conditions of volume preservation, $V = V_0$, from which the radius $R_0 = (3V_0/4\pi)^{1/3}$ and the associated area $S_0 = 4\pi R_0^2$ can be computed. Under these conditions, the radius of the equator needs only to be measured once, say after adhesion, and the surface growth can be monitored by simply capturing the successive values for r. Expressing all lengths in units of R_0, for the radius R of a vesicle adhering under constant volume, we have $R = \frac{1}{16}\left(\frac{r^8}{\Delta}+r^4+\Delta\right)$ with Δ given by $\Delta = \left(r^{12}+128r^6+16\left(\left((r^6+16)(r^6+32)^2\right)^{1/2}+128\right)\right)^{1/3}$. The excess area S/S_0-1 is shown in (d) as a function of r/R_0. Irradiation of DOPC vesicles in a solution of 50 mM erythrosine at three different illumination powers leads to variations of the excess area as depicted in (e). All figures in this box are reproduced from Aoki et al. (2015) with permission from the Royal Society of Chemistry.

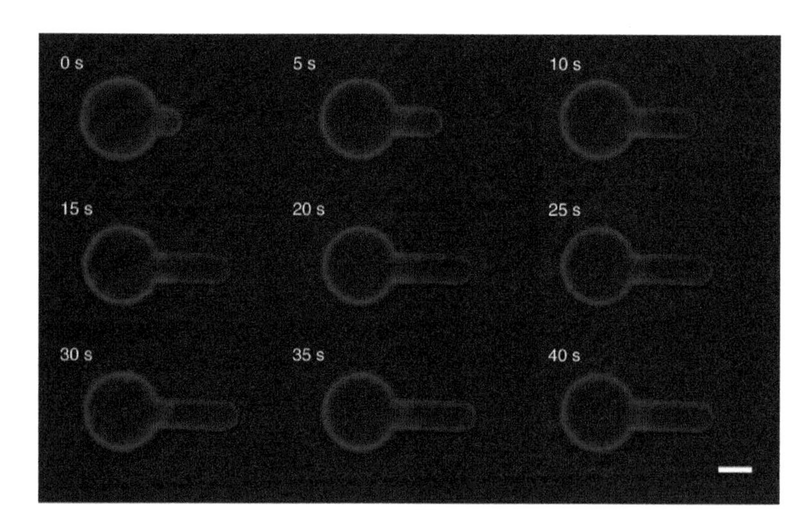

Figure 22.6 Fluorescence snapshots of a POPC GUV aspirated by a micropipette at constant tension 0.7 mN/m. The vesicle carries anchored PhSs of chlorin-12 (Weber et al., 2014). Area increase induced by hydroperoxidation can be measured from the different geometrical parameters of the images: vesicle diameter, micropipette inner diameter (equal to the "tongue" diameter) and the "tongue" length L_p (see also Chapter 11). Scale bar: 10 μm.

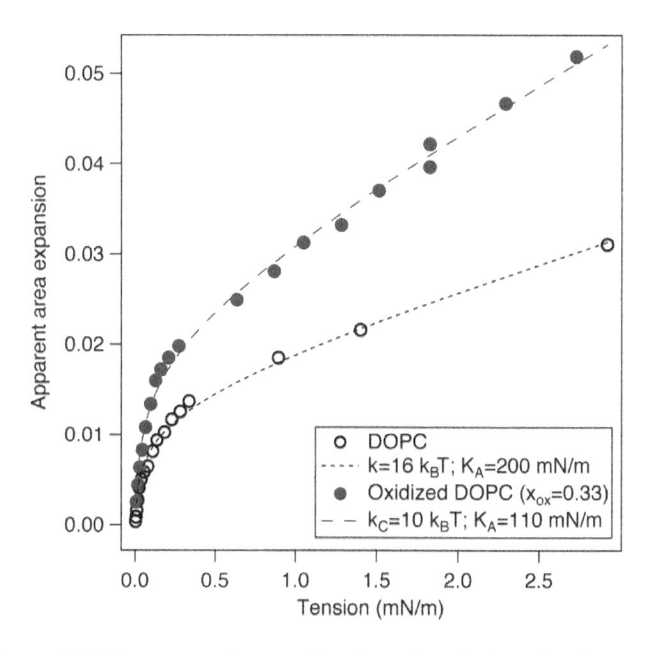

Figure 22.7 Area expansion as a function of applied tension determined from GUVs composed of nonoxidized (o) and 33% of oxidized (33%) DOPC. (From Weber, G., Photoinduced modifications of model membranes, PhD thesis, Université de Strasbourg, Strasbourg, France, 2012.) Dashed lines are standards fittings to the data that allows obtaining the bending K and elastic K_A membrane moduli.

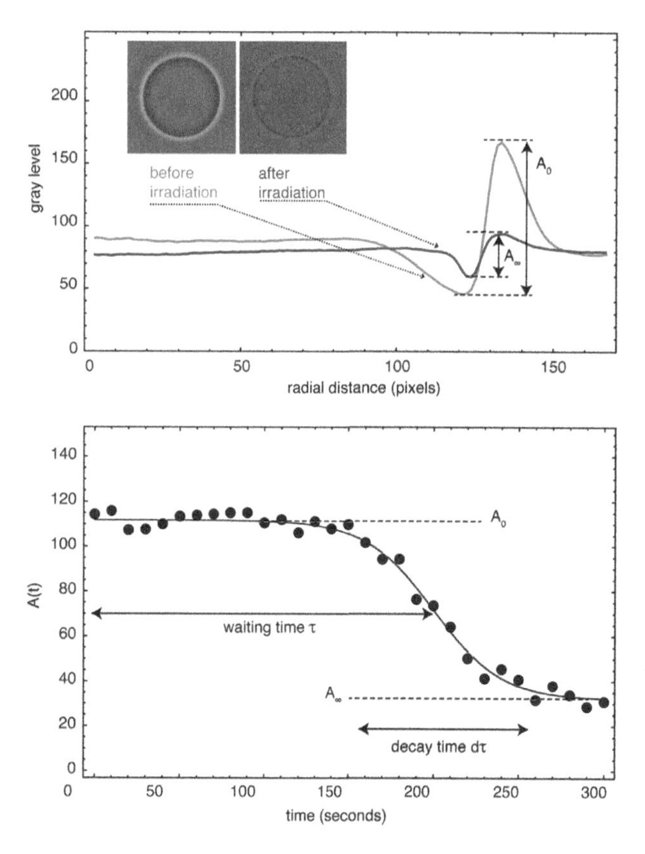

Figure 22.8 Membrane permeability increase measured by the contrast loss method. Above: GUVs observed under phase-contrast microscopy, before and after being exposed to a continuous irradiation at 630 nm in a methylene blue solution (10 μM) and the corresponding radial profiles with contrast amplitude. Below: a sketch of a typical time evolution of the contrast amplitude as a function of time and a sigmoidal fit with a Boltzmann function. (Non-published data from the authors acquired under conditions similar to Mertins, O. et al. *Biophys. J.*, 106, 162–171, 2014.)

gray level profiles of the phase-contrast microscopy images. Figure 22.8b shows two half-GUV images, one before (image above) and the other after (image below) the complete loss of contrast by irradiation. The gray level intensity measured along the equator is also displayed for both cases; it shows larger peaks for the vesicle before irradiation as compared with after irradiation. These peaks are symmetric with respect to the vesicle center and the gray level differences between the maxima and minima values allow monitoring the increased permeability of the membrane with respect to sucrose and glucose under irradiation as shown in Figure 22.8c (Heuvingh and Bonneau, 2009; Mertins et al., 2014). The amplitude of phase-contrast dependence on time has a sigmoidal decay shape, with a waiting time, τ, and a decay time, $d\tau$. It can be well fitted, for instance, by the Boltzmann function $I(t) = A_\infty + (A_0 - A_\infty)/(1 + \exp((t-\tau)/d\tau))$, with A_0 and A_∞ the initial and long-time values of the contrast.

The lag time, τ, has been shown (Mertins et al., 2014) to display a power law dependence on [PhS] (i.e., the concentration of the PhS), $\tau \sim [\mathrm{PhS}]^m$, with $m = -½$ for DOPC and $m = -1$ for POPC. The exponents can be explained by assuming that the oxidized lipids generated in the membrane diffuse until they eventually aggregate into a pore nucleus; exponent $m = -1$ suggests a reaction-limited aggregation between oxidized lipids, and points thus to a weak tendency of oxidized POPC species to aggregate, whereas exponent $m = -1/2$ can be understood from a diffusion-limited aggregation mechanism and suggests instead a strong aggregation tendency from oxidized DOPC species.

The decay time, $d\tau$, can be roughly described as the time needed for the vesicle to lose contrast once the pores are formed. The simplest estimation of this time is provided by a diffusive process where the inner and outer sugar concentrations equilibrate through a pore of size, r_{por}. In this limit, one refers to the effusion time (Heuvingh and Bonneau, 2009), written as $d\tau = V/(2D\, r_{por})$, where V is the volume of the vesicle; and D, the sugar's diffusion coefficient. For typical characteristic times, $d\tau$, observed between 10 and 50 s (Mertins et al., 2014), the average diameter of the photoinduced pores can be estimated between 10 and 50 nm. Interestingly, these pore sizes are in the same range as those measured in the electroporation of red blood cell ghosts or those calculated for stretched giant vesicles (Heuvingh and Bonneau, 2009). Care should, however, be exerted in this simple interpretation because membrane tension effects are also likely to play an important role in the decaying process.

22.4.4 LIPID REORGANIZATION TRIGGERED BY LIGHT

The chemical transformations of lipids under irradiation lead to fundamental changes in the bilayer composition: Even a bilayer made of a pure lipid will contain, after some irradiation time, not only the pure lipid but also its oxidized forms: such as hydroperoxides, aldehydes, and alcohols. This offers the exciting possibility of controlling the chemical composition of the bilayer by light and thus to finely tune the set of molecular interactions that determine lipid organization in the bilayer, domain formation and phase transition phenomena in general, as further described in Chapter 18.

The formation of lipid domains induced by light was noticed while observing GUVs composed of ternary lipid mixtures by fluorescence techniques (Ayuyan and Cohen, 2006; Zhao et al., 2007), introducing the possibility that lipid peroxidation may lead to serious artifacts in the study of GUV domains. Such effects can be at least partially prevented by using antioxidant agents such as ascorbic acid or *n*-propyl gallate (Morales-Penningston et al., 2010; Runas and Malmstadt, 2015). They can also be used to mimic the formation of membrane domains in GUVs and their modulation by specific macromolecules as, for instance, by the ganglioside GM1 (Staneva et al., 2011).

Therefore, it is important to understand whether the photoinduced processes change the organization of the lipids in a membrane as well as the mechanisms of these changes. On this ground, GUVs composed of POPC, 1,2-dipalmitoyl-*sn*-glycero-3-phosphocholine (DPPC), cholesterol and 1 mol% of porphyrin-based PhS incorporated into the lipid bilayer can be investigated by fluorescence microscopy (Haluska et al., 2012). Suitable ternary mixtures can be chosen from the POPC:DPPC:Chol phase diagram (Figure 2 in Haluska et al., 2012) with the cholesterol content varying from 9% to 23%. Under these conditions, all samples exhibit initially homogeneous fluorescence distribution under low-light nitrobenzoxadiazole (NBD)-probe fluorescence. Enhanced fluctuations and surface area increase are observed as described in Section 22.4.1. For particular compositions, domains typical of liquid-disordered/liquid-ordered (L_d/L_o) phase coexistence can be seen to grow with irradiation time (Haluska et al., 2012). Figure 22.9 shows an example of POPC:DPPC:Chol GUVs exposed to porphyrin-based PhS that is solubilized in the outer glucose solution. Under irradiation, GUVs initially displaying homogeneous fluorescence evolve from small to large L_d domains in the L_o phase due to continuous lipid photooxidation as well as to the coalescence of small domains. The oxidative side effect caused by irradiation of the fluorescent probe is minimal in comparison to the extensive lipid peroxidation caused by PhS photoactivation. The explanation for this is that the main relaxation from the excited state in the probe occurs via fluorescent

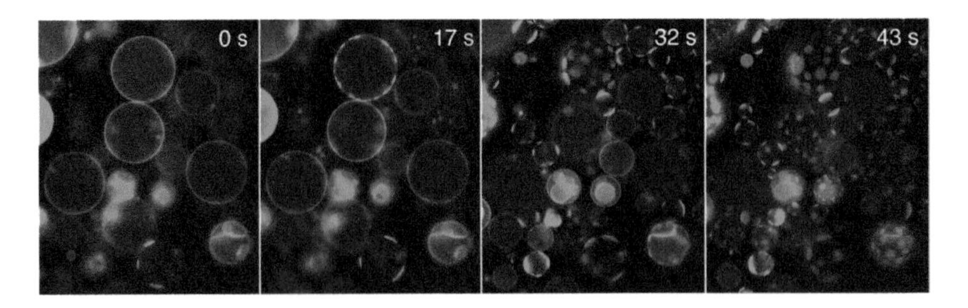

Figure 22.9 Fluorescence microscopy images of GUVs composed of POPC:DPPC 1:2 and cholesterol content of 23% (0.2 mol% rhodamine-labeled 1,2-dipalmitoyl-*sn*-glycero-3-phosphoethanolamine [DPPE-Rh]) dispersed in a glucose solution containing 6 mM of bis-metilperidil-difenilporphyrin (CisDiMPyP) PhS, under irradiation (400 nm). The beginning of irradiation is set at time 0 s. In the temporal sequence of images, GUVs initially homogeneous by fluorescence (thus not evidencing lipid demixing) quickly respond to light exposure, exhibiting L_o/L_d phase coexistence.

emission and not by intersystem crossing to the triplet state, as in the case of PhS molecules. Due to the observed initial expansion in the membrane area, the formation of the lipid domains was mainly attributed to *in situ* generated POPC hydroperoxide (Haluska et al., 2012). Indeed, part of the photoinduced POPC:DPPC:Chol phase diagram (Figure 2 in Haluska et al., 2012) could be reproduced in a purely chemical experimental setup by replacing POPC by its hydroperoxide counterpart in the membrane (Itri et al., 2014; Tsubone et al., 2018). These results emphasize the importance of POPC-hydroperoxide in lipid organization of membranes. Hydroperoxidation decreases the miscibility of lipids (saturated and unsaturated ones), thus inducing phase separation. In particular, for the hybrid POPC lipid, which is considered as a weak lineactant (Brewster et al., 2009), its chemical transformation to a hydroperoxide leads to distinct ordering and lipid packing properties, thereby favoring its action as a strong lineactant.

Cholesterol is also a target of $^1O_2$ and cholesterol hydroperoxides must also be formed concomitantly with POPC hydroperoxide during the photooxidative process. Note that oxysterols have been also recognized to affect membrane properties (Kulig et al., 2015). They may promote changes in L_o domains, impacting neurodegenerative diseases (Mitomo et al., 2009). Nevertheless, we are still lacking a better understanding of the coupling between both oxidized cholesterol and unsaturated lipids on the cell membranes. Using GUVs to better explore this issue is an exciting new path still to be trailed.

Furthermore, the exact mechanism that drives the oxidative-induced lipid phase separation has not yet been elucidated and represents a challenge for future work. Thus, employing GUVs as a tool to better interpret the membrane response to oxidative stress opens a number of important pathways for biophysical studies. For example, it could be hypothesized that domain formation after oxidation might facilitate oxidized lipid detection by signaling for recovery or elimination by the cell. Moreover, changes in the biophysical properties of the bilayer have been demonstrated to trigger signaling pathways (Cremesti et al., 2002). As a consequence, selective (dis)association of essential proteins within lipid domain scaffolds may take place in response to oxidative stress promoted in the cell membranes.

22.5 MEMBRANES AND LIGHT IN MEDICINE, DERMO-COSMETICS AND PHARMACEUTICS

22.5.1 PROTECTION FROM PHOTODAMAGE

Skin is the interface between the body and the environment being exposed to endogenous and environmental oxidants, which can cause molecular damage to nucleic acids, lipids and proteins, leading to cell and tissue damage and to the release of pro-inflammatory mediators, such as cytokines. This scenario leads to premature aging of the skin and often to cancer. To remain viable, various protection systems were selected during the evolution of living organisms, consisting of a complex antioxidant network that includes the sacrifice of antioxidant molecules, suppressors of excited states and enzymatic antioxidants working synergistically. The balance between damage and protection is shifted to damage if the individual has habits of excessive exposure to environmental oxidants, such as light. Administration of either natural or synthetic antioxidants, which can interact and neutralize free radicals or excited species (triplets and $^1O_2$), has been frequently suggested as a preventive therapy for skin photoaging, even with the hope of avoiding cancer (Briganti and Picardo, 2003; Uchoa et al., 2013).

Several methods were developed to quantify the antioxidant capacity of certain compounds or complex mixtures of natural compounds (plant extract). Given its importance in skin photoaging as well as in food oxidation, quantifying the inhibition of lipid oxidation is an important part of this task (Huang et al., 2005). Results obtained in liposomes have shown that both the intrinsic antioxidant capacity and the efficiency of membrane binding are important aspects of the membrane protection (Zhang et al., 2006). However, a direct evaluation of how various antioxidants protect membranes was not available. Liang et al. (2012) developed a GUV-based model membrane to monitor the effect of photosensitized oxidation in the presence and absence of potential antioxidants. The authors tested the integrity of GUVs challenged by oxidative damage induced by $^1O_2$ generated by photosensitization of chlorophyll-a and observed, as expected, that antioxidants are able to protect membrane integrity in GUVs (Liang et al., 2012). As mentioned in Sections 22.4.3 and 22.4.4, not only membrane structure but also other membrane properties are affected by photooxidative damage. One important feature of the lipid oxidation is the possibility to cause lipid demixing and domain formation. In fact, Georgieva et al. (2013) have shown that *n*-propyl galate, a hydrophobic derivative of gallic acid widely used in the food industry, pharmaceutics and cosmetics is able to decrease the liquid-ordered L_o/L_d miscibility transition temperature, protecting the membrane from the effects of photooxidation.

Another concept of membrane protection that has been tested with GUVs is the physical protection of the bilayer (Figure 22.10a). That is, by physically keeping the lipid cohesion on the membranes to avoid membrane leakage. Rodrigues et al. (2016) have shown that *Aloe vera* extract protects cells having a very poor antioxidant activity against UVA photodamage. Cell protection was mainly due to the physical protection in the lysosomal membranes. This physical effect was proved in experiments with GUVs. The concept of physical protection of membranes was further improved by Mertins et al. (2015), who prepared a polymer of gallic acid that had both antioxidant and mechanical protection properties. As we have seen above (Figure 22.8), GUVs challenged by photosensitized oxidations quickly lose contrast. In the presence of soluble gallic acid, there is a certain level of protection because there is a threefold increase in the waiting time to observe GUV contrast loss. Interestingly, in the presence of the polymer corona, no contrast loss was observed during 30 min of irradiation, indicating a high level of membrane protection (Mertins et al., 2015).

22.5.2 PHOTOINDUCED CONTROL OF MEMBRANE PERMEABILITY FOR DRUG RELEASE

The development of liposomes that can be destabilized in response to light has challenged the scientific community for several decades, and it is considered a promising strategy in the development of technologies for intelligent drug delivery (Alvarez-Lorenzo et al., 2009; Leung and Romanowski, 2012).

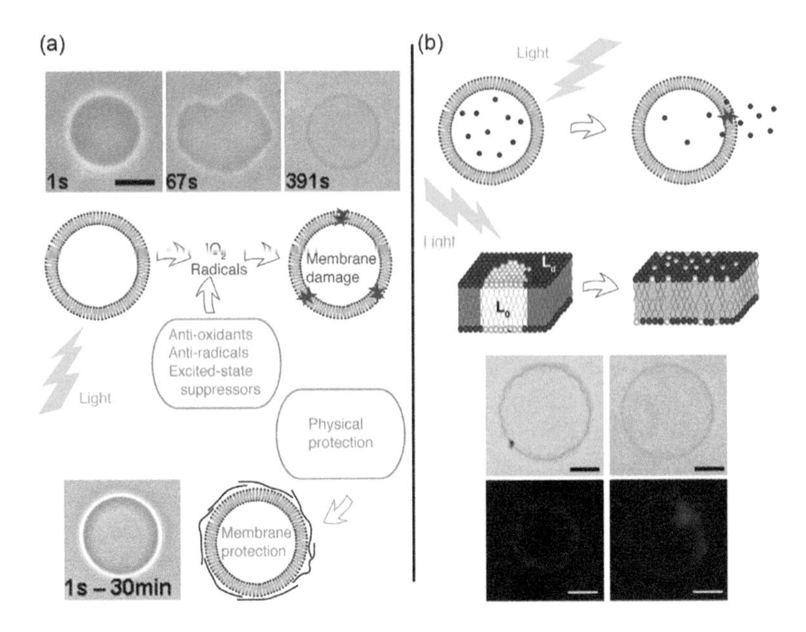

Figure 22.10 (a) Generic scheme of photoinduced damage in membranes, emphasizing the two most important mechanisms of membrane protection: antioxidant-based and physical protection. Above: time course, in seconds, of phase-contrast images from POPC GUVs under photo activation of methylene blue (40 µM at 665 nm). The image below shows the GUV covered with gallic acid–labeled chitosan with up to 30 min of irradiation. Scale bar: 20 µm. (With kind permission from Springer Science+Business Media: *Biochim Biophys Acta*, Effective protection of biological membranes against photo oxidative damage: Polymeric antioxidant forming a protecting shield over the membrane, 1848, 2015, 2180–2187, Mertins, O. et al.) (b) Upper scheme: photoinduced drug release from vesicles; lower scheme: representation of photoinduced disorganization of a liquid-ordered domain. Upper and lower images show phase-contrast and fluorescent microscopic images, respectively, of GUV composed of DPPC:DOPC:Chol:AzCh (38:38:19:5 in molar ratio) before and after UV-light irradiation. L_d phase in the GUV was stained with rhodamine-labeled 1,2-dioleoyl-*sn*-glycero-3-phosphoethanolamine (DOPE-Rho) (0.01 mol% of the total lipids). Scale bar: 10 µm for all. (With kind permission from Springer Science+Business Media: *Colloid. Polym. Sci.*, A photo-responsive cholesterol capable of inducing a morphological transformation of the liquid-ordered microdomain in lipid bilayers, 286, 2008, 1675–1680, Yasuhara, K. et al.)

The most common approaches involve the production of reactive species such as $^1O_2$ (Thompson et al., 1996) photopolymerization membrane components (Bondurant and O'Brien, 1998) and *cis-trans* photoisomerization in the double layer (Pidgeon and Hunt, 1983; Bisby et al., 2000). Although the release of internal material has been clearly identified and maximized by using liposomes, the mechanistic explanation and, consequently, the rational improvement of the technology are more difficult to be realized in liposome systems than in GUVs. The release induced by $^1O_2$ generation induces changes in the lipid membranes that were addressed in detail in Section 22.4 of this chapter.

The fact that both small liposomes and GUVs can be used to follow photoinduced drug release was clearly shown by Sebai et al. (2010) who used a light-responsive (*cis-trans*) copolymer to induce membrane destabilization in the *trans* configuration. An important mechanistic explanation for the membrane leakage related *cis-trans* photoisomerization was obtained with GUVs. An azobenzene-modified cholesterol (AzCh) is known to cause the release of material from the intravesicular compartment to the external medium. The work of Yasuhara et al. (2008) showed that the release was mainly due to the mixing of previously segregated L_o and L_d phases (Figure 22.10b). Phase-contrast and fluorescence microscopy observations revealed that the photo-isomerization reactions occurred in both phases and were mainly correlated with the disappearance of the L_o domain and appearance of a small L_d domain. There is a large increase in membrane permeability during this transition. Transient changes in lipid organization in a bilayer are known to affect considerably the membrane integrity, favoring leakage. GUV experiments have raised evidence that this phenomenon seems to be a key mechanism in photoinduced drug release methods developed in liposome systems.

22.6 THE CHALLENGES AHEAD

The experimental model constituted of photoreceptors in GUVs is quite close to reality in biology compared with isolated photoreceptors dispersed in micelles but still simple enough to allow for a better control of experimental parameters and theoretical modeling. The possibility of incorporating photoreceptors in GUVs shall continue to contribute to the understanding of fundamentally important processes such as vision and photosynthesis. Improvement in photosynthetic yields can contribute to biomass production and biofuels, which hold great promises in terms of energy production. In terms of photoreceptor applications, several degenerative diseases can be controlled if humans learn how to avoid generation of photoinduced reactive species after activation of light receptors. Opsin photoreceptors have become a key tool in the revolutionary field of optogetics (Zhang et al., 2011). Although GUVs could provide an interesting system to study the mechanism of selected new opsins, there is hardly any study in the literature with that aim. Based on this fact, we conjecture that in the upcoming years there will be a substantial increase in the studies using GUVs decorated with different type of opsins.

When membranes carry molecules that can be isomerized by exposure to light, the configuration changes lead to different kinds of stress that can open pores in the membrane or fully destabilize it. One can speculate that research in this area is just starting, and that an infinite amount of interesting possibilities are still to be explored as new isomerizable membrane-forming molecules will become available.

Much has been learned about photoinduced oxidative damage in membranes in the past eleven years after the first experiment proposing the use of GUVs to study physical changes induced by photoactivated methylene blue (Caetano et al., 2007). Nowadays, the main physical changes caused by the formation of the first product of this reaction, that is, lipid hydroperoxide are known quantitatively. The increase in membrane area can be as large as 15% by the formation of a single chain hydroperoxide in POPC and 19% by the formation of hydroperoxides in the two double bonds of DOPC. Both the bending and the stretching moduli of the membrane decreased considerably upon oxidation. Interestingly, in hydroperoxidized bilayers, the stretching modulus decreases four times from 200 to 50 mN m^{-1}. Even the intrinsic reaction efficiency between $^1O_2$ and the double bond of a lipid could be estimated to be 1 in 5 (Weber et al., 2014). The oxidation of the membrane also facilitates leakage of internal materials, an effect that has been correlated mainly with the formation of shorter chain lipids formed during oxidation. Another interesting aspects of the formation of hydroperoxides in GUVs are the promotion of domains and phase separation in membranes previously made of homogeneous distributed lipids (Haluska et al., 2012) and the influence on nanoparticle translocation (Su et al., 2019). The increase in the line tension was attributed to the increase in the hydrophobic mismatch upon formation of hydroperoxides.

The understanding of the physical changes in lipid membranes by the photoinduced generation of chemical species within the membranes has opened a window of opportunity to use this knowledge in new technologies. In this chapter, we emphasized the development of new ways to protect membranes from the photooxidation damages, which can lead to new strategies to protect the skin against sun exposition, that is, better sun screens. Furthermore a better control of this process allows the development of new strategies for drug delivery.

LIST OF ABBREVIATIONS

AzCh	azobenzene-modified cholesterol
azo-TAB	azobenzene-modified trimethylammonium bromide surfactant
BR	bacteriorhodopsin
CisDiMPyP	bis-metilperidil-difenilporphyrin
DMPC	1,2-dimyristoyl-*sn*-glycero-3-phosphocholine
DOPC	1,2-dioleoyl-*sn*-glycero-3-phosphocholine
DOPE	1,2-dioleoyl-*sn*-glycero-3-phosphoethanolamine
DOPE-Rho	rhodamine-labeled DOPE
DPPC	1,2-dipalmitoyl-*sn*-glycero-3-phosphocholine
DPPE	1,2-dipalmitoyl-*sn*-glycero-3-phosphoethanolamine
DPPE-Rh	rhodamine-labeled DPPE
FCS	fluorescence correlation spectroscopy
HtrII	transducer protein of SRII
MB	methylene blue
NMR	nuclear magnetic resonance
NpSRII	SRII of *Natronomonas pharaonis*
NpHtrII	HtrII of *Natronomonas pharaonis*
$^1O_2$	singlet oxygen
PDT	photodynamic therapy
PLinPC	1-palmitoyl-2-linoleoyl-*sn*-glycero-3-phosphocholine
POPC	1-palmitoyl-2-oleoyl-*sn*-glycero-3-phosphocholine
POxnoPC	oxidized form of PLinPC
PhS	photosensitizer
RICM	reflection interference contrast microscopy
SRII	sensory rhodopsin II

GLOSSARY OF SYMBOLS

A_0	Initial value of contrast amplitude
A_∞	Final value of contrast amplitude
α	fractional change in membrane area
ΔA	vesicle area change
c	solute concentration
ΔL	change in length of the vesicle projection in the pipette
D	diffusion coefficient
K_A	stretching or elastic modulus
k_B	Boltzmann constant
k	bending modulus
L_d	liquid disorder phase
L_o	liquid order phase
L_p	projection length of the vesicle inside the pipette
r_{por}	pore radius
τ	waiting time for optical contrast loss
$d\tau$	the characteristic time of optical contrast loss
V	volume of the GUV

REFERENCES

Alvarez-Lorenzo C, Bromberg L, Concheiro A (2009) Light-sensitive intelligent drug delivery systems. *Photochem Photobiol* 85:848–860.

Ames BN, Shigenaga MK, Hagen TM (1993). Oxidants, antioxidants, and the degenerative diseases of aging. *PNAS* 90:7915–7922.

Aoki PHB, Schroder AP, Constantino CJL, Marques CM (2015). Bioadhesive giant vesicles for monitoring hydroperoxidation in lipid membranes. *Soft Matter* 11:5995–5998.

Aranda S, Riske KA, Lipowsky R, Dimova R (2008) Morphological transitions of vesicles induced by alternating electric fields. *Biophys J* 95:L19–L21.

Ayuyan AG, Cohen FS (2006) Lipid peroxides promote large rafts: Effects of excitation of probes in fluorescence microscopy and electrochemical reactions during vesicle formation. *Biophys J* 91:2172–2183.

Bacellar IOL, Oliveira MC, Dantas LS, Costa EB, Junqueira HC, Martins WK, Durantini AM, Cosa G, Di Mascio P, Wainwright M, Miotto R, Cordeiro RM, Miyamoto S, Baptista MS (2018) Photosensitized membrane permeabilization requires contact-dependent reactions between photosensitizer and lipids. *Journal of the American Chemical Society* 140:9606–9615.

Bacellar IOL, Pavani C, Sales EM, Itri R, Wainwright M, Baptista MS (2014) Membrane damage efficiency of phenothiazinium photosensitizers. *Photochem Photobiol* 90:801–813.

Bacellar IOL, Tsubone TM, Pavani C, Baptista MS (2015) Photodynamic efficiency: From molecular photochemistry to cell death. *Int J Mol Sci* 16:20523–20559.

Bisby RH, Mead C, Morgan CG (2000) Active uptake of drugs into photosensitive liposomes and rapid release on UV photolysis. *Photochem Photobiol* 72:57–61.

Bondurant B, O'Brien DF (1998) Photoinduced destabilization of sterically stabilized liposomes. *J Am Chem Soc* 120:13541–13542.

Brewster R, Pincus PA, Safran SA (2009) Hybrid lipids as a biological surface-active component. *Biophys J* 97:1087–1094.

Briganti S, Picardo M (2003) Antioxidant activity, lipid peroxidation and skin diseases. What's new. *J Eur Acad Dermatol Venereol* 17:663–669.

Caetano W, Haddad PS, Itri R, Severino D, Vieira VC, Baptista MS, Schroder A, Marques C (2007) Photo-induced destruction of giant vesicles in methylene blue solutions. *Langmuir* 23:1307–1314.

Chiarelli-Neto O, Ferreira AS, Martins WK, Pavani C, Severino D, Faião-Flores F, Maria-Engler SS et al. (2014) Melanin photosensitization and the effect of visible light on epithelial cells. *PLoS One* 9:e113266.

Cremesti AE, Goni FM, Kolesnick R (2002) Role of sphingomyelinase and ceramide in modulating rafts: Do biophysical properties determine biologic outcome? *FEBS Lett* 531:47–53.

Cwiklik L, Jungwirth P (2010) Massive oxidation of phospholipid membranes leads to pore creation and bilayer disintegration. *Chem Phys Lett* 486:99–103.

De Rosa R, Spinozzi F, Itri R (2018) Hydroperoxide and carboxyl groups preferential location in oxidized biomembranes experimentally determined by small angle X-ray scattering: Implications in membrane structure. *Biochimica et Biophysica Acta (BBA)-Biomembranes* 1860:2299–2307.

Dencher NA, Heyn, MP (1979) Bacteriorhodopsin monomers pump protons. *FEBS Lett* 108: 307–310.

Dezi M, Di Cicco A, Bassereau P, Lévy D (2013) Detergent-mediated incorporation of transmembrane proteins in GUVs with controlled physiological contents. *Proc Natl Acad Sci USA* 110:7276–7281.

Diguet A, Yanagisawa M, Liu YJ, Brun E, Abadie S, Rudiuk S, Baigl D (2012) UV-induced bursting of cell-sized multicomponent lipid vesicles in a photosensitive surfactant solution. *J Am Chem Soc* 134: 4898–4904.

Fenz SF, Sengupta K (2012) Giant vesicles as cell models. *Integr Biol* 4:982–995.

Foote, CS (1968) Mechanisms of photosensitized oxidation. *Science* 162:963–970.

Fournier JB, Ajdari A, Peliti L (2001) Effective-area elasticity and tension of micromanipulated membranes. *Phys Rev Lett* 86:4970–4973.

Georgiev VN, Grafmüller A, Bléger D, Hecht S, Kunstmann S, Barbirz S, Lipowsky R, Dimova R (2017) Area increase and budding in giant vesicles triggered by light: Behind the scene. *Adv Sci* 5(8):1800432.

Georgieva R, Albena Momchilova A, Petkova D, Koumanov K, Galya Staneva G (2013) Effect of n-propyl gallate on lipid peroxidation in heterogenous model membranes. *Biotechnol Biotechnol* 27(5):4145–4149.

Guo Y, Baulin VA, Thalmann F (2016) Peroxidised phospholipid bilayers: Insight from coarse-grained molecular dynamics simulations. *Soft Matter* 12:263–271.

Haluska C, Baptista MS, Schroder AP, Marques CM, Itri R (2012) Photoactivated phase separation in giant vesicles made from different lipid mixtures. *Biochim Biophys Acta: Biomembranes* 1818:666–672.

Heuvingh J, Bonneau S (2009) Asymmetric oxidation of giant vesicles triggers curvature-associated shape transition and permeabilization. *Biophys J* 97:2904–2912.

Huang D, Ou B, Prior RL (2005) The chemistry behind antioxidant capacity assays. *J Agric Food Chem* 53:1841–1856.

Itri R, Junqueira, HC, Mertins O, Baptista MS (2014) Membrane changes under oxidative stress: The impact of oxidized lipids. *Biophys Rev* 6:47–61.

Jurkiewicz, P, Olzynskan A, Cwiklik L, Conte E, Jungwirth P, Megli FM, Hof M (2012) Biophysics of lipid bilayers containing oxidatively modified phospholipids: Insights from fluorescence and EPR experiments and from MD simulations. *Biochim Biophys Acta (BBA)-Biomembr* 1818:2388–2402.

Kahya N (2010) Protein–protein and protein–lipid interactions in domain-assembly: Lessons from giant unilamellar vesicles. *Biochim Biophys Acta* 1798:1392–1398.

Kahya N, Pécheur EI, de Boeij WP, Wiersma DA, Hoekstra D (2001) Reconstitution of membrane proteins into giant unilamellar vesicles via peptide-induced fusion. *Biophys J* 81:1464–1474.

Kahya N, Wiersma DA, Poolman B, Hoekstra D (2002) Spatial organization of bacteriorhodopsin in model membranes light-induced mobility changes. *J Biol Chem* 277:39304–39311.

Karatekin E, Sandre O, Brochard-Wyart F (2003b). Transient pores in vesicles. *Polym Int* 52:486–493.

Karatekin E, Sandre O, Guitouni H, Borghi N, Puech PH, Brochard-Wyart F (2003a) Cascades of transient pores in giant vesicles: Line tension and transport. *Biophys J* 84, 1734–1749.

Kerdous R, Heuvingh J, Bonneau S (2011) Photo-dynamic induction of oxidative stress within cholesterol-containing membranes: Shape transitions and permeabilization. *Biochim Biophys Acta* 1808:2965–2972.

Kriegsmann J, Brehs M, Klare JP, Engelhard M, Fitter J (2009a) Sensory rhodopsin II/transducer complex formation in detergent and in lipid bilayers studied with FRET. *Biochim Biophys Acta* 1788:522–531.

Kriegsmann J, Gregor I, von der Hocht I, Klare J, Engelhard M, Enderlein J, Fitter J (2009b) Translational diffusion and interaction of a photoreceptor and its cognate transducer observed in giant unilamellar vesicles by using dual-focus FCS. *Chem Bio Chem* 10:1823–1829.

Kuksis A (2012) Separation and determination of the structure of fatty acids. In: *Handbook of Lipid Research, Vol. 1.Fatty acids and glycerides*. New York, Springer Science & Business Media.

Kulig W, Olzynska A, Jurkiewicz P, Kantola AM, Komulainen S, Manna M, Pourmousa M et al. (2015) Cholesterol under oxidative stress—How lipid membranes sense oxidation as cholesterol is being replaced by oxysterols. *Free Radic Biol Med* 84:30–41.

Leung SJ, Romanowski M (2012) Light-activated content release from liposomes, *Theranostics* 2:1020–1036.

Liang R, Liu Y, Fu L-M, Ai X-C, Zhang J-P, Leif H. Skibsted LH (2012) Antioxidants and physical integrity of lipid bilayers under oxidative stress. *J Agric Food Chem* 60:10331–10336.

Long SP, Humphries S, Falkowski PG (1994) Photoinhibition of photosynthesis in nature. *Annu Rev Plant Physiol Plant Mol Biol* 45:633–662.

Mabrouk E, Bonneau S, Jia I, Cuvelier D, Li, M-H, Nassoy P (2010) Photosensitization of polymer vesicles: A multistep chemical process deciphered by micropipette manipulation. *Soft Matter* 6:4863–4875.

Mabrouk E, Cuvelier D, Brochard-Wyart F, Nassoy P, Li MH (2009). Bursting of sensitive polymersomes induced by curling. *PNAS* 106:7294–7298.

Mertins O, Bacellar IOL, Thalmann F, Marques CM, Baptista MS, Itri R (2014) Physical damage on giant vesicles membrane as a result of methylene blue photoirradiation. *Biophys J* 106:162–171.

Mertins O, Mathews PD, Gomide AB, Baptista MS, Itri R (2015) Effective protection of biological membranes against photo oxidative damage: Polymeric antioxidant forming a protecting shield over the membrane. *Biochim Biophys Acta* 1848:2180–2187.

Mitomo H, Chen W-H, Steven L, Regen SL (2009) Oxysterol-induced rearrangement of the liquid-ordered phase: A possible link to Alzheimer's disease? *J Am Chem Soc* 131:12354–12357.

Morales-Penningston BF, Wu J, Farkas ER, Goh SL, Konyakhina TM, Zheng JY, Webb WW, Feigenson GW (2010) GUV preparation and imaging: Minimizing artifacts. *Biochim Biophys Acta (Biomembranes)* 1798:1324–1332.

Pidgeon C, Hunt CA (1983) Light sensitive liposomes. *Photochem Photobiol* 37:491–494.

Poo MM, Cone RA (1974) Lateral diffusion of rhodopsin in the photoreceptor membrane. *Nature* 247:438–441.

Rawicz W, Olbrich K, McIntosh T, Needham D, Evans E (2000) Effect of chain length and unsaturation on elasticity of lipid bilayers. *Biophys J* 79:328–339.

Reis A, Domingues MR, Amado FM, Ferrer-Correia AJ, Domingues P (2005). Separation of peroxidation products of diacyl-phosphatidylcholines by reversed-phase liquid chromatography–mass spectrometry. *Biomed Chromatogr* 19: 129–137.

Riske KA, Sudbrack TP, Archilla NL, Uchoa AF, Schroder AP, Marques C, Baptista MS, Itri R (2009) Giant vesicles under oxidative stress induced by a membrane-anchored photosensitizer. *Biophys J* 97:1362–1370.

Rodrigues D, Viotto AC, Checchia R, Gomide A, Severino D, Itri R, Martins WK (2016) Mechanism of aloe vera extract protection against UVA: Shelter of lysosomal membrane avoids photodamage. *Photochem Photobio Sci* 15:334–350.

Runas KA, Malmstadt N (2015) Low levels of lipid oxidation radically increase the passive permeability of lipid bilayers. *Soft Matter* 11:599–505.

Sandre O, Moreaux L, Brochard-Wyart F (1999) Dynamics of transient pores in stretched vesicles. *PNAS* 96:10591–10596.

Sankhagowit S, Wu S-H, Biswas R, Riche CT, Povinelli ML, Malmstadt (2014) The dynamics of giant unilamellar vesicle oxidation probed by morphological changes. *Biochim Biophys Acta* 1838:2615–2624.

Sebai SC, Cribier S, Karimi A, Massotte D, Tribet C (2010) Permeabilization of lipid membranes and cells by a light-responsive copolymer. *Langmuir* 26:14135–14141.

Siani P, Souza RM, Dias LG, Itri R, Kandhelia H (2016) An overview of molecular dynamic simulations of oxidized lipid systems, with a comparison of ELBA and MARTINI force fields for coarse grained lipid simulations. *Biochim Biophys Acta* 1858(10):2498–2511.

Staneva G, Seigneuret M, Conjeaud H, Puff N, Angelova MI (2011) Making a tool of an artifact: the application of photoinduced Lo domains in giant unilamellar vesicles to the study of Lo/Ld phase spinodal decomposition and its modulation by the ganglioside GM1. *Langmuir* 27:15074–15082.

Staneva G, Seigneuret M, Conjeaud H, Puff N, Angelova MI (2011) Making a tool of an artifact: the application of photoinduced Lo domains in giant unilamellar vesicles to the study of Lo/Ld phase spinodal decomposition and its modulation by the ganglioside GM1. *Langmuir* 27:15074–15082.

Su C-F, Merlitz H, Thalmann F, Marques CM, Sommer J-U (2019) Coarse-Grained Model of Oxidized Membranes and Their Interactions with Nanoparticles of Various Degrees of Hydrophobicity *J. Phys. Chem. C* 123:6839–6848.

Thompson DH, Gerasimov OV, Wheeler JJ, Anderson VC (1996) Triggerable plasmalogen liposomes: Improvement of system efficiency. *Biochim Biophys Acta* 1279:25–34.

Tsubone TM, Junqueira HC, Baptista MS, Itri R (2019) Contrasting roles of oxidized lipids in modulating membrane microdomains. *Biochimica et Biophysica Acta (BBA)-Biomembranes* 1861: 660–669.

Uchoa AF, Knox PP, Turchielle R, Seifullina NK, Baptista MS (2008) Singlet oxygen generation in the reaction centers of Rhodobacter sphaeroides. *Eur Biophys J* 37:843–850.

Uchoa AF, Severino D, Baptista MS (2013) Antioxidant properties of singlet oxygen suppressors. *in Natural Antioxidants and Biocides from Wild Medicinal Plants.* CL Céspedes, DA Sampietro, DS Seigler, M Rai, editors. Boston, CABI, p. 65.

van der Horst MA, Hellingwerf KJ (2004) Photoreceptor proteins, "star actors of modern times": a review of the functional dynamics in the structure of representative members of six different photoreceptor families. *Acc Chem Res* 37:13–20.

Weber G (2012) Photo-induced modifications of model membranes. PhD thesis. Université de Strasbourg, Strasbourg, France.

Weber G, Charitat T, Baptista MS, Uchoa AF, Pavani C, Junqueira HC, Gu Y, Baulin VA, Itri R, Marques CM, Schroder AP (2014). Lipid oxidation induces structural changes in biomimetic membranes. *Soft Matter* 10:4241–4247.

Wong-ekkabut J, Xu Z, Triampo W, Tang I-M, Tieleman P, Monticelli L (2007) Effect of lipid peroxidation on the properties of lipid bilayers: A molecular dynamics study. *Biophys J* 93:4225–4236.

Yasuhara K, Sasaki Y, Kikuchi J-I (2008) A photo-responsive cholesterol capable of inducing a morphological transformation of the liquid-ordered microdomain in lipid bilayers. *Colloid Polym Sci* 286:1675–1680.

Zhang F, Vierock J, Yizhar O, Fenno LE, Tsunoda S, Kianianmomeni A, Prigge M et al. (2011) The microbial Opsin family of optogenetic tools. *Cell* 147:1446–1457.

Zhang J, Stanley RA, Melton LD (2006) Lipid peroxidation inhibition capacity assay for antioxidants based on liposomal membranes. *Mol Nutr Food Res* 50:714–724.

Zhao J, Wu J, Shao H, Kong F, Jain N, Hunt G, Feigenson GW (2007) Phase studies of model biomembranes: macroscopic coexistence of $L\alpha + L\beta$, with light induced coexistence of $L\alpha$ + Lo Phases. *Biochim Biophys Acta–Biomembr* 1768:2777–2786.

23 Protein–membrane interactions

Eva M. Schmid and Daniel A. Fletcher

Give me an organic vesicle endowed with life and
I will give you back the whole of the organized world.

(François Raspail 1794–1878)

Contents

23.1 INTRODUCTION

Cellular life critically depends on the presence of membranes to separate inside from outside and compartmentalize the cytoplasm into specific biochemical environments. A multitude of biochemical activities crucial for homeostasis take place on membrane surfaces, and abnormalities at the membrane level can have dramatic effects on cellular function. Through the interaction of proteins with lipid bilayers, cellular membranes actively guide cellular processes by, for example, (i) organizing cytoskeletal filament networks to give the cell structure, shape and allow for motility; (ii) transporting material in and out of the cell, between cells, and within intracellular organelles to allow for communication; and (iii) conferring selective sensing via membrane receptors during neuronal function, development, and the immune system.

Despite the fact that we can observe dynamic cellular phenomena involving the membrane in live cells, we cannot fully explain, predict, or replicate them from component molecules. There remain many open questions regarding the organization of membranes and the assembly of macromolecular structures on them. For example, how are membrane proteins spatially and temporally organized, and what effect does this organization have on organization of lipids in the membrane? How are membrane proteins concentrated in discreet clusters for efficient signaling? What are the physical consequences of protein organization in and on membranes? What are the protein parameters involved in membrane bending and fusion, and can we use this knowledge to modify and control cellular membranes? How is membrane organization affected at cell–cell junctions? Because the great complexity of even simple organisms like bacteria, yeast, and worms has made it difficult to isolate and

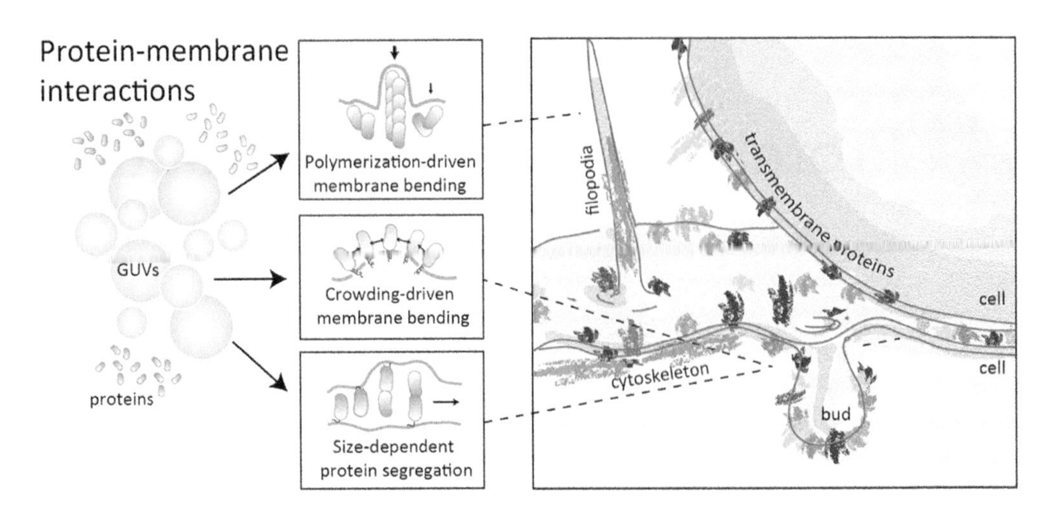

Figure 23.1 Schematic illustration of protein–membrane interactions discussed in this chapter. The reconstitution of peripheral proteins on synthetic membranes (illustrated on the left) reveals basic principles of complex cellular membrane behaviors and structures such as filopodia, endocytic bud, or cell–cell interface formation (schematized on the right).

understand the individual mechanisms by which cellular membranes perform their functions, *in vitro* reconstitution of cellular processes from their component parts has led to key insights into the physical principles underlying the function of all biological membranes (Liu and Fletcher, 2009; Loose and Schwille, 2009). Here the focus will be on *in vitro* reconstitution of protein–membrane processes using synthetic membranes and purified proteins.

This chapter will discuss three examples of reconstituted membrane–protein interactions that use synthetic membranes in the form of giant unilamellar vesicles (GUVs) in combination with purified proteins (Figure 23.1). These examples illustrate how protein organization on membranes can influence the shape of membranes, and how in turn the shape of the membrane can influence protein organization. Experimental details as well as theoretical considerations will be described, and the ways in which *in vitro* experiments can inform cell biological pathways will be discussed.

23.2 BACKGROUND

Membranes play a dual role in the life of the cell; they must not only separate cells and their organelles from their surroundings to permit critical biochemical reactions, but they must also facilitate communication of cells and their organelles with their environment through signaling and exchange of material. This dual role has necessitated the evolution of the highly complex protein–lipid composites that are found in cells. In the following, principles applicable to all cellular membranes will be discussed, but the focus will be on processes happening on the plasma membrane.

23.2.1 THE PLASMA MEMBRANE—LIPIDS AND PROTEINS (AND SUGARS)

The invention of the microscope in the seventeenth century revealed that all organisms were made of individual compartments separated from each other by some sort of barrier. It took almost two centuries for researchers to understand that the semipermeable barrier surrounding animal cells is organized as a bilayer of fatty molecules (Gorter and Grendel, 1925) and that proteins—attached and integrated—are a major constituent of membranes. Electron microscopy studies revealed the beautifully complex shapes of internal membranes, structures that still challenge our understanding of membrane–protein organization (Glauert, 1968; Robertson, 1981).

The modern view of biological membranes is one of a vastly crowded protein landscape in and on lipid bilayers of highly complex, asymmetric lipid composition. In addition, glycans (sugars) are often attached to proteins and lipids facing the extracellular milieu. This glycocalyx provides protection and is critical for cell recognition and cell adhesion. How the glycocalyx influences cell behavior is an area of growing interest but won't be discussed here.

The protein component of cellular membranes renders passive lipid sheets into active materials that animate cellular processes. Proteins inserted in membranes allow for transport across the otherwise mostly impermeable bilayer, facilitating cells to send and receive messages and to respond accordingly, take up nutrients and shed waste products. Membrane proteins (attached to the membrane or inserted) are also responsible for cellular shape and shape change, motility, and the formation of multicellular structures. Research over the past decades has updated—and complicated—the original fluid mosaic model of cellular membranes described by Singer and Nicolson in 1972 (Singer and Nicolson, 1972; Engelman, 2005). In this chapter, the description of membrane proteins will be focused on so-called peripheral membrane proteins, which are here defined as proteins that dynamically associate and dissociate with the membrane. Information on integral membrane proteins can be found in Chapter 3 of this book.

A complete understanding of the complex plasma membrane organization including both lipids and proteins is still missing, but the critical effects that proteins have on membranes, and vice-versa, are being revealed by increasingly sophisticated reconstitution experiments. Indeed, it is now clear that the biophysical behavior of the plasma membrane depends not only on the lipid composition but also on the protein content. The field of membrane reconstitution is continuing to develop new techniques for re-building protein–membrane interactions found in cells (Liu and Fletcher, 2009; Loose and Schwille, 2009; Lagny and Bassereau, 2015), which will continue to advance understanding of basic mechanisms that control organization of cell membrane. The following sections introduce interactions of peripheral membrane proteins with the plasma membrane (from hereon "protein–membrane interactions") and the remarkable structures they are able to form.

23.2.2 TYPES OF PROTEIN–MEMBRANE INTERACTIONS

Peripheral membrane proteins reversibly associate with the membrane in order to regulate its composition, dynamics, and morphology. Proteins become peripheral membrane proteins by directly interacting with the membrane in one of several ways (Figure 23.2).

First, proteins can interact with membranes via lipid headgroups, an interaction established either via electrostatic or ionic engagement between proteins and membrane lipids, through divalent cations (such as Ca^{2+}), or due to specific binding pockets for lipid heads on the protein surface (Mulgrew-Nesbitt et al., 2006). The internal leaflet of plasma membranes contains a variety of signaling lipids [such as phosphatidylinositol 4,5-bisphosphate (PIP_2), see Appendix 1 of the book for structure and data on this lipid], whose local concentration is a critical way to organize cytoplasmic proteins for biological function (McLaughlin and Murray, 2005; Lemmon, 2008). Alternatively, proteins can interact with the hydrophobic part of the bilayer by insertion of an amphipathic α-helix, a hydrophobic loop, or a lipid moiety, into the lipid bilayer, anchoring the protein peripherally at the membrane (Campelo et al., 2008). Some proteins contain unusually long amphipathic helixes (such as the ALPS motif) and have been shown to bind membranes of specific membrane curvature. In general it is understood that curved membranes provide spatial information to proteins, which can interpret the geometrical information and signal accordingly (for an in depth review on this topic see (Antonny, 2011) and for techniques to study these proteins see Chapter 16). In addition, proteins can get covalently attached to membrane lipids via a process called lipidation (Nadolski and Linder, 2007). Last, peripheral membrane proteins can interact with transmembrane proteins present in the membranes.

The attachment mechanism is an important determinant for biological function as it defines location, specificity, and affinity of the interaction, which in turn can influence enzymatic reactions, membrane shape and dynamics as well as downstream cellular signaling.

23.2.3 COLLECTIVE PROTEIN–MEMBRANE INTERACTIONS

Protein–membrane interactions must be thought of collectively as well as individually, as lipid-lipid and protein-protein interactions on membranes can result in behavior that can drive large-scale changes to cell membranes. These changes include alterations of biophysical properties of the membrane—tension, viscosity, elasticity, and bending rigidity—that have effects on membrane protein organization, dynamics and function (Janmey and Kinnunen, 2006). The interplay of lipids and proteins in cells leads to remarkably complex behaviors, and their roles in cell behavior are only beginning to be understood.

One striking example of how the molecular interaction between proteins and lipids lead to large-scale behavior is cell motility. Directional movement of a cell begins with proteins embedded in the plasma membrane that recognize chemical cues. After signal recognition, the cell polarizes (defines a front and a back), a process that involves localization of membrane bound signaling proteins to a defined membrane patch. Next, the actin cytoskeleton assembles at this newly defined site and generates force to push the membrane in the direction of the cues. The formed protrusions assemble and disassemble continuously in response to changes in the environment, allowing the cell to react to its changing surroundings, such as when a neutrophil tracks chemicals released by invading bacteria. Finally, in order to generate forward movement of the entire cell, contraction of the rear of the cell is triggered, again involving the cytoskeleton and the plasma membrane (Lauffenburger and Horwitz, 1996; Horwitz and Parsons, 1999; Pollard and Borisy, 2003; Ridley et al., 2003; Li et al., 2005). Both the biochemical and the mechanical characteristics of the plasma membrane and associated proteins are critical for each step in cell motility (Keren, 2011). Indeed, membrane rigidity, tension and curvature are all affected by the biochemical composition of the membrane, and the dynamic localization of proteins to the bilayer is in turn affected by these mechanical characteristics.

The dynamic interplay and feedback of cause and effect is likely critical for biological function, but it is very difficult to understand in live cells due to their complexity. This is where membrane reconstitution experiments, by minimizing and simplifying the system to isolate key mechanisms, have significantly improved understanding of isolated individual processes (Loisel et al., 1999; Upadhyaya and van Oudenaarden, 2003; Liu and Fletcher, 2009; Loose and Schwille, 2009).

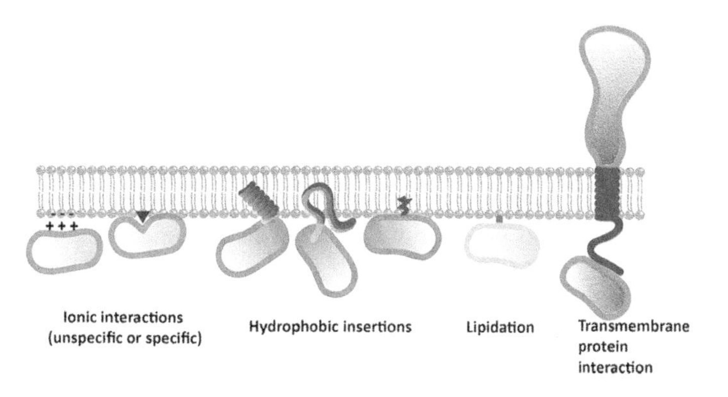

Figure 23.2 Types of protein–membrane interactions. Peripheral membrane proteins interact with lipid headgroups via electrostatic or ionic engagement, or bind to specific signaling lipids (such as PIP_2). Alternatively, peripheral membrane proteins are attached to the lipid bilayer by inserting hydrophobic helices, loops, or lipid moieties into one leaflet. Covalent attachment of a lipid (lipidation) anchors proteins to the membrane. Last, peripheral membrane proteins can gain access to the membrane by interacting with integral membrane proteins.

23.3 EXPERIMENTAL METHODS FOR INVESTIGATING PROTEIN–MEMBRANE INTERACTIONS

The field of membrane biology has benefitted tremendously from reducing the complexity of some membrane-based processes and reconstituting them from the bottom-up. The challenge for this field is to find the right balance between simplifying and over-simplifying the process of interest, so that key mechanistic insight can be obtained that advances understanding of the real process. In this way, membrane reconstitution is not unlike "course-grained" simulations, where the right level of detail must be included to obtain informative and nontrivial results. A variety of tools and assays have been developed to capture different subsets of biological membrane processes.

Before addressing a biological question with a reductionist approach, parameters important for a meaningful reconstitution experiment need to be considered and matched with the available tools. Parameters that should be taken into account for reconstitution include (i) the lipid composition of the membrane of interest (e.g., plasma membrane vs. intracellular membranes, inner vs. outer leaflet, symmetric or asymmetric lipid distribution); (ii) its membrane morphology (e.g., high or low curvature, supported or unsupported bilayer); and (iii) its protein composition (e.g., single or multiple species of proteins, full lengths vs protein domains). In addition, protein–membrane attachment possibilities (physiological vs. artificial) have to be decided on, and, last, appropriate tools for the visualization and characterization of the generated protein–membrane composites need to be at hand. Once these factors have been considered, the experimenter can choose the best techniques for the experiment.

Of the available model membrane systems, GUVs are often chosen for experiments reconstituting protein–membrane interactions (Monzel et al., 2009 and references therein, as well as Chapter 1). Their large (10–100 μm) size makes them amenable for light microscopy, see Chapter 10. The membranes are freestanding and, therefore, fluid and deformable. They are compatible with mechanical perturbations (see Chapter 16) and can be made with a large variety of lipid composition, which also allows for decoration in proteins, as seen later in this chapter. The reconstituted membranes successfully mimic the fluidity, tension and deformability of cellular membranes (see Chapters 11 through 16 and 21).

23.3.1 GIANT UNILAMELLAR VESICLE FORMATION

GUVs can be produced by a variety of different techniques, each with specific advantages and disadvantages (for a comprehensive review see Chapter 1). Probably the most widely used method for GUV generation is a process often referred to as electroformation (Angelova and Dimitrov, 1986; see Schmid et al., 2015 for detailed instructional videos). This chapter describes examples of where attachment of proteins peripherally to electroformed GUVs leads to remarkable behaviors and insights into biological mechanisms.

23.3.2 ADDITION OF PROTEINS

Controlling the composition of proteins on membranes is an essential part of any membrane reconstitution experiment. It is possible to emulate peripheral membrane proteins on synthetic membranes using a variety of protein attachment strategies. If the protein of interest has a specific lipid binding modality [e.g., PH, epsin N-terminal homology (ENTH), bin-amphyphysin-rvs (BAR) domains with specificity for phosphoinositides such as PIP_2 or other charged lipids], the binding lipid can simply be incorporated into the GUVs and the proteins will bind to their natural partner, linking the proteins to the synthetic membrane with physiological affinity. Lipid-anchored proteins, or proteins containing hydrophobic helices, will self-insert into a presented membrane spontaneously, yet it is often challenging to recombinantly express and purify such proteins in the first place, so that replacing the natural interaction with an artificial one is often favored. For example, individual amino acids in proteins can be attached to chemically modified lipids (e.g., cysteine–maleimide interactions, lysine–succinyl interactions, or histidine–Ni-NTA interactions). In another approach, the high-affinity interaction of streptavidin–biotin can be used to sandwich biotinylated proteins to biotinylated lipids via streptavidin (Schmid et al., 2015). Last, the lipidation of proteins to lipids in synthetic membranes can also be reproduced (Zens et al., 2015).

The membrane–protein anchoring strategies described above vary in affinity and so should be chosen with full appreciation for the role that affinity can have in experiments. In some cases, affinity can be tuned dynamically to meet the experiment's needs. The most versatile attachment strategy in this regard is the use of Ni-NTA functionalized lipid headgroups in combination with a multi-His tag cloned either N-or C-terminally onto the desired binding protein. *Practical note*: The affinity can be varied by changing the metal ion in the NTA group or the length of the His-tag (Kent et al., 2004; Nye and Groves, 2008) and the interaction can even be reversed by the addition of EDTA. Once a strategy for binding a protein of interest to a membrane has been selected, the next step in experimental design is to determine how best to characterize protein–membrane interactions.

23.3.3 QUANTIFICATION OF MEMBRANE RECONSTITUTIONS

Because GUVs with diameters 10–100 μm are amenable for light microscopy, they permit dynamic information to be obtained and quantified much as it would be for live cell microscopy. Protein organization on GUV membranes can be visualized by confocal microscopy, and biophysical parameters can be extracted with a variety of techniques. For example, membrane fluidity and protein diffusion on the membrane can be measured with fluorescence correlation spectroscopy or fluorescence recovery after photobleaching (FCS or FRAP, respectively; see Chapter 21); protein density and resulting membrane coverage can be determined by methods such as fluorescence lifetime imaging microscopy combined with Förster resonance energy transfer (FLIM-FRET), and distances between two membranes (established by adhesion proteins, for example) can be identified by reflection interference contrast microscopy (RICM) (Parthasarathy and Groves, 2004; Groves et al., 2008; Ries and Schwille, 2012; Basit et al., 2014; De Los Santos et al., 2015), see Chapter 17. The reconstituted membranes can also be mechanically manipulated by micropipette aspiration, tether pulling with an optical trap, or deformation by atomic force microscopy, see Chapter 12. All of these techniques allow for a quantification of collective protein–membrane behavior on model membranes.

Membrane reconstitution combined with these approaches for quantitation have been used to uncover the remarkable molecular mechanisms that animate biological membranes. For example, over the past few years, membrane reconstitution experiments have uncovered how the collective binding and unbinding of proteins to a supported membrane can lead to wave-like oscillations of protein assemblies, explaining decision making during symmetric cell division (Loose and Schwille, 2009), how mechanically induced membrane shape changes influence protein binding (Sorre et al., 2012), and how protein binding can influence membrane phase behavior (Liu and Fletcher, 2006). The following three subchapters describe different examples of specific protein–membrane reconstitution approaches and explain the insight gained from simplifying complex behaviors in these particular instances.

23.4 POLYMERIZATION-DRIVEN MEMBRANE BENDING

One prominent feature of cellular membranes is their extraordinary diversity when it comes to shape. Although the plasma membrane is often flat at the scale of a protein (tens of nanometers), it can be locally shaped into very high curvature regions, such as protrusions or intracellular buds. Intracellular membranes, such as the endoplasmic reticulum (ER) or the Golgi apparatus, exhibit highly curved structures, which get dynamically remodeled and from which high curvature vesicles and tubules are generated. Deviations from the energetically most favorable flat membrane conformation require energy input in the range of 10^{-19} J or ≈ 6 kBT (Helfrich, 1973). This can be provided by the application of force, or by the generation of asymmetry in the bilayer, which increases the spontaneous curvature of membranes (Kozlov et al., 2014; McMahon and Gallop, 2005), see Chapter 5. Among the most prominent shape changes of cells are those driven by the actin cytoskeleton, such as filopodia.

23.4.1 EXAMPLE: ACTIN-DRIVEN MEMBRANE TUBULATION

Filopodia are thin, spike-like protrusions of the plasma membrane that are filled with parallel bundled actin filaments. They play an important role in the guidance of crawling cells by probing the extracellular matrix and forming nascent adhesions with integrins found at their tips. Filopodia have also been implicated in the formation of cell–cell contacts and may also serve as sensors of other stimuli. How filopodia are formed at cell membranes has been the subject of extensive study, largely focusing on live cell studies and electron microscopy studies to identify the key molecular players and their organization (Gupton and Gertler, 2007; Mattila and Lappalainen, 2008). One prominent model, the convergent elongation model (Svitkina et al., 2003), proposed that filopodia emerge from branched actin networks that have been locally bundled to create parallel actin filaments and protected from filament capping by a tip complex. While a large number of proteins in addition to actin are associated with filopodia *in vivo*, two play a critical role in the convergent elongation model, formin and fascin. Formin, a processive actin filament elongation factor, is known to sit at the tips of filopodia and promote filament growth (Goode and Eck, 2007). Fascin, an actin bundling protein that forms parallel filament bundles *in vitro*, is thought to align filaments from branched actin networks into rigid structures that could protrude and bend the membrane into a filopodia's characteristic spike-like protrusion (Adams, 2004). In short, a system of proteins was believed to orchestrate the assembly of filopodia and deform the membrane during assembly. Was the plasma membrane just a passive player in the process?

23.4.2 METHODS AND FINDINGS

To investigate what role membranes might play in filopodia formation, branched actin network assembly was reconstituted on GUVs from pure proteins, and the membrane tubules produced by the growing actin network were analyzed (Liu et al., 2008). In brief, branched actin networks are assembled in a series of steps that create dense networks that push with actin filaments oriented, on average, orthogonal to the membrane. In addition to actin subunits and profilin, a protein that binds to actin subunits and prevents their spontaneous nucleation, branched actin network assembly requires a nucleation promoting factor localized to the membrane that activates the soluble protein Arp2/3, which is a seven-subunit protein complex that regulates the actin cytoskeleton, causing it to bind to the side of existing filaments and initiate a new filament. In cells, branched actin assembly is regulated by the presence of capping protein, which stops filament elongation, and is recycled by members of the ADF/cofilin family of proteins, which sever actin filaments. According to the convergent elongation model, the bundling factors like fascin and elongation factors like formins are also required in order to achieve filopodia-like protrusions.

Reconstitution of branched actin network assembly on GUVs provides the opportunity to test which components are necessary and sufficient for filopodia-like membrane tubule formation. To simplify the system, both capping protein and ADF/cofilin were left out of the reconstitution, and neither fascin nor formin were included to test whether tubules could form in the absence of these proteins. To drive actin polymerization, PIP_2 lipids were incorporated into the electroformed GUVs [75% L-α-phosphatidylcholine from chicken egg (eggPC), 20% 1,2-dioleoyl-*sn*-glycero-3-phospho-L-serine (DOPS), and 5% brain PIP_2, see Appendix 1 of the book for structure and data on lipids] (Box 23.1, COMPONENTS). This lipid composition leads to the localization and activation of N-WASP (a protein complex that regulates actin polymerization), which drives network assembly when added with actin and Arp2/3 as soluble proteins on the outside of the GUVs (~8.5 μM G-actin, ~160 nM Arp2/3 complex, and 400 nM N-WASP in actin polymerization buffer. Osmolarity of the final protein mixtures needs to be matched to the vesicles to within 5%) (Box 23.1, ASSEMBLY), creating an inverted geometry whereby the filopodial protrusions, should they occur, would extend into the lumen of the GUV. *Practical note*: By incorporating a small fraction (~10%) of labeled actin in the branched actin network mix, network assembly on the GUVs can be monitored with spinning disk microscopy until completion (~30 min) (Box 23.1, IMAGING).

Even without the presence of fascin or formins, it was observed that spike-like membrane protrusions containing long actin filaments extended into the lumen of the GUV. This result, produced with only the minimal set of factors necessary for branched actin network assembly, was unexpected because filament bundling by fascin was thought to be important. Single polymerizing filaments

Box 23.1 Actin-driven tubulation (Liu et al., 2008)

COMPONENTS

GUVs

Egg PC
PIP_2
lipid dye

purified proteins

Actin	cdc42/RhoGDI
Arp2/3	DH-PH
N-WASP	CP

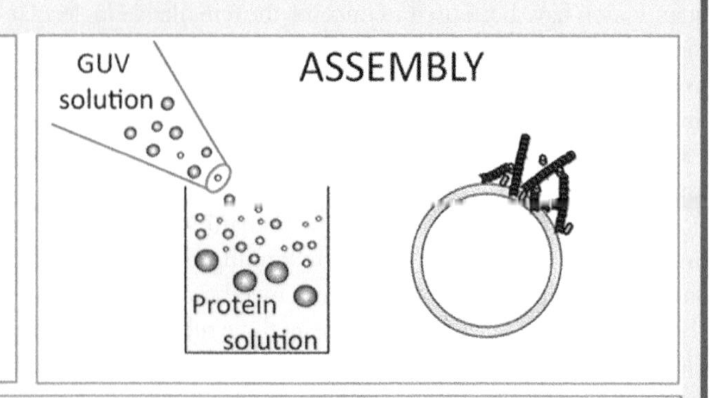

IMAGING

phase contrast and confocal microscopy

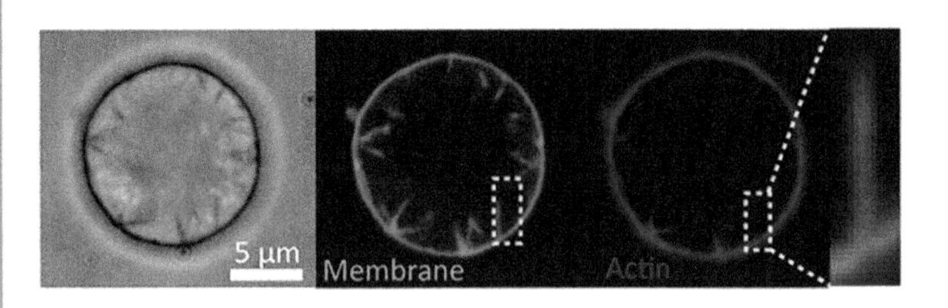

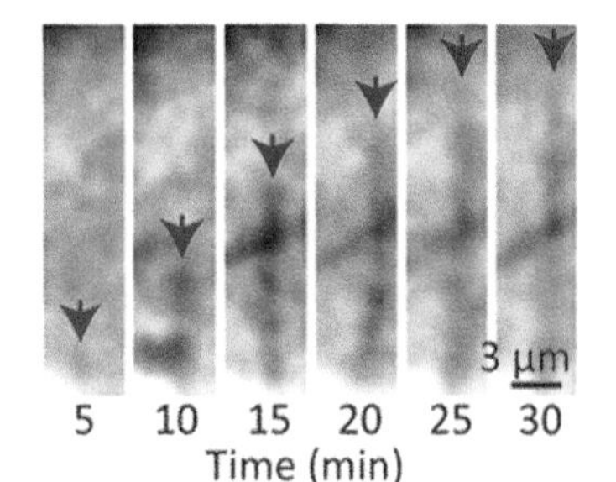

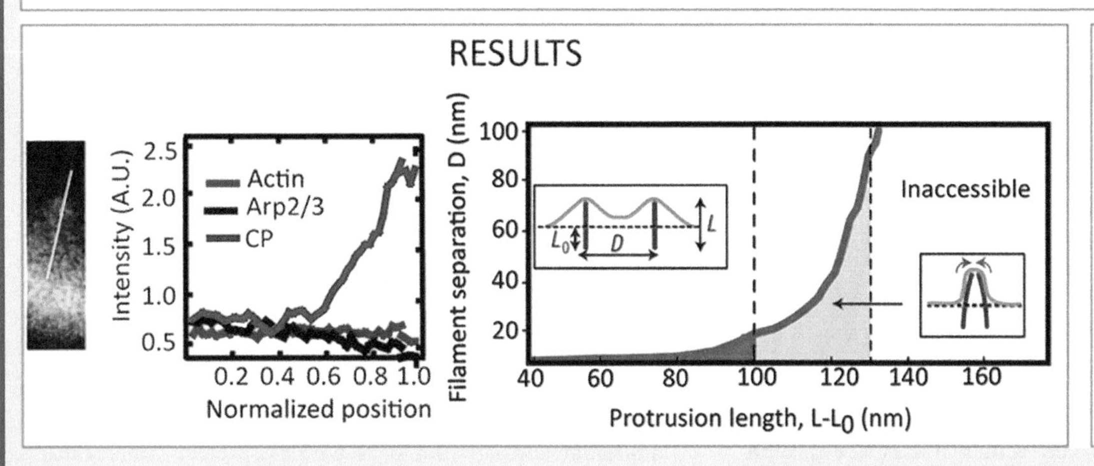

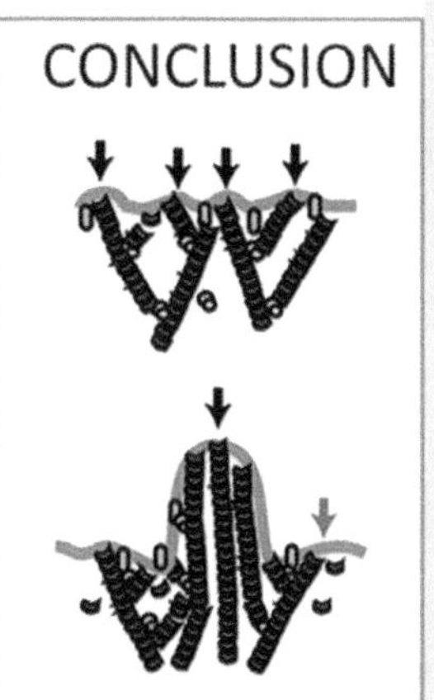

Electroformed GUVs are incubated with purified proteins (COMPONENTS). All components are introduced in solution on the outside of the GUV (ASSEMBLY).

Network assembly and protrusion formation are tracked by phase-contrast and spinning–disc confocal imaging of membrane and actin. Overlay of the fluorescence images confirms that the membrane protrusions are supported by actin filaments (IMAGING, left). Elongation of a thin protrusion can be visualized by phase-contrast microscopy. The length of the protrusion is measured over time and shows that growth initially occurs quickly but slows down as the protrusion elongates (IMAGING, right).

The localization of actin, Arp2/3 complex and capping protein along thin actin filament protrusions is shown in confocal images (RESULTS, left). The normalized Arp2/3 complex and capping protein traces are divided by the normalized actin line scans. A simulation of two filaments pushing against a membrane shows how bundling by the membrane is favorable under some conditions (RESULTS, right). In the simulation shown, two actin filaments are anchored 100 nm below the membrane with protrusion length $L–L_0$ and separation D. The lightly shaded region under the curve represents the set of thermodynamically accessible states that will lead to filament bundling by the membrane. The darkly shaded region represents a subset of these states that are likely to be accessible in a branched actin network with branches formed every ~100 nm.

These experiments and simulations show that it is energetically favorable for the membrane to have nearby filament-induced deformations merge into a single deformation. Once enough pushing filaments are present in the deformation, they can overcome the tube formation barrier and generate a spike-like protrusion (CONCLUSION).

do not generate enough force individually to deform flat membranes into stable cylindrical tubes, so bundling of multiple parallel filaments with fascin was believed to be necessary to generate sufficient force and stiffness to produce the spike-like projections associated with filopodia. Because formins prevent filament capping and promote continued filament elongation, the absence of capping protein emulated the anti-capping activity of formin.

Because no bundling protein was included in the reconstitution, the only component that could potentially contribute to parallel filament bundling in this scenario was the membrane itself. The reason is due to the energetic trade-off between filament bending and membrane bending. If two actin filaments pushing on the membrane are close enough together, they will merge together if the energetic cost of bending the filaments is less than the energetic gain of having one large membrane deformation rather than two small ones (Box 23.1, RESULTS). Once enough pushing filaments are present in the deformation, they can overcome the tube formation barrier and generate a spike-like protrusion—all without a bundling protein (Box 23.1, CONCLUSION).

23.4.3 INSIGHTS INTO FILOPODIA FORMATION FROM GUV EXPERIMENTS

These membrane reconstitution experiments demonstrate that actin filament protrusions depend intimately on the membrane—in this case its deformability—and not only on the soluble protein constituents. While there is no doubt that other proteins, fascin and formins included, are present on filopodia in live cells and play important roles, the described experiments suggest that membrane-driven filament bundling could contribute to clustering of filaments at early stages of filopodia formation in a manner that is sensitive to membrane tension (Liu et al., 2008). More broadly, this demonstration points to the need to consider the role of a deformable membrane on other protein assembly processes in cells. This reconstituted protein–membrane platform provides an opportunity to explore additional experimental conditions, such as the effect of varying protein concentrations, addition of recycling proteins, and incorporation of asymmetric bilayers. Encapsulation techniques could also be used to load proteins inside of GUVs rather than on the outside, though higher protein concentrations would be needed to generate the same actin networks from the smaller solution volume in the lumen.

23.5 CROWDING-DRIVEN MEMBRANE BENDING

In many cases, membrane bending is generated in the absence of actin filament protrusions. Multiple mechanisms by which protein binding to membranes can drive bending have been identified and described. The next example shows how the diffusion of proteins on the membrane, and resulting collisions, can turn out to be a remarkably powerful bending force.

23.5.1 EXAMPLE: EPSIN1-DRIVEN TUBULATION

Endocytosis is the process by which cells take up nutrients from their surroundings, recycle membrane components, regulate receptor numbers, and it is also often hijacked by pathogens to gain entry to cells (Doherty and McMahon, 2009). The process involves shaping the flat plasma membrane into highly curved transport vesicles, which then traffic to their destination. The energy that has to be overcome to bend a flat membrane is thought to be provided by peripheral proteins that are attached firmly to the membrane. However, in clathrin-mediated endocytosis, the entry route for most receptors, involves a highly complex protein network of over 60 different proteins—many of which bind the inner leaflet of the plasma membrane—which makes it very difficult to understand individual mechanistic steps in the formation of a coated vesicles (Schmid and McMahon, 2007). Investigation of individual proteins with *in vitro* experiments, combined with computational modeling, has provided valuable insight and identified two main mechanisms of how membrane bound proteins can deform a membrane: (i) coats or scaffolds of intrinsically curved proteins can imprint their curvature onto the membrane and (ii) the insertion of amphipathic helices can bend membranes in a wedge-like mechanism (McMahon and Gallop, 2005; Kozlov et al., 2014). To these models, a third mechanism of membrane bending can be added, one that is unspecific to individual proteins, but is a consequence of the crowded environment of cellular membranes (Stachowiak et al., 2012).

23.5.2 METHODS AND FINDINGS

To investigate how individual proteins bend membranes, it is useful to visualize the process by light microscopy. To this aim a GUV-based membrane bending/tubulation assay can be developed. Proteins that have previously been implicated in membrane bending in endocytosis (wild type epsin1-ENTH domain, wtENTH) (Ford et al., 2002), proteins that have been shown to be involved in endocytosis but without assigned membrane bending function [AP180 N-terminal homology domain (AP180-ANTH domain)] (Ford et al., 2001), and proteins that have no connection to either endocytosis or membrane bending (hexa-his Green Fluorescent Protein, his-GFP) can be compared in their capacity to bend membranes. To this end the proteins need to be purified from *E. coli* and chemically labeled with fluorescent dyes for fluorescent visualization. The labeling reactions can be performed using amine-reactive, NHS-ester functionalized dyes, whose concentration need to be adjusted experimentally for each individual protein to achieve a desired labeling ratio of approximately 1:1. After labeling, the proteins can then be incubated with electroformed GUVs containing binding lipids, which enable the proteins to bind to the synthetic membranes (PIP_2 for Epsin1-ENTH and AP180-ANTH or 1,2-dioleoyl-sn-glycero-3-[(*N*-(5-amino-1-carboxypentyl) iminodiacetic acid)succinyl] (nickel salt) (DOGS-Ni-NTA) lipids, and his-GFP). Protein density (and the extent of protein crowding) on the membranes can be controlled by varying the binding lipid concentration from 0 to 20 mol%. Because GUV yield decreases dramatically with the incorporation of high percentages of charged lipids, a practical trick can be applied to reach PIP_2 or DOGS-Ni-NTA concentrations of up to 20 mol%. One can choose a lipid composition that yields GUVs which phase separate (see also Chapter 18) at room temperature into a liquid-ordered majority phase and a small liquid-disordered domain. The composition can be tuned such that the binding lipid partitions only into the small liquid-disordered domain. The specificity of partitioning can be verified by protein addition—the protein only binds to the binding lipid and should therefore be absent from the liquid-ordered domain, but should be visualized on the liquid-disordered domain. (Box 23.2,

Box 23.2 Crowding-driven membrane bending (Stachowiak et al., 2012)

COMPONENTS

phase-separated GUVs

liquid disordered
DPhPC
Cholesterol
PIP_2 / DOGS-NiNTA

liquid ordered
DPPC
Cholesterol
Lipid dye

purified proteins

Helix0
helix inserting protein
(wt ENTH, wt Sar1)

6x-His
Ni-NTA lipid binding protein
(His-ENTH, His-GFP)

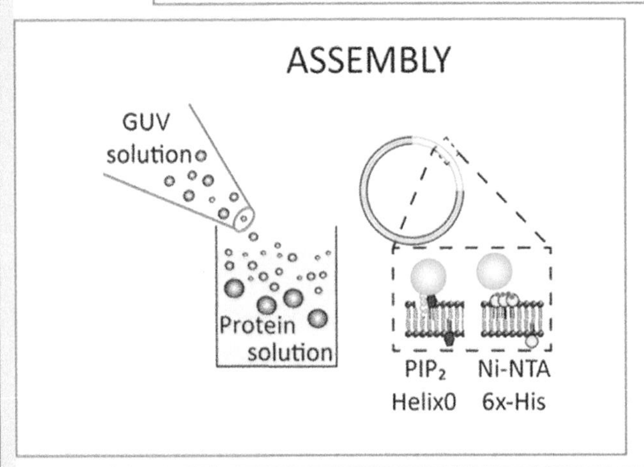

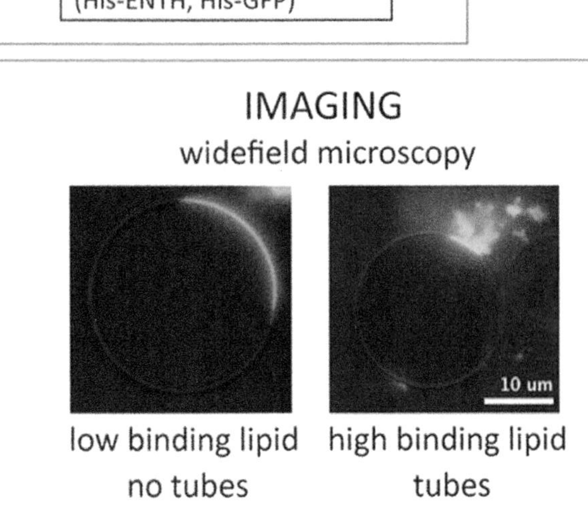

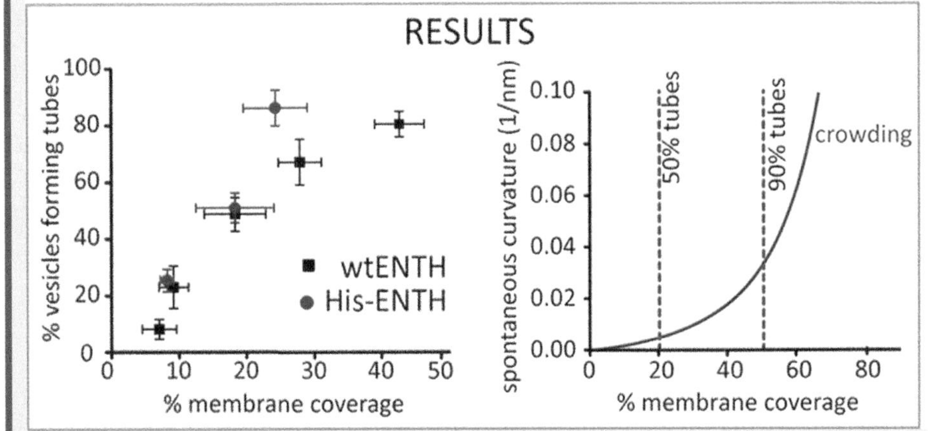

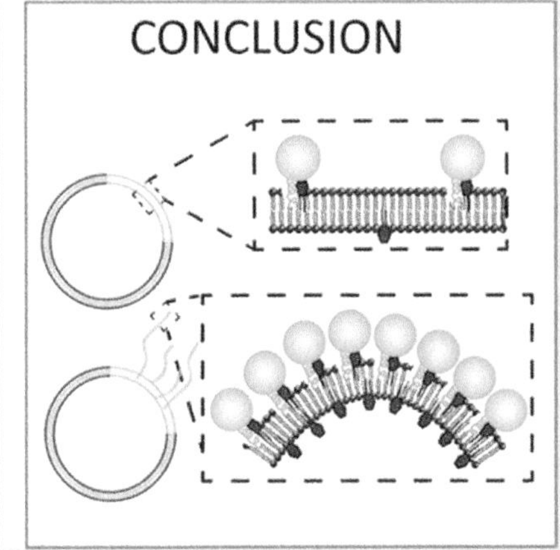

Electroformed, phase-separated GUVs are incubated with purified proteins. For binding of His-tagged proteins DOGS-Ni-NTA lipids are incorporated into the GUVs and wild-type proteins bind to PIP_2 (COMPONENTS). Proteins are introduced in solution on the outside of the GUV where they interacted with the binding lipid in the liquid disordered domain (ASSEMBLY).

Confocal cross sections of GUVs containing low or high (IMAGING, left vs. right) concentration of PtdIns(4,5)P_2. High binding lipid leads to high Epsin1 ENTH concentration on the membrane, resulting in tubulation of the membrane (red channel, lipid dye; green channel, Atto488 wtENTH).

Data analysis and quantification allows for correlation of the frequency of tubulated vesicles with protein coverage on the membranes (RESULTS, left). Protein coverage is quantified using a FRET-FLIM method that is not altered by membrane deformations (Stachowiak et al., 2012). Mathematical modeling helps to explain how high protein density can drive membrane bending. Predictions of spontaneous curvature (1/radius) are plotted as a function of coverage for the protein–protein crowding model (RESULTS, right, blue line). The red dashed lines represent the coverage for which 50% and 90% of vesicles form tubules in experiments with wtENTH. It can be concluded that diffusing, colliding proteins on a membrane create a lateral pressure big enough to bend a membrane (CONCLUSION).

COMPONENTS and ASSEMBLY). After protein binding, fluctuating tubules emerge from the protein–covered domain, a clear indication of membrane bending (Box 23.2, IMAGING). The frequency of tubulation can be determined by counting the tubulated GUVs and this number can be correlated with the amount of binding lipid in the domain. In addition, FLIM-FRET measurements determine the protein coverage on the membranes directly, and identify the density/coverage threshold at which the proteins induce tubulation. Last, mathematical modeling can be used to explain how high protein density can drive membrane bending. Estimating the pressure generated by colliding proteins using the Carnahan-Starling equation of state (Carnahan, 1969) for hard discs and balancing it against the mechanical resistance of the membrane to bending (Box 23.2, RESULTS) reinforces the experimental results. It predicts that 20%–50% protein coverage is sufficient to create curvatures resembling endocytic vesicles, independent of which protein is used or how the protein is attached to the membrane. It can be concluded: high local protein concentration can induce membrane bending! (Box 23.2, CONCLUSION).

23.5.3 INSIGHTS INTO ENDOCYTOSIS FROM GUV EXPERIMENTS

The main take home message from this reconstitution example is the general idea that diffusing membrane-bound proteins collide, and these collisions create a lateral pressure big enough to bend a membrane. In other words, proteins that attach to the membrane take up space, leading to a decrease in the free area that each protein can explore. At a certain protein density (surprisingly only 20% coverage!), protein collisions on the membrane lead to high enough pressures to overcome the energetic cost of membrane bending, effectively introducing a spontaneous curvature. The membrane bends to increase the area available for adhered proteins, thereby reducing the pressure between the colliding proteins. Proteins not previously implicated in membrane bending like AP180-ANTH or even GFP are able to induce membrane curvature when attached tightly to the membranes at high enough densities. Curvature inducing proteins such as Epsin1 and Sar1 are still able to bend membranes even after the putative membrane-bending helix is removed, which challenges the previously described helix-insertion mechanism of proteins.

This work is an example of how reconstitution experiments can reveal simple physical principles that must underlie biological mechanisms. Cellular membranes are densely covered in proteins, trans-membrane and peripheral, and it is tempting to speculate how lateral pressure in this crowded environment could influence membrane shape in cells.

23.6 SIZE-DEPENDENT PROTEIN SORTING AT MEMBRANE INTERFACES

The previous sections discussed two different scenarios whereby protein polymerization or the crowded protein landscape on membranes influenced membrane behavior. In this section, the opposite scenario will be explored, namely how the presence of lipid bilayers influences protein density on membranes and their localization. Specifically the physical underpinnings of how proteins are sorted at membrane interfaces will be described, and how their segregation is dependent on the relationship between their own size and the gap between the two apposing membranes forming the interface.

Biological membranes are frequently in close contact. Intracellular contact sites are formed between the ER and the plasma membrane, mitochondria or endosomes, and intracellular vesicles establish an adhesion (docking) site with their target membrane prior to fusion (Martens and McMahon, 2008; Helle et al., 2013; Kornmann, 2013). The extracellular leaflets of the plasma membrane also often come into close contact when cells adhere to each other to form tissues or to communicate from one to another (Bell, 1978; Rochlin et al., 2010). These interfaces are established by adhesion proteins, such as receptor/ligand pairs, on the apposing membranes, and the interface distance is set by the size of these adhesions. It has been observed that proteins not partaking directly in the adhesion formation undergo specific sorting behaviors, which influence cell-signaling (Adams et al., 1998; Grakoui et al., 1999; Goodridge et al., 2011).

23.6.1 EXAMPLE: CD45 EXCLUSION

One well-studied example of a cell–cell interface is the immunological synapse (Dustin, 2002) whereby antigen-presenting cells interact with T cells by forming a bond between peptide-bound major histocompatibility complexes (pMHCs) and T-cell receptors (TCRs). The initial adhesion is followed by a maturation step, in which the protein landscape undergoes significant reorganization. Most strikingly, the transmembrane phosphatase CD45, which has a large extracellular domain, becomes excluded from the pMHC/TCR interface, permitting stable TCR phosphorylation, which in turn leads to T-cell activation (Bunnell, 2002; Varma et al., 2006; James and Vale, 2012).

Multiple mechanisms have been suggested to be involved in the organization of proteins at membrane interfaces, including the idea of size-dependent protein segregation. CD45 isoforms can have extended conformations that are 15–40 nm larger than the space between apposing membranes established by pMHC/TCR pairs at the membrane interface. For the CD45 to reside in this short interface space, the flexible membrane needs to bend, which is energetically unfavorable and may lead to protein exclusion. This hypothesis can be tested in a purely synthetic membrane interface system using GUVs.

23.6.2 METHODS AND FINDINGS

To experimentally isolate the role of protein size on segregation, electroformed GUVs decorated with synthetic adhesion proteins (which are here called binding proteins or BP) or nonadhesion proteins (nonbinding proteins, NBP) can be generated. *Practical note*: GFPuv (Phillips, 1997) is an anti-parallel homodimer with a Kd = 20–100 μM, which is similar to physiological Kds of TCR–pMHC interactions and, therefore, a good stand-in for biological adhesion proteins. The structurally similar, but monomeric mCherry can be used as NBP (Schmid et al., 2016). Using mutated, nonfluorescent NBP proteins as modular building blocks, sets of adhesion and nonadhesive proteins of different heights (ranging from ~5 to ~20 nm) with constant lateral footprints (4 nm^2) can be generated. The membrane-proximal building block is engineered with a deca-His-tag for membrane attachment (Box 23.3, COMPONENTS).

All proteins are expressed and purified with an N-terminal deca-His tag, enabling fluid protein attachment to DOGS-Ni-NTA lipids onto the synthetic GUV membranes. Incubation of GUVs with BP led to interface formation between GUVs (Box 23.3, ASSEMBLY). To quantify the relative proportion of proteins at membrane interfaces containing both BPs and NBPs, confocal images at the GUV equator (Box 23.3, IMAGING) are taken. To describe the distribution of BP and NBP proteins on the vesicles, an "enrichment index" (EI)—the intensity ratio between the interface (I) and the sum of the individual vesicle intensities ($V_1 + V_2$)—can be calculated. For this the fluorescence intensity of GFP or mCherry is measured along a line bisecting the GUV-GUV pair. An intensity trace across the length of the linescan can be extracted from the confocal images after background subtraction. The three intensity peaks, corresponding to each of the two vesicles and the vesicle interface, are then used to compute the EI.

To determine the interface distance established by different lengths adhesion proteins, RICM between a GUV and a supported lipid bilayer can be used, and protein densities can be measured by FCS. To generalize and extend the obtained experimental results Monte Carlo Simulations are useful to resolve basic features of fluctuating protein arrangements and membrane topography. Such a reduced microscopic model of protein segregation at membrane interfaces can address both the properties identified as important in the experimental work, as well as features that are not easily accessible experimentally. The simulations in this example show that changes in binding protein affinity determine the maximum inclusion of nonbinding proteins, as binding proteins with high affinity enrich at the interface, subsequently laterally crowd out and exclude nonbinding proteins. In addition, the model reveals that changes in membrane bending rigidity have only a modest effect on protein segregation (Box 23.3, RESULTS).

It can be concluded that protein size (both their height and the lateral space they take up) is a highly effective means of altering local protein concentrations at membrane interfaces. Nanometer-scale changes in the height of nonadhesive proteins can dramatically change their densities at membrane interfaces, to avoid membrane bending. Membrane fluctuations further alter the interface gap size on a short time scale, leading to an added small effect in protein exclusion. Last, protein crowding at the interface is a critical factor for exclusion. As adhesion protein density goes up with maturation of the adhesion, nonadhesive proteins are pushed out of the interface, even when the vertical interface gap size is not limiting (Box 23.3, CONCLUSION).

23.6.3 INSIGHTS INTO T-CELL ACTIVATION FROM GUV EXPERIMENTS

These reconstitution study supports the "kinetic segregation model" that was postulated for protein sorting at the immunological synapse (van der Merwe, 2000). It shows how protein height influences protein segregation, but goes beyond that by describing the importance of the lateral space competition on the membrane. Furthermore, the reconstitution approach combined with mathematical modeling also allows for taking physical membrane characteristics such as rigidity and membrane fluctuations into account, something that is very hard to manipulate in cells.

Maybe the most striking finding of this work is that protein segregation at interfaces is surprisingly sensitive to changes in size of the protein. An only 2–5 nm increase in protein height leads to full exclusion of the protein from the interface! The described findings should hold true for all membrane–membrane interfaces between and WITHIN cells. Moreover, they open up even more general questions of protein size at membrane surfaces. It is striking to note that higher organisms like mice and humans have a significantly higher diversity of extracellular membrane protein sizes, with extracellular domains ranging from 20 to 250 nm, compared with single cell organisms (Teichmann and Chothia, 2000; Vogel, 2003). Has extracellular protein size increased in parallel with the increasing complexity of membrane junctions during the evolution of multicellular organisms to allow for more complex signaling pathways? These questions will need a more integrated approach, bringing together biophysics with bioinformatics and cell biology.

23.7 SUMMARY AND OUTLOOK

This chapter introduced the basics of membrane reconstitution with peripheral membrane proteins and described examples of the fundamental mechanisms that can be revealed by studying membrane–protein interactions *in vitro*. Membrane reconstitutions are by design incomplete—otherwise they would be as complex and difficult to interpret as the cellular process that inspires them—but they expose the fundamental physical and biochemical principles that are at work in the reconstituted system. Those principles can be harnessed by cells or they can be suppressed by cells, but they cannot be ignored.

A bottom-up approach to understanding membrane processes will most likely remain a powerful tool for understanding how biological systems are built. In the future, such bottom-up strategies may even help to develop strategies for therapeutic intervention. The latter goal, however, will require additional membrane-reconstitution capabilities that to date are still difficult to achieve. A few common challenges to routine techniques are brought about by the physiological composition of cellular membranes (high percentage of charged lipids, asymmetric lipid distribution), the desire to encapsulate biological content within membranous vesicles, reliable size control of these containers, and last but not least the oriented incorporation of transmembrane proteins. While efforts to provide solutions to each of these challenges are underway (Limozin et al., 2003; Pautot et al., 2003; Tan et al., 2006; Richmond et al., 2011; Dezi et al., 2013; Kang et al., 2013), see also Chapter 3, an all-in-one synthetic-cell generator is still a dream for the future. But one worth dreaming.

Box 23.3 Size-dependent protein sorting at membrane interfaces (Schmid et al., 2016)

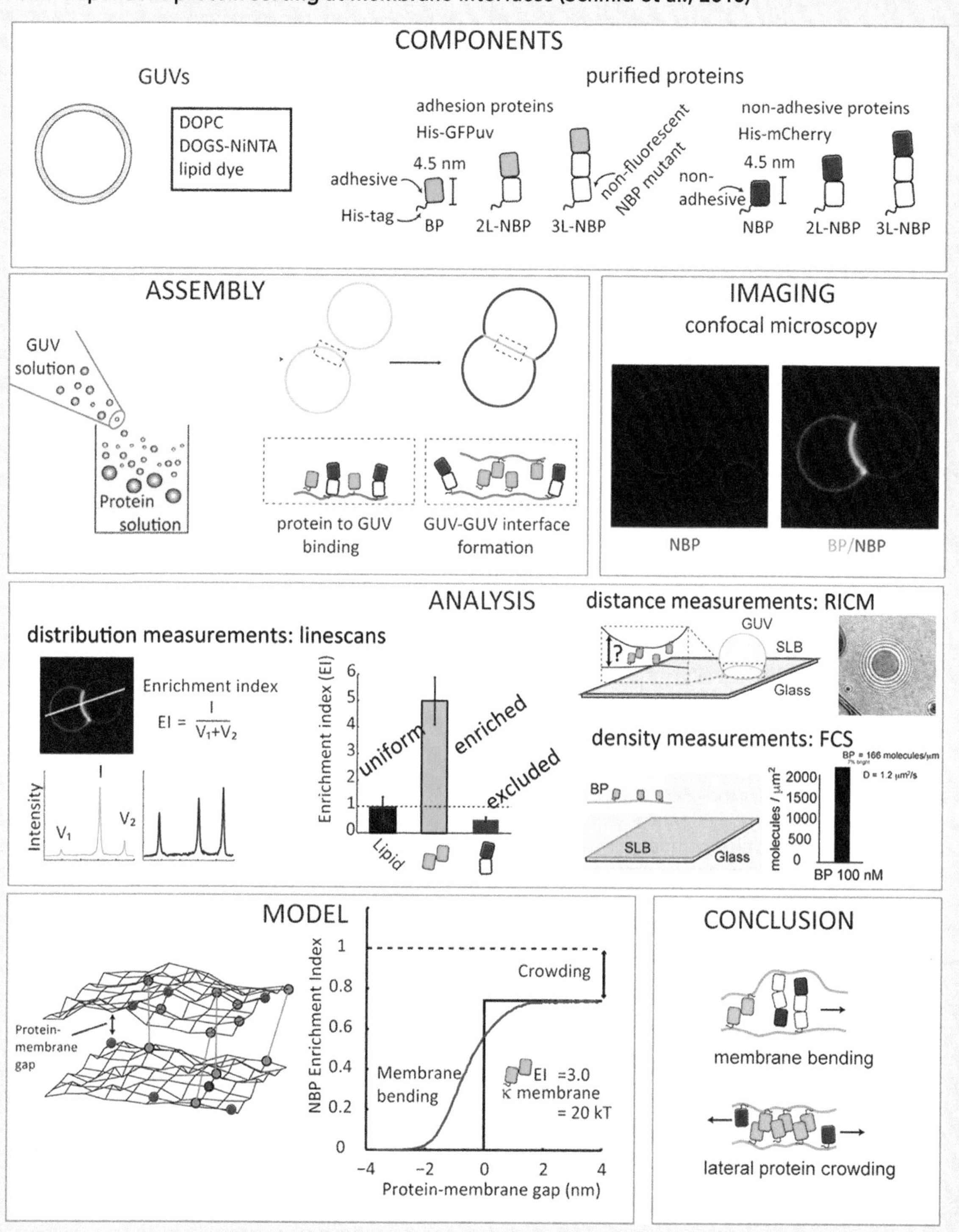

Electroformed GUVs containing the binding lipid DOGS-Ni-NTA are incubated with purified proteins. A synthetic tool-kit of binding (BP) and nonbinding (NBP) proteins of different sizes can be generated (COMPONENTS).

Proteins are introduced in solution on the outside of the GUV where they interact with the binding lipid (ASSEMBLY). Membrane interfaces are formed after the synthetic proteins bind to GUVs that then come into contact, which can be monitored fluorescently (IMAGING).

Linescans through vesicles dimers allow for quantification of fluorescence intensity at outside vesicle membranes (V_1 and V_2) and interface (I). An "Enrichment index" (EI) can be calculated by taking the ratio between (I) and the sum of (V_1 and V_2) and reveal uniform distribution of the fluorescently labeled lipid (EI = 1), enrichment of BP (EI > 1) and exclusion of 2L-NBP (EI < 1) at the interface. Simulation of protein enrichment for a nonbinding protein as a function of the gap between the protein and the apposing membrane (protein–membrane gap) characterizes mechanisms of protein sorting at membrane interfaces (RESULTS, right). Membrane bending and lateral protein crowding at interfaces leads to NBP sorting at interfaces (CONCLUSION).

LIST OF ABBREVIATIONS

AFM	atomic force microscopy
ANTH domain	AP180 *N*-terminal homology domain
Arp2/3	seven-subunit protein complex regulates the actin cytoskeleton
BAR domain	bin-amphyphysin rvs domain
BP	binding protein
CD45	a transmembrane phosphatase
CP	capping protein
DOGS Ni-NTA	1,2-dioleoyl-*sn*-glycero-3-[(*N*-(5-amino-1-carboxypentyl)iminodiacetic acid) succinyl] (nickel salt)
DOPC	1,2-dioleoyl-*sn*-glycero-3-phosphocholine
DOPS	1,2-dioleoyl-*sn*-glycero-3-phospho-*L*-serine
DPPC	1,2-dipalmitoyl-*sn*-glycero-3-phosphocholine
DPhPC	1,2-diphytanoyl-*sn*-glycero-3-phosphocholine
EDTA	ethylenediaminetetraacetic acid
EggPC	L-α-phosphatidylcholine from chicken egg
ENTH domain	epsin N-terminal homology domain
ER	endoplasmic reticulum
FCS	fluorescence correlation spectroscopy
FLIM-FRET	fluorescence lifetime imaging microscopy combined with Förster resonance energy transfer
FRAP	fluorescence recovery after photobleaching
GFP	green fluorescent protein
IS	immunological synapse
NBP	nonbinding protein
N-WASP	a protein complex that regulates actin polymerization
PIP_2 or $PI(4,5)P_2$	phosphatidylinositol 4,5-bisphosphate
PH domain	pleckstin homology domain
pMHC	peptide major histocompatibility complex
RICM	reflection interference contrast microscopy
TCR	T-cell receptor

REFERENCES

Adams, C.L., Chen, Y.T., Smith, S.J., and Nelson, W.J. (1998). Mechanisms of epithelial cell-cell adhesion and cell compaction revealed by high-resolution tracking of E-cadherin-green fluorescent protein. *J. Cell Biol. 142*, 1105–1119.

Adams, J.C. (2004). Roles of fascin in cell adhesion and motility. *Curr. Opin. Cell Biol. 16*, 590–596.

Angelova, M.I., and Dimitrov, D.S. (1986). Liposome electroformation. *Faraday Discuss. Chem. Soc. 81*, 303.

Antonny, B. (2011). Mechanisms of membrane curvature sensing. *Annu. Rev. Biochem. 80*, 101–123.

Basit, H., Lopez, S.G., and Keyes, T.E. (2014). Fluorescence correlation and lifetime correlation spectroscopy applied to the study of supported lipid bilayer models of the cell membrane. *Methods 68*, 286–299.

Bell, G.I. (1978). Models for the specific adhesion of cells to cells. *Science 200*, 618–627.

Bunnell, S.C. (2002). Determining the destiny of NF-kappa B after TCR ligation: It's CARMA1. *Mol. Interv. 2*, 356–360.

Campelo, F., McMahon, H.T., and Kozlov, M.M. (2008). The hydrophobic insertion mechanism of membrane curvature generation by proteins. *Biophys. J. 95*, 2325–2339.

Carnahan, N.F. (1969). Equation of state for nonattracting rigid spheres. *J. Chem. Phys. 51*, 635–636.

De Los Santos, C., Chang, C.-W., Mycek, M.-A., and Cardullo, R.A. (2015). FRAP, FLIM, and FRET: Detection and analysis of cellular dynamics on a molecular scale using fluorescence microscopy. *Mol. Reprod. Dev. 82*, 587–604.

Dezi, M., Di Cicco, A., Bassereau, P., and Lévy, D. (2013). Detergent-mediated incorporation of transmembrane proteins in giant unilamellar vesicles with controlled physiological contents. *Proc. Natl. Acad. Sci. 110*, 7276–7281.

Doherty, G.J., and McMahon, H.T. (2009). Mechanisms of endocytosis. *Annu. Rev. Biochem. 78*, 857–902.

Dustin, M.L. (2002). The immunological synapse. *Arthritis Res. Ther. 4*, S119–S125.

Engelman, D.M. (2005). Membranes are more mosaic than fluid. *Nature 438*, 578–580.

Ford, M.G., Pearse, B.M., Higgins, M.K., Vallis, Y., Owen, D.J., Gibson, A., Hopkins, C.R., Evans, P.R., and McMahon, H.T. (2001). Simultaneous binding of PtdIns(4,5)P_2 and clathrin by AP180 in the nucleation of clathrin lattices on membranes. *Science 291*, 1051–1055.

Ford, M.G.J., Mills, I.G., Peter, B.J., Vallis, Y., Praefcke, G.J.K., Evans, P.R., and McMahon, H.T. (2002). Curvature of clathrin-coated pits driven by epsin. *Nature 419*, 361–366.

Glauert, A.M. (1968). Electron microscopy of lipids and membranes. *J. R. Microsc. Soc. 88*, 49–70.

Goode, B.L., and Eck, M.J. (2007). Mechanism and function of formins in the control of actin assembly. *Annu. Rev. Biochem. 76*, 593–627.

Goodridge, H.S., Reyes, C.N., Becker, C.A., Katsumoto, T.R., Ma, J., Wolf, A.J., Bose, N. et al. (2011). Activation of the innate immune receptor Dectin-1 upon formation of a "phagocytic synapse." *Nature 472*, 471–475.

Gorter, E., and Grendel, F. (1925). On bimolecular layers of lipoids on the chromocytes of the blood. *J. Exp. Med. 41*, 439–443.

Grakoui, A., Bromley, S.K., Sumen, C., Davis, M.M., Shaw, A.S., Allen, P.M., and Dustin, M.L. (1999). The immunological synapse: A molecular machine controlling T cell activation. *Science 285*, 221–227.

Groves, J.T., Parthasarathy, R., and Forstner, M.B. (2008). Fluorescence imaging of membrane dynamics. *Annu. Rev. Biomed. Eng. 10*, 311–338.

Gupton, S.L., and Gertler, F.B. (2007). Filopodia: The fingers that do the walking. *Sci. Signal Transduct. Knowl. Environ. 2007*, re5–re5.

Helfrich, W. (1973). Elastic properties of lipid bilayers: Theory and possible experiments. *Z. Für Naturforschung Teil C Biochem. Biophys. Biol. Virol. 28*, 693–703.

Helle, S.C.J., Kanfer, G., Kolar, K., Lang, A., Michel, A.H., and Kornmann, B. (2013). Organization and function of membrane contact sites. *BBA Mol. Cell Res. 1833*, 2526–2541.

Horwitz, A.R., and Parsons, J.T. (1999). Cell migration–movin' on. *Science 286*, 1102–1103.

James, J.R., and Vale, R.D. (2012). Supplement to biophysical mechanism of T-cell receptor triggering in a reconstituted system. *Nature 487*, 64–69.

Janmey, P.A., and Kinnunen, P.K.J. (2006). Biophysical properties of lipids and dynamic membranes. *Trends Cell Biol. 16*, 538–546.

Kang, Y.J., Wostein, H.S., and Majd, S. (2013). A simple and versatile method for the formation of arrays of giant vesicles with controlled size and composition. *Adv. Mater. 25*, 6834–6838.

Kent, M.S., Yim, H., Sasaki, D.Y., Satija, S., Majewski, J., and Gog, T. (2004). Analysis of myoglobin adsorption to Cu(II)-IDA and Ni(II)-IDA functionalized Langmuir monolayers by grazing incidence neutron and X-ray techniques. *Langmuir 20*, 2819–2829.

Keren, K. (2011). Cell motility: The integrating role of the plasma membrane. *Eur. Biophys. J. 40*, 1013–1027.

Kornmann, B. (2013). The molecular hug between the ER and the mitochondria. *Curr. Opin. Cell Biol. 25*, 443–448.

Kozlov, M.M., Campelo, F., Liska, N., Chernomordik, L.V., Marrink, S.J., and McMahon, H.T. (2014). Mechanisms shaping cell membranes. *Curr. Opin. Cell Biol. 29*, 53–60.

Lagny, T.J., and Bassereau, P. (2015). Bioinspired membrane-based systems for a physical approach of cell organization and dynamics: Usefulness and limitations. *Interface Focus 5*, 20150038.

Lauffenburger, D.A., and Horwitz, A.F. (1996). Cell migration: A physically integrated molecular process. *Cell 84*, 359–369.

Lemmon, M.A. (2008). Membrane recognition by phospholipid-binding domains. *Nat. Rev. Mol. Cell Biol. 9*, 99–111.

Li, S., Guan, J.-L., and Chien, S. (2005). Biochemistry and biomechanics of cell motility. *Annu. Rev. Biomed. Eng. 7*, 105–150.

Limozin, L., Bärmann, M., and Sackmann, E. (2003). On the organization of self-assembled actin networks in giant vesicles. *Eur. Phys. J. E Soft Matter 10*, 319–330.

Liu, A., and Fletcher, D. (2009). Biology under construction: In vitro reconstitution of cellular function. *Nat. Rev. Mol. Cell Biol. 10*, 644.

Liu, A., Richmond, D., Maibaum, L., Pronk, S., Geissler, P., and Fletcher, D. (2008). Membrane-induced bundling of actin filaments. *Nat. Phys. 4*, 789–793.

Liu, A.P., and Fletcher, D.A. (2006). Actin polymerization serves as a membrane domain switch in model lipid bilayers. *Biophys. J. 91*, 4064–4070.

Loisel, T.P., Boujemaa, R., Pantaloni, D., and Carlier, M.-F. (1999). Reconstitution of actin-based motility of Listeria and Shigella using pure proteins. *Nature 401*, 613–616.

Loose, M., and Schwille, P. (2009). Biomimetic membrane systems to study cellular organization. *J. Struct. Biol. 168*, 143–151.

Martens, S., and McMahon, H.T. (2008). Mechanisms of membrane fusion: Disparate players and common principles. *Nat. Rev. Mol. Cell Biol. 9*, 543–556.

Mattila, P.K., and Lappalainen, P. (2008). Filopodia: Molecular architecture and cellular functions. *Nat. Rev. Mol. Cell Biol. 9*, 446–454.

McLaughlin, S., and Murray, D. (2005). Plasma membrane phosphoinositide organization by protein electrostatics. *Nature 438*, 605–611.

McMahon, H.T., and Gallop, J.L. (2005). Membrane curvature and mechanisms of dynamic cell membrane remodelling. *Nature 438*, 590–596.

Monzel, C., Fenz, S.F., Merkel, R., and Sengupta, K. (2009). Probing biomembrane dynamics by dual-wavelength reflection interference contrast microscopy. *Chemphyschem 10*, 2828–2838.

Mulgrew-Nesbitt, A., Diraviyam, K., Wang, J., Singh, S., Murray, P., Li, Z., Rogers, L., Mirkovic, N., and Murray, D. (2006). The role of electrostatics in protein-membrane interactions. *Biochim. Biophys. Acta 1761*, 812–826.

Nadolski, M.J., and Linder, M.E. (2007). Protein lipidation. *FEBS J. 274*, 5202–5210.

Nye, J.A., and Groves, J.T. (2008). Kinetic control of histidine-tagged protein surface density on supported lipid bilayers. *Langmuir 24*, 4145–4149.

Parthasarathy, R., and Groves, J.T. (2004). Optical techniques for imaging membrane topography. *Cell Biochem. Biophys. 41*, 391–414.

Pautot, S., Frisken, B., and Weitz, D. (2003). Production of unilamellar vesicles using an inverted emulsion. *Langmuir 19*, 2870–2879.

Phillips, G.N. (1997). Structure and dynamics of green fluorescent protein. *Curr. Opin. Struct. Biol. 7*, 821–827.

Pollard, T.D., and Borisy, G.G. (2003). Cellular motility driven by assembly and disassembly of actin filaments. *Cell 112*, 453–465.

Richmond, D.L., Schmid, E.M., Martens, S., Stachowiak, J.C., Liska, N., and Fletcher, D.A. (2011). Forming giant vesicles with controlled membrane composition, asymmetry, and contents. *Proc. Natl. Acad. Sci. USA 108*, 9431–9436.

Ridley, A.J., Schwartz, M.A., Burridge, K., Firtel, R.A., Ginsberg, M.H., Borisy, G., Parsons, J.T., and Horwitz, A.R. (2003). Cell migration: Integrating signals from front to back. *Science 302*, 1704–1709.

Ries, J., and Schwille, P. (2012). Fluorescence correlation spectroscopy. *BioEssays 34*, 361–368.

Robertson, J.D. (1981). Membrane structure. *J. Cell Biol. 91*, 189s–204s.

Rochlin, K., Yu, S., Roy, S., and Baylies, M.K. (2010). Myoblast fusion: When it takes more to make one. *Dev. Biol. 341*, 66–83.

Schmid, E.M., Bakalar, M.H., Choudhuri, K., Weichsel, J., Ann, H.S., Geissler, P.L., Dustin, M.L., and Fletcher, D.A. (2016). Size-dependent protein segregation at membrane interfaces. *Nat. Phys. 12*, 704–711.

Schmid, E.M., and McMahon, H.T. (2007). Integrating molecular and network biology to decode endocytosis. *Nature 448*, 883–888.

Schmid, E.M., Richmond, D.L., and Fletcher, D.A. (2015). Reconstitution of proteins on electroformed giant unilamellar vesicles. *Method Cell Biol. 128*, 319–338.

Singer, S.J., and Nicolson, G.L. (1972). The fluid mosaic model of the structure of cell membranes. *Science 175*, 720–731.

Sorre, B., Callan-Jones, A., Manzi, J., Goud, B., Prost, J., Bassereau, P., and Roux, A. (2012). Nature of curvature coupling of amphiphysin with membranes depends on its bound density. *Proc. Natl. Acad. Sci. USA 109*, 173–178.

Stachowiak, J.C., Schmid, E.M., Ryan, C.J., Ann, H.S., Sasaki, D.Y., Sherman, M.B., Geissler, P.L., Fletcher, D.A., and Hayden, C.C. (2012). Membrane bending by protein–protein crowding. *Nat. Cell Biol. 14*, 944–949.

Svitkina, T.M., Bulanova, E.A., Chaga, O.Y., Vignjevic, D.M., Kojima, S., Vasiliev, J.M., and Borisy, G.G. (2003). Mechanism of filopodia initiation by reorganization of a dendritic network. *J. Cell Biol. 160*, 409–421.

Tan, Y.-C., Hettiarachchi, K., Siu, M., Pan, Y.-R., and Lee, A.P. (2006). Controlled microfluidic encapsulation of cells, proteins, and microbeads in lipid vesicles. *J. Am. Chem. Soc. 128*, 5656–5658.

Teichmann, S.A., Chothia, C. (2000). Immunoglobulin superfamily proteins in Caenorhabditis elegans. *J. Mol. Biol. 296(5)*,1367–1383.

Upadhyaya, A., and van Oudenaarden, A. (2003). Biomimetic systems for studying actin-based motility. *Curr. Biol. 13*, R734–R744.

van der Merwe, P. Davis, S.J., Shaw, A.S., and Dustin, M.L. (2000). Cytoskeletal polarization and redistribution of cell-surface molecules during T cell antigen recognition. *Semin. Immunol. 12*, 5–21.

Varma, R., Campi, G., Yokosuka, T., Saito, T., and Dustin, M.L. (2006). T cell receptor-proximal signals are sustained in peripheral microclusters and terminated in the central supramolecular activation cluster. *Immunity 25*, 117–127.

Vogel, C., Teichmann, S.A., Chothia, C. (2003). The immunoglobulin superfamily in Drosophila melanogaster and Caenorhabditis elegans and the evolution of complexity. *Development. 130(25)*, 6317–6328.

Zens, B., Sawa-Makarska, J., and Martens, S. (2015). In vitro systems for Atg8 lipidation. *Methods 75*, 37–43.

Effects of antimicrobial peptides and detergents on giant unilamellar vesicles

Karin A. Riske

Tear down the wall!

Pink Floyd
The Wall

Contents

24.1 INTRODUCTORY WORDS

Amphipathic compounds usually have a high affinity for biological membranes, and several are able to (partially or fully) insert into membranes and alter their properties, such as packing, permeability and cohesiveness. Such compounds can therefore be classified as membrane-active agents. Among these are detergents, routinely used to solubilize and extract membrane components, and several natural and bio-inspired synthetic peptides, such as antimicrobial peptides, which exert their activity by interacting in a nonspecific way with biological membranes (i.e., independent of specific binding sites), mainly with the lipid matrix.

The effects of antimicrobial peptides and detergents on membranes have been widely studied using lipid vesicles, mainly small (~100 nm) liposomes, as biomimetic models. Results obtained with small liposomes represent an average over the whole vesicle population, and relevant details might be lost. More recently, giant unilamellar vesicles (GUVs) have also been employed to study the mechanistic details of the interaction of such membrane-active agents with lipid membranes. GUVs offer the unique advantage of being cell-sized objects that can be individually observed and the membrane response followed in real time with optical microscopy. Therefore, their use allows access to spatially and temporally resolved information otherwise inaccessible with conventional bulk assays with small vesicles. In this chapter, experimental protocols using GUVs to investigate the interaction of lipid bilayers with membrane-active agents, focusing on antimicrobial peptides and detergents, are given and discussed. The chapter is organized as follows. First, different experimental protocols relevant for this topic are described and discussed. Then, ways to measure area increase and membrane permeability using GUVs are introduced. Finally, specific aspects related to the activity of antimicrobial peptides and to the solubilization of membranes by detergents are discussed in detail.

24.2 EXPERIMENTAL METHODOLOGY

Membrane-active agents, especially antimicrobial peptides and detergents, are amphipathic molecules usually soluble in water. Thus, the first challenge is to mix them with the GUVs in appropriate conditions. Eventually they insert in and perturb the membrane, usually causing changes in membrane permeability and/or surface area. Here, protocols for assessing the interaction of membrane-active agents with GUVs and for quantifying changes in membrane area and permeability are introduced and discussed.

24.2.1 HOW TO MIX THE MOLECULES WITH THE GUVs

The first challenge to observe the interaction of GUVs with peptides and detergents is how and when to mix them. Ideally, it would be desirable to follow the same GUV, first in the absence of the studied molecules, and then when in contact with a known concentration of the molecules. This cannot be easily and simply accomplished and depends on the available facilities in the lab. The methods most often used are described in Box 24.1.

Box 24.1 Mixing protocols

- *Dilution protocol.* The simplest way to mix the GUVs with the molecules under investigation is to add an aliquot of the GUV suspension to an observation chamber filled with a defined concentration of the molecule. This method can be done in any lab without the need of any special apparatus. Simple observation chambers consist of two coverslips and/or glass slides sandwiching a thin spacer in between sealed with silicon grease. The chamber is filled with the solution to be observed and a small aliquot of the vesicle suspension is added, see figure (the preformed GUVs are indicated as spheres with black contour and gray interior; the shaded regions indicate the solution with the molecule of interest, more specifically peptides and detergents here). Thus, the effects caused as a function of the concentration of the molecules can be directly assessed. The typical active concentrations are usually in the micromolar range for antimicrobial peptides and around the critical micelle concentration (CMC) for detergents. The disadvantage of this method is that the GUVs cannot be imaged before the contact and a few seconds after mixing with the molecules. However, depending on the concentration chosen and the dynamics of the process, one can find a suitable GUV for observation before a substantial (i.e., detectable) effect has been induced on it (for instance, it is still spherical and with preserved contrast) (Apellániz et al., 2010; Domingues et al., 2010a; Wheaten et al., 2013a, 2013b). Alternatively, an aliquot (few microliters) of the molecules can be added to a chamber filled with GUVs (volumes usually around 0.1–1 mL). In that case, a GUV before coming in contact with the molecules can be frequently found, but the effective molecule concentration with time changes by diffusion (and/or convection) and is not precisely known during observation (Ambroggio et al., 2005).

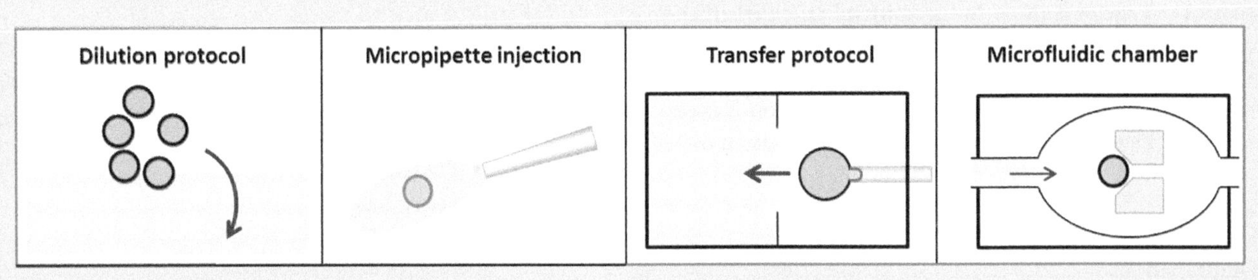

- *Local injection with a micropipette.* Injection with a glass micropipette is one of the preferred ways to add the molecules of interest close to selected GUVs. Micropipettes are easily prepared from glass capillaries using pullers and their diameter is usually around 5–10 μm. They are filled with the solution of interest (few microliters) and attached to an injection system and controlled with a micromanipulator (details on how to make and manipulate micropipettes are given in Section 11.2.1 of Chapter 11). Injection systems can be purchased from Eppendorf and Sutter Instruments, for instance. Typical injected volumes are in the picoliter range with pressures below 100 psi. The advantage of this method is that preselected GUVs can be exposed to the solution coming out of the pipette allowing observation of the effects caused by the molecule of interest on single GUVs. The injection is temporally and locally well-defined and affects only the GUVs close to the pipette, so that the effect on several GUVs can be observed from the beginning to the end in the same chamber. However, control of the injection flux is difficult and the effective concentration of the molecule on the GUV surface is in most cases not well defined, varying not only with time but also spatially around the vesicle. Furthermore, the injection flux can impose a deformation on the GUV membrane as well as a vesicle drift, which can be circumvented by holding the vesicle with a second micropipette connected to a pressure system. Nonetheless, this method usually allows following, at least qualitatively, the dynamic effects caused by membrane-active agents on GUVs (Tamba and Yamazaki, 2005; Cabrera et al., 2011; Alam et al., 2012).
- *Transfer protocol.* Special chambers composed of two compartments with a gated connection between them can be used to aspirate a selected GUV with a micropipette in one compartment and to transfer it to the adjacent compartment containing the molecule of interest. This procedure allows following a single vesicle throughout the whole process with control of the molecule concentration, but requires building specialized chambers and micromanipulating the vesicles (Longo et al., 1998; Mally et al., 2007; Lee et al., 2008, 2011).
- *Microfluidic chamber.* Microfluidic chambers offer the unique possibility to trap GUVs and then quickly and fully exchange the external solution. Thus, the molecule concentration and time after contact are easily controlled. Furthermore, different molecules can be added at different times, in a sequential mode. However, this method requires building chambers with high precision and connecting them with external pumps (Robinson et al., 2013). In addition, the geometry of the posts allows trapping GUVs with certain sizes only and their presence might hinder the imaging of the vesicle shape.

24.2.2 HOW TO QUANTIFY MEMBRANE PERMEABILITY

Most membrane-active agents are able to significantly perturb the membrane with accompanying increase in membrane permeability. Antimicrobial peptides, for instance, generally exert their biological activity by drastically altering membrane permeability (Brogden, 2005). Different mechanisms of permeabilization have been described in the literature, and the two most accepted and best described mechanisms are those of the toroidal pore and the carpet mode, as will be better discussed ahead. When sub-micrometer pores are formed, the membrane becomes permeable but the GUV is basically preserved. Detergents are usually cone-shaped molecules and therefore prefer a micellar structure. Thus, their presence in the bilayer can cause opening of pores at sub-solubilizing conditions, that is, at concentrations below the onset of solubilization, usually below the detergent CMC (Ahyayauch et al., 2010; Mattei et al., 2015).

A detailed description on ways to quantify membrane permeability to different molecules is presented in Chapter 20. Here, we will focus on the use of phase-contrast and confocal microscopy to visualize and quantify increased membrane permeability induced by peptides and detergents. For a detailed description of the microscopy techniques for imaging GUVs, the reader is referred to Chapter 10. For that, the GUVs are usually prepared with asymmetric distribution of marker molecules inside and outside the vesicle compartment. For observation under phase-contrast, GUVs are commonly grown in sucrose solution and dispersed in iso-osmolar glucose solution to create a sugar asymmetry that induces sedimentation of the GUVs onto the chamber bottom and increases their optical contrast due to differences in the refractive indexes of the enclosed and dispersing solutions. GUVs with sugar asymmetry exhibit a large halo effect, a feature of the phase-contrast technique due to spurious bright light around the phase object. The intensity profile across the membrane consists of upward and downward peaks relative to the bright and dark regions around the vesicle (Figure 24.1a, top). The membrane location is usually assigned to the point of the maximal gradient of the intensity profile, that is, around the midpoint between the two peaks. The intensity of the bright/dark peaks depends mainly on the refractive index difference between the two media. The high optical contrast is preserved as long as the membrane remains impermeable. If pores open across the membrane, their size is usually large enough to allow the passage of sugar molecules, and therefore the asymmetry is lost and the vesicle contrast decreases. The contrast loss can be easily quantified from the peak-to-peak height in the intensity profile across the GUV, as shown in Figure 24.1a. This allows following the extent and dynamics of permeabilization.

The use of fluorescence microscopy, and especially confocal microscopy, offers a broad range of possibilities to probe membrane permeability. Aqueous fluorescent probes exist in a wide choice of sizes and colors. Large macromolecules, such as dextran, linked to fluorescent dyes represent probes of high molecular weight. Invitrogen, for instance, offers a long list of fluorophores with different chemical structure, molecular size and spectroscopic properties. Depending on the size of the fluorescent probe used, the pore size can be distinguished (e.g., Ambroggio et al., 2005; Tamba et al., 2010). Small fluorescent dyes (~0.5 kDa) can usually permeate across small hydrophilic pores (around 1 nm). However, larger dyes (10–50 kDa) can only permeate through the membrane if relatively large pores open (of sizes of at least a few nanometers). The probes can be added to the external medium (Figure 24.1b) or be encapsulated in the vesicle lumen (Figure 24.1c). Probes of different sizes/colors can be followed with confocal microscopy and quantified simultaneously. Typical concentrations of aqueous soluble probes that yield good fluorescence microscopy images are around 10 μM, a concentration low enough not to cause any

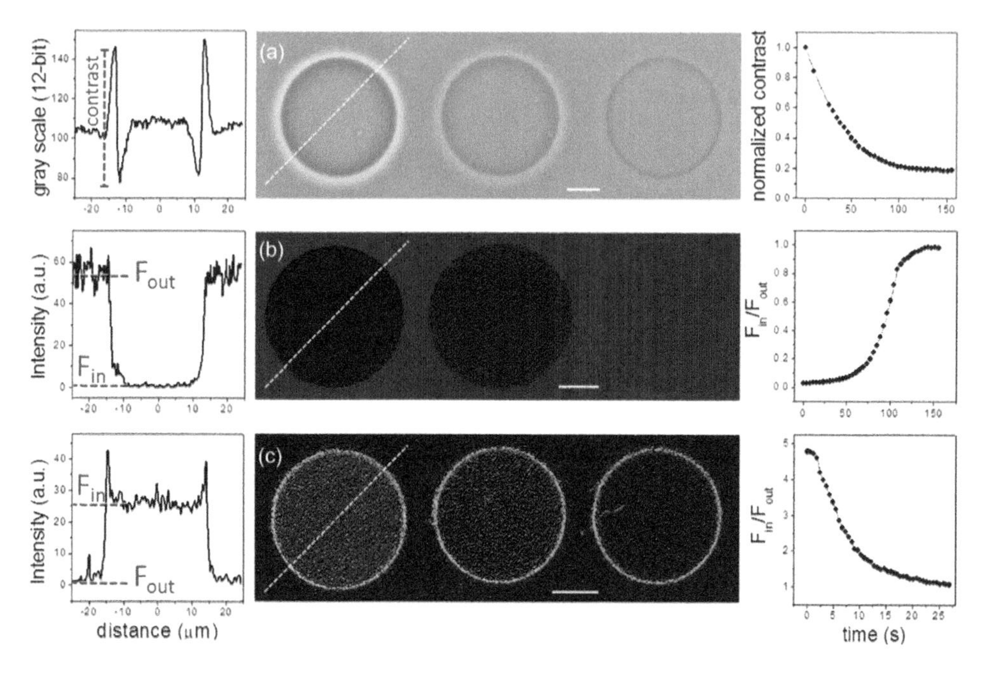

Figure 24.1 Quantification of membrane permeability. Vesicle permeabilization observed with (a) phase-contrast (sucrose inside/glucose outside) and (b, c) confocal microscopy (influx (b) and efflux (c) of fluorescent dyes—red; the membrane is labeled in green). The graphs on the left show the intensity profiles measured along the corresponding dashed yellow lines. The graphs on the right show the time dependence of the parameters defined for each experimental method. Scale bars: 10 μm.

osmotic effects on the GUVs. Both protocols (influx and efflux of fluorescent dyes) are similar and allow quantification of membrane permeability. However, when probes are added to the external medium, different probes (of different sizes/colors) can be added at defined times, to check long-term stability of pores, for instance. These approaches enable estimating the pore size, the permeation kinetics and the permeabilized areas. In addition, by fluorescently labeling the membrane (Figure 24.1c), macropores (with sizes in the micrometer range) can also be distinguished (Lira et al., 2014).

24.2.3 HOW TO MEASURE AREA INCREASE

When amphipathic molecules insert into the membrane, they can cause increase in vesicle surface area. This is particularly the case for detergents, which in some cases can be incorporated in relatively large fractions (Mattei et al., 2015). It should be mentioned, however, that the molecule inserted should be able to flip-flop across the bilayer in order to cause a detectable increase in area, otherwise mainly changes in spontaneous curvature (imbalance between the areas of the two monolayers) will occur (Sudbrack et al., 2011), as will be discussed in Section 24.4 of this chapter. Quantification of the area increase can be directly correlated with the partition coefficient of the molecule, provided the area per molecule is known, as will be shown ahead.

Increase in area at constant volume causes shape changes in GUVs that are mapped in a morphological phase diagram (Döbereiner, 2000, and Chapter 5). However, observation of the vesicle projection or cross section does not reveal the whole three-dimensional vesicle shape, unless one performs an image reconstruction from confocal slices (see Chapter 10), which is difficult to achieve and not always complete. Therefore, vesicle manipulation has to be done to quantify area change. Here, two methods used to quantify vesicle excess area without stretching the membrane at a molecular level are shown. The most commonly used method is to aspirate the GUV with a glass micropipette at controlled (relatively low) suction pressure, so that the relative area change ($\Delta A_{ve} / A_o$) can be calculated from the length of the projected tongue inside the pipette. Chapter 11 provides a complete and detailed discussion on the micropipette manipulation technique (see Section 11.3.2 of that chapter for the area expansivity measurement) and Figure 24.2a highlights the use of the aspiration technique to quantify changes in vesicle area.

Another way to manipulate GUVs and to quantify their area is by deforming them with moderate alternating current (AC) fields (i.e., amplitude of around 200 V/cm and frequency of about 200 kHz), which is experimentally easier compared with aspiration with micropipettes. The effects of electric fields on GUVs are extensively discussed in Chapter 15 and this approach is presented in detail in Section 15.4.1 of that chapter. Briefly, a chamber with electrodes that can be connected to a function generator is required (see Section 15.5 in Chapter 15). Under appropriate conditions (higher solution conductivity inside the vesicle), GUVs with excess area exposed to AC fields deform into a prolate shape. The extent of deformation, quantified by the ratio between the two vesicle semi-axes a / b, depends on the field strength and on the excess area available. Figure 24.2b shows GUVs with different excess area, and therefore extent of deformation, in the presence of an AC field, and how the area increase can be quantified from the aspect ratio a / b.

Other ways to measure area increase have also been reported. One such method is based on adhesion of biotinylated GUVs onto streptavidin-coated substrates (Aoki et al., 2015). By measuring the area of the adhered membrane patch, the molecular area increase can be directly obtained from simple geometrical considerations.

24.3 MODE OF ACTION OF ANTIMICROBIAL PEPTIDES

Antimicrobial peptides are key components of the immune defense system of organisms with the ability to kill a wide spectrum of pathogens, including bacteria and fungi (for reviews the reader is referred to Hwang and Vogel, 1998; Shai, 2002; Zasloff, 2002; Brogden, 2005; Jenssen et al., 2006; Wimley, 2010; Nguyen et al., 2011; Laverty et al., 2011; Sani and Separovic, 2016). This class of molecules is widely diverse in their amino acid sequences and secondary structures. Yet, most antimicrobial peptides are rich in

(a) **Micropipette aspiration**

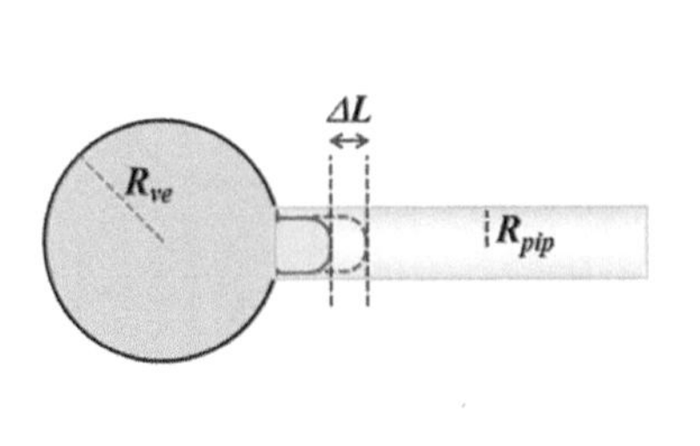

$$\frac{\Delta A_{ve}}{A_o} = 2\pi R_{pip}\left(1 - \frac{R_{pip}}{R_{ve}}\right)\Delta L$$

(b) **AC field deformation**

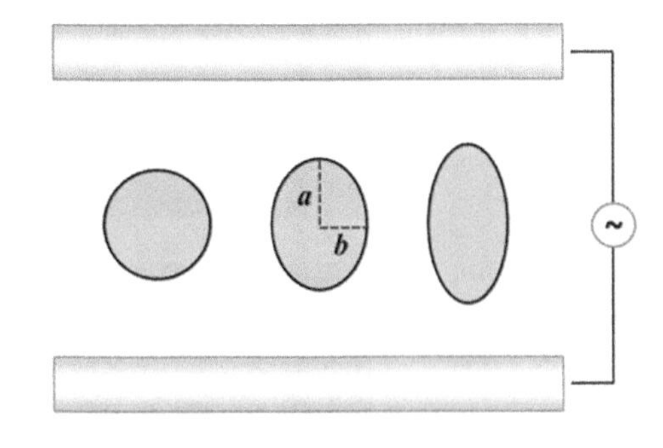

$$\frac{\Delta A_{ve}}{A_o} = \frac{1}{2}\sqrt[3]{\frac{a}{b}}\left(\frac{b}{a} + \frac{\sin^{-1}\varepsilon}{\varepsilon}\right) \qquad \varepsilon^2 = 1 - \left(\frac{b}{a}\right)^2$$

Figure 24.2 Quantification of relative area increase ($\Delta A_{ve} / A_o$, where ΔA_{ve} is the change in vesicle surface area and A_o is the initial vesicle surface area) using (a) the micropipette aspiration technique or (b) prolate deformation under an AC field. The expressions to determine the relative area increase are shown for each method.

cationic and hydrophobic residues and their active conformation has an amphipathic character. These properties warrant them a large affinity for the membranes of microorganisms, which are rich in anionic lipids (Matsuzaki, 1999). The ability of antimicrobial peptides to kill pathogens does not involve ligand-mediated pathways as most conventional antibiotics. They usually affect the integrity of the pathogen membrane barrier through nonspecific interactions with the lipid matrix. Therefore, antimicrobial peptides are a promising class of antibiotic agents in the ever increasing problem of antibiotic resistance.

The structural motifs of antimicrobial peptides are very diverse and a simple structure-activity relationship cannot be envisioned. Many antimicrobial peptides are found in a random conformation in solution but acquire an α-helix structure at the membrane surface. Another important group of antimicrobial peptides are β-sheet peptides, which quite often have their structure stabilized by disulfide bonds in a β-hairpin motif. Finally, there are also cyclic peptides and peptides with an extended structure, which exert their activity without acquiring any conventional secondary structure (these structural motifs can be viewed, e.g., in Nguyen et al., 2011). In common, these conformations are almost always amphipathic, a key feature to facilitate their "interfacial activity" as introduced by W. Wimley: "the ability of a molecule to bind to a membrane, partition into the membrane-water interface, and to alter the packing and organization of the lipids" (Rathinakumar and Wimley, 2008; Wimley, 2010).

Several mechanisms of action have been proposed in the literature to describe how antimicrobial peptides alter the membrane barrier. Even though each antimicrobial peptide interacts in a unique way with their target membranes and their mode of action can be also modulated by membrane composition and relative concentration, the perturbation they cause can be roughly divided into two major effects: pore formation and membrane disintegration (Shai, 2002; Brogden, 2005). In the former, the peptides accumulate on the membrane surface up to a threshold concentration when they insert perpendicularly into the membrane and, usually together with lipids, form the pore rims (Matsuzaki, 1998). Such pores are often called toroidal pores. In the latter mechanism, frequently called carpet mode of action, the peptides cover the membrane surface until the membrane eventually ruptures/disintegrates and in some cases a detergent-like solubilization occurs (Shai and Oren, 2001). Box 24.2 shows a schematic representation of these two membrane-perturbation mechanisms.

The mechanistic details on the mode of action of antimicrobial peptides were mostly obtained from studies using lipid vesicles as biomimetic systems; typically small vesicles with sizes below 1 μm were employed. Such studies have been traditionally performed with several biophysical techniques, such as fluorescence spectroscopy (Schibli et al., 2002), circular dichroism (Wieprecht et al., 1999), nuclear magnetic resonance (Mandard et al., 2002), calorimetry (Seelig, 2004) and X-ray and neutron small angle scattering (Wu et al., 1995), and allow correlation of peptide activity with membrane composition (especially with the fraction of anionic lipids) and quantification of the perturbations caused on the membrane structure.

During the last decade, studies with GUVs started to contribute to the antimicrobial peptide field as well. Observation and micromanipulation of GUVs in the presence of antimicrobial

Box 24.2 Models for the mode of action of antimicrobial peptides

- **Toroidal pore mechanism.** The peptides accumulate on the membrane surface up to a threshold concentration when they insert perpendicularly into the membrane and, usually together with lipids, form the pore rims. The toroidal pore mechanism induces membrane permeabilization but preserves the vesicle entity.
  The permeabilization induced by the toroidal pore mode of action follows either a graded or an all-or-none mechanism, which can be easily distinguished in GUV experiments, as shown below (the two mechanisms are illustrated with phase-contrast microscopy sequences of the effects caused on GUVs; the scale bars represent 10 μm; these sequences are shown in Movies 24.1 and 24.2; movies captions at the end of Chapter).
- **Carpet-like mechanism.** The peptides cover the membrane surface until the membrane eventually ruptures/disintegrates in a detergent-like mode. Membrane rupture is usually accompanied by vesicle collapse/burst.

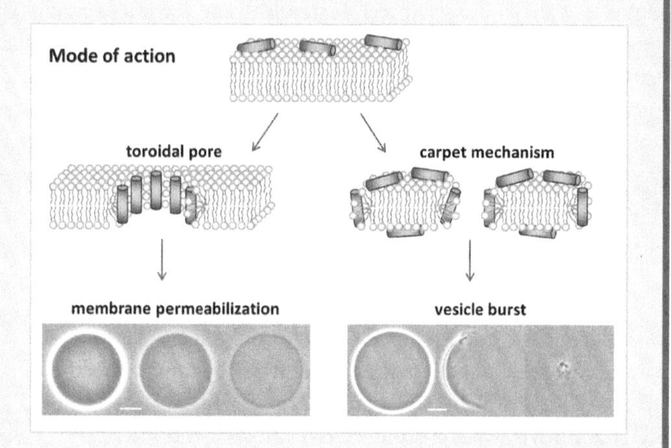

- The two main mechanisms of membrane permeabilization are illustrated below. The sketches show the distribution of dye content in a population of vesicles after addition of the antimicrobial peptide:
  1. **Graded mechanism.** The whole vesicle population has the same average dye content, which gradually changes with time as a unimodal distribution.
  2. **All-or-none mechanism.** Vesicles are either empty or full with dye and the distribution is bimodal.

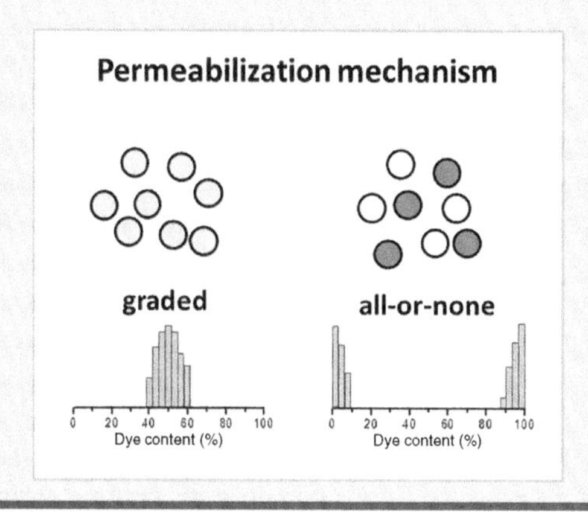

peptides revealed important aspects of the mode of action of such membrane-active molecules, not easily achieved with the aforementioned conventional approaches.

24.3.1 TYPE OF PEPTIDE-INDUCED MEMBRANE PERTURBATION

A very important characteristic of antimicrobial peptides is their lytic ability, which in model systems has been traditionally assessed with fluorescence-based leakage assays from small liposomes (Rex and Schwarz, 1988). However, these conventional leakage assays quantify the total amount of dye released from the internal volume of liposomes, but the result is a bulk average on the entire population. Information at the single vesicle level is lost so that leakage due to vesicle disruption or pore formation, for instance, cannot be distinguished. On the contrary, these two main mechanisms of membrane perturbation can be easily distinguished when observing GUVs: peptides inducing toroidal pores lead to GUVs with permeable membranes, whereas those acting via the carpet mode cause vesicle collapse/burst, as illustrated in Box 24.2 and in the online movies (Movies 24.1 and 24.2; movies captions at the end of chapter). It should be noticed, however, that vesicle burst can be observed also for pore-forming peptides, usually at high peptide concentration. In addition, micropipette injection can occasionally lead to vesicle burst also at moderate concentrations, mainly due to inhomogeneous peptide distribution around the vesicle. On the other hand, some peptides like gomesin (Domingues et al., 2010a) cause only vesicle burst and were not found to induce membrane permeabilization at any condition.

The different modes of action (toroidal pore and carpet-like) were described in one of the first studies on the interaction of antimicrobial peptides with GUVs (Ambroggio et al., 2005). The effects of three different peptides from Australian tree frogs, maculatin 1.1, citropin 1.1 and aurein 1.2, on GUVs composed of 1-palmitoyl-2-oleoyl-sn-glycero-3-phosphocholine (POPC, see Appendix 1 of the book for structure and data on this and other lipids) encapsulating dyes of different sizes (Alexa^{546}-maleimide and Alexa^{488}-dextran, of 1.3 and 10 kDa, respectively) were followed with confocal microscopy. The longer peptide maculatin 1.1 induced opening of pores that allowed the passage of the small dye only and the vesicle structure was preserved. On the other hand, the small peptides citropin 1.1 and aurein 1.2 caused sudden release of both dyes after membrane destabilization and vesicle burst. Therefore, the authors concluded that maculatin 1.1 acts via the toroidal pore formation mechanism, whereas the other two peptides follow the carpet mode of action.

The group of Yamasaki was also among the pioneers to use GUVs to investigate the mode of action of antimicrobial peptides, especially of magainin 2, a peptide first isolated from the skin of the Australian frog *Xenopus laevis* and one of the most studied antimicrobial peptides that acts via the toroidal pore mechanism (Matsuzaki, 1998). In their first work (Tamba and Yamasaki, 2005), they studied the permeabilization effect of magainin 2 on GUVs composed of 1:1 1,2-dioleoyl-sn-glycero-3-phosphocholine/1,2-dioleoyl-sn-glycero-3-phospho-(1′-rac-glycerol) (sodium salt) (DOPC:DOPG) encapsulating the fluorescent probe calcein (0.6 kDa) when exposed to injection of 3–10 μM magainin 2 with a micropipette (see Box 24.1). As expected from a pore-forming peptide, magainin 2 induced membrane permeabilization with preservation of the vesicle structure (see Box 24.2). Leakage of the encapsulated calcein started stochastically and once initiated, it proceeded until full release in less than 1 min, therefore, in an all-or-none permeabilization mechanism (see Box 24.2). By quantifying the fluorescence intensity inside the GUVs, in a similar way as shown in Figure 24.1c, the leakage kinetics was analyzed and was shown to become faster with increasing peptide concentration. In addition, the fraction of intact GUVs was also quantified as a function of time and peptide concentration. In parallel, the permeability of a population of large unilamellar vesicles (LUVs) was also quantified with conventional leakage assays. The single-GUV method revealed important mechanistic details of the peptide–lipid interaction that remain hidden in the bulk experiments with LUVs: leakage proceeds through pores in the membrane and pore formation is the rate limiting step, suggesting that the average leakage percentage measured with LUVs represent the fraction of leaked vesicles rather than the extent of leakage per vesicle. In following studies, the same group used the single-GUV method to gain further insight into the mode of action of magainin 2. By varying the DOPG fraction in the membrane and the magainin 2 concentration in the micropipette, it was shown that the rate of pore formation is determined by the surface concentration of magainin 2, which is modulated by the DOPG fraction in the membrane (Tamba and Yamazaki, 2009). Then, by encapsulating fluorescent probes of different sizes in GUVs of 1:1 DOPC:DOPG, the kinetic pathway of pore formation was also determined (Tamba et al., 2010). It was shown that magainin 2 initially induces opening of large transient pores that allow the passage of all probes, but later on, the pore size decreases to smaller and stable sizes selective for the small probes only. The proposed hypothesis is that accumulation of the peptides on the external monolayer causes an increase in surface tension that is released by opening of large pores (of up to 80 nm radius and scaling with the GUV radius). Then, migration of peptide to the pore rims and to the internal monolayer leads to a new equilibrium in which small pores (~2 nm) are stabilized.

Such behavior has also been reported for the α5 fragment from the proapoptotic protein Baxα5 (Fuertes et al., 2010). Influx of dyes into GUVs (Figure 24.1b) showed again the stochastic nature of the pores first formed, which rapidly proceeded to complete filling of the internal compartment. Then, by serial addition of new dyes, it was determined that long-term pores were smaller and corresponded to an equilibrium state.

Observation of GUVs was used by other groups to distinguish if the peptide-induced flux of dyes across the membrane followed a graded or an all-or-none mechanism (Almeida and Pokorny, 2009), as illustrated in Box 24.2. The type of flux is basically determined by the rate constants of pore opening/closing and dye flux. When pore opening is the rate limiting step, as shown in the example above for magainin 2 and Baxα5 fragment, then an all-or-none mechanism follows: once pores open across the membrane, dye leakage proceeds rapidly until equilibration. If the rate constant of pore formation increases and/or the dye flux becomes slower, then a graded mechanism is eventually observed. These two mechanisms can be discriminated in LUVs experiments with specially designed leakage assays based on requenching of a dye originally encapsulated in the LUVs together with its quencher

(Ladokhin et al., 1995, 1997; Gregory et al., 2008; Almeida and Pokorny, 2009). Clearly, observation of a population of GUVs directly reveals whether a graded or all-or-none mechanism prevails (see Box 24.2). The group of García-Saéz studied the mechanism of permeabilization of two peptides, CpreTM and NpreTM, derived from the HIV fusion glycoprotein gp41 subunit. Initially, with the LUVs requenching assay, they showed that CpreTM induced all-or-none flux, whereas NpreTM followed the graded mechanism (Apellániz et al., 2009). To gain further insight into the differences between these two mechanisms and directly visualize heterogeneities in a vesicle population, GUVs were used in a following study (Apellániz et al., 2010). A fluorescent dye (Alexa Fluor 488, 0.6 kDa) was added to the external medium of GUVs in the presence of the peptides and the degree of GUV filling with time was quantified using confocal microscopy (Figure 24.1b). Usually, membrane permeabilization started after a lag time (up to 20 min) after contact with peptide, evidencing the stochastic nature of the membrane permeabilization process at the single-vesicle level. The kinetic of dye influx was markedly different for both peptides. CpreTM induced a fast entrance of dyes: almost complete filling was reached within ~100 s, in accordance with the all-or-none mechanism. On the other hand, the permeation rate of NpreTM was much slower and only partial filling (~20%–40%) was achieved even after 30 min, validating the graded mechanism for this peptide. By fluorescently labeling both peptides, it was shown that they were both bound to the membrane, but with ensuing different permeabilization mechanisms. To assess the role of cholesterol on the permeabilization mechanism of both peptides, the degree of filling of vesicle populations (up to 500 GUVs) was quantified and revealed that the presence of cholesterol stabilized the NpreTM-induced pores. By adding a second fluorescent probe (Alexa Fluor 555, also 0.6 kDa) after incubation with the peptides, it was shown that the pores opened by CpreTM remain stable and allow entry of the second probe. In contrast, pores induced by NpreTM were transient and did not allow passage of the second probe, even in the vesicles that were partially filled with the first dye. Automated analysis of GUV images are also available (Hermann et al., 2014).

Alternatively, flow cytometry of GUVs can also be employed to analyze a population of GUVs (Nishimura et al., 2009). And so, by using GUVs, it is not only possible to visualize the effects of molecules on overall GUV integrity, but also study pore size and stability and extract kinetics parameters from permeation behavior.

In another study, the permeabilization mechanism of four different membrane-active peptides (CE-2, an analogue of cecropin A, TPW-3, δ-lysin and an analogue of δ-lysin, DL-1) was assessed by investigating the distribution of dye content in populations of GUVs (unimodal or bimodal, see Box 24.2) and by the requenching assay with LUVs (Wheaten et al., 2013b). The results showed that for three peptides the same mechanism was observed with LUVs and GUVs, but for one of the peptides (CE-2), vesicle size (also within the GUV population) modulated dye flux. Such experiments exemplify how observation at the single GUV level can improve our understanding of the mechanism of action of membrane-active peptides.

Quantitative information on the efficiency of antimicrobial peptides acting via the carpet mode using GUVs has been less explored in the literature. One example is the study of the mode of action of gomesin, a β-hairpin peptide from the Brazilian spider *Acanthoscurria gomesiana* (T Domingues et al., 2010a). This peptide causes sudden burst of GUVs without any previous membrane permeabilization, suggesting therefore a carpet mode of action. The burst efficiency of gomesin was quantified as follows. The number of intact GUVs on a representative large field of view (visualized with a low magnification objective) was counted for increasing peptide concentration (Figure 24.3a). A new parameter termed minimum bursting concentration (MBC) was introduced, representing the minimum peptide concentration necessary to induce extensive bursting of an ensemble of vesicles (>90%). The introduction of this parameter was inspired by the MIC (minimum inhibitory concentration) value, widely used in antimicrobial activity assays to represent the minimum peptide concentration necessary to inhibit growth of a microorganism colony after incubation. The MBC parameter was quantified for gomesin and its linear analogue, which lacks

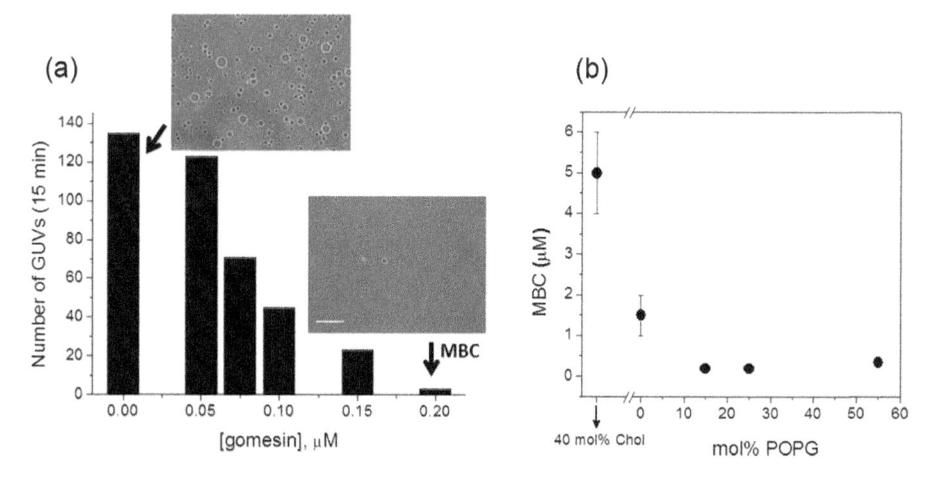

Figure 24.3 (a) Number of intact GUVs after 15 min incubation with the antimicrobial peptide gomesin. Representative images obtained with a low magnification objective (10×) are shown in the absence of peptide and at the minimum bursting concentration (MBC). Scale bar: 200 μm. (b) MBC as a function of membrane composition (POPC with 40 mol% cholesterol or with different mol% POPG). (Adapted from Domingues, T.M. et al., *Langmuir*, 26, 11077–11084, 2010a.)

the disulfide bonds as a function of the membrane composition of the GUVs (POPC with 40 mol% cholesterol (Chol) or with increasing concentration of 1-palmitoyl-2-oleoyl-*sn*-glycero-3-phospho-(1′-rac-glycerol) (sodium salt) (POPG) (Figure 24.3b). The bursting activity of both peptides was highly sensitive to the membrane composition, and increased with the fraction of POPG, whereas the presence of cholesterol protected the membrane from the peptide attack, an important result in the search for a nonhemolytic peptide.

24.3.2 PEPTIDE-INDUCED CHANGES IN AREA/VOLUME AND MEMBRANE FOLDING

The micropipette aspiration technique provides an accurate way to measure changes in vesicle surface area and/or volume (Figure 24.2a and Chapter 11). This method has been applied to investigate the effects of the pore-forming peptide melittin, the main toxin from the venom of the bee *Apis mellifera*, on GUVs made of PC or DOPC/DOPG 7:3 (Lee et al., 2008, 2013). GUVs held by a micropipette with controlled pressure were exposed to melittin, and the vesicle projection length inside the micropipette was measured as a function of time for different peptide concentration. It was shown that melittin initially induced an increase in the vesicle projection length while the membrane was still impermeable, thus showing an increase in relative area (up to 10%) at constant volume. Then, above a critical peptide concentration, the membrane became permeable (contrast loss and leakage of encapsulated fluorescent probes) and the projection length decreased. This result was interpreted as an increase in vesicle volume at constant area because of opening of pores of sizes that allowed the influx of glucose but not the efflux of sucrose, therefore causing an osmotic imbalance with consequent water influx. As discussed in Riske (2015), this hypothesis is inconsistent with the fact that leakage of the encapsulated fluorescent probe (Texas Red, 0.6 kDa), which is bigger than sucrose, occurred before the retraction of the vesicle projection length (Lee et al., 2013). Alternatively, this retraction can be explained by area decrease because of peptide-mediated membrane folding, as will be discussed ahead (Figure 24.4c). In fact, closer inspection of videos and images presented in Lee et al. (2008) and other studies (Sun et al., 2009; Chen et al., 2014), revealed that retraction of the projection length occurs concomitantly with the formation of dense spots on the vesicle surface, consistent with the hypothesis of membrane folding.

Peptide-induced formation of dense regions on the GUV surface has been observed for several membrane-active peptides and agents (Tamba et al., 2007; Sun et al., 2009; Domingues et al., 2010a; Cabrera et al., 2011; Lee et al., 2011; Manzini et al., 2014; Chen et al., 2014). Some representative images of GUVs exhibiting such spots and a sequence that culminates with vesicle collapse are shown in Figure 23.4a and b. This effect is usually overlooked and not deeply discussed in the literature. However, it seems to be a rather universal behavior induced by cationic membrane-active molecules. Furthermore, dense peptide–lipid lumps are often observed after GUVs burst/collapse induced by carpeting agents (Figure 23.4a, 8 s). The cartoon in Figure 23.4c shows the proposed peptide–lipid arrangement that could lead to membrane-folding with consequent local accumulation of lipids and peptides (Riske, 2015). Interestingly, cationic peptides that induce such arrangements were often found to induce aggregation of LUVs containing negatively charged lipids (Cabrera et al., 2011; Domingues et al., 2013; Manzini et al., 2014), suggesting that these phenomena are related.

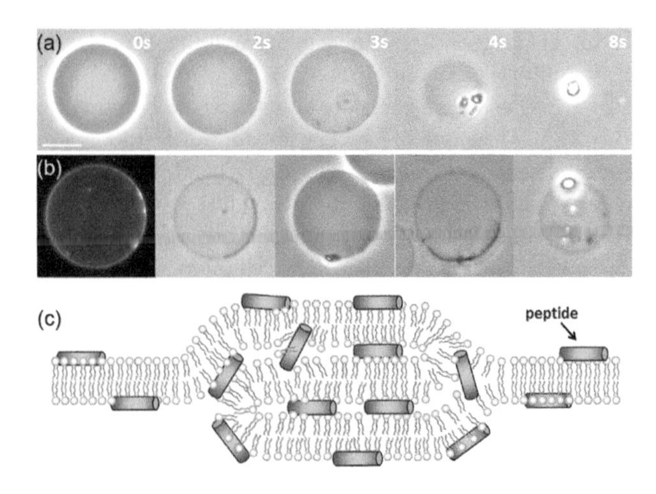

Figure 24.4 Peptide-induced membrane folding. (a) Representative sequence and (b) images of peptide-induced membrane folding resulting in the formation of dense spots on the vesicle surface with eventual total collapse into a peptide–lipid lump (8 s in a). Scale bar: 10 μm. (c) Cartoon illustrating the peptide–lipid organization that gives rise to peptide-induced membrane folding. (Adapted from Riske, K.A., *Adv. Planar Lipid Bilayers Liposomes*, 21, 99–129, 2015.)

GUVs have also been used as platforms to study the interaction of other membrane-active peptides with membranes, most notably cell penetrating peptides (CPPs), which are able to cross the cellular membrane and transport cargos into the intracellular environment (Sani and Separovic, 2016). They share some resemblance with antimicrobial peptides and also exhibit a high occurrence of the cationic amino-acid residues lysine and arginine. Usually, CPPs are able to permeate GUVs without affecting their permeability, although, at higher concentrations, they quite often cause membrane permeabilization/disruption (Thorén et al., 2000; Henriques et al., 2007; Ciobanasu et al., 2010; Säälika et al., 2011). Another class of peptides, the lantibiotic peptides, which are cyclic peptides with special affinity for the Lipid II of the bacterial cell wall, was found to induce clusters of Lipid II on the GUV surface (Hasper et al., 2006; Scherer et al., 2015) and eventually lead to increased membrane permeability (Scherer et al., 2015).

24.4 SOLUBILIZATION OF GUVs BY DETERGENTS

Detergents are amphiphilic molecules with the ability to intercalate into biological membranes and separate their hydrophobic components, both lipids and membrane proteins, under appropriate conditions (Jones, 1999; Seddon et al., 2004). They are routinely used in a great number of biochemical and molecular biology protocols and have been especially useful in the process of isolating and purifying membrane proteins and reconstitution of membrane proteins into liposomes. Most detergents self-assemble into micelles above a CMC, which basically depends on the hydrophobic-hydrophilic balance of the molecule. The solubilization process of lipid bilayers can be viewed as a competition between two distinct self-assembled structures: lamellar versus micellar, and the process is depicted in the classical three-stage model (see Box 24.3) (Helenius and Simon, 1975; Lichtenberg, 1985). During incorporation of detergents into membranes (stage I), the effects caused on the vesicle can be quite different depending on the detergent flip-flop rate, as will be discussed in more detail with examples in the following section.

Biological membranes treated with detergents often exhibit insoluble fragments, termed detergent resistant membranes (DRMs) (Brown and Rose, 1992). Analysis of the lipid composition of DRMs revealed a large fraction of sphingomyelin (SM), long-tail saturated lipids, and cholesterol. In addition, specific membrane proteins (e.g., GPI-anchors) were also found enriched in the insoluble fragments. In parallel, the so-called lipid raft hypothesis (Simons and Ikonen, 1997) postulated the existence of membrane microdomains (also rich in SM, cholesterol and specific proteins), that would transiently form to accomplish physiological tasks related to membrane traffic, cell signaling, immune response and apoptosis. In particular, the lipid composition of DRMs and that of the postulated rafts are both prone to forming a liquid-ordered (L_o) phase in mimetic lipid systems. Because of that, treatment with detergents has been initially interpreted as a means to isolate and extract lipid rafts and the terms DRMs, membrane rafts and L_o phase were in many examples misleadingly used as synonyms (Lichtenberg et al., 2005). However, it is now clear that detergents have the ability to reshape membrane architecture (Ingelmo-Torres et al., 2009) and that the composition and features of DRMs are strongly dependent on the detergent used and protocol details, such as temperature treatment (Schuck et al., 2003; Domingues et al., 2010b). Therefore, DRMs do not represent lipid rafts (Lichtenberg et al., 2013). Nonetheless, the interaction of detergents with biological and model membranes of different compositions shed light on many aspects of membrane properties and architecture.

Box 24.3 Solubilization of membranes by detergents

- *Three-stage model.* The solubilization process of membranes by detergents can be described with a three-stage model as the bound detergent-to-lipid molar ratio (X_b) increases: (I) At low X_b, detergent monomers insert into the lipid bilayers up to a saturation limit that sets the onset of solubilization. (II) Then, lipid molecules are continuously transferred to micelles and bilayers saturated with detergents coexist with mixed micelles. (III) Finally, above the solubilization threshold, no bilayer remains and the lipids are fully solubilized into micellar structures.

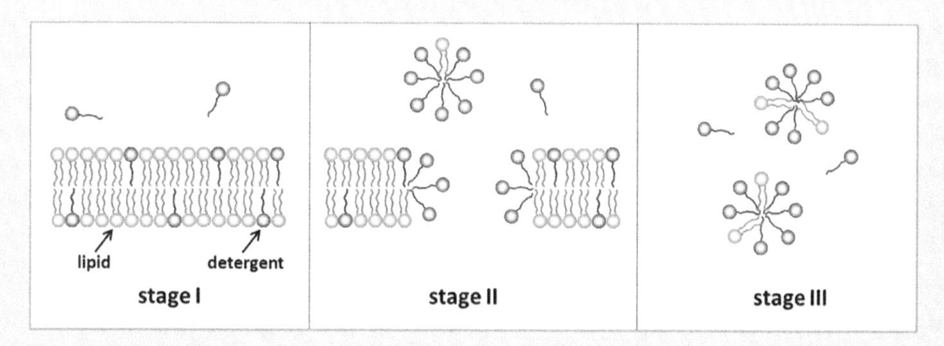

- *Role of detergent flip-flop.* During stage I, the detergent monomers insert initially unilaterally and their ability to equilibrate between both leaflets depends on the detergent flip-flop rate. Fast flip-flop (usually for nonionic detergents) leads to rapid equilibration and consequently the membrane area increases to accommodate the incorporated molecules. Detergents with slow flip-flop (with a bulky polar head and/or ionic detergents) are trapped in one of the leaflets and the membrane has to bend to relieve the stress of area difference between the monolayers and an increase in vesicle spontaneous curvature follows. These two effects (area and spontaneous curvature increase) can be clearly distinguished by observing GUVs (see images below and Figure 24.5a for the whole sequences).

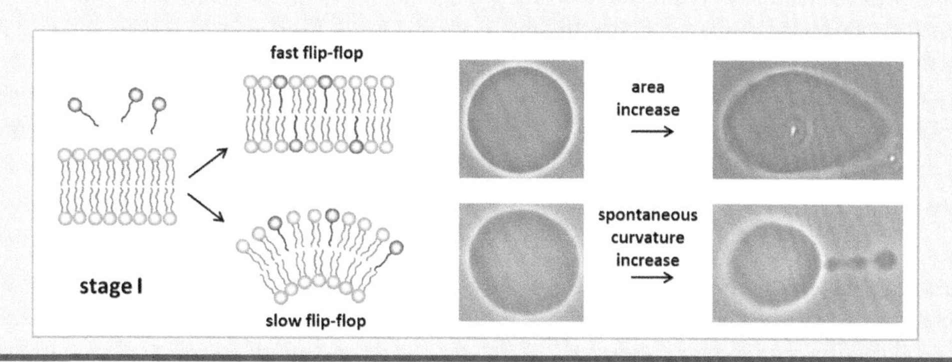

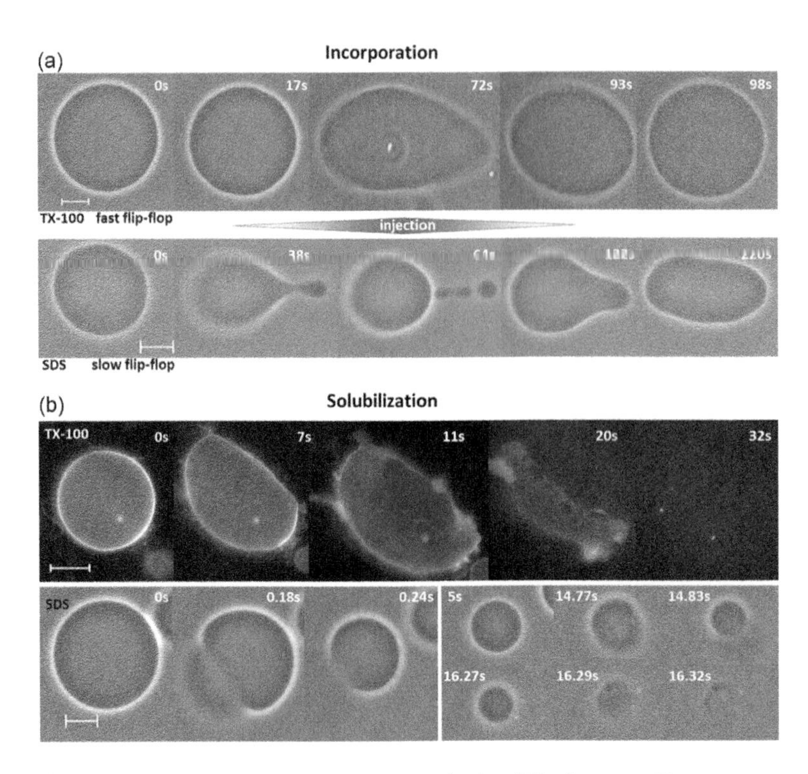

Figure 24.5 Effects of TX-100 and SDS in POPC GUVs. (a) Incorporation and role of flip-flop rate. Two representative sequences of GUVs experiencing a micropipette injection of low detergent concentration (0.1 mM TX-100 and 10 mM SDS). The micropipettes were initially brought close to the vesicle and then removed (see gradient between the two sequences). (b) Solubilization. Representative sequences of the solubilization process of TX-100 (fluorescence microscopy, 1 mM in the micropipette) and SDS (phase-contrast, 40 mM in the micropipette). Scale bars: 10 µm or 20 µm (b, TX-100). These sequences are also shown as online movies (Movies 24.3 and 24.4; movies captions at the end of chapter). (Adapted from Sudbrack, T.P. et al., *J. Phys. Chem. B*, 115, 269–277, 2011.)

The solubilization of lipid vesicles by a wide variety of detergents has provided a great wealth of information on the mechanistic details of the process. The extent and characteristics of the solubilization are strongly modulated by the lipid composition/phase and each detergent interacts in a unique way with membranes; thus detergent-membrane interactions are highly polymorphic (Heerklotz and Seelig, 2000; Heerklotz et al., 2003; Tsamaloukas et al., 2007; Arnulphi et al., 2007; Ahyayauch et al., 2010, 2012; Lichtenberg et al., 2013). Of special interest is the effect of membrane phase and presence of cholesterol on the resistance of the membrane to treatment with detergents. The incorporation of detergents into membranes (stage I) is highly dependent on the lipid phase. For instance, fluid-phase membranes were found to exhibit a significantly higher saturation threshold as compared with gel-phase membranes (Ahyayauch et al., 2012) and the presence of cholesterol was reported to usually diminish the affinity of the membranes to detergents (Sot et al., 2002; Tsamaloukas et al., 2006). In fact, insoluble lipid bilayers were also reported, mostly associated with the presence of cholesterol and with the L_o phase, that is, raft-like mixtures also similar to the composition of DRMs (Ahmed et al., 1997; Sot et al., 2002; Mattei et al., 2015).

24.4.1 MEMBRANE INSERTION AND SOLUBILIZATION PROCESS

GUVs have also been used to explore the effects of detergents on membranes, but to a lesser extent as compared with the field of antimicrobial peptides. The first report focused on the capability of liposomes for topological transformation induced by detergents and other molecules (Nomura et al., 2001). GUVs with and without charged lipids were mixed with a wide variety of detergents, including the widely used Triton X-100 (TX-100), sodium dodecyl sulfate (SDS), C12E8 and CHAPS; see Appendix 3 of the book for structure and data on these detergents. Different topological responses were reported: shrinkage, burst, opening up and inside-out inversion. The different transformation pathways were discussed in terms of membrane curvature and ability of the detergent to perform flip-flop, as discussed ahead.

Detergents are usually added to the external medium of vesicles, so that initially they insert into the external vesicle monolayer. Their transfer to the inner leaflet depends on the detergent flip-flop rate across the membrane. For most nonionic detergents, flip-flop rate is fast so that the detergent molecules rapidly equilibrate between both monolayers. On the other hand, detergents with bulky polar heads or ionic detergents will be mostly trapped in the external leaflet. The initial morphological transformations caused by the detergents on vesicles will then depend on the ability of the detergent to perform flip-flop (see Box 24.3). The effects of TX-100 and SDS on GUVs composed of POPC were investigated (Sudbrack et al., 2011). TX-100 is a nonionic detergent that exhibits a fast flip-flop rate across the membrane, whereas SDS is anionic and is therefore not able to easily traverse the hydrophobic membrane core. The CMC values of TX-100 and SDS are very different due to their distinct hydrophobic/hydrophilic balance: 0.27 mM for TX-100 and 8 mM for SDS in 0.2 M glucose (Sudbrack et al., 2011). For experiments with low lipid concentration (in the micromolar range), as is the case with GUVs, solubilization is expected to occur close to the CMC of the detergent (Heerklotz and Seelig, 2000; Lichtenberg et al., 2013). The situation is different for experiments with higher lipid concentration (e.g., as done with small liposomes), for which higher concentration of detergent is required to induce similar effects. GUVs were exposed to a detergent flux from a micropipette (see Box 24.1). For low detergent concentration (below or around the CMC of each detergent), only stage I was reached and the solubilization was not initiated. The effects on GUV morphology as a consequence of insertion of detergent monomers in the membrane were quite distinct for both detergents and could be well explained by their different flip-flop rates, as shown in Box 24.3 and Figure 24.5a. TX-100 induced an increase in vesicle surface area, whereas SDS caused an increase in spontaneous curvature. Both effects were partially reversed when the micropipette was removed and the detergent concentration around the vesicle decreased, showing also that detergents are in equilibrium between the membrane and water phase. For higher detergent concentrations (above the CMC of each detergent), different responses ensued for TX-100 and SDS (Figure 24.5b and online Movies 24.3 and 24.4; movies captions at the end of chapter). The three-stage model is clearly visualized in the solubilization of POPC by TX-100: Incorporation of TX-100 in the membrane during stage I causes a significant increase in vesicle surface area while the membrane is still impermeable (0–7 s in Figure 24.5b, top). The membrane becomes permeable and then large pores open (11 s) and the solubilization of lipids starts (20 s). Eventually, the whole vesicle is solubilized (32 s). Differently, SDS causes opening of large holes (Figure 24.5b, bottom; 0.18, 14.77 s) and eventually vesicle burst (16.29 s) to relieve the strain of unilateral incorporation of SDS. However, it is not resolved whether vesicle collapse coincides with the onset of solubilization because the small fragments that remain might be still bilayers (16.32 s). Therefore, the whole solubilization process induced by SDS could not be entirely followed with optical microscopy.

In a following study, the role of membrane composition and phase on the solubilization process induced by TX-100 was investigated (Mattei et al., 2015). The compositions/phases studied were POPC (fluid, also called liquid-disordered phase [L_d]), SM (gel), and binary mixtures of these lipids with 30 mol% cholesterol: POPC/Chol (fluid) and SM/Chol (L_o). It was shown that pure phospholipid membranes were fully solubilized by TX-100, whereas the presence of cholesterol rendered the mixture either partially (POPC/Chol) or completely(SM/Chol) insoluble. Clearly, the area increase caused by TX-100 incorporation was modulated by membrane composition. To quantify the relative area increase of GUVs during stage I, the AC field deformation method shown in Figure 24.2b was used. GUVs were placed in a chamber with two electrodes connected to a function generator (2 kV/cm, 200 kHz). An aliquot of a concentrated TX-100 solution was added to the corner of the chamber, which was then sealed with a coverslip and recording of initially spherical GUVs between the electrodes was started. Eventually, TX-100 diffused through the chamber and reached the examined vesicle, which then started to elongate as a response of the amount of TX-100 incorporated (Figure 24.6a; see also online Movie 24.5; movies captions at the end of chapter). Elongation proceeded until the membrane became permeable (last snapshot of each sequence) and solubilization started (found only for POPC and POPC/Chol; SM/Chol was not solubilized). The maximum vesicle aspect ratio a / b was measured for several GUVs of POPC, POPC/Chol and SM/Chol and the relative area increase was estimated as

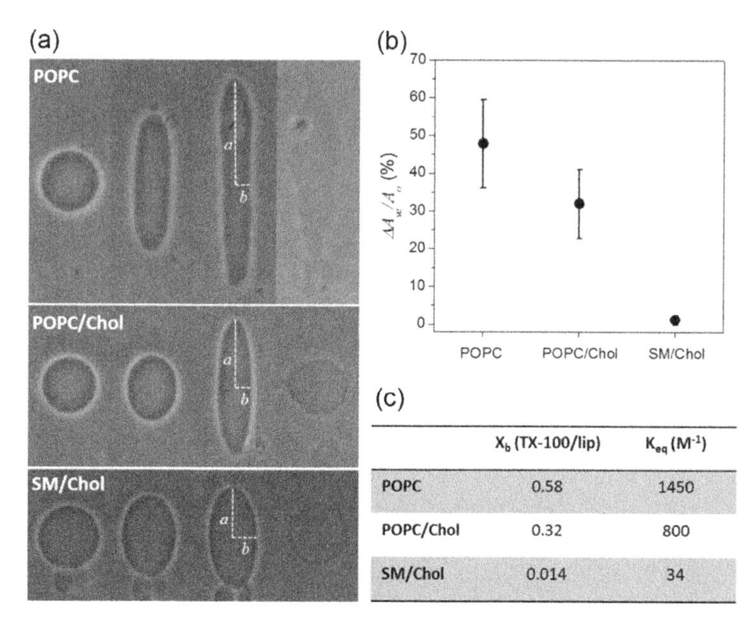

	X_b (TX-100/lip)	K_{eq} (M^{-1})
POPC	0.58	1450
POPC/Chol	0.32	800
SM/Chol	0.014	34

Figure 24.6 Quantification of TX-100 incorporation in GUVs. (a) Representative sequences of GUVs (POPC, POPC/Chol 7:3, SM/Chol 7:3) under an AC field (2 kV/cm, 200 kHz) in contact with TX-100. The sequence of POPC/Chol is also shown as online movie (Movie 24.5; movies captions at the end of chapter). (b) Maximum relative area increase ($\Delta A_{ve} / A_o$) measured from the aspect ratio a/b (Figure 24.2b) at the maximum deformation before the onset of solubilization. (c) TX-100/lipid molar ratio in the membrane (X_b) calculated from $\Delta A_{ve} / A_o = 1 + X_b A_{TX} / A_{lip}$, where A_{TX} is the area occupied by TX-100 in the membrane (54 Å^2, Nyholm and Slotte, 2001) and A_{lip} is the area per lipid (65 Å^2 for POPC, 27 Å^2 for cholesterol and 51 Å^2 for SM in the L_o phase; Maulik et al., 1991; Hofsäß et al., 2003; Kučerca et al., 2005, 2011). The binding constant K_{eq} was estimated from $X_b = K_{eq} c_{fr}$, where c_{fr} = 0.4 mM TX-100 (the concentration of TX-100 at the onset of solubilization). (Adapted from Mattei, B. et al., *Langmuir*, 31, 378–386, 2015.)

shown in Figure 24.2b and the results are shown in Figure 24.6b. It should be noticed, however, that for vesicles that gained a lot of excess area, as is the case of POPC (see last snapshot in Figure 24.2a), a perfect prolate shape is no longer maintained and this procedure gives only an approximate estimation of the real area increase. This procedure could not be applied to gel phase membranes, because of the high bending rigidity (Dimova et al., 2000) that prevents vesicle deformation. Because data on the area per lipid and TX-100 are available in the literature (see legend of Figure 24.6 for details), the relative area increase could be directly converted into TX-100/lipid molar ratio in the membrane (X_b) at the onset of solubilization. In parallel experiments, GUVs were added to increasing concentrations of TX-100. The onset of solubilization for POPC, POPC/Chol and SM occurred around the CMC of TX-100 (CMC ~0.3 mM; Sudbrack et al., 2011). Thus, the binding constant K_{eq} of TX-100 to the membrane could be inferred from $X_b = K_{eq}c_{fr}$, where c_{fr} is the concentration of detergent in solution, which can be assumed to be the total TX-100 concentration (0.4 mM at the onset of solubilization) because the bound TX-100 represents a small fraction due to the very low lipid concentration in GUV experiments (on the order of several μM). The results are shown in a Table in Figure 24.6c and reveal that K_{eq} is strongly modulated by membrane composition: it decreases when cholesterol is added to POPC and it becomes negligible in the L_o phase. Thus, the origin of detergent (total or partial) resistance of cholesterol-rich membranes arises from the limited incorporation of TX-100 in such mixtures.

Before the onset of solubilization, membranes become permeable in the presence of TX-100. Dilution experiments of GUVs in a detergent solution shown in Mattei et al. (2015) reveal that membrane permeabilization occurs at 0.3 mM TX-100, whereas solubilization starts at 0.4 mM TX-100. To resolve the nature of the pores opened before the onset of solubilization, GUVs of the same compositions reported in Mattei et al. (2015) were dispersed in a solution containing dyes of different sizes: sulforhodamine (0.6 kDa) and cascade blue-Dextran (10 kDa) (Mattei et al., 2015). For POPC, POPC/Chol and SM, pores allowing the passage of the small dye only open first and grow afterward to allow the influx of the large dye as well. For the insoluble SM/Chol, some GUVs exhibited high permeability to both dyes, in a graded mechanism, showing that the perturbations caused by TX-100 in the insoluble L_o phase are of different nature. The ability of TX-100 to permeabilize membranes at sub-solubilizing concentrations was interpreted as a result of reduction in the membrane edge tension, measured from electroporation experiments as discussed in Chapter 15 and in Portet and Dimova (2010).

24.4.2 DETERGENT-INDUCED LATERAL REORGANIZATION OF MEMBRANES

The detergent TX-100 was also found to reshape cellular (Ingelmo-Torres et al., 2009) and model membranes (El Kirat and Morandat, 2007; Garner et al., 2008). In model systems, TX-100 induced lateral phase separation (L_d/L_o) of mixtures containing cholesterol and selective solubilization of the L_d phase. In addition, selective solubilization was also found in mixtures of SM/ceramide, which exhibit lateral phase separation and the formation gel-phase domain (ceramide-rich and SM-rich). Confocal microscopy experiments with GUVs revealed that whereas SM-rich bilayers were preferentially solubilized by TX-100, ceramide-rich membranes were resistant to TX-100 (Sot et al., 2006).

The effects of TX-100 on GUVs composed of raft-like mixtures were explored in Casadei et al. (2014). GUVs made of the lipid extract of erythrocytes (termed erythro-GUVs, representing a model system that retains the complexity of the lipid composition of a biological membrane; see also Chapter 2) and of the ternary mixture POPC/SM/Chol 2:1:2 were exposed to TX-100, with either micropipette injection or dilution in a defined TX-100 concentration (see Box 24.1). Interestingly, both compositions gave rise to the same qualitative behavior and quantitative data. A rather unexpected effect was observed as a result of incorporation of TX-100 in the membrane (stage I). Erythro-GUVs and GUVs of the ternary mixture labeled with the membrane probe 1,1′-dioctadecyl-3,3,3′,3′-tetramethylindotricarbocyanine iodide (DiIC18; see Appendix 2 of the book for structure and data on this and other lipid dyes) appeared homogeneous under fluorescence microscopy, showing that no macroscopic phase separation existed in the absence of TX-100. As soon as the GUVs entered in contact with TX-100 (either by micropipette injection or with the dilution protocol), visible domains with enriched fluorescence appeared and coalesced on the vesicle surface, as shown in Figure 24.7b and c and online Movie 24.6 (movies captions at the end of chapter). The fact that the domains were round, diffused in the dye-depleted matrix and coalesced show that both phases were liquid, and therefore represented L_d/L_o phase coexistence (see Chapter 18, especially Figures 18.1 through 18.3). The fluorescent probe DiIC18 is known to prefer the L_d phase, thus the domains were in the L_d phase in an L_o matrix. Incorporation of TX-100 in such membranes did not cause significant area increase, showing that the partition of TX-100 in these mixtures is lower, but nonetheless induce lateral reorganization of lipids. However, observation with optical microscopy cannot discriminate whether sub-microscopic domains existing before contact with TX-100 coalesced and became visible or were induced by the detergent (Pathak and London, 2011). TX-100-induced domain formation was also occasionally seen for POPC/Chol GUVs (Mattei et al., 2015). If the concentration of TX-100 was above the CMC, selective solubilization of the L_d phase followed: the domains were removed from the vesicle and fully solubilized (Figure 24.7a–c). The remaining vesicle in the L_o phase, visibly smaller than the original one, was insoluble to TX-100 and exhibited reduced fluorescence (see last snapshots of each sequence in Figure 24.7a–c). By measuring the GUV diameter before (D_{ini}) and after (D_{end}) contact with TX-100 (Figure 24.7a), the insoluble area fraction X_{ins} was quantified as $X_{ins} = (D_{end}/D_{ini})^2$. For both erythro-GUVs and GUVs of the ternary lipid mixture $X_{ins} \sim 0.68 \pm 0.10$ (at room temperature), as shown in Figure 24.7d. This value decreased for both systems if the temperature was increased to 37°C: $X_{ins} \sim 0.54 \pm 0.10$. If it is assumed that POPC is enriched in the domain and is therefore solubilized from the ternary mixture at room temperature, and knowing the area per lipid of POPC, SM and cholesterol (Maulik et al., 1991; Hofsäß et al., 2003; Bezlyepkina et al., 2013), this would mean that the insoluble

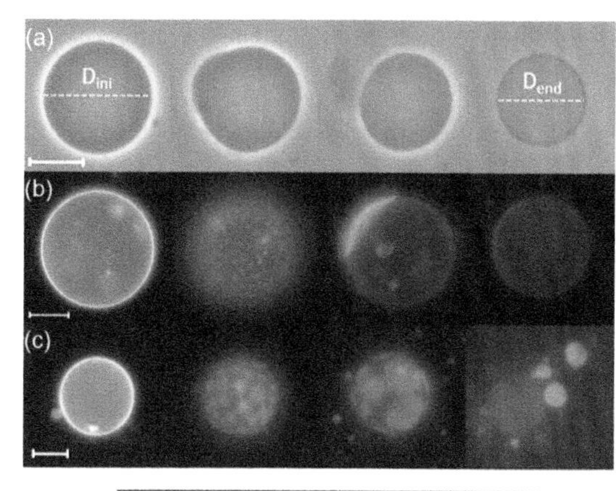

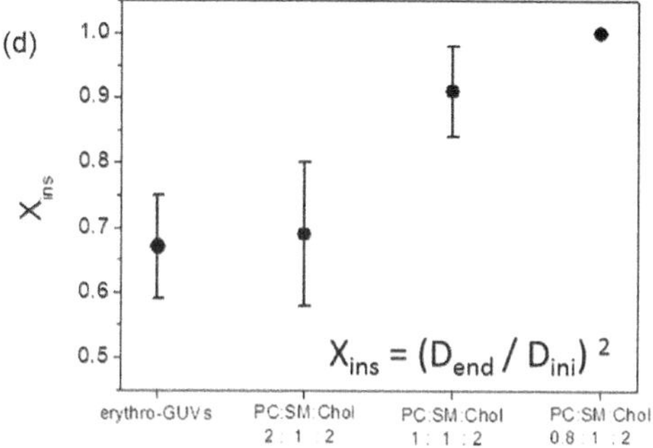

Figure 24.7 Domain formation induced by TX-100. (a–c) Representative sequences of erythro-GUVs experiencing a micropipette flux of TX-100 (5 mM, a, b) and diluted into 0.3 mM TX-100 (c). Scale bars: 10 (a, b) and 20 μm. The sequence shown in (c) is available as online movie (Movie 24.6). (d) Insoluble area fraction measured from the GUV diameter before and after contact with TX-100 for erythro-GUVs and GUVs composed of different ternary mixtures of POPC/SM/Chol as indicated. (Adapted from Casadei, B.R. et al., *Biophys. J.*, 106, 2417–2425, 2014.)

composition left is composed of POPC/SM/Chol 0.8:1:2 (see discussions in Casadei et al. (2014) for details). In fact, GUVs of that composition were found to be insoluble in TX-100 (Figure 24.7d). Taken together, these results show that TX-100 is able to alter the lateral distribution of lipids and selectively extract and solubilize L_d phase patches.

24.5 FINAL WORDS

In this chapter, the unique advantages of using GUVs to reveal mechanistic details of the interaction of membrane-active agents with model membranes were discussed in detail, with special attention to antimicrobial peptides and detergents. Some future directions on the field are envisaged. GUVs are mostly prepared from synthetic lipids, but GUVs of more complex membranes, as discussed in detail in Chapter 2, can also be prepared allowing the investigation of the role of integral proteins and bacterial membrane components on the activity of antimicrobial peptides and detergents. The effects of incorporation of membrane-active agents on lipid lateral diffusion, membrane edge tension and elastic properties of membranes (bending modulus, area compressibility modulus) can be achieved and will certainly reveal interesting aspects to the field.

MOVIES

The Movies can be found under http://www.crcpress.com/9781498752176.

Movie 24.1: Phase-contrast sequence of a vesicle burst event induced by an antimicrobial peptide. POPC/POPG 1:1, total elapsed time: 12 s.

Movie 24.2: Phase-contrast sequence of a membrane permeabilization event induced by a pore-forming antimicrobial peptide. POPC/POPG 1:1, total elapsed time: 25 s.

Movie 24.3: Fluorescence microscopy sequence of a POPC GUV experiencing a flux of TX-100 from a micropipette. Fluorescence dye DiIC18, total elapsed time: 32 s.

Movie 24.4: Phase-contrast sequence of a POPC GUV experiencing a flux of SDS from a micropipette. Total elapsed time: 18 s.

Movie 24.5: Phase-contrast sequence of a POPC/Chol 7:3 GUV in the presence of an AC field experiencing a flux of TX-100. Total elapsed time: 50 s. Mattei et al. (2015).

Movie 24.6: Fluorescence microscopy sequence of an erythro-GUV diluted into a 0.3 mM TX-100 solution. Fluorescence dye DiIC18, total elapsed time: 200 s. Casadei et al. (2014).

ACKNOWLEDGMENTS

I am greatly thankful to all students and co-workers involved in the studies presented and discussed here: Tatiana M. Domingues, Bruno Mattei, Rafael B. Lira, Ana D. C. França, Amanda C. Caritá, Tatiane P. Sudbrack, Nathaly L. Archilha, Bruna R. Casadei, Dayane S. Alvares, Mariana C. Manzini, Cleyton C. Domingues, Antonio Miranda, Katia R. Perez, Eneida de Paula, Rosangela Itri, Marcia Cabrera, João Ruggiero Neto and Iolanda M. Cuccovia. The financial support of FAPESP, CNPq and INCT-FCx is acknowledged.

LIST OF ABBREVIATIONS

AC	alternating current
CMC	critical micelle concentration
CPP	cell penetrating peptide
DOPC	1,2-dioleoyl-*sn*-glycero-3-phosphocholine
DOPG	1,2-dioleoyl-*sn*-glycero-3-phospho-(1′-*rac*-glycerol) (sodium salt)
Chol	cholesterol
DiIC18	1,1′-dioctadecyl-3,3,3′,3′-tetramethylindotricarbocyanine iodide
DRMs	detergent resistant membranes
L_d	liquid-disordered phase
L_o	liquid-ordered phase
LUVs	large unilamellar vesicles
MBC	minimum bursting concentration
POPC	1-palmitoyl-2-oleoyl-*sn*-glycero-3-phosphocholine
POPG	1-palmitoyl-2-oleoyl-*sn*-glycero-3-phospho-(1′-*rac*-glycerol) (sodium salt)
TX-100	Triton X-100
SDS	sodium dodecyl sulfate
SM	sphingomyelin (egg chicken)

GLOSSARY OF SYMBOLS

a / b	ratio between the two prolate shape semi-axes
A_{lip}	area per lipid
A_{TX}	area per TX-100
c_{fr}	free detergent concentration
D_{end}	final vesicle diameter after contact with detergent
D_{ini}	initial vesicle diameter before contact with detergent
$\Delta A_{ve} / A_o$	relative area change
K_{eq}	binding constant
R_{pip}	pipette radius
R_{ve}	vesicle radius
X_b	bound detergent-to-lipid molar ratio
X_{ins}	insoluble area fraction of vesicles

REFERENCES

Ahmed SN, DA Brown, E London (1997) On the origin of sphingolipid/cholesterol-rich detergent-insoluble cell membranes: physiological concentrations of cholesterol and sphingolipid induce formation of a detergent-insoluble, liquid-ordered lipid phase in model membranes. *Biochemistry* 36:10944–10953.

Ahyayauch H, M Bennouna, A Alonso, FM Goñi (2010) Detergent effects on membranes at subsolubilizing concentrations: Transmembrane lipid motion, bilayer permeabilization, and vesicle lysis/reassembly are independent phenomena. *Langmuir* 26:7307–7313.

Ahyayauch H, M Collado, A Alonso, FM Goñi (2012) Lipid bilayers in the gel phase become saturated by triton X-100 at lower surfactant concentrations than those in the fluid phase. *Biophys. J.* 102:2510–2516.

Alam JM, T Kobayashi, M Yamazaki (2012) The single-giant unilamellar vesicle method reveals lysenin-induced pore formation in lipid membranes containing sphingomyelin. *Biochemistry* 51:5160–5172.

Almeida PF, A Pokorny (2009) Mechanisms of antimicrobial, cytolytic, and cell-penetrating peptides: From kinetics to thermodynamics. *Biochemistry* 48:8083–8093.

Ambroggio EE, F Separovic, JH Bowie, GD Fidelio, LA Bagatolli (2005) Direct visualization of membrane leakage induced by the antibiotic peptides: Maculatin, citropin, and aurein. *Biophys. J.* 89:1874–1881.

Aoki PHB, AP Schroder, CJL Constantino, CM Marques (2015) Bioadhesive giant vesicles for monitoring hydroperoxidation in lipid membranes. *Soft Matter* 11:5995–5998.

Apellániz B, S Nir, JL Nieva (2009) Distinct mechanisms of lipid bilayer perturbation induced by peptides derived from the membrane-proximal external region of HIV-1 gp41. *Biochemistry* 48:5320–5331.

Apellániz B, JL Nieva, P Schwille, AJ García-Saéz (2010) All-or-none versus graded: single-vesicle analysis reveals lipid composition effects on membrane permeabilization. *Biophys. J.* 99:3619–3628.

Arnulphi C, J Sot, M García-Pacios, JL Arrondo, A Alonso, FM Goñi (2007) Triton X-100 partitioning into sphingomyelin bilayers at subsolubilizing detergent concentrations: Effect of lipid phase and a comparison with dipalmitoylphosphatidylcholine. *Biophys. J.* 93:3504–3514.

Bezlyepkina N, RS Gracià, P Shchelokovskyy, R Lipowsky, R Dimova (2013) Phase diagram and tie-line determination for the ternary mixture DOPC/eSM/cholesterol. *Biophys. J.* 104:1456–1464.

Brogden KA (2005) Antimicrobial peptides: Pore formers or metabolic inhibitors in bacteria. *Nat. Rev. Microbiol.* 3:238–250.

Brown DA, JK Rose (1992) Sorting of GPI-anchored proteins to glycolipid-enriched membrane subdomains during transport to the apical cell surface. *Cell.* 68:533–544.

Cabrera MPS, DS Alvares, NB Leite, BM Souza, MS Palma, KA Riske, JR Neto (2011) New insight into the mechanism of action of wasp mastoparan peptides: Lytic activity and clustering observed with giant vesicles. *Langmuir* 27:10805–10813.

Casadei BR, CC Domingues, E de Paula, KA Riske (2014) Direct visualization of the action of Triton X-100 on giant vesicles of erythrocyte membrane lipids. *Biophys. J.* 106:2417–2425.

Chen Y-F, T-L Sun, Y Sun, HW Huang (2014) Interaction of daptomycin with lipid bilayers: A lipid extracting effect. *Biochemistry* 53:5384–5392.

Ciobanasu C, JP Siebrasse, U Kubitscheck (2010) Cell-penetrating HIV1 TAT peptides can generate pores in model membranes. *Biophys. J.* 99:153–162.

Dimova R, B Pouligny, C Dietrich (2000) Pretransitional effects in dimyristoylphosphatidylcholine vesicle membranes: Optical dynamometry study. *Biophys. J.* 79:340–356.

Döbereiner H-G (2000) Properties of giant vesicles. *Opin. Colloid Interface Sci.* 5:2560263.

Domingues CC, A Ciana, A Buttafava, BR Casadei, C Balduini, E de Paula, G Minetti (2010b) Effect of cholesterol depletion and temperature on the isolation of detergent-resistant membranes from human erythrocytes. *J. Membr. Biol.* 234:195–205.

Domingues TM, B Mattei, J Seelig, KR Perez, A Miranda, KA Riske (2013) Interaction of the antimicrobial peptide gomesin with model membranes: A calorimetric study. *Langmuir* 29:8609–8618.

Domingues TM, KA Riske, A Miranda (2010a) Revealing the lytic mechanism of the antimicrobial peptide gomesin by observing giant unilamellar vesicles. *Langmuir* 26:11077–11084.

El Kirat K, S Morandat (2007) Cholesterol modulation of membrane resistance to Triton X-100 explored by atomic force microscopy. *Biochim. Biophys. Acta* 1768:2300–2309.

Fuertes G, AJ García-Sáez, S Esteban-Martín, D Giménez, OL Sánchez-Muñoz, P Schwille, J Salgado (2010) Pores formed by Baxα5 relax to a smaller size and keep at equilibrium. *Biophys. J.* 99:2917–2925.

Garner AE, DA Smith, NM Hooper (2008) Visualization of detergent solubilization of membranes: Implications for the isolation of rafts. *Biophys. J.* 94:1326–1340.

Gregory SM, A Cavenaugh, V Journigan, A Pokorny, PFF Almeida (2008) A quantitative model for the all-or-none permeabilization of phospholipid vesicles by the antimicrobial peptide cecropin A. *Biophys. J.* 94:1667–1680.

Hasper HE, NE Kramer, JL Smith, JD Hillman, C Zachariah, OP Kuipers, B de Kruijff, E Breukink (2006) An alternative bactericidal mechanism of action for lantibiotic peptides that target lipid II. *Science* 313:1636–1637.

Heerklotz H, H Szadkowska, T Anderson, J Seelig (2003) The sensitivity of lipid domains to small perturbations demonstrated by the effect of triton. *J. Mol. Biol.* 329:793–799.

Heerklotz H, J Seelig (2000) Titration calorimetry of surfactant-membrane partitioning and membrane solubilization. *Biochim. Biophys. Acta* 1508:69–85.

Helenius A., K Simons (1975) Solubilization of membranes by detergents. *Biochim. Biophys. Acta* 415:29–79.

Henriques ST, A Quintas, LA Bagatolli, F Homblé, MARB Castanho (2007) Energy-independent translocation of cell-penetrating peptides occurs without formation of pores. A biophysical study with pep-1. *Mol. Membr. Biol.* 24:282–293.

Hermann E, S Bleicken, Y Subburaj, AJ García-Sáez (2014) Automated analysis of giant unilamellar vesicles using circular Hough transformation. *Bioinformatics* 30:1747–1754.
Hofsäß C, E Lindahl, O Edholm (2003) Molecular dynamics simulations of phospholipid bilayers with cholesterol. *Biophys. J.* 84:2192–2206.
Hwang PM, J Vogel (1998) Structure–function relationships of antimicrobial peptides. *Biochem. Cell Biol.* 76:235–246.
Ingelmo-Torres M, K Gaus, A Herms, E González-Moreno, A Kassan, M Bosch, T Grewal, F Tebar, C Enrich, A Pol (2009) Triton X-100 promotes a cholesterol-dependent condensation of the plasma membrane. *Biochem. J.* 420:373–381.
Jenssen H, P Hamill, REW Hancock (2006) Peptide antimicrobial agents. *Clin. Microbiol. Rev.* 19:491–511.
Jones MN (1999) Surfactants in membrane solubilisation. *Int. J. Pharm.* 177:137–159.
Kučerca N, S Tristram-Nagle, JF Nagle (2005) Structure of fully hydrated fluid phase lipid bilayers with monounsaturated chains. *J. Membr. Biol.* 208:193–202.
Kučerca N. M Nieh, J Katsaras (2011) Fluid phase lipid areas and bilayer thicknesses of commonly used phosphatidylcholines as a function of temperature. *Biochim. Biophys. Acta* 1808:2761–2771.
Ladokhin AS, ME Selsted, SH White (1997) Sizing membrane pores in lipid vesicles by leakage of coencapsulated markers: Pore formation by melittin. *Biophys. J.* 72:1762–1766.
Ladokhin AS, WC Wimley, SH White (1995) Leakage of membrane vesicle contents: Determination of mechanism using fluorescence requenching. *Biophys. J.* 69:1964–1971.
Laverty G, SP Gorman, BF Gilmore (2011) The potential of antimicrobial peptides as biocides. *Int. J. Mol. Sci.* 12:6566–6596.
Lee C-C, Y Sun, S Qian, HW Huang (2011) Transmembrane pores formed by human antimicrobial peptide LL-37. *Biophys. J.* 100:1688–1696.
Lee M-T, T-L Sun, W-C Hung, HW Huang (2013) Process of inducing pores in membranes by melittin. *Proc. Natl. Acad. Sci.* 110:14243–14248.
Lee M-T, W-C Hung, F-Y Chen, HW Huang (2008) Mechanism and kinetics of pore formation in membranes by water-soluble amphipathic peptides. *Proc. Natl. Acad. Sci.* 105:5087–5092.
Lichtenberg D (1985) Characterization of the solubilization of lipid bilayers by surfactants. *Biochim. Biophys. Acta* 821:470–478.
Lichtenberg D, H Ahyaauch, A Alonson, FM Goñi (2013) Detergent solubilization of lipid bilayers: A balance of driving forces. *Trends Biochem. Sci.* 38:85–93.
Lira, RB, R Dimova, KA Riske (2014) Giant unilamellar vesicles formed by hybrid films of agarose and lipids display altered mechanical properties. *Biophys. J.* 107:1609–1619.
Longo ML, AJ Waring, LM Gordon, DA Hammer (1998) Area expansion and permeation of phospholipid membrane bilayers by influenza fusion peptides and melittin. *Langmuir* 14:2385–2395.
Mally M, J Majhenc, S Svetina, B Žekš (2007) The response of giant phospholipid vesicles to pore-forming peptide melittin. *Biochim. Biophys. Acta* 1768:1179–1189.
Mandard N, P Bulet, A Caile, S Daffre, F Vovelle (2002) The solution structure of gomesin, an antimicrobial cysteine-rich peptide from the spider. *Eur. J. Biochem.* 269:1190–1198.
Manzini MC, KR Perez, KA Riske, JC Bozelli Jr, TL Santos, MA da Silva, GKV Saraiva et al. (2014) Peptide: Lipid ratio and membrane surface charge determine the mechanism of action of the antimicrobial peptide BP100. Conformational and functional studies. *Biochim. Biophys. Acta* 1838:1985–1999.
Matsuzaki K (1998) Magainins as paradigm for the mode of action of pore forming polypeptide. *Biochim. Biophys. Acta* 1376:391–400.
Matsuzaki K (1999) Why and how are peptide-lipid interactions utilized for self-defense? Magainins and tachyplesins as archetypes. *Biochim. Biophys. Acta* 1462:1–10.
Mattei B, ADC França, KA Riske (2015) Solubilization of binary lipid mixtures by the detergent triton X-100: The role of cholesterol. *Langmuir* 31:378–386.
Maulik PR, PK Sripada, GG Shipley (1991) Structure and thermotropic properties of hydrated N-stearoyl sphingomyelin bilayer membranes. *Biochim. Biophys. Acta* 1062:211–219.
Nguyen LT, EF Haney, HJ Vogel (2011) The expanding scope of antimicrobial peptide structures and their modes of action. *Trends Biotechnol.* 29:464–472.
Nishimura K, T Hosoi, T Sunami, T Toyota, M Fujinami, K Oguma, T Matsuura, H Suzuki, T Yomo (2009). Population analysis of structural properties of giant liposomes by flow cytometry. *Langmuir* 25:10439–10443.
Nomura F, M Nagata, T Inaba, H Hiramatsu, H Hotani, K Takiguchi (2001) Capabilities of liposomes for topological transformation. *Proc. Natl. Acad. Sci. USA* 98:2340–2345.
Nyholm T, JP Slotte (2001) Comparison of Triton X-100 penetration into phosphatidylcholine and sphingomyelin mono-and bilayers. *Langmuir* 17:4724–4730.
Pathak P, E London (2011) Measurement of lipid nanodomain (raft) formation and size in sphingomyelin/POPC/cholesterol vesicles shows TX-100 and transmembrane helices increase domain size by coalescing preexisting nanodomains but do not induce domain formation. *Biophys. J.* 101:2417–2425.
Portet T, R Dimova (2010) A new method for measuring edge tensions and stability of lipid bilayers: Effect of membrane composition. *Biophys. J.* 99:3264–3273.
Rathinakumar R, WC Wimley (2008) Biomolecular engineering by combinatorial design and high-throughput screening: small, soluble peptides that permeabilize membranes. *J. Am. Chem. Soc.* 130:9849–9858.
Rex S, G Schwarz (1998) Quantitative studies on the melittin-induced leakage mechanism of lipid vesicles. *Biochemistry* 37 2336–2345.
Riske KA (2015) Optical microscopy of giant vesicles as a tool to reveal the mechanism of action of antimicrobial peptides and the specific case of gomesin. *Adv. Planar Lipid Bilayers Liposomes* 21:99–129.
Robinson T, P Kuhn, K Eyer, PS Dittrich (2013) Microfluidic trapping of giant unilamellar vesicles to study transport through a membrane pore. *Biomicrofluidics* 26:44105.
Säälika P, A Niinep, J Pae, M Hansen, D Lubenets, Ü Langel, M Pooga (2011) Penetration without cells: Membrane translocation of cell-penetrating peptides in the model giant plasma membrane vesicles. *J. Controlled Release* 153:117–125.
Sani M-A, F Separovic (2016) How membrane-active peptides get into lipid membranes. *Acc. Chem. Res.* 49:1130–1138.
Scherer KM, J-H Spille, H-G Sahl, F Grein, U Kubitscheck (2015) The lantibiotic nisin induces lipid ii aggregation, causing membrane instability and vesicle budding. *Biophys. J.* 108:1114–1124.
Schibli DJ, RF Epand, HJ Vogel, RM Epand (2002) Tryptophan-rich antimicrobial peptides: Comparative properties and membrane interactions. *Biochem. Cell Biol.* 80:667–677.
Schuck S, M Honsho, K Ekroos, A Shevchenko, K Simons (2003) Resistance of cell membranes to different detergents. *Proc. Natl. Acad. Sci. USA* 100:5795–5800.
Seddon AM, P Curnow, PJ Booth (2004) Membrane proteins, lipids and detergents: Not just a soap opera. *Biochim. Biophys. Acta* 1666:105–117.
Seelig J (2004) Thermodynamics of lipid–peptide interaction. *Biochim. Biophys. Acta* 1666:40–50.
Shai Y (2002) Mode of action of membrane active antimicrobial peptides. *Biopolymers* 66:236–248.

Shai Y, Z Oren (2001). From "Carpet" mechanism to de-novo designed diastereomeric cell-selective antimicrobial peptides. *Peptides* 22:1629–1641.

Simons K, E Ikonen (1997) Functional rafts in cell membranes. *Nature* 387:569–572.

Sot J, LA Bagatolli, FM Goñi, A Alonso (2006) Detergent-resistant, ceramide-enriched domains in sphingomyelin/ceramide bilayers. *Biophys. J.* 90:903–914.

Sot J, MI Collado, JLR Arrondo, A Alonso, FM Goñi (2002) Triton X-100-resistant bilayers: Effect of lipid composition and relevance to the raft phenomenon. *Langmuir* 18:2828–2835.

Sudbrack TP, NL Archilha, R Itri, KA Riske (2011) Observing the solubilization of lipid bilayers by detergents with optical microscopy of GUVs. *J. Phys. Chem. B* 115:269–277.

Sun Y, W-C Hung, F-Y Chen, C-C Lee, HW Huang (2009) Interaction of tea catechin ()-epigallocatechin gallate with lipid bilayers. *Biophys. J.* 96:1026–1035.

Tamba Y, H Ariyama, V Levadny, M Yamazaki (2010) Kinetic pathway of antimicrobial peptide magainin 2-induced pore formation in lipid membranes. *J. Phys. Chem. B* 114:12018–12026.

Tamba Y, M Yamazaki (2005) Single giant unilamellar vesicle method reveals effect of antimicrobial peptide magainin 2 on membrane permeability. *Biochemistry* 44:15823–15833.

Tamba Y, M Yamazaki (2009) Magainin 2-induced pore formation in the lipid membranes depends on its concentration in the membrane interface. *J. Phys. Chem. B* 113:4846–4852.

Tamba Y, S Ohba, M Kubota, H Yoshioka, H Yoshioka, M Yamazaki (2007) Single GUV method reveals interaction of tea catechin (2)-epigallocatechin gallate with lipid membranes. *Biophys. J.* 92 3178–3194.

Thorén PEG, D Persson, M Karlsson, B Nordén (2000) The antennapedia peptide penetratin translocates across lipid bilayers—The first direct observation. *FEBS Lett* 482:265–268.

Tsamaloukas A, H Szadkowska, H Heerklotz (2006) Nonideal mixing in multicomponent lipid/detergent systems. *J. Phys. Condens. Matter* 18:S1125–S1138.

Tsamaloukas AD, S Keller, H Heerklotz (2007) Uptake and release protocol for assessing membrane binding and permeation by way of isothermal titration calorimetry. *Nat Protoc* 2:695–704.

Wheaten SA, A Lakshmanan, PF Almeida (2013b) Statistical analysis of peptide-induced graded and all-or-none fluxes in giant vesicles. *Biophys. J.* 105:432–443.

Wheaten SA, FDO Ablan, BL Spaller, JM Trieu, PF Almeida (2013a) Translocation of cationic amphipathic peptides across the membranes of pure phospholipid giant vesicles. *J. Am. Chem. Soc.* 135:16517–16525.

Wieprecht T, M Beyermann, J Seelig (1999) Binding of antibacterial magainin peptides to electrically neutral membranes: Thermodynamics and structure. *Biochemistry* 38:10377–10387.

Wimley WC (2010) Describing the mechanism of antimicrobial peptide action with the interfacial activity model. *ACS Chem. Biol.* 5:905–917.

Wu Y, K He, SJ Ludtke, HW Huang (1995) X-ray diffraction study of lipid bilayer membranes interacting with amphiphilic helical peptides: Diphytanoyl phosphatidylcholine with alamethicin at low concentrations. *Biophys. J.* 68:2361–2369.

Zasloff M (2002) Antimicrobial peptides of multicellular organisms. *Nature* 415:389–395.

Lipid-polymer interactions: Effect on giant unilamellar vesicle shape and behavior

Brigitte Pépin-Donat, François Quemeneur, and Clément Campillo

Beyond the mountain is another mountain.

Haitian proverb

Contents

25.1 INTRODUCTION

Cell membranes, made of self-assembled lipids, define the boundary of living cells and act as a selective barrier regulating transport between the cell and the exterior. According to the temperature, which modulates interactions between self-assembled lipids, membranes are either in liquid or gel phase. In addition, they are often "decorated" on their outer side by macromolecules, such as proteins and glycol-conjugates (Singer and Nicolson, 1972) and interact on their inner side with the cytoskeleton, a dynamic macromolecular network, which dictates cell architecture. Interactions between lipids constituting the membrane as well as interactions between lipids and macromolecules drastically change structural and mechanical properties of membranes and, consequently, affect cell shape and cell behavior under various stresses. For example, the carbohydrate-rich layer called the glycocalyx (Sabri et al., 2000), which surrounds some living cells, modifies cell–cell or cell–surface interaction, and the cytoskeleton controls cell shape changes and therefore membrane deformations in processes such as motility, division or filopodia formation (Blanchoin et al., 2014).

Giant unilamellar vesicles (GUVs), constituted by a model biological membrane of controlled composition, are passive simple models of cell, see also Chapter 28. It is then of interest to use GUVs as platforms to study some aspects of cells behavior, especially at the cell interface, under various external stresses such as osmotic shocks, adhesion on a substrate or interaction with other GUVs. Therefore, developing composite polymer-coated GUVs (pGUVs) provides a way to improve the relevance of GUVs as cell models by (i) forming a controlled structure with tunable visco-elasticity in the GUV interior, (ii) mimicking cell membrane–internal cytoskeleton interaction and (iii) mimicking the interaction between cell membrane and macromolecules in contact with its external surface.

The present chapter is organized as follows: first, we deal with the various types of interactions between polymers and lipid GUVs before showing how these interactions can affect membrane structure at different length scales. Then, we focus on pGUVs with membranes in liquid phase interacting with neutral polymers exhibiting a low critical solubility temperature (LCST) and with charged polymers. After a general conclusion, we will present some perspectives.

25.2 INTERACTIONS BETWEEN POLYMERS AND LIPID MEMBRANES

Polymers are made of repeating monomeric units, connected by covalent chemical bonds. Homopolymers are composed of single repeating monomeric unit, whereas copolymer chains are composed of two (or more) different monomer units. Their structures are linear, branched or dendritic. The properties of polymers depend on the nature of monomeric units (e.g., hydrophobic, hydrophilic, charged) and, for copolymers, on their distribution along the chain (Fleer, 1993). Their properties in solution are defined both by the intrinsic polymer characteristics (e.g., chemical structure, molecular weight and concentration) and by external parameters (e.g., nature of the solvent, temperature, ionic concentration) (de Gennes, 1979; Flory, 1979).

In contrast with non-charged polymers whose properties are relatively well understood, polyelectrolytes (charged polymers) still raise many questions. Difficulties arise from the simultaneous action of short-range (excluded volume with electrostatic contribution) interactions between monomers on one chain and of long-range (Coulombic) interactions. This coupling of different length scales leads to a severe influence of local chain properties on the properties of the whole system.

Polymer/membrane interactions depend on both polymer and lipid bilayer chemical structures and charge properties, which can be varied in some cases by external parameters such as temperature, salt concentration and pH. Various polymer/lipid membrane interactions can be considered and classified into four categories (Tribet and Vial, 2008) detailed below. Structure and behavior of pGUVs usually result from the joint action of several of these interactions (Thomas et al., 1996).

Hydrophobic interactions can occur between the tails of the lipid molecules and the polymer that may result in its penetration into the lipid bilayer (Polozova and Winnik, 1997). For example, polymers containing hydrophobic moieties, such as alkyl or cholesterol groups, may exhibit stable anchoring in the lipid membrane. Polymer/lipid hydrophobic interactions have been extensively studied because of their crucial importance in the context of *in vivo* injection and circulation of liposomal drug carriers.

Hydrogen bonding interactions between lipids are at least partly responsible for the lipid self-assembly into bilayers (Boggs, 1987). Polymer/lipid associations via hydrogen bonds remain poorly described in the literature, likely because of their weak strength when compared with other interaction types (Sanderson and Whelan, 2004). In the case of polymers showing a LCST like poly(*N*-isopropylacrylamide) (polyNIPAM) (Schild, 1992) (see Table 25.1) and pluronics, dramatic changes

Table 25.1 **Chemical structures and properties of the principal polymers considered in this chapter: polyNIPAM, chitosan, hyaluronan**

POLYMER	CHEMICAL STRUCTURE	PROPERTIES	REFERENCES
poly(*N*-isopropylacrylamide) or (polyNIPAM)	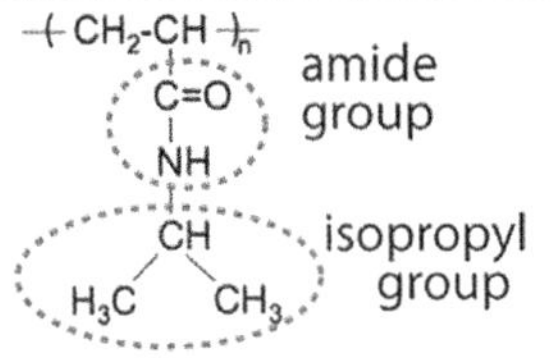 $-(CH_2-CH)_n-$; C=O; NH; CH; H_3C CH_3 — amide group; isopropyl group	• Thermoresponsive: LCST at 32°C (adjustable when copolymerized) • Nontoxic and biocompatible when polymerized	Schild (1992)
Chitosan	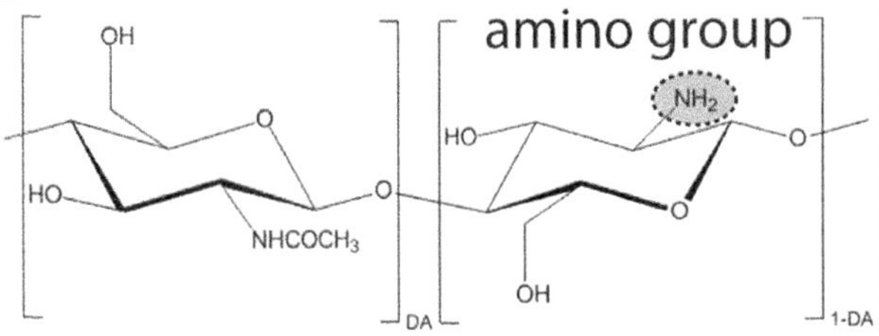 OH; HO; O; $NHCOCH_3$; DA; amino group; NH_2; HO; OH; O; 1-DA	• Pseudo-natural linear random copolymer • Obtained from chitin of crustaceous shells, cuticles of insects and cell walls of some fungi • Positively charged, pK=6.0 • Biocompatible • Biodegradable • Mucoadhesive • Nontoxic • Wound healing promoter	Rinaudo (2006)
Hyaluronan	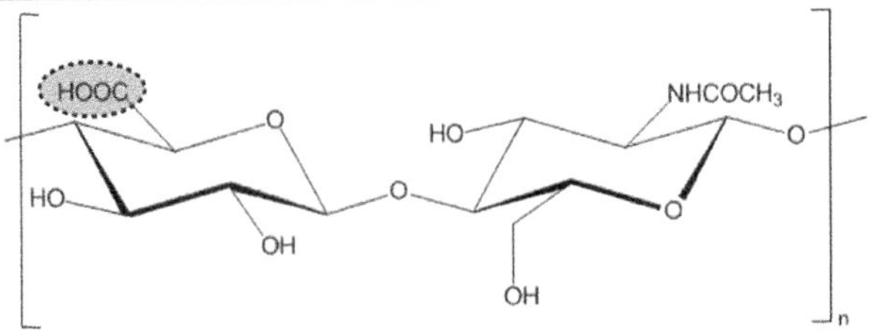 HOOC; HO; O; OH; $NHCOCH_3$; HO; OH; O; n	• Natural linear alternated copolymer • Component of synovial fluid, cartilage, vitreous humor and extracellular matrices • Negatively charged, pKa = 2.9 • Biocompatible • Viscosupplementation • Preserve tissue hydration • Wound healing promoter	Rinaudo (2008); Kennedy et al. (2002)

in the intrachain hydrophilic–hydrophobic balance and in solubility are observed above LCST; this will be discussed in more detail in Section 25.4.

Coulombic interactions. Lipid membranes are commonly made of zwitterionic phospholipids (i.e., phosphatidylcholine [PC]; phosphatidylethanolamine), sometimes associated with negatively charged phospholipids (i.e., phosphatidic acid [PA]; phosphatidylglycerol; phosphatidylserine; phosphatidylinositol). Zwitterionic phospholipids have polar headgroups with at least two local charges of opposite signs, which can be controlled by pH (Quemeneur et al., 2010). The average membrane surface charge depends on all these local charges. Whatever the global charge of the membrane (positive, negative or neutral), these local charges remain and consequently, electrostatic interactions can always occur with charged polymers. A polyelectrolyte (PE1) adsorbed on lipid surface can be completely removed from the membrane by a simple increase of salt concentration or by the addition of another soluble charged polymer (PE2) of opposite charge, which is able to associate with PE1 (Kabanov and Yaroslavov, 2002). Several examples of polyelectrolyte/membrane interactions are reported by Tribet and Vial (2008) and Schulz et al. (2012) and references therein. The interaction of chitosan and hyaluronan (see Table 25.1) with zwitterionic 1,2-dioleoyl-*sn*-glycero-3-phosphocholine (DOPC, see Appendix 1 of the book for structure and data on this lipid) membranes will be extensively discussed in Section 25.5.

Crosslinking by ions and polar groups. Multivalent ions or complexing agents (e.g., biotin-streptavidin) can mediate lipid/polymer interactions. In such cases, a complex balance between effects of polymer length, screening of Coulombic interaction and mediator bridging is involved and modulates the association. For instance, in the absence of multivalent ions, dextran sulfate does not bind to phosphocholine lipids but it does so in the presence of Ca^{2+}, Mg^{2+}, Mn^{2+} or La^{3+} (Huster et al., 1999).

25.3 POLYMERS AFFECT MEMBRANE STRUCTURES AT DIFFERENT SCALES

When macromolecules interact with lipid membranes, a large variety of effects are observed at different length and time scales (Tribet and Vial, 2008; Schulz et al., 2012).

First, at the molecular scale membranes coupled to polymers can adopt various configurations as illustrated in Figure 25.1. Polymers can be (a) **adsorbed** on the internal or external leaflet with various possible conformations (flat on the surface, with loops, or freely moving chain ends, (b) **embedded** in the hydrophobic part of the membrane, or (c) **grafted** either covalently or via a mediator. From a mechanical point of view, adsorbed and anchored polymers modify membrane mechanical properties by inducing a spontaneous curvature C_0 and changing their bending rigidity (see Chapters 5 and 9). From a dynamical point of view, the polymer can induce formation of lipid patterns (Ladavière et al., 2002; Kato et al., 2010; Slochower et al., 2014) and changes lipid lateral mobility (Deverall et al., 2005) or flip-flop (Yaroslavov et al., 2006). Besides, the amount of adsorbed polymer as well as its conformation may change over time (Xie and Granick, 2002; Slochower et al., 2014).

At the scale of the whole pGUV, different structural and morphological changes have been observed depending on the nature of the interactions with the polymer: surface adhesion (Shimobayashi et al., 2015), peeling (Luan and Ramos, 2007) (Figure 25.2a), pore opening (Vial et al., 2007) (Figure 25.2b), spontaneous curvature (Simon et al., 1995) (Figure 25.2c), budding (Tsafrir et al., 2003; Nikolov et al., 2007) (Figure 25.2d), invaginations (Angelova and Tsoneva, 1999), tube ejection and instabilities (Tsafrir et al., 2001) (Figure 25.2e), and even disruption leading to planar composite bilayers (Chung et al., 2009) or micelle formation (Schalchli-Plaszczynski and Auvray, 2002), some examples are illustrated in Figure 25.2. When polymer chains or gels are encapsulated inside GUVs, they also modify its mechanical properties (Viallat et al., 2004).

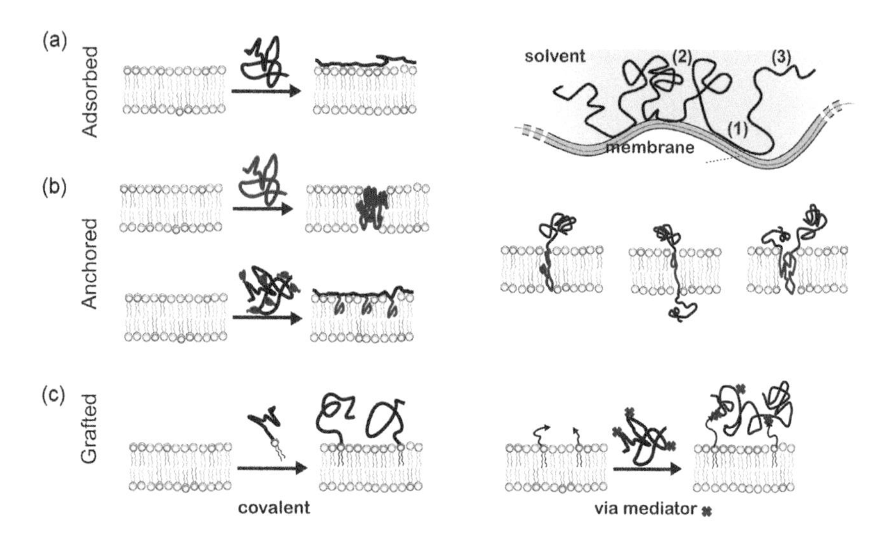

Figure 25.1 Some of the various scenarios that can occur upon interaction between polymers and lipid bilayers: (a) adsorption, (b) anchoring and (c) grafting. (Adapted from Schulz, M. et al., *Soft Matter*, 8, 4849, 2012.)

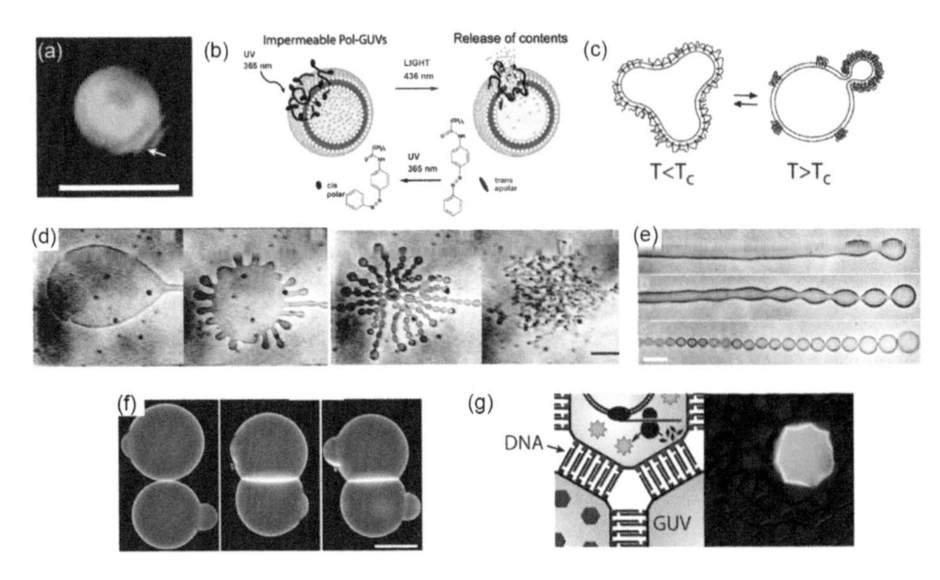

Figure 25.2 Morphological changes observed on GUVs for various types of interactions with polymers. (a) Pore formation and bilayer peeling on multilamellar vesicles due to a strong electrostatic attraction between oppositely charged lipid membrane and polyelectrolytes. (Reprinted with permission from Luan, Y. and Ramos, L. *J. Am. Chem. Soc.*, 129, 14619–14624, 2007. Copyright 2007 American Chemical Society.) (b) GUVs with partially embedded polymer release their internal content upon stimulation by light. (Reprinted with permission from Sebai, S.C. et al., *Langmuir*, 26, 14135–14141, 2010. Copyright 2010 American Chemical Society.) (c) Modified polyNIPAM chains grafted to the membrane collapse, induce spontaneous curvature at the LCST and finally get segregated. (Reprinted from *Chem. Phys. Lipids*, 76, Simon J, Kuhner, M. et al., Polymer-induced shape changes and capping in giant liposomes, 241–258, Copyright 1995, with permission from Elsevier.) (d) Budding and tube growth upon multianchor polymer injection on GUV. Scale = 10 μm. (Reprinted with permission from Tsafrir, I. et al., *Phys. Rev. Lett.*, 91, 138102, 2003. Copyright 2003 by the American Physical Society.) (e) Multilamellar tubular vesicle undergoing a pearling instability upon functionalized dextran injection, at 0, 70 and 150 s. Scale bar: 20 μm. (Reprinted with permission from Tsafrir, I. et al., *Phys. Rev. Lett.*, 86, 1138–1141, 2001. Copyright 2003 by the American Physical Society.) (f) Contact and hemifusion of two GUVs. Scale bar: 25 μm. (Reproduced from Sun, Y. et al., *Biophys. J.*, 100, 987–995, 2011. With permission from Cell press, Copyright 2011.) (g) Higher-order DNA-mediated assemblies of functional GUVs with tissue like architectures. (Reprinted with permission from Hadorn, M. et al., *Langmuir*, 29, 15309–15319, 2013. Copyright 2013 American Chemical Society.)

Finally, at the scale of **assemblies of GUVs**: hemifusion (Heuvingh et al., 2004; Sun et al., 2011) (Figure 25.2f), fusion (Terasawa et al., 2012), GUVs aggregation (Hadorn et al., 2013) (Figure 25.2g), and even transition from GUV to planar multilayer lipid/polymer assemblies (Chiappisi et al., 2013) were reported.

A large panel of methods can be used to characterize polymer/membrane interactions at different scales: at the molecular scale, see (Deleu et al., 2014) and references therein, and at the micrometer scale, that is, at the level of GUVs, see Chapters 10 through 22 of the present book.

25.4 GUVs WITH FLUID MEMBRANE IN INTERACTION WITH LCST NEUTRAL POLYMERS

25.4.1 INTRODUCTION

In this section, we consider fluid lipid membranes interacting with polymers presenting a LCST. At the LCST, such polymers undergo a very sharp reversible transition from an expanded coil to a collapsed globule configuration, thus with an abrupt decrease of their radius of gyration. This transition is easily observed because it induces demixing of the polymer solution that can be monitored optically by a “cloud point” measurement (Heskins and Guillet, 1968), see also Chapter 29. These polymers can be coupled with lipid membranes to produce pGUVs, which physical properties are controlled by temperature. Some of the numerous thermo-responsive polymers presenting an accessible LCST in water, which have been used to design composite membrane systems are: poly(ethylene oxide) (Chandaroy et al., 2001), poly(oligo(ethyleneglycol)methacrylate) (Saaka et al., 2012), poly(organophosphazene) (Couffin-Hoarau and Leroux, 2004) and, the most common of the LCST polymers, polyNIPAM. Here we will focus on membrane–polyNIPAM systems.

PolyNIPAM is widely used for biotechnology applications because its LCST can be adjusted to be close to 37°C by copolymerizing NIPAM with other polymers (Park and Hoffman, 1994). The balance between monomer-monomer and monomer-solvent interactions determines its behavior in solution. The chemical structure of a NIPAM monomer, presented on Table 25.1, explains the thermoresponsive behavior of polyNIPAM chains: below the LCST, the amide groups form hydrogen bonds with water molecules, whereas a hydrophobic cage surrounds the nonpolar isopropyl groups. Above the LCST, hydrogen bonds and the hydrophobic cage are destabilized; therefore the balance between monomer-monomer and monomer-water interactions is shifted toward hydrophobicity. This leads to the collapse of polyNIPAM chains and thus to phase separation in polyNIPAM solutions, and to the global shrinkage of crosslinked polyNIPAM gels. In this section, we present work on membrane-polyNIPAM systems, summarizing the motivations to design such systems, the strategies to couple polyNIPAM chains to the membrane, and the observed changes in pGUVs shape or behavior as the system is driven across the LCST.

Some of these studies were devoted to the reconstitution of an artificial polyNIPAM scaffold inside vesicles to mimic the cytoskeleton. Ringsdorf et al. (1993) prepared small unilamellar vesicles (SUVs) made of 1,2-dimyristoyl-*sn*-glycero-3-phosphocholine (DMPC), DSPC (1,2-distearoyl-*sn*-glycero-3-phosphocholine) and Egg-PA (negatively charged) interacting with chains of hydrophobically modified poly(NIPAM) (HM-polyNIPAM). HM-polyNIPAM is a fluorescently labeled copolymer of NIPAM, comprising octadecyl chains that represent hydrophobic anchors in the lipid membrane (Polozova and Winnik, 1997), see Appendix 1 of the book for structure and data on these lipids. At the LCST, HM-polyNIPAM forms hydrophobic globules, which aggregate in a separated hydrophobic phase. The interaction of these chains with the membrane was characterized by non-radiative energy transfer fluorescence spectroscopy, differential scanning calorimetry and dynamic light scattering. At the LCST, the alkyl chains of HM-polyNIPAM remain anchored in the liposome membrane and their collapse induces a change in the shape of HM-polyNIPAM covered liposomes (Simon et al., 1995). When individual chains collapse, their 2D diffusion coefficient is reduced, as demonstrated by fluorescence recovery after photobleaching experiments in supported bilayers. In the case of GUVs, the collapse of chains inserted in the membrane induces spontaneous curvature, and the collapsed chains tend to accumulate in the zones of high curvature (Simon et al., 1995) (Figure 25.2c). This eventually induces a global segregation of the collapsed chains. Therefore, a collapse of individual HM-polyNIPAM chains at the macromolecular scale affects the membrane bending properties and finally leads to a global phase separation at the scale of the GUVs.

An important motivation to develop composite membranes coupled to polyNIPAM is the design of thermoresponsive liposomes that release their content upon thermal stimulation, constituting *in vivo* drug carriers. Liposomes coupled with NIPAM/Acrylamide/*N,N'*-didodecylacrylamide copolymers, which are able to anchor in the membrane because of their alkylated chains, can indeed release their content when the temperature reaches the LCST (Hayashi et al., 1999). The ratio of NIPAM and acrylamide allows adjusting the LCST around 37°C for *in vivo* applications. Pursuing the same objective, (Stauch et al., 2002b) prepared egg phosphatidylcholine SUVs containing a reticulated network of polyNIPAM/tetraethylene glycol dimethacrylate (TEGDM). These copolymers are attached to the internal leaflet of the bilayer by a hydrophobic anchor of 1,2-distearyl-3-octaethylene glycol ether methacrylate (DOGM, see Appendix 1 of the book for structure and data on this lipid). DOGM possesses two hydrophobic tails that insert in the bilayer, a hydrophilic chain that prevents contact between the copolymer chains and the bilayer and a domain that copolymerizes with NIPAM and TEGDM. The polymerization of the polyNIPAM/TEGDM gel is triggered by UV light in the presence of dietoxyacetophenone (DEAP), an initiator of radical polymerization that is either inside the vesicles before polymerization or enters the vesicle though the lipid membrane. These two cases yield either polyNIPAM/TEGDM gels that fill the whole vesicle internal medium or bidimensional gels close to the membrane (Stauch et al., 2002a).

The encapsulation of mixtures of polyNIPAM and other hydrophilic polymers also allows forming hydrophilic/hydrophobic domains in GUVs when the LCST is reached, and thus to mimic the compartmentalization observed in living cells (Helfrich et al., 2002; Jesorka et al., 2005), see also Chapter 29.

25.4.2 PREPARATION OF GUVs ENCLOSING THERMORESPONSIVE polyNIPAM SOLUTIONS AND GELS

Following the work presented above, GUVs enclosing polyNIPAM solution or gels were prepared in order to obtain biomimetic systems with controlled mechanical properties, which behavior under stress could be compared with the one of living cells. The electroformation technique (Angelova and Dimitrov, 1988) (see Chapter 1) was used to prepare GUVs made of DOPC enclosing a "presol" or "pregel" solution, that contains NIPAM monomers (between 100 and 900 mM), DEAP as a polymerization photoinitiator and various concentrations of methylen-bis-acrylamide as a crosslinking agent (0% for a "presol" medium, 3%–9% molar of the NIPAM concentration for a "pregel" medium). After electroformation, the *in situ* polymerization of NIPAM was triggered by UV light (wavelength of 360 nm). In the absence of crosslinkers, GUVs enclosing a solution of non-reticulated polyNIPAM chains are obtained (these GUVs are called "sol–GUVs" in analogy to the sol/gel terminology). With crosslinkers, the GUVs are filled with a polyNIPAM gel; these GUVs are called "gel–GUVs" (Faivre et al., 2006). It is to be stressed that, with the electroformation technique, the internal and external mediums of the GUVs are identical. To prepare sol–GUVs, the polymerization reaction was triggered just after completing the electroformation of GUVs containing the "presol" medium, and therefore polyNIPAM polymerizes similarly inside and outside the GUVs. Sol–GUVs are floating in a medium containing polyNIPAM chains, but these external chains can easily be washed out. In the case of gel–GUVs, triggering the polymerization just after the electroformation of GUVs containing the "pregel" solution would lead to gel–GUVs trapped in a macroscopic polyNIPAM gel. Therefore, the "pre-gel" GUVs were dispersed in a medium containing no crosslinkers and the polymerization reaction was immediately triggered. Even though crosslinkers diffuse through the membrane, the internal gel polymerizes and gel–GUVs free in solution are finally obtained. This protocol allowed preparing various types of composite GUVs enclosing polyNIPAM solutions or gels, the thermoresponsive behavior and physical properties of which are presented in the following section.

25.4.3 THERMORESPONSIVE PROPERTIES OF GUV/polyNIPAM SYSTEMS

Thermoresponsive properties of GUVs enclosing polyNIPAM systems are summarized in Box 25.1, which presents the structure of standard "bare" GUVs (a, b), sol–GUVs (c, d; e, f; g, h) and gel–GUVs (i, j) below or above the LCST. Their structure modification at the LCST (observed by optical microscopy) depends on the volume fraction of chains and on their crosslink ratio. In the case of sol–GUVs, at low polyNIPAM volume fraction (lower than 600 mM), the chains inside the vesicle locally collapse at the LCST (Box 25.1c and d), and one can observe demixing of their internal medium (Campillo et al., 2007). At high volume fraction

Box 25.1 GUVs with encapsulated polyNIPAM

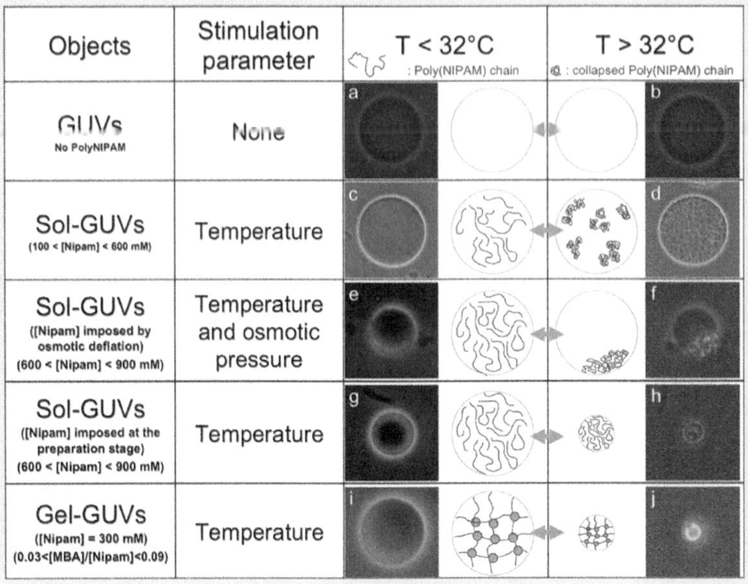

Objects	Stimulation parameter	T < 32°C (: Poly(NIPAM) chain)	T > 32°C (: collapsed Poly(NIPAM) chain)
GUVs No PolyNIPAM	None	a	b
Sol-GUVs (100 < [Nipam] < 600 mM)	Temperature	c	d
Sol-GUVs ([Nipam] imposed by osmotic deflation) (600 < [Nipam] < 900 mM)	Temperature and osmotic pressure	e	f
Sol-GUVs ([Nipam] imposed at the preparation stage) (600 < [Nipam] < 900 mM)	Temperature	g	h
Gel-GUVs ([Nipam] = 300 mM) (0.03<[MBA]/[Nipam]<0.09)	Temperature	i	j

Behavior of GUVs enclosing polyNIPAM above and below the LCST as obtained by Campillo et al. (2008): Bare GUVs show no thermoresponsive behavior (a, b). Sol–GUVs at low polyNIPAM volume fraction can exhibit local chain demixing (c, d). Sol–GUVs at high polyNIPAM volume fraction obtained by osmotic deflation of Sol–GUVs as presented in (c) show global demixing of the entangled chains (e, f). Sol–GUVs at high polyNIPAM volume fraction set during the preparation stage induce global collapse of the whole GUV (g, h). Gel–GUVs show global contraction when raising the temperature above the LCST (I, j).

(higher than 600 mM), two different cases are observed depending on how this concentration is attained (Campillo et al., 2008): the concentration can be either established by osmotically deflating sol–GUVs prepared at lower NIPAM concentration or it can be fixed at the preparation step. When the high concentration is achieved by osmotic deflation (by changing the external osmolarity from 300 mOsm/L to 600 or 900 mOsm/L using concentrated sucrose solutions), the entangled chains globally collapse and get segregated in one side of the GUV without changing the shape of the vesicle (Box 25.1e and f). For sol–GUVs initially prepared with a high polymer concentration, the internal medium globally collapses and deforms the vesicle (Box 25.1g and h). The difference between these two cases shows that the membrane and the internal polyNIPAM solution are interacting and that the magnitude of this interaction depends on the NIPAM concentration at the preparation stage. Indeed, the osmotic deflation changes the final polyNIPAM concentration inside pGUVs without changing the magnitude of polyNIPAM-membrane interaction. This also shows that the response to temperature changes of these GUV systems can be modulated by the conditions of polymerization and crosslinking. Finally, gel-vesicles globally collapse by expelling their internal fluid while their membrane crumples around the shrunk gel. This phenomenon is fully reversible, the gel–GUV returns to its initial state when temperature decreases below LCST without any apparent sign of damage. This deformation of the membrane further demonstrates the strong interaction with the internal polyNIPAM gel or solution, which will be further explored by mechanical measurements in the next section. It is to be emphasized that the global collapse of polyNIPAM solutions at high concentration has not been investigated in detail and remains to be explored. Moreover, because the transition process inside GUVs is similar to what is observed in macroscopic samples, we can conclude that there is not any confinement effect at this length scale. The ability to respond in very different ways to an external stimulus is a very preliminary step toward mimicking cytoskeleton contractility (see Chapter 4) or dynamic cellular compartmentalization, and allows using these objects for controlled drug delivery.

25.4.4 BEHAVIOR UNDER STRESS OF GUV/polyNIPAM SYSTEMS: MECHANICAL PROPERTIES AND STRUCTURE

In order to further inspect the coupling between the lipid bilayer and the polymer, and to demonstrate the relevance of these systems as mechanical models of living cells, various mechanical parameters have been measured on GUVs enclosing polyNIPAM systems as summarized in Box 25.2.

The application of a point force on a membrane allows forming a membrane nanotube of diameter ranging from 10 to 100 nm (Figure 25.3a) (see also Chapter 16). Hydrodynamic nanotube extrusion was used to determine mechanical characteristics of polyNIPAM pGUVs: the vesicle was attached to a glass micro needle and a flow with velocity U, and therefore a Stokes force, is applied on the vesicle (Borghi et al., 2003). In this experiment, standard tethered GUVs behave as entropic springs of stiffness K_{eff}: their membrane resists to the formation of the nanotube that increases their tension, and consequently, the force necessary to hold the nanotube varies with its length. In the case of sol–GUVs,

Box 25.2 Mechanical properties of GUVs with encapsulated polyNIPAM

	Point force (membrane nanotube)	Pressure (micropipette)	k_b (k_BT)	E (kPa)	η (Pa.s)	Control parameter
GUVs	Entropic spring $K_{eff\,GUV}$ F	Fluctuating Membrane ΔP	20	0	10^{-3}	none
Sol-GUVs (100 < [Nipam] < 600 mM)	$K_{eff\,solGUV} \neq K_{eff\,GUV}$ pNIPAM-membrane interaction	Fluctuating Membrane + pNIPAM	< 20	0	10^{-3}-10^{-1}	[Nipam]
Sol-GUVs ([Nipam] imposed by osmotic deflation) (600 < [Nipam] < 900 mM)	F	ΔP		0	10^{-3} - 3	$[\text{Glucose}]_{ext}$
Sol-GUVs ([Nipam] imposed at the preparation stage) (600 < [Nipam] < 900 mM)				0	10^{-1} - 3	[Nipam]
Gel-GUVs ([Nipam] = 300 mM) ($0.03 < [MBA]/[Nipam] < 0.09$)	$K_{eff} >> K_{eff\,solGUV}$ « patches » model F	Elastic sphere ΔP		0.5 - 25	0	[MBA]/[Nipam]

Synopsis of mechanical studies performed on composite GUVs encapsulating polyNIPAM systems. The approaches used to deduce the membrane properties are based on (i) the application of a point force on the GUV membrane to form membrane nanotubes, (ii) aspirating the GUV with a micropipette (see Chapter 11 for description of the technique), (iii) bulk viscosimetry of polyNIPAM solutions prepared in the same way as the internal medium of the sol–GUVs (Campillo et al., 2007). As a result, one can deduce the values of the bending modulus, κ, the elastic modulus, E and the viscosity, η. In the table above, K_{eff} is the effective spring constant of the tethered GUV, ΔP is the suction pressure applied via the micropipette.

K_{eff} is lowered compared with bare GUVs, because the bending rigidity of their membrane is decreased by polyNIPAM adsorption (Kremer et al., 2008), in agreement with theoretical predictions (Brooks et al., 1991) (see Chapter 9). In the case of gel–GUVs, K_{eff} is strongly increased, because the lipids have to be detached from the internal polyNIPAM gel to form a nanotube. This is also the case in cells, where, to form a tube, one has to pay the adhesion energy W required to detach the lipids from the underlying cytoskeleton (Hochmuth and Marcus, 2002). In the case of gel-vesicles, W is on the order of 10^{-5} J (Kremer et al., 2008), and thus close to the values measured on cells. Moreover, it was observed that shorter tubes were formed on gel–GUVS, suggesting that only a fraction of the lipids freely flows inside the nanotube. This led us to propose a model where "patches" of freely diffusing lipids are confined inside corrals that are fixed by the spots where the gel is anchored in the membrane (Kremer et al., 2008). It would be interesting to probe in detail how the crosslink ratio of the gel affects the gel-membrane interaction and therefore the forces required for the formation of membrane nanotubes.

Applying a suction pressure with a micropipette on the fluctuating membrane of standard GUVs allows measuring the membrane bending rigidity at low tensions (Fournier et al., 2001) (see Chapter 11). In sol–GUVs, preliminary results seem to confirm that, as in nanotube experiments, the bending rigidity is decreased by polyNIPAM adsorption. This result needs to be confirmed and compared in detail to theoretical predictions. On the contrary, gel–GUVs behave as elastic homogenous spheres: as the membrane is strongly adhered to the internal gel, micropipette aspiration under a controlled aspiration pressure ΔP leads to the deformation of the whole GUV without any observed detachment of their membrane (Figure 25.3b, insert) and therefore provides an *in situ* measurement of the enclosed gel elasticity. This experiment further confirms the strong gel-membrane interaction that we deduce from the membrane crumpling observed at the LCST. The deformation of the GUV surface scales linearly with the value of the aspiration pressure (Campillo et al., 2009), the slopes of deformation versus pressure curves (Figure 25.3b) thus allow calculating the elastic modulus E of the internal gel using the elastic model developed by (Theret et al., 1988): the aspired portion of the pGUV is Lp $= 6.2R_p\Delta P / 2\pi E$ where R_p is the micropipette internal radius. The elastic modulus is controlled by the crosslink ratio of the internal gel and ranges between 0.5 and 25 kPa (Figure 25.3c).

Finally, in the case of sol–GUVs, the internal and external media of polyNIPAM pGUVs are identical after the polymerization reaction. Therefore, to probe the internal viscosity of sol–GUVs, the viscosity of bulk polyNIPAM solutions that are prepared in identical conditions was measured. These solutions, and thus the internal medium of sol–GUVs have a viscosity that can be adjusted by the internal NIPAM concentration (50 to 800 mM) between 10^{-3} and 1 Pa.s (Figure 25.3d). This internal viscosity can also be varied by osmotically deflating the GUVs, which induces changes of the internal polymer concentration.

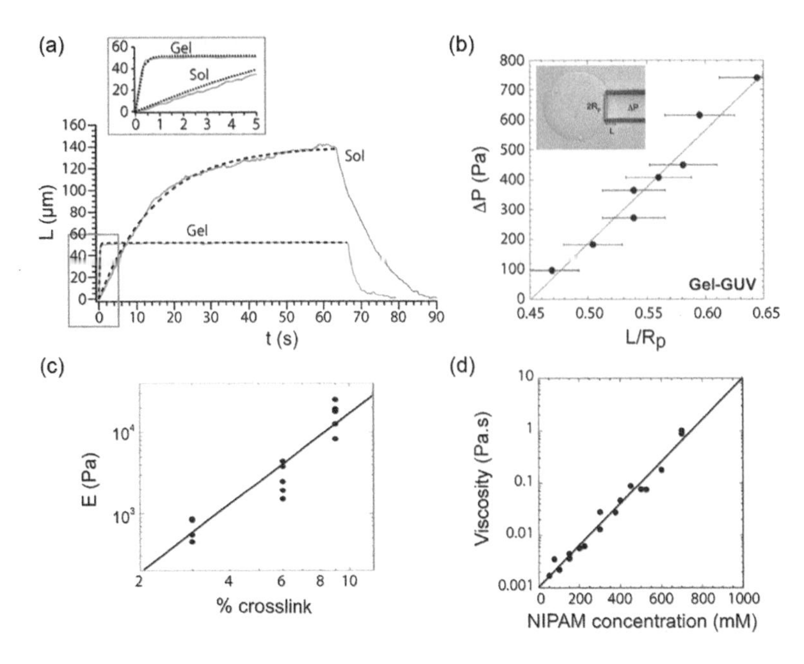

Figure 25.3 (a) Length of nanotubes hydrodynamically extruded from a sol–GUV and a gel–GUV. (Reprinted with permission Kremer, S. et al., Nanotubes from gelly vesicles, *Europhys. Lett.*, 82, 48002, 2008. Copyright 2008, Institute of Physics.) (b) Aspiration of a gel–GUV in a micropipette of radius Rp with an aspiration pressure ΔP. The part of the gel–GUV aspirated in the pipet (see insert) has the length L. Plotting L normalized by Rp versus the aspiration pressure ΔP yields the value of the elastic modulus E, see text for details. (Reprinted from *Mater. Sci. Eng. C*, 29, Campillo, C.C. et al., Composite gel-filled giant vesicles: Membrane homogeneity and mechanical properties, 393–397, Copyright 2009, with permission from Elsevier.) (c) Elastic moduli E measured on gel–GUVs from samples prepared with different crosslink ratios (% molar). (Reprinted from *Mater. Sci. Eng. C*, 29, Campillo, C.C. et al., Composite gel-filled giant vesicles: Membrane homogeneity and mechanical properties, 393–397, Copyright 2009, with permission from Elsevier.) (d) Internal viscosity of sol–GUVs enclosing non crosslinked polyNIPAM solution as a function of the initial NIPAM concentration. (Campillo, C. et al., *Soft Matter*, 3, 1421–1427, 2007. Reproduced by permission of the Royal Society of Chemistry.)

Altogether, these mechanical measurements show that the viscosity of sol–GUVs and the elasticity of gel–GUVs are controlled by the NIPAM and cross linker concentrations and can be adjusted in the range measured on various types of living cells (Campillo et al., 2007). Moreover, as in living cells, the internal gel adheres to the GUV membrane and its adhesion energy is close to the one coupling the cytoskeleton to the plasma membrane of living cells. Therefore, these mechanical measurements show the relevance of polyNIPAM pGUVs as basic mechanical models of living cells.

The behavior of composite GUVs upon specific adhesion on a surface decorated with adhesion molecules and upon application of a global isotropic pressure is summarized in Box 25.3. Sol–and gel–GUVs do not adhere to specifically coated substrates. This reveals that, in both cases, polyNIPAM chains coat the outer lipid membrane and prevent their adhesion. Changing the preparation technique of the pGUVs, for example by using the inverted emulsion method (Pautot et al., 2003), provided external membranes devoid of polyNIPAM, and therefore allowed studying how composite GUVs spread as a function of their internal viscosity and elasticity.

Notably, the shape of composite GUVs under osmotic deflation is very different from that of bare GUVs, which is well-described by the area difference elasticity (ADE) model (Mui et al., 1995) (see Chapter 5). In the case of sol–GUVs, ruffled shapes that resemble echinocyte red blood cells transiently form (Gerald Lim et al., 2002). These shapes relax to ADE shapes within minutes. It would be interesting to study these morphologies and their evolution as a function of the internal viscosity. As far as gel–GUVs are concerned, they keep their spherical shape upon deflation: their membrane is probably crumpled around the gel as when the gel collapses at the LCST, showing again the strong interaction that exists between the membrane and the internal gel, even though some membrane tubules appear at high deflation ratio.

These experiments, together with the mechanical measurements discussed above, led us to propose tentative structures for the polyNIPAM pGUVs, which are presented in Box 25.3.

As a conclusion, polyNIPAM, probably the most common LCST polymer, and copolymers based on polyNIPAM, allow preparing a family of thermoresponsive pGUVs. The development of such thermoresponsive systems has motivated several groups to use polyNIPAM to emphasize the importance of pGUVs as a mechanical model of living cells. Indeed, polyNIPAM gels can be used to form a scaffold inside GUVs. The mechanical properties of this model "cytoskeleton" are controlled by the chemistry of the NIPAM solution or gel, and can be adjusted to mimic the passive mechanical behavior of living cells. Moreover, the internal medium of composite pGUVs can form separated compartments and thus mimic the compartmentalization observed in living cells. Besides, obtaining thermoresponsive vesicles also provides stimuli-responsive drug carriers that can release the drugs they encapsulate upon a temperature variation. Finally, polymers gels were also used to form GUVs, in this case the polymeric scaffold is outside the vesicle and directs GUVs growth from partially hydrated agarose and lipids films (Horger et al., 2009). Even though this technique affects the mechanical properties of the GUVs (Lira et al., 2014), it offers novel possibilities for encapsulation and imaging of immobilized GUVs (Lira et al., 2016).

Box 25.3 Response of GUVs with encapsulated polyNIPAM coupled to the bilayer

	Specific adhesion A	Pressure (osmotic deflation) B	Structure C
GUVs	Adhesion	ADE shapes	$k_b = 20\ k_BT$
Sol-GUVs	No adhesion: pNIPAM on surface	Transient « echinocyte »	$k_b < 20\ k_BT$ $0.001 < \eta < 1$ Pa.s
Gel-GUVs	No adhesion: poly(Nipam) on surface	Spherical deflation	$0.5 < E < 25$ Pa

Response of composite GUVs coupled to polyNIPAM systems to specific adhesion and osmotic shocks. Proposed structures of these systems based on all the performed mechanical measurements are illustrated in the last column. The sol–GUVs are composed of a lipid membrane decorated by polyNIPAM chains on both sides enclosing a viscous internal medium (Campillo et al., 2007). Gel–GUVs are homogeneous gel spheres surrounded by a lipid membrane coupled to the internal gel (Campillo et al., 2009). Their membrane is made of patches of freely diffusing lipids and of lipids stuck to the internal gel (Kremer et al., 2008).

25.5 GUVs WITH FLUID MEMBRANE INTERACTING WITH CHARGED POLYMERS

25.5.1 INTRODUCTION

In this section we address the interaction of lipid membranes with charged polymers. Polyelectrolyte adsorption on various charged surfaces has been extensively discussed in the literature (Fleer, 1993; Szilagyi et al., 2014), especially on lipid assemblies (Xie and Granick, 2002; Tribet and Vial, 2008; Schulz et al., 2012). In particular, studies on GUVs report membrane interaction with cationic polymers such as poly(L-lysine), chitosan (Mertins and Dimova, 2013b), or anionic polyelectrolytes such as DNA (Hristova et al., 2002; Hisette et al., 2008).

Here, we illustrate the zwitterionic lipid/polyelectrolyte association considering DOPC membranes interacting with two natural polyelectrolytes: chitosan, a positively charged polymer in acidic conditions (pH < 6.5), and hyaluronic acid (hyaluronan), a negatively charged polymer at pH > 2.0 (Table 25.1). Both polyelectrolytes find many biomedical applications in association with lipids, such as tissue engineering and regenerative medicine (Kennedy et al., 2002; Rinaudo, 2006) due to their biocompatibility, biodegradability, and non-toxicity.

In this section, we present the physicochemical characterization of these systems using GUVs as a test platform. We discuss the adsorption mechanism. Then we deal with the structures of the composite membranes at different scales and finally we study the coating effect on the mechanics and resistance of these composite GUVs under various stresses.

25.5.2 DECORATION OF GUVs BY POLYELECTROLYTES: MECHANISMS AND STRUCTURE

For these experiments, bare DOPC GUVs suspensions were added to the polyelectrolyte solution, homogenized and then left to rest during 30 min at room temperature for incubation (Quemeneur et al., 2010). As already discussed in Section 25.3, DOPC zwitterionic membranes exhibit local charges tuned by pH, whatever its global charge. Figure 25.4a shows that, whatever

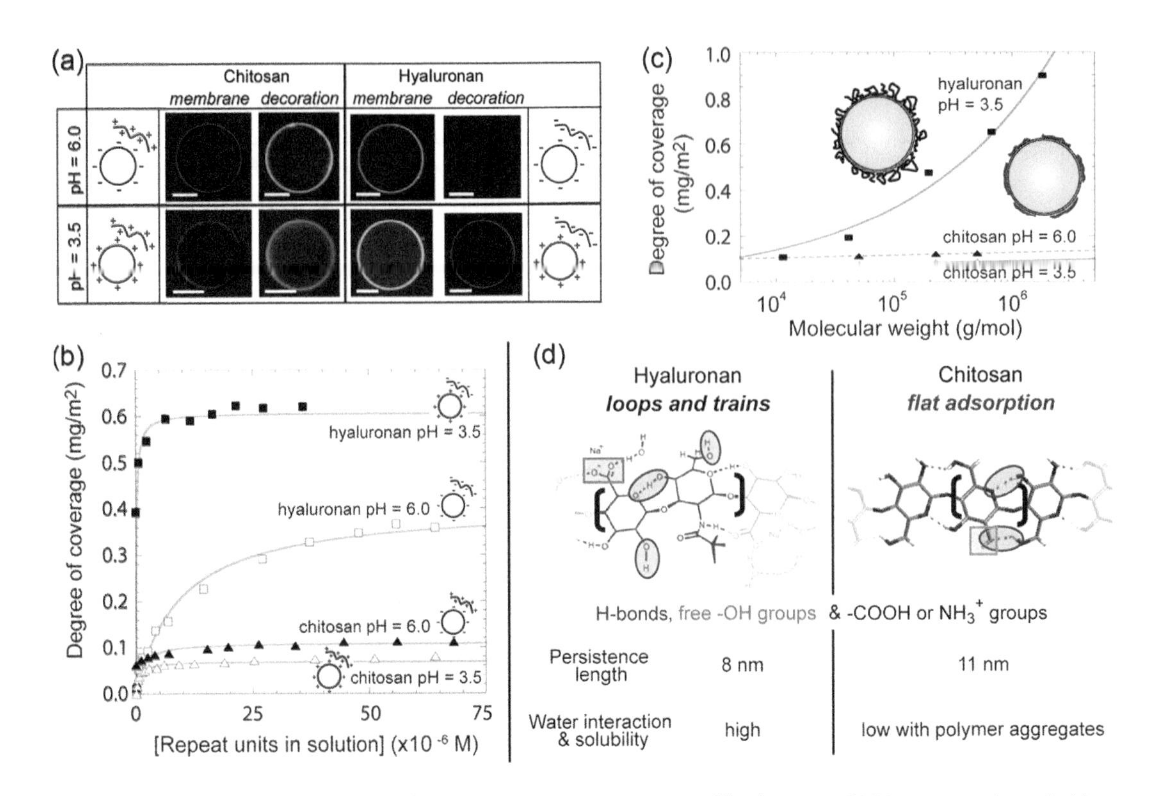

Figure 25.4 (a) Fluorescence confocal observations of GUVs coated with chitosan and hyaluronan. GUVs were incubated either at pH 3.5 or at pH 6.0. For the same GUV, the lipid membrane and polyelectrolyte coating are imaged successively. Scale bars: 10 µm. For each pH, sketches illustrate the respective charge signs of the membrane and polyelectrolytes. (Quemeneur, F. et al., *Soft Matter*, 6, 4471–4481, 2010. Reproduced by permission of the Royal Society of Chemistry.) (b) Degree of coverage (in mg/m2) for chitosan (triangles) and hyaluronan (squares) as a function of the repeat unit concentration of polyelectrolyte free in solution. The dotted lines are added to guide the eye. (Quemeneur, F. et al., *Soft Matter*, 6, 4471–4481, 2010. Reproduced by permission of the Royal Society of Chemistry.) (c) Influence of the molecular weight on the maximum amount of chitosan and hyaluronan adsorbed on the DOPC membrane. (Reproduced from Quemeneur, F., Relation entre les paramètres mécaniques et le comportement sous contraintes externes de vésicules lipidiques à membrane modifiée, PhD Dissertation, Université Joseph Fourier, Grenoble, France, 2010.) (d) Interpretation on the chitosan and hyaluronan conformation difference at the membrane surface based on their chemical structures (H-bonds, free –OH groups and –COOH or NH_3^+-groups are highlighted), chain rigidity and water interaction/solubility. (Reproduced from Quemeneur, F. Relation entre les paramètres mécaniques et le comportement sous contraintes externes de vésicules lipidiques à membrane modifiée, PhD Dissertation, Université Joseph Fourier, Grenoble, France, 2010.)

the pH of the solution (i.e., whatever the average net charge of the vesicle), positively charged chitosan or negatively charged hyaluronan adsorb and form homogeneous layers at the external lipid interface of DOPC GUVs at the microscopic scale. Because of the zwitterionic nature of DOPC lipids, polymer chains adsorb irrespectively of the respective global sign of the membrane and the polyelectrolyte. Electrostatic origin of the interaction between both polyelectrolytes and zwitterionic lipid membranes could be revealed, showing the role of their respective charge densities (both tunable by pH) on the amount of adsorbed polymer (Quemeneur et al., 2008). The amount of adsorbed polymer is higher in the case of membrane and polymer with opposite charges (Figure 25.4b). The adsorbed amount is also affected by the salt concentration (Quemeneur et al., 2010). These results were confirmed by other studies (Taglienti et al., 2006; Mertins et al., 2010; Mertins and Dimova, 2013a).

Adsorption isotherms obtained with polymers having different molecular weights, shown in Figure 25.4c, demonstrate that the polymer chemical structure controls the adsorbed polyelectrolyte chain conformation at the vesicle surface (i.e., chitosan adsorbs flat, whereas hyaluronan forms loops and trains), and consequently the amount of adsorbed polymer. Actually, as reported in Figure 25.4d, intrachain hydrogen bonds are more numerous for chitosan, which implies a higher rigidity of the chains illustrated by a higher persistence length: 11 nm for chitosan (Rinaudo, 2006) and 8 nm for hyaluronan (Kennedy et al., 2002). Moreover, chemical groups interacting with water molecules, such as hydroxyl, carboxyl and amino groups, are more numerous in hyaluronan, which can explain its higher water solubility compared with chitosan: hyaluronan tends to form loops at the membrane surface. Finally, considering the charge densities per repeat unit being equivalent for both polyelectrolytes (0.3 and 0.2 for chitosan at pH 6.0 and hyaluronan at pH 3.5 respectively), it was concluded that adsorbed chain conformations mainly depend on the polymer chemical structure. Global charge of the pGUV can be modified in a controlled way by changing the adsorbed amount of polymer, the pH, or the nature of the polyelectrolyte.

The behavior of a collection of GUVs in solution was also observed. Figure 25.5a describes the role of polyelectrolyte sorption on GUV suspensions state. GUVs reversibly aggregate upon progressive addition of both polyelectrolytes when membrane and polyelectrolyte charges are of opposite sign. The maximal aggregation is reached when the global charge of the lipid vesicles is compensated

(a)

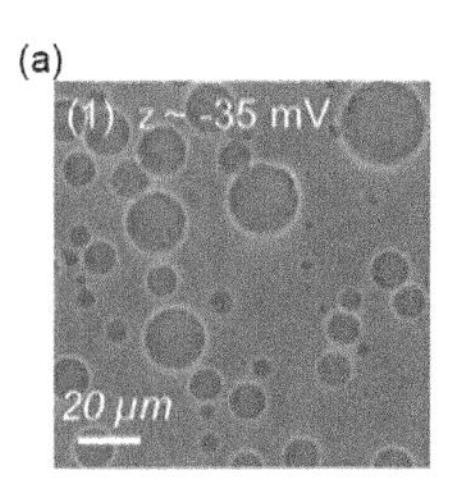

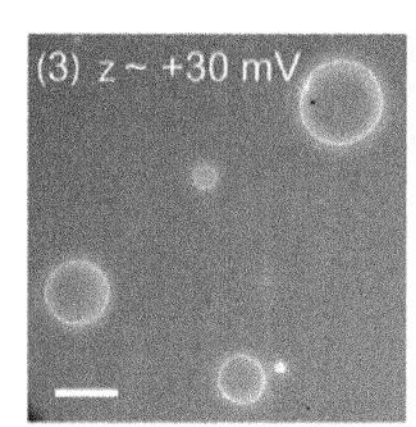

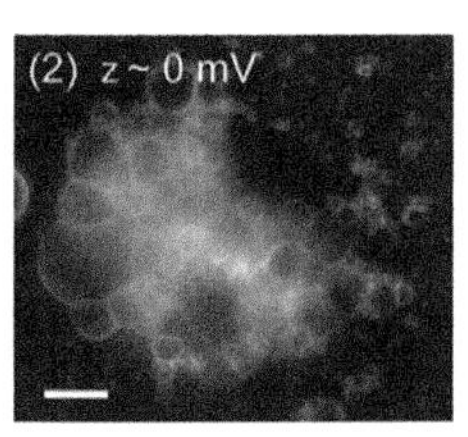

(b)

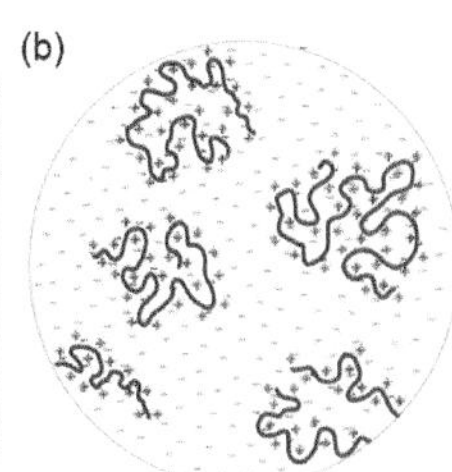

Figure 25.5 (a) Illustration of the aggregation-dissociation process for DOPC GUVs incubated, at pH 6.0, with chitosan at different molar ratios. The respective zeta potential of the membrane is indicated on each snapshot. Scale bars: 10 μm. (Quemeneur, F. et al., *Soft Matter*, 6, 4471–4481, 2010. Reproduced by of the Royal Society of Chemistry.) (b) Possible arrangement (patch-like) of adsorbed polycations on the surface of particle (e.g., a GUV) with low negative surface charge density.

by the adsorbed oppositely charged polyelectrolyte (i.e., at the isoelectric point). Then upon further polymer addition, adsorption still occurs but vesicle aggregates are progressively dissociated with charge inversion to finally obtain isolated overcharged vesicles exhibiting the same dimension as that of vesicles in the bare state.

This aggregation-disaggregation process was interpreted assuming a patch-like model (Figure 25.5b) of the coated membrane where domains of stuck charged polymer (i.e., with a local charge excess) alternate with domains of bare surface (Dobrynin et al., 2001). Within this model, particles association mainly results from short range electrostatic attraction between oppositely charged patches of two approaching vesicles, whereas aggregate dissociation, upon further polyelectrolyte addition, is attributed to long-range electrostatic repulsion between overcharged vesicles; the resulting dispersion of chitosan coated GUVs is stabilized and they are isolated in solution. Note that we never observed hemifusion or fusion processes in our experimental conditions as reported for other lipid membrane/polyelectrolyte interactions (Sun et al., 2011).

Using a combination of zeta-potential measurements on large unilamellar vesicles (LUVs) and fluorescence microscopy observations on GUVs, it was demonstrated that chitosan coating is stable over 4 days when the polyelectrolyte concentration in the external solution remains constant. When the polyelectrolyte concentration is decreased (upon dilution), desorption slowly occurs. Coating is reversible upon dilution: the degree of coverage obtained at the equilibrium (typically reached after 2 days) corresponds to the one expected by the direct incubation of bare vesicles at this polyelectrolyte concentration. Nevertheless, due to slow desorption kinetics, coating is stable during several hours (Quemeneur et al., 2010), which justifies the use of non-grafted polyelectrolyte coated vesicles as drug carriers.

To sum up, interactions of chitosan and hyaluronan with DOPC GUVs are mainly of electrostatic origins. Adsorbed amounts of chitosan and hyaluronan depend on pH and range between 0.07–0.12 and 0.11–0.90 mg/m^2 respectively. Chitosan adsorbs flat, whereas hyaluronan forms loops and trains on the membrane. Both polyelectrolyte adsorptions give raise to a patch-like structure at the GUV surface and are reversible upon dilution.

25.5.3 BEHAVIOR OF POLYELECTROLYTE-COATED VESICLES UNDER EXTERNAL STRESS

Polyelectrolyte GUVs were submitted to various external stimuli. Results are discussed in comparison with those obtained for bare vesicles in order to assess polyelectrolyte coating effect on membrane properties.

The response of bare and polyelectrolyte GUVs submitted to changes in osmotic pressure induced by glucose shocks (Figure 25.6a) was observed. For bare vesicles, a large panel of deflated shapes can be obtained (Mui et al., 1995), whereas for polyelectrolyte GUVs only spherical deflation with ejection of chitosan coated tubes are observed (Kremer et al., 2011). As discussed in Chapter 5, this behavior can be interpreted taking into account the spontaneous curvature (C_0) induced by the asymmetric polyelectrolyte coating on the external surface of the membrane (Figure 25.6b). This interpretation was confirmed by pulling membrane tethers (protocols are detailed in Chapter 16) from vesicles coated by chitosan at 0.90 mg/m^2. The spontaneous curvature of the chitosan coated membrane was measured to be $C_0 = 9.4 \pm 0.6\ 10^{-3}\ \text{nm}^{-1}$ (Kremer et al., 2011).

Salt shocks (Figure 25.6a) have also been applied: bare GUV were observed to systematically explode within 2 min, whereas polyelectrolyte GUVs deflated spherically with invaginations (Quemeneur et al., 2007). The formation of invaginations under salt shocks can be also explained by an increase in the magnitude of the spontaneous curvature (Figure 25.6b). Indeed, when a polyelectrolyte adsorbs on the membrane, it tends to bend it, inducing the previously measured positive spontaneous curvature. Upon salt addition, the intra and interchains electrostatic repulsions are suppressed, inducing the polymer collapse, hence causing a negative spontaneous curvature and consequently invaginations. At this point, it is to be stressed that the sign of the spontaneous curvature and consequently the response of coated vesicles submitted to stresses can be controlled by modifying the electrostatic screening as illustrated in Figure 25.6b: when osmotic deflation is applied, some excess membrane area becomes available and, corresponding to the sign of the spontaneous curvature, can lead to the ejection of tubes (positive curvature) or to the appearance of invaginations (negative curvature).

Finally, anisotropic compression of isolated GUV was studied, combining atomic force microscopy (AFM) with epifluorescence microscopy observations, as detailed in Chapter 12. This leads to reproducible and reversible force-deformation curves for individual vesicles in the range of small deformation (relative deformation up to 0.3). These curves were analyzed using a simple elastic model (Lulevich et al., 2004) for the deformation of impermeable, fluid-filled spherical elastic thin shells made of homogeneous isotropic material. This model describes well the observed radius-dependency of the force response (Figure 25.6c). The chitosan coating increases the effective stretching modulus of the lipid membranes (Rinaudo et al., 2013) and leads us to assume

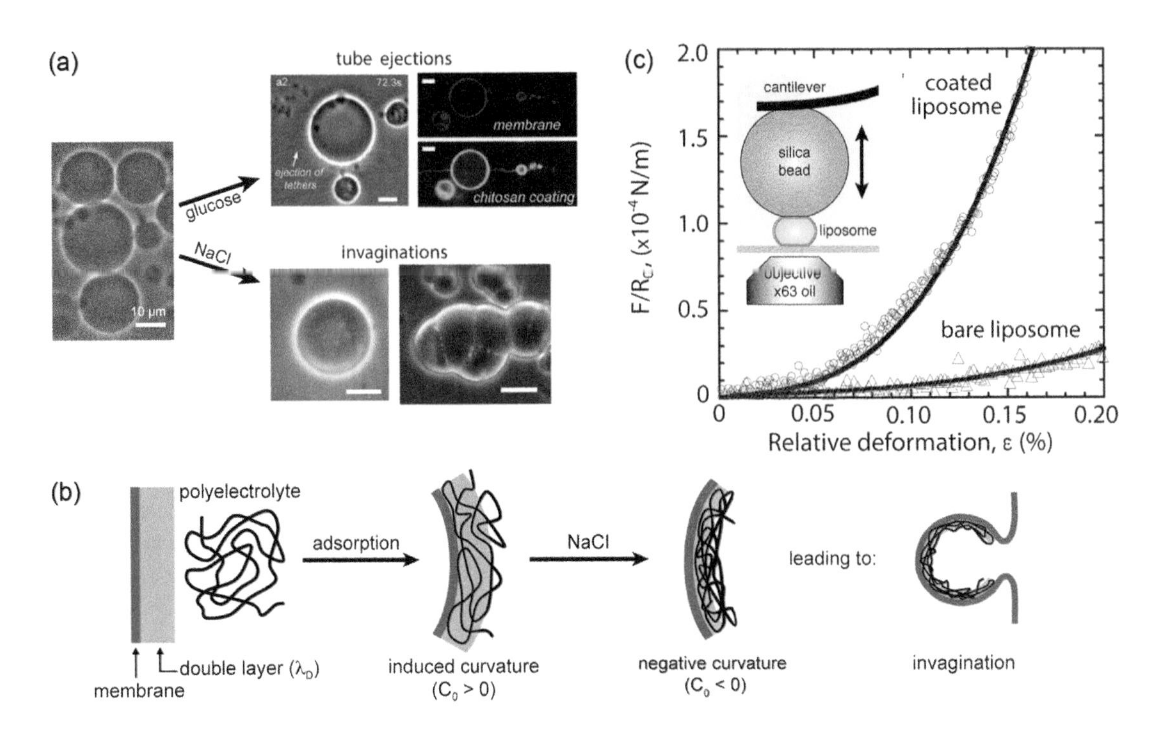

Figure 25.6 (a) Chitosan-coated vesicles submitted to a change in osmotic pressure induced by glucose shocks and NaCl shocks at pH 6.0. The lipid membrane and the polyelectrolyte coating are successively visualized by fluorescence confocal observations. Scale bars: 10 µm. (Reprinted with permission from Quemeneur, F. et al., *Biomacromolecules*, 8, 2512–2519, 2007. Copyright 2007 American Chemical Society; Reprinted with permission from Quemeneur, F. et al., *Biomacromolecules*, 9, 396–402, 2008. Copyright 2007 American Chemical Society; Reprinted with permission from Kremer, S. et al., Nanotubes from asymmetrically decorated vesicles, *Soft Matter*, 7, 946–951, 2011. Copyright 2011, Institute of Physics.) (b) Schematic figures of membrane bending induced by polyelectrolyte adsorption, invaginations formed upon addition of salt, and related spontaneous curvatures C_0. (c) Mechanical characterization of micrometric chitosan-coated vesicle by AFM. Experimental force-response/radius data as a function of the relative deformation ε for bare vesicles (triangles) and chitosan-coated vesicles (circles) and associated fit curves according to the elastic model from Lulevich et al. (solid line) (From Lulevich, V.V. et al., *J. Chem. Phys.*, 120, 3822, 2004; With kind permission from Taylor & Francis: *Int. J. Polym. Anal. Charact.*, Experimental characterization of liposomes stabilized by polyelectrolytes and mechanism of interaction involved 17, 2012, 1–10, Rinaudo, M. et al.)

that chitosan patches (already suggested in the aggregation-disaggregation process) are connected to form a physical network. (Mertins and Dimova, 2013b) also demonstrate membrane stiffening due to an increase of the bending modulus of the chitosan coated-membranes. Chitosan also stabilizes zwitterionic membranes increasing their resistance to surfactants (Mady et al., 2009), but it alters the permeability of the highly negatively charged membranes and may cause leakage.

As a conclusion, pGUVs coated with biocompatible and biodegradable polyelectrolytes such as chitosan and hyaluronan are promising as drug carriers because they present enhanced resistance to various stresses of biological interest, interesting structural and mechanical properties, and responsive properties under mechanical and chemical stresses if compared with standard lipid GUVs. They also hold promise for new potential applications such as multi-compartment carriers because of their ability to form well-defined finite aggregates when suspended in external polyelectrolyte solution at specific concentration.

25.6 VISION

In this chapter, composite pGUVs with controlled mechanical properties were presented. In this context, two future directions can be envisioned. First, it may be of interest to study the behavior of composite GUVs under stress (for example in confined shear flows, osmotic deflation or adhesion to a substrate) for different values of one mechanical parameter, as internal elasticity or viscosity. Then, this would allow comparing their behavior to the one of living cells to separate passive and active cellular response under stress. Second, developing GUVs with the internal membrane surface covered with a 2D layer of responsive gel may also be of interest. Indeed, in living cells, the actin cortex controls the shape of the cell and its modification during biological processes. This dynamical and contractile 2D structure lying beneath the plasma membrane can be remodeled in response to external cues and controls for cell reorganization for motility or division. Therefore, it would be interesting to design GUVs where a thin 2D layer of responsive gel is attached beneath the membrane in order to study how such composite objects respond at the LCST and under mechanical stresses.

In the context of designing various promising systems such as smart objects with sites exhibiting different reactivity for chemical applications or optimized drug carriers, it seems of interest to develop composite GUVs with controlled shapes. It has been recently demonstrated that GUVs with membrane in the gel phase (see Chapters 15 and 27), submitted to osmotic pressure exhibit original structures. Some reproducible and stable shapes as illustrated in Figure 25.7. Their characteristic length is directly linked to the initial vesicle size and the membrane mechanical properties (Quemeneur et al., 2012).

These results open new opportunities for designing stable anisotropic colloids. Therefore, it seems important to further

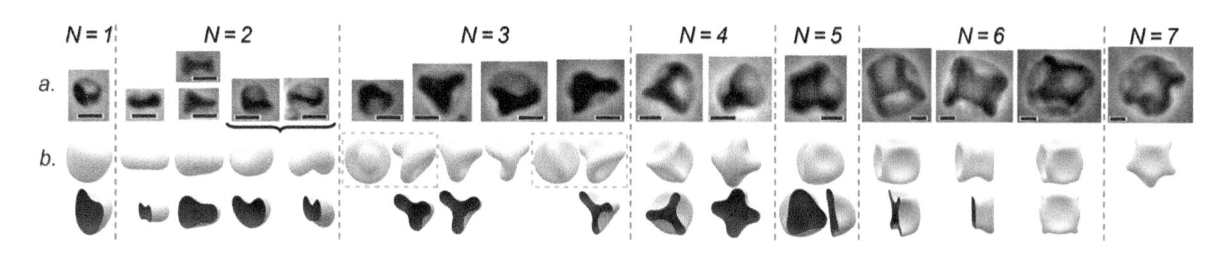

Figure 25.7 (a) Experimental shapes for deflated (reduced volume = 0.6) DMPC bare GUVs at 15°C (in the P_{β}gel-phase) for increasing radii. Black scale bars: 5 µm. (b) Numerical simulations: Each shape is characterized by the number of depressions N. (Reprinted with permission from Quemeneur, F. et al., *Phys. Rev. Lett.*, 108, 1–5, 2012. Copyright 2012 by the American Physical Society.)

investigate the role of polymer interaction with gel phase membrane systems. For example, we wonder if interaction with a polymer can induce alternative shapes for a GUV of specific size, and if introducing a polymer gel inside the GUV (see Chapter 29) allows maintaining its specific shape independently of the osmotic pressure. On the other hand, we can envision using such specific lipid shapes as a template to produce three-dimensional polymeric anisotropic structures.

LIST OF ABBREVIATIONS

ADE model	area difference elasticity model
AFM	atomic force microscopy
DEAP	dietoxyacetophenone
DOGM	1,2-distearyl-3-octaethylene glycol ether methacrylate
DOPC	1,2-dioleoyl-*sn*-glycero-3-phosphocholine
DMPC	1,2-dimyristoyl-*sn*-glycero-3-phosphocholine
HM-polyNIPAM	hydrophobically modified poly(NIPAM)
LCST	low critical solubility temperature
LUV	large unilamellar vesicle
MBA	methylen-bis-acrylamide
NIPAM	*N*-isopropylacrylamide
PA	phosphatidic acid
pGUVs	polymer-coated GUVs
SUV	small unilamellar vesicle
TEGDM	tetraethylene glycol dimethacrylate

GLOSSARY OF SYMBOLS

C_0	spontaneous curvature
E	elastic modulus
ΔP	suction pressure
K_{eff}	effective stiffness of a tethered vesicle
κ	bending modulus
L	nanotube length
R_p	the micropipette internal radius
H	viscosity
U	flow velocity
W	adhesion energy

REFERENCES

Angelova MI, Dimitrov DS (1988) A mechanism of liposome electroformation. *Prog Colloid Polym Sci* 76:59–67.

Angelova MI, Tsoneva I (1999) Interactions of DNA with giant liposomes. *Chem Phys Lipids* 101:123–137.

Blanchoin L, Boujemaa-Paterski R, Sykes C, Plastino J (2014) Actin dynamics, architecture, and mechanics in cell motility. *Physiol Rev* 94:235–263.

Boggs JM (1987) Lipid intermolecular hydrogen bonding: Influence on structural organization and membrane function. *Biochim Biophys Acta* 906:353–404.

Borghi N, Rossier O, Brochard-Wyart F (2003) Hydrodynamic extrusion of tubes from giant vesicles. *Europhys Lett* 64:837–843.

Brooks JT, Marques CM, Cates ME (1991) Role of adsorbed polymer in bilayer elasticity. *Europhys Lett* 14:713–718.

Campillo C, Pépin-Donat B, Viallat A (2007) Responsive viscoelastic giant lipid vesicles filled with a poly(N-isopropylacrylamide) artificial cytoskeleton. *Soft Matter* 3:1421–1427.

Campillo C, Schroder AP, Marques CM, Pépin-Donat B (2008) Volume transition in composite poly(NIPAM)–giant unilamellar vesicles. *Soft Matter* 4:2486–2491.

Campillo CC, Schroder AP, Marques CM, Pépin-Donat B (2009) Composite gel-filled giant vesicles: Membrane homogeneity and mechanical properties. *Mater Sci Eng C* 29:393–397.

Chandaroy P, Sen A, Hui SW (2001) Temperature-controlled content release from liposomes encapsulating pluronic F127. *J Control Release* 76:27–37.

Chiappisi L, Hoffmann I, Gradzielski M (2013) Complexes of oppositely charged polyelectrolytes and surfactants—Recent developments in the field of biologically derived polyelectrolytes. *Soft Matter* 9:3896.

Chung M, Lowe RD, Chan Y-HM, Ganesan PV, Boxer SG (2009) DNA-tethered membranes formed by giant vesicle rupture. *J Struct Biol* 168:190–199.

Couffin-Hoarau A-C, Leroux J-C (2004) Report on the use of poly(organophosphazenes) for the design of stimuli-responsive vesicles. *Biomacromolecules* 5:2082–2087.

de Gennes P-G (1979) *Scaling Concepts in Polymer Physics*. Ithaca, New York: Cornell University Press.

Deleu M, Crowet J-M, Nasir MN, Lins L (2014) Complementary biophysical tools to investigate lipid specificity in the interaction between bioactive molecules and the plasma membrane: A review. *Biochim Biophys Acta—Biomembr* 1838:3171–3190.

Deverall MA, Gindl E, Sinner E-K, Besir H, Ruehe J, Saxton MJ, Naumann CA (2005) Membrane lateral mobility obstructed by polymer-tethered lipids studied at the single molecule level. *Biophys J* 88:1875–1886.

Dobrynin AV, Deshkovski A, Rubinstein M (2001) Adsorption of polyelectrolytes at oppositely charged surfaces. *Macromolecules* 34:3421–3436.

Faivre M, Campillo C, Pépin-Donat B, Viallat A (2006) Responsive giant vesicles filled with Poly(N-isopropylacrylamide) sols or gels. *Prog Colloid Polym Sci* 133:41–44.

Fleer GJ (Gerard J. (1993) *Polymers at Interfaces*. London: Chapman & Hall.

Flory P (1979) *Statistical Mechanics of Chain Molecules*. New York: Wiley.

Fournier J, Ajdari A, Peliti L (2001) Effective-area elasticity and tension of micromanipulated membranes. *Phys Rev Lett* 86:4970–4973.

Gerald Lim HW, Wortis M, Mukhopadhyay R (2002) Stomatocyte–discocyte–echinocyte sequence of the human red blood cell: Evidence for the bilayer–couple hypothesis from membrane mechanics. *Proc Natl Acad Sci USA* 99:16766–16769.

Hadorn M, Boenzli E, Sørensen KT, De Lucrezia D, Hanczyc MM, Yomo T (2013) Defined DNA-mediated assemblies of gene-expressing giant unilamellar vesicles. *Langmuir* 29:15309–15319.

Hayashi H, Kono K, Takfagishi T (1999) Temperature sensitization of liposomes using copolymers of *N*-isopropylacrylamide. *Bioconjug Chem* 10:412–418.

Helfrich MR, Mangeney-slavin LK, Scott Long M, Djoko KY, Keating CD (2002) Aqueous phase separation in giant vesicles. *JACS* 124:13374–13375.

Heskins M, Guillet JE (1968) Solution properties of Poly(N-isopropylacrylamide). *J Macromol Sci Part A—Chem* 2:1441–1455.

Heuvingh J, Pincet F, Cribier S (2004) Hemifusion and fusion of giant vesicles induced by reduction of inter-membrane distance. *Eur Phys J E* 14:269–276.

Hisette M-L, Haddad P, Gisler T, Marques CM, Schröder AP (2008) Spreading of bio-adhesive vesicles on DNA carpets. *Soft Matter* 4:828.

Hochmuth RM, Marcus WD (2002) Membrane tethers formed from blood cells with available area and determination of their adhesion energy. *Biophys J* 82:2964–2969.

Horger KS, Estes DJ, Capone R, Mayer M (2009) Films of agarose enable rapid formation of giant liposomes in solutions of physiologic ionic strength. *JACS* 131:1810–1819.

Hristova NI, Angelova MI, Tsoneva I (2002) An experimental approach for direct observation of the interaction of polyanions with sphingosine-containing giant vesicles. *Bioelectrochemistry* 58:65–73.

Huster D, Paasche G, Dietrich U, Zschörnig O, Gutberlet T, Gawrisch K, Arnold K (1999) Investigation of phospholipid area compression induced by calcium-mediated dextran sulfate interaction. *Biophys J* 77:879–887.

Jesorka A, Markström M, Orwar O (2005) Controlling the internal structure of giant unilamellar vesicles by means of reversible temperature dependent sol-gel transition of internalized poly(N-isopropyl acrylamide). *Langmuir* 21:1230–1237.

Kabanov V., Yaroslavov A. (2002) What happens to negatively charged lipid vesicles upon interacting with polycation species? *J Control Release* 78:267–271.

Kato A, Tsuji A, Yanagisawa M, Saeki D, Juni K, Morimoto Y, Yoshikawa K (2010) Phase Separation on a phospholipid membrane inducing a characteristic localization of DNA accompanied by its structural transition. *J Phys Chem Lett* 1:3391–3395.

Kennedy JF (John F, Phillips GO, Williams PA, Hascall VC (2002) *Hyaluronan*. Cambridge: Woodhead Publishing Limited.

Kremer S, Campillo C, Pépin-Donat B, Viallat A, Brochard-Wyart F (2008) Nanotubes from gelly vesicles. *Europhys Lett* 82:48002.

Kremer S, Campillo C, Quemeneur F, Rinaudo M, Pépin-Donat B, Brochard-Wyart F (2011) Nanotubes from asymmetrically decorated vesicles. *Soft Matter* 7:946–951.

Ladavière C, Tribet C, Cribier S (2002) Lateral organization of lipid membranes induced by amphiphilic polymer inclusions. *Langmuir* 18:7320–7327.

Lira RB, Dimova R, Riske KA (2014) Giant unilamellar vesicles formed by hybrid films of agarose and lipids display altered mechanical properties. *Biophys J* 107:1609–1619.

Lira RB, Steinkühler J, Knorr RL, Dimova R, Riske KA (2016) Posing for a picture: Vesicle immobilization in agarose gel. *Sci Rep* 6:25254.

Luan Y, Ramos L (2007) Real-time observation of polyelectrolyte-induced binding of charged bilayers. *J Am Chem Soc* 129:14619–14624.

Lulevich VV, Andrienko D, Vinogradova OI (2004) Elasticity of polyelectrolyte multilayer microcapsules. *J Chem Phys* 120:3822.

Mady MM, Darwish MM, Khalil S, Khalil WM (2009) Biophysical studies on chitosan-coated liposomes. *Eur Biophys J* 38:1127–1133.

Mertins O, Dimova R (2013a) Insights on the interactions of chitosan with phospholipid vesicles. Part I: Effect of polymer deprotonation. *Langmuir* 29:14545–14551.

Mertins O, Dimova R (2013b) Insights on the interactions of chitosan with phospholipid vesicles. Part II: Membrane stiffening and pore formation. *Langmuir* 29:14552–14559.

Mertins O, Schneider PH, Pohlmann AR, Pesce N (2010) Interaction between phospholipids bilayer and chitosan in liposomes investigated by 31P NMR spectroscopy. *Colloids Surf B Biointerfaces* 75:294–299.

Mui BL-S, Dobereiner H-G, Madden TD, Cullis PR (1995) Influence of transbilayer area asymmetry on the morphology of large unilamellar vesicles. *Biophys J* 69:930–941.

Nikolov V, Lipowsky R, Dimova R (2007) Behavior of giant vesicles with anchored DNA molecules. *Biophys J* 92:4356–4368.

Park TG, Hoffman a S (1994) Estimation of temperature-dependent pore size in poly(N-isopropylacrylamide) hydrogel beads. *Biotechnol Prog* 10:82–86.

Pautot S, Frisken BJ, Weitz DA (2003) Production of unilamellar vesicles using an inverted emulsion. *Langmuir* 19:2870–2879.

Polozova A, Winnik FM (1997) Mechanism of the interaction of hydrophobically-modified poly-(N-isopropylacrylamides) with liposomes. *Biochim Biophys Acta* 1326:213–224.

Quemeneur F (2010) Relation entre les paramètres mécaniques et le comportement sous contraintes externes de vésicules lipidiques à membrane modifiée. PhD Dissertation, Université Joseph Fourier, Grenoble, France.

Quemeneur F, Quilliet C, Faivre M, Viallat A, Pépin-Donat B (2012) Gel phase vesicles buckle into specific shapes. *Phys Rev Lett* 108:1–5.

Quemeneur F, Rammal A, Rinaudo M, Pépin-Donat B (2007) Large and giant vesicles "Decorated" with chitosan: Effects of pH, salt or glucose stress, and surface adhesion. *Biomacromolecules* 8:2512–2519.

Quemeneur F, Rinaudo M, Maret G, Pépin-Donat B (2010) Decoration of lipid vesicles by polyelectrolytes: Mechanism and structure. *Soft Matter* 6:4471–4481.

Quemeneur F, Rinaudo M, Pépin-Donat B (2008) Influence of molecular weight and pH on adsorption of chitosan at the surface of large and giant vesicles. *Biomacromolecules* 9:396–402.

Rinaudo M (2006) Chitin and chitosan: Properties and applications. *Progress in Polymer Science* 31:603–632.

Rinaudo M, Quemeneur F, Dubreuil F, Fery A, Pépin-Donat B (2013) Mechanical characterization of micrometric chitosan-coated vesicle by atomic force microscopy. *Int J Polym Anal Charact* 18:617–626.

Rinaudo M, Quemeneur F, Pépin-Donat B (2012) Experimental characterization of liposomes stabilized by polyelectrolytes and mechanism of interaction involved. *Int J Polym Anal Charact* 17:1–10.

Ringsdorf H, Sackmann E, Simon J, Winnik F (1993) Interactions of liposomes and hydrophobically-modified poly-(N-isopropylacrylamides) an attempt to model the cytoskeleton. *Biochim Biophys Acta* 1153:335–344.

Saaka Y, Deller RC, Rodger A, Gibson MI (2012) Exploiting thermo-responsive polymers to modulate lipophilicity: Interactions with model membranes. *Macromol Rapid Commun* 33:779–784.

Sabri S, Soler M, Foa C, Pierres A, Benoliel A, Bongrand P (2000) Glycocalyx modulation is a physiological means of regulating cell adhesion. *J Cell Sci* 113:1589–1600.

Sanderson JM, Whelan EJ (2004) Characterisation of the interactions of aromatic amino acids with diacetyl phosphatidylcholine. *Phys Chem Chem Phys* 6(5):1012–1017.

Schalchli-Plaszczynski A, Auvray L (2002) Vesicle-to-micelle transition induced by grafted diblock copolymers. *Eur Phys J E* 7:339–344.

Schild HG (1992) Poly(N-Isopropylacrylamide): Experiment, theory and application. *Prog Polym Sci* 17:163–249.

Schulz M, Olubummo A, Binder WH (2012) Beyond the lipid-bilayer: Interaction of polymers and nanoparticles with membranes. *Soft Matter* 8:4849.

Sebai SC, Cribier S, Karimi A, Massotte D, Tribet C (2010) Permeabilization of lipid membranes and cells by a light-responsive copolymer. *Langmuir* 26:14135–14141.

Shimobayashi SF, Mognetti BM, Parolini L, Orsi D, Cicuta P, Di Michele L (2015) Direct measurement of DNA-mediated adhesion between lipid bilayers. *Phys Chem Chem Phys* 17:15615–15628.

Simon J, Kuhner M, Ringsdorf H, Sackmann E (1995) Polymer-induced shape changes and capping in giant liposomes. *Chem Phys Lipids* 76:241–258.

Singer SJ, Nicolson, GL (1972) The fluid mosaic model of the structure of cell membranes. *Science* (80) 175:720–731.

Slochower DR, Wang Y-H, Tourdot RW, Radhakrishnan R, Janmey PA (2014) Counterion-mediated pattern formation in membranes containing anionic lipids. *Adv Colloid Interface Sci* 208:177–188.

Stauch O, Schubert R, Savin G, Burchard W (2002a) Structure of artificial cytoskeleton containing liposomes in aqueous solution studied by static and dynamic light scattering. *Biomacromolecules* 3:565–578.

Stauch O, Uhlmann T, Fröhlich M, Thomann R, El-badry M, Kim Y, Schubert R (2002b) Mimicking a cytoskeleton by coupling Poly(N-isopropylacrylamide) to the inner leaflet of liposomal membranes: Effects of photopolymerization on vesicle shape and polymer architecture. *Biomacromolecules* 3:324–332.

Sun Y, Lee C-C, Huang HW (2011) Adhesion and merging of lipid bilayers: A method for measuring the free energy of adhesion and hemifusion. *Biophys J* 100:987–995.

Szilagyi I, Trefalt G, Tiraferri A, Maroni P, Borkovec M (2014) Polyelectrolyte adsorption, interparticle forces, and colloidal aggregation. *Soft Matter* 10:2479.

Taglienti A, Cellesi F, Crescenzi V, Sequi P, Valentini M, Tirelli N (2006) Investigating the interactions of hyaluronan derivatives with biomolecules. The use of diffusional NMR techniques. *Macromol Biosci* 6:611–622.

Terasawa H, Nishimura K, Suzuki H, Matsuura T, Yomo T (2012) Coupling of the fusion and budding of giant phospholipid vesicles containing macromolecules. *Proc Natl Acad Sci USA* 109:5942–5947.

Theret DP, Levesque MJ, Sato M, Nerem RM, Wheeler LT (1988) The Application of a homogeneous half-space model in the analysis of endothelial cell micropipette measurements. *Trans ASME* 110:190–199.

Thomas JL, Borden KA, Tirrell DA (1996) Modulation of mobilities of fluorescent membrane probes by adsorption of a hydrophobic polyelectrolyte. *Macromolecules* 29:2570–2576.

Tribet C, Vial F (2008) Flexible macromolecules attached to lipid bilayers: Impact on fluidity, curvature, permeability and stability of the membranes. *Soft Matter* 4:68–81.

Tsafrir I, Caspi Y, Arzi T, Stavans J (2003) Budding and tubulation in highly oblate vesicles by anchored amphiphilic molecules. *Phys Rev Lett* 91:138102.

Tsafrir I, Sagi D, Arzi T, Guedeau-Boudeville M-A, Frette V, Kandel D, Stavans J (2001) Pearling instabilities of membrane tubes with anchored polymers. *Phys Rev Lett* 86:1138–1141.

Vial F, Oukhaled A, Auvray L, Tribet C (2007) Long-living channels of well defined radius opened in lipid bilayers by polydisperse, hydrophobically-modified polyacrylic acids. *Soft Matter* 3:75–78.

Viallat a., Dalous J, Abkarian M (2004) Giant lipid vesicles filled with a gel: Shape instability induced by osmotic shrinkage. *Biophys J* 86:2179–2187.

Xie AF, Granick S (2002) Phospholipid membranes as substrates for polymer adsorption. *Nat Mater* 1:129–133.

Yaroslavov AA, Melik-Nubarov NS, Menger FM (2006) Polymer-induced flip-flop in biomembranes. *Acc Chem Res* 39:702–710.

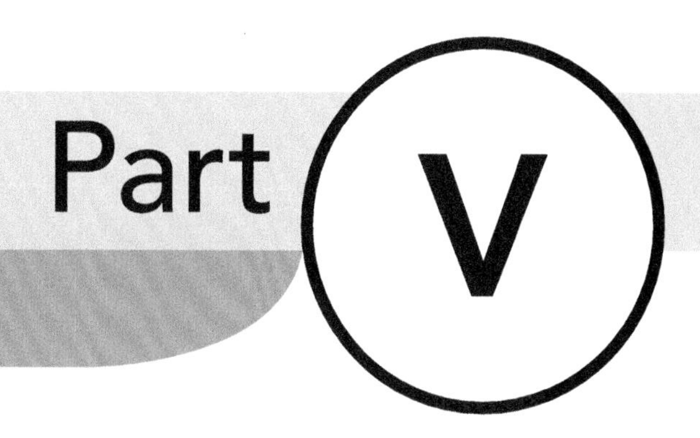

GUVs as complex membrane containers

26 Polymersomes

Praful Nair, David Christian, and Dennis E. Discher

GUVs with broadly controlled properties

Contents

26.1 INTRODUCTION

Phospholipids are the defining component of a cell's outermost plasma membrane, its nuclear " double" membrane, and many other vesicular bodies in cells such as endosomes and lysosomes. Liposomes assembled from purified lipids have been utilized for several decades to encapsulate drugs in their lumen for drug delivery. However, a lack of stability for many types of liposomes results in poor control over payload retention. This, in addition to other vesicle properties, provided some practical motivation nearly 20 year ago for the development of polymersomes, a family of vesicular structures self-assembled from block copolymers with lipid-like amphiphilicity (Discher and Eisenberg, 2002). It was also found from the onset that polymer giant unilamellar vesicles (pGUVs) could be assembled from this family of bilayer-forming polymer architectures. They have since not only provided a platform to study the physical properties of polymer bilayers, but also developed as a new class of low permeability, mechanically tough cell-sized vesicles that cannot be made from standard lipids.

The great variety in types and sizes of polymers provides mechanisms for tuning membrane properties for applications in fields that range from drug delivery to devices. For example, in drug delivery to tumors, the high toxicity of most chemotherapeutics can often be better controlled with nano-carriers such as polymer vesicles (polymersomes) by tuning the kinetics of drug release such that less of a toxic drug is delivered to healthy tissue (Chidambaram et al., 2011). Slow drug release from a nano-carrier will delay release into the bloodstream or to a disease site and limit drug excretion into urine, whereas very slow release will lead to drug accumulation in immune cells that generally clear nanoparticles from the bloodstream. Such design criteria are merely illustrative of properties that might be tuned with polymer based nano-carriers such as polymersomes. While such applications often require the use of nano-sized polymersomes, fundamental studies of GUVs has provided much of the foundation for understanding how to control membrane properties such as stiffness, phase separation, surface charge and degradation rate in order to make polymersomes useful in a broad range of dynamic applications.

26.2 BLOCK COPOLYMER AMPHIPHILES AND ASSEMBLIES

A hydrophobic polymer linked covalently to a hydrophilic polymer yields an amphiphilic diblock copolymer and, like lipids, the polymers will tend to self-assemble in aqueous solutions in order to minimize exposure of the hydrophobic block to water. The morphology of these assemblies includes among others spherical micelles, cylindrical micelles, and vesicles—all of which have received substantial attention for drug delivery. Spherical micelles have long been used for delivery of hydrophobic drugs integrated into the hydrophobic block to solid tumors (Kwon and Okano, 1996). Polymer vesicles are more recent and have been loaded not only with hydrophobic drugs into the hydrophobic core of the membrane, but also with hydrophilic drugs into the vesicle lumen in order to deliver two anti-cancer drugs simultaneously to tumors (Ahmed et al., 2006a, 2006b). Vesicles have also been used for the co-delivery of siRNA and antisense (AON) oligonucleotides to treat genetic diseases (Kim et al., 2009a).

Unlike acyl chains in lipids, a sufficiently long hydrophobic block can also contain oxygen among other polar or water-soluble groups. Polyesters are one type of oxygen-containing, medically approved polymers that are particularly useful as they degrade via hydrolysis. Controlled release from biodegradable materials has formed the basis for many types of drug delivery systems (Leong et al., 1985) among other applications (Middleton and Tipton, 2000). Polyester-based, rate-controlled release from degradable polymersomes have already been used to shrink tumors and treat genetic diseases (Ahmed et al., 2006a, 2006b; Kim et al., 2009a).

26.3 GENERAL BACKGROUND OF STRUCTURES

The self-directed assembly of amphiphilic molecules into a highly curved spherical micelle, a less curved cylindrical micelle, or a relatively flat vesicle morphology is primarily dictated by the ratio of the hydrophilic and hydrophobic fractions. In contrast with lipids, the structure of polymeric amphiphiles can be tuned across a wide range of properties including molecular weight (MW), polydispersity, charge, and crystallinity. Therefore, understanding the fundamental basis by which these block copolymers self-assemble into different morphologies is crucial.

26.3.1 MICROPHASE STRUCTURES: POLYMERSOMES, WORMS, AND SPHERES

A simple calculation of amphiphile geometry sheds light into its tendency to assemble into different morphologies. This is often obtained in terms of a "packing parameter," that indicates the curvature of the molecular structure (Figure 26.1) per (Discher and Eisenberg, 2002). Mathematically, the packing parameter p is expressed as:

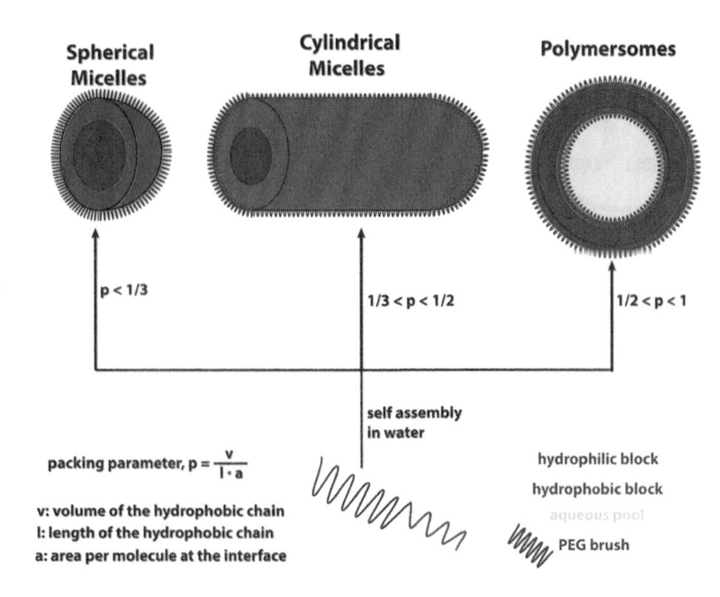

Figure 26.1 Aggregation of amphiphilic block copolymers into various morphologies. Copolymers with different values of packing parameter (p) form different assemblies as shown.

$$p = \frac{v}{al_c} \tag{26.1}$$

where v is the volume of the hydrophobic chain; a is the cross-sectional area of the hydrophilic chain at the interface; and l_c is the hydrophobic chain length (Israelachvili, 1991). The difference in structure from lipids leads to key differences in the calculation of p. The value of a is taken as the area of the chain at energetic minimum, whereas l_c is the average chain length over all conformations. These differences and have been the subject of extensive simulation and theoretical research. For spherical micelles $p < 1/3$, whereas cylindrical micelles (or worms) have a p value between 1/3 and 1/2, and vesicles (or polymersomes) result from a value of p between 1/2 and 1. While p values above 1 exist, their utilization in drug delivery systems has been limited. Differences in the shapes and sizes of assemblies or aggregates influence their properties as nano-carriers *in vitro* as well as *in vivo*, including properties such as drug loading capacity, and mode of clearance from bloodstream.

Spherical micelles can be created with a wide distribution of polymer weights and block ratios because a suitable packing parameter allows for very large hydrophilic blocks. Spherical micelles are also kinetically trapped states of the larger cylindrical micelle or vesicles assemblies if the latter are disrupted by excess energy (such as from ultrasound, shearing, heat, electrical fields) (Discher and Ahmed, 2006). For these reasons and more, spherical micelles are the most widely studied polymer assembly for drug delivery (Kwon and Okano, 1996). However, they load only hydrophobic drugs and have lower drug loading capacity per particle compared with other morphologies.

Cylindrical micelles exist only in a narrow range of the packing parameter, and it is challenging to synthesize block copolymers that generate such micelles. Synthesis needs to be precisely controlled, and strong shear (among other physical perturbations) needs to be avoided to obtain cylindrical micelles. These factors make it difficult to make this shape, and hence, they are less commonly studied. However, benefits of this elongated shape for drug delivery (Geng et al., 2007) have led to significant interest in these nano-carriers of unusual shape, as reviewed by Oltra et al. (2013). Higher drug loading capacity and evasion of uptake by phagocytes (cells that eat foreign substances) are particularly interesting in application of these assemblies.

Polymersomes have attracted considerable attention for their potential in drug delivery (Xu et al., 2005; Ahmed et al., 2006a; Onaca et al., 2009; Kim and Lee, 2010) among other applications. Unlike their micellar counterparts, polymersomes allow hydrophobic drugs to be loaded in the highly stable membrane core, whereas hydrophilic therapeutics can be encapsulated in the aqueous lumen.

Dual drug loading in one carrier allows the co-delivery of drugs that simultaneously target two different pathways (Ahmed et al., 2006b). Polymersomes have been used also as nanoreactors (Vriezema et al., 2007) among other diverse applications. Importantly, giant polymersomes (pGUVs) have been essential to measuring and understanding the properties of these novel polymer membranes (Discher et al., 1999; Aranda-Espinoza et al. 2001; Bermudez et al., 2002, 2003, 2004; Dimova et al., 2002; Itel et al., 2014).

26.3.2 STRUCTURE: THICKNESS SCALING WITH MOLECULAR WEIGHT DESPITE POLYDISPERSITY

Whereas lipid molecules are smaller than 1 kDa, polymers with mean MWs from 2 to 20 kDa can assemble into bilayers with thickness that range from 8 to 21 nm based on cryogenic electron microscopy (Aranda-Espinoza et al., 2001; Bermudez et al., 2002). Such experimental measurements agree with coarse-grained molecular dynamics simulations of the same diblock copolymers (Srinivas et al., 2004b) in establishing a power law relationship between the MW of the hydrophobic block (M) and bilayer core thickness (d) as

$$d \sim M^b, \text{ with } b = 0.55 \qquad (26.2)$$

This approach was found to be generic and could be used to study different polymers. Other segregating copolymer systems seem to give a 20% higher exponent of $b = 0.67$ (Battaglia and Ryan, 2005) which is consistent with a simple balance of interfacial energy and chain elasticity (Discher et al., 2007). Regardless, such nonlinear scaling is typical of the fractal nature of polymer physics.

26.3.3 SIMULATIONS ON A SMALL SCALE

The chemical simplicity of block copolymers has allowed the use of simulation studies to lend insight into the properties of block copolymer assemblies—including bilayer membranes. Block copolymer simulations have been performed using parameter sets that have been optimized for biological systems around 300 K. Coarse-grained models (that group a certain number of atoms together as one sphere, see also Chapter 6) sacrifice atomistic detail for computational efficiency and have become more complex and realistic with time. The first coarse-grained molecular dynamics simulations of PEG-PEE in computational water replicated the phase behavior experiments on PEG-PEE assemblies (Srinivas et al., 2004a, 2004b; Srinivas and Klein, 2004; Srinivas et al., 2005). The first studies also matched some properties measured on polymer GUVs (pGUVs) such as the area expansion modulus, K_a, which proves nearly independent of polymer MW. Dissipative-particle dynamics (see Chapter 6) allows longer simulation time scales (Warren, 1998; Peters, 2004; Ortiz et al., 2005), and has been used, for example, to understand how phase separated domains induced by differences in the hydrophilic block could register across the bilayer (Pantano et al., 2011). Finally, simulations were used to study how vesicles could be used for drug delivery by determining the release of drug payload following osmotic swelling and vesicle rupture (Ortiz et al., 2005).

26.4 MOTIVATION FOR POLYMER APPROACHES

While liposomes have already been used in drug delivery applications in the clinic, they have been limited by their low stability and rapid disintegration *in vivo* that prevents the controlled release of their cargo (Lasic and Papahadjopoulos, 1998; Semple et al., 1998). Liposomes are also quickly recognized and cleared by the immune system and must be modified to include an outer PEG brush to delay clearance *in vivo*. The tunability of block copolymer MW and chemistry while maintaining a vesicle morphology allows polymersomes to improve on the shortcomings of liposomes (Discher et al., 2007). Box 26.1 summarizes the key advantages of polymersomes as well as some of the differences from liposomes.

For drug release *in vivo*, the time scale for decay should be long enough for the vesicles to reach the target. However, if it is too long, the vesicles will be cleared by the body's immune system. Liposomes exhibit limited circulation time *in vivo*, and the addition of a PEG (or PEO) to <10% of lipid headgroups creates "stealthy" liposomes with circulation times extended from minutes or hours to 10–15 h (Klibanov et al., 1990; Woodle, 1993). PEG is repulsive in interactions with cells, and is the most common hydrophilic fraction of block copolymers to make polymersomes inherently stealthy (Photos et al., 2003).

26.4.1 TUNING PHYSICAL PROPERTIES WITH MOLECULAR WEIGHT FOR GIVEN CHEMISTRY

Liposomes are made up of constituent lipid molecules smaller than 1 kDa in size (Discher and Eisenberg, 2002). As mentioned above, polymersomes are made of copolymers that are up to 20 kDa in size, and this difference in MW manifests itself in membrane thickness, d. Parameters such as permeability, viscosity, and elasticity, as measured by experiments on single pGUVs as well as on nano-vesicles, are controlled by membrane thickness d (Dimova et al., 2002; Discher and Eisenberg, 2002; Hamley, 2005; Battaglia and Ryan, 2006), and pGUV composed of membranes with much more variance in d (8 to 21 nm) than liposomes (3 to 5 nm) were used to measure these membrane properties. The thinner membrane of liposomes leads to the low stability (leakiness) and short circulation time mentioned above. Measurement of water permeation through membranes confirmed that permeation rate was much higher for liposome membranes than polymersome membranes (Discher et al., 1999). Thus, liposomes are assembled such that there is much more fluidity and permeability versus stability in the bilayer at the expense of stability. In contrast, the fluidity, permeability, and stability of polymersome membranes can be tuned by controlling the MW.

Box 26.1 Comparison between polymersomes and liposomes.

Illustration and summary of the key differences between polymersomes and liposomes. Because liposomes are self-assembled from natural phospholipids, they are fully biocompatible. However, they exhibit low encapsulation efficiency and stability (Lasic and Papahadjopoulos, 1998). Polymersomes are tougher, and improve upon some of the aforementioned shortcomings. Some of these key differences are listed below (note that only lipid membranes in the fluid state are considered). A table comparing these properties of liposomes and polymersomes can be found in Chapters 15 and 27 and in the paper by Le Meins et al. (2011).

	POLYMERSOMES	LIPOSOMES
Structure	Structure of a polymersome. A polymersome can be dual loaded, with hydrophobic drugs in the core of the membrane (red) and hydrophilic drugs in the water pool in the center (blue). (Adapted from LoPresti et al., [2009] with permission of the Royal Society of Chemistry.)	Bottom: schematic diagram of a liposome. Top: hydrophilic cargo can be loaded in the aqueous pool in the interior, whereas hydrophobic cargo is stored in the membrane. (Courtesy from C.M. Marques.)
Permeability	Polymer membranes are much less permeable, thus making polymersomes slow to deflate; see also Chapter 27.	Permeability of lipid vesicles to water is high and GUV deflation is fast; for permeability values see Chapter 20.
Viscosity, diffusion	One study estimated the shear viscosity of polymersome membrane to be 500 times higher than that of lipid membranes (Dimova et al., 2002). Molecular mobility in the polymersome membrane is much less compared with liposomes. See Chapter 27 for comparison of diffusion between pure lipid and hybrid polymer-lipid membranes.	Membrane viscosity is much lower and diffusion of membrane probes is fast (see Chapter 21) which leads to faster domain coarsening in phase separated fluid vesicles.
Fluctuations	Polymersomes relax back very slowly after deformation; e.g., after aspiration with micropipettes, they may retain their shape for a while.	Brownian motion induces visible membrane undulation in lipid GUVs with excess area (low tension), see Chapter 14.
Thickness	Polymersome membranes are generally thicker and their thickness can range in a larger interval (e.g., 8–21 nm). This translates to higher loading of hydrophobic drugs (Ahmed et al., 2006b).	Lipid membranes are typically 4–5 nm thick.
Bilayer nature	For triblock copolymers, the bilayer (or two-leaflet) nature of the membrane may be lost and there could be polymers structured as loops (in one membrane half) or spanning across the membrane.	Membranes from natural lipid are characterized by two-layer (bilayer) structure, although interdigitation is possible; membranes made of bola-lipids cannot be separated in two leaflets.
Other features	Polydispersity of polymers can create variance between batches, and can pose some challenges to reproducing results. Synthesis of a series of block copolymers across a targeted range will increase the likelihood of generating some useful polymer.	Because liposomes are self-assembled from natural phospholipids, they are fully biocompatible. However, they exhibit low encapsulation efficiency and stability (Lasic and Papahadjopoulos, 1998).

Although polymersome membranes are generally thicker than liposomes, the customizability of polymers has allowed for membranes with thicknesses more similar to liposomes to be made (Battaglia and Ryan, 2005; LoPresti et al., 2009). Amphiphilic diblock copolymers of polyethylene glycol-b-polybutylene oxide self-assembled into membranes with thickness reportedly ranging from 2.4 to 4.5 nm. Triblock copolymers (polyethylene glycol-b-polybutylene oxide-b-polyethylene glycol) assembled into thicker membranes ranging from 3.4 to 6.2 nm. The authors attributed the thinner membranes to the higher flexibility of the polyether hydrophobic block. The permeability of the membrane was found to vary with pH, and was found to be higher than that of phosphatidylcholine membranes at lower pH (Battaglia et al., 2006). Such thin membranes might be particularly useful in the reconstitution of proteins into polymersomes (as discussed below in Section 26.8.2).

26.4.2 TUNING PROPERTIES WITH CHEMISTRY

Changing the chemical properties of either the hydrophilic or hydrophobic portion of the block copolymer is a means to control the characteristics of polymersomes. Three commonly employed hydrophobic chains used to make polymersomes are, in the order of increasing hydrophobicity, poly-L-lactic acid (PLA), polycaprolactone (PCL) and polybutadiene (PBD). Table 26.1 summarizes polymer abbreviations used in the chapter. PLA has a high degree of hydrophilicity due to the presence of an ester bond, leading to a rapid rate of hydrolysis. Thus, PEG-PLA (also abbreviated as OL) assemblies degrade soon after they self-assemble, with vesicles stable for a few days (Lee et al., 2001, 2002; Photos et al., 2003). PBD on the other hand is very hydrophobic and does not undergo any appreciable hydrolysis. Consequently, assemblies comprised of PEG-PBD (OB) are stable for up to years.

PEG-PCL (OCL) assemblies have stabilities between those of OL and OB and are thus more suitable for applications in drug delivery when used in pure form. Blends of polymers that vary the percentage of degradable OL or OCL blended with inert OB were utilized to produce vesicles that have controllable release times (Ahmed et al., 2004). Such principles of blend-controlled release apply also to filomicelles (Kim et al., 2005). Blending of polymers with varying chemistry provides a powerful method to create nano-vehicles with highly customizable release times. The main mode of drug release is the hydrolytic degradation of the copolymer, specifically, via "end-chain cleavage" and not at random points in the hydrophobic block ("random scission") (Geng and Discher, 2005). This changes the phase from vesicles to worms, destabilizing the assembly and releasing the contents (Ahmed et al., 2006b).

Environmental triggers have also been utilized to induce degradation of polymersomes, such as polymersomes that were designed to be sensitive to oxidative byproducts like H_2O_2 (Napoli et al., 2004). The hydrophobic core in PEG-(propylene sulfide)-PEG triblock copolymers was oxidized to hydrophilic groups, that destabilized the vesicle and favored micelles. pH responsive polymersomes from PEG-poly(2-vinylpyridine) (PEG-P2VP) have also been created, that fall apart after protonation of the hydrophobic core at low pH (Borchert et al., 2006). Stimuli can also be utilized to change the size of vesicles (Checot et al., 2003). Vesicles assembled from

Table 26.1 **Summary of abbreviations of some typical block copolymers and example references where they were used to form pGUVs**

DIBLOCK COPOLYMER	ABBREVIATION	pGUV FORMATION
Polyethylene oxide poly(ε-caprolactone)	OCL or PEO-PCL or PEG-PCL	Ahmed and Discher (2004)
Polyethylene oxide polybutadiene	OB or PEO-PBD or PEG-PBD	Discher et al. (1999); Dimova et al. (2002); Christian et al. (2009); Rodriguez-Garcia et al. (2011); Meeuwissen et al. (2014); Mabrouk et al. (2009); Nuss et al. (2012)
Polyethylene oxide polylactic acid	OL or PEO-PLA or PEG-PLA	Ahmed and Discher (2004); Shum et al. (2008)
Polyethylene oxide poly(ε-caprolactone co-lactide)	OCLA or PEO-PCLA or PEG-PCLA	Ghoroghchian et al. (2006)
Polyacrylic acid polybutadiene	AB or PAA-PBD	Christian et al. (2009); Meeuwissen et al. (2014)
Poly(normal-butyl acrylate) poly(acrylic acid)	PBA-PAA	Lorenceau et al. (2005)
Polyethylene oxide poly(2-vinylpyridine)	PEO-P2VP	Borchert et al. (2006)
Poly(ethylene oxide) poly(butylene oxide)	PEO-PBO	Howse et al. (2009)
TRIBLOCK COPOLYMER	**ABBREVIATION**	**pGUV FORMATION**
Poly(methyloxazoline) Poly(dimethylsiloxane) Poly(methyloxazoline)	PMOXA-PDMS-PMOXA	Kita-Tokarczyk et al. (2005); Itel et al. (2014)
Polyethylene oxide Polypropylene oxide Polyethylene oxide	PEO-PPO-PEO	Rodriguez-Garcia et al. (2011)

Table 26.2 **Summary of abbreviations of some typical block copolymers and example references where they were used to form hGUVs, see Appendix 1 of the book for structure and data on the lipids**

DIBLOCK COPOLYMER AND LIPID	ABBREVIATION	hGUV FORMATION
Polyethylene oxide polybutadiene and 1,2-dipalmitoyl-*sn*-glycero-3-phosphocholine	PEO-PBD and DPPC	Nam et al. (2012)
Polyethylene oxide polybutadiene and 1-palmitoyl 2 oleoyl *sn* glycero 3 phosphocholine	PEO-PBD and POPC	Nam et al. (2010, 2012)
Polyethylene oxide polydimethyl siloxane and 1-palmitoyl-2-oleoyl-*sn*-glycero-3-phosphocholine	PEO-PDMS and POPC	Chemin et al. (2012)
Polyethylene oxide polydimethyl siloxane and 1,2-dipalmitoyl-*sn*-glycero-3-phosphocholine	PEO-PDMS and DPPC	Chemin et al. (2012)
Polyethylene oxide polyisobutylene and 1,2-dipalmitoyl-*sn*-glycero-3-phosphocholine	PEO-PIB and DPPC	Schulz et al. (2011, 2013)

poly(butadiene)-b-poly(γ-L-glutamic acid) were shown to reversibly change size by varying the pH and ionic strength. Further, the 1,2-vinyl double bonds were crosslinked by exposure to UV, creating systems in which the release of encapsulated drug can be controlled.

Such broad customization options are not available with lipids. However, it must be noted that, because of polydispersity inherent to synthetic polymers, each batch can vary significantly, which poses a major challenge for reproducible results.

26.5 OTHER COMMON BLOCK COPOLYMERS

Assemblies made with PEG-based copolymers have been found to be "stealthy" (Photos et al., 2003). These properties in water can be attributed to the hydrogen bonding to ether oxygen atoms in PEG and high mobility in water (Discher et al., 2007). This stealthiness imparted by the PEG brush helped to delay phagocytosis, extending circulation time *in vivo*, thus making PEG an almost universal choice for the hydrophilic block. Charged hydrophilic blocks like polyacrylic acid (Schmaljohann, 2006) and polymethyloxazoline (Kim et al., 2009b) can be mixed with PEG-based block copolymers for various purposes. For example, such mixtures tune the lateral segregation of pGUV membranes (Christian et al., 2009, 2010) as discussed below.

The hydrophobic block dictates a number of properties of the polymersome membrane, and consequently there exist a number of polymers that have been utilized. Nondegradable membranes can be made to be fluid by using polybutadiene (Ahmed and Discher, 2004), polyethylethylene (Meng et al., 2009) or polydimethylsiloxane (Li et al., 2003), or membranes can be rigid by using strongly interacting glassy or semi-crystalline hydrophobic polymers such as polycaprolactone (Ahmed and Discher, 2004; Cai et al., 2007; Rajagopal et al., 2010). Other popular hydrophobic polymers are polylactic acid (Kim et al., 2005) and polystyrene (Kazunori et al., 1993). The list of polymers can be much longer, and block ratios as well as MW are all variables that can influence physical properties measured for pGUVs, which in turn often translate to application even of nano-vesicles. Another dimension of customization can be achieved by blending lipids and polymers to create hybrid GUVs (hGUVs)(Chemin et al., 2012; Le Meins et al., 2013). Benefits of these hGUVs (Table 26.2) include the scope for surface functionalization as well as the tunability of surface properties. Further in-depth discussion regarding these composite vesicles can be found in Chapter 27.

26.6 METHODS FOR SYNTHESIZING POLYMERS

Diblock copolymers can be created either by using one block as a macro-initiator for the polymerization (chain extension) of the other block or by covalently linking the two block polymers after they have been formed (conjugation).

26.6.1 EXAMPLE SYNTHESIS

Chain extension requires a polymer block (initiator) with an end-group that can trigger polymerization to be reacted with precise amounts of polymer (calculated based on the desired MW of the second block) in the presence of a catalyst and at high temperature. An example of this type of copolymer synthesis is the polymerization of caprolactone by methoxy-capped PEO to form PEO-PCL (OCL). Prior to this "Ring opening" polymerization, ε-Caprolactone has to be purified by distilling it under vacuum. This purification removes trace amounts of water in the system that also possesses hydroxyl group required to initiate polymerization. The reactants along with the catalyst, stannous octoate are then sealed under vacuum and reacted for 6 h at 140°C. A schematic representation of this reaction is shown in Figure 26.2.

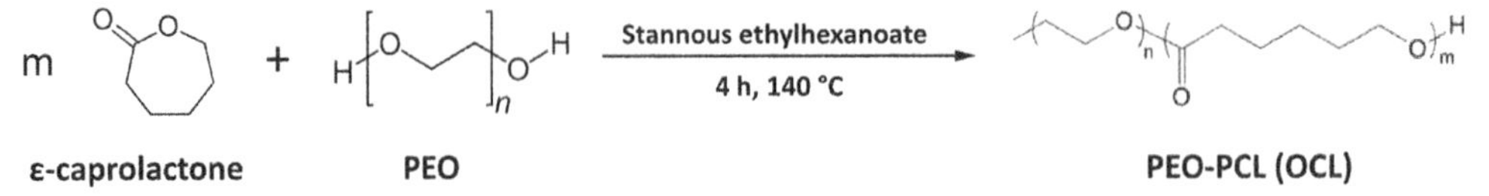

m ε-caprolactone + PEO —Stannous ethylhexanoate, 4 h, 140 °C→ PEO-PCL (OCL)

Figure 26.2 Reaction between PEO and ε-caprolactone. (From Carrot, G. et al., *Macromolecules*, 32, 5264–5269, 1999.)

Ratio of blocks can be confirmed by ^{1}H NMR spectroscopy and the size distribution of the polymer (or polydispersity index) is characterized using gel permeation chromatography.

Another example of chain extension is the reaction between PEO and a mixture of ε-caprolactone and D, L-lactide to form OCLA. Similar to OCL, this polymer has a polyester hydrophobic block, but has a faster hydrolysis time than OCL.

The other approach (conjugation), involves functionalizing the ends of the polymer after synthesizing them. These end groups must be capable of reacting with each other, and linking the blocks together when they do so. Cerritelli et al. (2007) utilized this strategy to link together blocks of PEG and polypropylene sulfide (PPS), with a reduction sensitive disulfide bond linking the two blocks. Thiolate terminated PPS was synthesized by living ring-opening polymerization of propylene sulfide using benzyl mercaptan as initiator. Thioacetate PEG was deprotected to form thiolate terminated PEG, which was reacted with PPS to obtain the final copolymer. Vesicles formed were found to be sensitive to various reducing agents (like cysteine, glutathione, and dithiothreitol or DTT), aiming to simulate conditions found early in endolysosomal processing. With the disulfide bridge getting reduced and the blocks no longer linked, the assemblies destabilize and release the encapsulated payload.

26.6.2 AGGREGATE PHASE DIAGRAM

Although calculation of the packing parameter, *p*, can be useful to predicting the expected morphology of self-assembled amphiphilic block copolymers in water, mapping phase diagrams for a given polymer remains essential. These thermodynamic phase diagrams indicate which shapes are dominant in the indicated regions. The variables are hydrophilic mass fraction (*f*) and the MW of the hydrophobic block (M_{CH2}) in *x* and *y* axis respectively. A phase diagram of OCL is shown in Figure 26.3. It is important to note that the temperature is held constant (25°C in Figure 26.3). Another point to be noted is that the oxygen atoms in the hydrophobic

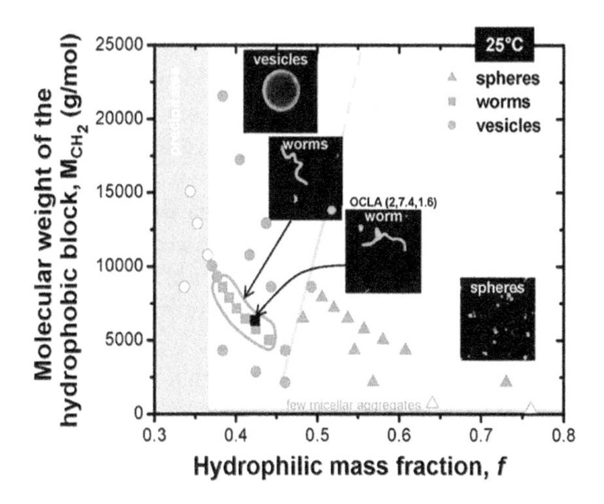

Figure 26.3 Phase diagram of OCL assemblies prepared at 25°C. Three morphologies: vesicles (shown as circles), worms (squares) and spheres (triangles) are represented here. Filled gray region to the left of the diagram indicates the region where stable colloidal assemblies are not found: open circles denote precipitates, whereas solid denote stable aggregates. Insets show representative images of fluorescently labeled aggregates. Concentration of polymer in water was 0.1 mM/mL. (Reprinted with permission from Rajagopal, K. et al., *Macromolecules*, 43,9736–9746, 2010. Copyright 2010 American Chemical Society.)

block contribute to the hydrophilicity of the polymer. Hence, the MW of the hydrophobic block is calculated by subtracting the weight of the oxygen atoms from the PCL block and the hydrophilic mass fraction f is calculated by adding the weight of the oxygen atoms to that of PEO (Rajagopal et al., 2010).

One prominent point to note is that f has to be at least 0.36 for any stable colloidal aggregates to form in water. Below this value, the PEO chains are too short to shield the hydrophobic block from water and minimize the energy of the system. This region corresponds to the higher end of packing parameter ($p > 1$). At f values just above 0.36, the polymer exhibits low curvature (large hydrophobic block), and self-assembles into vesicles (polymersomes). Increasing *f* increases the curvature and we obtain cylinder micelles also called worms (Rajagopal et al., 2010) and spherical micelles simply referred to elsewhere as spheres. It is interesting to note that worms exist in a narrow region that is embedded within the vesicle region.

It is important to note that controlling polydispersity is critical to obtaining the indicated phases. The presence of a wide distribution of different chain lengths may destabilize and/or favor the formation of other morphologies. Due to the narrow region in which they exist, cylindrical micelles are particularly prone to this. Further, due to different value of parameters for different polymers, phase diagrams are unique for a given copolymer. The regions at given values of M_{CH2} and f might be different for different polymers. However, OCLA worms exist in a similar region as OCL worms.

26.7 METHODS FOR MAKING AND OBSERVING pGUVs

Polymer GUVs or pGUVs are synthesized and prepared using film rehydration methods that are commonly utilized in the synthesis of liposomes. A more detailed explanation and discussion of GUV preparation has been described in Chapter 1. Several methods of preparing pGUVs have also been summarized by Walde et al., 2010. One essential difference is that the time for swelling of pGUVs is typically longer compared with lipid GUVs and electric fields applied in electroformation are typically higher. The polymer films used to make pGUVs can be a mixture of block copolymers to impart tunable properties of the membranes, and other fluorescent membrane labeling dyes such as lipophilic tracers (e.g., DiO, DiI, DiD, DiR, PKH67, PKH26, see Appendix 2 of the book for structure and data on some lipid dyes) and hydrophobic dyes (e.g., Nile Red, Laurdan) can be incorporated to aid in the imaging of the membrane. The hydrating solution can also be varied by altering the pH, concentration of ions, therapeutic molecules, or density, which provides another variable that can be used to change the properties of the polymersome membrane.

26.7.1 MIXING POLYMERS TO TUNE STRUCTURES AND PROPERTIES

Heterogeneous pGUV membranes can be made by the addition of a fluorescent membrane labeling dye that allows membrane visualization or the mixing of multiple block copolymers that enhance the tunability of the polymersome membrane architecture and functionality (e.g. degradability). Different block copolymers and fluorescent amphiphiles are typically mixed in an organic solvent like chloroform and subsequently dried to form a mixed polymer film. Rehydration of this film in an aqueous solution results in pGUVs with heterogeneous membranes.

These mixtures allow the visualization of the polymersome membrane through the addition of fluorescent amphiphilic dyes or block copolymers that are directly conjugated to such commercially available fluorescent molecules as tetramethylrhodamine-5-isothiocyanate (5-TRITC). Although amphiphilic dyes enable imaging of the entire membrane by fluorescence microscopy, fluorescent block copolymers provide the ability to track a specific copolymer within the mixed membrane. Fluorescently labeled pGUVs are an important tool for analyzing membrane properties such as fluidity and phase separation of different polymer species (Lee et al., 2001; Christian et al., 2009).

Mixing of block copolymers of varying chemistry in either the hydrophobic or hydrophilic block changes the properties of the membrane. As mentioned above, degradable polymersomes have been investigated for use in drug delivery applications, and the ability to tune the release kinetics of both hydrophilic and hydrophobic therapeutic cargo has been achieved by making polymersome membranes composed of inert OB copolymer with degradable OL or OCL copolymers (Ahmed et al., 2006a, 2006b). Initial characterization of the cargo release from these degradable polymersomes was performed using pGUVs loaded with either sucrose or fluorescent dextrans of increasing MW (Ahmed and Discher, 2004) (Figure 26.4).

pGUVs can also be made using mixtures of block copolymers with different electrostatic charge in the hydrophilic block. By dosing in increasing amounts of the polyanionic PAA-PBD (AB) block copolymer with OB copolymer that has a slight negative charge at neutral pH, the surface charge of the polymersome can be decreased. This change in polymersome surface charge subsequently results in changes in the biodistribution of nano-vesicles *in vivo* (Christian et al., 2010). Mixtures of AB and OB copolymers can also be induced under certain hydration pH and divalent cation concentrations to form pGUVs with laterally phase segregated domains enriched in either AB or OB copolymer. This phase separation results from electrostatic interactions between the polyanionic AB chains and the divalent cation (Christian et al., 2009).

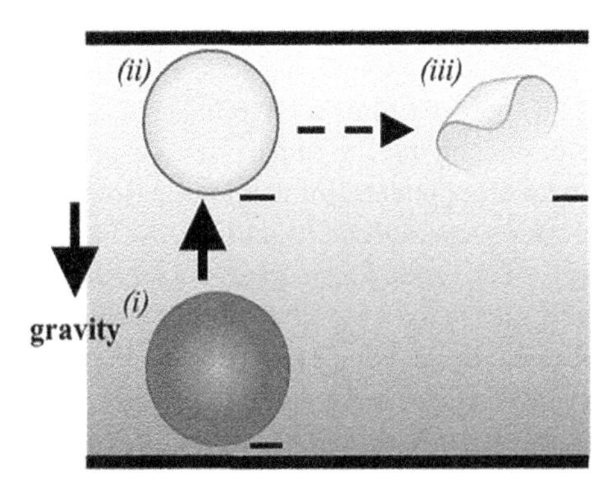

Figure 26.4 Release from polymer vesicles. Phase-contrast microscopy images of degradable polymersome carriers in a sealed chamber. Vesicles of 25 mol% blends of OL in OB are loaded with sucrose (300 mOsm) and suspended in an isotonic buffer. The vesicles are initially dense and phase dark (i). Over time (i.e., hours), vesicles become phase light—losing their encapsulant—and rise to the top of the chamber (ii). Over longer times (i.e., days), vesicles exhibit altered morphology and finally disintegrate (iii). (*J. Control Release*, 96, Ahmed F, and Discher DE, Self-porating polymersomes of PEG–PLA and PEG–PCL: Hydrolysis-triggered controlled release vesicles, 37–53, Copyright 2004, with permission from Elsevier.)

26.7.2 HYDRATING FILMS

pGUVs are synthesized using a variety of film rehydration techniques. Most commonly, amphiphilic block copolymers are dissolved in an organic solvent such as chloroform in a glass vial, and polymer films are created by first evaporating the solvent under an inert gas such as nitrogen or argon and then drying the polymer film under vacuum for 8–24 h to remove any residual solvent. A hydrating solution is then added to the dry polymer film at high temperature (commonly 60°C) for at least 12–16 h, and pGUV can then be observed to spontaneously bud off of the polymer film. Alternatively, pGUV can be formed via electroformation from polymer films. In this method, thin polymer films are formed on two parallel platinum wires by solvent evaporation. These polymer-covered wires are then placed in a hydrating solution and an oscillating voltage of 10 V, 10 Hz is applied to drive the budding of pGUV from the wires (Lee et al., 2001). Finally, pGUV of controlled size can be produced by drying polymer films on uniform micrometer-sized squares of glass created by lithography methods (Howse et al., 2009). These films are then hydrated as described in Chapter 1.

Film rehydration offers another variable to control the architecture and content of pGUV. As described more below, therapeutic molecules can be added to the aqueous hydrating solution and will subsequently be encapsulated inside the lumen of the polymersomes. The conditions of the hydrating solution can also be used to control the properties of polymersome membranes containing charged block copolymers like the polyanionic AB copolymer. When films of AB copolymer are hydrated in the presence of neutral pH and higher concentrations of multivalent cations like calcium, the chains become highly charged and then electrostatic crosslinking between the AB chains via the calcium ions causes a rigid, solid polymersome membrane (Christian et al., 2009). More explicitly, the acrylic acid groups of the AB polymer chains deprotonate in neutral pH solutions, resulting in a high concentration of anionic groups. In hydrating solutions containing multivalent cations like calcium, these cations are present among the polyanionic chains where a single divalent calcium ion can form a strong electrostatic bond between two different acrylic acid groups (i.e., electrostatic crosslink). In self-assembled AB polymersomes, these cation-induced electrostatic crosslinks between chains induce a solid polymersome membrane despite the fluid PBD hydrophobic core of the membrane (Figure 26.5a, inset). In contrast, when AB films are hydrated at low pH, the AB chains become neutral and the polymersomes are composed of a fluid membrane regardless of counter ion concentration. In addition to manipulating AB polymersome membranes, the architecture of polymersomes composed of neutral OB and polyanionic AB copolymers can be tuned with the hydrating solution. By carefully tuning the pH and calcium concentration chains of AB copolymer are attracted to one another such that they phase separate from the neutral OB copolymer, which results in the formation of spotted polymersomes (Figure 26.5) (Christian et al., 2009). Contrary to phase separation in lipid GUVs exhibiting

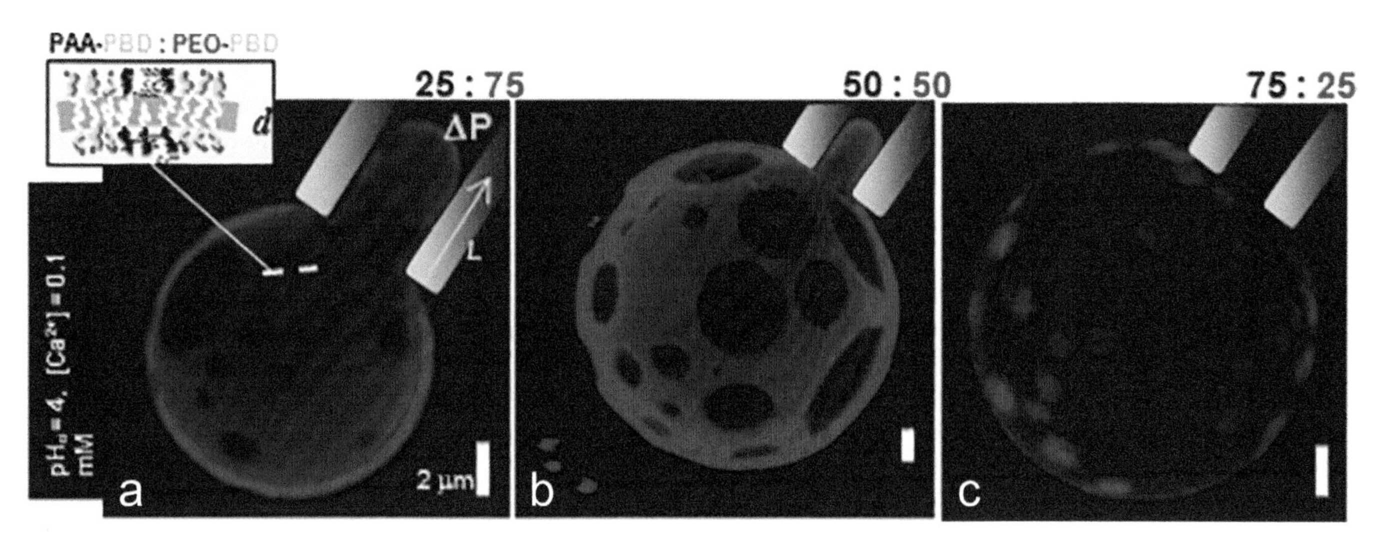

Figure 26.5 Cation-induced, lateral phase segregation of charged AB and neutral, tetramethylrhodamine-labeled OB diblock copolymers (Christian et al., 2009) formed at pH_0 4, 0.1 mM calcium at (a) 25% AB, (b) 50% AB, and (c) 75% AB. (a, inset): Schematic of phase-separated membrane of thickness *d*.

micrometer-sized domains of liquid-ordered or liquid-disordered phase (see Chapter 18), the domains here do not diffuse at a detectable speed.

26.8 COMMON TYPES OF EXPERIMENTS

26.8.1 CHARACTERIZING MECHANICS AND STABILITY OF SOLID VESICLES CROSSLINKED COVALENTLY OR BY DIVALENT IONS BY MICROPIPETTES AND FRAP

Two common methods used to characterize the physical properties of pGUV membranes are micropipette aspiration (described in Chapter 11) and fluorescence recovery after photobleaching (FRAP) as described in more detail in Chapter 21. Micropipette aspiration was used to first characterize the increased toughness and resistance to rupture of polymersome membranes compared with their liposome counterparts as it easily measures the membrane elasticity, area expansion modulus, and bending modulus (Discher et al., 1999). Polymersomes composed of copolymer with a PBD hydrophobic block can be covalently crosslinked to form pGUV with a solid membrane. Similarly, AB copolymer pGUV can also be made solid using pH and counter ion concentration as described above. Micropipette aspiration can be used to determine the stability of solid polymersome membranes as they are highly resistant to rupture and at high aspiration pressures are only capable of plastic deformation (Figure 26.6) (Discher et al., 2002; Christian et al., 2009).

A second method to measure the fluidity of pGUV membranes is using FRAP. In these experiments, an amphiphilic, fluorescent molecular tracer is added to the copolymer film prior to hydration and pGUV formation. The fluidity of pGUV membranes is then measured by analyzing the rate at which these fluorescent tracers diffuse through the membrane using established FRAP methods (Chapter 21) while making the fluorescently labeled pGUV stationary using micropipette aspiration. However, when membranes are made solid either by covalent crosslinking (OB polymersomes) or electrostatic crosslinking (AB polymersomes), the fluorophore is not capable of diffusing through the membrane and there is no recovery of fluorescence

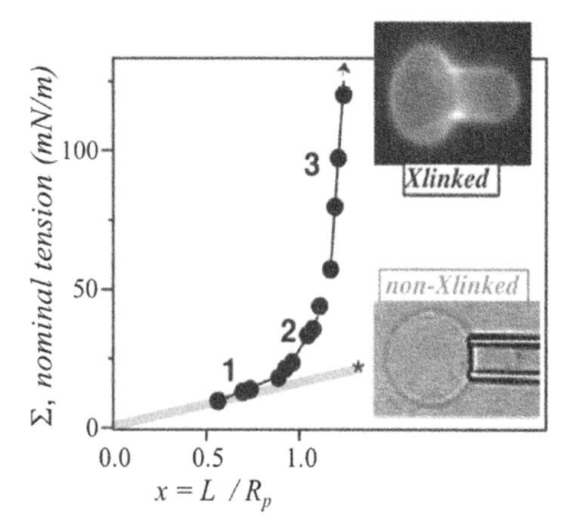

Figure 26.6 Comparing the deformability of crosslinked and non-crosslinked OB vesicles. Aspiration of similar vesicles generates a nominal membrane tension, S, in the aspirated direction which is proportional to the applied pressure but nonlinear with the extension of the membrane, *L*, into the pipette (here, rescaled by the pipette radius R_p). Three regimes of aspiration are typically found for a flaccid vesicle ($V < V_{MAX}$). At small extension (denoted regime 1), the membrane appears to seal against the micropipette as large "dents," typically smaller than the discocyte's dimple, are smoothed out. At intermediate extension (regime 2: $L \approx R_p$), the outer vesicle contour appears more axisymmetric with the membrane forced to constrict upon entering the pipette; this is the regime most readily identified with membrane shearing. In a final high pressure regime 3, extension into the micropipette is minimal while wrinkles appear locked in by contact with the micropipette wall. The upper inset image highlights such wrinkles emanating from the mouth of the pipette on a fluorescently labeled, crosslinked vesicle. In comparison, non-crosslinked vesicles exhibit more linear aspiration curves (gray line); and the tensions at rupture for such fluid membranes are orders of magnitude smaller than the ≈1,000 mN/m tensions sustainable by crosslinked membranes. The lower inset image shows a non-crosslinked vesicle in aspiration and illustrates the spherical contour of membrane outside of the micropipette, consistent with an isotropically tensed fluid membrane. (Reprinted with permission from Discher DE, Eisenberg A [2002] Polymer vesicles. *Science* 297:967–973. Copyright 2002 American Chemical Society.)

at the site of photobleaching (Christian et al., 2009). All of these membrane characterization techniques have been informative about the properties of polymer membranes and have required the use of pGUVs to allow the visualization and manipulation of the membranes.

26.8.2 ENCAPSULATION (DRUGS, OLIGONUCLEOTIDES) FOR FRAGMENTATION TO NANO-VESICLES AND DELIVERY *IN VIVO*

Due to the presence of a hydrophobic core in the bilayers as well as an aqueous lumen, polymersomes can load both hydrophobic and hydrophilic drugs. Doxorubicin (a drug that intercalates in the DNA and damages it (Swift et al., 2006) has been a commonly used hydrophilic anti-cancer drug and has been successfully loaded into the core of polymersomes using the same pH-gradient method used to load liposomes (Ahmed and Discher, 2004). This method utilizes a pH and ammonium sulfate gradient to load the drug, although Choucair et al. (2005) achieved loading without ammonium sulfate and induced drug release using dioxane. Doxorubicin has also been loaded in the membranes of polymersomes using a solvent-displacement (nanoprecipitation) method (Upadhyay et al., 2009, 2010). Drug loading efficiency is dependent on the pH of the buffer, with pH 10.5 leading to higher drug integration than pH of 7.4 (Sanson et al., 2010). Nanoprecipitation was also used for the simultaneous loading of doxorubicin and maghemite nanoparticles to create theranostic nanocarriers (Sanson et al., 2011).

Proteins (Lee et al., 2001; Arifin and Palmer, 2005) and anti-sense oligonucleotides (Kim et al., 2009a) have also been encapsulated in the center of the polymersome. Remarkably, hemoglobin encapsulated polymersomes displayed affinities similar to red blood cells (Arifin and Palmer, 2005). Preintegration of channel proteins into the bilayer has made it possible to encapsulate DNA into the core as well (Graff et al., 2002). Proteins (such as porins and other channel proteins) have also been incorporated in the membrane (assembled from PMOXA-b-PDMS-b-PMOXA) to facilitate loading of ions and macromolecules inside vesicles (Le Meins et al., 2011). Please refer to Chapter 27 for a more in depth discussion on composite vesicles. The reconstitution of proteins in the polymersome membrane is a more difficult task due to the thicker membrane (Mecke et al., 2006). Simulations reveal the existence of an energetic penalty for the insertion of proteins into a thicker membrane (Pata and Dan, 2003). Thus, tailoring of membrane thickness to match that of the hydrophobic regions of the protein is an important consideration. 10 nm membranes have been used to incorporate channel proteins (Meier et al., 2000). An alternative approach has been to modify the protein for easier reconstitution into the membrane (Muhammad et al., 2011).

Paclitaxel is a hydrophobic chemotherapeutic that is widely used in the clinic, and it works by stabilizing microtubules to block mitosis at the metaphase-anaphase transition (Jordan et al., 1996). This drug has been loaded into the bilayer core of polymersome membranes by injecting a solution of the drug (in methanol) into the polymersome suspension. This mixture is then allowed to stand to permit the drug to diffuse into the core of the membrane, following which, the excess free drug is dialyzed away into PBS (Ahmed et al., 2006a, 2006b). Using this procedure, efficiencies higher than those for liposomes (Ahmed and Discher, 2004) have been obtained. The membrane can also be utilized to incorporate fluorescent dyes for imaging and tracking *in vivo* (Lin et al., 1994, Photos et al., 2003). Delivery of two such drugs acting via separate cytotoxic pathways might reduce incidence of drug resistance by cancer cells, and is thus one of the most promising applications of polymersomes. Polymersomes loaded with both Paclitaxel and Doxorubicin led to tumor shrinkage *in vivo* (Ahmed et al., 2006a, 2006b). The maximum tolerated dose of drugs was several-fold increased by loading in polymersomes (compared with free drug), with unloaded polymersomes themselves being benign (Ahmed et al., 2006a, 2006b). In nude mice implanted with human derived tumor xenografts, intravenous injections of Doxorubicin and Paclitaxel encapsulated within polymersomes led to higher drug fluorescence in tumors, while producing consistent shrinkage. Dual loaded polymersomes also led to enhanced tumor cell apoptosis compared with free drug, with levels remaining steady with time (Ahmed et al., 2006a, 2006b).

Although polymersomes and liposomes rely on leaky tumor vasculature (leading to enhanced permeation and retention effect, EPR) (Maeda, 2002; Iyer et al., 2006; Cho et al., 2008), highly stable polymersomes allow attachment of targeting moieties to the surface of these vesicles without compromising their architecture. Active targeting may be particularly necessary with assemblies that have a PEG-coated surface, which reduces uptake by cells (that have negatively charged cell membrane) (Discher et al., 2007). Biotinylation (Dalhaimer et al., 2004) and attachment of anti-human IgG (Meng et al., 2005; Lin et al., 2006) to the surface have been explored as viable options. However, the presence of active targeting groups on the surface can also reduce circulation time (Discher et al., 2007), and this tradeoff is one that has to be carefully weighed before implementing.

Another important application of polymersomes has been as microreactors (Price et al., 2009; Serra et al., 2013). Giant polymer vesicles composed of PEG-12 dimethicone were shown to be permeable to calcium ions in the presence of an ionophore (and impermeable in the absence) (Picker et al., 2011). Multicompartmentalized microreactors have been synthesized to enclose different conditions in different compartments (Chandrawati and Caruso, 2012; Elani et al., 2014).

26.9 CONCLUSIONS

Polymersomes possess most features of liposomes, but broaden the possible range of properties through tuning polymer chemistry and MW. pGUV have been particularly crucial for characterization of polymer membranes composed of polymers with a wide range of chemical and physical properties. pGUV have also been useful in lending insight into the interactions with and mimicry of biological membranes, especially with respect to the role of charge and counter-ions in the formation of laterally segregated domains within the bilayer membrane. Since their conception, polymersomes have served as an important tool for the field of membrane physics while also being developed for a wide range of applications.

LIST OF ABBREVIATIONS

FRAP	fluorescence recovery after photobleaching
MW	molecular weight
OB	polyethylene oxide polybutadiene
OCL	polyethylene oxide poly(ε-caprolactone)
OCLA	polyethylene oxide poly(ε-caprolactoneco-lactide)
OL	polyethylene oxide polylactic acid
P2VP	poly(2-vinylpyridine)
PCL	poly(ε-caprolactone)
PCLA	poly(ε-caprolactoneco-lactide)
PBD	polybutadiene
PDMS	poly(dimethylsiloxane)
PEG	polyethylene glycol (same as PEO)
PEG-PBD	polyethylene oxide polybutadiene
PEG-PCL	polyethylene oxide poly(ε-caprolactone)
PEG-PEE	polyethylene oxide polybutadiene
PEG-PLA	polyethylene oxide polylactic acid
PEG-P2VP	PEG-poly(2-vinylpyridine)
PEO	polyethylene oxide (same as PEG)
PEO-PBD	polyethylene oxide polybutadiene
PEO-PCL	polyethylene oxide poly(ε-caprolactone)
PEO-PCLA	polyethylene oxide poly(ε-caprolactoneco-lactide)
PEO-PLA	polyethylene oxide polylactic acid
pGUVs	polymer giant unilamellar vesicles
PLA	poly-L-lactic acid
PMOXA	poly-2-methyl-2-oxazoline
PPS	polypropylene sulfide

GLOSSARY OF SYMBOLS

Different subscripts

c	chain
CH_2	hydrophobic block

Physical quantities:

a	the cross-sectional area of the hydrophilic group
d	polymersome membrane core thickness
f	hydrophilic mass fraction
l_c	the hydrophobic block chain length
M_{CH2}	molecular weight of the hydrophobic block
M	molecular weight of hydrophobic block
K_a	area expansion modulus
p	packing parameter
v	the volume of the hydrophobic part

REFERENCES

Ahmed F, and Discher DE (2004) Self-porating polymersomes of PEG–PLA and PEG–PCL: Hydrolysis-triggered controlled release vesicles. *J. Control Release* 96:37–53.

Ahmed F, Pakunlu RI, Srinivas G, Brannan A, Bates FS, Klein ML, Minko T, Discher DE (2006a) Shrinkage of a rapidly growing tumor by drug-loaded polymersomes: pH-triggered release through copolymer degradation. *Mol. Pharm.* 3:340–350.

Ahmed F, Pakunlu RI, Srinivas G, Brannan A, Bates FS, Klein ML, Minko T, Discher DE (2006b) Biodegradable polymersomes loaded with both paclitaxel and doxorubicin permeate and shrink tumors, inducing apoptosis in proportion to accumulated drug. *J. Control Release* 116:150–158.

Aranda-Espinoza H, Bermudez H, Bates FS, Discher DE (2001) Electromechanical limits of polymersomes. *Phys Rev Lett* 87:208301.

Arifin DR, and Palmer AF (2005) Polymersome encapsulated hemoglobin: A novel type of oxygen carrier. *Biomacromolecules* 6:2172–2181.

Battaglia G, and Ryan AJ (2005) Bilayers and interdigitation in block copolymer vesicles. *J. Am. Chem. Soc.* 127:8757–8764.

Battaglia G, Ryan AJ, Tomas S (2006) Polymeric vesicle permeability: A facile chemical assay. *Langmuir* 22:4910–4913.

Bermudez H, Aranda-Espinoza H, Hammer DA, Discher DE (2003) Pore stability and dynamics in polymer membranes. *Europhys. Lett.* 64:550.

Bermudez H, Brannan AK, Hammer DA, Bates FS, Discher DE (2002) Molecular weight dependence of polymersome membrane structure, elasticity, and stability. *Macromolecules* 35:8203–8208.

Bermudez H, Hammer DA, Discher DE (2004) Effect of bilayer thickness on membrane bending rigidity. *Langmuir* 20:540–543.

Blanazs A, Armes SP, Ryan AJ (2009). Self-assembled block copolymer aggregates: From micelles to vesicles and their biological applications. *Macromol. Rapid Commun.* 30:267–277.

Borchert U, Lipprandt U, Bilang M, Kimpfler A, Rank A, Peschka-Süss R, Schubert R, Lindner P, Förster, S (2006) pH-induced release from P2VP-PEO block copolymer vesicles. *Langmuir* 22:5843–5847.

Cai S, Vijayan K, Cheng D, Lima EM, Discher DE (2007) Micelles of different morphologies: Advantages of worm-like filomicelles of PEO-PCL in paclitaxel delivery. *Pharm Res* 24:2099–2109.

Carrot G, Hilborn JG, Trollsås M, Hedrick JL (1999) Two general methods for the synthesis of thiol-functional polycaprolactones. *Macromolecules* 32:5264–5269.

Cerritelli S, Velluto D, Hubbell JA (2007) PEG-SS-PPS: Reduction-sensitive disulfide block copolymer vesicles for intracellular drug delivery. *Biomacromolecules* 8:1966–1972.

Chandrawati R, and Caruso F (2012) Biomimetic liposome-and polymersome-based multicompartmentalized assemblies. *Langmuir* 28:13798–13807.

Checot F, Lecommandoux S, Klok HA, Gnanou Y (2003) From supramolecular polymersomes to stimuli-responsive nano-capsules based on poly (diene-b-peptide) diblock copolymers. *Eur. Phys. J. E.* 10:25–35.

Chemin M, Brun PM, Lecommandoux S, Sandre O, Le Meins JF (2012) Hybrid polymer/lipid vesicles: Fine control of the lipid and polymer distribution in the binary membrane. *Soft Matter* 8:2867–2874.

Chidambaram M, Manavalan R, Kathiresan K (2011) Nanotherapeutics to overcome conventional cancer chemotherapy limitations. *J. Pharm. Pharm. Sci.* 14:67–77.

Cho K, Wang XU, Nie S, Shin DM (2008) Therapeutic nanoparticles for drug delivery in cancer. *Clin. Cancer Res.* 14:1310–1316.

Choucair A, Lim Soo P, Eisenberg A (2005) Active loading and tunable release of doxorubicin from block copolymer vesicles. *Langmuir* 21:9308–9313.

Christian DA, Tian A, Ellenbroek WG, Levental I, Rajagopal K, Janmey PA, Liu AJ, Baumgart T, Discher DE (2009) Spotted vesicles, striped micelles and Janus assemblies induced by ligand binding. *Nat Mater* 8:843–849.

Christian DA, Garbuzenko OB, Minko T, Discher DE (2010) Polymer vesicles with a red cell-like surface charge: Microvascular imaging and in vivo tracking with near-infrared fluorescence. *Macromol. Rapid Commun.* 31:135–141.

Dalhaimer P, Engler AJ, Parthasarathy R, Discher DE (2004) Targeted worm micelles. *Biomacromolecules* 5:1714–1719.

Dimova R, Seifert U, Pouligny B, Förster S, Döbereiner HG (2002) Hyperviscous diblock copolymer vesicles. *Eur. Phys. J. E.* 7:241–250.

Discher BM, Bermudez H, Hammer DA, Discher DE, Won YY, Bates FS (2002) Cross-linked polymersome membranes: Vesicles with broadly adjustable properties. *J. Phys. Chem. B.* 106:2848–2854.

Discher BM, Won YY, Ege DS, Lee JC, Bates FS, Discher DE, Hammer DA (1999) Polymersomes: Tough vesicles made from diblock copolymers. *Science* 284:1143–1146.

Discher DE, and Ahmed F (2006) Polymersomes. *Annu. Rev. Biomed. Eng.* 8:323–341.

Discher DE, and Eisenberg A (2002) Polymer vesicles. *Science* 297:967–973.

Discher DE, Ortiz V, Srinivas G, Klein ML, Kim Y, Christian D, Cai S, Photos P, Ahmed F (2007) Emerging applications of polymersomes in delivery: From molecular dynamics to shrinkage of tumors. *Prog. Polym. Sci.* 32:838–857.

Elani Y, Law RV, Ces O (2014) Vesicle-based artificial cells as chemical microreactors with spatially segregated reaction pathways. *Nat. Commun.* 5:5305.

Geng Y, Dalhaimer P, Cai S, Tsai R, Tewari M, Minko T, Discher DE (2007) Shape effects of filaments versus spherical particles in flow and drug delivery. *Nat. Nanotechnol* 2:249–255.

Geng Y, and Discher DE (2005) Hydrolytic degradation of poly (ethylene oxide)-block-polycaprolactone worm micelles. *J. Am. Chem. Soc.* 127:12780–12781.

Ghoroghchian PP, Frail PR, Susumu K, Blessington D, Brannan AK, Bates FS, Chance B, Hammer DA, Therien MJ (2005) Near-infrared-emissive polymersomes: Self-assembled soft matter for in vivo optical imaging. *Proc. Natl. Acad. Sci. USA* 102:2922–2927.

Ghoroghchian PP, Li G, Levine DH, Davis KP, Bates FS, Hammer DA, Therien MJ (2006) Bioresorbable vesicles formed through spontaneous self-assembly of amphiphilic poly (ethylene oxide)-block-polycaprolactone. *Macromolecules* 39:1673–1675.

Graff A, Sauer M, Van Gelder P, Meier W (2002) Virus-assisted loading of polymer nanocontainer. *Proc. Natl. Acad. Sci. USA* 99:5064–5068.

Gutteridge WE (1985) Existing chemotherapy and its limitations. *Br. Med. Bull.* 41:162–168.

Hamley IW (2005). Nanoshells and nanotubes from block copolymers. *Soft Matter* 1:36–43.

Howse JR, Jones RA, Battaglia G, Ducker RE, Leggett GJ, Ryan AJ (2009) Templated formation of giant polymer vesicles with controlled size distributions. *Nat Mater* 8:507–511.

Israelachvili JN (1991) *Intermolecular and Surface Forces*, 2nd Ed, London, UK: Academic.

Itel F, Chami M, Najer A, Lörcher S, Wu D, Dinu IA, Meier W (2014) Molecular organization and dynamics in polymersome membranes: A lateral diffusion study. *Macromolecules* 47:7588–7596.

Iyer AK, Khaled G, Fang J, Maeda H (2006) Exploiting the enhanced permeability and retention effect for tumor targeting. *Drug. Discov. Today* 11:812–818.

Jordan MA, Wendell K, Gardiner S, Derry WB, Copp H, Wilson L (1996) Mitotic block induced in HeLa cells by low concentrations of paclitaxel (Taxol) results in abnormal mitotic exit and apoptotic cell death. *Cancer Res.* 56:816–825.

Kazunori K, Masayuki Y, Teruo O, Yasuhisa S (1993) Block copolymer micelles as vehicles for drug delivery. *J. Control Release* 24:119–132.

Kim MS, and Lee DS (2010) Biodegradable and pH-sensitive polymersome with tuning permeable membrane for drug delivery carrier. *Chem. Commun* 46:4481–4483.

Kim S, Kim JH, Jeon O, Kwon IC, Park K (2009a) Engineered polymers for advanced drug delivery. *Eur. J. Pharm. Biopharm.* 71:420–430.

Kim Y, Dalhaimer P, Christian DA, Discher DE (2005) Polymeric worm micelles as nano-carriers for drug delivery. *Nanotechnol* 16:S484.

Kim Y, Tewari M, Pajerowski JD, Cai S, Sen S, Williams J, Sirsi S, Lutz G, Discher DE (2009b) Polymersome delivery of siRNA and antisense oligonucleotides. *J. Control Release* 134:132–140.

Kita-Tokarczyk K, Grumelard J, Haefele T, Meier W (2005) Block copolymer vesicles—Using concepts from polymer chemistry to mimic biomembranes. *Polymer* 46:3540–3563.

Klibanov AL, Maruyama K, Torchilin VP, Huang L (1990) Amphipathic polyethyleneglycols effectively prolong the circulation time of liposomes. *FEBS Lett* 268:235–237.

Kwon GS, and Okano T (1996) Polymeric micelles as new drug carriers. *Adv. Drug. Deliv. Rev.* 21:107–116.

Lasic DD, and Papahadjopoulos D (1998) Medical applications of liposomes. Elsevier.

Lee JCM, Bermudez H, Discher BM, Sheehan MA, Won YY, Bates FS, Discher DE (2001) Preparation, stability, and in vitro performance of vesicles made with diblock copolymers. *Biotechnol. Bioeng.* 73:135–145.

Lee JCM, Santore M, Bates FS, Discher DE (2002) From membranes to melts, rouse to reptation: Diffusion in polymersome versus lipid bilayers. *Macromolecules* 35:323–326.

Le Meins JF, Sandre O, Lecommandoux S (2011) Recent trends in the tuning of polymersomes' membrane properties. *Eur. Phys. J. E.* 34:1–17.

Le Meins JF, Schatz C, Lecommandoux S, Sandre O (2013) Hybrid polymer/lipid vesicles: State of the art and future perspectives. *Mater Today* 16:397–402.

Leong KW, Brott BC, Langer R (1985) Bioerodible polyanhydrides as drug-carrier matrices. I: Characterization, degradation, and release characteristics. *J. Biomed. Mat. Res.* 19:941–955.

Li YY, Cunin F, Link JR, Gao T, Betts RE, Reiver SH, Chin V, Bhatia SN, Sailor MJ (2003) Polymer replicas of photonic porous silicon for sensing and drug delivery applications. *Science* 299:2045–2047.

Lian T, and Ho RJ (2001) Trends and developments in liposome drug delivery systems. *J. Pharm. Sci.* 90:667–680.

Lin JJ, Ghoroghchian PP, Zhang Y, Hammer DA (2006) Adhesion of antibody-functionalized polymersomes. *Langmuir* 22:3975–3979.

Lin VS, Di Magno SG, Therien MJ (1994) Highly conjugated, acetyleneyl bridged porphyrins: New models for light-harvesting antenna systems. *Science* 264:1105–1111.

LoPresti C, Lomas H, Massignani M, Smart T, Battaglia G (2009) Polymersomes: Nature inspired nanometer sized compartments. *J. Mater. Chem.* 19:3576–3590.

Lorenceau E, Utada AS, Link DR, Cristobal G, Joanicot M, Weitz DA (2005) Generation of polymerosomes from double-emulsions. *Langmuir* 21:9183–9186.

Mabrouk E, Cuvelier D, Pontani LL, Xu B, Lévy D, Keller P, Brochard-Wyart F, Nassoy P, Li MH (2009) Formation and material properties of giant liquid crystal polymersomes. *Soft Matter* 5:1870–1878.

Maeda H (2002) Enhanced permeability and retention (EPR) effect: Basis for drug targeting to tumor. In V. Muzykantov et al. (eds.), *Biomedical Aspects of Drug Targeting*. New York, Springer Science+Business Media.

Mecke A, Dittrich C, Meier W (2006) Biomimetic membranes designed from amphiphilic block copolymers. *Soft Matter* 2:751–759.

Meeuwissen SA, Bruekers SM, Chen Y, Pochan DJ, van Hest JC (2014) Spontaneous shape changes in polymersomes via polymer/polymer segregation. *Polym. Chem.* 5:489–501.

Meier W, Nardin C, Winterhalter M (2000) Reconstitution of channel proteins in (polymerized) ABA triblock copolymer membranes. *Angew. Chem. Int. Ed.* 39:4599–4602.

Meng F, Engbers GH, Feijen J (2005) Biodegradable polymersomes as a basis for artificial cells: Encapsulation, release and targeting. *J. Control Release* 101:187–198.

Meng F, Zhong Z, Feijen J (2009) Stimuli-responsive polymersomes for programmed drug delivery. *Biomacromolecules* 10:197–209.

Messager L, Gaitzsch J, Chierico L, Battaglia G (2014) Novel aspects of encapsulation and delivery using polymersomes. *Curr. Opin. Pharmacol.* 18:104–111.

Middleton JC, and Tipton AJ (2000) Synthetic biodegradable polymers as orthopedic devices. *Biomaterials* 21:2335–2346.

Muhammad N, Dworeck T, Fioroni M, Schwaneberg U (2011) Engineering of the *E. coli* outer membrane protein FhuA to overcome the hydrophobic mismatch in thick polymeric membranes. *J. Nanobiotechnology* 9:1:8.

Nam J, Beales PA, Vanderlick TK (2010) Giant phospholipid/block copolymer hybrid vesicles: Mixing behavior and domain formation. *Langmuir* 27:1–6.

Nam J, Vanderlick TK, Beales PA (2012) Formation and dissolution of phospholipid domains with varying textures in hybrid lipo-polymersomes. *Soft Matter* 8:7982–7988

Napoli A, Boerakker MJ, Tirelli N, Nolte RJ, Sommerdijk NA, Hubbell JA (2004) Glucose-oxidase based self-destructing polymeric vesicles. *Langmuir* 20:3487–3491.

Nuss H, Chevallard C, Guenoun P, Malloggi F (2012) Microfluidic trap-and-release system for lab-on-a-chip-based studies on giant vesicles. *Lab on a Chip* 12:5257–5261.

Oltra NS, Swift J, Mahmud A, Rajagopal K, Loverde SM, Discher DE (2013) Filomicelles in nanomedicine–from flexible, fragmentable, and ligand-targetable drug carrier designs to combination therapy for brain tumors. *J. Mater. Chem. B* 1:5177–5185.

Onaca O, Enea R, Hughes DW, Meier W (2009) Stimuli-responsive polymersomes as nanocarriers for drug and gene delivery. *Macromol. Biosci.* 9:129–139.

Ortiz V, Nielsen SO, Discher DE, Klein ML, Lipowsky R, Shillcock J (2005) Dissipative particle dynamics simulations of polymersomes. *J. Phys. Chem. B* 109:17708–17714.

Pantano DA, Moore PB, Klein ML, Discher DE (2011) Raft registration across bilayers in a molecularly detailed model. *Soft Matter* 7:8182–8191.

Pata V, and Dan N (2003) The effect of chain length on protein solubilization in polymer-based vesicles (polymersomes). *Biophys. J.* 85:2111–2118.

Peters EAJF (2004). Elimination of time step effects in DPD. *Europhys. Lett.* 66:311.

Photos PJ, Bacakova L, Discher BM, Bates FS, Discher DE (2003). Polymer vesicles in vivo: Correlations with PEG molecular weight. *J. Control Release* 90:323–334.

Picker A, Nuss H, Guenoun P, Chevallard C (2011) Polymer vesicles as microreactors for bioinspired calcium carbonate precipitation. *Langmuir* 27:3213–3218.

Price AD, Zelikin AN, Wang Y, Caruso F (2009) Triggered enzymatic degradation of DNA within selectively permeable polymer capsule microreactors. *Angew. Chem. Int. Ed.* 48:329–332.

Rajagopal K, Mahmud A, Christian DA, Pajerowski JD, Brown AE, Loverde SM, Discher DE (2010) Curvature-coupled hydration of semicrystalline polymer amphiphiles yields flexible worm micelles but favors rigid vesicles: Polycaprolactone-based block copolymers. *Macromolecules* 43:9736–9746.

Rodriguez-Garcia R, Mell M, López-Montero I, Netzel J, Hellweg T, Monroy F (2011) Polymersomes: Smart vesicles of tunable rigidity and permeability. *Soft Matter* 7:1532–1542.

Sanson C, Diou O, Thevenot J, Ibarboure E, Soum A, Brûlet A, Miraux S et al. (2011) Doxorubicin loaded magnetic polymersomes: theranostic nanocarriers for MR imaging and magneto-chemotherapy. *ACS Nano* 5:1122–1140.

Sanson C, Schatz C, Le Meins JF, Soum A, Thévenot J, Garanger E, Lecommandoux S (2010) A simple method to achieve high doxorubicin loading in biodegradable polymersomes. *J. Control Release* 147:428–435.

Schmaljohann D (2006) Thermo-and pH-responsive polymers in drug delivery. *Adv. Drug Deliv. Rev.* 58:1655–1670.

Schulz M, Glatte D, Meister A, Scholtysek P, Kerth A, Blume A, Bacia K, Binder WH (2011) Hybrid lipid/polymer giant unilamellar vesicles: Effects of incorporated biocompatible PIB–PEO block copolymers on vesicle properties. *Soft Matter* 7:8100–8110.

Schulz M, Werner S, Bacia K, Binder WH (2013) Controlling molecular recognition with lipid/polymer domains in vesicle membranes. *Angew. Chem. Int. Ed.* 52:1829–1833.

Semple SC, Chonn A, Cullis PR (1998) Interactions of liposomes and lipid-based carrier systems with blood proteins: Relation to clearance behavior in vivo. *Adv. Drug. Deliv. Rev.* 32:3–17.

Serra CA, Cortese B, Khan IU, Anton N, de Croon MH, Hessel V, Ono T, Vandamme T (2013) Coupling microreaction technologies, polymer chemistry, and processing to produce polymeric micro and nanoparticles with controlled size, morphology, and composition. *Macromol. React. Eng.* 7:414–439.

Sharma A, and Sharma US (1997) Liposomes in drug delivery: Progress and limitations. *Int. J. Pharm.* 154:123–140.

Shum HC, Kim JW, Weitz DA (2008) Microfluidic fabrication of monodisperse biocompatible and biodegradable polymersomes with controlled permeability. *J. Am. Chem. Soc.* 130:9543–9549.

Srinivas G, Discher DE, Klein ML (2004a). Self-assembly and properties of diblock copolymers by coarse-grain molecular dynamics. *Nat. Mater* 3:638–644.

Srinivas G, Discher DE, Klein ML (2005). Key roles for chain flexibility in block copolymer membranes that contain pores or make tubes. *Nano. Lett.* 5:2343–2349.

Srinivas G, and Klein ML (2004). Coarse-grain molecular dynamics simulations of diblock copolymer surfactants interacting with a lipid bilayer. *Mol. Phys.* 102:883–889.

Srinivas G, Shelley JC, Nielsen SO, Discher DE, Klein ML (2004b). Simulation of diblock copolymer self-assembly, using a coarse-grain model. *J. Phys. Chem. B* 108:8153–8160.

Swift LP, Rephaeli A, Nudelman A, Phillips DR, Cutts SM (2006). Doxorubicin-DNA adducts induce a non-topoisomerase II–mediated form of cell death. *Cancer Res.* 66:4863–4871.

Smart T, Lomas H, Massignani M, Flores-Merino MV, Perez LR, Battaglia G (2008). Block copolymer nanostructures. *Nano Today* 3:38–46.

Upadhyay KK, Bhatt AN, Mishra AK, Dwarakanath BS, Jain S, Schatz C, Le Meins JFL et al. (2010) The intracellular drug delivery and anti tumor activity of doxorubicin loaded poly (γ-benzyl l-glutamate)-b-hyaluronanpolymersomes. *Biomaterials* 31:2882–2892.

Upadhyay KK, Meins JFL, Misra A, Voisin P, Bouchaud V, Ibarboure E, Schatz C, Lecommandoux S (2009) Biomimetic doxorubicin loaded polymersomes from hyaluronan-block-poly (γ-benzyl glutamate) copolymers. *Biomacromolecules* 10:2802–2808.

Vriezema DM, Garcia PM, Sancho Oltra N, Hatzakis NS, Kuiper SM, Nolte RJ, Rowan AE, van Hest J (2007) Positional assembly of enzymes in polymersome nanoreactors for cascade reactions. *Angew. Chem.* 119:7522–7526.

Walde P, Cosentino K, Engel H, Stano P (2010) Giant vesicles: preparations and applications. *ChemBioChem* 11:848–865.

Warren PB (1998) Dissipative particle dynamics. *Curr. Opin. Colloid Interface Sci.* 3:620–624.

Woodle MC (1993) Surface-modified liposomes: Assessment and characterization for increased stability and prolonged blood circulation. *Chem. Phys. Lipids* 64:249–262.

Xu JP, Ji J, Chen WD, Shen JC (2005) Novel biomimetic polymersomes as polymer therapeutics for drug delivery. *J. Control Release* 107:502–512.

Giant hybrid polymer/lipid vesicles

Thi Phuong Tuyen Dao, Khalid Ferji, Fabio Fernandes, Manuel Prieto, Sébastien Lecommandoux, Emmanuel Ibarboure, Olivier Sandre, and Jean-François Le Meins

A compromise is the art of dividing a cake in such a way
that everyone believes he has the biggest piece.

Ludwig Erhard

Contents

27.1 INTRODUCTION

About 15 years ago, vesicles resulting from the self-assembly of amphiphilic copolymers referred to as "polymersomes" emerged as a potential alternative to liposomes for fundamental research and applications, such as cell membrane models, nano/microreactors or drug delivery systems (DDS) (Discher et al., 1999, 2000; Discher and Ahmed, 2006; LoPresti et al., 2009; Le Meins et al., 2011; Liao et al., 2012; Thevenot et al., 2013). Their major characteristics are described and discussed in Chapter 26, to be consulted by readers who are non-specialist in the field before going through the present chapter. This one deals with hybrid, that is, intimately mixed polymer/lipid vesicles that can be viewed as advanced vesicular structures as compared with their liposome and polymersome forerunners, as they potentially marry in a single membrane the best characteristics of the two separate components. Ideally, these structures could present biocompatibility and bio-functionality of liposomes, as well as robustness, low permeability and functional variability conferred by the copolymer chains. This should be of great interest in pharmaceutical applications for which only a few formulations based on liposomes are commercially available despite decades of research, (e.g., DaunoXome®, Doxil®/Caelyx®, Visudyne®), but also in personal care. In particular such moderate use of liposomes in clinics could be due to their lack of mechanical stability in the high shear rate of blood circulation through tiny vessels. Liposomal DDS can also often exhibit uncontrolled leakage phenomena (seen as a "burst release" effect on their pharmacokinetic profiles). As a consequence, the controlled release of encapsulated molecules at the predetermined biological target (e.g., a tumor site) remains a difficult challenge. Besides the obvious interest of the association of lipids and amphiphilic copolymers into a single membrane of LUVs for biomedical applications, hybrid giant unilamellar vesicles (hGUVs) can also be an excellent tool to get more insight into molecular and macroscopic parameters that govern the membrane domain formation, fusion or fission. Literature on the subject is still relatively limited (Le Meins et al., 2013; Schulz and Binder, 2015) although the scientific output is increasing with growing interest from different scientific communities (biophysicists, biologists, physicochemists). Promising results have been obtained regarding their drug targeting ability and biomolecular recognition properties (Cheng et al., 2011; Schulz et al., 2013).

To date, the physical and molecular factors governing the phase separation in these hybrid copolymer/lipid membranes are only partially understood. In addition to the expected chemical incompatibility between copolymer block chains and phospholipids, one also has to consider the respective dimensions of the molecules as well as those of the corresponding bilayers. In order to perfectly benefit from the potential of such systems, the membrane structure must be tuned either toward homogeneous mixing of the molecular components or, on the contrary, toward lateral phase separation leading to the presence of nano/micrometric domains. The relationship between membrane structure and physical and bio-functional properties must then be better understood in order to eventually optimize and validate the use of hybrid vesicles in future applications like drug delivery, tumor targeting, bio-recognition or bio-adhesion.

In the present chapter, an overview of hybrid copolymer/phospholipid vesicles will be given with a particular emphasis on hGUVs. The molecular and macroscopic parameters necessary to obtain stable hybrid vesicles with different membrane structural levels (homogenous distribution of the components, nanodomain or microdomain formation) will be first summarized, followed by a description of the different preparation methods used to obtain hGUVs. It has to be noted that hybrid vesicles reported over the last 10 years were exclusively prepared by a one-step process by which a film composed of the desired amount of copolymer and phospholipid is hydrated. We will not describe previous approaches based on the modulation of lipid membrane properties by adsorption of amphiphilic polymers onto preformed giant liposomes. Although such a method leads in some cases to a reorganization of the lipid membrane and to the induction of polymer-rich domains—see for instance (Ladavière et al., 2002; Tribet and Vial, 2007), readers interested by these approaches will find relevant information in Chapters 9 and 25 that describe membrane-polymer interactions. Finally, an overview of what is known about membrane properties of hybrid vesicles and especially hGUVs will be proposed. Tips and advices will be given in the two last sections about the preparation protocols of hGUVs and techniques used to characterize their membrane properties.

27.2 CRITERIA TO BE FULFILLED TO OBTAIN HYBRID GIANT VESICLES

27.2.1 A BRIEF OVERVIEW OF MULTICOMPONENT LIPID VESICLES

The existing work on hybrid copolymer/lipid vesicles is obviously inspired from all previously acquired knowledge on multicomponent lipid vesicles. These systems have been proposed as tools to understand the structure-properties relationship of biological cell membranes (both of the plasma membrane and that of the internal organelles), which are constituted of different lipids and membrane proteins ensuring part of the many biological functions of the cell: transport of matter, energy, cell division, signaling pathways…). The lipid composition in the membrane strongly depends on the nature of the cell (eukaryote, prokaryote or archaea) and comprises several classes: glycerophospholipids, sphyngolipids, sterols, saccharolipids, and so on. This subtle association allows flexibility and fluidity of the membrane and the formation of lipid raft like domains, which can arise from the aggregation of proteins such as clathrin but can also be driven by lipid segregation. Lipid rafts are mainly composed of sphingolipid and cholesterol-rich domains and contain a variety of signaling proteins. It has been established that the lipid rafts play an important role in health and disease (Michel and Bakovic, 2007).

Numerous studies have been realized on model GUVs to understand the role of lipid segregation in domain formation (Binder et al., 2003; Lipowsky and Dimova, 2003). This has been extensively discussed in Chapter 18. Here, we briefly recall some molecular aspects of domain formation. Basically, two types of phase separation in lipid membrane can occur: Lateral phase separation of two lipids into different areas or orthogonal phase separation between the two leaflets of the lipid bilayer. Orthogonal phase separation can be triggered *via* addition of an external compound (e.g., adsorption of polymer chains), whereas lateral phase separation can occur through several mechanisms that are mainly through interaction between lipid headgroups or tails within the membrane, but also by recruiting mobile "binders" among the lipids into the adhesion area with a substrate (Brochard-Wyart and de Gennes, 2002).

Nonideal mixing or even demixing (phase separation) can occur between lipids with similar structure (e.g., phosphocholine headgroup with two saturated acyl chains) provided that a sufficient length difference in lipid tails is present (typically four CH_2 groups). In that case, phase separation is obtained below a given temperature of fluid/solid or solid/solid transition. A different nature of the headgroups (e.g., charged/neutral) can also lead to phase separation, but in that case the ionic content of the solution is important (e.g., added Ca^{2+} ions, [Cevc and Richardsen, 1999]). A strong difference in melting temperatures is generally associated with a strong difference in chemical structures (e.g., sphingolipids and phospholipids) leading to solid/solid or fluid/solid phase separations versus temperature. Fluid-fluid phase separation can also occur through weak attractive forces. Cholesterol has been largely employed to modulate the fluidity of membranes and to create phase separation above the main transition temperature of a phosphocholine lipid (e.g., 1,2-di-palmitoyl-*sn*-glycero-3-phosphocholine [DPPC], 1,2-dimyristoyl-*sn*-glycero-3-phosphocholine [DMPC] see Appendix 1 of the book for structure and data on these lipids) leading to liquid-ordered and liquid-disordered phase coexistence (Garcia-Saez and Schwille, 2010), as seen in Chapter 18. Phase separation leads to lipid/lipid boundaries and possibly to a height mismatch between both phases. Consequently, the membrane elastically deforms at the domain interface to minimize the exposure of hydrophobic tails to water. The height mismatch has an energetic cost proportional to the length of the boundary line, thus defining the line tension. Thermodynamically, the line tension tends to favor domain coalescence (once a nucleation size is reached) to minimize the boundary length. As a consequence, fluid lipid domains should grow with time into one single large-circular domain in the membrane. However, some distribution of domain sizes can be found in model GUVs and biological membranes. This is due to the fact that the line tension is balanced by other mechanisms such as an "entropic trap" (Frolov et al., 2006) stabilizing the domains at a nano-metric size, the "elastic interaction" between dimpled domains due to deformation of the surrounding membrane (Ursell et al., 2009), the long range electrostatic dipolar interaction (Travesset, 2006), and the natural vesicle spontaneous curvature c_0 and bending rigidity κ of the membrane (Semrau et al., 2009; Ursell et al., 2009; Hu et al., 2011).

27.2.2 THE EFFECT OF KEY MOLECULAR PARAMETERS: LIPID FLUIDITY, CHEMICAL INCOMPATIBILITY AND HYDROPHOBIC MISMATCH

In the case of copolymer and lipid mixtures, a very important parameter controlling the formation of stable hybrid vesicles is the discrepancy of chemical composition and size of hydrophobic segments between polymers and lipids. In the case of lipid mixtures, one has to consider interactions between lipid tails, always constituted of saturated or unsaturated fatty acid chains. However, in the case of polymer/lipid mixture, the nature of monomeric unit may lead to a stronger immiscibility between the hydrophobic copolymer blocks and the lipid tails. In addition, a characteristic thickness of lipid membrane is around 3 to 5 nm, well below those commonly observed for polymersomes (~10 nm or more) although this parameter is directly controlled by the polymerization degree (see Chapter 26), and may lead to strong geometric differences between the molecules constituting the membrane and large entropic driving force toward demixing. This most often results in phase separation, leading to separate populations of liposomes and polymersomes.

A relatively limited number of amphiphilic copolymers have been used so far to form hGUVs. Hydrophobic blocks were based on poly(dimethyl siloxane) (PDMS) (Chemin et al., 2012; Chen and Santore, 2015), poly(isobutylene) (PIB) (Schulz et al., 2011, 2013, 2014; Olubummo et al., 2014) or poly(butadiene) (PBd) (Cheng et al., 2011; Nam et al., 2011, 2012; Lim et al., 2013), whereas hydrophilic blocks were either made of poly(ethylene oxide) (PEO) or poly(2-methyl oxazoline) (PMOXA). All these polymer blocks possess a low glass transition temperature (T_g), allowing dynamic exchanges of the chains and leading to the formation of membrane with a structure at thermal equilibrium. The low T_g is a criterion that appeared so far as essential, but not unique, to the successful formation of hGUVs. Concerning the choice of lipids, most studies were performed with phosphatidylethanolamine or phosphatidylcholine headgroups like 1-palmitoyl-2-oleoyl-*sn*-glycero-3-phosphocholine (POPC) (Nam et al., 2011, 2012; Lim et al., 2013), HSPC (Cheng et al., 2011), 1,2-dioleoyl-*sn*-glycero-3-phosphocholine (DOPC) and 1,2-dilauroyl-*sn*-glycero-3-phosphocholine (DLPC) (Olubummo et al., 2014; Schulz et al., 2014), and the most often used DPPC (Chemin et al., 2012; Nam et al., 2012; Schulz et al., 2013, 2014; Chen and Santore, 2015). An extensive list of systems used so far is given in Table 27.1.

It is interesting to note that the Hildebrand solubility parameters, δ, (which are derived from the heat of vaporization of a molecule and which reflect its cohesive energy density) of hydrocarbon moieties in hydrophobic polymer blocks and phospholipids are relatively close, that is δ = 9.1 cal$^{1/2}$/cm$^{3/2}$ for the fatty acid tail in lipids and δ = 7.3 cal$^{1/2}$/cm$^{3/2}$, 7.7 cal$^{1/2}$/

Table 27.1 **Composition and membrane structure of hGUVs-see Appendix 1 of the book for structure and data on the lipids in this stable**

COPOLYMER, MOLAR MASS M_n	LIPID	MASS COMPOSITION COPOLYMER/LIPID	MEMBRANE STRUCTURE	REFERENCE
(PMOXA-*b*-PDMS-*b*-PMOXA) M_n = 9,000 g·mol^{-1}	PC PE	Copolymer/PC/PE: 98%/1%/1%	Homogeneous membranes	Ruysschaert et al. (2005)
(PBd$_{46}$-*b*-PEO$_{30}$), M_n = 3,800 g·mol^{-1}	POPC	Copolymer/POPC: 100%-92%/0%-8%	Homogeneous membranes	Nam et al. (2011)
		Copolymer/POPC: 90%-73%/10%-27%	No hGUV formation	
		Copolymer/POPC: 0%-68%/100%-32%	Separated vesicles: liposomes + polymersomes	
		Copolymer/POPC/Biotinyl DSPE: 89%/6%/5%	Heterogeneous vesicles with small lipid domains	
		Copolymer/POPC/Biotinyl DSPE: 64%/28%/8%	Heterogeneous vesicles with large lipid domains	
		Copolymer/POPC/Chol: 84.5%/10.3%/5.2% 76.8%/15.4%/7.8% 71.4%/23.8%/4.8%	Heterogeneous vesicles with micrometer-sized lipid domains	Nam et al. (2012)
	DPPC	Copolymer/DPPC: 88.6%/11.4%	• At 50°C: homogeneous membranes • At room temperature (RT): heterogeneous membranes: lipid domain size and shape depend on cooling rate	
		Copolymer/DPPC/Chol: 86.5%/10%/3.5%	Round DPPC rich domains at RT	

(Continued)

Table 27.1 (*Continued*) **Composition and membrane structure of hGUVs-see Appendix 1 of the book for structure and data on the lipids in this stable**

COPOLYMER, MOLAR MASS M_n	LIPID	MASS COMPOSITION COPOLYMER/LIPID	MEMBRANE STRUCTURE	REFERENCE
(PBd_{22}-*b*-PEO_{14}) M_n = 1,800 g·mol^{-1}	HSPC	Copolymer/HSPC: 95%/5%	Homogeneous vesicles	Cheng et al. (2011)
		Copolymer/HSPC: 87%/13%		
	POPC	Copolymer/POPC: 70%/30%	Homogeneous vesicles	Lim et al. (2013)
(PIB_{87}-*b*-PEO_{17}) M_n = 5,350 g·mol^{-1}	DOPC	Copolymer/DOPC 74%/26% 63%/37%	Homogeneous and stable vesicles	Schulz et al. (2014)
		Copolymer/DOPC 43%/57%	Homogeneous membranes turning into heterogeneous membranes. Budding and fission leading to formation of separated polymersomes and liposomes	
	DPPC	Copolymer/DPPC 100%/0%	No hGUVs were formed	Schulz et al. (2011, 2013, 2014)
		Copolymer/DPPC 92%-99.85%/8%-0.15%	Homogeneous and small vesicles	
		Copolymer/DPPC 74%-91.6%/26%-8.4%	Homogeneous vesicles with smooth surface	
		Copolymer/DPPC 64.6%-74%/35.4%-26%	Heterogeneous vesicles with domains	
		Copolymer/DPPC 7%-54%/93%-46%	Homogeneous membranes with "DPPC aspect": facetted surface and hole defects	
(PIB_{37}-*b*-PEO_{48}) M_n = 3,970 g·mol^{-1}	DPPC	Copolymer/DPPC 57%/43%	Homogeneous vesicle with large holes and "lacerated edges"	Schulz et al. (2011)
		Copolymer/DPPC 37.6%/62.4%	Homogeneous membranes but vesicles were neither smooth nor round	
Cholesteryl-poly(ethylene oxide)-*b*-poly(glycerol) Ch-PEG_{30}-*b*-$hbPG_{23}$ M_n = 1,100 g·mol^{-1}	POPC DLPC DOPC	Copolymer/Lipid 1.5%/98.5% 7%/93% 23%/77% 59%/41%	Homogeneous membranes and stable vesicles	Scholtysek et al. (2015)
$PDMS_{22}$-*g*-$(PEO_{12})_2$ M_n = 3,200 g·mol^{-1}	POPC	Copolymer/POPC: 86%-97%/14%-3%	Homogeneous vesicles (stable during many days).	Chemin et al. (2012)
		Copolymer/POPC: 58%-81%/42%-19%	Unstable biphasic vesicles leading to separated liposomes and polymersomes after a budding and fission process	
	DPPC	Copolymer/DPPC: 96%/4%	Homogeneous vesicles membranes	
		Copolymer/DPPC: 93%-81%/7%-19%	Stable heterogeneous vesicles with lipid domains	
		Copolymer/DPPC: 59%/41%	Separated liposomes and polymersomes and some heterogeneous vesicles containing polymer domains	

(*Continued*)

Table 27.1 (*Continued*) **Composition and membrane structure of hGUVs-see Appendix 1 of the book for structure and data on the lipids in this stable**

COPOLYMER, MOLAR MASS M_n	LIPID	MASS COMPOSITION COPOLYMER/LIPID	MEMBRANE STRUCTURE	REFERENCE
		Copolymer/DPPC: 99%-93%/1%-7%	Homogeneous vesicle membranes	Chen and Santore (2015)
		Copolymer/DPPC: 93%-70%/7%-30%	Heterogeneous membranes with hexagonal facetted domains. Shape of domains is independent on cooling rate	
		Copolymer/DPPC: 70%-99%/30%-1%/	Heterogeneous membranes at RT. Domain morphologies depend on cooling rate: at 1°C/min: hexagonal facetted domains at 5°C/min: striped domains	

Note: Biotinyl-DSPE, 1,2-distearoyl-*sn*-glycero-3-phosphoethanolamine-*N*-[biotinyl(poly(ethylene glycol)]-2000.

$cm^{3/2}$ and 8.32 $cal^{1/2}/cm^{3/2}$ respectively for PDMS, PIB and PBd blocks (Roth, 1990; King, 2002). These relatively close values suggest that the chemical compatibility between the components is indeed a parameter of uppermost importance to enable the formation of such hybrid vesicles even though the lateral phase separation of components inside the membrane still can occur for other reasons, as it will be commented in the following.

In each of the abovementioned contributions, there is no real systematic investigation allowing a clear extraction of molecular and macroscopic parameters necessary to intimately mix the components into stable hGUVs presenting homogeneous distribution of both components, or on the contrary to induce formation of heterogeneous membranes patterned with domains. Moreover, another difficulty arises from the fact that the molar composition of lipid and copolymer in the final hybrid vesicles can be different from the starting composition, as evidenced by fluorescence microscopy, which complicates the analysis of the results. This is inherent to the experimental procedures used so far for the formation of hGUVs, which are described in Section 27.3.

The physical state of the lipids, which depends on their main transition temperature (from gel state at $T < T_m$ to fluid state at $T > T_m$, where T_m is the melting temperature) as well as the composition of the lipid/copolymer mixture are among the most relevant parameters. It seems that at high copolymer content (>70% weight), the formation of homogeneous hybrid vesicles is favored when using a lipid with phosphocholine as headgroup and fatty chains in a fluid state at room temperature (Cheng et al., 2011; Nam et al., 2011, 2012; Lim et al., 2013), except in one case where no hGUVs were obtained between 90% and 73% weight fraction of a polybutadiene-block-poly(ethylene oxide) (PBd-*b*-PEO) copolymer with a number average molecular weight of 3800 $g{\cdot}mol^{-1}$ (Nam et al., 2011).

Above a critical lipid weight fraction, one generally observes the formation of heterogeneous vesicles presenting lipid-rich micrometric domains, that progressively evolve through a budding and fission phenomenon toward separated liposomes and polymersomes (Chemin et al., 2012; Schulz et al., 2014). This ultimate phase separation into two pure GUVs occurs for fluid domains in fluid membranes, and is directly linked to a sufficiently high line tension. When the line tension is large enough, the energetic barrier induced by the larger curvature energy associated with membrane budding can be overcome by decreasing the boundary length between the lipid and copolymer domains and the associated excess energy. To get rid of the line energy implies a cost in bending energy, as the curvature of the membrane will increase through the formation of the bud. Therefore line tension between the domains and the bending rigidity of the membrane are two parameters of prime importance.

In copolymer/lipid hybrid vesicles, line tension and bending rigidities can be different to a large extent as compared with their values for lipid/lipid mixtures. The usual membrane (bilayer) thickness is indeed 3–5 nm for liposomes, whereas it may vary from 5 to 50 nm for polymersomes. In the case of a large size gap, the formation of a lipid domain would result in a high line tension at the lipid/copolymer boundaries arising from the exposure of hydrophobic polymer segments to water ("hydrophobic mismatch"). To reduce this exposure and the resulting energetic cost of the boundary lines, the two opposite plausible scenarios can be considered. The first one (i) consists in a conformational adaptation through elastic deformation of the polymer chains at the boundary to decrease the line tension (Figure 27.1) in analogy to elastic deformation of membrane at lipid/lipid domain boundary in lipid bilayers (Kuzmin et al., 2005). Another possibility (ii) is to decrease the interfacial length and therefore the interfacial energy, by coalescence into fewer domains of a larger area.

In Figure 27.1, which illustrates different properties resulting from the membrane structure and potential applications of hybrid vesicles, it is shown that the conformational adaptation of the polymer implies a collapse of the hydrophobic polymer chains near the lipid interface, therefore reducing the total number of conformations and opposing the entropic elasticity of chains. Therefore, it is clear that the molar mass (or chain length) and the rigidity (or Kühn length) of the hydrophobic polymer backbone also play a major role. If this adaptation cannot be achieved, then the domain formation is improbable (spontaneously nucleated domains eventually collapse) and a homogeneous mixture of the components is expected (Figure 27.1).

A large hydrophobic mismatch is met in most of the studies performed so far, as the diblock or triblock copolymers most often

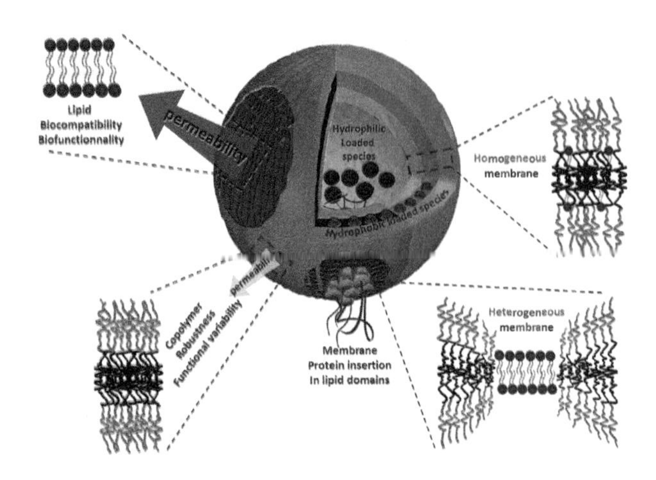

Figure 27.1 Illustration of the different membrane structural organization of the membrane in hGUVs and its principal characteristics.

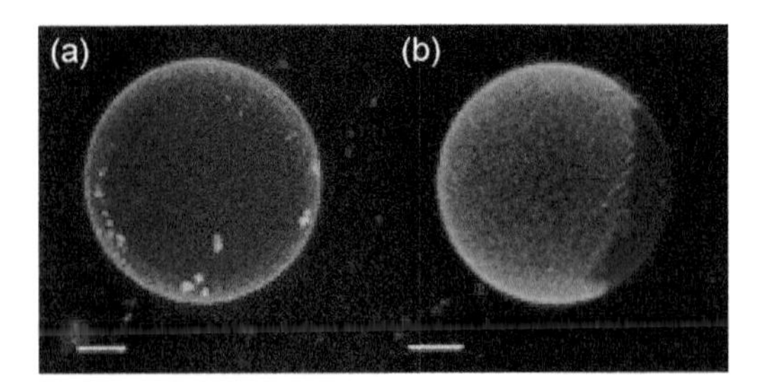

Figure 27.2 Overlay of maximum intensity 3D projection images taken for (a) homogeneous hGUV with the homogeneously distribution of both signals; (b) demixed hGUV with green polymer-rich phase and red lipid-rich phase. Scale bars: 5 µm. (Dao, T.P.T. et al., *Soft Matter*, 13, 627–637, 2017. Reproduced by permission of the Royal Society of Chemistry.)

used, form membranes whose thickness is at least 7 nm, and the line tension, although not yet quantified experimentally, is expected to be high and driving the budding and fission of existing fluid domains. However, the hydrophobic mismatch is certainly not the only parameter, as the budding and fission of lipid domains have been observed also in a study in which a grafted copolymer, PDMS-*g*-$(PEO)_2$ well-known to form vesicles with a membrane thickness close to liposomes (~5 nm), was used (Chemin et al., 2012). In addition to the chemical nature of the hydrophobic block, which obviously plays a role in the miscibility with the lipid phase and consequently on the interfacial energy, the architecture of the copolymer may also be an important factor to consider (e.g., *block* copolymer where the different polymers are linearly covalently coupled, versus *graft* copolymer where polymers chain are distributed along the backbone of another polymer).

The fluidity of the lipid phase is also of significance in the membrane structure of hGUVs. In the case of lipid in the gel state at room temperature, and using a formation protocol described in Section 27.3, the spontaneous formation of micrometer-sized domains was reported only as a rare event. For instance in one of these studies (Schulz et al., 2011), hGUVs presenting stable micrometric domains were spontaneously obtained using DPPC and PIB_{87}-*b*-PEO_{17}, but only in a narrow composition range (65–74 mass% polymer). It is supposed that the large hydrophobic block in that case limited the conformational adaptation at the copolymer–lipid boundary. The large hydrophobic thickness (~10 nm) plays in favor of a statistic distribution of the lipid in the copolymer phase as sketched on Figure 27.1. Interestingly, homogeneous vesicles, at least at the micrometric scale, were observed for all copolymer contents larger than 30 mol% (or 75 weight %). Below 60 weight%, homogeneous vesicles presenting faceted surfaces were obtained, which is the signature of the DPPC gel phase. As described above, using a copolymer presenting a membrane thickness close to that of liposomes, allows the spontaneous formation of micrometric lipid domains in hGUVs in a large polymer content range (from 10% to 93% in weight). Above 93%, the lipid is apparently dispersed in the polymer phase and homogenous vesicles can be observed microscopically. Ideally the homogenous or heterogeneous character of hGUVs can be easily observed under fluorescence microscopy by including a small amount of fluorescently tagged polymer and lipid in the initial mixture, such as fluorescein modified polymer and L-α-egg-phosphatidylethanolamine-*N*-(lissamine rhodamine B-sulfonyl) (Egg Liss Rhod PE) for instance. In homogeneous hybrid vesicles, both signals of the fluorescent probes are observed within the entire membrane, whereas in heterogeneous hybrid vesicles, the probes preferentially partition into the polymer and lipid rich phases, as illustrated in Figure 27.2.

27.2.3 THE EFFECT OF THE CHEMICAL MODIFICATIONS AND ADDITIVES

Obtaining polymer-rich vesicles with stable lipid-rich domains of controlled micrometric or nanometric size would be of great interest for different kinds of functions in drug-delivery systems, biotargeting or biophysical fundamental studies, such as the modeling of nanoparticles/membrane interactions and cellular internalization ("artificial endocytosis"). The use of lipids in the fluid state, which is probably the most interesting, regarding the abovementioned applications, leads only to homogeneous hGUVs (i.e., at the micrometric scale, as appearing under an optical microscope). Other approaches involving the use of additional components have been used to generate a stable phase-separated membrane from a mixture of a copolymer with a mixture of several fluid lipids. For instance, with lipids presenting biotinylated headgroups, stable micrometric domains can be generated in hybrid vesicles by reacting them with streptavidin (in solution). This protein with multivalent binding sites for biotin works as a "zipper" to gather the lipid molecules together in pure lipid phases or "monodomains." However, such a protein coating prevents further bio-functionalization of the domains (Schulz et al., 2011). Another approach consists in using a lipid mixture containing a given amount of cholesterol, which is well-known to promote lateral phase separation into "raft-like" domains within liposomes for particular lipid compositions such as cholesterol/egg-phosphatidylcholine/sphingomyelin (Chol/PC/SM) mixtures (Bagatolli and Kumar, 2009). This induces the same effects on hybrid copolymer/lipid vesicles in which round-shaped micrometric domains can be obtained with various phospholipids of low melting temperature Tm (e.g., DLPC, POPC, 2-ditridecanoyl-sn-glycerophosphocholine [DC13PC]). Interestingly, the domain size could be modulated *via* the polymer/lipid/cholesterol composition. However, no domain formation can be obtained when using DOPC, even at high cholesterol content, in agreement with what is commonly observed for DOPC/cholesterol mixtures that are not known to form liquid ordered phase bilayers (Mills et al., 2008). Regarding the hydrophobic mismatch, it has also been observed *via* cryo-TEM experiments on LUVs made of mixtures of PDMS-*b*-PMOXA and cholesterol that the vesicles obtained present lower membrane thickness compared with pure polymersomes. This was explained by an increase of the packing density in the membrane through the support of static light scattering measurements (Winzen et al., 2013). This approach could be useful to control the hydrophobic mismatch in hybrid lipid/copolymer giant vesicles.

27.2.4 THE EFFECT OF MACROSCOPIC PARAMETERS: TEMPERATURE, MEMBRANE TENSION

Forming hybrid copolymer/lipid vesicles containing a lipid with a melting temperature above room temperature implies a particular attention toward the process of vesicle formation. Generally, the film hydration method to prepare hGUVs is conducted at a temperature above the melting transition of the lipid. Then the vesicle suspension is cooled down to room temperature. It has been shown that the cooling rate plays a role on the resulting membrane structuration in hGUVs made of PBd-*b*-PEO and DPPC (Nam et al., 2012). A fast cooling rate results in the formation of numerous small domains, whereas a slow cooling rate favors formation of less numerous domains but larger in size, in agreement with the classical nucleation-growth theory.

In two-component phosphatidylcholine GUVs, it has been shown that membrane tension could affect the fluid-solid phase transition of lipid and consequently the membrane structuration (Chen and Santore, 2014). The same effect has been observed recently on hybrid vesicles made of PDMS-*g*-$(PEO)_2$ and DPPC (Chen and Santore, 2015). Using vesicles conditioned in a way that they did not present excess membrane area, the authors were able to tune the membrane tension by using different cooling rates after vesicle formation by electroformation at high temperature (see Section 27.3). The cooling causes a contraction of both the water solution and the membrane. As the membrane contracts more than the water in the compartment of the vesicle, a membrane surface tension appears, as estimated by a micropipette pulling experiment. Thereafter the membrane stress obviously relaxes as water diffuses progressively out of the vesicle. The authors were able to measure the membrane tension immediately after cooling to the temperature of interest, and a clear influence of cooling rate was shown on the membrane tension. Actually the membrane tension measured is maximal at intermediate cooling speed: A too small cooling rate allows membrane tension to relax by water diffusion by permeability, whereas a too fast quench induces stress relaxation through the membrane lysis and leaking by transient pore formation (Figure 27.3).

At low surface tension, patchy domains were generally obtained whereas higher tension lead to striped domains. The physical state of the DPPC phase (ripple phase P_β' solid or gel phase L_β) inside these stripes or patchy domains is still under debate in literature (Gordon et al., 2006; Bernchou et al., 2009; Chen and Santore, 2015).

Finally, the structure of hGUVs containing high melting temperature lipids (e.g., DPPC) has been rarely commented above the transition temperature T_m. Our team has started to evaluate the behavior of hGUVs composed of PDMS-*g*-$(PEO)_2$/DPPC previously studied at room temperature (Chemin et al., 2012) following an electroformation process realized at 55°C and storage conditions of the sample at 55°C. The sample was transferred as quickly as possible in a temperature-controlled stage at 55°C, in order to maintain the lipid phase in a continuous fluid (liquid disordered) state. Observations were made after different incubation times at 55°C, from 4 to 20 h (Figure 27.4). L-α-phosphatidylethanolamine-*N*-(lissamine rhodamine B sulfonyl) (Liss-Rhod PE) was used to reveal the lipid-rich phase (0.2% mol), and copolymer grafted with fluorescein was added to reveal the polymer-rich phase (1% mol). It appeared that the fluorescent lipid (Liss-Rhod PE) was excluded from the DPPC domains, although

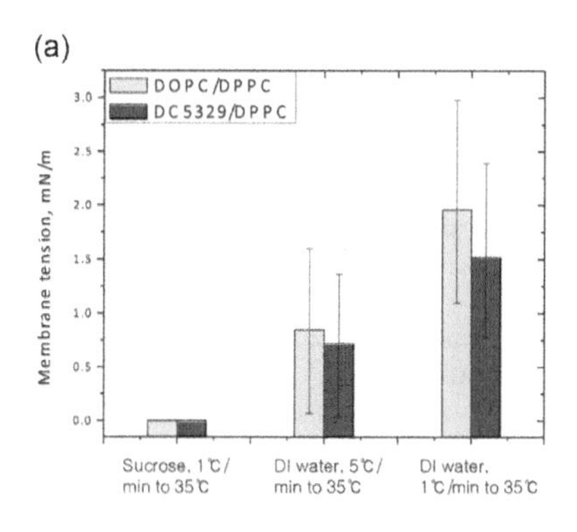

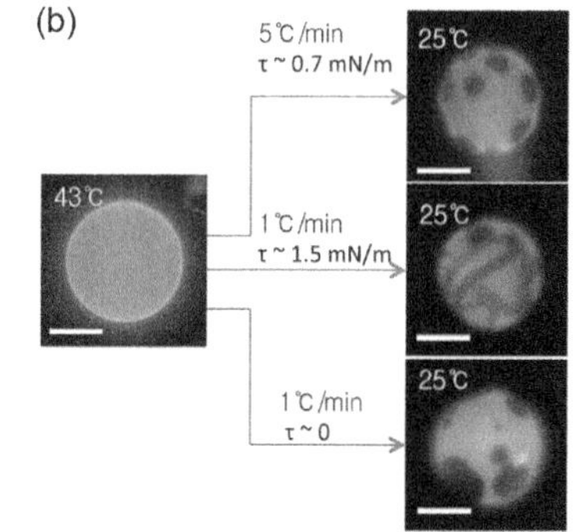

Figure 27.3 (a) Illustration of the variation of membrane tension resulting from different cooling rates after preparation onto the membrane structuration: examples on vesicles containing 70% mol (39 mass%) DPPC and 30% mol (61% mass) PDMS-*g*-$(PEO)_2$. (b) Images of vesicles submitted to different cooling and osmotic control. (Chen, D. and Santore, M.M., *Soft Matter*, 11, 2617–2626, 2015. Reproduced by permission of the Royal Society of Chemistry.)

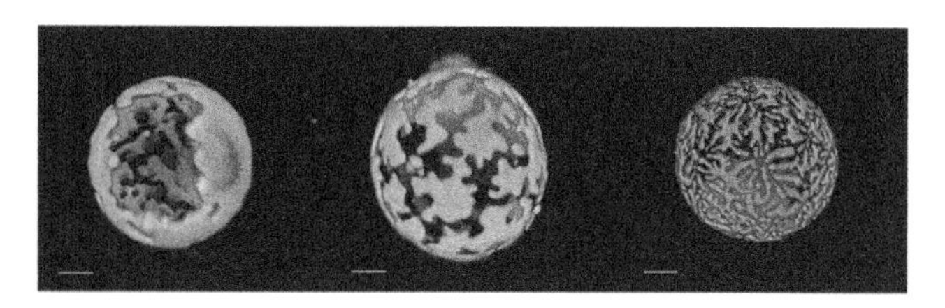

Figure 27.4 3D reconstructions of confocal microscopy images of PDMS-*g*-$(PEO)_2$/DPPC (90/10 mass) vesicles loaded with Liss-Rhod PE (red) and Fluorescein-PDMS-*g*-$(PEO)_2$ (green) observed at 55°C after different incubation times at 55°C; see Appendix 2 of the book for structures of fluorescent dyes. Scale bar: 10 μm.

they were in the fluid state as attested by the rounded boundary between the polymer and lipid domains. After a long incubation time, the Liss-Rhod PE repartitioned entirely into the more disordered polymer phase, which turned from green to yellow-orange. The most interesting feature is that the interface length between the lipid and polymer phases clearly increased, suggesting that a kind of compatibility increase occurred with time, and that such structures are strongly out-of-equilibrium in their early stage of formation.

27.3 SPECIFIC ASPECTS OF THE FORMATION OF GIANT HYBRID UNILAMELLAR VESICLES

Most of the knowledge acquired in the past 10 years about hybrid copolymer/lipid vesicles results from studies performed on GUVs formed by the two well-known methods of rehydration or electroformation (see Chapter 1) from copolymer/lipid mixtures as applied by several investigators (Ruysschaert et al., 2005; Cheng et al., 2011; Nam et al., 2011, 2012; Schulz et al., 2011, 2013, 2014; Chemin et al., 2012; Lim et al., 2013; Chen and Santore, 2015; Scholtysek et al., 2015) and summarized in Table 27.2. A few examples report the preparation of hGUVs using mixed suspensions of giant phospholipid liposomes and polymersomes and the triggering of their adhesion followed by membrane fusion, by using temperature-responsive copolymers, or salt addition (Henderson and Paxton, 2014; Morimoto et al., 2014). The ability of these methods to be extended to different polymer/lipid mixtures has not been confirmed so far. It has to be noted that simple film rehydration was used by Tsourkas and colleagues. (Cheng and Tsourkas, 2008; Cheng et al., 2011) to obtain hGUVs of HSPC phospholipid and PBd_{22}-*b*-PEO_{14} block copolymer, with long incubation time (24 h) in an aqueous solution of sucrose at 65°C, but electroformation is definitely the most commonly used technique. Practical tips about the preparation and collecting of hGUVs are described in Box 27.1.

Table 27.2 **Preparation methods for obtaining hGUVs-see Appendix 1 of the book for structure and data on the lipids in this stable**

COPOLYMER, MOLECULAR WEGHT, MEMBRANE THICKNESS	LIPID (PHASE STATE)	PREPARATION TECHNIQUE	REFERENCE
PMOXA-*b*-PDMS-*b*-PMOXA M_n = 9,000 g·mol^{-1} l_{me} = 14 nm	PC (fluid) PE (fluid) l_{me} = 3–4 nm	Electroformation method ITO coated glass: 5 V, 10 Hz, 3 h followed by 5 V, 0.5 Hz 30 min	Ruysschaert et al. (2005)
PBd_{46}-*b*-PEO_{30} M_n = 3,800 g·mol^{-1} l_{me} = 9 nm	POPC (fluid)	Electroformation method platinum electrodes: 3 V, 11 Hz several hours, at RT in sucrose solution (260 mOsm)	Nam et al. (2011)
	DPPC (gel)	Electroformation method platinum electrodes: 3 V, 15 Hz, at 50°C, in sucrose solution (260 mOsm)	Nam et al. (2012)
PBd_{22}-*b*-PEO_{14} M_n = 1,800 g·mol^{-1} l_{me} ~ 7 nm	HSPC (fluid)	Film rehydration in sucrose solution (285 mM), incubating at 65°C during 24 h	Cheng et al. (2011)
	POPC (fluid)	Electroformation method ITO-coated glass: 3 V, 10 Hz, at 45°C in sucrose solution (at 300 mM) during 120 min	Lim et al. (2013)
PIB_{87}-*b*-PEO_{17} M_n = 5,350 g·mol^{-1} l_{me} = 10 nm	DOPC (fluid) DPPC (gel)	Electroformation method ITO-coated glass electrodes: 1.3 V, 10 Hz at 70°C during 4 h	Schulz et al. (2011, 2013, 2014)
Poly(isobutylene)-*b*-poly(ethylene oxide) (PIB_{37}-*b*-PEO_{48}) M_n = 3,970 g·mol^{-1}			
Poly(dimethylsiloxane)-graft-poly(ethylene oxide) $PDMS_{12}$-*g*-$(PEO)_2$ M_n = 2,700 g·mol^{-1} l_{me} = 5 nm	POPC (fluid)	Electroformation method ITO-coated glass: 2 V, 10 Hz, at RT during 20 min	Chemin et al. (2012)
	DPPC (gel)	Electroformation method ITO-coated glass: 2 V, 10 Hz at 50°C during 20 min	
		Electroformation method platinum wire electrodes: 3 V, 10 Hz during 1 h at 52°C	Chen and Santore (2014)

Note: l_{me}: hydrophobic membrane thickness.

Box 27.1 One example for a practical method for obtaining hGUVs *via* electroformation

Preparation of hGUVs using this method is made in three steps:

- *Step 1*: Solution preparation
  - Organic solution of lipids and copolymers are prepared and then mixed in order to reach the desired composition. Typically, chloroform or a mixture of chloroform and MeOH (2/1 v/v) are the solvents most often used. The concentration range (copolymer+lipid) is typically around 1 mg·mL^{-1}. If hybrid vesicles are meant to be studied with fluorescence or confocal microscopy analyses, a tagged lipid should be included to reveal the lipid phase. Typically the use of Liss-Rhod PE gives good results at 0.2% molar ratio. Co-localization can also be performed using fluorescently tagged copolymers that have to be synthesized. If fluorescein or nitrobenzoxadiazole (NBD) is used as fluorescent moiety, a larger amount of copolymer probe should be used for visualization, such as 1% molar, because of the lower quantum yields and tendency of photobleaching of these dyes compared with rhodamine.
- *Step 2*: Film deposition
  - Films are prepared by spreading around 10 µL of the lipid/copolymer solution slowly over an area of about 2 cm diameter of an electrically conductive surface. Most often used are platinum electrodes and glass plates coated with indium tin oxide (ITO glass plates of surface resistivity 15–25 Ω/square). Surfaces must be cleaned with ethanol and chloroform before use. In the case of ITO plates, a *Hamilton* syringe with tapered needle is used to deposit the solution and also to gently scratch the film and generate irregularities that will help film swelling and vesicle formation. No additional benefit is observed on the amount of vesicle obtained, size distribution or homogeneity of composition by forming the film by spin-coating. Then the film is dried out under dynamic vacuum for at least 5 h. Too short drying times can lead to the formation of vesicles presenting "lipid filaments" on their surface.
- *Step 3*: Electroswelling
  - Film rehydration under an alternative electric field is performed most often in a glucose or sucrose solution to set the osmolarity of the buffer, defined as the total amount of soluble species per kg of solution, and checked for instance by a freezing point osmometry measurement. As illustrated in Table 27.2, there is no real tendency that can be extracted in terms of voltage applied or duration of electroformation. However, from our personal experience, it seems that at least 5 V is needed for molar masses of hydrophobic block above 5,000 g·mol^{-1}. Temperature is also a parameter of great importance. Obviously electroformation must be realized above the main chain transition temperature of the lipid used, but even in the case of lipids fluid at room temperature (e.g., POPC), temperature have to be controlled and slightly increased (~30°C–40°C) when using polymers with long hydrophobic blocks >5,000 g·mol^{-1}. To extract a solution containing giant vesicles, it is recommended to use a syringe with at least a 0.8 mm internal diameter (gauge ≤ 21 G) of the needle in order to minimize shear stress. During electroformation, some vesicles remain stuck on the conductive surfaces. To detach them, two methods are used depending on the type of surfaces employed. In case of platinum electrodes, gently manually shaking the electrode in the hydration medium is sufficient to detach vesicles. However, when ITO glass plates are used, GUVs are detached by turning (or rolling) around a bubble created in the chamber by taking out a few drops of solution (Figure B27.1.1).

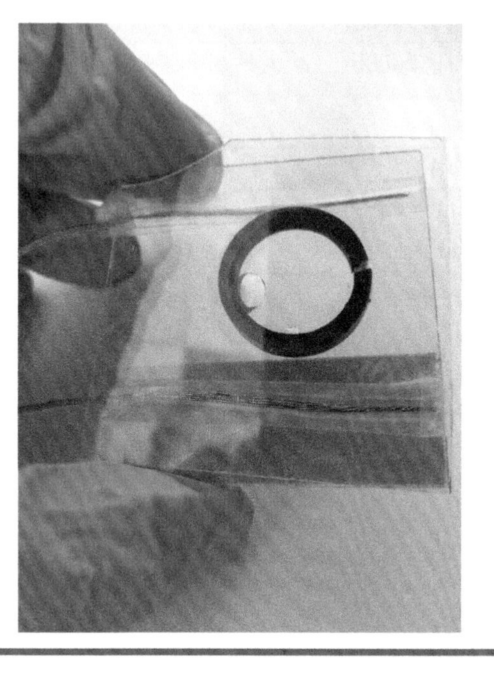

27.4 UNDERSTANDING MEMBRANE PROPERTIES FROM MEMBRANE STRUCTURE

The purpose of this section is to describe some of the properties of the hGUVs that were characterized so far, and try to point out a correlation between those properties and their membrane structure at macro- or nanoscale. Considering the intrinsic differences between lipid and copolymer membranes as illustrated in Chapter 26, in terms of bio-functionality and physical properties, it can be reasonably expected that lipid polymer mixtures provide numerous ways to modulate the membrane properties, provided that a good control of membrane composition and structure is achieved. In this part, the mechanical properties, permeability, fluidity, and bio-functionality will be discussed in priority, whereas stability and deformability will also be cursory mentioned.

27.4.1 MECHANICAL PROPERTIES, FLUIDITY

Two essential features are often analyzed on giant vesicles because of their importance in biological events (e.g., resistance of cells to osmotic shock, cell fission and fusion, cell motility) or drug delivery (resistance in the high shear rate of blood circulation through tiny vessels such as capillaries). They are related to their ability to resist to isotropic area dilation, and to deform from an initially flat surface into a curved structure, see Chapter 5. They are respectively quantified through the area compressibility modulus K_A, which is linked to the interfacial tension at the junction between hydrophilic and hydrophobic moieties of the membrane, and the bending rigidity of the membrane κ (also called bending modulus) that appears in the Helfrich expression of the curvature energy that becomes non-negligible for nearly zero surface tension systems like micro-emulsions and lipid bilayers. As illustrated in Chapter 15 and Chapter 26, K_A and κ from liposomes and polymersomes are rather different (see also Table 15.1 in Chapter 15) and they can be modulated to some extent by the hydrophobic block length and membrane thickness for polymersomes. This is clearly illustrated in Tables 27.3 and 27.4. Another important point to mention is the much larger lysis strain of the polymersomes, ascribed to a higher cohesive energy density between the molecules constituting the membrane. On the contrary, the excess area defined by subtracting the projected area A_p from the total true surface area of the membrane seems to be much lower for giant polymersomes than for giant liposomes, which explains why it is very difficult to extract the bending modulus κ from the very limited "entropic regime" in a micropipette pulling experiment.

Therefore, when mixing copolymer and lipid in a single membrane, a modulation of these properties can be expected, obviously by playing on the lipid/copolymer composition. Subtle modifications can also probably be obtained through precise membrane structuration. Interestingly, the area compressibility modulus has only been measured on hybrid vesicles through the help of the micropipette aspiration technique in recent studies. In their early work on hybrid vesicles in 2011, Cheng et al. reported that homogeneous hybrid vesicles composed

Table 27.3 **Area compressibility modulus (K_A) of different liposomes and polymersomes**

LIPID	K_A (mN/m)	COPOLYMER	K_A (mN/m)
POPC, RT	198 ± 8 (Shoemaker and Vanderlick, 2003)	PEO-*b*-PBBdt, RT	102 ± 10 (Bermudez et al., 2002)
DOPC, 15°C	310 ± 20 (Rawicz et al., 2008)	PEO-*g*-PDMS, RT	95 ± 9 (Chen and Santore, 2015)
SOPC, 15°C	290 ± 6 (Rawicz et al., 2008)	PEO-*b*-PBd, RT	102 ± 10 (Bermúdez et al., 2004) 470 ± 15 (Dimova et al., 2002) 112 ± 20 (Nam et al., 2011)
HSPC, >T_m	206 (Cheng et al., 2011)	PEO-*b*-PEE, RT	120 ± 20 (Discher et al., 1999)
DMPC, 29°C	234 (Bermúdez et al., 2004)		
DPPC, >T_m	~200 (Chen and Santore, 2015)		

Note: SOPC, 1-stearoyl-2-oleoyl-*sn*-glycero-3-phosphocholine.

Table 27.4 **Bending modulus (κ) of different liposomes and polymersomes**

LIPID	BENDING MODULUS (κ) (10^{-19} J)	POLYMER	BENDING MODULUS (κ) (10^{-19} J)
POPC, 25°C	1.58 ± 0.03 (Henriksen et al., 2004)	$PDMS_{60}$-*b*-$PMOXA_{21}$,	101 ± 23 (Winzen et al., 2013)
DOPC, 23°C	1.08 ± 0.1 (Gracia et al., 2010)	$PDMS_{68}$-*b*-$PMOXA_{11}$, RT	70 ± 50 (Jaskiewicz et al., 2012)
SOPC, 18°C	0.9 ± 0.06 (Evans and Rawicz, 1990)	PEO_{26}-*b*-PBd_{46}, RT	1.02 ± 0.46 (Bermúdez et al., 2004)
Egg PC, RT	0.3 ± 0.1 (Evans and Rawicz, 1990)	PEO_{80}-*b*-PBd_{125}, RT	19.1 ± 6.4 (Bermúdez et al., 2004)
DPPC, RT	10–15 (Jaskiewicz et al., 2012)	PEO_{40}-*b*-PEE_{37}, RT	1.37 ± 0.29 (Bermúdez et al., 2004)
DMPC, 29°C	0.56 ± 0.06 (Evans and Rawicz, 1990)	PS_{115}-*b*-PAA_{15}, RT	716 ± 103 (Chen et al., 2009)

of poly(butadiene)-*b*-poly(oxyethylene) (PBD_{22}-*b*-PEO_{14}) and hydrogenated soy phosphatidylcholine (HSPC) exhibited an intermediate elastic stretching modulus between the values of pure lipid and pure polymer vesicles (Cheng et al., 2011). Similar results for hybrid vesicles composed of PBD_{46}-*b*-PEO_{30} and POPC were also indicated (Nam et al., 2011). Although there was a rather large composition range where hybrid vesicles could not be formed, the remaining fractions showed distinctly a gradual decrease in K_A with increasing copolymer content (Figure 27.5). It has to be noted that in the above-mentioned contributions, the hGUVs obtained presented a homogeneous membrane structure, at least at the micrometer scale studied by optical microscopy. A recent study has focused on hGUVs composed of poly(dimethylsiloxane)-*graft*-poly(ethyleneoxide) $PDMS_{22}$-*g*-$(PEO_{12})_2$ and DPPC, which is at the gel state at room temperature. In this system, results from our group showed that heterogeneous membranes (presence of micrometric domains) could be obtained at room temperature in a large composition range (Chemin et al., 2012). The main results of the Santore group have been described in Section 27.2.4 of this chapter. However, they also performed some mechanical measurements through micropipette aspiration on these heterogeneous membranes presenting lipid gel domains. In the case of data obtained without having lipid micrometric domains entering in the micropipettes (therefore by stretching the copolymer part of the membrane), the obtained stretching modulus was similar to that of the pure copolymer membrane within the uncertainty of the measurement for hybrid vesicles containing 70 mol% of DPPC (39 mass%) (Chen and Santore, 2015). Interestingly, the lysis strain was still very high and similar to the lysis strain of pure

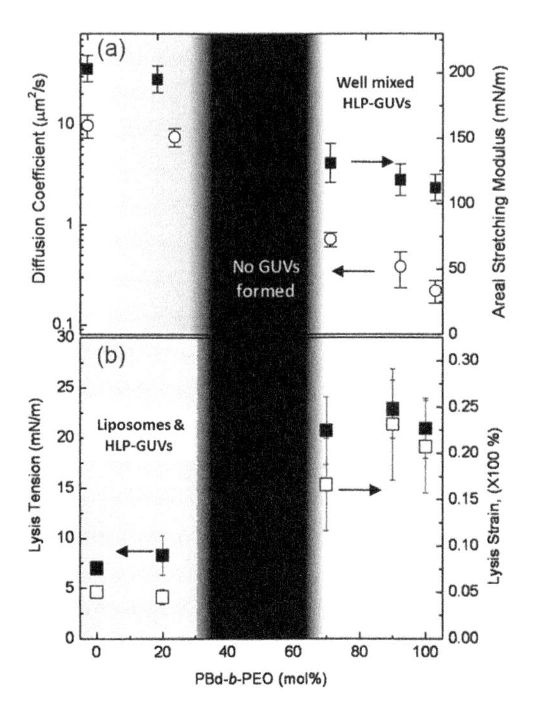

Figure 27.5 (a) Areal stretching modulus K_A (■) and lipid lateral diffusion coefficient D (○) for varying compositions of hGUVs. (b) Critical lysis tension (■) and lysis strain (□) of POPC/PBd-b-PEO GUVs. HLP-GUVs: Hybrid lipid-diblock copolymer GUVs. (Reprinted with permission from Nam, J. et al., *Langmuir*, 27, 1–6, 2011. Copyright 2011 American Chemical Society.)

polymersomes, despite the potential fragility that could result from the interfaces at the copolymer/lipid boundaries. This shows clearly that the mechanical properties of hybrid membranes can be directly linked to their lateral structure.

Regarding the bending modulus (κ) of hGUVs, quantitative data reported so far in the literature are scarce compared with the abundant values for lipid GUVs, which are floppier and much more flexible (see previous remark on the reduced excess area of polymersomes). There is only one alternate approach by atomic force microscopy to measure the κ of hybrid large unilamellar vesicles (hLUVs) composed of poly(dimethylsiloxane)-*block*-poly(methyloxazoline) and 1,2-dimyristoyl-*sn*-glycero-3-phosphocholine ($PDMS_{60}$-*b*-$PMOXA_{21}$/DMPC) with a diameter of ~200 nm. The mechanical properties of the vesicles were measured using quantitative forces versus distance curves. An intermediate value of κ between liposomes and polymersomes ($62 \cdot 10^{-19}$ J for hybrid $PDMS_{60}$-*b*-$PMOXA_{21}$/DMPC at 50:50 mol ratio or 90:10 mass ratio) was reported (Winzen et al., 2013). The authors investigated also the modification of the membrane properties using cholesterol instead of DMPC. Interestingly, they observed that the bending modulus of a membrane containing only ~5% in mass of cholesterol was more than four times larger than that of pure polymersomes. It was suggested that this could result from an increase of the packing density in the membrane induced by cholesterol. From the preliminary results, it is clear that the flexibility of hybrid membranes can be tuned on a very large range and in a subtle way. As this parameter and other mechanical properties control many aspects of the behavior of vesicles such as their formation, stability, fusion and budding processes, further studies need to be performed to concomitantly elucidate the membrane structure and the resulting properties.

Although there are many types of molecular individual and collective motions within membranes, the mobility of molecules inside a membrane has been mostly evaluated through the measurement of the lateral diffusion coefficient, which is directly linked to the surface shear viscosity of the membrane. Such measurement is commonly made by fluorescence recovery after photobleaching (FRAP) experiments, see also Chapter 21. In the case of hybrid copolymer lipid hGUVs, one can access either the mobility of the lipid molecules or the copolymer chains depending on the localization of the fluorescent probes. Numerous studies and reviews on liposomal membranes pointed out a strong influence of the lipid physical state and membrane composition on the molecular mobility. Therefore it is expected that the mobility and the fluidity of the membrane could be finely tuned by the association of copolymers and lipids in the membrane. Lateral diffusion coefficients for lipids are indeed in the range 3–5 $\mu m^2 \cdot s^{-1}$ and those of copolymer chains in the lower range 0.0024–0.12 $\mu m^2 \cdot s^{-1}$ (Lee et al., 2002). Large differences in surface shear viscosity have also been reported between PBd-*b*-PEO and lipid membranes (Dimova et al., 2002; Evans et al., 2003), with even higher values reported for liquid crystalline block copolymers (Mabrouk et al., 2009). Therefore large variations of such parameters are anticipated depending on the lipid/copolymer composition, but subtle modifications could result also from peculiar membrane lateral structuration (presence of domains) (Loura et al., 2005, 2010). Other types of effects can also be invoked, such as in the work

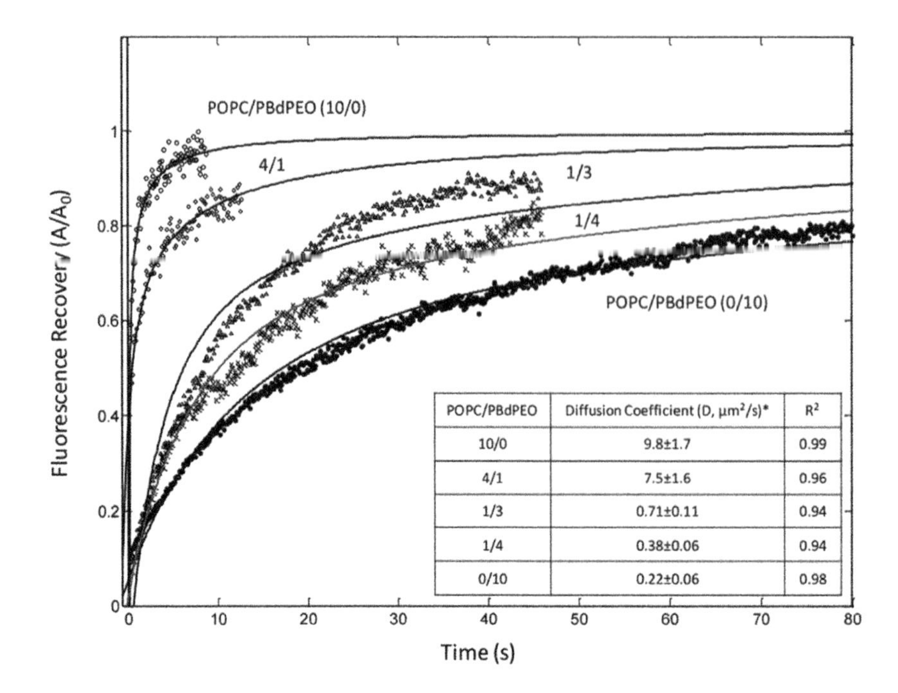

POPC/PBdPEO	Diffusion Coefficient (D, μm²/s)*	R²
10/0	9.8±1.7	0.99
4/1	7.5±1.6	0.96
1/3	0.71±0.11	0.94
1/4	0.38±0.06	0.94
0/10	0.22±0.06	0.98

Figure 27.6 Diffusion coefficient measurements by FRAP for different hGUVs of POPC/PBd$_{46}$-b-PEO$_{30}$. (Reprinted with permission from Nam, J. et al., *Langmuir*, 27, 1–6, 2011. Copyright 2011 American Chemical Society.)

of Nam et al. (2011) who first indicated that with increasing copolymer content, the diffusion of lipid molecules in hGUVs of POPC/PBd$_{46}$-*b*-PEO$_{30}$ becomes slower and in proportion of the copolymer amount incorporated, except at low copolymer fractions where only a weak dependence was observed (Figure 27.6). Because for this system, none of the hybrid vesicles showed macroscopic domains, the homogeneous insertion of copolymer chains into the lipid membrane may somewhat hamper the motion of lipid chains. It is interesting to note that the authors observed a fluorescence recovery shape for hybrid vesicles different from that of the pure components (polymersomes and liposomes), suggesting that dynamics in these hybrid membranes is not ideally described by a standard diffusion model with a single diffusion coefficient.

Other authors (Schulz et al., 2013) obtained information about the mobility of lipid molecules in hybrid membranes composed of DPPC, in the gel state at room temperature, and PIB$_{87}$-*b*-PEO. For that purpose, they used 1,2-dihexadecanoyl-*sn*-glycero-3-phosphothanolamine-*N*-(lissamine rhodamine B sulfonyl) (Rh-DHPE) as a diffusion probe. This molecule fails to incorporate ordered membranes and should be localized in the most disordered phase of the hybrid vesicles (Bagatolli and Gratton, 2000). Whereas no fluorescence recovery was detected for pure DPPC, above a threshold in copolymer fraction for hGUVs, the data revealed a clear increase in mobility of Rh-DHPE in the hybrid membranes with the copolymer content. The authors interpreted these observations by the breaking up of the rigid DPPC densely packed phase by the copolymer chains. The interpretation of such data is, however, not obvious as Rh-DHPE is known to exhibit a large preference for disordered domains. Therefore, it is likely that Rh-DHPE inserts into the copolymer phase, which presents a higher mobility compared with the DPPC gel phase.

In these preliminary results, the authors probed the lipid mobility in the entire hybrid systems, where there was no clear evidence of macroscopic domain presence. This does not rule out the existence of domains below the optical resolution, which can act as obstacles hindering the diffusion, in a similar manner as lipid rafts or membrane proteins in the so-called "mosaic membrane" model (Vaz et al., 1984; Saxton, 1987). Liquid ordered phases within liposomal membranes are indeed suspected for a long time as a factor that significantly impacts on the dynamic of molecules in natural and reconstructed membranes.

We studied the consequence of the presence of POPC on the mobility of copolymer chains. From our data shown in Figure 27.7, mixtures containing either 30% or 50% of POPC in mass demonstrated a significant reduction of the copolymer lateral diffusion compared with the pure copolymer membranes.

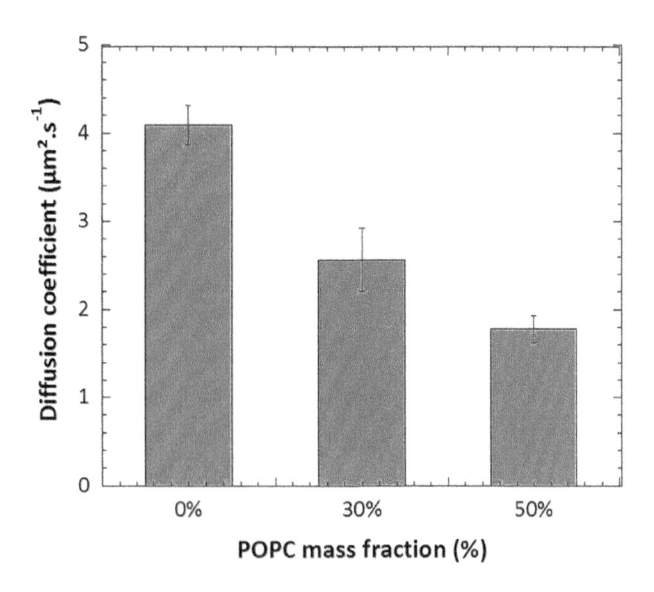

Figure 27.7 Diffusion coefficients of FITC-PDMS$_{22}$-*g*-(PEO$_{12}$)$_2$ in hybrid PDMS$_{22}$-*g*-(PEO$_{12}$)$_2$: POPC GUVs were plotted versus the POPC fraction in the membranes. (Data originating from Dao, T.P.T., PhD thesis, University of Bordeaux, 2016.)

It should be noted that 30% and 50% contents of POPC in $PDMS_{22}$-*g*-$(PEO_{12})_2$, POPC mixtures were enough to drive microdomain formation within the hybrid membranes, and in this situation, the tracer molecule used (FITC-$PDMS_{22}$-*g*-$(PEO_{12})_2$) was localized almost exclusively in the copolymer phase. Apart from the decrease of diffusion coefficient value, the FRAP curve of hybrid $PDMS_{22}$-*g*-$(PEO_{12})_2$: POPC vesicles also clearly indicated the presence of an immobile fraction as shown in Figure 27.8. Indeed, even though the measurement was carried out in polymer-rich domain, we rarely got a full recovery for all fractions of POPC incorporated. These results could be explained by the presence of nanoscale lipid domains, which were below the resolution of confocal microscopy yet hamper the mobility of the copolymer chains. In Figure 27.8, data were fitted with the circular spot model to extract diffusion coefficient values (Axelrod et al., 1976; Soumpasis, 1983; Ishikawa-Ankerlhod et al., 2012). Box 27.2 summarizes the important points that should be carefully considered for a successful FRAP measurement.

27.4.2 PERMEABILITY

It is also important to evaluate and control how easily the soluble molecules can pass through the semipermeable membrane of hybrid vesicles in the presence of concentration gradients of osmotically active solutes (i.e., that cannot pass through the membrane). Permeability is essential for regulation of transmembrane exchange of substances, such as drug release and water transport. The permeability of polymersomes lies far below the value encountered for liposomes (Le Meins et al., 2011) therefore the hybrid membrane permeability is expected to strongly depend on the lipid/copolymer fraction; see also Chapters 20 and 26. Again, so far there are no quantitative data available obtained on hGUVs, but information has been obtained on sub-micrometric hybrid vesicles (i.e., hLUVs) whose size was followed versus time by dynamic light scattering. Shen et al. examined the water permeability of hybrid triblock copolymer PMOXA-*b*-PDMS-*b*-PMOXA/DOPC vesicles with different copolymer/lipid ratios, from 0 to 10 copolymer chains per

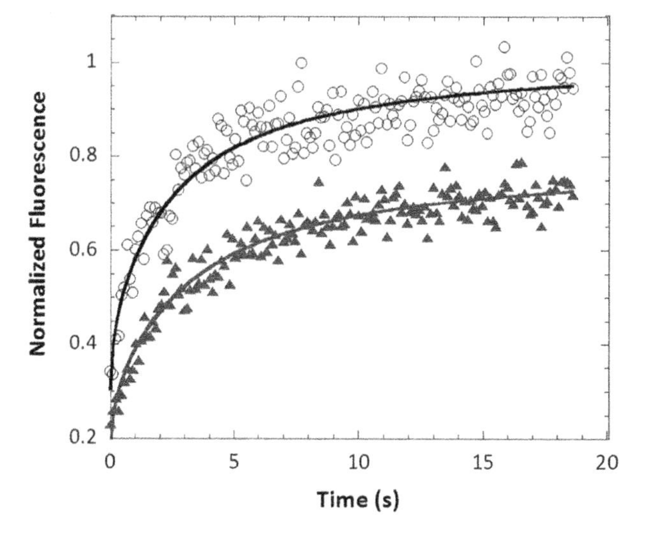

Figure 27.8 Normalized fluorescence versus time of pure $PDMS_{22}$-*g*-$(PEO_{12})_2$ (O) and hybrid $PDMS_{22}$-*g*-$(PEO_{12})_2$: POPC with 30% POPC in mass (→) The arrow indicates the immobile fraction. (Data originating from Dao, T.P.T., PhD thesis, University of Bordeaux, 2016.)

Box 27.2 Practical tips for FRAP measurements on hGUVs

1. *Fluorescent probes*: The probes should preferably have intermediate photostability, because, on the one hand, they should be not too difficult to bleach and, on the other hand, they should be sufficiently photostable to allow imaging of the recovery process. Furthermore, the fluorophore concentration should also be considered. Indeed, the concentration should not be too high in order to avoid possible self-quenching effects, whereas a too low concentration will result in very weak fluorescence, poor signal/noise ratio and number fluctuations will start to dominate the signal. From our experience, fluorescein, NBD, dichlorotriazinylamino fluorescein are normally good choices of fluorescent probes for FRAP.
2. *hGUV sample*: The vesicles chosen to be measured should be sufficiently large to have a significant excess number of molecules outside the circular bleached spot (region of interest [ROI]) to diffuse from. Indeed, because at least 3 μm is usually required for the minimum ROI diameter to get meaningful data, and because the precision of the measurement is improved when a larger bleached ROI is used, we recommend the use of hGUVs larger than 10 μm diameter for FRAP measurement. Another important point is that the vesicles must be fully immobilized during acquisition. The immobilization of hGUVs can be achieved by including a low ratio of biotinylated lipids (1:10^6 mol) into the vesicles, which allows for high affinity adhesion to an avidin-coated surface. This low amount is sufficient to fully immobilize the vesicles on the substrate, and larger amounts can lead to collapse of the vesicle on the surface. The avidin coating is realized by incubating the glass slide in an avidin solution at 0.1 mg/mL. After removing all the excess amount of non-attached avidin to the glass slide, GUVs can be added in the chamber, and they will adhere to the surface in about 15 min. Interestingly, this method initially developed for liposomes (Puech et al., 2006; Sarmento et al., 2012) was also effective for hGUVs. Another approach consists in diluting the vesicle suspension that has been hydrated in sucrose 0.1 M, in a glucose solution of the same concentration, to induce vesicle sedimentation.
3. *Instrument setup*: FRAP acquisition is processed in different stages, during which the excitation light intensity acquisition rate, and other parameters have to be tuned individually for each stage. The bleaching duration has to be as short as possible to prevent significant diffusion during this process, which biases the recovery intensity to lower values. On the other hand, the duration of the measurement of fluorescence recovery must be around 7 to 10 τ_D (time for the probe to diffuse over the bleached zone), in order to have statistical confidence in the different parameters recovered from the fit (Klein and Waharte, 2010), Furthermore, the illumination during fluorescence recovery has to be as low as possible to avoid photobleaching at this stage, whereas the power laser should be set at maximum for the bleach phase.

The rate of data acquisition during fluorescence recovery should also be relatively high in regard to τ_D in order to acquire a sufficient number of points during the initial stage of the recovery fluorescence curve.

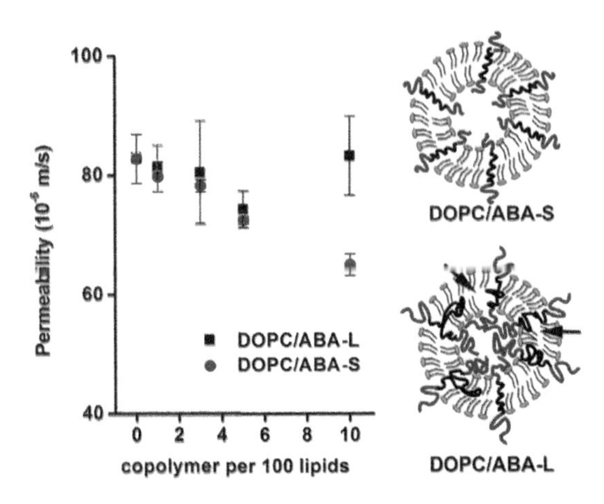

Figure 27.9 Water permeability of hybrid vesicles and schematic of different packing of DOPC/$PMOXA_6$-$PDMS_{33}$-$PMOXA_6$ (ABA-S) and DOPC/$PMOXA_{15}$-$PDMS_{67}$-$PMOXA_{15}$ (ABA-L) on hybrid vesicles. Arrows represent water flow in the vesicle DOPC/ABA-L. (Reprinted from *Chem. Phys. Lett.*, 600, Shen, W. et al., Impact of amphiphilic triblock copolymers on stability and permeability of phospholipid/polymer hybrid vesicles, 56–61, Copyright 2014, with permission from Elsevier.)

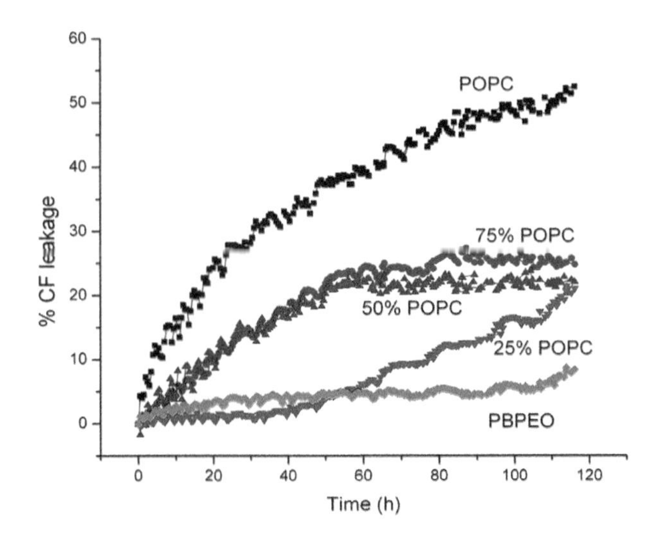

Figure 27.10 Hydrophilic dye carboxy-fluorescein (CF) release from different vesicles PB-PEO/POPC containing 0%, 25%, 50%, 75%, or 100% POPC. (Reproduced from Lim, S. et al., *Polymers*, 5, 1102–1114, 2013.)

100 lipid molecules (Shen et al., 2014). By using two copolymers with different molar masses ($PMOXA_6$-*b*-$PDMS_{33}$-*b*-$PMOXA_6$ and $PMOXA_{15}$-*b*-$PDMS_{67}$-*b*-$PMOXA_{15}$), they obtained different variation profiles of the permeability with the copolymer/lipid content (Figure 27.9). The hybrid system with the shorter triblock copolymer showed a regular decrease of permeability with an increasing copolymer ratio. The permeability of the hybrid systems obtained with the longer triblock copolymer, according to the authors, would decrease up to 5% of copolymer and then re-increase as the polymer ratio was further increased. These results were explained by the authors by the incorporation of copolymer chains into the defects and voids existing in DOPC membranes, partly due to steric hindrance of unsaturated chains, therefore changing the packing density and decreasing the permeability. In the case of long copolymer chains, due to a large thickness mismatch between the alkyl tails of lipids and the hydrophobic block of the copolymer, an excessive amount of copolymer in the hybrid membranes could cause over-saturated occupation of the filler space and create new void spaces, leading to a reduced packing density again, which would explain the re-increase of permeability at 10 copolymer chains/100 lipids. It is important to mention that according to the error bars, this nonmonotonous variation is not obvious to observe. Another possible explanation is that the high hydrophobic mismatch in this system plays a different role, in the sense that too long copolymer chains could not incorporate in lipid LUVs because of a high curvature energy of the membrane at such size and a large steric hindrance between the hydrophilic blocks confined in the inner compartment of the LUVs (as sketched on Figure 27.9), and therefore would not significantly modify permeability. Anyway, these results once again emphasized the impact of copolymer/lipid thickness mismatch onto the resulting membrane properties that was proven as a key point for structuration of hybrid membranes (Chemin et al., 2012; Le Meins et al., 2013; Dao et al., 2015). Besides the water permeability, we would like also to mention another work (Lim et al., 2013) in which the release profile of encapsulated hydrophilic molecules in hybrid LUVs has been measured. As shown in Figure 27.10, it was found that the permeability of the hybrid vesicles prepared from PBd_{22}-PEO_{14}/POPC changed clearly with the fraction of copolymer/lipid. The higher the copolymer content introduced, the more significantly delayed molecule release was detected.

27.4.3 SPECIFIC PROPERTIES, BIOFUNCTIONALITY

Besides the basic membrane physical properties mentioned above, the specific properties of hybrid polymer/lipid vesicles will be discussed in this section to illustrate their potential interest regarding biophysics and nanomedicine applications. One of the major advantages of such hybrid lipid-polymer vesicles is the formation of controlled lipid domains. These domains can be exploited to specifically incorporate active biomacromolecules, especially membrane proteins, thus providing specific transport of biorecognition properties to the hybrid systems.

Different successful reconstitutions either of membrane proteins or of specific ligand molecules in hybrid copolymer/lipid membranes shows that bio-functional vesicles can be obtained. For instance, the work of Binder et al. proved that hGUVs composed of poly(isobutene)-*b*-poly(ethylene oxide) (PIB_{87}-b-PEO_{17}) and DOPC, decorated with ganglioside GM1 receptors, were able to maintain their biological function in binding with cholera toxin B (Schulz et al., 2013). It should be noted that the recognition of the surface functionalized hGUVs by the protein was observed to be independent on the mixing state of the hybrid membrane. In addition, protein binding to the membranes resulted also in a lateral reorganization, leading to the creation of lipid rich phase domains in the polymer continuous phase of the membrane. However, the binding cannot be achieved at polymer ratios above 40 mol% (~85 mass%) due to a shielding effect of the PEO hydrophilic chains, preventing the binding process.

Hybrid copolymer/lipid vesicles offer new routes to obtain more complex and efficient bio-inspired and bio-functional nano- or micrometric structures. Liposomes and more recently polymersomes have been widely used for the reconstitution of membrane proteins and for developing complex artificial membranes embedding proteins or new bio-inspired reactors. Hybrid copolymer/lipid vesicles could enable developing stable and versatile platforms that could be easily modified to adapt to different membrane proteins, thanks to the robustness of copolymer membranes and to the "natural" (non-denaturizing) protein environment provided by the lipid phase. Although the results were not obtained on "real" membranes or hGUVs, but on copolymer/lipid films deposited onto a surface, successful reconstitution of cyclic nucleotide-modulated potassium channel from MloK1 has recently been achieved (Justyna et al., 2015). In this work, the authors clearly showed the possibility to control the protein insertion. By varying the copolymer choice and the degree of saturation of the hydrophobic lipid tails, they could favor a selective localization of proteins within the more fluid domains of the hybrid membranes.

27.5 CONCLUSION AND PERSPECTIVES

The design of hybrid copolymer/lipid giant vesicles with desired and controlled membrane structure at the micro- and nanoscales remains highly challenging. In particular, it would be of utmost interest to build a vesicular membrane with lipid micrometric domains but also "raft-like" nano-domains (from a few tens to hundreds of nanometers). Such structures could present bio-functional properties ("patchy surface" effect) and the membrane physical properties could be fine-tuned. Lipid rafts naturally occurring in the biological cell membrane, with relatively small size, are indeed well known to play a crucial role in numerous biological processes (Simons and Ikonen, 1997; Sens and Turner, 2006). As illustrated in Section 27.2.1 of this chapter, biophysical mechanisms that maintain small lipid domains in equilibrium are not yet completely understood. They involve several parameters that have been briefly described in Section 27.2.1 and have been the subject of many discussions in recent reviews focusing on the stability of lipid domains (Garcia-Saez and Schwille, 2010), and phase separation in biological membranes (Elson et al., 2010). Only few experimental examples of lipid nanodomains (i.e., below the optical microscopy resolution) in model liposomes (De Almeida et al., 2005; Brown et al., 2007; Loura et al., 2010; Heberle and Feigenson, 2011; Suga and Umakoshi, 2013) are available and their existence has been proven on hGUVs or hLUVs only very recently by our group (Dao et al., 2015, 2017). It is believed that amphiphilic block copolymers with hydrophobic blocks of low T_g, because of their flexibility and of their molar mass tuning during the synthesis, play a role in adjusting the line tension at the lipid–copolymer boundaries.

Finally, an emphasis should be given on the control of the membrane structure through physicochemical routes, such as playing with entropic parameters (copolymer/lipid adaptation at the boundaries), interactions (chemical nature of the copolymer blocks), incorporation of additional reactants and environmental parameters (temperature, pH, redox potential, electric or magnetic fields). In particular, lipids in gel phase state certainly play a role on the domain morphologies of copolymer/lipid vesicles, as they do for mixed lipid vesicles (Gordon et al., 2006). In addition, new processes of formation should be developed. Indeed, most of the processes used so far led to compositional variability in the membranes of different vesicles within the same batch, which prevents a fine characterization of the membrane structuration and the understanding of their structure-properties relationship.

Therefore progresses in understanding hGUVs thermodynamics and developing theoretical models for hybrid membranes will probably be necessary to develop new approaches, and could open the way to the design of advanced nano- and micrometric structures with specific functionalities (selective permeability, magnetic properties by incorporating magnetic nanoparticles, membrane disruption with stimuli-responsive copolymer block or lipids, incorporation of channel proteins...) that could be very useful in drug delivery (Cheng et al., 2011; Lim et al., 2013) and as templating agents (Cheng and Tsourkas, 2008), or fundamental systems to understand complex biological behaviors in cells.

LIST OF ABBREVIATIONS

AFM	atomic-force microscopy
Biotinyl DSPE	1,2-distearoyl-*sn*-glycero-3-phosphoethanolamine-*N*-[biotinyl(poly(ethylene glycol)]-2000
Chol	cholesterol
Ch-PEG-*b*-hbPG	cholesteryl-poly(ethylene oxide)-*b*-poly(glycerol)
DC13PC	2-ditridecanoyl-*sn*-glycerophosphocholine
DLPC	1,2-dilauroyl-*sn*-glycero-3-phosphocholine
DMPC	1,2-dimyristoyl-*sn*-glycero-3-phosphocholine
DOPC	1,2-dioleoyl-*sn*-glycero-3-phosphocholine
DPPC	1,2-di-palmitoyl-*sn*-glycero-3-phosphocholine
DDS	drug delivery system
Egg Liss Rhod PE	L-α-egg-phosphatidylethanolamine-*N*-(lissamine rhodamine B-sulfonyl)
hLUVs	hybrid large unilamellar vesicles
FITC	fluorescein isothiocyanate
FRAP	fluorescence recovery after photobleaching
FRET	fluorescence resonance energy transfer
hGUVs	hybrid giant unilamellar vesicles
HSPC	L-α-phosphatidylcholine, hydrogenated (Soy)
Liss-Rhod PE	1,2-dioleoyl-*sn*-glycero-3-phosphoethanolamine-*N*-(lissamine rhodamine B sulfonyl)
NBD	nitrobenzoxadiazole
PBd	poly(butadiene)
PBd-*b*-PEO	polybutadiene-*block*-poly(ethylene oxide)

PC	egg-phosphatidylcholine
PDMS	poly(dimethyl siloxane)
PDMS-*b*-PMOXA	poly(dimethylsiloxane)-*block*-poly(methyloxazoline)
PDMS-*g*-PEO	poly(dimethylsiloxane)-*graft*-poly(ethylene oxide)
PE	egg-phosphatidylethanolamine
PEO	poly(ethylene oxide)
PIB	poly(isobutylene)
PIB-*b*-PEO	poly(isobutylene)-*block*-poly(ethylene oxide)
PMOXA	poly(2-methyloxazoline)
PMOXA-*b*-PDMS-*b*-PMOXA	poly(methyloxazoline)-*block*-poly(dimethylsiloxane)-*block*-poly(methyloxazoline)
POPC	1-palmitoyl-2-oleoyl-*sn*-glycero-3-phosphocholine
ROI	region of interest
RT	room temperature
Rh-DHPE	1,2-dihexadecanoyl-*sn*-glycero-3-phosphothanolamine-*N*-(lissamine rhodamine B sulfonyl)
SLS	static light scattering
SOPC	1-stearoyl-2-oleoyl-*sn*-glycero-3-phosphocholine

GLOSSARY OF SYMBOLS

δ	Hildebrand solubility parameter
K_A	area compressibility modulus
κ	bending rigidity of the membrane
c_0	spontaneous curvature
l_{me}	hydrophobic membrane thickness
T_g	glass transition temperature
T_m	melting temperature
τ_D	time for a probe to diffuse over bleached zone

REFERENCES

Axelrod D, Koppel DE, Schlessingern J, Elson E, Webb WW (1976) Mobility measurement by analysis of fluorescence photobleaching recovery kinetics. *Biophys J* 16:1055–1069.

Bagatolli L, Kumar PBS (2009) Phase behavior of multicomponent membranes: Experimental and computational techniques. *Soft Matter* 5:3234–3248.

Bagatolli LA, Gratton E (2000) A correlation between lipid domain shape and binary phospholipid mixture composition in free standing bilayers: A two-photon fluorescence microscopy study. *Biophys J* 79:434–447.

Bermudez H, Brannan AK, Hammer DA, Bates FS, Discher DE (2002) Molecular weight dependence of polymersome membrane structure, elasticity, and stability. *Macromolecules* 35:8203–8208.

Bermúdez H, Hammer DA, Discher DE (2004) Effect of bilayer thickness on membrane bending rigidity. *Langmuir* 20:540–543.

Bernchou U, Brewer J, Midtiby HS, Ipsen JH, Bagatolli LA, Simonsen AC (2009) Texture of lipid bilayer domains. *J Am Chem Soc* 131:14130–14131.

Binder WH, Barragan V, Menger FM (2003) Domains and rafts in lipid membranes. *Angew Chem Int Edit* 42:5802–5827.

Brochard-Wyart F, de Gennes PG (2002) Adhesion induced by mobile binders: Dynamics. *Proc Natl Acad Sci* 99:7854–7859.

Brown AC, Towles KB, Wrenn SP (2007) Measuring raft size as a function of membrane composition in PC-Based systems: Part 1—Binary systems. *Langmuir* 23:11180–11187.

Cevc G, Richardsen H (1999) Lipid vesicles and membrane fusion. *Adv Drug Deliv Rev* 38:207–232.

Chemin M, Brun PM, Lecommandoux S, Sandre O, Le Meins JF (2012) Hybrid polymer/lipid vesicles: Fine control of the lipid and polymer distribution in the binary membrane. *Soft Matter* 8:2867–2874.

Chen D, Santore MM (2014) Large effect of membrane tension on the fluid–solid phase transitions of two-component phosphatidylcholine vesicles. *Proc Natl Acad Sci USA* 111:179–184.

Chen D, Santore MM (2015) Hybrid copolymer-phospholipid vesicles: Phase separation resembling mixed phospholipid lamellae, but with mechanical stability and control. *Soft Matter* 11:2617–2626.

Chen Q, Schonher H, Vancso GJ (2009) Mechanical properties of block copolymer vesicle membranes by atomic force microscopy. *Soft Matter* 5:4944–4950.

Cheng Z, Elias DR, Kamat NP, Johnston ED, Poloukhtine A, Popik V, Hammer DA, Tsourkas A (2011) Improved tumor targeting of polymer-based nanovesicles using polymer-lipid blends. *Bioconjug Chem* 22:2021–2029.

Cheng Z, Tsourkas A (2008) Paramagnetic porous polymersomes. *Langmuir* 24:8169–8173.

Dao TPT (2016) Hybrid Polymer/Lipid vesicle as new particles for drug delivery and cell mimics, PhD thesis, University of Bordeaux.

Dao TPT, Fernandes F, Er-Rafik M, Salva R, Schmutz M, Brûlet A, Prieto M, Sandre O, Le Meins JF (2015) Phase separation and nanodomain formation in hybrid polymer/lipid vesicles. *ACS Macro Lett* 4:182–186.

Dao TPT, Fernandes F, Ibarboure E, Ferji K, Prieto M, Sandre O, Le Meins J-F (2017) Modulation of phase separation at the micron scale and nanoscale in giant polymer/lipid hybrid unilamellar vesicles (GHUVs). *Soft Matter* 13:627–637.

De Almeida RFM, Loura LMS, Fedorov A, Prieto M (2005) Lipid rafts have different sizes depending on membrane composition: A time-resolved fluorescence resonance energy transfer study. *J Mol Biol* 346:1109–1120.

Dimova R, Seifert U, Pouligny B, Forster S, Dobereiner HG (2002) Hyperviscous diblock copolymer vesicles. *Eur Phys J E* 7:241–250.

Discher BM, Hammer DA, Frank SB, Discher DE (2000) Polymer vesicles in various media. *Curr Opin Colloid Interface Sci* 5:125–131.

Discher BM, Won YY, Ege DS, Lee JC, Bates FS, Discher DE, Hammer DA (1999) Polymersomes: Tough vesicles made from diblock copolymers. *Science* 284:1143–1146.

Discher DE, Ahmed F (2006) Polymersomes. *Annu Rev Biomed Eng* 8:323–341.

Elson EL, Fried E, Dolbow JE, Genin GM (2010) Phase separation in biological membranes: Integration of theory and experiment. In: *Annual Review of Biophysics* (Rees DC, Dill KA, Williamson JR, Eds.), pp. 207–226. Palo Alto, CA: Annual Reviews.

Evans E, Heinrich V, Ludwig F, Rawiczy W (2003) Dynamic tension spectroscopy and strength of biomembranes. *Biophys J* 85:2342–2350.

Evans E, Rawicz W (1990) Entropy-driven tension and bending elasticity in condensed-fluid membranes. *Phys Rev Lett* 64:2094–2097.

Frolov VA, Chizmadzhev YA, Cohen FS, Zimmerberg J (2006) “Entropic traps” in the kinetics of phase separation in multicomponent membranes stabilize nanodomains. *Biophys J* 91:189–205.

Garcia-Saez AJ, Schwille P (2010) Stability of lipid domains. *FEBS Lett* 584:1653–1658.

Gordon VD, Beales PA, Zhao Z, Blake C, MacKintosh FC, Olmsted PD, Cates ME, Egelhaaf SU, Poon WCK (2006) Lipid organization and the morphology of solid-like domains in phase-separating binary lipid membranes. *J Phys Condens Matter* 18:L415–L420.

Gracia RS, Bezlyepkina N, Knorr RL, Lipowsky R, Dimova R (2010) Effect of cholesterol on the rigidity of saturated and unsaturated membranes: Fluctuation and electrodeformation analysis of giant vesicles. *Soft Matter* 6:1472–1482.

Heberle FA, Feigenson GW (2011) Phase separation in lipid membranes. *Cold Spring Harb Perspect Biol* 3:a004630.

Henderson IM, Paxton WF (2014) Salt, shake, fuse-giant hybrid polymer/lipid vesicles through mechanically activated fusion. *Angew Chem Int Edit* 53:3372–3376.

Henriksen J, Rowat AC, Ipsen JH (2004) Vesicle fluctuation analysis of the effects of sterols on membrane bending rigidity. *Eur Biophys J* 33:732–741.

Hu J, Weikl T, Lipowsky R (2011) Vesicles with multiple membrane domains. *Soft Matter* 7:6092–6102.

Ishikawa-Ankerlhod HC, Ankerhold R, Drummen GPC (2012) Advanced fluorescence microscopy techniques—FRAP, FLIP, FLAP, FRET and FLIM. *Molecules* 17:4047–4132.

Jaskiewicz K, Makowski M, Kappl M, Landfester K, Kroeger A (2012) Mechanical properties of poly(dimethylsiloxane)-block-poly(2-methyloxazoline) polymersomes probed by atomic force microscopy. *Langmuir* 28:12629–12636.

Justyna K, Dalin W, Viktoria M, Cornelia GP, Wolfgang M (2015) Hybrid polymer–lipid films as platforms for directed membrane protein insertion. *Langmuir* 31:4868–4877.

King JW (2002) Supercritical fluid technology for lipid extraction, fractionation and reactions. In: *Lipid Biotechnology* (Kuo TM, Gardner HW, Eds.), pp. 663–687. New York: Marcel Dekker.

Klein C, Waharte F (2010) Analysis of molecular mobility by fluorescence recovery after photobleaching in living cells. In *Microscopy: Science, Technology, Applications and Education*, A. Méndez-Vilas and J. Díaz (Eds.), pp. 772–783. Badajoz, Formatex Research Center.

Kuzmin PI, Akimov SA, Chizmadzhev YA, Zimmerberg J, Cohen FS (2005) Line tension and interaction energies of membrane rafts calculated from lipid splay and tilt. *Biophys J* 88:1120–1133.

Ladavière C, Tribet C, Cribier S (2002) Lateral organization of lipid membranes induced by amphiphilic polymer inclusions. *Langmuir* 18:7320–7327.

Le Meins JF, Sandre O, Lecommandoux S (2011) Recent trends in the tuning of polymersomes' membrane properties. *Eur Phys J E* 34:14.

Le Meins JF, Schatz C, Lecommandoux S, Sandre O (2013) Hybrid polymer/lipid vesicles: State of the art and future perspectives. *Mater Today* 16:397–402.

Lee JCM, Santore M, Bates FS, Discher DE (2002) From membranes to melts, rouse to reptation: Diffusion in polymersome versus lipid bilayers. *Macromolecules* 35:323–326.

Liao JF, Wang C, Wang YJ, Luo F, Qian ZY (2012) Recent advances in formation, properties, and applications of polymersomes. *Curr Pharm Des* 18:3432–3441.

Lim S, de Hoog H-P, Parikh A, Nallani M, Liedberg B (2013) Hybrid, nanoscale phospholipid/block copolymer vesicles. *Polymers* 5:1102–1114.

Lipowsky R, Dimova R (2003) Domains in membranes and vesicles. *J Phys* 15:S31–S45.

LoPresti C, Lomas H, Massignani M, Smart T, Battaglia G (2009) Polymersomes: Nature inspired nanometer sized compartments. *J Mater Chem* 19:3576–3590.

Loura LMS, Fernandes F, Prieto M (2010) Membrane microheterogeneity: Förster resonance energy transfer characterization of lateral membrane domains. *Eur Biophys J* 39:589–607.

Loura MS, Fedorov A, Almeida RFMD, Prieto M (2005) Lipid rafts have different sizes depending on membrane composition: A time-resolved fluorescence resonance energy transfer study. *J Mol Biol* 346:1109–1120.

Mabrouk E, Cuvelier D, Pontani LL, Xu B, Levy D, Keller P, Brochard-Wyart F, Nassoy P, Li MH (2009) Formation and material properties of giant liquid crystal polymersomes. *Soft Matter* 5:1870–1878.

Michel V, Bakovic M (2007) Lipid rafts in health and disease. *Biol Cell* 99:129–140.

Mills TT, Toombes GES, Tristram-Nagle S, Smilgies D-M, Feigenson GW, Nagle JF (2008) Order parameters and areas in fluid-phase oriented lipid membranes using wide angle x-ray scattering. *Biophys J* 95:669–681.

Morimoto N, Sasaki Y, Mitsunushi K, Korchagina E, Wazawa T, Qiu X-P, Nomura S-iM, Suzuki M, Winnik FM (2014) Temperature-responsive telechelic dipalmitoylglyceryl poly(N-isopropylacrylamide) vesicles: Real-time morphology observation in aqueous suspension and in the presence of giant liposomes. *Chem Commun* 50:8350–8352.

Nam J, Beales PA, Vanderlick TK (2011) Giant phospholipid/block copolymer hybrid vesicles: Mixing behavior and domain formation. *Langmuir* 27:1–6.

Nam J, Vanderlick TK, Beales PA (2012) Formation and dissolution of phospholipid domains with varying textures in hybrid lipo-polymersomes. *Soft Matter* 8:7982–7988.

Olubummo A, Schulz M, Schöps R, Kressler Jr, Binder WH (2014) Phase changes in mixed lipid/polymer membranes by multivalent nanoparticle recognition. *Langmuir* 30:259–267.

Puech P-H, Askovic V, De Gennes P-G, Brochard-Wyart F (2006) Dynamics of vesicle adhesion: Spreading versus dewetting coupled to binder diffusion. *Biophys Rev Lett* 1:85–95.

Rawicz W, Smith BA, McIntosh TJ, Simon SA, Evans E (2008) Elasticity, strength, and water permeability of bilayers that contain raft microdomain-forming lipids. *Biophys J* 94:4725–4736.

Roth M (1990) Solubility parameter of poly(dimethyl siloxane) as a function of temperature and chain length. *J Polym Sci B* 28:2719.

Ruysschaert T, Sonnen AFP, Haefele T, Meier W, Winterhalter M, Fournier D (2005) Hybrid nanocapsules: Interactions of ABA block copolymers with liposomes. *J Am Chem Soc* 127:6242–6247.

Sarmento MJ, Prieto M, Fernandes F (2012) Reorganization of lipid domain distribution in giant unilamellar vesicles upon immobilization with different membrane tethers. *Biochim Biophys Acta* 1818:2605–2615.

Saxton MJ (1987) Lateral diffusion in an archipelago: The effect of mobile obstacles. *Biophys J* 52:989–997.

Scholtysek P, Shah SWH, Müller SS, Schöps R, Frey H, Blume A, Kressler J (2015) Unusual triskelion patterns and dye-labelled GUVs: Consequences of the interaction of cholesterol-containing linear-hyperbranched block copolymers with phospholipids. *Soft Matter* 11:6106–6117.

Schulz M, Binder WH (2015) Mixed hybrid lipid/polymer vesicles as a novel membrane platform. *Macromol Rapid Commun.* doi:10.1002/marc.201500344.

Schulz M, Glatte D, Meister A, Scholtysek P, Kerth A, Blume A, Bacia K, Binder WH (2011) Hybrid lipid/polymer giant unilamellar vesicles: Effects of incorporated biocompatible PIB-PEO block copolymers on vesicle properties. *Soft Matter* 7:8100–8110.

Schulz M, Olubummo A, Bacia K, Binder WH (2014) Lateral surface engineering of hybrid lipid-BCP vesicles and selective nanoparticle embedding. *Soft Matter* 10:831–839.

Schulz M, Werner S, Bacia K, Binder WH (2013) Controlling molecular recognition with lipid/polymer domains in vesicle membranes. *Angew Chem Int Ed* 52:1829–1833.

Semrau S, Idema T, Schmidt T, Storm C (2009) Membrane-mediated interactions measured using membrane domains. *Biophys J* 96:4906–4915.

Sens P, Turner MS (2006) Budded membrane microdomains as tension regulators. *Phys Rev E* 73:031918.

Shen W, Hu J, Hu X (2014) Impact of amphiphilic triblock copolymers on stability and permeability of phospholipid/polymer hybrid vesicles. *Chem Phys Lett* 600:56–61.

Shoemaker SD, Vanderlick KT (2003) Material studies of lipid vesicles in the La and La-gel coexistence regimes. *Biophys J* 84:998–1009.

Simons K, Ikonen E (1997) Functional rafts in cell membranes. *Nature* 387:569–572.

Soumpasis DM (1983) Theoretical analysis of fluorescence photobleaching recovery experiments. *Biophys J* 41:95–97.

Suga K, Umakoshi H (2013) Detection of nanosized ordered domains in DOPC/DPPC and DOPC/Ch binary lipid mixture systems of large unilamellar vesicles using a TEMPO quenching method. *Langmuir* 29:4830–4838.

Thevenot J, Oliveira H, Lecommandoux S (2013) Polymersomes for theranostics. *J Drug Deliv Sci Technol* 23:38–46.

Travesset A (2006) Effect of dipolar moments in domain sizes of lipid bilayers and monolayers. *J Chem Phys* 125:084905.

Tribet C, Vial F (2007) Flexible macromolecules attached to lipid bilayers: Impact on fluidity, curvature, permeability and stability of the membranes. *Soft Matter* 4:68–81.

Ursell TS, Klug WS, Phillips R (2009) Morphology and interaction between lipid domains. *Proc Natl Acad Sci USA* 106:13301–13306.

Vaz WLC, Goodsaid-Zalduondo F, Jacobson K (1984) Lateral diffusion of lipids and proteins in bilayer-membranes. *FEBS Lett* 174:199–207.

Winzen S, Bernhardt M, Schaeffel D, Koch A, Kappl M, Koynov K, Landfester K, Kroeger A (2013) Submicron hybrid vesicles consisting of polymer-lipid and polymer-cholesterol blends. *Soft Matter* 9:5883–5890.

Giant unilamellar vesicles: From protocell models to the construction of minimal cells

Masayuki Imai and Peter Walde

Did life start from vesicles?

Contents

28.1 INTRODUCTORY WORDS

Giant unilamellar vesicles (GUVs) are often used for the preparation of model systems of "protocells"—the hypothetical precursor structures of the first cells at the origin of life—and of "minimal cells" (Box 28.1). In minimal cells, replicating informational molecules, a simple metabolic reaction network and a vesicle self-reproduction mechanism are assembled with the aim of obtaining a system that has the essential properties of living cells. Some of the ideas and experimental approaches are summarized here as a motivation to enter into this fascinating and emerging field of research.

Unraveling how molecular assemblies emerged to the first forms of life has been one of the greatest challenges in science since decades. A crucial turning point was Schrödinger's famous book "What is Life?" (Schrödinger, 1944). In this book, two important concepts of any forms of life were suggested. One is the existence of a *genetic information molecule* and the other the formation of *ordered structures in an open system*. The former concept was later confirmed by the discovery of DNA (Watson and Crick, 1953) and the subsequent paradigm, which is known as the "central dogma of molecular biology" (DNA → RNA → protein) (Crick, 1970). This added up to the formulation of the "RNA world hypothesis," where the emergence of RNA molecules with genetic information and enzymatic abilities is considered an important early step toward the origin of life (Gilbert, 1986). On the other hand, the latter concept prompted the development of the physics of dissipative structures (Nicolis and Prigogine, 1977) with the aim of better understanding some of the fundamental principles of living systems (Kaneko, 2006).

More recently, a more comprehensive (and less theoretical) physicochemical approach toward understanding the origin of life has been presented, both as hypothesis and as an "invitation" for experimental investigations; it is the concept of "protocells" (Morowitz et al., 1988; Luisi, 2006; Deamer, 2011). Protocells are cell-like compartment systems that are thought to have formed in prebiotic times, before the first forms of life existed on Earth. Linked to the concept of protocells are those of "minimal cells" or "artificial cell-like systems" (or just "artificial cells") (Noireaux, 2015; Luisi, 2006). Because this field of research is still rather new, there are not yet commonly accepted definitions of the various terms used. Here, we consider "minimal cells" as assembled systems that have autonomous reproduction and Darwinian evolution abilities (Szostak et al., 2001). Thus, a minimal cell contains three important properties: (i) self-maintenance (metabolism) that extracts usable energy and chemical resources from the environment through processes that bring nutrients inside and allow excretion of waste products, (ii) self-reproduction that is growth and division of a compartment using metabolism products, and (iii) evolvability that requires the essential biological aspects of genetic variation and its phenotypic expression. If all these properties would be integrated successfully into a synthetic system, then one may claim that life has been synthesized, possibly in its most simple form, as minimal cells (in today's environment on the Earth or under specific laboratory conditions). This has not been achieved yet. The components used toward the construction of minimal cells are contemporary molecules of usually high chemical complexity, for example, enzymes or nucleic acids. This is in clear contrast with the preparation of models

Box 28.1 Suggested definitions of "protocells" and "minimal cells"

Protocells: Hypothetical compartment systems that formed by nonbiological processes *before* living forms of matter existed and which had many properties of living cells but were not yet living. There are many different types of models of protocells, for example, coacervates or vesicles (Oparin, 1965; Luisi, 2006; Deamer, 2011; Monnard and Walde, 2015).

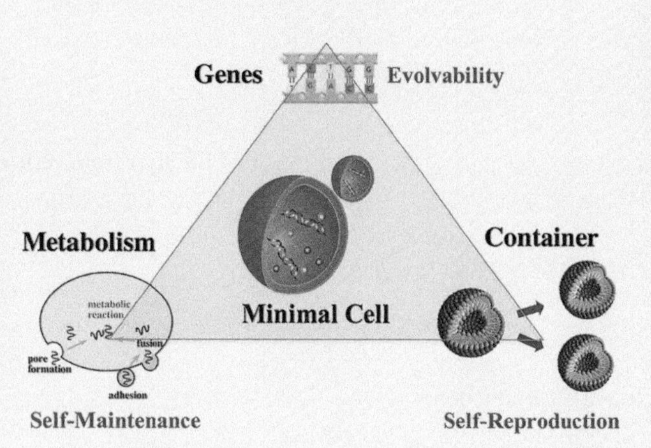

Minimal cells: Chemical systems that consist of self-reproducing cell-like compartments, replicating informational molecules and a simple metabolic reaction network so that the systems represent *minimal forms of living cells* with all the essential features so that the systems can be called living. Because all known forms of life are based on cells, which can be viewed as highly functionalized giant vesicles, the compartment structure of a minimal cell is a vesicle (Luisi and Stano, 2011).

of protocells, where the constituting molecules should be prebiotically plausible compounds.

The concepts of minimal cells and protocells have been reviewed in several books (Luisi, 2006; Rasmussen et al., 2008; Deamer and Szostak, 2010; Deamer, 2011; Luisi and Stano, 2011), although the use of the terms "protocells" and "minimal cells" may sometimes deviate from the one suggested here (Box 28.1). In any case, vesicles, including GUVs, have attracted much attention during the last years as the most cell-like compartment systems for the preparation of minimal cells and protocell models. In the latter case, one tries to combine the concepts of an early "RNA world" and the concept of "dissipative structures" in one and the same system. In the following, some of the original ideas and some of the current developments in the two interconnected fields of vesicle-based protocell model systems and of minimal cells are summarized.

28.2 FATTY ACID VESICLES AS PROTOCELL MODELS

It is currently assumed by many scientists—and also depicted in biology textbooks (Campbell et al., 2015)—that an important step during the transformation of organic molecules to cellular life on Earth was the formation of compartmentalized systems at some stage (Luisi, 2006; Ruiz-Mirazo et al., 2014; Tang et al., 2014; Damer and Deamer, 2015; Monnard and Walde, 2015; Deamer 2017), possibly as dispersed vesicles in an aqueous environment, as a result of the self-assembly of amphiphilic molecules. Vesicle-forming amphiphiles may have emerged from simple organic molecules by chemical evolution in the prebiotic environment on the primitive Earth (Deamer, 1997). Indeed, the prebiotic availability of simple organic compounds has been established by the discovery of a variety of amino acids in the Murchison meteorite (Kvenvolden et al., 1970). Furthermore, monocarboxylic acids ranging from 2 to 12 carbons in length (fatty acids) are abundant components of the organic mixture extracted from the Murchison meteorite (Lawless and Yuen, 1979), and have the capacity to assemble into membranous vesicles as shown in Figure 28.1 (Deamer, 1985; Monnard and Deamer, 2002). This suggests that simple lipidic amphiphiles like monocarboxylic acids (fatty acids) were present on the early Earth already at prebiotic times and that they had the capacity to assemble into vesicles and other types of aggregates, if in contact with an aqueous environment. It may well be that such and other types of amphiphilic lipid assemblies played various crucial roles in prebiotic times, not only as compartment for separating an internal volume from the surrounding environment ("lipid world" scenario [Segre et al., 2001; Walde, 2006]). In any case, it is reasonable to assume that self-assembled compartmentalization was once equally important as the formation of informational molecules and the development of metabolic reaction networks. Therefore, it is likely that a chemically coordinated interplay between a compartment-centered "lipid world," the informational molecule-centered "RNA world," and metabolism was an important step on the way from simple chemical compounds, informational molecules and molecular assemblies to the first forms of cellular life about 3.8 billion years ago (Mojzsis et al., 1996).

Based on these general considerations and the possible prebiotic importance of fatty acids as compartment-forming amphiphiles, investigations on the physicochemical properties of self-assembled structures of fatty acids—and other types of potentially prebiotic amphiphiles (Walde, 2006; Albertsen et al., 2014; Ruiz-Mirazo et al., 2014)—were initiated and put into the context of the origin of life (Gebicki and Hicks, 1973), to reveal the characteristics of these simple vesicular compartment systems

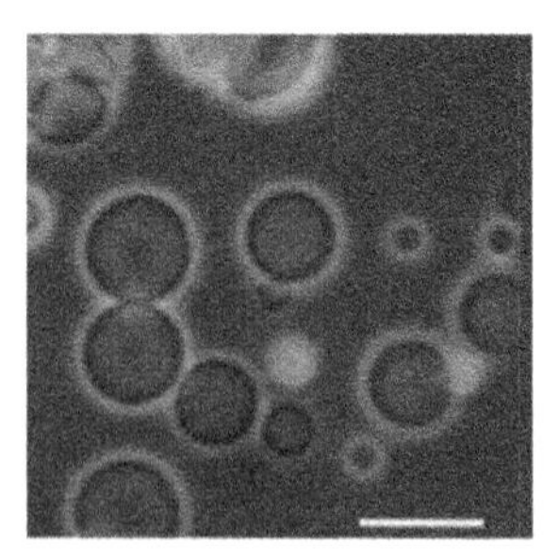

Figure 28.1 Amphiphilic compounds extracted from the Murchison meteorite form membranous vesicles when exposed to dilute aqueous salt solutions at pH 7.0. The probable components of the vesicles are monocarboxylic acids ranging from 8 to 11 carbons in length together with admixtures of polycyclic aromatic hydrocarbon derivatives. Scale bar: 10 μm. (From Monnard, P.A. and Deamer, D.: Membrane self-assembly processes: Steps toward the first cellular life. *Anatomical Rec.*, 2002, 268, 196–207. Copyright John Wiley & Sons. Reproduced with permission.)

Box 28.2 Preparation of fatty acid vesicles

Fatty acid vesicles can be prepared either by adjusting the pH of an aqueous fatty acid solution or by swelling a fatty acid film (Monnard and Deamer, 2003). In the former case, it is convenient first to completely deprotonate the acid by adding a strong base (1M NaOH). At pH 9–11, fatty acids typically form micelles, and these solutions are totally transparent. The pH of such micellar solution is then adjusted with acid (HCl) addition to a pH range near the apparent pK_a of the fatty acid (pH 7.3 for decanoic acid/decanoate vesicles). At this pH the solution appears slightly opalescent, indicating the formation of large and giant vesicles. Further lowering of the pH below the pK_a value results in protonation of the fatty acid molecules, which then become insoluble and form droplets, or precipitates, depending on the chemical structure of the fatty acid. It is essential to maintain the aqueous medium at a temperature (*T*) above the bulk melting point during vesicle formation to avoid crystallization.

In the swelling method, fatty acid films are prepared by first dissolving the fatty acid in an organic solvent, for example, oleic acid in chloroform-methanol (9:1 v/v), followed by solvent evaporation to produce a thin film on the interior surface of a glass vessel. Then the fatty acid film is hydrated at a temperature above the bulk melting point by a pH adjusted aqueous solution, the pH being close to the fatty acid apparent pK_a.

Example protocol for the preparation of 5.0 mL of a decanoic acid vesicle suspension following (Monnard and Deamer, 2003):

1. Heat the decanoic acid to 43°C to melt the crystals.
2. Add 72 μL of the melted decanoic acid to 4.5 mL of a 10 mM Tris-HCl buffer (pH = 7.4) heated at 43°C. The decanoic acid will form a droplet on the top of the aqueous medium.
3. Add aqueous NaOH (1.0 M) to the solution in 20 μL aliquots, with vortexing between additions, until the acid is completely dissolved near pH = 11.0 (i.e., after adding of ca. 330 μL). The solution becomes transparent with abundant foam formation.
4. Add aqueous HCl (1.0 M) in 10 μL aliquots (ca. 120 μL) with vortexing and checking the pH value after each addition, until pH = 7.4 is reached, at which point an opalescent suspension of decanoic acid vesicles is present (pK_a of decanoic acid is 7.1–7.3).

as hypothetical precursor structures (protocells) of the first cells. In the following, some of the properties of fatty acid vesicles will be summarized first, before experiments of the self-reproduction of vesicles from fatty acids—or other types of amphiphiles—are discussed (Section 28.3). The preparation of fatty acid giant vesicles is described in Box 28.2.

Fatty acids are carboxylic acids with long aliphatic tails. Due to the fact that they are weak acids, fatty acids can exist in neutral, slightly acidic or basic aqueous solutions in two forms, the nonionized (neutral) form and the ionized (negatively charged soap) form. Therefore, compared with bilayer-forming phosphatidylcholines (PCs), fatty acids have unique pH-dependent aggregation properties (Hargreaves and Deamer, 1978; Morigaki and Walde, 2007; Chen and Walde, 2010). At low pH, the carboxylic acid is in its undissociated neutral form, which results in the formation of a separate oil phase or the formation of fatty acid crystals, depending on the melting temperature of the fatty acids. At high pH values, the molecules are deprotonated and negatively charged, resulting in the formation of micelles, if the concentration is above the critical micellization concentration. The formation of fatty acid vesicles is restricted to a rather narrow pH range (ca. 7–9), where approximately half of the carboxylic groups are ionized and the fatty acid/soap molecules form transient hydrogen bonded dimers (Haines, 1983). The relationship between the assembly morphologies and the protonation degree for oleic acid/sodium oleate is shown in Figure 28.2. Although the pK_a of a carboxylic acid monomer is typically 4–5 (Smith and Tanford, 1973), a fatty acid within a bilayer membrane has a pK_a of 7–9 due to the condensation of counter ions on the membrane (Haines, 1983). Depending on the composition of the aqueous solution (i.e., the presence of guanidine hydrochloride), at room temperature fatty acid vesicles not only form from short and long-chain *unsaturated* fatty acids, like oleic acid/oleate (Figure 28.2), but also from long-chain *saturated* ones due to a decrease in the chain melting temperature (Douliez et al., 2016). Saturated fatty acids are prebiotically more plausible molecules than the unsaturated counter parts.

The second important feature of fatty acids is that they have a single aliphatic tail only, in contrast to contemporary double-chain phospholipids, like 1,2-dipalmitoyl-*sn*-glycero-3-phosphocholine (DPPC, see Appendix 1 of the book for structure and data on this and other lipids of this chapter). This means that the critical aggregation concentration (CAC) for fatty acids is considerably higher than in the case of double-chain phospholipids of comparable chain length. For example, whereas the monomer concentration in equilibrium with DPPC bilayers (vesicles) is around 10^{-10} M (the approximate CAC value of DPPC) (Smith and Tanford, 1972),

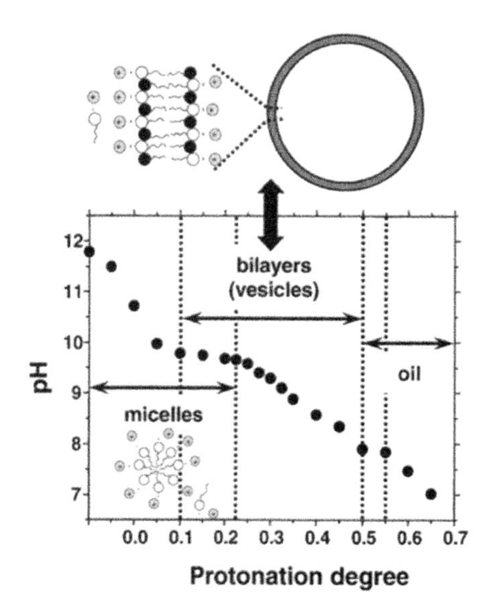

Figure 28.2 Titration curve for 80 mM oleic acid/sodium oleate. The regions for the formation of micelles in water at 25°C, vesicles and oil droplets are indicated. The schematic drawing of the amphiphile with the empty head represents the ionized form of oleic acid (oleate), the neutral form (oleic acid) is represented with a filled headgroup. (Reprinted from *Curr. Opin. Colloid Interface Sci.*, 12, Morigaki, K. and Walde, P. Fatty acid vesicles, 75–80, Copyright 2007, with permission from Elsevier.)

the CAC value for the formation of oleic acid/oleate vesicles is about 4–7 × 10^{-4} M (Walde et al., 1994b) and for decanoic acid/decanoate it varies from 14 mM at pH = 6.8 (vesicles) to 102 mM at pH = 11.8 (micelles) (Morigaki et al., 2003). A consequence of the high water solubility of single-chain amphiphiles is that they exchange more rapidly between the aggregates (vesicles or micelles) and the bulk solution, if compared with phospholipids like DPPC. This dynamic exchange property of fatty acids has particular consequences if fatty acid vesicles are used as protocell model systems (Blain and Szostak, 2014) because in contrast to phospholipids, fatty acid/soap monomers in a solution form vesicles spontaneously by supplying additional fatty acid monomers (Morigaki et al., 2003). Furthermore, fatty acid vesicles will rapidly transform upon changing the external conditions, or quickly grow through external addition of fatty acid micelles (Morigaki et al., 2003; Chen and Szostak 2004a), or by absorbing membrane-forming molecules from other vesicles (Chen et al., 2004; Budin and Szostak, 2011). In addition, as a consequence of the highly dynamic nature of fatty acid systems above their melting temperatures, the flip-flop of molecules between two monolayer leaflets of a fatty acid bilayer is much faster than in the case of a comparable fluid phospholipid membrane (Chen and Szostak, 2004b; Mansy, 2009).

Another interesting property of fatty acid vesicles is their ability to "self-reproduce" after they are fed with precursor molecules from the environment. Because compartment self-reproduction is considered as an essential feature of all cells, the self-reproduction ability of potentially prebiotic compartment systems is of particular relevance for the origin of the first cells. Fatty acid vesicles demonstrate that compartment reproduction is possible on the basis of simple physicochemical processes, without the involvement of proteins and other complex molecules. The differences between fatty acid and phospholipid systems are summarized in Box 28.3.

Box 28.3 Fatty acid vesicles as compared with phospholipid vesicles

- *Tails:* Compared with vesicles formed from double-chain phospholipids, fatty acid vesicles are built from single-chain amphiphiles (e.g., oleic acid or decanoic acid).
- *Headgroup:* The headgroup of fatty acids has to be partially ionized for vesicles to form.
- *pH range:* Fatty acid vesicles only exist in a certain pH range, which depends to some extend on the type of fatty acid and on the presence of other single-chain amphiphiles (e.g., decanol or 1-decanoyl-*rac*-glycerol in the case of decanoic acid).
- *Counter-ions:* The nature of the counter ion is important for obtaining stable fatty acid vesicles.
- *Concentration:* Compared with phospholipid vesicles, the concentration of nonassociated amphiphiles is much higher in the case of fatty acid vesicles.
- *Dynamics:* The dynamics of the molecules within the vesicle membrane is higher as well as the exchange kinetics of fatty acid molecules between the vesicle and the solution, and stabilizing headgroup interactions are more important, which endows distinct properties, such as fast flip-flop motion, growth of vesicles, and self-reproduction of vesicle (Walde et al., 1994b; Chen et al., 2004; Morigaki and Walde, 2007; Maurer et al., 2009; Douliez et al., 2016).
- *Permeability:* The permeability of fatty acid vesicles in general is also higher if compared with conventional phospholipid vesicles (see Section 28.5.2).

28.3 SELF-REPRODUCTION OF VESICLES

From a conceptual point of view, the first experiments on the self-reproduction of vesicles can be seen as an important step in the development of compartment systems as protocell models because, with these experiments, not only the cell-like compartment features of vesicles were taken into account but also the possibility of increasing the amount of compartments due to chemical and/or physicochemical processes taking place within the compartment structure. In one approach, preformed vesicles (large unilamellar vesicles [LUVs] or GUVs) take up precursors of the membrane molecules from the environment, which are then transformed by a chemical reaction into the very same membrane molecules the preexisting vesicles are already composed of. The increase in the amount of membrane molecules must lead to vesicle growth and/or to vesicle shape changes that may lead to vesicle budding or even division into independent vesicles having similar properties as the mother vesicles. Such simple self-reproducing vesicle systems were initially developed by Luisi's group, by mainly using LUVs (Walde et al., 1994b; Wick et al., 1995). Oleic acid/oleate vesicles were first prepared in buffer solution at pH 8.5. Afterward, a drop of oleic anhydride was added. The hydrolysis of the anhydride to oleic acid/oleate was found to be accelerated in the presence of the vesicles, most likely because the reaction took place—at least partially—at the membranous boundary of the vesicles. As shown in experiments with GUVs, the hydrolysis of oleic anhydride in the presence of preformed oleic acid/oleate vesicles resulted in various vesicle transformations that could be monitored by light microscopy (Figure 28.3). Although it is difficult to control the outcome of the reaction due to the heterogeneity of the system, at least two types of vesicle transformations were observed; one, although a very rare event, was "vesicle birthing" see also Menger and Gabrielson (1994), where a mother vesicle formed an inclusion

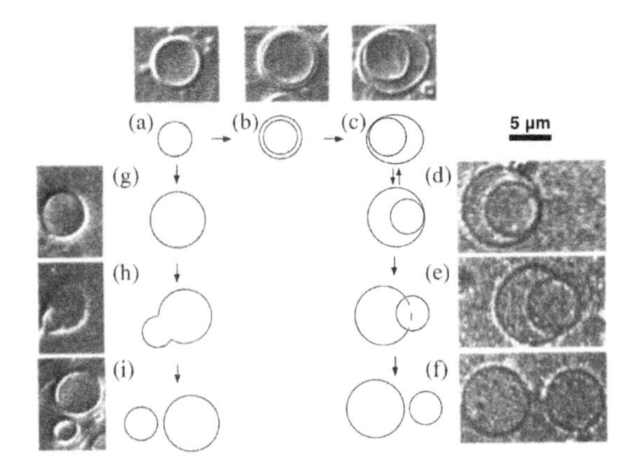

Figure 28.3 Two self-reproduction processes observed in giant oleic acid/oleate vesicles following oleic anhydride hydrolysis: (a–f) "birthing" and (g–i) "budding." (Reprinted with permission from Wick, R. et al., *J. Am. Chem. Soc.*, 117, 1435–1436, 1995. Copyright 1995 American Chemical Society.)

vesicle that was then expelled from the mother vesicle, and the other one was a budding pathway where the mother vesicle deformed to a pear-like shape and then divided into two vesicles, which were still connected to each other. Although the mechanism of vesicle birthing remains unclear, these initial findings were interesting because they strongly suggested that amphiphilic molecules themselves have the capability to perform vesicular compartment self-reproduction without the aid of proteins.

A conceptually different way of increasing the size of preformed unilamellar oleic acid/oleate vesicles—which eventually results in an increase in the number of vesicles due to vesicle divisions—is to add oleate micelles, as originally demonstrated by using suspensions of LUVs (Berclaz et al., 2001). Related to these investigations are the experiments described by the group of Szostak in which the cyclic self-reproduction of *multilamellar* fatty acid vesicles was demonstrated (Zhu and Szostak, 2009; Budin et al., 2012). This type of self-reproduction process starts from spherical multilamellar oleic acid/oleate vesicles as shown in Figure 28.4. By addition of oleate micelles, thin tubular structures begin to protrude from the mother vesicle and finally the initially spherical vesicles completely transform into long thread-like vesicles. By applying mild agitation (shear force), the thread-like vesicles divide into multiple smaller, spherical multilamellar daughter vesicles. The energy source for the division of thread-like vesicles is simply the kinetic energy of mildly agitated liquid water that can be considered as a likely event in prebiotically plausible environments. Upon further micelle addition, the daughter vesicles grow to the sphere-tube intermediate stage, and the growth-division cycle continues in a cyclic manner.

A unique feature of another self-reproducing fatty acid vesicle system developed by the group of Szostak is the one that exhibits resemblance to Darwinian evolution (a type of evolvable vesicles). The system studied consists of fatty-acid vesicles containing a dipeptide catalyst, which catalyzes the formation of a second dipeptide (Adamala and Szostak, 2013). The newly formed dipeptide binds to the vesicle membranes, which imparts enhanced affinity for fatty acids and decreases desorption rate of fatty acids from fatty acid vesicles, resulting in the promotion of vesicle growth. The catalyzed dipeptide synthesis proceeds with higher efficiency in vesicles than in bulk solution, which further enhances fitness, that is, has a selective advantage. This rapid competitive vesicle growth leads to the development of thread-like filamentous vesicles, which can subsequently divide into small daughter vesicles as a result of gentle agitation.

A chemically completely different self-reproducing giant vesicle system was studied by Sugawara's group. Followed by a unique preceding self-reproductive vesicle system (Takakura et al., 2003), a revised system was reported. It is based on the design of a non-natural membrane-forming amphiphile, abbreviated as *V*, and its precursor molecule V^* (Figure 28.5) (Takakura and Sugawara, 2004; Toyota et al., 2008). V^* is a bola-amphiphile containing two polar headgroups that are connected *via* a hydrophobic linker containing in the middle of the linker a chemically cleavable imine bond. Hydrolysis of this bond yields *V* and an electrolyte *E* through the aid of the catalyst *C*, as shown in Figure 28.5a. When the precursor V^* is added to a suspension of giant vesicles composed of the membrane molecule *V* and catalyst *C*, V^* is hydrolyzed within the vesicular membrane. The mother vesicle undergoes

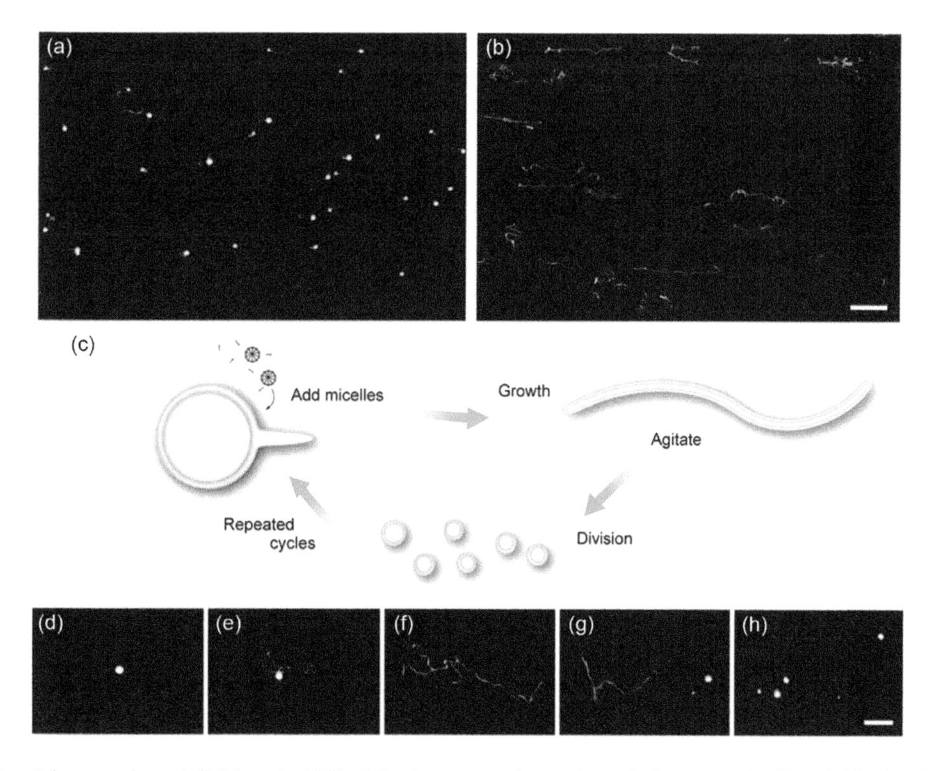

Figure 28.4 Fatty acid vesicle growth and division. (a, b) Vesicle shape transformations during growth, 10 and 30 min after the addition of 5 equiv of oleate micelles to multilamellar oleic acid/oleate vesicles, respectively. Scale bar: 50 μm. (c) Schematic diagram of cyclic multilamellar vesicle growth and division: vesicles remain multilamellar before and after division (shown as, but not limited to, two layers). (d–f) Growth of a single multilamellar vesicle, 3, 10, and 25 min after the addition of 5 equiv of oleate micelles, respectively. (g, h) In response to mild fluid agitation, this thread-like vesicle divided into multiple smaller daughter vesicles. Scale bar for d–h: 20 μm. (Reprinted with permission from Zhu, T.F. and Szostak, J.W., *J. Am. Chem. Soc.*, 131, 5705–5713, 2009. Copyright 2009 American Chemical Society.)

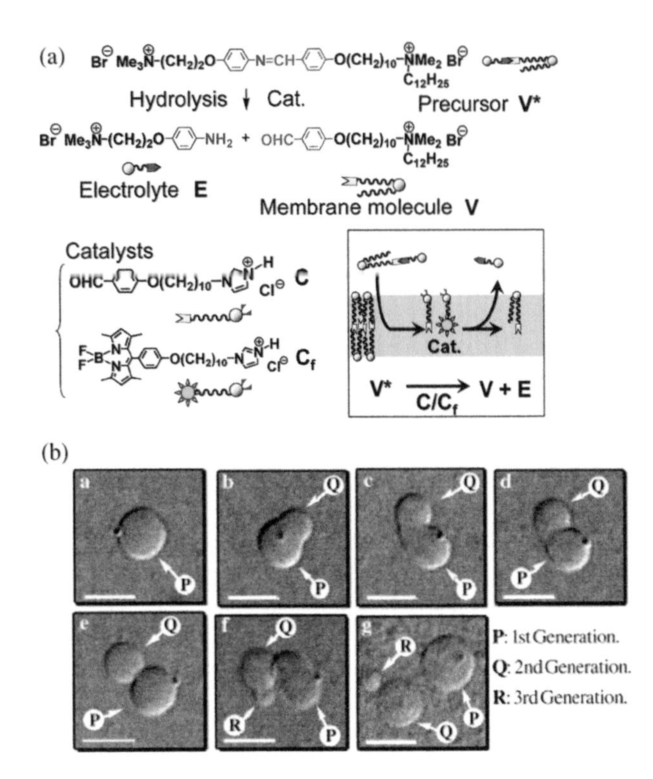

Figure 28.5 (a) Schematic representation of the self-reproduction of giant multilamellar vesicles (GMVs). The membrane molecule *V* and electrolyte *E* are formed by the hydrolysis of the membrane precursor *V** in the presence of the catalyst *C* and the fluorescence probe *Cf*, which are anchored within the vesicular membrane (panel at the bottom right). (Reprinted with permission from Toyota, T. et al., *Langmuir*, 24, 3037–3044, 2008. Copyright 2008 American Chemical Society.) (b) Morphological changes in a GMV composed of *V* and 10 mol%C: (a–g) images obtained at different times after mixing of a dispersion of GMV and a solution of precursor *V**. Scale bar: 10 μm. (Reprinted with permission from Takakura, K. and Sugawara, T., *Langmuir*, 20, 3832–3834, 2004. Copyright 2004 American Chemical Society.)

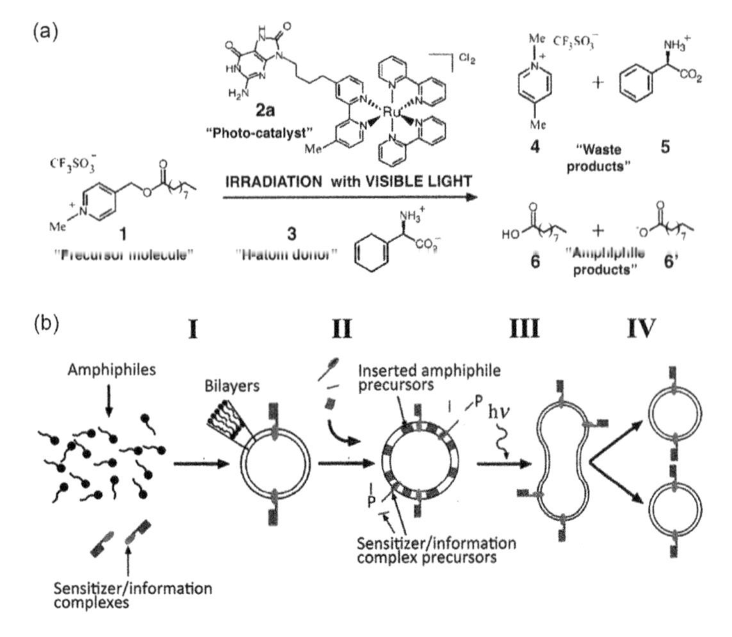

Figure 28.6 (a) Summary of the nucleobase-mediated photochemical production of decanoic acid from its picolyl ester precursor **1**. The precursor **1** is dispersed in an aqueous buffered solution containing the electron donor linked photocatalyst **2a** and the hydrogen donor **3**. Upon irradiation, decanoic acids 6 and deprotonated 6′ are formed and self-assemble into vesicles. Two "waste" compounds, N-methyl picolinium **4** and phenylglycine **5**, are also produced. (b) Co-location protocell model using the ruthenium-based photocatalyst in scheme (a). (With kind permission from Springer Science+Business Media: *The Minimal Cell–The Biophysics of Cell Compartment and the Origin of Cell Functionality*, Membrane self-assembly processes: Steps toward the first cellular life, 2011, pp. 123–151, Monnard, P.A. and Deamer, D.W., Springer, Heidelberg, Germany.)

division using the generated membrane molecule *V* as shown in Figure 28.5b. Although the molecules used in this work are not plausible prebiotic compounds, the experiments are conceptually interesting and related to the oleic anhydride hydrolysis in oleic acid/oleate vesicles. The advantage of this system is that the membrane molecule *V* is definitely formed within the vesicle, whereas the formation site of oleic acid/oleate system is not clearly defined. All in all, the experiments with *V*, *V**, and *C* suggest that simple chemical reactions occurring within preexisting vesicle membranes can lead to an increase in the size of the vesicles and to a division of the grown vesicles. The challenge remains to explaining the physical basis of the experimentally observed transformations.

Another more complex example was presented by Rasmussen's group. It is based on the generation of fatty acid amphiphiles from precursor molecules by visible light photolysis (Rasmussen et al., 2004; DeClue et al., 2009). This system consists of a decanoic acid ester precursor, hydrogen donor molecules, and a ruthenium-based photocatalyst that employs a linked electron donor that generates decanoic acid, as shown in Figure 28.6. The precursor is dispersed in an aqueous buffered solution containing the electron donor linked photocatalyst and the hydrogen donor. Upon irradiation, the ruthenium-based photocatalyst generates a metal-to-ligand charge transfer state, which can then transfer charge to the picolinium ester providing carboxylic acid in the presence of a hydrogen source. The obtained fatty acids self-assemble into membranous structures (Figure 28.6a). In the proposed co-location protocell model (Figure 28.6b), all chemical compounds that constitute the protocell are embedded in the vesicle membrane. The molecules that are required to synthesize more building blocks with hydrophobic character can—once added to the external medium—spontaneously insert into the protocell compartment, where they will be transformed using light energy into either amphiphiles or photocatalyst molecules. This production of new building blocks leads to the growth of the systems and once a threshold size is reached, should induce the division of the grown protocell into offspring with similar or altered properties.

28.4 TOWARD ENDOVESICULAR ENZYMATIC REACTIONS THAT PROMOTE VESICLE SELF-REPRODUCTION

One possible scenario for the construction of a minimal cell involves creating a system where genetic informational molecules replicate and express proteins inside a membranous vesicular compartment, which promote the synthesis of boundary molecules that build the membrane. In this way, offspring compartments may be obtained due to endovesicular reactions, that is, reactions occurring inside the vesicles (Stano and Luisi, 2010; Noireaux et al., 2011; Stano et al., 2011; Blain and Szostak, 2014), which would also link—at least conceptually—an RNA-centered view of the origin of life ("RNA-world" scenario) with compartmentalization and which represents a straightforward approach toward the construction of minimal cells.

A very first and modest attempt toward building minimal cells was the examination of the enzymatic synthesis of ribonucleic acids (RNA) from mononucleotides (adenosine diphosphate, ADP) inside LUVs. In these early studies the aim was to explore whether such enzyme-catalyzed endovesicular polymerization reactions were possible without any dependency between the reaction inside the vesicles and the vesicle membrane. These initial studies were carried out by the groups of Luisi and Deamer (Chakrabarti et al., 1994; Walde et al., 1994b). Using oleic anhydride-driven self-reproducing fatty acid LUVs (Walde et al., 1994b; Wick et al., 1995), the polymerization of ADP to poly (adenosine phosphate), poly(A), catalyzed by vesicle-trapped polynucleotide phosphorylase, and in other experiments the replication of a template RNA catalyzed by Qβ replicase (Oberholzer et al., 1995b) were established. These studies revealed (i) that by externally triggered vesicle divisions, the produced encapsulated chemical compounds may be delivered to their offspring vesicles, and (ii) that low molar mass polar nutrients can penetrate inside the fatty acid vesicle, which enables the enzymatic reactions to continue inside the new vesicles. Furthermore, the amplification of DNA using the polymerase chain reaction (PCR) inside phospholipid LUVs was reported, whereby 1-palmitoyl-2-oleoyl-*sn*-glycero-3-phosphocholine (POPC) was used and the vesicles formed therefrom were found to be stable at the high temperature condition of the PCR (up to 95°C) (Oberholzer et al., 1995a). These successful works encouraged the researchers to go one step further, namely to try a cell-free protein expression inside vesicles using a nucleic acid template and ribosomes. Luisi's group succeeded in synthesizing the polypeptide poly(Phe) inside POPC LUVs by using encapsulated ribosomes, messenger-RNA (poly(U)), transfer-RNA, phenylalanyl-tRNA synthetase, elongation factors, and the substrate phenylalanine (Oberholzer et al., 1999). Later, Yomo's group reported the synthesis of green fluorescent protein (GFP) in giant vesicles composed of phospholipid mixtures using cell extracts from *Escherichia coli* (Yu et al., 2001). Similar protein expressions inside giant vesicles were reported by other groups as well (Nomura et al., 2003; Nourian and Danelon, 2013).

One key requirement for the development of protocell model compartments as microreactors is the selective uptake of nutrients from the environment and the excretion of waste from the interior of the vesicles to the environment. Libchaber's group tackled this problem by expressing a pore protein, α-hemolysin, inside egg PC giant vesicles to create a size-selective permeability for nutrients (Noireaux and Libchaber, 2004). Under the chosen experimental conditions, the expression of GFP in the vesicles with this pore lasted for up to 4 days, whereas without α-hemolysin, the expression of GFP was observed only for 5 h.

A disadvantage of protein synthesis using cell extracts is that the exact composition of the cell extract is not known. Ueda's group established a cell-free protein synthesis kit from individually purified components called the Protein synthesis Using Recombinant Elements (PURE) system (Shimizu et al., 2001). This kit is composed of four subsystems, (i) transcription, (ii) translation, (iii) amino acid charging onto tRNA (aminoacylation), and (iv) energy regeneration. Using this PURE system Yomo's group developed a "self-encoding system," in which the genetic information was replicated by self-encoded replicase inside phospholipid giant vesicles (Kita et al., 2008). In these experiments, the catalytic subunit of Qβ replicase was synthesized from the template RNA that encoded the protein and the replicase, then replicated the template RNA that was used for its production. Thus the genetic material (RNA) was encoded for a function that ultimately led to its own replication. This work demonstrated that sophisticated template-controlled enzymatic reactions are possible within vesicular compartments, which was an important finding if one aims at constructing minimal cell systems.

An important goal in the field of minimal cell research is to express inside vesicular compartments membrane proteins that spontaneously insert into the vesicle boundary and then catalyze the synthesis of the vesicle membrane-forming lipids. Such lipid synthesis may result in vesicle self-reproduction. Luisi's group demonstrated the endovesicular synthesis and activity of two membrane proteins involved in the phospholipid biosynthesis pathway by using the PURE system encapsulated inside phospholipid LUVs (Kuruma et al., 2009). The two proteins catalyzed the synthesis of the two phospholipids, lysophosphatidic acid and phosphatidic acid. Unfortunately due to the low reaction yields and due to reaction network discontinuities, no morphological changes caused by the newly synthesized lipids could be observed. The amount of phospholipid products appeared to be restricted by the insufficient nutrient passage through the phospholipid membrane. In order to overcome such limitation in future, the vesicles must be a semi-open system as it is the case of living cells, that is, vesicle with chemically selective transport system.

An advanced system was developed by Sugawara's group (Sugawara, 2009; Sugawara et al., 2012). It stands somewhat between the simple fatty acid-based systems and the more sophisticated phospholipid vesicles with endovesicular protein expressions (Kurihara et al., 2011). The vesicle membrane-forming amphiphile used was the same as shown in Figure 28.5, the cationic membrane molecule *V*, but this time the components in the giant vesicles were more complex than discussed above. They contained a DNA amplification system. The vesicle-trapped DNA was amplified by the PCR technique and the vesicles also showed growth and division behavior after the addition of a vesicular membrane precursor. The amplified DNA was distributed among the daughter giant vesicles. In particular, the amplification of the DNA accelerated the division of the giant vesicles, which means that the self-replication of vesicle-trapped informational compounds were linked to the self-reproduction of a compartment through the interplay between DNA and the cationic vesicular membrane. This system was further advanced by investigating the recursive self-proliferation ability of the vesicles as protocell model system, which was, however, constructed from molecules that are not prebiotically plausible (Kurihara et al., 2015). This proliferation cycle is composed of four phases (Figure 28.7), ingestion, replication, maturity and division. In the ingestion phase, the model protocell ingests depleted nutrients (deoxyribonucleoside triphosphates: dNTPs) by a pH-induced fusion of conveyer vesicles filled with depleted substances (Suzuki et al., 2012). In the

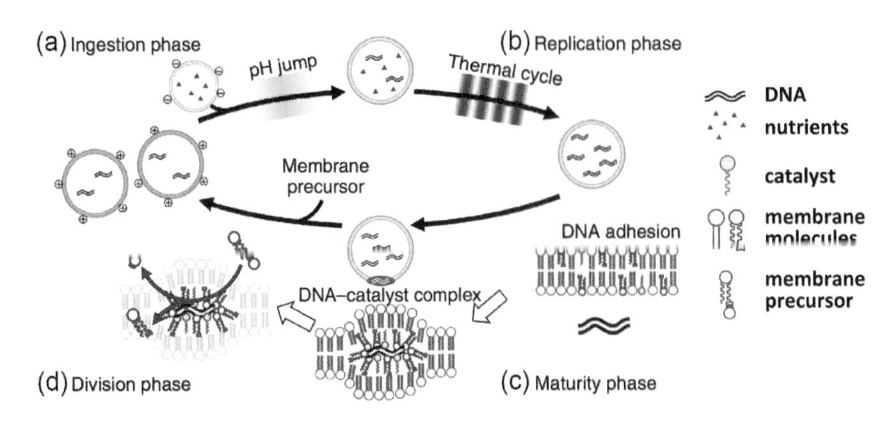

Figure 28.7 Primitive model cell cycle of a self-proliferative model protocell with four discrete phases. (a) In the ingestion phase, the giant vesicle (GV) of the next generation ingests substrates through vesicular fusion with conveyer GV containing dNTP, triggered by a *pH* jump. (b) In the replication phase, the replication of DNA in the next-generation GV proceeds using ingested dNTP. (c) In the maturity phase, the catalytic ability of the vesicular membrane matures in a sense that a complex between amplified DNA, amphiphilic catalyst C and cationic lipids V intrudes into the vesicular membrane, forming an active site for converting membrane precursor V^* to lipid membrane V. (d) In the division phase, the self-proliferative GV grows and exhibits a budding deformation and an equivolume division when the precursor V^* of the membrane lipid is added to the exterior of GVs. (Reprinted by permission from Macmillan Publishers Ltd. *Nat. Commun.*, Kurihara, K. et al., 2015, Copyright 2015.)

replication phase, DNA is replicated using the ingested dNTPs by periodic change of the temperature (PCR). The maturity phase is characterized by the formation of DNA–catalyst complexes within the membrane, which promotes the catalytic activity. The division phase is the phase of proliferation of the giant vesicles containing the amplified DNA after the addition of membrane precursor, where the DNA–catalyst complexes not only serve as the pseudo-enzyme to produce membrane lipid from its precursor, but also act as a scaffold for giant vesicle division.

All the efforts described here—and others described in the literature—are aimed at constructing minimal cells. If one is not too skeptical, then one may think that this type of experimental work and the general idea behind it may at some stage not only lead to the construction of synthetic minimal cells but it may also contribute to an understanding of the origin of the first cells. At the moment, we still do not know how the first cells formed. The first cells can be viewed as a particular type of minimal cells, but composed of chemically simpler molecules than contemporary cells are built from. The first cells must have contained only those molecules that were present on the early Earth due to nonbiological reactions (e.g., no sophisticated enzymes as we know them today).

28.5 THE MEMBRANE PHYSICS OF GIANT VESICLES AS PROTOCELL MODELS

In protocell research, like in any other field of scientific investigations, it is desired to provide a theory for experimental observations. In the case of "simple" vesicle systems (not containing complex endovesicular reactions), an understanding of the physics of the vesicles and their morphological changes may provide some useful insight (Sakuma and Imai, 2015), see also Chapter 5. This especially concerns aspects of (i) the self-reproduction of vesicles, and (ii) the exchange of nutrients from the outside to the inside of vesicles. These two topics are discussed in the following on the basis of a series of specifically designed experiments—not with vesicles from prebiotically plausible amphiphiles (see Section 28.2)—but with giant phospholipid vesicles.

28.5.1 ON THE SELF-REPRODUCTION OF VESICLES BUILT FROM PHOSPHOLIPIDS

In the self-reproduction of vesicles, by incorporating additional membrane molecules, mother vesicles grow and deform to a limiting shape consisting of a pair of spheres connected by a narrow neck and then divide to independent daughter vesicles by breaking the neck. If the daughter vesicles have the same properties as the mother vesicle, the self-reproduction process can be repeated. Although many attempts have been made to realize the self-reproduction of vesicles using simple amphiphiles, the successes are still very limited, as described in Section 28.3. This fact suggests that the self-reproduction of vesicles is not straightforward from a membrane physics point of view (Svetina, 2012). The requirements to attain the self-reproduction of vesicles—independent of whether they are made from phospholipids, fatty acids, or other types of amphiphiles—are (i) vesicle deformation from a sphere to the limiting shape, (ii) breaking the neck of the limiting shape vesicle to produce daughter vesicles, and (iii) a recursive nature of the vesicles after division.

The deformation of vesicles is well understood on the basis of the membrane elasticity energy model (curvature model) (Seifert, 1997), see also Chapter 5. The most successful model to describe vesicle shapes is the area-difference-elasticity model (Božič et al., 1992; Seifert et al., 1992; Wiese et al., 1992). In this area difference elasticity (ADE) model, the vesicle shape is determined by minimization of the total elastic energy composed of a bending energy and an area difference elastic energy for given vesicle area, A_{ve}, and volume, V. The area difference elastic energy originates from relative stretching of a monolayer. The bilayer has a preferred area difference given by $\Delta A_0 = (N^+ - N^-)A_{lip}$, where A_{lip} is the equilibrium area per membrane molecule and N^+ and N^- are the numbers of molecules in the outer and inner monolayer, respectively. On the other hand, the vesicle has the geometrical area difference, $\Delta A = A^{+} - A^{-}$, where A^{+} and A^{-} are the areas of the outer and inner monolayer. If the monolayer area difference ΔA deviates from the preferred value ΔA_0, the monolayers should be stretched relative to one another, which gives the area difference elastic energy. In this model the vesicle shapes can be mapped by two geometrical parameters, the reduced volume expressed as:

$$v = \frac{V}{4\pi R_s^3 / 3} \tag{28.1}$$

where $R_s = \sqrt{A_{\mathrm{ve}} / 4\pi}$, and the reduced preferred area difference

$$\Delta a_0 = \frac{\Delta A_0}{8\pi l_{me} R_s}, \tag{28.2}$$

where l_{me} is the distance between the monolayer's neutral planes, that is, roughly half of the bilayer thickness, as shown in Figure 28.8. In this phase diagram characteristic equilibrium shapes (pear, prolate, oblate, stomatocyte and sphere) and the limiting shape, are illustrated for each phase. The limiting shape line (L) and first-order discontinuous transitions, stomatocyte-oblate transition ($D^{sto/obl}$), oblate-prolate transition ($D^{obl/pro}$), and prolate-pear transition ($D^{pro/pea}$) lines are indicated by solid lines. These transitions take place when the deformation trajectory $\Delta a_0(v)$ crosses the shape boundary lines. This phase diagram clearly shows that to attain the deformation from a sphere with $\Delta a_0 = 1$ and $v = 1$ to the symmetric limiting shape (× on line L in Figure 28.8) with $\Delta a_0 \approx 1.7$ and $v \approx 0.7$, the vesicles have to not only decrease the reduced volume by increasing the membrane area but also increase the reduced preferred area difference. Thus incorporated membrane molecules should be delivered to the inner and outer leaflets to attain $\Delta a_0 \approx 1.7$. Svetina's group analyzed the growth and deformation process of self-reproducing vesicles and found that vesicle self-reproduction occurs only when the membrane hydraulic permeabilities and the membrane growth rate satisfy a geometrical condition (Božič and Svetina, 2004, 2007).

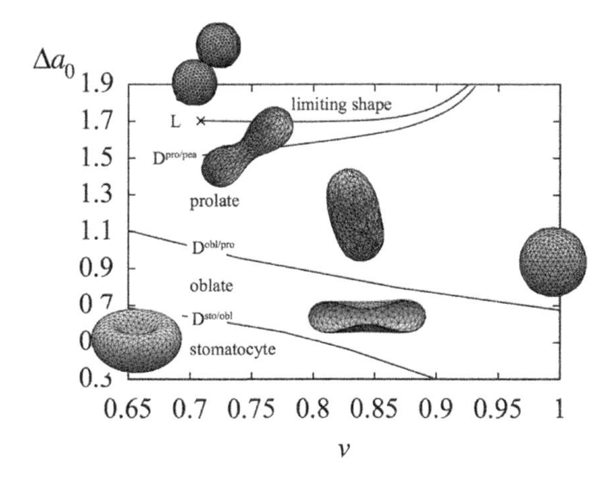

Figure 28.8 Phase diagram, Δa_0 versus v, as predicted by the ADE model. Localization of the characteristic equilibrium shapes—sphere, pear, prolate, oblate, stomatocyte—and the limiting shape are illustrated for each phase. Solid lines indicate discontinuous transitions between the stomatocyte and the oblate shape ($D^{sto/obl}$), the oblate and the prolate shape ($D^{obl/pro}$), and the prolate and the pear shape ($D^{pro/pea}$). The budded limiting shape line is labeled by L.

A unique technique to attain deformations to the limiting shape is to encapsulate colloidal particles (Natsume et al., 2010) or polymer chains (Terasawa, et al., 2012). By confining colloids densely, nonspherical vesicles deform to the limiting shape, see Chapters 8, 9, and 25. This deformation is caused by the maximization of the free volume of the confined colloids in a vesicle, that is, overlapping of the depletion zone next to the vesicle membrane (Asakura and Oosawa, 1954).

The second issue concerns the mechanism of breaking the neck after the limiting shape has been reached (division process). It is well known that single-component vesicles can deform to the budded limiting shape in response to external stimuli (e.g., a change of temperature or osmotic pressure) but breaking the narrow neck between the buds is hard (Käs et al., 1993) because the neck is stabilized by the spontaneous curvature (Miao et al., 1991; Fourcade et al., 1994). In such a case, vesicle division must involve a mechanism of neck destabilization, such as mechanical agitation (Zhu and Szostak, 2009; Adamala and Szostak, 2013). On the other hand, in multicomponent vesicles division is observed more readily (Döbereiner et al., 1993; Bailey and Cullis, 1997; Sakuma and Imai, 2011). These experiments suggest that a coupling between Gaussian curvature and local lipid composition is important for such a division process. The free energy analysis for the limiting shape and the two-vesicle states shows that this coupling can destabilize the narrow neck in a limiting shape of binary vesicles, if the minor component lipids prefer to stay at regions with large positive Gaussian curvature (Chen et al., 1997). Thus the vesicle division is facilitated by mixing two types of amphiphiles having different Gaussian bending rigidities. In addition, after the neck has been broken, offspring vesicles should have the same properties (composition) as the mother vesicle so that the self-reproduction process can be repeated. Thus to attain vesicle self-reproduction, the vesicles should satisfy several physical requirements.

To address the mechanism of vesicle self-reproduction, Imai's group developed experiments where a phospholipid mother vesicle produces twin offspring vesicles by cyclic heating and cooling (Sakuma and Imai, 2011). The applied protocol relies on binary vesicles that are composed of cylinder-shape phospholipids with a high melting temperature T_m (DPPC, $T_m = 41°C$) and of inverse-cone-shape lipids with a low T_m, (1,2-dilauroyl-*sn*-glycero-3-phosphoethanolamine [DLPE], $T_m = 29°C$), where PC lipids and PE lipid are assumed to have different Gaussian bending rigidities (Siegel and Kozlov, 2004) because bilayers composed of PE lipids tend to form an hourglass shape with a negative Gaussian curvature (rhombohedral phase) (Yang and Huang, 2002; Sakuma et al., 2008). The solution in/around the vesicles was pure water. By heating the vesicles to a temperature above T_m of DPPC, binary GUVs spontaneously deformed to a budded limiting shape using up the excess area produced by the chain melting of DPPC and then the neck was completely broken as shown in Figure 28.9. After cooling back to the initial temperature, offspring vesicles recovered the original spherical shape of the mother vesicle. In the next cycle, the process was repeated in both offspring vesicles and then again, eventually yielding more than 10 generations of vesicles. This model vesicle system captures the essence of self-reproduction, although the membrane area does not grow by addition of molecules and

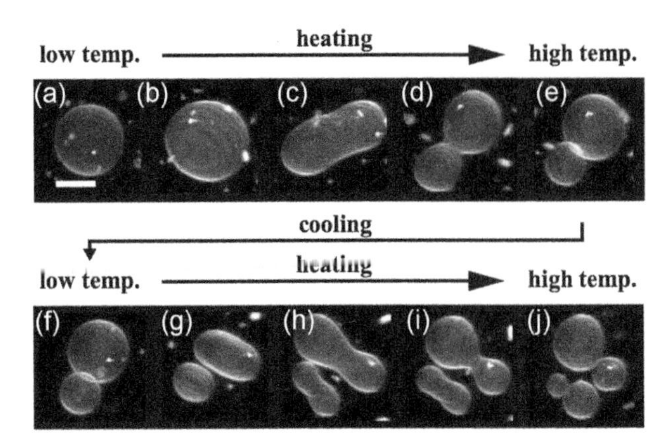

Figure 28.9 Sequence of z-projection of 3D images of a DPPC/DLPE = 8/2 binary GUV in the vesicle division process driven by a heating-cooling cycle between 30°C and 50°C. (a–e) Deformations of a first-generation vesicle, where the mother vesicle produced daughter vesicles. (f–j) Second-generation deformations, where the daughter vesicles produced granddaughter vesicles. Scale bar: 10 μm.

thus the total area is fixed. Three-dimensional analysis of this deformation-division process reveals that the spontaneous coupling between the membrane curvature and PE lipid shape might be responsible for the deformation, division and recursive nature (Jimbo et al., 2016).

28.5.2 ON THE NUTRIENT TRANSPORT THROUGH VESICLE MEMBRANES WITHOUT PROTEIN PORES

An important requirement to continue chemical reactions occurring inside vesicles is the uptake of nutrients from the environment and the excretion of waste through the vesicle membrane (Mansy, 2010). The main paths for this transportation are (i) passive diffusion, (ii) pore formation, and (iii) fusion of cargo vesicles.

Passive diffusion: The passive diffusion is governed by the permeability of the vesicle membrane. In the case of phospholipid vesicles, the permeability of small, uncharged molecules—such as water, oxygen, and carbon dioxide—is greater than the permeability of ionic compounds (Deamer and Dworkin, 2005; Monnard and Deamer, 2011) see also Chapter 20. For instance, the permeability coefficient of water in the case of phospholipid membranes ($T > T_m$) is approximately 10^{-3} cm/s (Fettiplace and Haydon, 1980), and that of phosphate and charged amino acids are found to be in the range of 10^{-11}–10^{-12} cm/s (Chakrabarti and Deamer, 1992). This permeability is interpreted by the solubility-diffusion model, where the permeability is simply governed by the hydrophobicity of the solute and the diffusion of the solute in the hydrophobic interior of the membrane. However, it was found that the permeability is strongly dependent on the chain length (Paula et al., 1996). For instance, shortening phospholipid chains from 18 to 14 carbons increases the permeability to ions by a thousand fold. To account for this observation, a transient pore model has been proposed (Paula and Deamer, 1999). The thinner membranes have increasing numbers of transient defects that open and close pores on nanosecond time scales. Then ionic solutes can diffuse across a bilayer composed of 1,2-dimyristoyl-*sn*-glycero-3-phospho (DMPC) with 14-carbon chains (Monnard and Deamer, 2001). The formation of transient pores significantly increases at T_m of the lipid (Kanehisa and Tsong, 1978), which enables to transport ionized substrate molecules as large as nucleoside triphosphates (NTPs) through a DMPC membrane (Monnard et al., 2007). For instance, the permeability coefficient of NTP for pure DMPC vesicles at $T_m = 23°C$ is about 1.3×10^{-9} cm/s, whereas the same vesicles at 37°C exhibit a permeability about two orders of magnitude lower. Using this property, vesicle systems with a temperature-controlled solute uptake (and release) can be designed.

Compared with double-chain phospholipid vesicles, fatty acid vesicles have different membrane permeation properties because the neutral form of a single-chain fatty acid crosses a bilayer membrane with a half-time of milliseconds (flip-flop time), much faster than in the case of phospholipids (Mansy, 2009). The more dynamic fatty acid membranes were shown to be permeable remarkably to nucleoside mono- and diphosphates (Apel et al., 2002; Mansy et al., 2008), whereas the less dynamic phospholipid membranes are typically impermeable to nucleotides at $T \neq T_m$, consult also Chapter 20. Then the fatty acid membrane can serve both as an energy source and as substrate for endovesicular RNA polymerization catalyzed by polynucleotide phosphorylase (Walde et al., 1994a). It should be noted that for small, uncharged solutes both fatty acid and phospholipid vesicles have similar permeability, for example, permeability coefficients of ribose for POPC and oleic acid/oleate membrane are 20×10^{-8} and 11×10^{-8} cm/s, respectively (Sacerdote and Szostak, 2005). Raising the temperature or adding low millimolar concentrations of Mg^{2+} further increases the permeability of fatty acid vesicles (Mansy and Szostak, 2008). In prebiotic times, where vesicular protocells may have existed, the passive diffusion might have been important for the uptake of critical nutrients.

Pore formation: The formation of protein pores is a powerful means to transport chemical substances through vesicle membranes, like in the case of contemporary biomembranes. To form such pores, several pore-forming proteins and peptides can be utilized such as α-hemolysin (Noireaux and Libchaber, 2004), porins (Graff et al., 2001), and melittin (Mally et al., 2007). Pore formation can be achieved without the use of any proteins (or peptides) by coupling the spontaneous curvature of the lipids with lipid phase separation (Sakuma et al., 2010). This was demonstrated by using binary phospholipid GUVs composed of the cylinder-shaped lipid, DPPC and the cone-shaped 1,2-dihexanoyl-*sn*-glycero-3-phosphocholine (DHPC). When the temperature was set below T_m of DPPC, the GUVs formed a single large pore with a radius of ~2 μm, whereby the cone shape lipids formed a cap at the edge of the bilayer to stabilize the pore. Nutrients can be transported into the vesicles through this pore. The pore closes when the temperature increases above the transition temperature. The opening and closing of pores in a vesicle results from a competition between two forces: surface tension and line tension. The membrane surface tension derives from the stretching of the lipids, which is caused by several stimuli, such as chain ordering (Sakuma et al., 2010), adhesion (Sandre et al., 1999), illumination of light (Karatekin et al., 2003; Rodriguez et al., 2006), and addition of detergent (Edwards and Almgren, 1991). This surface tension can rupture the membrane, leading to the formation of open pores. On the other hand, at the rim of the pore, the lipids are forced to reorganize for reducing the interface between their hydrophobic parts and the solvent. This reorganization leads to a line tension, which forces the pore to close up. Such, or similar, mechanism may have been in operation in protocells as wells.

Fusion of cargo vesicle: Another way of supplying nutrients to the interior of vesicles is by fusion of cargo vesicles filled with the nutrients (Kurihara et al., 2015). Indeed, the fusion between cationic vesicles and anionic vesicles has been reported for many systems (Caschera et al., 2010; Sunami et al., 2010; Suzuki et al., 2012). Fusion of lipid vesicles appears to proceed in several steps (Jahn and Grubmüller, 2002; Tamm et al., 2003). First two vesicles approach each other and come into contact. Then, the bilayers of the two vesicles form an intermediate hemifusion state, that is, a neck-like connection between the two bilayers. Subsequent expansion of these necks results in the completion of the fusion. Because the size of the fusion neck is about 10 nm and the corresponding time scale is less than 100 μs (Haluska et al., 2006), computer simulation techniques (Noguchi and Takasu, 2001; Grafmüller et al., 2007) have been used to obtain insight into the dynamics of fusion necks. A rate determinant process of the fusion is the flip of lipids between the membranes in contact to form hemifused patches, where the outer leaflets of both bilayers merged whereas the inner leaflets remained independent. Computer simulations show that the energy barrier for the flipping of lipid molecules is about 10 $k_B T$ (Grafmüller et al., 2007). This explains why the fusion of vesicles has to be induced by one of the various stimuli, such as by addition of fusogenic agents of opposite charge (Caschera et al., 2010; Sunami et al., 2010; Suzuki et al., 2012), by an osmotic swelling (Finkelstein et al., 1986), by electroporation (Riske and Dimova, 2005) (see also Chapter 15), by bridging vesicles (Heuvingh et al., 2004) or by addition of specific ions (La^{3+} or Eu^{3+}) (Tanaka and Yamazaki, 2004; Haluska et al., 2006). These stimuli increase the membrane tension, which results in a reduction of the energy barrier for the flipping of lipid molecules because such an increase moves the headgroups further apart.

Overall, from a more conceptual point of view, there are means by which organic nutrient molecules and inorganic ions from the environment can move into the interior of vesicles without the need of sophisticated pore proteins. Some are discussed here, others are emphasized in the literature (Mosgaard and Heimburg, 2013). Of course, all depends on the details of the vesicle membrane composition and the general experimental conditions. It is possible that such, or similar, mechanisms were in operation in prebiotic times when protocellular vesicular system might have existed without any sophisticated biomolecules.

28.6 SOME OF THE REMAINING BIG CHALLENGES

There is no doubt that an understanding of the origin of the first cells, that is, the origin of life on Earth, might only be possible if in any type of scenario the role of prebiotic compartmentalization is included. Nevertheless, we still know very little about this transition from the nonliving form of matter—likely to be compartmentalized systems (protocells)—to the first forms of life. Among other possible prebiological compartments (Mann, 2012; Monnard and Walde, 2015), for example, coacervates, giant (unilamellar) vesicles are particularly attractive because they are the most cell-like systems. Furthermore, numerous experimental studies have shown that nonenzymatic as well as complex enzymatic reactions are possible inside giant vesicles making them also attractive for the construction of artificial cell-like systems (Rasmussen et al., 2016; Engelhart et al., 2016), see also Chapters 4, 29 and 30. Despite the progress made in this field by fundamental research carried out over the last years in different groups, it is still too early to conclude, whether the synthesis of minimal cells—which one would generally accept to be living—will ever be possible. One particular analytical challenge is the quantification of the experiments and a statistical analysis, including chemical conversions and changes in the physical properties (vesicle size, shape and amounts). Because the concepts used for the experiments toward the construction of minimal cells are directly related to the general ideas about protocells, the two fields of research have much in common. It is hoped that more researchers will enter these fields, hopefully with completely new ideas. At least, the fundamentals about the preparation and the properties of giant vesicles have been elaborated. The challenge is to work with complex vesicles as chemical compartment systems—and not as individual containers—which are not at thermodynamic equilibrium but rather are characterized by constant chemical and physical transformations. Finally, it is worth emphasizing that vesicles not only offer a trapped aqueous volume that is separated from the bulk medium for volume-confined reactions, but they also offer soft—or semi-solid—interfaces (the vesicle membrane) for promoting and controlling surface-confined chemical or enzymatic reactions (Walde, 2006; Tessera, 2011; Walde et al., 2014c; Monnard and Walde, 2015). This possibility adds to the complexity of vesicle system and it may be beneficial, if it is included in scenarios dealing with vesicular models of protocells and minimal cells systems.

LIST OF ABBREVIATIONS

ADE model	area difference elasticity model
CAC	critical aggregation concentration
CMC	critical micellization concentration
DHPC	1,2-dihexanoyl-*sn*-glycero-3-phosphocholine
DLPE	1,2-dilauroyl-*sn*-glycero-3-phosphoethanolamine
DMPC	1,2-dimyristoyl-*sn*-glycero-3-phosphocholine
DPPC	1,2-dipalmitoyl-*sn*-glycero-3-phosphocholine
GFP	green fluorescent protein
GMV	giant multilamellar vesicle
GV	giant vesicle
LUV	large unilamellar vesicle
NTP	nucleoside triphosphate
dNTP	deoxyribonucleoside triphosphate
PC	phosphatidylcholine
PCR	polymerase chain reaction
PE	phosphatidylethanolamine
POPC	1-palmitoyl-2-oleoyl-*sn*-glycero-3-phosphocholine
PURE system	protein synthesis using recombinant elements system

GLOSSARY OF SYMBOLS

$A^{\pm}$	area of the outer (+)/inner (−) monolayer
A_{lip}	equilibrium area per membrane molecule
A_{ve}	total membrane/surface area of the vesicle
ΔA	geometrical area difference between inner and outer monolayer
ΔA_0	preferred area difference between inner and outer monolayer
Δa_0	normalized preferred area difference between inner and outer monolayer
l_{me}	distance between the monolayer's neutral planes
$N^{\pm}$	number of molecules in the outer (+)/inner (−) monolayer
R_s	radius of spherical vesicle with total membrane area A_{ve}
T	temperature
T_m	phase transition temperature
v	reduced volume
V	volume of the vesicle

REFERENCES

Adamala K, Szostak JW (2013) Competition between model protocells driven by an encapsulated catalyst. *Nature Chem.* 5: 495–501.

Albertsen AN, Duffy CD, Sutherland JD, Monnard PA (2014) Self-assembly of phosphate amphiphiles in mixtures of prebiotically plausible surfactants. *Astrobiology* 14: 462–472.

Apel CL, Deamer D, Mautner MN (2002) Self-assembled vesicles of monocarboxylic acids and alcohols: Conditions for stability and for the encapsulation of biopolymers. *Biochim. Biophys. Acta* 1559: 1–9.

Asakura S, Oosawa F (1954) On interaction between two bodies immersed in a solution of macromolecules. *J. Chem. Phys.* 22: 1255–1256.

Bailey AL, Cullis PR (1997) Membrane fusion with cationic liposomes: Effects of target membrane lipid composition. *Biochemistry* 36: 1628–1634.

Berclaz N, Müller M, Walde P, Luisi PL (2001) Growth and transformation of vesicles studied by ferritin labeling and cryotransmission electron microscopy. *J. Phys. Chem. B* 105: 1056–1064.

Blain JC, Szostak JW (2014) Progress toward synthetic cells. *Annu. Rev. Biochem.* 83: 11.1–11.26.

Božič B, Svetina S (2004) A relationship between membrane properties forms the basis of a selectivity mechanism for vesicle self-reproduction. *Eur. Biophys. J.* 33: 565–571.

Božič B, Svetina S (2007) Vesicle self-reproduction: The involvement of membrane hydraulic and solute permeabilities. *Eur. Phys. J. E* 24: 79–90.

Božič B, Svetina S, Žekš B, Waugh RE (1992) Role of lamellar membrane structure in tether formation from bilayer vesicles. *Biophys. J.* 61: 963–973.

Budin I, Debnath A, Szostak JW (2012) Concentration-driven growth of model protocell membranes. *J. Am. Chem. Soc.* 134: 20812–20819.

Budin I, Szostak JW (2011) Physical effects underlying the transition from primitive to modern cell membranes. *Proc. Natl. Acad. Sci. USA* 108: 5249–5254.

Campbell NA, Reece JB, Urry LA, Cain ML, Wasserman SA, Minorsky PV, Jackson RB (2015) *Biology: A Global Approach*, Essex: Pearson Education Ltd.

Caschera F, Stano P, Luisi PL (2010) Reactivity and fusion between cationic vesicles and fatty acid anionic vesicles. *J. Colloid Interface Sci.* 345: 561–565.

Chakrabarti AC, Breaker RR, Joyce GF, Deamer DW (1994) Production of RNA by a polymerase protein encapsulated within phospholipid vesicles. *J. Mol. Evol.* 39: 555–559.

Chakrabarti AC, Deamer D (1992) Permeability of lipid bilayers to amino acids and phosphate. *Biochim. Biophys. Acta* 1111: 171–177.

Chen CM, Higgs PG, MacKintosh FC (1997) Theory of fission for two-component lipid vesicles. *Phys. Rev. Lett.* 79: 1579–1582.

Chen IA, Roberts RW, Szostak JW (2004) The emergence of competition between model protocells. *Science* 305: 1474–1476.

Chen IA, Szostak JW (2004a) A kinetic study of the growth of fatty acid vesicles. *Biophys. J.* 87: 988–998.

Chen IA, Szostak JW (2004b) Membrane growth can generate a transmembrane pH gradient in fatty acid vesicles. *Proc. Natl. Acad. Sci. USA* 101: 7965–7970.

Chen IA, Walde P (2010) From self-assembled vesicles to protocells. *Cold Spring Harb. Perspect. Biol.* 2: a002170.

Crick F (1970) Central dogma of molecular biology. *Nature* 227: 561–563.

Damer B, Deamer D (2015) Coupled phases and combinatorial selection in fluctuating hydrothermal pools: A scenario to guide experimental approaches to the origin of cellular life. *Life* 5: 872–887.

Deamer D (1985) Boundary structures are formed by organic components of the Murchison carbonaceous chondrite. *Nature* 317:792–794.

Deamer D (1997) The first living systems: A bioenergetic perspective. *Microbiol. Mol. Biol. Rev.* 61: 239–261.

Deamer D (2011) *First Life-Discovering the Connections Between Stars, Cells, and How Life Began.* Berkeley, CA: University of California Press.

Deamer D (2017) The role of lipid membranes in life's origin. *Life* 7: 5.

Deamer D, Dworkin JP (2005) Chemistry and physics of primitive membranes. *Top Curr. Chem.* 259: 1–27.

Deamer D, Szostak JW (eds.) (2010) *The Origins of Life.* New York: Cold Spring Harbor Laboratory Press.

DeClue MS, Monnard PA, Bailey JA, Maurer SE, Collis GE, Ziock H-J, Rasmussen S, Boncella JM (2009) Nucleobase mediated, photocatalytic vesicle formation from an ester precursor. *J. Am. Chem. Soc.* 131: 931–933.

Döbereiner HG, Käs J, Noppl D, Sprenger I, Sackmann E (1993) Budding and fission of vesicles. *Biophys. J.* 65: 1396–1403.

Douliez JP, Houinsou Houssou B, Fameau A-L, Navailles L, Nallet F, Grélard A, Dufourc EJ, Gaillard C (2016) Self-assembly of bilayer vesicles made from saturated long chain fatty acids. *Langmuir* 32: 401–410.

Edwards K, Almgren M (1991) Solubilization of lecithin vesicles by C12E8: Structural transitions and temperature effects. *J. Colloid Interface Sci.* 147: 1–21.

Engelhart AE, Adamala KP, Szostak JW (2016) A simple physical mechanism enables homeostasis in primitive cells. *Nat. Chem.* 8: 448–453.

Fettiplace R, Haydon DA (1980) Water permeability of lipid membranes. *Physiol. Rev.* 60: 510–550.

Finkelstein A, Zimmerberg J, Cohen FS (1986) Osmotic swelling of vesicles: Its role in the fusion of vesicles with planar phospholipid bilayer membranes and its possible role in exocytosis. *Ann. Rev. Physiol.* 48: 163–174.

Fourcade B, Miao L, Rao M, Wortis M (1994) Scaling analysis of narrow necks in curvature models of fluid lipid-bilayer vesicles. *Phys. Rev. E* 49: 5276–5286.

Gebicki JM, Hicks M (1973) Ufasomes are stable particles surrounded by unsaturated fatty acid membranes. *Nature* 243: 232–234.

Gilbert W (1986) Origin of life: The RNA world. *Nature* 319: 618.

Graff A, Winterhalter M, Meier W (2001) Nanoreactors from polymer-stabilized liposomes. *Langmuir* 17: 919–923.

Grafmüller A, Shillcock J, Lipowsky R (2007) Pathway of membrane fusion with two tension-dependent energy barriers. *Phys. Rev. Lett.* 98: 218101.

Haines TH (1983) Anionic lipid headgroups as a proton-conducting pathway along the surface of membranes: A hypothesis. *Proc. Nat. Acad. Sci.* 80: 160–164.

Haluska CK, Riske KA, Marchi-Artzner V, Lehn J-M, Lipowsky R, Dimova R (2006) Time scales of membrane fusion revealed by direct imaging of vesicle fusion with high temporal resolution. *Proc. Natl. Acad. Sci. USA* 103: 15841–15846.

Hargreaves WP, Deamer DW (1978) Liposomes from Ionic, Single-Chain Amphiphiles. Biochemistry 17: 3759–3768.

Heuvingh J, Pincet F, Cribier S (2004) Hemifusion and fusion of giant vesicles induced by reduction of inter-membrane distance. *Eur. Phys. J. E* 14: 269–276.

Jahn R, Grubmüller H (2002) Membrane fusion. *Curr. Opin. Cell Biol.* 14: 488–495.

Jimbo T, Sakuma Y, Urakami N, Ziherl P, Imai M (2016) Role of inverse-cone-shape lipids in temperature-controlled self-reproduction of binary vesicles. *Biophys. J.* 110: 1151–1562.

Kanehisa MI, Tsong TY (1978) Cluster model of lipid phase transitions with application to passive permeation of molecules and structure relaxations in lipid bilayers. *J. Am. Chem. Soc.* 100: 424–432.

Kaneko K (2006) Life: *An Introduction to Complex Systems Biology.* Berlin, Germany: Springer.

Karatekin E, Sandre O, Guitouni H, Borghi N, Puech P-H, Brochard-Wyart F (2003) Cascades of transient pores in giant vesicles: Line tension and transport. *Biophys. J.* 84: 1734–1749.

Käs J, Sackmann E, Podgornik R, Svetina S, Žekš B (1993) Thermally induced budding of phospholipid vesicles—A discontinuous process. *J. Phys. II France* 3: 631–645.

Kita H, Matsuura T, Sunami T, Hosoda K, Ichihashi N, Tsukada K, Urabe I, Yomo T (2008) Replication of genetic information with self-encoded replicase in liposomes. *ChemBioChem* 9: 2403–2410.

Kurihara K, Okura Y, Matsuo M, Toyota T, Suzuki K, Sugawara T (2015) A recursive vesicle-based model protocell with a primitive model cell cycle. *Nat. Commun.* 6: 8352.

Kurihara K, Tamura M, Shoda K, Yoyota T, Suzuki K, Sugawara T (2011) Self-reproduction of supramolecular giant vesicles combined with the amplification of encapsulated DNA. *Nat. Chem.* 3: 775–781.

Kuruma Y, Stano P, T Ueda, Luisi PL (2009) A synthetic biology approach to the construction of membrane proteins in semi-synthetic minimal cells. *Biochim. Biophys. Acta Biomembr.* 1788: 567–574.

Kvenvolden KA, Lawless J, Pering K, Peterson E, Flores J, Ponnameruma C, Kaplan IR, Moore C (1970) Evidence for extraterrestrial amino-acids and hydrocarbons in the Murchison meteorite. *Nature* 228: 923–926.

Lawless LG, Yuen GU, (1979) Quantification of monocarboxylic acids in the Murchison carbonaceous meteorite. *Nature* 282: 396–398.

Luisi PL (2006) *The Emergence of Life–From Chemical Origins to Synthetic Biology.* Cambridge, UK: Cambridge University Press.

Luisi PL, Stano P (eds.) (2011) *The Minimal Cell: The Biophysics of Cell Compartment and the Origin of Cell Functionality.* Heidelberg, Germany: Springer.

Mally M, Majhenc J, Svetina S, Žekš B (2007) The response of giant phospholipid vesicles to pore-forming peptide melittin. *Biochim. Biophys. Acta* 1768: 1179–1189.

Mann S (2012) Systems of creation: The emergence of life from nonliving matter. *Acc. Chem. Res.* 45: 2131–2141.

Mansy SS (2009) Model protocells from single-chain lipids. *Int. J. Mol. Sci.* 10: 835–843.

Mansy SS (2010) Membrane transport in primitive cells. In: *The Origin of Life* (Deamer D, Szostak JW, eds.) pp. 193–206. New York: Cold Spring Harbor Laboratory Press.

Mansy SS, Schrum JP, Krishnamurthy M, Tobé S, Treco DA, Szostak JW (2008) Template-directed synthesis of a genetic polymer in a model protocell. *Nature* 454: 122–125.

Mansy SS, Szostak JW (2008) Thermostability of model protocell membranes. *Proc. Natl. Acad. Sci. USA.* 105: 13351–13355.

Maurer SE, Deamer DW, Boncella JM, Monnard P-A (2009) Chemical evolution of amphiphiles: Glycerol monoacyl derivatives stabilize plausible prebiotic membranes. *Astrobiology* 9: 979–987.

Menger FM, Gabrielson K (1994) Chemically-induced birthing and foraging in vesicle systems. *J. Am. Chem. Soc.* 116: 1567–1568.

Miao L, Fourcade B, Rao M, Wortis M (1991) Equilibrium budding and vesiculation in the curvature model of fluid lipid vesicles. *Phys. Rev. A* 43: 6843–6856.

Mojzsis SJ, Arrhenius G, McKeegan KD, Harrison TM, Nutman AP, Friend CRL (1996) Evidence for life on earth before 3,800 Million years ago. *Nature* 384: 55–59.

Monnard PA, Deamer D (2001) Loading of DMPC based liposomes with nucleotide triphosphates by passive diffusion: A plausible model for nutrient uptake by the protocell. *Orig. Life Evol. Biosph.* 31: 147–155.

Monnard PA, Deamer D (2002) Membrane self-assembly processes: Steps toward the first cellular life. *Anatomical Rec.* 268: 196–207.

Monnard PA, Deamer D (2003) Preparation of vesicles from nonphospholipid amphiphiles. *Methods Enzymol.* 372: 133–151.

Monnard PA, Deamer DW (2011) Membrane self-assembly processes: Steps toward the first cellular life. In: *The Minimal Cell–The Biophysics of Cell Compartment and the Origin of Cell Functionality* (Luisi L, Stano P eds.) pp. 123–151. Heidelberg, Germany: Springer.

Monnard PA, Luptak A, Deamer DW (2007) Models of primitive cellular life: Polymerases and templates in liposomes. *Phil. Trans. R Soc. B* 362: 1741–1750.

Monnard PA, Walde P (2015) Current ideas about prebiological compartmentalization. *Life* 5: 1239–1263.

Morigaki K, Walde P (2007) Fatty acid vesicles. *Curr. Opin. Colloid Interface Sci.* 12: 75–80.

Morigaki K, Walde P, Misran M, Robinson BH (2003) Thermodynamic and kinetic stability. Properties of micelles and vesicles formed by the decanoic acid/decanoate system. *Coll. Surf. A: Physicochem. Eng. Aspects* 213: 37–44.

Morowitz HJ, Heinz B, Deamer DW (1988) The chemical logic of a minimum protocell. *Orig. Life Evol. Biosph.* 18: 281–287.

Mosgaard LD, Heimburg T (2013) Lipid ion channels and the role of proteins. *Acc. Chem. Res.* 46: 2966–2976.

Natsume Y, Pravaz O, Yoshida H, Imai M (2010) Shape deformation of giant vesicles encapsulating charged colloidal particles. *Soft Matter* 6: 5359–6366.

Nicolis G, Prigogine I (1977) Self-organization in non-equilibrium systems. New York: John Wiley & Sons.

Noguchi H, Takasu M (2001) Fusion pathways of vesicles: A Brownian dynamics simulation. *J. Chem. Phys.* 115: 9547–9551.

Noireaux V (2015) Construction de cellules synthétiques. Du fondamental au pratique. *Med. Sci.* (Paris) 31: 1126–1132.

Noireaux V, Libchaber A (2004) A vesicle bioreactor as a step toward an artificial cell assembly. *Proc. Natl. Acad. Sci. USA* 101: 17669–17674.

Noireaux V, Maeda YT, Libchaber A (2011) Development of an artificial cell, from self-organization to computation and self-reproduction. *Proc. Natl. Acad. Sci. USA* 108: 3473–3480.

Nomura SM, Tsumoto K, Hamada T, Akiyoshi K, Nakatani Y, Yoshikawa K (2003) Gene Expression within cell-sized lipid vesicles. *ChemBioChem* 4: 1172–1175.

Nourian Z, Danelon C (2013) Linking genotype and phenotype in protein synthesizing liposomes with external supply of resources. *ACS Synth. Biol.* 2: 186–193.

Oberholzer T, Albrizio M, Luisi PL (1995a) Polymerase chain reaction in liposomes. *Chem. Bio.* 2: 677–682.

Oberholzer T, Nierhaus KH, Luisi PL (1999) Protein expression in liposomes. *Biochem. Biophys. Res. Commun.* 261: 238–241.

Oberholzer T, Wick R, Luisi PL, Biebricher CK (1995b) Enzymatic RNA replication in self-reproducing vesicles: an approach to a minimal cell. *Biochem. Biophys. Res. Commum.* 207: 250–257.

Oparin AI (1965) The origin of life and the origin of enzymes. *Adv. Enzymol.* 27: 347–380.

Paula S, Deamer D (1999) Membrane permeability barriers to ionic and polar solites. In: *Membrane Permeability* (Deamer D, Kleinzeller A, et al. eds.) pp. 77–96, San Diego, CA: Academic Press.

Paula S, Volkov AG, Van Hoek AN, Haines TH, Deamer DW (1996) Permeation of protons, potassium ions, and small polar molecules through phospholipid bilayers as a function of membrane thickness. *Biophys. J.* 70: 339–348.

Rasmussen S, Bedau MA, Chen L, Deamer D, Krakauer DC, Packard NH, Stadler PF (eds.) (2008) *Protocells–Bridging Nonliving and Living Matter.* Cambridge, UK: MIT Press.

Rasmussen S, Chen L, Deamer D, Krakauer DC, Packard NH, Stadler PF, Bedau MA (2004) Transitions from nonliving to living matter. *Science* 303: 963–965.

Rasmussen S, Constantinescu A, Svaneborg C (2016) Generating minimal living systems from non-living materials and increasing their evolutionary abilities. *Phil. Trans. R Soc. B* 371: 20150440.

Riske KA, Dimova R (2005) Electro-deformation and poration of giant vesicles viewed with high temporal resolution. *Biophys. J.* 88: 1143–1155.

Rodriguez N, Cribier S, Pincet F (2006) Transition from long-to short-lived transient pores in giant vesicles in an aqueous medium. *Phys. Rev. E* 74: 061902.

Ruiz-Mirazo K, Briones C, de la Escosura A (2014) Prebiotic systems chemistry: New perspectives for the origins of life. *Chem. Rev.* 114: 285–366.

Sacerdote MG, Szostak JW (2005) Semipermeable lipid bilayers exhibit diastereoselectivity favoring ribose. *Proc. Nat. Acad. Sci. USA* 102: 6004–6008.

Sakuma Y, Imai M (2011) Model system of self-reproducing vesicles. *Phys. Rev. Lett.* 107: 198101.

Sakuma Y, Imai M (2015) From vesicles to protocells: The roles of amphiphilic molecules. *Life* 5: 651–675.

Sakuma Y, Imai M, Yanagisawa M, Komura S (2008) Adhesion of binary giant vesicles containing negative spontaneous curvature lipids induced by phase separation. *Eur. Phys. J. E* 25: 403–413.

Sakuma Y, Taniguchi T, Imai M (2010) Pore formation in a binary giant vesicle induced by cone-shaped lipids. *Biophys. J.* 99: 472–479.

Sandre O, Moreaux L, Brochard-Wyart F (1999) Dynamics of transient pores in stretched vesicles. *Proc. Natl. Acad. Sci. USA* 96: 10591–10596.

Schrödinger E (1944) *What is Life?* Cambridge, UK: Cambridge University Press.

Segre D, Ben-Eli D, Deamer DW, Lancet D (2001) The lipid world. *Orig. Life Evol. Biosph.* 31: 119–145.

Shimizu Y, Inoue A, Tomari Y, Suzuki T, Yokogawa T, Nishikawa K, Ueda T (2001) Cell-free translation reconstituted with purified components. *Nat. Biotechnol.* 19: 751–755.

Seifert U (1997) Configurations of fluid membranes and vesicles. *Adv. Phys.* 46: 13–137.

Seifert U, Miao L, Döbereiner HG, Wortis M (1992) The structure and conformation of amphiphilic membranes. *Springer Proc. Phys.* 66: 93–96.

Siegel DP, Kozlov MM (2004) The Gaussian curvature elastic modulus of N-monomethylated dioleoylphosphatidylethanolamine: Relevance to membrane fusion and lipid phase behavior. *Biophys. J.* 87: 366–374.

Smith R, Tanford C (1972) The critical micelle concentration of l-α-dipalmitoylphosphatidyl-choline in water and water/methanol solutions. *J. Mol. Biol.* 67: 75–83.

Smith R, Tanford C (1973) Hydrophobicity of long chain *n*-Alkyl carboxylic acids, as measured by their distribution between heptane and aqueous solutions. *Proc. Nat. Acad. Sci. USA* 70: 289–293.

Stano P, Carrara P, Kuruma Y, de Souza TP, Luisi PL (2011) Compartmentalized reactions as a case of soft-matter biotechnology: Synthesis of proteins and nucleic acids inside lipid vesicles. *J. Mater. Chem.* 21: 18887–18902.

Stano P, Luisi PL, (2010) Achievements and open questions in the self-reproduction of vesicles and synthetic minimal cells. *Chem. Commun.* 46: 3639–3653.

Sugawara T (2009) Minimal cell model to understand origin of life and evolution. In: *Evolutionary Biology from Concept to Application II* (Pontarotti P, ed) pp23–50. Berlin, Germany: Springer.

Sugawara T, Kurihara K, Suzuki K (2012) Constructive approach towards protocells. In: *Engineering of Chemical Complexity* (Mikhailov A, Ertl G eds.) pp. 359–374. Berlin, Germany: World Scientific.

Sunami T, Caschera F, Morita Y, Toyota T, Nishimura K, Matsuura T, Suzuki H, Hanczyc MM, Yomo T (2010) Detection of association and fusion of giant vesicles using a fluorescence-activated cell sorter. *Langmuir* 26: 15098–15103.

Suzuki K, Aboshi R, Kurihara K, Sugawara T (2012) Adhesion and fusion of two kinds of phospholipid hybrid vesicles controlled by surface charges of vesicular membranes. *Chem. Lett.* 41: 789–791.

Svetina S, (2012) Cellular life could have emerged from properties of vesicles. *Orig. Life Evol. Biosph.* 42: 483–486.

Szostak JW, Bartel DP, Luisi PL (2001) Synthesizing life. *Nature* 409: 387–390.

Takakura K, Sugawara T (2004) Membrane dynamics of a myelin-like giant multilamellar vesicle applicable to a self-reproducing system. *Langmuir* 20: 3832–3834.

Takakura K, Toyota T, Sugawara T (2003). A novel system of self-reproducing giant vesicles. *J. Am. Chem. Soc.* 125: 8134–8140.

Tamm LK, Crane J, Kiessling V (2003) Membrane fusion: A structural perspective on the interplay of lipids and proteins. *Curr. Opin. Struct. Biol.* 13: 453–466.

Tanaka T, Yamazaki M (2004) Membrane fusion of giant unilamellar vesicles of neutral phospholipid membranes induced by La^{3+}. *Langmuir* 20: 5160–5164.

Tang T-YD, Hak CRC, Thompson AJ, Kuimova MK, Williams DS, Perriman AW, Mann S (2014) Fatty acid membrane assembly on coacervate microdroplets as a step towards a hybrid protocell model. *Nat. Chem.* 6: 527–533.

Terasawa H, Nishimura K, Suzuki H, Matsuura T, Yomo T (2012) Coupling of the fusion and budding of giant phospholipid vesicles containing macromolecules. *Proc. Natl. Acad. Sci. USA* 109: 5942–5947.

Tessera M (2011) Origin of evolution *versus* origin of life: A shift of paradigm. *Int. J. Mol. Sci.* 12: 3445–3458.

Toyota T, Takakura K, Kageyama Y, Kurihara K, Maru N, Ohnuma K, Kaneko K, Sugawara T (2008) Population study of sizes and components of self-reproducing giant multilamellar vesicles. *Langmuir* 24: 3037–3044.

Walde P (2006) Surfactant assemblies and their various possible role for the origin(s) of life. *Orig. Life Evol. Biosph.* 36: 109–150.

Walde P, Goto A, Monnard P-A, Wessicken M, Luisi PL (1994a) Oparin's reactions revisited: Enzymatic synthesis of poly(adenylic acid) in micelles and self-reproducing vesicles. *J. Am. Chem. Soc.* 116: 7541–7547.

Walde P, Umakoshi H, Stano P, Mavelli F (2014) Emergent properties arising from the assembly of amphiphiles. Artificial vesicle membranes as reaction promoters and regulators. *Chem. Commun.* 50: 10177–10197.

Walde P, Wick R, Fresta M, Mangone A, Luisi PL (1994b) Autopoietic self-reproduction of fatty acid vesicles. *J. Am. Chem. Soc.* 116: 11649–11654.

Watson JD, Crick FHC (1953) Molecular structure of nucleic acids. *Nature* 171: 737–738.

Wick R, Walde P, Luisi PL (1995) Light microscopic investigations of the autocatalytic self-reproduction of giant vesicles. *J. Am. Chem. Soc.* 117: 1435–1436.

Wiese W, Harbich W, Helfrich W (1992) Budding of lipid bilayer vesicles and flat membranes. *J. Phys. Condens. Matter* 4: 1647–1657.

Yang L, Huang HW (2002) Observation of a membrane fusion intermediate structure. *Science* 297: 1877–1879.

Yu W, Sato K, Wakabayashi M, Nakaishi T, Ko-Mitamura EP, Shima Y, Urabe I, Yomo T (2001) Synthesis of functional protein in liposome. *J. Biosci. Bioeng.* 92: 590–593.

Zhu TF, Szostak JW (2009) Coupled growth and division of model protocell membranes. *J. Am. Chem. Soc.* 131: 5705–5713.

Encapsulation of aqueous two-phase systems and gels within giant lipid vesicles

Allyson M. Marianelli and Christine D. Keating

From the point of view of intermolecular interactions, the cytoplasmic space is more like a crowded party in a house full of furniture than a game of tag in an empty field.

Kate Luby-Phelps (2013)

Contents

29.1 INTRODUCTORY WORDS

The aqueous interior of giant unilamellar vesicles (GUVs) can be subdivided into two or more distinct regions by encapsulating polymeric solutes to render the interior solution capable of either liquid–liquid phase separation or gelation. GUVs that incorporate coexisting microcompartments of distinct composition and properties are of interest as simple models for compartmentalization in biological cells. This chapter provides insight and advice for the encapsulation of aqueous two-phase systems (ATPS) and gels within GUVs. The phase behavior of polymer and gel systems is discussed, as well as the motivation for containing these systems within GUVs. Approaches to encapsulation are presented along with method-specific guidance. In order to highlight the range of information that can be gathered from studying these compartmentalized GUV systems, common types of experiments are reviewed.

29.2 AQUEOUS TWO-PHASE SYSTEMS

Aqueous polymer solutions above certain concentrations can undergo phase separation to form two or more chemically distinct phases that are in equilibrium with one another. Bulk ATPS have been used for partitioning studies and the separation of biomolecules for many years (Albertsson, 1986). For mixtures of neutral polymers such as polyethylene glycol (PEG) and dextran, phase separation results in an upper, PEG-rich phase and a lower, dextran-rich phase. Although the PEG/dextran system is the most frequently used ATPS of this general class, aqueous-aqueous phase coexistence can be observed with a wide variety of synthetic and biopolymers across a range of molecular weights (Albertsson, 1986; Wohlfarth, 2013; Aumiller and Keating, 2017). Bulk and microscopic images of a PEG/dextran ATPS are shown in Figure 29.1. The simplicity of this

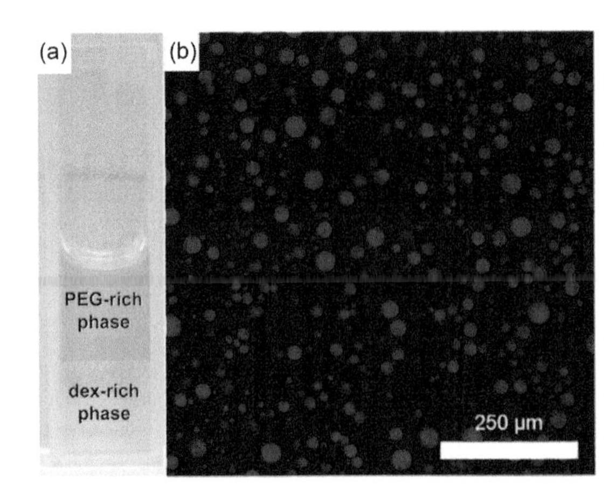

Figure 29.1 A PEG/dextran ATPS (a) in a cuvette and (b) viewed using fluorescence confocal microscopy after solution agitation. The upper PEG-rich phase is labeled with Alexa 647-PEG 5kDa and is the dispersed droplet phase in the fluorescence image. (Reprinted from *New Models of the Cell Nucleus: Crowding, Entropic Forces, Phase Separation, and Fractals*, Aumiller, W.M. et al., Phase separation as a possible means of nuclear compartmentalization, In, Hancock, R. Jeon, K. (Eds.), Academic Press, San Diego, CA, 109–149, Copyright 2014, with permission from Elsevier.)

polymer system, in which both aqueous phases are crowded with polymers, as a means of colocalizing biomolecules and biochemical reactions is attractive for mimicking microcompartmentalization and macromolecular crowding within cells.

Although the phases have different chemical properties (e.g., density, viscosity, polymer composition), all phases have high water content, making them suitable for compartmentalizing biomolecules and biochemical processes (Albertsson, 1986). In addition, these systems have low and tunable interfacial tensions. For example, the interfacial tension of a PEG 8 kDa/dextran 500 kDa system can range from 0.21 μN/m for a system with 8.28 wt% total polymer to 769 μN/m for a system with 36 wt% total polymer; for comparison, the air/water interfacial tension is 72.86 mM/m (Pallas and Harrison, 1990; Liu et al., 2012). The selective distribution of soluble molecules within an ATPS can be represented by the partition coefficient, K, where C_{top} and C_{bot} are the concentrations of solute in the top and bottom phases respectively (Eq. 29.1).

$$K = \frac{C_{top}}{C_{bot}} \tag{29.1}$$

In the PEG/dextran system, native, globular proteins generally partition into the dextran-rich phase ($K < 1$), whereas denatured proteins favor the PEG-rich phase ($K > 1$) and particulates such as chloroplasts accumulate at the interface (Albertsson, 1986; Di Nucci et al., 2001; Dominak et al., 2010). Affinity labels can be included to provide control over the distribution of a specific molecule of interest (Walter and Kopperschlager, 1994). Hence incorporation of this ATPS can be used to control the localization of biomolecules within the GUV interior.

29.2.1 MOTIVATION FOR ATPS ENCAPSULATION

Development of cellular mimics by a bottom-up approach where one begins with a lipid bilayer membrane and adds additional complexity in a stepwise manner, is attractive because it allows study of one or more particular aspects of living biological cells without their full complexity. The cellular environment is crowded with macromolecules and spatially segregated, with both membranous and non-membranous compartments sequestering biological molecules and activities. The resulting microenvironments are thought to play important roles in cell biology and to help explain many differences between typical *in vitro* biochemical studies conducted in dilute buffer solutions (<1 mg/mL of nucleic acids, proteins and small molecules) (Minton, 2001) and observations in living cells. The cellular environment can contain up to 300–400 mg/mL of macromolecules, resulting in a crowded solution that behaves non-ideally and can have a significant effect on biological activity (Sarkar et al., 2013). Recently, a number of membraneless intracellular bodies have been shown to exist as liquid microcompartments. For example, nucleoli from *Xenopus laevis* oocytes and cytoplasmic P granules from *Caenorhabditis elegans* both exhibit liquid-like behavior, such as dripping, droplet fusion and spherical shape (Brangwynne et al., 2009, 2011; Brangwynne, 2011). Observations such as these, and *in vitro* studies showing that major components of these "liquid organelles" are capable of phase separation, suggest that many cytoplasmic and nucleoplasmic membraneless organelles may be the result of liquid–liquid phase coexistence (Hyman et al., 2014). The importance of more closely mimicking the intracellular milieu has been increasingly realized (Ellis, 2001; Zhou et al., 2008; Luby-Phelps, 2013; Minton, 2013). Encapsulation of an ATPS within GUVs is an appealing means of capturing key biophysical features of this crowded, compartmentalized environment. Macromolecular crowding is supplied by the relatively high polymer concentrations, and the phases serve as models for membraneless organelles.

29.2.2 ATPS PHASE BEHAVIOR

It is important to understand the bulk phase behavior of your polymer system prior to attempting to encapsulate it within a GUV. A phase diagram illustrates the range of polymer concentrations that will result in the formation of either a single phase or two phases (Figure 29.2). When plotting the phase diagram, the total amount of the denser polymer (and the majority component of the bottom phase) is conventionally plotted on the x-axis and the less dense polymer (and majority component of the top phase) on the y-axis. In the PEG/dextran ATPS, the dextran-rich phase is the denser phase. Each point on the phase diagram represents the total polymer composition

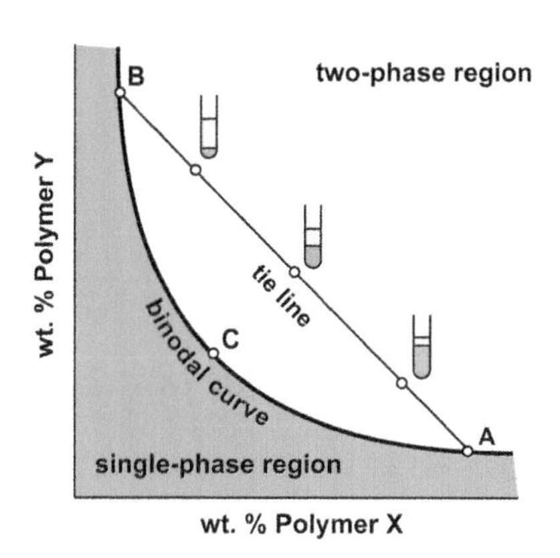

Figure 29.2 Schematic phase diagram for an ATPS composed of neutral polymers such as PEG and dextran in water. All points in the gray area below the binodal curve represent compositions that exist as a single phase, whereas all points in the white area above the binodal curve represent compositions that exist as two phases. Each point on the phase diagram represents the total polymer concentration of a solution. Point A represents the composition of the dextran-rich phase for any point along the given tie line, and point B represents the composition of the PEG-rich phase for any point along the tie line. Point C represents the critical point.

in the system. The binodal curve divides the phase diagram into two regions; all points on or below the curve represent compositions that exist as a single phase, whereas points above the curve represent compositions that exist as two phases. The binodal curve of a phase diagram can be experimentally determined by cloud point titration, as an ATPS will appear turbid when mixed (Box 29.1) (Albertsson and Tjerneld, 1994; Hatti-Kaul, 2000). It is important to note that binodal curves determined by cloud point titration for an ATPS containing polymers with high polydispersity (e.g., high molecular weight dextran) may not be as precise due to fractionation of different molecular weights of the polymer between the two phases (Zhao et al., 2016).

For compositions that lie in the two-phase region of the phase diagram, tie lines give the equilibrium polymer concentrations of each phase. Along a given tie line only the volume ratio of the two phases varies. The composition of each phase is represented by the point at which the tie line intersects the binodal curve (in Figure 29.2, point A represents the composition of the bottom phase at any point on the tie line and point B represents the composition of the top phase at any point on the tie line). Hence, longer tie lines suggest systems in which the two phases are more different in composition and consequently may show greater partitioning capabilities (Albertsson and Tjerneld, 1994; Hatti-Kaul, 2000).

Methods for the determination of equilibrium polymer composition of each phase vary based on the polymer system utilized. When only one of the polymers is optically active, as dextran is in the PEG/dextran ATPS, the concentration of that polymer can be determined using polarimetry. A standard curve is generated using known concentrations of dextran and their measured specific rotation, and can then be used to calculate the concentration of dextran in each of the phases based on specific rotation of the given phase. The concentration of the other polymer, PEG in this case, can then be determined from refractometry data. Two standard curves are generated, one with known dextran concentrations and one with known PEG concentrations. Due to the additive nature of refractive indices, and because the concentration of dextran in each phase has already been determined, the contribution to the total refractive index from dextran (RI_{dex}) can be subtracted (Eq. 29.2). The PEG contribution to the refractive index (RI_{PEG}) can then be used to calculate the concentration of PEG in each of the phases (Albertsson and Tjerneld, 1994).

$$RI_{tot} = RI_{PEG} + RI_{dex} \tag{29.2}$$

An alternative method for tie line determination is to use the intersection of the binodal curve and isopycnic (constant density) lines to deduce the compositions of the coexisting phases, which serve as the end points of the tie line (Liu et al. 2012). Eqs. 29.3a and 29.3b represent isopycnic lines for constant densities of each phase, which can be experimentally measured. w_p^D and w_p^P represent the weight fractions of PEG in the dextran-rich and PEG-rich phases, and w_d^D and w_d^P represent the weight fractions of dextran in the dextran-rich and PEG-rich phases, respectively. ρ^D and ρ^P represent the measured densities of the dextran-rich and PEG-rich phases, and v_s, v_p, and v_d represent the specific volumes of water, PEG and dextran, respectively (Liu et al. 2012).

$$w_p{}^D = \frac{1}{v_p - v_s}\left[\frac{1}{\rho^D} - v_s - (v_d - v_s)w_d{}^D\right] \tag{29.3a}$$

$$w_p{}^P = \frac{1}{v_p - v_s}\left[\frac{1}{\rho^P} - v_s - (v_d - v_s)w_d{}^P\right] \tag{29.3b}$$

The critical point represents the total polymer composition at which both the volumes and compositions of the individual phases are theoretically equal. The critical point can be determined by extrapolating the middle points of multiple tie lines to the binodal curve. A protocol for the construction of a phase diagram is highlighted in Box 29.1 (Albertsson and Tjerneld, 1994; Hatti-Kaul, 2000).

Preparation of polymer solutions is straightforward, and can be carried out by dissolving the desired mass of dry polymer in deionized water or a buffer solution of interest. All concentrations should be kept as weight polymer/total weight percent (w/w %), as it cannot be assumed that molar volumes in these solutions are additive. In addition, polymer solutions and the resulting phases formed from combining them often have a high viscosity, making it difficult to accurately measure their volumes. Solutions can be stored for 2–3 days at or below 4°C, after which time new solutions should be prepared. Alternatively, a small amount of sodium azide (~0.03 wt%) can be added to prevent bacterial growth for up to 1 month and allow longer-term storage (Brooks and Norris-Jones, 1994).

Box 29.1 Phase diagram protocol

Cloud point titration for binodal curve determination (illustrated in figure below)

1. Begin with concentrated stock solutions of each polymer (~25–30 wt%).
2. Add a few grams of concentrated dextran stock solution to a pre-weighed glass vial with a stir bar. Record the mass of dextran solution added.
3. Add PEG stock solution (**slowly**, dropwise) to the vial until the stirring solution appears persistently turbid.
4. Record the mass of the solution at this point (i.e., cloud point).
5. Add distilled water dropwise to the solution, until it returns to clear. Then add a few more drops of water.
6. Record the mass of the solution at this point.
7. Repeat Steps 3–6 until between 10–15 cloud points have been achieved.
8. Use the concentration of the stock solutions and the masses of solution present/added at each cloud point to calculate the weight percent of each polymer at each cloud point, and plot the resulting points with the wt% dextran as the x-axis and the wt% PEG as the *y*-axis.
9. Check the accuracy of your results by preparing samples at several compositions and allowing them to equilibrate for many hours in order to verify that compositions predicted to be one or two phases actually are. If necessary you can repeat the cloud point titration or produce a phase diagram by simply preparing many compositions and observing them after equilibration.

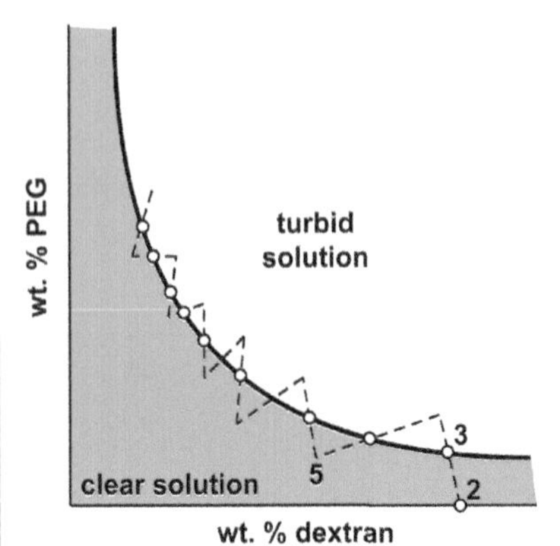

Schematic phase diagram illustrating the experimental determination of the binodal curve for a PEG/dextran ATPS. Numbers correspond to the step described above.

Tie line determination

Step 1. Choose a composition that will result in a two-phase system (by choosing a point that lies above the binodal curve).

Step 2. Make 10 g solution (to ensure enough of each phase for analysis) with this composition by using the *x*-coordinate of the point to determine the total wt% of dextran and the y-coordinate to determine the total wt% of PEG.

Step 3. Allow the system to phase separate. This can be done by centrifugation or allowing the solution to sit overnight.

Step 4. Separate the phases and analyze them to determine the composition of each phase (analysis techniques discussed in detail in the text).

Step 5. Plot the composition of each phase as points on the phase diagram; each of the two points should lie on the binodal curve. In the figure below, point A represents the composition of the dextran-rich phase and point B represents the composition of the PEG-rich phase.

Step 6. Draw a straight line connecting the three points (those representing the compositions of each phase, and the point representing the total composition of the solution made in Step 2).

29.2.3 ATPS ENCAPSULATION WITHIN GIANT LIPID VESICLES (GUVs)

Encapsulation of an ATPS is achieved by production of GUVs using a single-phase aqueous polymer solution as the hydrating medium, followed by subsequent induction of phase separation through a change in temperature or osmotic pressure (Helfrich et al., 2002; Li et al., 2008). ATPS have been encapsulated using either spontaneous swelling or electroformation (described and compared in Chapter 1) to produce the GUVs. These procedures are illustrated in Figure 29.3 and outlined in Box 29.2. Although a wide range of water-soluble polymers can be combined to form aqueous phase systems, the most commonly used polymers for encapsulation in GUVs are PEG 8 kDa with either dextran 10 kDa or dextran 500 kDa.

When using a temperature change to drive phase separation, two binodal curves should be determined; one at the temperature where encapsulation will occur and one at the temperature where experiments will occur. Compositions represented by points between these two binodal curves will exist as a single phase during encapsulation and phase separate upon temperature change. Heating is often used to aid spontaneous swelling, therefore it can be practical to choose a composition that exists as a single phase at a higher temperature and as two phases at room temperature (or lower temperatures) for ATPS encapsulation via spontaneous swelling (Helfrich et al., 2002; Long et al., 2005; Cans et al., 2008; Long et al., 2008; Dominak et al., 2010; Andes-Koback and Keating, 2011; Keating, 2012). Because the shift in binodal is not large between encapsulation at, for example, 37°C and examination at 4°C, care must be taken in generating the phase diagram. It is recommended that a new phase diagram be generated for each new batch of polymer to account for changes in molecular weight distribution, particularly for high molecular weight dextrans.

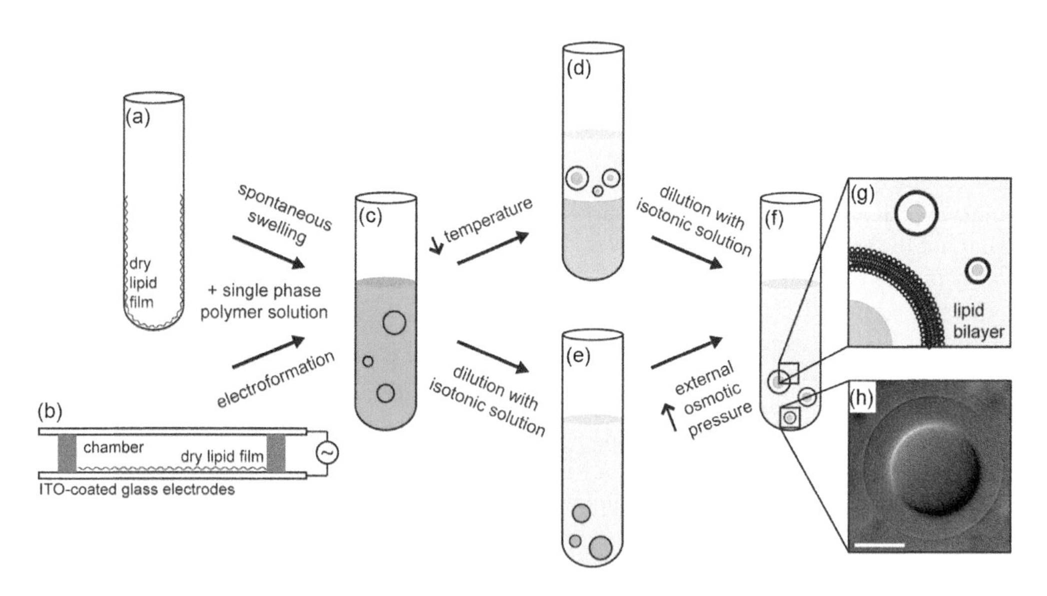

Figure 29.3 Scheme illustrating methods for generating GUVs containing an ATPS. (a) Shows a test tube containing a dried lipid film, whereas (b) shows an indium tin oxide (ITO)–coated glass electrode chamber with a lipid film dried on the bottom electrode, which are the experimental setups for spontaneous swelling and electroformation, respectively. (c) Illustrates a population of GUVs filled with a polymer solution existing as a single phase. The use of temperature change to induce phase separation and subsequent dilution of the external polymer solution is shown in (c → d → f), whereas (c → e → f) illustrates external polymer solution dilution followed by osmotic GUV deflation to induce phase separation. (g) Is a cartoon illustration of GUVs containing ATPS. (h) Is a DIC microscope image of a PEG/dextran ATPS encapsulated within a GUV. The PEG-rich phase is in contact with the lipid bilayer membrane, whereas the dextran-rich phase is completely enclosed within the PEG-rich phase. Scale bar: 10 μm. (Reprinted with permission from Long, M.S. et al., 2005. Dynamic microcompartmentation in synthetic cells. *Proc. Natl. Acad. Sci. USA*, 102, 5920–5925. Copyright 2005, National Academy of Sciences, U.S.A.)

Box 29.2 ATPS encapsulation protocols

Method 1: Temperature-driven phase separation

1. Choose polymer composition that will exist as a single phase at the encapsulation temperature and two phases at the experimental temperature. For example, we have used a composition of 7 wt% PEG 8 kDa/10 wt% dextran 10 kDa, which exists as a single phase above 37°C, but as two phases below 5°C (Andes-Koback and Keating, 2011).
2. Follow spontaneous swelling or electroformation protocol in Chapter 1, using the polymer solution as the hydrating medium.
3. After vesicle formation, allow solution to stand at the experimental temperature to drive phase separation.
4. Collect GUVs from the ATPS interface and transfer to a bulk solution of the top phase of the ATPS.

TIP: To increase yield of ATPS within the GUV and enhance partitioning, the interior polymer concentrations can be increased slightly after preparation. Add a small amount of sucrose (or other osmolyte) to the external solution to remove water by osmosis. Determine osmolality of your ATPS using vapor phase osmometry and then increase the external osmolality above this in a stepwise fashion. Note that large increases in external osmolality will lead to budding.

Method 2: Osmotic stress-driven phase separation

1. Choose polymer composition represented by a point below the binodal curve, so that it will exist as a single phase during encapsulation, but can be driven to phase separate upon osmotic deflation of vesicles. For example, the Dimova research group typically uses a composition of 4.05 wt% PEG 8 kDa/2.22 wt% dextran 500 kDa (Li et al., 2008, 2011, 2012).
2. Follow spontaneous swelling or electroformation protocol in Chapter 1, using the polymer solution as the hydrating medium.
3. After vesicle formation, dilute the external medium with an isotonic polymer solution (single phase) with a lower density than that of the encapsulated solution to ensure that vesicles sediment to the bottom of the chamber.
4. To drive phase separation, slowly add a hypertonic polymer solution (contains sucrose or other osmolyte) to osmotically deflate vesicles. This can be done in multiple steps to avoid rapid changes to the osmolality.

Osmotic deflation can also be used to drive phase separation (Li et al., 2008, 2011; Dimova and Lipowsky, 2012; Li et al., 2012). By increasing the osmotic pressure of the external solution (through the addition of sucrose or another osmolyte), water is drawn out of the vesicles in order to equilibrate the internal and external osmotic pressures. Because the lipid bilayer membrane is permeable to water but not the polymer components, this concentrates the polymers in the GUV interior and drives phase separation. When using osmotic deflation to drive phase separation, a solution composition represented by a point below the binodal curve should be chosen, so the solution exists as a single phase during encapsulation, and phase separates as the polymer concentration inside the vesicles increases in response to osmotic deflation. The external osmotic pressure should be changed gradually to avoid osmotic stress-induced vesicle rupture. This method is more versatile because it does not require a temperature-dependent phase transition in a temperature range acceptable for biomolecule inclusion, which will not be feasible for all polymer phase systems.

It is important to note that, for both methods described above, although all vesicles are formed in the presence of the same single-phase aqueous polymer solution, there will be vesicle-to-vesicle variations in the encapsulated concentration of each polymer across the population. Therefore, some vesicles may not contain an ATPS upon bulk phase separation in the exterior solution, and vesicles that do have ATPS will show variability in phase volume ratios and compositions (Long et al., 2005). Consequently, encapsulated ATPS may exhibit some differences in phase behavior when compared to bulk phase ATPS, for example, increased (or decreased) phase transition temperatures. This is due to the difficulty of passively (i.e., without using any attractive interactions to encourage encapsulation) encapsulating macromolecules within small volumes. Dextrans of higher molecular weights have been shown to exhibit lower encapsulation efficiencies than those of lower molecular weights, leading to lower concentrations encapsulated within GUVs (Dominak and Keating, 2007). Partitioning may also vary between encapsulated and bulk ATPS due to variations in the volume ratios and compositions of the encapsulated phases (Long et al., 2005).

A consideration when encapsulating polymers within GUVs is that macromolecular crowding may inhibit lipid lamella swelling and decrease the yield of giant unilamellar vesicles. Including lipids with electrostatically or sterically repulsive headgroups can increase GUV yield by aiding in the separation of lamella during hydration (Lasic, 1988; Sriwongsitanont and Ueno, 2004; Pozo Navas et al., 2005). Lipid composition can also be used to control the position of the polymer phases within a GUV. For example, the presence of PEGylated lipid in the GUV membrane can result in the PEG-rich phase contacting the membrane due to favorable interactions between the PEGylated lipids and the PEG-rich phase (Long et al., 2008).

GUVs containing ATPS can be observed using phase-contrast or differential interference contrast (DIC) optical microscopy. The difference in refractive indices between the phases results in one phase appearing thicker than the other (in a PEG/dextran ATPS, dextran appears as the thicker phase). However, due to sample complexity when unencapsulated phase droplets and/or multilamellar vesicles are also present, fluorescent labeling should be used to identify the membrane and each aqueous phase. Labeling at least one of the component polymers and the lipid membrane is recommended to alleviate confusion.

29.3 GELS

A wide variety of polymers in solution can also be cross-linked to form gels, or polymeric networks composed largely of solvent. Cross-linking can occur simultaneously with polymerization of monomer units or with existing polymer chains, and can be reversible in some cases. A wide variety of factors can be used to control cross-linking, including temperature changes, presence of divalent cations, or exposure to UV light (Osada and Kajiwara, 2001). Hydrogels, or gels with high water content, have been used in the development of drug delivery systems due to their biocompatibility (Hoare and Kohane, 2008; Vashist et al., 2014). When encapsulated in GUV, these materials can serve as microcompartments for a simple model for the cytoskeletal network.

29.3.1 MOTIVATION FOR GEL ENCAPSULATION

As mentioned previously, the bottom-up development of a more accurate cell mimic (see Chapter 28 for a comprehensive discussion of protocells) is important for better understanding the cellular environment, and GUVs encapsulating a dilute buffer solution lack important cellular features, such as cytoplasmic and cytoskeletal mimics. Incorporation of a gel within a GUV can provide shape control and stabilization (typically provided by the cytoskeleton—see Chapter 4) (Viallat et al., 2004; Campillo et al., 2007, 2008), as well as control over particle mobility as an alternate way to model the crowded intracellular milieu (Jesorka et al., 2005a, 2005b; Markstrom et al., 2007). Red blood cells have a unique discoid shape that can be modeled by deflation of vesicles (Seifert, 2000), but without shape stabilization these vesicles are difficult to handle experimentally as they are sensitive to small experimental stresses (Abkarian et al., 2002). Encapsulation of a gel within a GUV prior to vesicle deflation may therefore lead to an increased understanding of the mechanics relating to red blood cell shape. Microcompartmentalization within the cell controls the location of specific biomolecules during different periods of activity and can also be mimicked by encapsulating gels within GUVs. A variety of polymer gels have been incorporated into GUVs in order to immobilize or control the diffusion of particles (Viallat et al., 2004; Jesorka et al., 2005b). Encapsulation of gels within GUVs provides another route toward a greater understanding of intracellular organization.

29.3.2 GEL PHASE BEHAVIOR

Gel formation is possible with a wide variety of polymers through a number of different cross-linking methods. Cross-linking can occur via the formation of covalent bonds or intermolecular interactions, such as hydrogen bonding or van der Waals interactions, with greater mechanical strength and chemical stability in gels formed by covalent bond cross-linking. A wide variety of stimuli can induce cross-linking, ranging from simple mixing of polymer components to exposure to plasma or an electric field (Osada and Kajiwara, 2001). The encapsulated gels discussed in this chapter undergo cross-linking as a result of temperature change, exposure to divalent cations, or UV radiation.

The sol-gel transition, or the point at which a solution changes from liquid to gel, varies based on the gel system utilized and the polymer (or monomer) concentration in the pre-gel solution. For temperature-dependent gels, the upper (UCST) and lower (LCST) critical solution temperatures are important parameters for gelation as they represent the temperatures below and above which gelation occur, respectively. These temperature limits depend on the polydispersity and degree of polymerization of the given polymer system (Klouda and Mikos, 2008). Because gel phase behavior can greatly differ based on the identity of the gel, the behavior of the specific gels discussed in this chapter will be reviewed as they are introduced.

29.3.3 GEL ENCAPSULATION WITHIN GUVs

The encapsulation of gels within GUVs allows researchers to combine the phase behavior of gels with the small volumes and semipermeable lipid bilayer membrane of GUVs to produce cellular mimics. A variety of polymers can be encapsulated within vesicles, and there are a number of different methods for generating these cell mimics (illustrated in Figure 29.4, outlined in Box 29.3, and compiled in Table 29.1). Pre-gel polymer solutions can be encapsulated in GUVs during vesicle formation via spontaneous swelling or electroformation (Viallat et al., 2004; Campillo et al., 2007, 2008), or in some cases microinjected into existing vesicles (Jesorka et al., 2005a, 2005b; Markstrom et al., 2007; Węgrzyn et al., 2011). Microinjection requires additional equipment and must be done one vesicle at a time, but can allow more control over the final vesicle contents than the passive encapsulation occurring with spontaneous swelling or electroformation. Gelation can be induced by a change in temperature (Viallat et al., 2004; Jesorka et al., 2005b; Markstrom et al., 2007; Campillo et al., 2008), addition of divalent cation (Jesorka et al., 2005a) or exposure to UV light (Campillo et al., 2007), depending on the specific polymer being utilized.

Polymer concentrations necessary for gelation depend on polymer identity. When using a temperature change to drive gelation, a polymer concentration should be chosen that exists as a liquid at the encapsulation temperature and a gel at the experimental temperature. Depending on the polymer, gelation can occur in response to either an increase or decrease in temperature. Poly(*N*-isopropyl acrylamide) (PNIPAAm) forms a gel when the temperature increases above its LCST because the polymer undergoes a coil-to-globule transition and increases hydrophobicity (Markstrom et al., 2007), whereas agarose undergoes gelation upon a decrease in temperature due to increased hydrogen bonding (Tako and Nakamura, 1988). Gelation triggered by a temperature change is often reversible, which is a useful feature for a cellular mimic because the cellular environment undergoes structural changes during different periods of biological activity (Mao et al., 2011; Dundr, 2012).

Using divalent cation addition to trigger gelation is more challenging, because cations cannot cross the lipid bilayer membrane. Therefore, the cations have been microinjected into the vesicles, inserted via electroformation, or encapsulated within another vesicle and introduced to the original vesicle via fusion (Jesorka et al., 2005a). These methods add additional experimental steps and often more specialized equipment. Cation-driven gelation is typically irreversible, limiting the types of experiments that can be conducted using these systems. Polymer and cation concentrations required for gelation should be experimentally determined as they change based on the system being utilized. In a poly(ethylene dioxythiophene) (PEDOT)/poly(styrene sulfonate) (PSS) system, Ca^{+2} ions induce cross-linking of the PSS by binding to excess negative charges on the polyanionic PSS in order to form a gel (Jesorka et al., 2005a).

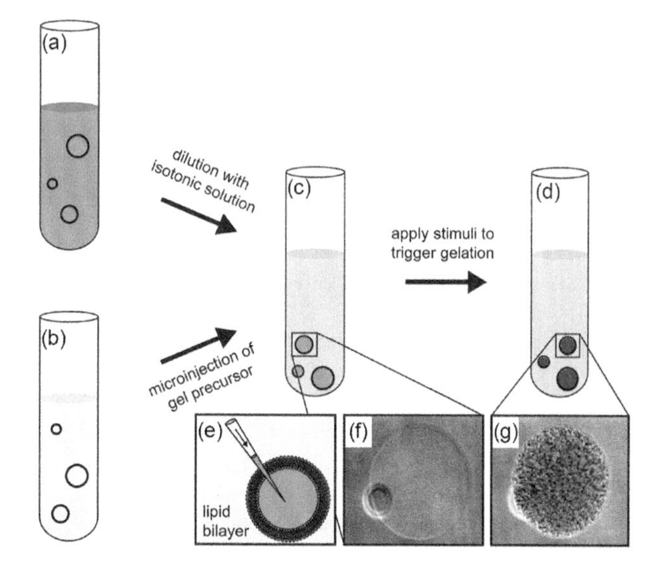

Figure 29.4 Scheme illustrating methods for encapsulating gels within GUVs. (a) GUVs with gel precursor encapsulated during vesicle formation through either spontaneous swelling or electroformation. (b) GUVs without gel precursor encapsulated, the hydrating media for spontaneous swelling or electroformation is water or a dilute buffer solution. (c) GUVs containing a gel precursor in a dilute external solution achieved by either dilution of the external solution after GUVs containing gel precursors were formed or microinjection of the gel precursor into already formed GUVs. An illustration of the microinjection process is shown in (e). (d) GUVs containing a gel; gelation has been triggered by exposure to a stimulus such as a temperature change, UV light, or divalent cations. (f, g) Microscope images of GUVs containing PNIPAAm gel precursor solution and a gel, respectively, with gelation being triggered by a temperature increase. (Panels f and g: Reprinted with permission from Jesorka, A. et al., *Langmuir* 21:1230–1237, 2005a. Copyright 2005a American Chemical Society.)

Table 29.1 **Commonly used polymers for encapsulating gels**

POLYMER NAME (ABBREVIATION)	MONOMER UNIT	ENCAPSULATION METHOD	GELATION TRIGGER	REFERENCES
Poly(*N*-isopropyl acrylamine) (PNIPAAm)	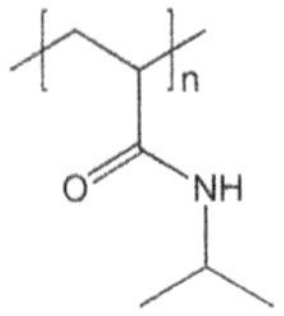	Electroformation Microinjection into gentle hydration GUVs	UV radiation Temperature	Campillo et al. (2007, 2008, 2009); Jesorka et al. (2005a), Markstrom et al. (2007)
PNIPAAm-vinylferrocene	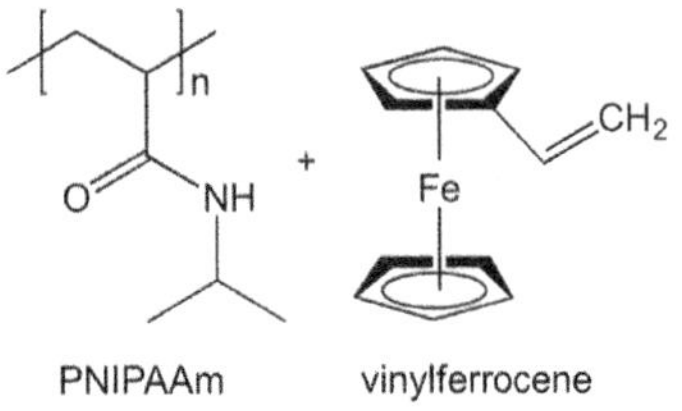 PNIPAAm + vinylferrocene	Microinjection into gentle hydration GUVs	Temperature	Węgrzyn et al. (2011)
Agarose	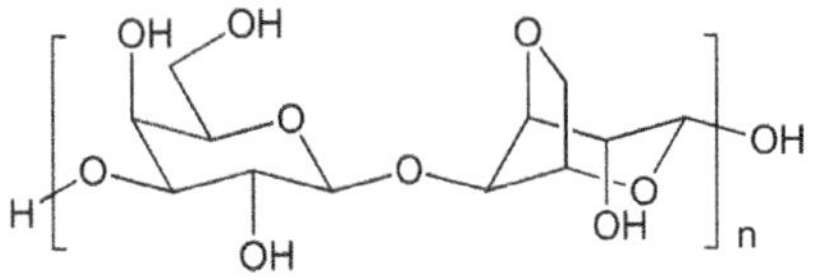	Electroformation	Temperature	Viallat et al. (2004)
Poly(ethylene dioxythiophene)/poly(styrene sulfonate) (PEDOT/PSS)	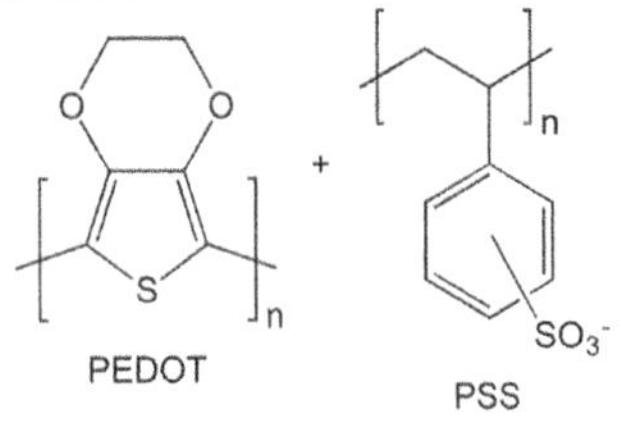 PEDOT + PSS	Microinjection into gentle hydration GUVs	Divalent cation exposure	Jesorka et al. (2005b)

Box 29.3 Gel encapsulation protocols

Encapsulation methods

Encapsulation Method 1: Encapsulation during vesicle formation

1. Choose polymer composition based on method for driving gelation.
2. Follow spontaneous swelling or electroformation protocol in Chapter 1, using the polymer solution as the hydrating medium.
3. Dilute the external solution to a concentration below that required for gelation.

Encapsulation Method 2: Microinjection (requires the use of specialized equipment)

1. Follow spontaneous swelling or electroformation protocol in Chapter 1, using the desired buffer solution as the hydrating medium.
2. Transfer a small amount of vesicle-containing solution to a glass slide for observation.
3. Beginning with a single GUV immobilized on the glass surface, use a tapered capillary micropipette and a microelectrode controlled by micromanipulators and a pulse generator to inject the liquid phase gel-forming solution.

Induction of gelation

Gelation Method 1: Temperature (Viallat et al., 2004; Jesorka et al., 2005b; Markstrom et al., 2007; Węgrzyn et al., 2011)

1. Choose polymer composition that will exist as a single phase at the encapsulation temperature and two phases at the experimental temperature.
2. Follow spontaneous swelling or electroformation protocol in Chapter 1, either using the polymer solution as the hydrating medium or using a buffer as the hydrating solution and microinjecting the polymer solution into GUVs.
3. Increase/decrease the solution temperature in the direction that induces gel formation.

Gelation Method 2: Divalent cation (Jesorka et al., 2005a)

1. Choose a polymer composition that will undergo cross-linking to form a gel when introduced to divalent cations.
2. Follow spontaneous swelling or electroformation protocol in Chapter 1, either using the polymer solution as the hydrating medium or using a buffer as the hydrating solution and microinjecting the polymer solution into GUVs.
3. Introduce divalent cation via microinjection, vesicle fusion (controlled with micromanipulators), or electroporation (see Chapter 15 on electro-effects).

Gelation Method 3: UV radiation (Campillo et al., 2007; Campillo et al., 2008, 2009)

1. Choose a polymer composition that will undergo cross-linking to form a gel when exposed to UV radiation.
2. Follow spontaneous swelling or electroformation protocol in Chapter 1, either using the polymer solution as the hydrating medium or using a buffer as the hydrating solution and microinjecting the polymer solution into GUVs.
3. Induce gel formation by irradiating GUVs with UV light.

UV light-triggered gelation is also an irreversible process, though it does not require addition of ions or molecules across the bilayer membrane. However, due to the exothermic nature of some UV-induced radical polymerization reactions, samples may need to be cooled during photopolymerization. In the presence of a cross-linker and a photoinitiator, NIPAAm monomers can polymerize to form PNIPAAm gel when exposed to UV light, and when encapsulated within a GUV, hydrophobic areas of the polymer can interact with the lipid bilayer membrane (see Chapter 25 on polymer-vesicle interactions), allowing researchers to mimic the attachment of a cytoskeleton to the cell membrane (Campillo et al., 2007). In addition, because PNIPAAm gels are reversibly temperature responsive, vesicle and gel size can be tuned by changing the solution temperature.

As with ATPS encapsulated using spontaneous swelling or electroformation, vesicles will likely contain varying amounts of gel precursor, leading to GUVs with slightly varying gel content and phase behavior. Although microinjection provides more control over the interior contents of the vesicle, it requires special instrumentation and must be done one vesicle at a time.

29.4 COMMON TYPES OF EXPERIMENTS WITH COMPARTMENTALIZED GUVs

The most common types of experiments investigate solute compartmentalization and morphological changes such as vesicle budding and lipid tubule formation (Dimova and Lipowsky, 2012; Keating, 2012). Complete budding, in which a vesicle splits at the ATPS phase boundary into asymmetric daughter vesicles, has also been observed (Andes-Koback and Keating, 2011). The experiments discussed in the following section are not an exhaustive list, but instead meant to provide suggestions for the range of phenomena that can be studied with encapsulated systems.

29.4.1 SOLUTE COMPARTMENTALIZATION

Living cells can sequester biomolecules in varying microcompartments throughout the cell during different periods of activity. As such, solute compartmentalization is an important goal in modeling cellular behavior. Partitioning of a number of different biomolecules within an encapsulated PEG/dextran ATPS has been studied, including soybean agglutinin (Long et al., 2008), phytohemagglutinin, concanavalin A, streptavidin (Long et al., 2005), and human serum albumin (Dominak et al., 2010). Proteins in their native conformations were observed to partition more strongly to the more hydrophilic dextran-rich phase, although partitioning was not as strong in encapsulated microvolumes as in bulk solutions, and varied from vesicle-to-vesicle due to differences in the composition of the encapsulated ATPS introduced during vesicle preparation. Incorporation of affinity labels provides a means of changing the partitioning, for example, streptavidin will accumulate in the PEG-rich phase when biotinylated PEG is present (Long et al., 2005). Partitioning could be tuned in response to variations in the external osmotic pressure. An increase in the external osmotic pressure concentrated the ATPS further, leading to improved partitioning (Long et al., 2005). Conformational changes induced by changes in pH led to re-partitioning from the PEG-rich phase (for denatured protein at low or high pH) to the dextran-rich phase (for native protein near neutral pH (Figure 29.5a) (Dominak et al., 2010).

GUVs encapsulating gels can also be used to generate microenvironments within GUVs. Gels have been shown to immobilize or restrict diffusion of small particles within the GUV interior. Carboxylate-modified fluorescent latex beads were reversibly immobilized within PNIPAAm gel compartments inside GUVs as a function of solution temperature (Jesorka et al., 2005b; Markstrom et al., 2007). When fluorescently labeled dextran was injected into these GUVs, the dextran remained in the aqueous phase surrounding the gel (Jesorka et al., 2005b). In the cation-induced PEDOT/PSS polymer gel system, gelation within a vesicle visibly reduced Brownian motion of latex beads (Jesorka et al., 2005a). The regions of different size-dependent diffusion can be used to model similar behavior in living cells.

29.4.2 MORPHOLOGICAL CHANGES

Morphological changes (highlighted in Box 29.4) can also be studied using encapsulated systems, as the interior microenvironments contact and interact with the lipid membrane. The wetting properties of an encapsulated PEG/dextran ATPS in relationship to the GUV membrane were studied and provided insight into a dewetting-to-wetting transition that the dextran-rich phase undergoes as an increase in external osmotic pressure deflates vesicles (Li et al., 2008). As water is removed from the vesicles, phase separation occurs in an initially homogeneous PEG/dextran interior solution, followed by changes in the wetting of the newly formed dextran-rich phase droplet, which is surrounded by the PEG-rich phase and also in contact with the membrane (Figure 29.5b). The wettability of the phases was determined by analyzing contact angle measurements, which were made using images captured from a horizontally aligned confocal microscope. The external solution density was tailored to ensure vesicles sedimented to the bottom of the observation chamber with a vertical orientation in which the dextran-rich phase was always at the bottom. In this way, more accurate measurements of the wetting transitions could be collected (Li et al., 2008).

Osmotic deflation of ATPS-containing vesicles ultimately leads to vesicle budding as the decrease in internal volume leads to excess membrane area that can allow this shape transformation (Figure 29.5c). As their concentration increases, the two aqueous phases become more chemically distinct and capable of improved partitioning. The interfacial tension between them also increases (Li et al., 2012; Liu et al., 2012). The vesicle shape change from spherical to budded decreases the contact area between the aqueous phases, reducing the interfacial energy (Dimova and Lipowsky, 2012; Li et al., 2012). In some cases, the excess membrane area resulting from vesicle deflation is stored in lipid nanotubes instead of adopting a budded shape (Li et al., 2011). These nanotubes are significant because eukaryotic cells can contain tubular membrane structures (Lee and Chen, 1988; De Matteis and Luini, 2008). Nanotubes formed by liquid-ordered and liquid-disordered membranes have been studied with varying vesicle morphologies and display different patterns based on vesicle morphology as a result of two competing kinetic pathways (Figure 29.5d) (Liu et al., 2016). Specific mechanical

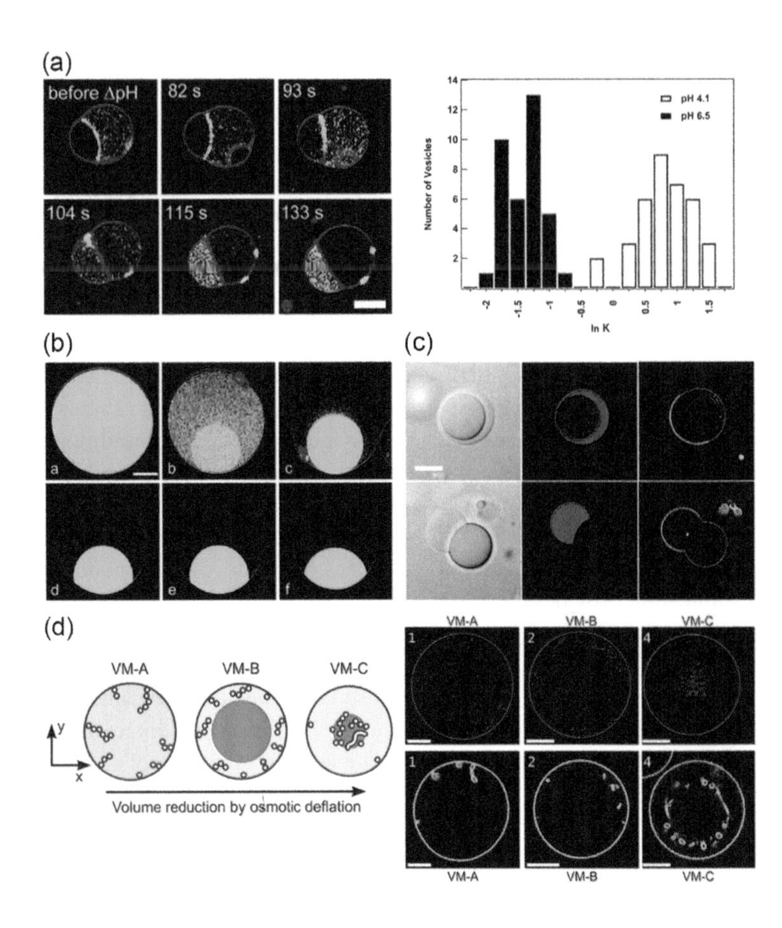

Figure 29.5 (a) Fluorescence microscope images showing the relocalization of a protein (human serum albumin) within a GUV from the PEG-rich phase to the dextran-rich phase as the pH changes from 4.1 to 6.5 (left) and histograms comparing the partitioning of the protein as a function of pH (right). A lipid component of the membrane was fluorescently labeled with rhodamine and false colored red, whereas human serum albumin was labeled with Alexa 588 and false colored green. Scale bar: 5 μm. (b) Fluorescence confocal images (vertical cross sections) of a GUV containing a polymer solution undergoing phase separation followed by a wetting transition as a result of an increase in external solution osmolarity. A membrane component was labeled with rhodamine (false colored red) and dextran was labeled with fluorescein isothiocyanate (false colored green). Scale bar: 20 μm. (c) Microscope images (DIC and fluorescence) of a GUV with two different lipid domains encapsulating a PEG/dextran ATPS before (top) and after (bottom) osmotic stress-induced budding. The PEG-rich phase was labeled with Alexa 647 (false colored blue), a lipid component of the liquid-disordered domain was labeled with rhodamine (false colored red), and a lipid component of the liquid-ordered domain was labeled with carboxyfluorescein (false colored green). Scale bar: 10 μm. (d) Illustration of three different nanotube patterns and the corresponding vesicle morphologies as the system is exposed to osmotic stress (left) and the fluorescence confocal microscope images showing those same nanotube patterns occurring experimentally (right). Top right panel shows vesicles composed of a liquid-disordered membrane (labeled with Texas Red and false colored red, see Appendix 2 of the book for fluorescently labeled lipids), bottom right panel shows vesicles composed of a liquid-ordered membrane (labeled with carboxyfluorescein and false colored green). Scale bar: 10 μm. (Panel a: Reprinted with permission from Dominak, L.M. et al., *Langmuir*, 26, 5697–5705, 2010. Copyright 2010 American Chemical Society.) (Panel b: Reprinted with permission from Li, Y.H. et al., *J. Am. Chem. Soc.*, 130, 12252–12253, 2008. Copyright 2008 American Chemical Society.) (Panel c: Reprinted with permission from Cans, A.S. et al., *J. Am. Chem. Soc.*, 130, 7400–7406, 2008. Copyright 2008 American Chemical Society.) (Panel d: Reprinted with permission from Liu et al., 2016 Copyright 2016 American Chemical Society.)

properties such as membrane tension and rigidity, wetting, and spontaneous curvature as well as the two-phase interfacial tension dictate which phenomenon will occur; spontaneous curvature favors nanotube formation, whereas partial wetting of both internal aqueous phases favors budding (Li et al., 2012).

Nanotube formation has also been observed in gel-containing GUVs when excess membrane area is present. GUVs containing the thermoresponsive polymer mix PNIPAAm/vinylferrocene underwent membrane nanotube formation and membrane protrusions as a result of gel shrinkage and the attachment between the encapsulated gel and the bilayer membrane (Figure 29.6) (Węgrzyn et al., 2011). Attachments between the encapsulated gel and the lipid membrane are reminiscent of membrane-cytoskeletal contacts in biological cells.

When budded vesicles contain multiple lipid membrane domains (liquid-ordered vs. liquid-disordered), nonspecific interactions between the aqueous phases and the membrane can result in localization of lipid domains to preferentially coat one of the interior aqueous phases. For example, in Figure 29.5c, lipids with PEGylated headgroups partitioned to the liquid-ordered domain of the membrane, causing that domain to localize around the PEG-rich aqueous bud. The resulting budded ATPS-containing GUVs are polarized both in their interior contents and their membrane, with different molecules localized at different ends of the vesicle (Cans et al., 2008).

GUVs encapsulating gels have also been used to investigate vesicle behavior under osmotic stress. Agarose gel-containing

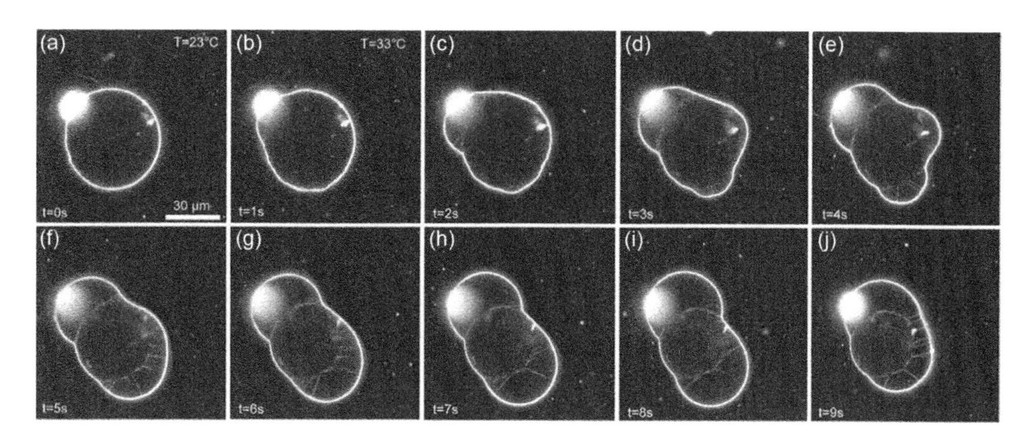

Figure 29.6 Membrane nanotube formation and membrane protrusions in a GUV containing a PNIPAAm/vinylferrocene gel as a result of temperature increase. (a) GUV at room temperature. (b-j) Formation of nanotubes and protrusions over time following an increase in temperature above the LCST. (Reprinted with permission from Węgrzyn et al., 2011. Copyright 2011 American Chemical Society.)

Box 29.4 Effects of osmotic deflation on morphology of ATPS-containing GUVs

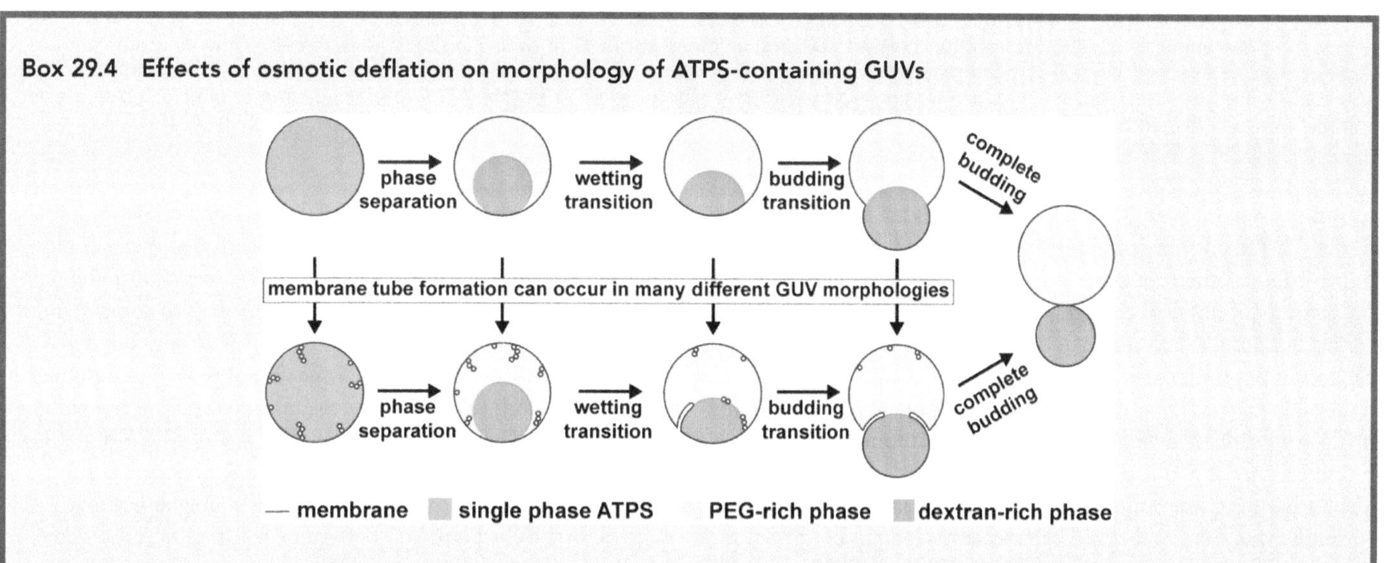

The phase behavior of an ATPS encapsulated within a GUV affects the morphology of the GUV itself, leading to a number of unique vesicle morphologies. As external stress is applied to an ATPS-containing vesicle, the single-phase solution undergoes phase separation, with one aqueous polymer solution completely wetting the membrane (the PEG-rich phase in the above scheme) and engulfing the second internal aqueous phase (the dextran-rich phase above). Under further external stress, the membrane becomes partially wet by each of the internal aqueous phases as the dextran-rich phase undergoes a dewetting-to-wetting transition (Li et al., 2008). External stress in the form of osmotic vesicle deflation leads to GUV budding as a result of excess membrane area, which decreases the contact area between encapsulated aqueous phases and reduces their interfacial tension (Dimova and Lipowsky, 2012; Li et al., 2012). Complete budding is possible through additional osmotic deflation, producing two daughter vesicles; one containing the PEG-rich phase and one containing the dextran-rich phase (Andes-Koback and Keating, 2011). Excess membrane area in any of the GUV morphologies can also result in membrane nanotube formation, in which thin tubules of membrane containing external solution form. When the encapsulated phase exists as a single phase, or when the membrane is wetted completely by one internal aqueous phase, the nanotubes exist throughout the entirety of the wetting phase. Once partial wetting of the membrane by both phases occurs, the nanotubes remain at the ATPS interface (Liu et al., 2016). The occurrence of vesicle budding, membrane nanotube formation, or a combination of both depend on properties such as the membrane tension and rigidity, wetting, spontaneous curvature, and the ATPS interfacial tension. GUV budding is favored by partial membrane wetting of both internal aqueous phases, whereas nanotube formation is favored by spontaneous curvature (Li et al., 2011, 2012).

vesicles were able to retain more water than vesicles filled with an aqueous solution when exposed to similar osmotic conditions, leading to a lower degree of vesicle shrinkage. In addition, agarose concentration affected vesicle shape after osmotic deflation due to a higher elastic modulus in gels with higher concentrations of agarose (Viallat et al., 2004).

29.4.3 MICROPIPETTE ASPIRATION

Micropipette aspiration can also be used to investigate morphological changes in GUVs. A full discussion on micropipette aspiration, including experimental protocols and a wider range of experiments that can be carried out using this

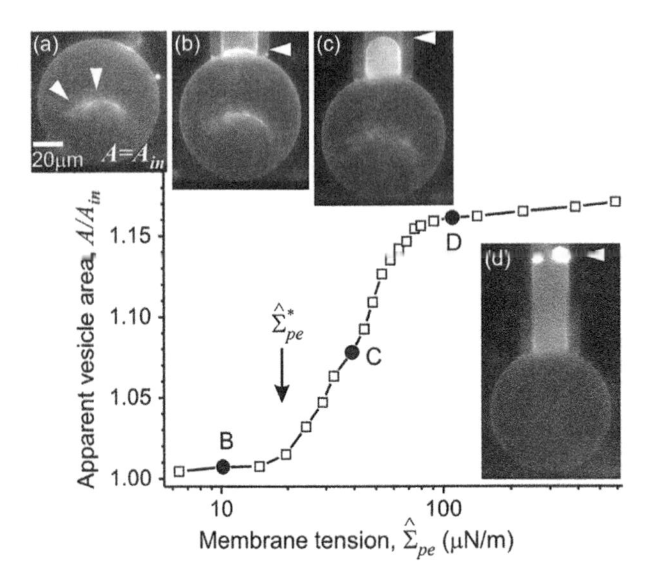

Figure 29.7 Micropipette aspiration is used to increase the membrane tension of a GUV containing a PEG/dextran ATPS and measure the apparent vesicle area as a function of membrane tension. (a) Before applying a suction pressure via micropipette a network of nanotubes is visible at the ATPS interface. (b–c) As the membrane tension is increased, the nanotube network at the ATPS interface decreases until (d) it disappears completely at a high membrane tension. (Reprinted with permission from Li, Y.H. et al., *Proc. Natl. Acad. Sci. USA*, 108, 4731–4736, 2011.)

technique, can be found in Chapter 11. Applying suction pressure to the GUV using a micropipette increases its membrane tension to provide a method for observing the effects of this tension on a variety of physical changes. Nanotube formation and vesicle budding have both been investigated using micropipette aspiration of pre-deflated vesicles containing ATPS. Membrane nanotubes formed by aqueous phase separation could be retracted using aspiration, shown in Figure 29.7, and in some cases were found to store 15% of the total membrane area (Li et al., 2011). Vesicle budding was controlled by varying the membrane tension through aspiration, and could be reversed as long as both phases of the ATPS partially wet the membrane (Li et al., 2012).

GUVs containing PNIPAAm cross-linked by UV irradiation were exposed to suction pressure by aspiration, revealing a strong coupling between the bilayer membrane and the polymer gel. Upon gel shrinkage in response to a temperature change, the membrane did not retain its spherical shape and instead remained attached to the shrunken gel. Further aspiration experiments revealed a tunable range for the elastic modulus of the gel between 0.5 and 25 kPa, which is within the observed range for biological cells. The coupling between bilayer membrane and internal vesicle components combined with the biologically relevant elastic modulus makes these systems an attractive model for the mechanical behavior of biological cells (Campillo et al., 2007, 2009).

29.4.4 GUV DIVISION

Asymmetric cell division is another important aspect of cellular biology that is critical for cell differentiation. Building on vesicle budding studies discussed above, complete vesicle budding has been achieved to produce two asymmetric daughter vesicles with distinct interior compositions. Beginning with a GUV encapsulating a PEG/dextran ATPS, osmotic stress-induced vesicle deflation first prompted vesicle budding then led to complete budding with one daughter vesicle containing the PEG-rich phase and the other containing the dextran-rich phase. A lipid membrane nanotube tether persisted between daughter vesicles initially following budding, which in some cases ruptured due to fluid flow-driven collisions with other liposomes to completely separate the daughter vesicles. This process was also conducted using GUVs with varying membrane domains, so that daughter vesicles with both distinct lipid compositions and interior contents were generated (Figure 29.8) (Andes-Koback and Keating, 2011). The division plane in ATPS-containing vesicles occurs at the aqueous/aqueous phase boundary regardless of whether that coincides with the boundary between coexisting liquid-ordered and liquid-disordered lipid membrane domains (Figure 29.8). This makes it possible to achieve daughter vesicles in which the membrane already contains a patch that breaks symmetry and can set up a new polarity axis upon a second aqueous phase separation event on the interior (Andes-Koback and Keating, 2011).

29.5 LOOKING AHEAD

Encapsulating ATPS and gels within GUVs is an attractive means of controlling intracellular organization, biomolecule colocalization and vesicle morphology. Because biological systems can contain many different aqueous compartments, the encapsulation of more than two aqueous polymer phases within GUVs is one potential future direction for this area of research. An aqueous three-phase system composed of PEG, dextran and Ficoll® has been contained within water-in-oil emulsion droplets to investigate cell-free protein expression (Torre et al., 2014). GUVs containing three or more aqueous phases could provide a better model for understanding intracellular organization. Another interesting direction for future research is encapsulation of ATPS made from biopolymers such as proteins and nucleic acids. This would allow formation of microcompartments more closely mimicking the membraneless organelles in cytoplasm and nucleoplasm of eukaryotic cells.

(a)

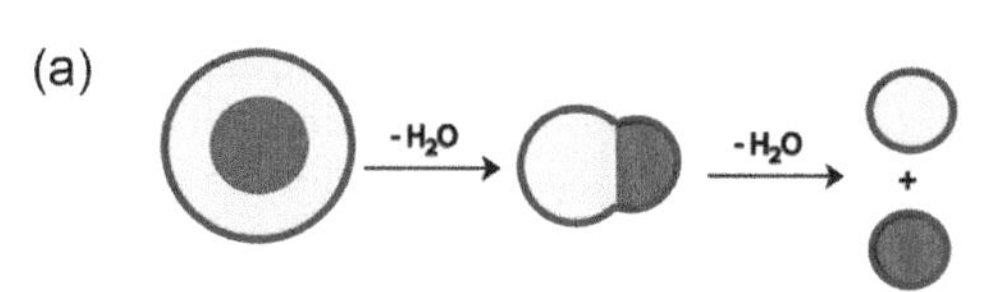

ATPS-containing GVs divide to produce daughter vesicles with different internal compositions

(b)

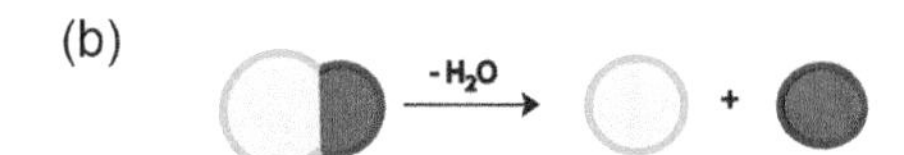

When membrane domains are present, they preferentially coat different aqueous phases; upon division, daughter vesicles differ in both interior and lipid compositions.

(c)

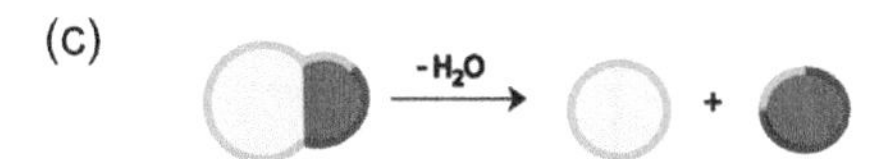

Mismatched areas/volumes between interior and membrane phase domains result in daughter vesicles where one inherits both L_d and L_o membrane.

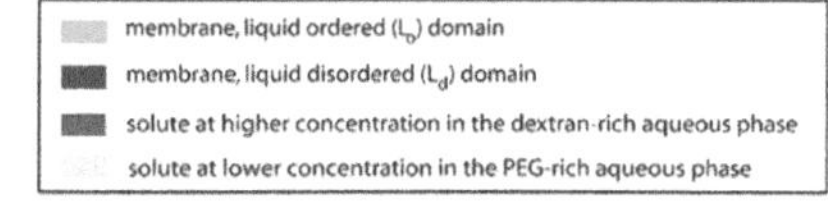

(d)

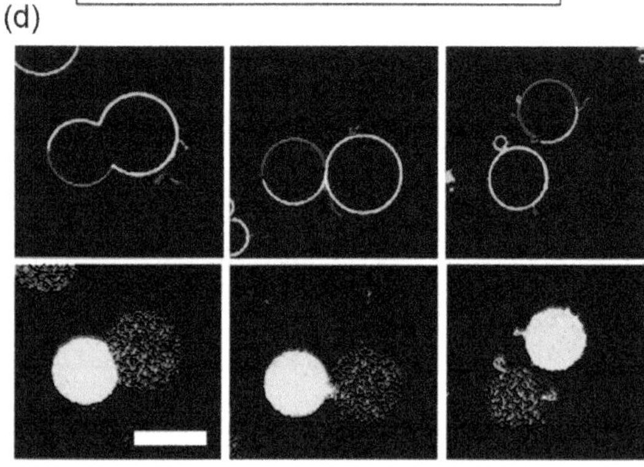

Figure 29.8 (a–c) Scheme illustrating the various types of asymmetric division of model cells. (d) Complete budding into asymmetric daughter vesicles of a PEG/dextran ATPS-containing GUV with coexisting lipid domains. External solution osmolality increases from left to right, causing budding followed by complete division. A lipid component of the liquid-disordered domain was labeled with rhodamine (false colored red), a lipid component of the liquid-ordered domain was labeled with Alexa 488 (false colored green), and soybean agglutinin, a protein added for greater biomolecular complexity, was labeled with Alexa 647 (false colored blue). Scale bar: 10 μm. (Reprinted with permission from Andes-Koback, M and Keating C.D. *J. Am. Chem. Soc.*, 133:9545–9555, 2011. Copyright 2011 American Chemical Society.)

Current experimental methods are limited by the difficulty of passively encapsulating macromolecules in GUVs, which leads to varying internal contents between vesicles within a batch. Spontaneous swelling and electroformation produce vesicles that vary widely in both size and lamellarity, creating another layer of heterogeneity in the system (see Chapter 1 for discussion on GUV preparation methods) (Reeves and Dowben, 1969; Bagatolli et al., 2000). A potential way to address this issue is to couple microfluidic vesicle production with macromolecule encapsulation. Because microfluidic devices disperse one phase within another and produce highly uniform droplets (Yamada et al., 2006; Teh et al., 2011; Nishimura et al., 2012), a much higher degree of control over the vesicle contents and the resulting properties of the gels and/or phases would be possible (see Chapter 30 on GUVs within droplets generated through microfluidics). The increased control over both GUV size and content would provide further opportunity to use these systems as simple cell mimics.

ACKNOWLEDGMENTS

The authors were supported by the National Science Foundation grant MCB-1244180 during the writing of this chapter.

LIST OF ABBREVIATIONS

ATPS	aqueous two-phase system(s)
DIC	differential interference contrast
LCST	lower critical solution temperature
PEDOT	poly(ethylene dioxythiophene)
PEG	polyethylene glycol
PNIPAAm	poly(*N*-isopropyl acrylamide)
PSS	poly(styrene sulfonate)
UCST	upper critical solution temperature

GLOSSARY OF SYMBOLS

C_{bot} concentration of solute in bottom phase of an ATPS
C_{top} concentration of solute in top phase of an ATPS
K partition coefficient
RI_{dex} dextran contribution to refractive index
RI_{PEG} PEG contribution to refractive index
RI_{tot} total refractive index
ρ^D measured density of dextran-rich phase
ρ^P measured density of PEG-rich phase
v_d specific volume of dextran
v_p specific volume of PEG
v_s specific volume of water
w_d^D weight fraction of dextran in dextran-rich phase
w_d^P weight fraction of dextran in PEG-rich phase
w_p^D weight fraction of PEG in dextran-rich phase
w_p^P weight fraction of PEG in PEG-rich phase

REFERENCES

Abkarian M, Lartigue C, Viallat A (2002) Tank treading and unbinding of deformable vesicles in shear flow: Determination of the lift force. *Phys Rev Lett* 88:068103.

Albertsson P-Å (1986) *Partition of Cell Particles and Macromolecules.* New York: Wiley.

Albertsson PA, Tjerneld F (1994) Phase diagrams. In: *Methods in Enzymology*, Walter, H. Johansson, G. (Eds.), pp 3–13. San Diego, CA: Academic Press.

Andes-Koback M, Keating CD (2011) Complete budding and asymmetric division of primitive model cells to produce daughter vesicles with different interior and membrane compositions. *J Am Chem Soc* 133:9545–9555.

Aumiller WM, Davis BW, Keating CD (2014) Phase separation as a possible means of nuclear compartmentalization. In: *New Models of the Cell Nucleus: Crowding, Entropic Forces, Phase Separation, and Fractals*, Hancock, R. Jeon, K. (Eds.), pp 109–149. San Diego, CA: Academic Press.

Aumiller WM, Keating CD (2017) Experimental models for dynamic compartmentalization of biomolecules in liquid organelles: Reversible formation and partitioning in aqueous biphasic systems. *Adv Colloid Interface Sci* 239:75–87.

Bagatolli LA, Parasassi T, Gratton E (2000) Giant phospholipid vesicles: Comparison among the whole lipid sample characteristics using different preparation methods: A two photon fluorescence microscopy study. *Chem Phys Lipids* 105:135–147.

Brangwynne CP (2011) Soft active aggregates: Mechanics, dynamics and self-assembly of liquid-like intracellular protein bodies. *Soft Matter* 7:3052–3059.

Brangwynne CP, Eckmann CR, Courson DS, Rybarska A, Hoege C, Gharakhani J, Jülicher F, Hyman AA (2009) Germline P granules are liquid droplets that localize by controlled dissolution/condensation. *Science* 324:1729–1732.

Brangwynne CP, Mitchison TJ, Hyman AA (2011) Active liquid-like behavior of nucleoli determines their size and shape in *Xenopus laevis* oocytes. *Proc Natl Acad Sci USA* 108:4334–4339.

Brooks DE, Norris-Jones R (1994) Preparation and analysis of two-phase systems. In: *Methods in Enzymology*, Walter, H. Johansson, G. (Eds.), San Diego, CA: Academic Press.

Campillo C, Pepin-Donat B, Viallat A (2007) Responsive viscoelastic giant lipid vesicles filled with a poly(N-isopropylacrylamide) artificial cytoskeleton. *Soft Matter* 3:1421–1427.

Campillo CC, Schroder AP, Marques CM, Pepin-Donat B (2008) Volume transition in composite poly(NIPAM)-giant unilamellar vesicles. *Soft Matter* 4:2486–2491.

Campillo CC, Schroder AP, Marques CM, Pepin-Donat B (2009) Composite gel-filled giant vesicles: Membrane homogeneity and mechanical properties. *Mat Sci Eng C-Bio S* 29:393–397.

Cans AS, Andes-Koback M, Keating CD (2008) Positioning lipid membrane domains in giant vesicles by micro-organization of aqueous cytoplasm mimic. *J Am Chem Soc* 130:7400–7406.

De Matteis MA, Luini A (2008) Exiting the Golgi complex. *Nat Rev Mol Cell Biol* 9:273–284.

Di Nucci H, Nerli B, Picó G (2001) Comparison between the thermodynamic features of α1-antitrypsin and human albumin partitioning in aqueous two-phase systems of polyethyleneglycol-dextran. *Biophys Chem* 89:219–229.

Dimova R, Lipowsky R (2012) Lipid membranes in contact with aqueous phases of polymer solutions. *Soft Matter* 8:6409–6415.

Dominak LM, Gundermann EL, Keating CD (2010) Microcompartmentation in artificial cells: pH-Induced conformational changes alter protein localization. *Langmuir* 26:5697–5705.

Dominak LM, Keating CD (2007) Polymer encapsulation within giant lipid vesicles. *Langmuir* 23:7148–7154.

Dundr M (2012) Nuclear bodies: Multifunctional companions of the genome. *Curr Opin Cell Biol* 24:415–422.

Ellis RJ (2001) Macromolecular crowding: Obvious but underappreciated. *Trends Biochem Sci* 26:597–604.

Hatti-Kaul R (2000) *Aqueous Two-Phase Systems: Methods and Protocols.* Totowa, NJ: Humana Press.

Helfrich MR, Mangeney-Slavin LK, Long MS, Djoko KY, Keating CD (2002) Aqueous phase separation in giant vesicles. *J Am Chem Soc* 124:13374–13375.

Hoare TR, Kohane DS (2008) Hydrogels in drug delivery: Progress and challenges. *Polymer* 49:1993–2007.

Hyman AA, Weber CA, Jülicher F (2014) Liquid-liquid phase separation in biology. *Annu Rev Cell Dev Biol* 30:39–58.

Jesorka A, Markstrom M, Karlsson M, Orwar O (2005a) Controlled hydrogel formation in the internal compartment of giant unilamellar vesicles. *J Phys Chem B* 109:14759–14763.

Jesorka A, Markstrom M, Orwar O (2005b) Controlling the internal structure of giant unilamellar vesicles by means of reversible temperature dependent sol-gel transition of internalized poly (N-isopropyl acrylamide). *Langmuir* 21:1230–1237.

Keating CD (2012) Aqueous phase separation as a possible route to compartmentalization of biological molecules. *Acc Chem Res* 45:2114–2124.

Klouda L, Mikos AG (2008) Thermoresponsive hydrogels in biomedical applications: A review. *Eur J Pharm Biopharm* 68:34–45.

Kopperschlager G (1994) Affinity extraction with dye ligands. In *Methods in Enzymoology*, Walter H, Johansson G (Eds.), San Diego, CA: Academic Press.

Lasic DD (1988) The mechanism of vesicle formation. *Biochem J* 256:1–11.

Lee C, Chen LB (1988) Dynamic behavior of endoplasmic reticulum in living cells. *Cell* 54:37–46.

Li YH, Kusumaatmaja H, Lipowsky R, Dimova R (2012) Wetting-induced budding of vesicles in contact with several aqueous phases. *J Phys Chem B* 116:1819–1823.

Li YH, Lipowsky R, Dimova R (2008) Transition from complete to partial wetting within membrane compartments. *J Am Chem Soc* 130:12252–12253.

Li YH, Lipowsky R, Dimova R (2011) Membrane nanotubes induced by aqueous phase separation and stabilized by spontaneous curvature. *Proc Natl Acad Sci USA* 108:4731–4736.

Liu Y, Agudo-Canalejo J, Grafmüller A, Dimova R, Lipowsky R (2016) Patterns of flexible nanotubes formed by liquid-ordered and liquid-disordered membranes. *ACS Nano* 10:463–474.

Liu Y, Lipowsky R, Dimova R (2012) Concentration dependence of the interfacial tension for aqueous two-phase polymer solutions of dextran and polyethylene glycol. *Langmuir* 28:3831–3839.

Long MS, Cans A-S, Keating CD (2008) Budding and asymmetric protein microcompartmentation in giant vesicles containing two aqueous phases. *J Am Chem Soc* 130:756–762.

Long MS, Jones CD, Helfrich MR, Mangeney-Slavin LK, Keating CD (2005) Dynamic microcompartmentation in synthetic cells. *Proc Natl Acad Sci USA* 102:5920–5925.

Luby-Phelps K (2013) The physical chemistry of cytoplasm and its influence on cell function: An update. *Mol Biol Cell* 24:2593–2596.

Mao YTS, Zhang B, Spector DL (2011) Biogenesis and function of nuclear bodies. *Trends Genet* 27:295–306.

Markstrom M, Gunnarsson A, Orwar O, Jesorka A (2007) Dynamic microcompartmentalization of giant unilamellar vesicles by sol gel transition and temperature induced shrinking/swelling of poly(N-isopropyl acrylamide). *Soft Matter* 3:587–595.

Minton AP (2013) Quantitative assessment of the relative contributions of steric repulsion and chemical interactions to macromolecular crowding. *Biopolymers* 99:239–244.

Nishimura K, Suzuki H, Toyota T, Yomo T (2012) Size control of giant unilamellar vesicles prepared from inverted emulsion droplets. *J Colloid Interf Sci* 376:119–125.

Osada Y, Kajiwara K (2001) *Gels Handbook*. San Diego, CA: Academic Press.

Pallas NR, Harrison Y (1990) An automated drop shape apparatus and the surface tension of pure water. *Colloids and Surfaces* 43:169–194.

Pozo Navas B, Lohner K, Deutsch G, Sevcsik E, Riske KA, Dimova R, Garidel P, Pabst G (2005) Composition dependence of vesicle morphology and mixing properties in a bacterial model membrane system. *Biochim Biophys Acta* 1716:40–48.

Reeves JP, Dowben RM (1969) Formation and properties of thin-walled phospholipid vesicles. *J Cell Physiol* 73:49–60.

Sarkar M, Smith AE, Pielak GJ (2013) Impact of reconstituted cytosol on protein stability. *Proc Natl Acad Sci USA* 110:19342–19347.

Seifert U (2000) Giant vesicles: A theoretical perspective. In: *Giant Vesicles*, Luisi, P.L. Walde, P. (Eds.), Chichester, UK: John Wiley & Sons.

Sriwongsitanont S, Ueno M (2004) Effect of a PEG lipid (DSPE-PEG2000) and freeze-thawing process on phospholipid vesicle size and lamellarity. *Colloid Polym Sci* 282:753–760.

Tako M, Nakamura S (1988) Gelation mechanism of agarose. *Carbohydr Res* 180:277–284.

Teh SY, Khnouf R, Fan H, Lee AP (2011) Stable, biocompatible lipid vesicle generation by solvent extraction-based droplet microfluidics. *Biomicrofluidics* 5:44113–44112.

Torre P, Keating CD, Mansy SS (2014) Multiphase water-in-oil emulsion droplets for cell-free transcription–translation. *Langmuir* 30:5695–5699.

Vashist A, Vashist A, Gupta YK, Ahmad S (2014) Recent advances in hydrogel based drug delivery systems for the human body. *J Mater Chem B* 2:147–166.

Viallat A, Dalous J, Abkarian M (2004) Giant lipid vesicles filled with a gel: Shape instability induced by osmotic shrinkage. *Biophys J* 86:2179–2187.

Węgrzyn I, Jeffries GDM, Nagel B, Katterle M, Gerrard SR, Brown T, Orwar O, Jesorka A (2011) Membrane protrusion coarsening and nanotubulation within giant unilamellar vesicles. *J Am Chem Soc* 133:18046–18049.

Wohlfarth C (2013) *CRC Handbook of Phase Equilibria and Thermodynamic Data of Aqueous Polymer Solutions*. Boca Raton, FL: Taylor & Francis Group.

Yamada A, Yamanaka T, Hamada T, Hase M, Yoshikawa K, Baigl D (2006) Spontaneous transfer of phospholipid-coated oil-in-oil and water-in-oil micro-droplets through an oil/water interface. *Langmuir* 22:9824–9828.

Zhao Z, Li Q, Ji X, Dimova R, Lipowsky R, Liu Y (2016) Molar mass fractionation in aqueous two-phase polymer solutions of dextran and poly(ethylene glycol). *J Chromatogr A* 1452:107–115.

Zhou H-X, Rivas GN, Minton AP (2008) Macromolecular crowding and confinement: Biochemical, biophysical, and potential physiological consequences. *Ann Rev Biophys* 37:375–397.

Droplet-stabilized giant lipid vesicles as compartments for synthetic biology

Johannes P. Frohnmayer, Marian Weiss, Lucia T. Benk, Jan-Willi Janiesch, Barbara Haller, Rafael B. Lira, Rumiana Dimova, Ilia Platzman, and Joachim P. Spatz

Wherever you go, whatever you do, I'll always be there, supporting you.

Microfluidic droplet

Contents

30.1 INTRODUCTORY WORDS

Compartmentalization, the formation of lipid-membrane compartments in which specific metabolic activity takes place, is one of the distinguishing features of eukaryotic cells (Agapakis et al., 2012; Diekmann and Pereira-Leal, 2013). Inspired by the important functions of cellular compartments, synthetic biologists have concentrated on an attempt to develop cell like compartments for biochemical reactions (Elani et al., 2014; Li et al., 2014). In this context, the most common studies have focused on the development of enclosed cell-size volumes of aqueous space with organic lipid-based membranes as in giant unilamellar vesicles (GUVs) (see also Chapter 28), or polymer-based membranes as in polymersomes and polymer-stabilized emulsion droplets.

GUV-based compartments provide a suitable model system for mimicking the cell membrane. The lipid matrix with reconstituted proteins resembles the *in vivo* frame (Ramadurai et al., 2009). Therefore, free-standing GUVs have been utilized in various synthetic biology applications (Carrara et al., 2012; Elani et al., 2014). However, the GUV-based compartment system has several drawbacks that are related mostly to its poor chemical and mechanical stability, and therefore limited potential for manipulation.

Due to their increased stability and lifetime, polymersomes (see also Chapter 26) made of amphiphilic block-copolymers in a continuous water phase are commonly used as an alternative in synthetic biology applications (Discher and Eisenberg, 2002). By adjusting the molecular properties of the block-copolymers, the thickness and properties such as the bending and stretching moduli of the membrane can be finely tuned (Bermudez et al., 2002). Despite their increased mechanical and chemical stability, the encapsulation of biomolecules and further manipulation of such bio-containing polymersomes still represent big challenges. These drawbacks are mainly caused by the lack of technological means that allow efficient incorporation or loading of bio-ingredients for biochemical activity.

Compartments based on block copolymer-stabilized water-in-oil droplets show potential to overcome the drawbacks related to technological limitations associated with "traditional" polymersomes. In the context of synthetic biology, droplet-based compartment systems possess the advantages of polymersomes and, in addition, area easily integrated into microfluidic technologies for controlled and precise loading with biologically relevant materials (Abate et al., 2010). Despite these advantages, the ability of these droplets or polymersomes to serve as optimal cell-like compartments is mainly hindered by their inability to mimic the biophysical properties of cellular membranes (Itel et al., 2014). To overcome these obstacles, we aimed at combining the biophysical properties of cellular lipid membranes, the stability of copolymer-stabilized droplets and the ability to be adapted for high-throughput manipulation and high encapsulation efficiency.

This chapter describes our recently developed approach that merges lipid vesicle formation and droplet microfluidics for the generation of stable, uniformly sized and easily manipulated droplet-stabilized GUV (dsGUV) compartments as schematically illustrated in Figure 30.1 (Weiss et al., 2018; Haller et al., 2019). In Section 30.2, we describe the microfluidics-based production

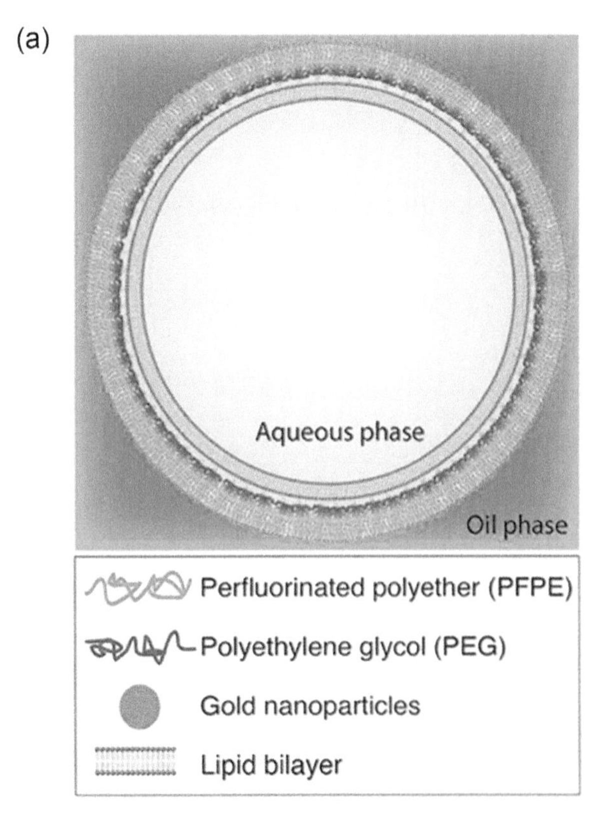

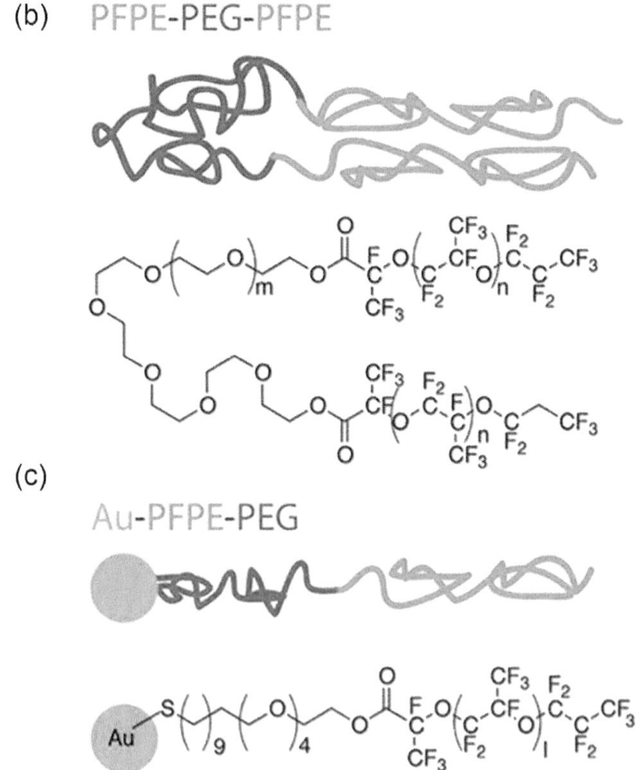

Figure 30.1 (a) Schematic representation of a droplet-supported GUV compartment. The water-in-oil droplet is stabilized by triblock-copolymer surfactants. Optional copolymer surfactants are the gold-linked diblock-copolymers. These gold-linked surfactants can be employed to biofunctionalize the droplets inner interface. The lipid bilayer is supported on the copolymer-stabilized oil-water interface of the droplet. (b) and (c) present chemical structures of PFPE-PEG-PFPE triblock- and PFPE-PEG-Gold diblock-copolymer surfactants, respectively. (Adapted with permission from Platzman et al., *J. Am. Chem. Soc.*, 135, 3339–3342, 2013.)

of the droplets and their stabilization by the copolymer surfactants. To form dsGUVs, liposome-containing aqueous solution is encapsulated within the copolymer-stabilized droplets. In the presence of divalent ions, the liposomes adsorb and fuse to the inner droplet surface forming a spherical supported bilayer (see Section 30.3). The process is reminiscent of the formation of conventional supported lipid bilayers (SLB), however, the support here is not a solid surface as in SLBs, but the fluid copolymer-stabilized droplet interface. In Section 30.3.5, we demonstrate that lipid mobility in dsGUVs is similar to that in the free-standing bilayer of GUVs with the same lipid composition. An advantage of the dsGUV system is that microfluidic pico-injection technology (Abate et al., 2010) can be applied to deliver adjustable amounts of biomaterials in their interior. For example, proteoliposomes can be injected into the copolymer-stabilized emulsion droplets and thus form biofunctionalized dsGUV compartments with reconstituted proteins (see Section 30.4). In contrast to conventional GUVs, the stability of the dsGUVs imparted by the supporting droplet interface is enhanced and their size can be finely tuned. Furthermore, we show that it is possible to release GUVs from dsGUVs (see Section 30.5). Therefore, the technique can also be used as an intermediate step, allowing the assembly of synthetic cells and later observing them in a physiological environment. This novel system is poised to overcome the fundamental limitations associated with manipulation of currently employed techniques to form cell-like compartments and possesses great potential for enabling efficient bottom-up assembly of minimal synthetic cells.

30.2 FORMATION, FUNCTIONALIZATION AND CHARACTERIZATION OF MICROFLUIDIC DROPLETS: NECESSARY INGREDIENTS AND EQUIPMENT

Droplet-based microfluidics combines principles of science and technology, and enables the user to handle, process and manipulate droplets of very small volumes (down to less than a few picoliters) via microchannels. This technology permits the integration of multiple laboratory functions into one single microfabricated chip, requires minimal manual user intervention and sample consumption, and allows for enhanced data acquisition, analysis speed and precision. Due to such advantages, the potential for this technology to be applied in biological (Pearce and Williams, 2007; Rowat et al., 2009; Platzman et al., 2013), chemical (Hung and Lee, 2007; deMello and Wootton, 2009) and medical (Huang et al., 2011) research is vast. This section will describe the preparation and characterization of copolymer-stabilized droplets required for the dsGUVs formation. The droplet dimension sets the size of the dsGUVs. Thus, droplet-based microfluidics offers (i) unprecedented control on the size and monodispersity of the resulting stable vesicles, a feature not yet available for conventional GUVs, and (ii) precise control and flexibility of the encapsulated content and the constituting membrane of the dsGUVs at high yield.

30.2.1 AMPHIPHILIC TRIBLOCK- AND GOLD-LINKED DIBLOCK-COPOLYMER SURFACTANTS

The most important key factor for the stability of water-in-oil emulsion droplets is the block copolymer surfactants at their interface. Nonionic fluorosurfactants made of perfluorinated polyether (PFPE) hydrophobic blocks attribute long-term stability to the droplets by preventing their coalescence, whereas polyethylene glycol (PEG) hydrophilic blocks serve as a biocompatible, inert droplet interface (Holtze et al., 2008). Optionally, to provide active sites for biochemical interactions within the droplets, a new type of surfactants that is covalently linked to gold nanoparticles (~ 5 nm in diameter) can be employed (Platzman et al., 2013). Such gold-linked surfactants are mixed with gold-free surfactants (at mixing molar ratios ranging from 1:1000–1:2000) to create stable droplets functionalized with gold nanoparticles (Figure 30.1).

The synthesis of the PFPE (7,000 g/mol)–PEG(1,400 g/mol)–PFPE(7,000 g/mol) (TRI7000) triblock-copolymer surfactants is based on a protocol reported earlier (Holtze et al., 2008), but with several modifications, as schematically illustrated in Figure 30.2 (Janiesch et al., 2015). Briefly, the synthesis is carried out under argon atmosphere in dry tetrahydrofuran (THF) solvent in a heated Schlenk-flask. PEG (1.4 g, 1 mmol) is dissolved in dry THF (90 mL) and cooled to -78°C. *N*-butyl lithium (1.25 mL of a 1.6 M solution in hexane, 2 mmol) is added dropwise over a period of 60 min to the PEG solution and stirred for additional 30 min at -78°C. Subsequently, under continuous stirring, the reaction is slowly heated to room temperature before PFPE-carboxylic acid (Krytox FSH, 14 g, 2 mmol, molecular weight 7,000 g/mol) is added dropwise over a period of 30 min and stirred for additional 2 h. A separatory funnel is used to remove the THF solvent with unreacted PEG traces. The product is dissolved in methanol (99.8%) to separate it from unreacted PFPE(7000)-carboxylic acid. The PFPE-PEG-PFPE product, soluble in methanol, is transferred to a clean flask and dried with a rotary evaporator at 40°C. The same synthesis procedure can be applied for PFPE (2,500 g/mol)–PEG(600 g/mol)–PFPE(2,500 g/mol) (TRI2500) surfactants (Table 30.1). The synthesis of the surfactants can be performed in standard chemical labs and requires basic vacuum and inert gas lines. The PEG polymers can be purchased from Fluka, Germany, and PFPE is available from DuPont, Netherlands. The product should be a clear viscous liquid.

Synthesis of the gold-linked diblock-copolymer PFPE(7000)-PEG(350)-Gold surfactants can be performed in a one-step

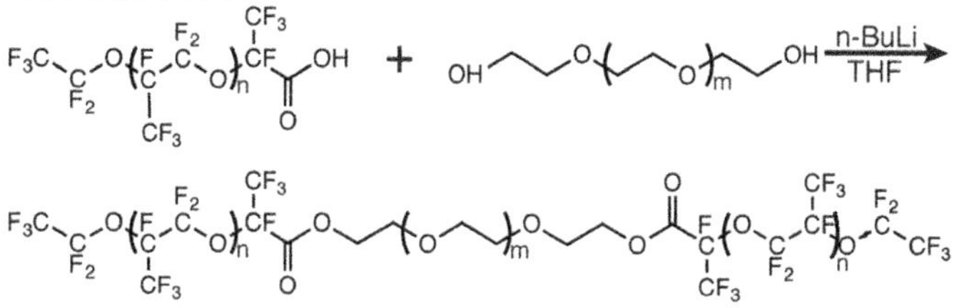

PFPE–PEG–PFPE

Figure 30.2 Synthesis of PFPE-PEG-PFPE triblock-copolymer surfactants from PFPE carboxylic acid and PEG. The protocol can be applied for polymers with various molecular weights of block copolymers. (Adapted from Platzman et al., *J. Am. Chem. Soc.*, 135, 3339–3342, 2013. With permission.)

Table 30.1 **Molecular characteristics of the copolymer surfactants employed for droplet stabilization and functionalization**

COPOLYMER SURFACTANT	FULL NAME	PEG MOIETY	PFPE MOIETIES
TRI2500	PFPE(2,500 g/mol)-PEG (600 g/mol)-PFPE(2,500 g/mol)	MW = 600 g/mol, m = 12	MW = 2,500 g/mol, n = 17
TRI7000	PFPE(7,000 g/mol)- PEG (1,400 g/mol)-PFPE(7,000 g/mol)	MW = 1,400 g/mol, m= 30	MW = 7,000 g/mol, nm = 48
PFPE-PEG-Gold	PFPE(7,000 g/mol)-PEG (350 g/mol)-Gold	MW = 350 g/mol, m= 4	MW = 7,000 g/mol, n = 48

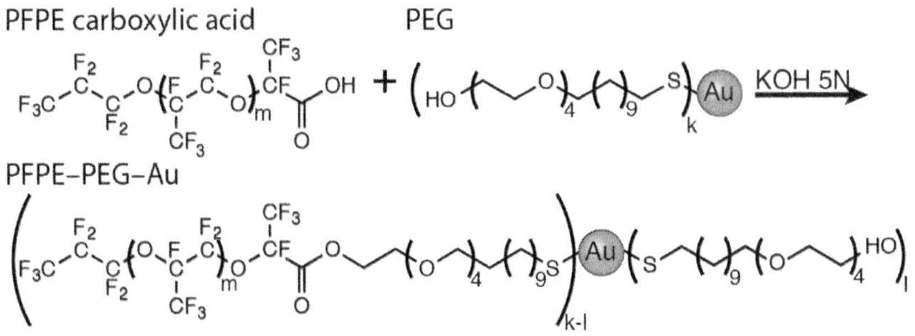

Figure 30.3 Synthesis of PFPE-PEG-Gold diblock-copolymer surfactants. (Adapted from Platzman et al., *J. Am. Chem. Soc.*, 135, 3339–3342, 2013. With permission.)

process as schematically illustrated in Figure 30.3 (Platzman et al., 2013). PFPE(7000)-carboxylic acid (9.3 mg, 1.3 μmol) and KOH (10 μL of a 5 N solution) are added to 5 mL functionalized gold nanoparticle solution (11-Mercaptoundecyl)tetra(ethylene glycol), 2% (w/w) (Sigma Aldrich, Germany, nanoparticle diameter ~ 5 nm) and stirred for 1 h to achieve flocculation of the PFPE-PEG-Gold product and of unreacted PFPE. At the end of the reaction, water is removed by freeze-drying for 24 h. PFPE(7000)-PEG(350)-Gold surfactants are dissolved in 1 mL of fluorinated oil FC-40 and filtered with a hydrophobic filter (PTFE 0.2 μm) to remove unreacted, hydrophilic (11-Mercaptoundecyl) tetra(ethylene glycol)-functionalized gold nanoparticles.

Following synthesis, TRI7000 and TRI2500 surfactants are mixed separately with gold-linked surfactants and are dissolved in FC-40 oil to final concentrations of 2.5 mM and 3 μM for triblock and gold-linked surfactants, respectively. These mixtures are used as an oil phase for droplet creation in the droplet-based microfluidic device (see the following section).

General notes: The fluorinated oil, FC-40 (3M, USA) is the standard oil in droplet-based microfluidics. Its physical properties such as viscosity and hydrophobicity facilitate droplet formation and prevent leakage of encapsulated bio-molecules into the oil phase. Moreover, FC-40 is highly permeable for gases. For droplet formation, the final copolymer surfactant concentration in FC-40 oil should be in the range between 1 and 20 mM. Exact concentration within this range has to be optimized according to the encapsulated biochemical system (Janiesch et al., 2015).

30.2.2 MICROFLUIDIC DEVICES FOR DROPLETS FORMATION AND THEIR MANIPULATION

Simple shaking of an aqueous phase and the oil with dissolved copolymer surfactants will generate the formation of droplets. However, droplet-based microfluidic devices are required to create monodisperse droplets of a defined size. Moreover, microfluidic devices are necessary for further droplet manipulation (e.g., injection, sorting and time-lapse analysis). All microfluidic devices used in this research are fabricated from poly(dimethylsiloxane) (PDMS) using photo- and soft-lithography methods (Duffy et al., 1998; Xia and Whitesides, 1998). PDMS is a common material in microfluidic devices due to its low price, good biocompatibility and permeability to gasses, high transparency and low fluorescent background.

Following is a concise description of the experimental steps required for the preparation of microfluidic devices (Figure 30.4). As a first step, a 2D version of the desired device structures is drawn with a computer-aided design software ("QCAD-pro"). A negative of the design structure is printed on a chrome-coated soda lime glass via ion-etching and is used as a mask for standard contact photolithography to obtain a positive relief/master on the silicon wafer. A replica is generated by pouring PDMS onto a wafer containing the master, degassing and heat curing the PDMS. The part of the PDMS containing the replica of the structure is cut out using disposable scalpels and removed from the master. Inlet and outlet holes are punched through the PDMS with tissue punchers. Finally, the PDMS molds undergo short oxygen plasma treatment to achieve better bonding to the glass coverslips. Note: Both glass and PDMS mold should be cleaned before the plasma treatment. The standard procedure for glass cleaning consists of three sonication steps, once in 20% Extrane and twice in MiliQ water. The molds are cleaned with 70% ethanol and dried with N_2.

Droplets are generated in a flow-focusing geometry junction, in which an aqueous phase is cut off by a surfactant-containing oil phase (Figure 30.5a and b). Following the formation,

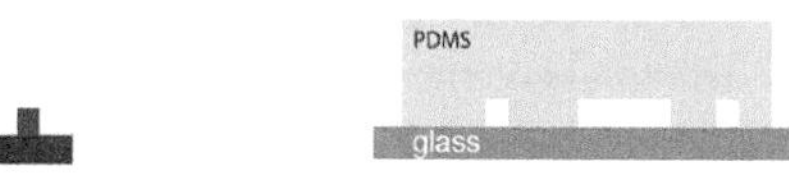

Figure 30.4 Schematic overview of the microfluidic device fabrication. The first three experimental steps(1 - 3) describe the standard contact photolithography methods required for silicon wafer master production. Following master production, soft-lithography method is applied to develop a PDMS mold (4 + 5). At the final step, the PDMS mold is bound to the glass coverslip to produce a microfluidic device (6). (From Frohnmayer, J.P., Bottom-Up assembly of synthetic model systems for cellular adhesion, in *Online-Ressource* (238 Seiten), Heidelberg University, Heidelberg, Germany, 1, 2017.)

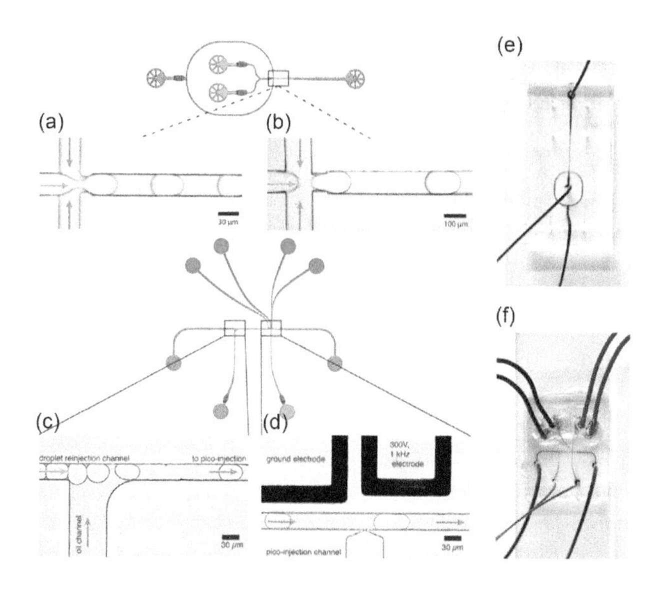

Figure 30.5 Presentation of two major droplet-based microfluidic operating units used in this research. (a) and (b) show the bright-field images of the flow-focusing junctions in which the droplets with diameters of 40 and 100 µm are generated, respectively. Bright-field images (c) and (d) represent the pico-injection microfluidic unit for controlled introduction of different biomaterials into the droplets. (c) The spacing between the droplets carrying different biological components is controlled through addition of oil via the second oil channel. (d) An AC field (1 kHz, 250 V) reduces the stability (poration) of the surfactant layer at the droplet surface and allows the injection of an aqueous solution of biological reagents from the pico-injection channel. The injection process can be visualized by comparing the droplets sizes before and after the injection. Photographs of (e) a droplet production and (f) a pico-injection device. For better visualization, channel and tubing are filled with ink. (Reproduced from Frohnmayer (2017) with permission of the author.)

water-in-oil droplets are stabilized by accretion of block copolymer surfactants at the water-oil interface leading to reduction of the oil/water interfacial tension (Mazutis et al., 2009) from 52.1 mN/m to a value of 19.5 mN/m for TRI7000 and 3.1 mN/m Tri2500 (Janiesch, 2015). The droplet diameter is mainly controlled by the channel dimensions, but can also be regulated to some extend by the variation of flow rates of the aqueous and oil phase. A channel height and width of 30 µm at the flow-focusing junction is used for the production of droplets of 40 µm in diameter (d) (Figure 30.5a). Syringe pumps (World Precision Instruments, Harvard Apparatus) are used to apply flow rates of 120 µL/h for the aqueous phase and 160 µL/h for the oil phase to achieve stable droplets formation at the rate of 1 kHz. For large droplets ($d = 100$ µm), the channel dimensions of 80 µm height and 100 µm width can be employed (Figure 30.5b). To achieve a stable formation of the large droplets at the rate of 1 kHz, the flow rates are adjusted to 650 and 850 µL/h for the aqueous and oil phase, respectively.

General notes: Nearly every laboratory has most of the requirements needed for microfludic experiments. A simple inverted microscope and standard syringe pumps are sufficient. The most likely hindrance is access to clean room facilities for photolithography. Another bottleneck might be the availability of a high-speed camera. An experienced user can produce droplets without the help of such a camera by observing the inlet and outlet channels. To monitor the formation of droplets, a camera with temporal resolution in the range 4,000–8,000 fps is recommended.

To allow precise delivery of various biological components into preformed droplets, the microfluidic devices can be integrated with small and compact electrodes to apply electric fields in the microchannels. These electric fields induce destabilization (poration) of the surfactant (mono)layer and facilitate controlled injection (pico-injection) of aqueous phase into the droplets. The design of our droplet-based pico-injection unit is adapted from Abate et al. (2010). A microfluidic flow control system (MFCS™-EZ, France) was used to introduce droplets into the pico-injection unit. The spacing between the droplets was controlled through addition of oil with surfactants via the second oil channel as presented in Figure 30.5c. Following the separation step, isolated droplets passed an electric alternating current (AC) field (frequency of 1 kHz, voltage of 250 V) generated by a HM 8150 signal generator (HAMEG®, Germany) and amplified by a 623B-H-CE amplifier (TREK®, USA) and two electrodes made of Indalloy 19 (51% indium, 32.5% bismuth, 16.5% tin). This process destabilizes the droplet interface and allows introduction of biological reagents via a pressurized injection channel as presented in Figure 30.5d. The injection volume can be controlled precisely between 1 and 100 pL dependent on the applied pressure in the injection channel.

30.2.3 BIOFUNCTIONALIZATION OF THE COPOLYMER-STABILIZED DROPLETS

Two different approaches can be implemented to achieve efficient biofunctionalization of the nanostructured copolymer-stabilized droplets (Platzman et al., 2013). The first approach is based on the functionalization of the created droplets containing gold-linked copolymer surfactants (see Section 30.2) using a nitrilotriacetic acid (NTA) and hexahistidine-tagged (His6-tag) protein chemistry. The second approach involves two experimental steps: (1)

synthesis of gold-linked surfactants coupled to bioactive molecules; and (2) formation of biofunctionalized droplets.

In the following, we describe the experimental steps required for biofunctionalization of gold-nanostructured droplets with (His6)-tag proteins via an NTA-thiol chemistry. NTA-thiol (300 μL, 1 mM dissolved in 99.9% ethanol, ProChimia, Poland) and $NiCl_2$ (9 μL, 100 mM) are mixed by stirring for 20 min. Ethanol is removed partly to a final volume of 50 μL by a nitrogen flow. Following this procedure, the (His6)-GFP (300 μL, 30 μM, GFP was a gift of S. Gardia, Addgene, plasmid #29663; comparable products are commercially available from ThermoFischer and BSP Bioscience) is added and mixed for another 1 h. The product is diluted in MilliQ water to a final (His6)-GFP-NTA-thiol concentration of 10 μM, as determined using extinction coefficient, $\varepsilon = 30000\ cm^{-1}M^{-1}$ at 395 nm.

Two types of droplets loaded with His6-GFP-NTA-thiol were investigated: (1) those stabilized only by TRI7000 PFPE-PEG-PFPE (2.5 mM) surfactants and (2) droplets stabilized by a mixture of TRI7000 PFPE-PEG-PFPE (2.5 mM) and PEG-PFPE-Gold (3 μM) surfactants. Figure 30.6 shows fluorescence images of the (His6)-GFP-NTA-thiol within the gold-nanostructured droplets (Figure 30.6a) and within the droplets containing no gold-linked surfactants (Figure 30.6b). It can easily be observed that the fluorescent signal is confined to the surface of the nanostructured droplets, whereas it is homogeneously distributed in the entire volume of the non-nanostructured droplets.

The two-step biofunctionalization approach was used to provide integrin interactions with nanostructured droplets (see Section 30.4.5). For this purpose, SN528 RGD peptides (Figure 30.7, kindly provided by Prof. Kessler's group from TU, Munich) were immobilized on gold-linked surfactants via thiol linker (Frohnmayer et al., 2015). For this, the freeze-dried PEG-PFPE-Gold diblock-copolymer surfactants (see Section 30.2.1) were dissolved in 100 μL fluorinated oil FC-40 to a concentration of 25 μM and then the aqueous solution of RGD peptides (50 μM, 100 μL) was added and stirred for 1 h. To remove the unbound RGD peptides, the crude product solution was centrifuged. The supernatant aqueous solution was removed and the precipitant was freeze-dried for 24 h to remove the entire remaining water. Finally, the product was dissolved in 1 mL of (the oil) FC-40 and filtered with a hydrophobic filter (PTFE 0.2 μm) to remove unreacted peptide traces. A mixture of TRI7000 PFPE-PEG-PFPE (2.5 mM) and PEG-PFPE-Gold-RGD (3 μM) copolymer surfactants in FC-40 oil was used for stable droplets creation.

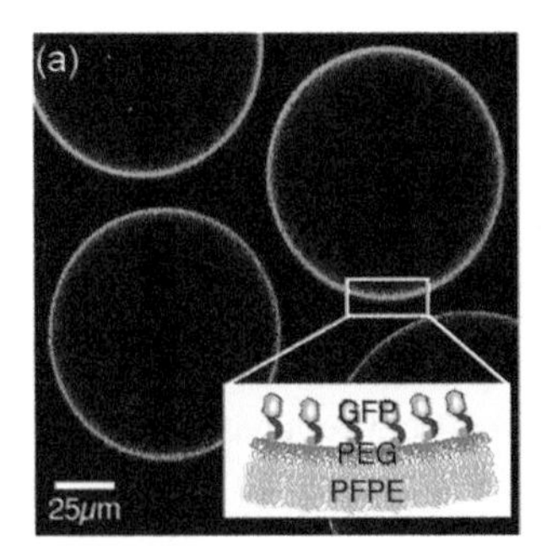

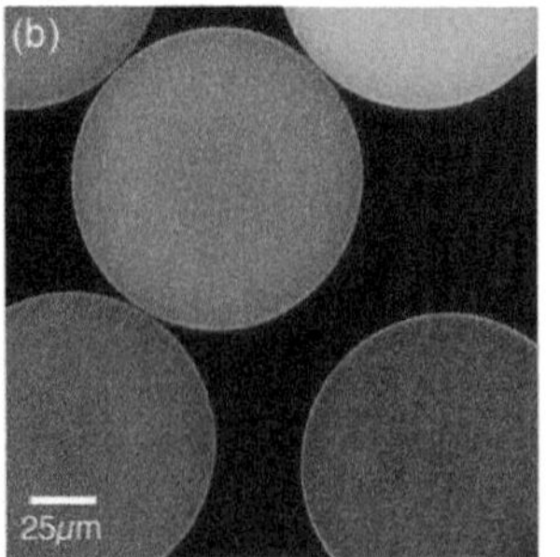

Figure 30.6 Fluorescence images of the (His6)-GFP-NTA-thiol within (a) gold-nanostructured and (b) gold-free polymer-stabilized droplets measured 1 day after formation. (From Weiss et al., 2018.)

30.2.4 MOBILITY OF THE COPOLYMER SURFACTANTS AT THE DROPLET INTERFACE

To acquire mobility values of molecules confined to the interface of the nanostructured droplet in comparison to the mobility of lipids in a dsGUV, one can perform fluorescence recovery after photobleaching (FRAP) measurements. Here, the data was collected with a Leica SP5 confocal microscope equipped with an argon as well as a white-light laser. The examined fluorophore was GFP. Images were recorded at a constant temperature of 25°C. Using a 63x oil objective (HCX PL APO 63x/1.40–0.60), GFP was exited at 488 nm, and the detection window was set to 495-540 nm. The pinhole for data acquisition was set to 1 Airy unit, which corresponds to the diameter of the Airy disk of 96 μm and 0.9 μm thickness of the optical section.

Following production, the droplets are collected at the outlet of the microfluidic device and transferred to an analysis chamber (Figure 30.8a). The chamber can be made of two coverslips glued by two or four double-face sticky tapes (thickness ≈ 80 μm or 160 μm for analysis of small or large droplets, respectively) or shards of coverslips (thicknesses #0 ≈ 130 μm, #1 ≈ 150 μm and #1.5 ≈ 170 μm). After filling with the droplets emulsion, the chamber is sealed by two-component Twinsil® glue (Picodent, Germany).

Due to the water/oil density difference (1.0 vs. 1.9 g/mL) the droplets tend to ascend toward the ceiling of the observation chamber (Figure 30.8b). In this position, a confocal microscope is used to scan the droplets in z-direction to locate the bottom slice of the droplet. This particular focal plane is chosen to exclude the influence of the surface (of the coverslip) on the droplets membrane dynamics during the FRAP measurements.

A circular spot with a diameter of 5 μm is selected as the bleaching area (Figure 30.8c). The time course of each FRAP experiment includes 10 pre-bleaching images, 2–10 bleach cycles to fully bleach the fluorescent signal and 50–200 post-bleaching images to record the fluorescence recovery.

FRAP analysis followed a protocol proposed by (Axelrod et al., 1976) and (Soumpasis, 1983); for details regarding FRAP analysis see Chapter 21. To correct for the background noise, I_{bg}, the fluorescence signal was measured in the oil phase using the same settings as for FRAP measurements. I_{bg} was subtracted from all the measured values. Intensity values for the bleaching spot, *I(t)*, and the whole droplet base, *T(t)*, were extracted. *I(t)* and *T(t)* were normalized by the averages of the prebleaching values, I_{pb} and T_{pb}. To correct for photofading, the intensities of the bleached spot were multiplied with the reciprocal, normalized intensities of the droplet base, *T(t)*. Thus, the normalized and corrected intensities, I_{nor}, were calculated as follows

$$I_{nor}(t) = \frac{I(t) - I_{bg}}{I_{pb} - I_{bg}} \frac{T_{pb} - I_{bg}}{T(t) - I_{bg}}. \tag{30.1}$$

A nonlinear least-square fit $f(t)$ was then applied using MATLAB R2015a SP1 to fit the following function to the normalized intensities, I_{nor}

$$f(t) = a(1 - \exp(\lambda t)). \tag{30.2}$$

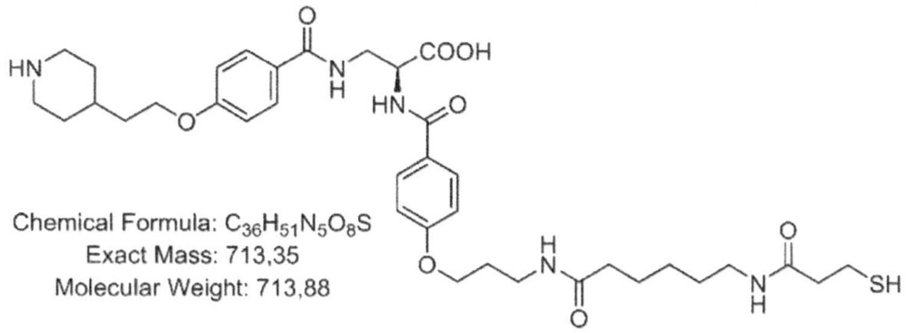

Chemical Formula: $C_{36}H_{51}N_5O_8S$
Exact Mass: 713,35
Molecular Weight: 713,88

Figure 30.7 Chemical structure of SN528 RGD mimetic peptide specific for $\alpha_{IIb}\beta_3$ integrin binding. (Frohnmayer et al., *Angew. Chem-Int. Edit.*, 54, 12472–12478, 2015.)

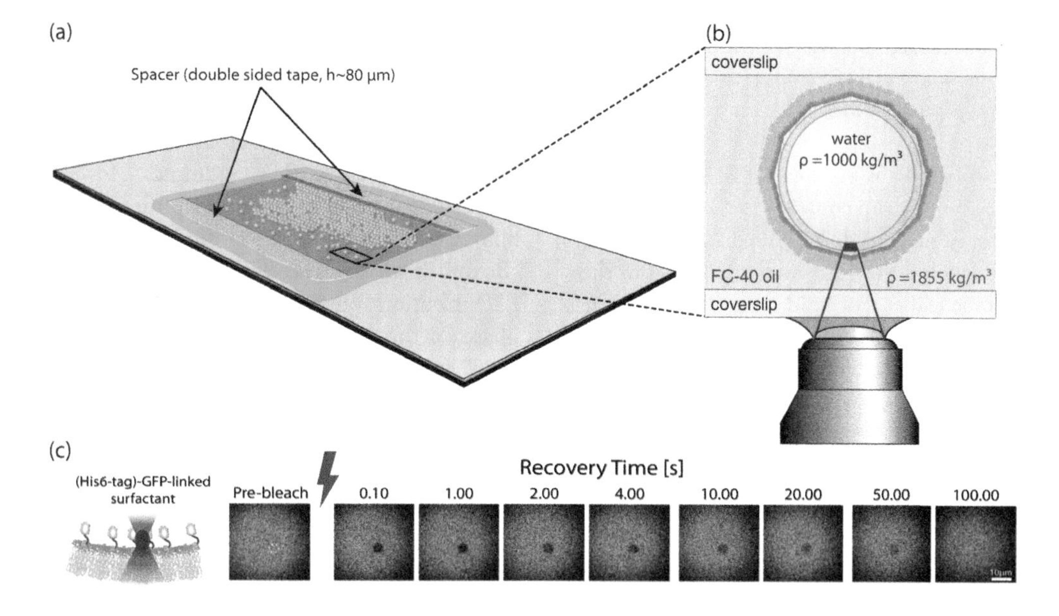

Figure 30.8 FRAP measurements of the mobility of (His6)-GFP-NTA-thiol immobilized to the PEG-PFPE-Gold copolymer surfactants. (a) Sketch of the observation chamber for droplet analysis. (b) Magnified vertical cross section representation of the analysis chamber. Due to the density differences, droplets are localized close to the top coverslip. (c) Example of a FRAP experiment on the droplet membrane surfactant mobility. FRAP was performed on droplets containing (His6)-GFP immobilized to the gold-linked copolymer surfactants. The bleached area is encircled in the pre-bleached frame and the recovery time (seconds after bleaching completion) is indicated to the top of the fluorescence recovery frames. (From Weiss et al., 2018.)

The resulting values of the coefficient λ were then used to calculate the half-recovery time $\tau_{1/2}$ for each bleaching experiment,

$$\tau_{1/2} = \frac{-\log(0.5)}{\lambda}. \tag{30.3}$$

The diffusion coefficient D is related to the half-recovery time $\tau_{1/2}$ via the square radius of the bleaching spot, assuming a Gaussian bleaching profile:

$$D = 0.32\frac{r^2}{\tau_{1/2}}. \tag{30.4}$$

The average diffusion coefficient for each experiment and its standard error were calculated from at least 20 measurements on different droplets.

FRAP analysis of the GFP-labeled gold-linked surfactants revealed similar diffusion coefficients of 0.21 ± 0.05 and 0.20 ± 0.05 µm²/s when mixed with TRI7000 and TRI2500, respectively. These values are slightly lower than the values obtained from the "traditional" water-in-water polymersomes (Itel et al., 2014). Lower diffusion coefficients values in case of water-in-oil droplets can be explained by the fact that the oil, FC-40, viscosity is 3.8 times higher than the viscosity of water. Oil viscosity might be also a potential reason for the similar diffusion coefficients, independent of the surfactant molecular weight. It is worth mentioning here that different concentrations of GFP and the surfactant in the range between 1 and 10 µM and 1 and 10 mM, respectively, did not affect the surfactant mobility.

30.3 FORMATION OF DROPLET-STABILIZED GUVs (dsGUVs)

In this section, we describe the approach that merges lipid bilayer vesicle formation and droplet-based microfluidics for the generation of stable and easily amenable to manipulation dsGUV compartments as schematically illustrated in Figure 30.1. To achieve the formation of dsGUV compartments, aqueous solution containing liposomes was encapsulated within droplets. To create a lipid bilayer within the droplets, we adapted the key factors that are necessary for an efficient formation of planar supported lipid bilayers (e.g., ion type and its concentration). In the following sections, we will describe the essential theoretical and experimental steps that are necessary for the development and analysis of the novel compartment system.

30.3.1 LIPID CONCENTRATION

Droplet-based microfluidics allows for a high-throughput generation of monodisperse droplets (diameter difference <1%) with precise volume and surface area (Thorsen et al., 2001). Therefore, it is possible to estimate the necessary amount of lipid required for formation of continuous lipid bilayer at the droplet surface. Because droplets are spherical, simple arithmetic gives for the necessary lipid concentration needed to cover the droplet by a bilayer

$$C_{lip} = 6/\left(rN_A A_{lip}\right) \tag{30.5}$$

where r is the droplet radius; A_{lip} is the area per lipid in lipid membrane; and N_A is the Avogadro constant. Because the radius of a droplet is two orders of magnitude larger than the thickness of a lipid layer, we have ignored the difference in the radii of the inner and outer membrane leaflet. For lipid area, $A_{lip} \approx 0.7\ \text{nm}^2$ (Petrache et al., 2000) the following lipid concentrations for different droplet dimensions were derived: $C_{lip}(r = 20\ \mu\text{m}) = 712\ \mu\text{M}$; $C_{lip}(r = 50\ \mu\text{m}) = 285\ \mu\text{M}$; and $C_{lip}(r = 75\ \mu\text{m}) = 190\ \mu\text{M}$. These concentrations are essential when preparing dsGUVs. Working with concentrations lower than the critical value will result in a partial lipid bilayer formation. In case of working with concentrations slightly higher than the calculated threshold will lead to accumulation of liposomes in the aqueous phase.

Note: working with lipid concentrations similar to the surfactant concentrations will potentially lead to destabilization of the droplets due to lipid/copolymer surfactant competition on the droplets interface. In these conditions, partial leakage of lipids to the oil phase can be observed.

30.3.2 FORMATION OF LARGE AND GIANT UNILAMELLAR VESICLES

Large unilamellar vesicles (LUVs) can be prepared following the protocols reported earlier (Johnson et al., 2002). In this chapter, we have used the following membrane compositions 1,2-dioleoyl-*sn*-glycero-3-phosphocholine/1,2-dioleoyl-*sn*-glycero-3-phosphoethanolamine/1,2-dioleoyl-*sn*-glycero-3-phospho-L-serine (DOPC:DOPE:DOPS) 8:1:1, including 1% ATTO 488-labeled DOPE. To prepare the LUVs, we first prepare a stock of the desired lipid mixture blended at a high concentration (5 mM) in chloroform and stored at −20°C. The desired amount of lipids is transferred to a tinted glass vial and dried under a gentle stream of nitrogen. To remove traces of the solvent, the lipids are kept under vacuum in a desiccator for roughly 1 h. The dried lipids are then resuspended in MilliQ water or in the water containing 10 mM $MgCl_2$ (see the following section) and vortexed for 1 h. Monodisperse LUVs can be formed by extruding the solution 7 times through a polycarbonate filter with a pore size of 50 nm. The mean LUVs diameter is typically 100 ± 10 nm and can be measured using dynamic light scattering (DLS). Solutions containing LUVs are stored at 4°C for up to 48 h or used immediately for their encapsulation into the droplets (see the following section).

General Note: A similar protocol was successfully applied by us on a range of other lipid compositions such as egg PC:egg PG 1:1 and pure DOPC, see Appendix 1 of the book for structure and data on these and other lipids.

GUVs were formed either in MilliQ water or glucose/sucrose conditions using the electroformation protocol, see Chapter 1. In addition to the lipid mixtures used for the LUVs, a mixture of DOPC/sphingomyelin/cholesterol (DOPC:SM:Chol), with a molar ratio of 7:7:6, was used to examine for phase separation. For observing GUV encapsulated in droplets, a lipid concentration of 100 μM is sufficient. To form dsGUVs, the concentration has to be calculated according to Section 30.3.1.

30.3.3 FORMATION OF dsGUV COMPARTMENTS BY LUVs ENCAPSULATION

To create dsGUV compartments, LUVs in MilliQ water are encapsulated into copolymer-stabilized water-in-oil droplets by means of droplet-based microfluidics. However, no transfer of the encapsulated LUVs to the droplet interface in the form of lipid bilayer was observed, if no additional divalent ions were added (Box 30.1a and b). Therefore, to create dsGUV compartments, LUV solution containing 10 mM $MgCl_2$ was used as aqueous phase for the droplet creation. Box 30.1c shows that the fluorescence intensity was localized at the droplet interface, in comparison to the homogeneous distribution of the fluorescence signal seen in Box 30.1a where no Mg^{2+} ions were applied.

From studies on planar supported lipid bilayers formation, Mg^{2+} ions are known to be the most efficient mediators of lipid vesicle rupturing due to promotion of adhesion to the substrate (Seantier and Kasemo, 2009; Bhatia et al., 2014). It is worth mentioning here that the addition of other divalent ions, such as Ca^{2+} (10 mM) or lower concentrations of $MgCl_2$ (1 mM) did not seem sufficient for dsGUV formation. Using the same lipid concentration (820 μM) only partial LUV fusion to the droplet interface was observed (Figure 30.9).

30.3.4 FORMATION OF dsGUV COMPARTMENTS BY MEANS OF PICO-INJECTION

Microfluidic pico-injection technology (see Section 30.2.2) can be used as an alternative approach to create dsGUVs. Toward this end, two experimental steps are required: the first is the formation

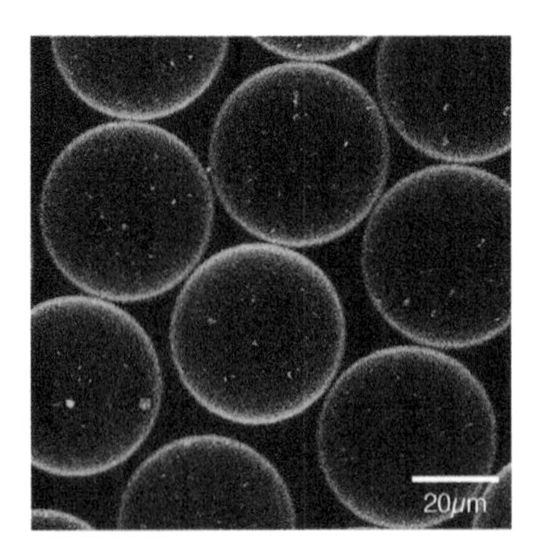

Figure 30.9 Fluorescence cross-section image of the lipid distribution (with the LUV composition DOPC:DOPE:DOPS 8:1:1 including 1% ATTO 488-labeled DOPE, see Appendix 2 of the book for structure and data on this fluorescent dye) within the droplets containing 10 mM Ca^{2+} ions, measured 1 h after formation.

Box 30.1 Formation process of dsGUVs by means of droplet-based microfluidic technology

To create dsGUVs, in the first step, lipids in the form of (a) LUVs or (b) GUVs are encapsulated into water-in-oil droplets. Mg^{2+} ions (10 mM) are introduced into the droplets during droplet creation, Pathway 1, or via pico-injection microfluidic technology, Pathway 2, in order to transfer the encapsulated LUVs or GUVs into the form of supported lipid bilayer at the droplet interface (c). Fluorescence signal in the droplets is due to ATTO 488-labeled DOPE, which is part (1%) of the lipids mixture consisting of DOPC:DOPE:DOPS 8:1:1 (Weiss et al., 2018).

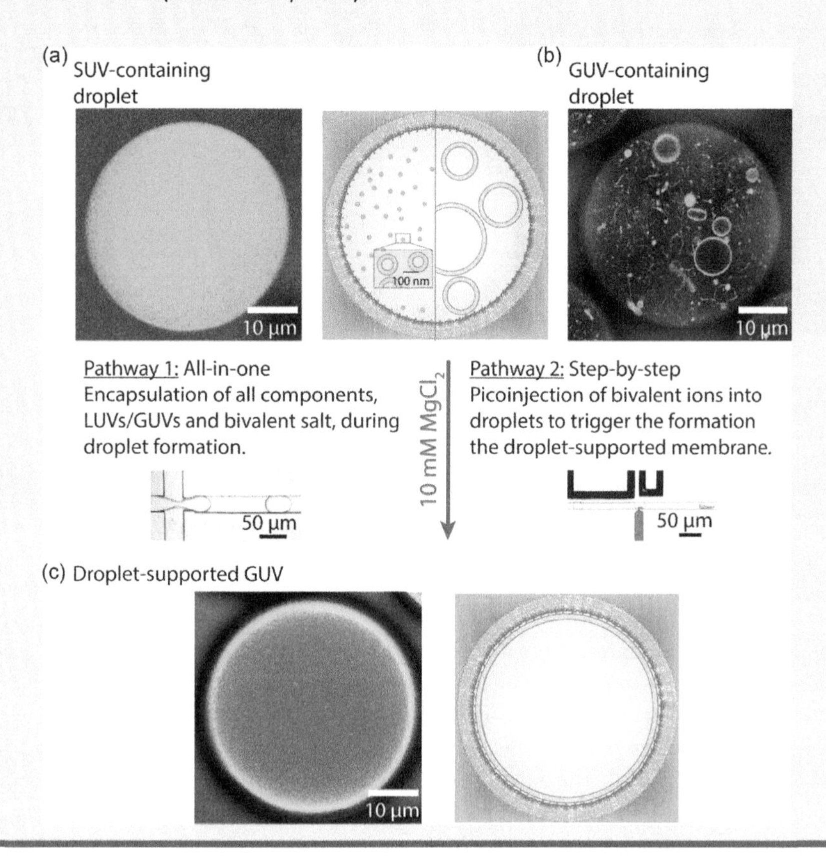

of copolymer-stabilized droplets containing LUVs or GUVs (Box 30.1); the second step is the injection of an ionic aqueous solution of $MgCl_2$ into these droplets to a final ionic concentration of 10 mM. Moreover, by using the pico-injection approach, one can analyze the dsGUV formation process, especially when free-standing GUVs are used for supported lipid bilayer formation (see the following paragraph).

Following the pico-injection step, time-lapse imaging can be used to observe the dynamics of the dsGUVs formation. In case of LUVs-containing droplets, the fusion process of the LUVs to the droplet interface is observed immediately and lasts no longer than a minute. In contrast to LUVs, fusion dynamics of GUVs to the droplet interface is significantly slower—in the range of half an hour (see Movie 30.1 (movies captions at the end of chapter)). These observations can be supported by the following considerations regarding the diffusion of LUV and GUV species: In order to form a lipid bilayer at the droplet interface, the vesicles have to diffuse toward the surface before adsorbing and rupturing to form the supported layer (Hamai et al., 2006). Small vesicles diffuse faster than large ones. The typical time of a particle to travel some mean distance is

$$<t> = \frac{r^2(t)}{4D}, \tag{30.6}$$

where $r^2(t)$ is the squared displacement per time; and $D = k_B T / (6\pi\eta R)$ is the diffusion coefficient of a spherical particle with the radius R in the medium with the known viscosity η (Perrin, 1909). Therefore, for a droplet with a radius $r = 20$ µm, the average time for a vesicle with radius R to fuse from the middle of the droplet to the interface is $<t>$ ($R = 50$ nm) ≈ 15 s for LUVs and $<t>$($R = 5$ µm) ≈ 0.5 h for GUVs.

The pico-injection approach was also used to test the ability of the droplet compartments to support the formation of dsGUVs with lipid phase separation. For this purpose we have probed phase separated GUVs (DOPC:SM:Chol 7:7:6 including 1% Rhodamine B-labeled DOPE, see Appendix 2 of the book for data and structure of fluorescence dyes) formed at 60°C using the electroformation method (see Section 30.3.2) and encapsulated immediately into the droplets (Figure 30.10a). Vesicles of this composition exhibit macroscopic domains of coexisting liquid-ordered and liquid-disordered phases (Bezlyepkina et al., 2013). To achieve the formation of dsGUVs (Figure 30.10b), $MgCl_2$ solution was subsequently injected into these droplets by pico-injection to a final concentration of 10 mM.

As can be observed in Figure 30.10d, no lipid phase separation in the dsGUV is detected. The possible explanation to the absence of domains can mainly be related to the perturbations from the

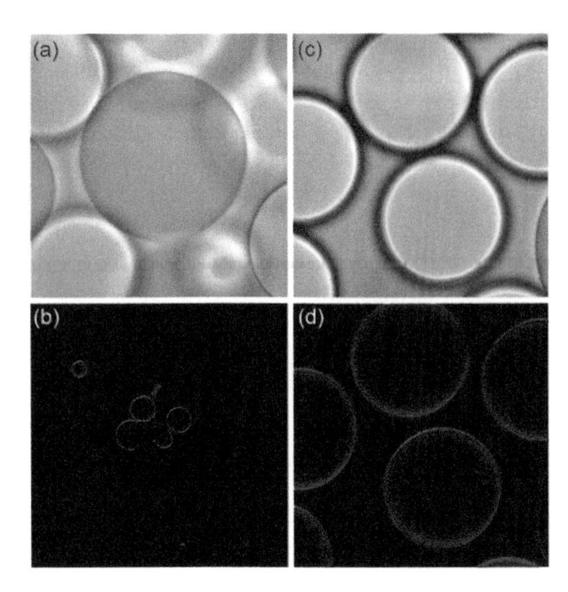

Figure 30.10 Representative (a) phase-contrast and (b) fluorescence images of phase separated GUVs (DOPC:SM:Chol, 3.5:3.5:3 including 1% Rhodamine B DOPE) encapsulated within a droplet; note that only half of some vesicle appear labeled in the cross section in (b) because of preferential partitioning of the fluorescent dye in the liquid disordered domain of the phase separated GUVs. (c) and (d) show representative phase-contrast and fluorescence images of dsGUV observed 0.5 h after injection of $MgCl_2$ into the droplets (10 mM) by means of pico-injection.

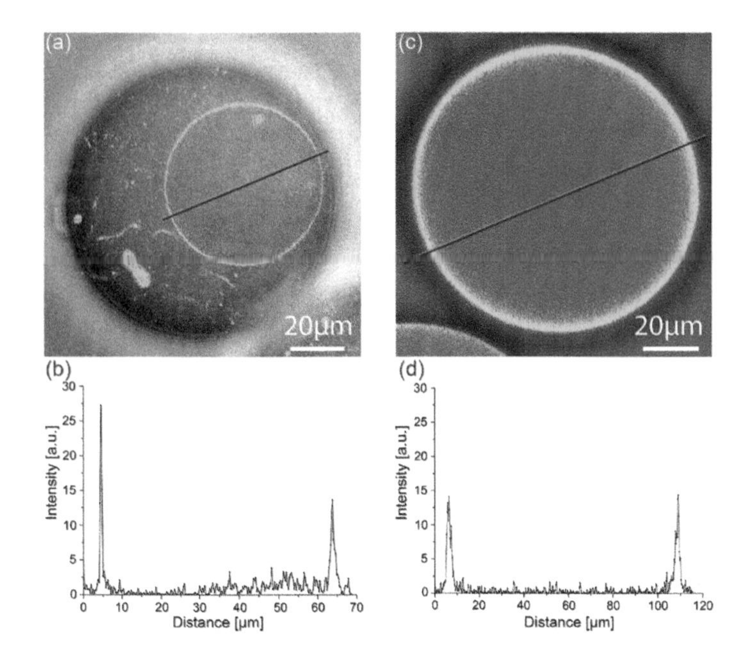

Figure 30.11 (a) Phase-contrast and (c) fluorescence images of the encapsulated GUVs and dsGUVs (egg PC:egg PG, 9:1, 0.5% ATTO 488 DOPE), respectively. (b) and (d) Fluorescence intensity profiles along the indicated lines as presented in (a) and (c), respectively. (From Weiss et al., 2018.)

droplet interface. Whereas bilayers, which are spread from a phase-separated GUV onto a solid support, preserve their phase separation (Bhatia et al., 2014), here the organization of the lipids in the bilayer might constantly be perturbed by the dynamic nature of the droplet support (see Section 30.2.4). The issue of lack of phase separation in the dsGUV membranes remains an open question for further studies. FRAP and fluorescence intensity measurements to confirm the bilayer nature of the lipid membranes in the droplets are presented in the following section.

General notes: The easiest way to generate dsGUVs is by the "all-in-one" approach (Box 30.1), that is, encapsulation of LUVs solution containing 10mM of Mg^{2+} in to the copolymer stabilized droplets. "Step-by-step" - pico-injection approach (Box 30.1) should be applied in case an analysis of dsGUVs formation is required.

30.3.5 FLUORESCENCE INTENSITY ANALYSIS AND FRAP MEASUREMENTS OF dsGUVs

To compare the membrane fluorescence intensity of encapsulated GUVs to that of dsGUVs as shown in Figure 30.11, GUVs (egg PC:egg PG, 9:1, including 0.5% ATTO 488-labeled DOPE) were encapsulated into the droplets and dsGUVs were produced from the same lipid composition using the "all-in-one" approach (see Box 30.1). Both types of droplets were evaluated with identical settings of the confocal microscope (see Section 30.2.4). At least twenty intensity profiles were extracted for each droplet type.

From both, microscope images as well as evaluated data, a weak blur close to the droplet interface is observed. This is caused by refraction and diffraction at the water-oil interface due to a slight difference in the refractive indices of water (1.333) and FC-40 oil (1.290). This effect causes a widening of the intensity profile (Figure 30.11b) and a reduction of the fluorescence intensity amplitude of the GUV part close to the droplet interface. Therefore, to compare fluorescence intensities, a Gaussian function with a background correction was fitted to the intensity peak profiles using a nonlinear least-square fit (MATLAB 2015 SP1). The fitting revealed similar integrated intensity values of 42 ± 8 and 44 ± 4 a.u. × µm for dsGUVs and encapsulated GUVs, respectively. These findings suggest that structurally, freely suspended GUVs and dsGUVs consist of the same lipid bilayers.

FRAP measurements were performed to compare the lipid diffusion coefficients in encapsulated GUVs and dsGUVs. Toward this end, the same microscope settings and analysis approach were applied as presented in Section 30.2.4. It is worth mentioning here that full recovery of the bleaching spot was observed in all measurements. Table 30.2 presents the summary of diffusion coefficients of encapsulated GUVs and dsGUVs, consisting of various

Table 30.2 **Summary of the diffusion coefficients obtained by FRAP measurements**

DIFFUSION COEFFICIENT LIPID COMPOSITION	ENCAPSULATED GUV [µm²/s]	dsGUV [µm²/s]
DOPC:DOPE:DOPS (8:1:1) + 1% ATTO488-labeled DOPE	3.52 ± 0.26	3.31 ± 0.77
Egg PC:Egg PG (1:1) +1% ATTO488-labeled DOPE	3.96 ± 0.51	2.88 ± 0.06
DOPC + 1% rhodamine B-labeled DOPE	4.42 ± 0.65	4.11 ± 0.59

lipid composition and fluorophore types. The data shows a weak slowdown of diffusion in the dsGUV membrane. This outcome can be related to the fact that supported lipid membranes are subject to pertubation from the copolymer shell of the droplet, whose mobility is an order of magnitude lower (see Section 30.2.4). Moreover, it is known that divalent ions bind to phosphatidylcholine membranes (Sinn et al., 2006) and lead to a decrease in the self-diffusion of lipids in the membrane (Böckmann and Grubmüller, 2004). Furthermore, FRAP and fluorescence correlation spectroscopy (FCS) measurements performed in other studies with similar lipid compositions revealed diffusion coefficients in the same range as well as a similar tendency to lower values in the case of supported lipid membranes (Machan and Hof, 2010; Bhatia et al., 2014).

30.4 dsGUV BIOFUNCTIONALIZATION

In order to dissect complex cellular sensory machinery by means of an automated droplet-based microfluidic approach, dsGUVs have to be adapted to allow the functional bottom-up assembly of various sub-cellular functional units. Thus, we focused on developing high-throughput strategies for incorporating transmembrane proteins, such as integrin and ATP synthase, and for immobilizing proteins to the biofunctionalized dsGUVs.

30.4.1 INTEGRIN LUVs

Integrin $\alpha_{IIb}\beta_3$ was purified from human blood platelets using TBS and Triton® X-100 according to a previously reported protocol (Eberhard, 2012). Affinity chromatography over Concanavalin A and Heparin columns was followed by gel filtration over a Superdex 200 Prep Grade column. The biological activity of the purified integrin was analyzed by an enzyme-linked immunosorbent assay (ELISA) using AB1967 anti-integrin α_{IIb} antibodies. Following purification, integrin were stored at –80°C in TRIS storage buffer, consisting of 20 mM TRIS/HCl pH 7.4, 150 mM NaCl, 1 mM $CaCl_2$, 1 mM $MgCl_2$, 0.1% (w/v) Triton X-100, 0.02% (w/v) NaN_3 and 2 mg/mL Aprotinin. Integrin $\alpha_{IIb}\beta_3$ was labeled with (5-(and-6)-carboxytetramethylrodamine, succinimidyl ester, #C1171 TAMRA (Eberhard, 2012). The protein/dye ratio was measured from molecular dye extinction ($\varepsilon_{555} = 80000\ M^{-1}cm^{-1}$) and the functionality was assessed with an ELISA binding essay.

Integrin $\alpha_{IIb}\beta_3$ were reconstituted into the LUVs according to a previously published protocol (Erb et al., 1997). Briefly, a chloroform lipid mixture of egg PC to egg PG 1:1 molar ratio was dried under a gentle flow of nitrogen and then placed in a desiccator overnight. The dried lipids were dissolved in integrin reconstitution buffer, consisting of 20 mM TRIS/HCl, pH 7.4, 50 mM NaCl, 0.5 mM $CaCl_2$, and 0.1% (w/v) Triton X-100. Integrin proteins in a storage buffer (see previous paragraph) were added to the final 1:1000 integrin/lipid ratio. The solution was incubated in a shaker at 600 rpm at 37°C for 2 h. Triton® X-100 was removed in two subsequent washing steps of 3.5 h each using 50 mg/mL BT Bio-Beads® SM-2. The size distribution of the integrin-containing proteoliposomes (100 ± 10 nm) was measured by DLS. For experiments with droplets containing integrin (see Section 30.4.3) activation buffer, consisting of 20 mM TRIS/HCl, pH 7.4, 50 mM NaCl, 0.5 mM $CaCl_2$, 1 mM $MnCl_2$, 1 mM $MgCl_2$ was used.

30.4.2 F_0F_1-ATP SYNTHASE LUVs

Alexa488-labeled F_0F_1-ATP synthase was kindly provided by Prof. Dr. Michael Börsch (Friedrich Schiller University Jena). Purification and labeling of F_0F_1-ATP synthase from *E. coli* were performed according to the protocol described by Zimmermann et al. (2005) and Heitkamp et al. (2016). Labeled ATP synthase was reconstituted into preformed LUVs (≈ 120 nm) as described by Fischer and Graber (1999) and stored at –80°C in tricine buffer, consisting of 20 mM tricine–NaOH (pH 8.0), 20 mM succinic acid and 0.6 mM KCl. For experiments with droplets containing F_0F_1-ATP synthase (see Section 30.4.3), ATP synthase working buffer, consisting of 20 mM tricine–NaOH (pH 8.0), 20 mM succinic acid, 0.6 mM KCl, 50 mM NaCl and 2.5 mM $MgCl_2$, was used.

30.4.3 PROTEIN RECONSTITUTION IN dsGUVs

Protein reconstitution in free-standing GUVs is addressed in detail in Chapter 3. Here, we describe an approach applicable to dsGUVs. Two experimental steps are required for proteoliposomes fusion (i.e., liposomes containing TAMRA-labeled $\alpha_{IIb}\beta_3$ integrin or ATTO 488-labeled F_0F_1-ATP synthase) with the preformed dsGUVs consisting of DOPC:DOPE:DOPS (8:1:1), including 1% ATTO488-labeled DOPE or 1% Rhodamine B-labeled DOPE, respectively. The first step is the creation of dsGUVs with the diameter of 40 μm using a solution of 800 μM liposomes and integrin activation buffer (Section 30.4.1) or ATP synthase working buffer (Section 30.4.2) following the all-in-one approach (Box 30.1). The second step is injection of proteoliposome solution into these droplets by means of pico-injection technology as schematically shown in Figure 30.12a. Colocalization of protein and lipid fluorescence signals is observed (Figure 30.12b–e), indicating successful fusion of proteoliposomes with the dsGUVs. Provided proteoliposomes are readily obtained; we expect this approach to be applicable for reconstitution of other types of transmembrane proteins into dsGUVs.

30.4.4 BIOFUNCTIONALIZATION OF dsGUVs WITH (His6)-PROTEINS

In order to link (His6)-GFP to the dsGUV lipid bilayer, the following procedure based on the use of DGS-NTA lipids can be employed. Toward this end, to form DGS-NTA(Ni) complex, $NiCl_2$ (9 μL, 100 mM) is mixed with water solution of DGS-NTA (300 μL, 1 mM) and stirred for 20 min. Following the DGS-NTA(Ni) complex formation, DOPC lipids are mixed with the DGS-NTA(Ni) to a final ratio of 9:1. LUVs composed of DOPC:DGS-NTA(Ni), 9:1 (220 μM) are encapsulated at a lipid concentration of 200 mM into the droplets with a diameter of 150 μm. Following encapsulation, 350 pL water solution containing (His6)-GFP (10 μM) and 40 mM $MgCl_2$ is pico-injected into each LUVs-loaded droplet in order to form dsGUVs with immobilized (His6)-GFP. Figure 30.13a, b shows fluorescence signal from (His6)-GFP within the DGS-NTA(Ni)-containing dsGUVs and within droplets, containing no NTA-lipids, respectively. It can be observed that the fluorescent signal is localized at the membrane of the DGS-NTA(Ni)-containing dsGUV in comparison to an equally distributed signal within the droplets

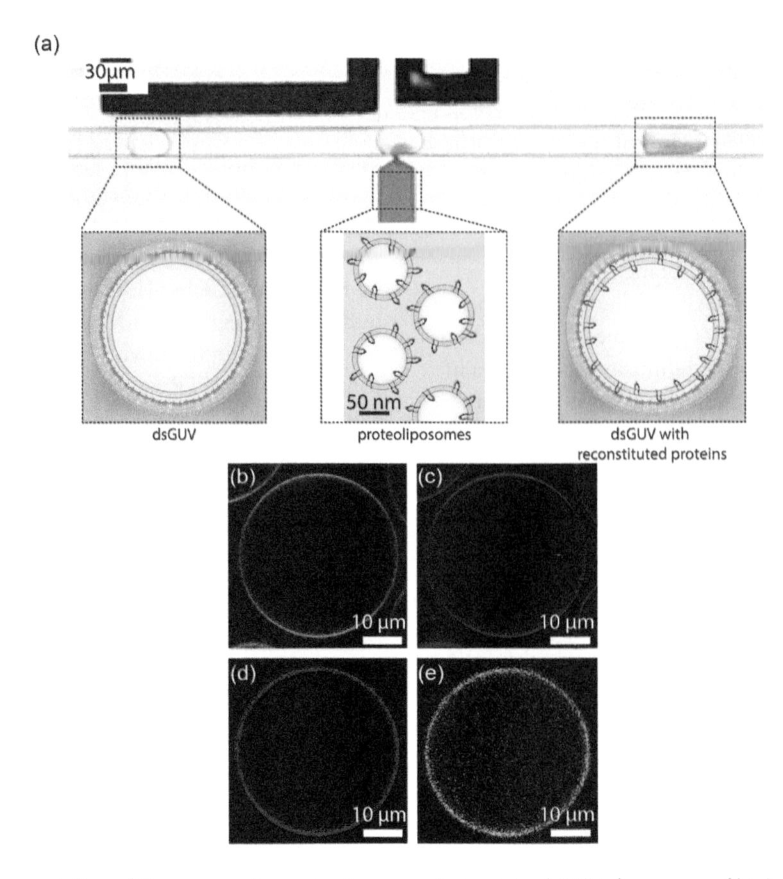

Figure 30.12 (a) Schematic representation of the process for proteoliposome fusion into dsGUVs by means of high-throughput pico-injection microfluidics. (b) and (d) are representative fluorescence images of the dsGUVs (DOPC:DOPE:DOPS, 8:1:1) 10 minutes after pico-injection, containing 1% ATTO 488- or Rhodamine B-labeled DOPE, respectively. (c) and (e) are representative fluorescence images of TAMRA-labeled $\alpha_{IIb}\beta_3$ integrin and ATTO 488-labeled F_0F_1 ATP synthase incorporated in the dsGUVs presented in (b) and (d), respectively. (From Weiss et al., 2018.)

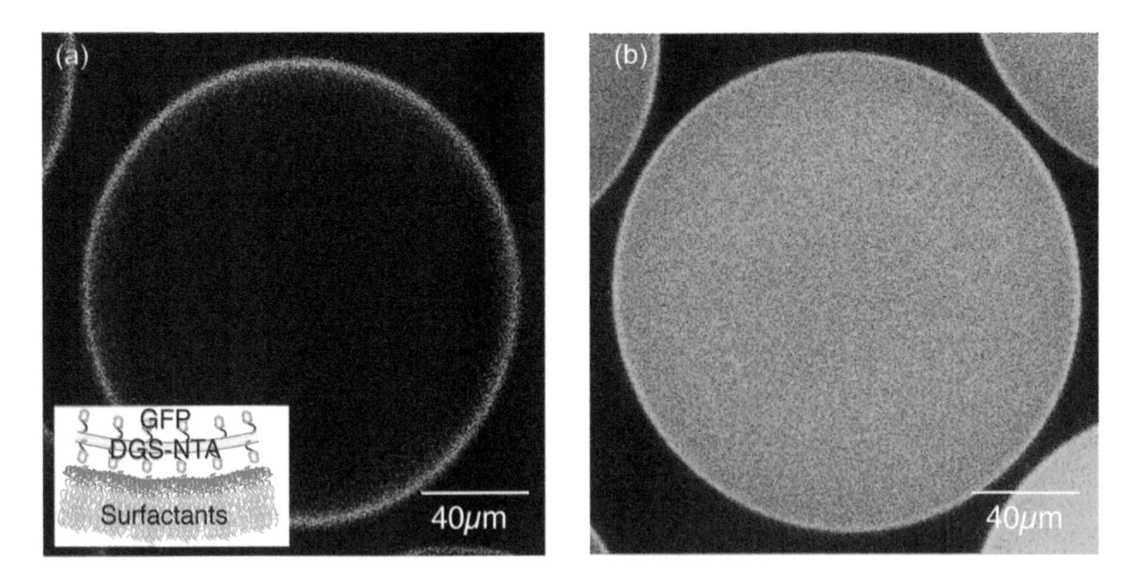

Figure 30.13 (a) A fluorescence cross-section image of (His6)-GFP linked to DGS-NTA(Ni)-containing dsGUV. (b) In contrast to the dsGUV containing NTA-lipids, the fluorescence intensity in the dsGUV containing DOPC only is distributed equally. (From Weiss, M. et al., 2018.)

containing no functionalized lipids. FRAP measurements of the GFP-decorated dsGUVs revealed a diffusion coefficient of 1.22 ± 0.03 μm²/s. Successful functionalization of the dsGUVs with (His6)-GFP is particularly important because the chemistry behind the immobilization of these proteins is the same as required for immobilization of many other proteins in synthetic biology applications.

30.4.5 MOBILITY OF TRANSMEMBRANE PROTEINS IN dsGUVs

FRAP measurements were performed to investigate the mobility of transmembrane proteins reconstituted into the dsGUVs with a diameter of 150 μm. The same microscopy settings and analysis approach were applied as described in

Table 30.3 **Summary of the diffusion coefficients of lipids and proteins reconstituted in protein-decorated dsGUVs obtained by FRAP measurements, see Appendices 1 and 2 of the book for structure and data on the lipids and lipid dyes**

LIPID COMPOSITION	PROTEIN	D_{Lip} [µm²/s]	D_{Pro} [µm²/s]
DOPC:DGS-NTA (9:1)	GFP	---	1.22 ± 0.03
DOPC:DOPE:DOPS (8:1:1) +1% Rhodamine B-labeled DOPE	F_0F_1 ATP synthase	2.80 ± 0.39	1.15 ± 0.76
DOPC:DOPE:DOPS (8:1:1) + 1% ATTO 488-labeled DOPE	Integrin (Proteoliposome)		0.67 ± 0.10
DOPC:DOPE:DOPS (8:1:1) + 1% ATTO 488-labeled DOPE	Integrin (pure)	2.27 ± 0.22	0.70 ± 0.06
DOPC:DOPE:DOPS (8:1:1) + ATTO 488-labeled DOPE	Integrin (Proteoliposome) + RGD	2.27 ± 0.16	0.13 ± 0.03

Sections 2.4 and 3.5. Table 30.3 presents the summary of diffusion coefficients of lipids D_{Lip} in protein-decorated dsGUVs and of the corresponding incorporated proteins D_{Pro}. In all cases, the measured D_{Lip} values were lower in comparison to the diffusion coefficients of the dsGUVs containing no proteins (see Table 30.2). Lower diffusion coefficient values can be attributed to the fact that the lipid lateral diffusion is a subject to steric and charge-related perturbations from the incorporated proteins (May et al., 2000). As can be observed from Table 30.3, the FRAP measurements indicated similar diffusion coefficient values of $D_{Pro} \approx 0.7$ µm²/s for integrin, independently on whether they were introduced as pure proteins or as proteoliposomes. These values are in good agreement with previously published studies on integrin $\alpha_{IIb}\beta_3$ lateral mobility in planar supported lipid bilayers or in the cellular membranes as obtained by FRAP (Erb et al., 1997; Goennenwein et al., 2003) and FCS measurements (Edel et al., 2005), respectively.

To test the functionality of the incorporated integrin proteins, nanostructured droplets-containing RGD-linked surfactants (see Section 30.2.3) were used to provide binding sites for integrin adhesion. In this case (see Table 30.3), the diffusion coefficient of integrin dropped significantly ($D_{Pro} = 0.13$ µm²/s) to values that represent the mobility of the surfactant layer. This observation might indicate a successful establishment of a linkage between the transmembrane integrin and RGD-peptides on the copolymer droplet interface. It also reveals that at least some of the integrin proteins are oriented correctly, that is, that the extracellular part points toward the copolymer-stabilized droplet inner interface.

30.5 APPROACHES FOR RELEASE OF GUVs FROM dsGUVs

The polymer-based surfactant shell provides great stability to the dsGUV and, therefore allows the sequential loading of the compartment with biomolecules. However, it greatly restricts the possibility to study the behavior of the GUV-based model system in a physiological environment. Toward this end, we developed approaches to recover/extract the synthetic protocells from the dsGUVs. Note that in the context of synthetic biology, protocells are synthetic, biomolecules-containing lipid-based compartments.

Several successful methods have been published to release content of polymer-stabilized droplets into continuous aqueous phase. However, most of these methods were developed for the release of cells that present high mechanical resistance to the forces during droplet coalescence (Platzman et al., 2013). When designing the methods for GUV release, one has to take in to account the fragile nature of cell-size lipid vesicles due to their low mechanical and chemical stability. Therefore, we decided to use copolymer-based destabilizing surfactants, or de-emulsifiers, for the gentle GUV release. These de-emulsifiers partially replace their stable counterparts at the droplet interface and by doing so reduce the energy barrier to allow droplet coalescence. In the following paragraphs two developed methods for GUVs release will be described.

30.5.1 BULK RELEASE

Using destabilizing surfactants, bulk de-emulsification is probably the most intuitive approach to release the content of droplets. Efficient release can be achieved by applying the destabilizing surfactant perfluoro-1-octanol (370533, Sigma-Aldrich, USA). The steps of our protocol to achieve bulk release are sketched in Figure 30.14.

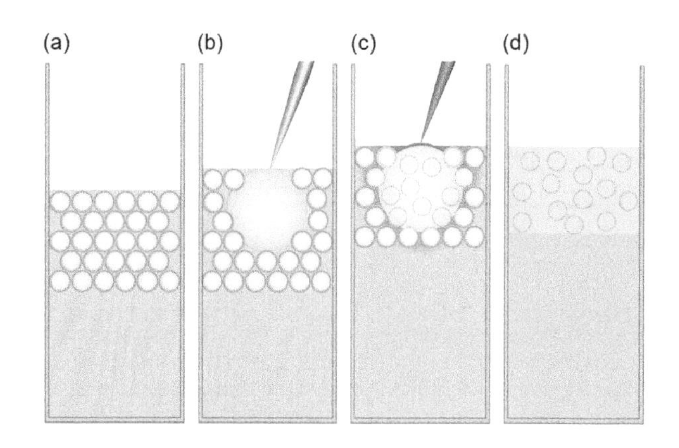

Figure 30.14 Schematic representation of experimental steps necessary for bulk release of GUVs from dsGUVs. (a) Droplets are stored in an Eppendorf tube. (b) To provide an aqueous phase for release, 100 µL of the buffer are added to the droplet layer. (c) 100 µL of the de-emulsifier solution is added dropwise on the buffer drop. (d) The emulsion brakes and the released GUVs go to the aqueous phase. (Reproduced from Fronmayer (2017) with permission of the author.)

For the bulk release approach, 100 μL of formed dsGUVs were collected in an Eppendorf tube. Due to the density differences between the FC-40 oil and water, the dsGUVs form a dense layer at the top of the tube. To provide an aqueous phase for release, 100 μL of buffer was placed as a one large drop in the center of the droplet layer. To reduce osmotic pressure effect, it is preferable that the buffer ionic content will be identical to the buffer content within the dsGUVs. Following the addition of buffer, a 20 vol% of de-emulsifier FC-40 oil solution was gently dripped on top of the buffer drop. After applying the complete volume of de-emulsifier, the tube was tilted to increase the interface area and slowly rotated about its longitudinal axis. In that conditions breaking the emulsion takes less than five minutes. Residual oil drops in the aqueous phase can be centrifuged down by briefly spinning at low speed with a table-top centrifuge. The aqueous solution containing GUV can be carefully removed with a pipette and immediately used for the analysis in the observation chamber (Figure 30.8) or for the following experiments.

30.5.2 MICROFLUIDIC SETUP FOR GUV RELEASE

A microfluidic device for GUVs release (Figure 30.15) was designed to allow monitoring of the GUVs release process under controlled high-throughput conditions. Observation of the release process allowed us to optimize the parameters necessary for high release yield. Moreover, findings obtained from microfluidic release setup can be in turn used to improve bulk release conditions.

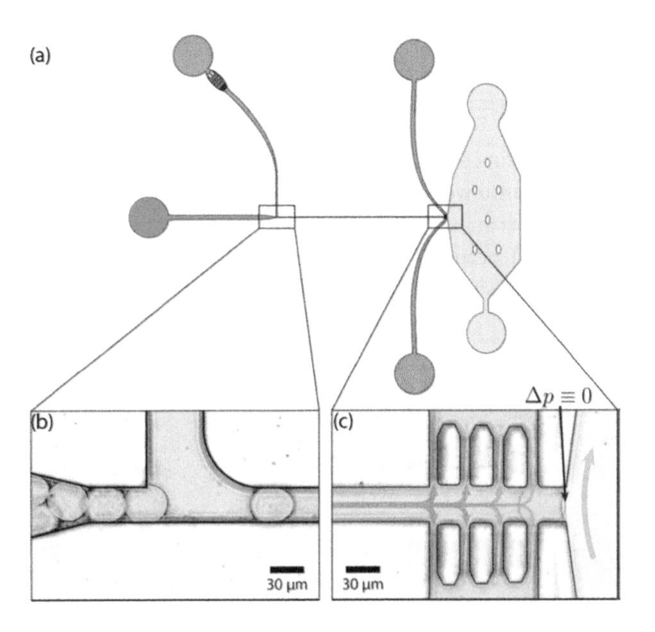

Figure 30.15 Schematics of the microfluidic device for release of GUVs from dsGUVs. (a) For the sake of presentation clarity, the droplet injection channel is marked in green, the droplet separating channel (introduction of de-emulsifier-containing oil) and the outlet oil channels are marked in orange and the aqueous wide channel in blue. (b) Bright-field microscopy image of the spacing T-junction. The displacement of each phase without droplets is indicated in a colored overlay—orange for de-emulsifier-containing oil surfactant and green for oil from the reinjection channel. (c) Bright-field microscopy image of the passive trapping structures (i.e., rows of pillars separated by distances smaller than the representative droplets dimensions) and the recovery area. (From Weiss et al., 2018.)

In the microfluidic device for GUVs release, a flow control system (MFCS-EZ, Fluigent, France) was used to control the pressure in the aqueous and oil inlet channels. To minimize stress on the droplets (e.g., shear forces), pressure levels on all inlets were set below 20 mbar with minor corrections for individual setups and experimental conditions. Moreover, the channel heights were designed to exceed the droplets diameter.

Preformed dsGUVs were reinjected into the release chip and separated at the T-junction (Figure 30.15b) by an additional oil flow containing 20 vol% perfluoro-1-octanol de-emulsifier. This T-junction is identical to the structure used in pico-injection devices for droplets separation (see Section 30.2.2). The difference here is that following the T-junction, the stabilizing surfactants are replaced by the de-emulsifier from the spacing oil. At similar pressure on both channels, the spacing channel displaces around 80%–90% of the oil from the droplet channel. The total flow was adjusted to allow efficient time (~300 ms) to replace the stabilizing surfactants by the de-emulsifier prior to reaching the release unit where dsGUV encounter the aqueous phase in a wide perpendicular channel. To minimize the mechanical impact on the droplets at the oil/water junction, passive trapping structures within the microfluidic channels (i.e., rows of pillars separated by slits) have been designed and used to decelerate the droplets before coming in contact with the aqueous phase (Figure 30.15c). On both sides, the trapping unit is connected to the adjacent outlet oil channels. To allow only the oil phase to flow to the outlet channels, the width of these slits was designed to be smaller than the representative droplets dimensions. In addition, to avoid net flow between the oil and aqueous channels, the aqueous flow was adjusted to achieve zero pressure gradient, $\Delta p \equiv 0$, at the oil/water junction when no droplet is in the trapping structures. When a droplet enters the trapping zone, it blocks the slits on both sides, therefore reducing the channel total cross section connecting the main and adjacent oil channels. According to the Hagen-Poiseuille equation, this increases the pressure. As the droplet traverses along the trapping area, it passes pairs of slits, opening them up again for oil flow. With each open pair of slits, the channel cross section for the oil to flow to the adjacent channels increases, subsequently decreasing pressure pushing the droplet along the channel. The droplet decelerates as it approaches the oil-water interface. Provided the concentration of destabilizing surfactant is sufficient, the residual surfactant layer peels off the droplet at contact with the water phase and its content is released into the water phase.

30.5.3 OBSERVATION OF RELEASED GUVs

Released GUVs can be observed in similar observation chambers as in Figure 30.8. GUVs are mechanically less stable than droplets and can rupture when coming into contact with bare glass surfaces. Therefore, prior to the assembly of the chamber, the glass slides were passivated with bovine serum albumin (BSA): the glasses were first activated in an oxygen plasma (0.4 mbar O_2, 150 W, 10 min) and then incubated for 2 h in 10 mg/mL BSA (SERVA, Germany) dissolved in PBS (Gibco, USA) on a see-saw rocker. Next, the glass slides were rinsed twice with PBS and washed in PBS for 5 minutes. The washing process was repeated with MilliQ water. Finally, the glass slides were dried with nitrogen.

30.5.4 LIPID COMPOSITIONS FOR RELEASE

The release efficiency of GUVs was found to be strongly dependent on their lipid composition. A few general key factors for the stability of lipid bilayers have been identified: (1) In accordance to the studies on the free-standing GUVs, cholesterol in the range of 10–20 mol% has a positive effect on the stability (durability) of the lipid membranes. Therefore, for the release experiments we included 10 mol% cholesterol (C8667, Sigma-Aldrich, USA); (2) Another important outcome of the release experiments was that GUVs consisting of lipid compositions exceeding a net concentration of 10 mol% of negatively charged lipids yielded only low release efficiency; (3) In addition, it was found that replacing part of DOPC by POPC enhanced membrane stability and, therefore better release yield. Finally, the lipid composition used for the results presented here, consisted of a molar ratio of 4:4:2 of DOPC, POPC and cholesterol, respectively. For release experiments 1 mM of LUVs were prepared through extrusion.

30.5.5 RELEASE OF INTEGRIN-RECONSTITUTED GUVs

For the release of $\alpha_{IIb}\beta_3$ integrin-reconstituted GUVs, the protocol for detergent removal presented in Section 30.4.1 was slightly modified. Instead of a 50 mol%: 50 mol% of egg PC: egg PG lipid mixture, pure egg PC was used. Note that the major constituent in egg PC is POPC. Thus, we omitted lipids with a negative net charge completely. The rest of the protocol was kept as described earlier. A second solution of liposomes was produced through extrusion with a total lipid concentration of 1.8 mM. The lipid composition for release listed in Section 30.4 was changed to a molar ratio of 4:3:2 of DOPC, POPC and cholesterol.

For droplet production both liposome solutions were mixed in a 1:9 ratio of proteoliposomes and liposomes with a total concentration of 1 mM in activation buffer resulting in dsGUV consisting of lipid composition with a molar ratio of 4:4:2 of DOPC, POPC/egg PC and cholesterol (Figure 30.16).

30.6 SUMMARY AND OUTLOOK FOR THE FUTURE

In this chapter, we explored the capacity of droplet-based microfluidics for high-throughput bottom-up assembly of cell-like compartments for synthetic biology applications. To illustrate the necessity of our novel compartment system, we addressed the drawbacks of the currently available protocell systems and described technological limitations related to their manipulation. Thereafter, we summarized in detail our recently developed approach that merges lipid vesicle formation and droplet-based microfluidics for the generation of stable dsGUV compartments. By applying several droplet-based microfluidic functional techniques, including droplet generation and pico-injection, the versatility and robustness of the dsGUV compartment system for high-throughput manipulations was presented. The combination of these various technologies allows the sequential assembly of lipid-based model systems with a high complexity. Moreover, by presenting methods to recover the constructed protocells from the dsGUV, we show that this method allows studying their interaction with a physiological environment.

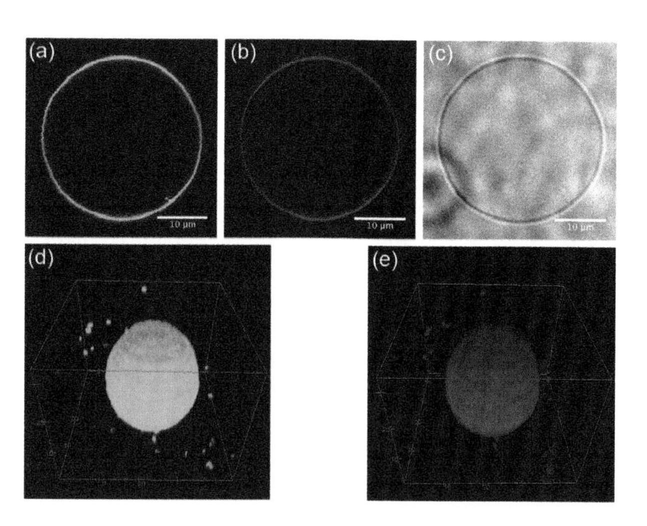

Figure 30.16 Representative fluorescence and bright-field images of $\alpha_{IIb}\beta_3$ integrin-reconstituted GUVs released from dsGUV. Fluorescence signal from (a) ATTO488-DOPE, (b) integrin $\alpha_{IIb}\beta_3$ labeled with TAMRA and (c) a bright-field image of the released GUV. Representative 3D reconstitution of confocal images of released GUV. Fluorescence signals are (d) ATTO 488-labeled DOPE lipids and (e) TAMRA-labeled $\alpha_{IIb}\beta_3$ integrin. The GUVs shown here preserved the size (30 μm) of the droplets they were released from. In the projection volume, fragments of broken GUVs are visible around the intact GUV. (Reproduced from Fronmayer (2017) with permission of the author.)

It is our expectation that the pliable biophysical properties of the dsGUV compartments and their integration to microfluidic technology might provide a system with superior properties for the assembly of a wide range of subcellular functional units. A potential example of application is adhesion-associated complexes and cytoskeleton filament organization as schematically illustrated in Figure 30.17.

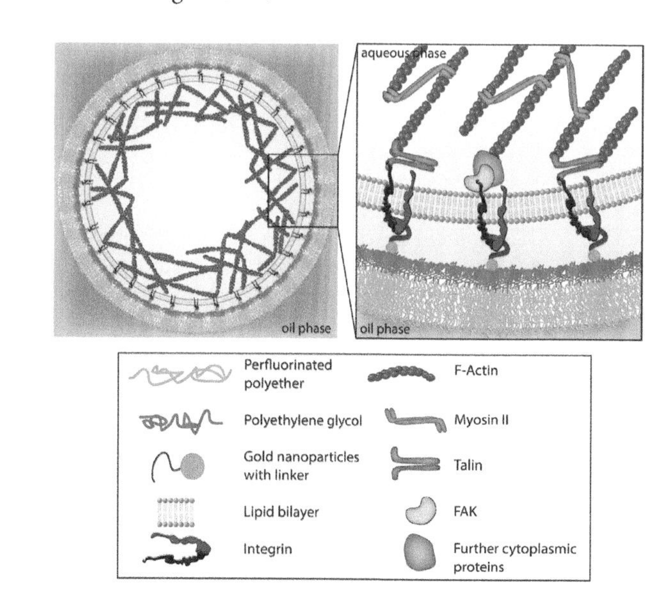

Figure 30.17 Schematic representation of a bio-inspired "dsGUV minimal synthetic cell." To provide bioactivity in terms of integrin adhesion, RGD peptides are immobilized on the gold nanoparticles via a thiol linker. G-actin and other proteins can be subsequently introduced into the dsGUVs by means of pico-injection technology. The dsGUV are ready for optical microscopy and force investigations and can be employed as cell-sized compartments within which interactions between different adhesion-associated proteins are systematically analyzed by means of a high-throughput screening platform. (Reproduced from Fronmayer (2017) with permission of the author.)

MOVIE

The Movies can be found under http://www.crcpress.com/9781498752176.

Movie 30.1: Formation process of dsGUVs out of encapsulated GUVs encapsulated into microfluidic droplets. Time lapse fluorescence images showing the equatorial plane of a microfluidic droplet with encapsulated, ATTO 488 labeled DOPE GUVs. After the addition of $MgCl_2$, GUVs adsorb, rupture and fuse to form a continuous lipid bilayer, supported by the droplet interface.

ACKNOWLEDGMENTS

Parts of the research leading to these results have received funding from the European Research Council/ERC Grant Agreement no. 294852, SynAd. This work is also part of the MaxSynBio consortium, which is jointly funded by the Federal Ministry of Education and Research of Germany and the Max Planck Society. The work was also partly supported by the SFB 1129 of the German Science Foundation and the Volkswagen Stiftung (priority call "Life?") I.P. gratefully acknowledges the support of the Alexander von Humboldt Foundation. J.P.S. is the Weston Visiting Professor at the Weizmann Institute of Science. The authors acknowledge Ms. Mollenhauer for her support with protein purification. The Max Planck Society is appreciated for its general support in all aspects of our research.

LIST OF ABBREVIATIONS

AC	alternating current
AU	arbitrary unit
Chol	cholesterol
DGS	1,2-dioleoyl-*sn*-glycero-3-succinate
DLS	dynamic light scattering
DOPC	1,2-dioleoyl-*sn*-glycero-3-phosphocholine
DOPE	1,2-dioleoyl-*sn*-glycero-3-phosphoethanolamine
DOPS	1,2-dioleoyl-*sn*-glycero-3-phospho-L-serine
DSB	droplet supported lipid-bilayer
ELISA	enzyme-linked immunosorbent assay
FCS	fluorescence correlation spectroscopy
FRAP	fluorescence recovery after photobleaching
dsGUV	droplet-stabilized giant unilamellar vesicle
GUV	giant unilamellar vesicle
ITO	indium tin oxide
LUV	large unilamellar vesicle
NTA	nitrilotriacetic acid
PDMS	poly(dimethylsiloxane)
PEG	polyethylene glycol
PFPE	perfluorinated polyether
SM	sphingomyelin
THF	tetrahydrofuran
TRI2500	PFPE (2,500 g/mol)–PEG(600 g/mol)–PFPE(2,500 g/mol)
TRI7000	PFPE (7,000 g/mol)–PEG(1,400 g/mol)–PFPE(7,000 g/mol)

GLOSSARY OF SYMBOLS

A_{lip}	lipid area
c_{lip}	lipid concentration
D	diffusion coefficient
$I(t)$	fluorescence intensity value of the bleaching spot
I_{bg}	fluorescence intensity value of the oil phase background
I_{nor}	normalized fluorescence intensity value
I_{pb}	prebleaching fluorescence intensity value of the bleaching spot
k_B	boltzmann constant
N_A	avogadro constant
N_{lip}	number of lipids
r	radius of droplet
R	radius of vesicle
t	time
$T(t)$	fluorescence intensity value of the whole droplet base
T_{pb}	prebleaching value of the whole droplet base
ε	extinction coefficient
η	viscosity
$\tau_{1/2}$	half-recovery time of the bleaching experiment

REFERENCES

Abate AR, Hung T, Mary P, Agresti JJ, Weitz DA (2010) High-throughput injection with microfluidics using picoinjectors. *Proc Natl Acad Sci U S A* 107:19163–19166.

Agapakis CM, Boyle PM, Silver PA (2012) Natural strategies for the spatial optimization of metabolism in synthetic biology. *Nat Chem Biol* 8:527–535.

Axelrod D, Koppel DE, Schlessinger J, Elson E, Webb WW (1976) Mobility measurement by analysis of fluorescence photobleaching recovery kinetics. *Biophys J* 16:1055–1069.

Bermudez H, Brannan AK, Hammer DA, Bates FS, Discher DE (2002) Molecular weight dependence of olymersome membrane structure, elasticity, and stability. *Macromolecules* 35:8203–8208.

Bezlyepkina N, Gracia RS, Shchelokovskyy P, Lipowsky R, Dimova R (2013) Phase diagram and tie-line determination for the ternary mixture DOPC/eSM/Cholesterol. *Biophys J* 104:1456–1464.

Bhatia T, Husen P, Ipsen JH, Bagatolli LA, Simonsen AC (2014) Fluid domain patterns in free-standing membranes captured on a solid support. *Biochim Biophys Acta-Biomembr* 1838:2503–2510.

Böckmann RA, Grubmüller H (2004) Multistep binding of divalent cations to phospholipid bilayers: A molecular dynamics study. *Angew Chem Int Ed Engl* 43:1021–1024.

Carrara P, Stano P, Luisi PL (2012) Giant vesicles "colonies": A model for primitive cell communities. *ChemBioChem* 13:1497–1502.

deMello AJ, Wootton RCR (2009) Chemistry at the crossroads. *Nat Chem* 1:28–29.

Diekmann Y, Pereira-Leal JB (2013) Evolution of intracellular compartmentalization. *Biochem J* 449:319–331.

Discher DE, Eisenberg A (2002) Polymer vesicles. *Science* 297:967–973.

Duffy DC, McDonald JC, Schueller OJA, Whitesides GM (1998) Rapid prototyping of microfluidic systems in poly(dimethylsiloxane). *Anal Chem* 70:4974–4984.

Eberhard C (2012) Development of a model system to study cell adhesion and cell mechanics. Dissertation, Ruperto-Carola University of Heidelberg, DOI:10.11588/heidok.00014257.

Edel JB, Wu M, Baird B, Craighead HG (2005) High spatial resolution observation of single-molecule dynamics in living cell membranes. *Biophys J* 88:L43–L45.

Elani Y, Law RV, Ces O (2014) Vesicle-based artificial cells as chemical microreactors with spatially segregated reaction pathways. *Nat Commun* 5:5.

Erb EM, Tangemann K, Bohrmann B, Muller B, Engel J (1997) Integrin alpha IIb beta 3 reconstituted into lipid bilayers is nonclustered in its activated state but clusters after fibrinogen binding. *Biochemistry* 36:7395–7402.

Fischer S, Graber P (1999) Comparison of delta pH- and delta phi-driven ATP synthesis catalyzed by the H+-ATPases from *Escherichia coli* or chloroplasts reconstituted into liposomes. *FEBS Lett* 457:327–332.

Frohnmayer JP (2017) Bottom-up assembly of synthetic model systems for cellular adhesion.

Frohnmayer JP, Bruggemann D, Eberhard C, Neubauer S, Mollenhauer C, Boehm H, Kessler H, Geiger B, Spatz JP (2015) Minimal synthetic cells to study integrin-mediated adhesion. *Angew Chem-Int Edit* 54:12472–12478.

Goennenwein S, Tanaka M, Hu B, Moroder L, Sackmann E (2003) Functional incorporation of integrins into solid supported membranes on ultrathin films of cellulose: Impact on adhesion. *Biophys J* 85:646–655.

Haller B, Gopfrich K, Schroter M, Janiesch JW, Platzman I, Spatz JP (2018) Charge-controlled microfluidic formation of lipid-based single- and multicompartment systems. Lab Chip 18:2665–2674.

Hamai C, Yang T, Kataoka S, Cremer PS, Musser SM (2006) Effect of average phospholipid curvature on supported bilayer formation on glass by vesicle fusion. *Biophys J* 90:1241–1248.

Heitkamp T, Deckers-Hebestreit G, Borsch M (2016) Observing single FoF1-ATP synthase at work using an improved fluorescent protein mNeonGreen as FRET donor. *SPIE Proceedings* 9714.

Holtze C, Rowat AC, Agresti JJ, Hutchison JB, Angile FE, Schmitz CHJ, Koster S et al., (2008) Biocompatible surfactants for water-in-fluorocarbon emulsions. *Lab Chip* 8:1632–1639.

Huang Y, Agrawal B, Sun DD, Kuo JS, Williams JC (2011) Microfluidics-based devices: New tools for studying cancer and cancer stem cell migration. *Biomicrofluidics* 5:013412.

Hung L-H, Lee AP (2007) Microfluidic devices for the synthesis of nanoparticles and biomaterials. *J. Med. Biol. Eng.*, 27:1–6.

Itel F, Chami M, Najer A, Lörcher S, Wu D, Dinu IA, Meier W (2014) Molecular organization and dynamics in polymersome membranes: A lateral diffusion study. *Macromolecules* 47:7588–7596.

Janiesch J-W (2015) Development of droplet-based microfluidics for synthetic biology applications. Dissertation, Ruperto-Carola University of Heidelberg, DOI:10.11588/heidok.00019990.

Janiesch JW, Weiss M, Kannenberg G, Hannabuss J, Surrey T, Platzman I, Spatz JP (2015) Key factors for stable retention of fluorophores and labeled biomolecules in droplet-based microfluidics. *Anal Chem* 87:2063–2067.

Johnson JM, Ha T, Chu S, Boxer SG (2002) Early steps of supported bilayer formation probed by single vesicle fluorescence assays. *Biophys J* 83:3371–3379.

Li M, Huang X, Tang TYD, Mann S (2014) Synthetic cellularity based on non-lipid micro-compartments and protocell models. *Curr Opin Chem Biol* 22:1–11.

Machan R, Hof M (2010) Lipid diffusion in planar membranes investigated by fluorescence correlation spectroscopy. *Biochim Biophys Acta* 1798:1377–1391.

May S, Harries D, Ben-Shaul A (2000) Lipid demixing and protein-protein interactions in the adsorption of charged proteins on mixed membranes. *Biophys J* 79:1747–1760.

Mazutis L, Baret JC, Griffiths AD (2009) A fast and efficient microfluidic system for highly selective one-to-one droplet fusion. *Lab Chip* 9:2665–2672.

Pearce TM, Williams JC (2007) Microtechnology: Meet neurobiology. *Lab Chip* 7:30–40.

Perrin J (1909) Brownian motion and molecular reality. *Ann Chim Phys* 18:5–114.

Petrache HI, Dodd SW, Brown MF (2000) Area per lipid and acyl length distributions in fluid phosphatidylcholines determined by H-2 NMR spectroscopy. *Biophys J* 79:3172–3192.

Platzman I, Janiesch J-W, Spatz JP (2013) Synthesis of nanostructured and biofunctionalized water-in-oil droplets as tools for homing T cells. *J Am Chem Soc* 135:3339–3342.

Ramadurai S, Holt A, Krasnikov V, van den Bogaart G, Killian JA, Poolman B (2009) Lateral diffusion of membrane proteins. *J Am Chem Soc* 131:12650–12656.

Rowat AC, Bird JC, Agresti JJ, Rando OJ, Weitz DA (2009) Tracking lineages of single cells in lines using a microfluidic device. *PNAS* 106:18149–18154.

Seantier B, Kasemo B (2009) Influence of mono-and divalent ions on the formation of supported phospholipid bilayers via vesicle adsorption. *Langmuir* 25:5767–5772.

Sinn CG, Antonietti M, Dimova R (2006) Binding of calcium to phosphatidylcholine-phosphatidylserine membranes. *Colloid Surf A-Physicochem Eng Asp* 282:410–419.

Soumpasis DM (1983) Theoretical-analysis of fluorescence photobleaching recovery experiments. *Biophys J* 41:95–97.

Thorsen T, Roberts RW, Arnold FH, Quake SR (2001) Dynamic pattern formation in a vesicle-generating microfluidic device. *Phys Rev Lett* 86:4163–4166.

Weiss M, Frohnmayer JP, Benk LT, Haller B, Janiesch J-W, Heitkamp T, Borsch M et al. (2018) Sequential bottom-up assembly of mechanically stabilized synthetic cells by microfluidics. *Nat Mater* 17:89–96.

Xia Y, Whitesides GM (1998) Soft lithography. *Annu Rev Mater Res Sci* 28:153–184.

Zimmermann B, Diez M, Zarrabi N, Graber P, Borsch M (2005) Movements of the epsilon-subunit during catalysis and activation in single membrane-bound H+-ATP synthase. *Embo J* 24:2053–2063.

Appendix 1: List of lipids and physical constants of lipids bilayers

Molecular weight, MW
Main phase transition temperature, T_m
Area per lipid, A_{lip}
Hydrophobic thickness, l_{hyd}
Head-head distance, l_{hh}

Data and structures are by courtesy of Walt Shaw, Kacee Sims, and Shengrong Li from Avanti Polar Lipids (https://avantilipids.com/). We thank the Tristram-Nagle/Nagel group for making available a table with their own data. We used extensively the *Handbook of Lipid Bilayers* (Marsh, 2013) for data mining.

13:0 PC (Chapter 27)
CAS: 71242-28-9; MW = 649.88; 1,2-ditridecanoyl-*sn*-glycero-3-phosphocholine
3,5,9-Trioxa-4-phosphadocosan-1-aminium, 4-hydroxy-*N*,*N*,*N*-trimethyl-10-oxo-7-[(1-oxotridecyl)oxy]-, inner salt, 4-oxide, (R)-; 1,2-Ditridecanoyl-*sn*-glycero-3-phosphatidylcholine; 1,2-Ditridecanoyl-*sn*-glycero-3-phosphocholine; PC(13:0/13:0)
T_m = 14°C
18:1 PI(4,5)P_2 (Chapters 2, 16, 20, 23, 26)
CAS: 799268-56-7; MW = 1074.16; 1,2-dioleoyl-*sn*-glycero-3-phospho-(1′-myo-inositol-4′,5′-bisphosphate) (ammonium salt)
1,2-di-(9Z-octadecenoyl)-*sn*-glycero-3-[phosphoinositol-4,5-bisphosphate] (ammonium salt); PIP2[4′,5′](18:1(9Z)/18:1(9Z))
Biotinyl DSPE (Chapters 4, 16, 27, 29)
CAS: 385437-57-0; MW = 3016.781; 1,2-distearoyl-*sn*-glycero-3-phosphoethanolamine-*N*-[biotinyl(polyethylene glycol)-2000] (ammonium salt); Poly(oxy-1,2-ethanediyl), α-[(9R)-6-hydroxy-6-oxido-1,12-dioxo-9-[(1-oxooctadecyl)oxy]-5,7,11-trioxa-2-aza-6-phosphanonacos-1-yl]-ω-[2-[[5-[(3aS,4S,6aR)-hexahydro-2-oxo-1H-thieno[3,4-d]imidazol-4-yl]-1-oxopentyl]amino]ethoxy]-, ammonium salt
Chol (Chapters 2, 4, 10, 11, 15, 16, 18, 20, 21, 22, 27, 28, 30)
CAS: 57-88-5; MW = 386.65; Cholesterol; Cholest-5-en-3-ol (3β)
Cholesterol (8CI); (3β)-Cholest-5-en-3-ol; (–)-Cholesterol; 3β-Hydroxycholest-5-ene; 5:6-Cholesten-3β-ol; Cholest-5-en-3β-ol; Cholesterin; Cholesteryl alcohol; Dythol; Lidinit; Lidinite; Marine Cholesterol; NSC 8798; Provitamin D; SyntheChol; Δ^5-Cholesten-3β-ol
for molecular areas of cholesterol in palmitoyl SM/Chol and DPPC/Chol bilayers see e.g. (Hofsäss et al., 2003; Khelashvili and Scott, 2004)

(*Continued*)

DAPC (Chapter 11)

CAS: 17688-29-8; MW = 830.12; 1,2-diarachidonoyl-*sn*-glycero-3-phosphocholine

Diarachidonoyl-phosphatidylcholine; 1,2-di-(5Z,8Z,11Z,14Z-eicosatetraenoyl)-*sn*-glycero-3-phosphocholine; PC(20:4(5Z,8Z,11Z,14Z)/20:4(5Z,8Z,11Z,14Z)); 1,2-diarachidonin, L-(8CI); 1,2-Diarachidonyl phosphatidylcholine; 1,2-Diarachidonyl-L-α-glycerophosphorylcholine; Diarachidonoyl phosphatidylcholine; Diarachidonyl lecithin; Diarachidonyl phosphatidylcholine; L-Diarachidonoyl lecithin

T_m = –69°C

DEiPC (Chapter 20)

CAS: 56782-46-8; MW = 786.11; 1,2-dielaidoyl-*sn*-glycero-3-phosphocholine

1,2-di-(9E-octadecenoyl)-*sn*-glycero-3-phosphocholine; PC(18:1(9E)/18:1(9E));1,2-Dielaidoyl-*sn*-glycero-3-phosphocholine; Dielaidoyl-L-α-glycerophosphorylcholine; L-α-Dielaidoylphosphatidylcholine

T_m = 12°C

DEPC (Chapters 11, 20)

CAS: 51779-95-4; MW = 898.33; 1,2-dierucoyl-*sn*-glycero-3-phosphocholine

Dierucoyl-phosphatidylcholine; 1,2-di-(13Z-docosenoyl)-*sn*-glycero-3-phosphocholine; PC(22:1(13Z)/22:1(13Z)); 1,2-Dierucoyl-*sn*-glycero-3-phosphocholine; Dierucoyl-L-α-glycerophosphorylcholine

T_m = 13°C
A_{lip} = 69.3 Å^2; l_{hyd} = 34.4 Å; l_{hh} = 44.3 Å at 30°C (Kučerka et al., 2006b)
A_{lip} = 65.7 Å^2; l_{hyd} = 36.30 Å; l_{hh} = 45.5 Å at 30°C (Kučerka et al., 2009)

DGDG (Chapters 9, 11)

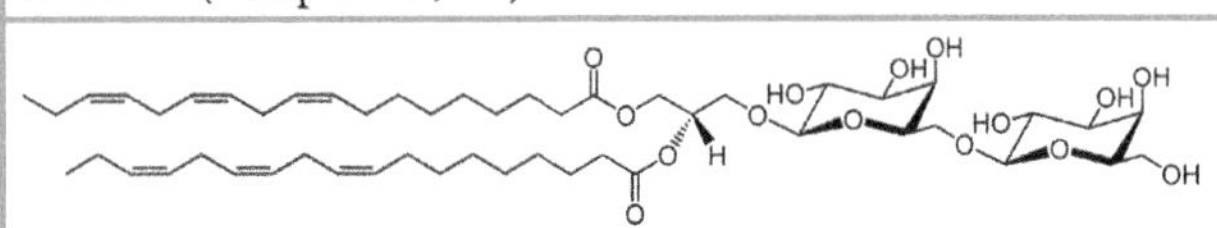

Representative structure only. This structure is only one of many possible structures in the product.

CAS: 63142-69-8; MW = 926.767; Digalactosyldiacylglycerol

Digalactosyldiglyceride; 1,2-diacyl-3-O-(α-D-galactosyl1-6)-β-D-galactosyl-*sn*-glycerol; (2S)-2,3-Bis[[(9Z,12Z,15Z)-1-oxo-9,12,15-octadecatrien-1-yl]oxy]propyl 6-O-α-D-galactopyranosyl-β-D-galactopyranoside

DGS-NTA(Ni) (Chapters 13, 30)

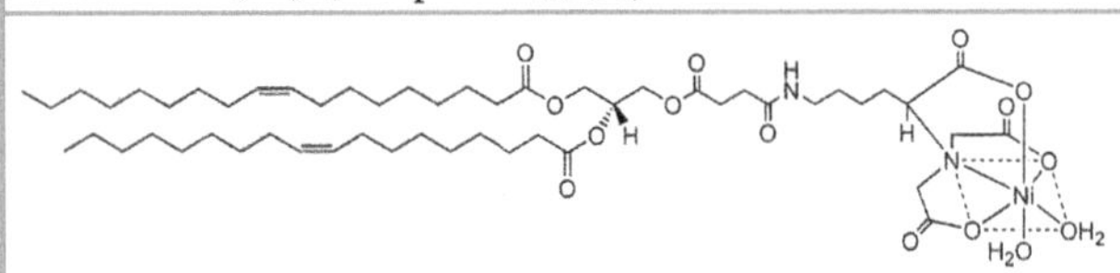

CAS: 231615-77-3; MW = 1021.98; 1,2-dioleoyl-*sn*-glycero-3-[(*N*-(5-amino-1-carboxypentyl)iminodiacetic acid)succinyl] (nickel salt)

1,2-di-(9Z-octadecenoyl)-*sn*-glycero-3-[(*N*-(5-amino-1-carboxypentyl)iminodiacetic acid)succinyl] (nickel salt); DOGS-NTA-Ni; DOGS-NiNTA; DOGS NTA

(Continued)

DLoPC (Chapter 11)
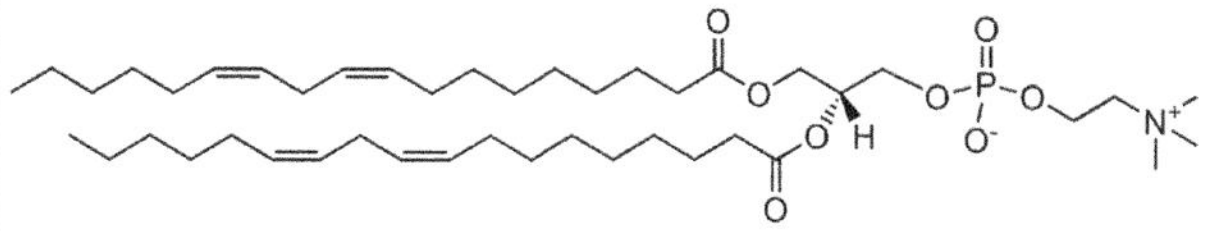
CAS: 998-06-1; $MW = 782.08$; 1,2-dilinoleoyl-*sn*-glycero-3-phosphocholine
Dilinoleoyl-phosphatidylcholine; 1,2-di-(9Z,12Z-octadecadienoyl)-*sn*-glycero-3-phosphocholine; PC(18:2(9Z,12Z)/18:2(9Z,12Z)); L-Dilinoleoyllecithin; L-α-Dilinoleoyl phosphatidylcholine
$T_m = -57°C$
DLPC (Chapters 10, 20, 21, 27)
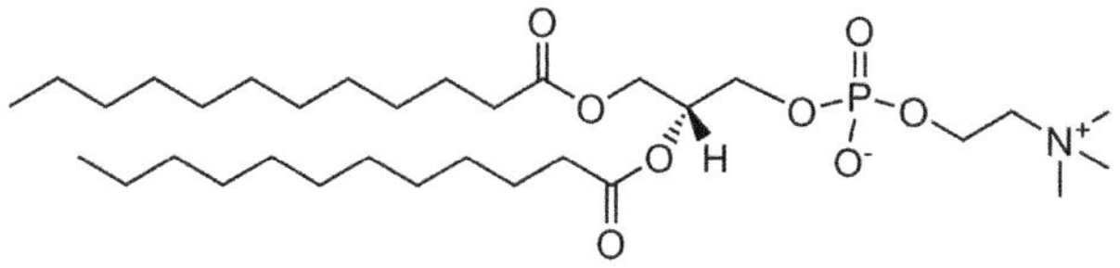
CAS: 18194-25-7; $MW = 621.8$; 1,2-dilauroyl-*sn*-glycero-3-phosphocholine
1,2-didodecanoyl-*sn*-glycero-3-phosphocholine; PC(12:0/12:0); Dilauroyl-L-α-glycerophosphocholine; 1,2-Dilauroyl-L-phosphatidylcholine; Coatsome MC-1212
$T_m = -2°C$ $A_{lip} = 63.2 \pm 0.5$ Å^2; $l_{hyd} = 20.9$ Å; $l_{hh} = 30.8$ Å at 30°C (Kučerka et al., 2005)
DLPE (Chapter 28)
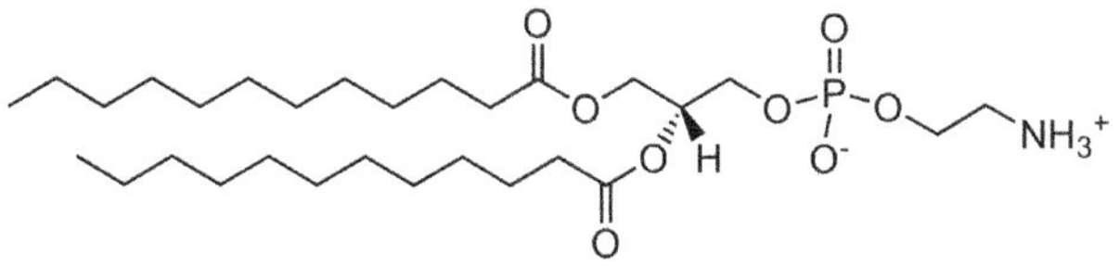
CAS: 59752-57-7; $MW = 579.75$; 1,2-dilauroyl-*sn*-glycero-3-phosphoethanolamine
1,2-didodecanoyl-*sn*-glycero-3-phosphoethanolamine; PE(12:0/12:0); 1,2-Dilauroyl-*sn*-glycero-3-phosphoethanolamine; L-α-Dilauroylphosphatidylethanolamine
$T_m = 29°C$ $A_{lip} = 51.2 \pm 0.5$ Å^2; $l_{hyd} = 25.8$ Å; $l_{hh} = 35.6$ Å at 35°C (Nagle and Tristram-Nagle, 2000)
DMPC (Chapters 8, 11, 15, 22, 25, 27)
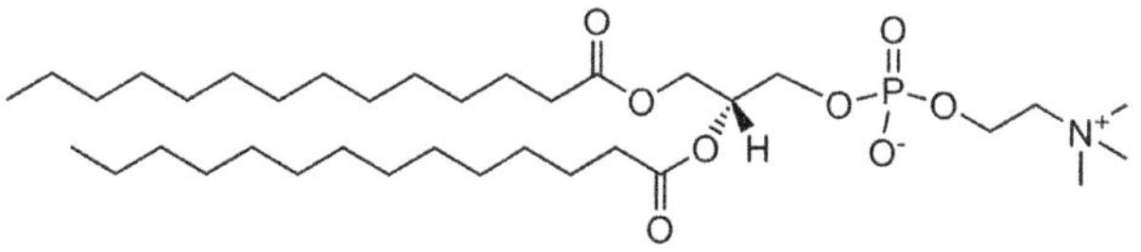
CAS: 18194-24-6; $MW = 677.93$; 1,2-dimyristoyl-*sn*-glycero-3-phosphocholine; Dimyristoyl-phosphatidylcholine; 1,2-ditetradecanoyl-*sn*-glycero-3-phosphocholine; PC(14:0/14:0); L-α-Dimyristoylphosphatidylcholine; 1,2-L-α-Dimyristoylphosphatidylcholine
$T_m = 24°C$ $A_{lip} = 59.7 \pm 0.5$ Å^2; $l_{hyd} = 26.2$ Å; $l_{hh} = 34.4$ Å at 30°C (Petrache et al., 1998) $A_{lip} = 60.6 \pm 0.5$ Å^2; $l_{hyd} = 25.4$ Å; $l_{hh} = 35.3$ Å at 30°C (Kučerka et al., 2005)
DOGM (Chapter 25)
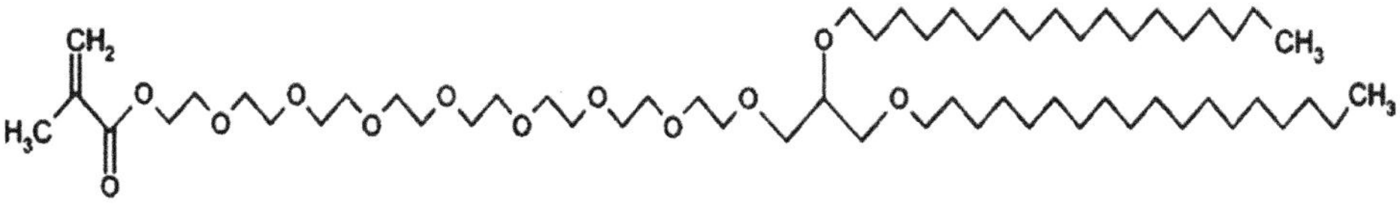
1,2-distearyl-3-octaethylene glycol glycerol ether methacrylate

(*Continued*)

DOPC (Chapters 1, 2, 4, 8–11, 15, 18–24, 27, 30)

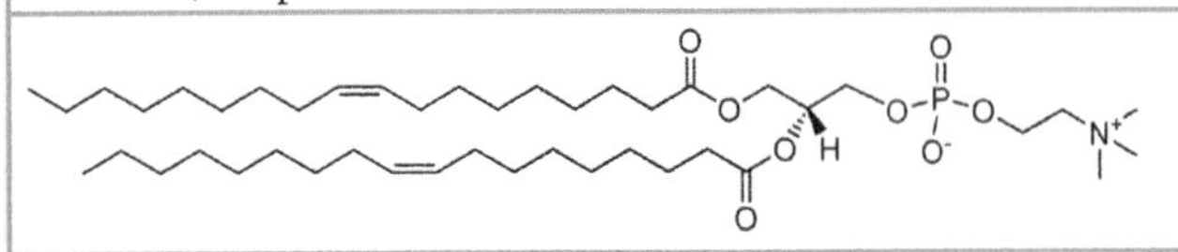

CAS: 4235-95-4; $MW = 786.11$; 1,2-dioleoyl-*sn*-glycero-3-phosphocholine; 1,2-dioctadecenoyl-*sn*-glycero-3-phosphocholine

Dioleoyl-phosphatidylcholine; PC(18:1(9Z)/18:1(9Z));1,2-di-(9Z-octadecenoyl)-*sn*-glycero-3-phosphocholine; L-α-Dioleylphosphatidylcholine; Dioleoyl-L-α-phosphatidylcholine

$T_m = -17°C$
$A_{lip} = 72.2 \pm 0.5$ Å^2; $l_{hyd} = 27.2$ Å; $l_{hh} = 35.3$ Å at 30°C (Tristram-Nagle et al., 1998)
$A_{lip} = 72.5 \pm 0.5$ Å^2; $l_{hyd} = 27.1$ Å; $l_{hh} = 36.9$ Å at 30°C (Nagle and Tristram-Nagle, 2000)
$A_{lip} = 72.1 \pm 0.5$ Å^2; $l_{hyd} = 27.2$ Å; $l_{hh} = 37.1$ Å at 30°C (Liu and Nagle, 2004)
$A_{lip} = 72.4 \pm 0.5$ Å^2; $l_{hyd} = 26.8$ Å; $l_{hh} = 36.7$ Å at 30°C (Kučerka et al., 2006b)
$A_{lip} = 67.4 \pm 0.5$ Å^2; $l_{hyd} = 28.8$ Å; $l_{hh} = 36.7$ Å at 30°C (Kučerka et al., 2008)
$A_{lip} = 69.1 \pm 0.5$ Å^2; $l_{hyd} = 27.7$ Å; $l_{hh} = 37.6$ Å at 15°C (Pan et al., 2008)
$A_{lip} = 75.5 \pm 0.5$ Å^2; $l_{hyd} = 26.2$ Å; $l_{hh} = 36.1$ Å at 45°C (Pan et al., 2008)

DOPE (Chapters 4, 15, 16, 20, 22, 30)

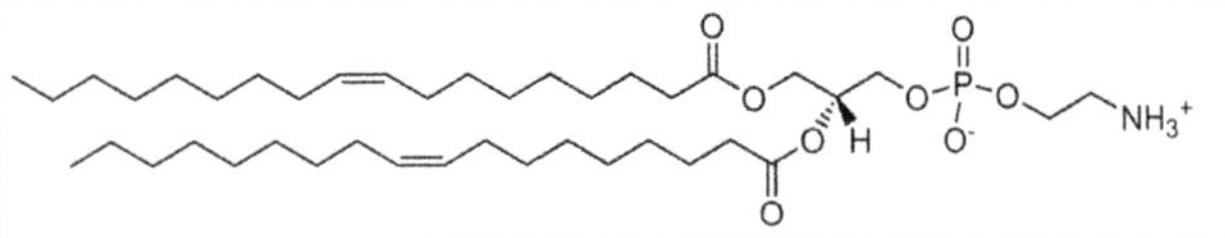

CAS: 4004-05-1; $MW = 744.03$; 1,2-dioleoyl-*sn*-glycero-3-phosphoethanolamine

Dioleoyl-phosphoethanolamine;1,2-di-(9Z-octadecenoyl)-*sn*-glycero-3-phosphoethanolamine;PE(18:1(9Z)/18:1(9Z)); Coatsome MC 8181; Dioleoyl-L-α-phosphatidylethanolamine

$T_m = -16°C$
$A_{lip} \approx 60$ Å^2 at 22.5°C (Gawrisch et al., 1992); $l_{hyd} = 33.9$ Å; $l_{hh} = 40.3$ Å see (Jambeck and Lyubartsev, 2012)

DOPG (Chapters 4, 8, 20, 24)

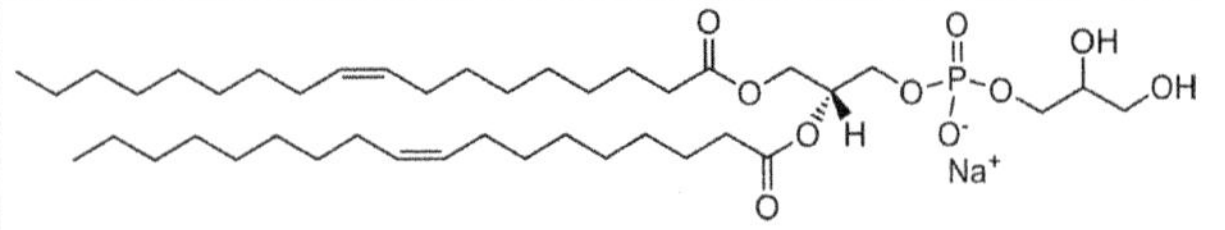

CAS: 67254-28-8; $MW = 797.03$; 1,2-di-oleoyl-*sn*-glycero-3-phospho-(1′-*rac*-glycerol) (sodium salt)

1,2-di-(9Z-octadecenoyl)-*sn*-glycero-3-phospho-(1′-*rac*-glycerol) (sodium salt); PG(18:1(9Z)/18:1(9Z)); 1,2-Dioleoyl-*sn*-glycero-3-phosphoglycerol

$T_m = -18°C$
$A_{lip} = 70.8 \pm 0.5$ Å^2; $l_{hyd} = 27.5$ Å at 30°C (Pan et al., 2012)

DPhPC (Chapter 23)

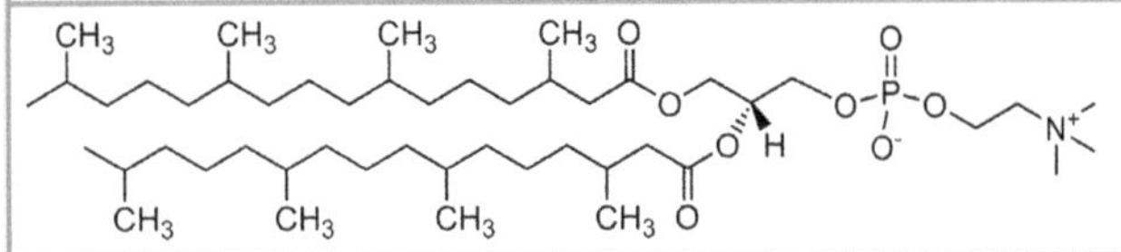

CAS: 207131-40-6; $MW = 846.25$; 1,2-diphytanoyl-*sn*-glycero-3-phosphocholine

1,2-di-(3,7,11,15-tetramethylhexadecanoyl)-*sn*-glycero-3-phosphocholine; PC(16:0(3me,7me,11me,15me)/16:0(3me,7me,11me,15me)); Diphytanoylphosphatidylcholine; L-Diphytanoylphosphatidylcholine

$A_{lip} = 80.5 \pm 1.5$ Å^2; $l_{hyd} = 27.2$ Å; $l_{hh} = 36.4$ Å at 30°C (Tristram-Nagle et al., 2010)

(Continued)

DPPC (Chapters 1, 4, 8, 10, 11, 13, 15, 20–23, 26, 27, 28)
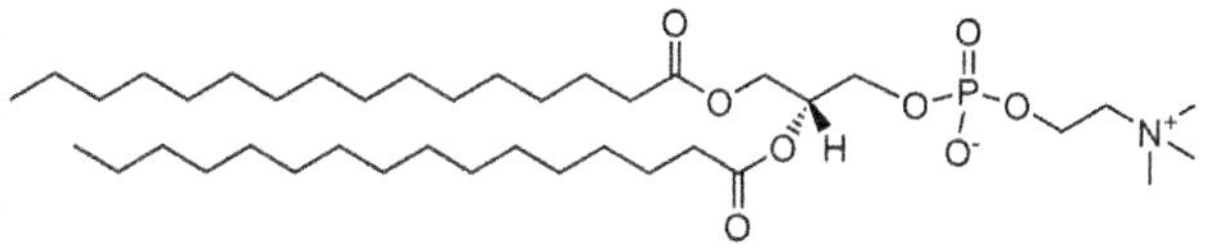
CAS: 63-89-8; MW = 734.04; 1,2-dipalmitoyl-*sn*-glycero-3-phosphocholine; 1,2-dihexadecanoyl-*sn*-glycero-3-phosphocholine; Dipalmitoyl-phosphocholine; PC(16:0/16:0); L-α-Dipalmitoylphosphatidylcholine; L-Dipalmitoylphosphatidylcholine
T_m = 41.5°C A_{lip} = 62.9 ± 1.3 Å^2; l_{hyd} = 29.2 Å; l_{hh} = 39.6 Å at 50°C (Nagle et al., 1996) A_{lip} = 64.0 ± 0.5 Å^2; l_{hyd} = 28.5 Å; l_{hh} = 38.3 Å at 50°C (Nagle and Tristram-Nagle, 2000) A_{lip} = 64.3 ± 0.5 Å^2; l_{hyd} = 27.9 Å; l_{hh} = 37.8 Å at 50°C (Kučerka et al., 2006a) A_{lip} = 63.1 ± 0.5 Å^2; l_{hyd} = 28.4 Å; l_{hh} = 38.0 Å at 50°C (Kučerka et al., 2008)
DPPE (Chapters 4, 10, 15, 18, 22)
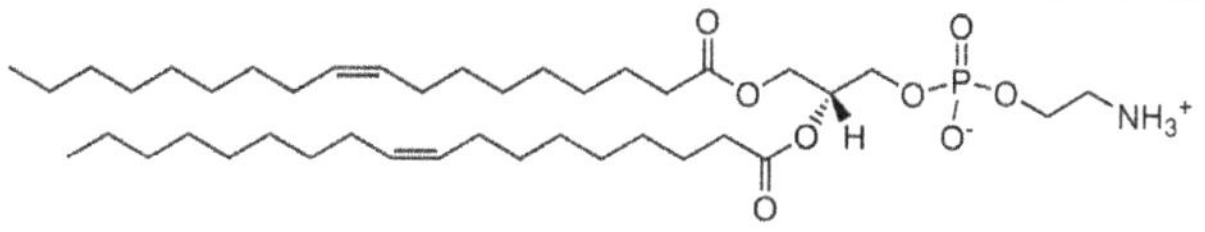
CAS: 923-61-5; MW = 691.96; 1,2-dipalmitoyl-*sn*-glycero-3-phosphoethanolamine
Dipalmitoylphosphatidyl-ethanolamine; 1,2-dihexadecanoyl-*sn*-glycero-3-phosphoethanolamine; PE(16:0/16:0); L-α-Dipalmitoylpho sphatidylethanolamine; DHPE; Coatsome ME 6060
T_m = 63°C A_{lip} = 60.5 Å^2; l_{hyd} = 30.4 Å see (Jambeck and Lyubartsev, 2012)
DPSPC (Chapter 20)
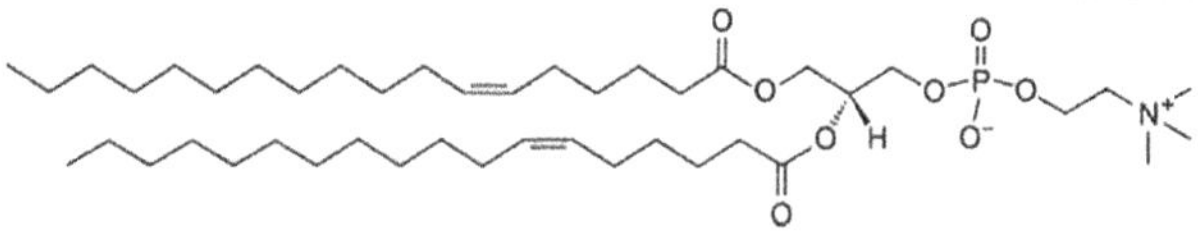
CAS: 56391-91-4; MW = 786.11; 1,2-dipetroselinoleoyl-*sn*-glycero-3-phosphocholine
PC(18:1(6Z)/18:1(6Z))
T_m = 1°C see (Barton and Gunstone, 1975)
DSPC (Chapters 1, 9, 13, 18, 21)
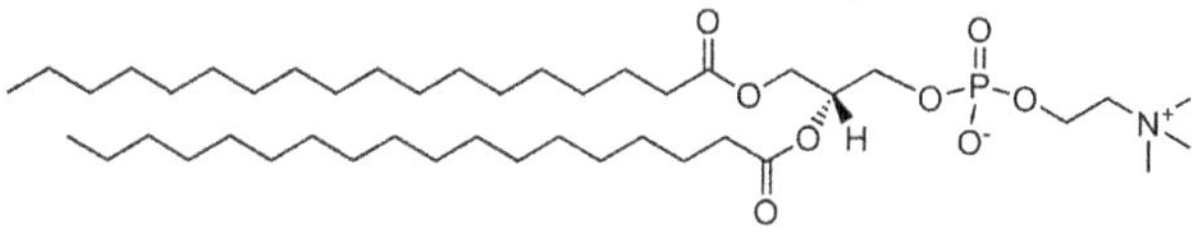
CAS: 816-94-4; MW = 790.15; 1,2-distearoyl-*sn*-glycero-3-phosphocholine
1,2-dioctadecanoyl-*sn*-glycero-3-phosphocholine; PC(18:0/18:0); L-α-Distearoylphosphatidylcholine; Coatsome MC 8080; 1,2-distearin, L-(8CI)
T_m = 55°C
DSPE-PEG biotin (Chapter 16)
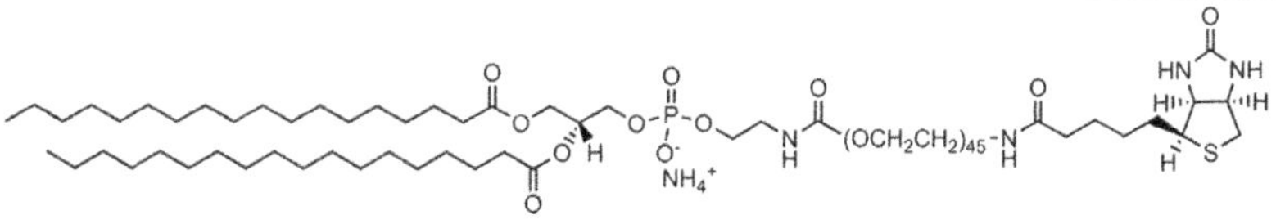
CAS: 385437-57-0; MW = 3016.78; 1,2-distearoyl-*sn*-glycero-3-phosphoethanolamine-*N*-[biotinyl(polyethylene glycol)-2000] (ammonium salt)
DSPE-PEG(2000) Biotin; Laysan Bio DSPE-PEG-biotin

(*Continued*)

Egg PA (Chapters 3, 25)
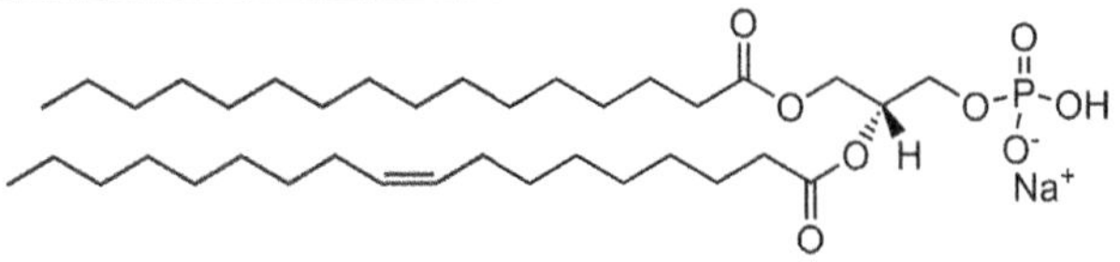
CAS: 383907-53-7; $MW = 696.91$; egg phosphatidic acid
L-α-phosphatidic acid (Egg, Chicken) (sodium salt)
Egg PC (Chapters 2, 4, 10, 11, 15, 18, 20, 21, 24, 27, 28, 30)
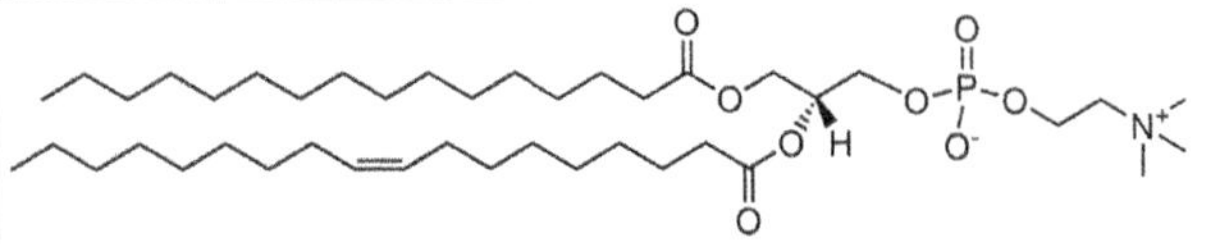
CAS: 97281-44-2; $MW = 770.12$; egg phosphatidylcholine
L-α-phosphatidylcholine; Egg PC 99%
$A_{lip} = 69.4 \pm 0.5$ Å^2; $l_{hyd} = 27.2$ Å; $l_{hh} = 35.4$ Å at 30°C (Petrache et al., 1998) $A_{lip} = 69.4 \pm 0.5$ Å^2; $l_{hyd} = 27.15$ Å; $l_{hh} = 36.9$ Å at 30°C (Nagle and Tristram-Nagle, 2000)
eSM, **egg SM** (Chapters 4, 10, 11, 15, 18, 24, 30)
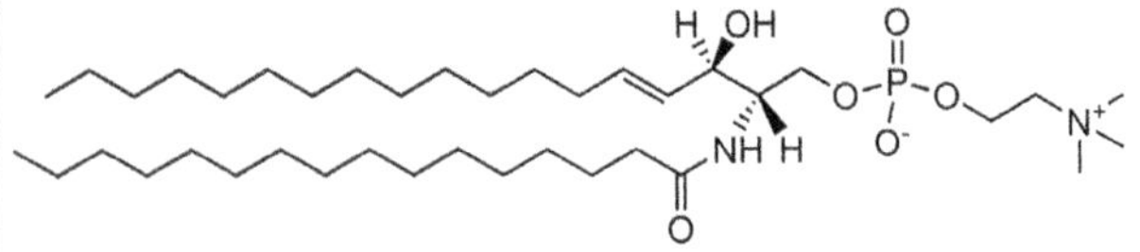
CAS: 383907-87-7; $MW = 703.03$; *N*-hexadecanoyl-D-*erythro*-sphingosylphosphorylcholine; Egg sphingomyelin; Hexadecanoyl Sphingomyelin; *N*-hexadecanoyl-D-*erythro*-sphingosylphosphorylcholine; *N*-(hexadecanoyl)-sphing-4-enine-1-phosphocholine
For eSM $A_{lip} = 64 \pm 2$ Å^2; $l_{hh} = 38.4$ Å at 45°C (Arsov et al., 2018) For palmitoyl (16:0) SM, the main component of eSM: $T_m = 41$°C $A_{lip} = 45$ Å^2; $l_{hh} = 48$ Å at 29°C $A_{lip} = 55$ Å^2; $l_{hh} = 42$ Å at 55°C (Maulik et al., 1991; Maulik and Shipley, 1996) $A_{lip} = 64 \pm 2$ Å^2; $l_{hh} = 37.6$ Å at 45°C (Arsov et al., 2018)
HSPC (Chapter 27)
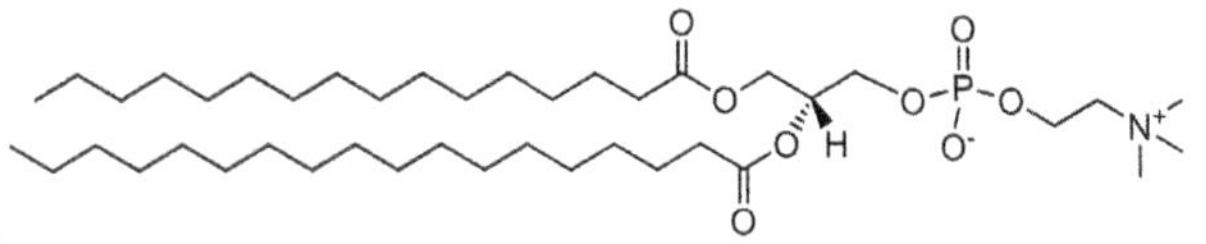
CAS: 97281-48-6; $MW = 783.77$; Hydrogenated Soybean L-α-phosphatidylcholine
Phospholipon 90N; Hydrogenated soy phosphatidylcholines; NC 21E
MOPC or **18:1 LysoPC** (Chapter 11)
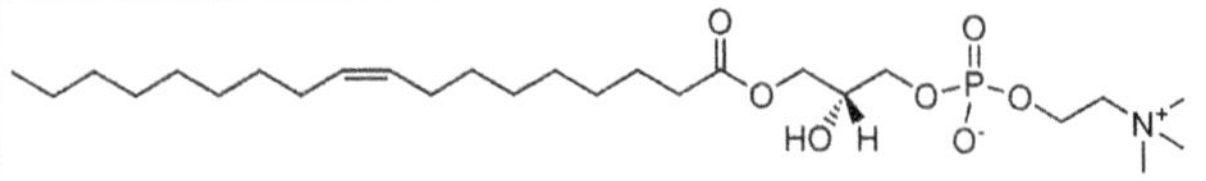
CAS: 19420-56-5; $MW = 521.67$; 1-oleoyl-2-hydroxy-*sn*-glycero-3-phosphocholine
Monooeloyl-phosphatidylcholine; 1-(9Z-octadecenoyl)-*sn*-glycero-3-phosphocholine; PC(18:1(9Z)/0:0); Choline, hydroxide, dihydrogen phosphate, inner salt, 3-ester with 1-monoolein, L-(8CI)

(Continued)

OSPC (Chapter 20)
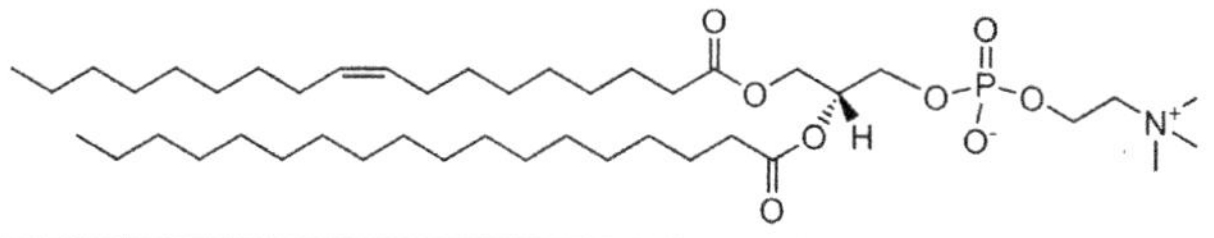
CAS: 7276-38-2; MW = 788.13; 1-oleoyl-2-stearoyl-*sn*-glycero-3-phosphatidylcholine
1(9Z-octadecenoyl)-2-octadecanoyl-*sn*-glycero-3-phosphocholine; PC(18:1(9Z)/18:0); L-1-Oleoyl-2-stearoyl-3-phosphatidylcholine;
T_m = 9°C
PL in **PC** (Chapter 22)
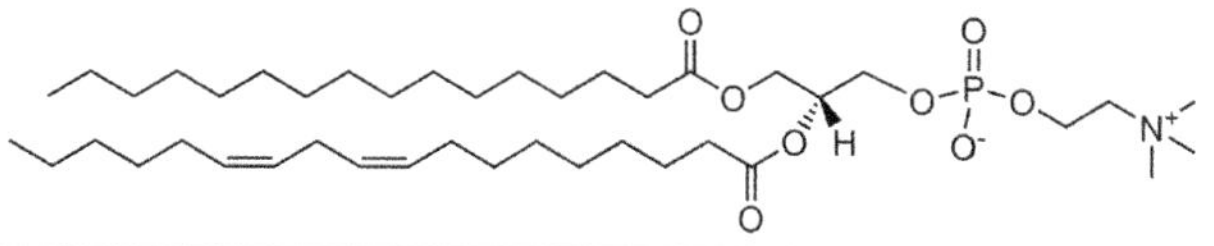
CAS: 159701-21-0; MW = 758.06; 1-palmitoyl-2-linoleoyl-*sn*-glycero-3-phosphocholine
PLPC, 1-hexadecanoyl-2-(9Z,12Z-octadecadienoyl)-*sn*-glycero-3-phosphocholine; PC(16:0/18:2(9Z,12Z));
POPC (Chapters 1, 2, 10, 11, 15, 20–22, 24, 26–28, 30)
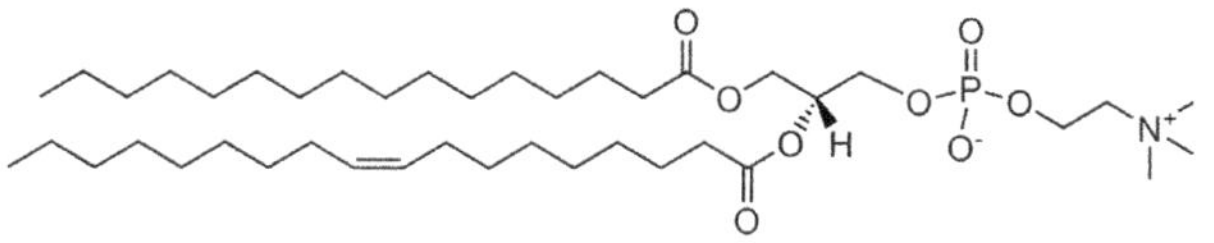
CAS: 26853-31-6; MW = 760.08; 1-palmitoyl-2-oleoyl-*sn*-glycero-3-phosphocholine
Palmitoyloleoyl-phosphatidylcholine; 1-hexadecanoyl-2-(9Z-octadecenoyl)-*sn*-glycero-3-phosphocholine; PC(16:0/18:1(9Z)); Palmitoyloleoylphosphatidylcholine; L-α-1-Palmitoyl-2-oleoylphosphatidylcholine
T_m = -2°C A_{lip} = 68.3 ± 1.5 Å^2; l_{hyd} = 27.1 Å; l_{hh} = 37.0 Å at 30°C (Kučerka et al., 2006b)
POPE (Chapter 11)
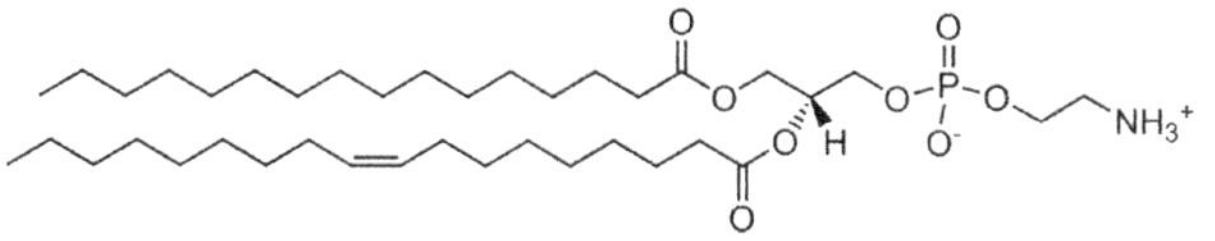
CAS: 26662-94-2; MW = 718.00; 1-palmitoyl-2-oleoyl-*sn*-glycero-3-phosphoethanolamine
Palmitoyloleoyl-phosphatidyl-ethanolamine; 1-hexadecanoyl-2-(9Z-octadecenoyl)-*sn*-glycero-3-phosphoethanolamine; PE(16:0/18:1(9Z)); L-α-1-Palmitoyl-2-oleoylglycerophosphoethanolamine; Palmitin, 2-oleo-1-, dihydrogen phosphate mono(2-aminoethyl) ester, L-(8CI); Olein, 1-palmito-2-, dihydrogen phosphate mono(2-aminoethyl) ester, L-(8CI)
T_m = 25°C A_{lip} = 56.6 Å^2; l_{hyd} = 32.6 Å; l_{hh} = 41.4 Å see (Jambeck and Lyubartsev, 2012)
POPG (Chapters 1, 15, 20, 22, 24)
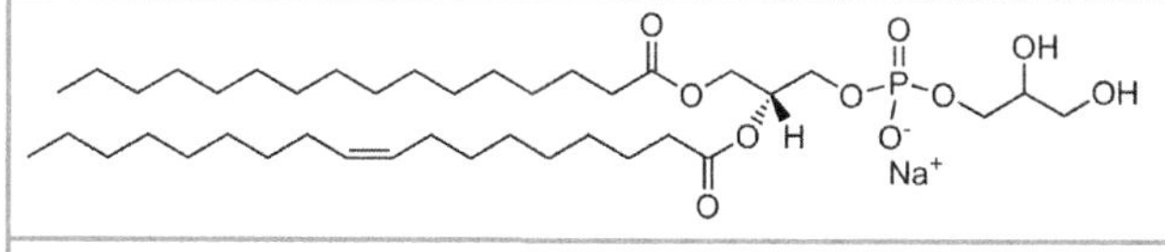
CAS: 268550-95-4; MW = 770.99; 1-palmitoyl-2-oleoyl-*sn*-glycero-3-phospho-(1′-*rac*-glycerol) (sodium salt)
Palmitoyloleoyl-phosphatidylglycerol; 1-hexadecanoyl-2-(9Z-octadecenoyl)-*sn*-glycero-3-phospho-(1′-*rac*-glycerol) (sodium salt); PG(16:0/18:1(9Z))
T_m = -2°C A_{lip} = 66.1 ± 0.5 Å^2; l_{hyd} = 27.8 Å at 30°C (Pan et al., 2012)

(*Continued*)

POPS (Chapter 15)
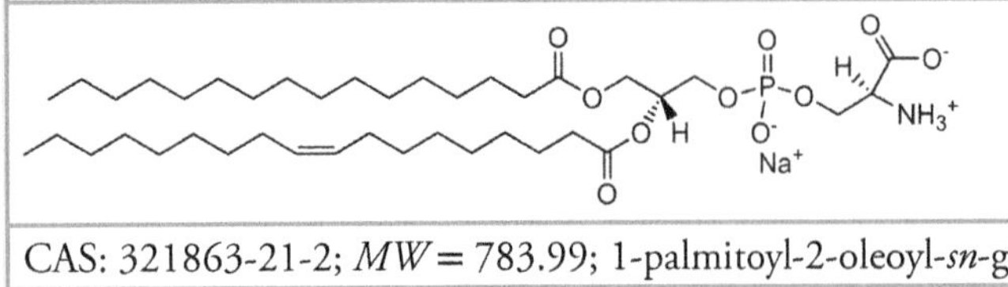
CAS: 321863-21-2; MW = 783.99; 1-palmitoyl-2-oleoyl-*sn*-glycero-3-phospho-L-serine (sodium salt)
Palmitoyloleoyl-phosphatidylglycerol; 1-hexadecanoyl-2--(9Z-octadecenoyl)-*sn*-glycero-3-phospho-L-serine (sodium salt); PS(16:0/18:1(9Z))
T_m = -11°C A_{lip} = 55 ± 1 Å^2 see (Mukhopadhyay et al., 2004)
SLPC or **SLoPC 18:0-18:2 PC** (Chapters 11, 20)
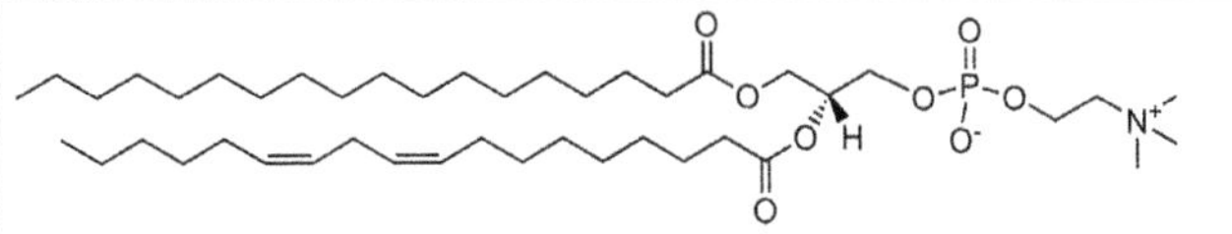
CAS: 27098-24-4; MW = 786.12; 1-stearoyl-2-linoleoyl-*sn*-glycero-3-phosphocholine
Stearoyllinoleoyl-phosphatidylcholine; 1-octadecanoyl-2-(9Z,12Z-octadecadienoyl)-*sn*-glycero-3-phosphocholine; PC(18:0/18:2(9Z,12Z)); 1-stearoyl-2-linolean, L-(8CI)
SOPC (Chapters 9, 11, 20, 27)
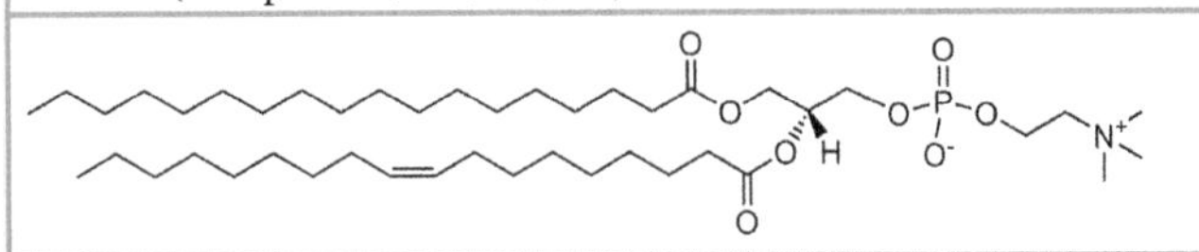
CAS: 56421-10-4; MW = 788.13; 1-stearoyl-2-oleoyl-*sn*-glycero-3-phosphocholine
Stearoyloeloyl-phosphatidylcholine; 1-octadecanoyl-2-(9Z-octadecenoyl)-*sn*-glycero-3-phosphocholine; PC(18:0/18:1(9Z)); L-α-1-Stearoyl-2-oleoylphosphatidylcholine; Choline phosphate, 3-ester with L-2-oleo-1-stearin (6CI); L-α-1-Stearoyl-2-oleoyl lecithin
T_m = 6°C A_{lip} = 67.0 ± 0.9 Å^2; l_{hyd} = 29.2 ± 0.4 Å; l_{hh} = 39 Å at 30°C (Greenwood et al., 2008)

REFERENCES

Arsov Z, González-Ramírez EJ, Goñi FM, Tristram-Nagle S, Nagle JF (2018) Phase behavior of palmitoyl and egg sphingomyelin. *Chemistry and Physics of Lipids*, 213, 102–110.

Barton PG, Gunstone FD (1975) Hydrocarbon chain packing and molecular motion in phospholipid bilayers formed from unsaturated lecithins. Synthesis and properties of sixteen positional isomers of 1, 2-dioctadecenoyl-*sn*-glycero-3-phosphorylcholine. *Journal of Biological Chemistry*, 250(12), 4470–4476.

Gawrisch K, Parsegian VA, Hajduk DA, Tate MW, Gruner SM, Fuller NL, Rand RP (1992) Energetics of a hexagonal lamellar hexagonal-phase transition sequence in dioleoylphosphatidylethanolamine membranes. *Biochemistry*, 31, 2856–2864.

Greenwood AI, Pan J, Mills TT, Nagle JF, Epand RM, Tristram-Nagle S (2008) CRAC motif peptide of the HIV-1 gp41 protein thins SOPC membranes and interacts with cholesterol. *Biochimica et Biophysica Acta (BBA)-Biomembranes*, 1778(4), 1120–1130.

Hofsäss C, Lindahl E, Edholm O (2003) Molecular dynamics simulations of phospholipid bilayers with cholesterol. *Biophysical Journal*, 84, 2192–2206.

Jambeck JPM, Lyubartsev AP (2012) An extension and further validation of an all-atomistic force field for biological membranes. *Journal of Chemical Theory and Computation*, 8, 2938–2948.

Khelashvili GA, Scott HL (2004) Combined monte carlo and molecular dynamics simulation of hydrated 18:0 sphingomyelin-cholesterol lipid bilayers. *Journal of Chemical Physics*, 120, 9841–9847.

Kučerka N, Gallová J, Uhríková D, Balgavý P, Bulacu M, Marrink SJ, Katsaras J (2009) Areas of monounsaturated diacylphosphatidylcholines. *Biophysical Journal*, 97(7), 1926–1932.

Kučerka N, Liu Y, Chu N, Petrache HI, Tristram-Nagle S, Nagle JF (2005) Structure of fully hydrated fluid phase DMPC and DLPC lipid bilayers using X-ray scattering from oriented multilamellar arrays and from unilamellar vesicles. *Biophysical Journal*, 88(4), 2626–2637.

Kučerka N, Nagle JF, Sachs JN, Feller SE, Pencer J, Jackson A, Katsaras J (2008) Lipid bilayer structure determined by the simultaneous analysis of neutron and X-ray scattering data. *Biophysical Journal*, 95(5), 2356–2367.

Kučerka N, Tristram-Nagle S, Nagle JF (2006a) Closer look at structure of fully hydrated fluid phase DPPC bilayers. *Biophysical Journal*, 90, L83–L85.

Kučerka N, Tristram-Nagle S, Nagle JF (2006b) Structure of fully hydrated fluid phase lipid bilayers with monounsaturated chains. *The Journal of Membrane Biology*, 208(3), 193–202.

Liu Y, Nagle JF (2004) Diffuse scattering provides material parameters and electron density profiles of biomembranes. *Physical Review E*, 69(4), 040901.

Marsh D (2013) *Handbook of Lipid Bilayers*, CRC press. Boca Raton
Maulik PR, Shipley GG (1996) N-Palmitoyl sphingomyelin bilayers: Structure and interactions with cholesterol and dipalmitoylphosphatidylcholine. *Biochemistry*, 35, 8025–8034.
Maulik PR, Sripada PK, Shipley GG (1991) Structure and thermotropic properties of hydrated N-stearoyl sphingomyelin bilayer membranes. *Biochimica et Biophysica Acta (BBA)-Biomembranes*, 1062, 211–219.
Mukhopadhyay P, Monticelli L, Tieleman DP (2004) Molecular dynamics simulation of a palmitoyl-oleoyl phosphatidylserine bilayer with Na^+ counterions and NaCl. *Biophysical Journal*, 86, 1601–1609.
Nagle JF, Tristram-Nagle S (2000) Structure of lipid bilayers. *Biochimica et Biophysica Acta (BBA)-Reviews on Biomembranes*, 1469(3), 159–195.
Nagle JF, Zhang R, Tristram-Nagle S, Sun W, Petrache HI, Suter RM (1996) X-ray structure determination of fully hydrated L alpha phase dipalmitoylphosphatidylcholine bilayers. *Biophysical Journal*, 70(3), 1419–1431.
Pan J, Heberle FA, Tristram-Nagle S, Szymanski M, Koepfinger M, Katsaras J, Kučerka N (2012) Molecular structures of fluid phase phosphatidylglycerol bilayers as determined by small angle neutron and X-ray scattering. *Biochimica et Biophysica Acta (BBA)-Biomembranes*, 1818(9), 2135–2148.
Pan J, Tristram-Nagle S, Kučerka N, Nagle JF (2008) Temperature dependence of structure, bending rigidity, and bilayer interactions of dioleoylphosphatidylcholine bilayers. *Biophysical Journal*, 94(1), 117–124.
Petrache HI, Tristram-Nagle S, Nagle JF (1998) Fluid phase structure of EPC and DMPC bilayers. *Chemistry and Physics of Lipids*, 95(1), 83–94.
Tristram-Nagle S, Kim DJ, Akhunzada N, Kučerka N, Mathai JC, Katsaras J, Zeidel M and Nagle JF (2010) Structure and water permeability of fully hydrated diphytanoylPC. *Chemistry and Physics of Lipids*, 163(6), 630–637.
Tristram-Nagle S, Petrache HI, Nagle JF (1998) Structure and interactions of fully hydrated dioleoylphosphatidylcholine bilayers. *Biophysical Journal*, 75(2), 917–925.

Appendix 2: List of membrane dyes and fluorescent groups conjugated to lipids

DYE ABBREVIATION IN THE BOOK; ABSORPTION/ EMISSION MAXIMA (nm); OCCURRENCE IN THE BOOK	FULL NAME; ALTERNATIVE NAME; CAS NUMBER	STRUCTURE OF DYE OR FLUORESCENT GROUP
Alexa 488-NHS Ester 496/520 (Chapters 3, 20, 22)	Alexa Fluor™ 488 NHS Ester (Succinimidyl Ester)	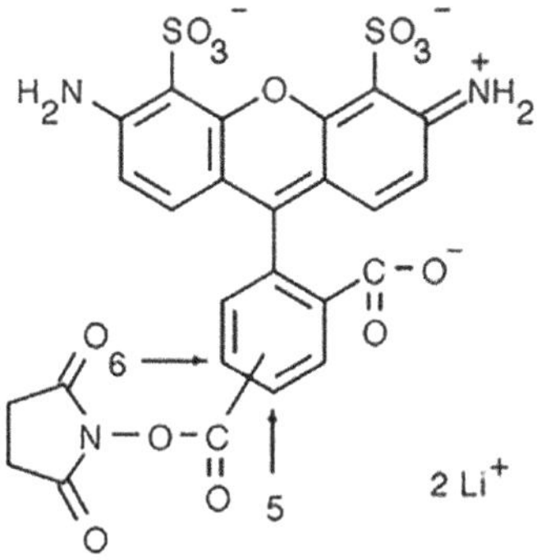
ATTO 488 DOPE 507/527 (Chapter 30)	ATTO 488 labeled DOPE	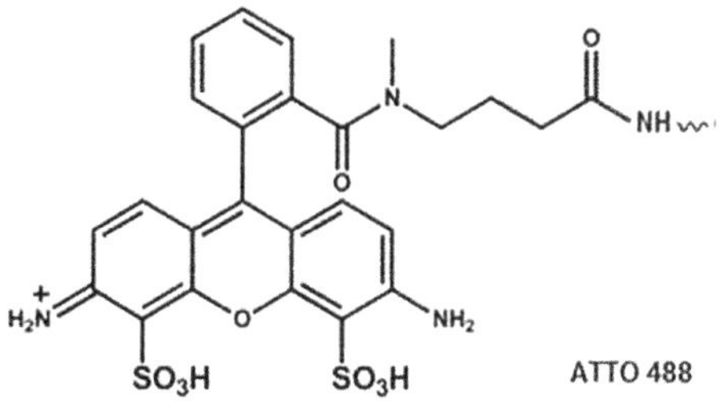
Atto647N-DOPE 643/662 (Chapter 4)	1,2-Dioleoyl-*sn*-glycero-3-phosphoethanolamine labeled with Atto 647N	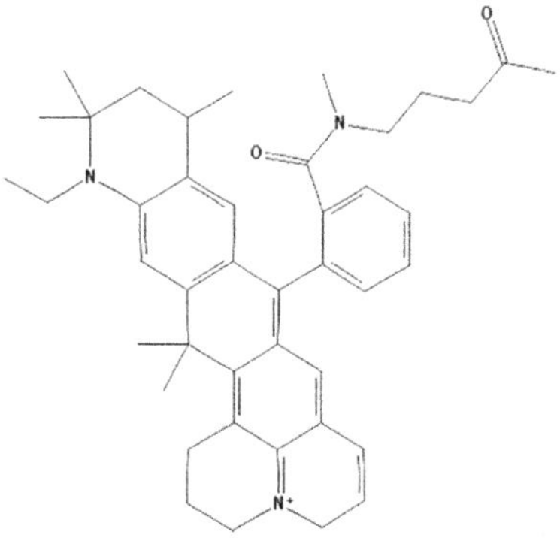

(*Continued*)

DYE ABBREVIATION IN THE BOOK; ABSORPTION/ EMISSION MAXIMA (nm); OCCURRENCE IN THE BOOK	FULL NAME; ALTERNATIVE NAME; CAS NUMBER	STRUCTURE OF DYE OR FLUORESCENT GROUP
BODIPY FLC5-GM1 502/511 (Chapter 16)	BODIPY® FL C5-ganglioside GM1	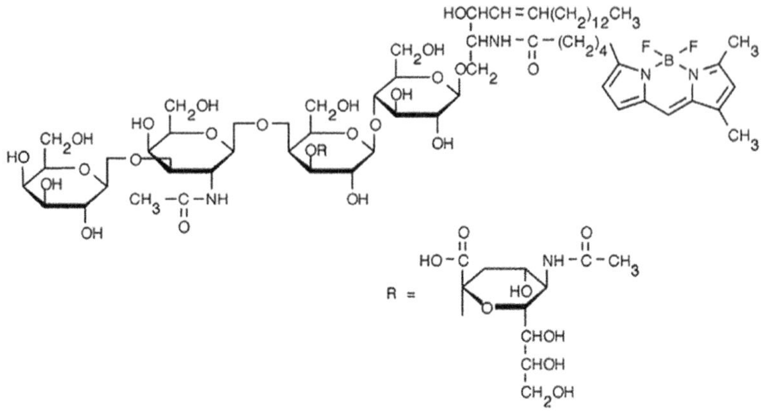
BODIPY-PC 545/570 (Chapter 10)		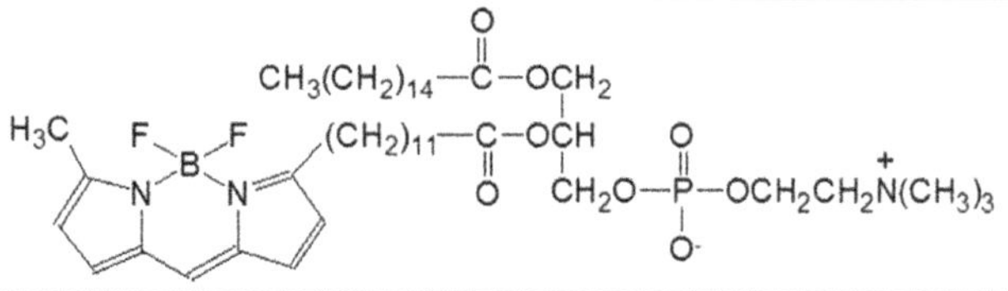
BODIPY TMR PIP_2 542/574 (Chapter 4)	BODIPY® TMR Phosphatidylinositol 4,5-bisphosphate	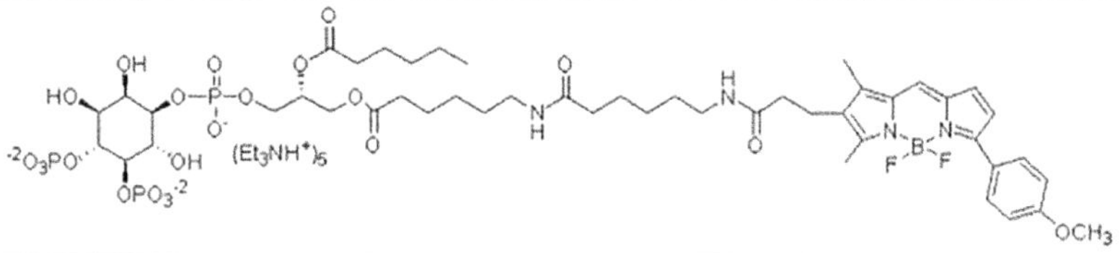
Bodipy-TRCer 590/621 (Chapter 16)	BODIPY® TR ceramide	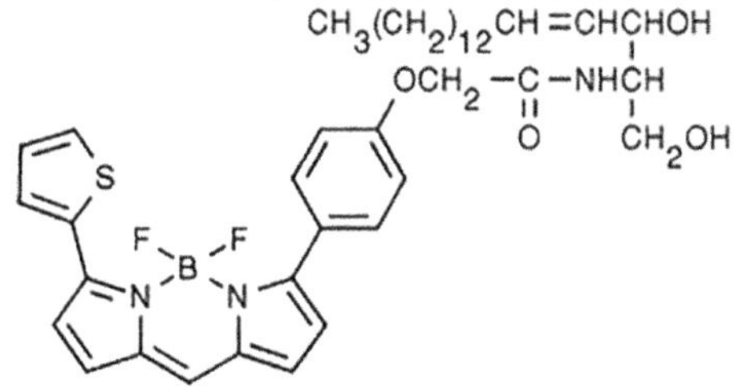
Carboxyfluorescein 492/517 (Chapter 29)	6-FAM; CAS: 3301-79-9	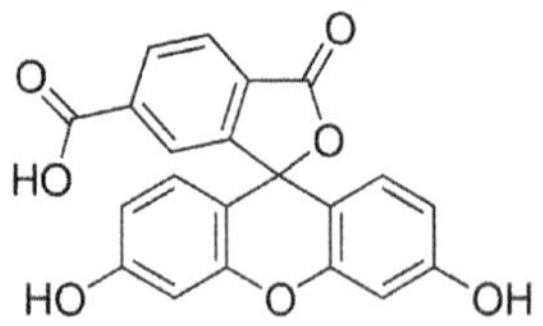
di6ASP-BS 495/613 (Chapter 22)	Di-6-ASP-BS [N-(3-Sulfobutyl)-4-(4-dihexylaMinostyryl) pyridinium, inner salt] CAS: 90133-77-0	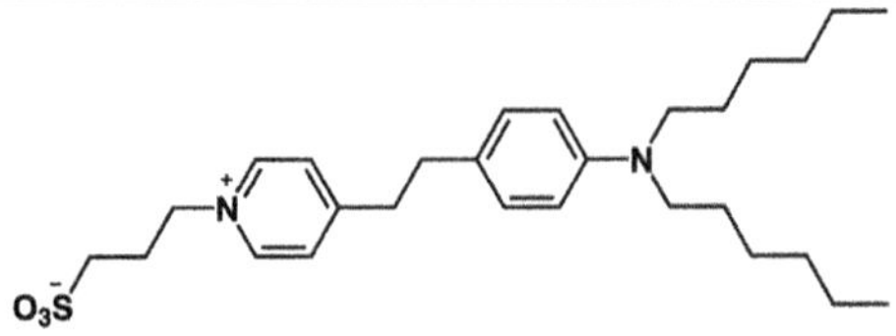

(Continued)

DYE ABBREVIATION IN THE BOOK; ABSORPTION/ EMISSION MAXIMA (nm); OCCURRENCE IN THE BOOK	FULL NAME; ALTERNATIVE NAME; CAS NUMBER	STRUCTURE OF DYE OR FLUORESCENT GROUP
$DiIC_{12}$ 550/567 (Chapter 10)	1,1′-Didodecyl-3,3,3′,3′-tetramethylindocarbocyanine perchlorate; CAS: 75664-01-6	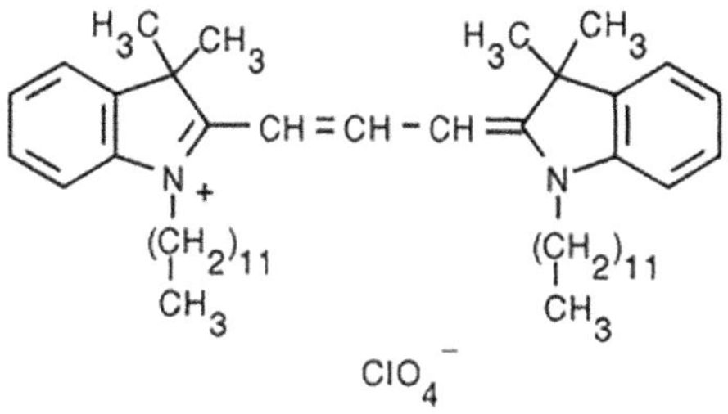 H_3C CH_3 H_3C CH_3 CH = CH – CH = N+ N $(CH_2)_{11}$ CH_3 $(CH_2)_{11}$ CH_3 ClO_4^-
$DiIC_{18}$ 550/567 (Chapter 4, 10, 21, 24)	1,1′-Dioctadecyl-3,3,3′,3′-tetramethylindocarbocyanine perchlorate; There is also a iodide form; CAS: 41085-99-8	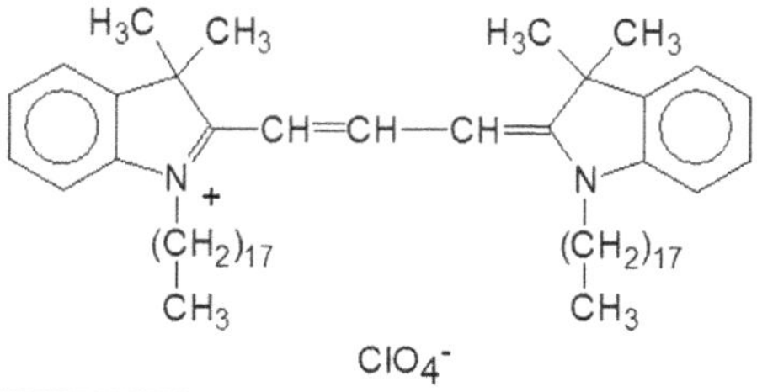 H_3C CH_3 H_3C CH_3 CH=CH—CH= N+ N $(CH_2)_{17}$ CH_3 $(CH_2)_{17}$ CH_3 ClO_4^-
$DiIC_{20}$ (Chapter 21)	1,1′-Dieicosanyl-3,3,3′,3′-tetramethylindocarbocyanine perchlorate	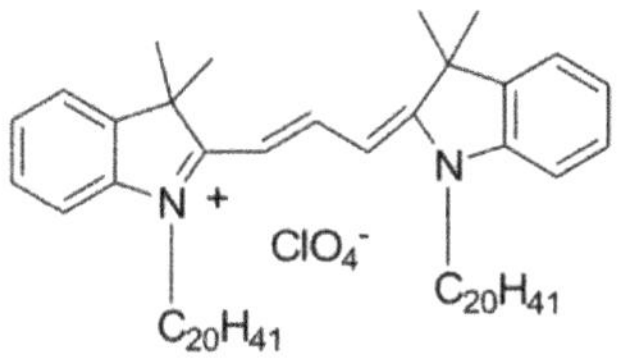 N+ N ClO_4^- $C_{20}H_{41}$ $C_{20}H_{41}$
DiD 550/567 (Chapters 20, 26)	1,1′-Dioctadecyl-3,3,3′,3′-tetramethylindodicarbocyanine (lipophilic dye); CAS: 41085-99-8	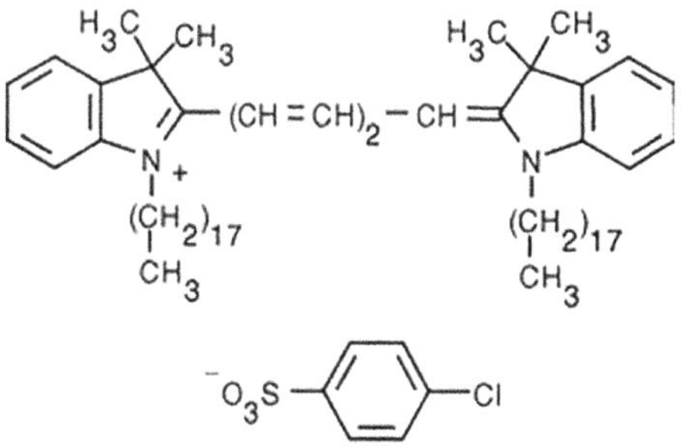 H_3C CH_3 H_3C CH_3 $(CH = CH)_2$ – CH = N+ N $(CH_2)_{17}$ CH_3 $(CH_2)_{17}$ CH_3 ^-O_3S Cl
$DiOC_{18}$ 484/501 (Chapters 10)	3,3′-Dioctadecyloxacarbocyanine perchlorate; CAS: 34215-57-1	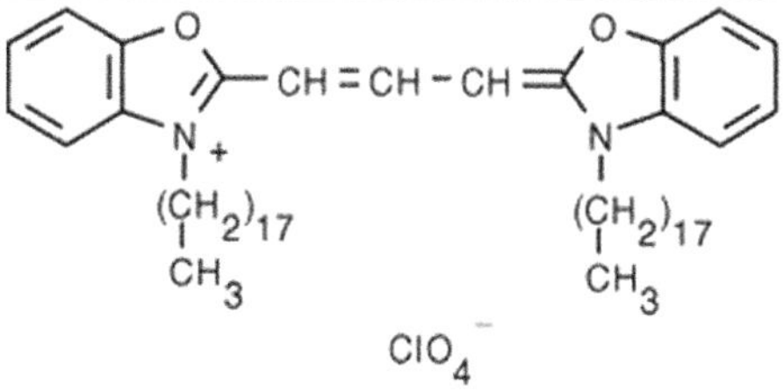 O O CH = CH – CH = N+ N $(CH_2)_{17}$ CH_3 $(CH_2)_{17}$ CH_3 ClO_4^-
DPH 352/452 (Chapters 10)	1,6-Diphenyl-hexa-1,3,5-triene; CAS: 1720-32-7	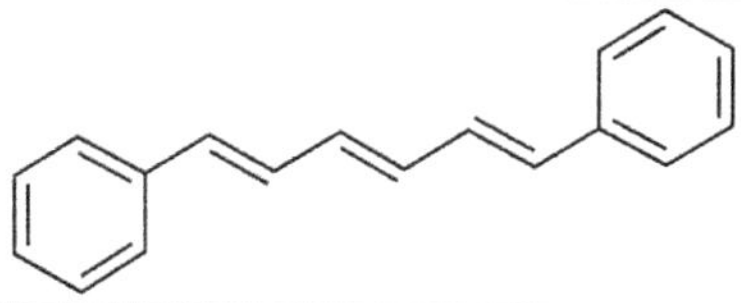

(*Continued*)

DYE ABBREVIATION IN THE BOOK; ABSORPTION/ EMISSION MAXIMA (nm); OCCURRENCE IN THE BOOK	FULL NAME; ALTERNATIVE NAME; CAS NUMBER	STRUCTURE OF DYE OR FLUORESCENT GROUP
FITC 495/519 (Chapter 27)	Fluorescein isothiocyanate; Fluorescein 5-isothiocyanate; CAS: 3326-32-7	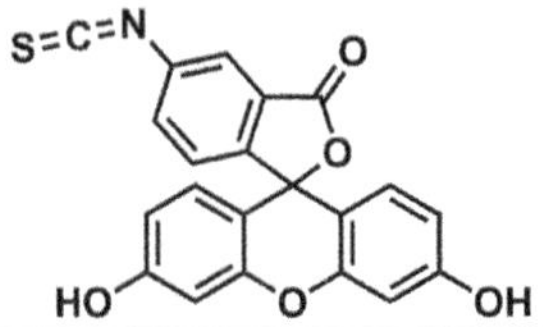 S=C=N, O, O, HO, O, OH
Fluorescein 492/512 (Chapter 22)	CAS: 2321-07-5	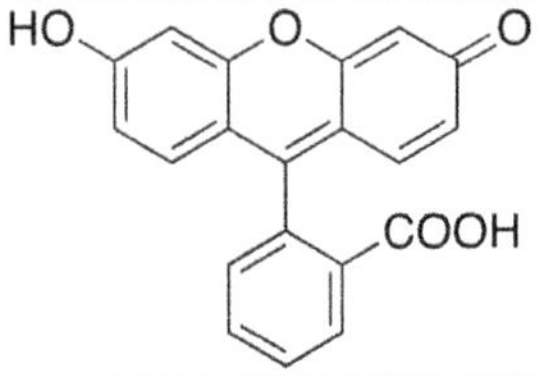 HO, O, O, COOH
LAURDAN 366/497 (in methanol; values depend on probe environment) (Chapters 10, 18, 21, 26)	6-Dodecanoyl-2-dimethylaminonaphthalene; CAS: 74515-25-6	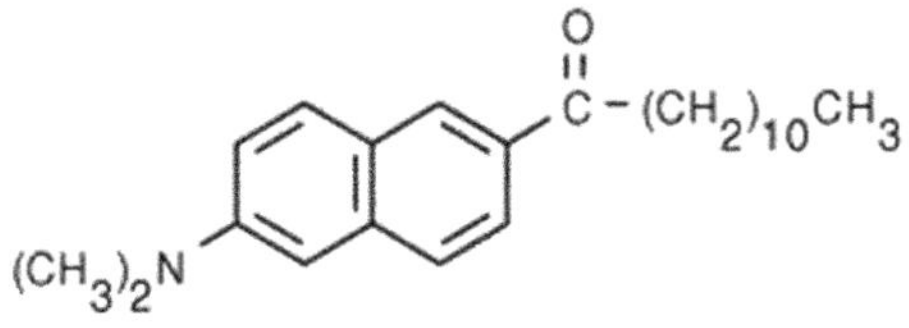 O, C – $(CH_2)_{10}CH_3$, $(CH_3)_2N$
Liss-Rhod PE 560/583 (Chapters 27)	1,2-Dioleoyl-*sn*-glycero-3-phosphoethanolamine-*N*-(lissamine rhodamine B sulfonyl) CAS: 384833-00-5	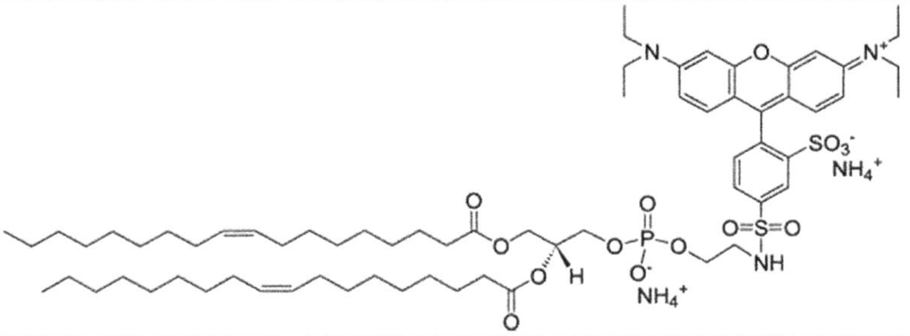 N, O, N^+, SO_3^-, NH_4^+, O, O, O=S=O, O, P, O, NH, O, H, O^-, NH_4^+, O
Nile Red 552/636 (in methanol) (Chapters 21, 26)	CAS: 7385-67-3	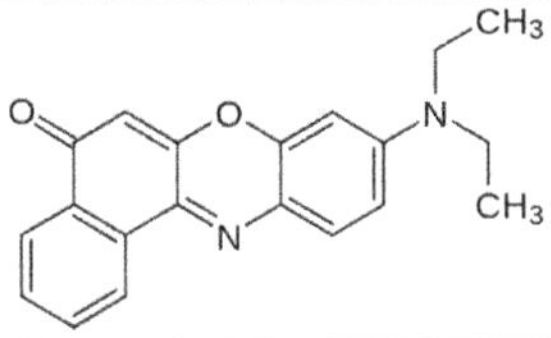 CH_3, O, O, N, N, CH_3
NBD-PE 460/535 (Chapter 19)	NBD-DPPE; 16:0 NBD PE; 1,2-dipalmitoyl-*sn*-glycero-3-phosphoethanolamine-*N*-(7-nitro-2-1,3-benzoxadiazol-4-yl) (ammonium salt)	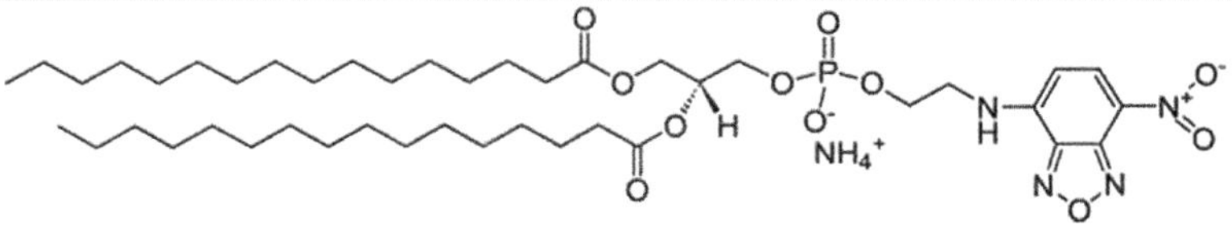 O, O, O, P, O, H, O^-, NH_4^+, N, H, N^+, O^-, O, N, O, N
NBD-sphingomyelin 460/534 (Chapter 2)	N-C6:0-NBD-Sphingomyelin; N-C6:0-NBD-Sphingosylphosphorylcholine; NBD C_6-sphingomyelin; CAS: 94885-04-8	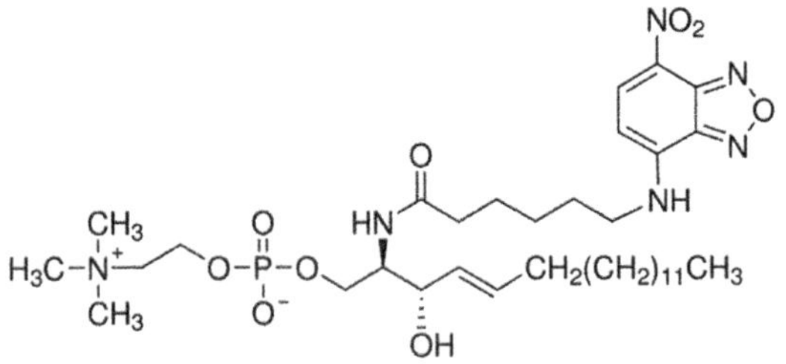 NO_2, N, O, N, O, NH, CH_3, O, HN, H_3C-N^+, O-P-O, $CH_2(CH_2)_{11}CH_3$, CH_3, O^-, OH

(Continued)

DYE ABBREVIATION IN THE BOOK; ABSORPTION/ EMISSION MAXIMA (nm); OCCURRENCE IN THE BOOK	FULL NAME; ALTERNATIVE NAME; CAS NUMBER	STRUCTURE OF DYE OR FLUORESCENT GROUP
Parinaric acid 320/430 (Chapter 10)	(9*Z*,11*E*,13*E*,15*Z*)-Octadeca-9,11,13,15-tetraenoic acid; CAS: 18427-44-6	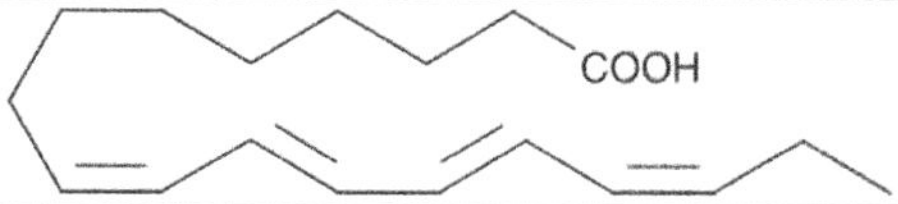 COOH
PRODAN 361/498 (Chapters 10)	*N, N*-Dimethyl-6-propionyl-2-naphthylamine dimethylaminonaphthalene; CAS: 70504-01-7	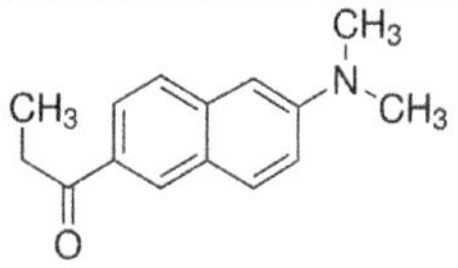 CH_3 N CH_3 CH_3 O
Pyrene 335/360 (in cyclohexane) (Chapter 10)	CAS: 129-00-0	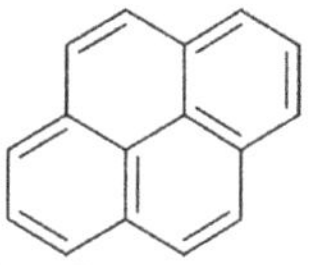
Rhodamine-DOPE (DOPE-Rh) 560/583 (Chapter 20, 22, 27, 30)	1,2-Dioleoyl-*sn*-glycero-3-phosphoethanolamine-*N*-(lissamine rhodamine B sulfonyl) (ammonium salt); 18:1 Liss Rhod PE; Rhodamine B-labeled DOPE; CAS: 384833-00-5	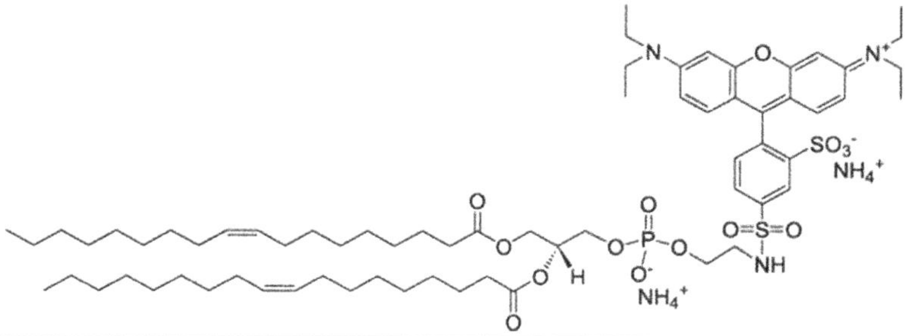 N O N^+ SO_3^- NH_4^+ O O O=S=O O O P O NH H O^- NH_4^+ O
Rh-DPPE (DPPE-Rh, Rh-DHPE) 560/583 (Chapters 10, 15, 22, 27)	1,2-Dipalmitoyl-*sn*-glycero-3-phosphoethanolamine-*N*-(lissamine rhodamine B); 1,2-Dihexadecanoyl-*sn*-glycero-3-phosphoethanolamine-N-(lissamine rhodamine B sulfonyl); 16:0 Liss Rhod PE; CAS: 384833-01-6	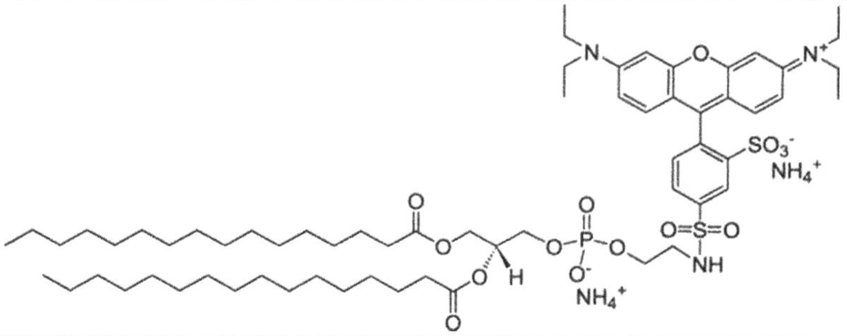 N O N^+ SO_3^- NH_4^+ O O O=S=O O O P O NH H O^- NH_4^+ O
Texas Red 589/615 (Chapter 29)	Sulforhodamine 101 acid chloride; CAS: 82354-19-6	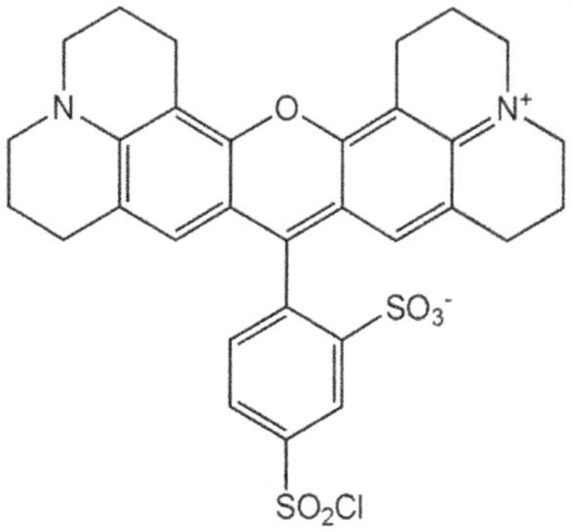 N O N^+ SO_3^- SO_2Cl

(*Continued*)

DYE ABBREVIATION IN THE BOOK; ABSORPTION/ EMISSION MAXIMA (nm); OCCURRENCE IN THE BOOK	FULL NAME; ALTERNATIVE NAME; CAS NUMBER	STRUCTURE OF DYE OR FLUORESCENT GROUP
Texas Red DPPE (Texas Red DHPE) 595/615 (Chapter 3, 16, 18)	Texas Red® 1,2-dihexadecanoyl-*sn*-glycero-3-phosphoethanolamine; Texas Red 1,2-dipalmitoyl-sn-glycerophosphoethanolamine; CAS: 187099-99-6	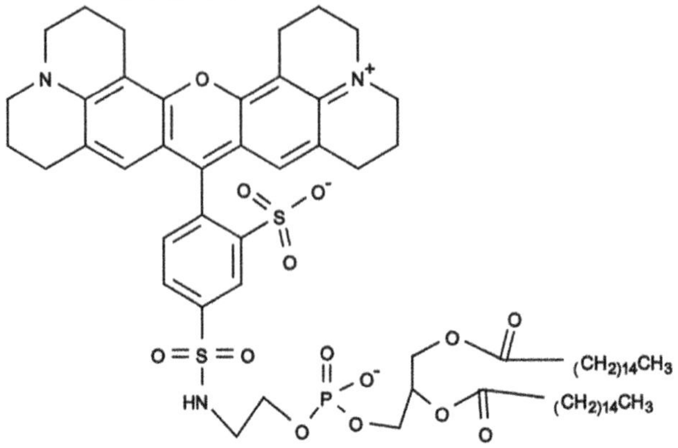 N, O, N^+, O, S, O^-, O, O=S=O, HN, O, P, O^-, O, O, O, O, O, $(CH_2)_{14}CH_3$, $(CH_2)_{14}CH_3$
TMA-DPH (Chapter 10)	(1-(4-Trimethylammoniumphenyl)-6-Phenyl-1,3,5-Hexatriene p-Toluenesulfonate)	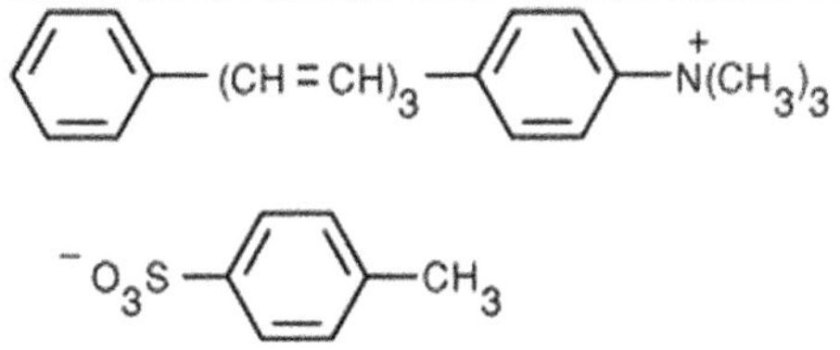 $(CH=CH)_3$, $\overset{+}{N}(CH_3)_3$; ^-O_3S, CH_3

Appendix 3: List of detergents

DETERGENT ABBREVIATION	FULL NAME	CAS NUMBER	STRUCTURE
$C_{12}E_8$ (Chapters 24, 28)	Dodecyloctaethyleneglycol monoether	3055-98-9	
CHAPS (Chapter 24)	3-[(3-Cholamidopropyl) dimethylammonio]-1-propanesulfonate	75621-03-3	
DDM (Chapter 3)	*n*-Dodecyl β-D-maltoside Lauryl-β-D-maltoside	69227-93-6	
DOTM (Chapter 3)	*n*-Dodecyl-β-D-thiomaltoside	148565-58-6	
OG (Chapter 3)	*n*-Octylglucoside	29836-26-8	
Tween® 20 (Chapters 4, 15)	Polyethylene glycol sorbitan monolaurate, Polyoxyethylenesorbitan monolaurate	9005-64-5	
TX-100 TRITON X-100 (Chapters 2, 3, 15, 24, 30)	Polyethylene glycol *p*-(1,1,3,3-tetramethylbutyl)-phenyl ether	9002-93-1	n = 9–10
SDS (Chapters 4, 24)	Sodium dodecyl sulfate	151-21-3	$CH_3(CH_2)_{10}CH_2O{-}S(=O)_2{-}ONa$

Appendix 4: List of water-soluble dyes or their fluorescent groups and their structures

DYE NAME; EXCITATION/ EMISSION MAXIMA (nm), OCCURRENCE IN THE BOOK	OTHER NAMES; CAS NUMBER	STRUCTURE
Alexa 488 490/525 (Chapters 16, 21; 23, 24, 30)		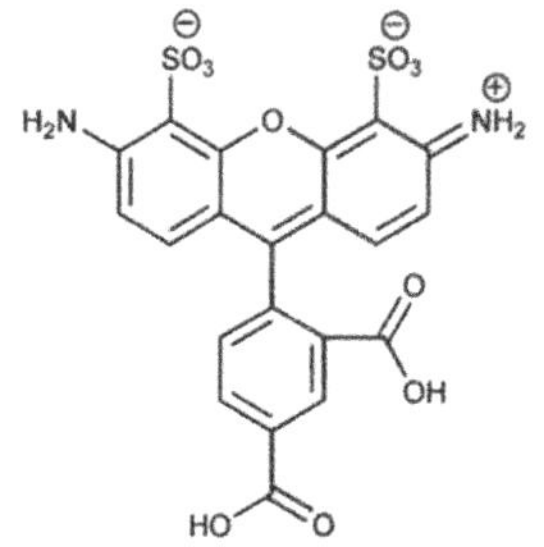 SO_3^{-} SO_3^{-} H_2N O NH_2^{+} O OH HO O
Alexa 546 556/573 (Chapters 21, 24)		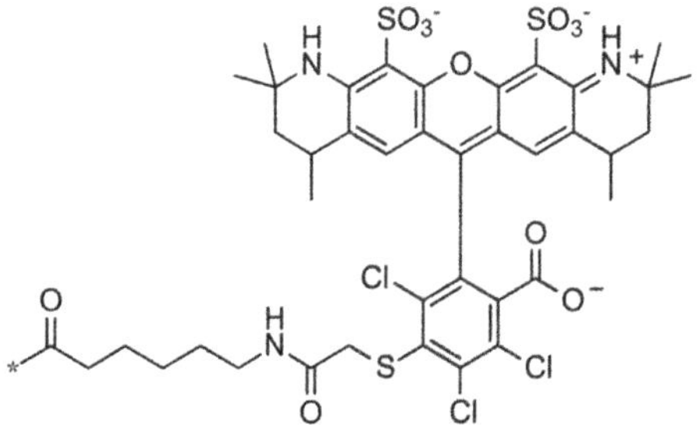 H N SO_3^- O SO_3^- H N^+ Cl O O^- O H N S Cl O Cl *
Alexa 555 555/580 (Chapter 24)		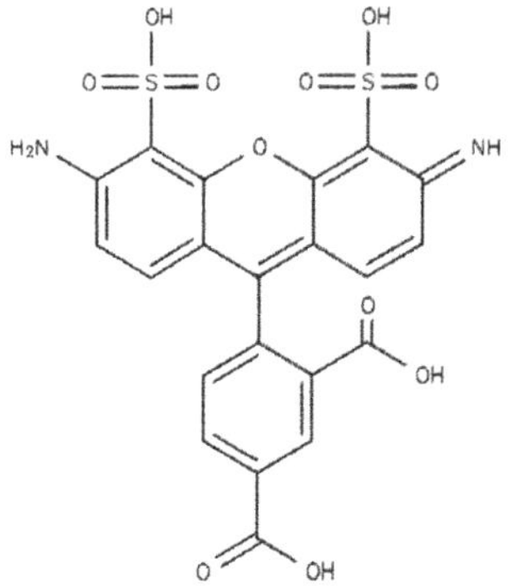 OH O=S=O OH O=S=O H_2N O NH O OH O OH

(Continued)

DYE NAME; EXCITATION/ EMISSION MAXIMA (nm), OCCURRENCE IN THE BOOK	OTHER NAMES; CAS NUMBER	STRUCTURE
Alexa 647 650/665 (Chapters 21, 29)		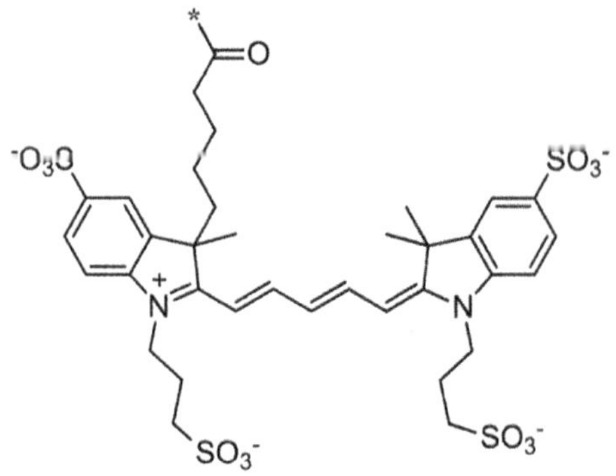
Atto 655 663/680 (Chapter 21)		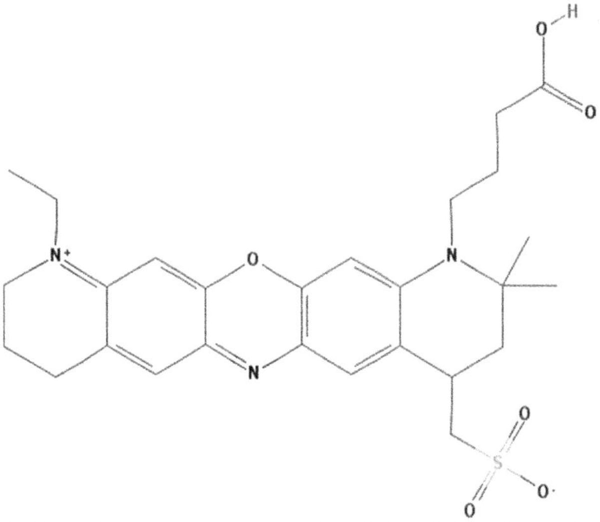
BODIPY 545/570 (Chapter 21)	4,4-Difluoro-5,7dimethyl-4-bora-3a,4a-diaza-*s*-indacene-3-pentanoyl (boron-dipyrromethene)	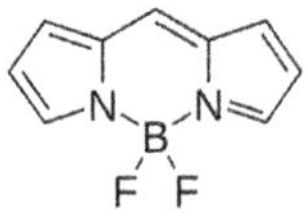
Calcein 495/515 (Chapter 24)	Fluorexon, fluorescein complex CAS: 1461-15-0	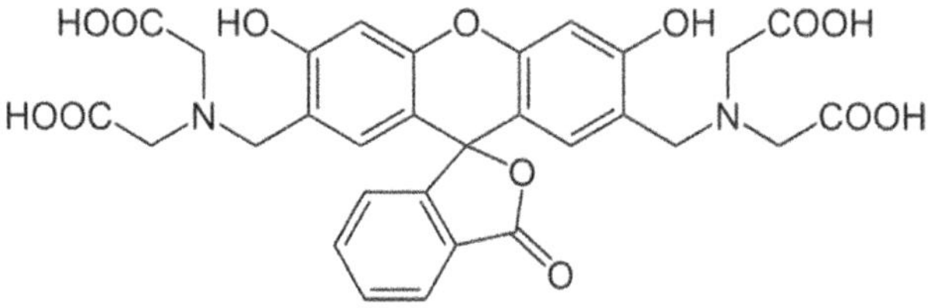
Carboxyfluorescein 495/517 (Chapter 27)	6-Carboxyfluorescein; 6-FAM CAS: 3301-79-9	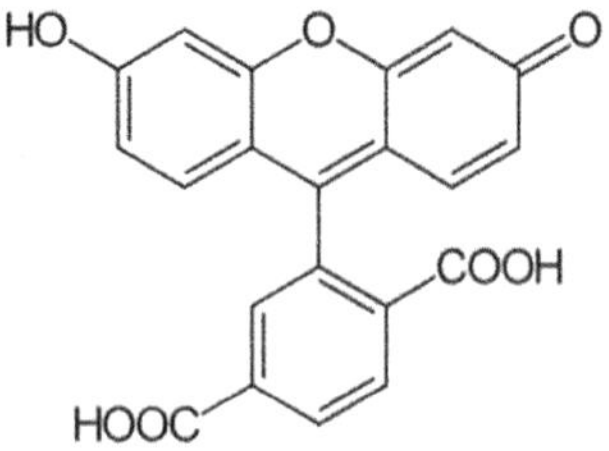
eGFP 488/507 (Chapter 21)	Enhanced green fluorescent protein	

(Continued)

DYE NAME; EXCITATION/ EMISSION MAXIMA (nm), OCCURRENCE IN THE BOOK	OTHER NAMES; CAS NUMBER	STRUCTURE
Fluorescein 494/512 (Chapters 20, 21, 22)	CAS: 2321-07-5	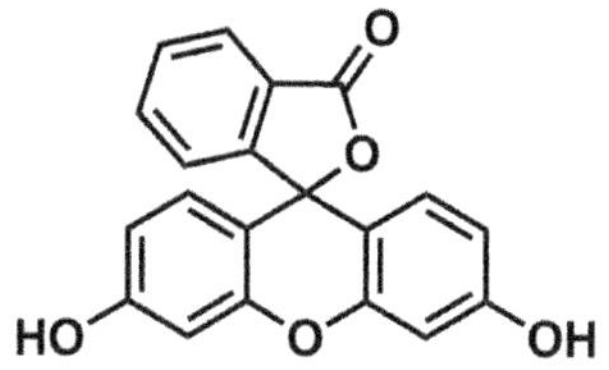 O O HO O OH
Fluorescein diphosphate 490/514 (Chapter 20)	Fluorescein diphosphate, tetraammonium salt; spiro[isobenzofuran-1(3H),9′-[9H]xanthen]-3-one, 3′,6′-bis(phosphonooxy)-, tetraammonium salt; CAS: 217305-49-2	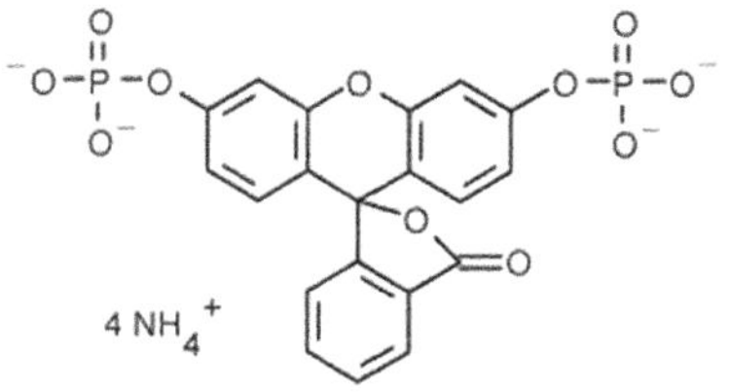 $4\ NH_4^+$
Fluorescein isothiocyanate 495/519 (Chapter 29)	FITC; fluorescein isothiocyanate; fluorescein 5-isothiocyanate; CAS: 3326-32-7	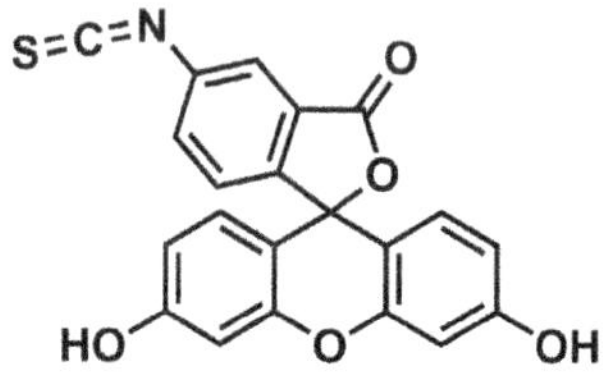 S=C=N O O HO O OH
Methylene Blue (292, 665)/675 (Chapter 22)	Methylthioninium chloride; CAS: 122965-43-9	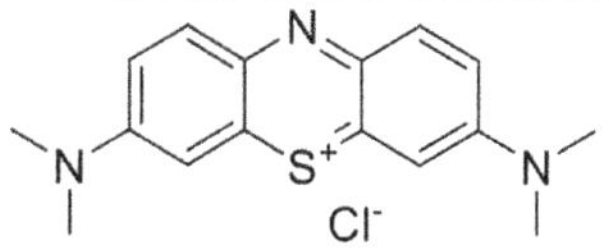 N N S^+ N Cl^-
Rhodamine 6G ~530/~560 (Chapter 21)	Basic red 1; CAS: 989-38-8	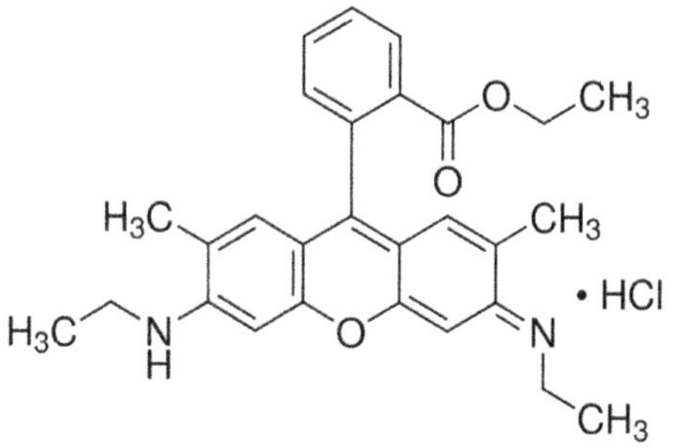 O CH_3 O H_3C CH_3 • HCl H_3C N H O N CH_3
Rhodamine B 553/627 (in methanol) (Chapters 4, 21)	Basic violet 10, brilliant pink B, rhodamine O, tetraethylrhodamine; CAS: 81-88-9	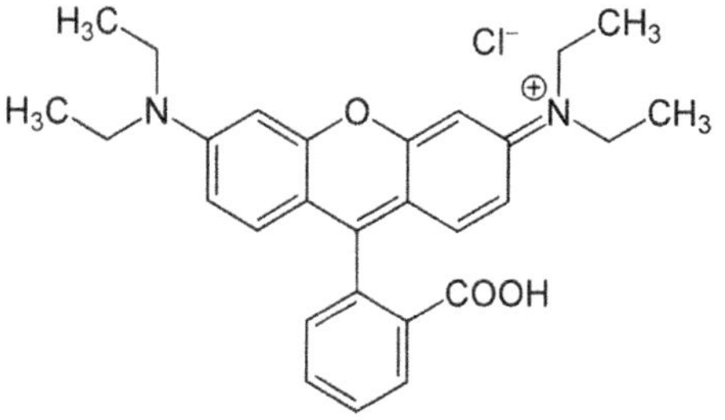 H_3C Cl^- CH_3 H_3C N O ⊕N CH_3 COOH

(*Continued*)

DYE NAME; EXCITATION/ EMISSION MAXIMA (nm), OCCURRENCE IN THE BOOK	OTHER NAMES; CAS NUMBER	STRUCTURE
Sulphorhodamine B 565/586	SRB; CAS: 3520-42-1	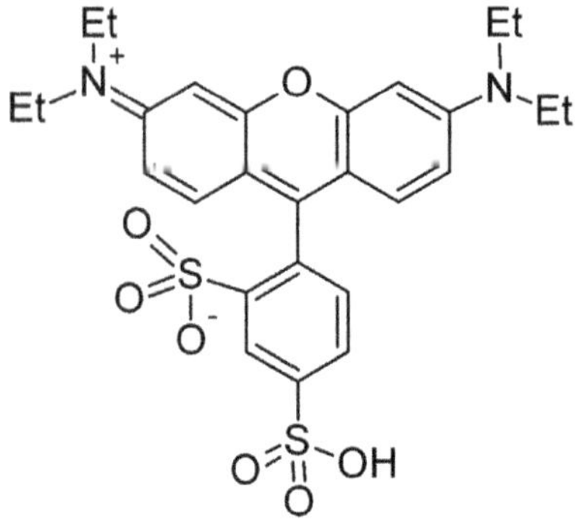
Texas Red 589/615 (Chapter 24)	Sulforhodamine 101 acid chloride; CAS: 82354-19-6	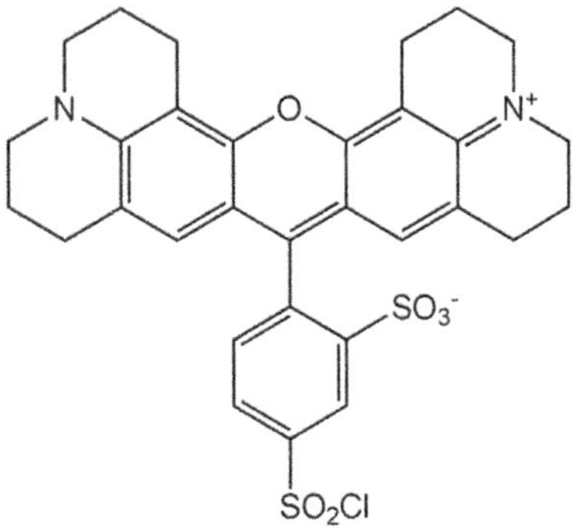
TMR 557/576 (Chapter 4)	Tetramethylrhodamine; CAS: 70281-37-7	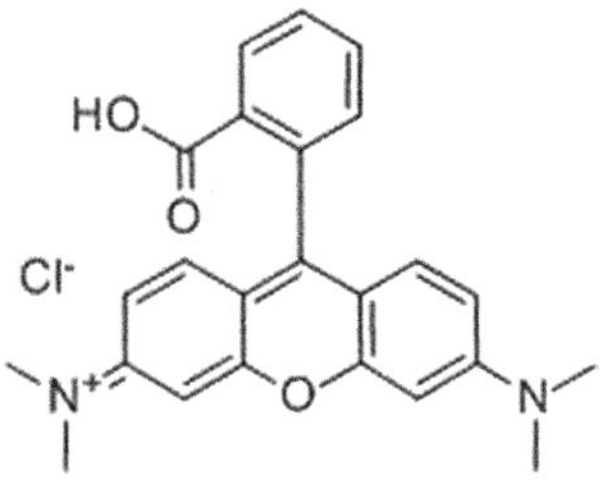

Index

Note: Page numbers in italic refer to figures and bold refer to boxes and tables.

N

Q

R

S

T

U

V

W

X

Y

Z

Printed and bound by CPI Group (UK) Ltd, Croydon, CR0 4YY
23/10/2024
01778381-0001